STATISTICS/STATISTIQUES

Basic Science and Technology Statistics

Statistiques de base de la science et de la technologie

AF795438

2001

ORGANISATION FOR ECONOMIC CO-OPERATION AND DEVELOPMENT

Pursuant to Article 1 of the Convention signed in Paris on 14th December 1960, and which came into force on 30th September 1961, the Organisation for Economic Co-operation and Development (OECD) shall promote policies designed:

- to achieve the highest sustainable economic growth and employment and a rising standard of living in Member countries, while maintaining financial stability, and thus to contribute to the development of the world economy;
- to contribute to sound economic expansion in Member as well as non-member countries in the process of economic development; and
- to contribute to the expansion of world trade on a multilateral, non-discriminatory basis in accordance with international obligations.

The original Member countries of the OECD are Austria, Belgium, Canada, Denmark, France, Germany, Greece, Iceland, Ireland, Italy, Luxembourg, the Netherlands, Norway, Portugal, Spain, Sweden, Switzerland, Turkey, the United Kingdom and the United States. The following countries became Members subsequently through accession at the dates indicated hereafter: Japan (28th April 1964), Finland (28th January 1969), Australia (7th June 1971), New Zealand (29th May 1973), Mexico (18th May 1994), the Czech Republic (21st December 1995), Hungary (7th May 1996), Poland (22nd November 1996), Korea (12th December 1996) and the Slovak Republic (14th December 2000). The Commission of the European Communities takes part in the work of the OECD (Article 13 of the OECD Convention).

ORGANISATION DE COOPÉRATION ET DE DÉVELOPPEMENT ÉCONOMIQUES

En vertu de l'article 1er de la Convention signée le 14 décembre 1960, à Paris, et entrée en vigueur le 30 septembre 1961, l'Organisation de Coopération et de Développement Économiques (OCDE) a pour objectif de promouvoir des politiques visant :

- à réaliser la plus forte expansion de l'économie et de l'emploi et une progression du niveau de vie dans les pays Membres, tout en maintenant la stabilité financière, et à contribuer ainsi au développement de l'économie mondiale ;
- à contribuer à une saine expansion économique dans les pays Membres, ainsi que les pays non membres, en voie de développement économique ;
- à contribuer à l'expansion du commerce mondial sur une base multilatérale et non discriminatoire conformément aux obligations internationales.

Les pays Membres originaires de l'OCDE sont : l'Allemagne, l'Autriche, la Belgique, le Canada, le Danemark, l'Espagne, les États-Unis, la France, la Grèce, l'Irlande, l'Islande, l'Italie, le Luxembourg, la Norvège, les Pays-Bas, le Portugal, le Royaume-Uni, la Suède, la Suisse et la Turquie. Les pays suivants sont ultérieurement devenus Membres par adhésion aux dates indiquées ci-après : le Japon (28 avril 1964), la Finlande (28 janvier 1969), l'Australie (7 juin 1971), la Nouvelle-Zélande (29 mai 1973), le Mexique (18 mai 1994), la République tchèque (21 décembre 1995), la Hongrie (7 mai 1996), la Pologne (22 novembre 1996), la Corée (12 décembre 1996) et la République slovaque (14 décembre 2000). La Commission des Communautés européennes participe aux travaux de l'OCDE (article 13 de la Convention de l'OCDE).

FOREWORD

This publication is prepared by the Economic Analysis and Statistics (EAS) Division of the Directorate for Science, Technology and Industry (DSTI) of the OECD Secretariat, working with the Working Party of National Experts on Science and Technology Indicators (NESTI). It contains 19 country tables giving the main basic statistical series selected from the OECD "Science and Technology Databases", of which 18 concern resources devoted to research and experimental development (R&D) and one the Technology balance of payments . These statistics are published on the responsibility of the Secretary-General of the OECD.

R&D data

The OECD has been collecting R&D data on a regular basis since the early 1960s. Tables 1 to 8 present basic statistics on ***total national resources devoted to R&D***. The first four and the seventh deal with the standard expenditure measure, the ***Gross Domestic Expenditure on Research and Experimental Development (GERD)*** which covers all R&D carried out on national territory in the year concerned by sector of performance, giving information on its breakdown by source of funds, type of costs, type of activity, socio-economic objective and by main scientific field.

Resources devoted to R&D are also measured in labour terms as shown in Tables 5, 6 and 8, where ***total R&D personnel*** by sector of employment is broken down by occupation, formal qualification and by main scientific field. Such R&D personnel data should be expressed in full-time equivalent (FTE) on R&D or as person-years (a person working half-time on R&D therefore counts as 0.5 person-years).

Similar sets of data are provided for R&D carried out in the ***Business Enterprise Sector*** with detailed data ***by industry*** (Tables 9 to 13) and the ***Higher Education Sector*** with detailed data ***by field of science*** (Tables 14 to 17).

All of these tables are essentially based on retrospective surveys of the units carrying out the R&D, though national forecasts have been included when available.

More up-to-date information on government intentions or objectives when committing money to R&D can be derived from budgets. These data are shown in Table 18, which give ***government R&D appropriations or outlays*** and their breakdown ***by socio-economic objective***. Users are warned that these budget R&D data vary in coverage from those in Tables 1 to 17, and percentages from the two types of table should not be combined.

Measures of the output of S&T

This publication does not include any direct measures of the output of Scientific and Technological (S&T) activities, but only one proxy indicator based on data collected for other purposes: ***the technology balance of payments***. Individually, these data are not reliable, but taken together with other kinds of indicators in a more global context, they may throw light on countries' technological performance.

A second indicator of S&T output, *the number of patent applications*, was previously included in this publication but is no longer. Due to methodological developments, new more reliable indicators presenting patent families have been developed and are included in the twice-yearly publication "*Main Science and Technology Indicators*".

The ***Technology Balance of Payments (TBP)*** series are data extracted from national balance of payments sources, with the aim of measuring the flow of proprietary technology and know-how into and out of the country concerned. Data collected to date are not always compiled according to international standards, but the series quoted in Table 19 generally comprise money paid or received for the acquisition or use of patents, licences, trademarks, designs, inventions, know-how and closely related technical services.

AVANT-PROPOS

Cette publication est préparée par la Division des analyses économiques et des statistiques (AES) de la Direction de la science, de la technologie et de l'industrie (DSTI) du Secrétariat de l'OCDE, qui travaille avec le Groupe de travail des experts nationaux sur les indicateurs de la science et de la technologie (ENIST). Elle contient 19 tableaux par pays présentant les principales séries de base choisies dans les "Bases de données Science et Technologie" de l'OCDE, dont 18 se rapportent aux ressources consacrées à la recherche et au développement expérimental (R-D) et un à la Balance des paiements technologiques. Ces statistiques sont publiées sous la responsabilité du Secrétaire général de l'OCDE.

Données sur les ressources consacrées à la R-D

L'OCDE recueille régulièrement des données sur la R-D depuis le début des années 1960. Les tableaux 1 à 8 présentent les données de bases concernant ***les ressources nationales totales consacrées à la R-D***. Les quatre premiers ainsi que le septième se rapportent à la mesure type des ressources financières : ***la dépense intérieure brute de recherche et développement expérimental (DIRD)*** qui comprend toute la R-D réalisée sur le territoire national pendant l'année concernée par secteur d'exécution, et fournissent les informations sur sa ventilation selon la source de financement, le type de coûts, le type d'activité, l'objectif socio-économique ou le domaine scientifique principal.

Les ressources consacrées à la R-D se mesurent aussi en terme de personnel comme indiqué dans les tableaux 5, 6 et 8, où ***le personnel total de R-D*** par secteur d'emploi est réparti par profession, par qualification et selon le domaine scientifique principal. Les données concernant le personnel de R-D sont exprimées en équivalence plein-temps (EPT) à la R-D, qui peut être assimilée à une année de travail d'une personne (une personne qui travaille à mi-temps dans le domaine de la R-D compte donc pour 0.5 année d'une personne EPT).

Des jeux de données similaires sont fournis pour la R-D réalisée dans ***les secteurs des entreprises*** avec des détails ***par industrie*** (tableaux 9 à 13) et de ***l'enseignement supérieur*** avec des détails ***par discipline scientifique*** (tableaux 14 à 17).

Tous ces tableaux sont fondés sur des enquêtes rétrospectives auprès des unités qui ont réalisé les travaux de R-D ; cependant, des prévisions nationales ont été incluses quand elles étaient disponibles.

Des informations plus à jour sur les intentions ou buts poursuivis par les gouvernements quand ils votent des crédits pour la R-D peuvent être tirées des budgets. Ces données sont indiquées dans le tableau 18 qui présente ***les crédits budgétaires publics de R-D*** et leur répartition ***par objectif socio-économique***. Les lecteurs sont avertis que ces données budgétaires sur la R-D couvrent un champ différent de celui des informations qui figurent dans les tableaux 1 à 17, et que les pourcentages calculés à partir de ces deux catégories de données ne doivent pas être combinés.

Mesure des résultats de la S-T

Cette publication n'inclue pas de mesure directe des résultats des activités scientifiques et technologiques (S-T), mais seulement un indicateur de remplacement fondé sur des données recueillies à d'autres fins : ***la balance des paiements technologiques***. Prises seules, ces données ne sont pas fiables, mais considérées avec d'autres sortes indicateurs dans un contexte plus global, elles peuvent fournir des indications sur la performance technologique des pays.

Un deuxième indicateur de résultats des activités scientifiques et technologiques, se rapportant au *nombre de demandes de brevets déposées*, était précédemment inclus dans cette publication et a été supprimé. À la suite de travaux méthodologiques, de nouveaux indicateurs plus pertinents, se rapportant à des familles de brevets, ont été développés et sont maintenant inclus dans la publication semestrielle « *Principaux indicateurs de la science et de la technologie* ».

Les séries de ***la balance des paiements technologiques*** (BPT) sont des données extraites des sources nationales sur la balance des paiements, dans le but de mesurer les opérations d'un pays se rapportant à l'achat et à la vente d'informations et de savoir-faire technologiques. Les données recueillies jusqu'ici n'ont pas toujours été compilées selon des normes internationales, mais les séries présentées dans le tableau 19 comprennent en général les fonds payés ou reçus pour l'acquisition ou l'utilisation de brevets, licences, marques de fabrique, designs, inventions, savoir-faire et l'assistance technique qui y est étroitement liée.

TABLE OF CONTENTS - TABLE DES MATIÈRES

LIST OF TABLES BY COUNTRY

LISTE DES TABLEAUX PAR PAYS

ABBREVIATIONS

BE	Business enterprise sector
BEMP	Business enterprise R&D personnel
BERD	Expenditure on R&D in the business enterprise sector
FTE	Full-time equivalent (on R&D)
GBAORD	Government budget appropriations or outlays for R&D
GDP	Gross domestic product
GERD	Gross domestic expenditure on R&D
GOVERD	Government intramural R&D expenditure
GUF	General university funds
HE	Higher education
HEMP	Higher education R&D personnel
HERD	Expenditure on R&D in the higher education sector
ISIC	International standard industrial classification
N.E.C.	Not elsewhere classified
NSE	Natural sciences and engineering
PNP	Private non-profit sector
PPP	Purchasing power parities
R&D	Research and experimental development
RSE	R&D scientists and engineers, researchers
SSH	Social sciences and humanities
TBP	Technology balance of payments
..	Data not available

For further explanations of the above terms, see the standard OECD methodology for the collection of R&D statistics entitled, *The Measurement of Scientific and Technical Activities: Proposed Standard Practice for Surveys of Research and Experimental Development -- Frascati Manual 1993*, known simply as the "*Frascati Manual*".

STANDARD FOOTNOTES

a) Break in series with previous year for which data is available.
b) Secretariat estimate or projection based on national sources.
c) National estimate or projection adjusted, if necessary, by the Secretariat to meet OECD norms.
d) Defence excluded (all or mostly).
e) National results adjusted by the Secretariat to meet OECD norms.
f) Including R&D in the social sciences and humanities.
g) Excluding R&D in the social sciences and humanities.
h) Federal or central government only.
i) Excludes data for the R&D content of general payment to the higher education sector for combined education and research (public GUF).
j) Excludes most or all capital expenditure.
k) Total intramural R&D expenditure instead of current intramural R&D expenditure.
l) Overestimated or based on overestimated data (see General Methodology).
m) Underestimated or based on underestimated data (see General Methodology).
n) Included elsewhere.
o) Includes other classes.
p) Provisional.
q) At current exchange rate and not at current purchasing power parities.
r) Note not currently used.
s) Unrevised breakdown not adding to the revised total.
t) Do not correspond exactly to the OECD recommendations (see General Methodology).
u) Including extramural R&D expenditure.
v) Incomplete breakdown. The sum of the elements does not add to the total (see General Methodology).

ABRÉVIATIONS

BPT	Balance des paiements technologiques
CBPRD	Crédits budgétaires publics de R-D
CITI	Classification internationale type par industrie
DIRD	Dépense intérieure brute de R-D
DIRDE	Dépense intérieure brute de R-D du secteur des entreprises
DIRDES	Dépense intérieure brute de R-D du secteur de l'enseignement supérieur
DIRDET	Dépense intérieure brute de R-D du secteur de l'État
EPT	Équivalence plein-temps (pour la R-D)
ES	Enseignement supérieur
FGU	Fonds généraux des universités
ISBL	Institutions sans but lucratif
N.C.A.	Non classé ailleurs
PIB	Produit intérieur brut
PPA	Parités de pouvoir d'achat
R-D	Recherche et développement expérimental
SENI	Sciences exactes, naturelles et de l'ingénieur
SSH	Sciences sociales et humaines
..	Donnée non disponible

Pour plus de détails sur les termes ci-dessus, voir la méthodologie type de l'OCDE pour la collecte des statistiques de R-D intitulée *La mesure des activités scientifiques et techniques : méthode type proposée pour les enquêtes sur la recherche et le développement expérimental, Manuel de Frascati 1993*, méthode connue sous le nom de "*Manuel de Frascati*".

NOTES TYPES

a) Discontinuité dans la série avec l'année précédente pour laquelle les données sont disponibles.
b) Estimation ou projection du Secrétariat fondée sur des sources nationales.
c) Estimation ou projection nationale ajustée si nécessaire par le Secrétariat pour correspondre aux normes de l'OCDE.
d) Défense exclue (toute ou principalement).
e) Résultats nationaux ajustés par le Secrétariat selon les normes de l'OCDE.
f) SSH (sciences sociales et humaines) comprises.
g) SSH exclues.
h) Gouvernement fédéral ou central seulement.
i) À l'exclusion de la part de R-D des paiements généraux au secteur de l'enseignement supérieur pour l'enseignement et la recherche (FGU d'origine publique).
j) Dépenses en capital exclues (toutes ou en partie).
k) Dépenses totales intra-muros de R-D au lieu des dépenses courantes intra-muros de R-D.
l) Surestimé ou fondé sur des données surestimées (voir Méthodologie générale).
m) Sous-estimé ou fondé sur des données sous-estimées (voir Méthodologie générale).
n) Compris ailleurs.
o) Y compris d'autres catégories.
p) Provisoire.
q) Aux taux de change courants et non aux parités de pouvoir d'achat (PPA).
r) Lettre non utilisée en note actuellement.
s) Sous-ventilation non révisée, ne correspondant plus tout à fait au total révisé.
t) Ne correspond pas tout à fait aux normes recommandées par l'OCDE (voir méthodologie générale).
u) Y compris les dépenses extra-muros de R-D.
v) Ventilation incomplète, la somme des éléments n'ajoute pas au total (voir méthodologie générale).

AVAILABILITY OF DATA BY COUNTRY - DISPONIBILITE DES DONNÉES PAR PAYS

	Table - Tableau																						
	1	2	3	4	5	6	7	8	9	10				11		12	13	14	15	16	17	18	19
Country / Pays										10.1	10.2	10.3	10.4	11.1	11.2								
Australia / Australie	x	x	x	x	x	-	x	x	x	x	x	x	x	x	x	x	x	x	x	x	x	x	/
Austria / Autriche	x	x	x	-	x	-	x	x	x	x	x	x	x	x	x	x	x	x	x	-	x	x	/
Belgium / Belgique	x	-	-	-	x	x	-	-	x	-	-	-	-	-	-	-	-	-	-	-	-	x	/
Canada	x	/	-	-	x	/	x	x	x	-	-	-	-	-	-	-	-	/	-	/	/	x	/
Czech R. / R. Tchèque	x	x	x	x	x	x	x	x	x	x	x	x	x	x	x	x	x	x	x	x	x	-	/
Denmark / Danemark	x	x	x	/	x	/	x	x	x	x	x	x	x	x	x	/	/	x	x	x	x	x	/
Finland / Finlande	/	x	-	-	-	x	/	x	x	x	x	x	x	x	x	x	x	x	x	x	/	x	/
France	x	x	x	-	/	-	-	-	x	x	x	x	x	x	x	x	x	-	-	-	-	x	/
Germany / Allemagne	x	x	/	-	x	x	x	x	x	x	x	x	x	x	x	x	x	x	x	-	x	x	/
Greece / Grèce	x	x	-	-	x	x	-	-	x	-	-	-	-	-	-	-	-	-	-	-	-	x	/
Hungary / Hongrie	x	x	x	-	x	-	x	/	x	-	-	-	-	-	-	x	x	x	x	/	x	-	/
Iceland / Islande	x	x	x	x	x	x	x	x	x	x	x	x	x	x	x	x	x	x	x	x	x	x	-
Ireland / Irlande	x	/	/	/	x	-	/	/	x	-	-	-	-	-	-	-	-	/	/	/	/	x	-
Italy / Italie	x	-	/	-	x	x	-	-	x	-	-	-	-	-	-	-	/	-	-	-	-	x	/
Japan / Japon	x	x	x	/	x	-	x	x	x	x	x	x	x	x	x	x	x	x	x	x	x	x	/
Korea / Corée	x	x	x	x	x	-	/	/	x	x	x	x	x	x	x	x	x	x	-	x	-	x	-
Mexico / Mexique	x	/	/	/	/	/	/	/	x	/	/	/	/	/	/	/	/	/	-	/	/	x	/
Netherlands / Pays-Bas	x	-	-	-	x	-	-	-	x	-	-	-	-	-	-	-	-	-	-	-	-	x	/
N. Zealand / N-Zélande	x	-	-	-	x	/	-	-	x	-	-	-	-	-	-	-	-	-	-	-	-	x	/
Norway / Norvège	x	x	x	x	/	/	x	x	x	x	x	x	x	x	x	x	x	x	x	x	/	x	/
Portugal	x	x	x	/	x	x	x	/	x	x	x	x	x	x	x	x	x	x	x	x	/	x	/
Poland / Pologne	x	x	x	-	x	-	x	x	x	x	x	x	x	x	x	x	x	x	x	x	x	/	/
Slovak R./ R. Slovaque	x	x	x	x	x	x	x	x	x	x	x	x	x	x	x	x	x	x	x	x	x	x	/
Spain / Espagne	x	x	x	x	x	x	x	x	x	x	x	x	x	x	x	x	x	x	x	x	x	x	/
Sweden / Suède	x	/	/	/	/	x	/	/	x	x	x	x	x	x	x	x	x	x	/	x	x	x	/
Switzerland / Suisse	/	/	/	/	/	/	/	/	/	/	/	/	/	/	/	/	/	-	-	-	-	x	/
Turkey / Turquie	x	/	/	/	x	x	/	-	x	/	/	/	/	/	/	/	/	/	/	/	/	-	-
United Kingd. / R. Uni	x	/	/	/	/	-	/	/	x	x	x	-	x	-	-	-	-	-	-	-	-	x	/
USA / États-Unis	x	/	x	-	/	/	/	-	x	x	x	-	-	x	-	-	x	x	-	-	-	x	x

x Data available (for 1997, 1998 or 1999 at least).
Données disponibles (pour 1997, 1998 ou 1999 au moins).

/ Data partially available (incomplete breakdown or data previous 1997).
Données partiellement disponibles (ventilation incomplète ou données antérieures à 1997).

- Data not available. The corresponding table has been suppressed if there is no other information on the page.
Données non disponibles. Le tableau correspondant a été supprimé si la page ne contient pas d'autre information.

AUSTRALIA Table 1 - Tableau 1 **AUSTRALIE**

Gross domestic expenditure on R&D (GERD) - Dépenses intérieures brutes de R-D (DIRD)

by sector of performance and source of funds - par secteur d'exécution et par source de financement

	1994	1995	1996	1997	1998	1999	2000	
	1.1 Million AUD - Millions AUD							
Business enterprise								**Entreprises**
financed by								*financé par*
Business enterprise	3 256.4	4 045.0	3 936.0	3 966.1	3 769.7	3 686.9	..	Entreprises
Direct government	107.2	104.5	108.4	99.6	121.5	134.6	..	État direct
Higher education	8.0	6.3	5.6	5.5	6.2	8.2	..	Enseignement supérieur
Private non-profit	26.6	68.0	51.7	37.0	29.9	25.5	..	Institutions sans but lucratif
Funds from abroad	110.2	132.9	135.4	111.0	140.4	190.2	..	Fonds de l'étranger
Total	3 508.3	4 356.6	4 237.2	4 219.1	4 067.6	4 045.3	..	*Total*
Government								**État**
financed by								*financé par*
Business enterprise	117.2	..	118.8	..	129.7	..	..	Entreprises
Direct government	1 690.8	..	1 767.0	..	1 705.6	..	..	État direct
Higher education	6.1	..	8.1	..	7.3	..	..	Enseignement supérieur
Private non-profit	149.4	..	167.6	..	198.1	..	..	Institutions sans but lucratif
Funds from abroad	12.6	..	15.3	..	30.9	..	..	Fonds de l'étranger
Total	1 976.0	..	2 076.9	..	2 071.6	..	..	*Total*
Higher education								**Enseignement supérieur**
financed by								*financé par*
Business enterprise	63.9	95.1	120.7	..	135.8	..	..	Entreprises
Direct government	510.1	505.8	575.8	..	673.7	..	..	État direct
General university funds	1 165.8	1 347.6	1 508.3	..	1 661.1	..	..	Fonds généraux des universités
Sub-total government	1 675.9	1 853.4	2 084.1	..	2 334.8	..	..	Sous-Total État
Higher education	..	..	..	..	..	..	..	Enseignement supérieur
Private non-profit	71.2	67.5	78.2	..	90.9	..	..	Institutions sans but lucratif
Funds from abroad	18.5	23.1	24.6	..	41.2	..	..	Fonds de l'étranger
Total	1 829.6	2 039.1	2 307.6	..	2 602.7	..	..	*Total*
Private non-profit								**Institutions sans but lucratif**
financed by								*financé par*
Business enterprise	13.4	..	29.7	..	27.3	..	..	Entreprises
Direct government	68.3	..	68.6	..	70.3	..	..	État direct
Higher education	2.1	..	2.4	..	3.4	..	..	Enseignement supérieur
Private non-profit	64.9	..	67.4	..	69.8	..	..	Institutions sans but lucratif
Funds from abroad	4.0	..	5.3	..	13.1	..	..	Fonds de l'étranger
Total	152.7	..	173.4	..	183.9	..	..	*Total*
GERD								**DIRD**
financed by								*financé par*
Business enterprise	3 450.9	..	4 205.2	..	4 062.5	..	..	Entreprises
Direct government	2 376.4	..	2 519.8	..	2 571.1	..	..	État direct
General university funds	1 165.8	..	1 508.3	..	1 661.1	..	..	Fonds généraux des universités
Sub-total government	3 542.2	..	4 028.1	..	4 232.2	..	..	Sous-Total État
Higher education	16.2	..	16.1	..	16.9	..	..	Enseignement supérieur
Private non-profit	312.1	..	364.9	..	388.7	..	..	Institutions sans but lucratif
Funds from abroad	145.3	..	180.6	..	225.6	..	..	Fonds de l'étranger
Total	7 466.6	..	8 795.1	..	8 925.8	..	..	*Total*
	1.2 Million constant USD (1995 prices and PPPs) - Millions de USD constants (prix et PPA de 1995)							
Business enterprise								**Entreprises**
financed by								*financé par*
Business enterprise	2 596.6	3 142.8	3 018.7	3 005.3	2 854.1	2 743.5	..	Entreprises
Direct government	85.5	81.2	83.1	75.5	92.0	100.2	..	État direct
Higher education	6.3	4.9	4.3	4.2	4.7	6.1	..	Enseignement supérieur
Private non-profit	21.2	52.8	39.7	28.0	22.6	19.0	..	Institutions sans but lucratif
Funds from abroad	87.8	103.2	103.8	84.1	106.3	141.5	..	Fonds de l'étranger
Total	2 797.4	3 384.9	3 249.7	3 197.0	3 079.6	3 010.1	..	*Total*
Government								**État**
financed by								*financé par*
Business enterprise	93.5	..	91.1	..	98.2	..	..	Entreprises
Direct government	1 348.2	..	1 355.2	..	1 291.3	..	..	État direct
Higher education	4.9	..	6.2	..	5.5	..	..	Enseignement supérieur
Private non-profit	119.1	..	128.5	..	150.0	..	..	Institutions sans but lucratif
Funds from abroad	10.0	..	11.7	..	23.4	..	..	Fonds de l'étranger
Total	1 575.6	..	1 592.9	..	1 568.4	..	..	*Total*
Higher education								**Enseignement supérieur**
financed by								*financé par*
Business enterprise	51.0	73.9	92.6	..	102.8	..	..	Entreprises
Direct government	406.7	393.0	441.6	..	510.1	..	..	État direct
General university funds	929.6	1 047.0	1 156.8	..	1 257.6	..	..	Fonds généraux des universités
Sub-total government	1 336.3	1 440.0	1 598.4	..	1 767.7	..	..	Sous-Total État
Higher education	..	..	..	..	..	..	..	Enseignement supérieur
Private non-profit	56.8	52.4	60.0	..	68.8	..	..	Institutions sans but lucratif
Funds from abroad	14.8	17.9	18.9	..	31.2	..	..	Fonds de l'étranger
Total	1 458.9	1 584.3	1 769.8	..	1 970.5	..	..	*Total*
Private non-profit								**Institutions sans but lucratif**
financed by								*financé par*
Business enterprise	10.7	..	22.8	..	20.7	..	..	Entreprises
Direct government	54.5	..	52.6	..	53.2	..	..	État direct
Higher education	1.7	..	1.8	..	2.6	..	..	Enseignement supérieur
Private non-profit	51.7	..	51.7	..	52.8	..	..	Institutions sans but lucratif
Funds from abroad	3.2	..	4.1	..	9.9	..	..	Fonds de l'étranger
Total	121.8	..	133.0	..	139.2	..	..	*Total*
GERD								**DIRD**
financed by								*financé par*
Business enterprise	2 751.6	..	3 225.2	..	3 075.7	..	..	Entreprises
Direct government	1 894.8	..	1 932.6	..	1 946.6	..	..	État direct
General university funds	929.6	..	1 156.8	..	1 257.6	..	..	Fonds généraux des universités
Sub-total government	2 824.4	..	3 089.3	..	3 204.2	..	..	Sous-Total État
Higher education	12.9	..	12.3	..	12.8	..	..	Enseignement supérieur
Private non-profit	248.9	..	279.9	..	294.3	..	..	Institutions sans but lucratif
Funds from abroad	115.8	..	138.5	..	170.8	..	..	Fonds de l'étranger
Total	5 953.6	..	6 745.4	..	6 757.8	..	..	*Total*

Source: OECD, R&D database (DSTI/EAS Division), January 2002.

Source : OCDE, base de données R-D (DSTI/Division AES), janvier 2002.

Gross domestic expenditure on R&D (GERD) - Dépenses intérieures brutes de R-D (DIRD)

by sector of performance and type of costs - par secteur d'exécution et par type de coûts

Million AUD *Millions AUD*

	1994	1995	1996	1997	1998	1999	2000	
Business enterprise								**Entreprises**
Labour costs	1 444.9	1 601.5	1 591.4	1 598.8	1 647.6	1 796.7	..	Coûts du personnel
Other current costs	1 631.2	2 224.4	2 068.2	2 083.5	1 965.7	1 854.7	..	Autres coûts courants
Total current expenditure	3 076.1	3 825.9	3 659.6	3 682.3	3 613.2	3 651.4	..	*Total dépenses courantes*
Land and buildings	78.9	59.0	34.7	68.5	80.4	83.3	..	Terrains et constructions
Instruments and equipment	353.3	471.7	542.9	468.4	374.1	310.6	..	Instruments et équipements
Total capital expenditure	432.2	530.7	577.6	536.8	454.4	393.9	..	*Total dépenses en capital*
Total	3 508.3	4 356.6	4 237.2	4 219.1	4 067.6	4 045.3	..	*Total*
Government								**État**
Labour costs	915.7	..	1 013.1	..	1 075.8	..	..	Coûts du personnel
Other current costs	796.5	..	790.3	..	835.4	..	..	Autres coûts courants
Total current expenditure	1 712.2	..	1 803.4	..	1 911.2	..	..	*Total dépenses courantes*
Land and buildings	123.2	..	160.0	..	57.6	..	..	Terrains et constructions
Instruments and equipment	140.6	..	113.5	..	102.8	..	..	Instruments et équipements
Total capital expenditure	263.8	..	273.5	..	160.4	..	..	*Total dépenses en capital*
Total	1 976.0	..	2 076.9	..	2 071.6	..	..	*Total*
Higher education								**Enseignement supérieur**
Labour costs	913.2	976.3	1 049.1	..	1 219.3	..	..	Coûts du personnel
Other current costs	751.4	919.0	1 080.1	..	1 203.2	..	..	Autres coûts courants
Total current expenditure	1 664.6	1 895.3	2 129.2	..	2 422.5	..	..	*Total dépenses courantes*
Land and buildings	47.9	27.9	47.4	..	34.8	..	..	Terrains et constructions
Instruments and equipment	117.0	115.9	131.0	..	145.5	..	..	Instruments et équipements
Total capital expenditure	165.0	143.8	178.4	..	180.3	..	..	*Total dépenses en capital*
Total	1 829.6	2 039.1	2 307.6	..	2 602.7	..	..	*Total*
Private non-profit								**Institutions sans but lucratif**
Labour costs	67.9	..	90.0	..	92.2	..	..	Coûts du personnel
Other current costs	54.9	..	62.3	..	73.5	..	..	Autres coûts courants
Total current expenditure	122.8	..	152.3	..	165.6	..	..	*Total dépenses courantes*
Land and buildings	16.6	..	8.1	..	3.5	..	..	Terrains et constructions
Instruments and equipment	13.3	..	13.0	..	14.8	..	..	Instruments et équipements
Total capital expenditure	29.9	..	21.1	..	18.3	..	..	*Total dépenses en capital*
Total	152.7	..	173.4	..	183.9	..	..	*Total*
GERD								**DIRD**
Labour costs	3 341.7	..	3 743.6	..	4 034.9	..	..	Coûts du personnel
Other current costs	3 234.0	..	4 000.9	..	4 077.8	..	..	Autres coûts courants
Total current expenditure	6 575.7	..	7 744.5	..	8 112.5	..	..	*Total dépenses courantes*
Land and buildings	266.6	..	250.2	..	176.3	..	..	Terrains et constructions
Instruments and equipment	624.2	..	800.4	..	637.2	..	..	Instruments et équipements
Total capital expenditure	890.9	..	1 050.6	..	813.4	..	..	*Total dépenses en capital*
Total	7 466.6	..	8 795.1	..	8 925.8	..	..	*Total*

Table 3 - Tableau 3

Current domestic expenditure on R&D - Dépenses intérieures courantes de R-D

by sector of performance and type of activity - par secteur d'exécution et par type d'activités

Million AUD *Millions AUD*

	1994	1995	1996	1997	1998	1999	2000	
Business enterprise								**Entreprises**
Basic research	215.5 k	..	200.9 k	..	134.5 k	..	..	Recherche fondamentale
Applied research	873.5 k	..	995.9 k	..	784.6 k	..	..	Recherche appliquée
Experimental development	2 419.4 k	..	3 040.4 k	..	3 148.5 k	..	..	Développement expérimental
Not specified	..	4 356.6 k	..	4 219.1 k	..	4 045.3 k	..	Non spécifié
Total	3 508.3 k	4 356.6 k	4 237.2 k	4 219.1 k	4 067.6 k	4 045.3 k	..	*Total*
Government								**État**
Basic research	579.0 k	..	588.7 k	..	594.3 k	..	..	Recherche fondamentale
Applied research	1 119.6 k	..	1 226.8 k	..	1 156.2 k	..	..	Recherche appliquée
Experimental development	277.4 k	..	261.4 k	..	321.1 k	..	..	Développement expérimental
Not specified	..	..	..	..	..	..	..	Non spécifié
Total	1 976.0 k	..	2 076.9 k	..	2 071.6 k	..	..	*Total*
Higher education								**Enseignement supérieur**
Basic research	1 113.4 k	1 261.6 k	1 363.4 k	..	1 523.7 k	..	..	Recherche fondamentale
Applied research	602.8 k	654.5 k	800.7 k	..	920.4 k	..	..	Recherche appliquée
Experimental development	113.4 k	123.0 k	143.6 k	..	158.6 k	..	..	Développement expérimental
Not specified	..	..	..	..	..	..	..	Non spécifié
Total	1 829.6 k	2 039.1 k	2 307.6 k	..	2 602.7 k	..	..	*Total*
Private non-profit								**Institutions sans but lucratif**
Basic research	110.7 k	..	106.8 k	..	121.7 k	..	..	Recherche fondamentale
Applied research	30.0 k	..	48.2 k	..	48.9 k	..	..	Recherche appliquée
Experimental development	12.0 k	..	18.4 k	..	13.3 k	..	..	Développement expérimental
Not specified	..	..	..	..	..	..	..	Non spécifié
Total	152.7 k	..	173.4 k	..	183.9 k	..	..	*Total*
GERD								**DIRD**
Basic research	2 018.6 k	..	2 259.8 k	..	2 374.2 k	..	..	Recherche fondamentale
Applied research	2 625.9 k	..	3 071.6 k	..	2 910.1 k	..	..	Recherche appliquée
Experimental development	2 822.2 k	..	3 463.8 k	..	3 641.5 k	..	..	Développement expérimental
Not specified	..	..	..	..	..	..	..	Non spécifié
Total	7 466.6 k	..	8 795.1 k	..	8 925.8 k	..	..	*Total*

Source: OECD, R&D database (DSTI/EAS Division), January 2002.

Source : OCDE, base de données R-D (DSTI/Division AES), janvier 2002.

AUSTRALIA — Table 4 - Tableau 4 — AUSTRALIE

Gross domestic expenditure on R&D (GERD) - Dépenses intérieures brutes de R-D (DIRD)

by sector of performance and socio-economic objective - par secteur d'exécution et par objectif socio-économique

Million AUD — *Millions AUD*

	1994	1995	1996	1997	1998	1999	2000	
Business enterprise								**Entreprises**
Agriculture, forestry and fishing	69.1	78.6	59.5	..	77.4	78.9	..	Agriculture, sylviculture et pêche
Industrial development	2 726.5	3 437.9	3 347.5	..	3 225.1	3 144.1	..	Développement industriel
Energy	267.5	316.7	297.9	..	233.3	213.6	..	Energie
Transport and telecommunications	63.0	146.9	93.8	..	92.8	125.5	..	Transports et télécommunications
Urban and rural planning	4.4	7.1	3.0	..	3.2	5.1	..	Aménagement urbain et rural
Prevention of pollution	.. n	.. n	.. n	..	.. n	.. n	..	Prévention de la pollution
Identification & treatment of pollution	51.3 o	48.9 o	47.3 o	..	99.3 o	62.1 o	..	Détection pollu. & mesures antipollu.
Health	86.4	112.2	104.0	..	132.0	202.3	..	Santé
Social development and services	18.3	14.8	15.2	..	24.8	24.1	..	Développement et services sociaux
Earth and atmosphere	28.1	20.1	25.8	..	16.8	31.5	..	Milieu terrestre et atmosphère
Advancement of knowledge	51.6	51.3	47.5	..	24.9	4.1	..	Promotion générale des connaissances
Civil space	..	..	..	..	..	..	..	Espace civil
Defence	142.1	121.9	195.9	..	138.0	154.0	..	Défense
Not elsewhere classified	..	..	..	4 219.1	..	..	..	Non classé
Total	3 508.3	4 356.6	4 237.2	4 219.1	4 067.6	4 045.3	..	*Total*
Government								**État**
Agriculture, forestry and fishing	537.4	..	574.0	..	590.2	..	..	Agriculture, sylviculture et pêche
Industrial development	388.6	..	413.7	..	435.6	..	..	Développement industriel
Energy	99.6	..	117.8	..	75.3	..	..	Energie
Transport and telecommunications	23.4	..	20.4	..	27.9	..	..	Transports et télécommunications
Urban and rural planning	8.0	..	6.2	..	10.4	..	..	Aménagement urbain et rural
Prevention of pollution	.. n	..	.. n	..	.. n	..	..	Prévention de la pollution
Identification & treatment of pollution	182.0 o	..	155.7 o	..	156.0 o	..	..	Détection pollu. & mesures antipollu.
Health	207.5	..	194.1	..	197.2	..	..	Santé
Social development and services	48.6	..	50.0	..	40.5	..	..	Développement et services sociaux
Earth and atmosphere	188.2	..	256.6	..	272.8	..	..	Milieu terrestre et atmosphère
Advancement of knowledge	66.6	..	54.4	..	60.7	..	..	Promotion générale des connaissances
Civil space	..	..	..	..	..	..	..	Espace civil
Defence	226.2	..	234.0	..	205.1	..	..	Défense
Not elsewhere classified	..	..	..	..	..	..	..	Non classé
Total	1 976.0	..	2 076.9	..	2 071.6	..	..	*Total*
Higher education								**Enseignement supérieur**
Agriculture, forestry and fishing	93.2	107.6	120.3	..	158.9	..	..	Agriculture, sylviculture et pêche
Industrial development	235.1	256.9	295.3	..	364.3	..	..	Développement industriel
Energy	29.5	26.6	39.9	..	38.3	..	..	Energie
Transport and telecommunications	7.0	11.9	16.4	..	26.1	..	..	Transports et télécommunications
Urban and rural planning	9.3	11.0	13.4	..	16.5	..	..	Aménagement urbain et rural
Prevention of pollution	.. n	.. n	.. n	..	.. n	..	..	Prévention de la pollution
Identification & treatment of pollution	46.3 o	54.2 o	63.7 o	..	84.7 o	..	..	Détection pollu. & mesures antipollu.
Health	305.8	356.9	413.5	..	506.7	..	..	Santé
Social development and services	142.8	146.0	168.6	..	203.5	..	..	Développement et services sociaux
Earth and atmosphere	74.8	91.0	107.0	..	103.9	..	..	Milieu terrestre et atmosphère
Advancement of knowledge	881.1	970.4	1 062.3	..	1 094.1	..	..	Promotion générale des connaissances
Civil space	..	..	..	..	..	..	..	Espace civil
Defence	4.8	6.5	7.2	..	5.8	..	..	Défense
Not elsewhere classified	..	..	..	..	..	..	..	Non classé
Total	1 829.6	2 039.1	2 307.6	..	2 602.7	..	..	*Total*
Private non-profit								**Institutions sans but lucratif**
Agriculture, forestry and fishing	0.0	..	0.0	..	1.9	..	..	Agriculture, sylviculture et pêche
Industrial development	6.1	..	4.2	..	1.8	..	..	Développement industriel
Energy	0.3	..	0.0	..	0.0	..	..	Energie
Transport and telecommunications	0.2	..	0.0	..	0.0	..	..	Transports et télécommunications
Urban and rural planning	0.0	..	..	..	..	..	..	Aménagement urbain et rural
Prevention of pollution	.. n	..	.. n	..	.. n	..	..	Prévention de la pollution
Identification & treatment of pollution	1.0 o	..	0.0 o	..	0.4 o	..	..	Détection pollu. & mesures antipollu.
Health	121.8	..	149.4	..	155.6	..	..	Santé
Social development and services	10.9	..	9.6	..	14.8	..	..	Développement et services sociaux
Earth and atmosphere	1.3	..	1.2	..	1.7	..	..	Milieu terrestre et atmosphère
Advancement of knowledge	11.3	..	8.9	..	7.6	..	..	Promotion générale des connaissances
Civil space	..	..	..	..	..	..	..	Espace civil
Defence	0.0	..	..	..	..	..	..	Défense
Not elsewhere classified	..	..	..	..	..	..	..	Non classé
Total	152.7	..	173.4	..	183.9	..	..	*Total*
GERD								**DIRD**
Agriculture, forestry and fishing	699.7	..	753.8	..	828.4	..	..	Agriculture, sylviculture et pêche
Industrial development	3 356.3	..	4 060.7	..	4 026.8	..	..	Développement industriel
Energy	396.9	..	455.6	..	346.9	..	..	Energie
Transport and telecommunications	93.6	..	130.6	..	146.8	..	..	Transports et télécommunications
Urban and rural planning	21.7	..	22.6	..	30.1	..	..	Aménagement urbain et rural
Prevention of pollution	.. n	..	.. n	..	.. n	..	..	Prévention de la pollution
Identification & treatment of pollution	280.6 o	..	266.7 o	..	340.4 o	..	..	Détection pollu. & mesures antipollu.
Health	721.5	..	861.0	..	991.5	..	..	Santé
Social development and services	220.6	..	243.4	..	283.6	..	..	Développement et services sociaux
Earth and atmosphere	292.4	..	390.6	..	395.2	..	..	Milieu terrestre et atmosphère
Advancement of knowledge	1 010.6	..	1 173.1	..	1 187.3	..	..	Promotion générale des connaissances
Civil space	..	..	..	..	..	..	..	Espace civil
Defence	373.1	..	437.1	..	348.9	..	..	Défense
Not elsewhere classified	..	..	..	..	..	..	..	Non classé
Total	7 466.6	..	8 795.1	..	8 925.8	..	..	*Total*

Source: OECD, R&D database (DSTI/EAS Division), January 2002.

Source : OCDE, base de données R-D (DSTI/Division AES), janvier 2002.

Table 5 - Tableau 5

Total R&D personnel - Personnel total de R-D

by sector of employment and occupation - par secteur d'emploi et par profession

In full-time equivalent — *En équivalence plein-temps*

	1994	1995	1996	1997	1998	1999	2000	
Business enterprise								**Entreprises**
RSE	14 903	15 589	15 282	14 577	14 626	15 283	..	Chercheurs
Technicians	7 057	7 443	7 483	6 777	7 119	7 471	..	Techniciens
Other occupations	3 852	4 091	3 681	3 432	3 229	3 354	..	Autres professions
Total	25 812	27 123	26 446	24 786	24 974	26 109	..	*Total*
Government								**État**
RSE	8 806	..	9 130	..	8 860	..	..	Chercheurs
Technicians	6 843	..	6 641	..	7 034	..	..	Techniciens
Other occupations	3 659	..	3 617	..	3 051	..	..	Autres professions
Total	19 309	..	19 388	..	18 946	..	..	*Total*
Higher education								**Enseignement supérieur**
RSE	32 272	..	35 472	..	38 137	..	..	Chercheurs
Technicians	.. [n]	..	.. [n]	..	.. [n]	..	..	Techniciens
Other occupations	7 824 [o]	..	7 266 [o]	..	7 365 [o]	..	..	Autres professions
Total	40 096	..	42 739	..	45 502	..	..	*Total*
Private non-profit								**Institutions sans but lucratif**
RSE	892	..	1 189	..	1 167	..	..	Chercheurs
Technicians	545	..	699	..	664	..	..	Techniciens
Other occupations	229	..	283	..	238	..	..	Autres professions
Total	1 666	..	2 171	..	2 068	..	..	*Total*
TOTAL								**TOTAL**
RSE	56 873	..	61 073	..	62 790	..	..	Chercheurs
Technicians	.. [n]	..	.. [n]	..	.. [n]	..	..	Techniciens
Other occupations	30 010 [o]	..	29 671 [o]	..	28 700 [o]	..	..	Autres professions
Total	86 883	..	90 744	..	91 490	..	..	*Total*

Table 6 - Tableau 6

Total R&D personnel - Personnel total de R-D

by sector of employment and formal qualification - par secteur d'emploi et par niveau de qualification

In full-time equivalent — *En équivalence plein-temps*

	1994	1995	1996	1997	1998	1999	2000	
Business enterprise								**Entreprises**
University PhD level degrees	..	..	..	..	..	..	..	Diplômes univ. de niv. doctorat
Other university degrees	..	..	..	..	..	..	..	Autres diplômes universitaires
Sub-total university degrees	..	..	..	..	..	..	..	Sous-total Universitaires
Other post-secondary diplomas	..	..	..	..	..	..	..	Autres diplômes post-secondaires
Secondary diplomas	..	..	..	..	..	..	..	Diplômes de fin d'études secondaires
Other qualifications	..	..	..	..	..	..	..	Autres qualifications
Total	25 812	27 123	26 446	24 786	24 974	26 109	..	*Total*
Government								**État**
University PhD level degrees	..	..	..	..	..	..	..	Diplômes univ. de niv. doctorat
Other university degrees	..	..	..	..	..	..	..	Autres diplômes universitaires
Sub-total university degrees	..	..	..	..	..	..	..	Sous-total Universitaires
Other post-secondary diplomas	..	..	..	..	..	..	..	Autres diplômes post-secondaires
Secondary diplomas	..	..	..	..	..	..	..	Diplômes de fin d'études secondaires
Other qualifications	..	..	..	..	..	..	..	Autres qualifications
Total	19 309	..	19 388	..	18 946	..	..	*Total*
Higher education								**Enseignement supérieur**
University PhD level degrees	..	..	..	..	..	..	..	Diplômes univ. de niv. doctorat
Other university degrees	..	..	..	..	..	..	..	Autres diplômes universitaires
Sub-total university degrees	..	..	..	..	..	..	..	Sous-total Universitaires
Other post-secondary diplomas	..	..	..	..	..	..	..	Autres diplômes post-secondaires
Secondary diplomas	..	..	..	..	..	..	..	Diplômes de fin d'études secondaires
Other qualifications	..	..	..	..	..	..	..	Autres qualifications
Total	40 096	..	42 739	..	45 502	..	..	*Total*
Private non-profit								**Institutions sans but lucratif**
University PhD level degrees	..	..	..	..	..	..	..	Diplômes univ. de niv. doctorat
Other university degrees	..	..	..	..	..	..	..	Autres diplômes universitaires
Sub-total university degrees	..	..	..	..	..	..	..	Sous-total Universitaires
Other post-secondary diplomas	..	..	..	..	..	..	..	Autres diplômes post-secondaires
Secondary diplomas	..	..	..	..	..	..	..	Diplômes de fin d'études secondaires
Other qualifications	..	..	..	..	..	..	..	Autres qualifications
Total	1 666	..	2 171	..	2 068	..	..	*Total*
TOTAL								**TOTAL**
University PhD level degrees	..	..	..	..	..	..	..	Diplômes univ. de niv. doctorat
Other university degrees	..	..	..	..	..	..	..	Autres diplômes universitaires
Sub-total university degrees	..	..	..	..	..	..	..	Sous-total Universitaires
Other post-secondary diplomas	..	..	..	..	..	..	..	Autres diplômes post-secondaires
Secondary diplomas	..	..	..	..	..	..	..	Diplômes de fin d'études secondaires
Other qualifications	..	..	..	..	..	..	..	Autres qualifications
Total	86 883	..	90 744	..	91 490	..	..	*Total*

Source: OECD, R&D database (DSTI/EAS Division), January 2002.

Source : OCDE, base de données R-D (DSTI/Division AES), janvier 2002.

Table 7 - Tableau 7

Gross domestic expenditure on R&D (GERD) - Dépenses intérieures brutes de R-D (DIRD)

by sector of performance and main scientific field - par secteur d'exécution et par grand domaine scientifique

Million AUD — *Millions AUD*

	1994	1995	1996	1997	1998	1999	2000	
Business enterprise								**Entreprises**
Natural sciences and engineering	3 508.3	4 347.7	4 229.4	4 210.7	4 059.9	4 031.3	..	Sciences exactes, nat. et de l'ingénieur
Social sciences and humanities	..	8.8	7.8	8.5	7.7	14.1	..	Sciences sociales et humaines
All fields of science	3 508.3	4 356.6	4 237.2	4 219.1	4 067.6	4 045.3	..	*Tous domaines scientifiques*
Government								**État**
Natural sciences and engineering	1 892.2	..	1 985.8	..	1 997.3	..	..	Sciences exactes, nat. et de l'ingénieur
Social sciences and humanities	83.8	..	91.1	..	74.3	..	..	Sciences sociales et humaines
All fields of science	1 976.0	..	2 076.9	..	2 071.6	..	..	*Tous domaines scientifiques*
Higher education								**Enseignement supérieur**
Natural sciences and engineering	1 312.5	1 488.4	1 678.0	..	1 898.7	..	..	Sciences exactes, nat. et de l'ingénieur
Social sciences and humanities	517.1	550.7	629.5	..	704.0	..	..	Sciences sociales et humaines
All fields of science	1 829.6	2 039.1	2 307.6	..	2 602.7	..	..	*Tous domaines scientifiques*
Private non-profit								**Institutions sans but lucratif**
Natural sciences and engineering	139.8	..	162.5	..	170.7	..	..	Sciences exactes, nat. et de l'ingénieur
Social sciences and humanities	12.9	..	10.9	..	13.2	..	..	Sciences sociales et humaines
All fields of science	152.7	..	173.4	..	183.9	..	..	*Tous domaines scientifiques*
TOTAL								**TOTAL**
Natural sciences and engineering	6 862.4	..	8 055.7	..	8 126.6	..	..	Sciences exactes, nat. et de l'ingénieur
Social sciences and humanities	613.8	..	739.4	..	799.3	..	..	Sciences sociales et humaines
All fields of science	7 466.6	..	8 795.1	..	8 925.8	..	..	*Tous domaines scientifiques*

Table 8 - Tableau 8

Total R&D personnel - Personnel total de R-D

by sector of employment and main scientific field - par secteur d'emploi et par grand domaine scientifique

In full-time equivalent — *En équivalence plein-temps*

	1994	1995	1996	1997	1998	1999	2000	
Business enterprise								**Entreprises**
Natural sciences and engineering	25 744	27 030	26 365	24 697	24 897	25 968	..	Sciences exactes, nat. et de l'ingénieur
Social sciences and humanities	..	93	81	89	77	141	..	Sciences sociales et humaines
All fields of science	25 812	27 123	26 446	24 786	24 974	26 109	..	*Tous domaines scientifiques*
Government								**État**
Natural sciences and engineering	18 459	..	18 457	..	18 147	..	..	Sciences exactes, nat. et de l'ingénieur
Social sciences and humanities	850	..	930	..	799	..	..	Sciences sociales et humaines
All fields of science	19 309	..	19 388	..	18 946	..	..	*Tous domaines scientifiques*
Higher education								**Enseignement supérieur**
Natural sciences and engineering	26 098	..	27 684	..	28 985	..	..	Sciences exactes, nat. et de l'ingénieur
Social sciences and humanities	13 998	..	15 055	..	16 517	..	..	Sciences sociales et humaines
All fields of science	40 096	..	42 739	..	45 502	..	..	*Tous domaines scientifiques*
Private non-profit								**Institutions sans but lucratif**
Natural sciences and engineering	1 541	..	2 084	..	1 978	..	..	Sciences exactes, nat. et de l'ingénieur
Social sciences and humanities	126	..	87	..	90	..	..	Sciences sociales et humaines
All fields of science	1 666	..	2 171	..	2 068	..	..	*Tous domaines scientifiques*
TOTAL								**TOTAL**
Natural sciences and engineering	71 929	..	74 590	..	74 007	..	..	Sciences exactes, nat. et de l'ingénieur
Social sciences and humanities	14 974	..	16 154	..	17 483	..	..	Sciences sociales et humaines
All fields of science	86 883	..	90 744	..	91 490	..	..	*Tous domaines scientifiques*

Source: OECD, R&D database (DSTI/EAS Division), January 2002.

Source : OCDE, base de données R-D (DSTI/Division AES), janvier 2002.

AUSTRALIA Table 9 - Tableau 9 **AUSTRALIE**

Total business enterprise intramural expenditure on R&D - Dépenses intra-muros totales de R-D des entreprises
(BERD) by industry - (DIRDE) par industrie

	1994	1995	1996	1997	1998	1999	2000	
	9.1 Million AUD - Millions AUD							
1 AGRICULTURE, HUNTING & FORESTRY	..	..	..	..	..	..	..	1 AGRICULTURE, CHASSE & SYLVICULTURE
2 MINING	302.8	521.1	562.1	536.6	478.2	272.7	..	2 ACTIVITES EXTRACTIVES
3 MANUFACTURING	1 932.1	2 392.2	2 289.1	2 229.0	2 054.8	2 052.1	..	3 ACTIVITES DE FABRICATION
4 Food, beverages & tobacco	140.1	290.7	231.0	181.1	208.2	186.0	..	**4 Produits alimentaires, boissons & tabac**
5 Food products & beverages	135.5	..	..	..	..	..	..	5 Produits alimentaires & boissons
6 Tobacco products	4.7	..	..	..	..	..	..	6 Produits à base de tabac
7 Textiles, wear app., fur & leather	28.4	25.7	21.4	21.3	20.4	16.8	..	**7 Textiles, habill., fourrures & cuirs**
8 Textiles	15.9	11.7	10.7	12.2	11.8	10.1	..	8 Textiles
9 Wearing apparel & fur	4.5	4.2	3.9	4.9	3.9	3.5	..	9 Articles d'habillement & fourrures
10 Leather products & footwear	8.0	9.8	6.7	4.2	4.8	3.2	..	10 Articles en cuir & chaussures
11 Wood, paper, printing, publishing	94.3	206.1	206.8	134.3	105.8	118.6	..	**11 Bois, papier, imprimerie, édition**
12 Wood & cork (not furniture)	22.1	23.3	22.3	23.4	11.9	13.7	..	12 Bois & liège (sauf meubles)
13 Pulp, paper & paper products	57.2	161.9	..	..	77.0	90.2	..	13 Carton, papier & articles
14 Publ., print. & repro. of rec. media	15.1	21.0	14.8	17.3	20.1	14.7	..	14 Edition, impress. & repro. enregis.
15 Coke, petroleum, nuclear fuel, chemicals, rubber plastics	317.5	326.8	309.8	327.9	351.4	413.3	..	**15 Coke, pétrole, combust. nucl., prod. chimi., caout. & mat. plast.**
16 Coke, ref. petrol. prod. & nuclear fuel	5.5	8.3	5.3	16.1	16.5	17.8	..	16 Coke, raffin. pétr. & combust. nucl.
17 Chemicals & chemical products	278.8	282.5	274.6	282.4	309.2	361.1	..	17 Produits chimiques
18 *Chemical prod.(less pharmaceuticals)*	144.0	159.7	152.9	152.3	151.4	178.7	..	18 *Prod. chimi. (moins prod. pharmaceutiques)*
19 *Pharmaceuticals*	134.8	122.8	121.7	130.2	157.8	182.5	..	19 *Produits pharmaceutiques*
20 Rubber & plastic products	33.3	36.1	29.9	29.4	25.6	34.4	..	20 Caoutchouc et matières plastiques
21 Non-metallic mineral products	45.3	80.4	62.7	69.8	52.9	48.9	..	**21 Produits minéraux non métalliques**
22 Basic metals	274.1	331.7	359.5	328.2	257.0	179.2	..	**22 Produits métallurgiques de base**
23 Basic metals, ferrous	157.0	223.4	210.4	194.1	159.0	104.4	..	23 Produits métallur. de base, ferreux
24 Basic metals, non-ferrous	117.0	108.3	149.1	134.1	97.9	74.7	..	24 Prod. métallur. de base, non ferreux
25 Fabricated metal products	50.1	50.4	43.9	40.3	38.6	47.6	..	**25 Ouvrages en métaux (sauf mach.& mat.)**
26 Machinery equipment, instruments and transport equipment	963.1	1 063.9	1 008.2	1 090.1	1 001.4	1 021.6	..	**26 Machines & matériel, instruments & mat. de transport**
27 Machinery n.e.c.	93.3	128.4	142.7	126.4	115.2	131.9	..	27 Machines, n.c.a.
28 Office, account. & computing machinery	43.3	27.8	25.5	48.8	34.5	39.6	..	28 Mach. de bur., compta. & trait. inform.
29 Electrical machinery	78.6	78.4	69.9	80.2	70.6	76.1	..	29 Machines électriques
30 Electro. equip. (radio, TV & commun.)	302.0	282.6	277.9	298.1	294.5	226.7	..	30 Equip. électro. (radio, tv & communic.)
31 *Electro. comp. (inc. semi-conductors)*	115.0	124.0	125.1	112.3	120.1	95.3	..	31 *Compos. électro. (semi-cond. incl.)*
32 *TV, radio & communications equip.*	187.0	158.6	152.8	185.8	174.4	131.4	..	32 *Appareils radio, tv et de communi.*
33 Instruments	114.4	131.4	91.0	95.2	106.8	127.4	..	33 Instruments
34 Motor vehicles	238.5	317.9	313.1	359.5	317.4	348.0	..	34 Véhicules automobiles
35 Other transport equipment	93.1	97.4	88.1	81.9	62.3	72.0	..	35 Autres matériels de transport
36 *Ships*	..	58.1	42.7	61.9	39.9	47.9	..	36 *Navires*
37 *Aerospace*	10.2	13.9	9.4	..	..	..	..	37 *Aérospatiale*
38 *Other transport equipment n.e.c.*	11.3	25.4	35.9	..	..	..	..	38 *Autres matériels de transport n.c.a.*
39 Furniture, other manufacturing n.e.c.	19.2	16.4	45.8	36.1	19.2	20.0	..	**39 Meubles, autres activ. de fabricat. n.c.a.**
40 Furniture	6.8	5.9	4.7	5.6	6.5	7.6	..	40 Meubles
41 Other manufacturing n.e.c.	12.4	10.5	41.1	30.5	12.7	12.4	..	41 Autres activités de fabrication n.c.a.
42 Recycling	0.0	0.0	0.0	0.0	0.0	0.0	..	**42 Récupération**
43 ELECTRICITY, GAS & WATER SUPPLY	129.0	38.0	26.2	24.8	29.3	24.8	..	43 DISTRIBUTION D'ELECTRICITE, GAZ & EAU
44 CONSTRUCTION	12.7	44.5	56.3	43.5	37.7	64.3	..	44 CONSTRUCTION
45 SERVICES SECTOR	1 131.8	1 360.7	1 303.6	1 385.2	1 467.6	1 631.5	..	45 SECTEUR DES SERVICES
46 Wholesale, retail trade & repair of motor vehicles, etc.	192.9	329.2	335.7	315.2	337.2	352.4	..	**46 Commerce (gros & détail), réparation de véhicules, etc.**
47 Hotels & restaurants	..	0.0	0.0	0.0	0.0	0.0	..	**47 Hôtels et restaurants**
48 Transport & storage	..	26.0	18.6	46.7	30.5	..	..	**48 Transports, entreposage**
49 Communications	..	96.4	..	..	..	..	..	**49 Communications**
50 Post	..	11.9	..	..	..	..	..	50 Postes
51 Telecommunications	..	84.5	67.2	125.2	196.9	115.3	..	51 Télécommunications
52 Financial intermediation (inc. insur.)	100.9	123.9	96.1	84.9	81.4	125.7	..	**52 Intermédiation financ. (assurances inc.)**
53 Real estate, renting & busin. activity	718.6	766.4	742.3	755.8	773.3	933.2	..	**53 Immobil., locat. & serv. aux entreprises**
54 Computer & related activities	318.3	363.6	392.6	368.5	425.6	509.4	..	54 Informatique et activités connexes
55 *Software consultancy*	300.2	343.8	362.7	334.0	401.0	489.0	..	55 *Production de logiciels*
56 *Other computer services n.e.c.*	18.1	19.8	30.0	34.5	24.6	20.4	..	56 *Autres activités informatiques*
57 Research & development	118.4	139.3	137.4	154.0	161.0	210.1	..	57 Recherche & développement expérimental
58 Other business activities n.e.c.	281.8	263.5	212.3	233.2	186.7	213.7	..	58 Autres activ. & serv. aux entreprises
59 Comm., soc. & pers. serv. activ., etc.	18.1	18.9	21.6	28.2	29.6	48.8	..	**59 Adm. pub., serv. soc. & collec., etc**
TOTAL BUSINESS ENTERPRISE SECTOR	**3 508.3**	**4 356.6**	**4 237.2**	**4 219.1**	**4 067.6**	**4 045.3**	..	TOTAL SECTEUR DES ENTREPRISES

Source: OECD, R&D database (DSTI/EAS Division), January 2002.

Source : OCDE, base de données R-D (DSTI/Division AES), janvier 2002.

Total business enterprise intramural expenditure on R&D - Dépenses intra-muros totales de R-D des entreprises
(BERD) by industry - (DIRDE) par industrie

	1994	1995	1996	1997	1998	1999	2000	
	9.2 Million constant USD(1995 prices and PPPs) - Millions de USD constants (prix et PPA de 1995)							
1 AGRICULTURE, HUNTING & FORESTRY	..	..	..	..	..	..	..	**1 AGRICULTURE, CHASSE & SYLVICULTURE**
2 MINING	241.4	404.9	431.1	406.6	362.1	202.9	..	**2 ACTIVITES EXTRACTIVES**
3 MANUFACTURING	1 540.6	1 858.6	1 755.6	1 689.0	1 555.7	1 527.0	..	**3 ACTIVITES DE FABRICATION**
4 Food, beverages & tobacco	111.7	225.9	177.1	137.2	157.6	138.4	..	**4 Produits alimentaires, boissons & tabac**
5 Food products & beverages	108.0	..	..	..	..	..	..	5 Produits alimentaires & boissons
6 Tobacco products	3.7	..	..	..	..	..	..	6 Produits à base de tabac
7 Textiles, wear app., fur & leather	22.7	20.0	16.4	16.1	15.5	12.5	..	**7 Textiles, habill., fourrures & cuirs**
8 Textiles	12.6	9.1	8.2	9.2	8.9	7.5	..	8 Textiles
9 Wearing apparel & fur	3.6	3.3	3.0	3.7	2.9	2.6	..	9 Articles d'habillement & fourrures
10 Leather products & footwear	6.4	7.6	5.1	3.2	3.6	2.4	..	10 Articles en cuir & chaussures
11 Wood, paper, printing, publishing	75.2	160.1	158.6	101.7	80.1	88.2	..	**11 Bois, papier, imprimerie, édition**
12 Wood & cork (not furniture)	17.6	18.1	17.1	17.7	9.0	10.2	..	12 Bois & liège (sauf meubles)
13 Pulp, paper & paper products	45.6	125.8	..	..	58.3	67.1	..	13 Carton, papier & articles
14 Publ., print. & repro. of rec. media	12.0	16.3	11.4	13.1	15.2	11.0	..	14 Edition, impress. & repro. enregis.
15 Coke, petroleum, nuclear fuel, chemicals, rubber plastics	253.2	253.9	237.6	248.5	266.0	307.6	..	**15 Coke, pétrole, combust. nucl., prod. chimi., caout. & mat. plast.**
16 Coke, ref. petrol. prod. & nuclear fuel	4.4	6.4	4.1	12.2	12.5	13.3	..	16 Coke, raffin. pétr. & combust. nucl.
17 Chemicals & chemical products	222.3	219.5	210.6	214.0	234.1	268.7	..	17 Produits chimiques
18 *Chemical prod.(less pharmaceuticals)*	114.8	124.1	117.3	115.4	114.6	132.9	..	18 *Prod. chimi. (moins prod. pharmaceutiques)*
19 *Pharmaceuticals*	107.5	95.4	93.4	98.6	119.5	135.8	..	19 *Produits pharmaceutiques*
20 Rubber & plastic products	26.5	28.0	22.9	22.3	19.4	25.6	..	20 Caoutchouc et matières plastiques
21 Non-metallic mineral products	36.1	62.4	48.1	52.9	40.1	36.4	..	**21 Produits minéraux non métalliques**
22 Basic metals	218.5	257.7	275.8	248.7	194.6	133.3	..	**22 Produits métallurgiques de base**
23 Basic metals, ferrous	125.2	173.6	161.4	147.1	120.4	77.7	..	23 Produits métallur. de base, ferreux
24 Basic metals, non-ferrous	93.3	84.2	114.4	101.6	74.2	55.6	..	24 Prod. métallur. de base, non ferreux
25 Fabricated metal products	39.9	39.2	33.7	30.5	29.2	35.4	..	**25 Ouvrages en métaux (sauf mach.& mat.)**
26 Machinery equipment, instruments and transport equipment	768.0	826.6	773.2	826.0	758.2	760.2	..	**26 Machines & matériel, instruments & mat. de transport**
27 Machinery n.e.c.	74.4	99.7	109.5	95.8	87.2	98.1	..	27 Machines, n.c.a.
28 Office, account. & computing machinery	34.5	21.6	19.6	37.0	26.1	29.5	..	28 Mach. de bur., compta. & trait. inform.
29 Electrical machinery	62.7	60.9	53.6	60.8	53.5	56.7	..	29 Machines électriques
30 Electro. equip. (radio, TV & commun.)	240.8	219.6	213.1	225.9	223.0	168.7	..	30 Equip. électro. (radio, tv & communic.)
31 *Electro. comp. (inc. semi-conductors)*	91.7	96.4	96.0	85.1	90.9	70.9	..	31 *Compos. électro. (semi-cond. incl.)*
32 *TV, radio & communications equip.*	149.1	123.2	117.2	140.8	132.1	97.8	..	32 *Appareils radio, tv et de communi.*
33 Instruments	91.3	102.1	69.8	72.1	80.9	94.8	..	33 Instruments
34 Motor vehicles	190.1	247.0	240.1	272.4	240.3	258.9	..	34 Véhicules automobiles
35 Other transport equipment	74.2	75.7	67.5	62.0	47.1	53.6	..	35 Autres matériels de transport
36 *Ships*	..	45.1	32.8	46.9	30.2	35.6	..	36 *Navires*
37 *Aerospace*	8.1	10.8	7.2	..	..	..	..	37 *Aérospatiale*
38 *Other transport equipment n.e.c.*	9.0	19.8	27.6	..	..	..	..	38 *Autres matériels de transport n.c.a.*
39 Furniture, other manufacturing n.e.c.	15.3	12.8	35.1	27.4	14.5	14.9	..	**39 Meubles, autres activ. de fabricat. n.c.a.**
40 Furniture	5.4	4.6	3.6	4.3	4.9	5.6	..	40 Meubles
41 Other manufacturing n.e.c.	9.9	8.2	31.5	23.1	9.6	9.2	..	41 Autres activités de fabrication n.c.a.
42 Recycling	0.0	0.0	0.0	0.0	0.0	0.0	..	**42 Récupération**
43 ELECTRICITY, GAS & WATER SUPPLY	102.8	29.5	20.1	18.8	22.2	18.4	..	**43 DISTRIBUTION D'ELECTRICITE, GAZ & EAU**
44 CONSTRUCTION	10.1	34.6	43.2	32.9	28.5	47.8	..	**44 CONSTRUCTION**
45 SERVICES SECTOR	902.5	1 057.2	999.8	1 049.6	1 111.1	1 214.0	..	**45 SECTEUR DES SERVICES**
46 Wholesale, retail trade & repair of motor vehicles, etc.	153.8	255.8	257.5	238.8	255.3	262.2	..	**46 Commerce (gros & détail), réparation de véhicules, etc.**
47 Hotels & restaurants	..	0.0	0.0	0.0	0.0	0.0	..	**47 Hôtels et restaurants**
48 Transport & storage	..	20.2	14.3	35.4	23.1	..	..	**48 Transports, entreposage**
49 Communications	..	74.9	..	..	..	..	..	**49 Communications**
50 Post	..	9.2	..	..	..	..	..	50 Postes
51 Telecommunications	..	65.6	51.5	94.9	149.1	85.8	..	51 Télécommunications
52 Financial intermediation (inc. insur.)	80.4	96.3	73.7	64.4	61.6	93.5	..	**52 Intermédiation financ. (assurances inc.)**
53 Real estate, renting & busin. activity	573.0	595.4	569.3	572.7	585.5	694.4	..	**53 Immobil., locat. & serv. aux entreprises**
54 Computer & related activities	253.8	282.5	301.1	279.3	322.2	379.0	..	54 Informatique et activités connexes
55 *Software consultancy*	239.4	267.1	278.2	253.1	303.6	363.8	..	55 *Production de logiciels*
56 *Other computer services n.e.c.*	14.4	15.4	23.0	26.1	18.7	15.2	..	56 *Autres activités informatiques*
57 Research & development	94.4	108.2	105.4	116.7	121.9	156.4	..	57 Recherche & développement expérimental
58 Other business activities n.e.c.	224.7	204.7	162.8	176.7	141.3	159.0	..	58 Autres activ. & serv. aux entreprises
59 Comm., soc. & pers. serv. activ., etc.	14.4	14.7	16.6	21.4	22.4	36.3	..	**59 Adm. pub., serv. soc. & collec., etc**
TOTAL BUSINESS ENTERPRISE SECTOR	**2 797.4**	**3 384.9**	**3 249.7**	**3 197.0**	**3 079.6**	**3 010.1**	..	**TOTAL SECTEUR DES ENTREPRISES**

Source: OECD, R&D database (DSTI/EAS Division), January 2002.

Source : OCDE, base de données R-D (DSTI/Division AES), janvier 2002.

AUSTRALIA — Table 10 - Tableau 10 — AUSTRALIE

Business enterprise intramural expenditure on R&D (BERD) - Dépenses intra-muros de R-D des entreprises (DIRDE)

by industry and source of funds - par industrie et par source de financement

Million AUD — *Millions AUD*

	1994	1995	1996	1997	1998	1999	2000	
	10.1 Business enterprise - Entreprises							
1 AGRICULTURE, HUNTING & FORESTRY	..	..	..	..	..	..	..	**1 AGRICULTURE, CHASSE & SYLVICULTURE**
2 MINING	302.5	520.1	561.4	531.9	473.0	267.3	..	**2 ACTIVITES EXTRACTIVES**
3 MANUFACTURING	1 843.2	2 296.1	2 177.2	2 149.4	1 959.5	1 931.5	..	**3 ACTIVITES DE FABRICATION**
4 Food, beverages & tobacco	138.2	288.7	228.2	179.1	203.2	182.0	..	**4 Produits alimentaires, boissons & tabac**
5 Food products & beverages	133.5	..	..	..	..	..	..	5 Produits alimentaires & boissons
6 Tobacco products	4.7	..	..	..	..	..	..	6 Produits à base de tabac
7 Textiles, wear app., fur & leather	..	..	21.3	21.1	..	..	..	**7 Textiles, habill., fourrures & cuirs**
8 Textiles	15.9	11.7	10.7	..	11.6	..	..	8 Textiles
9 Wearing apparel & fur	..	4.2	3.9	4.9	3.9	3.5	..	9 Articles d'habillement & fourrures
10 Leather products & footwear	..	..	6.7	..	..	..	..	10 Articles en cuir & chaussures
11 Wood, paper, printing, publishing	93.2	..	206.4	133.7	105.1	116.4	..	**11 Bois, papier, imprimerie, édition**
12 Wood & cork (not furniture)	..	23.3	..	23.9	..	13.6	..	12 Bois & liège (sauf meubles)
13 Pulp, paper & paper products	57.1	..	..	..	..	..	..	13 Carton, papier & articles
14 Publ., print. & repro. of rec. media	..	20.7	14.6	..	19.5	..	..	14 Edition, impress. & repro. enregis.
15 Coke, petroleum, nuclear fuel, chemicals, rubber plastics	298.2	315.1	299.9	305.2	322.3	382.1	..	**15 Coke, pétrole, combust. nucl., prod. chimi., caout. & mat. plast.**
16 Coke, ref. petrol. prod. & nuclear fuel	5.5	8.3	..	16.1	16.5	..	..	16 Coke, raffin. pétr. & combust. nucl.
17 Chemicals & chemical products	259.9	271.1	264.8	..	280.7	330.8	..	17 Produits chimiques
18 *Chemical prod.(less pharmaceuticals)*	141.8	157.4	150.6	147.2	146.7	177.8	..	18 *Prod. chimi. (moins prod. pharmaceutiques)*
19 *Pharmaceuticals*	118.0	113.6	114.3	..	134.0	153.1	..	19 *Produits pharmaceutiques*
20 Rubber & plastic products	32.8	35.8	..	..	25.1	..	..	20 Caoutchouc et matières plastiques
21 Non-metallic mineral products	44.8	..	62.7	69.5	52.7	48.5	..	**21 Produits minéraux non métalliques**
22 Basic metals	273.2	330.1	357.2	320.1	255.8	177.7	..	**22 Produits métallurgiques de base**
23 Basic metals, ferrous	..	222.6	209.7	..	..	..	..	23 Produits métallur. de base, ferreux
24 Basic metals, non-ferrous	..	107.5	..	..	..	..	..	24 Prod. métallur. de base, non ferreux
25 Fabricated metal products	49.1	50.2	..	40.2	38.2	..	..	**25 Ouvrages en métaux (sauf mach.& mat.)**
26 Machinery equipment, instruments and transport equipment	899.7	984.6	912.6	1 046.3	943.5	946.1	..	**26 Machines & matériel, instruments & mat. de transport**
27 Machinery n.e.c.	92.2	127.3	140.7	121.4	107.0	119.6	..	27 Machines, n.c.a.
28 Office, account. & computing machinery	42.1	24.4	21.4	46.9	31.4	..	..	28 Mach. de bur., compta. & trait. inform.
29 Electrical machinery	77.5	77.6	69.0	79.6	70.4	74.8	..	29 Machines électriques
30 Electro. equip. (radio, TV & commun.)	278.7	253.7	244.0	289.9	284.4	221.2	..	30 Equip. électro. (radio, tv & communic.)
31 *Electro. comp. (inc. semi-conductors)*	93.4	99.2	93.4	106.8	112.4	91.0	..	31 *Compos. électro. (semi-cond. incl.)*
32 *TV, radio & communications equip.*	185.3	154.5	150.6	183.2	172.0	130.2	..	32 *Appareils radio, tv et de communi.*
33 Instruments	93.0	108.0	83.6	86.3	90.6	110.0	..	33 Instruments
34 Motor vehicles	223.3	303.1	300.3	340.3	298.1	318.6	..	34 Véhicules automobiles
35 Other transport equipment	92.9	90.6	53.7	81.8	61.6	..	..	35 Autres matériels de transport
36 *Ships*	..	..	9.0	61.8	39.2	..	..	36 *Navires*
37 *Aerospace*	..	..	8.8	..	..	..	..	37 *Aérospatiale*
38 *Other transport equipment n.e.c.*	11.3	25.4	35.9	..	..	..	..	38 *Autres matériels de transport n.c.a.*
39 Furniture, other manufacturing n.e.c.	18.6	15.7	45.2	34.2	..	..	..	**39 Meubles, autres activ. de fabricat. n.c.a.**
40 Furniture	..	5.7	..	..	..	7.0	..	40 Meubles
41 Other manufacturing n.e.c.	..	10.0	..	..	..	11.3	..	41 Autres activités de fabrication n.c.a.
42 Recycling	0.0	0.0	0.0	0.0	0.0	..	..	**42 Récupération**
43 ELECTRICITY, GAS & WATER SUPPLY	117.2	37.4	..	24.3	28.7	..	..	**43 DISTRIBUTION D'ELECTRICITE, GAZ & EAU**
44 CONSTRUCTION	12.4	19.5	..	33.4	36.6	..	..	**44 CONSTRUCTION**
45 SERVICES SECTOR	981.2	1 171.9	1 144.1	1 227.1	1 271.8	1 400.5	..	**45 SECTEUR DES SERVICES**
46 Wholesale, retail trade & repair of motor vehicles, etc.	163.1	292.3	310.5	288.0	312.6	317.5	..	**46 Commerce (gros & détail), réparation de véhicules, etc.**
47 Hotels & restaurants	0.0	0.0	0.0	0.0	..	..	..	**47 Hôtels et restaurants**
48 Transport & storage	13.0	..	..	..	..	29.6	..	**48 Transports, entreposage**
49 Communications	..	..	..	..	..	141.3	..	**49 Communications**
50 Post	..	..	..	..	..	..	..	50 Postes
51 Telecommunications	..	..	..	..	..	..	..	51 Télécommunications
52 Financial intermediation (inc. insur.)	..	117.6	88.4	..	81.1	125.3	..	**52 Intermédiation financ. (assurances inc.)**
53 Real estate, renting & busin. activity	606.2	622.1	616.4	627.3	604.5	739.7	..	**53 Immobil., locat. & serv. aux entreprises**
54 Computer & related activities	277.0	319.3	344.1	341.1	371.3	445.2	..	54 Informatique et activités connexes
55 *Software consultancy*	259.0	300.0	314.7	306.8	347.0	..	..	55 *Production de logiciels*
56 *Other computer services n.e.c.*	..	19.3	29.5	34.4	24.3	..	..	56 *Autres activités informatiques*
57 Research & development	75.3	87.8	91.9	97.7	106.3	137.3	..	57 Recherche & développement expérimental
58 Other business activities n.e.c.	253.9	215.0	180.3	188.4	126.8	157.1	..	58 Autres activ. & serv. aux entreprises
59 Comm., soc. & pers. serv. activ., etc.	16.4	17.9	21.2	27.6	28.2	47.1	..	**59 Adm. pub., serv. soc. & collec., etc**
TOTAL BUSINESS ENTERPRISE SECTOR	**3 256.4**	**4 045.0**	**3 936.0**	**3 966.1**	**3 769.7**	**3 686.9**	**..**	**TOTAL SECTEUR DES ENTREPRISES**

Source: OECD, R&D database (DSTI/EAS Division), January 2002.

Source : OCDE, base de données R-D (DSTI/Division AES), janvier 2002.

Business enterprise intramural expenditure on R&D (BERD) - Dépenses intra-muros de R-D des entreprises (DIRDE)

by industry and source of funds - par industrie et par source de financement

Million AUD — *Millions AUD*

10.2 Government - État

	1994	1995	1996	1997	1998	1999	2000	
1 AGRICULTURE, HUNTING & FORESTRY	..	..	..	..	..	..	..	**1 AGRICULTURE, CHASSE & SYLVICULTURE**
2 MINING	..	..	..	0.5	2.2	1.6	..	**2 ACTIVITES EXTRACTIVES**
3 MANUFACTURING	51.6	53.5	61.5	36.6	39.1	48.7	..	**3 ACTIVITES DE FABRICATION**
4 Food, beverages & tobacco	..	..	..	..	..	2.6	..	**4 Produits alimentaires, boissons & tabac**
5 Food products & beverages	..	..	..	..	..	2.6	..	5 Produits alimentaires & boissons
6 Tobacco products	0.0	0.0	0.0	0.0	0.0	..	..	6 Produits à base de tabac
7 Textiles, wear app., fur & leather	..	..	0.0	..	..	..	..	**7 Textiles, habill., fourrures & cuirs**
8 Textiles	0.0	0.0	0.0	0.0	0.2	..	..	8 Textiles
9 Wearing apparel & fur	..	0.0	0.0	0.0	0.0	..	..	9 Articles d'habillement & fourrures
10 Leather products & footwear	..	..	0.0	..	..	..	..	10 Articles en cuir & chaussures
11 Wood, paper, printing, publishing	..	0.3	..	0.6	..	0.6	..	**11 Bois, papier, imprimerie, édition**
12 Wood & cork (not furniture)	0.0	0.0	..	0.0	..	..	..	12 Bois & liège (sauf meubles)
13 Pulp, paper & paper products	..	0.0	..	0.0	..	..	..	13 Carton, papier & articles
14 Publ., print. & repro. of rec. media	..	0.3	0.3	0.6	..	0.4	..	14 Edition, impress. & repro. enregis.
15 Coke, petroleum, nuclear fuel, chemicals, rubber plastics	2.1	2.8	3.1	7.1	7.7	5.5	..	**15 Coke, pétrole, combust. nucl., prod. chimi., caout. & mat. plast.**
16 Coke, ref. petrol. prod. & nuclear fuel	0.0	0.0	0.0	0.0	0.0	..	..	16 Coke, raffin. pétr. & combust. nucl.
17 Chemicals & chemical products	1.7	2.5	..	..	7.2	4.5	..	17 Produits chimiques
18 *Chemical prod.(less pharmaceuticals)*	..	2.1	2.2	..	..	0.4	..	18 *Prod. chimi. (moins prod. pharmaceutiques)*
19 *Pharmaceuticals*	..	0.4	0.7	..	..	4.1	..	19 *Produits pharmaceutiques*
20 Rubber & plastic products	0.4	0.3	..	..	0.5	..	..	20 Caoutchouc et matières plastiques
21 Non-metallic mineral products	..	..	0.0	0.2	0.2	..	..	**21 Produits minéraux non métalliques**
22 Basic metals	0.8	..	..	..	..	1.5	..	**22 Produits métallurgiques de base**
23 Basic metals, ferrous	..	..	..	0.6	..	..	..	23 Produits métallur. de base, ferreux
24 Basic metals, non-ferrous	..	..	..	..	..	..	..	24 Prod. métallur. de base, non ferreux
25 Fabricated metal products	1.0	0.2	..	0.1	..	..	..	**25 Ouvrages en métaux (sauf mach.& mat.)**
26 Machinery equipment, instruments and transport equipment	44.7	47.2	53.9	18.7	27.9	..	..	**26 Machines & matériel, instruments & mat. de transport**
27 Machinery n.e.c.	1.0	0.4	1.2	2.0	4.6	5.6	..	27 Machines, n.c.a.
28 Office, account. & computing machinery	0.4	..	..	1.9	3.1	4.1	..	28 Mach. de bur., compta. & trait. inform.
29 Electrical machinery	1.0	0.6	..	0.6	0.3	1.3	..	29 Machines électriques
30 Electro. equip. (radio, TV & commun.)	22.9	27.6	..	7.0	..	..	..	30 Equip. électro. (radio, tv & communic.)
31 *Electro. comp. (inc. semi-conductors)*	..	..	..	4.4	..	..	..	31 *Compos. électro. (semi-cond. incl.)*
32 *TV, radio & communications equip.*	..	..	..	2.6	..	1.2	..	32 *Appareils radio, tv et de communi.*
33 Instruments	..	..	4.5	6.5	11.1	..	..	33 Instruments
34 Motor vehicles	..	0.3	..	0.6	..	..	..	34 Véhicules automobiles
35 Other transport equipment	..	0.3	..	0.1	0.7	..	..	35 Autres matériels de transport
36 *Ships*	..	0.3	..	0.1	0.7	..	..	36 *Navires*
37 *Aerospace*	0.0	0.0	..	0.0	0.0	..	..	37 *Aérospatiale*
38 *Other transport equipment n.e.c.*	0.0	0.0	0.0	0.0	0.0	..	..	38 *Autres matériels de transport n.c.a.*
39 Furniture, other manufacturing n.e.c.	0.6	0.7	0.6	1.9	..	..	..	**39 Meubles, autres activ. de fabricat. n.c.a.**
40 Furniture	..	0.2	..	..	..	..	..	40 Meubles
41 Other manufacturing n.e.c.	..	0.6	..	..	0.6	..	..	41 Autres activités de fabrication n.c.a.
42 Recycling	0.0	0.0	0.0	0.0	0.0	..	..	**42 Récupération**
43 ELECTRICITY, GAS & WATER SUPPLY	11.6	0.5	..	0.1	..	..	..	**43 DISTRIBUTION D'ELECTRICITE, GAZ & EAU**
44 CONSTRUCTION	..	..	..	1.3	0.9	0.5	..	**44 CONSTRUCTION**
45 SERVICES SECTOR	43.8	49.7	46.2	61.2	79.4	83.8	..	**45 SECTEUR DES SERVICES**
46 Wholesale, retail trade & repair of motor vehicles, etc.	7.1	2.8	10.4	11.5	11.2	..	..	**46 Commerce (gros & détail), réparation de véhicules, etc.**
47 Hotels & restaurants	0.0	0.0	0.0	0.0	0.0	..	..	**47 Hôtels et restaurants**
48 Transport & storage	0.0	..	..	..	..	..	..	**48 Transports, entreposage**
49 Communications	0.0	..	0.0	..	..	..	..	**49 Communications**
50 Post	0.0	0.0	0.0	0.0	0.0	..	..	50 Postes
51 Telecommunications	0.0	..	0.0	..	..	..	..	51 Télécommunications
52 Financial intermediation (inc. insur.)	0.0	..	..	..	0.3	0.3	..	**52 Intermédiation financ. (assurances inc.)**
53 Real estate, renting & busin. activity	35.8	46.4	35.5	47.3	66.3	75.6	..	**53 Immobil., locat. & serv. aux entreprises**
54 Computer & related activities	5.7	5.0	3.9	6.6	15.6	22.1	..	54 Informatique et activités connexes
55 *Software consultancy*	5.6	4.5	3.5	6.5	15.2	..	..	55 *Production de logiciels*
56 *Other computer services n.e.c.*	..	0.5	0.5	0.1	0.4	..	..	56 *Autres activités informatiques*
57 Research & development	22.8	28.5	22.4	26.7	27.8	34.6	..	57 Recherche & développement expérimental
58 Other business activities n.e.c.	7.3	13.0	9.2	13.9	22.9	18.9	..	58 Autres activ. & serv. aux entreprises
59 Comm., soc. & pers. serv. activ., etc.	0.9	0.1	..	0.5	1.1	1.4	..	**59 Adm. pub., serv. soc. & collec., etc**
TOTAL BUSINESS ENTERPRISE SECTOR	**107.2**	**104.5**	**108.4**	**99.6**	**121.5**	**134.6**	..	**TOTAL SECTEUR DES ENTREPRISES**

Source: OECD, R&D database (DSTI/EAS Division), January 2002.

Source : OCDE, base de données R-D (DSTI/Division AES), janvier 2002.

Business enterprise intramural expenditure on R&D (BERD) - Dépenses intra-muros de R-D des entreprises (DIRDE)

by industry and source of funds - par industrie et par source de financement

Million AUD *Millions AUD*

	1994	1995	1996	1997	1998	1999	2000	
	10.3 Other national funds - Autres sources nationales							
1 AGRICULTURE, HUNTING & FORESTRY	..	..	..	..	..	..	..	**1 AGRICULTURE, CHASSE & SYLVICULTURE**
2 MINING	..	..	..	..	1.5	..	..	**2 ACTIVITES EXTRACTIVES**
3 MANUFACTURING	..	23.2	20.7	11.7	10.3	2.9	..	**3 ACTIVITES DE FABRICATION**
4 Food, beverages & tobacco	..	..	..	..	3.7	..	..	**4 Produits alimentaires, boissons & tabac**
5 Food products & beverages	..	..	..	..	3.7	..	..	5 Produits alimentaires & boissons
6 Tobacco products	0.0	0.0	0.0	0.0	0.0	..	..	6 Produits à base de tabac
7 Textiles, wear app., fur & leather	0.0	0.0	0.0	0.0	0.0	..	..	**7 Textiles, habill., fourrures & cuirs**
8 Textiles	0.0	0.0	0.0	0.0	0.0	..	..	8 Textiles
9 Wearing apparel & fur	0.0	0.0	0.0	0.0	0.0	..	..	9 Articles d'habillement & fourrures
10 Leather products & footwear	0.0	0.0	0.0	0.0	0.0	..	..	10 Articles en cuir & chaussures
11 Wood, paper, printing, publishing	..	..	..	0.0	..	..	..	**11 Bois, papier, imprimerie, édition**
12 Wood & cork (not furniture)	..	0.0	0.0	0.0	0.0	..	..	12 Bois & liège (sauf meubles)
13 Pulp, paper & paper products	..	..	..	0.0	0.0	..	..	13 Carton, papier & articles
14 Publ., print. & repro. of rec. media	..	0.0	0.0	0.0	..	..	..	14 Edition, impress. & repro. enregis.
15 Coke, petroleum, nuclear fuel, chemicals, rubber plastics	..	..	..	0.4	0.3	0.7	..	**15 Coke, pétrole, combust. nucl., prod. chimi., caout. & mat. plast.**
16 Coke, ref. petrol. prod. & nuclear fuel	0.0	0.0	0.0	0.0	0.0	..	..	16 Coke, raffin. pétr. & combust. nucl.
17 Chemicals & chemical products	..	..	..	0.4	0.3	0.7	..	17 Produits chimiques
18 *Chemical prod.(less pharmaceuticals)*	..	..	..	..	..	..	..	18 *Prod. chimi. (moins prod. pharmaceutiques)*
19 *Pharmaceuticals*	..	..	..	..	..	..	..	19 *Produits pharmaceutiques*
20 Rubber & plastic products	0.0	0.0	0.0	0.0	0.0	..	..	20 Caoutchouc et matières plastiques
21 Non-metallic mineral products	..	0.0	0.0	0.1	0.0	..	..	**21 Produits minéraux non métalliques**
22 Basic metals	0.0	..	..	..	0.0	..	..	**22 Produits métallurgiques de base**
23 Basic metals, ferrous	0.0	0.0	..	..	0.0	..	..	23 Produits métallur. de base, ferreux
24 Basic metals, non-ferrous	0.0	..	0.0	0.0	0.0	..	..	24 Prod. métallur. de base, non ferreux
25 Fabricated metal products	0.0	0.0	0.0	0.0	..	..	..	**25 Ouvrages en métaux (sauf mach.& mat.)**
26 Machinery equipment, instruments and transport equipment	..	21.9	..	9.6	5.8	..	..	**26 Machines & matériel, instruments & mat. de transport**
27 Machinery n.e.c.	0.1	..	0.4	..	..	..	..	27 Machines, n.c.a.
28 Office, account. & computing machinery	..	..	..	0.0	0.0	..	..	28 Mach. de bur., compta. & trait. inform.
29 Electrical machinery	..	..	0.0	..	0.0	..	..	29 Machines électriques
30 Electro. equip. (radio, TV & commun.)	..	..	2.3	..	1.6	..	..	30 Equip. électro. (radio, tv & communic.)
31 *Electro. comp. (inc. semi-conductors)*	0.0	0.6	..	..	..	..	..	31 *Compos. électro. (semi-cond. incl.)*
32 *TV, radio & communications equip.*	..	..	..	0.0	..	..	..	32 *Appareils radio, tv et de communi.*
33 Instruments	2.1	5.4	..	..	..	..	..	33 Instruments
34 Motor vehicles	..	..	10.2	..	3.4	..	..	34 Véhicules automobiles
35 Other transport equipment	0.0	..	0.0	0.0	0.0	..	..	35 Autres matériels de transport
36 *Ships*	0.0	0.0	0.0	0.0	0.0	..	..	36 *Navires*
37 *Aerospace*	0.0	..	0.0	0.0	0.0	..	..	37 *Aérospatiale*
38 *Other transport equipment n.e.c.*	0.0	0.0	0.0	0.0	0.0	..	..	38 *Autres matériels de transport n.c.a.*
39 Furniture, other manufacturing n.e.c.	0.0	0.0	0.0	0.0	..	..	..	**39 Meubles, autres activ. de fabricat. n.c.a.**
40 Furniture	0.0	0.0	0.0	0.0	0.0	..	..	40 Meubles
41 Other manufacturing n.e.c.	0.0	0.0	0.0	0.0	..	..	..	41 Autres activités de fabrication n.c.a.
42 Recycling	..	0.0	0.0	0.0	0.0	..	..	**42 Récupération**
43 ELECTRICITY, GAS & WATER SUPPLY	0.2	0.2	0.0	0.0	0.0	..	..	**43 DISTRIBUTION D'ELECTRICITE, GAZ & EAU**
44 CONSTRUCTION	..	..	..	..	0.2	0.2	..	**44 CONSTRUCTION**
45 SERVICES SECTOR	..	50.4	36.1	28.9	24.2	..	..	**45 SECTEUR DES SERVICES**
46 Wholesale, retail trade & repair of motor vehicles, etc.	0.5	0.7	..	..	..	..	..	**46 Commerce (gros & détail), réparation de véhicules, etc.**
47 Hotels & restaurants	..	0.0	0.0	0.0	0.0	..	..	**47 Hôtels et restaurants**
48 Transport & storage	0.0	0.0	0.0	..	0.0	..	..	**48 Transports, entreposage**
49 Communications	0.0	0.0	..	0.0	0.0	..	..	**49 Communications**
50 Post	0.0	0.0	0.0	0.0	0.0	..	..	50 Postes
51 Telecommunications	0.0	0.0	..	0.0	0.0	..	..	51 Télécommunications
52 Financial intermediation (inc. insur.)	..	..	..	0.0	0.0	..	..	**52 Intermédiation financ. (assurances inc.)**
53 Real estate, renting & busin. activity	..	42.9	27.7	28.7	23.8	28.1	..	**53 Immobil., locat. & serv. aux entreprises**
54 Computer & related activities	1.6	2.3	1.7	..	1.0	0.3	..	54 Informatique et activités connexes
55 *Software consultancy*	1.6	2.3	1.7	..	1.0	0.3	..	55 *Production de logiciels*
56 *Other computer services n.e.c.*	0.0	0.0	0.0	0.0	..	..	..	56 *Autres activités informatiques*
57 Research & development	17.2	..	..	18.0	..	..	..	57 Recherche & développement expérimental
58 Other business activities n.e.c.	..	..	..	..	..	..	..	58 Autres activ. & serv. aux entreprises
59 Comm., soc. & pers. serv. activ., etc.	0.1	..	0.2	0.1	..	..	..	**59 Adm. pub., serv. soc. & collec., etc**
TOTAL BUSINESS ENTERPRISE SECTOR	**34.6**	**74.3**	**57.3**	**42.5**	**36.1**	**33.7**	**..**	**TOTAL SECTEUR DES ENTREPRISES**

Source: OECD, R&D database (DSTI/EAS Division), January 2002. Source : OCDE, base de données R-D (DSTI/Division AES), janvier 2002.

AUSTRALIA — Table 10 - Tableau 10 — AUSTRALIE

Business enterprise intramural expenditure on R&D (BERD) - Dépenses intra-muros de R-D des entreprises (DIRDE)

by industry and source of funds - par industrie et par source de financement

Million AUD — *Millions AUD*

10.4 Funds from abroad - Fonds de l'étranger

	1994	1995	1996	1997	1998	1999	2000	
1 AGRICULTURE, HUNTING & FORESTRY	..	..	..	..	..	..	..	**1 AGRICULTURE, CHASSE & SYLVICULTURE**
2 MINING	0.0	0.0	..	..	..	..	..	**2 ACTIVITES EXTRACTIVES**
3 MANUFACTURING	..	19.3	29.6	31.4	46.0	69.1	..	**3 ACTIVITES DE FABRICATION**
4 Food, beverages & tobacco	0.0	0.0	0.0	0.0	..	..	..	**4 Produits alimentaires, boissons & tabac**
5 Food products & beverages	0.0	0.0	0.0	0.0	..	..	..	5 Produits alimentaires & boissons
6 Tobacco products	0.0	0.0	0.0	0.0	0.0	..	..	6 Produits à base de tabac
7 Textiles, wear app., fur & leather	0.0	0.0	0.0	..	0.0	..	..	**7 Textiles, habill., fourrures & cuirs**
8 Textiles	0.0	0.0	0.0	..	0.0	..	..	8 Textiles
9 Wearing apparel & fur	0.0	0.0	0.0	0.0	0.0	..	..	9 Articles d'habillement & fourrures
10 Leather products & footwear	0.0	0.0	0.0	0.0	0.0	..	..	10 Articles en cuir & chaussures
11 Wood, paper, printing, publishing	0.0	0.0	0.0	0.0	0.0	..	..	**11 Bois, papier, imprimerie, édition**
12 Wood & cork (not furniture)	0.0	0.0	0.0	0.0	0.0	..	..	12 Bois & liège (sauf meubles)
13 Pulp, paper & paper products	0.0	0.0	0.0	0.0	0.0	..	..	13 Carton, papier & articles
14 Publ., print. & repro. of rec. media	0.0	0.0	0.0	0.0	0.0	..	..	14 Edition, impress. & repro. enregis.
15 Coke, petroleum, nuclear fuel, chemicals, rubber plastics	..	..	..	15.2	21.1	25.1	..	**15 Coke, pétrole, combust. nucl., prod. chimi., caout. & mat. plast.**
16 Coke, ref. petrol. prod. & nuclear fuel	0.0	0.0	0.0	0.0	0.0	..	..	16 Coke, raffin. pétr. & combust. nucl.
17 Chemicals & chemical products	..	..	..	15.2	21.1	25.1	..	17 Produits chimiques
18 *Chemical prod.(less pharmaceuticals)*	0.0	..	..	0.9	0.9	..	..	18 *Prod. chimi. (moins prod. pharmaceutiques)*
19 *Pharmaceuticals*	..	..	..	14.3	20.2	..	..	19 *Produits pharmaceutiques*
20 Rubber & plastic products	0.0	0.0	0.0	0.0	0.0	..	..	20 Caoutchouc et matières plastiques
21 Non-metallic mineral products	0.0	0.0	0.0	0.0	0.0	..	..	**21 Produits minéraux non métalliques**
22 Basic metals	0.0	..	..	..	..	..	..	**22 Produits métallurgiques de base**
23 Basic metals, ferrous	0.0	..	..	..	..	..	..	23 Produits métallur. de base, ferreux
24 Basic metals, non-ferrous	0.0	0.0	0.0	0.0	0.0	..	..	24 Prod. métallur. de base, non ferreux
25 Fabricated metal products	0.0	0.0	0.0	0.0	0.0	..	..	**25 Ouvrages en métaux (sauf mach.& mat.)**
26 Machinery equipment, instruments and transport equipment	..	10.2	..	15.5	24.2	42.2	..	**26 Machines & matériel, instruments & mat. de transport**
27 Machinery n.e.c.	0.0	..	0.4	..	..	..	..	27 Machines, n.c.a.
28 Office, account. & computing machinery	..	..	0.0	0.0	0.0	..	..	28 Mach. de bur., compta. & trait. inform.
29 Electrical machinery	..	..	..	..	0.0	..	..	29 Machines électriques
30 Electro. equip. (radio, TV & commun.)	..	..	..	..	..	..	..	30 Equip. électro. (radio, tv & communic.)
31 *Electro. comp. (inc. semi-conductors)*	..	..	..	..	..	..	..	31 *Compos. électro. (semi-cond. incl.)*
32 *TV, radio & communications equip.*	0.0	0.0	0.0	0.0	0.0	..	..	32 *Appareils radio, tv et de communi.*
33 Instruments	..	..	0.0	..	..	..	..	33 Instruments
34 Motor vehicles	..	..	..	..	..	..	..	34 Véhicules automobiles
35 Other transport equipment	..	..	..	0.0	0.0	..	..	35 Autres matériels de transport
36 *Ships*	..	..	..	0.0	0.0	..	..	36 *Navires*
37 *Aerospace*	0.0	0.0	..	0.0	0.0	..	..	37 *Aérospatiale*
38 *Other transport equipment n.e.c.*	0.0	0.0	0.0	0.0	0.0	..	..	38 *Autres matériels de transport n.c.a.*
39 Furniture, other manufacturing n.e.c.	0.0	0.0	0.0	0.0	0.0	..	..	**39 Meubles, autres activ. de fabricat. n.c.a.**
40 Furniture	0.0	0.0	0.0	0.0	0.0	..	..	40 Meubles
41 Other manufacturing n.e.c.	0.0	0.0	0.0	0.0	0.0	..	..	41 Autres activités de fabrication n.c.a.
42 Recycling	0.0	0.0	0.0	0.0	0.0	..	..	**42 Récupération**
43 ELECTRICITY, GAS & WATER SUPPLY	0.0	0.0	0.0	..	0.0	..	..	**43 DISTRIBUTION D'ELECTRICITE, GAZ & EAU**
44 CONSTRUCTION	0.0	24.9	0.0	..	0.0	..	..	**44 CONSTRUCTION**
45 SERVICES SECTOR	..	88.6	77.3	68.1	92.1	..	..	**45 SECTEUR DES SERVICES**
46 Wholesale, retail trade & repair of motor vehicles, etc.	22.3	..	..	..	..	..	..	**46 Commerce (gros & détail), réparation de véhicules, etc.**
47 Hotels & restaurants	0.0	0.0	0.0	0.0	0.0	..	..	**47 Hôtels et restaurants**
48 Transport & storage	0.0	0.0	0.0	0.0	0.0	..	..	**48 Transports, entreposage**
49 Communications	0.0	0.0	0.0	0.0	0.0	..	..	**49 Communications**
50 Post	0.0	0.0	0.0	0.0	0.0	..	..	50 Postes
51 Telecommunications	0.0	0.0	0.0	0.0	0.0	..	..	51 Télécommunications
52 Financial intermediation (inc. insur.)	0.0	0.0	0.0	0.0	0.0	..	..	**52 Intermédiation financ. (assurances inc.)**
53 Real estate, renting & busin. activity	..	54.9	62.7	52.6	78.8	89.9	..	**53 Immobil., locat. & serv. aux entreprises**
54 Computer & related activities	34.0	37.0	42.9	..	37.7	41.8	..	54 Informatique et activités connexes
55 *Software consultancy*	34.0	37.0	42.9	..	37.7	41.8	..	55 *Production de logiciels*
56 *Other computer services n.e.c.*	0.0	0.0	0.0	0.0	0.0	..	..	56 *Autres activités informatiques*
57 Research & development	3.2	..	..	11.5	..	..	..	57 Recherche & développement expérimental
58 Other business activities n.e.c.	..	..	..	..	..	..	..	58 Autres activ. & serv. aux entreprises
59 Comm., soc. & pers. serv. activ., etc.	0.7	..	..	..	..	..	..	**59 Adm. pub., serv. soc. & collec., etc**
TOTAL BUSINESS ENTERPRISE SECTOR	**110.2**	**132.9**	**135.4**	**111.0**	**140.4**	**190.2**	**..**	**TOTAL SECTEUR DES ENTREPRISES**

Source: OECD, R&D database (DSTI/EAS Division), January 2002.

Source : OCDE, base de données R-D (DSTI/Division AES), janvier 2002.

Business enterprise intramural expenditure on R&D (BERD) - Dépenses intra-muros de R-D des entreprises (DIRDE)

by industry and type of costs - par industrie et par type de coûts

Million AUD *Millions AUD*

	1994	1995	1996	1997	1998	1999	2000	
				11.1 Current expenditure - Dépenses courantes				
1 AGRICULTURE, HUNTING & FORESTRY	..	..	..	..	..	..	..	**1 AGRICULTURE, CHASSE & SYLVICULTURE**
2 MINING	238.0	441.9	443.0	402.0	371.2	244.9	..	**2 ACTIVITES EXTRACTIVES**
3 MANUFACTURING	1 758.8	2 097.5	1 969.4	1 960.4	1 831.9	1 848.4	..	**3 ACTIVITES DE FABRICATION**
4 Food, beverages & tobacco	123.7	217.0	207.0	157.2	162.4	161.2	..	**4 Produits alimentaires, boissons & tabac**
5 Food products & beverages	..	..	..	..	..	..	..	5 Produits alimentaires & boissons
6 Tobacco products	..	..	..	..	..	..	..	6 Produits à base de tabac
7 Textiles, wear app., fur & leather	23.6	23.9	19.3	..	19.5	15.6	..	**7 Textiles, habill., fourrures & cuirs**
8 Textiles	11.4	10.4	9.2	10.5	11.2	9.2	..	8 Textiles
9 Wearing apparel & fur	4.5	4.1	3.6	..	..	..	..	9 Articles d'habillement & fourrures
10 Leather products & footwear	7.7	9.3	6.4	..	..	..	..	10 Articles en cuir & chaussures
11 Wood, paper, printing, publishing	81.9	179.6	..	102.5	92.1	104.4	..	**11 Bois, papier, imprimerie, édition**
12 Wood & cork (not furniture)	19.8	21.8	19.6	16.6	..	13.4	..	12 Bois & liège (sauf meubles)
13 Pulp, paper & paper products	..	..	..	..	..	77.0	..	13 Carton, papier & articles
14 Publ., print. & repro. of rec. media	..	..	..	..	19.1	14.0	..	14 Edition, impress. & repro. enregis.
15 Coke, petroleum, nuclear fuel, chemicals, rubber plastics	278.0	279.7	282.6	301.7	318.3	337.6	..	**15 Coke, pétrole, combust. nucl., prod. chimi., caout. & mat. plast.**
16 Coke, ref. petrol. prod. & nuclear fuel	4.9	7.3	3.9	..	..	..	..	16 Coke, raffin. pétr. & combust. nucl.
17 Chemicals & chemical products	244.1	241.2	253.4	262.1	280.1	294.7	..	17 Produits chimiques
18 *Chemical prod.(less pharmaceuticals)*	125.7	143.4	141.5	141.8	132.8	139.3	..	18 *Prod. chimi. (moins prod. pharmaceutiques)*
19 *Pharmaceuticals*	118.4	97.8	111.9	120.3	147.3	155.4	..	19 *Produits pharmaceutiques*
20 Rubber & plastic products	29.0	31.1	25.3	..	..	..	..	20 Caoutchouc et matières plastiques
21 Non-metallic mineral products	37.7	56.8	58.3	60.1	47.6	43.3	..	**21 Produits minéraux non métalliques**
22 Basic metals	252.0	287.4	270.9	260.2	200.3	168.6	..	**22 Produits métallurgiques de base**
23 Basic metals, ferrous	145.0	193.9	172.3	181.7	..	97.9	..	23 Produits métallur. de base, ferreux
24 Basic metals, non-ferrous	107.0	93.5	98.6	78.6	..	70.7	..	24 Prod. métallur. de base, non ferreux
25 Fabricated metal products	45.2	45.9	40.1	35.3	35.3	43.7	..	**25 Ouvrages en métaux (sauf mach.& mat.)**
26 Machinery equipment, instruments and transport equipment	898.8	992.4	936.4	1 006.3	939.8	956.8	..	**26 Machines & matériel, instruments & mat. de transport**
27 Machinery n.e.c.	86.9	121.6	130.1	118.9	108.0	120.0	..	27 Machines, n.c.a.
28 Office, account. & computing machinery	41.2	26.0	..	47.5	32.3	36.8	..	28 Mach. de bur., compta. & trait. inform.
29 Electrical machinery	73.7	71.8	63.2	72.0	67.2	73.7	..	29 Machines électriques
30 Electro. equip. (radio, TV & commun.)	276.3	263.3	264.1	285.5	277.5	210.1	..	30 Equip. électro. (radio, tv & communic.)
31 *Electro. comp. (inc. semi-conductors)*	102.6	113.4	117.9	108.5	113.2	91.0	..	31 *Compos. électro. (semi-cond. incl.)*
32 *TV, radio & communications equip.*	173.7	149.9	146.1	177.0	164.3	119.1	..	32 *Appareils radio, tv et de communi.*
33 Instruments	105.5	122.6	85.0	90.2	100.2	120.0	..	33 Instruments
34 Motor vehicles	223.7	290.1	294.7	311.2	294.6	330.9	..	34 Véhicules automobiles
35 Other transport equipment	91.6	97.0	..	81.0	59.9	65.3	..	35 Autres matériels de transport
36 *Ships*	..	57.9	..	61.4	..	43.8	..	36 *Navires*
37 *Aerospace*	..	..	..	..	..	..	..	37 *Aérospatiale*
38 *Other transport equipment n.e.c.*	..	..	35.9	..	18.3	..	..	38 *Autres matériels de transport n.c.a.*
39 Furniture, other manufacturing n.e.c.	17.8	14.9	..	..	16.6	17.2	..	**39 Meubles, autres activ. de fabricat. n.c.a.**
40 Furniture	6.6	5.7	4.4	..	6.2	6.7	..	40 Meubles
41 Other manufacturing n.e.c.	11.2	9.2	..	..	10.5	10.5	..	41 Autres activités de fabrication n.c.a.
42 Recycling	0.0	0.0	0.0	0.0	0.0	..	..	**42 Récupération**
43 ELECTRICITY, GAS & WATER SUPPLY	..	..	19.0	20.2	25.4	23.3	..	**43 DISTRIBUTION D'ELECTRICITE, GAZ & EAU**
44 CONSTRUCTION	..	..	41.5	33.5	34.4	42.8	..	**44 CONSTRUCTION**
45 SERVICES SECTOR	984.9	1 233.1	1 186.7	1 266.2	1 350.4	1 492.1	..	**45 SECTEUR DES SERVICES**
46 Wholesale, retail trade & repair of motor vehicles, etc.	176.2	296.8	313.1	290.6	292.0	315.5	..	**46 Commerce (gros & détail), réparation de véhicules, etc.**
47 Hotels & restaurants	0.0	0.0	0.0	0.0	..	..	..	**47 Hôtels et restaurants**
48 Transport & storage	12.6	..	..	39.2	..	20.8	..	**48 Transports, entreposage**
49 Communications	..	..	..	..	..	135.0	..	**49 Communications**
50 Post	..	11.4	..	..	..	..	..	50 Postes
51 Telecommunications	..	..	..	..	..	..	..	51 Télécommunications
52 Financial intermediation (inc. insur.)	88.1	105.5	86.3	74.4	79.5	..	..	**52 Intermédiation financ. (assurances inc.)**
53 Real estate, renting & busin. activity	615.1	703.6	666.1	689.2	718.4	869.4	..	**53 Immobil., locat. & serv. aux entreprises**
54 Computer & related activities	301.6	344.2	372.5	349.5	403.5	486.0	..	54 Informatique et activités connexes
55 *Software consultancy*	285.8	326.1	..	..	380.1	467.0	..	55 *Production de logiciels*
56 *Other computer services n.e.c.*	15.9	18.1	..	..	23.4	19.0	..	56 *Autres activités informatiques*
57 Research & development	108.2	123.2	121.5	141.6	144.6	188.8	..	57 Recherche & développement expérimental
58 Other business activities n.e.c.	205.2	236.1	172.1	198.1	170.3	194.5	..	58 Autres activ. & serv. aux entreprises
59 Comm., soc. & pers. serv. activ., etc.	14.0	15.0	17.9	..	24.9	..	..	**59 Adm. pub., serv. soc. & collec., etc**
TOTAL BUSINESS ENTERPRISE SECTOR	**3 076.1**	**3 825.9**	**3 659.6**	**3 682.3**	**3 613.2**	**3 651.4**	**..**	**TOTAL SECTEUR DES ENTREPRISES**

Source: OECD, R&D database (DSTI/EAS Division), January 2002.

Source : OCDE, base de données R-D (DSTI/Division AES), janvier 2002.

Business enterprise intramural expenditure on R&D (BERD) - Dépenses intra-muros de R-D des entreprises (DIRDE)

by industry and type of costs - par industrie et par type de coûts

Million AUD — *Millions AUD*

	1994	1995	1996	1997	1998	1999	2000	
	11.2 Capital expenditure - Dépenses en capital							
1 AGRICULTURE, HUNTING & FORESTRY	..	..	..	..	..	..	..	1 AGRICULTURE, CHASSE & SYLVICULTURE
2 MINING	64.8	79.2	119.1	134.7	107.1	27.8	..	2 ACTIVITES EXTRACTIVES
3 MANUFACTURING	173.3	294.7	319.7	268.6	222.9	203.6	..	3 ACTIVITES DE FABRICATION
4 Food, beverages & tobacco	16.4	73.7	24.0	23.8	45.8	24.8	..	**4 Produits alimentaires, boissons & tabac**
5 Food products & beverages	..	..	24.0	..	45.8	24.8	..	5 Produits alimentaires & boissons
6 Tobacco products	..	..	0.0	..	0.0	..	..	6 Produits à base de tabac
7 Textiles, wear app., fur & leather	4.9	1.8	2.1	..	1.0	1.2	..	**7 Textiles, habill., fourrures & cuirs**
8 Textiles	4.5	1.3	1.5	1.7	0.6	0.9	..	8 Textiles
9 Wearing apparel & fur	0.1	0.1	0.3	..	..	..	..	9 Articles d'habillement & fourrures
10 Leather products & footwear	0.3	0.4	0.3	..	..	..	..	10 Articles en cuir & chaussures
11 Wood, paper, printing, publishing	12.4	26.6	..	31.8	13.6	14.2	..	**11 Bois, papier, imprimerie, édition**
12 Wood & cork (not furniture)	..	1.5	3.3	7.3	..	0.3	..	12 Bois & liège (sauf meubles)
13 Pulp, paper & paper products	..	..	..	..	..	13.3	..	13 Carton, papier & articles
14 Publ., print. & repro. of rec. media	2.3	..	2.0	..	1.0	0.7	..	14 Edition, impress. & repro. enregis.
15 Coke, petroleum, nuclear fuel, chemicals, rubber plastics	39.5	47.2	27.2	26.2	33.0	75.7	..	**15 Coke, pétrole, combust. nucl., prod. chimi., caout. & mat. plast.**
16 Coke, ref. petrol. prod. & nuclear fuel	0.6	0.9	..	..	..	..	..	16 Coke, raffin. pétr. & combust. nucl.
17 Chemicals & chemical products	34.7	41.3	21.2	20.3	29.2	66.4	..	17 Produits chimiques
18 *Chemical prod.(less pharmaceuticals)*	18.3	16.3	11.4	10.5	18.6	39.4	..	18 *Prod. chimi. (moins prod. pharmaceutiques)*
19 *Pharmaceuticals*	16.4	25.0	9.9	9.8	10.5	27.0	..	19 *Produits pharmaceutiques*
20 Rubber & plastic products	4.2	4.9	..	..	..	..	..	20 Caoutchouc et matières plastiques
21 Non-metallic mineral products	7.6	..	4.3	9.7	5.3	5.6	..	**21 Produits minéraux non métalliques**
22 Basic metals	22.0	44.3	88.7	67.9	56.7	10.6	..	**22 Produits métallurgiques de base**
23 Basic metals, ferrous	12.0	29.5	38.2	12.4	..	6.5	..	23 Produits métallur. de base, ferreux
24 Basic metals, non-ferrous	10.0	14.8	50.5	55.5	..	4.0	..	24 Prod. métallur. de base, non ferreux
25 Fabricated metal products	4.8	..	3.8	5.0	3.3	3.9	..	**25 Ouvrages en métaux (sauf mach.& mat.)**
26 Machinery equipment, instruments and transport equipment	64.3	71.5	71.7	83.8	61.7	64.8	..	**26 Machines & matériel, instruments & mat. de transport**
27 Machinery n.e.c.	6.4	6.7	12.6	7.5	7.1	11.9	..	27 Machines, n.c.a.
28 Office, account. & computing machinery	2.2	1.9	..	1.3	2.2	2.8	..	28 Mach. de bur., compta. & trait. inform.
29 Electrical machinery	4.9	6.6	6.7	8.2	3.4	2.4	..	29 Machines électriques
30 Electro. equip. (radio, TV & commun.)	25.6	19.3	13.8	12.6	17.1	16.5	..	30 Equip. électro. (radio, tv & communic.)
31 *Electro. comp. (inc. semi-conductors)*	12.4	10.6	7.2	3.9	6.9	4.3	..	31 *Compos. électro. (semi-cond. incl.)*
32 *TV, radio & communications equip.*	13.3	8.7	6.6	8.8	10.1	12.3	..	32 *Appareils radio, tv et de communi.*
33 Instruments	9.0	8.7	6.1	5.0	6.6	7.4	..	33 Instruments
34 Motor vehicles	14.8	27.8	18.4	48.3	22.8	17.1	..	34 Véhicules automobiles
35 Other transport equipment	1.5	0.4	..	0.9	2.4	6.8	..	35 Autres matériels de transport
36 *Ships*	1.3	0.2	..	0.5	..	..	..	36 *Navires*
37 *Aerospace*	0.0	..	..	..	..	..	..	37 *Aérospatiale*
38 *Other transport equipment n.e.c.*	..	..	0.0	..	..	..	..	38 *Autres matériels de transport n.c.a.*
39 Furniture, other manufacturing n.e.c.	1.4	1.5	..	..	2.6	2.8	..	**39 Meubles, autres activ. de fabricat. n.c.a.**
40 Furniture	0.2	0.2	0.3	..	0.3	0.9	..	40 Meubles
41 Other manufacturing n.e.c.	1.2	1.3	..	..	2.2	1.9	..	41 Autres activités de fabrication n.c.a.
42 Recycling	0.0	0.0	0.0	0.0	0.0	..	..	**42 Récupération**
43 ELECTRICITY, GAS & WATER SUPPLY	..	..	7.2	4.6	3.9	1.5	..	43 DISTRIBUTION D'ELECTRICITE, GAZ & EAU
44 CONSTRUCTION	..	..	14.8	10.0	3.3	21.5	..	44 CONSTRUCTION
45 SERVICES SECTOR	146.9	127.5	116.9	119.0	117.2	139.4	..	45 SECTEUR DES SERVICES
46 Wholesale, retail trade & repair of motor vehicles, etc.	16.7	32.4	22.7	24.5	45.3	36.8	..	**46 Commerce (gros & détail), réparation de véhicules, etc.**
47 Hotels & restaurants	0.0	0.0	0.0	0.0	0.0	..	..	**47 Hôtels et restaurants**
48 Transport & storage	0.5	..	0.5	7.6	7.8	9.3	..	**48 Transports, entreposage**
49 Communications	..	..	4.0	6.3	2.7	6.4	..	**49 Communications**
50 Post	..	..	..	..	0.0	..	..	50 Postes
51 Telecommunications	..	..	..	..	2.7	..	..	51 Télécommunications
52 Financial intermediation (inc. insur.)	12.8	18.4	9.8	10.6	1.9	..	..	**52 Intermédiation financ. (assurances inc.)**
53 Real estate, renting & busin. activity	103.5	62.8	76.2	66.6	54.9	63.8	..	**53 Immobil., locat. & serv. aux entreprises**
54 Computer & related activities	16.7	19.3	20.2	19.0	22.2	23.4	..	54 Informatique et activités connexes
55 *Software consultancy*	14.4	17.6	..	..	20.9	22.0	..	55 *Production de logiciels*
56 *Other computer services n.e.c.*	2.3	1.7	..	..	1.3	1.5	..	56 *Autres activités informatiques*
57 Research & development	10.1	16.1	15.8	12.4	16.4	21.3	..	57 Recherche & développement expérimental
58 Other business activities n.e.c.	76.6	27.4	40.2	35.2	16.3	19.1	..	58 Autres activ. & serv. aux entreprises
59 Comm., soc. & pers. serv. activ., etc.	4.1	3.9	3.7	3.5	4.6	..	..	**59 Adm. pub., serv. soc. & collec., etc**
TOTAL BUSINESS ENTERPRISE SECTOR	**432.2**	**530.7**	**577.6**	**536.8**	**454.4**	**393.9**	**..**	**TOTAL SECTEUR DES ENTREPRISES**

Source: OECD, R&D database (DSTI/EAS Division), January 2002. — Source : OCDE, base de données R-D (DSTI/Division AES), janvier 2002.

AUSTRALIA — Table 12 - Tableau 12 — AUSTRALIE

Total business enterprise R&D personnel (BEMP) - Personnel total de R-D du secteur des entreprises (PRDE)

by industry - par industrie

In full time equivalent — *En équivalence plein-temps*

	1994	1995	1996	1997	1998	1999	2000	
1 AGRICULTURE, HUNTING & FORESTRY	..	..	..	..	..	..	..	1 AGRICULTURE, CHASSE & SYLVICULTURE
2 MINING	1 017	1 079	1 069	952	974	749	..	2 ACTIVITES EXTRACTIVES
3 MANUFACTURING	14 869	14 860	14 397	13 768	13 492	14 111	..	3 ACTIVITES DE FABRICATION
4 Food, beverages & tobacco	1 108	1 321	1 330	1 057	1 131	1 118	..	**4 Produits alimentaires, boissons & tabac**
5 Food products & beverages	1 078	..	..	..	..	..	..	5 Produits alimentaires & boissons
6 Tobacco products	30	..	..	..	..	..	..	6 Produits à base de tabac
7 Textiles, wear app., fur & leather	215	215	193	172	178	171	..	**7 Textiles, habill., fourrures & cuirs**
8 Textiles	93	99	92	86	92	90	..	8 Textiles
9 Wearing apparel & fur	72	63	56	55	39	36	..	9 Articles d'habillement & fourrures
10 Leather products & footwear	50	53	46	31	47	45	..	10 Articles en cuir & chaussures
11 Wood, paper, printing, publishing	430	483	400	437	399	553	..	**11 Bois, papier, imprimerie, édition**
12 Wood & cork (not furniture)	88	113	123	126	99	111	..	12 Bois & liège (sauf meubles)
13 Pulp, paper & paper products	168	162	146	148	136	275	..	13 Carton, papier & articles
14 Publ., print. & repro. of rec. media	174	209	132	163	164	166	..	14 Edition, impress. & repro. enregis.
15 Coke, petroleum, nuclear fuel, chemicals, rubber plastics	2 425	2 384	2 364	2 363	2 290	2 347	..	**15 Coke, pétrole, combust. nucl., prod. chimi., caout. & mat. plast.**
16 Coke, ref. petrol. prod. & nuclear fuel	36	58	35	44	31	25	..	16 Coke, raffin. pétr. & combust. nucl.
17 Chemicals & chemical products	2 061	2 007	2 063	2 057	2 040	2 079	..	17 Produits chimiques
18 *Chemical prod.(less pharmaceuticals)*	1 179	1 225	1 224	1 185	1 032	967	..	18 *Prod. chimi. (moins prod. pharmaceutiques)*
19 *Pharmaceuticals*	882	781	839	872	1 008	1 111	..	19 *Produits pharmaceutiques*
20 Rubber & plastic products	328	319	266	262	219	243	..	20 Caoutchouc et matières plastiques
21 Non-metallic mineral products	427	357	456	434	369	352	..	**21 Produits minéraux non métalliques**
22 Basic metals	1 466	1 531	1 265	1 086	839	725	..	**22 Produits métallurgiques de base**
23 Basic metals, ferrous	1 007	1 071	885	815	511	375	..	23 Produits métallur. de base, ferreux
24 Basic metals, non-ferrous	459	460	380	271	328	350	..	24 Prod. métallur. de base, non ferreux
25 Fabricated metal products	554	506	446	350	385	397	..	**25 Ouvrages en métaux (sauf mach.& mat.)**
26 Machinery equipment, instruments and transport equipment	8 011	7 864	7 744	7 660	7 690	8 211	..	**26 Machines & matériel, instruments & mat. de transport**
27 Machinery n.e.c.	1 092	1 270	1 284	1 185	1 100	1 270	..	27 Machines, n.c.a.
28 Office, account. & computing machinery	468	334	280	318	246	226	..	28 Mach. de bur., compta. & trait. inform.
29 Electrical machinery	892	797	711	676	636	602	..	29 Machines électriques
30 Electro. equip. (radio, TV & commun.)	2 561	1 871	1 891	1 844	1 994	1 935	..	30 Equip. électro. (radio, tv & communic.)
31 *Electro. comp. (inc. semi-conductors)*	1 129	1 001	998	897	1 000	903	..	31 *Compos. électro. (semi-cond. incl.)*
32 *TV, radio & communications equip.*	1 433	870	892	947	994	1 032	..	32 *Appareils radio, tv et de communi.*
33 Instruments	973	1 134	872	920	941	1 104	..	33 Instruments
34 Motor vehicles	1 568	1 913	2 192	2 276	2 335	2 675	..	34 Véhicules automobiles
35 Other transport equipment	457	544	514	440	438	399	..	35 Autres matériels de transport
36 *Ships*	254	247	160	248	233	179	..	36 *Navires*
37 *Aerospace*	66	110	52	66	..	..	..	37 *Aérospatiale*
38 *Other transport equipment n.e.c.*	138	188	302	126	..	..	..	38 *Autres matériels de transport n.c.a.*
39 Furniture, other manufacturing n.e.c.	233	201	199	210	212	238	..	**39 Meubles, autres activ. de fabricat. n.c.a.**
40 Furniture	102	94	76	74	90	89	..	40 Meubles
41 Other manufacturing n.e.c.	131	106	123	136	122	149	..	41 Autres activités de fabrication n.c.a.
42 Recycling	0	0	0	..	..	..	..	**42 Récupération**
43 ELECTRICITY, GAS & WATER SUPPLY	367	195	124	114	180	69	..	43 DISTRIBUTION D'ELECTRICITE, GAZ & EAU
44 CONSTRUCTION	108	163	215	216	188	241	..	44 CONSTRUCTION
45 SERVICES SECTOR	9 451	10 826	10 641	9 737	10 141	10 939	..	45 SECTEUR DES SERVICES
46 Wholesale, retail trade & repair of motor vehicles, etc.	1 590	2 588	2 662	2 469	2 273	2 475	..	**46 Commerce (gros & détail), réparation de véhicules, etc.**
47 Hotels & restaurants	0	0	0	..	..	..	..	**47 Hôtels et restaurants**
48 Transport & storage	..	..	147	267	104	113	..	**48 Transports, entreposage**
49 Communications	..	..	250	126	..	114	..	**49 Communications**
50 Post	..	17	..	..	..	..	..	50 Postes
51 Telecommunications	..	..	..	..	101	..	..	51 Télécommunications
52 Financial intermediation (inc. insur.)	979	1 219	1 138	430	664	844	..	**52 Intermédiation financ. (assurances inc.)**
53 Real estate, renting & busin. activity	5 966	6 032	6 240	6 160	6 621	7 085	..	**53 Immobil., locat. & serv. aux entreprises**
54 Computer & related activities	3 158	3 456	3 532	3 449	3 910	4 237	..	54 Informatique et activités connexes
55 *Software consultancy*	2 925	3 196	3 236	3 129	3 690	4 050	..	55 *Production de logiciels*
56 *Other computer services n.e.c.*	233	260	296	320	221	187	..	56 *Autres activités informatiques*
57 Research & development	933	974	944	1 065	1 184	1 289	..	57 Recherche & développement expérimental
58 Other business activities n.e.c.	1 875	1 602	1 765	1 647	1 527	1 559	..	58 Autres activ. & serv. aux entreprises
59 Comm., soc. & pers. serv. activ., etc.	188	195	203	286	309	308	..	**59 Adm. pub., serv. soc. & collec., etc**
TOTAL BUSINESS ENTERPRISE SECTOR	**25 812**	**27 123**	**26 446**	**24 786**	**24 974**	**26 109**	..	**TOTAL SECTEUR DES ENTREPRISES**

Source: OECD, R&D database (DSTI/EAS Division), January 2002.

Source : OCDE, base de données R-D (DSTI/Division AES), janvier 2002.

AUSTRALIA Table 13 - Tableau 13 **AUSTRALIE**

RSE in the business enterprise sector - Chercheurs dans le secteur des entreprises

by industry - par industrie

In full time equivalent / *En équivalence plein-temps*

	1994	1995	1996	1997	1998	1999	2000	
1 AGRICULTURE, HUNTING & FORESTRY	..	..	..	..	..	..	..	**1 AGRICULTURE, CHASSE & SYLVICULTURE**
2 MINING	422	437	639	424	421	423	..	**2 ACTIVITES EXTRACTIVES**
3 MANUFACTURING	8 123	7 910	7 538	7 505	7 548	7 930	..	**3 ACTIVITES DE FABRICATION**
4 Food, beverages & tobacco	630	670	691	595	622	597	..	**4 Produits alimentaires, boissons & tabac**
5 Food products & beverages	619	..	..	..	..	..	..	5 Produits alimentaires & boissons
6 Tobacco products	10	..	..	..	..	..	..	6 Produits à base de tabac
7 Textiles, wear app., fur & leather	103	87	79	75	72	65	..	**7 Textiles, habill., fourrures & cuirs**
8 Textiles	44	37	40	38	42	41	..	8 Textiles
9 Wearing apparel & fur	40	32	24	25	14	14	..	9 Articles d'habillement & fourrures
10 Leather products & footwear	19	18	15	12	16	10	..	10 Articles en cuir & chaussures
11 Wood, paper, printing, publishing	203	246	197	215	215	321	..	**11 Bois, papier, imprimerie, édition**
12 Wood & cork (not furniture)	37	45	58	53	40	46	..	12 Bois & liège (sauf meubles)
13 Pulp, paper & paper products	76	77	58	59	71	172	..	13 Carton, papier & articles
14 Publ., print. & repro. of rec. media	90	124	82	102	104	104	..	14 Edition, impress. & repro. enregis.
15 Coke, petroleum, nuclear fuel, chemicals, rubber plastics	1 272	1 341	1 331	1 289	1 215	1 292	..	**15 Coke, pétrole, combust. nucl., prod. chimi., caout. & mat. plast.**
16 Coke, ref. petrol. prod. & nuclear fuel	20	36	26	21	15	10	..	16 Coke, raffin. pétr. & combust. nucl.
17 Chemicals & chemical products	1 130	1 176	1 181	1 153	1 101	1 165	..	17 Produits chimiques
18 *Chemical prod.(less pharmaceuticals)*	697	720	719	695	572	546	..	18 *Prod. chimi. (moins prod. pharmaceutiques)*
19 *Pharmaceuticals*	433	456	462	458	529	619	..	19 *Produits pharmaceutiques*
20 Rubber & plastic products	122	130	123	115	100	116	..	20 Caoutchouc et matières plastiques
21 Non-metallic mineral products	161	143	211	211	185	187	..	**21 Produits minéraux non métalliques**
22 Basic metals	855	820	662	558	408	364	..	**22 Produits métallurgiques de base**
23 Basic metals, ferrous	604	562	431	411	258	183	..	23 Produits métallur. de base, ferreux
24 Basic metals, non-ferrous	251	259	231	146	150	181	..	24 Prod. métallur. de base, non ferreux
25 Fabricated metal products	240	232	221	176	194	201	..	**25 Ouvrages en métaux (sauf mach.& mat.)**
26 Machinery equipment, instruments and transport equipment	4 553	4 273	4 054	4 282	4 526	4 766	..	**26 Machines & matériel, instruments & mat. de transport**
27 Machinery n.e.c.	526	527	577	600	599	704	..	27 Machines, n.c.a.
28 Office, account. & computing machinery	331	232	169	222	156	162	..	28 Mach. de bur., compta. & trait. inform.
29 Electrical machinery	421	421	322	329	349	299	..	29 Machines électriques
30 Electro. equip. (radio, TV & commun.)	1 786	1 343	1 364	1 342	1 495	1 390	..	30 Equip. électro. (radio, tv & communic.)
31 *Electro. comp. (inc. semi-conductors)*	718	665	637	561	693	620	..	31 *Compos. électro. (semi-cond. incl.)*
32 *TV, radio & communications equip.*	1 069	678	727	782	802	770	..	32 *Appareils radio, tv et de communi.*
33 Instruments	621	741	526	589	632	759	..	33 Instruments
34 Motor vehicles	615	852	940	995	1 119	1 285	..	34 Véhicules automobiles
35 Other transport equipment	254	157	158	204	176	167	..	35 Autres matériels de transport
36 *Ships*	152	96	82	130	112	71	..	36 *Navires*
37 *Aerospace*	34	20	19	41	..	..	..	37 *Aérospatiale*
38 *Other transport equipment n.e.c.*	68	42	57	32	..	..	..	38 *Autres matériels de transport n.c.a.*
39 Furniture, other manufacturing n.e.c.	107	97	93	105	110	137	..	**39 Meubles, autres activ. de fabricat. n.c.a.**
40 Furniture	40	38	28	29	40	43	..	40 Meubles
41 Other manufacturing n.e.c.	67	59	65	76	71	94	..	41 Autres activités de fabrication n.c.a.
42 Recycling	0	0	0	..	..	..	..	**42 Récupération**
43 ELECTRICITY, GAS & WATER SUPPLY	203	115	71	62	88	46	..	**43 DISTRIBUTION D'ELECTRICITE, GAZ & EAU**
44 CONSTRUCTION	45	73	109	113	85	96	..	**44 CONSTRUCTION**
45 SERVICES SECTOR	6 110	7 054	6 925	6 473	6 484	6 790	..	**45 SECTEUR DES SERVICES**
46 Wholesale, retail trade & repair of motor vehicles, etc.	1 070	1 712	1 752	1 621	1 243	1 230	..	**46 Commerce (gros & détail), réparation de véhicules, etc.**
47 Hotels & restaurants	0	..	0	..	..	..	..	**47 Hôtels et restaurants**
48 Transport & storage	..	..	84	135	77	67	..	**48 Transports, entreposage**
49 Communications	..	..	149	89	..	53	..	**49 Communications**
50 Post	..	14	..	..	..	..	..	50 Postes
51 Telecommunications	..	..	..	..	70	..	..	51 Télécommunications
52 Financial intermediation (inc. insur.)	542	620	574	273	446	590	..	**52 Intermédiation financ. (assurances inc.)**
53 Real estate, renting & busin. activity	3 986	4 142	4 237	4 138	4 387	4 608	..	**53 Immobil., locat. & serv. aux entreprises**
54 Computer & related activities	2 365	2 498	2 558	2 393	2 758	2 869	..	54 Informatique et activités connexes
55 *Software consultancy*	2 192	2 314	2 374	2 181	2 609	2 739	..	55 *Production de logiciels*
56 *Other computer services n.e.c.*	173	184	184	211	149	131	..	56 *Autres activités informatiques*
57 Research & development	584	600	564	644	733	768	..	57 Recherche & développement expérimental
58 Other business activities n.e.c.	1 038	1 044	1 115	1 101	897	971	..	58 Autres activ. & serv. aux entreprises
59 Comm., soc. & pers. serv. activ., etc.	119	123	129	218	251	242	..	**59 Adm. pub., serv. soc. & collec., etc**
TOTAL BUSINESS ENTERPRISE SECTOR	**14 903**	**15 589**	**15 282**	**14 577**	**14 626**	**15 283**	**..**	**TOTAL SECTEUR DES ENTREPRISES**

Source: OECD, R&D database (DSTI/EAS Division), January 2002.

Source : OCDE, base de données R-D (DSTI/Division AES), janvier 2002.

Table 14 - Tableau 14

Higher education intramural expenditure on R&D - Dépenses intra-muros de R-D de l'enseignement supérieur
(HERD) by field of science - (DIRDES) par discipline scientifique

	1994	1995	1996	1997	1998	1999	2000	
	14.1 Million AUD - Millions AUD							
Natural sciences	534.3	591.0	664.1	..	715.4	..	..	Sciences exactes et naturelles
Engineering	292.5	320.2	394.8	..	420.3	..	..	Sciences de l'ingénieur
Medical sciences	375.9	453.8	491.4	..	592.5	..	..	Sciences médicales
Agricultural sciences	109.8	123.4	127.8	..	170.6	..	..	Sciences agricoles
Sub-total NSE	1 312.5	1 488.4	1 678.0	..	1 898.7	..	..	Sous-total SENI
Social sciences	372.4	388.3	446.0	..	506.5	..	..	Sciences sociales
Humanities	144.7	162.4	183.6	..	197.5	..	..	Sciences humaines
Sub-total SSH	517.1	550.7	629.5	..	704.0	..	..	Sous-total SSH
Total NSE+SSH	1 829.6	2 039.1	2 307.6	..	2 602.7	..	..	*Total SENI+SSH*
	14.2 Million constant USD (1995 prices and PPPs) - Millions de USD constants (prix et PPA de 1995)							
Natural sciences	426.0	459.2	509.3	..	541.6	..	..	Sciences exactes et naturelles
Engineering	233.2	248.8	302.8	..	318.2	..	..	Sciences de l'ingénieur
Medical sciences	299.7	352.6	376.9	..	448.6	..	..	Sciences médicales
Agricultural sciences	87.6	95.9	98.0	..	129.2	..	..	Sciences agricoles
Sub-total NSE	1 046.5	1 156.4	1 287.0	..	1 437.5	..	..	Sous-total SENI
Social sciences	296.9	301.7	342.0	..	383.5	..	..	Sciences sociales
Humanities	115.4	126.2	140.8	..	149.5	..	..	Sciences humaines
Sub-total SSH	412.3	427.9	482.8	..	533.0	..	..	Sous-total SSH
Total NSE+SSH	1 458.9	1 584.3	1 769.8	..	1 970.5	..	..	*Total SENI+SSH*

Table 15 - Tableau 15

Higher education intramural expenditure on R&D - Dépenses intra-muros de R-D de l'enseignement supérieur
(HERD) by field of science and type of costs - (DIRDES) par discipline scientifique et par type de coûts

Million AUD | *Millions AUD*

	1994	1995	1996	1997	1998	1999	2000	
Current expenditure								**Dépenses courantes**
Natural sciences	479.5	540.6	..	..	648.9	..	..	Sciences exactes et naturelles
Engineering	248.0	283.4	..	..	378.8	..	..	Sciences de l'ingénieur
Medical sciences	353.8	430.9	..	..	558.7	..	..	Sciences médicales
Agricultural sciences	93.7	113.2	..	..	160.8	..	..	Sciences agricoles
Sub-total NSE	1 175.1	1 368.2	..	..	1 747.2	..	..	Sous-total SENI
Social sciences	352.8	374.2	..	..	485.3	..	..	Sciences sociales
Humanities	136.8	152.9	..	..	190.0	..	..	Sciences humaines
Sub-total SSH	489.6	527.1	..	..	675.2	..	..	Sous-total SSH
Total NSE+SSH	1 664.6	1 895.3	2 129.2	..	2 422.5	..	..	*Total SENI+SSH*
Capital expenditure								**Dépenses en capital**
Natural sciences	54.8	50.4	..	..	66.5	..	..	Sciences exactes et naturelles
Engineering	44.5	36.8	..	..	41.5	..	..	Sciences de l'ingénieur
Medical sciences	22.1	22.8	..	..	33.7	..	..	Sciences médicales
Agricultural sciences	16.1	10.2	..	..	9.9	..	..	Sciences agricoles
Sub-total NSE	137.5	120.3	..	..	151.5	..	..	Sous-total SENI
Social sciences	19.6	14.1	..	..	21.3	..	..	Sciences sociales
Humanities	7.9	9.4	..	..	7.5	..	..	Sciences humaines
Sub-total SSH	27.5	23.5	..	..	28.8	..	..	Sous-total SSH
Total NSE+SSH	165.0	143.8	178.4	..	180.3	..	..	*Total SENI+SSH*

Source: OECD, R&D database (DSTI/EAS Division), January 2002.

Source : OCDE, base de données R-D (DSTI/Division AES), janvier 2002.

AUSTRALIA Table 16 - Tableau 16 AUSTRALIE

Higher education intramural expenditure on R&D - Dépenses intra-muros de R-D de l'enseignement supérieur

(HERD) by field of science and source of funds - (DIRDES) par discipline scientifique et par source de financement

Million AUD *Millions AUD*

	1994	1995	1996	1997	1998	1999	2000	
Direct government								**État direct**
Natural sciences	165.1	157.9	..	..	200.4	..	..	Sciences exactes et naturelles
Engineering	89.4	90.8	..	..	117.3	..	..	Sciences de l'ingénieur
Medical sciences	136.8	123.6	..	..	164.8	..	..	Sciences médicales
Agricultural sciences	43.2	42.7	..	..	68.1	..	..	Sciences agricoles
Sub-total NSE	434.6	415.0	..	..	550.6	..	..	Sous-total SENI
Social sciences	64.5	64.1	..	..	89.3	..	..	Sciences sociales
Humanities	23.9	26.7	..	..	33.8	..	..	Sciences humaines
Sub-total SSH	88.4	90.8	..	..	123.1	..	..	Sous-total SSH
Total NSE+SSH	510.1	505.8	575.8	..	673.7	..	..	*Total SENI+SSH*
General university funds								**Fonds généraux des universités**
Natural sciences	334.0	386.1	..	..	448.3	..	..	Sciences exactes et naturelles
Engineering	170.7	191.0	..	..	251.1	..	..	Sciences de l'ingénieur
Medical sciences	192.8	260.8	..	..	331.3	..	..	Sciences médicales
Agricultural sciences	59.0	73.1	..	..	81.6	..	..	Sciences agricoles
Sub-total NSE	756.5	911.0	..	..	1 112.3	..	..	Sous-total SENI
Social sciences	291.5	304.8	..	..	390.7	..	..	Sciences sociales
Humanities	117.7	131.8	..	..	158.1	..	..	Sciences humaines
Sub-total SSH	409.2	436.6	..	..	548.8	..	..	Sous-total SSH
Total NSE+SSH	1 165.8	1 347.6	1 508.3	..	1 661.1	..	..	*Total SENI+SSH*
Sub-total government								**Sous-total État**
Natural sciences	499.1	544.0	..	..	648.7	..	..	Sciences exactes et naturelles
Engineering	260.1	281.8	..	..	368.4	..	..	Sciences de l'ingénieur
Medical sciences	329.6	384.5	..	..	496.1	..	..	Sciences médicales
Agricultural sciences	102.2	115.8	..	..	149.7	..	..	Sciences agricoles
Sub-total NSE	1 191.1	1 326.0	..	..	1 662.9	..	..	Sous-total SENI
Social sciences	356.0	368.9	..	..	480.0	..	..	Sciences sociales
Humanities	141.6	158.5	..	..	191.9	..	..	Sciences humaines
Sub-total SSH	497.6	527.4	..	..	671.9	..	..	Sous-total SSH
Total NSE+SSH	1 675.9	1 853.4	2 084.1	..	2 334.8	..	..	*Total SENI+SSH*
Higher education								**Enseignement supérieur**
Natural sciences	..	..	..	..	..	..	..	Sciences exactes et naturelles
Engineering	..	..	..	..	..	..	..	Sciences de l'ingénieur
Medical sciences	..	..	..	..	..	..	..	Sciences médicales
Agricultural sciences	..	..	..	..	..	..	..	Sciences agricoles
Sub-total NSE	..	..	..	..	..	..	..	Sous-total SENI
Social sciences	..	..	..	..	..	..	..	Sciences sociales
Humanities	..	..	..	..	..	..	..	Sciences humaines
Sub-total SSH	..	..	..	..	..	..	..	Sous-total SSH
Total NSE+SSH	..	..	..	..	..	..	..	*Total SENI+SSH*
Private non-profit								**Institutions sans but lucratif**
Natural sciences	10.9	14.3	..	..	23.4	..	..	Sciences exactes et naturelles
Engineering	9.2	8.5	..	..	7.9	..	..	Sciences de l'ingénieur
Medical sciences	26.8	30.5	..	..	43.0	..	..	Sciences médicales
Agricultural sciences	2.8	3.7	..	..	7.7	..	..	Sciences agricoles
Sub-total NSE	49.6	57.0	..	..	82.0	..	..	Sous-total SENI
Social sciences	7.3	8.9	..	..	6.9	..	..	Sciences sociales
Humanities	1.4	1.6	..	..	2.1	..	..	Sciences humaines
Sub-total SSH	8.7	10.5	..	..	9.0	..	..	Sous-total SSH
Total NSE+SSH	71.2	67.5	78.2	..	90.9	..	..	*Total SENI+SSH*
Business entreprise								**Entreprises**
Natural sciences	20.1	24.8	..	..	35.5	..	..	Sciences exactes et naturelles
Engineering	19.2	26.9	..	..	37.6	..	..	Sciences de l'ingénieur
Medical sciences	14.1	31.3	..	..	34.9	..	..	Sciences médicales
Agricultural sciences	3.8	3.5	..	..	11.9	..	..	Sciences agricoles
Sub-total NSE	57.1	86.6	..	..	120.0	..	..	Sous-total SENI
Social sciences	5.7	7.0	..	..	13.4	..	..	Sciences sociales
Humanities	1.1	1.6	..	..	2.4	..	..	Sciences humaines
Sub-total SSH	6.8	8.6	..	..	15.8	..	..	Sous-total SSH
Total NSE+SSH	63.9	95.1	120.7	..	135.8	..	..	*Total SENI+SSH*
Funds from abroad								**Fonds de l'étranger**
Natural sciences	4.1	7.9	..	..	7.8	..	..	Sciences exactes et naturelles
Engineering	4.0	3.0	..	..	6.4	..	..	Sciences de l'ingénieur
Medical sciences	5.4	7.6	..	..	18.5	..	..	Sciences médicales
Agricultural sciences	1.1	0.3	..	..	1.3	..	..	Sciences agricoles
Sub-total NSE	14.6	18.8	..	..	33.9	..	..	Sous-total SENI
Social sciences	3.4	3.5	..	..	6.2	..	..	Sciences sociales
Humanities	0.5	0.7	..	..	1.1	..	..	Sciences humaines
Sub-total SSH	3.9	4.2	..	..	7.3	..	..	Sous-total SSH
Total NSE+SSH	18.5	23.1	24.6	..	41.2	..	..	*Total SENI+SSH*

Source: OECD, R&D database (DSTI/EAS Division), January 2002.

Source : OCDE, base de données R-D (DSTI/Division AES), janvier 2002.

AUSTRALIA — Table 17 - Tableau 17 — AUSTRALIE

Higher education R&D personnel (HEMP) - Personnel de R-D du secteur de l'enseignement supérieur

by field of science and occupation - par discipline scientifique et par profession

In full-time equivalent — *En équivalence plein-temps*

	1994	1995	1996	1997	1998	1999	2000	
Total R&D personnel								**Personnel total de R-D**
Natural sciences	10 387	..	..	..	10 979	..	..	Sciences exactes et naturelles
Engineering	6 369	..	..	..	6 865	..	..	Sciences de l'ingénieur
Medical sciences	7 103	..	..	..	8 482	..	..	Sciences médicales
Agricultural sciences	2 239	..	..	..	2 659	..	..	Sciences agricoles
Sub-total NSE	26 098	..	27 684	..	28 985	..	..	Sous-total SENI
Social sciences	9 473	..	..	..	10 939	..	..	Sciences sociales
Humanities	4 525	..	..	..	5 578	..	..	Sciences humaines
Sub-total SSH	13 998	..	15 055	..	16 517	..	..	Sous-total SSH
Total NSE+SSH	40 096	..	42 739	..	45 502	..	..	*Total SENI+SSH*
Researchers								**Chercheurs**
Natural sciences	8 059	..	..	..	8 708	..	..	Sciences exactes et naturelles
Engineering	5 071	..	..	..	5 604	..	..	Sciences de l'ingénieur
Medical sciences	4 989	..	..	..	6 710	..	..	Sciences médicales
Agricultural sciences	1 650	..	..	..	2 087	..	..	Sciences agricoles
Sub-total NSE	19 770	..	..	..	23 110	..	..	Sous-total SENI
Social sciences	8 329	..	..	..	9 796	..	..	Sciences sociales
Humanities	4 173	..	..	..	5 231	..	..	Sciences humaines
Sub-total SSH	12 502	..	..	..	15 027	..	..	Sous-total SSH
Total NSE+SSH	32 272	..	35 472	..	38 137	..	..	*Total SENI+SSH*
Technicians								**Technicians**
Natural sciences	.. [n]	..	..	..	.. [n]	..	..	Sciences exactes et naturelles
Engineering	.. [n]	..	..	..	.. [n]	..	..	Sciences de l'ingénieur
Medical sciences	.. [n]	..	..	..	.. [n]	..	..	Sciences médicales
Agricultural sciences	.. [n]	..	..	..	.. [n]	..	..	Sciences agricoles
Sub-total NSE	.. [n]	..	..	..	.. [n]	..	..	Sous-total SENI
Social sciences	.. [n]	..	..	..	.. [n]	..	..	Sciences sociales
Humanities	.. [n]	..	..	..	.. [n]	..	..	Sciences humaines
Sub-total SSH	.. [n]	..	..	..	.. [n]	..	..	Sous-total SSH
Total NSE+SSH	.. [n]	..	.. [n]	..	.. [n]	..	..	*Total SENI+SSH*
Other								**Autres**
Natural sciences	2 328 [o]	..	..	..	2 271 [o]	..	..	Sciences exactes et naturelles
Engineering	1 298 [o]	..	..	..	1 261 [o]	..	..	Sciences de l'ingénieur
Medical sciences	2 114 [o]	..	..	..	1 772 [o]	..	..	Sciences médicales
Agricultural sciences	589 [o]	..	..	..	572 [o]	..	..	Sciences agricoles
Sub-total NSE	6 328 [o]	..	..	..	5 875 [o]	..	..	Sous-total SENI
Social sciences	1 144 [o]	..	..	..	1 143 [o]	..	..	Sciences sociales
Humanities	352 [o]	..	..	..	347 [o]	..	..	Sciences humaines
Sub-total SSH	1 495 [o]	..	..	..	1 489 [o]	..	..	Sous-total SSH
Total NSE+SSH	7 824 [o]	..	7 266 [o]	..	7 365 [o]	..	..	*Total SENI+SSH*

Table 18 - Tableau 18

Government budget appropriations or outlays for R&D (GBAORD) - Crédits budgétaires publics de R-D (CBPRD)

by socio-economic objective - par objectif socio-économique

Million AUD — *Millions AUD*

	1995	1996	1997	1998	1999	2000	2001	
Agriculture, forestry and fishing	265.1 [h]	272.0 [h]	293.4 [h]	305.3 [h]	330.3 [a,h]	330.3 [h,p]	335.4 [h,p]	Agriculture, sylviculture et pêche
Industrial development	351.0 [h]	351.2 [h]	370.9 [h]	469.2 [h]	574.0 [a,h]	532.4 [h,p]	613.5 [h,p]	Développement industriel
Energy	62.7 [h]	54.1 [h]	72.4 [h]	55.7 [h]	69.7 [a,h]	94.6 [h,p]	109.5 [h,p]	Energie
Transport & telecommunications	21.8 [h]	24.5 [h]	26.2 [h]	26.6 [h]	33.8 [a,h]	32.9 [h,p]	35.9 [h,p]	*Transports & télécommunications*
Urban & rural planning	..	..	..	..	..	..	..	*Aménagement urbain et rural*
ST Infrastructure	21.8 [h]	24.5 [h]	26.2 [h]	26.6 [h]	33.8 [a,h]	32.9 [h,p]	35.9 [h,p]	ST Infrastructure
Prevention of pollution	56.8 [h]	52.5 [h]	57.3 [h]	58.7 [h]	72.8 [a,h]	75.1 [h,p]	74.9 [h,p]	*Prévention de la pollution*
Identificat. & treatment of pollution	28.7 [h]	27.2 [h]	28.2 [h]	29.1 [h]	39.8 [a,h]	52.0 [h,p]	56.2 [h,p]	*Détect. pollut. & mesures antipollut.*
ST Environment	85.5 [h]	79.7 [h]	85.4 [h]	87.8 [h]	112.5 [a,h]	127.1 [h,p]	131.1 [h,p]	ST Environnement
Health	204.7 [h]	209.4 [h]	221.7 [h]	244.4 [h]	241.4 [a,h]	317.2 [h,p]	450.9 [h,p]	Santé
Social development & services	8.9 [h]	8.7 [h]	9.4 [h]	9.6 [h]	12.4 [a,h]	12.3 [h,p]	12.3 [h,p]	Développement et services sociaux
Earth and atmosphere	125.8 [h]	186.1 [h]	156.4 [h]	138.3 [h]	158.3 [a,h]	161.1 [h,p]	164.4 [h,p]	Milieu terrestre et atmosphère
Advancement of research	613.8 [h]	673.6 [h]	706.3 [h]	730.2 [h]	781.0 [a,h]	..	..	*Promotion de la recherche*
General university funds	1 006.0 [h]	1 062.0 [h]	1 104.0 [h]	1 135.0 [h]	1 182.0 [a,h]	..	..	*Fonds généraux des universités*
ST Advancement of knowledge	1 619.8 [h]	1 735.6 [h]	1 810.3 [h]	1 865.2 [h]	1 963.0 [a,h]	2 015.7 [h,p]	2 089.8 [h,p]	ST promotion gén. des connaiss.
Civil space	.. [n]	.. [n]	.. [n]	.. [n]	.. [n]	.. [n]	.. [n]	Espace civil
Defence	269.2 [h]	257.0 [h]	247.8 [h]	251.5 [h]	272.3 [a,h]	290.8 [h,p]	287.3 [h,p]	Défense
Not elsewhere classified	..	..	..	..	..	..	..	Non classé
Total	**3 014.4 [h]**	**3 178.3 [h]**	**3 293.8 [h]**	**3 453.5 [h]**	**3 767.7 [a,h]**	**3 914.4 [h,p]**	**4 230.1 [h,p]**	**Total**

Source: OECD, R&D database (DSTI/EAS Division), January 2002.

Source : OCDE, base de données R-D (DSTI/Division AES), janvier 2002.

AUSTRALIA Table 19 - Tableau 19 **AUSTRALIE**

Technology balance of payments - Balance des paiements technologiques

by type of transfer - selon la nature des transferts

Million AUD *Millions AUD*

	1983	1984	1985	1986	1987	1988	1989	1990	1991	
Receipts										**Recettes**
Related firms	..	..	..	..	..	..	..	..	..	Firmes affiliées
Unrelated firms	..	..	..	..	..	..	..	..	..	Firmes non affiliées
Other	..	..	..	..	..	..	..	..	..	Autre
Total	..	37	..	102	..	135	..	134	..	*Total*
Payments										**Paiements**
Related firms	..	..	..	..	..	..	..	..	..	Firmes affiliées
Unrelated firms	..	..	..	..	..	..	..	..	..	Firmes non affiliées
Other	..	..	..	..	..	..	..	..	..	Autre
Total	..	167	..	281	..	376	..	374	..	*Total*

	1992	1993	1994	1995	1996	1997	1998	1999	2000	
Receipts										**Recettes**
Related firms	..	..	..	..	..	..	..	..	..	Firmes affiliées
Unrelated firms	..	..	..	..	..	..	..	..	..	Firmes non affiliées
Other	..	..	..	..	..	..	..	..	..	Autre
Total	272	..	214	..	164	..	164	..	..	*Total*
Payments										**Paiements**
Related firms	..	..	..	..	..	..	..	..	..	Firmes affiliées
Unrelated firms	..	..	..	..	..	..	..	..	..	Firmes non affiliées
Other	..	..	..	..	..	..	..	..	..	Autre
Total	504	..	451	..	439	..	358	..	..	*Total*

Source: OECD, TBPdatabase (DSTI/EAS Division), Januay 2002.

Source : OCDE, base de données TBP (DSTI/Division AES), janvier 2002.

Table 1 - Tableau 1

Gross domestic expenditure on R&D (GERD) - Dépenses intérieures brutes de R-D (DIRD)

by sector of performance and source of funds - par secteur d'exécution et par source de financement

	1994	1995	1996	1997	1998	1999	2000	
	1.1 Million EUR (1999 ATS euro, EUR from 1999) - Millions EUR (1999 ATS euro, EUR à partir de 1999)							
Business enterprise								**Entreprises**
financed by								*financé par*
Business enterprise	..	..	..	..	1 390.6	..	..	Entreprises
Direct government	..	..	..	..	..	..	..	État direct
Higher education	..	..	..	..	..	..	..	Enseignement supérieur
Private non-profit	..	..	..	..	..	..	..	Institutions sans but lucratif
Funds from abroad	..	..	..	..	649.5	..	..	Fonds de l'étranger
Total	..	..	..	..	2 160.7	..	..	*Total*
Government								**État**
financed by								*financé par*
Business enterprise	..	..	..	..	..	..	..	Entreprises
Direct government	..	..	..	..	..	..	..	État direct
Higher education	..	..	..	..	..	..	..	Enseignement supérieur
Private non-profit	..	..	..	..	..	..	..	Institutions sans but lucratif
Funds from abroad	..	..	..	..	..	..	..	Fonds de l'étranger
Total	..	..	..	..	218.9	..	..	*Total*
Higher education								**Enseignement supérieur**
financed by								*financé par*
Business enterprise	..	..	..	..	..	..	..	Entreprises
Direct government	..	..	..	..	..	..	..	État direct
General university funds	..	..	..	..	..	..	..	Fonds généraux des universités
Sub-total government	..	..	..	..	..	..	..	Sous-Total État
Higher education	..	..	..	..	..	..	..	Enseignement supérieur
Private non-profit	..	..	..	..	..	..	..	Institutions sans but lucratif
Funds from abroad	..	..	..	..	..	..	..	Fonds de l'étranger
Total	..	..	..	..	1 009.7	..	..	*Total*
Private non-profit								**Institutions sans but lucratif**
financed by								*financé par*
Business enterprise	..	..	..	..	3.5	..	..	Entreprises
Direct government	..	..	..	..	..	..	..	État direct
Higher education	..	..	..	..	..	..	..	Enseignement supérieur
Private non-profit	..	..	..	..	2.5	..	..	Institutions sans but lucratif
Funds from abroad	..	..	..	..	2.1	..	..	Fonds de l'étranger
Total	..	..	..	..	10.5	..	..	*Total*
GERD								**DIRD**
financed by								*financé par*
Business enterprise	1 168.1 [c]	1 212.8 [c]	1 262.1 [c]	1 315.9 [c]	1 389.9 [c,s]	1 444.1 [c]	1 511.9 [c,p]	Entreprises
Direct government	524.6 [c]	522.2 [c]	519.1 [c]	546.6 [c]	579.2 [c,s]	622.6 [c]	668.7 [c,p]	État direct
General university funds	729.8 [c]	745.0 [c]	727.5 [c]	734.3 [c]	766.7 [c,s]	805.4 [c]	762.9 [c,p]	Fonds généraux des universités
Sub-total government	1 254.4 [c]	1 267.3 [c]	1 246.6 [c]	1 280.9 [c]	1 346.0 [c,s]	1 428.0 [c]	1 431.6 [c,p]	Sous-Total État
Higher education	..	..	..	..	..	..	..	Enseignement supérieur
Private non-profit	10.4 [c]	10.8 [c]	11.2 [c]	11.5 [c]	11.9 [c,s]	12.4 [c]	12.9 [c,p]	Institutions sans but lucratif
Funds from abroad	106.5 [c]	189.1 [c]	335.5 [c]	476.7 [c]	686.6 [c,s]	715.3 [c]	731.0 [c,p]	Fonds de l'étranger
Total	2 539.4 [c]	2 679.9 [c]	2 855.4 [c]	3 085.0 [c]	3 399.8	3 599.8 [c]	3 687.5 [c,p]	*Total*
	1.2 Million constant USD (1995 prices and PPPs) - Millions de USD constants (prix et PPA de 1995)							
Business enterprise								**Entreprises**
financed by								*financé par*
Business enterprise	..	..	..	..	1 348.3	..	..	Entreprises
Direct government	..	..	..	..	..	..	..	État direct
Higher education	..	..	..	..	..	..	..	Enseignement supérieur
Private non-profit	..	..	..	..	..	..	..	Institutions sans but lucratif
Funds from abroad	..	..	..	..	629.7	..	..	Fonds de l'étranger
Total	..	..	..	..	2 095.0	..	..	*Total*
Government								**État**
financed by								*financé par*
Business enterprise	..	..	..	..	..	..	..	Entreprises
Direct government	..	..	..	..	..	..	..	État direct
Higher education	..	..	..	..	..	..	..	Enseignement supérieur
Private non-profit	..	..	..	..	..	..	..	Institutions sans but lucratif
Funds from abroad	..	..	..	..	..	..	..	Fonds de l'étranger
Total	..	..	..	..	212.3	..	..	*Total*
Higher education								**Enseignement supérieur**
financed by								*financé par*
Business enterprise	..	..	..	..	..	..	..	Entreprises
Direct government	..	..	..	..	..	..	..	État direct
General university funds	..	..	..	..	..	..	..	Fonds généraux des universités
Sub-total government	..	..	..	..	..	..	..	Sous-Total État
Higher education	..	..	..	..	..	..	..	Enseignement supérieur
Private non-profit	..	..	..	..	..	..	..	Institutions sans but lucratif
Funds from abroad	..	..	..	..	..	..	..	Fonds de l'étranger
Total	..	..	..	..	979.0	..	..	*Total*
Private non-profit								**Institutions sans but lucratif**
financed by								*financé par*
Business enterprise	..	..	..	..	3.4	..	..	Entreprises
Direct government	..	..	..	..	..	..	..	État direct
Higher education	..	..	..	..	..	..	..	Enseignement supérieur
Private non-profit	..	..	..	..	2.5	..	..	Institutions sans but lucratif
Funds from abroad	..	..	..	..	2.1	..	..	Fonds de l'étranger
Total	..	..	..	..	10.2	..	..	*Total*
GERD								**DIRD**
financed by								*financé par*
Business enterprise	1 199.6 [c]	1 215.2 [c]	1 248.2 [c]	1 285.4 [c]	1 347.7 [c,s]	1 388.0 [c]	1 431.6 [b,p]	Entreprises
Direct government	538.8 [c]	523.3 [c]	513.4 [c]	534.0 [c]	561.6 [c,s]	598.4 [c]	633.2 [b,p]	État direct
General university funds	749.5 [c]	746.5 [c]	719.5 [c]	717.3 [c]	743.4 [c,s]	774.2 [c]	722.4 [b,p]	Fonds généraux des universités
Sub-total government	1 288.2 [c]	1 269.8 [c]	1 232.9 [c]	1 251.3 [c]	1 305.0 [c,s]	1 372.6 [c]	1 355.6 [b,p]	Sous-Total État
Higher education	..	..	..	..	..	..	..	Enseignement supérieur
Private non-profit	10.7 [c]	10.9 [c]	11.1 [c]	11.2 [c]	11.6 [c,s]	11.9 [c]	12.2 [b,p]	Institutions sans but lucratif
Funds from abroad	109.3 [c]	189.5 [c]	331.8 [c]	465.6 [c]	665.7 [c,s]	687.5 [c]	692.2 [b,p]	Fonds de l'étranger
Total	2 607.9 [c]	2 685.3 [c]	2 823.9 [c]	3 013.6 [c]	3 296.4	3 460.0 [c]	3 491.6 [b,p]	*Total*

Source: OECD, R&D database (DSTI/EAS Division), January 2002.

Source : OCDE, base de données R-D (DSTI/Division AES), janvier 2002.

Table 2 - Tableau 2

Gross domestic expenditure on R&D (GERD) - Dépenses intérieures brutes de R-D (DIRD)

by sector of performance and type of costs - par secteur d'exécution et par type de coûts

Million EUR (1999 ATS euro, EUR from 1999) — *Millions EUR (1999 ATS euro, EUR à partir de 1999)*

	1994	1995	1996	1997	1998	1999	2000	
Business enterprise								**Entreprises**
Labour costs	..	..	..	..	1 202.0	..	..	Coûts du personnel
Other current costs	..	..	..	..	764.4	..	..	Autres coûts courants
Total current expenditure	..	..	..	..	1 966.4	..	..	*Total dépenses courantes*
Land and buildings	..	..	..	..	21.4	..	..	Terrains et constructions
Instruments and equipment	..	..	..	..	172.9	..	..	Instruments et équipements
Total capital expenditure	..	..	..	..	194.3	..	..	*Total dépenses en capital*
Total	..	..	..	..	2 160.7	..	..	*Total*
Government								**État**
Labour costs	..	..	..	..	119.4	..	..	Coûts du personnel
Other current costs	..	..	..	..	72.5	..	..	Autres coûts courants
Total current expenditure	..	..	..	..	191.9	..	..	*Total dépenses courantes*
Land and buildings	..	..	..	..	13.8	..	..	Terrains et constructions
Instruments and equipment	..	..	..	..	13.3	..	..	Instruments et équipements
Total capital expenditure	..	..	..	..	27.1	..	..	*Total dépenses en capital*
Total	..	..	..	..	218.9	..	..	*Total*
Higher education								**Enseignement supérieur**
Labour costs	..	..	..	..	387.2	..	..	Coûts du personnel
Other current costs	..	..	..	..	473.8	..	..	Autres coûts courants
Total current expenditure	..	..	..	..	860.9	..	..	*Total dépenses courantes*
Land and buildings	..	..	..	..	80.0	..	..	Terrains et constructions
Instruments and equipment	..	..	..	..	68.8	..	..	Instruments et équipements
Total capital expenditure	..	..	..	..	148.8	..	..	*Total dépenses en capital*
Total	..	..	..	..	1 009.7	..	..	*Total*
Private non-profit								**Institutions sans but lucratif**
Labour costs	..	..	..	..	5.5	..	..	Coûts du personnel
Other current costs	..	..	..	..	4.6	..	..	Autres coûts courants
Total current expenditure	..	..	..	..	10.1	..	..	*Total dépenses courantes*
Land and buildings	..	..	..	..	0.0	..	..	Terrains et constructions
Instruments and equipment	..	..	..	..	0.4	..	..	Instruments et équipements
Total capital expenditure	..	..	..	..	0.4	..	..	*Total dépenses en capital*
Total	..	..	..	..	10.5	..	..	*Total*
GERD								**DIRD**
Labour costs	..	..	..	..	1 714.1	..	..	Coûts du personnel
Other current costs	..	..	..	..	1 315.2	..	..	Autres coûts courants
Total current expenditure	..	..	..	..	3 029.3	..	..	*Total dépenses courantes*
Land and buildings	..	..	..	..	115.2	..	..	Terrains et constructions
Instruments and equipment	..	..	..	..	255.3	..	..	Instruments et équipements
Total capital expenditure	..	..	..	..	370.5	..	..	*Total dépenses en capital*
Total	2 539.4 [c]	2 679.9 [c]	2 855.4 [c]	3 085.0 [c]	3 399.8	3 599.8 [c]	3 687.5 [c,p]	*Total*

Table 3 - Tableau 3

Current domestic expenditure on R&D - Dépenses intérieures courantes de R-D

by sector of performance and type of activity - par secteur d'exécution et par type d'activités

Million EUR (1999 ATS euro, EUR from 1999) — *Millions EUR (1999 ATS euro, EUR à partir de 1999)*

	1994	1995	1996	1997	1998	1999	2000	
Business enterprise								**Entreprises**
Basic research	..	..	..	..	73.0 [a]	..	..	Recherche fondamentale
Applied research	..	..	..	..	716.4 [a]	..	..	Recherche appliquée
Experimental development	..	..	..	..	1 177.0 [a]	..	..	Développement expérimental
Not specified	..	..	..	..	0.0	..	..	Non spécifié
Total	..	..	..	..	1 966.4 [a]	..	..	*Total*
Government								**État**
Basic research	..	..	..	..	41.4 [a,m]	..	..	Recherche fondamentale
Applied research	..	..	..	..	71.4 [a,m]	..	..	Recherche appliquée
Experimental development	..	..	..	..	13.6 [a,m]	..	..	Développement expérimental
Not specified	..	..	..	..	65.4	..	..	Non spécifié
Total	..	..	..	..	191.9 [a]	..	..	*Total*
Higher education								**Enseignement supérieur**
Basic research	..	..	..	..	398.2 [a]	..	..	Recherche fondamentale
Applied research	..	..	..	..	353.3 [a]	..	..	Recherche appliquée
Experimental development	..	..	..	..	109.5 [a]	..	..	Développement expérimental
Not specified	..	..	..	..	0.0	..	..	Non spécifié
Total	..	..	..	..	860.9 [a]	..	..	*Total*
Private non-profit								**Institutions sans but lucratif**
Basic research	..	..	..	..	2.3 [a]	..	..	Recherche fondamentale
Applied research	..	..	..	..	6.4 [a]	..	..	Recherche appliquée
Experimental development	..	..	..	..	1.4 [a]	..	..	Développement expérimental
Not specified	..	..	..	..	0.0	..	..	Non spécifié
Total	..	..	..	..	10.1 [a]	..	..	*Total*
GERD								**DIRD**
Basic research	..	..	..	..	515.0 [a,m]	..	..	Recherche fondamentale
Applied research	..	..	..	..	1 147.5 [a,m]	..	..	Recherche appliquée
Experimental development	..	..	..	..	1 301.4 [a,m]	..	..	Développement expérimental
Not specified	..	..	..	..	65.4	..	..	Non spécifié
Total	..	..	..	..	3 029.3 [a]	..	..	*Total*

Source: OECD, R&D database (DSTI/EAS Division), January 2002.

Source : OCDE, base de données R-D (DSTI/Division AES), janvier 2002.

AUSTRIA **AUTRICHE**

Table 5 - Tableau 5

Total R&D personnel - Personnel total de R-D

by sector of employment and occupation - par secteur d'emploi et par profession

In full-time equivalent *En équivalence plein-temps*

	1994	1995	1996	1997	1998	1999	2000	
Business enterprise								**Entreprises**
RSE	..	..	..	..	11 716	..	..	Chercheurs
Technicians	..	..	..	..	6 319	..	..	Techniciens
Other occupations	..	..	..	..	2 350	..	..	Autres professions
Total	..	..	..	..	20 385	..	..	*Total*
Government								**État**
RSE	..	..	..	..	954 m	..	..	Chercheurs
Technicians	..	..	..	..	326 m	..	..	Techniciens
Other occupations	..	..	..	..	825 m	..	..	Autres professions
Total	..	..	..	..	2 104 m	..	..	*Total*
Higher education								**Enseignement supérieur**
RSE	..	..	..	..	5 955	..	..	Chercheurs
Technicians	..	..	..	..	1 242	..	..	Techniciens
Other occupations	..	..	..	..	1 473	..	..	Autres professions
Total	..	..	..	..	8 670	..	..	*Total*
Private non-profit								**Institutions sans but lucratif**
RSE	..	..	..	..	90	..	..	Chercheurs
Technicians	..	..	..	..	33	..	..	Techniciens
Other occupations	..	..	..	..	26	..	..	Autres professions
Total	..	..	..	..	148	..	..	*Total*
TOTAL								**TOTAL**
RSE	..	..	..	..	18 715 m	..	..	Chercheurs
Technicians	..	..	..	..	7 919 m	..	..	Techniciens
Other occupations	..	..	..	..	4 674 m	..	..	Autres professions
Total	..	..	..	..	31 308 m	..	..	*Total*

Table 6 - Tableau 6

Total R&D personnel - Personnel total de R-D

by sector of employment and formal qualification - par secteur d'emploi et par niveau de qualification

In full-time equivalent *En équivalence plein-temps*

	1994	1995	1996	1997	1998	1999	2000	
Business enterprise								**Entreprises**
University PhD level degrees	..	..	..	..	..	..	..	Diplômes univ. de niv. doctorat
Other university degrees	..	..	..	..	..	..	..	Autres diplômes universitaires
Sub-total university degrees	..	..	..	..	..	..	..	Sous-total Universitaires
Other post-secondary diplomas	..	..	..	..	..	..	..	Autres diplômes post-secondaires
Secondary diplomas	..	..	..	..	..	..	..	Diplômes de fin d'études secondaires
Other qualifications	..	..	..	..	..	..	..	Autres qualifications
Total	..	..	..	..	20 385	..	..	*Total*
Government								**État**
University PhD level degrees	..	..	..	..	..	..	..	Diplômes univ. de niv. doctorat
Other university degrees	..	..	..	..	..	..	..	Autres diplômes universitaires
Sub-total university degrees	..	..	..	..	..	..	..	Sous-total Universitaires
Other post-secondary diplomas	..	..	..	..	..	..	..	Autres diplômes post-secondaires
Secondary diplomas	..	..	..	..	..	..	..	Diplômes de fin d'études secondaires
Other qualifications	..	..	..	..	..	..	..	Autres qualifications
Total	..	..	..	..	2 104 m	..	..	*Total*
Higher education								**Enseignement supérieur**
University PhD level degrees	..	..	..	..	..	..	..	Diplômes univ. de niv. doctorat
Other university degrees	..	..	..	..	..	..	..	Autres diplômes universitaires
Sub-total university degrees	..	..	..	..	..	..	..	Sous-total Universitaires
Other post-secondary diplomas	..	..	..	..	..	..	..	Autres diplômes post-secondaires
Secondary diplomas	..	..	..	..	..	..	..	Diplômes de fin d'études secondaires
Other qualifications	..	..	..	..	..	..	..	Autres qualifications
Total	..	..	..	..	8 670	..	..	*Total*
Private non-profit								**Institutions sans but lucratif**
University PhD level degrees	..	..	..	..	..	..	..	Diplômes univ. de niv. doctorat
Other university degrees	..	..	..	..	..	..	..	Autres diplômes universitaires
Sub-total university degrees	..	..	..	..	..	..	..	Sous-total Universitaires
Other post-secondary diplomas	..	..	..	..	..	..	..	Autres diplômes post-secondaires
Secondary diplomas	..	..	..	..	..	..	..	Diplômes de fin d'études secondaires
Other qualifications	..	..	..	..	..	..	..	Autres qualifications
Total	..	..	..	..	148	..	..	*Total*
TOTAL								**TOTAL**
University PhD level degrees	..	..	..	..	..	..	..	Diplômes univ. de niv. doctorat
Other university degrees	..	..	..	..	..	..	..	Autres diplômes universitaires
Sub-total university degrees	..	..	..	..	..	..	..	Sous-total Universitaires
Other post-secondary diplomas	..	..	..	..	..	..	..	Autres diplômes post-secondaires
Secondary diplomas	..	..	..	..	..	..	..	Diplômes de fin d'études secondaires
Other qualifications	..	..	..	..	..	..	..	Autres qualifications
Total	..	..	..	..	31 308 m	..	..	*Total*

Source: OECD, R&D database (DSTI/EAS Division), January 2002.

Source : OCDE, base de données R-D (DSTI/Division AES), janvier 2002.

Table 7 - Tableau 7

Gross domestic expenditure on R&D (GERD) - Dépenses intérieures brutes de R-D (DIRD)

by sector of performance and main scientific field - par secteur d'exécution et par grand domaine scientifique

Million EUR (1999 ATS euro, EUR from 1999) *Millions EUR (1999 ATS euro, EUR à partir de 1999)*

	1994	1995	1996	1997	1998	1999	2000	
Business enterprise								**Entreprises**
Natural sciences and engineering	..	..	..	..	2 160.7	..	..	Sciences exactes, nat. et de l'ingénieur
Social sciences and humanities	..	..	..	..	..	..	..	Sciences sociales et humaines
All fields of science	..	..	..	..	2 160.7	..	..	*Tous domaines scientifiques*
Government								**État**
Natural sciences and engineering	..	..	..	..	140.3	..	..	Sciences exactes, nat. et de l'ingénieur
Social sciences and humanities	..	..	..	..	78.6	..	..	Sciences sociales et humaines
All fields of science	..	..	..	..	218.9	..	..	*Tous domaines scientifiques*
Higher education								**Enseignement supérieur**
Natural sciences and engineering	..	..	..	..	775.2	..	..	Sciences exactes, nat. et de l'ingénieur
Social sciences and humanities	..	..	..	..	234.5	..	..	Sciences sociales et humaines
All fields of science	..	..	..	..	1 009.7	..	..	*Tous domaines scientifiques*
Private non-profit								**Institutions sans but lucratif**
Natural sciences and engineering	..	..	..	..	4.8	..	..	Sciences exactes, nat. et de l'ingénieur
Social sciences and humanities	..	..	..	..	5.7	..	..	Sciences sociales et humaines
All fields of science	..	..	..	..	10.5	..	..	*Tous domaines scientifiques*
TOTAL								**TOTAL**
Natural sciences and engineering	..	..	..	..	3 081.0	..	..	Sciences exactes, nat. et de l'ingénieur
Social sciences and humanities	..	..	..	..	318.9	..	..	Sciences sociales et humaines
All fields of science	2 539.4 *c*	2 679.9 *c*	2 855.4 *c*	3 085.0 *c*	3 399.8	3 599.8 *c*	3 687.5 *c,p*	*Tous domaines scientifiques*

Table 8 - Tableau 8

Total R&D personnel - Personnel total de R-D

by sector of employment and main scientific field - par secteur d'emploi et par grand domaine scientifique

In full-time equivalent *En équivalence plein-temps*

	1994	1995	1996	1997	1998	1999	2000	
Business enterprise								**Entreprises**
Natural sciences and engineering	..	..	..	..	20 385	..	..	Sciences exactes, nat. et de l'ingénieur
Social sciences and humanities	..	..	..	..	..	..	..	Sciences sociales et humaines
All fields of science	..	..	..	..	20 385	..	..	*Tous domaines scientifiques*
Government								**État**
Natural sciences and engineering	..	..	..	..	1 046 *m*	..	..	Sciences exactes, nat. et de l'ingénieur
Social sciences and humanities	..	..	..	..	1 058	..	..	Sciences sociales et humaines
All fields of science	..	..	..	..	2 104 *m*	..	..	*Tous domaines scientifiques*
Higher education								**Enseignement supérieur**
Natural sciences and engineering	..	..	..	..	6 655	..	..	Sciences exactes, nat. et de l'ingénieur
Social sciences and humanities	..	..	..	..	2 016	..	..	Sciences sociales et humaines
All fields of science	..	..	..	..	8 670	..	..	*Tous domaines scientifiques*
Private non-profit								**Institutions sans but lucratif**
Natural sciences and engineering	..	..	..	..	76	..	..	Sciences exactes, nat. et de l'ingénieur
Social sciences and humanities	..	..	..	..	71	..	..	Sciences sociales et humaines
All fields of science	..	..	..	..	148	..	..	*Tous domaines scientifiques*
TOTAL								**TOTAL**
Natural sciences and engineering	..	..	..	..	28 162 *m*	..	..	Sciences exactes, nat. et de l'ingénieur
Social sciences and humanities	..	..	..	..	3 145	..	..	Sciences sociales et humaines
All fields of science	..	..	..	..	31 308 *m*	..	..	*Tous domaines scientifiques*

Source: OECD, R&D database (DSTI/EAS Division), January 2002.

Source : OCDE, base de données R-D (DSTI/Division AES), janvier 2002.

Total business enterprise intramural expenditure on R&D - Dépenses intra-muros totales de R-D des entreprises
(BERD) by industry - (DIRDE) par industrie

9.1 Million EUR (1999 ATS euro, EUR from 1999) - Millions EUR (1999 ATS euro, EUR à partir de 1999)

	1994	1995	1996	1997	1998	1999	2000	
1 AGRICULTURE, HUNTING & FORESTRY	..	..	..	..	.. n	..	..	**1 AGRICULTURE, CHASSE & SYLVICULTURE**
2 MINING	..	..	..	..	.. n	..	..	**2 ACTIVITES EXTRACTIVES**
3 MANUFACTURING	..	..	..	..	1 652.7 o	..	..	**3 ACTIVITES DE FABRICATION**
4 Food, beverages & tobacco	..	..	..	..	.. n	..	..	**4 Produits alimentaires, boissons & tabac**
5 Food products & beverages	..	..	..	..	18.2	..	..	5 Produits alimentaires & boissons
6 Tobacco products	..	..	..	..	.. n	..	..	6 Produits à base de tabac
7 Textiles, wear app., fur & leather	..	..	..	..	.. n	..	..	**7 Textiles, habill., fourrures & cuirs**
8 Textiles	..	..	..	..	17.9	..	..	8 Textiles
9 Wearing apparel & fur	..	..	..	..	.. n	..	..	9 Articles d'habillement & fourrures
10 Leather products & footwear	..	..	..	..	2.4	..	..	10 Articles en cuir & chaussures
11 Wood, paper, printing, publishing	..	..	..	..	25.4	..	..	**11 Bois, papier, imprimerie, édition**
12 Wood & cork (not furniture)	..	..	..	..	9.9	..	..	12 Bois & liège (sauf meubles)
13 Pulp, paper & paper products	..	..	..	..	11.9	..	..	13 Carton, papier & articles
14 Publ., print. & repro. of rec. media	..	..	..	..	3.7	..	..	14 Edition, impress. & repro. enregis.
15 Coke, petroleum, nuclear fuel, chemicals, rubber plastics	..	..	..	..	.. n	..	..	**15 Coke, pétrole, combust. nucl., prod. chimi., caout. & mat. plast.**
16 Coke, ref. petrol. prod. & nuclear fuel	..	..	..	..	.. n	..	..	16 Coke, raffin. pétr. & combust. nucl.
17 Chemicals & chemical products	..	..	..	..	202.3	..	..	17 Produits chimiques
18 *Chemical prod.(less pharmaceuticals)*	..	..	..	..	78.7	..	..	18 *Prod. chimi. (moins prod. pharmaceutiques)*
19 *Pharmaceuticals*	..	..	..	..	123.7	..	..	19 *Produits pharmaceutiques*
20 Rubber & plastic products	..	..	..	..	44.6	..	..	20 Caoutchouc et matières plastiques
21 Non-metallic mineral products	..	..	..	..	43.9	..	..	**21 Produits minéraux non métalliques**
22 Basic metals	..	..	..	..	57.1	..	..	**22 Produits métallurgiques de base**
23 Basic metals, ferrous	..	..	..	..	36.5	..	..	23 Produits métallur. de base, ferreux
24 Basic metals, non-ferrous	..	..	..	..	20.6	..	..	24 Prod. métallur. de base, non ferreux
25 Fabricated metal products	..	..	..	..	49.5	..	..	**25 Ouvrages en métaux (sauf mach.& mat.)**
26 Machinery equipment, instruments and transport equipment	..	..	..	..	1 151.2	..	..	**26 Machines & matériel, instruments & mat. de transport**
27 Machinery n.e.c.	..	..	..	..	185.8	..	..	27 Machines, n.c.a.
28 Office, account. & computing machinery	..	..	..	..	4.6	..	..	28 Mach. de bur., compta. & trait. inform.
29 Electrical machinery	..	..	..	..	96.6	..	..	29 Machines électriques
30 Electro. equip. (radio, TV & commun.)	..	..	..	..	575.6	..	..	30 Equip. électro. (radio, tv & communic.)
31 *Electro. comp. (inc. semi-conductors)*	..	..	..	..	73.3	..	..	31 *Compos. électro. (semi-cond. incl.)*
32 *TV, radio & communications equip.*	..	..	..	..	502.3	..	..	32 *Appareils radio, tv et de communi.*
33 Instruments	..	..	..	..	48.5	..	..	33 Instruments
34 Motor vehicles	..	..	..	..	183.2	..	..	34 Véhicules automobiles
35 Other transport equipment	..	..	..	..	56.9 o	..	..	35 Autres matériels de transport
36 *Ships*	..	..	..	..	.. n	..	..	36 *Navires*
37 *Aerospace*	..	..	..	..	.. n	..	..	37 *Aérospatiale*
38 *Other transport equipment n.e.c.*	..	..	..	..	49.8	..	..	38 *Autres matériels de transport n.c.a.*
39 Furniture, other manufacturing n.e.c.	..	..	..	..	20.1	..	..	**39 Meubles, autres activ. de fabricat. n.c.a.**
40 Furniture	..	..	..	..	6.8	..	..	40 Meubles
41 Other manufacturing n.e.c.	..	..	..	..	13.2	..	..	41 Autres activités de fabrication n.c.a.
42 Recycling	..	..	..	..	0.0	..	..	**42 Récupération**
43 ELECTRICITY, GAS & WATER SUPPLY	..	..	..	..	8.8	..	..	**43 DISTRIBUTION D'ELECTRICITE, GAZ & EAU**
44 CONSTRUCTION	..	..	..	..	13.8	..	..	**44 CONSTRUCTION**
45 SERVICES SECTOR	..	..	..	..	483.8 o	..	..	**45 SECTEUR DES SERVICES**
46 Wholesale, retail trade & repair of motor vehicles, etc.	..	..	..	..	55.4	..	..	**46 Commerce (gros & détail), réparation de véhicules, etc.**
47 Hotels & restaurants	..	..	..	..	0.0	..	..	**47 Hôtels et restaurants**
48 Transport & storage	..	..	..	..	.. n	..	..	**48 Transports, entreposage**
49 Communications	..	..	..	..	.. n	..	..	**49 Communications**
50 Post	..	..	..	..	0.0	..	..	50 Postes
51 Telecommunications	..	..	..	..	.. n	..	..	51 Télécommunications
52 Financial intermediation (inc. insur.)	..	..	..	..	27.2	..	..	**52 Intermédiation financ. (assurances inc.)**
53 Real estate, renting & busin. activity	..	..	..	..	376.9	..	..	**53 Immobil., locat. & serv. aux entreprises**
54 Computer & related activities	..	..	..	..	33.1	..	..	54 Informatique et activités connexes
55 *Software consultancy*	..	..	..	..	19.7	..	..	55 *Production de logiciels*
56 *Other computer services n.e.c.*	..	..	..	..	13.5	..	..	56 *Autres activités informatiques*
57 Research & development	..	..	..	..	156.2	..	..	57 Recherche & développement expérimental
58 Other business activities n.e.c.	..	..	..	..	187.6	..	..	58 Autres activ. & serv. aux entreprises
59 Comm., soc. & pers. serv. activ., etc.	..	..	..	..	1.3	..	..	**59 Adm. pub., serv. soc. & collec., etc**
TOTAL BUSINESS ENTERPRISE SECTOR	**..**	**..**	**..**	**..**	**2 160.7**	**..**	**..**	**TOTAL SECTEUR DES ENTREPRISES**

Source: OECD, R&D database (DSTI/EAS Division), January 2002.

Source : OCDE, base de données R-D (DSTI/Division AES), janvier 2002.

AUSTRIA — Table 9 - Tableau 9 — AUTRICHE

Total business enterprise intramural expenditure on R&D - Dépenses intra-muros totales de R-D des entreprises
(BERD) by industry - (DIRDE) par industrie

9.2 Million constant USD(1995 prices and PPPs) - Millions de USD constants (prix et PPA de 1995)

	1994	1995	1996	1997	1998	1999	2000	
1 AGRICULTURE, HUNTING & FORESTRY	..	..	..	..	..[n]	..	..	**1 AGRICULTURE, CHASSE & SYLVICULTURE**
2 MINING	..	..	..	..	..[n]	..	..	**2 ACTIVITES EXTRACTIVES**
3 MANUFACTURING	..	..	..	..	1 602.5[o]	..	..	**3 ACTIVITES DE FABRICATION**
4 Food, beverages & tobacco	..	..	..	..	..[n]	..	..	**4 Produits alimentaires, boissons & tabac**
5 Food products & beverages	..	..	..	..	17.7	..	..	5 Produits alimentaires & boissons
6 Tobacco products	..	..	..	..	..[n]	..	..	6 Produits à base de tabac
7 Textiles, wear app., fur & leather	..	..	..	..	..[n]	..	..	**7 Textiles, habill., fourrures & cuirs**
8 Textiles	..	..	..	..	17.3	..	..	8 Textiles
9 Wearing apparel & fur	..	..	..	..	..[n]	..	..	9 Articles d'habillement & fourrures
10 Leather products & footwear	..	..	..	..	2.3	..	..	10 Articles en cuir & chaussures
11 Wood, paper, printing, publishing	..	..	..	..	24.7	..	..	**11 Bois, papier, imprimerie, édition**
12 Wood & cork (not furniture)	..	..	..	..	9.6	..	..	12 Bois & liège (sauf meubles)
13 Pulp, paper & paper products	..	..	..	..	11.5	..	..	13 Carton, papier & articles
14 Publ., print. & repro. of rec. media	..	..	..	..	3.5	..	..	14 Edition, impress. & repro. enregis.
15 Coke, petroleum, nuclear fuel, chemicals, rubber plastics	..	..	..	..	..[n]	..	..	**15 Coke, pétrole, combust. nucl., prod. chimi., caout. & mat. plast.**
16 Coke, ref. petrol. prod. & nuclear fuel	..	..	..	..	..[n]	..	..	16 Coke, raffin. pétr. & combust. nucl.
17 Chemicals & chemical products	..	..	..	..	196.2	..	..	17 Produits chimiques
18 *Chemical prod.(less pharmaceuticals)*	..	..	..	..	76.3	..	..	18 *Prod. chimi. (moins prod. pharmaceutiques)*
19 *Pharmaceuticals*	..	..	..	..	119.9	..	..	19 *Produits pharmaceutiques*
20 Rubber & plastic products	..	..	..	..	43.3	..	..	20 Caoutchouc et matières plastiques
21 Non-metallic mineral products	..	..	..	..	42.5	..	..	**21 Produits minéraux non métalliques**
22 Basic metals	..	..	..	..	55.3	..	..	**22 Produits métallurgiques de base**
23 Basic metals, ferrous	..	..	..	..	35.4	..	..	23 Produits métallur. de base, ferreux
24 Basic metals, non-ferrous	..	..	..	..	19.9	..	..	24 Prod. métallur. de base, non ferreux
25 Fabricated metal products	..	..	..	..	48.0	..	..	**25 Ouvrages en métaux (sauf mach.& mat.)**
26 Machinery equipment, instruments and transport equipment	..	..	..	..	1 116.2	..	..	**26 Machines & matériel, instruments & mat. de transport**
27 Machinery n.e.c.	..	..	..	..	180.1	..	..	27 Machines, n.c.a.
28 Office, account. & computing machinery	..	..	..	..	4.5	..	..	28 Mach. de bur., compta. & trait. inform.
29 Electrical machinery	..	..	..	..	93.7	..	..	29 Machines électriques
30 Electro. equip. (radio, TV & commun.)	..	..	..	..	558.1	..	..	30 Equip. électro. (radio, tv & communic.)
31 *Electro. comp. (inc. semi-conductors)*	..	..	..	..	71.1	..	..	31 *Compos. électro. (semi-cond. incl.)*
32 *TV, radio & communications equip.*	..	..	..	..	487.0	..	..	32 *Appareils radio, tv et de communi.*
33 Instruments	..	..	..	..	47.0	..	..	33 Instruments
34 Motor vehicles	..	..	..	..	177.6	..	..	34 Véhicules automobiles
35 Other transport equipment	..	..	..	..	55.1[o]	..	..	35 Autres matériels de transport
36 *Ships*	..	..	..	..	..[n]	..	..	36 *Navires*
37 *Aerospace*	..	..	..	..	..[n]	..	..	37 *Aérospatiale*
38 *Other transport equipment n.e.c.*	..	..	..	..	48.3	..	..	38 *Autres matériels de transport n.c.a.*
39 Furniture, other manufacturing n.e.c.	..	..	..	..	19.5	..	..	**39 Meubles, autres activ. de fabricat. n.c.a.**
40 Furniture	..	..	..	..	6.6	..	..	40 Meubles
41 Other manufacturing n.e.c.	..	..	..	..	12.8	..	..	41 Autres activités de fabrication n.c.a.
42 Recycling	..	..	..	..	0.0	..	..	**42 Récupération**
43 ELECTRICITY, GAS & WATER SUPPLY	..	..	..	..	8.5	..	..	**43 DISTRIBUTION D'ELECTRICITE, GAZ & EAU**
44 CONSTRUCTION	..	..	..	..	13.4	..	..	**44 CONSTRUCTION**
45 SERVICES SECTOR	..	..	..	..	469.0[o]	..	..	**45 SECTEUR DES SERVICES**
46 Wholesale, retail trade & repair of motor vehicles, etc.	..	..	..	..	53.7	..	..	**46 Commerce (gros & détail), réparation de véhicules, etc.**
47 Hotels & restaurants	..	..	..	..	0.0	..	..	**47 Hôtels et restaurants**
48 Transport & storage	..	..	..	..	..[n]	..	..	**48 Transports, entreposage**
49 Communications	..	..	..	..	..[n]	..	..	**49 Communications**
50 Post	..	..	..	..	0.0	..	..	50 Postes
51 Telecommunications	..	..	..	..	..[n]	..	..	51 Télécommunications
52 Financial intermediation (inc. insur.)	..	..	..	..	26.4	..	..	**52 Intermédiation financ. (assurances inc.)**
53 Real estate, renting & busin. activity	..	..	..	..	365.5	..	..	**53 Immobil., locat. & serv. aux entreprises**
54 Computer & related activities	..	..	..	..	32.1	..	..	54 Informatique et activités connexes
55 *Software consultancy*	..	..	..	..	19.1	..	..	55 *Production de logiciels*
56 *Other computer services n.e.c.*	..	..	..	..	13.0	..	..	56 *Autres activités informatiques*
57 Research & development	..	..	..	..	151.5	..	..	57 Recherche & développement expérimental
58 Other business activities n.e.c.	..	..	..	..	181.9	..	..	58 Autres activ. & serv. aux entreprises
59 Comm., soc. & pers. serv. activ., etc.	..	..	..	..	1.2	..	..	**59 Adm. pub., serv. soc. & collec., etc**
TOTAL BUSINESS ENTERPRISE SECTOR	**..**	**..**	**..**	**..**	**2 095.0**	**..**	**..**	**TOTAL SECTEUR DES ENTREPRISES**

Source: OECD, R&D database (DSTI/EAS Division), January 2002.

Source : OCDE, base de données R-D (DSTI/Division AES), janvier 2002.

Business enterprise intramural expenditure on R&D (BERD) - Dépenses intra-muros de R-D des entreprises (DIRDE)

by industry and source of funds - par industrie et par source de financement

Million EUR (1999 ATS euro, EUR from 1999) — *Millions EUR (1999 ATS euro, EUR à partir de 1999)*

10.1 Business enterprise - Entreprises

	1994	1995	1996	1997	1998	1999	2000	
1 AGRICULTURE, HUNTING & FORESTRY	..	..	..	..	..[n]	..	..	**1 AGRICULTURE, CHASSE & SYLVICULTURE**
2 MINING	..	..	..	..	..[n]	..	..	**2 ACTIVITES EXTRACTIVES**
3 MANUFACTURING	..	..	..	..	1 145.1[o]	..	..	**3 ACTIVITES DE FABRICATION**
4 Food, beverages & tobacco	..	..	..	..	..[n]	..	..	**4 Produits alimentaires, boissons & tabac**
5 Food products & beverages	..	..	..	..	16.9	..	..	5 Produits alimentaires & boissons
6 Tobacco products	..	..	..	..	..[n]	..	..	6 Produits à base de tabac
7 Textiles, wear app., fur & leather	..	..	..	..	..[n]	..	..	**7 Textiles, habill., fourrures & cuirs**
8 Textiles	..	..	..	..	17.2	..	..	8 Textiles
9 Wearing apparel & fur	..	..	..	..	..[n]	..	..	9 Articles d'habillement & fourrures
10 Leather products & footwear	..	..	..	..	2.2	..	..	10 Articles en cuir & chaussures
11 Wood, paper, printing, publishing	..	..	..	..	23.1	..	..	**11 Bois, papier, imprimerie, édition**
12 Wood & cork (not furniture)	..	..	..	..	8.3	..	..	12 Bois & liège (sauf meubles)
13 Pulp, paper & paper products	..	..	..	..	11.3	..	..	13 Carton, papier & articles
14 Publ., print. & repro. of rec. media	..	..	..	..	3.5	..	..	14 Edition, impress. & repro. enregis.
15 Coke, petroleum, nuclear fuel, chemicals, rubber plastics	..	..	..	..	..[n]	..	..	**15 Coke, pétrole, combust. nucl., prod. chimi., caout. & mat. plast.**
16 Coke, ref. petrol. prod. & nuclear fuel	..	..	..	..	..[n]	..	..	16 Coke, raffin. pétr. & combust. nucl.
17 Chemicals & chemical products	..	..	..	..	153.2	..	..	17 Produits chimiques
18 *Chemical prod.(less pharmaceuticals)*	..	..	..	..	67.3	..	..	18 *Prod. chimi. (moins prod. pharmaceutiques)*
19 *Pharmaceuticals*	..	..	..	..	85.9	..	..	19 *Produits pharmaceutiques*
20 Rubber & plastic products	..	..	..	..	43.2	..	..	20 Caoutchouc et matières plastiques
21 Non-metallic mineral products	..	..	..	..	42.0	..	..	**21 Produits minéraux non métalliques**
22 Basic metals	..	..	..	..	50.3	..	..	**22 Produits métallurgiques de base**
23 Basic metals, ferrous	..	..	..	..	33.2	..	..	23 Produits métallur. de base, ferreux
24 Basic metals, non-ferrous	..	..	..	..	17.1	..	..	24 Prod. métallur. de base, non ferreux
25 Fabricated metal products	..	..	..	..	45.8	..	..	**25 Ouvrages en métaux (sauf mach.& mat.)**
26 Machinery equipment, instruments and transport equipment	..	..	..	..	713.5	..	..	**26 Machines & matériel, instruments & mat. de transport**
27 Machinery n.e.c.	..	..	..	..	170.7	..	..	27 Machines, n.c.a.
28 Office, account. & computing machinery	..	..	..	..	4.0	..	..	28 Mach. de bur., compta. & trait. inform.
29 Electrical machinery	..	..	..	..	88.0	..	..	29 Machines électriques
30 Electro. equip. (radio, TV & commun.)	..	..	..	..	256.7	..	..	30 Equip. électro. (radio, tv & communic.)
31 *Electro. comp. (inc. semi-conductors)*	..	..	..	..	69.2	..	..	31 *Compos. électro. (semi-cond. incl.)*
32 *TV, radio & communications equip.*	..	..	..	..	187.5	..	..	32 *Appareils radio, tv et de communi.*
33 Instruments	..	..	..	..	40.6	..	..	33 Instruments
34 Motor vehicles	..	..	..	..	104.7	..	..	34 Véhicules automobiles
35 Other transport equipment	..	..	..	..	48.7[o]	..	..	35 Autres matériels de transport
36 *Ships*	..	..	..	..	..[n]	..	..	36 *Navires*
37 *Aerospace*	..	..	..	..	..[n]	..	..	37 *Aérospatiale*
38 *Other transport equipment n.e.c.*	..	..	..	..	45.7	..	..	38 *Autres matériels de transport n.c.a.*
39 Furniture, other manufacturing n.e.c.	..	..	..	..	18.1	..	..	**39 Meubles, autres activ. de fabricat. n.c.a.**
40 Furniture	..	..	..	..	6.6	..	..	40 Meubles
41 Other manufacturing n.e.c.	..	..	..	..	11.5	..	..	41 Autres activités de fabrication n.c.a.
42 Recycling	..	..	..	..	0.0	..	..	**42 Récupération**
43 ELECTRICITY, GAS & WATER SUPPLY	..	..	..	..	6.3	..	..	**43 DISTRIBUTION D'ELECTRICITE, GAZ & EAU**
44 CONSTRUCTION	..	..	..	..	13.1	..	..	**44 CONSTRUCTION**
45 SERVICES SECTOR	..	..	..	..	224.7[o]	..	..	**45 SECTEUR DES SERVICES**
46 Wholesale, retail trade & repair of motor vehicles, etc.	..	..	..	..	50.8	..	..	**46 Commerce (gros & détail), réparation de véhicules, etc.**
47 Hotels & restaurants	..	..	..	..	0.0	..	..	**47 Hôtels et restaurants**
48 Transport & storage	..	..	..	..	..[n]	..	..	**48 Transports, entreposage**
49 Communications	..	..	..	..	..[n]	..	..	**49 Communications**
50 Post	..	..	..	..	0.0	..	..	50 Postes
51 Telecommunications	..	..	..	..	..[n]	..	..	51 Télécommunications
52 Financial intermediation (inc. insur.)	..	..	..	..	27.1	..	..	**52 Intermédiation financ. (assurances inc.)**
53 Real estate, renting & busin. activity	..	..	..	..	123.3	..	..	**53 Immobil., locat. & serv. aux entreprises**
54 Computer & related activities	..	..	..	..	28.7	..	..	54 Informatique et activités connexes
55 *Software consultancy*	..	..	..	..	16.8	..	..	55 *Production de logiciels*
56 *Other computer services n.e.c.*	..	..	..	..	11.9	..	..	56 *Autres activités informatiques*
57 Research & development	..	..	..	..	23.2	..	..	57 Recherche & développement expérimental
58 Other business activities n.e.c.	..	..	..	..	71.3	..	..	58 Autres activ. & serv. aux entreprises
59 Comm., soc. & pers. serv. activ., etc.	..	..	..	..	0.9	..	..	**59 Adm. pub., serv. soc. & collec., etc**
TOTAL BUSINESS ENTERPRISE SECTOR	**..**	**..**	**..**	**..**	**1 390.6**	**..**	**..**	**TOTAL SECTEUR DES ENTREPRISES**

Source: OECD, R&D database (DSTI/EAS Division), January 2002.

Source : OCDE, base de données R-D (DSTI/Division AES), janvier 2002.

AUSTRIA — Table 10 - Tableau 10 — AUTRICHE

Business enterprise intramural expenditure on R&D (BERD) - Dépenses intra-muros de R-D des entreprises (DIRDE)

by industry and source of funds - par industrie et par source de financement

Million EUR (1999 ATS euro, EUR from 1999) — *Millions EUR (1999 ATS euro, EUR à partir de 1999)*

10.2 Government - État

	1994	1995	1996	1997	1998	1999	2000	
1 AGRICULTURE, HUNTING & FORESTRY	..	..	..	..	.. [n]	..	..	**1 AGRICULTURE, CHASSE & SYLVICULTURE**
2 MINING	..	..	..	..	.. [n]	..	..	**2 ACTIVITES EXTRACTIVES**
3 MANUFACTURING	..	..	..	..	53.8 [o]	..	..	**3 ACTIVITES DE FABRICATION**
4 Food, beverages & tobacco	..	..	..	..	.. [n]	..	..	**4 Produits alimentaires, boissons & tabac**
5 Food products & beverages	..	..	..	..	1.3	..	..	5 Produits alimentaires & boissons
6 Tobacco products	..	..	..	..	.. [n]	..	..	6 Produits à base de tabac
7 Textiles, wear app., fur & leather	..	..	..	..	.. [n]	..	..	**7 Textiles, habill., fourrures & cuirs**
8 Textiles	..	..	..	..	0.7	..	..	8 Textiles
9 Wearing apparel & fur	..	..	..	..	.. [n]	..	..	9 Articles d'habillement & fourrures
10 Leather products & footwear	..	..	..	..	0.2	..	..	10 Articles en cuir & chaussures
11 Wood, paper, printing, publishing	..	..	..	..	1.6	..	..	**11 Bois, papier, imprimerie, édition**
12 Wood & cork (not furniture)	..	..	..	..	1.2	..	..	12 Bois & liège (sauf meubles)
13 Pulp, paper & paper products	..	..	..	..	0.3	..	..	13 Carton, papier & articles
14 Publ., print. & repro. of rec. media	..	..	..	..	0.2	..	..	14 Edition, impress. & repro. enregis.
15 Coke, petroleum, nuclear fuel, chemicals, rubber plastics	..	..	..	..	.. [n]	..	..	**15 Coke, pétrole, combust. nucl., prod. chimi., caout. & mat. plast.**
16 Coke, ref. petrol. prod. & nuclear fuel	..	..	..	..	.. [n]	..	..	16 Coke, raffin. pétr. & combust. nucl.
17 Chemicals & chemical products	..	..	..	..	8.5	..	..	17 Produits chimiques
18 *Chemical prod.(less pharmaceuticals)*	..	..	..	..	3.2	..	..	18 *Prod. chimi. (moins prod. pharmaceutiques)*
19 *Pharmaceuticals*	..	..	..	..	5.3	..	..	19 *Produits pharmaceutiques*
20 Rubber & plastic products	..	..	..	..	1.3	..	..	20 Caoutchouc et matières plastiques
21 Non-metallic mineral products	..	..	..	..	1.3	..	..	**21 Produits minéraux non métalliques**
22 Basic metals	..	..	..	..	3.2	..	..	**22 Produits métallurgiques de base**
23 Basic metals, ferrous	..	..	..	..	2.1	..	..	23 Produits métallur. de base, ferreux
24 Basic metals, non-ferrous	..	..	..	..	1.2	..	..	24 Prod. métallur. de base, non ferreux
25 Fabricated metal products	..	..	..	..	2.7	..	..	**25 Ouvrages en métaux (sauf mach.& mat.)**
26 Machinery equipment, instruments and transport equipment	..	..	..	..	32.2	..	..	**26 Machines & matériel, instruments & mat. de transport**
27 Machinery n.e.c.	..	..	..	..	8.1	..	..	27 Machines, n.c.a.
28 Office, account. & computing machinery	..	..	..	..	0.6	..	..	28 Mach. de bur., compta. & trait. inform.
29 Electrical machinery	..	..	..	..	4.4	..	..	29 Machines électriques
30 Electro. equip. (radio, TV & commun.)	..	..	..	..	6.2	..	..	30 Equip. électro. (radio, tv & communic.)
31 *Electro. comp. (inc. semi-conductors)*	..	..	..	..	2.8	..	..	31 *Compos. électro. (semi-cond. incl.)*
32 *TV, radio & communications equip.*	..	..	..	..	3.4	..	..	32 *Appareils radio, tv et de communi.*
33 Instruments	..	..	..	..	4.4	..	..	33 Instruments
34 Motor vehicles	..	..	..	..	7.1	..	..	34 Véhicules automobiles
35 Other transport equipment	..	..	..	..	1.4 [o]	..	..	35 Autres matériels de transport
36 *Ships*	..	..	..	..	.. [n]	..	..	36 *Navires*
37 *Aerospace*	..	..	..	..	.. [n]	..	..	37 *Aérospatiale*
38 *Other transport equipment n.e.c.*	..	..	..	..	0.9	..	..	38 *Autres matériels de transport n.c.a.*
39 Furniture, other manufacturing n.e.c.	..	..	..	..	0.4	..	..	**39 Meubles, autres activ. de fabricat. n.c.a.**
40 Furniture	..	..	..	..	0.2	..	..	40 Meubles
41 Other manufacturing n.e.c.	..	..	..	..	0.1	..	..	41 Autres activités de fabrication n.c.a.
42 Recycling	..	..	..	..	0.0	..	..	**42 Récupération**
43 ELECTRICITY, GAS & WATER SUPPLY	..	..	..	..	0.0	..	..	**43 DISTRIBUTION D'ELECTRICITE, GAZ & EAU**
44 CONSTRUCTION	..	..	..	..	0.7	..	..	**44 CONSTRUCTION**
45 SERVICES SECTOR	..	..	..	..	64.8 [o]	..	..	**45 SECTEUR DES SERVICES**
46 Wholesale, retail trade & repair of motor vehicles, etc.	..	..	..	..	1.2	..	..	**46 Commerce (gros & détail), réparation de véhicules, etc.**
47 Hotels & restaurants	..	..	..	..	0.0	..	..	**47 Hôtels et restaurants**
48 Transport & storage	..	..	..	..	.. [n]	..	..	**48 Transports, entreposage**
49 Communications	..	..	..	..	.. [n]	..	..	**49 Communications**
50 Post	..	..	..	..	0.0	..	..	50 Postes
51 Telecommunications	..	..	..	..	.. [n]	..	..	51 Télécommunications
52 Financial intermediation (inc. insur.)	..	..	..	..	0.1	..	..	**52 Intermédiation financ. (assurances inc.)**
53 Real estate, renting & busin. activity	..	..	..	..	63.0	..	..	**53 Immobil., locat. & serv. aux entreprises**
54 Computer & related activities	..	..	..	..	2.4	..	..	54 Informatique et activités connexes
55 *Software consultancy*	..	..	..	..	1.6	..	..	55 *Production de logiciels*
56 *Other computer services n.e.c.*	..	..	..	..	0.7	..	..	56 *Autres activités informatiques*
57 Research & development	..	..	..	..	50.6	..	..	57 Recherche & développement expérimental
58 Other business activities n.e.c.	..	..	..	..	10.0	..	..	58 Autres activ. & serv. aux entreprises
59 Comm., soc. & pers. serv. activ., etc.	..	..	..	..	0.4	..	..	**59 Adm. pub., serv. soc. & collec., etc**
TOTAL BUSINESS ENTERPRISE SECTOR	**..**	**..**	**..**	**..**	**119.5**	**..**	**..**	**TOTAL SECTEUR DES ENTREPRISES**

Source: OECD, R&D database (DSTI/EAS Division), January 2002.

Source : OCDE, base de données R-D (DSTI/Division AES), janvier 2002.

Business enterprise intramural expenditure on R&D (BERD) - Dépenses intra-muros de R-D des entreprises (DIRDE)

by industry and source of funds - par industrie et par source de financement

Million EUR (1999 ATS euro, EUR from 1999) — *Millions EUR (1999 ATS euro, EUR à partir de 1999)*

10.3 Other national funds - Autres sources nationales

	1994	1995	1996	1997	1998	1999	2000	
1 **AGRICULTURE, HUNTING & FORESTRY**	..	..	..	..	.. n	..	..	1 **AGRICULTURE, CHASSE & SYLVICULTURE**
2 **MINING**	..	..	..	..	.. n	..	..	2 **ACTIVITES EXTRACTIVES**
3 **MANUFACTURING**	..	..	..	..	0.2 o	..	..	3 **ACTIVITES DE FABRICATION**
4 **Food, beverages & tobacco**	..	..	..	..	.. n	..	..	4 **Produits alimentaires, boissons & tabac**
5 Food products & beverages	..	..	..	..	0.0	..	..	5 Produits alimentaires & boissons
6 Tobacco products	..	..	..	..	.. n	..	..	6 Produits à base de tabac
7 **Textiles, wear app., fur & leather**	..	..	..	..	.. n	..	..	7 **Textiles, habill., fourrures & cuirs**
8 Textiles	..	..	..	..	0.0	..	..	8 Textiles
9 Wearing apparel & fur	..	..	..	..	.. n	..	..	9 Articles d'habillement & fourrures
10 Leather products & footwear	..	..	..	..	0.0	..	..	10 Articles en cuir & chaussures
11 **Wood, paper, printing, publishing**	..	..	..	..	0.1	..	..	11 **Bois, papier, imprimerie, édition**
12 Wood & cork (not furniture)	..	..	..	..	0.1	..	..	12 Bois & liège (sauf meubles)
13 Pulp, paper & paper products	..	..	..	..	0.0	..	..	13 Carton, papier & articles
14 Publ., print. & repro. of rec. media	..	..	..	..	0.0	..	..	14 Edition, impress. & repro. enregis.
15 **Coke, petroleum, nuclear fuel, chemicals, rubber plastics**	..	..	..	..	.. n	..	..	15 **Coke, pétrole, combust. nucl., prod. chimi., caout. & mat. plast.**
16 Coke, ref. petrol. prod. & nuclear fuel	..	..	..	..	.. n	..	..	16 Coke, raffin. pétr. & combust. nucl.
17 Chemicals & chemical products	..	..	..	..	0.0	..	..	17 Produits chimiques
18 *Chemical prod.(less pharmaceuticals)*	..	..	..	..	0.0	..	..	18 *Prod. chimi. (moins prod. pharmaceutiques)*
19 *Pharmaceuticals*	..	..	..	..	0.0	..	..	19 *Produits pharmaceutiques*
20 Rubber & plastic products	..	..	..	..	0.0	..	..	20 Caoutchouc et matières plastiques
21 **Non-metallic mineral products**	..	..	..	..	0.0	..	..	21 **Produits minéraux non métalliques**
22 **Basic metals**	..	..	..	..	0.0	..	..	22 **Produits métallurgiques de base**
23 Basic metals, ferrous	..	..	..	..	0.0	..	..	23 Produits métallur. de base, ferreux
24 Basic metals, non-ferrous	..	..	..	..	0.0	..	..	24 Prod. métallur. de base, non ferreux
25 **Fabricated metal products**	..	..	..	..	0.0	..	..	25 **Ouvrages en métaux (sauf mach.& mat.)**
26 **Machinery equipment, instruments and transport equipment**	..	..	..	..	0.1	..	..	26 **Machines & matériel, instruments & mat. de transport**
27 Machinery n.e.c.	..	..	..	..	0.0	..	..	27 Machines, n.c.a.
28 Office, account. & computing machinery	..	..	..	..	0.0	..	..	28 Mach. de bur., compta. & trait. inform.
29 Electrical machinery	..	..	..	..	0.0	..	..	29 Machines électriques
30 Electro. equip. (radio, TV & commun.)	..	..	..	..	0.0	..	..	30 Equip. électro. (radio, tv & communic.)
31 *Electro. comp. (inc. semi-conductors)*	..	..	..	..	0.0	..	..	31 *Compos. électro. (semi-cond. incl.)*
32 *TV, radio & communications equip.*	..	..	..	..	0.0	..	..	32 *Appareils radio, tv et de communi.*
33 Instruments	..	..	..	..	0.1	..	..	33 Instruments
34 Motor vehicles	..	..	..	..	0.0	..	..	34 Véhicules automobiles
35 Other transport equipment	..	..	..	..	0.0 o	..	..	35 Autres matériels de transport
36 *Ships*	..	..	..	..	.. n	..	..	36 *Navires*
37 *Aerospace*	..	..	..	..	.. n	..	..	37 *Aérospatiale*
38 *Other transport equipment n.e.c.*	..	..	..	..	0.0	..	..	38 *Autres matériels de transport n.c.a.*
39 **Furniture, other manufacturing n.e.c.**	..	..	..	..	0.0	..	..	39 **Meubles, autres activ. de fabricat. n.c.a.**
40 Furniture	..	..	..	..	0.0	..	..	40 Meubles
41 Other manufacturing n.e.c.	..	..	..	..	0.0	..	..	41 Autres activités de fabrication n.c.a.
42 **Recycling**	..	..	..	..	0.0	..	..	42 **Récupération**
43 **ELECTRICITY, GAS & WATER SUPPLY**	..	..	..	..	0.0	..	..	43 **DISTRIBUTION D'ELECTRICITE, GAZ & EAU**
44 **CONSTRUCTION**	..	..	..	..	0.0	..	..	44 **CONSTRUCTION**
45 **SERVICES SECTOR**	..	..	..	..	0.9 o	..	..	45 **SECTEUR DES SERVICES**
46 **Wholesale, retail trade & repair of motor vehicles, etc.**	..	..	..	..	0.0	..	..	46 **Commerce (gros & détail), réparation de véhicules, etc.**
47 **Hotels & restaurants**	..	..	..	..	0.0	..	..	47 **Hôtels et restaurants**
48 **Transport & storage**	..	..	..	..	.. n	..	..	48 **Transports, entreposage**
49 **Communications**	..	..	..	..	.. n	..	..	49 **Communications**
50 Post	..	..	..	..	0.0	..	..	50 Postes
51 Telecommunications	..	..	..	..	.. n	..	..	51 Télécommunications
52 **Financial intermediation (inc. insur.)**	..	..	..	..	0.0	..	..	52 **Intermédiation financ. (assurances inc.)**
53 **Real estate, renting & busin. activity**	..	..	..	..	0.9	..	..	53 **Immobil., locat. & serv. aux entreprises**
54 Computer & related activities	..	..	..	..	0.0	..	..	54 Informatique et activités connexes
55 *Software consultancy*	..	..	..	..	0.0	..	..	55 *Production de logiciels*
56 *Other computer services n.e.c.*	..	..	..	..	0.0	..	..	56 *Autres activités informatiques*
57 Research & development	..	..	..	..	0.1	..	..	57 Recherche & développement expérimental
58 Other business activities n.e.c.	..	..	..	..	0.8	..	..	58 Autres activ. & serv. aux entreprises
59 **Comm., soc. & pers. serv. activ., etc.**	..	..	..	..	0.0	..	..	59 **Adm. pub., serv. soc. & collec., etc**
TOTAL BUSINESS ENTERPRISE SECTOR	..	..	..	..	1.2	..	..	**TOTAL SECTEUR DES ENTREPRISES**

Source: OECD, R&D database (DSTI/EAS Division), January 2002.

Source : OCDE, base de données R-D (DSTI/Division AES), janvier 2002.

Business enterprise intramural expenditure on R&D (BERD) - Dépenses intra-muros de R-D des entreprises (DIRDE)

by industry and source of funds - par industrie et par source de financement

Million EUR (1999 ATS euro, EUR from 1999) — *Millions EUR (1999 ATS euro, EUR à partir de 1999)*

10.4 Funds from abroad - Fonds de l'étranger

	1994	1995	1996	1997	1998	1999	2000	
1 AGRICULTURE, HUNTING & FORESTRY	..	..	..	..	.. n	..	..	**1 AGRICULTURE, CHASSE & SYLVICULTURE**
2 MINING	..	..	..	..	.. n	..	..	**2 ACTIVITES EXTRACTIVES**
3 MANUFACTURING	..	..	..	..	453.6 o	..	..	**3 ACTIVITES DE FABRICATION**
4 Food, beverages & tobacco	..	..	..	..	.. n	..	..	**4 Produits alimentaires, boissons & tabac**
5 Food products & beverages	..	..	..	..	0.1	..	..	5 Produits alimentaires & boissons
6 Tobacco products	..	..	..	..	.. n	..	..	6 Produits à base de tabac
7 Textiles, wear app., fur & leather	..	..	..	..	.. n	..	..	**7 Textiles, habill., fourrures & cuirs**
8 Textiles	..	..	..	..	0.0	..	..	8 Textiles
9 Wearing apparel & fur	..	..	..	..	.. n	..	..	9 Articles d'habillement & fourrures
10 Leather products & footwear	..	..	..	..	0.0	..	..	10 Articles en cuir & chaussures
11 Wood, paper, printing, publishing	..	..	..	..	0.6	..	..	**11 Bois, papier, imprimerie, édition**
12 Wood & cork (not furniture)	..	..	..	..	0.3	..	..	12 Bois & liège (sauf meubles)
13 Pulp, paper & paper products	..	..	..	..	0.3	..	..	13 Carton, papier & articles
14 Publ., print. & repro. of rec. media	..	..	..	..	0.0	..	..	14 Edition, impress. & repro. enregis.
15 Coke, petroleum, nuclear fuel, chemicals, rubber plastics	..	..	..	..	.. n	..	..	**15 Coke, pétrole, combust. nucl., prod. chimi., caout. & mat. plast.**
16 Coke, ref. petrol. prod. & nuclear fuel	..	..	..	..	.. n	..	..	16 Coke, raffin. pétr. & combust. nucl.
17 Chemicals & chemical products	..	..	..	..	40.6	..	..	17 Produits chimiques
18 *Chemical prod.(less pharmaceuticals)*	..	..	..	..	8.2	..	..	18 *Prod. chimi. (moins prod. pharmaceutiques)*
19 *Pharmaceuticals*	..	..	..	..	32.5	..	..	19 *Produits pharmaceutiques*
20 Rubber & plastic products	..	..	..	..	0.1	..	..	20 Caoutchouc et matières plastiques
21 Non-metallic mineral products	..	..	..	..	0.6	..	..	**21 Produits minéraux non métalliques**
22 Basic metals	..	..	..	..	3.5	..	..	**22 Produits métallurgiques de base**
23 Basic metals, ferrous	..	..	..	..	1.2	..	..	23 Produits métallur. de base, ferreux
24 Basic metals, non-ferrous	..	..	..	..	2.3	..	..	24 Prod. métallur. de base, non ferreux
25 Fabricated metal products	..	..	..	..	1.0	..	..	**25 Ouvrages en métaux (sauf mach.& mat.)**
26 Machinery equipment, instruments and transport equipment	..	..	..	..	405.5	..	..	**26 Machines & matériel, instruments & mat. de transport**
27 Machinery n.e.c.	..	..	..	..	7.0	..	..	27 Machines, n.c.a.
28 Office, account. & computing machinery	..	..	..	..	0.0	..	..	28 Mach. de bur., compta. & trait. inform.
29 Electrical machinery	..	..	..	..	4.2	..	..	29 Machines électriques
30 Electro. equip. (radio, TV & commun.)	..	..	..	..	312.7	..	..	30 Equip. électro. (radio, tv & communic.)
31 *Electro. comp. (inc. semi-conductors)*	..	..	..	..	1.3	..	..	31 *Compos. électro. (semi-cond. incl.)*
32 *TV, radio & communications equip.*	..	..	..	..	311.4	..	..	32 *Appareils radio, tv et de communi.*
33 Instruments	..	..	..	..	3.4	..	..	33 Instruments
34 Motor vehicles	..	..	..	..	71.4	..	..	34 Véhicules automobiles
35 Other transport equipment	..	..	..	..	6.8 o	..	..	35 Autres matériels de transport
36 *Ships*	..	..	..	..	.. n	..	..	36 *Navires*
37 *Aerospace*	..	..	..	..	.. n	..	..	37 *Aérospatiale*
38 *Other transport equipment n.e.c.*	..	..	..	..	3.2	..	..	38 *Autres matériels de transport n.c.a.*
39 Furniture, other manufacturing n.e.c.	..	..	..	..	1.6	..	..	**39 Meubles, autres activ. de fabricat. n.c.a.**
40 Furniture	..	..	..	..	0.0	..	..	40 Meubles
41 Other manufacturing n.e.c.	..	..	..	..	1.6	..	..	41 Autres activités de fabrication n.c.a.
42 Recycling	..	..	..	..	0.0	..	..	**42 Récupération**
43 ELECTRICITY, GAS & WATER SUPPLY	..	..	..	..	2.4	..	..	**43 DISTRIBUTION D'ELECTRICITE, GAZ & EAU**
44 CONSTRUCTION	..	..	..	..	0.0	..	..	**44 CONSTRUCTION**
45 SERVICES SECTOR	..	..	..	..	193.4 o	..	..	**45 SECTEUR DES SERVICES**
46 Wholesale, retail trade & repair of motor vehicles, etc.	..	..	..	..	3.5	..	..	**46 Commerce (gros & détail), réparation de véhicules, etc.**
47 Hotels & restaurants	..	..	..	..	0.0	..	..	**47 Hôtels et restaurants**
48 Transport & storage	..	..	..	..	.. n	..	..	**48 Transports, entreposage**
49 Communications	..	..	..	..	.. n	..	..	**49 Communications**
50 Post	..	..	..	..	0.0	..	..	50 Postes
51 Telecommunications	..	..	..	..	.. n	..	..	51 Télécommunications
52 Financial intermediation (inc. insur.)	..	..	..	..	0.0	..	..	**52 Intermédiation financ. (assurances inc.)**
53 Real estate, renting & busin. activity	..	..	..	..	189.8	..	..	**53 Immobil., locat. & serv. aux entreprises**
54 Computer & related activities	..	..	..	..	2.0	..	..	54 Informatique et activités connexes
55 *Software consultancy*	..	..	..	..	1.2	..	..	55 *Production de logiciels*
56 *Other computer services n.e.c.*	..	..	..	..	0.8	..	..	56 *Autres activités informatiques*
57 Research & development	..	..	..	..	82.3	..	..	57 Recherche & développement expérimental
58 Other business activities n.e.c.	..	..	..	..	105.5	..	..	58 Autres activ. & serv. aux entreprises
59 Comm., soc. & pers. serv. activ., etc.	..	..	..	..	0.0	..	..	**59 Adm. pub., serv. soc. & collec., etc**
TOTAL BUSINESS ENTERPRISE SECTOR	**..**	**..**	**..**	**..**	**649.5**	**..**	**..**	**TOTAL SECTEUR DES ENTREPRISES**

Source: OECD, R&D database (DSTI/EAS Division), January 2002.

Source : OCDE, base de données R-D (DSTI/Division AES), janvier 2002.

AUSTRIA Table 11 - Tableau 11 AUTRICHE

Business enterprise intramural expenditure on R&D (BERD) - Dépenses intra-muros de R-D des entreprises (DIRDE)

by industry and type of costs - par industrie et par type de coûts

Million EUR (1999 ATS euro, EUR from 1999) *Millions EUR (1999 ATS euro, EUR à partir de 1999)*

	1994	1995	1996	1997	1998	1999	2000	
	11.1 Current expenditure - Dépenses courantes							
1 AGRICULTURE, HUNTING & FORESTRY	..	..	..	..	..	..	..	**1 AGRICULTURE, CHASSE & SYLVICULTURE**
2 MINING	..	..	..	..	..	..	..	**2 ACTIVITES EXTRACTIVES**
3 MANUFACTURING	..	..	..	..	1 497.4	..	..	**3 ACTIVITES DE FABRICATION**
4 Food, beverages & tobacco	..	..	..	..	..	..	..	**4 Produits alimentaires, boissons & tabac**
5 Food products & beverages	..	..	..	..	13.9	..	..	5 Produits alimentaires & boissons
6 Tobacco products	..	..	..	..	..	..	..	6 Produits à base de tabac
7 Textiles, wear app., fur & leather	..	..	..	..	..	..	..	**7 Textiles, habill., fourrures & cuirs**
8 Textiles	..	..	..	..	16.5	..	..	8 Textiles
9 Wearing apparel & fur	..	..	..	..	..	..	..	9 Articles d'habillement & fourrures
10 Leather products & footwear	..	..	..	..	2.2	..	..	10 Articles en cuir & chaussures
11 Wood, paper, printing, publishing	..	..	..	..	16.0	..	..	**11 Bois, papier, imprimerie, édition**
12 Wood & cork (not furniture)	..	..	..	..	6.0	..	..	12 Bois & liège (sauf meubles)
13 Pulp, paper & paper products	..	..	..	..	8.0	..	..	13 Carton, papier & articles
14 Publ., print. & repro. of rec. media	..	..	..	..	2.0	..	..	14 Edition, impress. & repro. enregis.
15 Coke, petroleum, nuclear fuel, chemicals, rubber plastics	..	..	..	..	..	..	..	**15 Coke, pétrole, combust. nucl., prod. chimi., caout. & mat. plast.**
16 Coke, ref. petrol. prod. & nuclear fuel	..	..	..	..	..	..	..	16 Coke, raffin. pétr. & combust. nucl.
17 Chemicals & chemical products	..	..	..	..	175.1	..	..	17 Produits chimiques
18 *Chemical prod.(less pharmaceuticals)*	..	..	..	..	70.5	..	..	18 *Prod. chimi. (moins prod. pharmaceutiques)*
19 *Pharmaceuticals*	..	..	..	..	104.7	..	..	19 *Produits pharmaceutiques*
20 Rubber & plastic products	..	..	..	..	37.6	..	..	20 Caoutchouc et matières plastiques
21 Non-metallic mineral products	..	..	..	..	39.8	..	..	**21 Produits minéraux non métalliques**
22 Basic metals	..	..	..	..	50.9	..	..	**22 Produits métallurgiques de base**
23 Basic metals, ferrous	..	..	..	..	33.4	..	..	23 Produits métallur. de base, ferreux
24 Basic metals, non-ferrous	..	..	..	..	17.5	..	..	24 Prod. métallur. de base, non ferreux
25 Fabricated metal products	..	..	..	..	43.5	..	..	**25 Ouvrages en métaux (sauf mach.& mat.)**
26 Machinery equipment, instruments and transport equipment	..	..	..	..	1 066.4	..	..	**26 Machines & matériel, instruments & mat. de transport**
27 Machinery n.e.c.	..	..	..	..	162.0	..	..	27 Machines, n.c.a.
28 Office, account. & computing machinery	..	..	..	..	4.4	..	..	28 Mach. de bur., compta. & trait. inform.
29 Electrical machinery	..	..	..	..	87.8	..	..	29 Machines électriques
30 Electro. equip. (radio, TV & commun.)	..	..	..	..	552.0	..	..	30 Equip. électro. (radio, tv & communic.)
31 *Electro. comp. (inc. semi-conductors)*	..	..	..	..	67.1	..	..	31 *Compos. électro. (semi-cond. incl.)*
32 *TV, radio & communications equip.*	..	..	..	..	484.9	..	..	32 *Appareils radio, tv et de communi.*
33 Instruments	..	..	..	..	42.9	..	..	33 Instruments
34 Motor vehicles	..	..	..	..	163.3	..	..	34 Véhicules automobiles
35 Other transport equipment	..	..	..	..	54.1	..	..	35 Autres matériels de transport
36 *Ships*	..	..	..	..	..	..	..	36 *Navires*
37 *Aerospace*	..	..	..	..	..	..	..	37 *Aérospatiale*
38 *Other transport equipment n.e.c.*	..	..	..	..	47.7	..	..	38 *Autres matériels de transport n.c.a.*
39 Furniture, other manufacturing n.e.c.	..	..	..	..	17.4	..	..	**39 Meubles, autres activ. de fabricat. n.c.a.**
40 Furniture	..	..	..	..	5.5	..	..	40 Meubles
41 Other manufacturing n.e.c.	..	..	..	..	12.0	..	..	41 Autres activités de fabrication n.c.a.
42 Recycling	..	..	..	..	0.0	..	..	**42 Récupération**
43 ELECTRICITY, GAS & WATER SUPPLY	..	..	..	..	7.5	..	..	**43 DISTRIBUTION D'ELECTRICITE, GAZ & EAU**
44 CONSTRUCTION	..	..	..	..	12.0	..	..	**44 CONSTRUCTION**
45 SERVICES SECTOR	..	..	..	..	447.9	..	..	**45 SECTEUR DES SERVICES**
46 Wholesale, retail trade & repair of motor vehicles, etc.	..	..	..	..	50.6	..	..	**46 Commerce (gros & détail), réparation de véhicules, etc.**
47 Hotels & restaurants	..	..	..	..	0.0	..	..	**47 Hôtels et restaurants**
48 Transport & storage	..	..	..	..	..	..	..	**48 Transports, entreposage**
49 Communications	..	..	..	..	..	..	..	**49 Communications**
50 Post	..	..	..	..	0.0	..	..	50 Postes
51 Telecommunications	..	..	..	..	..	..	..	51 Télécommunications
52 Financial intermediation (inc. insur.)	..	..	..	..	27.1	..	..	**52 Intermédiation financ. (assurances inc.)**
53 Real estate, renting & busin. activity	..	..	..	..	349.9	..	..	**53 Immobil., locat. & serv. aux entreprises**
54 Computer & related activities	..	..	..	..	31.5	..	..	54 Informatique et activités connexes
55 *Software consultancy*	..	..	..	..	18.5	..	..	55 *Production de logiciels*
56 *Other computer services n.e.c.*	..	..	..	..	13.0	..	..	56 *Autres activités informatiques*
57 Research & development	..	..	..	..	139.2	..	..	57 Recherche & développement expérimental
58 Other business activities n.e.c.	..	..	..	..	179.2	..	..	58 Autres activ. & serv. aux entreprises
59 Comm., soc. & pers. serv. activ., etc.	..	..	..	..	1.0	..	..	**59 Adm. pub., serv. soc. & collec., etc**
TOTAL BUSINESS ENTERPRISE SECTOR	**..**	**..**	**..**	**..**	**1 966.4**	**..**	**..**	**TOTAL SECTEUR DES ENTREPRISES**

Source: OECD, R&D database (DSTI/EAS Division), January 2002.

Source : OCDE, base de données R-D (DSTI/Division AES), janvier 2002.

Business enterprise intramural expenditure on R&D (BERD) - Dépenses intra-muros de R-D des entreprises (DIRDE)

by industry and type of costs - par industrie et par type de coûts

Million EUR (1999 ATS euro, EUR from 1999) — *Millions EUR (1999 ATS euro, EUR à partir de 1999)*

11.2 Capital expenditure - Dépenses en capital

	1994	1995	1996	1997	1998	1999	2000	
1 AGRICULTURE, HUNTING & FORESTRY	..	..	..	..	..	..	..	**1 AGRICULTURE, CHASSE & SYLVICULTURE**
2 MINING	..	..	..	..	..	..	..	**2 ACTIVITES EXTRACTIVES**
3 MANUFACTURING	..	..	..	..	155.3	..	..	**3 ACTIVITES DE FABRICATION**
4 Food, beverages & tobacco	..	..	..	..	..	..	..	**4 Produits alimentaires, boissons & tabac**
5 Food products & beverages	..	..	..	..	4.4	..	..	5 Produits alimentaires & boissons
6 Tobacco products	..	..	..	..	..	..	..	6 Produits à base de tabac
7 Textiles, wear app., fur & leather	..	..	..	..	..	..	..	**7 Textiles, habill., fourrures & cuirs**
8 Textiles	..	..	..	..	1.4	..	..	8 Textiles
9 Wearing apparel & fur	..	..	..	..	..	..	..	9 Articles d'habillement & fourrures
10 Leather products & footwear	..	..	..	..	0.2	..	..	10 Articles en cuir & chaussures
11 Wood, paper, printing, publishing	..	..	..	..	9.4	..	..	**11 Bois, papier, imprimerie, édition**
12 Wood & cork (not furniture)	..	..	..	..	3.9	..	..	12 Bois & liège (sauf meubles)
13 Pulp, paper & paper products	..	..	..	..	3.9	..	..	13 Carton, papier & articles
14 Publ., print. & repro. of rec. media	..	..	..	..	1.6	..	..	14 Edition, impress. & repro. enregis.
15 Coke, petroleum, nuclear fuel, chemicals, rubber plastics	..	..	..	..	..	..	..	**15 Coke, pétrole, combust. nucl., prod. chimi., caout. & mat. plast.**
16 Coke, ref. petrol. prod. & nuclear fuel	..	..	..	..	..	..	..	16 Coke, raffin. pétr. & combust. nucl.
17 Chemicals & chemical products	..	..	..	..	27.2	..	..	17 Produits chimiques
18 *Chemical prod.(less pharmaceuticals)*	..	..	..	..	8.2	..	..	18 *Prod. chimi. (moins prod. pharmaceutiques)*
19 *Pharmaceuticals*	..	..	..	..	19.0	..	..	19 *Produits pharmaceutiques*
20 Rubber & plastic products	..	..	..	..	7.0	..	..	20 Caoutchouc et matières plastiques
21 Non-metallic mineral products	..	..	..	..	4.1	..	..	**21 Produits minéraux non métalliques**
22 Basic metals	..	..	..	..	6.2	..	..	**22 Produits métallurgiques de base**
23 Basic metals, ferrous	..	..	..	..	3.1	..	..	23 Produits métallur. de base, ferreux
24 Basic metals, non-ferrous	..	..	..	..	3.1	..	..	24 Prod. métallur. de base, non ferreux
25 Fabricated metal products	..	..	..	..	6.1	..	..	**25 Ouvrages en métaux (sauf mach.& mat.)**
26 Machinery equipment, instruments and transport equipment	..	..	..	..	84.7	..	..	**26 Machines & matériel, instruments & mat. de transport**
27 Machinery n.e.c.	..	..	..	..	23.8	..	..	27 Machines, n.c.a.
28 Office, account. & computing machinery	..	..	..	..	0.3	..	..	28 Mach. de bur., compta. & trait. inform.
29 Electrical machinery	..	..	..	..	8.8	..	..	29 Machines électriques
30 Electro. equip. (radio, TV & commun.)	..	..	..	..	23.6	..	..	30 Equip. électro. (radio, tv & communic.)
31 *Electro. comp. (inc. semi-conductors)*	..	..	..	..	6.2	..	..	31 *Compos. électro. (semi-cond. incl.)*
32 *TV, radio & communications equip.*	..	..	..	..	17.4	..	..	32 *Appareils radio, tv et de communi.*
33 Instruments	..	..	..	..	5.6	..	..	33 Instruments
34 Motor vehicles	..	..	..	..	19.9	..	..	34 Véhicules automobiles
35 Other transport equipment	..	..	..	..	2.8 [o]	..	..	35 Autres matériels de transport
36 *Ships*	..	..	..	..	..	..	..	36 *Navires*
37 *Aerospace*	..	..	..	..	..	..	..	37 *Aérospatiale*
38 *Other transport equipment n.e.c.*	..	..	..	..	2.1	..	..	38 *Autres matériels de transport n.c.a.*
39 Furniture, other manufacturing n.e.c.	..	..	..	..	2.6	..	..	**39 Meubles, autres activ. de fabricat. n.c.a.**
40 Furniture	..	..	..	..	1.4	..	..	40 Meubles
41 Other manufacturing n.e.c.	..	..	..	..	1.3	..	..	41 Autres activités de fabrication n.c.a.
42 Recycling	..	..	..	..	0.0	..	..	**42 Récupération**
43 ELECTRICITY, GAS & WATER SUPPLY	..	..	..	..	1.3	..	..	**43 DISTRIBUTION D'ELECTRICITE, GAZ & EAU**
44 CONSTRUCTION	..	..	..	..	1.8	..	..	**44 CONSTRUCTION**
45 SERVICES SECTOR	..	..	..	..	35.8	..	..	**45 SECTEUR DES SERVICES**
46 Wholesale, retail trade & repair of motor vehicles, etc.	..	..	..	..	4.8	..	..	**46 Commerce (gros & détail), réparation de véhicules, etc.**
47 Hotels & restaurants	..	..	..	..	0.0	..	..	**47 Hôtels et restaurants**
48 Transport & storage	..	..	..	..	..	..	..	**48 Transports, entreposage**
49 Communications	..	..	..	..	..	..	..	**49 Communications**
50 Post	..	..	..	..	0.0	..	..	50 Postes
51 Telecommunications	..	..	..	..	..	..	..	51 Télécommunications
52 Financial intermediation (inc. insur.)	..	..	..	..	0.1	..	..	**52 Intermédiation financ. (assurances inc.)**
53 Real estate, renting & busin. activity	..	..	..	..	27.0	..	..	**53 Immobil., locat. & serv. aux entreprises**
54 Computer & related activities	..	..	..	..	1.6	..	..	54 Informatique et activités connexes
55 *Software consultancy*	..	..	..	..	1.2	..	..	55 *Production de logiciels*
56 *Other computer services n.e.c.*	..	..	..	..	0.5	..	..	56 *Autres activités informatiques*
57 Research & development	..	..	..	..	17.1	..	..	57 Recherche & développement expérimental
58 Other business activities n.e.c.	..	..	..	..	8.4	..	..	58 Autres activ. & serv. aux entreprises
59 Comm., soc. & pers. serv. activ., etc.	..	..	..	..	0.2	..	..	**59 Adm. pub., serv. soc. & collec., etc**
TOTAL BUSINESS ENTERPRISE SECTOR	**..**	**..**	**..**	**..**	**194.3**	**..**	**..**	**TOTAL SECTEUR DES ENTREPRISES**

Source: OECD, R&D database (DSTI/EAS Division), January 2002.

Source : OCDE, base de données R-D (DSTI/Division AES), janvier 2002.

AUSTRIA — Table 12 - Tableau 12 — **AUTRICHE**

Total business enterprise R&D personnel (BEMP) - Personnel total de R-D du secteur des entreprises (PRDE)

by industry - par industrie

In full time equivalent — *En équivalence plein-temps*

	1994	1995	1996	1997	1998	1999	2000	
1 AGRICULTURE, HUNTING & FORESTRY	..	..	..	..	..	..	..	**1 AGRICULTURE, CHASSE & SYLVICULTURE**
2 MINING	..	..	..	..	..	..	..	**2 ACTIVITES EXTRACTIVES**
3 MANUFACTURING	..	..	..	..	15 437	..	..	**3 ACTIVITES DE FABRICATION**
4 Food, beverages & tobacco	..	..	..	..	..	..	..	**4 Produits alimentaires, boissons & tabac**
5 Food products & beverages	..	..	..	..	202	..	..	5 Produits alimentaires & boissons
6 Tobacco products	..	..	..	..	..	..	..	6 Produits à base de tabac
7 Textiles, wear app., fur & leather	..	..	..	..	..	..	..	**7 Textiles, habill., fourrures & cuirs**
8 Textiles	..	..	..	..	181	..	..	8 Textiles
9 Wearing apparel & fur	..	..	..	..	..	..	..	9 Articles d'habillement & fourrures
10 Leather products & footwear	..	..	..	..	27	..	..	10 Articles en cuir & chaussures
11 Wood, paper, printing, publishing	..	..	..	..	179	..	..	**11 Bois, papier, imprimerie, édition**
12 Wood & cork (not furniture)	..	..	..	..	71	..	..	12 Bois & liège (sauf meubles)
13 Pulp, paper & paper products	..	..	..	..	87	..	..	13 Carton, papier & articles
14 Publ., print. & repro. of rec. media	..	..	..	..	22	..	..	14 Edition, impress. & repro. enregis.
15 Coke, petroleum, nuclear fuel, chemicals, rubber plastics	..	..	..	..	..	..	..	**15 Coke, pétrole, combust. nucl., prod. chimi., caout. & mat. plast.**
16 Coke, ref. petrol. prod. & nuclear fuel	..	..	..	..	..	..	..	16 Coke, raffin. pétr. & combust. nucl.
17 Chemicals & chemical products	..	..	..	..	1 613	..	..	17 Produits chimiques
18 *Chemical prod.(less pharmaceuticals)*	..	..	..	..	749	..	..	18 *Prod. chimi. (moins prod. pharmaceutiques)*
19 *Pharmaceuticals*	..	..	..	..	863	..	..	19 *Produits pharmaceutiques*
20 Rubber & plastic products	..	..	..	..	502	..	..	20 Caoutchouc et matières plastiques
21 Non-metallic mineral products	..	..	..	..	504	..	..	**21 Produits minéraux non métalliques**
22 Basic metals	..	..	..	..	467	..	..	**22 Produits métallurgiques de base**
23 Basic metals, ferrous	..	..	..	..	278	..	..	23 Produits métallur. de base, ferreux
24 Basic metals, non-ferrous	..	..	..	..	189	..	..	24 Prod. métallur. de base, non ferreux
25 Fabricated metal products	..	..	..	..	588	..	..	**25 Ouvrages en métaux (sauf mach.& mat.)**
26 Machinery equipment, instruments and transport equipment	..	..	..	..	10 764	..	..	**26 Machines & matériel, instruments & mat. de transport**
27 Machinery n.e.c.	..	..	..	..	1 897	..	..	27 Machines, n.c.a.
28 Office, account. & computing machinery	..	..	..	..	71	..	..	28 Mach. de bur., compta. & trait. inform.
29 Electrical machinery	..	..	..	..	1 119	..	..	29 Machines électriques
30 Electro. equip. (radio, TV & commun.)	..	..	..	..	5 172	..	..	30 Equip. électro. (radio, tv & communic.)
31 *Electro. comp. (inc. semi-conductors)*	..	..	..	..	622	..	..	31 *Compos. électro. (semi-cond. incl.)*
32 *TV, radio & communications equip.*	..	..	..	..	4 550	..	..	32 *Appareils radio, tv et de communi.*
33 Instruments	..	..	..	..	598	..	..	33 Instruments
34 Motor vehicles	..	..	..	..	1 544	..	..	34 Véhicules automobiles
35 Other transport equipment	..	..	..	..	362	..	..	35 Autres matériels de transport
36 *Ships*	..	..	..	..	..	..	..	36 *Navires*
37 *Aerospace*	..	..	..	..	..	..	..	37 *Aérospatiale*
38 *Other transport equipment n.e.c.*	..	..	..	..	265	..	..	38 *Autres matériels de transport n.c.a.*
39 Furniture, other manufacturing n.e.c.	..	..	..	..	230	..	..	**39 Meubles, autres activ. de fabricat. n.c.a.**
40 Furniture	..	..	..	..	73	..	..	40 Meubles
41 Other manufacturing n.e.c.	..	..	..	..	156	..	..	41 Autres activités de fabrication n.c.a.
42 Recycling	..	..	..	..	0	..	..	**42 Récupération**
43 ELECTRICITY, GAS & WATER SUPPLY	..	..	..	..	86	..	..	**43 DISTRIBUTION D'ELECTRICITE, GAZ & EAU**
44 CONSTRUCTION	..	..	..	..	120	..	..	**44 CONSTRUCTION**
45 SERVICES SECTOR	..	..	..	..	4 718	..	..	**45 SECTEUR DES SERVICES**
46 Wholesale, retail trade & repair of motor vehicles, etc.	..	..	..	..	546	..	..	**46 Commerce (gros & détail), réparation de véhicules, etc.**
47 Hotels & restaurants	..	..	..	..	0	..	..	**47 Hôtels et restaurants**
48 Transport & storage	..	..	..	..	..	..	..	**48 Transports, entreposage**
49 Communications	..	..	..	..	..	..	..	**49 Communications**
50 Post	..	..	..	..	0	..	..	50 Postes
51 Telecommunications	..	..	..	..	..	..	..	51 Télécommunications
52 Financial intermediation (inc. insur.)	..	..	..	..	196	..	..	**52 Intermédiation financ. (assurances inc.)**
53 Real estate, renting & busin. activity	..	..	..	..	3 579	..	..	**53 Immobil., locat. & serv. aux entreprises**
54 Computer & related activities	..	..	..	..	407	..	..	54 Informatique et activités connexes
55 *Software consultancy*	..	..	..	..	288	..	..	55 *Production de logiciels*
56 *Other computer services n.e.c.*	..	..	..	..	118	..	..	56 *Autres activités informatiques*
57 Research & development	..	..	..	..	1 422	..	..	57 Recherche & développement expérimental
58 Other business activities n.e.c.	..	..	..	..	1 751	..	..	58 Autres activ. & serv. aux entreprises
59 Comm., soc. & pers. serv. activ., etc.	..	..	..	..	14	..	..	**59 Adm. pub., serv. soc. & collec., etc**
TOTAL BUSINESS ENTERPRISE SECTOR	**..**	**..**	**..**	**..**	**20 385**	**..**	**..**	**TOTAL SECTEUR DES ENTREPRISES**

Source: OECD, R&D database (DSTI/EAS Division), January 2002.

Source : OCDE, base de données R-D (DSTI/Division AES), janvier 2002.

AUSTRIA Table 13 - Tableau 13 **AUTRICHE**

RSE in the business enterprise sector - Chercheurs dans le secteur des entreprises

by industry - par industrie

In full time equivalent *En équivalence plein-temps*

	1994	1995	1996	1997	1998	1999	2000	
1 **AGRICULTURE, HUNTING & FORESTRY**	..	..	..	..	..	..	..	1 **AGRICULTURE, CHASSE & SYLVICULTURE**
2 **MINING**	..	..	..	..	..	..	..	2 **ACTIVITES EXTRACTIVES**
3 **MANUFACTURING**	..	..	..	..	8 886	..	..	3 **ACTIVITES DE FABRICATION**
4 **Food, beverages & tobacco**	..	..	..	..	..	..	..	4 **Produits alimentaires, boissons & tabac**
5 Food products & beverages	..	..	..	..	102	..	..	5 Produits alimentaires & boissons
6 Tobacco products	..	..	..	..	..	..	..	6 Produits à base de tabac
7 **Textiles, wear app., fur & leather**	..	..	..	..	..	..	..	7 **Textiles, habill., fourrures & cuirs**
8 Textiles	..	..	..	..	61	..	..	8 Textiles
9 Wearing apparel & fur	..	..	..	..	..	..	..	9 Articles d'habillement & fourrures
10 Leather products & footwear	..	..	..	..	9	..	..	10 Articles en cuir & chaussures
11 **Wood, paper, printing, publishing**	..	..	..	..	94	..	..	11 **Bois, papier, imprimerie, édition**
12 Wood & cork (not furniture)	..	..	..	..	38	..	..	12 Bois & liège (sauf meubles)
13 Pulp, paper & paper products	..	..	..	..	45	..	..	13 Carton, papier & articles
14 Publ., print. & repro. of rec. media	..	..	..	..	11	..	..	14 Edition, impress. & repro. enregis.
15 **Coke, petroleum, nuclear fuel, chemicals, rubber plastics**	..	..	..	..	..	..	..	15 **Coke, pétrole, combust. nucl., prod. chimi., caout. & mat. plast.**
16 Coke, ref. petrol. prod. & nuclear fuel	..	..	..	..	..	..	..	16 Coke, raffin. pétr. & combust. nucl.
17 Chemicals & chemical products	..	..	..	..	779	..	..	17 Produits chimiques
18 *Chemical prod.(less pharmaceuticals)*	..	..	..	..	309	..	..	18 *Prod. chimi. (moins prod. pharmaceutiques)*
19 *Pharmaceuticals*	..	..	..	..	470	..	..	19 *Produits pharmaceutiques*
20 Rubber & plastic products	..	..	..	..	165	..	..	20 Caoutchouc et matières plastiques
21 **Non-metallic mineral products**	..	..	..	..	251	..	..	21 **Produits minéraux non métalliques**
22 **Basic metals**	..	..	..	..	228	..	..	22 **Produits métallurgiques de base**
23 Basic metals, ferrous	..	..	..	..	156	..	..	23 Produits métallur. de base, ferreux
24 Basic metals, non-ferrous	..	..	..	..	72	..	..	24 Prod. métallur. de base, non ferreux
25 **Fabricated metal products**	..	..	..	..	264	..	..	25 **Ouvrages en métaux (sauf mach.& mat.)**
26 **Machinery equipment, instruments and transport equipment**	..	..	..	..	6 789	..	..	26 **Machines & matériel, instruments & mat. de transport**
27 Machinery n.e.c.	..	..	..	..	888	..	..	27 Machines, n.c.a.
28 Office, account. & computing machinery	..	..	..	..	58	..	..	28 Mach. de bur., compta. & trait. inform.
29 Electrical machinery	..	..	..	..	686	..	..	29 Machines électriques
30 Electro. equip. (radio, TV & commun.)	..	..	..	..	3 889	..	..	30 Equip. électro. (radio, tv & communic.)
31 *Electro. comp. (inc. semi-conductors)*	..	..	..	..	435	..	..	31 *Compos. électro. (semi-cond. incl.)*
32 *TV, radio & communications equip.*	..	..	..	..	3 454	..	..	32 *Appareils radio, tv et de communi.*
33 Instruments	..	..	..	..	381	..	..	33 Instruments
34 Motor vehicles	..	..	..	..	697	..	..	34 Véhicules automobiles
35 Other transport equipment	..	..	..	..	190	..	..	35 Autres matériels de transport
36 *Ships*	..	..	..	..	..	..	..	36 *Navires*
37 *Aerospace*	..	..	..	..	..	..	..	37 *Aérospatiale*
38 *Other transport equipment n.e.c.*	..	..	..	..	118	..	..	38 *Autres matériels de transport n.c.a.*
39 **Furniture, other manufacturing n.e.c.**	..	..	..	..	86	..	..	39 **Meubles, autres activ. de fabricat. n.c.a.**
40 Furniture	..	..	..	..	21	..	..	40 Meubles
41 Other manufacturing n.e.c.	..	..	..	..	64	..	..	41 Autres activités de fabrication n.c.a.
42 **Recycling**	..	..	..	..	0	..	..	42 **Récupération**
43 **ELECTRICITY, GAS & WATER SUPPLY**	..	..	..	..	65	..	..	43 **DISTRIBUTION D'ELECTRICITE, GAZ & EAU**
44 **CONSTRUCTION**	..	..	..	..	80	..	..	44 **CONSTRUCTION**
45 **SERVICES SECTOR**	..	..	..	..	2 675	..	..	45 **SECTEUR DES SERVICES**
46 **Wholesale, retail trade & repair of motor vehicles, etc.**	..	..	..	..	334	..	..	46 **Commerce (gros & détail), réparation de véhicules, etc.**
47 **Hotels & restaurants**	..	..	..	..	0	..	..	47 **Hôtels et restaurants**
48 **Transport & storage**	..	..	..	..	..	..	..	48 **Transports, entreposage**
49 **Communications**	..	..	..	..	..	..	..	49 **Communications**
50 Post	..	..	..	..	0	..	..	50 Postes
51 Telecommunications	..	..	..	..	..	..	..	51 Télécommunications
52 **Financial intermediation (inc. insur.)**	..	..	..	..	89	..	..	52 **Intermédiation financ. (assurances inc.)**
53 **Real estate, renting & busin. activity**	..	..	..	..	1 943	..	..	53 **Immobil., locat. & serv. aux entreprises**
54 Computer & related activities	..	..	..	..	293	..	..	54 Informatique et activités connexes
55 *Software consultancy*	..	..	..	..	212	..	..	55 *Production de logiciels*
56 *Other computer services n.e.c.*	..	..	..	..	82	..	..	56 *Autres activités informatiques*
57 Research & development	..	..	..	..	919	..	..	57 Recherche & développement expérimental
58 Other business activities n.e.c.	..	..	..	..	731	..	..	58 Autres activ. & serv. aux entreprises
59 **Comm., soc. & pers. serv. activ., etc.**	..	..	..	..	11	..	..	59 **Adm. pub., serv. soc. & collec., etc**
TOTAL BUSINESS ENTERPRISE SECTOR	..	..	..	..	**11 716**	..	..	**TOTAL SECTEUR DES ENTREPRISES**

Source: OECD, R&D database (DSTI/EAS Division), January 2002.

Source : OCDE, base de données R-D (DSTI/Division AES), janvier 2002.

Table 14 - Tableau 14

Higher education intramural expenditure on R&D - Dépenses intra-muros de R-D de l'enseignement supérieur
(HERD) by field of science - (DIRDES) par discipline scientifique

	1994	1995	1996	1997	1998	1999	2000	
	14.1 Million EUR (1999 ATS euro, EUR from 1999) - Millions EUR (1999 ATS euro, EUR à partir de 1999)							
Natural sciences	..	..	..	..	298.7	..	..	Sciences exactes et naturelles
Engineering	..	..	..	..	127.0	..	..	Sciences de l'ingénieur
Medical sciences	..	..	..	..	284.4	..	..	Sciences médicales
Agricultural sciences	..	..	..	..	65.1	..	..	Sciences agricoles
Sub-total NSE	..	..	..	..	775.2	..	..	Sous-total SENI
Social sciences	..	..	..	..	129.1	..	..	Sciences sociales
Humanities	..	..	..	..	105.5	..	..	Sciences humaines
Sub-total SSH	..	..	..	..	234.5	..	..	Sous-total SSH
Total NSE+SSH	..	..	..	..	1 009.7	..	..	*Total SENI+SSH*
	14.2 Million constant USD (1995 prices and PPPs) - Millions de USD constants (prix et PPA de 1995)							
Natural sciences	..	..	..	..	289.6	..	..	Sciences exactes et naturelles
Engineering	..	..	..	..	123.1	..	..	Sciences de l'ingénieur
Medical sciences	..	..	..	..	275.8	..	..	Sciences médicales
Agricultural sciences	..	..	..	..	63.1	..	..	Sciences agricoles
Sub-total NSE	..	..	..	..	751.6	..	..	Sous-total SENI
Social sciences	..	..	..	..	125.1	..	..	Sciences sociales
Humanities	..	..	..	..	102.3	..	..	Sciences humaines
Sub-total SSH	..	..	..	..	227.4	..	..	Sous-total SSH
Total NSE+SSH	..	..	..	..	979.0	..	..	*Total SENI+SSH*

Table 15 - Tableau 15

Higher education intramural expenditure on R&D - Dépenses intra-muros de R-D de l'enseignement supérieur
(HERD) by field of science and type of costs - (DIRDES) par discipline scientifique et par type de coûts

Million EUR (1999 ATS euro, EUR from 1999) — *Millions EUR (1999 ATS euro, EUR à partir de 1999)*

	1994	1995	1996	1997	1998	1999	2000	
Current expenditure								**Dépenses courantes**
Natural sciences	..	..	..	..	260.6	..	..	Sciences exactes et naturelles
Engineering	..	..	..	..	107.0	..	..	Sciences de l'ingénieur
Medical sciences	..	..	..	..	228.8	..	..	Sciences médicales
Agricultural sciences	..	..	..	..	55.7	..	..	Sciences agricoles
Sub-total NSE	..	..	..	..	652.1	..	..	Sous-total SENI
Social sciences	..	..	..	..	114.6	..	..	Sciences sociales
Humanities	..	..	..	..	94.3	..	..	Sciences humaines
Sub-total SSH	..	..	..	..	208.9	..	..	Sous-total SSH
Total NSE+SSH	..	..	..	..	860.9	..	..	*Total SENI+SSH*
Capital expenditure								**Dépenses en capital**
Natural sciences	..	..	..	..	38.2	..	..	Sciences exactes et naturelles
Engineering	..	..	..	..	20.0	..	..	Sciences de l'ingénieur
Medical sciences	..	..	..	..	55.6	..	..	Sciences médicales
Agricultural sciences	..	..	..	..	9.4	..	..	Sciences agricoles
Sub-total NSE	..	..	..	..	123.1	..	..	Sous-total SENI
Social sciences	..	..	..	..	14.5	..	..	Sciences sociales
Humanities	..	..	..	..	11.2	..	..	Sciences humaines
Sub-total SSH	..	..	..	..	25.7	..	..	Sous-total SSH
Total NSE+SSH	..	..	..	..	148.8	..	..	*Total SENI+SSH*

Source: OECD, R&D database (DSTI/EAS Division), January 2002.

Source : OCDE, base de données R-D (DSTI/Division AES), janvier 2002.

AUSTRIA — Table 17 - Tableau 17 — AUTRICHE

Higher education R&D personnel (HEMP) - Personnel de R-D du secteur de l'enseignement supérieur

by field of science and occupation - par discipline scientifique et par profession

In full-time equivalent — *En équivalence plein-temps*

	1994	1995	1996	1997	1998	1999	2000	
Total R&D personnel								**Personnel total de R-D**
Natural sciences	..	..	..	..	2 732	..	..	Sciences exactes et naturelles
Engineering	..	..	..	..	1 211	..	..	Sciences de l'ingénieur
Medical sciences	..	..	..	..	2 285	..	..	Sciences médicales
Agricultural sciences	..	..	..	..	427	..	..	Sciences agricoles
Sub-total NSE	..	..	..	..	6 655	..	..	Sous-total SENI
Social sciences	..	..	..	..	1 115	..	..	Sciences sociales
Humanities	..	..	..	..	900	..	..	Sciences humaines
Sub-total SSH	..	..	..	..	2 016	..	..	Sous-total SSH
Total NSE+SSH	..	..	..	..	8 670	..	..	*Total SENI+SSH*
Researchers								**Chercheurs**
Natural sciences	..	..	..	..	1 951	..	..	Sciences exactes et naturelles
Engineering	..	..	..	..	818	..	..	Sciences de l'ingénieur
Medical sciences	..	..	..	..	1 368	..	..	Sciences médicales
Agricultural sciences	..	..	..	..	216	..	..	Sciences agricoles
Sub-total NSE	..	..	..	..	4 352	..	..	Sous-total SENI
Social sciences	..	..	..	..	876	..	..	Sciences sociales
Humanities	..	..	..	..	728	..	..	Sciences humaines
Sub-total SSH	..	..	..	..	1 603	..	..	Sous-total SSH
Total NSE+SSH	..	..	..	..	5 955	..	..	*Total SENI+SSH*
Technicians								**Technicians**
Natural sciences	..	..	..	..	341	..	..	Sciences exactes et naturelles
Engineering	..	..	..	..	144	..	..	Sciences de l'ingénieur
Medical sciences	..	..	..	..	537	..	..	Sciences médicales
Agricultural sciences	..	..	..	..	92	..	..	Sciences agricoles
Sub-total NSE	..	..	..	..	1 114	..	..	Sous-total SENI
Social sciences	..	..	..	..	74	..	..	Sciences sociales
Humanities	..	..	..	..	54	..	..	Sciences humaines
Sub-total SSH	..	..	..	..	128	..	..	Sous-total SSH
Total NSE+SSH	..	..	..	..	1 242	..	..	*Total SENI+SSH*
Other								**Autres**
Natural sciences	..	..	..	..	440	..	..	Sciences exactes et naturelles
Engineering	..	..	..	..	249	..	..	Sciences de l'ingénieur
Medical sciences	..	..	..	..	380	..	..	Sciences médicales
Agricultural sciences	..	..	..	..	118	..	..	Sciences agricoles
Sub-total NSE	..	..	..	..	1 188	..	..	Sous-total SENI
Social sciences	..	..	..	..	166	..	..	Sciences sociales
Humanities	..	..	..	..	119	..	..	Sciences humaines
Sub-total SSH	..	..	..	..	285	..	..	Sous-total SSH
Total NSE+SSH	..	..	..	..	1 473	..	..	*Total SENI+SSH*

Table 18 - Tableau 18

Government budget appropriations or outlays for R&D (GBAORD) - Crédits budgétaires publics de R-D (CBPRD)

by socio-economic objective - par objectif socio-économique

Million EUR (1999 ATS euro, EUR from 1999) — *Millions EUR (1999 ATS euro, EUR à partir de 1999)*

	1995	1996	1997	1998	1999	2000	2001	
Agriculture, forestry and fishing	37.1 [h]	35.4 [h]	37.1 [h]	38.7 [h]	43.2 [h]	39.9 [h,p]	40.5 [h,p]	Agriculture, sylviculture et pêche
Industrial development	88.8 [h]	82.4 [h]	73.6 [h]	80.0 [h]	87.2 [h]	76.2 [h,p]	114.0 [h,p]	Développement industriel
Energy	5.3 [h]	4.5 [h]	10.1 [h]	7.0 [h]	9.0 [h]	5.5 [h,p]	7.9 [h,p]	Energie
Transport & telecommunications	23.9 [h]	19.2 [h]	21.5 [h]	21.7 [h]	19.4 [h]	22.1 [h,p]	23.0 [h,p]	*Transports & télécommunications*
Urban & rural planning	1.6 [h]	1.2 [h]	1.5 [h]	1.7 [h]	1.3 [h]	1.4 [h,p]	1.5 [h,p]	*Aménagement urbain et rural*
ST Infrastructure	25.6 [h]	20.4 [h]	23.0 [h]	23.5 [h]	20.7 [h]	23.5 [h,p]	24.4 [h,p]	ST Infrastructure
Prevention of pollution	..	..	..	..	..	..	..	*Prévention de la pollution*
Identificat. & treatment of pollution	..	..	..	..	..	..	..	*Détect. pollut. & mesures antipollut.*
ST Environment	28.3 [h]	24.9 [h]	23.8 [h]	22.6 [h]	22.4 [h]	24.9 [h,p]	24.6 [h,p]	ST Environnement
Health	28.9 [h]	26.0 [h]	34.8 [h]	30.2 [h]	33.2 [h]	30.0 [h,p]	40.0 [h,p]	Santé
Social development & services	20.2 [h]	24.3 [h]	24.4 [h]	25.7 [h]	27.0 [h]	26.4 [h,p]	28.1 [h,p]	Développement et services sociaux
Earth and atmosphere	19.5 [h]	18.6 [h]	20.1 [h]	28.1 [h]	30.9 [h]	29.4 [h,p]	31.0 [h,p]	Milieu terrestre et atmosphère
Advancement of research	149.9 [h]	157.5 [h]	149.5 [h]	166.8 [h]	181.2 [h]	176.7 [h,p]	201.1 [h,p]	*Promotion de la recherche*
General university funds	745.0 [h]	727.5 [h]	734.3 [h]	766.7 [h]	805.4 [h]	762.9 [h,p]	810.6 [h,p]	*Fonds généraux des universités*
ST Advancement of knowledge	894.9 [h]	885.0 [h]	883.8 [h]	933.5 [h]	986.6 [h]	939.5 [h,p]	1 011.7 [h,p]	ST promotion gén. des connaiss.
Civil space	0.3 [h]	0.9 [h]	0.2 [h]	0.3 [h]	0.5 [h]	0.9 [h,p]	1.4 [h,p]	Espace civil
Defence	0.1 [h]	0.1 [h]	0.0 [h]	0.1 [h]	0.0 [h]	0.3 [h,p]	0.3 [h,p]	Défense
Not elsewhere classified	1.6 [h]	1.3 [h]	1.7 [h]	1.2 [h]	0.3 [h]	0.3 [h,p]	59.9 [h,p]	Non classé
Total	**1 150.4 [h]**	**1 123.7 [h]**	**1 132.9 [h]**	**1 190.8 [h]**	**1 260.8 [h]**	**1 196.8 [h,p]**	**1 383.8 [h,p]**	**Total**

Source: OECD, R&D database (DSTI/EAS Division), January 2002.

Source : OCDE, base de données R-D (DSTI/Division AES), janvier 2002.

Table 19 - Tableau 19

Technology balance of payments - Balance des paiements technologiques

by type of transfer - selon la nature des transferts

Million EUR (1999 ATS euro, EUR from 1999) — *Millions EUR (1999 ATS euro, EUR à partir de 1999)*

	1983	1984	1985	1986	1987	1988	1989	1990	1991	
Receipts										**Recettes**
Related firms	..	..	..	..	..	..	..	..	..	Firmes affiliées
Unrelated firms	..	..	..	..	..	..	..	..	..	Firmes non affiliées
Other	..	..	..	..	..	..	..	..	..	Autre
Total	39 *m*	46 *m*	45 *m*	48 *m*	51 *m*	63 *m*	73 *m*	74 *m*	67 *m*	*Total*
Payments										**Paiements**
Related firms	..	..	..	..	..	..	..	..	..	Firmes affiliées
Unrelated firms	..	..	..	..	..	..	..	..	..	Firmes non affiliées
Other	..	..	..	..	..	..	..	..	..	Autre
Total	193 *m*	156 *m*	171 *m*	183 *m*	199 *m*	238 *m*	223 *m*	235 *m*	255 *m*	*Total*

	1992	1993	1994	1995	1996	1997	1998	1999	2000	
Receipts										**Recettes**
Related firms	..	..	..	..	..	..	..	..	..	Firmes affiliées
Unrelated firms	..	..	..	..	..	..	..	..	..	Firmes non affiliées
Other	..	..	..	..	..	..	..	..	..	Autre
Total	816 *a*	892	1 127	1 397	1 657	1 735	2 120	2 142	2 637	*Total*
Payments										**Paiements**
Related firms	..	..	..	..	..	..	..	..	..	Firmes affiliées
Unrelated firms	..	..	..	..	..	..	..	..	..	Firmes non affiliées
Other	..	..	..	..	..	..	..	..	..	Autre
Total	1 175 *a*	1 211	1 285	1 568	1 992	2 234	2 699	2 470	2 633	*Total*

Source: OECD, TBPdatabase (DSTI/EAS Division), Januay 2002.

Source : OCDE, base de données TBP (DSTI/Division AES), janvier 2002.

BELGIUM — Table 1 - Tableau 1 — BELGIQUE

Gross domestic expenditure on R&D (GERD) - Dépenses intérieures brutes de R-D (DIRD)

by sector of performance and source of funds - par secteur d'exécution et par source de financement

	1994	1995	1996	1997	1998	1999	2000	
	1.1 Million EUR (1999 BEF euro, EUR from 1999) - Millions EUR (1999 BEF euro, EUR à partir de 1999)							
Business enterprise								**Entreprises**
financed by								*financé par*
Business enterprise	2 100.2	2 204.6	2 400.8	2 595.5	2 682.9	2 924.4	3 190.0 p	Entreprises
Direct government	103.4	106.1	137.1	155.9	190.4	207.1	224.5 p	État direct
Higher education	8.8	9.2	14.0	8.1	0.0	0.0	0.0 p	Enseignement supérieur
Private non-profit	0.8	0.5	0.6	0.6	0.6	0.7	0.7 p	Institutions sans but lucratif
Funds from abroad	144.8	151.0	121.2	144.5	161.2	175.0	189.9 p	Fonds de l'étranger
Total	2 357.9	2 471.4	2 673.7	2 904.6	3 035.1	3 307.1	3 605.0 p	*Total*
Government								**État**
financed by								*financé par*
Business enterprise	2.1	2.5	2.1	2.6	1.8	1.7	..	Entreprises
Direct government	107.2	110.7	114.5	125.1	133.2	139.0	..	État direct
Higher education	0.0	0.0	0.0	0.0	0.0	0.0	..	Enseignement supérieur
Private non-profit	0.6	0.3	0.3	0.3	0.2	0.2	..	Institutions sans but lucratif
Funds from abroad	6.4	6.2	5.6	5.7	11.8	12.2	..	Fonds de l'étranger
Total	116.3	119.6	122.5	133.7	147.1	153.1	..	*Total*
Higher education								**Enseignement supérieur**
financed by								*financé par*
Business enterprise	96.7	109.1	113.0	116.9	113.8	120.7	..	Entreprises
Direct government	..	..	..	..	..	..	..	État direct
General university funds	..	..	..	..	..	..	..	Fonds généraux des universités
Sub-total government	545.4	581.0	606.4	626.2	680.9	721.2	..	Sous-Total État
Higher education	56.5	58.4	77.3	114.5	127.3	145.4	..	Enseignement supérieur
Private non-profit	8.2	8.4	6.2	6.4	5.1	6.7	..	Institutions sans but lucratif
Funds from abroad	75.6	71.6	85.3	101.7	113.9	109.4	..	Fonds de l'étranger
Total	782.3	828.5	888.3	965.6	1 041.1	1 103.4	..	*Total*
Private non-profit								**Institutions sans but lucratif**
financed by								*financé par*
Business enterprise	9.4	9.8	9.5	9.7	9.8	10.0	..	Entreprises
Direct government	2.8	2.9	2.3	2.0	2.0	2.1	..	État direct
Higher education	1.2	1.2	0.3	0.3	0.3	0.3	..	Enseignement supérieur
Private non-profit	2.1	2.6	1.0	1.2	0.9	0.8	..	Institutions sans but lucratif
Funds from abroad	33.4	32.0	37.9	39.0	40.5	41.3	..	Fonds de l'étranger
Total	48.9	48.4	51.0	52.2	53.5	54.5	..	*Total*
GERD								**DIRD**
financed by								*financé par*
Business enterprise	2 208.4	2 326.0	2 525.3	2 724.7	2 808.3	3 056.7	..	Entreprises
Direct government	..	..	..	..	..	..	..	État direct
General university funds	..	..	..	..	..	..	..	Fonds généraux des universités
Sub-total government	758.8	800.6	860.3	909.3	1 006.5	1 069.4	..	Sous-Total État
Higher education	66.4	68.8	91.7	122.9	127.7	145.8	..	Enseignement supérieur
Private non-profit	11.6	11.7	8.2	8.5	7.0	8.4	..	Institutions sans but lucratif
Funds from abroad	260.3	260.7	250.0	290.8	327.4	337.9	..	Fonds de l'étranger
Total	3 305.5	3 467.9	3 735.5	4 056.2	4 276.8	4 618.1	..	*Total*
	1.2 Million constant USD (1995 prices and PPPs) - Millions de USD constants (prix et PPA de 1995)							
Business enterprise								**Entreprises**
financed by								*financé par*
Business enterprise	2 346.6	2 420.5	2 604.9	2 779.1	2 827.5	3 051.4	3 288.4 p	Entreprises
Direct government	115.5	116.5	148.7	167.0	200.6	216.1	231.4 p	État direct
Higher education	9.8	10.1	15.2	8.6	0.0	0.0	0.0 p	Enseignement supérieur
Private non-profit	0.9	0.5	0.7	0.7	0.7	0.7	0.7 p	Institutions sans but lucratif
Funds from abroad	161.8	165.8	131.5	154.7	169.9	182.6	195.7 p	Fonds de l'étranger
Total	2 634.6	2 713.3	2 901.0	3 110.0	3 198.7	3 450.8	3 716.3 p	*Total*
Government								**État**
financed by								*financé par*
Business enterprise	2.3	2.8	2.3	2.8	1.9	1.8	..	Entreprises
Direct government	119.8	121.5	124.2	134.0	140.3	145.0	..	État direct
Higher education	0.0	0.0	0.0	0.0	0.0	0.0	..	Enseignement supérieur
Private non-profit	0.7	0.3	0.4	0.3	0.3	0.2	..	Institutions sans but lucratif
Funds from abroad	7.2	6.8	6.1	6.1	12.5	12.7	..	Fonds de l'étranger
Total	130.0	131.3	133.0	143.1	155.0	159.7	..	*Total*
Higher education								**Enseignement supérieur**
financed by								*financé par*
Business enterprise	108.1	119.7	122.6	125.1	119.9	125.9	..	Entreprises
Direct government	..	..	..	..	..	..	..	État direct
General university funds	..	..	..	..	..	..	..	Fonds généraux des universités
Sub-total government	609.4	637.9	658.0	670.4	717.6	752.5	..	Sous-Total État
Higher education	63.1	64.2	83.9	122.6	134.2	151.8	..	Enseignement supérieur
Private non-profit	9.1	9.2	6.8	6.8	5.4	7.0	..	Institutions sans but lucratif
Funds from abroad	84.5	78.6	92.6	108.9	120.1	114.1	..	Fonds de l'étranger
Total	874.1	909.7	963.8	1 033.9	1 097.2	1 151.3	..	*Total*
Private non-profit								**Institutions sans but lucratif**
financed by								*financé par*
Business enterprise	10.5	10.7	10.3	10.4	10.3	10.4	..	Entreprises
Direct government	3.1	3.2	2.5	2.2	2.2	2.2	..	État direct
Higher education	1.3	1.3	0.4	0.3	0.3	0.3	..	Enseignement supérieur
Private non-profit	2.3	2.8	1.1	1.3	1.0	0.9	..	Institutions sans but lucratif
Funds from abroad	37.4	35.1	41.1	41.7	42.6	43.1	..	Fonds de l'étranger
Total	54.7	53.1	55.3	55.9	56.4	56.9	..	*Total*
GERD								**DIRD**
financed by								*financé par*
Business enterprise	2 467.5	2 553.7	2 740.0	2 917.3	2 959.6	3 189.5	..	Entreprises
Direct government	..	..	..	..	..	..	..	État direct
General university funds	..	..	..	..	..	..	..	Fonds généraux des universités
Sub-total government	847.8	879.0	933.4	973.6	1 060.7	1 115.9	..	Sous-Total État
Higher education	74.2	75.6	99.5	131.6	134.6	152.1	..	Enseignement supérieur
Private non-profit	13.0	12.8	8.9	9.1	7.3	8.7	..	Institutions sans but lucratif
Funds from abroad	290.8	286.3	271.3	311.4	345.0	352.5	..	Fonds de l'étranger
Total	3 693.4	3 807.4	4 053.1	4 343.0	4 507.3	4 818.7	..	*Total*

Source: OECD, R&D database (DSTI/EAS Division), January 2002.

Source : OCDE, base de données R-D (DSTI/Division AES), janvier 2002.

Table 5 - Tableau 5

Total R&D personnel - Personnel total de R-D

by sector of employment and occupation - par secteur d'emploi et par profession

In full-time equivalent — *En équivalence plein-temps*

	1994	1995	1996	1997	1998	1999	2000	
Business enterprise								**Entreprises**
RSE	11 603	12 174	13 340	13 848	15 573	16 476	18 031 p	Chercheurs
Technicians	5 108	5 481	8 758	9 007	9 771	10 331	10 891 p	Techniciens
Other occupations	6 691	6 692	5 113	5 305	3 919	4 061	4 226 p	Autres professions
Total	23 402	24 346	27 211	28 161	29 264	30 868	33 148 p	*Total*
Government								**État**
RSE	1 007	1 014	1 045	1 118	1 120	1 210	..	Chercheurs
Technicians	111	110	159	165	135	179	..	Techniciens
Other occupations	908	896	867	862	816	841	..	Autres professions
Total	2 026	2 019	2 072	2 144	2 071	2 229	..	*Total*
Higher education								**Enseignement supérieur**
RSE	9 900	10 027	9 789	10 305	11 148	12 209	..	Chercheurs
Technicians	694	686	887	960	1 108	1 207	..	Techniciens
Other occupations	2 345	2 332	2 106	2 161	2 344	2 456	..	Autres professions
Total	12 939	13 045	12 782	13 426	14 600	15 871	..	*Total*
Private non-profit								**Institutions sans but lucratif**
RSE	262	277	303	308	308	324	..	Chercheurs
Technicians	103	112	128	128	129	127	..	Techniciens
Other occupations	47	47	53	54	57	57	..	Autres professions
Total	412	435	483	489	494	509	..	*Total*
TOTAL								**TOTAL**
RSE	22 773	23 491	24 478	25 579	28 149	30 219	..	Chercheurs
Technicians	6 016	6 388	9 931	10 259	11 142	11 843	..	Techniciens
Other occupations	9 990	9 966	8 139	8 382	7 136	7 415	..	Autres professions
Total	38 779	39 846	42 548	44 220	46 428	49 477	..	*Total*

Table 6 - Tableau 6

Total R&D personnel - Personnel total de R-D

by sector of employment and formal qualification - par secteur d'emploi et par niveau de qualification

In full-time equivalent — *En équivalence plein-temps*

	1994	1995	1996	1997	1998	1999	2000	
Business enterprise								**Entreprises**
University PhD level degrees	..	..	..	..	..	..	..	Diplômes univ. de niv. doctorat
Other university degrees	..	..	..	..	..	..	..	Autres diplômes universitaires
Sub-total university degrees	11 657	12 208	13 763	14 271	15 963	16 989	18 650 p	Sous-total Universitaires
Other post-secondary diplomas	5 108	5 481	7 420	7 715	7 221	7 488	7 847 p	Autres diplômes post-secondaires
Secondary diplomas	..	..	..	..	..	..	..	Diplômes de fin d'études secondaires
Other qualifications	6 637	6 657	6 028	6 174	6 080	6 392	6 651 p	Autres qualifications
Total	23 402	24 346	27 211	28 161	29 264	30 868	33 148 p	*Total*
Government								**État**
University PhD level degrees	..	..	..	..	..	..	..	Diplômes univ. de niv. doctorat
Other university degrees	..	..	..	..	..	..	..	Autres diplômes universitaires
Sub-total university degrees	1 007	1 014	1 045	1 118	1 120	1 210	..	Sous-total Universitaires
Other post-secondary diplomas	111	110	159	165	135	179	..	Autres diplômes post-secondaires
Secondary diplomas	..	..	..	..	..	..	..	Diplômes de fin d'études secondaires
Other qualifications	908	896	867	862	816	841	..	Autres qualifications
Total	2 026	2 019	2 072	2 144	2 071	2 229	..	*Total*
Higher education								**Enseignement supérieur**
University PhD level degrees	..	..	..	..	..	..	..	Diplômes univ. de niv. doctorat
Other university degrees	..	..	..	..	..	..	..	Autres diplômes universitaires
Sub-total university degrees	9 900	10 027	9 789	10 305	11 148	12 209	..	Sous-total Universitaires
Other post-secondary diplomas	694	686	887	960	1 108	1 207	..	Autres diplômes post-secondaires
Secondary diplomas	..	..	..	..	..	..	..	Diplômes de fin d'études secondaires
Other qualifications	2 345	2 332	2 106	2 161	2 344	2 456	..	Autres qualifications
Total	12 939	13 045	12 782	13 426	14 600	15 871	..	*Total*
Private non-profit								**Institutions sans but lucratif**
University PhD level degrees	64	71	..	..	..	..	..	Diplômes univ. de niv. doctorat
Other university degrees	198	206	..	..	..	..	..	Autres diplômes universitaires
Sub-total university degrees	262	277	303	308	308	324	..	Sous-total Universitaires
Other post-secondary diplomas	103	112	128	128	129	127	..	Autres diplômes post-secondaires
Secondary diplomas	..	..	..	..	..	..	..	Diplômes de fin d'études secondaires
Other qualifications	47	47	53	54	57	57	..	Autres qualifications
Total	412	435	483	489	494	509	..	*Total*
TOTAL								**TOTAL**
University PhD level degrees	..	..	..	..	..	..	..	Diplômes univ. de niv. doctorat
Other university degrees	..	..	..	..	..	..	..	Autres diplômes universitaires
Sub-total university degrees	22 827	23 525	24 900	26 001	28 539	30 732	..	Sous-total Universitaires
Other post-secondary diplomas	6 016	6 388	8 593	8 967	8 592	9 000	..	Autres diplômes post-secondaires
Secondary diplomas	..	..	..	..	..	..	..	Diplômes de fin d'études secondaires
Other qualifications	9 936	9 932	9 054	9 251	9 297	9 745	..	Autres qualifications
Total	38 779	39 846	42 548	44 220	46 428	49 477	..	*Total*

Source: OECD, R&D database (DSTI/EAS Division), January 2002.

Source : OCDE, base de données R-D (DSTI/Division AES), janvier 2002.

BELGIUM Table 9 - Tableau 9 **BELGIQUE**

Total business enterprise intramural expenditure on R&D - Dépenses intra-muros totales de R-D des entreprises

(BERD) by industry - (DIRDE) par industrie

	1994	1995	1996	1997	1998	1999	2000	
	9.1 Million EUR (1999 BEF euro, EUR from 1999) - Millions EUR (1999 BEF euro, EUR à partir de 1999)							
1 AGRICULTURE, HUNTING & FORESTRY	5.7	5.9	6.3	6.2	10.4	11.3	11.8 P	**1 AGRICULTURE, CHASSE & SYLVICULTURE**
2 MINING	0.9	1.0	1.0	1.0	2.8	3.0	3.3 P	**2 ACTIVITES EXTRACTIVES**
3 MANUFACTURING	1 992.2	2 077.4	2 221.1	2 402.4	2 446.3	2 655.6	2 878.6 P	**3 ACTIVITES DE FABRICATION**
4 Food, beverages & tobacco	61.4	61.7	73.8	75.0	83.0	87.3	89.3 P	**4 Produits alimentaires, boissons & tabac**
5 Food products & beverages	60.3	60.7	72.7	73.9	82.2	86.3	88.3 P	5 Produits alimentaires & boissons
6 Tobacco products	1.1	1.0	1.1	1.1	0.8	1.1	1.1 P	6 Produits à base de tabac
7 Textiles, wear app., fur & leather	25.4	24.6	34.9	35.7	44.2	48.8	54.0 P	**7 Textiles, habill., fourrures & cuirs**
8 Textiles	22.7	22.1	32.6	33.6	39.1	42.7	48.7 P	8 Textiles
9 Wearing apparel & fur	0.8	0.7	0.3	0.3	3.5	4.5	3.7 P	9 Articles d'habillement & fourrures
10 Leather products & footwear	1.9	1.8	1.9	1.7	1.6	1.6	1.7 P	10 Articles en cuir & chaussures
11 Wood, paper, printing, publishing	27.0	27.1	26.3	23.8	26.7	32.2	34.5 P	**11 Bois, papier, imprimerie, édition**
12 Wood & cork (not furniture)	1.2	1.3	1.5	1.5	1.7	2.1	2.3 P	12 Bois & liège (sauf meubles)
13 Pulp, paper & paper products	7.6	7.6	7.4	7.8	10.7	12.8	13.9 P	13 Carton, papier & articles
14 Publ., print. & repro. of rec. media	18.2	18.2	17.4	14.6	14.3	17.4	18.4 P	14 Edition, impress. & repro. enregis.
15 Coke, petroleum, nuclear fuel, chemicals, rubber plastics	930.3	984.2	1 089.1	1 226.4	1 192.3	1 280.9	1 356.0 P	**15 Coke, pétrole, combust. nucl., prod. chimi., caout. & mat. plast.**
16 Coke, ref. petrol. prod. & nuclear fuel	69.3	69.2	68.4	69.3	62.4	63.4	65.3 P	16 Coke, raffin. pétr. & combust. nucl.
17 Chemicals & chemical products	804.3	857.8	954.9	1 092.0	1 053.1	1 139.6	1 210.5 P	17 Produits chimiques
18 *Chemical prod.(less pharmaceuticals)*	494.0	521.2	552.4	580.7	550.5	565.1	581.1 P	18 *Prod. chimi. (moins prod. pharmaceutiques)*
19 *Pharmaceuticals*	310.3	336.6	402.4	511.4	502.6	574.5	629.5 P	19 *Produits pharmaceutiques*
20 Rubber & plastic products	56.7	57.2	65.8	65.1	76.8	77.8	80.2 P	20 Caoutchouc et matières plastiques
21 Non-metallic mineral products	36.1	38.7	57.0	56.5	63.2	67.6	72.5 P	**21 Produits minéraux non métalliques**
22 Basic metals	98.2	105.7	103.0	103.7	101.4	107.6	115.9 P	**22 Produits métallurgiques de base**
23 Basic metals, ferrous	72.9	77.1	70.6	71.2	70.8	75.4	82.7 P	23 Produits métallur. de base, ferreux
24 Basic metals, non-ferrous	25.3	28.6	32.4	32.5	30.6	32.2	33.2 P	24 Prod. métallur. de base, non ferreux
25 Fabricated metal products	36.2	37.6	36.1	37.1	41.8	42.8	48.7 P	**25 Ouvrages en métaux (sauf mach.& mat.)**
26 Machinery equipment, instruments and transport equipment	744.6	764.8	776.4	821.1	877.0	972.1	1 090.7 P	**26 Machines & matériel, instruments & mat. de transport**
27 Machinery n.e.c.	134.4	130.6	136.3	143.2	166.1	170.5	179.4 P	27 Machines, n.c.a.
28 Office, account. & computing machinery	5.6	5.1	5.4	5.9	7.1	6.4	8.3 P	28 Mach. de bur., compta. & trait. inform.
29 Electrical machinery	89.0	90.4	93.5	98.6	99.7	111.3	123.5 P	29 Machines électriques
30 Electro. equip. (radio, TV & commun.)	367.4	377.4	375.3	406.9	424.5	481.4	560.4 P	30 Equip. électro. (radio, tv & communic.)
31 *Electro. comp. (inc. semi-conductors)*	34.5	38.4	43.1	45.3	50.6	55.4	66.3 P	31 *Compos. électro. (semi-cond. incl.)*
32 *TV, radio & communications equip.*	332.9	339.0	332.1	361.6	373.9	426.0	494.1 P	32 *Appareils radio, tv et de communi.*
33 Instruments	53.9	55.6	61.4	63.8	42.7	42.6	44.7 P	33 Instruments
34 Motor vehicles	50.1	55.1	58.7	57.9	110.2	126.5	129.6 P	34 Véhicules automobiles
35 Other transport equipment	44.1	50.5	45.8	44.9	26.7	33.5	44.7 P	35 Autres matériels de transport
36 *Ships*	0.4	0.5	1.1	0.7	0.6	0.6	0.8 P	36 *Navires*
37 *Aerospace*	34.8	37.8	34.5	34.4	21.3	29.8	40.5 P	37 *Aérospatiale*
38 *Other transport equipment n.e.c.*	8.9	12.2	10.1	9.8	4.8	3.0	3.4 P	38 *Autres matériels de transport n.c.a.*
39 Furniture, other manufacturing n.e.c.	31.8	31.4	23.5	21.9	15.3	14.4	15.0 P	**39 Meubles, autres activ. de fabricat. n.c.a.**
40 Furniture	28.2	28.3	14.4	13.4	5.5	7.7	7.9 P	40 Meubles
41 Other manufacturing n.e.c.	3.6	3.1	9.1	8.5	9.8	6.7	7.1 P	41 Autres activités de fabrication n.c.a.
42 Recycling	1.2	1.6	1.1	1.2	1.4	1.9	1.9 P	**42 Récupération**
43 ELECTRICITY, GAS & WATER SUPPLY	3.8	4.8	2.6	2.5	2.0	2.3	1.9 P	**43 DISTRIBUTION D'ELECTRICITE, GAZ & EAU**
44 CONSTRUCTION	32.1	30.8	39.4	41.7	36.9	38.3	39.5 P	**44 CONSTRUCTION**
45 SERVICES SECTOR	323.2	351.6	403.5	450.8	536.8	596.6	669.9 P	**45 SECTEUR DES SERVICES**
46 Wholesale, retail trade & repair of motor vehicles, etc.	34.9	40.4	36.0	37.9	40.0	46.2	45.6 P	**46 Commerce (gros & détail), réparation de véhicules, etc.**
47 Hotels & restaurants	0.0	0.0	0.0	0.0	0.0	0.0	0.0 P	**47 Hôtels et restaurants**
48 Transport & storage	3.2	3.5	4.9	5.3	6.0	9.4	9.7 P	**48 Transports, entreposage**
49 Communications	6.1	13.7	10.0	10.7	12.9	13.1	13.2 P	**49 Communications**
50 Post	0.0	0.0	0.0	0.0	0.0	0.2	0.2 P	50 Postes
51 Telecommunications	6.1	13.7	10.0	10.7	12.9	12.8	13.0 P	51 Télécommunications
52 Financial intermediation (inc. insur.)	55.3	59.5	70.6	74.7	77.8	81.8	91.4 P	**52 Intermédiation financ. (assurances inc.)**
53 Real estate, renting & busin. activity	219.5	229.2	273.5	314.5	390.7	433.8	499.1 P	**53 Immobil., locat. & serv. aux entreprises**
54 Computer & related activities	102.5	107.4	129.8	158.7	183.4	213.7	254.2 P	54 Informatique et activités connexes
55 *Software consultancy*	17.9	20.8	23.0	28.2	34.3	39.7	41.8 P	55 *Production de logiciels*
56 *Other computer services n.e.c.*	84.6	86.6	106.9	130.5	149.1	174.0	212.4 P	56 *Autres activités informatiques*
57 Research & development	11.7	10.5	11.0	12.6	21.0	21.1	20.5 P	57 Recherche & développement expérimental
58 Other business activities n.e.c.	105.3	111.3	132.7	143.2	186.2	199.0	224.4	58 Autres activ. & serv. aux entreprises
59 Comm., soc. & pers. serv. activ., etc.	4.3	5.3	8.5	7.8	9.4	12.3	10.8 P	**59 Adm. pub., serv. soc. & collec., etc**
TOTAL BUSINESS ENTERPRISE SECTOR	**2 357.9**	**2 471.4**	**2 673.7**	**2 904.6**	**3 035.1**	**3 307.1**	**3 605.0 P**	**TOTAL SECTEUR DES ENTREPRISES**

Source: OECD, R&D database (DSTI/EAS Division), January 2002.

Source : OCDE, base de données R-D (DSTI/Division AES), janvier 2002.

BELGIUM — Table 9 - Tableau 9 — BELGIQUE

Total business enterprise intramural expenditure on R&D - Dépenses intra-muros totales de R-D des entreprises

(BERD) by industry - (DIRDE) par industrie

9.2 Million constant USD(1995 prices and PPPs) - Millions de USD constants (prix et PPA de 1995)

	1994	1995	1996	1997	1998	1999	2000	
1 AGRICULTURE, HUNTING & FORESTRY	6.4	6.5	6.8	6.6	10.9	11.8	12.2 p	**1 AGRICULTURE, CHASSE & SYLVICULTURE**
2 MINING	1.0	1.1	1.0	1.0	3.0	3.2	3.4 p	**2 ACTIVITES EXTRACTIVES**
3 MANUFACTURING	2 226.0	2 280.7	2 409.9	2 572.3	2 578.1	2 771.0	2 967.4 p	**3 ACTIVITES DE FABRICATION**
4 Food, beverages & tobacco	68.6	67.7	80.1	80.3	87.5	91.1	92.1 p	**4 Produits alimentaires, boissons & tabac**
5 Food products & beverages	67.4	66.6	78.9	79.1	86.6	90.0	91.0 p	5 Produits alimentaires & boissons
6 Tobacco products	1.2	1.1	1.2	1.1	0.9	1.1	1.1 p	6 Produits à base de tabac
7 Textiles, wear app., fur & leather	28.4	27.0	37.8	38.2	46.6	50.9	55.7 p	**7 Textiles, habill., fourrures & cuirs**
8 Textiles	25.4	24.2	35.4	36.0	41.2	44.6	50.2 p	8 Textiles
9 Wearing apparel & fur	0.9	0.7	0.3	0.3	3.7	4.7	3.8 p	9 Articles d'habillement & fourrures
10 Leather products & footwear	2.1	2.0	2.1	1.9	1.7	1.7	1.7 p	10 Articles en cuir & chaussures
11 Wood, paper, printing, publishing	30.2	29.7	28.5	25.5	28.2	33.6	35.6 p	**11 Bois, papier, imprimerie, édition**
12 Wood & cork (not furniture)	1.3	1.4	1.6	1.6	1.8	2.2	2.4 p	12 Bois & liège (sauf meubles)
13 Pulp, paper & paper products	8.5	8.3	8.0	8.3	11.3	13.3	14.3 p	13 Carton, papier & articles
14 Publ., print. & repro. of rec. media	20.3	20.0	18.9	15.6	15.0	18.1	18.9 p	14 Edition, impress. & repro. enregis.
15 Coke, petroleum, nuclear fuel, chemicals, rubber plastics	1 039.4	1 080.6	1 181.6	1 313.1	1 256.5	1 336.5	1 397.9 p	**15 Coke, pétrole, combust. nucl., prod. chimi., caout. & mat. plast.**
16 Coke, ref. petrol. prod. & nuclear fuel	77.4	76.0	74.2	74.2	65.7	66.1	67.3 p	16 Coke, raffin. pétr. & combust. nucl.
17 Chemicals & chemical products	898.6	941.8	1 036.0	1 169.2	1 109.8	1 189.2	1 247.9 p	17 Produits chimiques
18 *Chemical prod.(less pharmaceuticals)*	552.0	572.2	599.4	621.7	580.2	589.7	599.0 p	18 *Prod. chimi. (moins prod. pharmaceutiques)*
19 *Pharmaceuticals*	346.7	369.6	436.6	547.5	529.6	599.5	648.9 p	19 *Produits pharmaceutiques*
20 Rubber & plastic products	63.3	62.8	71.4	69.7	80.9	81.2	82.7 p	20 Caoutchouc et matières plastiques
21 Non-metallic mineral products	40.4	42.5	61.8	60.5	66.6	70.6	74.7 p	**21 Produits minéraux non métalliques**
22 Basic metals	109.8	116.1	111.8	111.0	106.8	112.2	119.5 p	**22 Produits métallurgiques de base**
23 Basic metals, ferrous	81.5	84.7	76.6	76.2	74.6	78.6	85.3 p	23 Produits métallur. de base, ferreux
24 Basic metals, non-ferrous	28.3	31.4	35.2	34.8	32.2	33.6	34.3 p	24 Prod. métallur. de base, non ferreux
25 Fabricated metal products	40.5	41.3	39.2	39.7	44.1	44.6	50.2 p	**25 Ouvrages en métaux (sauf mach.& mat.)**
26 Machinery equipment, instruments and transport equipment	832.0	839.7	842.3	879.2	924.3	1 014.4	1 124.3 p	**26 Machines & matériel, instruments & mat. de transport**
27 Machinery n.e.c.	150.2	143.4	147.9	153.3	175.1	177.9	185.0 p	27 Machines, n.c.a.
28 Office, account. & computing machinery	6.3	5.6	5.9	6.3	7.5	6.7	8.6 p	28 Mach. de bur., compta. & trait. inform.
29 Electrical machinery	99.5	99.3	101.4	105.5	105.0	116.1	127.3 p	29 Machines électriques
30 Electro. equip. (radio, TV & commun.)	410.5	414.3	407.2	435.7	447.4	502.4	577.6 p	30 Equip. électro. (radio, tv & communic.)
31 *Electro. comp. (inc. semi-conductors)*	38.5	42.1	46.8	48.5	53.3	57.8	68.3 p	31 *Compos. électro. (semi-cond. incl.)*
32 *TV, radio & communications equip.*	372.0	372.2	360.4	387.2	394.1	444.5	509.3 p	32 *Appareils radio, tv et de communi.*
33 Instruments	60.3	61.1	66.6	68.4	45.0	44.4	46.1 p	33 Instruments
34 Motor vehicles	56.0	60.5	63.7	61.9	116.1	132.0	133.6 p	34 Véhicules automobiles
35 Other transport equipment	49.3	55.5	49.6	48.1	28.1	34.9	46.1 p	35 Autres matériels de transport
36 *Ships*	0.5	0.5	1.2	0.8	0.7	0.7	0.8 p	36 *Navires*
37 *Aerospace*	38.9	41.5	37.4	36.8	22.4	31.1	41.8 p	37 *Aérospatiale*
38 *Other transport equipment n.e.c.*	9.9	13.4	11.0	10.5	5.0	3.1	3.5 p	38 *Autres matériels de transport n.c.a.*
39 Furniture, other manufacturing n.e.c.	35.5	34.4	25.5	23.5	16.1	15.0	15.5 p	**39 Meubles, autres activ. de fabricat. n.c.a.**
40 Furniture	31.5	31.1	15.6	14.4	5.8	8.1	8.2 p	40 Meubles
41 Other manufacturing n.e.c.	4.0	3.4	9.9	9.1	10.4	7.0	7.4 p	41 Autres activités de fabrication n.c.a.
42 Recycling	1.3	1.7	1.2	1.3	1.5	1.9	2.0 p	**42 Récupération**
43 ELECTRICITY, GAS & WATER SUPPLY	4.3	5.2	2.8	2.7	2.1	2.4	1.9 p	**43 DISTRIBUTION D'ELECTRICITE, GAZ & EAU**
44 CONSTRUCTION	35.8	33.8	42.7	44.7	38.9	39.9	40.7 p	**44 CONSTRUCTION**
45 SERVICES SECTOR	361.1	386.0	437.8	482.7	565.7	622.5	690.6 p	**45 SECTEUR DES SERVICES**
46 Wholesale, retail trade & repair of motor vehicles, etc.	38.9	44.3	39.1	40.6	42.2	48.2	47.0 p	**46 Commerce (gros & détail), réparation de véhicules, etc.**
47 Hotels & restaurants	0.0	0.0	0.0	0.0	0.0	0.0	0.0 p	**47 Hôtels et restaurants**
48 Transport & storage	3.6	3.8	5.3	5.6	6.3	9.8	10.0 p	**48 Transports, entreposage**
49 Communications	6.8	15.0	10.9	11.5	13.6	13.7	13.6 p	**49 Communications**
50 Post	0.0	0.0	0.0	0.0	0.0	0.3	0.3 p	50 Postes
51 Telecommunications	6.8	15.0	10.9	11.5	13.6	13.4	13.4 p	51 Télécommunications
52 Financial intermediation (inc. insur.)	61.8	65.3	76.6	79.9	82.0	85.3	94.3 p	**52 Intermédiation financ. (assurances inc.)**
53 Real estate, renting & busin. activity	245.2	251.6	296.7	336.7	411.7	452.6	514.5 p	**53 Immobil., locat. & serv. aux entreprises**
54 Computer & related activities	114.5	117.9	140.9	170.0	193.3	223.0	262.0 p	54 Informatique et activités connexes
55 *Software consultancy*	20.0	22.8	24.9	30.2	36.2	41.4	43.1 p	55 *Production de logiciels*
56 *Other computer services n.e.c.*	94.5	95.1	116.0	139.7	157.1	181.6	218.9 p	56 *Autres activités informatiques*
57 Research & development	13.0	11.6	11.9	13.5	22.2	22.0	21.2 p	57 Recherche & développement expérimental
58 Other business activities n.e.c.	117.7	122.1	143.9	153.3	196.3	207.6	231.3	58 Autres activ. & serv. aux entreprises
59 Comm., soc. & pers. serv. activ., etc.	4.8	5.9	9.2	8.4	9.9	12.9	11.2 p	**59 Adm. pub., serv. soc. & collec., etc**
TOTAL BUSINESS ENTERPRISE SECTOR	**2 634.6**	**2 713.3**	**2 901.0**	**3 110.0**	**3 198.7**	**3 450.8**	**3 716.3 p**	**TOTAL SECTEUR DES ENTREPRISES**

Source: OECD, R&D database (DSTI/EAS Division), January 2002.

Source : OCDE, base de données R-D (DSTI/Division AES), janvier 2002.

BELGIUM — Table 17 - Tableau 17 — BELGIQUE

Higher education R&D personnel (HEMP) - Personnel de R-D du secteur de l'enseignement supérieur

by field of science and occupation - par discipline scientifique et par profession

In full-time equivalent — *En équivalence plein-temps*

	1994	1995	1996	1997	1998	1999	2000	
Total R&D personnel								**Personnel total de R-D**
Natural sciences	..	..	..	..	..	..	..	Sciences exactes et naturelles
Engineering	..	..	..	..	..	..	..	Sciences de l'ingénieur
Medical sciences	..	..	..	..	..	..	..	Sciences médicales
Agricultural sciences	..	..	..	..	..	..	..	Sciences agricoles
Sub-total NSE	..	..	..	..	..	..	..	Sous-total SENI
Social sciences	..	..	..	..	..	..	..	Sciences sociales
Humanities	..	..	..	..	..	..	..	Sciences humaines
Sub-total SSH	..	..	..	..	..	..	..	Sous-total SSH
Total NSE+SSH	12 939	13 045	12 782	13 426	14 600	15 871	..	*Total SENI+SSH*
Researchers								**Chercheurs**
Natural sciences	..	..	..	..	..	..	..	Sciences exactes et naturelles
Engineering	..	..	..	..	..	..	..	Sciences de l'ingénieur
Medical sciences	..	..	..	..	..	..	..	Sciences médicales
Agricultural sciences	..	..	..	..	..	..	..	Sciences agricoles
Sub-total NSE	..	..	..	..	..	..	..	Sous-total SENI
Social sciences	..	..	..	..	..	..	..	Sciences sociales
Humanities	..	..	..	..	..	..	..	Sciences humaines
Sub-total SSH	..	..	..	..	..	..	..	Sous-total SSH
Total NSE+SSH	9 900	10 027	9 789	10 305	11 148	12 209	..	*Total SENI+SSH*
Technicians								**Technicians**
Natural sciences	..	..	..	..	..	..	..	Sciences exactes et naturelles
Engineering	..	..	..	..	..	..	..	Sciences de l'ingénieur
Medical sciences	..	..	..	..	..	..	..	Sciences médicales
Agricultural sciences	..	..	..	..	..	..	..	Sciences agricoles
Sub-total NSE	..	..	..	..	..	..	..	Sous-total SENI
Social sciences	..	..	..	..	..	..	..	Sciences sociales
Humanities	..	..	..	..	..	..	..	Sciences humaines
Sub-total SSH	..	..	..	..	..	..	..	Sous-total SSH
Total NSE+SSH	694	686	887	960	1 108	1 207	..	*Total SENI+SSH*
Other								**Autres**
Natural sciences	..	..	..	..	..	..	..	Sciences exactes et naturelles
Engineering	..	..	..	..	..	..	..	Sciences de l'ingénieur
Medical sciences	..	..	..	..	..	..	..	Sciences médicales
Agricultural sciences	..	..	..	..	..	..	..	Sciences agricoles
Sub-total NSE	..	..	..	..	..	..	..	Sous-total SENI
Social sciences	..	..	..	..	..	..	..	Sciences sociales
Humanities	..	..	..	..	..	..	..	Sciences humaines
Sub-total SSH	..	..	..	..	..	..	..	Sous-total SSH
Total NSE+SSH	2 345	2 332	2 106	2 161	2 344	2 456	..	*Total SENI+SSH*

Table 18 - Tableau 18

Government budget appropriations or outlays for R&D (GBAORD) - Crédits budgétaires publics de R-D (CBPRD)

by socio-economic objective - par objectif socio-économique

Million EUR (1999 BEF euro, EUR from 1999) — *Millions EUR (1999 BEF euro, EUR à partir de 1999)*

	1995	1996	1997	1998	1999	2000	2001	
Agriculture, forestry and fishing	42.8	42.8	42.4	42.3	42.7	42.2 p	..	Agriculture, sylviculture et pêche
Industrial development	158.7	195.7	246.5	292.8	330.6	321.1 p	..	Développement industriel
Energy	33.0	32.7	38.9	35.9	36.1	38.9 p	..	Energie
Transport & telecommunications	..	..	..	..	..	..	..	*Transports & télécommunications*
Urban & rural planning	..	..	..	..	..	..	..	*Aménagement urbain et rural*
ST Infrastructure	8.5	11.4	11.9	13.4	12.9	15.6 p	..	ST Infrastructure
Prevention of pollution	..	..	..	..	..	..	..	*Prévention de la pollution*
Identificat. & treatment of pollution	..	..	..	..	..	..	..	*Détect. pollut. & mesures antipollut.*
ST Environment	18.6	29.0	30.0	25.1	36.3	51.4 p	..	ST Environnement
Health	28.4	19.8	18.5	20.6	19.0	22.6 p	..	Santé
Social development & services	34.5	34.7	38.1	56.8	60.2	63.4 p	..	Développement et services sociaux
Earth and atmosphere	15.0	14.4	9.0	9.6	13.6	13.8 p	..	Milieu terrestre et atmosphère
Advancement of research	233.7	243.7	289.0	302.9	315.4	340.2 p	..	*Promotion de la recherche*
General university funds	252.5	282.9	261.3	258.5	268.1	273.5 p	..	*Fonds généraux des universités*
ST Advancement of knowledge	486.2	526.6	550.3	561.4	583.5	613.7 p	..	ST promotion gén. des connaiss.
Civil space	163.1	165.3	167.8	166.1	167.8	168.9 p	..	Espace civil
Defence	4.1	5.9	6.7	6.3	5.5	5.1 p	..	Défense
Not elsewhere classified	70.4	63.3	60.6	65.3	73.8	66.4 p	..	Non classé
Total	**1 063.2**	**1 141.5**	**1 220.6**	**1 295.7**	**1 382.1**	**1 423.2 p**	..	**Total**

Source: OECD, R&D database (DSTI/EAS Division), January 2002.

Source : OCDE, base de données R-D (DSTI/Division AES), janvier 2002.

BELGIUM — BELGIQUE

Table 19 - Tableau 19

Technology balance of payments - Balance des paiements technologiques

by type of transfer - selon la nature des transferts

Million EUR (1999 BEF euro, EUR from 1999) — *Millions EUR (1999 BEF euro, EUR à partir de 1999)*

	1983	1984	1985	1986	1987	1988	1989	1990	1991	
Receipts										**Recettes**
Related firms	..	..	..	..	..	..	..	..	..	Firmes affiliées
Unrelated firms	..	..	..	..	..	..	..	..	..	Firmes non affiliées
Other	..	..	..	..	..	..	..	..	..	Autre
Total	757	907	1 022	1 124	1 023	1 150	1 578	1 562	1 647	*Total*
Payments										**Paiements**
Related firms	..	..	..	..	..	..	..	..	..	Firmes affiliées
Unrelated firms	..	..	..	..	..	..	..	..	..	Firmes non affiliées
Other	..	..	..	..	..	..	..	..	..	Autre
Total	879	1 106	1 178	1 300	1 438	1 652	2 052	2 090	2 015	*Total*

	1992	1993	1994	1995	1996	1997	1998	1999	2000	
Receipts										**Recettes**
Related firms	..	..	..	..	..	..	..	..	..	Firmes affiliées
Unrelated firms	..	..	..	..	..	..	..	..	..	Firmes non affiliées
Other	..	..	..	..	..	..	..	..	..	Autre
Total	1 902	2 144	2 309	2 765 [a]	3 268	3 965	4 387	4 786	..	*Total*
Payments										**Paiements**
Related firms	..	..	..	..	..	..	..	..	..	Firmes affiliées
Unrelated firms	..	..	..	..	..	..	..	..	..	Firmes non affiliées
Other	..	..	..	..	..	..	..	..	..	Autre
Total	2 121	2 321	2 456	2 271 [a]	2 480	3 078	3 609	3 978	..	*Total*

Source: OECD, TBPdatabase (DSTI/EAS Division), Januay 2002.

Source : OCDE, base de données TBP (DSTI/Division AES), janvier 2002.

Gross domestic expenditure on R&D (GERD) - Dépenses intérieures brutes de R-D (DIRD)

by sector of performance and source of funds - par secteur d'exécution et par source de financement

	1994	1995	1996	1997	1998	1999	2000	
	1.1 Million CAD - Millions CAD							
Business enterprise								**Entreprises**
financed by								*financé par*
Business enterprise	5 535	5 946	5 963	6 592	6 795	6 805	7 527 p	Entreprises
Direct government	566	498	395	439	329	367	402 p	Etat direct
Higher education	0	0	0	0	0	0	0	Enseignement supérieur
Private non-profit	0	0	0	0	0	0	0	Institutions sans but lucratif
Funds from abroad	1 466	1 555	1 661	1 750	2 499	2 648	2 933 p	Fonds de l'étranger
Total	7 567	7 999	8 019	8 781	9 623	9 820	10 862 p	*Total*
Government								**État**
financed by								*financé par*
Business enterprise	29	36	111	82	70	60	66 p	Entreprises
Direct government	1 980	1 940	1 916	1 848	1 886	2 031	2 087 p	État direct
Higher education	0	0	0	0	0	0	0	Enseignement supérieur
Private non-profit	0	0	0	0	0	0	0	Institutions sans but lucratif
Funds from abroad	5	5	7	4	3	3	4 p	Fonds de l'étranger
Total	2 014	1 981	2 034	1 934	1 959	2 094	2 157 p	*Total*
Higher education								**Enseignement supérieur**
financed by								*financé par*
Business enterprise	296	297	335	381	408	456	525 p	Entreprises
Direct government	1 185	1 178	1 107	1 163	1 231	1 562	1 797 p	État direct
General university funds	1 305 c	1 297 c	1 223 c	1 244 c	1 416 c	1 569 c	1 753 c,p	Fonds généraux des universités
Sub-total government	2 490 c	2 475 c	2 330 c	2 407 c	2 647 c	3 131 c	3 550 c,p	Sous-Total État
Higher education	577 c	638 c	704 c	788 c	931 c	1 160 c	1 385 c,p	Enseignement supérieur
Private non-profit	259	266	313	324	334	350	401 p	Institutions sans but lucratif
Funds from abroad	21	24	37	40	50	57	65 p	Fonds de l'étranger
Total	3 643	3 700	3 719	3 940	4 370	5 154	5 926 p	*Total*
Private non-profit								**Institutions sans but lucratif**
financed by								*financé par*
Business enterprise	29	27	21	22	26	21	22 p	Entreprises
Direct government	47	46	48	51	55	48	51 p	État direct
Higher education	0	0	0	0	0	0	0	Enseignement supérieur
Private non-profit	66	70	73	85	88	94	98 p	Institutions sans but lucratif
Funds from abroad	9	10	12	11	12	12	13 p	Fonds de l'étranger
Total	151	153	154	169	181	175	184 p	*Total*
GERD								**DIRD**
financed by								*financé par*
Business enterprise	5 889	6 306	6 430	7 077	7 299	7 342	8 140 p	Entreprises
Direct government	3 778	3 662	3 466	3 501	3 501	4 008	4 337 p	État direct
General university funds	1 305 c	1 297 c	1 223 c	1 244 c	1 416 c	1 569 c	1 753 c,p	Fonds généraux des universités
Sub-total government	5 083 c	4 959 c	4 689 c	4 745 c	4 917 c	5 577 c	6 090 c,p	Sous-Total État
Higher education	577 c	638 c	704 c	788 c	931 c	1 160 c	1 385 c,p	Enseignement supérieur
Private non-profit	325	336	386	409	422	444	499 p	Institutions sans but lucratif
Funds from abroad	1 501	1 594	1 717	1 805	2 564	2 720	3 015 p	Fonds de l'étranger
Total	13 375	13 833	13 926	14 824	16 133	17 243	19 129 p	*Total*
	1.2 Million constant USD (1995 prices and PPPs) - Millions de USD constants (prix et PPA de 1995)							
Business enterprise								**Entreprises**
financed by								*financé par*
Business enterprise	4 822.4	5 028.1	4 996.6	5 468.4	5 703.5	5 650.0	6 120.6 b,p	Entreprises
Direct government	493.1	421.1	331.0	364.2	276.2	304.7	326.9 b,p	État direct
Higher education	0.0	0.0	0.0	0.0	0.0	0.0	0.0 b	Enseignement supérieur
Private non-profit	0.0	0.0	0.0	0.0	0.0	0.0	0.0 b	Institutions sans but lucratif
Funds from abroad	1 277.3	1 314.9	1 391.8	1 451.7	2 097.6	2 198.5	2 385.0 b,p	Fonds de l'étranger
Total	6 592.8	6 764.1	6 719.3	7 284.3	8 077.3	8 153.2	8 832.4 b,p	*Total*
Government								**État**
financed by								*financé par*
Business enterprise	25.3	30.4	93.0	68.0	58.8	49.8	53.7 b,p	Entreprises
Direct government	1 725.1	1 640.5	1 605.5	1 533.0	1 583.1	1 686.3	1 697.0 b,p	État direct
Higher education	0.0	0.0	0.0	0.0	0.0	0.0	0.0 b	Enseignement supérieur
Private non-profit	0.0	0.0	0.0	0.0	0.0	0.0	0.0 b	Institutions sans but lucratif
Funds from abroad	4.4	4.2	5.9	3.3	2.5	2.5	3.3 b,p	Fonds de l'étranger
Total	1 754.7	1 675.2	1 704.3	1 604.3	1 644.3	1 738.6	1 754.0 b,p	*Total*
Higher education								**Enseignement supérieur**
financed by								*financé par*
Business enterprise	257.9	251.2	280.7	316.1	342.5	378.6	426.9 b,p	Entreprises
Direct government	1 032.4	996.1	927.6	964.8	1 033.3	1 296.9	1 461.2 b,p	État direct
General university funds	1 137.0 c	1 096.8 c	1 024.8 c	1 032.0 c	1 188.5 c	1 302.7 c	1 425.5 b,p	Fonds généraux des universités
Sub-total government	2 169.4 c	2 092.9 c	1 952.4 c	1 996.7 c	2 221.8 c	2 599.6 c	2 886.7 b,p	Sous-Total État
Higher education	502.7 c	539.5 c	589.9 c	653.7 c	781.5 c	963.1 c	1 126.2 b,p	Enseignement supérieur
Private non-profit	225.7	224.9	262.3	268.8	280.4	290.6	326.1 b,p	Institutions sans but lucratif
Funds from abroad	18.3	20.3	31.0	33.2	42.0	47.3	52.9 b,p	Fonds de l'étranger
Total	3 174.0	3 128.8	3 116.3	3 268.4	3 668.1	4 279.2	4 818.7 b,p	*Total*
Private non-profit								**Institutions sans but lucratif**
financed by								*financé par*
Business enterprise	25.3	22.8	17.6	18.3	21.8	17.4	17.9 b,p	Entreprises
Direct government	40.9	38.9	40.2	42.3	46.2	39.9	41.5 b,p	État direct
Higher education	0.0	0.0	0.0	0.0	0.0	0.0	0.0 b	Enseignement supérieur
Private non-profit	57.5	59.2	61.2	70.5	73.9	78.0	79.7 b,p	Institutions sans but lucratif
Funds from abroad	7.8	8.5	10.1	9.1	10.1	10.0	10.6 b,p	Fonds de l'étranger
Total	131.6	129.4	129.0	140.2	151.9	145.3	149.6 b,p	*Total*
GERD								**DIRD**
financed by								*financé par*
Business enterprise	5 130.8	5 332.5	5 387.9	5 870.7	6 126.6	6 095.8	6 619.0 b,p	Entreprises
Direct government	3 291.6	3 096.7	2 904.3	2 904.3	2 938.6	3 327.7	3 526.6 b,p	État direct
General university funds	1 137.0 c	1 096.8 c	1 024.8 c	1 032.0 c	1 188.5 c	1 302.7 c	1 425.5 b,p	Fonds généraux des universités
Sub-total government	4 428.6 c	4 193.4 c	3 929.0 c	3 936.2 c	4 127.2 c	4 630.4 c	4 952.1 b,p	Sous-Total État
Higher education	502.7 c	539.5 c	589.9 c	653.7 c	781.5 c	963.1 c	1 126.2 b,p	Enseignement supérieur
Private non-profit	283.2	284.1	323.4	339.3	354.2	368.6	405.8 b,p	Institutions sans but lucratif
Funds from abroad	1 307.8	1 347.9	1 438.7	1 497.3	2 152.1	2 258.3	2 451.7 b,p	Fonds de l'étranger
Total	11 653.1	11 697.5	11 669.0	12 297.2	13 541.6	14 316.3	15 554.8 b,p	*Total*

Source: OECD, R&D database (DSTI/EAS Division), January 2002. *Source* : OCDE, base de données R-D (DSTI/Division AES), janvier 2002.

Table 2 - Tableau 2

Gross domestic expenditure on R&D (GERD) - Dépenses intérieures brutes de R-D (DIRD)

by sector of performance and type of costs - par secteur d'exécution et par type de coûts

Million CAD *Millions CAD*

	1994	1995	1996	1997	1998	1999	2000	
Business enterprise								**Entreprises**
Labour costs	3 562 [s]	..	..	..	..	..	..	Coûts du personnel
Other current costs	2 866 [s]	..	..	..	..	..	..	Autres coûts courants
Total current expenditure	6 428 [s]	..	..	..	..	..	..	*Total dépenses courantes*
Land and buildings	..	..	..	..	..	..	..	Terrains et constructions
Instruments and equipment	..	..	..	..	..	..	..	Instruments et équipements
Total capital expenditure	590 [s]	..	..	..	..	..	..	*Total dépenses en capital*
Total	7 567	7 999	8 019	8 781	9 623	9 820	10 862 [p]	*Total*
Government								**État**
Labour costs	..	..	..	..	..	..	..	Coûts du personnel
Other current costs	..	..	..	..	..	..	..	Autres coûts courants
Total current expenditure	1 682 [s]	1 632 [s]	..	..	..	..	..	*Total dépenses courantes*
Land and buildings	..	..	..	..	..	..	..	Terrains et constructions
Instruments and equipment	..	..	..	..	..	..	..	Instruments et équipements
Total capital expenditure	212 [s]	195 [s]	..	..	..	..	..	*Total dépenses en capital*
Total	2 014	1 981	2 034	1 934	1 959	2 094	2 157 [p]	*Total*
Higher education								**Enseignement supérieur**
Labour costs	..	..	..	..	..	..	..	Coûts du personnel
Other current costs	..	..	..	..	..	..	..	Autres coûts courants
Total current expenditure	..	..	..	..	..	..	..	*Total dépenses courantes*
Land and buildings	..	..	..	..	..	..	..	Terrains et constructions
Instruments and equipment	..	..	..	..	..	..	..	Instruments et équipements
Total capital expenditure	..	..	..	..	..	..	..	*Total dépenses en capital*
Total	3 643	3 700	3 719	3 940	4 370	5 154	5 926 [p]	*Total*
Private non-profit								**Institutions sans but lucratif**
Labour costs	..	..	..	..	..	..	..	Coûts du personnel
Other current costs	..	..	..	..	..	..	..	Autres coûts courants
Total current expenditure	..	148	..	..	..	..	..	*Total dépenses courantes*
Land and buildings	..	..	..	..	..	..	..	Terrains et constructions
Instruments and equipment	..	..	..	..	..	..	..	Instruments et équipements
Total capital expenditure	..	6	..	..	..	..	..	*Total dépenses en capital*
Total	151	153	154	169	181	175	184 [p]	*Total*
GERD								**DIRD**
Labour costs	..	..	..	..	..	..	..	Coûts du personnel
Other current costs	..	..	..	..	..	..	..	Autres coûts courants
Total current expenditure	..	..	..	..	..	..	..	*Total dépenses courantes*
Land and buildings	..	..	..	..	..	..	..	Terrains et constructions
Instruments and equipment	..	..	..	..	..	..	..	Instruments et équipements
Total capital expenditure	..	..	..	..	..	..	..	*Total dépenses en capital*
Total	13 375	13 833	13 926	14 824	16 133	17 243	19 129 [p]	*Total*

Table 3 - Tableau 3

Current domestic expenditure on R&D - Dépenses intérieures courantes de R-D

by sector of performance and type of activity - par secteur d'exécution et par type d'activités

Million CAD *Millions CAD*

	1994	1995	1996	1997	1998	1999	2000	
Business enterprise								**Entreprises**
Basic research	..	..	..	..	..	..	..	Recherche fondamentale
Applied research	..	..	..	..	..	..	..	Recherche appliquée
Experimental development	..	..	..	..	..	..	..	Développement expérimental
Not specified	..	..	..	..	..	..	..	Non spécifié
Total	..	..	..	..	..	..	..	*Total*
Government								**État**
Basic research	..	..	..	..	..	..	..	Recherche fondamentale
Applied research	..	..	..	..	..	..	..	Recherche appliquée
Experimental development	..	..	..	..	..	..	..	Développement expérimental
Not specified	..	..	..	..	..	..	..	Non spécifié
Total	..	..	..	..	..	..	..	*Total*
Higher education								**Enseignement supérieur**
Basic research	..	..	..	..	..	..	..	Recherche fondamentale
Applied research	..	..	..	..	..	..	..	Recherche appliquée
Experimental development	..	..	..	..	..	..	..	Développement expérimental
Not specified	..	..	..	..	..	..	..	Non spécifié
Total	..	..	..	..	..	..	..	*Total*
Private non-profit								**Institutions sans but lucratif**
Basic research	..	..	..	..	..	..	..	Recherche fondamentale
Applied research	..	..	..	..	..	..	..	Recherche appliquée
Experimental development	..	..	..	..	..	..	..	Développement expérimental
Not specified	..	..	..	..	..	..	..	Non spécifié
Total	..	..	..	..	..	..	..	*Total*
GERD								**DIRD**
Basic research	..	..	..	..	..	..	..	Recherche fondamentale
Applied research	..	..	..	..	..	..	..	Recherche appliquée
Experimental development	..	..	..	..	..	..	..	Développement expérimental
Not specified	..	..	..	..	..	..	..	Non spécifié
Total	..	..	..	..	..	..	..	*Total*

Source: OECD, R&D database (DSTI/EAS Division), January 2002.

Source : OCDE, base de données R-D (DSTI/Division AES), janvier 2002.

Table 5 - Tableau 5

Total R&D personnel - Personnel total de R-D

by sector of employment and occupation - par secteur d'emploi et par profession

In full-time equivalent / *En équivalence plein-temps*

	1994	1995	1996	1997	1998	1999	2000	
Business enterprise								**Entreprises**
RSE	46 880	49 120	49 050	51 510	48 980	49 500 [p]	..	Chercheurs
Technicians	22 750	23 360	21 940	21 570	19 310	19 690 [p]	..	Techniciens
Other occupations	9 290	9 810	9 340	9 080	8 200	8 340 [p]	..	Autres professions
Total	78 920	82 290	80 330	82 160	76 490	77 530 [p]	..	*Total*
Government								**État**
RSE	8 250	7 740	7 700	7 300	7 290	7 420 [p]	..	Chercheurs
Technicians	5 660	5 260	4 870	4 770	4 650	4 520 [p]	..	Techniciens
Other occupations	6 150	5 660	5 030	4 770	4 590	4 550 [p]	..	Autres professions
Total	20 060	18 660	17 600	16 840	16 530	16 490 [p]	..	*Total*
Higher education								**Enseignement supérieur**
RSE	30 930	30 840	34 220	33 960	33 250 [c]	33 250 [c]	..	Chercheurs
Technicians	6 660	6 500	6 320	6 200	6 220 [c]	6 220 [c]	..	Techniciens
Other occupations	5 210	5 020	4 810	4 780	4 690 [c]	4 690 [c]	..	Autres professions
Total	42 800	42 360	45 350	44 940	44 160 [c]	44 160 [c]	..	*Total*
Private non-profit								**Institutions sans but lucratif**
RSE	680	630	630	670	680	640	..	Chercheurs
Technicians	860	970	1 010	1 120	1 200	1 140	..	Techniciens
Other occupations	320	320	380	460	510	480	..	Autres professions
Total	1 860	1 920	2 020	2 250	2 390	2 260	..	*Total*
TOTAL								**TOTAL**
RSE	86 740	88 330	91 600	93 440	90 200 [c]	90 810 [c,p]	..	Chercheurs
Technicians	35 930	36 090	34 140	33 660	31 380 [c]	31 570 [c,p]	..	Techniciens
Other occupations	20 970	20 810	19 560	19 090	17 990 [c]	18 060 [c,p]	..	Autres professions
Total	143 640	145 230	145 300	146 190	139 570 [c]	140 440 [c,p]	..	*Total*

Table 6 - Tableau 6

Total R&D personnel - Personnel total de R-D

by sector of employment and formal qualification - par secteur d'emploi et par niveau de qualification

In full-time equivalent / *En équivalence plein-temps*

	1994	1995	1996	1997	1998	1999	2000	
Business enterprise								**Entreprises**
University PhD level degrees	3 987 [s]	3 840	3 830	3 800	3 770	3 810 [p]	..	Diplômes univ. de niv. doctorat
Other university degrees	41 735 [s]	45 280	45 220	47 710	45 210	45 690 [p]	..	Autres diplômes universitaires
Sub-total university degrees	45 722 [s]	49 120	49 050	51 510	48 980	49 500 [p]	..	Sous-total Universitaires
Other post-secondary diplomas	22 038 [s]	23 360	21 940	21 570	19 310	19 690 [p]	..	Autres diplômes post-secondaires
Secondary diplomas	..	..	..	..	..	..	..	Diplômes de fin d'études secondaires
Other qualifications	8 995 [s]	9 810	9 340	9 080	8 200	8 340 [p]	..	Autres qualifications
Total	78 920	82 290	80 330	82 160	76 490	77 530 [p]	..	*Total*
Government								**État**
University PhD level degrees	..	..	..	..	..	..	..	Diplômes univ. de niv. doctorat
Other university degrees	..	..	..	..	..	..	..	Autres diplômes universitaires
Sub-total university degrees	..	..	..	..	..	..	..	Sous-total Universitaires
Other post-secondary diplomas	..	..	..	..	..	..	..	Autres diplômes post-secondaires
Secondary diplomas	..	..	..	..	..	..	..	Diplômes de fin d'études secondaires
Other qualifications	..	..	..	..	..	..	..	Autres qualifications
Total	20 060	18 660	17 600	16 840	16 530	16 490 [p]	..	*Total*
Higher education								**Enseignement supérieur**
University PhD level degrees	..	..	..	..	..	..	..	Diplômes univ. de niv. doctorat
Other university degrees	..	..	..	..	..	..	..	Autres diplômes universitaires
Sub-total university degrees	30 930 [s]	30 840 [s]	..	..	..	..	..	Sous-total Universitaires
Other post-secondary diplomas	..	..	..	..	..	..	..	Autres diplômes post-secondaires
Secondary diplomas	..	..	..	..	..	..	..	Diplômes de fin d'études secondaires
Other qualifications	..	..	..	..	..	..	..	Autres qualifications
Total	42 800	42 360	45 350	44 940	44 160 [c]	44 160 [c]	..	*Total*
Private non-profit								**Institutions sans but lucratif**
University PhD level degrees	..	..	..	..	..	..	..	Diplômes univ. de niv. doctorat
Other university degrees	..	..	..	..	..	..	..	Autres diplômes universitaires
Sub-total university degrees	..	..	..	..	..	..	..	Sous-total Universitaires
Other post-secondary diplomas	..	..	..	..	..	..	..	Autres diplômes post-secondaires
Secondary diplomas	..	..	..	..	..	..	..	Diplômes de fin d'études secondaires
Other qualifications	..	..	..	..	..	..	..	Autres qualifications
Total	1 860	1 920	2 020	2 250	2 390	2 260	..	*Total*
TOTAL								**TOTAL**
University PhD level degrees	..	..	..	..	..	..	..	Diplômes univ. de niv. doctorat
Other university degrees	..	..	..	..	..	..	..	Autres diplômes universitaires
Sub-total university degrees	..	..	..	..	..	..	..	Sous-total Universitaires
Other post-secondary diplomas	..	..	..	..	..	..	..	Autres diplômes post-secondaires
Secondary diplomas	..	..	..	..	..	..	..	Diplômes de fin d'études secondaires
Other qualifications	..	..	..	..	..	..	..	Autres qualifications
Total	143 640	145 230	145 300	146 190	139 570 [c]	140 440 [c,p]	..	*Total*

Source: OECD, R&D database (DSTI/EAS Division), January 2002.

Source : OCDE, base de données R-D (DSTI/Division AES), janvier 2002.

Table 7 - Tableau 7

Gross domestic expenditure on R&D (GERD) - Dépenses intérieures brutes de R-D (DIRD)

by sector of performance and main scientific field - par secteur d'exécution et par grand domaine scientifique

Million CAD — *Millions CAD*

	1994	1995	1996	1997	1998	1999	2000	
Business enterprise								**Entreprises**
Natural sciences and engineering	7 567	7 999	8 019	8 781	9 623	9 820	10 862 p	Sciences exactes, nat. et de l'ingénieur
Social sciences and humanities	..	..	..	..	..	..	..	Sciences sociales et humaines
All fields of science	7 567	7 999	8 019	8 781	9 623	9 820	10 862 p	*Tous domaines scientifiques*
Government								**État**
Natural sciences and engineering	1 933	1 905	1 950	1 849	1 867	1 996	2 056 p	Sciences exactes, nat. et de l'ingénieur
Social sciences and humanities	81	76	84	85	92	98	101 p	Sciences sociales et humaines
All fields of science	2 014	1 981	2 034	1 934	1 959	2 094	2 157 p	*Tous domaines scientifiques*
Higher education								**Enseignement supérieur**
Natural sciences and engineering	2 918	2 975	3 014	3 208	3 470	4 098	4 712 p	Sciences exactes, nat. et de l'ingénieur
Social sciences and humanities	725	725	705	732	900	1 056	1 214 p	Sciences sociales et humaines
All fields of science	3 643	3 700	3 719	3 940	4 370	5 154	5 926 p	*Tous domaines scientifiques*
Private non-profit								**Institutions sans but lucratif**
Natural sciences and engineering	142	142	143	158	171	166	174 p	Sciences exactes, nat. et de l'ingénieur
Social sciences and humanities	9	11	11	11	10	9	10 p	Sciences sociales et humaines
All fields of science	151	153	154	169	181	175	184 p	*Tous domaines scientifiques*
TOTAL								**TOTAL**
Natural sciences and engineering	12 560	13 021	13 126	13 996	15 131	16 080	17 804 p	Sciences exactes, nat. et de l'ingénieur
Social sciences and humanities	815	812	800	828	1 003	1 163	1 325 p	Sciences sociales et humaines
All fields of science	13 375	13 833	13 926	14 824	16 133	17 243	19 129 p	*Tous domaines scientifiques*

Table 8 - Tableau 8

Total R&D personnel - Personnel total de R-D

by sector of employment and main scientific field - par secteur d'emploi et par grand domaine scientifique

In full-time equivalent — *En équivalence plein-temps*

	1994	1995	1996	1997	1998	1999	2000	
Business enterprise								**Entreprises**
Natural sciences and engineering	78 920	82 290	80 330	82 160	76 490	77 530 p	..	Sciences exactes, nat. et de l'ingénieur
Social sciences and humanities	..	..	..	..	..	..	..	Sciences sociales et humaines
All fields of science	78 920	82 290	80 330	82 160	76 490	77 530 p	..	*Tous domaines scientifiques*
Government								**État**
Natural sciences and engineering	19 180	17 850	16 820	16 110	15 840	15 770 p	..	Sciences exactes, nat. et de l'ingénieur
Social sciences and humanities	880	810	780	730	690	720 p	..	Sciences sociales et humaines
All fields of science	20 060	18 660	17 600	16 840	16 530	16 490 p	..	*Tous domaines scientifiques*
Higher education								**Enseignement supérieur**
Natural sciences and engineering	24 260	23 810	24 260	23 740	23 420 c	23 420 c	..	Sciences exactes, nat. et de l'ingénieur
Social sciences and humanities	18 540	18 550	21 090	21 200	20 740 c	20 740 c	..	Sciences sociales et humaines
All fields of science	42 800	42 360	45 350	44 940	44 160 c	44 160 c	..	*Tous domaines scientifiques*
Private non-profit								**Institutions sans but lucratif**
Natural sciences and engineering	1 860	1 920	2 020	2 250	2 390	2 260	..	Sciences exactes, nat. et de l'ingénieur
Social sciences and humanities	..	..	..	..	..	..	..	Sciences sociales et humaines
All fields of science	1 860	1 920	2 020	2 250	2 390	2 260	..	*Tous domaines scientifiques*
TOTAL								**TOTAL**
Natural sciences and engineering	124 220	125 870	123 430	124 260	118 140 c	118 980 c,p	..	Sciences exactes, nat. et de l'ingénieur
Social sciences and humanities	19 420	19 360	21 870	21 930	21 430 c	21 460 c,p	..	Sciences sociales et humaines
All fields of science	143 640	145 230	145 300	146 190	139 570 c	140 440 c,p	..	*Tous domaines scientifiques*

Source: OECD, R&D database (DSTI/EAS Division), January 2002.

Source : OCDE, base de données R-D (DSTI/Division AES), janvier 2002.

Total business enterprise intramural expenditure on R&D - Dépenses intra-muros totales de R-D des entreprises
(BERD) by industry - (DIRDE) par industrie

	1994	1995	1996	1997	1998	1999	2000	
	9.1 Million CAD - Millions CAD							
1 AGRICULTURE, HUNTING & FORESTRY	60.4	63.6	70.0	63.6	66.2	58.4	61.7 [p]	**1 AGRICULTURE, CHASSE & SYLVICULTURE**
2 MINING	192.3	206.1	213.0	199.2	155.1	120.2	139.6 [p]	**2 ACTIVITES EXTRACTIVES**
3 MANUFACTURING	4 523.9	4 912.9	5 048.8	5 720.6	6 424.8	6 557.9	7 313.1 [p]	**3 ACTIVITES DE FABRICATION**
4 Food, beverages & tobacco	104.1	114.9	106.9	99.9	100.3	113.1	106.1 [p]	**4 Produits alimentaires, boissons & tabac**
5 Food products & beverages	..	..	..	..	..	..	..	5 Produits alimentaires & boissons
6 Tobacco products	..	..	..	..	..	..	..	6 Produits à base de tabac
7 Textiles, wear app., fur & leather	58.7	65.1	57.8	62.2	67.1	68.1	75.4 [p]	**7 Textiles, habill., fourrures & cuirs**
8 Textiles	..	..	..	..	..	..	..	8 Textiles
9 Wearing apparel & fur	..	..	..	..	..	..	..	9 Articles d'habillement & fourrures
10 Leather products & footwear	..	..	..	..	..	..	..	10 Articles en cuir & chaussures
11 Wood, paper, printing, publishing	153.3	182.1	166.9	182.9	191.6	147.4	143.4 [p]	**11 Bois, papier, imprimerie, édition**
12 Wood & cork (not furniture)	30.6	32.5	29.1	35.1	33.0	32.9	33.2 [p]	12 Bois & liège (sauf meubles)
13 Pulp, paper & paper products	105.3	130.4	125.0	135.3	147.3	103.5	98.4 [p]	13 Carton, papier & articles
14 Publ., print. & repro. of rec. media	17.5	19.2	12.8	12.6	11.3	11.1	11.8 [p]	14 Edition, impress. & repro. enregis.
15 Coke, petroleum, nuclear fuel, chemicals, rubber plastics	705.5	785.2	822.9	811.9	847.5	952.4	907.4 [p]	**15 Coke, pétrole, combust. nucl., prod. chimi., caout. & mat. plast.**
16 Coke, ref. petrol. prod. & nuclear fuel	89.8	88.3	113.0	101.0	77.7	57.9	51.7 [p]	16 Coke, raffin. pétr. & combust. nucl.
17 Chemicals & chemical products	566.3	644.3	656.8	660.8	712.7	853.5	814.3 [p]	17 Produits chimiques
18 *Chemical prod.(less pharmaceuticals)*	181.8	203.7	159.0	159.0	164.8	228.1	149.4 [p]	18 *Prod. chimi. (moins prod. pharmaceutiques)*
19 *Pharmaceuticals*	384.5	440.5	497.9	501.8	547.9	625.5	664.9 [p]	19 *Produits pharmaceutiques*
20 Rubber & plastic products	49.5	52.6	53.0	50.1	57.2	40.9	41.3 [p]	20 Caoutchouc et matières plastiques
21 Non-metallic mineral products	16.0	13.3	12.3	10.7	12.8	12.2	12.9 [p]	**21 Produits minéraux non métalliques**
22 Basic metals	164.8	149.2	145.4	149.1	144.9	134.6	148.9 [p]	**22 Produits métallurgiques de base**
23 Basic metals, ferrous	34.9	21.2	18.5	17.7	15.1	10.0	9.3 [p]	23 Produits métallur. de base, ferreux
24 Basic metals, non-ferrous	129.9	128.0	126.9	131.5	129.8	124.6	139.6 [p]	24 Prod. métallur. de base, non ferreux
25 Fabricated metal products	109.8	104.5	100.0	107.4	128.3	115.6	120.8 [p]	**25 Ouvrages en métaux (sauf mach.& mat.)**
26 Machinery equipment, instruments and transport equipment	3 147.5	3 433.6	3 568.9	4 216.6	4 864.8	4 957.9	5 733.9 [p]	**26 Machines & matériel, instruments & mat. de transport**
27 Machinery n.e.c.	186.4	201.4	181.8	206.1	226.4	239.2	236.4 [p]	27 Machines, n.c.a.
28 Office, account. & computing machinery	331.3	338.3	329.9	356.0	423.4	471.4	526.6 [p]	28 Mach. de bur., compta. & trait. inform.
29 Electrical machinery	81.5	87.8	91.9	114.0	122.9	109.9	149.2 [p]	29 Machines électriques
30 Electro. equip. (radio, TV & commun.)	1 628.1	1 797.0	1 917.0	2 157.5	2 619.3	2 663.5	3 133.2 [p]	30 Equip. électro. (radio, tv & communic.)
31 *Electro. comp. (inc. semi-conductors)*	516.0	418.0	399.8	408.6	489.8	550.1	635.1 [p]	31 *Compos. électro. (semi-cond. incl.)*
32 *TV, radio & communications equip.*	1 112.2	1 379.0	1 517.1	1 748.8	2 129.5	2 113.4	2 498.1 [p]	32 *Appareils radio, tv et de communi.*
33 Instruments	93.0	106.2	106.5	114.6	144.2	125.6	144.0 [p]	33 Instruments
34 Motor vehicles	211.5	183.9	165.6	197.1	183.1	195.8	202.7 [p]	34 Véhicules automobiles
35 Other transport equipment	615.7	718.9	776.3	1 071.3	1 145.5	1 152.4	1 341.8 [p]	35 Autres matériels de transport
36 *Ships*	..	..	..	..	..	..	..	36 *Navires*
37 *Aerospace*	612.0	713.8	770.9	1 057.0	1 129.5	1 143.1	1 332.6 [p]	37 *Aérospatiale*
38 *Other transport equipment n.e.c.*	3.7	5.2	5.3	14.4	16.0	9.4	9.2 [p]	38 *Autres matériels de transport n.c.a.*
39 Furniture, other manufacturing n.e.c.	64.1	65.0	67.7	79.9	67.5	56.6	64.2 [p]	**39 Meubles, autres activ. de fabricat. n.c.a.**
40 Furniture	10.6	8.6	8.5	6.9	5.9	6.6	6.5 [p]	40 Meubles
41 Other manufacturing n.e.c.	53.5	56.4	59.2	73.0	61.7	50.0	57.7 [p]	41 Autres activités de fabrication n.c.a.
42 Recycling	..	..	..	..	..	..	..	**42 Récupération**
43 ELECTRICITY, GAS & WATER SUPPLY	224.4	205.2	233.3	186.4	216.3	185.1	175.1 [p]	**43 DISTRIBUTION D'ELECTRICITE, GAZ & EAU**
44 CONSTRUCTION	30.3	28.6	25.7	38.3	24.3	26.4	26.9 [p]	**44 CONSTRUCTION**
45 SERVICES SECTOR	2 535.9	2 582.6	2 429.1	2 572.8	2 736.2	2 871.8	3 145.6 [p]	**45 SECTEUR DES SERVICES**
46 Wholesale, retail trade & repair of motor vehicles, etc.	444.1	556.6	568.4	677.9	642.0	714.6	792.0 [p]	**46 Commerce (gros & détail), réparation de véhicules, etc.**
47 Hotels & restaurants	..	..	..	..	..	..	..	**47 Hôtels et restaurants**
48 Transport & storage	22.5	18.3	9.2	12.3	18.8	20.4	26.1 [p]	**48 Transports, entreposage**
49 Communications	267.4	216.0	158.8	138.0	136.1	88.3	93.0 [p]	**49 Communications**
50 Post	..	..	..	..	..	..	..	50 Postes
51 Telecommunications	..	..	..	..	..	..	..	51 Télécommunications
52 Financial intermediation (inc. insur.)	374.4	283.5	217.5	196.2	204.0	189.5	206.1 [p]	**52 Intermédiation financ. (assurances inc.)**
53 Real estate, renting & busin. activity	1 427.6	1 508.2	1 475.2	1 548.4	1 735.2	1 859.1	2 028.4 [p]	**53 Immobil., locat. & serv. aux entreprises**
54 Computer & related activities	503.1	543.3	551.3	596.5	652.3	614.5	678.6 [p]	54 Informatique et activités connexes
55 *Software consultancy*	..	..	..	..	..	..	..	55 *Production de logiciels*
56 *Other computer services n.e.c.*	..	..	..	..	..	..	..	56 *Autres activités informatiques*
57 Research & development	716.7	740.7	703.4	731.5	898.5	1 040.4	1 140.5 [p]	57 Recherche & développement expérimental
58 Other business activities n.e.c.	207.9	224.3	220.4	220.5	184.4	204.2	209.4 [p]	58 Autres activ. & serv. aux entreprises
59 Comm., soc. & pers. serv. activ., etc.	..	..	..	..	..	..	..	**59 Adm. pub., serv. soc. & collec., etc**
TOTAL BUSINESS ENTERPRISE SECTOR	**7 567.0**	**7 999.0**	**8 019.0**	**8 781.0**	**9 623.0**	**9 820.0**	**10 862.0 [p]**	**TOTAL SECTEUR DES ENTREPRISES**

Source: OECD, R&D database (DSTI/EAS Division), January 2002.

Source : OCDE, base de données R-D (DSTI/Division AES), janvier 2002.

Total business enterprise intramural expenditure on R&D - Dépenses intra-muros totales de R-D des entreprises
(BERD) by industry - (DIRDE) par industrie

9.2 Million constant USD(1995 prices and PPPs) - Millions de USD constants (prix et PPA de 1995)

	1994	1995	1996	1997	1998	1999	2000	
1 AGRICULTURE, HUNTING & FORESTRY	52.7	53.8	58.7	52.8	55.5	48.5	50.1 [b,p]	1 AGRICULTURE, CHASSE & SYLVICULTURE
2 MINING	167.5	174.3	178.5	165.3	130.2	99.8	113.5 [b,p]	2 ACTIVITES EXTRACTIVES
3 MANUFACTURING	3 941.5	4 154.5	4 230.5	4 745.5	5 392.8	5 444.9	5 946.7 [b,p]	3 ACTIVITES DE FABRICATION
4 Food, beverages & tobacco	90.7	97.2	89.6	82.9	84.2	93.9	86.3 [b,p]	**4 Produits alimentaires, boissons & tabac**
5 Food products & beverages	..	..	..	..	..	..	..	5 Produits alimentaires & boissons
6 Tobacco products	..	..	..	..	..	..	..	6 Produits à base de tabac
7 Textiles, wear app., fur & leather	51.2	55.1	48.5	51.6	56.3	56.5	61.3 [b,p]	**7 Textiles, habill., fourrures & cuirs**
8 Textiles	..	..	..	..	..	..	..	8 Textiles
9 Wearing apparel & fur	..	..	..	..	..	..	..	9 Articles d'habillement & fourrures
10 Leather products & footwear	..	..	..	..	..	..	..	10 Articles en cuir & chaussures
11 Wood, paper, printing, publishing	133.6	154.0	139.9	151.8	160.8	122.4	116.6 [b,p]	**11 Bois, papier, imprimerie, édition**
12 Wood & cork (not furniture)	26.7	27.4	24.4	29.1	27.7	27.3	27.0 [b,p]	12 Bois & liège (sauf meubles)
13 Pulp, paper & paper products	91.7	110.2	104.7	112.2	123.7	85.9	80.0 [b,p]	13 Carton, papier & articles
14 Publ., print. & repro. of rec. media	15.2	16.3	10.8	10.5	9.5	9.2	9.6 [b,p]	14 Edition, impress. & repro. enregis.
15 Coke, petroleum, nuclear fuel, chemicals, rubber plastics	614.7	664.0	689.5	673.5	711.4	790.7	737.8 [b,p]	**15 Coke, pétrole, combust. nucl., prod. chimi., caout. & mat. plast.**
16 Coke, ref. petrol. prod. & nuclear fuel	78.2	74.7	94.7	83.8	65.2	48.1	42.1 [b,p]	16 Coke, raffin. pétr. & combust. nucl.
17 Chemicals & chemical products	493.4	544.8	550.4	548.2	598.2	708.7	662.2 [b,p]	17 Produits chimiques
18 *Chemical prod.(less pharmaceuticals)*	158.4	172.3	133.2	131.9	138.3	189.4	121.5 [b,p]	18 *Prod. chimi. (moins prod. pharmaceutiques)*
19 *Pharmaceuticals*	335.0	372.5	417.2	416.3	459.9	519.3	540.6 [b,p]	19 *Produits pharmaceutiques*
20 Rubber & plastic products	43.1	44.5	44.4	41.5	48.0	33.9	33.6 [b,p]	20 Caoutchouc et matières plastiques
21 Non-metallic mineral products	13.9	11.3	10.3	8.8	10.8	10.2	10.5 [b,p]	**21 Produits minéraux non métalliques**
22 Basic metals	143.6	126.2	121.8	123.7	121.6	111.8	121.1 [b,p]	**22 Produits métallurgiques de base**
23 Basic metals, ferrous	30.4	18.0	15.5	14.7	12.7	8.3	7.6 [b,p]	23 Produits métallur. de base, ferreux
24 Basic metals, non-ferrous	113.2	108.2	106.3	109.0	109.0	103.4	113.5 [b,p]	24 Prod. métallur. de base, non ferreux
25 Fabricated metal products	95.6	88.4	83.8	89.1	107.7	96.0	98.2 [b,p]	**25 Ouvrages en métaux (sauf mach.& mat.)**
26 Machinery equipment, instruments and transport equipment	2 742.3	2 903.5	2 990.5	3 497.8	4 083.4	4 116.4	4 662.6 [b,p]	**26 Machines & matériel, instruments & mat. de transport**
27 Machinery n.e.c.	162.4	170.3	152.3	170.9	190.0	198.6	192.3 [b,p]	27 Machines, n.c.a.
28 Office, account. & computing machinery	288.7	286.0	276.4	295.3	355.4	391.4	428.2 [b,p]	28 Mach. de bur., compta. & trait. inform.
29 Electrical machinery	71.0	74.3	77.0	94.6	103.2	91.3	121.3 [b,p]	29 Machines électriques
30 Electro. equip. (radio, TV & commun.)	1 418.5	1 519.6	1 606.3	1 789.7	2 198.6	2 211.4	2 547.8 [b,p]	30 Equip. électro. (radio, tv & communic.)
31 *Electro. comp. (inc. semi-conductors)*	449.6	353.5	335.0	339.0	411.1	456.7	516.4 [b,p]	31 *Compos. électro. (semi-cond. incl.)*
32 *TV, radio & communications equip.*	969.0	1 166.1	1 271.3	1 450.8	1 787.5	1 754.7	2 031.3 [b,p]	32 *Appareils radio, tv et de communi.*
33 Instruments	81.0	89.8	89.2	95.1	121.0	104.3	117.1 [b,p]	33 Instruments
34 Motor vehicles	184.2	155.5	138.8	163.5	153.7	162.6	164.8 [b,p]	34 Véhicules automobiles
35 Other transport equipment	536.4	608.0	650.4	888.7	961.5	956.8	1 091.1 [b,p]	35 Autres matériels de transport
36 *Ships*	..	..	..	..	..	..	..	36 *Navires*
37 *Aerospace*	533.2	603.6	646.0	876.8	948.1	949.1	1 083.6 [b,p]	37 *Aérospatiale*
38 *Other transport equipment n.e.c.*	3.2	4.4	4.5	11.9	13.4	7.8	7.5 [b,p]	38 *Autres matériels de transport n.c.a.*
39 Furniture, other manufacturing n.e.c.	55.9	55.0	56.7	66.3	56.7	47.0	52.2 [b,p]	**39 Meubles, autres activ. de fabricat. n.c.a.**
40 Furniture	9.3	7.3	7.1	5.7	4.9	5.5	5.3 [b,p]	40 Meubles
41 Other manufacturing n.e.c.	46.6	47.7	49.6	60.6	51.8	41.5	46.9 [b,p]	41 Autres activités de fabrication n.c.a.
42 Recycling	..	..	..	..	..	..	..	**42 Récupération**
43 ELECTRICITY, GAS & WATER SUPPLY	195.5	173.5	195.5	154.6	181.5	153.7	142.4 [b,p]	43 DISTRIBUTION D'ELECTRICITE, GAZ & EAU
44 CONSTRUCTION	26.4	24.2	21.5	31.8	20.4	21.9	21.9 [b,p]	44 CONSTRUCTION
45 SERVICES SECTOR	2 209.5	2 183.9	2 035.4	2 134.3	2 296.6	2 384.3	2 557.8 [b,p]	45 SECTEUR DES SERVICES
46 Wholesale, retail trade & repair of motor vehicles, etc.	386.9	470.6	476.3	562.3	538.9	593.3	644.0 [b,p]	**46 Commerce (gros & détail), réparation de véhicules, etc.**
47 Hotels & restaurants	..	..	..	..	..	..	..	**47 Hôtels et restaurants**
48 Transport & storage	19.6	15.5	7.7	10.2	15.8	16.9	21.2 [b,p]	**48 Transports, entreposage**
49 Communications	233.0	182.6	133.1	114.4	114.3	73.3	75.6 [b,p]	**49 Communications**
50 Post	..	..	..	..	..	..	..	50 Postes
51 Telecommunications	..	..	..	..	..	..	..	51 Télécommunications
52 Financial intermediation (inc. insur.)	326.2	239.8	182.2	162.8	171.3	157.3	167.6 [b,p]	**52 Intermédiation financ. (assurances inc.)**
53 Real estate, renting & busin. activity	1 243.8	1 275.4	1 236.1	1 284.5	1 456.5	1 543.5	1 649.4 [b,p]	**53 Immobil., locat. & serv. aux entreprises**
54 Computer & related activities	438.3	459.4	461.9	494.8	547.5	510.2	551.8 [b,p]	54 Informatique et activités connexes
55 *Software consultancy*	..	..	..	..	..	..	..	55 *Production de logiciels*
56 *Other computer services n.e.c.*	..	..	..	..	..	..	..	56 *Autres activités informatiques*
57 Research & development	624.4	626.3	589.4	606.8	754.2	863.8	927.4 [b,p]	57 Recherche & développement expérimental
58 Other business activities n.e.c.	181.1	189.7	184.7	182.9	154.8	169.5	170.3 [b,p]	58 Autres activ. & serv. aux entreprises
59 Comm., soc. & pers. serv. activ., etc.	..	..	..	..	..	..	..	**59 Adm. pub., serv. soc. & collec., etc**
TOTAL BUSINESS ENTERPRISE SECTOR	**6 592.8**	**6 764.1**	**6 719.3**	**7 284.3**	**8 077.3**	**8 153.2**	**8 832.4 [b,p]**	**TOTAL SECTEUR DES ENTREPRISES**

Source: OECD, R&D database (DSTI/EAS Division), January 2002.

Source : OCDE, base de données R-D (DSTI/Division AES), janvier 2002.

Table 14 - Tableau 14

Higher education intramural expenditure on R&D - Dépenses intra-muros de R-D de l'enseignement supérieur (HERD) by field of science - (DIRDES) par discipline scientifique

	1994	1995	1996	1997	1998	1999	2000	
	14.1 Million CAD - Millions CAD							
Natural sciences	..	..	..	..	..	..	..	Sciences exactes et naturelles
Engineering	..	..	..	..	..	..	..	Sciences de l'ingénieur
Medical sciences	..	..	..	..	..	..	..	Sciences médicales
Agricultural sciences	..	..	..	..	..	..	..	Sciences agricoles
Sub-total NSE	2 918	2 975	3 014	3 208	3 470	4 098	4 712 p	Sous-total SENI
Social sciences	..	..	..	..	..	..	..	Sciences sociales
Humanities	..	..	..	..	..	..	..	Sciences humaines
Sub-total SSH	725	725	705	732	900	1 056	1 214 p	Sous-total SSH
Total NSE+SSH	3 643	3 700	3 719	3 940	4 370	5 154	5 926 p	*Total SENI+SSH*
	14.2 Million constant USD (1995 prices and PPPs) - Millions de USD constants (prix et PPA de 1995)							
Natural sciences	..	..	..	..	..	..	..	Sciences exactes et naturelles
Engineering	..	..	..	..	..	..	..	Sciences de l'ingénieur
Medical sciences	..	..	..	..	..	..	..	Sciences médicales
Agricultural sciences	..	..	..	..	..	..	..	Sciences agricoles
Sub-total NSE	2 542.3	2 515.7	2 525.5	2 661.2	2 912.6	3 402.4	3 831.6 b,p	Sous-total SENI
Social sciences	..	..	..	..	..	..	..	Sciences sociales
Humanities	..	..	..	..	..	..	..	Sciences humaines
Sub-total SSH	631.7	613.1	590.7	607.2	755.4	876.8	987.2 b,p	Sous-total SSH
Total NSE+SSH	3 174.0	3 128.8	3 116.3	3 268.4	3 668.1	4 279.2	4 818.7 b,p	*Total SENI+SSH*

Table 15 - Tableau 15

Higher education intramural expenditure on R&D - Dépenses intra-muros de R-D de l'enseignement supérieur (HERD) by field of science and type of costs - (DIRDES) par discipline scientifique et par type de coûts

Million CAD *Millions CAD*

	1994	1995	1996	1997	1998	1999	2000	
Current expenditure								**Dépenses courantes**
Natural sciences	..	..	..	..	..	..	..	Sciences exactes et naturelles
Engineering	..	..	..	..	..	..	..	Sciences de l'ingénieur
Medical sciences	..	..	..	..	..	..	..	Sciences médicales
Agricultural sciences	..	..	..	..	..	..	..	Sciences agricoles
Sub-total NSE	..	..	..	..	..	..	..	Sous-total SENI
Social sciences	..	..	..	..	..	..	..	Sciences sociales
Humanities	..	..	..	..	..	..	..	Sciences humaines
Sub-total SSH	..	..	..	..	..	..	..	Sous-total SSH
Total NSE+SSH	..	..	..	..	..	..	..	*Total SENI+SSH*
Capital expenditure								**Dépenses en capital**
Natural sciences	..	..	..	..	..	..	..	Sciences exactes et naturelles
Engineering	..	..	..	..	..	..	..	Sciences de l'ingénieur
Medical sciences	..	..	..	..	..	..	..	Sciences médicales
Agricultural sciences	..	..	..	..	..	..	..	Sciences agricoles
Sub-total NSE	..	..	..	..	..	..	..	Sous-total SENI
Social sciences	..	..	..	..	..	..	..	Sciences sociales
Humanities	..	..	..	..	..	..	..	Sciences humaines
Sub-total SSH	..	..	..	..	..	..	..	Sous-total SSH
Total NSE+SSH	..	..	..	..	..	..	..	*Total SENI+SSH*

Source: OECD, R&D database (DSTI/EAS Division), January 2002.

Source : OCDE, base de données R-D (DSTI/Division AES), janvier 2002.

Higher education intramural expenditure on R&D - Dépenses intra-muros de R-D de l'enseignement supérieur

(HERD) by field of science and source of funds - (DIRDES) par discipline scientifique et par source de financement

Million CAD | *Millions CAD*

	1994	1995	1996	1997	1998	1999	2000	
Direct government								**État direct**
Natural sciences	..	..	..	..	..	..	..	Sciences exactes et naturelles
Engineering	..	..	..	..	..	..	..	Sciences de l'ingénieur
Medical sciences	..	..	..	..	..	..	..	Sciences médicales
Agricultural sciences	..	..	..	..	..	..	..	Sciences agricoles
Sub-total NSE	..	..	..	..	1 045	1 217 [s]	..	Sous-total SENI
Social sciences	..	..	..	..	..	..	..	Sciences sociales
Humanities	..	..	..	..	..	..	..	Sciences humaines
Sub-total SSH	..	..	..	..	186	202 [s]	..	Sous-total SSH
Total NSE+SSH	1 185	1 178	1 107	1 163	1 231	1 562	1 797 [p]	*Total SENI+SSH*
General university funds								**Fonds généraux des universités**
Natural sciences	..	..	..	..	..	..	..	Sciences exactes et naturelles
Engineering	..	..	..	..	..	..	..	Sciences de l'ingénieur
Medical sciences	..	..	..	..	..	..	..	Sciences médicales
Agricultural sciences	..	..	..	..	..	..	..	Sciences agricoles
Sub-total NSE	..	..	..	..	902 [c,s]	882 [c,s]	..	Sous-total SENI
Social sciences	..	..	..	..	..	..	..	Sciences sociales
Humanities	..	..	..	..	..	..	..	Sciences humaines
Sub-total SSH	..	..	..	..	269 [c,s]	285 [c,s]	..	Sous-total SSH
Total NSE+SSH	1 305 [c]	1 297 [c]	1 223 [c]	1 244 [c]	1 416 [c]	1 569 [c]	1 753 [c,p]	*Total SENI+SSH*
Sub-total government								**Sous-total État**
Natural sciences	..	..	..	..	..	..	..	Sciences exactes et naturelles
Engineering	..	..	..	..	..	..	..	Sciences de l'ingénieur
Medical sciences	..	..	..	..	..	..	..	Sciences médicales
Agricultural sciences	..	..	..	..	..	..	..	Sciences agricoles
Sub-total NSE	..	..	..	..	1 947 [c,s]	2 099 [c,s]	..	Sous-total SENI
Social sciences	..	..	..	..	..	..	..	Sciences sociales
Humanities	..	..	..	..	..	..	..	Sciences humaines
Sub-total SSH	..	..	..	..	455 [c,s]	487 [c,s]	..	Sous-total SSH
Total NSE+SSH	2 490 [c]	2 475 [c]	2 330 [c]	2 407 [c]	2 647 [c]	3 131 [c]	3 550 [c,p]	*Total SENI+SSH*
Higher education								**Enseignement supérieur**
Natural sciences	..	..	..	..	..	..	..	Sciences exactes et naturelles
Engineering	..	..	..	..	..	..	..	Sciences de l'ingénieur
Medical sciences	..	..	..	..	..	..	..	Sciences médicales
Agricultural sciences	..	..	..	..	..	..	..	Sciences agricoles
Sub-total NSE	..	..	..	..	592 [c,s]	580 [c,s]	..	Sous-total SENI
Social sciences	..	..	..	..	..	..	..	Sciences sociales
Humanities	..	..	..	..	..	..	..	Sciences humaines
Sub-total SSH	..	..	..	..	177 [c,s]	187 [c,s]	..	Sous-total SSH
Total NSE+SSH	577 [c]	638 [c]	704 [c]	788 [c]	931 [c]	1 160 [c,p]	1 385 [c,p]	*Total SENI+SSH*
Private non-profit								**Institutions sans but lucratif**
Natural sciences	..	..	..	..	..	..	..	Sciences exactes et naturelles
Engineering	..	..	..	..	..	..	..	Sciences de l'ingénieur
Medical sciences	..	..	..	..	..	..	..	Sciences médicales
Agricultural sciences	..	..	..	..	..	..	..	Sciences agricoles
Sub-total NSE	..	..	..	..	277 [s]	275 [s]	..	Sous-total SENI
Social sciences	..	..	..	..	..	..	..	Sciences sociales
Humanities	..	..	..	..	..	..	..	Sciences humaines
Sub-total SSH	..	..	..	..	57 [s]	69 [s]	..	Sous-total SSH
Total NSE+SSH	259	266	313	324	334	350	401 [p]	*Total SENI+SSH*
Business entreprise								**Entreprises**
Natural sciences	..	..	..	..	..	..	..	Sciences exactes et naturelles
Engineering	..	..	..	..	..	..	..	Sciences de l'ingénieur
Medical sciences	..	..	..	..	..	..	..	Sciences médicales
Agricultural sciences	..	..	..	..	..	..	..	Sciences agricoles
Sub-total NSE	..	..	..	..	391 [s]	396 [s]	..	Sous-total SENI
Social sciences	..	..	..	..	..	..	..	Sciences sociales
Humanities	..	..	..	..	..	..	..	Sciences humaines
Sub-total SSH	..	..	..	..	17 [s]	44 [s]	..	Sous-total SSH
Total NSE+SSH	296	297	335	381	408	456	525 [p]	*Total SENI+SSH*
Funds from abroad								**Fonds de l'étranger**
Natural sciences	..	..	..	..	..	..	..	Sciences exactes et naturelles
Engineering	..	..	..	..	..	..	..	Sciences de l'ingénieur
Medical sciences	..	..	..	..	..	..	..	Sciences médicales
Agricultural sciences	..	..	..	..	..	..	..	Sciences agricoles
Sub-total NSE	..	..	..	..	50 [s]	54 [s]	..	Sous-total SENI
Social sciences	..	..	..	..	..	..	..	Sciences sociales
Humanities	..	..	..	..	..	..	..	Sciences humaines
Sub-total SSH	..	..	..	..	..	..	..	Sous-total SSH
Total NSE+SSH	21	24	37	40	50	57	65 [p]	*Total SENI+SSH*

Source: OECD, R&D database (DSTI/EAS Division), January 2002.

Source : OCDE, base de données R-D (DSTI/Division AES), janvier 2002.

Higher education R&D personnel (HEMP) - Personnel de R-D du secteur de l'enseignement supérieur

by field of science and occupation - par discipline scientifique et par profession

In full-time equivalent — *En équivalence plein-temps*

	1994	1995	1996	1997	1998	1999	2000	
Total R&D personnel								**Personnel total de R-D**
Natural sciences	..	..	..	..	..	..	..	Sciences exactes et naturelles
Engineering	..	..	..	..	..	..	..	Sciences de l'ingénieur
Medical sciences	..	..	..	..	..	..	..	Sciences médicales
Agricultural sciences	..	..	..	..	..	..	..	Sciences agricoles
Sub-total NSE	24 260	23 810	24 260	23 740	23 420 [c]	23 420 [c]	..	Sous-total SENI
Social sciences	..	..	..	..	..	..	..	Sciences sociales
Humanities	..	..	..	..	..	..	..	Sciences humaines
Sub-total SSH	18 540	18 550	21 090	21 200	20 740 [c]	20 740 [c]	..	Sous-total SSH
Total NSE+SSH	42 800	42 360	45 350	44 940	44 160 [c]	44 160 [c]	..	*Total SENI+SSH*
Researchers								**Chercheurs**
Natural sciences	..	..	..	..	..	..	..	Sciences exactes et naturelles
Engineering	..	..	..	..	..	..	..	Sciences de l'ingénieur
Medical sciences	..	..	..	..	..	..	..	Sciences médicales
Agricultural sciences	..	..	..	..	..	..	..	Sciences agricoles
Sub-total NSE	..	..	..	..	16 350 [c]	16 350 [c]	..	Sous-total SENI
Social sciences	..	..	..	..	..	..	..	Sciences sociales
Humanities	..	..	..	..	..	..	..	Sciences humaines
Sub-total SSH	..	..	..	..	16 900 [c]	16 900 [c]	..	Sous-total SSH
Total NSE+SSH	30 930	30 840	34 220	33 960	33 250 [c]	33 250 [c]	..	*Total SENI+SSH*
Technicians								**Technicians**
Natural sciences	..	..	..	..	..	..	..	Sciences exactes et naturelles
Engineering	..	..	..	..	..	..	..	Sciences de l'ingénieur
Medical sciences	..	..	..	..	..	..	..	Sciences médicales
Agricultural sciences	..	..	..	..	..	..	..	Sciences agricoles
Sub-total NSE	..	..	..	..	6 220 [c]	6 220 [c]	..	Sous-total SENI
Social sciences	..	..	..	..	..	..	..	Sciences sociales
Humanities	..	..	..	..	..	..	..	Sciences humaines
Sub-total SSH	..	..	..	..	..	..	..	Sous-total SSH
Total NSE+SSH	6 660	6 500	6 320	6 200	6 220 [c]	6 220 [c]	..	*Total SENI+SSH*
Other								**Autres**
Natural sciences	..	..	..	..	..	..	..	Sciences exactes et naturelles
Engineering	..	..	..	..	..	..	..	Sciences de l'ingénieur
Medical sciences	..	..	..	..	..	..	..	Sciences médicales
Agricultural sciences	..	..	..	..	..	..	..	Sciences agricoles
Sub-total NSE	..	..	..	..	850 [c]	850 [c]	..	Sous-total SENI
Social sciences	..	..	..	..	..	..	..	Sciences sociales
Humanities	..	..	..	..	..	..	..	Sciences humaines
Sub-total SSH	..	..	..	..	3 840 [c]	3 840 [c]	..	Sous-total SSH
Total NSE+SSH	5 210	5 020	4 810	4 780	4 690 [c]	4 690 [c]	..	*Total SENI+SSH*

Table 18 - Tableau 18

Government budget appropriations or outlays for R&D (GBAORD) - Crédits budgétaires publics de R-D (CBPRD)

by socio-economic objective - par objectif socio-économique

Million CAD — *Millions CAD*

	1995	1996	1997	1998	1999	2000	2001	
Agriculture, forestry and fishing	532 [a,h]	529 [h]	489 [h,p]	502 [h,p]	577 [h,p]	..	..	Agriculture, sylviculture et pêche
Industrial development	453 [a,h]	437 [h]	548 [h,p]	529 [h,p]	535 [h,p]	..	..	Développement industriel
Energy	268 [a,h]	337 [h]	266 [h,p]	235 [h,p]	239 [h,p]	..	..	Energie
Transport & telecommunications	147 [a,h]	106 [h]	120 [h,p]	133 [h,p]	123 [h,p]	..	..	*Transports & télécommunications*
Urban & rural planning	83 [h]	76 [h]	67 [h]	65 [h]	58 [h]	..	..	*Aménagement urbain et rural*
ST Infrastructure	230 [a,h]	182 [h]	187 [h,p]	198 [h,p]	181 [h,p]	..	..	ST Infrastructure
Prevention of pollution	..	..	..	..	..	..	..	*Prévention de la pollution*
Identificat. & treatment of pollution	..	..	..	..	..	..	..	*Détect. pollut. & mesures antipollut.*
ST Environment	173 [a,h]	146 [h]	170 [h,p]	181 [h,p]	210 [h,p]	..	..	ST Environnement
Health	388 [a,h]	389 [h]	362 [h,p]	405 [h,p]	493 [h,p]	..	..	Santé
Social development & services	167 [a,h]	142 [h]	141 [h,p]	215 [h,p]	137 [h,p]	..	..	Développement et services sociaux
Earth and atmosphere	204 [a,h]	229 [h]	203 [h,p]	208 [h,p]	285 [h,p]	..	..	Milieu terrestre et atmosphère
Advancement of research	274 [a,h]	253 [h]	288 [h,p]	283 [h,p]	406 [h,p]	..	..	*Promotion de la recherche*
General university funds	1 297 [c]	1 223 [c]	1 244 [c]	1 416 [c]	1 569 [c]	..	..	*Fonds généraux des universités*
ST Advancement of knowledge	1 571 [c]	1 476 [c]	1 532 [c]	1 699 [c]	1 975 [c]	..	..	ST promotion gén. des connaiss.
Civil space	329 [a,h]	296 [h]	265 [h,p]	377 [h,p]	352 [h,p]	..	..	Espace civil
Defence	218 [a,h]	212 [h]	251 [h,p]	256 [h,p]	288 [h,p]	..	..	Défense
Not elsewhere classified	111 [a,h]	101 [h]	77 [h,p]	72 [h,p]	62 [h,p]	..	..	Non classé
Total	**4 644 [a,h]**	**4 476 [h]**	**4 491 [h,p]**	**4 877 [h,p]**	**5 334 [h,p]**	**..**	**..**	**Total**

Source: OECD, R&D database (DSTI/EAS Division), January 2002.

Source : OCDE, base de données R-D (DSTI/Division AES), janvier 2002.

Table 19 - Tableau 19

Technology balance of payments - Balance des paiements technologiques

by type of transfer - selon la nature des transferts

Million CAD / *Millions CAD*

	1983	1984	1985	1986	1987	1988	1989	1990	1991	
Receipts										**Recettes**
Related firms	..	..	..	..	..	..	..	..	..	Firmes affiliées
Unrelated firms	..	..	..	..	..	..	..	..	..	Firmes non affiliées
Other	..	..	..	..	..	..	..	..	..	Autre
Total	459	546	545	587	767	893	885	987	1 064	*Total*
Payments										**Paiements**
Related firms	..	..	..	..	..	..	..	..	..	Firmes affiliées
Unrelated firms	..	..	..	..	..	..	..	..	..	Firmes non affiliées
Other	..	..	..	..	..	..	..	..	..	Autre
Total	584	640	751	788	785	861	931	988	1 063	*Total*

	1992	1993	1994	1995	1996	1997	1998	1999	2000	
Receipts										**Recettes**
Related firms	..	..	..	..	..	..	..	..	..	Firmes affiliées
Unrelated firms	..	..	..	..	..	..	..	..	..	Firmes non affiliées
Other	..	..	..	..	..	..	..	..	..	Autre
Total	1 109	1 276	1 627	1 761	1 903	1 934	2 794	2 964	..	*Total*
Payments										**Paiements**
Related firms	..	..	..	..	..	..	..	..	..	Firmes affiliées
Unrelated firms	..	..	..	..	..	..	..	..	..	Firmes non affiliées
Other	..	..	..	..	..	..	..	..	..	Autre
Total	1 032	1 125	1 251	1 383	1 396	1 625	1 771	2 094	..	*Total*

Source: OECD, TBPdatabase (DSTI/EAS Division), Januay 2002.

Source : OCDE, base de données TBP (DSTI/Division AES), janvier 2002.

CZECH REPUBLIC — Table 1 - Tableau 1 — RÉPUBLIQUE TCHEQUE

Gross domestic expenditure on R&D (GERD) - Dépenses intérieures brutes de R-D (DIRD)

by sector of performance and source of funds - par secteur d'exécution et par source de financement

	1994	1995	1996	1997	1998	1999	2000	
				1.1 Million CZK - Millions CZK				
Business enterprise								**Entreprises**
financed by								*financé par*
Business enterprise	..	8 384.5	8 840.1	11 095.7	13 197.0	11 943.8	12 807.5	Entreprises
Direct government	260.0 [d,t]	406.7 [a]	714.6	962.3	1 207.8	2 101.8	2 341.0	État direct
Higher education	..	2.9	2.9	3.8	0.0 [a]	95.0	0.2	Enseignement supérieur
Private non-profit	..	17.9	22.6	3.1	9.7	3.4	159.5	Institutions sans but lucratif
Funds from abroad	..	286.6	167.0	167.2	345.3	718.3	573.7	Fonds de l'étranger
Total	8 699.0 [d,t]	9 098.6 [a]	9 747.2	12 232.1	14 759.8	14 862.3	15 881.9	*Total*
Government								**État**
financed by								*financé par*
Business enterprise	..	416.4	851.1	504.7	481.1	394.5	645.6	Entreprises
Direct government	2 818.0 [d,t]	3 128.5 [a]	4 107.7	4 503.9	5 265.4	5 210.7	5 855.4	État direct
Higher education	..	21.4	17.8	46.5	5.2 [a]	7.6	69.4	Enseignement supérieur
Private non-profit	..	19.4	3.8	41.4	2.4	0.6	12.7	Institutions sans but lucratif
Funds from abroad	..	112.4	82.6	91.1	123.7	123.2	123.9	Fonds de l'étranger
Total	3 704.0 [d,t]	3 698.1 [a]	5 063.0	5 187.6	5 877.8	5 736.6	6 707.0	*Total*
Higher education								**Enseignement supérieur**
financed by								*financé par*
Business enterprise	..	23.2	6.2	26.5	44.4	36.5	40.0	Entreprises
Direct government	549.0 [l]	977.8 [a]	820.3	526.5	..	..	..	État direct
General university funds	..	..	..	..	..	..	..	Fonds généraux des universités
Sub-total government	..	977.8 [a,m]	820.3 [m]	526.5 [m]	1 933.4 [a]	2 709.0	3 534.2	Sous-Total État
Higher education	..	113.0 [i]	549.3 [i]	1 110.5 [i]	63.9 [a]	65.3	57.6	Enseignement supérieur
Private non-profit	..	3.3	0.9	1.9	1.7	7.0	0.4	Institutions sans but lucratif
Funds from abroad	..	68.6	63.6	112.6	126.9	100.0	131.8	Fonds de l'étranger
Total	580.0 [l]	1 185.9 [a]	1 440.3	1 778.0	2 170.3	2 917.8	3 764.0	*Total*
Private non-profit								**Institutions sans but lucratif**
financed by								*financé par*
Business enterprise	..	..	3.7	24.7	40.4	64.7	70.9	Entreprises
Direct government	..	..	8.8	4.0	16.7	55.8	58.0	État direct
Higher education	..	..	0.3	249.6	0.0	0.6	..	Enseignement supérieur
Private non-profit	..	..	0.8	1.4	0.0	5.7	1.8	Institutions sans but lucratif
Funds from abroad	..	..	0.2	0.0	0.0	3.2	3.6	Fonds de l'étranger
Total	..	..	13.8	279.7	57.1	130.0	134.3	*Total*
GERD								**DIRD**
financed by								*financé par*
Business enterprise	..	8 824.1	9 701.1	11 651.6	13 762.9	12 439.5	13 564.0	Entreprises
Direct government	3 627.0 [d,t]	4 513.0 [a]	5 651.4	5 996.7	8 423.3 [a]	10 077.3	11 788.6	État direct
General university funds	..	..	..	..	..	..	..	Fonds généraux des universités
Sub-total government	3 627.0 [d,m,t]	4 513.0 [a,m]	5 651.4 [m]	5 996.7 [m]	8 423.3 [a]	10 077.3	11 788.6	Sous-Total État
Higher education	..	137.3 [i]	570.3 [i]	1 410.4 [i]	69.1 [a]	168.5	127.2	Enseignement supérieur
Private non-profit	..	40.6	28.1	47.8	13.8	16.7	174.4	Institutions sans but lucratif
Funds from abroad	..	467.6	313.4	370.9	595.9	944.7	833.0	Fonds de l'étranger
Total	12 983.0 [d,t]	13 982.6 [a]	16 264.3	19 477.4	22 865.0	23 646.7	26 487.2	*Total*
				1.2 Million constant USD (1995 prices and PPPs) - Millions de USD constants (prix et PPA de 1995)				
Business enterprise								**Entreprises**
financed by								*financé par*
Business enterprise	..	775.5	751.6	873.2	938.4	823.4	875.5	Entreprises
Direct government	26.5 [d,t]	37.6 [a]	60.8	75.7	85.9	144.9	160.0	État direct
Higher education	..	0.3	0.2	0.3	0.0 [a]	6.5	0.0	Enseignement supérieur
Private non-profit	..	1.7	1.9	0.2	0.7	0.2	10.9	Institutions sans but lucratif
Funds from abroad	..	26.5	14.2	13.2	24.6	49.5	39.2	Fonds de l'étranger
Total	886.8 [d,t]	841.6 [a]	828.7	962.6	1 049.5	1 024.6	1 085.6	*Total*
Government								**État**
financed by								*financé par*
Business enterprise	..	38.5	72.4	39.7	34.2	27.2	44.1	Entreprises
Direct government	287.3 [d,t]	289.4 [a]	349.2	354.4	374.4	359.2	400.3	État direct
Higher education	..	2.0	1.5	3.7	0.4 [a]	0.5	4.7	Enseignement supérieur
Private non-profit	..	1.8	0.3	3.3	0.2	0.0	0.9	Institutions sans but lucratif
Funds from abroad	..	10.4	7.0	7.2	8.8	8.5	8.5	Fonds de l'étranger
Total	377.6 [d,t]	342.1 [a]	430.5	408.3	417.9	395.5	458.5	*Total*
Higher education								**Enseignement supérieur**
financed by								*financé par*
Business enterprise	..	2.1	0.5	2.1	3.2	2.5	2.7	Entreprises
Direct government	56.0 [l]	90.4 [a]	69.7	41.4	..	..	..	État direct
General university funds	..	..	..	..	..	..	..	Fonds généraux des universités
Sub-total government	..	90.4 [a,m]	69.7 [m]	41.4 [m]	137.5 [a]	186.8	241.6	Sous-Total État
Higher education	..	10.5 [i]	46.7 [i]	87.4 [i]	4.5 [a]	4.5	3.9	Enseignement supérieur
Private non-profit	..	0.3	0.1	0.2	0.1	0.5	0.0	Institutions sans but lucratif
Funds from abroad	..	6.3	5.4	8.9	9.0	6.9	9.0	Fonds de l'étranger
Total	59.1 [l]	109.7 [a]	122.5	139.9	154.3	201.2	257.3	*Total*
Private non-profit								**Institutions sans but lucratif**
financed by								*financé par*
Business enterprise	..	..	0.3	1.9	2.9	4.5	4.8	Entreprises
Direct government	..	..	0.7	0.3	1.2	3.8	4.0	État direct
Higher education	..	..	0.0	19.6	0.0	0.0	..	Enseignement supérieur
Private non-profit	..	..	0.1	0.1	0.0	0.4	0.1	Institutions sans but lucratif
Funds from abroad	..	..	0.0	0.0	0.0	0.2	0.2	Fonds de l'étranger
Total	..	..	1.2	22.0	4.1	9.0	9.2	*Total*
GERD								**DIRD**
financed by								*financé par*
Business enterprise	..	816.2	824.8	917.0	978.6	857.6	927.2	Entreprises
Direct government	369.8 [d,t]	417.4 [a]	480.5	471.9	598.9 [a]	694.7	805.8	État direct
General university funds	..	..	..	..	..	..	..	Fonds généraux des universités
Sub-total government	369.8 [d,m,t]	417.4 [a,m]	480.5 [m]	471.9 [m]	598.9 [a]	694.7	805.8	Sous-Total État
Higher education	..	12.7 [i]	48.5 [i]	111.0 [i]	4.9 [a]	11.6	8.7	Enseignement supérieur
Private non-profit	..	3.8	2.4	3.8	1.0	1.2	11.9	Institutions sans but lucratif
Funds from abroad	..	43.3	26.6	29.2	42.4	65.1	56.9	Fonds de l'étranger
Total	1 323.6 [d,t]	1 293.3 [a]	1 382.8	1 532.8	1 625.8	1 630.2	1 810.6	*Total*

Source: OECD, R&D database (DSTI/EAS Division), January 2002.

Source : OCDE, base de données R-D (DSTI/Division AES), janvier 2002.

Gross domestic expenditure on R&D (GERD) - Dépenses intérieures brutes de R-D (DIRD)

by sector of performance and type of costs - par secteur d'exécution et par type de coûts

Million CZK — *Millions CZK*

	1994	1995	1996	1997	1998	1999	2000	
Business enterprise								**Entreprises**
Labour costs	..	2 419.3	2 807.3	3 122.5	3 406.9	4 025.5	..	Coûts du personnel
Other current costs	..	5 884.5	5 894.3	7 637.2	9 897.0	9 246.6	..	Autres coûts courants
Total current expenditure	..	8 303.9	8 701.6	10 759.7	13 303.9	13 272.1	..	*Total dépenses courantes*
Land and buildings	..	124.6	123.3	251.5	284.1	280.3	..	Terrains et constructions
Instruments and equipment	..	670.2	922.3	1 220.9	1 171.8	1 309.9	..	Instruments et équipements
Total capital expenditure	..	794.7	1 045.6	1 472.4	1 455.9	1 590.2	..	*Total dépenses en capital*
Total	8 699.0 [d,t]	9 098.6 [a]	9 747.2	12 232.1	14 759.8	14 862.3	15 881.9	*Total*
Government								**État**
Labour costs	..	1 480.7	1 847.5	1 714.5	1 764.4	2 118.8	..	Coûts du personnel
Other current costs	..	1 712.8	2 524.9	2 739.0	3 244.7	2 818.4	..	Autres coûts courants
Total current expenditure	..	3 193.5	4 372.4	4 453.5	5 009.1	4 937.2	..	*Total dépenses courantes*
Land and buildings	..	120.5	135.8	175.5	242.9	189.4	..	Terrains et constructions
Instruments and equipment	..	384.1	554.8	558.6	625.8	610.0	..	Instruments et équipements
Total capital expenditure	..	504.6	690.5	734.1	868.7	799.4	..	*Total dépenses en capital*
Total	3 704.0 [d,t]	3 698.1 [a]	5 063.0	5 187.6	5 877.8	5 736.6	6 707.0	*Total*
Higher education								**Enseignement supérieur**
Labour costs	..	439.1	390.0	562.2	593.0	964.5	..	Coûts du personnel
Other current costs	..	494.8	562.8	841.9	1 178.5	1 525.8	..	Autres coûts courants
Total current expenditure	..	933.9	952.8	1 404.1	1 771.5	2 490.3	..	*Total dépenses courantes*
Land and buildings	..	34.7	14.0	5.8	10.6	2.7	..	Terrains et constructions
Instruments and equipment	..	217.2	473.5	368.1	388.2	424.8	..	Instruments et équipements
Total capital expenditure	..	252.0	487.5	373.9	398.8	427.5	..	*Total dépenses en capital*
Total	580.0 [t]	1 185.9 [a]	1 440.3	1 778.0	2 170.3	2 917.8	3 764.0	*Total*
Private non-profit								**Institutions sans but lucratif**
Labour costs	..	..	3.1	16.9	14.3	33.6	..	Coûts du personnel
Other current costs	..	..	10.2	236.1	37.1	82.5	..	Autres coûts courants
Total current expenditure	..	..	13.3	253.0	51.4	116.1	..	*Total dépenses courantes*
Land and buildings	..	..	0.0	0.5	0.0	4.8	..	Terrains et constructions
Instruments and equipment	..	..	0.5	26.1	5.7	9.1	..	Instruments et équipements
Total capital expenditure	..	..	0.5	26.6	5.7	13.9	..	*Total dépenses en capital*
Total	..	..	13.8	279.7	57.1	130.0	134.3	*Total*
GERD								**DIRD**
Labour costs	..	4 339.1	5 047.9	5 416.2	5 778.6	7 142.4	..	Coûts du personnel
Other current costs	..	8 092.1	899.2	11 454.1	14 357.3	13 673.3	..	Autres coûts courants
Total current expenditure	..	12 431.3	14 040.1	16 870.3	20 135.9	20 815.7	..	*Total dépenses courantes*
Land and buildings	..	279.8	273.1	433.3	537.6	477.2	..	Terrains et constructions
Instruments and equipment	..	1 271.6	1 951.1	2 173.8	2 191.5	2 353.8	..	Instruments et équipements
Total capital expenditure	..	1 551.3	2 224.1	2 607.1	2 729.1	2 831.0	..	*Total dépenses en capital*
Total	12 983.0 [d,t]	13 982.6 [a]	16 264.3	19 477.4	22 865.0	23 646.7	26 487.2	*Total*

Table 3 - Tableau 3

Current domestic expenditure on R&D - Dépenses intérieures courantes de R-D

by sector of performance and type of activity - par secteur d'exécution et par type d'activités

Million CZK — *Millions CZK*

	1994	1995	1996	1997	1998	1999	2000	
Business enterprise								**Entreprises**
Basic research	..	105.2 [k]	54.2 [k]	87.7 [k]	52.6 [k]	266.3 [k]	3 819.7 [k]	Recherche fondamentale
Applied research	..	2 720.0 [k]	2 711.4 [k]	2 906.6 [k]	3 236.1 [k]	3 655.7 [k]	4 954.0 [k]	Recherche appliquée
Experimental development	..	6 273.4 [k]	6 981.6 [k]	9 237.8 [k]	11 471.1 [k]	10 940.3 [k]	7 108.2 [k]	Développement expérimental
Not specified	..	..	..	..	..	..	..	Non spécifié
Total	..	9 098.6 [a,k]	9 747.2 [k]	12 232.1 [k]	14 759.8 [k]	14 862.3 [k]	15 881.9 [k]	*Total*
Government								**État**
Basic research	..	1 777.9 [k]	2 389.3 [k]	2 571.7 [k]	2 815.5 [k]	2 924.5 [k]	3 949.9 [k]	Recherche fondamentale
Applied research	..	1 403.6 [k]	1 737.7 [k]	1 940.1 [k]	1 973.8 [k]	1 547.2 [k]	1 598.2 [k]	Recherche appliquée
Experimental development	..	516.7 [k]	936.0 [k]	675.8 [k]	1 088.5 [k]	1 264.9 [k]	1 158.9 [k]	Développement expérimental
Not specified	..	..	..	..	..	..	..	Non spécifié
Total	..	3 698.1 [a,k]	5 063.0 [k]	5 187.6 [k]	5 877.8 [k]	5 736.6 [k]	6 707.0 [k]	*Total*
Higher education								**Enseignement supérieur**
Basic research	..	492.0 [k]	605.9 [k]	848.1 [k]	1 187.2 [k]	1 637.6 [k]	1 891.4 [k]	Recherche fondamentale
Applied research	..	598.3 [k]	679.9 [k]	719.9 [k]	736.7 [k]	1 100.3 [k]	1 569.8 [k]	Recherche appliquée
Experimental development	..	95.5 [k]	154.5 [k]	210.0 [k]	246.4 [k]	179.9 [k]	302.8 [k]	Développement expérimental
Not specified	..	..	..	..	..	..	..	Non spécifié
Total	..	1 185.9 [a,k]	1 440.3 [k]	1 778.0 [k]	2 170.3 [k]	2 917.8 [k]	3 764.0 [k]	*Total*
Private non-profit								**Institutions sans but lucratif**
Basic research	..	..	0.4 [k]	0.0 [k]	0.0 [k]	11.1 [k]	43.4 [k]	Recherche fondamentale
Applied research	..	..	2.9 [k]	11.9 [k]	10.9 [k]	38.6 [k]	62.7 [k]	Recherche appliquée
Experimental development	..	..	10.5 [k]	267.8 [k]	46.2 [k]	80.3 [k]	28.2 [k]	Développement expérimental
Not specified	..	..	..	..	..	..	..	Non spécifié
Total	..	..	13.8 [k]	279.7 [k]	57.1 [k]	130.0 [k]	134.3 [k]	*Total*
GERD								**DIRD**
Basic research	..	2 375.1 [k]	3 049.8 [k]	3 507.5 [k]	4 055.3 [k]	4 839.5 [k]	9 704.4 [k]	Recherche fondamentale
Applied research	..	4 722.0 [k]	5 131.9 [k]	5 578.5 [k]	5 957.5 [k]	6 341.8 [k]	8 184.7 [k]	Recherche appliquée
Experimental development	..	6 885.6 [k]	8 082.6 [k]	10 391.4 [k]	12 852.2 [k]	12 465.4 [k]	8 598.1 [k]	Développement expérimental
Not specified	..	..	..	..	..	..	..	Non spécifié
Total	..	13 982.6 [a,k]	16 264.3 [k]	19 477.4 [k]	22 865.0 [k]	23 646.7 [k]	26 487.2 [k]	*Total*

Source: OECD, R&D database (DSTI/EAS Division), January 2002.

Source : OCDE, base de données R-D (DSTI/Division AES), janvier 2002.

Gross domestic expenditure on R&D (GERD) - Dépenses intérieures brutes de R-D (DIRD)

by sector of performance and socio-economic objective - par secteur d'exécution et par objectif socio-économique

Million CZK / *Millions CZK*

	1994	1995	1996	1997	1998	1999	2000	
Business enterprise								**Entreprises**
Agriculture, forestry and fishing	..	..	..	..	355.4	470.3	..	Agriculture, sylviculture et pêche
Industrial development	..	..	..	..	10 401.7	10 101.6	..	Développement industriel
Energy	..	..	..	..	798.4	810.5	..	Energie
Transport and telecommunications	..	..	..	..	1 340.8	1 406.5	..	Transports et télécommunications
Urban and rural planning	..	..	..	..	44.7	54.7	..	Aménagement urbain et rural
Prevention of pollution	..	..	..	..	242.4	290.6	..	Prévention de la pollution
Identification & treatment of pollution	..	..	..	..	135.7	162.8	..	Détection pollu. & mesures antipollu.
Health	..	..	..	..	495.7	1 030.1	..	Santé
Social development and services	..	..	..	..	233.3	127.8	..	Développement et services sociaux
Earth and atmosphere	..	..	..	..	3.1	12.4	..	Milieu terrestre et atmosphère
Advancement of knowledge	..	..	..	..	345.7	72.2	..	Promotion générale des connaissances
Civil space	..	..	..	..	0.0	5.8	..	Espace civil
Defence	..	..	..	..	358.3	316.6	..	Défense
Not elsewhere classified	8 699.0 [d,t]	9 098.6 [a]	9 747.2	12 232.1	4.6	0.4	15 881.9	Non classé
Total	8 699.0 [d,t]	9 098.6 [a]	9 747.2	12 232.1	14 759.8	14 862.3	15 881.9	*Total*
Government								**État**
Agriculture, forestry and fishing	..	..	..	..	355.6	327.2	..	Agriculture, sylviculture et pêche
Industrial development	..	..	..	..	263.3	225.6	..	Développement industriel
Energy	..	..	..	..	30.6	152.1	..	Energie
Transport and telecommunications	..	..	..	..	128.9	79.1	..	Transports et télécommunications
Urban and rural planning	..	..	..	..	14.1	6.1	..	Aménagement urbain et rural
Prevention of pollution	..	..	..	..	117.6	150.2	..	Prévention de la pollution
Identification & treatment of pollution	..	..	..	..	188.8	110.2	..	Détection pollu. & mesures antipollu.
Health	..	..	..	..	711.1	324.3	..	Santé
Social development and services	..	..	..	..	138.4	306.1	..	Développement et services sociaux
Earth and atmosphere	..	..	..	..	207.9	267.0	..	Milieu terrestre et atmosphère
Advancement of knowledge	..	..	..	..	2 418.5	2 450.6	..	Promotion générale des connaissances
Civil space	..	..	..	..	62.7	65.6	..	Espace civil
Defence	..	..	..	..	1 185.1	1 258.6	..	Défense
Not elsewhere classified	3 704.0 [d,t]	3 698.1 [a]	5 063.0	5 187.6	55.2	13.9	6 707.0	Non classé
Total	3 704.0 [d,t]	3 698.1 [a]	5 063.0	5 187.6	5 877.8	5 736.6	6 707.0	*Total*
Higher education								**Enseignement supérieur**
Agriculture, forestry and fishing	..	..	..	..	144.7	185.5	..	Agriculture, sylviculture et pêche
Industrial development	..	..	..	..	205.3	605.8	..	Développement industriel
Energy	..	..	..	..	31.0	33.0	..	Energie
Transport and telecommunications	..	..	..	..	55.7	45.7	..	Transports et télécommunications
Urban and rural planning	..	..	..	..	0.7	14.0	..	Aménagement urbain et rural
Prevention of pollution	..	..	..	..	76.9	76.4	..	Prévention de la pollution
Identification & treatment of pollution	..	..	..	..	52.6	44.9	..	Détection pollu. & mesures antipollu.
Health	..	..	..	..	343.9	200.2	..	Santé
Social development and services	..	..	..	..	9.2	22.7	..	Développement et services sociaux
Earth and atmosphere	..	..	..	..	48.1	66.5	..	Milieu terrestre et atmosphère
Advancement of knowledge	..	..	..	..	1 198.1	1 620.4	..	Promotion générale des connaissances
Civil space	..	..	..	..	0.9	0.0	..	Espace civil
Defence	..	..	..	..	0.5	1.9	..	Défense
Not elsewhere classified	580.0	1 185.9 [a]	1 440.3	1 778.0	2.7	0.8	3 764.0	Non classé
Total	580.0 [t]	1 185.9 [a]	1 440.3	1 778.0	2 170.3	2 917.8	3 764.0	*Total*
Private non-profit								**Institutions sans but lucratif**
Agriculture, forestry and fishing	..	..	..	..	0.0	1.2	..	Agriculture, sylviculture et pêche
Industrial development	..	..	..	..	49.9	76.1	..	Développement industriel
Energy	..	..	..	..	1.7	3.2	..	Energie
Transport and telecommunications	..	..	..	..	0.0	7.2	..	Transports et télécommunications
Urban and rural planning	..	..	..	..	0.0	1.7	..	Aménagement urbain et rural
Prevention of pollution	..	..	..	..	1.7	5.8	..	Prévention de la pollution
Identification & treatment of pollution	..	..	..	..	0.0	4.1	..	Détection pollu. & mesures antipollu.
Health	..	..	..	..	2.2	3.8	..	Santé
Social development and services	..	..	..	..	0.0	6.1	..	Développement et services sociaux
Earth and atmosphere	..	..	..	..	0.0	1.6	..	Milieu terrestre et atmosphère
Advancement of knowledge	..	..	..	..	0.0	12.6	..	Promotion générale des connaissances
Civil space	..	..	..	..	0.0	4.6	..	Espace civil
Defence	..	..	..	..	0.0	0.3	..	Défense
Not elsewhere classified	..	..	13.8	279.7	1.6	1.7	134.0	Non classé
Total	..	..	13.8	279.7	57.1	130.0	134.3	*Total*
GERD								**DIRD**
Agriculture, forestry and fishing	..	..	..	..	855.7	984.2	..	Agriculture, sylviculture et pêche
Industrial development	..	..	..	..	10 920.2	11 009.1	..	Développement industriel
Energy	..	..	..	..	861.7	998.8	..	Energie
Transport and telecommunications	..	..	..	..	1 525.4	1 538.5	..	Transports et télécommunications
Urban and rural planning	..	..	..	..	59.5	76.5	..	Aménagement urbain et rural
Prevention of pollution	..	..	..	..	438.6	523.0	..	Prévention de la pollution
Identification & treatment of pollution	..	..	..	..	377.1	322.0	..	Détection pollu. & mesures antipollu.
Health	..	..	..	..	1 552.9	1 558.4	..	Santé
Social development and services	..	..	..	..	380.9	462.7	..	Développement et services sociaux
Earth and atmosphere	..	..	..	..	259.1	347.5	..	Milieu terrestre et atmosphère
Advancement of knowledge	..	..	..	..	3 962.3	4 155.8	..	Promotion générale des connaissances
Civil space	..	..	..	..	63.6	76.0	..	Espace civil
Defence	..	..	..	..	1 543.9	1 577.4	..	Défense
Not elsewhere classified	12 983.0 [d,t]	13 982.6 [a]	16 264.3	19 477.4	64.1	16.8	26 487.2	Non classé
Total	12 983.0 [d,t]	13 982.6 [a]	16 264.3	19 477.4	22 865.0	23 646.7	26 487.2	*Total*

Source: OECD, R&D database (DSTI/EAS Division), January 2002.

Source : OCDE, base de données R-D (DSTI/Division AES), janvier 2002.

Table 5 - Tableau 5

Total R&D personnel - Personnel total de R-D

by sector of employment and occupation - par secteur d'emploi et par profession

In full-time equivalent — *En équivalence plein-temps*

	1994	1995	1996	1997	1998	1999	2000	
Business enterprise								**Entreprises**
RSE	..	4 936	4 863	5 120	5 067	5 811	5 533	Chercheurs
Technicians	..	4 412	4 351	4 475	4 267	4 592	4 430	Techniciens
Other occupations	..	1 998	1 953	1 869	1 953	1 880	1 564	Autres professions
Total	23 114 [d,l,t]	11 346 [a]	11 167	11 464	11 287	12 283	11 527	*Total*
Government								**État**
RSE	..	4 315	4 574	4 597	4 596	4 281	4 424	Chercheurs
Technicians	..	2 076	2 071	2 034	1 923	1 757	1 765	Techniciens
Other occupations	..	1 252	1 261	1 120	871	925	959	Autres professions
Total	11 268 [d,l,t]	7 643 [a]	7 906	7 751	7 390	6 963	7 148	*Total*
Higher education								**Enseignement supérieur**
RSE	..	2 685	3 505	2 848	2 884	3 380	3 768	Chercheurs
Technicians	..	665	601	772	812	1 009	1 073	Techniciens
Other occupations	..	339	298	361	330	347	490	Autres professions
Total	3 397 [l,t]	3 689 [a]	4 404	3 981	4 026	4 736	5 331	*Total*
Private non-profit								**Institutions sans but lucratif**
RSE	..	..	21	15	19	63	127	Chercheurs
Technicians	..	..	3	17	14	45	51	Techniciens
Other occupations	..	..	0	2	4	16	14	Autres professions
Total	..	..	24	34	37	124	192	*Total*
TOTAL								**TOTAL**
RSE	..	11 936	12 963	12 580	12 566	13 535	13 852	Chercheurs
Technicians	..	7 153	7 026	7 298	7 016	7 403	7 319	Techniciens
Other occupations	..	3 589	3 512	3 352	3 158	3 168	3 027	Autres professions
Total	37 779 [d,l,t]	22 678 [a]	23 501	23 230	22 740	24 106	24 198	*Total*

Table 6 - Tableau 6

Total R&D personnel - Personnel total de R-D

by sector of employment and formal qualification - par secteur d'emploi et par niveau de qualification

In full-time equivalent — *En équivalence plein-temps*

	1994	1995	1996	1997	1998	1999	2000	
Business enterprise								**Entreprises**
University PhD level degrees	1 459 [l,t]	781 [l,t]	..	614 [a]	564	792	721	Diplômes univ. de niv. doctorat
Other university degrees	5 169 [l,t]	6 161 [l,t]	..	4 502 [a]	4 346	5 028	4 939	Autres diplômes universitaires
Sub-total university degrees	6 628 [d,l,t]	6 942 [l,t]	..	5 116 [a]	4 910	5 820	5 660	Sous-total Universitaires
Other post-secondary diplomas	..	..	..	..	810	614	509	Autres diplômes post-secondaires
Secondary diplomas	6 103 [d,l,t]	8 085 [l,t]	..	4 236 [a]	3 848	4 277	4 037	Diplômes de fin d'études secondaires
Other qualifications	880 [d,l,t]	5 148 [l,t]	..	2 112 [a]	1 719	1 572	1 321	Autres qualifications
Total	23 114 [d,l,t]	11 346 [a]	11 167	11 464	11 287	12 283	11 527	*Total*
Government								**État**
University PhD level degrees	2 746 [l,t]	3 031 [l,t]	..	2 388 [a]	2 291	2 176	2 065	Diplômes univ. de niv. doctorat
Other university degrees	2 220 [l,t]	3 565 [l,t]	..	2 437 [a]	2 433	2 292	2 517	Autres diplômes universitaires
Sub-total university degrees	4 966 [d,l,t]	6 596 [l,t]	..	4 825 [a]	4 724	4 468	4 582	Sous-total Universitaires
Other post-secondary diplomas	..	..	..	..	243	151	111	Autres diplômes post-secondaires
Secondary diplomas	2 507 [d,l,t]	3 526 [l,t]	..	2 010 [a]	1 930	1 912	2 069	Diplômes de fin d'études secondaires
Other qualifications	94 [d,l,t]	2 435 [l,t]	..	916 [a]	493	432	386	Autres qualifications
Total	11 268 [d,l,t]	7 643 [a]	7 906	7 751	7 390	6 963	7 148	*Total*
Higher education								**Enseignement supérieur**
University PhD level degrees	1 115 [l,t]	4 587 [l,t]	..	1 698 [a]	1 558	2 005	2 494	Diplômes univ. de niv. doctorat
Other university degrees	616 [l,t]	5 089 [l,t]	..	1 485 [a]	1 692	1 783	1 912	Autres diplômes universitaires
Sub-total university degrees	1 731 [l,t]	9 676 [l,t]	..	3 183 [a]	3 250	3 788	4 406	Sous-total Universitaires
Other post-secondary diplomas	..	..	..	..	78	59	62	Autres diplômes post-secondaires
Secondary diplomas	187 [l,t]	1 763 [l,t]	..	596 [a]	568	779	782	Diplômes de fin d'études secondaires
Other qualifications	0 [l,t]	545 [l,t]	..	202 [a]	130	110	81	Autres qualifications
Total	3 397 [l,t]	3 689 [a]	4 404	3 981	4 026	4 736	5 331	*Total*
Private non-profit								**Institutions sans but lucratif**
University PhD level degrees	..	..	..	2 [a]	0	34	82	Diplômes univ. de niv. doctorat
Other university degrees	..	..	..	13 [a]	7	38	60	Autres diplômes universitaires
Sub-total university degrees	..	..	..	15 [a]	7	72	142	Sous-total Universitaires
Other post-secondary diplomas	..	..	..	..	4	7	11	Autres diplômes post-secondaires
Secondary diplomas	..	..	..	16 [a]	14	34	32	Diplômes de fin d'études secondaires
Other qualifications	..	..	..	3 [a]	12	11	7	Autres qualifications
Total	..	..	24	34	37	124	192	*Total*
TOTAL								**TOTAL**
University PhD level degrees	5 320 [l,t]	8 399 [l,t]	..	4 702 [a]	4 413	5 007	5 362	Diplômes univ. de niv. doctorat
Other university degrees	8 005 [l,t]	14 815 [l,t]	..	8 437 [a]	8 478	9 141	9 428	Autres diplômes universitaires
Sub-total university degrees	13 325 [d,l,t]	23 214 [l,t]	..	13 139 [a]	12 891	14 148	14 790	Sous-total Universitaires
Other post-secondary diplomas	0 [d,l,t]	..	..	..	1 135	831	693	Autres diplômes post-secondaires
Secondary diplomas	8 797 [d,l,t]	13 374 [l,t]	..	6 858 [a]	6 360	7 002	6 920	Diplômes de fin d'études secondaires
Other qualifications	974 [d,l,t]	8 128 [l,t]	..	3 233 [a]	2 354	2 125	1 795	Autres qualifications
Total	37 779 [d,l,t]	22 678 [a]	23 501	23 230	22 740	24 106	24 198	*Total*

Source: OECD, R&D database (DSTI/EAS Division), January 2002.

Source : OCDE, base de données R-D (DSTI/Division AES), janvier 2002.

Gross domestic expenditure on R&D (GERD) - Dépenses intérieures brutes de R-D (DIRD)

by sector of performance and main scientific field - par secteur d'exécution et par grand domaine scientifique

Million CZK *Millions CZK*

	1994	1995	1996	1997	1998	1999	2000	
Business enterprise								**Entreprises**
Natural sciences and engineering	8 699.0 [d,t]	9 098.2	9 731.6	12 179.0	14 754.0	14 792.1	..	Sciences exactes, nat. et de l'ingénieur
Social sciences and humanities	..	0.4	15.6	53.1	5.8	70.2	..	Sciences sociales et humaines
All fields of science	8 699.0 [d,t]	9 098.6 [a]	9 747.2	12 232.1	14 759.8	14 862.3	15 881.9	*Tous domaines scientifiques*
Government								**État**
Natural sciences and engineering	..	3 230.5	4 679.8	4 567.3	5 283.4	5 018.1	..	Sciences exactes, nat. et de l'ingénieur
Social sciences and humanities	..	467.6	383.2	620.3	594.4	718.5	..	Sciences sociales et humaines
All fields of science	3 704.0 [d,t]	3 698.1 [a]	5 063.0	5 187.6	5 877.8	5 736.6	6 707.0	*Tous domaines scientifiques*
Higher education								**Enseignement supérieur**
Natural sciences and engineering	..	1 072.4	1 281.6	1 470.8	1 999.4	2 662.6	..	Sciences exactes, nat. et de l'ingénieur
Social sciences and humanities	..	113.5 [m]	158.7 [m]	307.2 [m]	170.9	255.2	..	Sciences sociales et humaines
All fields of science	580.0 [t]	1 185.9 [a]	1 440.3	1 778.0	2 170.3	2 917.8	3 764.0	*Tous domaines scientifiques*
Private non-profit								**Institutions sans but lucratif**
Natural sciences and engineering	..	..	10.6	269.7	57.1	117.1	..	Sciences exactes, nat. et de l'ingénieur
Social sciences and humanities	..	..	3.2 [m]	10.0 [m]	0.0	12.9	..	Sciences sociales et humaines
All fields of science	..	..	13.8	279.7	57.1	130.0	134.3	*Tous domaines scientifiques*
TOTAL								**TOTAL**
Natural sciences and engineering	..	13 401.1	15 703.6	18 487.2	22 093.9	22 589.9	..	Sciences exactes, nat. et de l'ingénieur
Social sciences and humanities	..	581.5	560.7	990.5	771.1	1 056.8	..	Sciences sociales et humaines
All fields of science	12 983.0 [d,t]	13 982.6 [a]	16 264.3	19 477.4	22 865.0	23 646.7	26 487.2	*Tous domaines scientifiques*

Table 8 - Tableau 8

Total R&D personnel - Personnel total de R-D

by sector of employment and main scientific field - par secteur d'emploi et par grand domaine scientifique

In full-time equivalent *En équivalence plein-temps*

	1994	1995	1996	1997	1998	1999	2000	
Business enterprise								**Entreprises**
Natural sciences and engineering	23 114 [d,l,t]	11 341	11 037	11 382	11 274	12 148	..	Sciences exactes, nat. et de l'ingénieur
Social sciences and humanities	..	5	11	82	13	135	..	Sciences sociales et humaines
All fields of science	23 114 [d,l,t]	11 346 [a]	11 167	11 464	11 287	12 283	11 527	*Tous domaines scientifiques*
Government								**État**
Natural sciences and engineering	..	6 677	6 917	6 509	6 258	5 818	..	Sciences exactes, nat. et de l'ingénieur
Social sciences and humanities	..	966	973	1 242	1 132	1 145	..	Sciences sociales et humaines
All fields of science	11 268 [d,l,t]	7 643 [a]	7 906	7 751	7 390	6 963	7 148	*Tous domaines scientifiques*
Higher education								**Enseignement supérieur**
Natural sciences and engineering	..	3 309	2 682	2 994	3 598	4 192	..	Sciences exactes, nat. et de l'ingénieur
Social sciences and humanities	..	380 [m]	1 721 [m]	987 [m]	428	544	..	Sciences sociales et humaines
All fields of science	3 397 [l,t]	3 689 [a]	4 404	3 981	4 026	4 736	5 331	*Tous domaines scientifiques*
Private non-profit								**Institutions sans but lucratif**
Natural sciences and engineering	..	..	28	16	37	102	..	Sciences exactes, nat. et de l'ingénieur
Social sciences and humanities	..	..	4 [m]	18 [m]	0	22	..	Sciences sociales et humaines
All fields of science	..	..	24	34	37	124	192	*Tous domaines scientifiques*
TOTAL								**TOTAL**
Natural sciences and engineering	..	21 327	20 664	20 901	21 167	22 260	..	Sciences exactes, nat. et de l'ingénieur
Social sciences and humanities	..	1 351	2 709	2 329	1 573	1 846	..	Sciences sociales et humaines
All fields of science	37 779 [d,l,t]	22 678 [a]	23 501	23 230	22 740	24 106	24 198	*Tous domaines scientifiques*

Source: OECD, R&D database (DSTI/EAS Division), January 2002.

Source : OCDE, base de données R-D (DSTI/Division AES), janvier 2002.

Total business enterprise intramural expenditure on R&D - Dépenses intra-muros totales de R-D des entreprises
(BERD) by industry - (DIRDE) par industrie

	1994	1995	1996	1997	1998	1999	2000	
				9.1 Million CZK - Millions CZK				
1 AGRICULTURE, HUNTING & FORESTRY	..	20.0 [a]	52.3	37.0	65.4	70.0	74.2	**1 AGRICULTURE, CHASSE & SYLVICULTURE**
2 MINING	37.0 [d,t]	123.6 [a]	102.9	87.3	95.4	65.1	40.6	**2 ACTIVITES EXTRACTIVES**
3 MANUFACTURING	5 615.0 [d,t]	6 877.8 [a]	7 027.7	9 175.4	11 210.2	10 408.1	10 600.9	**3 ACTIVITES DE FABRICATION**
4 Food, beverages & tobacco	26.0 [d,t]	31.8 [a]	35.5	62.7	172.8	63.0	88.0	**4 Produits alimentaires, boissons & tabac**
5 Food products & beverages	26.0 [d,t]	31.8 [a]	35.5	62.7	172.8	63.0	88.0	5 Produits alimentaires & boissons
6 Tobacco products	0.0 [d,t]	0.0	0.0	0.0	0.0	0.0	0.0	6 Produits à base de tabac
7 Textiles, wear app., fur & leather	79.0 [d,t]	58.8 [a]	41.2	84.4	150.8	121.6	129.1	**7 Textiles, habill., fourrures & cuirs**
8 Textiles	24.0 [d,t]	26.5 [a]	30.1	73.1	137.2	85.6	78.7	8 Textiles
9 Wearing apparel & fur	1.0 [d,t]	1.9 [a]	2.0	1.7	1.6	21.9	35.3	9 Articles d'habillement & fourrures
10 Leather products & footwear	54.0 [d,t]	30.4 [a]	9.1	9.6	12.0	14.1	15.1	10 Articles en cuir & chaussures
11 Wood, paper, printing, publishing	23.0 [d]	20.2 [a]	25.8	13.5	22.6	12.5	7.2	**11 Bois, papier, imprimerie, édition**
12 Wood & cork (not furniture)	1.0 [d,t]	2.6 [a]	2.1	0.9	7.1	1.4	0.0	12 Bois & liège (sauf meubles)
13 Pulp, paper & paper products	22.0 [d,t]	17.6 [a]	21.5	12.4	12.5	10.7	3.2	13 Carton, papier & articles
14 Publ., print. & repro. of rec. media	0.0 [d,t]	0.0	2.2	0.2	3.0	0.4	4.0	14 Edition, impress. & repro. enregis.
15 Coke, petroleum, nuclear fuel, chemicals, rubber plastics	994.0 [d,t]	1 031.1 [a]	1 152.9	1 094.0	1 677.3	1 272.4	1 330.7	**15 Coke, pétrole, combust. nucl., prod. chimi., caout. & mat. plast.**
16 Coke, ref. petrol. prod. & nuclear fuel	110.0 [d,t]	131.7 [a]	18.4	15.6	13.7	15.5	12.5	16 Coke, raffin. pétr. & combust. nucl.
17 Chemicals & chemical products	589.0 [d,t]	726.2 [a]	891.7	810.5	934.0	992.8	1 092.8	17 Produits chimiques
18 *Chemical prod.(less pharmaceuticals)*	322.0 [d,t]	549.9 [a]	640.9	636.8	714.0	681.4	634.7	18 *Prod. chimi. (moins prod. pharmaceutiques)*
19 *Pharmaceuticals*	267.0 [d,t]	176.3 [a]	250.8	173.7	220.0	311.4	458.1	19 *Produits pharmaceutiques*
20 Rubber & plastic products	295.0 [d,t]	173.2 [a]	242.8	267.9	729.6	264.1	225.4	20 Caoutchouc et matières plastiques
21 Non-metallic mineral products	136.0 [d,t]	94.4 [a]	103.4	152.5	259.8	212.1	232.8	**21 Produits minéraux non métalliques**
22 Basic metals	529.0 [d,t]	436.4 [a]	509.2	498.0	604.4	447.5	462.1	**22 Produits métallurgiques de base**
23 Basic metals, ferrous	518.0 [d,t]	411.2 [a]	504.5	485.4	581.1	419.8	433.5	23 Produits métallur. de base, ferreux
24 Basic metals, non-ferrous	11.0 [d,t]	25.2 [a]	4.7	12.6	23.3	27.7	28.6	24 Prod. métallur. de base, non ferreux
25 Fabricated metal products	150.0 [d,t]	179.0 [a]	194.4	284.7	238.6	404.8	326.4	**25 Ouvrages en métaux (sauf mach.& mat.)**
26 Machinery equipment, instruments and transport equipment	3 576.0 [d,t]	4 899.3 [a]	4 801.1	6 807.5	8 033.7	7 652.8	7 807.5	**26 Machines & matériel, instruments & mat. de transport**
27 Machinery n.e.c.	855.0 [d,t]	976.5 [a]	1 172.8	1 340.9	1 392.9	1 259.1	1 377.8	27 Machines, n.c.a.
28 Office, account. & computing machinery	3.0 [d,t]	10.8 [a]	1.8	24.6	3.0	3.9	6.9	28 Mach. de bur., compta. & trait. inform.
29 Electrical machinery	211.0 [d,t]	272.7 [a]	210.5	200.4	375.9	365.1	393.8	29 Machines électriques
30 Electro. equip. (radio, TV & commun.)	268.0 [d,t]	254.8 [a]	207.1	283.8	377.2	316.9	376.4	30 Equip. électro. (radio, tv & communic.)
31 *Electro. comp. (inc. semi-conductors)*	214.0 [d,t]	183.1 [a]	122.8	116.1	50.4	53.0	69.5	31 *Compos. électro. (semi-cond. incl.)*
32 *TV, radio & communications equip.*	54.0 [d,t]	71.7 [a]	84.3	167.7	326.8	263.9	306.9	32 *Appareils radio, tv et de communi.*
33 Instruments	140.0 [d,t]	93.1 [a]	75.5	118.4	111.4	279.9	205.5	33 Instruments
34 Motor vehicles	1 602.0 [d,t]	2 256.1 [a]	1 995.8	2 921.7	4 119.4	4 030.9	4 703.5	34 Véhicules automobiles
35 Other transport equipment	497.0 [d,t]	1 035.4 [a]	1 137.6	1 917.7	1 654.0	1 397.0	743.6	35 Autres matériels de transport
36 *Ships*	0.0 [d,t]	0.0	0.0	0.0	0.0	0.0	0.0	36 *Navires*
37 *Aerospace*	340.0 [d,t]	714.3 [a]	779.1	1 013.2	878.6	774.9	549.3	37 *Aérospatiale*
38 *Other transport equipment n.e.c.*	157.0 [d,t]	321.1 [a]	358.5	904.5	775.4	622.1	194.3	38 *Autres matériels de transport n.c.a.*
39 Furniture, other manufacturing n.e.c.	102.0 [d,t]	126.8 [a]	164.2	178.1	50.2	190.9	214.2	**39 Meubles, autres activ. de fabricat. n.c.a.**
40 Furniture	5.0 [d,t]	2.5 [a]	14.5	15.7	30.7	22.2	17.3	40 Meubles
41 Other manufacturing n.e.c.	97.0 [d,t]	124.3 [a]	149.7	162.4	19.5	168.7	196.9	41 Autres activités de fabrication n.c.a.
42 Recycling	0.0 [d,t]	0.0	0.0	0.0	0.0	30.5	2.9	**42 Récupération**
43 ELECTRICITY, GAS & WATER SUPPLY	9.0 [d,t]	9.4 [a]	4.9	4.9	1.5	17.8	1.3	**43 DISTRIBUTION D'ELECTRICITE, GAZ & EAU**
44 CONSTRUCTION	41.0 [d,t]	63.8 [a]	85.2	100.7	118.9	174.0	191.1	**44 CONSTRUCTION**
45 SERVICES SECTOR	2 997.0 [d,t]	2 004.0 [a]	2 474.2	2 826.8	3 268.4	4 127.3	4 973.8	**45 SECTEUR DES SERVICES**
46 Wholesale, retail trade & repair of motor vehicles, etc.	..	0.0	26.1	75.9	41.8	111.1	141.7	**46 Commerce (gros & détail), réparation de véhicules, etc.**
47 Hotels & restaurants	..	0.0	0.0	3.2	0.7	0.0	0.0	**47 Hôtels et restaurants**
48 Transport & storage	..	0.0	94.4	90.7	104.1	85.7	106.0	**48 Transports, entreposage**
49 Communications	..	0.0	25.6	18.6	8.1	13.1	11.7	**49 Communications**
50 Post	..	0.0	10.4	3.2	7.3	8.7	11.7	50 Postes
51 Telecommunications	..	0.0	15.2	15.4	0.8	4.4	0.0	51 Télécommunications
52 Financial intermediation (inc. insur.)	..	0.0	0.0	0.8	0.0	10.6	5.1	**52 Intermédiation financ. (assurances inc.)**
53 Real estate, renting & busin. activity	2 997.0 [d,t]	2 004.0 [a]	2 327.2	2 635.3	2 881.8	3 279.3	3 988.2	**53 Immobil., locat. & serv. aux entreprises**
54 Computer & related activities	..	0.0	29.1	170.7	180.1	273.8	426.1	54 Informatique et activités connexes
55 *Software consultancy*	..	0.0	24.0	126.8	174.7	247.0	390.2	55 *Production de logiciels*
56 *Other computer services n.e.c.*	..	0.0	5.1	43.9	5.4	26.8	35.9	56 *Autres activités informatiques*
57 Research & development	2 997.0 [d,t]	1 925.4 [a]	2 105.2	2 319.2	2 548.5	2 801.8	3 320.4	57 Recherche & développement expérimental
58 Other business activities n.e.c.	..	78.6 [a]	193.0	145.4	153.2	203.7	241.7	58 Autres activ. & serv. aux entreprises
59 Comm., soc. & pers. serv. activ., etc.	..	0.0	0.9	2.3	231.9	627.5	721.1	**59 Adm. pub., serv. soc. & collec., etc**
TOTAL BUSINESS ENTERPRISE SECTOR	**8 699.0 [d,t]**	**9 098.6 [a]**	**9 747.2**	**12 232.1**	**14 759.8**	**14 862.3**	**15 881.9**	**TOTAL SECTEUR DES ENTREPRISES**

Source: OECD, R&D database (DSTI/EAS Division), January 2002.

Source : OCDE, base de données R-D (DSTI/Division AES), janvier 2002.

Total business enterprise intramural expenditure on R&D - Dépenses intra-muros totales de R-D des entreprises
(BERD) by industry - (DIRDE) par industrie

	1994	1995	1996	1997	1998	1999	2000	
	9.2 Million constant USD(1995 prices and PPPs) - Millions de USD constants (prix et PPA de 1995)							
1 AGRICULTURE, HUNTING & FORESTRY	..	1.9 [a]	4.4	2.9	4.7	4.8	5.1	**1 AGRICULTURE, CHASSE & SYLVICULTURE**
2 MINING	3.8 [d,t]	11.4 [a]	8.7	6.9	6.8	4.5	2.8	**2 ACTIVITES EXTRACTIVES**
3 MANUFACTURING	572.4 [d,t]	636.2 [a]	597.5	722.1	797.1	717.6	724.6	**3 ACTIVITES DE FABRICATION**
4 Food, beverages & tobacco	2.7 [d,t]	2.9 [a]	3.0	4.9	12.3	4.3	6.0	**4 Produits alimentaires, boissons & tabac**
5 Food products & beverages	2.7 [d,t]	2.9 [a]	3.0	4.9	12.3	4.3	6.0	5 Produits alimentaires & boissons
6 Tobacco products	0.0 [d,t]	0.0	0.0	0.0	0.0	0.0	0.0	6 Produits à base de tabac
7 Textiles, wear app., fur & leather	8.1 [d,t]	5.4 [a]	3.5	6.6	10.7	8.4	8.8	**7 Textiles, habill., fourrures & cuirs**
8 Textiles	2.4 [d,t]	2.5 [a]	2.6	5.8	9.8	5.9	5.4	8 Textiles
9 Wearing apparel & fur	0.1 [d,t]	0.2 [a]	0.2	0.1	0.1	1.5	2.4	9 Articles d'habillement & fourrures
10 Leather products & footwear	5.5 [d,t]	2.8 [a]	0.8	0.8	0.9	1.0	1.0	10 Articles en cuir & chaussures
11 Wood, paper, printing, publishing	2.3 [d]	1.9 [a]	2.2	1.1	1.6	0.9	0.5	**11 Bois, papier, imprimerie, édition**
12 Wood & cork (not furniture)	0.1 [d,t]	0.2 [a]	0.2	0.1	0.5	0.1	0.0	12 Bois & liège (sauf meubles)
13 Pulp, paper & paper products	2.2 [d,t]	1.6 [a]	1.8	1.0	0.9	0.7	0.2	13 Carton, papier & articles
14 Publ., print. & repro. of rec. media	0.0 [d,t]	0.0	0.2	0.0	0.2	0.0	0.3	14 Edition, impress. & repro. enregis.
15 Coke, petroleum, nuclear fuel, chemicals, rubber plastics	101.3 [d,t]	95.4 [a]	98.0	86.1	119.3	87.7	91.0	**15 Coke, pétrole, combust. nucl., prod. chimi., caout. & mat. plast.**
16 Coke, ref. petrol. prod. & nuclear fuel	11.2 [d,t]	12.2 [a]	1.6	1.2	1.0	1.1	0.9	16 Coke, raffin. pétr. & combust. nucl.
17 Chemicals & chemical products	60.0 [d,t]	67.2 [a]	75.8	63.8	66.4	68.4	74.7	17 Produits chimiques
18 *Chemical prod.(less pharmaceuticals)*	32.8 [d,t]	50.9 [a]	54.5	50.1	50.8	47.0	43.4	18 *Prod. chimi. (moins prod. pharmaceutiques)*
19 *Pharmaceuticals*	27.2 [d,t]	16.3 [a]	21.3	13.7	15.6	21.5	31.3	19 *Produits pharmaceutiques*
20 Rubber & plastic products	30.1 [d,t]	16.0 [a]	20.6	21.1	51.9	18.2	15.4	20 Caoutchouc et matières plastiques
21 Non-metallic mineral products	13.9 [d,t]	8.7 [a]	8.8	12.0	18.5	14.6	15.9	**21 Produits minéraux non métalliques**
22 Basic metals	53.9 [d,t]	40.4 [a]	43.3	39.2	43.0	30.9	31.6	**22 Produits métallurgiques de base**
23 Basic metals, ferrous	52.8 [d,t]	38.0 [a]	42.9	38.2	41.3	28.9	29.6	23 Produits métallur. de base, ferreux
24 Basic metals, non-ferrous	1.1 [d,t]	2.3 [a]	0.4	1.0	1.7	1.9	2.0	24 Prod. métallur. de base, non ferreux
25 Fabricated metal products	15.3 [d,t]	16.6 [a]	16.5	22.4	17.0	27.9	22.3	**25 Ouvrages en métaux (sauf mach.& mat.)**
26 Machinery equipment, instruments and transport equipment	364.6 [d,t]	453.2 [a]	408.2	535.7	571.2	527.6	533.7	**26 Machines & matériel, instruments & mat. de transport**
27 Machinery n.e.c.	87.2 [d,t]	90.3 [a]	99.7	105.5	99.0	86.8	94.2	27 Machines, n.c.a.
28 Office, account. & computing machinery	0.3 [d,t]	1.0 [a]	0.2	1.9	0.2	0.3	0.5	28 Mach. de bur., compta. & trait. inform.
29 Electrical machinery	21.5 [d,t]	25.2 [a]	17.9	15.8	26.7	25.2	26.9	29 Machines électriques
30 Electro. equip. (radio, TV & commun.)	27.3 [d,t]	23.6 [a]	17.6	22.3	26.8	21.8	25.7	30 Equip. électro. (radio, tv & communic.)
31 *Electro. comp. (inc. semi-conductors)*	21.8 [d,t]	16.9 [a]	10.4	9.1	3.6	3.7	4.8	31 *Compos. électro. (semi-cond. incl.)*
32 *TV, radio & communications equip.*	5.5 [d,t]	6.6 [a]	7.2	13.2	23.2	18.2	21.0	32 *Appareils radio, tv et de communi.*
33 Instruments	14.3 [d,t]	8.6 [a]	6.4	9.3	7.9	19.3	14.0	33 Instruments
34 Motor vehicles	163.3 [d,t]	208.7 [a]	169.7	229.9	292.9	277.9	321.5	34 Véhicules automobiles
35 Other transport equipment	50.7 [d,t]	95.8 [a]	96.7	150.9	117.6	96.3	50.8	35 Autres matériels de transport
36 *Ships*	0.0 [d,t]	0.0	0.0	0.0	0.0	0.0	0.0	36 *Navires*
37 *Aerospace*	34.7 [d,t]	66.1 [a]	66.2	79.7	62.5	53.4	37.5	37 *Aérospatiale*
38 *Other transport equipment n.e.c.*	16.0 [d,t]	29.7 [a]	30.5	71.2	55.1	42.9	13.3	38 *Autres matériels de transport n.c.a.*
39 Furniture, other manufacturing n.e.c.	10.4 [d,t]	11.7 [a]	14.0	14.0	3.6	13.2	14.6	**39 Meubles, autres activ. de fabricat. n.c.a.**
40 Furniture	0.5 [d,t]	0.2 [a]	1.2	1.2	2.2	1.5	1.2	40 Meubles
41 Other manufacturing n.e.c.	9.9 [d,t]	11.5 [a]	12.7	12.8	1.4	11.6	13.5	41 Autres activités de fabrication n.c.a.
42 Recycling	0.0 [d,t]	0.0	0.0	0.0	0.0	2.1	0.2	**42 Récupération**
43 ELECTRICITY, GAS & WATER SUPPLY	0.9 [d,t]	0.9 [a]	0.4	0.4	0.1	1.2	0.1	**43 DISTRIBUTION D'ELECTRICITE, GAZ & EAU**
44 CONSTRUCTION	4.2 [d,t]	5.9 [a]	7.2	7.9	8.5	12.0	13.1	**44 CONSTRUCTION**
45 SERVICES SECTOR	305.5 [d,t]	185.4 [a]	210.4	222.5	232.4	284.5	340.0	**45 SECTEUR DES SERVICES**
46 Wholesale, retail trade & repair of motor vehicles, etc.	..	0.0	2.2	6.0	3.0	7.7	9.7	**46 Commerce (gros & détail), réparation de véhicules, etc.**
47 Hotels & restaurants	..	0.0	0.0	0.3	0.1	0.0	0.0	**47 Hôtels et restaurants**
48 Transport & storage	..	0.0	8.0	7.1	7.4	5.9	7.2	**48 Transports, entreposage**
49 Communications	..	0.0	2.2	1.5	0.6	0.9	0.8	**49 Communications**
50 Post	..	0.0	0.9	0.3	0.5	0.6	0.8	50 Postes
51 Telecommunications	..	0.0	1.3	1.2	0.1	0.3	0.0	51 Télécommunications
52 Financial intermediation (inc. insur.)	..	0.0	0.0	0.1	0.0	0.7	0.3	**52 Intermédiation financ. (assurances inc.)**
53 Real estate, renting & busin. activity	305.5 [d,t]	185.4 [a]	197.9	207.4	204.9	226.1	272.6	**53 Immobil., locat. & serv. aux entreprises**
54 Computer & related activities	..	0.0	2.5	13.4	12.8	18.9	29.1	54 Informatique et activités connexes
55 *Software consultancy*	..	0.0	2.0	10.0	12.4	17.0	26.7	55 *Production de logiciels*
56 *Other computer services n.e.c.*	..	0.0	0.4	3.5	0.4	1.8	2.5	56 *Autres activités informatiques*
57 Research & development	305.5 [d,t]	178.1 [a]	179.0	182.5	181.2	193.2	227.0	57 Recherche & développement expérimental
58 Other business activities n.e.c.	..	7.3 [a]	16.4	11.4	10.9	14.0	16.5	58 Autres activ. & serv. aux entreprises
59 Comm., soc. & pers. serv. activ., etc.	..	0.0	0.1	0.2	16.5	43.3	49.3	**59 Adm. pub., serv. soc. & collec., etc**
TOTAL BUSINESS ENTERPRISE SECTOR	**886.8** [d,t]	**841.6** [a]	**828.7**	**962.6**	**1 049.5**	**1 024.6**	**1 085.6**	**TOTAL SECTEUR DES ENTREPRISES**

Source: OECD, R&D database (DSTI/EAS Division), January 2002.

Source : OCDE, base de données R-D (DSTI/Division AES), janvier 2002.

Business enterprise intramural expenditure on R&D (BERD) - Dépenses intra-muros de R-D des entreprises (DIRDE)

by industry and source of funds - par industrie et par source de financement

Million CZK *Millions CZK*

	1994	1995	1996	1997	1998	1999	2000	
	10.1 Business enterprise - Entreprises							
1 AGRICULTURE, HUNTING & FORESTRY	..	6.8	19.4	14.4	22.1	22.2	..	**1 AGRICULTURE, CHASSE & SYLVICULTURE**
2 MINING	..	122.6	102.9	87.3	95.4	63.9	..	**2 ACTIVITES EXTRACTIVES**
3 MANUFACTURING	..	6 412.2	6 584.6	8 563.0	10 429.2	9 275.1	..	**3 ACTIVITES DE FABRICATION**
4 Food, beverages & tobacco	..	20.1	28.9	55.0	167.7	58.2	..	**4 Produits alimentaires, boissons & tabac**
5 Food products & beverages	..	20.1	28.9	55.0	167.7	58.2	..	5 Produits alimentaires & boissons
6 Tobacco products	..	0.0	0.0	0.0	0.0	0.0	..	6 Produits à base de tabac
7 Textiles, wear app., fur & leather	..	58.8	41.2	81.7	147.6	117.7	..	**7 Textiles, habill., fourrures & cuirs**
8 Textiles	..	26.5	30.1	70.4	134.0	81.7	..	8 Textiles
9 Wearing apparel & fur	..	1.9	2.0	1.7	1.6	21.9	..	9 Articles d'habillement & fourrures
10 Leather products & footwear	..	30.4	9.1	9.6	12.0	14.1	..	10 Articles en cuir & chaussures
11 Wood, paper, printing, publishing	..	20.1	25.8	13.4	22.6	11.4	..	**11 Bois, papier, imprimerie, édition**
12 Wood & cork (not furniture)	..	2.5	2.1	0.9	7.1	0.7	..	12 Bois & liège (sauf meubles)
13 Pulp, paper & paper products	..	17.6	21.5	12.3	12.5	10.7	..	13 Carton, papier & articles
14 Publ., print. & repro. of rec. media	..	0.0	2.2	0.2	3.0	0.0	..	14 Edition, impress. & repro. enregis.
15 Coke, petroleum, nuclear fuel, chemicals, rubber plastics	..	992.7	1 105.4	1 050.0	1 599.0	1 182.8	..	**15 Coke, pétrole, combust. nucl., prod. chimi., caout. & mat. plast.**
16 Coke, ref. petrol. prod. & nuclear fuel	..	131.7	18.4	15.6	13.7	15.5	..	16 Coke, raffin. pétr. & combust. nucl.
17 Chemicals & chemical products	..	687.8	844.2	769.9	857.3	912.4	..	17 Produits chimiques
18 *Chemical prod.(less pharmaceuticals)*	..	511.5	595.7	598.1	639.8	609.2	..	18 *Prod. chimi. (moins prod. pharmaceutiques)*
19 *Pharmaceuticals*	..	176.3	248.5	171.8	217.5	303.2	..	19 *Produits pharmaceutiques*
20 Rubber & plastic products	..	173.2	242.8	264.4	728.0	254.9	..	20 Caoutchouc et matières plastiques
21 Non-metallic mineral products	..	90.8	100.4	96.0	246.1	204.1	..	**21 Produits minéraux non métalliques**
22 Basic metals	..	426.6	492.9	483.1	578.7	416.7	..	**22 Produits métallurgiques de base**
23 Basic metals, ferrous	..	401.4	488.2	471.9	557.7	390.9	..	23 Produits métallur. de base, ferreux
24 Basic metals, non-ferrous	..	25.2	4.7	11.2	21.0	25.8	..	24 Prod. métallur. de base, non ferreux
25 Fabricated metal products	..	169.8	192.1	269.6	223.8	328.2	..	**25 Ouvrages en métaux (sauf mach.& mat.)**
26 Machinery equipment, instruments and transport equipment	..	4 506.5	4 440.1	6 339.7	7 396.0	6 743.1	..	**26 Machines & matériel, instruments & mat. de transport**
27 Machinery n.e.c.	..	930.5	1 021.4	1 205.6	1 284.9	1 146.3	..	27 Machines, n.c.a.
28 Office, account. & computing machinery	..	10.8	1.8	24.6	3.0	3.9	..	28 Mach. de bur., compta. & trait. inform.
29 Electrical machinery	..	268.2	202.8	189.4	335.5	324.5	..	29 Machines électriques
30 Electro. equip. (radio, TV & commun.)	..	154.6	148.6	241.6	362.9	294.8	..	30 Equip. électro. (radio, tv & communic.)
31 *Electro. comp. (inc. semi-conductors)*	..	112.4	84.3	73.9	49.3	50.1	..	31 *Compos. électro. (semi-cond. incl.)*
32 *TV, radio & communications equip.*	..	42.2	64.3	167.7	313.6	244.7	..	32 *Appareils radio, tv et de communi.*
33 Instruments	..	66.4	65.7	102.9	105.1	243.2	..	33 Instruments
34 Motor vehicles	..	2 077.7	1 995.8	2 915.0	4 115.9	3 798.3	..	34 Véhicules automobiles
35 Other transport equipment	..	998.3	1 004.0	1 660.6	1 188.7	932.1	..	35 Autres matériels de transport
36 *Ships*	..	0.0	0.0	0.0	0.0	0.0	..	36 *Navires*
37 *Aerospace*	..	686.5	707.6	871.3	582.6	425.3	..	37 *Aérospatiale*
38 *Other transport equipment n.e.c.*	..	311.8	296.4	789.3	606.1	506.8	..	38 *Autres matériels de transport n.c.a.*
39 Furniture, other manufacturing n.e.c.	..	126.8	157.8	174.7	47.7	188.9	..	**39 Meubles, autres activ. de fabricat. n.c.a.**
40 Furniture	..	2.5	10.5	14.7	28.2	20.3	..	40 Meubles
41 Other manufacturing n.e.c.	..	124.3	147.3	160.0	19.5	168.6	..	41 Autres activités de fabrication n.c.a.
42 Recycling	..	0.0	0.0	0.0	0.0	24.0	..	**42 Récupération**
43 ELECTRICITY, GAS & WATER SUPPLY	..	9.4	4.9	3.6	0.1	17.8	..	**43 DISTRIBUTION D'ELECTRICITE, GAZ & EAU**
44 CONSTRUCTION	..	63.8	82.6	89.9	109.7	167.9	..	**44 CONSTRUCTION**
45 SERVICES SECTOR	..	1 769.7	2 045.7	2 337.6	2 540.5	2 396.9	..	**45 SECTEUR DES SERVICES**
46 Wholesale, retail trade & repair of motor vehicles, etc.	..	0.0	7.8	22.7	27.5	57.3	..	**46 Commerce (gros & détail), réparation de véhicules, etc.**
47 Hotels & restaurants	..	0.0	0.0	1.0	0.7	0.0	..	**47 Hôtels et restaurants**
48 Transport & storage	..	0.0	90.7	84.6	97.7	79.9	..	**48 Transports, entreposage**
49 Communications	..	0.0	25.4	18.0	7.3	12.0	..	**49 Communications**
50 Post	..	0.0	10.4	3.2	7.3	8.7	..	50 Postes
51 Telecommunications	..	0.0	15.0	14.8	0.0	3.3	..	51 Télécommunications
52 Financial intermediation (inc. insur.)	..	0.0	0.0	0.2	0.0	9.5	..	**52 Intermédiation financ. (assurances inc.)**
53 Real estate, renting & busin. activity	..	1 769.7	1 921.8	2 210.8	2 384.2	2 221.7	..	**53 Immobil., locat. & serv. aux entreprises**
54 Computer & related activities	..	0.0	25.3	164.4	171.3	245.6	..	54 Informatique et activités connexes
55 *Software consultancy*	..	0.0	24.0	125.6	170.0	231.7	..	55 *Production de logiciels*
56 *Other computer services n.e.c.*	..	0.0	1.3	38.8	1.3	13.9	..	56 *Autres activités informatiques*
57 Research & development	..	1 699.9	1 726.3	1 924.1	2 111.1	1 852.9	..	57 Recherche & développement expérimental
58 Other business activities n.e.c.	..	69.8	170.3	122.2	101.8	123.1	..	58 Autres activ. & serv. aux entreprises
59 Comm., soc. & pers. serv. activ., etc.	..	0.0	0.0	0.3	23.1	16.5	..	**59 Adm. pub., serv. soc. & collec., etc**
TOTAL BUSINESS ENTERPRISE SECTOR	**..**	**8 384.5**	**8 840.1**	**11 095.7**	**13 197.0**	**11 943.8**	**12 807.5**	**TOTAL SECTEUR DES ENTREPRISES**

Source: OECD, R&D database (DSTI/EAS Division), January 2002.

Source : OCDE, base de données R-D (DSTI/Division AES), janvier 2002.

Business enterprise intramural expenditure on R&D (BERD) - Dépenses intra-muros de R-D des entreprises (DIRDE)

by industry and source of funds - par industrie et par source de financement

Million CZK — *Millions CZK*

10.2 Government - État

	1994	1995	1996	1997	1998	1999	2000	
1 AGRICULTURE, HUNTING & FORESTRY	..	13.2	32.8	22.6	43.3	47.8	..	**1 AGRICULTURE, CHASSE & SYLVICULTURE**
2 MINING	..	1.0	0.0	0.0	0.0	1.2	..	**2 ACTIVITES EXTRACTIVES**
3 MANUFACTURING	..	228.5	404.1	514.7	485.6	555.4	..	**3 ACTIVITES DE FABRICATION**
4 Food, beverages & tobacco	..	4.5	5.8	7.4	5.1	4.8	..	**4 Produits alimentaires, boissons & tabac**
5 Food products & beverages	..	4.5	5.8	7.4	5.1	4.8	..	5 Produits alimentaires & boissons
6 Tobacco products	..	0.0	0.0	0.0	0.0	0.0	..	6 Produits à base de tabac
7 Textiles, wear app., fur & leather	..	0.0	0.0	2.7	3.1	3.2	..	**7 Textiles, habill., fourrures & cuirs**
8 Textiles	..	0.0	0.0	2.7	3.1	3.2	..	8 Textiles
9 Wearing apparel & fur	..	0.0	0.0	0.0	0.0	0.0	..	9 Articles d'habillement & fourrures
10 Leather products & footwear	..	0.0	0.0	0.0	0.0	0.0	..	10 Articles en cuir & chaussures
11 Wood, paper, printing, publishing	..	0.0	0.0	0.0	0.0	1.1	..	**11 Bois, papier, imprimerie, édition**
12 Wood & cork (not furniture)	..	0.0	0.0	0.0	0.0	0.7	..	12 Bois & liège (sauf meubles)
13 Pulp, paper & paper products	..	0.0	0.0	0.0	0.0	0.0	..	13 Carton, papier & articles
14 Publ., print. & repro. of rec. media	..	0.0	0.0	0.0	0.0	0.4	..	14 Edition, impress. & repro. enregis.
15 Coke, petroleum, nuclear fuel, chemicals, rubber plastics	..	28.1	39.0	41.5	77.2	86.6	..	**15 Coke, pétrole, combust. nucl., prod. chimi., caout. & mat. plast.**
16 Coke, ref. petrol. prod. & nuclear fuel	..	0.0	0.0	0.0	0.0	0.0	..	16 Coke, raffin. pétr. & combust. nucl.
17 Chemicals & chemical products	..	28.1	39.0	38.1	75.6	77.4	..	17 Produits chimiques
18 *Chemical prod.(less pharmaceuticals)*	..	28.1	36.7	36.2	73.1	69.2	..	18 *Prod. chimi. (moins prod. pharmaceutiques)*
19 *Pharmaceuticals*	..	0.0	2.3	1.9	2.5	8.2	..	19 *Produits pharmaceutiques*
20 Rubber & plastic products	..	0.0	0.0	3.5	1.6	9.2	..	20 Caoutchouc et matières plastiques
21 Non-metallic mineral products	..	3.6	3.0	16.0	0.7	8.0	..	**21 Produits minéraux non métalliques**
22 Basic metals	..	3.5	1.1	8.5	19.1	29.9	..	**22 Produits métallurgiques de base**
23 Basic metals, ferrous	..	3.5	1.1	7.1	16.8	28.0	..	23 Produits métallur. de base, ferreux
24 Basic metals, non-ferrous	..	0.0	0.0	1.4	2.3	1.9	..	24 Prod. métallur. de base, non ferreux
25 Fabricated metal products	..	0.0	1.6	15.1	14.8	71.9	..	**25 Ouvrages en métaux (sauf mach.& mat.)**
26 Machinery equipment, instruments and transport equipment	..	188.8	349.6	422.5	363.1	341.4	..	**26 Machines & matériel, instruments & mat. de transport**
27 Machinery n.e.c.	..	40.1	146.3	131.3	101.8	107.3	..	27 Machines, n.c.a.
28 Office, account. & computing machinery	..	0.0	0.0	0.0	0.0	0.0	..	28 Mach. de bur., compta. & trait. inform.
29 Electrical machinery	..	0.0	2.6	7.9	28.8	31.1	..	29 Machines électriques
30 Electro. equip. (radio, TV & commun.)	..	99.8	57.7	39.9	13.7	19.8	..	30 Equip. électro. (radio, tv & communic.)
31 *Electro. comp. (inc. semi-conductors)*	..	70.3	37.7	39.9	0.5	0.6	..	31 *Compos. électro. (semi-cond. incl.)*
32 *TV, radio & communications equip.*	..	29.5	20.0	0.0	13.2	19.2	..	32 *Appareils radio, tv et de communi.*
33 Instruments	..	20.2	9.4	14.8	6.1	36.6	..	33 Instruments
34 Motor vehicles	..	0.9	0.0	6.7	3.5	6.1	..	34 Véhicules automobiles
35 Other transport equipment	..	27.8	133.6	221.9	209.3	140.5	..	35 Autres matériels de transport
36 *Ships*	..	0.0	0.0	0.0	0.0	0.0	..	36 *Navires*
37 *Aerospace*	..	27.8	71.5	106.7	40.0	25.2	..	37 *Aérospatiale*
38 *Other transport equipment n.e.c.*	..	0.0	62.1	115.2	169.3	115.3	..	38 *Autres matériels de transport n.c.a.*
39 Furniture, other manufacturing n.e.c.	..	0.0	4.0	1.0	2.5	2.0	..	**39 Meubles, autres activ. de fabricat. n.c.a.**
40 Furniture	..	0.0	4.0	1.0	2.5	1.9	..	40 Meubles
41 Other manufacturing n.e.c.	..	0.0	0.0	0.0	0.0	0.1	..	41 Autres activités de fabrication n.c.a.
42 Recycling	..	0.0	0.0	0.0	0.0	6.5	..	**42 Récupération**
43 ELECTRICITY, GAS & WATER SUPPLY	..	0.0	0.0	1.3	1.4	0.0	..	**43 DISTRIBUTION D'ELECTRICITE, GAZ & EAU**
44 CONSTRUCTION	..	0.0	2.6	10.9	9.2	6.1	..	**44 CONSTRUCTION**
45 SERVICES SECTOR	..	164.0	275.1	412.8	668.3	1 491.3	..	**45 SECTEUR DES SERVICES**
46 Wholesale, retail trade & repair of motor vehicles, etc.	..	0.0	17.9	48.2	14.3	52.3	..	**46 Commerce (gros & détail), réparation de véhicules, etc.**
47 Hotels & restaurants	..	0.0	0.0	2.2	0.0	0.0	..	**47 Hôtels et restaurants**
48 Transport & storage	..	0.0	3.7	6.2	6.4	5.8	..	**48 Transports, entreposage**
49 Communications	..	0.0	0.0	0.1	0.0	0.0	..	**49 Communications**
50 Post	..	0.0	0.0	0.0	0.0	0.0	..	50 Postes
51 Telecommunications	..	0.0	0.0	0.1	0.0	0.0	..	51 Télécommunications
52 Financial intermediation (inc. insur.)	..	0.0	0.0	0.5	0.0	1.1	..	**52 Intermédiation financ. (assurances inc.)**
53 Real estate, renting & busin. activity	..	164.0	252.6	354.0	439.1	916.2	..	**53 Immobil., locat. & serv. aux entreprises**
54 Computer & related activities	..	0.0	3.8	6.2	5.1	25.2	..	54 Informatique et activités connexes
55 *Software consultancy*	..	0.0	0.0	1.1	1.0	12.3	..	55 *Production de logiciels*
56 *Other computer services n.e.c.*	..	0.0	3.8	5.1	4.1	12.9	..	56 *Autres activités informatiques*
57 Research & development	..	162.3	226.1	326.0	382.6	817.6	..	57 Recherche & développement expérimental
58 Other business activities n.e.c.	..	1.7	22.7	21.8	51.4	73.4	..	58 Autres activ. & serv. aux entreprises
59 Comm., soc. & pers. serv. activ., etc.	..	0.0	0.9	1.7	208.5	515.9	..	**59 Adm. pub., serv. soc. & collec., etc**
TOTAL BUSINESS ENTERPRISE SECTOR	**260.0** [d,t]	**406.7** [a]	**714.6**	**962.3**	**1 207.8**	**2 101.8**	**2 341.0**	**TOTAL SECTEUR DES ENTREPRISES**

Source: OECD, R&D database (DSTI/EAS Division), January 2002.

Source : OCDE, base de données R-D (DSTI/Division AES), janvier 2002.

CZECH REPUBLIC — Table 10 - Tableau 10 — RÉPUBLIQUE TCHEQUE

Business enterprise intramural expenditure on R&D (BERD) - Dépenses intra-muros de R-D des entreprises (DIRDE)

by industry and source of funds - par industrie et par source de financement

Million CZK — *Millions CZK*

	1994	1995	1996	1997	1998	1999	2000	
	10.3 Other national funds - Autres sources nationales							
1 AGRICULTURE, HUNTING & FORESTRY	..	0.0	0.1	0.0	0.0	0.0	..	**1 AGRICULTURE, CHASSE & SYLVICULTURE**
2 MINING	..	0.0	0.0	0.0	0.0	0.0	..	**2 ACTIVITES EXTRACTIVES**
3 MANUFACTURING	..	0.6	1.1	0.5	5.7	0.5	..	**3 ACTIVITES DE FABRICATION**
4 Food, beverages & tobacco	..	0.0	0.3	0.1	0.0	0.0	..	**4 Produits alimentaires, boissons & tabac**
5 Food products & beverages	..	0.0	0.3	0.1	0.0	0.0	..	5 Produits alimentaires & boissons
6 Tobacco products	..	0.0	0.0	0.0	0.0	0.0	..	6 Produits à base de tabac
7 Textiles, wear app., fur & leather	..	0.0	0.0	0.0	0.0	0.0	..	**7 Textiles, habill., fourrures & cuirs**
8 Textiles	..	0.0	0.0	0.0	0.0	0.0	..	8 Textiles
9 Wearing apparel & fur	..	0.0	0.0	0.0	0.0	0.0	..	9 Articles d'habillement & fourrures
10 Leather products & footwear	..	0.0	0.0	0.0	0.0	0.0	..	10 Articles en cuir & chaussures
11 Wood, paper, printing, publishing	..	0.0	0.0	0.0	0.0	0.0	..	**11 Bois, papier, imprimerie, édition**
12 Wood & cork (not furniture)	..	0.0	0.0	0.0	0.0	0.0	..	12 Bois & liège (sauf meubles)
13 Pulp, paper & paper products	..	0.0	0.0	0.0	0.0	0.0	..	13 Carton, papier & articles
14 Publ., print. & repro. of rec. media	..	0.0	0.0	0.0	0.0	0.0	..	14 Edition, impress. & repro. enregis.
15 Coke, petroleum, nuclear fuel, chemicals, rubber plastics	..	0.0	0.0	0.0	0.0	0.4	..	**15 Coke, pétrole, combust. nucl., prod. chimi., caout. & mat. plast.**
16 Coke, ref. petrol. prod. & nuclear fuel	..	0.0	0.0	0.0	0.0	0.0	..	16 Coke, raffin. pétr. & combust. nucl.
17 Chemicals & chemical products	..	0.0	0.0	0.0	0.0	0.4	..	17 Produits chimiques
18 *Chemical prod.(less pharmaceuticals)*	..	0.0	0.0	0.0	0.0	0.4	..	18 *Prod. chimi. (moins prod. pharmaceutiques)*
19 *Pharmaceuticals*	..	0.0	0.0	0.0	0.0	0.0	..	19 *Produits pharmaceutiques*
20 Rubber & plastic products	..	0.0	0.0	0.0	0.0	0.0	..	20 Caoutchouc et matières plastiques
21 Non-metallic mineral products	..	0.0	0.0	0.0	0.0	0.0	..	**21 Produits minéraux non métalliques**
22 Basic metals	..	0.4	0.6	0.0	0.0	0.0	..	**22 Produits métallurgiques de base**
23 Basic metals, ferrous	..	0.4	0.6	0.0	0.0	0.0	..	23 Produits métallur. de base, ferreux
24 Basic metals, non-ferrous	..	0.0	0.0	0.0	0.0	0.0	..	24 Prod. métallur. de base, non ferreux
25 Fabricated metal products	..	0.0	0.0	0.0	0.0	0.0	..	**25 Ouvrages en métaux (sauf mach.& mat.)**
26 Machinery equipment, instruments and transport equipment	..	0.2	0.2	0.3	5.7	0.1	..	**26 Machines & matériel, instruments & mat. de transport**
27 Machinery n.e.c.	..	0.2	0.1	0.3	5.7	0.0	..	27 Machines, n.c.a.
28 Office, account. & computing machinery	..	0.0	0.0	0.0	0.0	0.0	..	28 Mach. de bur., compta. & trait. inform.
29 Electrical machinery	..	0.0	0.0	0.0	0.0	0.0	..	29 Machines électriques
30 Electro. equip. (radio, TV & commun.)	..	0.0	0.0	0.0	0.0	0.0	..	30 Equip. électro. (radio, tv & communic.)
31 *Electro. comp. (inc. semi-conductors)*	..	0.0	0.0	0.0	0.0	0.0	..	31 *Compos. électro. (semi-cond. incl.)*
32 *TV, radio & communications equip.*	..	0.0	0.0	0.0	0.0	0.0	..	32 *Appareils radio, tv et de communi.*
33 Instruments	..	0.0	0.1	0.0	0.0	0.1	..	33 Instruments
34 Motor vehicles	..	0.0	0.0	0.0	0.0	0.0	..	34 Véhicules automobiles
35 Other transport equipment	..	0.0	0.0	0.0	0.0	0.0	..	35 Autres matériels de transport
36 *Ships*	..	0.0	0.0	0.0	0.0	0.0	..	36 *Navires*
37 *Aerospace*	..	0.0	0.0	0.0	0.0	0.0	..	37 *Aérospatiale*
38 *Other transport equipment n.e.c.*	..	0.0	0.0	0.0	0.0	0.0	..	38 *Autres matériels de transport n.c.a.*
39 Furniture, other manufacturing n.e.c.	..	0.0	0.0	0.0	0.0	0.0	..	**39 Meubles, autres activ. de fabricat. n.c.a.**
40 Furniture	..	0.0	0.0	0.0	0.0	0.0	..	40 Meubles
41 Other manufacturing n.e.c.	..	0.0	0.0	0.0	0.0	0.0	..	41 Autres activités de fabrication n.c.a.
42 Recycling	..	0.0	0.0	0.0	0.0	0.0	..	**42 Récupération**
43 ELECTRICITY, GAS & WATER SUPPLY	..	0.0	0.0	0.0	0.0	0.0	..	**43 DISTRIBUTION D'ELECTRICITE, GAZ & EAU**
44 CONSTRUCTION	..	0.0	0.0	0.0	0.0	0.0	..	**44 CONSTRUCTION**
45 SERVICES SECTOR	..	20.2	24.3	6.4	4.0	97.9	..	**45 SECTEUR DES SERVICES**
46 Wholesale, retail trade & repair of motor vehicles, etc.	..	0.0	0.0	2.0	0.0	0.0	..	**46 Commerce (gros & détail), réparation de véhicules, etc.**
47 Hotels & restaurants	..	0.0	0.0	0.0	0.0	0.0	..	**47 Hôtels et restaurants**
48 Transport & storage	..	0.0	0.0	0.0	0.0	0.0	..	**48 Transports, entreposage**
49 Communications	..	0.0	0.0	0.0	0.0	0.0	..	**49 Communications**
50 Post	..	0.0	0.0	0.0	0.0	0.0	..	50 Postes
51 Telecommunications	..	0.0	0.0	0.0	0.0	0.0	..	51 Télécommunications
52 Financial intermediation (inc. insur.)	..	0.0	0.0	0.0	0.0	0.0	..	**52 Intermédiation financ. (assurances inc.)**
53 Real estate, renting & busin. activity	..	20.2	24.3	4.0	3.7	3.7	..	**53 Immobil., locat. & serv. aux entreprises**
54 Computer & related activities	..	0.0	0.0	0.0	3.7	3.0	..	54 Informatique et activités connexes
55 *Software consultancy*	..	0.0	0.0	0.0	3.7	3.0	..	55 *Production de logiciels*
56 *Other computer services n.e.c.*	..	0.0	0.0	0.0	0.0	0.0	..	56 *Autres activités informatiques*
57 Research & development	..	20.2	24.3	2.8	0.0	0.7	..	57 Recherche & développement expérimental
58 Other business activities n.e.c.	..	0.0	0.0	1.3	0.0	0.0	..	58 Autres activ. & serv. aux entreprises
59 Comm., soc. & pers. serv. activ., etc.	..	0.0	0.0	0.3	0.3	94.2	..	**59 Adm. pub., serv. soc. & collec., etc**
TOTAL BUSINESS ENTERPRISE SECTOR	**..**	**20.8**	**25.5**	**6.9**	**9.7**	**98.4**	**159.7**	**TOTAL SECTEUR DES ENTREPRISES**

Source: OECD, R&D database (DSTI/EAS Division), January 2002.

Source : OCDE, base de données R-D (DSTI/Division AES), janvier 2002.

Business enterprise intramural expenditure on R&D (BERD) - Dépenses intra-muros de R-D des entreprises (DIRDE)

by industry and source of funds - par industrie et par source de financement

Million CZK *Millions CZK*

10.4 Funds from abroad - Fonds de l'étranger

	1994	1995	1996	1997	1998	1999	2000	
1 AGRICULTURE, HUNTING & FORESTRY	..	0.0	0.0	0.0	0.0	0.0	..	**1 AGRICULTURE, CHASSE & SYLVICULTURE**
2 MINING	..	0.0	0.0	0.0	0.0	0.0	..	**2 ACTIVITES EXTRACTIVES**
3 MANUFACTURING	..	236.4	37.9	97.2	289.7	577.1	..	**3 ACTIVITES DE FABRICATION**
4 Food, beverages & tobacco	..	7.2	0.5	0.2	0.0	0.0	..	**4 Produits alimentaires, boissons & tabac**
5 Food products & beverages	..	7.2	0.5	0.2	0.0	0.0	..	5 Produits alimentaires & boissons
6 Tobacco products	..	0.0	0.0	0.0	0.0	0.0	..	6 Produits à base de tabac
7 Textiles, wear app., fur & leather	..	0.0	0.0	0.0	0.1	0.7	..	**7 Textiles, habill., fourrures & cuirs**
8 Textiles	..	0.0	0.0	0.0	0.1	0.7	..	8 Textiles
9 Wearing apparel & fur	..	0.0	0.0	0.0	0.0	0.0	..	9 Articles d'habillement & fourrures
10 Leather products & footwear	..	0.0	0.0	0.0	0.0	0.0	..	10 Articles en cuir & chaussures
11 Wood, paper, printing, publishing	..	0.0	0.0	0.1	0.0	0.0	..	**11 Bois, papier, imprimerie, édition**
12 Wood & cork (not furniture)	..	0.0	0.0	0.0	0.0	0.0	..	12 Bois & liège (sauf meubles)
13 Pulp, paper & paper products	..	0.0	0.0	0.1	0.0	0.0	..	13 Carton, papier & articles
14 Publ., print. & repro. of rec. media	..	0.0	0.0	0.0	0.0	0.0	..	14 Edition, impress. & repro. enregis.
15 Coke, petroleum, nuclear fuel, chemicals, rubber plastics	..	10.2	8.5	2.5	1.1	2.6	..	**15 Coke, pétrole, combust. nucl., prod. chimi., caout. & mat. plast.**
16 Coke, ref. petrol. prod. & nuclear fuel	..	0.0	0.0	0.0	0.0	0.0	..	16 Coke, raffin. pétr. & combust. nucl.
17 Chemicals & chemical products	..	10.2	8.5	2.5	1.1	2.6	..	17 Produits chimiques
18 *Chemical prod.(less pharmaceuticals)*	..	10.2	8.5	2.5	1.1	2.6	..	18 *Prod. chimi. (moins prod. pharmaceutiques)*
19 *Pharmaceuticals*	..	0.0	0.0	0.0	0.0	0.0	..	19 *Produits pharmaceutiques*
20 Rubber & plastic products	..	0.0	0.0	0.0	0.0	0.0	..	20 Caoutchouc et matières plastiques
21 Non-metallic mineral products	..	0.0	0.0	40.5	13.0	0.0	..	**21 Produits minéraux non métalliques**
22 Basic metals	..	6.0	14.6	6.4	6.6	0.9	..	**22 Produits métallurgiques de base**
23 Basic metals, ferrous	..	6.0	14.6	6.4	6.6	0.9	..	23 Produits métallur. de base, ferreux
24 Basic metals, non-ferrous	..	0.0	0.0	0.0	0.0	0.0	..	24 Prod. métallur. de base, non ferreux
25 Fabricated metal products	..	9.2	0.7	0.0	0.0	4.7	..	**25 Ouvrages en métaux (sauf mach.& mat.)**
26 Machinery equipment, instruments and transport equipment	..	203.8	11.2	45.0	268.9	568.2	..	**26 Machines & matériel, instruments & mat. de transport**
27 Machinery n.e.c.	..	5.7	5.0	3.7	0.5	5.5	..	27 Machines, n.c.a.
28 Office, account. & computing machinery	..	0.0	0.0	0.0	0.0	0.0	..	28 Mach. de bur., compta. & trait. inform.
29 Electrical machinery	..	4.5	5.1	3.1	11.6	9.5	..	29 Machines électriques
30 Electro. equip. (radio, TV & commun.)	..	0.4	0.8	2.3	0.6	2.3	..	30 Equip. électro. (radio, tv & communic.)
31 *Electro. comp. (inc. semi-conductors)*	..	0.4	0.8	2.3	0.6	2.3	..	31 *Compos. électro. (semi-cond. incl.)*
32 *TV, radio & communications equip.*	..	0.0	0.0	0.0	0.0	0.0	..	32 *Appareils radio, tv et de communi.*
33 Instruments	..	6.5	0.3	0.7	0.2	0.0	..	33 Instruments
34 Motor vehicles	..	177.4	0.0	0.0	0.0	226.5	..	34 Véhicules automobiles
35 Other transport equipment	..	9.3	0.0	35.2	256.0	324.4	..	35 Autres matériels de transport
36 *Ships*	..	0.0	0.0	0.0	0.0	0.0	..	36 *Navires*
37 *Aerospace*	..	0.0	0.0	35.2	256.0	324.4	..	37 *Aérospatiale*
38 *Other transport equipment n.e.c.*	..	9.3	0.0	0.0	0.0	0.0	..	38 *Autres matériels de transport n.c.a.*
39 Furniture, other manufacturing n.e.c.	..	0.0	2.4	2.4	0.0	0.0	..	**39 Meubles, autres activ. de fabricat. n.c.a.**
40 Furniture	..	0.0	0.0	0.0	0.0	0.0	..	40 Meubles
41 Other manufacturing n.e.c.	..	0.0	2.4	2.4	0.0	0.0	..	41 Autres activités de fabrication n.c.a.
42 Recycling	..	0.0	0.0	0.0	0.0	0.0	..	**42 Récupération**
43 ELECTRICITY, GAS & WATER SUPPLY	..	0.0	0.0	0.0	0.0	0.0	..	**43 DISTRIBUTION D'ELECTRICITE, GAZ & EAU**
44 CONSTRUCTION	..	0.0	0.0	0.0	0.0	0.0	..	**44 CONSTRUCTION**
45 SERVICES SECTOR	..	50.2	129.1	70.0	55.6	141.2	..	**45 SECTEUR DES SERVICES**
46 Wholesale, retail trade & repair of motor vehicles, etc.	..	0.0	0.4	3.0	0.0	1.5	..	**46 Commerce (gros & détail), réparation de véhicules, etc.**
47 Hotels & restaurants	..	0.0	0.0	0.0	0.0	0.0	..	**47 Hôtels et restaurants**
48 Transport & storage	..	0.0	0.0	0.0	0.0	0.0	..	**48 Transports, entreposage**
49 Communications	..	0.0	0.2	0.5	0.8	1.1	..	**49 Communications**
50 Post	..	0.0	0.0	0.0	0.0	0.0	..	50 Postes
51 Telecommunications	..	0.0	0.2	0.5	0.8	1.1	..	51 Télécommunications
52 Financial intermediation (inc. insur.)	..	0.0	0.0	0.0	0.0	0.0	..	**52 Intermédiation financ. (assurances inc.)**
53 Real estate, renting & busin. activity	..	50.2	128.5	66.5	54.8	137.7	..	**53 Immobil., locat. & serv. aux entreprises**
54 Computer & related activities	..	0.0	0.0	0.0	0.0	0.0	..	54 Informatique et activités connexes
55 *Software consultancy*	..	0.0	0.0	0.0	0.0	0.0	..	55 *Production de logiciels*
56 *Other computer services n.e.c.*	..	0.0	0.0	0.0	0.0	0.0	..	56 *Autres activités informatiques*
57 Research & development	..	43.2	128.5	66.4	54.8	130.6	..	57 Recherche & développement expérimental
58 Other business activities n.e.c.	..	7.0	0.0	0.1	0.0	7.2	..	58 Autres activ. & serv. aux entreprises
59 Comm., soc. & pers. serv. activ., etc.	..	0.0	0.0	0.0	0.0	0.9	..	**59 Adm. pub., serv. soc. & collec., etc**
TOTAL BUSINESS ENTERPRISE SECTOR	..	**286.6**	**167.0**	**167.2**	**345.3**	**718.3**	**573.7**	**TOTAL SECTEUR DES ENTREPRISES**

Source: OECD, R&D database (DSTI/EAS Division), January 2002.

Source : OCDE, base de données R-D (DSTI/Division AES), janvier 2002.

Business enterprise intramural expenditure on R&D (BERD) - Dépenses intra-muros de R-D des entreprises (DIRDE)

by industry and type of costs - par industrie et par type de coûts

Million CZK *Millions CZK*

	1994	1995	1996	1997	1998	1999	2000	
			11.1 Current expenditure - Dépenses courantes					
1 AGRICULTURE, HUNTING & FORESTRY	..	20.0	51.0	34.8	62.1	64.9	..	**1 AGRICULTURE, CHASSE & SYLVICULTURE**
2 MINING	..	102.0	62.8	49.8	50.6	34.6	..	**2 ACTIVITES EXTRACTIVES**
3 MANUFACTURING	..	6 278.2	6 422.8	8 105.6	10 173.4	9 411.0	..	**3 ACTIVITES DE FABRICATION**
4 Food, beverages & tobacco	..	26.0	27.3	57.6	85.1	61.7	..	**4 Produits alimentaires, boissons & tabac**
5 Food products & beverages	..	26.0	27.3	57.6	85.1	61.7	..	5 Produits alimentaires & boissons
6 Tobacco products	..	0.0	0.0	0.0	0.0	0.0	..	6 Produits à base de tabac
7 Textiles, wear app., fur & leather	..	57.0	40.9	54.9	110.3	113.6	..	**7 Textiles, habill., fourrures & cuirs**
8 Textiles	..	25.1	30.1	43.7	97.5	77.7	..	8 Textiles
9 Wearing apparel & fur	..	1.9	2.0	1.6	1.6	21.9	..	9 Articles d'habillement & fourrures
10 Leather products & footwear	..	30.0	8.8	9.6	11.2	14.0	..	10 Articles en cuir & chaussures
11 Wood, paper, printing, publishing	..	16.7	23.6	12.9	15.7	12.4	..	**11 Bois, papier, imprimerie, édition**
12 Wood & cork (not furniture)	..	2.5	1.7	0.9	1.6	1.4	..	12 Bois & liège (sauf meubles)
13 Pulp, paper & paper products	..	14.2	19.7	12.0	12.4	10.6	..	13 Carton, papier & articles
14 Publ., print. & repro. of rec. media	..	0.0	2.2	0.1	1.7	0.4	..	14 Edition, impress. & repro. enregis.
15 Coke, petroleum, nuclear fuel, chemicals, rubber plastics	..	844.2	996.8	943.4	1 594.9	1 143.5	..	**15 Coke, pétrole, combust. nucl., prod. chimi., caout. & mat. plast.**
16 Coke, ref. petrol. prod. & nuclear fuel	..	113.5	16.6	14.3	12.5	12.6	..	16 Coke, raffin. pétr. & combust. nucl.
17 Chemicals & chemical products	..	570.9	743.0	697.3	864.6	891.1	..	17 Produits chimiques
18 *Chemical prod.(less pharmaceuticals)*	..	413.6	535.3	545.2	657.5	595.2	..	18 *Prod. chimi. (moins prod. pharmaceutiques)*
19 *Pharmaceuticals*	..	157.3	207.7	152.1	207.1	295.9	..	19 *Produits pharmaceutiques*
20 Rubber & plastic products	..	159.9	237.2	231.8	717.8	239.8	..	20 Caoutchouc et matières plastiques
21 Non-metallic mineral products	..	86.4	101.1	138.6	243.6	161.7	..	**21 Produits minéraux non métalliques**
22 Basic metals	..	389.6	451.0	473.6	559.6	427.7	..	**22 Produits métallurgiques de base**
23 Basic metals, ferrous	..	364.8	446.3	461.1	538.1	401.3	..	23 Produits métallur. de base, ferreux
24 Basic metals, non-ferrous	..	24.8	4.7	12.5	21.5	26.4	..	24 Prod. métallur. de base, non ferreux
25 Fabricated metal products	..	170.3	182.2	262.9	212.5	368.2	..	**25 Ouvrages en métaux (sauf mach.& mat.)**
26 Machinery equipment, instruments and transport equipment	..	4 568.5	4 453.8	6 000.4	7 312.2	6 935.9	..	**26 Machines & matériel, instruments & mat. de transport**
27 Machinery n.e.c.	..	903.3	1 103.2	1 195.3	1 213.5	1 073.4	..	27 Machines, n.c.a.
28 Office, account. & computing machinery	..	10.6	1.8	24.0	2.9	3.8	..	28 Mach. de bur., compta. & trait. inform.
29 Electrical machinery	..	253.7	199.6	181.6	342.5	328.0	..	29 Machines électriques
30 Electro. equip. (radio, TV & commun.)	..	249.9	201.8	243.4	294.4	297.9	..	30 Equip. électro. (radio, tv & communic.)
31 *Electro. comp. (inc. semi-conductors)*	..	179.5	122.6	115.9	39.3	50.8	..	31 *Compos. électro. (semi-cond. incl.)*
32 *TV, radio & communications equip.*	..	70.4	79.2	127.5	255.1	247.1	..	32 *Appareils radio, tv et de communi.*
33 Instruments	..	91.5	71.3	112.0	98.2	253.3	..	33 Instruments
34 Motor vehicles	..	2 087.4	1 809.8	2 634.1	3 878.0	3 627.1	..	34 Véhicules automobiles
35 Other transport equipment	..	972.1	1 066.3	1 610.0	1 482.7	1 352.4	..	35 Autres matériels de transport
36 *Ships*	..	0.0	0.0	0.0	0.0	0.0	..	36 *Navires*
37 *Aerospace*	..	689.6	731.1	934.6	850.3	758.5	..	37 *Aérospatiale*
38 *Other transport equipment n.e.c.*	..	282.5	335.2	675.4	632.4	593.9	..	38 *Autres matériels de transport n.c.a.*
39 Furniture, other manufacturing n.e.c.	..	119.2	146.1	161.3	39.5	170.4	..	**39 Meubles, autres activ. de fabricat. n.c.a.**
40 Furniture	..	2.3	8.9	10.3	21.3	10.4	..	40 Meubles
41 Other manufacturing n.e.c.	..	116.9	137.2	151.0	18.2	160.0	..	41 Autres activités de fabrication n.c.a.
42 Recycling	..	0.0	0.0	0.0	0.0	15.9	..	**42 Récupération**
43 ELECTRICITY, GAS & WATER SUPPLY	..	8.6	2.5	4.7	1.4	10.2	..	**43 DISTRIBUTION D'ELECTRICITE, GAZ & EAU**
44 CONSTRUCTION	..	58.7	78.9	92.7	108.0	168.6	..	**44 CONSTRUCTION**
45 SERVICES SECTOR	..	1 836.4	2 083.6	2 472.2	2 903.9	3 582.8	..	**45 SECTEUR DES SERVICES**
46 Wholesale, retail trade & repair of motor vehicles, etc.	..	0.0	21.7	70.3	28.4	98.4	..	**46 Commerce (gros & détail), réparation de véhicules, etc.**
47 Hotels & restaurants	..	0.0	0.0	3.2	0.7	0.0	..	**47 Hôtels et restaurants**
48 Transport & storage	..	0.0	84.7	71.3	72.8	78.0	..	**48 Transports, entreposage**
49 Communications	..	0.0	16.8	17.4	5.4	10.9	..	**49 Communications**
50 Post	..	0.0	6.5	2.5	4.6	7.7	..	50 Postes
51 Telecommunications	..	0.0	10.3	14.9	0.8	3.2	..	51 Télécommunications
52 Financial intermediation (inc. insur.)	..	0.0	0.0	0.7	0.0	10.1	..	**52 Intermédiation financ. (assurances inc.)**
53 Real estate, renting & busin. activity	..	1 836.4	1 959.5	2 306.9	2 595.9	2 847.4	..	**53 Immobil., locat. & serv. aux entreprises**
54 Computer & related activities	..	0.0	28.0	151.7	164.5	249.6	..	54 Informatique et activités connexes
55 *Software consultancy*	..	0.0	22.9	116.1	160.5	223.5	..	55 *Production de logiciels*
56 *Other computer services n.e.c.*	..	0.0	5.0	35.6	4.0	26.1	..	56 *Autres activités informatiques*
57 Research & development	..	1 763.8	1 793.2	2 022.3	2 289.8	2 413.1	..	57 Recherche & développement expérimental
58 Other business activities n.e.c.	..	72.6	138.3	133.0	141.6	184.7	..	58 Autres activ. & serv. aux entreprises
59 Comm., soc. & pers. serv. activ., etc.	..	0.0	0.9	2.3	200.7	538.0	..	**59 Adm. pub., serv. soc. & collec., etc**
TOTAL BUSINESS ENTERPRISE SECTOR	..	**8 303.9**	**8 701.6**	**10 759.7**	**13 303.9**	**13 272.1**	..	**TOTAL SECTEUR DES ENTREPRISES**

Source: OECD, R&D database (DSTI/EAS Division), January 2002.

Source : OCDE, base de données R-D (DSTI/Division AES), janvier 2002.

Business enterprise intramural expenditure on R&D (BERD) - Dépenses intra-muros de R-D des entreprises (DIRDE)

by industry and type of costs - par industrie et par type de coûts

Million CZK — *Millions CZK*

11.2 Capital expenditure - Dépenses en capital

	1994	1995	1996	1997	1998	1999	2000	
1 AGRICULTURE, HUNTING & FORESTRY	..	0.0	1.3	2.2	3.3	5.1	..	1 AGRICULTURE, CHASSE & SYLVICULTURE
2 MINING	..	21.6	40.1	37.5	44.8	30.5	..	2 ACTIVITES EXTRACTIVES
3 MANUFACTURING	..	599.6	604.9	1 069.8	1 036.8	997.1	..	3 ACTIVITES DE FABRICATION
4 Food, beverages & tobacco	..	5.8	8.2	5.1	87.7	1.3	..	4 Produits alimentaires, boissons & tabac
5 Food products & beverages	..	5.8	8.2	5.1	87.7	1.3	..	5 Produits alimentaires & boissons
6 Tobacco products	..	0.0	0.0	0.0	0.0	0.0	..	6 Produits à base de tabac
7 Textiles, wear app., fur & leather	..	1.7	0.3	29.5	40.5	8.0	..	7 Textiles, habill., fourrures & cuirs
8 Textiles	..	1.4	0.0	29.4	39.7	7.9	..	8 Textiles
9 Wearing apparel & fur	..	0.0	0.0	0.1	0.0	0.0	..	9 Articles d'habillement & fourrures
10 Leather products & footwear	..	0.3	0.3	0.0	0.8	0.1	..	10 Articles en cuir & chaussures
11 Wood, paper, printing, publishing	..	3.4	2.2	0.6	6.9	0.1	..	11 Bois, papier, imprimerie, édition
12 Wood & cork (not furniture)	..	0.0	0.4	0.0	5.5	0.0	..	12 Bois & liège (sauf meubles)
13 Pulp, paper & paper products	..	3.4	1.8	0.4	0.1	0.1	..	13 Carton, papier & articles
14 Publ., print. & repro. of rec. media	..	0.0	0.0	0.1	1.3	0.0	..	14 Edition, impress. & repro. enregis.
15 Coke, petroleum, nuclear fuel, chemicals, rubber plastics	..	186.9	156.1	150.6	82.4	128.9	..	15 Coke, pétrole, combust. nucl., prod. chimi., caout. & mat. plast.
16 Coke, ref. petrol. prod. & nuclear fuel	..	18.2	1.8	1.3	1.2	2.9	..	16 Coke, raffin. pétr. & combust. nucl.
17 Chemicals & chemical products	..	155.4	148.7	113.2	69.4	101.7	..	17 Produits chimiques
18 *Chemical prod.(less pharmaceuticals)*	..	136.3	105.6	91.6	56.5	86.2	..	18 *Prod. chimi. (moins prod. pharmaceutiques)*
19 *Pharmaceuticals*	..	19.1	43.1	21.6	12.9	15.5	..	19 *Produits pharmaceutiques*
20 Rubber & plastic products	..	13.3	5.6	36.1	11.8	24.3	..	20 Caoutchouc et matières plastiques
21 Non-metallic mineral products	..	8.0	2.3	13.9	16.2	50.4	..	21 Produits minéraux non métalliques
22 Basic metals	..	46.8	58.2	24.4	44.8	19.8	..	22 Produits métallurgiques de base
23 Basic metals, ferrous	..	46.4	58.2	24.3	43.0	18.5	..	23 Produits métallur. de base, ferreux
24 Basic metals, non-ferrous	..	0.4	0.0	0.1	1.8	1.3	..	24 Prod. métallur. de base, non ferreux
25 Fabricated metal products	..	8.7	12.2	21.8	26.1	36.6	..	25 Ouvrages en métaux (sauf mach.& mat.)
26 Machinery equipment, instruments and transport equipment	..	330.8	347.3	807.1	721.5	716.9	..	26 Machines & matériel, instruments & mat. de transport
27 Machinery n.e.c.	..	73.4	69.6	145.6	179.4	185.7	..	27 Machines, n.c.a.
28 Office, account. & computing machinery	..	0.1	0.0	0.6	0.1	0.1	..	28 Mach. de bur., compta. & trait. inform.
29 Electrical machinery	..	18.9	10.9	18.8	33.4	37.1	..	29 Machines électriques
30 Electro. equip. (radio, TV & commun.)	..	4.9	5.3	40.4	82.8	19.0	..	30 Equip. électro. (radio, tv & communic.)
31 *Electro. comp. (inc. semi-conductors)*	..	3.6	0.2	0.2	11.1	2.2	..	31 *Compos. électro. (semi-cond. incl.)*
32 *TV, radio & communications equip.*	..	1.3	5.1	40.2	71.7	16.8	..	32 *Appareils radio, tv et de communi.*
33 Instruments	..	1.7	4.2	6.4	13.1	26.6	..	33 Instruments
34 Motor vehicles	..	168.7	186.0	287.6	241.4	403.8	..	34 Véhicules automobiles
35 Other transport equipment	..	63.2	71.3	307.7	171.3	44.6	..	35 Autres matériels de transport
36 *Ships*	..	0.0	0.0	0.0	0.0	0.0	..	36 *Navires*
37 *Aerospace*	..	24.8	48.0	78.6	28.3	16.4	..	37 *Aérospatiale*
38 *Other transport equipment n.e.c.*	..	38.4	23.3	229.1	143.0	28.2	..	38 *Autres matériels de transport n.c.a.*
39 Furniture, other manufacturing n.e.c.	..	7.5	18.1	16.8	10.7	20.5	..	39 Meubles, autres activ. de fabricat. n.c.a.
40 Furniture	..	0.2	5.6	5.4	9.4	11.8	..	40 Meubles
41 Other manufacturing n.e.c.	..	7.3	12.5	11.4	1.3	8.7	..	41 Autres activités de fabrication n.c.a.
42 Recycling	..	0.0	0.0	0.0	0.0	14.6	..	42 Récupération
43 ELECTRICITY, GAS & WATER SUPPLY	..	0.8	2.4	0.2	0.1	7.6	..	43 DISTRIBUTION D'ELECTRICITE, GAZ & EAU
44 CONSTRUCTION	..	5.1	6.3	8.1	6.4	5.4	..	44 CONSTRUCTION
45 SERVICES SECTOR	..	167.6	390.6	354.6	364.5	544.5	..	45 SECTEUR DES SERVICES
46 Wholesale, retail trade & repair of motor vehicles, etc.	..	0.0	4.4	5.6	13.4	12.7	..	46 Commerce (gros & détail), réparation de véhicules, etc.
47 Hotels & restaurants	..	0.0	0.0	0.0	0.0	0.0	..	47 Hôtels et restaurants
48 Transport & storage	..	0.0	9.7	19.5	31.3	7.7	..	48 Transports, entreposage
49 Communications	..	0.0	8.8	1.2	2.7	2.2	..	49 Communications
50 Post	..	0.0	3.9	0.7	2.7	1.0	..	50 Postes
51 Telecommunications	..	0.0	4.9	0.5	0.0	1.2	..	51 Télécommunications
52 Financial intermediation (inc. insur.)	..	0.0	0.0	0.0	0.0	0.5	..	52 Intermédiation financ. (assurances inc.)
53 Real estate, renting & busin. activity	..	167.6	367.7	328.4	285.9	431.9	..	53 Immobil., locat. & serv. aux entreprises
54 Computer & related activities	..	0.0	1.1	18.9	15.6	24.2	..	54 Informatique et activités connexes
55 *Software consultancy*	..	0.0	1.1	10.6	14.2	23.5	..	55 *Production de logiciels*
56 *Other computer services n.e.c.*	..	0.0	0.0	8.3	1.4	0.7	..	56 *Autres activités informatiques*
57 Research & development	..	161.6	312.0	297.0	258.7	388.7	..	57 Recherche & développement expérimental
58 Other business activities n.e.c.	..	6.0	54.7	12.4	11.6	19.0	..	58 Autres activ. & serv. aux entreprises
59 Comm., soc. & pers. serv. activ., etc.	..	0.0	0.0	0.0	31.2	89.5	..	59 Adm. pub., serv. soc. & collec., etc
TOTAL BUSINESS ENTERPRISE SECTOR	..	**794.7**	**1 045.6**	**1 472.4**	**1 455.9**	**1 590.2**	..	**TOTAL SECTEUR DES ENTREPRISES**

Source: OECD, R&D database (DSTI/EAS Division), January 2002.

Source : OCDE, base de données R-D (DSTI/Division AES), janvier 2002.

CZECH REPUBLIC — Table 12 - Tableau 12 — RÉPUBLIQUE TCHEQUE

Total business enterprise R&D personnel (BEMP) - Personnel total de R-D du secteur des entreprises (PRDE)

by industry - par industrie

In full time equivalent / *En équivalence plein-temps*

	1994	1995	1996	1997	1998	1999	2000	
1 AGRICULTURE, HUNTING & FORESTRY	..	37	127	99	130	118	..	**1 AGRICULTURE, CHASSE & SYLVICULTURE**
2 MINING	..	43	17	17	17	15	..	**2 ACTIVITES EXTRACTIVES**
3 MANUFACTURING	..	7 812	7 533	7 793	7 776	7 963	..	**3 ACTIVITES DE FABRICATION**
4 Food, beverages & tobacco	..	75	52	83	47	66	..	**4 Produits alimentaires, boissons & tabac**
5 Food products & beverages	..	75	52	83	47	66	..	5 Produits alimentaires & boissons
6 Tobacco products	..	0	0	0	0	0	..	6 Produits à base de tabac
7 Textiles, wear app., fur & leather	..	149	84	90	170	193	..	**7 Textiles, habill., fourrures & cuirs**
8 Textiles	..	59	58	69	144	98	..	8 Textiles
9 Wearing apparel & fur	..	8	7	5	7	75	..	9 Articles d'habillement & fourrures
10 Leather products & footwear	..	82	19	16	19	20	..	10 Articles en cuir & chaussures
11 Wood, paper, printing, publishing	..	30	29	18	22	20	..	**11 Bois, papier, imprimerie, édition**
12 Wood & cork (not furniture)	..	2	1	4	7	1	..	12 Bois & liège (sauf meubles)
13 Pulp, paper & paper products	..	28	26	13	12	18	..	13 Carton, papier & articles
14 Publ., print. & repro. of rec. media	..	0	2	1	3	1	..	14 Edition, impress. & repro. enregis.
15 Coke, petroleum, nuclear fuel, chemicals, rubber plastics	..	1 311	1 294	1 231	1 262	1 276	..	**15 Coke, pétrole, combust. nucl., prod. chimi., caout. & mat. plast.**
16 Coke, ref. petrol. prod. & nuclear fuel	..	211	18	7	7	11	..	16 Coke, raffin. pétr. & combust. nucl.
17 Chemicals & chemical products	..	844	1 027	920	965	1 019	..	17 Produits chimiques
18 *Chemical prod.(less pharmaceuticals)*	..	653	738	714	708	770	..	18 *Prod. chimi. (moins prod. pharmaceutiques)*
19 *Pharmaceuticals*	..	191	289	206	257	249	..	19 *Produits pharmaceutiques*
20 Rubber & plastic products	..	256	249	304	290	246	..	20 Caoutchouc et matières plastiques
21 Non-metallic mineral products	..	107	111	125	197	137	..	**21 Produits minéraux non métalliques**
22 Basic metals	..	322	310	289	337	282	..	**22 Produits métallurgiques de base**
23 Basic metals, ferrous	..	275	299	269	306	251	..	23 Produits métallur. de base, ferreux
24 Basic metals, non-ferrous	..	47	11	20	31	31	..	24 Prod. métallur. de base, non ferreux
25 Fabricated metal products	..	340	355	386	319	437	..	**25 Ouvrages en métaux (sauf mach.& mat.)**
26 Machinery equipment, instruments and transport equipment	..	5 311	5 094	5 406	5 355	5 370	..	**26 Machines & matériel, instruments & mat. de transport**
27 Machinery n.e.c.	..	1 824	1 764	1 819	1 774	1 638	..	27 Machines, n.c.a.
28 Office, account. & computing machinery	..	56	6	62	11	13	..	28 Mach. de bur., compta. & trait. inform.
29 Electrical machinery	..	418	396	397	489	409	..	29 Machines électriques
30 Electro. equip. (radio, TV & commun.)	..	236	170	188	330	385	..	30 Equip. électro. (radio, tv & communic.)
31 *Electro. comp. (inc. semi-conductors)*	..	87	61	41	55	70	..	31 *Compos. électro. (semi-cond. incl.)*
32 *TV, radio & communications equip.*	..	149	109	147	275	315	..	32 *Appareils radio, tv et de communi.*
33 Instruments	..	231	158	274	238	367	..	33 Instruments
34 Motor vehicles	..	1 646	1 699	1 739	1 685	1 941	..	34 Véhicules automobiles
35 Other transport equipment	..	899	901	927	828	617	..	35 Autres matériels de transport
36 *Ships*	..	0	0	0	0	0	..	36 *Navires*
37 *Aerospace*	..	494	515	532	541	376	..	37 *Aérospatiale*
38 *Other transport equipment n.e.c.*	..	405	386	395	287	241	..	38 *Autres matériels de transport n.c.a.*
39 Furniture, other manufacturing n.e.c.	..	167	204	165	67	179	..	**39 Meubles, autres activ. de fabricat. n.c.a.**
40 Furniture	..	9	28	22	24	13	..	40 Meubles
41 Other manufacturing n.e.c.	..	158	176	143	43	166	..	41 Autres activités de fabrication n.c.a.
42 Recycling	..	0	0	0	0	3	..	**42 Récupération**
43 ELECTRICITY, GAS & WATER SUPPLY	..	10	4	4	2	13	..	**43 DISTRIBUTION D'ELECTRICITE, GAZ & EAU**
44 CONSTRUCTION	..	93	124	112	119	116	..	**44 CONSTRUCTION**
45 SERVICES SECTOR	..	3 351	3 243	3 439	3 243	4 058	..	**45 SECTEUR DES SERVICES**
46 Wholesale, retail trade & repair of motor vehicles, etc.	..	0	18	72	43	145	..	**46 Commerce (gros & détail), réparation de véhicules, etc.**
47 Hotels & restaurants	..	0	0	3	1	0	..	**47 Hôtels et restaurants**
48 Transport & storage	..	0	159	141	112	71	..	**48 Transports, entreposage**
49 Communications	..	0	26	31	6	9	..	**49 Communications**
50 Post	..	0	5	4	5	4	..	50 Postes
51 Telecommunications	..	0	21	27	1	5	..	51 Télécommunications
52 Financial intermediation (inc. insur.)	..	0	0	6	0	13	..	**52 Intermédiation financ. (assurances inc.)**
53 Real estate, renting & busin. activity	..	3 351	3 038	3 178	3 065	3 563	..	**53 Immobil., locat. & serv. aux entreprises**
54 Computer & related activities	..	0	44	188	193	375	..	54 Informatique et activités connexes
55 *Software consultancy*	..	0	37	153	187	365	..	55 *Production de logiciels*
56 *Other computer services n.e.c.*	..	0	7	35	6	10	..	56 *Autres activités informatiques*
57 Research & development	..	3 286	2 762	2 808	2 682	2 946	..	57 Recherche & développement expérimental
58 Other business activities n.e.c.	..	65	232	182	190	241	..	58 Autres activ. & serv. aux entreprises
59 Comm., soc. & pers. serv. activ., etc.	..	0	2	8	16	257	..	**59 Adm. pub., serv. soc. & collec., etc**
TOTAL BUSINESS ENTERPRISE SECTOR	**23 114** [d,l,t]	**11 346** [a]	**11 167**	**11 464**	**11 287**	**12 283**	**11 527**	**TOTAL SECTEUR DES ENTREPRISES**

Source: OECD, R&D database (DSTI/EAS Division), January 2002.

Source : OCDE, base de données R-D (DSTI/Division AES), janvier 2002.

CZECH REPUBLIC Table 13 - Tableau 13 **RÉPUBLIQUE TCHEQUE**

RSE in the business enterprise sector - Chercheurs dans le secteur des entreprises

by industry - par industrie

In full time equivalent *En équivalence plein-temps*

	1994	1995	1996	1997	1998	1999	2000	
1 AGRICULTURE, HUNTING & FORESTRY	..	12	40	38	43	39	..	**1 AGRICULTURE, CHASSE & SYLVICULTURE**
2 MINING	..	6	0	0	0	0	..	**2 ACTIVITES EXTRACTIVES**
3 MANUFACTURING	..	3 280	3 138	3 435	3 388	3 626	..	**3 ACTIVITES DE FABRICATION**
4 Food, beverages & tobacco	..	25	27	55	20	33	..	**4 Produits alimentaires, boissons & tabac**
5 Food products & beverages	..	25	27	55	20	33	..	5 Produits alimentaires & boissons
6 Tobacco products	..	0	0	0	0	0	..	6 Produits à base de tabac
7 Textiles, wear app., fur & leather	..	74	41	49	57	38	..	**7 Textiles, habill., fourrures & cuirs**
8 Textiles	..	29	27	39	45	25	..	8 Textiles
9 Wearing apparel & fur	..	0	0	0	0	0	..	9 Articles d'habillement & fourrures
10 Leather products & footwear	..	45	14	10	12	13	..	10 Articles en cuir & chaussures
11 Wood, paper, printing, publishing	..	16	8	5	3	7	..	**11 Bois, papier, imprimerie, édition**
12 Wood & cork (not furniture)	..	1	0	1	0	1	..	12 Bois & liège (sauf meubles)
13 Pulp, paper & paper products	..	15	7	4	3	5	..	13 Carton, papier & articles
14 Publ., print. & repro. of rec. media	..	0	1	0	0	1	..	14 Edition, impress. & repro. enregis.
15 Coke, petroleum, nuclear fuel, chemicals, rubber plastics	..	636	601	585	610	583	..	**15 Coke, pétrole, combust. nucl., prod. chimi., caout. & mat. plast.**
16 Coke, ref. petrol. prod. & nuclear fuel	..	90	9	3	3	6	..	16 Coke, raffin. pétr. & combust. nucl.
17 Chemicals & chemical products	..	423	489	444	473	472	..	17 Produits chimiques
18 *Chemical prod.(less pharmaceuticals)*	..	324	363	323	330	342	..	18 *Prod. chimi. (moins prod. pharmaceutiques)*
19 *Pharmaceuticals*	..	99	126	121	143	130	..	19 *Produits pharmaceutiques*
20 Rubber & plastic products	..	123	103	138	134	105	..	20 Caoutchouc et matières plastiques
21 Non-metallic mineral products	..	34	32	27	91	18	..	**21 Produits minéraux non métalliques**
22 Basic metals	..	153	143	135	164	127	..	**22 Produits métallurgiques de base**
23 Basic metals, ferrous	..	139	135	117	144	110	..	23 Produits métallur. de base, ferreux
24 Basic metals, non-ferrous	..	14	8	18	20	17	..	24 Prod. métallur. de base, non ferreux
25 Fabricated metal products	..	129	124	144	98	190	..	**25 Ouvrages en métaux (sauf mach.& mat.)**
26 Machinery equipment, instruments and transport equipment	..	2 136	2 063	2 346	2 329	2 524	..	**26 Machines & matériel, instruments & mat. de transport**
27 Machinery n.e.c.	..	792	767	753	726	750	..	27 Machines, n.c.a.
28 Office, account. & computing machinery	..	19	0	32	0	0	..	28 Mach. de bur., compta. & trait. inform.
29 Electrical machinery	..	111	118	157	165	146	..	29 Machines électriques
30 Electro. equip. (radio, TV & commun.)	..	84	89	90	160	211	..	30 Equip. électro. (radio, tv & communic.)
31 *Electro. comp. (inc. semi-conductors)*	..	15	31	13	29	35	..	31 *Compos. électro. (semi-cond. incl.)*
32 *TV, radio & communications equip.*	..	69	58	77	131	176	..	32 *Appareils radio, tv et de communi.*
33 Instruments	..	126	67	130	113	215	..	33 Instruments
34 Motor vehicles	..	624	647	732	737	880	..	34 Véhicules automobiles
35 Other transport equipment	..	380	375	452	428	322	..	35 Autres matériels de transport
36 *Ships*	..	0	0	0	0	0	..	36 *Navires*
37 *Aerospace*	..	254	253	286	297	209	..	37 *Aérospatiale*
38 *Other transport equipment n.e.c.*	..	127	122	166	131	113	..	38 *Autres matériels de transport n.c.a.*
39 Furniture, other manufacturing n.e.c.	..	77	99	89	16	103	..	**39 Meubles, autres activ. de fabricat. n.c.a.**
40 Furniture	..	4	4	4	6	5	..	40 Meubles
41 Other manufacturing n.e.c.	..	73	95	85	10	98	..	41 Autres activités de fabrication n.c.a.
42 Recycling	..	0	0	0	0	3	..	**42 Récupération**
43 ELECTRICITY, GAS & WATER SUPPLY	..	4	2	1	1	6	..	**43 DISTRIBUTION D'ELECTRICITE, GAZ & EAU**
44 CONSTRUCTION	..	43	47	57	64	59	..	**44 CONSTRUCTION**
45 SERVICES SECTOR	..	1 591	1 595	1 589	1 571	2 081	..	**45 SECTEUR DES SERVICES**
46 Wholesale, retail trade & repair of motor vehicles, etc.	..	0	9	20	10	68	..	**46 Commerce (gros & détail), réparation de véhicules, etc.**
47 Hotels & restaurants	..	0	0	1	0	0	..	**47 Hôtels et restaurants**
48 Transport & storage	..	0	53	38	77	40	..	**48 Transports, entreposage**
49 Communications	..	0	19	24	1	3	..	**49 Communications**
50 Post	..	0	0	0	0	0	..	50 Postes
51 Telecommunications	..	0	19	24	1	3	..	51 Télécommunications
52 Financial intermediation (inc. insur.)	..	0	0	6	0	2	..	**52 Intermédiation financ. (assurances inc.)**
53 Real estate, renting & busin. activity	..	1 591	1 514	1 494	1 471	1 818	..	**53 Immobil., locat. & serv. aux entreprises**
54 Computer & related activities	..	0	16	108	106	221	..	54 Informatique et activités connexes
55 *Software consultancy*	..	0	11	83	101	215	..	55 *Production de logiciels*
56 *Other computer services n.e.c.*	..	0	5	25	5	6	..	56 *Autres activités informatiques*
57 Research & development	..	1 542	1 352	1 268	1 247	1 460	..	57 Recherche & développement expérimental
58 Other business activities n.e.c.	..	49	146	118	118	138	..	58 Autres activ. & serv. aux entreprises
59 Comm., soc. & pers. serv. activ., etc.	..	0	0	6	12	150	..	**59 Adm. pub., serv. soc. & collec., etc**
TOTAL BUSINESS ENTERPRISE SECTOR	..	**4 936**	**4 863**	**5 120**	**5 067**	**5 811**	**5 533**	**TOTAL SECTEUR DES ENTREPRISES**

Source: OECD, R&D database (DSTI/EAS Division), January 2002.

Source : OCDE, base de données R-D (DSTI/Division AES), janvier 2002.

Table 14 - Tableau 14

Higher education intramural expenditure on R&D - Dépenses intra-muros de R-D de l'enseignement supérieur

(HERD) by field of science - (DIRDES) par discipline scientifique

	1994	1995	1996	1997	1998	1999	2000	
	14.1 Million CZK - Millions CZK							
Natural sciences	..	203.9	463.2	295.1	787.2 [a]	861.7	..	Sciences exactes et naturelles
Engineering	..	600.8	393.3	856.7	770.6 [a]	1 245.4	..	Sciences de l'ingénieur
Medical sciences	..	206.8	331.7	177.7	290.4 [a]	330.9	..	Sciences médicales
Agricultural sciences	..	61.9	93.4	141.3	151.2 [a]	224.6	..	Sciences agricoles
Sub-total NSE	..	1 073.2	1 281.6	1 470.8	1 999.4 [a]	2 662.6	..	Sous-total SENI
Social sciences	..	113.0	158.7	307.2	161.3 [a]	250.9	..	Sciences sociales
Humanities	..	..	..	..	9.6 [a]	4.3	..	Sciences humaines
Sub-total SSH	..	113.0 [m]	158.7 [m]	307.2 [m]	170.9 [a]	255.2	..	Sous-total SSH
Total NSE+SSH	580.0 [l]	1 185.9 [a]	1 440.3	1 778.0	2 170.3	2 917.8	3 764.0	*Total SENI+SSH*
	14.2 Million constant USD (1995 prices and PPPs) - Millions de USD constants (prix et PPA de 1995)							
Natural sciences	..	18.9	39.4	23.2	56.0 [a]	59.4	..	Sciences exactes et naturelles
Engineering	..	55.6	33.4	67.4	54.8 [a]	85.9	..	Sciences de l'ingénieur
Medical sciences	..	19.1	28.2	14.0	20.7 [a]	22.8	..	Sciences médicales
Agricultural sciences	..	5.7	7.9	11.1	10.8 [a]	15.5	..	Sciences agricoles
Sub-total NSE	..	99.3	109.0	115.8	142.2 [a]	183.6	..	Sous-total SENI
Social sciences	..	10.5	13.5	24.2	11.5 [a]	17.3	..	Sciences sociales
Humanities	..	..	..	..	0.7 [a]	0.3	..	Sciences humaines
Sub-total SSH	..	10.5 [m]	13.5 [m]	24.2 [m]	12.2 [a]	17.6	..	Sous-total SSH
Total NSE+SSH	59.1 [l]	109.7 [a]	122.5	139.9	154.3	201.2	257.3	*Total SENI+SSH*

Table 15 - Tableau 15

Higher education intramural expenditure on R&D - Dépenses intra-muros de R-D de l'enseignement supérieur

(HERD) by field of science and type of costs - (DIRDES) par discipline scientifique et par type de coûts

Million CZK *Millions CZK*

	1994	1995	1996	1997	1998	1999	2000	
Current expenditure								**Dépenses courantes**
Natural sciences	..	139.7	267.0	174.9	585.5 [a]	708.7	..	Sciences exactes et naturelles
Engineering	..	465.1	258.2	714.3	647.0 [a]	1 072.6	..	Sciences de l'ingénieur
Medical sciences	..	179.3	234.7	135.4	252.0 [a]	281.4	..	Sciences médicales
Agricultural sciences	..	45.3	54.5	128.3	134.6 [a]	189.9	..	Sciences agricoles
Sub-total NSE	..	829.5	814.4	1 152.9	1 619.1 [a]	2 252.6	..	Sous-total SENI
Social sciences	..	104.4	138.4	251.2	147.9 [a]	233.9	..	Sciences sociales
Humanities	..	..	..	..	4.5 [a]	3.8	..	Sciences humaines
Sub-total SSH	..	104.4	138.4	251.2	152.4 [a]	237.7	..	Sous-total SSH
Total NSE+SSH	..	933.9	952.8	1 404.1	1 771.5	2 490.3	..	*Total SENI+SSH*
Capital expenditure								**Dépenses en capital**
Natural sciences	..	64.3	196.1	120.2	201.7 [a]	153.0	..	Sciences exactes et naturelles
Engineering	..	133.6	135.0	142.4	123.6 [a]	172.8	..	Sciences de l'ingénieur
Medical sciences	..	27.9	97.2	42.3	38.4 [a]	49.5	..	Sciences médicales
Agricultural sciences	..	17.0	38.9	13.0	16.6 [a]	34.7	..	Sciences agricoles
Sub-total NSE	..	242.8	467.2	317.9	380.3 [a]	410.0	..	Sous-total SENI
Social sciences	..	9.2	20.3	56.0	13.4 [a]	17.0	..	Sciences sociales
Humanities	..	..	..	..	5.1 [a]	0.5	..	Sciences humaines
Sub-total SSH	..	9.2	20.3	56.0	18.5 [a]	17.5	..	Sous-total SSH
Total NSE+SSH	..	252.0	487.5	373.9	398.8	427.5	..	*Total SENI+SSH*

Source: OECD, R&D database (DSTI/EAS Division), January 2002.

Source : OCDE, base de données R-D (DSTI/Division AES), janvier 2002.

CZECH REPUBLIC — Table 16 - Tableau 16 — RÉPUBLIQUE TCHEQUE

Higher education intramural expenditure on R&D - Dépenses intra-muros de R-D de l'enseignement supérieur

(HERD) by field of science and source of funds - (DIRDES) par discipline scientifique et par source de financement

Million CZK — *Millions CZK*

	1994	1995	1996	1997	1998	1999	2000	
Direct government								**État direct**
Natural sciences	..	180.0	249.5	29.7	..	..	..	Sciences exactes et naturelles
Engineering	..	453.0	265.9	351.7	..	..	..	Sciences de l'ingénieur
Medical sciences	..	172.9	265.2	25.3	..	..	..	Sciences médicales
Agricultural sciences	..	61.9	74.0	43.4	..	..	..	Sciences agricoles
Sub-total NSE	..	867.8	854.6	450.1	..	..	..	Sous-total SENI
Social sciences	..	110.0	91.0	76.4	..	..	..	Sciences sociales
Humanities	..	..	..	..	..	..	..	Sciences humaines
Sub-total SSH	..	110.0	91.0	76.4	..	..	..	Sous-total SSH
Total NSE+SSH	549.0 [l]	977.8 [a]	820.3	526.5	..	..	..	*Total SENI+SSH*
General university funds								**Fonds généraux des universités**
Natural sciences	..	..	..	..	..	..	..	Sciences exactes et naturelles
Engineering	..	..	..	..	..	..	..	Sciences de l'ingénieur
Medical sciences	..	..	..	..	..	..	..	Sciences médicales
Agricultural sciences	..	..	..	..	..	..	..	Sciences agricoles
Sub-total NSE	..	..	..	..	..	..	..	Sous-total SENI
Social sciences	..	..	..	..	..	..	..	Sciences sociales
Humanities	..	..	..	..	..	..	..	Sciences humaines
Sub-total SSH	..	..	..	..	..	..	..	Sous-total SSH
Total NSE+SSH	..	..	..	..	..	..	..	*Total SENI+SSH*
Sub-total government								**Sous-total État**
Natural sciences	..	180.0 [m]	249.5 [m]	29.7 [m]	707.6 [a]	816.9	..	Sciences exactes et naturelles
Engineering	..	453.0 [m]	265.9 [m]	351.7 [m]	691.3 [a]	1 136.5	..	Sciences de l'ingénieur
Medical sciences	..	172.9 [m]	265.2 [m]	25.3 [m]	250.4 [a]	314.5	..	Sciences médicales
Agricultural sciences	..	61.9 [m]	74.0 [m]	43.4 [m]	134.5 [a]	204.5	..	Sciences agricoles
Sub-total NSE	..	867.8 [m]	854.6 [m]	450.1 [m]	1 783.8 [a]	2 472.4	..	Sous-total SENI
Social sciences	..	110.0 [m]	91.0 [m]	76.4 [m]	140.0 [a]	232.3	..	Sciences sociales
Humanities	..	..	..	..	9.6 [a]	4.3	..	Sciences humaines
Sub-total SSH	..	110.0 [m]	91.0 [m]	76.4 [m]	149.6 [a]	236.6	..	Sous-total SSH
Total NSE+SSH	..	977.8 [a,m]	820.3 [m]	526.5 [m]	1 933.4 [a]	2 709.0	3 534.2	*Total SENI+SSH*
Higher education								**Enseignement supérieur**
Natural sciences	..	1.0 [l]	185.0 [l]	243.6 [l]	52.0 [a]	13.5	..	Sciences exactes et naturelles
Engineering	..	95.0 [l]	102.2 [l]	418.7 [l]	5.6 [a]	44.9	..	Sciences de l'ingénieur
Medical sciences	..	16.0 [l]	55.6 [l]	140.1 [l]	0.0 [a]	0.0	..	Sciences médicales
Agricultural sciences	..	0.0	19.1 [l]	86.0 [l]	2.5 [a]	3.6	..	Sciences agricoles
Sub-total NSE	..	112.0 [l]	361.9 [l]	888.4 [l]	60.1 [a]	62.0	..	Sous-total SENI
Social sciences	..	1.0 [l]	62.1 [l]	222.1 [l]	3.8 [a]	3.3	..	Sciences sociales
Humanities	..	..	..	..	0.0 [a]	..	..	Sciences humaines
Sub-total SSH	..	1.0 [l]	62.1 [l]	222.1 [l]	3.8 [a]	3.3	..	Sous-total SSH
Total NSE+SSH	..	113.0 [l]	424.0 [l]	1 110.5 [l]	63.9 [a]	65.3	57.6	*Total SENI+SSH*
Private non-profit								**Institutions sans but lucratif**
Natural sciences	..	0.0	0.5	0.4	1.0	0.5	..	Sciences exactes et naturelles
Engineering	..	3.3	0.0	0.0	0.2	6.1	..	Sciences de l'ingénieur
Medical sciences	..	0.0	0.3	0.0	0.0	..	..	Sciences médicales
Agricultural sciences	..	0.0	0.0	0.0	0.0	..	..	Sciences agricoles
Sub-total NSE	..	3.3	0.8	0.4	1.2	6.6	..	Sous-total SENI
Social sciences	..	0.0	0.1	1.5	0.5	0.4	..	Sciences sociales
Humanities	..	..	..	..	0.0	..	..	Sciences humaines
Sub-total SSH	..	0.0	0.1	1.5	0.5	0.4	..	Sous-total SSH
Total NSE+SSH	..	3.3	0.9	1.9	1.7	7.0	0.4	*Total SENI+SSH*
Business entreprise								**Entreprises**
Natural sciences	..	0.0	3.5	3.5	2.1	11.0	..	Sciences exactes et naturelles
Engineering	..	8.1	0.0	11.0	10.4	10.2	..	Sciences de l'ingénieur
Medical sciences	..	15.1	2.5	3.3	27.2	6.6	..	Sciences médicales
Agricultural sciences	..	0.0	0.0	6.9	4.4	8.7	..	Sciences agricoles
Sub-total NSE	..	23.2	6.0	24.7	44.1	36.5	..	Sous-total SENI
Social sciences	..	0.0	0.2	1.8	0.3	..	..	Sciences sociales
Humanities	..	..	..	..	0.0	..	..	Sciences humaines
Sub-total SSH	..	0.0	0.2	1.8	0.3	..	..	Sous-total SSH
Total NSE+SSH	..	23.2	6.2	26.5	44.4	36.5	40.0	*Total SENI+SSH*
Funds from abroad								**Fonds de l'étranger**
Natural sciences	..	22.9	24.7	17.9	24.5	19.8	..	Sciences exactes et naturelles
Engineering	..	40.9	25.2	75.3	63.1	47.7	..	Sciences de l'ingénieur
Medical sciences	..	2.8	8.1	9.0	12.8	9.8	..	Sciences médicales
Agricultural sciences	..	0.0	0.3	5.0	9.8	7.8	..	Sciences agricoles
Sub-total NSE	..	66.6	58.3	107.2	110.2	85.1	..	Sous-total SENI
Social sciences	..	2.0	5.3	5.4	16.7	14.9	..	Sciences sociales
Humanities	..	..	..	..	0.0	..	..	Sciences humaines
Sub-total SSH	..	2.0	5.3	5.4	16.7	14.9	..	Sous-total SSH
Total NSE+SSH	..	68.6	63.6	112.6	126.9	100.0	131.8	*Total SENI+SSH*

Source: OECD, R&D database (DSTI/EAS Division), January 2002.

Source : OCDE, base de données R-D (DSTI/Division AES), janvier 2002.

Table 17 - Tableau 17

Higher education R&D personnel (HEMP) - Personnel de R-D du secteur de l'enseignement supérieur

by field of science and occupation - par discipline scientifique et par profession

In full-time equivalent — *En équivalence plein-temps*

	1994	1995	1996	1997	1998	1999	2000	
Total R&D personnel								**Personnel total de R-D**
Natural sciences	..	828	823	574	1 411	..	..	Sciences exactes et naturelles
Engineering	..	1 375	927	1 842	1 497	..	..	Sciences de l'ingénieur
Medical sciences	..	705	665	368	388	..	..	Sciences médicales
Agricultural sciences	..	401	267	210	302	..	..	Sciences agricoles
Sub-total NSE	..	3 309	2 682	2 994	3 598	..	..	Sous-total SENI
Social sciences	..	380	1 721	987	421	..	..	Sciences sociales
Humanities	..	..	..	..	7	..	..	Sciences humaines
Sub-total SSH	..	380	1 721	987	428	..	..	Sous-total SSH
Total NSE+SSH	3 397 [l,t]	3 689 [a]	4 404	3 981	4 026	4 736	5 331	*Total SENI+SSH*
Researchers								**Chercheurs**
Natural sciences	..	673	669	478	1 067	..	..	Sciences exactes et naturelles
Engineering	..	1 027	687	1 334	1 098	..	..	Sciences de l'ingénieur
Medical sciences	..	364	339	201	189	..	..	Sciences médicales
Agricultural sciences	..	293	157	129	174	..	..	Sciences agricoles
Sub-total NSE	..	2 357	1 852	2 142	2 528	..	..	Sous-total SENI
Social sciences	..	328	1 652	706	354	..	..	Sciences sociales
Humanities	..	..	..	..	2	..	..	Sciences humaines
Sub-total SSH	..	328	1 652	706	356	..	..	Sous-total SSH
Total NSE+SSH	..	2 685	3 505	2 848	2 884	3 380	3 768	*Total SENI+SSH*
Technicians								**Technicians**
Natural sciences	..	109	116	71	272	..	..	Sciences exactes et naturelles
Engineering	..	225	148	332	237	..	..	Sciences de l'ingénieur
Medical sciences	..	216	207	91	151	..	..	Sciences médicales
Agricultural sciences	..	97	103	79	120	..	..	Sciences agricoles
Sub-total NSE	..	647	574	573	780	..	..	Sous-total SENI
Social sciences	..	19	27	199	27	..	..	Sciences sociales
Humanities	..	..	..	..	5	..	..	Sciences humaines
Sub-total SSH	..	19	27	199	32	..	..	Sous-total SSH
Total NSE+SSH	..	665	601	772	812	1 009	1 073	*Total SENI+SSH*
Other								**Autres**
Natural sciences	..	46	38	25	72	..	..	Sciences exactes et naturelles
Engineering	..	123	92	176	162	..	..	Sciences de l'ingénieur
Medical sciences	..	125	119	76	48	..	..	Sciences médicales
Agricultural sciences	..	12	7	2	8	..	..	Sciences agricoles
Sub-total NSE	..	306	256	279	290	..	..	Sous-total SENI
Social sciences	..	32	42	82	40	..	..	Sciences sociales
Humanities	..	..	..	..	0	..	..	Sciences humaines
Sub-total SSH	..	32	42	82	40	..	..	Sous-total SSH
Total NSE+SSH	..	339	298	361	330	347	490	*Total SENI+SSH*

Table 18 - Tableau 18

Government budget appropriations or outlays for R&D (GBAORD) - Crédits budgétaires publics de R-D (CBPRD)

by socio-economic objective - par objectif socio-économique

Million CZK — *Millions CZK*

	1995	1996	1997	1998	1999	2000	2001	
Agriculture, forestry and fishing	..	..	..	..	..	..	..	Agriculture, sylviculture et pêche
Industrial development	..	..	..	..	..	..	..	Développement industriel
Energy	..	..	..	..	..	..	..	Energie
Transport & telecommunications	..	..	..	..	..	..	..	*Transports & télécommunications*
Urban & rural planning	..	..	..	..	..	..	..	*Aménagement urbain et rural*
ST Infrastructure	..	..	..	..	..	..	..	ST Infrastructure
Prevention of pollution	..	..	..	..	..	..	..	*Prévention de la pollution*
Identificat. & treatment of pollution	..	..	..	..	..	..	..	*Détect. pollut. & mesures antipollut.*
ST Environment	..	..	..	..	..	..	..	ST Environnement
Health	..	..	..	..	..	..	..	Santé
Social development & services	..	..	..	..	..	..	..	Développement et services sociaux
Earth and atmosphere	..	..	..	..	..	..	..	Milieu terrestre et atmosphère
Advancement of research	..	..	..	..	..	..	..	*Promotion de la recherche*
General university funds	..	..	..	..	..	..	..	*Fonds généraux des universités*
ST Advancement of knowledge	..	..	..	..	..	..	..	ST promotion gén. des connaiss.
Civil space	..	..	..	..	..	..	..	Espace civil
Defence	..	..	..	..	..	..	..	Défense
Not elsewhere classified	..	..	..	..	..	..	..	Non classé
Total	**..**	**..**	**..**	**..**	**..**	**..**	**..**	**Total**

Source: OECD, R&D database (DSTI/EAS Division), January 2002.

Source : OCDE, base de données R-D (DSTI/Division AES), janvier 2002.

CZECH REPUBLIC — Table 19 - Tableau 19 — RÉPUBLIQUE TCHEQUE

Technology balance of payments - Balance des paiements technologiques

by type of transfer - selon la nature des transferts

Million CZK — *Millions CZK*

	1983	1984	1985	1986	1987	1988	1989	1990	1991	
Receipts										**Recettes**
Related firms	..	..	..	..	..	..	..	..	..	Firmes affiliées
Unrelated firms	..	..	..	..	..	..	..	..	..	Firmes non affiliées
Other	..	..	..	..	..	..	..	..	..	Autre
Total	..	..	..	..	..	..	..	..	..	*Total*
Payments										**Paiements**
Related firms	..	..	..	..	..	..	..	..	..	Firmes affiliées
Unrelated firms	..	..	..	..	..	..	..	..	..	Firmes non affiliées
Other	..	..	..	..	..	..	..	..	..	Autre
Total	..	..	..	..	..	..	..	..	..	*Total*

	1992	1993	1994	1995	1996	1997	1998	1999	2000	
Receipts										**Recettes**
Related firms	..	..	..	..	..	..	..	..	..	Firmes affiliées
Unrelated firms	..	..	..	..	..	..	..	..	..	Firmes non affiliées
Other	..	..	..	..	..	..	..	..	..	Autre
Total	..	..	..	..	..	7 082	9 189	9 935	..	*Total*
Payments										**Paiements**
Related firms	..	..	..	..	..	..	..	..	..	Firmes affiliées
Unrelated firms	..	..	..	..	..	..	..	..	..	Firmes non affiliées
Other	..	..	..	..	..	..	..	..	..	Autre
Total	..	..	..	..	..	9 694	16 492	19 849	..	*Total*

Source: OECD, TBPdatabase (DSTI/EAS Division), Januay 2002.

Source : OCDE, base de données TBP (DSTI/Division AES), janvier 2002.

DENMARK — Table 1 - Tableau 1 — DANEMARK

Gross domestic expenditure on R&D (GERD) - Dépenses intérieures brutes de R-D (DIRD)

by sector of performance and source of funds - par secteur d'exécution et par source de financement

	1994	1995	1996	1997	1998	1999	2000	
	1.1 Million DKK - Millions DKK							
Business enterprise								**Entreprises**
financed by								*financé par*
Business enterprise	..	8 182.9	9 752.9 [c]	11 322.9	13 853.7	14 348.3	..	Entreprises
Direct government	..	652.3	677.5 [c]	702.6	646.6	700.5	..	État direct
Higher education	..	..	..	..	..	..	..	Enseignement supérieur
Private non-profit	..	157.2	154.3 [c]	151.3	90.7	95.8	..	Institutions sans but lucratif
Funds from abroad	..	1 648.6	1 388.4 [c]	1 128.2	803.2	909.2	..	Fonds de l'étranger
Total	..	10 641.0	11 973.0 [c]	13 305.0	15 394.1	16 053.7	..	*Total*
Government								**État**
financed by								*financé par*
Business enterprise	..	110.0	90.5 [c]	71.0	..	213.0 [v]	..	Entreprises
Direct government	..	2 596.0	2 589.5 [c]	2 740.0	..	2 999.0 [v]	..	État direct
Higher education	..	..	..	..	..	..	..	Enseignement supérieur
Private non-profit	..	266.0	330.5 [c]	396.0	..	293.0 [v]	..	Institutions sans but lucratif
Funds from abroad	..	182.0	188.0 [c]	130.0	..	213.0 [v]	..	Fonds de l'étranger
Total	..	3 154.0	3 198.5 [c]	3 337.0	3 410.0 [c]	3 839.0	..	*Total*
Higher education								**Enseignement supérieur**
financed by								*financé par*
Business enterprise	..	84.0	80.2 [c]	164.0	..	107.0 [v]	..	Entreprises
Direct government	..	1 029.0	1 079.7 [c]	1 070.0	..	635.0 [v]	..	État direct
General university funds	..	3 040.0	2 649.5 [c]	3 310.0	..	3 883.0 [v]	..	Fonds généraux des universités
Sub-total government	..	4 069.0	3 729.2 [c]	4 380.0	..	4 518.0 [v]	..	Sous-Total État
Higher education	..	..	..	..	..	..	..	Enseignement supérieur
Private non-profit	..	205.0	223.0 [c]	130.0	..	247.0 [v]	..	Institutions sans but lucratif
Funds from abroad	..	190.0	220.2 [c]	130.0	..	187.0 [v]	..	Fonds de l'étranger
Total	..	4 548.0	4 252.6 [c]	4 802.0	4 773.0 [c]	5 133.0	..	*Total*
Private non-profit								**Institutions sans but lucratif**
financed by								*financé par*
Business enterprise	..	2.0	3.0 [c]	2.0	..	4.0	..	Entreprises
Direct government	..	24.0	23.0 [c]	3.0	..	24.0	..	État direct
Higher education	..	..	..	..	..	..	..	Enseignement supérieur
Private non-profit	..	159.0	191.9 [c]	201.0	..	242.0	..	Institutions sans but lucratif
Funds from abroad	..	15.0	15.0 [c]	2.0	..	21.0	..	Fonds de l'étranger
Total	..	200.0	232.9 [c]	208.0	216.0 [c]	292.0	..	*Total*
GERD								**DIRD**
financed by								*financé par*
Business enterprise	..	8 378.9	9 926.6 [c]	11 559.9	..	14 672.3 [v]	..	Entreprises
Direct government	..	4 301.3	4 369.7 [c]	4 515.6	..	4 358.5 [v]	..	État direct
General university funds	..	3 040.0	2 649.5 [c]	3 310.0	..	3 883.0 [v]	..	Fonds généraux des universités
Sub-total government	..	7 341.3	7 019.2 [c]	7 825.6	..	8 241.5 [v]	..	Sous-Total État
Higher education	..	..	..	..	..	..	..	Enseignement supérieur
Private non-profit	..	787.2	899.7 [c]	878.3	..	877.8 [v]	..	Institutions sans but lucratif
Funds from abroad	..	2 035.6	1 811.6 [c]	1 390.0	..	1 330.2 [v]	..	Fonds de l'étranger
Total	..	18 543.0	19 657.0 [c]	21 652.0	23 793.1 [c]	25 317.7	..	*Total*
	1.2 Million constant USD (1995 prices and PPPs) - Millions de USD constants (prix et PPA de 1995)							
Business enterprise								**Entreprises**
financed by								*financé par*
Business enterprise	..	972.2	1 130.7 [c]	1 284.6	1 542.2	1 550.9	..	Entreprises
Direct government	..	77.5	78.5 [c]	79.7	72.0	75.7	..	État direct
Higher education	..	..	..	..	..	..	..	Enseignement supérieur
Private non-profit	..	18.7	17.9 [c]	17.2	10.1	10.4	..	Institutions sans but lucratif
Funds from abroad	..	195.9	161.0 [c]	128.0	89.4	98.3	..	Fonds de l'étranger
Total	..	1 264.3	1 388.1 [c]	1 509.4	1 713.6	1 735.2	..	*Total*
Government								**État**
financed by								*financé par*
Business enterprise	..	13.1	10.5 [c]	8.1	..	23.0 [v]	..	Entreprises
Direct government	..	308.4	300.2 [c]	310.8	..	324.2 [v]	..	État direct
Higher education	..	..	..	..	..	..	..	Enseignement supérieur
Private non-profit	..	31.6	38.3 [c]	44.9	..	31.7 [v]	..	Institutions sans but lucratif
Funds from abroad	..	21.6	21.8 [c]	14.7	..	23.0 [v]	..	Fonds de l'étranger
Total	..	374.7	370.8 [c]	378.6	379.6 [c]	415.0	..	*Total*
Higher education								**Enseignement supérieur**
financed by								*financé par*
Business enterprise	..	10.0	9.3 [c]	18.6	..	11.6 [v]	..	Entreprises
Direct government	..	122.3	125.2 [c]	121.4	..	68.6 [v]	..	État direct
General university funds	..	361.2	307.2 [c]	375.5	..	419.7 [v]	..	Fonds généraux des universités
Sub-total government	..	483.4	432.3 [c]	496.9	..	488.3 [v]	..	Sous-Total État
Higher education	..	..	..	..	..	..	..	Enseignement supérieur
Private non-profit	..	24.4	25.9 [c]	14.7	..	26.7 [v]	..	Institutions sans but lucratif
Funds from abroad	..	22.6	25.5 [c]	14.7	..	20.2 [v]	..	Fonds de l'étranger
Total	..	540.4	493.0 [c]	544.8	531.3 [c]	554.8	..	*Total*
Private non-profit								**Institutions sans but lucratif**
financed by								*financé par*
Business enterprise	..	0.2	0.3 [c]	0.2	..	0.4	..	Entreprises
Direct government	..	2.9	2.7 [c]	0.3	..	2.6	..	État direct
Higher education	..	..	..	..	..	..	..	Enseignement supérieur
Private non-profit	..	18.9	22.2 [c]	22.8	..	26.2	..	Institutions sans but lucratif
Funds from abroad	..	1.8	1.7 [c]	0.2	..	2.3	..	Fonds de l'étranger
Total	..	23.8	27.0 [c]	23.6	24.0 [c]	31.6	..	*Total*
GERD								**DIRD**
financed by								*financé par*
Business enterprise	..	995.5	1 150.8 [c]	1 311.5	..	1 585.9 [v]	..	Entreprises
Direct government	..	511.0	506.6 [c]	512.3	..	471.1 [v]	..	État direct
General university funds	..	361.2	307.2 [c]	375.5	..	419.7 [v]	..	Fonds généraux des universités
Sub-total government	..	872.2	813.8 [c]	887.8	..	890.8 [v]	..	Sous-Total État
Higher education	..	..	..	..	..	..	..	Enseignement supérieur
Private non-profit	..	93.5	104.3 [c]	99.6	..	94.9 [v]	..	Institutions sans but lucratif
Funds from abroad	..	241.9	210.0 [c]	157.7	..	143.8 [v]	..	Fonds de l'étranger
Total	..	2 203.1	2 278.9 [c]	2 456.4	2 648.6 [c]	2 736.6	..	*Total*

Source: OECD, R&D database (DSTI/EAS Division), January 2002.

Source : OCDE, base de données R-D (DSTI/Division AES), janvier 2002.

Gross domestic expenditure on R&D (GERD) - Dépenses intérieures brutes de R-D (DIRD)

by sector of performance and type of costs - par secteur d'exécution et par type de coûts

Million DKK *Millions DKK*

	1994	1995	1996	1997	1998	1999	2000	
Business enterprise								**Entreprises**
Labour costs	..	5 957.0	..	7 513.0	8 423.0	..	..	Coûts du personnel
Other current costs	..	3 593.0	..	4 535.0	5 169.0	..	..	Autres coûts courants
Total current expenditure	..	9 550.0	..	12 047.0	13 593.0	13 949.0	..	*Total dépenses courantes*
Land and buildings	..	285.0	..	349.0	672.0	..	..	Terrains et constructions
Instruments and equipment	..	806.0	..	908.0	1 130.0	..	..	Instruments et équipements
Total capital expenditure	..	1 091.0	..	1 258.0	1 802.0	2 104.7	..	*Total dépenses en capital*
Total	..	10 641.0	11 973.0 [c]	13 305.0	15 394.1	16 053.7	..	*Total*
Government								**État**
Labour costs	..	1 613.0	..	1 963.0	..	2 420.0	..	Coûts du personnel
Other current costs	..	1 373.0	..	1 156.0	..	1 166.0	..	Autres coûts courants
Total current expenditure	..	2 985.0	..	3 119.0	..	3 586.0	..	*Total dépenses courantes*
Land and buildings	..	45.0	..	16.0	..	39.0	..	Terrains et constructions
Instruments and equipment	..	125.0	..	202.0	..	214.0	..	Instruments et équipements
Total capital expenditure	..	170.0	..	218.0	..	253.0	..	*Total dépenses en capital*
Total	..	3 154.0	3 198.5 [c]	3 337.0	3 410.0 [c]	3 839.0	..	*Total*
Higher education								**Enseignement supérieur**
Labour costs	..	2 101.0	..	2 524.0	..	2 830.0	..	Coûts du personnel
Other current costs	..	1 986.0	..	1 886.0	..	1 874.0	..	Autres coûts courants
Total current expenditure	..	4 087.0	..	4 410.0	..	4 704.0	..	*Total dépenses courantes*
Land and buildings	..	261.0	..	241.0	..	305.0	..	Terrains et constructions
Instruments and equipment	..	200.0	..	151.0	..	124.0	..	Instruments et équipements
Total capital expenditure	..	461.0	..	392.0	..	429.0	..	*Total dépenses en capital*
Total	..	4 548.0	4 252.6 [c]	4 802.0	4 773.0 [c]	5 133.0	..	*Total*
Private non-profit								**Institutions sans but lucratif**
Labour costs	..	104.0	..	116.0	..	166.0	..	Coûts du personnel
Other current costs	..	81.0	..	77.0	..	116.0	..	Autres coûts courants
Total current expenditure	..	185.0	..	193.0	..	283.0	..	*Total dépenses courantes*
Land and buildings	..	4.0	..	3.0	..	4.0	..	Terrains et constructions
Instruments and equipment	..	11.0	..	12.0	..	5.0	..	Instruments et équipements
Total capital expenditure	..	15.0	..	15.0	..	9.0	..	*Total dépenses en capital*
Total	..	200.0	232.9 [c]	208.0	216.0 [c]	292.0	..	*Total*
GERD								**DIRD**
Labour costs	..	9 774.0	..	12 116.0	..	..	..	Coûts du personnel
Other current costs	..	7 032.0	..	7 654.0	..	..	..	Autres coûts courants
Total current expenditure	..	16 806.0	..	19 769.0	..	22 522.0	..	*Total dépenses courantes*
Land and buildings	..	595.0	..	609.0	..	..	..	Terrains et constructions
Instruments and equipment	..	1 142.0	..	1 273.0	..	..	..	Instruments et équipements
Total capital expenditure	..	1 736.0	..	1 883.0	..	2 795.7	..	*Total dépenses en capital*
Total	..	18 543.0	19 657.0 [c]	21 652.0	23 793.1 [c]	25 317.7	..	*Total*

Table 3 - Tableau 3

Current domestic expenditure on R&D - Dépenses intérieures courantes de R-D

by sector of performance and type of activity - par secteur d'exécution et par type d'activités

Million DKK *Millions DKK*

	1994	1995	1996	1997	1998	1999	2000	
Business enterprise								**Entreprises**
Basic research	..	..	..	..	616.0 [k]	837.0 [a]	..	Recherche fondamentale
Applied research	..	..	..	..	2 617.0 [k]	3 208.0 [a]	..	Recherche appliquée
Experimental development	..	..	..	..	12 162.0 [k]	9 904.0 [a]	..	Développement expérimental
Not specified	..	9 550.0	..	13 305.0 [a,k]	..	..	..	Non spécifié
Total	..	9 550.0	..	13 305.0 [a,k]	15 394.1 [k]	13 949.0 [a]	..	*Total*
Government								**État**
Basic research	..	857.6	..	917.7 [a,k]	..	1 152.0 [k]	..	Recherche fondamentale
Applied research	..	1 502.1	..	1 670.2 [a,k]	..	1 958.0 [k]	..	Recherche appliquée
Experimental development	..	625.9	..	752.3 [a,k]	..	729.0 [k]	..	Développement expérimental
Not specified	..	..	..	..	..	..	..	Non spécifié
Total	..	2 985.0	..	3 337.0 [a,k]	3 410.0 [c,k]	3 839.0 [k]	..	*Total*
Higher education								**Enseignement supérieur**
Basic research	..	2 533.3	..	2 976.4 [a,k]	..	3 234.0 [k]	..	Recherche fondamentale
Applied research	..	1 184.9	..	1 388.5 [a,k]	..	1 437.0 [k]	..	Recherche appliquée
Experimental development	..	367.7	..	438.3 [a,k]	..	462.0 [k]	..	Développement expérimental
Not specified	..	..	..	..	..	..	..	Non spécifié
Total	..	4 087.0	..	4 802.0 [a,k]	4 773.0 [c,k]	5 133.0 [k]	..	*Total*
Private non-profit								**Institutions sans but lucratif**
Basic research	..	94.4	..	125.6 [a,k]	..	131.0 [k]	..	Recherche fondamentale
Applied research	..	70.3	..	64.0 [a,k]	..	126.0 [k]	..	Recherche appliquée
Experimental development	..	22.2	..	16.7 [a,k]	..	35.0 [k]	..	Développement expérimental
Not specified	..	..	..	..	..	..	..	Non spécifié
Total	..	185.0	..	208.0 [a,k]	216.0 [c,k]	292.0 [k]	..	*Total*
GERD								**DIRD**
Basic research	..	..	..	..	..	5 354.0 [o]	..	Recherche fondamentale
Applied research	..	..	..	..	..	6 729.0 [o]	..	Recherche appliquée
Experimental development	..	..	..	..	..	11 130.0 [o]	..	Développement expérimental
Not specified	..	16 806.0	..	21 652.0 [a,k]	..	..	..	Non spécifié
Total	..	16 806.0	..	21 652.0 [a,k]	23 793.1 [c,k]	23 213.0 [a,o]	..	*Total*

Source: OECD, R&D database (DSTI/EAS Division), January 2002.

Source : OCDE, base de données R-D (DSTI/Division AES), janvier 2002.

Gross domestic expenditure on R&D (GERD) - Dépenses intérieures brutes de R-D (DIRD)

by sector of performance and socio-economic objective - par secteur d'exécution et par objectif socio-économique

Million DKK — *Millions DKK*

	1994	1995	1996	1997	1998	1999	2000	
Business enterprise								**Entreprises**
Agriculture, forestry and fishing	..	..	..	..	..	..	..	Agriculture, sylviculture et pêche
Industrial development	..	..	..	..	..	..	..	Développement industriel
Energy	..	..	..	491.1	2 336.8	..	..	Energie
Transport and telecommunications	..	..	..	..	..	..	..	Transports et télécommunications
Urban and rural planning	..	..	..	..	..	..	..	Aménagement urbain et rural
Prevention of pollution	..	..	..	..	..	..	..	Prévention de la pollution
Identification & treatment of pollution	..	..	..	..	..	..	..	Détection pollu. & mesures antipollu.
Health	..	..	..	477.1	611.0	..	..	Santé
Social development and services	..	..	..	..	..	..	..	Développement et services sociaux
Earth and atmosphere	..	..	..	..	..	..	..	Milieu terrestre et atmosphère
Advancement of knowledge	..	..	..	..	..	..	..	Promotion générale des connaissances
Civil space	..	..	..	..	..	..	..	Espace civil
Defence	..	..	..	48.2	87.3	..	..	Défense
Not elsewhere classified	..	10 641.0	11 973.0 c	12 288.6	12 359.9	16 053.0	..	Non classé
Total	..	10 641.0	11 973.0 c	13 305.0	15 394.1	16 053.7	..	*Total*
Government								**État**
Agriculture, forestry and fishing	..	559.0	..	677.5	..	697.0	..	Agriculture, sylviculture et pêche
Industrial development	..	77.0	..	122.1	..	146.0	..	Développement industriel
Energy	..	158.0	..	209.9	..	157.0	..	Energie
Transport and telecommunications	..	31.0	..	35.0	..	38.0	..	Transports et télécommunications
Urban and rural planning	..	45.0	..	30.4	..	62.0	..	Aménagement urbain et rural
Prevention of pollution	..	62.0	..	174.6	..	162.0	..	Prévention de la pollution
Identification & treatment of pollution	..	216.0	..	110.5	..	156.0	..	Détection pollu. & mesures antipollu.
Health	..	1 078.0	..	1 111.7	..	1 380.0	..	Santé
Social development and services	..	407.0	..	369.2	..	426.0	..	Développement et services sociaux
Earth and atmosphere	..	110.0	..	148.2	..	118.0	..	Milieu terrestre et atmosphère
Advancement of knowledge	..	345.0	..	220.6	..	411.0	..	Promotion générale des connaissances
Civil space	..	31.0	..	38.7	..	56.0	..	Espace civil
Defence	..	38.0	..	88.8	..	30.0	..	Défense
Not elsewhere classified	..	0.0	3 198.5 c	0.0	3 410.0 c	0.0	..	Non classé
Total	..	3 154.0	3 198.5 c	3 337.0	3 410.0 c	3 839.0	..	*Total*
Higher education								**Enseignement supérieur**
Agriculture, forestry and fishing	..	373.0	..	236.8	..	263.0	..	Agriculture, sylviculture et pêche
Industrial development	..	359.0	..	414.8	..	503.0	..	Développement industriel
Energy	..	91.0	..	85.5	..	129.0	..	Energie
Transport and telecommunications	..	93.0	..	123.7	..	186.0	..	Transports et télécommunications
Urban and rural planning	..	72.0	..	94.1	..	87.0	..	Aménagement urbain et rural
Prevention of pollution	..	86.0	..	126.0	..	84.0	..	Prévention de la pollution
Identification & treatment of pollution	..	87.0	..	119.8	..	91.0	..	Détection pollu. & mesures antipollu.
Health	..	595.0	..	548.5	..	714.0	..	Santé
Social development and services	..	569.0	..	730.1	..	781.0	..	Développement et services sociaux
Earth and atmosphere	..	139.0	..	190.7	..	139.0	..	Milieu terrestre et atmosphère
Advancement of knowledge	..	2 061.0	..	1 905.6	..	2 129.0	..	Promotion générale des connaissances
Civil space	..	21.0	..	59.6	..	25.0	..	Espace civil
Defence	..	0.0	..	22.3	..	1.0	..	Défense
Not elsewhere classified	..	0.0	4 252.6 c	147.7	4 773.0 c	0.0	..	Non classé
Total	..	4 548.0	4 252.6 c	4 802.0	4 773.0 c	5 133.0	..	*Total*
Private non-profit								**Institutions sans but lucratif**
Agriculture, forestry and fishing	..	0.0	..	0.0	..	8.0	..	Agriculture, sylviculture et pêche
Industrial development	..	1.0	..	1.0	..	5.0	..	Développement industriel
Energy	..	0.0	..	0.0	..	0.0	..	Energie
Transport and telecommunications	..	2.0	..	0.6	..	2.0	..	Transports et télécommunications
Urban and rural planning	..	2.0	..	0.0	..	1.0	..	Aménagement urbain et rural
Prevention of pollution	..	1.0	..	0.7	..	4.0	..	Prévention de la pollution
Identification & treatment of pollution	..	25.0	..	0.0	..	8.0	..	Détection pollu. & mesures antipollu.
Health	..	128.0	..	81.2	..	224.0	..	Santé
Social development and services	..	20.0	..	2.6	..	18.0	..	Développement et services sociaux
Earth and atmosphere	..	0.0	..	1.1	..	0.0	..	Milieu terrestre et atmosphère
Advancement of knowledge	..	20.0	..	15.5	..	23.0	..	Promotion générale des connaissances
Civil space	..	0.0	..	0.0	..	0.0	..	Espace civil
Defence	..	0.0	..	0.0	..	0.0	..	Défense
Not elsewhere classified	..	0.0	232.9 c	105.3	216.0 c	0.0	..	Non classé
Total	..	200.0	232.9 c	208.0	216.0 c	292.0	..	*Total*
GERD								**DIRD**
Agriculture, forestry and fishing	..	..	..	..	..	..	..	Agriculture, sylviculture et pêche
Industrial development	..	..	..	..	..	..	..	Développement industriel
Energy	..	..	..	786.5	..	..	..	Energie
Transport and telecommunications	..	..	..	..	..	..	..	Transports et télécommunications
Urban and rural planning	..	..	..	..	..	..	..	Aménagement urbain et rural
Prevention of pollution	..	..	..	..	..	..	..	Prévention de la pollution
Identification & treatment of pollution	..	..	..	..	..	..	..	Détection pollu. & mesures antipollu.
Health	..	..	..	2 218.5	..	..	..	Santé
Social development and services	..	..	..	..	..	..	..	Développement et services sociaux
Earth and atmosphere	..	..	..	..	..	..	..	Milieu terrestre et atmosphère
Advancement of knowledge	..	..	..	..	..	..	..	Promotion générale des connaissances
Civil space	..	..	..	..	..	..	..	Espace civil
Defence	..	..	..	159.3	..	..	..	Défense
Not elsewhere classified	..	18 543.0	19 657.0 c	12 541.5	23 793.1 c	25 317.7	..	Non classé
Total	..	18 543.0	19 657.0 c	21 652.0	23 793.1 c	25 317.7	..	*Total*

Source: OECD, R&D database (DSTI/EAS Division), January 2002.

Source : OCDE, base de données R-D (DSTI/Division AES), janvier 2002.

Table 5 - Tableau 5

Total R&D personnel - Personnel total de R-D

by sector of employment and occupation - par secteur d'emploi et par profession

In full-time equivalent · *En équivalence plein-temps*

	1994	1995	1996	1997	1998	1999	2000	
Business enterprise								**Entreprises**
RSE	..	6 674	7 098 [c]	7 522	8 009	8 575	..	Chercheurs
Technicians	..	7 427	8 306 [c]	9 185	9 649	8 993	..	Techniciens
Other occupations	..	3 094	3 212 [c]	3 330	3 539	3 454	..	Autres professions
Total	..	17 195	18 615 [c]	20 037	21 198	21 022	..	*Total*
Government								**État**
RSE	..	3 575	3 590 [c]	3 673	..	3 918	..	Chercheurs
Technicians	..	..	..	..	..	..	..	Techniciens
Other occupations	..	..	..	..	..	..	..	Autres professions
Total	..	5 439	5 506 [c]	5 675	..	6 236	..	*Total*
Higher education								**Enseignement supérieur**
RSE	..	5 520	5 832 [c]	6 143	..	5 722	..	Chercheurs
Technicians	..	..	..	..	..	..	..	Techniciens
Other occupations	..	..	..	..	..	..	..	Autres professions
Total	..	7 213	7 676 [c]	8 138	..	8 017	..	*Total*
Private non-profit								**Institutions sans but lucratif**
RSE	..	185	179 [c]	173	..	223	..	Chercheurs
Technicians	..	..	..	..	..	..	..	Techniciens
Other occupations	..	..	..	..	..	..	..	Autres professions
Total	..	366	351 [c]	335	..	375	..	*Total*
TOTAL								**TOTAL**
RSE	..	15 954	16 699 [c]	17 511	..	18 438	..	Chercheurs
Technicians	..	..	..	13 344	..	13 758	..	Techniciens
Other occupations	..	..	..	3 330	..	3 454	..	Autres professions
Total	..	30 213	32 148 [c]	34 187	..	35 650	..	*Total*

Table 6 - Tableau 6

Total R&D personnel - Personnel total de R-D

by sector of employment and formal qualification - par secteur d'emploi et par niveau de qualification

In full-time equivalent · *En équivalence plein-temps*

	1994	1995	1996	1997	1998	1999	2000	
Business enterprise								**Entreprises**
University PhD level degrees	..	..	..	..	..	..	..	Diplômes univ. de niv. doctorat
Other university degrees	..	..	..	..	..	..	..	Autres diplômes universitaires
Sub-total university degrees	..	6 674	7 098 [c]	7 522	8 009	8 575	..	Sous-total Universitaires
Other post-secondary diplomas	..	..	..	..	..	..	..	Autres diplômes post-secondaires
Secondary diplomas	..	..	..	..	..	..	..	Diplômes de fin d'études secondaires
Other qualifications	..	..	..	..	..	..	..	Autres qualifications
Total	..	17 195	18 615 [c]	20 037	21 198	21 022	..	*Total*
Government								**État**
University PhD level degrees	..	..	..	..	..	..	..	Diplômes univ. de niv. doctorat
Other university degrees	..	..	..	..	..	..	..	Autres diplômes universitaires
Sub-total university degrees	..	3 575	3 590 [c]	3 673	..	3 918	..	Sous-total Universitaires
Other post-secondary diplomas	..	..	..	..	..	..	..	Autres diplômes post-secondaires
Secondary diplomas	..	..	..	..	..	..	..	Diplômes de fin d'études secondaires
Other qualifications	..	..	..	..	..	..	..	Autres qualifications
Total	..	5 439	5 506 [c]	5 675	..	6 236	..	*Total*
Higher education								**Enseignement supérieur**
University PhD level degrees	..	..	..	..	..	..	..	Diplômes univ. de niv. doctorat
Other university degrees	..	..	..	..	..	..	..	Autres diplômes universitaires
Sub-total university degrees	..	5 520	5 832 [c]	6 143	..	5 722	..	Sous-total Universitaires
Other post-secondary diplomas	..	..	..	..	..	..	..	Autres diplômes post-secondaires
Secondary diplomas	..	..	..	..	..	..	..	Diplômes de fin d'études secondaires
Other qualifications	..	..	..	..	..	..	..	Autres qualifications
Total	..	7 213	7 676 [c]	8 138	..	8 017	..	*Total*
Private non-profit								**Institutions sans but lucratif**
University PhD level degrees	..	..	..	..	..	..	..	Diplômes univ. de niv. doctorat
Other university degrees	..	..	..	..	..	..	..	Autres diplômes universitaires
Sub-total university degrees	..	185	179 [c]	173	..	223	..	Sous-total Universitaires
Other post-secondary diplomas	..	..	..	..	..	..	..	Autres diplômes post-secondaires
Secondary diplomas	..	..	..	..	..	..	..	Diplômes de fin d'études secondaires
Other qualifications	..	..	..	..	..	..	..	Autres qualifications
Total	..	366	351 [c]	335	..	375	..	*Total*
TOTAL								**TOTAL**
University PhD level degrees	..	..	..	..	..	..	..	Diplômes univ. de niv. doctorat
Other university degrees	..	..	..	..	..	..	..	Autres diplômes universitaires
Sub-total university degrees	..	15 954	16 699 [c]	17 511	..	18 438	..	Sous-total Universitaires
Other post-secondary diplomas	..	..	..	..	..	..	..	Autres diplômes post-secondaires
Secondary diplomas	..	..	..	..	..	..	..	Diplômes de fin d'études secondaires
Other qualifications	..	..	..	..	..	..	..	Autres qualifications
Total	..	30 213	32 148 [c]	34 187	..	35 650	..	*Total*

Source: OECD, R&D database (DSTI/EAS Division), January 2002.

Source : OCDE, base de données R-D (DSTI/Division AES), janvier 2002.

DENMARK — Table 7 - Tableau 7 — DANEMARK

Gross domestic expenditure on R&D (GERD) - Dépenses intérieures brutes de R-D (DIRD)

by sector of performance and main scientific field - par secteur d'exécution et par grand domaine scientifique

Million DKK — *Millions DKK*

	1994	1995	1996	1997	1998	1999	2000	
Business enterprise								**Entreprises**
Natural sciences and engineering	..	10 641.0	11 973.0 [c]	13 305.0	15 394.1	16 053.7	..	Sciences exactes, nat. et de l'ingénieur
Social sciences and humanities	..	..	..	..	..	..	..	Sciences sociales et humaines
All fields of science	..	10 641.0	11 973.0 [c]	13 305.0	15 394.1	16 053.7	..	*Tous domaines scientifiques*
Government								**État**
Natural sciences and engineering	..	2 745.0	..	2 919.0	..	3 329.0	..	Sciences exactes, nat. et de l'ingénieur
Social sciences and humanities	..	409.0	..	418.0	..	510.0	..	Sciences sociales et humaines
All fields of science	..	3 154.0	3 198.5 [c]	3 337.0	3 410.0 [c]	3 839.0	..	*Tous domaines scientifiques*
Higher education								**Enseignement supérieur**
Natural sciences and engineering	..	3 295.0	..	3 436.0	..	3 380.0	..	Sciences exactes, nat. et de l'ingénieur
Social sciences and humanities	..	1 253.0	..	1 366.0	..	1 752.0	..	Sciences sociales et humaines
All fields of science	..	4 548.0	4 252.6 [c]	4 802.0	4 773.0 [c]	5 133.0	..	*Tous domaines scientifiques*
Private non-profit								**Institutions sans but lucratif**
Natural sciences and engineering	..	182.0	..	203.0	..	281.0	..	Sciences exactes, nat. et de l'ingénieur
Social sciences and humanities	..	18.0	..	5.0	..	8.0	..	Sciences sociales et humaines
All fields of science	..	200.0	232.9 [c]	208.0	216.0 [c]	292.0	..	*Tous domaines scientifiques*
TOTAL								**TOTAL**
Natural sciences and engineering	..	16 863.0	..	19 863.0	..	23 043.0	..	Sciences exactes, nat. et de l'ingénieur
Social sciences and humanities	..	1 680.0	..	1 789.0	..	2 270.0	..	Sciences sociales et humaines
All fields of science	..	18 543.0	19 657.0 [c]	21 652.0	23 793.1 [c]	25 317.7	..	*Tous domaines scientifiques*

Table 8 - Tableau 8

Total R&D personnel - Personnel total de R-D

by sector of employment and main scientific field - par secteur d'emploi et par grand domaine scientifique

In full-time equivalent — *En équivalence plein-temps*

	1994	1995	1996	1997	1998	1999	2000	
Business enterprise								**Entreprises**
Natural sciences and engineering	..	17 195	18 614 [c]	20 037	21 198	21 022	..	Sciences exactes, nat. et de l'ingénieur
Social sciences and humanities	..	..	..	..	..	..	..	Sciences sociales et humaines
All fields of science	..	17 195	18 615 [c]	20 037	21 198	21 022	..	*Tous domaines scientifiques*
Government								**État**
Natural sciences and engineering	..	4 707	..	4 920	..	5 516	..	Sciences exactes, nat. et de l'ingénieur
Social sciences and humanities	..	732	..	755	..	721	..	Sciences sociales et humaines
All fields of science	..	5 439	5 506 [c]	5 675	..	6 236	..	*Tous domaines scientifiques*
Higher education								**Enseignement supérieur**
Natural sciences and engineering	..	5 208	..	5 838	..	5 573	..	Sciences exactes, nat. et de l'ingénieur
Social sciences and humanities	..	2 005	..	2 302	..	2 445	..	Sciences sociales et humaines
All fields of science	..	7 213	7 676 [c]	8 138	..	8 017	..	*Tous domaines scientifiques*
Private non-profit								**Institutions sans but lucratif**
Natural sciences and engineering	..	333	..	326	..	358	..	Sciences exactes, nat. et de l'ingénieur
Social sciences and humanities	..	34	..	9	..	16	..	Sciences sociales et humaines
All fields of science	..	366	351 [c]	335	..	375	..	*Tous domaines scientifiques*
TOTAL								**TOTAL**
Natural sciences and engineering	..	27 443	..	31 121	..	32 469	..	Sciences exactes, nat. et de l'ingénieur
Social sciences and humanities	..	2 770	..	3 066	..	3 182	..	Sciences sociales et humaines
All fields of science	..	30 213	32 148 [c]	34 187	..	35 650	..	*Tous domaines scientifiques*

Source: OECD, R&D database (DSTI/EAS Division), January 2002.

Source : OCDE, base de données R-D (DSTI/Division AES), janvier 2002.

DENMARK | Table 9 - Tableau 9 | DANEMARK

Total business enterprise intramural expenditure on R&D - Dépenses intra-muros totales de R-D des entreprises
(BERD) by industry - (DIRDE) par industrie

	1994	1995	1996	1997	1998	1999	2000	
				9.1 Million DKK - Millions DKK				
1 AGRICULTURE, HUNTING & FORESTRY	..	23.0 [o]	27.3 [c,o]	29.3 [c,o]	.. [o]	20.0 [o]	..	1 AGRICULTURE, CHASSE & SYLVICULTURE
2 MINING	..	.. [n]	.. [n]	.. [n]	.. [n]	.. [n]	..	2 ACTIVITES EXTRACTIVES
3 MANUFACTURING	..	7 227.0	7 812.5 [c]	8 399.0 [c]	9 938.0	10 294.0	..	3 ACTIVITES DE FABRICATION
4 Food, beverages & tobacco	..	445.0	443.5 [c]	441.0 [c]	503.0	423.0	..	**4 Produits alimentaires, boissons & tabac**
5 Food products & beverages	..	..	..	..	..	..	..	5 Produits alimentaires & boissons
6 Tobacco products	..	..	..	..	..	..	..	6 Produits à base de tabac
7 Textiles, wear app., fur & leather	..	10.0	12.0 [c]	14.0 [c]	.. [n]	42.0	..	**7 Textiles, habill., fourrures & cuirs**
8 Textiles	..	..	..	..	..	..	..	8 Textiles
9 Wearing apparel & fur	..	..	..	..	..	..	..	9 Articles d'habillement & fourrures
10 Leather products & footwear	..	..	..	..	..	..	..	10 Articles en cuir & chaussures
11 Wood, paper, printing, publishing	..	.. [n]	.. [n]	.. [n]	.. [n]	81.0	..	**11 Bois, papier, imprimerie, édition**
12 Wood & cork (not furniture)	..	..	..	..	..	..	..	12 Bois & liège (sauf meubles)
13 Pulp, paper & paper products	..	..	..	..	..	..	..	13 Carton, papier & articles
14 Publ., print. & repro. of rec. media	..	..	..	..	..	..	..	14 Edition, impress. & repro. enregis.
15 Coke, petroleum, nuclear fuel, chemicals, rubber plastics	..	2 574.0	2 943.0 [c]	3 312.0 [c]	4 291.0	4 926.0	..	**15 Coke, pétrole, combust. nucl., prod. chimi., caout. & mat. plast.**
16 Coke, ref. petrol. prod. & nuclear fuel	..	.. [n]	.. [n]	.. [n]	.. [n]	.. [n]	..	16 Coke, raffin. pétr. & combust. nucl.
17 Chemicals & chemical products	..	2 453.0	2 798.0 [c]	3 143.0 [c]	4 109.0	4 550.0	..	17 Produits chimiques
18 *Chemical prod.(less pharmaceuticals)*	..	321.0	405.5 [c]	490.0 [c]	522.0	598.0	..	18 *Prod. chimi. (moins prod. pharmaceutiques)*
19 *Pharmaceuticals*	..	2 132.0	2 392.5 [c]	2 653.0 [c]	3 588.0	3 952.0	..	19 *Produits pharmaceutiques*
20 Rubber & plastic products	..	121.0	145.0 [c]	169.0 [c]	182.0	376.0	..	20 Caoutchouc et matières plastiques
21 Non-metallic mineral products	..	48.0	55.0 [c]	62.0 [c]	77.0	106.0	..	**21 Produits minéraux non métalliques**
22 Basic metals	..	95.0 [o]	96.0 [c,o]	97.0 [c,o]	130.0 [o]	192.0 [o]	..	**22 Produits métallurgiques de base**
23 Basic metals, ferrous	..	..	..	..	..	..	..	23 Produits métallur. de base, ferreux
24 Basic metals, non-ferrous	..	..	..	..	..	..	..	24 Prod. métallur. de base, non ferreux
25 Fabricated metal products	..	.. [n]	.. [n]	.. [n]	.. [n]	.. [n]	..	**25 Ouvrages en métaux (sauf mach.& mat.)**
26 Machinery equipment, instruments and transport equipment	..	3 481.0	3 700.5 [c]	3 923.0 [c]	4 165.0	4 366.0	..	**26 Machines & matériel, instruments & mat. de transport**
27 Machinery n.e.c.	..	1 798.0	1 868.0 [c]	1 939.0 [c]	1 801.0	1 895.0	..	27 Machines, n.c.a.
28 Office, account. & computing machinery	..	98.0	95.0 [c]	92.0 [c]	.. [n]	141.0	..	28 Mach. de bur., compta. & trait. inform.
29 Electrical machinery	..	187.0	188.5 [c]	190.0 [c]	310.0	504.0	..	29 Machines électriques
30 Electro. equip. (radio, TV & commun.)	..	670.0	791.5 [c]	915.0 [c]	1 096.0	684.0	..	30 Equip. électro. (radio, tv & communic.)
31 *Electro. comp. (inc. semi-conductors)*	..	..	..	..	..	..	..	31 *Compos. électro. (semi-cond. incl.)*
32 *TV, radio & communications equip.*	..	..	..	..	..	..	..	32 *Appareils radio, tv et de communi.*
33 Instruments	..	636.0	661.5 [c]	687.0 [c]	852.0	985.0	..	33 Instruments
34 Motor vehicles	..	.. [n]	.. [n]	.. [n]	.. [n]	.. [n]	..	34 Véhicules automobiles
35 Other transport equipment	..	92.0 [o]	96.0 [c,o]	100.0 [c,o]	106.0 [o]	156.0 [o]	..	35 Autres matériels de transport
36 *Ships*	..	..	..	..	..	..	..	36 *Navires*
37 *Aerospace*	..	..	..	..	..	..	..	37 *Aérospatiale*
38 *Other transport equipment n.e.c.*	..	..	..	..	..	..	..	38 *Autres matériels de transport n.c.a.*
39 Furniture, other manufacturing n.e.c.	..	575.0 [o]	562.0 [c,o]	549.0 [c,o]	770.0 [o]	157.0 [o]	..	**39 Meubles, autres activ. de fabricat. n.c.a.**
40 Furniture	..	..	..	..	..	..	..	40 Meubles
41 Other manufacturing n.e.c.	..	..	..	..	..	..	..	41 Autres activités de fabrication n.c.a.
42 Recycling	..	.. [n]	.. [n]	.. [n]	.. [n]	.. [n]	..	**42 Récupération**
43 ELECTRICITY, GAS & WATER SUPPLY	..	41.0	56.5 [c]	72.0 [c]	53.0	.. [n]	..	43 DISTRIBUTION D'ELECTRICITE, GAZ & EAU
44 CONSTRUCTION	..	24.0	25.5 [c]	27.0 [c]	29.0	38.0	..	44 CONSTRUCTION
45 SERVICES SECTOR	..	3 327.0	4 067.5 [c]	4 808.0 [c]	5 374.0 [o]	5 702.0 [o]	..	45 SECTEUR DES SERVICES
46 Wholesale, retail trade & repair of motor vehicles, etc.	..	597.0	822.5 [c]	1 188.0 [c]	1 076.0	1 238.0	..	**46 Commerce (gros & détail), réparation de véhicules, etc.**
47 Hotels & restaurants	..	..	..	..	..	..	..	**47 Hôtels et restaurants**
48 Transport & storage	..	.. [n]	.. [n]	.. [n]	.. [n]	.. [n]	..	**48 Transports, entreposage**
49 Communications	..	259.0 [o]	243.0 [c,o]	227.0 [c,o]	.. [n]	525.0 [o]	..	**49 Communications**
50 Post	..	..	..	..	..	.. [n]	..	50 Postes
51 Telecommunications	..	..	..	..	..	.. [n]	..	51 Télécommunications
52 Financial intermediation (inc. insur.)	..	..	..	..	..	222.0	..	**52 Intermédiation financ. (assurances inc.)**
53 Real estate, renting & busin. activity	..	..	2 962.0 [c]	3 393.0 [c]	4 298.0	3 717.0	..	**53 Immobil., locat. & serv. aux entreprises**
54 Computer & related activities	..	809.0	1 020.5 [c]	1 232.0 [c]	1 797.0	1 512.0	..	54 Informatique et activités connexes
55 *Software consultancy*	..	..	..	..	1 292.0	1 221.0	..	55 *Production de logiciels*
56 *Other computer services n.e.c.*	..	..	..	..	504.0	291.0	..	56 *Autres activités informatiques*
57 Research & development	..	320.0	386.0 [c]	452.0 [c]	483.0	1 206.0	..	57 Recherche & développement expérimental
58 Other business activities n.e.c.	..	1 342.0 [o]	1 555.5 [c,o]	1 709.0 [c,o]	2 018.0 [o]	999.0 [o]	..	58 Autres activ. & serv. aux entreprises
59 Comm., soc. & pers. serv. activ., etc.	..	.. [n]	.. [n]	.. [n]	.. [n]	.. [n]	..	**59 Adm. pub., serv. soc. & collec., etc**
TOTAL BUSINESS ENTERPRISE SECTOR	..	**10 641.0**	**11 973.0** [c]	**13 305.0**	**15 394.1**	**16 053.7**	..	**TOTAL SECTEUR DES ENTREPRISES**

Source: OECD, R&D database (DSTI/EAS Division), January 2002.

Source : OCDE, base de données R-D (DSTI/Division AES), janvier 2002.

Total business enterprise intramural expenditure on R&D - Dépenses intra-muros totales de R-D des entreprises
(BERD) by industry - (DIRDE) par industrie

	1994	1995	1996	1997	1998	1999	2000	
	9.2 Million constant USD(1995 prices and PPPs) - Millions de USD constants (prix et PPA de 1995)							
1 AGRICULTURE, HUNTING & FORESTRY	..	2.7 [o]	3.2 [c,o]	3.3 [c,o]	..	2.2 [o]	..	**1 AGRICULTURE, CHASSE & SYLVICULTURE**
2 MINING	..	.. [n]	.. [n]	.. [n]	.. [n]	.. [n]	..	**2 ACTIVITES EXTRACTIVES**
3 MANUFACTURING	..	858.7	905.7 [c]	952.9 [c]	1 106.3	1 112.7	..	**3 ACTIVITES DE FABRICATION**
4 Food, beverages & tobacco	..	52.9	51.4 [c]	50.0 [c]	56.0	45.7	..	**4 Produits alimentaires, boissons & tabac**
5 Food products & beverages	..	..	..	..	..	..	..	5 Produits alimentaires & boissons
6 Tobacco products	..	..	..	..	..	..	..	6 Produits à base de tabac
7 Textiles, wear app., fur & leather	..	1.2	1.4 [c]	1.6 [c]	.. [n]	4.5	..	**7 Textiles, habill., fourrures & cuirs**
8 Textiles	..	..	..	..	..	..	..	8 Textiles
9 Wearing apparel & fur	..	..	..	..	..	..	..	9 Articles d'habillement & fourrures
10 Leather products & footwear	..	..	..	..	..	..	..	10 Articles en cuir & chaussures
11 Wood, paper, printing, publishing	..	.. [n]	.. [n]	.. [n]	.. [n]	8.8	..	**11 Bois, papier, imprimerie, édition**
12 Wood & cork (not furniture)	..	..	..	..	..	..	..	12 Bois & liège (sauf meubles)
13 Pulp, paper & paper products	..	..	..	..	..	..	..	13 Carton, papier & articles
14 Publ., print. & repro. of rec. media	..	..	..	..	..	..	..	14 Edition, impress. & repro. enregis.
15 Coke, petroleum, nuclear fuel, chemicals, rubber plastics	..	305.8	341.2 [c]	375.7 [c]	477.7	532.4	..	**15 Coke, pétrole, combust. nucl., prod. chimi., caout. & mat. plast.**
16 Coke, ref. petrol. prod. & nuclear fuel	..	.. [n]	.. [n]	.. [n]	.. [n]	.. [n]	..	16 Coke, raffin. pétr. & combust. nucl.
17 Chemicals & chemical products	..	291.4	324.4 [c]	356.6 [c]	457.4	491.8	..	17 Produits chimiques
18 *Chemical prod.(less pharmaceuticals)*	..	38.1	47.0 [c]	55.6 [c]	58.1	64.6	..	18 *Prod. chimi. (moins prod. pharmaceutiques)*
19 *Pharmaceuticals*	..	253.3	277.4 [c]	301.0 [c]	399.4	427.2	..	19 *Produits pharmaceutiques*
20 Rubber & plastic products	..	14.4	16.8 [c]	19.2 [c]	20.3	40.6	..	20 Caoutchouc et matières plastiques
21 Non-metallic mineral products	..	5.7	6.4 [c]	7.0 [c]	8.6	11.5	..	**21 Produits minéraux non métalliques**
22 Basic metals	..	11.3 [o]	11.1 [c,o]	11.0 [c,o]	14.5 [o]	20.8 [o]	..	**22 Produits métallurgiques de base**
23 Basic metals, ferrous	..	..	..	..	..	..	..	23 Produits métallur. de base, ferreux
24 Basic metals, non-ferrous	..	..	..	..	..	..	..	24 Prod. métallur. de base, non ferreux
25 Fabricated metal products	..	.. [n]	.. [n]	.. [n]	.. [n]	.. [n]	..	**25 Ouvrages en métaux (sauf mach.& mat.)**
26 Machinery equipment, instruments and transport equipment	..	413.6	429.0 [c]	445.1 [c]	463.6	471.9	..	**26 Machines & matériel, instruments & mat. de transport**
27 Machinery n.e.c.	..	213.6	216.6 [c]	220.0 [c]	200.5	204.8	..	27 Machines, n.c.a.
28 Office, account. & computing machinery	..	11.6	11.0 [c]	10.4 [c]	.. [n]	15.2	..	28 Mach. de bur., compta. & trait. inform.
29 Electrical machinery	..	22.2	21.9 [c]	21.6 [c]	34.5	54.5	..	29 Machines électriques
30 Electro. equip. (radio, TV & commun.)	..	79.6	91.8 [c]	103.8 [c]	122.0	73.9	..	30 Equip. électro. (radio, tv & communic.)
31 *Electro. comp. (inc. semi-conductors)*	..	..	..	..	..	..	..	31 *Compos. électro. (semi-cond. incl.)*
32 *TV, radio & communications equip.*	..	..	..	..	..	..	..	32 *Appareils radio, tv et de communi.*
33 Instruments	..	75.6	76.7 [c]	77.9 [c]	94.8	106.5	..	33 Instruments
34 Motor vehicles	..	.. [n]	.. [n]	.. [n]	.. [n]	.. [n]	..	34 Véhicules automobiles
35 Other transport equipment	..	10.9 [o]	11.1 [c,o]	11.3 [c,o]	11.8 [o]	16.9 [o]	..	35 Autres matériels de transport
36 *Ships*	..	..	..	..	..	..	..	36 *Navires*
37 *Aerospace*	..	..	..	..	..	..	..	37 *Aérospatiale*
38 *Other transport equipment n.e.c.*	..	..	..	..	..	..	..	38 *Autres matériels de transport n.c.a.*
39 Furniture, other manufacturing n.e.c.	..	68.3 [o]	65.2 [c,o]	62.3 [c,o]	85.7 [o]	17.0 [o]	..	**39 Meubles, autres activ. de fabricat. n.c.a.**
40 Furniture	..	..	..	..	..	..	..	40 Meubles
41 Other manufacturing n.e.c.	..	..	..	..	..	..	..	41 Autres activités de fabrication n.c.a.
42 Recycling	..	.. [n]	.. [n]	.. [n]	.. [n]	.. [n]	..	**42 Récupération**
43 ELECTRICITY, GAS & WATER SUPPLY	..	4.9	6.6 [c]	8.2 [c]	5.9	.. [n]	..	**43 DISTRIBUTION D'ELECTRICITE, GAZ & EAU**
44 CONSTRUCTION	..	2.9	3.0 [c]	3.1 [c]	3.2	4.1	..	**44 CONSTRUCTION**
45 SERVICES SECTOR	..	395.3	471.6 [c]	545.5 [c]	598.2 [o]	616.3 [o]	..	**45 SECTEUR DES SERVICES**
46 Wholesale, retail trade & repair of motor vehicles, etc.	..	70.9	95.4 [c]	134.8 [c]	119.8	133.8	..	**46 Commerce (gros & détail), réparation de véhicules, etc.**
47 Hotels & restaurants	..	..	..	..	..	..	..	**47 Hôtels et restaurants**
48 Transport & storage	..	.. [n]	.. [n]	.. [n]	.. [n]	.. [n]	..	**48 Transports, entreposage**
49 Communications	..	30.8 [o]	28.2 [c,o]	25.8 [c,o]	.. [n]	56.7 [o]	..	**49 Communications**
50 Post	..	..	..	..	..	.. [n]	..	50 Postes
51 Telecommunications	..	..	..	..	..	.. [n]	..	51 Télécommunications
52 Financial intermediation (inc. insur.)	..	..	..	..	..	24.0	..	**52 Intermédiation financ. (assurances inc.)**
53 Real estate, renting & busin. activity	..	..	343.4 [c]	384.9 [c]	478.4	401.8	..	**53 Immobil., locat. & serv. aux entreprises**
54 Computer & related activities	..	96.1	118.3 [c]	139.8 [c]	200.0	163.4	..	54 Informatique et activités connexes
55 *Software consultancy*	..	..	..	..	143.8	132.0	..	55 *Production de logiciels*
56 *Other computer services n.e.c.*	..	..	..	..	56.1	31.5	..	56 *Autres activités informatiques*
57 Research & development	..	38.0	44.8 [c]	51.3 [c]	53.8	130.4	..	57 Recherche & développement expérimental
58 Other business activities n.e.c.	..	159.4 [o]	180.3 [c,o]	193.9 [c,o]	224.6 [o]	108.0 [o]	..	58 Autres activ. & serv. aux entreprises
59 Comm., soc. & pers. serv. activ., etc.	..	.. [n]	.. [n]	.. [n]	.. [n]	.. [n]	..	**59 Adm. pub., serv. soc. & collec., etc**
TOTAL BUSINESS ENTERPRISE SECTOR	..	**1 264.3**	**1 388.1** [c]	**1 509.4**	**1 713.6**	**1 735.2**	..	**TOTAL SECTEUR DES ENTREPRISES**

Source: OECD, R&D database (DSTI/EAS Division), January 2002.

Source : OCDE, base de données R-D (DSTI/Division AES), janvier 2002.

DENMARK — Table 10 - Tableau 10 — DANEMARK

Business enterprise intramural expenditure on R&D (BERD) - Dépenses intra-muros de R-D des entreprises (DIRDE)

by industry and source of funds - par industrie et par source de financement

Million DKK — *Millions DKK*

10.1 Business enterprise - Entreprises

	1994	1995	1996	1997	1998	1999	2000	
1 AGRICULTURE, HUNTING & FORESTRY	..	20.0 [o]	..	.. [n]	.. [n]	.. [n]	..	**1 AGRICULTURE, CHASSE & SYLVICULTURE**
2 MINING	..	.. [n]	..	..	..	..	..	**2 ACTIVITES EXTRACTIVES**
3 MANUFACTURING	..	5 851.0	..	7 805.7	9 331.2	..	..	**3 ACTIVITES DE FABRICATION**
4 Food, beverages & tobacco	..	349.0	..	409.5	429.6	..	..	**4 Produits alimentaires, boissons & tabac**
5 Food products & beverages	..	..	..	..	..	..	..	5 Produits alimentaires & boissons
6 Tobacco products	..	..	..	..	..	..	..	6 Produits à base de tabac
7 Textiles, wear app., fur & leather	..	10.0	..	13.5	.. [n]	.. [n]	..	**7 Textiles, habill., fourrures & cuirs**
8 Textiles	..	..	..	..	..	..	..	8 Textiles
9 Wearing apparel & fur	..	..	..	..	..	..	..	9 Articles d'habillement & fourrures
10 Leather products & footwear	..	..	..	..	..	..	..	10 Articles en cuir & chaussures
11 Wood, paper, printing, publishing	..	..	..	..	..	..	..	**11 Bois, papier, imprimerie, édition**
12 Wood & cork (not furniture)	..	..	..	..	..	..	..	12 Bois & liège (sauf meubles)
13 Pulp, paper & paper products	..	..	..	..	..	..	..	13 Carton, papier & articles
14 Publ., print. & repro. of rec. media	..	..	..	..	..	..	..	14 Edition, impress. & repro. enregis.
15 Coke, petroleum, nuclear fuel, chemicals, rubber plastics	..	1 696.0	..	3 160.1	4 226.0	..	..	**15 Coke, pétrole, combust. nucl., prod. chimi., caout. & mat. plast.**
16 Coke, ref. petrol. prod. & nuclear fuel	..	.. [n]	..	.. [n]	.. [n]	..	..	16 Coke, raffin. pétr. & combust. nucl.
17 Chemicals & chemical products	..	1 579.0	..	2 993.2	4 047.5	..	..	17 Produits chimiques
18 *Chemical prod.(less pharmaceuticals)*	..	299.0	..	475.6	513.1	..	..	18 *Prod. chimi. (moins prod. pharmaceutiques)*
19 *Pharmaceuticals*	..	1 280.0	..	2 517.6	3 534.4	..	..	19 *Produits pharmaceutiques*
20 Rubber & plastic products	..	117.0	..	166.9	178.4	..	..	20 Caoutchouc et matières plastiques
21 Non-metallic mineral products	..	39.0	..	57.8	66.7	..	..	**21 Produits minéraux non métalliques**
22 Basic metals	..	90.0 [o]	..	87.0 [o]	102.2 [o]	..	..	**22 Produits métallurgiques de base**
23 Basic metals, ferrous	..	..	..	..	..	..	..	23 Produits métallur. de base, ferreux
24 Basic metals, non-ferrous	..	..	..	..	..	..	..	24 Prod. métallur. de base, non ferreux
25 Fabricated metal products	..	.. [n]	..	.. [n]	.. [n]	..	..	**25 Ouvrages en métaux (sauf mach.& mat.)**
26 Machinery equipment, instruments and transport equipment	..	3 149.0	..	3 592.2	3 762.7	..	..	**26 Machines & matériel, instruments & mat. de transport**
27 Machinery n.e.c.	..	1 616.0	..	1 740.0	1 533.0	..	..	27 Machines, n.c.a.
28 Office, account. & computing machinery	..	98.0	..	92.4	.. [n]	..	..	28 Mach. de bur., compta. & trait. inform.
29 Electrical machinery	..	174.0	..	153.7	273.6	..	..	29 Machines électriques
30 Electro. equip. (radio, TV & commun.)	..	584.0	..	854.2	1 033.9	..	..	30 Equip. électro. (radio, tv & communic.)
31 *Electro. comp. (inc. semi-conductors)*	..	..	..	..	..	..	..	31 *Compos. électro. (semi-cond. incl.)*
32 *TV, radio & communications equip.*	..	..	..	..	..	..	..	32 *Appareils radio, tv et de communi.*
33 Instruments	..	610.0	..	668.3	825.3	..	..	33 Instruments
34 Motor vehicles	..	.. [n]	..	.. [n]	.. [n]	..	..	34 Véhicules automobiles
35 Other transport equipment	..	67.0 [o]	..	81.8 [o]	97.0 [o]	..	..	35 Autres matériels de transport
36 *Ships*	..	..	..	..	..	..	..	36 *Navires*
37 *Aerospace*	..	..	..	..	..	..	..	37 *Aérospatiale*
38 *Other transport equipment n.e.c.*	..	..	..	..	..	..	..	38 *Autres matériels de transport n.c.a.*
39 Furniture, other manufacturing n.e.c.	..	520.0 [o]	..	485.6 [o]	744.1 [o]	..	..	**39 Meubles, autres activ. de fabricat. n.c.a.**
40 Furniture	..	..	..	..	..	..	..	40 Meubles
41 Other manufacturing n.e.c.	..	..	..	..	..	..	..	41 Autres activités de fabrication n.c.a.
42 Recycling	..	.. [n]	..	.. [n]	.. [n]	..	..	**42 Récupération**
43 ELECTRICITY, GAS & WATER SUPPLY	..	..	..	69.7	..	..	..	**43 DISTRIBUTION D'ELECTRICITE, GAZ & EAU**
44 CONSTRUCTION	..	17.0	..	27.0	..	..	..	**44 CONSTRUCTION**
45 SERVICES SECTOR	..	2 324.0 [o]	..	3 556.2 [o]	4 522.5 [o,s]	..	..	**45 SECTEUR DES SERVICES**
46 Wholesale, retail trade & repair of motor vehicles, etc.	..	514.0	..	845.6	986.2	..	..	**46 Commerce (gros & détail), réparation de véhicules, etc.**
47 Hotels & restaurants	..	..	..	..	..	..	..	**47 Hôtels et restaurants**
48 Transport & storage	..	.. [n]	..	.. [n]	.. [n]	..	..	**48 Transports, entreposage**
49 Communications	..	238.0 [o]	..	219.7 [o]	.. [n]	..	..	**49 Communications**
50 Post	..	..	..	..	..	..	..	50 Postes
51 Telecommunications	..	..	..	..	..	..	..	51 Télécommunications
52 Financial intermediation (inc. insur.)	..	..	..	..	..	..	..	**52 Intermédiation financ. (assurances inc.)**
53 Real estate, renting & busin. activity	..	..	..	2 490.9 [o]	3 536.3 [o,s]	..	..	**53 Immobil., locat. & serv. aux entreprises**
54 Computer & related activities	..	688.0	..	1 098.6	1 717.5	..	..	54 Informatique et activités connexes
55 *Software consultancy*	..	..	..	..	..	..	..	55 *Production de logiciels*
56 *Other computer services n.e.c.*	..	..	..	..	..	..	..	56 *Autres activités informatiques*
57 Research & development	..	195.0	..	344.1	385.7	..	..	57 Recherche & développement expérimental
58 Other business activities n.e.c.	..	689.0 [o]	..	1 048.2 [o]	1 433.0 [o,s]	..	..	58 Autres activ. & serv. aux entreprises
59 Comm., soc. & pers. serv. activ., etc.	..	.. [n]	..	.. [n]	.. [n]	..	..	**59 Adm. pub., serv. soc. & collec., etc**
TOTAL BUSINESS ENTERPRISE SECTOR	**..**	**8 182.9**	**9 752.9** [c]	**11 322.9**	**13 853.7**	**14 348.3**	**..**	**TOTAL SECTEUR DES ENTREPRISES**

Source: OECD, R&D database (DSTI/EAS Division), January 2002.

Source : OCDE, base de données R-D (DSTI/Division AES), janvier 2002.

DENMARK — Table 10 - Tableau 10 — DANEMARK

Business enterprise intramural expenditure on R&D (BERD) - Dépenses intra-muros de R-D des entreprises (DIRDE)

by industry and source of funds - par industrie et par source de financement

Million DKK — *Millions DKK*

10.2 Government - État

	1994	1995	1996	1997	1998	1999	2000	
1 AGRICULTURE, HUNTING & FORESTRY	..	2.0 [o]	..	.. [n]	.. [n]	..	..	**1 AGRICULTURE, CHASSE & SYLVICULTURE**
2 MINING	..	.. [n]	..	.. [n]	.. [n]	..	..	**2 ACTIVITES EXTRACTIVES**
3 MANUFACTURING	..	185.0	..	110.5	92.9	..	..	**3 ACTIVITES DE FABRICATION**
4 Food, beverages & tobacco	..	40.0	..	12.3	18.2	..	..	**4 Produits alimentaires, boissons & tabac**
5 Food products & beverages	..	..	..	..	..	..	..	5 Produits alimentaires & boissons
6 Tobacco products	..	..	..	..	..	..	..	6 Produits à base de tabac
7 Textiles, wear app., fur & leather	..	0.0	..	0.0	.. [n]	..	..	**7 Textiles, habill., fourrures & cuirs**
8 Textiles	..	..	..	..	..	..	..	8 Textiles
9 Wearing apparel & fur	..	..	..	..	..	..	..	9 Articles d'habillement & fourrures
10 Leather products & footwear	..	..	..	..	..	..	..	10 Articles en cuir & chaussures
11 Wood, paper, printing, publishing	..	..	..	..	..	..	..	**11 Bois, papier, imprimerie, édition**
12 Wood & cork (not furniture)	..	..	..	..	..	..	..	12 Bois & liège (sauf meubles)
13 Pulp, paper & paper products	..	..	..	..	..	..	..	13 Carton, papier & articles
14 Publ., print. & repro. of rec. media	..	..	..	..	..	..	..	14 Edition, impress. & repro. enregis.
15 Coke, petroleum, nuclear fuel, chemicals, rubber plastics	..	17.0	..	17.4	7.8	..	..	**15 Coke, pétrole, combust. nucl., prod. chimi., caout. & mat. plast.**
16 Coke, ref. petrol. prod. & nuclear fuel	..	.. [n]	..	.. [n]	.. [n]	..	..	16 Coke, raffin. pétr. & combust. nucl.
17 Chemicals & chemical products	..	15.0	..	15.7	6.6	..	..	17 Produits chimiques
18 *Chemical prod.(less pharmaceuticals)*	..	3.0	..	4.1	2.2	..	..	18 *Prod. chimi. (moins prod. pharmaceutiques)*
19 *Pharmaceuticals*	..	12.0	..	11.7	4.4	..	..	19 *Produits pharmaceutiques*
20 Rubber & plastic products	..	2.0	..	1.6	1.2	..	..	20 Caoutchouc et matières plastiques
21 Non-metallic mineral products	..	2.0	..	2.0	1.5	..	..	**21 Produits minéraux non métalliques**
22 Basic metals	..	2.0 [o]	..	0.9 [o]	1.8 [o]	..	..	**22 Produits métallurgiques de base**
23 Basic metals, ferrous	..	..	..	..	..	..	..	23 Produits métallur. de base, ferreux
24 Basic metals, non-ferrous	..	..	..	..	..	..	..	24 Prod. métallur. de base, non ferreux
25 Fabricated metal products	..	.. [n]	..	.. [n]	.. [n]	..	..	**25 Ouvrages en métaux (sauf mach.& mat.)**
26 Machinery equipment, instruments and transport equipment	..	77.0	..	42.6	42.4	..	..	**26 Machines & matériel, instruments & mat. de transport**
27 Machinery n.e.c.	..	25.0	..	18.5	29.7	..	..	27 Machines, n.c.a.
28 Office, account. & computing machinery	..	0.0	..	0.0	.. [n]	..	..	28 Mach. de bur., compta. & trait. inform.
29 Electrical machinery	..	5.0	..	2.5	1.9	..	..	29 Machines électriques
30 Electro. equip. (radio, TV & commun.)	..	29.0	..	12.7	5.8	..	..	30 Equip. électro. (radio, tv & communic.)
31 *Electro. comp. (inc. semi-conductors)*	..	..	..	..	..	..	..	31 *Compos. électro. (semi-cond. incl.)*
32 *TV, radio & communications equip.*	..	..	..	..	..	..	..	32 *Appareils radio, tv et de communi.*
33 Instruments	..	16.0	..	6.3	4.8	..	..	33 Instruments
34 Motor vehicles	..	.. [n]	..	.. [n]	.. [n]	..	..	34 Véhicules automobiles
35 Other transport equipment	..	2.0 [o]	..	0.8	0.2	..	..	35 Autres matériels de transport
36 *Ships*	..	..	..	..	..	..	..	36 *Navires*
37 *Aerospace*	..	..	..	..	..	..	..	37 *Aérospatiale*
38 *Other transport equipment n.e.c.*	..	..	..	..	..	..	..	38 *Autres matériels de transport n.c.a.*
39 Furniture, other manufacturing n.e.c.	..	47.0 [o]	..	35.4 [o]	21.2 [o]	..	..	**39 Meubles, autres activ. de fabricat. n.c.a.**
40 Furniture	..	..	..	..	..	..	..	40 Meubles
41 Other manufacturing n.e.c.	..	..	..	..	..	..	..	41 Autres activités de fabrication n.c.a.
42 Recycling	..	.. [n]	..	.. [n]	.. [n]	..	..	**42 Récupération**
43 ELECTRICITY, GAS & WATER SUPPLY	..	..	..	1.0	.. [n]	..	..	**43 DISTRIBUTION D'ELECTRICITE, GAZ & EAU**
44 CONSTRUCTION	..	6.0	..	0.0	.. [n]	..	..	**44 CONSTRUCTION**
45 SERVICES SECTOR	..	432.0	..	557.4 [o]	553.7 [o,s]	..	..	**45 SECTEUR DES SERVICES**
46 Wholesale, retail trade & repair of motor vehicles, etc.	..	2.0	..	2.6	1.9	..	..	**46 Commerce (gros & détail), réparation de véhicules, etc.**
47 Hotels & restaurants	..	..	..	..	..	..	..	**47 Hôtels et restaurants**
48 Transport & storage	..	..	..	..	..	..	..	**48 Transports, entreposage**
49 Communications	..	0.0	..	0.0	.. [n]	..	..	**49 Communications**
50 Post	..	..	..	..	..	..	..	50 Postes
51 Telecommunications	..	..	..	..	..	..	..	51 Télécommunications
52 Financial intermediation (inc. insur.)	..	..	..	..	..	..	..	**52 Intermédiation financ. (assurances inc.)**
53 Real estate, renting & busin. activity	..	..	..	554.8 [o]	551.8 [o,s]	..	..	**53 Immobil., locat. & serv. aux entreprises**
54 Computer & related activities	..	16.0	..	22.2	22.3	..	..	54 Informatique et activités connexes
55 *Software consultancy*	..	..	..	..	..	..	..	55 *Production de logiciels*
56 *Other computer services n.e.c.*	..	..	..	..	..	..	..	56 *Autres activités informatiques*
57 Research & development	..	23.0	..	30.0	33.4	..	..	57 Recherche & développement expérimental
58 Other business activities n.e.c.	..	391.0 [o]	..	502.7 [o]	496.0 [o,s]	..	..	58 Autres activ. & serv. aux entreprises
59 Comm., soc. & pers. serv. activ., etc.	..	.. [n]	..	.. [n]	.. [n]	..	..	**59 Adm. pub., serv. soc. & collec., etc**
TOTAL BUSINESS ENTERPRISE SECTOR	..	**652.3**	**677.5** [c]	**702.6**	**646.6**	**700.5**	..	**TOTAL SECTEUR DES ENTREPRISES**

Source: OECD, R&D database (DSTI/EAS Division), January 2002.

Source : OCDE, base de données R-D (DSTI/Division AES), janvier 2002.

Business enterprise intramural expenditure on R&D (BERD) - Dépenses intra-muros de R-D des entreprises (DIRDE)

by industry and source of funds - par industrie et par source de financement

Million DKK *Millions DKK*

	1994	1995	1996	1997	1998	1999	2000	
	10.3 Other national funds - Autres sources nationales							
1 AGRICULTURE, HUNTING & FORESTRY	..	0.0	..	.. n	.. n	..	..	**1 AGRICULTURE, CHASSE & SYLVICULTURE**
2 MINING	..	..	..	..	..	..	..	**2 ACTIVITES EXTRACTIVES**
3 MANUFACTURING	..	49.0	..	22.3	45.9	..	..	**3 ACTIVITES DE FABRICATION**
4 Food, beverages & tobacco	..	48.0	..	13.6	16.6	..	..	**4 Produits alimentaires, boissons & tabac**
5 Food products & beverages	..	..	..	..	..	..	..	5 Produits alimentaires & boissons
6 Tobacco products	..	..	..	..	..	..	..	6 Produits à base de tabac
7 Textiles, wear app., fur & leather	..	0.0	..	0.0	.. n	..	..	**7 Textiles, habill., fourrures & cuirs**
8 Textiles	..	..	..	..	..	..	..	8 Textiles
9 Wearing apparel & fur	..	..	..	..	..	..	..	9 Articles d'habillement & fourrures
10 Leather products & footwear	..	..	..	..	..	..	..	10 Articles en cuir & chaussures
11 Wood, paper, printing, publishing	..	..	..	..	..	..	..	**11 Bois, papier, imprimerie, édition**
12 Wood & cork (not furniture)	..	..	..	..	..	..	..	12 Bois & liège (sauf meubles)
13 Pulp, paper & paper products	..	..	..	..	..	..	..	13 Carton, papier & articles
14 Publ., print. & repro. of rec. media	..	..	..	..	..	..	..	14 Edition, impress. & repro. enregis.
15 Coke, petroleum, nuclear fuel, chemicals, rubber plastics	..	0.0	..	0.2	0.4	..	..	**15 Coke, pétrole, combust. nucl., prod. chimi., caout. & mat. plast.**
16 Coke, ref. petrol. prod. & nuclear fuel	..	..	..	..	..	..	..	16 Coke, raffin. pétr. & combust. nucl.
17 Chemicals & chemical products	..	0.0	..	0.0	0.0	..	..	17 Produits chimiques
18 *Chemical prod.(less pharmaceuticals)*	..	0.0	..	0.0	0.0	..	..	18 *Prod. chimi. (moins prod. pharmaceutiques)*
19 *Pharmaceuticals*	..	0.0	..	0.0	0.0	..	..	19 *Produits pharmaceutiques*
20 Rubber & plastic products	..	0.0	..	0.2	0.4	..	..	20 Caoutchouc et matières plastiques
21 Non-metallic mineral products	..	0.0	..	0.0	0.0	..	..	**21 Produits minéraux non métalliques**
22 Basic metals	..	0.0	..	0.0	25.5	..	..	**22 Produits métallurgiques de base**
23 Basic metals, ferrous	..	..	..	..	..	..	..	23 Produits métallur. de base, ferreux
24 Basic metals, non-ferrous	..	..	..	..	..	..	..	24 Prod. métallur. de base, non ferreux
25 Fabricated metal products	..	.. n	..	.. n	.. n	..	..	**25 Ouvrages en métaux (sauf mach.& mat.)**
26 Machinery equipment, instruments and transport equipment	..	0.0	..	2.7	3.4	..	..	**26 Machines & matériel, instruments & mat. de transport**
27 Machinery n.e.c.	..	0.0	..	0.2	0.8	..	..	27 Machines, n.c.a.
28 Office, account. & computing machinery	..	0.0	..	0.0	.. n	..	..	28 Mach. de bur., compta. & trait. inform.
29 Electrical machinery	..	0.0	..	0.0	0.7	..	..	29 Machines électriques
30 Electro. equip. (radio, TV & commun.)	..	0.0	..	0.4	1.8	..	..	30 Equip. électro. (radio, tv & communic.)
31 *Electro. comp. (inc. semi-conductors)*	..	..	..	..	..	..	..	31 *Compos. électro. (semi-cond. incl.)*
32 *TV, radio & communications equip.*	..	..	..	..	..	..	..	32 *Appareils radio, tv et de communi.*
33 Instruments	..	0.0	..	2.1	0.1	..	..	33 Instruments
34 Motor vehicles	..	..	..	..	..	..	..	34 Véhicules automobiles
35 Other transport equipment	..	0.0	..	0.0	0.0	..	..	35 Autres matériels de transport
36 *Ships*	..	..	..	..	..	..	..	36 *Navires*
37 *Aerospace*	..	..	..	..	..	..	..	37 *Aérospatiale*
38 *Other transport equipment n.e.c.*	..	..	..	..	..	..	..	38 *Autres matériels de transport n.c.a.*
39 Furniture, other manufacturing n.e.c.	..	0.0 o	..	5.8 o	0.0 o	..	..	**39 Meubles, autres activ. de fabricat. n.c.a.**
40 Furniture	..	..	..	..	..	..	..	40 Meubles
41 Other manufacturing n.e.c.	..	..	..	..	..	..	..	41 Autres activités de fabrication n.c.a.
42 Recycling	..	..	..	..	..	..	..	**42 Récupération**
43 ELECTRICITY, GAS & WATER SUPPLY	..	..	..	1.1	.. n	..	..	**43 DISTRIBUTION D'ELECTRICITE, GAZ & EAU**
44 CONSTRUCTION	..	2.0	..	0.0	.. n	..	..	**44 CONSTRUCTION**
45 SERVICES SECTOR	..	105.0 o	..	123.4 o	44.8 o,s	..	..	**45 SECTEUR DES SERVICES**
46 Wholesale, retail trade & repair of motor vehicles, etc.	..	2.0	..	0.0	1.1	..	..	**46 Commerce (gros & détail), réparation de véhicules, etc.**
47 Hotels & restaurants	..	..	..	..	..	..	..	**47 Hôtels et restaurants**
48 Transport & storage	..	.. n	..	.. n	.. n	..	..	**48 Transports, entreposage**
49 Communications	..	0.0	..	0.2	.. n	..	..	**49 Communications**
50 Post	..	..	..	..	..	..	..	50 Postes
51 Telecommunications	..	..	..	..	..	..	..	51 Télécommunications
52 Financial intermediation (inc. insur.)	..	..	..	..	..	..	..	**52 Intermédiation financ. (assurances inc.)**
53 Real estate, renting & busin. activity	..	..	..	123.2 o	43.7 o,s	..	..	**53 Immobil., locat. & serv. aux entreprises**
54 Computer & related activities	..	66.0	..	86.1	13.9	..	..	54 Informatique et activités connexes
55 *Software consultancy*	..	..	..	..	..	..	..	55 *Production de logiciels*
56 *Other computer services n.e.c.*	..	..	..	..	..	..	..	56 *Autres activités informatiques*
57 Research & development	..	17.0	..	12.8	3.9	..	..	57 Recherche & développement expérimental
58 Other business activities n.e.c.	..	20.0 o	..	24.4 o	25.8 o,s	..	..	58 Autres activ. & serv. aux entreprises
59 Comm., soc. & pers. serv. activ., etc.	..	.. n	..	.. n	.. n	..	..	**59 Adm. pub., serv. soc. & collec., etc**
TOTAL BUSINESS ENTERPRISE SECTOR	..	**157.2**	**154.3 c**	**151.3**	**90.7**	**95.8**	..	**TOTAL SECTEUR DES ENTREPRISES**

Source: OECD, R&D database (DSTI/EAS Division), January 2002.

Source : OCDE, base de données R-D (DSTI/Division AES), janvier 2002.

Business enterprise intramural expenditure on R&D (BERD) - Dépenses intra-muros de R-D des entreprises (DIRDE)

by industry and source of funds - par industrie et par source de financement

Million DKK / *Millions DKK*

	1994	1995	1996	1997	1998	1999	2000	
	10.4 Funds from abroad - Fonds de l'étranger							
1 **AGRICULTURE, HUNTING & FORESTRY**	..	1.0 [o]	..	.. [n]	.. [n]	..	..	1 **AGRICULTURE, CHASSE & SYLVICULTURE**
2 **MINING**	..	.. [n]	..	..	..	..	..	2 **ACTIVITES EXTRACTIVES**
3 **MANUFACTURING**	..	1 142.0	..	460.7	468.0	..	..	3 **ACTIVITES DE FABRICATION**
4 **Food, beverages & tobacco**	..	9.0	..	5.8	38.5	..	..	4 **Produits alimentaires, boissons & tabac**
5 Food products & beverages	..	..	..	..	..	..	..	5 Produits alimentaires & boissons
6 Tobacco products	..	..	..	..	..	..	..	6 Produits à base de tabac
7 **Textiles, wear app., fur & leather**	..	0.0	..	0.2	.. [n]	..	..	7 **Textiles, habill., fourrures & cuirs**
8 Textiles	..	..	..	..	..	..	..	8 Textiles
9 Wearing apparel & fur	..	..	..	..	..	..	..	9 Articles d'habillement & fourrures
10 Leather products & footwear	..	..	..	..	..	..	..	10 Articles en cuir & chaussures
11 **Wood, paper, printing, publishing**	..	..	..	..	..	..	..	11 **Bois, papier, imprimerie, édition**
12 Wood & cork (not furniture)	..	..	..	..	..	..	..	12 Bois & liège (sauf meubles)
13 Pulp, paper & paper products	..	..	..	..	..	..	..	13 Carton, papier & articles
14 Publ., print. & repro. of rec. media	..	..	..	..	..	..	..	14 Edition, impress. & repro. enregis.
15 **Coke, petroleum, nuclear fuel, chemicals, rubber plastics**	..	861.0	..	133.4	58.2	..	..	15 **Coke, pétrole, combust. nucl., prod. chimi., caout. & mat. plast.**
16 Coke, ref. petrol. prod. & nuclear fuel	..	.. [n]	..	.. [n]	.. [n]	..	..	16 Coke, raffin. pétr. & combust. nucl.
17 Chemicals & chemical products	..	859.0	..	133.1	55.8	..	..	17 Produits chimiques
18 *Chemical prod.(less pharmaceuticals)*	..	19.0	..	9.8	6.2	..	..	18 *Prod. chimi. (moins prod. pharmaceutiques)*
19 *Pharmaceuticals*	..	840.0	..	123.3	49.5	..	..	19 *Produits pharmaceutiques*
20 Rubber & plastic products	..	2.0	..	0.3	2.4	..	..	20 Caoutchouc et matières plastiques
21 **Non-metallic mineral products**	..	8.0	..	2.5	8.8	..	..	21 **Produits minéraux non métalliques**
22 **Basic metals**	..	2.0 [o]	..	9.0	0.6	..	..	22 **Produits métallurgiques de base**
23 Basic metals, ferrous	..	..	..	..	..	..	..	23 Produits métallur. de base, ferreux
24 Basic metals, non-ferrous	..	..	..	..	..	..	..	24 Prod. métallur. de base, non ferreux
25 **Fabricated metal products**	..	.. [n]	..	.. [n]	.. [n]	..	..	25 **Ouvrages en métaux (sauf mach.& mat.)**
26 **Machinery equipment, instruments and transport equipment**	..	255.0	..	287.1	356.9	..	..	26 **Machines & matériel, instruments & mat. de transport**
27 Machinery n.e.c.	..	157.0	..	176.9	237.5	..	..	27 Machines, n.c.a.
28 Office, account. & computing machinery	..	0.0	..	0.0	.. [n]	..	..	28 Mach. de bur., compta. & trait. inform.
29 Electrical machinery	..	8.0	..	33.8	33.7	..	..	29 Machines électriques
30 Electro. equip. (radio, TV & commun.)	..	58.0	..	48.6	54.6	..	..	30 Equip. électro. (radio, tv & communic.)
31 *Electro. comp. (inc. semi-conductors)*	..	..	..	..	..	..	..	31 *Compos. électro. (semi-cond. incl.)*
32 *TV, radio & communications equip.*	..	..	..	..	..	..	..	32 *Appareils radio, tv et de communi.*
33 Instruments	..	10.0	..	10.3	22.2	..	..	33 Instruments
34 Motor vehicles	..	.. [n]	..	.. [n]	.. [n]	..	..	34 Véhicules automobiles
35 Other transport equipment	..	22.0 [o]	..	17.6 [o]	8.9 [o]	..	..	35 Autres matériels de transport
36 *Ships*	..	..	..	..	..	..	..	36 *Navires*
37 *Aerospace*	..	..	..	..	..	..	..	37 *Aérospatiale*
38 *Other transport equipment n.e.c.*	..	..	..	..	..	..	..	38 *Autres matériels de transport n.c.a.*
39 **Furniture, other manufacturing n.e.c.**	..	8.0 [o]	..	22.7 [o]	4.9 [o]	..	..	39 **Meubles, autres activ. de fabricat. n.c.a.**
40 Furniture	..	..	..	..	..	..	..	40 Meubles
41 Other manufacturing n.e.c.	..	..	..	..	..	..	..	41 Autres activités de fabrication n.c.a.
42 **Recycling**	..	.. [n]	..	..	..	..	..	42 **Récupération**
43 **ELECTRICITY, GAS & WATER SUPPLY**	..	..	..	0.0	.. [n]	..	..	43 **DISTRIBUTION D'ELECTRICITE, GAZ & EAU**
44 **CONSTRUCTION**	..	0.0	..	0.0	.. [n]	..	..	44 **CONSTRUCTION**
45 **SERVICES SECTOR**	..	464.0 [o]	..	569.9 [o]	335.2 [o,s]	..	..	45 **SECTEUR DES SERVICES**
46 **Wholesale, retail trade & repair of motor vehicles, etc.**	..	80.0	..	339.9	87.2	..	..	46 **Commerce (gros & détail), réparation de véhicules, etc.**
47 **Hotels & restaurants**	..	..	..	..	..	..	..	47 **Hôtels et restaurants**
48 **Transport & storage**	..	.. [n]	..	.. [n]	.. [n]	..	..	48 **Transports, entreposage**
49 **Communications**	..	20.0 [o]	..	6.9 [o]	.. [n]	..	..	49 **Communications**
50 Post	..	..	..	..	..	..	..	50 Postes
51 Telecommunications	..	..	..	..	..	..	..	51 Télécommunications
52 **Financial intermediation (inc. insur.)**	..	..	..	..	..	..	..	52 **Intermédiation financ. (assurances inc.)**
53 **Real estate, renting & busin. activity**	..	..	..	223.2 [o]	248.0 [o,s]	..	..	53 **Immobil., locat. & serv. aux entreprises**
54 Computer & related activities	..	38.0	..	24.6	42.3	..	..	54 Informatique et activités connexes
55 *Software consultancy*	..	..	..	..	..	..	..	55 *Production de logiciels*
56 *Other computer services n.e.c.*	..	..	..	..	..	..	..	56 *Autres activités informatiques*
57 Research & development	..	85.0	..	65.6	60.3	..	..	57 Recherche & développement expérimental
58 Other business activities n.e.c.	..	241.0 [o]	..	133.0 [o]	145.4 [o,s]	..	..	58 Autres activ. & serv. aux entreprises
59 **Comm., soc. & pers. serv. activ., etc.**	..	.. [n]	..	.. [n]	.. [n]	..	..	59 **Adm. pub., serv. soc. & collec., etc**
TOTAL BUSINESS ENTERPRISE SECTOR	..	**1 648.6**	**1 388.4** [c]	**1 128.2**	**803.2**	**909.2**	..	**TOTAL SECTEUR DES ENTREPRISES**

Source: OECD, R&D database (DSTI/EAS Division), January 2002.

Source : OCDE, base de données R-D (DSTI/Division AES), janvier 2002.

Business enterprise intramural expenditure on R&D (BERD) - Dépenses intra-muros de R-D des entreprises (DIRDE)

by industry and type of costs - par industrie et par type de coûts

Million DKK — *Millions DKK*

11.1 Current expenditure - Dépenses courantes

	1994	1995	1996	1997	1998	1999	2000	
1 AGRICULTURE, HUNTING & FORESTRY	..	20.0 [o]	..	.. [n]	.. [n]	..	..	**1 AGRICULTURE, CHASSE & SYLVICULTURE**
2 MINING	..	.. [n]	..	.. [n]	.. [n]	..	..	**2 ACTIVITES EXTRACTIVES**
3 MANUFACTURING	..	6 368.0	..	7 463.0	8 549.0	..	..	**3 ACTIVITES DE FABRICATION**
4 Food, beverages & tobacco	..	418.0	..	391.0	464.0	..	..	**4 Produits alimentaires, boissons & tabac**
5 Food products & beverages	..	..	..	..	..	..	..	5 Produits alimentaires & boissons
6 Tobacco products	..	..	..	..	..	..	..	6 Produits à base de tabac
7 Textiles, wear app., fur & leather	..	10.0	..	13.0	.. [n]	..	..	**7 Textiles, habill., fourrures & cuirs**
8 Textiles	..	..	..	..	..	..	..	8 Textiles
9 Wearing apparel & fur	..	..	..	..	..	..	..	9 Articles d'habillement & fourrures
10 Leather products & footwear	..	..	..	..	..	..	..	10 Articles en cuir & chaussures
11 Wood, paper, printing, publishing	..	..	..	..	..	..	..	**11 Bois, papier, imprimerie, édition**
12 Wood & cork (not furniture)	..	..	..	..	..	..	..	12 Bois & liège (sauf meubles)
13 Pulp, paper & paper products	..	..	..	..	..	..	..	13 Carton, papier & articles
14 Publ., print. & repro. of rec. media	..	..	..	..	..	..	..	14 Edition, impress. & repro. enregis.
15 Coke, petroleum, nuclear fuel, chemicals, rubber plastics	..	2 189.0	..	2 933.0	3 569.0	..	..	**15 Coke, pétrole, combust. nucl., prod. chimi., caout. & mat. plast.**
16 Coke, ref. petrol. prod. & nuclear fuel	..	.. [n]	..	.. [n]	.. [n]	..	..	16 Coke, raffin. pétr. & combust. nucl.
17 Chemicals & chemical products	..	2 090.0	..	2 776.0	3 419.0	..	..	17 Produits chimiques
18 *Chemical prod.(less pharmaceuticals)*	..	281.0	..	440.0	441.0	..	..	18 *Prod. chimi. (moins prod. pharmaceutiques)*
19 *Pharmaceuticals*	..	1 809.0	..	2 336.0	2 978.0	..	..	19 *Produits pharmaceutiques*
20 Rubber & plastic products	..	109.0	..	157.0	150.0	..	..	20 Caoutchouc et matières plastiques
21 Non-metallic mineral products	..	45.0	..	57.0	69.0	..	..	**21 Produits minéraux non métalliques**
22 Basic metals	..	69.0 [o]	..	91.0 [o]	92.0 [o]	..	..	**22 Produits métallurgiques de base**
23 Basic metals, ferrous	..	..	..	..	..	..	..	23 Produits métallur. de base, ferreux
24 Basic metals, non-ferrous	..	..	..	..	..	..	..	24 Prod. métallur. de base, non ferreux
25 Fabricated metal products	..	.. [n]	..	.. [n]	.. [n]	..	..	**25 Ouvrages en métaux (sauf mach.& mat.)**
26 Machinery equipment, instruments and transport equipment	..	2 489.0	..	3 489.0	3 666.0	..	..	**26 Machines & matériel, instruments & mat. de transport**
27 Machinery n.e.c.	..	1 557.0	..	1 680.0	1 642.0	..	..	27 Machines, n.c.a.
28 Office, account. & computing machinery	..	93.0	..	89.0	.. [n]	..	..	28 Mach. de bur., compta. & trait. inform.
29 Electrical machinery	..	171.0	..	160.0	269.0	..	..	29 Machines électriques
30 Electro. equip. (radio, TV & commun.)	..	584.0	..	830.0	866.0	..	..	30 Equip. électro. (radio, tv & communic.)
31 *Electro. comp. (inc. semi-conductors)*	..	..	..	..	..	..	..	31 *Compos. électro. (semi-cond. incl.)*
32 *TV, radio & communications equip.*	..	..	..	..	..	..	..	32 *Appareils radio, tv et de communi.*
33 Instruments	..	595.0	..	636.0	792.0	..	..	33 Instruments
34 Motor vehicles	..	.. [n]	..	.. [n]	.. [n]	..	..	34 Véhicules automobiles
35 Other transport equipment	..	84.0 [o]	..	94.0 [o]	97.0 [o]	..	..	35 Autres matériels de transport
36 *Ships*	..	..	..	..	..	..	..	36 *Navires*
37 *Aerospace*	..	..	..	..	..	..	..	37 *Aérospatiale*
38 *Other transport equipment n.e.c.*	..	..	..	..	..	..	..	38 *Autres matériels de transport n.c.a.*
39 Furniture, other manufacturing n.e.c.	..	542.0 [o]	..	489.0 [o]	690.0 [o]	..	..	**39 Meubles, autres activ. de fabricat. n.c.a.**
40 Furniture	..	..	..	..	..	..	..	40 Meubles
41 Other manufacturing n.e.c.	..	..	..	..	..	..	..	41 Autres activités de fabrication n.c.a.
42 Recycling	..	.. [n]	..	.. [n]	.. [n]	..	..	**42 Récupération**
43 ELECTRICITY, GAS & WATER SUPPLY	..	36.0	..	64.0	41.0	..	..	**43 DISTRIBUTION D'ELECTRICITE, GAZ & EAU**
44 CONSTRUCTION	..	22.0	..	26.0	28.0	..	..	**44 CONSTRUCTION**
45 SERVICES SECTOR	..	3 105.0 [o]	..	4 496.0 [o]	4 974.0 [o]	..	..	**45 SECTEUR DES SERVICES**
46 Wholesale, retail trade & repair of motor vehicles, etc.	..	509.0	..	1 115.0	1 017.0	..	..	**46 Commerce (gros & détail), réparation de véhicules, etc.**
47 Hotels & restaurants	..	..	..	..	..	..	..	**47 Hôtels et restaurants**
48 Transport & storage	..	.. [n]	..	.. [n]	.. [n]	..	..	**48 Transports, entreposage**
49 Communications	..	229.0 [o]	..	218.0 [o]	.. [n]	..	..	**49 Communications**
50 Post	..	..	..	..	..	..	..	50 Postes
51 Telecommunications	..	..	..	..	..	..	..	51 Télécommunications
52 Financial intermediation (inc. insur.)	..	.. [n]	..	.. [n]	.. [n]	..	..	**52 Intermédiation financ. (assurances inc.)**
53 Real estate, renting & busin. activity	..	2 365.0 [o]	..	3 163.0 [o]	3 957.0 [o]	..	..	**53 Immobil., locat. & serv. aux entreprises**
54 Computer & related activities	..	763.0	..	1 184.0	1 703.0	..	..	54 Informatique et activités connexes
55 *Software consultancy*	..	..	..	..	..	..	..	55 *Production de logiciels*
56 *Other computer services n.e.c.*	..	..	..	..	..	..	..	56 *Autres activités informatiques*
57 Research & development	..	297.0	..	419.0	448.0	..	..	57 Recherche & développement expérimental
58 Other business activities n.e.c.	..	1 307.0 [o]	..	1 560.0 [o]	1 806.0 [o]	..	..	58 Autres activ. & serv. aux entreprises
59 Comm., soc. & pers. serv. activ., etc.	..	.. [n]	..	.. [n]	.. [n]	..	..	**59 Adm. pub., serv. soc. & collec., etc**
TOTAL BUSINESS ENTERPRISE SECTOR	..	**9 550.0**	..	**12 047.0**	**13 593.0**	**13 949.0**	..	**TOTAL SECTEUR DES ENTREPRISES**

Source: OECD, R&D database (DSTI/EAS Division), January 2002.

Source : OCDE, base de données R-D (DSTI/Division AES), janvier 2002.

Business enterprise intramural expenditure on R&D (BERD) - Dépenses intra-muros de R-D des entreprises (DIRDE)

by industry and type of costs - par industrie et par type de coûts

Million DKK — *Millions DKK*

11.2 Capital expenditure - Dépenses en capital

	1994	1995	1996	1997	1998	1999	2000	
1 AGRICULTURE, HUNTING & FORESTRY	..	3.0 [o]	..	.. [n]	.. [n]	..	..	**1 AGRICULTURE, CHASSE & SYLVICULTURE**
2 MINING	..	.. [n]	..	.. [n]	.. [n]	..	..	**2 ACTIVITES EXTRACTIVES**
3 MANUFACTURING	..	857.0	..	936.0	1 389.0	..	..	**3 ACTIVITES DE FABRICATION**
4 Food, beverages & tobacco	..	12.0	..	49.0	38.0	..	..	**4 Produits alimentaires, boissons & tabac**
5 Food products & beverages	..	..	..	..	..	..	..	5 Produits alimentaires & boissons
6 Tobacco products	..	..	..	..	..	..	..	6 Produits à base de tabac
7 Textiles, wear app., fur & leather	..	1.0	..	1.0	.. [n]	..	..	**7 Textiles, habill., fourrures & cuirs**
8 Textiles	..	..	..	..	..	..	..	8 Textiles
9 Wearing apparel & fur	..	..	..	..	..	..	..	9 Articles d'habillement & fourrures
10 Leather products & footwear	..	..	..	..	..	..	..	10 Articles en cuir & chaussures
11 Wood, paper, printing, publishing	..	..	..	..	..	..	..	**11 Bois, papier, imprimerie, édition**
12 Wood & cork (not furniture)	..	..	..	..	..	..	..	12 Bois & liège (sauf meubles)
13 Pulp, paper & paper products	..	..	..	..	..	..	..	13 Carton, papier & articles
14 Publ., print. & repro. of rec. media	..	..	..	..	..	..	..	14 Edition, impress. & repro. enregis.
15 Coke, petroleum, nuclear fuel, chemicals, rubber plastics	..	373.0	..	378.0	725.0	..	..	**15 Coke, pétrole, combust. nucl., prod. chimi., caout. & mat. plast.**
16 Coke, ref. petrol. prod. & nuclear fuel	..	.. [n]	..	.. [n]	.. [n]	..	..	16 Coke, raffin. pétr. & combust. nucl.
17 Chemicals & chemical products	..	361.0	..	366.0	692.0	..	..	17 Produits chimiques
18 *Chemical prod.(less pharmaceuticals)*	..	38.0	..	49.0	81.0	..	..	18 *Prod. chimi. (moins prod. pharmaceutiques)*
19 *Pharmaceuticals*	..	323.0	..	317.0	611.0	..	..	19 *Produits pharmaceutiques*
20 Rubber & plastic products	..	12.0	..	12.0	33.0	..	..	20 Caoutchouc et matières plastiques
21 Non-metallic mineral products	..	3.0	..	5.0	8.0	..	..	**21 Produits minéraux non métalliques**
22 Basic metals	..	27.0 [o]	..	6.0 [o]	38.0 [o]	..	..	**22 Produits métallurgiques de base**
23 Basic metals, ferrous	..	..	..	..	..	..	..	23 Produits métallur. de base, ferreux
24 Basic metals, non-ferrous	..	..	..	..	..	..	..	24 Prod. métallur. de base, non ferreux
25 Fabricated metal products	..	.. [n]	..	.. [n]	.. [n]	..	..	**25 Ouvrages en métaux (sauf mach.& mat.)**
26 Machinery equipment, instruments and transport equipment	..	353.0	..	436.0	500.0	..	..	**26 Machines & matériel, instruments & mat. de transport**
27 Machinery n.e.c.	..	241.0	..	260.0	159.0	..	..	27 Machines, n.c.a.
28 Office, account. & computing machinery	..	5.0	..	4.0	.. [n]	..	..	28 Mach. de bur., compta. & trait. inform.
29 Electrical machinery	..	16.0	..	30.0	41.0	..	..	29 Machines électriques
30 Electro. equip. (radio, TV & commun.)	..	84.0	..	85.0	230.0	..	..	30 Equip. électro. (radio, tv & communic.)
31 *Electro. comp. (inc. semi-conductors)*	..	..	..	..	..	..	..	31 *Compos. électro. (semi-cond. incl.)*
32 *TV, radio & communications equip.*	..	..	..	..	..	..	..	32 *Appareils radio, tv et de communi.*
33 Instruments	..	41.0	..	51.0	61.0	..	..	33 Instruments
34 Motor vehicles	..	.. [n]	..	.. [n]	.. [n]	..	..	34 Véhicules automobiles
35 Other transport equipment	..	7.0 [o]	..	6.0 [o]	9.0 [o]	..	..	35 Autres matériels de transport
36 *Ships*	..	..	..	..	..	..	..	36 *Navires*
37 *Aerospace*	..	..	..	..	..	..	..	37 *Aérospatiale*
38 *Other transport equipment n.e.c.*	..	..	..	..	..	..	..	38 *Autres matériels de transport n.c.a.*
39 Furniture, other manufacturing n.e.c.	..	34.0 [o]	..	60.0 [o]	80.0 [o]	..	..	**39 Meubles, autres activ. de fabricat. n.c.a.**
40 Furniture	..	..	..	..	..	..	..	40 Meubles
41 Other manufacturing n.e.c.	..	..	..	..	..	..	..	41 Autres activités de fabrication n.c.a.
42 Recycling	..	.. [n]	..	.. [n]	.. [n]	..	..	**42 Récupération**
43 ELECTRICITY, GAS & WATER SUPPLY	..	6.0	..	8.0	13.0	..	..	**43 DISTRIBUTION D'ELECTRICITE, GAZ & EAU**
44 CONSTRUCTION	..	2.0	..	1.0	1.0	..	..	**44 CONSTRUCTION**
45 SERVICES SECTOR	..	222.0 [o]	..	312.0 [o]	400.0 [n]	..	..	**45 SECTEUR DES SERVICES**
46 Wholesale, retail trade & repair of motor vehicles, etc.	..	28.0	..	73.0	59.0	..	..	**46 Commerce (gros & détail), réparation de véhicules, etc.**
47 Hotels & restaurants	..	..	..	..	..	..	..	**47 Hôtels et restaurants**
48 Transport & storage	..	.. [n]	..	.. [n]	.. [n]	..	..	**48 Transports, entreposage**
49 Communications	..	30.0 [o]	..	9.0	.. [n]	..	..	**49 Communications**
50 Post	..	..	..	..	..	..	..	50 Postes
51 Telecommunications	..	..	..	..	..	..	..	51 Télécommunications
52 Financial intermediation (inc. insur.)	..	.. [n]	..	.. [n]	.. [n]	..	..	**52 Intermédiation financ. (assurances inc.)**
53 Real estate, renting & busin. activity	..	167.0 [o]	..	230.0 [o]	341.0 [o]	..	..	**53 Immobil., locat. & serv. aux entreprises**
54 Computer & related activities	..	46.0	..	48.0	94.0	..	..	54 Informatique et activités connexes
55 *Software consultancy*	..	..	..	..	..	..	..	55 *Production de logiciels*
56 *Other computer services n.e.c.*	..	..	..	..	..	..	..	56 *Autres activités informatiques*
57 Research & development	..	24.0	..	34.0	36.0	..	..	57 Recherche & développement expérimental
58 Other business activities n.e.c.	..	97.0 [o]	..	148.0 [o]	211.0 [o]	..	..	58 Autres activ. & serv. aux entreprises
59 Comm., soc. & pers. serv. activ., etc.	..	.. [n]	..	.. [n]	.. [n]	..	..	**59 Adm. pub., serv. soc. & collec., etc**
TOTAL BUSINESS ENTERPRISE SECTOR	..	**1 091.0**	..	**1 258.0**	**1 802.0**	**2 104.7**	..	**TOTAL SECTEUR DES ENTREPRISES**

Source: OECD, R&D database (DSTI/EAS Division), January 2002.

Source : OCDE, base de données R-D (DSTI/Division AES), janvier 2002.

Total business enterprise R&D personnel (BEMP) - Personnel total de R-D du secteur des entreprises (PRDE)

by industry - par industrie

In full time equivalent — *En équivalence plein-temps*

	1994	1995	1996	1997	1998	1999	2000	
1 AGRICULTURE, HUNTING & FORESTRY	..	41 [o]	..	..	..	..	..	**1 AGRICULTURE, CHASSE & SYLVICULTURE**
2 MINING	..	.. [n]	..	..	..	..	..	**2 ACTIVITES EXTRACTIVES**
3 MANUFACTURING	..	11 726	..	..	..	..	..	**3 ACTIVITES DE FABRICATION**
4 Food, beverages & tobacco	..	712	..	..	..	..	..	**4 Produits alimentaires, boissons & tabac**
5 Food products & beverages	..	..	..	..	..	..	..	5 Produits alimentaires & boissons
6 Tobacco products	..	..	..	..	..	..	..	6 Produits à base de tabac
7 Textiles, wear app., fur & leather	..	21	..	..	..	..	..	**7 Textiles, habill., fourrures & cuirs**
8 Textiles	..	..	..	..	..	..	..	8 Textiles
9 Wearing apparel & fur	..	..	..	..	..	..	..	9 Articles d'habillement & fourrures
10 Leather products & footwear	..	..	..	..	..	..	..	10 Articles en cuir & chaussures
11 Wood, paper, printing, publishing	..	.. [n]	..	..	..	..	..	**11 Bois, papier, imprimerie, édition**
12 Wood & cork (not furniture)	..	..	..	..	..	..	..	12 Bois & liège (sauf meubles)
13 Pulp, paper & paper products	..	..	..	..	..	..	..	13 Carton, papier & articles
14 Publ., print. & repro. of rec. media	..	..	..	..	..	..	..	14 Edition, impress. & repro. enregis.
15 Coke, petroleum, nuclear fuel, chemicals, rubber plastics	..	4 175	..	..	..	..	..	**15 Coke, pétrole, combust. nucl., prod. chimi., caout. & mat. plast.**
16 Coke, ref. petrol. prod. & nuclear fuel	..	.. [n]	..	..	..	..	..	16 Coke, raffin. pétr. & combust. nucl.
17 Chemicals & chemical products	..	3 954	..	..	..	..	..	17 Produits chimiques
18 *Chemical prod.(less pharmaceuticals)*	..	650	..	..	..	..	..	18 *Prod. chimi. (moins prod. pharmaceutiques)*
19 *Pharmaceuticals*	..	3 304	..	..	..	..	..	19 *Produits pharmaceutiques*
20 Rubber & plastic products	..	221	..	..	..	..	..	20 Caoutchouc et matières plastiques
21 Non-metallic mineral products	..	97	..	..	..	..	..	**21 Produits minéraux non métalliques**
22 Basic metals	..	147 [o]	..	..	..	..	..	**22 Produits métallurgiques de base**
23 Basic metals, ferrous	..	..	..	..	..	..	..	23 Produits métallur. de base, ferreux
24 Basic metals, non-ferrous	..	..	..	..	..	..	..	24 Prod. métallur. de base, non ferreux
25 Fabricated metal products	..	.. [n]	..	..	..	..	..	**25 Ouvrages en métaux (sauf mach.& mat.)**
26 Machinery equipment, instruments and transport equipment	..	5 306	..	..	..	..	..	**26 Machines & matériel, instruments & mat. de transport**
27 Machinery n.e.c.	..	2 314	..	..	..	..	..	27 Machines, n.c.a.
28 Office, account. & computing machinery	..	201	..	..	..	..	..	28 Mach. de bur., compta. & trait. inform.
29 Electrical machinery	..	359	..	..	..	..	..	29 Machines électriques
30 Electro. equip. (radio, TV & commun.)	..	1 121	..	..	..	..	..	30 Equip. électro. (radio, tv & communic.)
31 *Electro. comp. (inc. semi-conductors)*	..	..	..	..	..	..	..	31 *Compos. électro. (semi-cond. incl.)*
32 *TV, radio & communications equip.*	..	..	..	..	..	..	..	32 *Appareils radio, tv et de communi.*
33 Instruments	..	1 159	..	..	..	..	..	33 Instruments
34 Motor vehicles	..	.. [n]	..	..	..	..	..	34 Véhicules automobiles
35 Other transport equipment	..	152 [o]	..	..	..	..	..	35 Autres matériels de transport
36 *Ships*	..	..	..	..	..	..	..	36 *Navires*
37 *Aerospace*	..	..	..	..	..	..	..	37 *Aérospatiale*
38 *Other transport equipment n.e.c.*	..	..	..	..	..	..	..	38 *Autres matériels de transport n.c.a.*
39 Furniture, other manufacturing n.e.c.	..	1 268 [o]	..	..	..	..	..	**39 Meubles, autres activ. de fabricat. n.c.a.**
40 Furniture	..	..	..	..	..	..	..	40 Meubles
41 Other manufacturing n.e.c.	..	..	..	..	..	..	..	41 Autres activités de fabrication n.c.a.
42 Recycling	..	.. [n]	..	..	..	..	..	**42 Récupération**
43 ELECTRICITY, GAS & WATER SUPPLY	..	53	..	..	..	..	..	**43 DISTRIBUTION D'ELECTRICITE, GAZ & EAU**
44 CONSTRUCTION	..	39	..	..	..	..	..	**44 CONSTRUCTION**
45 SERVICES SECTOR	..	5 336	..	..	..	..	..	**45 SECTEUR DES SERVICES**
46 Wholesale, retail trade & repair of motor vehicles, etc.	..	929	..	..	..	..	..	**46 Commerce (gros & détail), réparation de véhicules, etc.**
47 Hotels & restaurants	..	..	..	..	..	..	..	**47 Hôtels et restaurants**
48 Transport & storage	..	.. [n]	..	..	..	..	..	**48 Transports, entreposage**
49 Communications	..	440 [o]	..	..	..	..	..	**49 Communications**
50 Post	..	..	..	..	..	..	..	50 Postes
51 Telecommunications	..	..	..	..	..	..	..	51 Télécommunications
52 Financial intermediation (inc. insur.)	..	..	..	..	..	..	..	**52 Intermédiation financ. (assurances inc.)**
53 Real estate, renting & busin. activity	..	3 965	..	..	..	..	..	**53 Immobil., locat. & serv. aux entreprises**
54 Computer & related activities	..	1 479	..	..	..	..	..	54 Informatique et activités connexes
55 *Software consultancy*	..	..	..	..	..	..	..	55 *Production de logiciels*
56 *Other computer services n.e.c.*	..	..	..	..	..	..	..	56 *Autres activités informatiques*
57 Research & development	..	447	..	..	..	..	..	57 Recherche & développement expérimental
58 Other business activities n.e.c.	..	2 040 [o]	..	..	..	..	..	58 Autres activ. & serv. aux entreprises
59 Comm., soc. & pers. serv. activ., etc.	..	.. [n]	..	..	..	..	..	**59 Adm. pub., serv. soc. & collec., etc**
TOTAL BUSINESS ENTERPRISE SECTOR	**..**	**17 195**	**18 615 [c]**	**20 037**	**21 198**	**21 022**	**..**	**TOTAL SECTEUR DES ENTREPRISES**

Source: OECD, R&D database (DSTI/EAS Division), January 2002. — *Source* : OCDE, base de données R-D (DSTI/Division AES), janvier 2002.

RSE in the business enterprise sector - Chercheurs dans le secteur des entreprises

by industry - par industrie

In full time equivalent — *En équivalence plein-temps*

	1994	1995	1996	1997	1998	1999	2000	
1 AGRICULTURE, HUNTING & FORESTRY	..	24 [o]	..	..	..	..	..	**1 AGRICULTURE, CHASSE & SYLVICULTURE**
2 MINING	..	.. [n]	..	..	..	..	..	**2 ACTIVITES EXTRACTIVES**
3 MANUFACTURING	..	3 961	..	..	..	..	..	**3 ACTIVITES DE FABRICATION**
4 Food, beverages & tobacco	..	268	..	..	..	..	..	**4 Produits alimentaires, boissons & tabac**
5 Food products & beverages	..	..	..	..	..	..	..	5 Produits alimentaires & boissons
6 Tobacco products	..	..	..	..	..	..	..	6 Produits à base de tabac
7 Textiles, wear app., fur & leather	..	5	..	..	..	..	..	**7 Textiles, habill., fourrures & cuirs**
8 Textiles	..	..	..	..	..	..	..	8 Textiles
9 Wearing apparel & fur	..	..	..	..	..	..	..	9 Articles d'habillement & fourrures
10 Leather products & footwear	..	..	..	..	..	..	..	10 Articles en cuir & chaussures
11 Wood, paper, printing, publishing	..	.. [n]	..	..	..	..	..	**11 Bois, papier, imprimerie, édition**
12 Wood & cork (not furniture)	..	..	..	..	..	..	..	12 Bois & liège (sauf meubles)
13 Pulp, paper & paper products	..	..	..	..	..	..	..	13 Carton, papier & articles
14 Publ., print. & repro. of rec. media	..	..	..	..	..	..	..	14 Edition, impress. & repro. enregis.
15 Coke, petroleum, nuclear fuel, chemicals, rubber plastics	..	1 548	..	..	..	..	..	**15 Coke, pétrole, combust. nucl., prod. chimi., caout. & mat. plast.**
16 Coke, ref. petrol. prod. & nuclear fuel	..	.. [n]	..	..	..	..	..	16 Coke, raffin. pétr. & combust. nucl.
17 Chemicals & chemical products	..	1 484	..	..	..	..	..	17 Produits chimiques
18 *Chemical prod.(less pharmaceuticals)*	..	229	..	..	..	..	..	18 *Prod. chimi. (moins prod. pharmaceutiques)*
19 *Pharmaceuticals*	..	1 255	..	..	..	..	..	19 *Produits pharmaceutiques*
20 Rubber & plastic products	..	64	..	..	..	..	..	20 Caoutchouc et matières plastiques
21 Non-metallic mineral products	..	37	..	..	..	..	..	**21 Produits minéraux non métalliques**
22 Basic metals	..	35 [o]	..	..	..	..	..	**22 Produits métallurgiques de base**
23 Basic metals, ferrous	..	..	..	..	..	..	..	23 Produits métallur. de base, ferreux
24 Basic metals, non-ferrous	..	..	..	..	..	..	..	24 Prod. métallur. de base, non ferreux
25 Fabricated metal products	..	.. [n]	..	..	..	..	..	**25 Ouvrages en métaux (sauf mach.& mat.)**
26 Machinery equipment, instruments and transport equipment	..	1 699	..	..	..	..	..	**26 Machines & matériel, instruments & mat. de transport**
27 Machinery n.e.c.	..	509	..	..	..	..	..	27 Machines, n.c.a.
28 Office, account. & computing machinery	..	102	..	..	..	..	..	28 Mach. de bur., compta. & trait. inform.
29 Electrical machinery	..	102	..	..	..	..	..	29 Machines électriques
30 Electro. equip. (radio, TV & commun.)	..	491	..	..	..	..	..	30 Equip. électro. (radio, tv & communic.)
31 *Electro. comp. (inc. semi-conductors)*	..	..	..	..	..	..	..	31 *Compos. électro. (semi-cond. incl.)*
32 *TV, radio & communications equip.*	..	..	..	..	..	..	..	32 *Appareils radio, tv et de communi.*
33 Instruments	..	472	..	..	..	..	..	33 Instruments
34 Motor vehicles	..	.. [n]	..	..	..	..	..	34 Véhicules automobiles
35 Other transport equipment	..	23 [o]	..	..	..	..	..	35 Autres matériels de transport
36 *Ships*	..	..	..	..	..	..	..	36 *Navires*
37 *Aerospace*	..	..	..	..	..	..	..	37 *Aérospatiale*
38 *Other transport equipment n.e.c.*	..	..	..	..	..	..	..	38 *Autres matériels de transport n.c.a.*
39 Furniture, other manufacturing n.e.c.	..	369 [o]	..	..	..	..	..	**39 Meubles, autres activ. de fabricat. n.c.a.**
40 Furniture	..	..	..	..	..	..	..	40 Meubles
41 Other manufacturing n.e.c.	..	..	..	..	..	..	..	41 Autres activités de fabrication n.c.a.
42 Recycling	..	.. [n]	..	..	..	..	..	**42 Récupération**
43 ELECTRICITY, GAS & WATER SUPPLY	..	35	..	..	..	..	..	**43 DISTRIBUTION D'ELECTRICITE, GAZ & EAU**
44 CONSTRUCTION	..	20	..	..	..	..	..	**44 CONSTRUCTION**
45 SERVICES SECTOR	..	2 638	..	..	..	..	..	**45 SECTEUR DES SERVICES**
46 Wholesale, retail trade & repair of motor vehicles, etc.	..	284	..	..	..	..	..	**46 Commerce (gros & détail), réparation de véhicules, etc.**
47 Hotels & restaurants	..	..	..	..	..	..	..	**47 Hôtels et restaurants**
48 Transport & storage	..	.. [n]	..	..	..	..	..	**48 Transports, entreposage**
49 Communications	..	270 [o]	..	..	..	..	..	**49 Communications**
50 Post	..	..	..	..	..	..	..	50 Postes
51 Telecommunications	..	..	..	..	..	..	..	51 Télécommunications
52 Financial intermediation (inc. insur.)	..	..	..	..	..	..	..	**52 Intermédiation financ. (assurances inc.)**
53 Real estate, renting & busin. activity	..	2 084	..	..	..	..	..	**53 Immobil., locat. & serv. aux entreprises**
54 Computer & related activities	..	716	..	..	..	..	..	54 Informatique et activités connexes
55 *Software consultancy*	..	..	..	..	..	..	..	55 *Production de logiciels*
56 *Other computer services n.e.c.*	..	..	..	..	..	..	..	56 *Autres activités informatiques*
57 Research & development	..	233	..	..	..	..	..	57 Recherche & développement expérimental
58 Other business activities n.e.c.	..	1 135 [o]	..	..	..	..	..	58 Autres activ. & serv. aux entreprises
59 Comm., soc. & pers. serv. activ., etc.	..	.. [n]	..	..	..	..	..	**59 Adm. pub., serv. soc. & collec., etc**
TOTAL BUSINESS ENTERPRISE SECTOR	**..**	**6 674**	**7 098** [c]	**7 522**	**8 009**	**8 575**	**..**	**TOTAL SECTEUR DES ENTREPRISES**

Source: OECD, R&D database (DSTI/EAS Division), January 2002.

Source : OCDE, base de données R-D (DSTI/Division AES), janvier 2002.

Higher education intramural expenditure on R&D - Dépenses intra-muros de R-D de l'enseignement supérieur
(HERD) by field of science - (DIRDES) par discipline scientifique

	1994	1995	1996	1997	1998	1999	2000	
				14.1 Million DKK - Millions DKK				
Natural sciences	..	1 364.0	..	1 657.0	..	1 616.0	..	Sciences exactes et naturelles
Engineering	..	718.0	..	786.0	..	792.0	..	Sciences de l'ingénieur
Medical sciences	..	753.0	..	688.0	..	651.0	..	Sciences médicales
Agricultural sciences	..	461.0	..	306.0	..	321.0	..	Sciences agricoles
Sub-total NSE	..	3 295.0	..	3 437.0	..	3 380.0	..	Sous-total SENI
Social sciences	..	553.0	..	662.0	..	876.0	..	Sciences sociales
Humanities	..	700.0	..	705.0	..	876.0	..	Sciences humaines
Sub-total SSH	..	1 253.0	..	1 367.0	..	1 752.0	..	Sous-total SSH
Total NSE+SSH	..	4 548.0	4 252.6 c	4 802.0	4 773.0 c	5 133.0	..	*Total SENI+SSH*
				14.2 Million constant USD (1995 prices and PPPs) - Millions de USD constants (prix et PPA de 1995)				
Natural sciences	..	162.1	..	188.0	..	174.7	..	Sciences exactes et naturelles
Engineering	..	85.3	..	89.2	..	85.6	..	Sciences de l'ingénieur
Medical sciences	..	89.5	..	78.1	..	70.4	..	Sciences médicales
Agricultural sciences	..	54.8	..	34.7	..	34.7	..	Sciences agricoles
Sub-total NSE	..	391.5	..	389.9	..	365.3	..	Sous-total SENI
Social sciences	..	65.7	..	75.1	..	94.7	..	Sciences sociales
Humanities	..	83.2	..	80.0	..	94.7	..	Sciences humaines
Sub-total SSH	..	148.9	..	155.1	..	189.4	..	Sous-total SSH
Total NSE+SSH	..	540.4	493.0 c	544.8	531.3 c	554.8	..	*Total SENI+SSH*

Table 15 - Tableau 15

Higher education intramural expenditure on R&D - Dépenses intra-muros de R-D de l'enseignement supérieur
(HERD) by field of science and type of costs - (DIRDES) par discipline scientifique et par type de coûts

Million DKK *Millions DKK*

	1994	1995	1996	1997	1998	1999	2000	
Current expenditure								**Dépenses courantes**
Natural sciences	..	1 204.0	..	1 506.0	..	1 429.0	..	Sciences exactes et naturelles
Engineering	..	642.0	..	720.0	..	748.0	..	Sciences de l'ingénieur
Medical sciences	..	707.0	..	657.0	..	608.0	..	Sciences médicales
Agricultural sciences	..	347.0	..	258.0	..	294.0	..	Sciences agricoles
Sub-total NSE	..	2 899.0	..	3 141.0	..	3 079.0	..	Sous-total SENI
Social sciences	..	522.0	..	642.0	..	832.0	..	Sciences sociales
Humanities	..	665.0	..	627.0	..	795.0	..	Sciences humaines
Sub-total SSH	..	1 187.0	..	1 269.0	..	1 627.0	..	Sous-total SSH
Total NSE+SSH	..	4 087.0	..	4 410.0	..	4 704.0	..	*Total SENI+SSH*
Capital expenditure								**Dépenses en capital**
Natural sciences	..	160.0	..	152.0	..	188.0	..	Sciences exactes et naturelles
Engineering	..	76.0	..	65.0	..	44.0	..	Sciences de l'ingénieur
Medical sciences	..	46.0	..	30.0	..	44.0	..	Sciences médicales
Agricultural sciences	..	114.0	..	48.0	..	27.0	..	Sciences agricoles
Sub-total NSE	..	396.0	..	295.0	..	303.0	..	Sous-total SENI
Social sciences	..	30.0	..	19.0	..	44.0	..	Sciences sociales
Humanities	..	35.0	..	78.0	..	81.0	..	Sciences humaines
Sub-total SSH	..	65.0	..	97.0	..	125.0	..	Sous-total SSH
Total NSE+SSH	..	461.0	..	392.0	..	429.0	..	*Total SENI+SSH*

Source: OECD, R&D database (DSTI/EAS Division), January 2002.

Source : OCDE, base de données R-D (DSTI/Division AES), janvier 2002.

Higher education intramural expenditure on R&D - Dépenses intra-muros de R-D de l'enseignement supérieur

(HERD) by field of science and source of funds - (DIRDES) par discipline scientifique et par source de financement

Million DKK *Millions DKK*

	1994	1995	1996	1997	1998	1999	2000	
Direct government								**État direct**
Natural sciences	..	361.0	..	422.0	..	339.0 [s]	..	Sciences exactes et naturelles
Engineering	..	243.0	..	187.0	..	204.0 [s]	..	Sciences de l'ingénieur
Medical sciences	..	113.0	..	122.0	..	98.0 [s]	..	Sciences médicales
Agricultural sciences	..	120.0	..	100.0	..	98.0 [s]	..	Sciences agricoles
Sub-total NSE	..	836.0	..	831.0	..	739.0 [s]	..	Sous-total SENI
Social sciences	..	94.0	..	113.0	..	158.0 [s]	..	Sciences sociales
Humanities	..	99.0	..	81.0	..	122.0 [s]	..	Sciences humaines
Sub-total SSH	..	193.0	..	194.0	..	280.0 [s]	..	Sous-total SSH
Total NSE+SSH	..	1 029.0	1 079.7 [c]	1 070.0	..	635.0 [v]	..	*Total SENI+SSH*
General university funds								**Fonds généraux des universités**
Natural sciences	..	866.0	..	1 082.0	..	1 053.0 [s]	..	Sciences exactes et naturelles
Engineering	..	373.0	..	516.0	..	464.0 [s]	..	Sciences de l'ingénieur
Medical sciences	..	502.0	..	427.0	..	419.0 [s]	..	Sciences médicales
Agricultural sciences	..	308.0	..	176.0	..	187.0 [s]	..	Sciences agricoles
Sub-total NSE	..	2 050.0	..	2 201.0	..	2 123.0 [s]	..	Sous-total SENI
Social sciences	..	426.0	..	521.0	..	671.0 [s]	..	Sciences sociales
Humanities	..	564.0	..	588.0	..	705.0 [s]	..	Sciences humaines
Sub-total SSH	..	990.0	..	1 109.0	..	1 376.0 [s]	..	Sous-total SSH
Total NSE+SSH	..	3 040.0	2 649.5 [c]	3 310.0	..	3 883.0 [v]	..	*Total SENI+SSH*
Sub-total government								**Sous-total État**
Natural sciences	..	1 227.0	..	1 504.0	..	1 392.0	..	Sciences exactes et naturelles
Engineering	..	616.0	..	703.0	..	668.0	..	Sciences de l'ingénieur
Medical sciences	..	615.0	..	549.0	..	517.0	..	Sciences médicales
Agricultural sciences	..	428.0	..	276.0	..	285.0	..	Sciences agricoles
Sub-total NSE	..	2 886.0	..	3 032.0	..	2 862.0	..	Sous-total SENI
Social sciences	..	520.0	..	634.0	..	829.0	..	Sciences sociales
Humanities	..	663.0	..	669.0	..	827.0	..	Sciences humaines
Sub-total SSH	..	1 183.0	..	1 303.0	..	1 656.0	..	Sous-total SSH
Total NSE+SSH	..	4 069.0	3 729.2 [c]	4 380.0	..	4 518.0 [v]	..	*Total SENI+SSH*
Higher education								**Enseignement supérieur**
Natural sciences	..	..	..	..	..	..	..	Sciences exactes et naturelles
Engineering	..	..	..	..	..	..	..	Sciences de l'ingénieur
Medical sciences	..	..	..	..	..	..	..	Sciences médicales
Agricultural sciences	..	..	..	..	..	..	..	Sciences agricoles
Sub-total NSE	..	..	..	..	..	..	..	Sous-total SENI
Social sciences	..	..	..	..	..	..	..	Sciences sociales
Humanities	..	..	..	..	..	..	..	Sciences humaines
Sub-total SSH	..	..	..	..	..	..	..	Sous-total SSH
Total NSE+SSH	..	..	..	..	..	..	..	*Total SENI+SSH*
Private non-profit								**Institutions sans but lucratif**
Natural sciences	..	51.0	..	54.0	..	85.0	..	Sciences exactes et naturelles
Engineering	..	11.0	..	11.0	..	26.0	..	Sciences de l'ingénieur
Medical sciences	..	97.0	..	39.0	..	80.0	..	Sciences médicales
Agricultural sciences	..	13.0	..	10.0	..	24.0	..	Sciences agricoles
Sub-total NSE	..	173.0	..	114.0	..	215.0	..	Sous-total SENI
Social sciences	..	9.0	..	8.0	..	11.0	..	Sciences sociales
Humanities	..	23.0	..	8.0	..	22.0	..	Sciences humaines
Sub-total SSH	..	32.0	..	16.0	..	33.0	..	Sous-total SSH
Total NSE+SSH	..	205.0	223.0 [c]	130.0	..	247.0 [v]	..	*Total SENI+SSH*
Business entreprise								**Entreprises**
Natural sciences	..	13.0	..	40.0	..	31.0	..	Sciences exactes et naturelles
Engineering	..	31.0	..	24.0	..	40.0	..	Sciences de l'ingénieur
Medical sciences	..	24.0	..	73.0	..	19.0	..	Sciences médicales
Agricultural sciences	..	6.0	..	6.0	..	1.0	..	Sciences agricoles
Sub-total NSE	..	75.0	..	143.0	..	91.0	..	Sous-total SENI
Social sciences	..	6.0	..	8.0	..	9.0	..	Sciences sociales
Humanities	..	3.0	..	13.0	..	7.0	..	Sciences humaines
Sub-total SSH	..	9.0	..	21.0	..	16.0	..	Sous-total SSH
Total NSE+SSH	..	84.0	80.2 [c]	164.0	..	107.0 [v]	..	*Total SENI+SSH*
Funds from abroad								**Fonds de l'étranger**
Natural sciences	..	72.0	..	56.0	..	66.0	..	Sciences exactes et naturelles
Engineering	..	60.0	..	35.0	..	56.0	..	Sciences de l'ingénieur
Medical sciences	..	16.0	..	10.0	..	27.0	..	Sciences médicales
Agricultural sciences	..	13.0	..	13.0	..	11.0	..	Sciences agricoles
Sub-total NSE	..	161.0	..	114.0	..	160.0	..	Sous-total SENI
Social sciences	..	18.0	..	12.0	..	20.0	..	Sciences sociales
Humanities	..	11.0	..	4.0	..	6.0	..	Sciences humaines
Sub-total SSH	..	29.0	..	16.0	..	26.0	..	Sous-total SSH
Total NSE+SSH	..	190.0	220.2 [c]	130.0	..	187.0 [v]	..	*Total SENI+SSH*

Source: OECD, R&D database (DSTI/EAS Division), January 2002.

Source : OCDE, base de données R-D (DSTI/Division AES), janvier 2002.

Table 17 - Tableau 17

Higher education R&D personnel (HEMP) - Personnel de R-D du secteur de l'enseignement supérieur

by field of science and occupation - par discipline scientifique et par profession

In full-time equivalent — *En équivalence plein-temps*

	1994	1995	1996	1997	1998	1999	2000	
Total R&D personnel								**Personnel total de R-D**
Natural sciences	..	2 213	..	2 791	..	2 666	..	Sciences exactes et naturelles
Engineering	..	1 112	..	1 235	..	1 149	..	Sciences de l'ingénieur
Medical sciences	..	1 193	..	1 086	..	1 099	..	Sciences médicales
Agricultural sciences	..	690	..	726	..	658	..	Sciences agricoles
Sub-total NSE	..	5 208	..	5 838	..	5 572	..	Sous-total SENI
Social sciences	..	835	..	1 024	..	1 107	..	Sciences sociales
Humanities	..	1 170	..	1 278	..	1 339	..	Sciences humaines
Sub-total SSH	..	2 005	..	2 302	..	2 446	..	Sous-total SSH
Total NSE+SSH	..	7 213	7 676 [c]	8 138	..	8 017	..	*Total SENI+SSH*
Researchers								**Chercheurs**
Natural sciences	..	1 603	..	1 941	..	1 751	..	Sciences exactes et naturelles
Engineering	..	946	..	1 003	..	914	..	Sciences de l'ingénieur
Medical sciences	..	670	..	638	..	553	..	Sciences médicales
Agricultural sciences	..	443	..	506	..	378	..	Sciences agricoles
Sub-total NSE	..	3 662	..	4 088	..	3 596	..	Sous-total SENI
Social sciences	..	749	..	877	..	921	..	Sciences sociales
Humanities	..	1 109	..	1 179	..	1 205	..	Sciences humaines
Sub-total SSH	..	1 858	..	2 056	..	2 126	..	Sous-total SSH
Total NSE+SSH	..	5 520	5 832 [c]	6 143	..	5 722	..	*Total SENI+SSH*
Technicians								**Techniciens**
Natural sciences	..	..	..	..	..	..	..	Sciences exactes et naturelles
Engineering	..	..	..	..	..	..	..	Sciences de l'ingénieur
Medical sciences	..	..	..	..	..	..	..	Sciences médicales
Agricultural sciences	..	..	..	..	..	..	..	Sciences agricoles
Sub-total NSE	..	..	..	..	..	..	..	Sous-total SENI
Social sciences	..	..	..	..	..	..	..	Sciences sociales
Humanities	..	..	..	..	..	..	..	Sciences humaines
Sub-total SSH	..	..	..	..	..	..	..	Sous-total SSH
Total NSE+SSH	..	..	..	..	..	..	..	*Total SENI+SSH*
Other								**Autres**
Natural sciences	..	..	..	..	..	..	..	Sciences exactes et naturelles
Engineering	..	..	..	..	..	..	..	Sciences de l'ingénieur
Medical sciences	..	..	..	..	..	..	..	Sciences médicales
Agricultural sciences	..	..	..	..	..	..	..	Sciences agricoles
Sub-total NSE	..	..	..	..	..	..	..	Sous-total SENI
Social sciences	..	..	..	..	..	..	..	Sciences sociales
Humanities	..	..	..	..	..	..	..	Sciences humaines
Sub-total SSH	..	..	..	..	..	..	..	Sous-total SSH
Total NSE+SSH	..	..	..	..	..	..	..	*Total SENI+SSH*

Table 18 - Tableau 18

Government budget appropriations or outlays for R&D (GBAORD) - Crédits budgétaires publics de R-D (CBPRD)

by socio-economic objective - par objectif socio-économique

Million DKK — *Millions DKK*

	1995	1996	1997	1998	1999	2000	2001	
Agriculture, forestry and fishing	460.1	526.0	626.1	720.6	787.9	1 100.8	919.3	Agriculture, sylviculture et pêche
Industrial development	777.9	823.6	842.3	859.4	843.5	565.9	769.2	Développement industriel
Energy	151.5	175.6	213.3	173.1	182.2	176.4	197.4	Energie
Transport & telecommunications	95.5	81.4	68.4	60.5	77.7	84.0	95.9	*Transports & télécommunications*
Urban & rural planning	38.6	44.3	54.6	60.8	89.1	78.6	100.8	*Aménagement urbain et rural*
ST Infrastructure	134.1	125.7	123.0	121.3	166.8	162.6	196.7	ST Infrastructure
Prevention of pollution	..	..	..	..	..	..	..	*Prévention de la pollution*
Identificat. & treatment of pollution	..	..	..	..	..	..	..	*Détect. pollut. & mesures antipollut.*
ST Environment	294.0	330.0	231.1	313.1	301.7	261.3	229.0	ST Environnement
Health	110.3	124.6	130.7	156.7	157.3	544.0 [a]	565.0 [a]	Santé
Social development & services	582.5	763.3	714.5	807.5	819.5	1 015.7	1 000.7	Développement et services sociaux
Earth and atmosphere	114.1	101.8	109.5	106.2	113.2	113.6	115.6	Milieu terrestre et atmosphère
Advancement of research	1 472.3	1 498.9	1 654.4	1 738.3	1 720.6	1 579.5	1 539.3	*Promotion de la recherche*
General university funds	2 405.4	2 893.2	3 155.0	3 295.7	3 426.0	3 451.7	3 516.3	*Fonds généraux des universités*
ST Advancement of knowledge	3 877.7	4 392.1	4 809.4	5 034.0	5 146.6	5 031.2	5 055.6	ST promotion gén. des connaiss.
Civil space	175.9	139.0	234.0	219.8	229.8	247.2	240.1	Espace civil
Defence	41.8	42.9	45.9	47.9	49.7	50.4	49.2	Défense
Not elsewhere classified	..	..	..	..	..	385.5 [a]	319.0 [a]	Non classé
Total	**6 719.9**	**7 544.6**	**8 079.8**	**8 559.6**	**8 798.2**	**9 652.6 [a]**	**9 656.8 [a]**	**Total**

Source: OECD, R&D database (DSTI/EAS Division), January 2002.

Source : OCDE, base de données R-D (DSTI/Division AES), janvier 2002.

DENMARK | Table 19 - Tableau 19 | **DANEMARK**

Technology balance of payments - Balance des paiements technologiques

by type of transfer - selon la nature des transferts

Million DKK | *Millions DKK*

	1983	1984	1985	1986	1987	1988	1989	1990	1991	
Receipts										**Recettes**
Related firms	..	..	..	..	..	..	..	..	..	Firmes affiliées
Unrelated firms	..	..	..	..	..	..	..	..	..	Firmes non affiliées
Other	..	..	..	..	..	..	..	..	..	Autre
Total	1 219	1 330	1 945	..	..	..	..	..	..	*Total*
Payments										**Paiements**
Related firms	..	..	..	..	..	..	..	..	..	Firmes affiliées
Unrelated firms	..	..	..	..	..	..	..	..	..	Firmes non affiliées
Other	..	..	..	..	..	..	..	..	..	Autre
Total	1 247	1 471	1 706	..	..	..	..	..	..	*Total*

	1992	1993	1994	1995	1996	1997	1998	1999	2000	
Receipts										**Recettes**
Related firms	..	..	..	..	..	..	..	..	..	Firmes affiliées
Unrelated firms	..	..	..	..	..	..	..	..	..	Firmes non affiliées
Other	..	..	..	..	..	..	..	..	..	Autre
Total	..	..	..	..	..	..	..	..	..	*Total*
Payments										**Paiements**
Related firms	..	..	..	..	..	..	..	..	..	Firmes affiliées
Unrelated firms	..	..	..	..	..	..	..	..	..	Firmes non affiliées
Other	..	..	..	..	..	..	..	..	..	Autre
Total	..	..	..	..	..	..	..	..	..	*Total*

Source: OECD, TBPdatabase (DSTI/EAS Division), Januay 2002.

Source : OCDE, base de données TBP (DSTI/Division AES), janvier 2002.

FINLAND Table 1 - Tableau 1 **FINLANDE**

Gross domestic expenditure on R&D (GERD) - Dépenses intérieures brutes de R-D (DIRD)

by sector of performance and source of funds - par secteur d'exécution et par source de financement

	1994	1995	1996	1997	1998	1999	2000	
	1.1 Million EUR (1999 FIM euro, EUR from 1999) - Millions EUR (1999 FIM euro, EUR à partir de 1999)							
Business enterprise								**Entreprises**
financed by								*financé par*
Business enterprise	..	1 223.3	..	1 738.7	2 048.5 [a]	2 494.8	..	Entreprises
Direct government	..	76.6	..	78.8	99.4	110.9	..	État direct
Higher education	..	..	..	..	..	..	..	Enseignement supérieur
Private non-profit	..	0.6	..	0.8	3.6	0.6	..	Institutions sans but lucratif
Funds from abroad	..	72.8	..	98.5	101.3 [a]	37.6	..	Fonds de l'étranger
Total	1 249.8	1 373.4	1 656.7 [c]	1 916.7	2 252.8	2 643.9	3 096.5 [c,p]	*Total*
Government								**État**
financed by								*financé par*
Business enterprise	..	42.9	..	55.7	62.6	62.6	..	Entreprises
Direct government	..	304.1	..	313.7	322.5	342.4	..	État direct
Higher education	..	..	..	..	..	..	..	Enseignement supérieur
Private non-profit	..	3.3	..	2.9	5.4	4.4	..	Institutions sans but lucratif
Funds from abroad	..	11.3	..	22.8	31.8	32.3	..	Fonds de l'étranger
Total	379.9 [a,o]	361.6	394.6 [o]	395.1	422.3	441.8	484.0 [c,o,p]	*Total*
Higher education								**Enseignement supérieur**
financed by								*financé par*
Business enterprise	..	24.3	..	30.3	29.7	36.0	..	Entreprises
Direct government	..	136.4	..	209.3 [a]	262.8	313.1	..	État direct
General university funds	..	241.5	..	290.6 [a]	316.5	354.2	..	Fonds généraux des universités
Sub-total government	..	377.9	..	499.8 [a]	579.2	667.3	..	Sous-Total État
Higher education	..	4.8	..	8.3	7.3	5.9	..	Enseignement supérieur
Private non-profit	..	5.2	..	7.3	7.3	12.1	..	Institutions sans but lucratif
Funds from abroad	..	12.4	..	33.8	34.3	43.5	..	Fonds de l'étranger
Total	378.7	424.6	452.4 [c]	579.5 [a]	657.9	764.8	773.8 [c,p]	*Total*
Private non-profit								**Institutions sans but lucratif**
financed by								*financé par*
Business enterprise	..	1.5	..	2.4 [c]	2.0	3.4	..	Entreprises
Direct government	..	3.7	..	4.1 [c]	6.7	11.3	..	État direct
Higher education	..	..	..	..	..	..	..	Enseignement supérieur
Private non-profit	..	7.3	..	6.7 [c]	9.7	12.1	..	Institutions sans but lucratif
Funds from abroad	..	0.3	..	0.4 [c]	3.1	1.7	..	Fonds de l'étranger
Total	.. [n]	12.8	.. [n]	13.5 [c]	21.6	28.4	.. [n]	*Total*
GERD								**DIRD**
financed by								*financé par*
Business enterprise	..	1 292.0	..	1 827.0	2 142.7 [a]	2 596.8	..	Entreprises
Direct government	..	520.8	..	605.8	691.4	777.7	..	État direct
General university funds	..	241.5	..	290.6	316.5	354.2	..	Fonds généraux des universités
Sub-total government	..	762.3	..	896.4	1 007.9	1 131.8	..	Sous-Total État
Higher education	..	4.8	..	8.3	7.3	5.9	..	Enseignement supérieur
Private non-profit	..	16.4	..	17.7	26.0	29.2	..	Institutions sans but lucratif
Funds from abroad	..	96.9	..	155.4	170.5 [a]	115.0	..	Fonds de l'étranger
Total	2 008.3	2 172.4	2 503.6 [c]	2 904.9	3 354.5	3 878.8	4 354.4 [c,p]	*Total*
	1.2 Million constant USD (1995 prices and PPPs) - Millions de USD constants (prix et PPA de 1995)							
Business enterprise								**Entreprises**
financed by								*financé par*
Business enterprise	..	1 240.9	..	1 732.1	1 981.2 [a]	2 415.7	..	Entreprises
Direct government	..	77.7	..	78.5	96.2	107.3	..	État direct
Higher education	..	..	..	..	..	..	..	Enseignement supérieur
Private non-profit	..	0.6	..	0.8	3.5	0.6	..	Institutions sans but lucratif
Funds from abroad	..	73.9	..	98.1	98.0 [a]	36.4	..	Fonds de l'étranger
Total	1 320.0	1 393.1	1 684.3 [c]	1 909.4	2 178.9	2 560.1	2 900.2 [c,p]	*Total*
Government								**État**
financed by								*financé par*
Business enterprise	..	43.5	..	55.5	60.5	60.6	..	Entreprises
Direct government	..	308.5	..	312.5	311.9	331.6	..	État direct
Higher education	..	..	..	..	..	..	..	Enseignement supérieur
Private non-profit	..	3.3	..	2.9	5.3	4.2	..	Institutions sans but lucratif
Funds from abroad	..	11.5	..	22.7	30.8	31.3	..	Fonds de l'étranger
Total	401.2 [a,o]	366.8	401.2 [o]	393.6	408.4	427.8	453.4 [c,o,p]	*Total*
Higher education								**Enseignement supérieur**
financed by								*financé par*
Business enterprise	..	24.6	..	30.2	28.7	34.9	..	Entreprises
Direct government	..	138.4	..	208.5 [a]	254.2	303.2	..	État direct
General university funds	..	245.0	..	289.5 [a]	306.1	342.9	..	Fonds généraux des universités
Sub-total government	..	383.3	..	497.9 [a]	560.2	646.1	..	Sous-Total État
Higher education	..	4.8	..	8.3	7.1	5.7	..	Enseignement supérieur
Private non-profit	..	5.3	..	7.3	7.0	11.7	..	Institutions sans but lucratif
Funds from abroad	..	12.6	..	33.6	33.2	42.1	..	Fonds de l'étranger
Total	399.9	430.7	460.0 [c]	577.3 [a]	636.3	740.5	724.8 [c,p]	*Total*
Private non-profit								**Institutions sans but lucratif**
financed by								*financé par*
Business enterprise	..	1.5	..	2.3 [c]	1.9	3.3	..	Entreprises
Direct government	..	3.7	..	4.0 [c]	6.5	10.9	..	État direct
Higher education	..	..	..	..	..	..	..	Enseignement supérieur
Private non-profit	..	7.4	..	6.7 [c]	9.4	11.7	..	Institutions sans but lucratif
Funds from abroad	..	0.3	..	0.4 [c]	3.0	1.6	..	Fonds de l'étranger
Total	.. [n]	13.0	.. [n]	13.5 [c]	20.9	27.5	.. [n]	*Total*
GERD								**DIRD**
financed by								*financé par*
Business enterprise	..	1 310.5	..	1 820.1	2 072.4 [a]	2 514.4	..	Entreprises
Direct government	..	528.3	..	603.5	668.7	753.0	..	État direct
General university funds	..	245.0	..	289.5	306.1	342.9	..	Fonds généraux des universités
Sub-total government	..	773.3	..	893.0	974.8	1 095.9	..	Sous-Total État
Higher education	..	4.8	..	8.3	7.1	5.7	..	Enseignement supérieur
Private non-profit	..	16.7	..	17.7	25.2	28.3	..	Institutions sans but lucratif
Funds from abroad	..	98.3	..	154.8	164.9 [a]	111.4	..	Fonds de l'étranger
Total	2 121.2	2 203.6	2 545.4 [c]	2 893.9	3 244.4	3 755.8	4 078.4 [c,p]	*Total*

Source: OECD, R&D database (DSTI/EAS Division), January 2002.

Source : OCDE, base de données R-D (DSTI/Division AES), janvier 2002.

Gross domestic expenditure on R&D (GERD) - Dépenses intérieures brutes de R-D (DIRD)

by sector of performance and type of costs - par secteur d'exécution et par type de coûts

Million EUR (1999 FIM euro, EUR from 1999) — *Millions EUR (1999 FIM euro, EUR à partir de 1999)*

	1994	1995	1996	1997	1998	1999	2000	
Business enterprise								**Entreprises**
Labour costs	640.7	707.1	..	932.6	1 060.1	1 350.1	..	Coûts du personnel
Other current costs	481.2	500.3	..	780.1	956.1	1 083.8	..	Autres coûts courants
Total current expenditure	1 121.9	1 207.4	..	1 712.7	2 016.2	2 433.8	..	*Total dépenses courantes*
Land and buildings	6.2	7.4	..	14.3	15.8	21.0	..	Terrains et constructions
Instruments and equipment	121.7	158.6	..	189.9	220.8	189.1	..	Instruments et équipements
Total capital expenditure	127.9	166.0	..	204.2	236.6	210.1	..	*Total dépenses en capital*
Total	1 249.8	1 373.4	1 656.7 [c]	1 916.7	2 252.8	2 643.9	3 096.5 [c,p]	*Total*
Government								**État**
Labour costs	..	210.5	..	231.1	251.7	257.8	..	Coûts du personnel
Other current costs	..	118.7	..	140.4	149.4	163.0	..	Autres coûts courants
Total current expenditure	..	329.2	..	371.5	401.1	420.8	..	*Total dépenses courantes*
Land and buildings	..	10.3	..	0.0	0.1	..	..	Terrains et constructions
Instruments and equipment	..	22.2	..	23.5	21.1	21.0	..	Instruments et équipements
Total capital expenditure	..	32.4	..	23.5	21.2	21.0	..	*Total dépenses en capital*
Total	379.9 [a,o]	361.6	394.6 [o]	395.1	422.3	441.8	484.0 [c,o,p]	*Total*
Higher education								**Enseignement supérieur**
Labour costs	..	256.3	..	345.0 [a]	388.8	434.3	..	Coûts du personnel
Other current costs	..	148.1	..	224.5 [a]	254.8	313.2	..	Autres coûts courants
Total current expenditure	..	404.3	..	569.5 [a]	643.6	747.5	..	*Total dépenses courantes*
Land and buildings	..	13.2	..	..	..	..	..	Terrains et constructions
Instruments and equipment	..	7.0	..	10.1 [a]	14.3	17.3	..	Instruments et équipements
Total capital expenditure	..	20.2	..	10.1 [a]	14.3	17.3	..	*Total dépenses en capital*
Total	378.7	424.6	452.4 [c]	579.5 [a]	657.9	764.8	773.8 [c,p]	*Total*
Private non-profit								**Institutions sans but lucratif**
Labour costs	..	8.2	..	8.4 [c]	13.1	17.7	..	Coûts du personnel
Other current costs	..	4.3	..	4.5 [c]	7.9	10.3	..	Autres coûts courants
Total current expenditure	..	12.6	..	13.0 [c]	20.9	27.9	..	*Total dépenses courantes*
Land and buildings	..	..	..	0.3 [c]	0.3	..	..	Terrains et constructions
Instruments and equipment	..	0.2	..	0.3 [c]	0.4	0.5	..	Instruments et équipements
Total capital expenditure	..	0.2	..	0.7 [c]	0.7	0.5	..	*Total dépenses en capital*
Total	.. [n]	12.8	.. [n]	13.5 [c]	21.6	28.4	.. [n]	*Total*
GERD								**DIRD**
Labour costs	..	1 182.2	..	1 517.1	1 713.6	2 059.8	..	Coûts du personnel
Other current costs	..	771.3	..	1 149.6	1 368.1	1 570.2	..	Autres coûts courants
Total current expenditure	..	1 953.5	..	2 666.6	3 081.7	3 630.1	..	*Total dépenses courantes*
Land and buildings	..	30.8	..	14.6	16.1	21.0	..	Terrains et constructions
Instruments and equipment	..	188.0	..	223.9	256.6	227.9	..	Instruments et équipements
Total capital expenditure	..	218.9	..	238.5	272.8	248.9	..	*Total dépenses en capital*
Total	2 008.3	2 172.4	2 503.6 [c]	2 904.9	3 354.5	3 878.8	4 354.4 [c,p]	*Total*

Table 3 - Tableau 3

Current domestic expenditure on R&D - Dépenses intérieures courantes de R-D

by sector of performance and type of activity - par secteur d'exécution et par type d'activités

Million EUR (1999 FIM euro, EUR from 1999) — *Millions EUR (1999 FIM euro, EUR à partir de 1999)*

	1994	1995	1996	1997	1998	1999	2000	
Business enterprise								**Entreprises**
Basic research	..	..	..	..	..	..	..	Recherche fondamentale
Applied research	..	..	..	..	..	..	..	Recherche appliquée
Experimental development	..	..	..	..	..	..	..	Développement expérimental
Not specified	..	..	..	..	..	..	..	Non spécifié
Total	..	..	..	..	..	..	..	*Total*
Government								**État**
Basic research	..	..	..	..	..	..	..	Recherche fondamentale
Applied research	..	..	..	..	..	..	..	Recherche appliquée
Experimental development	..	..	..	..	..	..	..	Développement expérimental
Not specified	..	..	..	..	..	..	..	Non spécifié
Total	..	..	..	..	..	..	..	*Total*
Higher education								**Enseignement supérieur**
Basic research	..	..	..	..	..	..	..	Recherche fondamentale
Applied research	..	..	..	..	..	..	..	Recherche appliquée
Experimental development	..	..	..	..	..	..	..	Développement expérimental
Not specified	..	..	..	..	..	..	..	Non spécifié
Total	..	..	..	..	..	..	..	*Total*
Private non-profit								**Institutions sans but lucratif**
Basic research	..	..	..	..	..	..	..	Recherche fondamentale
Applied research	..	..	..	..	..	..	..	Recherche appliquée
Experimental development	..	..	..	..	..	..	..	Développement expérimental
Not specified	..	..	..	..	..	..	..	Non spécifié
Total	..	..	..	..	..	..	..	*Total*
GERD								**DIRD**
Basic research	..	..	..	..	..	..	..	Recherche fondamentale
Applied research	..	..	..	..	..	..	..	Recherche appliquée
Experimental development	..	..	..	..	..	..	..	Développement expérimental
Not specified	..	..	..	..	..	..	..	Non spécifié
Total	..	..	..	..	..	..	..	*Total*

Source: OECD, R&D database (DSTI/EAS Division), January 2002. *Source* : OCDE, base de données R-D (DSTI/Division AES), janvier 2002.

FINLAND — FINLANDE

Table 5 - Tableau 5

Total R&D personnel - Personnel total de R-D

by sector of employment and occupation - par secteur d'emploi et par profession

In full-time equivalent — *En équivalence plein-temps*

	1994	1995	1996	1997	1998	1999	2000	
Business enterprise								**Entreprises**
RSE	..	..	..	..	..	..	..	Chercheurs
Technicians	..	..	..	..	..	..	..	Techniciens
Other occupations	..	..	..	..	..	..	..	Autres professions
Total	16 900	17 798	20 756	22 302	25 010	27 818	..	*Total*
Government								**État**
RSE	..	..	..	..	..	..	..	Chercheurs
Technicians	..	..	..	..	..	..	..	Techniciens
Other occupations	..	..	..	..	..	..	..	Autres professions
Total	6 849 [a,o]	6 430 [a]	..	6 826	7 498	7 454	..	*Total*
Higher education								**Enseignement supérieur**
RSE	..	..	..	..	..	..	..	Chercheurs
Technicians	..	..	..	..	..	..	..	Techniciens
Other occupations	..	..	..	..	..	..	..	Autres professions
Total	8 582	9 146	..	11 855 [a]	13 653	14 840	..	*Total*
Private non-profit								**Institutions sans but lucratif**
RSE	..	..	..	..	..	..	..	Chercheurs
Technicians	..	..	..	..	..	..	..	Techniciens
Other occupations	..	..	..	..	..	..	..	Autres professions
Total	.. [n]	261	..	273	357	492	..	*Total*
TOTAL								**TOTAL**
RSE	..	..	..	..	..	..	..	Chercheurs
Technicians	..	..	..	..	..	..	..	Techniciens
Other occupations	..	..	..	..	..	..	..	Autres professions
Total	32 331	33 634	..	41 256	46 517	50 604	..	*Total*

Table 6 - Tableau 6

Total R&D personnel - Personnel total de R-D

by sector of employment and formal qualification - par secteur d'emploi et par niveau de qualification

In full-time equivalent — *En équivalence plein-temps*

	1994	1995	1996	1997	1998	1999	2000	
Business enterprise								**Entreprises**
University PhD level degrees	..	777	..	999	1 116	1 236	..	Diplômes univ. de niv. doctorat
Other university degrees	..	5 906	..	7 676	8 573	9 319	..	Autres diplômes universitaires
Sub-total university degrees	..	6 683	..	8 675	9 689	10 555	..	Sous-total Universitaires
Other post-secondary diplomas	..	6 460	..	8 661	7 758 [a]	8 717	..	Autres diplômes post-secondaires
Secondary diplomas	..	3 298	..	3 584	5 662 [a]	6 268	..	Diplômes de fin d'études secondaires
Other qualifications	..	1 356	..	1 383	1 902	2 278	..	Autres qualifications
Total	16 900	17 798	20 756	22 302	25 010	27 818	..	*Total*
Government								**État**
University PhD level degrees	..	1 176	..	1 313	1 460	1 411	..	Diplômes univ. de niv. doctorat
Other university degrees	..	2 323	..	2 448	2 657	2 704	..	Autres diplômes universitaires
Sub-total university degrees	..	3 499	..	3 761	4 117	4 115	..	Sous-total Universitaires
Other post-secondary diplomas	..	646	..	614	822 [a]	602	..	Autres diplômes post-secondaires
Secondary diplomas	..	1 268	..	1 430	1 452 [a]	1 782	..	Diplômes de fin d'études secondaires
Other qualifications	..	1 018	..	1 022	1 107	956	..	Autres qualifications
Total	6 849 [a,o]	6 430 [a]	..	6 826	7 498	7 454	..	*Total*
Higher education								**Enseignement supérieur**
University PhD level degrees	..	3 114	..	3 892 [a]	4 475	4 839	..	Diplômes univ. de niv. doctorat
Other university degrees	..	3 367	..	4 626 [a]	5 208	5 555	..	Autres diplômes universitaires
Sub-total university degrees	..	6 481	..	8 518 [a]	9 683	10 395	..	Sous-total Universitaires
Other post-secondary diplomas	..	991	..	1 147 [a]	923 [a]	1 629	..	Autres diplômes post-secondaires
Secondary diplomas	..	1 003	..	1 670 [a]	2 055 [a]	1 909	..	Diplômes de fin d'études secondaires
Other qualifications	..	671	..	520 [a]	991	907	..	Autres qualifications
Total	8 582	9 146	..	11 855 [a]	13 653	14 840	..	*Total*
Private non-profit								**Institutions sans but lucratif**
University PhD level degrees	..	87	..	81	106	113	..	Diplômes univ. de niv. doctorat
Other university degrees	..	114	..	114	151	220	..	Autres diplômes universitaires
Sub-total university degrees	..	200	..	195	257	333	..	Sous-total Universitaires
Other post-secondary diplomas	..	12	..	32	35 [a]	42	..	Autres diplômes post-secondaires
Secondary diplomas	..	43	..	38	57 [a]	91	..	Diplômes de fin d'études secondaires
Other qualifications	..	6	..	6	8	27	..	Autres qualifications
Total	.. [n]	261	..	273	357	492	..	*Total*
TOTAL								**TOTAL**
University PhD level degrees	..	5 153	..	6 284	7 156	7 599	..	Diplômes univ. de niv. doctorat
Other university degrees	..	11 710	..	14 865	16 589	17 798	..	Autres diplômes universitaires
Sub-total university degrees	..	16 863	..	21 149	23 745	25 398	..	Sous-total Universitaires
Other post-secondary diplomas	..	8 108	..	10 454	9 538 [a]	10 990	..	Autres diplômes post-secondaires
Secondary diplomas	..	5 611	..	6 723	9 226 [a]	10 050	..	Diplômes de fin d'études secondaires
Other qualifications	..	3 050	..	2 931	4 009	4 168	..	Autres qualifications
Total	32 331	33 634	..	41 256	46 517	50 604	..	*Total*

Source: OECD, R&D database (DSTI/EAS Division), January 2002.

Source : OCDE, base de données R-D (DSTI/Division AES), janvier 2002.

Table 7 - Tableau 7

Gross domestic expenditure on R&D (GERD) - Dépenses intérieures brutes de R-D (DIRD)

by sector of performance and main scientific field - par secteur d'exécution et par grand domaine scientifique

Million EUR (1999 FIM euro, EUR from 1999) *Millions EUR (1999 FIM euro, EUR à partir de 1999)*

	1994	1995	1996	1997	1998	1999	2000	
Business enterprise								**Entreprises**
Natural sciences and engineering	1 249.8	1 373.4	1 656.7	1 916.7	2 252.8	2 643.9	3 096.5 [c]	Sciences exactes, nat. et de l'ingénieur
Social sciences and humanities	..	..	..	..	..	..	..	Sciences sociales et humaines
All fields of science	1 249.8	1 373.4	1 656.7 [c]	1 916.7	2 252.8	2 643.9	3 096.5 [c,p]	*Tous domaines scientifiques*
Government								**État**
Natural sciences and engineering	..	..	..	..	..	..	..	Sciences exactes, nat. et de l'ingénieur
Social sciences and humanities	..	..	..	..	..	..	..	Sciences sociales et humaines
All fields of science	379.9 [a,o]	361.6	394.6 [o]	395.1	422.3	441.8	484.0 [c,o,p]	*Tous domaines scientifiques*
Higher education								**Enseignement supérieur**
Natural sciences and engineering	..	288.5 [v]	..	422.2 [a]	479.5	562.8	..	Sciences exactes, nat. et de l'ingénieur
Social sciences and humanities	..	122.8 [v]	..	157.3	178.3	202.0	..	Sciences sociales et humaines
All fields of science	378.7	424.6	452.4 [c]	579.5 [a]	657.9	764.8	773.8 [c,p]	*Tous domaines scientifiques*
Private non-profit								**Institutions sans but lucratif**
Natural sciences and engineering	..	..	..	..	..	..	..	Sciences exactes, nat. et de l'ingénieur
Social sciences and humanities	..	..	..	..	..	..	..	Sciences sociales et humaines
All fields of science	.. [n]	12.8	.. [n]	13.5 [c]	21.6	28.4	.. [n]	*Tous domaines scientifiques*
TOTAL								**TOTAL**
Natural sciences and engineering	..	..	..	..	..	..	..	Sciences exactes, nat. et de l'ingénieur
Social sciences and humanities	..	..	..	..	..	..	..	Sciences sociales et humaines
All fields of science	2 008.3	2 172.4	2 503.6 [c]	2 904.9	3 354.5	3 878.8	4 354.4 [c,p]	*Tous domaines scientifiques*

Table 8 - Tableau 8

Total R&D personnel - Personnel total de R-D

by sector of employment and main scientific field - par secteur d'emploi et par grand domaine scientifique

In full-time equivalent *En équivalence plein-temps*

	1994	1995	1996	1997	1998	1999	2000	
Business enterprise								**Entreprises**
Natural sciences and engineering	16 900	17 798	20 756	22 302	25 010	27 817	..	Sciences exactes, nat. et de l'ingénieur
Social sciences and humanities	..	..	..	..	..	..	..	Sciences sociales et humaines
All fields of science	16 900	17 798	20 756	22 302	25 010	27 818	..	*Tous domaines scientifiques*
Government								**État**
Natural sciences and engineering	..	..	..	6 349 [o]	..	7 100 [o]	..	Sciences exactes, nat. et de l'ingénieur
Social sciences and humanities	..	..	..	750 [o]	..	846 [o]	..	Sciences sociales et humaines
All fields of science	6 849 [a,o]	6 430 [a]	..	6 826	7 498	7 454	..	*Tous domaines scientifiques*
Higher education								**Enseignement supérieur**
Natural sciences and engineering	..	6 601	..	8 562 [a]	9 983	10 933	..	Sciences exactes, nat. et de l'ingénieur
Social sciences and humanities	..	2 545	..	3 294	3 670	3 907	..	Sciences sociales et humaines
All fields of science	8 582	9 146	..	11 855 [a]	13 653	14 840	..	*Tous domaines scientifiques*
Private non-profit								**Institutions sans but lucratif**
Natural sciences and engineering	..	..	..	.. [n]	..	.. [n]	..	Sciences exactes, nat. et de l'ingénieur
Social sciences and humanities	..	..	..	.. [n]	..	.. [n]	..	Sciences sociales et humaines
All fields of science	.. [n]	261	..	273	357	492	..	*Tous domaines scientifiques*
TOTAL								**TOTAL**
Natural sciences and engineering	..	..	..	37 213	..	45 851	..	Sciences exactes, nat. et de l'ingénieur
Social sciences and humanities	..	..	..	4 044	..	4 753	..	Sciences sociales et humaines
All fields of science	32 331	33 634	..	41 256	46 517	50 604	..	*Tous domaines scientifiques*

Source: OECD, R&D database (DSTI/EAS Division), January 2002.

Source : OCDE, base de données R-D (DSTI/Division AES), janvier 2002.

Total business enterprise intramural expenditure on R&D - Dépenses intra-muros totales de R-D des entreprises
(BERD) by industry - (DIRDE) par industrie

	1994	1995	1996	1997	1998	1999	2000	
	9.1 Million EUR (1999 FIM euro, EUR from 1999) - Millions EUR (1999 FIM euro, EUR à partir de 1999)							
1 AGRICULTURE, HUNTING & FORESTRY	..	2.7	..	0.7	2.5	2.1	..	**1 AGRICULTURE, CHASSE & SYLVICULTURE**
2 MINING	5.3	4.9	..	5.2	8.2	8.9	..	**2 ACTIVITES EXTRACTIVES**
3 MANUFACTURING	1 053.6	1 128.3	1 361.7	1 540.4	1 852.2	2 161.9	..	**3 ACTIVITES DE FABRICATION**
4 Food, beverages & tobacco	49.0	43.7	50.9	50.0	56.8	53.8	..	**4 Produits alimentaires, boissons & tabac**
5 Food products & beverages	..	..	..	..	..	..	..	5 Produits alimentaires & boissons
6 Tobacco products	..	..	..	..	..	..	..	6 Produits à base de tabac
7 Textiles, wear app., fur & leather	5.2	7.3	9.1	10.2	12.2	12.6	..	**7 Textiles, habill., fourrures & cuirs**
8 Textiles	4.3	6.5	..	9.3	10.2	10.9	..	8 Textiles
9 Wearing apparel & fur	.. [n]	0.5	..	0.5	1.1	0.5	..	9 Articles d'habillement & fourrures
10 Leather products & footwear	0.9 [o]	0.3 [a]	..	0.4	0.9	1.2	..	10 Articles en cuir & chaussures
11 Wood, paper, printing, publishing	68.4	68.7	74.2	79.8	81.7	81.7	..	**11 Bois, papier, imprimerie, édition**
12 Wood & cork (not furniture)	4.5	17.3	3.8	5.7	7.0	7.9	..	12 Bois & liège (sauf meubles)
13 Pulp, paper & paper products	58.5	45.7	61.6	67.7	68.5	68.4	..	13 Carton, papier & articles
14 Publ., print. & repro. of rec. media	5.4	5.7	8.8	6.4	6.1	5.3	..	14 Edition, impress. & repro. enregis.
15 Coke, petroleum, nuclear fuel, chemicals, rubber plastics	186.3	188.8	183.0	196.8	221.6	222.5	..	**15 Coke, pétrole, combust. nucl., prod. chimi., caout. & mat. plast.**
16 Coke, ref. petrol. prod. & nuclear fuel	.. [n]	.. [n]	.. [n]	.. [n]	.. [n]	.. [n]	..	16 Coke, raffin. pétr. & combust. nucl.
17 Chemicals & chemical products	168.0	143.9	139.6	142.8	157.3	177.8	..	17 Produits chimiques
18 *Chemical prod.(less pharmaceuticals)*	..	..	..	..	..	..	..	18 *Prod. chimi. (moins prod. pharmaceutiques)*
19 *Pharmaceuticals*	..	..	..	..	..	..	..	19 *Produits pharmaceutiques*
20 Rubber & plastic products	18.3 [o]	45.0 [o]	43.4 [o]	54.0 [o]	64.4 [o]	27.2 [o]	..	20 Caoutchouc et matières plastiques
21 Non-metallic mineral products	15.5	13.3	6.7	12.6	13.5	17.8	..	**21 Produits minéraux non métalliques**
22 Basic metals	54.5	32.6 [a]	36.5	37.6	34.4	36.1	..	**22 Produits métallurgiques de base**
23 Basic metals, ferrous	..	..	..	..	..	..	..	23 Produits métallur. de base, ferreux
24 Basic metals, non-ferrous	..	..	..	..	..	..	..	24 Prod. métallur. de base, non ferreux
25 Fabricated metal products	23.6	28.2	14.8	23.2	26.8	27.5	..	**25 Ouvrages en métaux (sauf mach.& mat.)**
26 Machinery equipment, instruments and transport equipment	639.2	740.7	981.2	1 123.2	1 395.4	1 696.3	..	**26 Machines & matériel, instruments & mat. de transport**
27 Machinery n.e.c.	119.0	134.3	152.2	189.1	233.9	248.3	..	27 Machines, n.c.a.
28 Office, account. & computing machinery	.. [n]	38.4	36.9	.. [n]	.. [n]	.. [n]	..	28 Mach. de bur., compta. & trait. inform.
29 Electrical machinery	.. [n]	64.8	113.6	.. [n]	.. [n]	.. [n]	..	29 Machines électriques
30 Electro. equip. (radio, TV & commun.)	478.9 [o]	406.1 [a]	584.9	798.9 [o]	1 018.5 [o]	1 278.0 [o]	..	30 Equip. électro. (radio, tv & communic.)
31 *Electro. comp. (inc. semi-conductors)*	..	..	..	..	..	..	..	31 *Compos. électro. (semi-cond. incl.)*
32 *TV, radio & communications equip.*	..	..	..	..	..	..	..	32 *Appareils radio, tv et de communi.*
33 Instruments	.. [n]	75.9	67.9	103.7	117.3	143.1	..	33 Instruments
34 Motor vehicles	.. [n]	11.9	19.4	14.6	11.8	12.5	..	34 Véhicules automobiles
35 Other transport equipment	41.3 [o]	9.3 [a]	6.4	16.9	13.8	14.4	..	35 Autres matériels de transport
36 *Ships*	..	..	..	..	..	..	..	36 *Navires*
37 *Aerospace*	..	..	..	..	..	..	..	37 *Aérospatiale*
38 *Other transport equipment n.e.c.*	..	..	..	..	..	..	..	38 *Autres matériels de transport n.c.a.*
39 Furniture, other manufacturing n.e.c.	11.9	4.9	5.1	7.0	9.9	13.6	..	**39 Meubles, autres activ. de fabricat. n.c.a.**
40 Furniture	3.3	..	..	..	..	..	..	40 Meubles
41 Other manufacturing n.e.c.	8.6	..	..	..	..	..	..	41 Autres activités de fabrication n.c.a.
42 Recycling	..	..	..	..	..	..	..	**42 Récupération**
43 ELECTRICITY, GAS & WATER SUPPLY	25.5	12.6	31.5	29.7	31.1	28.0	..	**43 DISTRIBUTION D'ELECTRICITE, GAZ & EAU**
44 CONSTRUCTION	3.5	9.5	..	16.4	16.9	24.7	..	**44 CONSTRUCTION**
45 SERVICES SECTOR	161.9	215.3	263.5	324.4	341.9	418.4	..	**45 SECTEUR DES SERVICES**
46 Wholesale, retail trade & repair of motor vehicles, etc.	4.1	12.3	..	25.9	34.2	42.1	..	**46 Commerce (gros & détail), réparation de véhicules, etc.**
47 Hotels & restaurants	..	..	..	..	..	..	..	**47 Hôtels et restaurants**
48 Transport & storage	4.6	4.2	8.5	7.6	6.2	13.2	..	**48 Transports, entreposage**
49 Communications	30.9 [a]	40.0	41.7	62.5	76.3	97.3	..	**49 Communications**
50 Post	23.9	..	..	..	..	..	..	50 Postes
51 Telecommunications	7.1	..	..	..	..	..	..	51 Télécommunications
52 Financial intermediation (inc. insur.)	..	..	..	..	..	..	..	**52 Intermédiation financ. (assurances inc.)**
53 Real estate, renting & busin. activity	119.6	147.4	..	198.1	213.9	256.2	..	**53 Immobil., locat. & serv. aux entreprises**
54 Computer & related activities	59.9	39.5	31.6	41.7	56.9	94.0	..	54 Informatique et activités connexes
55 *Software consultancy*	..	..	..	..	..	..	..	55 *Production de logiciels*
56 *Other computer services n.e.c.*	..	..	..	..	..	..	..	56 *Autres activités informatiques*
57 Research & development	20.8	34.3 [a]	..	91.1	104.3	111.1	..	57 Recherche & développement expérimental
58 Other business activities n.e.c.	38.9	73.5	48.0	65.3	52.6	51.0	..	58 Autres activ. & serv. aux entreprises
59 Comm., soc. & pers. serv. activ., etc.	2.6	11.4	..	30.3	11.3	9.5	..	**59 Adm. pub., serv. soc. & collec., etc**
TOTAL BUSINESS ENTERPRISE SECTOR	**1 249.8**	**1 373.4**	**1 656.7** [c]	**1 916.7**	**2 252.8**	**2 643.9**	**3 096.5** [c,p]	**TOTAL SECTEUR DES ENTREPRISES**

Source: OECD, R&D database (DSTI/EAS Division), January 2002.

Source : OCDE, base de données R-D (DSTI/Division AES), janvier 2002.

Total business enterprise intramural expenditure on R&D - Dépenses intra-muros totales de R-D des entreprises
(BERD) by industry - (DIRDE) par industrie

	1994	1995	1996	1997	1998	1999	2000	
	9.2 Million constant USD(1995 prices and PPPs) - Millions de USD constants (prix et PPA de 1995)							
1 **AGRICULTURE, HUNTING & FORESTRY**	..	2.7	..	0.7	2.4	2.0	..	1 **AGRICULTURE, CHASSE & SYLVICULTURE**
2 **MINING**	5.6	4.9	..	5.1	8.0	8.6	..	2 **ACTIVITES EXTRACTIVES**
3 **MANUFACTURING**	1 112.8	1 144.5	1 384.4	1 534.6	1 791.4	2 093.3	..	3 **ACTIVITES DE FABRICATION**
4 **Food, beverages & tobacco**	51.7	44.3	51.8	49.8	54.9	52.1	..	4 **Produits alimentaires, boissons & tabac**
5 Food products & beverages	..	..	..	..	..	..	..	5 Produits alimentaires & boissons
6 Tobacco products	..	..	..	..	..	..	..	6 Produits à base de tabac
7 **Textiles, wear app., fur & leather**	5.5	7.4	9.3	10.1	11.8	12.2	..	7 **Textiles, habill., fourrures & cuirs**
8 Textiles	4.5	6.6	..	9.3	9.9	10.5	..	8 Textiles
9 Wearing apparel & fur	.. [n]	0.5	..	0.5	1.1	0.5	..	9 Articles d'habillement & fourrures
10 Leather products & footwear	0.9 [o]	0.3 [a]	..	0.4	0.9	1.2	..	10 Articles en cuir & chaussures
11 **Wood, paper, printing, publishing**	72.3	69.7	75.4	79.5	79.0	79.1	..	11 **Bois, papier, imprimerie, édition**
12 Wood & cork (not furniture)	4.8	17.5	3.8	5.7	6.8	7.7	..	12 Bois & liège (sauf meubles)
13 Pulp, paper & paper products	61.8	46.4	62.6	67.4	66.3	66.3	..	13 Carton, papier & articles
14 Publ., print. & repro. of rec. media	5.7	5.7	9.0	6.4	5.9	5.2	..	14 Edition, impress. & repro. enregis.
15 **Coke, petroleum, nuclear fuel, chemicals, rubber plastics**	196.8	191.6	186.0	196.0	214.4	215.4	..	15 **Coke, pétrole, combust. nucl., prod. chimi., caout. & mat. plast.**
16 Coke, ref. petrol. prod. & nuclear fuel	.. [n]	.. [n]	.. [n]	.. [n]	.. [n]	.. [n]	..	16 Coke, raffin. pétr. & combust. nucl.
17 Chemicals & chemical products	177.5	146.0	142.0	142.2	152.1	172.1	..	17 Produits chimiques
18 *Chemical prod.(less pharmaceuticals)*	..	..	..	..	..	..	..	18 *Prod. chimi. (moins prod. pharmaceutiques)*
19 *Pharmaceuticals*	..	..	..	..	..	..	..	19 *Produits pharmaceutiques*
20 Rubber & plastic products	19.3 [o]	45.6 [o]	44.1 [o]	53.8 [o]	62.3 [o]	26.3 [o]	..	20 Caoutchouc et matières plastiques
21 **Non-metallic mineral products**	16.3	13.5	6.8	12.6	13.0	17.2	..	21 **Produits minéraux non métalliques**
22 **Basic metals**	57.6	33.1 [a]	37.1	37.5	33.2	35.0	..	22 **Produits métallurgiques de base**
23 Basic metals, ferrous	..	..	..	..	..	..	..	23 Produits métallur. de base, ferreux
24 Basic metals, non-ferrous	..	..	..	..	..	..	..	24 Prod. métallur. de base, non ferreux
25 **Fabricated metal products**	24.9	28.6	15.1	23.1	25.9	26.6	..	25 **Ouvrages en métaux (sauf mach.& mat.)**
26 **Machinery equipment, instruments and transport equipment**	675.1	751.4	997.6	1 119.0	1 349.6	1 642.5	..	26 **Machines & matériel, instruments & mat. de transport**
27 Machinery n.e.c.	125.7	136.2	154.7	188.3	226.2	240.5	..	27 Machines, n.c.a.
28 Office, account. & computing machinery	.. [n]	38.9	37.6	.. [n]	.. [n]	.. [n]	..	28 Mach. de bur., compta. & trait. inform.
29 Electrical machinery	.. [n]	65.8	115.5	.. [n]	.. [n]	.. [n]	..	29 Machines électriques
30 Electro. equip. (radio, TV & commun.)	505.8 [o]	411.9 [a]	594.6	795.9 [o]	985.1 [o]	1 237.5 [o]	..	30 Equip. électro. (radio, tv & communic.)
31 *Electro. comp. (inc. semi-conductors)*	..	..	..	..	..	..	..	31 *Compos. électro. (semi-cond. incl.)*
32 *TV, radio & communications equip.*	..	..	..	..	..	..	..	32 *Appareils radio, tv et de communi.*
33 Instruments	.. [n]	77.0	69.0	103.3	113.5	138.6	..	33 Instruments
34 Motor vehicles	.. [n]	12.1	19.7	14.6	11.4	12.1	..	34 Véhicules automobiles
35 Other transport equipment	43.6 [o]	9.4 [a]	6.5	16.8	13.4	13.9	..	35 Autres matériels de transport
36 *Ships*	..	..	..	..	..	..	..	36 *Navires*
37 *Aerospace*	..	..	..	..	..	..	..	37 *Aérospatiale*
38 *Other transport equipment n.e.c.*	..	..	..	..	..	..	..	38 *Autres matériels de transport n.c.a.*
39 **Furniture, other manufacturing n.e.c.**	12.6	5.0	5.2	7.0	9.6	13.2	..	39 **Meubles, autres activ. de fabricat. n.c.a.**
40 Furniture	3.5	..	..	..	..	..	..	40 Meubles
41 Other manufacturing n.e.c.	9.1	..	..	..	..	..	..	41 Autres activités de fabrication n.c.a.
42 **Recycling**	..	..	..	..	..	..	..	42 **Récupération**
43 **ELECTRICITY, GAS & WATER SUPPLY**	26.9	12.8	32.0	29.6	30.1	27.1	..	43 **DISTRIBUTION D'ELECTRICITE, GAZ & EAU**
44 **CONSTRUCTION**	3.7	9.7	..	16.4	16.3	23.9	..	44 **CONSTRUCTION**
45 **SERVICES SECTOR**	171.0	218.4	267.9	323.1	330.7	405.1	..	45 **SECTEUR DES SERVICES**
46 **Wholesale, retail trade & repair of motor vehicles, etc.**	4.4	12.5	..	25.8	33.1	40.8	..	46 **Commerce (gros & détail), réparation de véhicules, etc.**
47 **Hotels & restaurants**	..	..	..	..	..	..	..	47 **Hôtels et restaurants**
48 **Transport & storage**	4.9	4.2	8.6	7.5	6.0	12.8	..	48 **Transports, entreposage**
49 **Communications**	32.7 [a]	40.6	42.4	62.2	73.8	94.3	..	49 **Communications**
50 Post	25.2	..	..	..	..	..	..	50 Postes
51 Telecommunications	7.5	..	..	..	..	..	..	51 Télécommunications
52 **Financial intermediation (inc. insur.)**	..	..	..	..	..	..	..	52 **Intermédiation financ. (assurances inc.)**
53 **Real estate, renting & busin. activity**	126.3	149.5	..	197.3	206.9	248.0	..	53 **Immobil., locat. & serv. aux entreprises**
54 Computer & related activities	63.2	40.1	32.1	41.5	55.1	91.1	..	54 Informatique et activités connexes
55 *Software consultancy*	..	..	..	..	..	..	..	55 *Production de logiciels*
56 *Other computer services n.e.c.*	..	..	..	..	..	..	..	56 *Autres activités informatiques*
57 Research & development	22.0	34.8 [a]	..	90.8	100.9	107.6	..	57 Recherche & développement expérimental
58 Other business activities n.e.c.	41.1	74.6	48.8	65.0	50.9	49.4	..	58 Autres activ. & serv. aux entreprises
59 **Comm., soc. & pers. serv. activ., etc.**	2.8	11.6	..	30.2	10.9	9.2	..	59 **Adm. pub., serv. soc. & collec., etc**
TOTAL BUSINESS ENTERPRISE SECTOR	**1 320.0**	**1 393.1**	**1 684.3** [c]	**1 909.4**	**2 178.9**	**2 560.1**	**2 900.2** [c,p]	**TOTAL SECTEUR DES ENTREPRISES**

Source: OECD, R&D database (DSTI/EAS Division), January 2002.

Source : OCDE, base de données R-D (DSTI/Division AES), janvier 2002.

Business enterprise intramural expenditure on R&D (BERD) - Dépenses intra-muros de R-D des entreprises (DIRDE)

by industry and source of funds - par industrie et par source de financement

Million EUR (1999 FIM euro, EUR from 1999) — *Millions EUR (1999 FIM euro, EUR à partir de 1999)*

10.1 Business enterprise - Entreprises

	1994	1995	1996	1997	1998	1999	2000	
1 AGRICULTURE, HUNTING & FORESTRY	..	2.6	..	0.5	2.2	2.0	..	**1 AGRICULTURE, CHASSE & SYLVICULTURE**
2 MINING	..	4.4	..	4.7	6.7	8.1	..	**2 ACTIVITES EXTRACTIVES**
3 MANUFACTURING	..	1 011.0	..	1 446.9	1 751.2	2 067.7	..	**3 ACTIVITES DE FABRICATION**
4 Food, beverages & tobacco	..	42.6	..	45.3	52.0	50.5	..	**4 Produits alimentaires, boissons & tabac**
5 Food products & beverages	..	..	..	..	..	..	..	5 Produits alimentaires & boissons
6 Tobacco products	..	..	..	..	..	..	..	6 Produits à base de tabac
7 Textiles, wear app., fur & leather	..	7.0	..	9.5	11.6	12.1	..	**7 Textiles, habill., fourrures & cuirs**
8 Textiles	..	6.2	..	8.7	9.7	10.4	..	8 Textiles
9 Wearing apparel & fur	..	0.5	..	0.4	0.9	0.5	..	9 Articles d'habillement & fourrures
10 Leather products & footwear	..	0.3 [a]	..	0.4	0.9	1.2	..	10 Articles en cuir & chaussures
11 Wood, paper, printing, publishing	..	66.4	..	77.2	78.5	78.8	..	**11 Bois, papier, imprimerie, édition**
12 Wood & cork (not furniture)	..	16.7	..	5.2	5.6	6.8	..	12 Bois & liège (sauf meubles)
13 Pulp, paper & paper products	..	44.3	..	66.0	66.9	66.8	..	13 Carton, papier & articles
14 Publ., print. & repro. of rec. media	..	5.4	..	6.0	6.0	5.1	..	14 Edition, impress. & repro. enregis.
15 Coke, petroleum, nuclear fuel, chemicals, rubber plastics	..	174.2	..	184.4	202.5	209.6	..	**15 Coke, pétrole, combust. nucl., prod. chimi., caout. & mat. plast.**
16 Coke, ref. petrol. prod. & nuclear fuel	..	.. [n]	..	.. [n]	.. [n]	.. [n]	..	16 Coke, raffin. pétr. & combust. nucl.
17 Chemicals & chemical products	..	133.7	..	135.3	149.2	169.2	..	17 Produits chimiques
18 *Chemical prod.(less pharmaceuticals)*	..	..	..	..	..	..	..	18 *Prod. chimi. (moins prod. pharmaceutiques)*
19 *Pharmaceuticals*	..	..	..	..	..	..	..	19 *Produits pharmaceutiques*
20 Rubber & plastic products	..	40.5 [o]	..	49.1 [o]	53.3 [o]	24.3 [o]	..	20 Caoutchouc et matières plastiques
21 Non-metallic mineral products	..	12.1	..	11.7	12.4	16.6	..	**21 Produits minéraux non métalliques**
22 Basic metals	..	25.8 [a]	..	30.4	29.3	32.5	..	**22 Produits métallurgiques de base**
23 Basic metals, ferrous	..	..	..	..	..	..	..	23 Produits métallur. de base, ferreux
24 Basic metals, non-ferrous	..	..	..	..	..	..	..	24 Prod. métallur. de base, non ferreux
25 Fabricated metal products	..	21.2	..	18.6	21.9	21.4	..	**25 Ouvrages en métaux (sauf mach.& mat.)**
26 Machinery equipment, instruments and transport equipment	..	657.4	..	1 063.4	1 333.7	1 634.0	..	**26 Machines & matériel, instruments & mat. de transport**
27 Machinery n.e.c.	..	127.4	..	173.9	212.9	233.6	..	27 Machines, n.c.a.
28 Office, account. & computing machinery	..	37.9	..	.. [n]	.. [n]	.. [n]	..	28 Mach. de bur., compta. & trait. inform.
29 Electrical machinery	..	61.3	..	.. [n]	.. [n]	.. [n]	..	29 Machines électriques
30 Electro. equip. (radio, TV & commun.)	..	341.6 [a]	..	770.7 [o]	994.7 [o]	1 247.5 [o]	..	30 Equip. électro. (radio, tv & communic.)
31 *Electro. comp. (inc. semi-conductors)*	..	..	..	..	..	..	..	31 *Compos. électro. (semi-cond. incl.)*
32 *TV, radio & communications equip.*	..	..	..	..	..	..	..	32 *Appareils radio, tv et de communi.*
33 Instruments	..	71.0	..	95.5	104.6	131.1	..	33 Instruments
34 Motor vehicles	..	10.1	..	13.7	10.5	11.4	..	34 Véhicules automobiles
35 Other transport equipment	..	8.0 [a]	..	9.6	10.8	10.5	..	35 Autres matériels de transport
36 *Ships*	..	..	..	..	..	..	..	36 *Navires*
37 *Aerospace*	..	..	..	..	..	..	..	37 *Aérospatiale*
38 *Other transport equipment n.e.c.*	..	..	..	..	..	..	..	38 *Autres matériels de transport n.c.a.*
39 Furniture, other manufacturing n.e.c.	..	4.4	..	6.4	9.2	12.2	..	**39 Meubles, autres activ. de fabricat. n.c.a.**
40 Furniture	..	..	..	..	..	..	..	40 Meubles
41 Other manufacturing n.e.c.	..	..	..	..	..	..	..	41 Autres activités de fabrication n.c.a.
42 Recycling	..	..	..	..	..	..	..	**42 Récupération**
43 ELECTRICITY, GAS & WATER SUPPLY	..	12.5	..	28.6	29.3	23.0	..	**43 DISTRIBUTION D'ELECTRICITE, GAZ & EAU**
44 CONSTRUCTION	..	8.7	..	12.9	13.9	21.9	..	**44 CONSTRUCTION**
45 SERVICES SECTOR	..	184.0	..	245.2	245.2	372.2	..	**45 SECTEUR DES SERVICES**
46 Wholesale, retail trade & repair of motor vehicles, etc.	..	11.8	..	24.5	29.8	38.6	..	**46 Commerce (gros & détail), réparation de véhicules, etc.**
47 Hotels & restaurants	..	..	..	..	..	..	..	**47 Hôtels et restaurants**
48 Transport & storage	..	.. [n]	..	7.4	5.9	12.7	..	**48 Transports, entreposage**
49 Communications	..	39.4 [o]	..	57.3	71.3	91.3	..	**49 Communications**
50 Post	..	..	..	..	..	..	..	50 Postes
51 Telecommunications	..	..	..	..	..	..	..	51 Télécommunications
52 Financial intermediation (inc. insur.)	..	..	..	..	..	..	..	**52 Intermédiation financ. (assurances inc.)**
53 Real estate, renting & busin. activity	..	122.7	..	127.4	128.6	220.5	..	**53 Immobil., locat. & serv. aux entreprises**
54 Computer & related activities	..	35.0	..	37.0	46.1	84.8	..	54 Informatique et activités connexes
55 *Software consultancy*	..	..	..	..	..	..	..	55 *Production de logiciels*
56 *Other computer services n.e.c.*	..	..	..	..	..	..	..	56 *Autres activités informatiques*
57 Research & development	..	26.9 [a]	..	34.9	42.8	95.9	..	57 Recherche & développement expérimental
58 Other business activities n.e.c.	..	60.7	..	55.4	39.8	39.8	..	58 Autres activ. & serv. aux entreprises
59 Comm., soc. & pers. serv. activ., etc.	..	10.1	..	28.6	9.5	9.0	..	**59 Adm. pub., serv. soc. & collec., etc**
TOTAL BUSINESS ENTERPRISE SECTOR	**..**	**1 223.3**	**..**	**1 738.7**	**2 048.5** [a]	**2 494.8**	**..**	**TOTAL SECTEUR DES ENTREPRISES**

Source: OECD, R&D database (DSTI/EAS Division), January 2002.

Source : OCDE, base de données R-D (DSTI/Division AES), janvier 2002.

Business enterprise intramural expenditure on R&D (BERD) - Dépenses intra-muros de R-D des entreprises (DIRDE)

by industry and source of funds - par industrie et par source de financement

Million EUR (1999 FIM euro, EUR from 1999) — *Millions EUR (1999 FIM euro, EUR à partir de 1999)*

10.2 Government - État

	1994	1995	1996	1997	1998	1999	2000	
1 AGRICULTURE, HUNTING & FORESTRY	..	0.1	..	0.2	0.3	0.1	..	**1 AGRICULTURE, CHASSE & SYLVICULTURE**
2 MINING	..	0.4	..	0.4	1.4	0.7	..	**2 ACTIVITES EXTRACTIVES**
3 MANUFACTURING	..	54.5	..	51.0	65.5	71.6	..	**3 ACTIVITES DE FABRICATION**
4 Food, beverages & tobacco	..	0.8	..	2.6	3.0	2.9	..	**4 Produits alimentaires, boissons & tabac**
5 Food products & beverages	..	..	..	..	..	..	..	5 Produits alimentaires & boissons
6 Tobacco products	..	..	..	..	..	..	..	6 Produits à base de tabac
7 Textiles, wear app., fur & leather	..	0.3	..	0.6	0.7	0.5	..	**7 Textiles, habill., fourrures & cuirs**
8 Textiles	..	0.3	..	0.6	0.5	0.5	..	8 Textiles
9 Wearing apparel & fur	..	0.0	..	0.1	0.2	..	..	9 Articles d'habillement & fourrures
10 Leather products & footwear	..	..	..	0.0	0.0	0.0	..	10 Articles en cuir & chaussures
11 Wood, paper, printing, publishing	..	2.2	..	2.5	3.1	2.8	..	**11 Bois, papier, imprimerie, édition**
12 Wood & cork (not furniture)	..	0.5	..	0.5	1.4	1.1	..	12 Bois & liège (sauf meubles)
13 Pulp, paper & paper products	..	1.4	..	1.6	1.5	1.5	..	13 Carton, papier & articles
14 Publ., print. & repro. of rec. media	..	0.3	..	0.4	0.2	0.2	..	14 Edition, impress. & repro. enregis.
15 Coke, petroleum, nuclear fuel, chemicals, rubber plastics	..	13.8	..	6.4	10.2	10.6	..	**15 Coke, pétrole, combust. nucl., prod. chimi., caout. & mat. plast.**
16 Coke, ref. petrol. prod. & nuclear fuel	..	..[n]	..	..[n]	..[n]	..[n]	..	16 Coke, raffin. pétr. & combust. nucl.
17 Chemicals & chemical products	..	9.4	..	5.0	6.0	7.7	..	17 Produits chimiques
18 *Chemical prod.(less pharmaceuticals)*	..	..	..	..	..	..	..	18 *Prod. chimi. (moins prod. pharmaceutiques)*
19 *Pharmaceuticals*	..	..	..	..	..	..	..	19 *Produits pharmaceutiques*
20 Rubber & plastic products	..	4.4[o]	..	1.4[o]	4.2[o]	1.6[o]	..	20 Caoutchouc et matières plastiques
21 Non-metallic mineral products	..	1.0	..	0.8	0.9	1.1	..	**21 Produits minéraux non métalliques**
22 Basic metals	..	5.9[a]	..	4.6	3.7	2.3	..	**22 Produits métallurgiques de base**
23 Basic metals, ferrous	..	..	..	..	..	..	..	23 Produits métallur. de base, ferreux
24 Basic metals, non-ferrous	..	..	..	..	..	..	..	24 Prod. métallur. de base, non ferreux
25 Fabricated metal products	..	3.1	..	2.0	2.5	2.8	..	**25 Ouvrages en métaux (sauf mach.& mat.)**
26 Machinery equipment, instruments and transport equipment	..	26.7	..	30.9	40.9	47.2	..	**26 Machines & matériel, instruments & mat. de transport**
27 Machinery n.e.c.	..	5.8	..	9.9	12.9	12.7	..	27 Machines, n.c.a.
28 Office, account. & computing machinery	..	0.5	..	..[n]	..[n]	..[n]	..	28 Mach. de bur., compta. & trait. inform.
29 Electrical machinery	..	1.8	..	..[n]	..[n]	..[n]	..	29 Machines électriques
30 Electro. equip. (radio, TV & commun.)	..	12.7[a]	..	9.2[o]	18.0[o]	23.8[o]	..	30 Equip. électro. (radio, tv & communic.)
31 *Electro. comp. (inc. semi-conductors)*	..	..	..	..	..	..	..	31 *Compos. électro. (semi-cond. incl.)*
32 *TV, radio & communications equip.*	..	..	..	..	..	..	..	32 *Appareils radio, tv et de communi.*
33 Instruments	..	4.7	..	6.4	7.0	7.4	..	33 Instruments
34 Motor vehicles	..	0.6	..	0.8	0.3	0.2	..	34 Véhicules automobiles
35 Other transport equipment	..	0.6[a]	..	4.6	2.7	3.1	..	35 Autres matériels de transport
36 *Ships*	..	..	..	..	..	..	..	36 *Navires*
37 *Aerospace*	..	..	..	..	..	..	..	37 *Aérospatiale*
38 *Other transport equipment n.e.c.*	..	..	..	..	..	..	..	38 *Autres matériels de transport n.c.a.*
39 Furniture, other manufacturing n.e.c.	..	0.5	..	0.6	0.7	1.4	..	**39 Meubles, autres activ. de fabricat. n.c.a.**
40 Furniture	..	..	..	..	..	..	..	40 Meubles
41 Other manufacturing n.e.c.	..	..	..	..	..	..	..	41 Autres activités de fabrication n.c.a.
42 Recycling	..	..	..	..	..	..	..	**42 Récupération**
43 ELECTRICITY, GAS & WATER SUPPLY	..	0.2	..	0.9	1.6	3.0	..	**43 DISTRIBUTION D'ELECTRICITE, GAZ & EAU**
44 CONSTRUCTION	..	0.8	..	3.4	2.8	2.4	..	**44 CONSTRUCTION**
45 SERVICES SECTOR	..	20.6	..	22.8	27.8	33.0	..	**45 SECTEUR DES SERVICES**
46 Wholesale, retail trade & repair of motor vehicles, etc.	..	0.5	..	1.1	1.8	3.3	..	**46 Commerce (gros & détail), réparation de véhicules, etc.**
47 Hotels & restaurants	..	..	..	..	..	..	..	**47 Hôtels et restaurants**
48 Transport & storage	..	..[n]	..	0.2	0.3	0.5	..	**48 Transports, entreposage**
49 Communications	..	1.2[o]	..	1.0	0.8	1.3	..	**49 Communications**
50 Post	..	..	..	..	..	..	..	50 Postes
51 Telecommunications	..	..	..	..	..	..	..	51 Télécommunications
52 Financial intermediation (inc. insur.)	..	..	..	..	..	..	..	**52 Intermédiation financ. (assurances inc.)**
53 Real estate, renting & busin. activity	..	17.6	..	19.1	23.7	27.4	..	**53 Immobil., locat. & serv. aux entreprises**
54 Computer & related activities	..	4.0	..	3.9	6.4	7.1	..	54 Informatique et activités connexes
55 *Software consultancy*	..	..	..	..	..	..	..	55 *Production de logiciels*
56 *Other computer services n.e.c.*	..	..	..	..	..	..	..	56 *Autres activités informatiques*
57 Research & development	..	5.7[a]	..	6.8	7.5	10.7	..	57 Recherche & développement expérimental
58 Other business activities n.e.c.	..	7.9	..	8.4	9.8	9.7	..	58 Autres activ. & serv. aux entreprises
59 Comm., soc. & pers. serv. activ., etc.	..	1.3	..	1.4	1.3	0.5	..	**59 Adm. pub., serv. soc. & collec., etc**
TOTAL BUSINESS ENTERPRISE SECTOR	**..**	**76.6**	**..**	**78.8**	**99.4**	**110.9**	**..**	**TOTAL SECTEUR DES ENTREPRISES**

Source: OECD, R&D database (DSTI/EAS Division), January 2002.

Source : OCDE, base de données R-D (DSTI/Division AES), janvier 2002.

FINLAND	Table 10 - Tableau 10							FINLANDE

Business enterprise intramural expenditure on R&D (BERD) - Dépenses intra-muros de R-D des entreprises (DIRDE)

by industry and source of funds - par industrie et par source de financement

Million EUR (1999 FIM euro, EUR from 1999) — *Millions EUR (1999 FIM euro, EUR à partir de 1999)*

10.3 Other national funds - Autres sources nationales

	1994	1995	1996	1997	1998	1999	2000	
1 AGRICULTURE, HUNTING & FORESTRY	..	..	..	..	..	..	..	**1 AGRICULTURE, CHASSE & SYLVICULTURE**
2 MINING	..	..	..	..	..	..	..	**2 ACTIVITES EXTRACTIVES**
3 MANUFACTURING	..	0.5	..	0.4	0.3	0.3	..	**3 ACTIVITES DE FABRICATION**
4 Food, beverages & tobacco	..	0.0	..	..	..	..	..	**4 Produits alimentaires, boissons & tabac**
5 Food products & beverages	..	..	..	..	..	..	..	5 Produits alimentaires & boissons
6 Tobacco products	..	..	..	..	..	..	..	6 Produits à base de tabac
7 Textiles, wear app., fur & leather	..	..	..	..	..	..	..	**7 Textiles, habill., fourrures & cuirs**
8 Textiles	..	..	..	..	..	..	..	8 Textiles
9 Wearing apparel & fur	..	..	..	..	..	..	..	9 Articles d'habillement & fourrures
10 Leather products & footwear	..	..	..	..	..	..	..	10 Articles en cuir & chaussures
11 Wood, paper, printing, publishing	..	..	..	..	0.0	..	..	**11 Bois, papier, imprimerie, édition**
12 Wood & cork (not furniture)	..	..	..	..	..	..	..	12 Bois & liège (sauf meubles)
13 Pulp, paper & paper products	..	..	..	..	..	..	..	13 Carton, papier & articles
14 Publ., print. & repro. of rec. media	..	..	..	..	0.0	..	..	14 Edition, impress. & repro. enregis.
15 Coke, petroleum, nuclear fuel, chemicals, rubber plastics	..	0.0	..	..	..	..	..	**15 Coke, pétrole, combust. nucl., prod. chimi., caout. & mat. plast.**
16 Coke, ref. petrol. prod. & nuclear fuel	..	.. [n]	..	..	..	..	..	16 Coke, raffin. pétr. & combust. nucl.
17 Chemicals & chemical products	..	0.0	..	..	..	..	..	17 Produits chimiques
18 *Chemical prod.(less pharmaceuticals)*	..	..	..	..	..	..	..	18 *Prod. chimi. (moins prod. pharmaceutiques)*
19 *Pharmaceuticals*	..	..	..	..	..	..	..	19 *Produits pharmaceutiques*
20 Rubber & plastic products	..	0.0 [o]	..	..	..	..	..	20 Caoutchouc et matières plastiques
21 Non-metallic mineral products	..	0.1	..	0.0	0.0	..	..	**21 Produits minéraux non métalliques**
22 Basic metals	..	..	..	..	0.2	0.1	..	**22 Produits métallurgiques de base**
23 Basic metals, ferrous	..	..	..	..	..	..	..	23 Produits métallur. de base, ferreux
24 Basic metals, non-ferrous	..	..	..	..	..	..	..	24 Prod. métallur. de base, non ferreux
25 Fabricated metal products	..	..	..	..	..	..	..	**25 Ouvrages en métaux (sauf mach.& mat.)**
26 Machinery equipment, instruments and transport equipment	..	0.4	..	0.3	0.1	0.2	..	**26 Machines & matériel, instruments & mat. de transport**
27 Machinery n.e.c.	..	0.0	..	..	0.1	0.1	..	27 Machines, n.c.a.
28 Office, account. & computing machinery	..	..	..	.. [n]	..	..	..	28 Mach. de bur., compta. & trait. inform.
29 Electrical machinery	..	0.2	..	.. [n]	..	..	..	29 Machines électriques
30 Electro. equip. (radio, TV & commun.)	..	0.1 [a]	..	0.3 [o]	..	..	..	30 Equip. électro. (radio, tv & communic.)
31 *Electro. comp. (inc. semi-conductors)*	..	..	..	..	..	..	..	31 *Compos. électro. (semi-cond. incl.)*
32 *TV, radio & communications equip.*	..	..	..	..	..	..	..	32 *Appareils radio, tv et de communi.*
33 Instruments	..	..	..	..	0.0	0.1	..	33 Instruments
34 Motor vehicles	..	..	..	..	..	..	..	34 Véhicules automobiles
35 Other transport equipment	..	..	..	..	..	..	..	35 Autres matériels de transport
36 *Ships*	..	..	..	..	..	..	..	36 *Navires*
37 *Aerospace*	..	..	..	..	..	..	..	37 *Aérospatiale*
38 *Other transport equipment n.e.c.*	..	..	..	..	..	..	..	38 *Autres matériels de transport n.c.a.*
39 Furniture, other manufacturing n.e.c.	..	0.0	..	0.0	..	..	..	**39 Meubles, autres activ. de fabricat. n.c.a.**
40 Furniture	..	..	..	..	..	..	..	40 Meubles
41 Other manufacturing n.e.c.	..	..	..	..	..	..	..	41 Autres activités de fabrication n.c.a.
42 Recycling	..	..	..	..	..	..	..	**42 Récupération**
43 ELECTRICITY, GAS & WATER SUPPLY	..	..	..	..	..	..	..	**43 DISTRIBUTION D'ELECTRICITE, GAZ & EAU**
44 CONSTRUCTION	..	..	..	..	0.1	0.1	..	**44 CONSTRUCTION**
45 SERVICES SECTOR	..	0.2	..	0.5	3.3	0.3	..	**45 SECTEUR DES SERVICES**
46 Wholesale, retail trade & repair of motor vehicles, etc.	..	..	..	..	..	0.0	..	**46 Commerce (gros & détail), réparation de véhicules, etc.**
47 Hotels & restaurants	..	..	..	..	..	..	..	**47 Hôtels et restaurants**
48 Transport & storage	..	..	..	..	..	..	..	**48 Transports, entreposage**
49 Communications	..	..	..	..	..	..	..	**49 Communications**
50 Post	..	..	..	..	..	..	..	50 Postes
51 Telecommunications	..	..	..	..	..	..	..	51 Télécommunications
52 Financial intermediation (inc. insur.)	..	..	..	..	..	..	..	**52 Intermédiation financ. (assurances inc.)**
53 Real estate, renting & busin. activity	..	0.1	..	0.4	3.3	0.3	..	**53 Immobil., locat. & serv. aux entreprises**
54 Computer & related activities	..	..	..	0.2	..	0.0	..	54 Informatique et activités connexes
55 *Software consultancy*	..	..	..	..	..	..	..	55 *Production de logiciels*
56 *Other computer services n.e.c.*	..	..	..	..	..	..	..	56 *Autres activités informatiques*
57 Research & development	..	0.1 [a]	..	0.2	3.0	0.1	..	57 Recherche & développement expérimental
58 Other business activities n.e.c.	..	..	..	0.0	0.3	0.2	..	58 Autres activ. & serv. aux entreprises
59 Comm., soc. & pers. serv. activ., etc.	..	0.0	..	0.1	..	..	..	**59 Adm. pub., serv. soc. & collec., etc**
TOTAL BUSINESS ENTERPRISE SECTOR	..	**0.6**	..	**0.8**	**3.6**	**0.6**	..	**TOTAL SECTEUR DES ENTREPRISES**

Source: OECD, R&D database (DSTI/EAS Division), January 2002.

Source : OCDE, base de données R-D (DSTI/Division AES), janvier 2002.

Business enterprise intramural expenditure on R&D (BERD) - Dépenses intra-muros de R-D des entreprises (DIRDE)

by industry and source of funds - par industrie et par source de financement

Million EUR (1999 FIM euro, EUR from 1999) — *Millions EUR (1999 FIM euro, EUR à partir de 1999)*

10.4 Funds from abroad - Fonds de l'étranger

	1994	1995	1996	1997	1998	1999	2000	
1 AGRICULTURE, HUNTING & FORESTRY	..	..	..	0.0	..	0.0	..	1 AGRICULTURE, CHASSE & SYLVICULTURE
2 MINING	..	..	..	0.0	0.1	0.0	..	2 ACTIVITES EXTRACTIVES
3 MANUFACTURING	..	62.3	..	42.1	35.2	22.3	..	3 ACTIVITES DE FABRICATION
4 Food, beverages & tobacco	..	0.2	..	2.0	1.8	0.4	..	**4 Produits alimentaires, boissons & tabac**
5 Food products & beverages	..	..	..	..	..	..	..	5 Produits alimentaires & boissons
6 Tobacco products	..	..	..	..	..	..	..	6 Produits à base de tabac
7 Textiles, wear app., fur & leather	..	..	..	0.0	..	..	..	**7 Textiles, habill., fourrures & cuirs**
8 Textiles	..	..	..	0.0	..	..	..	8 Textiles
9 Wearing apparel & fur	..	..	..	0.0	..	..	..	9 Articles d'habillement & fourrures
10 Leather products & footwear	..	..	..	0.0	..	..	..	10 Articles en cuir & chaussures
11 Wood, paper, printing, publishing	..	0.1	..	0.1	0.1	0.1	..	**11 Bois, papier, imprimerie, édition**
12 Wood & cork (not furniture)	..	0.1	..	0.0	..	..	..	12 Bois & liège (sauf meubles)
13 Pulp, paper & paper products	..	..	..	0.1	0.0	0.1	..	13 Carton, papier & articles
14 Publ., print. & repro. of rec. media	..	..	..	0.0	0.0	..	..	14 Edition, impress. & repro. enregis.
15 Coke, petroleum, nuclear fuel, chemicals, rubber plastics	..	0.8	..	5.9	9.0	2.3	..	**15 Coke, pétrole, combust. nucl., prod. chimi., caout. & mat. plast.**
16 Coke, ref. petrol. prod. & nuclear fuel	..	..	..	.. [n]	.. [n]	.. [n]	..	16 Coke, raffin. pétr. & combust. nucl.
17 Chemicals & chemical products	..	0.8	..	2.4	2.1	0.8	..	17 Produits chimiques
18 *Chemical prod.(less pharmaceuticals)*	..	..	..	..	..	..	..	18 *Prod. chimi. (moins prod. pharmaceutiques)*
19 *Pharmaceuticals*	..	..	..	..	..	..	..	19 *Produits pharmaceutiques*
20 Rubber & plastic products	..	..	..	3.5 [o]	6.9 [o]	1.3 [o]	..	20 Caoutchouc et matières plastiques
21 Non-metallic mineral products	..	0.1	..	0.1	0.2	0.1	..	**21 Produits minéraux non métalliques**
22 Basic metals	..	1.0	..	2.6	1.1	1.2	..	**22 Produits métallurgiques de base**
23 Basic metals, ferrous	..	..	..	..	..	..	..	23 Produits métallur. de base, ferreux
24 Basic metals, non-ferrous	..	..	..	..	..	..	..	24 Prod. métallur. de base, non ferreux
25 Fabricated metal products	..	3.9	..	2.7	2.3	3.3	..	**25 Ouvrages en métaux (sauf mach.& mat.)**
26 Machinery equipment, instruments and transport equipment	..	56.4	..	28.7	20.8	14.9	..	**26 Machines & matériel, instruments & mat. de transport**
27 Machinery n.e.c.	..	1.2	..	5.3	8.0	1.9	..	27 Machines, n.c.a.
28 Office, account. & computing machinery	..	..	..	.. [n]	.. [n]	.. [n]	..	28 Mach. de bur., compta. & trait. inform.
29 Electrical machinery	..	1.5	..	.. [n]	.. [n]	.. [n]	..	29 Machines électriques
30 Electro. equip. (radio, TV & commun.)	..	51.7 [a]	..	18.7 [o]	5.8 [o]	6.7 [o]	..	30 Equip. électro. (radio, tv & communic.)
31 *Electro. comp. (inc. semi-conductors)*	..	..	..	..	..	..	..	31 *Compos. électro. (semi-cond. incl.)*
32 *TV, radio & communications equip.*	..	..	..	..	..	..	..	32 *Appareils radio, tv et de communi.*
33 Instruments	..	0.1	..	1.8	5.7	4.5	..	33 Instruments
34 Motor vehicles	..	1.2	..	0.2	1.0	0.9	..	34 Véhicules automobiles
35 Other transport equipment	..	0.7 [a]	..	2.7	0.3	0.8	..	35 Autres matériels de transport
36 *Ships*	..	..	..	..	..	..	..	36 *Navires*
37 *Aerospace*	..	..	..	..	..	..	..	37 *Aérospatiale*
38 *Other transport equipment n.e.c.*	..	..	..	..	..	..	..	38 *Autres matériels de transport n.c.a.*
39 Furniture, other manufacturing n.e.c.	..	0.0	..	0.0	..	..	..	**39 Meubles, autres activ. de fabricat. n.c.a.**
40 Furniture	..	..	..	..	..	..	..	40 Meubles
41 Other manufacturing n.e.c.	..	..	..	..	..	..	..	41 Autres activités de fabrication n.c.a.
42 Recycling	..	..	..	..	..	..	..	**42 Récupération**
43 ELECTRICITY, GAS & WATER SUPPLY	..	..	..	0.2	0.2	2.0	..	43 DISTRIBUTION D'ELECTRICITE, GAZ & EAU
44 CONSTRUCTION	..	..	..	0.2	0.1	0.4	..	44 CONSTRUCTION
45 SERVICES SECTOR	..	10.5	..	56.0	65.6	12.9	..	45 SECTEUR DES SERVICES
46 Wholesale, retail trade & repair of motor vehicles, etc.	..	..	..	0.3	2.6	0.1	..	**46 Commerce (gros & détail), réparation de véhicules, etc.**
47 Hotels & restaurants	..	..	..	..	..	..	..	**47 Hôtels et restaurants**
48 Transport & storage	..	.. [n]	..	0.0	0.0	0.1	..	**48 Transports, entreposage**
49 Communications	..	3.5 [o]	..	4.2	4.2	4.7	..	**49 Communications**
50 Post	..	..	..	..	..	..	..	50 Postes
51 Telecommunications	..	..	..	..	..	..	..	51 Télécommunications
52 Financial intermediation (inc. insur.)	..	..	..	..	..	..	..	**52 Intermédiation financ. (assurances inc.)**
53 Real estate, renting & busin. activity	..	7.0	..	51.2	58.3	8.0	..	**53 Immobil., locat. & serv. aux entreprises**
54 Computer & related activities	..	0.6	..	0.6	4.4	2.1	..	54 Informatique et activités connexes
55 *Software consultancy*	..	..	..	..	..	..	..	55 *Production de logiciels*
56 *Other computer services n.e.c.*	..	..	..	..	..	..	..	56 *Autres activités informatiques*
57 Research & development	..	1.5 [a]	..	49.2	51.1	4.4	..	57 Recherche & développement expérimental
58 Other business activities n.e.c.	..	4.9	..	1.4	2.8	1.4	..	58 Autres activ. & serv. aux entreprises
59 Comm., soc. & pers. serv. activ., etc.	..	0.0	..	0.3	0.5	0.0	..	**59 Adm. pub., serv. soc. & collec., etc**
TOTAL BUSINESS ENTERPRISE SECTOR	**..**	**72.8**	**..**	**98.5**	**101.3 [a]**	**37.6**	**..**	**TOTAL SECTEUR DES ENTREPRISES**

Source: OECD, R&D database (DSTI/EAS Division), January 2002.

Source : OCDE, base de données R-D (DSTI/Division AES), janvier 2002.

Business enterprise intramural expenditure on R&D (BERD) - Dépenses intra-muros de R-D des entreprises (DIRDE)

by industry and type of costs - par industrie et par type de coûts

Million EUR (1999 FIM euro, EUR from 1999) — *Millions EUR (1999 FIM euro, EUR à partir de 1999)*

11.1 Current expenditure - Dépenses courantes

	1994	1995	1996	1997	1998	1999	2000	
1 AGRICULTURE, HUNTING & FORESTRY	..	2.5	..	0.7	2.3	2.0	..	**1 AGRICULTURE, CHASSE & SYLVICULTURE**
2 MINING	5.2	4.7	..	5.0	7.9	8.3	..	**2 ACTIVITES EXTRACTIVES**
3 MANUFACTURING	954.9	982.4	..	1 361.2	1 643.3	1 977.4	..	**3 ACTIVITES DE FABRICATION**
4 Food, beverages & tobacco	46.7	40.7	..	46.2	53.2	50.7	..	**4 Produits alimentaires, boissons & tabac**
5 Food products & beverages	..	..	..	..	..	..	..	5 Produits alimentaires & boissons
6 Tobacco products	..	..	..	..	..	..	..	6 Produits à base de tabac
7 Textiles, wear app., fur & leather	5.1	6.4	..	8.5	10.3	11.5	..	**7 Textiles, habill., fourrures & cuirs**
8 Textiles	4.2	5.6	..	7.7	8.2	9.8	..	8 Textiles
9 Wearing apparel & fur	.. [n]	0.5	..	0.4	1.1	0.5	..	9 Articles d'habillement & fourrures
10 Leather products & footwear	0.8 [o]	0.3 [a]	..	0.4	0.9	1.2	..	10 Articles en cuir & chaussures
11 Wood, paper, printing, publishing	64.1	63.5	..	71.8	73.6	74.6	..	**11 Bois, papier, imprimerie, édition**
12 Wood & cork (not furniture)	4.0	15.6	..	5.6	6.6	7.6	..	12 Bois & liège (sauf meubles)
13 Pulp, paper & paper products	55.3	42.2	..	60.1	60.8	61.7	..	13 Carton, papier & articles
14 Publ., print. & repro. of rec. media	4.8	5.6	..	6.1	6.1	5.3	..	14 Edition, impress. & repro. enregis.
15 Coke, petroleum, nuclear fuel, chemicals, rubber plastics	168.5	169.8	..	172.7	200.0	189.2	..	**15 Coke, pétrole, combust. nucl., prod. chimi., caout. & mat. plast.**
16 Coke, ref. petrol. prod. & nuclear fuel	.. [n]	.. [n]	..	.. [n]	.. [n]	.. [n]	..	16 Coke, raffin. pétr. & combust. nucl.
17 Chemicals & chemical products	152.8	131.0	..	130.0	140.2	149.2	..	17 Produits chimiques
18 *Chemical prod.(less pharmaceuticals)*	..	..	..	..	..	..	..	18 *Prod. chimi. (moins prod. pharmaceutiques)*
19 *Pharmaceuticals*	..	..	..	..	..	..	..	19 *Produits pharmaceutiques*
20 Rubber & plastic products	15.6 [o]	38.8 [o]	..	42.7 [o]	59.8 [o]	23.5 [o]	..	20 Caoutchouc et matières plastiques
21 Non-metallic mineral products	14.4	12.5	..	12.4	13.3	17.5	..	**21 Produits minéraux non métalliques**
22 Basic metals	51.2	30.6 [a]	..	35.1	33.6	35.5	..	**22 Produits métallurgiques de base**
23 Basic metals, ferrous	..	..	..	..	..	..	..	23 Produits métallur. de base, ferreux
24 Basic metals, non-ferrous	..	..	..	..	..	..	..	24 Prod. métallur. de base, non ferreux
25 Fabricated metal products	22.9	25.6	..	22.2	26.1	26.8	..	**25 Ouvrages en métaux (sauf mach.& mat.)**
26 Machinery equipment, instruments and transport equipment	571.5	628.8	..	985.7	1 224.1	1 558.3	..	**26 Machines & matériel, instruments & mat. de transport**
27 Machinery n.e.c.	104.3	110.2	..	173.9	219.7	235.7	..	27 Machines, n.c.a.
28 Office, account. & computing machinery	.. [n]	37.2	..	.. [n]	.. [n]	.. [n]	..	28 Mach. de bur., compta. & trait. inform.
29 Electrical machinery	.. [n]	61.4	..	.. [n]	.. [n]	.. [n]	..	29 Machines électriques
30 Electro. equip. (radio, TV & commun.)	426.8 [o]	328.4 [a]	..	682.2 [o]	868.2 [o]	1 158.2 [o]	..	30 Equip. électro. (radio, tv & communic.)
31 *Electro. comp. (inc. semi-conductors)*	..	..	..	..	..	..	..	31 *Compos. électro. (semi-cond. incl.)*
32 *TV, radio & communications equip.*	..	..	..	..	..	..	..	32 *Appareils radio, tv et de communi.*
33 Instruments	.. [n]	72.3	..	99.6	111.4	138.3	..	33 Instruments
34 Motor vehicles	.. [n]	10.5	..	13.9	11.3	12.2	..	34 Véhicules automobiles
35 Other transport equipment	40.3 [o]	8.9 [a]	..	16.2	13.4	13.9	..	35 Autres matériels de transport
36 *Ships*	..	..	..	..	..	..	..	36 *Navires*
37 *Aerospace*	..	..	..	..	..	..	..	37 *Aérospatiale*
38 *Other transport equipment n.e.c.*	..	..	..	..	..	..	..	38 *Autres matériels de transport n.c.a.*
39 Furniture, other manufacturing n.e.c.	10.6	4.6	..	6.6	9.2	13.2	..	**39 Meubles, autres activ. de fabricat. n.c.a.**
40 Furniture	3.1	..	..	..	..	..	..	40 Meubles
41 Other manufacturing n.e.c.	7.5	..	..	..	..	..	..	41 Autres activités de fabrication n.c.a.
42 Recycling	..	..	..	..	..	..	..	**42 Récupération**
43 ELECTRICITY, GAS & WATER SUPPLY	22.7	12.1	..	28.5	30.2	27.2	..	**43 DISTRIBUTION D'ELECTRICITE, GAZ & EAU**
44 CONSTRUCTION	3.4	9.4	..	15.9	16.3	23.9	..	**44 CONSTRUCTION**
45 SERVICES SECTOR	135.8	196.3	..	301.4	316.2	395.1	..	**45 SECTEUR DES SERVICES**
46 Wholesale, retail trade & repair of motor vehicles, etc.	3.5	11.0	..	23.2	31.4	39.2	..	**46 Commerce (gros & détail), réparation de véhicules, etc.**
47 Hotels & restaurants	..	..	..	..	..	..	..	**47 Hôtels et restaurants**
48 Transport & storage	4.6	.. [n]	..	7.3	6.2	12.9	..	**48 Transports, entreposage**
49 Communications	27.1 [a]	40.9 [a,o]	..	54.2	69.8	88.3	..	**49 Communications**
50 Post	20.4	..	..	..	..	..	..	50 Postes
51 Telecommunications	6.7	..	..	..	..	..	..	51 Télécommunications
52 Financial intermediation (inc. insur.)	..	..	..	..	..	..	..	**52 Intermédiation financ. (assurances inc.)**
53 Real estate, renting & busin. activity	98.2	133.4	..	188.5	198.0	245.6	..	**53 Immobil., locat. & serv. aux entreprises**
54 Computer & related activities	57.2	37.7	..	40.2	55.2	92.0	..	54 Informatique et activités connexes
55 *Software consultancy*	..	..	..	..	..	..	..	55 *Production de logiciels*
56 *Other computer services n.e.c.*	..	..	..	..	..	..	..	56 *Autres activités informatiques*
57 Research & development	20.0	31.2 [a]	..	85.9	94.5	104.4	..	57 Recherche & développement expérimental
58 Other business activities n.e.c.	20.9	64.5	..	62.4	48.4	49.2	..	58 Autres activ. & serv. aux entreprises
59 Comm., soc. & pers. serv. activ., etc.	2.4	11.1	..	28.2	10.9	9.1	..	**59 Adm. pub., serv. soc. & collec., etc**
TOTAL BUSINESS ENTERPRISE SECTOR	**1 121.9**	**1 207.4**	**..**	**1 712.7**	**2 016.2**	**2 433.8**	**..**	**TOTAL SECTEUR DES ENTREPRISES**

Source: OECD, R&D database (DSTI/EAS Division), January 2002.

Source : OCDE, base de données R-D (DSTI/Division AES), janvier 2002.

Business enterprise intramural expenditure on R&D (BERD) - Dépenses intra-muros de R-D des entreprises (DIRDE)

by industry and type of costs - par industrie et par type de coûts

Million EUR (1999 FIM euro, EUR from 1999) — *Millions EUR (1999 FIM euro, EUR à partir de 1999)*

11.2 Capital expenditure - Dépenses en capital

	1994	1995	1996	1997	1998	1999	2000	
1 AGRICULTURE, HUNTING & FORESTRY	..	0.2	..	0.0	0.2	0.1	..	**1 AGRICULTURE, CHASSE & SYLVICULTURE**
2 MINING	0.1	0.1	..	0.2	0.3	0.6	..	**2 ACTIVITES EXTRACTIVES**
3 MANUFACTURING	98.7	145.9	..	179.2	208.9	184.5	..	**3 ACTIVITES DE FABRICATION**
4 Food, beverages & tobacco	2.3	3.0	..	3.9	3.5	3.1	..	**4 Produits alimentaires, boissons & tabac**
5 Food products & beverages	..	..	..	..	..	..	..	5 Produits alimentaires & boissons
6 Tobacco products	..	..	..	..	..	..	..	6 Produits à base de tabac
7 Textiles, wear app., fur & leather	0.1	0.9	..	1.7	2.0	1.1	..	**7 Textiles, habill., fourrures & cuirs**
8 Textiles	0.1	0.9	..	1.6	2.0	1.1	..	8 Textiles
9 Wearing apparel & fur	.. [n]	0.0	..	0.0	..	..	..	9 Articles d'habillement & fourrures
10 Leather products & footwear	0.1 [o]	..	..	0.0	..	..	..	10 Articles en cuir & chaussures
11 Wood, paper, printing, publishing	4.3	5.2	..	8.0	8.1	7.1	..	**11 Bois, papier, imprimerie, édition**
12 Wood & cork (not furniture)	0.5	1.6	..	0.2	0.4	0.3	..	12 Bois & liège (sauf meubles)
13 Pulp, paper & paper products	3.2	3.5	..	7.6	7.7	6.7	..	13 Carton, papier & articles
14 Publ., print. & repro. of rec. media	0.6	0.1	..	0.2	0.0	0.0	..	14 Edition, impress. & repro. enregis.
15 Coke, petroleum, nuclear fuel, chemicals, rubber plastics	17.8	19.0	..	24.1	21.7	33.3	..	**15 Coke, pétrole, combust. nucl., prod. chimi., caout. & mat. plast.**
16 Coke, ref. petrol. prod. & nuclear fuel	.. [n]	.. [n]	..	.. [n]	.. [n]	.. [n]	..	16 Coke, raffin. pétr. & combust. nucl.
17 Chemicals & chemical products	15.2	12.9	..	12.8	17.1	28.6	..	17 Produits chimiques
18 *Chemical prod.(less pharmaceuticals)*	..	..	..	..	..	..	..	18 *Prod. chimi. (moins prod. pharmaceutiques)*
19 *Pharmaceuticals*	..	..	..	..	..	..	..	19 *Produits pharmaceutiques*
20 Rubber & plastic products	2.6 [o]	6.1 [o]	..	11.3 [o]	4.6 [o]	3.7 [o]	..	20 Caoutchouc et matières plastiques
21 Non-metallic mineral products	1.1	0.8	..	0.2	0.1	0.3	..	**21 Produits minéraux non métalliques**
22 Basic metals	3.3	2.1 [a]	..	2.5	0.8	0.6	..	**22 Produits métallurgiques de base**
23 Basic metals, ferrous	..	..	..	..	..	..	..	23 Produits métallur. de base, ferreux
24 Basic metals, non-ferrous	..	..	..	..	..	..	..	24 Prod. métallur. de base, non ferreux
25 Fabricated metal products	0.8	2.6	..	1.0	0.7	0.7	..	**25 Ouvrages en métaux (sauf mach.& mat.)**
26 Machinery equipment, instruments and transport equipment	67.7	112.0	..	137.5	171.3	138.0	..	**26 Machines & matériel, instruments & mat. de transport**
27 Machinery n.e.c.	14.7	24.1	..	15.2	14.1	12.6	..	27 Machines, n.c.a.
28 Office, account. & computing machinery	.. [n]	1.2	..	.. [n]	.. [n]	.. [n]	..	28 Mach. de bur., compta. & trait. inform.
29 Electrical machinery	.. [n]	3.5	..	.. [n]	.. [n]	.. [n]	..	29 Machines électriques
30 Electro. equip. (radio, TV & commun.)	52.1 [o]	77.7 [a]	..	116.8 [o]	150.4 [o]	119.9 [o]	..	30 Equip. électro. (radio, tv & communic.)
31 *Electro. comp. (inc. semi-conductors)*	..	..	..	..	..	..	..	31 *Compos. électro. (semi-cond. incl.)*
32 *TV, radio & communications equip.*	..	..	..	..	..	..	..	32 *Appareils radio, tv et de communi.*
33 Instruments	.. [n]	3.7	..	4.2	5.9	4.8	..	33 Instruments
34 Motor vehicles	.. [n]	1.4	..	0.7	0.5	0.2	..	34 Véhicules automobiles
35 Other transport equipment	1.0 [o]	0.4 [a]	..	0.7	0.4	0.5	..	35 Autres matériels de transport
36 *Ships*	..	..	..	..	..	..	..	36 *Navires*
37 *Aerospace*	..	..	..	..	..	..	..	37 *Aérospatiale*
38 *Other transport equipment n.e.c.*	..	..	..	..	..	..	..	38 *Autres matériels de transport n.c.a.*
39 Furniture, other manufacturing n.e.c.	1.3	0.3	..	0.4	0.7	0.4	..	**39 Meubles, autres activ. de fabricat. n.c.a.**
40 Furniture	0.2	..	..	..	..	..	..	40 Meubles
41 Other manufacturing n.e.c.	1.1	..	..	..	..	..	..	41 Autres activités de fabrication n.c.a.
42 Recycling	..	..	..	..	..	..	..	**42 Récupération**
43 ELECTRICITY, GAS & WATER SUPPLY	2.8	0.6	..	1.2	0.9	0.8	..	**43 DISTRIBUTION D'ELECTRICITE, GAZ & EAU**
44 CONSTRUCTION	0.2	0.2	..	0.6	0.6	0.8	..	**44 CONSTRUCTION**
45 SERVICES SECTOR	26.1	19.0	..	22.9	25.7	23.3	..	**45 SECTEUR DES SERVICES**
46 Wholesale, retail trade & repair of motor vehicles, etc.	0.6	1.3	..	2.7	2.8	2.9	..	**46 Commerce (gros & détail), réparation de véhicules, etc.**
47 Hotels & restaurants	..	..	..	..	..	..	..	**47 Hôtels et restaurants**
48 Transport & storage	0.1	.. [n]	..	0.2	..	0.4	..	**48 Transports, entreposage**
49 Communications	3.8 [a]	3.3 [a,o]	..	8.3	6.5	9.1	..	**49 Communications**
50 Post	3.5	..	..	..	..	..	..	50 Postes
51 Telecommunications	0.3	..	..	..	..	..	..	51 Télécommunications
52 Financial intermediation (inc. insur.)	..	..	..	..	..	..	..	**52 Intermédiation financ. (assurances inc.)**
53 Real estate, renting & busin. activity	21.4	14.0	..	9.6	15.9	10.6	..	**53 Immobil., locat. & serv. aux entreprises**
54 Computer & related activities	2.7	1.8	..	1.5	1.8	2.0	..	54 Informatique et activités connexes
55 *Software consultancy*	..	..	..	..	..	..	..	55 *Production de logiciels*
56 *Other computer services n.e.c.*	..	..	..	..	..	..	..	56 *Autres activités informatiques*
57 Research & development	0.8	3.1 [a]	..	5.2	9.9	6.7	..	57 Recherche & développement expérimental
58 Other business activities n.e.c.	17.9	9.0	..	2.8	4.3	1.9	..	58 Autres activ. & serv. aux entreprises
59 Comm., soc. & pers. serv. activ., etc.	0.2	0.4	..	2.1	0.4	0.5	..	**59 Adm. pub., serv. soc. & collec., etc**
TOTAL BUSINESS ENTERPRISE SECTOR	**127.9**	**166.0**	..	**204.2**	**236.6**	**210.1**	..	**TOTAL SECTEUR DES ENTREPRISES**

Source: OECD, R&D database (DSTI/EAS Division), January 2002.

Source : OCDE, base de données R-D (DSTI/Division AES), janvier 2002.

FINLAND Table 12 - Tableau 12 **FINLANDE**

Total business enterprise R&D personnel (BEMP) - Personnel total de R-D du secteur des entreprises (PRDE)

by industry - par industrie

In full time equivalent — *En équivalence plein-temps*

	1994	1995	1996	1997	1998	1999	2000	
1 **AGRICULTURE, HUNTING & FORESTRY**	..	16	..	7	28	15	..	1 **AGRICULTURE, CHASSE & SYLVICULTURE**
2 **MINING**	60	66	..	51	59	75	..	2 **ACTIVITES EXTRACTIVES**
3 **MANUFACTURING**	14 443	14 417	17 093	17 673	20 268	22 334	..	3 **ACTIVITES DE FABRICATION**
4 **Food, beverages & tobacco**	674	542	736	578	633	607	..	4 **Produits alimentaires, boissons & tabac**
5 Food products & beverages	..	..	..	..	..	..	..	5 Produits alimentaires & boissons
6 Tobacco products	..	..	..	..	..	..	..	6 Produits à base de tabac
7 **Textiles, wear app., fur & leather**	117	117	57	141	165	169	..	7 **Textiles, habill., fourrures & cuirs**
8 Textiles	100	106	..	133	146	143	..	8 Textiles
9 Wearing apparel & fur	.. [n]	5	..	5	6	10	..	9 Articles d'habillement & fourrures
10 Leather products & footwear	17 [o]	6 [a]	..	2	13	15	..	10 Articles en cuir & chaussures
11 **Wood, paper, printing, publishing**	978	955	979	932	907	986	..	11 **Bois, papier, imprimerie, édition**
12 Wood & cork (not furniture)	60	179	55	81	93	81	..	12 Bois & liège (sauf meubles)
13 Pulp, paper & paper products	841	681	797	755	728	836	..	13 Carton, papier & articles
14 Publ., print. & repro. of rec. media	77	95	127	97	86	69	..	14 Edition, impress. & repro. enregis.
15 **Coke, petroleum, nuclear fuel, chemicals, rubber plastics**	2 673	2 616	2 908	2 743	3 038	2 864	..	15 **Coke, pétrole, combust. nucl., prod. chimi., caout. & mat. plast.**
16 Coke, ref. petrol. prod. & nuclear fuel	.. [n]	.. [n]	.. [n]	.. [n]	.. [n]	.. [n]	..	16 Coke, raffin. pétr. & combust. nucl.
17 Chemicals & chemical products	2 431	1 994	2 241	2 024	2 206	2 206	..	17 Produits chimiques
18 *Chemical prod.(less pharmaceuticals)*	..	..	..	..	..	..	..	18 *Prod. chimi. (moins prod. pharmaceutiques)*
19 *Pharmaceuticals*	..	..	..	..	..	..	..	19 *Produits pharmaceutiques*
20 Rubber & plastic products	242 [o]	622 [o]	667 [o]	719 [o]	832 [o]	359 [o]	..	20 Caoutchouc et matières plastiques
21 **Non-metallic mineral products**	192	166	162	150	155	177	..	21 **Produits minéraux non métalliques**
22 **Basic metals**	719	430 [a]	579	362	447	421	..	22 **Produits métallurgiques de base**
23 Basic metals, ferrous	..	..	..	..	..	..	..	23 Produits métallur. de base, ferreux
24 Basic metals, non-ferrous	..	..	..	..	..	..	..	24 Prod. métallur. de base, non ferreux
25 **Fabricated metal products**	306	382	198	295	323	316	..	25 **Ouvrages en métaux (sauf mach.& mat.)**
26 **Machinery equipment, instruments and transport equipment**	8 588	9 138	11 393	12 376	14 469	16 570	..	26 **Machines & matériel, instruments & mat. de transport**
27 Machinery n.e.c.	1 375	1 408	1 729	1 900	2 204	2 265	..	27 Machines, n.c.a.
28 Office, account. & computing machinery	..	500	424	.. [n]	.. [n]	.. [n]	..	28 Mach. de bur., compta. & trait. inform.
29 Electrical machinery	.. [n]	806	1 423	.. [n]	.. [n]	.. [n]	..	29 Machines électriques
30 Electro. equip. (radio, TV & commun.)	6 650 [o]	5 151 [a]	6 553	8 703 [o]	10 447 [o]	12 188 [o]	..	30 Equip. électro. (radio, tv & communic.)
31 *Electro. comp. (inc. semi-conductors)*	..	..	..	..	..	..	..	31 *Compos. électro. (semi-cond. incl.)*
32 *TV, radio & communications equip.*	..	..	..	..	..	..	..	32 *Appareils radio, tv et de communi.*
33 Instruments	..	1 000	962	1 344	1 472	1 786	..	33 Instruments
34 Motor vehicles	.. [n]	158	203	241	178	156	..	34 Véhicules automobiles
35 Other transport equipment	563 [o]	115 [a]	99	189	168	176	..	35 Autres matériels de transport
36 *Ships*	..	..	..	..	..	..	..	36 *Navires*
37 *Aerospace*	..	..	..	..	..	..	..	37 *Aérospatiale*
38 *Other transport equipment n.e.c.*	..	..	..	..	..	..	..	38 *Autres matériels de transport n.c.a.*
39 **Furniture, other manufacturing n.e.c.**	196	71	81	96	131	224	..	39 **Meubles, autres activ. de fabricat. n.c.a.**
40 Furniture	53	..	..	..	..	..	..	40 Meubles
41 Other manufacturing n.e.c.	143	..	..	..	..	..	..	41 Autres activités de fabrication n.c.a.
42 **Recycling**	..	..	..	..	..	..	..	42 **Récupération**
43 **ELECTRICITY, GAS & WATER SUPPLY**	260	159	275	344	355	157	..	43 **DISTRIBUTION D'ELECTRICITE, GAZ & EAU**
44 **CONSTRUCTION**	60	125	..	179	183	275	..	44 **CONSTRUCTION**
45 **SERVICES SECTOR**	2 077	3 017	3 388	4 049	6 724	4 963	..	45 **SECTEUR DES SERVICES**
46 **Wholesale, retail trade & repair of motor vehicles, etc.**	64	179	..	323	380	400	..	46 **Commerce (gros & détail), réparation de véhicules, etc.**
47 **Hotels & restaurants**	..	..	..	..	..	..	..	47 **Hôtels et restaurants**
48 **Transport & storage**	70	.. [n]	73	85	106	157	..	48 **Transports, entreposage**
49 **Communications**	421 [a]	601 [o]	484	723	877	1 119	..	49 **Communications**
50 Post	310	..	..	..	..	..	..	50 Postes
51 Telecommunications	111	..	..	..	..	..	..	51 Télécommunications
52 **Financial intermediation (inc. insur.)**	..	..	..	..	..	..	..	52 **Intermédiation financ. (assurances inc.)**
53 **Real estate, renting & busin. activity**	1 483	2 101	..	2 617	2 607	3 174	..	53 **Immobil., locat. & serv. aux entreprises**
54 Computer & related activities	910	675	580	699	890	1 243	..	54 Informatique et activités connexes
55 *Software consultancy*	..	..	..	..	..	..	..	55 *Production de logiciels*
56 *Other computer services n.e.c.*	..	..	..	..	..	..	..	56 *Autres activités informatiques*
57 Research & development	316	560 [a]	..	1 114	1 104	1 275	..	57 Recherche & développement expérimental
58 Other business activities n.e.c.	257	866	505	804	612	657	..	58 Autres activ. & serv. aux entreprises
59 **Comm., soc. & pers. serv. activ., etc.**	39	136	..	302	147	114	..	59 **Adm. pub., serv. soc. & collec., etc**
TOTAL BUSINESS ENTERPRISE SECTOR	**16 900**	**17 798**	**20 756**	**22 302**	**25 010**	**27 818**	**..**	**TOTAL SECTEUR DES ENTREPRISES**

Source: OECD, R&D database (DSTI/EAS Division), January 2002.

Source : OCDE, base de données R-D (DSTI/Division AES), janvier 2002.

Business enterprise university graduates - Diplômés universitaires des entreprises

by industry - par industrie

In full time equivalent | *En équivalence plein-temps*

	1994	1995	1996	1997	1998	1999	2000	
1 AGRICULTURE, HUNTING & FORESTRY	..	4	..	3	13	3	..	**1 AGRICULTURE, CHASSE & SYLVICULTURE**
2 MINING	..	25	..	28	33	36	..	**2 ACTIVITES EXTRACTIVES**
3 MANUFACTURING	..	4 920	..	6 455	7 357	7 921	..	**3 ACTIVITES DE FABRICATION**
4 Food, beverages & tobacco	..	218	..	257	264	255	..	**4 Produits alimentaires, boissons & tabac**
5 Food products & beverages	..	..	..	..	..	..	..	5 Produits alimentaires & boissons
6 Tobacco products	..	..	..	..	..	..	..	6 Produits à base de tabac
7 Textiles, wear app., fur & leather	..	30	..	36	57	52	..	**7 Textiles, habill., fourrures & cuirs**
8 Textiles	..	28	..	34	53	46	..	8 Textiles
9 Wearing apparel & fur	..	1	..	1	1	2	..	9 Articles d'habillement & fourrures
10 Leather products & footwear	..	1 [a]	..	0	4	4	..	10 Articles en cuir & chaussures
11 Wood, paper, printing, publishing	..	347	..	382	367	413	..	**11 Bois, papier, imprimerie, édition**
12 Wood & cork (not furniture)	..	89	..	19	28	24	..	12 Bois & liège (sauf meubles)
13 Pulp, paper & paper products	..	225	..	322	305	343	..	13 Carton, papier & articles
14 Publ., print. & repro. of rec. media	..	33	..	41	34	46	..	14 Edition, impress. & repro. enregis.
15 Coke, petroleum, nuclear fuel, chemicals, rubber plastics	..	1 043	..	1 082	1 224	1 213	..	**15 Coke, pétrole, combust. nucl., prod. chimi., caout. & mat. plast.**
16 Coke, ref. petrol. prod. & nuclear fuel	..	.. [n]	..	.. [n]	.. [n]	.. [n]	..	16 Coke, raffin. pétr. & combust. nucl.
17 Chemicals & chemical products	..	835	..	854	957	998	..	17 Produits chimiques
18 *Chemical prod.(less pharmaceuticals)*	..	..	..	..	..	..	..	18 *Prod. chimi. (moins prod. pharmaceutiques)*
19 *Pharmaceuticals*	..	..	..	..	..	..	..	19 *Produits pharmaceutiques*
20 Rubber & plastic products	..	208 [o]	..	229 [o]	267 [o]	96 [o]	..	20 Caoutchouc et matières plastiques
21 Non-metallic mineral products	..	63	..	66	51	63	..	**21 Produits minéraux non métalliques**
22 Basic metals	..	179 [a]	..	197	238	227	..	**22 Produits métallurgiques de base**
23 Basic metals, ferrous	..	..	..	..	..	..	..	23 Produits métallur. de base, ferreux
24 Basic metals, non-ferrous	..	..	..	..	..	..	..	24 Prod. métallur. de base, non ferreux
25 Fabricated metal products	..	91	..	56	72	85	..	**25 Ouvrages en métaux (sauf mach.& mat.)**
26 Machinery equipment, instruments and transport equipment	..	2 939	..	4 368	5 049	5 555	..	**26 Machines & matériel, instruments & mat. de transport**
27 Machinery n.e.c.	..	430	..	705	756	750	..	27 Machines, n.c.a.
28 Office, account. & computing machinery	..	51	..	.. [n]	.. [n]	.. [n]	..	28 Mach. de bur., compta. & trait. inform.
29 Electrical machinery	..	279	..	.. [n]	.. [n]	.. [n]	..	29 Machines électriques
30 Electro. equip. (radio, TV & commun.)	..	1 665 [a]	..	2 967 [o]	3 587 [o]	3 979 [o]	..	30 Equip. électro. (radio, tv & communic.)
31 *Electro. comp. (inc. semi-conductors)*	..	..	..	..	..	..	..	31 *Compos. électro. (semi-cond. incl.)*
32 *TV, radio & communications equip.*	..	..	..	..	..	..	..	32 *Appareils radio, tv et de communi.*
33 Instruments	..	443	..	590	630	751	..	33 Instruments
34 Motor vehicles	..	32	..	40	30	24	..	34 Véhicules automobiles
35 Other transport equipment	..	39 [a]	..	66	47	50	..	35 Autres matériels de transport
36 *Ships*	..	..	..	..	..	..	..	36 *Navires*
37 *Aerospace*	..	..	..	..	..	..	..	37 *Aérospatiale*
38 *Other transport equipment n.e.c.*	..	..	..	..	..	..	..	38 *Autres matériels de transport n.c.a.*
39 Furniture, other manufacturing n.e.c.	..	7	..	12	35	60	..	**39 Meubles, autres activ. de fabricat. n.c.a.**
40 Furniture	..	..	..	..	..	..	..	40 Meubles
41 Other manufacturing n.e.c.	..	..	..	..	..	..	..	41 Autres activités de fabrication n.c.a.
42 Recycling	..	..	..	..	..	..	..	**42 Récupération**
43 ELECTRICITY, GAS & WATER SUPPLY	..	97	..	251	255	90	..	**43 DISTRIBUTION D'ELECTRICITE, GAZ & EAU**
44 CONSTRUCTION	..	36	..	57	54	68	..	**44 CONSTRUCTION**
45 SERVICES SECTOR	..	1 601	..	1 882	3 224	2 437	..	**45 SECTEUR DES SERVICES**
46 Wholesale, retail trade & repair of motor vehicles, etc.	..	76	..	115	145	189	..	**46 Commerce (gros & détail), réparation de véhicules, etc.**
47 Hotels & restaurants	..	..	..	..	..	..	..	**47 Hôtels et restaurants**
48 Transport & storage	..	.. [n]	..	21	33	55	..	**48 Transports, entreposage**
49 Communications	..	341 [o]	..	355	467	580	..	**49 Communications**
50 Post	..	..	..	..	..	..	..	50 Postes
51 Telecommunications	..	..	..	..	..	..	..	51 Télécommunications
52 Financial intermediation (inc. insur.)	..	..	..	..	..	..	..	**52 Intermédiation financ. (assurances inc.)**
53 Real estate, renting & busin. activity	..	1 097	..	1 174	1 248	1 555	..	**53 Immobil., locat. & serv. aux entreprises**
54 Computer & related activities	..	368	..	333	387	572	..	54 Informatique et activités connexes
55 *Software consultancy*	..	..	..	..	..	..	..	55 *Production de logiciels*
56 *Other computer services n.e.c.*	..	..	..	..	..	..	..	56 *Autres activités informatiques*
57 Research & development	..	261 [a]	..	481	528	622	..	57 Recherche & développement expérimental
58 Other business activities n.e.c.	..	468	..	360	333	361	..	58 Autres activ. & serv. aux entreprises
59 Comm., soc. & pers. serv. activ., etc.	..	87	..	216	84	57	..	**59 Adm. pub., serv. soc. & collec., etc**
TOTAL BUSINESS ENTERPRISE SECTOR	**..**	**6 683**	**..**	**8 675**	**9 689**	**10 555**	**..**	**TOTAL SECTEUR DES ENTREPRISES**

Source: OECD, R&D database (DSTI/EAS Division), January 2002.

Source : OCDE, base de données R-D (DSTI/Division AES), janvier 2002.

Table 14 - Tableau 14

Higher education intramural expenditure on R&D - Dépenses intra-muros de R-D de l'enseignement supérieur

(HERD) by field of science - (DIRDES) par discipline scientifique

	1994	1995	1996	1997	1998	1999	2000	
	14.1 Million EUR (1999 FIM euro, EUR from 1999) - Millions EUR (1999 FIM euro, EUR à partir de 1999)							
Natural sciences	..	93.2	..	154.4	187.8	216.4	..	Sciences exactes et naturelles
Engineering	..	101.7	..	106.1	118.3	146.0	..	Sciences de l'ingénieur
Medical sciences	..	76.7	..	149.6 [a]	159.9	181.7	..	Sciences médicales
Agricultural sciences	..	16.8	..	12.2	13.5	18.7	..	Sciences agricoles
Sub-total NSE	..	288.5	..	422.2 [a]	479.5	562.8	..	Sous-total SENI
Social sciences	..	76.1	..	105.8	117.8	136.6	..	Sciences sociales
Humanities	..	46.7	..	51.5	60.6	65.4	..	Sciences humaines
Sub-total SSH	..	122.9	..	157.3	178.4	202.0	..	Sous-total SSH
Total NSE+SSH	378.7	424.6	452.4 [c]	579.5 [a]	657.9	764.8	773.8 [c,p]	*Total SENI+SSH*
	14.2 Million constant USD (1995 prices and PPPs) - Millions de USD constants (prix et PPA de 1995)							
Natural sciences	..	94.6	..	153.8	181.6	209.6	..	Sciences exactes et naturelles
Engineering	..	103.2	..	105.7	114.4	141.4	..	Sciences de l'ingénieur
Medical sciences	..	77.8	..	149.0 [a]	154.7	175.9	..	Sciences médicales
Agricultural sciences	..	17.1	..	12.1	13.1	18.1	..	Sciences agricoles
Sub-total NSE	..	292.6	..	420.6 [a]	463.8	545.0	..	Sous-total SENI
Social sciences	..	77.2	..	105.4	113.9	132.3	..	Sciences sociales
Humanities	..	47.4	..	51.3	58.6	63.3	..	Sciences humaines
Sub-total SSH	..	124.6	..	156.7	172.5	195.6	..	Sous-total SSH
Total NSE+SSH	399.9	430.7	460.0 [c]	577.3 [a]	636.3	740.5	724.8 [c,p]	*Total SENI+SSH*

Table 15 - Tableau 15

Higher education intramural expenditure on R&D - Dépenses intra-muros de R-D de l'enseignement supérieur

(HERD) by field of science and type of costs - (DIRDES) par discipline scientifique et par type de coûts

Million EUR (1999 FIM euro, EUR from 1999) — *Millions EUR (1999 FIM euro, EUR à partir de 1999)*

	1994	1995	1996	1997	1998	1999	2000	
Current expenditure								**Dépenses courantes**
Natural sciences	..	..	..	150.5	182.1	209.0	..	Sciences exactes et naturelles
Engineering	..	..	..	102.7	113.8	140.5	..	Sciences de l'ingénieur
Medical sciences	..	..	..	148.0 [a]	157.3	178.8	..	Sciences médicales
Agricultural sciences	..	..	..	12.0	13.1	18.6	..	Sciences agricoles
Sub-total NSE	..	..	..	413.2 [a]	466.3	546.8	..	Sous-total SENI
Social sciences	..	..	..	104.9	117.2	135.7	..	Sciences sociales
Humanities	..	..	..	51.5	60.1	64.9	..	Sciences humaines
Sub-total SSH	..	..	..	156.4	177.3	200.7	..	Sous-total SSH
Total NSE+SSH	..	404.3	..	569.5 [a]	643.6	747.5	..	*Total SENI+SSH*
Capital expenditure								**Dépenses en capital**
Natural sciences	..	..	..	3.9	5.7	7.4	..	Sciences exactes et naturelles
Engineering	..	..	..	3.4	4.5	5.5	..	Sciences de l'ingénieur
Medical sciences	..	..	..	1.6 [a]	2.7	2.9	..	Sciences médicales
Agricultural sciences	..	..	..	0.2	0.4	0.2	..	Sciences agricoles
Sub-total NSE	..	..	..	9.1 [a]	13.3	16.0	..	Sous-total SENI
Social sciences	..	..	..	0.8	0.6	0.9	..	Sciences sociales
Humanities	..	..	..	0.1	0.4	0.5	..	Sciences humaines
Sub-total SSH	..	..	..	0.9	1.0	1.3	..	Sous-total SSH
Total NSE+SSH	..	20.2	..	10.1 [a]	14.3	17.3	..	*Total SENI+SSH*

Source: OECD, R&D database (DSTI/EAS Division), January 2002.

Source : OCDE, base de données R-D (DSTI/Division AES), janvier 2002.

FINLAND — Table 16 - Tableau 16 — FINLANDE

Higher education intramural expenditure on R&D - Dépenses intra-muros de R-D de l'enseignement supérieur

(HERD) by field of science and source of funds - (DIRDES) par discipline scientifique et par source de financement

Million EUR (1999 FIM euro, EUR from 1999) — *Millions EUR (1999 FIM euro, EUR à partir de 1999)*

	1994	1995	1996	1997	1998	1999	2000	
Direct government								**État direct**
Natural sciences	..	31.0	..	53.4	79.0	91.9	..	Sciences exactes et naturelles
Engineering	..	42.0	..	37.0	47.7	61.9	..	Sciences de l'ingénieur
Medical sciences	..	15.3	..	75.3 [a]	83.5	90.3	..	Sciences médicales
Agricultural sciences	..	7.0	..	5.0	5.4	7.8	..	Sciences agricoles
Sub-total NSE	..	95.3	..	170.6 [a]	215.7	251.9	..	Sous-total SENI
Social sciences	..	18.8	..	27.6	33.1	44.4	..	Sciences sociales
Humanities	..	9.2	..	11.1	14.0	16.8	..	Sciences humaines
Sub-total SSH	..	28.0	..	38.7	47.1	61.2	..	Sous-total SSH
Total NSE+SSH	..	136.4	..	209.3 [a]	262.8	313.1	..	*Total SENI+SSH*
General university funds								**Fonds généraux des universités**
Natural sciences	..	54.4	..	78.5	87.4	100.5	..	Sciences exactes et naturelles
Engineering	..	37.7	..	44.6	48.7	58.5	..	Sciences de l'ingénieur
Medical sciences	..	51.9	..	58.2 [a]	58.9	67.2	..	Sciences médicales
Agricultural sciences	..	8.1	..	5.2	5.8	6.5	..	Sciences agricoles
Sub-total NSE	..	152.0	..	186.6 [a]	200.8	232.7	..	Sous-total SENI
Social sciences	..	53.4	..	66.3	72.8	76.9	..	Sciences sociales
Humanities	..	36.0	..	37.7	42.8	44.6	..	Sciences humaines
Sub-total SSH	..	89.4	..	104.0	115.7	121.5	..	Sous-total SSH
Total NSE+SSH	..	241.5	..	290.6 [a]	316.5	354.2	..	*Total SENI+SSH*
Sub-total government								**Sous-total État**
Natural sciences	..	85.3	..	131.9	166.5	192.4	..	Sciences exactes et naturelles
Engineering	..	79.7	..	81.6	96.5	120.4	..	Sciences de l'ingénieur
Medical sciences	..	67.2	..	133.5 [a]	142.4	157.5	..	Sciences médicales
Agricultural sciences	..	15.1	..	10.2	11.2	14.3	..	Sciences agricoles
Sub-total NSE	..	247.3	..	357.2 [a]	416.5	484.6	..	Sous-total SENI
Social sciences	..	72.2	..	93.9	105.9	121.3	..	Sciences sociales
Humanities	..	45.2	..	48.8	56.9	61.4	..	Sciences humaines
Sub-total SSH	..	117.4	..	142.7	162.8	182.6	..	Sous-total SSH
Total NSE+SSH	..	377.9	..	499.8 [a]	579.2	667.3	..	*Total SENI+SSH*
Higher education								**Enseignement supérieur**
Natural sciences	..	1.1	..	1.9	1.6	1.2	..	Sciences exactes et naturelles
Engineering	..	0.6	..	2.0	1.7	1.5	..	Sciences de l'ingénieur
Medical sciences	..	0.9	..	1.5	1.0	0.8	..	Sciences médicales
Agricultural sciences	..	0.4	..	0.2	0.2	0.2	..	Sciences agricoles
Sub-total NSE	..	3.0	..	5.5	4.4	3.8	..	Sous-total SENI
Social sciences	..	0.8	..	1.7	1.6	1.0	..	Sciences sociales
Humanities	..	1.1	..	1.1	1.3	1.1	..	Sciences humaines
Sub-total SSH	..	1.8	..	2.8	2.9	2.2	..	Sous-total SSH
Total NSE+SSH	..	4.8	..	8.3	7.3	5.9	..	*Total SENI+SSH*
Private non-profit								**Institutions sans but lucratif**
Natural sciences	..	0.5	..	0.9	1.0	3.5	..	Sciences exactes et naturelles
Engineering	..	1.9	..	1.3	1.4	1.1	..	Sciences de l'ingénieur
Medical sciences	..	1.8	..	2.6	2.5	3.6	..	Sciences médicales
Agricultural sciences	..	0.1	..	0.1	0.2	1.1	..	Sciences agricoles
Sub-total NSE	..	4.3	..	4.9	5.1	9.3	..	Sous-total SENI
Social sciences	..	0.8	..	2.3	2.0	2.0	..	Sciences sociales
Humanities	..	0.1	..	0.2	0.3	0.8	..	Sciences humaines
Sub-total SSH	..	0.9	..	2.5	2.2	2.8	..	Sous-total SSH
Total NSE+SSH	..	5.2	..	7.3	7.3	12.1	..	*Total SENI+SSH*
Business entreprise								**Entreprises**
Natural sciences	..	4.0	..	8.8	7.6	8.5	..	Sciences exactes et naturelles
Engineering	..	16.8	..	14.9	13.3	15.4	..	Sciences de l'ingénieur
Medical sciences	..	1.9	..	3.1 [a]	4.1	6.0	..	Sciences médicales
Agricultural sciences	..	0.4	..	0.3	0.5	0.3	..	Sciences agricoles
Sub-total NSE	..	23.1	..	27.0 [a]	25.5	30.2	..	Sous-total SENI
Social sciences	..	1.2	..	3.3	4.1	5.7	..	Sciences sociales
Humanities	..	0.0	..	0.1	0.1	0.2	..	Sciences humaines
Sub-total SSH	..	1.2	..	3.3	4.2	5.9	..	Sous-total SSH
Total NSE+SSH	..	24.3	..	30.3	29.7	36.0	..	*Total SENI+SSH*
Funds from abroad								**Fonds de l'étranger**
Natural sciences	..	2.2	..	11.0	11.2	10.8	..	Sciences exactes et naturelles
Engineering	..	2.9	..	6.3	5.5	7.6	..	Sciences de l'ingénieur
Medical sciences	..	5.0	..	9.0	9.9	13.8	..	Sciences médicales
Agricultural sciences	..	0.9	..	1.4	1.4	2.8	..	Sciences agricoles
Sub-total NSE	..	10.9	..	27.7	28.0	34.9	..	Sous-total SENI
Social sciences	..	1.2	..	4.7	4.3	6.6	..	Sciences sociales
Humanities	..	0.3	..	1.4	2.1	1.9	..	Sciences humaines
Sub-total SSH	..	1.5	..	6.1	6.3	8.5	..	Sous-total SSH
Total NSE+SSH	..	12.5	..	33.8	34.3	43.5	..	*Total SENI+SSH*

Source: OECD, R&D database (DSTI/EAS Division), January 2002.

Source : OCDE, base de données R-D (DSTI/Division AES), janvier 2002.

Higher education R&D personnel (HEMP) - Personnel de R-D du secteur de l'enseignement supérieur

by field of science and formal education - par discipline scientifique et niveau de qualification

In full-time equivalent — *En équivalence plein-temps*

	1994	1995	1996	1997	1998	1999	2000	
Total R&D personnel								**Personnel total de R-D**
Natural sciences	..	2 182	..	3 353	4 240	4 506	..	Sciences exactes et naturelles
Engineering	..	2 324	..	2 363	2 736	3 076	..	Sciences de l'ingénieur
Medical sciences	..	1 684	..	2 572 [a]	2 684	2 950	..	Sciences médicales
Agricultural sciences	..	410	..	274	322	402	..	Sciences agricoles
Sub-total NSE	..	6 601	..	8 562 [a]	9 982	10 933	..	Sous-total SENI
Social sciences	..	1 609	..	2 264	2 508	2 721	..	Sciences sociales
Humanities	..	936	..	1 030	1 163	1 186	..	Sciences humaines
Sub-total SSH	..	2 545	..	3 294	3 670	3 907	..	Sous-total SSH
Total NSE+SSH	8 582	9 146	..	11 855 [a]	13 653	14 840	..	*Total SENI+SSH*
University degrees								**Diplômes universitaires**
Natural sciences	..	..	..	..	..	..	..	Sciences exactes et naturelles
Engineering	..	..	..	..	..	..	..	Sciences de l'ingénieur
Medical sciences	..	..	..	..	..	..	..	Sciences médicales
Agricultural sciences	..	..	..	..	..	..	..	Sciences agricoles
Sub-total NSE	..	..	..	..	..	..	..	Sous-total SENI
Social sciences	..	..	..	..	..	..	..	Sciences sociales
Humanities	..	..	..	..	..	..	..	Sciences humaines
Sub-total SSH	..	..	..	..	..	..	..	Sous-total SSH
Total NSE+SSH	..	6 481	..	8 518 [a]	9 683	10 395	..	*Total SENI+SSH*
Other post-secondary								**Autres post-secondaires**
Natural sciences	..	..	..	..	..	..	..	Sciences exactes et naturelles
Engineering	..	..	..	..	..	..	..	Sciences de l'ingénieur
Medical sciences	..	..	..	..	..	..	..	Sciences médicales
Agricultural sciences	..	..	..	..	..	..	..	Sciences agricoles
Sub-total NSE	..	..	..	..	..	..	..	Sous-total SENI
Social sciences	..	..	..	..	..	..	..	Sciences sociales
Humanities	..	..	..	..	..	..	..	Sciences humaines
Sub-total SSH	..	..	..	..	..	..	..	Sous-total SSH
Total NSE+SSH	..	991	..	1 147 [a]	923 [a]	1 629	..	*Total SENI+SSH*
Other								**Autres**
Natural sciences	..	..	..	..	..	..	..	Sciences exactes et naturelles
Engineering	..	..	..	..	..	..	..	Sciences de l'ingénieur
Medical sciences	..	..	..	..	..	..	..	Sciences médicales
Agricultural sciences	..	..	..	..	..	..	..	Sciences agricoles
Sub-total NSE	..	..	..	..	..	..	..	Sous-total SENI
Social sciences	..	..	..	..	..	..	..	Sciences sociales
Humanities	..	..	..	..	..	..	..	Sciences humaines
Sub-total SSH	..	..	..	..	..	..	..	Sous-total SSH
Total NSE+SSH	..	1 674	..	2 190 [a]	3 046 [a]	2 816	..	*Total SENI+SSH*

Table 18 - Tableau 18

Government budget appropriations or outlays for R&D (GBAORD) - Crédits budgétaires publics de R-D (CBPRD)

by socio-economic objective - par objectif socio-économique

Million EUR (1999 FIM euro, EUR from 1999) — *Millions EUR (1999 FIM euro, EUR à partir de 1999)*

	1995	1996	1997	1998	1999	2000	2001	
Agriculture, forestry and fishing	66.4 [a]	67.1	75.3	78.1	72.8	69.8	75.2 [p]	Agriculture, sylviculture et pêche
Industrial development	294.9 [a]	267.4	323.7	334.6	356.6	367.5	370.0 [p]	Développement industriel
Energy	32.8 [a]	31.1	48.1	80.0	79.3	68.2	77.0 [p]	Energie
Transport & telecommunications	..	26.6	21.1	21.9	17.4	18.0	18.1 [p]	*Transports & télécommunications*
Urban & rural planning	..	5.1	13.5	14.0	13.1	10.0	10.8 [p]	*Aménagement urbain et rural*
ST Infrastructure	32.3 [a]	31.8	34.6	35.9	30.5	28.0	28.9 [p]	ST Infrastructure
Prevention of pollution	..	..	..	..	..	..	..	*Prévention de la pollution*
Identificat. & treatment of pollution	..	..	..	..	..	..	..	*Détect. pollut. & mesures antipollut.*
ST Environment	23.7 [a]	24.1	27.2	27.9	28.0	29.4	29.4 [p]	ST Environnement
Health	32.8 [a]	32.1	93.9 [a]	94.5	89.1	88.4	86.9 [p]	Santé
Social development & services	59.0 [a]	68.5	63.5	68.8	65.4	70.3	75.1 [p]	Développement et services sociaux
Earth and atmosphere	11.3 [a]	11.0	13.6	15.9	21.4	19.8	17.3 [p]	Milieu terrestre et atmosphère
Advancement of research	92.6 [a]	100.1	143.1	148.7	158.7	159.2	189.7 [p]	*Promotion de la recherche*
General university funds	245.0 [a]	258.6	313.0	324.1	323.3	346.4	349.9 [p]	*Fonds généraux des universités*
ST Advancement of knowledge	337.5 [a]	358.6	456.1	472.8	482.0	505.6	539.6 [p]	ST promotion gén. des connaiss.
Civil space	20.5 [a]	27.9	29.8	24.0	32.6	26.8	29.2 [p]	Espace civil
Defence	19.4 [a]	19.2	18.2	17.1	17.5	16.8	21.1 [p]	Défense
Not elsewhere classified	..	..	..	..	..	..	..	Non classé
Total	**930.4 [a]**	**938.8**	**1 183.9 [a]**	**1 249.7**	**1 275.2**	**1 290.6**	**1 349.7 [p]**	**Total**

Source: OECD, R&D database (DSTI/EAS Division), January 2002. — Source : OCDE, base de données R-D (DSTI/Division AES), janvier 2002.

FINLAND | Table 19 - Tableau 19 | **FINLANDE**

Technology balance of payments - Balance des paiements technologiques

by type of transfer - selon la nature des transferts

Million EUR (1999 FIM euro, EUR from 1999) — *Millions EUR (1999 FIM euro, EUR à partir de 1999)*

	1983	1984	1985	1986	1987	1988	1989	1990	1991	
Receipts										**Recettes**
Related firms	..	..	..	..	..	..	..	..	..	Firmes affiliées
Unrelated firms	..	..	..	..	..	..	..	..	..	Firmes non affiliées
Other	..	..	..	..	..	..	..	..	..	Autre
Total	5	3	5	6	18	25	33	32	37	*Total*
Payments										**Paiements**
Related firms	..	..	..	..	..	..	..	..	..	Firmes affiliées
Unrelated firms	..	..	..	..	..	..	..	..	..	Firmes non affiliées
Other	..	..	..	..	..	..	..	..	..	Autre
Total	80	95	111	120	143	159	194	203	212	*Total*

	1992	1993	1994	1995	1996	1997	1998	1999	2000	
Receipts										**Recettes**
Related firms	..	..	..	..	..	..	..	..	..	Firmes affiliées
Unrelated firms	..	..	..	..	..	..	..	..	..	Firmes non affiliées
Other	..	..	..	..	..	..	..	..	..	Autre
Total	47	86	67	43	51	82	97	..	..	*Total*
Payments										**Paiements**
Related firms	..	..	..	..	..	..	..	..	..	Firmes affiliées
Unrelated firms	..	..	..	..	..	..	..	..	..	Firmes non affiliées
Other	..	..	..	..	..	..	..	..	..	Autre
Total	203	314	283	286	359	439	371	..	..	*Total*

Source: OECD, TBPdatabase (DSTI/EAS Division), Januay 2002.

Source : OCDE, base de données TBP (DSTI/Division AES), janvier 2002.

Table 1 - Tableau 1

Gross domestic expenditure on R&D (GERD) - Dépenses intérieures brutes de R-D (DIRD)

by sector of performance and source of funds - par secteur d'exécution et par source de financement

	1994	1995	1996	1997	1998	1999	2000	
	1.1 Million EUR (1999 FRF euro, EUR from 1999) - Millions EUR (1999 FRF euro, EUR à partir de 1999)							
Business enterprise								**Entreprises**
financed by								*financé par*
Business enterprise	12 541.3	12 676.6	12 943.1	13 708.4 [a]	14 384.8	15 142.1	..	Entreprises
Direct government	2 143.8	2 117.8	2 236.3	1 799.7 [a]	1 584.9	1 857.3	..	État direct
Higher education	3.2	1.0	1.8	5.0	1.7	4.4	..	Enseignement supérieur
Private non-profit	5.0	4.0	4.7	9.9	9.5	5.1	..	Institutions sans but lucratif
Funds from abroad	1 857.8	1 850.0	1 945.2	1 834.0	1 650.6	1 646.3	..	Fonds de l'étranger
Total	16 551.1	16 649.3	17 131.2	17 356.9 [a]	17 631.6	18 655.1	19 308.0 [p]	*Total*
Government								**État**
financed by								*financé par*
Business enterprise	292.1	308.3	349.4	392.1	494.9	580.7	..	Entreprises
Direct government	4 981.3	5 158.1	5 048.5	4 540.7 [a]	4 498.6	4 500.4	..	État direct
Higher education	8.5	9.7	10.2	10.4	11.9	13.9	..	Enseignement supérieur
Private non-profit	4.1	5.6	5.5	6.7	5.9	6.4	..	Institutions sans but lucratif
Funds from abroad	235.2	249.1	228.2	230.8	267.5	255.7	..	Fonds de l'étranger
Total	5 521.2	5 730.7	5 641.8	5 180.7 [a]	5 278.9	5 357.1	5 357.1 [p]	*Total*
Higher education								**Enseignement supérieur**
financed by								*financé par*
Business enterprise	136.6	152.2	149.4	147.3	167.4	173.5	..	Entreprises
Direct government	2 017.1	2 100.2	2 121.5	2 213.7	2 136.0	2 140.1	..	État direct
General university funds	1 953.6	2 032.6	2 096.5	2 175.1	2 297.8	2 346.2	..	Fonds généraux des universités
Sub-total government	3 970.7	4 132.8	4 218.0	4 388.9	4 433.8	4 486.3	..	Sous-Total État
Higher education	109.2	182.4	179.9	174.6	224.6	263.3	..	Enseignement supérieur
Private non-profit	15.9	22.1	24.7	16.2	18.6	14.8	..	Institutions sans but lucratif
Funds from abroad	98.5	71.5	115.4	106.7	141.9	130.0	..	Fonds de l'étranger
Total	4 330.6	4 561.1	4 687.4	4 833.5 [a]	4 986.3	5 067.9	5 040.3 [p]	*Total*
Private non-profit								**Institutions sans but lucratif**
financed by								*financé par*
Business enterprise	58.5	62.3	63.4	80.8	91.9	82.8	..	Entreprises
Direct government	40.6	41.2	47.6	50.0	55.0	64.3	..	État direct
Higher education	30.9	18.8	20.3	14.8	13.9	21.6	..	Enseignement supérieur
Private non-profit	204.1	214.9	211.1	209.0	225.5	241.0	..	Institutions sans but lucratif
Funds from abroad	27.1	24.1	32.6	30.2	35.8	38.6	..	Fonds de l'étranger
Total	361.2	361.4	375.0	384.6	422.1	448.4	447.4 [p]	*Total*
GERD								**DIRD**
financed by								*financé par*
Business enterprise	13 028.6	13 199.5	13 505.3	14 328.5 [a]	15 139.0	15 979.0	..	Entreprises
Direct government	9 182.7	9 417.2	9 453.8	8 604.1 [a]	8 274.6	8 562.2	..	État direct
General university funds	1 953.6	2 032.6	2 096.5	2 175.1	2 297.8	2 346.2	..	Fonds généraux des universités
Sub-total government	11 136.3	11 449.9	11 550.3	10 779.2 [a]	10 572.4	10 908.3	..	Sous-Total État
Higher education	151.8	211.8	212.2	204.7	252.1	303.2	..	Enseignement supérieur
Private non-profit	229.1	246.6	246.1	241.8	259.5	267.3	..	Institutions sans but lucratif
Funds from abroad	2 218.7	2 194.8	2 321.5	2 201.7	2 095.9	2 070.5	..	Fonds de l'étranger
Total	26 764.4	27 302.5	27 835.4	27 755.8 [a]	28 318.9	29 528.4	30 152.7 [p]	*Total*
	1.2 Million constant USD (1995 prices and PPPs) - Millions de USD constants (prix et PPA de 1995)							
Business enterprise								**Entreprises**
financed by								*financé par*
Business enterprise	12 947.2	12 871.7	12 954.7	13 546.3 [a]	14 082.8	14 755.8	..	Entreprises
Direct government	2 213.2	2 150.3	2 238.3	1 778.4 [a]	1 551.7	1 809.9	..	État direct
Higher education	3.3	1.0	1.8	5.0	1.7	4.3	..	Enseignement supérieur
Private non-profit	5.2	4.1	4.7	9.8	9.3	5.0	..	Institutions sans but lucratif
Funds from abroad	1 917.9	1 878.4	1 947.0	1 812.3	1 615.9	1 604.3	..	Fonds de l'étranger
Total	17 086.7	16 905.5	17 146.6	17 151.8 [a]	17 261.4	18 179.2	18 644.3 [p]	*Total*
Government								**État**
financed by								*financé par*
Business enterprise	301.5	313.0	349.7	387.5	484.5	565.9	..	Entreprises
Direct government	5 142.5	5 237.4	5 053.0	4 487.0 [a]	4 404.1	4 385.6	..	État direct
Higher education	8.8	9.8	10.2	10.2	11.6	13.5	..	Enseignement supérieur
Private non-profit	4.2	5.7	5.5	6.6	5.8	6.2	..	Institutions sans but lucratif
Funds from abroad	242.8	253.0	228.4	228.1	261.9	249.1	..	Fonds de l'étranger
Total	5 699.9	5 818.9	5 646.9	5 119.4 [a]	5 168.0	5 220.4	5 172.9 [p]	*Total*
Higher education								**Enseignement supérieur**
financed by								*financé par*
Business enterprise	141.0	154.6	149.5	145.5	163.9	169.1	..	Entreprises
Direct government	2 082.3	2 132.5	2 123.4	2 187.5	2 091.1	2 085.5	..	État direct
General university funds	2 016.9	2 063.9	2 098.4	2 149.4	2 249.5	2 286.3	..	Fonds généraux des universités
Sub-total government	4 099.2	4 196.4	4 221.8	4 337.0	4 340.7	4 371.8	..	Sous-Total État
Higher education	112.7	185.2	180.1	172.5	219.9	256.5	..	Enseignement supérieur
Private non-profit	16.4	22.4	24.7	16.0	18.2	14.4	..	Institutions sans but lucratif
Funds from abroad	101.7	72.6	115.5	105.5	139.0	126.7	..	Fonds de l'étranger
Total	4 470.8	4 631.2	4 691.6	4 776.4 [a]	4 881.6	4 938.6	4 867.0 [p]	*Total*
Private non-profit								**Institutions sans but lucratif**
financed by								*financé par*
Business enterprise	60.4	63.3	63.5	79.8	90.0	80.7	..	Entreprises
Direct government	41.9	41.9	47.6	49.4	53.9	62.7	..	État direct
Higher education	31.9	19.1	20.3	14.6	13.6	21.1	..	Enseignement supérieur
Private non-profit	210.7	218.2	211.3	206.5	220.7	234.9	..	Institutions sans but lucratif
Funds from abroad	28.0	24.5	32.7	29.8	35.1	37.6	..	Fonds de l'étranger
Total	372.8	366.9	375.4	380.1	413.3	436.9	432.1 [p]	*Total*
GERD								**DIRD**
financed by								*financé par*
Business enterprise	13 450.2	13 402.6	13 517.5	14 159.2 [a]	14 821.0	15 571.4	..	Entreprises
Direct government	9 479.8	9 562.1	9 462.3	8 502.4 [a]	8 100.8	8 343.7	..	État direct
General university funds	2 016.9	2 063.9	2 098.4	2 149.4	2 249.5	2 286.3	..	Fonds généraux des universités
Sub-total government	11 496.7	11 626.1	11 560.7	10 651.8 [a]	10 350.3	10 630.0	..	Sous-Total État
Higher education	156.7	215.0	212.4	202.3	246.8	295.4	..	Enseignement supérieur
Private non-profit	236.5	250.4	246.3	238.9	254.1	260.5	..	Institutions sans but lucratif
Funds from abroad	2 290.4	2 228.6	2 323.6	2 175.6	2 051.9	2 017.7	..	Fonds de l'étranger
Total	27 630.5	27 722.6	27 860.4	27 427.7 [a]	27 724.2	28 775.1	29 116.2 [p]	*Total*

Source: OECD, R&D database (DSTI/EAS Division), January 2002.

Source : OCDE, base de données R-D (DSTI/Division AES), janvier 2002.

Gross domestic expenditure on R&D (GERD) - Dépenses intérieures brutes de R-D (DIRD)

by sector of performance and type of costs - par secteur d'exécution et par type de coûts

Million EUR (1999 FRF euro, EUR from 1999) — *Millions EUR (1999 FRF euro, EUR à partir de 1999)*

	1994	1995	1996	1997	1998	1999	2000	
Business enterprise								**Entreprises**
Labour costs	8 334.8	8 518.4	8 776.2	9 219.0 [a]	9 379.1	9 758.8	..	Coûts du personnel
Other current costs	7 095.7	6 908.5	7 090.1	6 802.2 [a]	6 983.0	7 628.0	..	Autres coûts courants
Total current expenditure	15 430.6	15 426.9	15 866.3	16 021.3 [a]	16 362.2	17 386.8	..	*Total dépenses courantes*
Land and buildings	184.7	201.5	203.5	216.7 [a]	195.8	189.2	..	Terrains et constructions
Instruments and equipment	935.9	1 020.9	1 061.4	1 118.9 [a]	1 073.7	1 079.1	..	Instruments et équipements
Total capital expenditure	1 120.6	1 222.4	1 264.9	1 335.6 [a]	1 269.5	1 268.3	..	*Total dépenses en capital*
Total	16 551.1	16 649.3	17 131.2	17 356.9 [a]	17 631.6	18 655.1	19 308.0 [p]	*Total*
Government								**État**
Labour costs	..	..	..	..	..	..	..	Coûts du personnel
Other current costs	..	..	..	..	..	..	..	Autres coûts courants
Total current expenditure	4 843.7	4 859.0	4 773.4	4 346.6 [a]	4 185.7	..	..	*Total dépenses courantes*
Land and buildings	..	..	..	..	..	..	..	Terrains et constructions
Instruments and equipment	..	..	..	..	..	..	..	Instruments et équipements
Total capital expenditure	677.6	871.7	868.6	833.9 [a]	1 093.3	..	..	*Total dépenses en capital*
Total	5 521.2	5 730.7	5 641.8	5 180.7 [a]	5 278.9	5 357.1	5 357.1 [p]	*Total*
Higher education								**Enseignement supérieur**
Labour costs	3 229.9	3 417.5	3 521.7	3 463.7	3 608.8	..	..	Coûts du personnel
Other current costs	749.1	766.7	811.5	934.0	954.5	..	..	Autres coûts courants
Total current expenditure	3 979.0	4 184.1	4 333.2	4 397.8	4 563.3	..	..	*Total dépenses courantes*
Land and buildings	72.2	93.4	73.6	110.4	118.8	..	..	Terrains et constructions
Instruments and equipment	279.5	283.6	280.5	325.3	304.4	..	..	Instruments et équipements
Total capital expenditure	351.7	376.9	354.1	435.7	423.2	..	..	*Total dépenses en capital*
Total	4 330.6	4 561.1	4 687.4	4 833.5 [a]	4 986.3	5 067.9	5 040.3 [p]	*Total*
Private non-profit								**Institutions sans but lucratif**
Labour costs	187.4	191.9	209.5	228.8	236.4	..	..	Coûts du personnel
Other current costs	127.6	129.4	133.5	124.5	143.0	..	..	Autres coûts courants
Total current expenditure	315.0	321.3	343.0	353.3	379.4	..	..	*Total dépenses courantes*
Land and buildings	19.8	12.5	10.4	7.9	12.8	..	..	Terrains et constructions
Instruments and equipment	26.3	27.5	21.6	23.5	29.9	..	..	Instruments et équipements
Total capital expenditure	46.1	40.0	32.0	31.4	42.7	..	..	*Total dépenses en capital*
Total	361.2	361.4	375.0	384.6	422.1	448.4	447.4 [p]	*Total*
GERD								**DIRD**
Labour costs	..	..	..	..	..	..	..	Coûts du personnel
Other current costs	..	..	..	..	..	..	..	Autres coûts courants
Total current expenditure	24 568.3	24 791.3	25 315.9	25 119.0 [a]	25 490.6	..	..	*Total dépenses courantes*
Land and buildings	..	..	..	..	..	..	..	Terrains et constructions
Instruments and equipment	..	..	..	..	..	..	..	Instruments et équipements
Total capital expenditure	2 196.0	2 511.1	2 519.6	2 636.6 [a]	2 828.7	..	..	*Total dépenses en capital*
Total	26 764.4	27 302.5	27 835.4	27 755.8 [a]	28 318.9	29 528.4	30 152.7 [p]	*Total*

Table 3 - Tableau 3

Current domestic expenditure on R&D - Dépenses intérieures courantes de R-D

by sector of performance and type of activity - par secteur d'exécution et par type d'activités

Million EUR (1999 FRF euro, EUR from 1999) — *Millions EUR (1999 FRF euro, EUR à partir de 1999)*

	1994	1995	1996	1997	1998	1999	2000	
Business enterprise								**Entreprises**
Basic research	790.1 [k]	699.3 [k]	716.6 [k]	765.3 [a,k]	784.7 [k]	843.7 [k]	..	Recherche fondamentale
Applied research	4 159.4 [k]	4 245.6 [k]	4 306.7 [k]	4 737.9 [a,k]	4 948.1 [k]	5 473.1 [k]	..	Recherche appliquée
Experimental development	11 601.5 [k]	11 704.5 [k]	12 108.0 [k]	11 853.7 [a,k]	11 898.9 [k]	12 338.3 [k]	..	Développement expérimental
Not specified	..	..	..	..	..	..	..	Non spécifié
Total	16 551.1 [k]	16 649.3 [k]	17 131.2 [k]	17 356.9 [a,k]	17 631.6 [k]	18 655.1 [k]	19 308.0 [k,p]	*Total*
Government								**État**
Basic research	1 173.6 [k]	1 243.1 [k]	1 167.5 [k]	1 051.4 [a,k]	1 865.1 [k]	1 801.2 [k]	..	Recherche fondamentale
Applied research	2 965.6 [k]	3 080.5 [k]	2 880.2 [k]	2 664.4 [a,k]	1 782.0 [k]	1 898.6 [k]	..	Recherche appliquée
Experimental development	1 382.1 [k]	1 407.0 [k]	1 594.2 [k]	1 464.7 [a,k]	1 632.0 [k]	1 657.5 [k]	..	Développement expérimental
Not specified	..	..	..	..	..	..	..	Non spécifié
Total	5 521.2 [k]	5 730.7 [k]	5 641.8 [k]	5 180.7 [a,k]	5 278.9 [k]	5 357.1 [k]	5 357.1 [k,p]	*Total*
Higher education								**Enseignement supérieur**
Basic research	3 806.8 [k]	3 949.4 [k]	4 072.2 [k]	4 154.8 [k]	4 288.5 [k]	4 384.9 [k]	..	Recherche fondamentale
Applied research	495.6 [k]	582.3 [k]	574.4 [k]	652.0 [k]	663.6 [k]	551.4 [k]	..	Recherche appliquée
Experimental development	28.3 [k]	29.4 [k]	40.7 [k]	26.6 [k]	34.3 [k]	131.6 [k]	..	Développement expérimental
Not specified	..	..	..	..	..	..	..	Non spécifié
Total	4 330.6 [k]	4 561.1 [k]	4 687.4 [k]	4 833.5 [a,k]	4 986.3 [k]	5 067.9 [k]	5 040.3 [k,p]	*Total*
Private non-profit								**Institutions sans but lucratif**
Basic research	175.5 [k]	169.2 [k]	161.3 [k]	139.5 [k]	164.0 [k]	179.8 [k]	..	Recherche fondamentale
Applied research	157.6 [k]	161.0 [k]	171.5 [k]	172.5 [k]	183.2 [k]	203.2 [k]	..	Recherche appliquée
Experimental development	28.1 [k]	31.2 [k]	42.2 [k]	72.7 [k]	74.9 [k]	65.4 [k]	..	Développement expérimental
Not specified	..	..	..	..	..	..	..	Non spécifié
Total	361.2 [k]	361.4 [k]	375.0 [k]	384.6 [k]	422.1 [k]	448.4 [k]	447.4 [k,p]	*Total*
GERD								**DIRD**
Basic research	5 946.0 [k]	6 060.9 [k]	6 117.5 [k]	6 111.1 [a,k]	7 102.4 [k]	7 209.4 [k]	..	Recherche fondamentale
Applied research	7 778.2 [k]	8 069.4 [k]	7 932.8 [k]	8 226.8 [a,k]	7 576.9 [k]	8 126.2 [k]	..	Recherche appliquée
Experimental development	13 040.0 [k]	13 172.1 [k]	13 785.0 [k]	13 417.7 [a]	13 639.9 [k]	14 192.8 [k]	..	Développement expérimental
Not specified	..	..	..	..	..	..	..	Non spécifié
Total	26 764.4 [k]	27 302.5 [k]	27 835.4 [k]	27 755.8 [a,k]	28 318.9 [k]	29 528.4 [k]	30 152.7 [k,p]	*Total*

Source: OECD, R&D database (DSTI/EAS Division), January 2002. *Source* : OCDE, base de données R-D (DSTI/Division AES), janvier 2002.

FRANCE

Table 5 - Tableau 5

Total R&D personnel - Personnel total de R-D

by sector of employment and occupation - par secteur d'emploi et par profession

In full-time equivalent — *En équivalence plein-temps*

	1994	1995	1996	1997	1998	1999	2000	
Business enterprise								**Entreprises**
RSE	66 713	66 618	68 487	72 023 [a]	71 717	75 390	..	Chercheurs
Technicians	69 082	69 756	69 097	68 899 [a]	71 071	71 858	..	Techniciens
Other occupations	26 160	25 669	25 006	25 387 [a]	24 977	24 315	..	Autres professions
Total	161 955	162 042	162 590	166 310 [a]	167 765	171 564	..	*Total*
Government								**État**
RSE	26 403	27 195	27 803	24 249 [a,o]	24 216 [o]	25 187 [o]	..	Chercheurs
Technicians	..	..	..	..	..	..	..	Techniciens
Other occupations	..	..	..	..	..	..	..	Autres professions
Total	68 082	68 539	69 184	52 693 [a]	52 082	53 452	..	*Total*
Higher education								**Enseignement supérieur**
RSE	52 119	53 726	54 592	54 916 [a]	56 288	56 717	..	Chercheurs
Technicians	..	..	..	..	..	..	..	Techniciens
Other occupations	..	..	..	..	..	..	..	Autres professions
Total	78 132	80 678	81 538	79 926 [a]	82 177	83 254	..	*Total*
Private non-profit								**Institutions sans but lucratif**
RSE	3 958	3 710	3 945	3 554 [a]	3 506	3 130	..	Chercheurs
Technicians	..	..	..	..	..	..	..	Techniciens
Other occupations	..	..	..	..	..	..	..	Autres professions
Total	6 990	7 125	7 493	7 249 [a]	7 137	6 182	..	*Total*
TOTAL								**TOTAL**
RSE	149 193	151 249	154 827	154 742 [a]	155 727	160 424	..	Chercheurs
Technicians	..	..	..	..	..	..	..	Techniciens
Other occupations	..	..	..	..	..	..	..	Autres professions
Total	315 159	318 384	320 805	306 178 [a]	309 161	314 452	..	*Total*

Table 6 - Tableau 6

Total R&D personnel - Personnel total de R-D

by sector of employment and formal qualification - par secteur d'emploi et par niveau de qualification

In full-time equivalent — *En équivalence plein-temps*

	1994	1995	1996	1997	1998	1999	2000	
Business enterprise								**Entreprises**
University PhD level degrees	..	..	..	..	..	..	..	Diplômes univ. de niv. doctorat
Other university degrees	..	..	..	..	..	..	..	Autres diplômes universitaires
Sub-total university degrees	..	..	..	..	..	..	..	Sous-total Universitaires
Other post-secondary diplomas	..	..	..	..	..	..	..	Autres diplômes post-secondaires
Secondary diplomas	..	..	..	..	..	..	..	Diplômes de fin d'études secondaires
Other qualifications	..	..	..	..	..	..	..	Autres qualifications
Total	161 955	162 042	162 590	166 310 [a]	167 765	171 564	..	*Total*
Government								**État**
University PhD level degrees	..	..	..	..	..	..	..	Diplômes univ. de niv. doctorat
Other university degrees	..	..	..	..	..	..	..	Autres diplômes universitaires
Sub-total university degrees	..	..	..	..	..	..	..	Sous-total Universitaires
Other post-secondary diplomas	..	..	..	..	..	..	..	Autres diplômes post-secondaires
Secondary diplomas	..	..	..	..	..	..	..	Diplômes de fin d'études secondaires
Other qualifications	..	..	..	..	..	..	..	Autres qualifications
Total	68 082	68 539	69 184	52 693 [a]	52 082	53 452	..	*Total*
Higher education								**Enseignement supérieur**
University PhD level degrees	..	..	..	..	..	..	..	Diplômes univ. de niv. doctorat
Other university degrees	..	..	..	..	..	..	..	Autres diplômes universitaires
Sub-total university degrees	..	..	..	..	..	..	..	Sous-total Universitaires
Other post-secondary diplomas	..	..	..	..	..	..	..	Autres diplômes post-secondaires
Secondary diplomas	..	..	..	..	..	..	..	Diplômes de fin d'études secondaires
Other qualifications	..	..	..	..	..	..	..	Autres qualifications
Total	78 132	80 678	81 538	79 926 [a]	82 177	83 254	..	*Total*
Private non-profit								**Institutions sans but lucratif**
University PhD level degrees	..	..	..	..	..	..	..	Diplômes univ. de niv. doctorat
Other university degrees	..	..	..	..	..	..	..	Autres diplômes universitaires
Sub-total university degrees	..	..	..	..	..	..	..	Sous-total Universitaires
Other post-secondary diplomas	..	..	..	..	..	..	..	Autres diplômes post-secondaires
Secondary diplomas	..	..	..	..	..	..	..	Diplômes de fin d'études secondaires
Other qualifications	..	..	..	..	..	..	..	Autres qualifications
Total	6 990	7 125	7 493	7 249 [a]	7 137	6 182	..	*Total*
TOTAL								**TOTAL**
University PhD level degrees	..	..	..	..	..	..	..	Diplômes univ. de niv. doctorat
Other university degrees	..	..	..	..	..	..	..	Autres diplômes universitaires
Sub-total university degrees	..	..	..	..	..	..	..	Sous-total Universitaires
Other post-secondary diplomas	..	..	..	..	..	..	..	Autres diplômes post-secondaires
Secondary diplomas	..	..	..	..	..	..	..	Diplômes de fin d'études secondaires
Other qualifications	..	..	..	..	..	..	..	Autres qualifications
Total	315 159	318 384	320 805	306 178 [a]	309 161	314 452	..	*Total*

Source: OECD, R&D database (DSTI/EAS Division), January 2002.

Source : OCDE, base de données R-D (DSTI/Division AES), janvier 2002.

Total business enterprise intramural expenditure on R&D - Dépenses intra-muros totales de R-D des entreprises
(BERD) by industry - (DIRDE) par industrie

	1994	1995	1996	1997	1998	1999	2000	
	9.1 Million EUR (1999 FRF euro, EUR from 1999) - Millions EUR (1999 FRF euro, EUR à partir de 1999)							
1 AGRICULTURE, HUNTING & FORESTRY	190.3	191.3	201.1	211.4 [a]	256.2	247.6	..	**1 AGRICULTURE, CHASSE & SYLVICULTURE**
2 MINING	103.4	95.9	93.3	88.5 [a]	84.3	76.8	..	**2 ACTIVITES EXTRACTIVES**
3 MANUFACTURING	14 683.6	14 705.2	15 016.4	14 812.5 [a]	15 069.7	15 990.7	..	**3 ACTIVITES DE FABRICATION**
4 Food, beverages & tobacco	289.4	292.0	304.8	308.2 [a]	328.9	328.0	..	**4 Produits alimentaires, boissons & tabac**
5 Food products & beverages	..	..	..	..	306.8	304.6	..	5 Produits alimentaires & boissons
6 Tobacco products	..	..	..	..	22.1	23.4	..	6 Produits à base de tabac
7 Textiles, wear app., fur & leather	108.4	110.4	107.5	106.8 [a]	104.5	101.1	..	**7 Textiles, habill., fourrures & cuirs**
8 Textiles	70.8	69.3	76.0	69.6 [a]	76.6	75.2	..	8 Textiles
9 Wearing apparel & fur	26.3	29.0	21.2	26.0 [a]	16.4	19.9	..	9 Articles d'habillement & fourrures
10 Leather products & footwear	11.4	12.1	10.3	11.2 [a]	11.6	6.1	..	10 Articles en cuir & chaussures
11 Wood, paper, printing, publishing	69.3	68.9	67.3	72.8 [a]	67.0	79.7	..	**11 Bois, papier, imprimerie, édition**
12 Wood & cork (not furniture)	9.4	12.5	12.6	11.8 [a]	13.1	15.4	..	12 Bois & liège (sauf meubles)
13 Pulp, paper & paper products	53.2	48.6	51.0	54.4 [a]	51.0	59.7	..	13 Carton, papier & articles
14 Publ., print. & repro. of rec. media	6.6	7.8	3.7	6.5 [a]	2.9	4.6	..	14 Edition, impress. & repro. enregis.
15 Coke, petroleum, nuclear fuel, chemicals, rubber plastics	3 398.3	3 576.8	3 838.2	3 953.0 [a]	4 114.9	4 379.6	..	**15 Coke, pétrole, combust. nucl., prod. chimi., caout. & mat. plast.**
16 Coke, ref. petrol. prod. & nuclear fuel	214.8	205.3	225.0	245.8 [a]	263.6	261.2	..	16 Coke, raffin. pétr. & combust. nucl.
17 Chemicals & chemical products	2 865.2	3 025.8	3 180.7	3 254.2 [a]	3 363.2	3 594.0	..	17 Produits chimiques
18 *Chemical prod.(less pharmaceuticals)*	974.3	1 030.1	1 080.4	1 074.8 [a]	1 113.2	1 136.4	..	18 *Prod. chimi. (moins prod. pharmaceutiques)*
19 *Pharmaceuticals*	1 890.9	1 995.7	2 100.3	2 179.3 [a]	2 250.0	2 457.6	..	19 *Produits pharmaceutiques*
20 Rubber & plastic products	318.3	345.7	432.5	453.1 [a]	488.1	524.4	..	20 Caoutchouc et matières plastiques
21 Non-metallic mineral products	178.7	192.3	212.2	212.0 [a]	218.2	236.4	..	**21 Produits minéraux non métalliques**
22 Basic metals	278.7	305.5	287.9	283.9 [a]	293.4	254.9	..	**22 Produits métallurgiques de base**
23 Basic metals, ferrous	199.1	216.0	202.0	207.7 [a]	216.0	180.4	..	23 Produits métallur. de base, ferreux
24 Basic metals, non-ferrous	79.5	89.5	85.9	76.2 [a]	77.4	74.5	..	24 Prod. métallur. de base, non ferreux
25 Fabricated metal products	209.5	225.4	209.9	205.1 [a]	180.7	191.8	..	**25 Ouvrages en métaux (sauf mach.& mat.)**
26 Machinery equipment, instruments and transport equipment	10 078.0	9 835.9	9 879.1	9 556.4 [a]	9 637.8	10 256.2	..	**26 Machines & matériel, instruments & mat. de transport**
27 Machinery n.e.c.	873.2	807.8	794.8	766.1 [a]	810.0	830.6	..	27 Machines, n.c.a.
28 Office, account. & computing machinery	493.3	445.5	436.9	409.5 [a]	417.1	353.6	..	28 Mach. de bur., compta. & trait. inform.
29 Electrical machinery	573.1	597.8	582.4	614.4 [a]	645.0	689.7	..	29 Machines électriques
30 Electro. equip. (radio, TV & commun.)	1 886.8	1 828.8	1 972.7	2 005.8 [a]	2 270.0	2 334.7	..	30 Equip. électro. (radio, tv & communic.)
31 *Electro. comp. (inc. semi-conductors)*	..	..	..	..	596.2	656.1	..	31 *Compos. électro. (semi-cond. incl.)*
32 *TV, radio & communications equip.*	..	..	..	..	1 673.8	1 678.6	..	32 *Appareils radio, tv et de communi.*
33 Instruments	1 789.7	1 723.0	1 618.9	1 671.6 [a]	1 249.0	1 256.4	..	33 Instruments
34 Motor vehicles	2 169.9	2 148.3	2 030.4	2 045.0 [a]	2 099.7	2 491.0	..	34 Véhicules automobiles
35 Other transport equipment	2 292.0	2 284.8	2 443.0	2 043.8 [a]	2 147.0	2 300.2	..	35 Autres matériels de transport
36 *Ships*	21.6	13.6	21.7	16.2 [a]	20.0	14.2	..	36 *Navires*
37 *Aerospace*	2 197.4	2 197.6	2 343.6	1 949.5 [a]	2 047.5	2 194.1	..	37 *Aérospatiale*
38 *Other transport equipment n.e.c.*	73.0	73.5	77.6	78.1 [a]	79.5	91.8	..	38 *Autres matériels de transport n.c.a.*
39 Furniture, other manufacturing n.e.c.	73.3 [o]	97.9 [o]	109.6 [o]	106.2 [a]	117.6	157.2	..	**39 Meubles, autres activ. de fabricat. n.c.a.**
40 Furniture	..	..	..	..	96.0	133.8	..	40 Meubles
41 Other manufacturing n.e.c.	..	..	..	..	21.5	23.3	..	41 Autres activités de fabrication n.c.a.
42 Recycling	.. [n]	.. [n]	.. [n]	8.1 [a]	6.8	5.7	..	**42 Récupération**
43 ELECTRICITY, GAS & WATER SUPPLY	332.5	344.3	524.6	503.3 [a]	477.8	472.4	..	**43 DISTRIBUTION D'ELECTRICITE, GAZ & EAU**
44 CONSTRUCTION	118.3	122.6	122.3	163.9 [a]	171.1	176.9	..	**44 CONSTRUCTION**
45 SERVICES SECTOR	1 123.0	1 190.1	1 173.5	1 577.3 [a]	1 572.5	1 690.7	..	**45 SECTEUR DES SERVICES**
46 Wholesale, retail trade & repair of motor vehicles, etc.	..	..	..	..	0.0	0.0	..	**46 Commerce (gros & détail), réparation de véhicules, etc.**
47 Hotels & restaurants	..	..	..	..	0.0	0.0	..	**47 Hôtels et restaurants**
48 Transport & storage	479.0 [o]	505.4 [o]	499.2 [o]	864.8 [a,o]	777.9 [o]	673.9 [o]	..	**48 Transports, entreposage**
49 Communications	.. [n]	.. [n]	.. [n]	.. [n]	.. [n]	.. [n]	..	**49 Communications**
50 Post	..	..	..	..	..	..	..	50 Postes
51 Telecommunications	..	..	..	..	..	..	..	51 Télécommunications
52 Financial intermediation (inc. insur.)	..	..	..	..	..	..	..	**52 Intermédiation financ. (assurances inc.)**
53 Real estate, renting & busin. activity	644.0	684.7	674.3	712.5 [a]	794.6	1 016.9	..	**53 Immobil., locat. & serv. aux entreprises**
54 Computer & related activities	409.0	423.6	394.0	404.9 [a]	375.8	464.2	..	54 Informatique et activités connexes
55 *Software consultancy*	..	..	..	..	..	..	..	55 *Production de logiciels*
56 *Other computer services n.e.c.*	..	..	..	..	..	..	..	56 *Autres activités informatiques*
57 Research & development	..	..	..	..	..	..	..	57 Recherche & développement expérimental
58 Other business activities n.e.c.	235.0	261.1	280.4	307.6 [a]	418.8	552.6	..	58 Autres activ. & serv. aux entreprises
59 Comm., soc. & pers. serv. activ., etc.	..	..	..	..	..	..	..	**59 Adm. pub., serv. soc. & collec., etc**
TOTAL BUSINESS ENTERPRISE SECTOR	**16 551.1**	**16 649.3**	**17 131.2**	**17 356.9 [a]**	**17 631.6**	**18 655.1**	**19 308.0 [p]**	**TOTAL SECTEUR DES ENTREPRISES**

Source: OECD, R&D database (DSTI/EAS Division), January 2002.

Source : OCDE, base de données R-D (DSTI/Division AES), janvier 2002.

Total business enterprise intramural expenditure on R&D - Dépenses intra-muros totales de R-D des entreprises
(BERD) by industry - (DIRDE) par industrie

	1994	1995	1996	1997	1998	1999	2000	
	9.2 Million constant USD(1995 prices and PPPs) - Millions de USD constants (prix et PPA de 1995)							
1 AGRICULTURE, HUNTING & FORESTRY	196.5	194.2	201.3	208.9 [a]	250.8	241.3	..	**1 AGRICULTURE, CHASSE & SYLVICULTURE**
2 MINING	106.7	97.3	93.4	87.4 [a]	82.6	74.9	..	**2 ACTIVITES EXTRACTIVES**
3 MANUFACTURING	15 158.7	14 931.4	15 030.0	14 637.4 [a]	14 753.3	15 582.8	..	**3 ACTIVITES DE FABRICATION**
4 Food, beverages & tobacco	298.8	296.5	305.1	304.6 [a]	322.0	319.7	..	**4 Produits alimentaires, boissons & tabac**
5 Food products & beverages	..	..	..	..	300.4	296.8	..	5 Produits alimentaires & boissons
6 Tobacco products	..	..	..	..	21.7	22.8	..	6 Produits à base de tabac
7 Textiles, wear app., fur & leather	111.9	112.1	107.6	105.5 [a]	102.3	98.5	..	**7 Textiles, habill., fourrures & cuirs**
8 Textiles	73.1	70.4	76.0	68.8 [a]	75.0	73.2	..	8 Textiles
9 Wearing apparel & fur	27.1	29.5	21.2	25.7 [a]	16.0	19.4	..	9 Articles d'habillement & fourrures
10 Leather products & footwear	11.7	12.3	10.3	11.1 [a]	11.3	5.9	..	10 Articles en cuir & chaussures
11 Wood, paper, printing, publishing	71.5	70.0	67.4	71.9 [a]	65.6	77.7	..	**11 Bois, papier, imprimerie, édition**
12 Wood & cork (not furniture)	9.7	12.7	12.6	11.7 [a]	12.8	15.0	..	12 Bois & liège (sauf meubles)
13 Pulp, paper & paper products	54.9	49.3	51.0	53.8 [a]	49.9	58.1	..	13 Carton, papier & articles
14 Publ., print. & repro. of rec. media	6.8	8.0	3.7	6.4 [a]	2.9	4.5	..	14 Edition, impress. & repro. enregis.
15 Coke, petroleum, nuclear fuel, chemicals, rubber plastics	3 508.3	3 631.8	3 841.7	3 906.3 [a]	4 028.5	4 267.9	..	**15 Coke, pétrole, combust. nucl., prod. chimi., caout. & mat. plast.**
16 Coke, ref. petrol. prod. & nuclear fuel	221.8	208.4	225.2	242.8 [a]	258.1	254.6	..	16 Coke, raffin. pétr. & combust. nucl.
17 Chemicals & chemical products	2 957.9	3 072.3	3 183.6	3 215.7 [a]	3 292.5	3 502.3	..	17 Produits chimiques
18 *Chemical prod.(less pharmaceuticals)*	1 005.8	1 045.9	1 081.4	1 062.1 [a]	1 089.8	1 107.4	..	18 *Prod. chimi. (moins prod. pharmaceutiques)*
19 *Pharmaceuticals*	1 952.1	2 026.4	2 102.2	2 153.6 [a]	2 202.7	2 394.9	..	19 *Produits pharmaceutiques*
20 Rubber & plastic products	328.6	351.1	432.9	447.7 [a]	477.8	511.1	..	20 Caoutchouc et matières plastiques
21 Non-metallic mineral products	184.5	195.3	212.4	209.5 [a]	213.6	230.4	..	**21 Produits minéraux non métalliques**
22 Basic metals	287.7	310.2	288.2	280.5 [a]	287.2	248.4	..	**22 Produits métallurgiques de base**
23 Basic metals, ferrous	205.6	219.3	202.1	205.3 [a]	211.5	175.8	..	23 Produits métallur. de base, ferreux
24 Basic metals, non-ferrous	82.1	90.8	86.0	75.3 [a]	75.8	72.6	..	24 Prod. métallur. de base, non ferreux
25 Fabricated metal products	216.2	228.9	210.1	202.7 [a]	176.9	186.9	..	**25 Ouvrages en métaux (sauf mach.& mat.)**
26 Machinery equipment, instruments and transport equipment	10 404.1	9 987.3	9 888.0	9 443.4 [a]	9 435.4	9 994.5	..	**26 Machines & matériel, instruments & mat. de transport**
27 Machinery n.e.c.	901.5	820.2	795.5	757.1 [a]	793.0	809.4	..	27 Machines, n.c.a.
28 Office, account. & computing machinery	509.3	452.3	437.3	404.7 [a]	408.4	344.5	..	28 Mach. de bur., compta. & trait. inform.
29 Electrical machinery	591.6	607.0	582.9	607.1 [a]	631.4	672.1	..	29 Machines électriques
30 Electro. equip. (radio, TV & commun.)	1 947.8	1 856.9	1 974.5	1 982.1 [a]	2 222.3	2 275.2	..	30 Equip. électro. (radio, tv & communic.)
31 *Electro. comp. (inc. semi-conductors)*	..	..	..	..	583.7	639.4	..	31 *Compos. électro. (semi-cond. incl.)*
32 *TV, radio & communications equip.*	..	..	..	..	1 638.6	1 635.8	..	32 *Appareils radio, tv et de communi.*
33 Instruments	1 847.6	1 749.5	1 620.4	1 651.8 [a]	1 222.7	1 224.3	..	33 Instruments
34 Motor vehicles	2 240.1	2 181.4	2 032.3	2 020.9 [a]	2 055.6	2 427.5	..	34 Véhicules automobiles
35 Other transport equipment	2 366.2	2 319.9	2 445.2	2 019.7 [a]	2 101.9	2 241.5	..	35 Autres matériels de transport
36 *Ships*	22.3	13.8	21.7	16.0 [a]	19.6	13.8	..	36 *Navires*
37 *Aerospace*	2 268.5	2 231.5	2 345.8	1 926.5 [a]	2 004.5	2 138.1	..	37 *Aérospatiale*
38 *Other transport equipment n.e.c.*	75.4	74.6	77.7	77.2 [a]	77.8	89.5	..	38 *Autres matériels de transport n.c.a.*
39 Furniture, other manufacturing n.e.c.	75.7 [o]	99.4 [o]	109.7 [o]	105.0 [a]	115.1	153.2	..	**39 Meubles, autres activ. de fabricat. n.c.a.**
40 Furniture	..	..	..	..	94.0	130.4	..	40 Meubles
41 Other manufacturing n.e.c.	..	..	..	..	21.1	22.8	..	41 Autres activités de fabrication n.c.a.
42 Recycling	.. [n]	.. [n]	.. [n]	8.1 [a]	6.7	5.6	..	**42 Récupération**
43 ELECTRICITY, GAS & WATER SUPPLY	343.3	349.6	525.0	497.4 [a]	467.7	460.3	..	**43 DISTRIBUTION D'ELECTRICITE, GAZ & EAU**
44 CONSTRUCTION	122.1	124.5	122.4	162.0 [a]	167.5	172.4	..	**44 CONSTRUCTION**
45 SERVICES SECTOR	1 159.3	1 208.4	1 174.6	1 558.6 [a]	1 539.5	1 647.6	..	**45 SECTEUR DES SERVICES**
46 Wholesale, retail trade & repair of motor vehicles, etc.	..	..	..	..	0.0	0.0	..	**46 Commerce (gros & détail), réparation de véhicules, etc.**
47 Hotels & restaurants	..	..	..	..	0.0	0.0	..	**47 Hôtels et restaurants**
48 Transport & storage	494.5 [o]	513.2 [o]	499.6 [o]	854.6 [a,o]	761.5 [o]	656.7 [o]	..	**48 Transports, entreposage**
49 Communications	.. [n]	.. [n]	.. [n]	.. [n]	.. [n]	.. [n]	..	**49 Communications**
50 Post	..	..	..	..	..	..	..	50 Postes
51 Telecommunications	..	..	..	..	..	..	..	51 Télécommunications
52 Financial intermediation (inc. insur.)	..	..	..	..	..	..	..	**52 Intermédiation financ. (assurances inc.)**
53 Real estate, renting & busin. activity	664.8	695.2	674.9	704.1 [a]	778.0	990.9	..	**53 Immobil., locat. & serv. aux entreprises**
54 Computer & related activities	422.2	430.1	394.3	400.1 [a]	367.9	452.4	..	54 Informatique et activités connexes
55 *Software consultancy*	..	..	..	..	..	..	..	55 *Production de logiciels*
56 *Other computer services n.e.c.*	..	..	..	..	..	..	..	56 *Autres activités informatiques*
57 Research & development	..	..	..	..	..	..	..	57 Recherche & développement expérimental
58 Other business activities n.e.c.	242.6	265.1	280.6	303.9 [a]	410.1	538.5	..	58 Autres activ. & serv. aux entreprises
59 Comm., soc. & pers. serv. activ., etc.	..	..	..	..	..	..	..	**59 Adm. pub., serv. soc. & collec., etc**
TOTAL BUSINESS ENTERPRISE SECTOR	**17 086.7**	**16 905.5**	**17 146.6**	**17 151.8** [a]	**17 261.4**	**18 179.2**	**18 644.3** [p]	**TOTAL SECTEUR DES ENTREPRISES**

Source: OECD, R&D database (DSTI/EAS Division), January 2002.

Source : OCDE, base de données R-D (DSTI/Division AES), janvier 2002.

Business enterprise intramural expenditure on R&D (BERD) - Dépenses intra-muros de R-D des entreprises (DIRDE)

by industry and source of funds - par industrie et par source de financement

Million EUR (1999 FRF euro, EUR from 1999) — *Millions EUR (1999 FRF euro, EUR à partir de 1999)*

10.1 Business enterprise - Entreprises

	1994	1995	1996	1997	1998	1999	2000	
1 AGRICULTURE, HUNTING & FORESTRY	152.9	153.7	165.5	174.6 [a]	214.0	177.3	..	**1 AGRICULTURE, CHASSE & SYLVICULTURE**
2 MINING	74.5	70.2	69.5	64.2 [a]	68.5	61.4	..	**2 ACTIVITES EXTRACTIVES**
3 MANUFACTURING	10 906.6	10 936.7	11 030.3	11 376.3 [a]	12 034.9	12 834.7	..	**3 ACTIVITES DE FABRICATION**
4 Food, beverages & tobacco	243.1	245.2	254.0	252.5 [a]	270.8	265.7	..	**4 Produits alimentaires, boissons & tabac**
5 Food products & beverages	..	..	..	..	250.2	242.3	..	5 Produits alimentaires & boissons
6 Tobacco products	..	..	..	..	20.6	23.4	..	6 Produits à base de tabac
7 Textiles, wear app., fur & leather	103.2	106.3	104.3	100.3 [a]	99.4	96.1	..	**7 Textiles, habill., fourrures & cuirs**
8 Textiles	66.5	65.8	74.2	64.8 [a]	73.1	71.3	..	8 Textiles
9 Wearing apparel & fur	25.8	28.5	19.9	24.4 [a]	14.8	18.8	..	9 Articles d'habillement & fourrures
10 Leather products & footwear	11.0	12.0	10.2	11.0 [a]	11.5	6.0	..	10 Articles en cuir & chaussures
11 Wood, paper, printing, publishing	65.5	64.7	62.2	66.2 [a]	61.4	71.7	..	**11 Bois, papier, imprimerie, édition**
12 Wood & cork (not furniture)	7.6	10.5	10.5	9.2 [a]	12.0	13.4	..	12 Bois & liège (sauf meubles)
13 Pulp, paper & paper products	51.7	46.5	48.2	51.8 [a]	46.9	53.7	..	13 Carton, papier & articles
14 Publ., print. & repro. of rec. media	6.3	7.6	3.5	5.3 [a]	2.4	4.5	..	14 Edition, impress. & repro. enregis.
15 Coke, petroleum, nuclear fuel, chemicals, rubber plastics	3 085.4	3 218.0	3 485.6	3 576.6 [a]	3 734.6	3 888.9	..	**15 Coke, pétrole, combust. nucl., prod. chimi., caout. & mat. plast.**
16 Coke, ref. petrol. prod. & nuclear fuel	174.0	162.3	182.0	197.9 [a]	219.9	216.2	..	16 Coke, raffin. pétr. & combust. nucl.
17 Chemicals & chemical products	2 599.3	2 715.1	2 874.6	2 930.3 [a]	3 029.1	3 153.0	..	17 Produits chimiques
18 *Chemical prod.(less pharmaceuticals)*	858.2	900.5	965.6	972.6 [a]	1 017.2	1 032.7	..	18 *Prod. chimi. (moins prod. pharmaceutiques)*
19 *Pharmaceuticals*	1 741.1	1 814.6	1 909.0	1 957.7 [a]	2 011.9	2 120.2	..	19 *Produits pharmaceutiques*
20 Rubber & plastic products	312.2	340.6	428.9	448.4 [a]	485.6	519.7	..	20 Caoutchouc et matières plastiques
21 Non-metallic mineral products	172.6	185.5	204.4	206.4 [a]	211.3	231.3	..	**21 Produits minéraux non métalliques**
22 Basic metals	264.9	290.0	273.8	267.7 [a]	277.3	234.5	..	**22 Produits métallurgiques de base**
23 Basic metals, ferrous	189.4	205.0	194.0	197.2 [a]	206.4	167.3	..	23 Produits métallur. de base, ferreux
24 Basic metals, non-ferrous	75.5	85.0	79.8	70.6 [a]	70.9	67.2	..	24 Prod. métallur. de base, non ferreux
25 Fabricated metal products	198.9	213.3	197.6	189.5 [a]	166.3	180.8	..	**25 Ouvrages en métaux (sauf mach.& mat.)**
26 Machinery equipment, instruments and transport equipment	6 700.6	6 517.0	6 341.7	6 607.1 [a]	7 092.5	7 707.7	..	**26 Machines & matériel, instruments & mat. de transport**
27 Machinery n.e.c.	658.7	597.2	601.3	569.1 [a]	624.2	605.0	..	27 Machines, n.c.a.
28 Office, account. & computing machinery	268.4	226.9	281.9	266.8 [a]	280.9	299.2	..	28 Mach. de bur., compta. & trait. inform.
29 Electrical machinery	536.3	558.9	554.3	580.7 [a]	621.0	645.5	..	29 Machines électriques
30 Electro. equip. (radio, TV & commun.)	1 470.4	1 490.0	1 557.5	1 595.8 [a]	1 670.8	1 777.2	..	30 Equip. électro. (radio, tv & communic.)
31 *Electro. comp. (inc. semi-conductors)*	..	..	..	..	498.4	545.0	..	31 *Compos. électro. (semi-cond. incl.)*
32 *TV, radio & communications equip.*	..	..	..	..	1 172.4	1 232.3	..	32 *Appareils radio, tv et de communi.*
33 Instruments	959.0	874.8	798.0	897.9 [a]	778.1	634.7	..	33 Instruments
34 Motor vehicles	2 125.4	2 094.2	2 000.1	2 013.5 [a]	2 078.3	2 471.2	..	34 Véhicules automobiles
35 Other transport equipment	682.4	674.9	548.6	683.2 [a]	1 039.2	1 274.9	..	35 Autres matériels de transport
36 *Ships*	15.6	11.5	19.1	13.5 [a]	17.3	12.4	..	36 *Navires*
37 *Aerospace*	598.6	595.7	459.3	594.0 [a]	949.8	1 178.0	..	37 *Aérospatiale*
38 *Other transport equipment n.e.c.*	68.2	67.7	70.2	75.7 [a]	72.1	84.6	..	38 *Autres matériels de transport n.c.a.*
39 Furniture, other manufacturing n.e.c.	72.3 [o]	96.7 [o]	106.7 [o]	103.6 [a]	115.8	153.4	..	**39 Meubles, autres activ. de fabricat. n.c.a.**
40 Furniture	..	..	..	..	94.8	131.3	..	40 Meubles
41 Other manufacturing n.e.c.	..	..	..	..	21.1	22.1	..	41 Autres activités de fabrication n.c.a.
42 Recycling	.. [n]	.. [n]	.. [n]	6.3 [a]	5.7	4.6	..	**42 Récupération**
43 ELECTRICITY, GAS & WATER SUPPLY	325.8	339.9	516.5	495.8 [a]	473.3	465.1	..	**43 DISTRIBUTION D'ELECTRICITE, GAZ & EAU**
44 CONSTRUCTION	108.5	110.1	117.1	152.7 [a]	163.7	168.6	..	**44 CONSTRUCTION**
45 SERVICES SECTOR	973.1	1 066.1	1 044.0	1 444.8 [a]	1 430.5	1 435.1	..	**45 SECTEUR DES SERVICES**
46 Wholesale, retail trade & repair of motor vehicles, etc.	..	..	..	..	0.0	0.0	..	**46 Commerce (gros & détail), réparation de véhicules, etc.**
47 Hotels & restaurants	..	..	..	..	0.0	0.0	..	**47 Hôtels et restaurants**
48 Transport & storage	457.8 [o]	488.2 [o]	475.6 [o]	837.9 [a,o]	753.7 [o]	653.6 [o]	..	**48 Transports, entreposage**
49 Communications	.. [n]	.. [n]	.. [n]	.. [n]	.. [n]	.. [n]	..	**49 Communications**
50 Post	..	..	..	..	..	..	..	50 Postes
51 Telecommunications	..	..	..	..	..	..	..	51 Télécommunications
52 Financial intermediation (inc. insur.)	..	..	..	..	0.0	0.0	..	**52 Intermédiation financ. (assurances inc.)**
53 Real estate, renting & busin. activity	515.3	578.0	568.4	606.9 [a]	676.7	781.5	..	**53 Immobil., locat. & serv. aux entreprises**
54 Computer & related activities	357.8	393.2	364.4	372.1 [a]	348.4	395.9	..	54 Informatique et activités connexes
55 *Software consultancy*	..	..	..	..	..	..	..	55 *Production de logiciels*
56 *Other computer services n.e.c.*	..	..	..	..	..	..	..	56 *Autres activités informatiques*
57 Research & development	..	..	..	..	0.0	0.0	..	57 Recherche & développement expérimental
58 Other business activities n.e.c.	157.5	184.8	204.0	234.8 [a]	328.4	385.6	..	58 Autres activ. & serv. aux entreprises
59 Comm., soc. & pers. serv. activ., etc.	..	..	..	..	0.0	0.0	..	**59 Adm. pub., serv. soc. & collec., etc**
TOTAL BUSINESS ENTERPRISE SECTOR	**12 541.3**	**12 676.6**	**12 943.1**	**13 708.4** [a]	**14 384.8**	**15 142.1**	..	**TOTAL SECTEUR DES ENTREPRISES**

Source: OECD, R&D database (DSTI/EAS Division), January 2002.

Source : OCDE, base de données R-D (DSTI/Division AES), janvier 2002.

Business enterprise intramural expenditure on R&D (BERD) - Dépenses intra-muros de R-D des entreprises (DIRDE)

by industry and source of funds - par industrie et par source de financement

Million EUR (1999 FRF euro, EUR from 1999) — *Millions EUR (1999 FRF euro, EUR à partir de 1999)*

10.2 Government - État

	1994	1995	1996	1997	1998	1999	2000	
1 AGRICULTURE, HUNTING & FORESTRY	18.6	20.1	16.6	16.8	19.7	25.8	..	**1 AGRICULTURE, CHASSE & SYLVICULTURE**
2 MINING	14.4	12.0	13.2	7.9	7.7	9.9	..	**2 ACTIVITES EXTRACTIVES**
3 MANUFACTURING	1 997.4	1 995.2	2 121.8	1 693.1 [a]	1 482.5	1 731.5	..	**3 ACTIVITES DE FABRICATION**
4 Food, beverages & tobacco	6.6	6.4	7.1	8.2 [a]	8.2	10.4	..	**4 Produits alimentaires, boissons & tabac**
5 Food products & beverages	..	..	..	..	8.2	10.4	..	5 Produits alimentaires & boissons
6 Tobacco products	..	..	..	..	0.0	0.0	..	6 Produits à base de tabac
7 Textiles, wear app., fur & leather	3.9	2.9	1.7	4.9 [a]	3.6	3.2	..	**7 Textiles, habill., fourrures & cuirs**
8 Textiles	3.2	2.5	0.7	3.7 [a]	2.6	2.4	..	8 Textiles
9 Wearing apparel & fur	0.2	0.4	0.9	1.0 [a]	1.0	0.8	..	9 Articles d'habillement & fourrures
10 Leather products & footwear	0.4	0.1	0.1	0.2 [a]	0.1	0.0	..	10 Articles en cuir & chaussures
11 Wood, paper, printing, publishing	1.9	2.1	2.4	4.0 [a]	1.1	2.2	..	**11 Bois, papier, imprimerie, édition**
12 Wood & cork (not furniture)	0.9	1.1	1.2	2.0 [a]	0.7	1.6	..	12 Bois & liège (sauf meubles)
13 Pulp, paper & paper products	0.6	1.0	1.1	1.0 [a]	0.2	0.5	..	13 Carton, papier & articles
14 Publ., print. & repro. of rec. media	0.4	0.1	0.2	1.1 [a]	0.3	0.1	..	14 Edition, impress. & repro. enregis.
15 Coke, petroleum, nuclear fuel, chemicals, rubber plastics	100.1	121.7	104.0	72.0 [a]	69.2	81.4	..	**15 Coke, pétrole, combust. nucl., prod. chimi., caout. & mat. plast.**
16 Coke, ref. petrol. prod. & nuclear fuel	13.6	13.9	12.4	10.9 [a]	10.2	11.2	..	16 Coke, raffin. pétr. & combust. nucl.
17 Chemicals & chemical products	84.2	104.9	89.2	57.6 [a]	56.9	67.0	..	17 Produits chimiques
18 *Chemical prod.(less pharmaceuticals)*	59.3	66.1	52.0	40.6 [a]	38.1	52.6	..	18 *Prod. chimi. (moins prod. pharmaceutiques)*
19 *Pharmaceuticals*	24.9	38.8	37.3	17.0 [a]	18.7	14.4	..	19 *Produits pharmaceutiques*
20 Rubber & plastic products	2.3	3.0	2.3	3.5 [a]	2.2	3.3	..	20 Caoutchouc et matières plastiques
21 Non-metallic mineral products	4.6	4.7	5.8	3.7 [a]	4.7	4.3	..	**21 Produits minéraux non métalliques**
22 Basic metals	3.9	3.2	3.4	4.0 [a]	2.8	3.6	..	**22 Produits métallurgiques de base**
23 Basic metals, ferrous	2.4	2.6	2.2	3.1 [a]	2.1	2.0	..	23 Produits métallur. de base, ferreux
24 Basic metals, non-ferrous	1.5	0.5	1.2	0.9 [a]	0.7	1.7	..	24 Prod. métallur. de base, non ferreux
25 Fabricated metal products	6.5	6.7	7.6	9.6 [a]	8.0	6.8	..	**25 Ouvrages en métaux (sauf mach.& mat.)**
26 Machinery equipment, instruments and transport equipment	1 868.9	1 846.3	1 987.3	1 584.3 [a]	1 383.3	1 617.0	..	**26 Machines & matériel, instruments & mat. de transport**
27 Machinery n.e.c.	179.6	164.3	151.4	144.8 [a]	138.5	209.1	..	27 Machines, n.c.a.
28 Office, account. & computing machinery	87.6	90.5	56.6	45.7 [a]	52.6	27.1	..	28 Mach. de bur., compta. & trait. inform.
29 Electrical machinery	19.6	20.0	13.4	14.7 [a]	8.7	10.1	..	29 Machines électriques
30 Electro. equip. (radio, TV & commun.)	287.8	231.5	276.5	258.2 [a]	427.0	406.6	..	30 Equip. électro. (radio, tv & communic.)
31 *Electro. comp. (inc. semi-conductors)*	..	..	..	..	80.3	76.2	..	31 *Compos. électro. (semi-cond. incl.)*
32 *TV, radio & communications equip.*	..	..	..	..	346.7	330.5	..	32 *Appareils radio, tv et de communi.*
33 Instruments	619.6	598.5	604.9	503.8 [a]	260.9	333.3	..	33 Instruments
34 Motor vehicles	24.6	17.2	16.7	16.7 [a]	10.2	8.9	..	34 Véhicules automobiles
35 Other transport equipment	650.0	724.2	867.8	600.4 [a]	485.4	621.8	..	35 Autres matériels de transport
36 *Ships*	5.9	1.8	1.9	1.6 [a]	2.6	1.6	..	36 *Navires*
37 *Aerospace*	639.8	716.6	859.0	596.5 [a]	475.4	613.0	..	37 *Aérospatiale*
38 *Other transport equipment n.e.c.*	4.2	5.8	6.9	2.3 [a]	7.3	7.2	..	38 *Autres matériels de transport n.c.a.*
39 Furniture, other manufacturing n.e.c.	1.0 [o]	1.1 [o]	2.5 [o]	0.8 [a]	0.5	1.4	..	**39 Meubles, autres activ. de fabricat. n.c.a.**
40 Furniture	..	..	..	..	0.1	0.1	..	40 Meubles
41 Other manufacturing n.e.c.	..	..	..	..	0.4	1.2	..	41 Autres activités de fabrication n.c.a.
42 Recycling	.. [n]	.. [n]	.. [n]	1.6 [a]	1.1	1.1	..	**42 Récupération**
43 ELECTRICITY, GAS & WATER SUPPLY	3.7	2.0	5.0	4.5	1.7	2.5	..	**43 DISTRIBUTION D'ELECTRICITE, GAZ & EAU**
44 CONSTRUCTION	7.0	10.9	3.3	6.8	5.1	6.6	..	**44 CONSTRUCTION**
45 SERVICES SECTOR	102.8	77.5	76.3	70.6	68.2	81.0	..	**45 SECTEUR DES SERVICES**
46 Wholesale, retail trade & repair of motor vehicles, etc.	..	..	..	..	0.0	0.0	..	**46 Commerce (gros & détail), réparation de véhicules, etc.**
47 Hotels & restaurants	..	..	..	..	0.0	0.0	..	**47 Hôtels et restaurants**
48 Transport & storage	6.1 [o]	2.4 [o]	5.4 [o]	3.3 [o]	3.6 [o]	3.5 [o]	..	**48 Transports, entreposage**
49 Communications	.. [n]	.. [n]	.. [n]	.. [n]	.. [n]	.. [n]	..	**49 Communications**
50 Post	..	..	..	..	..	..	..	50 Postes
51 Telecommunications	..	..	..	..	..	..	..	51 Télécommunications
52 Financial intermediation (inc. insur.)	..	..	..	..	0.0	0.0	..	**52 Intermédiation financ. (assurances inc.)**
53 Real estate, renting & busin. activity	96.7	75.1	70.9	67.3	64.6	77.5	..	**53 Immobil., locat. & serv. aux entreprises**
54 Computer & related activities	33.4	18.8	16.4	19.6	12.8	19.9	..	54 Informatique et activités connexes
55 *Software consultancy*	..	..	..	..	..	..	..	55 *Production de logiciels*
56 *Other computer services n.e.c.*	..	..	..	..	..	..	..	56 *Autres activités informatiques*
57 Research & development	..	..	..	..	0.0	0.0	..	57 Recherche & développement expérimental
58 Other business activities n.e.c.	63.3	56.3	54.5	47.6	51.8	57.6	..	58 Autres activ. & serv. aux entreprises
59 Comm., soc. & pers. serv. activ., etc.	..	..	..	..	0.0	0.0	..	**59 Adm. pub., serv. soc. & collec., etc**
TOTAL BUSINESS ENTERPRISE SECTOR	**2 143.8**	**2 117.8**	**2 236.3**	**1 799.7 [a]**	**1 584.9**	**1 857.3**	**..**	**TOTAL SECTEUR DES ENTREPRISES**

Source: OECD, R&D database (DSTI/EAS Division), January 2002.

Source : OCDE, base de données R-D (DSTI/Division AES), janvier 2002.

Business enterprise intramural expenditure on R&D (BERD) - Dépenses intra-muros de R-D des entreprises (DIRDE)

by industry and source of funds - par industrie et par source de financement

Million EUR (1999 FRF euro, EUR from 1999) — *Millions EUR (1999 FRF euro, EUR à partir de 1999)*

10.3 Other national funds - Autres sources nationales

	1994	1995	1996	1997	1998	1999	2000	
1 AGRICULTURE, HUNTING & FORESTRY	0.4	0.0	0.0	0.0	0.1	3.1	..	**1 AGRICULTURE, CHASSE & SYLVICULTURE**
2 MINING	0.1	0.0	0.0	0.0	0.0	0.0	..	**2 ACTIVITES EXTRACTIVES**
3 MANUFACTURING	6.0	4.2	5.8	10.4	8.4	4.3	..	**3 ACTIVITES DE FABRICATION**
4 Food, beverages & tobacco	2.2	1.4	1.2	1.9	2.1	0.8	..	**4 Produits alimentaires, boissons & tabac**
5 Food products & beverages	..	..	..	..	0.6	0.8	..	5 Produits alimentaires & boissons
6 Tobacco products	..	..	..	..	1.4	0.0	..	6 Produits à base de tabac
7 Textiles, wear app., fur & leather	0.1	0.1	0.0	0.0	0.0	0.0	..	**7 Textiles, habill., fourrures & cuirs**
8 Textiles	0.1	0.1	0.0	0.0	0.0	0.0	..	8 Textiles
9 Wearing apparel & fur	0.0	0.0	0.0	0.0	0.0	0.0	..	9 Articles d'habillement & fourrures
10 Leather products & footwear	0.0	0.0	0.0	0.0	0.0	0.0	..	10 Articles en cuir & chaussures
11 Wood, paper, printing, publishing	0.0	0.0	0.0	0.3	0.1	0.0	..	**11 Bois, papier, imprimerie, édition**
12 Wood & cork (not furniture)	0.0	0.0	0.0	0.1	0.0	0.0	..	12 Bois & liège (sauf meubles)
13 Pulp, paper & paper products	0.0	0.0	0.0	0.0	0.0	0.0	..	13 Carton, papier & articles
14 Publ., print. & repro. of rec. media	0.0	0.0	0.0	0.1	0.0	0.0	..	14 Edition, impress. & repro. enregis.
15 Coke, petroleum, nuclear fuel, chemicals, rubber plastics	0.8	0.9	0.5	4.9	0.6	0.1	..	**15 Coke, pétrole, combust. nucl., prod. chimi., caout. & mat. plast.**
16 Coke, ref. petrol. prod. & nuclear fuel	0.1	0.0	0.2	0.0	0.0	0.0	..	16 Coke, raffin. pétr. & combust. nucl.
17 Chemicals & chemical products	0.7	0.9	0.3	4.8	0.6	0.1	..	17 Produits chimiques
18 *Chemical prod.(less pharmaceuticals)*	0.1	0.0	0.1	0.0	0.1	0.0	..	18 *Prod. chimi. (moins prod. pharmaceutiques)*
19 *Pharmaceuticals*	0.7	0.9	0.2	4.8	0.5	0.1	..	19 *Produits pharmaceutiques*
20 Rubber & plastic products	0.0	0.0	0.0	0.0	0.0	0.0	..	20 Caoutchouc et matières plastiques
21 Non-metallic mineral products	0.0	0.0	0.0	0.0	0.0	0.1	..	**21 Produits minéraux non métalliques**
22 Basic metals	0.0	0.0	0.0	0.0	0.0	0.0	..	**22 Produits métallurgiques de base**
23 Basic metals, ferrous	0.0	0.0	0.0	0.0	0.0	0.0	..	23 Produits métallur. de base, ferreux
24 Basic metals, non-ferrous	0.0	0.0	0.0	0.0	0.0	0.0	..	24 Prod. métallur. de base, non ferreux
25 Fabricated metal products	0.5	0.5	0.6	0.6	0.7	0.0	..	**25 Ouvrages en métaux (sauf mach.& mat.)**
26 Machinery equipment, instruments and transport equipment	2.4	1.3	3.2	2.7	4.9	3.2	..	**26 Machines & matériel, instruments & mat. de transport**
27 Machinery n.e.c.	0.2	0.3	0.6	0.4	1.2	0.4	..	27 Machines, n.c.a.
28 Office, account. & computing machinery	0.0	0.2	0.0	0.0	0.0	0.1	..	28 Mach. de bur., compta. & trait. inform.
29 Electrical machinery	0.0	0.0	0.0	0.2	0.4	0.0	..	29 Machines électriques
30 Electro. equip. (radio, TV & commun.)	1.2	0.3	1.3	0.4	0.5	0.7	..	30 Equip. électro. (radio, tv & communic.)
31 *Electro. comp. (inc. semi-conductors)*	..	..	..	..	0.0	0.0	..	31 *Compos. électro. (semi-cond. incl.)*
32 *TV, radio & communications equip.*	..	..	..	..	0.5	0.7	..	32 *Appareils radio, tv et de communi.*
33 Instruments	0.9	0.5	0.4	0.4	2.7	1.6	..	33 Instruments
34 Motor vehicles	0.1	0.1	0.2	0.0	0.0	0.0	..	34 Véhicules automobiles
35 Other transport equipment	0.0	0.0	0.7	1.2	0.1	0.4	..	35 Autres matériels de transport
36 *Ships*	0.0	0.0	0.0	0.0	0.0	0.0	..	36 *Navires*
37 *Aerospace*	0.0	0.0	0.7	1.2	0.1	0.4	..	37 *Aérospatiale*
38 *Other transport equipment n.e.c.*	0.0	0.0	0.0	0.0	0.0	0.0	..	38 *Autres matériels de transport n.c.a.*
39 Furniture, other manufacturing n.e.c.	0.0 [o]	0.0 [o]	0.1 [o]	0.0	0.0	0.0	..	**39 Meubles, autres activ. de fabricat. n.c.a.**
40 Furniture	..	..	..	..	0.0	0.0	..	40 Meubles
41 Other manufacturing n.e.c.	..	..	..	..	0.0	0.0	..	41 Autres activités de fabrication n.c.a.
42 Recycling	.. [n]	.. [n]	.. [n]	0.1	0.0	0.0	..	**42 Récupération**
43 ELECTRICITY, GAS & WATER SUPPLY	0.0	0.0	0.0	0.2	0.0	0.0	..	**43 DISTRIBUTION D'ELECTRICITE, GAZ & EAU**
44 CONSTRUCTION	0.0	0.0	0.0	1.3	0.0	0.0	..	**44 CONSTRUCTION**
45 SERVICES SECTOR	1.6	0.8	0.8	2.9	2.7	2.1	..	**45 SECTEUR DES SERVICES**
46 Wholesale, retail trade & repair of motor vehicles, etc.	..	..	..	..	0.0	0.0	..	**46 Commerce (gros & détail), réparation de véhicules, etc.**
47 Hotels & restaurants	..	..	..	..	0.0	0.0	..	**47 Hôtels et restaurants**
48 Transport & storage	0.0 [o]	0.0 [o]	0.1 [o]	0.0 [o]	1.0 [o]	0.1 [o]	..	**48 Transports, entreposage**
49 Communications	.. [n]	.. [n]	.. [n]	.. [n]	0.0 [n]	.. [n]	..	**49 Communications**
50 Post	..	..	..	..	..	..	..	50 Postes
51 Telecommunications	..	..	..	..	..	..	..	51 Télécommunications
52 Financial intermediation (inc. insur.)	..	..	..	..	..	..	..	**52 Intermédiation financ. (assurances inc.)**
53 Real estate, renting & busin. activity	1.6	0.8	0.7	2.9	1.8	2.0	..	**53 Immobil., locat. & serv. aux entreprises**
54 Computer & related activities	0.2	0.3	0.0	0.0	0.8	0.1	..	54 Informatique et activités connexes
55 *Software consultancy*	..	..	..	..	..	..	..	55 *Production de logiciels*
56 *Other computer services n.e.c.*	..	..	..	..	..	..	..	56 *Autres activités informatiques*
57 Research & development	..	..	..	..	0.0	0.0	..	57 Recherche & développement expérimental
58 Other business activities n.e.c.	1.5	0.5	0.7	2.9	0.9	1.9	..	58 Autres activ. & serv. aux entreprises
59 Comm., soc. & pers. serv. activ., etc.	..	..	..	..	0.0	0.0	..	**59 Adm. pub., serv. soc. & collec., etc**
TOTAL BUSINESS ENTERPRISE SECTOR	**8.2**	**5.0**	**6.6**	**14.9**	**11.3**	**9.5**	**..**	**TOTAL SECTEUR DES ENTREPRISES**

Source: OECD, R&D database (DSTI/EAS Division), January 2002.

Source : OCDE, base de données R-D (DSTI/Division AES), janvier 2002.

Business enterprise intramural expenditure on R&D (BERD) - Dépenses intra-muros de R-D des entreprises (DIRDE)

by industry and source of funds - par industrie et par source de financement

Million EUR (1999 FRF euro, EUR from 1999) — Millions EUR (1999 FRF euro, EUR à partir de 1999)

10.4 Funds from abroad - Fonds de l'étranger

	1994	1995	1996	1997	1998	1999	2000	
1 AGRICULTURE, HUNTING & FORESTRY	18.5	17.5	19.0	19.9	22.4	41.4	..	**1 AGRICULTURE, CHASSE & SYLVICULTURE**
2 MINING	14.5	13.7	10.5	16.3	8.1	5.6	..	**2 ACTIVITES EXTRACTIVES**
3 MANUFACTURING	1 773.6	1 769.1	1 858.6	1 732.7	1 543.9	1 420.2	..	**3 ACTIVITES DE FABRICATION**
4 Food, beverages & tobacco	37.5	39.1	42.5	45.6	47.8	51.1	..	**4 Produits alimentaires, boissons & tabac**
5 Food products & beverages	..	..	..	..	47.8	51.1	..	5 Produits alimentaires & boissons
6 Tobacco products	..	..	..	..	0.1	0.0	..	6 Produits à base de tabac
7 Textiles, wear app., fur & leather	1.3	1.1	1.4	1.7	1.5	1.8	..	**7 Textiles, habill., fourrures & cuirs**
8 Textiles	1.0	0.9	1.0	1.1	0.9	1.5	..	8 Textiles
9 Wearing apparel & fur	0.3	0.2	0.4	0.5	0.6	0.4	..	9 Articles d'habillement & fourrures
10 Leather products & footwear	0.0	0.0	0.0	0.0	0.0	0.0	..	10 Articles en cuir & chaussures
11 Wood, paper, printing, publishing	1.8	2.1	2.7	2.3	4.4	5.8	..	**11 Bois, papier, imprimerie, édition**
12 Wood & cork (not furniture)	0.9	1.0	1.0	0.6	0.4	0.4	..	12 Bois & liège (sauf meubles)
13 Pulp, paper & paper products	0.9	1.0	1.8	1.7	3.8	5.4	..	13 Carton, papier & articles
14 Publ., print. & repro. of rec. media	0.0	0.1	0.0	0.0	0.2	0.0	..	14 Edition, impress. & repro. enregis.
15 Coke, petroleum, nuclear fuel, chemicals, rubber plastics	212.0	236.1	248.1	299.5	310.5	409.1	..	**15 Coke, pétrole, combust. nucl., prod. chimi., caout. & mat. plast.**
16 Coke, ref. petrol. prod. & nuclear fuel	27.2	29.1	30.3	37.0	33.5	33.8	..	16 Coke, raffin. pétr. & combust. nucl.
17 Chemicals & chemical products	181.0	204.8	216.6	261.4	276.6	373.9	..	17 Produits chimiques
18 *Chemical prod.(less pharmaceuticals)*	56.7	63.5	62.7	61.6	57.8	51.0	..	18 *Prod. chimi. (moins prod. pharmaceutiques)*
19 *Pharmaceuticals*	124.2	141.4	153.8	199.8	218.9	322.8	..	19 *Produits pharmaceutiques*
20 Rubber & plastic products	3.8	2.2	1.2	1.1	0.3	1.4	..	20 Caoutchouc et matières plastiques
21 Non-metallic mineral products	1.6	2.0	1.9	1.9	2.3	0.7	..	**21 Produits minéraux non métalliques**
22 Basic metals	9.9	12.3	10.6	12.2	13.3	16.8	..	**22 Produits métallurgiques de base**
23 Basic metals, ferrous	7.3	8.3	5.7	7.5	7.5	11.1	..	23 Produits métallur. de base, ferreux
24 Basic metals, non-ferrous	2.6	3.9	4.9	4.7	5.8	5.7	..	24 Prod. métallur. de base, non ferreux
25 Fabricated metal products	3.5	4.8	4.1	5.4	5.7	4.1	..	**25 Ouvrages en métaux (sauf mach.& mat.)**
26 Machinery equipment, instruments and transport equipment	1 506.1	1 471.4	1 547.0	1 362.3	1 157.2	928.3	..	**26 Machines & matériel, instruments & mat. de transport**
27 Machinery n.e.c.	34.7	45.9	41.5	51.8	46.1	16.1	..	27 Machines, n.c.a.
28 Office, account. & computing machinery	137.3	127.9	98.5	97.0	83.6	27.2	..	28 Mach. de bur., compta. & trait. inform.
29 Electrical machinery	17.2	18.9	14.6	18.9	14.9	34.1	..	29 Machines électriques
30 Electro. equip. (radio, TV & commun.)	127.4	107.0	137.4	151.4	171.7	150.2	..	30 Equip. électro. (radio, tv & communic.)
31 *Electro. comp. (inc. semi-conductors)*	..	..	..	..	17.5	35.0	..	31 *Compos. électro. (semi-cond. incl.)*
32 *TV, radio & communications equip.*	..	..	..	..	154.2	115.2	..	32 *Appareils radio, tv et de communi.*
33 Instruments	210.2	249.2	215.6	269.4	207.3	286.7	..	33 Instruments
34 Motor vehicles	19.8	36.9	13.5	14.9	11.3	10.9	..	34 Véhicules automobiles
35 Other transport equipment	959.7	885.7	1 025.9	759.0	622.4	403.0	..	35 Autres matériels de transport
36 *Ships*	0.1	0.4	0.7	1.1	0.1	0.2	..	36 *Navires*
37 *Aerospace*	959.0	885.3	1 024.7	757.8	622.2	402.7	..	37 *Aérospatiale*
38 *Other transport equipment n.e.c.*	0.6	0.0	0.5	0.1	0.1	0.1	..	38 *Autres matériels de transport n.c.a.*
39 Furniture, other manufacturing n.e.c.	0.0 [o]	0.1 [o]	0.2 [o]	1.8	1.2	2.4	..	**39 Meubles, autres activ. de fabricat. n.c.a.**
40 Furniture	..	..	..	..	1.2	2.4	..	40 Meubles
41 Other manufacturing n.e.c.	..	..	..	..	0.0	0.0	..	41 Autres activités de fabrication n.c.a.
42 Recycling	.. [n]	.. [n]	.. [n]	0.1	0.0	0.0	..	**42 Récupération**
43 ELECTRICITY, GAS & WATER SUPPLY	3.0	2.3	3.0	2.9	2.8	4.9	..	**43 DISTRIBUTION D'ELECTRICITE, GAZ & EAU**
44 CONSTRUCTION	2.8	1.6	1.9	3.1	2.3	1.7	..	**44 CONSTRUCTION**
45 SERVICES SECTOR	45.4	45.7	52.4	59.1	71.1	172.5	..	**45 SECTEUR DES SERVICES**
46 Wholesale, retail trade & repair of motor vehicles, etc.	..	..	..	..	0.0	0.0	..	**46 Commerce (gros & détail), réparation de véhicules, etc.**
47 Hotels & restaurants	..	..	..	..	0.0	0.0	..	**47 Hôtels et restaurants**
48 Transport & storage	15.1 [o]	14.9 [o]	18.1 [o]	23.6 [o]	19.5 [o]	16.7 [o]	..	**48 Transports, entreposage**
49 Communications	.. [n]	.. [n]	.. [n]	.. [n]	.. [n]	.. [n]	..	**49 Communications**
50 Post	..	..	..	..	..	..	..	50 Postes
51 Telecommunications	..	..	..	..	..	..	..	51 Télécommunications
52 Financial intermediation (inc. insur.)	..	..	0.0	..	0.0	0.0	..	**52 Intermédiation financ. (assurances inc.)**
53 Real estate, renting & busin. activity	30.3	30.9	34.3	35.5	51.6	155.9	..	**53 Immobil., locat. & serv. aux entreprises**
54 Computer & related activities	17.6	11.3	13.2	13.2	13.8	48.4	..	54 Informatique et activités connexes
55 *Software consultancy*	..	..	..	..	..	..	..	55 *Production de logiciels*
56 *Other computer services n.e.c.*	..	..	..	..	..	..	..	56 *Autres activités informatiques*
57 Research & development	..	..	..	..	0.0	0.0	..	57 Recherche & développement expérimental
58 Other business activities n.e.c.	12.8	19.6	21.1	22.3	37.8	107.5	..	58 Autres activ. & serv. aux entreprises
59 Comm., soc. & pers. serv. activ., etc.	..	..	..	..	0.0	0.0	..	**59 Adm. pub., serv. soc. & collec., etc**
TOTAL BUSINESS ENTERPRISE SECTOR	**1 857.8**	**1 850.0**	**1 945.2**	**1 834.0**	**1 650.6**	**1 646.3**	**..**	**TOTAL SECTEUR DES ENTREPRISES**

Source: OECD, R&D database (DSTI/EAS Division), January 2002.

Source : OCDE, base de données R-D (DSTI/Division AES), janvier 2002.

Business enterprise intramural expenditure on R&D (BERD) - Dépenses intra-muros de R-D des entreprises (DIRDE)

by industry and type of costs - par industrie et par type de coûts

Million EUR (1999 FRF euro, EUR from 1999) — *Millions EUR (1999 FRF euro, EUR à partir de 1999)*

11.1 Current expenditure - Dépenses courantes

	1994	1995	1996	1997	1998	1999	2000	
1 AGRICULTURE, HUNTING & FORESTRY	179.4	180.6	188.1	194.5	230.3	220.7	..	**1 AGRICULTURE, CHASSE & SYLVICULTURE**
2 MINING	99.7	92.6	89.7	85.4	81.6	74.5	..	**2 ACTIVITES EXTRACTIVES**
3 MANUFACTURING	13 744.4	13 672.8	13 971.6	13 760.9	14 060.8	14 969.7	..	**3 ACTIVITES DE FABRICATION**
4 Food, beverages & tobacco	265.7	267.4	280.4	277.5	299.8	296.1	..	**4 Produits alimentaires, boissons & tabac**
5 Food products & beverages	..	..	..	..	280.3	273.4	..	5 Produits alimentaires & boissons
6 Tobacco products	..	..	..	..	19.4	22.7	..	6 Produits à base de tabac
7 Textiles, wear app., fur & leather	99.5	100.8	94.3	97.7	96.0	91.6	..	**7 Textiles, habill., fourrures & cuirs**
8 Textiles	64.7	63.1	65.6	63.1	69.1	66.5	..	8 Textiles
9 Wearing apparel & fur	24.0	26.7	20.3	24.8	15.9	19.6	..	9 Articles d'habillement & fourrures
10 Leather products & footwear	10.8	11.0	8.4	9.8	11.0	5.5	..	10 Articles en cuir & chaussures
11 Wood, paper, printing, publishing	64.4	61.7	62.7	67.5	58.7	73.4	..	**11 Bois, papier, imprimerie, édition**
12 Wood & cork (not furniture)	8.1	11.0	11.2	10.9	9.2	13.6	..	12 Bois & liège (sauf meubles)
13 Pulp, paper & paper products	50.1	43.5	47.9	51.0	47.1	55.6	..	13 Carton, papier & articles
14 Publ., print. & repro. of rec. media	6.1	7.2	3.6	5.6	2.4	4.3	..	14 Edition, impress. & repro. enregis.
15 Coke, petroleum, nuclear fuel, chemicals, rubber plastics	3 131.5	3 260.1	3 508.1	3 616.8	3 783.4	4 044.4	..	**15 Coke, pétrole, combust. nucl., prod. chimi., caout. & mat. plast.**
16 Coke, ref. petrol. prod. & nuclear fuel	187.7	183.7	191.1	214.8	229.3	223.2	..	16 Coke, raffin. pétr. & combust. nucl.
17 Chemicals & chemical products	2 649.3	2 761.9	2 919.2	2 989.5	3 101.8	3 338.1	..	17 Produits chimiques
18 *Chemical prod.(less pharmaceuticals)*	905.8	953.8	1 003.0	998.9	1 035.0	1 061.3	..	18 *Prod. chimi. (moins prod. pharmaceutiques)*
19 *Pharmaceuticals*	1 743.5	1 808.0	1 916.1	1 990.6	2 066.8	2 276.8	..	19 *Produits pharmaceutiques*
20 Rubber & plastic products	294.5	314.5	397.8	412.5	452.3	483.0	..	20 Caoutchouc et matières plastiques
21 Non-metallic mineral products	163.4	173.4	189.0	191.1	199.7	203.2	..	**21 Produits minéraux non métalliques**
22 Basic metals	256.4	261.8	264.9	265.8	274.3	238.1	..	**22 Produits métallurgiques de base**
23 Basic metals, ferrous	181.9	180.7	187.3	194.7	202.4	170.8	..	23 Produits métallur. de base, ferreux
24 Basic metals, non-ferrous	74.5	81.1	77.6	71.2	72.0	67.3	..	24 Prod. métallur. de base, non ferreux
25 Fabricated metal products	197.0	213.9	194.6	193.4	169.3	182.3	..	**25 Ouvrages en métaux (sauf mach.& mat.)**
26 Machinery equipment, instruments and transport equipment	9 499.0	9 245.2	9 275.6	8 944.4	9 062.7	9 685.5	..	**26 Machines & matériel, instruments & mat. de transport**
27 Machinery n.e.c.	821.9	762.7	745.3	722.9	765.1	784.5	..	27 Machines, n.c.a.
28 Office, account. & computing machinery	469.7	424.4	410.2	385.9	394.7	340.0	..	28 Mach. de bur., compta. & trait. inform.
29 Electrical machinery	528.5	558.8	542.7	562.8	605.4	641.5	..	29 Machines électriques
30 Electro. equip. (radio, TV & commun.)	1 712.4	1 657.0	1 800.2	1 837.7	2 089.5	2 149.6	..	30 Equip. électro. (radio, tv & communic.)
31 *Electro. comp. (inc. semi-conductors)*	..	..	..	..	525.1	578.7	..	31 *Compos. électro. (semi-cond. incl.)*
32 *TV, radio & communications equip.*	..	..	..	..	1 564.4	1 570.9	..	32 *Appareils radio, tv et de communi.*
33 Instruments	1 709.0	1 644.5	1 550.1	1 598.9	1 197.8	1 197.5	..	33 Instruments
34 Motor vehicles	2 010.4	1 950.6	1 834.4	1 847.1	1 920.6	2 315.3	..	34 Véhicules automobiles
35 Other transport equipment	2 246.9	2 247.2	2 392.6	1 989.0	2 089.6	2 257.0	..	35 Autres matériels de transport
36 *Ships*	20.9	13.4	20.9	15.6	17.7	13.0	..	36 *Navires*
37 *Aerospace*	2 154.2	2 161.7	2 294.8	1 896.3	1 993.2	2 153.3	..	37 *Aérospatiale*
38 *Other transport equipment n.e.c.*	71.8	72.2	76.9	77.1	78.8	90.7	..	38 *Autres matériels de transport n.c.a.*
39 Furniture, other manufacturing n.e.c.	67.7 [o]	88.5 [o]	102.0 [o]	99.0	110.7	149.6	..	**39 Meubles, autres activ. de fabricat. n.c.a.**
40 Furniture	..	..	..	62.2	90.8	128.9	..	40 Meubles
41 Other manufacturing n.e.c.	..	..	..	36.8	19.9	20.7	..	41 Autres activités de fabrication n.c.a.
42 Recycling	.. [n]	.. [n]	.. [n]	7.6	6.3	5.3	..	**42 Récupération**
43 ELECTRICITY, GAS & WATER SUPPLY	303.8	314.4	494.3	477.1	459.9	450.8	..	**43 DISTRIBUTION D'ELECTRICITE, GAZ & EAU**
44 CONSTRUCTION	111.3	115.7	116.2	156.1	164.2	168.9	..	**44 CONSTRUCTION**
45 SERVICES SECTOR	992.0	1 050.9	1 006.3	1 347.4	1 365.3	1 502.2	..	**45 SECTEUR DES SERVICES**
46 Wholesale, retail trade & repair of motor vehicles, etc.	..	..	..	..	0.0	0.0	..	**46 Commerce (gros & détail), réparation de véhicules, etc.**
47 Hotels & restaurants	..	..	..	..	0.0	0.0	..	**47 Hôtels et restaurants**
48 Transport & storage	383.0 [o]	407.5 [o]	380.5 [o]	677.6 [o]	621.1 [o]	543.0 [o]	..	**48 Transports, entreposage**
49 Communications	.. [n]	.. [n]	.. [n]	.. [n]	.. [n]	.. [n]	..	**49 Communications**
50 Post	..	..	..	..	..	..	..	50 Postes
51 Telecommunications	..	..	..	..	..	..	..	51 Télécommunications
52 Financial intermediation (inc. insur.)	..	..	..	..	0.0	0.0	..	**52 Intermédiation financ. (assurances inc.)**
53 Real estate, renting & busin. activity	609.0	643.3	625.8	669.7	744.2	959.2	..	**53 Immobil., locat. & serv. aux entreprises**
54 Computer & related activities	391.3	408.7	374.5	385.3	354.9	446.1	..	54 Informatique et activités connexes
55 *Software consultancy*	..	..	..	..	..	..	..	55 *Production de logiciels*
56 *Other computer services n.e.c.*	..	..	..	..	..	..	..	56 *Autres activités informatiques*
57 Research & development	..	..	..	..	0.0	0.0	..	57 Recherche & développement expérimental
58 Other business activities n.e.c.	217.7	234.6	251.3	284.5	389.3	513.1	..	58 Autres activ. & serv. aux entreprises
59 Comm., soc. & pers. serv. activ., etc.	..	..	..	..	0.0	0.0	..	**59 Adm. pub., serv. soc. & collec., etc**
TOTAL BUSINESS ENTERPRISE SECTOR	**15 430.6**	**15 426.9**	**15 866.3**	**16 021.3** [a]	**16 362.2**	**17 386.8**	..	**TOTAL SECTEUR DES ENTREPRISES**

Source: OECD, R&D database (DSTI/EAS Division), January 2002.

Source : OCDE, base de données R-D (DSTI/Division AES), janvier 2002.

Business enterprise intramural expenditure on R&D (BERD) - Dépenses intra-muros de R-D des entreprises (DIRDE)

by industry and type of costs - par industrie et par type de coûts

Million EUR (1999 FRF euro, EUR from 1999) — *Millions EUR (1999 FRF euro, EUR à partir de 1999)*

11.2 Capital expenditure - Dépenses en capital

	1994	1995	1996	1997	1998	1999	2000	
1 AGRICULTURE, HUNTING & FORESTRY	11.0	10.7	13.0	16.9	25.9	26.9	..	**1 AGRICULTURE, CHASSE & SYLVICULTURE**
2 MINING	3.7	3.2	3.5	3.0	2.7	2.3	..	**2 ACTIVITES EXTRACTIVES**
3 MANUFACTURING	939.1	1 032.4	1 044.8	1 051.6	1 008.9	1 021.0	..	**3 ACTIVITES DE FABRICATION**
4 Food, beverages & tobacco	23.7	24.6	24.4	30.7	29.2	31.9	..	**4 Produits alimentaires, boissons & tabac**
5 Food products & beverages	..	..	..	..	26.5	31.2	..	5 Produits alimentaires & boissons
6 Tobacco products	..	..	..	..	2.7	0.7	..	6 Produits à base de tabac
7 Textiles, wear app., fur & leather	8.9	9.7	13.2	9.1	8.5	9.5	..	**7 Textiles, habill., fourrures & cuirs**
8 Textiles	6.0	6.2	10.3	6.6	7.5	8.6	..	8 Textiles
9 Wearing apparel & fur	2.3	2.4	0.9	1.1	0.5	0.3	..	9 Articles d'habillement & fourrures
10 Leather products & footwear	0.6	1.1	1.9	1.4	0.6	0.6	..	10 Articles en cuir & chaussures
11 Wood, paper, printing, publishing	4.9	7.2	4.6	5.2	8.3	6.3	..	**11 Bois, papier, imprimerie, édition**
12 Wood & cork (not furniture)	1.3	1.5	1.4	0.9	3.8	1.9	..	12 Bois & liège (sauf meubles)
13 Pulp, paper & paper products	3.1	5.0	3.1	3.4	3.9	4.1	..	13 Carton, papier & articles
14 Publ., print. & repro. of rec. media	0.5	0.6	0.1	0.9	0.5	0.3	..	14 Edition, impress. & repro. enregis.
15 Coke, petroleum, nuclear fuel, chemicals, rubber plastics	266.8	316.7	330.1	336.2	331.4	335.3	..	**15 Coke, pétrole, combust. nucl., prod. chimi., caout. & mat. plast.**
16 Coke, ref. petrol. prod. & nuclear fuel	27.1	21.6	33.9	31.0	34.3	38.0	..	16 Coke, raffin. pétr. & combust. nucl.
17 Chemicals & chemical products	215.9	263.9	261.5	264.7	261.4	255.8	..	17 Produits chimiques
18 *Chemical prod.(less pharmaceuticals)*	68.4	76.2	77.3	75.9	78.2	75.1	..	18 *Prod. chimi. (moins prod. pharmaceutiques)*
19 *Pharmaceuticals*	147.5	187.6	184.2	188.8	183.2	180.8	..	19 *Produits pharmaceutiques*
20 Rubber & plastic products	23.8	31.2	34.8	40.6	35.8	41.4	..	20 Caoutchouc et matières plastiques
21 Non-metallic mineral products	15.4	18.9	23.2	20.9	18.5	33.2	..	**21 Produits minéraux non métalliques**
22 Basic metals	22.3	43.7	23.0	18.0	19.1	16.8	..	**22 Produits métallurgiques de base**
23 Basic metals, ferrous	17.2	35.3	14.7	13.1	13.6	9.6	..	23 Produits métallur. de base, ferreux
24 Basic metals, non-ferrous	5.1	8.4	8.3	5.0	5.4	7.2	..	24 Prod. métallur. de base, non ferreux
25 Fabricated metal products	12.4	11.5	15.2	11.7	11.4	9.5	..	**25 Ouvrages en métaux (sauf mach.& mat.)**
26 Machinery equipment, instruments and transport equipment	579.0	590.7	603.5	612.0	575.1	570.6	..	**26 Machines & matériel, instruments & mat. de transport**
27 Machinery n.e.c.	51.3	45.1	49.5	43.2	44.9	46.1	..	27 Machines, n.c.a.
28 Office, account. & computing machinery	23.6	21.1	26.7	23.6	22.5	13.6	..	28 Mach. de bur., compta. & trait. inform.
29 Electrical machinery	44.5	39.0	39.7	51.6	39.6	48.2	..	29 Machines électriques
30 Electro. equip. (radio, TV & commun.)	174.3	171.8	172.5	168.1	180.5	185.1	..	30 Equip. électro. (radio, tv & communic.)
31 *Electro. comp. (inc. semi-conductors)*	..	..	..	..	71.1	77.4	..	31 *Compos. électro. (semi-cond. incl.)*
32 *TV, radio & communications equip.*	..	..	..	..	109.4	107.7	..	32 *Appareils radio, tv et de communi.*
33 Instruments	80.6	78.5	68.8	72.7	51.2	58.9	..	33 Instruments
34 Motor vehicles	159.5	197.8	196.0	197.9	179.1	175.7	..	34 Véhicules automobiles
35 Other transport equipment	45.2	37.6	50.4	54.8	57.4	43.1	..	35 Autres matériels de transport
36 *Ships*	0.7	0.3	0.8	0.6	2.4	1.2	..	36 *Navires*
37 *Aerospace*	43.2	36.0	48.8	53.2	54.3	40.8	..	37 *Aérospatiale*
38 *Other transport equipment n.e.c.*	1.2	1.3	0.7	1.0	0.7	1.1	..	38 *Autres matériels de transport n.c.a.*
39 Furniture, other manufacturing n.e.c.	5.7 [o]	9.4 [o]	7.6 [o]	7.2	6.9	7.6	..	**39 Meubles, autres activ. de fabricat. n.c.a.**
40 Furniture	..	..	..	..	5.3	4.9	..	40 Meubles
41 Other manufacturing n.e.c.	..	..	..	..	1.6	2.6	..	41 Autres activités de fabrication n.c.a.
42 Recycling	.. [n]	.. [n]	.. [n]	0.5	0.5	0.4	..	**42 Récupération**
43 ELECTRICITY, GAS & WATER SUPPLY	28.8	29.9	30.2	26.3	17.9	21.6	..	**43 DISTRIBUTION D'ELECTRICITE, GAZ & EAU**
44 CONSTRUCTION	6.9	7.0	6.0	7.8	6.9	8.0	..	**44 CONSTRUCTION**
45 SERVICES SECTOR	131.0	139.2	167.3	229.9	207.2	188.5	..	**45 SECTEUR DES SERVICES**
46 Wholesale, retail trade & repair of motor vehicles, etc.	..	..	..	..	0.0	0.0	..	**46 Commerce (gros & détail), réparation de véhicules, etc.**
47 Hotels & restaurants	..	..	..	..	0.0	0.0	..	**47 Hôtels et restaurants**
48 Transport & storage	96.0 [o]	97.9 [o]	118.7 [o]	187.1 [o]	156.7 [o]	130.9 [o]	..	**48 Transports, entreposage**
49 Communications	.. [n]	.. [n]	.. [n]	.. [n]	.. [n]	.. [n]	..	**49 Communications**
50 Post	..	..	..	..	..	..	..	50 Postes
51 Telecommunications	..	..	..	..	..	..	..	51 Télécommunications
52 Financial intermediation (inc. insur.)	..	..	..	..	0.0	0.0	..	**52 Intermédiation financ. (assurances inc.)**
53 Real estate, renting & busin. activity	35.0	41.3	48.5	42.8	50.4	57.7	..	**53 Immobil., locat. & serv. aux entreprises**
54 Computer & related activities	17.7	14.9	19.5	19.7	20.9	18.2	..	54 Informatique et activités connexes
55 *Software consultancy*	..	..	..	..	..	..	..	55 *Production de logiciels*
56 *Other computer services n.e.c.*	..	..	..	..	..	..	..	56 *Autres activités informatiques*
57 Research & development	..	..	..	..	0.0	0.0	..	57 Recherche & développement expérimental
58 Other business activities n.e.c.	17.4	26.4	29.1	23.1	29.5	39.5	..	58 Autres activ. & serv. aux entreprises
59 Comm., soc. & pers. serv. activ., etc.	..	..	..	..	0.0	0.0	..	**59 Adm. pub., serv. soc. & collec., etc**
TOTAL BUSINESS ENTERPRISE SECTOR	**1 120.6**	**1 222.4**	**1 264.9**	**1 335.6** [a]	**1 269.5**	**1 268.3**	**..**	**TOTAL SECTEUR DES ENTREPRISES**

Source: OECD, R&D database (DSTI/EAS Division), January 2002.

Source : OCDE, base de données R-D (DSTI/Division AES), janvier 2002.

FRANCE — Table 12 - Tableau 12 — FRANCE

Total business enterprise R&D personnel (BEMP) - Personnel total de R-D du secteur des entreprises (PRDE)

by industry - par industrie

In full time equivalent — *En équivalence plein-temps*

	1994	1995	1996	1997	1998	1999	2000	
1 AGRICULTURE, HUNTING & FORESTRY	2 654	2 749	2 588	2 565	2 916	2 820	..	**1 AGRICULTURE, CHASSE & SYLVICULTURE**
2 MINING	762	558	591	571	540	515	..	**2 ACTIVITES EXTRACTIVES**
3 MANUFACTURING	142 599	142 140	142 015	142 260	143 168	144 475	..	**3 ACTIVITES DE FABRICATION**
4 Food, beverages & tobacco	3 506	3 501	3 507	3 393	3 572	3 557	..	**4 Produits alimentaires, boissons & tabac**
5 Food products & beverages	..	..	..	..	3 335	3 262	..	5 Produits alimentaires & boissons
6 Tobacco products	..	..	..	..	237	295	..	6 Produits à base de tabac
7 Textiles, wear app., fur & leather	1 585	1 504	1 328	1 522	1 528	1 438	..	**7 Textiles, habill., fourrures & cuirs**
8 Textiles	992	872	830	980	1 051	1 020	..	8 Textiles
9 Wearing apparel & fur	407	456	357	385	303	314	..	9 Articles d'habillement & fourrures
10 Leather products & footwear	186	176	141	157	174	105	..	10 Articles en cuir & chaussures
11 Wood, paper, printing, publishing	846	864	878	911	739	899	..	**11 Bois, papier, imprimerie, édition**
12 Wood & cork (not furniture)	117	172	166	113	101	190	..	12 Bois & liège (sauf meubles)
13 Pulp, paper & paper products	634	585	652	702	592	618	..	13 Carton, papier & articles
14 Publ., print. & repro. of rec. media	95	107	60	96	46	91	..	14 Edition, impress. & repro. enregis.
15 Coke, petroleum, nuclear fuel, chemicals, rubber plastics	35 296	36 080	36 923	36 650	37 043	36 974	..	**15 Coke, pétrole, combust. nucl., prod. chimi., caout. & mat. plast.**
16 Coke, ref. petrol. prod. & nuclear fuel	2 161	2 170	2 125	2 080	2 063	2 064	..	16 Coke, raffin. pétr. & combust. nucl.
17 Chemicals & chemical products	28 378	28 986	29 400	29 368	29 709	29 364	..	17 Produits chimiques
18 *Chemical prod.(less pharmaceuticals)*	11 065	11 238	11 440	11 221	11 533	11 100	..	18 *Prod. chimi. (moins prod. pharmaceutiques)*
19 *Pharmaceuticals*	17 313	17 748	17 960	18 147	18 176	18 264	..	19 *Produits pharmaceutiques*
20 Rubber & plastic products	4 757	4 924	5 397	5 202	5 270	5 546	..	20 Caoutchouc et matières plastiques
21 Non-metallic mineral products	2 046	2 053	2 178	2 151	2 050	2 014	..	**21 Produits minéraux non métalliques**
22 Basic metals	3 399	3 244	3 192	3 207	3 177	2 804	..	**22 Produits métallurgiques de base**
23 Basic metals, ferrous	2 445	2 230	2 267	2 314	2 339	2 010	..	23 Produits métallur. de base, ferreux
24 Basic metals, non-ferrous	953	1 015	926	893	838	793	..	24 Prod. métallur. de base, non ferreux
25 Fabricated metal products	2 554	2 773	2 672	2 665	2 391	2 345	..	**25 Ouvrages en métaux (sauf mach.& mat.)**
26 Machinery equipment, instruments and transport equipment	92 241	90 889	89 974	90 349	91 258	92 894	..	**26 Machines & matériel, instruments & mat. de transport**
27 Machinery n.e.c.	10 175	10 077	9 935	9 748	10 109	10 421	..	27 Machines, n.c.a.
28 Office, account. & computing machinery	5 329	4 456	4 255	3 899	3 679	2 926	..	28 Mach. de bur., compta. & trait. inform.
29 Electrical machinery	6 504	6 578	6 391	6 845	6 967	7 784	..	29 Machines électriques
30 Electro..equip. (radio, TV & commun.)	14 582	14 927	16 596	17 800	19 896	20 042	..	30 Equip. électro. (radio, tv & communic.)
31 *Electro. comp. (inc. semi-conductors)*	..	..	..	..	4 378	4 923	..	31 *Compos. électro. (semi-cond. incl.)*
32 *TV, radio & communications equip.*	..	..	..	..	15 518	15 119	..	32 *Appareils radio, tv et de communi.*
33 Instruments	17 808	16 470	15 511	15 922	12 484	11 931	..	33 Instruments
34 Motor vehicles	18 042	20 249	19 964	20 103	22 006	23 832	..	34 Véhicules automobiles
35 Other transport equipment	19 801	18 133	17 322	16 031	16 118	15 960	..	35 Autres matériels de transport
36 *Ships*	323	225	347	300	269	271	..	36 *Navires*
37 *Aerospace*	18 637	17 074	16 184	14 931	14 992	14 738	..	37 *Aérospatiale*
38 *Other transport equipment n.e.c.*	841	834	792	800	857	950	..	38 *Autres matériels de transport n.c.a.*
39 Furniture, other manufacturing n.e.c.	1 127 [o]	1 231 [o]	1 364	1 308	1 355	1 489	..	**39 Meubles, autres activ. de fabricat. n.c.a.**
40 Furniture	..	..	..	804	1 009	1 184	..	40 Meubles
41 Other manufacturing n.e.c.	..	..	..	504	346	305	..	41 Autres activités de fabrication n.c.a.
42 Recycling	.. [n]	.. [n]	..	104	54	62	..	**42 Récupération**
43 ELECTRICITY, GAS & WATER SUPPLY	2 160	2 276	3 501	3 494	3 318	3 299	..	**43 DISTRIBUTION D'ELECTRICITE, GAZ & EAU**
44 CONSTRUCTION	1 085	1 192	1 138	1 391	1 238	1 237	..	**44 CONSTRUCTION**
45 SERVICES SECTOR	12 696	13 127	12 757	16 030	16 586	19 218	..	**45 SECTEUR DES SERVICES**
46 Wholesale, retail trade & repair of motor vehicles, etc.	..	..	..	..	0	0	..	**46 Commerce (gros & détail), réparation de véhicules, etc.**
47 Hotels & restaurants	..	..	..	..	0	0	..	**47 Hôtels et restaurants**
48 Transport & storage	5 247 [o]	5 280 [o]	5 171 [o]	7 558 [o]	6 925 [o]	6 993 [o]	..	**48 Transports, entreposage**
49 Communications	.. [n]	.. [n]	.. [n]	.. [n]	.. [n]	.. [n]	..	**49 Communications**
50 Post	..	..	..	..	..	..	..	50 Postes
51 Telecommunications	..	..	..	..	..	..	..	51 Télécommunications
52 Financial intermediation (inc. insur.)	..	..	..	..	0	0	..	**52 Intermédiation financ. (assurances inc.)**
53 Real estate, renting & busin. activity	7 449	7 848	7 586	8 471	9 661	12 225	..	**53 Immobil., locat. & serv. aux entreprises**
54 Computer & related activities	4 857	5 030	4 688	4 940	4 713	5 990	..	54 Informatique et activités connexes
55 *Software consultancy*	..	..	..	..	..	..	..	55 *Production de logiciels*
56 *Other computer services n.e.c.*	..	..	..	..	..	..	..	56 *Autres activités informatiques*
57 Research & development	..	..	..	..	..	..	..	57 Recherche & développement expérimental
58 Other business activities n.e.c.	2 592	2 817	2 898	3 532	4 948	6 235	..	58 Autres activ. & serv. aux entreprises
59 Comm., soc. & pers. serv. activ., etc.	..	..	..	..	0	0	..	**59 Adm. pub., serv. soc. & collec., etc**
TOTAL BUSINESS ENTERPRISE SECTOR	**161 955**	**162 042**	**162 590**	**166 310** [a]	**167 765**	**171 564**	..	**TOTAL SECTEUR DES ENTREPRISES**

Source: OECD, R&D database (DSTI/EAS Division), January 2002.

Source : OCDE, base de données R-D (DSTI/Division AES), janvier 2002.

FRANCE

Table 13 - Tableau 13

RSE in the business enterprise sector - Chercheurs dans le secteur des entreprises

by industry - par industrie

In full time equivalent — *En équivalence plein-temps*

	1994	1995	1996	1997	1998	1999	2000	
1 **AGRICULTURE, HUNTING & FORESTRY**	857	874	835	828	1 006	982	..	1 **AGRICULTURE, CHASSE & SYLVICULTURE**
2 **MINING**	320	217	236	237	221	219	..	2 **ACTIVITES EXTRACTIVES**
3 **MANUFACTURING**	55 857	55 744	57 017	58 287	58 686	59 978	..	3 **ACTIVITES DE FABRICATION**
4 **Food, beverages & tobacco**	1 336	1 346	1 398	1 365	1 496	1 488	..	4 **Produits alimentaires, boissons & tabac**
5 Food products & beverages	..	..	..	..	1 461	1 444	..	5 Produits alimentaires & boissons
6 Tobacco products	..	..	..	..	35	44	..	6 Produits à base de tabac
7 **Textiles, wear app., fur & leather**	512	480	449	540	492	436	..	7 **Textiles, habill., fourrures & cuirs**
8 Textiles	355	305	286	381	376	317	..	8 Textiles
9 Wearing apparel & fur	121	139	135	131	83	94	..	9 Articles d'habillement & fourrures
10 Leather products & footwear	37	36	29	28	33	24	..	10 Articles en cuir & chaussures
11 **Wood, paper, printing, publishing**	378	339	345	388	344	397	..	11 **Bois, papier, imprimerie, édition**
12 Wood & cork (not furniture)	64	81	77	72	70	97	..	12 Bois & liège (sauf meubles)
13 Pulp, paper & paper products	243	203	223	262	245	247	..	13 Carton, papier & articles
14 Publ., print. & repro. of rec. media	72	55	45	55	29	54	..	14 Edition, impress. & repro. enregis.
15 **Coke, petroleum, nuclear fuel, chemicals, rubber plastics**	11 855	12 022	12 296	12 103	12 415	12 856	..	15 **Coke, pétrole, combust. nucl., prod. chimi., caout. & mat. plast.**
16 Coke, ref. petrol. prod. & nuclear fuel	881	938	914	905	939	973	..	16 Coke, raffin. pétr. & combust. nucl.
17 Chemicals & chemical products	9 778	9 820	10 122	10 058	10 225	10 542	..	17 Produits chimiques
18 *Chemical prod.(less pharmaceuticals)*	3 789	3 764	3 904	3 852	3 933	4 025	..	18 *Prod. chimi. (moins prod. pharmaceutiques)*
19 *Pharmaceuticals*	5 988	6 056	6 218	6 206	6 291	6 517	..	19 *Produits pharmaceutiques*
20 Rubber & plastic products	1 196	1 264	1 260	1 139	1 251	1 341	..	20 Caoutchouc et matières plastiques
21 **Non-metallic mineral products**	737	738	792	813	786	813	..	21 **Produits minéraux non métalliques**
22 **Basic metals**	1 110	1 134	1 110	1 148	1 063	938	..	22 **Produits métallurgiques de base**
23 Basic metals, ferrous	831	795	796	833	791	667	..	23 Produits métallur. de base, ferreux
24 Basic metals, non-ferrous	280	338	313	315	272	271	..	24 Prod. métallur. de base, non ferreux
25 **Fabricated metal products**	1 100	1 179	1 066	1 201	1 063	993	..	25 **Ouvrages en métaux (sauf mach.& mat.)**
26 **Machinery equipment, instruments and transport equipment**	38 444	38 120	39 110	40 277	40 670	41 680	..	26 **Machines & matériel, instruments & mat. de transport**
27 Machinery n.e.c.	3 347	3 272	3 274	3 486	3 473	3 639	..	27 Machines, n.c.a.
28 Office, account. & computing machinery	3 523	3 258	3 053	2 885	2 794	2 023	..	28 Mach. de bur., compta. & trait. inform.
29 Electrical machinery	2 423	2 521	2 618	2 810	2 872	3 654	..	29 Machines électriques
30 Electro. equip. (radio, TV & commun.)	7 056	7 646	8 420	9 151	11 120	11 476	..	30 Equip. électro. (radio, tv & communic.)
31 *Electro. comp. (inc. semi-conductors)*	..	..	..	..	2 835	3 442	..	31 *Compos. électro. (semi-cond. incl.)*
32 *TV, radio & communications equip.*	..	..	..	..	8 285	8 035	..	32 *Appareils radio, tv et de communi.*
33 Instruments	10 206	9 682	9 440	10 031	7 472	7 471	..	33 Instruments
34 Motor vehicles	4 448	4 868	5 305	5 590	6 432	7 074	..	34 Véhicules automobiles
35 Other transport equipment	7 441	6 873	7 000	6 323	6 507	6 343	..	35 Autres matériels de transport
36 *Ships*	95	56	93	74	85	76	..	36 *Navires*
37 *Aerospace*	7 029	6 497	6 599	5 917	6 125	5 938	..	37 *Aérospatiale*
38 *Other transport equipment n.e.c.*	317	320	309	332	297	329	..	38 *Autres matériels de transport n.c.a.*
39 **Furniture, other manufacturing n.e.c.**	386 [o]	387 [o]	452	374	311	328	..	39 **Meubles, autres activ. de fabricat. n.c.a.**
40 Furniture	..	..	..	213	214	241	..	40 Meubles
41 Other manufacturing n.e.c.	..	..	..	162	97	87	..	41 Autres activités de fabrication n.c.a.
42 **Recycling**	.. [n]	.. [n]	..	78	46	48	..	42 **Récupération**
43 **ELECTRICITY, GAS & WATER SUPPLY**	1 187	1 235	1 932	1 969	1 878	1 914	..	43 **DISTRIBUTION D'ELECTRICITE, GAZ & EAU**
44 **CONSTRUCTION**	608	699	642	758	659	694	..	44 **CONSTRUCTION**
45 **SERVICES SECTOR**	7 885	7 848	7 825	9 945	9 267	11 604	..	45 **SECTEUR DES SERVICES**
46 **Wholesale, retail trade & repair of motor vehicles, etc.**	..	..	..	..	0	0	..	46 **Commerce (gros & détail), réparation de véhicules, etc.**
47 **Hotels & restaurants**	..	..	..	..	0	0	..	47 **Hôtels et restaurants**
48 **Transport & storage**	2 480 [o]	2 539 [o]	2 627 [o]	4 008	2 720 [o]	2 814 [o]	..	48 **Transports, entreposage**
49 **Communications**	.. [n]	.. [n]	.. [n]	.. [n]	.. [n]	.. [n]	..	49 **Communications**
50 Post	..	..	..	..	..	..	..	50 Postes
51 Telecommunications	..	..	..	..	..	..	..	51 Télécommunications
52 **Financial intermediation (inc. insur.)**	..	..	..	..	0	0	..	52 **Intermédiation financ. (assurances inc.)**
53 **Real estate, renting & busin. activity**	5 405	5 310	5 197	5 937	6 546	8 790	..	53 **Immobil., locat. & serv. aux entreprises**
54 Computer & related activities	4 002	3 801	3 675	3 940	3 844	4 997	..	54 Informatique et activités connexes
55 *Software consultancy*	..	..	..	..	..	..	..	55 *Production de logiciels*
56 *Other computer services n.e.c.*	..	..	..	..	..	..	..	56 *Autres activités informatiques*
57 Research & development	..	..	..	..	..	..	..	57 Recherche & développement expérimental
58 Other business activities n.e.c.	1 402	1 509	1 522	1 996	2 702	3 793	..	58 Autres activ. & serv. aux entreprises
59 **Comm., soc. & pers. serv. activ., etc.**	..	..	..	..	..	0	..	59 **Adm. pub., serv. soc. & collec., etc**
TOTAL BUSINESS ENTERPRISE SECTOR	**66 713**	**66 618**	**68 487**	**72 023** [a]	**71 717**	**75 390**	**..**	**TOTAL SECTEUR DES ENTREPRISES**

Source: OECD, R&D database (DSTI/EAS Division), January 2002.

Source : OCDE, base de données R-D (DSTI/Division AES), janvier 2002.

Table 17 - Tableau 17

FRANCE

Higher education R&D personnel (HEMP) - Personnel de R-D du secteur de l'enseignement supérieur

by field of science and occupation - par discipline scientifique et par profession

In full-time equivalent | *En équivalence plein-temps*

	1994	1995	1996	1997	1998	1999	2000	
Total R&D personnel								**Personnel total de R-D**
Natural sciences	..	..	..	..	..	..	..	Sciences exactes et naturelles
Engineering	..	..	..	..	..	..	..	Sciences de l'ingénieur
Medical sciences	..	..	..	..	..	..	..	Sciences médicales
Agricultural sciences	..	..	..	..	..	..	..	Sciences agricoles
Sub-total NSE	..	..	..	..	..	..	..	Sous-total SENI
Social sciences	..	..	..	..	..	..	..	Sciences sociales
Humanities	..	..	..	..	..	..	..	Sciences humaines
Sub-total SSH	..	..	..	..	..	..	..	Sous-total SSH
Total NSE+SSH	78 132	80 678	81 538	79 926 [a]	82 177	83 254	..	*Total SENI+SSH*
Researchers								**Chercheurs**
Natural sciences	..	..	..	..	..	..	..	Sciences exactes et naturelles
Engineering	..	..	..	..	..	..	..	Sciences de l'ingénieur
Medical sciences	..	..	..	..	..	..	..	Sciences médicales
Agricultural sciences	..	..	..	..	..	..	..	Sciences agricoles
Sub-total NSE	..	..	..	..	..	..	..	Sous-total SENI
Social sciences	..	..	..	..	..	..	..	Sciences sociales
Humanities	..	..	..	..	..	..	..	Sciences humaines
Sub-total SSH	..	..	..	..	..	..	..	Sous-total SSH
Total NSE+SSH	52 119	53 726	54 592	54 916 [a]	56 288	56 717	..	*Total SENI+SSH*
Technicians								**Technicians**
Natural sciences	..	..	..	..	..	..	..	Sciences exactes et naturelles
Engineering	..	..	..	..	..	..	..	Sciences de l'ingénieur
Medical sciences	..	..	..	..	..	..	..	Sciences médicales
Agricultural sciences	..	..	..	..	..	..	..	Sciences agricoles
Sub-total NSE	..	..	..	..	..	..	..	Sous-total SENI
Social sciences	..	..	..	..	..	..	..	Sciences sociales
Humanities	..	..	..	..	..	..	..	Sciences humaines
Sub-total SSH	..	..	..	..	..	..	..	Sous-total SSH
Total NSE+SSH	..	..	..	..	..	..	..	*Total SENI+SSH*
Other								**Autres**
Natural sciences	..	..	..	..	..	..	..	Sciences exactes et naturelles
Engineering	..	..	..	..	..	..	..	Sciences de l'ingénieur
Medical sciences	..	..	..	..	..	..	..	Sciences médicales
Agricultural sciences	..	..	..	..	..	..	..	Sciences agricoles
Sub-total NSE	..	..	..	..	..	..	..	Sous-total SENI
Social sciences	..	..	..	..	..	..	..	Sciences sociales
Humanities	..	..	..	..	..	..	..	Sciences humaines
Sub-total SSH	..	..	..	..	..	..	..	Sous-total SSH
Total NSE+SSH	..	..	..	..	..	..	..	*Total SENI+SSH*

Table 18 - Tableau 18

Government budget appropriations or outlays for R&D (GBAORD) - Crédits budgétaires publics de R-D (CBPRD)

by socio-economic objective - par objectif socio-économique

Million EUR (1999 FRF euro, EUR from 1999) | *Millions EUR (1999 FRF euro, EUR à partir de 1999)*

	1995	1996	1997	1998	1999	2000	2001	
Agriculture, forestry and fishing	452.4	468.9	477.0	490.2	383.3	321.0 [p]	..	Agriculture, sylviculture et pêche
Industrial development	794.3	628.7	593.0	726.2	789.5	837.7 [p]	..	Développement industriel
Energy	583.4	609.8	604.6	653.7	630.0	663.9 [p]	..	Energie
Transport & telecommunications	..	..	..	..	..	..	..	*Transports & télécommunications*
Urban & rural planning	..	..	..	..	..	..	..	*Aménagement urbain et rural*
ST Infrastructure	82.5	66.5	84.3	81.1	82.9	96.4 [p]	..	ST Infrastructure
Prevention of pollution	..	..	..	..	..	..	..	*Prévention de la pollution*
Identificat. & treatment of pollution	..	..	..	..	..	..	..	*Détect. pollut. & mesures antipollut.*
ST Environment	255.4	260.3	260.9	278.5	202.8	238.8 [p]	..	ST Environnement
Health	655.5	670.2	700.4	708.0	707.6	734.7 [p]	..	Santé
Social development & services	101.1	112.3	118.0	134.6	124.2	98.2 [p]	..	Développement et services sociaux
Earth and atmosphere	104.7	92.9	96.8	112.8	93.4	75.6 [p]	..	Milieu terrestre et atmosphère
Advancement of research	2 531.0	2 451.5	2 510.4	2 553.1	2 808.3	2 941.7 [p]	..	*Promotion de la recherche*
General university funds	2 047.5	2 097.4	2 175.2	2 286.0	2 346.2	2 344.5 [p]	..	*Fonds généraux des universités*
ST Advancement of knowledge	4 578.4	4 548.9	4 685.6	4 839.1	5 154.5	5 286.2 [p]	..	ST promotion gén. des connaiss.
Civil space	1 381.2	1 433.7	1 460.8	1 403.8	1 417.2	1 436.6 [p]	..	Espace civil
Defence	3 955.0	3 897.3	3 162.7 [a]	2 970.1	2 931.5	2 959.9 [p]	..	Défense
Not elsewhere classified	248.6	315.6	313.0	305.3	375.1	342.9 [p]	..	Non classé
Total	**13 192.5**	**13 105.0**	**12 557.0 [a]**	**12 703.3**	**12 891.9**	**13 092.0 [p]**	**..**	**Total**

Source: OECD, R&D database (DSTI/EAS Division), January 2002.

Source : OCDE, base de données R-D (DSTI/Division AES), janvier 2002.

Table 19 - Tableau 19

Technology balance of payments - Balance des paiements technologiques

by type of transfer - selon la nature des transferts

Million EUR (1999 FRF euro, EUR from 1999) — *Millions EUR (1999 FRF euro, EUR à partir de 1999)*

	1983	1984	1985	1986	1987	1988	1989	1990	1991	
Receipts										**Recettes**
Related firms	..	..	..	..	..	..	..	..	..	Firmes affiliées
Unrelated firms	..	..	..	..	..	..	..	..	..	Firmes non affiliées
Other	..	..	..	..	..	..	..	..	..	Autre
Total	952	1 207	1 224	1 203	1 235	1 343	1 507	1 574	1 498	*Total*
Payments										**Paiements**
Related firms	..	..	..	..	..	..	..	..	..	Firmes affiliées
Unrelated firms	..	..	..	..	..	..	..	..	..	Firmes non affiliées
Other	..	..	..	..	..	..	..	..	..	Autre
Total	1 055	1 347	1 457	1 446	1 587	1 689	1 817	2 081	2 109	*Total*

	1992	1993	1994	1995	1996	1997	1998	1999	2000	
Receipts										**Recettes**
Related firms	..	..	..	..	..	..	..	..	..	Firmes affiliées
Unrelated firms	..	..	..	..	..	..	..	..	..	Firmes non affiliées
Other	..	..	..	..	..	..	..	..	..	Autre
Total	1 624	1 568	1 577	1 651	1 867	1 930	2 330	2 586	..	*Total*
Payments										**Paiements**
Related firms	..	..	..	..	..	..	..	..	..	Firmes affiliées
Unrelated firms	..	..	..	..	..	..	..	..	..	Firmes non affiliées
Other	..	..	..	..	..	..	..	..	..	Autre
Total	2 253	2 202	2 153	2 274	2 473	2 699	2 810	2 975	..	*Total*

Source: OECD, TBPdatabase (DSTI/EAS Division), Januay 2002.

Source : OCDE, base de données TBP (DSTI/Division AES), janvier 2002.

GERMANY — Table 1 - Tableau 1 — ALLEMAGNE

Gross domestic expenditure on R&D (GERD) - Dépenses intérieures brutes de R-D (DIRD)

by sector of performance and source of funds - par secteur d'exécution et par source de financement

	1994	1995	1996	1997	1998	1999	2000	
	1.1 Million EUR (1999 DEM euro, EUR from 1999) - Millions EUR (1999 DEM euro, EUR à partir de 1999)							
Business enterprise								**Entreprises**
financed by								*financé par*
Business enterprise	23 258.7 c	24 033.3	24 317.0 c	25 428.6	26 878.1 c	30 266.4	31 677.1 c	Entreprises
Direct government	2 351.9 c	2 376.0 o	2 454.2 c	2 656.7	2 607.6 c	2 564.6	2 607.6 c	État direct
Higher education	..	..	..	..	..	..	..	Enseignement supérieur
Private non-profit	51.1 c	20.5	20.5 c	28.6	30.7 c	77.7	80.7 c	Institutions sans but lucratif
Funds from abroad	511.3 c	584.4	613.6 c	795.6	818.1 c	713.8	744.3 c	Fonds de l'étranger
Total	26 173.0 c	27 014.1 o	27 405.2 c	28 910.0	30 334.4 c	33 622.6	35 110.0 c	*Total*
Government								**État**
financed by								*financé par*
Business enterprise	202.3	213.7	116.0	123.7	136.1	140.0	141.0 c	Entreprises
Direct government	5 576.4	5 889.9	5 980.4	5 930.6	6 166.5	6 218.2	6 261.7 c	État direct
Higher education	..	..	..	..	..	..	..	Enseignement supérieur
Private non-profit	78.9	83.3	105.4	112.6	123.6	127.3	128.2 c	Institutions sans but lucratif
Funds from abroad	74.8	78.9	103.4	105.4	120.6	146.0	147.0 c	Fonds de l'étranger
Total	5 932.3 o	6 265.8 o	6 305.1 o	6 272.3 o	6 546.9 o	6 631.6 o	6 678.0 c,o	*Total*
Higher education								**Enseignement supérieur**
financed by								*financé par*
Business enterprise	585.0	605.2	705.3	747.5	818.5	897.6	908.7 c	Entreprises
Direct government	..	1 490.7	1 636.5	1 659.4	1 673.2	1 757.6	1 779.2 c	État direct
General university funds	..	5 204.0	5 198.6	5 139.1	5 118.7	5 144.6	5 208.0 c	Fonds généraux des universités
Sub-total government	6 385.5 o	6 694.7 o	6 835.1 o	6 798.5 o	6 792.0 o	6 902.2 o	6 987.3 c,o	Sous-Total État
Higher education	.. n	.. n	.. n	.. n	.. n	.. n	.. n	Enseignement supérieur
Private non-profit	.. n	.. n	.. n	.. n	.. n	.. n	.. n	Institutions sans but lucratif
Funds from abroad	89.0	78.0	111.9	130.8	157.6	136.8	138.5 c	Fonds de l'étranger
Total	7 059.4	7 377.8	7 652.3	7 676.8	7 768.1	7 936.6	8 034.4 c	*Total*
Private non-profit								**Institutions sans but lucratif**
financed by								*financé par*
Business enterprise	..	..	..	..	..	..	..	Entreprises
Direct government	..	..	..	..	..	..	..	État direct
Higher education	..	..	..	..	..	..	..	Enseignement supérieur
Private non-profit	..	..	..	..	..	..	..	Institutions sans but lucratif
Funds from abroad	..	..	..	..	..	..	..	Fonds de l'étranger
Total	..	..	..	..	..	..	..	*Total*
GERD								**DIRD**
financed by								*financé par*
Business enterprise	24 045.9 c	24 852.1	25 138.4 c	26 299.8	27 832.7 c	31 304.1	32 726.8 c	Entreprises
Direct government	..	9 756.5	10 071.1	10 246.6	10 447.2	10 540.4	10 648.5 c	État direct
General university funds	..	5 204.0	5 198.6	5 139.1	5 118.7	5 144.6	5 208.0 c	Fonds généraux des universités
Sub-total government	14 314.1 c	14 960.6	15 269.7 c	15 385.7	15 566.0 c	15 685.0	15 927.9 c	Sous-Total État
Higher education	..	..	..	..	..	..	..	Enseignement supérieur
Private non-profit	130.0 c	103.8	125.8 c	141.3	154.3 c	205.0	209.0 c	Institutions sans but lucratif
Funds from abroad	675.1 c	741.3	828.8 c	1 031.8	1 096.3 c	996.6	1 029.9 c	Fonds de l'étranger
Total	39 165.0 c,o	40 657.8 c,o	41 362.7 c,o	42 859.1 o	44 649.3 c,o	48 190.7	49 822.4 c	*Total*
	1.2 Million constant USD (1995 prices and PPPs) - Millions de USD constants (prix et PPA de 1995)							
Business enterprise								**Entreprises**
financed by								*financé par*
Business enterprise	23 026.0 c	23 320.2	23 355.4 c	24 263.0	25 364.6 c	28 429.8	29 872.1 c	Entreprises
Direct government	2 328.4 c	2 305.5 o	2 357.1 c	2 534.9	2 460.8 c	2 409.0	2 459.0 c	État direct
Higher education	..	..	..	..	..	..	..	Enseignement supérieur
Private non-profit	50.6 c	19.8	19.6 c	27.3	29.0 c	73.0	76.1 c	Institutions sans but lucratif
Funds from abroad	506.2 c	567.1	589.3 c	759.1	772.0 c	670.5	701.9 c	Fonds de l'étranger
Total	25 911.2 c	26 212.6 o	26 321.5 c	27 584.8	28 626.3 c	31 582.3	33 109.4 c	*Total*
Government								**État**
financed by								*financé par*
Business enterprise	200.2	207.3	111.4	118.1	128.4	131.5	133.0 c	Entreprises
Direct government	5 520.6	5 715.1	5 743.9	5 658.7	5 819.3	5 840.9	5 904.9 c	État direct
Higher education	..	..	..	..	..	..	..	Enseignement supérieur
Private non-profit	78.1	80.9	101.2	107.5	116.7	119.6	120.9 c	Institutions sans but lucratif
Funds from abroad	74.0	76.6	99.3	100.5	113.8	137.2	138.7 c	Fonds de l'étranger
Total	5 873.0 o	6 079.9 o	6 055.8 o	5 984.8 o	6 178.2 o	6 229.1 o	6 297.5 c,o	*Total*
Higher education								**Enseignement supérieur**
financed by								*financé par*
Business enterprise	579.1	587.2	677.4	713.2	772.4	843.2	856.9 c	Entreprises
Direct government	..	1 446.4	1 571.8	1 583.3	1 578.9	1 650.9	1 677.9 c	État direct
General university funds	..	5 049.6	4 993.0	4 903.5	4 830.5	4 832.4	4 911.3 c	Fonds généraux des universités
Sub-total government	6 321.7 o	6 496.1 o	6 564.8 o	6 486.9 o	6 409.5 o	6 483.4 o	6 589.1 c,o	Sous-Total État
Higher education	.. n	.. n	.. n	.. n	.. n	.. n	.. n	Enseignement supérieur
Private non-profit	.. n	.. n	.. n	.. n	.. n	.. n	.. n	Institutions sans but lucratif
Funds from abroad	88.1	75.7	107.4	124.8	148.8	128.5	130.6 c	Fonds de l'étranger
Total	6 988.8	7 158.9	7 349.7	7 324.9	7 330.6	7 455.0	7 576.6 c	*Total*
Private non-profit								**Institutions sans but lucratif**
financed by								*financé par*
Business enterprise	..	..	..	..	..	..	..	Entreprises
Direct government	..	..	..	..	..	..	..	État direct
Higher education	..	..	..	..	..	..	..	Enseignement supérieur
Private non-profit	..	..	..	..	..	..	..	Institutions sans but lucratif
Funds from abroad	..	..	..	..	..	..	..	Fonds de l'étranger
Total	..	..	..	..	..	..	..	*Total*
GERD								**DIRD**
financed by								*financé par*
Business enterprise	23 805.4 c	24 114.7	24 144.3 c	25 094.2	26 265.4 c	29 404.5	30 862.0 c	Entreprises
Direct government	..	9 467.0	9 672.9	9 776.9	9 858.9	9 900.8	10 041.8 c	État direct
General university funds	..	5 049.6	4 993.0	4 903.5	4 830.5	4 832.4	4 911.3 c	Fonds généraux des universités
Sub-total government	14 170.9 c	14 516.7	14 665.8 c	14 680.5	14 689.5 c	14 733.2	15 020.3 c	Sous-Total État
Higher education	..	..	..	..	..	..	..	Enseignement supérieur
Private non-profit	128.7 c	100.7	120.9 c	134.8	145.6 c	192.6	197.1 c	Institutions sans but lucratif
Funds from abroad	668.3 c	719.3	796.0 c	984.5	1 034.6 c	936.1	971.2 c	Fonds de l'étranger
Total	38 773.2 c,o	39 451.5 c,o	39 727.0 c,o	40 894.4 o	42 135.2 c,o	45 266.5	46 983.5 c	*Total*

Source: OECD, R&D database (DSTI/EAS Division), January 2002.

Source : OCDE, base de données R-D (DSTI/Division AES), janvier 2002.

Gross domestic expenditure on R&D (GERD) - Dépenses intérieures brutes de R-D (DIRD)

by sector of performance and type of costs - par secteur d'exécution et par type de coûts

Million EUR (1999 DEM euro, EUR from 1999) — *Millions EUR (1999 DEM euro, EUR à partir de 1999)*

	1994	1995	1996	1997	1998	1999	2000	
Business enterprise								**Entreprises**
Labour costs	..	16 044.9	..	17 786.8	..	19 925.0	..	Coûts du personnel
Other current costs	..	8 955.8	..	9 004.9	..	10 802.1	..	Autres coûts courants
Total current expenditure	..	25 000.6 [v]	..	26 791.2	..	30 727.1	..	*Total dépenses courantes*
Land and buildings	..	..	..	..	..	..	..	Terrains et constructions
Instruments and equipment	..	..	..	..	..	..	..	Instruments et équipements
Total capital expenditure	..	1 816.1 [v]	..	2 118.3	..	2 895.4	..	*Total dépenses en capital*
Total	26 173.0 [c]	27 014.1 [o]	27 405.2 [c]	28 910.0	30 334.4 [c]	33 622.6	35 110.0 [c]	*Total*
Government								**État**
Labour costs	3 016.9	3 209.0	3 287.0	3 308.8	3 427.2	3 458.8	..	Coûts du personnel
Other current costs	1 858.9	1 948.7	1 865.9	1 870.8	1 943.0	1 959.9	..	Autres coûts courants
Total current expenditure	4 875.8	5 157.7	5 153.0	5 179.6	5 370.2	5 418.7	..	*Total dépenses courantes*
Land and buildings	378.6	384.1	463.7	449.0	494.2	507.6	..	Terrains et constructions
Instruments and equipment	678.0	724.1	688.5	643.8	682.4	705.3	..	Instruments et équipements
Total capital expenditure	1 056.5	1 108.2	1 152.1	1 092.7	1 176.6	1 212.9	..	*Total dépenses en capital*
Total	5 932.3 [o]	6 265.8 [o]	6 305.1 [o]	6 272.3 [o]	6 546.9 [o]	6 631.6 [o]	6 678.0 [c,o]	*Total*
Higher education								**Enseignement supérieur**
Labour costs	4 196.1	4 382.5	4 542.1	4 644.9	4 730.4	4 864.3	..	Coûts du personnel
Other current costs	2 113.4	2 247.0	2 340.1	2 326.7	2 301.4	2 326.7	..	Autres coûts courants
Total current expenditure	6 309.5 [v]	6 629.6 [v]	6 882.1 [v]	6 971.7 [v]	7 031.8 [v]	7 191.0 [v]	..	*Total dépenses courantes*
Land and buildings	291.4	336.0	384.6	373.0	395.1	377.1	..	Terrains et constructions
Instruments and equipment	303.5	305.1	322.3	292.1	272.2	282.6	..	Instruments et équipements
Total capital expenditure	594.9 [v]	641.1 [v]	706.9 [v]	665.2 [v]	667.3 [v]	659.7 [v]	..	*Total dépenses en capital*
Total	7 059.4	7 377.8	7 652.3	7 676.8	7 768.1	7 936.6	8 034.4 [c]	*Total*
Private non-profit								**Institutions sans but lucratif**
Labour costs	..	..	..	..	..	..	..	Coûts du personnel
Other current costs	..	..	..	..	..	..	..	Autres coûts courants
Total current expenditure	..	..	..	..	..	..	..	*Total dépenses courantes*
Land and buildings	..	..	..	..	..	..	..	Terrains et constructions
Instruments and equipment	..	..	..	..	..	..	..	Instruments et équipements
Total capital expenditure	..	..	..	..	..	..	..	*Total dépenses en capital*
Total	..	..	..	..	..	..	..	*Total*
GERD								**DIRD**
Labour costs	..	23 636.4	..	25 740.6	..	28 248.1	..	Coûts du personnel
Other current costs	..	13 151.6	..	13 202.4	..	15 088.6	..	Autres coûts courants
Total current expenditure	..	36 787.9 [v]	..	38 942.4 [v]	..	43 336.7 [v]	..	*Total dépenses courantes*
Land and buildings	..	..	..	..	..	..	..	Terrains et constructions
Instruments and equipment	..	..	..	..	..	..	..	Instruments et équipements
Total capital expenditure	..	3 565.4 [v]	..	3 876.2 [v]	..	4 768.0 [v]	..	*Total dépenses en capital*
Total	39 165.0 [c,o]	40 657.8 [c,o]	41 362.7 [c,o]	42 859.1 [o]	44 649.3 [c,o]	48 190.7	49 822.4 [c]	*Total*

Table 3 - Tableau 3

Current domestic expenditure on R&D - Dépenses intérieures courantes de R-D

by sector of performance and type of activity - par secteur d'exécution et par type d'activités

Million EUR (1999 DEM euro, EUR from 1999) — *Millions EUR (1999 DEM euro, EUR à partir de 1999)*

	1994	1995	1996	1997	1998	1999	2000	
Business enterprise								**Entreprises**
Basic research	..	1 273.6	..	1 404.5	..	1 579.4	..	Recherche fondamentale
Applied research	..	..	..	..	..	..	..	Recherche appliquée
Experimental development	..	..	..	..	..	..	..	Développement expérimental
Not specified	..	23 727.0	..	25 387.2	..	29 147.7	..	Non spécifié
Total	..	25 000.6 [v]	..	26 791.2	..	30 727.1	..	*Total*
Government								**État**
Basic research	..	..	..	..	..	..	..	Recherche fondamentale
Applied research	..	..	..	..	..	..	..	Recherche appliquée
Experimental development	..	..	..	..	..	..	..	Développement expérimental
Not specified	4 875.8	5 157.4	5 153.0	5 179.6	5 370.2	5 418.7	..	Non spécifié
Total	4 875.8	5 157.7	5 153.0	5 179.6	5 370.2	5 418.7	..	*Total*
Higher education								**Enseignement supérieur**
Basic research	..	..	..	..	..	..	..	Recherche fondamentale
Applied research	..	..	..	..	..	..	..	Recherche appliquée
Experimental development	..	..	..	..	..	..	..	Développement expérimental
Not specified	6 309.5 [v]	6 629.4 [v]	6 882.1 [v]	6 971.7 [v]	7 031.8 [v]	7 191.0 [v]	..	Non spécifié
Total	6 309.5 [v]	6 629.6 [v]	6 882.1 [v]	6 971.7 [v]	7 031.8 [v]	7 191.0 [v]	..	*Total*
Private non-profit								**Institutions sans but lucratif**
Basic research	..	..	..	..	..	..	..	Recherche fondamentale
Applied research	..	..	..	..	..	..	..	Recherche appliquée
Experimental development	..	..	..	..	..	..	..	Développement expérimental
Not specified	..	..	..	..	..	..	..	Non spécifié
Total	..	..	..	..	..	..	..	*Total*
GERD								**DIRD**
Basic research	..	..	..	..	..	..	..	Recherche fondamentale
Applied research	..	..	..	..	..	..	..	Recherche appliquée
Experimental development	..	..	..	..	..	..	..	Développement expérimental
Not specified	..	36 787.5 [v]	..	38 942.4 [v]	..	43 336.7 [v]	..	Non spécifié
Total	..	36 787.9 [v]	..	38 942.4 [v]	..	43 336.7 [v]	..	*Total*

Source: OECD, R&D database (DSTI/EAS Division), January 2002.

Source : OCDE, base de données R-D (DSTI/Division AES), janvier 2002.

Table 5 - Tableau 5

Total R&D personnel - Personnel total de R-D

by sector of employment and occupation - par secteur d'emploi et par profession

In full-time equivalent | *En équivalence plein-temps*

	1994	1995	1996	1997	1998	1999	2000	
Business enterprise								**Entreprises**
RSE	..	129 370	126 392 [c]	132 687	133 529 [c]	150 150	..	Chercheurs
Technicians	..	78 155	76 356 [c]	79 016	79 518 [c]	80 165	..	Techniciens
Other occupations	..	75 791	74 046 [c]	74 567	75 043 [c]	76 378	..	Autres professions
Total	..	283 316	276 794 [c]	286 270	288 090 [c]	306 693	..	*Total*
Government								**État**
RSE	35 383	37 324	37 687	37 402	38 210	38 415	38 500 [c]	Chercheurs
Technicians	20 179	20 380	20 220	19 364	18 618	17 533	17 300 [c]	Techniciens
Other occupations	17 263	17 444	16 818	16 729	16 541	16 303	16 200 [c]	Autres professions
Total	72 825	75 148	74 725	73 495	73 369	72 251	72 000 [c]	*Total*
Higher education								**Enseignement supérieur**
RSE	..	64 434	66 110	65 704	65 973	66 695	67 504 [c]	Chercheurs
Technicians	..	13 636	13 578	13 369	12 929	12 666	..	Techniciens
Other occupations	..	22 604	22 472	21 573	21 179	22 110	..	Autres professions
Total	..	100 674	102 160	100 646	100 080	101 471	103 171 [c]	*Total*
Private non-profit								**Institutions sans but lucratif**
RSE	..	..	..	..	..	..	..	Chercheurs
Technicians	..	..	..	..	..	..	..	Techniciens
Other occupations	..	..	..	..	..	..	..	Autres professions
Total	..	..	..	..	..	..	..	*Total*
TOTAL								**TOTAL**
RSE	..	231 128	230 189 [c]	235 793	237 712 [c]	255 260	..	Chercheurs
Technicians	..	112 171	110 154 [c]	111 749	111 065 [c]	110 364	..	Techniciens
Other occupations	..	115 839	113 336 [c]	112 869	112 763 [c]	114 791	..	Autres professions
Total	..	459 138	453 679 [c]	460 411	461 539 [c]	480 415	..	*Total*

Table 6 - Tableau 6

Total R&D personnel - Personnel total de R-D

by sector of employment and formal qualification - par secteur d'emploi et par niveau de qualification

In full-time equivalent | *En équivalence plein-temps*

	1994	1995	1996	1997	1998	1999	2000	
Business enterprise								**Entreprises**
University PhD level degrees	..	..	..	..	..	..	..	Diplômes univ. de niv. doctorat
Other university degrees	..	..	..	..	..	..	..	Autres diplômes universitaires
Sub-total university degrees	..	129 370 [o]	126 392 [c,o]	132 687 [o]	133 529 [c,o]	150 150 [o]	..	Sous-total Universitaires
Other post-secondary diplomas	..	.. [n]	.. [n]	.. [n]	.. [n]	.. [n]	..	Autres diplômes post-secondaires
Secondary diplomas	..	78 155	76 356 [c]	79 016	79 518 [c]	80 165	..	Diplômes de fin d'études secondaires
Other qualifications	..	75 791	74 046 [c]	74 567	75 043 [c]	76 378	..	Autres qualifications
Total	..	283 316	276 794 [c]	286 270	288 090 [c]	306 693	..	*Total*
Government								**État**
University PhD level degrees	..	..	..	..	..	..	..	Diplômes univ. de niv. doctorat
Other university degrees	..	..	..	..	..	..	..	Autres diplômes universitaires
Sub-total university degrees	35 383 [o]	37 324 [o]	37 687 [o]	37 402 [o]	38 210 [o]	38 415 [o]	38 500 [c,o]	Sous-total Universitaires
Other post-secondary diplomas	.. [n]	.. [n]	.. [n]	.. [n]	.. [n]	.. [n]	.. [n]	Autres diplômes post-secondaires
Secondary diplomas	20 179	20 380	20 220	19 364	18 618	17 533	17 300 [c]	Diplômes de fin d'études secondaires
Other qualifications	17 263	17 444	16 818	16 729	16 541	16 303	16 200 [c]	Autres qualifications
Total	72 825	75 148	74 725	73 495	73 369	72 251	72 000 [c]	*Total*
Higher education								**Enseignement supérieur**
University PhD level degrees	..	..	..	..	..	..	..	Diplômes univ. de niv. doctorat
Other university degrees	..	..	..	..	..	..	..	Autres diplômes universitaires
Sub-total university degrees	..	64 434 [o]	66 110 [o]	65 704 [o]	65 973 [o]	66 695 [o]	67 504 [c,o]	Sous-total Universitaires
Other post-secondary diplomas	..	.. [n]	.. [n]	.. [n]	.. [n]	.. [n]	.. [n]	Autres diplômes post-secondaires
Secondary diplomas	..	13 636	13 578	13 369	12 929	12 666	..	Diplômes de fin d'études secondaires
Other qualifications	..	22 604	22 472	21 573	21 179	22 110	35 668 [c,o]	Autres qualifications
Total	..	100 674	102 160	100 646	100 080	101 471	103 171 [c]	*Total*
Private non-profit								**Institutions sans but lucratif**
University PhD level degrees	..	..	..	..	..	..	..	Diplômes univ. de niv. doctorat
Other university degrees	..	..	..	..	..	..	..	Autres diplômes universitaires
Sub-total university degrees	..	..	..	..	..	..	..	Sous-total Universitaires
Other post-secondary diplomas	..	..	..	..	..	..	..	Autres diplômes post-secondaires
Secondary diplomas	..	..	..	..	..	..	..	Diplômes de fin d'études secondaires
Other qualifications	..	..	..	..	..	..	..	Autres qualifications
Total	..	..	..	..	..	..	..	*Total*
TOTAL								**TOTAL**
University PhD level degrees	..	..	..	..	..	..	..	Diplômes univ. de niv. doctorat
Other university degrees	..	..	..	..	..	..	..	Autres diplômes universitaires
Sub-total university degrees	..	231 128 [o]	230 189 [c,o]	235 793 [o]	237 712 [c,o]	255 260 [o]	..	Sous-total Universitaires
Other post-secondary diplomas	..	.. [n]	.. [n]	.. [n]	.. [n]	.. [n]	..	Autres diplômes post-secondaires
Secondary diplomas	..	112 171	110 154 [c]	111 749	111 065 [c]	110 364	..	Diplômes de fin d'études secondaires
Other qualifications	..	115 839	113 336 [c]	112 869	112 763 [c]	114 791	..	Autres qualifications
Total	..	459 138	453 679 [c]	460 411	461 539 [c]	480 415	..	*Total*

Source: OECD, R&D database (DSTI/EAS Division), January 2002.

Source : OCDE, base de données R-D (DSTI/Division AES), janvier 2002.

Table 7 - Tableau 7

Gross domestic expenditure on R&D (GERD) - Dépenses intérieures brutes de R-D (DIRD)

by sector of performance and main scientific field - par secteur d'exécution et par grand domaine scientifique

Million EUR (1999 DEM euro, EUR from 1999) — *Millions EUR (1999 DEM euro, EUR à partir de 1999)*

	1994	1995	1996	1997	1998	1999	2000	
Business enterprise								**Entreprises**
Natural sciences and engineering	26 173.0 [c]	27 014.1 [o]	27 405.2 [c,v]	28 910.0 [c]	30 334.4 [c]	33 622.6	35 110.0 [c]	Sciences exactes, nat. et de l'ingénieur
Social sciences and humanities	..	..	..	..	..	..	..	Sciences sociales et humaines
All fields of science	26 173.0 [c]	27 014.1 [o]	27 405.2 [c]	28 910.0	30 334.4 [c]	33 622.6	35 110.0 [c]	*Tous domaines scientifiques*
Government								**État**
Natural sciences and engineering	5 211.5	5 500.5	5 550.8	5 511.2	5 756.8	5 831.3	..	Sciences exactes, nat. et de l'ingénieur
Social sciences and humanities	720.8	765.4	754.3	761.1	790.0	800.3	..	Sciences sociales et humaines
All fields of science	5 932.3 [o]	6 265.8 [o]	6 305.1 [o]	6 272.3 [o]	6 546.9 [o]	6 631.6 [o]	6 678.0 [c,o]	*Tous domaines scientifiques*
Higher education								**Enseignement supérieur**
Natural sciences and engineering	5 505.9 [v]	5 747.0 [v]	5 965.4 [v]	5 980.6 [v]	6 041.0 [v]	6 219.6 [v]	..	Sciences exactes, nat. et de l'ingénieur
Social sciences and humanities	1 398.5 [v]	1 523.7 [v]	1 623.6 [v]	1 656.3 [v]	1 658.1 [v]	1 631.0 [v]	..	Sciences sociales et humaines
All fields of science	7 059.4	7 377.8	7 652.3	7 676.8	7 768.1	7 936.6	8 034.4 [c]	*Tous domaines scientifiques*
Private non-profit								**Institutions sans but lucratif**
Natural sciences and engineering	..	..	..	..	..	..	..	Sciences exactes, nat. et de l'ingénieur
Social sciences and humanities	..	..	..	..	..	..	..	Sciences sociales et humaines
All fields of science	..	..	..	..	..	..	..	*Tous domaines scientifiques*
TOTAL								**TOTAL**
Natural sciences and engineering	36 890.4 [v]	38 261.7 [v]	38 921.5 [v]	40 401.8 [v]	42 132.2 [v]	45 673.4 [v]	..	Sciences exactes, nat. et de l'ingénieur
Social sciences and humanities	2 119.3 [v]	2 289.0 [v]	2 377.8 [v]	2 417.4 [v]	2 448.1 [v]	2 431.3 [v]	..	Sciences sociales et humaines
All fields of science	39 165.0 [c,o]	40 657.8 [c,o]	41 362.7 [c,o]	42 859.1 [o]	44 649.3 [c,o]	48 190.7	49 822.4 [c]	*Tous domaines scientifiques*

Table 8 - Tableau 8

Total R&D personnel - Personnel total de R-D

by sector of employment and main scientific field - par secteur d'emploi et par grand domaine scientifique

In full-time equivalent — *En équivalence plein-temps*

	1994	1995	1996	1997	1998	1999	2000	
Business enterprise								**Entreprises**
Natural sciences and engineering	..	283 316	276 794 [c]	286 270	288 090 [c]	306 693	..	Sciences exactes, nat. et de l'ingénieur
Social sciences and humanities	..	..	..	..	..	..	..	Sciences sociales et humaines
All fields of science	..	283 316	276 794 [c]	286 270	288 090 [c]	306 693	..	*Tous domaines scientifiques*
Government								**État**
Natural sciences and engineering	62 656	64 111	63 981	62 716	63 139	62 729	..	Sciences exactes, nat. et de l'ingénieur
Social sciences and humanities	10 169	11 037	10 744	10 779	10 231	9 522	..	Sciences sociales et humaines
All fields of science	72 825	75 148	74 725	73 495	73 369	72 251	72 000 [c]	*Tous domaines scientifiques*
Higher education								**Enseignement supérieur**
Natural sciences and engineering	..	78 471	79 409	78 039	77 335	78 667	..	Sciences exactes, nat. et de l'ingénieur
Social sciences and humanities	..	22 204	22 751	22 605	22 744	22 801	..	Sciences sociales et humaines
All fields of science	..	100 674	102 160	100 646	100 080	101 471	103 171 [c]	*Tous domaines scientifiques*
Private non-profit								**Institutions sans but lucratif**
Natural sciences and engineering	..	..	..	..	..	..	..	Sciences exactes, nat. et de l'ingénieur
Social sciences and humanities	..	..	..	..	..	..	..	Sciences sociales et humaines
All fields of science	..	..	..	..	..	..	..	*Tous domaines scientifiques*
TOTAL								**TOTAL**
Natural sciences and engineering	..	425 898	420 184 [c]	427 025	428 564 [c]	448 089	..	Sciences exactes, nat. et de l'ingénieur
Social sciences and humanities	..	33 241	33 495	33 384	32 975	32 323	..	Sciences sociales et humaines
All fields of science	..	459 138	453 679 [c]	460 411	461 539 [c]	480 415	..	*Tous domaines scientifiques*

Source: OECD, R&D database (DSTI/EAS Division), January 2002. *Source* : OCDE, base de données R-D (DSTI/Division AES), janvier 2002.

GERMANY — Table 9 - Tableau 9 — ALLEMAGNE

Total business enterprise intramural expenditure on R&D - Dépenses intra-muros totales de R-D des entreprises

(BERD) by industry - (DIRDE) par industrie

	1994	1995	1996	1997	1998	1999	2000	
	9.1 Million EUR (1999 DEM euro, EUR from 1999) - Millions EUR (1999 DEM euro, EUR à partir de 1999)							
1 AGRICULTURE, HUNTING & FORESTRY	..	57.8 [v]	..	65.4	..	81.3	..	**1 AGRICULTURE, CHASSE & SYLVICULTURE**
2 MINING	..	72.6 [v]	..	73.1	..	55.2	..	**2 ACTIVITES EXTRACTIVES**
3 MANUFACTURING	..	25 544.7 [v]	..	27 020.8	..	30 550.2	32 120.0 [c]	**3 ACTIVITES DE FABRICATION**
4 Food, beverages & tobacco	..	217.3 [v]	..	190.2	..	217.8	..	**4 Produits alimentaires, boissons & tabac**
5 Food products & beverages	..	187.6 [v]	..	173.3	..	188.7	..	5 Produits alimentaires & boissons
6 Tobacco products	..	29.7 [v]	..	16.9	..	29.1	..	6 Produits à base de tabac
7 Textiles, wear app., fur & leather	..	161.6 [v]	..	190.2	..	210.7	..	**7 Textiles, habill., fourrures & cuirs**
8 Textiles	..	139.6 [v]	..	153.4	..	173.3	..	8 Textiles
9 Wearing apparel & fur	..	14.8 [v]	..	30.7	..	33.7	..	9 Articles d'habillement & fourrures
10 Leather products & footwear	..	7.2 [v]	..	6.1	..	3.6	..	10 Articles en cuir & chaussures
11 Wood, paper, printing, publishing	..	134.5 [v]	..	111.5	..	131.4	..	**11 Bois, papier, imprimerie, édition**
12 Wood & cork (not furniture)	..	30.2 [v]	..	26.1	..	26.1	..	12 Bois & liège (sauf meubles)
13 Pulp, paper & paper products	..	78.2 [v]	..	60.3	..	80.3	..	13 Carton, papier & articles
14 Publ., print. & repro. of rec. media	..	26.1 [v]	..	24.5	..	24.5	..	14 Edition, impress. & repro. enregis.
15 Coke, petroleum, nuclear fuel, chemicals, rubber plastics	..	5 306.7 [v]	..	5 987.2	..	6 357.9	..	**15 Coke, pétrole, combust. nucl., prod. chimi., caout. & mat. plast.**
16 Coke, ref. petrol. prod. & nuclear fuel	..	66.0 [v]	..	75.7	..	47.6	..	16 Coke, raffin. pétr. & combust. nucl.
17 Chemicals & chemical products	..	4 847.0 [v]	..	5 429.9	..	5 694.3	..	17 Produits chimiques
18 *Chemical prod.(less pharmaceuticals)*	..	3 596.9 [v]	..	3 536.6	..	3 604.6	..	18 *Prod. chimi. (moins prod. pharmaceutiques)*
19 *Pharmaceuticals*	..	1 250.1 [v]	..	1 892.8	..	2 089.7	..	19 *Produits pharmaceutiques*
20 Rubber & plastic products	..	393.7 [v]	..	481.6	..	616.1	..	20 Caoutchouc et matières plastiques
21 Non-metallic mineral products	..	275.6 [v]	..	265.9	..	335.4	..	**21 Produits minéraux non métalliques**
22 Basic metals	..	270.0 [v]	..	275.6	..	259.2	..	**22 Produits métallurgiques de base**
23 Basic metals, ferrous	..	190.7 [v]	..	188.2	..	179.0	..	23 Produits métallur. de base, ferreux
24 Basic metals, non-ferrous	..	79.3 [v]	..	87.9	..	80.3	..	24 Prod. métallur. de base, non ferreux
25 Fabricated metal products	..	386.0 [v]	..	432.6	..	516.9	..	**25 Ouvrages en métaux (sauf mach.& mat.)**
26 Machinery equipment, instruments and transport equipment	..	18 632.0 [v]	..	19 427.0	..	22 345.5	..	**26 Machines & matériel, instruments & mat. de transport**
27 Machinery n.e.c.	..	3 062.1 [v]	..	3 191.0	..	3 403.2	..	27 Machines, n.c.a.
28 Office, account. & computing machinery	..	1 052.2 [v]	..	673.4	..	633.5	..	28 Mach. de bur., compta. & trait. inform.
29 Electrical machinery	..	1 942.4 [v]	..	876.9	..	1 021.1	..	29 Machines électriques
30 Electro. equip. (radio, TV & commun.)	..	2 707.3 [v]	..	3 276.4	..	3 642.4	..	30 Equip. électro. (radio, tv & communic.)
31 *Electro. comp. (inc. semi-conductors)*	..	654.5 [v]	..	1 034.9	..	1 207.2	..	31 *Compos. électro. (semi-cond. incl.)*
32 *TV, radio & communications equip.*	..	2 052.8 [v]	..	2 241.5	..	2 435.3	..	32 *Appareils radio, tv et de communi.*
33 Instruments	..	1 616.2 [v]	..	1 499.6	..	1 664.8	..	33 Instruments
34 Motor vehicles	..	5 723.9 [v]	..	6 991.9	..	9 420.0	..	34 Véhicules automobiles
35 Other transport equipment	..	2 527.8 [v]	..	2 919.0	..	2 560.6	..	35 Autres matériels de transport
36 *Ships*	..	75.2 [v]	..	65.4	..	36.3	..	36 *Navires*
37 *Aerospace*	..	2 196.0 [v]	..	2 466.0	..	2 234.9	..	37 *Aérospatiale*
38 *Other transport equipment n.e.c.*	..	256.7 [v]	..	387.6	..	289.4	..	38 *Autres matériels de transport n.c.a.*
39 Furniture, other manufacturing n.e.c.	..	155.4 [v]	..	136.5	..	171.3	..	**39 Meubles, autres activ. de fabricat. n.c.a.**
40 Furniture	..	96.1 [v]	..	77.7	..	112.0	..	40 Meubles
41 Other manufacturing n.e.c.	..	59.3 [v]	..	58.8	..	59.3	..	41 Autres activités de fabrication n.c.a.
42 Recycling	..	5.1 [v]	..	4.1	..	4.6	..	**42 Récupération**
43 ELECTRICITY, GAS & WATER SUPPLY	..	114.0 [v]	..	91.5	..	106.9	..	**43 DISTRIBUTION D'ELECTRICITE, GAZ & EAU**
44 CONSTRUCTION	..	74.1 [v]	..	93.1	..	87.9	..	**44 CONSTRUCTION**
45 SERVICES SECTOR	..	953.0 [v]	..	1 566.1	..	2 741.0	..	**45 SECTEUR DES SERVICES**
46 Wholesale, retail trade & repair of motor vehicles, etc.	..	33.2 [v]	..	38.3	..	130.9	..	**46 Commerce (gros & détail), réparation de véhicules, etc.**
47 Hotels & restaurants	..	..	..	..	..	0.0	..	**47 Hôtels et restaurants**
48 Transport & storage	..	67.0 [v]	..	68.0	..	..	..	**48 Transports, entreposage**
49 Communications	..	..	..	..	..	..	..	**49 Communications**
50 Post	..	..	..	..	..	..	..	50 Postes
51 Telecommunications	..	..	..	..	..	..	..	51 Télécommunications
52 Financial intermediation (inc. insur.)	..	27.1 [v]	..	7.7	..	12.8	..	**52 Intermédiation financ. (assurances inc.)**
53 Real estate, renting & busin. activity	..	683.1 [v]	..	1 307.4	..	2 173.5	..	**53 Immobil., locat. & serv. aux entreprises**
54 Computer & related activities	..	112.0 [v]	..	496.5	..	865.6	..	54 Informatique et activités connexes
55 *Software consultancy*	..	88.5 [v]	..	460.7	..	649.9	..	55 *Production de logiciels*
56 *Other computer services n.e.c.*	..	23.5 [v]	..	35.8	..	216.3	..	56 *Autres activités informatiques*
57 Research & development	..	182.5 [v]	..	418.7	..	861.0	..	57 Recherche & développement expérimental
58 Other business activities n.e.c.	..	388.6 [v]	..	391.7	..	446.9	..	58 Autres activ. & serv. aux entreprises
59 Comm., soc. & pers. serv. activ., etc.	..	22.5 [v]	..	15.3	..	12.3	..	**59 Adm. pub., serv. soc. & collec., etc**
TOTAL BUSINESS ENTERPRISE SECTOR	**26 173.0 [c]**	**27 014.1 [o]**	**27 405.2 [c]**	**28 910.0**	**30 334.4 [c]**	**33 622.6**	**35 110.0 [c]**	**TOTAL SECTEUR DES ENTREPRISES**

Source: OECD, R&D database (DSTI/EAS Division), January 2002.

Source : OCDE, base de données R-D (DSTI/Division AES), janvier 2002.

Total business enterprise intramural expenditure on R&D - Dépenses intra-muros totales de R-D des entreprises
(BERD) by industry - (DIRDE) par industrie

	1994	1995	1996	1997	1998	1999	2000	
	9.2 Million constant USD(1995 prices and PPPs) - Millions de USD constants (prix et PPA de 1995)							
1 AGRICULTURE, HUNTING & FORESTRY	..	56.1 v	..	62.4	..	76.4	..	**1 AGRICULTURE, CHASSE & SYLVICULTURE**
2 MINING	..	70.4 v	..	69.8	..	51.9	..	**2 ACTIVITES EXTRACTIVES**
3 MANUFACTURING	..	24 786.7 v	..	25 782.1	..	28 696.4	30 289.8 c	**3 ACTIVITES DE FABRICATION**
4 Food, beverages & tobacco	..	210.9 v	..	181.5	..	204.6	..	**4 Produits alimentaires, boissons & tabac**
5 Food products & beverages	..	182.1 v	..	165.4	..	177.2	..	5 Produits alimentaires & boissons
6 Tobacco products	..	28.8 v	..	16.1	..	27.4	..	6 Produits à base de tabac
7 Textiles, wear app., fur & leather	..	156.8 v	..	181.5	..	197.9	..	**7 Textiles, habill., fourrures & cuirs**
8 Textiles	..	135.4 v	..	146.4	..	162.8	..	8 Textiles
9 Wearing apparel & fur	..	14.4 v	..	29.3	..	31.7	..	9 Articles d'habillement & fourrures
10 Leather products & footwear	..	6.9 v	..	5.9	..	3.4	..	10 Articles en cuir & chaussures
11 Wood, paper, printing, publishing	..	130.5 v	..	106.4	..	123.4	..	**11 Bois, papier, imprimerie, édition**
12 Wood & cork (not furniture)	..	29.3 v	..	24.9	..	24.5	..	12 Bois & liège (sauf meubles)
13 Pulp, paper & paper products	..	75.9 v	..	57.6	..	75.4	..	13 Carton, papier & articles
14 Publ., print. & repro. of rec. media	..	25.3 v	..	23.4	..	23.1	..	14 Edition, impress. & repro. enregis.
15 Coke, petroleum, nuclear fuel, chemicals, rubber plastics	..	5 149.2 v	..	5 712.8	..	5 972.1	..	**15 Coke, pétrole, combust. nucl., prod. chimi., caout. & mat. plast.**
16 Coke, ref. petrol. prod. & nuclear fuel	..	64.0 v	..	72.2	..	44.7	..	16 Coke, raffin. pétr. & combust. nucl.
17 Chemicals & chemical products	..	4 703.2 v	..	5 181.0	..	5 348.7	..	17 Produits chimiques
18 *Chemical prod.(less pharmaceuticals)*	..	3 490.2 v	..	3 374.5	..	3 385.9	..	18 *Prod. chimi. (moins prod. pharmaceutiques)*
19 *Pharmaceuticals*	..	1 213.0 v	..	1 806.0	..	1 962.8	..	19 *Produits pharmaceutiques*
20 Rubber & plastic products	..	382.0 v	..	459.6	..	578.7	..	20 Caoutchouc et matières plastiques
21 Non-metallic mineral products	..	267.4 v	..	253.7	..	315.1	..	**21 Produits minéraux non métalliques**
22 Basic metals	..	262.0 v	..	263.0	..	243.5	..	**22 Produits métallurgiques de base**
23 Basic metals, ferrous	..	185.1 v	..	179.5	..	168.1	..	23 Produits métallur. de base, ferreux
24 Basic metals, non-ferrous	..	76.9 v	..	83.9	..	75.4	..	24 Prod. métallur. de base, non ferreux
25 Fabricated metal products	..	374.6 v	..	412.7	..	485.5	..	**25 Ouvrages en métaux (sauf mach.& mat.)**
26 Machinery equipment, instruments and transport equipment	..	18 079.2 v	..	18 536.5	..	20 989.6	..	**26 Machines & matériel, instruments & mat. de transport**
27 Machinery n.e.c.	..	2 971.3 v	..	3 044.7	..	3 196.7	..	27 Machines, n.c.a.
28 Office, account. & computing machinery	..	1 021.0 v	..	642.5	..	595.1	..	28 Mach. de bur., compta. & trait. inform.
29 Electrical machinery	..	1 884.8 v	..	836.7	..	959.1	..	29 Machines électriques
30 Electro. equip. (radio, TV & commun.)	..	2 627.0 v	..	3 126.2	..	3 421.4	..	30 Equip. électro. (radio, tv & communic.)
31 *Electro. comp. (inc. semi-conductors)*	..	635.0 v	..	987.4	..	1 133.9	..	31 *Compos. électro. (semi-cond. incl.)*
32 *TV, radio & communications equip.*	..	1 991.9 v	..	2 138.8	..	2 287.5	..	32 *Appareils radio, tv et de communi.*
33 Instruments	..	1 568.2 v	..	1 430.9	..	1 563.7	..	33 Instruments
34 Motor vehicles	..	5 554.1 v	..	6 671.4	..	8 848.4	..	34 Véhicules automobiles
35 Other transport equipment	..	2 452.8 v	..	2 785.2	..	2 405.2	..	35 Autres matériels de transport
36 *Ships*	..	72.9 v	..	62.4	..	34.1	..	36 *Navires*
37 *Aerospace*	..	2 130.8 v	..	2 352.9	..	2 099.2	..	37 *Aérospatiale*
38 *Other transport equipment n.e.c.*	..	249.1 v	..	369.8	..	271.8	..	38 *Autres matériels de transport n.c.a.*
39 Furniture, other manufacturing n.e.c.	..	150.8 v	..	130.3	..	160.9	..	**39 Meubles, autres activ. de fabricat. n.c.a.**
40 Furniture	..	93.3 v	..	74.2	..	105.2	..	40 Meubles
41 Other manufacturing n.e.c.	..	57.6 v	..	56.1	..	55.7	..	41 Autres activités de fabrication n.c.a.
42 Recycling	..	5.0 v	..	3.9	..	4.3	..	**42 Récupération**
43 ELECTRICITY, GAS & WATER SUPPLY	..	110.6 v	..	87.3	..	100.4	..	**43 DISTRIBUTION D'ELECTRICITE, GAZ & EAU**
44 CONSTRUCTION	..	71.9 v	..	88.8	..	82.6	..	**44 CONSTRUCTION**
45 SERVICES SECTOR	..	924.8 v	..	1 494.3	..	2 574.7	..	**45 SECTEUR DES SERVICES**
46 Wholesale, retail trade & repair of motor vehicles, etc.	..	32.2 v	..	36.6	..	122.9	..	**46 Commerce (gros & détail), réparation de véhicules, etc.**
47 Hotels & restaurants	..	..	..	..	..	0.0	..	**47 Hôtels et restaurants**
48 Transport & storage	..	65.0 v	..	64.9	..	..	..	**48 Transports, entreposage**
49 Communications	..	..	..	..	..	..	..	**49 Communications**
50 Post	..	..	..	..	..	..	..	50 Postes
51 Telecommunications	..	..	..	..	..	..	..	51 Télécommunications
52 Financial intermediation (inc. insur.)	..	26.3 v	..	7.3	..	12.0	..	**52 Intermédiation financ. (assurances inc.)**
53 Real estate, renting & busin. activity	..	662.8 v	..	1 247.4	..	2 041.6	..	**53 Immobil., locat. & serv. aux entreprises**
54 Computer & related activities	..	108.7 v	..	473.7	..	813.1	..	54 Informatique et activités connexes
55 *Software consultancy*	..	85.8 v	..	439.6	..	610.4	..	55 *Production de logiciels*
56 *Other computer services n.e.c.*	..	22.8 v	..	34.2	..	203.2	..	56 *Autres activités informatiques*
57 Research & development	..	177.1 v	..	399.6	..	808.8	..	57 Recherche & développement expérimental
58 Other business activities n.e.c.	..	377.1 v	..	373.7	..	419.8	..	58 Autres activ. & serv. aux entreprises
59 Comm., soc. & pers. serv. activ., etc.	..	21.8 v	..	14.6	..	11.5	..	**59 Adm. pub., serv. soc. & collec., etc**
TOTAL BUSINESS ENTERPRISE SECTOR	25 911.2 c	26 212.6 o	26 321.5 c	27 584.8	28 626.3 c	31 582.3	33 109.4 c	**TOTAL SECTEUR DES ENTREPRISES**

Source: OECD, R&D database (DSTI/EAS Division), January 2002.

Source : OCDE, base de données R-D (DSTI/Division AES), janvier 2002.

Business enterprise intramural expenditure on R&D (BERD) - Dépenses intra-muros de R-D des entreprises (DIRDE)

by industry and source of funds - par industrie et par source de financement

Million EUR (1999 DEM euro, EUR from 1999) — *Millions EUR (1999 DEM euro, EUR à partir de 1999)*

10.1 Business enterprise - Entreprises

	1994	1995	1996	1997	1998	1999	2000	
1 AGRICULTURE, HUNTING & FORESTRY	..	54.7	..	62.9	..	77.2	..	**1 AGRICULTURE, CHASSE & SYLVICULTURE**
2 MINING	..	66.0	..	71.1	..	52.2	..	**2 ACTIVITES EXTRACTIVES**
3 MANUFACTURING	..	23 129.8	..	24 015.9	..	27 664.0	..	**3 ACTIVITES DE FABRICATION**
4 Food, beverages & tobacco	..	200.4	..	182.0	..	203.5	..	**4 Produits alimentaires, boissons & tabac**
5 Food products & beverages	..	170.8	..	165.1	..	175.4	..	5 Produits alimentaires & boissons
6 Tobacco products	..	29.1	..	16.9	..	28.1	..	6 Produits à base de tabac
7 Textiles, wear app., fur & leather	..	127.3	..	153.4	..	167.7	..	**7 Textiles, habill., fourrures & cuirs**
8 Textiles	..	108.4	..	119.1	..	134.5	..	8 Textiles
9 Wearing apparel & fur	..	12.8	..	29.1	..	31.2	..	9 Articles d'habillement & fourrures
10 Leather products & footwear	..	6.1	..	5.6	..	2.6	..	10 Articles en cuir & chaussures
11 Wood, paper, printing, publishing	..	125.3	..	102.3	..	118.1	..	**11 Bois, papier, imprimerie, édition**
12 Wood & cork (not furniture)	..	26.6	..	23.0	..	21.0	..	12 Bois & liège (sauf meubles)
13 Pulp, paper & paper products	..	75.7	..	57.8	..	76.7	..	13 Carton, papier & articles
14 Publ., print. & repro. of rec. media	..	23.0	..	21.5	..	20.5	..	14 Edition, impress. & repro. enregis.
15 Coke, petroleum, nuclear fuel, chemicals, rubber plastics	..	5 191.1	..	5 776.6	..	6 117.1	..	**15 Coke, pétrole, combust. nucl., prod. chimi., caout. & mat. plast.**
16 Coke, ref. petrol. prod. & nuclear fuel	..	48.1	..	55.2	..	45.5	..	16 Coke, raffin. pétr. & combust. nucl.
17 Chemicals & chemical products	..	4 769.8	..	5 262.7	..	5 498.9	..	17 Produits chimiques
18 *Chemical prod.(less pharmaceuticals)*	..	3 532.0	..	3 407.8	..	3 460.9	..	18 *Prod. chimi. (moins prod. pharmaceutiques)*
19 *Pharmaceuticals*	..	1 237.3	..	1 855.0	..	2 037.5	..	19 *Produits pharmaceutiques*
20 Rubber & plastic products	..	373.2	..	458.6	..	572.6	..	20 Caoutchouc et matières plastiques
21 Non-metallic mineral products	..	257.2	..	244.9	..	312.9	..	**21 Produits minéraux non métalliques**
22 Basic metals	..	241.8	..	237.8	..	219.9	..	**22 Produits métallurgiques de base**
23 Basic metals, ferrous	..	170.8	..	157.5	..	148.8	..	23 Produits métallur. de base, ferreux
24 Basic metals, non-ferrous	..	71.6	..	80.3	..	71.1	..	24 Prod. métallur. de base, non ferreux
25 Fabricated metal products	..	353.3	..	404.9	..	481.1	..	**25 Ouvrages en métaux (sauf mach.& mat.)**
26 Machinery equipment, instruments and transport equipment	..	16 473.3	..	16 775.0	..	19 871.9	..	**26 Machines & matériel, instruments & mat. de transport**
27 Machinery n.e.c.	..	2 952.7	..	3 055.5	..	3 229.3	..	27 Machines, n.c.a.
28 Office, account. & computing machinery	..	1 043.0	..	497.5	..	624.8	..	28 Mach. de bur., compta. & trait. inform.
29 Electrical machinery	..	1 907.6	..	853.9	..	990.4	..	29 Machines électriques
30 Electro. equip. (radio, TV & commun.)	..	2 445.0	..	2 933.8	..	3 237.5	..	30 Equip. électro. (radio, tv & communic.)
31 *Electro. comp. (inc. semi-conductors)*	..	538.9	..	896.3	..	1 024.1	..	31 *Compos. électro. (semi-cond. incl.)*
32 *TV, radio & communications equip.*	..	1 906.1	..	2 037.5	..	2 213.4	..	32 *Appareils radio, tv et de communi.*
33 Instruments	..	1 430.6	..	1 412.2	..	1 522.6	..	33 Instruments
34 Motor vehicles	..	5 672.3	..	6 909.6	..	9 328.0	..	34 Véhicules automobiles
35 Other transport equipment	..	1 022.1	..	1 112.6	..	939.2	..	35 Autres matériels de transport
36 *Ships*	..	59.3	..	54.7	..	27.6	..	36 *Navires*
37 *Aerospace*	..	712.2	..	677.0	..	627.4	..	37 *Aérospatiale*
38 *Other transport equipment n.e.c.*	..	250.5	..	380.9	..	284.3	..	38 *Autres matériels de transport n.c.a.*
39 Furniture, other manufacturing n.e.c.	..	154.4	..	135.0	..	168.2	..	**39 Meubles, autres activ. de fabricat. n.c.a.**
40 Furniture	..	96.1	..	77.7	..	110.4	..	40 Meubles
41 Other manufacturing n.e.c.	..	58.8	..	57.8	..	57.8	..	41 Autres activités de fabrication n.c.a.
42 Recycling	..	4.6	..	4.1	..	3.6	..	**42 Récupération**
43 ELECTRICITY, GAS & WATER SUPPLY	..	101.2	..	74.1	..	94.1	..	**43 DISTRIBUTION D'ELECTRICITE, GAZ & EAU**
44 CONSTRUCTION	..	72.6	..	89.0	..	80.3	..	**44 CONSTRUCTION**
45 SERVICES SECTOR	..	609.5	..	1 115.6	..	2 299.3	..	**45 SECTEUR DES SERVICES**
46 Wholesale, retail trade & repair of motor vehicles, etc.	..	30.7	..	35.8	..	129.9	..	**46 Commerce (gros & détail), réparation de véhicules, etc.**
47 Hotels & restaurants	..	..	..	..	..	0.0	..	**47 Hôtels et restaurants**
48 Transport & storage	..	16.4	..	44.0	..	44.0	..	**48 Transports, entreposage**
49 Communications	..	..	..	..	..	354.3	..	**49 Communications**
50 Post	..	..	..	..	..	..	..	50 Postes
51 Telecommunications	..	..	..	..	..	..	..	51 Télécommunications
52 Financial intermediation (inc. insur.)	..	27.1	..	7.7	..	12.8	..	**52 Intermédiation financ. (assurances inc.)**
53 Real estate, renting & busin. activity	..	394.7	..	892.2	..	1 757.8	..	**53 Immobil., locat. & serv. aux entreprises**
54 Computer & related activities	..	95.6	..	468.9	..	823.2	..	54 Informatique et activités connexes
55 *Software consultancy*	..	78.2	..	446.4	..	621.2	..	55 *Production de logiciels*
56 *Other computer services n.e.c.*	..	16.9	..	22.0	..	202.0	..	56 *Autres activités informatiques*
57 Research & development	..	94.6	..	204.0	..	638.6	..	57 Recherche & développement expérimental
58 Other business activities n.e.c.	..	204.5	..	219.3	..	295.5	..	58 Autres activ. & serv. aux entreprises
59 Comm., soc. & pers. serv. activ., etc.	..	20.5	..	8.2	..	1.0	..	**59 Adm. pub., serv. soc. & collec., etc**
TOTAL BUSINESS ENTERPRISE SECTOR	**23 258.7 [c]**	**24 033.3**	**24 317.0 [c]**	**25 428.6**	**26 878.1 [c]**	**30 266.4**	**31 677.1 [c]**	**TOTAL SECTEUR DES ENTREPRISES**

Source: OECD, R&D database (DSTI/EAS Division), January 2002.

Source : OCDE, base de données R-D (DSTI/Division AES), janvier 2002.

Business enterprise intramural expenditure on R&D (BERD) - Dépenses intra-muros de R-D des entreprises (DIRDE)

by industry and source of funds - par industrie et par source de financement

Million EUR (1999 DEM euro, EUR from 1999) — *Millions EUR (1999 DEM euro, EUR à partir de 1999)*

10.2 Government - État

	1994	1995	1996	1997	1998	1999	2000	
1 AGRICULTURE, HUNTING & FORESTRY	..	3.1 [v]	..	2.6 [v]	..	4.1	..	1 AGRICULTURE, CHASSE & SYLVICULTURE
2 MINING	..	..	..	..	..	2.0	..	2 ACTIVITES EXTRACTIVES
3 MANUFACTURING	..	1 843.2 [v]	..	2 214.4 [v]	..	2 146.4	..	3 ACTIVITES DE FABRICATION
4 Food, beverages & tobacco	..	6.6 [v]	..	6.1 [v]	..	11.2	..	4 Produits alimentaires, boissons & tabac
5 Food products & beverages	..	6.6 [v]	..	6.1 [v]	..	..	..	5 Produits alimentaires & boissons
6 Tobacco products	..	0.0 [v]	..	0.0	..	..	..	6 Produits à base de tabac
7 Textiles, wear app., fur & leather	..	33.7 [v]	..	36.8 [v]	..	41.9	..	7 Textiles, habill., fourrures & cuirs
8 Textiles	..	31.2 [v]	..	34.3 [v]	..	38.3	..	8 Textiles
9 Wearing apparel & fur	..	..	..	..	..	2.6	..	9 Articles d'habillement & fourrures
10 Leather products & footwear	..	..	..	..	..	0.5	..	10 Articles en cuir & chaussures
11 Wood, paper, printing, publishing	..	7.7 [v]	..	7.7 [v]	..	11.2	..	11 Bois, papier, imprimerie, édition
12 Wood & cork (not furniture)	..	..	..	..	..	5.1	..	12 Bois & liège (sauf meubles)
13 Pulp, paper & paper products	..	..	..	..	..	3.6	..	13 Carton, papier & articles
14 Publ., print. & repro. of rec. media	..	..	..	..	..	3.1	..	14 Edition, impress. & repro. enregis.
15 Coke, petroleum, nuclear fuel, chemicals, rubber plastics	..	79.8 [v]	..	89.5 [v]	..	102.8	..	15 Coke, pétrole, combust. nucl., prod. chimi., caout. & mat. plast.
16 Coke, ref. petrol. prod. & nuclear fuel	..	..	..	..	..	2.0	..	16 Coke, raffin. pétr. & combust. nucl.
17 Chemicals & chemical products	..	47.6 [v]	..	56.8 [v]	..	68.5	..	17 Produits chimiques
18 *Chemical prod.(less pharmaceuticals)*	..	..	..	..	..	38.9	..	18 *Prod. chimi. (moins prod. pharmaceutiques)*
19 *Pharmaceuticals*	..	12.3 [v]	..	22.0 [v]	..	29.7	..	19 *Produits pharmaceutiques*
20 Rubber & plastic products	..	..	..	..	..	32.2	..	20 Caoutchouc et matières plastiques
21 Non-metallic mineral products	..	12.3 [v]	..	14.3 [v]	..	13.8	..	21 Produits minéraux non métalliques
22 Basic metals	..	22.5 [v]	..	29.7 [v]	..	25.6	..	22 Produits métallurgiques de base
23 Basic metals, ferrous	..	16.4 [v]	..	23.5 [v]	..	18.4	..	23 Produits métallur. de base, ferreux
24 Basic metals, non-ferrous	..	6.1 [v]	..	5.6 [v]	..	7.2	..	24 Prod. métallur. de base, non ferreux
25 Fabricated metal products	..	32.2 [v]	..	27.6 [v]	..	32.7	..	25 Ouvrages en métaux (sauf mach.& mat.)
26 Machinery equipment, instruments and transport equipment	..	1 647.4 [v]	..	2 001.2 [v]	..	1 903.5	..	26 Machines & matériel, instruments & mat. de transport
27 Machinery n.e.c.	..	93.6 [v]	..	106.9 [v]	..	124.2	..	27 Machines, n.c.a.
28 Office, account. & computing machinery	..	8.7 [v]	..	9.2 [v]	..	8.7	..	28 Mach. de bur., compta. & trait. inform.
29 Electrical machinery	..	30.2 [v]	..	21.5 [v]	..	27.6	..	29 Machines électriques
30 Electro. equip. (radio, TV & commun.)	..	84.4 [v]	..	101.2 [v]	..	143.2	..	30 Equip. électro. (radio, tv & communic.)
31 *Electro. comp. (inc. semi-conductors)*	..	33.2 [v]	..	65.4 [v]	..	61.9	..	31 *Compos. électro. (semi-cond. incl.)*
32 *TV, radio & communications equip.*	..	51.1 [v]	..	36.3 [v]	..	81.3	..	32 *Appareils radio, tv et de communi.*
33 Instruments	..	158.0 [v]	..	78.7 [v]	..	112.0	..	33 Instruments
34 Motor vehicles	..	51.6 [v]	..	63.4 [v]	..	70.6	..	34 Véhicules automobiles
35 Other transport equipment	..	1 220.5 [v]	..	1 620.3 [v]	..	1 417.3	..	35 Autres matériels de transport
36 *Ships*	..	14.3 [v]	..	10.7 [v]	..	8.2	..	36 *Navires*
37 *Aerospace*	..	1 201.5 [v]	..	1 604.9 [v]	..	1 405.0	..	37 *Aérospatiale*
38 *Other transport equipment n.e.c.*	..	5.1 [v]	..	4.6 [v]	..	4.1	..	38 *Autres matériels de transport n.c.a.*
39 Furniture, other manufacturing n.e.c.	..	..	..	..	..	3.1	..	39 Meubles, autres activ. de fabricat. n.c.a.
40 Furniture	..	..	..	..	..	1.5	..	40 Meubles
41 Other manufacturing n.e.c.	..	..	..	..	..	1.5	..	41 Autres activités de fabrication n.c.a.
42 Recycling	..	..	..	..	..	1.0	..	42 Récupération
43 ELECTRICITY, GAS & WATER SUPPLY	..	..	..	15.9 [v]	..	11.8	..	43 DISTRIBUTION D'ELECTRICITE, GAZ & EAU
44 CONSTRUCTION	..	..	..	..	..	7.2	..	44 CONSTRUCTION
45 SERVICES SECTOR	..	317.5 [v]	..	419.3 [v]	..	393.2	..	45 SECTEUR DES SERVICES
46 Wholesale, retail trade & repair of motor vehicles, etc.	..	0.0 [v]	..	0.5 [v]	..	0.5	..	46 Commerce (gros & détail), réparation de véhicules, etc.
47 Hotels & restaurants	..	0.0 [v]	..	0.0 [v]	..	0.0	..	47 Hôtels et restaurants
48 Transport & storage	..	50.6 [v]	..	23.0 [v]	..	..	..	48 Transports, entreposage
49 Communications	..	0.0 [v]	..	1.0 [v]	..	..	..	49 Communications
50 Post	..	0.0 [v]	..	0.0 [v]	..	..	..	50 Postes
51 Telecommunications	..	0.0 [v]	..	1.0 [v]	..	..	..	51 Télécommunications
52 Financial intermediation (inc. insur.)	..	0.0 [v]	..	0.0	..	0.0	..	52 Intermédiation financ. (assurances inc.)
53 Real estate, renting & busin. activity	..	264.8 [v]	..	387.6 [v]	..	369.7	..	53 Immobil., locat. & serv. aux entreprises
54 Computer & related activities	..	15.3 [v]	..	26.6 [v]	..	28.6	..	54 Informatique et activités connexes
55 *Software consultancy*	..	8.7 [v]	..	13.3 [v]	..	17.9	..	55 *Production de logiciels*
56 *Other computer services n.e.c.*	..	6.6 [v]	..	13.8 [v]	..	10.7	..	56 *Autres activités informatiques*
57 Research & development	..	77.2 [v]	..	199.9 [v]	..	199.9	..	57 Recherche & développement expérimental
58 Other business activities n.e.c.	..	172.3 [v]	..	160.5 [v]	..	141.1	..	58 Autres activ. & serv. aux entreprises
59 Comm., soc. & pers. serv. activ., etc.	..	2.0 [v]	..	..	..	..	..	59 Adm. pub., serv. soc. & collec., etc
TOTAL BUSINESS ENTERPRISE SECTOR	2 351.9 [c]	2 376.0	2 454.2 [c]	2 656.7	2 607.6 [c]	2 564.6	2 607.6 [c]	TOTAL SECTEUR DES ENTREPRISES

Source: OECD, R&D database (DSTI/EAS Division), January 2002.

Source : OCDE, base de données R-D (DSTI/Division AES), janvier 2002.

Business enterprise intramural expenditure on R&D (BERD) - Dépenses intra-muros de R-D des entreprises (DIRDE)

by industry and source of funds - par industrie et par source de financement

Million EUR (1999 DEM euro, EUR from 1999) — *Millions EUR (1999 DEM euro, EUR à partir de 1999)*

10.3 Other national funds - Autres sources nationales

	1994	1995	1996	1997	1998	1999	2000	
1 AGRICULTURE, HUNTING & FORESTRY	..	..	..	..	..	0.0	..	1 AGRICULTURE, CHASSE & SYLVICULTURE
2 MINING	..	..	..	..	..	0.0	..	2 ACTIVITES EXTRACTIVES
3 MANUFACTURING	..	15.3	..	20.5	..	69.5	..	3 ACTIVITES DE FABRICATION
4 Food, beverages & tobacco	..	..	..	..	..	0.0	..	**4 Produits alimentaires, boissons & tabac**
5 Food products & beverages	..	..	..	..	..	..	..	5 Produits alimentaires & boissons
6 Tobacco products	..	..	..	..	..	..	..	6 Produits à base de tabac
7 Textiles, wear app., fur & leather	..	..	..	..	..	0.5	..	**7 Textiles, habill., fourrures & cuirs**
8 Textiles	..	..	..	..	..	0.5	..	8 Textiles
9 Wearing apparel & fur	..	..	..	..	..	0.0	..	9 Articles d'habillement & fourrures
10 Leather products & footwear	..	..	..	..	..	..	..	10 Articles en cuir & chaussures
11 Wood, paper, printing, publishing	..	..	..	..	..	1.0	..	**11 Bois, papier, imprimerie, édition**
12 Wood & cork (not furniture)	..	..	..	..	..	0.0	..	12 Bois & liège (sauf meubles)
13 Pulp, paper & paper products	..	..	..	..	..	0.0	..	13 Carton, papier & articles
14 Publ., print. & repro. of rec. media	..	..	..	..	..	1.0	..	14 Edition, impress. & repro. enregis.
15 Coke, petroleum, nuclear fuel, chemicals, rubber plastics	..	0.5	..	7.7	..	9.7	..	**15 Coke, pétrole, combust. nucl., prod. chimi., caout. & mat. plast.**
16 Coke, ref. petrol. prod. & nuclear fuel	..	0.0	..	..	..	..	..	16 Coke, raffin. pétr. & combust. nucl.
17 Chemicals & chemical products	..	0.5	..	3.6	..	6.6	..	17 Produits chimiques
18 *Chemical prod.(less pharmaceuticals)*	..	..	..	..	..	3.1	..	18 *Prod. chimi. (moins prod. pharmaceutiques)*
19 *Pharmaceuticals*	..	0.0	..	3.1	..	3.1	..	19 *Produits pharmaceutiques*
20 Rubber & plastic products	..	..	..	..	..	3.1	..	20 Caoutchouc et matières plastiques
21 Non-metallic mineral products	..	..	..	..	..	2.6	..	**21 Produits minéraux non métalliques**
22 Basic metals	..	1.0	..	2.0	..	2.0	..	**22 Produits métallurgiques de base**
23 Basic metals, ferrous	..	..	..	..	..	2.0	..	23 Produits métallur. de base, ferreux
24 Basic metals, non-ferrous	..	..	..	..	..	0.5	..	24 Prod. métallur. de base, non ferreux
25 Fabricated metal products	..	..	..	..	..	0.5	..	**25 Ouvrages en métaux (sauf mach.& mat.)**
26 Machinery equipment, instruments and transport equipment	..	9.7	..	6.6	..	52.7	..	**26 Machines & matériel, instruments & mat. de transport**
27 Machinery n.e.c.	..	1.5	..	2.0	..	13.3	..	27 Machines, n.c.a.
28 Office, account. & computing machinery	..	..	..	..	..	0.0	..	28 Mach. de bur., compta. & trait. inform.
29 Electrical machinery	..	..	..	..	..	0.5	..	29 Machines électriques
30 Electro. equip. (radio, TV & commun.)	..	..	..	1.5	..	0.5	..	30 Equip. électro. (radio, tv & communic.)
31 *Electro. comp. (inc. semi-conductors)*	..	..	..	..	..	0.5	..	31 *Compos. électro. (semi-cond. incl.)*
32 *TV, radio & communications equip.*	..	..	..	..	..	0.0	..	32 *Appareils radio, tv et de communi.*
33 Instruments	..	7.7	..	2.0	..	16.4	..	33 Instruments
34 Motor vehicles	..	..	..	..	..	0.5	..	34 Véhicules automobiles
35 Other transport equipment	..	..	..	0.5	..	22.0	..	35 Autres matériels de transport
36 *Ships*	..	..	..	..	..	0.0	..	36 *Navires*
37 *Aerospace*	..	..	..	..	..	..	..	37 *Aérospatiale*
38 *Other transport equipment n.e.c.*	..	..	..	..	..	0.5	..	38 *Autres matériels de transport n.c.a.*
39 Furniture, other manufacturing n.e.c.	..	..	..	..	..	..	..	**39 Meubles, autres activ. de fabricat. n.c.a.**
40 Furniture	..	..	..	..	..	..	..	40 Meubles
41 Other manufacturing n.e.c.	..	..	..	..	..	..	..	41 Autres activités de fabrication n.c.a.
42 Recycling	..	..	..	..	..	..	..	**42 Récupération**
43 ELECTRICITY, GAS & WATER SUPPLY	..	..	..	..	..	0.0	..	43 DISTRIBUTION D'ELECTRICITE, GAZ & EAU
44 CONSTRUCTION	..	..	..	..	..	0.0	..	44 CONSTRUCTION
45 SERVICES SECTOR	..	4.6	..	8.2	..	8.2	..	45 SECTEUR DES SERVICES
46 Wholesale, retail trade & repair of motor vehicles, etc.	..	..	..	..	..	0.0	..	**46 Commerce (gros & détail), réparation de véhicules, etc.**
47 Hotels & restaurants	..	0.0	..	0.0	..	0.0	..	**47 Hôtels et restaurants**
48 Transport & storage	..	..	..	..	..	..	..	**48 Transports, entreposage**
49 Communications	..	0.0	..	..	..	..	..	**49 Communications**
50 Post	..	0.0	..	..	..	..	..	50 Postes
51 Telecommunications	..	0.0	..	..	..	..	..	51 Télécommunications
52 Financial intermediation (inc. insur.)	..	..	..	0.0	..	0.0	..	**52 Intermédiation financ. (assurances inc.)**
53 Real estate, renting & busin. activity	..	4.1	..	7.7	..	7.7	..	**53 Immobil., locat. & serv. aux entreprises**
54 Computer & related activities	..	..	..	0.5	..	0.5	..	54 Informatique et activités connexes
55 *Software consultancy*	..	..	..	..	..	0.0	..	55 *Production de logiciels*
56 *Other computer services n.e.c.*	..	..	..	..	..	0.0	..	56 *Autres activités informatiques*
57 Research & development	..	..	..	..	..	6.1	..	57 Recherche & développement expérimental
58 Other business activities n.e.c.	..	..	..	..	..	1.0	..	58 Autres activ. & serv. aux entreprises
59 Comm., soc. & pers. serv. activ., etc.	..	..	..	..	..	..	..	**59 Adm. pub., serv. soc. & collec., etc**
TOTAL BUSINESS ENTERPRISE SECTOR	51.1 c	20.5	20.5 c	28.6	30.7 c	77.7	80.7 c	**TOTAL SECTEUR DES ENTREPRISES**

Source: OECD, R&D database (DSTI/EAS Division), January 2002.

Source : OCDE, base de données R-D (DSTI/Division AES), janvier 2002.

Business enterprise intramural expenditure on R&D (BERD) - Dépenses intra-muros de R-D des entreprises (DIRDE)

by industry and source of funds - par industrie et par source de financement

Million EUR (1999 DEM euro, EUR from 1999) — *Millions EUR (1999 DEM euro, EUR à partir de 1999)*

10.4 Funds from abroad - Fonds de l'étranger

	1994	1995	1996	1997	1998	1999	2000	
1 **AGRICULTURE, HUNTING & FORESTRY**	..	..	..	..	..	0.5	..	1 **AGRICULTURE, CHASSE & SYLVICULTURE**
2 **MINING**	..	..	..	..	..	1.0	..	2 **ACTIVITES EXTRACTIVES**
3 **MANUFACTURING**	..	556.3	..	770.0	..	670.3	..	3 **ACTIVITES DE FABRICATION**
4 **Food, beverages & tobacco**	..	..	..	..	..	3.1	..	4 **Produits alimentaires, boissons & tabac**
5 Food products & beverages	..	..	..	..	..	..	..	5 Produits alimentaires & boissons
6 Tobacco products	..	..	..	..	..	..	..	6 Produits à base de tabac
7 **Textiles, wear app., fur & leather**	..	..	..	..	..	1.0	..	7 **Textiles, habill., fourrures & cuirs**
8 Textiles	..	..	..	..	..	0.5	..	8 Textiles
9 Wearing apparel & fur	..	..	..	..	..	0.0	..	9 Articles d'habillement & fourrures
10 Leather products & footwear	..	..	..	..	..	..	..	10 Articles en cuir & chaussures
11 **Wood, paper, printing, publishing**	..	..	..	..	..	0.5	..	11 **Bois, papier, imprimerie, édition**
12 Wood & cork (not furniture)	..	..	..	..	..	0.0	..	12 Bois & liège (sauf meubles)
13 Pulp, paper & paper products	..	..	..	..	..	0.0	..	13 Carton, papier & articles
14 Publ., print. & repro. of rec. media	..	..	..	..	..	0.5	..	14 Edition, impress. & repro. enregis.
15 **Coke, petroleum, nuclear fuel, chemicals, rubber plastics**	..	35.3	..	113.5	..	128.3	..	15 **Coke, pétrole, combust. nucl., prod. chimi., caout. & mat. plast.**
16 Coke, ref. petrol. prod. & nuclear fuel	..	..	..	..	..	..	..	16 Coke, raffin. pétr. & combust. nucl.
17 Chemicals & chemical products	..	29.1	..	106.3	..	120.7	..	17 Produits chimiques
18 *Chemical prod.(less pharmaceuticals)*	..	..	..	..	..	101.2	..	18 *Prod. chimi. (moins prod. pharmaceutiques)*
19 *Pharmaceuticals*	..	0.5	..	12.8	..	18.9	..	19 *Produits pharmaceutiques*
20 Rubber & plastic products	..	..	..	..	..	7.7	..	20 Caoutchouc et matières plastiques
21 **Non-metallic mineral products**	..	..	..	..	..	5.6	..	21 **Produits minéraux non métalliques**
22 **Basic metals**	..	5.1	..	6.1	..	11.2	..	22 **Produits métallurgiques de base**
23 Basic metals, ferrous	..	..	..	..	..	9.7	..	23 Produits métallur. de base, ferreux
24 Basic metals, non-ferrous	..	..	..	..	..	2.0	..	24 Prod. métallur. de base, non ferreux
25 **Fabricated metal products**	..	..	..	..	..	2.6	..	25 **Ouvrages en métaux (sauf mach.& mat.)**
26 **Machinery equipment, instruments and transport equipment**	..	501.6	..	644.2	..	517.4	..	26 **Machines & matériel, instruments & mat. de transport**
27 Machinery n.e.c.	..	13.8	..	25.6	..	36.3	..	27 Machines, n.c.a.
28 Office, account. & computing machinery	..	..	..	..	..	0.0	..	28 Mach. de bur., compta. & trait. inform.
29 Electrical machinery	..	..	..	..	..	2.6	..	29 Machines électriques
30 Electro. equip. (radio, TV & commun.)	..	..	..	239.3	..	261.3	..	30 Equip. électro. (radio, tv & communic.)
31 *Electro. comp. (inc. semi-conductors)*	..	..	..	..	..	120.7	..	31 *Compos. électro. (semi-cond. incl.)*
32 *TV, radio & communications equip.*	..	..	..	..	..	140.6	..	32 *Appareils radio, tv et de communi.*
33 Instruments	..	20.5	..	7.2	..	13.8	..	33 Instruments
34 Motor vehicles	..	..	..	..	..	21.0	..	34 Véhicules automobiles
35 Other transport equipment	..	..	..	186.1	..	182.0	..	35 Autres matériels de transport
36 *Ships*	..	..	..	..	..	0.5	..	36 *Navires*
37 *Aerospace*	..	..	..	..	..	..	..	37 *Aérospatiale*
38 *Other transport equipment n.e.c.*	..	..	..	..	..	1.0	..	38 *Autres matériels de transport n.c.a.*
39 **Furniture, other manufacturing n.e.c.**	..	..	..	..	..	..	..	39 **Meubles, autres activ. de fabricat. n.c.a.**
40 Furniture	..	..	..	..	..	..	..	40 Meubles
41 Other manufacturing n.e.c.	..	..	..	..	..	..	..	41 Autres activités de fabrication n.c.a.
42 **Recycling**	..	..	..	..	..	..	..	42 **Récupération**
43 **ELECTRICITY, GAS & WATER SUPPLY**	..	..	..	..	..	1.0	..	43 **DISTRIBUTION D'ELECTRICITE, GAZ & EAU**
44 **CONSTRUCTION**	..	..	..	..	..	0.5	..	44 **CONSTRUCTION**
45 **SERVICES SECTOR**	..	22.0	..	23.5	..	40.4	..	45 **SECTEUR DES SERVICES**
46 **Wholesale, retail trade & repair of motor vehicles, etc.**	..	..	..	..	..	0.0	..	46 **Commerce (gros & détail), réparation de véhicules, etc.**
47 **Hotels & restaurants**	..	0.0	..	0.0	..	0.0	..	47 **Hôtels et restaurants**
48 **Transport & storage**	..	..	..	..	..	..	..	48 **Transports, entreposage**
49 **Communications**	..	..	..	..	..	..	..	49 **Communications**
50 Post	..	..	..	..	..	..	..	50 Postes
51 Telecommunications	..	..	..	..	..	..	..	51 Télécommunications
52 **Financial intermediation (inc. insur.)**	..	..	..	0.0	..	0.0	..	52 **Intermédiation financ. (assurances inc.)**
53 **Real estate, renting & busin. activity**	..	19.9	..	19.9	..	38.9	..	53 **Immobil., locat. & serv. aux entreprises**
54 Computer & related activities	..	..	..	1.0	..	13.3	..	54 Informatique et activités connexes
55 *Software consultancy*	..	..	..	..	..	10.2	..	55 *Production de logiciels*
56 *Other computer services n.e.c.*	..	..	..	..	..	3.6	..	56 *Autres activités informatiques*
57 Research & development	..	..	..	..	..	16.4	..	57 Recherche & développement expérimental
58 Other business activities n.e.c.	..	..	..	..	..	8.7	..	58 Autres activ. & serv. aux entreprises
59 **Comm., soc. & pers. serv. activ., etc.**	..	..	..	..	..	..	..	59 **Adm. pub., serv. soc. & collec., etc**
TOTAL BUSINESS ENTERPRISE SECTOR	511.3 c	584.4	613.6 c	795.6	818.1 c	713.8	744.3 c	**TOTAL SECTEUR DES ENTREPRISES**

Source: OECD, R&D database (DSTI/EAS Division), January 2002.

Source : OCDE, base de données R-D (DSTI/Division AES), janvier 2002.

Business enterprise intramural expenditure on R&D (BERD) - Dépenses intra-muros de R-D des entreprises (DIRDE)

by industry and type of costs - par industrie et par type de coûts

Million EUR (1999 DEM euro, EUR from 1999) — *Millions EUR (1999 DEM euro, EUR à partir de 1999)*

	1994	1995	1996	1997	1998	1999	2000	
				11.1 Current expenditure - Dépenses courantes				
1 AGRICULTURE, HUNTING & FORESTRY	..	53.2	..	58.8	..	71.6	..	1 AGRICULTURE, CHASSE & SYLVICULTURE
2 MINING	..	71.1	..	70.6	..	52.2	..	2 ACTIVITES EXTRACTIVES
3 MANUFACTURING	..	23 911.6	..	25 145.8	..	27 991.2	..	3 ACTIVITES DE FABRICATION
4 Food, beverages & tobacco	..	189.2	..	162.1	..	186.1	..	**4 Produits alimentaires, boissons & tabac**
5 Food products & beverages	..	161.6	..	146.7	..	160.5	..	5 Produits alimentaires & boissons
6 Tobacco products	..	27.6	..	14.8	..	25.6	..	6 Produits à base de tabac
7 Textiles, wear app., fur & leather	..	141.6	..	168.2	..	187.6	..	**7 Textiles, habill., fourrures & cuirs**
8 Textiles	..	122.2	..	135.5	..	154.9	..	8 Textiles
9 Wearing apparel & fur	..	12.8	..	27.1	..	29.7	..	9 Articles d'habillement & fourrures
10 Leather products & footwear	..	6.1	..	5.6	..	3.1	..	10 Articles en cuir & chaussures
11 Wood, paper, printing, publishing	..	100.2	..	85.9	..	108.9	..	**11 Bois, papier, imprimerie, édition**
12 Wood & cork (not furniture)	..	..	..	..	..	22.5	..	12 Bois & liège (sauf meubles)
13 Pulp, paper & paper products	..	..	..	..	..	65.4	..	13 Carton, papier & articles
14 Publ., print. & repro. of rec. media	..	..	..	..	..	21.0	..	14 Edition, impress. & repro. enregis.
15 Coke, petroleum, nuclear fuel, chemicals, rubber plastics	..	4 894.6	..	5 416.6	..	5 737.7	..	**15 Coke, pétrole, combust. nucl., prod. chimi., caout. & mat. plast.**
16 Coke, ref. petrol. prod. & nuclear fuel	..	..	..	..	..	..	..	16 Coke, raffin. pétr. & combust. nucl.
17 Chemicals & chemical products	..	4 488.6	..	4 981.5	..	5 149.2	..	17 Produits chimiques
18 *Chemical prod.(less pharmaceuticals)*	..	3 324.9	..	3 211.4	..	3 240.6	..	18 *Prod. chimi. (moins prod. pharmaceutiques)*
19 *Pharmaceuticals*	..	1 163.7	..	1 770.1	..	1 909.2	..	19 *Produits pharmaceutiques*
20 Rubber & plastic products	..	349.2	..	..	..	545.0	..	20 Caoutchouc et matières plastiques
21 Non-metallic mineral products	..	243.9	..	233.7	..	281.2	..	**21 Produits minéraux non métalliques**
22 Basic metals	..	242.4	..	246.4	..	239.3	..	**22 Produits métallurgiques de base**
23 Basic metals, ferrous	..	171.3	..	168.7	..	166.2	..	23 Produits métallur. de base, ferreux
24 Basic metals, non-ferrous	..	71.1	..	77.7	..	72.6	..	24 Prod. métallur. de base, non ferreux
25 Fabricated metal products	..	328.8	..	369.7	..	452.5	..	**25 Ouvrages en métaux (sauf mach.& mat.)**
26 Machinery equipment, instruments and transport equipment	..	17 624.2	..	18 339.0	..	20 642.4	..	**26 Machines & matériel, instruments & mat. de transport**
27 Machinery n.e.c.	..	2 854.5	..	2 953.7	..	3 151.6	..	27 Machines, n.c.a.
28 Office, account. & computing machinery	..	1 028.7	..	639.1	..	604.3	..	28 Mach. de bur., compta. & trait. inform.
29 Electrical machinery	..	1 798.7	..	808.9	..	933.1	..	29 Machines électriques
30 Electro. equip. (radio, TV & commun.)	..	2 555.4	..	3 107.1	..	3 452.2	..	30 Equip. électro. (radio, tv & communic.)
31 *Electro. comp. (inc. semi-conductors)*	..	610.0	..	981.7	..	1 143.8	..	31 *Compos. électro. (semi-cond. incl.)*
32 *TV, radio & communications equip.*	..	1 945.5	..	2 125.4	..	2 308.5	..	32 *Appareils radio, tv et de communi.*
33 Instruments	..	1 503.2	..	1 397.9	..	1 551.8	..	33 Instruments
34 Motor vehicles	..	5 416.6	..	6 558.3	..	8 577.9	..	34 Véhicules automobiles
35 Other transport equipment	..	2 466.5	..	2 874.0	..	2 371.4	..	35 Autres matériels de transport
36 *Ships*	..	71.1	..	61.4	..	33.7	..	36 *Navires*
37 *Aerospace*	..	2 153.6	..	2 429.1	..	2 051.3	..	37 *Aérospatiale*
38 *Other transport equipment n.e.c.*	..	241.8	..	383.5	..	286.3	..	38 *Autres matériels de transport n.c.a.*
39 Furniture, other manufacturing n.e.c.	..	143.2	..	122.2	..	..	..	**39 Meubles, autres activ. de fabricat. n.c.a.**
40 Furniture	..	89.0	..	69.5	..	..	..	40 Meubles
41 Other manufacturing n.e.c.	..	53.7	..	52.7	..	..	..	41 Autres activités de fabrication n.c.a.
42 Recycling	..	3.6	..	2.6	..	..	..	**42 Récupération**
43 ELECTRICITY, GAS & WATER SUPPLY	..	80.3	..	57.8	..	59.3	..	43 DISTRIBUTION D'ELECTRICITE, GAZ & EAU
44 CONSTRUCTION	..	57.3	..	75.2	..	..	..	44 CONSTRUCTION
45 SERVICES SECTOR	..	827.3	..	1 383.0	..	2 478.7	..	45 SECTEUR DES SERVICES
46 Wholesale, retail trade & repair of motor vehicles, etc.	..	29.7	..	33.7	..	115.6	..	**46 Commerce (gros & détail), réparation de véhicules, etc.**
47 Hotels & restaurants	..	..	..	..	..	0.0	..	**47 Hôtels et restaurants**
48 Transport & storage	..	55.2	..	57.8	..	..	..	**48 Transports, entreposage**
49 Communications	..	..	..	..	..	..	..	**49 Communications**
50 Post	..	..	..	..	..	..	..	50 Postes
51 Telecommunications	..	..	..	..	..	..	..	51 Télécommunications
52 Financial intermediation (inc. insur.)	..	..	..	6.6	..	8.7	..	**52 Intermédiation financ. (assurances inc.)**
53 Real estate, renting & busin. activity	..	609.5	..	1 169.3	..	1 959.8	..	**53 Immobil., locat. & serv. aux entreprises**
54 Computer & related activities	..	100.7	..	456.1	..	784.3	..	54 Informatique et activités connexes
55 *Software consultancy*	..	79.8	..	424.4	..	594.6	..	55 *Production de logiciels*
56 *Other computer services n.e.c.*	..	21.0	..	31.7	..	189.7	..	56 *Autres activités informatiques*
57 Research & development	..	155.9	..	354.3	..	762.8	..	57 Recherche & développement expérimental
58 Other business activities n.e.c.	..	352.3	..	358.4	..	412.6	..	58 Autres activ. & serv. aux entreprises
59 Comm., soc. & pers. serv. activ., etc.	..	..	..	13.8	..	10.2	..	**59 Adm. pub., serv. soc. & collec., etc**
TOTAL BUSINESS ENTERPRISE SECTOR	..	**25 000.6** v	..	**26 791.2**	..	**30 727.1**	..	**TOTAL SECTEUR DES ENTREPRISES**

Source: OECD, R&D database (DSTI/EAS Division), January 2002.

Source : OCDE, base de données R-D (DSTI/Division AES), janvier 2002.

Business enterprise intramural expenditure on R&D (BERD) - Dépenses intra-muros de R-D des entreprises (DIRDE)

by industry and type of costs - par industrie et par type de coûts

Million EUR (1999 DEM euro, EUR from 1999) — *Millions EUR (1999 DEM euro, EUR à partir de 1999)*

11.2 Capital expenditure - Dépenses en capital

	1994	1995	1996	1997	1998	1999	2000	
1 AGRICULTURE, HUNTING & FORESTRY	..	4.6	..	6.6	..	10.2	..	**1 AGRICULTURE, CHASSE & SYLVICULTURE**
2 MINING	..	1.5	..	2.6	..	3.1	..	**2 ACTIVITES EXTRACTIVES**
3 MANUFACTURING	..	1 633.1	..	1 874.9	..	2 559.5	..	**3 ACTIVITES DE FABRICATION**
4 Food, beverages & tobacco	..	28.1	..	28.1	..	31.7	..	**4 Produits alimentaires, boissons & tabac**
5 Food products & beverages	..	26.1	..	26.1	..	28.6	..	5 Produits alimentaires & boissons
6 Tobacco products	..	2.0	..	1.5	..	3.6	..	6 Produits à base de tabac
7 Textiles, wear app., fur & leather	..	19.9	..	22.0	..	23.0	..	**7 Textiles, habill., fourrures & cuirs**
8 Textiles	..	17.4	..	17.9	..	18.4	..	8 Textiles
9 Wearing apparel & fur	..	2.0	..	3.6	..	4.1	..	9 Articles d'habillement & fourrures
10 Leather products & footwear	..	0.5	..	1.0	..	0.5	..	10 Articles en cuir & chaussures
11 Wood, paper, printing, publishing	..	34.3	..	25.6	..	22.0	..	**11 Bois, papier, imprimerie, édition**
12 Wood & cork (not furniture)	..	..	..	..	..	3.6	..	12 Bois & liège (sauf meubles)
13 Pulp, paper & paper products	..	..	..	..	..	14.8	..	13 Carton, papier & articles
14 Publ., print. & repro. of rec. media	..	..	..	..	..	3.6	..	14 Edition, impress. & repro. enregis.
15 Coke, petroleum, nuclear fuel, chemicals, rubber plastics	..	412.1	..	570.6	..	620.2	..	**15 Coke, pétrole, combust. nucl., prod. chimi., caout. & mat. plast.**
16 Coke, ref. petrol. prod. & nuclear fuel	..	..	..	..	..	..	..	16 Coke, raffin. pétr. & combust. nucl.
17 Chemicals & chemical products	..	..	..	448.4	..	545.0	..	17 Produits chimiques
18 *Chemical prod.(less pharmaceuticals)*	..	272.0	..	325.7	..	364.6	..	18 *Prod. chimi. (moins prod. pharmaceutiques)*
19 *Pharmaceuticals*	..	86.4	..	122.7	..	181.0	..	19 *Produits pharmaceutiques*
20 Rubber & plastic products	..	44.5	..	..	..	71.1	..	20 Caoutchouc et matières plastiques
21 Non-metallic mineral products	..	31.7	..	32.2	..	53.7	..	**21 Produits minéraux non métalliques**
22 Basic metals	..	27.6	..	29.7	..	19.9	..	**22 Produits métallurgiques de base**
23 Basic metals, ferrous	..	19.9	..	19.4	..	12.3	..	23 Produits métallur. de base, ferreux
24 Basic metals, non-ferrous	..	8.2	..	10.2	..	7.7	..	24 Prod. métallur. de base, non ferreux
25 Fabricated metal products	..	57.8	..	62.9	..	64.4	..	**25 Ouvrages en métaux (sauf mach.& mat.)**
26 Machinery equipment, instruments and transport equipment	..	1 007.8	..	1 088.0	..	1 703.1	..	**26 Machines & matériel, instruments & mat. de transport**
27 Machinery n.e.c.	..	207.6	..	236.7	..	251.6	..	27 Machines, n.c.a.
28 Office, account. & computing machinery	..	23.5	..	34.3	..	29.1	..	28 Mach. de bur., compta. & trait. inform.
29 Electrical machinery	..	143.7	..	68.0	..	87.9	..	29 Machines électriques
30 Electro. equip. (radio, TV & commun.)	..	151.9	..	168.7	..	190.2	..	30 Equip. électro. (radio, tv & communic.)
31 *Electro. comp. (inc. semi-conductors)*	..	44.5	..	53.2	..	63.4	..	31 *Compos. électro. (semi-cond. incl.)*
32 *TV, radio & communications equip.*	..	107.4	..	116.1	..	126.8	..	32 *Appareils radio, tv et de communi.*
33 Instruments	..	112.5	..	101.7	..	113.0	..	33 Instruments
34 Motor vehicles	..	307.3	..	433.6	..	842.1	..	34 Véhicules automobiles
35 Other transport equipment	..	61.4	..	45.0	..	188.7	..	35 Autres matériels de transport
36 *Ships*	..	4.1	..	4.1	..	2.0	..	36 *Navires*
37 *Aerospace*	..	42.4	..	36.8	..	183.0	..	37 *Aérospatiale*
38 *Other transport equipment n.e.c.*	..	14.3	..	4.1	..	3.1	..	38 *Autres matériels de transport n.c.a.*
39 Furniture, other manufacturing n.e.c.	..	12.8	..	13.8	..	..	..	**39 Meubles, autres activ. de fabricat. n.c.a.**
40 Furniture	..	7.2	..	8.2	..	..	..	40 Meubles
41 Other manufacturing n.e.c.	..	5.6	..	6.1	..	..	..	41 Autres activités de fabrication n.c.a.
42 Recycling	..	1.5	..	1.5	..	..	..	**42 Récupération**
43 ELECTRICITY, GAS & WATER SUPPLY	..	33.7	..	33.7	..	47.6	..	**43 DISTRIBUTION D'ELECTRICITE, GAZ & EAU**
44 CONSTRUCTION	..	16.9	..	17.9	..	..	..	**44 CONSTRUCTION**
45 SERVICES SECTOR	..	125.8	..	182.5	..	262.3	..	**45 SECTEUR DES SERVICES**
46 Wholesale, retail trade & repair of motor vehicles, etc.	..	3.6	..	4.1	..	15.3	..	**46 Commerce (gros & détail), réparation de véhicules, etc.**
47 Hotels & restaurants	..	..	..	..	..	0.0	..	**47 Hôtels et restaurants**
48 Transport & storage	..	11.8	..	10.2	..	..	..	**48 Transports, entreposage**
49 Communications	..	..	..	..	..	..	..	**49 Communications**
50 Post	..	..	..	..	..	..	..	50 Postes
51 Telecommunications	..	..	..	..	..	..	..	51 Télécommunications
52 Financial intermediation (inc. insur.)	..	..	..	1.0	..	3.6	..	**52 Intermédiation financ. (assurances inc.)**
53 Real estate, renting & busin. activity	..	74.1	..	138.0	..	214.2	..	**53 Immobil., locat. & serv. aux entreprises**
54 Computer & related activities	..	11.2	..	40.4	..	81.3	..	54 Informatique et activités connexes
55 *Software consultancy*	..	8.2	..	36.3	..	55.2	..	55 *Production de logiciels*
56 *Other computer services n.e.c.*	..	2.6	..	4.1	..	26.1	..	56 *Autres activités informatiques*
57 Research & development	..	26.6	..	64.9	..	98.2	..	57 Recherche & développement expérimental
58 Other business activities n.e.c.	..	36.3	..	33.2	..	34.3	..	58 Autres activ. & serv. aux entreprises
59 Comm., soc. & pers. serv. activ., etc.	..	..	..	1.5	..	2.0	..	**59 Adm. pub., serv. soc. & collec., etc**
TOTAL BUSINESS ENTERPRISE SECTOR	..	**1 816.1** v	..	**2 118.3**	..	**2 895.4**	..	**TOTAL SECTEUR DES ENTREPRISES**

Source: OECD, R&D database (DSTI/EAS Division), January 2002.

Source : OCDE, base de données R-D (DSTI/Division AES), janvier 2002.

Total business enterprise R&D personnel (BEMP) - Personnel total de R-D du secteur des entreprises (PRDE)

by industry - par industrie

In full time equivalent | *En équivalence plein-temps*

	1994	1995	1996	1997	1998	1999	2000	
1 AGRICULTURE, HUNTING & FORESTRY	..	1 037	..	1 085	..	1 202	..	**1 AGRICULTURE, CHASSE & SYLVICULTURE**
2 MINING	..	375	..	458	..	363	..	**2 ACTIVITES EXTRACTIVES**
3 MANUFACTURING	..	267 478	..	265 766	..	272 965	..	**3 ACTIVITES DE FABRICATION**
4 Food, beverages & tobacco	..	2 888	..	2 621	..	2 481	..	**4 Produits alimentaires, boissons & tabac**
5 Food products & beverages	..	2 600	..	2 441	..	2 267	..	5 Produits alimentaires & boissons
6 Tobacco products	..	288	..	180	..	214	..	6 Produits à base de tabac
7 Textiles, wear app., fur & leather	..	2 512	..	2 678	..	2 449	..	**7 Textiles, habill., fourrures & cuirs**
8 Textiles	..	2 108	..	2 122	..	1 969	..	8 Textiles
9 Wearing apparel & fur	..	286	..	446	..	398	..	9 Articles d'habillement & fourrures
10 Leather products & footwear	..	118	..	110	..	82	..	10 Articles en cuir & chaussures
11 Wood, paper, printing, publishing	..	1 403	..	1 372	..	1 606	..	**11 Bois, papier, imprimerie, édition**
12 Wood & cork (not furniture)	..	468	..	..	..	460	..	12 Bois & liège (sauf meubles)
13 Pulp, paper & paper products	..	752	..	..	..	795	..	13 Carton, papier & articles
14 Publ., print. & repro. of rec. media	..	184	..	..	..	352	..	14 Edition, impress. & repro. enregis.
15 Coke, petroleum, nuclear fuel, chemicals, rubber plastics	..	..	..	53 086	..	51 437	..	**15 Coke, pétrole, combust. nucl., prod. chimi., caout. & mat. plast.**
16 Coke, ref. petrol. prod. & nuclear fuel	..	..	..	..	..	..	..	16 Coke, raffin. pétr. & combust. nucl.
17 Chemicals & chemical products	..	49 081	..	47 297	..	44 158	..	17 Produits chimiques
18 *Chemical prod.(less pharmaceuticals)*	..	36 277	..	30 289	..	28 922	..	18 *Prod. chimi. (moins prod. pharmaceutiques)*
19 *Pharmaceuticals*	..	12 804	..	17 008	..	15 236	..	19 *Produits pharmaceutiques*
20 Rubber & plastic products	..	5 314	..	..	..	6 890	..	20 Caoutchouc et matières plastiques
21 Non-metallic mineral products	..	3 385	..	2 974	..	2 788	..	**21 Produits minéraux non métalliques**
22 Basic metals	..	2 686	..	2 881	..	2 553	..	**22 Produits métallurgiques de base**
23 Basic metals, ferrous	..	1 786	..	1 900	..	1 663	..	23 Produits métallur. de base, ferreux
24 Basic metals, non-ferrous	..	901	..	981	..	890	..	24 Prod. métallur. de base, non ferreux
25 Fabricated metal products	..	4 885	..	5 392	..	6 443	..	**25 Ouvrages en métaux (sauf mach.& mat.)**
26 Machinery equipment, instruments and transport equipment	..	192 441	..	192 537	..	200 822	..	**26 Machines & matériel, instruments & mat. de transport**
27 Machinery n.e.c.	..	39 323	..	39 251	..	37 584	..	27 Machines, n.c.a.
28 Office, account. & computing machinery	..	10 433	..	7 093	..	6 364	..	28 Mach. de bur., compta. & trait. inform.
29 Electrical machinery	..	21 189	..	10 739	..	11 662	..	29 Machines électriques
30 Electro. equip. (radio, TV & commun.)	..	28 875	..	35 427	..	35 491	..	30 Equip. électro. (radio, tv & communic.)
31 *Electro. comp. (inc. semi-conductors)*	..	7 524	..	11 813	..	12 215	..	31 *Compos. électro. (semi-cond. incl.)*
32 *TV, radio & communications equip.*	..	21 351	..	23 615	..	23 276	..	32 *Appareils radio, tv et de communi.*
33 Instruments	..	21 800	..	18 498	..	18 752	..	33 Instruments
34 Motor vehicles	..	51 254	..	59 717	..	72 761	..	34 Véhicules automobiles
35 Other transport equipment	..	19 567	..	21 812	..	18 208	..	35 Autres matériels de transport
36 *Ships*	..	933	..	740	..	383	..	36 *Navires*
37 *Aerospace*	..	14 953	..	16 783	..	14 991	..	37 *Aérospatiale*
38 *Other transport equipment n.e.c.*	..	3 681	..	4 288	..	2 834	..	38 *Autres matériels de transport n.c.a.*
39 Furniture, other manufacturing n.e.c.	..	2 127	..	2 160	..	..	..	**39 Meubles, autres activ. de fabricat. n.c.a.**
40 Furniture	..	1 280	..	1 288	..	..	..	40 Meubles
41 Other manufacturing n.e.c.	..	847	..	872	..	..	..	41 Autres activités de fabrication n.c.a.
42 Recycling	..	..	..	66	..	..	..	**42 Récupération**
43 ELECTRICITY, GAS & WATER SUPPLY	..	729	..	607	..	601	..	**43 DISTRIBUTION D'ELECTRICITE, GAZ & EAU**
44 CONSTRUCTION	..	..	..	937	..	..	..	**44 CONSTRUCTION**
45 SERVICES SECTOR	..	12 840	..	17 417	..	30 521	..	**45 SECTEUR DES SERVICES**
46 Wholesale, retail trade & repair of motor vehicles, etc.	..	375	..	495	..	1 919	..	**46 Commerce (gros & détail), réparation de véhicules, etc.**
47 Hotels & restaurants	..	..	..	..	..	0	..	**47 Hôtels et restaurants**
48 Transport & storage	..	1 720	..	813	..	..	..	**48 Transports, entreposage**
49 Communications	..	..	..	..	..	..	..	**49 Communications**
50 Post	..	..	..	..	..	..	..	50 Postes
51 Telecommunications	..	..	..	..	..	..	..	51 Télécommunications
52 Financial intermediation (inc. insur.)	..	..	..	33	..	80	..	**52 Intermédiation financ. (assurances inc.)**
53 Real estate, renting & busin. activity	..	9 020	..	14 522	..	23 099	..	**53 Immobil., locat. & serv. aux entreprises**
54 Computer & related activities	..	1 794	..	5 033	..	9 065	..	54 Informatique et activités connexes
55 *Software consultancy*	..	1 532	..	4 705	..	6 764	..	55 *Production de logiciels*
56 *Other computer services n.e.c.*	..	262	..	327	..	2 301	..	56 *Autres activités informatiques*
57 Research & development	..	2 585	..	5 175	..	9 300	..	57 Recherche & développement expérimental
58 Other business activities n.e.c.	..	4 642	..	4 315	..	4 735	..	58 Autres activ. & serv. aux entreprises
59 Comm., soc. & pers. serv. activ., etc.	..	..	..	206	..	187	..	**59 Adm. pub., serv. soc. & collec., etc**
TOTAL BUSINESS ENTERPRISE SECTOR	**..**	**283 316**	**276 794** c	**286 270**	**288 090** c	**306 693**	**..**	**TOTAL SECTEUR DES ENTREPRISES**

Source: OECD, R&D database (DSTI/EAS Division), January 2002.

Source : OCDE, base de données R-D (DSTI/Division AES), janvier 2002.

RSE in the business enterprise sector - Chercheurs dans le secteur des entreprises

by industry - par industrie

In full time equivalent | *En équivalence plein-temps*

	1994	1995	1996	1997	1998	1999	2000	
1 AGRICULTURE, HUNTING & FORESTRY	..	291	..	301	..	374	..	**1 AGRICULTURE, CHASSE & SYLVICULTURE**
2 MINING	..	159	..	174	..	171	..	**2 ACTIVITES EXTRACTIVES**
3 MANUFACTURING	..	120 759	..	120 368	..	129 911	..	**3 ACTIVITES DE FABRICATION**
4 Food, beverages & tobacco	..	1 004	..	885	..	969	..	**4 Produits alimentaires, boissons & tabac**
5 Food products & beverages	..	918	..	..	..	896	..	5 Produits alimentaires & boissons
6 Tobacco products	..	86	..	..	..	73	..	6 Produits à base de tabac
7 Textiles, wear app., fur & leather	..	908	..	1 095	..	1 116	..	**7 Textiles, habill., fourrures & cuirs**
8 Textiles	..	770	..	872	..	916	..	8 Textiles
9 Wearing apparel & fur	..	108	..	177	..	164	..	9 Articles d'habillement & fourrures
10 Leather products & footwear	..	30	..	46	..	36	..	10 Articles en cuir & chaussures
11 Wood, paper, printing, publishing	..	515	..	529	..	724	..	**11 Bois, papier, imprimerie, édition**
12 Wood & cork (not furniture)	..	164	..	167	..	212	..	12 Bois & liège (sauf meubles)
13 Pulp, paper & paper products	..	261	..	292	..	359	..	13 Carton, papier & articles
14 Publ., print. & repro. of rec. media	..	91	..	69	..	154	..	14 Edition, impress. & repro. enregis.
15 Coke, petroleum, nuclear fuel, chemicals, rubber plastics	..	15 301	..	15 079	..	15 717	..	**15 Coke, pétrole, combust. nucl., prod. chimi., caout. & mat. plast.**
16 Coke, ref. petrol. prod. & nuclear fuel	..	255	..	259	..	112	..	16 Coke, raffin. pétr. & combust. nucl.
17 Chemicals & chemical products	..	12 811	..	12 560	..	12 325	..	17 Produits chimiques
18 *Chemical prod.(less pharmaceuticals)*	..	8 682	..	7 339	..	6 986	..	18 *Prod. chimi. (moins prod. pharmaceutiques)*
19 *Pharmaceuticals*	..	4 129	..	5 221	..	5 340	..	19 *Produits pharmaceutiques*
20 Rubber & plastic products	..	2 235	..	2 260	..	3 280	..	20 Caoutchouc et matières plastiques
21 Non-metallic mineral products	..	1 310	..	1 118	..	1 069	..	**21 Produits minéraux non métalliques**
22 Basic metals	..	975	..	1 139	..	1 019	..	**22 Produits métallurgiques de base**
23 Basic metals, ferrous	..	613	..	724	..	623	..	23 Produits métallur. de base, ferreux
24 Basic metals, non-ferrous	..	362	..	415	..	396	..	24 Prod. métallur. de base, non ferreux
25 Fabricated metal products	..	1 986	..	2 136	..	2 747	..	**25 Ouvrages en métaux (sauf mach.& mat.)**
26 Machinery equipment, instruments and transport equipment	..	98 032	..	97 734	..	105 559	..	**26 Machines & matériel, instruments & mat. de transport**
27 Machinery n.e.c.	..	19 113	..	19 217	..	18 652	..	27 Machines, n.c.a.
28 Office, account. & computing machinery	..	7 076	..	5 158	..	4 630	..	28 Mach. de bur., compta. & trait. inform.
29 Electrical machinery	..	11 439	..	5 475	..	6 401	..	29 Machines électriques
30 Electro. equip. (radio, TV & commun.)	..	17 204	..	22 059	..	22 697	..	30 Equip. électro. (radio, tv & communic.)
31 *Electro. comp. (inc. semi-conductors)*	..	4 028	..	6 523	..	7 215	..	31 *Compos. électro. (semi-cond. incl.)*
32 *TV, radio & communications equip.*	..	13 176	..	15 536	..	15 483	..	32 *Appareils radio, tv et de communi.*
33 Instruments	..	13 379	..	11 403	..	11 525	..	33 Instruments
34 Motor vehicles	..	20 642	..	23 878	..	31 379	..	34 Véhicules automobiles
35 Other transport equipment	..	9 178	..	10 545	..	10 275	..	35 Autres matériels de transport
36 *Ships*	..	579	..	440	..	230	..	36 *Navires*
37 *Aerospace*	..	6 387	..	7 577	..	8 331	..	37 *Aérospatiale*
38 *Other transport equipment n.e.c.*	..	2 213	..	2 528	..	1 714	..	38 *Autres matériels de transport n.c.a.*
39 Furniture, other manufacturing n.e.c.	..	688	..	622	..	960	..	**39 Meubles, autres activ. de fabricat. n.c.a.**
40 Furniture	..	394	..	372	..	637	..	40 Meubles
41 Other manufacturing n.e.c.	..	294	..	249	..	323	..	41 Autres activités de fabrication n.c.a.
42 Recycling	..	39	..	30	..	32	..	**42 Récupération**
43 ELECTRICITY, GAS & WATER SUPPLY	..	418	..	407	..	343	..	**43 DISTRIBUTION D'ELECTRICITE, GAZ & EAU**
44 CONSTRUCTION	..	457	..	492	..	..	..	**44 CONSTRUCTION**
45 SERVICES SECTOR	..	7 286	..	10 944	..	18 707	..	**45 SECTEUR DES SERVICES**
46 Wholesale, retail trade & repair of motor vehicles, etc.	..	172	..	268	..	875	..	**46 Commerce (gros & détail), réparation de véhicules, etc.**
47 Hotels & restaurants	..	..	..	..	..	0	..	**47 Hôtels et restaurants**
48 Transport & storage	..	334	..	290	..	..	..	**48 Transports, entreposage**
49 Communications	..	..	..	..	..	..	..	**49 Communications**
50 Post	..	..	..	..	..	..	..	50 Postes
51 Telecommunications	..	..	..	..	..	..	..	51 Télécommunications
52 Financial intermediation (inc. insur.)	..	..	..	18	..	54	..	**52 Intermédiation financ. (assurances inc.)**
53 Real estate, renting & busin. activity	..	5 634	..	9 285	..	14 028	..	**53 Immobil., locat. & serv. aux entreprises**
54 Computer & related activities	..	1 268	..	3 481	..	5 682	..	54 Informatique et activités connexes
55 *Software consultancy*	..	1 112	..	3 259	..	4 563	..	55 *Production de logiciels*
56 *Other computer services n.e.c.*	..	156	..	222	..	1 119	..	56 *Autres activités informatiques*
57 Research & development	..	1 411	..	2 979	..	5 302	..	57 Recherche & développement expérimental
58 Other business activities n.e.c.	..	2 955	..	2 825	..	3 044	..	58 Autres activ. & serv. aux entreprises
59 Comm., soc. & pers. serv. activ., etc.	..	147	..	107	..	116	..	**59 Adm. pub., serv. soc. & collec., etc**
TOTAL BUSINESS ENTERPRISE SECTOR	**..**	**129 370**	**126 392** c	**132 687**	**133 529** c	**150 150**	**..**	**TOTAL SECTEUR DES ENTREPRISES**

Source: OECD, R&D database (DSTI/EAS Division), January 2002.

Source : OCDE, base de données R-D (DSTI/Division AES), janvier 2002.

Table 14 - Tableau 14

Higher education intramural expenditure on R&D - Dépenses intra-muros de R-D de l'enseignement supérieur
(HERD) by field of science - (DIRDES) par discipline scientifique

	1994	1995	1996	1997	1998	1999	2000	
	14.1 Million EUR (1999 DEM euro, EUR from 1999) - Millions EUR (1999 DEM euro, EUR à partir de 1999)							
Natural sciences	2 066.7	2 146.5	2 200.6	2 232.7	2 333.5	2 317.4	..	Sciences exactes et naturelles
Engineering	1 444.4	1 497.7	1 579.7	1 591.1	1 576.4	1 608.4	..	Sciences de l'ingénieur
Medical sciences	1 661.5	1 728.0	1 844.0	1 825.2	1 791.2	1 961.1	..	Sciences médicales
Agricultural sciences	333.3	374.9	341.1	331.5	339.8	332.7	..	Sciences agricoles
Sub-total NSE	5 505.9 v	5 747.0 v	5 965.5 v	5 980.6 v	6 041.0 v	6 219.6 v	..	Sous-total SENI
Social sciences	534.4	616.1	647.2	661.7	662.9	673.2	..	Sciences sociales
Humanities	864.1	907.5	976.3	994.6	995.2	957.9	..	Sciences humaines
Sub-total SSH	1 398.5 v	1 523.7 v	1 623.6 v	1 656.3 v	1 658.1 v	1 631.0 v	..	Sous-total SSH
Total NSE+SSH	7 059.4	7 377.8	7 652.3	7 676.8	7 768.1	7 936.6	8 034.4 c	*Total SENI+SSH*
	14.2 Million constant USD (1995 prices and PPPs) - Millions de USD constants (prix et PPA de 1995)							
Natural sciences	2 046.0	2 082.8	2 113.5	2 130.4	2 202.1	2 176.8	..	Sciences exactes et naturelles
Engineering	1 430.0	1 453.2	1 517.3	1 518.2	1 487.7	1 510.8	..	Sciences de l'ingénieur
Medical sciences	1 644.9	1 676.7	1 771.1	1 741.5	1 690.3	1 842.1	..	Sciences médicales
Agricultural sciences	330.0	363.8	327.6	316.3	320.6	312.5	..	Sciences agricoles
Sub-total NSE	5 450.8 v	5 576.5 v	5 729.5 v	5 706.4 v	5 700.9 v	5 842.2 v	..	Sous-total SENI
Social sciences	529.1	597.8	621.7	631.3	625.6	632.3	..	Sciences sociales
Humanities	855.4	880.6	937.7	949.0	939.1	899.7	..	Sciences humaines
Sub-total SSH	1 384.5 v	1 478.4 v	1 559.4 v	1 580.4 v	1 564.7 v	1 532.1 v	..	Sous-total SSH
Total NSE+SSH	6 988.8	7 158.9	7 349.7	7 324.9	7 330.6	7 455.0	7 576.6 c	*Total SENI+SSH*

Table 15 - Tableau 15

Higher education intramural expenditure on R&D - Dépenses intra-muros de R-D de l'enseignement supérieur
(HERD) by field of science and type of costs - (DIRDES) par discipline scientifique et par type de coûts

Million EUR (1999 DEM euro, EUR from 1999) — *Millions EUR (1999 DEM euro, EUR à partir de 1999)*

	1994	1995	1996	1997	1998	1999	2000	
Current expenditure								**Dépenses courantes**
Natural sciences	..	1 919.9	..	..	2 055.0	2 057.0	..	Sciences exactes et naturelles
Engineering	..	1 331.2	..	..	1 431.7	1 460.4	..	Sciences de l'ingénieur
Medical sciences	..	1 615.8	..	..	1 678.2	1 841.3	..	Sciences médicales
Agricultural sciences	..	347.9	..	..	309.1	300.1	..	Sciences agricoles
Sub-total NSE	..	5 214.8	..	..	5 473.9	5 658.8	..	Sous-total SENI
Social sciences	..	565.4	..	..	620.9	629.3	..	Sciences sociales
Humanities	..	849.3	..	..	936.8	902.9	..	Sciences humaines
Sub-total SSH	..	1 414.7	..	..	1 557.8	1 532.2	..	Sous-total SSH
Total NSE+SSH	6 309.5 v	6 629.6 v	6 882.1 v	6 971.7 v	7 031.8 v	7 191.0 v	..	*Total SENI+SSH*
Capital expenditure								**Dépenses en capital**
Natural sciences	..	226.6	..	..	278.5	260.4	..	Sciences exactes et naturelles
Engineering	..	166.4	..	..	144.9	148.1	..	Sciences de l'ingénieur
Medical sciences	..	112.2	..	..	112.9	119.9	..	Sciences médicales
Agricultural sciences	..	27.0	..	..	30.6	32.5	..	Sciences agricoles
Sub-total NSE	..	532.2	..	..	566.9	560.8	..	Sous-total SENI
Social sciences	..	50.7	..	..	42.0	43.9	..	Sciences sociales
Humanities	..	58.2	..	..	58.3	55.0	..	Sciences humaines
Sub-total SSH	..	108.9	..	..	100.3	98.9	..	Sous-total SSH
Total NSE+SSH	594.9 v	641.1 v	706.9 v	665.2 v	667.3 v	659.7 v	..	*Total SENI+SSH*

Source: OECD, R&D database (DSTI/EAS Division), January 2002.

Source : OCDE, base de données R-D (DSTI/Division AES), janvier 2002.

GERMANY — Table 17 - Tableau 17 — ALLEMAGNE

Higher education R&D personnel (HEMP) - Personnel de R-D du secteur de l'enseignement supérieur

by field of science and occupation - par discipline scientifique et par profession

In full-time equivalent — *En équivalence plein-temps*

	1994	1995	1996	1997	1998	1999	2000	
Total R&D personnel								**Personnel total de R-D**
Natural sciences	..	29 085	..	..	28 650	28 402	..	Sciences exactes et naturelles
Engineering	..	19 782	..	..	20 157	19 764	..	Sciences de l'ingénieur
Medical sciences	..	24 595	..	..	23 928	25 945	..	Sciences médicales
Agricultural sciences	..	5 009	..	..	4 600	4 556	..	Sciences agricoles
Sub-total NSE	..	78 471	79 409	78 039	77 335	78 667	..	Sous-total SENI
Social sciences	..	8 235	..	..	8 675	8 762	..	Sciences sociales
Humanities	..	13 969	..	..	14 069	14 039	..	Sciences humaines
Sub-total SSH	..	22 204	22 751	22 605	22 744	22 801	..	Sous-total SSH
Total NSE+SSH	..	100 674	102 160	100 646	100 080	101 471	103 171 c	*Total SENI+SSH*
Researchers								**Chercheurs**
Natural sciences	..	20 888	..	..	20 964	20 884	..	Sciences exactes et naturelles
Engineering	..	13 743	..	..	14 086	13 901	..	Sciences de l'ingénieur
Medical sciences	..	8 072	..	..	8 738	9 415	..	Sciences médicales
Agricultural sciences	..	2 970	..	..	2 753	2 830	..	Sciences agricoles
Sub-total NSE	..	45 673	..	..	46 541	47 030	..	Sous-total SENI
Social sciences	..	6 888	..	..	7 273	7 430	..	Sciences sociales
Humanities	..	11 872	..	..	12 159	12 232	..	Sciences humaines
Sub-total SSH	..	18 760	..	..	19 432	19 662	..	Sous-total SSH
Total NSE+SSH	..	64 434	66 110	65 704	65 973	66 695	67 504 c	*Total SENI+SSH*
Technicians								**Technicians**
Natural sciences	..	4 737	..	..	4 395	4 246	..	Sciences exactes et naturelles
Engineering	..	3 590	..	..	3 614	3 486	..	Sciences de l'ingénieur
Medical sciences	..	3 471	..	..	3 113	3 271	..	Sciences médicales
Agricultural sciences	..	1 143	..	..	1 083	963	..	Sciences agricoles
Sub-total NSE	..	12 941	..	..	12 205	11 966	..	Sous-total SENI
Social sciences	..	212	..	..	265	257	..	Sciences sociales
Humanities	..	483	..	..	459	443	..	Sciences humaines
Sub-total SSH	..	695	..	..	724	700	..	Sous-total SSH
Total NSE+SSH	..	13 636	13 578	13 369	12 929	12 666	..	*Total SENI+SSH*
Other								**Autres**
Natural sciences	..	3 460	..	..	3 291	3 272	..	Sciences exactes et naturelles
Engineering	..	2 449	..	..	2 457	2 377	..	Sciences de l'ingénieur
Medical sciences	..	13 051	..	..	12 077	13 259	..	Sciences médicales
Agricultural sciences	..	897	..	..	764	763	..	Sciences agricoles
Sub-total NSE	..	19 857	..	..	18 589	19 671	..	Sous-total SENI
Social sciences	..	1 134	..	..	1 138	1 075	..	Sciences sociales
Humanities	..	1 614	..	..	1 452	1 364	..	Sciences humaines
Sub-total SSH	..	2 748	..	..	2 590	2 439	..	Sous-total SSH
Total NSE+SSH	..	22 604	22 472	21 573	21 179	22 110	..	*Total SENI+SSH*

Table 18 - Tableau 18

Government budget appropriations or outlays for R&D (GBAORD) - Crédits budgétaires publics de R-D (CBPRD)

by socio-economic objective - par objectif socio-économique

Million EUR (1999 DEM euro, EUR from 1999) — *Millions EUR (1999 DEM euro, EUR à partir de 1999)*

	1995	1996	1997	1998	1999	2000	2001	
Agriculture, forestry and fishing	420.8	424.6	432.1 s	433.0	416.9	410.1 p	409.2 p	Agriculture, sylviculture et pêche
Industrial development	2 157.1	2 225.3	2 040.3 s	1 997.8	2 091.5	1 999.4 p	2 071.4 p	Développement industriel
Energy	556.1	578.1	562.9 s	588.7	593.8	569.8 p	607.8 p	Energie
Transport & telecommunications	86.3	86.2	128.1 s	130.5	142.5	130.1 p	136.0 p	*Transports & télécommunications*
Urban & rural planning	160.0	155.7	144.7 s	144.5	137.7	138.6 p	142.2 p	*Aménagement urbain et rural*
ST Infrastructure	246.2	241.9	272.8 s	275.0	280.2	268.6 p	278.3 p	ST Infrastructure
Prevention of pollution	..	..	..	..	..	..	..	*Prévention de la pollution*
Identificat. & treatment of pollution	..	..	..	..	..	..	..	*Détect. pollut. & mesures antipollut.*
ST Environment	580.3	614.9	562.6 s	551.8	562.7	560.1 p	559.0 p	ST Environnement
Health	520.4	557.0	530.1 s	518.6	543.4	558.4 p	652.7 p	Santé
Social development & services	387.8	393.3	395.9 s	411.6	549.5	592.1 p	770.3 p	Développement et services sociaux
Earth and atmosphere	370.1	356.0	325.3 s	293.8	293.1	296.9 p	313.3 p	Milieu terrestre et atmosphère
Advancement of research	2 430.9	2 569.4	2 476.6 s	2 531.0	2 605.7	2 712.2 p	2 841.6 p	*Promotion de la recherche*
General university funds	6 104.4	6 009.1	6 160.7 s	6 209.4	6 252.5	6 274.0 p	6 399.0 p	*Fonds généraux des universités*
ST Advancement of knowledge	8 535.3	8 578.5	8 637.3 s	8 740.4	8 858.2	8 986.2 p	9 240.6 p	ST promotion gén. des connaiss.
Civil space	833.0	809.3	762.7 s	759.7	736.4	741.3 p	767.8 p	Espace civil
Defence	1 465.3	1 637.6	1 532.8 s	1 402.6	1 359.4	1 308.1 p	1 255.4 p	Défense
Not elsewhere classified	104.5	44.3	..	43.9	37.3	17.0 p	9.3 p	Non classé
Total	**16 176.9**	**16 460.6**	**16 009.4 a**	**16 017.0**	**16 322.3**	**16 308.1 p**	**16 935.1 p**	**Total**

Source: OECD, R&D database (DSTI/EAS Division), January 2002.

Source : OCDE, base de données R-D (DSTI/Division AES), janvier 2002.

Table 19 - Tableau 19

Technology balance of payments - Balance des paiements technologiques

by type of transfer - selon la nature des transferts

Million EUR (1999 DEM euro, EUR from 1999) *Millions EUR (1999 DEM euro, EUR à partir de 1999)*

	1983	1984	1985	1986	1987	1988	1989	1990	1991	
Receipts										**Recettes**
Related firms	153 m	146 m	124 m	135 m	134 m	108 m	102 m	195 m	293 m	Firmes affiliées
Unrelated firms	518 m	607 m	698 m	663 m	641 m	793 m	..	..	..	Firmes non affiliées
Other	..	..	..	..	..	..	..	..	..	Autre
Total	1 487	1 603	1 763	3 191 a	3 541	3 633	4 220	5 234 a	5 331	*Total*
Payments										**Paiements**
Related firms	1 046 m	1 056 m	1 202 m	1 360 m	1 395 m	1 581 m	1 618 m	1 876 m	2 037 m	Firmes affiliées
Unrelated firms	223 m	269 m	301 m	276 m	250 m	278 m	..	..	..	Firmes non affiliées
Other	..	..	..	..	..	..	..	..	..	Autre
Total	2 010	2 081	2 484	3 584 a	3 972	4 297	5 382	5 735 a	6 771	*Total*

	1992	1993	1994	1995	1996	1997	1998	1999	2000	
Receipts										**Recettes**
Related firms	377 m	421 m	405 m	452 m	503 m	454 m	673 m	588 m	..	Firmes affiliées
Unrelated firms	..	..	..	..	..	..	..	..	..	Firmes non affiliées
Other	..	..	..	..	..	..	..	..	..	Autre
Total	5 825	6 115	6 792	7 827	8 334	10 968	11 923	11 746	12 995 p	*Total*
Payments										**Paiements**
Related firms	2 022 m	2 017 m	1 897 m	1 957 m	1 800 m	1 632 m	1 817 m	1 386 m	..	Firmes affiliées
Unrelated firms	..	..	..	..	..	..	..	..	..	Firmes non affiliées
Other	..	..	..	..	..	..	..	..	..	Autre
Total	8 077	8 696	8 505	9 773	11 008	13 124	14 084	15 224	17 803 p	*Total*

Source: OECD, TBPdatabase (DSTI/EAS Division), Januay 2002.

Source : OCDE, base de données TBP (DSTI/Division AES), janvier 2002.

Table 1 - Tableau 1

Gross domestic expenditure on R&D (GERD) - Dépenses intérieures brutes de R-D (DIRD)

by sector of performance and source of funds - par secteur d'exécution et par source de financement

	1994	1995	1996	1997	1998	1999	2000	
	1.1 Million EUR (2001 GRD euro, EUR from 2001) - Millions EUR (2001 GRD euro, EUR à partir de 2001)							
Business enterprise								**Entreprises**
financed by								*financé par*
Business enterprise	..	87.2	75.1	90.3	..	161.8 [p]	..	Entreprises
Direct government	..	1.6	1.9	2.3	..	8.1 [p]	..	État direct
Higher education	..	0.0	0.0	0.0	..	0.0 [p]	..	Enseignement supérieur
Private non-profit	..	0.0	0.0	0.0	..	0.0 [p]	..	Institutions sans but lucratif
Funds from abroad	..	25.9	29.9	33.2	..	46.8 [p]	..	Fonds de l'étranger
Total	..	114.6	106.9	125.8	..	216.7 [p]	..	*Total*
Government								**État**
financed by								*financé par*
Business enterprise	..	2.3	..	2.0	..	2.0	..	Entreprises
Direct government	..	75.9	..	81.1	..	108.3	..	État direct
Higher education	..	0.0	..	0.0	..	0.0	..	Enseignement supérieur
Private non-profit	..	0.4	..	0.1	..	0.1	..	Institutions sans but lucratif
Funds from abroad	..	20.5	..	32.0	..	54.6	..	Fonds de l'étranger
Total	..	99.0	..	115.3	..	165.0	..	*Total*
Higher education								**Enseignement supérieur**
financed by								*financé par*
Business enterprise	..	9.6	..	13.8	..	18.8	..	Entreprises
Direct government	..	22.9	..	36.0	..	92.1	..	État direct
General university funds	..	101.8 [a]	..	143.1	..	161.4	..	Fonds généraux des universités
Sub-total government	..	124.7 [a]	..	179.1	..	253.5	..	Sous-Total État
Higher education	..	7.0	..	7.0	..	16.2	..	Enseignement supérieur
Private non-profit	..	1.6	..	0.6	..	2.1	..	Institutions sans but lucratif
Funds from abroad	..	29.3	..	48.5	..	85.8	..	Fonds de l'étranger
Total	..	172.2 [a]	..	249.0	..	376.3	..	*Total*
Private non-profit								**Institutions sans but lucratif**
financed by								*financé par*
Business enterprise	..	0.0	..	0.0	..	0.0	..	Entreprises
Direct government	..	1.0	..	0.7	..	0.8	..	État direct
Higher education	..	0.0	..	0.0	..	0.0	..	Enseignement supérieur
Private non-profit	..	0.6	..	0.3	..	0.4	..	Institutions sans but lucratif
Funds from abroad	..	1.2	..	1.1	..	1.2	..	Fonds de l'étranger
Total	..	2.8	..	2.1	..	2.3	..	*Total*
GERD								**DIRD**
financed by								*financé par*
Business enterprise	..	99.1	..	106.2	..	182.6 [p]	..	Entreprises
Direct government	..	101.3	..	120.1	..	209.3 [p]	..	État direct
General university funds	..	101.8 [a]	..	143.1	..	161.4	..	Fonds généraux des universités
Sub-total government	..	203.1 [a]	..	263.2	..	370.7 [p]	..	Sous-Total État
Higher education	..	7.0	..	7.0	..	16.2 [p]	..	Enseignement supérieur
Private non-profit	..	2.6	..	1.1	..	2.5 [p]	..	Institutions sans but lucratif
Funds from abroad	..	76.8	..	114.8	..	188.4 [p]	..	Fonds de l'étranger
Total	..	388.6 [a]	..	492.2	..	760.4 [p]	..	*Total*
	1.2 Million constant USD (1995 prices and PPPs) - Millions de USD constants (prix et PPA de 1995)							
Business enterprise								**Entreprises**
financed by								*financé par*
Business enterprise	..	146.2	117.4	132.1	..	218.8 [p]	..	Entreprises
Direct government	..	2.6	2.9	3.3	..	10.9 [p]	..	État direct
Higher education	..	0.0	0.0	0.0	..	0.0 [p]	..	Enseignement supérieur
Private non-profit	..	0.0	0.0	0.0	..	0.0 [p]	..	Institutions sans but lucratif
Funds from abroad	..	43.4	46.7	48.6	..	63.3 [p]	..	Fonds de l'étranger
Total	..	192.2	167.0	184.0	..	293.0 [p]	..	*Total*
Government								**État**
financed by								*financé par*
Business enterprise	..	3.9	..	3.0	..	2.7	..	Entreprises
Direct government	..	127.3	..	118.7	..	146.4	..	État direct
Higher education	..	0.0	..	0.0	..	0.0	..	Enseignement supérieur
Private non-profit	..	0.7	..	0.2	..	0.1	..	Institutions sans but lucratif
Funds from abroad	..	34.4	..	46.8	..	73.9	..	Fonds de l'étranger
Total	..	166.2	..	168.6	..	223.1	..	*Total*
Higher education								**Enseignement supérieur**
financed by								*financé par*
Business enterprise	..	16.1	..	20.3	..	25.4	..	Entreprises
Direct government	..	38.4	..	52.6	..	124.5	..	État direct
General university funds	..	170.8 [a]	..	209.4	..	218.2	..	Fonds généraux des universités
Sub-total government	..	209.2 [a]	..	262.0	..	342.7	..	Sous-Total État
Higher education	..	11.7	..	10.2	..	21.9	..	Enseignement supérieur
Private non-profit	..	2.8	..	0.9	..	2.8	..	Institutions sans but lucratif
Funds from abroad	..	49.1	..	70.9	..	115.9	..	Fonds de l'étranger
Total	..	288.9 [a]	..	364.3	..	508.8	..	*Total*
Private non-profit								**Institutions sans but lucratif**
financed by								*financé par*
Business enterprise	..	0.1	..	0.0	..	0.0	..	Entreprises
Direct government	..	1.7	..	1.0	..	1.0	..	État direct
Higher education	..	0.0	..	0.0	..	0.0	..	Enseignement supérieur
Private non-profit	..	1.0	..	0.5	..	0.5	..	Institutions sans but lucratif
Funds from abroad	..	2.0	..	1.6	..	1.6	..	Fonds de l'étranger
Total	..	4.7	..	3.1	..	3.1	..	*Total*
GERD								**DIRD**
financed by								*financé par*
Business enterprise	..	166.2	..	155.4	..	246.9 [p]	..	Entreprises
Direct government	..	169.9	..	175.6	..	282.9 [p]	..	État direct
General university funds	..	170.8 [a]	..	209.4	..	218.2	..	Fonds généraux des universités
Sub-total government	..	340.8 [a]	..	385.0	..	501.1 [p]	..	Sous-Total État
Higher education	..	11.7	..	10.2	..	21.9 [p]	..	Enseignement supérieur
Private non-profit	..	4.4	..	1.5	..	3.4 [p]	..	Institutions sans but lucratif
Funds from abroad	..	128.9	..	167.9	..	254.7 [p]	..	Fonds de l'étranger
Total	..	652.0 [a]	..	720.0	..	1 028.0 [p]	..	*Total*

Source: OECD, R&D database (DSTI/EAS Division), January 2002.

Source : OCDE, base de données R-D (DSTI/Division AES), janvier 2002.

Table 2 - Tableau 2

Gross domestic expenditure on R&D (GERD) - Dépenses intérieures brutes de R-D (DIRD)

by sector of performance and type of costs - par secteur d'exécution et par type de coûts

Million EUR (2001 GRD euro, EUR from 2001) — *Millions EUR (2001 GRD euro, EUR à partir de 2001)*

	1994	1995	1996	1997	1998	1999	2000	
Business enterprise								**Entreprises**
Labour costs	..	63.2	54.3	65.2	..	99.2 [p]	..	Coûts du personnel
Other current costs	..	26.6	27.5	31.1	..	68.1 [p]	..	Autres coûts courants
Total current expenditure	..	89.8	81.8	96.3	..	167.3 [p]	..	*Total dépenses courantes*
Land and buildings	..	..	..	..	..	..	..	Terrains et constructions
Instruments and equipment	..	..	..	..	..	..	..	Instruments et équipements
Total capital expenditure	..	24.8	25.1	29.5	..	49.4 [p]	..	*Total dépenses en capital*
Total	..	114.6	106.9	125.8	..	216.7 [p]	..	*Total*
Government								**État**
Labour costs	..	55.6	..	65.9	..	89.8	..	Coûts du personnel
Other current costs	..	24.9	..	29.7	..	44.7	..	Autres coûts courants
Total current expenditure	..	80.5	..	95.6	..	134.5	..	*Total dépenses courantes*
Land and buildings	..	..	..	..	..	..	..	Terrains et constructions
Instruments and equipment	..	..	..	..	..	..	..	Instruments et équipements
Total capital expenditure	..	18.5	..	19.7	..	30.5	..	*Total dépenses en capital*
Total	..	99.0	..	115.3	..	165.0	..	*Total*
Higher education								**Enseignement supérieur**
Labour costs	..	103.0 [a]	..	159.1	..	225.2	..	Coûts du personnel
Other current costs	..	41.2	..	61.9	..	93.1	..	Autres coûts courants
Total current expenditure	..	144.2 [a]	..	221.0	..	318.3	..	*Total dépenses courantes*
Land and buildings	..	..	..	..	..	..	..	Terrains et constructions
Instruments and equipment	..	..	..	..	..	..	..	Instruments et équipements
Total capital expenditure	..	28.0	..	28.1	..	58.0	..	*Total dépenses en capital*
Total	..	172.2 [a]	..	249.0	..	376.3	..	*Total*
Private non-profit								**Institutions sans but lucratif**
Labour costs	..	1.0	..	0.8	..	0.9	..	Coûts du personnel
Other current costs	..	0.9	..	0.6	..	0.6	..	Autres coûts courants
Total current expenditure	..	1.9	..	1.4	..	1.6	..	*Total dépenses courantes*
Land and buildings	..	..	..	..	..	..	..	Terrains et constructions
Instruments and equipment	..	..	..	..	..	..	..	Instruments et équipements
Total capital expenditure	..	0.9	..	0.7	..	0.7	..	*Total dépenses en capital*
Total	..	2.8	..	2.1	..	2.3	..	*Total*
GERD								**DIRD**
Labour costs	..	222.7 [a]	..	291.0	..	415.2 [p]	..	Coûts du personnel
Other current costs	..	93.6	..	123.3	..	206.6 [p]	..	Autres coûts courants
Total current expenditure	..	316.4 [a]	..	414.3	..	621.8 [p]	..	*Total dépenses courantes*
Land and buildings	..	..	..	..	..	..	..	Terrains et constructions
Instruments and equipment	..	..	..	..	..	..	..	Instruments et équipements
Total capital expenditure	..	72.2	..	78.0	..	138.6 [p]	..	*Total dépenses en capital*
Total	..	388.6 [a]	..	492.2	..	760.4 [p]	..	*Total*

Table 3 - Tableau 3

Current domestic expenditure on R&D - Dépenses intérieures courantes de R-D

by sector of performance and type of activity - par secteur d'exécution et par type d'activités

Million EUR (2001 GRD euro, EUR from 2001) — *Millions EUR (2001 GRD euro, EUR à partir de 2001)*

	1994	1995	1996	1997	1998	1999	2000	
Business enterprise								**Entreprises**
Basic research	..	..	..	..	..	..	..	Recherche fondamentale
Applied research	..	..	..	..	..	..	..	Recherche appliquée
Experimental development	..	..	..	..	..	..	..	Développement expérimental
Not specified	..	114.6 [k]	106.9 [k]	125.8 [k]	..	216.7 [k,p]	..	Non spécifié
Total	..	114.6 [k]	106.9 [k]	125.8 [k]	..	216.7 [k,p]	..	*Total*
Government								**État**
Basic research	..	..	..	..	..	..	..	Recherche fondamentale
Applied research	..	..	..	..	..	..	..	Recherche appliquée
Experimental development	..	..	..	..	..	..	..	Développement expérimental
Not specified	..	99.0 [k]	..	115.3 [k]	..	165.0 [k]	..	Non spécifié
Total	..	99.0 [k]	..	115.3 [k]	..	165.0 [k]	..	*Total*
Higher education								**Enseignement supérieur**
Basic research	..	..	..	..	..	..	..	Recherche fondamentale
Applied research	..	..	..	..	..	..	..	Recherche appliquée
Experimental development	..	..	..	..	..	..	..	Développement expérimental
Not specified	..	172.2 [a,k]	..	249.0 [k]	..	376.3 [k]	..	Non spécifié
Total	..	172.2 [a,k]	..	249.0 [k]	..	376.3 [k]	..	*Total*
Private non-profit								**Institutions sans but lucratif**
Basic research	..	..	..	..	..	..	..	Recherche fondamentale
Applied research	..	..	..	..	..	..	..	Recherche appliquée
Experimental development	..	..	..	..	..	..	..	Développement expérimental
Not specified	..	2.8 [k]	..	2.1 [k]	..	2.3 [k]	..	Non spécifié
Total	..	2.8 [k]	..	2.1 [k]	..	2.3 [k]	..	*Total*
GERD								**DIRD**
Basic research	..	..	..	..	..	..	..	Recherche fondamentale
Applied research	..	..	..	..	..	..	..	Recherche appliquée
Experimental development	..	..	..	..	..	..	..	Développement expérimental
Not specified	..	388.6 [a,k]	..	492.2 [k]	..	760.4 [k,p]	..	Non spécifié
Total	..	388.6 [a,k]	..	492.2 [k]	..	760.4 [k,p]	..	*Total*

Source: OECD, R&D database (DSTI/EAS Division), January 2002.

Source : OCDE, base de données R-D (DSTI/Division AES), janvier 2002.

Table 5 - Tableau 5

Total R&D personnel - Personnel total de R-D

by sector of employment and occupation - par secteur d'emploi et par profession

In full-time equivalent / *En équivalence plein-temps*

	1994	1995	1996	1997	1998	1999	2000	
Business enterprise								**Entreprises**
RSE	..	1 554	1 538	1 815	..	2 315	..	Chercheurs
Technicians	..	911	881	956	..	1 428	..	Techniciens
Other occupations	..	635	479	520	..	946	..	Autres professions
Total	..	3 100	2 898	3 291	..	4 690	..	*Total*
Government								**État**
RSE	..	2 012	..	1 985	..	2 000	..	Chercheurs
Technicians	..	1 243	..	1 126	..	1 111	..	Techniciens
Other occupations	..	1 653	..	1 370	..	1 320	..	Autres professions
Total	..	4 908	..	4 481	..	4 431	..	*Total*
Higher education								**Enseignement supérieur**
RSE	..	6 068 [a]	..	7 127	..	10 471	..	Chercheurs
Technicians	..	1 573 [a]	..	2 175	..	3 320	..	Techniciens
Other occupations	..	1 774 [a]	..	3 006	..	3 502	..	Autres professions
Total	..	9 415 [a]	..	12 309	..	17 294	..	*Total*
Private non-profit								**Institutions sans but lucratif**
RSE	..	71	..	45	..	41	..	Chercheurs
Technicians	..	13	..	9	..	7	..	Techniciens
Other occupations	..	65	..	39	..	32	..	Autres professions
Total	..	148	..	92	..	80	..	*Total*
TOTAL								**TOTAL**
RSE	..	9 705 [a]	..	10 972	..	14 828	..	Chercheurs
Technicians	..	3 740 [a]	..	4 265	..	5 867	..	Techniciens
Other occupations	..	4 126 [a]	..	4 935	..	5 800	..	Autres professions
Total	..	17 571 [a]	..	20 173	..	26 495	..	*Total*

Table 6 - Tableau 6

Total R&D personnel - Personnel total de R-D

by sector of employment and formal qualification - par secteur d'emploi et par niveau de qualification

In full-time equivalent / *En équivalence plein-temps*

	1994	1995	1996	1997	1998	1999	2000	
Business enterprise								**Entreprises**
University PhD level degrees	..	319	277	308	..	332	..	Diplômes univ. de niv. doctorat
Other university degrees	..	1 632	1 662	1 962	..	2 329	..	Autres diplômes universitaires
Sub-total university degrees	..	1 950	1 939	2 270	..	2 661	..	Sous-total Universitaires
Other post-secondary diplomas	..	473	433	490	..	830	..	Autres diplômes post-secondaires
Secondary diplomas	..	578	452	456	..	1 101	..	Diplômes de fin d'études secondaires
Other qualifications	..	98	73	74	..	96	..	Autres qualifications
Total	..	3 100	2 898	3 291	..	4 690	..	*Total*
Government								**État**
University PhD level degrees	..	1 071	..	1 103	..	1 175	..	Diplômes univ. de niv. doctorat
Other university degrees	..	1 709	..	1 485	..	1 473	..	Autres diplômes universitaires
Sub-total university degrees	..	2 780	..	2 589	..	2 648	..	Sous-total Universitaires
Other post-secondary diplomas	..	271	..	275	..	294	..	Autres diplômes post-secondaires
Secondary diplomas	..	1 213	..	1 151	..	1 033	..	Diplômes de fin d'études secondaires
Other qualifications	..	644	..	467	..	455	..	Autres qualifications
Total	..	4 908	..	4 481	..	4 431	..	*Total*
Higher education								**Enseignement supérieur**
University PhD level degrees	..	3 566 [a]	..	4 176	..	5 374	..	Diplômes univ. de niv. doctorat
Other university degrees	..	3 500 [a]	..	4 332	..	8 250	..	Autres diplômes universitaires
Sub-total university degrees	..	7 066 [a]	..	8 508	..	13 623	..	Sous-total Universitaires
Other post-secondary diplomas	..	850 [a]	..	1 296	..	581	..	Autres diplômes post-secondaires
Secondary diplomas	..	1 359 [a]	..	2 239	..	2 774	..	Diplômes de fin d'études secondaires
Other qualifications	..	140 [a]	..	266	..	316	..	Autres qualifications
Total	..	9 415 [a]	..	12 309	..	17 294	..	*Total*
Private non-profit								**Institutions sans but lucratif**
University PhD level degrees	..	24	..	14	..	13	..	Diplômes univ. de niv. doctorat
Other university degrees	..	62	..	40	..	36	..	Autres diplômes universitaires
Sub-total university degrees	..	85	..	54	..	49	..	Sous-total Universitaires
Other post-secondary diplomas	..	2	..	2	..	1	..	Autres diplômes post-secondaires
Secondary diplomas	..	55	..	32	..	26	..	Diplômes de fin d'études secondaires
Other qualifications	..	6	..	5	..	4	..	Autres qualifications
Total	..	148	..	92	..	80	..	*Total*
TOTAL								**TOTAL**
University PhD level degrees	..	4 980 [a]	..	5 602	..	6 894	..	Diplômes univ. de niv. doctorat
Other university degrees	..	6 902 [a]	..	7 819	..	12 088	..	Autres diplômes universitaires
Sub-total university degrees	..	11 882 [a]	..	13 421	..	18 981	..	Sous-total Universitaires
Other post-secondary diplomas	..	1 596 [a]	..	2 063	..	1 707	..	Autres diplômes post-secondaires
Secondary diplomas	..	3 205 [a]	..	3 877	..	4 934	..	Diplômes de fin d'études secondaires
Other qualifications	..	888 [a]	..	812	..	871	..	Autres qualifications
Total	..	17 571 [a]	..	20 173	..	26 495	..	*Total*

Source: OECD, R&D database (DSTI/EAS Division), January 2002.

Source : OCDE, base de données R-D (DSTI/Division AES), janvier 2002.

GREECE Table 9 - Tableau 9 **GRECE**

Total business enterprise intramural expenditure on R&D - Dépenses intra-muros totales de R-D des entreprises
(BERD) by industry - (DIRDE) par industrie

	1994	1995	1996	1997	1998	1999	2000	
	9.1 Million EUR (2001 GRD euro, EUR from 2001) - Millions EUR (2001 GRD euro, EUR à partir de 2001)							
1 AGRICULTURE, HUNTING & FORESTRY	..	1.7	1.7	1.8	..	2.0 p	..	**1 AGRICULTURE, CHASSE & SYLVICULTURE**
2 MINING	..	3.4	3.6	3.0	..	2.7 p	..	**2 ACTIVITES EXTRACTIVES**
3 MANUFACTURING	..	64.7	59.5	72.3	..	115.4 p	..	**3 ACTIVITES DE FABRICATION**
4 Food, beverages & tobacco	..	6.2	6.3	6.5	..	9.2 p	..	**4 Produits alimentaires, boissons & tabac**
5 Food products & beverages	..	6.2	6.3	6.5	..	9.2 p	..	5 Produits alimentaires & boissons
6 Tobacco products	..	0.0	0.0	0.0	..	0.0 p	..	6 Produits à base de tabac
7 Textiles, wear app., fur & leather	..	1.5	2.0	2.9	..	3.4 p	..	**7 Textiles, habill., fourrures & cuirs**
8 Textiles	..	0.6	0.6	0.6	..	2.2 p	..	8 Textiles
9 Wearing apparel & fur	..	0.6	0.5	2.2	..	1.2 p	..	9 Articles d'habillement & fourrures
10 Leather products & footwear	..	0.3	0.8	0.1	..	0.0 p	..	10 Articles en cuir & chaussures
11 Wood, paper, printing, publishing	..	3.5	3.6	4.0	..	2.8 p	..	**11 Bois, papier, imprimerie, édition**
12 Wood & cork (not furniture)	..	1.3	1.0	1.5	..	0.2 p	..	12 Bois & liège (sauf meubles)
13 Pulp, paper & paper products	..	0.7	0.0	0.0	..	0.8 p	..	13 Carton, papier & articles
14 Publ., print. & repro. of rec. media	..	1.5	2.5	2.4	..	1.7 p	..	14 Edition, impress. & repro. enregis.
15 Coke, petroleum, nuclear fuel, chemicals, rubber plastics	..	14.5	12.1	14.4	..	23.1 p	..	**15 Coke, pétrole, combust. nucl., prod. chimi., caout. & mat. plast.**
16 Coke, ref. petrol. prod. & nuclear fuel	..	1.8	2.1	2.5	..	6.6 p	..	16 Coke, raffin. pétr. & combust. nucl.
17 Chemicals & chemical products	..	9.7	7.8	9.6	..	14.2 p	..	17 Produits chimiques
18 *Chemical prod.(less pharmaceuticals)*	..	7.1	5.4	6.1	..	5.1 p	..	18 *Prod. chimi. (moins prod. pharmaceutiques)*
19 *Pharmaceuticals*	..	2.6	2.4	3.5	..	9.1 p	..	19 *Produits pharmaceutiques*
20 Rubber & plastic products	..	3.0	2.3	2.2	..	2.4 p	..	20 Caoutchouc et matières plastiques
21 Non-metallic mineral products	..	3.4	3.8	2.6	..	3.3 p	..	**21 Produits minéraux non métalliques**
22 Basic metals	..	2.3	1.0	1.1	..	5.1 p	..	**22 Produits métallurgiques de base**
23 Basic metals, ferrous	..	0.4	0.2	0.4	..	4.5 p	..	23 Produits métallur. de base, ferreux
24 Basic metals, non-ferrous	..	1.9	0.9	0.6	..	0.6 p	..	24 Prod. métallur. de base, non ferreux
25 Fabricated metal products	..	2.8	2.8	3.5	..	2.3 p	..	**25 Ouvrages en métaux (sauf mach.& mat.)**
26 Machinery equipment, instruments and transport equipment	..	30.1	27.4	36.5	..	63.5 p	..	**26 Machines & matériel, instruments & mat. de transport**
27 Machinery n.e.c.	..	5.7	5.1	6.6	..	12.9 p	..	27 Machines, n.c.a.
28 Office, account. & computing machinery	..	0.1	0.0	0.0	..	0.2 p	..	28 Mach. de bur., compta. & trait. inform.
29 Electrical machinery	..	2.3	1.5	2.2	..	5.7 p	..	29 Machines électriques
30 Electro. equip. (radio, TV & commun.)	..	18.3	16.3	21.7	..	42.0 p	..	30 Equip. électro. (radio, tv & communic.)
31 *Electro. comp. (inc. semi-conductors)*	..	0.8	0.4	1.1	..	0.6 p	..	31 *Compos. électro. (semi-cond. incl.)*
32 *TV, radio & communications equip.*	..	17.5	15.9	20.5	..	41.4 p	..	32 *Appareils radio, tv et de communi.*
33 Instruments	..	1.8	3.9	4.7	..	0.8 p	..	33 Instruments
34 Motor vehicles	..	1.4	0.0	0.2	..	0.2 p	..	34 Véhicules automobiles
35 Other transport equipment	..	0.4	0.6	1.2	..	1.6 p	..	35 Autres matériels de transport
36 *Ships*	..	0.1	0.2	0.4	..	0.9 p	..	36 *Navires*
37 *Aerospace*	..	0.3	0.4	0.8	..	0.7 p	..	37 *Aérospatiale*
38 *Other transport equipment n.e.c.*	..	0.1	0.0	0.0	..	0.0 p	..	38 *Autres matériels de transport n.c.a.*
39 Furniture, other manufacturing n.e.c.	..	0.4	0.6	0.8	..	2.6 p	..	**39 Meubles, autres activ. de fabricat. n.c.a.**
40 Furniture	..	0.3	0.2	0.5	..	0.1 p	..	40 Meubles
41 Other manufacturing n.e.c.	..	0.1	0.4	0.3	..	2.5 p	..	41 Autres activités de fabrication n.c.a.
42 Recycling	..	0.0	0.0	0.0	..	0.0 p	..	**42 Récupération**
43 ELECTRICITY, GAS & WATER SUPPLY	..	4.4	4.4	4.8	..	1.6 p	..	**43 DISTRIBUTION D'ELECTRICITE, GAZ & EAU**
44 CONSTRUCTION	..	0.3	1.0	0.9	..	24.9 p	..	**44 CONSTRUCTION**
45 SERVICES SECTOR	..	39.9	36.5	42.9	..	70.1 p	..	**45 SECTEUR DES SERVICES**
46 Wholesale, retail trade & repair of motor vehicles, etc.	..	4.5	3.9	4.2	..	8.8 p	..	**46 Commerce (gros & détail), réparation de véhicules, etc.**
47 Hotels & restaurants	..	0.2	0.1	0.0	..	0.0 p	..	**47 Hôtels et restaurants**
48 Transport & storage	..	0.2	0.2	0.3	..	0.3 p	..	**48 Transports, entreposage**
49 Communications	..	1.0	1.2	1.2	..	7.0 p	..	**49 Communications**
50 Post	..	0.0	0.0	0.0	..	0.0 p	..	50 Postes
51 Telecommunications	..	1.0	1.2	1.2	..	7.0 p	..	51 Télécommunications
52 Financial intermediation (inc. insur.)	..	4.0	0.2	1.0	..	6.6 p	..	**52 Intermédiation financ. (assurances inc.)**
53 Real estate, renting & busin. activity	..	26.6	27.9	32.9	..	43.6 p	..	**53 Immobil., locat. & serv. aux entreprises**
54 Computer & related activities	..	15.0	17.3	19.8	..	27.9 p	..	54 Informatique et activités connexes
55 *Software consultancy*	..	13.2	13.9	14.7	..	24.1 p	..	55 *Production de logiciels*
56 *Other computer services n.e.c.*	..	1.8	3.4	5.1	..	3.8 p	..	56 *Autres activités informatiques*
57 Research & development	..	5.2	5.2	5.7	..	2.9 p	..	57 Recherche & développement expérimental
58 Other business activities n.e.c.	..	6.5	5.5	7.3	..	12.9 p	..	58 Autres activ. & serv. aux entreprises
59 Comm., soc. & pers. serv. activ., etc.	..	3.5	2.9	3.4	..	3.9 p	..	**59 Adm. pub., serv. soc. & collec., etc**
TOTAL BUSINESS ENTERPRISE SECTOR	**..**	**114.6**	**106.9**	**125.8**	**..**	**216.7 p**	**..**	**TOTAL SECTEUR DES ENTREPRISES**

Source: OECD, R&D database (DSTI/EAS Division), January 2002.

Source : OCDE, base de données R-D (DSTI/Division AES), janvier 2002.

Total business enterprise intramural expenditure on R&D - Dépenses intra-muros totales de R-D des entreprises
(BERD) by industry - (DIRDE) par industrie

9.2 Million constant USD(1995 prices and PPPs) - Millions de USD constants (prix et PPA de 1995)

	1994	1995	1996	1997	1998	1999	2000	
1 **AGRICULTURE, HUNTING & FORESTRY**	..	2.9	2.7	2.7	..	2.7 p	..	1 **AGRICULTURE, CHASSE & SYLVICULTURE**
2 **MINING**	..	5.7	5.7	4.4	..	3.6 p	..	2 **ACTIVITES EXTRACTIVES**
3 **MANUFACTURING**	..	108.5	93.0	105.7	..	156.0 p	..	3 **ACTIVITES DE FABRICATION**
4 **Food, beverages & tobacco**	..	10.4	9.8	9.5	..	12.5 p	..	4 **Produits alimentaires, boissons & tabac**
5 Food products & beverages	..	10.4	9.8	9.5	..	12.5 p	..	5 Produits alimentaires & boissons
6 Tobacco products	..	0.0	0.0	0.0	..	0.0 p	..	6 Produits à base de tabac
7 **Textiles, wear app., fur & leather**	..	2.6	3.1	4.3	..	4.6 p	..	7 **Textiles, habill., fourrures & cuirs**
8 Textiles	..	1.1	0.9	0.9	..	3.0 p	..	8 Textiles
9 Wearing apparel & fur	..	1.1	0.8	3.2	..	1.6 p	..	9 Articles d'habillement & fourrures
10 Leather products & footwear	..	0.4	1.3	0.2	..	0.0 p	..	10 Articles en cuir & chaussures
11 **Wood, paper, printing, publishing**	..	5.8	5.6	5.9	..	3.8 p	..	11 **Bois, papier, imprimerie, édition**
12 Wood & cork (not furniture)	..	2.2	1.6	2.2	..	0.3 p	..	12 Bois & liège (sauf meubles)
13 Pulp, paper & paper products	..	1.2	0.1	0.0	..	1.1 p	..	13 Carton, papier & articles
14 Publ., print. & repro. of rec. media	..	2.4	3.9	3.6	..	2.4 p	..	14 Edition, impress. & repro. enregis.
15 **Coke, petroleum, nuclear fuel, chemicals, rubber plastics**	..	24.4	18.9	21.0	..	31.3 p	..	15 **Coke, pétrole, combust. nucl., prod. chimi., caout. & mat. plast.**
16 Coke, ref. petrol. prod. & nuclear fuel	..	3.0	3.2	3.7	..	8.9 p	..	16 Coke, raffin. pétr. & combust. nucl.
17 Chemicals & chemical products	..	16.3	12.1	14.0	..	19.2 p	..	17 Produits chimiques
18 *Chemical prod.(less pharmaceuticals)*	..	11.9	8.4	8.9	..	6.9 p	..	18 *Prod. chimi. (moins prod. pharmaceutiques)*
19 *Pharmaceuticals*	..	4.4	3.7	5.2	..	12.3 p	..	19 *Produits pharmaceutiques*
20 Rubber & plastic products	..	5.1	3.6	3.3	..	3.2 p	..	20 Caoutchouc et matières plastiques
21 **Non-metallic mineral products**	..	5.7	5.9	3.7	..	4.4 p	..	21 **Produits minéraux non métalliques**
22 **Basic metals**	..	3.8	1.6	1.6	..	6.9 p	..	22 **Produits métallurgiques de base**
23 Basic metals, ferrous	..	0.6	0.3	0.6	..	6.1 p	..	23 Produits métallur. de base, ferreux
24 Basic metals, non-ferrous	..	3.2	1.4	0.9	..	0.7 p	..	24 Prod. métallur. de base, non ferreux
25 **Fabricated metal products**	..	4.8	4.3	5.1	..	3.1 p	..	25 **Ouvrages en métaux (sauf mach.& mat.)**
26 **Machinery equipment, instruments and transport equipment**	..	50.5	42.8	53.4	..	85.8 p	..	26 **Machines & matériel, instruments & mat. de transport**
27 Machinery n.e.c.	..	9.6	8.0	9.7	..	17.4 p	..	27 Machines, n.c.a.
28 Office, account. & computing machinery	..	0.1	0.0	0.0	..	0.3 p	..	28 Mach. de bur., compta. & trait. inform.
29 Electrical machinery	..	3.9	2.3	3.2	..	7.8 p	..	29 Machines électriques
30 Electro. equip. (radio, TV & commun.)	..	30.7	25.4	31.7	..	56.8 p	..	30 Equip. électro. (radio, tv & communic.)
31 *Electro. comp. (inc. semi-conductors)*	..	1.3	0.6	1.6	..	0.8 p	..	31 *Compos. électro. (semi-cond. incl.)*
32 *TV, radio & communications equip.*	..	29.4	24.8	30.0	..	56.0 p	..	32 *Appareils radio, tv et de communi.*
33 Instruments	..	3.0	6.1	6.9	..	1.1 p	..	33 Instruments
34 Motor vehicles	..	2.4	0.1	0.2	..	0.3 p	..	34 Véhicules automobiles
35 Other transport equipment	..	0.7	1.0	1.7	..	2.2 p	..	35 Autres matériels de transport
36 *Ships*	..	0.1	0.3	0.6	..	1.2 p	..	36 *Navires*
37 *Aerospace*	..	0.4	0.7	1.2	..	1.0 p	..	37 *Aérospatiale*
38 *Other transport equipment n.e.c.*	..	0.1	0.0	0.0	..	0.0 p	..	38 *Autres matériels de transport n.c.a.*
39 **Furniture, other manufacturing n.e.c.**	..	0.7	0.9	1.2	..	3.6 p	..	39 **Meubles, autres activ. de fabricat. n.c.a.**
40 Furniture	..	0.4	0.3	0.7	..	0.2 p	..	40 Meubles
41 Other manufacturing n.e.c.	..	0.2	0.6	0.5	..	3.4 p	..	41 Autres activités de fabrication n.c.a.
42 **Recycling**	..	0.0	0.0	0.0	..	0.0 p	..	42 **Récupération**
43 **ELECTRICITY, GAS & WATER SUPPLY**	..	7.5	6.9	7.0	..	2.2 p	..	43 **DISTRIBUTION D'ELECTRICITE, GAZ & EAU**
44 **CONSTRUCTION**	..	0.6	1.6	1.4	..	33.7 p	..	44 **CONSTRUCTION**
45 **SERVICES SECTOR**	..	67.0	57.0	62.8	..	94.8 p	..	45 **SECTEUR DES SERVICES**
46 **Wholesale, retail trade & repair of motor vehicles, etc.**	..	7.6	6.1	6.1	..	11.9 p	..	46 **Commerce (gros & détail), réparation de véhicules, etc.**
47 **Hotels & restaurants**	..	0.3	0.1	0.0	..	0.0 p	..	47 **Hôtels et restaurants**
48 **Transport & storage**	..	0.3	0.3	0.5	..	0.4 p	..	48 **Transports, entreposage**
49 **Communications**	..	1.7	1.9	1.8	..	9.5 p	..	49 **Communications**
50 Post	..	0.0	0.0	0.0	..	0.0 p	..	50 Postes
51 Telecommunications	..	1.7	1.9	1.8	..	9.5 p	..	51 Télécommunications
52 **Financial intermediation (inc. insur.)**	..	6.7	0.4	1.4	..	8.9 p	..	52 **Intermédiation financ. (assurances inc.)**
53 **Real estate, renting & busin. activity**	..	44.7	43.6	48.1	..	58.9 p	..	53 **Immobil., locat. & serv. aux entreprises**
54 Computer & related activities	..	25.1	27.0	29.0	..	37.7 p	..	54 Informatique et activités connexes
55 *Software consultancy*	..	22.1	21.7	21.5	..	32.6 p	..	55 *Production de logiciels*
56 *Other computer services n.e.c.*	..	3.0	5.3	7.5	..	5.1 p	..	56 *Autres activités informatiques*
57 Research & development	..	8.7	8.1	8.4	..	3.9 p	..	57 Recherche & développement expérimental
58 Other business activities n.e.c.	..	10.9	8.6	10.7	..	17.4 p	..	58 Autres activ. & serv. aux entreprises
59 **Comm., soc. & pers. serv. activ., etc.**	..	5.8	4.5	4.9	..	5.2 p	..	59 **Adm. pub., serv. soc. & collec., etc**
TOTAL BUSINESS ENTERPRISE SECTOR	..	**192.2**	**167.0**	**184.0**	..	**293.0 p**	..	**TOTAL SECTEUR DES ENTREPRISES**

Source: OECD, R&D database (DSTI/EAS Division), January 2002.

Source : OCDE, base de données R-D (DSTI/Division AES), janvier 2002.

Table 17 - Tableau 17

Higher education R&D personnel (HEMP) - Personnel de R-D du secteur de l'enseignement supérieur

by field of science and occupation - par discipline scientifique et par profession

In full-time equivalent — *En équivalence plein-temps*

	1994	1995	1996	1997	1998	1999	2000	
Total R&D personnel								**Personnel total de R-D**
Natural sciences	..	..	..	..	..	..	..	Sciences exactes et naturelles
Engineering	..	..	..	..	..	..	..	Sciences de l'ingénieur
Medical sciences	..	..	..	..	..	..	..	Sciences médicales
Agricultural sciences	..	..	..	..	..	..	..	Sciences agricoles
Sub-total NSE	..	..	..	..	..	..	..	Sous-total SENI
Social sciences	..	..	..	..	..	..	..	Sciences sociales
Humanities	..	..	..	..	..	..	..	Sciences humaines
Sub-total SSH	..	..	..	..	..	..	..	Sous-total SSH
Total NSE+SSH	..	9 415	..	12 309	..	17 294	..	*Total SENI+SSH*
Researchers								**Chercheurs**
Natural sciences	..	..	..	..	..	..	..	Sciences exactes et naturelles
Engineering	..	..	..	..	..	..	..	Sciences de l'ingénieur
Medical sciences	..	..	..	..	..	..	..	Sciences médicales
Agricultural sciences	..	..	..	..	..	..	..	Sciences agricoles
Sub-total NSE	..	..	..	..	..	..	..	Sous-total SENI
Social sciences	..	..	..	..	..	..	..	Sciences sociales
Humanities	..	..	..	..	..	..	..	Sciences humaines
Sub-total SSH	..	..	..	..	..	..	..	Sous-total SSH
Total NSE+SSH	..	6 068	..	7 127	..	10 471	..	*Total SENI+SSH*
Technicians								**Technicians**
Natural sciences	..	..	..	..	..	..	..	Sciences exactes et naturelles
Engineering	..	..	..	..	..	..	..	Sciences de l'ingénieur
Medical sciences	..	..	..	..	..	..	..	Sciences médicales
Agricultural sciences	..	..	..	..	..	..	..	Sciences agricoles
Sub-total NSE	..	..	..	..	..	..	..	Sous-total SENI
Social sciences	..	..	..	..	..	..	..	Sciences sociales
Humanities	..	..	..	..	..	..	..	Sciences humaines
Sub-total SSH	..	..	..	..	..	..	..	Sous-total SSH
Total NSE+SSH	..	1 573	..	2 175	..	3 320	..	*Total SENI+SSH*
Other								**Autres**
Natural sciences	..	..	..	..	..	..	..	Sciences exactes et naturelles
Engineering	..	..	..	..	..	..	..	Sciences de l'ingénieur
Medical sciences	..	..	..	..	..	..	..	Sciences médicales
Agricultural sciences	..	..	..	..	..	..	..	Sciences agricoles
Sub-total NSE	..	..	..	..	..	..	..	Sous-total SENI
Social sciences	..	..	..	..	..	..	..	Sciences sociales
Humanities	..	..	..	..	..	..	..	Sciences humaines
Sub-total SSH	..	..	..	..	..	..	..	Sous-total SSH
Total NSE+SSH	..	1 774	..	3 006	..	3 502	..	*Total SENI+SSH*

Table 18 - Tableau 18

Government budget appropriations or outlays for R&D (GBAORD) - Crédits budgétaires publics de R-D (CBPRD)

by socio-economic objective - par objectif socio-économique

Million EUR (2001 GRD euro, EUR from 2001) — *Millions EUR (2001 GRD euro, EUR à partir de 2001)*

	1995	1996	1997	1998	1999	2000	2001	
Agriculture, forestry and fishing	24.8	24.8	25.6	26.6	24.9	26.8 p	..	Agriculture, sylviculture et pêche
Industrial development	24.8	23.8	26.9	24.5	39.7	53.2 p	..	Développement industriel
Energy	7.1	5.4	5.9	5.4	5.5	5.7 p	..	Energie
Transport & telecommunications	..	..	..	..	..	..	..	*Transports & télécommunications*
Urban & rural planning	..	..	..	..	..	..	..	*Aménagement urbain et rural*
ST Infrastructure	4.2	7.3	7.3	11.4	12.8	19.6 p	..	ST Infrastructure
Prevention of pollution	..	..	..	..	..	..	..	*Prévention de la pollution*
Identificat. & treatment of pollution	..	..	..	..	..	..	..	*Détect. pollut. & mesures antipollut.*
ST Environment	8.4	9.9	10.0	9.9	11.3	12.9 p	..	ST Environnement
Health	12.3	15.7	12.4	14.8	19.0	22.4 p	..	Santé
Social development & services	5.1	10.1	10.6	11.9	15.3	15.7 p	..	Développement et services sociaux
Earth and atmosphere	12.3	14.4	13.8	13.0	16.5	14.9 p	..	Milieu terrestre et atmosphère
Advancement of research	22.0	19.3	26.0	21.2	27.2	30.7 p	..	*Promotion de la recherche*
General university funds	105.0	126.9	149.7	156.8	170.8	176.9 p	..	*Fonds généraux des universités*
ST Advancement of knowledge	127.0	146.1	175.6	178.0	198.1	207.6 p	..	ST promotion gén. des connaiss.
Civil space	0.9	1.2	1.7	3.0	2.2	3.5 p	..	Espace civil
Defence	2.9	3.3	3.5	3.9	3.1	3.0 p	..	Défense
Not elsewhere classified	0.2	0.7	0.6	0.7	1.1	1.1 p	..	Non classé
Total	**229.9**	**262.7**	**293.9**	**302.9**	**349.4**	**386.4 p**	**..**	**Total**

Source: OECD, R&D database (DSTI/EAS Division), January 2002.

Source : OCDE, base de données R-D (DSTI/Division AES), janvier 2002.

Table 19 - Tableau 19

Technology balance of payments - Balance des paiements technologiques

by type of transfer - selon la nature des transferts

Million EUR (2001 GRD euro, EUR from 2001) — *Millions EUR (2001 GRD euro, EUR à partir de 2001)*

	1983	1984	1985	1986	1987	1988	1989	1990	1991	
Receipts										**Recettes**
Related firms	..	..	..	..	..	..	..	..	..	Firmes affiliées
Unrelated firms	..	..	..	..	..	..	..	..	..	Firmes non affiliées
Other	..	..	..	..	..	..	..	..	..	Autre
Total	..	..	..	..	..	..	..	..	..	*Total*
Payments										**Paiements**
Related firms	..	..	..	..	..	..	..	..	..	Firmes affiliées
Unrelated firms	..	..	..	..	..	..	..	..	..	Firmes non affiliées
Other	..	..	..	..	..	..	..	..	..	Autre
Total	4	4	3	4	4	5	..	..	..	*Total*

	1992	1993	1994	1995	1996	1997	1998	1999	2000	
Receipts										**Recettes**
Related firms	..	..	..	..	..	..	..	..	..	Firmes affiliées
Unrelated firms	..	..	..	..	..	..	..	..	..	Firmes non affiliées
Other	..	..	..	..	..	..	..	..	..	Autre
Total	..	..	..	..	..	..	..	..	..	*Total*
Payments										**Paiements**
Related firms	..	..	..	..	..	..	..	..	..	Firmes affiliées
Unrelated firms	..	..	..	..	..	..	..	..	..	Firmes non affiliées
Other	..	..	..	..	..	..	..	..	..	Autre
Total	..	..	..	..	..	..	..	..	..	*Total*

Source: OECD, TBPdatabase (DSTI/EAS Division), Januay 2002.

Source : OCDE, base de données TBP (DSTI/Division AES), janvier 2002.

Gross domestic expenditure on R&D (GERD) - Dépenses intérieures brutes de R-D (DIRD)

by sector of performance and source of funds - par secteur d'exécution et par source de financement

	1994	1995	1996	1997	1998	1999	2000	
	1.1 Million HUF - Millions HUF							
Business enterprise								**Entreprises**
financed by								*financé par*
Business enterprise	9 402.0 [a,v]	14 011.4 [v]	15 181.8 [v]	20 195.0 [v]	22 141.5 [v]	27 158.0 [v]	35 413.9 [v]	Entreprises
Direct government	3 143.0 [a,v]	2 907.1 [v]	2 664.8 [v]	3 744.0 [v]	2 479.4 [v]	1 781.6 [v]	2 837.4 [v]	État direct
Higher education	..	..	..	..	..	..	..	Enseignement supérieur
Private non-profit	..	..	4.3 [v]	1.5 [v]	6.3	6.1	15.9	Institutions sans but lucratif
Funds from abroad	446.0 [v]	733.4 [v]	544.7 [v]	742.7 [v]	1 479.3	1 991.1	8 026.8	Fonds de l'étranger
Total	13 704.0 [a,v]	17 895.9 [v]	19 385.0 [v]	25 628.5 [v]	26 380.3 [v]	31 458.2 [v]	46 703.5 [v]	*Total*
Government								**État**
financed by								*financé par*
Business enterprise	1 414.0 [a,v]	1 587.2 [v]	1 934.6 [v]	2 103.6 [v]	1 693.6 [v]	1 847.3 [v]	2 992.4 [v]	Entreprises
Direct government	7 769.0 [a,d,v]	7 247.4 [d,v]	8 679.7 [d,v]	11 448.5 [d,v]	17 959.1 [d,v]	20 864.3 [d,v]	21 764.4 [d,v]	État direct
Higher education	..	..	..	..	..	..	..	Enseignement supérieur
Private non-profit	..	..	26.8 [v]	26.1 [v]	66.8	58.4	50.1	Institutions sans but lucratif
Funds from abroad	705.0 [v]	875.2 [v]	993.4 [v]	1 230.4 [v]	916.1	1 297.1	1 811.3	Fonds de l'étranger
Total	10 579.0 [a,d,v]	10 537.3 [d,v]	12 723.9 [d,v]	15 515.5 [d,v]	21 391.5 [d,v]	25 247.1 [d,v]	27 494.1 [d,v]	*Total*
Higher education								**Enseignement supérieur**
financed by								*financé par*
Business enterprise	241.0 [a,v]	212.7 [v]	325.4 [v]	271.7 [v]	933.2 [v]	1 064.3 [v]	1 383.6 [v]	Entreprises
Direct government	..	..	..	..	..	..	..	État direct
General university funds	..	..	..	..	..	..	..	Fonds généraux des universités
Sub-total government	9 224.0 [a,v]	9 158.2 [v]	9 460.4 [v]	12 222.6 [v]	14 579.2 [v]	14 581.3 [v]	21 725.7 [v]	Sous-Total État
Higher education	..	..	..	..	..	..	..	Enseignement supérieur
Private non-profit	..	..	209.8 [v]	264.9 [v]	228.7	142.5	257.6	Institutions sans but lucratif
Funds from abroad	282.0 [v]	388.1 [v]	537.5 [v]	681.5 [v]	979.1	1 074.5	1 363.9	Fonds de l'étranger
Total	10 271.0 [a,v]	10 200.5 [v]	11 127.2 [v]	14 198.8 [v]	17 283.2 [v]	17 472.0 [v]	25 310.4 [v]	*Total*
Private non-profit								**Institutions sans but lucratif**
financed by								*financé par*
Business enterprise	..	..	..	..	..	..	..	Entreprises
Direct government	..	..	..	..	..	..	..	État direct
Higher education	..	..	..	..	..	..	..	Enseignement supérieur
Private non-profit	..	..	..	..	..	..	..	Institutions sans but lucratif
Funds from abroad	..	..	..	..	..	..	..	Fonds de l'étranger
Total	..	..	..	..	..	..	..	*Total*
GERD								**DIRD**
financed by								*financé par*
Business enterprise	14 760.0 [a,o,v]	15 811.3 [v]	17 441.8 [v]	22 570.3 [v]	24 768.3 [v]	30 069.6 [v]	39 789.9 [v]	Entreprises
Direct government	..	..	..	..	..	..	..	État direct
General university funds	..	..	..	..	..	..	..	Fonds généraux des universités
Sub-total government	20 731.0 [a,d,o]	21 880.0 [d,v]	22 458.2 [d,v]	33 851.6 [d,v]	38 583.1 [d,v]	41 624.2 [d,v]	52 207.1 [d,v]	Sous-Total État
Higher education	..	..	..	..	..	..	..	Enseignement supérieur
Private non-profit	57.0 [v]	192.9 [v]	240.9 [v]	292.5 [v]	301.8	207.0	323.6	Institutions sans but lucratif
Funds from abroad	1 433.0 [v]	1 996.7 [v]	2 075.6 [v]	2 654.6 [v]	3 374.5	4 362.7	11 202.0	Fonds de l'étranger
Total	38 852.0 [a,d]	41 201.0 [d]	44 889.4 [d]	61 749.3 [d]	68 620.4 [d]	78 187.9 [d]	105 387.6 [d]	*Total*
	1.2 Million constant USD (1995 prices and PPPs) - Millions de USD constants (prix et PPA de 1995)							
Business enterprise								**Entreprises**
financed by								*financé par*
Business enterprise	195.0 [a,b,v]	231.4 [v]	206.9 [v]	232.3 [v]	226.1 [v]	255.8 [v]	310.5 [v]	Entreprises
Direct government	65.2 [a,b,v]	48.0 [v]	36.3 [v]	43.1 [v]	25.3 [v]	16.8 [v]	24.9 [v]	État direct
Higher education	..	..	..	..	..	..	..	Enseignement supérieur
Private non-profit	..	..	0.1 [v]	0.0 [v]	0.1	0.1	0.1	Institutions sans but lucratif
Funds from abroad	9.2 [b,v]	12.1 [v]	7.4 [v]	8.5 [v]	15.1	18.8	70.4	Fonds de l'étranger
Total	284.2 [a,b,v]	295.5 [v]	264.2 [v]	294.8 [v]	269.4 [v]	296.3 [v]	409.4 [v]	*Total*
Government								**État**
financed by								*financé par*
Business enterprise	29.3 [a,b,v]	26.2 [v]	26.4 [v]	24.2 [v]	17.3 [v]	17.4 [v]	26.2 [v]	Entreprises
Direct government	161.1 [a,b,d]	119.7 [d,v]	118.3 [d,v]	131.7 [d,v]	183.4 [d,v]	196.5 [d,v]	190.8 [d,v]	État direct
Higher education	..	..	..	..	..	..	..	Enseignement supérieur
Private non-profit	..	..	0.4 [v]	0.3 [v]	0.7	0.6	0.4	Institutions sans but lucratif
Funds from abroad	14.6 [b,v]	14.5 [v]	13.5 [v]	14.2 [v]	9.4	12.2	15.9	Fonds de l'étranger
Total	219.4 [a,b,d]	174.0 [d,v]	173.4 [d,v]	178.5 [d,v]	218.5 [d,v]	237.8 [d,v]	241.0 [d,v]	*Total*
Higher education								**Enseignement supérieur**
financed by								*financé par*
Business enterprise	5.0 [a,b,v]	3.5 [v]	4.4 [v]	3.1 [v]	9.5 [v]	10.0 [v]	12.1 [v]	Entreprises
Direct government	..	..	..	..	..	..	..	État direct
General university funds	..	..	..	..	..	..	..	Fonds généraux des universités
Sub-total government	191.3 [a,b,v]	151.2 [v]	128.9 [v]	140.6 [v]	148.9 [v]	137.4 [v]	190.5 [v]	Sous-Total État
Higher education	..	..	..	..	..	..	..	Enseignement supérieur
Private non-profit	..	..	2.9 [v]	3.0 [v]	2.3	1.3	2.3	Institutions sans but lucratif
Funds from abroad	5.8 [b,v]	6.4 [v]	7.3 [v]	7.8 [v]	10.0	10.1	12.0	Fonds de l'étranger
Total	213.0 [a,b,v]	168.5 [v]	151.7 [v]	163.3 [v]	176.5 [v]	164.6 [v]	221.9 [v]	*Total*
Private non-profit								**Institutions sans but lucratif**
financed by								*financé par*
Business enterprise	..	..	..	..	..	..	..	Entreprises
Direct government	..	..	..	..	..	..	..	État direct
Higher education	..	..	..	..	..	..	..	Enseignement supérieur
Private non-profit	..	..	..	..	..	..	..	Institutions sans but lucratif
Funds from abroad	..	..	..	..	..	..	..	Fonds de l'étranger
Total	..	..	..	..	..	..	..	*Total*
GERD								**DIRD**
financed by								*financé par*
Business enterprise	306.0 [a,b,o]	261.1 [v]	237.7 [v]	259.7 [v]	253.0 [v]	283.2 [v]	348.8 [v]	Entreprises
Direct government	..	..	..	..	..	..	..	État direct
General university funds	..	..	..	..	..	..	..	Fonds généraux des universités
Sub-total government	429.9 [a,b,d]	361.3 [d,v]	306.1 [d,v]	389.4 [d,v]	394.1 [d,v]	392.1 [d,v]	457.7 [d,v]	Sous-Total État
Higher education	..	..	..	..	..	..	..	Enseignement supérieur
Private non-profit	1.2 [b,v]	3.2 [v]	3.3 [v]	3.4 [v]	3.1	2.0	2.8	Institutions sans but lucratif
Funds from abroad	29.7 [b,v]	33.0 [v]	28.3 [v]	30.5 [v]	34.5	41.1	98.2	Fonds de l'étranger
Total	805.6 [a,b,d]	680.4 [d]	611.8 [d]	710.4 [d]	700.8 [d]	736.5 [d]	923.9 [d]	*Total*

Source: OECD, R&D database (DSTI/EAS Division), January 2002.

Source : OCDE, base de données R-D (DSTI/Division AES), janvier 2002.

Table 2 - Tableau 2

Gross domestic expenditure on R&D (GERD) - Dépenses intérieures brutes de R-D (DIRD)

by sector of performance and type of costs - par secteur d'exécution et par type de coûts

Million HUF · *Millions HUF*

	1994	1995	1996	1997	1998	1999	2000	
Business enterprise								**Entreprises**
Labour costs	3 940.0	4 785.0	5 489.4	6 874.0	..	..	..	Coûts du personnel
Other current costs	7 952.0 [a]	10 722.0	10 713.1	13 488.9	..	..	..	Autres coûts courants
Total current expenditure	11 892.0 [a]	15 507.0	16 202.5	20 362.9	20 254.4	23 457.4	33 760.1	*Total dépenses courantes*
Land and buildings	243.0	688.0	1 217.0	1 931.3	..	..	..	Terrains et constructions
Instruments and equipment	1 569.0	1 701.0	1 965.5	3 334.3	..	..	..	Instruments et équipements
Total capital expenditure	1 812.0	2 389.0	3 182.5	5 265.6	6 125.9	8 000.8	12 943.4	*Total dépenses en capital*
Total	13 704.0 [a,v]	17 895.9 [v]	19 385.0 [v]	25 628.5 [v]	26 380.3 [v]	31 458.2 [v]	46 703.5 [v]	*Total*
Government								**État**
Labour costs	3 413.0	3 955.0	4 768.7	5 529.0	..	..	..	Coûts du personnel
Other current costs	5 674.0	5 668.0	6 962.8	8 244.0	..	..	..	Autres coûts courants
Total current expenditure	9 087.0	9 623.0	11 731.5	13 773.0	18 103.9	22 139.9	24 472.2	*Total dépenses courantes*
Land and buildings	393.0	146.0	186.8	372.5	..	..	..	Terrains et constructions
Instruments and equipment	1 099.0	768.0	805.6	1 370.0	..	..	..	Instruments et équipements
Total capital expenditure	1 492.0	914.0	992.4	1 742.5	3 287.6	3 107.2	3 021.9	*Total dépenses en capital*
Total	10 579.0 [a,d,v]	10 537.3 [d,v]	12 723.9 [d,v]	15 515.5 [d,v]	21 391.5 [d,v]	25 247.1 [d,v]	27 494.1 [d,v]	*Total*
Higher education								**Enseignement supérieur**
Labour costs	3 847.0	3 250.0	3 483.4	4 552.0	..	..	..	Coûts du personnel
Other current costs	5 048.0	5 541.0	6 486.4	8 514.1	..	..	..	Autres coûts courants
Total current expenditure	8 895.0	8 790.9	9 969.8	13 066.1	15 316.2	15 869.2	23 123.6	*Total dépenses courantes*
Land and buildings	151.0	132.0	125.3	94.4	..	..	..	Terrains et constructions
Instruments and equipment	1 225.0	1 277.0	1 032.1	1 038.3	..	..	..	Instruments et équipements
Total capital expenditure	1 376.0	1 409.6	1 157.4	1 132.7	1 966.9	1 602.8	2 186.8	*Total dépenses en capital*
Total	10 271.0 [a,v]	10 200.5 [v]	11 127.2 [v]	14 198.8 [v]	17 283.2 [v]	17 472.0 [v]	25 310.4 [v]	*Total*
Private non-profit								**Institutions sans but lucratif**
Labour costs	..	..	..	..	..	..	..	Coûts du personnel
Other current costs	..	..	..	..	..	..	..	Autres coûts courants
Total current expenditure	..	..	..	..	..	..	..	*Total dépenses courantes*
Land and buildings	..	..	..	..	..	..	..	Terrains et constructions
Instruments and equipment	..	..	..	..	..	..	..	Instruments et équipements
Total capital expenditure	..	..	..	..	..	..	..	*Total dépenses en capital*
Total	..	..	..	..	..	..	..	*Total*
GERD								**DIRD**
Labour costs	11 200.0 [v]	11 990.0 [v]	13 741.5 [v]	16 955.0 [v]	..	..	..	Coûts du personnel
Other current costs	18 674.0 [a,v]	21 931.0 [v]	24 162.2 [v]	30 247.0 [v]	..	..	..	Autres coûts courants
Total current expenditure	34 172.0 [a]	36 488.0	39 557.1	53 608.5	57 240.0	65 477.1	81 355.9	*Total dépenses courantes*
Land and buildings	787.0	966.0	1 529.1	2 398.2	..	..	..	Terrains et constructions
Instruments and equipment	3 893.0	3 746.0	3 803.2	5 742.6	..	..	..	Instruments et équipements
Total capital expenditure	4 680.0	4 713.0	5 332.3	8 140.8	11 380.4	12 710.8	24 031.7	*Total dépenses en capital*
Total	38 852.0 [a,d]	41 201.0 [d]	44 889.4 [d]	61 749.3 [d]	68 620.4 [d]	78 187.9 [d]	105 387.6 [d]	*Total*

Table 3 - Tableau 3

Current domestic expenditure on R&D - Dépenses intérieures courantes de R-D

by sector of performance and type of activity - par secteur d'exécution et par type d'activités

Million HUF · *Millions HUF*

	1994	1995	1996	1997	1998	1999	2000	
Business enterprise								**Entreprises**
Basic research	734.0 [a]	649.0	758.1	904.0	581.0	512.5	1 014.3	Recherche fondamentale
Applied research	5 324.0 [a]	6 021.0	5 317.0	6 214.0	8 135.0	6 213.3	8 892.7	Recherche appliquée
Experimental development	5 834.0 [a]	8 837.0	10 127.4	13 245.0	11 537.9	16 731.6	23 853.1	Développement expérimental
Not specified	0.0	0.0	0.0	0.0	0.0	0.0	0.0	Non spécifié
Total	11 892.0 [a]	15 507.0	16 202.5	20 362.9	20 254.4	23 457.4	33 760.1	*Total*
Government								**État**
Basic research	5 546.0	5 559.0	6 672.5	8 039.0	9 836.9	12 050.5	13 881.8	Recherche fondamentale
Applied research	2 712.0	3 285.0	4 087.5	4 420.0	5 993.2	7 353.8	8 076.6	Recherche appliquée
Experimental development	829.0	779.0	971.5	1 314.0	2 273.8	2 735.6	2 513.8	Développement expérimental
Not specified	0.0	0.0	0.0	0.0	0.0	0.0	0.0	Non spécifié
Total	9 087.0	9 623.0	11 731.5	13 773.0	18 103.9	22 139.9	24 472.2	*Total*
Higher education								**Enseignement supérieur**
Basic research	4 254.0	3 987.0	4 736.3	5 867.0	6 701.5	7 027.4	10 647.1	Recherche fondamentale
Applied research	3 486.0	3 604.0	3 894.0	5 076.0	6 082.0	5 972.2	8 605.4	Recherche appliquée
Experimental development	1 155.0	1 200.0	1 339.5	2 123.0	2 532.8	2 869.6	3 871.1	Développement expérimental
Not specified	0.0	0.0	0.0	0.0	0.0	0.0	0.0	Non spécifié
Total	8 895.0	8 790.9	9 969.8	13 066.1	15 316.3	15 869.2	23 123.6	*Total*
Private non-profit								**Institutions sans but lucratif**
Basic research	..	..	..	..	..	..	..	Recherche fondamentale
Applied research	..	..	..	..	..	..	..	Recherche appliquée
Experimental development	..	..	..	..	..	..	..	Développement expérimental
Not specified	..	..	..	..	..	..	..	Non spécifié
Total	..	..	..	..	..	..	..	*Total*
GERD								**DIRD**
Basic research	10 534.0 [a]	10 195.0	12 166.9	14 810.0	17 119.4	19 590.4	25 543.2	Recherche fondamentale
Applied research	11 522.0 [a]	12 910.0	13 298.5	15 710.0	20 210.7	19 539.3	25 574.7	Recherche appliquée
Experimental development	7 818.0 [a]	10 816.0	12 438.4	16 682.0	16 344.4	22 336.8	30 238.0	Développement expérimental
Not specified	4 298.0	2 567.0	1 653.0	6 406.5	0.0	0.0	0.0	Non spécifié
Total	34 172.0 [a]	36 488.0	39 557.1	53 608.5	57 240.0	65 477.1	81 355.9	*Total*

Source: OECD, R&D database (DSTI/EAS Division), January 2002.

Source : OCDE, base de données R-D (DSTI/Division AES), janvier 2002.

Table 5 - Tableau 5

Total R&D personnel - Personnel total de R-D

by sector of employment and occupation - par secteur d'emploi et par profession

In full-time equivalent *En équivalence plein-temps*

	1994	1995	1996	1997	1998	1999	2000	
Business enterprise								**Entreprises**
RSE	3 330	2 926	2 626	3 049	3 044	3 261	3 901	Chercheurs
Technicians	2 918	2 516	2 224	2 378	1 787	1 899	1 839	Techniciens
Other occupations	1 534	1 350	1 262	1 118	762	739	731	Autres professions
Total	7 782	6 792	6 112	6 545	5 593	5 899	6 471	*Total*
Government								**État**
RSE	3 833 [a]	3 529 [a]	3 925 [a]	3 911 [a]	4 289 [a]	4 550 [a]	4 653 [a]	Chercheurs
Technicians	1 500 [a]	1 542 [a]	1 661 [a]	1 605 [a]	1 839 [a]	2 034 [a]	2 086 [a]	Techniciens
Other occupations	1 282 [a]	1 412 [a]	1 520 [a]	1 487 [a]	1 033 [a]	1 394 [a]	1 465 [a]	Autres professions
Total	6 615 [a]	6 483 [a]	7 106 [a]	7 003 [a]	7 161 [a]	7 978 [a]	8 204 [a]	*Total*
Higher education								**Enseignement supérieur**
RSE	4 589	4 044	3 857	4 194	4 398	4 768	5 852	Chercheurs
Technicians	1 504	1 149	1 229	1 222	1 281	1 104	1 241	Techniciens
Other occupations	1 518	1 117	1 472	1 794	1 882	1 580	1 766	Autres professions
Total	7 611	6 310	6 558	7 210	7 561	7 452	8 859	*Total*
Private non-profit								**Institutions sans but lucratif**
RSE	..	..	..	..	..	..	..	Chercheurs
Technicians	..	..	..	..	..	..	..	Techniciens
Other occupations	..	..	..	..	..	..	..	Autres professions
Total	..	..	..	..	..	..	..	*Total*
TOTAL								**TOTAL**
RSE	11 752 [a]	10 499 [a]	10 408 [a]	11 154 [a]	11 731 [a]	12 579 [a]	14 406 [a]	Chercheurs
Technicians	5 922 [a]	5 207 [a]	5 114 [a]	5 205 [a]	4 907 [a]	5 037 [a]	5 166 [a]	Techniciens
Other occupations	4 334 [a]	3 879 [a]	4 254 [a]	4 399 [a]	3 677 [a]	3 713 [a]	3 962 [a]	Autres professions
Total	22 008 [a]	19 585 [a]	19 776 [a]	20 758 [a]	20 315 [a]	21 329 [a]	23 534 [a]	*Total*

Table 6 - Tableau 6

Total R&D personnel - Personnel total de R-D

by sector of employment and formal qualification - par secteur d'emploi et par niveau de qualification

In full-time equivalent *En équivalence plein-temps*

	1994	1995	1996	1997	1998	1999	2000	
Business enterprise								**Entreprises**
University PhD level degrees	..	..	..	..	..	..	..	Diplômes univ. de niv. doctorat
Other university degrees	..	..	..	..	..	..	..	Autres diplômes universitaires
Sub-total university degrees	..	..	..	..	..	..	..	Sous-total Universitaires
Other post-secondary diplomas	..	..	..	..	..	..	..	Autres diplômes post-secondaires
Secondary diplomas	..	..	..	..	..	..	..	Diplômes de fin d'études secondaires
Other qualifications	..	..	..	..	..	..	..	Autres qualifications
Total	7 782	6 792	6 112	6 545	5 593	5 899	6 471	*Total*
Government								**État**
University PhD level degrees	..	..	..	..	..	..	..	Diplômes univ. de niv. doctorat
Other university degrees	..	..	..	..	..	..	..	Autres diplômes universitaires
Sub-total university degrees	..	..	..	..	..	..	..	Sous-total Universitaires
Other post-secondary diplomas	..	..	..	..	..	..	..	Autres diplômes post-secondaires
Secondary diplomas	..	..	..	..	..	..	..	Diplômes de fin d'études secondaires
Other qualifications	..	..	..	..	..	..	..	Autres qualifications
Total	6 615 [a]	6 483 [a]	7 106 [a]	7 003 [a]	7 161 [a]	7 978 [a]	8 204 [a]	*Total*
Higher education								**Enseignement supérieur**
University PhD level degrees	..	..	..	..	..	..	..	Diplômes univ. de niv. doctorat
Other university degrees	..	..	..	..	..	..	..	Autres diplômes universitaires
Sub-total university degrees	..	..	..	..	..	..	..	Sous-total Universitaires
Other post-secondary diplomas	..	..	..	..	..	..	..	Autres diplômes post-secondaires
Secondary diplomas	..	..	..	..	..	..	..	Diplômes de fin d'études secondaires
Other qualifications	..	..	..	..	..	..	..	Autres qualifications
Total	7 611	6 310	6 558	7 210	7 561	7 452	8 859	*Total*
Private non-profit								**Institutions sans but lucratif**
University PhD level degrees	..	..	..	..	..	..	..	Diplômes univ. de niv. doctorat
Other university degrees	..	..	..	..	..	..	..	Autres diplômes universitaires
Sub-total university degrees	..	..	..	..	..	..	..	Sous-total Universitaires
Other post-secondary diplomas	..	..	..	..	..	..	..	Autres diplômes post-secondaires
Secondary diplomas	..	..	..	..	..	..	..	Diplômes de fin d'études secondaires
Other qualifications	..	..	..	..	..	..	..	Autres qualifications
Total	..	..	..	..	..	..	..	*Total*
TOTAL								**TOTAL**
University PhD level degrees	..	..	..	..	..	..	..	Diplômes univ. de niv. doctorat
Other university degrees	..	..	..	..	..	..	..	Autres diplômes universitaires
Sub-total university degrees	..	..	..	..	..	..	..	Sous-total Universitaires
Other post-secondary diplomas	..	..	..	..	..	..	..	Autres diplômes post-secondaires
Secondary diplomas	..	..	..	..	..	..	..	Diplômes de fin d'études secondaires
Other qualifications	..	..	..	..	..	..	..	Autres qualifications
Total	22 008 [a]	19 585 [a]	19 776 [a]	20 758 [a]	20 315 [a]	21 329 [a]	23 534 [a]	*Total*

Source: OECD, R&D database (DSTI/EAS Division), January 2002.

Source : OCDE, base de données R-D (DSTI/Division AES), janvier 2002.

Table 7 - Tableau 7

Gross domestic expenditure on R&D (GERD) - Dépenses intérieures brutes de R-D (DIRD)

by sector of performance and main scientific field - par secteur d'exécution et par grand domaine scientifique

Million HUF | *Millions HUF*

	1994	1995	1996	1997	1998	1999	2000	
Business enterprise								**Entreprises**
Natural sciences and engineering	13 480.0 [a]	17 740.3	19 172.1	25 280.0	..	..	..	Sciences exactes, nat. et de l'ingénieur
Social sciences and humanities	..	155.6	212.9	348.5	..	..	..	Sciences sociales et humaines
All fields of science	13 704.0 [a,v]	17 895.9 [v]	19 385.0 [v]	25 628.5 [v]	26 380.3 [v]	31 458.2 [v]	46 703.5 [v]	*Tous domaines scientifiques*
Government								**État**
Natural sciences and engineering	8 105.0	8 019.5	9 292.0	10 738.9	..	..	..	Sciences exactes, nat. et de l'ingénieur
Social sciences and humanities	2 474.0	2 517.8	2 711.4	3 891.6	..	..	..	Sciences sociales et humaines
All fields of science	10 579.0 [a,d,v]	10 537.3 [d,v]	12 723.9 [d,v]	15 515.5 [d,v]	21 391.5 [d,v]	25 247.1 [d,v]	27 494.1 [d,v]	*Tous domaines scientifiques*
Higher education								**Enseignement supérieur**
Natural sciences and engineering	8 375.0	8 173.8	8 945.6	10 705.1	..	..	..	Sciences exactes, nat. et de l'ingénieur
Social sciences and humanities	1 896.0	2 026.7	2 181.6	3 371.1	..	..	..	Sciences sociales et humaines
All fields of science	10 271.0 [a,v]	10 200.5 [v]	11 127.2 [v]	14 198.8 [v]	17 283.2 [v]	17 472.0 [v]	25 310.4 [v]	*Tous domaines scientifiques*
Private non-profit								**Institutions sans but lucratif**
Natural sciences and engineering	..	..	..	..	..	..	..	Sciences exactes, nat. et de l'ingénieur
Social sciences and humanities	..	..	..	..	..	..	..	Sciences sociales et humaines
All fields of science	..	..	..	..	..	..	..	*Tous domaines scientifiques*
TOTAL								**TOTAL**
Natural sciences and engineering	33 430.0 [a,v]	33 933.6 [v]	37 409.7 [v]	46 724.0 [v]	..	..	..	Sciences exactes, nat. et de l'ingénieur
Social sciences and humanities	4 594.0 [v]	4 700.1 [v]	5 105.9 [v]	7 611.2 [v]	..	..	..	Sciences sociales et humaines
All fields of science	38 852.0 [a,d]	41 201.0 [d]	44 889.4 [d]	61 749.3 [d]	68 620.4 [d]	78 187.9 [d]	105 387.6 [d]	*Tous domaines scientifiques*

Table 8 - Tableau 8

Total R&D personnel - Personnel total de R-D

by sector of employment and main scientific field - par secteur d'emploi et par grand domaine scientifique

In full-time equivalent | *En équivalence plein-temps*

	1994	1995	1996	1997	1998	1999	2000	
Business enterprise								**Entreprises**
Natural sciences and engineering	7 622	6 773	6 040	..	..	..	..	Sciences exactes, nat. et de l'ingénieur
Social sciences and humanities	160	19	72	..	..	..	..	Sciences sociales et humaines
All fields of science	7 782	6 792	6 112	6 545	5 593	5 899	6 471	*Tous domaines scientifiques*
Government								**État**
Natural sciences and engineering	4 629	4 455	4 768	..	..	..	..	Sciences exactes, nat. et de l'ingénieur
Social sciences and humanities	1 986	2 028	2 338	..	..	..	..	Sciences sociales et humaines
All fields of science	6 615 [d]	6 483 [d]	7 106 [d]	7 003 [d]	7 161 [d]	7 978 [d]	8 204 [d]	*Tous domaines scientifiques*
Higher education								**Enseignement supérieur**
Natural sciences and engineering	6 141	5 035	5 284	5 469	..	..	..	Sciences exactes, nat. et de l'ingénieur
Social sciences and humanities	1 470	1 275	1 274	1 688	..	..	..	Sciences sociales et humaines
All fields of science	7 611	6 310	6 558	7 210	7 561	7 452	8 859	*Tous domaines scientifiques*
Private non-profit								**Institutions sans but lucratif**
Natural sciences and engineering	..	..	..	..	..	..	..	Sciences exactes, nat. et de l'ingénieur
Social sciences and humanities	..	..	..	..	..	..	..	Sciences sociales et humaines
All fields of science	..	..	..	..	..	..	..	*Tous domaines scientifiques*
TOTAL								**TOTAL**
Natural sciences and engineering	18 392	16 263	16 092	..	..	..	..	Sciences exactes, nat. et de l'ingénieur
Social sciences and humanities	3 616	3 322	3 684	..	..	..	..	Sciences sociales et humaines
All fields of science	22 008 [d]	19 585 [d]	19 776 [d]	20 758 [d]	20 315 [d]	21 329 [d]	23 534 [d]	*Tous domaines scientifiques*

Source: OECD, R&D database (DSTI/EAS Division), January 2002.

Source : OCDE, base de données R-D (DSTI/Division AES), janvier 2002.

Total business enterprise intramural expenditure on R&D - Dépenses intra-muros totales de R-D des entreprises

(BERD) by industry - (DIRDE) par industrie

	1994	1995	1996	1997	1998	1999	2000	
	9.1 Million HUF - Millions HUF							
1 AGRICULTURE, HUNTING & FORESTRY	2 678.0 [a]	2 918.0	2 457.1	2 388.4	1 239.2	690.7	555.0	**1 AGRICULTURE, CHASSE & SYLVICULTURE**
2 MINING	14.0 [a]	39.0	3.4	5.5	0.0	11.0	1.0	**2 ACTIVITES EXTRACTIVES**
3 MANUFACTURING	9 813.0 [a]	13 650.0	14 534.3	20 013.0	22 185.0	25 718.4	36 879.1	**3 ACTIVITES DE FABRICATION**
4 Food, beverages & tobacco	229.0 [a]	281.0	616.2	830.3	682.4	900.9	987.2	**4 Produits alimentaires, boissons & tabac**
5 Food products & beverages	229.0 [a]	271.0	594.4	792.6	637.9	669.0	759.6	5 Produits alimentaires & boissons
6 Tobacco products	0.0	10.0	21.8	37.7	44.4	231.9	227.6	6 Produits à base de tabac
7 Textiles, wear app., fur & leather	47.0 [a]	67.0	42.4	51.2	136.3	103.4	39.1	**7 Textiles, habill., fourrures & cuirs**
8 Textiles	25.0 [a]	51.0	27.2	12.6	97.9	53.9	39.1	8 Textiles
9 Wearing apparel & fur	6.0 [a]	0.0	0.0	29.6	38.3	49.1	0.0	9 Articles d'habillement & fourrures
10 Leather products & footwear	16.0 [a]	16.0	15.2	9.0	0.0	0.0	0.0	10 Articles en cuir & chaussures
11 Wood, paper, printing, publishing	37.0 [a]	37.0	14.9	96.8	74.9	186.6	223.8	**11 Bois, papier, imprimerie, édition**
12 Wood & cork (not furniture)	..	0.0	0.0	0.0	0.0	10.0	11.5	12 Bois & liège (sauf meubles)
13 Pulp, paper & paper products	..	37.0	..	88.8	52.2	86.2	129.7	13 Carton, papier & articles
14 Publ., print. & repro. of rec. media	..	0.0	..	8.0	22.7	90.4	82.6	14 Edition, impress. & repro. enregis.
15 Coke, petroleum, nuclear fuel, chemicals, rubber plastics	7 232.0 [a]	10 084.0	11 104.1	13 899.8	14 696.8	17 185.3	19 699.8	**15 Coke, pétrole, combust. nucl., prod. chimi., caout. & mat. plast.**
16 Coke, ref. petrol. prod. & nuclear fuel	961.0 [a]	1 414.0	824.7	887.9	1 746.4	1 198.7	362.4	16 Coke, raffin. pétr. & combust. nucl.
17 Chemicals & chemical products	6 036.0 [a]	8 307.0	9 894.3	12 344.5	12 193.1	15 298.9	18 695.4	17 Produits chimiques
18 *Chemical prod.(less pharmaceuticals)*	564.0 [a]	1 413.0	1 577.4	1 393.8	1 288.8	1 106.7	1 382.6	18 *Prod. chimi. (moins prod. pharmaceutiques)*
19 *Pharmaceuticals*	5 472.0 [a]	6 894.0	8 316.9	10 950.7	10 904.3	14 192.2	17 312.8	19 *Produits pharmaceutiques*
20 Rubber & plastic products	235.0 [a]	363.0	385.1	667.5	757.3	687.7	641.9	20 Caoutchouc et matières plastiques
21 Non-metallic mineral products	57.0 [a]	211.0	109.9	67.9	111.8	91.5	101.6	**21 Produits minéraux non métalliques**
22 Basic metals	198.0 [a]	477.0	298.2	59.7	45.2	124.3	196.9	**22 Produits métallurgiques de base**
23 Basic metals, ferrous	..	135.0	155.3	0.0	0.0	14.4	132.5	23 Produits métallur. de base, ferreux
24 Basic metals, non-ferrous	..	342.0	142.9	59.7	45.2	109.9	64.4	24 Prod. métallur. de base, non ferreux
25 Fabricated metal products	164.0 [a]	163.0	50.6	123.7	48.7	130.3	111.3	**25 Ouvrages en métaux (sauf mach.& mat.)**
26 Machinery equipment, instruments and transport equipment	1 837.0 [a]	2 315.0	2 280.3	4 878.1	6 381.8	6 979.9	15 437.6	**26 Machines & matériel, instruments & mat. de transport**
27 Machinery n.e.c.	240.0 [a]	600.0	541.8	1 838.9	2 293.7	1 730.0	1 472.8	27 Machines, n.c.a.
28 Office, account. & computing machinery	3.0 [a]	35.0	0.0	164.7	68.3	125.1	140.9	28 Mach. de bur., compta. & trait. inform.
29 Electrical machinery	338.0 [a]	191.0	240.4	274.2	861.0	1 470.0	7 003.2	29 Machines électriques
30 Electro. equip. (radio, TV & commun.)	545.0 [a]	568.0	355.8	780.1	1 011.0	1 225.3	4 236.7	30 Equip. électro. (radio, tv & communic.)
31 *Electro. comp. (inc. semi-conductors)*	8.0 [a]	24.0	9.0	78.0	119.1	161.4	321.2	31 *Compos. électro. (semi-cond. incl.)*
32 *TV, radio & communications equip.*	537.0 [a]	544.0	346.8	702.1	891.9	1 063.9	3 915.5	32 *Appareils radio, tv et de communi.*
33 Instruments	354.0 [a]	506.0	488.8	895.2	826.8	550.7	974.2	33 Instruments
34 Motor vehicles	357.0 [a]	402.0	645.6	925.0	1 320.6	1 878.7	1 609.8	34 Véhicules automobiles
35 Other transport equipment	0.0	13.0	7.9	0.0	0.0	..	..	35 Autres matériels de transport
36 *Ships*	0.0	0.0	0.0	0.0	0.0	..	..	36 *Navires*
37 *Aerospace*	0.0	0.0	0.0	0.0	0.0	..	..	37 *Aérospatiale*
38 *Other transport equipment n.e.c.*	0.0	13.0	7.9	0.0	0.0	..	..	38 *Autres matériels de transport n.c.a.*
39 Furniture, other manufacturing n.e.c.	12.0 [a]	15.0	17.7	5.5	7.5	16.1	69.6	**39 Meubles, autres activ. de fabricat. n.c.a.**
40 Furniture	6.0 [a]	..	..	0.9	4.2	15.1	61.5	40 Meubles
41 Other manufacturing n.e.c.	6.0 [a]	..	..	4.6	3.3	1.0	8.1	41 Autres activités de fabrication n.c.a.
42 Recycling	0.0	0.0	0.0	0.0	0.0	..	12.1	**42 Récupération**
43 ELECTRICITY, GAS & WATER SUPPLY	417.0 [a]	401.0	711.5	759.7	315.6	339.9	406.2	**43 DISTRIBUTION D'ELECTRICITE, GAZ & EAU**
44 CONSTRUCTION	65.0 [a]	84.0	26.1	21.1	25.3	39.9	53.2	**44 CONSTRUCTION**
45 SERVICES SECTOR	717.0 [a]	804.0	1 652.6	2 440.8	2 615.2	4 658.3	8 809.0	**45 SECTEUR DES SERVICES**
46 Wholesale, retail trade & repair of motor vehicles, etc.	81.0 [a]	43.0	31.7	168.1	137.8	386.7	3 794.4	**46 Commerce (gros & détail), réparation de véhicules, etc.**
47 Hotels & restaurants	0.0	0.0	0.0	0.0	0.0	..	..	**47 Hôtels et restaurants**
48 Transport & storage	..	247.0	159.8	125.4	87.4	103.6	130.2	**48 Transports, entreposage**
49 Communications	..	263.0	989.8	1 030.9	620.2	1 329.0	1 394.3	**49 Communications**
50 Post	0.0	0.0	0.0	0.0	0.0	54.3	113.4	50 Postes
51 Telecommunications	..	263.0	989.8	1 030.9	620.2	1 274.7	1 280.9	51 Télécommunications
52 Financial intermediation (inc. insur.)	0.0	0.0	0.0	46.9	0.0	..	26.3	**52 Intermédiation financ. (assurances inc.)**
53 Real estate, renting & busin. activity	465.0 [a]	232.0	439.4	1 010.9	1 710.4	2 755.7	3 209.0	**53 Immobil., locat. & serv. aux entreprises**
54 Computer & related activities	191.0 [a]	62.0	193.6	485.0	707.4	843.1	1 228.5	54 Informatique et activités connexes
55 *Software consultancy*	32.0 [a]	33.0	41.3	448.9	470.6	639.1	878.5	55 *Production de logiciels*
56 *Other computer services n.e.c.*	159.0 [a]	29.0	152.2	36.1	236.7	204.0	350.0	56 *Autres activités informatiques*
57 Research & development	172.0 [a]	101.0	240.3	134.2	295.5	320.3	335.8	57 Recherche & développement expérimental
58 Other business activities n.e.c.	102.0 [a]	69.0	5.5	391.7	707.6	1 592.3	1 644.6	58 Autres activ. & serv. aux entreprises
59 Comm., soc. & pers. serv. activ., etc.	17.0 [a]	19.0	31.9	58.6	59.4	83.3	254.8	**59 Adm. pub., serv. soc. & collec., etc**
TOTAL BUSINESS ENTERPRISE SECTOR	**13 704.0** [a,v]	**17 895.9** [v]	**19 385.0** [v]	**25 628.5** [v]	**26 380.3** [v]	**31 458.2** [v]	**46 703.5** [v]	**TOTAL SECTEUR DES ENTREPRISES**

Source: OECD, R&D database (DSTI/EAS Division), January 2002.

Source : OCDE, base de données R-D (DSTI/Division AES), janvier 2002.

Total business enterprise intramural expenditure on R&D - Dépenses intra-muros totales de R-D des entreprises (BERD) by industry - (DIRDE) par industrie

9.2 Million constant USD(1995 prices and PPPs) - Millions de USD constants (prix et PPA de 1995)

	1994	1995	1996	1997	1998	1999	2000	
1 AGRICULTURE, HUNTING & FORESTRY	55.5 [a,b]	48.2	33.5	27.5	12.7	6.5	4.9	**1 AGRICULTURE, CHASSE & SYLVICULTURE**
2 MINING	0.3 [a,b]	0.6	0.0	0.1	0.0	0.1	0.0	**2 ACTIVITES EXTRACTIVES**
3 MANUFACTURING	203.5 [a,b]	225.4	198.1	230.2	226.6	242.3	323.3	**3 ACTIVITES DE FABRICATION**
4 Food, beverages & tobacco	4.7 [a,b]	4.6	8.4	9.6	7.0	8.5	8.7	**4 Produits alimentaires, boissons & tabac**
5 Food products & beverages	4.7 [a,b]	4.5	8.1	9.1	6.5	6.3	6.7	5 Produits alimentaires & boissons
6 Tobacco products	0.0 [b]	0.2	0.3	0.4	0.5	2.2	2.0	6 Produits à base de tabac
7 Textiles, wear app., fur & leather	1.0 [a,b]	1.1	0.6	0.6	1.4	1.0	0.3	**7 Textiles, habill., fourrures & cuirs**
8 Textiles	0.5 [a,b]	0.8	0.4	0.1	1.0	0.5	0.3	8 Textiles
9 Wearing apparel & fur	0.1 [a,b]	0.0	0.0	0.3	0.4	0.5	0.0	9 Articles d'habillement & fourrures
10 Leather products & footwear	0.3 [a,b]	0.3	0.2	0.1	0.0	0.0	0.0	10 Articles en cuir & chaussures
11 Wood, paper, printing, publishing	0.8 [a,b]	0.6	0.2	1.1	0.8	1.8	2.0	**11 Bois, papier, imprimerie, édition**
12 Wood & cork (not furniture)	..	0.0	0.0	0.0	0.0	0.1	0.1	12 Bois & liège (sauf meubles)
13 Pulp, paper & paper products	..	0.6	..	1.0	0.5	0.8	1.1	13 Carton, papier & articles
14 Publ., print. & repro. of rec. media	..	0.0	..	0.1	0.2	0.9	0.7	14 Edition, impress. & repro. enregis.
15 Coke, petroleum, nuclear fuel, chemicals, rubber plastics	150.0 [a,b]	166.5	151.3	159.9	150.1	161.9	172.7	**15 Coke, pétrole, combust. nucl., prod. chimi., caout. & mat. plast.**
16 Coke, ref. petrol. prod. & nuclear fuel	19.9 [a,b]	23.4	11.2	10.2	17.8	11.3	3.2	16 Coke, raffin. pétr. & combust. nucl.
17 Chemicals & chemical products	125.2 [a,b]	137.2	134.8	142.0	124.5	144.1	163.9	17 Produits chimiques
18 *Chemical prod.(less pharmaceuticals)*	11.7 [a,b]	23.3	21.5	16.0	13.2	10.4	12.1	18 *Prod. chimi. (moins prod. pharmaceutiques)*
19 *Pharmaceuticals*	113.5 [a,b]	113.9	113.3	126.0	111.4	133.7	151.8	19 *Produits pharmaceutiques*
20 Rubber & plastic products	4.9 [a,b]	6.0	5.2	7.7	7.7	6.5	5.6	20 Caoutchouc et matières plastiques
21 Non-metallic mineral products	1.2 [a,b]	3.5	1.5	0.8	1.1	0.9	0.9	**21 Produits minéraux non métalliques**
22 Basic metals	4.1 [a,b]	7.9	4.1	0.7	0.5	1.2	1.7	**22 Produits métallurgiques de base**
23 Basic metals, ferrous	..	2.2	2.1	0.0	0.0	0.1	1.2	23 Produits métallur. de base, ferreux
24 Basic metals, non-ferrous	..	5.6	1.9	0.7	0.5	1.0	0.6	24 Prod. métallur. de base, non ferreux
25 Fabricated metal products	3.4 [a,b]	2.7	0.7	1.4	0.5	1.2	1.0	**25 Ouvrages en métaux (sauf mach.& mat.)**
26 Machinery equipment, instruments and transport equipment	38.1 [a,b]	38.2	31.1	56.1	65.2	65.7	135.3	**26 Machines & matériel, instruments & mat. de transport**
27 Machinery n.e.c.	5.0 [a,b]	9.9	7.4	21.2	23.4	16.3	12.9	27 Machines, n.c.a.
28 Office, account. & computing machinery	0.1 [a,b]	0.6	0.0	1.9	0.7	1.2	1.2	28 Mach. de bur., compta. & trait. inform.
29 Electrical machinery	7.0 [a,b]	3.2	3.3	3.2	8.8	13.8	61.4	29 Machines électriques
30 Electro. equip. (radio, TV & commun.)	11.3 [a,b]	9.4	4.8	9.0	10.3	11.5	37.1	30 Equip. électro. (radio, tv & communic.)
31 *Electro. comp. (inc. semi-conductors)*	0.2 [a,b]	0.4	0.1	0.9	1.2	1.5	2.8	31 *Compos. électro. (semi-cond. incl.)*
32 *TV, radio & communications equip.*	11.1 [a,b]	9.0	4.7	8.1	9.1	10.0	34.3	32 *Appareils radio, tv et de communi.*
33 Instruments	7.3 [a,b]	8.4	6.7	10.3	8.4	5.2	8.5	33 Instruments
34 Motor vehicles	7.4 [a,b]	6.6	8.8	10.6	13.5	17.7	14.1	34 Véhicules automobiles
35 Other transport equipment	0.0 [b]	0.2	0.1	0.0	0.0	..	..	35 Autres matériels de transport
36 *Ships*	0.0 [b]	0.0	0.0	0.0	0.0	..	..	36 *Navires*
37 *Aerospace*	0.0 [b]	0.0	0.0	0.0	0.0	..	..	37 *Aérospatiale*
38 *Other transport equipment n.e.c.*	0.0 [b]	0.2	0.1	0.0	0.0	..	..	38 *Autres matériels de transport n.c.a.*
39 Furniture, other manufacturing n.e.c.	0.2 [a,b]	0.2	0.2	0.1	0.1	0.2	0.6	**39 Meubles, autres activ. de fabricat. n.c.a.**
40 Furniture	0.1 [a,b]	..	..	0.0	0.0	0.1	0.5	40 Meubles
41 Other manufacturing n.e.c.	0.1 [a,b]	..	..	0.1	0.0	0.0	0.1	41 Autres activités de fabrication n.c.a.
42 Recycling	0.0 [b]	0.0	0.0	0.0	0.0	..	0.1	**42 Récupération**
43 ELECTRICITY, GAS & WATER SUPPLY	8.6 [a,b]	6.6	9.7	8.7	3.2	3.2	3.6	**43 DISTRIBUTION D'ELECTRICITE, GAZ & EAU**
44 CONSTRUCTION	1.3 [a,b]	1.4	0.4	0.2	0.3	0.4	0.5	**44 CONSTRUCTION**
45 SERVICES SECTOR	14.9 [a,b]	13.3	22.5	28.1	26.7	43.9	77.2	**45 SECTEUR DES SERVICES**
46 Wholesale, retail trade & repair of motor vehicles, etc.	1.7 [a,b]	0.7	0.4	1.9	1.4	3.6	33.3	**46 Commerce (gros & détail), réparation de véhicules, etc.**
47 Hotels & restaurants	0.0 [b]	0.0	0.0	0.0	0.0	..	..	**47 Hôtels et restaurants**
48 Transport & storage	..	4.1	2.2	1.4	0.9	1.0	1.1	**48 Transports, entreposage**
49 Communications	..	4.3	13.5	11.9	6.3	12.5	12.2	**49 Communications**
50 Post	0.0 [b]	0.0	0.0	0.0	0.0	0.5	1.0	50 Postes
51 Telecommunications	..	4.3	13.5	11.9	6.3	12.0	11.2	51 Télécommunications
52 Financial intermediation (inc. insur.)	0.0 [b]	0.0	0.0	0.5	0.0	..	0.2	**52 Intermédiation financ. (assurances inc.)**
53 Real estate, renting & busin. activity	9.6 [a,b]	3.8	6.0	11.6	17.5	26.0	28.1	**53 Immobil., locat. & serv. aux entreprises**
54 Computer & related activities	4.0 [a,b]	1.0	2.6	5.6	7.2	7.9	10.8	54 Informatique et activités connexes
55 *Software consultancy*	0.7 [a,b]	0.5	0.6	5.2	4.8	6.0	7.7	55 *Production de logiciels*
56 *Other computer services n.e.c.*	3.3 [a,b]	0.5	2.1	0.4	2.4	1.9	3.1	56 *Autres activités informatiques*
57 Research & development	3.6 [a,b]	1.7	3.3	1.5	3.0	3.0	2.9	57 Recherche & développement expérimental
58 Other business activities n.e.c.	2.1 [a,b]	1.1	0.1	4.5	7.2	15.0	14.4	58 Autres activ. & serv. aux entreprises
59 Comm., soc. & pers. serv. activ., etc.	0.4 [a,b]	0.3	0.4	0.7	0.6	0.8	2.2	**59 Adm. pub., serv. soc. & collec., etc**
TOTAL BUSINESS ENTERPRISE SECTOR	**284.2 [a,b,]**	**295.5 [v]**	**264.2 [v]**	**294.8 [v]**	**269.4 [v]**	**296.3 [v]**	**409.4 [v]**	**TOTAL SECTEUR DES ENTREPRISES**

Source: OECD, R&D database (DSTI/EAS Division), January 2002.

Source : OCDE, base de données R-D (DSTI/Division AES), janvier 2002.

Total business enterprise R&D personnel (BEMP) - Personnel total de R-D du secteur des entreprises (PRDE)

by industry - par industrie

In full time equivalent / *En équivalence plein-temps*

	1994	1995	1996	1997	1998	1999	2000	
1 AGRICULTURE, HUNTING & FORESTRY	1 864	1 613	1 235	1 075	..	..	..	**1 AGRICULTURE, CHASSE & SYLVICULTURE**
2 MINING	6	8	4	4	..	..	..	**2 ACTIVITES EXTRACTIVES**
3 MANUFACTURING	5 036	4 492	4 133	4 545	..	..	..	**3 ACTIVITES DE FABRICATION**
4 Food, beverages & tobacco	80	109	250	226	..	..	..	**4 Produits alimentaires, boissons & tabac**
5 Food products & beverages	80	106	232	209	..	..	..	5 Produits alimentaires & boissons
6 Tobacco products	0	3	18	17	..	..	..	6 Produits à base de tabac
7 Textiles, wear app., fur & leather	21	34	29	20	..	..	..	**7 Textiles, habill., fourrures & cuirs**
8 Textiles	8	18	19	9	..	..	..	8 Textiles
9 Wearing apparel & fur	6	0	0	6	..	..	..	9 Articles d'habillement & fourrures
10 Leather products & footwear	7	16	10	5	..	..	..	10 Articles en cuir & chaussures
11 Wood, paper, printing, publishing	21	9	12	24	..	..	..	**11 Bois, papier, imprimerie, édition**
12 Wood & cork (not furniture)	..	0	0	0	..	..	..	12 Bois & liège (sauf meubles)
13 Pulp, paper & paper products	..	9	..	22	..	..	..	13 Carton, papier & articles
14 Publ., print. & repro. of rec. media	..	0	..	2	..	..	..	14 Edition, impress. & repro. enregis.
15 Coke, petroleum, nuclear fuel, chemicals, rubber plastics	3 271	2 734	2 596	2 793	..	..	..	**15 Coke, pétrole, combust. nucl., prod. chimi., caout. & mat. plast.**
16 Coke, ref. petrol. prod. & nuclear fuel	194	203	210	209	..	..	..	16 Coke, raffin. pétr. & combust. nucl.
17 Chemicals & chemical products	2 981	2 354	2 248	2 415	..	..	..	17 Produits chimiques
18 *Chemical prod.(less pharmaceuticals)*	601	694	627	524	..	..	..	18 *Prod. chimi. (moins prod. pharmaceutiques)*
19 *Pharmaceuticals*	2 380	1 660	1 621	1 891	..	..	..	19 *Produits pharmaceutiques*
20 Rubber & plastic products	96	177	138	169	..	..	..	20 Caoutchouc et matières plastiques
21 Non-metallic mineral products	14	58	36	25	..	..	..	**21 Produits minéraux non métalliques**
22 Basic metals	99	153	130	44	..	..	..	**22 Produits métallurgiques de base**
23 Basic metals, ferrous	..	42	78	0	..	..	..	23 Produits métallur. de base, ferreux
24 Basic metals, non-ferrous	..	111	52	44	..	..	..	24 Prod. métallur. de base, non ferreux
25 Fabricated metal products	112	96	54	64	..	..	..	**25 Ouvrages en métaux (sauf mach.& mat.)**
26 Machinery equipment, instruments and transport equipment	1 394	1 284	1 020	1 346	..	..	..	**26 Machines & matériel, instruments & mat. de transport**
27 Machinery n.e.c.	165	399	353	402	..	..	..	27 Machines, n.c.a.
28 Office, account. & computing machinery	5	31	0	33	..	..	..	28 Mach. de bur., compta. & trait. inform.
29 Electrical machinery	180	153	94	82	..	..	..	29 Machines électriques
30 Electro. equip. (radio, TV & commun.)	489	306	119	155	..	..	..	30 Equip. électro. (radio, tv & communic.)
31 *Electro. comp. (inc. semi-conductors)*	117	15	6	36	..	..	..	31 *Compos. électro. (semi-cond. incl.)*
32 *TV, radio & communications equip.*	372	291	113	119	..	..	..	32 *Appareils radio, tv et de communi.*
33 Instruments	221	187	177	265	..	..	..	33 Instruments
34 Motor vehicles	334	195	271	409	..	..	..	34 Véhicules automobiles
35 Other transport equipment	0	13	6	0	..	..	..	35 Autres matériels de transport
36 *Ships*	0	0	0	0	..	..	..	36 *Navires*
37 *Aerospace*	0	0	0	0	..	..	..	37 *Aérospatiale*
38 *Other transport equipment n.e.c.*	0	13	6	0	..	..	..	38 *Autres matériels de transport n.c.a.*
39 Furniture, other manufacturing n.e.c.	24	15	6	3	..	..	..	**39 Meubles, autres activ. de fabricat. n.c.a.**
40 Furniture	14	..	0	..	..	..	..	40 Meubles
41 Other manufacturing n.e.c.	10	..	6	..	..	..	..	41 Autres activités de fabrication n.c.a.
42 Recycling	0	0	0	0	..	..	..	**42 Récupération**
43 ELECTRICITY, GAS & WATER SUPPLY	267	194	227	254	..	..	..	**43 DISTRIBUTION D'ELECTRICITE, GAZ & EAU**
44 CONSTRUCTION	49	63	70	7	..	..	..	**44 CONSTRUCTION**
45 SERVICES SECTOR	560	422	443	660	..	..	..	**45 SECTEUR DES SERVICES**
46 Wholesale, retail trade & repair of motor vehicles, etc.	48	24	15	57	..	..	..	**46 Commerce (gros & détail), réparation de véhicules, etc.**
47 Hotels & restaurants	0	0	0	0	..	..	..	**47 Hôtels et restaurants**
48 Transport & storage	..	233	53	36	..	..	..	**48 Transports, entreposage**
49 Communications	..	85	163	215	..	..	..	**49 Communications**
50 Post	0	0	0	0	..	..	..	50 Postes
51 Telecommunications	..	..	163	215	..	..	..	51 Télécommunications
52 Financial intermediation (inc. insur.)	0	0	0	7	..	..	..	**52 Intermédiation financ. (assurances inc.)**
53 Real estate, renting & busin. activity	165	76	205	316	..	..	..	**53 Immobil., locat. & serv. aux entreprises**
54 Computer & related activities	30	20	39	121	..	..	..	54 Informatique et activités connexes
55 *Software consultancy*	10	12	14	111	..	..	..	55 *Production de logiciels*
56 *Other computer services n.e.c.*	20	8	25	10	..	..	..	56 *Autres activités informatiques*
57 Research & development	80	41	58	22	..	..	..	57 Recherche & développement expérimental
58 Other business activities n.e.c.	55	15	108	173	..	..	..	58 Autres activ. & serv. aux entreprises
59 Comm., soc. & pers. serv. activ., etc.	13	4	7	29	..	..	..	**59 Adm. pub., serv. soc. & collec., etc**
TOTAL BUSINESS ENTERPRISE SECTOR	**7 782**	**6 792**	**6 112**	**6 545**	**5 593**	**5 899**	**6 471**	**TOTAL SECTEUR DES ENTREPRISES**

Source: OECD, R&D database (DSTI/EAS Division), January 2002.

Source : OCDE, base de données R-D (DSTI/Division AES), janvier 2002.

RSE in the business enterprise sector - Chercheurs dans le secteur des entreprises

by industry - par industrie

In full time equivalent / *En équivalence plein-temps*

	1994	1995	1996	1997	1998	1999	2000	
1 AGRICULTURE, HUNTING & FORESTRY	505	454	372	332	..	..	..	**1 AGRICULTURE, CHASSE & SYLVICULTURE**
2 MINING	5	7	3	2	..	..	..	**2 ACTIVITES EXTRACTIVES**
3 MANUFACTURING	2 325	2 069	1 871	2 161	..	..	..	**3 ACTIVITES DE FABRICATION**
4 Food, beverages & tobacco	36	60	113	85	..	..	..	**4 Produits alimentaires, boissons & tabac**
5 Food products & beverages	36	58	110	81	..	..	..	5 Produits alimentaires & boissons
6 Tobacco products	0	2	3	4	..	..	..	6 Produits à base de tabac
7 Textiles, wear app., fur & leather	17	22	19	12	..	..	..	**7 Textiles, habill., fourrures & cuirs**
8 Textiles	6	14	14	4	..	..	..	8 Textiles
9 Wearing apparel & fur	6	0	0	4	..	..	..	9 Articles d'habillement & fourrures
10 Leather products & footwear	5	8	5	4	..	..	..	10 Articles en cuir & chaussures
11 Wood, paper, printing, publishing	13	6	3	10	..	..	..	**11 Bois, papier, imprimerie, édition**
12 Wood & cork (not furniture)	..	0	0	0	..	..	..	12 Bois & liège (sauf meubles)
13 Pulp, paper & paper products	..	6	..	8	..	..	..	13 Carton, papier & articles
14 Publ., print. & repro. of rec. media	..	0	..	2	..	..	..	14 Edition, impress. & repro. enregis.
15 Coke, petroleum, nuclear fuel, chemicals, rubber plastics	1 284	1 126	1 085	1 244	..	..	..	**15 Coke, pétrole, combust. nucl., prod. chimi., caout. & mat. plast.**
16 Coke, ref. petrol. prod. & nuclear fuel	64	76	79	77	..	..	..	16 Coke, raffin. pétr. & combust. nucl.
17 Chemicals & chemical products	1 180	949	948	1 083	..	..	..	17 Produits chimiques
18 *Chemical prod.(less pharmaceuticals)*	209	286	251	218	..	..	..	18 *Prod. chimi. (moins prod. pharmaceutiques)*
19 *Pharmaceuticals*	971	663	697	865	..	..	..	19 *Produits pharmaceutiques*
20 Rubber & plastic products	40	101	58	84	..	..	..	20 Caoutchouc et matières plastiques
21 Non-metallic mineral products	7	16	14	10	..	..	..	**21 Produits minéraux non métalliques**
22 Basic metals	46	70	47	11	..	..	..	**22 Produits métallurgiques de base**
23 Basic metals, ferrous	..	16	22	0	..	..	..	23 Produits métallur. de base, ferreux
24 Basic metals, non-ferrous	..	54	25	11	..	..	..	24 Prod. métallur. de base, non ferreux
25 Fabricated metal products	59	45	24	22	..	..	..	**25 Ouvrages en métaux (sauf mach.& mat.)**
26 Machinery equipment, instruments and transport equipment	857	719	564	765	..	..	..	**26 Machines & matériel, instruments & mat. de transport**
27 Machinery n.e.c.	88	181	141	182	..	..	..	27 Machines, n.c.a.
28 Office, account. & computing machinery	3	16	0	18	..	..	..	28 Mach. de bur., compta. & trait. inform.
29 Electrical machinery	127	92	56	54	..	..	..	29 Machines électriques
30 Electro. equip. (radio, TV & commun.)	323	191	90	105	..	..	..	30 Equip. électro. (radio, tv & communic.)
31 *Electro. comp. (inc. semi-conductors)*	117	8	3	22	..	..	..	31 *Compos. électro. (semi-cond. incl.)*
32 *TV, radio & communications equip.*	206	183	87	83	..	..	..	32 *Appareils radio, tv et de communi.*
33 Instruments	146	116	115	156	..	..	..	33 Instruments
34 Motor vehicles	170	114	158	250	..	..	..	34 Véhicules automobiles
35 Other transport equipment	0	9	4	0	..	..	..	35 Autres matériels de transport
36 *Ships*	0	0	0	0	..	..	..	36 *Navires*
37 *Aerospace*	0	0	0	0	..	..	..	37 *Aérospatiale*
38 *Other transport equipment n.e.c.*	0	9	4	0	..	..	..	38 *Autres matériels de transport n.c.a.*
39 Furniture, other manufacturing n.e.c.	6	5	2	2	..	..	..	**39 Meubles, autres activ. de fabricat. n.c.a.**
40 Furniture	2	..	0	..	..	..	..	40 Meubles
41 Other manufacturing n.e.c.	4	..	2	..	..	..	..	41 Autres activités de fabrication n.c.a.
42 Recycling	0	0	0	0	..	..	..	**42 Récupération**
43 ELECTRICITY, GAS & WATER SUPPLY	180	131	118	125	..	..	..	**43 DISTRIBUTION D'ELECTRICITE, GAZ & EAU**
44 CONSTRUCTION	18	40	14	5	..	..	..	**44 CONSTRUCTION**
45 SERVICES SECTOR	297	225	248	424	..	..	..	**45 SECTEUR DES SERVICES**
46 Wholesale, retail trade & repair of motor vehicles, etc.	18	11	7	27	..	..	..	**46 Commerce (gros & détail), réparation de véhicules, etc.**
47 Hotels & restaurants	0	0	0	0	..	..	..	**47 Hôtels et restaurants**
48 Transport & storage	..	95	26	19	..	..	..	**48 Transports, entreposage**
49 Communications	..	60	126	176	..	..	..	**49 Communications**
50 Post	0	0	0	0	..	..	..	50 Postes
51 Telecommunications	..	..	126	176	..	..	..	51 Télécommunications
52 Financial intermediation (inc. insur.)	0	0	0	7	..	..	..	**52 Intermédiation financ. (assurances inc.)**
53 Real estate, renting & busin. activity	109	56	84	177	..	..	..	**53 Immobil., locat. & serv. aux entreprises**
54 Computer & related activities	28	19	28	116	..	..	..	54 Informatique et activités connexes
55 *Software consultancy*	9	11	10	106	..	..	..	55 *Production de logiciels*
56 *Other computer services n.e.c.*	19	8	18	10	..	..	..	56 *Autres activités informatiques*
57 Research & development	33	22	41	15	..	..	..	57 Recherche & développement expérimental
58 Other business activities n.e.c.	48	15	15	46	..	..	..	58 Autres activ. & serv. aux entreprises
59 Comm., soc. & pers. serv. activ., etc.	11	3	5	18	..	..	..	**59 Adm. pub., serv. soc. & collec., etc**
TOTAL BUSINESS ENTERPRISE SECTOR	**3 330**	**2 926**	**2 626**	**3 049**	**3 044**	**3 261**	**3 901**	**TOTAL SECTEUR DES ENTREPRISES**

Source: OECD, R&D database (DSTI/EAS Division), January 2002.

Source : OCDE, base de données R-D (DSTI/Division AES), janvier 2002.

Higher education intramural expenditure on R&D - Dépenses intra-muros de R-D de l'enseignement supérieur
(HERD) by field of science - (DIRDES) par discipline scientifique

	1994	1995	1996	1997	1998	1999	2000	
	14.1 Million HUF - Millions HUF							
Natural sciences	2 834.0	2 708.7	2 789.9	3 290.4	..	..	..	Sciences exactes et naturelles
Engineering	1 893.0	2 073.9	2 229.6	2 802.0	..	..	..	Sciences de l'ingénieur
Medical sciences	1 914.0	1 736.0	2 063.6	2 430.1	..	..	..	Sciences médicales
Agricultural sciences	1 734.0	1 655.3	1 862.5	2 182.6	..	..	..	Sciences agricoles
Sub-total NSE	8 375.0	8 173.8	8 945.6	10 705.1	..	..	..	Sous-total SENI
Social sciences	..	..	..	..	..	..	..	Sciences sociales
Humanities	..	..	..	..	..	..	..	Sciences humaines
Sub-total SSH	1 896.0	2 026.6	2 181.6	3 371.1	..	..	..	Sous-total SSH
Total NSE+SSH	10 271.0 [a,v]	10 200.5 [v]	11 127.2 [v]	14 198.8 [v]	17 283.2 [v]	17 472.0 [v]	25 310.4 [v]	*Total SENI+SSH*
	14.2 Million constant USD (1995 prices and PPPs) - Millions de USD constants (prix et PPA de 1995)							
Natural sciences	58.8 [b]	44.7	38.0	37.9	..	..	..	Sciences exactes et naturelles
Engineering	39.3 [b]	34.3	30.4	32.2	..	..	..	Sciences de l'ingénieur
Medical sciences	39.7 [b]	28.7	28.1	28.0	..	..	..	Sciences médicales
Agricultural sciences	36.0 [b]	27.3	25.4	25.1	..	..	..	Sciences agricoles
Sub-total NSE	173.7 [b]	135.0	121.9	123.2	..	..	..	Sous-total SENI
Social sciences	..	..	..	..	..	..	..	Sciences sociales
Humanities	..	..	..	..	..	..	..	Sciences humaines
Sub-total SSH	39.3 [b]	33.5	29.7	38.8	..	..	..	Sous-total SSH
Total NSE+SSH	213.0 [a,b,v]	168.5 [v]	151.7 [v]	163.3 [v]	176.5 [v]	164.6 [v]	221.9 [v]	*Total SENI+SSH*

Table 15 - Tableau 15

Higher education intramural expenditure on R&D - Dépenses intra-muros de R-D de l'enseignement supérieur
(HERD) by field of science and type of costs - (DIRDES) par discipline scientifique et par type de coûts

Million HUF *Millions HUF*

	1994	1995	1996	1997	1998	1999	2000	
Current expenditure								**Dépenses courantes**
Natural sciences	..	2 155.7	..	2 999.3	..	..	..	Sciences exactes et naturelles
Engineering	..	1 720.2	..	2 440.6	..	..	..	Sciences de l'ingénieur
Medical sciences	..	1 596.4	..	2 325.5	..	..	..	Sciences médicales
Agricultural sciences	..	1 414.1	..	2 017.4	..	..	..	Sciences agricoles
Sub-total NSE	..	6 886.3	..	9 782.8	..	..	..	Sous-total SENI
Social sciences	..	..	..	..	..	..	..	Sciences sociales
Humanities	..	..	..	..	..	..	..	Sciences humaines
Sub-total SSH	..	1 904.5	..	3 175.7	..	..	..	Sous-total SSH
Total NSE+SSH	8 895.0	8 790.9	9 969.8	13 066.1	15 316.2	15 869.2	23 123.6	*Total SENI+SSH*
Capital expenditure								**Dépenses en capital**
Natural sciences	..	553.0	..	291.1	..	..	..	Sciences exactes et naturelles
Engineering	..	353.7	..	361.4	..	..	..	Sciences de l'ingénieur
Medical sciences	..	139.5	..	104.6	..	..	..	Sciences médicales
Agricultural sciences	..	241.3	..	165.2	..	..	..	Sciences agricoles
Sub-total NSE	..	1 287.5	..	922.3	..	..	..	Sous-total SENI
Social sciences	..	..	..	..	..	..	..	Sciences sociales
Humanities	..	..	..	..	..	..	..	Sciences humaines
Sub-total SSH	..	122.1	..	195.4	..	..	..	Sous-total SSH
Total NSE+SSH	1 376.0	1 409.6	1 157.4	1 132.7	1 966.9	1 602.8	2 186.8	*Total SENI+SSH*

Source: OECD, R&D database (DSTI/EAS Division), January 2002.

Source : OCDE, base de données R-D (DSTI/Division AES), janvier 2002.

Higher education intramural expenditure on R&D - Dépenses intra-muros de R-D de l'enseignement supérieur

(HERD) by field of science and source of funds - (DIRDES) par discipline scientifique et par source de financement

Million HUF *Millions HUF*

	1994	1995	1996	1997	1998	1999	2000	
Direct government								**État direct**
Natural sciences	..	..	..	..	..	..	..	Sciences exactes et naturelles
Engineering	..	..	..	..	..	..	..	Sciences de l'ingénieur
Medical sciences	..	..	..	..	..	..	..	Sciences médicales
Agricultural sciences	..	..	..	..	..	..	..	Sciences agricoles
Sub-total NSE	..	..	..	..	..	..	..	Sous-total SENI
Social sciences	..	..	..	..	..	..	..	Sciences sociales
Humanities	..	..	..	..	..	..	..	Sciences humaines
Sub-total SSH	..	..	..	..	..	..	..	Sous-total SSH
Total NSE+SSH	..	..	..	..	..	..	..	*Total SENI+SSH*
General university funds								**Fonds généraux des universités**
Natural sciences	..	..	..	..	..	..	..	Sciences exactes et naturelles
Engineering	..	..	..	..	..	..	..	Sciences de l'ingénieur
Medical sciences	..	..	..	..	..	..	..	Sciences médicales
Agricultural sciences	..	..	..	..	..	..	..	Sciences agricoles
Sub-total NSE	..	..	..	..	..	..	..	Sous-total SENI
Social sciences	..	..	..	..	..	..	..	Sciences sociales
Humanities	..	..	..	..	..	..	..	Sciences humaines
Sub-total SSH	..	..	..	..	..	..	..	Sous-total SSH
Total NSE+SSH	..	..	..	..	..	..	..	*Total SENI+SSH*
Sub-total government								**Sous-total État**
Natural sciences	..	..	..	2 886.3	..	..	..	Sciences exactes et naturelles
Engineering	..	..	..	2 295.0	..	..	..	Sciences de l'ingénieur
Medical sciences	..	..	..	2 294.1	..	..	..	Sciences médicales
Agricultural sciences	..	..	..	1 589.6	..	..	..	Sciences agricoles
Sub-total NSE	..	..	..	9 065.0	..	..	..	Sous-total SENI
Social sciences	..	..	..	..	..	..	..	Sciences sociales
Humanities	..	..	..	..	..	..	..	Sciences humaines
Sub-total SSH	..	..	..	3 035.0	..	..	..	Sous-total SSH
Total NSE+SSH	9 224.0 [a,v]	9 158.2 [v]	9 460.4 [v]	12 222.6 [v]	14 579.2 [v]	14 581.3 [v]	21 725.7 [v]	*Total SENI+SSH*
Higher education								**Enseignement supérieur**
Natural sciences	..	..	..	..	..	..	..	Sciences exactes et naturelles
Engineering	..	..	..	..	..	..	..	Sciences de l'ingénieur
Medical sciences	..	..	..	..	..	..	..	Sciences médicales
Agricultural sciences	..	..	..	..	..	..	..	Sciences agricoles
Sub-total NSE	..	..	..	..	..	..	..	Sous-total SENI
Social sciences	..	..	..	..	..	..	..	Sciences sociales
Humanities	..	..	..	..	..	..	..	Sciences humaines
Sub-total SSH	..	..	..	..	..	..	..	Sous-total SSH
Total NSE+SSH	..	..	..	..	..	..	..	*Total SENI+SSH*
Private non-profit								**Institutions sans but lucratif**
Natural sciences	..	..	..	..	..	..	..	Sciences exactes et naturelles
Engineering	..	..	..	..	..	..	..	Sciences de l'ingénieur
Medical sciences	..	..	..	..	..	..	..	Sciences médicales
Agricultural sciences	..	..	..	..	..	..	..	Sciences agricoles
Sub-total NSE	..	..	..	..	..	..	..	Sous-total SENI
Social sciences	..	..	..	..	..	..	..	Sciences sociales
Humanities	..	..	..	..	..	..	..	Sciences humaines
Sub-total SSH	..	..	..	..	..	..	..	Sous-total SSH
Total NSE+SSH	..	..	209.8 [v]	264.9 [v]	228.7	142.5	257.6	*Total SENI+SSH*
Business entreprise								**Entreprises**
Natural sciences	..	..	..	71.7	..	..	..	Sciences exactes et naturelles
Engineering	..	..	..	98.6	..	..	..	Sciences de l'ingénieur
Medical sciences	..	..	..	56.0	..	..	..	Sciences médicales
Agricultural sciences	..	..	..	22.0	..	..	..	Sciences agricoles
Sub-total NSE	..	..	..	248.3	..	..	..	Sous-total SENI
Social sciences	..	..	..	..	..	..	..	Sciences sociales
Humanities	..	..	..	..	..	..	..	Sciences humaines
Sub-total SSH	..	..	..	23.4	..	..	..	Sous-total SSH
Total NSE+SSH	241.0 [a,v]	212.7 [v]	325.4 [v]	271.7 [v]	933.2 [v]	1 064.3 [v]	1 383.6 [v]	*Total SENI+SSH*
Funds from abroad								**Fonds de l'étranger**
Natural sciences	..	..	..	191.3	..	..	..	Sciences exactes et naturelles
Engineering	..	..	..	247.8	..	..	..	Sciences de l'ingénieur
Medical sciences	..	..	..	37.4	..	..	..	Sciences médicales
Agricultural sciences	..	..	..	101.9	..	..	..	Sciences agricoles
Sub-total NSE	..	..	..	578.4	..	..	..	Sous-total SENI
Social sciences	..	..	..	..	..	..	..	Sciences sociales
Humanities	..	..	..	..	..	..	..	Sciences humaines
Sub-total SSH	..	..	..	103.1	..	..	..	Sous-total SSH
Total NSE+SSH	282.0 [v]	388.1 [v]	537.5 [v]	681.5 [v]	979.1	1 074.5	1 363.9	*Total SENI+SSH*

Source: OECD, R&D database (DSTI/EAS Division), January 2002.

Source : OCDE, base de données R-D (DSTI/Division AES), janvier 2002.

Table 17 - Tableau 17

Higher education R&D personnel (HEMP) - Personnel de R-D du secteur de l'enseignement supérieur

by field of science and occupation - par discipline scientifique et par profession

In full-time equivalent — *En équivalence plein-temps*

	1994	1995	1996	1997	1998	1999	2000	
Total R&D personnel								**Personnel total de R-D**
Natural sciences	..	1 383	..	1 172	..	..	..	Sciences exactes et naturelles
Engineering	..	826	..	793	..	..	..	Sciences de l'ingénieur
Medical sciences	..	1 679	..	2 110	..	..	..	Sciences médicales
Agricultural sciences	..	1 147	..	1 394	..	..	..	Sciences agricoles
Sub-total NSE	6 141	5 035	5 284	5 469	..	..	..	Sous-total SENI
Social sciences	..	..	..	..	..	..	..	Sciences sociales
Humanities	..	..	..	..	..	..	..	Sciences humaines
Sub-total SSH	1 470	1 275	1 274	1 688	..	..	..	Sous-total SSH
Total NSE+SSH	7 611	6 310	6 558	7 210	7 561	7 452	8 859	*Total SENI+SSH*
Researchers								**Chercheurs**
Natural sciences	..	845	..	710	..	..	..	Sciences exactes et naturelles
Engineering	..	584	..	543	..	..	..	Sciences de l'ingénieur
Medical sciences	..	965	..	976	..	..	..	Sciences médicales
Agricultural sciences	..	476	..	468	..	..	..	Sciences agricoles
Sub-total NSE	..	2 870	..	2 697	..	..	..	Sous-total SENI
Social sciences	..	..	..	..	..	..	..	Sciences sociales
Humanities	..	..	..	..	..	..	..	Sciences humaines
Sub-total SSH	..	1 174	..	1 455	..	..	..	Sous-total SSH
Total NSE+SSH	4 589	4 044	3 857	4 194	4 398	4 768	5 852	*Total SENI+SSH*
Technicians								**Technicians**
Natural sciences	..	334	..	272	..	..	..	Sciences exactes et naturelles
Engineering	..	135	..	114	..	..	..	Sciences de l'ingénieur
Medical sciences	..	431	..	537	..	..	..	Sciences médicales
Agricultural sciences	..	203	..	228	..	..	..	Sciences agricoles
Sub-total NSE	..	1 103	..	1 151	..	..	..	Sous-total SENI
Social sciences	..	..	..	..	..	..	..	Sciences sociales
Humanities	..	..	..	..	..	..	..	Sciences humaines
Sub-total SSH	..	46	..	69	..	..	..	Sous-total SSH
Total NSE+SSH	1 504	1 149	1 229	1 222	1 281	1 104	1 241	*Total SENI+SSH*
Other								**Autres**
Natural sciences	..	204	..	190	..	..	..	Sciences exactes et naturelles
Engineering	..	107	..	136	..	..	..	Sciences de l'ingénieur
Medical sciences	..	283	..	597	..	..	..	Sciences médicales
Agricultural sciences	..	468	..	698	..	..	..	Sciences agricoles
Sub-total NSE	..	1 062	..	1 621	..	..	..	Sous-total SENI
Social sciences	..	..	..	..	..	..	..	Sciences sociales
Humanities	..	..	..	..	..	..	..	Sciences humaines
Sub-total SSH	..	55	..	164	..	..	..	Sous-total SSH
Total NSE+SSH	1 518	1 117	1 472	1 794	1 882	1 580	1 766	*Total SENI+SSH*

Table 18 - Tableau 18

Government budget appropriations or outlays for R&D (GBAORD) - Crédits budgétaires publics de R-D (CBPRD)

by socio-economic objective - par objectif socio-économique

Million HUF — *Millions HUF*

	1995	1996	1997	1998	1999	2000	2001	
Agriculture, forestry and fishing	..	..	..	..	..	..	..	Agriculture, sylviculture et pêche
Industrial development	..	..	..	..	..	..	..	Développement industriel
Energy	..	..	..	..	..	..	..	Energie
Transport & telecommunications	..	..	..	..	..	..	..	*Transports & télécommunications*
Urban & rural planning	..	..	..	..	..	..	..	*Aménagement urbain et rural*
ST Infrastructure	..	..	..	..	..	..	..	ST Infrastructure
Prevention of pollution	..	..	..	..	..	..	..	*Prévention de la pollution*
Identificat. & treatment of pollution	..	..	..	..	..	..	..	*Détect. pollut. & mesures antipollut.*
ST Environment	..	..	..	..	..	..	..	ST Environnement
Health	..	..	..	..	..	..	..	Santé
Social development & services	..	..	..	..	..	..	..	Développement et services sociaux
Earth and atmosphere	..	..	..	..	..	..	..	Milieu terrestre et atmosphère
Advancement of research	..	..	..	..	..	..	..	*Promotion de la recherche*
General university funds	..	..	..	..	..	..	..	*Fonds généraux des universités*
ST Advancement of knowledge	..	..	..	..	..	..	..	ST promotion gén. des connaiss.
Civil space	..	..	..	..	..	..	..	Espace civil
Defence	..	..	..	..	..	..	..	Défense
Not elsewhere classified	..	..	..	..	..	..	..	Non classé
Total	..	..	..	..	..	..	..	**Total**

Source: OECD, R&D database (DSTI/EAS Division), January 2002.

Source : OCDE, base de données R-D (DSTI/Division AES), janvier 2002.

Table 19 - Tableau 19

Technology balance of payments - Balance des paiements technologiques

by type of transfer - selon la nature des transferts

Million HUF — *Millions HUF*

	1983	1984	1985	1986	1987	1988	1989	1990	1991	
Receipts										**Recettes**
Related firms	..	..	..	..	..	..	..	..	..	Firmes affiliées
Unrelated firms	..	..	..	..	..	..	..	..	..	Firmes non affiliées
Other	..	..	..	..	..	..	..	..	..	Autre
Total	..	..	..	..	..	..	..	..	..	*Total*
Payments										**Paiements**
Related firms	..	..	..	..	..	..	..	..	..	Firmes affiliées
Unrelated firms	..	..	..	..	..	..	..	..	..	Firmes non affiliées
Other	..	..	..	..	..	..	..	..	..	Autre
Total	..	..	..	..	..	..	..	..	..	*Total*

	1992	1993	1994	1995	1996	1997	1998	1999	2000	
Receipts										**Recettes**
Related firms	..	..	..	..	..	..	..	..	..	Firmes affiliées
Unrelated firms	..	..	..	..	..	..	..	..	..	Firmes non affiliées
Other	..	..	..	..	..	..	..	..	..	Autre
Total	..	..	..	..	27 598	36 896	32 750	51 241	..	*Total*
Payments										**Paiements**
Related firms	..	..	..	..	..	..	..	..	..	Firmes affiliées
Unrelated firms	..	..	..	..	..	..	..	..	..	Firmes non affiliées
Other	..	..	..	..	..	..	..	..	..	Autre
Total	..	..	..	..	32 891	58 731	74 800	119 451	..	*Total*

Source: OECD, TBPdatabase (DSTI/EAS Division), Januay 2002.

Source : OCDE, base de données TBP (DSTI/Division AES), janvier 2002.

Table 1 - Tableau 1

Gross domestic expenditure on R&D (GERD) - Dépenses intérieures brutes de R-D (DIRD)

by sector of performance and source of funds - par secteur d'exécution et par source de financement

	1994	1995	1996	1997	1998	1999	2000	
	1.1 Million ISK - Millions ISK							
Business enterprise								**Entreprises**
financed by								*financé par*
Business enterprise	1 577.6	2 116.4	..	3 577.3	3 935.1 c	5 205.3	..	Entreprises
Direct government	270.7	72.6	..	196.3	215.9 c	163.9	..	État direct
Higher education	0.0	0.0	..	..	..	..	..	Enseignement supérieur
Private non-profit	0.0	0.0	..	0.4	0.4 c	0.0	..	Institutions sans but lucratif
Funds from abroad	33.3	27.4	..	144.0	158.4 c	1 407.7	..	Fonds de l'étranger
Total	1 881.6	2 216.4	..	3 918.0	4 309.8 c	6 776.8	..	*Total*
Government								**État**
financed by								*financé par*
Business enterprise	273.8	186.8	..	218.3	234.7 c	970.9	..	Entreprises
Direct government	2 091.3	2 316.2	..	2 409.1	3 889.8 c	3 102.9	..	État direct
Higher education	..	..	..	..	..	..	..	Enseignement supérieur
Private non-profit	41.0	..	..	22.8	24.5 c	51.9	..	Institutions sans but lucratif
Funds from abroad	65.6	102.7	..	225.1	241.9 c	260.3	..	Fonds de l'étranger
Total	2 471.7	2 605.7	..	2 875.3	4 391.0 c	4 386.0	..	*Total*
Higher education								**Enseignement supérieur**
financed by								*financé par*
Business enterprise	61.9	103.3	..	252.0	270.9 c	121.8	..	Entreprises
Direct government	1 162.4	1 200.3	..	1 586.4	1 705.4 c	2 126.8	..	État direct
General university funds	131.2	269.6	..	668.4	718.5 c	447.9	..	Fonds généraux des universités
Sub-total government	1 293.6	1 470.0	..	2 254.9	2 424.0 c	2 574.7	..	Sous-Total État
Higher education	..	..	..	..	..	..	..	Enseignement supérieur
Private non-profit	4.7	164.3	..	12.3	13.2 c	0.5	..	Institutions sans but lucratif
Funds from abroad	92.1	177.7	..	212.0	227.9 c	336.2	..	Fonds de l'étranger
Total	1 452.3	1 915.3	..	2 731.2	2 936.0 c	3 033.2	..	*Total*
Private non-profit								**Institutions sans but lucratif**
financed by								*financé par*
Business enterprise	..	..	..	0.6	0.6 c	6.6	..	Entreprises
Direct government	150.4	130.5	..	51.0	54.9 c	136.0	..	État direct
Higher education	..	..	..	..	..	..	..	Enseignement supérieur
Private non-profit	91.4	90.0	..	54.1	58.1 c	164.0	..	Institutions sans but lucratif
Funds from abroad	..	..	..	20.5	22.1 c	19.0	..	Fonds de l'étranger
Total	241.8	220.5	..	126.2	135.7 c	325.6	..	*Total*
GERD								**DIRD**
financed by								*financé par*
Business enterprise	1 913.4	2 406.5	..	4 048.3	4 441.3 c	6 304.5	..	Entreprises
Direct government	3 674.8	3 719.7	..	4 242.9	5 866.1 c	5 529.6	..	État direct
General university funds	131.2	269.6	..	668.4	718.5 c	447.9	..	Fonds généraux des universités
Sub-total government	3 806.0	3 989.3	..	4 911.3	6 584.6 c	5 977.5	..	Sous-Total État
Higher education	..	..	..	..	..	..	..	Enseignement supérieur
Private non-profit	137.1	254.3	..	89.5	96.2 c	216.4	..	Institutions sans but lucratif
Funds from abroad	191.0	307.8	..	601.6	650.3 c	2 023.2	..	Fonds de l'étranger
Total	6 047.4	6 957.9	..	9 650.7	11 772.5 c	14 521.6	..	*Total*
	1.2 Million constant USD (1995 prices and PPPs) - Millions de USD constants (prix et PPA de 1995)							
Business enterprise								**Entreprises**
financed by								*financé par*
Business enterprise	21.4	27.9	..	44.7	46.7 c	59.5	..	Entreprises
Direct government	3.7	1.0	..	2.5	2.6 c	1.9	..	État direct
Higher education	0.0	0.0	..	..	..	..	..	Enseignement supérieur
Private non-profit	0.0	0.0	..	0.0	0.0 c	0.0	..	Institutions sans but lucratif
Funds from abroad	0.5	0.4	..	1.8	1.9 c	16.1	..	Fonds de l'étranger
Total	25.5	29.2	..	48.9	51.1 c	77.5	..	*Total*
Government								**État**
financed by								*financé par*
Business enterprise	3.7	2.5	..	2.7	2.8 c	11.1	..	Entreprises
Direct government	28.3	30.5	..	30.1	46.2 c	35.5	..	État direct
Higher education	..	..	..	..	..	..	..	Enseignement supérieur
Private non-profit	0.6	..	..	0.3	0.3 c	0.6	..	Institutions sans but lucratif
Funds from abroad	0.9	1.4	..	2.8	2.9 c	3.0	..	Fonds de l'étranger
Total	33.5	34.3	..	35.9	52.1 c	50.2	..	*Total*
Higher education								**Enseignement supérieur**
financed by								*financé par*
Business enterprise	0.8	1.4	..	3.1	3.2 c	1.4	..	Entreprises
Direct government	15.7	15.8	..	19.8	20.2 c	24.3	..	État direct
General university funds	1.8	3.6	..	8.3	8.5 c	5.1	..	Fonds généraux des universités
Sub-total government	17.5	19.4	..	28.2	28.8 c	29.4	..	Sous-Total État
Higher education	..	..	..	..	..	..	..	Enseignement supérieur
Private non-profit	0.1	2.2	..	0.2	0.2 c	0.0	..	Institutions sans but lucratif
Funds from abroad	1.2	2.3	..	2.6	2.7 c	3.8	..	Fonds de l'étranger
Total	19.7	25.2	..	34.1	34.8 c	34.7	..	*Total*
Private non-profit								**Institutions sans but lucratif**
financed by								*financé par*
Business enterprise	..	..	..	0.0	0.0 c	0.1	..	Entreprises
Direct government	2.0	1.7	..	0.6	0.7 c	1.6	..	État direct
Higher education	..	..	..	..	..	..	..	Enseignement supérieur
Private non-profit	1.2	1.2	..	0.7	0.7 c	1.9	..	Institutions sans but lucratif
Funds from abroad	..	..	..	0.3	0.3 c	0.2	..	Fonds de l'étranger
Total	3.3	2.9	..	1.6	1.6 c	3.7	..	*Total*
GERD								**DIRD**
financed by								*financé par*
Business enterprise	25.9	31.7	..	50.6	52.7 c	72.1	..	Entreprises
Direct government	49.8	49.0	..	53.0	69.6 c	63.2	..	État direct
General university funds	1.8	3.6	..	8.3	8.5 c	5.1	..	Fonds généraux des universités
Sub-total government	51.5	52.6	..	61.3	78.1 c	68.4	..	Sous-Total État
Higher education	..	..	..	..	..	..	..	Enseignement supérieur
Private non-profit	1.9	3.4	..	1.1	1.1 c	2.5	..	Institutions sans but lucratif
Funds from abroad	2.6	4.1	..	7.5	7.7 c	23.1	..	Fonds de l'étranger
Total	81.9	91.7	..	120.6	139.7 c	166.1	..	*Total*

Source: OECD, R&D database (DSTI/EAS Division), January 2002.

Source : OCDE, base de données R-D (DSTI/Division AES), janvier 2002.

Table 2 - Tableau 2

Gross domestic expenditure on R&D (GERD) - Dépenses intérieures brutes de R-D (DIRD)

by sector of performance and type of costs - par secteur d'exécution et par type de coûts

Million ISK — *Millions ISK*

	1994	1995	1996	1997	1998	1999	2000	
Business enterprise								**Entreprises**
Labour costs	..	1 279.9 [s]	..	2 349.2	..	4 094.4	..	Coûts du personnel
Other current costs	..	600.0 [s]	..	1 107.5	..	1 813.5	..	Autres coûts courants
Total current expenditure	..	1 879.8 [s]	..	3 456.7	..	5 907.9	..	*Total dépenses courantes*
Land and buildings	..	209.7 [s]	..	351.1	..	815.7	..	Terrains et constructions
Instruments and equipment	..	91.1 [s]	..	110.3	..	53.3	..	Instruments et équipements
Total capital expenditure	..	300.8 [s]	..	461.4	..	869.0	..	*Total dépenses en capital*
Total	1 881.6	2 216.4	..	3 918.0	4 309.8 [c]	6 776.8	..	*Total*
Government								**État**
Labour costs	..	1 321.8 [s]	..	1 559.0	..	2 156.7	..	Coûts du personnel
Other current costs	..	961.1 [s]	..	1 094.0	..	1 375.5	..	Autres coûts courants
Total current expenditure	..	2 282.8 [s]	..	2 653.0	..	3 532.2	..	*Total dépenses courantes*
Land and buildings	..	155.9 [s]	..	182.5	..	834.8	..	Terrains et constructions
Instruments and equipment	..	82.8 [s]	..	39.8	..	19.0	..	Instruments et équipements
Total capital expenditure	..	238.7 [s]	..	222.3	..	853.8	..	*Total dépenses en capital*
Total	2 471.7	2 605.7	..	2 875.3	4 391.0 [c]	4 386.0	..	*Total*
Higher education								**Enseignement supérieur**
Labour costs	..	1 152.2 [s]	..	1 494.3	..	1 842.9	..	Coûts du personnel
Other current costs	..	474.1 [s]	..	770.1	..	889.5	..	Autres coûts courants
Total current expenditure	..	1 626.3 [s]	..	2 264.4	..	2 732.4	..	*Total dépenses courantes*
Land and buildings	..	140.8 [s]	..	271.9	..	186.3	..	Terrains et constructions
Instruments and equipment	..	268.2 [s]	..	194.8	..	114.5	..	Instruments et équipements
Total capital expenditure	..	409.0 [s]	..	466.8	..	300.8	..	*Total dépenses en capital*
Total	1 452.3	1 915.3	..	2 731.2	2 936.0 [c]	3 033.2	..	*Total*
Private non-profit								**Institutions sans but lucratif**
Labour costs	..	123.5	..	75.8	..	216.9	..	Coûts du personnel
Other current costs	..	97.0	..	47.4	..	96.3	..	Autres coûts courants
Total current expenditure	..	220.5	..	123.2	..	313.2	..	*Total dépenses courantes*
Land and buildings	..	..	..	3.0	..	8.9	..	Terrains et constructions
Instruments and equipment	..	..	..	..	..	3.5	..	Instruments et équipements
Total capital expenditure	..	..	..	3.0	..	12.4	..	*Total dépenses en capital*
Total	241.8	220.5	..	126.2	135.7 [c]	325.6	..	*Total*
GERD								**DIRD**
Labour costs	..	3 877.3	..	5 478.3	..	8 310.9	..	Coûts du personnel
Other current costs	..	2 132.1	..	3 019.0	..	4 174.8	..	Autres coûts courants
Total current expenditure	..	6 009.5	..	8 497.3	..	12 485.7	..	*Total dépenses courantes*
Land and buildings	..	506.4	..	808.5	..	1 845.7	..	Terrains et constructions
Instruments and equipment	..	442.1	..	344.9	..	190.3	..	Instruments et équipements
Total capital expenditure	..	948.5	..	1 153.4	..	2 036.0	..	*Total dépenses en capital*
Total	6 047.4	6 957.9	..	9 650.7	11 772.5 [c]	14 521.6	..	*Total*

Table 3 - Tableau 3

Current domestic expenditure on R&D - Dépenses intérieures courantes de R-D

by sector of performance and type of activity - par secteur d'exécution et par type d'activités

Million ISK — *Millions ISK*

	1994	1995	1996	1997	1998	1999	2000	
Business enterprise								**Entreprises**
Basic research	..	..	..	..	..	0.0	..	Recherche fondamentale
Applied research	..	468.8 [k,s]	..	1 374.8 [k]	..	3 719.8 [k]	..	Recherche appliquée
Experimental development	..	1 711.8 [k,s]	..	2 543.2 [k]	..	3 057.0 [k]	..	Développement expérimental
Not specified	1 881.6 [k]	..	..	..	4 309.8 [c,k]	..	..	Non spécifié
Total	1 881.6 [k]	2 216.4 [k]	..	3 918.0 [k]	4 309.8 [c,k]	6 776.8 [k]	..	*Total*
Government								**État**
Basic research	..	540.1 [k,s]	..	640.2 [k]	..	971.1 [k]	..	Recherche fondamentale
Applied research	..	1 361.7 [k,s]	..	1 601.8 [k]	..	2 572.5 [k]	..	Recherche appliquée
Experimental development	..	619.7 [k,s]	..	633.3 [k]	..	842.5 [k]	..	Développement expérimental
Not specified	2 471.7 [k]	..	..	..	4 391.0 [c,k]	..	..	Non spécifié
Total	2 471.7 [k]	2 605.7 [k]	..	2 875.3 [k]	4 391.0 [c,k]	4 386.0 [k]	..	*Total*
Higher education								**Enseignement supérieur**
Basic research	..	1 050.8 [k,s]	..	1 367.6 [k]	..	1 453.3 [k]	..	Recherche fondamentale
Applied research	..	712.3 [k,s]	..	947.3 [k]	..	1 131.8 [k]	..	Recherche appliquée
Experimental development	..	272.2 [k,s]	..	416.3 [k]	..	448.1 [k]	..	Développement expérimental
Not specified	1 452.3 [k]	..	..	..	2 936.0 [c,k]	..	..	Non spécifié
Total	1 452.3 [k]	1 915.3 [k]	..	2 731.2 [k]	2 936.0 [c,k]	3 033.2 [k]	..	*Total*
Private non-profit								**Institutions sans but lucratif**
Basic research	..	110.3 [k]	..	57.9 [k]	..	155.9 [k]	..	Recherche fondamentale
Applied research	..	110.3 [k]	..	67.4 [k]	..	155.1 [k]	..	Recherche appliquée
Experimental development	..	..	..	1.0 [k]	..	14.7 [k]	..	Développement expérimental
Not specified	241.8 [k]	..	..	..	135.7 [c,k]	..	..	Non spécifié
Total	241.8 [k]	220.5 [k]	..	126.2 [k]	135.7 [c,k]	325.6 [k]	..	*Total*
GERD								**DIRD**
Basic research	..	1 701.1 [k]	..	2 065.7 [k]	..	2 580.2 [k]	..	Recherche fondamentale
Applied research	..	2 653.1 [k]	..	3 991.2 [k]	..	7 579.2 [k]	..	Recherche appliquée
Experimental development	..	2 603.7 [k]	..	3 593.8 [k]	..	4 362.3 [k]	..	Développement expérimental
Not specified	6 047.4 [k]	..	..	..	11 772.5 [c,k]	..	..	Non spécifié
Total	6 047.4 [k]	6 957.9 [k]	..	9 650.7 [k]	11 772.5 [c,k]	14 521.6 [k]	..	*Total*

Source: OECD, R&D database (DSTI/EAS Division), January 2002.

Source : OCDE, base de données R-D (DSTI/Division AES), janvier 2002.

ICELAND | Table 4 - Tableau 4 | ISLANDE

Gross domestic expenditure on R&D (GERD) - Dépenses intérieures brutes de R-D (DIRD)

by sector of performance and socio-economic objective - par secteur d'exécution et par objectif socio-économique

Million ISK | *Millions ISK*

	1994	1995	1996	1997	1998	1999	2000	
Business enterprise								**Entreprises**
Agriculture, forestry and fishing	..	237.1	..	524.1	..	248.3	..	Agriculture, sylviculture et pêche
Industrial development	..	1 534.7	..	2 207.1	..	3 128.3	..	Développement industriel
Energy	..	123.7	..	69.9	..	72.9	..	Energie
Transport and telecommunications	..	110.8	..	184.4	..	218.6	..	Transports et télécommunications
Urban and rural planning	..	23.6	..	6.0	..	23.7	..	Aménagement urbain et rural
Prevention of pollution	..	4.8 [o]	..	14.7 [o]	..	29.3 [o]	..	Prévention de la pollution
Identification & treatment of pollution	..	.. [n]	..	.. [n]	..	.. [n]	..	Détection pollu. & mesures antipollu.
Health	..	163.0	..	803.5	..	2 896.7	..	Santé
Social development and services	..	5.4	..	23.9	..	113.0	..	Développement et services sociaux
Earth and atmosphere	..	13.4	..	84.4	..	46.0	..	Milieu terrestre et atmosphère
Advancement of knowledge	..	..	..	..	..	0.0	..	Promotion générale des connaissances
Civil space	..	..	..	..	..	0.0	..	Espace civil
Defence	..	..	..	..	..	0.0	..	Défense
Not elsewhere classified	1 881.6	..	..	..	4 309.8 [c]	0.0	..	Non classé
Total	1 881.6	2 216.4	..	3 918.0	4 309.8 [c]	6 776.8	..	*Total*
Government								**État**
Agriculture, forestry and fishing	..	895.1	..	1 032.4	..	1 802.7	..	Agriculture, sylviculture et pêche
Industrial development	..	426.8	..	479.4	..	570.6	..	Développement industriel
Energy	..	229.1	..	170.6	..	171.1	..	Energie
Transport and telecommunications	..	53.2	..	5.6	..	33.8	..	Transports et télécommunications
Urban and rural planning	..	7.3	..	35.7	..	31.1	..	Aménagement urbain et rural
Prevention of pollution	..	113.1 [o]	..	140.5 [o]	..	185.5 [o]	..	Prévention de la pollution
Identification & treatment of pollution	..	.. [n]	..	.. [n]	..	.. [n]	..	Détection pollu. & mesures antipollu.
Health	..	77.8	..	129.4	..	254.3	..	Santé
Social development and services	..	102.0	..	63.8	..	165.7	..	Développement et services sociaux
Earth and atmosphere	..	123.8	..	182.1	..	198.6	..	Milieu terrestre et atmosphère
Advancement of knowledge	..	577.5	..	635.8	..	972.5	..	Promotion générale des connaissances
Civil space	..	..	..	..	..	0.0	..	Espace civil
Defence	..	..	..	..	..	0.0	..	Défense
Not elsewhere classified	2 471.7	..	..	..	4 391.0 [c]	0.0	..	Non classé
Total	2 471.7	2 605.7	..	2 875.3	4 391.0 [c]	4 386.0	..	*Total*
Higher education								**Enseignement supérieur**
Agriculture, forestry and fishing	..	73.4	..	126.8	..	112.3	..	Agriculture, sylviculture et pêche
Industrial development	..	121.3	..	162.3	..	176.3	..	Développement industriel
Energy	..	55.3	..	72.3	..	75.3	..	Energie
Transport and telecommunications	..	1.5	..	1.9	..	2.0	..	Transports et télécommunications
Urban and rural planning	..	..	..	2.1	..	0.0	..	Aménagement urbain et rural
Prevention of pollution	..	15.7 [o]	..	23.6 [o]	..	33.4 [o]	..	Prévention de la pollution
Identification & treatment of pollution	..	.. [n]	..	.. [n]	..	.. [n]	..	Détection pollu. & mesures antipollu.
Health	..	207.5	..	357.1	..	350.4	..	Santé
Social development and services	..	389.2	..	605.8	..	817.1	..	Développement et services sociaux
Earth and atmosphere	..	2.3	..	11.7	..	13.0	..	Milieu terrestre et atmosphère
Advancement of knowledge	..	1 049.2	..	1 367.6	..	1 453.3	..	Promotion générale des connaissances
Civil space	..	..	..	..	..	0.0	..	Espace civil
Defence	..	..	..	..	..	0.0	..	Défense
Not elsewhere classified	1 452.3	..	..	..	2 936.0 [c]	0.0	..	Non classé
Total	1 452.3	1 915.3	..	2 731.2	2 936.0 [c]	3 033.2	..	*Total*
Private non-profit								**Institutions sans but lucratif**
Agriculture, forestry and fishing	..	..	..	..	..	0.0	..	Agriculture, sylviculture et pêche
Industrial development	..	..	..	..	..	0.0	..	Développement industriel
Energy	..	..	..	..	..	0.0	..	Energie
Transport and telecommunications	..	..	..	..	..	1.2	..	Transports et télécommunications
Urban and rural planning	..	..	..	2.0	..	2.7	..	Aménagement urbain et rural
Prevention of pollution	..	..	..	2.0 [o]	..	2.7 [o]	..	Prévention de la pollution
Identification & treatment of pollution	..	..	..	.. [n]	..	.. [n]	..	Détection pollu. & mesures antipollu.
Health	..	110.3	..	61.4	..	107.7	..	Santé
Social development and services	..	..	..	1.0	..	41.9	..	Développement et services sociaux
Earth and atmosphere	..	..	..	2.0	..	13.5	..	Milieu terrestre et atmosphère
Advancement of knowledge	..	110.3	..	57.9	..	155.9	..	Promotion générale des connaissances
Civil space	..	..	..	..	..	0.0	..	Espace civil
Defence	..	..	..	..	..	0.0	..	Défense
Not elsewhere classified	241.8	..	..	..	135.7 [c]	0.0	..	Non classé
Total	241.8	220.5	..	126.2	135.7 [c]	325.6	..	*Total*
GERD								**DIRD**
Agriculture, forestry and fishing	..	1 205.6	..	1 683.2	..	2 163.3	..	Agriculture, sylviculture et pêche
Industrial development	..	2 082.8	..	2 848.7	..	3 875.1	..	Développement industriel
Energy	..	408.0	..	312.9	..	319.3	..	Energie
Transport and telecommunications	..	165.5	..	191.9	..	255.7	..	Transports et télécommunications
Urban and rural planning	..	30.8	..	45.9	..	57.5	..	Aménagement urbain et rural
Prevention of pollution	..	133.6 [o]	..	180.8 [o]	..	250.9 [o]	..	Prévention de la pollution
Identification & treatment of pollution	..	.. [n]	..	.. [n]	..	0.0 [n]	..	Détection pollu. & mesures antipollu.
Health	..	558.5	..	1 351.3	..	3 609.1	..	Santé
Social development and services	..	496.5	..	694.5	..	1 137.8	..	Développement et services sociaux
Earth and atmosphere	..	139.6	..	280.3	..	271.2	..	Milieu terrestre et atmosphère
Advancement of knowledge	..	1 736.9	..	2 061.2	..	2 581.6	..	Promotion générale des connaissances
Civil space	..	..	..	..	..	0.0	..	Espace civil
Defence	..	..	..	..	..	0.0	..	Défense
Not elsewhere classified	6 047.4	..	..	..	11 772.5 [c]	0.0	..	Non classé
Total	6 047.4	6 957.9	..	9 650.7	11 772.5 [c]	14 521.6	..	*Total*

Source: OECD, R&D database (DSTI/EAS Division), January 2002.

Source : OCDE, base de données R-D (DSTI/Division AES), janvier 2002.

Table 5 - Tableau 5

Total R&D personnel - Personnel total de R-D

by sector of employment and occupation - par secteur d'emploi et par profession

In full-time equivalent — *En équivalence plein-temps*

	1994	1995	1996	1997	1998	1999	2000	
Business enterprise								**Entreprises**
RSE	287	359	..	474	521 c	626	..	Chercheurs
Technicians	94	146	..	243	267 c	223	..	Techniciens
Other occupations	38	46	..	115	126 c	112	..	Autres professions
Total	418	551	..	832	915 c	961	..	*Total*
Government								**État**
RSE	330	324	..	390	401 c	422	..	Chercheurs
Technicians	88	129	..	125	128 c	119	..	Techniciens
Other occupations	135	110	..	114	118 c	105	..	Autres professions
Total	554	563	..	629	647 c	645	..	*Total*
Higher education								**Enseignement supérieur**
RSE	211	380	..	462	476 c	480	..	Chercheurs
Technicians	109	102	..	115	118 c	108	..	Techniciens
Other occupations	65	48	..	80	83 c	125	..	Autres professions
Total	385	530	..	656	676 c	712	..	*Total*
Private non-profit								**Institutions sans but lucratif**
RSE	18	13	..	16	16 c	50	..	Chercheurs
Technicians	17	17	..	7	7 c	11	..	Techniciens
Other occupations	21	20	..	11	12 c	11	..	Autres professions
Total	55	50	..	34	35 c	73	..	*Total*
TOTAL								**TOTAL**
RSE	845	1 076	..	1 341	1 414 c	1 578	..	Chercheurs
Technicians	308	393	..	489	520 c	461	..	Techniciens
Other occupations	259	225	..	321	338 c	352	..	Autres professions
Total	1 412	1 694	..	2 150	2 273 c	2 390	..	*Total*

Table 6 - Tableau 6

Total R&D personnel - Personnel total de R-D

by sector of employment and formal qualification - par secteur d'emploi et par niveau de qualification

In full-time equivalent — *En équivalence plein-temps*

	1994	1995	1996	1997	1998	1999	2000	
Business enterprise								**Entreprises**
University PhD level degrees	26	32	..	61	67 c	96	..	Diplômes univ. de niv. doctorat
Other university degrees	221	311	..	419	461 c	540	..	Autres diplômes universitaires
Sub-total university degrees	246	343	..	481	529 c	635	..	Sous-total Universitaires
Other post-secondary diplomas	119	29	..	77	85 c	84	..	Autres diplômes post-secondaires
Secondary diplomas	24	150	..	199	219 c	196	..	Diplômes de fin d'études secondaires
Other qualifications	29	29	..	76	83 c	45	..	Autres qualifications
Total	418	551	..	832	915 c	961	..	*Total*
Government								**État**
University PhD level degrees	92	86	..	115	118 c	108	..	Diplômes univ. de niv. doctorat
Other university degrees	272	276	..	297	306 c	329	..	Autres diplômes universitaires
Sub-total university degrees	364	362	..	412	424 c	437	..	Sous-total Universitaires
Other post-secondary diplomas	49	45	..	47	49 c	75	..	Autres diplômes post-secondaires
Secondary diplomas	61	80	..	77	79 c	51	..	Diplômes de fin d'études secondaires
Other qualifications	80	77	..	93	96 c	82	..	Autres qualifications
Total	554	563	..	629	647 c	645	..	*Total*
Higher education								**Enseignement supérieur**
University PhD level degrees	132	234	..	263	271 c	222	..	Diplômes univ. de niv. doctorat
Other university degrees	177	192	..	284	292 c	370	..	Autres diplômes universitaires
Sub-total university degrees	309	427	..	547	564 c	592	..	Sous-total Universitaires
Other post-secondary diplomas	33	29	..	29	30 c	67	..	Autres diplômes post-secondaires
Secondary diplomas	16	58	..	19	20 c	34	..	Diplômes de fin d'études secondaires
Other qualifications	27	17	..	61	63 c	19	..	Autres qualifications
Total	385	530	..	656	676 c	712	..	*Total*
Private non-profit								**Institutions sans but lucratif**
University PhD level degrees	5	3	..	5	5 c	12	..	Diplômes univ. de niv. doctorat
Other university degrees	22	9	..	11	11 c	40	..	Autres diplômes universitaires
Sub-total university degrees	28	12	..	16	16 c	52	..	Sous-total Universitaires
Other post-secondary diplomas	11	9	..	4	4 c	6	..	Autres diplômes post-secondaires
Secondary diplomas	9	19	..	14	15 c	9	..	Diplômes de fin d'études secondaires
Other qualifications	8	9	..	0	0 c	5	..	Autres qualifications
Total	55	50	..	34	35 c	73	..	*Total*
TOTAL								**TOTAL**
University PhD level degrees	255	355	..	444	462 c	438	..	Diplômes univ. de niv. doctorat
Other university degrees	692	788	..	1 011	1 070 c	1 279	..	Autres diplômes universitaires
Sub-total university degrees	947	1 144	..	1 455	1 532 c	1 717	..	Sous-total Universitaires
Other post-secondary diplomas	211	112	..	157	167 c	232	..	Autres diplômes post-secondaires
Secondary diplomas	110	306	..	309	332 c	291	..	Diplômes de fin d'études secondaires
Other qualifications	144	132	..	230	242 c	152	..	Autres qualifications
Total	1 412	1 694	..	2 150	2 273 c	2 390	..	*Total*

Source: OECD, R&D database (DSTI/EAS Division), January 2002.

Source : OCDE, base de données R-D (DSTI/Division AES), janvier 2002.

Table 7 - Tableau 7

Gross domestic expenditure on R&D (GERD) - Dépenses intérieures brutes de R-D (DIRD)

by sector of performance and main scientific field - par secteur d'exécution et par grand domaine scientifique

Million ISK / *Millions ISK*

	1994	1995	1996	1997	1998	1999	2000	
Business enterprise								**Entreprises**
Natural sciences and engineering	1 881.6	2 216.4	..	3 918.0	4 309.8 [c]	6 696.7	..	Sciences exactes, nat. et de l'ingénieur
Social sciences and humanities	..	..	..	..	..	80.1	..	Sciences sociales et humaines
All fields of science	1 881.6	2 216.4	..	3 918.0	4 309.8 [c]	6 776.8	..	*Tous domaines scientifiques*
Government								**État**
Natural sciences and engineering	..	2 326.6 [s]	..	2 663.0	..	4 008.4	..	Sciences exactes, nat. et de l'ingénieur
Social sciences and humanities	..	194.8 [s]	..	212.3	..	377.6	..	Sciences sociales et humaines
All fields of science	2 471.7	2 605.7	..	2 875.3	4 391.0 [c]	4 386.0	..	*Tous domaines scientifiques*
Higher education								**Enseignement supérieur**
Natural sciences and engineering	..	1 668.9 [s]	..	2 176.7	..	2 277.8	..	Sciences exactes, nat. et de l'ingénieur
Social sciences and humanities	..	366.4 [s]	..	554.5	..	755.4	..	Sciences sociales et humaines
All fields of science	1 452.3	1 915.3	..	2 731.2	2 936.0 [c]	3 033.2	..	*Tous domaines scientifiques*
Private non-profit								**Institutions sans but lucratif**
Natural sciences and engineering	..	220.5	..	116.2	..	178.8	..	Sciences exactes, nat. et de l'ingénieur
Social sciences and humanities	..	0.0	..	10.0	..	146.8	..	Sciences sociales et humaines
All fields of science	241.8	220.5	..	126.2	135.7 [c]	325.6	..	*Tous domaines scientifiques*
TOTAL								**TOTAL**
Natural sciences and engineering	..	6 396.7	..	8 873.9	..	13 161.0	..	Sciences exactes, nat. et de l'ingénieur
Social sciences and humanities	..	561.3	..	776.8	..	1 359.9	..	Sciences sociales et humaines
All fields of science	6 047.4	6 957.9	..	9 650.7	11 772.5 [c]	14 521.6	..	*Tous domaines scientifiques*

Table 8 - Tableau 8

Total R&D personnel - Personnel total de R-D

by sector of employment and main scientific field - par secteur d'emploi et par grand domaine scientifique

In full-time equivalent / *En équivalence plein-temps*

	1994	1995	1996	1997	1998	1999	2000	
Business enterprise								**Entreprises**
Natural sciences and engineering	418	551	..	831	915 [c]	947	..	Sciences exactes, nat. et de l'ingénieur
Social sciences and humanities	..	..	..	..	..	14	..	Sciences sociales et humaines
All fields of science	418	551	..	832	915 [c]	961	..	*Tous domaines scientifiques*
Government								**État**
Natural sciences and engineering	..	532	..	591	..	598	..	Sciences exactes, nat. et de l'ingénieur
Social sciences and humanities	..	31	..	38	..	46	..	Sciences sociales et humaines
All fields of science	554	563	..	629	647 [c]	645	..	*Tous domaines scientifiques*
Higher education								**Enseignement supérieur**
Natural sciences and engineering	..	414	..	505	..	557	..	Sciences exactes, nat. et de l'ingénieur
Social sciences and humanities	..	116	..	152	..	155	..	Sciences sociales et humaines
All fields of science	385	530	..	656	676 [c]	712	..	*Tous domaines scientifiques*
Private non-profit								**Institutions sans but lucratif**
Natural sciences and engineering	..	50	..	31	..	41	..	Sciences exactes, nat. et de l'ingénieur
Social sciences and humanities	..	0	..	3	..	32	..	Sciences sociales et humaines
All fields of science	55	50	..	34	35 [c]	73	..	*Tous domaines scientifiques*
TOTAL								**TOTAL**
Natural sciences and engineering	..	1 547	..	1 957	..	2 144	..	Sciences exactes, nat. et de l'ingénieur
Social sciences and humanities	..	147	..	193	..	246	..	Sciences sociales et humaines
All fields of science	1 412	1 694	..	2 150	2 273 [c]	2 390	..	*Tous domaines scientifiques*

Source: OECD, R&D database (DSTI/EAS Division), January 2002.

Source : OCDE, base de données R-D (DSTI/Division AES), janvier 2002.

ICELAND — Table 9 - Tableau 9 — ISLANDE

Total business enterprise intramural expenditure on R&D - Dépenses intra-muros totales de R-D des entreprises

(BERD) by industry - (DIRDE) par industrie

	1994	1995	1996	1997	1998	1999	2000	
				9.1 Million ISK - Millions ISK				
1 AGRICULTURE, HUNTING & FORESTRY	155.7	151.6 [s]	..	212.2	218.0 [c]	199.9	..	**1 AGRICULTURE, CHASSE & SYLVICULTURE**
2 MINING	..	10.8 [s]	..	4.3	4.8 [c]	5.0	..	**2 ACTIVITES EXTRACTIVES**
3 MANUFACTURING	1 137.9	1 067.5 [s]	..	1 773.8	1 951.2 [c]	1 703.6	..	**3 ACTIVITES DE FABRICATION**
4 Food, beverages & tobacco	426.3	249.3	..	665.9	732.5 [c]	254.2	..	**4 Produits alimentaires, boissons & tabac**
5 Food products & beverages	..	249.3	..	665.9	732.5 [c]	254.2	..	5 Produits alimentaires & boissons
6 Tobacco products	..	0.0	..	0.0	0.0 [c]	0.0	..	6 Produits à base de tabac
7 Textiles, wear app., fur & leather	24.3	48.2	..	81.6	89.8 [c]	75.9	..	**7 Textiles, habill., fourrures & cuirs**
8 Textiles	..	24.7	..	62.9	53.8 [c]	66.0	..	8 Textiles
9 Wearing apparel & fur	..	23.5 [o]	..	18.7 [o]	36.0 [c,o]	..	..	9 Articles d'habillement & fourrures
10 Leather products & footwear	..	.. [n]	..	.. [n]	.. [n]	9.9	..	10 Articles en cuir & chaussures
11 Wood, paper, printing, publishing	3.9	23.2	..	..	..	5.0	..	**11 Bois, papier, imprimerie, édition**
12 Wood & cork (not furniture)	..	7.5	..	..	..	..	..	12 Bois & liège (sauf meubles)
13 Pulp, paper & paper products	3.9	15.7	..	..	..	..	..	13 Carton, papier & articles
14 Publ., print. & repro. of rec. media	..	0.0	..	..	..	5.0	..	14 Edition, impress. & repro. enregis.
15 Coke, petroleum, nuclear fuel, chemicals, rubber plastics	205.6	233.2	..	326.9	359.6 [c]	350.1	..	**15 Coke, pétrole, combust. nucl., prod. chimi., caout. & mat. plast.**
16 Coke, ref. petrol. prod. & nuclear fuel	..	..	..	..	..	..	..	16 Coke, raffin. pétr. & combust. nucl.
17 Chemicals & chemical products	156.1	185.4	..	282.8	311.0 [c]	306.6	..	17 Produits chimiques
18 *Chemical prod.(less pharmaceuticals)*	66.7	54.2	..	127.1	139.8 [c]	135.0	..	18 *Prod. chimi. (moins prod. pharmaceutiques)*
19 *Pharmaceuticals*	89.4	131.2	..	155.7	171.3 [c]	171.6	..	19 *Produits pharmaceutiques*
20 Rubber & plastic products	49.5	47.8	..	44.2	48.6 [c]	43.5	..	20 Caoutchouc et matières plastiques
21 Non-metallic mineral products	..	37.1	..	24.7	27.2 [c]	36.1	..	**21 Produits minéraux non métalliques**
22 Basic metals	36.0	68.9	..	69.3	43.2 [c]	37.1	..	**22 Produits métallurgiques de base**
23 Basic metals, ferrous	26.4	42.1	..	40.9	0.0 [c]	0.0	..	23 Produits métallur. de base, ferreux
24 Basic metals, non-ferrous	9.6	26.9	..	28.4	43.2 [c]	37.1	..	24 Prod. métallur. de base, non ferreux
25 Fabricated metal products	23.1	..	..	..	41.9 [c]	..	..	**25 Ouvrages en métaux (sauf mach.& mat.)**
26 Machinery equipment, instruments and transport equipment	380.4	403.4	..	598.9	649.9 [c]	935.7	..	**26 Machines & matériel, instruments & mat. de transport**
27 Machinery n.e.c.	33.3	14.5	..	10.2	188.0 [c]	450.3	..	27 Machines, n.c.a.
28 Office, account. & computing machinery	..	..	..	..	..	..	..	28 Mach. de bur., compta. & trait. inform.
29 Electrical machinery	305.2	266.1	..	356.0	233.1 [c]	16.2	..	29 Machines électriques
30 Electro. equip. (radio, TV & commun.)	..	2.7	..	2.0	..	..	..	30 Equip. électro. (radio, tv & communic.)
31 *Electro. comp. (inc. semi-conductors)*	..	..	..	..	..	..	..	31 *Compos. électro. (semi-cond. incl.)*
32 *TV, radio & communications equip.*	..	2.7	..	2.0	..	..	..	32 *Appareils radio, tv et de communi.*
33 Instruments	18.3	101.4	..	196.0	200.1 [c]	392.2	..	33 Instruments
34 Motor vehicles	..	..	..	..	..	..	..	34 Véhicules automobiles
35 Other transport equipment	23.7	18.8	..	34.6	28.7 [c]	76.9	..	35 Autres matériels de transport
36 *Ships*	9.5	9.0	..	12.3	13.6 [c]	..	..	36 *Navires*
37 *Aerospace*	10.4	0.0	..	0.0	0.0 [c]	0.0	..	37 *Aérospatiale*
38 *Other transport equipment n.e.c.*	3.7	9.9	..	22.3	15.2 [c]	76.9	..	38 *Autres matériels de transport n.c.a.*
39 Furniture, other manufacturing n.e.c.	38.2	3.5	..	6.5	7.2 [c]	9.3	..	**39 Meubles, autres activ. de fabricat. n.c.a.**
40 Furniture	6.0	3.5	..	6.5	7.2 [c]	9.3	..	40 Meubles
41 Other manufacturing n.e.c.	32.2	0.0	..	0.0	0.0 [c]	0.0	..	41 Autres activités de fabrication n.c.a.
42 Recycling	..	1.5	..	..	..	..	..	**42 Récupération**
43 ELECTRICITY, GAS & WATER SUPPLY	87.3	128.2 [s]	..	11.6	12.8 [c]	45.7	..	**43 DISTRIBUTION D'ELECTRICITE, GAZ & EAU**
44 CONSTRUCTION	92.0	22.8 [s]	..	21.3	23.4 [c]	26.0	..	**44 CONSTRUCTION**
45 SERVICES SECTOR	408.7	835.5 [s]	..	1 894.7	2 099.7 [c]	4 796.5	..	**45 SECTEUR DES SERVICES**
46 Wholesale, retail trade & repair of motor vehicles, etc.	..	144.2	..	146.9	165.0 [c]	90.9	..	**46 Commerce (gros & détail), réparation de véhicules, etc.**
47 Hotels & restaurants	..	..	..	..	..	..	..	**47 Hôtels et restaurants**
48 Transport & storage	66.0	103.0	..	139.7	153.7 [c]	161.6	..	**48 Transports, entreposage**
49 Communications	..	0.0	..	0.0	..	95.0	..	**49 Communications**
50 Post	..	..	..	..	..	..	..	50 Postes
51 Telecommunications	..	..	..	..	..	95.0	..	51 Télécommunications
52 Financial intermediation (inc. insur.)	..	..	..	..	..	..	..	**52 Intermédiation financ. (assurances inc.)**
53 Real estate, renting & busin. activity	339.7	588.3	..	1 608.1	1 781.0 [c]	4 449.1	..	**53 Immobil., locat. & serv. aux entreprises**
54 Computer & related activities	333.0	398.9	..	756.3	832.1 [c]	997.8	..	54 Informatique et activités connexes
55 *Software consultancy*	246.4	397.4	..	752.9	820.0 [c]	997.8	..	55 *Production de logiciels*
56 *Other computer services n.e.c.*	86.6	1.5	..	3.4	12.1 [c]	..	..	56 *Autres activités informatiques*
57 Research & development	6.7	11.4	..	587.9	646.7 [c]	3 081.2	..	57 Recherche & développement expérimental
58 Other business activities n.e.c.	..	178.0	..	263.9	302.2 [c]	370.1	..	58 Autres activ. & serv. aux entreprises
59 Comm., soc. & pers. serv. activ., etc.	3.0	..	..	..	..	..	..	**59 Adm. pub., serv. soc. & collec., etc**
TOTAL BUSINESS ENTERPRISE SECTOR	**1 881.6**	**2 216.4**	**..**	**3 918.0**	**4 309.8 [c]**	**6 776.8**	**..**	**TOTAL SECTEUR DES ENTREPRISES**

Source: OECD, R&D database (DSTI/EAS Division), January 2002.

Source : OCDE, base de données R-D (DSTI/Division AES), janvier 2002.

ICELAND — Table 9 - Tableau 9 — ISLANDE

Total business enterprise intramural expenditure on R&D - Dépenses intra-muros totales de R-D des entreprises
(BERD) by industry - (DIRDE) par industrie

9.2 Million constant USD(1995 prices and PPPs) - Millions de USD constants (prix et PPA de 1995)

	1994	1995	1996	1997	1998	1999	2000	
1 AGRICULTURE, HUNTING & FORESTRY	2.1	2.0 [s]	..	2.7	2.6 [c]	2.3	..	**1 AGRICULTURE, CHASSE & SYLVICULTURE**
2 MINING	..	0.1 [s]	..	0.1	0.1 [c]	0.1	..	**2 ACTIVITES EXTRACTIVES**
3 MANUFACTURING	15.4	14.1 [s]	..	22.2	23.2 [c]	19.5	..	**3 ACTIVITES DE FABRICATION**
4 Food, beverages & tobacco	5.8	3.3	..	8.3	8.7 [c]	2.9	..	**4 Produits alimentaires, boissons & tabac**
5 Food products & beverages	..	3.3	..	8.3	8.7 [c]	2.9	..	5 Produits alimentaires & boissons
6 Tobacco products	..	0.0	..	0.0	0.0 [c]	0.0	..	6 Produits à base de tabac
7 Textiles, wear app., fur & leather	0.3	0.6	..	1.0	1.1 [c]	0.9	..	**7 Textiles, habill., fourrures & cuirs**
8 Textiles	..	0.3	..	0.8	0.6 [c]	0.8	..	8 Textiles
9 Wearing apparel & fur	..	0.3 [o]	..	0.2 [o]	0.4 [c,o]	..	..	9 Articles d'habillement & fourrures
10 Leather products & footwear	..	.. [n]	..	.. [n]	.. [n]	0.1	..	10 Articles en cuir & chaussures
11 Wood, paper, printing, publishing	0.1	0.3	..	..	..	0.1	..	**11 Bois, papier, imprimerie, édition**
12 Wood & cork (not furniture)	..	0.1	..	..	..	..	..	12 Bois & liège (sauf meubles)
13 Pulp, paper & paper products	0.1	0.2	..	..	..	..	..	13 Carton, papier & articles
14 Publ., print. & repro. of rec. media	..	0.0	..	..	..	0.1	..	14 Edition, impress. & repro. enregis.
15 Coke, petroleum, nuclear fuel, chemicals, rubber plastics	2.8	3.1	..	4.1	4.3 [c]	4.0	..	**15 Coke, pétrole, combust. nucl., prod. chimi., caout. & mat. plast.**
16 Coke, ref. petrol. prod. & nuclear fuel	..	..	..	..	..	..	..	16 Coke, raffin. pétr. & combust. nucl.
17 Chemicals & chemical products	2.1	2.4	..	3.5	3.7 [c]	3.5	..	17 Produits chimiques
18 *Chemical prod.(less pharmaceuticals)*	0.9	0.7	..	1.6	1.7 [c]	1.5	..	18 *Prod. chimi. (moins prod. pharmaceutiques)*
19 *Pharmaceuticals*	1.2	1.7	..	1.9	2.0 [c]	2.0	..	19 *Produits pharmaceutiques*
20 Rubber & plastic products	0.7	0.6	..	0.6	0.6 [c]	0.5	..	20 Caoutchouc et matières plastiques
21 Non-metallic mineral products	..	0.5	..	0.3	0.3 [c]	0.4	..	**21 Produits minéraux non métalliques**
22 Basic metals	0.5	0.9	..	0.9	0.5 [c]	0.4	..	**22 Produits métallurgiques de base**
23 Basic metals, ferrous	0.4	0.6	..	0.5	0.0 [c]	0.0	..	23 Produits métallur. de base, ferreux
24 Basic metals, non-ferrous	0.1	0.4	..	0.4	0.5 [c]	0.4	..	24 Prod. métallur. de base, non ferreux
25 Fabricated metal products	0.3	..	..	..	0.5 [c]	..	..	**25 Ouvrages en métaux (sauf mach.& mat.)**
26 Machinery equipment, instruments and transport equipment	5.2	5.3	..	7.5	7.7 [c]	10.7	..	**26 Machines & matériel, instruments & mat. de transport**
27 Machinery n.e.c.	0.5	0.2	..	0.1	2.2 [c]	5.2	..	27 Machines, n.c.a.
28 Office, account. & computing machinery	..	..	..	..	..	..	..	28 Mach. de bur., compta. & trait. inform.
29 Electrical machinery	4.1	3.5	..	4.4	2.8 [c]	0.2	..	29 Machines électriques
30 Electro. equip. (radio, TV & commun.)	..	0.0	..	0.0	..	..	..	30 Equip. électro. (radio, tv & communic.)
31 *Electro. comp. (inc. semi-conductors)*	..	..	..	..	..	..	..	31 *Compos. électro. (semi-cond. incl.)*
32 *TV, radio & communications equip.*	..	0.0	..	0.0	..	..	..	32 *Appareils radio, tv et de communi.*
33 Instruments	0.2	1.3	..	2.4	2.4 [c]	4.5	..	33 Instruments
34 Motor vehicles	..	..	..	..	..	..	..	34 Véhicules automobiles
35 Other transport equipment	0.3	0.2	..	0.4	0.3 [c]	0.9	..	35 Autres matériels de transport
36 *Ships*	0.1	0.1	..	0.2	0.2 [c]	..	..	36 *Navires*
37 *Aerospace*	0.1	0.0	..	0.0	0.0 [c]	0.0	..	37 *Aérospatiale*
38 *Other transport equipment n.e.c.*	0.1	0.1	..	0.3	0.2 [c]	0.9	..	38 *Autres matériels de transport n.c.a.*
39 Furniture, other manufacturing n.e.c.	0.5	0.0	..	0.1	0.1 [c]	0.1	..	**39 Meubles, autres activ. de fabricat. n.c.a.**
40 Furniture	0.1	0.0	..	0.1	0.1 [c]	0.1	..	40 Meubles
41 Other manufacturing n.e.c.	0.4	0.0	..	0.0	0.0 [c]	0.0	..	41 Autres activités de fabrication n.c.a.
42 Recycling	..	0.0	..	..	..	..	..	**42 Récupération**
43 ELECTRICITY, GAS & WATER SUPPLY	1.2	1.7 [s]	..	0.1	0.2 [c]	0.5	..	**43 DISTRIBUTION D'ELECTRICITE, GAZ & EAU**
44 CONSTRUCTION	1.2	0.3 [s]	..	0.3	0.3 [c]	0.3	..	**44 CONSTRUCTION**
45 SERVICES SECTOR	5.5	11.0 [s]	..	23.7	24.9 [c]	54.9	..	**45 SECTEUR DES SERVICES**
46 Wholesale, retail trade & repair of motor vehicles, etc.	..	1.9	..	1.8	2.0 [c]	1.0	..	**46 Commerce (gros & détail), réparation de véhicules, etc.**
47 Hotels & restaurants	..	..	..	..	..	..	..	**47 Hôtels et restaurants**
48 Transport & storage	0.9	1.4	..	1.7	1.8 [c]	1.8	..	**48 Transports, entreposage**
49 Communications	..	0.0	..	0.0	..	1.1	..	**49 Communications**
50 Post	..	..	..	..	..	..	..	50 Postes
51 Telecommunications	..	..	..	..	..	1.1	..	51 Télécommunications
52 Financial intermediation (inc. insur.)	..	..	..	..	..	..	..	**52 Intermédiation financ. (assurances inc.)**
53 Real estate, renting & busin. activity	4.6	7.8	..	20.1	21.1 [c]	50.9	..	**53 Immobil., locat. & serv. aux entreprises**
54 Computer & related activities	4.5	5.3	..	9.4	9.9 [c]	11.4	..	54 Informatique et activités connexes
55 *Software consultancy*	3.3	5.2	..	9.4	9.7 [c]	11.4	..	55 *Production de logiciels*
56 *Other computer services n.e.c.*	1.2	0.0	..	0.0	0.1 [c]	..	..	56 *Autres activités informatiques*
57 Research & development	0.1	0.2	..	7.3	7.7 [c]	35.2	..	57 Recherche & développement expérimental
58 Other business activities n.e.c.	..	2.3	..	3.3	3.6 [c]	4.2	..	58 Autres activ. & serv. aux entreprises
59 Comm., soc. & pers. serv. activ., etc.	0.0	..	..	..	..	..	..	**59 Adm. pub., serv. soc. & collec., etc**
TOTAL BUSINESS ENTERPRISE SECTOR	**25.5**	**29.2**	**..**	**48.9**	**51.1 [c]**	**77.5**	**..**	**TOTAL SECTEUR DES ENTREPRISES**

Source: OECD, R&D database (DSTI/EAS Division), January 2002.

Source : OCDE, base de données R-D (DSTI/Division AES), janvier 2002.

Business enterprise intramural expenditure on R&D (BERD) - Dépenses intra-muros de R-D des entreprises (DIRDE)

by industry and source of funds - par industrie et par source de financement

Million ISK — *Millions ISK*

	1994	1995	1996	1997	1998	1999	2000	
				10.1 Business enterprise - Entreprises				
1 AGRICULTURE, HUNTING & FORESTRY	..	123.0	..	155.8	..	141.1	..	**1 AGRICULTURE, CHASSE & SYLVICULTURE**
2 MINING	..	10.8	..	4.3	..	5.0	..	**2 ACTIVITES EXTRACTIVES**
3 MANUFACTURING	..	1 016.4	..	1 644.9	..	1 644.9	..	**3 ACTIVITES DE FABRICATION**
4 Food, beverages & tobacco	..	242.6	..	614.8	..	241.0	..	**4 Produits alimentaires, boissons & tabac**
5 Food products & beverages	..	242.6	..	614.8	..	241.0	..	5 Produits alimentaires & boissons
6 Tobacco products	..	..	..	..	..	..	..	6 Produits à base de tabac
7 Textiles, wear app., fur & leather	..	48.2	..	79.0	..	71.3	..	**7 Textiles, habill., fourrures & cuirs**
8 Textiles	..	24.7	..	60.7	..	61.4	..	8 Textiles
9 Wearing apparel & fur	..	23.5	..	18.3	..	..	..	9 Articles d'habillement & fourrures
10 Leather products & footwear	..	..	..	..	..	9.9	..	10 Articles en cuir & chaussures
11 Wood, paper, printing, publishing	..	23.2	..	..	..	5.0	..	**11 Bois, papier, imprimerie, édition**
12 Wood & cork (not furniture)	..	7.5	..	..	..	..	..	12 Bois & liège (sauf meubles)
13 Pulp, paper & paper products	..	15.7	..	..	..	..	..	13 Carton, papier & articles
14 Publ., print. & repro. of rec. media	..	..	..	..	..	5.0	..	14 Edition, impress. & repro. enregis.
15 Coke, petroleum, nuclear fuel, chemicals, rubber plastics	..	222.8	..	307.1	..	342.9	..	**15 Coke, pétrole, combust. nucl., prod. chimi., caout. & mat. plast.**
16 Coke, ref. petrol. prod. & nuclear fuel	..	..	..	..	..	..	..	16 Coke, raffin. pétr. & combust. nucl.
17 Chemicals & chemical products	..	177.4	..	264.1	..	300.6	..	17 Produits chimiques
18 *Chemical prod.(less pharmaceuticals)*	..	47.9	..	108.4	..	131.3	..	18 *Prod. chimi. (moins prod. pharmaceutiques)*
19 *Pharmaceuticals*	..	129.5	..	155.7	..	169.2	..	19 *Produits pharmaceutiques*
20 Rubber & plastic products	..	45.4	..	43.0	..	42.4	..	20 Caoutchouc et matières plastiques
21 Non-metallic mineral products	..	36.9	..	24.7	..	36.1	..	**21 Produits minéraux non métalliques**
22 Basic metals	..	66.7	..	66.3	..	37.0	..	**22 Produits métallurgiques de base**
23 Basic metals, ferrous	..	39.8	..	37.9	..	..	..	23 Produits métallur. de base, ferreux
24 Basic metals, non-ferrous	..	26.9	..	28.4	..	37.0	..	24 Prod. métallur. de base, non ferreux
25 Fabricated metal products	..	..	..	..	..	..	..	**25 Ouvrages en métaux (sauf mach.& mat.)**
26 Machinery equipment, instruments and transport equipment	..	371.0	..	546.4	..	902.2	..	**26 Machines & matériel, instruments & mat. de transport**
27 Machinery n.e.c.	..	13.9	..	9.3	..	423.9	..	27 Machines, n.c.a.
28 Office, account. & computing machinery	..	..	..	..	..	..	..	28 Mach. de bur., compta. & trait. inform.
29 Electrical machinery	..	240.1	..	329.5	..	16.2	..	29 Machines électriques
30 Electro. equip. (radio, TV & commun.)	..	2.7	..	2.0	..	..	..	30 Equip. électro. (radio, tv & communic.)
31 *Electro. comp. (inc. semi-conductors)*	..	..	..	..	..	..	..	31 *Compos. électro. (semi-cond. incl.)*
32 *TV, radio & communications equip.*	..	2.7	..	2.0	..	..	..	32 *Appareils radio, tv et de communi.*
33 Instruments	..	95.5	..	177.2	..	385.2	..	33 Instruments
34 Motor vehicles	..	..	..	..	..	..	..	34 Véhicules automobiles
35 Other transport equipment	..	18.8	..	28.5	..	76.9	..	35 Autres matériels de transport
36 *Ships*	..	9.0	..	12.0	..	..	..	36 *Navires*
37 *Aerospace*	..	..	..	..	..	..	..	37 *Aérospatiale*
38 *Other transport equipment n.e.c.*	..	9.9	..	16.5	..	76.9	..	38 *Autres matériels de transport n.c.a.*
39 Furniture, other manufacturing n.e.c.	..	3.5	..	6.5	..	9.3	..	**39 Meubles, autres activ. de fabricat. n.c.a.**
40 Furniture	..	3.5	..	6.5	..	9.3	..	40 Meubles
41 Other manufacturing n.e.c.	..	..	..	..	..	..	..	41 Autres activités de fabrication n.c.a.
42 Recycling	..	1.5	..	..	..	..	..	**42 Récupération**
43 ELECTRICITY, GAS & WATER SUPPLY	..	128.2	..	11.6	..	45.7	..	**43 DISTRIBUTION D'ELECTRICITE, GAZ & EAU**
44 CONSTRUCTION	..	22.3	..	21.0	..	25.5	..	**44 CONSTRUCTION**
45 SERVICES SECTOR	..	815.9	..	1 739.7	..	3 343.0	..	**45 SECTEUR DES SERVICES**
46 Wholesale, retail trade & repair of motor vehicles, etc.	..	142.2	..	144.9	..	81.6	..	**46 Commerce (gros & détail), réparation de véhicules, etc.**
47 Hotels & restaurants	..	..	..	..	..	..	..	**47 Hôtels et restaurants**
48 Transport & storage	..	103.0	..	131.0	..	161.6	..	**48 Transports, entreposage**
49 Communications	..	..	..	..	..	86.6	..	**49 Communications**
50 Post	..	..	..	..	..	..	..	50 Postes
51 Telecommunications	..	..	..	..	..	86.6	..	51 Télécommunications
52 Financial intermediation (inc. insur.)	..	..	..	..	..	..	..	**52 Intermédiation financ. (assurances inc.)**
53 Real estate, renting & busin. activity	..	570.7	..	1 463.8	..	3 013.2	..	**53 Immobil., locat. & serv. aux entreprises**
54 Computer & related activities	..	398.9	..	671.7	..	834.3	..	54 Informatique et activités connexes
55 *Software consultancy*	..	397.4	..	668.4	..	834.3	..	55 *Production de logiciels*
56 *Other computer services n.e.c.*	..	1.5	..	3.3	..	..	..	56 *Autres activités informatiques*
57 Research & development	..	7.4	..	578.8	..	1 886.0	..	57 Recherche & développement expérimental
58 Other business activities n.e.c.	..	164.4	..	213.2	..	292.9	..	58 Autres activ. & serv. aux entreprises
59 Comm., soc. & pers. serv. activ., etc.	..	..	..	..	..	..	..	**59 Adm. pub., serv. soc. & collec., etc**
TOTAL BUSINESS ENTERPRISE SECTOR	**1 577.6**	**2 116.4**	**..**	**3 577.3**	**3 935.1** [c]	**5 205.3**	**..**	**TOTAL SECTEUR DES ENTREPRISES**

Source: OECD, R&D database (DSTI/EAS Division), January 2002.

Source : OCDE, base de données R-D (DSTI/Division AES), janvier 2002.

Business enterprise intramural expenditure on R&D (BERD) - Dépenses intra-muros de R-D des entreprises (DIRDE)

by industry and source of funds - par industrie et par source de financement

Million ISK — *Millions ISK*

10.2 Government - État

	1994	1995	1996	1997	1998	1999	2000	
1 AGRICULTURE, HUNTING & FORESTRY	..	23.4	..	34.5	..	30.8	..	**1 AGRICULTURE, CHASSE & SYLVICULTURE**
2 MINING	..	..	..	..	..	..	..	**2 ACTIVITES EXTRACTIVES**
3 MANUFACTURING	..	31.2	..	87.4	..	44.4	..	**3 ACTIVITES DE FABRICATION**
4 Food, beverages & tobacco	..	6.7	..	30.3	..	9.2	..	**4 Produits alimentaires, boissons & tabac**
5 Food products & beverages	..	6.7	..	30.3	..	9.2	..	5 Produits alimentaires & boissons
6 Tobacco products	..	..	..	..	..	..	..	6 Produits à base de tabac
7 Textiles, wear app., fur & leather	..	0.0	..	2.3	..	4.6	..	**7 Textiles, habill., fourrures & cuirs**
8 Textiles	..	..	..	2.3	..	4.6	..	8 Textiles
9 Wearing apparel & fur	..	..	..	..	..	..	..	9 Articles d'habillement & fourrures
10 Leather products & footwear	..	..	..	..	..	..	..	10 Articles en cuir & chaussures
11 Wood, paper, printing, publishing	..	0.0	..	0.0	..	..	..	**11 Bois, papier, imprimerie, édition**
12 Wood & cork (not furniture)	..	..	..	..	..	..	..	12 Bois & liège (sauf meubles)
13 Pulp, paper & paper products	..	..	..	..	..	..	..	13 Carton, papier & articles
14 Publ., print. & repro. of rec. media	..	..	..	..	..	..	..	14 Edition, impress. & repro. enregis.
15 Coke, petroleum, nuclear fuel, chemicals, rubber plastics	..	10.4	..	13.5	..	7.1	..	**15 Coke, pétrole, combust. nucl., prod. chimi., caout. & mat. plast.**
16 Coke, ref. petrol. prod. & nuclear fuel	..	..	..	..	..	..	..	16 Coke, raffin. pétr. & combust. nucl.
17 Chemicals & chemical products	..	8.0	..	12.4	..	6.0	..	17 Produits chimiques
18 *Chemical prod.(less pharmaceuticals)*	..	6.3	..	12.4	..	3.6	..	18 *Prod. chimi. (moins prod. pharmaceutiques)*
19 *Pharmaceuticals*	..	1.7	..	..	..	2.4	..	19 *Produits pharmaceutiques*
20 Rubber & plastic products	..	2.4	..	1.1	..	1.1	..	20 Caoutchouc et matières plastiques
21 Non-metallic mineral products	..	0.2	..	..	..	..	..	**21 Produits minéraux non métalliques**
22 Basic metals	..	1.1	..	2.9	..	0.1	..	**22 Produits métallurgiques de base**
23 Basic metals, ferrous	..	1.1	..	2.9	..	..	..	23 Produits métallur. de base, ferreux
24 Basic metals, non-ferrous	..	..	..	..	..	0.1	..	24 Prod. métallur. de base, non ferreux
25 Fabricated metal products	..	..	..	..	..	..	..	**25 Ouvrages en métaux (sauf mach.& mat.)**
26 Machinery equipment, instruments and transport equipment	..	12.8	..	37.1	..	23.4	..	**26 Machines & matériel, instruments & mat. de transport**
27 Machinery n.e.c.	..	0.6	..	..	..	16.3	..	27 Machines, n.c.a.
28 Office, account. & computing machinery	..	..	..	..	..	..	..	28 Mach. de bur., compta. & trait. inform.
29 Electrical machinery	..	6.3	..	14.9	..	..	..	29 Machines électriques
30 Electro. equip. (radio, TV & commun.)	..	0.0	..	0.0	..	..	..	30 Equip. électro. (radio, tv & communic.)
31 *Electro. comp. (inc. semi-conductors)*	..	..	..	..	..	..	..	31 *Compos. électro. (semi-cond. incl.)*
32 *TV, radio & communications equip.*	..	..	..	..	..	..	..	32 *Appareils radio, tv et de communi.*
33 Instruments	..	5.9	..	16.1	..	7.0	..	33 Instruments
34 Motor vehicles	..	..	..	..	..	..	..	34 Véhicules automobiles
35 Other transport equipment	..	0.0	..	6.2	..	..	..	35 Autres matériels de transport
36 *Ships*	..	..	..	0.4	..	..	..	36 *Navires*
37 *Aerospace*	..	..	..	..	..	..	..	37 *Aérospatiale*
38 *Other transport equipment n.e.c.*	..	..	..	5.8	..	..	..	38 *Autres matériels de transport n.c.a.*
39 Furniture, other manufacturing n.e.c.	..	0.0	..	0.0	..	..	..	**39 Meubles, autres activ. de fabricat. n.c.a.**
40 Furniture	..	..	..	..	..	..	..	40 Meubles
41 Other manufacturing n.e.c.	..	..	..	..	..	..	..	41 Autres activités de fabrication n.c.a.
42 Recycling	..	..	..	..	..	..	..	**42 Récupération**
43 ELECTRICITY, GAS & WATER SUPPLY	..	..	..	..	..	..	..	**43 DISTRIBUTION D'ELECTRICITE, GAZ & EAU**
44 CONSTRUCTION	..	0.5	..	0.3	..	0.5	..	**44 CONSTRUCTION**
45 SERVICES SECTOR	..	17.4	..	74.6	..	53.4	..	**45 SECTEUR DES SERVICES**
46 Wholesale, retail trade & repair of motor vehicles, etc.	..	2.0	..	2.0	..	8.3	..	**46 Commerce (gros & détail), réparation de véhicules, etc.**
47 Hotels & restaurants	..	..	..	..	..	..	..	**47 Hôtels et restaurants**
48 Transport & storage	..	..	..	5.8	..	..	..	**48 Transports, entreposage**
49 Communications	..	0.0	..	0.0	..	3.4	..	**49 Communications**
50 Post	..	..	..	..	..	..	..	50 Postes
51 Telecommunications	..	..	..	..	..	3.4	..	51 Télécommunications
52 Financial intermediation (inc. insur.)	..	..	..	..	..	..	..	**52 Intermédiation financ. (assurances inc.)**
53 Real estate, renting & busin. activity	..	15.4	..	66.8	..	41.7	..	**53 Immobil., locat. & serv. aux entreprises**
54 Computer & related activities	..	0.0	..	17.8	..	19.6	..	54 Informatique et activités connexes
55 *Software consultancy*	..	..	..	17.7	..	19.6	..	55 *Production de logiciels*
56 *Other computer services n.e.c.*	..	..	..	0.1	..	..	..	56 *Autres activités informatiques*
57 Research & development	..	4.0	..	9.1	..	..	..	57 Recherche & développement expérimental
58 Other business activities n.e.c.	..	11.4	..	39.8	..	22.1	..	58 Autres activ. & serv. aux entreprises
59 Comm., soc. & pers. serv. activ., etc.	..	..	..	..	..	..	..	**59 Adm. pub., serv. soc. & collec., etc**
TOTAL BUSINESS ENTERPRISE SECTOR	**270.7**	**72.6**	**..**	**196.3**	**215.9** [c]	**163.9**	**..**	**TOTAL SECTEUR DES ENTREPRISES**

Source: OECD, R&D database (DSTI/EAS Division), January 2002.

Source : OCDE, base de données R-D (DSTI/Division AES), janvier 2002.

Business enterprise intramural expenditure on R&D (BERD) - Dépenses intra-muros de R-D des entreprises (DIRDE)

by industry and source of funds - par industrie et par source de financement

Million ISK / *Millions ISK*

	1994	1995	1996	1997	1998	1999	2000	
	10.3 Other national funds - Autres sources nationales							
1 **AGRICULTURE, HUNTING & FORESTRY**	..	..	..	..	..	..	..	1 **AGRICULTURE, CHASSE & SYLVICULTURE**
2 **MINING**	..	..	..	..	..	..	..	2 **ACTIVITES EXTRACTIVES**
3 **MANUFACTURING**	..	..	..	0.4	..	..	..	3 **ACTIVITES DE FABRICATION**
4 **Food, beverages & tobacco**	..	..	..	..	..	..	..	4 **Produits alimentaires, boissons & tabac**
5 Food products & beverages	..	..	..	..	..	..	..	5 Produits alimentaires & boissons
6 Tobacco products	..	..	..	..	..	..	..	6 Produits à base de tabac
7 **Textiles, wear app., fur & leather**	..	..	..	..	..	..	..	7 **Textiles, habill., fourrures & cuirs**
8 Textiles	..	..	..	..	..	..	..	8 Textiles
9 Wearing apparel & fur	..	..	..	..	..	..	..	9 Articles d'habillement & fourrures
10 Leather products & footwear	..	..	..	..	..	..	..	10 Articles en cuir & chaussures
11 **Wood, paper, printing, publishing**	..	..	..	..	..	..	..	11 **Bois, papier, imprimerie, édition**
12 Wood & cork (not furniture)	..	..	..	..	..	..	..	12 Bois & liège (sauf meubles)
13 Pulp, paper & paper products	..	..	..	..	..	..	..	13 Carton, papier & articles
14 Publ., print. & repro. of rec. media	..	..	..	..	..	..	..	14 Edition, impress. & repro. enregis.
15 **Coke, petroleum, nuclear fuel, chemicals, rubber plastics**	..	..	..	..	..	..	..	15 **Coke, pétrole, combust. nucl., prod. chimi., caout. & mat. plast.**
16 Coke, ref. petrol. prod. & nuclear fuel	..	..	..	..	..	..	..	16 Coke, raffin. pétr. & combust. nucl.
17 Chemicals & chemical products	..	..	..	..	..	..	..	17 Produits chimiques
18 *Chemical prod.(less pharmaceuticals)*	..	..	..	..	..	..	..	18 *Prod. chimi. (moins prod. pharmaceutiques)*
19 *Pharmaceuticals*	..	..	..	0.4	..	..	..	19 *Produits pharmaceutiques*
20 Rubber & plastic products	..	..	..	..	..	..	..	20 Caoutchouc et matières plastiques
21 **Non-metallic mineral products**	..	..	..	..	..	..	..	21 **Produits minéraux non métalliques**
22 **Basic metals**	..	..	..	..	..	..	..	22 **Produits métallurgiques de base**
23 Basic metals, ferrous	..	..	..	..	..	..	..	23 Produits métallur. de base, ferreux
24 Basic metals, non-ferrous	..	..	..	..	..	..	..	24 Prod. métallur. de base, non ferreux
25 **Fabricated metal products**	..	..	..	..	..	..	..	25 **Ouvrages en métaux (sauf mach.& mat.)**
26 **Machinery equipment, instruments and transport equipment**	..	..	..	..	..	..	..	26 **Machines & matériel, instruments & mat. de transport**
27 Machinery n.e.c.	..	..	..	..	..	..	..	27 Machines, n.c.a.
28 Office, account. & computing machinery	..	..	..	..	..	..	..	28 Mach. de bur., compta. & trait. inform.
29 Electrical machinery	..	..	..	..	..	..	..	29 Machines électriques
30 Electro. equip. (radio, TV & commun.)	..	..	..	..	..	..	..	30 Equip. électro. (radio, tv & communic.)
31 *Electro. comp. (inc. semi-conductors)*	..	..	..	..	..	..	..	31 *Compos. électro. (semi-cond. incl.)*
32 *TV, radio & communications equip.*	..	..	..	..	..	..	..	32 *Appareils radio, tv et de communi.*
33 Instruments	..	..	..	..	..	..	..	33 Instruments
34 Motor vehicles	..	..	..	..	..	..	..	34 Véhicules automobiles
35 Other transport equipment	..	..	..	..	..	..	..	35 Autres matériels de transport
36 *Ships*	..	..	..	..	..	..	..	36 *Navires*
37 *Aerospace*	..	..	..	..	..	..	..	37 *Aérospatiale*
38 *Other transport equipment n.e.c.*	..	..	..	..	..	..	..	38 *Autres matériels de transport n.c.a.*
39 **Furniture, other manufacturing n.e.c.**	..	..	..	..	..	..	..	39 **Meubles, autres activ. de fabricat. n.c.a.**
40 Furniture	..	..	..	..	..	..	..	40 Meubles
41 Other manufacturing n.e.c.	..	..	..	..	..	..	..	41 Autres activités de fabrication n.c.a.
42 **Recycling**	..	..	..	..	..	..	..	42 **Récupération**
43 **ELECTRICITY, GAS & WATER SUPPLY**	..	..	..	..	..	..	..	43 **DISTRIBUTION D'ELECTRICITE, GAZ & EAU**
44 **CONSTRUCTION**	..	..	..	..	..	..	..	44 **CONSTRUCTION**
45 **SERVICES SECTOR**	..	..	..	..	..	..	..	45 **SECTEUR DES SERVICES**
46 **Wholesale, retail trade & repair of motor vehicles, etc.**	..	..	..	..	..	..	..	46 **Commerce (gros & détail), réparation de véhicules, etc.**
47 **Hotels & restaurants**	..	..	..	..	..	..	..	47 **Hôtels et restaurants**
48 **Transport & storage**	..	..	..	..	..	..	..	48 **Transports, entreposage**
49 **Communications**	..	..	..	..	..	..	..	49 **Communications**
50 Post	..	..	..	..	..	..	..	50 Postes
51 Telecommunications	..	..	..	..	..	..	..	51 Télécommunications
52 **Financial intermediation (inc. insur.)**	..	..	..	..	..	..	..	52 **Intermédiation financ. (assurances inc.)**
53 **Real estate, renting & busin. activity**	..	..	..	..	..	34.7	..	53 **Immobil., locat. & serv. aux entreprises**
54 Computer & related activities	..	..	..	..	..	..	..	54 Informatique et activités connexes
55 *Software consultancy*	..	..	..	..	..	..	..	55 *Production de logiciels*
56 *Other computer services n.e.c.*	..	..	..	..	..	..	..	56 *Autres activités informatiques*
57 Research & development	..	..	..	..	..	..	..	57 Recherche & développement expérimental
58 Other business activities n.e.c.	..	..	..	..	..	34.7	..	58 Autres activ. & serv. aux entreprises
59 **Comm., soc. & pers. serv. activ., etc.**	..	..	..	..	..	..	..	59 **Adm. pub., serv. soc. & collec., etc**
TOTAL BUSINESS ENTERPRISE SECTOR	**0.0**	**0.0**	**..**	**0.4**	**0.4** [c]	**..**	**..**	**TOTAL SECTEUR DES ENTREPRISES**

Source: OECD, R&D database (DSTI/EAS Division), January 2002.

Source : OCDE, base de données R-D (DSTI/Division AES), janvier 2002.

ICELAND Table 10 - Tableau 10 ISLANDE

Business enterprise intramural expenditure on R&D (BERD) - Dépenses intra-muros de R-D des entreprises (DIRDE)

by industry and source of funds - par industrie et par source de financement

Million ISK *Millions ISK*

10.4 Funds from abroad - Fonds de l'étranger

	1994	1995	1996	1997	1998	1999	2000	
1 AGRICULTURE, HUNTING & FORESTRY	..	5.2	..	21.9	..	28.0	..	**1 AGRICULTURE, CHASSE & SYLVICULTURE**
2 MINING	..	..	..	..	..	..	..	**2 ACTIVITES EXTRACTIVES**
3 MANUFACTURING	..	20.0	..	41.6	..	4.1	..	**3 ACTIVITES DE FABRICATION**
4 Food, beverages & tobacco	..	0.0	..	20.8	..	4.1	..	**4 Produits alimentaires, boissons & tabac**
5 Food products & beverages	..	..	..	20.8	..	4.1	..	5 Produits alimentaires & boissons
6 Tobacco products	..	..	..	..	..	..	..	6 Produits à base de tabac
7 Textiles, wear app., fur & leather	..	0.0	..	0.0	..	..	..	**7 Textiles, habill., fourrures & cuirs**
8 Textiles	..	..	..	..	..	..	..	8 Textiles
9 Wearing apparel & fur	..	..	..	..	..	..	..	9 Articles d'habillement & fourrures
10 Leather products & footwear	..	..	..	..	..	..	..	10 Articles en cuir & chaussures
11 Wood, paper, printing, publishing	..	0.0	..	0.0	..	..	..	**11 Bois, papier, imprimerie, édition**
12 Wood & cork (not furniture)	..	..	..	..	..	..	..	12 Bois & liège (sauf meubles)
13 Pulp, paper & paper products	..	..	..	..	..	..	..	13 Carton, papier & articles
14 Publ., print. & repro. of rec. media	..	..	..	..	..	..	..	14 Edition, impress. & repro. enregis.
15 Coke, petroleum, nuclear fuel, chemicals, rubber plastics	..	..	..	6.3	..	..	..	**15 Coke, pétrole, combust. nucl., prod. chimi., caout. & mat. plast.**
16 Coke, ref. petrol. prod. & nuclear fuel	..	..	..	..	..	..	..	16 Coke, raffin. pétr. & combust. nucl.
17 Chemicals & chemical products	..	0.0	..	6.3	..	..	..	17 Produits chimiques
18 *Chemical prod.(less pharmaceuticals)*	..	..	..	6.3	..	..	..	18 *Prod. chimi. (moins prod. pharmaceutiques)*
19 *Pharmaceuticals*	..	..	..	..	..	..	..	19 *Produits pharmaceutiques*
20 Rubber & plastic products	..	..	..	..	..	..	..	20 Caoutchouc et matières plastiques
21 Non-metallic mineral products	..	..	..	..	..	..	..	**21 Produits minéraux non métalliques**
22 Basic metals	..	0.3	..	0.0	..	..	..	**22 Produits métallurgiques de base**
23 Basic metals, ferrous	..	0.3	..	..	..	..	..	23 Produits métallur. de base, ferreux
24 Basic metals, non-ferrous	..	..	..	..	..	..	..	24 Prod. métallur. de base, non ferreux
25 Fabricated metal products	..	..	..	..	..	..	..	**25 Ouvrages en métaux (sauf mach.& mat.)**
26 Machinery equipment, instruments and transport equipment	..	19.7	..	14.5	..	..	..	**26 Machines & matériel, instruments & mat. de transport**
27 Machinery n.e.c.	..	..	..	..	..	..	..	27 Machines, n.c.a.
28 Office, account. & computing machinery	..	..	..	..	..	..	..	28 Mach. de bur., compta. & trait. inform.
29 Electrical machinery	..	19.7	..	11.7	..	..	..	29 Machines électriques
30 Electro. equip. (radio, TV & commun.)	..	0.0	..	0.0	..	..	..	30 Equip. électro. (radio, tv & communic.)
31 *Electro. comp. (inc. semi-conductors)*	..	..	..	..	..	..	..	31 *Compos. électro. (semi-cond. incl.)*
32 *TV, radio & communications equip.*	..	..	..	..	..	..	..	32 *Appareils radio, tv et de communi.*
33 Instruments	..	..	..	2.8	..	..	..	33 Instruments
34 Motor vehicles	..	..	..	..	..	..	..	34 Véhicules automobiles
35 Other transport equipment	..	0.0	..	0.0	..	..	..	35 Autres matériels de transport
36 *Ships*	..	..	..	..	..	..	..	36 *Navires*
37 *Aerospace*	..	..	..	..	..	..	..	37 *Aérospatiale*
38 *Other transport equipment n.e.c.*	..	..	..	..	..	..	..	38 *Autres matériels de transport n.c.a.*
39 Furniture, other manufacturing n.e.c.	..	0.0	..	0.0	..	..	..	**39 Meubles, autres activ. de fabricat. n.c.a.**
40 Furniture	..	..	..	..	..	..	..	40 Meubles
41 Other manufacturing n.e.c.	..	..	..	..	..	..	..	41 Autres activités de fabrication n.c.a.
42 Recycling	..	..	..	..	..	..	..	**42 Récupération**
43 ELECTRICITY, GAS & WATER SUPPLY	..	..	..	..	..	..	..	**43 DISTRIBUTION D'ELECTRICITE, GAZ & EAU**
44 CONSTRUCTION	..	..	..	..	..	..	..	**44 CONSTRUCTION**
45 SERVICES SECTOR	..	2.2	..	80.5	..	1 365.5	..	**45 SECTEUR DES SERVICES**
46 Wholesale, retail trade & repair of motor vehicles, etc.	..	..	..	..	..	1.0	..	**46 Commerce (gros & détail), réparation de véhicules, etc.**
47 Hotels & restaurants	..	..	..	..	..	..	..	**47 Hôtels et restaurants**
48 Transport & storage	..	..	..	2.9	..	..	..	**48 Transports, entreposage**
49 Communications	..	0.0	..	0.0	..	5.0	..	**49 Communications**
50 Post	..	..	..	..	..	..	..	50 Postes
51 Telecommunications	..	..	..	..	..	5.0	..	51 Télécommunications
52 Financial intermediation (inc. insur.)	..	..	..	..	..	..	..	**52 Intermédiation financ. (assurances inc.)**
53 Real estate, renting & busin. activity	..	2.2	..	77.5	..	1 359.5	..	**53 Immobil., locat. & serv. aux entreprises**
54 Computer & related activities	..	0.0	..	66.8	..	143.9	..	54 Informatique et activités connexes
55 *Software consultancy*	..	..	..	66.8	..	143.9	..	55 *Production de logiciels*
56 *Other computer services n.e.c.*	..	..	..	..	..	..	..	56 *Autres activités informatiques*
57 Research & development	..	..	..	..	..	1 195.3	..	57 Recherche & développement expérimental
58 Other business activities n.e.c.	..	2.2	..	10.7	..	20.3	..	58 Autres activ. & serv. aux entreprises
59 Comm., soc. & pers. serv. activ., etc.	..	..	..	..	..	..	..	**59 Adm. pub., serv. soc. & collec., etc**
TOTAL BUSINESS ENTERPRISE SECTOR	**33.3**	**27.4**	..	**144.0**	**158.4** c	**1 407.7**	..	**TOTAL SECTEUR DES ENTREPRISES**

Source: OECD, R&D database (DSTI/EAS Division), January 2002.

Source : OCDE, base de données R-D (DSTI/Division AES), janvier 2002.

Business enterprise intramural expenditure on R&D (BERD) - Dépenses intra-muros de R-D des entreprises (DIRDE)

by industry and type of costs - par industrie et par type de coûts

Million ISK *Millions ISK*

	1994	1995	1996	1997	1998	1999	2000	
				11.1 Current expenditure - Dépenses courantes				
1 AGRICULTURE, HUNTING & FORESTRY	..	65.2	..	146.3	..	141.1	..	**1 AGRICULTURE, CHASSE & SYLVICULTURE**
2 MINING	..	8.4	..	3.7	..	5.0	..	**2 ACTIVITES EXTRACTIVES**
3 MANUFACTURING	..	937.7	..	1 503.0	..	1 554.6	..	**3 ACTIVITES DE FABRICATION**
4 Food, beverages & tobacco	..	214.5	..	495.7	..	191.9	..	**4 Produits alimentaires, boissons & tabac**
5 Food products & beverages	..	214.5	..	495.7	..	191.9	..	5 Produits alimentaires & boissons
6 Tobacco products	..	..	..	..	..	..	..	6 Produits à base de tabac
7 Textiles, wear app., fur & leather	..	44.7	..	71.3	..	72.8	..	**7 Textiles, habill., fourrures & cuirs**
8 Textiles	..	23.7	..	54.2	..	63.0	..	8 Textiles
9 Wearing apparel & fur	..	21.0	..	17.1	..	..	..	9 Articles d'habillement & fourrures
10 Leather products & footwear	..	..	..	..	..	9.9	..	10 Articles en cuir & chaussures
11 Wood, paper, printing, publishing	..	19.0	..	..	..	5.0	..	**11 Bois, papier, imprimerie, édition**
12 Wood & cork (not furniture)	..	6.0	..	..	..	..	..	12 Bois & liège (sauf meubles)
13 Pulp, paper & paper products	..	13.0	..	..	..	..	..	13 Carton, papier & articles
14 Publ., print. & repro. of rec. media	..	..	..	..	..	5.0	..	14 Edition, impress. & repro. enregis.
15 Coke, petroleum, nuclear fuel, chemicals, rubber plastics	..	198.5	..	300.7	..	318.3	..	**15 Coke, pétrole, combust. nucl., prod. chimi., caout. & mat. plast.**
16 Coke, ref. petrol. prod. & nuclear fuel	..	..	..	..	..	..	..	16 Coke, raffin. pétr. & combust. nucl.
17 Chemicals & chemical products	..	156.6	..	260.6	..	280.1	..	17 Produits chimiques
18 *Chemical prod.(less pharmaceuticals)*	..	49.7	..	113.1	..	129.6	..	18 *Prod. chimi. (moins prod. pharmaceutiques)*
19 *Pharmaceuticals*	..	106.9	..	147.5	..	150.4	..	19 *Produits pharmaceutiques*
20 Rubber & plastic products	..	41.9	..	40.1	..	38.2	..	20 Caoutchouc et matières plastiques
21 Non-metallic mineral products	..	35.2	..	24.4	..	35.0	..	**21 Produits minéraux non métalliques**
22 Basic metals	..	65.7	..	62.5	..	37.1	..	**22 Produits métallurgiques de base**
23 Basic metals, ferrous	..	38.9	..	34.1	..	..	..	23 Produits métallur. de base, ferreux
24 Basic metals, non-ferrous	..	26.9	..	28.4	..	37.1	..	24 Prod. métallur. de base, non ferreux
25 Fabricated metal products	..	..	..	..	..	..	..	**25 Ouvrages en métaux (sauf mach.& mat.)**
26 Machinery equipment, instruments and transport equipment	..	355.0	..	542.4	..	885.1	..	**26 Machines & matériel, instruments & mat. de transport**
27 Machinery n.e.c.	..	14.2	..	9.2	..	417.3	..	27 Machines, n.c.a.
28 Office, account. & computing machinery	..	..	..	..	..	..	..	28 Mach. de bur., compta. & trait. inform.
29 Electrical machinery	..	237.4	..	327.3	..	15.0	..	29 Machines électriques
30 Electro. equip. (radio, TV & commun.)	..	2.4	..	1.7	..	..	..	30 Equip. électro. (radio, tv & communic.)
31 *Electro. comp. (inc. semi-conductors)*	..	..	..	..	..	..	..	31 *Compos. électro. (semi-cond. incl.)*
32 *TV, radio & communications equip.*	..	2.4	..	1.7	..	..	..	32 *Appareils radio, tv et de communi.*
33 Instruments	..	85.5	..	181.5	..	375.9	..	33 Instruments
34 Motor vehicles	..	..	..	..	..	..	..	34 Véhicules automobiles
35 Other transport equipment	..	15.5	..	22.7	..	76.9	..	35 Autres matériels de transport
36 *Ships*	..	7.9	..	6.2	..	..	..	36 *Navires*
37 *Aerospace*	..	..	..	..	..	..	..	37 *Aérospatiale*
38 *Other transport equipment n.e.c.*	..	7.6	..	16.5	..	76.9	..	38 *Autres matériels de transport n.c.a.*
39 Furniture, other manufacturing n.e.c.	..	3.5	..	6.0	..	9.3	..	**39 Meubles, autres activ. de fabricat. n.c.a.**
40 Furniture	..	3.5	..	6.0	..	9.3	..	40 Meubles
41 Other manufacturing n.e.c.	..	..	..	..	..	..	..	41 Autres activités de fabrication n.c.a.
42 Recycling	..	1.5	..	..	..	..	..	**42 Récupération**
43 ELECTRICITY, GAS & WATER SUPPLY	..	116.0	..	7.6	..	44.0	..	**43 DISTRIBUTION D'ELECTRICITE, GAZ & EAU**
44 CONSTRUCTION	..	21.7	..	19.8	..	25.0	..	**44 CONSTRUCTION**
45 SERVICES SECTOR	..	765.4	..	1 776.3	..	4 132.9	..	**45 SECTEUR DES SERVICES**
46 Wholesale, retail trade & repair of motor vehicles, etc.	..	131.8	..	136.6	..	84.0	..	**46 Commerce (gros & détail), réparation de véhicules, etc.**
47 Hotels & restaurants	..	..	..	..	..	..	..	**47 Hôtels et restaurants**
48 Transport & storage	..	87.0	..	122.5	..	136.1	..	**48 Transports, entreposage**
49 Communications	..	..	..	..	..	85.0	..	**49 Communications**
50 Post	..	..	..	..	..	..	..	50 Postes
51 Telecommunications	..	..	..	..	..	85.0	..	51 Télécommunications
52 Financial intermediation (inc. insur.)	..	..	..	..	..	..	..	**52 Intermédiation financ. (assurances inc.)**
53 Real estate, renting & busin. activity	..	546.6	..	1 517.2	..	3 827.8	..	**53 Immobil., locat. & serv. aux entreprises**
54 Computer & related activities	..	368.4	..	718.0	..	920.6	..	54 Informatique et activités connexes
55 *Software consultancy*	..	366.9	..	715.6	..	920.6	..	55 *Production de logiciels*
56 *Other computer services n.e.c.*	..	1.5	..	2.4	..	..	..	56 *Autres activités informatiques*
57 Research & development	..	11.4	..	559.4	..	2 574.5	..	57 Recherche & développement expérimental
58 Other business activities n.e.c.	..	166.9	..	239.7	..	332.7	..	58 Autres activ. & serv. aux entreprises
59 Comm., soc. & pers. serv. activ., etc.	..	..	..	..	..	..	..	**59 Adm. pub., serv. soc. & collec., etc**
TOTAL BUSINESS ENTERPRISE SECTOR	**..**	**1 879.8**	**..**	**3 456.7**	**..**	**5 907.9**	**..**	**TOTAL SECTEUR DES ENTREPRISES**

Source: OECD, R&D database (DSTI/EAS Division), January 2002.

Source : OCDE, base de données R-D (DSTI/Division AES), janvier 2002.

Business enterprise intramural expenditure on R&D (BERD) - Dépenses intra-muros de R-D des entreprises (DIRDE)

by industry and type of costs - par industrie et par type de coûts

Million ISK | *Millions ISK*

11.2 Capital expenditure - Dépenses en capital

	1994	1995	1996	1997	1998	1999	2000	
1 AGRICULTURE, HUNTING & FORESTRY	..	86.4	..	65.9	..	58.8	..	**1 AGRICULTURE, CHASSE & SYLVICULTURE**
2 MINING	..	2.4	..	0.7	..	..	..	**2 ACTIVITES EXTRACTIVES**
3 MANUFACTURING	..	129.8	..	270.8	..	148.9	..	**3 ACTIVITES DE FABRICATION**
4 Food, beverages & tobacco	..	34.8	..	170.2	..	62.3	..	**4 Produits alimentaires, boissons & tabac**
5 Food products & beverages	..	34.8	..	170.2	..	62.3	..	5 Produits alimentaires & boissons
6 Tobacco products	..	..	..	..	..	..	..	6 Produits à base de tabac
7 Textiles, wear app., fur & leather	..	3.5	..	10.3	..	3.1	..	**7 Textiles, habill., fourrures & cuirs**
8 Textiles	..	1.0	..	8.7	..	3.1	..	8 Textiles
9 Wearing apparel & fur	..	2.5	..	1.6	..	..	..	9 Articles d'habillement & fourrures
10 Leather products & footwear	..	..	..	..	..	..	..	10 Articles en cuir & chaussures
11 Wood, paper, printing, publishing	..	4.2	..	..	..	..	..	**11 Bois, papier, imprimerie, édition**
12 Wood & cork (not furniture)	..	1.5	..	..	..	..	..	12 Bois & liège (sauf meubles)
13 Pulp, paper & paper products	..	2.7	..	..	..	..	..	13 Carton, papier & articles
14 Publ., print. & repro. of rec. media	..	..	..	..	..	..	..	14 Edition, impress. & repro. enregis.
15 Coke, petroleum, nuclear fuel, chemicals, rubber plastics	..	34.7	..	26.2	..	31.8	..	**15 Coke, pétrole, combust. nucl., prod. chimi., caout. & mat. plast.**
16 Coke, ref. petrol. prod. & nuclear fuel	..	..	..	..	..	..	..	16 Coke, raffin. pétr. & combust. nucl.
17 Chemicals & chemical products	..	28.8	..	22.2	..	26.5	..	17 Produits chimiques
18 *Chemical prod.(less pharmaceuticals)*	..	4.5	..	14.0	..	5.3	..	18 *Prod. chimi. (moins prod. pharmaceutiques)*
19 *Pharmaceuticals*	..	24.3	..	8.2	..	21.2	..	19 *Produits pharmaceutiques*
20 Rubber & plastic products	..	5.9	..	4.1	..	5.3	..	20 Caoutchouc et matières plastiques
21 Non-metallic mineral products	..	1.9	..	0.4	..	1.1	..	**21 Produits minéraux non métalliques**
22 Basic metals	..	2.3	..	6.8	..	..	..	**22 Produits métallurgiques de base**
23 Basic metals, ferrous	..	2.3	..	6.8	..	..	..	23 Produits métallur. de base, ferreux
24 Basic metals, non-ferrous	..	..	..	..	..	..	..	24 Prod. métallur. de base, non ferreux
25 Fabricated metal products	..	..	..	..	..	..	..	**25 Ouvrages en métaux (sauf mach.& mat.)**
26 Machinery equipment, instruments and transport equipment	..	48.4	..	56.4	..	50.6	..	**26 Machines & matériel, instruments & mat. de transport**
27 Machinery n.e.c.	..	0.3	..	1.0	..	33.0	..	27 Machines, n.c.a.
28 Office, account. & computing machinery	..	..	..	..	..	..	..	28 Mach. de bur., compta. & trait. inform.
29 Electrical machinery	..	28.7	..	28.7	..	1.3	..	29 Machines électriques
30 Electro. equip. (radio, TV & commun.)	..	0.3	..	0.3	..	..	..	30 Equip. électro. (radio, tv & communic.)
31 *Electro. comp. (inc. semi-conductors)*	..	..	..	..	..	..	..	31 *Compos. électro. (semi-cond. incl.)*
32 *TV, radio & communications equip.*	..	0.3	..	0.3	..	..	..	32 *Appareils radio, tv et de communi.*
33 Instruments	..	15.8	..	14.5	..	16.3	..	33 Instruments
34 Motor vehicles	..	..	..	..	..	..	..	34 Véhicules automobiles
35 Other transport equipment	..	3.3	..	11.9	..	..	..	35 Autres matériels de transport
36 *Ships*	..	1.1	..	6.2	..	..	..	36 *Navires*
37 *Aerospace*	..	..	..	..	..	..	..	37 *Aérospatiale*
38 *Other transport equipment n.e.c.*	..	2.3	..	5.8	..	..	..	38 *Autres matériels de transport n.c.a.*
39 Furniture, other manufacturing n.e.c.	..	..	..	0.5	..	..	..	**39 Meubles, autres activ. de fabricat. n.c.a.**
40 Furniture	..	..	..	0.5	..	..	..	40 Meubles
41 Other manufacturing n.e.c.	..	..	..	..	..	..	..	41 Autres activités de fabrication n.c.a.
42 Recycling	..	..	..	..	..	..	..	**42 Récupération**
43 ELECTRICITY, GAS & WATER SUPPLY	..	12.2	..	4.0	..	1.7	..	**43 DISTRIBUTION D'ELECTRICITE, GAZ & EAU**
44 CONSTRUCTION	..	1.1	..	1.5	..	1.0	..	**44 CONSTRUCTION**
45 SERVICES SECTOR	..	70.1	..	118.5	..	659.8	..	**45 SECTEUR DES SERVICES**
46 Wholesale, retail trade & repair of motor vehicles, etc.	..	12.4	..	10.3	..	3.0	..	**46 Commerce (gros & détail), réparation de véhicules, etc.**
47 Hotels & restaurants	..	..	..	..	..	..	..	**47 Hôtels et restaurants**
48 Transport & storage	..	16.0	..	17.2	..	25.5	..	**48 Transports, entreposage**
49 Communications	..	..	..	..	..	10.0	..	**49 Communications**
50 Post	..	..	..	..	..	..	..	50 Postes
51 Telecommunications	..	..	..	..	..	10.0	..	51 Télécommunications
52 Financial intermediation (inc. insur.)	..	..	..	..	..	..	..	**52 Intermédiation financ. (assurances inc.)**
53 Real estate, renting & busin. activity	..	41.7	..	90.9	..	621.3	..	**53 Immobil., locat. & serv. aux entreprises**
54 Computer & related activities	..	30.5	..	38.3	..	77.2	..	54 Informatique et activités connexes
55 *Software consultancy*	..	30.5	..	37.3	..	77.2	..	55 *Production de logiciels*
56 *Other computer services n.e.c.*	..	..	..	1.0	..	..	..	56 *Autres activités informatiques*
57 Research & development	..	..	..	28.5	..	506.7	..	57 Recherche & développement expérimental
58 Other business activities n.e.c.	..	11.1	..	24.2	..	37.4	..	58 Autres activ. & serv. aux entreprises
59 Comm., soc. & pers. serv. activ., etc.	..	..	..	..	..	..	..	**59 Adm. pub., serv. soc. & collec., etc**
TOTAL BUSINESS ENTERPRISE SECTOR	**..**	**300.8**	**..**	**461.4**	**..**	**869.0**	**..**	**TOTAL SECTEUR DES ENTREPRISES**

Source: OECD, R&D database (DSTI/EAS Division), January 2002.

Source : OCDE, base de données R-D (DSTI/Division AES), janvier 2002.

ICELAND Table 12 - Tableau 12 **ISLANDE**

Total business enterprise R&D personnel (BEMP) - Personnel total de R-D du secteur des entreprises (PRDE)

by industry - par industrie

In full time equivalent — *En équivalence plein-temps*

	1994	1995	1996	1997	1998	1999	2000	
1 AGRICULTURE, HUNTING & FORESTRY	..	25	..	42	..	33	..	**1 AGRICULTURE, CHASSE & SYLVICULTURE**
2 MINING	..	2	..	1	..	1	..	**2 ACTIVITES EXTRACTIVES**
3 MANUFACTURING	..	253	..	400	..	306	..	**3 ACTIVITES DE FABRICATION**
4 Food, beverages & tobacco	..	66	..	140	..	48	..	**4 Produits alimentaires, boissons & tabac**
5 Food products & beverages	..	66	..	140	..	48	..	5 Produits alimentaires & boissons
6 Tobacco products	..	..	..	..	..	..	..	6 Produits à base de tabac
7 Textiles, wear app., fur & leather	..	10	..	17	..	13	..	**7 Textiles, habill., fourrures & cuirs**
8 Textiles	..	4	..	11	..	10	..	8 Textiles
9 Wearing apparel & fur	..	6	..	5	..	0	..	9 Articles d'habillement & fourrures
10 Leather products & footwear	..	..	..	..	..	3	..	10 Articles en cuir & chaussures
11 Wood, paper, printing, publishing	..	7	..	..	..	2	..	**11 Bois, papier, imprimerie, édition**
12 Wood & cork (not furniture)	..	2	..	..	..	0	..	12 Bois & liège (sauf meubles)
13 Pulp, paper & paper products	..	5	..	..	..	0	..	13 Carton, papier & articles
14 Publ., print. & repro. of rec. media	..	..	..	..	..	2	..	14 Edition, impress. & repro. enregis.
15 Coke, petroleum, nuclear fuel, chemicals, rubber plastics	..	50	..	72	..	74	..	**15 Coke, pétrole, combust. nucl., prod. chimi., caout. & mat. plast.**
16 Coke, ref. petrol. prod. & nuclear fuel	..	..	..	..	..	0	..	16 Coke, raffin. pétr. & combust. nucl.
17 Chemicals & chemical products	..	41	..	64	..	67	..	17 Produits chimiques
18 *Chemical prod.(less pharmaceuticals)*	..	17	..	25	..	30	..	18 *Prod. chimi. (moins prod. pharmaceutiques)*
19 *Pharmaceuticals*	..	24	..	39	..	36	..	19 *Produits pharmaceutiques*
20 Rubber & plastic products	..	9	..	9	..	7	..	20 Caoutchouc et matières plastiques
21 Non-metallic mineral products	..	16	..	11	..	14	..	**21 Produits minéraux non métalliques**
22 Basic metals	..	10	..	13	..	7	..	**22 Produits métallurgiques de base**
23 Basic metals, ferrous	..	5	..	8	..	0	..	23 Produits métallur. de base, ferreux
24 Basic metals, non-ferrous	..	5	..	5	..	7	..	24 Prod. métallur. de base, non ferreux
25 Fabricated metal products	..	..	..	..	..	..	..	**25 Ouvrages en métaux (sauf mach.& mat.)**
26 Machinery equipment, instruments and transport equipment	..	92	..	144	..	146	..	**26 Machines & matériel, instruments & mat. de transport**
27 Machinery n.e.c.	..	5	..	3	..	83	..	27 Machines, n.c.a.
28 Office, account. & computing machinery	..	..	..	..	..	..	..	28 Mach. de bur., compta. & trait. inform.
29 Electrical machinery	..	50	..	76	..	4	..	29 Machines électriques
30 Electro. equip. (radio, TV & commun.)	..	1	..	1	..	..	..	30 Equip. électro. (radio, tv & communic.)
31 *Electro. comp. (inc. semi-conductors)*	..	..	..	..	..	..	..	31 *Compos. électro. (semi-cond. incl.)*
32 *TV, radio & communications equip.*	..	1	..	1	..	..	..	32 *Appareils radio, tv et de communi.*
33 Instruments	..	29	..	56	..	49	..	33 Instruments
34 Motor vehicles	..	..	..	..	..	0	..	34 Véhicules automobiles
35 Other transport equipment	..	7	..	10	..	10	..	35 Autres matériels de transport
36 *Ships*	..	3	..	2	..	..	..	36 *Navires*
37 *Aerospace*	..	..	..	..	..	..	..	37 *Aérospatiale*
38 *Other transport equipment n.e.c.*	..	4	..	8	..	10	..	38 *Autres matériels de transport n.c.a.*
39 Furniture, other manufacturing n.e.c.	..	1	..	2	..	2	..	**39 Meubles, autres activ. de fabricat. n.c.a.**
40 Furniture	..	1	..	2	..	2	..	40 Meubles
41 Other manufacturing n.e.c.	..	..	..	..	..	0	..	41 Autres activités de fabrication n.c.a.
42 Recycling	..	1	..	..	..	0	..	**42 Récupération**
43 ELECTRICITY, GAS & WATER SUPPLY	..	40	..	2	..	9	..	**43 DISTRIBUTION D'ELECTRICITE, GAZ & EAU**
44 CONSTRUCTION	..	3	..	4	..	4	..	**44 CONSTRUCTION**
45 SERVICES SECTOR	..	201	..	382	..	609	..	**45 SECTEUR DES SERVICES**
46 Wholesale, retail trade & repair of motor vehicles, etc.	..	21	..	23	..	14	..	**46 Commerce (gros & détail), réparation de véhicules, etc.**
47 Hotels & restaurants	..	..	..	..	..	0	..	**47 Hôtels et restaurants**
48 Transport & storage	..	19	..	28	..	27	..	**48 Transports, entreposage**
49 Communications	..	..	..	..	..	13	..	**49 Communications**
50 Post	..	..	..	..	..	0	..	50 Postes
51 Telecommunications	..	..	..	..	..	13	..	51 Télécommunications
52 Financial intermediation (inc. insur.)	..	..	..	..	..	0	..	**52 Intermédiation financ. (assurances inc.)**
53 Real estate, renting & busin. activity	..	161	..	331	..	555	..	**53 Immobil., locat. & serv. aux entreprises**
54 Computer & related activities	..	97	..	165	..	192	..	54 Informatique et activités connexes
55 *Software consultancy*	..	97	..	164	..	192	..	55 *Production de logiciels*
56 *Other computer services n.e.c.*	..	..	..	1	..	0	..	56 *Autres activités informatiques*
57 Research & development	..	6	..	114	..	296	..	57 Recherche & développement expérimental
58 Other business activities n.e.c.	..	58	..	52	..	67	..	58 Autres activ. & serv. aux entreprises
59 Comm., soc. & pers. serv. activ., etc.	..	..	..	..	..	0	..	**59 Adm. pub., serv. soc. & collec., etc**
TOTAL BUSINESS ENTERPRISE SECTOR	**418**	**551**	**..**	**832**	**915** [c]	**961**	**..**	**TOTAL SECTEUR DES ENTREPRISES**

Source: OECD, R&D database (DSTI/EAS Division), January 2002.

Source : OCDE, base de données R-D (DSTI/Division AES), janvier 2002.

ICELAND — Table 13 - Tableau 13 — ISLANDE

RSE in the business enterprise sector - Chercheurs dans le secteur des entreprises

by industry - par industrie

In full time equivalent — *En équivalence plein-temps*

	1994	1995	1996	1997	1998	1999	2000	
1 AGRICULTURE, HUNTING & FORESTRY	..	15	..	16	..	13	..	**1 AGRICULTURE, CHASSE & SYLVICULTURE**
2 MINING	..	1	..	0	..	0	..	**2 ACTIVITES EXTRACTIVES**
3 MANUFACTURING	..	170	..	217	..	204	..	**3 ACTIVITES DE FABRICATION**
4 Food, beverages & tobacco	..	35	..	58	..	22	..	**4 Produits alimentaires, boissons & tabac**
5 Food products & beverages	..	35	..	58	..	22	..	5 Produits alimentaires & boissons
6 Tobacco products	..	..	..	..	..	..	..	6 Produits à base de tabac
7 Textiles, wear app., fur & leather	..	5	..	6	..	6	..	**7 Textiles, habill., fourrures & cuirs**
8 Textiles	..	2	..	4	..	5	..	8 Textiles
9 Wearing apparel & fur	..	3	..	2	..	..	..	9 Articles d'habillement & fourrures
10 Leather products & footwear	..	..	..	..	..	1	..	10 Articles en cuir & chaussures
11 Wood, paper, printing, publishing	..	5	..	0	..	1	..	**11 Bois, papier, imprimerie, édition**
12 Wood & cork (not furniture)	..	1	..	..	..	..	..	12 Bois & liège (sauf meubles)
13 Pulp, paper & paper products	..	4	..	..	..	..	..	13 Carton, papier & articles
14 Publ., print. & repro. of rec. media	..	..	..	..	..	1	..	14 Edition, impress. & repro. enregis.
15 Coke, petroleum, nuclear fuel, chemicals, rubber plastics	..	33	..	33	..	37	..	**15 Coke, pétrole, combust. nucl., prod. chimi., caout. & mat. plast.**
16 Coke, ref. petrol. prod. & nuclear fuel	..	..	..	..	..	..	..	16 Coke, raffin. pétr. & combust. nucl.
17 Chemicals & chemical products	..	29	..	27	..	33	..	17 Produits chimiques
18 *Chemical prod.(less pharmaceuticals)*	..	11	..	14	..	18	..	18 *Prod. chimi. (moins prod. pharmaceutiques)*
19 *Pharmaceuticals*	..	18	..	13	..	15	..	19 *Produits pharmaceutiques*
20 Rubber & plastic products	..	4	..	6	..	5	..	20 Caoutchouc et matières plastiques
21 Non-metallic mineral products	..	8	..	6	..	5	..	**21 Produits minéraux non métalliques**
22 Basic metals	..	9	..	9	..	7	..	**22 Produits métallurgiques de base**
23 Basic metals, ferrous	..	4	..	4	..	..	..	23 Produits métallur. de base, ferreux
24 Basic metals, non-ferrous	..	5	..	5	..	7	..	24 Prod. métallur. de base, non ferreux
25 Fabricated metal products	..	..	..	..	..	..	..	**25 Ouvrages en métaux (sauf mach.& mat.)**
26 Machinery equipment, instruments and transport equipment	..	73	..	105	..	123	..	**26 Machines & matériel, instruments & mat. de transport**
27 Machinery n.e.c.	..	3	..	2	..	75	..	27 Machines, n.c.a.
28 Office, account. & computing machinery	..	..	..	..	..	..	..	28 Mach. de bur., compta. & trait. inform.
29 Electrical machinery	..	39	..	57	..	4	..	29 Machines électriques
30 Electro. equip. (radio, TV & commun.)	..	1	..	1	..	..	..	30 Equip. électro. (radio, tv & communic.)
31 *Electro. comp. (inc. semi-conductors)*	..	..	..	..	..	..	..	31 *Compos. électro. (semi-cond. incl.)*
32 *TV, radio & communications equip.*	..	1	..	1	..	..	..	32 *Appareils radio, tv et de communi.*
33 Instruments	..	24	..	42	..	38	..	33 Instruments
34 Motor vehicles	..	..	..	..	..	..	..	34 Véhicules automobiles
35 Other transport equipment	..	7	..	4	..	6	..	35 Autres matériels de transport
36 *Ships*	..	3	..	0	..	..	..	36 *Navires*
37 *Aerospace*	..	..	..	..	..	..	..	37 *Aérospatiale*
38 *Other transport equipment n.e.c.*	..	4	..	4	..	6	..	38 *Autres matériels de transport n.c.a.*
39 Furniture, other manufacturing n.e.c.	..	1	..	1	..	2	..	**39 Meubles, autres activ. de fabricat. n.c.a.**
40 Furniture	..	1	..	1	..	2	..	40 Meubles
41 Other manufacturing n.e.c.	..	..	..	..	..	..	..	41 Autres activités de fabrication n.c.a.
42 Recycling	..	1	..	..	..	..	..	**42 Récupération**
43 ELECTRICITY, GAS & WATER SUPPLY	..	25	..	1	..	9	..	**43 DISTRIBUTION D'ELECTRICITE, GAZ & EAU**
44 CONSTRUCTION	..	2	..	3	..	3	..	**44 CONSTRUCTION**
45 SERVICES SECTOR	..	136	..	235	..	400	..	**45 SECTEUR DES SERVICES**
46 Wholesale, retail trade & repair of motor vehicles, etc.	..	16	..	16	..	9	..	**46 Commerce (gros & détail), réparation de véhicules, etc.**
47 Hotels & restaurants	..	..	..	..	..	..	..	**47 Hôtels et restaurants**
48 Transport & storage	..	12	..	16	..	15	..	**48 Transports, entreposage**
49 Communications	..	0	..	0	..	10	..	**49 Communications**
50 Post	..	..	..	..	..	..	..	50 Postes
51 Telecommunications	..	..	..	..	..	10	..	51 Télécommunications
52 Financial intermediation (inc. insur.)	..	..	..	..	..	..	..	**52 Intermédiation financ. (assurances inc.)**
53 Real estate, renting & busin. activity	..	108	..	203	..	367	..	**53 Immobil., locat. & serv. aux entreprises**
54 Computer & related activities	..	75	..	126	..	131	..	54 Informatique et activités connexes
55 *Software consultancy*	..	75	..	125	..	131	..	55 *Production de logiciels*
56 *Other computer services n.e.c.*	..	..	..	1	..	..	..	56 *Autres activités informatiques*
57 Research & development	..	4	..	41	..	191	..	57 Recherche & développement expérimental
58 Other business activities n.e.c.	..	29	..	36	..	45	..	58 Autres activ. & serv. aux entreprises
59 Comm., soc. & pers. serv. activ., etc.	..	..	..	..	..	..	..	**59 Adm. pub., serv. soc. & collec., etc**
TOTAL BUSINESS ENTERPRISE SECTOR	**287**	**359**	**..**	**474**	**521** [c]	**626**	**..**	**TOTAL SECTEUR DES ENTREPRISES**

Source: OECD, R&D database (DSTI/EAS Division), January 2002.

Source : OCDE, base de données R-D (DSTI/Division AES), janvier 2002.

Higher education intramural expenditure on R&D - Dépenses intra-muros de R-D de l'enseignement supérieur
(HERD) by field of science - (DIRDES) par discipline scientifique

	1994	1995	1996	1997	1998	1999	2000	
	14.1 Million ISK - Millions ISK							
Natural sciences	..	505.8 s	..	633.1	..	665.9	..	Sciences exactes et naturelles
Engineering	..	613.8 s	..	927.6	..	1 080.8	..	Sciences de l'ingénieur
Medical sciences	..	425.7 s	..	438.9	..	346.1	..	Sciences médicales
Agricultural sciences	..	123.6 s	..	177.0	..	185.0	..	Sciences agricoles
Sub-total NSE	..	1 668.9 s	..	2 176.7	..	2 277.8	..	Sous-total SENI
Social sciences	..	209.4 s	..	359.7	..	506.6	..	Sciences sociales
Humanities	..	157.0 s	..	194.8	..	248.8	..	Sciences humaines
Sub-total SSH	..	366.4 s	..	554.5	..	755.4	..	Sous-total SSH
Total NSE+SSH	1 452.3	1 915.3	..	2 731.2	2 936.0 c	3 033.2	..	*Total SENI+SSH*
	14.2 Million constant USD (1995 prices and PPPs) - Millions de USD constants (prix et PPA de 1995)							
Natural sciences	..	6.7	..	7.9	..	7.6	..	Sciences exactes et naturelles
Engineering	..	8.1	..	11.6	..	12.4	..	Sciences de l'ingénieur
Medical sciences	..	5.6	..	5.5	..	4.0	..	Sciences médicales
Agricultural sciences	..	1.6	..	2.2	..	2.1	..	Sciences agricoles
Sub-total NSE	..	22.0	..	27.2	..	26.1	..	Sous-total SENI
Social sciences	..	2.8	..	4.5	..	5.8	..	Sciences sociales
Humanities	..	2.1	..	2.4	..	2.8	..	Sciences humaines
Sub-total SSH	..	4.8	..	6.9	..	8.6	..	Sous-total SSH
Total NSE+SSH	19.7	25.2	..	34.1	34.8 c	34.7	..	*Total SENI+SSH*

Table 15 - Tableau 15

Higher education intramural expenditure on R&D - Dépenses intra-muros de R-D de l'enseignement supérieur
(HERD) by field of science and type of costs - (DIRDES) par discipline scientifique et par type de coûts

Million ISK *Millions ISK*

	1994	1995	1996	1997	1998	1999	2000	
Current expenditure								**Dépenses courantes**
Natural sciences	..	..	..	562.7	..	612.2	..	Sciences exactes et naturelles
Engineering	..	..	..	640.6	..	911.2	..	Sciences de l'ingénieur
Medical sciences	..	..	..	400.8	..	320.2	..	Sciences médicales
Agricultural sciences	..	..	..	162.9	..	172.0	..	Sciences agricoles
Sub-total NSE	..	..	..	1 767.0	..	2 015.6	..	Sous-total SENI
Social sciences	..	..	..	315.6	..	476.3	..	Sciences sociales
Humanities	..	..	..	181.8	..	240.4	..	Sciences humaines
Sub-total SSH	..	..	..	497.4	..	716.8	..	Sous-total SSH
Total NSE+SSH	..	1 626.3	..	2 264.4	..	2 732.4	..	*Total SENI+SSH*
Capital expenditure								**Dépenses en capital**
Natural sciences	..	..	..	70.4	..	53.7	..	Sciences exactes et naturelles
Engineering	..	..	..	287.1	..	169.7	..	Sciences de l'ingénieur
Medical sciences	..	..	..	38.1	..	25.9	..	Sciences médicales
Agricultural sciences	..	..	..	14.1	..	13.0	..	Sciences agricoles
Sub-total NSE	..	..	..	409.7	..	262.2	..	Sous-total SENI
Social sciences	..	..	..	44.1	..	30.3	..	Sciences sociales
Humanities	..	..	..	13.1	..	8.3	..	Sciences humaines
Sub-total SSH	..	..	..	57.1	..	38.6	..	Sous-total SSH
Total NSE+SSH	..	409.0	..	466.8	..	300.8	..	*Total SENI+SSH*

Source: OECD, R&D database (DSTI/EAS Division), January 2002. *Source* : OCDE, base de données R-D (DSTI/Division AES), janvier 2002.

Higher education intramural expenditure on R&D - Dépenses intra-muros de R-D de l'enseignement supérieur

(HERD) by field of science and source of funds - (DIRDES) par discipline scientifique et par source de financement

Million ISK *Millions ISK*

	1994	1995	1996	1997	1998	1999	2000	
Direct government								**État direct**
Natural sciences	..	..	..	314.9	..	353.6	..	Sciences exactes et naturelles
Engineering	..	..	..	482.1	..	793.7	..	Sciences de l'ingénieur
Medical sciences	..	..	..	263.7	..	243.1	..	Sciences médicales
Agricultural sciences	..	..	..	130.8	..	144.0	..	Sciences agricoles
Sub-total NSE	..	..	..	1 191.5	..	1 534.4	..	Sous-total SENI
Social sciences	..	..	..	248.3	..	393.4	..	Sciences sociales
Humanities	..	..	..	146.7	..	199.1	..	Sciences humaines
Sub-total SSH	..	..	..	395.0	..	592.5	..	Sous-total SSH
Total NSE+SSH	1 162.4	1 200.3	..	1 586.4	1 705.4 c	2 126.8	..	*Total SENI+SSH*
General university funds								**Fonds généraux des universités**
Natural sciences	..	..	..	179.7	..	192.4	..	Sciences exactes et naturelles
Engineering	..	..	..	264.0	..	76.1	..	Sciences de l'ingénieur
Medical sciences	..	..	..	103.0	..	80.5	..	Sciences médicales
Agricultural sciences	..	..	..	23.6	..	24.0	..	Sciences agricoles
Sub-total NSE	..	..	..	570.8	..	373.0	..	Sous-total SENI
Social sciences	..	..	..	66.3	..	53.7	..	Sciences sociales
Humanities	..	..	..	31.3	..	21.2	..	Sciences humaines
Sub-total SSH	..	..	..	97.6	..	75.0	..	Sous-total SSH
Total NSE+SSH	131.2	269.6	..	668.4	718.6 c	447.9	..	*Total SENI+SSH*
Sub-total government								**Sous-total État**
Natural sciences	..	..	..	494.6	..	546.0	..	Sciences exactes et naturelles
Engineering	..	..	..	746.1	..	869.8	..	Sciences de l'ingénieur
Medical sciences	..	..	..	367.1	..	323.5	..	Sciences médicales
Agricultural sciences	..	..	..	154.4	..	168.0	..	Sciences agricoles
Sub-total NSE	..	..	..	1 762.2	..	1 907.3	..	Sous-total SENI
Social sciences	..	..	..	314.6	..	447.1	..	Sciences sociales
Humanities	..	..	..	178.0	..	220.3	..	Sciences humaines
Sub-total SSH	..	..	..	492.6	..	667.4	..	Sous-total SSH
Total NSE+SSH	1 293.6	1 470.0	..	2 254.9	2 424.0 c	2 574.7	..	*Total SENI+SSH*
Higher education								**Enseignement supérieur**
Natural sciences	..	..	..	..	..	..	..	Sciences exactes et naturelles
Engineering	..	..	..	..	..	..	..	Sciences de l'ingénieur
Medical sciences	..	..	..	..	..	..	..	Sciences médicales
Agricultural sciences	..	..	..	..	..	..	..	Sciences agricoles
Sub-total NSE	..	..	..	..	..	..	..	Sous-total SENI
Social sciences	..	..	..	..	..	..	..	Sciences sociales
Humanities	..	..	..	..	..	..	..	Sciences humaines
Sub-total SSH	..	..	..	..	..	..	..	Sous-total SSH
Total NSE+SSH	..	..	..	..	..	..	..	*Total SENI+SSH*
Private non-profit								**Institutions sans but lucratif**
Natural sciences	..	..	..	..	..	0.0	..	Sciences exactes et naturelles
Engineering	..	..	..	..	..	0.0	..	Sciences de l'ingénieur
Medical sciences	..	..	..	12.3	..	0.0	..	Sciences médicales
Agricultural sciences	..	..	..	..	..	0.0	..	Sciences agricoles
Sub-total NSE	..	..	..	12.3	..	0.0	..	Sous-total SENI
Social sciences	..	..	..	..	..	0.5	..	Sciences sociales
Humanities	..	..	..	..	..	0.0	..	Sciences humaines
Sub-total SSH	..	..	..	..	..	0.5	..	Sous-total SSH
Total NSE+SSH	4.7	164.4	..	12.3	13.2 c	0.5	..	*Total SENI+SSH*
Business entreprise								**Entreprises**
Natural sciences	..	..	..	12.8	..	1.6	..	Sciences exactes et naturelles
Engineering	..	..	..	126.2	..	68.1	..	Sciences de l'ingénieur
Medical sciences	..	..	..	53.6	..	8.1	..	Sciences médicales
Agricultural sciences	..	..	..	19.0	..	10.7	..	Sciences agricoles
Sub-total NSE	..	..	..	211.5	..	88.6	..	Sous-total SENI
Social sciences	..	..	..	23.7	..	16.8	..	Sciences sociales
Humanities	..	..	..	16.8	..	16.4	..	Sciences humaines
Sub-total SSH	..	..	..	40.5	..	33.2	..	Sous-total SSH
Total NSE+SSH	61.9	103.3	..	252.0	270.9 c	121.8	..	*Total SENI+SSH*
Funds from abroad								**Fonds de l'étranger**
Natural sciences	..	..	..	125.7	..	118.2	..	Sciences exactes et naturelles
Engineering	..	..	..	55.3	..	143.0	..	Sciences de l'ingénieur
Medical sciences	..	..	..	6.0	..	14.5	..	Sciences médicales
Agricultural sciences	..	..	..	3.7	..	6.3	..	Sciences agricoles
Sub-total NSE	..	..	..	190.7	..	281.9	..	Sous-total SENI
Social sciences	..	..	..	21.4	..	42.1	..	Sciences sociales
Humanities	..	..	..	..	..	12.1	..	Sciences humaines
Sub-total SSH	..	..	..	21.4	..	54.2	..	Sous-total SSH
Total NSE+SSH	92.1	177.8	..	212.0	227.9 c	336.2	..	*Total SENI+SSH*

Source: OECD, R&D database (DSTI/EAS Division), January 2002.

Source : OCDE, base de données R-D (DSTI/Division AES), janvier 2002.

ICELAND — Table 17 - Tableau 17 — ISLANDE

Higher education R&D personnel (HEMP) - Personnel de R-D du secteur de l'enseignement supérieur

by field of science and occupation - par discipline scientifique et par profession

In full-time equivalent — *En équivalence plein-temps*

	1994	1995	1996	1997	1998	1999	2000	
Total R&D personnel								**Personnel total de R-D**
Natural sciences	..	137	..	162	..	148	..	Sciences exactes et naturelles
Engineering	..	84	..	153	..	271	..	Sciences de l'ingénieur
Medical sciences	..	155	..	137	..	88	..	Sciences médicales
Agricultural sciences	..	38	..	53	..	50	..	Sciences agricoles
Sub-total NSE	..	414	..	505	..	557	..	Sous-total SENI
Social sciences	..	61	..	84	..	89	..	Sciences sociales
Humanities	..	55	..	68	..	66	..	Sciences humaines
Sub-total SSH	..	116	..	152	..	155	..	Sous-total SSH
Total NSE+SSH	385	530	..	656	676 [c]	712	..	*Total SENI+SSH*
Researchers								**Chercheurs**
Natural sciences	..	..	..	108	..	..	..	Sciences exactes et naturelles
Engineering	..	..	..	85	..	..	..	Sciences de l'ingénieur
Medical sciences	..	..	..	111	..	..	..	Sciences médicales
Agricultural sciences	..	..	..	31	..	..	..	Sciences agricoles
Sub-total NSE	..	..	..	334	..	..	..	Sous-total SENI
Social sciences	..	..	..	69	..	..	..	Sciences sociales
Humanities	..	..	..	58	..	..	..	Sciences humaines
Sub-total SSH	..	..	..	127	..	..	..	Sous-total SSH
Total NSE+SSH	211	380	..	462	476 [c]	480	..	*Total SENI+SSH*
Technicians								**Technicians**
Natural sciences	..	..	..	44	..	..	..	Sciences exactes et naturelles
Engineering	..	..	..	33	..	..	..	Sciences de l'ingénieur
Medical sciences	..	..	..	14	..	..	..	Sciences médicales
Agricultural sciences	..	..	..	12	..	..	..	Sciences agricoles
Sub-total NSE	..	..	..	102	..	..	..	Sous-total SENI
Social sciences	..	..	..	11	..	..	..	Sciences sociales
Humanities	..	..	..	2	..	..	..	Sciences humaines
Sub-total SSH	..	..	..	12	..	..	..	Sous-total SSH
Total NSE+SSH	109	102	..	115	118 [c]	108	..	*Total SENI+SSH*
Other								**Autres**
Natural sciences	..	..	..	11	..	..	..	Sciences exactes et naturelles
Engineering	..	..	..	35	..	..	..	Sciences de l'ingénieur
Medical sciences	..	..	..	12	..	..	..	Sciences médicales
Agricultural sciences	..	..	..	10	..	..	..	Sciences agricoles
Sub-total NSE	..	..	..	68	..	..	..	Sous-total SENI
Social sciences	..	..	..	4	..	..	..	Sciences sociales
Humanities	..	..	..	8	..	..	..	Sciences humaines
Sub-total SSH	..	..	..	12	..	..	..	Sous-total SSH
Total NSE+SSH	65	48	..	80	83 [c]	125	..	*Total SENI+SSH*

Table 18 - Tableau 18

Government budget appropriations or outlays for R&D (GBAORD) - Crédits budgétaires publics de R-D (CBPRD)

by socio-economic objective - par objectif socio-économique

Million ISK — *Millions ISK*

	1995	1996	1997	1998	1999	2000	2001	
Agriculture, forestry and fishing	1 339.1	1 376.2	1 454.6	3 032.8	2 520.0	1 649.1 [p]	1 758.9 [p]	Agriculture, sylviculture et pêche
Industrial development	384.8	397.8	453.5	429.0	562.8	460.4 [p]	473.6 [p]	Développement industriel
Energy	180.1	181.9	207.9	177.4	304.4	215.0 [p]	228.7 [p]	Energie
Transport & telecommunications	..	..	..	..	..	..	..	*Transports & télécommunications*
Urban & rural planning	..	..	..	..	..	..	..	*Aménagement urbain et rural*
ST Infrastructure	46.0	64.8	68.5	65.5	77.9	74.9 [p]	108.2 [p]	ST Infrastructure
Prevention of pollution	..	..	..	..	..	..	0.0	*Prévention de la pollution*
Identificat. & treatment of pollution	..	..	..	..	..	..	0.0	*Détect. pollut. & mesures antipollut.*
ST Environment	162.4	208.7	250.3	215.2	292.4	255.0 [p]	289.6 [p]	ST Environnement
Health	180.7	195.2	201.7	207.8	278.2	265.7 [p]	367.3 [p]	Santé
Social development & services	447.6	473.3	425.7	512.3	1 437.8	1 367.3 [p]	1 527.5 [p]	Développement et services sociaux
Earth and atmosphere	.. [n]	.. [n]	.. [n]	.. [n]	.. [n]	.. [n]	.. [n]	Milieu terrestre et atmosphère
Advancement of research	1 293.4	1 479.4	1 572.5	1 624.9	636.3	534.4 [p]	519.0 [p]	*Promotion de la recherche*
General university funds	624.9	716.6	793.6	996.1	2 006.7	2 013.7 [p]	2 025.1 [p]	*Fonds généraux des universités*
ST Advancement of knowledge	1 918.3	2 196.0	2 366.0	2 621.0	2 643.0	2 548.1 [p]	2 544.1 [p]	ST promotion gén. des connaiss.
Civil space	.. [n]	.. [n]	.. [n]	.. [n]	.. [n]	.. [n]	.. [n]	Espace civil
Defence	0.0	0.0	0.0	0.0	0.0	0.0	0.0	Défense
Not elsewhere classified	58.6 [o]	64.7 [o]	59.9 [o]	67.1 [o]	89.2 [o]	93.4 [p]	102.1 [p]	Non classé
Total	**4 717.6**	**5 158.6**	**5 488.1**	**7 328.1**	**8 205.8**	**6 928.9 [p]**	**7 400.1 [p]**	**Total**

Source: OECD, R&D database (DSTI/EAS Division), January 2002. — *Source* : OCDE, base de données R-D (DSTI/Division AES), janvier 2002.

IRELAND — Table 1 - Tableau 1 — IRLANDE

Gross domestic expenditure on R&D (GERD) - Dépenses intérieures brutes de R-D (DIRD)

by sector of performance and source of funds - par secteur d'exécution et par source de financement

	1994	1995	1996	1997	1998	1999	2000	
	1.1 Million EUR (1999 IEP euro, EUR from 1999) - Millions EUR (1999 IEP euro, EUR à partir de 1999)							
Business enterprise								**Entreprises**
financed by								*financé par*
Business enterprise	398.2 c	461.6	534.1 c	619.5	..	..	..	Entreprises
Direct government	9.0 c	22.9	31.0 c	35.9	..	..	..	État direct
Higher education	0.0	0.0	0.0	0.0	..	..	..	Enseignement supérieur
Private non-profit	1.8 c	2.1	3.3 c	3.8	..	..	..	Institutions sans but lucratif
Funds from abroad	15.5 c	17.9	17.0 c	19.8	..	..	..	Fonds de l'étranger
Total	424.5 c	504.5	585.3 c	679.1	..	..	..	*Total*
Government								**État**
financed by								*financé par*
Business enterprise	10.5	13.1	11.1	10.8	11.4	9.4	11.3 p	Entreprises
Direct government	43.2	43.4	49.3	51.3	54.1	52.2	53.9 p	État direct
Higher education	0.0	0.0	0.0	0.0	0.0	..	..	Enseignement supérieur
Private non-profit	0.1	0.0	0.0	0.0	0.0	0.0	0.0 p	Institutions sans but lucratif
Funds from abroad	4.4	3.6	3.7	3.2	4.4	2.3	3.2 p	Fonds de l'étranger
Total	58.1	60.2	64.2	65.4	70.0	63.9	68.5 p	*Total*
Higher education								**Enseignement supérieur**
financed by								*financé par*
Business enterprise	8.4	9.5 c	9.8	11.6 c	13.3	..	..	Entreprises
Direct government	24.1	27.4 c	32.8	40.6 c	48.4	..	..	État direct
General university funds	51.0	57.6 c	69.6	77.9 c	86.3	..	..	Fonds généraux des universités
Sub-total government	75.1	85.0 c	102.4	118.5 c	134.7	..	..	Sous-Total État
Higher education	5.5	6.2 c	5.9	10.4 c	15.0	..	..	Enseignement supérieur
Private non-profit	3.0	3.4 c	2.5	1.6 c	0.7	..	..	Institutions sans but lucratif
Funds from abroad	29.1	33.0 c	32.6	36.2 c	39.9	..	..	Fonds de l'étranger
Total	121.1	137.1 c	153.1	178.4 c	203.7	..	..	*Total*
Private non-profit								**Institutions sans but lucratif**
financed by								*financé par*
Business enterprise	1.3 c	1.3 c	1.3 c	1.3 c	..	..	..	Entreprises
Direct government	0.0	0.0	0.0	0.0	..	..	..	État direct
Higher education	0.0	0.0	0.0	0.0	..	..	..	Enseignement supérieur
Private non-profit	1.3 c	1.3 c	2.5 c	2.5 c	..	..	..	Institutions sans but lucratif
Funds from abroad	1.3 c	2.5 c	2.5 c	2.5 c	..	..	..	Fonds de l'étranger
Total	3.8 c	5.1 c	6.3 c	6.3 c	..	..	..	*Total*
GERD								**DIRD**
financed by								*financé par*
Business enterprise	418.4 c	485.5 c	556.2 c	643.1 c	..	..	..	Entreprises
Direct government	76.3 c	93.7 c	113.1 c	127.9 c	..	..	..	État direct
General university funds	51.0 c	57.6 c	69.6 c	77.9 c	..	..	..	Fonds généraux des universités
Sub-total government	127.2 c	151.3 c	182.7 c	205.8 c	..	..	..	Sous-Total État
Higher education	5.5 c	6.2 c	5.9 c	10.4 c	..	..	..	Enseignement supérieur
Private non-profit	6.2 c	6.8 c	8.3 c	8.0 c	..	..	..	Institutions sans but lucratif
Funds from abroad	50.2 c	57.0 c	55.9 c	61.8 c	..	..	..	Fonds de l'étranger
Total	607.5 c	706.9 c	808.9 c	929.2 c	..	..	..	*Total*
	1.2 Million constant USD (1995 prices and PPPs) - Millions de USD constants (prix et PPA de 1995)							
Business enterprise								**Entreprises**
financed by								*financé par*
Business enterprise	509.1 c	572.8	648.1 c	721.8	..	..	..	Entreprises
Direct government	11.5 c	28.4	37.6 c	41.9	..	..	..	État direct
Higher education	0.0	0.0	0.0	0.0	..	..	..	Enseignement supérieur
Private non-profit	2.3 c	2.6	4.0 c	4.4	..	..	..	Institutions sans but lucratif
Funds from abroad	19.8 c	22.2	20.6 c	23.1	..	..	..	Fonds de l'étranger
Total	542.7 c	626.1	710.3 c	791.3	..	..	..	*Total*
Government								**État**
financed by								*financé par*
Business enterprise	13.5	16.3	13.5	12.6	12.6	10.0	11.5 p	Entreprises
Direct government	55.2	53.9	59.8	59.8	59.5	55.1	54.6 p	État direct
Higher education	0.0	0.0	0.0	0.0	0.0	..	..	Enseignement supérieur
Private non-profit	0.1	0.0	0.0	0.0	0.0	0.0	0.0 p	Institutions sans but lucratif
Funds from abroad	5.6	4.5	4.5	3.8	4.9	2.4	3.3 p	Fonds de l'étranger
Total	74.3	74.7	77.9	76.2	77.0	67.5	69.3 p	*Total*
Higher education								**Enseignement supérieur**
financed by								*financé par*
Business enterprise	10.8	11.8 c	11.9	13.5 c	14.7	..	..	Entreprises
Direct government	30.8	34.0 c	39.8	47.3 c	53.2	..	..	État direct
General university funds	65.2	71.4 c	84.4	90.8 c	94.9	..	..	Fonds généraux des universités
Sub-total government	96.0	105.5 c	124.2	138.1 c	148.2	..	..	Sous-Total État
Higher education	7.0	7.7 c	7.1	12.2 c	16.5	..	..	Enseignement supérieur
Private non-profit	3.9	4.3 c	3.0	1.9 c	0.8	..	..	Institutions sans but lucratif
Funds from abroad	37.2	40.9 c	39.6	42.2 c	43.8	..	..	Fonds de l'étranger
Total	154.8	170.2 c	185.8	207.9 c	224.0	..	..	*Total*
Private non-profit								**Institutions sans but lucratif**
financed by								*financé par*
Business enterprise	1.6 c	1.6 c	1.5 c	1.5 c	..	..	..	Entreprises
Direct government	0.0	0.0	0.0	0.0	..	..	..	État direct
Higher education	0.0	0.0	0.0	0.0	..	..	..	Enseignement supérieur
Private non-profit	1.6 c	1.6 c	3.1 c	3.0 c	..	..	..	Institutions sans but lucratif
Funds from abroad	1.6 c	3.2 c	3.1 c	3.0 c	..	..	..	Fonds de l'étranger
Total	4.9 c	6.3 c	7.7 c	7.4 c	..	..	..	*Total*
GERD								**DIRD**
financed by								*financé par*
Business enterprise	535.0 c	602.5 c	675.0 c	749.4 c	..	..	..	Entreprises
Direct government	97.5 c	116.3 c	137.3 c	149.0 c	..	..	..	État direct
General university funds	65.2 c	71.4 c	84.4 c	90.8 c	..	..	..	Fonds généraux des universités
Sub-total government	162.7 c	187.7 c	221.7 c	239.8 c	..	..	..	Sous-Total État
Higher education	7.0 c	7.7 c	7.1 c	12.2 c	..	..	..	Enseignement supérieur
Private non-profit	7.9 c	8.5 c	10.1 c	9.3 c	..	..	..	Institutions sans but lucratif
Funds from abroad	64.2 c	70.8 c	67.8 c	72.0 c	..	..	..	Fonds de l'étranger
Total	776.7 c	877.2 c	981.7 c	1 082.7 c	..	..	..	*Total*

Source: OECD, R&D database (DSTI/EAS Division), January 2002.

Source : OCDE, base de données R-D (DSTI/Division AES), janvier 2002.

Table 2 - Tableau 2

Gross domestic expenditure on R&D (GERD) - Dépenses intérieures brutes de R-D (DIRD)

by sector of performance and type of costs - par secteur d'exécution et par type de coûts

Million EUR (1999 IEP euro, EUR from 1999) — *Millions EUR (1999 IEP euro, EUR à partir de 1999)*

	1994	1995	1996	1997	1998	1999	2000	
Business enterprise								**Entreprises**
Labour costs	..	..	..	..	..	..	..	Coûts du personnel
Other current costs	..	..	..	..	..	..	..	Autres coûts courants
Total current expenditure	..	..	..	..	..	..	..	*Total dépenses courantes*
Land and buildings	..	..	..	..	..	..	..	Terrains et constructions
Instruments and equipment	..	..	..	..	..	..	..	Instruments et équipements
Total capital expenditure	..	..	..	..	..	..	..	*Total dépenses en capital*
Total	424.5 c	504.5	585.3 c	679.1	..	..	..	*Total*
Government								**État**
Labour costs	34.7 s	..	..	..	..	..	..	Coûts du personnel
Other current costs	31.0 s	..	..	..	..	..	..	Autres coûts courants
Total current expenditure	65.6 s	..	..	..	..	..	..	*Total dépenses courantes*
Land and buildings	..	..	..	..	..	..	..	Terrains et constructions
Instruments and equipment	..	..	..	..	..	..	..	Instruments et équipements
Total capital expenditure	0.9 s	..	..	..	..	..	..	*Total dépenses en capital*
Total	58.1	60.2	64.2	65.4	70.0	63.9	68.5 p	*Total*
Higher education								**Enseignement supérieur**
Labour costs	78.7	..	..	..	..	..	..	Coûts du personnel
Other current costs	32.5	..	..	..	..	..	..	Autres coûts courants
Total current expenditure	111.2	..	..	..	..	..	..	*Total dépenses courantes*
Land and buildings	0.0	..	..	..	..	..	..	Terrains et constructions
Instruments and equipment	9.9	..	..	..	..	..	..	Instruments et équipements
Total capital expenditure	9.9	..	..	..	..	..	..	*Total dépenses en capital*
Total	121.1	137.1 c	153.1	178.4 c	203.7	..	..	*Total*
Private non-profit								**Institutions sans but lucratif**
Labour costs	2.4	..	..	..	..	..	..	Coûts du personnel
Other current costs	1.3	..	..	..	..	..	..	Autres coûts courants
Total current expenditure	3.7	..	..	..	..	..	..	*Total dépenses courantes*
Land and buildings	..	..	..	..	..	..	..	Terrains et constructions
Instruments and equipment	..	..	..	..	..	..	..	Instruments et équipements
Total capital expenditure	0.1	..	..	..	..	..	..	*Total dépenses en capital*
Total	3.8 c	5.1 c	6.3 c	6.3 c	..	..	..	*Total*
GERD								**DIRD**
Labour costs	..	..	..	..	..	..	..	Coûts du personnel
Other current costs	..	..	..	..	..	..	..	Autres coûts courants
Total current expenditure	..	..	..	..	..	..	..	*Total dépenses courantes*
Land and buildings	..	..	..	..	..	..	..	Terrains et constructions
Instruments and equipment	..	..	..	..	..	..	..	Instruments et équipements
Total capital expenditure	..	..	..	..	..	..	..	*Total dépenses en capital*
Total	607.5 c	706.9 c	808.9 c	929.2 c	..	..	..	*Total*

Table 3 - Tableau 3

Current domestic expenditure on R&D - Dépenses intérieures courantes de R-D

by sector of performance and type of activity - par secteur d'exécution et par type d'activités

Million EUR (1999 IEP euro, EUR from 1999) — *Millions EUR (1999 IEP euro, EUR à partir de 1999)*

	1994	1995	1996	1997	1998	1999	2000	
Business enterprise								**Entreprises**
Basic research	..	..	..	..	..	..	..	Recherche fondamentale
Applied research	..	..	..	..	..	..	..	Recherche appliquée
Experimental development	..	..	..	..	..	..	..	Développement expérimental
Not specified	..	..	..	..	..	..	..	Non spécifié
Total	..	..	..	..	..	..	..	*Total*
Government								**État**
Basic research	2.2	..	..	..	..	..	..	Recherche fondamentale
Applied research	31.0	..	..	..	..	..	..	Recherche appliquée
Experimental development	32.5	..	..	..	..	..	..	Développement expérimental
Not specified	0.0	..	..	..	..	..	..	Non spécifié
Total	65.6 s	..	..	..	..	..	..	*Total*
Higher education								**Enseignement supérieur**
Basic research	36.1	..	..	..	..	..	..	Recherche fondamentale
Applied research	57.8	..	..	..	..	..	..	Recherche appliquée
Experimental development	17.4	..	..	..	..	..	..	Développement expérimental
Not specified	0.0	..	..	..	..	..	..	Non spécifié
Total	111.2	..	..	..	..	..	..	*Total*
Private non-profit								**Institutions sans but lucratif**
Basic research	0.3	..	..	..	..	..	..	Recherche fondamentale
Applied research	2.7	..	..	..	..	..	..	Recherche appliquée
Experimental development	0.8	..	..	..	..	..	..	Développement expérimental
Not specified	0.0	..	..	..	..	..	..	Non spécifié
Total	3.7	..	..	..	..	..	..	*Total*
GERD								**DIRD**
Basic research	..	..	..	..	..	..	..	Recherche fondamentale
Applied research	..	..	..	..	..	..	..	Recherche appliquée
Experimental development	..	..	..	..	..	..	..	Développement expérimental
Not specified	..	..	..	..	..	..	..	Non spécifié
Total	..	..	..	..	..	..	..	*Total*

Source: OECD, R&D database (DSTI/EAS Division), January 2002.

Source : OCDE, base de données R-D (DSTI/Division AES), janvier 2002.

Gross domestic expenditure on R&D (GERD) - Dépenses intérieures brutes de R-D (DIRD)

by sector of performance and socio-economic objective - par secteur d'exécution et par objectif socio-économique

Million EUR (1999 IEP euro, EUR from 1999) — *Millions EUR (1999 IEP euro, EUR à partir de 1999)*

	1994	1995	1996	1997	1998	1999	2000	
Business enterprise								**Entreprises**
Agriculture, forestry and fishing	..	..	..	..	..	..	..	Agriculture, sylviculture et pêche
Industrial development	..	..	..	..	..	..	..	Développement industriel
Energy	..	..	..	..	..	..	..	Energie
Transport and telecommunications	..	..	..	..	..	..	..	Transports et télécommunications
Urban and rural planning	..	..	..	..	..	..	..	Aménagement urbain et rural
Prevention of pollution	..	..	..	..	..	..	..	Prévention de la pollution
Identification & treatment of pollution	..	..	..	..	..	..	..	Détection pollu. & mesures antipollu.
Health	..	..	..	..	..	..	..	Santé
Social development and services	..	..	..	..	..	..	..	Développement et services sociaux
Earth and atmosphere	..	..	..	..	..	..	..	Milieu terrestre et atmosphère
Advancement of knowledge	..	..	..	..	..	..	..	Promotion générale des connaissances
Civil space	..	..	..	..	..	..	..	Espace civil
Defence	..	..	..	..	..	..	..	Défense
Not elsewhere classified	424.5 [c]	504.5	585.3 [c]	679.1	..	..	..	Non classé
Total	424.5 [c]	504.5	585.3 [c]	679.1	..	..	..	*Total*
Government								**État**
Agriculture, forestry and fishing	22.0 [s]	..	..	..	..	..	..	Agriculture, sylviculture et pêche
Industrial development	14.0 [s]	..	..	..	..	..	..	Développement industriel
Energy	0.3 [s]	..	..	..	..	..	..	Energie
Transport and telecommunications	..	..	..	..	..	..	..	Transports et télécommunications
Urban and rural planning	5.6 [s]	..	..	..	..	..	..	Aménagement urbain et rural
Prevention of pollution	..	..	..	..	..	..	..	Prévention de la pollution
Identification & treatment of pollution	1.8 [s]	..	..	..	..	..	..	Détection pollu. & mesures antipollu.
Health	0.8 [s]	..	..	..	..	..	..	Santé
Social development and services	18.9 [s]	..	..	..	..	..	..	Développement et services sociaux
Earth and atmosphere	0.4 [s]	..	..	..	..	..	..	Milieu terrestre et atmosphère
Advancement of knowledge	2.9 [s]	..	..	..	..	..	..	Promotion générale des connaissances
Civil space	..	..	..	..	..	..	..	Espace civil
Defence	..	..	..	..	..	..	..	Défense
Not elsewhere classified	..	60.2	64.2	65.4	70.0	63.9	68.5 [p]	Non classé
Total	58.1	60.2	64.2	65.4	70.0	63.9	68.5 [p]	*Total*
Higher education								**Enseignement supérieur**
Agriculture, forestry and fishing	6.2	..	..	..	..	..	..	Agriculture, sylviculture et pêche
Industrial development	54.0	..	..	..	..	..	..	Développement industriel
Energy	2.7	..	..	..	..	..	..	Energie
Transport and telecommunications	..	..	..	..	..	..	..	Transports et télécommunications
Urban and rural planning	..	..	..	..	..	..	..	Aménagement urbain et rural
Prevention of pollution	7.2	..	..	..	..	..	..	Prévention de la pollution
Identification & treatment of pollution	..	..	..	..	..	..	..	Détection pollu. & mesures antipollu.
Health	13.0	..	..	..	..	..	..	Santé
Social development and services	11.6	..	..	..	..	..	..	Développement et services sociaux
Earth and atmosphere	..	..	..	..	..	..	..	Milieu terrestre et atmosphère
Advancement of knowledge	..	..	..	..	..	..	..	Promotion générale des connaissances
Civil space	..	..	..	..	..	..	..	Espace civil
Defence	..	..	..	..	..	..	..	Défense
Not elsewhere classified	26.5	137.1 [c]	153.1	178.4 [c]	203.7	..	..	Non classé
Total	121.1	137.1 [c]	153.1	178.4 [c]	203.7	..	..	*Total*
Private non-profit								**Institutions sans but lucratif**
Agriculture, forestry and fishing	..	..	..	..	..	..	..	Agriculture, sylviculture et pêche
Industrial development	..	..	..	..	..	..	..	Développement industriel
Energy	..	..	..	..	..	..	..	Energie
Transport and telecommunications	..	..	..	..	..	..	..	Transports et télécommunications
Urban and rural planning	..	..	..	..	..	..	..	Aménagement urbain et rural
Prevention of pollution	..	..	..	..	..	..	..	Prévention de la pollution
Identification & treatment of pollution	..	..	..	..	..	..	..	Détection pollu. & mesures antipollu.
Health	1.3	..	..	..	..	..	..	Santé
Social development and services	2.5	..	..	..	..	..	..	Développement et services sociaux
Earth and atmosphere	..	..	..	..	..	..	..	Milieu terrestre et atmosphère
Advancement of knowledge	..	..	..	..	..	..	..	Promotion générale des connaissances
Civil space	..	..	..	..	..	..	..	Espace civil
Defence	..	..	..	..	..	..	..	Défense
Not elsewhere classified	3.8 [c]	5.1 [c]	6.3 [c]	6.3 [c]	..	..	..	Non classé
Total	3.8 [c]	5.1 [c]	6.3 [c]	6.3 [c]	..	..	..	*Total*
GERD								**DIRD**
Agriculture, forestry and fishing	..	..	..	..	..	..	..	Agriculture, sylviculture et pêche
Industrial development	..	..	..	..	..	..	..	Développement industriel
Energy	..	..	..	..	..	..	..	Energie
Transport and telecommunications	..	..	..	..	..	..	..	Transports et télécommunications
Urban and rural planning	..	..	..	..	..	..	..	Aménagement urbain et rural
Prevention of pollution	..	..	..	..	..	..	..	Prévention de la pollution
Identification & treatment of pollution	..	..	..	..	..	..	..	Détection pollu. & mesures antipollu.
Health	..	..	..	..	..	..	..	Santé
Social development and services	..	..	..	..	..	..	..	Développement et services sociaux
Earth and atmosphere	..	..	..	..	..	..	..	Milieu terrestre et atmosphère
Advancement of knowledge	..	..	..	..	..	..	..	Promotion générale des connaissances
Civil space	..	..	..	..	..	..	..	Espace civil
Defence	..	..	..	..	..	..	..	Défense
Not elsewhere classified	607.5 [c]	706.9 [c]	808.9 [c]	929.2 [c]	..	..	..	Non classé
Total	607.5 [c]	706.9 [c]	808.9 [c]	929.2 [c]	..	..	..	*Total*

Source: OECD, R&D database (DSTI/EAS Division), January 2002.

Source : OCDE, base de données R-D (DSTI/Division AES), janvier 2002.

Table 5 - Tableau 5

Total R&D personnel - Personnel total de R-D

by sector of employment and occupation - par secteur d'emploi et par profession

In full-time equivalent *En équivalence plein-temps*

	1994	1995	1996	1997	1998	1999	2000	
Business enterprise								**Entreprises**
RSE	2 982 c	3 383	4 241 c	5 098	..	..	..	Chercheurs
Technicians	1 278 c	1 476	1 516 c	1 555	..	..	..	Techniciens
Other occupations	1 065 c	1 292	1 407 c	1 521	..	..	..	Autres professions
Total	5 325 c	6 151	7 164 c	8 174	..	..	..	*Total*
Government								**État**
RSE	291	283	314	302	310	300 p	306 p	Chercheurs
Technicians	279	315	290	288	293	276 p	269 p	Techniciens
Other occupations	372	362	341	348	338	308 p	307 p	Autres professions
Total	942	959	945	938	941	884 p	882 p	*Total*
Higher education								**Enseignement supérieur**
RSE	1 817	1 918 c	2 066	2 245 c	2 425	..	..	Chercheurs
Technicians	206	241 c	260	274 c	288	..	..	Techniciens
Other occupations	105	133 c	143	139 c	135	..	..	Autres professions
Total	2 127	2 292 c	2 469	2 658 c	2 847	..	..	*Total*
Private non-profit								**Institutions sans but lucratif**
RSE	180 c	180 c	180 c	180 c	..	..	..	Chercheurs
Technicians	40 c	40 c	40 c	40 c	..	..	..	Techniciens
Other occupations	40 c	40 c	40 c	40 c	..	..	..	Autres professions
Total	260 c	260 c	260 c	260 c	..	..	..	*Total*
TOTAL								**TOTAL**
RSE	5 270 c	5 764 c	6 801 c	7 825 c	..	..	..	Chercheurs
Technicians	1 803 c	2 072 c	2 106 c	2 157 c	..	..	..	Techniciens
Other occupations	1 582 c	1 827 c	1 931 c	2 048 c	..	..	..	Autres professions
Total	8 654 c	9 662 c	10 838 c	12 030 c	..	..	..	*Total*

Table 6 - Tableau 6

Total R&D personnel - Personnel total de R-D

by sector of employment and formal qualification - par secteur d'emploi et par niveau de qualification

In full-time equivalent *En équivalence plein-temps*

	1994	1995	1996	1997	1998	1999	2000	
Business enterprise								**Entreprises**
University PhD level degrees	..	..	..	..	..	..	..	Diplômes univ. de niv. doctorat
Other university degrees	..	..	..	..	..	..	..	Autres diplômes universitaires
Sub-total university degrees	..	..	..	..	..	..	..	Sous-total Universitaires
Other post-secondary diplomas	..	..	..	..	..	..	..	Autres diplômes post-secondaires
Secondary diplomas	..	..	..	..	..	..	..	Diplômes de fin d'études secondaires
Other qualifications	..	..	..	..	..	..	..	Autres qualifications
Total	5 325 c	6 151	7 164 c	8 174	..	..	..	*Total*
Government								**État**
University PhD level degrees	..	..	..	..	..	..	..	Diplômes univ. de niv. doctorat
Other university degrees	..	..	..	..	..	..	..	Autres diplômes universitaires
Sub-total university degrees	..	..	..	..	..	..	..	Sous-total Universitaires
Other post-secondary diplomas	..	..	..	..	..	..	..	Autres diplômes post-secondaires
Secondary diplomas	..	..	..	..	..	..	..	Diplômes de fin d'études secondaires
Other qualifications	..	..	..	..	..	..	..	Autres qualifications
Total	942	959	945	938	941	884 p	882 p	*Total*
Higher education								**Enseignement supérieur**
University PhD level degrees	..	..	..	..	..	..	..	Diplômes univ. de niv. doctorat
Other university degrees	..	..	..	..	..	..	..	Autres diplômes universitaires
Sub-total university degrees	..	..	..	..	..	..	..	Sous-total Universitaires
Other post-secondary diplomas	..	..	..	..	..	..	..	Autres diplômes post-secondaires
Secondary diplomas	..	..	..	..	..	..	..	Diplômes de fin d'études secondaires
Other qualifications	..	..	..	..	..	..	..	Autres qualifications
Total	2 127	2 292 c	2 469	2 658 c	2 847	..	..	*Total*
Private non-profit								**Institutions sans but lucratif**
University PhD level degrees	..	..	..	..	..	..	..	Diplômes univ. de niv. doctorat
Other university degrees	..	..	..	..	..	..	..	Autres diplômes universitaires
Sub-total university degrees	..	..	..	..	..	..	..	Sous-total Universitaires
Other post-secondary diplomas	..	..	..	..	..	..	..	Autres diplômes post-secondaires
Secondary diplomas	..	..	..	..	..	..	..	Diplômes de fin d'études secondaires
Other qualifications	..	..	..	..	..	..	..	Autres qualifications
Total	260 c	260 c	260 c	260 c	..	..	..	*Total*
TOTAL								**TOTAL**
University PhD level degrees	..	..	..	..	..	..	..	Diplômes univ. de niv. doctorat
Other university degrees	..	..	..	..	..	..	..	Autres diplômes universitaires
Sub-total university degrees	..	..	..	..	..	..	..	Sous-total Universitaires
Other post-secondary diplomas	..	..	..	..	..	..	..	Autres diplômes post-secondaires
Secondary diplomas	..	..	..	..	..	..	..	Diplômes de fin d'études secondaires
Other qualifications	..	..	..	..	..	..	..	Autres qualifications
Total	8 654 c	9 662 c	10 838 c	12 030 c	..	..	..	*Total*

Source: OECD, R&D database (DSTI/EAS Division), January 2002.

Source : OCDE, base de données R-D (DSTI/Division AES), janvier 2002.

Table 7 - Tableau 7

Gross domestic expenditure on R&D (GERD) - Dépenses intérieures brutes de R-D (DIRD)

by sector of performance and main scientific field - par secteur d'exécution et par grand domaine scientifique

Million EUR (1999 IEP euro, EUR from 1999) | *Millions EUR (1999 IEP euro, EUR à partir de 1999)*

	1994	1995	1996	1997	1998	1999	2000	
Business enterprise								**Entreprises**
Natural sciences and engineering	424.5 [c]	504.5	585.3 [c]	679.1	..	..	..	Sciences exactes, nat. et de l'ingénieur
Social sciences and humanities	0.0	0.0	0.0	0.0	..	..	..	Sciences sociales et humaines
All fields of science	424.5 [c]	504.5	585.3 [c]	679.1	..	..	..	*Tous domaines scientifiques*
Government								**État**
Natural sciences and engineering	47.6 [s]	..	..	..	..	..	..	Sciences exactes, nat. et de l'ingénieur
Social sciences and humanities	18.9 [s]	..	..	..	..	..	..	Sciences sociales et humaines
All fields of science	58.1	60.2	64.2	65.4	70.0	63.9	68.5 [p]	*Tous domaines scientifiques*
Higher education								**Enseignement supérieur**
Natural sciences and engineering	95.4	..	..	..	..	..	..	Sciences exactes, nat. et de l'ingénieur
Social sciences and humanities	25.8	..	..	..	..	..	..	Sciences sociales et humaines
All fields of science	121.1	137.1 [c]	153.1	178.4 [c]	203.7	..	..	*Tous domaines scientifiques*
Private non-profit								**Institutions sans but lucratif**
Natural sciences and engineering	1.4	..	..	..	..	..	..	Sciences exactes, nat. et de l'ingénieur
Social sciences and humanities	2.4	..	..	..	..	..	..	Sciences sociales et humaines
All fields of science	3.8 [c]	5.1 [c]	6.3 [c]	6.3 [c]	..	..	..	*Tous domaines scientifiques*
TOTAL								**TOTAL**
Natural sciences and engineering	568.8 [s]	..	..	..	..	..	..	Sciences exactes, nat. et de l'ingénieur
Social sciences and humanities	47.1 [s]	..	..	..	..	..	..	Sciences sociales et humaines
All fields of science	607.5 [c]	706.9 [c]	808.9 [c]	929.2 [c]	..	..	..	*Tous domaines scientifiques*

Table 8 - Tableau 8

Total R&D personnel - Personnel total de R-D

by sector of employment and main scientific field - par secteur d'emploi et par grand domaine scientifique

In full-time equivalent | *En équivalence plein-temps*

	1994	1995	1996	1997	1998	1999	2000	
Business enterprise								**Entreprises**
Natural sciences and engineering	5 325 [c]	6 151	7 164 [c]	8 174	..	..	..	Sciences exactes, nat. et de l'ingénieur
Social sciences and humanities	0	0	0	0	..	..	..	Sciences sociales et humaines
All fields of science	5 325 [c]	6 151	7 164 [c]	8 174	..	..	..	*Tous domaines scientifiques*
Government								**État**
Natural sciences and engineering	820 [s]	..	..	..	..	..	..	Sciences exactes, nat. et de l'ingénieur
Social sciences and humanities	259 [s]	..	..	..	..	..	..	Sciences sociales et humaines
All fields of science	942	959	945	938	941	884 [p]	882 [p]	*Tous domaines scientifiques*
Higher education								**Enseignement supérieur**
Natural sciences and engineering	3 524	..	..	..	..	..	..	Sciences exactes, nat. et de l'ingénieur
Social sciences and humanities	645	..	..	..	..	..	..	Sciences sociales et humaines
All fields of science	2 127	2 292 [c]	2 469	2 658 [c]	2 847	..	..	*Tous domaines scientifiques*
Private non-profit								**Institutions sans but lucratif**
Natural sciences and engineering	130	..	..	..	..	..	..	Sciences exactes, nat. et de l'ingénieur
Social sciences and humanities	130	..	..	..	..	..	..	Sciences sociales et humaines
All fields of science	260 [c]	260 [c]	260 [c]	260 [c]	..	..	..	*Tous domaines scientifiques*
TOTAL								**TOTAL**
Natural sciences and engineering	9 799 [s]	..	..	..	..	..	..	Sciences exactes, nat. et de l'ingénieur
Social sciences and humanities	1 034 [s]	..	..	..	..	..	..	Sciences sociales et humaines
All fields of science	8 654 [c]	9 662 [c]	10 838 [c]	12 030 [c]	..	..	..	*Tous domaines scientifiques*

Source: OECD, R&D database (DSTI/EAS Division), January 2002.

Source : OCDE, base de données R-D (DSTI/Division AES), janvier 2002.

Total business enterprise intramural expenditure on R&D - Dépenses intra-muros totales de R-D des entreprises
(BERD) by industry - (DIRDE) par industrie

	1994	1995	1996	1997	1998	1999	2000	
	9.1 Million EUR (1999 IEP euro, EUR from 1999) - Millions EUR (1999 IEP euro, EUR à partir de 1999)							
1 AGRICULTURE, HUNTING & FORESTRY	..	0.1	..	0.6 [c]	..	..	..	1 AGRICULTURE, CHASSE & SYLVICULTURE
2 MINING	..	0.3	..	0.6 [c]	..	..	..	2 ACTIVITES EXTRACTIVES
3 MANUFACTURING	..	455.5	..	589.6	..	..	..	3 ACTIVITES DE FABRICATION
4 Food, beverages & tobacco	..	58.7	..	77.2	..	..	..	**4 Produits alimentaires, boissons & tabac**
5 Food products & beverages	..	57.9	..	76.2	..	..	..	5 Produits alimentaires & boissons
6 Tobacco products	..	0.8	..	1.0	..	..	..	6 Produits à base de tabac
7 Textiles, wear app., fur & leather	..	19.4	..	11.1	..	..	..	**7 Textiles, habill., fourrures & cuirs**
8 Textiles	..	9.5	..	6.2	..	..	..	8 Textiles
9 Wearing apparel & fur	..	8.5	..	4.5	..	..	..	9 Articles d'habillement & fourrures
10 Leather products & footwear	..	1.4	..	0.4	..	..	..	10 Articles en cuir & chaussures
11 Wood, paper, printing, publishing	..	15.8	..	8.0	..	..	..	**11 Bois, papier, imprimerie, édition**
12 Wood & cork (not furniture)	..	7.3	..	1.2	..	..	..	12 Bois & liège (sauf meubles)
13 Pulp, paper & paper products	..	2.7	..	3.0	..	..	..	13 Carton, papier & articles
14 Publ., print. & repro. of rec. media	..	5.8	..	3.8	..	..	..	14 Edition, impress. & repro. enregis.
15 Coke, petroleum, nuclear fuel, chemicals, rubber plastics	..	97.0	..	131.2	..	..	..	**15 Coke, pétrole, combust. nucl., prod. chimi., caout. & mat. plast.**
16 Coke, ref. petrol. prod. & nuclear fuel	..	0.3	..	0.0	..	..	..	16 Coke, raffin. pétr. & combust. nucl.
17 Chemicals & chemical products	..	86.3	..	119.3	..	..	..	17 Produits chimiques
18 *Chemical prod.(less pharmaceuticals)*	..	17.0	..	21.7	..	..	..	18 *Prod. chimi. (moins prod. pharmaceutiques)*
19 *Pharmaceuticals*	..	69.4	..	97.6	..	..	..	19 *Produits pharmaceutiques*
20 Rubber & plastic products	..	10.3	..	11.9	..	..	..	20 Caoutchouc et matières plastiques
21 Non-metallic mineral products	..	12.2	..	9.7	..	..	..	**21 Produits minéraux non métalliques**
22 Basic metals	..	3.9	..	1.8	..	..	..	**22 Produits métallurgiques de base**
23 Basic metals, ferrous	..	2.9	..	1.3	..	..	..	23 Produits métallur. de base, ferreux
24 Basic metals, non-ferrous	..	1.0	..	0.6	..	..	..	24 Prod. métallur. de base, non ferreux
25 Fabricated metal products	..	13.8	..	7.4	..	..	..	**25 Ouvrages en métaux (sauf mach.& mat.)**
26 Machinery equipment, instruments and transport equipment	..	223.8	..	337.2	..	..	..	**26 Machines & matériel, instruments & mat. de transport**
27 Machinery n.e.c.	..	17.6	..	22.6	..	..	..	27 Machines, n.c.a.
28 Office, account. & computing machinery	..	25.0	..	34.3	..	..	..	28 Mach. de bur., compta. & trait. inform.
29 Electrical machinery	..	22.8	..	31.6	..	..	..	29 Machines électriques
30 Electro. equip. (radio, TV & commun.)	..	113.0	..	206.7	..	..	..	30 Equip. électro. (radio, tv & communic.)
31 *Electro. comp. (inc. semi-conductors)*	..	43.0	..	86.8	..	..	..	31 *Compos. électro. (semi-cond. incl.)*
32 *TV, radio & communications equip.*	..	70.0	..	119.9	..	..	..	32 *Appareils radio, tv et de communi.*
33 Instruments	..	36.8	..	34.3	..	..	..	33 Instruments
34 Motor vehicles	..	5.9	..	5.6	..	..	..	34 Véhicules automobiles
35 Other transport equipment	..	2.8	..	2.1	..	..	..	35 Autres matériels de transport
36 *Ships*	..	1.4	..	0.4	..	..	..	36 *Navires*
37 *Aerospace*	..	1.4	..	1.7	..	..	..	37 *Aérospatiale*
38 *Other transport equipment n.e.c.*	..	0.0	..	0.0	..	..	..	38 *Autres matériels de transport n.c.a.*
39 Furniture, other manufacturing n.e.c.	..	10.5	..	5.0	..	..	..	**39 Meubles, autres activ. de fabricat. n.c.a.**
40 Furniture	..	5.9	..	3.7	..	..	..	40 Meubles
41 Other manufacturing n.e.c.	..	4.5	..	1.3	..	..	..	41 Autres activités de fabrication n.c.a.
42 Recycling	..	0.5	..	1.1	..	..	..	**42 Récupération**
43 ELECTRICITY, GAS & WATER SUPPLY	..	0.0	..	0.0	..	..	..	43 DISTRIBUTION D'ELECTRICITE, GAZ & EAU
44 CONSTRUCTION	..	0.2	..	1.5	..	..	..	44 CONSTRUCTION
45 SERVICES SECTOR	..	48.4	..	86.7	..	..	..	45 SECTEUR DES SERVICES
46 Wholesale, retail trade & repair of motor vehicles, etc.	..	0.2	..	0.0	..	..	..	**46 Commerce (gros & détail), réparation de véhicules, etc.**
47 Hotels & restaurants	..	0.0	..	0.0	..	..	..	**47 Hôtels et restaurants**
48 Transport & storage	..	0.3	..	0.8	..	..	..	**48 Transports, entreposage**
49 Communications	..	16.2	..	29.2	..	..	..	**49 Communications**
50 Post	..	0.0	..	0.0	..	..	..	50 Postes
51 Telecommunications	..	16.2	..	29.2	..	..	..	51 Télécommunications
52 Financial intermediation (inc. insur.)	..	4.5	..	4.8	..	..	..	**52 Intermédiation financ. (assurances inc.)**
53 Real estate, renting & busin. activity	..	27.0	..	51.6	..	..	..	**53 Immobil., locat. & serv. aux entreprises**
54 Computer & related activities	..	22.6	..	36.4	..	..	..	54 Informatique et activités connexes
55 *Software consultancy*	..	21.1	..	31.9	..	..	..	55 *Production de logiciels*
56 *Other computer services n.e.c.*	..	1.5	..	4.5	..	..	..	56 *Autres activités informatiques*
57 Research & development	..	1.4	..	9.3	..	..	..	57 Recherche & développement expérimental
58 Other business activities n.e.c.	..	3.0	..	5.9	..	..	..	58 Autres activ. & serv. aux entreprises
59 Comm., soc. & pers. serv. activ., etc.	..	0.2	..	0.2	..	..	..	**59 Adm. pub., serv. soc. & collec., etc**
TOTAL BUSINESS ENTERPRISE SECTOR	**424.5 [c]**	**504.5**	**585.3 [c]**	**679.1**	**..**	**..**	**..**	**TOTAL SECTEUR DES ENTREPRISES**

Source: OECD, R&D database (DSTI/EAS Division), January 2002.

Source : OCDE, base de données R-D (DSTI/Division AES), janvier 2002.

Total business enterprise intramural expenditure on R&D - Dépenses intra-muros totales de R-D des entreprises (BERD) by industry - (DIRDE) par industrie

	1994	1995	1996	1997	1998	1999	2000	
	9.2 Million constant USD(1995 prices and PPPs) - Millions de USD constants (prix et PPA de 1995)							
1 AGRICULTURE, HUNTING & FORESTRY	..	0.1	..	0.7 [c]	..	..	..	**1 AGRICULTURE, CHASSE & SYLVICULTURE**
2 MINING	..	0.4	..	0.7 [c]	..	..	..	**2 ACTIVITES EXTRACTIVES**
3 MANUFACTURING	..	565.3	..	687.0	..	..	..	**3 ACTIVITES DE FABRICATION**
4 Food, beverages & tobacco	..	72.8	..	90.0	..	..	..	**4 Produits alimentaires, boissons & tabac**
5 Food products & beverages	..	71.8	..	88.8	..	..	..	5 Produits alimentaires & boissons
6 Tobacco products	..	1.0	..	1.2	..	..	..	6 Produits à base de tabac
7 Textiles, wear app., fur & leather	..	24.1	..	12.9	..	..	..	**7 Textiles, habill., fourrures & cuirs**
8 Textiles	..	11.8	..	7.2	..	..	..	8 Textiles
9 Wearing apparel & fur	..	10.6	..	5.2	..	..	..	9 Articles d'habillement & fourrures
10 Leather products & footwear	..	1.7	..	0.4	..	..	..	10 Articles en cuir & chaussures
11 Wood, paper, printing, publishing	..	19.6	..	9.3	..	..	..	**11 Bois, papier, imprimerie, édition**
12 Wood & cork (not furniture)	..	9.1	..	1.4	..	..	..	12 Bois & liège (sauf meubles)
13 Pulp, paper & paper products	..	3.3	..	3.5	..	..	..	13 Carton, papier & articles
14 Publ., print. & repro. of rec. media	..	7.2	..	4.4	..	..	..	14 Edition, impress. & repro. enregis.
15 Coke, petroleum, nuclear fuel, chemicals, rubber plastics	..	120.4	..	152.9	..	..	..	**15 Coke, pétrole, combust. nucl., prod. chimi., caout. & mat. plast.**
16 Coke, ref. petrol. prod. & nuclear fuel	..	0.4	..	0.0	..	..	..	16 Coke, raffin. pétr. & combust. nucl.
17 Chemicals & chemical products	..	107.1	..	139.0	..	..	..	17 Produits chimiques
18 *Chemical prod.(less pharmaceuticals)*	..	21.1	..	25.3	..	..	..	18 *Prod. chimi. (moins prod. pharmaceutiques)*
19 *Pharmaceuticals*	..	86.1	..	113.7	..	..	..	19 *Produits pharmaceutiques*
20 Rubber & plastic products	..	12.8	..	13.8	..	..	..	20 Caoutchouc et matières plastiques
21 Non-metallic mineral products	..	15.2	..	11.3	..	..	..	**21 Produits minéraux non métalliques**
22 Basic metals	..	4.9	..	2.2	..	..	..	**22 Produits métallurgiques de base**
23 Basic metals, ferrous	..	3.6	..	1.5	..	..	..	23 Produits métallur. de base, ferreux
24 Basic metals, non-ferrous	..	1.3	..	0.7	..	..	..	24 Prod. métallur. de base, non ferreux
25 Fabricated metal products	..	17.1	..	8.6	..	..	..	**25 Ouvrages en métaux (sauf mach.& mat.)**
26 Machinery equipment, instruments and transport equipment	..	277.7	..	392.9	..	..	..	**26 Machines & matériel, instruments & mat. de transport**
27 Machinery n.e.c.	..	21.8	..	26.3	..	..	..	27 Machines, n.c.a.
28 Office, account. & computing machinery	..	31.0	..	40.0	..	..	..	28 Mach. de bur., compta. & trait. inform.
29 Electrical machinery	..	28.2	..	36.9	..	..	..	29 Machines électriques
30 Electro. equip. (radio, TV & commun.)	..	140.2	..	240.8	..	..	..	30 Equip. électro. (radio, tv & communic.)
31 *Electro. comp. (inc. semi-conductors)*	..	53.3	..	101.1	..	..	..	31 *Compos. électro. (semi-cond. incl.)*
32 *TV, radio & communications equip.*	..	86.9	..	139.7	..	..	..	32 *Appareils radio, tv et de communi.*
33 Instruments	..	45.7	..	40.0	..	..	..	33 Instruments
34 Motor vehicles	..	7.3	..	6.5	..	..	..	34 Véhicules automobiles
35 Other transport equipment	..	3.5	..	2.4	..	..	..	35 Autres matériels de transport
36 *Ships*	..	1.7	..	0.4	..	..	..	36 *Navires*
37 *Aerospace*	..	1.8	..	2.0	..	..	..	37 *Aérospatiale*
38 *Other transport equipment n.e.c.*	..	0.0	..	0.0	..	..	..	38 *Autres matériels de transport n.c.a.*
39 Furniture, other manufacturing n.e.c.	..	13.0	..	5.8	..	..	..	**39 Meubles, autres activ. de fabricat. n.c.a.**
40 Furniture	..	7.3	..	4.3	..	..	..	40 Meubles
41 Other manufacturing n.e.c.	..	5.6	..	1.5	..	..	..	41 Autres activités de fabrication n.c.a.
42 Recycling	..	0.6	..	1.3	..	..	..	**42 Récupération**
43 ELECTRICITY, GAS & WATER SUPPLY	..	0.0	..	0.0	..	..	..	**43 DISTRIBUTION D'ELECTRICITE, GAZ & EAU**
44 CONSTRUCTION	..	0.2	..	1.8	..	..	..	**44 CONSTRUCTION**
45 SERVICES SECTOR	..	60.1	..	101.0	..	..	..	**45 SECTEUR DES SERVICES**
46 Wholesale, retail trade & repair of motor vehicles, etc.	..	0.2	..	0.0	..	..	..	**46 Commerce (gros & détail), réparation de véhicules, etc.**
47 Hotels & restaurants	..	0.0	..	0.0	..	..	..	**47 Hôtels et restaurants**
48 Transport & storage	..	0.3	..	1.0	..	..	..	**48 Transports, entreposage**
49 Communications	..	20.1	..	34.0	..	..	..	**49 Communications**
50 Post	..	0.0	..	0.0	..	..	..	50 Postes
51 Telecommunications	..	20.1	..	34.0	..	..	..	51 Télécommunications
52 Financial intermediation (inc. insur.)	..	5.6	..	5.6	..	..	..	**52 Intermédiation financ. (assurances inc.)**
53 Real estate, renting & busin. activity	..	33.5	..	60.1	..	..	..	**53 Immobil., locat. & serv. aux entreprises**
54 Computer & related activities	..	28.1	..	42.4	..	..	..	54 Informatique et activités connexes
55 *Software consultancy*	..	26.1	..	37.2	..	..	..	55 *Production de logiciels*
56 *Other computer services n.e.c.*	..	1.9	..	5.3	..	..	..	56 *Autres activités informatiques*
57 Research & development	..	1.7	..	10.8	..	..	..	57 Recherche & développement expérimental
58 Other business activities n.e.c.	..	3.7	..	6.9	..	..	..	58 Autres activ. & serv. aux entreprises
59 Comm., soc. & pers. serv. activ., etc.	..	0.2	..	0.2	..	..	..	**59 Adm. pub., serv. soc. & collec., etc**
TOTAL BUSINESS ENTERPRISE SECTOR	**542.7 [c]**	**626.1**	**710.3 [c]**	**791.3**	**..**	**..**	**..**	**TOTAL SECTEUR DES ENTREPRISES**

Source: OECD, R&D database (DSTI/EAS Division), January 2002.

Source : OCDE, base de données R-D (DSTI/Division AES), janvier 2002.

Table 14 - Tableau 14

Higher education intramural expenditure on R&D - Dépenses intra-muros de R-D de l'enseignement supérieur

(HERD) by field of science - (DIRDES) par discipline scientifique

	1994	1995	1996	1997	1998	1999	2000	
	14.1 Million EUR (1999 IEP euro, EUR from 1999) - Millions EUR (1999 IEP euro, EUR à partir de 1999)							
Natural sciences	53.0	..	..	..	..	..	..	Sciences exactes et naturelles
Engineering	26.8	..	..	..	..	..	..	Sciences de l'ingénieur
Medical sciences	12.7	..	..	..	..	..	..	Sciences médicales
Agricultural sciences	2.9	..	..	..	..	..	..	Sciences agricoles
Sub-total NSE	95.4	..	..	..	..	..	..	Sous-total SENI
Social sciences	16.0	..	..	..	..	..	..	Sciences sociales
Humanities	9.8	..	..	..	..	..	..	Sciences humaines
Sub-total SSH	25.8	..	..	..	..	..	..	Sous-total SSH
Total NSE+SSH	121.1	137.1 c	153.1	178.4 c	203.7	..	..	*Total SENI+SSH*
	14.2 Million constant USD (1995 prices and PPPs) - Millions de USD constants (prix et PPA de 1995)							
Natural sciences	67.7	..	..	..	..	..	..	Sciences exactes et naturelles
Engineering	34.3	..	..	..	..	..	..	Sciences de l'ingénieur
Medical sciences	16.2	..	..	..	..	..	..	Sciences médicales
Agricultural sciences	3.7	..	..	..	..	..	..	Sciences agricoles
Sub-total NSE	121.9	..	..	..	..	..	..	Sous-total SENI
Social sciences	20.5	..	..	..	..	..	..	Sciences sociales
Humanities	12.5	..	..	..	..	..	..	Sciences humaines
Sub-total SSH	33.0	..	..	..	..	..	..	Sous-total SSH
Total NSE+SSH	154.8	170.2 c	185.8	207.9 c	224.0	..	..	*Total SENI+SSH*

Table 15 - Tableau 15

Higher education intramural expenditure on R&D - Dépenses intra-muros de R-D de l'enseignement supérieur

(HERD) by field of science and type of costs - (DIRDES) par discipline scientifique et par type de coûts

Million EUR (1999 IEP euro, EUR from 1999) *Millions EUR (1999 IEP euro, EUR à partir de 1999)*

	1994	1995	1996	1997	1998	1999	2000	
Current expenditure								**Dépenses courantes**
Natural sciences	48.3	..	..	..	..	..	..	Sciences exactes et naturelles
Engineering	24.0	..	..	..	..	..	..	Sciences de l'ingénieur
Medical sciences	11.6	..	..	..	..	..	..	Sciences médicales
Agricultural sciences	2.7	..	..	..	..	..	..	Sciences agricoles
Sub-total NSE	86.5	..	..	..	..	..	..	Sous-total SENI
Social sciences	15.2	..	..	..	..	..	..	Sciences sociales
Humanities	9.5	..	..	..	..	..	..	Sciences humaines
Sub-total SSH	24.8	..	..	..	..	..	..	Sous-total SSH
Total NSE+SSH	111.2	..	..	..	..	..	..	*Total SENI+SSH*
Capital expenditure								**Dépenses en capital**
Natural sciences	4.7	..	..	..	..	..	..	Sciences exactes et naturelles
Engineering	2.8	..	..	..	..	..	..	Sciences de l'ingénieur
Medical sciences	1.1	..	..	..	..	..	..	Sciences médicales
Agricultural sciences	0.3	..	..	..	..	..	..	Sciences agricoles
Sub-total NSE	8.9	..	..	..	..	..	..	Sous-total SENI
Social sciences	0.8	..	..	..	..	..	..	Sciences sociales
Humanities	0.3	..	..	..	..	..	..	Sciences humaines
Sub-total SSH	1.0	..	..	..	..	..	..	Sous-total SSH
Total NSE+SSH	9.9	..	..	..	..	..	..	*Total SENI+SSH*

Source: OECD, R&D database (DSTI/EAS Division), January 2002.

Source : OCDE, base de données R-D (DSTI/Division AES), janvier 2002.

IRELAND — Table 16 - Tableau 16 — IRLANDE

Higher education intramural expenditure on R&D - Dépenses intra-muros de R-D de l'enseignement supérieur

(HERD) by field of science and source of funds - (DIRDES) par discipline scientifique et par source de financement

Million EUR (1999 IEP euro, EUR from 1999) — *Millions EUR (1999 IEP euro, EUR à partir de 1999)*

	1994	1995	1996	1997	1998	1999	2000	
Direct government								**État direct**
Natural sciences	9.9	..	..	..	..	..	..	Sciences exactes et naturelles
Engineering	9.4	..	..	..	..	..	..	Sciences de l'ingénieur
Medical sciences	1.7	..	..	..	..	..	..	Sciences médicales
Agricultural sciences	0.9	..	..	..	..	..	..	Sciences agricoles
Sub-total NSE	21.8	..	..	..	..	..	..	Sous-total SENI
Social sciences	1.5	..	..	..	..	..	..	Sciences sociales
Humanities	0.9	..	..	..	..	..	..	Sciences humaines
Sub-total SSH	2.4	..	..	..	..	..	..	Sous-total SSH
Total NSE+SSH	24.1	27.4 [c]	32.8	40.6 [c]	48.4	..	..	*Total SENI+SSH*
General university funds								**Fonds généraux des universités**
Natural sciences	18.3	..	..	..	..	..	..	Sciences exactes et naturelles
Engineering	6.9	..	..	..	..	..	..	Sciences de l'ingénieur
Medical sciences	6.0	..	..	..	..	..	..	Sciences médicales
Agricultural sciences	1.7	..	..	..	..	..	..	Sciences agricoles
Sub-total NSE	32.8	..	..	..	..	..	..	Sous-total SENI
Social sciences	10.2	..	..	..	..	..	..	Sciences sociales
Humanities	8.0	..	..	..	..	..	..	Sciences humaines
Sub-total SSH	18.2	..	..	..	..	..	..	Sous-total SSH
Total NSE+SSH	51.0	57.6 [c]	69.6	77.9 [c]	86.3	..	..	*Total SENI+SSH*
Sub-total government								**Sous-total État**
Natural sciences	28.2	..	..	..	..	..	..	Sciences exactes et naturelles
Engineering	16.3	..	..	..	..	..	..	Sciences de l'ingénieur
Medical sciences	7.6	..	..	..	..	..	..	Sciences médicales
Agricultural sciences	2.5	..	..	..	..	..	..	Sciences agricoles
Sub-total NSE	54.6	..	..	..	..	..	..	Sous-total SENI
Social sciences	11.7	..	..	..	..	..	..	Sciences sociales
Humanities	8.9	..	..	..	..	..	..	Sciences humaines
Sub-total SSH	20.6	..	..	..	..	..	..	Sous-total SSH
Total NSE+SSH	75.1	85.0 [c]	102.4	118.5 [c]	134.7	..	..	*Total SENI+SSH*
Higher education								**Enseignement supérieur**
Natural sciences	1.8	..	..	..	..	..	..	Sciences exactes et naturelles
Engineering	0.8	..	..	..	..	..	..	Sciences de l'ingénieur
Medical sciences	1.4	..	..	..	..	..	..	Sciences médicales
Agricultural sciences	0.1	..	..	..	..	..	..	Sciences agricoles
Sub-total NSE	4.1	..	..	..	..	..	..	Sous-total SENI
Social sciences	1.1	..	..	..	..	..	..	Sciences sociales
Humanities	0.4	..	..	..	..	..	..	Sciences humaines
Sub-total SSH	1.5	..	..	..	..	..	..	Sous-total SSH
Total NSE+SSH	5.5	6.2 [c]	5.9	10.4 [p]	15.0	..	..	*Total SENI+SSH*
Private non-profit								**Institutions sans but lucratif**
Natural sciences	1.9	..	..	..	..	..	..	Sciences exactes et naturelles
Engineering	0.3	..	..	..	..	..	..	Sciences de l'ingénieur
Medical sciences	0.6	..	..	..	..	..	..	Sciences médicales
Agricultural sciences	..	..	..	..	..	..	..	Sciences agricoles
Sub-total NSE	2.8	..	..	..	..	..	..	Sous-total SENI
Social sciences	0.1	..	..	..	..	..	..	Sciences sociales
Humanities	..	..	..	..	..	..	..	Sciences humaines
Sub-total SSH	0.1	..	..	..	..	..	..	Sous-total SSH
Total NSE+SSH	3.0	3.4 [c]	2.5	1.6 [c]	0.7	..	..	*Total SENI+SSH*
Business entreprise								**Entreprises**
Natural sciences	4.1	..	..	..	..	..	..	Sciences exactes et naturelles
Engineering	2.5	..	..	..	..	..	..	Sciences de l'ingénieur
Medical sciences	0.8	..	..	..	..	..	..	Sciences médicales
Agricultural sciences	0.1	..	..	..	..	..	..	Sciences agricoles
Sub-total NSE	7.5	..	..	..	..	..	..	Sous-total SENI
Social sciences	0.8	..	..	..	..	..	..	Sciences sociales
Humanities	0.1	..	..	..	..	..	..	Sciences humaines
Sub-total SSH	0.9	..	..	..	..	..	..	Sous-total SSH
Total NSE+SSH	8.4	9.5 [c]	9.8	11.6 [c]	13.3	..	..	*Total SENI+SSH*
Funds from abroad								**Fonds de l'étranger**
Natural sciences	17.0	..	..	..	..	..	..	Sciences exactes et naturelles
Engineering	7.0	..	..	..	..	..	..	Sciences de l'ingénieur
Medical sciences	2.3	..	..	..	..	..	..	Sciences médicales
Agricultural sciences	0.1	..	..	..	..	..	..	Sciences agricoles
Sub-total NSE	26.4	..	..	..	..	..	..	Sous-total SENI
Social sciences	2.3	..	..	..	..	..	..	Sciences sociales
Humanities	0.4	..	..	..	..	..	..	Sciences humaines
Sub-total SSH	2.7	..	..	..	..	..	..	Sous-total SSH
Total NSE+SSH	29.1	33.0 [c]	32.6	36.2 [c]	39.9	..	..	*Total SENI+SSH*

Source: OECD, R&D database (DSTI/EAS Division), January 2002.

Source : OCDE, base de données R-D (DSTI/Division AES), janvier 2002.

Higher education R&D personnel (HEMP) - Personnel de R-D du secteur de l'enseignement supérieur

by field of science and occupation - par discipline scientifique et par profession

In full-time equivalent — *En équivalence plein-temps*

	1994	1995	1996	1997	1998	1999	2000	
Total R&D personnel								**Personnel total de R-D**
Natural sciences	2 338	..	..	..	..	..	..	Sciences exactes et naturelles
Engineering	665	..	..	..	..	..	..	Sciences de l'ingénieur
Medical sciences	452	..	..	..	..	..	..	Sciences médicales
Agricultural sciences	69	..	..	..	..	..	..	Sciences agricoles
Sub-total NSE	3 524	..	..	..	..	..	..	Sous-total SENI
Social sciences	339	..	..	..	..	..	..	Sciences sociales
Humanities	306	..	..	..	..	..	..	Sciences humaines
Sub-total SSH	645	..	..	..	..	..	..	Sous-total SSH
Total NSE+SSH	2 127	2 292 c	2 469	2 658 c	2 847	..	..	*Total SENI+SSH*
Researchers								**Chercheurs**
Natural sciences	2 197	..	..	..	..	..	..	Sciences exactes et naturelles
Engineering	570	..	..	..	..	..	..	Sciences de l'ingénieur
Medical sciences	410	..	..	..	..	..	..	Sciences médicales
Agricultural sciences	52	..	..	..	..	..	..	Sciences agricoles
Sub-total NSE	3 229	..	..	..	..	..	..	Sous-total SENI
Social sciences	327	..	..	..	..	..	..	Sciences sociales
Humanities	302	..	..	..	..	..	..	Sciences humaines
Sub-total SSH	629	..	..	..	..	..	..	Sous-total SSH
Total NSE+SSH	1 817	1 918 c	2 066	2 245 c	2 425	..	..	*Total SENI+SSH*
Technicians								**Technicians**
Natural sciences	95	..	..	..	..	..	..	Sciences exactes et naturelles
Engineering	58	..	..	..	..	..	..	Sciences de l'ingénieur
Medical sciences	32	..	..	..	..	..	..	Sciences médicales
Agricultural sciences	15	..	..	..	..	..	..	Sciences agricoles
Sub-total NSE	200	..	..	..	..	..	..	Sous-total SENI
Social sciences	4	..	..	..	..	..	..	Sciences sociales
Humanities	2	..	..	..	..	..	..	Sciences humaines
Sub-total SSH	6	..	..	..	..	..	..	Sous-total SSH
Total NSE+SSH	206	241 c	260	274 c	288	..	..	*Total SENI+SSH*
Other								**Autres**
Natural sciences	46	..	..	..	..	..	..	Sciences exactes et naturelles
Engineering	37	..	..	..	..	..	..	Sciences de l'ingénieur
Medical sciences	10	..	..	..	..	..	..	Sciences médicales
Agricultural sciences	2	..	..	..	..	..	..	Sciences agricoles
Sub-total NSE	95	..	..	..	..	..	..	Sous-total SENI
Social sciences	8	..	..	..	..	..	..	Sciences sociales
Humanities	2	..	..	..	..	..	..	Sciences humaines
Sub-total SSH	10	..	..	..	..	..	..	Sous-total SSH
Total NSE+SSH	105	133 c	143	139 c	135	..	..	*Total SENI+SSH*

Table 18 - Tableau 18

Government budget appropriations or outlays for R&D (GBAORD) - Crédits budgétaires publics de R-D (CBPRD)

by socio-economic objective - par objectif socio-économique

Million EUR (1999 IEP euro, EUR from 1999) — *Millions EUR (1999 IEP euro, EUR à partir de 1999)*

	1995	1996	1997	1998	1999	2000	2001	
Agriculture, forestry and fishing	27.9	29.7	30.6	24.4	51.1	..	..	Agriculture, sylviculture et pêche
Industrial development	56.4	68.0	83.1	81.1	76.3	..	..	Développement industriel
Energy	0.3	0.3	0.3	0.2	0.0	..	..	Energie
Transport & telecommunications	0.0	0.0	0.0	0.0	0.0	..	..	*Transports & télécommunications*
Urban & rural planning	4.4	4.3	5.0	6.2	6.1	..	..	*Aménagement urbain et rural*
ST Infrastructure	4.4	4.3	5.0	6.2	6.1	..	..	ST Infrastructure
Prevention of pollution	..	..	..	..	..	..	..	*Prévention de la pollution*
Identificat. & treatment of pollution	..	..	..	..	..	..	..	*Détect. pollut. & mesures antipollut.*
ST Environment	2.4	3.3	3.2	2.3	3.6	..	..	ST Environnement
Health	3.8	4.7	5.2	6.5	7.4	..	..	Santé
Social development & services	18.5	14.0	11.2	13.7	17.6	..	..	Développement et services sociaux
Earth and atmosphere	0.6	0.6	0.8	1.0	1.0	..	..	Milieu terrestre et atmosphère
Advancement of research	6.2	10.2	13.4	26.4	32.4	..	..	*Promotion de la recherche*
General university funds	54.4	57.5	44.4	51.4	60.4	..	..	*Fonds généraux des universités*
ST Advancement of knowledge	60.7	67.8	57.7	77.8	92.7	..	..	ST promotion gén. des connaiss.
Civil space	0.1	0.0	0.0	0.0	0.0	..	..	Espace civil
Defence	0.0	0.0	0.0	0.0	0.0	..	..	Défense
Not elsewhere classified	0.0	0.0	0.0	0.0	0.0	..	..	Non classé
Total	**174.7**	**192.6**	**197.1**	**213.1**	**255.8**	**..**	**..**	**Total**

Source: OECD, R&D database (DSTI/EAS Division), January 2002.

Source : OCDE, base de données R-D (DSTI/Division AES), janvier 2002.

Table 1 - Tableau 1

Gross domestic expenditure on R&D (GERD) - Dépenses intérieures brutes de R-D (DIRD)

by sector of performance and source of funds - par secteur d'exécution et par source de financement

	1994	1995	1996	1997	1998	1999	2000	
	1.1 Million EUR (1999 ITL euro, EUR from 1999) - Millions EUR (1999 ITL euro, EUR à partir de 1999)							
Business enterprise								**Entreprises**
financed by								*financé par*
Business enterprise	3 771.8	3 705.4	4 104.1	4 166.9	4 578.9 p,s	4 837.8 p,s	..	Entreprises
Direct government	532.9	820.5	681.5	725.7	776.4 p,s	819.6 p,s	..	État direct
Higher education	..	..	..	..	..	..	..	Enseignement supérieur
Private non-profit	..	..	..	..	..	..	..	Institutions sans but lucratif
Funds from abroad	449.9	401.5	506.3	484.1	477.5 p,s	508.3 p,s	..	Fonds de l'étranger
Total	4 754.5	4 927.4		5 376.7	5 532.8 p	6 033.2 p	6 454.2 p	*Total*
Government								**État**
financed by								*financé par*
Business enterprise	35.2	35.4	47.3	85.9	58.6 p,s	54.9 p,s	..	Entreprises
Direct government	1 828.7	1 876.9	1 864.5	1 955.0	2 210.6 p,s	2 326.3 p,s	..	État direct
Higher education	..	..	..	..	..	..	..	Enseignement supérieur
Private non-profit	..	..	..	..	..	..	..	Institutions sans but lucratif
Funds from abroad	45.1	37.0	64.7	52.6	44.0 p,s	44.5 p,s	..	Fonds de l'étranger
Total	1 909.0	1 949.3	1 976.5	2 093.6	2 302.9 p	2 517.7 p	2 522.4 p	*Total*
Higher education								**Enseignement supérieur**
financed by								*financé par*
Business enterprise	121.4	109.9	100.5	125.7	129.2 p	136.7 p	..	Entreprises
Direct government	..	..	..	..	..	..	..	État direct
General university funds	..	..	..	..	..	..	..	Fonds généraux des universités
Sub-total government	2 143.6	2 192.2	2 480.9	2 493.8	2 564.3 p	2 713.6 p	..	Sous-Total État
Higher education	..	..	..	..	..	..	..	Enseignement supérieur
Private non-profit	..	..	..	..	..	..	..	Institutions sans but lucratif
Funds from abroad	52.0	47.1	43.1	22.5	23.1 p	24.5 p	..	Fonds de l'étranger
Total	2 317.1	2 349.3	2 624.4	2 641.9	2 716.7 p	2 874.8 p	..	*Total*
Private non-profit								**Institutions sans but lucratif**
financed by								*financé par*
Business enterprise	..	..	..	..	..	..	..	Entreprises
Direct government	..	..	..	..	..	..	..	État direct
Higher education	..	..	..	..	..	..	..	Enseignement supérieur
Private non-profit	..	..	..	..	..	..	..	Institutions sans but lucratif
Funds from abroad	..	..	..	..	..	..	..	Fonds de l'étranger
Total	..	..	..	..	..	..	..	*Total*
GERD								**DIRD**
financed by								*financé par*
Business enterprise	3 928.4	3 850.8	4 251.9	4 378.5	4 766.8 p,s	5 029.5 p,s	..	Entreprises
Direct government	..	..	..	..	..	..	..	État direct
General university funds	..	..	..	..	..	..	..	Fonds généraux des universités
Sub-total government	4 505.2	4 889.6	5 026.9	5 174.5	5 551.4 p,s	5 859.4 p,s	..	Sous-Total État
Higher education	..	..	..	..	..	..	..	Enseignement supérieur
Private non-profit	..	..	..	..	..	..	..	Institutions sans but lucratif
Funds from abroad	547.0	485.6	614.2	559.3	544.7 p,s	577.3 p,s	..	Fonds de l'étranger
Total	8 980.6	9 225.9	9 892.9	10 112.2	10 552.4 p	11 425.8 p	..	*Total*
	1.2 Million constant USD (1995 prices and PPPs) - Millions de USD constants (prix et PPA de 1995)							
Business enterprise								**Entreprises**
financed by								*financé par*
Business enterprise	4 947.5	4 627.9	4 868.7	4 827.9	5 166.8 p,s	5 372.1 p,s	..	Entreprises
Direct government	699.0	1 024.7	808.5	840.8	876.1 p,s	910.1 p,s	..	État direct
Higher education	..	..	..	..	..	..	..	Enseignement supérieur
Private non-profit	..	..	..	..	..	..	..	Institutions sans but lucratif
Funds from abroad	590.1	501.4	600.7	560.9	538.8 p,s	564.4 p,s	..	Fonds de l'étranger
Total	6 236.7	6 154.1	6 277.8	6 229.5	6 243.1 p	6 699.5 p	7 009.5 p	*Total*
Government								**État**
financed by								*financé par*
Business enterprise	46.2	44.3	56.1	99.6	66.1 p,s	61.0 p,s	..	Entreprises
Direct government	2 398.7	2 344.2	2 211.8	2 265.1	2 494.4 p,s	2 583.2 p,s	..	État direct
Higher education	..	..	..	..	..	..	..	Enseignement supérieur
Private non-profit	..	..	..	..	..	..	..	Institutions sans but lucratif
Funds from abroad	59.2	46.2	76.8	61.0	49.7 p,s	49.4 p,s	..	Fonds de l'étranger
Total	2 504.1	2 434.6	2 344.7	2 425.7	2 598.5 p	2 795.8 p	2 739.4 p	*Total*
Higher education								**Enseignement supérieur**
financed by								*financé par*
Business enterprise	159.3	137.3	119.2	145.6	145.8 p	151.9 p	..	Entreprises
Direct government	..	..	..	..	..	..	..	État direct
General university funds	..	..	..	..	..	..	..	Fonds généraux des universités
Sub-total government	2 811.9	2 738.0	2 943.0	2 889.3	2 893.5 p	3 013.3 p	..	Sous-Total Etat
Higher education	..	..	..	..	..	..	..	Enseignement supérieur
Private non-profit	..	..	..	..	..	..	..	Institutions sans but lucratif
Funds from abroad	68.3	58.8	51.1	26.1	26.1 p	27.2 p	..	Fonds de l'étranger
Total	3 039.4	2 934.1	3 113.3	3 061.0	3 065.5 p	3 192.3 p	..	*Total*
Private non-profit								**Institutions sans but lucratif**
financed by								*financé par*
Business enterprise	..	..	..	..	..	..	..	Entreprises
Direct government	..	..	..	..	..	..	..	État direct
Higher education	..	..	..	..	..	..	..	Enseignement supérieur
Private non-profit	..	..	..	..	..	..	..	Institutions sans but lucratif
Funds from abroad	..	..	..	..	..	..	..	Fonds de l'étranger
Total	..	..	..	..	..	..	..	*Total*
GERD								**DIRD**
financed by								*financé par*
Business enterprise	5 153.0	4 809.5	5 043.9	5 073.0	5 378.7 p,s	5 585.0 p,s	..	Entreprises
Direct government	..	..	..	..	..	..	..	État direct
General university funds	..	..	..	..	..	..	..	Fonds généraux des universités
Sub-total government	5 909.6	6 106.9	5 963.3	5 995.2	6 264.1 p,s	6 506.5 p,s	..	Sous-Total État
Higher education	..	..	..	..	..	..	..	Enseignement supérieur
Private non-profit	..	..	..	..	..	..	..	Institutions sans but lucratif
Funds from abroad	717.6	606.4	728.6	648.0	614.6 p,s	641.1 p,s	..	Fonds de l'étranger
Total	11 780.1	11 522.8	11 735.8	11 716.2	11 907.0 p	12 687.6 p	..	*Total*

Source: OECD, R&D database (DSTI/EAS Division), January 2002.

Source : OCDE, base de données R-D (DSTI/Division AES), janvier 2002.

Table 2 - Tableau 2

Gross domestic expenditure on R&D (GERD) - Dépenses intérieures brutes de R-D (DIRD)

by sector of performance and type of costs - par secteur d'exécution et par type de coûts

Million EUR (1999 ITL euro, EUR from 1999) — *Millions EUR (1999 ITL euro, EUR à partir de 1999)*

	1994	1995	1996	1997	1998	1999	2000	
Business enterprise								**Entreprises**
Labour costs	..	..	..	..	..	..	..	Coûts du personnel
Other current costs	..	..	..	..	..	..	..	Autres coûts courants
Total current expenditure	..	..	..	..	..	..	..	*Total dépenses courantes*
Land and buildings	..	..	..	..	..	..	..	Terrains et constructions
Instruments and equipment	..	..	..	..	..	..	..	Instruments et équipements
Total capital expenditure	..	..	..	..	..	..	..	*Total dépenses en capital*
Total	4 754.5	4 927.4	5 292.0	5 376.7	5 532.8 p	6 033.2 p	6 454.2 p	*Total*
Government								**État**
Labour costs	..	..	..	..	..	..	..	Coûts du personnel
Other current costs	..	..	..	..	..	..	..	Autres coûts courants
Total current expenditure	..	..	..	..	..	..	..	*Total dépenses courantes*
Land and buildings	..	..	..	..	..	..	..	Terrains et constructions
Instruments and equipment	..	..	..	..	..	..	..	Instruments et équipements
Total capital expenditure	..	..	..	..	..	..	..	*Total dépenses en capital*
Total	1 909.0	1 949.3	1 976.5	2 093.6	2 302.9 p	2 517.7 p	2 522.4 p	*Total*
Higher education								**Enseignement supérieur**
Labour costs	..	..	..	..	..	..	..	Coûts du personnel
Other current costs	..	..	..	..	..	..	..	Autres coûts courants
Total current expenditure	..	..	..	..	..	..	..	*Total dépenses courantes*
Land and buildings	..	..	..	..	..	..	..	Terrains et constructions
Instruments and equipment	..	..	..	..	..	..	..	Instruments et équipements
Total capital expenditure	..	..	..	..	..	..	..	*Total dépenses en capital*
Total	2 317.1	2 349.3	2 624.4	2 641.9	2 716.7 p	2 874.8 p	..	*Total*
Private non-profit								**Institutions sans but lucratif**
Labour costs	..	..	..	..	..	..	..	Coûts du personnel
Other current costs	..	..	..	..	..	..	..	Autres coûts courants
Total current expenditure	..	..	..	..	..	..	..	*Total dépenses courantes*
Land and buildings	..	..	..	..	..	..	..	Terrains et constructions
Instruments and equipment	..	..	..	..	..	..	..	Instruments et équipements
Total capital expenditure	..	..	..	..	..	..	..	*Total dépenses en capital*
Total	..	..	..	..	..	..	..	*Total*
GERD								**DIRD**
Labour costs	..	..	..	..	..	..	..	Coûts du personnel
Other current costs	..	..	..	..	..	..	..	Autres coûts courants
Total current expenditure	..	..	..	..	..	..	..	*Total dépenses courantes*
Land and buildings	..	..	..	..	..	..	..	Terrains et constructions
Instruments and equipment	..	..	..	..	..	..	..	Instruments et équipements
Total capital expenditure	..	..	..	..	..	..	..	*Total dépenses en capital*
Total	8 980.6	9 225.9	9 892.9	10 112.2	10 552.4 p	11 425.8 p	..	*Total*

Table 3 - Tableau 3

Current domestic expenditure on R&D - Dépenses intérieures courantes de R-D

by sector of performance and type of activity - par secteur d'exécution et par type d'activités

Million EUR (1999 ITL euro, EUR from 1999) — *Millions EUR (1999 ITL euro, EUR à partir de 1999)*

	1994	1995	1996	1997	1998	1999	2000	
Business enterprise								**Entreprises**
Basic research	131.6 k	140.7 k	142.2 k	147.4 k,s	173.5 k,p,s	..	..	Recherche fondamentale
Applied research	1 979.1 k	2 112.0 k	2 320.7 k	2 591.3 k,s	2 712.5 k,p,s	..	..	Recherche appliquée
Experimental development	2 643.8 k	2 674.7 k	2 829.1 k	3 133.2 k,s	3 353.5 k,p,s	..	..	Développement expérimental
Not specified	..	..	..	..	..	..	..	Non spécifié
Total	4 754.5 k	4 927.4 k	5 292.0 k	5 376.7 k	5 532.8 k,p	6 033.2 k,p	6 454.2 k,p	*Total*
Government								**État**
Basic research	662.9 k	722.9 k	741.5 k	854.3 k,s	926.8 k,p,s	..	..	Recherche fondamentale
Applied research	964.7 k	980.7 k	1 005.2 k	1 067.5 k,s	1 186.7 k,p,s	..	..	Recherche appliquée
Experimental development	281.4 k	245.8 k	229.7 k	307.7 k,s	317.6 k,p,s	..	..	Développement expérimental
Not specified	..	..	..	..	..	..	..	Non spécifié
Total	1 909.0 k	1 949.3 k	1 976.5 k	2 093.6 k	2 302.9 k,p	2 517.7 k,p	2 522.4 k,p	*Total*
Higher education								**Enseignement supérieur**
Basic research	1 158.5 k	1 174.6 k	1 312.2 k	1 411.3 k,s	1 475.3 k,p,s	..	..	Recherche fondamentale
Applied research	926.8 k	939.7 k	1 049.8 k	1 129.1 k,s	1 180.2 k,p,s	..	..	Recherche appliquée
Experimental development	231.7 k	234.9 k	262.4 k	282.3 k,s	295.1 k,p,s	..	..	Développement expérimental
Not specified	..	..	..	..	..	..	..	Non spécifié
Total	2 317.1 k	2 349.3 k	2 624.4 k	2 641.9 k	2 716.7 k,p	2 874.8 k,p	..	*Total*
Private non-profit								**Institutions sans but lucratif**
Basic research	..	..	..	..	..	..	..	Recherche fondamentale
Applied research	..	..	..	..	..	..	..	Recherche appliquée
Experimental development	..	..	..	..	..	..	..	Développement expérimental
Not specified	..	..	..	..	..	..	..	Non spécifié
Total	..	..	..	..	..	..	..	*Total*
GERD								**DIRD**
Basic research	1 953.0 k	2 038.2 k	2 196.0 k	2 413.1 k,s	2 575.5 k,p,s	..	..	Recherche fondamentale
Applied research	3 870.6 k	4 032.4 k	4 375.7 k	4 787.9 k,s	5 079.3 k,p,s	..	..	Recherche appliquée
Experimental development	3 156.9 k	3 155.4 k	3 321.3 k	3 723.1 k,s	3 966.2 k,p,s	..	..	Développement expérimental
Not specified	..	..	..	..	..	..	..	Non spécifié
Total	8 980.6 k	9 225.9 k	9 892.9 k	10 112.2 k	10 552.4 k,p	11 425.8 k,p	..	*Total*

Source: OECD, R&D database (DSTI/EAS Division), January 2002.

Source : OCDE, base de données R-D (DSTI/Division AES), janvier 2002.

Table 5 - Tableau 5

Total R&D personnel - Personnel total de R-D

by sector of employment and occupation - par secteur d'emploi et par profession

In full-time equivalent — *En équivalence plein-temps*

	1994	1995	1996	1997	1998	1999	2000	
Business enterprise								**Entreprises**
RSE	28 228	27 104	27 735	27 612	27 333	..	..	Chercheurs
Technicians	22 679	21 800	22 076	22 716	..	..	..	Techniciens
Other occupations	12 198	11 419	11 104	11 086	..	..	..	Autres professions
Total	63 105	60 323	60 915	61 414	61 117	..	..	*Total*
Government								**État**
RSE	13 587	13 916	13 637	13 685	12 175	..	..	Chercheurs
Technicians	13 797	13 544	13 168	11 953	..	..	..	Techniciens
Other occupations	5 384	5 579	5 420	5 654	..	..	..	Autres professions
Total	32 768	33 039	32 225	31 292	31 023	..	..	*Total*
Higher education								**Enseignement supérieur**
RSE	33 907	34 516	35 069	34 759	..	..	..	Chercheurs
Technicians	9 225	11 363	11 550	11 710	..	..	..	Techniciens
Other occupations	4 818	2 548	2 529	2 562	..	..	..	Autres professions
Total	47 950	48 427	49 148	49 031	..	..	..	*Total*
Private non-profit								**Institutions sans but lucratif**
RSE	..	..	..	..	..	..	..	Chercheurs
Technicians	..	..	..	..	..	..	..	Techniciens
Other occupations	..	..	..	..	..	..	..	Autres professions
Total	..	..	..	..	..	..	..	*Total*
TOTAL								**TOTAL**
RSE	75 722	75 536	76 441	76 056	..	..	..	Chercheurs
Technicians	45 701	46 707	46 794	46 379	..	..	..	Techniciens
Other occupations	22 400	19 546	19 053	19 302	..	..	..	Autres professions
Total	143 823	141 789	142 288	141 737	..	..	..	*Total*

Table 6 - Tableau 6

Total R&D personnel - Personnel total de R-D

by sector of employment and formal qualification - par secteur d'emploi et par niveau de qualification

In full-time equivalent — *En équivalence plein-temps*

	1994	1995	1996	1997	1998	1999	2000	
Business enterprise								**Entreprises**
University PhD level degrees	..	..	..	..	..	..	..	Diplômes univ. de niv. doctorat
Other university degrees	..	..	..	..	..	..	..	Autres diplômes universitaires
Sub-total university degrees	20 110	20 244	21 379	22 239	..	..	..	Sous-total Universitaires
Other post-secondary diplomas	..	..	..	..	..	..	..	Autres diplômes post-secondaires
Secondary diplomas	27 950	26 635	26 413	26 288	..	..	..	Diplômes de fin d'études secondaires
Other qualifications	15 045	13 444	13 123	12 887	..	..	..	Autres qualifications
Total	63 105	60 323	60 915	61 414	61 117	..	..	*Total*
Government								**État**
University PhD level degrees	..	..	..	..	..	..	..	Diplômes univ. de niv. doctorat
Other university degrees	..	..	..	..	..	..	..	Autres diplômes universitaires
Sub-total university degrees	14 629	15 034	15 589	15 681	..	..	..	Sous-total Universitaires
Other post-secondary diplomas	..	..	..	..	..	..	..	Autres diplômes post-secondaires
Secondary diplomas	10 858	11 293	10 418	10 561	..	..	..	Diplômes de fin d'études secondaires
Other qualifications	7 281	6 712	6 218	5 050	..	..	..	Autres qualifications
Total	32 768	33 039	32 225	31 292	31 023	..	..	*Total*
Higher education								**Enseignement supérieur**
University PhD level degrees	..	..	..	..	..	..	..	Diplômes univ. de niv. doctorat
Other university degrees	..	..	..	..	..	..	..	Autres diplômes universitaires
Sub-total university degrees	36 431	37 513	38 141	37 835	..	..	..	Sous-total Universitaires
Other post-secondary diplomas	..	..	..	..	..	..	..	Autres diplômes post-secondaires
Secondary diplomas	6 681	6 341	6 556	6 961	..	..	..	Diplômes de fin d'études secondaires
Other qualifications	4 838	4 573	4 451	4 235	..	..	..	Autres qualifications
Total	47 950	48 427	49 148	49 031	..	..	..	*Total*
Private non-profit								**Institutions sans but lucratif**
University PhD level degrees	..	..	..	..	..	..	..	Diplômes univ. de niv. doctorat
Other university degrees	..	..	..	..	..	..	..	Autres diplômes universitaires
Sub-total university degrees	..	..	..	..	..	..	..	Sous-total Universitaires
Other post-secondary diplomas	..	..	..	..	..	..	..	Autres diplômes post-secondaires
Secondary diplomas	..	..	..	..	..	..	..	Diplômes de fin d'études secondaires
Other qualifications	..	..	..	..	..	..	..	Autres qualifications
Total	..	..	..	..	..	..	..	*Total*
TOTAL								**TOTAL**
University PhD level degrees	..	..	..	..	..	..	..	Diplômes univ. de niv. doctorat
Other university degrees	..	..	..	..	..	..	..	Autres diplômes universitaires
Sub-total university degrees	71 170	72 791	75 109	75 755	..	..	..	Sous-total Universitaires
Other post-secondary diplomas	..	..	..	..	..	..	..	Autres diplômes post-secondaires
Secondary diplomas	45 489	44 269	43 387	43 810	..	..	..	Diplômes de fin d'études secondaires
Other qualifications	27 164	24 729	23 792	22 172	..	..	..	Autres qualifications
Total	143 823	141 789	142 288	141 737	..	..	..	*Total*

Source: OECD, R&D database (DSTI/EAS Division), January 2002.

Source : OCDE, base de données R-D (DSTI/Division AES), janvier 2002.

Total business enterprise intramural expenditure on R&D - Dépenses intra-muros totales de R-D des entreprises
(BERD) by industry - (DIRDE) par industrie

	1994	1995	1996	1997	1998	1999	2000	
	9.1 Million EUR (1999 ITL euro, EUR from 1999) - Millions EUR (1999 ITL euro, EUR à partir de 1999)							
1 AGRICULTURE, HUNTING & FORESTRY	..	..	..	..	..	..	..	**1 AGRICULTURE, CHASSE & SYLVICULTURE**
2 MINING	..	..	..	..	..	..	..	**2 ACTIVITES EXTRACTIVES**
3 MANUFACTURING	4 191.9	4 256.6	4 547.4	4 552.9	4 234.0	4 730.1 [p]	5 125.4 [p]	**3 ACTIVITES DE FABRICATION**
4 Food, beverages & tobacco	65.3	66.8	68.0	61.1	81.1	83.0 [p]	82.7 [p]	**4 Produits alimentaires, boissons & tabac**
5 Food products & beverages	65.3	65.7	67.2	60.5	80.5	82.3 [p]	82.0 [p]	5 Produits alimentaires & boissons
6 Tobacco products	0.0	1.1	0.8	0.6	0.7	0.7 [p]	0.7 [p]	6 Produits à base de tabac
7 Textiles, wear app., fur & leather	29.4 [a]	21.0	19.8	24.5	21.6	21.7 [p]	22.4 [p]	**7 Textiles, habill., fourrures & cuirs**
8 Textiles	15.2 [a]	13.3	11.4	16.3	13.0	14.1 [p]	14.8 [p]	8 Textiles
9 Wearing apparel & fur	1.0 [a]	0.8	0.8	0.0	0.8	0.7 [p]	0.8 [p]	9 Articles d'habillement & fourrures
10 Leather products & footwear	13.1 [a]	6.9	7.7	8.3	7.8	6.8 [p]	6.8 [p]	10 Articles en cuir & chaussures
11 Wood, paper, printing, publishing	15.0 [a]	18.9	13.7	19.5	19.0 [p,s]	19.9 [p,s]	..	**11 Bois, papier, imprimerie, édition**
12 Wood & cork (not furniture)	8.1 [a]	7.8	6.7	6.0	5.0 [p,s]	5.9 [p,s]	..	12 Bois & liège (sauf meubles)
13 Pulp, paper & paper products	5.4 [a]	8.7	5.3	10.6	12.4	12.3 [p]	15.0 [p]	13 Carton, papier & articles
14 Publ., print. & repro. of rec. media	1.6 [a]	2.4	1.7	2.9	4.5	3.6 [p]	3.9 [p]	14 Edition, impress. & repro. enregis.
15 Coke, petroleum, nuclear fuel, chemicals, rubber plastics	857.8	871.1	911.9	897.5	916.6	999.7 [p]	1 081.8 [p]	**15 Coke, pétrole, combust. nucl., prod. chimi., caout. & mat. plast.**
16 Coke, ref. petrol. prod. & nuclear fuel	31.5	31.4	29.1	25.6	13.4	17.8 [p]	29.3 [p]	16 Coke, raffin. pétr. & combust. nucl.
17 Chemicals & chemical products	735.6	738.3	777.0	769.5	807.9	881.7 [p]	938.2 [p]	17 Produits chimiques
18 *Chemical prod.(less pharmaceuticals)*	268.3	266.8	314.8	322.0	320.0 [p,s]	319.9 [p,s]	..	18 *Prod. chimi. (moins prod. pharmaceutiques)*
19 *Pharmaceuticals*	467.3	471.5	462.2	447.5	483.9 [p,s]	548.8 [p,s]	..	19 *Produits pharmaceutiques*
20 Rubber & plastic products	90.7	101.4	105.8	102.5	95.2	100.2 [p]	114.3 [p]	20 Caoutchouc et matières plastiques
21 Non-metallic mineral products	21.1	18.7	18.3	17.3	18.8	17.7 [p]	18.7 [p]	**21 Produits minéraux non métalliques**
22 Basic metals	62.0	60.1	55.7	65.8	28.7	22.1 [p]	25.7 [p]	**22 Produits métallurgiques de base**
23 Basic metals, ferrous	57.4	53.3	48.2	56.4	56.6 [p,s]	55.8 [p,s]	..	23 Produits métallur. de base, ferreux
24 Basic metals, non-ferrous	4.7	6.8	7.4	9.4	8.7 [p,s]	6.1 [p,s]	..	24 Prod. métallur. de base, non ferreux
25 Fabricated metal products	67.6	68.9	122.5	255.7	258.0 [p,s]	260.3 [p,s]	..	**25 Ouvrages en métaux (sauf mach.& mat.)**
26 Machinery equipment, instruments and transport equipment	3 054.9	3 108.7	3 315.0	3 190.9	3 570.9 [p,s]	3 795.4 [p,s]	..	**26 Machines & matériel, instruments & mat. de transport**
27 Machinery n.e.c.	301.7	292.1	300.0	312.4	329.8 [p,s]	343.0 [p,s]	..	27 Machines, n.c.a.
28 Office, account. & computing machinery	238.4	225.1	219.1	100.1	47.6	43.5 [p]	49.2 [p]	28 Mach. de bur., compta. & trait. inform.
29 Electrical machinery	226.5	237.8	264.6	280.8	322.5 [p,s]	359.4 [p,s]	..	29 Machines électriques
30 Electro. equip. (radio, TV & commun.)	994.6	982.5	1 033.8	1 116.2	1 158.5 [p,s]	1 268.3 [p,s]	..	30 Equip. électro. (radio, tv & communic.)
31 *Electro. comp. (inc. semi-conductors)*	174.0	200.8	229.1	277.5	329.3 [p,s]	429.6 [p,s]	..	31 *Compos. électro. (semi-cond. incl.)*
32 *TV, radio & communications equip.*	820.6	781.7	804.7	838.7	829.2 [p,s]	838.7 [p,s]	..	32 *Appareils radio, tv et de communi.*
33 Instruments	46.5	122.3	105.4	107.8	97.0 [p,s]	105.8 [p,s]	..	33 Instruments
34 Motor vehicles	635.5	720.6	769.0	745.0	720.2	875.0 [p]	951.7 [p]	34 Véhicules automobiles
35 Other transport equipment	611.7	528.3	623.0	528.6	585.4	690.9 [p]	744.8 [p]	35 Autres matériels de transport
36 *Ships*	35.1	47.8	55.4	46.8	46.6 [p,s]	45.7 [p,s]	..	36 *Navires*
37 *Aerospace*	527.9	429.5	507.0	436.1	574.9 [p,s]	593.3 [p,s]	..	37 *Aérospatiale*
38 *Other transport equipment n.e.c.*	48.7	51.0	60.5	45.8	42.0 [p,s]	43.6 [p,s]	..	38 *Autres matériels de transport n.c.a.*
39 Furniture, other manufacturing n.e.c.	18.6 [a]	22.2	22.4	20.6	18.7	19.3 [p]	20.4 [p]	**39 Meubles, autres activ. de fabricat. n.c.a.**
40 Furniture	13.4 [a]	15.7	15.3	13.5	15.3 [p,s]	16.3 [p,s]	..	40 Meubles
41 Other manufacturing n.e.c.	5.3 [a]	6.4	7.2	7.1	7.7 [p,s]	8.4 [p,s]	..	41 Autres activités de fabrication n.c.a.
42 Recycling	0.2	0.2	0.1	0.1	0.1 [n]	0.1 [n]	..	**42 Récupération**
43 ELECTRICITY, GAS & WATER SUPPLY	120.8	131.7	125.9	84.2	146.5 [o]	137.8 [o,p]	141.0 [p]	**43 DISTRIBUTION D'ELECTRICITE, GAZ & EAU**
44 CONSTRUCTION	10.0	11.1	13.1	12.7	21.1	17.0 [p]	15.7 [p]	**44 CONSTRUCTION**
45 SERVICES SECTOR	431.8	527.9	605.7	727.0	1 131.3	1 148.3 [p]	1 171.9 [p]	**45 SECTEUR DES SERVICES**
46 Wholesale, retail trade & repair of motor vehicles, etc.	6.1 [o]	11.8 [o]	14.1 [o]	28.3 [a,o]	32.1 [o]	30.0 [o,p]	..	**46 Commerce (gros & détail), réparation de véhicules, etc.**
47 Hotels & restaurants	.. [n]	.. [n]	.. [n]	.. [n]	.. [n]	.. [n]	..	**47 Hôtels et restaurants**
48 Transport & storage	6.5	7.6	9.5	6.7	5.4	7.0 [p]	8.6 [p]	**48 Transports, entreposage**
49 Communications	19.0	73.5	99.7	30.9	30.6	25.0 [p]	21.8 [p]	**49 Communications**
50 Post	0.9	0.9	0.9	0.9	13.3 [p,s]	7.4 [p,s]	..	50 Postes
51 Telecommunications	18.1	72.7	98.9	30.0	29.2 [p,s]	30.6 [p,s]	..	51 Télécommunications
52 Financial intermediation (inc. insur.)	.. [n]	.. [n]	.. [n]	45.6	61.9	63.3 [p]	66.1 [p]	**52 Intermédiation financ. (assurances inc.)**
53 Real estate, renting & busin. activity	389.3	425.4	471.0	602.7	984.7	1 006.5 [p]	1 029.5 [p]	**53 Immobil., locat. & serv. aux entreprises**
54 Computer & related activities	42.5	63.5	66.2	123.6	173.1	151.2 [p]	145.0 [p]	54 Informatique et activités connexes
55 *Software consultancy*	39.2	53.1	60.0	97.4	100.1 [p,s]	92.5 [p,s]	..	55 *Production de logiciels*
56 *Other computer services n.e.c.*	3.4	10.4	6.2	26.2	26.9 [p,s]	30.4 [p,s]	..	56 *Autres activités informatiques*
57 Research & development	314.0 [g]	318.4 [g]	330.9 [g]	348.6 [g]	705.1 [g]	753.3 [g,p]	771.3 [p]	57 Recherche & développement expérimental
58 Other business activities n.e.c.	32.8	43.5	74.0	130.5	106.5	102.0 [p]	113.2 [p]	58 Autres activ. & serv. aux entreprises
59 Comm., soc. & pers. serv. activ., etc.	10.9	9.5	11.3	12.8	14.5	14.2 [p]	14.3 [p]	**59 Adm. pub., serv. soc. & collec., etc**
TOTAL BUSINESS ENTERPRISE SECTOR	**4 754.5**	**4 927.4**	**5 292.0**	**5 376.7**	**5 532.8** [p]	**6 033.2** [p]	**6 454.2** [p]	**TOTAL SECTEUR DES ENTREPRISES**

Source: OECD, R&D database (DSTI/EAS Division), January 2002.

Source : OCDE, base de données R-D (DSTI/Division AES), janvier 2002.

Total business enterprise intramural expenditure on R&D - Dépenses intra-muros totales de R-D des entreprises
(BERD) by industry - (DIRDE) par industrie

9.2 Million constant USD(1995 prices and PPPs) - Millions de USD constants (prix et PPA de 1995)

	1994	1995	1996	1997	1998	1999	2000	
1 **AGRICULTURE, HUNTING & FORESTRY**	..	..	..	..	..	..	..	1 **AGRICULTURE, CHASSE & SYLVICULTURE**
2 **MINING**	..	..	..	..	..	..	..	2 **ACTIVITES EXTRACTIVES**
3 **MANUFACTURING**	5 498.6	5 316.4	5 394.5	5 275.1	4 777.5	5 252.5 [p]	5 566.4 [p]	3 **ACTIVITES DE FABRICATION**
4 **Food, beverages & tobacco**	85.6	83.5	80.7	70.8	91.5	92.2 [p]	89.8 [p]	4 **Produits alimentaires, boissons & tabac**
5 Food products & beverages	85.6	82.1	79.7	70.1	90.8	91.4 [p]	89.1 [p]	5 Produits alimentaires & boissons
6 Tobacco products	0.0	1.4	0.9	0.7	0.7	0.7 [p]	0.8 [p]	6 Produits à base de tabac
7 **Textiles, wear app., fur & leather**	38.5 [a]	26.2	23.5	28.4	24.3	24.0 [p]	24.3 [p]	7 **Textiles, habill., fourrures & cuirs**
8 Textiles	19.9 [a]	16.6	13.5	18.8	14.7	15.7 [p]	16.1 [p]	8 Textiles
9 Wearing apparel & fur	1.4 [a]	1.0	0.9	0.0	0.9	0.8 [p]	0.9 [p]	9 Articles d'habillement & fourrures
10 Leather products & footwear	17.2 [a]	8.6	9.1	9.6	8.8	7.5 [p]	7.3 [p]	10 Articles en cuir & chaussures
11 **Wood, paper, printing, publishing**	19.7 [a]	23.6	16.3	22.5	21.4 [p,s]	22.1 [p,s]	..	11 **Bois, papier, imprimerie, édition**
12 Wood & cork (not furniture)	10.6 [a]	9.8	7.9	6.9	5.6 [p,s]	6.5 [p,s]	..	12 Bois & liège (sauf meubles)
13 Pulp, paper & paper products	7.0 [a]	10.8	6.3	12.2	14.0	13.7 [p]	16.3 [p]	13 Carton, papier & articles
14 Publ., print. & repro. of rec. media	2.0 [a]	3.0	2.0	3.4	5.1	4.0 [p]	4.3 [p]	14 Edition, impress. & repro. enregis.
15 **Coke, petroleum, nuclear fuel, chemicals, rubber plastics**	1 125.2	1 088.0	1 081.8	1 039.9	1 034.2	1 110.1 [p]	1 174.8 [p]	15 **Coke, pétrole, combust. nucl., prod. chimi., caout. & mat. plast.**
16 Coke, ref. petrol. prod. & nuclear fuel	41.3	39.2	34.5	29.6	15.2	19.7 [p]	31.8 [p]	16 Coke, raffin. pétr. & combust. nucl.
17 Chemicals & chemical products	964.9	922.1	921.8	891.5	911.6	979.0 [p]	1 018.9 [p]	17 Produits chimiques
18 *Chemical prod.(less pharmaceuticals)*	351.9	333.2	373.4	373.1	361.1 [p,s]	355.2 [p,s]	..	18 *Prod. chimi. (moins prod. pharmaceutiques)*
19 *Pharmaceuticals*	613.0	588.9	548.3	518.4	546.0 [p,s]	609.4 [p,s]	..	19 *Produits pharmaceutiques*
20 Rubber & plastic products	118.9	126.7	125.5	118.7	107.4	111.3 [p]	124.2 [p]	20 Caoutchouc et matières plastiques
21 **Non-metallic mineral products**	27.7	23.4	21.7	20.0	21.2	19.6 [p]	20.3 [p]	21 **Produits minéraux non métalliques**
22 **Basic metals**	81.4	75.0	66.0	76.2	32.4	24.5 [p]	27.9 [p]	22 **Produits métallurgiques de base**
23 Basic metals, ferrous	75.2	66.6	57.2	65.3	63.8 [p,s]	61.9 [p,s]	..	23 Produits métallur. de base, ferreux
24 Basic metals, non-ferrous	6.1	8.5	8.8	10.9	9.8 [p,s]	6.7 [p,s]	..	24 Prod. métallur. de base, non ferreux
25 **Fabricated metal products**	88.7	86.0	145.3	296.3	291.1 [p,s]	289.1 [p,s]	..	25 **Ouvrages en métaux (sauf mach.& mat.)**
26 **Machinery equipment, instruments and transport equipment**	4 007.2	3 882.6	3 932.5	3 697.0	4 029.4 [p,s]	4 214.6 [p,s]	..	26 **Machines & matériel, instruments & mat. de transport**
27 Machinery n.e.c.	395.7	364.8	355.9	362.0	372.1 [p,s]	380.9 [p,s]	..	27 Machines, n.c.a.
28 Office, account. & computing machinery	312.7	281.1	259.9	115.9	53.8	48.3 [p]	53.5 [p]	28 Mach. de bur., compta. & trait. inform.
29 Electrical machinery	297.2	297.0	313.9	325.3	363.9 [p,s]	399.1 [p,s]	..	29 Machines électriques
30 Electro. equip. (radio, TV & commun.)	1 304.6	1 227.1	1 226.4	1 293.2	1 307.2 [p,s]	1 408.3 [p,s]	..	30 Equip. électro. (radio, tv & communic.)
31 *Electro. comp. (inc. semi-conductors)*	228.2	250.8	271.8	321.5	371.6 [p,s]	477.0 [p,s]	..	31 *Compos. électro. (semi-cond. incl.)*
32 *TV, radio & communications equip.*	1 076.4	976.3	954.6	971.7	935.6 [p,s]	931.3 [p,s]	..	32 *Appareils radio, tv et de communi.*
33 Instruments	61.0	152.8	125.1	124.9	109.5 [p,s]	117.5 [p,s]	..	33 Instruments
34 Motor vehicles	833.6	900.0	912.3	863.2	812.7	971.7 [p]	1 033.6 [p]	34 Véhicules automobiles
35 Other transport equipment	802.4	659.8	739.0	612.5	660.5	767.1 [p]	808.9 [p]	35 Autres matériels de transport
36 *Ships*	46.0	59.7	65.8	54.2	52.6 [p,s]	50.7 [p,s]	..	36 *Navires*
37 *Aerospace*	692.5	536.4	601.5	505.2	648.7 [p,s]	658.8 [p,s]	..	37 *Aérospatiale*
38 *Other transport equipment n.e.c.*	63.8	63.6	71.8	53.1	47.4 [p,s]	48.4 [p,s]	..	38 *Autres matériels de transport n.c.a.*
39 **Furniture, other manufacturing n.e.c.**	24.4 [a]	27.7	26.6	23.9	21.1	21.4 [p]	22.2 [p]	39 **Meubles, autres activ. de fabricat. n.c.a.**
40 Furniture	17.5 [a]	19.7	18.1	15.7	17.3 [p,s]	18.1 [p,s]	..	40 Meubles
41 Other manufacturing n.e.c.	6.9 [a]	8.1	8.5	8.2	8.6 [p,s]	9.3 [p,s]	..	41 Autres activités de fabrication n.c.a.
42 **Recycling**	0.2	0.2	0.1	0.1	0.1 [n]	0.1 [n]	..	42 **Récupération**
43 **ELECTRICITY, GAS & WATER SUPPLY**	158.5	164.5	149.3	97.5	165.3 [o]	153.0 [o,p]	153.1 [p]	43 **DISTRIBUTION D'ELECTRICITE, GAZ & EAU**
44 **CONSTRUCTION**	13.1	13.9	15.5	14.7	23.8	18.9 [p]	17.0 [p]	44 **CONSTRUCTION**
45 **SERVICES SECTOR**	566.4	659.3	718.5	842.3	1 276.5	1 275.2 [p]	1 272.8 [p]	45 **SECTEUR DES SERVICES**
46 **Wholesale, retail trade & repair of motor vehicles, etc.**	8.0 [o]	14.8 [o]	16.8 [o]	32.8 [a,o]	36.2 [o]	33.3 [o,p]	..	46 **Commerce (gros & détail), réparation de véhicules, etc.**
47 **Hotels & restaurants**	.. [n]	.. [n]	.. [n]	.. [n]	.. [n]	.. [n]	..	47 **Hôtels et restaurants**
48 **Transport & storage**	8.6	9.5	11.3	7.8	6.0	7.8 [p]	9.3 [p]	48 **Transports, entreposage**
49 **Communications**	24.9	91.9	118.3	35.8	34.5	27.8 [p]	23.7 [p]	49 **Communications**
50 Post	1.2	1.1	1.0	1.0	15.0 [p,s]	8.2 [p,s]	..	50 Postes
51 Telecommunications	23.7	90.8	117.3	34.8	33.0 [p,s]	34.0 [p,s]	..	51 Télécommunications
52 **Financial intermediation (inc. insur.)**	.. [n]	.. [n]	.. [n]	52.9	69.8	70.3 [p]	71.8 [p]	52 **Intermédiation financ. (assurances inc.)**
53 **Real estate, renting & busin. activity**	510.7	531.3	558.8	698.2	1 111.2	1 117.7 [p]	1 118.0 [p]	53 **Immobil., locat. & serv. aux entreprises**
54 Computer & related activities	55.8	79.3	78.5	143.2	195.4	167.9 [p]	157.4 [p]	54 Informatique et activités connexes
55 *Software consultancy*	51.4	66.3	71.2	112.8	112.9 [p,s]	102.7 [p,s]	..	55 *Production de logiciels*
56 *Other computer services n.e.c.*	4.4	13.0	7.3	30.4	30.4 [p,s]	33.8 [p,s]	..	56 *Autres activités informatiques*
57 Research & development	411.8 [g]	397.7 [g]	392.5 [g]	403.8 [g]	795.6 [g]	836.5 [g,p]	837.6 [p]	57 Recherche & développement expérimental
58 Other business activities n.e.c.	43.1	54.3	87.8	151.2	120.1	113.2 [p]	123.0 [p]	58 Autres activ. & serv. aux entreprises
59 **Comm., soc. & pers. serv. activ., etc.**	14.2	11.9	13.4	14.8	16.3	15.8 [p]	15.6 [p]	59 **Adm. pub., serv. soc. & collec., etc**
TOTAL BUSINESS ENTERPRISE SECTOR	**6 236.7**	**6 154.1**	**6 277.8**	**6 229.5**	**6 243.1** [p]	**6 699.5** [p]	**7 009.5** [p]	**TOTAL SECTEUR DES ENTREPRISES**

Source: OECD, R&D database (DSTI/EAS Division), January 2002.

Source : OCDE, base de données R-D (DSTI/Division AES), janvier 2002.

RSE in the business enterprise sector - Chercheurs dans le secteur des entreprises

by industry - par industrie

In full time equivalent / *En équivalence plein-temps*

	1994	1995	1996	1997	1998	1999	2000	
1 AGRICULTURE, HUNTING & FORESTRY	0	0	0	..	..	..	..	**1 AGRICULTURE, CHASSE & SYLVICULTURE**
2 MINING	0	0	0	..	..	..	..	**2 ACTIVITES EXTRACTIVES**
3 MANUFACTURING	24 136	22 854	23 076	..	..	..	..	**3 ACTIVITES DE FABRICATION**
4 Food, beverages & tobacco	308	304	296	..	..	..	..	**4 Produits alimentaires, boissons & tabac**
5 Food products & beverages	308	299	291	..	..	..	..	5 Produits alimentaires & boissons
6 Tobacco products	0	5	5	..	..	..	..	6 Produits à base de tabac
7 Textiles, wear app., fur & leather	88	98	96	..	..	..	..	**7 Textiles, habill., fourrures & cuirs**
8 Textiles	57	53	44	..	..	..	..	8 Textiles
9 Wearing apparel & fur	1	2	1	..	..	..	..	9 Articles d'habillement & fourrures
10 Leather products & footwear	30	43	51	..	..	..	..	10 Articles en cuir & chaussures
11 Wood, paper, printing, publishing	38	48	79	..	..	..	..	**11 Bois, papier, imprimerie, édition**
12 Wood & cork (not furniture)	16	15	13	..	..	..	..	12 Bois & liège (sauf meubles)
13 Pulp, paper & paper products	13	24	19	..	..	..	..	13 Carton, papier & articles
14 Publ., print. & repro. of rec. media	9	9	47	..	..	..	..	14 Edition, impress. & repro. enregis.
15 Coke, petroleum, nuclear fuel, chemicals, rubber plastics	5 234	4 872	4 736	..	..	..	..	**15 Coke, pétrole, combust. nucl., prod. chimi., caout. & mat. plast.**
16 Coke, ref. petrol. prod. & nuclear fuel	205	194	185	..	..	..	..	16 Coke, raffin. pétr. & combust. nucl.
17 Chemicals & chemical products	4 654	4 281	4 154	..	..	..	..	17 Produits chimiques
18 *Chemical prod.(less pharmaceuticals)*	1 836	1 652	1 604	..	..	..	..	18 *Prod. chimi. (moins prod. pharmaceutiques)*
19 *Pharmaceuticals*	2 818	2 629	2 550	..	..	..	..	19 *Produits pharmaceutiques*
20 Rubber & plastic products	375	397	397	..	..	..	..	20 Caoutchouc et matières plastiques
21 Non-metallic mineral products	121	109	109	..	..	..	..	**21 Produits minéraux non métalliques**
22 Basic metals	278	248	251	..	..	..	..	**22 Produits métallurgiques de base**
23 Basic metals, ferrous	245	198	200	..	..	..	..	23 Produits métallur. de base, ferreux
24 Basic metals, non-ferrous	33	50	51	..	..	..	..	24 Prod. métallur. de base, non ferreux
25 Fabricated metal products	321	312	503	..	..	..	..	**25 Ouvrages en métaux (sauf mach.& mat.)**
26 Machinery equipment, instruments and transport equipment	17 701	16 813	16 955	..	..	..	..	**26 Machines & matériel, instruments & mat. de transport**
27 Machinery n.e.c.	1 379	1 083	1 091	..	..	..	..	27 Machines, n.c.a.
28 Office, account. & computing machinery	2 110	1 510	1 467	..	..	..	..	28 Mach. de bur., compta. & trait. inform.
29 Electrical machinery	1 331	1 287	1 390	..	..	..	..	29 Machines électriques
30 Electro. equip. (radio, TV & commun.)	7 754	7 786	7 624	..	..	..	..	30 Equip. électro. (radio, tv & communic.)
31 *Electro. comp. (inc. semi-conductors)*	817	986	1 088	..	..	..	..	31 *Compos. électro. (semi-cond. incl.)*
32 *TV, radio & communications equip.*	6 937	6 800	6 536	..	..	..	..	32 *Appareils radio, tv et de communi.*
33 Instruments	267	749	638	..	..	..	..	33 Instruments
34 Motor vehicles	1 967	2 019	2 018	..	..	..	..	34 Véhicules automobiles
35 Other transport equipment	2 893	2 379	2 727	..	..	..	..	35 Autres matériels de transport
36 *Ships*	218	248	281	..	..	..	..	36 *Navires*
37 *Aerospace*	2 552	1 981	2 298	..	..	..	..	37 *Aérospatiale*
38 *Other transport equipment n.e.c.*	123	150	148	..	..	..	..	38 *Autres matériels de transport n.c.a.*
39 Furniture, other manufacturing n.e.c.	47 [m]	50 [m]	51 [m]	..	..	..	..	**39 Meubles, autres activ. de fabricat. n.c.a.**
40 Furniture	47	50	51	..	..	..	..	40 Meubles
41 Other manufacturing n.e.c.	.. [n]	.. [n]	.. [n]	..	..	..	..	41 Autres activités de fabrication n.c.a.
42 Recycling	0	0	0	..	..	..	..	**42 Récupération**
43 ELECTRICITY, GAS & WATER SUPPLY	475	451	489	..	..	..	..	**43 DISTRIBUTION D'ELECTRICITE, GAZ & EAU**
44 CONSTRUCTION	32	33	32	..	..	..	..	**44 CONSTRUCTION**
45 SERVICES SECTOR	3 585	3 766	4 138	..	..	..	..	**45 SECTEUR DES SERVICES**
46 Wholesale, retail trade & repair of motor vehicles, etc.	27 [o]	49 [o]	64 [o]	..	..	..	..	**46 Commerce (gros & détail), réparation de véhicules, etc.**
47 Hotels & restaurants	.. [n]	.. [n]	.. [n]	..	..	..	..	**47 Hôtels et restaurants**
48 Transport & storage	35	38	46	..	..	..	..	**48 Transports, entreposage**
49 Communications	251	281	377	..	..	..	..	**49 Communications**
50 Post	45	45	45	..	..	..	..	50 Postes
51 Telecommunications	206	236	332	..	..	..	..	51 Télécommunications
52 Financial intermediation (inc. insur.)	.. [n]	.. [n]	.. [n]	..	..	..	..	**52 Intermédiation financ. (assurances inc.)**
53 Real estate, renting & busin. activity	3 198	3 332	3 569	..	..	..	..	**53 Immobil., locat. & serv. aux entreprises**
54 Computer & related activities	594	708	756	..	..	..	..	54 Informatique et activités connexes
55 *Software consultancy*	517	603	666	..	..	..	..	55 *Production de logiciels*
56 *Other computer services n.e.c.*	77	105	90	..	..	..	..	56 *Autres activités informatiques*
57 Research & development	2 333	2 309	2 446	..	..	..	..	57 Recherche & développement expérimental
58 Other business activities n.e.c.	271	315	367	..	..	..	..	58 Autres activ. & serv. aux entreprises
59 Comm., soc. & pers. serv. activ., etc.	74	66	82	..	..	..	..	**59 Adm. pub., serv. soc. & collec., etc**
TOTAL BUSINESS ENTERPRISE SECTOR	**28 228**	**27 104**	**27 735**	**27 612**	**27 333**	..	..	**TOTAL SECTEUR DES ENTREPRISES**

Source: OECD, R&D database (DSTI/EAS Division), January 2002.

Source : OCDE, base de données R-D (DSTI/Division AES), janvier 2002.

Table 17 - Tableau 17

Higher education R&D personnel (HEMP) - Personnel de R-D du secteur de l'enseignement supérieur

by field of science and occupation - par discipline scientifique et par profession

In full-time equivalent | *En équivalence plein-temps*

	1994	1995	1996	1997	1998	1999	2000	
Total R&D personnel								**Personnel total de R-D**
Natural sciences	..	..	..	..	..	..	..	Sciences exactes et naturelles
Engineering	..	..	..	..	..	..	..	Sciences de l'ingénieur
Medical sciences	..	..	..	..	..	..	..	Sciences médicales
Agricultural sciences	..	..	..	..	..	..	..	Sciences agricoles
Sub-total NSE	..	..	..	..	..	..	..	Sous-total SENI
Social sciences	..	..	..	..	..	..	..	Sciences sociales
Humanities	..	..	..	..	..	..	..	Sciences humaines
Sub-total SSH	..	..	..	..	..	..	..	Sous-total SSH
Total NSE+SSH	47 950	48 427	49 148	49 031	..	..	..	*Total SENI+SSH*
Researchers								**Chercheurs**
Natural sciences	..	..	..	..	..	..	..	Sciences exactes et naturelles
Engineering	..	..	..	..	..	..	..	Sciences de l'ingénieur
Medical sciences	..	..	..	..	..	..	..	Sciences médicales
Agricultural sciences	..	..	..	..	..	..	..	Sciences agricoles
Sub-total NSE	..	..	..	..	..	..	..	Sous-total SENI
Social sciences	..	..	..	..	..	..	..	Sciences sociales
Humanities	..	..	..	..	..	..	..	Sciences humaines
Sub-total SSH	..	..	..	..	..	..	..	Sous-total SSH
Total NSE+SSH	33 907	34 516	35 069	34 759	..	..	..	*Total SENI+SSH*
Technicians								**Technicians**
Natural sciences	..	..	..	..	..	..	..	Sciences exactes et naturelles
Engineering	..	..	..	..	..	..	..	Sciences de l'ingénieur
Medical sciences	..	..	..	..	..	..	..	Sciences médicales
Agricultural sciences	..	..	..	..	..	..	..	Sciences agricoles
Sub-total NSE	..	..	..	..	..	..	..	Sous-total SENI
Social sciences	..	..	..	..	..	..	..	Sciences sociales
Humanities	..	..	..	..	..	..	..	Sciences humaines
Sub-total SSH	..	..	..	..	..	..	..	Sous-total SSH
Total NSE+SSH	9 225	11 363	11 550	11 710	..	..	..	*Total SENI+SSH*
Other								**Autres**
Natural sciences	..	..	..	..	..	..	..	Sciences exactes et naturelles
Engineering	..	..	..	..	..	..	..	Sciences de l'ingénieur
Medical sciences	..	..	..	..	..	..	..	Sciences médicales
Agricultural sciences	..	..	..	..	..	..	..	Sciences agricoles
Sub-total NSE	..	..	..	..	..	..	..	Sous-total SENI
Social sciences	..	..	..	..	..	..	..	Sciences sociales
Humanities	..	..	..	..	..	..	..	Sciences humaines
Sub-total SSH	..	..	..	..	..	..	..	Sous-total SSH
Total NSE+SSH	4 818	2 548	2 529	2 562	..	..	..	*Total SENI+SSH*

Table 18 - Tableau 18

Government budget appropriations or outlays for R&D (GBAORD) - Crédits budgétaires publics de R-D (CBPRD)

by socio-economic objective - par objectif socio-économique

Million EUR (1999 ITL euro, EUR from 1999) | *Millions EUR (1999 ITL euro, EUR à partir de 1999)*

	1995	1996	1997	1998	1999	2000	2001	
Agriculture, forestry and fishing	143.0	119.9	137.9	115.7	136.4	141.7 p	..	Agriculture, sylviculture et pêche
Industrial development	468.2	478.9	558.8	497.8	457.2	1 044.9 p	..	Développement industriel
Energy	165.8	203.8	252.9	305.7	278.0	302.4 p	..	Energie
Transport & telecommunications	..	..	..	..	..	..	..	*Transports & télécommunications*
Urban & rural planning	..	..	..	..	..	..	..	*Aménagement urbain et rural*
ST Infrastructure	24.0	22.2	28.3	38.9	21.9	18.4 p	..	ST Infrastructure
Prevention of pollution	..	..	..	..	..	..	..	*Prévention de la pollution*
Identificat. & treatment of pollution	..	..	..	..	..	..	..	*Détect. pollut. & mesures antipollut.*
ST Environment	128.5	136.2	151.6	210.7	166.8	168.3 p	..	ST Environnement
Health	467.7	431.2	436.7	343.8	429.9	456.9 p	..	Santé
Social development & services	147.7	258.4	253.9	223.3	278.5	237.7 p	..	Développement et services sociaux
Earth and atmosphere	75.2	74.4	84.6	97.9	111.0	110.1 p	..	Milieu terrestre et atmosphère
Advancement of research	427.4	806.7	675.5	713.4	710.1	755.2 p	..	*Promotion de la recherche*
General university funds	2 380.9	2 426.3	2 822.4	2 950.5	2 874.6	2 874.6 p	..	*Fonds généraux des universités*
ST Advancement of knowledge	2 808.3	3 233.1	3 498.0	3 663.9	3 584.7	3 629.8	..	ST promotion gén. des connaiss.
Civil space	460.0	577.3	544.5	511.0	538.8	586.5 p	..	Espace civil
Defence	250.7	174.8	274.5	163.0	76.2	59.2 p	..	Défense
Not elsewhere classified	176.1	..	..	..	..	..	..	Non classé
Total	**5 315.2**	**5 710.2**	**6 221.5**	**6 171.7**	**6 079.4**	**6 755.8 p**	**..**	**Total**

Source: OECD, R&D database (DSTI/EAS Division), January 2002. | *Source* : OCDE, base de données R-D (DSTI/Division AES), janvier 2002.

ITALY Table 19 - Tableau 19 ITALIE

Technology balance of payments - Balance des paiements technologiques

by type of transfer - selon la nature des transferts

Million EUR (1999 ITL euro, EUR from 1999) *Millions EUR (1999 ITL euro, EUR à partir de 1999)*

	1983	1984	1985	1986	1987	1988	1989	1990	1991	
Receipts										**Recettes**
Related firms	..	..	..	..	..	..	..	..	..	Firmes affiliées
Unrelated firms	..	..	..	..	..	..	..	..	..	Firmes non affiliées
Other	..	..	..	..	..	..	..	..	..	Autre
Total	116	150	142	169	201	429	365	437	904	*Total*
Payments										**Paiements**
Related firms	..	..	..	..	..	..	..	..	..	Firmes affiliées
Unrelated firms	..	..	..	..	..	..	..	..	..	Firmes non affiliées
Other	..	..	..	..	..	..	..	..	..	Autre
Total	471	511	538	546	527	792	736	759	1 516	*Total*

	1992	1993	1994	1995	1996	1997	1998	1999	2000	
Receipts										**Recettes**
Related firms	..	..	..	..	..	..	..	..	..	Firmes affiliées
Unrelated firms	..	..	..	..	..	..	..	..	..	Firmes non affiliées
Other	..	..	..	..	..	..	..	..	..	Autre
Total	2 086 [a]	2 171	2 123	2 570	2 540	3 000	2 724	3 161	3 044	*Total*
Payments										**Paiements**
Related firms	..	..	..	..	..	..	..	..	..	Firmes affiliées
Unrelated firms	..	..	..	..	..	..	..	..	..	Firmes non affiliées
Other	..	..	..	..	..	..	..	..	..	Autre
Total	2 678 [a]	2 855	2 877	2 897	3 086	3 206	3 249	3 976	3 802	*Total*

Source: OECD, TBPdatabase (DSTI/EAS Division), Januay 2002.

Source : OCDE, base de données TBP (DSTI/Division AES), janvier 2002.

Table 1 - Tableau 1

Gross domestic expenditure on R&D (GERD) - Dépenses intérieures brutes de R-D (DIRD)

by sector of performance and source of funds - par secteur d'exécution et par source de financement

	1994	1995	1996	1997	1998	1999	2000	
	1.1 Million JPY - Millions JPY							
Business enterprise								**Entreprises**
financed by								*financé par*
Business enterprise	8 850 586 [i]	9 222 996 [i]	9 916 024 [a]	10 464 078	10 513 491	10 369 702	..	Entreprises
Direct government	106 831	149 244	115 100 [a]	139 120	223 274	187 621	..	État direct
Higher education	365	173	249 [a]	1 948	272	307	..	Enseignement supérieur
Private non-profit	10 402	10 293	14 791 [a]	12 157	17 242	16 200	..	Institutions sans but lucratif
Funds from abroad	12 069	13 189	12 246	41 053	45 784	56 331	..	Fonds de l'étranger
Total	8 980 253 [i]	9 395 896 [i]	10 058 409 [a]	10 658 357	10 800 063	10 630 161	..	*Total*
Government								**État**
financed by								*financé par*
Business enterprise	6 623	10 231	23 244	12 309	20 551	27 113	..	Entreprises
Direct government	1 219 057	1 378 903	1 304 724	1 293 945	1 381 590	1 453 940	..	État direct
Higher education	499	578	323	389	368	384	..	Enseignement supérieur
Private non-profit	240	314	222	258	361	268	..	Institutions sans but lucratif
Funds from abroad	8	105	23	75	45	26	..	Fonds de l'étranger
Total	1 226 427	1 390 132	1 328 535	1 306 976	1 402 914	1 481 731	..	*Total*
Higher education								**Enseignement supérieur**
financed by								*financé par*
Business enterprise	63 840	70 272	49 393	51 305	50 975	50 442	..	Entreprises
Direct government	261 257 [e]	309 572 [e]	150 914 [a]	165 628	185 708	196 035	..	État direct
General university funds	1 124 194 [e,i]	1 251 593 [e,i]	874 603 [a,e]	877 331 [e]	933 407 [e]	918 903 [e]	..	Fonds généraux des universités
Sub-total government	1 385 451 [e,i]	1 561 165 [e,i]	1 025 517 [a,e]	1 042 959 [e]	1 119 115 [e]	1 114 938 [e]	..	Sous-Total État
Higher education	1 298 645 [e,i]	1 346 360 [e,i]	1 010 318 [a,e]	1 013 468 [e]	1 078 294 [e]	1 061 342 [e]	..	Enseignement supérieur
Private non-profit	4 383	4 032	3 260	3 389	3 229	3 889	..	Institutions sans but lucratif
Funds from abroad	231	357	373	608	544	547	..	Fonds de l'étranger
Total	2 752 551 [i]	2 982 186 [i]	2 088 861 [a]	2 111 730	2 252 158	2 231 159	..	*Total*
Private non-profit								**Institutions sans but lucratif**
financed by								*financé par*
Business enterprise	350 392	365 687	397 669	417 613	422 772	402 911	..	Entreprises
Direct government	205 281	201 350	200 132	213 392	209 032	196 249	..	État direct
Higher education	820	1 053	867	994	1 150	360	..	Enseignement supérieur
Private non-profit	78 631	70 217	79 221	83 117	79 510	88 583	..	Institutions sans but lucratif
Funds from abroad	1 676	1 715	1 362	1 851	1 604	1 506	..	Fonds de l'étranger
Total	636 800	640 021	679 251	716 967	714 068	689 609	..	*Total*
GERD								**DIRD**
financed by								*financé par*
Business enterprise	9 271 441 [i]	9 669 186 [i]	10 386 330 [a]	10 945 305	11 007 789	10 850 168	..	Entreprises
Direct government	1 792 426 [e]	2 039 069 [e]	1 770 870 [a]	1 812 085	1 999 604	2 033 845	..	État direct
General university funds	1 124 194 [e,i]	1 251 593 [e,i]	874 603 [a,e]	877 331 [e]	933 407 [e]	918 903 [e]	..	Fonds généraux des universités
Sub-total government	2 916 620 [e,i]	3 290 662 [e,i]	2 645 473 [a,e]	2 689 416 [e]	2 933 011 [e]	2 952 748 [e]	..	Sous-Total État
Higher education	1 300 329 [e,i]	1 348 164 [e,i]	1 011 757 [a,e]	1 016 799 [e]	1 080 084 [e]	1 062 393 [e]	..	Enseignement supérieur
Private non-profit	93 656	84 856	97 494	98 921	100 342	108 940	..	Institutions sans but lucratif
Funds from abroad	13 984	15 366	14 004	43 587	47 977	58 410	..	Fonds de l'étranger
Total	13 596 030 [i]	14 408 236 [i]	14 155 058 [a]	14 794 030	15 169 203	15 032 660	..	*Total*
	1.2 Million constant USD (1995 prices and PPPs) - Millions de USD constants (prix et PPA de 1995)							
Business enterprise								**Entreprises**
financed by								*financé par*
Business enterprise	51 890.7 [i]	54 271.5 [i]	58 830.1 [a]	61 860.1	62 186.8	62 198.7	..	Entreprises
Direct government	626.3	878.2	682.9 [a]	822.4	1 320.7	1 125.4	..	État direct
Higher education	2.1	1.0	1.5 [a]	11.5	1.6	1.8	..	Enseignement supérieur
Private non-profit	61.0	60.6	87.8 [a]	71.9	102.0	97.2	..	Institutions sans but lucratif
Funds from abroad	70.8	77.6	72.7	242.7	270.8	337.9	..	Fonds de l'étranger
Total	52 650.9 [i]	55 288.9 [i]	59 674.9 [a]	63 008.6	63 881.9	63 761.0	..	*Total*
Government								**État**
financed by								*financé par*
Business enterprise	38.8	60.2	137.9	72.8	121.6	162.6	..	Entreprises
Direct government	7 147.3	8 114.0	7 740.7	7 649.4	8 172.0	8 720.9	..	État direct
Higher education	2.9	3.4	1.9	2.3	2.2	2.3	..	Enseignement supérieur
Private non-profit	1.4	1.8	1.3	1.5	2.1	1.6	..	Institutions sans but lucratif
Funds from abroad	0.0	0.6	0.1	0.4	0.3	0.2	..	Fonds de l'étranger
Total	7 190.5	8 180.0	7 882.0	7 726.4	8 298.2	8 887.6	..	*Total*
Higher education								**Enseignement supérieur**
financed by								*financé par*
Business enterprise	374.3	413.5	293.0	303.3	301.5	302.6	..	Entreprises
Direct government	1 531.7 [e]	1 821.6 [e]	895.3 [a]	979.1	1 098.5	1 175.8	..	État direct
General university funds	6 591.1 [e,i]	7 364.8 [e,i]	5 188.9 [a,e]	5 186.5 [e]	5 521.1 [e]	5 511.7 [e]	..	Fonds généraux des universités
Sub-total government	8 122.9 [e,i]	9 186.5 [e,i]	6 084.2 [a,e]	6 165.6 [e]	6 619.5 [e]	6 687.5 [e]	..	Sous-Total État
Higher education	7 613.9 [e,i]	7 922.5 [e,i]	5 994.1 [a,e]	5 991.3 [e]	6 378.1 [e]	6 366.1 [e]	..	Enseignement supérieur
Private non-profit	25.7	23.7	19.3	20.0	19.1	23.3	..	Institutions sans but lucratif
Funds from abroad	1.4	2.1	2.2	3.6	3.2	3.3	..	Fonds de l'étranger
Total	16 138.1 [i]	17 548.3 [i]	12 392.9 [a]	12 483.8	13 321.4	13 382.8	..	*Total*
Private non-profit								**Institutions sans but lucratif**
financed by								*financé par*
Business enterprise	2 054.3	2 151.8	2 359.3	2 468.8	2 500.7	2 416.7	..	Entreprises
Direct government	1 203.6	1 184.8	1 187.4	1 261.5	1 236.4	1 177.1	..	État direct
Higher education	4.8	6.2	5.1	5.9	6.8	2.2	..	Enseignement supérieur
Private non-profit	461.0	413.2	470.0	491.4	470.3	531.3	..	Institutions sans but lucratif
Funds from abroad	9.8	10.1	8.1	10.9	9.5	9.0	..	Fonds de l'étranger
Total	3 733.5	3 766.1	4 029.9	4 238.5	4 223.7	4 136.4	..	*Total*
GERD								**DIRD**
financed by								*financé par*
Business enterprise	54 358.2 [i]	56 897.0 [i]	61 620.4 [a]	64 704.9	65 110.6	65 080.6	..	Entreprises
Direct government	10 508.9 [e]	11 998.6 [e]	10 506.3 [a]	10 712.4	11 827.6	12 199.2	..	État direct
General university funds	6 591.1 [e,i]	7 364.8 [e,i]	5 188.9 [a,e]	5 186.5 [e]	5 521.1 [e]	5 511.7 [e]	..	Fonds généraux des universités
Sub-total government	17 100.1 [e,i]	19 363.5 [e,i]	15 695.2 [a,e]	15 898.9 [e]	17 348.6 [e]	17 710.9 [e]	..	Sous-Total État
Higher education	7 623.8 [e,i]	7 933.1 [e,i]	6 002.6 [a,e]	6 011.0 [e]	6 388.6 [e]	6 372.4 [e]	..	Enseignement supérieur
Private non-profit	549.1	499.3	578.4	584.8	593.5	653.4	..	Institutions sans but lucratif
Funds from abroad	82.0	90.4	83.1	257.7	283.8	350.4	..	Fonds de l'étranger
Total	79 713.1 [i]	84 783.3 [i]	83 979.6 [a]	87 457.3	89 725.2	90 167.7	..	*Total*

Source: OECD, R&D database (DSTI/EAS Division), January 2002.

Source : OCDE, base de données R-D (DSTI/Division AES), janvier 2002.

JAPAN — JAPON

Table 2 - Tableau 2

Gross domestic expenditure on R&D (GERD) - Dépenses intérieures brutes de R-D (DIRD)

by sector of performance and type of costs - par secteur d'exécution et par type de coûts

Million JPY — *Millions JPY*

	1994	1995	1996	1997	1998	1999	2000	
Business enterprise								**Entreprises**
Labour costs	4 022 383 [l]	4 167 244 [l]	4 375 519 [a]	4 532 908	4 665 438	4 627 391	..	Coûts du personnel
Other current costs	4 023 547	4 215 127	4 627 918 [a]	4 968 355	5 069 873	5 015 770	..	Autres coûts courants
Total current expenditure	8 045 930 [l]	8 382 371 [l]	9 003 437 [a]	9 501 263	9 735 311	9 643 161	..	*Total dépenses courantes*
Land and buildings	183 381	162 271	170 737	199 520	146 813	146 737	..	Terrains et constructions
Instruments and equipment	750 941	851 255	884 236	957 574	917 939	840 263	..	Instruments et équipements
Total capital expenditure	934 322	1 013 525	1 054 973	1 157 094	1 064 752	987 000	..	*Total dépenses en capital*
Total	8 980 253 [l]	9 395 896 [l]	10 058 409 [a]	10 658 357	10 800 063	10 630 161	..	*Total*
Government								**État**
Labour costs	400 905	408 432	413 465	421 214	440 399	439 591	..	Coûts du personnel
Other current costs	473 951	554 550	603 642	611 351	617 743	633 241	..	Autres coûts courants
Total current expenditure	874 856	962 982	1 017 107	1 032 565	1 058 141	1 072 831	..	*Total dépenses courantes*
Land and buildings	72 171	73 321	67 926	87 352	44 642	90 550	..	Terrains et constructions
Instruments and equipment	279 400	353 829	243 501	187 059	300 131	318 350	..	Instruments et équipements
Total capital expenditure	351 571	427 150	311 427	274 411	344 773	408 900	..	*Total dépenses en capital*
Total	1 226 427	1 390 132	1 328 535	1 306 976	1 402 914	1 481 731	..	*Total*
Higher education								**Enseignement supérieur**
Labour costs	1 861 740 [l]	1 920 783 [l]	1 040 750 [a]	1 067 307	1 097 762	1 108 161	..	Coûts du personnel
Other current costs	571 274	625 958	628 286	648 704	688 026	713 434	..	Autres coûts courants
Total current expenditure	2 433 014 [l]	2 546 741 [l]	1 669 036 [a]	1 716 011	1 785 788	1 821 595	..	*Total dépenses courantes*
Land and buildings	111 860	130 498	159 105	140 435	162 014	150 869	..	Terrains et constructions
Instruments and equipment	207 677	304 948	260 721	255 284	304 358	258 694	..	Instruments et équipements
Total capital expenditure	319 537	435 446	419 826	395 719	466 372	409 563	..	*Total dépenses en capital*
Total	2 752 551 [l]	2 982 186 [l]	2 088 861 [a]	2 111 730	2 252 158	2 231 159	..	*Total*
Private non-profit								**Institutions sans but lucratif**
Labour costs	213 930	223 438	233 526	240 498	241 670	218 649	..	Coûts du personnel
Other current costs	322 241	322 096	363 626	406 490	409 955	412 053	..	Autres coûts courants
Total current expenditure	536 171	545 535	597 152	646 988	651 625	630 702	..	*Total dépenses courantes*
Land and buildings	7 274	5 427	6 717	7 998	5 702	7 196	..	Terrains et constructions
Instruments and equipment	93 354	89 059	75 382	61 980	56 741	51 712	..	Instruments et équipements
Total capital expenditure	100 628	94 486	82 099	69 979	62 443	58 907	..	*Total dépenses en capital*
Total	636 800	640 021	679 251	716 967	714 068	689 609	..	*Total*
GERD								**DIRD**
Labour costs	6 498 958 [l]	6 719 897 [l]	6 063 260 [a]	6 261 927	6 445 269	6 393 792	..	Coûts du personnel
Other current costs	5 391 013	5 717 732	6 223 472	6 634 900	6 785 597	6 774 498	..	Autres coûts courants
Total current expenditure	11 889 971 [l]	12 437 629 [l]	12 286 732 [a]	12 896 827	13 230 865	13 168 289	..	*Total dépenses courantes*
Land and buildings	374 687	371 516	404 486	435 305	359 171	395 352	..	Terrains et constructions
Instruments and equipment	1 331 372	1 599 091	1 463 840	1 461 897	1 579 169	1 469 019	..	Instruments et équipements
Total capital expenditure	1 706 059	1 970 607	1 868 326	1 897 203	1 938 340	1 864 370	..	*Total dépenses en capital*
Total	13 596 030 [l]	14 408 236 [l]	14 155 058 [a]	14 794 030	15 169 203	15 032 660	..	*Total*

Table 3 - Tableau 3

Current domestic expenditure on R&D - Dépenses intérieures courantes de R-D

by sector of performance and type of activity - par secteur d'exécution et par type d'activités

Million JPY — *Millions JPY*

	1994	1995	1996	1997	1998	1999	2000	
Business enterprise								**Entreprises**
Basic research	606 380 [k,l]	624 004 [k,l]	619 790 [a,k]	665 953 [k]	600 746 [k]	614 213 [k]	..	Recherche fondamentale
Applied research	1 991 048 [k,l]	2 071 564 [k,l]	2 218 713 [a,k]	2 298 707 [k]	2 360 466 [k]	2 183 879 [k]	..	Recherche appliquée
Experimental development	6 382 824 [k,l]	6 700 328 [k,l]	7 219 907 [a,k]	7 693 697 [k]	7 838 851 [k]	7 832 069 [k]	..	Développement expérimental
Not specified	..	..	..	..	..	..	..	Non spécifié
Total	8 980 253 [k,l]	9 395 896 [k,l]	10 058 409 [a,k]	10 658 357 [k]	10 800 063 [k]	10 630 161 [k]	..	*Total*
Government								**État**
Basic research	220 125 [k]	278 069 [k]	255 771 [k]	271 889 [k]	332 312 [k]	359 832 [k]	..	Recherche fondamentale
Applied research	330 713 [k]	374 448 [k]	346 064 [k]	405 650 [k]	406 167 [k]	402 642 [k]	..	Recherche appliquée
Experimental development	637 045 [k]	698 004 [k]	686 854 [k]	596 749 [k]	620 188 [k]	679 033 [k]	..	Développement expérimental
Not specified	38 544 [k]	39 611 [k]	39 845 [k]	32 688 [k]	44 247 [k]	40 224 [k]	..	Non spécifié
Total	1 226 427 [k]	1 390 132 [k]	1 328 535 [k]	1 306 976 [k]	1 402 914 [k]	1 481 731 [k]	..	*Total*
Higher education								**Enseignement supérieur**
Basic research	909 890 [k,l]	1 021 792 [k,l]	730 750 [a,k]	733 323 [k]	796 628 [k]	769 192 [k]	..	Recherche fondamentale
Applied research	622 961 [k,l]	682 230 [k,l]	476 991 [a,k]	488 614 [k]	520 144 [k]	516 614 [k]	..	Recherche appliquée
Experimental development	145 613 [k,l]	167 302 [k,l]	123 705 [a,k]	124 841 [k]	132 052 [k]	132 841 [k]	..	Développement expérimental
Not specified	1 074 087 [k,l]	1 110 863 [k,l]	757 417 [a,k]	764 952 [k]	803 337 [k]	812 511 [k]	..	Non spécifié
Total	2 752 551 [k,l]	2 982 186 [k,l]	2 088 861 [a,k]	2 111 730 [k]	2 252 158 [k]	2 231 159 [k]	..	*Total*
Private non-profit								**Institutions sans but lucratif**
Basic research	122 172 [k]	117 473 [k]	114 891 [k]	99 225 [k]	95 603 [k]	98 362 [k]	..	Recherche fondamentale
Applied research	108 057 [k]	110 353 [k]	119 471 [k]	140 950 [k]	145 310 [k]	141 159 [k]	..	Recherche appliquée
Experimental development	348 822 [k]	357 259 [k]	388 190 [k]	427 126 [k]	423 773 [k]	404 273 [k]	..	Développement expérimental
Not specified	57 749 [k]	54 936 [k]	56 699 [k]	49 666 [k]	49 382 [k]	45 815 [k]	..	Non spécifié
Total	636 800 [k]	640 021 [k]	679 251 [k]	716 967 [k]	714 068 [k]	689 609 [k]	..	*Total*
GERD								**DIRD**
Basic research	1 858 567 [k,l]	2 041 338 [k,l]	1 721 202 [a,k]	1 770 390 [k]	1 825 289 [k]	1 841 599 [k]	..	Recherche fondamentale
Applied research	3 052 779 [k,l]	3 238 595 [k,l]	3 161 239 [a,k]	3 333 921 [k]	3 432 087 [k]	3 244 294 [k]	..	Recherche appliquée
Experimental development	7 514 304 [k,l]	7 922 893 [k,l]	8 418 656 [a,k]	8 842 413 [k]	9 014 864 [k]	9 048 216 [k]	..	Développement expérimental
Not specified	1 170 380 [k,l]	1 205 410 [k,l]	853 961 [a,k]	847 306 [k]	896 966 [k]	898 550 [k]	..	Non spécifié
Total	13 596 030 [k,l]	14 408 236 [k,l]	14 155 058 [a,k]	14 794 030 [k]	15 169 203 [k]	15 032 660 [k]	..	*Total*

Source: OECD, R&D database (DSTI/EAS Division), January 2002.

Source : OCDE, base de données R-D (DSTI/Division AES), janvier 2002.

Gross domestic expenditure on R&D (GERD) - Dépenses intérieures brutes de R-D (DIRD)

by sector of performance and socio-economic objective - par secteur d'exécution et par objectif socio-économique

Million JPY — *Millions JPY*

	1994	1995	1996	1997	1998	1999	2000	
Business enterprise								**Entreprises**
Agriculture, forestry and fishing	..	..	..	..	..	..	..	Agriculture, sylviculture et pêche
Industrial development	..	..	..	..	..	..	..	Développement industriel
Energy	..	..	..	..	..	..	..	Energie
Transport and telecommunications	..	..	..	..	..	..	..	Transports et télécommunications
Urban and rural planning	..	..	..	..	..	..	..	Aménagement urbain et rural
Prevention of pollution	..	..	..	..	..	..	..	Prévention de la pollution
Identification & treatment of pollution	..	..	..	..	..	..	..	Détection pollu. & mesures antipollu.
Health	..	..	..	..	..	..	..	Santé
Social development and services	..	..	..	..	..	..	..	Développement et services sociaux
Earth and atmosphere	..	..	..	..	..	..	..	Milieu terrestre et atmosphère
Advancement of knowledge	..	..	..	..	..	..	..	Promotion générale des connaissances
Civil space	..	..	..	..	..	..	..	Espace civil
Defence	..	..	..	..	..	..	..	Défense
Not elsewhere classified	8 980 253 [i]	9 395 896 [i]	10 058 409 [a]	10 658 357	10 800 063	10 630 161	..	Non classé
Total	8 980 253 [i]	9 395 896 [i]	10 058 409 [a]	10 658 357	10 800 063	10 630 161	..	*Total*
Government								**État**
Agriculture, forestry and fishing	230 135	237 502	235 867	229 704	239 189	244 406	..	Agriculture, sylviculture et pêche
Industrial development	158 622	186 299	165 552	195 198	237 973	246 816	..	Développement industriel
Energy	317 601	341 939	339 522	296 091	322 938	333 041	..	Energie
Transport and telecommunications	41 757	56 540	39 778	42 738	49 770	51 858	..	Transports et télécommunications
Urban and rural planning	32 027	37 255	38 474	38 192	39 790	34 412	..	Aménagement urbain et rural
Prevention of pollution	24 157	25 693	28 418	26 232	27 782	27 983	..	Prévention de la pollution
Identification & treatment of pollution	..	..	..	..	..	..	..	Détection pollu. & mesures antipollu.
Health	57 095	61 686	56 896	58 216	64 066	63 475	..	Santé
Social development and services	22 207	25 212	23 562	23 210	30 568	29 820	..	Développement et services sociaux
Earth and atmosphere	31 640	60 029	50 187	41 682	58 064	80 071	..	Milieu terrestre et atmosphère
Advancement of knowledge	27 615	29 775	30 043	23 771	26 723	24 048	..	Promotion générale des connaissances
Civil space	141 526	176 499	155 650	156 569	161 703	197 444	..	Espace civil
Defence	142 046	151 704	164 587	175 375	144 347	148 357	..	Défense
Not elsewhere classified	..	..	..	..	..	..	..	Non classé
Total	1 226 427	1 390 132	1 328 535	1 306 976	1 402 914	1 481 731	..	*Total*
Higher education								**Enseignement supérieur**
Agriculture, forestry and fishing	..	..	..	..	..	..	..	Agriculture, sylviculture et pêche
Industrial development	..	..	..	..	..	..	..	Développement industriel
Energy	..	..	..	..	..	..	..	Energie
Transport and telecommunications	..	..	..	..	..	..	..	Transports et télécommunications
Urban and rural planning	..	..	..	..	..	..	..	Aménagement urbain et rural
Prevention of pollution	..	..	..	..	..	..	..	Prévention de la pollution
Identification & treatment of pollution	..	..	..	..	..	..	..	Détection pollu. & mesures antipollu.
Health	..	..	..	..	..	..	..	Santé
Social development and services	..	..	..	..	..	..	..	Développement et services sociaux
Earth and atmosphere	..	..	..	..	..	..	..	Milieu terrestre et atmosphère
Advancement of knowledge	..	..	..	..	..	..	..	Promotion générale des connaissances
Civil space	..	..	..	..	..	..	..	Espace civil
Defence	..	..	..	..	..	..	..	Défense
Not elsewhere classified	2 752 551 [i]	2 982 186 [i]	2 088 861 [a]	2 111 730	2 252 160	2 231 159	..	Non classé
Total	2 752 551 [i]	2 982 186 [i]	2 088 861 [a]	2 111 730	2 252 158	2 231 159	..	*Total*
Private non-profit								**Institutions sans but lucratif**
Agriculture, forestry and fishing	18 864	18 006	20 261	20 710	22 572	20 441	..	Agriculture, sylviculture et pêche
Industrial development	291 348	294 819	350 664	402 638	409 506	403 905	..	Développement industriel
Energy	102 989	101 822	94 665	92 370	73 722	72 260	..	Energie
Transport and telecommunications	70 285	71 224	58 804	46 162	45 105	44 489	..	Transports et télécommunications
Urban and rural planning	8 944	9 829	10 853	11 659	12 508	12 715	..	Aménagement urbain et rural
Prevention of pollution	22 670	18 503	22 793	18 548	21 492	24 915	..	Prévention de la pollution
Identification & treatment of pollution	..	..	..	..	..	..	..	Détection pollu. & mesures antipollu.
Health	48 573	47 561	48 215	50 996	56 800	55 018	..	Santé
Social development and services	19 582	19 791	18 952	21 505	19 515	18 590	..	Développement et services sociaux
Earth and atmosphere	5 121	5 893	5 999	6 406	5 161	5 440	..	Milieu terrestre et atmosphère
Advancement of knowledge	46 768	51 143	46 340	44 428	44 703	28 507	..	Promotion générale des connaissances
Civil space	1 656	1 430	1 704	1 545	2 983	3 329	..	Espace civil
Defence	..	..	..	..	..	..	..	Défense
Not elsewhere classified	..	..	..	..	..	..	..	Non classé
Total	636 800	640 021	679 251	716 967	714 068	689 609	..	*Total*
GERD								**DIRD**
Agriculture, forestry and fishing	..	..	..	..	..	..	..	Agriculture, sylviculture et pêche
Industrial development	..	..	..	..	..	..	..	Développement industriel
Energy	..	..	..	..	..	..	..	Energie
Transport and telecommunications	..	..	..	..	..	..	..	Transports et télécommunications
Urban and rural planning	..	..	..	..	..	..	..	Aménagement urbain et rural
Prevention of pollution	..	..	..	..	..	..	..	Prévention de la pollution
Identification & treatment of pollution	..	..	..	..	..	..	..	Détection pollu. & mesures antipollu.
Health	..	..	..	..	..	..	..	Santé
Social development and services	..	..	..	..	..	..	..	Développement et services sociaux
Earth and atmosphere	..	..	..	..	..	..	..	Milieu terrestre et atmosphère
Advancement of knowledge	..	..	..	..	..	..	..	Promotion générale des connaissances
Civil space	..	..	..	..	..	..	..	Espace civil
Defence	..	..	..	..	..	..	..	Défense
Not elsewhere classified	13 596 030 [i]	14 408 236 [i]	14 155 058 [a]	14 794 030	15 169 203	15 032 660	..	Non classé
Total	13 596 030 [i]	14 408 236 [i]	14 155 058 [a]	14 794 030	15 169 203	15 032 660	..	*Total*

Source: OECD, R&D database (DSTI/EAS Division), January 2002.

Source : OCDE, base de données R-D (DSTI/Division AES), janvier 2002.

Table 5 - Tableau 5

Total R&D personnel - Personnel total de R-D

by sector of employment and occupation - par secteur d'emploi et par profession

In full-time equivalent — *En équivalence plein-temps*

	1994	1995	1996	1997	1998	1999	2000	
Business enterprise								**Entreprises**
RSE	376 639 [l]	384 100 [l]	400 361 [a]	404 232	429 195	433 758	..	Chercheurs
Technicians	74 225	67 246	67 845 [a]	66 477	69 541	66 956	..	Techniciens
Other occupations	126 861	122 368	121 285 [a]	115 447	114 424	103 830	..	Autres professions
Total	577 725 [l]	573 714 [l]	589 491 [a]	586 156	613 160	604 544	..	*Total*
Government								**État**
RSE	30 263	30 346	30 241	30 212	30 910	30 987	..	Chercheurs
Technicians	2 681	2 911	3 138	3 318	4 158	4 487	..	Techniciens
Other occupations	22 689	22 733	22 797	23 024	23 694	23 551	..	Autres professions
Total	55 633	55 990	56 176	56 554	58 762	59 025	..	*Total*
Higher education								**Enseignement supérieur**
RSE	235 702 [l]	242 862 [l]	170 017 [a]	174 093	176 627	178 418	..	Chercheurs
Technicians	9 784	9 386	9 423	10 101	9 678	9 686	..	Techniciens
Other occupations	38 757	38 301	38 118	38 091	38 874	39 458	..	Autres professions
Total	284 243 [l]	290 549 [l]	217 558 [a]	222 285	225 179	227 562	..	*Total*
Private non-profit								**Institutions sans but lucratif**
RSE	16 262	16 113	16 746	16 905	16 113	15 747	..	Chercheurs
Technicians	3 382	3 308	3 500	3 643	3 445	3 398	..	Techniciens
Other occupations	8 578	8 414	8 312	8 460	8 910	8 856	..	Autres professions
Total	28 222	27 835	28 558	29 008	28 468	28 001	..	*Total*
TOTAL								**TOTAL**
RSE	658 866 [l]	673 421 [l]	617 365 [a]	625 442	652 845	658 910	..	Chercheurs
Technicians	90 072	82 851	83 906 [a]	83 539	86 822	84 527	..	Techniciens
Other occupations	196 885	191 816	190 512 [a]	185 022	185 902	175 695	..	Autres professions
Total	945 823 [l]	948 088 [l]	891 783 [a]	894 003	925 569	919 132	..	*Total*

Table 6 - Tableau 6

Total R&D personnel - Personnel total de R-D

by sector of employment and formal qualification - par secteur d'emploi et par niveau de qualification

In full-time equivalent — *En équivalence plein-temps*

	1994	1995	1996	1997	1998	1999	2000	
Business enterprise								**Entreprises**
University PhD level degrees	..	..	..	..	..	..	..	Diplômes univ. de niv. doctorat
Other university degrees	..	..	..	..	..	..	..	Autres diplômes universitaires
Sub-total university degrees	..	..	..	..	..	..	..	Sous-total Universitaires
Other post-secondary diplomas	..	..	..	..	..	..	..	Autres diplômes post-secondaires
Secondary diplomas	..	..	..	..	..	..	..	Diplômes de fin d'études secondaires
Other qualifications	..	..	..	..	..	..	..	Autres qualifications
Total	577 725 [l]	573 714 [l]	589 491 [a]	586 156	613 160	604 544	..	*Total*
Government								**État**
University PhD level degrees	..	..	..	..	..	..	..	Diplômes univ. de niv. doctorat
Other university degrees	..	..	..	..	..	..	..	Autres diplômes universitaires
Sub-total university degrees	..	..	..	..	..	..	..	Sous-total Universitaires
Other post-secondary diplomas	..	..	..	..	..	..	..	Autres diplômes post-secondaires
Secondary diplomas	..	..	..	..	..	..	..	Diplômes de fin d'études secondaires
Other qualifications	..	..	..	..	..	..	..	Autres qualifications
Total	55 633	55 990	56 176	56 554	58 762	59 025	..	*Total*
Higher education								**Enseignement supérieur**
University PhD level degrees	..	..	..	..	..	..	..	Diplômes univ. de niv. doctorat
Other university degrees	..	..	..	..	..	..	..	Autres diplômes universitaires
Sub-total university degrees	..	..	..	..	..	..	..	Sous-total Universitaires
Other post-secondary diplomas	..	..	..	..	..	..	..	Autres diplômes post-secondaires
Secondary diplomas	..	..	..	..	..	..	..	Diplômes de fin d'études secondaires
Other qualifications	..	..	..	..	..	..	..	Autres qualifications
Total	284 243 [l]	290 549 [l]	217 558 [a]	222 285	225 179	227 562	..	*Total*
Private non-profit								**Institutions sans but lucratif**
University PhD level degrees	..	..	..	..	..	..	..	Diplômes univ. de niv. doctorat
Other university degrees	..	..	..	..	..	..	..	Autres diplômes universitaires
Sub-total university degrees	..	..	..	..	..	..	..	Sous-total Universitaires
Other post-secondary diplomas	..	..	..	..	..	..	..	Autres diplômes post-secondaires
Secondary diplomas	..	..	..	..	..	..	..	Diplômes de fin d'études secondaires
Other qualifications	..	..	..	..	..	..	..	Autres qualifications
Total	28 222	27 835	28 558	29 008	28 468	28 001	..	*Total*
TOTAL								**TOTAL**
University PhD level degrees	..	..	..	..	..	..	..	Diplômes univ. de niv. doctorat
Other university degrees	..	..	..	..	..	..	..	Autres diplômes universitaires
Sub-total university degrees	..	..	..	..	..	..	..	Sous-total Universitaires
Other post-secondary diplomas	..	..	..	..	..	..	..	Autres diplômes post-secondaires
Secondary diplomas	..	..	..	..	..	..	..	Diplômes de fin d'études secondaires
Other qualifications	..	..	..	..	..	..	..	Autres qualifications
Total	945 823 [l]	948 088 [l]	891 783 [a]	894 003	925 569	919 132	..	*Total*

Source: OECD, R&D database (DSTI/EAS Division), January 2002.

Source : OCDE, base de données R-D (DSTI/Division AES), janvier 2002.

Table 7 - Tableau 7

Gross domestic expenditure on R&D (GERD) - Dépenses intérieures brutes de R-D (DIRD)

by sector of performance and main scientific field - par secteur d'exécution et par grand domaine scientifique

Million JPY *Millions JPY*

	1994	1995	1996	1997	1998	1999	2000	
Business enterprise								**Entreprises**
Natural sciences and engineering	8 980 253 [l]	9 395 896 [l]	10 058 409 [a]	10 658 357	10 800 063	10 630 161	..	Sciences exactes, nat. et de l'ingénieur
Social sciences and humanities	..	..	..	..	..	..	..	Sciences sociales et humaines
All fields of science	8 980 253 [l]	9 395 896 [l]	10 058 409 [a]	10 658 357	10 800 063	10 630 161	..	*Tous domaines scientifiques*
Government								**État**
Natural sciences and engineering	1 187 315	1 349 346	1 288 776	1 274 585	1 361 708	1 444 997	..	Sciences exactes, nat. et de l'ingénieur
Social sciences and humanities	39 112	40 786	39 759	32 391	41 206	36 734	..	Sciences sociales et humaines
All fields of science	1 226 427	1 390 132	1 328 535	1 306 976	1 402 914	1 481 731	..	*Tous domaines scientifiques*
Higher education								**Enseignement supérieur**
Natural sciences and engineering	1 685 526 [l]	1 874 668 [l]	1 384 344 [a]	1 394 427	1 487 748	1 463 547	..	Sciences exactes, nat. et de l'ingénieur
Social sciences and humanities	1 067 024 [l]	1 107 518 [l]	704 517 [a]	717 303	764 410	767 612	..	Sciences sociales et humaines
All fields of science	2 752 551 [l]	2 982 186 [l]	2 088 861 [a]	2 111 730	2 252 158	2 231 159	..	*Tous domaines scientifiques*
Private non-profit								**Institutions sans but lucratif**
Natural sciences and engineering	566 115	571 273	610 379	652 720	650 489	645 440	..	Sciences exactes, nat. et de l'ingénieur
Social sciences and humanities	70 684	68 748	68 872	64 247	63 579	44 169	..	Sciences sociales et humaines
All fields of science	636 800	640 021	679 251	716 967	714 068	689 609	..	*Tous domaines scientifiques*
TOTAL								**TOTAL**
Natural sciences and engineering	12 419 209 [l]	13 191 183 [l]	13 341 908 [a]	13 980 089	14 300 008	14 184 145	..	Sciences exactes, nat. et de l'ingénieur
Social sciences and humanities	1 176 820 [l]	1 217 052 [l]	813 148 [a]	813 941	869 195	848 515	..	Sciences sociales et humaines
All fields of science	13 596 030 [l]	14 408 236 [l]	14 155 058 [a]	14 794 030	15 169 203	15 032 660	..	*Tous domaines scientifiques*

Table 8 - Tableau 8

Total R&D personnel - Personnel total de R-D

by sector of employment and main scientific field - par secteur d'emploi et par grand domaine scientifique

In full-time equivalent *En équivalence plein-temps*

	1994	1995	1996	1997	1998	1999	2000	
Business enterprise								**Entreprises**
Natural sciences and engineering	577 725 [l]	573 714 [l]	589 491 [a]	586 156	613 160	604 544	..	Sciences exactes, nat. et de l'ingénieur
Social sciences and humanities	..	..	..	..	..	..	..	Sciences sociales et humaines
All fields of science	577 725 [l]	573 714 [l]	589 491 [a]	586 156	613 160	604 544	..	*Tous domaines scientifiques*
Government								**État**
Natural sciences and engineering	52 267	52 521	52 665	53 725	55 801	56 286	..	Sciences exactes, nat. et de l'ingénieur
Social sciences and humanities	3 366	3 469	3 511	2 829	2 961	2 739	..	Sciences sociales et humaines
All fields of science	55 633	55 990	56 176	56 554	58 762	59 025	..	*Tous domaines scientifiques*
Higher education								**Enseignement supérieur**
Natural sciences and engineering	189 881 [l]	194 087 [l]	153 558 [a]	156 853	156 078	156 555	..	Sciences exactes, nat. et de l'ingénieur
Social sciences and humanities	94 362 [l]	96 462 [l]	64 000 [a]	65 432	69 101	71 006	..	Sciences sociales et humaines
All fields of science	284 243 [l]	290 549 [l]	217 558 [a]	222 285	225 179	227 562	..	*Tous domaines scientifiques*
Private non-profit								**Institutions sans but lucratif**
Natural sciences and engineering	24 230	23 913	24 801	25 246	24 904	25 098	..	Sciences exactes, nat. et de l'ingénieur
Social sciences and humanities	3 992	3 922	3 757	3 762	3 564	2 903	..	Sciences sociales et humaines
All fields of science	28 222	27 835	28 558	29 008	28 468	28 001	..	*Tous domaines scientifiques*
TOTAL								**TOTAL**
Natural sciences and engineering	844 103 [l]	844 234 [l]	820 515 [a]	821 980	849 943	842 483	..	Sciences exactes, nat. et de l'ingénieur
Social sciences and humanities	101 720 [l]	103 853 [l]	71 268 [a]	72 023	75 626	76 648	..	Sciences sociales et humaines
All fields of science	945 823 [l]	948 088 [l]	891 783 [a]	894 003	925 569	919 132	..	*Tous domaines scientifiques*

Source: OECD, R&D database (DSTI/EAS Division), January 2002.

Source : OCDE, base de données R-D (DSTI/Division AES), janvier 2002.

Total business enterprise intramural expenditure on R&D - Dépenses intra-muros totales de R-D des entreprises
(BERD) by industry - (DIRDE) par industrie

9.1 Million JPY - Millions JPY

	1994	1995	1996	1997	1998	1999	2000	
1 AGRICULTURE, HUNTING & FORESTRY	7 794	8 741	8 981	10 839	11 936	10 975	..	**1 AGRICULTURE, CHASSE & SYLVICULTURE**
2 MINING	23 903	22 854	21 677	25 237	27 863	26 187	..	**2 ACTIVITES EXTRACTIVES**
3 MANUFACTURING	8 365 478	8 774 360	9 263 151	9 816 437	9 807 147	9 521 573	..	**3 ACTIVITES DE FABRICATION**
4 Food, beverages & tobacco	233 746	249 441	250 806	268 353	262 948	280 208	..	**4 Produits alimentaires, boissons & tabac**
5 Food products & beverages	..	..	..	..	..	..	..	5 Produits alimentaires & boissons
6 Tobacco products	..	..	..	..	..	..	..	6 Produits à base de tabac
7 Textiles, wear app., fur & leather	76 744 [m]	79 891 [m]	76 323 [m]	81 652 [m]	76 201 [m]	77 471 [m]	..	**7 Textiles, habill., fourrures & cuirs**
8 Textiles	..	..	..	..	..	..	..	8 Textiles
9 Wearing apparel & fur	..	..	..	..	..	..	..	9 Articles d'habillement & fourrures
10 Leather products & footwear	.. [n]	.. [n]	.. [n]	.. [n]	.. [n]	.. [n]	..	10 Articles en cuir & chaussures
11 Wood, paper, printing, publishing	105 328 [o]	115 825 [o]	115 863 [o]	121 016 [o]	127 331 [o]	123 359 [o]	..	**11 Bois, papier, imprimerie, édition**
12 Wood & cork (not furniture)	32 021 [o]	34 465 [o]	32 659 [o]	34 054 [o]	34 576 [o]	27 327 [o]	..	12 Bois & liège (sauf meubles)
13 Pulp, paper & paper products	73 307 [o]	81 360 [o]	83 204 [o]	86 962 [o]	92 755 [o]	96 032 [o]	..	13 Carton, papier & articles
14 Publ., print. & repro. of rec. media	.. [n]	.. [n]	.. [n]	.. [n]	.. [n]	.. [n]	..	14 Edition, impress. & repro. enregis.
15 Coke, petroleum, nuclear fuel, chemicals, rubber plastics	1 877 051	1 869 412	1 918 946	1 933 666	1 941 614	1 888 098	..	**15 Coke, pétrole, combust. nucl., prod. chimi., caout. & mat. plast.**
16 Coke, ref. petrol. prod. & nuclear fuel	79 256	67 750	62 509	65 351	50 331	40 150	..	16 Coke, raffin. pétr. & combust. nucl.
17 Chemicals & chemical products	1 548 794	1 554 884	1 593 250	1 609 251	1 630 928	1 588 074	..	17 Produits chimiques
18 *Chemical prod.(less pharmaceuticals)*	915 992	912 694	926 105	965 960	949 810	898 625	..	18 *Prod. chimi. (moins prod. pharmaceutiques)*
19 *Pharmaceuticals*	632 802	642 190	667 145	643 291	681 118	689 449	..	19 *Produits pharmaceutiques*
20 Rubber & plastic products	249 001	246 778	263 187	259 064	260 355	259 874	..	20 Caoutchouc et matières plastiques
21 Non-metallic mineral products	182 690	199 970	213 713	215 391	217 152	183 617	..	**21 Produits minéraux non métalliques**
22 Basic metals	377 594	367 135	354 794	379 581	329 230	306 616	..	**22 Produits métallurgiques de base**
23 Basic metals, ferrous	237 707	213 541	201 476	213 631	187 596	168 267	..	23 Produits métallur. de base, ferreux
24 Basic metals, non-ferrous	139 887	153 594	153 318	165 950	141 634	138 349	..	24 Prod. métallur. de base, non ferreux
25 Fabricated metal products	111 658	118 093	147 749	134 716	125 628	109 685	..	**25 Ouvrages en métaux (sauf mach.& mat.)**
26 Machinery equipment, instruments and transport equipment	5 315 004	5 695 094	6 104 882	6 589 724	6 630 184	6 448 364	..	**26 Machines & matériel, instruments & mat. de transport**
27 Machinery n.e.c.	696 736 [o]	705 222 [o]	733 707 [o]	790 057 [o]	811 653 [o]	811 364 [o]	..	27 Machines, n.c.a.
28 Office, account. & computing machinery	.. [n]	.. [n]	.. [n]	.. [n]	.. [n]	.. [n]	..	28 Mach. de bur., compta. & trait. inform.
29 Electrical machinery	1 007 125	1 032 887	1 093 333	1 141 846	1 124 146	1 009 639	..	29 Machines électriques
30 Electro. equip. (radio, TV & commun.)	2 057 642 [o]	2 240 714 [o]	2 400 304 [o]	2 577 601 [o]	2 588 619 [o]	2 606 237 [o]	..	30 Equip. électro. (radio, tv & communic.)
31 *Electro. comp. (inc. semi-conductors)*	..	..	..	..	..	..	..	31 *Compos. électro. (semi-cond. incl.)*
32 *TV, radio & communications equip.*	..	..	..	..	..	..	..	32 *Appareils radio, tv et de communi.*
33 Instruments	333 506	355 401	366 297	426 182	473 722	491 530	..	33 Instruments
34 Motor vehicles	1 021 291 [o]	1 169 269 [o]	1 315 445 [o]	1 445 220 [o]	1 427 331 [o]	1 335 053 [o]	..	34 Véhicules automobiles
35 Other transport equipment	198 704 [n]	191 601 [n]	195 796 [n]	208 818 [n]	204 713 [n]	194 541 [n]	..	35 Autres matériels de transport
36 *Ships*	158 674	150 917	155 163	167 136	162 066	156 374	..	36 *Navires*
37 *Aerospace*	1 978	2 092	2 188	2 281	2 204	1 379	..	37 *Aérospatiale*
38 *Other transport equipment n.e.c.*	38 052	38 592	38 445	39 401	40 443	36 788	..	38 *Autres matériels de transport n.c.a.*
39 Furniture, other manufacturing n.e.c.	85 664 [m,o]	79 497 [m,o]	80 075 [m,o]	92 339 [m,o]	96 860 [m,o]	104 156 [m,o]	..	**39 Meubles, autres activ. de fabricat. n.c.a.**
40 Furniture	.. [n]	.. [n]	.. [n]	.. [n]	.. [n]	.. [n]	..	40 Meubles
41 Other manufacturing n.e.c.	85 664 [o]	79 497 [o]	80 075 [o]	92 339 [o]	96 860 [o]	104 156 [o]	..	41 Autres activités de fabrication n.c.a.
42 Recycling	..	..	..	..	..	..	..	**42 Récupération**
43 ELECTRICITY, GAS & WATER SUPPLY	134 948	103 233	112 983	102 759	102 350	107 955	..	**43 DISTRIBUTION D'ELECTRICITE, GAZ & EAU**
44 CONSTRUCTION	221 999	204 363	224 514	225 162	176 690	199 453	..	**44 CONSTRUCTION**
45 SERVICES SECTOR	226 131	282 344	427 103 [a]	477 923	674 077	764 019	..	**45 SECTEUR DES SERVICES**
46 Wholesale, retail trade & repair of motor vehicles, etc.	..	..	..	..	..	..	..	**46 Commerce (gros & détail), réparation de véhicules, etc.**
47 Hotels & restaurants	..	..	..	..	..	..	..	**47 Hôtels et restaurants**
48 Transport & storage	12 853	16 079	11 384	14 613	14 514	16 986	..	**48 Transports, entreposage**
49 Communications	213 278	266 265	238 561	289 787	330 799	472 340	..	**49 Communications**
50 Post	..	..	..	..	..	..	..	50 Postes
51 Telecommunications	..	..	..	..	..	..	..	51 Télécommunications
52 Financial intermediation (inc. insur.)	..	..	..	..	..	..	..	**52 Intermédiation financ. (assurances inc.)**
53 Real estate, renting & busin. activity	..	..	177 158	173 523	328 764	274 693	..	**53 Immobil., locat. & serv. aux entreprises**
54 Computer & related activities	..	..	177 158	173 523	328 764	274 693	..	54 Informatique et activités connexes
55 *Software consultancy*	..	..	..	..	..	..	..	55 *Production de logiciels*
56 *Other computer services n.e.c.*	..	..	..	..	..	..	..	56 *Autres activités informatiques*
57 Research & development	..	..	..	..	..	..	..	57 Recherche & développement expérimental
58 Other business activities n.e.c.	..	..	..	..	..	..	..	58 Autres activ. & serv. aux entreprises
59 Comm., soc. & pers. serv. activ., etc.	..	..	..	..	..	..	..	**59 Adm. pub., serv. soc. & collec., etc**
TOTAL BUSINESS ENTERPRISE SECTOR	**8 980 253 [l]**	**9 395 896 [l]**	**10 058 409 [a]**	**10 658 357**	**10 800 063**	**10 630 161**	**..**	**TOTAL SECTEUR DES ENTREPRISES**

Source: OECD, R&D database (DSTI/EAS Division), January 2002.

Source : OCDE, base de données R-D (DSTI/Division AES), janvier 2002.

Total business enterprise intramural expenditure on R&D - Dépenses intra-muros totales de R-D des entreprises
(BERD) by industry - (DIRDE) par industrie

	1994	1995	1996	1997	1998	1999	2000	
	9.2 Million constant USD(1995 prices and PPPs) - Millions de USD constants (prix et PPA de 1995)							
1 **AGRICULTURE, HUNTING & FORESTRY**	45.7	51.4	53.3	64.1	70.6	65.8	..	1 **AGRICULTURE, CHASSE & SYLVICULTURE**
2 **MINING**	140.1	134.5	128.6	149.2	164.8	157.1	..	2 **ACTIVITES EXTRACTIVES**
3 **MANUFACTURING**	49 046.5	51 631.5	54 956.8	58 031.5	58 008.8	57 111.5	..	3 **ACTIVITES DE FABRICATION**
4 **Food, beverages & tobacco**	1 370.4	1 467.8	1 488.0	1 586.4	1 555.3	1 680.7	..	4 **Produits alimentaires, boissons & tabac**
5 Food products & beverages	..	..	..	..	..	..	..	5 Produits alimentaires & boissons
6 Tobacco products	..	..	..	..	..	..	..	6 Produits à base de tabac
7 **Textiles, wear app., fur & leather**	449.9 [m]	470.1 [m]	452.8 [m]	482.7 [m]	450.7 [m]	464.7 [m]	..	7 **Textiles, habill., fourrures & cuirs**
8 Textiles	..	..	..	..	..	..	..	8 Textiles
9 Wearing apparel & fur	..	..	..	..	..	..	..	9 Articles d'habillement & fourrures
10 Leather products & footwear	.. [n]	.. [n]	.. [n]	.. [n]	.. [n]	.. [n]	..	10 Articles en cuir & chaussures
11 **Wood, paper, printing, publishing**	617.5 [o]	681.6 [o]	687.4 [o]	715.4 [o]	753.2 [o]	739.9 [o]	..	11 **Bois, papier, imprimerie, édition**
12 Wood & cork (not furniture)	187.7 [o]	202.8 [o]	193.8 [o]	201.3 [o]	204.5 [o]	163.9 [o]	..	12 Bois & liège (sauf meubles)
13 Pulp, paper & paper products	429.8 [o]	478.8 [o]	493.6 [o]	514.1 [o]	548.6 [o]	576.0 [o]	..	13 Carton, papier & articles
14 Publ., print. & repro. of rec. media	.. [n]	.. [n]	.. [n]	.. [n]	.. [n]	.. [n]	..	14 Edition, impress. & repro. enregis.
15 **Coke, petroleum, nuclear fuel, chemicals, rubber plastics**	11 005.1	11 000.3	11 384.8	11 431.2	11 484.6	11 325.0	..	15 **Coke, pétrole, combust. nucl., prod. chimi., caout. & mat. plast.**
16 Coke, ref. petrol. prod. & nuclear fuel	464.7	398.7	370.9	386.3	297.7	240.8	..	16 Coke, raffin. pétr. & combust. nucl.
17 Chemicals & chemical products	9 080.5	9 149.5	9 452.5	9 513.3	9 646.9	9 525.5	..	17 Produits chimiques
18 *Chemical prod.(less pharmaceuticals)*	5 370.4	5 370.6	5 494.4	5 710.4	5 618.1	5 390.1	..	18 *Prod. chimi. (moins prod. pharmaceutiques)*
19 *Pharmaceuticals*	3 710.1	3 778.9	3 958.1	3 802.9	4 028.8	4 135.4	..	19 *Produits pharmaceutiques*
20 Rubber & plastic products	1 459.9	1 452.1	1 561.4	1 531.5	1 540.0	1 558.8	..	20 Caoutchouc et matières plastiques
21 **Non-metallic mineral products**	1 071.1	1 176.7	1 267.9	1 273.3	1 284.4	1 101.4	..	21 **Produits minéraux non métalliques**
22 **Basic metals**	2 213.8	2 160.4	2 104.9	2 244.0	1 947.4	1 839.1	..	22 **Produits métallurgiques de base**
23 Basic metals, ferrous	1 393.7	1 256.6	1 195.3	1 262.9	1 109.6	1 009.3	..	23 Produits métallur. de base, ferreux
24 Basic metals, non-ferrous	820.2	903.8	909.6	981.0	837.8	829.8	..	24 Prod. métallur. de base, non ferreux
25 **Fabricated metal products**	654.6	694.9	876.6	796.4	743.1	657.9	..	25 **Ouvrages en métaux (sauf mach.& mat.)**
26 **Machinery equipment, instruments and transport equipment**	31 161.7	33 512.0	36 219.3	38 956.2	39 217.2	38 678.1	..	26 **Machines & matériel, instruments & mat. de transport**
27 Machinery n.e.c.	4 084.9 [o]	4 149.8 [o]	4 353.0 [o]	4 670.5 [o]	4 800.9 [o]	4 866.7 [o]	..	27 Machines, n.c.a.
28 Office, account. & computing machinery	.. [n]	.. [n]	.. [n]	.. [n]	.. [n]	.. [n]	..	28 Mach. de bur., compta. & trait. inform.
29 Electrical machinery	5 904.7	6 077.9	6 486.6	6 750.2	6 649.3	6 055.9	..	29 Machines électriques
30 Electro. equip. (radio, TV & commun.)	12 063.9 [o]	13 185.2 [o]	14 240.6 [o]	15 237.9 [o]	15 311.6 [o]	15 632.5 [o]	..	30 Equip. électro. (radio, tv & communic.)
31 *Electro. comp. (inc. semi-conductors)*	..	..	..	..	..	..	..	31 *Compos. électro. (semi-cond. incl.)*
32 *TV, radio & communications equip.*	..	..	..	..	..	..	..	32 *Appareils radio, tv et de communi.*
33 Instruments	1 955.3	2 091.3	2 173.2	2 519.4	2 802.0	2 948.3	..	33 Instruments
34 Motor vehicles	5 987.8 [o]	6 880.4 [o]	7 804.3 [o]	8 543.7 [o]	8 442.6 [o]	8 007.8 [o]	..	34 Véhicules automobiles
35 Other transport equipment	1 165.0 [n]	1 127.5 [n]	1 161.6 [n]	1 234.5 [n]	1 210.9 [n]	1 166.9 [n]	..	35 Autres matériels de transport
36 *Ships*	930.3	888.1	920.6	988.1	958.6	938.0	..	36 *Navires*
37 *Aerospace*	11.6	12.3	13.0	13.5	13.0	8.3	..	37 *Aérospatiale*
38 *Other transport equipment n.e.c.*	223.1	227.1	228.1	232.9	239.2	220.7	..	38 *Autres matériels de transport n.c.a.*
39 **Furniture, other manufacturing n.e.c.**	502.2 [m,o]	467.8 [m,o]	475.1 [m,o]	545.9 [m,o]	572.9 [m,o]	624.7 [m,o]	..	39 **Meubles, autres activ. de fabricat. n.c.a.**
40 Furniture	.. [n]	.. [n]	.. [n]	.. [n]	.. [n]	.. [n]	..	40 Meubles
41 Other manufacturing n.e.c.	502.2 [o]	467.8 [o]	475.1 [o]	545.9 [o]	572.9 [o]	624.7 [o]	..	41 Autres activités de fabrication n.c.a.
42 **Recycling**	..	..	..	..	..	..	..	42 **Récupération**
43 **ELECTRICITY, GAS & WATER SUPPLY**	791.2	607.5	670.3	607.5	605.4	647.5	..	43 **DISTRIBUTION D'ELECTRICITE, GAZ & EAU**
44 **CONSTRUCTION**	1 301.6	1 202.5	1 332.0	1 331.1	1 045.1	1 196.3	..	44 **CONSTRUCTION**
45 **SERVICES SECTOR**	1 325.8	1 661.4	2 533.9 [a]	2 825.3	3 987.1	4 582.7	..	45 **SECTEUR DES SERVICES**
46 **Wholesale, retail trade & repair of motor vehicles, etc.**	..	..	..	..	..	..	..	46 **Commerce (gros & détail), réparation de véhicules, etc.**
47 **Hotels & restaurants**	..	..	..	..	..	..	..	47 **Hôtels et restaurants**
48 **Transport & storage**	75.4	94.6	67.5	86.4	85.9	101.9	..	48 **Transports, entreposage**
49 **Communications**	1 250.4	1 566.8	1 415.3	1 713.1	1 956.7	2 833.2	..	49 **Communications**
50 Post	..	..	..	..	..	..	..	50 Postes
51 Telecommunications	..	..	..	..	..	..	..	51 Télécommunications
52 **Financial intermediation (inc. insur.)**	..	..	..	..	..	..	..	52 **Intermédiation financ. (assurances inc.)**
53 **Real estate, renting & busin. activity**	..	..	1 051.0	1 025.8	1 944.6	1 647.6	..	53 **Immobil., locat. & serv. aux entreprises**
54 Computer & related activities	..	..	1 051.0	1 025.8	1 944.6	1 647.6	..	54 Informatique et activités connexes
55 *Software consultancy*	..	..	..	..	..	..	..	55 *Production de logiciels*
56 *Other computer services n.e.c.*	..	..	..	..	..	..	..	56 *Autres activités informatiques*
57 Research & development	..	..	..	..	..	..	..	57 Recherche & développement expérimental
58 Other business activities n.e.c.	..	..	..	..	..	..	..	58 Autres activ. & serv. aux entreprises
59 **Comm., soc. & pers. serv. activ., etc.**	..	..	..	..	..	..	..	59 **Adm. pub., serv. soc. & collec., etc**
TOTAL BUSINESS ENTERPRISE SECTOR	**52 650.9** [l]	**55 288.9** [l]	**59 674.9** [a]	**63 008.6**	**63 881.9**	**63 761.0**	..	**TOTAL SECTEUR DES ENTREPRISES**

Source: OECD, R&D database (DSTI/EAS Division), January 2002.

Source : OCDE, base de données R-D (DSTI/Division AES), janvier 2002.

Business enterprise intramural expenditure on R&D (BERD) - Dépenses intra-muros de R-D des entreprises (DIRDE)

by industry and source of funds - par industrie et par source de financement

Million JPY — *Millions JPY*

	1994	1995	1996	1997	1998	1999	2000	
	10.1 Business enterprise - Entreprises							
1 AGRICULTURE, HUNTING & FORESTRY	7 721	8 711	8 981	10 806	11 853	10 916	..	**1 AGRICULTURE, CHASSE & SYLVICULTURE**
2 MINING	15 897	17 785	16 219	18 593	20 332	19 860	..	**2 ACTIVITES EXTRACTIVES**
3 MANUFACTURING	8 250 363	8 658 117	9 142 721	9 650 262	9 647 204	9 331 351	..	**3 ACTIVITES DE FABRICATION**
4 Food, beverages & tobacco	233 236	248 813	250 188	266 906	261 625	273 406	..	**4 Produits alimentaires, boissons & tabac**
5 Food products & beverages	..	..	..	..	..	..	..	5 Produits alimentaires & boissons
6 Tobacco products	..	..	..	..	..	..	..	6 Produits à base de tabac
7 Textiles, wear app., fur & leather	76 195 m	79 772 m	76 042 m	81 263 m	75 916 m	76 069 m	..	**7 Textiles, habill., fourrures & cuirs**
8 Textiles	..	..	..	..	..	..	..	8 Textiles
9 Wearing apparel & fur	..	..	..	..	..	..	..	9 Articles d'habillement & fourrures
10 Leather products & footwear	.. n	.. n	.. n	.. n	.. n	.. n	..	10 Articles en cuir & chaussures
11 Wood, paper, printing, publishing	105 121 o	115 623 o	115 565 o	120 803 o	126 956 o	123 075 o	..	**11 Bois, papier, imprimerie, édition**
12 Wood & cork (not furniture)	31 856 o	34 303 o	32 440 o	33 946 o	34 462 o	27 272 o	..	12 Bois & liège (sauf meubles)
13 Pulp, paper & paper products	73 265 o	81 320 o	83 125 o	86 857 o	92 494 o	95 803 o	..	13 Carton, papier & articles
14 Publ., print. & repro. of rec. media	.. n	.. n	.. n	.. n	.. n	.. n	..	14 Edition, impress. & repro. enregis.
15 Coke, petroleum, nuclear fuel, chemicals, rubber plastics	1 857 567	1 850 039	1 900 802	1 911 528	1 921 888	1 868 597	..	**15 Coke, pétrole, combust. nucl., prod. chimi., caout. & mat. plast.**
16 Coke, ref. petrol. prod. & nuclear fuel	72 159	60 697	55 763	58 834	43 922	34 203	..	16 Coke, raffin. pétr. & combust. nucl.
17 Chemicals & chemical products	1 536 923	1 543 145	1 582 898	1 594 504	1 618 554	1 575 016	..	17 Produits chimiques
18 *Chemical prod.(less pharmaceuticals)*	910 528	906 888	921 827	960 000	944 861	893 192	..	18 *Prod. chimi. (moins prod. pharmaceutiques)*
19 *Pharmaceuticals*	626 395	636 257	661 071	634 504	673 693	681 824	..	19 *Produits pharmaceutiques*
20 Rubber & plastic products	248 485	246 197	262 141	258 190	259 412	259 378	..	20 Caoutchouc et matières plastiques
21 Non-metallic mineral products	179 289	197 723	211 629	213 529	215 315	180 837	..	**21 Produits minéraux non métalliques**
22 Basic metals	369 152	358 997	343 735	368 404	317 377	293 851	..	**22 Produits métallurgiques de base**
23 Basic metals, ferrous	231 506	207 638	193 429	205 834	179 318	159 364	..	23 Produits métallur. de base, ferreux
24 Basic metals, non-ferrous	137 646	151 359	150 306	162 570	138 059	134 487	..	24 Prod. métallur. de base, non ferreux
25 Fabricated metal products	111 283	117 589	146 760	134 246	124 411	109 095	..	**25 Ouvrages en métaux (sauf mach.& mat.)**
26 Machinery equipment, instruments and transport equipment	5 233 069	5 610 200	6 017 993	6 461 340	6 507 016	6 302 386	..	**26 Machines & matériel, instruments & mat. de transport**
27 Machinery n.e.c.	687 784 o	694 980 o	726 615 o	780 857 o	803 151 o	795 294 o	..	27 Machines, n.c.a.
28 Office, account. & computing machinery	.. n	.. n	.. n	.. n	.. n	.. n	..	28 Mach. de bur., compta. & trait. inform.
29 Electrical machinery	996 277	1 023 230	1 081 250	1 128 990	1 114 163	996 792	..	29 Machines électriques
30 Electro. equip. (radio, TV & commun.)	2 049 023 o	2 230 106 o	2 388 636 o	2 537 189 o	2 540 693 o	2 552 899 o	..	30 Equip. électro. (radio, tv & communic.)
31 *Electro. comp. (inc. semi-conductors)*	..	..	..	..	..	..	..	31 *Compos. électro. (semi-cond. incl.)*
32 *TV, radio & communications equip.*	..	..	..	..	..	..	..	32 *Appareils radio, tv et de communi.*
33 Instruments	332 680	353 648	364 621	424 877	471 930	489 057	..	33 Instruments
34 Motor vehicles	1 019 765	1 166 608	1 310 932	1 440 095	1 424 459	1 332 619	..	34 Véhicules automobiles
35 Other transport equipment	147 540	141 628	145 939	149 332	152 620	135 725	..	35 Autres matériels de transport
36 *Ships*	108 884	102 661	106 400	108 154	111 013	98 225	..	36 *Navires*
37 *Aerospace*	1 721	2 092	2 148	2 255	2 204	1 366	..	37 *Aérospatiale*
38 *Other transport equipment n.e.c.*	36 935	36 875	37 391	38 923	39 403	36 134	..	38 *Autres matériels de transport n.c.a.*
39 Furniture, other manufacturing n.e.c.	85 451 m,o	79 360 m,o	80 009 m,o	92 241 m,o	96 697 m	104 037 m	..	**39 Meubles, autres activ. de fabricat. n.c.a.**
40 Furniture	.. n	.. n	.. n	.. n	.. n	.. n	..	40 Meubles
41 Other manufacturing n.e.c.	85 451 o	79 360 o	80 009 o	92 241 o	96 697 o	104 037 o	..	41 Autres activités de fabrication n.c.a.
42 Recycling	..	..	..	..	..	..	..	**42 Récupération**
43 ELECTRICITY, GAS & WATER SUPPLY	131 730	98 786	107 600	95 669	98 468	104 916	..	**43 DISTRIBUTION D'ELECTRICITE, GAZ & EAU**
44 CONSTRUCTION	219 094	200 551	219 403	218 252	172 912	192 993	..	**44 CONSTRUCTION**
45 SERVICES SECTOR	225 779	239 047	421 102 a	470 498	562 723	709 666	..	**45 SECTEUR DES SERVICES**
46 Wholesale, retail trade & repair of motor vehicles, etc.	..	..	..	..	..	..	..	**46 Commerce (gros & détail), réparation de véhicules, etc.**
47 Hotels & restaurants	..	..	..	..	..	..	..	**47 Hôtels et restaurants**
48 Transport & storage	12 586	15 820	11 124	14 386	14 292	16 926	..	**48 Transports, entreposage**
49 Communications	213 193	223 227	234 862	285 313	222 222	418 979	..	**49 Communications**
50 Post	..	..	..	..	..	..	..	50 Postes
51 Telecommunications	..	..	..	..	..	..	..	51 Télécommunications
52 Financial intermediation (inc. insur.)	..	..	..	..	..	..	..	**52 Intermédiation financ. (assurances inc.)**
53 Real estate, renting & busin. activity	..	..	175 116	170 799	326 209	273 761	..	**53 Immobil., locat. & serv. aux entreprises**
54 Computer & related activities	..	..	175 116	170 799	326 209	273 761	..	54 Informatique et activités connexes
55 *Software consultancy*	..	..	..	..	..	..	..	55 *Production de logiciels*
56 *Other computer services n.e.c.*	..	..	..	..	..	..	..	56 *Autres activités informatiques*
57 Research & development	..	..	..	..	..	..	..	57 Recherche & développement expérimental
58 Other business activities n.e.c.	..	..	..	..	..	..	..	58 Autres activ. & serv. aux entreprises
59 Comm., soc. & pers. serv. activ., etc.	..	..	..	..	..	..	..	**59 Adm. pub., serv. soc. & collec., etc**
TOTAL BUSINESS ENTERPRISE SECTOR	**8 850 586 l**	**9 222 996 l**	**9 916 024 a**	**10 464 078**	**10 513 491**	**10 369 702**	**..**	**TOTAL SECTEUR DES ENTREPRISES**

Source: OECD, R&D database (DSTI/EAS Division), January 2002. — Source : OCDE, base de données R-D (DSTI/Division AES), janvier 2002.

Business enterprise intramural expenditure on R&D (BERD) - Dépenses intra-muros de R-D des entreprises (DIRDE)

by industry and source of funds - par industrie et par source de financement

Million JPY / *Millions JPY*

	1994	1995	1996	1997	1998	1999	2000	
				10.2 Government - État				
1 AGRICULTURE, HUNTING & FORESTRY	62	..	..	28	83	60	..	**1 AGRICULTURE, CHASSE & SYLVICULTURE**
2 MINING	7 647	5 020	5 450	6 640	7 527	6 327	..	**2 ACTIVITES EXTRACTIVES**
3 MANUFACTURING	93 477	94 102	95 231	112 512	97 591	119 606	..	**3 ACTIVITES DE FABRICATION**
4 Food, beverages & tobacco	480	519	521	1 259	1 114	1 980	..	**4 Produits alimentaires, boissons & tabac**
5 Food products & beverages	..	..	..	..	..	..	..	5 Produits alimentaires & boissons
6 Tobacco products	..	..	..	..	..	..	..	6 Produits à base de tabac
7 Textiles, wear app., fur & leather	549 [m]	119 [m]	281 [m]	377 [m]	284 [m]	1 402 [m]	..	**7 Textiles, habill., fourrures & cuirs**
8 Textiles	..	..	..	..	..	..	..	8 Textiles
9 Wearing apparel & fur	..	..	..	..	..	..	..	9 Articles d'habillement & fourrures
10 Leather products & footwear	.. [n]	.. [n]	.. [n]	.. [n]	.. [n]	.. [n]	..	10 Articles en cuir & chaussures
11 Wood, paper, printing, publishing	184 [o]	194 [o]	290 [o]	176 [o]	375 [o]	283 [o]	..	**11 Bois, papier, imprimerie, édition**
12 Wood & cork (not furniture)	153 [o]	155 [o]	211 [o]	108 [o]	114 [o]	55 [o]	..	12 Bois & liège (sauf meubles)
13 Pulp, paper & paper products	31 [o]	39 [o]	79 [o]	68 [o]	261 [o]	228 [o]	..	13 Carton, papier & articles
14 Publ., print. & repro. of rec. media	.. [n]	.. [n]	.. [n]	.. [n]	.. [n]	.. [n]	..	14 Edition, impress. & repro. enregis.
15 Coke, petroleum, nuclear fuel, chemicals, rubber plastics	9 713	9 752	9 415	9 886	9 601	10 688	..	**15 Coke, pétrole, combust. nucl., prod. chimi., caout. & mat. plast.**
16 Coke, ref. petrol. prod. & nuclear fuel	5 175	5 102	4 752	4 795	5 115	4 845	..	16 Coke, raffin. pétr. & combust. nucl.
17 Chemicals & chemical products	4 045	4 107	3 767	4 477	3 554	5 361	..	17 Produits chimiques
18 *Chemical prod.(less pharmaceuticals)*	3 888	3 916	3 102	4 211	3 332	3 154	..	18 *Prod. chimi. (moins prod. pharmaceutiques)*
19 *Pharmaceuticals*	157	191	665	266	222	2 207	..	19 *Produits pharmaceutiques*
20 Rubber & plastic products	493	543	896	614	932	482	..	20 Caoutchouc et matières plastiques
21 Non-metallic mineral products	3 233	2 123	2 008	1 862	1 734	2 743	..	**21 Produits minéraux non métalliques**
22 Basic metals	4 760	4 503	6 511	8 352	9 742	10 086	..	**22 Produits métallurgiques de base**
23 Basic metals, ferrous	2 848	2 466	3 973	5 126	6 399	6 499	..	23 Produits métallur. de base, ferreux
24 Basic metals, non-ferrous	1 912	2 037	2 538	3 226	3 343	3 587	..	24 Prod. métallur. de base, non ferreux
25 Fabricated metal products	352	500	704	452	1 217	581	..	**25 Ouvrages en métaux (sauf mach.& mat.)**
26 Machinery equipment, instruments and transport equipment	74 000	76 256	75 435	90 050	73 362	91 726	..	**26 Machines & matériel, instruments & mat. de transport**
27 Machinery n.e.c.	6 418 [o]	7 532 [o]	4 906 [o]	7 169 [o]	6 669 [o]	12 574 [o]	..	27 Machines, n.c.a.
28 Office, account. & computing machinery	.. [n]	.. [n]	.. [n]	.. [n]	.. [n]	.. [n]	..	28 Mach. de bur., compta. & trait. inform.
29 Electrical machinery	8 439	7 108	8 390	10 613	7 897	9 223	..	29 Machines électriques
30 Electro. equip. (radio, TV & commun.)	7 735 [o]	10 041 [o]	10 586 [o]	11 428 [o]	14 037 [o]	13 833 [o]	..	30 Equip. électro. (radio, tv & communic.)
31 *Electro. comp. (inc. semi-conductors)*	..	..	..	..	..	..	..	31 *Compos. électro. (semi-cond. incl.)*
32 *TV, radio & communications equip.*	..	..	..	..	..	..	..	32 *Appareils radio, tv et de communi.*
33 Instruments	794	1 428	1 330	1 109	1 467	2 045	..	33 Instruments
34 Motor vehicles	898	1 684	3 369	3 796	1 315	1 876	..	34 Véhicules automobiles
35 Other transport equipment	49 716	48 463	46 854	55 935	41 977	52 175	..	35 Autres matériels de transport
36 *Ships*	48 599	46 745	45 764	55 441	40 936	51 527	..	36 *Navires*
37 *Aerospace*	..	..	40	26	..	..	..	37 *Aérospatiale*
38 *Other transport equipment n.e.c.*	1 117	1 718	1 050	468	1 041	648	..	38 *Autres matériels de transport n.c.a.*
39 Furniture, other manufacturing n.e.c.	206 [m,o]	137 [m,o]	67 [m,o]	98 [m,o]	163 [m]	118 [m]	..	**39 Meubles, autres activ. de fabricat. n.c.a.**
40 Furniture	.. [n]	.. [n]	.. [n]	.. [n]	.. [n]	.. [n]	..	40 Meubles
41 Other manufacturing n.e.c.	206 [o]	137 [o]	67 [o]	98 [o]	163 [o]	118 [o]	..	41 Autres activités de fabrication n.c.a.
42 Recycling	..	..	..	..	..	..	..	**42 Récupération**
43 ELECTRICITY, GAS & WATER SUPPLY	3 019	4 409	5 363	6 994	3 877	3 039	..	**43 DISTRIBUTION D'ELECTRICITE, GAZ & EAU**
44 CONSTRUCTION	2 273	2 463	3 955	5 789	3 067	5 147	..	**44 CONSTRUCTION**
45 SERVICES SECTOR	352	43 250	5 101 [a]	7 158	111 129	53 442	..	**45 SECTEUR DES SERVICES**
46 Wholesale, retail trade & repair of motor vehicles, etc.	..	..	..	..	..	..	..	**46 Commerce (gros & détail), réparation de véhicules, etc.**
47 Hotels & restaurants	..	..	..	..	..	..	..	**47 Hôtels et restaurants**
48 Transport & storage	267	259	259	224	209	56	..	**48 Transports, entreposage**
49 Communications	85	42 991	3 699	4 474	108 577	52 533	..	**49 Communications**
50 Post	..	..	..	..	..	..	..	50 Postes
51 Telecommunications	..	..	..	..	..	..	..	51 Télécommunications
52 Financial intermediation (inc. insur.)	..	..	..	..	..	..	..	**52 Intermédiation financ. (assurances inc.)**
53 Real estate, renting & busin. activity	..	..	1 143	2 460	2 343	853	..	**53 Immobil., locat. & serv. aux entreprises**
54 Computer & related activities	..	..	1 143	2 460	2 343	853	..	54 Informatique et activités connexes
55 *Software consultancy*	..	..	..	..	..	..	..	55 *Production de logiciels*
56 *Other computer services n.e.c.*	..	..	..	..	..	..	..	56 *Autres activités informatiques*
57 Research & development	..	..	..	..	..	..	..	57 Recherche & développement expérimental
58 Other business activities n.e.c.	..	..	..	..	..	..	..	58 Autres activ. & serv. aux entreprises
59 Comm., soc. & pers. serv. activ., etc.	..	..	..	..	..	..	..	**59 Adm. pub., serv. soc. & collec., etc**
TOTAL BUSINESS ENTERPRISE SECTOR	**106 831**	**149 244**	**115 100 [a]**	**139 120**	**223 274**	**187 621**	**..**	**TOTAL SECTEUR DES ENTREPRISES**

Source: OECD, R&D database (DSTI/EAS Division), January 2002.

Source : OCDE, base de données R-D (DSTI/Division AES), janvier 2002.

Business enterprise intramural expenditure on R&D (BERD) - Dépenses intra-muros de R-D des entreprises (DIRDE)

by industry and source of funds - par industrie et par source de financement

Million JPY *Millions JPY*

10.3 Other national funds - Autres sources nationales

	1994	1995	1996	1997	1998	1999	2000	
1 AGRICULTURE, HUNTING & FORESTRY	10	30	..	5	..	..	..	**1 AGRICULTURE, CHASSE & SYLVICULTURE**
2 MINING	358	16	9	4	4	0	..	**2 ACTIVITES EXTRACTIVES**
3 MANUFACTURING	9 568	9 282	12 985	12 614	16 586	15 993	..	**3 ACTIVITES DE FABRICATION**
4 Food, beverages & tobacco	30	24	..	21	20	5	..	**4 Produits alimentaires, boissons & tabac**
5 Food products & beverages	..	..	..	..	..	..	..	5 Produits alimentaires & boissons
6 Tobacco products	..	..	..	..	..	..	..	6 Produits à base de tabac
7 Textiles, wear app., fur & leather	0	0	..	12	0	0	..	**7 Textiles, habill., fourrures & cuirs**
8 Textiles	..	..	..	..	..	..	..	8 Textiles
9 Wearing apparel & fur	..	..	..	..	..	..	..	9 Articles d'habillement & fourrures
10 Leather products & footwear	..	..	..	..	..	..	..	10 Articles en cuir & chaussures
11 Wood, paper, printing, publishing	23 [o]	10 [o]	9 [o]	36 [o]	0	..	..	**11 Bois, papier, imprimerie, édition**
12 Wood & cork (not furniture)	12 [o]	8 [o]	9 [o]	..	0	..	..	12 Bois & liège (sauf meubles)
13 Pulp, paper & paper products	11 [o]	2 [o]	..	36 [o]	..	..	..	13 Carton, papier & articles
14 Publ., print. & repro. of rec. media	.. [n]	.. [n]	..	..	..	..	..	14 Edition, impress. & repro. enregis.
15 Coke, petroleum, nuclear fuel, chemicals, rubber plastics	1 980	1 892	1 942	1 859	1 459	2 044	..	**15 Coke, pétrole, combust. nucl., prod. chimi., caout. & mat. plast.**
16 Coke, ref. petrol. prod. & nuclear fuel	1 681	1 678	1 707	1 433	1 002	799	..	16 Coke, raffin. pétr. & combust. nucl.
17 Chemicals & chemical products	275	176	205	403	447	1 230	..	17 Produits chimiques
18 *Chemical prod.(less pharmaceuticals)*	262	176	195	392	429	1 226	..	18 *Prod. chimi. (moins prod. pharmaceutiques)*
19 *Pharmaceuticals*	13	..	10	11	18	4	..	19 *Produits pharmaceutiques*
20 Rubber & plastic products	24	38	30	23	10	15	..	20 Caoutchouc et matières plastiques
21 Non-metallic mineral products	168	106	76	..	11	36	..	**21 Produits minéraux non métalliques**
22 Basic metals	3 637	3 634	4 122	2 766	2 015	2 614	..	**22 Produits métallurgiques de base**
23 Basic metals, ferrous	3 308	3 437	4 073	2 655	1 816	2 372	..	23 Produits métallur. de base, ferreux
24 Basic metals, non-ferrous	329	197	49	111	199	242	..	24 Prod. métallur. de base, non ferreux
25 Fabricated metal products	23	4	285	18	0	9	..	**25 Ouvrages en métaux (sauf mach.& mat.)**
26 Machinery equipment, instruments and transport equipment	3 710	3 614	6 553	7 901	13 080	11 286	..	**26 Machines & matériel, instruments & mat. de transport**
27 Machinery n.e.c.	569 [o]	593 [o]	611 [o]	698 [o]	413 [o]	418 [o]	..	27 Machines, n.c.a.
28 Office, account. & computing machinery	.. [n]	.. [n]	.. [n]	.. [n]	.. [n]	.. [n]	..	28 Mach. de bur., compta. & trait. inform.
29 Electrical machinery	827	357	1 528	2 086	2 056	3 623	..	29 Machines électriques
30 Electro. equip. (radio, TV & commun.)	858 [o]	525 [o]	893 [o]	1 056 [o]	438 [o]	450 [o]	..	30 Equip. électro. (radio, tv & communic.)
31 *Electro. comp. (inc. semi-conductors)*	..	..	..	..	..	..	..	31 *Compos. électro. (semi-cond. incl.)*
32 *TV, radio & communications equip.*	..	..	..	..	..	..	..	32 *Appareils radio, tv et de communi.*
33 Instruments	32	25	15	12	28	191	..	33 Instruments
34 Motor vehicles	231	602	645	544	377	86	..	34 Véhicules automobiles
35 Other transport equipment	1 193	1 512	2 861	3 505	9 768	6 518	..	35 Autres matériels de transport
36 *Ships*	1 122	1 512	2 858	3 505	9 768	6 498	..	36 *Navires*
37 *Aerospace*	71	..	..	..	..	13	..	37 *Aérospatiale*
38 *Other transport equipment n.e.c.*	..	..	3	..	..	7	..	38 *Autres matériels de transport n.c.a.*
39 Furniture, other manufacturing n.e.c.	..	..	..	..	..	..	..	**39 Meubles, autres activ. de fabricat. n.c.a.**
40 Furniture	.. [n]	.. [n]	.. [n]	.. [n]	.. [n]	.. [n]	..	40 Meubles
41 Other manufacturing n.e.c.	..	..	..	..	..	..	..	41 Autres activités de fabrication n.c.a.
42 Recycling	..	..	..	..	..	..	..	**42 Récupération**
43 ELECTRICITY, GAS & WATER SUPPLY	198	38	20	96	6	..	..	**43 DISTRIBUTION D'ELECTRICITE, GAZ & EAU**
44 CONSTRUCTION	633	1 099	1 145	1 117	700	445	..	**44 CONSTRUCTION**
45 SERVICES SECTOR	..	..	881 [a]	268	217	68	..	**45 SECTEUR DES SERVICES**
46 Wholesale, retail trade & repair of motor vehicles, etc.	..	..	..	..	..	..	..	**46 Commerce (gros & détail), réparation de véhicules, etc.**
47 Hotels & restaurants	..	..	..	..	..	..	..	**47 Hôtels et restaurants**
48 Transport & storage	..	..	2	4	13	4	..	**48 Transports, entreposage**
49 Communications	..	..	..	..	..	..	..	**49 Communications**
50 Post	..	..	..	..	..	..	..	50 Postes
51 Telecommunications	..	..	..	..	..	..	..	51 Télécommunications
52 Financial intermediation (inc. insur.)	..	..	..	..	..	..	..	**52 Intermédiation financ. (assurances inc.)**
53 Real estate, renting & busin. activity	..	..	879	264	204	64	..	**53 Immobil., locat. & serv. aux entreprises**
54 Computer & related activities	..	..	879	264	204	64	..	54 Informatique et activités connexes
55 *Software consultancy*	..	..	..	..	..	..	..	55 *Production de logiciels*
56 *Other computer services n.e.c.*	..	..	..	..	..	..	..	56 *Autres activités informatiques*
57 Research & development	..	..	..	..	..	..	..	57 Recherche & développement expérimental
58 Other business activities n.e.c.	..	..	..	..	..	..	..	58 Autres activ. & serv. aux entreprises
59 Comm., soc. & pers. serv. activ., etc.	..	..	..	..	..	..	..	**59 Adm. pub., serv. soc. & collec., etc**
TOTAL BUSINESS ENTERPRISE SECTOR	**10 767**	**10 466**	**15 040** [a]	**14 105**	**17 514**	**16 507**	..	**TOTAL SECTEUR DES ENTREPRISES**

Source: OECD, R&D database (DSTI/EAS Division), January 2002.

Source : OCDE, base de données R-D (DSTI/Division AES), janvier 2002.

Business enterprise intramural expenditure on R&D (BERD) - Dépenses intra-muros de R-D des entreprises (DIRDE)

by industry and source of funds - par industrie et par source de financement

Million JPY | *Millions JPY*

	1994	1995	1996	1997	1998	1999	2000	
	10.4 Funds from abroad - Fonds de l'étranger							
1 AGRICULTURE, HUNTING & FORESTRY	..	..	..	..	..	..	..	**1 AGRICULTURE, CHASSE & SYLVICULTURE**
2 MINING	..	33	..	..	..	..	..	**2 ACTIVITES EXTRACTIVES**
3 MANUFACTURING	12 069	12 858	12 214	41 049	45 765	54 622	..	**3 ACTIVITES DE FABRICATION**
4 Food, beverages & tobacco	..	85	97	166	190	4 817	..	**4 Produits alimentaires, boissons & tabac**
5 Food products & beverages	..	..	..	..	..	..	..	5 Produits alimentaires & boissons
6 Tobacco products	..	..	..	..	..	..	..	6 Produits à base de tabac
7 Textiles, wear app., fur & leather	..	..	..	..	..	..	..	**7 Textiles, habill., fourrures & cuirs**
8 Textiles	..	..	..	..	..	..	..	8 Textiles
9 Wearing apparel & fur	..	..	..	..	..	..	..	9 Articles d'habillement & fourrures
10 Leather products & footwear	..	..	..	..	..	..	..	10 Articles en cuir & chaussures
11 Wood, paper, printing, publishing	..	..	..	..	..	..	..	**11 Bois, papier, imprimerie, édition**
12 Wood & cork (not furniture)	..	..	..	..	..	..	..	12 Bois & liège (sauf meubles)
13 Pulp, paper & paper products	..	..	..	..	..	..	..	13 Carton, papier & articles
14 Publ., print. & repro. of rec. media	..	..	..	..	..	..	..	14 Edition, impress. & repro. enregis.
15 Coke, petroleum, nuclear fuel, chemicals, rubber plastics	7 792	7 730	6 787	10 393	8 664	6 769	..	**15 Coke, pétrole, combust. nucl., prod. chimi., caout. & mat. plast.**
16 Coke, ref. petrol. prod. & nuclear fuel	241	273	287	288	292	302	..	16 Coke, raffin. pétr. & combust. nucl.
17 Chemicals & chemical products	7 551	7 457	6 381	9 868	8 372	6 467	..	17 Produits chimiques
18 *Chemical prod.(less pharmaceuticals)*	1 315	1 715	982	1 358	1 188	1 054	..	18 *Prod. chimi. (moins prod. pharmaceutiques)*
19 *Pharmaceuticals*	6 236	5 742	5 399	8 510	7 184	5 413	..	19 *Produits pharmaceutiques*
20 Rubber & plastic products	..	..	119	237	..	..	..	20 Caoutchouc et matières plastiques
21 Non-metallic mineral products	..	18	..	..	92	..	..	**21 Produits minéraux non métalliques**
22 Basic metals	45	..	428	58	95	66	..	**22 Produits métallurgiques de base**
23 Basic metals, ferrous	45	..	2	16	62	33	..	23 Produits métallur. de base, ferreux
24 Basic metals, non-ferrous	..	..	426	42	33	33	..	24 Prod. métallur. de base, non ferreux
25 Fabricated metal products	..	..	..	..	..	..	..	**25 Ouvrages en métaux (sauf mach.& mat.)**
26 Machinery equipment, instruments and transport equipment	4 225	5 027	4 902	30 431	36 723	42 970	..	**26 Machines & matériel, instruments & mat. de transport**
27 Machinery n.e.c.	1 965 [o]	2 118 [o]	1 575 [o]	1 332 [o]	1 418 [o]	3 079 [o]	..	27 Machines, n.c.a.
28 Office, account. & computing machinery	.. [n]	.. [n]	.. [n]	.. [n]	.. [n]	.. [n]	..	28 Mach. de bur., compta. & trait. inform.
29 Electrical machinery	1 582	2 191	2 165	157	29	..	..	29 Machines électriques
30 Electro. equip. (radio, TV & commun.)	28 [o]	42 [o]	189 [o]	27 928 [o]	33 451 [o]	..	..	30 Equip. électro. (radio, tv & communic.)
31 *Electro. comp. (inc. semi-conductors)*	..	..	..	..	..	..	..	31 *Compos. électro. (semi-cond. incl.)*
32 *TV, radio & communications equip.*	..	..	..	..	..	..	..	32 *Appareils radio, tv et de communi.*
33 Instruments	0	301	331	183	297	237	..	33 Instruments
34 Motor vehicles	397	375	500	786	1 179	473	..	34 Véhicules automobiles
35 Other transport equipment	253	..	142	45	349	125	..	35 Autres matériels de transport
36 *Ships*	68	..	142	36	349	125	..	36 *Navires*
37 *Aerospace*	185	..	..	..	..	..	..	37 *Aérospatiale*
38 *Other transport equipment n.e.c.*	..	..	..	9	..	..	..	38 *Autres matériels de transport n.c.a.*
39 Furniture, other manufacturing n.e.c.	7	..	..	..	..	..	..	**39 Meubles, autres activ. de fabricat. n.c.a.**
40 Furniture	..	..	..	..	..	..	..	40 Meubles
41 Other manufacturing n.e.c.	7	..	..	..	..	..	..	41 Autres activités de fabrication n.c.a.
42 Recycling	..	..	..	..	..	..	..	**42 Récupération**
43 ELECTRICITY, GAS & WATER SUPPLY	..	..	..	..	..	..	..	**43 DISTRIBUTION D'ELECTRICITE, GAZ & EAU**
44 CONSTRUCTION	0	250	11	4	11	867	..	**44 CONSTRUCTION**
45 SERVICES SECTOR	..	48	21 [a]	..	8	842	..	**45 SECTEUR DES SERVICES**
46 Wholesale, retail trade & repair of motor vehicles, etc.	..	..	..	..	..	..	..	**46 Commerce (gros & détail), réparation de véhicules, etc.**
47 Hotels & restaurants	..	..	..	..	..	..	..	**47 Hôtels et restaurants**
48 Transport & storage	..	..	..	..	..	..	..	**48 Transports, entreposage**
49 Communications	..	48	..	..	..	827	..	**49 Communications**
50 Post	..	..	..	..	..	..	..	50 Postes
51 Telecommunications	..	..	..	..	..	..	..	51 Télécommunications
52 Financial intermediation (inc. insur.)	..	..	..	..	..	..	..	**52 Intermédiation financ. (assurances inc.)**
53 Real estate, renting & busin. activity	..	..	21	..	8	15	..	**53 Immobil., locat. & serv. aux entreprises**
54 Computer & related activities	..	..	21	..	8	15	..	54 Informatique et activités connexes
55 *Software consultancy*	..	..	..	..	..	..	..	55 *Production de logiciels*
56 *Other computer services n.e.c.*	..	..	..	..	..	..	..	56 *Autres activités informatiques*
57 Research & development	..	..	..	..	..	..	..	57 Recherche & développement expérimental
58 Other business activities n.e.c.	..	..	..	..	..	..	..	58 Autres activ. & serv. aux entreprises
59 Comm., soc. & pers. serv. activ., etc.	..	..	..	..	..	..	..	**59 Adm. pub., serv. soc. & collec., etc**
TOTAL BUSINESS ENTERPRISE SECTOR	**12 069**	**13 189**	**12 246**	**41 053**	**45 784**	**56 331**	..	**TOTAL SECTEUR DES ENTREPRISES**

Source: OECD, R&D database (DSTI/EAS Division), January 2002.

Source : OCDE, base de données R-D (DSTI/Division AES), janvier 2002.

Business enterprise intramural expenditure on R&D (BERD) - Dépenses intra-muros de R-D des entreprises (DIRDE)

by industry and type of costs - par industrie et par type de coûts

Million JPY — *Millions JPY*

	1994	1995	1996	1997	1998	1999	2000	
	11.1 Current expenditure - Dépenses courantes							
1 AGRICULTURE, HUNTING & FORESTRY	6 260	7 859	7 317	7 786	8 555	8 960	..	**1 AGRICULTURE, CHASSE & SYLVICULTURE**
2 MINING	21 219	20 630	19 632	20 752	23 333	23 765	..	**2 ACTIVITES EXTRACTIVES**
3 MANUFACTURING	7 625 905	7 962 155	8 384 739	8 862 679	8 955 799	8 763 810	..	**3 ACTIVITES DE FABRICATION**
4 Food, beverages & tobacco	208 632	222 561	219 175	233 540	234 546	255 295	..	**4 Produits alimentaires, boissons & tabac**
5 Food products & beverages	..	..	..	..	..	..	..	5 Produits alimentaires & boissons
6 Tobacco products	..	..	..	..	..	..	..	6 Produits à base de tabac
7 Textiles, wear app., fur & leather	72 276 [m]	73 774 [m]	71 278 [m]	72 240 [m]	67 225 [m]	72 381 [m]	..	**7 Textiles, habill., fourrures & cuirs**
8 Textiles	..	..	..	..	..	..	..	8 Textiles
9 Wearing apparel & fur	..	..	..	..	..	..	..	9 Articles d'habillement & fourrures
10 Leather products & footwear	.. [n]	.. [n]	.. [n]	.. [n]	.. [n]	.. [n]	..	10 Articles en cuir & chaussures
11 Wood, paper, printing, publishing	93 603 [o]	100 490 [o]	99 577 [o]	103 127 [o]	108 617 [o]	102 437 [o]	..	**11 Bois, papier, imprimerie, édition**
12 Wood & cork (not furniture)	29 533 [o]	32 365 [o]	29 175 [o]	30 966 [o]	32 890 [o]	25 581 [o]	..	12 Bois & liège (sauf meubles)
13 Pulp, paper & paper products	64 070 [o]	68 125 [o]	70 402 [o]	72 161 [o]	75 727 [o]	76 856 [o]	..	13 Carton, papier & articles
14 Publ., print. & repro. of rec. media	.. [n]	.. [n]	.. [n]	.. [n]	.. [n]	.. [n]	..	14 Edition, impress. & repro. enregis.
15 Coke, petroleum, nuclear fuel, chemicals, rubber plastics	1 680 829	1 652 099	1 685 160	1 705 605	1 730 992	1 692 030	..	**15 Coke, pétrole, combust. nucl., prod. chimi., caout. & mat. plast.**
16 Coke, ref. petrol. prod. & nuclear fuel	67 110	61 290	54 708	57 189	44 845	35 176	..	16 Coke, raffin. pétr. & combust. nucl.
17 Chemicals & chemical products	1 395 891	1 379 722	1 406 027	1 425 538	1 466 823	1 430 580	..	17 Produits chimiques
18 *Chemical prod.(less pharmaceuticals)*	814 748	818 450	820 715	853 339	840 920	799 955	..	18 *Prod. chimi. (moins prod. pharmaceutiques)*
19 *Pharmaceuticals*	581 143	561 272	585 312	572 199	625 903	630 625	..	19 *Produits pharmaceutiques*
20 Rubber & plastic products	217 828	211 087	224 425	222 878	219 324	226 274	..	20 Caoutchouc et matières plastiques
21 Non-metallic mineral products	164 183	178 958	187 436	185 226	186 775	163 609	..	**21 Produits minéraux non métalliques**
22 Basic metals	336 888	313 704	316 104	325 247	289 934	277 885	..	**22 Produits métallurgiques de base**
23 Basic metals, ferrous	213 706	188 701	182 920	188 290	166 926	158 341	..	23 Produits métallur. de base, ferreux
24 Basic metals, non-ferrous	123 182	125 003	133 184	136 957	123 008	119 544	..	24 Prod. métallur. de base, non ferreux
25 Fabricated metal products	102 861	107 466	132 233	119 241	114 422	103 538	..	**25 Ouvrages en métaux (sauf mach.& mat.)**
26 Machinery equipment, instruments and transport equipment	4 894 349	5 240 707	5 601 816	6 034 294	6 132 728	5 997 911	..	**26 Machines & matériel, instruments & mat. de transport**
27 Machinery n.e.c.	646 354 [o]	659 582 [o]	665 856 [o]	717 768 [o]	746 428 [o]	749 667 [o]	..	27 Machines, n.c.a.
28 Office, account. & computing machinery	.. [n]	.. [n]	.. [n]	.. [n]	.. [n]	.. [n]	..	28 Mach. de bur., compta. & trait. inform.
29 Electrical machinery	923 578	947 135	1 005 321	1 041 657	1 030 316	930 269	..	29 Machines électriques
30 Electro. equip. (radio, TV & commun.)	1 884 971 [o]	2 033 999 [o]	2 180 141 [o]	2 348 094 [o]	2 384 995 [o]	2 422 288 [o]	..	30 Equip. électro. (radio, tv & communic.)
31 *Electro. comp. (inc. semi-conductors)*	..	..	..	..	..	..	..	31 *Compos. électro. (semi-cond. incl.)*
32 *TV, radio & communications equip.*	..	..	..	..	..	..	..	32 *Appareils radio, tv et de communi.*
33 Instruments	290 100	327 461	335 145	373 174	426 912	439 019	..	33 Instruments
34 Motor vehicles	959 560	1 092 057	1 231 503	1 352 568	1 347 426	1 265 223	..	34 Véhicules automobiles
35 Other transport equipment	189 786	180 473	183 850	201 033	196 651	191 445	..	35 Autres matériels de transport
36 *Ships*	152 155	145 355	149 347	162 984	157 083	154 257	..	36 *Navires*
37 *Aerospace*	1 947	2 004	2 077	2 128	2 124	1 001	..	37 *Aérospatiale*
38 *Other transport equipment n.e.c.*	35 684	33 114	32 426	35 921	37 444	36 187	..	38 *Autres matériels de transport n.c.a.*
39 Furniture, other manufacturing n.e.c.	72 284 [m,o]	72 395 [m,o]	71 964 [m,o]	84 156 [m,o]	90 560 [m]	98 725 [m]	..	**39 Meubles, autres activ. de fabricat. n.c.a.**
40 Furniture	.. [n]	.. [n]	.. [n]	.. [n]	.. [n]	.. [n]	..	40 Meubles
41 Other manufacturing n.e.c.	72 284 [o]	72 395 [o]	71 964 [o]	84 156 [o]	90 560 [o]	98 725 [o]	..	41 Autres activités de fabrication n.c.a.
42 Recycling	..	..	..	..	..	..	..	**42 Récupération**
43 ELECTRICITY, GAS & WATER SUPPLY	72 909	75 557	75 352	77 195	71 090	89 247	..	**43 DISTRIBUTION D'ELECTRICITE, GAZ & EAU**
44 CONSTRUCTION	190 822	182 533	205 910	202 342	158 645	178 256	..	**44 CONSTRUCTION**
45 SERVICES SECTOR	128 816	133 638	310 487 [a]	330 509	517 889	579 123	..	**45 SECTEUR DES SERVICES**
46 Wholesale, retail trade & repair of motor vehicles, etc.	..	..	..	..	..	..	..	**46 Commerce (gros & détail), réparation de véhicules, etc.**
47 Hotels & restaurants	..	..	..	..	..	..	..	**47 Hôtels et restaurants**
48 Transport & storage	8 879	10 113	8 094	11 982	11 447	15 037	..	**48 Transports, entreposage**
49 Communications	119 937	123 525	129 371	148 403	183 134	294 496	..	**49 Communications**
50 Post	..	..	..	..	..	..	..	50 Postes
51 Telecommunications	..	..	..	..	..	..	..	51 Télécommunications
52 Financial intermediation (inc. insur.)	..	..	..	..	..	..	..	**52 Intermédiation financ. (assurances inc.)**
53 Real estate, renting & busin. activity	..	..	173 022	170 124	323 308	269 590	..	**53 Immobil., locat. & serv. aux entreprises**
54 Computer & related activities	..	..	173 022	170 124	323 308	269 590	..	54 Informatique et activités connexes
55 *Software consultancy*	..	..	..	..	..	..	..	55 *Production de logiciels*
56 *Other computer services n.e.c.*	..	..	..	..	..	..	..	56 *Autres activités informatiques*
57 Research & development	..	..	..	..	..	..	..	57 Recherche & développement expérimental
58 Other business activities n.e.c.	..	..	..	..	..	..	..	58 Autres activ. & serv. aux entreprises
59 Comm., soc. & pers. serv. activ., etc.	..	..	..	..	..	..	..	**59 Adm. pub., serv. soc. & collec., etc**
TOTAL BUSINESS ENTERPRISE SECTOR	**8 045 930** [l]	**8 382 371** [l]	**9 003 437** [a]	**9 501 263**	**9 735 311**	**9 643 161**	**..**	**TOTAL SECTEUR DES ENTREPRISES**

Source: OECD, R&D database (DSTI/EAS Division), January 2002.

Source : OCDE, base de données R-D (DSTI/Division AES), janvier 2002.

Business enterprise intramural expenditure on R&D (BERD) - Dépenses intra-muros de R-D des entreprises (DIRDE)

by industry and type of costs - par industrie et par type de coûts

Million JPY — *Millions JPY*

11.2 Capital expenditure - Dépenses en capital

	1994	1995	1996	1997	1998	1999	2000	
1 AGRICULTURE, HUNTING & FORESTRY	1 534	883	1 664	3 053	3 382	2 015	..	**1 AGRICULTURE, CHASSE & SYLVICULTURE**
2 MINING	2 683	2 224	2 045	4 485	4 530	2 422	..	**2 ACTIVITES EXTRACTIVES**
3 MANUFACTURING	739 573	812 205	878 412	953 759	851 348	757 763	..	**3 ACTIVITES DE FABRICATION**
4 Food, beverages & tobacco	25 114	26 879	31 632	34 812	28 402	24 913	..	**4 Produits alimentaires, boissons & tabac**
5 Food products & beverages	..	..	..	..	..	..	..	5 Produits alimentaires & boissons
6 Tobacco products	..	..	..	..	..	..	..	6 Produits à base de tabac
7 Textiles, wear app., fur & leather	4 468 [m]	6 117 [m]	5 045 [m]	9 412 [m]	8 976 [m]	5 090 [m]	..	**7 Textiles, habill., fourrures & cuirs**
8 Textiles	..	..	..	..	..	..	..	8 Textiles
9 Wearing apparel & fur	..	..	..	..	..	..	..	9 Articles d'habillement & fourrures
10 Leather products & footwear	.. [n]	.. [n]	.. [n]	.. [n]	.. [n]	.. [n]	..	10 Articles en cuir & chaussures
11 Wood, paper, printing, publishing	11 725 [o]	15 336 [o]	16 286 [o]	17 889 [o]	18 713 [o]	20 923 [o]	..	**11 Bois, papier, imprimerie, édition**
12 Wood & cork (not furniture)	2 488 [o]	2 100 [o]	3 484 [o]	3 088 [o]	1 685 [o]	1 747 [o]	..	12 Bois & liège (sauf meubles)
13 Pulp, paper & paper products	9 237 [o]	13 236 [o]	12 802 [o]	14 801 [o]	17 028 [o]	19 176 [o]	..	13 Carton, papier & articles
14 Publ., print. & repro. of rec. media	.. [n]	.. [n]	.. [n]	.. [n]	.. [n]	.. [n]	..	14 Edition, impress. & repro. enregis.
15 Coke, petroleum, nuclear fuel, chemicals, rubber plastics	196 222	217 313	233 787	228 061	210 623	196 068	..	**15 Coke, pétrole, combust. nucl., prod. chimi., caout. & mat. plast.**
16 Coke, ref. petrol. prod. & nuclear fuel	12 146	6 459	7 801	8 162	5 486	4 974	..	16 Coke, raffin. pétr. & combust. nucl.
17 Chemicals & chemical products	152 903	175 163	187 224	183 713	164 106	157 494	..	17 Produits chimiques
18 *Chemical prod.(less pharmaceuticals)*	101 244	94 245	105 391	112 621	108 890	98 669	..	18 *Prod. chimi. (moins prod. pharmaceutiques)*
19 *Pharmaceuticals*	51 659	80 918	81 833	71 092	55 216	58 825	..	19 *Produits pharmaceutiques*
20 Rubber & plastic products	31 173	35 691	38 762	36 186	41 031	33 600	..	20 Caoutchouc et matières plastiques
21 Non-metallic mineral products	18 507	21 012	26 278	30 164	30 377	20 008	..	**21 Produits minéraux non métalliques**
22 Basic metals	40 705	53 430	38 691	54 334	39 296	28 731	..	**22 Produits métallurgiques de base**
23 Basic metals, ferrous	24 001	24 840	18 557	25 341	20 670	9 927	..	23 Produits métallur. de base, ferreux
24 Basic metals, non-ferrous	16 704	28 590	20 134	28 993	18 626	18 804	..	24 Prod. métallur. de base, non ferreux
25 Fabricated metal products	8 798	10 627	15 516	15 475	11 206	6 147	..	**25 Ouvrages en métaux (sauf mach.& mat.)**
26 Machinery equipment, instruments and transport equipment	420 653	454 388	503 067	555 429	497 455	450 453	..	**26 Machines & matériel, instruments & mat. de transport**
27 Machinery n.e.c.	50 382 [o]	45 640 [o]	67 851 [o]	72 289 [o]	65 225 [o]	61 697 [o]	..	27 Machines, n.c.a.
28 Office, account. & computing machinery	.. [n]	.. [n]	.. [n]	.. [n]	.. [n]	.. [n]	..	28 Mach. de bur., compta. & trait. inform.
29 Electrical machinery	83 547	85 752	88 013	100 189	93 830	79 370	..	29 Machines électriques
30 Electro. equip. (radio, TV & commun.)	172 671 [o]	206 715 [o]	220 163 [o]	229 507 [o]	203 624 [o]	183 949 [o]	..	30 Equip. électro. (radio, tv & communic.)
31 *Electro. comp. (inc. semi-conductors)*	..	..	..	..	..	..	..	31 *Compos. électro. (semi-cond. incl.)*
32 *TV, radio & communications equip.*	..	..	..	..	..	..	..	32 *Appareils radio, tv et de communi.*
33 Instruments	43 406	27 940	31 152	53 007	46 809	52 511	..	33 Instruments
34 Motor vehicles	61 730	77 213	83 942	92 652	79 906	69 830	..	34 Véhicules automobiles
35 Other transport equipment	8 917	11 128	11 946	7 785	8 061	3 096	..	35 Autres matériels de transport
36 *Ships*	6 518	5 562	5 816	4 152	4 982	2 117	..	36 *Navires*
37 *Aerospace*	31	88	111	153	80	378	..	37 *Aérospatiale*
38 *Other transport equipment n.e.c.*	2 368	5 478	6 019	3 480	2 999	601	..	38 *Autres matériels de transport n.c.a.*
39 Furniture, other manufacturing n.e.c.	13 380 [m,o]	7 102 [m,o]	8 112 [m,o]	8 183 [m,o]	6 300 [m]	5 430 [m]	..	**39 Meubles, autres activ. de fabricat. n.c.a.**
40 Furniture	.. [n]	.. [n]	.. [n]	.. [n]	.. [n]	.. [n]	..	40 Meubles
41 Other manufacturing n.e.c.	13 380 [o]	7 102 [o]	8 112 [o]	8 183 [o]	6 300 [o]	5 430 [o]	..	41 Autres activités de fabrication n.c.a.
42 Recycling	..	..	..	..	..	..	..	**42 Récupération**
43 ELECTRICITY, GAS & WATER SUPPLY	62 039	27 677	37 631	25 564	31 260	18 708	..	**43 DISTRIBUTION D'ELECTRICITE, GAZ & EAU**
44 CONSTRUCTION	31 178	21 830	18 604	22 819	18 045	21 197	..	**44 CONSTRUCTION**
45 SERVICES SECTOR	97 315	148 707	116 617 [a]	147 414	156 188	184 894	..	**45 SECTEUR DES SERVICES**
46 Wholesale, retail trade & repair of motor vehicles, etc.	..	..	..	..	..	..	..	**46 Commerce (gros & détail), réparation de véhicules, etc.**
47 Hotels & restaurants	..	..	..	..	..	..	..	**47 Hôtels et restaurants**
48 Transport & storage	3 974	5 967	3 291	2 631	3 067	1 948	..	**48 Transports, entreposage**
49 Communications	93 341	142 740	109 190	141 384	147 666	177 843	..	**49 Communications**
50 Post	..	..	..	..	..	..	..	50 Postes
51 Telecommunications	..	..	..	..	..	..	..	51 Télécommunications
52 Financial intermediation (inc. insur.)	..	..	..	..	..	..	..	**52 Intermédiation financ. (assurances inc.)**
53 Real estate, renting & busin. activity	..	..	4 136	3 399	5 455	5 103	..	**53 Immobil., locat. & serv. aux entreprises**
54 Computer & related activities	..	..	4 136	3 399	5 455	5 103	..	54 Informatique et activités connexes
55 *Software consultancy*	..	..	..	..	..	..	..	55 *Production de logiciels*
56 *Other computer services n.e.c.*	..	..	..	..	..	..	..	56 *Autres activités informatiques*
57 Research & development	..	..	..	..	..	..	..	57 Recherche & développement expérimental
58 Other business activities n.e.c.	..	..	..	..	..	..	..	58 Autres activ. & serv. aux entreprises
59 Comm., soc. & pers. serv. activ., etc.	..	..	..	..	..	..	..	**59 Adm. pub., serv. soc. & collec., etc**
TOTAL BUSINESS ENTERPRISE SECTOR	**934 322**	**1 013 525**	**1 054 973**	**1 157 094**	**1 064 752**	**987 000**	**..**	**TOTAL SECTEUR DES ENTREPRISES**

Source: OECD, R&D database (DSTI/EAS Division), January 2002.

Source : OCDE, base de données R-D (DSTI/Division AES), janvier 2002.

Total business enterprise R&D personnel (BEMP) - Personnel total de R-D du secteur des entreprises (PRDE)

by industry - par industrie

In full time equivalent / *En équivalence plein-temps*

	1994	1995	1996	1997	1998	1999	2000	
1 AGRICULTURE, HUNTING & FORESTRY	597	643	647	709	743	811	..	**1 AGRICULTURE, CHASSE & SYLVICULTURE**
2 MINING	1 225	1 363	1 301	1 300	1 289	1 424	..	**2 ACTIVITES EXTRACTIVES**
3 MANUFACTURING	556 377	553 096	544 297	539 417	552 353	548 731	..	**3 ACTIVITES DE FABRICATION**
4 Food, beverages & tobacco	17 956	19 532	19 399	19 117	18 706	20 366	..	**4 Produits alimentaires, boissons & tabac**
5 Food products & beverages	..	..	..	..	..	..	..	5 Produits alimentaires & boissons
6 Tobacco products	..	..	..	..	..	..	..	6 Produits à base de tabac
7 Textiles, wear app., fur & leather	7 217 [m]	7 043 [m]	5 773 [m]	5 714 [m]	5 910 [m]	5 898 [m]	..	**7 Textiles, habill., fourrures & cuirs**
8 Textiles	..	..	..	..	..	..	..	8 Textiles
9 Wearing apparel & fur	..	..	..	..	..	..	..	9 Articles d'habillement & fourrures
10 Leather products & footwear	.. [n]	.. [n]	.. [n]	.. [n]	.. [n]	.. [n]	..	10 Articles en cuir & chaussures
11 Wood, paper, printing, publishing	8 293 [o]	9 001 [o]	8 765 [o]	8 785 [o]	8 957 [o]	8 111 [o]	..	**11 Bois, papier, imprimerie, édition**
12 Wood & cork (not furniture)	2 860 [o]	3 237 [o]	2 741 [o]	2 858 [o]	3 022 [o]	2 368 [o]	..	12 Bois & liège (sauf meubles)
13 Pulp, paper & paper products	5 433 [o]	5 764 [o]	6 024 [o]	5 927 [o]	5 935 [o]	5 743 [o]	..	13 Carton, papier & articles
14 Publ., print. & repro. of rec. media	.. [n]	.. [n]	.. [n]	.. [n]	.. [n]	.. [n]	..	14 Edition, impress. & repro. enregis.
15 Coke, petroleum, nuclear fuel, chemicals, rubber plastics	119 817	113 364	112 928	108 704	108 206	102 564	..	**15 Coke, pétrole, combust. nucl., prod. chimi., caout. & mat. plast.**
16 Coke, ref. petrol. prod. & nuclear fuel	3 978	3 509	3 213	3 331	2 715	2 174	..	16 Coke, raffin. pétr. & combust. nucl.
17 Chemicals & chemical products	97 187	93 448	92 559	88 070	88 467	84 725	..	17 Produits chimiques
18 *Chemical prod.(less pharmaceuticals)*	65 078	61 772	61 460	60 082	59 516	56 873	..	18 *Prod. chimi. (moins prod. pharmaceutiques)*
19 *Pharmaceuticals*	32 109	31 676	31 099	27 988	28 951	27 852	..	19 *Produits pharmaceutiques*
20 Rubber & plastic products	18 652	16 407	17 156	17 303	17 024	15 665	..	20 Caoutchouc et matières plastiques
21 Non-metallic mineral products	14 128	15 144	14 886	14 500	14 866	15 265	..	**21 Produits minéraux non métalliques**
22 Basic metals	21 042	19 924	19 212	19 298	17 266	15 609	..	**22 Produits métallurgiques de base**
23 Basic metals, ferrous	11 731	10 392	9 730	10 275	8 762	8 011	..	23 Produits métallur. de base, ferreux
24 Basic metals, non-ferrous	9 311	9 532	9 482	9 023	8 504	7 598	..	24 Prod. métallur. de base, non ferreux
25 Fabricated metal products	9 971	10 730	13 133	11 114	10 246	10 868	..	**25 Ouvrages en métaux (sauf mach.& mat.)**
26 Machinery equipment, instruments and transport equipment	350 837	351 866	343 515	345 616	361 006	361 844	..	**26 Machines & matériel, instruments & mat. de transport**
27 Machinery n.e.c.	49 254 [o]	49 172 [o]	46 433 [o]	49 455 [o]	49 600 [o]	49 727 [o]	..	27 Machines, n.c.a.
28 Office, account. & computing machinery	.. [n]	.. [n]	.. [n]	.. [n]	.. [n]	.. [n]	..	28 Mach. de bur., compta. & trait. inform.
29 Electrical machinery	58 923	55 945	56 623	54 542	57 717	54 763	..	29 Machines électriques
30 Electro. equip. (radio, TV & commun.)	144 414 [o]	144 014 [o]	142 214 [o]	141 660 [o]	146 708 [o]	152 630 [o]	..	30 Equip. électro. (radio, tv & communic.)
31 *Electro. comp. (inc. semi-conductors)*	..	..	..	..	..	..	..	31 *Compos. électro. (semi-cond. incl.)*
32 *TV, radio & communications equip.*	..	..	..	..	..	..	..	32 *Appareils radio, tv et de communi.*
33 Instruments	25 525	27 265	25 315	24 913	29 895	30 808	..	33 Instruments
34 Motor vehicles	64 143	66 428	65 142	67 138	69 010	67 056	..	34 Véhicules automobiles
35 Other transport equipment	8 578	9 042	7 788	7 908	8 076	6 860	..	35 Autres matériels de transport
36 *Ships*	4 524	5 059	4 351	4 208	4 102	3 266	..	36 *Navires*
37 *Aerospace*	120	158	120	167	160	210	..	37 *Aérospatiale*
38 *Other transport equipment n.e.c.*	3 934	3 825	3 317	3 533	3 814	3 384	..	38 *Autres matériels de transport n.c.a.*
39 Furniture, other manufacturing n.e.c.	7 116 [m,o]	6 491 [m,o]	6 686 [m,o]	6 569 [m,o]	7 190 [m]	8 206 [m]	..	**39 Meubles, autres activ. de fabricat. n.c.a.**
40 Furniture	.. [n]	.. [n]	.. [n]	.. [n]	.. [n]	.. [n]	..	40 Meubles
41 Other manufacturing n.e.c.	7 116 [o]	6 491 [o]	6 686 [o]	6 569 [o]	7 190 [o]	8 206 [o]	..	41 Autres activités de fabrication n.c.a.
42 Recycling	..	..	..	..	..	..	..	**42 Récupération**
43 ELECTRICITY, GAS & WATER SUPPLY	2 933	2 898	2 935	2 902	2 840	2 799	..	**43 DISTRIBUTION D'ELECTRICITE, GAZ & EAU**
44 CONSTRUCTION	11 138	10 242	13 679	18 406	10 369	10 905	..	**44 CONSTRUCTION**
45 SERVICES SECTOR	5 455	5 472	26 632 [a]	23 422	45 566	39 874	..	**45 SECTEUR DES SERVICES**
46 Wholesale, retail trade & repair of motor vehicles, etc.	..	..	..	..	..	..	..	**46 Commerce (gros & détail), réparation de véhicules, etc.**
47 Hotels & restaurants	..	..	..	..	..	..	..	**47 Hôtels et restaurants**
48 Transport & storage	715	712	662	600	549	750	..	**48 Transports, entreposage**
49 Communications	4 740	4 760	4 701	5 407	6 895	6 810	..	**49 Communications**
50 Post	..	..	..	..	..	..	..	50 Postes
51 Telecommunications	..	..	..	..	..	..	..	51 Télécommunications
52 Financial intermediation (inc. insur.)	..	..	..	..	..	..	..	**52 Intermédiation financ. (assurances inc.)**
53 Real estate, renting & busin. activity	..	..	21 269	17 415	38 122	32 314	..	**53 Immobil., locat. & serv. aux entreprises**
54 Computer & related activities	..	..	21 269	17 415	38 122	32 314	..	54 Informatique et activités connexes
55 *Software consultancy*	..	..	..	..	..	..	..	55 *Production de logiciels*
56 *Other computer services n.e.c.*	..	..	..	..	..	..	..	56 *Autres activités informatiques*
57 Research & development	..	..	..	..	..	..	..	57 Recherche & développement expérimental
58 Other business activities n.e.c.	..	..	..	..	..	..	..	58 Autres activ. & serv. aux entreprises
59 Comm., soc. & pers. serv. activ., etc.	..	..	..	..	..	..	..	**59 Adm. pub., serv. soc. & collec., etc**
TOTAL BUSINESS ENTERPRISE SECTOR	**577 725** [l]	**573 714** [l]	**589 491** [a]	**586 156**	**613 160**	**604 544**	**..**	**TOTAL SECTEUR DES ENTREPRISES**

Source: OECD, R&D database (DSTI/EAS Division), January 2002.

Source : OCDE, base de données R-D (DSTI/Division AES), janvier 2002.

RSE in the business enterprise sector - Chercheurs dans le secteur des entreprises

by industry - par industrie

In full time equivalent — *En équivalence plein-temps*

	1994	1995	1996	1997	1998	1999	2000	
1 AGRICULTURE, HUNTING & FORESTRY	360	405	416	453	498	482	..	**1 AGRICULTURE, CHASSE & SYLVICULTURE**
2 MINING	668	785	776	765	728	935	..	**2 ACTIVITES EXTRACTIVES**
3 MANUFACTURING	361 064	368 960	367 593	367 382	382 761	389 104	..	**3 ACTIVITES DE FABRICATION**
4 Food, beverages & tobacco	11 650	12 947	13 094	12 708	12 659	13 555	..	**4 Produits alimentaires, boissons & tabac**
5 Food products & beverages	..	..	..	..	..	..	..	5 Produits alimentaires & boissons
6 Tobacco products	..	..	..	..	..	..	..	6 Produits à base de tabac
7 Textiles, wear app., fur & leather	3 904 [m]	3 943 [m]	3 582 [m]	3 446 [m]	3 518 [m]	3 523 [m]	..	**7 Textiles, habill., fourrures & cuirs**
8 Textiles	..	..	..	..	..	..	..	8 Textiles
9 Wearing apparel & fur	..	..	..	..	..	..	..	9 Articles d'habillement & fourrures
10 Leather products & footwear	.. [n]	.. [n]	.. [n]	.. [n]	.. [n]	.. [n]	..	10 Articles en cuir & chaussures
11 Wood, paper, printing, publishing	5 705 [o]	6 219 [o]	5 965 [o]	6 180 [o]	6 099 [o]	5 771 [o]	..	**11 Bois, papier, imprimerie, édition**
12 Wood & cork (not furniture)	1 955 [o]	2 264 [o]	1 978 [o]	2 055 [o]	2 051 [o]	1 735 [o]	..	12 Bois & liège (sauf meubles)
13 Pulp, paper & paper products	3 750 [o]	3 955 [o]	3 987 [o]	4 125 [o]	4 048 [o]	4 036 [o]	..	13 Carton, papier & articles
14 Publ., print. & repro. of rec. media	.. [n]	.. [n]	.. [n]	.. [n]	.. [n]	.. [n]	..	14 Edition, impress. & repro. enregis.
15 Coke, petroleum, nuclear fuel, chemicals, rubber plastics	75 321	72 777	74 020	72 828	72 990	70 908	..	**15 Coke, pétrole, combust. nucl., prod. chimi., caout. & mat. plast.**
16 Coke, ref. petrol. prod. & nuclear fuel	2 097	1 946	1 803	2 086	1 739	1 308	..	16 Coke, raffin. pétr. & combust. nucl.
17 Chemicals & chemical products	61 124	60 056	61 068	58 782	59 573	58 751	..	17 Produits chimiques
18 *Chemical prod.(less pharmaceuticals)*	41 083	39 985	41 098	40 702	40 840	39 936	..	18 *Prod. chimi. (moins prod. pharmaceutiques)*
19 *Pharmaceuticals*	20 041	20 071	19 970	18 080	18 733	18 815	..	19 *Produits pharmaceutiques*
20 Rubber & plastic products	12 100	10 775	11 149	11 960	11 678	10 849	..	20 Caoutchouc et matières plastiques
21 Non-metallic mineral products	8 633	8 853	8 953	8 774	9 489	8 994	..	**21 Produits minéraux non métalliques**
22 Basic metals	11 313	11 080	11 072	11 192	9 846	9 218	..	**22 Produits métallurgiques de base**
23 Basic metals, ferrous	6 093	5 509	5 472	5 806	4 826	4 700	..	23 Produits métallur. de base, ferreux
24 Basic metals, non-ferrous	5 220	5 571	5 600	5 386	5 020	4 518	..	24 Prod. métallur. de base, non ferreux
25 Fabricated metal products	6 757	7 759	9 262	7 734	7 466	7 818	..	**25 Ouvrages en métaux (sauf mach.& mat.)**
26 Machinery equipment, instruments and transport equipment	232 752	240 893	237 027	239 845	255 115	262 518	..	**26 Machines & matériel, instruments & mat. de transport**
27 Machinery n.e.c.	33 973 [o]	34 728 [o]	33 201 [o]	35 636 [o]	36 492 [o]	37 632 [o]	..	27 Machines, n.c.a.
28 Office, account. & computing machinery	.. [n]	.. [n]	.. [n]	.. [n]	.. [n]	.. [n]	..	28 Mach. de bur., compta. & trait. inform.
29 Electrical machinery	41 051	40 129	41 240	39 598	42 248	39 960	..	29 Machines électriques
30 Electro. equip. (radio, TV & commun.)	103 959 [o]	106 686 [o]	106 555 [o]	106 686 [o]	113 478 [o]	120 078 [o]	..	30 Equip. électro. (radio, tv & communic.)
31 *Electro. comp. (inc. semi-conductors)*	..	..	..	..	..	..	..	31 *Compos. électro. (semi-cond. incl.)*
32 *TV, radio & communications equip.*	..	..	..	..	..	..	..	32 *Appareils radio, tv et de communi.*
33 Instruments	18 146	20 174	17 583	17 325	21 585	23 306	..	33 Instruments
34 Motor vehicles	29 942	33 218	33 063	35 143	35 606	36 934	..	34 Véhicules automobiles
35 Other transport equipment	5 681	5 958	5 385	5 457	5 706	4 608	..	35 Autres matériels de transport
36 *Ships*	3 159	3 413	3 104	2 958	2 973	2 274	..	36 *Navires*
37 *Aerospace*	82	84	83	102	117	159	..	37 *Aérospatiale*
38 *Other transport equipment n.e.c.*	2 440	2 461	2 198	2 397	2 616	2 175	..	38 *Autres matériels de transport n.c.a.*
39 Furniture, other manufacturing n.e.c.	5 029 [m,o]	4 489 [m,o]	4 618 [m,o]	4 675 [m,o]	5 579 [m]	6 799 [m]	..	**39 Meubles, autres activ. de fabricat. n.c.a.**
40 Furniture	.. [n]	.. [n]	.. [n]	.. [n]	.. [n]	.. [n]	..	40 Meubles
41 Other manufacturing n.e.c.	5 029 [o]	4 489 [o]	4 618 [o]	4 675 [o]	5 579 [o]	6 799 [o]	..	41 Autres activités de fabrication n.c.a.
42 Recycling	..	..	..	..	..	..	..	**42 Récupération**
43 ELECTRICITY, GAS & WATER SUPPLY	2 046	1 964	2 025	2 017	1 990	1 997	..	**43 DISTRIBUTION D'ELECTRICITE, GAZ & EAU**
44 CONSTRUCTION	8 082	7 553	9 652	14 813	7 797	9 023	..	**44 CONSTRUCTION**
45 SERVICES SECTOR	4 419	4 433	19 899 [a]	18 802	35 421	32 217	..	**45 SECTEUR DES SERVICES**
46 Wholesale, retail trade & repair of motor vehicles, etc.	..	..	..	..	..	..	..	**46 Commerce (gros & détail), réparation de véhicules, etc.**
47 Hotels & restaurants	..	..	..	..	..	..	..	**47 Hôtels et restaurants**
48 Transport & storage	296	309	268	400	355	538	..	**48 Transports, entreposage**
49 Communications	4 123	4 124	4 081	4 860	6 336	6 368	..	**49 Communications**
50 Post	..	..	..	..	..	..	..	50 Postes
51 Telecommunications	..	..	..	..	..	..	..	51 Télécommunications
52 Financial intermediation (inc. insur.)	..	..	..	..	..	..	..	**52 Intermédiation financ. (assurances inc.)**
53 Real estate, renting & busin. activity	..	..	15 550	13 542	28 730	25 311	..	**53 Immobil., locat. & serv. aux entreprises**
54 Computer & related activities	..	..	15 550	13 542	28 730	25 311	..	54 Informatique et activités connexes
55 *Software consultancy*	..	..	..	..	..	..	..	55 *Production de logiciels*
56 *Other computer services n.e.c.*	..	..	..	..	..	..	..	56 *Autres activités informatiques*
57 Research & development	..	..	..	..	..	..	..	57 Recherche & développement expérimental
58 Other business activities n.e.c.	..	..	..	..	..	..	..	58 Autres activ. & serv. aux entreprises
59 Comm., soc. & pers. serv. activ., etc.	..	..	..	..	..	..	..	**59 Adm. pub., serv. soc. & collec., etc**
TOTAL BUSINESS ENTERPRISE SECTOR	**376 639 [l]**	**384 100 [l]**	**400 361 [a]**	**404 232**	**429 195**	**433 758**	**..**	**TOTAL SECTEUR DES ENTREPRISES**

Source: OECD, R&D database (DSTI/EAS Division), January 2002.

Source : OCDE, base de données R-D (DSTI/Division AES), janvier 2002.

Higher education intramural expenditure on R&D - Dépenses intra-muros de R-D de l'enseignement supérieur
(HERD) by field of science - (DIRDES) par discipline scientifique

	1994	1995	1996	1997	1998	1999	2000	
				14.1 Million JPY - Millions JPY				
Natural sciences	..	300 440 [l]	261 634 [a]	258 052	258 587	253 774	..	Sciences exactes et naturelles
Engineering	..	673 989 [l]	507 149 [a]	509 804	550 153	558 455	..	Sciences de l'ingénieur
Medical sciences	..	776 988 [l]	528 930 [a]	537 036	574 335	548 208	..	Sciences médicales
Agricultural sciences	..	123 252 [l]	86 630 [a]	89 536	104 673	103 110	..	Sciences agricoles
Sub-total NSE	1 685 526 [l]	1 874 668 [l]	1 384 344 [a]	1 394 427	1 487 748	1 463 547	..	Sous-total SENI
Social sciences	..	..	..	..	..	..	..	Sciences sociales
Humanities	..	..	..	..	..	..	..	Sciences humaines
Sub-total SSH	1 067 024 [l]	1 107 518 [l]	704 517 [a]	717 303	764 410	767 612	..	Sous-total SSH
Total NSE+SSH	2 752 551 [l]	2 982 186 [l]	2 088 861 [a]	2 111 730	2 252 160	2 231 159	..	*Total SENI+SSH*
				14.2 Million constant USD (1995 prices and PPPs) - Millions de USD constants (prix et PPA de 1995)				
Natural sciences	..	1 767.9 [l]	1 552.2 [a]	1 525.5	1 529.5	1 522.2	..	Sciences exactes et naturelles
Engineering	..	3 966.0 [l]	3 008.8 [a]	3 013.8	3 254.1	3 349.7	..	Sciences de l'ingénieur
Medical sciences	..	4 572.1 [l]	3 138.1 [a]	3 174.8	3 397.2	3 288.2	..	Sciences médicales
Agricultural sciences	..	725.3 [l]	514.0 [a]	529.3	619.1	618.5	..	Sciences agricoles
Sub-total NSE	9 882.2 [l]	11 031.2 [l]	8 213.1 [a]	8 243.4	8 800.0	8 778.5	..	Sous-total SENI
Social sciences	..	..	..	..	..	..	..	Sciences sociales
Humanities	..	..	..	..	..	..	..	Sciences humaines
Sub-total SSH	6 255.9 [l]	6 517.0 [l]	4 179.8 [a]	4 240.5	4 521.5	4 604.2	..	Sous-total SSH
Total NSE+SSH	16 138.1 [l]	17 548.3 [l]	12 392.9 [a]	12 483.8	13 321.4	13 382.8	..	*Total SENI+SSH*

Table 15 - Tableau 15

Higher education intramural expenditure on R&D - Dépenses intra-muros de R-D de l'enseignement supérieur
(HERD) by field of science and type of costs - (DIRDES) par discipline scientifique et par type de coûts

Million JPY *Millions JPY*

	1994	1995	1996	1997	1998	1999	2000	
Current expenditure								**Dépenses courantes**
Natural sciences	..	220 452 [l]	184 308 [a]	188 128	189 612	197 825	..	Sciences exactes et naturelles
Engineering	..	536 889 [l]	374 327 [a]	382 204	402 764	412 247	..	Sciences de l'ingénieur
Medical sciences	..	664 713 [l]	422 203 [a]	438 658	460 237	464 779	..	Sciences médicales
Agricultural sciences	..	104 778 [l]	73 043 [a]	75 746	78 985	81 611	..	Sciences agricoles
Sub-total NSE	..	1 526 832 [l]	1 053 881 [a]	1 084 735	1 131 598	1 156 462	..	Sous-total SENI
Social sciences	..	..	..	..	..	..	..	Sciences sociales
Humanities	..	..	..	..	..	..	..	Sciences humaines
Sub-total SSH	..	1 019 910 [l]	615 154 [a]	631 276	654 190	665 134	..	Sous-total SSH
Total NSE+SSH	2 433 014 [l]	2 546 741 [l]	1 669 036 [a]	1 716 011	1 785 788	1 821 595	..	*Total SENI+SSH*
Capital expenditure								**Dépenses en capital**
Natural sciences	..	79 988	77 325	69 924	68 975	55 948	..	Sciences exactes et naturelles
Engineering	..	137 100	132 822	127 600	147 389	146 208	..	Sciences de l'ingénieur
Medical sciences	..	112 275	106 727	98 378	114 098	83 429	..	Sciences médicales
Agricultural sciences	..	18 474	13 588	13 790	25 688	21 499	..	Sciences agricoles
Sub-total NSE	..	347 837	330 462	309 692	356 151	307 084	..	Sous-total SENI
Social sciences	..	..	..	..	..	..	..	Sciences sociales
Humanities	..	..	..	..	..	..	..	Sciences humaines
Sub-total SSH	..	87 609	89 363	86 027	110 221	102 478	..	Sous-total SSH
Total NSE+SSH	319 537	435 446	419 826	395 719	466 372	409 563	..	*Total SENI+SSH*

Source: OECD, R&D database (DSTI/EAS Division), January 2002.

Source : OCDE, base de données R-D (DSTI/Division AES), janvier 2002.

JAPAN — Table 16 - Tableau 16 — JAPON

Higher education intramural expenditure on R&D - Dépenses intra-muros de R-D de l'enseignement supérieur

(HERD) by field of science and source of funds - (DIRDES) par discipline scientifique et par source de financement

Million JPY — *Millions JPY*

	1994	1995	1996	1997	1998	1999	2000	
Direct government								**État direct**
Natural sciences	..	23 281	22 447 [a]	24 763	24 862	29 596	..	Sciences exactes et naturelles
Engineering	..	51 769	39 889 [a]	45 250	52 801	54 747	..	Sciences de l'ingénieur
Medical sciences	..	57 740	39 598 [a]	43 494	50 318	52 508	..	Sciences médicales
Agricultural sciences	..	8 315	6 912 [a]	8 563	9 952	11 163	..	Sciences agricoles
Sub-total NSE	..	141 105	108 846 [a]	122 069	137 932	148 015	..	Sous-total SENI
Social sciences	..	..	..	..	..	..	..	Sciences sociales
Humanities	..	..	..	..	..	..	..	Sciences humaines
Sub-total SSH	..	68 953	42 068 [a]	43 558	47 777	48 020	..	Sous-total SSH
Total NSE+SSH	..	210 058	150 914 [a]	165 628	185 708	196 035	..	*Total SENI+SSH*
General university funds								**Fonds généraux des universités**
Natural sciences	..	.. [n]	.. [n]	.. [n]	.. [n]	..	..	Sciences exactes et naturelles
Engineering	..	.. [n]	.. [n]	.. [n]	.. [n]	..	..	Sciences de l'ingénieur
Medical sciences	..	.. [n]	.. [n]	.. [n]	.. [n]	..	..	Sciences médicales
Agricultural sciences	..	.. [n]	.. [n]	.. [n]	.. [n]	..	..	Sciences agricoles
Sub-total NSE	..	.. [n]	.. [n]	.. [n]	.. [n]	..	..	Sous-total SENI
Social sciences	..	..	..	..	..	..	..	Sciences sociales
Humanities	..	..	..	..	..	..	..	Sciences humaines
Sub-total SSH	..	.. [n]	.. [n]	.. [n]	.. [n]	..	..	Sous-total SSH
Total NSE+SSH	..	.. [n]	.. [n]	.. [n]	.. [n]	..	..	*Total SENI+SSH*
Sub-total government								**Sous-total État**
Natural sciences	..	..	..	..	..	..	..	Sciences exactes et naturelles
Engineering	..	..	..	..	..	..	..	Sciences de l'ingénieur
Medical sciences	..	..	..	..	..	..	..	Sciences médicales
Agricultural sciences	..	..	..	..	..	..	..	Sciences agricoles
Sub-total NSE	..	..	..	..	..	..	..	Sous-total SENI
Social sciences	..	..	..	..	..	..	..	Sciences sociales
Humanities	..	..	..	..	..	..	..	Sciences humaines
Sub-total SSH	..	..	..	..	..	..	..	Sous-total SSH
Total NSE+SSH	..	..	..	..	..	..	..	*Total SENI+SSH*
Higher education								**Enseignement supérieur**
Natural sciences	..	273 134 [l,o]	235 626 [a,o]	228 951 [o]	230 457 [o]	220 534	..	Sciences exactes et naturelles
Engineering	..	598 745 [l,o]	451 496 [a,o]	447 861 [o]	479 737 [o]	486 679	..	Sciences de l'ingénieur
Medical sciences	..	678 448 [l,o]	460 597 [a,o]	463 961 [o]	494 977 [o]	467 109	..	Sciences médicales
Agricultural sciences	..	111 946 [l,o]	76 866 [a,o]	78 487 [o]	92 325 [o]	88 782	..	Sciences agricoles
Sub-total NSE	..	1 662 273 [l,o]	1 224 586 [a,o]	1 219 261 [o]	1 297 496 [o]	1 263 105	..	Sous-total SENI
Social sciences	..	..	..	..	..	..	..	Sciences sociales
Humanities	..	..	..	..	..	..	..	Sciences humaines
Sub-total SSH	..	1 035 194 [l,o]	660 338 [a,o]	671 539 [o]	714 204 [o]	717 139	..	Sous-total SSH
Total NSE+SSH	..	2 697 467 [l,o]	1 884 924 [a,o]	1 890 799 [o]	2 011 700 [o]	1 980 244	..	*Total SENI+SSH*
Private non-profit								**Institutions sans but lucratif**
Natural sciences	..	123	191	291	336	399	..	Sciences exactes et naturelles
Engineering	..	1 247	1 312	1 199	1 231	1 553	..	Sciences de l'ingénieur
Medical sciences	..	2 241	1 459	1 490	1 365	1 522	..	Sciences médicales
Agricultural sciences	..	258	208	314	186	226	..	Sciences agricoles
Sub-total NSE	..	3 869	3 170	3 294	3 118	3 700	..	Sous-total SENI
Social sciences	..	..	..	..	..	..	..	Sciences sociales
Humanities	..	..	..	..	..	..	..	Sciences humaines
Sub-total SSH	..	163	90	95	111	189	..	Sous-total SSH
Total NSE+SSH	4 383	4 032	3 260	3 389	3 229	3 889	..	*Total SENI+SSH*
Business entreprise								**Entreprises**
Natural sciences	..	3 804	3 124	3 713	2 502	2 809	..	Sciences exactes et naturelles
Engineering	..	22 198	14 427	15 438	16 329	15 399	..	Sciences de l'ingénieur
Medical sciences	..	38 435	27 240	28 047	27 661	27 043	..	Sciences médicales
Agricultural sciences	..	2 725	2 638	2 169	2 209	2 939	..	Sciences agricoles
Sub-total NSE	..	67 162	47 429	49 367	48 701	48 190	..	Sous-total SENI
Social sciences	..	..	..	..	..	..	..	Sciences sociales
Humanities	..	..	..	..	..	..	..	Sciences humaines
Sub-total SSH	..	3 110	1 964	1 938	2 275	2 252	..	Sous-total SSH
Total NSE+SSH	63 840	70 271	49 393	51 305	50 975	50 442	..	*Total SENI+SSH*
Funds from abroad								**Fonds de l'étranger**
Natural sciences	..	98	247	333	431	435	..	Sciences exactes et naturelles
Engineering	..	30	24	57	55	77	..	Sciences de l'ingénieur
Medical sciences	..	125	37	45	16	25	..	Sciences médicales
Agricultural sciences	..	7	6	4	0	0	..	Sciences agricoles
Sub-total NSE	..	259	314	438	502	536	..	Sous-total SENI
Social sciences	..	..	..	..	..	..	..	Sciences sociales
Humanities	..	..	..	..	..	..	..	Sciences humaines
Sub-total SSH	..	98	59	170	42	11	..	Sous-total SSH
Total NSE+SSH	231	357	373	608	544	547	..	*Total SENI+SSH*

Source: OECD, R&D database (DSTI/EAS Division), January 2002.

Source : OCDE, base de données R-D (DSTI/Division AES), janvier 2002.

Table 17 - Tableau 17

Higher education R&D personnel (HEMP) - Personnel de R-D du secteur de l'enseignement supérieur

by field of science and occupation - par discipline scientifique et par profession

In full-time equivalent | *En équivalence plein-temps*

	1994	1995	1996	1997	1998	1999	2000	
Total R&D personnel								**Personnel total de R-D**
Natural sciences	21 460 [l]	22 626 [l]	20 318 [a]	20 997	19 019	19 303	..	Sciences exactes et naturelles
Engineering	52 897 [l]	54 445 [l]	41 600 [a]	42 910	43 956	45 062	..	Sciences de l'ingénieur
Medical sciences	103 314 [l]	104 920 [l]	81 566 [a]	82 668	82 466	81 386	..	Sciences médicales
Agricultural sciences	12 210 [l]	12 096 [l]	10 074 [a]	10 277	10 637	10 806	..	Sciences agricoles
Sub-total NSE	189 881 [l]	194 087 [l]	153 558 [a]	156 853	156 078	156 555	..	Sous-total SENI
Social sciences	..	..	..	..	..	..	..	Sciences sociales
Humanities	..	..	..	..	..	..	..	Sciences humaines
Sub-total SSH	94 362 [l]	96 462 [l]	64 000 [a]	65 432	69 101	71 006	..	Sous-total SSH
Total NSE+SSH	284 243 [l]	290 549 [l]	217 558 [a]	222 285	225 179	227 562	..	*Total SENI+SSH*
Researchers								**Chercheurs**
Natural sciences	..	17 544 [l]	15 076 [a]	15 723	13 945	13 992	..	Sciences exactes et naturelles
Engineering	..	43 521 [l]	30 504 [a]	31 447	32 416	33 527	..	Sciences de l'ingénieur
Medical sciences	..	90 595 [l]	67 313 [a]	67 877	67 408	65 933	..	Sciences médicales
Agricultural sciences	..	9 644 [l]	7 720 [a]	7 933	8 321	8 497	..	Sciences agricoles
Sub-total NSE	..	161 304 [l]	120 613 [a]	122 981	122 090	121 947	..	Sous-total SENI
Social sciences	..	..	..	..	..	..	..	Sciences sociales
Humanities	..	..	..	..	..	..	..	Sciences humaines
Sub-total SSH	..	81 558 [l]	49 404 [a]	51 112	54 537	56 470	..	Sous-total SSH
Total NSE+SSH	235 702 [l]	242 862 [l]	170 017 [a]	174 093	176 627	178 418	..	*Total SENI+SSH*
Technicians								**Technicians**
Natural sciences	..	566	492	719	695	741	..	Sciences exactes et naturelles
Engineering	..	1 457	1 568	1 565	1 551	1 454	..	Sciences de l'ingénieur
Medical sciences	..	4 741	4 688	5 118	4 831	4 964	..	Sciences médicales
Agricultural sciences	..	301	267	325	231	251	..	Sciences agricoles
Sub-total NSE	..	7 065	7 015	7 727	7 308	7 410	..	Sous-total SENI
Social sciences	..	..	..	..	..	..	..	Sciences sociales
Humanities	..	..	..	..	..	..	..	Sciences humaines
Sub-total SSH	..	2 321	2 408	2 374	2 370	2 276	..	Sous-total SSH
Total NSE+SSH	9 784	9 386	9 423	10 101	9 678	9 686	..	*Total SENI+SSH*
Other								**Autres**
Natural sciences	..	4 516	4 750	4 555	4 379	4 570	..	Sciences exactes et naturelles
Engineering	..	9 467	9 528	9 898	9 989	10 081	..	Sciences de l'ingénieur
Medical sciences	..	9 584	9 565	9 673	10 227	10 489	..	Sciences médicales
Agricultural sciences	..	2 151	2 087	2 019	2 085	2 058	..	Sciences agricoles
Sub-total NSE	..	25 718	25 930	26 145	26 680	27 198	..	Sous-total SENI
Social sciences	..	..	..	..	..	..	..	Sciences sociales
Humanities	..	..	..	..	..	..	..	Sciences humaines
Sub-total SSH	..	12 583	12 188	11 946	12 194	12 260	..	Sous-total SSH
Total NSE+SSH	38 757	38 301	38 118	38 091	38 874	39 458	..	*Total SENI+SSH*

Table 18 - Tableau 18

Government budget appropriations or outlays for R&D (GBAORD) - Crédits budgétaires publics de R-D (CBPRD)

by socio-economic objective - par objectif socio-économique

Million JPY | *Millions JPY*

	1995	1996	1997	1998	1999	2000	2001	
Agriculture, forestry and fishing	85 418.0	94 944.0	100 894.0	104 501.0	109 176.0	113 754.0	..	Agriculture, sylviculture et pêche
Industrial development	93 820.0	94 721.0	197 427.0	208 023.0	205 485.0	221 917.0	..	Développement industriel
Energy	510 077.0	653 634.0	605 329.0	603 474.0	608 322.0	593 330.0	..	Energie
Transport & telecommunications	39 085.0	32 376.0	41 585.0	44 575.0	70 492.0	78 595.0	..	*Transports & télécommunications*
Urban & rural planning	8 441.0	33 885.0	39 144.0	39 767.0	41 504.0	44 440.0	..	*Aménagement urbain et rural*
ST Infrastructure	47 526.0	66 261.0	80 729.0	84 342.0	111 996.0	123 035.0	..	ST Infrastructure
Prevention of pollution	3 167.0	3 206.0	4 031.0	4 347.0	5 436.0	6 132.0	..	*Prévention de la pollution*
Identificat. & treatment of pollution	10 805.0	12 990.0	13 047.0	14 179.0	16 993.0	20 115.0	..	*Détect. pollut. & mesures antipollut.*
ST Environment	13 972.0	16 196.0	17 078.0	18 526.0	22 429.0	26 247.0	..	ST Environnement
Health	70 813.0	99 367.0	120 756.0	110 471.0	117 102.0	127 527.0	..	Santé
Social development & services	29 160.0	28 872.0	28 162.0	28 891.0	27 146.0	29 933.0	..	Développement et services sociaux
Earth and atmosphere	31 646.0	37 875.0	40 486.0	40 827.0	46 750.0	55 289.0	..	Milieu terrestre et atmosphère
Advancement of research	241 570.0	270 851.0	324 263.0	362 857.0	405 774.0	459 690.0	..	*Promotion de la recherche*
General university funds	1 036 663.0 [g]	1 095 843.0 [g]	1 122 780.0 [g]	1 135 015.0 [g]	1 156 116.0 [g]	1 164 087.0 [g]	..	*Fonds généraux des universités*
ST Advancement of knowledge	1 278 233.0 [g]	1 366 694.0 [g]	1 447 043.0 [g]	1 497 872.0 [g]	1 561 890.0 [g]	1 623 777.0 [g]	..	ST promotion gén. des connaiss.
Civil space	184 386.0	186 367.0	189 126.0	190 976.0	198 419.0	182 650.0	..	Espace civil
Defence	154 499.0 [m]	165 279.0 [m]	175 340.0 [m]	144 176.0 [m]	146 529.0 [m]	136 081.0 [m]	..	Défense
Not elsewhere classified	0.0	243.0	242.0	101.0	1 482.0	50 780.0	..	Non classé
Total	**2 499 549.0 [g]**	**2 810 452.0 [g]**	**3 002 610.0 [g]**	**3 032 179.0 [g]**	**3 156 728.0 [g]**	**3 284 320.0 [g]**	**..**	**Total**

Source: OECD, R&D database (DSTI/EAS Division), January 2002.

Source : OCDE, base de données R-D (DSTI/Division AES), janvier 2002.

Table 19 - Tableau 19

Technology balance of payments - Balance des paiements technologiques

by type of transfer - selon la nature des transferts

Million JPY *Millions JPY*

	1983	1984	1985	1986	1987	1988	1989	1990	1991	
Receipts										**Recettes**
Related firms	..	..	..	..	..	..	..	..	..	Firmes affiliées
Unrelated firms	..	..	..	..	..	..	..	..	..	Firmes non affiliées
Other	..	..	..	..	..	..	..	..	..	Autre
Total	240 887	277 512	234 220	224 078	215 575	246 255	329 348	339 352	370 552	*Total*
Payments										**Paiements**
Related firms	..	..	..	..	..	..	..	..	..	Firmes affiliées
Unrelated firms	..	..	..	..	..	..	..	..	..	Firmes non affiliées
Other	..	..	..	..	..	..	..	..	..	Autre
Total	279 280	281 447	293 173	260 577	283 245	312 195	329 925	371 907	394 661	*Total*

	1992	1993	1994	1995	1996	1997	1998	1999	2000	
Receipts										**Recettes**
Related firms	..	..	..	..	..	..	..	..	..	Firmes affiliées
Unrelated firms	..	..	..	..	..	..	..	..	..	Firmes non affiliées
Other	..	..	..	..	..	..	..	..	..	Autre
Total	377 691	400 362	462 128	562 077	703 033	831 563	916 098	960 800	..	*Total*
Payments										**Paiements**
Related firms	..	..	..	..	..	..	..	..	..	Firmes affiliées
Unrelated firms	..	..	..	..	..	..	..	..	..	Firmes non affiliées
Other	..	..	..	..	..	..	..	..	..	Autre
Total	413 908	362 974	370 693	391 715	451 169	438 400	430 054	410 296	..	*Total*

Source: OECD, TBPdatabase (DSTI/EAS Division), Januay 2002.

Source : OCDE, base de données TBP (DSTI/Division AES), janvier 2002.

Gross domestic expenditure on R&D (GERD) - Dépenses intérieures brutes de R-D (DIRD)

by sector of performance and source of funds - par secteur d'exécution et par source de financement

	1994	1995	1996	1997	1998	1999	2000	
	1.1 Million KRW - Millions KRW							
Business enterprise								**Entreprises**
financed by								*financé par*
Business enterprise	..	6 700 235	7 544 915	8 381 794	7 494 610	7 993 094	..	Entreprises
Direct government	..	247 189	347 430	423 374	466 458	496 518	..	État direct
Higher education	..	1 948	603	1 600	453	415	..	Enseignement supérieur
Private non-profit	..	9 843	60 505	33 711	7 481	17 369	..	Institutions sans but lucratif
Funds from abroad	..	1 107	10 159	4 828	3 071	3 761	..	Fonds de l'étranger
Total	..	6 960 322	7 963 612	8 845 307	7 972 073	8 511 157	..	*Total*
Government								**État**
financed by								*financé par*
Business enterprise	..	264 260 g	309 683 g	190 288 g	133 095 g	118 620 g	..	Entreprises
Direct government	..	1 195 733 g	1 389 688 g	1 729 364 g	1 840 227 g	1 574 270 g	..	État direct
Higher education	..	574 g	363 g	368 g	1 378 g	466 g	..	Enseignement supérieur
Private non-profit	..	140 341 g	63 673 g	2 233 g	2 255 g	29 176 g	..	Institutions sans but lucratif
Funds from abroad	..	40 g	560 g	223 g	2 264 g	365 g	..	Fonds de l'étranger
Total	..	1 600 948 g	1 763 967 g	1 922 476 g	1 979 219 g	1 722 897 g	..	*Total*
Higher education								**Enseignement supérieur**
financed by								*financé par*
Business enterprise	..	173 012 g	514 721 g	189 540 g	165 159 g	154 500 g	..	Entreprises
Direct government	..	..	..	..	..	..	..	État direct
General university funds	..	..	..	..	..	..	..	Fonds généraux des universités
Sub-total government	..	338 727 g	448 063 g	611 477 g	585 675 g	806 828 g	..	Sous-Total État
Higher education	..	246 548 g	20 530 g	439 511 g	504 533 g	457 646 g	..	Enseignement supérieur
Private non-profit	..	12 467 g	32 591 g	24 337 g	6 875 g	9 237 g	..	Institutions sans but lucratif
Funds from abroad	..	158 g	2 916 g	6 734 g	2 832 g	3 210 g	..	Fonds de l'étranger
Total	..	770 912 g	1 018 821 g	1 271 599 g	1 265 074 g	1 431 421 g	..	*Total*
Private non-profit								**Institutions sans but lucratif**
financed by								*financé par*
Business enterprise	..	61 722 g	90 749 g	67 574 g	44 219 g	73 826 g	..	Entreprises
Direct government	..	15 846 g	17 639 g	28 719 g	47 281 g	91 081 g	..	État direct
Higher education	..	45 g	140 g	24 g	365 g	198 g	..	Enseignement supérieur
Private non-profit	..	30 785 g	23 122 g	50 098 g	28 121 g	91 166 g	..	Institutions sans but lucratif
Funds from abroad	..	26 g	..	10 g	265 g	6 g	..	Fonds de l'étranger
Total	..	108 424 g	131 650 g	146 425 g	120 251 g	256 277 g	..	*Total*
GERD								**DIRD**
financed by								*financé par*
Business enterprise	..	7 199 229 g	8 460 068 g	8 829 196 g	7 837 083 g	8 340 040 g	..	Entreprises
Direct government	..	..	..	..	..	..	..	État direct
General university funds	..	..	..	..	..	..	..	Fonds généraux des universités
Sub-total government	..	1 797 495 g	2 202 820 g	2 792 934 g	2 939 641 g	2 968 697 g	..	Sous-Total État
Higher education	..	249 115 g	21 636 g	441 503 g	506 729 g	458 725 g	..	Enseignement supérieur
Private non-profit	..	193 436 g	179 891 g	110 379 g	44 732 g	146 948 g	..	Institutions sans but lucratif
Funds from abroad	..	1 331 g	13 635 g	11 796 g	8 432 g	7 342 g	..	Fonds de l'étranger
Total	7 894 746 g	9 440 606 g	10 878 050 g	12 185 807 g	11 336 617 g	11 921 752 g	..	*Total*
	1.2 Million constant USD (1995 prices and PPPs) - Millions de USD constants (prix et PPA de 1995)							
Business enterprise								**Entreprises**
financed by								*financé par*
Business enterprise	..	10 891.2	11 805.3	12 714.7	10 821.1	11 780.7	..	Entreprises
Direct government	..	401.8	543.6	642.2	673.5	731.8	..	État direct
Higher education	..	3.2	0.9	2.4	0.7	0.6	..	Enseignement supérieur
Private non-profit	..	16.0	94.7	51.1	10.8	25.6	..	Institutions sans but lucratif
Funds from abroad	..	1.8	15.9	7.3	4.4	5.5	..	Fonds de l'étranger
Total	..	11 314.0	12 460.5	13 417.8	11 510.5	12 544.2	..	*Total*
Government								**État**
financed by								*financé par*
Business enterprise	..	429.6 g	484.6 g	288.7 g	192.2 g	174.8 g	..	Entreprises
Direct government	..	1 943.7 g	2 174.4 g	2 623.4 g	2 657.0 g	2 320.2 g	..	État direct
Higher education	..	0.9 g	0.6 g	0.6 g	2.0 g	0.7 g	..	Enseignement supérieur
Private non-profit	..	228.1 g	99.6 g	3.4 g	3.3 g	43.0 g	..	Institutions sans but lucratif
Funds from abroad	..	0.1 g	0.9 g	0.3 g	3.3 g	0.5 g	..	Fonds de l'étranger
Total	..	2 602.3 g	2 760.0 g	2 916.3 g	2 857.7 g	2 539.3 g	..	*Total*
Higher education								**Enseignement supérieur**
financed by								*financé par*
Business enterprise	..	281.2 g	805.4 g	287.5 g	238.5 g	227.7 g	..	Entreprises
Direct government	..	..	..	..	..	..	..	État direct
General university funds	..	..	..	..	..	..	..	Fonds généraux des universités
Sub-total government	..	550.6 g	701.1 g	927.6 g	845.6 g	1 189.1 g	..	Sous-Total État
Higher education	..	400.8 g	32.1 g	666.7 g	728.5 g	674.5 g	..	Enseignement supérieur
Private non-profit	..	20.3 g	51.0 g	36.9 g	9.9 g	13.6 g	..	Institutions sans but lucratif
Funds from abroad	..	0.3 g	4.6 g	10.2 g	4.1 g	4.7 g	..	Fonds de l'étranger
Total	..	1 253.1 g	1 594.1 g	1 928.9 g	1 826.6 g	2 109.7 g	..	*Total*
Private non-profit								**Institutions sans but lucratif**
financed by								*financé par*
Business enterprise	..	100.3 g	142.0 g	102.5 g	63.8 g	108.8 g	..	Entreprises
Direct government	..	25.8 g	27.6 g	43.6 g	68.3 g	134.2 g	..	État direct
Higher education	..	0.1 g	0.2 g	0.0 g	0.5 g	0.3 g	..	Enseignement supérieur
Private non-profit	..	50.0 g	36.2 g	76.0 g	40.6 g	134.4 g	..	Institutions sans but lucratif
Funds from abroad	..	0.0 g	..	0.0 g	0.4 g	0.0 g	..	Fonds de l'étranger
Total	..	176.2 g	206.0 g	222.1 g	173.6 g	377.7 g	..	*Total*
GERD								**DIRD**
financed by								*financé par*
Business enterprise	..	11 702.3 g	13 237.3 g	13 393.4 g	11 315.6 g	12 292.0 g	..	Entreprises
Direct government	..	..	..	..	..	..	..	État direct
General university funds	..	..	..	..	..	..	..	Fonds généraux des universités
Sub-total government	..	2 921.8 g	3 446.7 g	4 236.7 g	4 244.4 g	4 375.4 g	..	Sous-Total État
Higher education	..	404.9 g	33.9 g	669.7 g	731.6 g	676.1 g	..	Enseignement supérieur
Private non-profit	..	314.4 g	281.5 g	167.4 g	64.6 g	216.6 g	..	Institutions sans but lucratif
Funds from abroad	..	2.2 g	21.3 g	17.9 g	12.2 g	10.8 g	..	Fonds de l'étranger
Total	13 747.1 g	15 345.7 g	17 020.6 g	18 485.2 g	16 368.4 g	17 570.9 g	..	*Total*

Source: OECD, R&D database (DSTI/EAS Division), January 2002. *Source* : OCDE, base de données R-D (DSTI/Division AES), janvier 2002.

Table 2 - Tableau 2

Gross domestic expenditure on R&D (GERD) - Dépenses intérieures brutes de R-D (DIRD)

by sector of performance and type of costs - par secteur d'exécution et par type de coûts

Million KRW — *Millions KRW*

	1994	1995	1996	1997	1998	1999	2000	
Business enterprise								**Entreprises**
Labour costs	..	..	2 692 034	3 070 402	2 813 093	..	..	Coûts du personnel
Other current costs	..	..	3 198 055	3 752 081	3 250 649	..	..	Autres coûts courants
Total current expenditure	..	..	5 890 089	6 822 483	6 063 742	..	..	*Total dépenses courantes*
Land and buildings	..	..	478 156	344 658	261 875	..	..	Terrains et constructions
Instruments and equipment	..	..	1 595 366	1 678 166	1 646 456	..	..	Instruments et équipements
Total capital expenditure	..	..	2 073 522	2 022 824	1 908 331	..	..	*Total dépenses en capital*
Total	..	6 960 322	7 963 612	8 845 307	7 972 073	8 511 157	..	*Total*
Government								**État**
Labour costs	..	..	551 228 [g]	581 879 [g]	623 538 [g,s]	..	..	Coûts du personnel
Other current costs	..	..	853 455 [g]	917 518 [g]	977 713 [g,s]	..	..	Autres coûts courants
Total current expenditure	..	..	1 404 683 [g]	1 499 397 [g]	1 601 251 [g,s]	..	..	*Total dépenses courantes*
Land and buildings	..	..	174 977 [g]	181 447 [g]	176 181 [g,s]	..	..	Terrains et constructions
Instruments and equipment	..	..	184 308 [g]	241 631 [g]	217 062 [g,s]	..	..	Instruments et équipements
Total capital expenditure	..	..	359 285 [g]	423 078 [g]	393 243 [g,s]	..	..	*Total dépenses en capital*
Total	..	1 600 948 [g]	1 763 967 [g]	1 922 476 [g]	1 979 219 [g]	1 722 897 [g]	..	*Total*
Higher education								**Enseignement supérieur**
Labour costs	..	..	337 461 [g]	397 364 [g]	424 528 [g]	..	..	Coûts du personnel
Other current costs	..	..	495 648 [g]	595 311 [g]	590 106 [g]	..	..	Autres coûts courants
Total current expenditure	..	..	833 109 [g]	992 675 [g]	1 014 634 [g]	..	..	*Total dépenses courantes*
Land and buildings	..	..	31 892 [g]	69 148 [g]	60 474 [g]	..	..	Terrains et constructions
Instruments and equipment	..	..	153 821 [g]	209 777 [g]	189 966 [g]	..	..	Instruments et équipements
Total capital expenditure	..	..	185 713 [g]	278 925 [g]	250 440 [g]	..	..	*Total dépenses en capital*
Total	..	770 912 [g]	1 018 821 [g]	1 271 599 [g]	1 265 074 [g]	1 431 421 [g]	..	*Total*
Private non-profit								**Institutions sans but lucratif**
Labour costs	..	..	40 832 [g]	41 558 [g]	19 333 [g,s]	..	..	Coûts du personnel
Other current costs	..	..	64 641 [g]	71 208 [g]	76 423 [g,s]	..	..	Autres coûts courants
Total current expenditure	..	..	105 473 [g]	112 766 [g]	95 756 [g,s]	..	..	*Total dépenses courantes*
Land and buildings	..	..	9 132 [g]	6 251 [g]	273 [g,s]	..	..	Terrains et constructions
Instruments and equipment	..	..	17 045 [g]	27 408 [g]	8 947 [g,s]	..	..	Instruments et équipements
Total capital expenditure	..	..	26 177 [g]	33 659 [g]	9 220 [g,s]	..	..	*Total dépenses en capital*
Total	..	108 424 [g]	131 650 [g]	146 425 [g]	120 251 [g]	256 277 [g]	..	*Total*
GERD								**DIRD**
Labour costs	..	3 160 442 [g]	3 621 555 [g]	4 091 203 [g]	3 880 492 [g]	..	..	Coûts du personnel
Other current costs	..	4 001 205 [g]	4 611 799 [g]	5 336 118 [g]	4 894 891 [g]	..	..	Autres coûts courants
Total current expenditure	..	7 161 647 [g]	8 233 354 [g]	9 427 321 [g]	8 775 383 [g]	..	..	*Total dépenses courantes*
Land and buildings	..	603 126 [g]	694 157 [g]	601 504 [g]	498 803 [g]	..	..	Terrains et constructions
Instruments and equipment	..	1 675 833 [g]	1 950 540 [g]	2 156 982 [g]	2 062 431 [g]	..	..	Instruments et équipements
Total capital expenditure	..	2 278 959 [g]	2 644 697 [g]	2 758 486 [g]	2 561 234 [g]	..	..	*Total dépenses en capital*
Total	7 894 746 [g]	9 440 606 [g]	10 878 050 [g]	12 185 807 [g]	11 336 617 [g]	11 921 752 [g]	..	*Total*

Table 3 - Tableau 3

Current domestic expenditure on R&D - Dépenses intérieures courantes de R-D

by sector of performance and type of activity - par secteur d'exécution et par type d'activités

Million KRW — *Millions KRW*

	1994	1995	1996	1997	1998	1999	2000	
Business enterprise								**Entreprises**
Basic research	..	..	646 801 [k]	717 011 [k]	516 585 [k]	639 937 [k]	..	Recherche fondamentale
Applied research	..	..	2 094 968 [k]	2 353 388 [k]	1 679 720 [k]	1 829 348 [k]	..	Recherche appliquée
Experimental development	..	..	5 221 842 [k]	5 774 908 [k]	5 775 768 [k]	6 041 872 [k]	..	Développement expérimental
Not specified	..	6 960 322 [k]	..	..	..	..	..	Non spécifié
Total	..	6 960 322 [k]	7 963 612 [k]	8 845 307 [k]	7 972 073 [k]	8 511 157 [k]	..	*Total*
Government								**État**
Basic research	..	..	308 026 [g,k]	390 337 [g,k]	556 556 [g,k,s]	455 556 [g,k]	..	Recherche fondamentale
Applied research	..	..	529 375 [g,k]	675 444 [g,k]	670 270 [g,k,s]	629 109 [g,k]	..	Recherche appliquée
Experimental development	..	..	926 567 [g,k]	856 695 [g,k]	767 668 [g,k,s]	638 232 [g,k]	..	Développement expérimental
Not specified	..	1 600 948 [g,k]	..	..	..	..	..	Non spécifié
Total	..	1 600 948 [g,k]	1 763 967 [g,k]	1 922 476 [g,k]	1 979 219 [g,k]	1 722 897 [g,k]	..	*Total*
Higher education								**Enseignement supérieur**
Basic research	..	372 851 [g,k]	452 063 [g,k]	480 168 [g,k]	507 261 [g,k]	491 402 [g,k]	..	Recherche fondamentale
Applied research	..	205 618 [g,k]	270 913 [g,k]	400 962 [g,k]	427 539 [g,k]	479 440 [g,k]	..	Recherche appliquée
Experimental development	..	192 443 [g,k]	295 846 [g,k]	390 469 [g,k]	330 274 [g,k]	460 579 [g,k]	..	Développement expérimental
Not specified	..	..	..	..	..	..	..	Non spécifié
Total	..	770 912 [g,k]	1 018 821 [g,k]	1 271 599 [g,k]	1 265 074 [g,k]	1 431 421 [g,k]	..	*Total*
Private non-profit								**Institutions sans but lucratif**
Basic research	..	..	32 130 [g,k]	28 974 [g,k]	4 965 [g,k,s]	38 582 [g,k]	..	Recherche fondamentale
Applied research	..	..	32 014 [g,k]	40 766 [g,k]	70 929 [g,k,s]	127 300 [g,k]	..	Recherche appliquée
Experimental development	..	..	67 506 [g,k]	76 685 [g,k]	29 082 [g,k,s]	90 395 [g,k]	..	Développement expérimental
Not specified	..	108 424 [g,k]	..	..	..	..	..	Non spécifié
Total	..	108 424 [g,k]	131 650 [g,k]	146 425 [g,k]	120 251 [g,k]	256 277 [g,k]	..	*Total*
GERD								**DIRD**
Basic research	..	1 176 819 [g,k]	1 439 020 [g,k]	1 616 490 [g,k]	1 585 367 [g,k]	1 625 477 [g,k]	..	Recherche fondamentale
Applied research	..	2 362 110 [g,k]	2 927 270 [g,k]	3 470 560 [g,k]	2 848 458 [g,k]	3 065 197 [g,k]	..	Recherche appliquée
Experimental development	..	5 901 677 [g,k]	6 511 761 [g,k]	7 098 757 [g,k]	6 902 792 [g,k]	7 231 078 [g,k]	..	Développement expérimental
Not specified	7 894 746 [g,k]	..	..	..	..	..	..	Non spécifié
Total	7 894 746 [g,k]	9 440 606 [g,k]	10 878 050 [g,k]	12 185 807 [g,k]	11 336 617 [g,k]	11 921 752 [g,k]	..	*Total*

Source: OECD, R&D database (DSTI/EAS Division), January 2002.

Source : OCDE, base de données R-D (DSTI/Division AES), janvier 2002.

Gross domestic expenditure on R&D (GERD) - Dépenses intérieures brutes de R-D (DIRD)

by sector of performance and socio-economic objective - par secteur d'exécution et par objectif socio-économique

Million KRW — *Millions KRW*

	1994	1995	1996	1997	1998	1999	2000	
Business enterprise								**Entreprises**
Agriculture, forestry and fishing	..	75 369	..	93 982	77 337	..	..	Agriculture, sylviculture et pêche
Industrial development	..	4 204 359	..	3 424 794	4 038 823	..	..	Développement industriel
Energy	..	311 895	..	470 208	283 853	..	..	Energie
Transport and telecommunications	..	1 502 290	..	3 184 291	2 252 181	..	..	Transports et télécommunications
Urban and rural planning	..	93 676	..	73 726	37 890	..	..	Aménagement urbain et rural
Prevention of pollution	..	109 919	..	238 777	119 875	..	..	Prévention de la pollution
Identification & treatment of pollution	..	..	..	..	..	..	..	Détection pollu. & mesures antipollu.
Health	..	190 363	..	231 082	189 188	..	..	Santé
Social development and services	..	45 354	..	146 627	254 102	..	..	Développement et services sociaux
Earth and atmosphere	..	2 113	..	40 589	24 230	..	..	Milieu terrestre et atmosphère
Advancement of knowledge	..	15 252	..	83 956	37 051	..	..	Promotion générale des connaissances
Civil space	..	12 047	..	44 530	21 388	..	..	Espace civil
Defence	..	104 768	..	103 873	187 851	..	..	Défense
Not elsewhere classified	..	292 917	7 963 612	708 872	448 304	8 511 157	..	Non classé
Total	..	6 960 322	7 963 612	8 845 307	7 972 073	8 511 157	..	*Total*
Government								**État**
Agriculture, forestry and fishing	..	302 880 [g]	..	359 298 [g]	341 864 [g,s]	..	..	Agriculture, sylviculture et pêche
Industrial development	..	186 468 [g]	..	267 941 [g]	242 820 [g,s]	..	..	Développement industriel
Energy	..	259 114 [g]	..	245 235 [g]	236 187 [g,s]	..	..	Energie
Transport and telecommunications	..	244 673 [g]	..	255 515 [g]	371 918 [g,s]	..	..	Transports et télécommunications
Urban and rural planning	..	104 [g]	..	2 099 [g]	4 103 [g,s]	..	..	Aménagement urbain et rural
Prevention of pollution	..	50 769 [g]	..	61 720 [g]	50 033 [g,s]	..	..	Prévention de la pollution
Identification & treatment of pollution	..	..	..	..	..	..	..	Détection pollu. & mesures antipollu.
Health	..	63 537 [g]	..	83 735 [g]	77 825 [g,s]	..	..	Santé
Social development and services	..	10 794 [g]	..	12 464 [g]	5 665 [g,s]	..	..	Développement et services sociaux
Earth and atmosphere	..	11 312 [g]	..	29 622 [g]	36 941 [g,s]	..	..	Milieu terrestre et atmosphère
Advancement of knowledge	..	1 544 [g]	..	2 164 [g]	82 954 [g,s]	..	..	Promotion générale des connaissances
Civil space	..	59 128 [g]	..	51 821 [g]	59 827 [g,s]	..	..	Espace civil
Defence	..	327 897 [g]	..	418 202 [g]	403 452 [g,s]	..	..	Défense
Not elsewhere classified	..	82 728 [g]	1 763 967 [g]	132 660 [g]	80 905 [g,s]	1 722 897 [g]	..	Non classé
Total	..	1 600 948 [g]	1 763 967 [g]	1 922 476 [g]	1 979 219 [g]	1 722 897 [g]	..	*Total*
Higher education								**Enseignement supérieur**
Agriculture, forestry and fishing	..	52 110 [g]	..	80 469 [g]	87 940 [g]	..	..	Agriculture, sylviculture et pêche
Industrial development	..	197 875 [g]	..	292 843 [g]	268 098 [g]	..	..	Développement industriel
Energy	..	34 553 [g]	..	70 585 [g]	87 852 [g]	..	..	Energie
Transport and telecommunications	..	51 599 [g]	..	110 629 [g]	149 067 [g]	..	..	Transports et télécommunications
Urban and rural planning	..	26 893 [g]	..	45 218 [g]	48 621 [g]	..	..	Aménagement urbain et rural
Prevention of pollution	..	26 138 [g]	..	51 615 [g]	58 348 [g]	..	..	Prévention de la pollution
Identification & treatment of pollution	..	..	..	..	..	..	..	Détection pollu. & mesures antipollu.
Health	..	65 980 [g]	..	170 183 [g]	189 040 [g]	..	..	Santé
Social development and services	..	15 350 [g]	..	46 544 [g]	60 645 [g]	..	..	Développement et services sociaux
Earth and atmosphere	..	5 144 [g]	..	17 212 [g]	15 187 [g]	..	..	Milieu terrestre et atmosphère
Advancement of knowledge	..	168 235 [g]	..	274 734 [g]	159 224 [g]	..	..	Promotion générale des connaissances
Civil space	..	5 163 [g]	..	8 074 [g]	5 739 [g]	..	..	Espace civil
Defence	..	4 270 [g]	..	10 127 [g]	16 562 [g]	..	..	Défense
Not elsewhere classified	..	117 602 [g]	1 018 821 [g]	93 366 [g]	118 750 [g]	1 431 421 [g]	..	Non classé
Total	..	770 912 [g]	1 018 821 [g]	1 271 599 [g]	1 265 074 [g]	1 431 421 [g]	..	*Total*
Private non-profit								**Institutions sans but lucratif**
Agriculture, forestry and fishing	..	48 [g]	..	864 [g]	0 [g,s]	..	..	Agriculture, sylviculture et pêche
Industrial development	..	63 766 [g]	..	72 847 [g]	13 276 [g,s]	..	..	Développement industriel
Energy	..	5 356 [g]	..	12 296 [g]	724 [g,s]	..	..	Energie
Transport and telecommunications	..	3 653 [g]	..	398 [g]	13 905 [g,s]	..	..	Transports et télécommunications
Urban and rural planning	..	2 166 [g]	..	3 312 [g]	391 [g,s]	..	..	Aménagement urbain et rural
Prevention of pollution	..	1 244 [g]	..	12 934 [g]	1 174 [g,s]	..	..	Prévention de la pollution
Identification & treatment of pollution	..	..	..	..	..	..	..	Détection pollu. & mesures antipollu.
Health	..	28 399 [g]	..	29 879 [g]	13 452 [g,s]	..	..	Santé
Social development and services	..	0 [g]	..	2 937 [g]	603 [g,s]	..	..	Développement et services sociaux
Earth and atmosphere	..	0 [g]	..	3 077 [g]	0 [g,s]	..	..	Milieu terrestre et atmosphère
Advancement of knowledge	..	252 [g]	..	155 [g]	392 [g,s]	..	..	Promotion générale des connaissances
Civil space	..	0 [g]	..	..	0 [g,s]	..	..	Espace civil
Defence	..	0 [g]	..	..	262 [g,s]	..	..	Défense
Not elsewhere classified	..	3 540 [g]	131 650 [g]	7 726 [g]	60 797 [g,s]	256 277 [g]	..	Non classé
Total	..	108 424 [g]	131 650 [g]	146 425 [g]	120 251 [g]	256 277 [g]	..	*Total*
GERD								**DIRD**
Agriculture, forestry and fishing	..	430 407 [g]	..	534 613 [g]	507 142 [g]	..	..	Agriculture, sylviculture et pêche
Industrial development	..	4 652 468 [g]	..	4 058 425 [g]	4 563 016 [g]	..	..	Développement industriel
Energy	..	610 918 [g]	..	798 324 [g]	608 616 [g]	..	..	Energie
Transport and telecommunications	..	1 802 215 [g]	..	3 550 833 [g]	2 787 072 [g]	..	..	Transports et télécommunications
Urban and rural planning	..	122 839 [g]	..	124 355 [g]	91 005 [g]	..	..	Aménagement urbain et rural
Prevention of pollution	..	188 070 [g]	..	365 046 [g]	229 430 [g]	..	..	Prévention de la pollution
Identification & treatment of pollution	..	..	..	..	..	..	..	Détection pollu. & mesures antipollu.
Health	..	348 279 [g]	..	514 879 [g]	469 505 [g]	..	..	Santé
Social development and services	..	71 498 [g]	..	208 572 [g]	321 014 [g]	..	..	Développement et services sociaux
Earth and atmosphere	..	18 569 [g]	..	90 500 [g]	76 358 [g]	..	..	Milieu terrestre et atmosphère
Advancement of knowledge	..	185 283 [g]	..	361 009 [g]	279 621 [g]	..	..	Promotion générale des connaissances
Civil space	..	76 338 [g]	..	104 425 [g]	86 954 [g]	..	..	Espace civil
Defence	..	436 935 [g]	..	532 202 [g]	608 127 [g]	..	..	Défense
Not elsewhere classified	7 894 746 [g]	496 787 [g]	10 878 050 [g]	942 624 [g]	708 756 [g]	11 921 752 [g]	..	Non classé
Total	7 894 746 [g]	9 440 606 [g]	10 878 050 [g]	12 185 807 [g]	11 336 617 [g]	11 921 752 [g]	..	*Total*

Source: OECD, R&D database (DSTI/EAS Division), January 2002.

Source : OCDE, base de données R-D (DSTI/Division AES), janvier 2002.

Table 5 - Tableau 5

Total R&D personnel - Personnel total de R-D

by sector of employment and occupation - par secteur d'emploi et par profession

In full-time equivalent — *En équivalence plein-temps*

	1994	1995	1996	1997	1998	1999	2000	
Business enterprise								**Entreprises**
RSE	..	67 226	66 218	69 871	60 064	65 474	..	Chercheurs
Technicians	..	20 862	17 410	15 634	12 230	13 268	..	Techniciens
Other occupations	..	8 802	5 374	4 920	5 568	5 328	..	Autres professions
Total	..	96 890	89 002	90 425	77 862	84 070	..	*Total*
Government								**État**
RSE	..	12 711 g	12 322 g	12 323 g	10 098 g	11 745 g	..	Chercheurs
Technicians	..	5 635 g	3 214 g	2 948 g	2 793 g	2 565 g	..	Techniciens
Other occupations	..	3 510 g	872 g	743 g	1 858 g	1 049 g	..	Autres professions
Total	..	21 856 g	16 408 g	16 014 g	14 749 g	15 359 g	..	*Total*
Higher education								**Enseignement supérieur**
RSE	..	19 406 g	19 483 g	19 573 g	21 525 g	21 723 g	..	Chercheurs
Technicians	..	11 217 g	7 602 g	7 354 g	9 117 g	9 899 g	..	Techniciens
Other occupations	..	888 g	1 189 g	1 619 g	4 111 g	4 912 g	..	Autres professions
Total	..	31 511 g	28 274 g	28 546 g	34 753 g	36 534 g	..	*Total*
Private non-profit								**Institutions sans but lucratif**
RSE	..	1 113 g	1 410 g	893 g	854 g	1 268 g	..	Chercheurs
Technicians	..	589 g	374 g	449 g	304 g	428 g	..	Techniciens
Other occupations	..	288 g	235 g	232 g	147 g	215 g	..	Autres professions
Total	..	1 990 g	2 019 g	1 574 g	1 305 g	1 911 g	..	*Total*
TOTAL								**TOTAL**
RSE	..	100 456 g	99 433 g	102 660 g	92 541 g	100 210 g	..	Chercheurs
Technicians	..	38 303 g	28 600 g	26 385 g	24 444 g	26 160 g	..	Techniciens
Other occupations	..	13 488 g	7 670 g	7 514 g	11 684 g	11 504 g	..	Autres professions
Total	..	152 247 g	135 703 g	136 559 g	128 669 g	137 874 g	..	*Total*

Table 6 - Tableau 6

Total R&D personnel - Personnel total de R-D

by sector of employment and formal qualification - par secteur d'emploi et par niveau de qualification

In full-time equivalent — *En équivalence plein-temps*

	1994	1995	1996	1997	1998	1999	2000	
Business enterprise								**Entreprises**
University PhD level degrees	..	..	..	..	..	..	..	Diplômes univ. de niv. doctorat
Other university degrees	..	..	..	..	..	..	..	Autres diplômes universitaires
Sub-total university degrees	..	..	..	..	..	..	..	Sous-total Universitaires
Other post-secondary diplomas	..	..	..	..	..	..	..	Autres diplômes post-secondaires
Secondary diplomas	..	..	..	..	..	..	..	Diplômes de fin d'études secondaires
Other qualifications	..	..	..	..	..	..	..	Autres qualifications
Total	..	96 890	89 002	90 425	77 862	84 070	..	*Total*
Government								**État**
University PhD level degrees	..	..	..	..	..	..	..	Diplômes univ. de niv. doctorat
Other university degrees	..	..	..	..	..	..	..	Autres diplômes universitaires
Sub-total university degrees	..	..	..	..	..	..	..	Sous-total Universitaires
Other post-secondary diplomas	..	..	..	..	..	..	..	Autres diplômes post-secondaires
Secondary diplomas	..	..	..	..	..	..	..	Diplômes de fin d'études secondaires
Other qualifications	..	..	..	..	..	..	..	Autres qualifications
Total	..	21 856 g	16 408 g	16 014 g	14 749 g	15 359 g	..	*Total*
Higher education								**Enseignement supérieur**
University PhD level degrees	..	..	..	..	..	..	..	Diplômes univ. de niv. doctorat
Other university degrees	..	..	..	..	..	..	..	Autres diplômes universitaires
Sub-total university degrees	..	..	..	..	..	..	..	Sous-total Universitaires
Other post-secondary diplomas	..	..	..	..	..	..	..	Autres diplômes post-secondaires
Secondary diplomas	..	..	..	..	..	..	..	Diplômes de fin d'études secondaires
Other qualifications	..	..	..	..	..	..	..	Autres qualifications
Total	..	31 511 g	28 274 g	28 546 g	34 753 g	36 534 g	..	*Total*
Private non-profit								**Institutions sans but lucratif**
University PhD level degrees	..	..	..	..	..	..	..	Diplômes univ. de niv. doctorat
Other university degrees	..	..	..	..	..	..	..	Autres diplômes universitaires
Sub-total university degrees	..	..	..	..	..	..	..	Sous-total Universitaires
Other post-secondary diplomas	..	..	..	..	..	..	..	Autres diplômes post-secondaires
Secondary diplomas	..	..	..	..	..	..	..	Diplômes de fin d'études secondaires
Other qualifications	..	..	..	..	..	..	..	Autres qualifications
Total	..	1 990 g	2 019 g	1 574 g	1 305 g	1 911 g	..	*Total*
TOTAL								**TOTAL**
University PhD level degrees	..	..	..	..	..	..	..	Diplômes univ. de niv. doctorat
Other university degrees	..	..	..	..	..	..	..	Autres diplômes universitaires
Sub-total university degrees	..	..	..	..	..	..	..	Sous-total Universitaires
Other post-secondary diplomas	..	..	..	..	..	..	..	Autres diplômes post-secondaires
Secondary diplomas	..	..	..	..	..	..	..	Diplômes de fin d'études secondaires
Other qualifications	..	..	..	..	..	..	..	Autres qualifications
Total	..	152 247 g	135 703 g	136 559 g	128 669 g	137 874 g	..	*Total*

Source: OECD, R&D database (DSTI/EAS Division), January 2002.

Source : OCDE, base de données R-D (DSTI/Division AES), janvier 2002.

Table 7 - Tableau 7

Gross domestic expenditure on R&D (GERD) - Dépenses intérieures brutes de R-D (DIRD)

by sector of performance and main scientific field - par secteur d'exécution et par grand domaine scientifique

Million KRW — *Millions KRW*

	1994	1995	1996	1997	1998	1999	2000	
Business enterprise								**Entreprises**
Natural sciences and engineering	..	6 960 322	7 963 612	8 845 307	7 972 073	8 511 157	..	Sciences exactes, nat. et de l'ingénieur
Social sciences and humanities	..	..	..	..	..	..	..	Sciences sociales et humaines
All fields of science	..	6 960 322	7 963 612	8 845 307	7 972 073	8 511 157	..	*Tous domaines scientifiques*
Government								**État**
Natural sciences and engineering	..	1 600 948	1 763 967	1 922 476	1 979 219	1 722 897	..	Sciences exactes, nat. et de l'ingénieur
Social sciences and humanities	..	..	..	..	..	..	..	Sciences sociales et humaines
All fields of science	..	1 600 948 [g]	1 763 967 [g]	1 922 476 [g]	1 979 219 [g]	1 722 897 [g]	..	*Tous domaines scientifiques*
Higher education								**Enseignement supérieur**
Natural sciences and engineering	..	770 912 [g]	1 018 821	1 271 599	1 265 074	1 431 421	..	Sciences exactes, nat. et de l'ingénieur
Social sciences and humanities	..	..	..	..	..	..	..	Sciences sociales et humaines
All fields of science	..	770 912 [g]	1 018 821 [g]	1 271 599 [g]	1 265 074 [g]	1 431 421 [g]	..	*Tous domaines scientifiques*
Private non-profit								**Institutions sans but lucratif**
Natural sciences and engineering	..	108 424	131 650	146 425	120 251	256 277	..	Sciences exactes, nat. et de l'ingénieur
Social sciences and humanities	..	..	..	..	..	..	..	Sciences sociales et humaines
All fields of science	..	108 424 [g]	131 650 [g]	146 425 [g]	120 251 [g]	256 277 [g]	..	*Tous domaines scientifiques*
TOTAL								**TOTAL**
Natural sciences and engineering	7 894 746	9 440 606	10 878 050	12 185 807	11 336 617	11 921 752	..	Sciences exactes, nat. et de l'ingénieur
Social sciences and humanities	..	..	..	..	..	..	..	Sciences sociales et humaines
All fields of science	7 894 746 [g]	9 440 606 [g]	10 878 050 [g]	12 185 807 [g]	11 336 617 [g]	11 921 752 [g]	..	*Tous domaines scientifiques*

Table 8 - Tableau 8

Total R&D personnel - Personnel total de R-D

by sector of employment and main scientific field - par secteur d'emploi et par grand domaine scientifique

In full-time equivalent — *En équivalence plein-temps*

	1994	1995	1996	1997	1998	1999	2000	
Business enterprise								**Entreprises**
Natural sciences and engineering	..	96 890	89 002	90 425	77 862	84 070	..	Sciences exactes, nat. et de l'ingénieur
Social sciences and humanities	..	..	..	..	..	..	..	Sciences sociales et humaines
All fields of science	..	96 890	89 002	90 425	77 862	84 070	..	*Tous domaines scientifiques*
Government								**État**
Natural sciences and engineering	..	21 856	16 408	16 014	14 749	15 359	..	Sciences exactes, nat. et de l'ingénieur
Social sciences and humanities	..	..	..	..	..	..	..	Sciences sociales et humaines
All fields of science	..	21 856 [g]	16 408 [g]	16 014 [g]	14 749 [g]	15 359 [g]	..	*Tous domaines scientifiques*
Higher education								**Enseignement supérieur**
Natural sciences and engineering	..	31 511	28 274	28 546	34 753	36 534	..	Sciences exactes, nat. et de l'ingénieur
Social sciences and humanities	..	..	..	..	..	..	..	Sciences sociales et humaines
All fields of science	..	31 511 [g]	28 274 [g]	28 546 [g]	34 753 [g]	36 534 [g]	..	*Tous domaines scientifiques*
Private non-profit								**Institutions sans but lucratif**
Natural sciences and engineering	..	1 990	2 019	1 574	1 305	1 911	..	Sciences exactes, nat. et de l'ingénieur
Social sciences and humanities	..	..	..	..	..	..	..	Sciences sociales et humaines
All fields of science	..	1 990 [g]	2 019 [g]	1 574 [g]	1 305 [g]	1 911 [g]	..	*Tous domaines scientifiques*
TOTAL								**TOTAL**
Natural sciences and engineering	..	152 247	135 703	136 559	128 669	137 874	..	Sciences exactes, nat. et de l'ingénieur
Social sciences and humanities	..	..	..	..	..	..	..	Sciences sociales et humaines
All fields of science	..	152 247 [g]	135 703 [g]	136 559 [g]	128 669 [g]	137 874 [g]	..	*Tous domaines scientifiques*

Source: OECD, R&D database (DSTI/EAS Division), January 2002.

Source : OCDE, base de données R-D (DSTI/Division AES), janvier 2002.

Total business enterprise intramural expenditure on R&D - Dépenses intra-muros totales de R-D des entreprises
(BERD) by industry - (DIRDE) par industrie

9.1 Million KRW - Millions KRW

	1994	1995	1996	1997	1998	1999	2000	
1 **AGRICULTURE, HUNTING & FORESTRY**	..	16 854	15 809	14 599	19 913	18 893	..	1 **AGRICULTURE, CHASSE & SYLVICULTURE**
2 **MINING**	..	7 206	7 350	4 196	66 661	5 570	..	2 **ACTIVITES EXTRACTIVES**
3 **MANUFACTURING**	..	5 800 886	6 734 033	7 371 281	6 439 212	6 945 015	..	3 **ACTIVITES DE FABRICATION**
4 **Food, beverages & tobacco**	..	100 376	118 002	106 126	84 396	116 777	..	4 **Produits alimentaires, boissons & tabac**
5 Food products & beverages	..	100 084	117 629	105 772	84 142	116 487	..	5 Produits alimentaires & boissons
6 Tobacco products	..	292	373	354	254	290	..	6 Produits à base de tabac
7 **Textiles, wear app., fur & leather**	..	50 886	51 097	43 154	45 491	60 569	..	7 **Textiles, habill., fourrures & cuirs**
8 Textiles	..	44 452	45 429	37 462	37 228	46 262	..	8 Textiles
9 Wearing apparel & fur	..	5 258	4 630	4 341	3 329	6 003	..	9 Articles d'habillement & fourrures
10 Leather products & footwear	..	1 176	1 038	1 351	4 934	8 304	..	10 Articles en cuir & chaussures
11 **Wood, paper, printing, publishing**	..	35 960	54 661	46 985	38 143	28 943	..	11 **Bois, papier, imprimerie, édition**
12 Wood & cork (not furniture)	..	1 706	1 946	2 195	1 160	869	..	12 Bois & liège (sauf meubles)
13 Pulp, paper & paper products	..	20 403	26 093	26 305	17 296	11 158	..	13 Carton, papier & articles
14 Publ., print. & repro. of rec. media	..	13 851	26 622	18 485	19 687	16 916	..	14 Edition, impress. & repro. enregis.
15 **Coke, petroleum, nuclear fuel, chemicals, rubber plastics**	..	741 194	798 696	830 648	688 132	784 909	..	15 **Coke, pétrole, combust. nucl., prod. chimi., caout. & mat. plast.**
16 Coke, ref. petrol. prod. & nuclear fuel	..	87 640	106 999	125 029	86 755	102 636	..	16 Coke, raffin. pétr. & combust. nucl.
17 Chemicals & chemical products	..	560 871	600 806	604 299	462 924	521 501	..	17 Produits chimiques
18 *Chemical prod.(less pharmaceuticals)*	..	464 013	481 135	482 152	382 066	353 839	..	18 *Prod. chimi. (moins prod. pharmaceutiques)*
19 *Pharmaceuticals*	..	96 858	119 671	122 147	80 858	167 662	..	19 *Produits pharmaceutiques*
20 Rubber & plastic products	..	92 683	90 891	101 320	138 453	160 772	..	20 Caoutchouc et matières plastiques
21 **Non-metallic mineral products**	..	69 445	99 498	99 680	79 614	100 355	..	21 **Produits minéraux non métalliques**
22 **Basic metals**	..	215 028	164 174	128 817	117 885	112 856	..	22 **Produits métallurgiques de base**
23 Basic metals, ferrous	..	179 093	127 915	106 366	94 079	96 411	..	23 Produits métallur. de base, ferreux
24 Basic metals, non-ferrous	..	35 935	36 259	22 451	23 806	16 445	..	24 Prod. métallur. de base, non ferreux
25 **Fabricated metal products**	..	36 514	86 745	77 727	62 969	54 580	..	25 **Ouvrages en métaux (sauf mach.& mat.)**
26 **Machinery equipment, instruments and transport equipment**	..	4 538 227	5 326 187	6 023 344	5 280 528	5 639 028	..	26 **Machines & matériel, instruments & mat. de transport**
27 Machinery n.e.c.	..	352 353	291 155	311 166	436 596	247 538	..	27 Machines, n.c.a.
28 Office, account. & computing machinery	..	128 431	165 304	150 203	111 282	469 933	..	28 Mach. de bur., compta. & trait. inform.
29 Electrical machinery	..	133 679	143 770	160 095	191 561	202 942	..	29 Machines électriques
30 Electro. equip. (radio, TV & commun.)	..	2 196 489	2 614 398	3 045 477	2 785 202	3 383 986	..	30 Equip. électro. (radio, tv & communic.)
31 *Electro. comp. (inc. semi-conductors)*	..	1 440 781	2 103 033	2 543 038	970 770	1 425 252	..	31 *Compos. électro. (semi-cond. incl.)*
32 *TV, radio & communications equip.*	..	755 708	511 365	502 439	1 814 432	1 958 734	..	32 *Appareils radio, tv et de communi.*
33 Instruments	..	48 629	54 796	75 280	74 367	61 327	..	33 Instruments
34 Motor vehicles	..	1 466 866	1 841 587	1 969 551	1 468 998	1 124 193	..	34 Véhicules automobiles
35 Other transport equipment	..	211 780	215 177	311 572	212 522	149 109	..	35 Autres matériels de transport
36 *Ships*	..	99 417	104 605	177 839	176 580	125 906	..	36 *Navires*
37 *Aerospace*	..	103 049	102 425	121 598	20 672	182 m	..	37 *Aérospatiale*
38 *Other transport equipment n.e.c.*	..	9 314	8 147	12 135	15 270	23 021	..	38 *Autres matériels de transport n.c.a.*
39 **Furniture, other manufacturing n.e.c.**	..	13 256	34 044	14 175	34 531	39 457	..	39 **Meubles, autres activ. de fabricat. n.c.a.**
40 Furniture	..	..	..	..	2 561	1 712	..	40 Meubles
41 Other manufacturing n.e.c.	..	..	..	..	31 970	37 745	..	41 Autres activités de fabrication n.c.a.
42 **Recycling**	..	..	929	625	7 523	7 541	..	42 **Récupération**
43 **ELECTRICITY, GAS & WATER SUPPLY**	..	139 813	186 602	203 327	211 054	112 953	..	43 **DISTRIBUTION D'ELECTRICITE, GAZ & EAU**
44 **CONSTRUCTION**	..	467 684	437 683	433 548	282 160	294 332	..	44 **CONSTRUCTION**
45 **SERVICES SECTOR**	..	527 879	582 134	818 356	953 073	1 134 394	..	45 **SECTEUR DES SERVICES**
46 **Wholesale, retail trade & repair of motor vehicles, etc.**	..	..	..	..	7 638	12 645	..	46 **Commerce (gros & détail), réparation de véhicules, etc.**
47 **Hotels & restaurants**	..	..	..	..	0	0	..	47 **Hôtels et restaurants**
48 **Transport & storage**	..	..	..	..	4 961	29 883	..	48 **Transports, entreposage**
49 **Communications**	..	..	..	..	474 704	622 633	..	49 **Communications**
50 Post	..	..	..	..	0	0	..	50 Postes
51 Telecommunications	..	..	..	..	474 704	622 633	..	51 Télécommunications
52 **Financial intermediation (inc. insur.)**	..	..	..	..	8 105	3 409	..	52 **Intermédiation financ. (assurances inc.)**
53 **Real estate, renting & busin. activity**	..	..	..	..	415 716	424 290	..	53 **Immobil., locat. & serv. aux entreprises**
54 Computer & related activities	..	..	..	..	183 615	236 562	..	54 Informatique et activités connexes
55 *Software consultancy*	..	..	..	..	149 576	136 667	..	55 *Production de logiciels*
56 *Other computer services n.e.c.*	..	..	..	..	34 039	99 895	..	56 *Autres activités informatiques*
57 Research & development	..	..	..	..	2 854	1 903	..	57 Recherche & développement expérimental
58 Other business activities n.e.c.	..	90 629	44 076	73 003	66 004	185 825	..	58 Autres activ. & serv. aux entreprises
59 **Comm., soc. & pers. serv. activ., etc.**	..	..	..	..	41 949	41 534	..	59 **Adm. pub., serv. soc. & collec., etc**
TOTAL BUSINESS ENTERPRISE SECTOR	..	**6 960 322**	**7 963 612**	**8 845 307**	**7 972 073**	**8 511 157**	..	**TOTAL SECTEUR DES ENTREPRISES**

Source: OECD, R&D database (DSTI/EAS Division), January 2002.

Source : OCDE, base de données R-D (DSTI/Division AES), janvier 2002.

Total business enterprise intramural expenditure on R&D - Dépenses intra-muros totales de R-D des entreprises
(BERD) by industry - (DIRDE) par industrie

	1994	1995	1996	1997	1998	1999	2000	
	9.2 Million constant USD(1995 prices and PPPs) - Millions de USD constants (prix et PPA de 1995)							
1 **AGRICULTURE, HUNTING & FORESTRY**	..	27.4	24.7	22.1	28.8	27.8	..	1 **AGRICULTURE, CHASSE & SYLVICULTURE**
2 **MINING**	..	11.7	11.5	6.4	96.2	8.2	..	2 **ACTIVITES EXTRACTIVES**
3 **MANUFACTURING**	..	9 429.3	10 536.6	11 181.8	9 297.3	10 235.9	..	3 **ACTIVITES DE FABRICATION**
4 **Food, beverages & tobacco**	..	163.2	184.6	161.0	121.9	172.1	..	4 **Produits alimentaires, boissons & tabac**
5 Food products & beverages	..	162.7	184.1	160.5	121.5	171.7	..	5 Produits alimentaires & boissons
6 Tobacco products	..	0.5	0.6	0.5	0.4	0.4	..	6 Produits à base de tabac
7 **Textiles, wear app., fur & leather**	..	82.7	80.0	65.5	65.7	89.3	..	7 **Textiles, habill., fourrures & cuirs**
8 Textiles	..	72.3	71.1	56.8	53.8	68.2	..	8 Textiles
9 Wearing apparel & fur	..	8.5	7.2	6.6	4.8	8.8	..	9 Articles d'habillement & fourrures
10 Leather products & footwear	..	1.9	1.6	2.0	7.1	12.2	..	10 Articles en cuir & chaussures
11 **Wood, paper, printing, publishing**	..	58.5	85.5	71.3	55.1	42.7	..	11 **Bois, papier, imprimerie, édition**
12 Wood & cork (not furniture)	..	2.8	3.0	3.3	1.7	1.3	..	12 Bois & liège (sauf meubles)
13 Pulp, paper & paper products	..	33.2	40.8	39.9	25.0	16.4	..	13 Carton, papier & articles
14 Publ., print. & repro. of rec. media	..	22.5	41.7	28.0	28.4	24.9	..	14 Edition, impress. & repro. enregis.
15 **Coke, petroleum, nuclear fuel, chemicals, rubber plastics**	..	1 204.8	1 249.7	1 260.0	993.6	1 156.8	..	15 **Coke, pétrole, combust. nucl., prod. chimi., caout. & mat. plast.**
16 Coke, ref. petrol. prod. & nuclear fuel	..	142.5	167.4	189.7	125.3	151.3	..	16 Coke, raffin. pétr. & combust. nucl.
17 Chemicals & chemical products	..	911.7	940.1	916.7	668.4	768.6	..	17 Produits chimiques
18 *Chemical prod.(less pharmaceuticals)*	..	754.3	752.8	731.4	551.6	521.5	..	18 *Prod. chimi. (moins prod. pharmaceutiques)*
19 *Pharmaceuticals*	..	157.4	187.2	185.3	116.7	247.1	..	19 *Produits pharmaceutiques*
20 Rubber & plastic products	..	150.7	142.2	153.7	199.9	237.0	..	20 Caoutchouc et matières plastiques
21 **Non-metallic mineral products**	..	112.9	155.7	151.2	115.0	147.9	..	21 **Produits minéraux non métalliques**
22 **Basic metals**	..	349.5	256.9	195.4	170.2	166.3	..	22 **Produits métallurgiques de base**
23 Basic metals, ferrous	..	291.1	200.1	161.4	135.8	142.1	..	23 Produits métallur. de base, ferreux
24 Basic metals, non-ferrous	..	58.4	56.7	34.1	34.4	24.2	..	24 Prod. métallur. de base, non ferreux
25 **Fabricated metal products**	..	59.4	135.7	117.9	90.9	80.4	..	25 **Ouvrages en métaux (sauf mach.& mat.)**
26 **Machinery equipment, instruments and transport equipment**	..	7 376.9	8 333.8	9 137.1	7 624.3	8 311.1	..	26 **Machines & matériel, instruments & mat. de transport**
27 Machinery n.e.c.	..	572.7	455.6	472.0	630.4	364.8	..	27 Machines, n.c.a.
28 Office, account. & computing machinery	..	208.8	258.6	227.9	160.7	692.6	..	28 Mach. de bur., compta. & trait. inform.
29 Electrical machinery	..	217.3	225.0	242.9	276.6	299.1	..	29 Machines électriques
30 Electro. equip. (radio, TV & commun.)	..	3 570.4	4 090.7	4 619.8	4 021.4	4 987.5	..	30 Equip. électro. (radio, tv & communic.)
31 *Electro. comp. (inc. semi-conductors)*	..	2 342.0	3 290.6	3 857.6	1 401.6	2 100.6	..	31 *Compos. électro. (semi-cond. incl.)*
32 *TV, radio & communications equip.*	..	1 228.4	800.1	762.2	2 619.8	2 886.9	..	32 *Appareils radio, tv et de communi.*
33 Instruments	..	79.0	85.7	114.2	107.4	90.4	..	33 Instruments
34 Motor vehicles	..	2 384.4	2 881.5	2 987.7	2 121.0	1 656.9	..	34 Véhicules automobiles
35 Other transport equipment	..	344.2	336.7	472.6	306.9	219.8	..	35 Autres matériels de transport
36 *Ships*	..	161.6	163.7	269.8	255.0	185.6	..	36 *Navires*
37 *Aerospace*	..	167.5	160.3	184.5	29.8	0.3 m	..	37 *Aérospatiale*
38 *Other transport equipment n.e.c.*	..	15.1	12.7	18.4	22.0	33.9	..	38 *Autres matériels de transport n.c.a.*
39 **Furniture, other manufacturing n.e.c.**	..	21.5	53.3	21.5	49.9	58.2	..	39 **Meubles, autres activ. de fabricat. n.c.a.**
40 Furniture	..	..	..	..	3.7	2.5	..	40 Meubles
41 Other manufacturing n.e.c.	..	..	..	..	46.2	55.6	..	41 Autres activités de fabrication n.c.a.
42 **Recycling**	..	..	1.5	0.9	10.9	11.1	..	42 **Récupération**
43 **ELECTRICITY, GAS & WATER SUPPLY**	..	227.3	292.0	308.4	304.7	166.5	..	43 **DISTRIBUTION D'ELECTRICITE, GAZ & EAU**
44 **CONSTRUCTION**	..	760.2	684.8	657.7	407.4	433.8	..	44 **CONSTRUCTION**
45 **SERVICES SECTOR**	..	858.1	910.9	1 241.4	1 376.1	1 671.9	..	45 **SECTEUR DES SERVICES**
46 **Wholesale, retail trade & repair of motor vehicles, etc.**	..	..	..	..	11.0	18.6	..	46 **Commerce (gros & détail), réparation de véhicules, etc.**
47 **Hotels & restaurants**	..	..	..	..	0.0	0.0	..	47 **Hôtels et restaurants**
48 **Transport & storage**	..	..	..	..	7.2	44.0	..	48 **Transports, entreposage**
49 **Communications**	..	..	..	..	685.4	917.7	..	49 **Communications**
50 Post	..	..	..	..	0.0	0.0	..	50 Postes
51 Telecommunications	..	..	..	..	685.4	917.7	..	51 Télécommunications
52 **Financial intermediation (inc. insur.)**	..	..	..	..	11.7	5.0	..	52 **Intermédiation financ. (assurances inc.)**
53 **Real estate, renting & busin. activity**	..	..	..	..	600.2	625.3	..	53 **Immobil., locat. & serv. aux entreprises**
54 Computer & related activities	..	..	..	..	265.1	348.7	..	54 Informatique et activités connexes
55 *Software consultancy*	..	..	..	..	216.0	201.4	..	55 *Production de logiciels*
56 *Other computer services n.e.c.*	..	..	..	..	49.1	147.2	..	56 *Autres activités informatiques*
57 Research & development	..	..	..	..	4.1	2.8	..	57 Recherche & développement expérimental
58 Other business activities n.e.c.	..	147.3	69.0	110.7	95.3	273.9	..	58 Autres activ. & serv. aux entreprises
59 **Comm., soc. & pers. serv. activ., etc.**	..	..	..	..	60.6	61.2	..	59 **Adm. pub., serv. soc. & collec., etc**
TOTAL BUSINESS ENTERPRISE SECTOR	..	**11 314.0**	**12 460.5**	**13 417.8**	**11 510.5**	**12 544.2**	..	**TOTAL SECTEUR DES ENTREPRISES**

Source: OECD, R&D database (DSTI/EAS Division), January 2002.

Source : OCDE, base de données R-D (DSTI/Division AES), janvier 2002.

KOREA — Table 10 - Tableau 10 — **CORÉE**

Business enterprise intramural expenditure on R&D (BERD) - Dépenses intra-muros de R-D des entreprises (DIRDE)

by industry and source of funds - par industrie et par source de financement

Million KRW — *Millions KRW*

	1994	1995	1996	1997	1998	1999	2000	
				10.1 Business enterprise - Entreprises				
1 **AGRICULTURE, HUNTING & FORESTRY**	..	..	12 458	12 299	15 650	..	..	1 **AGRICULTURE, CHASSE & SYLVICULTURE**
2 **MINING**	..	..	7 350	4 196	65 108	..	..	2 **ACTIVITES EXTRACTIVES**
3 **MANUFACTURING**	..	..	6 371 362	6 985 603	6 038 176	..	..	3 **ACTIVITES DE FABRICATION**
4 **Food, beverages & tobacco**	..	..	114 722	103 537	80 523	..	..	4 **Produits alimentaires, boissons & tabac**
5 Food products & beverages	..	..	114 349	103 183	80 269	..	..	5 Produits alimentaires & boissons
6 Tobacco products	..	..	373	354	254	..	..	6 Produits à base de tabac
7 **Textiles, wear app., fur & leather**	..	..	46 806	39 911	42 671	..	..	7 **Textiles, habill., fourrures & cuirs**
8 Textiles	..	..	41 138	34 219	34 765	..	..	8 Textiles
9 Wearing apparel & fur	..	..	4 630	4 341	3 217	..	..	9 Articles d'habillement & fourrures
10 Leather products & footwear	..	..	1 038	1 351	4 690	..	..	10 Articles en cuir & chaussures
11 **Wood, paper, printing, publishing**	..	..	52 363	45 119	36 505	..	..	11 **Bois, papier, imprimerie, édition**
12 Wood & cork (not furniture)	..	..	1 936	2 188	1 142	..	..	12 Bois & liège (sauf meubles)
13 Pulp, paper & paper products	..	..	24 316	24 633	15 676	..	..	13 Carton, papier & articles
14 Publ., print. & repro. of rec. media	..	..	26 111	18 298	19 687	..	..	14 Edition, impress. & repro. enregis.
15 **Coke, petroleum, nuclear fuel, chemicals, rubber plastics**	..	..	746 407	781 983	642 984	..	..	15 **Coke, pétrole, combust. nucl., prod. chimi., caout. & mat. plast.**
16 Coke, ref. petrol. prod. & nuclear fuel	..	..	106 417	123 291	84 629	..	..	16 Coke, raffin. pétr. & combust. nucl.
17 Chemicals & chemical products	..	..	553 221	561 765	428 149	..	..	17 Produits chimiques
18 *Chemical prod.(less pharmaceuticals)*	..	..	453 872	453 998	359 379	..	..	18 *Prod. chimi. (moins prod. pharmaceutiques)*
19 *Pharmaceuticals*	..	..	99 349	107 767	68 770	..	..	19 *Produits pharmaceutiques*
20 Rubber & plastic products	..	..	86 769	96 927	130 205	..	..	20 Caoutchouc et matières plastiques
21 **Non-metallic mineral products**	..	..	95 067	93 965	75 821	..	..	21 **Produits minéraux non métalliques**
22 **Basic metals**	..	..	157 520	121 185	105 290	..	..	22 **Produits métallurgiques de base**
23 Basic metals, ferrous	..	..	124 368	101 278	84 066	..	..	23 Produits métallur. de base, ferreux
24 Basic metals, non-ferrous	..	..	33 152	19 907	21 224	..	..	24 Prod. métallur. de base, non ferreux
25 **Fabricated metal products**	..	..	77 345	70 270	52 059	..	..	25 **Ouvrages en métaux (sauf mach.& mat.)**
26 **Machinery equipment, instruments and transport equipment**	..	..	5 047 132	5 715 460	4 972 824	..	..	26 **Machines & matériel, instruments & mat. de transport**
27 Machinery n.e.c.	..	..	258 982	237 610	336 165	..	..	27 Machines, n.c.a.
28 Office, account. & computing machinery	..	..	151 777	129 842	97 175	..	..	28 Mach. de bur., compta. & trait. inform.
29 Electrical machinery	..	..	122 381	139 591	167 155	..	..	29 Machines électriques
30 Electro. equip. (radio, TV & commun.)	..	..	2 474 201	2 923 817	2 683 709	..	..	30 Equip. électro. (radio, tv & communic.)
31 *Electro. comp. (inc. semi-conductors)*	..	..	2 040 500	2 462 768	926 963	..	..	31 *Compos. électro. (semi-cond. incl.)*
32 *TV, radio & communications equip.*	..	..	433 701	461 049	1 756 746	..	..	32 *Appareils radio, tv et de communi.*
33 Instruments	..	..	43 043	57 379	57 602	..	..	33 Instruments
34 Motor vehicles	..	..	1 805 078	1 935 240	1 432 893	..	..	34 Véhicules automobiles
35 Other transport equipment	..	..	191 670	291 981	198 125	..	..	35 Autres matériels de transport
36 *Ships*	..	..	103 278	174 377	168 070	..	..	36 *Navires*
37 *Aerospace*	..	..	80 739	107 380	15 602	..	..	37 *Aérospatiale*
38 *Other transport equipment n.e.c.*	..	..	7 653	10 224	14 453	..	..	38 *Autres matériels de transport n.c.a.*
39 **Furniture, other manufacturing n.e.c.**	..	..	33 307	13 738	27 047	..	..	39 **Meubles, autres activ. de fabricat. n.c.a.**
40 Furniture	..	..	..	..	2 484	..	..	40 Meubles
41 Other manufacturing n.e.c.	..	..	..	..	24 563	..	..	41 Autres activités de fabrication n.c.a.
42 **Recycling**	..	..	691	436	2 452	..	..	42 **Récupération**
43 **ELECTRICITY, GAS & WATER SUPPLY**	..	..	183 107	199 485	200 523	..	..	43 **DISTRIBUTION D'ELECTRICITE, GAZ & EAU**
44 **CONSTRUCTION**	..	..	427 275	428 726	278 524	..	..	44 **CONSTRUCTION**
45 **SERVICES SECTOR**	..	..	543 362	751 485	896 628	..	..	45 **SECTEUR DES SERVICES**
46 **Wholesale, retail trade & repair of motor vehicles, etc.**	..	..	..	..	7 538	..	..	46 **Commerce (gros & détail), réparation de véhicules, etc.**
47 **Hotels & restaurants**	..	..	..	..	0	..	..	47 **Hôtels et restaurants**
48 **Transport & storage**	..	..	..	..	4 436	..	..	48 **Transports, entreposage**
49 **Communications**	..	..	..	..	465 923	..	..	49 **Communications**
50 Post	..	..	..	..	0	..	..	50 Postes
51 Telecommunications	..	..	..	..	465 923	..	..	51 Télécommunications
52 **Financial intermediation (inc. insur.)**	..	..	..	..	8 016	..	..	52 **Intermédiation financ. (assurances inc.)**
53 **Real estate, renting & busin. activity**	..	..	..	..	371 267	..	..	53 **Immobil., locat. & serv. aux entreprises**
54 Computer & related activities	..	..	..	..	144 095	..	..	54 Informatique et activités connexes
55 *Software consultancy*	..	..	..	..	119 332	..	..	55 *Production de logiciels*
56 *Other computer services n.e.c.*	..	..	..	..	24 763	..	..	56 *Autres activités informatiques*
57 Research & development	..	..	..	..	1 753	..	..	57 Recherche & développement expérimental
58 Other business activities n.e.c.	..	..	43 530	69 687	62 367	..	..	58 Autres activ. & serv. aux entreprises
59 **Comm., soc. & pers. serv. activ., etc.**	..	..	..	..	39 448	..	..	59 **Adm. pub., serv. soc. & collec., etc**
TOTAL BUSINESS ENTERPRISE SECTOR	..	6 700 235	7 544 915	8 381 794	7 494 610	7 993 094	..	**TOTAL SECTEUR DES ENTREPRISES**

Source: OECD, R&D database (DSTI/EAS Division), January 2002.

Source : OCDE, base de données R-D (DSTI/Division AES), janvier 2002.

Business enterprise intramural expenditure on R&D (BERD) - Dépenses intra-muros de R-D des entreprises (DIRDE)

by industry and source of funds - par industrie et par source de financement

Million KRW / Millions KRW

	1994	1995	1996	1997	1998	1999	2000	
				10.2 Government - État				
1 AGRICULTURE, HUNTING & FORESTRY	..	..	3 215	2 280	4 263	..	..	**1 AGRICULTURE, CHASSE & SYLVICULTURE**
2 MINING	..	..	..	..	1 400	..	..	**2 ACTIVITES EXTRACTIVES**
3 MANUFACTURING	..	..	307 462	352 250	392 865	..	..	**3 ACTIVITES DE FABRICATION**
4 Food, beverages & tobacco	..	..	3 165	2 524	3 824	..	..	**4 Produits alimentaires, boissons & tabac**
5 Food products & beverages	..	..	3 165	2 524	3 824	..	..	5 Produits alimentaires & boissons
6 Tobacco products	..	..	..	..	0	..	..	6 Produits à base de tabac
7 Textiles, wear app., fur & leather	..	..	4 098	3 230	2 808	..	..	**7 Textiles, habill., fourrures & cuirs**
8 Textiles	..	..	4 098	3 230	2 452	..	..	8 Textiles
9 Wearing apparel & fur	..	..	..	..	112	..	..	9 Articles d'habillement & fourrures
10 Leather products & footwear	..	..	..	..	244	..	..	10 Articles en cuir & chaussures
11 Wood, paper, printing, publishing	..	..	2 247	1 483	1 638	..	..	**11 Bois, papier, imprimerie, édition**
12 Wood & cork (not furniture)	..	..	10	5	18	..	..	12 Bois & liège (sauf meubles)
13 Pulp, paper & paper products	..	..	1 777	1 291	1 620	..	..	13 Carton, papier & articles
14 Publ., print. & repro. of rec. media	..	..	460	187	0	..	..	14 Edition, impress. & repro. enregis.
15 Coke, petroleum, nuclear fuel, chemicals, rubber plastics	..	..	47 423	45 868	43 811	..	..	**15 Coke, pétrole, combust. nucl., prod. chimi., caout. & mat. plast.**
16 Coke, ref. petrol. prod. & nuclear fuel	..	..	581	789	2 126	..	..	16 Coke, raffin. pétr. & combust. nucl.
17 Chemicals & chemical products	..	..	43 146	40 823	33 573	..	..	17 Produits chimiques
18 *Chemical prod.(less pharmaceuticals)*	..	..	23 509	26 883	21 564	..	..	18 *Prod. chimi. (moins prod. pharmaceutiques)*
19 *Pharmaceuticals*	..	..	19 637	13 940	12 009	..	..	19 *Produits pharmaceutiques*
20 Rubber & plastic products	..	..	3 696	4 256	8 112	..	..	20 Caoutchouc et matières plastiques
21 Non-metallic mineral products	..	..	4 374	5 171	3 787	..	..	**21 Produits minéraux non métalliques**
22 Basic metals	..	..	6 584	7 633	12 575	..	..	**22 Produits métallurgiques de base**
23 Basic metals, ferrous	..	..	3 516	5 088	10 013	..	..	23 Produits métallur. de base, ferreux
24 Basic metals, non-ferrous	..	..	3 068	2 545	2 563	..	..	24 Prod. métallur. de base, non ferreux
25 Fabricated metal products	..	..	9 262	6 681	9 599	..	..	**25 Ouvrages en métaux (sauf mach.& mat.)**
26 Machinery equipment, instruments and transport equipment	..	..	229 349	279 099	302 267	..	..	**26 Machines & matériel, instruments & mat. de transport**
27 Machinery n.e.c.	..	..	27 700	70 701	99 421	..	..	27 Machines, n.c.a.
28 Office, account. & computing machinery	..	..	13 477	20 361	14 107	..	..	28 Mach. de bur., compta. & trait. inform.
29 Electrical machinery	..	..	13 115	16 509	22 237	..	..	29 Machines électriques
30 Electro. equip. (radio, TV & commun.)	..	..	114 381	103 635	100 355	..	..	30 Equip. électro. (radio, tv & communic.)
31 *Electro. comp. (inc. semi-conductors)*	..	..	44 456	63 733	42 899	..	..	31 *Compos. électro. (semi-cond. incl.)*
32 *TV, radio & communications equip.*	..	..	69 925	39 902	57 456	..	..	32 *Appareils radio, tv et de communi.*
33 Instruments	..	..	11 057	16 178	16 500	..	..	33 Instruments
34 Motor vehicles	..	..	26 312	32 456	35 249	..	..	34 Véhicules automobiles
35 Other transport equipment	..	..	23 307	19 259	14 397	..	..	35 Autres matériels de transport
36 *Ships*	..	..	1 130	3 130	8 510	..	..	36 *Navires*
37 *Aerospace*	..	..	21 686	14 218	5 070	..	..	37 *Aérospatiale*
38 *Other transport equipment n.e.c.*	..	..	491	1 911	817	..	..	38 *Autres matériels de transport n.c.a.*
39 Furniture, other manufacturing n.e.c.	..	..	721	370	7 484	..	..	**39 Meubles, autres activ. de fabricat. n.c.a.**
40 Furniture	..	..	..	..	77	..	..	40 Meubles
41 Other manufacturing n.e.c.	..	..	..	..	7 407	..	..	41 Autres activités de fabrication n.c.a.
42 Recycling	..	..	238	189	5 071	..	..	**42 Récupération**
43 ELECTRICITY, GAS & WATER SUPPLY	..	..	3 464	3 834	10 512	..	..	**43 DISTRIBUTION D'ELECTRICITE, GAZ & EAU**
44 CONSTRUCTION	..	..	5 938	4 200	3 434	..	..	**44 CONSTRUCTION**
45 SERVICES SECTOR	..	..	27 352	60 809	53 984	..	..	**45 SECTEUR DES SERVICES**
46 Wholesale, retail trade & repair of motor vehicles, etc.	..	..	..	..	100	..	..	**46 Commerce (gros & détail), réparation de véhicules, etc.**
47 Hotels & restaurants	..	..	..	..	0	..	..	**47 Hôtels et restaurants**
48 Transport & storage	..	..	..	..	525	..	..	**48 Transports, entreposage**
49 Communications	..	..	..	..	8 781	..	..	**49 Communications**
50 Post	..	..	..	..	0	..	..	50 Postes
51 Telecommunications	..	..	..	..	8 781	..	..	51 Télécommunications
52 Financial intermediation (inc. insur.)	..	..	..	..	89	..	..	**52 Intermédiation financ. (assurances inc.)**
53 Real estate, renting & busin. activity	..	..	..	..	41 987	..	..	**53 Immobil., locat. & serv. aux entreprises**
54 Computer & related activities	..	..	..	..	38 630	..	..	54 Informatique et activités connexes
55 *Software consultancy*	..	..	..	..	29 971	..	..	55 *Production de logiciels*
56 *Other computer services n.e.c.*	..	..	..	..	8 659	..	..	56 *Autres activités informatiques*
57 Research & development	..	..	..	..	0	..	..	57 Recherche & développement expérimental
58 Other business activities n.e.c.	..	..	534	3 118	3 182	..	..	58 Autres activ. & serv. aux entreprises
59 Comm., soc. & pers. serv. activ., etc.	..	..	..	..	2 501	..	..	**59 Adm. pub., serv. soc. & collec., etc**
TOTAL BUSINESS ENTERPRISE SECTOR	..	**247 189**	**347 430**	**423 374**	**466 458**	**496 518**	..	**TOTAL SECTEUR DES ENTREPRISES**

Source: OECD, R&D database (DSTI/EAS Division), January 2002.

Source : OCDE, base de données R-D (DSTI/Division AES), janvier 2002.

Business enterprise intramural expenditure on R&D (BERD) - Dépenses intra-muros de R-D des entreprises (DIRDE)

by industry and source of funds - par industrie et par source de financement

Million KRW *Millions KRW*

10.3 Other national funds - Autres sources nationales

	1994	1995	1996	1997	1998	1999	2000	
1 AGRICULTURE, HUNTING & FORESTRY	..	..	135	20	0	..	..	**1 AGRICULTURE, CHASSE & SYLVICULTURE**
2 MINING	..	..	..	..	152	..	..	**2 ACTIVITES EXTRACTIVES**
3 MANUFACTURING	..	..	47 861	30 223	6 450	..	..	**3 ACTIVITES DE FABRICATION**
4 Food, beverages & tobacco	..	..	115	66	27	..	..	**4 Produits alimentaires, boissons & tabac**
5 Food products & beverages	..	..	115	66	27	..	..	5 Produits alimentaires & boissons
6 Tobacco products	..	..	0	0	0	..	..	6 Produits à base de tabac
7 Textiles, wear app., fur & leather	..	..	67	13	11	..	..	**7 Textiles, habill., fourrures & cuirs**
8 Textiles	..	..	67	13	11	..	..	8 Textiles
9 Wearing apparel & fur	..	..	0	0	0	..	..	9 Articles d'habillement & fourrures
10 Leather products & footwear	..	..	0	0	0	..	..	10 Articles en cuir & chaussures
11 Wood, paper, printing, publishing	..	..	51	383	0	..	..	**11 Bois, papier, imprimerie, édition**
12 Wood & cork (not furniture)	..	..	0	2	0	..	..	12 Bois & liège (sauf meubles)
13 Pulp, paper & paper products	..	..	0	381	0	..	..	13 Carton, papier & articles
14 Publ., print. & repro. of rec. media	..	..	51	0	0	..	..	14 Edition, impress. & repro. enregis.
15 Coke, petroleum, nuclear fuel, chemicals, rubber plastics	..	..	3 719	1 278	414	..	..	**15 Coke, pétrole, combust. nucl., prod. chimi., caout. & mat. plast.**
16 Coke, ref. petrol. prod. & nuclear fuel	..	..	..	1	0	..	..	16 Coke, raffin. pétr. & combust. nucl.
17 Chemicals & chemical products	..	..	3 293	1 141	278	..	..	17 Produits chimiques
18 *Chemical prod.(less pharmaceuticals)*	..	..	3 059	701	198	..	..	18 *Prod. chimi. (moins prod. pharmaceutiques)*
19 *Pharmaceuticals*	..	..	234	441	79	..	..	19 *Produits pharmaceutiques*
20 Rubber & plastic products	..	..	426	137	136	..	..	20 Caoutchouc et matières plastiques
21 Non-metallic mineral products	..	..	57	544	6	..	..	**21 Produits minéraux non métalliques**
22 Basic metals	..	..	71	0	20	..	..	**22 Produits métallurgiques de base**
23 Basic metals, ferrous	..	..	31	0	0	..	..	23 Produits métallur. de base, ferreux
24 Basic metals, non-ferrous	..	..	40	0	20	..	..	24 Prod. métallur. de base, non ferreux
25 Fabricated metal products	..	..	137	777	1 295	..	..	**25 Ouvrages en métaux (sauf mach.& mat.)**
26 Machinery equipment, instruments and transport equipment	..	..	43 629	27 163	4 679	..	..	**26 Machines & matériel, instruments & mat. de transport**
27 Machinery n.e.c.	..	..	4 473	2 791	811	..	..	27 Machines, n.c.a.
28 Office, account. & computing machinery	..	..	51	..	0	..	..	28 Mach. de bur., compta. & trait. inform.
29 Electrical machinery	..	..	7 756	2 484	1 689	..	..	29 Machines électriques
30 Electro. equip. (radio, TV & commun.)	..	..	20 309	18 018	1 138	..	..	30 Equip. électro. (radio, tv & communic.)
31 *Electro. comp. (inc. semi-conductors)*	..	..	17 785	16 536	908	..	..	31 *Compos. électro. (semi-cond. incl.)*
32 *TV, radio & communications equip.*	..	..	2 524	1 481	230	..	..	32 *Appareils radio, tv et de communi.*
33 Instruments	..	..	642	1 683	220	..	..	33 Instruments
34 Motor vehicles	..	..	10 197	1 855	820	..	..	34 Véhicules automobiles
35 Other transport equipment	..	..	201	332	0	..	..	35 Autres matériels de transport
36 *Ships*	..	..	198	332	0	..	..	36 *Navires*
37 *Aerospace*	..	..	0	0	0	..	..	37 *Aérospatiale*
38 *Other transport equipment n.e.c.*	..	..	3	0	0	..	..	38 *Autres matériels de transport n.c.a.*
39 Furniture, other manufacturing n.e.c.	..	..	16	0	0	..	..	**39 Meubles, autres activ. de fabricat. n.c.a.**
40 Furniture	..	..	..	..	0	..	..	40 Meubles
41 Other manufacturing n.e.c.	..	..	..	..	0	..	..	41 Autres activités de fabrication n.c.a.
42 Recycling	..	..	..	..	0	..	..	**42 Récupération**
43 ELECTRICITY, GAS & WATER SUPPLY	..	..	27	..	17	..	..	**43 DISTRIBUTION D'ELECTRICITE, GAZ & EAU**
44 CONSTRUCTION	..	..	4 470	622	179	..	..	**44 CONSTRUCTION**
45 SERVICES SECTOR	..	..	8 615	4 446	1 136	..	..	**45 SECTEUR DES SERVICES**
46 Wholesale, retail trade & repair of motor vehicles, etc.	..	..	..	..	0	..	..	**46 Commerce (gros & détail), réparation de véhicules, etc.**
47 Hotels & restaurants	..	..	..	..	0	..	..	**47 Hôtels et restaurants**
48 Transport & storage	..	..	..	..	0	..	..	**48 Transports, entreposage**
49 Communications	..	..	..	..	0	..	..	**49 Communications**
50 Post	..	..	..	..	0	..	..	50 Postes
51 Telecommunications	..	..	..	..	0	..	..	51 Télécommunications
52 Financial intermediation (inc. insur.)	..	..	..	..	0	..	..	**52 Intermédiation financ. (assurances inc.)**
53 Real estate, renting & busin. activity	..	..	..	..	1 136	..	..	**53 Immobil., locat. & serv. aux entreprises**
54 Computer & related activities	..	..	..	..	890	..	..	54 Informatique et activités connexes
55 *Software consultancy*	..	..	..	..	273	..	..	55 *Production de logiciels*
56 *Other computer services n.e.c.*	..	..	..	..	617	..	..	56 *Autres activités informatiques*
57 Research & development	..	..	..	..	0	..	..	57 Recherche & développement expérimental
58 Other business activities n.e.c.	..	..	13	198	231	..	..	58 Autres activ. & serv. aux entreprises
59 Comm., soc. & pers. serv. activ., etc.	..	..	..	..	0	..	..	**59 Adm. pub., serv. soc. & collec., etc**
TOTAL BUSINESS ENTERPRISE SECTOR	..	11 791	61 108	35 311	7 934	17 784	..	**TOTAL SECTEUR DES ENTREPRISES**

Source: OECD, R&D database (DSTI/EAS Division), January 2002.

Source : OCDE, base de données R-D (DSTI/Division AES), janvier 2002.

Business enterprise intramural expenditure on R&D (BERD) - Dépenses intra-muros de R-D des entreprises (DIRDE)

by industry and source of funds - par industrie et par source de financement

Million KRW *Millions KRW*

	1994	1995	1996	1997	1998	1999	2000	
	10.4 Funds from abroad - Fonds de l'étranger							
1 AGRICULTURE, HUNTING & FORESTRY	..	..	0	0	0	..	..	**1 AGRICULTURE, CHASSE & SYLVICULTURE**
2 MINING	..	..	0	0	0	..	..	**2 ACTIVITES EXTRACTIVES**
3 MANUFACTURING	..	..	7 350	3 206	1 721	..	..	**3 ACTIVITES DE FABRICATION**
4 Food, beverages & tobacco	..	..	0	0	22	..	..	**4 Produits alimentaires, boissons & tabac**
5 Food products & beverages	..	..	0	0	22	..	..	5 Produits alimentaires & boissons
6 Tobacco products	..	..	0	0	0	..	..	6 Produits à base de tabac
7 Textiles, wear app., fur & leather	..	..	125	0	0	..	..	**7 Textiles, habill., fourrures & cuirs**
8 Textiles	..	..	125	0	0	..	..	8 Textiles
9 Wearing apparel & fur	..	..	0	0	0	..	..	9 Articles d'habillement & fourrures
10 Leather products & footwear	..	..	0	0	0	..	..	10 Articles en cuir & chaussures
11 Wood, paper, printing, publishing	..	..	0	0	0	..	..	**11 Bois, papier, imprimerie, édition**
12 Wood & cork (not furniture)	..	..	0	0	0	..	..	12 Bois & liège (sauf meubles)
13 Pulp, paper & paper products	..	..	0	0	0	..	..	13 Carton, papier & articles
14 Publ., print. & repro. of rec. media	..	..	0	0	0	..	..	14 Edition, impress. & repro. enregis.
15 Coke, petroleum, nuclear fuel, chemicals, rubber plastics	..	..	1 146	1 518	924	..	..	**15 Coke, pétrole, combust. nucl., prod. chimi., caout. & mat. plast.**
16 Coke, ref. petrol. prod. & nuclear fuel	..	..	0	948	0	..	..	16 Coke, raffin. pétr. & combust. nucl.
17 Chemicals & chemical products	..	..	1 146	570	924	..	..	17 Produits chimiques
18 *Chemical prod.(less pharmaceuticals)*	..	..	695	570	924	..	..	18 *Prod. chimi. (moins prod. pharmaceutiques)*
19 *Pharmaceuticals*	..	..	451	0	0	..	..	19 *Produits pharmaceutiques*
20 Rubber & plastic products	..	..	0	0	0	..	..	20 Caoutchouc et matières plastiques
21 Non-metallic mineral products	..	..	0	0	0	..	..	**21 Produits minéraux non métalliques**
22 Basic metals	..	..	0	0	0	..	..	**22 Produits métallurgiques de base**
23 Basic metals, ferrous	..	..	0	0	0	..	..	23 Produits métallur. de base, ferreux
24 Basic metals, non-ferrous	..	..	0	0	0	..	..	24 Prod. métallur. de base, non ferreux
25 Fabricated metal products	..	..	0	0	16	..	..	**25 Ouvrages en métaux (sauf mach.& mat.)**
26 Machinery equipment, instruments and transport equipment	..	..	6 079	1 620	758	..	..	**26 Machines & matériel, instruments & mat. de transport**
27 Machinery n.e.c.	..	..	0	64	198	..	..	27 Machines, n.c.a.
28 Office, account. & computing machinery	..	..	0	0	0	..	..	28 Mach. de bur., compta. & trait. inform.
29 Electrical machinery	..	..	518	1 511	480	..	..	29 Machines électriques
30 Electro. equip. (radio, TV & commun.)	..	..	5 507	7	0	..	..	30 Equip. électro. (radio, tv & communic.)
31 *Electro. comp. (inc. semi-conductors)*	..	..	292	..	0	..	..	31 *Compos. électro. (semi-cond. incl.)*
32 *TV, radio & communications equip.*	..	..	5 215	7	0	..	..	32 *Appareils radio, tv et de communi.*
33 Instruments	..	..	54	38	44	..	..	33 Instruments
34 Motor vehicles	..	..	0	0	36	..	..	34 Véhicules automobiles
35 Other transport equipment	..	..	0	0	0	..	..	35 Autres matériels de transport
36 *Ships*	..	..	0	0	0	..	..	36 *Navires*
37 *Aerospace*	..	..	0	0	0	..	..	37 *Aérospatiale*
38 *Other transport equipment n.e.c.*	..	..	0	0	0	..	..	38 *Autres matériels de transport n.c.a.*
39 Furniture, other manufacturing n.e.c.	..	..	0	67	0	..	..	**39 Meubles, autres activ. de fabricat. n.c.a.**
40 Furniture	..	..	..	..	0	..	..	40 Meubles
41 Other manufacturing n.e.c.	..	..	..	..	0	..	..	41 Autres activités de fabrication n.c.a.
42 Recycling	..	..	..	..	0	..	..	**42 Récupération**
43 ELECTRICITY, GAS & WATER SUPPLY	..	..	4	7	2	..	..	**43 DISTRIBUTION D'ELECTRICITE, GAZ & EAU**
44 CONSTRUCTION	..	..	..	..	23	..	..	**44 CONSTRUCTION**
45 SERVICES SECTOR	..	..	2 805	1 616	1 326	..	..	**45 SECTEUR DES SERVICES**
46 Wholesale, retail trade & repair of motor vehicles, etc.	..	..	..	..	0	..	..	**46 Commerce (gros & détail), réparation de véhicules, etc.**
47 Hotels & restaurants	..	..	..	..	0	..	..	**47 Hôtels et restaurants**
48 Transport & storage	..	..	..	..	0	..	..	**48 Transports, entreposage**
49 Communications	..	..	..	..	0	..	..	**49 Communications**
50 Post	..	..	..	..	0	..	..	50 Postes
51 Telecommunications	..	..	..	..	0	..	..	51 Télécommunications
52 Financial intermediation (inc. insur.)	..	..	..	..	0	..	..	**52 Intermédiation financ. (assurances inc.)**
53 Real estate, renting & busin. activity	..	..	..	..	1 326	..	..	**53 Immobil., locat. & serv. aux entreprises**
54 Computer & related activities	..	..	..	..	0	..	..	54 Informatique et activités connexes
55 *Software consultancy*	..	..	..	..	0	..	..	55 *Production de logiciels*
56 *Other computer services n.e.c.*	..	..	..	..	0	..	..	56 *Autres activités informatiques*
57 Research & development	..	..	..	..	1 102	..	..	57 Recherche & développement expérimental
58 Other business activities n.e.c.	..	..	..	..	224	..	..	58 Autres activ. & serv. aux entreprises
59 Comm., soc. & pers. serv. activ., etc.	..	..	..	..	0	..	..	**59 Adm. pub., serv. soc. & collec., etc**
TOTAL BUSINESS ENTERPRISE SECTOR	..	1 107	10 159	4 828	3 071	3 761	..	**TOTAL SECTEUR DES ENTREPRISES**

Source: OECD, R&D database (DSTI/EAS Division), January 2002.

Source : OCDE, base de données R-D (DSTI/Division AES), janvier 2002.

Business enterprise intramural expenditure on R&D (BERD) - Dépenses intra-muros de R-D des entreprises (DIRDE)

by industry and type of costs - par industrie et par type de coûts

Million KRW — *Millions KRW*

	1994	1995	1996	1997	1998	1999	2000	
				11.1 Current expenditure - Dépenses courantes				
1 AGRICULTURE, HUNTING & FORESTRY	..	..	14 529	13 660	16 551	..	..	**1 AGRICULTURE, CHASSE & SYLVICULTURE**
2 MINING	..	..	6 881	3 961	58 019	..	..	**2 ACTIVITES EXTRACTIVES**
3 MANUFACTURING	..	..	4 987 391	5 682 645	4 899 804	..	..	**3 ACTIVITES DE FABRICATION**
4 Food, beverages & tobacco	..	..	99 439	93 411	73 779	..	..	**4 Produits alimentaires, boissons & tabac**
5 Food products & beverages	..	..	99 166	93 107	73 525	..	..	5 Produits alimentaires & boissons
6 Tobacco products	..	..	273	304	254	..	..	6 Produits à base de tabac
7 Textiles, wear app., fur & leather	..	..	42 214	35 420	38 291	..	..	**7 Textiles, habill., fourrures & cuirs**
8 Textiles	..	..	36 740	29 995	30 468	..	..	8 Textiles
9 Wearing apparel & fur	..	..	4 436	4 124	3 177	..	..	9 Articles d'habillement & fourrures
10 Leather products & footwear	..	..	1 038	1 301	4 646	..	..	10 Articles en cuir & chaussures
11 Wood, paper, printing, publishing	..	..	49 577	39 415	34 904	..	..	**11 Bois, papier, imprimerie, édition**
12 Wood & cork (not furniture)	..	..	1 836	1 998	1 004	..	..	12 Bois & liège (sauf meubles)
13 Pulp, paper & paper products	..	..	22 346	19 552	15 042	..	..	13 Carton, papier & articles
14 Publ., print. & repro. of rec. media	..	..	25 395	17 865	18 858	..	..	14 Edition, impress. & repro. enregis.
15 Coke, petroleum, nuclear fuel, chemicals, rubber plastics	..	..	588 859	642 686	549 026	..	..	**15 Coke, pétrole, combust. nucl., prod. chimi., caout. & mat. plast.**
16 Coke, ref. petrol. prod. & nuclear fuel	..	..	71 161	81 714	62 542	..	..	16 Coke, raffin. pétr. & combust. nucl.
17 Chemicals & chemical products	..	..	452 022	485 396	384 707	..	..	17 Produits chimiques
18 *Chemical prod.(less pharmaceuticals)*	..	..	357 839	392 608	316 476	..	..	18 *Prod. chimi. (moins prod. pharmaceutiques)*
19 *Pharmaceuticals*	..	..	94 183	92 788	68 231	..	..	19 *Produits pharmaceutiques*
20 Rubber & plastic products	..	..	65 676	75 576	101 777	..	..	20 Caoutchouc et matières plastiques
21 Non-metallic mineral products	..	..	70 150	71 919	42 479	..	..	**21 Produits minéraux non métalliques**
22 Basic metals	..	..	145 130	111 507	90 369	..	..	**22 Produits métallurgiques de base**
23 Basic metals, ferrous	..	..	117 543	92 865	68 998	..	..	23 Produits métallur. de base, ferreux
24 Basic metals, non-ferrous	..	..	27 587	18 642	21 371	..	..	24 Prod. métallur. de base, non ferreux
25 Fabricated metal products	..	..	44 026	55 588	40 804	..	..	**25 Ouvrages en métaux (sauf mach.& mat.)**
26 Machinery equipment, instruments and transport equipment	..	..	3 926 390	4 621 197	3 995 136	..	..	**26 Machines & matériel, instruments & mat. de transport**
27 Machinery n.e.c.	..	..	196 415	255 360	301 275	..	..	27 Machines, n.c.a.
28 Office, account. & computing machinery	..	..	133 040	120 050	84 541	..	..	28 Mach. de bur., compta. & trait. inform.
29 Electrical machinery	..	..	116 454	134 252	167 778	..	..	29 Machines électriques
30 Electro. equip. (radio, TV & commun.)	..	..	2 023 416	2 400 960	2 070 277	..	..	30 Equip. électro. (radio, tv & communic.)
31 *Electro. comp. (inc. semi-conductors)*	..	..	1 610 071	2 012 762	710 225	..	..	31 *Compos. électro. (semi-cond. incl.)*
32 *TV, radio & communications equip.*	..	..	413 345	388 198	1 360 052	..	..	32 *Appareils radio, tv et de communi.*
33 Instruments	..	..	44 261	63 838	57 120	..	..	33 Instruments
34 Motor vehicles	..	..	1 217 075	1 366 017	1 111 828	..	..	34 Véhicules automobiles
35 Other transport equipment	..	..	195 729	280 720	202 317	..	..	35 Autres matériels de transport
36 *Ships*	..	..	99 162	158 382	169 072	..	..	36 *Navires*
37 *Aerospace*	..	..	89 529	113 525	20 477	..	..	37 *Aérospatiale*
38 *Other transport equipment n.e.c.*	..	..	7 038	8 813	12 768	..	..	38 *Autres matériels de transport n.c.a.*
39 Furniture, other manufacturing n.e.c.	..	..	20 922	11 140	29 548	..	..	**39 Meubles, autres activ. de fabricat. n.c.a.**
40 Furniture	..	..	..	..	2 551	..	..	40 Meubles
41 Other manufacturing n.e.c.	..	..	..	..	26 997	..	..	41 Autres activités de fabrication n.c.a.
42 Recycling	..	..	684	362	5 468	..	..	**42 Récupération**
43 ELECTRICITY, GAS & WATER SUPPLY	..	..	166 998	179 374	142 630	..	..	**43 DISTRIBUTION D'ELECTRICITE, GAZ & EAU**
44 CONSTRUCTION	..	..	308 472	324 565	229 033	..	..	**44 CONSTRUCTION**
45 SERVICES SECTOR	..	..	405 818	618 278	717 705	..	..	**45 SECTEUR DES SERVICES**
46 Wholesale, retail trade & repair of motor vehicles, etc.	..	..	..	..	7 310	..	..	**46 Commerce (gros & détail), réparation de véhicules, etc.**
47 Hotels & restaurants	..	..	..	..	0	..	..	**47 Hôtels et restaurants**
48 Transport & storage	..	..	..	..	3 285	..	..	**48 Transports, entreposage**
49 Communications	..	..	..	..	339 647	..	..	**49 Communications**
50 Post	..	..	..	..	0	..	..	50 Postes
51 Telecommunications	..	..	..	..	339 647	..	..	51 Télécommunications
52 Financial intermediation (inc. insur.)	..	..	..	..	6 698	..	..	**52 Intermédiation financ. (assurances inc.)**
53 Real estate, renting & busin. activity	..	..	..	..	330 436	..	..	**53 Immobil., locat. & serv. aux entreprises**
54 Computer & related activities	..	..	..	..	146 443	..	..	54 Informatique et activités connexes
55 *Software consultancy*	..	..	..	..	120 671	..	..	55 *Production de logiciels*
56 *Other computer services n.e.c.*	..	..	..	..	25 772	..	..	56 *Autres activités informatiques*
57 Research & development	..	..	..	..	2 332	..	..	57 Recherche & développement expérimental
58 Other business activities n.e.c.	..	..	32 262	53 886	60 168	..	..	58 Autres activ. & serv. aux entreprises
59 Comm., soc. & pers. serv. activ., etc.	..	..	..	..	30 329	..	..	**59 Adm. pub., serv. soc. & collec., etc**
TOTAL BUSINESS ENTERPRISE SECTOR	..	..	**5 890 089**	**6 822 483**	**6 063 742**	..	..	**TOTAL SECTEUR DES ENTREPRISES**

Source: OECD, R&D database (DSTI/EAS Division), January 2002.

Source : OCDE, base de données R-D (DSTI/Division AES), janvier 2002.

Business enterprise intramural expenditure on R&D (BERD) - Dépenses intra-muros de R-D des entreprises (DIRDE)

by industry and type of costs - par industrie et par type de coûts

Million KRW *Millions KRW*

	1994	1995	1996	1997	1998	1999	2000	
				11.2 Capital expenditure - Dépenses en capital				
1 AGRICULTURE, HUNTING & FORESTRY	..	..	1 280	939	3 362	..	..	**1 AGRICULTURE, CHASSE & SYLVICULTURE**
2 MINING	..	..	469	235	8 642	..	..	**2 ACTIVITES EXTRACTIVES**
3 MANUFACTURING	..	..	1 746 642	1 688 636	1 539 408	..	..	**3 ACTIVITES DE FABRICATION**
4 Food, beverages & tobacco	..	..	18 563	12 715	10 617	..	..	**4 Produits alimentaires, boissons & tabac**
5 Food products & beverages	..	..	18 463	12 665	10 617	..	..	5 Produits alimentaires & boissons
6 Tobacco products	..	..	100	50	0	..	..	6 Produits à base de tabac
7 Textiles, wear app., fur & leather	..	..	8 883	7 734	7 200	..	..	**7 Textiles, habill., fourrures & cuirs**
8 Textiles	..	..	8 689	7 467	6 760	..	..	8 Textiles
9 Wearing apparel & fur	..	..	194	217	152	..	..	9 Articles d'habillement & fourrures
10 Leather products & footwear	..	..	0	50	288	..	..	10 Articles en cuir & chaussures
11 Wood, paper, printing, publishing	..	..	5 084	7 570	3 239	..	..	**11 Bois, papier, imprimerie, édition**
12 Wood & cork (not furniture)	..	..	110	197	156	..	..	12 Bois & liège (sauf meubles)
13 Pulp, paper & paper products	..	..	3 747	6 753	2 254	..	..	13 Carton, papier & articles
14 Publ., print. & repro. of rec. media	..	..	1 227	620	829	..	..	14 Edition, impress. & repro. enregis.
15 Coke, petroleum, nuclear fuel, chemicals, rubber plastics	..	..	209 837	187 962	139 106	..	..	**15 Coke, pétrole, combust. nucl., prod. chimi., caout. & mat. plast.**
16 Coke, ref. petrol. prod. & nuclear fuel	..	..	35 838	43 315	24 213	..	..	16 Coke, raffin. pétr. & combust. nucl.
17 Chemicals & chemical products	..	..	148 784	118 903	78 217	..	..	17 Produits chimiques
18 *Chemical prod.(less pharmaceuticals)*	..	..	123 296	89 544	65 590	..	..	18 *Prod. chimi. (moins prod. pharmaceutiques)*
19 *Pharmaceuticals*	..	..	25 488	29 359	12 627	..	..	19 *Produits pharmaceutiques*
20 Rubber & plastic products	..	..	25 215	25 744	36 676	..	..	20 Caoutchouc et matières plastiques
21 Non-metallic mineral products	..	..	29 348	27 761	37 135	..	..	**21 Produits minéraux non métalliques**
22 Basic metals	..	..	19 044	17 310	27 516	..	..	**22 Produits métallurgiques de base**
23 Basic metals, ferrous	..	..	10 372	13 501	25 081	..	..	23 Produits métallur. de base, ferreux
24 Basic metals, non-ferrous	..	..	8 672	3 809	2 435	..	..	24 Prod. métallur. de base, non ferreux
25 Fabricated metal products	..	..	42 719	22 139	22 165	..	..	**25 Ouvrages en métaux (sauf mach.& mat.)**
26 Machinery equipment, instruments and transport equipment	..	..	1 399 797	1 402 147	1 285 392	..	..	**26 Machines & matériel, instruments & mat. de transport**
27 Machinery n.e.c.	..	..	94 740	55 806	135 321	..	..	27 Machines, n.c.a.
28 Office, account. & computing machinery	..	..	32 264	30 153	26 741	..	..	28 Mach. de bur., compta. & trait. inform.
29 Electrical machinery	..	..	27 316	25 843	23 783	..	..	29 Machines électriques
30 Electro. equip. (radio, TV & commun.)	..	..	590 982	644 517	714 925	..	..	30 Equip. électro. (radio, tv & communic.)
31 *Electro. comp. (inc. semi-conductors)*	..	..	492 962	530 276	260 545	..	..	31 *Compos. électro. (semi-cond. incl.)*
32 *TV, radio & communications equip.*	..	..	98 020	114 241	454 380	..	..	32 *Appareils radio, tv et de communi.*
33 Instruments	..	..	10 535	11 442	17 247	..	..	33 Instruments
34 Motor vehicles	..	..	624 512	603 534	357 170	..	..	34 Véhicules automobiles
35 Other transport equipment	..	..	19 448	30 852	10 205	..	..	35 Autres matériels de transport
36 *Ships*	..	..	5 443	19 457	7 508	..	..	36 *Navires*
37 *Aerospace*	..	..	12 896	8 073	195	..	..	37 *Aérospatiale*
38 *Other transport equipment n.e.c.*	..	..	1 109	3 322	2 502	..	..	38 *Autres matériels de transport n.c.a.*
39 Furniture, other manufacturing n.e.c.	..	..	13 122	3 035	4 983	..	..	**39 Meubles, autres activ. de fabricat. n.c.a.**
40 Furniture	..	..	..	..	10	..	..	40 Meubles
41 Other manufacturing n.e.c.	..	..	..	..	4 973	..	..	41 Autres activités de fabrication n.c.a.
42 Recycling	..	..	245	263	2 055	..	..	**42 Récupération**
43 ELECTRICITY, GAS & WATER SUPPLY	..	..	19 604	23 953	68 424	..	..	**43 DISTRIBUTION D'ELECTRICITE, GAZ & EAU**
44 CONSTRUCTION	..	..	129 211	108 983	53 127	..	..	**44 CONSTRUCTION**
45 SERVICES SECTOR	..	..	176 316	200 078	235 368	..	..	**45 SECTEUR DES SERVICES**
46 Wholesale, retail trade & repair of motor vehicles, etc.	..	..	..	..	328	..	..	**46 Commerce (gros & détail), réparation de véhicules, etc.**
47 Hotels & restaurants	..	..	..	..	0	..	..	**47 Hôtels et restaurants**
48 Transport & storage	..	..	..	..	1 676	..	..	**48 Transports, entreposage**
49 Communications	..	..	..	..	135 057	..	..	**49 Communications**
50 Post	..	..	..	..	0	..	..	50 Postes
51 Telecommunications	..	..	..	..	135 057	..	..	51 Télécommunications
52 Financial intermediation (inc. insur.)	..	..	..	..	1 407	..	..	**52 Intermédiation financ. (assurances inc.)**
53 Real estate, renting & busin. activity	..	..	..	..	85 280	..	..	**53 Immobil., locat. & serv. aux entreprises**
54 Computer & related activities	..	..	..	..	37 172	..	..	54 Informatique et activités connexes
55 *Software consultancy*	..	..	..	..	28 905	..	..	55 *Production de logiciels*
56 *Other computer services n.e.c.*	..	..	..	..	8 267	..	..	56 *Autres activités informatiques*
57 Research & development	..	..	..	..	522	..	..	57 Recherche & développement expérimental
58 Other business activities n.e.c.	..	..	11 814	19 117	5 836	..	..	58 Autres activ. & serv. aux entreprises
59 Comm., soc. & pers. serv. activ., etc.	..	..	..	..	11 620	..	..	**59 Adm. pub., serv. soc. & collec., etc**
TOTAL BUSINESS ENTERPRISE SECTOR	**..**	**..**	**2 073 522**	**2 022 824**	**1 908 331**	**..**	**..**	**TOTAL SECTEUR DES ENTREPRISES**

Source: OECD, R&D database (DSTI/EAS Division), January 2002. Source : OCDE, base de données R-D (DSTI/Division AES), janvier 2002.

KOREA — Table 12 - Tableau 12 — CORÉE

Total business enterprise R&D personnel (BEMP) - Personnel total de R-D du secteur des entreprises (PRDE)

by industry - par industrie

In full time equivalent — *En équivalence plein-temps*

	1994	1995	1996	1997	1998	1999	2000	
1 AGRICULTURE, HUNTING & FORESTRY	..	..	351	297	279	..	..	**1 AGRICULTURE, CHASSE & SYLVICULTURE**
2 MINING	..	..	85	40	523	..	..	**2 ACTIVITES EXTRACTIVES**
3 MANUFACTURING	..	..	79 377	78 532	67 194	..	..	**3 ACTIVITES DE FABRICATION**
4 Food, beverages & tobacco	..	..	2 380	2 180	1 607	..	..	**4 Produits alimentaires, boissons & tabac**
5 Food products & beverages	..	..	2 373	2 172	1 597	..	..	5 Produits alimentaires & boissons
6 Tobacco products	..	..	7	8	10	..	..	6 Produits à base de tabac
7 Textiles, wear app., fur & leather	..	..	792	827	850	..	..	**7 Textiles, habill., fourrures & cuirs**
8 Textiles	..	..	675	679	631	..	..	8 Textiles
9 Wearing apparel & fur	..	..	93	109	133	..	..	9 Articles d'habillement & fourrures
10 Leather products & footwear	..	..	24	39	86	..	..	10 Articles en cuir & chaussures
11 Wood, paper, printing, publishing	..	..	820	616	503	..	..	**11 Bois, papier, imprimerie, édition**
12 Wood & cork (not furniture)	..	..	76	69	59	..	..	12 Bois & liège (sauf meubles)
13 Pulp, paper & paper products	..	..	397	339	273	..	..	13 Carton, papier & articles
14 Publ., print. & repro. of rec. media	..	..	347	208	170	..	..	14 Edition, impress. & repro. enregis.
15 Coke, petroleum, nuclear fuel, chemicals, rubber plastics	..	..	10 765	11 377	9 583	..	..	**15 Coke, pétrole, combust. nucl., prod. chimi., caout. & mat. plast.**
16 Coke, ref. petrol. prod. & nuclear fuel	..	..	863	1 033	997	..	..	16 Coke, raffin. pétr. & combust. nucl.
17 Chemicals & chemical products	..	..	8 696	9 117	6 962	..	..	17 Produits chimiques
18 *Chemical prod.(less pharmaceuticals)*	..	..	6 759	7 249	5 692	..	..	18 *Prod. chimi. (moins prod. pharmaceutiques)*
19 *Pharmaceuticals*	..	..	1 937	1 868	1 270	..	..	19 *Produits pharmaceutiques*
20 Rubber & plastic products	..	..	1 206	1 227	1 625	..	..	20 Caoutchouc et matières plastiques
21 Non-metallic mineral products	..	..	1 350	1 084	833	..	..	**21 Produits minéraux non métalliques**
22 Basic metals	..	..	1 761	1 226	1 002	..	..	**22 Produits métallurgiques de base**
23 Basic metals, ferrous	..	..	1 296	886	578	..	..	23 Produits métallur. de base, ferreux
24 Basic metals, non-ferrous	..	..	465	340	423	..	..	24 Prod. métallur. de base, non ferreux
25 Fabricated metal products	..	..	1 030	1 042	1 046	..	..	**25 Ouvrages en métaux (sauf mach.& mat.)**
26 Machinery equipment, instruments and transport equipment	..	..	59 993	59 850	50 644	..	..	**26 Machines & matériel, instruments & mat. de transport**
27 Machinery n.e.c.	..	..	4 219	4 839	4 443	..	..	27 Machines, n.c.a.
28 Office, account. & computing machinery	..	..	2 124	2 099	1 609	..	..	28 Mach. de bur., compta. & trait. inform.
29 Electrical machinery	..	..	2 497	2 474	2 062	..	..	29 Machines électriques
30 Electro. equip. (radio, TV & commun.)	..	..	30 326	31 722	26 911	..	..	30 Equip. électro. (radio, tv & communic.)
31 *Electro. comp. (inc. semi-conductors)*	..	..	23 284	25 127	10 434	..	..	31 *Compos. électro. (semi-cond. incl.)*
32 *TV, radio & communications equip.*	..	..	7 042	6 595	16 477	..	..	32 *Appareils radio, tv et de communi.*
33 Instruments	..	..	1 064	1 419	1 226	..	..	33 Instruments
34 Motor vehicles	..	..	17 582	14 602	12 752	..	..	34 Véhicules automobiles
35 Other transport equipment	..	..	2 181	2 695	1 641	..	..	35 Autres matériels de transport
36 *Ships*	..	..	775	1 449	1 124	..	..	36 *Navires*
37 *Aerospace*	..	..	1 212	1 052	291	..	..	37 *Aérospatiale*
38 *Other transport equipment n.e.c.*	..	..	194	194	226	..	..	38 *Autres matériels de transport n.c.a.*
39 Furniture, other manufacturing n.e.c.	..	..	469	313	1 076	..	..	**39 Meubles, autres activ. de fabricat. n.c.a.**
40 Furniture	..	..	..	..	65	..	..	40 Meubles
41 Other manufacturing n.e.c.	..	..	..	..	1 011	..	..	41 Autres activités de fabrication n.c.a.
42 Recycling	..	..	17	17	50	..	..	**42 Récupération**
43 ELECTRICITY, GAS & WATER SUPPLY	..	..	879	1 018	941	..	..	**43 DISTRIBUTION D'ELECTRICITE, GAZ & EAU**
44 CONSTRUCTION	..	..	1 990	1 601	1 186	..	..	**44 CONSTRUCTION**
45 SERVICES SECTOR	..	..	6 320	8 933	7 739	..	..	**45 SECTEUR DES SERVICES**
46 Wholesale, retail trade & repair of motor vehicles, etc.	..	..	..	..	137	..	..	**46 Commerce (gros & détail), réparation de véhicules, etc.**
47 Hotels & restaurants	..	..	..	..	0	..	..	**47 Hôtels et restaurants**
48 Transport & storage	..	..	..	..	113	..	..	**48 Transports, entreposage**
49 Communications	..	..	..	..	1 412	..	..	**49 Communications**
50 Post	..	..	..	..	0	..	..	50 Postes
51 Telecommunications	..	..	..	..	1 412	..	..	51 Télécommunications
52 Financial intermediation (inc. insur.)	..	..	..	..	111	..	..	**52 Intermédiation financ. (assurances inc.)**
53 Real estate, renting & busin. activity	..	..	..	..	5 481	..	..	**53 Immobil., locat. & serv. aux entreprises**
54 Computer & related activities	..	..	..	..	3 816	..	..	54 Informatique et activités connexes
55 *Software consultancy*	..	..	..	..	3 313	..	..	55 *Production de logiciels*
56 *Other computer services n.e.c.*	..	..	..	..	503	..	..	56 *Autres activités informatiques*
57 Research & development	..	..	..	..	65	..	..	57 Recherche & développement expérimental
58 Other business activities n.e.c.	..	..	423	785	1 309	..	..	58 Autres activ. & serv. aux entreprises
59 Comm., soc. & pers. serv. activ., etc.	..	..	..	..	485	..	..	**59 Adm. pub., serv. soc. & collec., etc**
TOTAL BUSINESS ENTERPRISE SECTOR	**..**	**96 890**	**89 002**	**90 425**	**77 862**	**84 070**	**..**	**TOTAL SECTEUR DES ENTREPRISES**

Source: OECD, R&D database (DSTI/EAS Division), January 2002.

Source : OCDE, base de données R-D (DSTI/Division AES), janvier 2002.

RSE in the business enterprise sector - Chercheurs dans le secteur des entreprises

by industry - par industrie

In full time equivalent | *En équivalence plein-temps*

	1994	1995	1996	1997	1998	1999	2000	
1 **AGRICULTURE, HUNTING & FORESTRY**	..	..	177	169	180	..	..	1 **AGRICULTURE, CHASSE & SYLVICULTURE**
2 **MINING**	..	..	57	22	276	..	..	2 **ACTIVITES EXTRACTIVES**
3 **MANUFACTURING**	..	..	58 103	59 741	51 418	..	..	3 **ACTIVITES DE FABRICATION**
4 **Food, beverages & tobacco**	..	..	1 929	1 724	1 274	..	..	4 **Produits alimentaires, boissons & tabac**
5 Food products & beverages	..	..	1 925	1 720	1 268	..	..	5 Produits alimentaires & boissons
6 Tobacco products	..	..	4	4	6	..	..	6 Produits à base de tabac
7 **Textiles, wear app., fur & leather**	..	..	495	530	453	..	..	7 **Textiles, habill., fourrures & cuirs**
8 Textiles	..	..	430	454	368	..	..	8 Textiles
9 Wearing apparel & fur	..	..	60	62	33	..	..	9 Articles d'habillement & fourrures
10 Leather products & footwear	..	..	5	14	53	..	..	10 Articles en cuir & chaussures
11 **Wood, paper, printing, publishing**	..	..	664	454	334	..	..	11 **Bois, papier, imprimerie, édition**
12 Wood & cork (not furniture)	..	..	50	48	39	..	..	12 Bois & liège (sauf meubles)
13 Pulp, paper & paper products	..	..	322	242	183	..	..	13 Carton, papier & articles
14 Publ., print. & repro. of rec. media	..	..	292	164	112	..	..	14 Edition, impress. & repro. enregis.
15 **Coke, petroleum, nuclear fuel, chemicals, rubber plastics**	..	..	8 083	8 728	7 223	..	..	15 **Coke, pétrole, combust. nucl., prod. chimi., caout. & mat. plast.**
16 Coke, ref. petrol. prod. & nuclear fuel	..	..	659	763	682	..	..	16 Coke, raffin. pétr. & combust. nucl.
17 Chemicals & chemical products	..	..	6 665	7 173	5 455	..	..	17 Produits chimiques
18 *Chemical prod.(less pharmaceuticals)*	..	..	5 184	5 682	4 463	..	..	18 *Prod. chimi. (moins prod. pharmaceutiques)*
19 *Pharmaceuticals*	..	..	1 481	1 491	992	..	..	19 *Produits pharmaceutiques*
20 Rubber & plastic products	..	..	759	792	1 086	..	..	20 Caoutchouc et matières plastiques
21 **Non-metallic mineral products**	..	..	916	705	519	..	..	21 **Produits minéraux non métalliques**
22 **Basic metals**	..	..	850	735	674	..	..	22 **Produits métallurgiques de base**
23 Basic metals, ferrous	..	..	540	491	380	..	..	23 Produits métallur. de base, ferreux
24 Basic metals, non-ferrous	..	..	310	244	294	..	..	24 Prod. métallur. de base, non ferreux
25 **Fabricated metal products**	..	..	602	605	724	..	..	25 **Ouvrages en métaux (sauf mach.& mat.)**
26 **Machinery equipment, instruments and transport equipment**	..	..	44 221	46 044	39 302	..	..	26 **Machines & matériel, instruments & mat. de transport**
27 Machinery n.e.c.	..	..	3 137	3 702	3 499	..	..	27 Machines, n.c.a.
28 Office, account. & computing machinery	..	..	1 775	1 805	1 365	..	..	28 Mach. de bur., compta. & trait. inform.
29 Electrical machinery	..	..	1 935	1 947	1 548	..	..	29 Machines électriques
30 Electro. equip. (radio, TV & commun.)	..	..	23 704	24 836	20 771	..	..	30 Equip. électro. (radio, tv & communic.)
31 *Electro. comp. (inc. semi-conductors)*	..	..	17 657	19 318	8 198	..	..	31 *Compos. électro. (semi-cond. incl.)*
32 *TV, radio & communications equip.*	..	..	6 047	5 518	12 573	..	..	32 *Appareils radio, tv et de communi.*
33 Instruments	..	..	803	1 053	944	..	..	33 Instruments
34 Motor vehicles	..	..	11 079	10 471	9 876	..	..	34 Véhicules automobiles
35 Other transport equipment	..	..	1 788	2 230	1 298	..	..	35 Autres matériels de transport
36 *Ships*	..	..	566	1 192	890	..	..	36 *Navires*
37 *Aerospace*	..	..	1 111	915	235	..	..	37 *Aérospatiale*
38 *Other transport equipment n.e.c.*	..	..	111	123	173	..	..	38 *Autres matériels de transport n.c.a.*
39 **Furniture, other manufacturing n.e.c.**	..	..	332	203	882	..	..	39 **Meubles, autres activ. de fabricat. n.c.a.**
40 Furniture	..	..	..	..	46	..	..	40 Meubles
41 Other manufacturing n.e.c.	..	..	..	..	836	..	..	41 Autres activités de fabrication n.c.a.
42 **Recycling**	..	..	11	13	32	..	..	42 **Récupération**
43 **ELECTRICITY, GAS & WATER SUPPLY**	..	..	789	907	799	..	..	43 **DISTRIBUTION D'ELECTRICITE, GAZ & EAU**
44 **CONSTRUCTION**	..	..	1 558	1 213	932	..	..	44 **CONSTRUCTION**
45 **SERVICES SECTOR**	..	..	5 534	7 819	6 460	..	..	45 **SECTEUR DES SERVICES**
46 **Wholesale, retail trade & repair of motor vehicles, etc.**	..	..	..	..	89	..	..	46 **Commerce (gros & détail), réparation de véhicules, etc.**
47 **Hotels & restaurants**	..	..	..	..	0	..	..	47 **Hôtels et restaurants**
48 **Transport & storage**	..	..	..	..	91	..	..	48 **Transports, entreposage**
49 **Communications**	..	..	..	..	1 184	..	..	49 **Communications**
50 Post	..	..	..	..	0	..	..	50 Postes
51 Telecommunications	..	..	..	..	1 184	..	..	51 Télécommunications
52 **Financial intermediation (inc. insur.)**	..	..	..	..	75	..	..	52 **Intermédiation financ. (assurances inc.)**
53 **Real estate, renting & busin. activity**	..	..	..	..	4 593	..	..	53 **Immobil., locat. & serv. aux entreprises**
54 Computer & related activities	..	..	..	..	3 300	..	..	54 Informatique et activités connexes
55 *Software consultancy*	..	..	..	..	2 889	..	..	55 *Production de logiciels*
56 *Other computer services n.e.c.*	..	..	..	..	411	..	..	56 *Autres activités informatiques*
57 Research & development	..	..	..	..	59	..	..	57 Recherche & développement expérimental
58 Other business activities n.e.c.	..	..	321	603	1 054	..	..	58 Autres activ. & serv. aux entreprises
59 **Comm., soc. & pers. serv. activ., etc.**	..	..	..	..	428	..	..	59 **Adm. pub., serv. soc. & collec., etc**
TOTAL BUSINESS ENTERPRISE SECTOR	**..**	**67 226**	**66 218**	**69 871**	**60 064**	**65 474**	**..**	**TOTAL SECTEUR DES ENTREPRISES**

Source: OECD, R&D database (DSTI/EAS Division), January 2002.

Source : OCDE, base de données R-D (DSTI/Division AES), janvier 2002.

Table 14 - Tableau 14

Higher education intramural expenditure on R&D - Dépenses intra-muros de R-D de l'enseignement supérieur
(HERD) by field of science - (DIRDES) par discipline scientifique

	1994	1995	1996	1997	1998	1999	2000	
	14.1 Million KRW - Millions KRW							
Natural sciences	..	..	..	..	234 294	..	..	Sciences exactes et naturelles
Engineering	..	..	..	..	620 861	..	..	Sciences de l'ingénieur
Medical sciences	..	..	..	..	214 495	..	..	Sciences médicales
Agricultural sciences	..	..	..	..	89 068	..	..	Sciences agricoles
Sub-total NSE	..	770 912	1 018 821	1 271 599	1 158 718	1 431 421	..	Sous-total SENI
Social sciences	..	..	..	..	..	..	..	Sciences sociales
Humanities	..	..	..	..	..	..	..	Sciences humaines
Sub-total SSH	..	..	..	..	..	..	..	Sous-total SSH
Total NSE+SSH	..	770 912 g	1 018 821 g	1 271 599 g	1 265 074 g	1 431 421 g	..	*Total SENI+SSH*
	14.2 Million constant USD (1995 prices and PPPs) - Millions de USD constants (prix et PPA de 1995)							
Natural sciences	..	..	..	..	338.3	..	..	Sciences exactes et naturelles
Engineering	..	..	..	..	896.4	..	..	Sciences de l'ingénieur
Medical sciences	..	..	..	..	309.7	..	..	Sciences médicales
Agricultural sciences	..	..	..	..	128.6	..	..	Sciences agricoles
Sub-total NSE	..	1 253.1	1 594.1	1 929.0	1 673.0	2 109.7	..	Sous-total SENI
Social sciences	..	..	..	..	..	..	..	Sciences sociales
Humanities	..	..	..	..	..	..	..	Sciences humaines
Sub-total SSH	..	..	..	..	..	..	..	Sous-total SSH
Total NSE+SSH	..	1 253.1 g	1 594.1 g	1 929.0 g	1 826.6 g	2 109.7 g	..	*Total SENI+SSH*

Table 15 - Tableau 15

Higher education intramural expenditure on R&D - Dépenses intra-muros de R-D de l'enseignement supérieur
(HERD) by field of science and type of costs - (DIRDES) par discipline scientifique et par type de coûts

Million KRW *Millions KRW*

	1994	1995	1996	1997	1998	1999	2000	
Current expenditure								**Dépenses courantes**
Natural sciences	..	..	..	..	..	..	..	Sciences exactes et naturelles
Engineering	..	..	..	..	..	..	..	Sciences de l'ingénieur
Medical sciences	..	..	..	..	..	..	..	Sciences médicales
Agricultural sciences	..	..	..	..	..	..	..	Sciences agricoles
Sub-total NSE	..	..	833 109	992 675	1 014 634	..	..	Sous-total SENI
Social sciences	..	..	..	..	..	..	..	Sciences sociales
Humanities	..	..	..	..	..	..	..	Sciences humaines
Sub-total SSH	..	..	..	..	..	..	..	Sous-total SSH
Total NSE+SSH	..	..	833 109 g	992 675 g	1 014 634 g	..	..	*Total SENI+SSH*
Capital expenditure								**Dépenses en capital**
Natural sciences	..	..	..	..	..	..	..	Sciences exactes et naturelles
Engineering	..	..	..	..	..	..	..	Sciences de l'ingénieur
Medical sciences	..	..	..	..	..	..	..	Sciences médicales
Agricultural sciences	..	..	..	..	..	..	..	Sciences agricoles
Sub-total NSE	..	..	185 713	278 925	250 440	..	..	Sous-total SENI
Social sciences	..	..	..	..	..	..	..	Sciences sociales
Humanities	..	..	..	..	..	..	..	Sciences humaines
Sub-total SSH	..	..	..	..	..	..	..	Sous-total SSH
Total NSE+SSH	..	..	185 713 g	278 925 g	250 440 g	..	..	*Total SENI+SSH*

Source: OECD, R&D database (DSTI/EAS Division), January 2002.

Source : OCDE, base de données R-D (DSTI/Division AES), janvier 2002.

KOREA Table 16 - Tableau 16 **CORÉE**

Higher education intramural expenditure on R&D - Dépenses intra-muros de R-D de l'enseignement supérieur

(HERD) by field of science and source of funds - (DIRDES) par discipline scientifique et par source de financement

Million KRW *Millions KRW*

	1994	1995	1996	1997	1998	1999	2000	
Direct government								**État direct**
Natural sciences	..	..	..	..	..	..	..	Sciences exactes et naturelles
Engineering	..	..	..	..	..	..	..	Sciences de l'ingénieur
Medical sciences	..	..	..	..	..	..	..	Sciences médicales
Agricultural sciences	..	..	..	..	..	..	..	Sciences agricoles
Sub-total NSE	..	..	..	..	..	..	..	Sous-total SENI
Social sciences	..	..	..	..	..	..	..	Sciences sociales
Humanities	..	..	..	..	..	..	..	Sciences humaines
Sub-total SSH	..	..	..	..	..	..	..	Sous-total SSH
Total NSE+SSH	..	..	..	..	..	..	..	*Total SENI+SSH*
General university funds								**Fonds généraux des universités**
Natural sciences	..	..	..	..	..	..	..	Sciences exactes et naturelles
Engineering	..	..	..	..	..	..	..	Sciences de l'ingénieur
Medical sciences	..	..	..	..	..	..	..	Sciences médicales
Agricultural sciences	..	..	..	..	..	..	..	Sciences agricoles
Sub-total NSE	..	..	..	..	..	..	..	Sous-total SENI
Social sciences	..	..	..	..	..	..	..	Sciences sociales
Humanities	..	..	..	..	..	..	..	Sciences humaines
Sub-total SSH	..	..	..	..	..	..	..	Sous-total SSH
Total NSE+SSH	..	..	..	..	..	..	..	*Total SENI+SSH*
Sub-total government								**Sous-total État**
Natural sciences	..	..	..	..	111 089	..	..	Sciences exactes et naturelles
Engineering	..	..	..	..	302 786	..	..	Sciences de l'ingénieur
Medical sciences	..	..	..	..	70 129	..	..	Sciences médicales
Agricultural sciences	..	..	..	..	46 935	..	..	Sciences agricoles
Sub-total NSE	..	338 727	448 063	611 477	530 939	806 828	..	Sous-total SENI
Social sciences	..	..	..	..	..	..	..	Sciences sociales
Humanities	..	..	..	..	..	..	..	Sciences humaines
Sub-total SSH	..	..	..	..	..	..	..	Sous-total SSH
Total NSE+SSH	..	338 727 [g]	448 063 [g]	611 477 [g]	585 675 [g]	806 828 [g]	..	*Total SENI+SSH*
Higher education								**Enseignement supérieur**
Natural sciences	..	..	..	..	93 738	..	..	Sciences exactes et naturelles
Engineering	..	..	..	..	221 942	..	..	Sciences de l'ingénieur
Medical sciences	..	..	..	..	125 360	..	..	Sciences médicales
Agricultural sciences	..	..	..	..	27 911	..	..	Sciences agricoles
Sub-total NSE	..	..	..	..	468 950	457 646	..	Sous-total SENI
Social sciences	..	..	..	..	..	..	..	Sciences sociales
Humanities	..	..	..	..	..	..	..	Sciences humaines
Sub-total SSH	..	..	..	..	..	..	..	Sous-total SSH
Total NSE+SSH	..	..	..	..	504 533 [g]	457 646 [g]	..	*Total SENI+SSH*
Private non-profit								**Institutions sans but lucratif**
Natural sciences	..	..	..	..	1 233	..	..	Sciences exactes et naturelles
Engineering	..	..	..	..	3 464	..	..	Sciences de l'ingénieur
Medical sciences	..	..	..	..	896	..	..	Sciences médicales
Agricultural sciences	..	..	..	..	907	..	..	Sciences agricoles
Sub-total NSE	..	12 467	32 591	24 337	6 499	9 237	..	Sous-total SENI
Social sciences	..	..	..	..	..	..	..	Sciences sociales
Humanities	..	..	..	..	..	..	..	Sciences humaines
Sub-total SSH	..	..	..	..	..	..	..	Sous-total SSH
Total NSE+SSH	..	12 467 [g]	32 591 [g]	24 337 [g]	6 875 [g]	9 237 [g]	..	*Total SENI+SSH*
Business entreprise								**Entreprises**
Natural sciences	..	..	..	..	27 789	..	..	Sciences exactes et naturelles
Engineering	..	..	..	..	91 123	..	..	Sciences de l'ingénieur
Medical sciences	..	..	..	..	17 722	..	..	Sciences médicales
Agricultural sciences	..	..	..	..	13 167	..	..	Sciences agricoles
Sub-total NSE	..	173 012	514 721	189 540	149 800	154 500	..	Sous-total SENI
Social sciences	..	..	..	..	..	..	..	Sciences sociales
Humanities	..	..	..	..	..	..	..	Sciences humaines
Sub-total SSH	..	..	..	..	..	..	..	Sous-total SSH
Total NSE+SSH	..	173 012 [g]	514 721 [g]	189 540 [g]	165 159 [g]	154 500 [g]	..	*Total SENI+SSH*
Funds from abroad								**Fonds de l'étranger**
Natural sciences	..	..	..	..	446	..	..	Sciences exactes et naturelles
Engineering	..	..	..	..	1 546	..	..	Sciences de l'ingénieur
Medical sciences	..	..	..	..	388	..	..	Sciences médicales
Agricultural sciences	..	..	..	..	148	..	..	Sciences agricoles
Sub-total NSE	..	158	2 916	6 734	2 529	3 210	..	Sous-total SENI
Social sciences	..	..	..	..	..	..	..	Sciences sociales
Humanities	..	..	..	..	..	..	..	Sciences humaines
Sub-total SSH	..	..	..	..	..	..	..	Sous-total SSH
Total NSE+SSH	..	158 [g]	2 916 [g]	6 734 [g]	2 832 [g]	3 210 [g]	..	*Total SENI+SSH*

Source: OECD, R&D database (DSTI/EAS Division), January 2002.

Source : OCDE, base de données R-D (DSTI/Division AES), janvier 2002.

Higher education R&D personnel (HEMP) - Personnel de R-D du secteur de l'enseignement supérieur

by field of science and occupation - par discipline scientifique et par profession

In full-time equivalent / *En équivalence plein-temps*

	1994	1995	1996	1997	1998	1999	2000	
Total R&D personnel								**Personnel total de R-D**
Natural sciences	..	..	..	..	..	..	..	Sciences exactes et naturelles
Engineering	..	..	..	..	..	..	..	Sciences de l'ingénieur
Medical sciences	..	..	..	..	..	..	..	Sciences médicales
Agricultural sciences	..	..	..	..	..	..	..	Sciences agricoles
Sub-total NSE	..	31 511	28 274	28 546	34 753	36 534	..	Sous-total SENI
Social sciences	..	..	..	..	..	..	..	Sciences sociales
Humanities	..	..	..	..	..	..	..	Sciences humaines
Sub-total SSH	..	..	..	..	..	..	..	Sous-total SSH
Total NSE+SSH	..	31 511 g	28 274 g	28 546 g	34 753 g	36 534 g	..	*Total SENI+SSH*
Researchers								**Chercheurs**
Natural sciences	..	..	..	..	..	..	..	Sciences exactes et naturelles
Engineering	..	..	..	..	..	..	..	Sciences de l'ingénieur
Medical sciences	..	..	..	..	..	..	..	Sciences médicales
Agricultural sciences	..	..	..	..	..	..	..	Sciences agricoles
Sub-total NSE	..	19 406	19 483	19 573	21 525	21 723	..	Sous-total SENI
Social sciences	..	..	..	..	..	..	..	Sciences sociales
Humanities	..	..	..	..	..	..	..	Sciences humaines
Sub-total SSH	..	..	..	..	..	..	..	Sous-total SSH
Total NSE+SSH	..	19 406 g	19 483 g	19 573 g	21 525 g	21 723 g	..	*Total SENI+SSH*
Technicians								**Technicians**
Natural sciences	..	..	..	..	..	..	..	Sciences exactes et naturelles
Engineering	..	..	..	..	..	..	..	Sciences de l'ingénieur
Medical sciences	..	..	..	..	..	..	..	Sciences médicales
Agricultural sciences	..	..	..	..	..	..	..	Sciences agricoles
Sub-total NSE	..	11 217	7 602	7 354	9 117	9 899	..	Sous-total SENI
Social sciences	..	..	..	..	..	..	..	Sciences sociales
Humanities	..	..	..	..	..	..	..	Sciences humaines
Sub-total SSH	..	..	..	..	..	..	..	Sous-total SSH
Total NSE+SSH	..	11 217 g	7 602 g	7 354 g	9 117 g	9 899 g	..	*Total SENI+SSH*
Other								**Autres**
Natural sciences	..	..	..	..	..	..	..	Sciences exactes et naturelles
Engineering	..	..	..	..	..	..	..	Sciences de l'ingénieur
Medical sciences	..	..	..	..	..	..	..	Sciences médicales
Agricultural sciences	..	..	..	..	..	..	..	Sciences agricoles
Sub-total NSE	..	888	1 189	1 619	4 111	4 912	..	Sous-total SENI
Social sciences	..	..	..	..	..	..	..	Sciences sociales
Humanities	..	..	..	..	..	..	..	Sciences humaines
Sub-total SSH	..	..	..	..	..	..	..	Sous-total SSH
Total NSE+SSH	..	888 g	1 189 g	1 619 g	4 111 g	4 912 g	..	*Total SENI+SSH*

Table 18 - Tableau 18

Government budget appropriations or outlays for R&D (GBAORD) - Crédits budgétaires publics de R-D (CBPRD)

by socio-economic objective - par objectif socio-économique

Million KRW / *Millions KRW*

	1995	1996	1997	1998	1999	2000	2001	
Agriculture, forestry and fishing	..	..	..	..	295 641	..	..	Agriculture, sylviculture et pêche
Industrial development	..	..	..	..	744 404	..	..	Développement industriel
Energy	..	..	..	..	184 779	..	..	Energie
Transport & telecommunications	..	..	..	..	88 389	..	..	*Transports & télécommunications*
Urban & rural planning	..	..	..	..	10 888	..	..	*Aménagement urbain et rural*
ST Infrastructure	..	..	..	..	99 277	..	..	ST Infrastructure
Prevention of pollution	..	..	..	..	..	..	..	*Prévention de la pollution*
Identificat. & treatment of pollution	..	..	..	..	..	..	..	*Détect. pollut. & mesures antipollut.*
ST Environment	..	..	..	..	107 181	..	..	ST Environnement
Health	..	..	..	..	161 760	..	..	Santé
Social development & services	..	..	..	..	94 331	..	..	Développement et services sociaux
Earth and atmosphere	..	..	..	..	52 890	..	..	Milieu terrestre et atmosphère
Advancement of research	..	..	..	..	..	..	..	*Promotion de la recherche*
General university funds	..	..	..	..	..	..	..	*Fonds généraux des universités*
ST Advancement of knowledge	..	..	..	..	730 277	..	..	ST promotion gén. des connaiss.
Civil space	..	..	..	..	69 722	..	..	Espace civil
Defence	..	..	..	..	733 751	..	..	Défense
Not elsewhere classified	..	..	..	..	0	..	..	Non classé
Total	**..**	**..**	**..**	**..**	**3 274 013**	**..**	**..**	**Total**

Source: OECD, R&D database (DSTI/EAS Division), January 2002. *Source* : OCDE, base de données R-D (DSTI/Division AES), janvier 2002.

Table 1 - Tableau 1

Gross domestic expenditure on R&D (GERD) - Dépenses intérieures brutes de R-D (DIRD)

by sector of performance and source of funds - par secteur d'exécution et par source de financement

	1994	1995	1996	1997	1998	1999	2000	
				1.1 Million MXN - Millions MXN				
Business enterprise								**Entreprises**
financed by								*financé par*
Business enterprise	690.4	899.4	1 102.3	1 366.8	3 020.3 p	4 053.2 p	..	Entreprises
Direct government	25.4	32.5	428.8	569.2	87.8 p	54.9 p	..	État direct
Higher education	0.0	0.8	0.4	1.4	..	0.3 p	..	Enseignement supérieur
Private non-profit	3.8	3.5	8.7	10.2	..	..	..	Institutions sans but lucratif
Funds from abroad	332.8	244.0	209.7	211.9	983.9 p	934.6 p	..	Fonds de l'étranger
Total	1 052.3	1 180.2	1 750.0	2 159.5	4 092.1 p	5 043.0 p	..	*Total*
Government								**État**
financed by								*financé par*
Business enterprise	64.6	61.4	317.9	364.1	..	..	..	Entreprises
Direct government	1 066.2	1 686.4	2 444.0	3 811.2	..	..	..	État direct
Higher education	1.7	8.0	0.7	3.3	..	..	..	Enseignement supérieur
Private non-profit	3.2	39.2	54.2	24.9	..	..	..	Institutions sans but lucratif
Funds from abroad	18.9	82.2	34.3	37.4	..	..	..	Fonds de l'étranger
Total	1 154.5	1 877.1	2 851.1	4 240.9	6 748.4 p	6 019.3 p	..	*Total*
Higher education								**Enseignement supérieur**
financed by								*financé par*
Business enterprise	34.0	35.6	99.1	116.2	..	..	..	Entreprises
Direct government	602.7	745.7	573.1	800.4	..	..	..	État direct
General university funds	956.6	1 297.5	1 628.4	2 479.7	..	..	..	Fonds généraux des universités
Sub-total government	1 559.3	2 043.2	2 201.5	3 280.1	..	..	..	Sous-Total État
Higher education	317.5	465.9	633.3	936.4	..	..	..	Enseignement supérieur
Private non-profit	11.9	12.3	10.2	11.7	..	..	..	Institutions sans but lucratif
Funds from abroad	23.7	48.9	22.1	20.7	..	..	..	Fonds de l'étranger
Total	1 946.4	2 605.9	2 966.3	4 365.2	6 513.8 p	7 145.6 p	..	*Total*
Private non-profit								**Institutions sans but lucratif**
financed by								*financé par*
Business enterprise	2.3	3.6	2.8	4.2	..	..	..	Entreprises
Direct government	1.5	2.0	154.8	116.9	..	..	..	État direct
Higher education	0.8	0.7	0.0	0.0	..	..	..	Enseignement supérieur
Private non-profit	5.7	9.6	98.0	49.5	..	..	..	Institutions sans but lucratif
Funds from abroad	5.2	8.2	5.8	8.3	..	..	..	Fonds de l'étranger
Total	15.5	24.0	261.4	178.8	275.4 p	318.7 p	..	*Total*
GERD								**DIRD**
financed by								*financé par*
Business enterprise	791.2	1 000.0	1 522.1	1 851.3	3 136.0 p	4 364.0 p	..	Entreprises
Direct government	1 695.7	2 466.5	3 600.8	5 297.7	..	..	..	État direct
General university funds	956.6	1 297.5	1 628.4	2 479.7	..	..	..	Fonds généraux des universités
Sub-total government	2 652.3	3 764.0	5 229.2	7 777.4	12 221.9 p	12 096.8 p	..	Sous-Total État
Higher education	320.0	475.3	634.5	941.1	784.3 p	843.7 p	..	Enseignement supérieur
Private non-profit	24.5	64.6	171.2	96.3	134.9 p	161.2 p	..	Institutions sans but lucratif
Funds from abroad	380.6	383.3	271.8	278.2	1 352.7 p	1 060.9 p	..	Fonds de l'étranger
Total	4 168.7	5 687.3	7 828.8	10 944.4	17 629.7 p	18 526.6 p	..	*Total*
				1.2 Million constant USD (1995 prices and PPPs) - Millions de USD constants (prix et PPA de 1995)				
Business enterprise								**Entreprises**
financed by								*financé par*
Business enterprise	321.8	304.1	285.1	300.3	575.2 p	672.0 p	..	Entreprises
Direct government	11.8	11.0	110.9	125.1	16.7 p	9.1 p	..	État direct
Higher education	0.0	0.3	0.1	0.3	..	0.0 p	..	Enseignement supérieur
Private non-profit	1.8	1.2	2.3	2.2	..	..	..	Institutions sans but lucratif
Funds from abroad	155.1	82.5	54.2	46.6	187.4 p	155.0 p	..	Fonds de l'étranger
Total	490.5	399.1	452.6	474.5	779.3 p	836.1 p	..	*Total*
Government								**État**
financed by								*financé par*
Business enterprise	30.1	20.7	82.2	80.0	..	..	..	Entreprises
Direct government	497.0	570.2	632.1	837.4	..	..	..	État direct
Higher education	0.8	2.7	0.2	0.7	..	..	..	Enseignement supérieur
Private non-profit	1.5	13.3	14.0	5.5	..	..	..	Institutions sans but lucratif
Funds from abroad	8.8	27.8	8.9	8.2	..	..	..	Fonds de l'étranger
Total	538.2	634.7	737.4	931.8	1 285.2 p	998.0 p	..	*Total*
Higher education								**Enseignement supérieur**
financed by								*financé par*
Business enterprise	15.8	12.0	25.6	25.5	..	..	..	Entreprises
Direct government	280.9	252.2	148.2	175.9	..	..	..	État direct
General university funds	445.9	438.7	421.2	544.8	..	..	..	Fonds généraux des universités
Sub-total government	726.9	690.9	569.4	720.7	..	..	..	Sous-Total État
Higher education	148.0	157.5	163.8	205.7	..	..	..	Enseignement supérieur
Private non-profit	5.5	4.2	2.6	2.6	..	..	..	Institutions sans but lucratif
Funds from abroad	11.1	16.5	5.7	4.5	..	..	..	Fonds de l'étranger
Total	907.3	881.2	767.2	959.1	1 240.5 p	1 184.8 p	..	*Total*
Private non-profit								**Institutions sans but lucratif**
financed by								*financé par*
Business enterprise	1.1	1.2	0.7	0.9	..	..	..	Entreprises
Direct government	0.7	0.7	40.0	25.7	..	..	..	État direct
Higher education	0.4	0.2	0.0	0.0	..	..	..	Enseignement supérieur
Private non-profit	2.7	3.2	25.3	10.9	..	..	..	Institutions sans but lucratif
Funds from abroad	2.4	2.8	1.5	1.8	..	..	..	Fonds de l'étranger
Total	7.2	8.1	67.6	39.3	52.5 p	52.8 p	..	*Total*
GERD								**DIRD**
financed by								*financé par*
Business enterprise	368.8	338.1	393.7	406.7	597.2 p	723.6 p	..	Entreprises
Direct government	790.5	834.0	931.3	1 164.0	..	..	..	État direct
General university funds	445.9	438.7	421.2	544.8	..	..	..	Fonds généraux des universités
Sub-total government	1 236.4	1 272.8	1 352.4	1 708.8	2 327.5 p	2 005.7 p	..	Sous-Total État
Higher education	149.2	160.7	164.1	206.8	149.4 p	139.9 p	..	Enseignement supérieur
Private non-profit	11.4	21.9	44.3	21.2	25.7 p	26.7 p	..	Institutions sans but lucratif
Funds from abroad	177.4	129.6	70.3	61.1	257.6 p	175.9 p	..	Fonds de l'étranger
Total	1 943.2	1 923.1	2 024.8	2 404.6	3 357.4 p	3 071.8 p	..	*Total*

Source: OECD, R&D database (DSTI/EAS Division), January 2002.

Source : OCDE, base de données R-D (DSTI/Division AES), janvier 2002.

Table 2 - Tableau 2

Gross domestic expenditure on R&D (GERD) - Dépenses intérieures brutes de R-D (DIRD)

by sector of performance and type of costs - par secteur d'exécution et par type de coûts

Million MXN | *Millions MXN*

	1994	1995	1996	1997	1998	1999	2000	
Business enterprise								**Entreprises**
Labour costs	429.0	547.8	..	..	..	..	..	Coûts du personnel
Other current costs	494.1	466.0	..	..	..	..	..	Autres coûts courants
Total current expenditure	923.1	1 013.8	..	..	..	..	..	*Total dépenses courantes*
Land and buildings	37.5	49.6	..	..	..	..	..	Terrains et constructions
Instruments and equipment	91.7	116.9	..	..	..	..	..	Instruments et équipements
Total capital expenditure	129.2	166.4	..	..	..	..	..	*Total dépenses en capital*
Total	1 052.3	1 180.2	1 750.0	2 159.5	4 092.1 p	5 043.0 p	..	*Total*
Government								**État**
Labour costs	687.2	1 151.1	..	..	..	..	..	Coûts du personnel
Other current costs	308.3	478.7	..	..	..	..	..	Autres coûts courants
Total current expenditure	995.6	1 629.7	..	..	..	..	..	*Total dépenses courantes*
Land and buildings	82.6	76.1	..	..	..	..	..	Terrains et constructions
Instruments and equipment	76.3	171.3	..	..	..	..	..	Instruments et équipements
Total capital expenditure	158.9	247.4	..	..	..	..	..	*Total dépenses en capital*
Total	1 154.5	1 877.1	2 851.1	4 240.9	6 748.4 p	6 019.3 p	..	*Total*
Higher education								**Enseignement supérieur**
Labour costs	1 088.7	1 461.5	..	..	..	..	..	Coûts du personnel
Other current costs	542.1	686.7	..	..	..	..	..	Autres coûts courants
Total current expenditure	1 630.8	2 148.2	..	..	..	..	..	*Total dépenses courantes*
Land and buildings	73.0	97.9	..	..	..	..	..	Terrains et constructions
Instruments and equipment	242.6	359.8	..	..	..	..	..	Instruments et équipements
Total capital expenditure	315.6	457.7	..	..	..	..	..	*Total dépenses en capital*
Total	1 946.4	2 605.9	2 966.3	4 365.2	6 513.8 p	7 145.6 p	..	*Total*
Private non-profit								**Institutions sans but lucratif**
Labour costs	7.8	13.8	..	..	..	..	..	Coûts du personnel
Other current costs	6.8	8.5	..	..	..	..	..	Autres coûts courants
Total current expenditure	14.6	22.3	..	..	..	..	..	*Total dépenses courantes*
Land and buildings	0.3	0.2	..	..	..	..	..	Terrains et constructions
Instruments and equipment	0.6	1.5	..	..	..	..	..	Instruments et équipements
Total capital expenditure	0.9	1.7	..	..	..	..	..	*Total dépenses en capital*
Total	15.5	24.0	261.4	178.8	275.4 p	318.7 p	..	*Total*
GERD								**DIRD**
Labour costs	2 212.8	3 174.2	..	..	..	..	..	Coûts du personnel
Other current costs	1 351.3	1 639.9	..	..	..	..	..	Autres coûts courants
Total current expenditure	3 564.1	4 814.0	..	..	..	..	..	*Total dépenses courantes*
Land and buildings	193.4	223.8	..	..	..	..	..	Terrains et constructions
Instruments and equipment	411.2	649.5	..	..	..	..	..	Instruments et équipements
Total capital expenditure	604.6	873.2	..	..	..	..	..	*Total dépenses en capital*
Total	4 168.7	5 687.3	7 828.8	10 944.4	17 629.7 p	18 526.6 p	..	*Total*

Table 3 - Tableau 3

Current domestic expenditure on R&D - Dépenses intérieures courantes de R-D

by sector of performance and type of activity - par secteur d'exécution et par type d'activités

Million MXN | *Millions MXN*

	1994	1995	1996	1997	1998	1999	2000	
Business enterprise								**Entreprises**
Basic research	47.3	67.6	..	..	..	..	..	Recherche fondamentale
Applied research	399.6	310.2	..	..	..	..	..	Recherche appliquée
Experimental development	476.2	635.9	..	..	..	..	..	Développement expérimental
Not specified	0.0	0.0	..	..	..	..	..	Non spécifié
Total	923.1	1 013.8	..	..	..	..	..	*Total*
Government								**État**
Basic research	260.9	766.9	..	..	..	..	..	Recherche fondamentale
Applied research	419.3	447.9	..	..	..	..	..	Recherche appliquée
Experimental development	315.4	415.0	..	..	..	..	..	Développement expérimental
Not specified	0.0	0.0	..	..	..	..	..	Non spécifié
Total	995.6	1 629.7	..	..	..	..	..	*Total*
Higher education								**Enseignement supérieur**
Basic research	639.2	884.6	..	..	..	..	..	Recherche fondamentale
Applied research	661.5	821.2	..	..	..	..	..	Recherche appliquée
Experimental development	330.1	442.4	..	..	..	..	..	Développement expérimental
Not specified	0.0	0.0	..	..	..	..	..	Non spécifié
Total	1 630.8	2 148.2	..	..	..	..	..	*Total*
Private non-profit								**Institutions sans but lucratif**
Basic research	2.9	6.5	..	..	..	..	..	Recherche fondamentale
Applied research	9.9	13.1	..	..	..	..	..	Recherche appliquée
Experimental development	1.8	2.7	..	..	..	..	..	Développement expérimental
Not specified	0.0	0.0	..	..	..	..	..	Non spécifié
Total	14.6	22.3	..	..	..	..	..	*Total*
GERD								**DIRD**
Basic research	950.3	1 725.6	..	..	..	..	..	Recherche fondamentale
Applied research	1 490.3	1 592.4	..	..	..	..	..	Recherche appliquée
Experimental development	1 123.5	1 496.0	..	..	..	..	..	Développement expérimental
Not specified	0.0	0.0	..	..	..	..	..	Non spécifié
Total	3 564.1	4 814.0	..	..	..	..	..	*Total*

Source: OECD, R&D database (DSTI/EAS Division), January 2002. *Source* : OCDE, base de données R-D (DSTI/Division AES), janvier 2002.

Gross domestic expenditure on R&D (GERD) - Dépenses intérieures brutes de R-D (DIRD)

by sector of performance and socio-economic objective - par secteur d'exécution et par objectif socio-économique

Million MXN — *Millions MXN*

	1994	1995	1996	1997	1998	1999	2000	
Business enterprise								**Entreprises**
Agriculture, forestry and fishing	189.4	236.7	..	..	..	..	..	Agriculture, sylviculture et pêche
Industrial development	464.3	628.2	..	..	..	..	..	Développement industriel
Energy	10.8	12.8	..	..	..	..	..	Energie
Transport and telecommunications	36.9	43.3	..	..	..	..	..	Transports et télécommunications
Urban and rural planning	1.5	2.0	..	..	..	..	..	Aménagement urbain et rural
Prevention of pollution	23.9	35.2	..	..	..	..	..	Prévention de la pollution
Identification & treatment of pollution	15.3	15.7	..	..	..	..	..	Détection pollu. & mesures antipollu.
Health	52.1	44.6	..	..	..	..	..	Santé
Social development and services	157.9	42.9	..	..	..	..	..	Développement et services sociaux
Earth and atmosphere	20.8	30.8	..	..	..	..	..	Milieu terrestre et atmosphère
Advancement of knowledge	11.0	6.7	..	..	..	..	..	Promotion générale des connaissances
Civil space	0.0	0.0	..	..	..	..	..	Espace civil
Defence	0.0	0.0	..	..	..	..	..	Défense
Not elsewhere classified	68.3	81.3	1 750.0	2 159.5	4 092.1 p	5 043.0 p	..	Non classé
Total	1 052.3	1 180.2	1 750.0	2 159.5	4 092.1 p	5 043.0 p	..	*Total*
Government								**État**
Agriculture, forestry and fishing	332.4	452.7	..	..	..	..	..	Agriculture, sylviculture et pêche
Industrial development	75.9	164.0	..	..	..	..	..	Développement industriel
Energy	135.3	160.7	..	..	..	..	..	Energie
Transport and telecommunications	22.8	71.2	..	..	..	..	..	Transports et télécommunications
Urban and rural planning	6.7	6.7	..	..	..	..	..	Aménagement urbain et rural
Prevention of pollution	10.3	17.8	..	..	..	..	..	Prévention de la pollution
Identification & treatment of pollution	8.8	40.9	..	..	..	..	..	Détection pollu. & mesures antipollu.
Health	269.6	443.3	..	..	..	..	..	Santé
Social development and services	69.2	78.1	..	..	..	..	..	Développement et services sociaux
Earth and atmosphere	37.5	21.3	..	..	..	..	..	Milieu terrestre et atmosphère
Advancement of knowledge	129.8	319.0	..	..	..	..	..	Promotion générale des connaissances
Civil space	0.0	0.0	..	..	..	..	..	Espace civil
Defence	0.0	2.1	..	..	..	..	..	Défense
Not elsewhere classified	56.1	99.3	2 851.1	4 240.9	6 748.4 p	6 019.3 p	..	Non classé
Total	1 154.5	1 877.1	2 851.1	4 240.9	6 748.4 p	6 019.3 p	..	*Total*
Higher education								**Enseignement supérieur**
Agriculture, forestry and fishing	..	..	..	..	..	..	..	Agriculture, sylviculture et pêche
Industrial development	..	..	..	..	..	..	..	Développement industriel
Energy	..	..	..	..	..	..	..	Energie
Transport and telecommunications	..	..	..	..	..	..	..	Transports et télécommunications
Urban and rural planning	..	..	..	..	..	..	..	Aménagement urbain et rural
Prevention of pollution	..	..	..	..	..	..	..	Prévention de la pollution
Identification & treatment of pollution	..	..	..	..	..	..	..	Détection pollu. & mesures antipollu.
Health	..	..	..	..	..	..	..	Santé
Social development and services	..	..	..	..	..	..	..	Développement et services sociaux
Earth and atmosphere	..	..	..	..	..	..	..	Milieu terrestre et atmosphère
Advancement of knowledge	..	..	..	..	..	..	..	Promotion générale des connaissances
Civil space	..	..	..	..	..	..	..	Espace civil
Defence	..	..	..	..	..	..	..	Défense
Not elsewhere classified	1 946.4	2 605.9	2 966.3	4 365.2	6 513.8 p	7 145.6 p	..	Non classé
Total	1 946.4	2 605.9	2 966.3	4 365.2	6 513.8 p	7 145.6 p	..	*Total*
Private non-profit								**Institutions sans but lucratif**
Agriculture, forestry and fishing	..	..	..	..	..	..	..	Agriculture, sylviculture et pêche
Industrial development	..	..	..	..	..	..	..	Développement industriel
Energy	..	..	..	..	..	..	..	Energie
Transport and telecommunications	..	..	..	..	..	..	..	Transports et télécommunications
Urban and rural planning	..	..	..	..	..	..	..	Aménagement urbain et rural
Prevention of pollution	..	..	..	..	..	..	..	Prévention de la pollution
Identification & treatment of pollution	..	..	..	..	..	..	..	Détection pollu. & mesures antipollu.
Health	..	..	..	..	..	..	..	Santé
Social development and services	..	..	..	..	..	..	..	Développement et services sociaux
Earth and atmosphere	..	..	..	..	..	..	..	Milieu terrestre et atmosphère
Advancement of knowledge	..	..	..	..	..	..	..	Promotion générale des connaissances
Civil space	..	..	..	..	..	..	..	Espace civil
Defence	..	..	..	..	..	..	..	Défense
Not elsewhere classified	15.5	24.1	261.4	178.8	275.4 p	318.7 p	..	Non classé
Total	15.5	24.0	261.4	178.8	275.4 p	318.7 p	..	*Total*
GERD								**DIRD**
Agriculture, forestry and fishing	..	..	..	..	..	..	..	Agriculture, sylviculture et pêche
Industrial development	..	..	..	..	..	..	..	Développement industriel
Energy	..	..	..	..	..	..	..	Energie
Transport and telecommunications	..	..	..	..	..	..	..	Transports et télécommunications
Urban and rural planning	..	..	..	..	..	..	..	Aménagement urbain et rural
Prevention of pollution	..	..	..	..	..	..	..	Prévention de la pollution
Identification & treatment of pollution	..	..	..	..	..	..	..	Détection pollu. & mesures antipollu.
Health	..	..	..	..	..	..	..	Santé
Social development and services	..	..	..	..	..	..	..	Développement et services sociaux
Earth and atmosphere	..	..	..	..	..	..	..	Milieu terrestre et atmosphère
Advancement of knowledge	..	..	..	..	..	..	..	Promotion générale des connaissances
Civil space	..	..	..	..	..	..	..	Espace civil
Defence	..	..	..	..	..	..	..	Défense
Not elsewhere classified	2 086.3	2 810.5	7 828.8	10 944.4	17 629.7 p	18 526.6 p	..	Non classé
Total	4 168.7	5 687.3	7 828.8	10 944.4	17 629.7 p	18 526.6 p	..	*Total*

Source: OECD, R&D database (DSTI/EAS Division), January 2002.

Source : OCDE, base de données R-D (DSTI/Division AES), janvier 2002.

Table 5 - Tableau 5

Total R&D personnel - Personnel total de R-D

by sector of employment and occupation - par secteur d'emploi et par profession

In full-time equivalent *En équivalence plein-temps*

	1994	1995	1996	1997	1998	1999	2000	
Business enterprise								**Entreprises**
RSE	1 754	2 006	..	..	..	..	..	Chercheurs
Technicians	1 250	1 405	..	..	..	..	..	Techniciens
Other occupations	872	1 055	..	..	..	..	..	Autres professions
Total	3 876	4 466	..	..	..	..	..	*Total*
Government								**État**
RSE	5 446	6 015	..	..	..	..	..	Chercheurs
Technicians	6 498	3 307	..	..	..	..	..	Techniciens
Other occupations	1 758	4 321	..	..	..	..	..	Autres professions
Total	13 702	13 643	..	..	..	..	..	*Total*
Higher education								**Enseignement supérieur**
RSE	9 735	11 233	..	..	..	..	..	Chercheurs
Technicians	1 639	1 902	..	..	..	..	..	Techniciens
Other occupations	1 329	1 754	..	..	..	..	..	Autres professions
Total	12 703	14 889	..	..	..	..	..	*Total*
Private non-profit								**Institutions sans but lucratif**
RSE	126	180	..	..	..	..	..	Chercheurs
Technicians	50	61	..	..	..	..	..	Techniciens
Other occupations	44	58	..	..	..	..	..	Autres professions
Total	220	299	..	..	..	..	..	*Total*
TOTAL								**TOTAL**
RSE	17 061	19 434	..	..	..	..	..	Chercheurs
Technicians	9 437	6 675	..	..	..	..	..	Techniciens
Other occupations	4 003	7 188	..	..	..	..	..	Autres professions
Total	30 501	33 297	..	..	..	..	..	*Total*

Table 6 - Tableau 6

Total R&D personnel - Personnel total de R-D

by sector of employment and formal qualification - par secteur d'emploi et par niveau de qualification

In full-time equivalent *En équivalence plein-temps*

	1994	1995	1996	1997	1998	1999	2000	
Business enterprise								**Entreprises**
University PhD level degrees	209	251	..	..	..	..	..	Diplômes univ. de niv. doctorat
Other university degrees	2 216	2 579	..	..	..	..	..	Autres diplômes universitaires
Sub-total university degrees	2 425	2 830	..	..	..	..	..	Sous-total Universitaires
Other post-secondary diplomas	932	985	..	..	..	..	..	Autres diplômes post-secondaires
Secondary diplomas	112	132	..	..	..	..	..	Diplômes de fin d'études secondaires
Other qualifications	407	519	..	..	..	..	..	Autres qualifications
Total	3 876	4 466	..	..	..	..	..	*Total*
Government								**État**
University PhD level degrees	1 143	1 291	..	..	..	..	..	Diplômes univ. de niv. doctorat
Other university degrees	6 095	6 423	..	..	..	..	..	Autres diplômes universitaires
Sub-total university degrees	7 238	7 714	..	..	..	..	..	Sous-total Universitaires
Other post-secondary diplomas	2 951	3 668	..	..	..	..	..	Autres diplômes post-secondaires
Secondary diplomas	1 559	556	..	..	..	..	..	Diplômes de fin d'études secondaires
Other qualifications	1 954	1 705	..	..	..	..	..	Autres qualifications
Total	13 702	13 643	..	..	..	..	..	*Total*
Higher education								**Enseignement supérieur**
University PhD level degrees	3 165	3 743	..	..	..	..	..	Diplômes univ. de niv. doctorat
Other university degrees	8 114	9 270	..	..	..	..	..	Autres diplômes universitaires
Sub-total university degrees	11 279	13 013	..	..	..	..	..	Sous-total Universitaires
Other post-secondary diplomas	934	1 199	..	..	..	..	..	Autres diplômes post-secondaires
Secondary diplomas	142	162	..	..	..	..	..	Diplômes de fin d'études secondaires
Other qualifications	348	515	..	..	..	..	..	Autres qualifications
Total	12 703	14 889	..	..	..	..	..	*Total*
Private non-profit								**Institutions sans but lucratif**
University PhD level degrees	35	33	..	..	..	..	..	Diplômes univ. de niv. doctorat
Other university degrees	121	193	..	..	..	..	..	Autres diplômes universitaires
Sub-total university degrees	156	226	..	..	..	..	..	Sous-total Universitaires
Other post-secondary diplomas	40	42	..	..	..	..	..	Autres diplômes post-secondaires
Secondary diplomas	0	11	..	..	..	..	..	Diplômes de fin d'études secondaires
Other qualifications	24	20	..	..	..	..	..	Autres qualifications
Total	220	299	..	..	..	..	..	*Total*
TOTAL								**TOTAL**
University PhD level degrees	4 552	5 317	..	..	..	..	..	Diplômes univ. de niv. doctorat
Other university degrees	16 546	18 466	..	..	..	..	..	Autres diplômes universitaires
Sub-total university degrees	21 098	23 783	..	..	..	..	..	Sous-total Universitaires
Other post-secondary diplomas	4 857	5 894	..	..	..	..	..	Autres diplômes post-secondaires
Secondary diplomas	1 813	861	..	..	..	..	..	Diplômes de fin d'études secondaires
Other qualifications	2 733	2 759	..	..	..	..	..	Autres qualifications
Total	30 501	33 297	..	..	..	..	..	*Total*

Source: OECD, R&D database (DSTI/EAS Division), January 2002. *Source* : OCDE, base de données R-D (DSTI/Division AES), janvier 2002.

Table 7 - Tableau 7

Gross domestic expenditure on R&D (GERD) - Dépenses intérieures brutes de R-D (DIRD)

by sector of performance and main scientific field - par secteur d'exécution et par grand domaine scientifique

Million MXN / *Millions MXN*

	1994	1995	1996	1997	1998	1999	2000	
Business enterprise								**Entreprises**
Natural sciences and engineering	1 010.9	1 158.3	..	..	..	..	..	Sciences exactes, nat. et de l'ingénieur
Social sciences and humanities	41.5	21.9	..	..	..	..	..	Sciences sociales et humaines
All fields of science	1 052.3	1 180.2	1 750.0	2 159.5	4 092.1 [p]	5 043.0 [p]	..	*Tous domaines scientifiques*
Government								**État**
Natural sciences and engineering	1 002.6	1 676.5	..	..	..	..	..	Sciences exactes, nat. et de l'ingénieur
Social sciences and humanities	151.9	200.6	..	..	..	..	..	Sciences sociales et humaines
All fields of science	1 154.5	1 877.1	2 851.1	4 240.9	6 748.4 [p]	6 019.3 [p]	..	*Tous domaines scientifiques*
Higher education								**Enseignement supérieur**
Natural sciences and engineering	1 412.8	1 907.0	..	..	..	..	..	Sciences exactes, nat. et de l'ingénieur
Social sciences and humanities	533.6	698.8	..	..	..	..	..	Sciences sociales et humaines
All fields of science	1 946.4	2 605.9	2 966.3	4 365.2	6 513.8 [p]	7 145.6 [p]	..	*Tous domaines scientifiques*
Private non-profit								**Institutions sans but lucratif**
Natural sciences and engineering	7.7	11.9	..	..	..	..	..	Sciences exactes, nat. et de l'ingénieur
Social sciences and humanities	7.7	12.1	..	..	..	..	..	Sciences sociales et humaines
All fields of science	15.5	24.0	261.4	178.8	275.4 [p]	318.7 [p]	..	*Tous domaines scientifiques*
TOTAL								**TOTAL**
Natural sciences and engineering	3 434.0	4 753.8	..	..	..	..	..	Sciences exactes, nat. et de l'ingénieur
Social sciences and humanities	734.7	933.5	..	..	..	..	..	Sciences sociales et humaines
All fields of science	4 168.7	5 687.3	7 828.8	10 944.4	17 629.7 [p]	18 526.6 [p]	..	*Tous domaines scientifiques*

Table 8 - Tableau 8

Total R&D personnel - Personnel total de R-D

by sector of employment and main scientific field - par secteur d'emploi et par grand domaine scientifique

In full-time equivalent / *En équivalence plein-temps*

	1994	1995	1996	1997	1998	1999	2000	
Business enterprise								**Entreprises**
Natural sciences and engineering	3 659	4 212	..	..	..	..	..	Sciences exactes, nat. et de l'ingénieur
Social sciences and humanities	217	254	..	..	..	..	..	Sciences sociales et humaines
All fields of science	3 876	4 466	..	..	..	..	..	*Tous domaines scientifiques*
Government								**État**
Natural sciences and engineering	12 282	12 371 [p]	..	..	..	..	..	Sciences exactes, nat. et de l'ingénieur
Social sciences and humanities	1 420	1 272 [p]	..	..	..	..	..	Sciences sociales et humaines
All fields of science	13 702	13 643	..	..	..	..	..	*Tous domaines scientifiques*
Higher education								**Enseignement supérieur**
Natural sciences and engineering	8 905	10 587	..	..	..	..	..	Sciences exactes, nat. et de l'ingénieur
Social sciences and humanities	3 798	4 302	..	..	..	..	..	Sciences sociales et humaines
All fields of science	12 703	14 889	..	..	..	..	..	*Tous domaines scientifiques*
Private non-profit								**Institutions sans but lucratif**
Natural sciences and engineering	102	160 [p]	..	..	..	..	..	Sciences exactes, nat. et de l'ingénieur
Social sciences and humanities	118	139 [p]	..	..	..	..	..	Sciences sociales et humaines
All fields of science	220	299	..	..	..	..	..	*Tous domaines scientifiques*
TOTAL								**TOTAL**
Natural sciences and engineering	24 948	27 330 [p]	..	..	..	..	..	Sciences exactes, nat. et de l'ingénieur
Social sciences and humanities	5 553	5 967 [p]	..	..	..	..	..	Sciences sociales et humaines
All fields of science	30 501	33 297	..	..	..	..	..	*Tous domaines scientifiques*

Source: OECD, R&D database (DSTI/EAS Division), January 2002.

Source : OCDE, base de données R-D (DSTI/Division AES), janvier 2002.

Total business enterprise intramural expenditure on R&D - Dépenses intra-muros totales de R-D des entreprises
(BERD) by industry - (DIRDE) par industrie

9.1 Million MXN - Millions MXN

	1994	1995	1996	1997	1998	1999	2000	
1 AGRICULTURE, HUNTING & FORESTRY	0.0	0.0	0.0	0.0	0.7 p	3.8 p	..	**1 AGRICULTURE, CHASSE & SYLVICULTURE**
2 MINING	8.0	13.0	35.5	77.9	392.6 p	274.3 p	..	**2 ACTIVITES EXTRACTIVES**
3 MANUFACTURING	595.5	782.4	1 119.2	1 356.0	3 289.9 p	4 011.9 p	..	**3 ACTIVITES DE FABRICATION**
4 Food, beverages & tobacco	56.8	69.6	135.2	164.7	225.3 p	258.6 p	..	**4 Produits alimentaires, boissons & tabac**
5 Food products & beverages	54.5	66.6	133.6	162.4	225.3 p	258.6 p	..	5 Produits alimentaires & boissons
6 Tobacco products	2.3	3.0	1.6	2.3	0.0 p	0.0 p	..	6 Produits à base de tabac
7 Textiles, wear app., fur & leather	5.4	5.2	7.4	12.1	27.7 p	143.6 p	..	**7 Textiles, habill., fourrures & cuirs**
8 Textiles	0.2	0.3	3.9	6.4	9.1 p	32.5 p	..	8 Textiles
9 Wearing apparel & fur	2.2	1.8	0.0	0.0	3.1 p	7.2 p	..	9 Articles d'habillement & fourrures
10 Leather products & footwear	3.0	3.2	3.5	5.7	15.5 p	103.9 p	..	10 Articles en cuir & chaussures
11 Wood, paper, printing, publishing	1.5	4.2	2.6	5.0	534.7 p	790.7 p	..	**11 Bois, papier, imprimerie, édition**
12 Wood & cork (not furniture)	..	..	0.0	0.0	0.7 p	0.1 p	..	12 Bois & liège (sauf meubles)
13 Pulp, paper & paper products	0.1	0.4	2.6	4.9	432.5 p	560.0 p	..	13 Carton, papier & articles
14 Publ., print. & repro. of rec. media	1.4	3.8	0.0	0.1	101.5 p	230.5 p	..	14 Edition, impress. & repro. enregis.
15 Coke, petroleum, nuclear fuel, chemicals, rubber plastics	160.3	219.9	562.7	715.1	700.7 p	769.9 p	..	**15 Coke, pétrole, combust. nucl., prod. chimi., caout. & mat. plast.**
16 Coke, ref. petrol. prod. & nuclear fuel	1.2	2.5	0.0	0.0	18.8 p	31.7 p	..	16 Coke, raffin. pétr. & combust. nucl.
17 Chemicals & chemical products	154.8	215.2	562.3	714.6	426.3 p	475.1 p	..	17 Produits chimiques
18 *Chemical prod.(less pharmaceuticals)*	136.0	174.6	509.9	640.4	297.3 p	315.6 p	..	18 *Prod. chimi. (moins prod. pharmaceutiques)*
19 *Pharmaceuticals*	18.8	40.7	52.4	74.2	128.9 p	159.5 p	..	19 *Produits pharmaceutiques*
20 Rubber & plastic products	4.3	2.3	0.4	0.5	255.6 p	263.1 p	..	20 Caoutchouc et matières plastiques
21 Non-metallic mineral products	16.3	17.1	30.3	33.7	155.2 p	164.4 p	..	**21 Produits minéraux non métalliques**
22 Basic metals	50.3	50.0	11.0	29.4	927.0 p	743.7 p	..	**22 Produits métallurgiques de base**
23 Basic metals, ferrous	17.9	25.3	6.9	17.4	804.5 p	574.6 p	..	23 Produits métallur. de base, ferreux
24 Basic metals, non-ferrous	32.3	24.7	4.0	11.9	122.4 p	169.1 p	..	24 Prod. métallur. de base, non ferreux
25 Fabricated metal products	1.3	1.3	144.6	101.9	50.6 p	89.1 p	..	**25 Ouvrages en métaux (sauf mach.& mat.)**
26 Machinery equipment, instruments and transport equipment	170.3	251.0	123.2	138.5	442.1 p	853.4 p	..	**26 Machines & matériel, instruments & mat. de transport**
27 Machinery n.e.c.	20.9	27.7	1.1	0.4	5.6 p	10.9 p	..	27 Machines, n.c.a.
28 Office, account. & computing machinery	23.0	62.3	0.0	0.0	73.6 p	47.6 p	..	28 Mach. de bur., compta. & trait. inform.
29 Electrical machinery	56.3	83.4	35.3	47.7	101.4 p	170.0 p	..	29 Machines électriques
30 Electro. equip. (radio, TV & commun.)	0.4	0.5	0.5	0.5	20.9 p	43.8 p	..	30 Equip. électro. (radio, tv & communic.)
31 *Electro. comp. (inc. semi-conductors)*	..	..	0.5	0.5	18.0 p	41.0 p	..	31 *Compos. électro. (semi-cond. incl.)*
32 *TV, radio & communications equip.*	0.4	0.5	0.0	0.0	2.9 p	2.7 p	..	32 *Appareils radio, tv et de communi.*
33 Instruments	..	..	0.0	0.0	9.7 p	14.0 p	..	33 Instruments
34 Motor vehicles	69.6	77.1	86.3	89.8	226.8 p	564.2 p	..	34 Véhicules automobiles
35 Other transport equipment	..	..	0.0	0.0	4.2 p	2.9 p	..	35 Autres matériels de transport
36 *Ships*	..	..	0.0	0.0	1.5 p	0.0 p	..	36 *Navires*
37 *Aerospace*	..	..	0.0	0.0	0.0 p	0.0 p	..	37 *Aérospatiale*
38 *Other transport equipment n.e.c.*	..	..	0.0	0.0	2.6 p	2.9 p	..	38 *Autres matériels de transport n.c.a.*
39 Furniture, other manufacturing n.e.c.	133.3	164.0	102.2	155.7	226.7 p	198.5 p	..	**39 Meubles, autres activ. de fabricat. n.c.a.**
40 Furniture	1.6	1.6	0.5	0.6	52.4 p	100.0 p	..	40 Meubles
41 Other manufacturing n.e.c.	131.7	162.4	101.7	155.1	174.3 p	98.5 p	..	41 Autres activités de fabrication n.c.a.
42 Recycling	..	..	0.0	0.0	0.0 p	0.0 p	..	**42 Récupération**
43 ELECTRICITY, GAS & WATER SUPPLY	..	..	0.0	0.0	92.0 p	167.3 p	..	**43 DISTRIBUTION D'ELECTRICITE, GAZ & EAU**
44 CONSTRUCTION	0.9	1.1	0.0	0.0	31.0 p	101.9 p	..	**44 CONSTRUCTION**
45 SERVICES SECTOR	447.9	383.6	595.3	725.6	285.8 p	483.7 p	..	**45 SECTEUR DES SERVICES**
46 Wholesale, retail trade & repair of motor vehicles, etc.	..	..	0.0	0.0	0.0 p	0.0 p	..	**46 Commerce (gros & détail), réparation de véhicules, etc.**
47 Hotels & restaurants	..	..	0.0	0.0	0.0 p	4.8 p	..	**47 Hôtels et restaurants**
48 Transport & storage	2.9	0.6	408.7	467.7	100.0 p	4.5 p	..	**48 Transports, entreposage**
49 Communications	2.5	3.6	1.0	3.3	154.4 p	169.9 p	..	**49 Communications**
50 Post	0.0	0.2	0.0	0.0	0.0 p	0.0 p	..	50 Postes
51 Telecommunications	2.5	3.4	1.0	3.3	154.4 p	169.9 p	..	51 Télécommunications
52 Financial intermediation (inc. insur.)	0.4	0.4	0.0	0.0	3.8 p	28.8 p	..	**52 Intermédiation financ. (assurances inc.)**
53 Real estate, renting & busin. activity	388.8	311.2	158.8	215.7	127.6 p	267.7 p	..	**53 Immobil., locat. & serv. aux entreprises**
54 Computer & related activities	14.5	18.7	0.0	0.0	0.0 p	0.0 p	..	54 Informatique et activités connexes
55 *Software consultancy*	14.5	18.7	0.0	0.0	0.0 p	0.0 p	..	55 *Production de logiciels*
56 *Other computer services n.e.c.*	..	..	0.0	0.0	0.0 p	0.0 p	..	56 *Autres activités informatiques*
57 Research & development	374.3	292.5	117.4	158.8	43.9 p	56.6 p	..	57 Recherche & développement expérimental
58 Other business activities n.e.c.	..	..	41.4	56.9	83.6 p	211.2 p	..	58 Autres activ. & serv. aux entreprises
59 Comm., soc. & pers. serv. activ., etc.	53.3	67.8	26.7	39.0	0.0 p	8.0 p	..	**59 Adm. pub., serv. soc. & collec., etc**
TOTAL BUSINESS ENTERPRISE SECTOR	**1 052.3**	**1 180.2**	**1 750.0**	**2 159.5**	**4 092.1 p**	**5 043.0 p**	..	**TOTAL SECTEUR DES ENTREPRISES**

Source: OECD, R&D database (DSTI/EAS Division), January 2002.

Source : OCDE, base de données R-D (DSTI/Division AES), janvier 2002.

Total business enterprise intramural expenditure on R&D - Dépenses intra-muros totales de R-D des entreprises
(BERD) by industry - (DIRDE) par industrie

	1994	1995	1996	1997	1998	1999	2000	
	9.2 Million constant USD(1995 prices and PPPs) - Millions de USD constants (prix et PPA de 1995)							
1 AGRICULTURE, HUNTING & FORESTRY	0.0	0.0	0.0	0.0	0.1 p	0.6 p	..	**1 AGRICULTURE, CHASSE & SYLVICULTURE**
2 MINING	3.7	4.4	9.2	17.1	74.8 p	45.5 p	..	**2 ACTIVITES EXTRACTIVES**
3 MANUFACTURING	277.6	264.6	289.5	297.9	626.5 p	665.2 p	..	**3 ACTIVITES DE FABRICATION**
4 Food, beverages & tobacco	26.5	23.5	35.0	36.2	42.9 p	42.9 p	..	**4 Produits alimentaires, boissons & tabac**
5 Food products & beverages	25.4	22.5	34.6	35.7	42.9 p	42.9 p	..	5 Produits alimentaires & boissons
6 Tobacco products	1.1	1.0	0.4	0.5	0.0 p	0.0 p	..	6 Produits à base de tabac
7 Textiles, wear app., fur & leather	2.5	1.7	1.9	2.7	5.3 p	23.8 p	..	**7 Textiles, habill., fourrures & cuirs**
8 Textiles	0.1	0.1	1.0	1.4	1.7 p	5.4 p	..	8 Textiles
9 Wearing apparel & fur	1.0	0.6	0.0	0.0	0.6 p	1.2 p	..	9 Articles d'habillement & fourrures
10 Leather products & footwear	1.4	1.1	0.9	1.3	3.0 p	17.2 p	..	10 Articles en cuir & chaussures
11 Wood, paper, printing, publishing	0.7	1.4	0.7	1.1	101.8 p	131.1 p	..	**11 Bois, papier, imprimerie, édition**
12 Wood & cork (not furniture)	..	..	0.0	0.0	0.1 p	0.0 p	..	12 Bois & liège (sauf meubles)
13 Pulp, paper & paper products	0.1	0.1	0.7	1.1	82.4 p	92.9 p	..	13 Carton, papier & articles
14 Publ., print. & repro. of rec. media	0.7	1.3	0.0	0.0	19.3 p	38.2 p	..	14 Edition, impress. & repro. enregis.
15 Coke, petroleum, nuclear fuel, chemicals, rubber plastics	74.7	74.4	145.5	157.1	133.4 p	127.7 p	..	**15 Coke, pétrole, combust. nucl., prod. chimi., caout. & mat. plast.**
16 Coke, ref. petrol. prod. & nuclear fuel	0.6	0.8	0.0	0.0	3.6 p	5.3 p	..	16 Coke, raffin. pétr. & combust. nucl.
17 Chemicals & chemical products	72.2	72.8	145.4	157.0	81.2 p	78.8 p	..	17 Produits chimiques
18 *Chemical prod.(less pharmaceuticals)*	63.4	59.0	131.9	140.7	56.6 p	52.3 p	..	18 *Prod. chimi. (moins prod. pharmaceutiques)*
19 *Pharmaceuticals*	8.8	13.7	13.6	16.3	24.6 p	26.4 p	..	19 *Produits pharmaceutiques*
20 Rubber & plastic products	2.0	0.8	0.1	0.1	48.7 p	43.6 p	..	20 Caoutchouc et matières plastiques
21 Non-metallic mineral products	7.6	5.8	7.8	7.4	29.6 p	27.3 p	..	**21 Produits minéraux non métalliques**
22 Basic metals	23.4	16.9	2.8	6.5	176.5 p	123.3 p	..	**22 Produits métallurgiques de base**
23 Basic metals, ferrous	8.4	8.6	1.8	3.8	153.2 p	95.3 p	..	23 Produits métallur. de base, ferreux
24 Basic metals, non-ferrous	15.1	8.4	1.0	2.6	23.3 p	28.0 p	..	24 Prod. métallur. de base, non ferreux
25 Fabricated metal products	0.6	0.4	37.4	22.4	9.6 p	14.8 p	..	**25 Ouvrages en métaux (sauf mach.& mat.)**
26 Machinery equipment, instruments and transport equipment	79.4	84.9	31.9	30.4	84.2 p	141.5 p	..	**26 Machines & matériel, instruments & mat. de transport**
27 Machinery n.e.c.	9.8	9.4	0.3	0.1	1.1 p	1.8 p	..	27 Machines, n.c.a.
28 Office, account. & computing machinery	10.7	21.1	0.0	0.0	14.0 p	7.9 p	..	28 Mach. de bur., compta. & trait. inform.
29 Electrical machinery	26.2	28.2	9.1	10.5	19.3 p	28.2 p	..	29 Machines électriques
30 Electro. equip. (radio, TV & commun.)	0.2	0.2	0.1	0.1	4.0 p	7.3 p	..	30 Equip. électro. (radio, tv & communic.)
31 *Electro. comp. (inc. semi-conductors)*	..	..	0.1	0.1	3.4 p	6.8 p	..	31 *Compos. électro. (semi-cond. incl.)*
32 *TV, radio & communications equip.*	0.2	0.2	0.0	0.0	0.6 p	0.5 p	..	32 *Appareils radio, tv et de communi.*
33 Instruments	..	..	0.0	0.0	1.9 p	2.3 p	..	33 Instruments
34 Motor vehicles	32.5	26.1	22.3	19.7	43.2 p	93.6 p	..	34 Véhicules automobiles
35 Other transport equipment	..	..	0.0	0.0	0.8 p	0.5 p	..	35 Autres matériels de transport
36 *Ships*	..	..	0.0	0.0	0.3 p	0.0 p	..	36 *Navires*
37 *Aerospace*	..	..	0.0	0.0	0.0 p	0.0 p	..	37 *Aérospatiale*
38 *Other transport equipment n.e.c.*	..	..	0.0	0.0	0.5 p	0.5 p	..	38 *Autres matériels de transport n.c.a.*
39 Furniture, other manufacturing n.e.c.	62.1	55.5	26.4	34.2	43.2 p	32.9 p	..	**39 Meubles, autres activ. de fabricat. n.c.a.**
40 Furniture	0.8	0.6	0.1	0.1	10.0 p	16.6 p	..	40 Meubles
41 Other manufacturing n.e.c.	61.4	54.9	26.3	34.1	33.2 p	16.3 p	..	41 Autres activités de fabrication n.c.a.
42 Recycling	..	..	0.0	0.0	0.0 p	0.0 p	..	**42 Récupération**
43 ELECTRICITY, GAS & WATER SUPPLY	..	..	0.0	0.0	17.5 p	27.7 p	..	**43 DISTRIBUTION D'ELECTRICITE, GAZ & EAU**
44 CONSTRUCTION	0.4	0.4	0.0	0.0	5.9 p	16.9 p	..	**44 CONSTRUCTION**
45 SERVICES SECTOR	208.8	129.7	154.0	159.4	54.4 p	80.2 p	..	**45 SECTEUR DES SERVICES**
46 Wholesale, retail trade & repair of motor vehicles, etc.	..	..	0.0	0.0	0.0 p	0.0 p	..	**46 Commerce (gros & détail), réparation de véhicules, etc.**
47 Hotels & restaurants	..	..	0.0	0.0	0.0 p	0.8 p	..	**47 Hôtels et restaurants**
48 Transport & storage	1.3	0.2	105.7	102.8	19.0 p	0.8 p	..	**48 Transports, entreposage**
49 Communications	1.2	1.2	0.3	0.7	29.4 p	28.2 p	..	**49 Communications**
50 Post	0.0	0.1	0.0	0.0	0.0 p	0.0 p	..	50 Postes
51 Telecommunications	1.2	1.2	0.3	0.7	29.4 p	28.2 p	..	51 Télécommunications
52 Financial intermediation (inc. insur.)	0.2	0.1	0.0	0.0	0.7 p	4.8 p	..	**52 Intermédiation financ. (assurances inc.)**
53 Real estate, renting & busin. activity	181.2	105.2	41.1	47.4	24.3 p	44.4 p	..	**53 Immobil., locat. & serv. aux entreprises**
54 Computer & related activities	6.7	6.3	0.0	0.0	0.0 p	0.0 p	..	54 Informatique et activités connexes
55 *Software consultancy*	6.7	6.3	0.0	0.0	0.0 p	0.0 p	..	55 *Production de logiciels*
56 *Other computer services n.e.c.*	..	..	0.0	0.0	0.0 p	0.0 p	..	56 *Autres activités informatiques*
57 Research & development	174.5	98.9	30.4	34.9	8.4 p	9.4 p	..	57 Recherche & développement expérimental
58 Other business activities n.e.c.	..	..	10.7	12.5	15.9 p	35.0 p	..	58 Autres activ. & serv. aux entreprises
59 Comm., soc. & pers. serv. activ., etc.	24.9	22.9	6.9	8.6	0.0 p	1.3 p	..	**59 Adm. pub., serv. soc. & collec., etc**
TOTAL BUSINESS ENTERPRISE SECTOR	**490.5**	**399.1**	**452.6**	**474.5**	**779.3 p**	**836.1 p**	..	**TOTAL SECTEUR DES ENTREPRISES**

Source: OECD, R&D database (DSTI/EAS Division), January 2002.

Source : OCDE, base de données R-D (DSTI/Division AES), janvier 2002.

Business enterprise intramural expenditure on R&D (BERD) - Dépenses intra-muros de R-D des entreprises (DIRDE)

by industry and source of funds - par industrie et par source de financement

Million MXN / *Millions MXN*

	1994	1995	1996	1997	1998	1999	2000	
				10.1 Business enterprise - Entreprises				
1 **AGRICULTURE, HUNTING & FORESTRY**	0.0	0.0	..	..	..	..	..	1 **AGRICULTURE, CHASSE & SYLVICULTURE**
2 **MINING**	8.0	13.0	..	..	..	..	..	2 **ACTIVITES EXTRACTIVES**
3 **MANUFACTURING**	594.6	778.3	..	..	..	..	..	3 **ACTIVITES DE FABRICATION**
4 **Food, beverages & tobacco**	56.8	69.6	..	..	..	..	..	4 **Produits alimentaires, boissons & tabac**
5 Food products & beverages	54.5	66.6	..	..	..	..	..	5 Produits alimentaires & boissons
6 Tobacco products	2.3	3.0	..	..	..	..	..	6 Produits à base de tabac
7 **Textiles, wear app., fur & leather**	5.4	5.2	..	..	..	..	..	7 **Textiles, habill., fourrures & cuirs**
8 Textiles	0.2	0.3	..	..	..	..	..	8 Textiles
9 Wearing apparel & fur	2.2	1.8	..	..	..	..	..	9 Articles d'habillement & fourrures
10 Leather products & footwear	3.0	3.2	..	..	..	..	..	10 Articles en cuir & chaussures
11 **Wood, paper, printing, publishing**	1.5	4.1	..	..	..	..	..	11 **Bois, papier, imprimerie, édition**
12 Wood & cork (not furniture)	0.0	0.0	..	..	..	..	..	12 Bois & liège (sauf meubles)
13 Pulp, paper & paper products	0.1	0.4	..	..	..	..	..	13 Carton, papier & articles
14 Publ., print. & repro. of rec. media	1.4	3.7	..	..	..	..	..	14 Edition, impress. & repro. enregis.
15 **Coke, petroleum, nuclear fuel, chemicals, rubber plastics**	160.3	219.9	..	..	..	..	..	15 **Coke, pétrole, combust. nucl., prod. chimi., caout. & mat. plast.**
16 Coke, ref. petrol. prod. & nuclear fuel	1.2	2.5	..	..	..	..	..	16 Coke, raffin. pétr. & combust. nucl.
17 Chemicals & chemical products	154.8	215.2	..	..	..	..	..	17 Produits chimiques
18 *Chemical prod.(less pharmaceuticals)*	136.0	174.6	..	..	..	..	..	18 *Prod. chimi. (moins prod. pharmaceutiques)*
19 *Pharmaceuticals*	18.8	40.7	..	..	..	..	..	19 *Produits pharmaceutiques*
20 Rubber & plastic products	4.3	2.3	..	..	..	..	..	20 Caoutchouc et matières plastiques
21 **Non-metallic mineral products**	16.3	17.1	..	..	..	..	..	21 **Produits minéraux non métalliques**
22 **Basic metals**	50.3	50.0	..	..	..	..	..	22 **Produits métallurgiques de base**
23 Basic metals, ferrous	17.9	25.3	..	..	..	..	..	23 Produits métallur. de base, ferreux
24 Basic metals, non-ferrous	32.3	24.7	..	..	..	..	..	24 Prod. métallur. de base, non ferreux
25 **Fabricated metal products**	1.3	1.3	..	..	..	..	..	25 **Ouvrages en métaux (sauf mach.& mat.)**
26 **Machinery equipment, instruments and transport equipment**	169.4	247.0	..	..	..	..	..	26 **Machines & matériel, instruments & mat. de transport**
27 Machinery n.e.c.	20.5	27.7	..	..	..	..	..	27 Machines, n.c.a.
28 Office, account. & computing machinery	23.0	62.3	..	..	..	..	..	28 Mach. de bur., compta. & trait. inform.
29 Electrical machinery	56.3	83.4	..	..	..	..	..	29 Machines électriques
30 Electro. equip. (radio, TV & commun.)	0.0	0.0	..	..	..	..	..	30 Equip. électro. (radio, tv & communic.)
31 *Electro. comp. (inc. semi-conductors)*	0.0	0.0	..	..	..	..	..	31 *Compos. électro. (semi-cond. incl.)*
32 *TV, radio & communications equip.*	0.0	0.0	..	..	..	..	..	32 *Appareils radio, tv et de communi.*
33 Instruments	0.0	0.0	..	..	..	..	..	33 Instruments
34 Motor vehicles	69.6	73.5	..	..	..	..	..	34 Véhicules automobiles
35 Other transport equipment	0.0	0.0	..	..	..	..	..	35 Autres matériels de transport
36 *Ships*	0.0	0.0	..	..	..	..	..	36 *Navires*
37 *Aerospace*	0.0	0.0	..	..	..	..	..	37 *Aérospatiale*
38 *Other transport equipment n.e.c.*	0.0	0.0	..	..	..	..	..	38 *Autres matériels de transport n.c.a.*
39 **Furniture, other manufacturing n.e.c.**	133.3	164.0	..	..	..	..	..	39 **Meubles, autres activ. de fabricat. n.c.a.**
40 Furniture	1.6	1.6	..	..	..	..	..	40 Meubles
41 Other manufacturing n.e.c.	131.7	162.4	..	..	..	..	..	41 Autres activités de fabrication n.c.a.
42 **Recycling**	0.0	0.0	..	..	..	..	..	42 **Récupération**
43 **ELECTRICITY, GAS & WATER SUPPLY**	0.0	0.0	..	..	..	..	..	43 **DISTRIBUTION D'ELECTRICITE, GAZ & EAU**
44 **CONSTRUCTION**	0.9	1.1	..	..	..	..	..	44 **CONSTRUCTION**
45 **SERVICES SECTOR**	86.9	107.0	..	..	..	..	..	45 **SECTEUR DES SERVICES**
46 **Wholesale, retail trade & repair of motor vehicles, etc.**	0.0	0.0	..	..	..	..	..	46 **Commerce (gros & détail), réparation de véhicules, etc.**
47 **Hotels & restaurants**	0.0	0.0	..	..	..	..	..	47 **Hôtels et restaurants**
48 **Transport & storage**	0.3	0.4	..	..	..	..	..	48 **Transports, entreposage**
49 **Communications**	0.8	1.0	..	..	..	..	..	49 **Communications**
50 Post	0.0	0.0	..	..	..	..	..	50 Postes
51 Telecommunications	0.8	1.0	..	..	..	..	..	51 Télécommunications
52 **Financial intermediation (inc. insur.)**	0.4	0.4	..	..	..	..	..	52 **Intermédiation financ. (assurances inc.)**
53 **Real estate, renting & busin. activity**	49.3	63.9	..	..	..	..	..	53 **Immobil., locat. & serv. aux entreprises**
54 Computer & related activities	14.5	18.7	..	..	..	..	..	54 Informatique et activités connexes
55 *Software consultancy*	14.5	18.7	..	..	..	..	..	55 *Production de logiciels*
56 *Other computer services n.e.c.*	0.0	0.0	..	..	..	..	..	56 *Autres activités informatiques*
57 Research & development	34.8	45.2	..	..	..	..	..	57 Recherche & développement expérimental
58 Other business activities n.e.c.	0.0	0.0	..	..	..	..	..	58 Autres activ. & serv. aux entreprises
59 **Comm., soc. & pers. serv. activ., etc.**	36.0	41.3	..	..	..	..	..	59 **Adm. pub., serv. soc. & collec., etc**
TOTAL BUSINESS ENTERPRISE SECTOR	**690.4**	**899.4**	**1 102.3**	**1 366.8**	**3 020.3** p	**4 053.2** p	..	**TOTAL SECTEUR DES ENTREPRISES**

Source: OECD, R&D database (DSTI/EAS Division), January 2002.

Source : OCDE, base de données R-D (DSTI/Division AES), janvier 2002.

Business enterprise intramural expenditure on R&D (BERD) - Dépenses intra-muros de R-D des entreprises (DIRDE)

by industry and source of funds - par industrie et par source de financement

Million MXN — *Millions MXN*

10.2 Government - État

	1994	1995	1996	1997	1998	1999	2000	
1 AGRICULTURE, HUNTING & FORESTRY	0.0	0.0	..	..	..	..	..	**1 AGRICULTURE, CHASSE & SYLVICULTURE**
2 MINING	0.0	0.0	..	..	..	..	..	**2 ACTIVITES EXTRACTIVES**
3 MANUFACTURING	0.5	0.1	..	..	..	..	..	**3 ACTIVITES DE FABRICATION**
4 Food, beverages & tobacco	0.0	0.0	..	..	..	..	..	**4 Produits alimentaires, boissons & tabac**
5 Food products & beverages	0.0	0.0	..	..	..	..	..	5 Produits alimentaires & boissons
6 Tobacco products	0.0	0.0	..	..	..	..	..	6 Produits à base de tabac
7 Textiles, wear app., fur & leather	0.0	0.0	..	..	..	..	..	**7 Textiles, habill., fourrures & cuirs**
8 Textiles	0.0	0.0	..	..	..	..	..	8 Textiles
9 Wearing apparel & fur	0.0	0.0	..	..	..	..	..	9 Articles d'habillement & fourrures
10 Leather products & footwear	0.0	0.0	..	..	..	..	..	10 Articles en cuir & chaussures
11 Wood, paper, printing, publishing	0.0	0.1	..	..	..	..	..	**11 Bois, papier, imprimerie, édition**
12 Wood & cork (not furniture)	0.0	0.0	..	..	..	..	..	12 Bois & liège (sauf meubles)
13 Pulp, paper & paper products	0.0	0.0	..	..	..	..	..	13 Carton, papier & articles
14 Publ., print. & repro. of rec. media	0.0	0.1	..	..	..	..	..	14 Edition, impress. & repro. enregis.
15 Coke, petroleum, nuclear fuel, chemicals, rubber plastics	0.0	0.0	..	..	..	..	..	**15 Coke, pétrole, combust. nucl., prod. chimi., caout. & mat. plast.**
16 Coke, ref. petrol. prod. & nuclear fuel	0.0	0.0	..	..	..	..	..	16 Coke, raffin. pétr. & combust. nucl.
17 Chemicals & chemical products	0.0	0.0	..	..	..	..	..	17 Produits chimiques
18 *Chemical prod.(less pharmaceuticals)*	0.0	0.0	..	..	..	..	..	18 *Prod. chimi. (moins prod. pharmaceutiques)*
19 *Pharmaceuticals*	0.0	0.0	..	..	..	..	..	19 *Produits pharmaceutiques*
20 Rubber & plastic products	0.0	0.0	..	..	..	..	..	20 Caoutchouc et matières plastiques
21 Non-metallic mineral products	0.0	0.0	..	..	..	..	..	**21 Produits minéraux non métalliques**
22 Basic metals	0.0	0.0	..	..	..	..	..	**22 Produits métallurgiques de base**
23 Basic metals, ferrous	0.0	0.0	..	..	..	..	..	23 Produits métallur. de base, ferreux
24 Basic metals, non-ferrous	0.0	0.0	..	..	..	..	..	24 Prod. métallur. de base, non ferreux
25 Fabricated metal products	0.0	0.0	..	..	..	..	..	**25 Ouvrages en métaux (sauf mach.& mat.)**
26 Machinery equipment, instruments and transport equipment	0.5	0.0	..	..	..	..	..	**26 Machines & matériel, instruments & mat. de transport**
27 Machinery n.e.c.	0.5	0.0	..	..	..	..	..	27 Machines, n.c.a.
28 Office, account. & computing machinery	0.0	0.0	..	..	..	..	..	28 Mach. de bur., compta. & trait. inform.
29 Electrical machinery	0.0	0.0	..	..	..	..	..	29 Machines électriques
30 Electro. equip. (radio, TV & commun.)	0.0	0.0	..	..	..	..	..	30 Equip. électro. (radio, tv & communic.)
31 *Electro. comp. (inc. semi-conductors)*	0.0	0.0	..	..	..	..	..	31 *Compos. électro. (semi-cond. incl.)*
32 *TV, radio & communications equip.*	0.0	0.0	..	..	..	..	..	32 *Appareils radio, tv et de communi.*
33 Instruments	0.0	0.0	..	..	..	..	..	33 Instruments
34 Motor vehicles	0.0	0.0	..	..	..	..	..	34 Véhicules automobiles
35 Other transport equipment	0.0	0.0	..	..	..	..	..	35 Autres matériels de transport
36 *Ships*	0.0	0.0	..	..	..	..	..	36 *Navires*
37 *Aerospace*	0.0	0.0	..	..	..	..	..	37 *Aérospatiale*
38 *Other transport equipment n.e.c.*	0.0	0.0	..	..	..	..	..	38 *Autres matériels de transport n.c.a.*
39 Furniture, other manufacturing n.e.c.	0.0	0.0	..	..	..	..	..	**39 Meubles, autres activ. de fabricat. n.c.a.**
40 Furniture	0.0	0.0	..	..	..	..	..	40 Meubles
41 Other manufacturing n.e.c.	0.0	0.0	..	..	..	..	..	41 Autres activités de fabrication n.c.a.
42 Recycling	0.0	0.0	..	..	..	..	..	**42 Récupération**
43 ELECTRICITY, GAS & WATER SUPPLY	0.0	0.0	..	..	..	..	..	**43 DISTRIBUTION D'ELECTRICITE, GAZ & EAU**
44 CONSTRUCTION	0.0	0.0	..	..	..	..	..	**44 CONSTRUCTION**
45 SERVICES SECTOR	24.9	32.4	..	..	..	..	..	**45 SECTEUR DES SERVICES**
46 Wholesale, retail trade & repair of motor vehicles, etc.	0.0	0.0	..	..	..	..	..	**46 Commerce (gros & détail), réparation de véhicules, etc.**
47 Hotels & restaurants	0.0	0.0	..	..	..	..	..	**47 Hôtels et restaurants**
48 Transport & storage	2.6	0.3	..	..	..	..	..	**48 Transports, entreposage**
49 Communications	1.7	2.6	..	..	..	..	..	**49 Communications**
50 Post	0.0	0.2	..	..	..	..	..	50 Postes
51 Telecommunications	1.7	2.4	..	..	..	..	..	51 Télécommunications
52 Financial intermediation (inc. insur.)	0.0	0.0	..	..	..	..	..	**52 Intermédiation financ. (assurances inc.)**
53 Real estate, renting & busin. activity	3.3	3.1	..	..	..	..	..	**53 Immobil., locat. & serv. aux entreprises**
54 Computer & related activities	0.0	0.0	..	..	..	..	..	54 Informatique et activités connexes
55 *Software consultancy*	0.0	0.0	..	..	..	..	..	55 *Production de logiciels*
56 *Other computer services n.e.c.*	0.0	0.0	..	..	..	..	..	56 *Autres activités informatiques*
57 Research & development	3.3	3.1	..	..	..	..	..	57 Recherche & développement expérimental
58 Other business activities n.e.c.	0.0	0.0	..	..	..	..	..	58 Autres activ. & serv. aux entreprises
59 Comm., soc. & pers. serv. activ., etc.	17.3	26.5	..	..	..	..	..	**59 Adm. pub., serv. soc. & collec., etc**
TOTAL BUSINESS ENTERPRISE SECTOR	**25.4**	**32.5**	**428.8**	**569.2**	**87.8** p	**54.9** p	..	**TOTAL SECTEUR DES ENTREPRISES**

Source: OECD, R&D database (DSTI/EAS Division), January 2002.

Source : OCDE, base de données R-D (DSTI/Division AES), janvier 2002.

Business enterprise intramural expenditure on R&D (BERD) - Dépenses intra-muros de R-D des entreprises (DIRDE)

by industry and source of funds - par industrie et par source de financement

Million MXN — *Millions MXN*

10.3 Other national funds - Autres sources nationales

	1994	1995	1996	1997	1998	1999	2000	
1 **AGRICULTURE, HUNTING & FORESTRY**	0.0	0.0	..	..	..	..	..	1 **AGRICULTURE, CHASSE & SYLVICULTURE**
2 **MINING**	0.0	0.0	..	..	..	..	..	2 **ACTIVITES EXTRACTIVES**
3 **MANUFACTURING**	0.0	0.0	..	..	..	..	..	3 **ACTIVITES DE FABRICATION**
4 **Food, beverages & tobacco**	0.0	0.0	..	..	..	..	..	4 **Produits alimentaires, boissons & tabac**
5 Food products & beverages	0.0	0.0	..	..	..	..	..	5 Produits alimentaires & boissons
6 Tobacco products	0.0	0.0	..	..	..	..	..	6 Produits à base de tabac
7 **Textiles, wear app., fur & leather**	0.0	0.0	..	..	..	..	..	7 **Textiles, habill., fourrures & cuirs**
8 Textiles	0.0	0.0	..	..	..	..	..	8 Textiles
9 Wearing apparel & fur	0.0	0.0	..	..	..	..	..	9 Articles d'habillement & fourrures
10 Leather products & footwear	0.0	0.0	..	..	..	..	..	10 Articles en cuir & chaussures
11 **Wood, paper, printing, publishing**	0.0	0.0	..	..	..	..	..	11 **Bois, papier, imprimerie, édition**
12 Wood & cork (not furniture)	0.0	0.0	..	..	..	..	..	12 Bois & liège (sauf meubles)
13 Pulp, paper & paper products	0.0	0.0	..	..	..	..	..	13 Carton, papier & articles
14 Publ., print. & repro. of rec. media	0.0	0.0	..	..	..	..	..	14 Edition, impress. & repro. enregis.
15 **Coke, petroleum, nuclear fuel, chemicals, rubber plastics**	0.0	0.0	..	..	..	..	..	15 **Coke, pétrole, combust. nucl., prod. chimi., caout. & mat. plast.**
16 Coke, ref. petrol. prod. & nuclear fuel	0.0	0.0	..	..	..	..	..	16 Coke, raffin. pétr. & combust. nucl.
17 Chemicals & chemical products	0.0	0.0	..	..	..	..	..	17 Produits chimiques
18 *Chemical prod.(less pharmaceuticals)*	0.0	0.0	..	..	..	..	..	18 *Prod. chimi. (moins prod. pharmaceutiques)*
19 *Pharmaceuticals*	0.0	0.0	..	..	..	..	..	19 *Produits pharmaceutiques*
20 Rubber & plastic products	0.0	0.0	..	..	..	..	..	20 Caoutchouc et matières plastiques
21 **Non-metallic mineral products**	0.0	0.0	..	..	..	..	..	21 **Produits minéraux non métalliques**
22 **Basic metals**	0.0	0.0	..	..	..	..	..	22 **Produits métallurgiques de base**
23 Basic metals, ferrous	0.0	0.0	..	..	..	..	..	23 Produits métallur. de base, ferreux
24 Basic metals, non-ferrous	0.0	0.0	..	..	..	..	..	24 Prod. métallur. de base, non ferreux
25 **Fabricated metal products**	0.0	0.0	..	..	..	..	..	25 **Ouvrages en métaux (sauf mach.& mat.)**
26 **Machinery equipment, instruments and transport equipment**	0.0	0.0	..	..	..	..	..	26 **Machines & matériel, instruments & mat. de transport**
27 Machinery n.e.c.	0.0	0.0	..	..	..	..	..	27 Machines, n.c.a.
28 Office, account. & computing machinery	0.0	0.0	..	..	..	..	..	28 Mach. de bur., compta. & trait. inform.
29 Electrical machinery	0.0	0.0	..	..	..	..	..	29 Machines électriques
30 Electro. equip. (radio, TV & commun.)	0.0	0.0	..	..	..	..	..	30 Equip. électro. (radio, tv & communic.)
31 *Electro. comp. (inc. semi-conductors)*	0.0	0.0	..	..	..	..	..	31 *Compos. électro. (semi-cond. incl.)*
32 *TV, radio & communications equip.*	0.0	0.0	..	..	..	..	..	32 *Appareils radio, tv et de communi.*
33 Instruments	0.0	0.0	..	..	..	..	..	33 Instruments
34 Motor vehicles	0.0	0.0	..	..	..	..	..	34 Véhicules automobiles
35 Other transport equipment	0.0	0.0	..	..	..	..	..	35 Autres matériels de transport
36 *Ships*	0.0	0.0	..	..	..	..	..	36 *Navires*
37 *Aerospace*	0.0	0.0	..	..	..	..	..	37 *Aérospatiale*
38 *Other transport equipment n.e.c.*	0.0	0.0	..	..	..	..	..	38 *Autres matériels de transport n.c.a.*
39 **Furniture, other manufacturing n.e.c.**	0.0	0.0	..	..	..	..	..	39 **Meubles, autres activ. de fabricat. n.c.a.**
40 Furniture	0.0	0.0	..	..	..	..	..	40 Meubles
41 Other manufacturing n.e.c.	0.0	0.0	..	..	..	..	..	41 Autres activités de fabrication n.c.a.
42 **Recycling**	0.0	0.0	..	..	..	..	..	42 **Récupération**
43 **ELECTRICITY, GAS & WATER SUPPLY**	0.0	0.0	..	..	..	..	..	43 **DISTRIBUTION D'ELECTRICITE, GAZ & EAU**
44 **CONSTRUCTION**	0.0	0.0	..	..	..	..	..	44 **CONSTRUCTION**
45 **SERVICES SECTOR**	3.8	4.3	..	..	..	..	..	45 **SECTEUR DES SERVICES**
46 **Wholesale, retail trade & repair of motor vehicles, etc.**	0.0	0.0	..	..	..	..	..	46 **Commerce (gros & détail), réparation de véhicules, etc.**
47 **Hotels & restaurants**	0.0	0.0	..	..	..	..	..	47 **Hôtels et restaurants**
48 **Transport & storage**	0.0	0.0	..	..	..	..	..	48 **Transports, entreposage**
49 **Communications**	0.0	0.0	..	..	..	..	..	49 **Communications**
50 Post	0.0	0.0	..	..	..	..	..	50 Postes
51 Telecommunications	0.0	0.0	..	..	..	..	..	51 Télécommunications
52 **Financial intermediation (inc. insur.)**	0.0	0.0	..	..	..	..	..	52 **Intermédiation financ. (assurances inc.)**
53 **Real estate, renting & busin. activity**	3.8	4.3	..	..	..	..	..	53 **Immobil., locat. & serv. aux entreprises**
54 Computer & related activities	0.0	0.0	..	..	..	..	..	54 Informatique et activités connexes
55 *Software consultancy*	0.0	0.0	..	..	..	..	..	55 *Production de logiciels*
56 *Other computer services n.e.c.*	0.0	0.0	..	..	..	..	..	56 *Autres activités informatiques*
57 Research & development	3.8	4.3	..	..	..	..	..	57 Recherche & développement expérimental
58 Other business activities n.e.c.	0.0	0.0	..	..	..	..	..	58 Autres activ. & serv. aux entreprises
59 **Comm., soc. & pers. serv. activ., etc.**	0.0	0.0	..	..	..	..	..	59 **Adm. pub., serv. soc. & collec., etc**
TOTAL BUSINESS ENTERPRISE SECTOR	**3.8**	**4.3**	**9.1**	**11.6**	**..**	**0.3** p	**..**	**TOTAL SECTEUR DES ENTREPRISES**

Source: OECD, R&D database (DSTI/EAS Division), January 2002.

Source : OCDE, base de données R-D (DSTI/Division AES), janvier 2002.

Business enterprise intramural expenditure on R&D (BERD) - Dépenses intra-muros de R-D des entreprises (DIRDE)

by industry and source of funds - par industrie et par source de financement

Million MXN *Millions MXN*

	1994	1995	1996	1997	1998	1999	2000	
	10.4 Funds from abroad - Fonds de l'étranger							
1 AGRICULTURE, HUNTING & FORESTRY	0.0	0.0	..	..	..	..	..	**1 AGRICULTURE, CHASSE & SYLVICULTURE**
2 MINING	0.0	0.0	..	..	..	..	..	**2 ACTIVITES EXTRACTIVES**
3 MANUFACTURING	0.4	4.1	..	..	..	..	..	**3 ACTIVITES DE FABRICATION**
4 Food, beverages & tobacco	0.0	0.0	..	..	..	..	..	**4 Produits alimentaires, boissons & tabac**
5 Food products & beverages	0.0	0.0	..	..	..	..	..	5 Produits alimentaires & boissons
6 Tobacco products	0.0	0.0	..	..	..	..	..	6 Produits à base de tabac
7 Textiles, wear app., fur & leather	0.0	0.0	..	..	..	..	..	**7 Textiles, habill., fourrures & cuirs**
8 Textiles	0.0	0.0	..	..	..	..	..	8 Textiles
9 Wearing apparel & fur	0.0	0.0	..	..	..	..	..	9 Articles d'habillement & fourrures
10 Leather products & footwear	0.0	0.0	..	..	..	..	..	10 Articles en cuir & chaussures
11 Wood, paper, printing, publishing	0.0	0.0	..	..	..	..	..	**11 Bois, papier, imprimerie, édition**
12 Wood & cork (not furniture)	0.0	0.0	..	..	..	..	..	12 Bois & liège (sauf meubles)
13 Pulp, paper & paper products	0.0	0.0	..	..	..	..	..	13 Carton, papier & articles
14 Publ., print. & repro. of rec. media	0.0	0.0	..	..	..	..	..	14 Edition, impress. & repro. enregis.
15 Coke, petroleum, nuclear fuel, chemicals, rubber plastics	0.0	0.0	..	..	..	..	..	**15 Coke, pétrole, combust. nucl., prod. chimi., caout. & mat. plast.**
16 Coke, ref. petrol. prod. & nuclear fuel	0.0	0.0	..	..	..	..	..	16 Coke, raffin. pétr. & combust. nucl.
17 Chemicals & chemical products	0.0	0.0	..	..	..	..	..	17 Produits chimiques
18 *Chemical prod.(less pharmaceuticals)*	0.0	0.0	..	..	..	..	..	18 *Prod. chimi. (moins prod. pharmaceutiques)*
19 *Pharmaceuticals*	0.0	0.0	..	..	..	..	..	19 *Produits pharmaceutiques*
20 Rubber & plastic products	0.0	0.0	..	..	..	..	..	20 Caoutchouc et matières plastiques
21 Non-metallic mineral products	0.0	0.0	..	..	..	..	..	**21 Produits minéraux non métalliques**
22 Basic metals	0.0	0.0	..	..	..	..	..	**22 Produits métallurgiques de base**
23 Basic metals, ferrous	0.0	0.0	..	..	..	..	..	23 Produits métallur. de base, ferreux
24 Basic metals, non-ferrous	0.0	0.0	..	..	..	..	..	24 Prod. métallur. de base, non ferreux
25 Fabricated metal products	0.0	0.0	..	..	..	..	..	**25 Ouvrages en métaux (sauf mach.& mat.)**
26 Machinery equipment, instruments and transport equipment	0.4	4.1	..	..	..	..	..	**26 Machines & matériel, instruments & mat. de transport**
27 Machinery n.e.c.	0.0	0.0	..	..	..	..	..	27 Machines, n.c.a.
28 Office, account. & computing machinery	0.0	0.0	..	..	..	..	..	28 Mach. de bur., compta. & trait. inform.
29 Electrical machinery	0.0	0.0	..	..	..	..	..	29 Machines électriques
30 Electro. equip. (radio, TV & commun.)	0.4	0.5	..	..	..	..	..	30 Equip. électro. (radio, tv & communic.)
31 *Electro. comp. (inc. semi-conductors)*	0.0	0.0	..	..	..	..	..	31 *Compos. électro. (semi-cond. incl.)*
32 *TV, radio & communications equip.*	0.4	0.5	..	..	..	..	..	32 *Appareils radio, tv et de communi.*
33 Instruments	0.0	0.0	..	..	..	..	..	33 Instruments
34 Motor vehicles	0.0	3.6	..	..	..	..	..	34 Véhicules automobiles
35 Other transport equipment	0.0	0.0	..	..	..	..	..	35 Autres matériels de transport
36 *Ships*	0.0	0.0	..	..	..	..	..	36 *Navires*
37 *Aerospace*	0.0	0.0	..	..	..	..	..	37 *Aérospatiale*
38 *Other transport equipment n.e.c.*	0.0	0.0	..	..	..	..	..	38 *Autres matériels de transport n.c.a.*
39 Furniture, other manufacturing n.e.c.	0.0	0.0	..	..	..	..	..	**39 Meubles, autres activ. de fabricat. n.c.a.**
40 Furniture	0.0	0.0	..	..	..	..	..	40 Meubles
41 Other manufacturing n.e.c.	0.0	0.0	..	..	..	..	..	41 Autres activités de fabrication n.c.a.
42 Recycling	0.0	0.0	..	..	..	..	..	**42 Récupération**
43 ELECTRICITY, GAS & WATER SUPPLY	0.0	0.0	..	..	..	..	..	**43 DISTRIBUTION D'ELECTRICITE, GAZ & EAU**
44 CONSTRUCTION	0.0	0.0	..	..	..	..	..	**44 CONSTRUCTION**
45 SERVICES SECTOR	332.4	239.9	..	..	..	..	..	**45 SECTEUR DES SERVICES**
46 Wholesale, retail trade & repair of motor vehicles, etc.	0.0	0.0	..	..	..	..	..	**46 Commerce (gros & détail), réparation de véhicules, etc.**
47 Hotels & restaurants	0.0	0.0	..	..	..	..	..	**47 Hôtels et restaurants**
48 Transport & storage	0.0	0.0	..	..	..	..	..	**48 Transports, entreposage**
49 Communications	0.0	0.0	..	..	..	..	..	**49 Communications**
50 Post	0.0	0.0	..	..	..	..	..	50 Postes
51 Telecommunications	0.0	0.0	..	..	..	..	..	51 Télécommunications
52 Financial intermediation (inc. insur.)	0.0	0.0	..	..	..	..	..	**52 Intermédiation financ. (assurances inc.)**
53 Real estate, renting & busin. activity	332.4	239.9	..	..	..	..	..	**53 Immobil., locat. & serv. aux entreprises**
54 Computer & related activities	0.0	0.0	..	..	..	..	..	54 Informatique et activités connexes
55 *Software consultancy*	0.0	0.0	..	..	..	..	..	55 *Production de logiciels*
56 *Other computer services n.e.c.*	0.0	0.0	..	..	..	..	..	56 *Autres activités informatiques*
57 Research & development	332.4	239.9	..	..	..	..	..	57 Recherche & développement expérimental
58 Other business activities n.e.c.	0.0	0.0	..	..	..	..	..	58 Autres activ. & serv. aux entreprises
59 Comm., soc. & pers. serv. activ., etc.	0.0	0.0	..	..	..	..	..	**59 Adm. pub., serv. soc. & collec., etc**
TOTAL BUSINESS ENTERPRISE SECTOR	**332.8**	**244.0**	**209.7**	**211.9**	**983.9 p**	**934.6 p**	..	**TOTAL SECTEUR DES ENTREPRISES**

Source: OECD, R&D database (DSTI/EAS Division), January 2002.

Source : OCDE, base de données R-D (DSTI/Division AES), janvier 2002.

MEXICO — Table 11 - Tableau 11 — MEXIQUE

Business enterprise intramural expenditure on R&D (BERD) - Dépenses intra-muros de R-D des entreprises (DIRDE)

by industry and type of costs - par industrie et par type de coûts

Million MXN — *Millions MXN*

	1994	1995	1996	1997	1998	1999	2000	
	11.1 Current expenditure - Dépenses courantes							
1 AGRICULTURE, HUNTING & FORESTRY	0.0	0.0	..	..	..	..	..	**1 AGRICULTURE, CHASSE & SYLVICULTURE**
2 MINING	4.7	7.2	..	..	..	..	..	**2 ACTIVITES EXTRACTIVES**
3 MANUFACTURING	489.5	656.4	..	..	..	..	..	**3 ACTIVITES DE FABRICATION**
4 Food, beverages & tobacco	54.8	64.1	..	..	..	..	..	**4 Produits alimentaires, boissons & tabac**
5 Food products & beverages	53.1	62.0	..	..	..	..	..	5 Produits alimentaires & boissons
6 Tobacco products	1.7	2.1	..	..	..	..	..	6 Produits à base de tabac
7 Textiles, wear app., fur & leather	4.3	4.6	..	..	..	..	..	**7 Textiles, habill., fourrures & cuirs**
8 Textiles	0.2	0.3	..	..	..	..	..	8 Textiles
9 Wearing apparel & fur	1.1	1.2	..	..	..	..	..	9 Articles d'habillement & fourrures
10 Leather products & footwear	3.0	3.2	..	..	..	..	..	10 Articles en cuir & chaussures
11 Wood, paper, printing, publishing	0.7	1.0	..	..	..	..	..	**11 Bois, papier, imprimerie, édition**
12 Wood & cork (not furniture)	0.0	0.0	..	..	..	..	..	12 Bois & liège (sauf meubles)
13 Pulp, paper & paper products	0.1	0.2	..	..	..	..	..	13 Carton, papier & articles
14 Publ., print. & repro. of rec. media	0.6	0.8	..	..	..	..	..	14 Edition, impress. & repro. enregis.
15 Coke, petroleum, nuclear fuel, chemicals, rubber plastics	152.6	205.8	..	..	..	..	..	**15 Coke, pétrole, combust. nucl., prod. chimi., caout. & mat. plast.**
16 Coke, ref. petrol. prod. & nuclear fuel	0.8	1.7	..	..	..	..	..	16 Coke, raffin. pétr. & combust. nucl.
17 Chemicals & chemical products	147.7	201.8	..	..	..	..	..	17 Produits chimiques
18 *Chemical prod.(less pharmaceuticals)*	129.5	162.7	..	..	..	..	..	18 *Prod. chimi. (moins prod. pharmaceutiques)*
19 *Pharmaceuticals*	18.2	39.1	..	..	..	..	..	19 *Produits pharmaceutiques*
20 Rubber & plastic products	4.0	2.3	..	..	..	..	..	20 Caoutchouc et matières plastiques
21 Non-metallic mineral products	13.0	15.0	..	..	..	..	..	**21 Produits minéraux non métalliques**
22 Basic metals	38.4	40.2	..	..	..	..	..	**22 Produits métallurgiques de base**
23 Basic metals, ferrous	16.5	22.5	..	..	..	..	..	23 Produits métallur. de base, ferreux
24 Basic metals, non-ferrous	21.9	17.8	..	..	..	..	..	24 Prod. métallur. de base, non ferreux
25 Fabricated metal products	0.7	0.7	..	..	..	..	..	**25 Ouvrages en métaux (sauf mach.& mat.)**
26 Machinery equipment, instruments and transport equipment	116.8	191.0	..	..	..	..	..	**26 Machines & matériel, instruments & mat. de transport**
27 Machinery n.e.c.	18.4	27.1	..	..	..	..	..	27 Machines, n.c.a.
28 Office, account. & computing machinery	21.3	58.7	..	..	..	..	..	28 Mach. de bur., compta. & trait. inform.
29 Electrical machinery	28.7	38.2	..	..	..	..	..	29 Machines électriques
30 Electro. equip. (radio, TV & commun.)	0.3	0.4	..	..	..	..	..	30 Equip. électro. (radio, tv & communic.)
31 *Electro. comp. (inc. semi-conductors)*	0.0	0.0	..	..	..	..	..	31 *Compos. électro. (semi-cond. incl.)*
32 *TV, radio & communications equip.*	0.3	0.4	..	..	..	..	..	32 *Appareils radio, tv et de communi.*
33 Instruments	0.0	0.0	..	..	..	..	..	33 Instruments
34 Motor vehicles	48.1	66.5	..	..	..	..	..	34 Véhicules automobiles
35 Other transport equipment	0.0	0.0	..	..	..	..	..	35 Autres matériels de transport
36 *Ships*	0.0	0.0	..	..	..	..	..	36 *Navires*
37 *Aerospace*	0.0	0.0	..	..	..	..	..	37 *Aérospatiale*
38 *Other transport equipment n.e.c.*	0.0	0.0	..	..	..	..	..	38 *Autres matériels de transport n.c.a.*
39 Furniture, other manufacturing n.e.c.	108.3	134.1	..	..	..	..	..	**39 Meubles, autres activ. de fabricat. n.c.a.**
40 Furniture	1.6	1.6	..	..	..	..	..	40 Meubles
41 Other manufacturing n.e.c.	106.7	132.5	..	..	..	..	..	41 Autres activités de fabrication n.c.a.
42 Recycling	0.0	0.0	..	..	..	..	..	**42 Récupération**
43 ELECTRICITY, GAS & WATER SUPPLY	0.0	0.0	..	..	..	..	..	**43 DISTRIBUTION D'ELECTRICITE, GAZ & EAU**
44 CONSTRUCTION	0.9	1.1	..	..	..	..	..	**44 CONSTRUCTION**
45 SERVICES SECTOR	428.0	349.0	..	..	..	..	..	**45 SECTEUR DES SERVICES**
46 Wholesale, retail trade & repair of motor vehicles, etc.	0.0	0.0	..	..	..	..	..	**46 Commerce (gros & détail), réparation de véhicules, etc.**
47 Hotels & restaurants	0.0	0.0	..	..	..	..	..	**47 Hôtels et restaurants**
48 Transport & storage	0.8	0.5	..	..	..	..	..	**48 Transports, entreposage**
49 Communications	2.5	3.6	..	..	..	..	..	**49 Communications**
50 Post	0.0	0.2	..	..	..	..	..	50 Postes
51 Telecommunications	2.5	3.4	..	..	..	..	..	51 Télécommunications
52 Financial intermediation (inc. insur.)	0.3	0.4	..	..	..	..	..	**52 Intermédiation financ. (assurances inc.)**
53 Real estate, renting & busin. activity	377.5	283.2	..	..	..	..	..	**53 Immobil., locat. & serv. aux entreprises**
54 Computer & related activities	12.3	9.2	..	..	..	..	..	54 Informatique et activités connexes
55 *Software consultancy*	12.3	9.2	..	..	..	..	..	55 *Production de logiciels*
56 *Other computer services n.e.c.*	0.0	0.0	..	..	..	..	..	56 *Autres activités informatiques*
57 Research & development	365.2	274.0	..	..	..	..	..	57 Recherche & développement expérimental
58 Other business activities n.e.c.	0.0	0.0	..	..	..	..	..	58 Autres activ. & serv. aux entreprises
59 Comm., soc. & pers. serv. activ., etc.	46.8	61.3	..	..	..	..	..	**59 Adm. pub., serv. soc. & collec., etc**
TOTAL BUSINESS ENTERPRISE SECTOR	**923.1**	**1 013.8**	**..**	**..**	**..**	**..**	**..**	**TOTAL SECTEUR DES ENTREPRISES**

Source: OECD, R&D database (DSTI/EAS Division), January 2002.

Source : OCDE, base de données R-D (DSTI/Division AES), janvier 2002.

Business enterprise intramural expenditure on R&D (BERD) - Dépenses intra-muros de R-D des entreprises (DIRDE)

by industry and type of costs - par industrie et par type de coûts

Million MXN *Millions MXN*

11.2 Capital expenditure - Dépenses en capital

	1994	1995	1996	1997	1998	1999	2000	
1 AGRICULTURE, HUNTING & FORESTRY	0.0	0.0	..	..	..	..	..	**1 AGRICULTURE, CHASSE & SYLVICULTURE**
2 MINING	3.3	5.8	..	..	..	..	..	**2 ACTIVITES EXTRACTIVES**
3 MANUFACTURING	106.0	126.0	..	..	..	..	..	**3 ACTIVITES DE FABRICATION**
4 Food, beverages & tobacco	2.1	5.5	..	..	..	..	..	**4 Produits alimentaires, boissons & tabac**
5 Food products & beverages	1.5	4.6	..	..	..	..	..	5 Produits alimentaires & boissons
6 Tobacco products	0.6	0.9	..	..	..	..	..	6 Produits à base de tabac
7 Textiles, wear app., fur & leather	1.1	0.5	..	..	..	..	..	**7 Textiles, habill., fourrures & cuirs**
8 Textiles	0.0	0.0	..	..	..	..	..	8 Textiles
9 Wearing apparel & fur	1.1	0.5	..	..	..	..	..	9 Articles d'habillement & fourrures
10 Leather products & footwear	0.0	0.0	..	..	..	..	..	10 Articles en cuir & chaussures
11 Wood, paper, printing, publishing	0.8	3.2	..	..	..	..	..	**11 Bois, papier, imprimerie, édition**
12 Wood & cork (not furniture)	0.0	0.0	..	..	..	..	..	12 Bois & liège (sauf meubles)
13 Pulp, paper & paper products	0.0	0.2	..	..	..	..	..	13 Carton, papier & articles
14 Publ., print. & repro. of rec. media	0.8	3.0	..	..	..	..	..	14 Edition, impress. & repro. enregis.
15 Coke, petroleum, nuclear fuel, chemicals, rubber plastics	7.7	14.2	..	..	..	..	..	**15 Coke, pétrole, combust. nucl., prod. chimi., caout. & mat. plast.**
16 Coke, ref. petrol. prod. & nuclear fuel	0.4	0.8	..	..	..	..	..	16 Coke, raffin. pétr. & combust. nucl.
17 Chemicals & chemical products	7.1	13.4	..	..	..	..	..	17 Produits chimiques
18 *Chemical prod.(less pharmaceuticals)*	6.5	11.8	..	..	..	..	..	18 *Prod. chimi. (moins prod. pharmaceutiques)*
19 *Pharmaceuticals*	0.6	1.6	..	..	..	..	..	19 *Produits pharmaceutiques*
20 Rubber & plastic products	0.2	0.0	..	..	..	..	..	20 Caoutchouc et matières plastiques
21 Non-metallic mineral products	3.3	2.2	..	..	..	..	..	**21 Produits minéraux non métalliques**
22 Basic metals	11.9	9.8	..	..	..	..	..	**22 Produits métallurgiques de base**
23 Basic metals, ferrous	1.5	2.9	..	..	..	..	..	23 Produits métallur. de base, ferreux
24 Basic metals, non-ferrous	10.4	6.9	..	..	..	..	..	24 Prod. métallur. de base, non ferreux
25 Fabricated metal products	0.7	0.7	..	..	..	..	..	**25 Ouvrages en métaux (sauf mach.& mat.)**
26 Machinery equipment, instruments and transport equipment	53.5	60.1	..	..	..	..	..	**26 Machines & matériel, instruments & mat. de transport**
27 Machinery n.e.c.	2.6	0.5	..	..	..	..	..	27 Machines, n.c.a.
28 Office, account. & computing machinery	1.7	3.6	..	..	..	..	..	28 Mach. de bur., compta. & trait. inform.
29 Electrical machinery	27.6	45.2	..	..	..	..	..	29 Machines électriques
30 Electro. equip. (radio, TV & commun.)	0.2	0.2	..	..	..	..	..	30 Equip. électro. (radio, tv & communic.)
31 *Electro. comp. (inc. semi-conductors)*	0.0	0.0	..	..	..	..	..	31 *Compos. électro. (semi-cond. incl.)*
32 *TV, radio & communications equip.*	0.2	0.2	..	..	..	..	..	32 *Appareils radio, tv et de communi.*
33 Instruments	0.0	0.0	..	..	..	..	..	33 Instruments
34 Motor vehicles	21.5	10.5	..	..	..	..	..	34 Véhicules automobiles
35 Other transport equipment	0.0	0.0	..	..	..	..	..	35 Autres matériels de transport
36 *Ships*	0.0	0.0	..	..	..	..	..	36 *Navires*
37 *Aerospace*	0.0	0.0	..	..	..	..	..	37 *Aérospatiale*
38 *Other transport equipment n.e.c.*	0.0	0.0	..	..	..	..	..	38 *Autres matériels de transport n.c.a.*
39 Furniture, other manufacturing n.e.c.	25.0	29.9	..	..	..	..	..	**39 Meubles, autres activ. de fabricat. n.c.a.**
40 Furniture	0.0	0.0	..	..	..	..	..	40 Meubles
41 Other manufacturing n.e.c.	25.0	29.9	..	..	..	..	..	41 Autres activités de fabrication n.c.a.
42 Recycling	0.0	0.0	..	..	..	..	..	**42 Récupération**
43 ELECTRICITY, GAS & WATER SUPPLY	0.0	0.0	..	..	..	..	..	**43 DISTRIBUTION D'ELECTRICITE, GAZ & EAU**
44 CONSTRUCTION	0.0	0.0	..	..	..	..	..	**44 CONSTRUCTION**
45 SERVICES SECTOR	19.9	34.6	..	..	..	..	..	**45 SECTEUR DES SERVICES**
46 Wholesale, retail trade & repair of motor vehicles, etc.	0.0	0.0	..	..	..	..	..	**46 Commerce (gros & détail), réparation de véhicules, etc.**
47 Hotels & restaurants	0.0	0.0	..	..	..	..	..	**47 Hôtels et restaurants**
48 Transport & storage	2.1	0.1	..	..	..	..	..	**48 Transports, entreposage**
49 Communications	0.0	0.0	..	..	..	..	..	**49 Communications**
50 Post	0.0	0.0	..	..	..	..	..	50 Postes
51 Telecommunications	0.0	0.0	..	..	..	..	..	51 Télécommunications
52 Financial intermediation (inc. insur.)	0.1	0.0	..	..	..	..	..	**52 Intermédiation financ. (assurances inc.)**
53 Real estate, renting & busin. activity	11.2	28.0	..	..	..	..	..	**53 Immobil., locat. & serv. aux entreprises**
54 Computer & related activities	2.1	9.5	..	..	..	..	..	54 Informatique et activités connexes
55 *Software consultancy*	2.1	9.5	..	..	..	..	..	55 *Production de logiciels*
56 *Other computer services n.e.c.*	0.0	0.0	..	..	..	..	..	56 *Autres activités informatiques*
57 Research & development	9.1	18.6	..	..	..	..	..	57 Recherche & développement expérimental
58 Other business activities n.e.c.	0.0	0.0	..	..	..	..	..	58 Autres activ. & serv. aux entreprises
59 Comm., soc. & pers. serv. activ., etc.	6.5	6.5	..	..	..	..	..	**59 Adm. pub., serv. soc. & collec., etc**
TOTAL BUSINESS ENTERPRISE SECTOR	**129.2**	**166.4**	**..**	**..**	**..**	**..**	**..**	**TOTAL SECTEUR DES ENTREPRISES**

Source: OECD, R&D database (DSTI/EAS Division), January 2002.

Source : OCDE, base de données R-D (DSTI/Division AES), janvier 2002.

Total business enterprise R&D personnel (BEMP) - Personnel total de R-D du secteur des entreprises (PRDE)

by industry - par industrie

In full time equivalent — *En équivalence plein-temps*

	1994	1995	1996	1997	1998	1999	2000	
1 AGRICULTURE, HUNTING & FORESTRY	0	0	..	..	..	..	..	**1 AGRICULTURE, CHASSE & SYLVICULTURE**
2 MINING	50	60	..	..	..	..	..	**2 ACTIVITES EXTRACTIVES**
3 MANUFACTURING	2 594	2 956	..	..	..	..	..	**3 ACTIVITES DE FABRICATION**
4 Food, beverages & tobacco	128	149	..	..	..	..	..	**4 Produits alimentaires, boissons & tabac**
5 Food products & beverages	106	128	..	..	..	..	..	5 Produits alimentaires & boissons
6 Tobacco products	22	21	..	..	..	..	..	6 Produits à base de tabac
7 Textiles, wear app., fur & leather	121	243	..	..	..	..	..	**7 Textiles, habill., fourrures & cuirs**
8 Textiles	3	3	..	..	..	..	..	8 Textiles
9 Wearing apparel & fur	90	211	..	..	..	..	..	9 Articles d'habillement & fourrures
10 Leather products & footwear	28	29	..	..	..	..	..	10 Articles en cuir & chaussures
11 Wood, paper, printing, publishing	19	28	..	..	..	..	..	**11 Bois, papier, imprimerie, édition**
12 Wood & cork (not furniture)	0	0	..	..	..	..	..	12 Bois & liège (sauf meubles)
13 Pulp, paper & paper products	1	2	..	..	..	..	..	13 Carton, papier & articles
14 Publ., print. & repro. of rec. media	18	26	..	..	..	..	..	14 Edition, impress. & repro. enregis.
15 Coke, petroleum, nuclear fuel, chemicals, rubber plastics	882	1 012	..	..	..	..	..	**15 Coke, pétrole, combust. nucl., prod. chimi., caout. & mat. plast.**
16 Coke, ref. petrol. prod. & nuclear fuel	6	7	..	..	..	..	..	16 Coke, raffin. pétr. & combust. nucl.
17 Chemicals & chemical products	826	992	..	..	..	..	..	17 Produits chimiques
18 *Chemical prod.(less pharmaceuticals)*	728	646	..	..	..	..	..	18 *Prod. chimi. (moins prod. pharmaceutiques)*
19 *Pharmaceuticals*	98	346	..	..	..	..	..	19 *Produits pharmaceutiques*
20 Rubber & plastic products	50	13	..	..	..	..	..	20 Caoutchouc et matières plastiques
21 Non-metallic mineral products	89	110	..	..	..	..	..	**21 Produits minéraux non métalliques**
22 Basic metals	182	182	..	..	..	..	..	**22 Produits métallurgiques de base**
23 Basic metals, ferrous	81	110	..	..	..	..	..	23 Produits métallur. de base, ferreux
24 Basic metals, non-ferrous	101	72	..	..	..	..	..	24 Prod. métallur. de base, non ferreux
25 Fabricated metal products	15	15	..	..	..	..	..	**25 Ouvrages en métaux (sauf mach.& mat.)**
26 Machinery equipment, instruments and transport equipment	409	414	..	..	..	..	..	**26 Machines & matériel, instruments & mat. de transport**
27 Machinery n.e.c.	161	132	..	..	..	..	..	27 Machines, n.c.a.
28 Office, account. & computing machinery	24	28	..	..	..	..	..	28 Mach. de bur., compta. & trait. inform.
29 Electrical machinery	130	138	..	..	..	..	..	29 Machines électriques
30 Electro. equip. (radio, TV & commun.)	8	8	..	..	..	..	..	30 Equip. électro. (radio, tv & communic.)
31 *Electro. comp. (inc. semi-conductors)*	0	0	..	..	..	..	..	31 *Compos. électro. (semi-cond. incl.)*
32 *TV, radio & communications equip.*	8	8	..	..	..	..	..	32 *Appareils radio, tv et de communi.*
33 Instruments	0	0	..	..	..	..	..	33 Instruments
34 Motor vehicles	86	108	..	..	..	..	..	34 Véhicules automobiles
35 Other transport equipment	0	0	..	..	..	..	..	35 Autres matériels de transport
36 *Ships*	0	0	..	..	..	..	..	36 *Navires*
37 *Aerospace*	0	0	..	..	..	..	..	37 *Aérospatiale*
38 *Other transport equipment n.e.c.*	0	0	..	..	..	..	..	38 *Autres matériels de transport n.c.a.*
39 Furniture, other manufacturing n.e.c.	749	803	..	..	..	..	..	**39 Meubles, autres activ. de fabricat. n.c.a.**
40 Furniture	17	18	..	..	..	..	..	40 Meubles
41 Other manufacturing n.e.c.	732	785	..	..	..	..	..	41 Autres activités de fabrication n.c.a.
42 Recycling	0	0	..	..	..	..	..	**42 Récupération**
43 ELECTRICITY, GAS & WATER SUPPLY	0	0	..	..	..	..	..	**43 DISTRIBUTION D'ELECTRICITE, GAZ & EAU**
44 CONSTRUCTION	15	18	..	..	..	..	..	**44 CONSTRUCTION**
45 SERVICES SECTOR	1 219	1 431	..	..	..	..	..	**45 SECTEUR DES SERVICES**
46 Wholesale, retail trade & repair of motor vehicles, etc.	0	0	..	..	..	..	..	**46 Commerce (gros & détail), réparation de véhicules, etc.**
47 Hotels & restaurants	0	0	..	..	..	..	..	**47 Hôtels et restaurants**
48 Transport & storage	10	5	..	..	..	..	..	**48 Transports, entreposage**
49 Communications	23	40	..	..	..	..	..	**49 Communications**
50 Post	1	18	..	..	..	..	..	50 Postes
51 Telecommunications	22	22	..	..	..	..	..	51 Télécommunications
52 Financial intermediation (inc. insur.)	3	3	..	..	..	..	..	**52 Intermédiation financ. (assurances inc.)**
53 Real estate, renting & busin. activity	760	811	..	..	..	..	..	**53 Immobil., locat. & serv. aux entreprises**
54 Computer & related activities	18	24	..	..	..	..	..	54 Informatique et activités connexes
55 *Software consultancy*	18	24	..	..	..	..	..	55 *Production de logiciels*
56 *Other computer services n.e.c.*	0	0	..	..	..	..	..	56 *Autres activités informatiques*
57 Research & development	742	787	..	..	..	..	..	57 Recherche & développement expérimental
58 Other business activities n.e.c.	0	0	..	..	..	..	..	58 Autres activ. & serv. aux entreprises
59 Comm., soc. & pers. serv. activ., etc.	423	572	..	..	..	..	..	**59 Adm. pub., serv. soc. & collec., etc**
TOTAL BUSINESS ENTERPRISE SECTOR	**3 876**	**4 466**	**..**	**..**	**..**	**..**	**..**	**TOTAL SECTEUR DES ENTREPRISES**

Source: OECD, R&D database (DSTI/EAS Division), January 2002.

Source : OCDE, base de données R-D (DSTI/Division AES), janvier 2002.

RSE in the business enterprise sector - Chercheurs dans le secteur des entreprises

by industry - par industrie

In full time equivalent — *En équivalence plein-temps*

	1994	1995	1996	1997	1998	1999	2000	
1 AGRICULTURE, HUNTING & FORESTRY	0	0	..	..	..	..	..	**1 AGRICULTURE, CHASSE & SYLVICULTURE**
2 MINING	20	25	..	..	..	..	..	**2 ACTIVITES EXTRACTIVES**
3 MANUFACTURING	1 200	1 339	..	..	..	..	..	**3 ACTIVITES DE FABRICATION**
4 Food, beverages & tobacco	75	81	..	..	..	..	..	**4 Produits alimentaires, boissons & tabac**
5 Food products & beverages	65	70	..	..	..	..	..	5 Produits alimentaires & boissons
6 Tobacco products	10	11	..	..	..	..	..	6 Produits à base de tabac
7 Textiles, wear app., fur & leather	8	9	..	..	..	..	..	**7 Textiles, habill., fourrures & cuirs**
8 Textiles	3	3	..	..	..	..	..	8 Textiles
9 Wearing apparel & fur	3	4	..	..	..	..	..	9 Articles d'habillement & fourrures
10 Leather products & footwear	2	2	..	..	..	..	..	10 Articles en cuir & chaussures
11 Wood, paper, printing, publishing	0	4	..	..	..	..	..	**11 Bois, papier, imprimerie, édition**
12 Wood & cork (not furniture)	0	0	..	..	..	..	..	12 Bois & liège (sauf meubles)
13 Pulp, paper & paper products	0	0	..	..	..	..	..	13 Carton, papier & articles
14 Publ., print. & repro. of rec. media	0	4	..	..	..	..	..	14 Edition, impress. & repro. enregis.
15 Coke, petroleum, nuclear fuel, chemicals, rubber plastics	522	587	..	..	..	..	..	**15 Coke, pétrole, combust. nucl., prod. chimi., caout. & mat. plast.**
16 Coke, ref. petrol. prod. & nuclear fuel	4	4	..	..	..	..	..	16 Coke, raffin. pétr. & combust. nucl.
17 Chemicals & chemical products	497	572	..	..	..	..	..	17 Produits chimiques
18 *Chemical prod.(less pharmaceuticals)*	442	426	..	..	..	..	..	18 *Prod. chimi. (moins prod. pharmaceutiques)*
19 *Pharmaceuticals*	55	146	..	..	..	..	..	19 *Produits pharmaceutiques*
20 Rubber & plastic products	21	11	..	..	..	..	..	20 Caoutchouc et matières plastiques
21 Non-metallic mineral products	56	68	..	..	..	..	..	**21 Produits minéraux non métalliques**
22 Basic metals	55	84	..	..	..	..	..	**22 Produits métallurgiques de base**
23 Basic metals, ferrous	38	67	..	..	..	..	..	23 Produits métallur. de base, ferreux
24 Basic metals, non-ferrous	17	17	..	..	..	..	..	24 Prod. métallur. de base, non ferreux
25 Fabricated metal products	7	7	..	..	..	..	..	**25 Ouvrages en métaux (sauf mach.& mat.)**
26 Machinery equipment, instruments and transport equipment	174	185	..	..	..	..	..	**26 Machines & matériel, instruments & mat. de transport**
27 Machinery n.e.c.	74	75	..	..	..	..	..	27 Machines, n.c.a.
28 Office, account. & computing machinery	22	25	..	..	..	..	..	28 Mach. de bur., compta. & trait. inform.
29 Electrical machinery	56	58	..	..	..	..	..	29 Machines électriques
30 Electro. equip. (radio, TV & commun.)	4	4	..	..	..	..	..	30 Equip. électro. (radio, tv & communic.)
31 *Electro. comp. (inc. semi-conductors)*	0	0	..	..	..	..	..	31 *Compos. électro. (semi-cond. incl.)*
32 *TV, radio & communications equip.*	4	4	..	..	..	..	..	32 *Appareils radio, tv et de communi.*
33 Instruments	0	0	..	..	..	..	..	33 Instruments
34 Motor vehicles	18	23	..	..	..	..	..	34 Véhicules automobiles
35 Other transport equipment	0	0	..	..	..	..	..	35 Autres matériels de transport
36 *Ships*	0	0	..	..	..	..	..	36 *Navires*
37 *Aerospace*	0	0	..	..	..	..	..	37 *Aérospatiale*
38 *Other transport equipment n.e.c.*	0	0	..	..	..	..	..	38 *Autres matériels de transport n.c.a.*
39 Furniture, other manufacturing n.e.c.	303	314	..	..	..	..	..	**39 Meubles, autres activ. de fabricat. n.c.a.**
40 Furniture	9	10	..	..	..	..	..	40 Meubles
41 Other manufacturing n.e.c.	294	304	..	..	..	..	..	41 Autres activités de fabrication n.c.a.
42 Recycling	0	0	..	..	..	..	..	**42 Récupération**
43 ELECTRICITY, GAS & WATER SUPPLY	0	0	..	..	..	..	..	**43 DISTRIBUTION D'ELECTRICITE, GAZ & EAU**
44 CONSTRUCTION	15	18	..	..	..	..	..	**44 CONSTRUCTION**
45 SERVICES SECTOR	521	624	..	..	..	..	..	**45 SECTEUR DES SERVICES**
46 Wholesale, retail trade & repair of motor vehicles, etc.	0	0	..	..	..	..	..	**46 Commerce (gros & détail), réparation de véhicules, etc.**
47 Hotels & restaurants	0	0	..	..	..	..	..	**47 Hôtels et restaurants**
48 Transport & storage	3	3	..	..	..	..	..	**48 Transports, entreposage**
49 Communications	19	20	..	..	..	..	..	**49 Communications**
50 Post	0	1	..	..	..	..	..	50 Postes
51 Telecommunications	19	19	..	..	..	..	..	51 Télécommunications
52 Financial intermediation (inc. insur.)	3	3	..	..	..	..	..	**52 Intermédiation financ. (assurances inc.)**
53 Real estate, renting & busin. activity	279	306	..	..	..	..	..	**53 Immobil., locat. & serv. aux entreprises**
54 Computer & related activities	18	24	..	..	..	..	..	54 Informatique et activités connexes
55 *Software consultancy*	18	24	..	..	..	..	..	55 *Production de logiciels*
56 *Other computer services n.e.c.*	0	0	..	..	..	..	..	56 *Autres activités informatiques*
57 Research & development	261	282	..	..	..	..	..	57 Recherche & développement expérimental
58 Other business activities n.e.c.	0	0	..	..	..	..	..	58 Autres activ. & serv. aux entreprises
59 Comm., soc. & pers. serv. activ., etc.	217	292	..	..	..	..	..	**59 Adm. pub., serv. soc. & collec., etc**
TOTAL BUSINESS ENTERPRISE SECTOR	**1 754**	**2 006**	**..**	**..**	**..**	**..**	**..**	**TOTAL SECTEUR DES ENTREPRISES**

Source: OECD, R&D database (DSTI/EAS Division), January 2002.

Source : OCDE, base de données R-D (DSTI/Division AES), janvier 2002.

Higher education intramural expenditure on R&D - Dépenses intra-muros de R-D de l'enseignement supérieur
(HERD) by field of science - (DIRDES) par discipline scientifique

	1994	1995	1996	1997	1998	1999	2000	
	14.1 Million MXN - Millions MXN							
Natural sciences	..	950.3	..	..	..	..	..	Sciences exactes et naturelles
Engineering	..	523.7	..	..	..	..	..	Sciences de l'ingénieur
Medical sciences	..	269.4	..	..	..	..	..	Sciences médicales
Agricultural sciences	..	163.6	..	..	..	..	..	Sciences agricoles
Sub-total NSE	1 412.8	1 907.0	..	..	..	..	..	Sous-total SENI
Social sciences	..	520.6	..	..	..	..	..	Sciences sociales
Humanities	..	178.2	..	..	..	..	..	Sciences humaines
Sub-total SSH	533.6	698.8	..	..	..	..	..	Sous-total SSH
Total NSE+SSH	1 946.4	2 605.9	2 966.3	4 365.2	6 513.8 p	7 145.6 p	..	*Total SENI+SSH*
	14.2 Million constant USD (1995 prices and PPPs) - Millions de USD constants (prix et PPA de 1995)							
Natural sciences	..	321.4	..	..	..	..	..	Sciences exactes et naturelles
Engineering	..	177.1	..	..	..	..	..	Sciences de l'ingénieur
Medical sciences	..	91.1	..	..	..	..	..	Sciences médicales
Agricultural sciences	..	55.3	..	..	..	..	..	Sciences agricoles
Sub-total NSE	658.6	644.9	..	..	..	..	..	Sous-total SENI
Social sciences	..	176.0	..	..	..	..	..	Sciences sociales
Humanities	..	60.3	..	..	..	..	..	Sciences humaines
Sub-total SSH	248.7	236.3	..	..	..	..	..	Sous-total SSH
Total NSE+SSH	907.3	881.2	767.2	959.1	1 240.5 p	1 184.8 p	..	*Total SENI+SSH*

Table 15 - Tableau 15

Higher education intramural expenditure on R&D - Dépenses intra-muros de R-D de l'enseignement supérieur
(HERD) by field of science and type of costs - (DIRDES) par discipline scientifique et par type de coûts

Million MXN *Millions MXN*

	1994	1995	1996	1997	1998	1999	2000	
Current expenditure								**Dépenses courantes**
Natural sciences	..	..	..	..	..	..	..	Sciences exactes et naturelles
Engineering	..	..	..	..	..	..	..	Sciences de l'ingénieur
Medical sciences	..	..	..	..	..	..	..	Sciences médicales
Agricultural sciences	..	..	..	..	..	..	..	Sciences agricoles
Sub-total NSE	..	..	..	..	..	..	..	Sous-total SENI
Social sciences	..	..	..	..	..	..	..	Sciences sociales
Humanities	..	..	..	..	..	..	..	Sciences humaines
Sub-total SSH	..	..	..	..	..	..	..	Sous-total SSH
Total NSE+SSH	1 630.8	2 148.2	..	..	..	..	..	*Total SENI+SSH*
Capital expenditure								**Dépenses en capital**
Natural sciences	..	..	..	..	..	..	..	Sciences exactes et naturelles
Engineering	..	..	..	..	..	..	..	Sciences de l'ingénieur
Medical sciences	..	..	..	..	..	..	..	Sciences médicales
Agricultural sciences	..	..	..	..	..	..	..	Sciences agricoles
Sub-total NSE	..	..	..	..	..	..	..	Sous-total SENI
Social sciences	..	..	..	..	..	..	..	Sciences sociales
Humanities	..	..	..	..	..	..	..	Sciences humaines
Sub-total SSH	..	..	..	..	..	..	..	Sous-total SSH
Total NSE+SSH	315.6	457.7	..	..	..	..	..	*Total SENI+SSH*

Source: OECD, R&D database (DSTI/EAS Division), January 2002.

Source : OCDE, base de données R-D (DSTI/Division AES), janvier 2002.

MEXICO — Table 16 - Tableau 16 — MEXIQUE

Higher education intramural expenditure on R&D - Dépenses intra-muros de R-D de l'enseignement supérieur

(HERD) by field of science and source of funds - (DIRDES) par discipline scientifique et par source de financement

Million MXN — *Millions MXN*

	1994	1995	1996	1997	1998	1999	2000	
Direct government								**État direct**
Natural sciences	..	228.3	..	..	..	..	..	Sciences exactes et naturelles
Engineering	..	161.9	..	..	..	..	..	Sciences de l'ingénieur
Medical sciences	..	46.7	..	..	..	..	..	Sciences médicales
Agricultural sciences	..	89.1	..	..	..	..	..	Sciences agricoles
Sub-total NSE	..	526.0	..	..	..	..	..	Sous-total SENI
Social sciences	..	163.4	..	..	..	..	..	Sciences sociales
Humanities	..	56.3	..	..	..	..	..	Sciences humaines
Sub-total SSH	..	219.7	..	..	..	..	..	Sous-total SSH
Total NSE+SSH	602.7	745.7	573.1	800.4	..	..	..	*Total SENI+SSH*
General university funds								**Fonds généraux des universités**
Natural sciences	..	518.6	..	..	..	..	..	Sciences exactes et naturelles
Engineering	..	250.5	..	..	..	..	..	Sciences de l'ingénieur
Medical sciences	..	172.2	..	..	..	..	..	Sciences médicales
Agricultural sciences	..	29.8	..	..	..	..	..	Sciences agricoles
Sub-total NSE	..	971.1	..	..	..	..	..	Sous-total SENI
Social sciences	..	226.2	..	..	..	..	..	Sciences sociales
Humanities	..	100.2	..	..	..	..	..	Sciences humaines
Sub-total SSH	..	326.4	..	..	..	..	..	Sous-total SSH
Total NSE+SSH	956.6	1 297.5	1 628.4	2 479.7	..	..	..	*Total SENI+SSH*
Sub-total government								**Sous-total État**
Natural sciences	..	746.9	..	..	..	..	..	Sciences exactes et naturelles
Engineering	..	412.3	..	..	..	..	..	Sciences de l'ingénieur
Medical sciences	..	218.9	..	..	..	..	..	Sciences médicales
Agricultural sciences	..	118.9	..	..	..	..	..	Sciences agricoles
Sub-total NSE	..	1 497.1	..	..	..	..	..	Sous-total SENI
Social sciences	..	389.6	..	..	..	..	..	Sciences sociales
Humanities	..	156.5	..	..	..	..	..	Sciences humaines
Sub-total SSH	..	546.1	..	..	..	..	..	Sous-total SSH
Total NSE+SSH	1 559.3	2 043.2	2 201.5	3 280.1	..	..	..	*Total SENI+SSH*
Higher education								**Enseignement supérieur**
Natural sciences	..	165.5	..	..	..	..	..	Sciences exactes et naturelles
Engineering	..	84.2	..	..	..	..	..	Sciences de l'ingénieur
Medical sciences	..	42.3	..	..	..	..	..	Sciences médicales
Agricultural sciences	..	36.5	..	..	..	..	..	Sciences agricoles
Sub-total NSE	..	328.4	..	..	..	..	..	Sous-total SENI
Social sciences	..	117.0	..	..	..	..	..	Sciences sociales
Humanities	..	20.5	..	..	..	..	..	Sciences humaines
Sub-total SSH	..	137.5	..	..	..	..	..	Sous-total SSH
Total NSE+SSH	317.5	465.9	633.3	936.4	..	..	..	*Total SENI+SSH*
Private non-profit								**Institutions sans but lucratif**
Natural sciences	..	4.9	..	..	..	..	..	Sciences exactes et naturelles
Engineering	..	1.9	..	..	..	..	..	Sciences de l'ingénieur
Medical sciences	..	0.9	..	..	..	..	..	Sciences médicales
Agricultural sciences	..	0.5	..	..	..	..	..	Sciences agricoles
Sub-total NSE	..	8.2	..	..	..	..	..	Sous-total SENI
Social sciences	..	3.8	..	..	..	..	..	Sciences sociales
Humanities	..	0.3	..	..	..	..	..	Sciences humaines
Sub-total SSH	..	4.1	..	..	..	..	..	Sous-total SSH
Total NSE+SSH	11.9	12.3	10.2	11.7	..	..	..	*Total SENI+SSH*
Business entreprise								**Entreprises**
Natural sciences	..	11.6	..	..	..	..	..	Sciences exactes et naturelles
Engineering	..	9.6	..	..	..	..	..	Sciences de l'ingénieur
Medical sciences	..	3.0	..	..	..	..	..	Sciences médicales
Agricultural sciences	..	3.8	..	..	..	..	..	Sciences agricoles
Sub-total NSE	..	28.1	..	..	..	..	..	Sous-total SENI
Social sciences	..	6.8	..	..	..	..	..	Sciences sociales
Humanities	..	0.7	..	..	..	..	..	Sciences humaines
Sub-total SSH	..	7.6	..	..	..	..	..	Sous-total SSH
Total NSE+SSH	34.0	35.6	99.1	116.2	..	..	..	*Total SENI+SSH*
Funds from abroad								**Fonds de l'étranger**
Natural sciences	..	21.5	..	..	..	..	..	Sciences exactes et naturelles
Engineering	..	15.7	..	..	..	..	..	Sciences de l'ingénieur
Medical sciences	..	4.2	..	..	..	..	..	Sciences médicales
Agricultural sciences	..	3.9	..	..	..	..	..	Sciences agricoles
Sub-total NSE	..	45.4	..	..	..	..	..	Sous-total SENI
Social sciences	..	3.4	..	..	..	..	..	Sciences sociales
Humanities	..	0.2	..	..	..	..	..	Sciences humaines
Sub-total SSH	..	3.6	..	..	..	..	..	Sous-total SSH
Total NSE+SSH	23.7	48.9	22.1	20.7	..	..	..	*Total SENI+SSH*

Source: OECD, R&D database (DSTI/EAS Division), January 2002.

Source : OCDE, base de données R-D (DSTI/Division AES), janvier 2002.

Table 17 - Tableau 17

Higher education R&D personnel (HEMP) - Personnel de R-D du secteur de l'enseignement supérieur

by field of science and occupation - par discipline scientifique et par profession

In full-time equivalent — *En équivalence plein-temps*

	1994	1995	1996	1997	1998	1999	2000	
Total R&D personnel								**Personnel total de R-D**
Natural sciences	..	..	..	..	..	..	..	Sciences exactes et naturelles
Engineering	..	..	..	..	..	..	..	Sciences de l'ingénieur
Medical sciences	..	..	..	..	..	..	..	Sciences médicales
Agricultural sciences	..	..	..	..	..	..	..	Sciences agricoles
Sub-total NSE	8 905	10 587	..	..	..	..	..	Sous-total SENI
Social sciences	..	..	..	..	..	..	..	Sciences sociales
Humanities	..	..	..	..	..	..	..	Sciences humaines
Sub-total SSH	3 798	4 302	..	..	..	..	..	Sous-total SSH
Total NSE+SSH	12 703	14 889	..	..	..	..	..	*Total SENI+SSH*
Researchers								**Chercheurs**
Natural sciences	..	..	..	..	..	..	..	Sciences exactes et naturelles
Engineering	..	..	..	..	..	..	..	Sciences de l'ingénieur
Medical sciences	..	..	..	..	..	..	..	Sciences médicales
Agricultural sciences	..	..	..	..	..	..	..	Sciences agricoles
Sub-total NSE	..	..	..	..	..	..	..	Sous-total SENI
Social sciences	..	..	..	..	..	..	..	Sciences sociales
Humanities	..	..	..	..	..	..	..	Sciences humaines
Sub-total SSH	..	..	..	..	..	..	..	Sous-total SSH
Total NSE+SSH	9 735	11 233	..	..	..	..	..	*Total SENI+SSH*
Technicians								**Technicians**
Natural sciences	..	..	..	..	..	..	..	Sciences exactes et naturelles
Engineering	..	..	..	..	..	..	..	Sciences de l'ingénieur
Medical sciences	..	..	..	..	..	..	..	Sciences médicales
Agricultural sciences	..	..	..	..	..	..	..	Sciences agricoles
Sub-total NSE	..	..	..	..	..	..	..	Sous-total SENI
Social sciences	..	..	..	..	..	..	..	Sciences sociales
Humanities	..	..	..	..	..	..	..	Sciences humaines
Sub-total SSH	..	..	..	..	..	..	..	Sous-total SSH
Total NSE+SSH	1 639	1 902	..	..	..	..	..	*Total SENI+SSH*
Other								**Autres**
Natural sciences	..	..	..	..	..	..	..	Sciences exactes et naturelles
Engineering	..	..	..	..	..	..	..	Sciences de l'ingénieur
Medical sciences	..	..	..	..	..	..	..	Sciences médicales
Agricultural sciences	..	..	..	..	..	..	..	Sciences agricoles
Sub-total NSE	..	..	..	..	..	..	..	Sous-total SENI
Social sciences	..	..	..	..	..	..	..	Sciences sociales
Humanities	..	..	..	..	..	..	..	Sciences humaines
Sub-total SSH	..	..	..	..	..	..	..	Sous-total SSH
Total NSE+SSH	1 329	1 754	..	..	..	..	..	*Total SENI+SSH*

Table 18 - Tableau 18

Government budget appropriations or outlays for R&D (GBAORD) - Crédits budgétaires publics de R-D (CBPRD)

by socio-economic objective - par objectif socio-économique

Million MXN — *Millions MXN*

	1995	1996	1997	1998	1999	2000	2001	
Agriculture, forestry and fishing	96.4 h	247.5 h	260.6	110.9 p	775.0 p	..	..	Agriculture, sylviculture et pêche
Industrial development	165.4 h	191.8 h	296.5	274.9 p	1 013.8 p	..	..	Développement industriel
Energy	410.4 h	739.2 h	2 928.3	2 160.1 p	1 517.7 p	..	..	Energie
Transport & telecommunications	17.1 h	26.5 h	31.1	46.9 p	66.1 p	..	..	*Transports & télécommunications*
Urban & rural planning	..	..	..	..	..	..	..	*Aménagement urbain et rural*
ST Infrastructure	17.1 h	26.5 h	31.1	46.9 p	66.1 p	..	..	ST Infrastructure
Prevention of pollution	..	..	..	..	..	..	..	*Prévention de la pollution*
Identificat. & treatment of pollution	..	..	..	..	..	..	..	*Détect. pollut. & mesures antipollut.*
ST Environment	21.8 h	39.0 h	70.9	87.9 p	117.9 p	..	..	ST Environnement
Health	195.0 h	260.8 h	332.8	498.7 p	735.4 p	..	..	Santé
Social development & services	157.9 h	136.3 h	329.6	376.4 p	706.3 p	..	..	Développement et services sociaux
Earth and atmosphere	162.7 h	226.3 h	306.2	411.6 p	720.9 p	..	..	Milieu terrestre et atmosphère
Advancement of research	963.3 h	967.2 h	1 216.9	1 344.9 p	1 425.5 p	..	..	*Promotion de la recherche*
General university funds	1 510.7 h	2 394.6 h	2 724.4	3 512.4 p	4 163.5 p	..	..	*Fonds généraux des universités*
ST Advancement of knowledge	2 474.0 h	3 361.8 h	3 941.3	4 857.3 p	5 588.9 p	..	..	ST promotion gén. des connaiss.
Civil space	0.0	0.0	0.0	0.0 p	0.0 p	..	..	Espace civil
Defence	0.0	0.0	0.0	0.0 p	0.0 p	..	..	Défense
Not elsewhere classified	..	..	..	..	..	..	..	Non classé
Total	**3 700.6 h**	**5 229.2 h**	**8 497.0**	**8 824.5 p**	**11 241.9 p**	**..**	**..**	**Total**

Source: OECD, R&D database (DSTI/EAS Division), January 2002. — *Source* : OCDE, base de données R-D (DSTI/Division AES), janvier 2002.

MEXICO

Table 19 - Tableau 19

MEXIQUE

Technology balance of payments - Balance des paiements technologiques

by type of transfer - selon la nature des transferts

Million MXN *Millions MXN*

	1983	1984	1985	1986	1987	1988	1989	1990	1991	
Receipts										**Recettes**
Related firms	..	..	27	..	..	..	..	..	..	Firmes affiliées
Unrelated firms	..	..	..	..	..	..	..	..	..	Firmes non affiliées
Other	..	..	..	..	..	..	..	..	..	Autre
Total	..	..	..	..	..	..	..	..	237	*Total*
Payments										**Paiements**
Related firms	..	..	149	..	..	..	..	..	..	Firmes affiliées
Unrelated firms	..	..	..	..	..	..	..	..	..	Firmes non affiliées
Other	..	..	..	..	..	..	..	..	..	Autre
Total	..	..	..	..	..	..	..	..	1 267	*Total*

	1992	1993	1994	1995	1996	1997	1998	1999	2000	
Receipts										**Recettes**
Related firms	..	133	..	..	..	..	..	..	..	Firmes affiliées
Unrelated firms	..	..	..	..	..	..	..	..	..	Firmes non affiliées
Other	..	..	..	..	..	..	..	..	..	Autre
Total	266	302	328	759	926	1 028	1 265	610	..	*Total*
Payments										**Paiements**
Related firms	..	916	..	..	..	..	..	..	..	Firmes affiliées
Unrelated firms	..	..	..	..	..	..	..	..	..	Firmes non affiliées
Other	..	..	..	..	..	..	..	..	..	Autre
Total	1 460	1 564	2 000	3 128	2 736	3 969	4 146	4 339	..	*Total*

Source: OECD, TBPdatabase (DSTI/EAS Division), Januay 2002.

Source : OCDE, base de données TBP (DSTI/Division AES), janvier 2002.

Gross domestic expenditure on R&D (GERD) - Dépenses intérieures brutes de R-D (DIRD)

by sector of performance and source of funds - par secteur d'exécution et par source de financement

	1994	1995	1996	1997	1998	1999	2000	
	1.1 Million EUR (1999 NLG euro, EUR from 1999) - Millions EUR (1999 NLG euro, EUR à partir de 1999)							
Business enterprise								**Entreprises**
financed by								*financé par*
Business enterprise	2 307.0 [a]	2 505.8	2 824.3 [a]	2 812.1	3 016.7	3 396.5	..	Entreprises
Direct government	246.4 [a]	206.0	187.9	199.2	162.0	217.8	..	État direct
Higher education	0.0	0.0	0.0	0.0	0.0	0.0	..	Enseignement supérieur
Private non-profit	6.8 [a]	4.5	2.7	3.2	3.2	3.2	..	Institutions sans but lucratif
Funds from abroad	356.7 [a]	414.3	327.2	700.2	538.6	645.7	..	Fonds de l'étranger
Total	2 916.9 [a]	3 130.6	3 342.1 [a]	3 714.6	3 720.5	4 263.3	..	*Total*
Government								**État**
financed by								*financé par*
Business enterprise	161.1 [a]	181.1	177.4	206.0	224.2	254.6	..	Entreprises
Direct government	789.1 [a]	800.0	833.1	831.3	855.8	814.5	..	État direct
Higher education	4.1 [a]	4.5	2.7	5.4	5.0	6.4	..	Enseignement supérieur
Private non-profit	21.3 [a]	20.0	21.8	22.2	20.9	50.4	..	Institutions sans but lucratif
Funds from abroad	80.3 [a]	81.7	90.3	101.2	110.7	124.3	..	Fonds de l'étranger
Total	1 055.9 [a]	1 087.3	1 125.4	1 166.2	1 216.6	1 250.2	..	*Total*
Higher education								**Enseignement supérieur**
financed by								*financé par*
Business enterprise	65.3	69.9	68.5	80.8	92.6	101.2	..	Entreprises
Direct government	126.6	109.8	121.2	118.0	120.7	121.2	..	État direct
General university funds	1 287.4	1 372.7	1 449.8	1 474.8	1 423.1	1 509.3	..	Fonds généraux des universités
Sub-total government	1 414.0	1 482.5	1 571.0	1 592.8	1 543.8	1 630.4	..	Sous-Total État
Higher education	5.9	5.0	0.0	5.9	5.0	5.0	..	Enseignement supérieur
Private non-profit	93.5	113.0	116.2	120.3	156.6	176.1	..	Institutions sans but lucratif
Funds from abroad	55.8	59.9	60.4	61.7	66.7	70.3	..	Fonds de l'étranger
Total	1 634.5	1 730.3	1 816.0	1 861.4	1 864.6	1 983.0	..	*Total*
Private non-profit								**Institutions sans but lucratif**
financed by								*financé par*
Business enterprise	5.0 [a]	4.1	4.1	3.6	3.6	3.6	..	Entreprises
Direct government	34.9 [a]	44.0	39.9	38.6	39.5	40.8	..	État direct
Higher education	0.9 [a]	0.9	0.9	0.9	0.9	0.9	..	Enseignement supérieur
Private non-profit	13.6 [a]	5.4	10.9	17.2	17.7	16.3	..	Institutions sans but lucratif
Funds from abroad	4.1 [a]	4.1	5.0	5.4	5.4	5.4	..	Fonds de l'étranger
Total	58.5 [a]	58.5	60.8	65.8	67.2	67.2	..	*Total*
GERD								**DIRD**
financed by								*financé par*
Business enterprise	2 538.4 [a]	2 760.8	3 074.4 [a]	3 102.5	3 337.1	3 755.9	..	Entreprises
Direct government	1 197.1	1 159.9	1 182.1	1 187.1	1 178.0	1 194.4	..	État direct
General university funds	1 287.4	1 372.7	1 449.8	1 474.8	1 423.1	1 509.3	..	Fonds généraux des universités
Sub-total government	2 484.4	2 532.5	2 631.9	2 661.9	2 601.1	2 703.6	..	Sous-Total État
Higher education	10.9	10.4	3.6	12.3	10.9	12.3	..	Enseignement supérieur
Private non-profit	135.2	142.9	151.6	162.9	198.3	245.9	..	Institutions sans but lucratif
Funds from abroad	496.9	560.0	482.8	868.5	721.5	845.8	..	Fonds de l'étranger
Total	5 665.9 [a]	6 006.7	6 344.3 [a]	6 808.1	6 868.9	7 563.6	..	*Total*
	1.2 Million constant USD (1995 prices and PPPs) - Millions de USD constants (prix et PPA de 1995)							
Business enterprise								**Entreprises**
financed by								*financé par*
Business enterprise	2 553.6 [a]	2 723.6	3 034.4 [a]	2 964.1	3 133.1	3 492.0	..	Entreprises
Direct government	272.7 [a]	223.9	201.8	210.0	168.3	223.9	..	État direct
Higher education	0.0	0.0	0.0	0.0	0.0	0.0	..	Enseignement supérieur
Private non-profit	7.5 [a]	4.9	2.9	3.3	3.3	3.3	..	Institutions sans but lucratif
Funds from abroad	394.8 [a]	450.3	351.5	738.0	559.4	663.9	..	Fonds de l'étranger
Total	3 228.7 [a]	3 402.8	3 590.6 [a]	3 915.4	3 864.1	4 383.1	..	*Total*
Government								**État**
financed by								*financé par*
Business enterprise	178.3 [a]	196.8	190.6	217.2	232.8	261.7	..	Entreprises
Direct government	873.5 [a]	869.6	895.1	876.3	888.9	837.4	..	État direct
Higher education	4.5 [a]	4.9	2.9	5.7	5.2	6.5	..	Enseignement supérieur
Private non-profit	23.6 [a]	21.7	23.4	23.4	21.7	51.8	..	Institutions sans but lucratif
Funds from abroad	88.9 [a]	88.8	97.0	106.7	115.0	127.8	..	Fonds de l'étranger
Total	1 168.8 [a]	1 181.8	1 209.1	1 229.2	1 263.5	1 285.3	..	*Total*
Higher education								**Enseignement supérieur**
financed by								*financé par*
Business enterprise	72.3	76.0	73.6	85.1	96.1	104.0	..	Entreprises
Direct government	140.1	119.4	130.2	124.4	125.4	124.6	..	État direct
General university funds	1 425.0	1 492.0	1 557.7	1 554.5	1 478.0	1 551.7	..	Fonds généraux des universités
Sub-total government	1 565.1	1 611.4	1 687.8	1 678.8	1 603.3	1 676.2	..	Sous-Total État
Higher education	6.5	5.4	0.0	6.2	5.2	5.1	..	Enseignement supérieur
Private non-profit	103.5	122.8	124.8	126.8	162.6	181.0	..	Institutions sans but lucratif
Funds from abroad	61.8	65.1	64.8	65.0	69.3	72.3	..	Fonds de l'étranger
Total	1 809.2	1 880.7	1 951.1	1 962.0	1 936.5	2 038.7	..	*Total*
Private non-profit								**Institutions sans but lucratif**
financed by								*financé par*
Business enterprise	5.5 [a]	4.4	4.4	3.8	3.8	3.7	..	Entreprises
Direct government	38.7 [a]	47.8	42.9	40.7	41.0	42.0	..	État direct
Higher education	1.0 [a]	1.0	1.0	1.0	0.9	0.9	..	Enseignement supérieur
Private non-profit	15.1 [a]	5.9	11.7	18.2	18.4	16.8	..	Institutions sans but lucratif
Funds from abroad	4.5 [a]	4.4	5.4	5.7	5.7	5.6	..	Fonds de l'étranger
Total	64.8 [a]	63.6	65.3	69.4	69.8	69.0	..	*Total*
GERD								**DIRD**
financed by								*financé par*
Business enterprise	2 809.8 [a]	3 000.8	3 303.0 [a]	3 270.2	3 465.9	3 861.5	..	Entreprises
Direct government	1 325.0	1 260.7	1 270.0	1 251.2	1 223.5	1 227.9	..	État direct
General university funds	1 425.0	1 492.0	1 557.7	1 554.5	1 478.0	1 551.7	..	Fonds généraux des universités
Sub-total government	2 750.0	2 752.7	2 827.7	2 805.7	2 701.4	2 779.6	..	Sous-Total État
Higher education	12.1	11.3	3.9	12.9	11.3	12.6	..	Enseignement supérieur
Private non-profit	149.7	155.4	162.8	171.7	206.0	252.9	..	Institutions sans but lucratif
Funds from abroad	550.0	608.6	518.7	915.5	749.4	869.6	..	Fonds de l'étranger
Total	6 271.5 [a]	6 528.9	6 816.1 [a]	7 176.0	7 133.9	7 776.1	..	*Total*

Source: OECD, R&D database (DSTI/EAS Division), January 2002.

Source : OCDE, base de données R-D (DSTI/Division AES), janvier 2002.

Table 5 - Tableau 5

Total R&D personnel - Personnel total de R-D

by sector of employment and occupation - par secteur d'emploi et par profession

In full-time equivalent — *En équivalence plein-temps*

	1994	1995	1996	1997	1998	1999	2000	
Business enterprise								**Entreprises**
RSE	13 140 [a]	13 245	13 797 [a]	17 302	18 164	19 359	..	Chercheurs
Technicians	15 640 [a]	16 673	19 044 [a]	17 760	17 517	17 226	..	Techniciens
Other occupations	7 260 [a]	7 538	6 620 [a]	7 347	8 191	8 595	..	Autres professions
Total	36 040 [a]	37 456	39 461 [a]	42 409	43 872	45 181	..	*Total*
Government								**État**
RSE	7 650 [a]	7 830	7 840	7 817	7 996	8 048	..	Chercheurs
Technicians	4 950 [a]	5 040	4 900	4 895	5 113	4 981	..	Techniciens
Other occupations	3 370 [a]	3 150	3 290	3 448	3 343	3 537	..	Autres professions
Total	15 970 [a]	16 020	16 030	16 160	16 451	16 565	..	*Total*
Higher education								**Enseignement supérieur**
RSE	12 920	12 483	12 375	12 427	12 407	12 740	..	Chercheurs
Technicians	..	..	..	..	..	..	..	Techniciens
Other occupations	..	..	..	..	..	..	..	Autres professions
Total	26 050	24 860	24 398	24 411	24 165	24 302	..	*Total*
Private non-profit								**Institutions sans but lucratif**
RSE	490 [a]	480	470	509	514	492	..	Chercheurs
Technicians	190 [a]	190	190	188	190	190	..	Techniciens
Other occupations	240 [a]	250	240	290	293	292	..	Autres professions
Total	920 [a]	920	900	987	997	974	..	*Total*
TOTAL								**TOTAL**
RSE	34 200 [a]	34 038	34 482 [a]	38 055	39 081	40 623	..	Chercheurs
Technicians	..	..	..	..	..	..	..	Techniciens
Other occupations	..	..	..	..	..	..	..	Autres professions
Total	78 980 [a]	79 256	80 789 [a]	83 967	85 486	87 006	..	*Total*

Table 6 - Tableau 6

Total R&D personnel - Personnel total de R-D

by sector of employment and formal qualification - par secteur d'emploi et par niveau de qualification

In full-time equivalent — *En équivalence plein-temps*

	1994	1995	1996	1997	1998	1999	2000	
Business enterprise								**Entreprises**
University PhD level degrees	..	..	..	..	..	..	..	Diplômes univ. de niv. doctorat
Other university degrees	..	..	..	..	..	..	..	Autres diplômes universitaires
Sub-total university degrees	..	..	..	..	..	..	..	Sous-total Universitaires
Other post-secondary diplomas	..	..	..	..	..	..	..	Autres diplômes post-secondaires
Secondary diplomas	..	..	..	..	..	..	..	Diplômes de fin d'études secondaires
Other qualifications	..	..	..	..	..	..	..	Autres qualifications
Total	36 040 [a]	37 456	39 461 [a]	42 409	43 872	45 181	..	*Total*
Government								**État**
University PhD level degrees	..	..	..	..	..	..	..	Diplômes univ. de niv. doctorat
Other university degrees	..	..	..	..	..	..	..	Autres diplômes universitaires
Sub-total university degrees	..	..	..	..	..	..	..	Sous-total Universitaires
Other post-secondary diplomas	..	..	..	..	..	..	..	Autres diplômes post-secondaires
Secondary diplomas	..	..	..	..	..	..	..	Diplômes de fin d'études secondaires
Other qualifications	..	..	..	..	..	..	..	Autres qualifications
Total	15 970 [a]	16 020	16 030	16 160	16 451	16 565	..	*Total*
Higher education								**Enseignement supérieur**
University PhD level degrees	..	..	..	..	..	..	..	Diplômes univ. de niv. doctorat
Other university degrees	..	..	..	..	..	..	..	Autres diplômes universitaires
Sub-total university degrees	..	..	..	..	..	..	..	Sous-total Universitaires
Other post-secondary diplomas	..	..	..	..	..	..	..	Autres diplômes post-secondaires
Secondary diplomas	..	..	..	..	..	..	..	Diplômes de fin d'études secondaires
Other qualifications	..	..	..	..	..	..	..	Autres qualifications
Total	26 050	24 860	24 398	24 411	24 165	24 302	..	*Total*
Private non-profit								**Institutions sans but lucratif**
University PhD level degrees	..	..	..	..	..	..	..	Diplômes univ. de niv. doctorat
Other university degrees	..	..	..	..	..	..	..	Autres diplômes universitaires
Sub-total university degrees	..	..	..	..	..	..	..	Sous-total Universitaires
Other post-secondary diplomas	..	..	..	..	..	..	..	Autres diplômes post-secondaires
Secondary diplomas	..	..	..	..	..	..	..	Diplômes de fin d'études secondaires
Other qualifications	..	..	..	..	..	..	..	Autres qualifications
Total	920 [a]	920	900	987	997	974	..	*Total*
TOTAL								**TOTAL**
University PhD level degrees	..	..	..	..	..	..	..	Diplômes univ. de niv. doctorat
Other university degrees	..	..	..	..	..	..	..	Autres diplômes universitaires
Sub-total university degrees	..	..	..	..	..	..	..	Sous-total Universitaires
Other post-secondary diplomas	..	..	..	..	..	..	..	Autres diplômes post-secondaires
Secondary diplomas	..	..	..	..	..	..	..	Diplômes de fin d'études secondaires
Other qualifications	..	..	..	..	..	..	..	Autres qualifications
Total	78 980 [a]	79 256	80 789 [a]	83 967	85 486	87 006	..	*Total*

Source: OECD, R&D database (DSTI/EAS Division), January 2002.

Source : OCDE, base de données R-D (DSTI/Division AES), janvier 2002.

Total business enterprise intramural expenditure on R&D - Dépenses intra-muros totales de R-D des entreprises

(BERD) by industry - (DIRDE) par industrie

	1994	1995	1996	1997	1998	1999	2000	
	9.1 Million EUR (1999 NLG euro, EUR from 1999) - Millions EUR (1999 NLG euro, EUR à partir de 1999)							
1 AGRICULTURE, HUNTING & FORESTRY	39.0	40.4	30.9 [a]	36.3 [a]	63.1 [a]	87.6 [a]	..	**1 AGRICULTURE, CHASSE & SYLVICULTURE**
2 MINING	120.3 [o]	97.1	57.2	120.7	151.1	86.2	..	**2 ACTIVITES EXTRACTIVES**
3 MANUFACTURING	2 389.2 [a]	2 563.0	2 641.9	2 812.1	2 786.2	3 242.3	..	**3 ACTIVITES DE FABRICATION**
4 Food, beverages & tobacco	182.0 [a]	216.9	198.8	196.9	191.5	249.6	..	**4 Produits alimentaires, boissons & tabac**
5 Food products & beverages	..	..	..	..	..	245.0	..	5 Produits alimentaires & boissons
6 Tobacco products	..	..	..	..	..	4.5	..	6 Produits à base de tabac
7 Textiles, wear app., fur & leather	12.7 [a]	12.3	12.7	13.6	10.4	17.2	..	**7 Textiles, habill., fourrures & cuirs**
8 Textiles	11.3 [a]	11.3	11.3	12.3	8.6	15.0	..	8 Textiles
9 Wearing apparel & fur	..	..	..	..	..	0.0	..	9 Articles d'habillement & fourrures
10 Leather products & footwear	1.4 [a]	0.9	0.9	0.9	1.4	2.3	..	10 Articles en cuir & chaussures
11 Wood, paper, printing, publishing	15.4 [a]	17.7	20.0	20.0	23.6	30.4	..	**11 Bois, papier, imprimerie, édition**
12 Wood & cork (not furniture)	1.4 [a]	1.8	1.8	3.6	1.4	0.9 [n]	..	12 Bois & liège (sauf meubles)
13 Pulp, paper & paper products	10.0 [a]	9.5	9.5	10.0	15.0	15.4	..	13 Carton, papier & articles
14 Publ., print. & repro. of rec. media	4.1 [a]	6.4	8.6	6.4	7.3	13.6	..	14 Edition, impress. & repro. enregis.
15 Coke, petroleum, nuclear fuel, chemicals, rubber plastics	857.2 [a]	841.3	880.8	910.7	925.7	996.0	..	**15 Coke, pétrole, combust. nucl., prod. chimi., caout. & mat. plast.**
16 Coke, ref. petrol. prod. & nuclear fuel	.. [n]	.. [n]	.. [n]	.. [n]	.. [n]	36.8	..	16 Coke, raffin. pétr. & combust. nucl.
17 Chemicals & chemical products	822.3 [a,o]	809.5 [o]	840.4 [o]	863.5	894.9 [o]	917.5	..	17 Produits chimiques
18 *Chemical prod.(less pharmaceuticals)*	623.9 [a,o]	597.2 [o]	534.6 [o]	556.3	567.7 [o]	498.7	..	18 *Prod. chimi. (moins prod. pharmaceutiques)*
19 *Pharmaceuticals*	198.3 [a]	212.4	305.8	307.7	327.2	418.8	..	19 *Produits pharmaceutiques*
20 Rubber & plastic products	34.9 [a]	31.8	40.4	46.7	30.9	41.7	..	20 Caoutchouc et matières plastiques
21 Non-metallic mineral products	12.3 [a]	13.6	14.1	17.2	13.2	15.9 [n]	..	**21 Produits minéraux non métalliques**
22 Basic metals	79.4 [a]	47.2	57.6	56.7	47.6	60.4	..	**22 Produits métallurgiques de base**
23 Basic metals, ferrous	..	..	..	..	..	..	..	23 Produits métallur. de base, ferreux
24 Basic metals, non-ferrous	..	..	..	..	..	..	..	24 Prod. métallur. de base, non ferreux
25 Fabricated metal products	.. [n]	.. [n]	.. [n]	.. [n]	32.2	54.0	..	**25 Ouvrages en métaux (sauf mach.& mat.)**
26 Machinery equipment, instruments and transport equipment	1 219.3 [a,o]	1 402.6 [o]	1 442.6 [o]	1 581.4 [o]	1 530.6 [o]	1 802.0 [o,p]	..	**26 Machines & matériel, instruments & mat. de transport**
27 Machinery n.e.c.	..	182.0	211.5	243.7	262.7	339.0	..	27 Machines, n.c.a.
28 Office, account. & computing machinery	..	.. [n]	.. [n]	.. [n]	.. [n]	.. [n]	..	28 Mach. de bur., compta. & trait. inform.
29 Electrical machinery	..	1 013.7 [o]	1 101.3 [o]	1 151.2 [o]	1 154.4 [o]	1 307.8 [o]	..	29 Machines électriques
30 Electro. equip. (radio, TV & commun.)	..	.. [n]	.. [n]	.. [n]	.. [n]	.. [n]	..	30 Equip. électro. (radio, tv & communic.)
31 *Electro. comp. (inc. semi-conductors)*	..	..	..	..	..	..	..	31 *Compos. électro. (semi-cond. incl.)*
32 *TV, radio & communications equip.*	..	..	..	..	..	..	..	32 *Appareils radio, tv et de communi.*
33 Instruments	..	.. [n]	.. [n]	.. [n]	.. [n]	.. [n]	..	33 Instruments
34 Motor vehicles	..	162.9 [o]	93.9 [o]	140.2 [o]	111.6 [o]	126.6 [o,p]	..	34 Véhicules automobiles
35 Other transport equipment	..	.. [n]	.. [n]	.. [n]	.. [n]	28.6 [n]	..	35 Autres matériels de transport
36 *Ships*	..	..	..	..	..	..	..	36 *Navires*
37 *Aerospace*	..	..	..	..	..	..	..	37 *Aérospatiale*
38 *Other transport equipment n.e.c.*	..	..	..	..	..	..	..	38 *Autres matériels de transport n.c.a.*
39 Furniture, other manufacturing n.e.c.	10.9 [a]	11.8	14.5	15.4	10.4	15.9 [o,p]	..	**39 Meubles, autres activ. de fabricat. n.c.a.**
40 Furniture	..	..	..	..	..	..	..	40 Meubles
41 Other manufacturing n.e.c.	.. [n]	.. [n]	.. [n]	.. [n]	.. [n]	..	..	41 Autres activités de fabrication n.c.a.
42 Recycling	..	..	0.5	0.9	0.9	0.9	..	**42 Récupération**
43 ELECTRICITY, GAS & WATER SUPPLY	5.0	8.2	9.5 [a]	20.4 [a]	27.2 [a]	20.9 [a]	..	**43 DISTRIBUTION D'ELECTRICITE, GAZ & EAU**
44 CONSTRUCTION	.. [n]	42.7	50.4 [a]	36.8 [a]	63.5 [a]	60.8 [a]	..	**44 CONSTRUCTION**
45 SERVICES SECTOR	363.0 [a]	379.4	548.6 [a]	687.9 [a]	629.4 [a]	766.0 [a]	..	**45 SECTEUR DES SERVICES**
46 Wholesale, retail trade & repair of motor vehicles, etc.	120.3	110.7	139.3 [a]	158.4 [a]	177.9 [a]	206.0 [a]	..	**46 Commerce (gros & détail), réparation de véhicules, etc.**
47 Hotels & restaurants	..	..	..	..	..	..	..	**47 Hôtels et restaurants**
48 Transport & storage	85.3 [o]	86.2 [o]	82.1 [a,o]	79.9 [a,o]	69.9 [a,o]	11.3 [a]	..	**48 Transports, entreposage**
49 Communications	.. [n]	.. [n]	.. [n]	.. [n]	.. [n]	93.5 [n]	..	**49 Communications**
50 Post	..	..	..	..	..	..	..	50 Postes
51 Telecommunications	..	..	..	..	..	..	..	51 Télécommunications
52 Financial intermediation (inc. insur.)	9.1	8.6	61.7 [a]	86.2 [a]	102.6 [a]	100.3 [a]	..	**52 Intermédiation financ. (assurances inc.)**
53 Real estate, renting & busin. activity	148.4	173.8	265.5 [a]	363.5 [a]	278.2 [a]	355.3 [a]	..	**53 Immobil., locat. & serv. aux entreprises**
54 Computer & related activities	..	29.0	83.5 [a]	111.2 [a]	96.2 [a]	107.5 [a]	..	54 Informatique et activités connexes
55 *Software consultancy*	..	..	..	..	..	..	..	55 *Production de logiciels*
56 *Other computer services n.e.c.*	..	..	..	..	..	..	..	56 *Autres activités informatiques*
57 Research & development	28.1	26.3	29.5	30.4	26.8	28.1 [p]	..	57 Recherche & développement expérimental
58 Other business activities n.e.c.	..	118.4	152.5 [a]	222.4 [a]	155.2 [a]	219.6 [a]	..	58 Autres activ. & serv. aux entreprises
59 Comm., soc. & pers. serv. activ., etc.	..	..	..	..	..	..	..	**59 Adm. pub., serv. soc. & collec., etc**
TOTAL BUSINESS ENTERPRISE SECTOR	**2 916.9** [a]	**3 130.6**	**3 342.1** [a]	**3 714.6**	**3 720.5**	**4 263.3**	..	**TOTAL SECTEUR DES ENTREPRISES**

Source: OECD, R&D database (DSTI/EAS Division), January 2002.

Source : OCDE, base de données R-D (DSTI/Division AES), janvier 2002.

Total business enterprise intramural expenditure on R&D - Dépenses intra-muros totales de R-D des entreprises
(BERD) by industry - (DIRDE) par industrie

9.2 Million constant USD(1995 prices and PPPs) - Millions de USD constants (prix et PPA de 1995)

	1994	1995	1996	1997	1998	1999	2000	
1 AGRICULTURE, HUNTING & FORESTRY	43.2	43.9	33.2 [a]	38.3 [a]	65.5 [a]	90.0 [a]	..	**1 AGRICULTURE, CHASSE & SYLVICULTURE**
2 MINING	133.1 [o]	105.6	61.4	127.2	156.9	88.6	..	**2 ACTIVITES EXTRACTIVES**
3 MANUFACTURING	2 644.5 [a]	2 785.8	2 838.4	2 964.1	2 893.7	3 333.4	..	**3 ACTIVITES DE FABRICATION**
4 Food, beverages & tobacco	201.4 [a]	235.8	213.5	207.6	198.9	256.6	..	**4 Produits alimentaires, boissons & tabac**
5 Food products & beverages	..	..	..	..	..	251.9	..	5 Produits alimentaires & boissons
6 Tobacco products	..	..	..	..	..	4.7	..	6 Produits à base de tabac
7 Textiles, wear app., fur & leather	14.1 [a]	13.3	13.7	14.3	10.8	17.7	..	**7 Textiles, habill., fourrures & cuirs**
8 Textiles	12.6 [a]	12.3	12.2	12.9	9.0	15.4	..	8 Textiles
9 Wearing apparel & fur	..	..	..	..	..	0.0	..	9 Articles d'habillement & fourrures
10 Leather products & footwear	1.5 [a]	1.0	1.0	1.0	1.4	2.3	..	10 Articles en cuir & chaussures
11 Wood, paper, printing, publishing	17.1 [a]	19.2	21.5	21.0	24.5	31.3	..	**11 Bois, papier, imprimerie, édition**
12 Wood & cork (not furniture)	1.5 [a]	2.0	2.0	3.8	1.4	0.9 [n]	..	12 Bois & liège (sauf meubles)
13 Pulp, paper & paper products	11.1 [a]	10.4	10.2	10.5	15.6	15.9	..	13 Carton, papier & articles
14 Publ., print. & repro. of rec. media	4.5 [a]	6.9	9.3	6.7	7.5	14.0	..	14 Edition, impress. & repro. enregis.
15 Coke, petroleum, nuclear fuel, chemicals, rubber plastics	948.8 [a]	914.5	946.3	960.0	961.4	1 024.0	..	**15 Coke, pétrole, combust. nucl., prod. chimi., caout. & mat. plast.**
16 Coke, ref. petrol. prod. & nuclear fuel	.. [n]	.. [n]	.. [n]	.. [n]	.. [n]	37.8	..	16 Coke, raffin. pétr. & combust. nucl.
17 Chemicals & chemical products	910.1 [a,o]	879.9 [o]	902.9 [o]	910.2	929.4 [o]	943.3	..	17 Produits chimiques
18 *Chemical prod.(less pharmaceuticals)*	690.6 [a,o]	649.1 [o]	574.3 [o]	586.4	589.6 [o]	512.7	..	18 *Prod. chimi. (moins prod. pharmaceutiques)*
19 *Pharmaceuticals*	219.5 [a]	230.8	328.6	324.3	339.8	430.6	..	19 *Produits pharmaceutiques*
20 Rubber & plastic products	38.7 [a]	34.5	43.4	49.3	32.0	42.9	..	20 Caoutchouc et matières plastiques
21 Non-metallic mineral products	13.6 [a]	14.8	15.1	18.2	13.7	16.3 [n]	..	**21 Produits minéraux non métalliques**
22 Basic metals	87.9 [a]	51.3	61.9	59.8	49.5	62.0	..	**22 Produits métallurgiques de base**
23 Basic metals, ferrous	..	..	..	..	..	..	..	23 Produits métallur. de base, ferreux
24 Basic metals, non-ferrous	..	..	..	..	..	..	..	24 Prod. métallur. de base, non ferreux
25 Fabricated metal products	.. [n]	.. [n]	.. [n]	.. [n]	33.5	55.5	..	**25 Ouvrages en métaux (sauf mach.& mat.)**
26 Machinery equipment, instruments and transport equipment	1 349.6 [a,o]	1 524.6 [o]	1 549.9 [o]	1 666.9 [o]	1 589.7 [o]	1 852.6 [o,p]	..	**26 Machines & matériel, instruments & mat. de transport**
27 Machinery n.e.c.	..	197.8	227.2	256.8	272.9	348.5	..	27 Machines, n.c.a.
28 Office, account. & computing machinery	..	.. [n]	.. [n]	.. [n]	.. [n]	.. [n]	..	28 Mach. de bur., compta. & trait. inform.
29 Electrical machinery	..	1 101.9 [o]	1 183.2 [o]	1 213.5 [o]	1 199.0 [o]	1 344.5 [o]	..	29 Machines électriques
30 Electro. equip. (radio, TV & commun.)	..	.. [n]	.. [n]	.. [n]	.. [n]	.. [n]	..	30 Equip. électro. (radio, tv & communic.)
31 *Electro. comp. (inc. semi-conductors)*	..	..	..	..	..	..	..	31 *Compos. électro. (semi-cond. incl.)*
32 *TV, radio & communications equip.*	..	..	..	..	..	..	..	32 *Appareils radio, tv et de communi.*
33 Instruments	..	.. [n]	.. [n]	.. [n]	.. [n]	.. [n]	..	33 Instruments
34 Motor vehicles	..	177.1 [o]	100.9 [o]	147.8 [o]	115.9 [o]	130.2 [o,p]	..	34 Véhicules automobiles
35 Other transport equipment	..	.. [n]	.. [n]	.. [n]	.. [n]	29.4 [n]	..	35 Autres matériels de transport
36 *Ships*	..	..	..	..	..	..	..	36 *Navires*
37 *Aerospace*	..	..	..	..	..	..	..	37 *Aérospatiale*
38 *Other transport equipment n.e.c.*	..	..	..	..	..	..	..	38 *Autres matériels de transport n.c.a.*
39 Furniture, other manufacturing n.e.c.	12.1 [a]	12.8	15.6	16.3	10.8	16.3 [o,p]	..	**39 Meubles, autres activ. de fabricat. n.c.a.**
40 Furniture	..	..	..	..	..	..	..	40 Meubles
41 Other manufacturing n.e.c.	.. [n]	.. [n]	.. [n]	.. [n]	.. [n]	..	..	41 Autres activités de fabrication n.c.a.
42 Recycling	..	..	0.5	1.0	0.9	0.9	..	**42 Récupération**
43 ELECTRICITY, GAS & WATER SUPPLY	5.5	8.9	10.2 [a]	21.5 [a]	28.3 [a]	21.5 [a]	..	**43 DISTRIBUTION D'ELECTRICITE, GAZ & EAU**
44 CONSTRUCTION	.. [n]	46.4	54.1 [a]	38.7 [a]	66.0 [a]	62.5 [a]	..	**44 CONSTRUCTION**
45 SERVICES SECTOR	401.8 [a]	412.3	589.4 [a]	725.1 [a]	653.7 [a]	787.5 [a]	..	**45 SECTEUR DES SERVICES**
46 Wholesale, retail trade & repair of motor vehicles, etc.	133.1	120.3	149.7 [a]	166.9 [a]	184.7 [a]	211.8 [a]	..	**46 Commerce (gros & détail), réparation de véhicules, etc.**
47 Hotels & restaurants	..	..	..	..	..	..	..	**47 Hôtels et restaurants**
48 Transport & storage	94.4 [o]	93.7 [o]	88.2 [a,o]	84.2 [a,o]	72.6 [a,o]	11.7 [a]	..	**48 Transports, entreposage**
49 Communications	.. [n]	.. [n]	.. [n]	.. [n]	.. [n]	96.1 [n]	..	**49 Communications**
50 Post	..	..	..	..	..	..	..	50 Postes
51 Telecommunications	..	..	..	..	..	..	..	51 Télécommunications
52 Financial intermediation (inc. insur.)	10.0	9.4	66.3 [a]	90.9 [a]	106.5 [a]	103.1 [a]	..	**52 Intermédiation financ. (assurances inc.)**
53 Real estate, renting & busin. activity	164.2	188.9	285.2 [a]	383.1 [a]	288.9 [a]	365.3 [a]	..	**53 Immobil., locat. & serv. aux entreprises**
54 Computer & related activities	..	31.6	89.7 [a]	117.2 [a]	99.9 [a]	110.6 [a]	..	54 Informatique et activités connexes
55 *Software consultancy*	..	..	..	..	..	..	..	55 *Production de logiciels*
56 *Other computer services n.e.c.*	..	..	..	..	..	..	..	56 *Autres activités informatiques*
57 Research & development	31.1	28.6	31.7	32.0	27.8	28.9 [p]	..	57 Recherche & développement expérimental
58 Other business activities n.e.c.	..	128.7	163.8 [a]	234.4 [a]	161.2 [a]	225.8 [a]	..	58 Autres activ. & serv. aux entreprises
59 Comm., soc. & pers. serv. activ., etc.	..	..	..	..	..	..	..	**59 Adm. pub., serv. soc. & collec., etc**
TOTAL BUSINESS ENTERPRISE SECTOR	**3 228.7** [a]	**3 402.8**	**3 590.6** [a]	**3 915.4**	**3 864.1**	**4 383.1**	..	**TOTAL SECTEUR DES ENTREPRISES**

Source: OECD, R&D database (DSTI/EAS Division), January 2002.

Source : OCDE, base de données R-D (DSTI/Division AES), janvier 2002.

Table 17 - Tableau 17

Higher education R&D personnel (HEMP) - Personnel de R-D du secteur de l'enseignement supérieur

by field of science and occupation - par discipline scientifique et par profession

In full-time equivalent — *En équivalence plein-temps*

	1994	1995	1996	1997	1998	1999	2000	
Total R&D personnel								**Personnel total de R-D**
Natural sciences	..	..	..	..	..	..	..	Sciences exactes et naturelles
Engineering	..	..	..	..	..	..	..	Sciences de l'ingénieur
Medical sciences	..	..	..	..	..	..	..	Sciences médicales
Agricultural sciences	..	..	..	..	..	..	..	Sciences agricoles
Sub-total NSE	..	..	..	..	..	..	..	Sous-total SENI
Social sciences	..	..	..	..	..	..	..	Sciences sociales
Humanities	..	..	..	..	..	..	..	Sciences humaines
Sub-total SSH	..	..	..	..	..	..	..	Sous-total SSH
Total NSE+SSH	26 050	24 860	24 398	24 411	24 165	24 302	..	*Total SENI+SSH*
Researchers								**Chercheurs**
Natural sciences	..	..	..	..	..	..	..	Sciences exactes et naturelles
Engineering	..	..	..	..	..	..	..	Sciences de l'ingénieur
Medical sciences	..	..	..	..	..	..	..	Sciences médicales
Agricultural sciences	..	..	..	..	..	..	..	Sciences agricoles
Sub-total NSE	..	..	..	..	..	..	..	Sous-total SENI
Social sciences	..	..	..	..	..	..	..	Sciences sociales
Humanities	..	..	..	..	..	..	..	Sciences humaines
Sub-total SSH	..	..	..	..	..	..	..	Sous-total SSH
Total NSE+SSH	12 920	12 483	12 375	12 427	12 407	12 740	..	*Total SENI+SSH*
Technicians								**Technicians**
Natural sciences	..	..	..	..	..	..	..	Sciences exactes et naturelles
Engineering	..	..	..	..	..	..	..	Sciences de l'ingénieur
Medical sciences	..	..	..	..	..	..	..	Sciences médicales
Agricultural sciences	..	..	..	..	..	..	..	Sciences agricoles
Sub-total NSE	..	..	..	..	..	..	..	Sous-total SENI
Social sciences	..	..	..	..	..	..	..	Sciences sociales
Humanities	..	..	..	..	..	..	..	Sciences humaines
Sub-total SSH	..	..	..	..	..	..	..	Sous-total SSH
Total NSE+SSH	..	..	..	..	..	..	..	*Total SENI+SSH*
Other								**Autres**
Natural sciences	..	..	..	..	..	..	..	Sciences exactes et naturelles
Engineering	..	..	..	..	..	..	..	Sciences de l'ingénieur
Medical sciences	..	..	..	..	..	..	..	Sciences médicales
Agricultural sciences	..	..	..	..	..	..	..	Sciences agricoles
Sub-total NSE	..	..	..	..	..	..	..	Sous-total SENI
Social sciences	..	..	..	..	..	..	..	Sciences sociales
Humanities	..	..	..	..	..	..	..	Sciences humaines
Sub-total SSH	..	..	..	..	..	..	..	Sous-total SSH
Total NSE+SSH	..	..	12 023	11 984	11 758	11 562	..	*Total SENI+SSH*

Table 18 - Tableau 18

Government budget appropriations or outlays for R&D (GBAORD) - Crédits budgétaires publics de R-D (CBPRD)

by socio-economic objective - par objectif socio-économique

Million EUR (1999 NLG euro, EUR from 1999) — *Millions EUR (1999 NLG euro, EUR à partir de 1999)*

	1995	1996	1997	1998	1999	2000	2001	
Agriculture, forestry and fishing	108.0	108.7	80.5	81.8 p	86.2 p	..	..	Agriculture, sylviculture et pêche
Industrial development	278.5	258.0	355.5	369.2 p	411.5 p	..	..	Développement industriel
Energy	85.4	82.6	76.3	71.0 p	80.9 p	..	..	Energie
Transport & telecommunications	35.4	40.3	49.8	60.9 p	50.0 p	..	..	*Transports & télécommunications*
Urban & rural planning	60.5	64.6	56.1	58.0 p	58.8 p	..	..	*Aménagement urbain et rural*
ST Infrastructure	95.8	104.9	106.0	118.9 p	108.8 p	..	..	ST Infrastructure
Prevention of pollution	..	..	..	..	..	..	..	*Prévention de la pollution*
Identificat. & treatment of pollution	..	..	..	..	..	..	..	*Détect. pollut. & mesures antipollut.*
ST Environment	95.0	88.2	108.9	114.0 p	115.2 p	..	..	ST Environnement
Health	46.8	49.9	77.0	103.7 p	108.5 p	..	..	Santé
Social development & services	56.2	61.4	76.1	76.7 p	76.0 p	..	..	Développement et services sociaux
Earth and atmosphere	6.7	11.0	23.1	24.5 p	23.5 p	..	..	Milieu terrestre et atmosphère
Advancement of research	279.3	294.2	297.0	313.0 p	320.0 p	..	..	*Promotion de la recherche*
General university funds	1 039.8	1 029.1	1 336.1 [a]	1 237.9 p	1 286.5 p	..	..	*Fonds généraux des universités*
ST Advancement of knowledge	1 319.1	1 323.3	1 633.0	1 550.9 p	1 606.5 p	..	..	ST promotion gén. des connaiss.
Civil space	106.9	96.5	99.0	92.3 p	84.0 p	..	..	Espace civil
Defence	78.5	91.6	86.5	95.3 p	88.9 p	..	..	Défense
Not elsewhere classified	129.4	122.7	110.9	116.5 p	121.3 p	..	..	Non classé
Total	**2 406.4**	**2 398.7**	**2 832.8**	**2 814.8 p**	**2 911.2 p**	**..**	**..**	**Total**

Source: OECD, R&D database (DSTI/EAS Division), January 2002.

Source : OCDE, base de données R-D (DSTI/Division AES), janvier 2002.

Table 19 - Tableau 19

Technology balance of payments - Balance des paiements technologiques

by type of transfer - selon la nature des transferts

Million EUR (1999 NLG euro, EUR from 1999) — *Millions EUR (1999 NLG euro, EUR à partir de 1999)*

	1983	1984	1985	1986	1987	1988	1989	1990	1991	
Receipts										**Recettes**
Related firms	..	..	..	..	..	..	..	..	..	Firmes affiliées
Unrelated firms	..	..	..	..	..	..	..	..	..	Firmes non affiliées
Other	..	..	..	..	..	..	..	..	..	Autre
Total	1 559	1 736	1 802	1 853	2 050	2 367	2 894	3 478	4 137	*Total*
Payments										**Paiements**
Related firms	..	..	..	..	..	..	..	..	..	Firmes affiliées
Unrelated firms	..	..	..	..	..	..	..	..	..	Firmes non affiliées
Other	..	..	..	..	..	..	..	..	..	Autre
Total	1 750	2 013	2 266	1 924	1 980	2 381	3 143	3 353	5 034	*Total*

	1992	1993	1994	1995	1996	1997	1998	1999	2000	
Receipts										**Recettes**
Related firms	..	..	..	..	..	..	..	..	..	Firmes affiliées
Unrelated firms	..	..	..	..	..	..	..	..	..	Firmes non affiliées
Other	..	..	..	..	..	..	..	..	..	Autre
Total	4 954	..	..	..	..	..	..	..	..	*Total*
Payments										**Paiements**
Related firms	..	..	..	..	..	..	..	..	..	Firmes affiliées
Unrelated firms	..	..	..	..	..	..	..	..	..	Firmes non affiliées
Other	..	..	..	..	..	..	..	..	..	Autre
Total	4 899	..	..	..	..	..	..	..	..	*Total*

Source: OECD, TBPdatabase (DSTI/EAS Division), Januay 2002.

Source : OCDE, base de données TBP (DSTI/Division AES), janvier 2002.

NEW ZEALAND — Table 1 - Tableau 1 — NOUVELLE-ZÉLANDE

Gross domestic expenditure on R&D (GERD) - Dépenses intérieures brutes de R-D (DIRD)

by sector of performance and source of funds - par secteur d'exécution et par source de financement

	1994	1995	1996	1997	1998	1999	2000	
				1.1 Million NZD - Millions NZD				
Business enterprise								**Entreprises**
financed by								*financé par*
Business enterprise	..	207.7	..	247.8	..	272.1	..	Entreprises
Direct government	..	16.5	..	27.1	..	31.0	..	État direct
Higher education	..	0.3	..	2.0	..	0.1	..	Enseignement supérieur
Private non-profit	..	2.1	..	0.6	..	3.7	..	Institutions sans but lucratif
Funds from abroad	..	13.8	..	35.0	..	17.2	..	Fonds de l'étranger
Total	..	240.3	..	312.5	..	324.1	..	*Total*
Government								**État**
financed by								*financé par*
Business enterprise	..	66.6	..	71.0	..	78.3	..	Entreprises
Direct government	..	299.1	..	308.7	..	301.5	..	État direct
Higher education	..	0.7	..	1.1	..	1.2	..	Enseignement supérieur
Private non-profit	..	2.0	..	1.0	..	0.9	..	Institutions sans but lucratif
Funds from abroad	..	7.2	..	9.6	..	11.1	..	Fonds de l'étranger
Total	..	375.7	..	391.3	..	393.0	..	*Total*
Higher education								**Enseignement supérieur**
financed by								*financé par*
Business enterprise	..	25.6	..	18.8	..	..	..	Entreprises
Direct government	..	53.4	..	99.2	..	..	..	État direct
General university funds	..	96.0	..	144.2	..	..	..	Fonds généraux des universités
Sub-total government	..	149.4	..	243.3	..	..	..	Sous-Total État
Higher education	..	74.5	..	117.6	..	..	..	Enseignement supérieur
Private non-profit	..	10.4	..	10.7	..	..	..	Institutions sans but lucratif
Funds from abroad	..	13.4	..	13.2	..	..	..	Fonds de l'étranger
Total	..	273.4	..	403.6	..	..	..	*Total*
Private non-profit								**Institutions sans but lucratif**
financed by								*financé par*
Business enterprise	..	..	..	..	..	..	..	Entreprises
Direct government	..	..	..	..	..	..	..	État direct
Higher education	..	..	..	..	..	..	..	Enseignement supérieur
Private non-profit	..	..	..	..	..	..	..	Institutions sans but lucratif
Funds from abroad	..	..	..	..	..	..	..	Fonds de l'étranger
Total	..	..	..	..	..	..	..	*Total*
GERD								**DIRD**
financed by								*financé par*
Business enterprise	..	300.0	..	337.5	..	..	..	Entreprises
Direct government	..	369.0	..	434.9	..	..	..	État direct
General university funds	..	96.0	..	144.2	..	..	..	Fonds généraux des universités
Sub-total government	..	465.0	..	579.1	..	..	..	Sous-Total État
Higher education	..	75.5	..	120.8	..	..	..	Enseignement supérieur
Private non-profit	..	14.4	..	12.2	..	..	..	Institutions sans but lucratif
Funds from abroad	..	34.4	..	57.8	..	..	..	Fonds de l'étranger
Total	..	889.3	..	1 107.4	..	..	..	*Total*
				1.2 Million constant USD (1995 prices and PPPs) - Millions de USD constants (prix et PPA de 1995)				
Business enterprise								**Entreprises**
financed by								*financé par*
Business enterprise	..	141.6	..	164.9	..	179.1	..	Entreprises
Direct government	..	11.2	..	18.0	..	20.4	..	Etat direct
Higher education	..	0.2	..	1.4	..	0.1	..	Enseignement supérieur
Private non-profit	..	1.4	..	0.4	..	2.4	..	Institutions sans but lucratif
Funds from abroad	..	9.4	..	23.3	..	11.3	..	Fonds de l'étranger
Total	..	163.8	..	207.9	..	213.3	..	*Total*
Government								**État**
financed by								*financé par*
Business enterprise	..	45.4	..	47.2	..	51.5	..	Entreprises
Direct government	..	203.9	..	205.4	..	198.4	..	État direct
Higher education	..	0.5	..	0.7	..	0.8	..	Enseignement supérieur
Private non-profit	..	1.3	..	0.7	..	0.6	..	Institutions sans but lucratif
Funds from abroad	..	4.9	..	6.4	..	7.3	..	Fonds de l'étranger
Total	..	256.0	..	260.3	..	258.6	..	*Total*
Higher education								**Enseignement supérieur**
financed by								*financé par*
Business enterprise	..	17.5	..	12.5	..	..	..	Entreprises
Direct government	..	36.4	..	66.0	..	..	..	État direct
General university funds	..	65.5	..	95.9	..	..	..	Fonds généraux des universités
Sub-total government	..	101.9	..	161.9	..	..	..	Sous-Total État
Higher education	..	50.8	..	78.3	..	..	..	Enseignement supérieur
Private non-profit	..	7.1	..	7.1	..	..	..	Institutions sans but lucratif
Funds from abroad	..	9.1	..	8.8	..	..	..	Fonds de l'étranger
Total	..	186.3	..	268.5	..	..	..	*Total*
Private non-profit								**Institutions sans but lucratif**
financed by								*financé par*
Business enterprise	..	..	..	..	..	..	..	Entreprises
Direct government	..	..	..	..	..	..	..	État direct
Higher education	..	..	..	..	..	..	..	Enseignement supérieur
Private non-profit	..	..	..	..	..	..	..	Institutions sans but lucratif
Funds from abroad	..	..	..	..	..	..	..	Fonds de l'étranger
Total	..	..	..	..	..	..	..	*Total*
GERD								**DIRD**
financed by								*financé par*
Business enterprise	..	204.5	..	224.6	..	..	..	Entreprises
Direct government	..	251.5	..	289.4	..	..	..	État direct
General university funds	..	65.5	..	95.9	..	..	..	Fonds généraux des universités
Sub-total government	..	316.9	..	385.3	..	..	..	Sous-Total État
Higher education	..	51.4	..	80.4	..	..	..	Enseignement supérieur
Private non-profit	..	9.8	..	8.1	..	..	..	Institutions sans but lucratif
Funds from abroad	..	23.5	..	38.4	..	..	..	Fonds de l'étranger
Total	..	606.2	..	736.8	..	..	..	*Total*

Source: OECD, R&D database (DSTI/EAS Division), January 2002.

Source : OCDE, base de données R-D (DSTI/Division AES), janvier 2002.

Table 5 - Tableau 5

Total R&D personnel - Personnel total de R-D

by sector of employment and occupation - par secteur d'emploi et par profession

In full-time equivalent *En équivalence plein-temps*

	1994	1995	1996	1997	1998	1999	2000	
Business enterprise								**Entreprises**
RSE	..	1 580	..	1 692	..	2 141	..	Chercheurs
Technicians	..	823	..	829	..	820	..	Techniciens
Other occupations	..	425	..	415	..	327	..	Autres professions
Total	..	2 828	..	2 937	..	3 288	..	*Total*
Government								**État**
RSE	..	1 498	..	1 765	..	1 631	..	Chercheurs
Technicians	..	1 518	..	1 211	..	1 118	..	Techniciens
Other occupations	..	968	..	840	..	696	..	Autres professions
Total	..	3 984	..	3 816	..	3 445	..	*Total*
Higher education								**Enseignement supérieur**
RSE	..	3 026	..	4 807	..	..	..	Chercheurs
Technicians	..	497	..	714	..	..	..	Techniciens
Other occupations	..	213	..	635	..	..	..	Autres professions
Total	..	3 735	..	6 156	..	..	..	*Total*
Private non-profit								**Institutions sans but lucratif**
RSE	..	..	..	..	..	..	..	Chercheurs
Technicians	..	..	..	..	..	..	..	Techniciens
Other occupations	..	..	..	..	..	..	..	Autres professions
Total	..	..	..	..	..	..	..	*Total*
TOTAL								**TOTAL**
RSE	..	6 104	..	8 264	..	..	..	Chercheurs
Technicians	..	2 838	..	2 754	..	..	..	Techniciens
Other occupations	..	1 606	..	1 890	..	..	..	Autres professions
Total	..	10 547	..	12 908	..	..	..	*Total*

Table 6 - Tableau 6

Total R&D personnel - Personnel total de R-D

by sector of employment and formal qualification - par secteur d'emploi et par niveau de qualification

In full-time equivalent *En équivalence plein-temps*

	1994	1995	1996	1997	1998	1999	2000	
Business enterprise								**Entreprises**
University PhD level degrees	..	..	..	..	..	392	..	Diplômes univ. de niv. doctorat
Other university degrees	..	..	..	..	..	1 657	..	Autres diplômes universitaires
Sub-total university degrees	..	..	..	..	..	2 049	..	Sous-total Universitaires
Other post-secondary diplomas	..	..	..	..	..	831	..	Autres diplômes post-secondaires
Secondary diplomas	..	..	..	..	..	233	..	Diplômes de fin d'études secondaires
Other qualifications	..	..	..	..	..	141	..	Autres qualifications
Total	..	2 828	..	2 937	..	3 288	..	*Total*
Government								**État**
University PhD level degrees	..	..	..	..	..	978	..	Diplômes univ. de niv. doctorat
Other university degrees	..	..	..	..	..	1 195	..	Autres diplômes universitaires
Sub-total university degrees	..	..	..	..	..	2 173	..	Sous-total Universitaires
Other post-secondary diplomas	..	..	..	..	..	677	..	Autres diplômes post-secondaires
Secondary diplomas	..	..	..	..	..	290	..	Diplômes de fin d'études secondaires
Other qualifications	..	..	..	..	..	294	..	Autres qualifications
Total	..	3 984	..	3 816	..	3 445	..	*Total*
Higher education								**Enseignement supérieur**
University PhD level degrees	..	..	..	..	..	..	..	Diplômes univ. de niv. doctorat
Other university degrees	..	..	..	..	..	..	..	Autres diplômes universitaires
Sub-total university degrees	..	..	..	..	..	..	..	Sous-total Universitaires
Other post-secondary diplomas	..	..	..	..	..	..	..	Autres diplômes post-secondaires
Secondary diplomas	..	..	..	..	..	..	..	Diplômes de fin d'études secondaires
Other qualifications	..	..	..	..	..	..	..	Autres qualifications
Total	..	3 735	..	6 156	..	..	..	*Total*
Private non-profit								**Institutions sans but lucratif**
University PhD level degrees	..	..	..	..	..	..	..	Diplômes univ. de niv. doctorat
Other university degrees	..	..	..	..	..	..	..	Autres diplômes universitaires
Sub-total university degrees	..	..	..	..	..	..	..	Sous-total Universitaires
Other post-secondary diplomas	..	..	..	..	..	..	..	Autres diplômes post-secondaires
Secondary diplomas	..	..	..	..	..	..	..	Diplômes de fin d'études secondaires
Other qualifications	..	..	..	..	..	..	..	Autres qualifications
Total	..	..	..	..	..	..	..	*Total*
TOTAL								**TOTAL**
University PhD level degrees	..	..	..	..	..	..	..	Diplômes univ. de niv. doctorat
Other university degrees	..	..	..	..	..	..	..	Autres diplômes universitaires
Sub-total university degrees	..	..	..	..	..	..	..	Sous-total Universitaires
Other post-secondary diplomas	..	..	..	..	..	..	..	Autres diplômes post-secondaires
Secondary diplomas	..	..	..	..	..	..	..	Diplômes de fin d'études secondaires
Other qualifications	..	..	..	..	..	..	..	Autres qualifications
Total	..	10 547	..	12 908	..	..	..	*Total*

Source: OECD, R&D database (DSTI/EAS Division), January 2002.

Source : OCDE, base de données R-D (DSTI/Division AES), janvier 2002.

Total business enterprise intramural expenditure on R&D - Dépenses intra-muros totales de R-D des entreprises
(BERD) by industry - (DIRDE) par industrie

	1994	1995	1996	1997	1998	1999	2000	
				9.1 Million NZD - Millions NZD				
1 AGRICULTURE, HUNTING & FORESTRY	..	12.6	..	11.7	..	10.4	..	**1 AGRICULTURE, CHASSE & SYLVICULTURE**
2 MINING	..	0.7	..	6.0	..	3.3	..	**2 ACTIVITES EXTRACTIVES**
3 MANUFACTURING	..	164.5	..	182.6	..	182.3	..	**3 ACTIVITES DE FABRICATION**
4 Food, beverages & tobacco	..	59.7	..	65.1	..	60.3	..	**4 Produits alimentaires, boissons & tabac**
5 Food products & beverages	..	59.7 [o]	..	65.1 [o]	..	60.3 [o]	..	5 Produits alimentaires & boissons
6 Tobacco products	..	.. [n]	..	.. [n]	..	.. [n]	..	6 Produits à base de tabac
7 Textiles, wear app., fur & leather	..	25.2	..	17.3	..	17.0	..	**7 Textiles, habill., fourrures & cuirs**
8 Textiles	..	22.1	..	15.0	..	14.8	..	8 Textiles
9 Wearing apparel & fur	..	0.6	..	0.5	..	0.6	..	9 Articles d'habillement & fourrures
10 Leather products & footwear	..	2.6	..	1.8	..	1.7	..	10 Articles en cuir & chaussures
11 Wood, paper, printing, publishing	..	7.0	..	4.3	..	1.6	..	**11 Bois, papier, imprimerie, édition**
12 Wood & cork (not furniture)	..	1.4	..	0.4	..	1.4	..	12 Bois & liège (sauf meubles)
13 Pulp, paper & paper products	..	5.6	..	3.8	..	..	..	13 Carton, papier & articles
14 Publ., print. & repro. of rec. media	..	0.0	..	0.1	..	0.3	..	14 Edition, impress. & repro. enregis.
15 Coke, petroleum, nuclear fuel, chemicals, rubber plastics	..	16.0	..	17.6	..	13.4	..	**15 Coke, pétrole, combust. nucl., prod. chimi., caout. & mat. plast.**
16 Coke, ref. petrol. prod. & nuclear fuel	..	..	..	..	..	..	..	16 Coke, raffin. pétr. & combust. nucl.
17 Chemicals & chemical products	..	12.3	..	13.2	..	8.8	..	17 Produits chimiques
18 *Chemical prod.(less pharmaceuticals)*	..	9.1	..	8.7	..	5.8	..	18 *Prod. chimi. (moins prod. pharmaceutiques)*
19 *Pharmaceuticals*	..	3.2	..	4.6	..	3.0	..	19 *Produits pharmaceutiques*
20 Rubber & plastic products	..	3.7	..	4.4	..	4.6	..	20 Caoutchouc et matières plastiques
21 Non-metallic mineral products	..	2.3	..	2.0	..	2.7	..	**21 Produits minéraux non métalliques**
22 Basic metals	..	1.6	..	0.4	..	0.9	..	**22 Produits métallurgiques de base**
23 Basic metals, ferrous	..	1.1	..	0.0	..	..	..	23 Produits métallur. de base, ferreux
24 Basic metals, non-ferrous	..	0.6	..	0.4	..	0.9	..	24 Prod. métallur. de base, non ferreux
25 Fabricated metal products	..	3.5	..	3.7	..	8.2	..	**25 Ouvrages en métaux (sauf mach.& mat.)**
26 Machinery equipment, instruments and transport equipment	..	47.0	..	72.1	..	76.7	..	**26 Machines & matériel, instruments & mat. de transport**
27 Machinery n.e.c.	..	7.4	..	7.0	..	8.6	..	27 Machines, n.c.a.
28 Office, account. & computing machinery	..	1.3	..	1.7	..	0.3	..	28 Mach. de bur., compta. & trait. inform.
29 Electrical machinery	..	22.0	..	21.8	..	20.6	..	29 Machines électriques
30 Electro. equip. (radio, TV & commun.)	..	14.1	..	33.2	..	37.0	..	30 Equip. électro. (radio, tv & communic.)
31 *Electro. comp. (inc. semi-conductors)*	..	..	..	..	..	..	..	31 *Compos. électro. (semi-cond. incl.)*
32 *TV, radio & communications equip.*	..	..	..	..	..	37.0	..	32 *Appareils radio, tv et de communi.*
33 Instruments	..	..	..	6.9	..	8.3	..	33 Instruments
34 Motor vehicles	..	0.2	..	0.7	..	1.2	..	34 Véhicules automobiles
35 Other transport equipment	..	1.0 [a]	..	0.9	..	0.7	..	35 Autres matériels de transport
36 *Ships*	..	..	..	..	..	0.4	..	36 *Navires*
37 *Aerospace*	..	..	..	..	..	..	..	37 *Aérospatiale*
38 *Other transport equipment n.e.c.*	..	..	..	..	..	..	..	38 *Autres matériels de transport n.c.a.*
39 Furniture, other manufacturing n.e.c.	..	2.0	..	0.1	..	1.5	..	**39 Meubles, autres activ. de fabricat. n.c.a.**
40 Furniture	..	0.8	..	0.1	..	0.2	..	40 Meubles
41 Other manufacturing n.e.c.	..	1.2	..	0.1	..	1.3	..	41 Autres activités de fabrication n.c.a.
42 Recycling	..	..	..	..	..	..	..	**42 Récupération**
43 ELECTRICITY, GAS & WATER SUPPLY	..	1.6	..	0.4	..	..	..	**43 DISTRIBUTION D'ELECTRICITE, GAZ & EAU**
44 CONSTRUCTION	..	0.5 [a]	..	2.9	..	5.8	..	**44 CONSTRUCTION**
45 SERVICES SECTOR	..	60.4	..	108.9	..	122.1	..	**45 SECTEUR DES SERVICES**
46 Wholesale, retail trade & repair of motor vehicles, etc.	..	13.9 [o]	..	40.8 [o]	..	25.1 [o]	..	**46 Commerce (gros & détail), réparation de véhicules, etc.**
47 Hotels & restaurants	..	.. [n]	..	.. [n]	..	.. [n]	..	**47 Hôtels et restaurants**
48 Transport & storage	..	0.2 [o]	..	0.1	..	4.2	..	**48 Transports, entreposage**
49 Communications	..	.. [n]	..	2.9	..	0.8	..	**49 Communications**
50 Post	..	..	..	..	..	..	..	50 Postes
51 Telecommunications	..	..	..	..	..	..	..	51 Télécommunications
52 Financial intermediation (inc. insur.)	..	0.4	..	0.1	..	6.5	..	**52 Intermédiation financ. (assurances inc.)**
53 Real estate, renting & busin. activity	..	42.9	..	62.7	..	84.2	..	**53 Immobil., locat. & serv. aux entreprises**
54 Computer & related activities	..	16.0	..	14.0	..	25.5	..	54 Informatique et activités connexes
55 *Software consultancy*	..	15.2	..	12.2	..	23.5	..	55 *Production de logiciels*
56 *Other computer services n.e.c.*	..	0.8	..	1.8	..	2.0	..	56 *Autres activités informatiques*
57 Research & development	..	12.7	..	21.7	..	34.4	..	57 Recherche & développement expérimental
58 Other business activities n.e.c.	..	14.3	..	26.9	..	24.4	..	58 Autres activ. & serv. aux entreprises
59 Comm., soc. & pers. serv. activ., etc.	..	1.9	..	2.3	..	1.3	..	**59 Adm. pub., serv. soc. & collec., etc**
TOTAL BUSINESS ENTERPRISE SECTOR	**..**	**240.3**	**..**	**312.5**	**..**	**324.1**	**..**	**TOTAL SECTEUR DES ENTREPRISES**

Source: OECD, R&D database (DSTI/EAS Division), January 2002.

Source : OCDE, base de données R-D (DSTI/Division AES), janvier 2002.

Total business enterprise intramural expenditure on R&D - Dépenses intra-muros totales de R-D des entreprises (BERD) by industry - (DIRDE) par industrie

9.2 Million constant USD(1995 prices and PPPs) - Millions de USD constants (prix et PPA de 1995)

	1994	1995	1996	1997	1998	1999	2000	
1 AGRICULTURE, HUNTING & FORESTRY	..	8.6	..	7.8	..	6.8	..	**1 AGRICULTURE, CHASSE & SYLVICULTURE**
2 MINING	..	0.5	..	4.0	..	2.2	..	**2 ACTIVITES EXTRACTIVES**
3 MANUFACTURING	..	112.1	..	121.5	..	120.0	..	**3 ACTIVITES DE FABRICATION**
4 Food, beverages & tobacco	..	40.7	..	43.3	..	39.7	..	**4 Produits alimentaires, boissons & tabac**
5 Food products & beverages	..	40.7 [o]	..	43.3 [o]	..	39.7 [o]	..	5 Produits alimentaires & boissons
6 Tobacco products	..	.. [n]	..	.. [n]	..	.. [n]	..	6 Produits à base de tabac
7 Textiles, wear app., fur & leather	..	17.2	..	11.5	..	11.2	..	**7 Textiles, habill., fourrures & cuirs**
8 Textiles	..	15.0	..	10.0	..	9.7	..	8 Textiles
9 Wearing apparel & fur	..	0.4	..	0.3	..	0.4	..	9 Articles d'habillement & fourrures
10 Leather products & footwear	..	1.7	..	1.2	..	1.1	..	10 Articles en cuir & chaussures
11 Wood, paper, printing, publishing	..	4.8	..	2.9	..	1.1	..	**11 Bois, papier, imprimerie, édition**
12 Wood & cork (not furniture)	..	0.9	..	0.3	..	0.9	..	12 Bois & liège (sauf meubles)
13 Pulp, paper & paper products	..	3.8	..	2.5	..	..	..	13 Carton, papier & articles
14 Publ., print. & repro. of rec. media	..	0.0	..	0.0	..	0.2	..	14 Edition, impress. & repro. enregis.
15 Coke, petroleum, nuclear fuel, chemicals, rubber plastics	..	10.9	..	11.7	..	8.8	..	**15 Coke, pétrole, combust. nucl., prod. chimi., caout. & mat. plast.**
16 Coke, ref. petrol. prod. & nuclear fuel	..	..	..	..	..	..	..	16 Coke, raffin. pétr. & combust. nucl.
17 Chemicals & chemical products	..	8.4	..	8.8	..	5.8	..	17 Produits chimiques
18 *Chemical prod.(less pharmaceuticals)*	..	6.2	..	5.8	..	3.8	..	18 *Prod. chimi. (moins prod. pharmaceutiques)*
19 *Pharmaceuticals*	..	2.2	..	3.0	..	2.0	..	19 *Produits pharmaceutiques*
20 Rubber & plastic products	..	2.5	..	2.9	..	3.0	..	20 Caoutchouc et matières plastiques
21 Non-metallic mineral products	..	1.6	..	1.3	..	1.8	..	**21 Produits minéraux non métalliques**
22 Basic metals	..	1.1	..	0.3	..	0.6	..	**22 Produits métallurgiques de base**
23 Basic metals, ferrous	..	0.7	..	0.0	..	..	..	23 Produits métallur. de base, ferreux
24 Basic metals, non-ferrous	..	0.4	..	0.3	..	0.6	..	24 Prod. métallur. de base, non ferreux
25 Fabricated metal products	..	2.4	..	2.5	..	5.4	..	**25 Ouvrages en métaux (sauf mach.& mat.)**
26 Machinery equipment, instruments and transport equipment	..	32.1	..	48.0	..	50.4	..	**26 Machines & matériel, instruments & mat. de transport**
27 Machinery n.e.c.	..	5.1	..	4.7	..	5.7	..	27 Machines, n.c.a.
28 Office, account. & computing machinery	..	0.9	..	1.2	..	0.2	..	28 Mach. de bur., compta. & trait. inform.
29 Electrical machinery	..	15.0	..	14.5	..	13.5	..	29 Machines électriques
30 Electro. equip. (radio, TV & commun.)	..	9.6	..	22.1	..	24.3	..	30 Equip. électro. (radio, tv & communic.)
31 *Electro. comp. (inc. semi-conductors)*	..	..	..	..	..	..	..	31 *Compos. électro. (semi-cond. incl.)*
32 *TV, radio & communications equip.*	..	..	..	..	..	24.3	..	32 *Appareils radio, tv et de communi.*
33 Instruments	..	..	..	4.6	..	5.5	..	33 Instruments
34 Motor vehicles	..	0.1	..	0.4	..	0.8	..	34 Véhicules automobiles
35 Other transport equipment	..	0.7 [a]	..	0.6	..	0.5	..	35 Autres matériels de transport
36 *Ships*	..	..	..	..	..	0.3	..	36 *Navires*
37 *Aerospace*	..	..	..	..	..	..	..	37 *Aérospatiale*
38 *Other transport equipment n.e.c.*	..	..	..	..	..	..	..	38 *Autres matériels de transport n.c.a.*
39 Furniture, other manufacturing n.e.c.	..	1.4	..	0.1	..	1.0	..	**39 Meubles, autres activ. de fabricat. n.c.a.**
40 Furniture	..	0.5	..	0.0	..	0.1	..	40 Meubles
41 Other manufacturing n.e.c.	..	0.8	..	0.0	..	0.9	..	41 Autres activités de fabrication n.c.a.
42 Recycling	..	..	..	..	..	..	..	**42 Récupération**
43 ELECTRICITY, GAS & WATER SUPPLY	..	1.1	..	0.3	..	..	..	**43 DISTRIBUTION D'ELECTRICITE, GAZ & EAU**
44 CONSTRUCTION	..	0.3 [a]	..	1.9	..	3.8	..	**44 CONSTRUCTION**
45 SERVICES SECTOR	..	41.2	..	72.4	..	80.4	..	**45 SECTEUR DES SERVICES**
46 Wholesale, retail trade & repair of motor vehicles, etc.	..	9.5 [o]	..	27.2 [o]	..	16.5 [o]	..	**46 Commerce (gros & détail), réparation de véhicules, etc.**
47 Hotels & restaurants	..	.. [n]	..	.. [n]	..	.. [n]	..	**47 Hôtels et restaurants**
48 Transport & storage	..	0.1 [o]	..	0.0	..	2.8	..	**48 Transports, entreposage**
49 Communications	..	.. [n]	..	1.9	..	0.5	..	**49 Communications**
50 Post	..	..	..	..	..	..	..	50 Postes
51 Telecommunications	..	..	..	..	..	..	..	51 Télécommunications
52 Financial intermediation (inc. insur.)	..	0.2	..	0.1	..	4.3	..	**52 Intermédiation financ. (assurances inc.)**
53 Real estate, renting & busin. activity	..	29.3	..	41.7	..	55.4	..	**53 Immobil., locat. & serv. aux entreprises**
54 Computer & related activities	..	10.9	..	9.3	..	16.8	..	54 Informatique et activités connexes
55 *Software consultancy*	..	10.3	..	8.1	..	15.5	..	55 *Production de logiciels*
56 *Other computer services n.e.c.*	..	0.6	..	1.2	..	1.3	..	56 *Autres activités informatiques*
57 Research & development	..	8.6	..	14.4	..	22.6	..	57 Recherche & développement expérimental
58 Other business activities n.e.c.	..	9.7	..	17.9	..	16.0	..	58 Autres activ. & serv. aux entreprises
59 Comm., soc. & pers. serv. activ., etc.	..	1.3	..	1.5	..	0.9	..	**59 Adm. pub., serv. soc. & collec., etc**
TOTAL BUSINESS ENTERPRISE SECTOR	**..**	**163.8**	**..**	**207.9**	**..**	**213.3**	**..**	**TOTAL SECTEUR DES ENTREPRISES**

Source: OECD, R&D database (DSTI/EAS Division), January 2002.

Source : OCDE, base de données R-D (DSTI/Division AES), janvier 2002.

Higher education R&D personnel (HEMP) - Personnel de R-D du secteur de l'enseignement supérieur

by field of science and occupation - par discipline scientifique et par profession

In full-time equivalent — *En équivalence plein-temps*

	1994	1995	1996	1997	1998	1999	2000	
Total R&D personnel								**Personnel total de R-D**
Natural sciences	..	..	..	..	..	..	..	Sciences exactes et naturelles
Engineering	..	..	..	..	..	..	..	Sciences de l'ingénieur
Medical sciences	..	..	..	..	..	..	..	Sciences médicales
Agricultural sciences	..	..	..	..	..	..	..	Sciences agricoles
Sub-total NSE	..	..	..	..	..	..	..	Sous-total SENI
Social sciences	..	..	..	..	..	..	..	Sciences sociales
Humanities	..	..	..	..	..	..	..	Sciences humaines
Sub-total SSH	..	..	..	..	..	..	..	Sous-total SSH
Total NSE+SSH	..	3 735	..	6 156	..	..	..	*Total SENI+SSH*
Researchers								**Chercheurs**
Natural sciences	..	..	..	..	..	..	..	Sciences exactes et naturelles
Engineering	..	..	..	..	..	..	..	Sciences de l'ingénieur
Medical sciences	..	..	..	..	..	..	..	Sciences médicales
Agricultural sciences	..	..	..	..	..	..	..	Sciences agricoles
Sub-total NSE	..	..	..	..	..	..	..	Sous-total SENI
Social sciences	..	..	..	..	..	..	..	Sciences sociales
Humanities	..	..	..	..	..	..	..	Sciences humaines
Sub-total SSH	..	..	..	..	..	..	..	Sous-total SSH
Total NSE+SSH	..	3 026	..	4 807	..	..	..	*Total SENI+SSH*
Technicians								**Technicians**
Natural sciences	..	..	..	..	..	..	..	Sciences exactes et naturelles
Engineering	..	..	..	..	..	..	..	Sciences de l'ingénieur
Medical sciences	..	..	..	..	..	..	..	Sciences médicales
Agricultural sciences	..	..	..	..	..	..	..	Sciences agricoles
Sub-total NSE	..	..	..	..	..	..	..	Sous-total SENI
Social sciences	..	..	..	..	..	..	..	Sciences sociales
Humanities	..	..	..	..	..	..	..	Sciences humaines
Sub-total SSH	..	..	..	..	..	..	..	Sous-total SSH
Total NSE+SSH	..	497	..	714	..	..	..	*Total SENI+SSH*
Other								**Autres**
Natural sciences	..	..	..	..	..	..	..	Sciences exactes et naturelles
Engineering	..	..	..	..	..	..	..	Sciences de l'ingénieur
Medical sciences	..	..	..	..	..	..	..	Sciences médicales
Agricultural sciences	..	..	..	..	..	..	..	Sciences agricoles
Sub-total NSE	..	..	..	..	..	..	..	Sous-total SENI
Social sciences	..	..	..	..	..	..	..	Sciences sociales
Humanities	..	..	..	..	..	..	..	Sciences humaines
Sub-total SSH	..	..	..	..	..	..	..	Sous-total SSH
Total NSE+SSH	..	213	..	635	..	..	..	*Total SENI+SSH*

Table 18 - Tableau 18

Government budget appropriations or outlays for R&D (GBAORD) - Crédits budgétaires publics de R-D (CBPRD)

by socio-economic objective - par objectif socio-économique

Million NZD — *Millions NZD*

	1995	1996	1997	1998	1999	2000	2001	
Agriculture, forestry and fishing	144.0	..	168.3	..	..	..	..	Agriculture, sylviculture et pêche
Industrial development	66.9	..	70.8	..	..	..	..	Développement industriel
Energy	8.3	..	7.8	..	..	..	..	Energie
Transport & telecommunications	6.2	..	5.0	..	..	..	..	*Transports & télécommunications*
Urban & rural planning	1.4	..	0.0	..	..	..	..	*Aménagement urbain et rural*
ST Infrastructure	7.7	..	5.0	..	..	..	..	ST Infrastructure
Prevention of pollution	..	..	..	..	..	..	..	*Prévention de la pollution*
Identificat. & treatment of pollution	..	..	..	..	..	..	..	*Détect. pollut. & mesures antipollut.*
ST Environment	14.9	..	4.3	..	..	..	..	ST Environnement
Health	26.7	..	33.9	..	..	..	..	Santé
Social development & services	18.9	..	15.9	..	..	..	..	Développement et services sociaux
Earth and atmosphere	53.7	..	94.3	..	..	..	..	Milieu terrestre et atmosphère
Advancement of research	7.9	..	26.2	..	..	..	..	*Promotion de la recherche*
General university funds	96.1	..	130.9	..	..	..	..	*Fonds généraux des universités*
ST Advancement of knowledge	104.0	..	157.2	..	..	..	..	ST promotion gén. des connaiss.
Civil space	0.1	..	0.3	..	..	..	..	Espace civil
Defence	5.4	..	3.8	..	..	..	..	Défense
Not elsewhere classified	0.1	..	0.4	..	..	..	..	Non classé
Total	**450.7**	**..**	**561.8**	**..**	**..**	**..**	**..**	**Total**

Source: OECD, R&D database (DSTI/EAS Division), January 2002.

Source : OCDE, base de données R-D (DSTI/Division AES), janvier 2002.

NEW ZEALAND — Table 19 - Tableau 19 — NOUVELLE-ZÉLANDE

Technology balance of payments - Balance des paiements technologiques

by type of transfer - selon la nature des transferts

Million NZD — *Millions NZD*

	1983	1984	1985	1986	1987	1988	1989	1990	1991	
Receipts										**Recettes**
Related firms	..	..	..	..	..	..	..	..	..	Firmes affiliées
Unrelated firms	..	..	..	..	..	..	..	..	..	Firmes non affiliées
Other	..	..	..	..	..	..	..	..	..	Autre
Total	..	..	..	..	..	..	45	36	36	*Total*
Payments										**Paiements**
Related firms	..	..	..	..	..	..	..	..	..	Firmes affiliées
Unrelated firms	..	..	..	..	..	..	..	..	..	Firmes non affiliées
Other	..	..	..	..	..	..	..	..	..	Autre
Total	..	..	..	..	..	..	41	34	27	*Total*

	1992	1993	1994	1995	1996	1997	1998	1999	2000	
Receipts										**Recettes**
Related firms	..	..	..	..	..	..	..	..	..	Firmes affiliées
Unrelated firms	..	..	..	..	..	..	..	..	..	Firmes non affiliées
Other	..	..	..	..	..	..	..	..	..	Autre
Total	37	36	..	31	..	8	..	..	..	*Total*
Payments										**Paiements**
Related firms	..	..	..	..	..	..	..	..	..	Firmes affiliées
Unrelated firms	..	..	..	..	..	..	..	..	..	Firmes non affiliées
Other	..	..	..	..	..	..	..	..	..	Autre
Total	15	13	..	12	..	14	..	..	..	*Total*

Source: OECD, TBPdatabase (DSTI/EAS Division), Januay 2002.

Source : OCDE, base de données TBP (DSTI/Division AES), janvier 2002.

NORWAY — Table 1 - Tableau 1 — NORVEGE

Gross domestic expenditure on R&D (GERD) - Dépenses intérieures brutes de R-D (DIRD)

by sector of performance and source of funds - par secteur d'exécution et par source de financement

	1994	1995	1996	1997	1998	1999	2000	
				1.1 Million NOK - Millions NOK				
Business enterprise								**Entreprises**
financed by								*financé par*
Business enterprise	..	7 441.5 [a]	..	8 427.4	..	9 446.0	..	Entreprises
Direct government	..	1 073.0 [a]	..	1 135.3	..	1 102.3	..	État direct
Higher education	..	0.0	..	0.0	..	0.0	..	Enseignement supérieur
Private non-profit	..	5.1 [a]	..	0.0	..	0.0	..	Institutions sans but lucratif
Funds from abroad	..	501.6 [a]	..	789.1	..	821.2	..	Fonds de l'étranger
Total	..	9 021.2 [a]	..	10 351.8	..	11 369.5	..	*Total*
Government								**État**
financed by								*financé par*
Business enterprise	..	274.5	..	301.9	..	322.8	..	Entreprises
Direct government	..	2 222.1	..	2 395.3	..	2 499.5	..	État direct
Higher education	..	0.0	..	0.0	..	0.0	..	Enseignement supérieur
Private non-profit	..	33.0	..	35.4	..	7.4	..	Institutions sans but lucratif
Funds from abroad	..	217.7	..	257.0	..	300.1	..	Fonds de l'étranger
Total	..	2 747.3	..	2 989.6	..	3 129.8	..	*Total*
Higher education								**Enseignement supérieur**
financed by								*financé par*
Business enterprise	..	219.8	..	254.3	..	295.6	..	Entreprises
Direct government	..	807.5	..	967.7	..	1 053.7	..	État direct
General university funds	..	2 894.7	..	3 305.7	..	3 989.1	..	Fonds généraux des universités
Sub-total government	..	3 702.2	..	4 273.4	..	5 042.8	..	Sous-Total État
Higher education	..	42.0	..	60.1	..	121.9	..	Enseignement supérieur
Private non-profit	..	110.6	..	129.2	..	191.0	..	Institutions sans but lucratif
Funds from abroad	..	64.5	..	128.8	..	168.1	..	Fonds de l'étranger
Total	..	4 139.1	..	4 845.8	..	5 819.4	..	*Total*
Private non-profit								**Institutions sans but lucratif**
financed by								*financé par*
Business enterprise	..	..	..	..	..	..	..	Entreprises
Direct government	..	..	..	..	..	..	..	État direct
Higher education	..	..	..	..	..	..	..	Enseignement supérieur
Private non-profit	..	..	..	..	..	..	..	Institutions sans but lucratif
Funds from abroad	..	..	..	..	..	..	..	Fonds de l'étranger
Total	..	..	..	..	..	..	..	*Total*
GERD								**DIRD**
financed by								*financé par*
Business enterprise	..	7 935.8 [a]	..	8 983.6	..	10 064.4	..	Entreprises
Direct government	..	4 102.6	..	4 498.3	..	4 655.5	..	État direct
General university funds	..	2 894.7	..	3 305.7	..	3 989.1	..	Fonds généraux des universités
Sub-total government	..	6 997.3	..	7 804.0	..	8 644.6	..	Sous-Total État
Higher education	..	42.0	..	60.1	..	121.9	..	Enseignement supérieur
Private non-profit	..	148.7	..	164.6	..	198.4	..	Institutions sans but lucratif
Funds from abroad	..	783.8	..	1 174.9	..	1 289.4	..	Fonds de l'étranger
Total	..	15 907.6 [a]	..	18 187.2	..	20 318.7	..	*Total*
				1.2 Million constant USD (1995 prices and PPPs) - Millions de USD constants (prix et PPA de 1995)				
Business enterprise								**Entreprises**
financed by								*financé par*
Business enterprise	..	813.8 [a]	..	878.6	..	930.6 [c]	..	Entreprises
Direct government	..	117.3 [a]	..	118.4	..	108.6 [c]	..	État direct
Higher education	..	0.0	..	0.0	..	0.0 [c]	..	Enseignement supérieur
Private non-profit	..	0.6 [a]	..	0.0	..	0.0 [c]	..	Institutions sans but lucratif
Funds from abroad	..	54.9 [a]	..	82.3	..	80.9 [c]	..	Fonds de l'étranger
Total	..	986.6 [a]	..	1 079.2	..	1 120.1 [c]	..	*Total*
Government								**État**
financed by								*financé par*
Business enterprise	..	30.0	..	31.5	..	31.8 [c]	..	Entreprises
Direct government	..	243.0	..	249.7	..	246.3 [c]	..	État direct
Higher education	..	0.0	..	0.0	..	0.0 [c]	..	Enseignement supérieur
Private non-profit	..	3.6	..	3.7	..	0.7 [c]	..	Institutions sans but lucratif
Funds from abroad	..	23.8	..	26.8	..	29.6 [c]	..	Fonds de l'étranger
Total	..	300.4	..	311.7	..	308.4 [c]	..	*Total*
Higher education								**Enseignement supérieur**
financed by								*financé par*
Business enterprise	..	24.0	..	26.5	..	29.1 [c]	..	Entreprises
Direct government	..	88.3	..	100.9	..	103.8 [c]	..	État direct
General university funds	..	316.6	..	344.6	..	393.0 [c]	..	Fonds généraux des universités
Sub-total government	..	404.9	..	445.5	..	496.8 [c]	..	Sous-Total État
Higher education	..	4.6	..	6.3	..	12.0 [c]	..	Enseignement supérieur
Private non-profit	..	12.1	..	13.5	..	18.8 [c]	..	Institutions sans but lucratif
Funds from abroad	..	7.1	..	13.4	..	16.6 [c]	..	Fonds de l'étranger
Total	..	452.6	..	505.2	..	573.3 [c]	..	*Total*
Private non-profit								**Institutions sans but lucratif**
financed by								*financé par*
Business enterprise	..	..	..	..	..	..	..	Entreprises
Direct government	..	..	..	..	..	..	..	État direct
Higher education	..	..	..	..	..	..	..	Enseignement supérieur
Private non-profit	..	..	..	..	..	..	..	Institutions sans but lucratif
Funds from abroad	..	..	..	..	..	..	..	Fonds de l'étranger
Total	..	..	..	..	..	..	..	*Total*
GERD								**DIRD**
financed by								*financé par*
Business enterprise	..	867.9 [a]	..	936.6	..	991.6 [c]	..	Entreprises
Direct government	..	448.7	..	469.0	..	458.7 [c]	..	État direct
General university funds	..	316.6	..	344.6	..	393.0 [c]	..	Fonds généraux des universités
Sub-total government	..	765.2	..	813.6	..	851.7 [c]	..	Sous-Total État
Higher education	..	4.6	..	6.3	..	12.0 [c]	..	Enseignement supérieur
Private non-profit	..	16.3	..	17.2	..	19.5 [c]	..	Institutions sans but lucratif
Funds from abroad	..	85.7	..	122.5	..	127.0 [c]	..	Fonds de l'étranger
Total	..	1 739.6 [a]	..	1 896.0	..	2 001.8 [c]	..	*Total*

Source: OECD, R&D database (DSTI/EAS Division), January 2002.

Source : OCDE, base de données R-D (DSTI/Division AES), janvier 2002.

Table 2 - Tableau 2

Gross domestic expenditure on R&D (GERD) - Dépenses intérieures brutes de R-D (DIRD)

by sector of performance and type of costs - par secteur d'exécution et par type de coûts

Million NOK — *Millions NOK*

	1994	1995	1996	1997	1998	1999	2000	
Business enterprise								**Entreprises**
Labour costs	..	4 703.5 [a]	..	5 837.1	..	6 616.0	..	Coûts du personnel
Other current costs	..	3 348.4 [a]	..	3 566.5	..	3 870.1	..	Autres coûts courants
Total current expenditure	..	8 051.9 [a]	..	9 403.6	..	10 486.1	..	*Total dépenses courantes*
Land and buildings	..	51.7 [a]	..	59.4	..	152.0	..	Terrains et constructions
Instruments and equipment	..	917.6 [a]	..	888.8	..	731.4	..	Instruments et équipements
Total capital expenditure	..	969.3 [a]	..	948.2	..	883.4	..	*Total dépenses en capital*
Total	..	9 021.2 [a]	..	10 351.8	..	11 369.5	..	*Total*
Government								**État**
Labour costs	..	1 618.5	..	1 775.3	..	1 902.5	..	Coûts du personnel
Other current costs	..	990.8	..	1 041.8	..	1 110.1	..	Autres coûts courants
Total current expenditure	..	2 609.3	..	2 817.1	..	3 012.6	..	*Total dépenses courantes*
Land and buildings	..	9.5	..	43.8	..	15.7	..	Terrains et constructions
Instruments and equipment	..	128.5	..	128.7	..	101.5	..	Instruments et équipements
Total capital expenditure	..	138.0	..	172.5	..	117.2	..	*Total dépenses en capital*
Total	..	2 747.3	..	2 989.6	..	3 129.8	..	*Total*
Higher education								**Enseignement supérieur**
Labour costs	..	2 109.9	..	2 418.4	..	2 737.7	..	Coûts du personnel
Other current costs	..	1 570.2	..	1 806.2	..	2 178.6	..	Autres coûts courants
Total current expenditure	..	3 680.1	..	4 224.6	..	4 916.3	..	*Total dépenses courantes*
Land and buildings	..	297.6	..	418.5	..	687.4	..	Terrains et constructions
Instruments and equipment	..	161.4	..	202.7	..	215.7	..	Instruments et équipements
Total capital expenditure	..	459.0	..	621.2	..	903.1	..	*Total dépenses en capital*
Total	..	4 139.1	..	4 845.8	..	5 819.4	..	*Total*
Private non-profit								**Institutions sans but lucratif**
Labour costs	..	..	..	..	..	..	..	Coûts du personnel
Other current costs	..	..	..	..	..	..	..	Autres coûts courants
Total current expenditure	..	..	..	..	..	..	..	*Total dépenses courantes*
Land and buildings	..	..	..	..	..	..	..	Terrains et constructions
Instruments and equipment	..	..	..	..	..	..	..	Instruments et équipements
Total capital expenditure	..	..	..	..	..	..	..	*Total dépenses en capital*
Total	..	..	..	..	..	..	..	*Total*
GERD								**DIRD**
Labour costs	..	8 431.9 [a]	..	10 030.8	..	11 256.2	..	Coûts du personnel
Other current costs	..	5 909.4 [a]	..	6 414.5	..	7 158.8	..	Autres coûts courants
Total current expenditure	..	14 341.3 [a]	..	16 445.3	..	18 415.0	..	*Total dépenses courantes*
Land and buildings	..	358.8 [a]	..	521.7	..	855.1	..	Terrains et constructions
Instruments and equipment	..	1 207.5 [a]	..	1 220.2	..	1 048.6	..	Instruments et équipements
Total capital expenditure	..	1 566.3 [a]	..	1 741.9	..	1 903.7	..	*Total dépenses en capital*
Total	..	15 907.6 [a]	..	18 187.2	..	20 318.7	..	*Total*

Table 3 - Tableau 3

Current domestic expenditure on R&D - Dépenses intérieures courantes de R-D

by sector of performance and type of activity - par secteur d'exécution et par type d'activités

Million NOK — *Millions NOK*

	1994	1995	1996	1997	1998	1999	2000	
Business enterprise								**Entreprises**
Basic research	..	170.2 [a]	..	251.3	..	287.0	..	Recherche fondamentale
Applied research	..	2 470.0 [a]	..	2 735.3	..	3 109.1	..	Recherche appliquée
Experimental development	..	5 411.7 [a]	..	6 417.0	..	7 090.0	..	Développement expérimental
Not specified	..	0.0	..	0.0	..	0.0	..	Non spécifié
Total	..	8 051.9 [a]	..	9 403.6	..	10 486.1	..	*Total*
Government								**État**
Basic research	..	378.2	..	430.9	..	408.0	..	Recherche fondamentale
Applied research	..	1 566.1	..	1 597.7	..	1 725.9	..	Recherche appliquée
Experimental development	..	665.0	..	788.5	..	878.7	..	Développement expérimental
Not specified	..	0.0	..	0.0	..	0.0	..	Non spécifié
Total	..	2 609.3	..	2 817.1	..	3 012.6	..	*Total*
Higher education								**Enseignement supérieur**
Basic research	..	1 766.9	..	2 001.9	..	2 352.8	..	Recherche fondamentale
Applied research	..	1 362.5	..	1 557.2	..	1 808.1	..	Recherche appliquée
Experimental development	..	550.7	..	665.5	..	755.4	..	Développement expérimental
Not specified	..	0.0	..	0.0	..	0.0	..	Non spécifié
Total	..	3 680.1	..	4 224.6	..	4 916.3	..	*Total*
Private non-profit								**Institutions sans but lucratif**
Basic research	..	..	..	..	..	..	..	Recherche fondamentale
Applied research	..	..	..	..	..	..	..	Recherche appliquée
Experimental development	..	..	..	..	..	..	..	Développement expérimental
Not specified	..	..	..	..	..	..	..	Non spécifié
Total	..	..	..	..	..	..	..	*Total*
GERD								**DIRD**
Basic research	..	2 315.3	..	2 684.1	..	3 047.8	..	Recherche fondamentale
Applied research	..	5 398.7 [a]	..	5 890.2	..	6 643.1	..	Recherche appliquée
Experimental development	..	6 627.4 [a]	..	7 871.0	..	8 724.1	..	Développement expérimental
Not specified	..	0.0	..	0.0	..	0.0	..	Non spécifié
Total	..	14 341.3 [a]	..	16 445.3	..	18 415.0	..	*Total*

Source: OECD, R&D database (DSTI/EAS Division), January 2002.

Source : OCDE, base de données R-D (DSTI/Division AES), janvier 2002.

NORWAY	Table 4 - Tableau 4	NORVEGE

Gross domestic expenditure on R&D (GERD) - Dépenses intérieures brutes de R-D (DIRD)

by sector of performance and socio-economic objective - par secteur d'exécution et par objectif socio-économique

Million NOK — *Millions NOK*

	1994	1995	1996	1997	1998	1999	2000	
Business enterprise								**Entreprises**
Agriculture, forestry and fishing	..	108.2 [a]	..	87.7	..	238.8	..	Agriculture, sylviculture et pêche
Industrial development	..	5 286.2 [a]	..	5 530.1	..	5 741.8	..	Développement industriel
Energy	..	127.0 [a]	..	804.9	..	1 052.8	..	Energie
Transport and telecommunications	..	696.4 [a]	..	796.9	..	892.0	..	Transports et télécommunications
Urban and rural planning	..	26.5 [a]	..	28.6	..	38.2	..	Aménagement urbain et rural
Prevention of pollution	..	..	..	..	..	..	..	Prévention de la pollution
Identification & treatment of pollution	..	80.0 [a]	..	87.3	..	91.7	..	Détection pollu. & mesures antipollu.
Health	..	9.9 [a]	..	13.3	..	49.9	..	Santé
Social development and services	..	100.6 [a]	..	107.0	..	72.8	..	Développement et services sociaux
Earth and atmosphere	..	88.8 [a]	..	44.9	..	30.8	..	Milieu terrestre et atmosphère
Advancement of knowledge	..	..	..	1.6	..	0.0	..	Promotion générale des connaissances
Civil space	..	8.6 [a]	..	7.3	..	11.0	..	Espace civil
Defence	..	2.3 [a]	..	2.9	..	14.8	..	Défense
Not elsewhere classified	..	2 486.7 [a]	..	2 839.3	..	3 134.9	..	Non classé
Total	..	9 021.2 [a]	..	10 351.8	..	11 369.5	..	*Total*
Government								**État**
Agriculture, forestry and fishing	..	577.0	..	593.1	..	663.0	..	Agriculture, sylviculture et pêche
Industrial development	..	219.9	..	260.1	..	255.8	..	Développement industriel
Energy	..	171.0	..	179.6	..	207.9	..	Energie
Transport and telecommunications	..	46.4	..	52.9	..	65.5	..	Transports et télécommunications
Urban and rural planning	..	8.4	..	14.2	..	18.9	..	Aménagement urbain et rural
Prevention of pollution	..	..	..	..	..	..	..	Prévention de la pollution
Identification & treatment of pollution	..	354.6	..	445.9	..	410.3	..	Détection pollu. & mesures antipollu.
Health	..	312.5	..	328.8	..	304.0	..	Santé
Social development and services	..	538.0	..	548.2	..	623.2	..	Développement et services sociaux
Earth and atmosphere	..	34.1	..	35.4	..	51.9	..	Milieu terrestre et atmosphère
Advancement of knowledge	..	75.0	..	66.8	..	82.3	..	Promotion générale des connaissances
Civil space	..	31.9	..	17.3	..	24.4	..	Espace civil
Defence	..	240.6	..	274.8	..	305.4	..	Défense
Not elsewhere classified	..	137.9	..	172.5	..	117.2	..	Non classé
Total	..	2 747.3	..	2 989.6	..	3 129.8	..	*Total*
Higher education								**Enseignement supérieur**
Agriculture, forestry and fishing	..	203.2	..	191.9	..	233.0	..	Agriculture, sylviculture et pêche
Industrial development	..	327.6	..	364.7	..	378.8	..	Développement industriel
Energy	..	58.5	..	56.0	..	79.8	..	Energie
Transport and telecommunications	..	65.2	..	66.8	..	105.2	..	Transports et télécommunications
Urban and rural planning	..	20.7	..	18.0	..	25.5	..	Aménagement urbain et rural
Prevention of pollution	..	..	..	..	..	..	..	Prévention de la pollution
Identification & treatment of pollution	..	165.6	..	205.9	..	205.8	..	Détection pollu. & mesures antipollu.
Health	..	699.4	..	863.1	..	1 129.9	..	Santé
Social development and services	..	526.5	..	799.7	..	953.5	..	Développement et services sociaux
Earth and atmosphere	..	145.3	..	161.1	..	184.6	..	Milieu terrestre et atmosphère
Advancement of knowledge	..	1 435.8	..	1 482.9	..	1 583.0	..	Promotion générale des connaissances
Civil space	..	23.2	..	11.7	..	28.3	..	Espace civil
Defence	..	9.1	..	2.8	..	8.9	..	Défense
Not elsewhere classified	..	459.0	..	621.2	..	903.1	..	Non classé
Total	..	4 139.1	..	4 845.8	..	5 819.4	..	*Total*
Private non-profit								**Institutions sans but lucratif**
Agriculture, forestry and fishing	..	..	..	..	..	..	..	Agriculture, sylviculture et pêche
Industrial development	..	..	..	..	..	..	..	Développement industriel
Energy	..	..	..	..	..	..	..	Energie
Transport and telecommunications	..	..	..	..	..	..	..	Transports et télécommunications
Urban and rural planning	..	..	..	..	..	..	..	Aménagement urbain et rural
Prevention of pollution	..	..	..	..	..	..	..	Prévention de la pollution
Identification & treatment of pollution	..	..	..	..	..	..	..	Détection pollu. & mesures antipollu.
Health	..	..	..	..	..	..	..	Santé
Social development and services	..	..	..	..	..	..	..	Développement et services sociaux
Earth and atmosphere	..	..	..	..	..	..	..	Milieu terrestre et atmosphère
Advancement of knowledge	..	..	..	..	..	..	..	Promotion générale des connaissances
Civil space	..	..	..	..	..	..	..	Espace civil
Defence	..	..	..	..	..	..	..	Défense
Not elsewhere classified	..	..	..	..	..	..	..	Non classé
Total	..	..	..	..	..	..	..	*Total*
GERD								**DIRD**
Agriculture, forestry and fishing	..	888.4 [j]	..	872.7 [j]	..	1 134.8 [j]	..	Agriculture, sylviculture et pêche
Industrial development	..	5 833.7 [a,j]	..	6 154.9 [j]	..	6 376.4 [j]	..	Développement industriel
Energy	..	356.5 [a,j]	..	1 040.5 [j]	..	1 340.5 [j]	..	Energie
Transport and telecommunications	..	808.0 [a,j]	..	916.6 [j]	..	1 062.7 [j]	..	Transports et télécommunications
Urban and rural planning	..	55.6 [j]	..	60.8 [j]	..	82.6 [j]	..	Aménagement urbain et rural
Prevention of pollution	..	0.0	..	0.0 [j]	..	0.0 [j]	..	Prévention de la pollution
Identification & treatment of pollution	..	600.2 [j]	..	739.1 [j]	..	707.8 [j]	..	Détection pollu. & mesures antipollu.
Health	..	1 021.8 [j]	..	1 205.2 [j]	..	1 483.8 [j]	..	Santé
Social development and services	..	1 165.1 [j]	..	1 454.9 [j]	..	1 649.5 [j]	..	Développement et services sociaux
Earth and atmosphere	..	268.2 [j]	..	241.4 [j]	..	267.3 [j]	..	Milieu terrestre et atmosphère
Advancement of knowledge	..	1 510.8 [j]	..	1 551.3 [j]	..	1 665.3 [j]	..	Promotion générale des connaissances
Civil space	..	63.7 [j]	..	36.3 [j]	..	63.7 [j]	..	Espace civil
Defence	..	252.0 [j]	..	280.5 [j]	..	329.1 [j]	..	Défense
Not elsewhere classified	..	3 083.6 [a]	..	3 633.0	..	4 155.2	..	Non classé
Total	..	15 907.6 [a]	..	18 187.2	..	20 318.7	..	*Total*

Source: OECD, R&D database (DSTI/EAS Division), January 2002.

Source : OCDE, base de données R-D (DSTI/Division AES), janvier 2002.

Table 5 - Tableau 5

Total R&D personnel - Personnel total de R-D

by sector of employment and occupation - par secteur d'emploi et par profession

In full-time equivalent *En équivalence plein-temps*

	1994	1995	1996	1997	1998	1999	2000	
Business enterprise								**Entreprises**
RSE	..	..	..	..	..	..	..	Chercheurs
Technicians	..	..	..	..	..	..	..	Techniciens
Other occupations	..	..	..	..	..	..	..	Autres professions
Total	..	12 090 [a]	..	12 942	..	13 308	..	*Total*
Government								**État**
RSE	..	..	..	..	..	..	..	Chercheurs
Technicians	..	..	..	..	..	..	..	Techniciens
Other occupations	..	..	..	..	..	..	..	Autres professions
Total	..	4 893	..	4 873	..	4 779	..	*Total*
Higher education								**Enseignement supérieur**
RSE	..	4 993	..	5 091	..	5 521	..	Chercheurs
Technicians	..	..	..	..	..	..	..	Techniciens
Other occupations	..	..	..	..	..	..	..	Autres professions
Total	..	6 955	..	7 062	..	7 313	..	*Total*
Private non-profit								**Institutions sans but lucratif**
RSE	..	..	..	..	..	..	..	Chercheurs
Technicians	..	..	..	..	..	..	..	Techniciens
Other occupations	..	..	..	..	..	..	..	Autres professions
Total	..	..	..	..	..	..	..	*Total*
TOTAL								**TOTAL**
RSE	..	15 931 [a]	..	17 490	..	18 295	..	Chercheurs
Technicians	..	..	..	..	..	..	..	Techniciens
Other occupations	..	..	..	..	..	..	..	Autres professions
Total	..	23 938 [a]	..	24 877	..	25 400	..	*Total*

Table 6 - Tableau 6

Total R&D personnel - Personnel total de R-D

by sector of employment and formal qualification - par secteur d'emploi et par niveau de qualification

In full-time equivalent *En équivalence plein-temps*

	1994	1995	1996	1997	1998	1999	2000	
Business enterprise								**Entreprises**
University PhD level degrees	..	..	..	..	..	..	..	Diplômes univ. de niv. doctorat
Other university degrees	..	..	..	..	..	..	..	Autres diplômes universitaires
Sub-total university degrees	..	7 921 [a]	..	9 348	..	9 737	..	Sous-total Universitaires
Other post-secondary diplomas	..	..	..	..	..	..	..	Autres diplômes post-secondaires
Secondary diplomas	..	..	..	..	..	..	..	Diplômes de fin d'études secondaires
Other qualifications	..	..	..	..	..	..	..	Autres qualifications
Total	..	12 090 [a]	..	12 942	..	13 308	..	*Total*
Government								**État**
University PhD level degrees	..	..	..	..	..	..	..	Diplômes univ. de niv. doctorat
Other university degrees	..	..	..	..	..	..	..	Autres diplômes universitaires
Sub-total university degrees	..	3 017	..	3 051	..	3 037	..	Sous-total Universitaires
Other post-secondary diplomas	..	..	..	..	..	..	..	Autres diplômes post-secondaires
Secondary diplomas	..	..	..	..	..	..	..	Diplômes de fin d'études secondaires
Other qualifications	..	..	..	..	..	..	..	Autres qualifications
Total	..	4 893	..	4 873	..	4 779	..	*Total*
Higher education								**Enseignement supérieur**
University PhD level degrees	..	..	..	..	..	..	..	Diplômes univ. de niv. doctorat
Other university degrees	..	..	..	..	..	..	..	Autres diplômes universitaires
Sub-total university degrees	..	..	..	..	..	..	..	Sous-total Universitaires
Other post-secondary diplomas	..	..	..	..	..	..	..	Autres diplômes post-secondaires
Secondary diplomas	..	..	..	..	..	..	..	Diplômes de fin d'études secondaires
Other qualifications	..	..	..	..	..	..	..	Autres qualifications
Total	..	6 955	..	7 062	..	7 313	..	*Total*
Private non-profit								**Institutions sans but lucratif**
University PhD level degrees	..	..	..	..	..	..	..	Diplômes univ. de niv. doctorat
Other university degrees	..	..	..	..	..	..	..	Autres diplômes universitaires
Sub-total university degrees	..	..	..	..	..	..	..	Sous-total Universitaires
Other post-secondary diplomas	..	..	..	..	..	..	..	Autres diplômes post-secondaires
Secondary diplomas	..	..	..	..	..	..	..	Diplômes de fin d'études secondaires
Other qualifications	..	..	..	..	..	..	..	Autres qualifications
Total	..	..	..	..	..	..	..	*Total*
TOTAL								**TOTAL**
University PhD level degrees	..	..	..	..	..	..	..	Diplômes univ. de niv. doctorat
Other university degrees	..	..	..	..	..	..	..	Autres diplômes universitaires
Sub-total university degrees	..	15 931 [a]	..	17 490	..	18 295	..	Sous-total Universitaires
Other post-secondary diplomas	..	..	..	..	..	..	..	Autres diplômes post-secondaires
Secondary diplomas	..	..	..	..	..	..	..	Diplômes de fin d'études secondaires
Other qualifications	..	..	..	..	..	..	..	Autres qualifications
Total	..	23 938 [a]	..	24 877	..	25 400	..	*Total*

Source: OECD, R&D database (DSTI/EAS Division), January 2002.

Source : OCDE, base de données R-D (DSTI/Division AES), janvier 2002.

Table 7 - Tableau 7

Gross domestic expenditure on R&D (GERD) - Dépenses intérieures brutes de R-D (DIRD)

by sector of performance and main scientific field - par secteur d'exécution et par grand domaine scientifique

Million NOK — *Millions NOK*

	1994	1995	1996	1997	1998	1999	2000	
Business enterprise								**Entreprises**
Natural sciences and engineering	..	8 966.3 [a]	..	10 270.7	..	11 259.6	..	Sciences exactes, nat. et de l'ingénieur
Social sciences and humanities	..	54.9 [a]	..	81.1	..	109.9	..	Sciences sociales et humaines
All fields of science	..	9 021.2 [a]	..	10 351.8	..	11 369.5	..	*Tous domaines scientifiques*
Government								**État**
Natural sciences and engineering	..	1 978.2	..	2 179.0	..	2 220.6	..	Sciences exactes, nat. et de l'ingénieur
Social sciences and humanities	..	769.1	..	810.6	..	909.2	..	Sciences sociales et humaines
All fields of science	..	2 747.3	..	2 989.6	..	3 129.8	..	*Tous domaines scientifiques*
Higher education								**Enseignement supérieur**
Natural sciences and engineering	..	2 820.0	..	3 347.0	..	4 061.4	..	Sciences exactes, nat. et de l'ingénieur
Social sciences and humanities	..	1 319.1	..	1 498.8	..	1 758.0	..	Sciences sociales et humaines
All fields of science	..	4 139.1	..	4 845.8	..	5 819.4	..	*Tous domaines scientifiques*
Private non-profit								**Institutions sans but lucratif**
Natural sciences and engineering	..	..	..	..	..	..	..	Sciences exactes, nat. et de l'ingénieur
Social sciences and humanities	..	..	..	..	..	..	..	Sciences sociales et humaines
All fields of science	..	..	..	..	..	..	..	*Tous domaines scientifiques*
TOTAL								**TOTAL**
Natural sciences and engineering	..	13 764.5 [a]	..	15 796.7	..	17 541.6	..	Sciences exactes, nat. et de l'ingénieur
Social sciences and humanities	..	2 143.1	..	2 390.5	..	2 777.1	..	Sciences sociales et humaines
All fields of science	..	15 907.6 [a]	..	18 187.2	..	20 318.7	..	*Tous domaines scientifiques*

Table 8 - Tableau 8

Total R&D personnel - Personnel total de R-D

by sector of employment and main scientific field - par secteur d'emploi et par grand domaine scientifique

In full-time equivalent — *En équivalence plein-temps*

	1994	1995	1996	1997	1998	1999	2000	
Business enterprise								**Entreprises**
Natural sciences and engineering	..	12 002 [a]	..	12 827	..	13 166	..	Sciences exactes, nat. et de l'ingénieur
Social sciences and humanities	..	88 [a]	..	115	..	142	..	Sciences sociales et humaines
All fields of science	..	12 090 [a]	..	12 942	..	13 308	..	*Tous domaines scientifiques*
Government								**État**
Natural sciences and engineering	..	3 427	..	3 446	..	3 321	..	Sciences exactes, nat. et de l'ingénieur
Social sciences and humanities	..	1 466	..	1 427	..	1 458	..	Sciences sociales et humaines
All fields of science	..	4 893	..	4 873	..	4 779	..	*Tous domaines scientifiques*
Higher education								**Enseignement supérieur**
Natural sciences and engineering	..	4 795	..	4 771	..	4 889	..	Sciences exactes, nat. et de l'ingénieur
Social sciences and humanities	..	2 160	..	2 291	..	2 424	..	Sciences sociales et humaines
All fields of science	..	6 955	..	7 062	..	7 313	..	*Tous domaines scientifiques*
Private non-profit								**Institutions sans but lucratif**
Natural sciences and engineering	..	..	..	..	..	..	..	Sciences exactes, nat. et de l'ingénieur
Social sciences and humanities	..	..	..	..	..	..	..	Sciences sociales et humaines
All fields of science	..	..	..	..	..	..	..	*Tous domaines scientifiques*
TOTAL								**TOTAL**
Natural sciences and engineering	..	20 224 [a]	..	21 044	..	21 376	..	Sciences exactes, nat. et de l'ingénieur
Social sciences and humanities	..	3 714	..	3 833	..	4 024	..	Sciences sociales et humaines
All fields of science	..	23 938 [a]	..	24 877	..	25 400	..	*Tous domaines scientifiques*

Source: OECD, R&D database (DSTI/EAS Division), January 2002.

Source : OCDE, base de données R-D (DSTI/Division AES), janvier 2002.

Total business enterprise intramural expenditure on R&D - Dépenses intra-muros totales de R-D des entreprises

(BERD) by industry - (DIRDE) par industrie

	1994	1995	1996	1997	1998	1999	2000	
				9.1 Million NOK - Millions NOK				
1 AGRICULTURE, HUNTING & FORESTRY	..	41.7 [a]	..	8.3	..	169.4	..	**1 AGRICULTURE, CHASSE & SYLVICULTURE**
2 MINING	..	634.6 [a]	..	625.6	..	791.3	..	**2 ACTIVITES EXTRACTIVES**
3 MANUFACTURING	..	4 286.8 [a]	..	4 469.6	..	4 731.9	..	**3 ACTIVITES DE FABRICATION**
4 Food, beverages & tobacco	..	267.9 [a]	..	190.5	..	171.5	..	**4 Produits alimentaires, boissons & tabac**
5 Food products & beverages	..	265.0 [a]	..	190.5	..	171.0	..	5 Produits alimentaires & boissons
6 Tobacco products	..	2.9 [a]	..	0.0	..	0.5	..	6 Produits à base de tabac
7 Textiles, wear app., fur & leather	..	39.3 [a]	..	38.7	..	24.7	..	**7 Textiles, habill., fourrures & cuirs**
8 Textiles	..	17.0 [a]	..	13.5	..	17.1	..	8 Textiles
9 Wearing apparel & fur	..	18.2 [a]	..	17.1	..	6.1	..	9 Articles d'habillement & fourrures
10 Leather products & footwear	..	4.1 [a]	..	8.0	..	1.6	..	10 Articles en cuir & chaussures
11 Wood, paper, printing, publishing	..	249.9 [a]	..	224.6	..	236.6	..	**11 Bois, papier, imprimerie, édition**
12 Wood & cork (not furniture)	..	23.9 [a]	..	28.7	..	24.5	..	12 Bois & liège (sauf meubles)
13 Pulp, paper & paper products	..	148.1 [a]	..	144.7	..	168.8	..	13 Carton, papier & articles
14 Publ., print. & repro. of rec. media	..	77.9 [a]	..	51.2	..	43.3	..	14 Edition, impress. & repro. enregis.
15 Coke, petroleum, nuclear fuel, chemicals, rubber plastics	..	1 016.8 [a]	..	952.1	..	1 013.1	..	**15 Coke, pétrole, combust. nucl., prod. chimi., caout. & mat. plast.**
16 Coke, ref. petrol. prod. & nuclear fuel	..	65.1 [a]	..	71.4	..	110.8	..	16 Coke, raffin. pétr. & combust. nucl.
17 Chemicals & chemical products	..	902.1 [a]	..	766.5	..	831.8	..	17 Produits chimiques
18 *Chemical prod.(less pharmaceuticals)*	..	501.7 [a]	..	436.2	..	286.8	..	18 *Prod. chimi. (moins prod. pharmaceutiques)*
19 *Pharmaceuticals*	..	400.4 [a]	..	330.3	..	545.0	..	19 *Produits pharmaceutiques*
20 Rubber & plastic products	..	49.6 [a]	..	114.2	..	70.4	..	20 Caoutchouc et matières plastiques
21 Non-metallic mineral products	..	50.4 [a]	..	83.9	..	61.2	..	**21 Produits minéraux non métalliques**
22 Basic metals	..	211.2 [a]	..	215.2	..	358.6	..	**22 Produits métallurgiques de base**
23 Basic metals, ferrous	..	16.2 [a]	..	28.0	..	139.9	..	23 Produits métallur. de base, ferreux
24 Basic metals, non-ferrous	..	195.0 [a]	..	187.1	..	218.7	..	24 Prod. métallur. de base, non ferreux
25 Fabricated metal products	..	52.0 [a]	..	52.9	..	73.8	..	**25 Ouvrages en métaux (sauf mach.& mat.)**
26 Machinery equipment, instruments and transport equipment	..	2 338.7 [a]	..	2 619.6	..	2 712.3	..	**26 Machines & matériel, instruments & mat. de transport**
27 Machinery n.e.c.	..	667.8 [a]	..	695.9	..	569.2	..	27 Machines, n.c.a.
28 Office, account. & computing machinery	..	98.5 [a]	..	77.9	..	119.5	..	28 Mach. de bur., compta. & trait. inform.
29 Electrical machinery	..	229.3 [a]	..	296.7	..	363.2	..	29 Machines électriques
30 Electro. equip. (radio, TV & commun.)	..	784.4 [a]	..	773.8	..	798.0	..	30 Equip. électro. (radio, tv & communic.)
31 *Electro. comp. (inc. semi-conductors)*	..	13.4 [a]	..	7.8	..	21.8	..	31 *Compos. électro. (semi-cond. incl.)*
32 *TV, radio & communications equip.*	..	771.0 [a]	..	766.0	..	776.2	..	32 *Appareils radio, tv et de communi.*
33 Instruments	..	309.7 [a]	..	420.7	..	572.3	..	33 Instruments
34 Motor vehicles	..	53.2 [a]	..	151.5	..	120.4	..	34 Véhicules automobiles
35 Other transport equipment	..	195.8 [a]	..	203.1	..	169.7	..	35 Autres matériels de transport
36 *Ships*	..	130.9 [a]	..	156.1	..	129.7	..	36 *Navires*
37 *Aerospace*	..	40.7 [a]	..	45.0	..	35.0	..	37 *Aérospatiale*
38 *Other transport equipment n.e.c.*	..	24.2 [a]	..	2.0	..	5.0	..	38 *Autres matériels de transport n.c.a.*
39 Furniture, other manufacturing n.e.c.	..	60.6 [a]	..	90.8	..	78.0	..	**39 Meubles, autres activ. de fabricat. n.c.a.**
40 Furniture	..	40.3 [a]	..	59.3	..	58.2	..	40 Meubles
41 Other manufacturing n.e.c.	..	20.3 [a]	..	31.5	..	19.8	..	41 Autres activités de fabrication n.c.a.
42 Recycling	..	0.0	..	1.7	..	2.1	..	**42 Récupération**
43 ELECTRICITY, GAS & WATER SUPPLY	..	18.7 [a]	..	69.0	..	80.3	..	**43 DISTRIBUTION D'ELECTRICITE, GAZ & EAU**
44 CONSTRUCTION	..	71.4 [a]	..	65.5	..	51.8	..	**44 CONSTRUCTION**
45 SERVICES SECTOR	..	3 968.0 [a]	..	5 113.8	..	5 544.8	..	**45 SECTEUR DES SERVICES**
46 Wholesale, retail trade & repair of motor vehicles, etc.	..	269.3 [a]	..	156.1	..	333.8	..	**46 Commerce (gros & détail), réparation de véhicules, etc.**
47 Hotels & restaurants	..	2.3 [a]	..	0.0	..	0.0	..	**47 Hôtels et restaurants**
48 Transport & storage	..	51.6 [a]	..	91.6	..	29.0	..	**48 Transports, entreposage**
49 Communications	..	521.5 [a]	..	586.7	..	718.8	..	**49 Communications**
50 Post	..	39.2 [a]	..	0.0	..	0.0	..	50 Postes
51 Telecommunications	..	482.3 [a]	..	586.7	..	718.8	..	51 Télécommunications
52 Financial intermediation (inc. insur.)	..	236.2 [a]	..	259.2	..	196.0	..	**52 Intermédiation financ. (assurances inc.)**
53 Real estate, renting & busin. activity	..	2 872.4 [a]	..	4 017.2	..	4 264.5	..	**53 Immobil., locat. & serv. aux entreprises**
54 Computer & related activities	..	566.2 [a]	..	1 372.3	..	1 560.1	..	54 Informatique et activités connexes
55 *Software consultancy*	..	415.5 [a]	..	1 056.1	..	1 083.4	..	55 *Production de logiciels*
56 *Other computer services n.e.c.*	..	150.7 [a]	..	316.2	..	476.7	..	56 *Autres activités informatiques*
57 Research & development	..	1 709.2 [a]	..	1 882.0	..	1 829.5	..	57 Recherche & développement expérimental
58 Other business activities n.e.c.	..	597.0 [a]	..	762.9	..	874.9	..	58 Autres activ. & serv. aux entreprises
59 Comm., soc. & pers. serv. activ., etc.	..	14.7 [a]	..	3.0	..	2.8	..	**59 Adm. pub., serv. soc. & collec., etc**
TOTAL BUSINESS ENTERPRISE SECTOR	..	**9 021.2 [a]**	..	**10 351.8**	..	**11 369.5**	..	**TOTAL SECTEUR DES ENTREPRISES**

Source: OECD, R&D database (DSTI/EAS Division), January 2002.

Source : OCDE, base de données R-D (DSTI/Division AES), janvier 2002.

NORWAY Table 9 - Tableau 9 **NORVEGE**

Total business enterprise intramural expenditure on R&D - Dépenses intra-muros totales de R-D des entreprises
(BERD) by industry - (DIRDE) par industrie

9.2 Million constant USD(1995 prices and PPPs) - Millions de USD constants (prix et PPA de 1995)

	1994	1995	1996	1997	1998	1999	2000	
1 **AGRICULTURE, HUNTING & FORESTRY**	..	4.6 [a]	..	0.9	..	16.7 [c]	..	1 **AGRICULTURE, CHASSE & SYLVICULTURE**
2 **MINING**	..	69.4 [a]	..	65.2	..	78.0 [c]	..	2 **ACTIVITES EXTRACTIVES**
3 **MANUFACTURING**	..	468.8 [a]	..	466.0	..	466.2 [c]	..	3 **ACTIVITES DE FABRICATION**
4 **Food, beverages & tobacco**	..	29.3 [a]	..	19.9	..	16.9 [c]	..	4 **Produits alimentaires, boissons & tabac**
5 Food products & beverages	..	29.0 [a]	..	19.9	..	16.8 [c]	..	5 Produits alimentaires & boissons
6 Tobacco products	..	0.3 [a]	..	0.0	..	0.0 [c]	..	6 Produits à base de tabac
7 **Textiles, wear app., fur & leather**	..	4.3 [a]	..	4.0	..	2.4 [c]	..	7 **Textiles, habill., fourrures & cuirs**
8 Textiles	..	1.9 [a]	..	1.4	..	1.7 [c]	..	8 Textiles
9 Wearing apparel & fur	..	2.0 [a]	..	1.8	..	0.6 [c]	..	9 Articles d'habillement & fourrures
10 Leather products & footwear	..	0.4 [a]	..	0.8	..	0.2 [c]	..	10 Articles en cuir & chaussures
11 **Wood, paper, printing, publishing**	..	27.3 [a]	..	23.4	..	23.3 [c]	..	11 **Bois, papier, imprimerie, édition**
12 Wood & cork (not furniture)	..	2.6 [a]	..	3.0	..	2.4 [c]	..	12 Bois & liège (sauf meubles)
13 Pulp, paper & paper products	..	16.2 [a]	..	15.1	..	16.6 [c]	..	13 Carton, papier & articles
14 Publ., print. & repro. of rec. media	..	8.5 [a]	..	5.3	..	4.3 [c]	..	14 Edition, impress. & repro. enregis.
15 **Coke, petroleum, nuclear fuel, chemicals, rubber plastics**	..	111.2 [a]	..	99.3	..	99.8 [c]	..	15 **Coke, pétrole, combust. nucl., prod. chimi., caout. & mat. plast.**
16 Coke, ref. petrol. prod. & nuclear fuel	..	7.1 [a]	..	7.4	..	10.9 [c]	..	16 Coke, raffin. pétr. & combust. nucl.
17 Chemicals & chemical products	..	98.7 [a]	..	79.9	..	82.0 [c]	..	17 Produits chimiques
18 *Chemical prod.(less pharmaceuticals)*	..	54.9 [a]	..	45.5	..	28.3 [c]	..	18 *Prod. chimi. (moins prod. pharmaceutiques)*
19 *Pharmaceuticals*	..	43.8 [a]	..	34.4	..	53.7 [c]	..	19 *Produits pharmaceutiques*
20 Rubber & plastic products	..	5.4 [a]	..	11.9	..	6.9 [c]	..	20 Caoutchouc et matières plastiques
21 **Non-metallic mineral products**	..	5.5 [a]	..	8.7	..	6.0 [c]	..	21 **Produits minéraux non métalliques**
22 **Basic metals**	..	23.1 [a]	..	22.4	..	35.3 [c]	..	22 **Produits métallurgiques de base**
23 Basic metals, ferrous	..	1.8 [a]	..	2.9	..	13.8 [c]	..	23 Produits métallur. de base, ferreux
24 Basic metals, non-ferrous	..	21.3 [a]	..	19.5	..	21.5 [c]	..	24 Prod. métallur. de base, non ferreux
25 **Fabricated metal products**	..	5.7 [a]	..	5.5	..	7.3 [c]	..	25 **Ouvrages en métaux (sauf mach.& mat.)**
26 **Machinery equipment, instruments and transport equipment**	..	255.8 [a]	..	273.1	..	267.2 [c]	..	26 **Machines & matériel, instruments & mat. de transport**
27 Machinery n.e.c.	..	73.0 [a]	..	72.5	..	56.1 [c]	..	27 Machines, n.c.a.
28 Office, account. & computing machinery	..	10.8 [a]	..	8.1	..	11.8 [c]	..	28 Mach. de bur., compta. & trait. inform.
29 Electrical machinery	..	25.1 [a]	..	30.9	..	35.8 [c]	..	29 Machines électriques
30 Electro. equip. (radio, TV & commun.)	..	85.8 [a]	..	80.7	..	78.6 [c]	..	30 Equip. électro. (radio, tv & communic.)
31 *Electro. comp. (inc. semi-conductors)*	..	1.5 [a]	..	0.8	..	2.1 [c]	..	31 *Compos. électro. (semi-cond. incl.)*
32 *TV, radio & communications equip.*	..	84.3 [a]	..	79.9	..	76.5 [c]	..	32 *Appareils radio, tv et de communi.*
33 Instruments	..	33.9 [a]	..	43.9	..	56.4 [c]	..	33 Instruments
34 Motor vehicles	..	5.8 [a]	..	15.8	..	11.9 [c]	..	34 Véhicules automobiles
35 Other transport equipment	..	21.4 [a]	..	21.2	..	16.7 [c]	..	35 Autres matériels de transport
36 *Ships*	..	14.3 [a]	..	16.3	..	12.8 [c]	..	36 *Navires*
37 *Aerospace*	..	4.5 [a]	..	4.7	..	3.4 [c]	..	37 *Aérospatiale*
38 *Other transport equipment n.e.c.*	..	2.6 [a]	..	0.2	..	0.5 [c]	..	38 *Autres matériels de transport n.c.a.*
39 **Furniture, other manufacturing n.e.c.**	..	6.6 [a]	..	9.5	..	7.7 [c]	..	39 **Meubles, autres activ. de fabricat. n.c.a.**
40 Furniture	..	4.4 [a]	..	6.2	..	5.7 [c]	..	40 Meubles
41 Other manufacturing n.e.c.	..	2.2 [a]	..	3.3	..	2.0 [c]	..	41 Autres activités de fabrication n.c.a.
42 **Recycling**	..	0.0	..	0.2	..	0.2 [c]	..	42 **Récupération**
43 **ELECTRICITY, GAS & WATER SUPPLY**	..	2.0 [a]	..	7.2	..	7.9 [c]	..	43 **DISTRIBUTION D'ELECTRICITE, GAZ & EAU**
44 **CONSTRUCTION**	..	7.8 [a]	..	6.8	..	5.1 [c]	..	44 **CONSTRUCTION**
45 **SERVICES SECTOR**	..	433.9 [a]	..	533.1	..	546.3 [c]	..	45 **SECTEUR DES SERVICES**
46 **Wholesale, retail trade & repair of motor vehicles, etc.**	..	29.5 [a]	..	16.3	..	32.9 [c]	..	46 **Commerce (gros & détail), réparation de véhicules, etc.**
47 **Hotels & restaurants**	..	0.3 [a]	..	0.0	..	0.0 [c]	..	47 **Hôtels et restaurants**
48 **Transport & storage**	..	5.6 [a]	..	9.5	..	2.9 [c]	..	48 **Transports, entreposage**
49 **Communications**	..	57.0 [a]	..	61.2	..	70.8 [c]	..	49 **Communications**
50 Post	..	4.3 [a]	..	0.0	..	0.0 [c]	..	50 Postes
51 Telecommunications	..	52.7 [a]	..	61.2	..	70.8 [c]	..	51 Télécommunications
52 **Financial intermediation (inc. insur.)**	..	25.8 [a]	..	27.0	..	19.3 [c]	..	52 **Intermédiation financ. (assurances inc.)**
53 **Real estate, renting & busin. activity**	..	314.1 [a]	..	418.8	..	420.1 [c]	..	53 **Immobil., locat. & serv. aux entreprises**
54 Computer & related activities	..	61.9 [a]	..	143.1	..	153.7 [c]	..	54 Informatique et activités connexes
55 *Software consultancy*	..	45.4 [a]	..	110.1	..	106.7 [c]	..	55 *Production de logiciels*
56 *Other computer services n.e.c.*	..	16.5 [a]	..	33.0	..	47.0 [c]	..	56 *Autres activités informatiques*
57 Research & development	..	186.9 [a]	..	196.2	..	180.2 [c]	..	57 Recherche & développement expérimental
58 Other business activities n.e.c.	..	65.3 [a]	..	79.5	..	86.2 [c]	..	58 Autres activ. & serv. aux entreprises
59 **Comm., soc. & pers. serv. activ., etc.**	..	1.6 [a]	..	0.3	..	0.3 [c]	..	59 **Adm. pub., serv. soc. & collec., etc**
TOTAL BUSINESS ENTERPRISE SECTOR	..	**986.6** [a]	..	**1 079.2**	..	**1 120.1** [c]	..	**TOTAL SECTEUR DES ENTREPRISES**

Source: OECD, R&D database (DSTI/EAS Division), January 2002.

Source : OCDE, base de données R-D (DSTI/Division AES), janvier 2002.

NORWAY Table 10 - Tableau 10 **NORVEGE**

Business enterprise intramural expenditure on R&D (BERD) - Dépenses intra-muros de R-D des entreprises (DIRDE)

by industry and source of funds - par industrie et par source de financement

Million NOK *Millions NOK*

10.1 Business enterprise - Entreprises

	1994	1995	1996	1997	1998	1999	2000	
1 AGRICULTURE, HUNTING & FORESTRY	..	37.5 [a]	..	5.9	..	105.7	..	**1 AGRICULTURE, CHASSE & SYLVICULTURE**
2 MINING	..	633.2 [a]	..	611.3	..	779.7	..	**2 ACTIVITES EXTRACTIVES**
3 MANUFACTURING	..	3 784.8 [a]	..	3 802.7	..	4 045.8	..	**3 ACTIVITES DE FABRICATION**
4 Food, beverages & tobacco	..	251.2 [a]	..	158.2	..	165.1	..	**4 Produits alimentaires, boissons & tabac**
5 Food products & beverages	..	248.3 [a]	..	158.2	..	164.6	..	5 Produits alimentaires & boissons
6 Tobacco products	..	2.9 [a]	..	0.0	..	0.5	..	6 Produits à base de tabac
7 Textiles, wear app., fur & leather	..	37.7 [a]	..	36.7	..	23.4	..	**7 Textiles, habill., fourrures & cuirs**
8 Textiles	..	15.7 [a]	..	11.9	..	16.2	..	8 Textiles
9 Wearing apparel & fur	..	18.0 [a]	..	16.9	..	6.1	..	9 Articles d'habillement & fourrures
10 Leather products & footwear	..	4.0 [a]	..	7.9	..	1.1	..	10 Articles en cuir & chaussures
11 Wood, paper, printing, publishing	..	164.2 [a]	..	168.9	..	165.8	..	**11 Bois, papier, imprimerie, édition**
12 Wood & cork (not furniture)	..	19.6 [a]	..	24.2	..	22.8	..	12 Bois & liège (sauf meubles)
13 Pulp, paper & paper products	..	74.8 [a]	..	95.0	..	103.9	..	13 Carton, papier & articles
14 Publ., print. & repro. of rec. media	..	69.8 [a]	..	49.7	..	39.1	..	14 Edition, impress. & repro. enregis.
15 Coke, petroleum, nuclear fuel, chemicals, rubber plastics	..	986.9 [a]	..	893.8	..	874.6	..	**15 Coke, pétrole, combust. nucl., prod. chimi., caout. & mat. plast.**
16 Coke, ref. petrol. prod. & nuclear fuel	..	65.1 [a]	..	68.2	..	96.5	..	16 Coke, raffin. pétr. & combust. nucl.
17 Chemicals & chemical products	..	890.2 [a]	..	728.1	..	712.2	..	17 Produits chimiques
18 *Chemical prod.(less pharmaceuticals)*	..	495.1 [a]	..	424.0	..	272.9	..	18 *Prod. chimi. (moins prod. pharmaceutiques)*
19 *Pharmaceuticals*	..	395.1 [a]	..	304.1	..	439.3	..	19 *Produits pharmaceutiques*
20 Rubber & plastic products	..	31.6 [a]	..	97.5	..	65.9	..	20 Caoutchouc et matières plastiques
21 Non-metallic mineral products	..	48.2 [a]	..	80.6	..	55.3	..	**21 Produits minéraux non métalliques**
22 Basic metals	..	187.2 [a]	..	201.3	..	344.0	..	**22 Produits métallurgiques de base**
23 Basic metals, ferrous	..	15.9 [a]	..	26.0	..	136.3	..	23 Produits métallur. de base, ferreux
24 Basic metals, non-ferrous	..	171.3 [a]	..	175.3	..	207.7	..	24 Prod. métallur. de base, non ferreux
25 Fabricated metal products	..	48.1 [a]	..	44.6	..	57.3	..	**25 Ouvrages en métaux (sauf mach.& mat.)**
26 Machinery equipment, instruments and transport equipment	..	2 003.6 [a]	..	2 127.0	..	2 281.6	..	**26 Machines & matériel, instruments & mat. de transport**
27 Machinery n.e.c.	..	423.4 [a]	..	361.1	..	334.6	..	27 Machines, n.c.a.
28 Office, account. & computing machinery	..	98.5 [a]	..	75.0	..	117.4	..	28 Mach. de bur., compta. & trait. inform.
29 Electrical machinery	..	212.0 [a]	..	261.0	..	330.1	..	29 Machines électriques
30 Electro. equip. (radio, TV & commun.)	..	751.0 [a]	..	732.2	..	784.9	..	30 Equip. électro. (radio, tv & communic.)
31 *Electro. comp. (inc. semi-conductors)*	..	7.0 [a]	..	3.8	..	10.1	..	31 *Compos. électro. (semi-cond. incl.)*
32 *TV, radio & communications equip.*	..	744.0 [a]	..	728.4	..	774.8	..	32 *Appareils radio, tv et de communi.*
33 Instruments	..	288.9 [a]	..	378.5	..	446.4	..	33 Instruments
34 Motor vehicles	..	47.6 [a]	..	124.7	..	109.1	..	34 Véhicules automobiles
35 Other transport equipment	..	182.2 [a]	..	194.5	..	159.1	..	35 Autres matériels de transport
36 *Ships*	..	123.0 [a]	..	147.7	..	121.2	..	36 *Navires*
37 *Aerospace*	..	38.8 [a]	..	45.0	..	32.9	..	37 *Aérospatiale*
38 *Other transport equipment n.e.c.*	..	20.4 [a]	..	1.8	..	5.0	..	38 *Autres matériels de transport n.c.a.*
39 Furniture, other manufacturing n.e.c.	..	57.7 [a]	..	90.2	..	76.6	..	**39 Meubles, autres activ. de fabricat. n.c.a.**
40 Furniture	..	37.5 [a]	..	58.7	..	57.4	..	40 Meubles
41 Other manufacturing n.e.c.	..	20.2 [a]	..	31.5	..	19.2	..	41 Autres activités de fabrication n.c.a.
42 Recycling	..	0.0 [a]	..	1.4	..	2.1	..	**42 Récupération**
43 ELECTRICITY, GAS & WATER SUPPLY	..	17.4 [a]	..	66.0	..	79.5	..	**43 DISTRIBUTION D'ELECTRICITE, GAZ & EAU**
44 CONSTRUCTION	..	60.7 [a]	..	54.4	..	48.4	..	**44 CONSTRUCTION**
45 SERVICES SECTOR	..	2 907.9 [a]	..	3 887.2	..	4 386.9	..	**45 SECTEUR DES SERVICES**
46 Wholesale, retail trade & repair of motor vehicles, etc.	..	241.7 [a]	..	140.0	..	298.2	..	**46 Commerce (gros & détail), réparation de véhicules, etc.**
47 Hotels & restaurants	..	2.3 [a]	..	0.0	..	0.0	..	**47 Hôtels et restaurants**
48 Transport & storage	..	50.5 [a]	..	90.3	..	27.6	..	**48 Transports, entreposage**
49 Communications	..	505.7 [a]	..	551.5	..	688.3	..	**49 Communications**
50 Post	..	39.2 [a]	..	0.0	..	0.0	..	50 Postes
51 Telecommunications	..	466.5 [a]	..	551.5	..	688.3	..	51 Télécommunications
52 Financial intermediation (inc. insur.)	..	236.2 [a]	..	259.2	..	196.0	..	**52 Intermédiation financ. (assurances inc.)**
53 Real estate, renting & busin. activity	..	1 858.0 [a]	..	2 843.1	..	3 174.0	..	**53 Immobil., locat. & serv. aux entreprises**
54 Computer & related activities	..	476.0 [a]	..	1 129.5	..	1 355.3	..	54 Informatique et activités connexes
55 *Software consultancy*	..	326.1 [a]	..	840.4	..	899.3	..	55 *Production de logiciels*
56 *Other computer services n.e.c.*	..	149.9 [a]	..	289.1	..	456.0	..	56 *Autres activités informatiques*
57 Research & development	..	909.4 [a]	..	1 018.9	..	1 026.8	..	57 Recherche & développement expérimental
58 Other business activities n.e.c.	..	472.6 [a]	..	694.8	..	792.0	..	58 Autres activ. & serv. aux entreprises
59 Comm., soc. & pers. serv. activ., etc.	..	13.5 [a]	..	3.0	..	2.8	..	**59 Adm. pub., serv. soc. & collec., etc**
TOTAL BUSINESS ENTERPRISE SECTOR	**..**	**7 441.5** [a]	**..**	**8 427.4**	**..**	**9 446.0**	**..**	**TOTAL SECTEUR DES ENTREPRISES**

Source: OECD, R&D database (DSTI/EAS Division), January 2002.

Source : OCDE, base de données R-D (DSTI/Division AES), janvier 2002.

Business enterprise intramural expenditure on R&D (BERD) - Dépenses intra-muros de R-D des entreprises (DIRDE)

by industry and source of funds - par industrie et par source de financement

Million NOK — *Millions NOK*

	1994	1995	1996	1997	1998	1999	2000	
				10.2 Government - État				
1 AGRICULTURE, HUNTING & FORESTRY	..	4.2 [a]	..	2.4	..	5.1	..	1 AGRICULTURE, CHASSE & SYLVICULTURE
2 MINING	..	1.3 [a]	..	9.2	..	3.0	..	2 ACTIVITES EXTRACTIVES
3 MANUFACTURING	..	323.6 [a]	..	415.5	..	378.2	..	3 ACTIVITES DE FABRICATION
4 Food, beverages & tobacco	..	16.7 [a]	..	14.8	..	5.0	..	**4 Produits alimentaires, boissons & tabac**
5 Food products & beverages	..	16.7 [a]	..	14.8	..	5.0	..	5 Produits alimentaires & boissons
6 Tobacco products	..	0.0 [a]	..	0.0	..	0.0	..	6 Produits à base de tabac
7 Textiles, wear app., fur & leather	..	1.6 [a]	..	2.0	..	1.4	..	**7 Textiles, habill., fourrures & cuirs**
8 Textiles	..	1.3 [a]	..	1.6	..	0.9	..	8 Textiles
9 Wearing apparel & fur	..	0.2 [a]	..	0.3	..	0.0	..	9 Articles d'habillement & fourrures
10 Leather products & footwear	..	0.1 [a]	..	0.1	..	0.5	..	10 Articles en cuir & chaussures
11 Wood, paper, printing, publishing	..	10.5 [a]	..	11.1	..	6.9	..	**11 Bois, papier, imprimerie, édition**
12 Wood & cork (not furniture)	..	3.3 [a]	..	4.5	..	1.7	..	12 Bois & liège (sauf meubles)
13 Pulp, paper & paper products	..	1.8 [a]	..	5.1	..	1.0	..	13 Carton, papier & articles
14 Publ., print. & repro. of rec. media	..	5.4 [a]	..	1.5	..	4.2	..	14 Edition, impress. & repro. enregis.
15 Coke, petroleum, nuclear fuel, chemicals, rubber plastics	..	11.8 [a]	..	20.8	..	40.0	..	**15 Coke, pétrole, combust. nucl., prod. chimi., caout. & mat. plast.**
16 Coke, ref. petrol. prod. & nuclear fuel	..	0.0 [a]	..	0.6	..	2.9	..	16 Coke, raffin. pétr. & combust. nucl.
17 Chemicals & chemical products	..	6.4 [a]	..	12.5	..	32.6	..	17 Produits chimiques
18 *Chemical prod.(less pharmaceuticals)*	..	6.4 [a]	..	10.7	..	4.4	..	18 *Prod. chimi. (moins prod. pharmaceutiques)*
19 *Pharmaceuticals*	..	0.0 [a]	..	1.8	..	28.2	..	19 *Produits pharmaceutiques*
20 Rubber & plastic products	..	5.4 [a]	..	7.7	..	4.5	..	20 Caoutchouc et matières plastiques
21 Non-metallic mineral products	..	2.2 [a]	..	2.5	..	4.0	..	**21 Produits minéraux non métalliques**
22 Basic metals	..	2.0 [a]	..	7.8	..	10.3	..	**22 Produits métallurgiques de base**
23 Basic metals, ferrous	..	0.3 [a]	..	2.0	..	0.3	..	23 Produits métallur. de base, ferreux
24 Basic metals, non-ferrous	..	1.7 [a]	..	5.8	..	10.0	..	24 Prod. métallur. de base, non ferreux
25 Fabricated metal products	..	3.8 [a]	..	8.2	..	16.4	..	**25 Ouvrages en métaux (sauf mach.& mat.)**
26 Machinery equipment, instruments and transport equipment	..	272.1 [a]	..	347.4	..	292.7	..	**26 Machines & matériel, instruments & mat. de transport**
27 Machinery n.e.c.	..	201.7 [a]	..	249.2	..	216.6	..	27 Machines, n.c.a.
28 Office, account. & computing machinery	..	0.0 [a]	..	2.9	..	2.1	..	28 Mach. de bur., compta. & trait. inform.
29 Electrical machinery	..	10.1 [a]	..	17.7	..	14.0	..	29 Machines électriques
30 Electro. equip. (radio, TV & commun.)	..	25.2 [a]	..	16.5	..	3.3	..	30 Equip. électro. (radio, tv & communic.)
31 *Electro. comp. (inc. semi-conductors)*	..	5.3 [a]	..	0.0	..	2.8	..	31 *Compos. électro. (semi-cond. incl.)*
32 *TV, radio & communications equip.*	..	19.9 [a]	..	16.5	..	0.5	..	32 *Appareils radio, tv et de communi.*
33 Instruments	..	20.8 [a]	..	31.8	..	41.4	..	33 Instruments
34 Motor vehicles	..	0.7 [a]	..	20.7	..	5.2	..	34 Véhicules automobiles
35 Other transport equipment	..	13.6 [a]	..	8.6	..	10.1	..	35 Autres matériels de transport
36 *Ships*	..	7.9 [a]	..	8.4	..	8.5	..	36 *Navires*
37 *Aerospace*	..	1.9 [a]	..	0.0	..	1.6	..	37 *Aérospatiale*
38 *Other transport equipment n.e.c.*	..	3.8 [a]	..	0.2	..	0.0	..	38 *Autres matériels de transport n.c.a.*
39 Furniture, other manufacturing n.e.c.	..	2.9 [a]	..	0.6	..	1.5	..	**39 Meubles, autres activ. de fabricat. n.c.a.**
40 Furniture	..	2.8 [a]	..	0.6	..	0.9	..	40 Meubles
41 Other manufacturing n.e.c.	..	0.1 [a]	..	0.0	..	0.6	..	41 Autres activités de fabrication n.c.a.
42 Recycling	..	0.0 [a]	..	0.3	..	0.0	..	**42 Récupération**
43 ELECTRICITY, GAS & WATER SUPPLY	..	1.3 [a]	..	3.0	..	0.8	..	43 DISTRIBUTION D'ELECTRICITE, GAZ & EAU
44 CONSTRUCTION	..	4.7 [a]	..	6.0	..	1.9	..	44 CONSTRUCTION
45 SERVICES SECTOR	..	737.9 [a]	..	699.2	..	713.3	..	45 SECTEUR DES SERVICES
46 Wholesale, retail trade & repair of motor vehicles, etc.	..	13.6 [a]	..	6.0	..	4.1	..	**46 Commerce (gros & détail), réparation de véhicules, etc.**
47 Hotels & restaurants	..	0.0 [a]	..	0.0	..	0.0	..	**47 Hôtels et restaurants**
48 Transport & storage	..	1.1 [a]	..	1.3	..	1.3	..	**48 Transports, entreposage**
49 Communications	..	4.0 [a]	..	1.7	..	4.2	..	**49 Communications**
50 Post	..	0.0 [a]	..	0.0	..	0.0	..	50 Postes
51 Telecommunications	..	4.0 [a]	..	1.7	..	4.2	..	51 Télécommunications
52 Financial intermediation (inc. insur.)	..	0.0 [a]	..	0.0	..	0.0	..	**52 Intermédiation financ. (assurances inc.)**
53 Real estate, renting & busin. activity	..	718.0 [a]	..	690.2	..	703.7	..	**53 Immobil., locat. & serv. aux entreprises**
54 Computer & related activities	..	26.5 [a]	..	10.8	..	45.2	..	54 Informatique et activités connexes
55 *Software consultancy*	..	25.7 [a]	..	10.1	..	24.5	..	55 *Production de logiciels*
56 *Other computer services n.e.c.*	..	0.8 [a]	..	0.7	..	20.7	..	56 *Autres activités informatiques*
57 Research & development	..	612.4 [a]	..	629.7	..	614.6	..	57 Recherche & développement expérimental
58 Other business activities n.e.c.	..	79.1 [a]	..	49.7	..	43.9	..	58 Autres activ. & serv. aux entreprises
59 Comm., soc. & pers. serv. activ., etc.	..	1.2 [a]	..	0.0	..	0.0	..	**59 Adm. pub., serv. soc. & collec., etc**
TOTAL BUSINESS ENTERPRISE SECTOR	..	**1 073.0 [a]**	..	**1 135.3**	..	**1 102.3**	..	**TOTAL SECTEUR DES ENTREPRISES**

Source: OECD, R&D database (DSTI/EAS Division), January 2002.

Source : OCDE, base de données R-D (DSTI/Division AES), janvier 2002.

Business enterprise intramural expenditure on R&D (BERD) - Dépenses intra-muros de R-D des entreprises (DIRDE)

by industry and source of funds - par industrie et par source de financement

Million NOK — *Millions NOK*

10.3 Other national funds - Autres sources nationales

	1994	1995	1996	1997	1998	1999	2000	
1 AGRICULTURE, HUNTING & FORESTRY	..	..	..	..	..	..	..	**1 AGRICULTURE, CHASSE & SYLVICULTURE**
2 MINING	..	..	..	..	..	..	..	**2 ACTIVITES EXTRACTIVES**
3 MANUFACTURING	..	..	..	..	..	..	..	**3 ACTIVITES DE FABRICATION**
4 Food, beverages & tobacco	..	..	..	..	..	..	..	**4 Produits alimentaires, boissons & tabac**
5 Food products & beverages	..	..	..	..	..	..	..	5 Produits alimentaires & boissons
6 Tobacco products	..	..	..	..	..	..	..	6 Produits à base de tabac
7 Textiles, wear app., fur & leather	..	..	..	..	..	..	..	**7 Textiles, habill., fourrures & cuirs**
8 Textiles	..	..	..	..	..	..	..	8 Textiles
9 Wearing apparel & fur	..	..	..	..	..	..	..	9 Articles d'habillement & fourrures
10 Leather products & footwear	..	..	..	..	..	..	..	10 Articles en cuir & chaussures
11 Wood, paper, printing, publishing	..	..	..	..	..	..	..	**11 Bois, papier, imprimerie, édition**
12 Wood & cork (not furniture)	..	..	..	..	..	..	..	12 Bois & liège (sauf meubles)
13 Pulp, paper & paper products	..	..	..	..	..	..	..	13 Carton, papier & articles
14 Publ., print. & repro. of rec. media	..	..	..	..	..	..	..	14 Edition, impress. & repro. enregis.
15 Coke, petroleum, nuclear fuel, chemicals, rubber plastics	..	..	..	..	..	..	..	**15 Coke, pétrole, combust. nucl., prod. chimi., caout. & mat. plast.**
16 Coke, ref. petrol. prod. & nuclear fuel	..	..	..	..	..	..	..	16 Coke, raffin. pétr. & combust. nucl.
17 Chemicals & chemical products	..	..	..	..	..	..	..	17 Produits chimiques
18 *Chemical prod.(less pharmaceuticals)*	..	..	..	..	..	..	..	18 *Prod. chimi. (moins prod. pharmaceutiques)*
19 *Pharmaceuticals*	..	..	..	..	..	..	..	19 *Produits pharmaceutiques*
20 Rubber & plastic products	..	..	..	..	..	..	..	20 Caoutchouc et matières plastiques
21 Non-metallic mineral products	..	..	..	..	..	..	..	**21 Produits minéraux non métalliques**
22 Basic metals	..	..	..	..	..	..	..	**22 Produits métallurgiques de base**
23 Basic metals, ferrous	..	..	..	..	..	..	..	23 Produits métallur. de base, ferreux
24 Basic metals, non-ferrous	..	..	..	..	..	..	..	24 Prod. métallur. de base, non ferreux
25 Fabricated metal products	..	..	..	..	..	..	..	**25 Ouvrages en métaux (sauf mach.& mat.)**
26 Machinery equipment, instruments and transport equipment	..	..	..	..	..	..	..	**26 Machines & matériel, instruments & mat. de transport**
27 Machinery n.e.c.	..	..	..	..	..	..	..	27 Machines, n.c.a.
28 Office, account. & computing machinery	..	..	..	..	..	..	..	28 Mach. de bur., compta. & trait. inform.
29 Electrical machinery	..	..	..	..	..	..	..	29 Machines électriques
30 Electro. equip. (radio, TV & commun.)	..	..	..	..	..	..	..	30 Equip. électro. (radio, tv & communic.)
31 *Electro. comp. (inc. semi-conductors)*	..	..	..	..	..	..	..	31 *Compos. électro. (semi-cond. incl.)*
32 *TV, radio & communications equip.*	..	..	..	..	..	..	..	32 *Appareils radio, tv et de communi.*
33 Instruments	..	..	..	..	..	..	..	33 Instruments
34 Motor vehicles	..	..	..	..	..	..	..	34 Véhicules automobiles
35 Other transport equipment	..	..	..	..	..	..	..	35 Autres matériels de transport
36 *Ships*	..	..	..	..	..	..	..	36 *Navires*
37 *Aerospace*	..	..	..	..	..	..	..	37 *Aérospatiale*
38 *Other transport equipment n.e.c.*	..	..	..	..	..	..	..	38 *Autres matériels de transport n.c.a.*
39 Furniture, other manufacturing n.e.c.	..	..	..	..	..	..	..	**39 Meubles, autres activ. de fabricat. n.c.a.**
40 Furniture	..	..	..	..	..	..	..	40 Meubles
41 Other manufacturing n.e.c.	..	..	..	..	..	..	..	41 Autres activités de fabrication n.c.a.
42 Recycling	..	..	..	..	..	..	..	**42 Récupération**
43 ELECTRICITY, GAS & WATER SUPPLY	..	..	..	..	..	..	..	**43 DISTRIBUTION D'ELECTRICITE, GAZ & EAU**
44 CONSTRUCTION	..	..	..	..	..	..	..	**44 CONSTRUCTION**
45 SERVICES SECTOR	..	5.1 [a]	..	..	..	..	..	**45 SECTEUR DES SERVICES**
46 Wholesale, retail trade & repair of motor vehicles, etc.	..	..	..	..	..	..	..	**46 Commerce (gros & détail), réparation de véhicules, etc.**
47 Hotels & restaurants	..	..	..	..	..	..	..	**47 Hôtels et restaurants**
48 Transport & storage	..	..	..	..	..	..	..	**48 Transports, entreposage**
49 Communications	..	..	..	..	..	..	..	**49 Communications**
50 Post	..	..	..	..	..	..	..	50 Postes
51 Telecommunications	..	..	..	..	..	..	..	51 Télécommunications
52 Financial intermediation (inc. insur.)	..	..	..	..	..	..	..	**52 Intermédiation financ. (assurances inc.)**
53 Real estate, renting & busin. activity	..	5.1 [a]	..	..	..	..	..	**53 Immobil., locat. & serv. aux entreprises**
54 Computer & related activities	..	..	..	..	..	..	..	54 Informatique et activités connexes
55 *Software consultancy*	..	..	..	..	..	..	..	55 *Production de logiciels*
56 *Other computer services n.e.c.*	..	..	..	..	..	..	..	56 *Autres activités informatiques*
57 Research & development	..	5.1 [a]	..	..	..	..	..	57 Recherche & développement expérimental
58 Other business activities n.e.c.	..	..	..	..	..	..	..	58 Autres activ. & serv. aux entreprises
59 Comm., soc. & pers. serv. activ., etc.	..	..	..	..	..	..	..	**59 Adm. pub., serv. soc. & collec., etc**
TOTAL BUSINESS ENTERPRISE SECTOR	..	5.1	..	0.0	..	0.0	..	**TOTAL SECTEUR DES ENTREPRISES**

Source: OECD, R&D database (DSTI/EAS Division), January 2002.

Source : OCDE, base de données R-D (DSTI/Division AES), janvier 2002.

NORWAY Table 10 - Tableau 10 NORVEGE

Business enterprise intramural expenditure on R&D (BERD) - Dépenses intra-muros de R-D des entreprises (DIRDE)

by industry and source of funds - par industrie et par source de financement

Million NOK *Millions NOK*

	1994	1995	1996	1997	1998	1999	2000	
	10.4 Funds from abroad - Fonds de l'étranger							
1 AGRICULTURE, HUNTING & FORESTRY	..	0.0 [a]	..	0.0	..	58.7	..	**1 AGRICULTURE, CHASSE & SYLVICULTURE**
2 MINING	..	0.1 [a]	..	5.1	..	8.5	..	**2 ACTIVITES EXTRACTIVES**
3 MANUFACTURING	..	178.4 [a]	..	251.5	..	308.1	..	**3 ACTIVITES DE FABRICATION**
4 Food, beverages & tobacco	..	0.0 [a]	..	17.5	..	1.4	..	**4 Produits alimentaires, boissons & tabac**
5 Food products & beverages	..	0.0 [a]	..	17.5	..	1.4	..	5 Produits alimentaires & boissons
6 Tobacco products	..	0.0 [a]	..	0.0	..	0.0	..	6 Produits à base de tabac
7 Textiles, wear app., fur & leather	..	0.0 [a]	..	0.0	..	0.0	..	**7 Textiles, habill., fourrures & cuirs**
8 Textiles	..	0.0 [a]	..	0.0	..	0.0	..	8 Textiles
9 Wearing apparel & fur	..	0.0 [a]	..	0.0	..	0.0	..	9 Articles d'habillement & fourrures
10 Leather products & footwear	..	0.0 [a]	..	0.0	..	0.0	..	10 Articles en cuir & chaussures
11 Wood, paper, printing, publishing	..	75.2 [a]	..	44.6	..	63.9	..	**11 Bois, papier, imprimerie, édition**
12 Wood & cork (not furniture)	..	1.0 [a]	..	0.0	..	0.0	..	12 Bois & liège (sauf meubles)
13 Pulp, paper & paper products	..	71.5 [a]	..	44.6	..	63.9	..	13 Carton, papier & articles
14 Publ., print. & repro. of rec. media	..	2.7 [a]	..	0.0	..	0.0	..	14 Edition, impress. & repro. enregis.
15 Coke, petroleum, nuclear fuel, chemicals, rubber plastics	..	18.1 [a]	..	37.5	..	98.5	..	**15 Coke, pétrole, combust. nucl., prod. chimi., caout. & mat. plast.**
16 Coke, ref. petrol. prod. & nuclear fuel	..	0.0 [a]	..	2.6	..	11.4	..	16 Coke, raffin. pétr. & combust. nucl.
17 Chemicals & chemical products	..	5.5 [a]	..	25.9	..	87.0	..	17 Produits chimiques
18 *Chemical prod.(less pharmaceuticals)*	..	0.2 [a]	..	1.5	..	9.5	..	18 *Prod. chimi. (moins prod. pharmaceutiques)*
19 *Pharmaceuticals*	..	5.3 [a]	..	24.4	..	77.5	..	19 *Produits pharmaceutiques*
20 Rubber & plastic products	..	12.6 [a]	..	9.0	..	0.1	..	20 Caoutchouc et matières plastiques
21 Non-metallic mineral products	..	0.0 [a]	..	0.8	..	1.9	..	**21 Produits minéraux non métalliques**
22 Basic metals	..	22.0 [a]	..	6.0	..	4.3	..	**22 Produits métallurgiques de base**
23 Basic metals, ferrous	..	0.0 [a]	..	0.0	..	3.3	..	23 Produits métallur. de base, ferreux
24 Basic metals, non-ferrous	..	22.0 [a]	..	6.0	..	1.0	..	24 Prod. métallur. de base, non ferreux
25 Fabricated metal products	..	0.1 [a]	..	0.0	..	0.0	..	**25 Ouvrages en métaux (sauf mach.& mat.)**
26 Machinery equipment, instruments and transport equipment	..	63.0 [a]	..	145.2	..	138.1	..	**26 Machines & matériel, instruments & mat. de transport**
27 Machinery n.e.c.	..	42.7 [a]	..	85.6	..	18.0	..	27 Machines, n.c.a.
28 Office, account. & computing machinery	..	0.0 [a]	..	0.0	..	0.0	..	28 Mach. de bur., compta. & trait. inform.
29 Electrical machinery	..	7.2 [a]	..	18.0	..	19.1	..	29 Machines électriques
30 Electro. equip. (radio, TV & commun.)	..	8.2 [a]	..	25.0	..	9.9	..	30 Equip. électro. (radio, tv & communic.)
31 *Electro. comp. (inc. semi-conductors)*	..	1.1 [a]	..	4.0	..	8.9	..	31 *Compos. électro. (semi-cond. incl.)*
32 *TV, radio & communications equip.*	..	7.1 [a]	..	21.0	..	1.0	..	32 *Appareils radio, tv et de communi.*
33 Instruments	..	0.0 [a]	..	10.3	..	84.5	..	33 Instruments
34 Motor vehicles	..	4.9 [a]	..	6.2	..	6.1	..	34 Véhicules automobiles
35 Other transport equipment	..	0.0 [a]	..	0.0	..	0.5	..	35 Autres matériels de transport
36 *Ships*	..	0.0 [a]	..	0.0	..	0.0	..	36 *Navires*
37 *Aerospace*	..	0.0 [a]	..	0.0	..	0.5	..	37 *Aérospatiale*
38 *Other transport equipment n.e.c.*	..	0.0 [a]	..	0.0	..	0.0	..	38 *Autres matériels de transport n.c.a.*
39 Furniture, other manufacturing n.e.c.	..	0.0 [a]	..	0.0	..	0.0	..	**39 Meubles, autres activ. de fabricat. n.c.a.**
40 Furniture	..	0.0 [a]	..	0.0	..	0.0	..	40 Meubles
41 Other manufacturing n.e.c.	..	0.0 [a]	..	0.0	..	0.0	..	41 Autres activités de fabrication n.c.a.
42 Recycling	..	0.0 [a]	..	0.0	..	0.0	..	**42 Récupération**
43 ELECTRICITY, GAS & WATER SUPPLY	..	0.0 [a]	..	0.0	..	0.0	..	**43 DISTRIBUTION D'ELECTRICITE, GAZ & EAU**
44 CONSTRUCTION	..	6.0 [a]	..	5.2	..	1.5	..	**44 CONSTRUCTION**
45 SERVICES SECTOR	..	317.1 [a]	..	527.4	..	444.4	..	**45 SECTEUR DES SERVICES**
46 Wholesale, retail trade & repair of motor vehicles, etc.	..	14.0 [a]	..	10.1	..	31.4	..	**46 Commerce (gros & détail), réparation de véhicules, etc.**
47 Hotels & restaurants	..	0.0 [a]	..	0.0	..	0.0	..	**47 Hôtels et restaurants**
48 Transport & storage	..	0.0 [a]	..	0.0	..	0.2	..	**48 Transports, entreposage**
49 Communications	..	11.8 [a]	..	33.5	..	26.2	..	**49 Communications**
50 Post	..	0.0 [a]	..	0.0	..	0.0	..	50 Postes
51 Telecommunications	..	11.8 [a]	..	33.5	..	26.2	..	51 Télécommunications
52 Financial intermediation (inc. insur.)	..	0.0 [a]	..	0.0	..	0.0	..	**52 Intermédiation financ. (assurances inc.)**
53 Real estate, renting & busin. activity	..	291.3 [a]	..	483.8	..	386.6	..	**53 Immobil., locat. & serv. aux entreprises**
54 Computer & related activities	..	63.7 [a]	..	232.0	..	159.6	..	54 Informatique et activités connexes
55 *Software consultancy*	..	63.7 [a]	..	205.6	..	159.6	..	55 *Production de logiciels*
56 *Other computer services n.e.c.*	..	0.0 [a]	..	26.4	..	0.0	..	56 *Autres activités informatiques*
57 Research & development	..	182.3 [a]	..	233.5	..	188.0	..	57 Recherche & développement expérimental
58 Other business activities n.e.c.	..	45.3 [a]	..	18.3	..	39.0	..	58 Autres activ. & serv. aux entreprises
59 Comm., soc. & pers. serv. activ., etc.	..	0.0 [a]	..	0.0	..	0.0	..	**59 Adm. pub., serv. soc. & collec., etc**
TOTAL BUSINESS ENTERPRISE SECTOR	..	**501.6** [a]	..	**789.1**	..	**821.2**	..	**TOTAL SECTEUR DES ENTREPRISES**

Source: OECD, R&D database (DSTI/EAS Division), January 2002.

Source : OCDE, base de données R-D (DSTI/Division AES), janvier 2002.

Business enterprise intramural expenditure on R&D (BERD) - Dépenses intra-muros de R-D des entreprises (DIRDE)

by industry and type of costs - par industrie et par type de coûts

Million NOK — *Millions NOK*

	1994	1995	1996	1997	1998	1999	2000	
	11.1 Current expenditure - Dépenses courantes							
1 AGRICULTURE, HUNTING & FORESTRY	..	36.4 [a]	..	7.4	..	151.1	..	1 AGRICULTURE, CHASSE & SYLVICULTURE
2 MINING	..	596.2 [a]	..	573.2	..	772.1	..	2 ACTIVITES EXTRACTIVES
3 MANUFACTURING	..	3 811.3 [a]	..	4 005.4	..	4 339.5	..	3 ACTIVITES DE FABRICATION
4 Food, beverages & tobacco	..	199.6 [a]	..	161.0	..	148.4	..	**4 Produits alimentaires, boissons & tabac**
5 Food products & beverages	..	197.2 [a]	..	161.0	..	148.0	..	5 Produits alimentaires & boissons
6 Tobacco products	..	2.4 [a]	..	0.0	..	0.4	..	6 Produits à base de tabac
7 Textiles, wear app., fur & leather	..	32.1 [a]	..	33.6	..	21.1	..	**7 Textiles, habill., fourrures & cuirs**
8 Textiles	..	12.5 [a]	..	11.3	..	14.4	..	8 Textiles
9 Wearing apparel & fur	..	16.1 [a]	..	15.9	..	5.1	..	9 Articles d'habillement & fourrures
10 Leather products & footwear	..	3.5 [a]	..	6.4	..	1.6	..	10 Articles en cuir & chaussures
11 Wood, paper, printing, publishing	..	207.5 [a]	..	172.0	..	199.2	..	**11 Bois, papier, imprimerie, édition**
12 Wood & cork (not furniture)	..	19.7 [a]	..	21.5	..	20.7	..	12 Bois & liège (sauf meubles)
13 Pulp, paper & paper products	..	125.3 [a]	..	133.1	..	159.9	..	13 Carton, papier & articles
14 Publ., print. & repro. of rec. media	..	62.5 [a]	..	17.4	..	18.6	..	14 Edition, impress. & repro. enregis.
15 Coke, petroleum, nuclear fuel, chemicals, rubber plastics	..	929.5 [a]	..	859.9	..	927.4	..	**15 Coke, pétrole, combust. nucl., prod. chimi., caout. & mat. plast.**
16 Coke, ref. petrol. prod. & nuclear fuel	..	53.6 [a]	..	68.8	..	109.8	..	16 Coke, raffin. pétr. & combust. nucl.
17 Chemicals & chemical products	..	834.9 [a]	..	704.8	..	758.5	..	17 Produits chimiques
18 *Chemical prod.(less pharmaceuticals)*	..	466.2 [a]	..	392.2	..	259.3	..	18 *Prod. chimi. (moins prod. pharmaceutiques)*
19 *Pharmaceuticals*	..	368.7 [a]	..	312.6	..	499.2	..	19 *Produits pharmaceutiques*
20 Rubber & plastic products	..	41.0 [a]	..	86.3	..	59.1	..	20 Caoutchouc et matières plastiques
21 Non-metallic mineral products	..	41.2 [a]	..	57.4	..	51.4	..	**21 Produits minéraux non métalliques**
22 Basic metals	..	195.0 [a]	..	194.5	..	339.4	..	**22 Produits métallurgiques de base**
23 Basic metals, ferrous	..	14.3 [a]	..	23.9	..	131.7	..	23 Produits métallur. de base, ferreux
24 Basic metals, non-ferrous	..	180.7 [a]	..	170.6	..	207.7	..	24 Prod. métallur. de base, non ferreux
25 Fabricated metal products	..	42.7 [a]	..	44.0	..	65.4	..	**25 Ouvrages en métaux (sauf mach.& mat.)**
26 Machinery equipment, instruments and transport equipment	..	2 117.5 [a]	..	2 417.5	..	2 533.0	..	**26 Machines & matériel, instruments & mat. de transport**
27 Machinery n.e.c.	..	590.8 [a]	..	657.4	..	540.7	..	27 Machines, n.c.a.
28 Office, account. & computing machinery	..	81.7 [a]	..	66.0	..	112.5	..	28 Mach. de bur., compta. & trait. inform.
29 Electrical machinery	..	207.8 [a]	..	273.8	..	338.4	..	29 Machines électriques
30 Electro. equip. (radio, TV & commun.)	..	746.4 [a]	..	725.7	..	749.0	..	30 Equip. électro. (radio, tv & communic.)
31 *Electro. comp. (inc. semi-conductors)*	..	11.0 [a]	..	7.4	..	20.9	..	31 *Compos. électro. (semi-cond. incl.)*
32 *TV, radio & communications equip.*	..	735.4 [a]	..	718.3	..	728.1	..	32 *Appareils radio, tv et de communi.*
33 Instruments	..	281.5 [a]	..	383.4	..	525.9	..	33 Instruments
34 Motor vehicles	..	48.4 [a]	..	144.6	..	113.3	..	34 Véhicules automobiles
35 Other transport equipment	..	160.9 [a]	..	166.6	..	153.2	..	35 Autres matériels de transport
36 *Ships*	..	110.4 [a]	..	139.3	..	125.4	..	36 *Navires*
37 *Aerospace*	..	27.1 [a]	..	25.5	..	23.6	..	37 *Aérospatiale*
38 *Other transport equipment n.e.c.*	..	23.4 [a]	..	1.8	..	4.2	..	38 *Autres matériels de transport n.c.a.*
39 Furniture, other manufacturing n.e.c.	..	46.2 [a]	..	63.8	..	52.1	..	**39 Meubles, autres activ. de fabricat. n.c.a.**
40 Furniture	..	28.0 [a]	..	44.0	..	36.7	..	40 Meubles
41 Other manufacturing n.e.c.	..	18.2 [a]	..	19.8	..	15.4	..	41 Autres activités de fabrication n.c.a.
42 Recycling	..	0.0 [a]	..	1.7	..	2.1	..	**42 Récupération**
43 ELECTRICITY, GAS & WATER SUPPLY	..	16.7 [a]	..	62.9	..	50.0	..	43 DISTRIBUTION D'ELECTRICITE, GAZ & EAU
44 CONSTRUCTION	..	63.7 [a]	..	54.8	..	38.3	..	44 CONSTRUCTION
45 SERVICES SECTOR	..	3 527.6 [a]	..	4 700.0	..	5 135.1	..	45 SECTEUR DES SERVICES
46 Wholesale, retail trade & repair of motor vehicles, etc.	..	184.7 [a]	..	110.8	..	318.8	..	**46 Commerce (gros & détail), réparation de véhicules, etc.**
47 Hotels & restaurants	..	2.0 [a]	..	0.0	..	0.0	..	**47 Hôtels et restaurants**
48 Transport & storage	..	25.8 [a]	..	78.8	..	26.4	..	**48 Transports, entreposage**
49 Communications	..	401.0 [a]	..	538.0	..	653.3	..	**49 Communications**
50 Post	..	8.0 [a]	..	0.0	..	0.0	..	50 Postes
51 Telecommunications	..	393.0 [a]	..	538.0	..	653.3	..	51 Télécommunications
52 Financial intermediation (inc. insur.)	..	196.9 [a]	..	227.4	..	177.8	..	**52 Intermédiation financ. (assurances inc.)**
53 Real estate, renting & busin. activity	..	2 703.4 [a]	..	3 742.1	..	3 956.0	..	**53 Immobil., locat. & serv. aux entreprises**
54 Computer & related activities	..	528.4 [a]	..	1 291.4	..	1 424.3	..	54 Informatique et activités connexes
55 *Software consultancy*	..	390.4 [a]	..	995.2	..	974.7	..	55 *Production de logiciels*
56 *Other computer services n.e.c.*	..	138.0 [a]	..	296.2	..	449.6	..	56 *Autres activités informatiques*
57 Research & development	..	1 639.9 [a]	..	1 755.7	..	1 713.8	..	57 Recherche & développement expérimental
58 Other business activities n.e.c.	..	535.1 [a]	..	695.0	..	817.9	..	58 Autres activ. & serv. aux entreprises
59 Comm., soc. & pers. serv. activ., etc.	..	13.8 [a]	..	2.9	..	2.8	..	**59 Adm. pub., serv. soc. & collec., etc**
TOTAL BUSINESS ENTERPRISE SECTOR	**..**	**8 051.9 [a]**	**..**	**9 403.6**	**..**	**10 486.1**	**..**	**TOTAL SECTEUR DES ENTREPRISES**

Source: OECD, R&D database (DSTI/EAS Division), January 2002.

Source : OCDE, base de données R-D (DSTI/Division AES), janvier 2002.

Business enterprise intramural expenditure on R&D (BERD) - Dépenses intra-muros de R-D des entreprises (DIRDE)

by industry and type of costs - par industrie et par type de coûts

Million NOK *Millions NOK*

11.2 Capital expenditure - Dépenses en capital

	1994	1995	1996	1997	1998	1999	2000	
1 AGRICULTURE, HUNTING & FORESTRY	..	5.3 [a]	..	0.9	..	18.3	..	**1 AGRICULTURE, CHASSE & SYLVICULTURE**
2 MINING	..	38.4 [a]	..	52.4	..	19.2	..	**2 ACTIVITES EXTRACTIVES**
3 MANUFACTURING	..	475.5 [a]	..	464.2	..	392.4	..	**3 ACTIVITES DE FABRICATION**
4 Food, beverages & tobacco	..	68.3 [a]	..	29.5	..	23.1	..	**4 Produits alimentaires, boissons & tabac**
5 Food products & beverages	..	67.8 [a]	..	29.5	..	23.0	..	5 Produits alimentaires & boissons
6 Tobacco products	..	0.5 [a]	..	0.0	..	0.1	..	6 Produits à base de tabac
7 Textiles, wear app., fur & leather	..	7.2 [a]	..	5.0	..	3.6	..	**7 Textiles, habill., fourrures & cuirs**
8 Textiles	..	4.5 [a]	..	2.3	..	2.7	..	8 Textiles
9 Wearing apparel & fur	..	2.1 [a]	..	1.2	..	0.9	..	9 Articles d'habillement & fourrures
10 Leather products & footwear	..	0.6 [a]	..	1.5	..	0.0	..	10 Articles en cuir & chaussures
11 Wood, paper, printing, publishing	..	42.4 [a]	..	52.4	..	37.3	..	**11 Bois, papier, imprimerie, édition**
12 Wood & cork (not furniture)	..	4.2 [a]	..	7.2	..	3.8	..	12 Bois & liège (sauf meubles)
13 Pulp, paper & paper products	..	22.8 [a]	..	11.5	..	8.9	..	13 Carton, papier & articles
14 Publ., print. & repro. of rec. media	..	15.4 [a]	..	33.7	..	24.6	..	14 Edition, impress. & repro. enregis.
15 Coke, petroleum, nuclear fuel, chemicals, rubber plastics	..	87.3 [a]	..	92.2	..	85.7	..	**15 Coke, pétrole, combust. nucl., prod. chimi., caout. & mat. plast.**
16 Coke, ref. petrol. prod. & nuclear fuel	..	11.5 [a]	..	2.6	..	1.0	..	16 Coke, raffin. pétr. & combust. nucl.
17 Chemicals & chemical products	..	67.2 [a]	..	61.8	..	73.3	..	17 Produits chimiques
18 *Chemical prod.(less pharmaceuticals)*	..	35.5 [a]	..	43.9	..	27.5	..	18 *Prod. chimi. (moins prod. pharmaceutiques)*
19 *Pharmaceuticals*	..	31.7 [a]	..	17.8	..	45.8	..	19 *Produits pharmaceutiques*
20 Rubber & plastic products	..	8.6 [a]	..	27.9	..	11.4	..	20 Caoutchouc et matières plastiques
21 Non-metallic mineral products	..	9.2 [a]	..	26.5	..	9.8	..	**21 Produits minéraux non métalliques**
22 Basic metals	..	16.2 [a]	..	20.6	..	19.2	..	**22 Produits métallurgiques de base**
23 Basic metals, ferrous	..	1.9 [a]	..	4.1	..	8.2	..	23 Produits métallur. de base, ferreux
24 Basic metals, non-ferrous	..	14.3 [a]	..	16.5	..	11.0	..	24 Prod. métallur. de base, non ferreux
25 Fabricated metal products	..	9.3 [a]	..	8.8	..	8.4	..	**25 Ouvrages en métaux (sauf mach.& mat.)**
26 Machinery equipment, instruments and transport equipment	..	221.2 [a]	..	202.1	..	179.2	..	**26 Machines & matériel, instruments & mat. de transport**
27 Machinery n.e.c.	..	77.0 [a]	..	38.4	..	28.5	..	27 Machines, n.c.a.
28 Office, account. & computing machinery	..	16.8 [a]	..	12.0	..	7.0	..	28 Mach. de bur., compta. & trait. inform.
29 Electrical machinery	..	21.5 [a]	..	22.8	..	24.8	..	29 Machines électriques
30 Electro. equip. (radio, TV & commun.)	..	38.0 [a]	..	48.2	..	48.9	..	30 Equip. électro. (radio, tv & communic.)
31 *Electro. comp. (inc. semi-conductors)*	..	2.4 [a]	..	0.4	..	0.9	..	31 *Compos. électro. (semi-cond. incl.)*
32 *TV, radio & communications equip.*	..	35.6 [a]	..	47.8	..	48.0	..	32 *Appareils radio, tv et de communi.*
33 Instruments	..	28.2 [a]	..	37.3	..	46.4	..	33 Instruments
34 Motor vehicles	..	4.8 [a]	..	6.9	..	7.1	..	34 Véhicules automobiles
35 Other transport equipment	..	34.9 [a]	..	36.5	..	16.5	..	35 Autres matériels de transport
36 *Ships*	..	20.5 [a]	..	16.8	..	4.3	..	36 *Navires*
37 *Aerospace*	..	13.6 [a]	..	19.5	..	11.4	..	37 *Aérospatiale*
38 *Other transport equipment n.e.c.*	..	0.8 [a]	..	0.2	..	0.8	..	38 *Autres matériels de transport n.c.a.*
39 Furniture, other manufacturing n.e.c.	..	14.4 [a]	..	27.1	..	26.0	..	**39 Meubles, autres activ. de fabricat. n.c.a.**
40 Furniture	..	12.3 [a]	..	15.4	..	21.5	..	40 Meubles
41 Other manufacturing n.e.c.	..	2.1 [a]	..	11.7	..	4.5	..	41 Autres activités de fabrication n.c.a.
42 Recycling	..	0.0 [a]	..	0.1	..	0.1	..	**42 Récupération**
43 ELECTRICITY, GAS & WATER SUPPLY	..	2.0 [a]	..	6.1	..	30.3	..	**43 DISTRIBUTION D'ELECTRICITE, GAZ & EAU**
44 CONSTRUCTION	..	7.7 [a]	..	10.7	..	13.5	..	**44 CONSTRUCTION**
45 SERVICES SECTOR	..	440.4 [a]	..	413.8	..	409.7	..	**45 SECTEUR DES SERVICES**
46 Wholesale, retail trade & repair of motor vehicles, etc.	..	84.6 [a]	..	45.3	..	15.0	..	**46 Commerce (gros & détail), réparation de véhicules, etc.**
47 Hotels & restaurants	..	0.3 [a]	..	0.0	..	0.0	..	**47 Hôtels et restaurants**
48 Transport & storage	..	25.8 [a]	..	12.8	..	2.6	..	**48 Transports, entreposage**
49 Communications	..	120.5 [a]	..	48.7	..	65.4	..	**49 Communications**
50 Post	..	31.2 [a]	..	0.0	..	0.0	..	50 Postes
51 Telecommunications	..	89.3 [a]	..	48.7	..	65.4	..	51 Télécommunications
52 Financial intermediation (inc. insur.)	..	39.3 [a]	..	31.8	..	18.2	..	**52 Intermédiation financ. (assurances inc.)**
53 Real estate, renting & busin. activity	..	169.0 [a]	..	275.1	..	308.5	..	**53 Immobil., locat. & serv. aux entreprises**
54 Computer & related activities	..	37.8 [a]	..	80.9	..	135.9	..	54 Informatique et activités connexes
55 *Software consultancy*	..	25.1 [a]	..	60.8	..	108.8	..	55 *Production de logiciels*
56 *Other computer services n.e.c.*	..	12.7 [a]	..	20.1	..	27.1	..	56 *Autres activités informatiques*
57 Research & development	..	69.3 [a]	..	126.5	..	115.6	..	57 Recherche & développement expérimental
58 Other business activities n.e.c.	..	61.9 [a]	..	67.7	..	57.0	..	58 Autres activ. & serv. aux entreprises
59 Comm., soc. & pers. serv. activ., etc.	..	0.9 [a]	..	0.1	..	0.0	..	**59 Adm. pub., serv. soc. & collec., etc**
TOTAL BUSINESS ENTERPRISE SECTOR	**..**	**969.3 [a]**	**..**	**948.2**	**..**	**883.4**	**..**	**TOTAL SECTEUR DES ENTREPRISES**

Source: OECD, R&D database (DSTI/EAS Division), January 2002.

Source : OCDE, base de données R-D (DSTI/Division AES), janvier 2002.

NORWAY — Table 12 - Tableau 12 — NORVEGE

Total business enterprise R&D personnel (BEMP) - Personnel total de R-D du secteur des entreprises (PRDE)

by industry - par industrie

In full time equivalent — *En équivalence plein-temps*

	1994	1995	1996	1997	1998	1999	2000	
1 AGRICULTURE, HUNTING & FORESTRY	..	78 [a]	..	14	..	164	..	**1 AGRICULTURE, CHASSE & SYLVICULTURE**
2 MINING	..	499 [a]	..	480	..	598	..	**2 ACTIVITES EXTRACTIVES**
3 MANUFACTURING	..	5 836 [a]	..	5 755	..	5 817	..	**3 ACTIVITES DE FABRICATION**
4 Food, beverages & tobacco	..	335 [a]	..	277	..	224	..	**4 Produits alimentaires, boissons & tabac**
5 Food products & beverages	..	331 [a]	..	277	..	223	..	5 Produits alimentaires & boissons
6 Tobacco products	..	4 [a]	..	0	..	1	..	6 Produits à base de tabac
7 Textiles, wear app., fur & leather	..	71 [a]	..	50	..	35	..	**7 Textiles, habill., fourrures & cuirs**
8 Textiles	..	35 [a]	..	17	..	23	..	8 Textiles
9 Wearing apparel & fur	..	29 [a]	..	20	..	9	..	9 Articles d'habillement & fourrures
10 Leather products & footwear	..	7 [a]	..	12	..	3	..	10 Articles en cuir & chaussures
11 Wood, paper, printing, publishing	..	282 [a]	..	215	..	246	..	**11 Bois, papier, imprimerie, édition**
12 Wood & cork (not furniture)	..	44 [a]	..	41	..	34	..	12 Bois & liège (sauf meubles)
13 Pulp, paper & paper products	..	131 [a]	..	137	..	168	..	13 Carton, papier & articles
14 Publ., print. & repro. of rec. media	..	107 [a]	..	37	..	44	..	14 Edition, impress. & repro. enregis.
15 Coke, petroleum, nuclear fuel, chemicals, rubber plastics	..	1 486 [a]	..	1 221	..	1 246	..	**15 Coke, pétrole, combust. nucl., prod. chimi., caout. & mat. plast.**
16 Coke, ref. petrol. prod. & nuclear fuel	..	40 [a]	..	52	..	76	..	16 Coke, raffin. pétr. & combust. nucl.
17 Chemicals & chemical products	..	1 371 [a]	..	1 054	..	1 075	..	17 Produits chimiques
18 *Chemical prod.(less pharmaceuticals)*	..	815 [a]	..	626	..	403	..	18 *Prod. chimi. (moins prod. pharmaceutiques)*
19 *Pharmaceuticals*	..	556 [a]	..	429	..	672	..	19 *Produits pharmaceutiques*
20 Rubber & plastic products	..	75 [a]	..	115	..	95	..	20 Caoutchouc et matières plastiques
21 Non-metallic mineral products	..	67 [a]	..	103	..	104	..	**21 Produits minéraux non métalliques**
22 Basic metals	..	296 [a]	..	329	..	545	..	**22 Produits métallurgiques de base**
23 Basic metals, ferrous	..	26 [a]	..	36	..	173	..	23 Produits métallur. de base, ferreux
24 Basic metals, non-ferrous	..	270 [a]	..	293	..	372	..	24 Prod. métallur. de base, non ferreux
25 Fabricated metal products	..	95 [a]	..	77	..	125	..	**25 Ouvrages en métaux (sauf mach.& mat.)**
26 Machinery equipment, instruments and transport equipment	..	3 105 [a]	..	3 363	..	3 203	..	**26 Machines & matériel, instruments & mat. de transport**
27 Machinery n.e.c.	..	833 [a]	..	896	..	790	..	27 Machines, n.c.a.
28 Office, account. & computing machinery	..	122 [a]	..	128	..	93	..	28 Mach. de bur., compta. & trait. inform.
29 Electrical machinery	..	315 [a]	..	377	..	444	..	29 Machines électriques
30 Electro. equip. (radio, TV & commun.)	..	1 002 [a]	..	914	..	816	..	30 Equip. électro. (radio, tv & communic.)
31 *Electro. comp. (inc. semi-conductors)*	..	20 [a]	..	12	..	23	..	31 *Compos. électro. (semi-cond. incl.)*
32 *TV, radio & communications equip.*	..	982 [a]	..	902	..	793	..	32 *Appareils radio, tv et de communi.*
33 Instruments	..	469 [a]	..	634	..	696	..	33 Instruments
34 Motor vehicles	..	58 [a]	..	186	..	132	..	34 Véhicules automobiles
35 Other transport equipment	..	306 [a]	..	228	..	232	..	35 Autres matériels de transport
36 *Ships*	..	204 [a]	..	183	..	190	..	36 *Navires*
37 *Aerospace*	..	54 [a]	..	41	..	34	..	37 *Aérospatiale*
38 *Other transport equipment n.e.c.*	..	48 [a]	..	4	..	8	..	38 *Autres matériels de transport n.c.a.*
39 Furniture, other manufacturing n.e.c.	..	99 [a]	..	117	..	84	..	**39 Meubles, autres activ. de fabricat. n.c.a.**
40 Furniture	..	68 [a]	..	80	..	55	..	40 Meubles
41 Other manufacturing n.e.c.	..	31 [a]	..	37	..	29	..	41 Autres activités de fabrication n.c.a.
42 Recycling	..	0 [a]	..	2	..	5	..	**42 Récupération**
43 ELECTRICITY, GAS & WATER SUPPLY	..	16 [a]	..	45	..	55	..	**43 DISTRIBUTION D'ELECTRICITE, GAZ & EAU**
44 CONSTRUCTION	..	90 [a]	..	91	..	52	..	**44 CONSTRUCTION**
45 SERVICES SECTOR	..	5 571 [a]	..	6 558	..	6 622	..	**45 SECTEUR DES SERVICES**
46 Wholesale, retail trade & repair of motor vehicles, etc.	..	323 [a]	..	164	..	433	..	**46 Commerce (gros & détail), réparation de véhicules, etc.**
47 Hotels & restaurants	..	7 [a]	..	0	..	0	..	**47 Hôtels et restaurants**
48 Transport & storage	..	45 [a]	..	110	..	43	..	**48 Transports, entreposage**
49 Communications	..	557 [a]	..	653	..	701	..	**49 Communications**
50 Post	..	27 [a]	..	0	..	0	..	50 Postes
51 Telecommunications	..	530 [a]	..	653	..	701	..	51 Télécommunications
52 Financial intermediation (inc. insur.)	..	292 [a]	..	435	..	176	..	**52 Intermédiation financ. (assurances inc.)**
53 Real estate, renting & busin. activity	..	4 313 [a]	..	5 187	..	5 262	..	**53 Immobil., locat. & serv. aux entreprises**
54 Computer & related activities	..	857 [a]	..	1 731	..	1 961	..	54 Informatique et activités connexes
55 *Software consultancy*	..	598 [a]	..	1 332	..	1 368	..	55 *Production de logiciels*
56 *Other computer services n.e.c.*	..	259 [a]	..	399	..	593	..	56 *Autres activités informatiques*
57 Research & development	..	2 693 [a]	..	2 675	..	2 313	..	57 Recherche & développement expérimental
58 Other business activities n.e.c.	..	763 [a]	..	780	..	988	..	58 Autres activ. & serv. aux entreprises
59 Comm., soc. & pers. serv. activ., etc.	..	34 [a]	..	9	..	7	..	**59 Adm. pub., serv. soc. & collec., etc**
TOTAL BUSINESS ENTERPRISE SECTOR	**..**	**12 090** [a]	**..**	**12 942**	**..**	**13 308**	**..**	**TOTAL SECTEUR DES ENTREPRISES**

Source: OECD, R&D database (DSTI/EAS Division), January 2002.

Source : OCDE, base de données R-D (DSTI/Division AES), janvier 2002.

Business enterprise university graduates - Diplômés universitaires des entreprises

by industry - par industrie

In full time equivalent / *En équivalence plein-temps*

	1994	1995	1996	1997	1998	1999	2000	
1 AGRICULTURE, HUNTING & FORESTRY	..	45 [a]	..	11	..	111	..	**1 AGRICULTURE, CHASSE & SYLVICULTURE**
2 MINING	..	365 [a]	..	353	..	481	..	**2 ACTIVITES EXTRACTIVES**
3 MANUFACTURING	..	3 620 [a]	..	4 005	..	4 037	..	**3 ACTIVITES DE FABRICATION**
4 Food, beverages & tobacco	..	190 [a]	..	137	..	136	..	**4 Produits alimentaires, boissons & tabac**
5 Food products & beverages	..	188 [a]	..	137	..	135	..	5 Produits alimentaires & boissons
6 Tobacco products	..	2 [a]	..	0	..	1	..	6 Produits à base de tabac
7 Textiles, wear app., fur & leather	..	24 [a]	..	22	..	14	..	**7 Textiles, habill., fourrures & cuirs**
8 Textiles	..	14 [a]	..	11	..	9	..	8 Textiles
9 Wearing apparel & fur	..	5 [a]	..	8	..	2	..	9 Articles d'habillement & fourrures
10 Leather products & footwear	..	5 [a]	..	4	..	3	..	10 Articles en cuir & chaussures
11 Wood, paper, printing, publishing	..	177 [a]	..	119	..	145	..	**11 Bois, papier, imprimerie, édition**
12 Wood & cork (not furniture)	..	29 [a]	..	19	..	19	..	12 Bois & liège (sauf meubles)
13 Pulp, paper & paper products	..	64 [a]	..	68	..	96	..	13 Carton, papier & articles
14 Publ., print. & repro. of rec. media	..	84 [a]	..	33	..	30	..	14 Edition, impress. & repro. enregis.
15 Coke, petroleum, nuclear fuel, chemicals, rubber plastics	..	839 [a]	..	767	..	700	..	**15 Coke, pétrole, combust. nucl., prod. chimi., caout. & mat. plast.**
16 Coke, ref. petrol. prod. & nuclear fuel	..	26 [a]	..	40	..	63	..	16 Coke, raffin. pétr. & combust. nucl.
17 Chemicals & chemical products	..	774 [a]	..	662	..	580	..	17 Produits chimiques
18 *Chemical prod.(less pharmaceuticals)*	..	465 [a]	..	372	..	184	..	18 *Prod. chimi. (moins prod. pharmaceutiques)*
19 *Pharmaceuticals*	..	309 [a]	..	290	..	396	..	19 *Produits pharmaceutiques*
20 Rubber & plastic products	..	39 [a]	..	65	..	57	..	20 Caoutchouc et matières plastiques
21 Non-metallic mineral products	..	28 [a]	..	60	..	63	..	**21 Produits minéraux non métalliques**
22 Basic metals	..	178 [a]	..	184	..	357	..	**22 Produits métallurgiques de base**
23 Basic metals, ferrous	..	19 [a]	..	24	..	138	..	23 Produits métallur. de base, ferreux
24 Basic metals, non-ferrous	..	159 [a]	..	160	..	219	..	24 Prod. métallur. de base, non ferreux
25 Fabricated metal products	..	38 [a]	..	40	..	46	..	**25 Ouvrages en métaux (sauf mach.& mat.)**
26 Machinery equipment, instruments and transport equipment	..	2 099 [a]	..	2 614	..	2 531	..	**26 Machines & matériel, instruments & mat. de transport**
27 Machinery n.e.c.	..	533 [a]	..	596	..	548	..	27 Machines, n.c.a.
28 Office, account. & computing machinery	..	76 [a]	..	86	..	77	..	28 Mach. de bur., compta. & trait. inform.
29 Electrical machinery	..	166 [a]	..	280	..	362	..	29 Machines électriques
30 Electro. equip. (radio, TV & commun.)	..	828 [a]	..	861	..	755	..	30 Equip. électro. (radio, tv & communic.)
31 *Electro. comp. (inc. semi-conductors)*	..	13 [a]	..	10	..	22	..	31 *Compos. électro. (semi-cond. incl.)*
32 *TV, radio & communications equip.*	..	815 [a]	..	852	..	733	..	32 *Appareils radio, tv et de communi.*
33 Instruments	..	277 [a]	..	446	..	535	..	33 Instruments
34 Motor vehicles	..	49 [a]	..	159	..	92	..	34 Véhicules automobiles
35 Other transport equipment	..	170 [a]	..	185	..	162	..	35 Autres matériels de transport
36 *Ships*	..	138 [a]	..	153	..	147	..	36 *Navires*
37 *Aerospace*	..	20 [a]	..	30	..	10	..	37 *Aérospatiale*
38 *Other transport equipment n.e.c.*	..	12 [a]	..	2	..	5	..	38 *Autres matériels de transport n.c.a.*
39 Furniture, other manufacturing n.e.c.	..	47 [a]	..	59	..	43	..	**39 Meubles, autres activ. de fabricat. n.c.a.**
40 Furniture	..	29 [a]	..	36	..	26	..	40 Meubles
41 Other manufacturing n.e.c.	..	18 [a]	..	23	..	17	..	41 Autres activités de fabrication n.c.a.
42 Recycling	..	0 [a]	..	2	..	2	..	**42 Récupération**
43 ELECTRICITY, GAS & WATER SUPPLY	..	12 [a]	..	38	..	46	..	**43 DISTRIBUTION D'ELECTRICITE, GAZ & EAU**
44 CONSTRUCTION	..	76 [a]	..	69	..	39	..	**44 CONSTRUCTION**
45 SERVICES SECTOR	..	3 803 [a]	..	4 873	..	5 023	..	**45 SECTEUR DES SERVICES**
46 Wholesale, retail trade & repair of motor vehicles, etc.	..	259 [a]	..	124	..	314	..	**46 Commerce (gros & détail), réparation de véhicules, etc.**
47 Hotels & restaurants	..	6 [a]	..	0	..	0	..	**47 Hôtels et restaurants**
48 Transport & storage	..	1 [a]	..	81	..	35	..	**48 Transports, entreposage**
49 Communications	..	340 [a]	..	494	..	535	..	**49 Communications**
50 Post	..	10 [a]	..	0	..	0	..	50 Postes
51 Telecommunications	..	330 [a]	..	494	..	535	..	51 Télécommunications
52 Financial intermediation (inc. insur.)	..	184 [a]	..	282	..	143	..	**52 Intermédiation financ. (assurances inc.)**
53 Real estate, renting & busin. activity	..	2 992 [a]	..	3 886	..	3 991	..	**53 Immobil., locat. & serv. aux entreprises**
54 Computer & related activities	..	582 [a]	..	1 387	..	1 466	..	54 Informatique et activités connexes
55 *Software consultancy*	..	447 [a]	..	1 114	..	1 027	..	55 *Production de logiciels*
56 *Other computer services n.e.c.*	..	135 [a]	..	273	..	439	..	56 *Autres activités informatiques*
57 Research & development	..	1 785 [a]	..	1 791	..	1 657	..	57 Recherche & développement expérimental
58 Other business activities n.e.c.	..	625 [a]	..	708	..	868	..	58 Autres activ. & serv. aux entreprises
59 Comm., soc. & pers. serv. activ., etc.	..	21 [a]	..	6	..	5	..	**59 Adm. pub., serv. soc. & collec., etc**
TOTAL BUSINESS ENTERPRISE SECTOR	**..**	**7 921 [a]**	**..**	**9 348**	**..**	**9 737**	**..**	**TOTAL SECTEUR DES ENTREPRISES**

Source: OECD, R&D database (DSTI/EAS Division), January 2002.

Source : OCDE, base de données R-D (DSTI/Division AES), janvier 2002.

Higher education intramural expenditure on R&D - Dépenses intra-muros de R-D de l'enseignement supérieur
(HERD) by field of science - (DIRDES) par discipline scientifique

	1994	1995	1996	1997	1998	1999	2000	
	14.1 Million NOK - Millions NOK							
Natural sciences	..	1 069.9	..	1 223.1	..	1 513.9	..	Sciences exactes et naturelles
Engineering	..	422.2	..	544.7	..	574.4	..	Sciences de l'ingénieur
Medical sciences	..	1 074.9	..	1 328.9	..	1 673.1	..	Sciences médicales
Agricultural sciences	..	253.0	..	250.3	..	300.0	..	Sciences agricoles
Sub-total NSE	..	2 820.0	..	3 347.0	..	4 061.4	..	Sous-total SENI
Social sciences	..	846.6	..	941.2	..	1 164.9	..	Sciences sociales
Humanities	..	472.5	..	557.6	..	593.1	..	Sciences humaines
Sub-total SSH	..	1 319.1	..	1 498.8	..	1 758.0	..	Sous-total SSH
Total NSE+SSH	..	4 139.1	..	4 845.8	..	5 819.4	..	*Total SENI+SSH*
	14.2 Million constant USD (1995 prices and PPPs) - Millions de USD constants (prix et PPA de 1995)							
Natural sciences	..	117.0	..	127.5	..	149.2 c	..	Sciences exactes et naturelles
Engineering	..	46.2	..	56.8	..	56.6 c	..	Sciences de l'ingénieur
Medical sciences	..	117.6	..	138.5	..	164.8 c	..	Sciences médicales
Agricultural sciences	..	27.7	..	26.1	..	29.6 c	..	Sciences agricoles
Sub-total NSE	..	308.4	..	348.9	..	400.1 c	..	Sous-total SENI
Social sciences	..	92.6	..	98.1	..	114.8 c	..	Sciences sociales
Humanities	..	51.7	..	58.1	..	58.4 c	..	Sciences humaines
Sub-total SSH	..	144.3	..	156.3	..	173.2 c	..	Sous-total SSH
Total NSE+SSH	..	452.7	..	505.2	..	573.3 c	..	*Total SENI+SSH*

Table 15 - Tableau 15

Higher education intramural expenditure on R&D - Dépenses intra-muros de R-D de l'enseignement supérieur
(HERD) by field of science and type of costs - (DIRDES) par discipline scientifique et par type de coûts

Million NOK *Millions NOK*

	1994	1995	1996	1997	1998	1999	2000	
Current expenditure								**Dépenses courantes**
Natural sciences	..	958.8	..	1 023.4	..	1 090.0	..	Sciences exactes et naturelles
Engineering	..	393.9	..	475.5	..	515.6	..	Sciences de l'ingénieur
Medical sciences	..	929.8	..	1 057.1	..	1 360.4	..	Sciences médicales
Agricultural sciences	..	229.0	..	235.0	..	284.3	..	Sciences agricoles
Sub-total NSE	..	2 511.5	..	2 791.0	..	3 250.3	..	Sous-total SENI
Social sciences	..	737.1	..	899.0	..	1 107.8	..	Sciences sociales
Humanities	..	431.5	..	534.6	..	558.2	..	Sciences humaines
Sub-total SSH	..	1 168.6	..	1 433.6	..	1 666.0	..	Sous-total SSH
Total NSE+SSH	..	3 680.1	..	4 224.6	..	4 916.3	..	*Total SENI+SSH*
Capital expenditure								**Dépenses en capital**
Natural sciences	..	111.1	..	199.7	..	423.9	..	Sciences exactes et naturelles
Engineering	..	28.3	..	69.2	..	58.8	..	Sciences de l'ingénieur
Medical sciences	..	145.1	..	271.8	..	312.7	..	Sciences médicales
Agricultural sciences	..	24.0	..	15.3	..	15.7	..	Sciences agricoles
Sub-total NSE	..	308.5	..	556.0	..	811.1	..	Sous-total SENI
Social sciences	..	109.5	..	42.2	..	57.1	..	Sciences sociales
Humanities	..	41.0	..	23.0	..	34.9	..	Sciences humaines
Sub-total SSH	..	150.5	..	65.2	..	92.0	..	Sous-total SSH
Total NSE+SSH	..	459.0	..	621.2	..	903.1	..	*Total SENI+SSH*

Source: OECD, R&D database (DSTI/EAS Division), January 2002.

Source : OCDE, base de données R-D (DSTI/Division AES), janvier 2002.

Higher education intramural expenditure on R&D - Dépenses intra-muros de R-D de l'enseignement supérieur

(HERD) by field of science and source of funds - (DIRDES) par discipline scientifique et par source de financement

Million NOK *Millions NOK*

	1994	1995	1996	1997	1998	1999	2000	
Direct government								**État direct**
Natural sciences	..	247.1	..	301.1	..	320.4	..	Sciences exactes et naturelles
Engineering	..	87.9	..	101.1	..	102.2	..	Sciences de l'ingénieur
Medical sciences	..	148.7	..	186.0	..	207.0	..	Sciences médicales
Agricultural sciences	..	90.2	..	86.8	..	87.9	..	Sciences agricoles
Sub-total NSE	..	573.9	..	675.0	..	717.5	..	Sous-total SENI
Social sciences	..	169.3	..	214.6	..	246.7	..	Sciences sociales
Humanities	..	64.3	..	78.1	..	89.5	..	Sciences humaines
Sub-total SSH	..	233.6	..	292.7	..	336.2	..	Sous-total SSH
Total NSE+SSH	..	807.5	..	967.7	..	1 053.7	..	*Total SENI+SSH*
General university funds								**Fonds généraux des universités**
Natural sciences	..	708.0	..	771.9	..	1 041.4	..	Sciences exactes et naturelles
Engineering	..	263.4	..	344.8	..	349.1	..	Sciences de l'ingénieur
Medical sciences	..	775.2	..	960.5	..	1 188.4	..	Sciences médicales
Agricultural sciences	..	142.0	..	134.4	..	161.0	..	Sciences agricoles
Sub-total NSE	..	1 888.6	..	2 211.6	..	2 739.9	..	Sous-total SENI
Social sciences	..	619.3	..	650.1	..	783.8	..	Sciences sociales
Humanities	..	386.8	..	444.0	..	465.4	..	Sciences humaines
Sub-total SSH	..	1 006.1	..	1 094.1	..	1 249.2	..	Sous-total SSH
Total NSE+SSH	..	2 894.7	..	3 305.7	..	3 989.1	..	*Total SENI+SSH*
Sub-total government								**Sous-total État**
Natural sciences	..	955.1	..	1 073.0	..	1 361.8	..	Sciences exactes et naturelles
Engineering	..	351.3	..	446.0	..	451.3	..	Sciences de l'ingénieur
Medical sciences	..	923.9	..	1 146.4	..	1 395.4	..	Sciences médicales
Agricultural sciences	..	232.2	..	221.2	..	248.9	..	Sciences agricoles
Sub-total NSE	..	2 462.5	..	2 886.6	..	3 457.4	..	Sous-total SENI
Social sciences	..	788.6	..	864.7	..	1 030.5	..	Sciences sociales
Humanities	..	451.1	..	522.1	..	554.9	..	Sciences humaines
Sub-total SSH	..	1 239.7	..	1 386.8	..	1 585.4	..	Sous-total SSH
Total NSE+SSH	..	3 702.2	..	4 273.4	..	5 042.8	..	*Total SENI+SSH*
Higher education								**Enseignement supérieur**
Natural sciences	..	2.0	..	4.3	..	14.5	..	Sciences exactes et naturelles
Engineering	..	1.1	..	0.6	..	4.2	..	Sciences de l'ingénieur
Medical sciences	..	10.6	..	15.2	..	20.9	..	Sciences médicales
Agricultural sciences	..	1.2	..	1.7	..	5.2	..	Sciences agricoles
Sub-total NSE	..	14.9	..	21.8	..	44.8	..	Sous-total SENI
Social sciences	..	16.3	..	23.0	..	60.0	..	Sciences sociales
Humanities	..	10.8	..	15.2	..	17.1	..	Sciences humaines
Sub-total SSH	..	27.1	..	38.3	..	77.1	..	Sous-total SSH
Total NSE+SSH	..	42.0	..	60.1	..	121.9	..	*Total SENI+SSH*
Private non-profit								**Institutions sans but lucratif**
Natural sciences	..	12.2	..	14.4	..	11.9	..	Sciences exactes et naturelles
Engineering	..	1.9	..	0.6	..	1.0	..	Sciences de l'ingénieur
Medical sciences	..	87.2	..	106.2	..	167.2	..	Sciences médicales
Agricultural sciences	..	1.8	..	0.4	..	0.0	..	Sciences agricoles
Sub-total NSE	..	103.1	..	121.6	..	180.1	..	Sous-total SENI
Social sciences	..	5.7	..	6.2	..	9.7	..	Sciences sociales
Humanities	..	1.8	..	1.4	..	1.2	..	Sciences humaines
Sub-total SSH	..	7.5	..	7.6	..	10.9	..	Sous-total SSH
Total NSE+SSH	..	110.6	..	129.2	..	191.0	..	*Total SENI+SSH*
Business entreprise								**Entreprises**
Natural sciences	..	67.3	..	63.1	..	61.3	..	Sciences exactes et naturelles
Engineering	..	63.7	..	85.2	..	101.5	..	Sciences de l'ingénieur
Medical sciences	..	40.8	..	43.5	..	50.1	..	Sciences médicales
Agricultural sciences	..	14.1	..	19.4	..	32.2	..	Sciences agricoles
Sub-total NSE	..	185.9	..	211.3	..	245.1	..	Sous-total SENI
Social sciences	..	27.8	..	30.8	..	40.6	..	Sciences sociales
Humanities	..	6.1	..	12.2	..	9.9	..	Sciences humaines
Sub-total SSH	..	33.9	..	43.0	..	50.5	..	Sous-total SSH
Total NSE+SSH	..	219.8	..	254.3	..	295.6	..	*Total SENI+SSH*
Funds from abroad								**Fonds de l'étranger**
Natural sciences	..	33.3	..	68.3	..	64.4	..	Sciences exactes et naturelles
Engineering	..	4.2	..	12.3	..	16.4	..	Sciences de l'ingénieur
Medical sciences	..	12.4	..	17.5	..	39.5	..	Sciences médicales
Agricultural sciences	..	3.7	..	7.6	..	13.7	..	Sciences agricoles
Sub-total NSE	..	53.6	..	105.7	..	134.0	..	Sous-total SENI
Social sciences	..	8.2	..	16.4	..	24.1	..	Sciences sociales
Humanities	..	2.7	..	6.7	..	10.0	..	Sciences humaines
Sub-total SSH	..	10.9	..	23.1	..	34.1	..	Sous-total SSH
Total NSE+SSH	..	64.5	..	128.8	..	168.1	..	*Total SENI+SSH*

Source: OECD, R&D database (DSTI/EAS Division), January 2002.

Source : OCDE, base de données R-D (DSTI/Division AES), janvier 2002.

Table 17 - Tableau 17

Higher education R&D personnel (HEMP) - Personnel de R-D du secteur de l'enseignement supérieur

by field of science and occupation - par discipline scientifique et par profession

In full-time equivalent — *En équivalence plein-temps*

	1994	1995	1996	1997	1998	1999	2000	
Total R&D personnel								**Personnel total de R-D**
Natural sciences	..	1 832	..	1 698	..	1 669	..	Sciences exactes et naturelles
Engineering	..	776	..	825	..	823	..	Sciences de l'ingénieur
Medical sciences	..	1 774	..	1 863	..	1 985	..	Sciences médicales
Agricultural sciences	..	413	..	385	..	412	..	Sciences agricoles
Sub-total NSE	..	4 795	..	4 771	..	4 889	..	Sous-total SENI
Social sciences	..	1 376	..	1 412	..	1 566	..	Sciences sociales
Humanities	..	784	..	879	..	858	..	Sciences humaines
Sub-total SSH	..	2 160	..	2 291	..	2 424	..	Sous-total SSH
Total NSE+SSH	..	6 955	..	7 062	..	7 313	..	*Total SENI+SSH*
Researchers								**Chercheurs**
Natural sciences	..	1 316	..	1 210	..	1 248	..	Sciences exactes et naturelles
Engineering	..	586	..	617	..	656	..	Sciences de l'ingénieur
Medical sciences	..	1 156	..	1 211	..	1 361	..	Sciences médicales
Agricultural sciences	..	224	..	228	..	260	..	Sciences agricoles
Sub-total NSE	..	3 282	..	3 266	..	3 525	..	Sous-total SENI
Social sciences	..	1 068	..	1 096	..	1 283	..	Sciences sociales
Humanities	..	643	..	729	..	713	..	Sciences humaines
Sub-total SSH	..	1 711	..	1 825	..	1 996	..	Sous-total SSH
Total NSE+SSH	..	4 993	..	5 091	..	5 521	..	*Total SENI+SSH*
Technicians								**Technicians**
Natural sciences	..	..	..	..	..	..	..	Sciences exactes et naturelles
Engineering	..	..	..	..	..	..	..	Sciences de l'ingénieur
Medical sciences	..	..	..	..	..	..	..	Sciences médicales
Agricultural sciences	..	..	..	..	..	..	..	Sciences agricoles
Sub-total NSE	..	..	..	..	..	..	..	Sous-total SENI
Social sciences	..	..	..	..	..	..	..	Sciences sociales
Humanities	..	..	..	..	..	..	..	Sciences humaines
Sub-total SSH	..	..	..	..	..	..	..	Sous-total SSH
Total NSE+SSH	..	..	..	..	..	..	..	*Total SENI+SSH*
Other								**Autres**
Natural sciences	..	..	..	..	..	..	..	Sciences exactes et naturelles
Engineering	..	..	..	..	..	..	..	Sciences de l'ingénieur
Medical sciences	..	..	..	..	..	..	..	Sciences médicales
Agricultural sciences	..	..	..	..	..	..	..	Sciences agricoles
Sub-total NSE	..	..	..	..	..	..	..	Sous-total SENI
Social sciences	..	..	..	..	..	..	..	Sciences sociales
Humanities	..	..	..	..	..	..	..	Sciences humaines
Sub-total SSH	..	..	..	..	..	..	..	Sous-total SSH
Total NSE+SSH	..	..	..	..	..	..	..	*Total SENI+SSH*

Table 18 - Tableau 18

Government budget appropriations or outlays for R&D (GBAORD) - Crédits budgétaires publics de R-D (CBPRD)

by socio-economic objective - par objectif socio-économique

Million NOK — *Millions NOK*

	1995	1996	1997	1998	1999	2000	2001	
Agriculture, forestry and fishing	693.8	747.7	729.0	752.8	786.2	857.1	897.1 p	Agriculture, sylviculture et pêche
Industrial development	1 197.8	1 160.8	1 129.6	1 123.5	1 113.2	1 168.0	1 428.1 p	Développement industriel
Energy	188.6	186.2	184.9	208.1	186.3	219.0	210.3 p	Energie
Transport & telecommunications	104.3	119.8	168.1	175.4	183.2	196.8	210.4 p	*Transports & télécommunications*
Urban & rural planning	17.7	21.2	23.5	25.7	25.6	25.9	26.6 p	*Aménagement urbain et rural*
ST Infrastructure	122.0	141.0	191.6	201.1	208.8	222.7	236.9 p	ST Infrastructure
Prevention of pollution	..	..	..	..	..	..	..	*Prévention de la pollution*
Identificat. & treatment of pollution	..	..	..	..	..	..	..	*Détect. pollut. & mesures antipollut.*
ST Environment	210.2	224.3	241.2	261.5	272.6	268.8	279.6 p	ST Environnement
Health	512.2	546.8	559.9	584.3	650.1	689.7	719.8 p	Santé
Social development & services	551.5	608.5	597.5	593.1	642.5	681.4	716.5 p	Développement et services sociaux
Earth and atmosphere	220.6	229.6	182.0	198.1	206.1	217.9	217.1 p	Milieu terrestre et atmosphère
Advancement of research	588.2	572.6	649.1	663.9	734.5	833.5	938.6 p	*Promotion de la recherche*
General university funds	2 593.2	2 827.6	3 064.5	3 367.1	3 570.4	3 811.4	3 851.9 p	*Fonds généraux des universités*
ST Advancement of knowledge	3 181.4	3 400.2	3 713.6	4 031.0	4 304.9	4 644.9	4 790.5 p	ST promotion gén. des connaiss.
Civil space	245.8	243.7	217.4	231.4	222.7	224.3	235.4 p	Espace civil
Defence	430.8	456.7	456.6	473.5	484.7	484.4	488.7 p	Défense
Not elsewhere classified	..	..	..	..	..	..	..	Non classé
Total	**7 554.7**	**7 945.5**	**8 203.3**	**8 658.4**	**9 078.0**	**9 678.3**	**10 219.9 p**	**Total**

Source: OECD, R&D database (DSTI/EAS Division), January 2002.

Source : OCDE, base de données R-D (DSTI/Division AES), janvier 2002.

NORWAY

Table 19 - Tableau 19

NORVEGE

Technology balance of payments - Balance des paiements technologiques

by type of transfer - selon la nature des transferts

Million NOK | *Millions NOK*

	1983	1984	1985	1986	1987	1988	1989	1990	1991	
Receipts										**Recettes**
Related firms	..	..	..	..	..	..	..	..	..	Firmes affiliées
Unrelated firms	..	..	..	..	..	..	..	..	..	Firmes non affiliées
Other	..	..	..	..	..	..	..	..	..	Autre
Total	342 [m]	235 [m]	243 [m]	316 [m]	455 [m]	2 114 [a]	2 614	2 821	3 073	*Total*
Payments										**Paiements**
Related firms	..	..	..	..	..	..	..	..	..	Firmes affiliées
Unrelated firms	..	..	..	..	..	..	..	..	..	Firmes non affiliées
Other	..	..	..	..	..	..	..	..	..	Autre
Total	638 [m]	639 [m]	657 [m]	818 [m]	1 325 [m]	2 720 [a]	2 868	3 412	4 219	*Total*

	1992	1993	1994	1995	1996	1997	1998	1999	2000	
Receipts										**Recettes**
Related firms	..	..	..	..	..	..	..	..	..	Firmes affiliées
Unrelated firms	..	..	..	..	..	..	..	..	..	Firmes non affiliées
Other	..	..	..	..	..	..	..	..	..	Autre
Total	3 189	3 751	3 758	3 438	4 240	4 102	6 546	7 153	..	*Total*
Payments										**Paiements**
Related firms	..	..	..	..	..	..	..	..	..	Firmes affiliées
Unrelated firms	..	..	..	..	..	..	..	..	..	Firmes non affiliées
Other	..	..	..	..	..	..	..	..	..	Autre
Total	5 419	6 564	6 453	6 706	5 088	6 194	9 629	9 676	..	*Total*

Source: OECD, TBPdatabase (DSTI/EAS Division), Januay 2002.

Source : OCDE, base de données TBP (DSTI/Division AES), janvier 2002.

Table 1 - Tableau 1

Gross domestic expenditure on R&D (GERD) - Dépenses intérieures brutes de R-D (DIRD)

by sector of performance and source of funds - par secteur d'exécution et par source de financement

	1994	1995	1996	1997	1998	1999	2000	
	1.1 Million PLN - Millions PLN							
Business enterprise								**Entreprises**
financed by								*financé par*
Business enterprise	461.7	534.1	798.9	921.6	1 195.6	1 369.3	1 147.2	Entreprises
Direct government	240.9 [a]	279.5 [a]	319.0	386.8	447.4	503.4	553.0	État direct
Higher education	0.6	1.4	0.8	1.3	1.0	1.3	1.6	Enseignement supérieur
Private non-profit	0.5	0.4	0.7	1.1	0.7	0.5	0.7	Institutions sans but lucratif
Funds from abroad	6.3	10.7	10.7	14.3	16.4	22.8	28.2	Fonds de l'étranger
Total	710.0 [a]	826.1 [a]	1 130.1	1 325.1	1 661.1	1 897.3	1 730.7	*Total*
Government								**État**
financed by								*financé par*
Business enterprise	171.7	168.6	186.7	157.1	210.7	256.0	296.1	Entreprises
Direct government	414.9 [a]	558.7 [a]	654.4	886.2	989.1	1 114.6	1 202.3	État direct
Higher education	1.0	0.7	1.3	1.5	1.5	4.2	1.6	Enseignement supérieur
Private non-profit	2.9	3.6	3.5	4.2	7.7	6.7	7.9	Institutions sans but lucratif
Funds from abroad	13.2	14.1	13.4	25.0	25.6	31.7	38.7	Fonds de l'étranger
Total	603.7 [a]	745.7 [a]	859.3	1 074.0	1 234.6	1 413.2	1 546.6	*Total*
Higher education								**Enseignement supérieur**
financed by								*financé par*
Business enterprise	46.2	64.2	86.8	100.5	107.6	124.9	118.7	Entreprises
Direct government	330.4	446.7	622.0	799.6	925.5	1 066.0	1 286.5	État direct
General university funds	0.0	0.0	0.0	0.0	0.0	0.0	0.0	Fonds généraux des universités
Sub-total government	330.4	446.7	622.0	799.6	925.5	1 066.0	1 286.5	Sous-Total État
Higher education	23.7	35.2	41.4	41.4	51.0	58.2	78.7	Enseignement supérieur
Private non-profit	3.0	3.7	3.9	5.0	4.7	4.5	10.4	Institutions sans but lucratif
Funds from abroad	3.7	11.2	14.4	15.4	17.9	20.7	18.1	Fonds de l'étranger
Total	407.0	561.0	768.5	961.9	1 106.7	1 274.3	1 512.4	*Total*
Private non-profit								**Institutions sans but lucratif**
financed by								*financé par*
Business enterprise	0.0	..	0.6	..	1.0	0.6	2.4	Entreprises
Direct government	0.3	..	0.2	..	1.0	2.7	0.8	État direct
Higher education	0.0	..	0.0	..	0.0	0.4	0.0	Enseignement supérieur
Private non-profit	0.0	..	1.4	..	0.5	0.9	1.0	Institutions sans but lucratif
Funds from abroad	0.0	..	1.3	..	0.2	1.1	2.2	Fonds de l'étranger
Total	0.3	..	3.5	..	2.7	5.7	6.4	*Total*
GERD								**DIRD**
financed by								*financé par*
Business enterprise	679.6	766.9	1 073.0	1 179.2	1 514.9	1 750.8	1 564.4	Entreprises
Direct government	986.5 [a]	1 284.9 [a]	1 595.6	2 072.6	2 363.0	2 686.7	3 042.6	État direct
General university funds	0.0	0.0	0.0	0.0	0.0	0.0	0.0	Fonds généraux des universités
Sub-total government	986.5 [a]	1 284.9 [a]	1 595.6	2 072.6	2 363.0	2 686.7	3 042.6	Sous-Total État
Higher education	25.3	37.3	43.5	44.2	53.5	64.1	81.9	Enseignement supérieur
Private non-profit	6.4	7.7	9.5	10.3	13.6	12.6	20.0	Institutions sans but lucratif
Funds from abroad	23.2	36.0	39.8	54.7	60.1	76.3	87.2	Fonds de l'étranger
Total	1 721.0 [a]	2 132.8 [a]	2 761.4	3 361.0	4 005.1	4 590.5	4 796.1	*Total*
	1.2 Million constant USD (1995 prices and PPPs) - Millions de USD constants (prix et PPA de 1995)							
Business enterprise								**Entreprises**
financed by								*financé par*
Business enterprise	519.3	469.7	591.8	598.8	694.9	745.3	582.9	Entreprises
Direct government	270.9 [a]	245.8 [a]	236.3	251.3	260.1	274.0	281.0	État direct
Higher education	0.7	1.2	0.6	0.8	0.6	0.7	0.8	Enseignement supérieur
Private non-profit	0.6	0.4	0.5	0.7	0.4	0.3	0.4	Institutions sans but lucratif
Funds from abroad	7.1	9.4	7.9	9.3	9.5	12.4	14.3	Fonds de l'étranger
Total	798.5 [a]	726.5 [a]	837.1	860.9	965.5	1 032.6	879.4	*Total*
Government								**État**
financed by								*financé par*
Business enterprise	193.1	148.3	138.3	102.1	122.5	139.3	150.4	Entreprises
Direct government	466.6 [a]	491.3 [a]	484.7	575.8	574.9	606.6	610.9	État direct
Higher education	1.1	0.6	1.0	1.0	0.9	2.3	0.8	Enseignement supérieur
Private non-profit	3.3	3.2	2.6	2.7	4.5	3.6	4.0	Institutions sans but lucratif
Funds from abroad	14.8	12.4	9.9	16.2	14.9	17.3	19.7	Fonds de l'étranger
Total	679.0 [a]	655.8 [a]	636.5	697.8	717.6	769.2	785.8	*Total*
Higher education								**Enseignement supérieur**
financed by								*financé par*
Business enterprise	52.0	56.5	64.3	65.3	62.5	68.0	60.3	Entreprises
Direct government	371.6	392.8	460.7	519.5	537.9	580.2	653.7	État direct
General university funds	0.0	0.0	0.0	0.0	0.0	0.0	0.0	Fonds généraux des universités
Sub-total government	371.6	392.8	460.7	519.5	537.9	580.2	653.7	Sous-Total État
Higher education	26.7	31.0	30.7	26.9	29.6	31.7	40.0	Enseignement supérieur
Private non-profit	3.4	3.3	2.9	3.2	2.7	2.4	5.3	Institutions sans but lucratif
Funds from abroad	4.2	9.8	10.7	10.0	10.4	11.3	9.2	Fonds de l'étranger
Total	457.7	493.3	569.3	625.0	643.3	693.6	768.4	*Total*
Private non-profit								**Institutions sans but lucratif**
financed by								*financé par*
Business enterprise	0.0	..	0.4	..	0.6	0.3	1.2	Entreprises
Direct government	0.3	..	0.1	..	0.6	1.5	0.4	État direct
Higher education	0.0	..	0.0	..	0.0	0.2	0.0	Enseignement supérieur
Private non-profit	0.0	..	1.0	..	0.3	0.5	0.5	Institutions sans but lucratif
Funds from abroad	0.0	..	1.0	..	0.1	0.6	1.1	Fonds de l'étranger
Total	0.3	..	2.6	..	1.6	3.1	3.3	*Total*
GERD								**DIRD**
financed by								*financé par*
Business enterprise	764.3	674.4	794.8	766.1	880.5	952.9	794.9	Entreprises
Direct government	1 109.5 [a]	1 129.9 [a]	1 181.9	1 346.6	1 373.5	1 462.3	1 545.9	État direct
General university funds	0.0	0.0	0.0	0.0	0.0	0.0	0.0	Fonds généraux des universités
Sub-total government	1 109.5 [a]	1 129.9 [a]	1 181.9	1 346.6	1 373.5	1 462.3	1 545.9	Sous-Total État
Higher education	28.5	32.8	32.2	28.7	31.1	34.9	41.6	Enseignement supérieur
Private non-profit	7.2	6.8	7.0	6.7	7.9	6.9	10.2	Institutions sans but lucratif
Funds from abroad	26.1	31.7	29.5	35.5	34.9	41.5	44.3	Fonds de l'étranger
Total	1 935.6 [a]	1 875.6 [a]	2 045.5	2 183.7	2 328.0	2 498.5	2 436.9	*Total*

Source: OECD, R&D database (DSTI/EAS Division), January 2002.

Source : OCDE, base de données R-D (DSTI/Division AES), janvier 2002.

Table 2 - Tableau 2

Gross domestic expenditure on R&D (GERD) - Dépenses intérieures brutes de R-D (DIRD)

by sector of performance and type of costs - par secteur d'exécution et par type de coûts

Million PLN *Millions PLN*

	1994	1995	1996	1997	1998	1999	2000	
Business enterprise								**Entreprises**
Labour costs	280.2 [d]	325.7 [a]	442.1	521.6	632.1	767.5	..	Coûts du personnel
Other current costs	300.7 [d]	353.2 [a]	412.2	497.8	576.3	628.6	..	Autres coûts courants
Total current expenditure	580.9 [d]	678.9 [a]	854.3	1 019.4	1 208.4	1 396.1	..	*Total dépenses courantes*
Land and buildings	20.9 [d]	22.3 [a]	136.7	55.8	77.8	111.3	..	Terrains et constructions
Instruments and equipment	108.2 [d]	124.9 [a]	139.1	249.9	374.9	389.9	..	Instruments et équipements
Total capital expenditure	129.1 [d]	147.2 [a]	275.8	305.7	452.7	501.2	..	*Total dépenses en capital*
Total	710.0 [d]	826.1 [a]	1 130.1	1 325.1	1 661.1	1 897.3	1 730.7	*Total*
Government								**État**
Labour costs	285.1 [d]	381.9 [a]	444.6	532.1	616.6	687.5	..	Coûts du personnel
Other current costs	216.2 [d]	244.2 [a]	264.4	393.5	462.0	514.7	..	Autres coûts courants
Total current expenditure	501.3 [d]	626.1 [a]	709.0	925.6	1 078.6	1 202.2	..	*Total dépenses courantes*
Land and buildings	38.9 [d]	37.0 [a]	30.5	41.6	47.0	70.3	..	Terrains et constructions
Instruments and equipment	63.5 [d]	82.6 [a]	119.8	106.8	109.0	140.7	..	Instruments et équipements
Total capital expenditure	102.4 [d]	119.6 [a]	150.3	148.4	156.0	211.0	..	*Total dépenses en capital*
Total	603.7 [d]	745.7 [a]	859.3	1 074.0	1 234.6	1 413.2	1 546.6	*Total*
Higher education								**Enseignement supérieur**
Labour costs	119.4	191.1	232.7	302.3	345.6	417.1	..	Coûts du personnel
Other current costs	159.5	241.5	338.0	397.9	428.3	524.3	..	Autres coûts courants
Total current expenditure	278.9	432.6	570.7	700.2	773.9	941.4	..	*Total dépenses courantes*
Land and buildings	17.1	8.9	69.4	70.2	108.6	75.2	..	Terrains et constructions
Instruments and equipment	111.0	119.5	128.4	191.5	224.2	257.7	..	Instruments et équipements
Total capital expenditure	128.1	128.4	197.8	261.7	332.8	332.9	..	*Total dépenses en capital*
Total	407.0	561.0	768.5	961.9	1 106.7	1 274.3	1 512.4	*Total*
Private non-profit								**Institutions sans but lucratif**
Labour costs	0.2	..	1.6	..	1.2	3.9	..	Coûts du personnel
Other current costs	0.1	..	1.1	..	1.2	1.6	..	Autres coûts courants
Total current expenditure	0.3	..	2.7	..	2.4	5.5	..	*Total dépenses courantes*
Land and buildings	0.0	..	0.1	..	0.0	0.0	..	Terrains et constructions
Instruments and equipment	0.0	..	0.7	..	0.3	0.2	..	Instruments et équipements
Total capital expenditure	0.0	..	0.8	..	0.3	0.2	..	*Total dépenses en capital*
Total	0.3	..	3.5	..	2.7	5.7	6.4	*Total*
GERD								**DIRD**
Labour costs	684.9 [d]	898.7 [a]	1 121.0	1 356.0	1 595.5	1 876.0	..	Coûts du personnel
Other current costs	676.5 [d]	838.9 [a]	1 015.7	1 289.3	1 467.8	1 669.2	..	Autres coûts courants
Total current expenditure	1 361.4 [d]	1 737.6 [a]	2 136.7	2 645.3	3 063.3	3 545.2	..	*Total dépenses courantes*
Land and buildings	76.9 [d]	68.2 [a]	236.7	167.6	233.4	256.8	..	Terrains et constructions
Instruments and equipment	282.7 [d]	327.0 [a]	388.0	548.1	708.4	788.5	..	Instruments et équipements
Total capital expenditure	359.6 [d]	395.2 [a]	624.7	715.7	941.8	1 045.3	..	*Total dépenses en capital*
Total	1 721.0 [d]	2 132.8 [a]	2 761.4	3 361.0	4 005.1	4 590.5	4 796.1	*Total*

Table 3 - Tableau 3

Current domestic expenditure on R&D - Dépenses intérieures courantes de R-D

by sector of performance and type of activity - par secteur d'exécution et par type d'activités

Million PLN *Millions PLN*

	1994	1995	1996	1997	1998	1999	2000	
Business enterprise								**Entreprises**
Basic research	46.7 [d,o]	32.3 [a,o]	51.6 [o]	41.8 [o]	61.2 [o]	75.1 [o]	118.4 [o]	Recherche fondamentale
Applied research	168.6 [d,o]	180.9 [a,o]	231.5 [o]	271.6 [o]	307.2 [o]	328.8 [o]	323.4 [o]	Recherche appliquée
Experimental development	381.4 [d,o]	484.2 [a,o]	585.3 [o]	724.4 [o]	898.9 [o]	1 013.2 [o]	976.8 [o]	Développement expérimental
Not specified	0.0	0.0	0.0	0.0	0.0	0.0	0.0	Non spécifié
Total	596.7 [d,o]	697.4 [a,o]	868.4 [o]	1 037.8 [o]	1 267.3 [o]	1 417.1 [o]	1 418.6 [o]	*Total*
Government								**État**
Basic research	258.7 [d,o]	337.2 [a,o]	323.9 [o]	454.7 [o]	557.6 [o]	669.7 [o]	708.5 [o]	Recherche fondamentale
Applied research	135.0 [d,o]	160.6 [a,o]	218.4 [o]	248.1 [o]	269.5 [o]	265.0 [o]	319.4 [o]	Recherche appliquée
Experimental development	124.2 [d,o]	146.4 [a,o]	192.3 [o]	241.2 [o]	270.9 [o]	294.6 [o]	327.4 [o]	Développement expérimental
Not specified	0.0	0.0	0.0	0.0	0.0	0.0	0.0	Non spécifié
Total	517.9 [d,o]	644.2 [a,o]	734.6 [o]	944.0 [o]	1 098.0 [o]	1 229.3 [o]	1 355.3 [o]	*Total*
Higher education								**Enseignement supérieur**
Basic research	172.5 [o]	298.5 [o]	370.5 [o]	441.4 [o]	492.5 [o]	594.4 [o]	707.3 [o]	Recherche fondamentale
Applied research	118.8 [o]	150.0 [o]	200.2 [o]	251.3 [o]	253.6 [o]	307.2 [o]	344.5 [o]	Recherche appliquée
Experimental development	55.6 [o]	44.8 [o]	81.0 [o]	90.9 [o]	106.0 [o]	141.4 [o]	149.6 [o]	Développement expérimental
Not specified	0.0	0.0	0.0	0.0	0.0	0.0	0.0	Non spécifié
Total	346.9 [o]	493.3 [o]	651.7 [o]	783.6 [o]	852.1 [o]	1 043.0 [o]	1 201.4 [o]	*Total*
Private non-profit								**Institutions sans but lucratif**
Basic research	0.1	..	0.0	..	0.0	0.0	0.0	Recherche fondamentale
Applied research	0.1	..	2.7	..	0.2	3.8	4.4	Recherche appliquée
Experimental development	0.1	..	0.0	..	2.2	1.7	1.8	Développement expérimental
Not specified	0.0	..	0.0	..	0.0	0.0	0.0	Non spécifié
Total	0.3	..	2.7	..	2.4	5.5	6.2	*Total*
GERD								**DIRD**
Basic research	478.0 [d,o]	668.0 [a,o]	746.0 [o]	937.9 [o]	1 111.3 [o]	1 339.2 [o]	1 534.2 [o]	Recherche fondamentale
Applied research	422.5 [d,o]	491.5 [a,o]	652.8 [o]	771.0 [o]	830.5 [o]	904.8 [o]	991.7 [o]	Recherche appliquée
Experimental development	561.3 [d,o]	675.4 [a,o]	858.6 [o]	1 056.5 [o]	1 278.0 [o]	1 450.9 [o]	1 455.6 [o]	Développement expérimental
Not specified	0.0	0.0	0.0	0.0	0.0	0.0	0.0	Non spécifié
Total	1 461.8 [d,o]	1 834.9 [a,o]	2 257.4 [o]	2 765.4 [o]	3 219.8 [o]	3 694.9 [o]	3 981.5 [o]	*Total*

Source: OECD, R&D database (DSTI/EAS Division), January 2002. *Source* : OCDE, base de données R-D (DSTI/Division AES), janvier 2002.

Table 5 - Tableau 5

Total R&D personnel - Personnel total de R-D

by sector of employment and occupation - par secteur d'emploi et par profession

In full-time equivalent / *En équivalence plein-temps*

	1994	1995	1996	1997	1998	1999	2000	
Business enterprise								**Entreprises**
RSE	10 629	11 155	10 365	11 039	10 173	10 327	9 821	Chercheurs
Technicians	8 309	8 593	7 585	7 614	6 735	5 756	5 051	Techniciens
Other occupations	6 593	6 534	5 586	4 528	5 006	4 231	3 714	Autres professions
Total	25 531	26 282	23 536	23 181	21 914	20 314	18 586	*Total*
Government								**État**
RSE	10 751	11 244	10 954	11 717	11 570	10 811	11 100	Chercheurs
Technicians	5 653	5 773	5 534	4 837	5 040	4 689	4 388	Techniciens
Other occupations	4 560	4 670	4 256	3 091	3 492	3 585	3 335	Autres professions
Total	20 964	21 687	20 744	19 645	20 102	19 085	18 823	*Total*
Higher education								**Enseignement supérieur**
RSE	26 053	28 026	31 133	32 846	34 425	35 284	34 246	Chercheurs
Technicians	4 974	5 287	5 717	5 444	5 162	4 915	4 205	Techniciens
Other occupations	1 718	2 308	2 196	2 687	2 891	2 749	3 048	Autres professions
Total	32 745	35 621	39 046	40 977	42 478	42 948	41 499	*Total*
Private non-profit								**Institutions sans but lucratif**
RSE	0	..	22	..	11	11	7	Chercheurs
Technicians	0	..	0	..	2	2	4	Techniciens
Other occupations	8	..	0	..	3	8	6	Autres professions
Total	8	..	22	..	16	21	17	*Total*
TOTAL								**TOTAL**
RSE	47 433	50 425	52 474	55 602	56 179	56 433	55 174	Chercheurs
Technicians	18 936	19 653	18 836	17 895	16 939	15 362	13 648	Techniciens
Other occupations	12 879	13 512	12 038	10 306	11 392	10 573	10 103	Autres professions
Total	79 248	83 590	83 348	83 803	84 510	82 368	78 925	*Total*

Table 6 - Tableau 6

Total R&D personnel - Personnel total de R-D

by sector of employment and formal qualification - par secteur d'emploi et par niveau de qualification

In full-time equivalent / *En équivalence plein-temps*

	1994	1995	1996	1997	1998	1999	2000	
Business enterprise								**Entreprises**
University PhD level degrees	..	..	..	..	..	..	..	Diplômes univ. de niv. doctorat
Other university degrees	..	..	..	..	..	..	..	Autres diplômes universitaires
Sub-total university degrees	..	..	..	..	..	..	..	Sous-total Universitaires
Other post-secondary diplomas	..	..	..	..	..	..	..	Autres diplômes post-secondaires
Secondary diplomas	..	..	..	..	..	..	..	Diplômes de fin d'études secondaires
Other qualifications	..	..	..	..	..	..	..	Autres qualifications
Total	25 531	26 282	23 536	23 181	21 914	20 314	18 586	*Total*
Government								**État**
University PhD level degrees	..	..	..	..	..	..	..	Diplômes univ. de niv. doctorat
Other university degrees	..	..	..	..	..	..	..	Autres diplômes universitaires
Sub-total university degrees	..	..	..	..	..	..	..	Sous-total Universitaires
Other post-secondary diplomas	..	..	..	..	..	..	..	Autres diplômes post-secondaires
Secondary diplomas	..	..	..	..	..	..	..	Diplômes de fin d'études secondaires
Other qualifications	..	..	..	..	..	..	..	Autres qualifications
Total	20 964	21 687	20 744	19 645	20 102	19 085	18 823	*Total*
Higher education								**Enseignement supérieur**
University PhD level degrees	..	..	..	..	..	..	..	Diplômes univ. de niv. doctorat
Other university degrees	..	..	..	..	..	..	..	Autres diplômes universitaires
Sub-total university degrees	..	..	..	..	..	..	..	Sous-total Universitaires
Other post-secondary diplomas	..	..	..	..	..	..	..	Autres diplômes post-secondaires
Secondary diplomas	..	..	..	..	..	..	..	Diplômes de fin d'études secondaires
Other qualifications	..	..	..	..	..	..	..	Autres qualifications
Total	32 745	35 621	39 046	40 977	42 478	42 948	41 499	*Total*
Private non-profit								**Institutions sans but lucratif**
University PhD level degrees	..	..	..	..	..	..	..	Diplômes univ. de niv. doctorat
Other university degrees	..	..	..	..	..	..	..	Autres diplômes universitaires
Sub-total university degrees	..	..	..	..	..	..	..	Sous-total Universitaires
Other post-secondary diplomas	..	..	..	..	..	..	..	Autres diplômes post-secondaires
Secondary diplomas	..	..	..	..	..	..	..	Diplômes de fin d'études secondaires
Other qualifications	..	..	..	..	..	..	..	Autres qualifications
Total	8	..	22	..	16	21	17	*Total*
TOTAL								**TOTAL**
University PhD level degrees	..	..	..	..	..	..	..	Diplômes univ. de niv. doctorat
Other university degrees	..	..	..	..	..	..	..	Autres diplômes universitaires
Sub-total university degrees	..	..	..	..	..	..	..	Sous-total Universitaires
Other post-secondary diplomas	..	..	..	..	..	..	..	Autres diplômes post-secondaires
Secondary diplomas	..	..	..	..	..	..	..	Diplômes de fin d'études secondaires
Other qualifications	..	..	..	..	..	..	..	Autres qualifications
Total	79 248	83 590	83 348	83 803	84 510	82 368	78 925	*Total*

Source: OECD, R&D database (DSTI/EAS Division), January 2002.

Source : OCDE, base de données R-D (DSTI/Division AES), janvier 2002.

Table 7 - Tableau 7

Gross domestic expenditure on R&D (GERD) - Dépenses intérieures brutes de R-D (DIRD)

by sector of performance and main scientific field - par secteur d'exécution et par grand domaine scientifique

Million PLN / *Millions PLN*

	1994	1995	1996	1997	1998	1999	2000	
Business enterprise								**Entreprises**
Natural sciences and engineering	705.9 [a]	822.2 [a]	1 126.8	1 320.9	1 657.2	1 893.2	..	Sciences exactes, nat. et de l'ingénieur
Social sciences and humanities	4.1 [a]	3.9 [a]	3.3	4.2	3.9	4.1	..	Sciences sociales et humaines
All fields of science	710.0 [a]	826.1 [a]	1 130.1	1 325.1	1 661.1	1 897.3	1 730.7	*Tous domaines scientifiques*
Government								**État**
Natural sciences and engineering	567.5 [a]	699.6 [a]	804.6	1 009.9	1 160.8	1 323.1	..	Sciences exactes, nat. et de l'ingénieur
Social sciences and humanities	36.2 [a]	46.1 [a]	54.7	64.1	73.8	90.1	..	Sciences sociales et humaines
All fields of science	603.7 [a]	745.7 [a]	859.3	1 074.0	1 234.6	1 413.2	1 546.6	*Tous domaines scientifiques*
Higher education								**Enseignement supérieur**
Natural sciences and engineering	..	487.0	655.0	762.2	894.2	1 070.1	..	Sciences exactes, nat. et de l'ingénieur
Social sciences and humanities	..	74.0	113.5	199.7	212.5	204.2	..	Sciences sociales et humaines
All fields of science	407.0	561.0	768.5	961.9	1 106.7	1 274.3	1 512.4	*Tous domaines scientifiques*
Private non-profit								**Institutions sans but lucratif**
Natural sciences and engineering	0.0	..	0.0	..	2.0	1.4	..	Sciences exactes, nat. et de l'ingénieur
Social sciences and humanities	0.0	..	3.5	..	0.7	4.3	..	Sciences sociales et humaines
All fields of science	0.3	..	3.5	..	2.7	5.7	6.4	*Tous domaines scientifiques*
TOTAL								**TOTAL**
Natural sciences and engineering	..	2 008.8	2 586.4	3 093.0	3 714.2	4 287.8	..	Sciences exactes, nat. et de l'ingénieur
Social sciences and humanities	..	124.0	175.0	268.0	290.9	302.7	..	Sciences sociales et humaines
All fields of science	1 721.0 [a]	2 132.8 [a]	2 761.4	3 361.0	4 005.1	4 590.5	4 796.1	*Tous domaines scientifiques*

Table 8 - Tableau 8

Total R&D personnel - Personnel total de R-D

by sector of employment and main scientific field - par secteur d'emploi et par grand domaine scientifique

In full-time equivalent / *En équivalence plein-temps*

	1994	1995	1996	1997	1998	1999	2000	
Business enterprise								**Entreprises**
Natural sciences and engineering	25 262	26 145	23 421	23 079	21 824	20 226	..	Sciences exactes, nat. et de l'ingénieur
Social sciences and humanities	269	137	115	102	90	88	..	Sciences sociales et humaines
All fields of science	25 531	26 282	23 536	23 181	21 914	20 314	18 586	*Tous domaines scientifiques*
Government								**État**
Natural sciences and engineering	19 185	19 560	18 845	17 921	18 296	17 346	..	Sciences exactes, nat. et de l'ingénieur
Social sciences and humanities	1 779	2 127	1 899	1 724	1 806	1 739	..	Sciences sociales et humaines
All fields of science	20 964	21 687	20 744	19 645	20 102	19 085	18 823	*Tous domaines scientifiques*
Higher education								**Enseignement supérieur**
Natural sciences and engineering	..	28 095	29 368	29 644	33 075 [s]	30 753	..	Sciences exactes, nat. et de l'ingénieur
Social sciences and humanities	..	7 526	9 678	11 333	12 190 [s]	12 195	..	Sciences sociales et humaines
All fields of science	32 745	35 621	39 046	40 977	42 478	42 948	41 499	*Tous domaines scientifiques*
Private non-profit								**Institutions sans but lucratif**
Natural sciences and engineering	5	..	0	..	13	9	..	Sciences exactes, nat. et de l'ingénieur
Social sciences and humanities	3	..	22	..	3	12	..	Sciences sociales et humaines
All fields of science	8	..	22	..	16	21	17	*Tous domaines scientifiques*
TOTAL								**TOTAL**
Natural sciences and engineering	..	73 800	71 634	70 644	73 208 [s]	68 334	..	Sciences exactes, nat. et de l'ingénieur
Social sciences and humanities	..	9 790	11 714	13 159	14 089 [s]	14 034	..	Sciences sociales et humaines
All fields of science	79 248	83 590	83 348	83 803	84 510	82 368	78 925	*Tous domaines scientifiques*

Source: OECD, R&D database (DSTI/EAS Division), January 2002.

Source : OCDE, base de données R-D (DSTI/Division AES), janvier 2002.

Total business enterprise intramural expenditure on R&D - Dépenses intra-muros totales de R-D des entreprises
(BERD) by industry - (DIRDE) par industrie

	1994	1995	1996	1997	1998	1999	2000	
				9.1 Million PLN - Millions PLN				
1 AGRICULTURE, HUNTING & FORESTRY	8.4	16.2	36.9	35.3	35.3	41.5	40.5	1 AGRICULTURE, CHASSE & SYLVICULTURE
2 MINING	54.3	39.4	51.6	60.4	65.8	66.6	73.6	2 ACTIVITES EXTRACTIVES
3 MANUFACTURING	510.5	610.1	803.9	970.1	1 276.7	1 451.9	1 185.6	3 ACTIVITES DE FABRICATION
4 Food, beverages & tobacco	10.5	10.6	13.9	31.4	26.8	33.0	34.6	**4 Produits alimentaires, boissons & tabac**
5 Food products & beverages	9.1	9.0	11.7	28.9	24.3	26.6	25.1	5 Produits alimentaires & boissons
6 Tobacco products	1.4	1.6	2.2	2.5	2.5	6.4	9.5	6 Produits à base de tabac
7 Textiles, wear app., fur & leather	26.6	34.4	26.3	25.3	29.8	29.4	34.2	**7 Textiles, habill., fourrures & cuirs**
8 Textiles	24.7	32.6	24.3	23.3	27.8	26.8	32.0	8 Textiles
9 Wearing apparel & fur	0.1	0.4	2.0	0.0	0.0	0.0	0.0	9 Articles d'habillement & fourrures
10 Leather products & footwear	1.8	1.4	0.0	2.0	2.0	2.6	2.2	10 Articles en cuir & chaussures
11 Wood, paper, printing, publishing	5.2	6.1	8.5	10.2	10.9	23.6	15.1	**11 Bois, papier, imprimerie, édition**
12 Wood & cork (not furniture)	3.3	4.5	6.0	7.3	7.5	19.2	7.0	12 Bois & liège (sauf meubles)
13 Pulp, paper & paper products	1.7	1.4	2.3	2.7	3.2	4.3	5.2	13 Carton, papier & articles
14 Publ., print. & repro. of rec. media	0.2	0.2	0.2	0.2	0.2	0.1	2.9	14 Edition, impress. & repro. enregis.
15 Coke, petroleum, nuclear fuel, chemicals, rubber plastics	115.0	139.6	263.6	255.9	299.6	280.3	259.4	**15 Coke, pétrole, combust. nucl., prod. chimi., caout. & mat. plast.**
16 Coke, ref. petrol. prod. & nuclear fuel	16.4	12.3	14.0	11.9	14.2	13.5	22.3	16 Coke, raffin. pétr. & combust. nucl.
17 Chemicals & chemical products	78.3	111.7	197.7	165.9	220.2	224.6	203.6	17 Produits chimiques
18 *Chemical prod.(less pharmaceuticals)*	62.3	71.7	99.6	106.4	155.9	133.6	124.7	18 *Prod. chimi. (moins prod. pharmaceutiques)*
19 *Pharmaceuticals*	16.0	40.0	98.1	59.5	64.3	91.0	78.9	19 *Produits pharmaceutiques*
20 Rubber & plastic products	20.4	15.6	51.9	78.1	65.2	42.2	33.5	20 Caoutchouc et matières plastiques
21 Non-metallic mineral products	6.1	8.2	20.3	9.5	18.3	21.4	16.9	**21 Produits minéraux non métalliques**
22 Basic metals	31.4	27.1	36.6	52.1	54.9	65.0	63.3	**22 Produits métallurgiques de base**
23 Basic metals, ferrous	24.0	16.6	24.9	33.3	37.5	43.5	40.4	23 Produits métallur. de base, ferreux
24 Basic metals, non-ferrous	7.4	10.5	11.7	18.8	17.4	21.5	22.9	24 Prod. métallur. de base, non ferreux
25 Fabricated metal products	6.8	9.5	8.0	7.5	10.4	15.8	12.3	**25 Ouvrages en métaux (sauf mach.& mat.)**
26 Machinery equipment, instruments and transport equipment	305.1	371.4	419.4	557.7	806.5	969.8	733.3	**26 Machines & matériel, instruments & mat. de transport**
27 Machinery n.e.c.	98.7	118.9	147.7	160.6	246.6	326.3	228.2	27 Machines, n.c.a.
28 Office, account. & computing machinery	0.3	0.3	0.8	1.1	2.2	1.3	11.1	28 Mach. de bur., compta. & trait. inform.
29 Electrical machinery	38.1	57.7	71.7	107.8	158.3	155.6	94.7	29 Machines électriques
30 Electro. equip. (radio, TV & commun.)	41.4	42.4	46.3	60.9	112.5	104.2	115.7	30 Equip. électro. (radio, tv & communic.)
31 *Electro. comp. (inc. semi-conductors)*	28.8	25.5	31.3	35.9	38.1	49.6	42.4	31 *Compos. électro. (semi-cond. incl.)*
32 *TV, radio & communications equip.*	12.6	16.9	15.0	25.0	74.4	54.6	73.3	32 *Appareils radio, tv et de communi.*
33 Instruments	9.0	9.1	12.8	15.8	31.0	27.3	26.8	33 Instruments
34 Motor vehicles	36.8	41.1	50.2	95.1	143.1	196.1	121.3	34 Véhicules automobiles
35 Other transport equipment	80.8	101.9	90.1	116.4	112.8	159.0	135.5	35 Autres matériels de transport
36 *Ships*	8.1	18.1	19.0	24.3	36.2	45.6	35.5	36 *Navires*
37 *Aerospace*	31.1	41.5	43.0	63.0	52.8	75.9	70.9	37 *Aérospatiale*
38 *Other transport equipment n.e.c.*	41.6	42.3	28.2	29.1	23.8	37.5	29.1	38 *Autres matériels de transport n.c.a.*
39 Furniture, other manufacturing n.e.c.	1.2	0.7	4.1	15.8	15.0	9.5	11.8	**39 Meubles, autres activ. de fabricat. n.c.a.**
40 Furniture	0.2	0.1	2.5	14.1	13.8	8.2	10.6	40 Meubles
41 Other manufacturing n.e.c.	1.0	0.6	1.6	1.7	1.2	1.3	1.2	41 Autres activités de fabrication n.c.a.
42 Recycling	2.7	2.6	3.4	4.7	4.5	4.1	4.7	**42 Récupération**
43 ELECTRICITY, GAS & WATER SUPPLY	4.4	8.4	9.1	15.1	19.9	19.7	30.4	43 DISTRIBUTION D'ELECTRICITE, GAZ & EAU
44 CONSTRUCTION	30.0	24.6	25.4	29.4	43.3	51.7	67.6	44 CONSTRUCTION
45 SERVICES SECTOR	102.4	127.4	203.2	214.8	220.1	265.9	333.0	45 SECTEUR DES SERVICES
46 Wholesale, retail trade & repair of motor vehicles, etc.	2.1	2.0	2.9	2.9	3.1	4.6	7.5	**46 Commerce (gros & détail), réparation de véhicules, etc.**
47 Hotels & restaurants	0.0	0.0	0.0	0.0	0.0	0.0	0.0	**47 Hôtels et restaurants**
48 Transport & storage	9.5	13.8	46.6	52.9	59.1	61.3	42.1	**48 Transports, entreposage**
49 Communications	19.4	24.6	25.4	26.6	25.0	30.9	85.2	**49 Communications**
50 Post	0.0	0.0	0.0	0.0	0.0	0.0	0.0	50 Postes
51 Telecommunications	19.4	24.6	25.4	26.6	25.0	30.9	85.2	51 Télécommunications
52 Financial intermediation (inc. insur.)	0.0	0.0	0.0	0.0	0.0	0.0	7.5	**52 Intermédiation financ. (assurances inc.)**
53 Real estate, renting & busin. activity	39.4	45.4	58.0	58.2	52.7	59.1	61.5	**53 Immobil., locat. & serv. aux entreprises**
54 Computer & related activities	7.7	3.0	1.4	4.6	0.1	0.1	3.4	54 Informatique et activités connexes
55 *Software consultancy*	5.9	0.2	1.3	0.1	0.1	0.1	0.9	55 *Production de logiciels*
56 *Other computer services n.e.c.*	1.8	2.8	0.1	4.5	0.0	0.0	2.5	56 *Autres activités informatiques*
57 Research & development	31.7	42.2	56.6	53.2	52.3	59.0	58.1	57 Recherche & développement expérimental
58 Other business activities n.e.c.	0.0	0.0	0.0	0.4	0.3	0.0	0.0	58 Autres activ. & serv. aux entreprises
59 Comm., soc. & pers. serv. activ., etc.	32.0	41.6	70.2	74.2	80.2	110.0	129.2	**59 Adm. pub., serv. soc. & collec., etc**
TOTAL BUSINESS ENTERPRISE SECTOR	**710.0** [d]	**826.1** [a]	**1 130.1**	**1 325.1**	**1 661.1**	**1 897.3**	**1 730.7**	**TOTAL SECTEUR DES ENTREPRISES**

Source: OECD, R&D database (DSTI/EAS Division), January 2002.

Source : OCDE, base de données R-D (DSTI/Division AES), janvier 2002.

POLAND Table 9 - Tableau 9 **POLOGNE**

Total business enterprise intramural expenditure on R&D - Dépenses intra-muros totales de R-D des entreprises
(BERD) by industry - (DIRDE) par industrie

	1994	1995	1996	1997	1998	1999	2000	
	9.2 Million constant USD(1995 prices and PPPs) - Millions de USD constants (prix et PPA de 1995)							
1 AGRICULTURE, HUNTING & FORESTRY	9.4	14.2	27.3	22.9	20.5	22.6	20.6	**1 AGRICULTURE, CHASSE & SYLVICULTURE**
2 MINING	61.1	34.6	38.2	39.2	38.2	36.2	37.4	**2 ACTIVITES EXTRACTIVES**
3 MANUFACTURING	574.2	536.5	595.5	630.3	742.1	790.2	602.4	**3 ACTIVITES DE FABRICATION**
4 Food, beverages & tobacco	11.8	9.3	10.3	20.4	15.6	18.0	17.6	**4 Produits alimentaires, boissons & tabac**
5 Food products & beverages	10.2	7.9	8.7	18.8	14.1	14.5	12.8	5 Produits alimentaires & boissons
6 Tobacco products	1.6	1.4	1.6	1.6	1.5	3.5	4.8	6 Produits à base de tabac
7 Textiles, wear app., fur & leather	29.9	30.3	19.5	16.4	17.3	16.0	17.4	**7 Textiles, habill., fourrures & cuirs**
8 Textiles	27.8	28.7	18.0	15.1	16.2	14.6	16.3	8 Textiles
9 Wearing apparel & fur	0.1	0.4	1.5	0.0	0.0	0.0	0.0	9 Articles d'habillement & fourrures
10 Leather products & footwear	2.0	1.2	0.0	1.3	1.2	1.4	1.1	10 Articles en cuir & chaussures
11 Wood, paper, printing, publishing	5.8	5.4	6.3	6.6	6.3	12.8	7.7	**11 Bois, papier, imprimerie, édition**
12 Wood & cork (not furniture)	3.7	4.0	4.4	4.7	4.4	10.5	3.6	12 Bois & liège (sauf meubles)
13 Pulp, paper & paper products	1.9	1.2	1.7	1.8	1.9	2.3	2.6	13 Carton, papier & articles
14 Publ., print. & repro. of rec. media	0.2	0.2	0.1	0.1	0.1	0.1	1.5	14 Edition, impress. & repro. enregis.
15 Coke, petroleum, nuclear fuel, chemicals, rubber plastics	129.3	122.8	195.3	166.3	174.1	152.6	131.8	**15 Coke, pétrole, combust. nucl., prod. chimi., caout. & mat. plast.**
16 Coke, ref. petrol. prod. & nuclear fuel	18.4	10.8	10.4	7.7	8.3	7.3	11.3	16 Coke, raffin. pétr. & combust. nucl.
17 Chemicals & chemical products	88.1	98.2	146.4	107.8	128.0	122.2	103.4	17 Produits chimiques
18 *Chemical prod.(less pharmaceuticals)*	70.1	63.1	73.8	69.1	90.6	72.7	63.4	18 *Prod. chimi. (moins prod. pharmaceutiques)*
19 *Pharmaceuticals*	18.0	35.2	72.7	38.7	37.4	49.5	40.1	19 *Produits pharmaceutiques*
20 Rubber & plastic products	22.9	13.7	38.4	50.7	37.9	23.0	17.0	20 Caoutchouc et matières plastiques
21 Non-metallic mineral products	6.9	7.2	15.0	6.2	10.6	11.6	8.6	**21 Produits minéraux non métalliques**
22 Basic metals	35.3	23.8	27.1	33.9	31.9	35.4	32.2	**22 Produits métallurgiques de base**
23 Basic metals, ferrous	27.0	14.6	18.4	21.6	21.8	23.7	20.5	23 Produits métallur. de base, ferreux
24 Basic metals, non-ferrous	8.3	9.2	8.7	12.2	10.1	11.7	11.6	24 Prod. métallur. de base, non ferreux
25 Fabricated metal products	7.6	8.4	5.9	4.9	6.0	8.6	6.3	**25 Ouvrages en métaux (sauf mach.& mat.)**
26 Machinery equipment, instruments and transport equipment	343.1	326.6	310.7	362.3	468.8	527.8	372.6	**26 Machines & matériel, instruments & mat. de transport**
27 Machinery n.e.c.	111.0	104.6	109.4	104.3	143.3	177.6	115.9	27 Machines, n.c.a.
28 Office, account. & computing machinery	0.3	0.3	0.6	0.7	1.3	0.7	5.6	28 Mach. de bur., compta. & trait. inform.
29 Electrical machinery	42.9	50.7	53.1	70.0	92.0	84.7	48.1	29 Machines électriques
30 Electro. equip. (radio, TV & commun.)	46.6	37.3	34.3	39.6	65.4	56.7	58.8	30 Equip. électro. (radio, tv & communic.)
31 *Electro. comp. (inc. semi-conductors)*	32.4	22.4	23.2	23.3	22.1	27.0	21.5	31 *Compos. électro. (semi-cond. incl.)*
32 *TV, radio & communications equip.*	14.2	14.9	11.1	16.2	43.2	29.7	37.2	32 *Appareils radio, tv et de communi.*
33 Instruments	10.1	8.0	9.5	10.3	18.0	14.9	13.6	33 Instruments
34 Motor vehicles	41.4	36.1	37.2	61.8	83.2	106.7	61.6	34 Véhicules automobiles
35 Other transport equipment	90.9	89.6	66.7	75.6	65.6	86.5	68.8	35 Autres matériels de transport
36 *Ships*	9.1	15.9	14.1	15.8	21.0	24.8	18.0	36 *Navires*
37 *Aerospace*	35.0	36.5	31.9	40.9	30.7	41.3	36.0	37 *Aérospatiale*
38 *Other transport equipment n.e.c.*	46.8	37.2	20.9	18.9	13.8	20.4	14.8	38 *Autres matériels de transport n.c.a.*
39 Furniture, other manufacturing n.e.c.	1.4	0.6	3.0	10.3	8.7	5.2	6.0	**39 Meubles, autres activ. de fabricat. n.c.a.**
40 Furniture	0.2	0.1	1.9	9.2	8.0	4.5	5.4	40 Meubles
41 Other manufacturing n.e.c.	1.1	0.5	1.2	1.1	0.7	0.7	0.6	41 Autres activités de fabrication n.c.a.
42 Recycling	3.0	2.3	2.5	3.1	2.6	2.2	2.4	**42 Récupération**
43 ELECTRICITY, GAS & WATER SUPPLY	4.9	7.4	6.7	9.8	11.6	10.7	15.4	**43 DISTRIBUTION D'ELECTRICITE, GAZ & EAU**
44 CONSTRUCTION	33.7	21.6	18.8	19.1	25.2	28.1	34.3	**44 CONSTRUCTION**
45 SERVICES SECTOR	115.2	112.0	150.5	139.6	127.9	144.7	169.2	**45 SECTEUR DES SERVICES**
46 Wholesale, retail trade & repair of motor vehicles, etc.	2.4	1.8	2.1	1.9	1.8	2.5	3.8	**46 Commerce (gros & détail), réparation de véhicules, etc.**
47 Hotels & restaurants	0.0	0.0	0.0	0.0	0.0	0.0	0.0	**47 Hôtels et restaurants**
48 Transport & storage	10.7	12.1	34.5	34.4	34.4	33.4	21.4	**48 Transports, entreposage**
49 Communications	21.8	21.6	18.8	17.3	14.5	16.8	43.3	**49 Communications**
50 Post	0.0	0.0	0.0	0.0	0.0	0.0	0.0	50 Postes
51 Telecommunications	21.8	21.6	18.8	17.3	14.5	16.8	43.3	51 Télécommunications
52 Financial intermediation (inc. insur.)	0.0	0.0	0.0	0.0	0.0	0.0	3.8	**52 Intermédiation financ. (assurances inc.)**
53 Real estate, renting & busin. activity	44.3	39.9	43.0	37.8	30.6	32.2	31.2	**53 Immobil., locat. & serv. aux entreprises**
54 Computer & related activities	8.7	2.6	1.0	3.0	0.1	0.1	1.7	54 Informatique et activités connexes
55 *Software consultancy*	6.6	0.2	1.0	0.1	0.1	0.1	0.5	55 *Production de logiciels*
56 *Other computer services n.e.c.*	2.0	2.5	0.1	2.9	0.0	0.0	1.3	56 *Autres activités informatiques*
57 Research & development	35.7	37.1	41.9	34.6	30.4	32.1	29.5	57 Recherche & développement expérimental
58 Other business activities n.e.c.	0.0	0.0	0.0	0.3	0.2	0.0	0.0	58 Autres activ. & serv. aux entreprises
59 Comm., soc. & pers. serv. activ., etc.	36.0	36.6	52.0	48.2	46.6	59.9	65.6	**59 Adm. pub., serv. soc. & collec., etc**
TOTAL BUSINESS ENTERPRISE SECTOR	**798.5** [d]	**726.5** [a]	**837.1**	**860.9**	**965.5**	**1 032.6**	**879.4**	**TOTAL SECTEUR DES ENTREPRISES**

Source: OECD, R&D database (DSTI/EAS Division), January 2002.

Source : OCDE, base de données R-D (DSTI/Division AES), janvier 2002.

Business enterprise intramural expenditure on R&D (BERD) - Dépenses intra-muros de R-D des entreprises (DIRDE)

by industry and source of funds - par industrie et par source de financement

Million PLN *Millions PLN*

10.1 Business enterprise - Entreprises

	1994	1995	1996	1997	1998	1999	2000	
1 AGRICULTURE, HUNTING & FORESTRY	5.0	7.5	21.7	20.5	19.0	24.7	..	1 AGRICULTURE, CHASSE & SYLVICULTURE
2 MINING	33.5	20.5	27.2	25.4	34.6	32.6	..	2 ACTIVITES EXTRACTIVES
3 MANUFACTURING	348.6	420.9	599.3	715.9	969.9	1 115.1	..	3 ACTIVITES DE FABRICATION
4 Food, beverages & tobacco	6.7	5.4	8.4	24.8	19.9	24.5	..	**4 Produits alimentaires, boissons & tabac**
5 Food products & beverages	5.4	3.9	6.2	22.4	17.5	18.2	..	5 Produits alimentaires & boissons
6 Tobacco products	1.3	1.5	2.2	2.4	2.4	6.3	..	6 Produits à base de tabac
7 Textiles, wear app., fur & leather	16.1	18.3	13.3	9.6	11.8	8.2	..	**7 Textiles, habill., fourrures & cuirs**
8 Textiles	15.3	17.2	12.7	8.6	11.4	7.3	..	8 Textiles
9 Wearing apparel & fur	0.1	0.4	0.6	0.0	0.0	0.0	..	9 Articles d'habillement & fourrures
10 Leather products & footwear	0.7	0.8	0.0	1.0	0.4	0.9	..	10 Articles en cuir & chaussures
11 Wood, paper, printing, publishing	3.3	3.6	3.9	5.1	6.3	18.3	..	**11 Bois, papier, imprimerie, édition**
12 Wood & cork (not furniture)	2.2	2.8	3.2	3.5	4.8	15.7	..	12 Bois & liège (sauf meubles)
13 Pulp, paper & paper products	1.0	0.6	0.5	1.4	1.3	2.5	..	13 Carton, papier & articles
14 Publ., print. & repro. of rec. media	0.1	0.2	0.2	0.2	0.2	0.1	..	14 Edition, impress. & repro. enregis.
15 Coke, petroleum, nuclear fuel, chemicals, rubber plastics	86.2	98.8	216.3	198.0	242.6	216.0	..	**15 Coke, pétrole, combust. nucl., prod. chimi., caout. & mat. plast.**
16 Coke, ref. petrol. prod. & nuclear fuel	12.6	11.5	12.3	10.2	12.5	11.4	..	16 Coke, raffin. pétr. & combust. nucl.
17 Chemicals & chemical products	56.1	74.3	157.1	115.6	173.0	167.2	..	17 Produits chimiques
18 *Chemical prod.(less pharmaceuticals)*	41.4	43.0	66.2	67.9	113.4	87.7	..	18 *Prod. chimi. (moins prod. pharmaceutiques)*
19 *Pharmaceuticals*	14.7	31.3	90.9	47.7	59.6	79.5	..	19 *Produits pharmaceutiques*
20 Rubber & plastic products	17.5	13.0	46.9	72.2	57.1	37.4	..	20 Caoutchouc et matières plastiques
21 Non-metallic mineral products	3.0	3.6	14.7	4.4	11.4	12.3	..	**21 Produits minéraux non métalliques**
22 Basic metals	17.5	18.3	23.7	38.3	33.6	43.9	..	**22 Produits métallurgiques de base**
23 Basic metals, ferrous	13.6	13.6	19.4	28.1	27.7	35.6	..	23 Produits métallur. de base, ferreux
24 Basic metals, non-ferrous	3.9	4.7	4.3	10.2	5.9	8.3	..	24 Prod. métallur. de base, non ferreux
25 Fabricated metal products	5.2	8.4	5.6	4.8	8.0	13.9	..	**25 Ouvrages en métaux (sauf mach.& mat.)**
26 Machinery equipment, instruments and transport equipment	208.3	263.2	308.9	413.5	620.3	767.1	..	**26 Machines & matériel, instruments & mat. de transport**
27 Machinery n.e.c.	76.5	87.1	112.8	126.4	215.2	285.9	..	27 Machines, n.c.a.
28 Office, account. & computing machinery	0.3	0.3	0.8	1.1	2.2	1.3	..	28 Mach. de bur., compta. & trait. inform.
29 Electrical machinery	24.7	42.3	52.2	82.9	132.7	123.5	..	29 Machines électriques
30 Electro. equip. (radio, TV & commun.)	17.8	12.7	24.8	31.5	54.7	59.9	..	30 Equip. électro. (radio, tv & communic.)
31 *Electro. comp. (inc. semi-conductors)*	10.6	7.5	9.8	6.5	7.0	5.8	..	31 *Compos. électro. (semi-cond. incl.)*
32 *TV, radio & communications equip.*	7.2	5.2	15.0	25.0	47.7	54.1	..	32 *Appareils radio, tv et de communi.*
33 Instruments	7.5	8.1	9.0	10.9	25.4	20.6	..	33 Instruments
34 Motor vehicles	33.8	35.9	47.0	90.2	136.1	189.9	..	34 Véhicules automobiles
35 Other transport equipment	47.7	76.8	62.3	70.5	54.0	86.0	..	35 Autres matériels de transport
36 *Ships*	4.4	12.9	14.2	18.4	14.8	15.8	..	36 *Navires*
37 *Aerospace*	11.8	24.9	22.7	27.6	20.3	42.6	..	37 *Aérospatiale*
38 *Other transport equipment n.e.c.*	31.5	39.0	25.4	24.5	18.9	27.6	..	38 *Autres matériels de transport n.c.a.*
39 Furniture, other manufacturing n.e.c.	1.1	0.3	3.4	15.3	14.0	8.8	..	**39 Meubles, autres activ. de fabricat. n.c.a.**
40 Furniture	0.2	0.1	1.8	13.7	13.3	7.7	..	40 Meubles
41 Other manufacturing n.e.c.	0.9	0.2	1.6	1.6	0.7	1.1	..	41 Autres activités de fabrication n.c.a.
42 Recycling	1.1	1.1	1.3	2.1	2.0	2.1	..	**42 Récupération**
43 ELECTRICITY, GAS & WATER SUPPLY	4.2	7.3	7.2	12.6	18.0	16.5	..	43 DISTRIBUTION D'ELECTRICITE, GAZ & EAU
44 CONSTRUCTION	14.8	14.2	15.1	18.2	30.8	38.3	..	44 CONSTRUCTION
45 SERVICES SECTOR	55.6	63.7	128.4	129.0	123.3	142.1	..	45 SECTEUR DES SERVICES
46 Wholesale, retail trade & repair of motor vehicles, etc.	1.7	1.7	2.4	2.7	2.9	4.4	..	**46 Commerce (gros & détail), réparation de véhicules, etc.**
47 Hotels & restaurants	0.0	0.0	0.0	0.0	0.0	0.0	..	**47 Hôtels et restaurants**
48 Transport & storage	7.2	8.6	38.2	43.0	47.1	46.9	..	**48 Transports, entreposage**
49 Communications	12.0	14.5	13.2	11.9	10.8	10.6	..	**49 Communications**
50 Post	0.0	0.0	0.0	0.0	0.0	0.0	..	50 Postes
51 Telecommunications	12.0	14.5	13.2	11.9	10.8	10.6	..	51 Télécommunications
52 Financial intermediation (inc. insur.)	0.0	0.0	0.0	0.0	0.0	0.0	..	**52 Intermédiation financ. (assurances inc.)**
53 Real estate, renting & busin. activity	24.7	25.8	31.8	32.6	19.8	25.0	..	**53 Immobil., locat. & serv. aux entreprises**
54 Computer & related activities	6.6	2.1	1.4	4.6	0.1	0.1	..	54 Informatique et activités connexes
55 *Software consultancy*	4.8	0.2	1.3	0.1	0.1	0.1	..	55 *Production de logiciels*
56 *Other computer services n.e.c.*	1.8	1.9	0.1	4.5	0.0	0.0	..	56 *Autres activités informatiques*
57 Research & development	18.1	23.7	30.4	27.6	19.4	24.9	..	57 Recherche & développement expérimental
58 Other business activities n.e.c.	0.0	0.0	0.0	0.4	0.3	0.0	..	58 Autres activ. & serv. aux entreprises
59 Comm., soc. & pers. serv. activ., etc.	10.0	13.1	42.8	38.8	42.7	55.2	..	**59 Adm. pub., serv. soc. & collec., etc**
TOTAL BUSINESS ENTERPRISE SECTOR	**461.7**	**534.1**	**798.9**	**921.6**	**1 195.6**	**1 369.3**	**1 147.2**	**TOTAL SECTEUR DES ENTREPRISES**

Source: OECD, R&D database (DSTI/EAS Division), January 2002.

Source : OCDE, base de données R-D (DSTI/Division AES), janvier 2002.

POLAND — Table 10 - Tableau 10 — POLOGNE

Business enterprise intramural expenditure on R&D (BERD) - Dépenses intra-muros de R-D des entreprises (DIRDE)

by industry and source of funds - par industrie et par source de financement

Million PLN — *Millions PLN*

	1994	1995	1996	1997	1998	1999	2000	
				10.2 Government - État				
1 AGRICULTURE, HUNTING & FORESTRY	3.4	7.1	12.8	12.5	13.7	14.6	..	**1 AGRICULTURE, CHASSE & SYLVICULTURE**
2 MINING	20.1	18.2	23.8	34.0	29.9	32.9	..	**2 ACTIVITES EXTRACTIVES**
3 MANUFACTURING	157.9	182.6	199.0	244.8	298.0	319.5	..	**3 ACTIVITES DE FABRICATION**
4 Food, beverages & tobacco	3.8	5.2	5.4	6.4	6.9	8.3	..	**4 Produits alimentaires, boissons & tabac**
5 Food products & beverages	3.7	5.1	5.4	6.3	6.8	8.2	..	5 Produits alimentaires & boissons
6 Tobacco products	0.1	0.1	0.0	0.1	0.1	0.1	..	6 Produits à base de tabac
7 Textiles, wear app., fur & leather	10.4	16.0	12.6	15.6	17.9	19.9	..	**7 Textiles, habill., fourrures & cuirs**
8 Textiles	9.3	15.4	11.4	14.6	16.3	18.2	..	8 Textiles
9 Wearing apparel & fur	0.0	0.0	1.2	0.0	0.0	0.0	..	9 Articles d'habillement & fourrures
10 Leather products & footwear	1.1	0.6	0.0	1.0	1.6	1.7	..	10 Articles en cuir & chaussures
11 Wood, paper, printing, publishing	1.8	2.4	4.1	4.8	4.4	5.1	..	**11 Bois, papier, imprimerie, édition**
12 Wood & cork (not furniture)	1.0	1.6	2.3	3.5	2.5	3.3	..	12 Bois & liège (sauf meubles)
13 Pulp, paper & paper products	0.7	0.8	1.8	1.3	1.9	1.8	..	13 Carton, papier & articles
14 Publ., print. & repro. of rec. media	0.1	0.0	0.0	0.0	0.0	0.0	..	14 Edition, impress. & repro. enregis.
15 Coke, petroleum, nuclear fuel, chemicals, rubber plastics	28.1	39.0	45.3	53.4	53.0	59.6	..	**15 Coke, pétrole, combust. nucl., prod. chimi., caout. & mat. plast.**
16 Coke, ref. petrol. prod. & nuclear fuel	3.8	0.8	1.7	1.7	1.6	2.1	..	16 Coke, raffin. pétr. & combust. nucl.
17 Chemicals & chemical products	21.6	35.6	38.7	45.9	43.4	53.0	..	17 Produits chimiques
18 *Chemical prod.(less pharmaceuticals)*	20.3	27.6	31.6	36.5	38.7	41.5	..	18 *Prod. chimi. (moins prod. pharmaceutiques)*
19 *Pharmaceuticals*	1.3	8.1	7.1	9.4	4.7	11.5	..	19 *Produits pharmaceutiques*
20 Rubber & plastic products	2.8	2.5	4.9	5.8	8.0	4.5	..	20 Caoutchouc et matières plastiques
21 Non-metallic mineral products	3.1	4.6	5.6	5.1	6.9	9.0	..	**21 Produits minéraux non métalliques**
22 Basic metals	12.9	8.7	12.9	13.5	20.9	20.7	..	**22 Produits métallurgiques de base**
23 Basic metals, ferrous	9.4	3.0	5.5	5.1	9.7	7.9	..	23 Produits métallur. de base, ferreux
24 Basic metals, non-ferrous	3.5	5.7	7.4	8.4	11.2	12.8	..	24 Prod. métallur. de base, non ferreux
25 Fabricated metal products	1.5	1.1	2.4	2.7	2.4	1.9	..	**25 Ouvrages en métaux (sauf mach.& mat.)**
26 Machinery equipment, instruments and transport equipment	94.9	104.0	108.3	140.2	182.2	192.3	..	**26 Machines & matériel, instruments & mat. de transport**
27 Machinery n.e.c.	21.6	30.4	34.0	33.4	30.3	38.2	..	27 Machines, n.c.a.
28 Office, account. & computing machinery	0.0	0.0	0.0	0.0	0.0	0.0	..	28 Mach. de bur., compta. & trait. inform.
29 Electrical machinery	13.1	14.8	19.4	23.0	25.2	31.9	..	29 Machines électriques
30 Electro. equip. (radio, TV & commun.)	22.8	28.1	20.7	28.6	57.5	43.3	..	30 Equip. électro. (radio, tv & communic.)
31 *Electro. comp. (inc. semi-conductors)*	17.4	16.4	20.7	28.6	30.8	43.3	..	31 *Compos. électro. (semi-cond. incl.)*
32 *TV, radio & communications equip.*	5.4	11.7	0.0	0.0	26.7	0.0	..	32 *Appareils radio, tv et de communi.*
33 Instruments	1.5	1.0	3.5	4.9	5.4	6.1	..	33 Instruments
34 Motor vehicles	3.0	5.2	3.2	4.9	7.0	6.2	..	34 Véhicules automobiles
35 Other transport equipment	32.9	24.5	27.5	45.4	56.8	66.6	..	35 Autres matériels de transport
36 *Ships*	3.6	4.7	4.6	5.7	21.3	28.7	..	36 *Navires*
37 *Aerospace*	19.3	16.6	20.3	35.2	30.6	28.7	..	37 *Aérospatiale*
38 *Other transport equipment n.e.c.*	10.0	3.2	2.6	4.5	4.9	9.2	..	38 *Autres matériels de transport n.c.a.*
39 Furniture, other manufacturing n.e.c.	0.1	0.4	0.4	0.5	1.0	0.7	..	**39 Meubles, autres activ. de fabricat. n.c.a.**
40 Furniture	0.0	0.0	0.4	0.4	0.5	0.5	..	40 Meubles
41 Other manufacturing n.e.c.	0.1	0.4	0.0	0.1	0.5	0.2	..	41 Autres activités de fabrication n.c.a.
42 Recycling	1.4	1.2	2.0	2.6	2.4	1.9	..	**42 Récupération**
43 ELECTRICITY, GAS & WATER SUPPLY	0.2	0.9	1.4	2.3	1.9	3.0	..	**43 DISTRIBUTION D'ELECTRICITE, GAZ & EAU**
44 CONSTRUCTION	15.1	10.4	10.1	11.2	12.5	13.4	..	**44 CONSTRUCTION**
45 SERVICES SECTOR	44.2	60.3	71.9	82.0	91.4	120.0	..	**45 SECTEUR DES SERVICES**
46 Wholesale, retail trade & repair of motor vehicles, etc.	0.4	0.3	0.4	0.2	0.2	0.2	..	**46 Commerce (gros & détail), réparation de véhicules, etc.**
47 Hotels & restaurants	0.0	0.0	0.0	0.0	0.0	0.0	..	**47 Hôtels et restaurants**
48 Transport & storage	1.9	5.0	8.3	9.2	11.1	14.0	..	**48 Transports, entreposage**
49 Communications	7.2	10.1	12.2	14.5	14.2	20.3	..	**49 Communications**
50 Post	0.0	0.0	0.0	0.0	0.0	0.0	..	50 Postes
51 Telecommunications	7.2	10.1	12.2	14.5	14.2	20.3	..	51 Télécommunications
52 Financial intermediation (inc. insur.)	0.0	0.0	0.0	0.0	0.0	0.0	..	**52 Intermédiation financ. (assurances inc.)**
53 Real estate, renting & busin. activity	13.7	17.4	24.2	24.2	30.2	32.1	..	**53 Immobil., locat. & serv. aux entreprises**
54 Computer & related activities	1.1	0.9	0.0	0.0	0.0	0.0	..	54 Informatique et activités connexes
55 *Software consultancy*	1.1	0.0	0.0	0.0	0.0	0.0	..	55 *Production de logiciels*
56 *Other computer services n.e.c.*	0.0	0.9	0.0	..	0.0	0.0	..	56 *Autres activités informatiques*
57 Research & development	12.6	16.5	24.2	24.2	30.2	32.1	..	57 Recherche & développement expérimental
58 Other business activities n.e.c.	0.0	0.0	0.0	0.0	0.0	0.0	..	58 Autres activ. & serv. aux entreprises
59 Comm., soc. & pers. serv. activ., etc.	21.0	27.5	26.8	33.9	35.7	53.4	..	**59 Adm. pub., serv. soc. & collec., etc**
TOTAL BUSINESS ENTERPRISE SECTOR	**240.9** [d]	**279.5** [a]	**319.0**	**386.8**	**447.4**	**503.4**	**553.0**	**TOTAL SECTEUR DES ENTREPRISES**

Source: OECD, R&D database (DSTI/EAS Division), January 2002.

Source : OCDE, base de données R-D (DSTI/Division AES), janvier 2002.

Business enterprise intramural expenditure on R&D (BERD) - Dépenses intra-muros de R-D des entreprises (DIRDE)

by industry and source of funds - par industrie et par source de financement

Million PLN / *Millions PLN*

10.3 Other national funds - Autres sources nationales

	1994	1995	1996	1997	1998	1999	2000	
1 AGRICULTURE, HUNTING & FORESTRY	0.0	0.0	0.0	0.0	0.0	0.0	..	**1 AGRICULTURE, CHASSE & SYLVICULTURE**
2 MINING	0.0	0.0	0.0	0.3	0.2	0.1	..	**2 ACTIVITES EXTRACTIVES**
3 MANUFACTURING	0.6	1.0	0.9	1.3	0.9	1.3	..	**3 ACTIVITES DE FABRICATION**
4 Food, beverages & tobacco	0.0	0.0	0.0	0.1	0.0	0.0	..	**4 Produits alimentaires, boissons & tabac**
5 Food products & beverages	0.0	0.0	0.0	0.1	0.0	0.0	..	5 Produits alimentaires & boissons
6 Tobacco products	0.0	0.0	0.0	0.0	0.0	0.0	..	6 Produits à base de tabac
7 Textiles, wear app., fur & leather	0.0	0.0	0.0	0.1	0.0	0.1	..	**7 Textiles, habill., fourrures & cuirs**
8 Textiles	0.0	0.0	0.0	0.1	0.0	0.1	..	8 Textiles
9 Wearing apparel & fur	0.0	0.0	0.0	0.0	0.0	0.0	..	9 Articles d'habillement & fourrures
10 Leather products & footwear	0.0	0.0	0.0	0.0	0.0	0.0	..	10 Articles en cuir & chaussures
11 Wood, paper, printing, publishing	0.1	0.1	0.4	0.2	0.2	0.2	..	**11 Bois, papier, imprimerie, édition**
12 Wood & cork (not furniture)	0.1	0.1	0.4	0.2	0.2	0.2	..	12 Bois & liège (sauf meubles)
13 Pulp, paper & paper products	0.0	0.0	0.0	0.0	0.0	0.0	..	13 Carton, papier & articles
14 Publ., print. & repro. of rec. media	0.0	0.0	0.0	0.0	0.0	0.0	..	14 Edition, impress. & repro. enregis.
15 Coke, petroleum, nuclear fuel, chemicals, rubber plastics	0.2	0.1	0.2	0.1	0.1	0.4	..	**15 Coke, pétrole, combust. nucl., prod. chimi., caout. & mat. plast.**
16 Coke, ref. petrol. prod. & nuclear fuel	0.0	0.0	0.0	0.0	0.0	0.0	..	16 Coke, raffin. pétr. & combust. nucl.
17 Chemicals & chemical products	0.1	0.1	0.1	0.0	0.1	0.2	..	17 Produits chimiques
18 *Chemical prod.(less pharmaceuticals)*	0.1	0.1	0.1	0.0	0.1	0.2	..	18 *Prod. chimi. (moins prod. pharmaceutiques)*
19 *Pharmaceuticals*	0.0	0.0	0.0	0.0	0.0	0.0	..	19 *Produits pharmaceutiques*
20 Rubber & plastic products	0.1	0.0	0.1	0.1	0.0	0.2	..	20 Caoutchouc et matières plastiques
21 Non-metallic mineral products	0.0	0.0	0.0	0.0	0.0	0.0	..	**21 Produits minéraux non métalliques**
22 Basic metals	0.0	0.1	0.0	0.0	0.2	0.1	..	**22 Produits métallurgiques de base**
23 Basic metals, ferrous	0.0	0.0	0.0	0.0	0.1	0.0	..	23 Produits métallur. de base, ferreux
24 Basic metals, non-ferrous	0.0	0.1	0.0	0.0	0.1	0.1	..	24 Prod. métallur. de base, non ferreux
25 Fabricated metal products	0.0	0.0	0.0	0.0	0.0	0.0	..	**25 Ouvrages en métaux (sauf mach.& mat.)**
26 Machinery equipment, instruments and transport equipment	0.2	0.7	0.3	0.8	0.4	0.5	..	**26 Machines & matériel, instruments & mat. de transport**
27 Machinery n.e.c.	0.1	0.4	0.2	0.2	0.2	0.2	..	27 Machines, n.c.a.
28 Office, account. & computing machinery	0.0	0.0	0.0	0.0	0.0	0.0	..	28 Mach. de bur., compta. & trait. inform.
29 Electrical machinery	0.0	0.1	0.1	0.5	0.1	0.0	..	29 Machines électriques
30 Electro. equip. (radio, TV & commun.)	0.1	0.1	0.1	0.1	0.1	0.0	..	30 Equip. électro. (radio, tv & communic.)
31 *Electro. comp. (inc. semi-conductors)*	0.1	0.1	0.1	0.1	0.1	0.0	..	31 *Compos. électro. (semi-cond. incl.)*
32 *TV, radio & communications equip.*	0.0	0.0	0.0	0.0	0.0	0.0	..	32 *Appareils radio, tv et de communi.*
33 Instruments	0.0	0.0	0.1	0.0	0.0	0.0	..	33 Instruments
34 Motor vehicles	0.0	0.0	0.0	0.0	0.0	0.0	..	34 Véhicules automobiles
35 Other transport equipment	0.0	0.1	0.0	0.0	0.0	0.3	..	35 Autres matériels de transport
36 *Ships*	0.0	0.1	0.0	0.0	0.0	0.0	..	36 *Navires*
37 *Aerospace*	0.0	0.0	0.0	0.0	0.0	0.3	..	37 *Aérospatiale*
38 *Other transport equipment n.e.c.*	0.0	0.0	0.0	0.0	0.0	0.0	..	38 *Autres matériels de transport n.c.a.*
39 Furniture, other manufacturing n.e.c.	0.0	0.0	0.0	0.0	0.0	0.0	..	**39 Meubles, autres activ. de fabricat. n.c.a.**
40 Furniture	0.0	0.0	0.0	0.0	0.0	0.0	..	40 Meubles
41 Other manufacturing n.e.c.	0.0	0.0	0.0	0.0	0.0	0.0	..	41 Autres activités de fabrication n.c.a.
42 Recycling	0.0	0.0	0.0	0.0	0.0	0.0	..	**42 Récupération**
43 ELECTRICITY, GAS & WATER SUPPLY	0.0	0.0	0.0	0.0	0.0	0.1	..	**43 DISTRIBUTION D'ELECTRICITE, GAZ & EAU**
44 CONSTRUCTION	0.0	0.1	0.2	0.0	0.0	0.0	..	**44 CONSTRUCTION**
45 SERVICES SECTOR	0.5	0.7	0.4	0.8	0.6	0.3	..	**45 SECTEUR DES SERVICES**
46 Wholesale, retail trade & repair of motor vehicles, etc.	0.0	0.0	0.0	0.0	0.0	0.0	..	**46 Commerce (gros & détail), réparation de véhicules, etc.**
47 Hotels & restaurants	0.0	0.0	0.0	0.0	0.0	0.0	..	**47 Hôtels et restaurants**
48 Transport & storage	0.0	0.1	0.0	0.1	0.1	0.0	..	**48 Transports, entreposage**
49 Communications	0.1	0.0	0.0	0.1	0.0	0.0	..	**49 Communications**
50 Post	0.0	0.0	0.0	0.0	0.0	0.0	..	50 Postes
51 Telecommunications	0.1	0.0	0.0	0.1	0.0	0.0	..	51 Télécommunications
52 Financial intermediation (inc. insur.)	0.0	0.0	0.0	0.0	0.0	0.0	..	**52 Intermédiation financ. (assurances inc.)**
53 Real estate, renting & busin. activity	0.3	0.6	0.2	0.2	0.4	0.2	..	**53 Immobil., locat. & serv. aux entreprises**
54 Computer & related activities	0.0	0.1	0.0	0.0	0.0	0.0	..	54 Informatique et activités connexes
55 *Software consultancy*	0.0	0.0	0.0	0.0	0.0	0.0	..	55 *Production de logiciels*
56 *Other computer services n.e.c.*	0.0	0.0	0.0	0.0	0.0	0.0	..	56 *Autres activités informatiques*
57 Research & development	0.3	0.5	0.2	0.2	0.4	0.2	..	57 Recherche & développement expérimental
58 Other business activities n.e.c.	0.0	0.0	0.0	0.0	0.0	0.0	..	58 Autres activ. & serv. aux entreprises
59 Comm., soc. & pers. serv. activ., etc.	0.1	0.0	0.0	0.4	0.1	0.1	..	**59 Adm. pub., serv. soc. & collec., etc**
TOTAL BUSINESS ENTERPRISE SECTOR	**1.1**	**1.8**	**1.5**	**2.4**	**1.7**	**1.8**	**2.3**	**TOTAL SECTEUR DES ENTREPRISES**

Source: OECD, R&D database (DSTI/EAS Division), January 2002.

Source : OCDE, base de données R-D (DSTI/Division AES), janvier 2002.

Business enterprise intramural expenditure on R&D (BERD) - Dépenses intra-muros de R-D des entreprises (DIRDE)

by industry and source of funds - par industrie et par source de financement

Million PLN *Millions PLN*

	1994	1995	1996	1997	1998	1999	2000	
	10.4 Funds from abroad - Fonds de l'étranger							
1 AGRICULTURE, HUNTING & FORESTRY	0.0	1.6	2.4	2.3	2.6	2.2	..	**1 AGRICULTURE, CHASSE & SYLVICULTURE**
2 MINING	0.7	0.6	0.6	0.7	1.1	1.0	..	**2 ACTIVITES EXTRACTIVES**
3 MANUFACTURING	3.4	5.6	4.7	8.1	7.9	16.0	..	**3 ACTIVITES DE FABRICATION**
4 Food, beverages & tobacco	0.0	0.0	0.1	0.1	0.0	0.2	..	**4 Produits alimentaires, boissons & tabac**
5 Food products & beverages	0.0	0.0	0.1	0.1	0.0	0.2	..	5 Produits alimentaires & boissons
6 Tobacco products	0.0	0.0	0.0	0.0	0.0	0.0	..	6 Produits à base de tabac
7 Textiles, wear app., fur & leather	0.1	0.1	0.4	0.0	0.1	1.2	..	**7 Textiles, habill., fourrures & cuirs**
8 Textiles	0.1	0.0	0.2	0.0	0.1	1.2	..	8 Textiles
9 Wearing apparel & fur	0.0	0.0	0.2	0.0	0.0	0.0	..	9 Articles d'habillement & fourrures
10 Leather products & footwear	0.0	0.0	0.0	0.0	0.0	0.0	..	10 Articles en cuir & chaussures
11 Wood, paper, printing, publishing	0.0	0.0	0.1	0.1	0.0	0.0	..	**11 Bois, papier, imprimerie, édition**
12 Wood & cork (not furniture)	0.0	0.0	0.1	0.1	0.0	0.0	..	12 Bois & liège (sauf meubles)
13 Pulp, paper & paper products	0.0	0.0	0.0	0.0	0.0	0.0	..	13 Carton, papier & articles
14 Publ., print. & repro. of rec. media	0.0	0.0	0.0	0.0	0.0	0.0	..	14 Edition, impress. & repro. enregis.
15 Coke, petroleum, nuclear fuel, chemicals, rubber plastics	0.5	1.7	1.8	4.4	3.9	4.3	..	**15 Coke, pétrole, combust. nucl., prod. chimi., caout. & mat. plast.**
16 Coke, ref. petrol. prod. & nuclear fuel	0.0	0.0	0.0	0.0	0.1	0.0	..	16 Coke, raffin. pétr. & combust. nucl.
17 Chemicals & chemical products	0.5	1.6	1.8	4.4	3.7	4.2	..	17 Produits chimiques
18 *Chemical prod.(less pharmaceuticals)*	0.5	1.0	1.7	2.0	3.7	4.2	..	18 *Prod. chimi. (moins prod. pharmaceutiques)*
19 *Pharmaceuticals*	0.0	0.6	0.1	2.4	0.0	0.0	..	19 *Produits pharmaceutiques*
20 Rubber & plastic products	0.0	0.1	0.0	0.0	0.1	0.1	..	20 Caoutchouc et matières plastiques
21 Non-metallic mineral products	0.0	0.0	0.0	0.0	0.0	0.1	..	**21 Produits minéraux non métalliques**
22 Basic metals	1.0	0.0	0.0	0.3	0.2	0.3	..	**22 Produits métallurgiques de base**
23 Basic metals, ferrous	1.0	0.0	0.0	0.1	0.0	0.0	..	23 Produits métallur. de base, ferreux
24 Basic metals, non-ferrous	0.0	0.0	0.0	0.2	0.2	0.3	..	24 Prod. métallur. de base, non ferreux
25 Fabricated metal products	0.1	0.0	0.0	0.0	0.0	0.0	..	**25 Ouvrages en métaux (sauf mach.& mat.)**
26 Machinery equipment, instruments and transport equipment	1.6	3.5	1.9	3.2	3.6	9.8	..	**26 Machines & matériel, instruments & mat. de transport**
27 Machinery n.e.c.	0.5	1.0	0.7	0.6	0.9	2.0	..	27 Machines, n.c.a.
28 Office, account. & computing machinery	0.0	0.0	0.0	..	0.0	0.0	..	28 Mach. de bur., compta. & trait. inform.
29 Electrical machinery	0.2	0.5	0.0	1.4	0.3	0.2	..	29 Machines électriques
30 Electro. equip. (radio, TV & commun.)	0.7	1.5	0.7	0.7	0.2	1.0	..	30 Equip. électro. (radio, tv & communic.)
31 *Electro. comp. (inc. semi-conductors)*	0.7	1.5	0.7	0.7	0.2	0.5	..	31 *Compos. électro. (semi-cond. incl.)*
32 *TV, radio & communications equip.*	0.0	0.0	0.0	0.0	0.0	0.5	..	32 *Appareils radio, tv et de communi.*
33 Instruments	0.0	0.0	0.2	0.0	0.2	0.5	..	33 Instruments
34 Motor vehicles	0.0	0.0	0.0	0.0	0.0	0.0	..	34 Véhicules automobiles
35 Other transport equipment	0.2	0.5	0.3	0.5	2.0	6.1	..	35 Autres matériels de transport
36 *Ships*	0.1	0.4	0.2	0.2	0.1	1.1	..	36 *Navires*
37 *Aerospace*	0.0	0.0	0.0	0.2	1.9	4.3	..	37 *Aérospatiale*
38 *Other transport equipment n.e.c.*	0.1	0.1	0.1	0.1	0.0	0.7	..	38 *Autres matériels de transport n.c.a.*
39 Furniture, other manufacturing n.e.c.	0.0	0.0	0.3	0.0	0.0	0.0	..	**39 Meubles, autres activ. de fabricat. n.c.a.**
40 Furniture	0.0	0.0	0.3	0.0	0.0	0.0	..	40 Meubles
41 Other manufacturing n.e.c.	0.0	0.0	0.0	0.0	0.0	0.0	..	41 Autres activités de fabrication n.c.a.
42 Recycling	0.1	0.3	0.1	0.0	0.1	0.1	..	**42 Récupération**
43 ELECTRICITY, GAS & WATER SUPPLY	0.0	0.2	0.5	0.2	0.0	0.1	..	**43 DISTRIBUTION D'ELECTRICITE, GAZ & EAU**
44 CONSTRUCTION	0.1	0.0	0.0	0.0	0.0	0.0	..	**44 CONSTRUCTION**
45 SERVICES SECTOR	2.1	2.7	2.5	3.0	4.8	3.5	..	**45 SECTEUR DES SERVICES**
46 Wholesale, retail trade & repair of motor vehicles, etc.	0.0	0.0	0.0	0.0	0.0	0.0	..	**46 Commerce (gros & détail), réparation de véhicules, etc.**
47 Hotels & restaurants	0.0	0.0	0.0	0.0	0.0	0.0	..	**47 Hôtels et restaurants**
48 Transport & storage	0.4	0.2	0.1	0.6	0.8	0.4	..	**48 Transports, entreposage**
49 Communications	0.1	0.0	0.0	0.1	0.0	0.0	..	**49 Communications**
50 Post	0.0	0.0	0.0	0.0	0.0	0.0	..	50 Postes
51 Telecommunications	0.1	0.0	0.0	0.1	0.0	0.0	..	51 Télécommunications
52 Financial intermediation (inc. insur.)	0.0	0.0	0.0	0.0	0.0	0.0	..	**52 Intermédiation financ. (assurances inc.)**
53 Real estate, renting & busin. activity	0.7	1.5	1.8	1.2	2.3	1.8	..	**53 Immobil., locat. & serv. aux entreprises**
54 Computer & related activities	0.0	0.0	0.0	0.0	0.0	0.0	..	54 Informatique et activités connexes
55 *Software consultancy*	0.0	0.0	0.0	0.0	0.0	0.0	..	55 *Production de logiciels*
56 *Other computer services n.e.c.*	0.0	0.0	0.0	0.0	0.0	0.0	..	56 *Autres activités informatiques*
57 Research & development	0.7	1.5	1.8	1.2	2.3	1.8	..	57 Recherche & développement expérimental
58 Other business activities n.e.c.	0.0	0.0	0.0	0.0	0.0	0.0	..	58 Autres activ. & serv. aux entreprises
59 Comm., soc. & pers. serv. activ., etc.	0.9	1.0	0.6	1.1	1.7	1.3	..	**59 Adm. pub., serv. soc. & collec., etc**
TOTAL BUSINESS ENTERPRISE SECTOR	**6.3**	**10.7**	**10.7**	**14.3**	**16.4**	**22.8**	**28.2**	**TOTAL SECTEUR DES ENTREPRISES**

Source: OECD, R&D database (DSTI/EAS Division), January 2002.

Source : OCDE, base de données R-D (DSTI/Division AES), janvier 2002.

Business enterprise intramural expenditure on R&D (BERD) - Dépenses intra-muros de R-D des entreprises (DIRDE)

by industry and type of costs - par industrie et par type de coûts

Million PLN *Millions PLN*

	1994	1995	1996	1997	1998	1999	2000	
	11.1 Current expenditure - Dépenses courantes							
1 **AGRICULTURE, HUNTING & FORESTRY**	8.2	15.4	30.0	30.0	31.8	31.9	..	1 **AGRICULTURE, CHASSE & SYLVICULTURE**
2 **MINING**	44.1	33.1	42.1	50.5	56.3	54.9	..	2 **ACTIVITES EXTRACTIVES**
3 **MANUFACTURING**	411.1	496.2	586.8	726.3	873.5	1 029.2	..	3 **ACTIVITES DE FABRICATION**
4 **Food, beverages & tobacco**	7.0	8.7	11.7	21.2	24.0	23.6	..	4 **Produits alimentaires, boissons & tabac**
5 Food products & beverages	5.7	7.1	9.5	18.8	21.5	21.9	..	5 Produits alimentaires & boissons
6 Tobacco products	1.3	1.6	2.2	2.4	2.5	1.7	..	6 Produits à base de tabac
7 **Textiles, wear app., fur & leather**	16.6	22.0	16.6	19.3	23.1	24.7	..	7 **Textiles, habill., fourrures & cuirs**
8 Textiles	15.4	20.6	15.1	17.5	21.2	22.2	..	8 Textiles
9 Wearing apparel & fur	0.6	1.0	1.5	0.0	0.0	0.0	..	9 Articles d'habillement & fourrures
10 Leather products & footwear	0.6	0.4	0.0	1.8	1.9	2.5	..	10 Articles en cuir & chaussures
11 **Wood, paper, printing, publishing**	4.3	5.3	7.8	8.6	9.5	9.0	..	11 **Bois, papier, imprimerie, édition**
12 Wood & cork (not furniture)	2.9	4.0	5.7	6.2	6.2	6.5	..	12 Bois & liège (sauf meubles)
13 Pulp, paper & paper products	1.2	1.1	1.9	2.2	3.1	2.4	..	13 Carton, papier & articles
14 Publ., print. & repro. of rec. media	0.2	0.2	0.2	0.2	0.2	0.1	..	14 Edition, impress. & repro. enregis.
15 **Coke, petroleum, nuclear fuel, chemicals, rubber plastics**	93.2	111.5	145.3	169.0	196.4	212.2	..	15 **Coke, pétrole, combust. nucl., prod. chimi., caout. & mat. plast.**
16 Coke, ref. petrol. prod. & nuclear fuel	14.2	9.3	6.4	9.1	11.4	10.2	..	16 Coke, raffin. pétr. & combust. nucl.
17 Chemicals & chemical products	65.3	93.0	124.4	134.8	165.9	181.2	..	17 Produits chimiques
18 *Chemical prod.(less pharmaceuticals)*	50.6	60.6	76.1	86.5	112.2	107.9	..	18 *Prod. chimi. (moins prod. pharmaceutiques)*
19 *Pharmaceuticals*	14.7	32.4	48.3	48.3	53.7	73.3	..	19 *Produits pharmaceutiques*
20 Rubber & plastic products	13.7	9.2	14.5	25.1	19.1	20.8	..	20 Caoutchouc et matières plastiques
21 **Non-metallic mineral products**	5.4	7.4	8.6	7.5	12.0	14.4	..	21 **Produits minéraux non métalliques**
22 **Basic metals**	25.8	23.3	32.0	43.1	48.1	56.9	..	22 **Produits métallurgiques de base**
23 Basic metals, ferrous	19.8	14.8	22.1	26.7	33.5	41.0	..	23 Produits métallur. de base, ferreux
24 Basic metals, non-ferrous	6.0	8.5	9.9	16.4	14.6	15.9	..	24 Prod. métallur. de base, non ferreux
25 **Fabricated metal products**	4.2	7.2	5.6	4.9	7.8	10.3	..	25 **Ouvrages en métaux (sauf mach.& mat.)**
26 **Machinery equipment, instruments and transport equipment**	251.0	307.9	353.6	445.6	540.2	665.1	..	26 **Machines & matériel, instruments & mat. de transport**
27 Machinery n.e.c.	82.7	101.8	120.2	130.8	135.6	168.9	..	27 Machines, n.c.a.
28 Office, account. & computing machinery	1.0	0.3	0.8	1.0	2.0	1.2	..	28 Mach. de bur., compta. & trait. inform.
29 Electrical machinery	28.8	45.3	58.8	72.1	80.4	133.5	..	29 Machines électriques
30 Electro. equip. (radio, TV & commun.)	34.7	38.1	38.7	52.2	89.4	77.5	..	30 Equip. électro. (radio, tv & communic.)
31 *Electro. comp. (inc. semi-conductors)*	23.6	21.3	24.7	30.3	34.0	32.1	..	31 *Compos. électro. (semi-cond. incl.)*
32 *TV, radio & communications equip.*	11.1	16.8	14.0	21.9	55.4	45.4	..	32 *Appareils radio, tv et de communi.*
33 Instruments	6.3	8.4	10.7	12.0	24.1	23.3	..	33 Instruments
34 Motor vehicles	33.1	35.8	40.3	78.5	103.0	120.9	..	34 Véhicules automobiles
35 Other transport equipment	65.1	78.2	84.1	99.0	105.6	139.8	..	35 Autres matériels de transport
36 *Ships*	7.3	16.8	17.8	22.0	33.9	40.0	..	36 *Navires*
37 *Aerospace*	26.0	32.5	40.6	52.5	48.9	64.8	..	37 *Aérospatiale*
38 *Other transport equipment n.e.c.*	31.8	28.9	25.7	24.5	22.8	35.0	..	38 *Autres matériels de transport n.c.a.*
39 **Furniture, other manufacturing n.e.c.**	1.1	0.7	2.8	3.4	8.2	9.0	..	39 **Meubles, autres activ. de fabricat. n.c.a.**
40 Furniture	0.1	0.1	2.2	1.9	7.1	7.7	..	40 Meubles
41 Other manufacturing n.e.c.	1.0	0.6	0.6	1.5	1.1	1.3	..	41 Autres activités de fabrication n.c.a.
42 **Recycling**	2.5	2.2	2.8	3.7	4.2	4.0	..	42 **Récupération**
43 **ELECTRICITY, GAS & WATER SUPPLY**	3.8	6.9	7.9	12.3	16.8	17.0	..	43 **DISTRIBUTION D'ELECTRICITE, GAZ & EAU**
44 **CONSTRUCTION**	25.1	22.4	23.1	26.8	40.1	47.9	..	44 **CONSTRUCTION**
45 **SERVICES SECTOR**	88.6	104.9	164.4	173.5	189.9	215.2	..	45 **SECTEUR DES SERVICES**
46 **Wholesale, retail trade & repair of motor vehicles, etc.**	1.7	1.6	2.5	2.2	2.4	3.0	..	46 **Commerce (gros & détail), réparation de véhicules, etc.**
47 **Hotels & restaurants**	0.0	0.0	0.0	0.0	0.0	0.0	..	47 **Hôtels et restaurants**
48 **Transport & storage**	8.1	11.7	32.3	39.6	48.8	54.3	..	48 **Transports, entreposage**
49 **Communications**	16.8	20.2	18.9	17.9	21.3	23.9	..	49 **Communications**
50 Post	0.0	0.0	0.0	0.0	0.0	0.0	..	50 Postes
51 Telecommunications	16.8	20.2	18.9	17.9	21.3	23.9	..	51 Télécommunications
52 **Financial intermediation (inc. insur.)**	0.0	0.0	0.0	0.0	0.0	0.0	..	52 **Intermédiation financ. (assurances inc.)**
53 **Real estate, renting & busin. activity**	36.4	40.5	49.1	52.4	46.9	46.5	..	53 **Immobil., locat. & serv. aux entreprises**
54 Computer & related activities	7.2	3.1	1.4	4.6	0.1	0.1	..	54 Informatique et activités connexes
55 *Software consultancy*	5.5	0.3	0.2	1.6	0.0	0.0	..	55 *Production de logiciels*
56 *Other computer services n.e.c.*	1.7	2.8	1.2	3.0	0.1	0.1	..	56 *Autres activités informatiques*
57 Research & development	29.2	37.4	47.7	47.6	46.6	46.4	..	57 Recherche & développement expérimental
58 Other business activities n.e.c.	0.0	0.0	0.0	0.2	0.2	0.0	..	58 Autres activ. & serv. aux entreprises
59 **Comm., soc. & pers. serv. activ., etc.**	25.6	30.9	61.6	61.4	70.5	87.5	..	59 **Adm. pub., serv. soc. & collec., etc**
TOTAL BUSINESS ENTERPRISE SECTOR	**580.9** [d]	**678.9** [a]	**854.3**	**1 019.4**	**1 208.4**	**1 396.1**	**..**	**TOTAL SECTEUR DES ENTREPRISES**

Source: OECD, R&D database (DSTI/EAS Division), January 2002.

Source : OCDE, base de données R-D (DSTI/Division AES), janvier 2002.

Business enterprise intramural expenditure on R&D (BERD) - Dépenses intra-muros de R-D des entreprises (DIRDE)

by industry and type of costs - par industrie et par type de coûts

Million PLN *Millions PLN*

	1994	1995	1996	1997	1998	1999	2000	
	11.2 Capital expenditure - Dépenses en capital							
1 AGRICULTURE, HUNTING & FORESTRY	0.3	0.8	6.9	5.3	3.5	9.6	..	**1 AGRICULTURE, CHASSE & SYLVICULTURE**
2 MINING	10.3	6.3	9.4	9.9	9.4	11.7	..	**2 ACTIVITES EXTRACTIVES**
3 MANUFACTURING	99.3	114.0	217.3	244.0	403.5	422.7	..	**3 ACTIVITES DE FABRICATION**
4 Food, beverages & tobacco	3.5	1.8	2.1	10.1	2.8	9.4	..	**4 Produits alimentaires, boissons & tabac**
5 Food products & beverages	3.4	1.8	2.1	10.1	2.8	4.7	..	5 Produits alimentaires & boissons
6 Tobacco products	0.1	0.0	0.0	0.0	0.0	4.7	..	6 Produits à base de tabac
7 Textiles, wear app., fur & leather	9.9	12.3	9.8	6.0	6.8	4.7	..	**7 Textiles, habill., fourrures & cuirs**
8 Textiles	9.4	12.1	9.2	5.8	6.7	4.6	..	8 Textiles
9 Wearing apparel & fur	0.5	0.3	0.6	0.0	0.0	0.0	..	9 Articles d'habillement & fourrures
10 Leather products & footwear	0.0	0.0	0.0	0.2	0.1	0.1	..	10 Articles en cuir & chaussures
11 Wood, paper, printing, publishing	0.8	0.8	0.8	1.7	1.3	14.6	..	**11 Bois, papier, imprimerie, édition**
12 Wood & cork (not furniture)	0.3	0.5	0.3	1.1	1.2	12.7	..	12 Bois & liège (sauf meubles)
13 Pulp, paper & paper products	0.5	0.3	0.5	0.6	0.1	1.9	..	13 Carton, papier & articles
14 Publ., print. & repro. of rec. media	0.0	0.0	0.0	0.0	0.0	0.0	..	14 Edition, impress. & repro. enregis.
15 Coke, petroleum, nuclear fuel, chemicals, rubber plastics	21.8	28.3	118.1	86.9	103.4	68.1	..	**15 Coke, pétrole, combust. nucl., prod. chimi., caout. & mat. plast.**
16 Coke, ref. petrol. prod. & nuclear fuel	2.3	3.0	7.4	2.8	2.9	3.3	..	16 Coke, raffin. pétr. & combust. nucl.
17 Chemicals & chemical products	12.8	18.8	73.3	31.1	54.3	43.4	..	17 Produits chimiques
18 *Chemical prod.(less pharmaceuticals)*	11.6	11.1	23.4	19.9	43.7	25.7	..	18 *Prod. chimi. (moins prod. pharmaceutiques)*
19 *Pharmaceuticals*	1.2	7.7	49.9	11.2	10.6	17.7	..	19 *Produits pharmaceutiques*
20 Rubber & plastic products	6.7	6.5	37.4	53.0	46.2	21.4	..	20 Caoutchouc et matières plastiques
21 Non-metallic mineral products	0.7	0.9	11.8	2.0	6.2	7.0	..	**21 Produits minéraux non métalliques**
22 Basic metals	5.6	3.9	4.6	9.0	6.9	8.1	..	**22 Produits métallurgiques de base**
23 Basic metals, ferrous	4.1	2.0	2.8	6.6	4.1	2.5	..	23 Produits métallur. de base, ferreux
24 Basic metals, non-ferrous	1.5	1.9	1.8	2.4	2.8	5.6	..	24 Prod. métallur. de base, non ferreux
25 Fabricated metal products	2.7	2.3	2.3	2.6	2.6	5.5	..	**25 Ouvrages en métaux (sauf mach.& mat.)**
26 Machinery equipment, instruments and transport equipment	54.1	63.5	65.9	112.1	266.5	304.7	..	**26 Machines & matériel, instruments & mat. de transport**
27 Machinery n.e.c.	16.0	17.1	27.4	29.9	111.1	157.4	..	27 Machines, n.c.a.
28 Office, account. & computing machinery	0.0	0.0	0.0	0.1	0.2	0.1	..	28 Mach. de bur., compta. & trait. inform.
29 Electrical machinery	9.3	12.4	12.8	35.6	78.0	22.1	..	29 Machines électriques
30 Electro. equip. (radio, TV & commun.)	6.6	4.4	7.5	8.6	22.9	26.7	..	30 Equip. électro. (radio, tv & communic.)
31 *Electro. comp. (inc. semi-conductors)*	5.1	4.3	6.7	5.5	4.0	17.5	..	31 *Compos. électro. (semi-cond. incl.)*
32 *TV, radio & communications equip.*	1.5	0.1	0.8	3.1	18.9	9.2	..	32 *Appareils radio, tv et de communi.*
33 Instruments	2.7	0.8	2.1	3.9	6.9	4.0	..	33 Instruments
34 Motor vehicles	3.8	5.2	10.0	16.6	40.1	75.2	..	34 Véhicules automobiles
35 Other transport equipment	15.7	23.6	6.1	17.4	7.3	19.2	..	35 Autres matériels de transport
36 *Ships*	0.9	1.3	1.2	2.3	2.4	5.6	..	36 *Navires*
37 *Aerospace*	5.1	8.9	2.4	10.4	3.9	11.1	..	37 *Aérospatiale*
38 *Other transport equipment n.e.c.*	9.7	13.4	2.5	4.7	1.0	2.5	..	38 *Autres matériels de transport n.c.a.*
39 Furniture, other manufacturing n.e.c.	0.0	0.0	1.3	12.5	6.6	0.5	..	**39 Meubles, autres activ. de fabricat. n.c.a.**
40 Furniture	0.0	0.0	0.3	12.3	6.6	0.5	..	40 Meubles
41 Other manufacturing n.e.c.	0.0	0.0	1.0	0.2	0.0	0.0	..	41 Autres activités de fabrication n.c.a.
42 Recycling	0.2	0.2	0.6	1.1	0.4	0.1	..	**42 Récupération**
43 ELECTRICITY, GAS & WATER SUPPLY	0.6	1.4	1.1	2.8	3.1	2.7	..	**43 DISTRIBUTION D'ELECTRICITE, GAZ & EAU**
44 CONSTRUCTION	4.9	2.3	2.4	2.6	3.2	3.8	..	**44 CONSTRUCTION**
45 SERVICES SECTOR	13.7	22.4	38.7	41.1	30.0	50.7	..	**45 SECTEUR DES SERVICES**
46 Wholesale, retail trade & repair of motor vehicles, etc.	0.3	0.4	0.4	0.7	0.6	1.6	..	**46 Commerce (gros & détail), réparation de véhicules, etc.**
47 Hotels & restaurants	0.0	0.0	0.0	0.0	0.0	0.0	..	**47 Hôtels et restaurants**
48 Transport & storage	1.4	2.3	14.3	13.3	10.3	7.0	..	**48 Transports, entreposage**
49 Communications	2.6	4.3	6.4	8.7	3.6	7.0	..	**49 Communications**
50 Post	0.0	0.0	0.0	0.0	0.0	0.0	..	50 Postes
51 Telecommunications	2.6	4.3	6.4	8.7	3.6	7.0	..	51 Télécommunications
52 Financial intermediation (inc. insur.)	0.0	0.0	0.0	0.0	0.0	0.0	..	**52 Intermédiation financ. (assurances inc.)**
53 Real estate, renting & busin. activity	2.9	4.8	9.0	5.6	5.7	12.6	..	**53 Immobil., locat. & serv. aux entreprises**
54 Computer & related activities	0.5	0.1	0.1	0.1	0.0	0.0	..	54 Informatique et activités connexes
55 *Software consultancy*	0.4	0.0	0.0	0.1	0.0	0.0	..	55 *Production de logiciels*
56 *Other computer services n.e.c.*	0.1	0.1	0.1	0.0	0.0	0.0	..	56 *Autres activités informatiques*
57 Research & development	2.4	4.7	8.9	5.5	5.7	12.6	..	57 Recherche & développement expérimental
58 Other business activities n.e.c.	0.0	0.0	0.0	0.0	0.0	0.0	..	58 Autres activ. & serv. aux entreprises
59 Comm., soc. & pers. serv. activ., etc.	6.5	10.6	8.6	12.8	9.8	22.5	..	**59 Adm. pub., serv. soc. & collec., etc**
TOTAL BUSINESS ENTERPRISE SECTOR	**129.1** [d]	**147.2** [a]	**275.8**	**305.7**	**452.7**	**501.2**	**..**	**TOTAL SECTEUR DES ENTREPRISES**

Source: OECD, R&D database (DSTI/EAS Division), January 2002.

Source : OCDE, base de données R-D (DSTI/Division AES), janvier 2002.

Total business enterprise R&D personnel (BEMP) - Personnel total de R-D du secteur des entreprises (PRDE)

by industry - par industrie

In full time equivalent — *En équivalence plein-temps*

	1994	1995	1996	1997	1998	1999	2000	
1 AGRICULTURE, HUNTING & FORESTRY	557	517	543	524	438	..	..	**1 AGRICULTURE, CHASSE & SYLVICULTURE**
2 MINING	1 805	1 352	1 501	1 463	1 264	..	..	**2 ACTIVITES EXTRACTIVES**
3 MANUFACTURING	18 811	20 495	17 090	17 045	16 653	..	..	**3 ACTIVITES DE FABRICATION**
4 Food, beverages & tobacco	351	364	335	423	435	..	..	**4 Produits alimentaires, boissons & tabac**
5 Food products & beverages	278	300	259	350	357	..	..	5 Produits alimentaires & boissons
6 Tobacco products	73	64	76	73	78	..	..	6 Produits à base de tabac
7 Textiles, wear app., fur & leather	753	680	666	612	629	..	..	**7 Textiles, habill., fourrures & cuirs**
8 Textiles	670	596	589	531	554	..	..	8 Textiles
9 Wearing apparel & fur	9	9	77	0	0	..	..	9 Articles d'habillement & fourrures
10 Leather products & footwear	74	75	0	81	75	..	..	10 Articles en cuir & chaussures
11 Wood, paper, printing, publishing	201	210	212	193	189	..	..	**11 Bois, papier, imprimerie, édition**
12 Wood & cork (not furniture)	143	142	142	132	128	..	..	12 Bois & liège (sauf meubles)
13 Pulp, paper & paper products	55	63	66	61	58	..	..	13 Carton, papier & articles
14 Publ., print. & repro. of rec. media	3	5	4	0	3	..	..	14 Edition, impress. & repro. enregis.
15 Coke, petroleum, nuclear fuel, chemicals, rubber plastics	3 808	6 197	4 213	4 192	4 142	..	..	**15 Coke, pétrole, combust. nucl., prod. chimi., caout. & mat. plast.**
16 Coke, ref. petrol. prod. & nuclear fuel	453	299	239	249	222	..	..	16 Coke, raffin. pétr. & combust. nucl.
17 Chemicals & chemical products	2 718	5 474	3 399	3 429	3 495	..	..	17 Produits chimiques
18 *Chemical prod.(less pharmaceuticals)*	2 329	4 720	2 465	2 445	2 675	..	..	18 *Prod. chimi. (moins prod. pharmaceutiques)*
19 *Pharmaceuticals*	389	754	934	984	820	..	..	19 *Produits pharmaceutiques*
20 Rubber & plastic products	637	424	575	514	425	..	..	20 Caoutchouc et matières plastiques
21 Non-metallic mineral products	389	431	324	301	288	..	..	**21 Produits minéraux non métalliques**
22 Basic metals	1 283	980	984	884	965	..	..	**22 Produits métallurgiques de base**
23 Basic metals, ferrous	883	540	558	562	603	..	..	23 Produits métallur. de base, ferreux
24 Basic metals, non-ferrous	400	440	426	322	362	..	..	24 Prod. métallur. de base, non ferreux
25 Fabricated metal products	220	440	183	150	159	..	..	**25 Ouvrages en métaux (sauf mach.& mat.)**
26 Machinery equipment, instruments and transport equipment	11 679	11 056	10 060	10 122	9 662	..	..	**26 Machines & matériel, instruments & mat. de transport**
27 Machinery n.e.c.	4 591	3 916	3 308	3 168	2 815	..	..	27 Machines, n.c.a.
28 Office, account. & computing machinery	24	13	29	13	16	..	..	28 Mach. de bur., compta. & trait. inform.
29 Electrical machinery	1 347	1 789	1 778	1 636	1 520	..	..	29 Machines électriques
30 Electro. equip. (radio, TV & commun.)	1 418	1 063	882	913	1 113	..	..	30 Equip. électro. (radio, tv & communic.)
31 *Electro. comp. (inc. semi-conductors)*	943	588	591	625	573	..	..	31 *Compos. électro. (semi-cond. incl.)*
32 *TV, radio & communications equip.*	475	475	291	288	540	..	..	32 *Appareils radio, tv et de communi.*
33 Instruments	385	445	525	445	449	..	..	33 Instruments
34 Motor vehicles	1 572	1 536	1 384	1 984	2 058	..	..	34 Véhicules automobiles
35 Other transport equipment	2 342	2 294	2 154	1 964	1 691	..	..	35 Autres matériels de transport
36 *Ships*	194	304	294	296	247	..	..	36 *Navires*
37 *Aerospace*	924	1 032	1 100	1 137	1 055	..	..	37 *Aérospatiale*
38 *Other transport equipment n.e.c.*	1 224	958	760	531	389	..	..	38 *Autres matériels de transport n.c.a.*
39 Furniture, other manufacturing n.e.c.	56	75	50	99	123	..	..	**39 Meubles, autres activ. de fabricat. n.c.a.**
40 Furniture	12	4	41	63	95	..	..	40 Meubles
41 Other manufacturing n.e.c.	44	71	9	36	28	..	..	41 Autres activités de fabrication n.c.a.
42 Recycling	71	62	63	69	61	..	..	**42 Récupération**
43 ELECTRICITY, GAS & WATER SUPPLY	226	253	253	350	384	..	..	**43 DISTRIBUTION D'ELECTRICITE, GAZ & EAU**
44 CONSTRUCTION	1 054	794	682	622	540	..	..	**44 CONSTRUCTION**
45 SERVICES SECTOR	3 078	2 871	3 467	3 177	2 635	..	..	**45 SECTEUR DES SERVICES**
46 Wholesale, retail trade & repair of motor vehicles, etc.	42	44	45	33	32	..	..	**46 Commerce (gros & détail), réparation de véhicules, etc.**
47 Hotels & restaurants	0	0	0	0	0	..	..	**47 Hôtels et restaurants**
48 Transport & storage	305	278	679	632	618	..	..	**48 Transports, entreposage**
49 Communications	437	452	437	484	493	..	..	**49 Communications**
50 Post	0	0	0	0	0	..	..	50 Postes
51 Telecommunications	437	452	437	484	493	..	..	51 Télécommunications
52 Financial intermediation (inc. insur.)	0	0	0	0	0	..	..	**52 Intermédiation financ. (assurances inc.)**
53 Real estate, renting & busin. activity	1 341	1 152	1 155	1 012	808	..	..	**53 Immobil., locat. & serv. aux entreprises**
54 Computer & related activities	324	33	49	38	9	..	..	54 Informatique et activités connexes
55 *Software consultancy*	294	16	13	15	9	..	..	55 *Production de logiciels*
56 *Other computer services n.e.c.*	30	17	36	23	0	..	..	56 *Autres activités informatiques*
57 Research & development	1 013	1 120	1 106	954	786	..	..	57 Recherche & développement expérimental
58 Other business activities n.e.c.	4	0	0	20	13	..	..	58 Autres activ. & serv. aux entreprises
59 Comm., soc. & pers. serv. activ., etc.	953	945	1 151	1 016	684	..	..	**59 Adm. pub., serv. soc. & collec., etc**
TOTAL BUSINESS ENTERPRISE SECTOR	**25 531**	**26 282**	**23 536**	**23 181**	**21 914**	**20 314**	**18 586**	**TOTAL SECTEUR DES ENTREPRISES**

Source: OECD, R&D database (DSTI/EAS Division), January 2002.

Source : OCDE, base de données R-D (DSTI/Division AES), janvier 2002.

POLAND Table 13 - Tableau 13 **POLOGNE**

RSE in the business enterprise sector - Chercheurs dans le secteur des entreprises

by industry - par industrie

In full time equivalent — *En équivalence plein-temps*

	1994	1995	1996	1997	1998	1999	2000	
1 AGRICULTURE, HUNTING & FORESTRY	177	178	372	357	304	..	..	**1 AGRICULTURE, CHASSE & SYLVICULTURE**
2 MINING	946	748	898	895	777	..	..	**2 ACTIVITES EXTRACTIVES**
3 MANUFACTURING	7 345	8 405	6 890	7 621	7 210	..	..	**3 ACTIVITES DE FABRICATION**
4 Food, beverages & tobacco	178	181	138	235	202	..	..	**4 Produits alimentaires, boissons & tabac**
5 Food products & beverages	147	155	122	207	182	..	..	5 Produits alimentaires & boissons
6 Tobacco products	31	26	16	28	20	..	..	6 Produits à base de tabac
7 Textiles, wear app., fur & leather	339	336	340	313	301	..	..	**7 Textiles, habill., fourrures & cuirs**
8 Textiles	292	288	289	277	264	..	..	8 Textiles
9 Wearing apparel & fur	1	1	51	0	0	..	..	9 Articles d'habillement & fourrures
10 Leather products & footwear	46	47	0	36	37	..	..	10 Articles en cuir & chaussures
11 Wood, paper, printing, publishing	97	108	106	101	101	..	..	**11 Bois, papier, imprimerie, édition**
12 Wood & cork (not furniture)	70	75	73	70	67	..	..	12 Bois & liège (sauf meubles)
13 Pulp, paper & paper products	24	29	29	31	31	..	..	13 Carton, papier & articles
14 Publ., print. & repro. of rec. media	3	4	4	0	3	..	..	14 Edition, impress. & repro. enregis.
15 Coke, petroleum, nuclear fuel, chemicals, rubber plastics	1 624	2 799	1 798	1 923	1 748	..	..	**15 Coke, pétrole, combust. nucl., prod. chimi., caout. & mat. plast.**
16 Coke, ref. petrol. prod. & nuclear fuel	259	122	106	106	97	..	..	16 Coke, raffin. pétr. & combust. nucl.
17 Chemicals & chemical products	1 110	2 524	1 492	1 610	1 471	..	..	17 Produits chimiques
18 *Chemical prod.(less pharmaceuticals)*	941	2 197	1 076	1 148	1 100	..	..	18 *Prod. chimi. (moins prod. pharmaceutiques)*
19 *Pharmaceuticals*	169	327	416	462	371	..	..	19 *Produits pharmaceutiques*
20 Rubber & plastic products	255	153	200	207	180	..	..	20 Caoutchouc et matières plastiques
21 Non-metallic mineral products	144	181	147	134	132	..	..	**21 Produits minéraux non métalliques**
22 Basic metals	527	367	364	340	362	..	..	**22 Produits métallurgiques de base**
23 Basic metals, ferrous	365	195	195	196	197	..	..	23 Produits métallur. de base, ferreux
24 Basic metals, non-ferrous	162	172	169	144	165	..	..	24 Prod. métallur. de base, non ferreux
25 Fabricated metal products	61	80	81	73	75	..	..	**25 Ouvrages en métaux (sauf mach.& mat.)**
26 Machinery equipment, instruments and transport equipment	4 318	4 323	3 848	4 400	4 209	..	..	**26 Machines & matériel, instruments & mat. de transport**
27 Machinery n.e.c.	1 480	1 507	1 124	1 265	1 039	..	..	27 Machines, n.c.a.
28 Office, account. & computing machinery	11	9	8	9	11	..	..	28 Mach. de bur., compta. & trait. inform.
29 Electrical machinery	546	491	568	632	582	..	..	29 Machines électriques
30 Electro. equip. (radio, TV & commun.)	687	579	515	582	720	..	..	30 Equip. électro. (radio, tv & communic.)
31 *Electro. comp. (inc. semi-conductors)*	447	367	353	354	324	..	..	31 *Compos. électro. (semi-cond. incl.)*
32 *TV, radio & communications equip.*	240	212	162	228	396	..	..	32 *Appareils radio, tv et de communi.*
33 Instruments	179	240	217	249	268	..	..	33 Instruments
34 Motor vehicles	520	512	514	809	862	..	..	34 Véhicules automobiles
35 Other transport equipment	895	985	902	854	727	..	..	35 Autres matériels de transport
36 *Ships*	96	151	131	139	119	..	..	36 *Navires*
37 *Aerospace*	370	378	398	465	389	..	..	37 *Aérospatiale*
38 *Other transport equipment n.e.c.*	429	456	373	250	219	..	..	38 *Autres matériels de transport n.c.a.*
39 Furniture, other manufacturing n.e.c.	11	2	21	63	37	..	..	**39 Meubles, autres activ. de fabricat. n.c.a.**
40 Furniture	2	2	12	36	18	..	..	40 Meubles
41 Other manufacturing n.e.c.	9	0	9	27	19	..	..	41 Autres activités de fabrication n.c.a.
42 Recycling	46	28	47	39	43	..	..	**42 Récupération**
43 ELECTRICITY, GAS & WATER SUPPLY	86	165	145	197	197	..	..	**43 DISTRIBUTION D'ELECTRICITE, GAZ & EAU**
44 CONSTRUCTION	511	384	321	358	302	..	..	**44 CONSTRUCTION**
45 SERVICES SECTOR	1 564	1 275	1 739	1 611	1 383	..	..	**45 SECTEUR DES SERVICES**
46 Wholesale, retail trade & repair of motor vehicles, etc.	15	22	22	11	6	..	..	**46 Commerce (gros & détail), réparation de véhicules, etc.**
47 Hotels & restaurants	0	0	0	0	0	..	..	**47 Hôtels et restaurants**
48 Transport & storage	148	87	378	358	361	..	..	**48 Transports, entreposage**
49 Communications	190	192	218	253	261	..	..	**49 Communications**
50 Post	0	0	..	0	0	..	..	50 Postes
51 Telecommunications	190	192	218	253	261	..	..	51 Télécommunications
52 Financial intermediation (inc. insur.)	0	0	0	0	0	..	..	**52 Intermédiation financ. (assurances inc.)**
53 Real estate, renting & busin. activity	721	525	528	523	443	..	..	**53 Immobil., locat. & serv. aux entreprises**
54 Computer & related activities	185	21	39	14	1	..	..	54 Informatique et activités connexes
55 *Software consultancy*	167	4	3	3	1	..	..	55 *Production de logiciels*
56 *Other computer services n.e.c.*	18	17	36	11	0	..	..	56 *Autres activités informatiques*
57 Research & development	536	504	489	502	439	..	..	57 Recherche & développement expérimental
58 Other business activities n.e.c.	0	0	0	7	3	..	..	58 Autres activ. & serv. aux entreprises
59 Comm., soc. & pers. serv. activ., etc.	490	449	593	466	312	..	..	**59 Adm. pub., serv. soc. & collec., etc**
TOTAL BUSINESS ENTERPRISE SECTOR	**10 629**	**11 155**	**10 365**	**11 039**	**10 173**	**10 327**	**9 821**	**TOTAL SECTEUR DES ENTREPRISES**

Source: OECD, R&D database (DSTI/EAS Division), January 2002.

Source : OCDE, base de données R-D (DSTI/Division AES), janvier 2002.

Higher education intramural expenditure on R&D - Dépenses intra-muros de R-D de l'enseignement supérieur
(HERD) by field of science - (DIRDES) par discipline scientifique

	1994	1995	1996	1997	1998	1999	2000	
	14.1 Million PLN - Millions PLN							
Natural sciences	..	150.2	169.9	258.8	255.3	297.3	..	Sciences exactes et naturelles
Engineering	..	202.3	318.0	307.2	403.6	485.3	..	Sciences de l'ingénieur
Medical sciences	..	72.9	89.2	118.6	147.1	173.4	..	Sciences médicales
Agricultural sciences	..	61.6	77.9	77.6	88.2	114.1	..	Sciences agricoles
Sub-total NSE	..	487.0	655.0	762.2	894.2	1 070.1	..	Sous-total SENI
Social sciences	..	41.9	65.4	134.8	152.2	129.2	..	Sciences sociales
Humanities	..	32.1	48.1	64.9	60.3	75.0	..	Sciences humaines
Sub-total SSH	..	74.0	113.5	199.7	212.5	204.2	..	Sous-total SSH
Total NSE+SSH	407.0	561.0	768.5	961.9	1 106.7	1 274.3	..	*Total SENI+SSH*
	14.2 Million constant USD (1995 prices and PPPs) - Millions de USD constants (prix et PPA de 1995)							
Natural sciences	..	132.1	125.9	168.2	148.4	161.8	..	Sciences exactes et naturelles
Engineering	..	177.9	235.6	199.6	234.6	264.1	..	Sciences de l'ingénieur
Medical sciences	..	64.1	66.1	77.1	85.5	94.4	..	Sciences médicales
Agricultural sciences	..	54.2	57.7	50.4	51.3	62.1	..	Sciences agricoles
Sub-total NSE	..	428.3	485.2	495.2	519.8	582.4	..	Sous-total SENI
Social sciences	..	36.9	48.4	87.6	88.5	70.3	..	Sciences sociales
Humanities	..	28.2	35.6	42.2	35.1	40.8	..	Sciences humaines
Sub-total SSH	..	65.1	84.1	129.8	123.5	111.1	..	Sous-total SSH
Total NSE+SSH	457.8	493.3	569.3	625.0	643.3	693.6	..	*Total SENI+SSH*

Table 15 - Tableau 15

Higher education intramural expenditure on R&D - Dépenses intra-muros de R-D de l'enseignement supérieur
(HERD) by field of science and type of costs - (DIRDES) par discipline scientifique et par type de coûts

Million PLN *Millions PLN*

	1994	1995	1996	1997	1998	1999	2000	
Current expenditure								**Dépenses courantes**
Natural sciences	..	114.0	125.9	187.6	175.3	203.7	..	Sciences exactes et naturelles
Engineering	..	155.1	240.2	221.7	305.4	373.6	..	Sciences de l'ingénieur
Medical sciences	..	48.8 s	53.3 s	86.3	102.7	122.1	..	Sciences médicales
Agricultural sciences	..	50.4	60.4	60.5	69.7	79.1	..	Sciences agricoles
Sub-total NSE	..	368.3 s	479.8 s	556.1	653.1	778.5	..	Sous-total SENI
Social sciences	..	37.2 s	55.1 s	99.2	76.3	100.2	..	Sciences sociales
Humanities	..	27.1 s	35.8 s	44.9	44.5	62.7	..	Sciences humaines
Sub-total SSH	..	64.3 s	90.9 s	144.1	120.8	162.9	..	Sous-total SSH
Total NSE+SSH	278.9	432.6	570.7	700.2	773.9	941.4	..	*Total SENI+SSH*
Capital expenditure								**Dépenses en capital**
Natural sciences	..	36.2	44.0	71.2	80.0	93.6	..	Sciences exactes et naturelles
Engineering	..	47.2	77.7	85.5	98.2	111.7	..	Sciences de l'ingénieur
Medical sciences	..	21.3 s	32.5 s	32.3	44.4	51.3	..	Sciences médicales
Agricultural sciences	..	11.3	17.6	17.1	18.5	35.0	..	Sciences agricoles
Sub-total NSE	..	116.0 s	171.8 s	206.1	241.1	291.6	..	Sous-total SENI
Social sciences	..	8.4 s	16.8 s	35.6	75.9	29.0	..	Sciences sociales
Humanities	..	4.0 s	9.2 s	20.0	15.8	12.3	..	Sciences humaines
Sub-total SSH	..	12.4 s	26.0 s	55.6	91.7	41.3	..	Sous-total SSH
Total NSE+SSH	128.1	128.4	197.8	261.7	332.8	332.9	..	*Total SENI+SSH*

Source: OECD, R&D database (DSTI/EAS Division), January 2002.

Source : OCDE, base de données R-D (DSTI/Division AES), janvier 2002.

POLAND — Table 16 - Tableau 16 — POLOGNE

Higher education intramural expenditure on R&D - Dépenses intra-muros de R-D de l'enseignement supérieur

(HERD) by field of science and source of funds - (DIRDES) par discipline scientifique et par source de financement

Million PLN — *Millions PLN*

	1994	1995	1996	1997	1998	1999	2000	
Direct government								**État direct**
Natural sciences	..	126.7	157.8	224.9	228.3	262.2	..	Sciences exactes et naturelles
Engineering	..	150.4	229.5	224.4	309.5	387.0	..	Sciences de l'ingénieur
Medical sciences	..	54.0 [s]	70.8 [s]	96.8	118.7	144.9	..	Sciences médicales
Agricultural sciences	..	53.9	62.2	67.6	79.6	95.7	..	Sciences agricoles
Sub-total NSE	..	385.0 [s]	520.3 [s]	613.7	736.1	889.8	..	Sous-total SENI
Social sciences	..	36.9 [s]	61.9 [s]	123.6	133.7	111.3	..	Sciences sociales
Humanities	..	24.8 [s]	39.8 [s]	62.3	55.7	64.9	..	Sciences humaines
Sub-total SSH	..	61.7 [s]	101.7 [s]	185.9	189.4	176.2	..	Sous-total SSH
Total NSE+SSH	330.4	446.7	622.0	799.6	925.5	1 066.0	..	*Total SENI+SSH*
General university funds								**Fonds généraux des universités**
Natural sciences	..	..	..	..	..	..	..	Sciences exactes et naturelles
Engineering	..	..	..	..	..	..	..	Sciences de l'ingénieur
Medical sciences	..	..	..	..	..	..	..	Sciences médicales
Agricultural sciences	..	..	..	..	..	..	..	Sciences agricoles
Sub-total NSE	..	..	..	..	..	..	..	Sous-total SENI
Social sciences	..	..	..	..	..	..	..	Sciences sociales
Humanities	..	..	..	..	..	..	..	Sciences humaines
Sub-total SSH	..	..	..	..	..	..	..	Sous-total SSH
Total NSE+SSH	0.0	0.0	0.0	0.0	0.0	0.0	..	*Total SENI+SSH*
Sub-total government								**Sous-total État**
Natural sciences	..	126.7	157.8	224.9	228.3	262.2	..	Sciences exactes et naturelles
Engineering	..	150.4	229.5	224.4	309.5	387.0	..	Sciences de l'ingénieur
Medical sciences	..	54.0 [s]	70.8 [s]	96.8	118.7	144.9	..	Sciences médicales
Agricultural sciences	..	53.9	62.2	67.6	79.6	95.7	..	Sciences agricoles
Sub-total NSE	..	385.0 [s]	520.3 [s]	613.7	736.1	889.8	..	Sous-total SENI
Social sciences	..	36.9 [s]	61.9 [s]	123.6	133.7	111.3	..	Sciences sociales
Humanities	..	24.8 [s]	39.8 [s]	62.3	55.7	64.9	..	Sciences humaines
Sub-total SSH	..	61.7 [s]	101.7 [s]	185.9	189.4	176.2	..	Sous-total SSH
Total NSE+SSH	330.4	446.7	622.0	799.6	925.5	1 066.0	..	*Total SENI+SSH*
Higher education								**Enseignement supérieur**
Natural sciences	..	4.4	2.2	4.6	7.6	16.4	..	Sciences exactes et naturelles
Engineering	..	11.1	18.7	23.6	16.6	11.3	..	Sciences de l'ingénieur
Medical sciences	..	11.1 [s]	6.0 [s]	6.9	10.9	11.2	..	Sciences médicales
Agricultural sciences	..	1.7	9.2	1.7	1.8	4.0	..	Sciences agricoles
Sub-total NSE	..	28.3 [s]	36.1 [s]	36.8	36.9	42.9	..	Sous-total SENI
Social sciences	..	2.2 [s]	2.4 [s]	3.4	11.7	11.2	..	Sciences sociales
Humanities	..	4.7 [s]	2.9 [s]	1.2	2.4	4.1	..	Sciences humaines
Sub-total SSH	..	6.9 [s]	5.3 [s]	4.6	14.1	15.3	..	Sous-total SSH
Total NSE+SSH	23.7	35.2	41.4	41.4	51.0	58.2	..	*Total SENI+SSH*
Private non-profit								**Institutions sans but lucratif**
Natural sciences	..	2.3	1.8	2.8	2.0	2.6	..	Sciences exactes et naturelles
Engineering	..	0.3	1.2	0.3	1.0	0.5	..	Sciences de l'ingénieur
Medical sciences	..	0.0 [s]	0.0 [s]	0.0	0.5	0.1	..	Sciences médicales
Agricultural sciences	..	0.2	0.0	0.4	0.3	0.4	..	Sciences agricoles
Sub-total NSE	..	2.8 [s]	3.0 [s]	3.5	3.8	3.6	..	Sous-total SENI
Social sciences	..	0.7 [s]	0.9 [s]	1.1	0.5	0.6	..	Sciences sociales
Humanities	..	0.2 [s]	0.0 [s]	0.4	0.4	0.3	..	Sciences humaines
Sub-total SSH	..	0.9 [s]	0.9 [s]	1.5	0.9	0.9	..	Sous-total SSH
Total NSE+SSH	3.0	3.7	3.9	5.0	4.7	4.5	..	*Total SENI+SSH*
Business entreprise								**Entreprises**
Natural sciences	..	14.1	5.6	21.6	13.5	12.8	..	Sciences exactes et naturelles
Engineering	..	36.9	65.7	53.9	68.9	75.7	..	Sciences de l'ingénieur
Medical sciences	..	4.9 [s]	6.1 [s]	14.1	16.3	15.8	..	Sciences médicales
Agricultural sciences	..	4.8	4.8	6.7	5.5	13.3	..	Sciences agricoles
Sub-total NSE	..	60.7 [s]	82.2 [s]	96.3	104.2	117.6	..	Sous-total SENI
Social sciences	..	2.4 [s]	2.8 [s]	3.4	2.5	2.2	..	Sciences sociales
Humanities	..	1.1 [s]	1.8 [s]	0.8	0.9	5.1	..	Sciences humaines
Sub-total SSH	..	3.5 [s]	4.6 [s]	4.2	3.4	7.3	..	Sous-total SSH
Total NSE+SSH	46.2	64.2	86.8	100.5	107.6	124.9	..	*Total SENI+SSH*
Funds from abroad								**Fonds de l'étranger**
Natural sciences	..	2.7	2.5	4.9	3.9	3.3	..	Sciences exactes et naturelles
Engineering	..	3.6	2.9	5.0	7.6	10.8	..	Sciences de l'ingénieur
Medical sciences	..	0.2 [s]	2.9 [s]	0.8	0.7	1.4	..	Sciences médicales
Agricultural sciences	..	1.0	1.7	1.2	1.0	0.7	..	Sciences agricoles
Sub-total NSE	..	7.5 [s]	10.0 [s]	11.9	13.2	16.2	..	Sous-total SENI
Social sciences	..	3.4 [s]	3.9 [s]	3.3	3.8	3.9	..	Sciences sociales
Humanities	..	0.3 [s]	0.5 [s]	0.2	0.9	0.6	..	Sciences humaines
Sub-total SSH	..	3.7 [s]	4.4 [s]	3.5	4.7	4.5	..	Sous-total SSH
Total NSE+SSH	3.7	11.2	14.4	15.4	17.9	20.7	..	*Total SENI+SSH*

Source: OECD, R&D database (DSTI/EAS Division), January 2002.

Source : OCDE, base de données R-D (DSTI/Division AES), janvier 2002.

Higher education R&D personnel (HEMP) - Personnel de R-D du secteur de l'enseignement supérieur

by field of science and occupation - par discipline scientifique et par profession

In full-time equivalent — *En équivalence plein-temps*

	1994	1995	1996	1997	1998	1999	2000	
Total R&D personnel								**Personnel total de R-D**
Natural sciences	..	8 163	8 104	10 027	9 077	9 244	..	Sciences exactes et naturelles
Engineering	..	10 285	11 749	9 166	10 902	10 699	..	Sciences de l'ingénieur
Medical sciences	..	6 240	5 725	7 308	8 562	6 557	..	Sciences médicales
Agricultural sciences	..	3 407	3 790	3 143	4 534	4 253	..	Sciences agricoles
Sub-total NSE	..	28 095	29 368	29 644	33 075	30 753	..	Sous-total SENI
Social sciences	..	4 299	5 257	6 112	6 708	6 813	..	Sciences sociales
Humanities	..	3 227	4 421	5 221	5 482	5 382	..	Sciences humaines
Sub-total SSH	..	7 526	9 678	11 333	12 190	12 195	..	Sous-total SSH
Total NSE+SSH	32 745	35 621	39 046	40 977	42 478	42 948	41 499	*Total SENI+SSH*
Researchers								**Chercheurs**
Natural sciences	..	6 036	6 017	7 511	7 037	7 226	..	Sciences exactes et naturelles
Engineering	..	7 950	9 140	7 143	8 546	8 481	..	Sciences de l'ingénieur
Medical sciences	..	4 334 [s]	3 991 [s]	5 785	6 473	5 353	..	Sciences médicales
Agricultural sciences	..	2 530	2 785	2 339	3 486	3 273	..	Sciences agricoles
Sub-total NSE	..	20 850 [s]	21 933 [s]	22 778	25 542	24 333	..	Sous-total SENI
Social sciences	..	3 886 [s]	4 982 [s]	5 344	5 923	5 998	..	Sciences sociales
Humanities	..	3 290 [s]	4 218 [s]	4 724	5 007	4 953	..	Sciences humaines
Sub-total SSH	..	7 176 [s]	9 200 [s]	10 068	10 930	10 951	..	Sous-total SSH
Total NSE+SSH	26 053	28 026	31 133	32 846	34 425	35 284	34 246	*Total SENI+SSH*
Technicians								**Technicians**
Natural sciences	..	1 477	1 442	1 862	1 559	1 526	..	Sciences exactes et naturelles
Engineering	..	1 760	2 006	1 209	1 463	1 333	..	Sciences de l'ingénieur
Medical sciences	..	767 [s]	733 [s]	1 046	1 128	801	..	Sciences médicales
Agricultural sciences	..	808	883	688	902	846	..	Sciences agricoles
Sub-total NSE	..	4 812 [s]	5 064 [s]	4 805	5 052	4 506	..	Sous-total SENI
Social sciences	..	302 [s]	307 [s]	392	327	239	..	Sciences sociales
Humanities	..	173 [s]	346 [s]	247	212	170	..	Sciences humaines
Sub-total SSH	..	475 [s]	653 [s]	639	539	409	..	Sous-total SSH
Total NSE+SSH	4 974	5 287	5 717	5 444	5 162	4 915	4 205	*Total SENI+SSH*
Other								**Autres**
Natural sciences	..	649	645	654	481	492	..	Sciences exactes et naturelles
Engineering	..	575	603	814	893	885	..	Sciences de l'ingénieur
Medical sciences	..	468 [s]	275 [s]	477	961	403	..	Sciences médicales
Agricultural sciences	..	69	122	116	146	134	..	Sciences agricoles
Sub-total NSE	..	1 761 [s]	1 645 [s]	2 061	2 481	1 914	..	Sous-total SENI
Social sciences	..	382 [s]	348 [s]	376	458	576	..	Sciences sociales
Humanities	..	165 [s]	203 [s]	250	263	259	..	Sciences humaines
Sub-total SSH	..	547 [s]	551 [s]	626	721	835	..	Sous-total SSH
Total NSE+SSH	1 718	2 308	2 196	2 687	2 891	2 749	3 048	*Total SENI+SSH*

Table 18 - Tableau 18

Government budget appropriations or outlays for R&D (GBAORD) - Crédits budgétaires publics de R-D (CBPRD)

by socio-economic objective - par objectif socio-économique

Million PLN — *Millions PLN*

	1995	1996	1997	1998	1999	2000	2001	
Agriculture, forestry and fishing	..	..	..	..	..	..	..	Agriculture, sylviculture et pêche
Industrial development	..	..	..	..	..	..	..	Développement industriel
Energy	..	..	..	..	..	..	..	Energie
Transport & telecommunications	..	..	..	..	..	..	..	*Transports & télécommunications*
Urban & rural planning	..	..	..	..	..	..	..	*Aménagement urbain et rural*
ST Infrastructure	..	..	..	..	..	..	..	ST Infrastructure
Prevention of pollution	..	..	..	..	..	..	..	*Prévention de la pollution*
Identificat. & treatment of pollution	..	..	..	..	..	..	..	*Détect. pollut. & mesures antipollut.*
ST Environment	..	..	..	..	..	..	..	ST Environnement
Health	..	..	..	..	..	..	..	Santé
Social development & services	..	..	..	..	..	..	..	Développement et services sociaux
Earth and atmosphere	..	..	..	..	..	..	..	Milieu terrestre et atmosphère
Advancement of research	..	..	..	..	..	..	..	*Promotion de la recherche*
General university funds	..	..	..	..	..	..	..	*Fonds généraux des universités*
ST Advancement of knowledge	..	..	..	..	..	..	..	ST promotion gén. des connaiss.
Civil space	..	..	..	..	..	..	..	Espace civil
Defence	..	..	..	..	..	..	..	Défense
Not elsewhere classified	..	..	..	..	..	..	..	Non classé
Total	**1 343.9**	**1 741.0**	**2 065.0**	**2 340.4**	**2 633.5**	**..**	**..**	**Total**

Source: OECD, R&D database (DSTI/EAS Division), January 2002. — *Source* : OCDE, base de données R-D (DSTI/Division AES), janvier 2002.

POLAND Table 19 - Tableau 19 **POLOGNE**

Technology balance of payments - Balance des paiements technologiques

by type of transfer - selon la nature des transferts

Million PLN *Millions PLN*

	1983	1984	1985	1986	1987	1988	1989	1990	1991	
Receipts										**Recettes**
Related firms	..	..	..	..	..	..	..	..	..	Firmes affiliées
Unrelated firms	..	..	..	..	..	..	..	..	..	Firmes non affiliées
Other	..	..	..	..	..	..	..	..	..	Autre
Total	..	..	..	..	..	..	..	..	..	*Total*
Payments										**Paiements**
Related firms	..	..	..	..	..	..	..	..	..	Firmes affiliées
Unrelated firms	..	..	..	..	..	..	..	..	..	Firmes non affiliées
Other	..	..	..	..	..	..	..	..	..	Autre
Total	..	..	..	..	..	..	..	..	..	*Total*

	1992	1993	1994	1995	1996	1997	1998	1999	2000	
Receipts										**Recettes**
Related firms	..	..	..	..	..	..	..	..	..	Firmes affiliées
Unrelated firms	..	..	..	..	..	..	..	..	..	Firmes non affiliées
Other	..	..	..	..	..	..	..	..	..	Autre
Total	..	..	301	560	543	641	497	512	..	*Total*
Payments										**Paiements**
Related firms	..	..	..	..	..	..	..	..	..	Firmes affiliées
Unrelated firms	..	..	..	..	..	..	..	..	..	Firmes non affiliées
Other	..	..	..	..	..	..	..	..	..	Autre
Total	..	..	306	568	955	1 349	1 421	2 651	..	*Total*

Source: OECD, TBPdatabase (DSTI/EAS Division), Januay 2002.

Source : OCDE, base de données TBP (DSTI/Division AES), janvier 2002.

Gross domestic expenditure on R&D (GERD) - Dépenses intérieures brutes de R-D (DIRD)

by sector of performance and source of funds - par secteur d'exécution et par source de financement

	1994	1995	1996	1997	1998	1999	2000	
	1.1 Million EUR (1999 PTE euro, EUR from 1999) - Millions EUR (1999 PTE euro, EUR à partir de 1999)							
Business enterprise								**Entreprises**
financed by								*financé par*
Business enterprise	..	75.6 [a]	..	107.2	..	156.0	..	Entreprises
Direct government	..	4.9 [a]	..	12.1 [a]	..	14.9	..	État direct
Higher education	..	..	..	..	..	..	..	Enseignement supérieur
Private non-profit	..	0.2 [a]	..	..	..	..	..	Institutions sans but lucratif
Funds from abroad	..	15.5 [a]	..	10.3 [a]	..	13.9	..	Fonds de l'étranger
Total	..	96.2 [a]	..	129.6	..	184.8	..	*Total*
Government								**État**
financed by								*financé par*
Business enterprise	..	0.3	..	5.8	..	8.6	..	Entreprises
Direct government	..	113.2	..	123.8 [a]	..	210.3	..	État direct
Higher education	..	..	..	..	..	..	..	Enseignement supérieur
Private non-profit	..	0.5	..	1.4	..	1.1	..	Institutions sans but lucratif
Funds from abroad	..	10.3	..	8.8 [a]	..	7.7	..	Fonds de l'étranger
Total	..	124.3	..	139.7	..	227.7	..	*Total*
Higher education								**Enseignement supérieur**
financed by								*financé par*
Business enterprise	..	1.6 [a]	..	4.0	..	3.9	..	Entreprises
Direct government	..	..	..	..	..	..	..	État direct
General university funds	..	..	..	..	..	..	..	Fonds généraux des universités
Sub-total government	..	147.1 [a]	..	207.5 [a]	..	279.0	..	Sous-Total État
Higher education	..	5.6 [a]	..	8.9	..	10.7	..	Enseignement supérieur
Private non-profit	..	3.1 [a]	..	3.1	..	10.3	..	Institutions sans but lucratif
Funds from abroad	..	13.0 [a]	..	7.5 [a]	..	10.5	..	Fonds de l'étranger
Total	..	170.4 [a]	..	231.0	..	314.4	..	*Total*
Private non-profit								**Institutions sans but lucratif**
financed by								*financé par*
Business enterprise	..	12.1 [a]	..	5.5	..	5.3	..	Entreprises
Direct government	..	35.2 [a]	..	50.2 [a]	..	63.5	..	État direct
Higher education	..	..	..	..	..	..	..	Enseignement supérieur
Private non-profit	..	5.9 [a]	..	12.2	..	7.7	..	Institutions sans but lucratif
Funds from abroad	..	15.9 [a]	..	8.8 [a]	..	11.4	..	Fonds de l'étranger
Total	..	69.1 [a]	..	76.6	..	87.9	..	*Total*
GERD								**DIRD**
financed by								*financé par*
Business enterprise	..	89.6	..	122.5	..	173.7	..	Entreprises
Direct government	..	..	..	..	..	..	..	État direct
General university funds	..	..	..	..	..	..	..	Fonds généraux des universités
Sub-total government	..	300.3	..	393.5 [a]	..	567.7	..	Sous-Total État
Higher education	..	5.6	..	8.9	..	10.7	..	Enseignement supérieur
Private non-profit	..	9.8	..	16.6	..	19.1	..	Institutions sans but lucratif
Funds from abroad	..	54.7	..	35.4 [a]	..	43.5	..	Fonds de l'étranger
Total	..	460.0	..	576.9	..	814.7	..	*Total*
	1.2 Million constant USD (1995 prices and PPPs) - Millions de USD constants (prix et PPA de 1995)							
Business enterprise								**Entreprises**
financed by								*financé par*
Business enterprise	..	127.3 [a]	..	168.9	..	229.3	..	Entreprises
Direct government	..	8.2 [a]	..	19.1 [a]	..	21.9	..	État direct
Higher education	..	..	..	..	..	..	..	Enseignement supérieur
Private non-profit	..	0.4 [a]	..	..	..	..	..	Institutions sans but lucratif
Funds from abroad	..	26.1 [a]	..	16.2 [a]	..	20.4	..	Fonds de l'étranger
Total	..	162.0 [a]	..	204.2	..	271.6	..	*Total*
Government								**État**
financed by								*financé par*
Business enterprise	..	0.5	..	9.2	..	12.6	..	Entreprises
Direct government	..	190.6	..	195.0 [a]	..	309.1	..	État direct
Higher education	..	..	..	..	..	..	..	Enseignement supérieur
Private non-profit	..	0.8	..	2.1	..	1.7	..	Institutions sans but lucratif
Funds from abroad	..	17.4	..	13.9 [a]	..	11.3	..	Fonds de l'étranger
Total	..	209.3	..	220.2	..	334.7	..	*Total*
Higher education								**Enseignement supérieur**
financed by								*financé par*
Business enterprise	..	2.7 [a]	..	6.3	..	5.7	..	Entreprises
Direct government	..	..	..	..	..	..	..	État direct
General university funds	..	..	..	..	..	..	..	Fonds généraux des universités
Sub-total government	..	247.7 [a]	..	327.0 [a]	..	410.1	..	Sous-Total État
Higher education	..	9.4 [a]	..	14.0	..	15.7	..	Enseignement supérieur
Private non-profit	..	5.3 [a]	..	4.9	..	15.1	..	Institutions sans but lucratif
Funds from abroad	..	22.0 [a]	..	11.8 [a]	..	15.5	..	Fonds de l'étranger
Total	..	286.9 [a]	..	364.0	..	462.1	..	*Total*
Private non-profit								**Institutions sans but lucratif**
financed by								*financé par*
Business enterprise	..	20.4 [a]	..	8.7	..	7.7	..	Entreprises
Direct government	..	59.2 [a]	..	79.0 [a]	..	93.4	..	État direct
Higher education	..	..	..	..	..	..	..	Enseignement supérieur
Private non-profit	..	10.0 [a]	..	19.2	..	11.3	..	Institutions sans but lucratif
Funds from abroad	..	26.7 [a]	..	13.8 [a]	..	16.8	..	Fonds de l'étranger
Total	..	116.3 [a]	..	120.8	..	129.2	..	*Total*
GERD								**DIRD**
financed by								*financé par*
Business enterprise	..	150.8	..	193.1	..	255.4	..	Entreprises
Direct government	..	..	..	..	..	..	..	État direct
General university funds	..	..	..	..	..	..	..	Fonds généraux des universités
Sub-total government	..	505.7	..	620.2 [a]	..	834.5	..	Sous-Total État
Higher education	..	9.4	..	14.0	..	15.7	..	Enseignement supérieur
Private non-profit	..	16.5	..	26.2	..	28.1	..	Institutions sans but lucratif
Funds from abroad	..	92.2	..	55.7 [a]	..	64.0	..	Fonds de l'étranger
Total	..	774.6	..	909.2	..	1 197.6	..	*Total*

Source: OECD, R&D database (DSTI/EAS Division), January 2002.

Source : OCDE, base de données R-D (DSTI/Division AES), janvier 2002.

Table 2 - Tableau 2

Gross domestic expenditure on R&D (GERD) - Dépenses intérieures brutes de R-D (DIRD)

by sector of performance and type of costs - par secteur d'exécution et par type de coûts

Million EUR (1999 PTE euro, EUR from 1999) — *Millions EUR (1999 PTE euro, EUR à partir de 1999)*

	1994	1995	1996	1997	1998	1999	2000	
Business enterprise								**Entreprises**
Labour costs	..	43.9 [s]	..	55.1	..	101.6	..	Coûts du personnel
Other current costs	..	29.1 [s]	..	31.7	..	48.8	..	Autres coûts courants
Total current expenditure	..	73.0 [s]	..	86.8	..	150.5	..	*Total dépenses courantes*
Land and buildings	..	3.2 [s]	..	14.9	..	3.9	..	Terrains et constructions
Instruments and equipment	..	14.7 [s]	..	27.9	..	30.5	..	Instruments et équipements
Total capital expenditure	..	17.9 [s]	..	42.8	..	34.3	..	*Total dépenses en capital*
Total	..	96.2 [a]	..	129.6	..	184.8	..	*Total*
Government								**État**
Labour costs	..	74.2 [s]	..	95.9	..	121.3	..	Coûts du personnel
Other current costs	..	29.5 [s]	..	27.2	..	40.4	..	Autres coûts courants
Total current expenditure	..	103.6 [s]	..	123.0	..	161.7	..	*Total dépenses courantes*
Land and buildings	..	6.8 [s]	..	3.0	..	46.2	..	Terrains et constructions
Instruments and equipment	..	12.2 [s]	..	13.7	..	19.8	..	Instruments et équipements
Total capital expenditure	..	19.0 [s]	..	16.7	..	65.9	..	*Total dépenses en capital*
Total	..	124.3	..	139.7	..	227.7	..	*Total*
Higher education								**Enseignement supérieur**
Labour costs	..	122.8 [s]	..	183.0	..	234.7	..	Coûts du personnel
Other current costs	..	22.0 [s]	..	33.4	..	56.8	..	Autres coûts courants
Total current expenditure	..	144.8 [s]	..	216.3	..	291.5	..	*Total dépenses courantes*
Land and buildings	..	2.1 [s]	..	1.2	..	2.9	..	Terrains et constructions
Instruments and equipment	..	7.6 [s]	..	13.5	..	20.0	..	Instruments et équipements
Total capital expenditure	..	9.6 [s]	..	14.7	..	22.8	..	*Total dépenses en capital*
Total	..	170.4 [a]	..	231.0	..	314.4	..	*Total*
Private non-profit								**Institutions sans but lucratif**
Labour costs	..	56.2 [s]	..	45.9	..	52.4	..	Coûts du personnel
Other current costs	..	24.6 [s]	..	20.3	..	24.5	..	Autres coûts courants
Total current expenditure	..	80.8 [s]	..	66.2	..	76.9	..	*Total dépenses courantes*
Land and buildings	..	3.6 [s]	..	3.8	..	2.2	..	Terrains et constructions
Instruments and equipment	..	5.9 [s]	..	6.6	..	8.9	..	Instruments et équipements
Total capital expenditure	..	9.6 [s]	..	10.4	..	11.1	..	*Total dépenses en capital*
Total	..	69.1 [a]	..	76.6	..	87.9	..	*Total*
GERD								**DIRD**
Labour costs	..	297.0 [s]	..	379.8	..	510.0	..	Coûts du personnel
Other current costs	..	105.1 [s]	..	112.5	..	170.6	..	Autres coûts courants
Total current expenditure	..	402.2 [s]	..	492.3	..	680.6	..	*Total dépenses courantes*
Land and buildings	..	15.7 [s]	..	22.9	..	55.1	..	Terrains et constructions
Instruments and equipment	..	40.4 [s]	..	61.7	..	79.1	..	Instruments et équipements
Total capital expenditure	..	56.1 [s]	..	84.6	..	134.2	..	*Total dépenses en capital*
Total	..	460.0	..	576.9	..	814.7	..	*Total*

Table 3 - Tableau 3

Current domestic expenditure on R&D - Dépenses intérieures courantes de R-D

by sector of performance and type of activity - par secteur d'exécution et par type d'activités

Million EUR (1999 PTE euro, EUR from 1999) — *Millions EUR (1999 PTE euro, EUR à partir de 1999)*

	1994	1995	1996	1997	1998	1999	2000	
Business enterprise								**Entreprises**
Basic research	..	1.4 [k,s]	..	3.8 [a]	..	4.8	..	Recherche fondamentale
Applied research	..	35.9 [k,s]	..	31.6 [a]	..	43.3	..	Recherche appliquée
Experimental development	..	53.7 [k,s]	..	51.4 [a]	..	102.4	..	Développement expérimental
Not specified	..	..	..	..	..	..	..	Non spécifié
Total	..	96.2 [a,k]	..	86.8 [a]	..	150.5	..	*Total*
Government								**État**
Basic research	..	7.4 [k,s]	..	11.7 [a]	..	10.8	..	Recherche fondamentale
Applied research	..	69.2 [k,s]	..	63.6 [a]	..	83.0	..	Recherche appliquée
Experimental development	..	46.0 [k,s]	..	47.7 [a]	..	68.0	..	Développement expérimental
Not specified	..	..	..	..	..	..	..	Non spécifié
Total	..	124.3 [k]	..	123.0 [a]	..	161.7	..	*Total*
Higher education								**Enseignement supérieur**
Basic research	..	82.2 [k,s]	..	108.5 [a]	..	145.7	..	Recherche fondamentale
Applied research	..	58.0 [k,s]	..	84.2 [a]	..	112.9	..	Recherche appliquée
Experimental development	..	14.3 [k,s]	..	23.6 [a]	..	32.9	..	Développement expérimental
Not specified	..	..	..	..	..	..	..	Non spécifié
Total	..	170.4 [a,k]	..	216.3 [a]	..	291.5	..	*Total*
Private non-profit								**Institutions sans but lucratif**
Basic research	..	23.8 [k,s]	..	21.9 [a]	..	27.9	..	Recherche fondamentale
Applied research	..	41.6 [k,s]	..	30.7 [a]	..	36.0	..	Recherche appliquée
Experimental development	..	25.0 [k,s]	..	13.6 [a]	..	12.9	..	Développement expérimental
Not specified	..	..	..	..	..	..	..	Non spécifié
Total	..	69.1 [a,k]	..	66.2 [a]	..	76.9	..	*Total*
GERD								**DIRD**
Basic research	..	114.7 [k,s]	..	145.9 [a]	..	189.2	..	Recherche fondamentale
Applied research	..	204.6 [k,s]	..	210.2 [a]	..	275.2	..	Recherche appliquée
Experimental development	..	139.0 [k,s]	..	136.2 [a]	..	216.2	..	Développement expérimental
Not specified	..	..	..	..	..	..	..	Non spécifié
Total	..	460.0 [k]	..	492.3 [a]	..	680.6	..	*Total*

Source: OECD, R&D database (DSTI/EAS Division), January 2002.

Source : OCDE, base de données R-D (DSTI/Division AES), janvier 2002.

Gross domestic expenditure on R&D (GERD) - Dépenses intérieures brutes de R-D (DIRD)

by sector of performance and socio-economic objective - par secteur d'exécution et par objectif socio-économique

Million EUR (1999 PTE euro, EUR from 1999) — *Millions EUR (1999 PTE euro, EUR à partir de 1999)*

	1994	1995	1996	1997	1998	1999	2000	
Business enterprise								**Entreprises**
Agriculture, forestry and fishing	..	..	..	..	..	..	..	Agriculture, sylviculture et pêche
Industrial development	..	..	..	..	..	..	..	Développement industriel
Energy	..	..	..	..	..	..	..	Energie
Transport and telecommunications	..	..	..	..	..	..	..	Transports et télécommunications
Urban and rural planning	..	..	..	..	..	..	..	Aménagement urbain et rural
Prevention of pollution	..	..	..	..	..	..	..	Prévention de la pollution
Identification & treatment of pollution	..	..	..	..	..	..	..	Détection pollu. & mesures antipollu.
Health	..	..	..	..	..	..	..	Santé
Social development and services	..	..	..	..	..	..	..	Développement et services sociaux
Earth and atmosphere	..	..	..	..	..	..	..	Milieu terrestre et atmosphère
Advancement of knowledge	..	..	..	..	..	..	..	Promotion générale des connaissances
Civil space	..	..	..	..	..	..	..	Espace civil
Defence	..	..	..	..	..	..	..	Défense
Not elsewhere classified	..	96.2 [a]	..	129.6	..	184.8	..	Non classé
Total	..	96.2 [a]	..	129.6	..	184.8	..	*Total*
Government								**État**
Agriculture, forestry and fishing	..	37.7 [s]	..	49.9	..	65.6	..	Agriculture, sylviculture et pêche
Industrial development	..	15.8 [s]	..	12.6	..	14.9	..	Développement industriel
Energy	..	..	..	2.6	..	3.0	..	Energie
Transport and telecommunications	..	..	..	..	..	..	..	Transports et télécommunications
Urban and rural planning	..	7.6 [s]	..	9.6	..	51.2	..	Aménagement urbain et rural
Prevention of pollution	..	..	..	..	..	..	..	Prévention de la pollution
Identification & treatment of pollution	..	..	..	12.1	..	20.6	..	Détection pollu. & mesures antipollu.
Health	..	14.7 [s]	..	16.4	..	26.6	..	Santé
Social development and services	..	13.2 [s]	..	9.5	..	12.3	..	Développement et services sociaux
Earth and atmosphere	..	26.4 [s]	..	11.2	..	8.0	..	Milieu terrestre et atmosphère
Advancement of knowledge	..	3.0 [s]	..	4.4	..	12.3	..	Promotion générale des connaissances
Civil space	..	..	..	1.0	..	2.5	..	Espace civil
Defence	..	4.2 [s]	..	3.5	..	2.9	..	Défense
Not elsewhere classified	..	..	..	6.9	..	7.8	..	Non classé
Total	..	124.3	..	139.7	..	227.7	..	*Total*
Higher education								**Enseignement supérieur**
Agriculture, forestry and fishing	..	10.8 [s]	..	23.8	..	33.2	..	Agriculture, sylviculture et pêche
Industrial development	..	18.0 [s]	..	29.7	..	39.0	..	Développement industriel
Energy	..	..	..	1.9	..	2.5	..	Energie
Transport and telecommunications	..	..	..	..	..	..	..	Transports et télécommunications
Urban and rural planning	..	5.3 [s]	..	8.1	..	13.3	..	Aménagement urbain et rural
Prevention of pollution	..	..	..	..	..	..	..	Prévention de la pollution
Identification & treatment of pollution	..	..	..	16.4	..	21.6	..	Détection pollu. & mesures antipollu.
Health	..	14.1 [s]	..	22.0	..	31.8	..	Santé
Social development and services	..	17.8 [s]	..	31.7	..	36.5	..	Développement et services sociaux
Earth and atmosphere	..	17.5 [s]	..	4.6	..	8.1	..	Milieu terrestre et atmosphère
Advancement of knowledge	..	69.8 [s]	..	90.3	..	124.3	..	Promotion générale des connaissances
Civil space	..	..	..	0.4	..	0.6	..	Espace civil
Defence	..	1.1 [s]	..	0.1	..	0.3	..	Défense
Not elsewhere classified	..	..	..	2.0	..	3.2	..	Non classé
Total	..	170.4 [a]	..	231.0	..	314.4	..	*Total*
Private non-profit								**Institutions sans but lucratif**
Agriculture, forestry and fishing	..	2.5 [s]	..	3.9	..	6.1	..	Agriculture, sylviculture et pêche
Industrial development	..	43.0 [s]	..	20.1	..	21.8	..	Développement industriel
Energy	..	..	..	4.3	..	2.6	..	Energie
Transport and telecommunications	..	..	..	..	..	..	..	Transports et télécommunications
Urban and rural planning	..	1.5 [s]	..	3.8	..	3.7	..	Aménagement urbain et rural
Prevention of pollution	..	..	..	..	..	..	..	Prévention de la pollution
Identification & treatment of pollution	..	..	..	6.2	..	6.0	..	Détection pollu. & mesures antipollu.
Health	..	4.7 [s]	..	7.3	..	13.9	..	Santé
Social development and services	..	2.8 [s]	..	7.7	..	6.1	..	Développement et services sociaux
Earth and atmosphere	..	4.3 [s]	..	1.9	..	1.4	..	Milieu terrestre et atmosphère
Advancement of knowledge	..	31.6 [s]	..	17.4	..	21.2	..	Promotion générale des connaissances
Civil space	..	..	..	0.5	..	1.6	..	Espace civil
Defence	..	..	..	0.6	..	0.6	..	Défense
Not elsewhere classified	..	..	..	2.9	..	2.8	..	Non classé
Total	..	69.1 [a]	..	76.6	..	87.9	..	*Total*
GERD								**DIRD**
Agriculture, forestry and fishing	..	..	..	77.6	..	104.9	..	Agriculture, sylviculture et pêche
Industrial development	..	..	..	62.4	..	75.7	..	Développement industriel
Energy	..	..	..	8.8	..	8.1	..	Energie
Transport and telecommunications	..	..	..	..	..	..	..	Transports et télécommunications
Urban and rural planning	..	..	..	21.5	..	68.2	..	Aménagement urbain et rural
Prevention of pollution	..	..	..	..	..	..	..	Prévention de la pollution
Identification & treatment of pollution	..	..	..	34.6	..	48.2	..	Détection pollu. & mesures antipollu.
Health	..	..	..	45.7	..	72.3	..	Santé
Social development and services	..	..	..	48.9	..	54.9	..	Développement et services sociaux
Earth and atmosphere	..	..	..	17.7	..	17.4	..	Milieu terrestre et atmosphère
Advancement of knowledge	..	..	..	112.2	..	157.9	..	Promotion générale des connaissances
Civil space	..	..	..	1.9	..	4.7	..	Espace civil
Defence	..	..	..	4.2	..	3.8	..	Défense
Not elsewhere classified	..	460.0	..	141.4	..	198.6	..	Non classé
Total	..	460.0	..	576.9	..	814.7	..	*Total*

Source: OECD, R&D database (DSTI/EAS Division), January 2002.

Source : OCDE, base de données R-D (DSTI/Division AES), janvier 2002.

PORTUGAL

Table 5 - Tableau 5

Total R&D personnel - Personnel total de R-D

by sector of employment and occupation - par secteur d'emploi et par profession

In full-time equivalent — *En équivalence plein-temps*

	1994	1995	1996	1997	1998	1999	2000	
Business enterprise								**Entreprises**
RSE	..	1 076 [a]	..	1 193	..	1 994	..	Chercheurs
Technicians	..	431 [a]	..	435	..	868	..	Techniciens
Other occupations	..	410 [a]	..	353	..	398	..	Autres professions
Total	..	1 917 [a]	..	1 981	..	3 260	..	*Total*
Government								**État**
RSE	..	2 741	..	2 930	..	3 445	..	Chercheurs
Technicians	..	743	..	915	..	971	..	Techniciens
Other occupations	..	1 232	..	1 385	..	1 486	..	Autres professions
Total	..	4 716	..	5 230	..	5 902	..	*Total*
Higher education								**Enseignement supérieur**
RSE	..	5 850 [a]	..	7 475	..	8 243	..	Chercheurs
Technicians	..	297 [a]	..	405	..	467	..	Techniciens
Other occupations	..	338 [a]	..	562	..	477	..	Autres professions
Total	..	6 484 [a]	..	8 442	..	9 187	..	*Total*
Private non-profit								**Institutions sans but lucratif**
RSE	..	1 933 [a]	..	2 045	..	2 070	..	Chercheurs
Technicians	..	189 [a]	..	150	..	173	..	Techniciens
Other occupations	..	227 [a]	..	188	..	214	..	Autres professions
Total	..	2 349 [a]	..	2 383	..	2 457	..	*Total*
TOTAL								**TOTAL**
RSE	..	11 599	..	13 642	..	15 752	..	Chercheurs
Technicians	..	1 660	..	1 906	..	2 479	..	Techniciens
Other occupations	..	2 207	..	2 487	..	2 575	..	Autres professions
Total	..	15 465	..	18 035	..	20 806	..	*Total*

Table 6 - Tableau 6

Total R&D personnel - Personnel total de R-D

by sector of employment and formal qualification - par secteur d'emploi et par niveau de qualification

In full-time equivalent — *En équivalence plein-temps*

	1994	1995	1996	1997	1998	1999	2000	
Business enterprise								**Entreprises**
University PhD level degrees	..	..	..	..	..	58	..	Diplômes univ. de niv. doctorat
Other university degrees	..	..	..	..	..	1 936 [o]	..	Autres diplômes universitaires
Sub-total university degrees	..	1 076 [a,o]	..	1 193 [o]	..	1 994 [o]	..	Sous-total Universitaires
Other post-secondary diplomas	..	.. [n]	..	.. [n]	..	.. [n]	..	Autres diplômes post-secondaires
Secondary diplomas	..	431 [a]	..	435	..	868	..	Diplômes de fin d'études secondaires
Other qualifications	..	410 [a]	..	353	..	398	..	Autres qualifications
Total	..	1 917 [a]	..	1 981	..	3 260	..	*Total*
Government								**État**
University PhD level degrees	..	..	..	..	..	671	..	Diplômes univ. de niv. doctorat
Other university degrees	..	..	..	..	..	2 773 [o]	..	Autres diplômes universitaires
Sub-total university degrees	..	2 741 [o]	..	2 930 [o]	..	3 444 [o]	..	Sous-total Universitaires
Other post-secondary diplomas	..	.. [n]	..	.. [n]	..	.. [n]	..	Autres diplômes post-secondaires
Secondary diplomas	..	743	..	915	..	971	..	Diplômes de fin d'études secondaires
Other qualifications	..	1 232	..	1 385	..	1 486	..	Autres qualifications
Total	..	4 716	..	5 230	..	5 902	..	*Total*
Higher education								**Enseignement supérieur**
University PhD level degrees	..	..	..	..	..	3 034	..	Diplômes univ. de niv. doctorat
Other university degrees	..	..	..	..	..	5 209 [o]	..	Autres diplômes universitaires
Sub-total university degrees	..	5 850 [a,o]	..	7 475 [o]	..	8 243 [o]	..	Sous-total Universitaires
Other post-secondary diplomas	..	.. [n]	..	.. [n]	..	.. [n]	..	Autres diplômes post-secondaires
Secondary diplomas	..	297 [a]	..	405	..	467	..	Diplômes de fin d'études secondaires
Other qualifications	..	338 [a]	..	562	..	477	..	Autres qualifications
Total	..	6 484 [a]	..	8 442	..	9 187	..	*Total*
Private non-profit								**Institutions sans but lucratif**
University PhD level degrees	..	..	..	..	..	647	..	Diplômes univ. de niv. doctorat
Other university degrees	..	..	..	..	..	1 423 [o]	..	Autres diplômes universitaires
Sub-total university degrees	..	1 933 [a,o]	..	2 045 [o]	..	2 070 [o]	..	Sous-total Universitaires
Other post-secondary diplomas	..	.. [n]	..	.. [n]	..	.. [n]	..	Autres diplômes post-secondaires
Secondary diplomas	..	189 [a]	..	150	..	173	..	Diplômes de fin d'études secondaires
Other qualifications	..	227 [a]	..	188	..	214	..	Autres qualifications
Total	..	2 349 [a]	..	2 383	..	2 457	..	*Total*
TOTAL								**TOTAL**
University PhD level degrees	..	..	..	..	..	4 409	..	Diplômes univ. de niv. doctorat
Other university degrees	..	..	..	..	..	11 342 [o]	..	Autres diplômes universitaires
Sub-total university degrees	..	11 599 [o]	..	13 642 [o]	..	15 751 [o]	..	Sous-total Universitaires
Other post-secondary diplomas	..	.. [n]	..	.. [n]	..	.. [n]	..	Autres diplômes post-secondaires
Secondary diplomas	..	1 660	..	1 906	..	2 479	..	Diplômes de fin d'études secondaires
Other qualifications	..	2 207	..	2 487	..	2 575	..	Autres qualifications
Total	..	15 465	..	18 035	..	20 806	..	*Total*

Source: OECD, R&D database (DSTI/EAS Division), January 2002.

Source : OCDE, base de données R-D (DSTI/Division AES), janvier 2002.

Table 7 - Tableau 7

Gross domestic expenditure on R&D (GERD) - Dépenses intérieures brutes de R-D (DIRD)

by sector of performance and main scientific field - par secteur d'exécution et par grand domaine scientifique

Million EUR (1999 PTE euro, EUR from 1999) — *Millions EUR (1999 PTE euro, EUR à partir de 1999)*

	1994	1995	1996	1997	1998	1999	2000	
Business enterprise								**Entreprises**
Natural sciences and engineering	..	96.2 [a]	..	129.6	..	184.8	..	Sciences exactes, nat. et de l'ingénieur
Social sciences and humanities	..	..	..	..	..	..	..	Sciences sociales et humaines
All fields of science	..	96.2 [a]	..	129.6	..	184.8	..	*Tous domaines scientifiques*
Government								**État**
Natural sciences and engineering	..	111.1 [s]	..	126.0	..	205.2	..	Sciences exactes, nat. et de l'ingénieur
Social sciences and humanities	..	11.5 [s]	..	13.7	..	22.5	..	Sciences sociales et humaines
All fields of science	..	124.3	..	139.7	..	227.7	..	*Tous domaines scientifiques*
Higher education								**Enseignement supérieur**
Natural sciences and engineering	..	115.0 [s]	..	162.5 [a]	..	214.1	..	Sciences exactes, nat. et de l'ingénieur
Social sciences and humanities	..	39.5 [s]	..	68.5 [a]	..	100.3	..	Sciences sociales et humaines
All fields of science	..	170.4 [a]	..	231.0	..	314.4	..	*Tous domaines scientifiques*
Private non-profit								**Institutions sans but lucratif**
Natural sciences and engineering	..	84.7 [s]	..	62.6 [a]	..	72.6	..	Sciences exactes, nat. et de l'ingénieur
Social sciences and humanities	..	5.7 [s]	..	14.0 [n]	..	15.3	..	Sciences sociales et humaines
All fields of science	..	69.1 [a]	..	76.6	..	87.9	..	*Tous domaines scientifiques*
TOTAL								**TOTAL**
Natural sciences and engineering	..	407.0 [s]	..	480.7	..	676.7	..	Sciences exactes, nat. et de l'ingénieur
Social sciences and humanities	..	56.6 [s]	..	96.2	..	138.1	..	Sciences sociales et humaines
All fields of science	..	460.0	..	576.9	..	814.7	..	*Tous domaines scientifiques*

Table 8 - Tableau 8

Total R&D personnel - Personnel total de R-D

by sector of employment and main scientific field - par secteur d'emploi et par grand domaine scientifique

In full-time equivalent — *En équivalence plein-temps*

	1994	1995	1996	1997	1998	1999	2000	
Business enterprise								**Entreprises**
Natural sciences and engineering	..	1 917 [a]	..	1 981	..	3 260	..	Sciences exactes, nat. et de l'ingénieur
Social sciences and humanities	..	..	..	..	..	..	..	Sciences sociales et humaines
All fields of science	..	1 917 [a]	..	1 981	..	3 260	..	*Tous domaines scientifiques*
Government								**État**
Natural sciences and engineering	..	1 984 [s]	..	..	..	..	..	Sciences exactes, nat. et de l'ingénieur
Social sciences and humanities	..	391 [s]	..	..	..	..	..	Sciences sociales et humaines
All fields of science	..	4 716	..	5 230	..	5 902	..	*Tous domaines scientifiques*
Higher education								**Enseignement supérieur**
Natural sciences and engineering	..	4 401 [s]	..	..	..	..	..	Sciences exactes, nat. et de l'ingénieur
Social sciences and humanities	..	1 493 [s]	..	..	..	..	..	Sciences sociales et humaines
All fields of science	..	6 484 [a]	..	8 442	..	9 187	..	*Tous domaines scientifiques*
Private non-profit								**Institutions sans but lucratif**
Natural sciences and engineering	..	2 431 [s]	..	..	..	..	..	Sciences exactes, nat. et de l'ingénieur
Social sciences and humanities	..	215 [s]	..	..	..	..	..	Sciences sociales et humaines
All fields of science	..	2 349 [a]	..	2 383	..	2 457	..	*Tous domaines scientifiques*
TOTAL								**TOTAL**
Natural sciences and engineering	..	10 732	..	..	..	..	..	Sciences exactes, nat. et de l'ingénieur
Social sciences and humanities	..	2 100	..	..	..	..	..	Sciences sociales et humaines
All fields of science	..	15 465	..	18 035	..	20 806	..	*Tous domaines scientifiques*

Source: OECD, R&D database (DSTI/EAS Division), January 2002.

Source : OCDE, base de données R-D (DSTI/Division AES), janvier 2002.

Total business enterprise intramural expenditure on R&D - Dépenses intra-muros totales de R-D des entreprises

(BERD) by industry - (DIRDE) par industrie

	1994	1995	1996	1997	1998	1999	2000	
	9.1 Million EUR (1999 PTE euro, EUR from 1999) - Millions EUR (1999 PTE euro, EUR à partir de 1999)							
1 AGRICULTURE, HUNTING & FORESTRY	..	0.1 [a]	..	0.3	..	0.9	..	**1 AGRICULTURE, CHASSE & SYLVICULTURE**
2 MINING	..	0.4 [a]	..	0.1	..	0.2	..	**2 ACTIVITES EXTRACTIVES**
3 MANUFACTURING	..	62.3 [a]	..	85.4	..	98.0	..	**3 ACTIVITES DE FABRICATION**
4 Food, beverages & tobacco	..	3.2 [a,o]	..	2.9 [o]	..	2.5 [o]	..	**4 Produits alimentaires, boissons & tabac**
5 Food products & beverages	..	.. [n]	..	.. [n]	..	.. [n]	..	5 Produits alimentaires & boissons
6 Tobacco products	..	.. [n]	..	.. [n]	..	.. [n]	..	6 Produits à base de tabac
7 Textiles, wear app., fur & leather	..	2.7 [a]	..	3.5	..	5.5	..	**7 Textiles, habill., fourrures & cuirs**
8 Textiles	..	2.0 [a]	..	3.1	..	3.9	..	8 Textiles
9 Wearing apparel & fur	..	0.0	..	0.0	..	0.3	..	9 Articles d'habillement & fourrures
10 Leather products & footwear	..	0.7 [a]	..	0.3	..	1.2	..	10 Articles en cuir & chaussures
11 Wood, paper, printing, publishing	..	4.7 [a]	..	3.6	..	8.2	..	**11 Bois, papier, imprimerie, édition**
12 Wood & cork (not furniture)	..	1.1 [a]	..	0.5	..	6.4	..	12 Bois & liège (sauf meubles)
13 Pulp, paper & paper products	..	3.4 [a]	..	1.1	..	1.3	..	13 Carton, papier & articles
14 Publ., print. & repro. of rec. media	..	0.2 [a]	..	2.1	..	0.4	..	14 Edition, impress. & repro. enregis.
15 Coke, petroleum, nuclear fuel, chemicals, rubber plastics	..	13.9 [a,o]	..	22.1 [o]	..	19.7 [o]	..	**15 Coke, pétrole, combust. nucl., prod. chimi., caout. & mat. plast.**
16 Coke, ref. petrol. prod. & nuclear fuel	..	.. [n]	..	.. [n]	..	.. [n]	..	16 Coke, raffin. pétr. & combust. nucl.
17 Chemicals & chemical products	..	.. [n]	..	.. [n]	..	.. [n]	..	17 Produits chimiques
18 *Chemical prod.(less pharmaceuticals)*	..	.. [n]	..	.. [n]	..	.. [n]	..	18 *Prod. chimi. (moins prod. pharmaceutiques)*
19 *Pharmaceuticals*	..	.. [n]	..	.. [n]	..	.. [n]	..	19 *Produits pharmaceutiques*
20 Rubber & plastic products	..	.. [n]	..	.. [n]	..	.. [n]	..	20 Caoutchouc et matières plastiques
21 Non-metallic mineral products	..	0.6 [a]	..	0.8	..	1.1	..	**21 Produits minéraux non métalliques**
22 Basic metals	..	0.1 [a,o]	..	0.1 [o]	..	0.8 [o]	..	**22 Produits métallurgiques de base**
23 Basic metals, ferrous	..	.. [n]	..	.. [n]	..	.. [n]	..	23 Produits métallur. de base, ferreux
24 Basic metals, non-ferrous	..	.. [n]	..	.. [n]	..	.. [n]	..	24 Prod. métallur. de base, non ferreux
25 Fabricated metal products	..	2.1 [a]	..	1.0	..	3.6	..	**25 Ouvrages en métaux (sauf mach.& mat.)**
26 Machinery equipment, instruments and transport equipment	..	34.6 [a]	..	51.2	..	56.1	..	**26 Machines & matériel, instruments & mat. de transport**
27 Machinery n.e.c.	..	5.0 [a]	..	6.4	..	11.7	..	27 Machines, n.c.a.
28 Office, account. & computing machinery	..	0.1 [a]	..	1.3	..	5.3	..	28 Mach. de bur., compta. & trait. inform.
29 Electrical machinery	..	3.9 [a]	..	9.9	..	10.7	..	29 Machines électriques
30 Electro. equip. (radio, TV & commun.)	..	18.2 [a]	..	17.2	..	18.0	..	30 Equip. électro. (radio, tv & communic.)
31 *Electro. comp. (inc. semi-conductors)*	..	..	..	..	..	..	..	31 *Compos. électro. (semi-cond. incl.)*
32 *TV, radio & communications equip.*	..	..	..	..	..	..	..	32 *Appareils radio, tv et de communi.*
33 Instruments	..	1.4 [a]	..	1.6	..	4.9	..	33 Instruments
34 Motor vehicles	..	3.3 [a]	..	1.4	..	3.8	..	34 Véhicules automobiles
35 Other transport equipment	..	2.7 [a]	..	13.4	..	1.7	..	35 Autres matériels de transport
36 *Ships*	..	..	..	..	..	..	..	36 *Navires*
37 *Aerospace*	..	0.0	..	0.0	..	..	..	37 *Aérospatiale*
38 *Other transport equipment n.e.c.*	..	..	..	..	..	..	..	38 *Autres matériels de transport n.c.a.*
39 Furniture, other manufacturing n.e.c.	..	0.3 [a]	..	0.2	..	0.7	..	**39 Meubles, autres activ. de fabricat. n.c.a.**
40 Furniture	..	..	..	..	..	..	..	40 Meubles
41 Other manufacturing n.e.c.	..	..	..	..	..	..	..	41 Autres activités de fabrication n.c.a.
42 Recycling	..	0.0 [a]	..	0.0	..	0.0	..	**42 Récupération**
43 ELECTRICITY, GAS & WATER SUPPLY	..	5.7 [a]	..	5.4	..	3.4	..	**43 DISTRIBUTION D'ELECTRICITE, GAZ & EAU**
44 CONSTRUCTION	..	0.1 [a]	..	0.9	..	1.1	..	**44 CONSTRUCTION**
45 SERVICES SECTOR	..	27.7 [a]	..	37.5	..	81.2	..	**45 SECTEUR DES SERVICES**
46 Wholesale, retail trade & repair of motor vehicles, etc.	..	1.5 [a]	..	4.8	..	6.8	..	**46 Commerce (gros & détail), réparation de véhicules, etc.**
47 Hotels & restaurants	..	0.0	..	0.0	..	0.0	..	**47 Hôtels et restaurants**
48 Transport & storage	..	1.2 [a]	..	4.8	..	2.6	..	**48 Transports, entreposage**
49 Communications	..	12.2 [a,o]	..	7.2 [o]	..	0.1 [o]	..	**49 Communications**
50 Post	..	.. [n]	..	.. [n]	..	.. [n]	..	50 Postes
51 Telecommunications	..	.. [n]	..	.. [n]	..	.. [n]	..	51 Télécommunications
52 Financial intermediation (inc. insur.)	..	1.0 [a]	..	2.3	..	23.1	..	**52 Intermédiation financ. (assurances inc.)**
53 Real estate, renting & busin. activity	..	11.5 [a]	..	17.2	..	46.1	..	**53 Immobil., locat. & serv. aux entreprises**
54 Computer & related activities	..	2.1 [a]	..	4.0	..	16.5	..	54 Informatique et activités connexes
55 *Software consultancy*	..	..	..	..	..	..	..	55 *Production de logiciels*
56 *Other computer services n.e.c.*	..	..	..	..	..	..	..	56 *Autres activités informatiques*
57 Research & development	..	3.9 [a]	..	9.7	..	7.4	..	57 Recherche & développement expérimental
58 Other business activities n.e.c.	..	5.5 [a]	..	3.6	..	22.2	..	58 Autres activ. & serv. aux entreprises
59 Comm., soc. & pers. serv. activ., etc.	..	0.3 [a]	..	1.2	..	2.4	..	**59 Adm. pub., serv. soc. & collec., etc**
TOTAL BUSINESS ENTERPRISE SECTOR	**..**	**96.2 [a]**	**..**	**129.6**	**..**	**184.8**	**..**	**TOTAL SECTEUR DES ENTREPRISES**

Source: OECD, R&D database (DSTI/EAS Division), January 2002.

Source : OCDE, base de données R-D (DSTI/Division AES), janvier 2002.

Total business enterprise intramural expenditure on R&D - Dépenses intra-muros totales de R-D des entreprises
(BERD) by industry - (DIRDE) par industrie

9.2 Million constant USD(1995 prices and PPPs) - Millions de USD constants (prix et PPA de 1995)

	1994	1995	1996	1997	1998	1999	2000	
1 **AGRICULTURE, HUNTING & FORESTRY**	..	0.2 [a]	..	0.5	..	1.3	..	1 **AGRICULTURE, CHASSE & SYLVICULTURE**
2 **MINING**	..	0.6 [a]	..	0.1	..	0.2	..	2 **ACTIVITES EXTRACTIVES**
3 **MANUFACTURING**	..	104.8 [a]	..	134.6	..	144.1	..	3 **ACTIVITES DE FABRICATION**
4 **Food, beverages & tobacco**	..	5.4 [a,o]	..	4.6 [o]	..	3.6 [o]	..	4 **Produits alimentaires, boissons & tabac**
5 Food products & beverages	..	.. [n]	..	.. [n]	..	.. [n]	..	5 Produits alimentaires & boissons
6 Tobacco products	..	.. [n]	..	.. [n]	..	.. [n]	..	6 Produits à base de tabac
7 **Textiles, wear app., fur & leather**	..	4.5 [a]	..	5.5	..	8.0	..	7 **Textiles, habill., fourrures & cuirs**
8 Textiles	..	3.4 [a]	..	4.9	..	5.8	..	8 Textiles
9 Wearing apparel & fur	..	0.0	..	0.1	..	0.5	..	9 Articles d'habillement & fourrures
10 Leather products & footwear	..	1.1 [a]	..	0.5	..	1.8	..	10 Articles en cuir & chaussures
11 **Wood, paper, printing, publishing**	..	7.9 [a]	..	5.6	..	12.0	..	11 **Bois, papier, imprimerie, édition**
12 Wood & cork (not furniture)	..	1.9 [a]	..	0.7	..	9.4	..	12 Bois & liège (sauf meubles)
13 Pulp, paper & paper products	..	5.7 [a]	..	1.7	..	2.0	..	13 Carton, papier & articles
14 Publ., print. & repro. of rec. media	..	0.3 [a]	..	3.2	..	0.7	..	14 Edition, impress. & repro. enregis.
15 **Coke, petroleum, nuclear fuel, chemicals, rubber plastics**	..	23.4 [a,o]	..	34.8 [o]	..	28.9 [o]	..	15 **Coke, pétrole, combust. nucl., prod. chimi., caout. & mat. plast.**
16 Coke, ref. petrol. prod. & nuclear fuel	..	.. [n]	..	.. [n]	..	.. [n]	..	16 Coke, raffin. pétr. & combust. nucl.
17 Chemicals & chemical products	..	.. [n]	..	.. [n]	..	.. [n]	..	17 Produits chimiques
18 *Chemical prod.(less pharmaceuticals)*	..	.. [n]	..	.. [n]	..	.. [n]	..	18 *Prod. chimi. (moins prod. pharmaceutiques)*
19 *Pharmaceuticals*	..	.. [n]	..	.. [n]	..	.. [n]	..	19 *Produits pharmaceutiques*
20 Rubber & plastic products	..	.. [n]	..	.. [n]	..	.. [n]	..	20 Caoutchouc et matières plastiques
21 **Non-metallic mineral products**	..	1.1 [a]	..	1.2	..	1.6	..	21 **Produits minéraux non métalliques**
22 **Basic metals**	..	0.1 [a,o]	..	0.2 [o]	..	1.2 [o]	..	22 **Produits métallurgiques de base**
23 Basic metals, ferrous	..	.. [n]	..	.. [n]	..	.. [n]	..	23 Produits métallur. de base, ferreux
24 Basic metals, non-ferrous	..	.. [n]	..	.. [n]	..	.. [n]	..	24 Prod. métallur. de base, non ferreux
25 **Fabricated metal products**	..	3.6 [a]	..	1.6	..	5.4	..	25 **Ouvrages en métaux (sauf mach.& mat.)**
26 **Machinery equipment, instruments and transport equipment**	..	58.2 [a]	..	80.8	..	82.4	..	26 **Machines & matériel, instruments & mat. de transport**
27 Machinery n.e.c.	..	8.4 [a]	..	10.1	..	17.2	..	27 Machines, n.c.a.
28 Office, account. & computing machinery	..	0.1 [a]	..	2.0	..	7.9	..	28 Mach. de bur., compta. & trait. inform.
29 Electrical machinery	..	6.5 [a]	..	15.6	..	15.7	..	29 Machines électriques
30 Electro. equip. (radio, TV & commun.)	..	30.7 [a]	..	27.2	..	26.4	..	30 Equip. électro. (radio, tv & communic.)
31 *Electro. comp. (inc. semi-conductors)*	..	..	..	..	..	..	..	31 *Compos. électro. (semi-cond. incl.)*
32 *TV, radio & communications equip.*	..	..	..	..	..	..	..	32 *Appareils radio, tv et de communi.*
33 Instruments	..	2.4 [a]	..	2.5	..	7.1	..	33 Instruments
34 Motor vehicles	..	5.6 [a]	..	2.2	..	5.6	..	34 Véhicules automobiles
35 Other transport equipment	..	4.5 [a]	..	21.1	..	2.5	..	35 Autres matériels de transport
36 *Ships*	..	..	..	..	..	..	..	36 *Navires*
37 *Aerospace*	..	0.0	..	0.0	..	..	..	37 *Aérospatiale*
38 *Other transport equipment n.e.c.*	..	..	..	..	..	..	..	38 *Autres matériels de transport n.c.a.*
39 **Furniture, other manufacturing n.e.c.**	..	0.5 [a]	..	0.3	..	1.0	..	39 **Meubles, autres activ. de fabricat. n.c.a.**
40 Furniture	..	..	..	..	..	..	..	40 Meubles
41 Other manufacturing n.e.c.	..	..	..	..	..	..	..	41 Autres activités de fabrication n.c.a.
42 **Recycling**	..	0.0 [a]	..	0.1	..	0.0	..	42 **Récupération**
43 **ELECTRICITY, GAS & WATER SUPPLY**	..	9.7 [a]	..	8.5	..	5.0	..	43 **DISTRIBUTION D'ELECTRICITE, GAZ & EAU**
44 **CONSTRUCTION**	..	0.1 [a]	..	1.4	..	1.7	..	44 **CONSTRUCTION**
45 **SERVICES SECTOR**	..	46.6 [a]	..	59.1	..	119.4	..	45 **SECTEUR DES SERVICES**
46 **Wholesale, retail trade & repair of motor vehicles, etc.**	..	2.6 [a]	..	7.6	..	10.0	..	46 **Commerce (gros & détail), réparation de véhicules, etc.**
47 **Hotels & restaurants**	..	0.0	..	0.0	..	0.0	..	47 **Hôtels et restaurants**
48 **Transport & storage**	..	2.0 [a]	..	7.6	..	3.9	..	48 **Transports, entreposage**
49 **Communications**	..	20.5 [a,o]	..	11.3 [o]	..	0.2 [o]	..	49 **Communications**
50 Post	..	.. [n]	..	.. [n]	..	.. [n]	..	50 Postes
51 Telecommunications	..	.. [n]	..	.. [n]	..	.. [n]	..	51 Télécommunications
52 **Financial intermediation (inc. insur.)**	..	1.7 [a]	..	3.7	..	34.0	..	52 **Intermédiation financ. (assurances inc.)**
53 **Real estate, renting & busin. activity**	..	19.3 [a]	..	27.1	..	67.8	..	53 **Immobil., locat. & serv. aux entreprises**
54 Computer & related activities	..	3.6 [a]	..	6.2	..	24.3	..	54 Informatique et activités connexes
55 *Software consultancy*	..	..	..	..	..	..	..	55 *Production de logiciels*
56 *Other computer services n.e.c.*	..	..	..	..	..	..	..	56 *Autres activités informatiques*
57 Research & development	..	6.5 [a]	..	15.2	..	10.9	..	57 Recherche & développement expérimental
58 Other business activities n.e.c.	..	9.2 [a]	..	5.7	..	32.6	..	58 Autres activ. & serv. aux entreprises
59 **Comm., soc. & pers. serv. activ., etc.**	..	0.6 [a]	..	1.8	..	3.6	..	59 **Adm. pub., serv. soc. & collec., etc**
TOTAL BUSINESS ENTERPRISE SECTOR	**..**	**162.0** [a]	**..**	**204.2**	**..**	**271.6**	**..**	**TOTAL SECTEUR DES ENTREPRISES**

Source: OECD, R&D database (DSTI/EAS Division), January 2002.

Source : OCDE, base de données R-D (DSTI/Division AES), janvier 2002.

PORTUGAL Table 10 - Tableau 10 **PORTUGAL**

Business enterprise intramural expenditure on R&D (BERD) - Dépenses intra-muros de R-D des entreprises (DIRDE)

by industry and source of funds - par industrie et par source de financement

Million EUR (1999 PTE euro, EUR from 1999) — *Millions EUR (1999 PTE euro, EUR à partir de 1999)*

	1994	1995	1996	1997	1998	1999	2000	
				10.1 Business enterprise - Entreprises				
1 AGRICULTURE, HUNTING & FORESTRY	..	0.0 [a]	..	0.2	..	0.4	..	**1 AGRICULTURE, CHASSE & SYLVICULTURE**
2 MINING	..	0.2 [a]	..	0.0	..	0.1	..	**2 ACTIVITES EXTRACTIVES**
3 MANUFACTURING	..	53.1 [a]	..	74.9	..	86.1	..	**3 ACTIVITES DE FABRICATION**
4 Food, beverages & tobacco	..	2.9 [a,o]	..	2.7 [o]	..	2.2 [o]	..	**4 Produits alimentaires, boissons & tabac**
5 Food products & beverages	..	.. [n]	..	.. [n]	..	.. [n]	..	5 Produits alimentaires & boissons
6 Tobacco products	..	.. [n]	..	.. [n]	..	.. [n]	..	6 Produits à base de tabac
7 Textiles, wear app., fur & leather	..	2.5 [a]	..	2.9	..	4.6	..	**7 Textiles, habill., fourrures & cuirs**
8 Textiles	..	1.9 [a]	..	2.7	..	3.2	..	8 Textiles
9 Wearing apparel & fur	..	0.0	..	0.0	..	0.3	..	9 Articles d'habillement & fourrures
10 Leather products & footwear	..	0.6 [a]	..	0.2	..	1.1	..	10 Articles en cuir & chaussures
11 Wood, paper, printing, publishing	..	4.2 [a]	..	3.1	..	6.8	..	**11 Bois, papier, imprimerie, édition**
12 Wood & cork (not furniture)	..	1.0 [a]	..	0.4	..	5.2	..	12 Bois & liège (sauf meubles)
13 Pulp, paper & paper products	..	3.0 [a]	..	1.0	..	1.1	..	13 Carton, papier & articles
14 Publ., print. & repro. of rec. media	..	0.2 [a]	..	1.6	..	0.4	..	14 Edition, impress. & repro. enregis.
15 Coke, petroleum, nuclear fuel, chemicals, rubber plastics	..	10.8 [a,o]	..	20.4 [o]	..	18.0 [o]	..	**15 Coke, pétrole, combust. nucl., prod. chimi., caout. & mat. plast.**
16 Coke, ref. petrol. prod. & nuclear fuel	..	.. [n]	..	.. [n]	..	.. [n]	..	16 Coke, raffin. pétr. & combust. nucl.
17 Chemicals & chemical products	..	.. [n]	..	.. [n]	..	.. [n]	..	17 Produits chimiques
18 *Chemical prod.(less pharmaceuticals)*	..	.. [n]	..	.. [n]	..	.. [n]	..	18 *Prod. chimi. (moins prod. pharmaceutiques)*
19 *Pharmaceuticals*	..	.. [n]	..	.. [n]	..	.. [n]	..	19 *Produits pharmaceutiques*
20 Rubber & plastic products	..	.. [n]	..	.. [n]	..	.. [n]	..	20 Caoutchouc et matières plastiques
21 Non-metallic mineral products	..	0.6 [a]	..	0.7	..	1.0	..	**21 Produits minéraux non métalliques**
22 Basic metals	..	0.0 [a,o]	..	0.1 [o]	..	0.6 [o]	..	**22 Produits métallurgiques de base**
23 Basic metals, ferrous	..	.. [n]	..	.. [n]	..	.. [n]	..	23 Produits métallur. de base, ferreux
24 Basic metals, non-ferrous	..	.. [n]	..	.. [n]	..	.. [n]	..	24 Prod. métallur. de base, non ferreux
25 Fabricated metal products	..	1.2 [a]	..	0.7	..	3.2	..	**25 Ouvrages en métaux (sauf mach.& mat.)**
26 Machinery equipment, instruments and transport equipment	..	30.6 [a]	..	44.2	..	49.1	..	**26 Machines & matériel, instruments & mat. de transport**
27 Machinery n.e.c.	..	3.8 [a]	..	5.2	..	9.0	..	27 Machines, n.c.a.
28 Office, account. & computing machinery	..	0.0 [a]	..	0.8	..	4.4	..	28 Mach. de bur., compta. & trait. inform.
29 Electrical machinery	..	3.8 [a]	..	8.8	..	9.9	..	29 Machines électriques
30 Electro. equip. (radio, TV & commun.)	..	17.0 [a,o]	..	13.8 [o]	..	17.3 [o]	..	30 Equip. électro. (radio, tv & communic.)
31 *Electro. comp. (inc. semi-conductors)*	..	.. [n]	..	.. [n]	..	.. [n]	..	31 *Compos. électro. (semi-cond. incl.)*
32 *TV, radio & communications equip.*	..	.. [n]	..	.. [n]	..	.. [n]	..	32 *Appareils radio, tv et de communi.*
33 Instruments	..	1.1 [a]	..	1.3	..	3.9	..	33 Instruments
34 Motor vehicles	..	3.3 [a]	..	1.3	..	3.6	..	34 Véhicules automobiles
35 Other transport equipment	..	1.5 [a,o]	..	13.0 [o]	..	1.0 [o]	..	35 Autres matériels de transport
36 *Ships*	..	.. [n]	..	.. [n]	..	.. [n]	..	36 *Navires*
37 *Aerospace*	..	0.0	..	0.0	..	.. [n]	..	37 *Aérospatiale*
38 *Other transport equipment n.e.c.*	..	.. [n]	..	.. [n]	..	.. [n]	..	38 *Autres matériels de transport n.c.a.*
39 Furniture, other manufacturing n.e.c.	..	0.3 [a,o]	..	0.1 [o]	..	0.6 [o]	..	**39 Meubles, autres activ. de fabricat. n.c.a.**
40 Furniture	..	.. [n]	..	.. [n]	..	.. [n]	..	40 Meubles
41 Other manufacturing n.e.c.	..	.. [n]	..	.. [n]	..	.. [n]	..	41 Autres activités de fabrication n.c.a.
42 Recycling	..	0.0 [a]	..	0.0	..	0.0	..	**42 Récupération**
43 ELECTRICITY, GAS & WATER SUPPLY	..	4.2 [a]	..	3.7	..	2.7	..	**43 DISTRIBUTION D'ELECTRICITE, GAZ & EAU**
44 CONSTRUCTION	..	0.1 [a]	..	0.9	..	0.8	..	**44 CONSTRUCTION**
45 SERVICES SECTOR	..	18.0 [a]	..	27.4	..	65.9	..	**45 SECTEUR DES SERVICES**
46 Wholesale, retail trade & repair of motor vehicles, etc.	..	1.2 [a]	..	4.3	..	5.6	..	**46 Commerce (gros & détail), réparation de véhicules, etc.**
47 Hotels & restaurants	..	0.0	..	0.0	..	0.0	..	**47 Hôtels et restaurants**
48 Transport & storage	..	1.1 [a]	..	4.8	..	2.4	..	**48 Transports, entreposage**
49 Communications	..	7.3 [a,o]	..	3.7 [o]	..	0.1 [o]	..	**49 Communications**
50 Post	..	.. [n]	..	.. [n]	..	.. [n]	..	50 Postes
51 Telecommunications	..	.. [n]	..	.. [n]	..	.. [n]	..	51 Télécommunications
52 Financial intermediation (inc. insur.)	..	1.0 [a]	..	2.3	..	23.0	..	**52 Intermédiation financ. (assurances inc.)**
53 Real estate, renting & busin. activity	..	7.2 [a]	..	11.9	..	33.7	..	**53 Immobil., locat. & serv. aux entreprises**
54 Computer & related activities	..	2.1 [a]	..	3.1	..	14.6 [o]	..	54 Informatique et activités connexes
55 *Software consultancy*	..	..	..	..	..	.. [n]	..	55 *Production de logiciels*
56 *Other computer services n.e.c.*	..	..	..	..	..	.. [n]	..	56 *Autres activités informatiques*
57 Research & development	..	2.6 [a]	..	6.6	..	4.6	..	57 Recherche & développement expérimental
58 Other business activities n.e.c.	..	2.5 [a]	..	2.2	..	14.4	..	58 Autres activ. & serv. aux entreprises
59 Comm., soc. & pers. serv. activ., etc.	..	0.1 [a]	..	0.4	..	1.1	..	**59 Adm. pub., serv. soc. & collec., etc**
TOTAL BUSINESS ENTERPRISE SECTOR	..	**75.6** [a]	..	**107.2**	..	**156.0**	..	**TOTAL SECTEUR DES ENTREPRISES**

Source: OECD, R&D database (DSTI/EAS Division), January 2002.

Source : OCDE, base de données R-D (DSTI/Division AES), janvier 2002.

Business enterprise intramural expenditure on R&D (BERD) - Dépenses intra-muros de R-D des entreprises (DIRDE)

by industry and source of funds - par industrie et par source de financement

Million EUR (1999 PTE euro, EUR from 1999) — *Millions EUR (1999 PTE euro, EUR à partir de 1999)*

10.2 Government - État

	1994	1995	1996	1997	1998	1999	2000	
1 AGRICULTURE, HUNTING & FORESTRY	..	0.0 [a]	..	0.1	..	0.2	..	**1 AGRICULTURE, CHASSE & SYLVICULTURE**
2 MINING	..	0.1 [a]	..	0.0	..	0.0	..	**2 ACTIVITES EXTRACTIVES**
3 MANUFACTURING	..	2.9 [a]	..	8.6	..	8.6	..	**3 ACTIVITES DE FABRICATION**
4 Food, beverages & tobacco	..	0.1 [a,o]	..	0.2 [o]	..	0.2 [o]	..	**4 Produits alimentaires, boissons & tabac**
5 Food products & beverages	..	.. [n]	..	.. [n]	..	.. [n]	..	5 Produits alimentaires & boissons
6 Tobacco products	..	.. [n]	..	.. [n]	..	.. [n]	..	6 Produits à base de tabac
7 Textiles, wear app., fur & leather	..	0.1 [a]	..	0.5	..	0.4	..	**7 Textiles, habill., fourrures & cuirs**
8 Textiles	..	0.0 [a]	..	0.4	..	0.3	..	8 Textiles
9 Wearing apparel & fur	..	0.0	..	0.0	..	0.0	..	9 Articles d'habillement & fourrures
10 Leather products & footwear	..	0.1 [a]	..	0.1	..	0.1	..	10 Articles en cuir & chaussures
11 Wood, paper, printing, publishing	..	0.2 [a]	..	0.5	..	1.2	..	**11 Bois, papier, imprimerie, édition**
12 Wood & cork (not furniture)	..	0.0 [a]	..	0.0	..	1.2	..	12 Bois & liège (sauf meubles)
13 Pulp, paper & paper products	..	0.2 [a]	..	0.0	..	0.1	..	13 Carton, papier & articles
14 Publ., print. & repro. of rec. media	..	0.0 [a]	..	0.4	..	0.0	..	14 Edition, impress. & repro. enregis.
15 Coke, petroleum, nuclear fuel, chemicals, rubber plastics	..	0.9 [a,o]	..	1.4 [o]	..	1.4 [o]	..	**15 Coke, pétrole, combust. nucl., prod. chimi., caout. & mat. plast.**
16 Coke, ref. petrol. prod. & nuclear fuel	..	.. [n]	..	.. [n]	..	.. [n]	..	16 Coke, raffin. pétr. & combust. nucl.
17 Chemicals & chemical products	..	.. [n]	..	.. [n]	..	.. [n]	..	17 Produits chimiques
18 *Chemical prod.(less pharmaceuticals)*	..	.. [n]	..	.. [n]	..	.. [n]	..	18 *Prod. chimi. (moins prod. pharmaceutiques)*
19 *Pharmaceuticals*	..	.. [n]	..	.. [n]	..	.. [n]	..	19 *Produits pharmaceutiques*
20 Rubber & plastic products	..	.. [n]	..	.. [n]	..	.. [n]	..	20 Caoutchouc et matières plastiques
21 Non-metallic mineral products	..	0.0 [a]	..	0.1	..	0.0	..	**21 Produits minéraux non métalliques**
22 Basic metals	..	0.0	..	0.0	..	0.2 [o]	..	**22 Produits métallurgiques de base**
23 Basic metals, ferrous	..	..	..	..	..	.. [n]	..	23 Produits métallur. de base, ferreux
24 Basic metals, non-ferrous	..	..	..	..	..	.. [n]	..	24 Prod. métallur. de base, non ferreux
25 Fabricated metal products	..	0.3 [a]	..	0.1	..	0.3	..	**25 Ouvrages en métaux (sauf mach.& mat.)**
26 Machinery equipment, instruments and transport equipment	..	1.3 [a]	..	5.8	..	4.7	..	**26 Machines & matériel, instruments & mat. de transport**
27 Machinery n.e.c.	..	0.4 [a]	..	1.1	..	2.3	..	27 Machines, n.c.a.
28 Office, account. & computing machinery	..	0.0 [a]	..	0.2	..	0.0	..	28 Mach. de bur., compta. & trait. inform.
29 Electrical machinery	..	0.0 [a]	..	1.1	..	0.8	..	29 Machines électriques
30 Electro. equip. (radio, TV & commun.)	..	0.5 [a,o]	..	2.9 [o]	..	0.5 [o]	..	30 Equip. électro. (radio, tv & communic.)
31 *Electro. comp. (inc. semi-conductors)*	..	.. [n]	..	.. [n]	..	.. [n]	..	31 *Compos. électro. (semi-cond. incl.)*
32 *TV, radio & communications equip.*	..	.. [n]	..	.. [n]	..	.. [n]	..	32 *Appareils radio, tv et de communi.*
33 Instruments	..	0.1 [a]	..	0.2	..	0.8	..	33 Instruments
34 Motor vehicles	..	0.0	..	0.1	..	0.0	..	34 Véhicules automobiles
35 Other transport equipment	..	0.2 [a,o]	..	0.2 [o]	..	0.3 [o]	..	35 Autres matériels de transport
36 *Ships*	..	.. [n]	..	.. [n]	..	.. [n]	..	36 *Navires*
37 *Aerospace*	..	0.0	..	0.0	..	.. [n]	..	37 *Aérospatiale*
38 *Other transport equipment n.e.c.*	..	.. [n]	..	.. [n]	..	.. [n]	..	38 *Autres matériels de transport n.c.a.*
39 Furniture, other manufacturing n.e.c.	..	0.0 [a,o]	..	0.0 [o]	..	0.0 [o]	..	**39 Meubles, autres activ. de fabricat. n.c.a.**
40 Furniture	..	.. [n]	..	.. [n]	..	.. [n]	..	40 Meubles
41 Other manufacturing n.e.c.	..	.. [n]	..	.. [n]	..	.. [n]	..	41 Autres activités de fabrication n.c.a.
42 Recycling	..	0.0 [a]	..	0.0	..	0.0	..	**42 Récupération**
43 ELECTRICITY, GAS & WATER SUPPLY	..	0.5 [a]	..	0.0	..	0.4	..	**43 DISTRIBUTION D'ELECTRICITE, GAZ & EAU**
44 CONSTRUCTION	..	0.0 [a]	..	0.0	..	0.3	..	**44 CONSTRUCTION**
45 SERVICES SECTOR	..	1.4 [a]	..	3.4	..	5.4	..	**45 SECTEUR DES SERVICES**
46 Wholesale, retail trade & repair of motor vehicles, etc.	..	0.1 [a]	..	0.4	..	0.4	..	**46 Commerce (gros & détail), réparation de véhicules, etc.**
47 Hotels & restaurants	..	0.0	..	0.0	..	0.0	..	**47 Hôtels et restaurants**
48 Transport & storage	..	0.0 [a]	..	0.0	..	0.2	..	**48 Transports, entreposage**
49 Communications	..	0.8 [a,o]	..	0.0 [o]	..	0.0 [o]	..	**49 Communications**
50 Post	..	.. [n]	..	.. [n]	..	.. [n]	..	50 Postes
51 Telecommunications	..	.. [n]	..	.. [n]	..	.. [n]	..	51 Télécommunications
52 Financial intermediation (inc. insur.)	..	0.0	..	0.1	..	0.1	..	**52 Intermédiation financ. (assurances inc.)**
53 Real estate, renting & busin. activity	..	0.5 [a]	..	2.6	..	4.2	..	**53 Immobil., locat. & serv. aux entreprises**
54 Computer & related activities	..	0.0 [a,o]	..	0.5 [o]	..	1.5 [o]	..	54 Informatique et activités connexes
55 *Software consultancy*	..	.. [n]	..	.. [n]	..	.. [n]	..	55 *Production de logiciels*
56 *Other computer services n.e.c.*	..	.. [n]	..	.. [n]	..	.. [n]	..	56 *Autres activités informatiques*
57 Research & development	..	0.1 [a]	..	1.4	..	1.0	..	57 Recherche & développement expérimental
58 Other business activities n.e.c.	..	0.4 [a]	..	0.7	..	1.8	..	58 Autres activ. & serv. aux entreprises
59 Comm., soc. & pers. serv. activ., etc.	..	0.0 [a]	..	0.3	..	0.5	..	**59 Adm. pub., serv. soc. & collec., etc**
TOTAL BUSINESS ENTERPRISE SECTOR	**..**	**4.9 [a]**	**..**	**12.1 [a]**	**..**	**14.9**	**..**	**TOTAL SECTEUR DES ENTREPRISES**

Source: OECD, R&D database (DSTI/EAS Division), January 2002.

Source : OCDE, base de données R-D (DSTI/Division AES), janvier 2002.

Business enterprise intramural expenditure on R&D (BERD) - Dépenses intra-muros de R-D des entreprises (DIRDE)

by industry and source of funds - par industrie et par source de financement

Million EUR (1999 PTE euro, EUR from 1999) — *Millions EUR (1999 PTE euro, EUR à partir de 1999)*

10.3 Other national funds - Autres sources nationales

	1994	1995	1996	1997	1998	1999	2000	
1 **AGRICULTURE, HUNTING & FORESTRY**	..	0.0 [a]	..	..	..	..	..	1 **AGRICULTURE, CHASSE & SYLVICULTURE**
2 **MINING**	..	0.0	..	..	..	..	..	2 **ACTIVITES EXTRACTIVES**
3 **MANUFACTURING**	..	0.2 [a]	..	..	..	..	..	3 **ACTIVITES DE FABRICATION**
4 **Food, beverages & tobacco**	..	0.0 [a]	..	..	..	..	..	4 **Produits alimentaires, boissons & tabac**
5 Food products & beverages	..	..	..	..	..	..	..	5 Produits alimentaires & boissons
6 Tobacco products	..	..	..	..	..	..	..	6 Produits à base de tabac
7 **Textiles, wear app., fur & leather**	..	0.0	..	..	..	..	..	7 **Textiles, habill., fourrures & cuirs**
8 Textiles	..	0.0	..	..	..	..	..	8 Textiles
9 Wearing apparel & fur	..	0.0	..	..	..	..	..	9 Articles d'habillement & fourrures
10 Leather products & footwear	..	0.0	..	..	..	..	..	10 Articles en cuir & chaussures
11 **Wood, paper, printing, publishing**	..	0.0	..	..	..	..	..	11 **Bois, papier, imprimerie, édition**
12 Wood & cork (not furniture)	..	0.0	..	..	..	..	..	12 Bois & liège (sauf meubles)
13 Pulp, paper & paper products	..	0.0	..	..	..	..	..	13 Carton, papier & articles
14 Publ., print. & repro. of rec. media	..	0.0	..	..	..	..	..	14 Edition, impress. & repro. enregis.
15 **Coke, petroleum, nuclear fuel, chemicals, rubber plastics**	..	0.0 [a]	..	..	..	..	..	15 **Coke, pétrole, combust. nucl., prod. chimi., caout. & mat. plast.**
16 Coke, ref. petrol. prod. & nuclear fuel	..	..	..	..	..	..	..	16 Coke, raffin. pétr. & combust. nucl.
17 Chemicals & chemical products	..	..	..	..	..	..	..	17 Produits chimiques
18 *Chemical prod.(less pharmaceuticals)*	..	..	..	..	..	..	..	18 *Prod. chimi. (moins prod. pharmaceutiques)*
19 *Pharmaceuticals*	..	..	..	..	..	..	..	19 *Produits pharmaceutiques*
20 Rubber & plastic products	..	..	..	..	..	..	..	20 Caoutchouc et matières plastiques
21 **Non-metallic mineral products**	..	0.0 [a]	..	..	..	..	..	21 **Produits minéraux non métalliques**
22 **Basic metals**	..	0.0 [a]	..	..	..	..	..	22 **Produits métallurgiques de base**
23 Basic metals, ferrous	..	..	..	..	..	..	..	23 Produits métallur. de base, ferreux
24 Basic metals, non-ferrous	..	..	..	..	..	..	..	24 Prod. métallur. de base, non ferreux
25 **Fabricated metal products**	..	0.0	..	..	..	..	..	25 **Ouvrages en métaux (sauf mach.& mat.)**
26 **Machinery equipment, instruments and transport equipment**	..	0.2 [a]	..	..	..	..	..	26 **Machines & matériel, instruments & mat. de transport**
27 Machinery n.e.c.	..	0.0 [a]	..	..	..	..	..	27 Machines, n.c.a.
28 Office, account. & computing machinery	..	0.0	..	..	..	..	..	28 Mach. de bur., compta. & trait. inform.
29 Electrical machinery	..	0.0	..	..	..	..	..	29 Machines électriques
30 Electro. equip. (radio, TV & commun.)	..	0.2 [a]	..	..	..	..	..	30 Equip. électro. (radio, tv & communic.)
31 *Electro. comp. (inc. semi-conductors)*	..	..	..	..	..	..	..	31 *Compos. électro. (semi-cond. incl.)*
32 *TV, radio & communications equip.*	..	..	..	..	..	..	..	32 *Appareils radio, tv et de communi.*
33 Instruments	..	0.0 [a]	..	..	..	..	..	33 Instruments
34 Motor vehicles	..	0.0	..	..	..	..	..	34 Véhicules automobiles
35 Other transport equipment	..	0.0 [a]	..	..	..	..	..	35 Autres matériels de transport
36 *Ships*	..	..	..	..	..	..	..	36 *Navires*
37 *Aerospace*	..	0.0	..	..	..	..	..	37 *Aérospatiale*
38 *Other transport equipment n.e.c.*	..	..	..	..	..	..	..	38 *Autres matériels de transport n.c.a.*
39 **Furniture, other manufacturing n.e.c.**	..	0.0	..	..	..	..	..	39 **Meubles, autres activ. de fabricat. n.c.a.**
40 Furniture	..	..	..	..	..	..	..	40 Meubles
41 Other manufacturing n.e.c.	..	..	..	..	..	..	..	41 Autres activités de fabrication n.c.a.
42 **Recycling**	..	0.0	..	..	..	..	..	42 **Récupération**
43 **ELECTRICITY, GAS & WATER SUPPLY**	..	0.0	..	..	..	..	..	43 **DISTRIBUTION D'ELECTRICITE, GAZ & EAU**
44 **CONSTRUCTION**	..	0.0	..	..	..	..	..	44 **CONSTRUCTION**
45 **SERVICES SECTOR**	..	0.0 [a]	..	..	..	..	..	45 **SECTEUR DES SERVICES**
46 **Wholesale, retail trade & repair of motor vehicles, etc.**	..	0.0	..	..	..	..	..	46 **Commerce (gros & détail), réparation de véhicules, etc.**
47 **Hotels & restaurants**	..	0.0	..	..	..	..	..	47 **Hôtels et restaurants**
48 **Transport & storage**	..	0.0	..	..	..	..	..	48 **Transports, entreposage**
49 **Communications**	..	0.0	..	..	..	..	..	49 **Communications**
50 Post	..	..	..	..	..	..	..	50 Postes
51 Telecommunications	..	..	..	..	..	..	..	51 Télécommunications
52 **Financial intermediation (inc. insur.)**	..	0.0	..	..	..	..	..	52 **Intermédiation financ. (assurances inc.)**
53 **Real estate, renting & busin. activity**	..	0.0	..	..	..	..	..	53 **Immobil., locat. & serv. aux entreprises**
54 Computer & related activities	..	0.0	..	..	..	..	..	54 Informatique et activités connexes
55 *Software consultancy*	..	..	..	..	..	..	..	55 *Production de logiciels*
56 *Other computer services n.e.c.*	..	..	..	..	..	..	..	56 *Autres activités informatiques*
57 Research & development	..	0.0 [a]	..	..	..	..	..	57 Recherche & développement expérimental
58 Other business activities n.e.c.	..	0.0	..	..	..	..	..	58 Autres activ. & serv. aux entreprises
59 **Comm., soc. & pers. serv. activ., etc.**	..	0.0	..	..	..	..	..	59 **Adm. pub., serv. soc. & collec., etc**
TOTAL BUSINESS ENTERPRISE SECTOR	**..**	**0.2**	**..**	**..**	**..**	**..**	**..**	**TOTAL SECTEUR DES ENTREPRISES**

Source: OECD, R&D database (DSTI/EAS Division), January 2002.

Source : OCDE, base de données R-D (DSTI/Division AES), janvier 2002.

Business enterprise intramural expenditure on R&D (BERD) - Dépenses intra-muros de R-D des entreprises (DIRDE)

by industry and source of funds - par industrie et par source de financement

Million EUR (1999 PTE euro, EUR from 1999) — *Millions EUR (1999 PTE euro, EUR à partir de 1999)*

10.4 Funds from abroad - Fonds de l'étranger

	1994	1995	1996	1997	1998	1999	2000	
1 AGRICULTURE, HUNTING & FORESTRY	..	0.0 [a]	..	0.0	..	0.3	..	**1 AGRICULTURE, CHASSE & SYLVICULTURE**
2 MINING	..	0.1 [a]	..	0.0	..	0.0	..	**2 ACTIVITES EXTRACTIVES**
3 MANUFACTURING	..	6.0 [a]	..	1.9	..	3.3	..	**3 ACTIVITES DE FABRICATION**
4 Food, beverages & tobacco	..	0.2 [a,o]	..	0.0 [o]	..	0.0 [o]	..	**4 Produits alimentaires, boissons & tabac**
5 Food products & beverages	..	.. [n]	..	.. [n]	..	.. [n]	..	5 Produits alimentaires & boissons
6 Tobacco products	..	.. [n]	..	.. [n]	..	.. [n]	..	6 Produits à base de tabac
7 Textiles, wear app., fur & leather	..	0.1 [a]	..	0.1	..	0.4	..	**7 Textiles, habill., fourrures & cuirs**
8 Textiles	..	0.1 [a]	..	0.1	..	0.4	..	8 Textiles
9 Wearing apparel & fur	..	0.0	..	0.0	..	0.0	..	9 Articles d'habillement & fourrures
10 Leather products & footwear	..	0.0 [a]	..	0.0	..	0.0	..	10 Articles en cuir & chaussures
11 Wood, paper, printing, publishing	..	0.3 [a]	..	0.0	..	0.2	..	**11 Bois, papier, imprimerie, édition**
12 Wood & cork (not furniture)	..	0.0 [a]	..	0.0	..	0.0	..	12 Bois & liège (sauf meubles)
13 Pulp, paper & paper products	..	0.2 [a]	..	0.0	..	0.2	..	13 Carton, papier & articles
14 Publ., print. & repro. of rec. media	..	0.0 [a]	..	0.0	..	0.0	..	14 Edition, impress. & repro. enregis.
15 Coke, petroleum, nuclear fuel, chemicals, rubber plastics	..	2.2 [a,o]	..	0.3 [o]	..	0.2 [o]	..	**15 Coke, pétrole, combust. nucl., prod. chimi., caout. & mat. plast.**
16 Coke, ref. petrol. prod. & nuclear fuel	..	.. [n]	..	.. [n]	..	.. [n]	..	16 Coke, raffin. pétr. & combust. nucl.
17 Chemicals & chemical products	..	.. [n]	..	.. [n]	..	.. [n]	..	17 Produits chimiques
18 *Chemical prod.(less pharmaceuticals)*	..	.. [n]	..	.. [n]	..	.. [n]	..	18 *Prod. chimi. (moins prod. pharmaceutiques)*
19 *Pharmaceuticals*	..	.. [n]	..	.. [n]	..	.. [n]	..	19 *Produits pharmaceutiques*
20 Rubber & plastic products	..	.. [n]	..	.. [n]	..	.. [n]	..	20 Caoutchouc et matières plastiques
21 Non-metallic mineral products	..	0.0 [a]	..	0.0	..	0.1	..	**21 Produits minéraux non métalliques**
22 Basic metals	..	0.0 [a,o]	..	0.0 [o]	..	0.0 [o]	..	**22 Produits métallurgiques de base**
23 Basic metals, ferrous	..	.. [n]	..	.. [n]	..	.. [n]	..	23 Produits métallur. de base, ferreux
24 Basic metals, non-ferrous	..	.. [n]	..	.. [n]	..	.. [n]	..	24 Prod. métallur. de base, non ferreux
25 Fabricated metal products	..	0.7 [a]	..	0.2	..	0.1	..	**25 Ouvrages en métaux (sauf mach.& mat.)**
26 Machinery equipment, instruments and transport equipment	..	2.4 [a]	..	1.2	..	2.3	..	**26 Machines & matériel, instruments & mat. de transport**
27 Machinery n.e.c.	..	0.8 [a]	..	0.1	..	0.5	..	27 Machines, n.c.a.
28 Office, account. & computing machinery	..	0.0 [a]	..	0.3	..	0.9	..	28 Mach. de bur., compta. & trait. inform.
29 Electrical machinery	..	0.0 [a]	..	0.0	..	0.0	..	29 Machines électriques
30 Electro. equip. (radio, TV & commun.)	..	0.4 [a,o]	..	0.5 [o]	..	0.2 [o]	..	30 Equip. électro. (radio, tv & communic.)
31 *Electro. comp. (inc. semi-conductors)*	..	.. [n]	..	.. [n]	..	.. [n]	..	31 *Compos. électro. (semi-cond. incl.)*
32 *TV, radio & communications equip.*	..	.. [n]	..	.. [n]	..	.. [n]	..	32 *Appareils radio, tv et de communi.*
33 Instruments	..	0.3 [a]	..	0.1	..	0.1	..	33 Instruments
34 Motor vehicles	..	0.0	..	0.0	..	0.2	..	34 Véhicules automobiles
35 Other transport equipment	..	0.9 [a,o]	..	0.2 [o]	..	0.5 [o]	..	35 Autres matériels de transport
36 *Ships*	..	.. [n]	..	.. [n]	..	.. [n]	..	36 *Navires*
37 *Aerospace*	..	0.0	..	0.0	..	.. [n]	..	37 *Aérospatiale*
38 *Other transport equipment n.e.c.*	..	.. [n]	..	.. [n]	..	.. [n]	..	38 *Autres matériels de transport n.c.a.*
39 Furniture, other manufacturing n.e.c.	..	0.0 [a,o]	..	0.0 [o]	..	0.1 [o]	..	**39 Meubles, autres activ. de fabricat. n.c.a.**
40 Furniture	..	.. [n]	..	.. [n]	..	.. [n]	..	40 Meubles
41 Other manufacturing n.e.c.	..	.. [n]	..	.. [n]	..	.. [n]	..	41 Autres activités de fabrication n.c.a.
42 Recycling	..	0.0 [a]	..	0.0	..	0.0	..	**42 Récupération**
43 ELECTRICITY, GAS & WATER SUPPLY	..	1.1 [a]	..	1.7	..	0.3	..	**43 DISTRIBUTION D'ELECTRICITE, GAZ & EAU**
44 CONSTRUCTION	..	0.0 [a]	..	0.0	..	0.1	..	**44 CONSTRUCTION**
45 SERVICES SECTOR	..	8.3 [a]	..	6.7	..	9.9	..	**45 SECTEUR DES SERVICES**
46 Wholesale, retail trade & repair of motor vehicles, etc.	..	0.2 [a]	..	0.1	..	0.8	..	**46 Commerce (gros & détail), réparation de véhicules, etc.**
47 Hotels & restaurants	..	0.0	..	0.0	..	0.0	..	**47 Hôtels et restaurants**
48 Transport & storage	..	0.0 [a]	..	0.0	..	0.0	..	**48 Transports, entreposage**
49 Communications	..	4.2 [a,o]	..	3.5 [o]	..	0.0 [o]	..	**49 Communications**
50 Post	..	.. [n]	..	.. [n]	..	.. [n]	..	50 Postes
51 Telecommunications	..	.. [n]	..	.. [n]	..	.. [n]	..	51 Télécommunications
52 Financial intermediation (inc. insur.)	..	0.0	..	0.0	..	0.1	..	**52 Intermédiation financ. (assurances inc.)**
53 Real estate, renting & busin. activity	..	3.8 [a]	..	2.7	..	8.2	..	**53 Immobil., locat. & serv. aux entreprises**
54 Computer & related activities	..	0.0 [a,o]	..	0.3 [o]	..	0.4 [o]	..	54 Informatique et activités connexes
55 *Software consultancy*	..	.. [n]	..	.. [n]	..	.. [n]	..	55 *Production de logiciels*
56 *Other computer services n.e.c.*	..	.. [n]	..	.. [n]	..	.. [n]	..	56 *Autres activités informatiques*
57 Research & development	..	1.2 [a]	..	1.7	..	1.8	..	57 Recherche & développement expérimental
58 Other business activities n.e.c.	..	2.5 [a]	..	0.6	..	5.9	..	58 Autres activ. & serv. aux entreprises
59 Comm., soc. & pers. serv. activ., etc.	..	0.2 [a]	..	0.4	..	0.9	..	**59 Adm. pub., serv. soc. & collec., etc**
TOTAL BUSINESS ENTERPRISE SECTOR	**..**	**15.5 [a]**	**..**	**10.3 [a]**	**..**	**13.9**	**..**	**TOTAL SECTEUR DES ENTREPRISES**

Source: OECD, R&D database (DSTI/EAS Division), January 2002. — Source : OCDE, base de données R-D (DSTI/Division AES), janvier 2002.

Business enterprise intramural expenditure on R&D (BERD) - Dépenses intra-muros de R-D des entreprises (DIRDE)

by industry and type of costs - par industrie et par type de coûts

Million EUR (1999 PTE euro, EUR from 1999) — *Millions EUR (1999 PTE euro, EUR à partir de 1999)*

	1994	1995	1996	1997	1998	1999	2000	
				11.1 Current expenditure - Dépenses courantes				
1 AGRICULTURE, HUNTING & FORESTRY	..	0.1 [a,s]	..	0.2	..	0.8	..	**1 AGRICULTURE, CHASSE & SYLVICULTURE**
2 MINING	..	0.4 [a,s]	..	0.1	..	0.1	..	**2 ACTIVITES EXTRACTIVES**
3 MANUFACTURING	..	48.0 [a,s]	..	59.9	..	77.1	..	**3 ACTIVITES DE FABRICATION**
4 Food, beverages & tobacco	..	2.9 [a,o,.]	..	2.3 [o]	..	2.3 [o]	..	**4 Produits alimentaires, boissons & tabac**
5 Food products & beverages	..	.. [n]	..	.. [n]	..	.. [n]	..	5 Food products & beverages
6 Tobacco products	..	.. [n]	..	.. [n]	..	.. [n]	..	6 Produits à base de tabac
7 Textiles, wear app., fur & leather	..	2.1 [a,s]	..	2.6	..	4.4	..	**7 Textiles, habill., fourrures & cuirs**
8 Textiles	..	1.8 [a,s]	..	2.4	..	3.0	..	8 Textiles
9 Wearing apparel & fur	..	0.0	..	0.0	..	0.3	..	9 Articles d'habillement & fourrures
10 Leather products & footwear	..	0.2 [a,s]	..	0.1	..	1.0	..	10 Articles en cuir & chaussures
11 Wood, paper, printing, publishing	..	2.8 [a,s]	..	1.9	..	3.1	..	**11 Bois, papier, imprimerie, édition**
12 Wood & cork (not furniture)	..	0.3 [a,s]	..	0.4	..	1.5	..	12 Bois & liège (sauf meubles)
13 Pulp, paper & paper products	..	2.4 [a,s]	..	1.0	..	1.2	..	13 Carton, papier & articles
14 Publ., print. & repro. of rec. media	..	0.2 [a,s]	..	0.4	..	0.4	..	14 Edition, impress. & repro. enregis.
15 Coke, petroleum, nuclear fuel, chemicals, rubber plastics	..	11.4 [a,o,.]	..	16.8 [o]	..	15.9 [o]	..	**15 Coke, pétrole, combust. nucl., prod. chimi., caout. & mat. plast.**
16 Coke, ref. petrol. prod. & nuclear fuel	..	.. [n]	..	.. [n]	..	.. [n]	..	16 Coke, raffin. pétr. & combust. nucl.
17 Chemicals & chemical products	..	.. [n]	..	.. [n]	..	.. [n]	..	17 Produits chimiques
18 *Chemical prod.(less pharmaceuticals)*	..	.. [n]	..	.. [n]	..	.. [n]	..	18 *Prod. chimi. (moins prod. pharmaceutiques)*
19 *Pharmaceuticals*	..	.. [n]	..	.. [n]	..	.. [n]	..	19 *Produits pharmaceutiques*
20 Rubber & plastic products	..	.. [n]	..	.. [n]	..	.. [n]	..	20 Caoutchouc et matières plastiques
21 Non-metallic mineral products	..	0.6 [a,s]	..	0.6	..	0.9	..	**21 Produits minéraux non métalliques**
22 Basic metals	..	0.1 [a,o,.]	..	0.1 [o]	..	0.6 [o]	..	**22 Produits métallurgiques de base**
23 Basic metals, ferrous	..	.. [n]	..	.. [n]	..	.. [n]	..	23 Produits métallur. de base, ferreux
24 Basic metals, non-ferrous	..	.. [n]	..	.. [n]	..	.. [n]	..	24 Prod. métallur. de base, non ferreux
25 Fabricated metal products	..	0.7 [a,s]	..	0.9	..	1.5	..	**25 Ouvrages en métaux (sauf mach.& mat.)**
26 Machinery equipment, instruments and transport equipment	..	27.2 [a,s]	..	34.7	..	47.9	..	**26 Machines & matériel, instruments & mat. de transport**
27 Machinery n.e.c.	..	4.3 [a,s]	..	4.8	..	8.8	..	27 Machines, n.c.a.
28 Office, account. & computing machinery	..	0.1 [a,s]	..	1.2	..	4.1	..	28 Mach. de bur., compta. & trait. inform.
29 Electrical machinery	..	3.6 [a,s]	..	8.6	..	9.3	..	29 Machines électriques
30 Electro. equip. (radio, TV & commun.)	..	15.2 [a,o,.]	..	15.6 [o]	..	16.8 [o]	..	30 Equip. électro. (radio, tv & communic.)
31 *Electro. comp. (inc. semi-conductors)*	..	.. [n]	..	.. [n]	..	.. [n]	..	31 *Compos. électro. (semi-cond. incl.)*
32 *TV, radio & communications equip.*	..	.. [n]	..	.. [n]	..	.. [n]	..	32 *Appareils radio, tv et de communi.*
33 Instruments	..	1.2 [a,s]	..	1.4	..	4.0	..	33 Instruments
34 Motor vehicles	..	0.9 [a,s]	..	1.4	..	3.3	..	34 Véhicules automobiles
35 Other transport equipment	..	1.8 [a,o,.]	..	1.7 [o]	..	1.5 [o]	..	35 Autres matériels de transport
36 *Ships*	..	.. [n]	..	.. [n]	..	.. [n]	..	36 *Navires*
37 *Aerospace*	..	0.0	..	0.0	..	.. [n]	..	37 *Aérospatiale*
38 *Other transport equipment n.e.c.*	..	.. [n]	..	.. [n]	..	.. [n]	..	38 *Autres matériels de transport n.c.a.*
39 Furniture, other manufacturing n.e.c.	..	0.3 [a,o,.]	..	0.1 [o]	..	0.6 [o]	..	**39 Meubles, autres activ. de fabricat. n.c.a.**
40 Furniture	..	.. [n]	..	.. [n]	..	.. [n]	..	40 Meubles
41 Other manufacturing n.e.c.	..	.. [n]	..	.. [n]	..	.. [n]	..	41 Autres activités de fabrication n.c.a.
42 Recycling	..	0.0 [a,s]	..	0.0	..	0.0	..	**42 Récupération**
43 ELECTRICITY, GAS & WATER SUPPLY	..	5.3 [a,s]	..	2.6	..	3.3	..	**43 DISTRIBUTION D'ELECTRICITE, GAZ & EAU**
44 CONSTRUCTION	..	0.1 [a,s]	..	0.8	..	0.6	..	**44 CONSTRUCTION**
45 SERVICES SECTOR	..	23.9 [a,s]	..	23.1	..	68.5	..	**45 SECTEUR DES SERVICES**
46 Wholesale, retail trade & repair of motor vehicles, etc.	..	1.4 [a,s]	..	4.5	..	5.3	..	**46 Commerce (gros & détail), réparation de véhicules, etc.**
47 Hotels & restaurants	..	0.0	..	0.0	..	0.0	..	**47 Hôtels et restaurants**
48 Transport & storage	..	1.2 [a,s]	..	0.8	..	1.8	..	**48 Transports, entreposage**
49 Communications	..	9.9 [a,o,.]	..	4.3 [o]	..	0.1 [o]	..	**49 Communications**
50 Post	..	.. [n]	..	.. [n]	..	.. [n]	..	50 Postes
51 Telecommunications	..	.. [n]	..	.. [n]	..	.. [n]	..	51 Télécommunications
52 Financial intermediation (inc. insur.)	..	0.7 [a,s]	..	1.0	..	18.7	..	**52 Intermédiation financ. (assurances inc.)**
53 Real estate, renting & busin. activity	..	10.5 [a,s]	..	11.6	..	40.9	..	**53 Immobil., locat. & serv. aux entreprises**
54 Computer & related activities	..	1.9 [a,o,.]	..	3.4 [o]	..	15.6 [o]	..	54 Informatique et activités connexes
55 *Software consultancy*	..	.. [n]	..	.. [n]	..	.. [n]	..	55 *Production de logiciels*
56 *Other computer services n.e.c.*	..	.. [n]	..	.. [n]	..	.. [n]	..	56 *Autres activités informatiques*
57 Research & development	..	3.4 [a,s]	..	5.3	..	6.6	..	57 Recherche & développement expérimental
58 Other business activities n.e.c.	..	5.3 [a,s]	..	2.9	..	18.8	..	58 Autres activ. & serv. aux entreprises
59 Comm., soc. & pers. serv. activ., etc.	..	0.3 [a,s]	..	1.0	..	1.7	..	**59 Adm. pub., serv. soc. & collec., etc**
TOTAL BUSINESS ENTERPRISE SECTOR	**..**	**73.0** [s]	**..**	**86.8**	**..**	**150.5**	**..**	**TOTAL SECTEUR DES ENTREPRISES**

Source: OECD, R&D database (DSTI/EAS Division), January 2002.

Source : OCDE, base de données R-D (DSTI/Division AES), janvier 2002.

Business enterprise intramural expenditure on R&D (BERD) - Dépenses intra-muros de R-D des entreprises (DIRDE)

by industry and type of costs - par industrie et par type de coûts

Million EUR (1999 PTE euro, EUR from 1999) — *Millions EUR (1999 PTE euro, EUR à partir de 1999)*

11.2 Capital expenditure - Dépenses en capital

	1994	1995	1996	1997	1998	1999	2000	
1 AGRICULTURE, HUNTING & FORESTRY	..	0.0 [a,s]	..	0.1	..	0.1	..	**1 AGRICULTURE, CHASSE & SYLVICULTURE**
2 MINING	..	0.0 [a,s]	..	0.0	..	0.0	..	**2 ACTIVITES EXTRACTIVES**
3 MANUFACTURING	..	14.2 [a,s]	..	25.5	..	20.9	..	**3 ACTIVITES DE FABRICATION**
4 Food, beverages & tobacco	..	0.3 [a,o,s]	..	0.6 [o]	..	0.2 [o]	..	**4 Produits alimentaires, boissons & tabac**
5 Food products & beverages	..	.. [n]	..	.. [n]	..	.. [n]	..	5 Produits alimentaires & boissons
6 Tobacco products	..	.. [n]	..	.. [n]	..	.. [n]	..	6 Produits à base de tabac
7 Textiles, wear app., fur & leather	..	0.6 [a,s]	..	0.9	..	1.1	..	**7 Textiles, habill., fourrures & cuirs**
8 Textiles	..	0.2 [a,s]	..	0.7	..	0.9	..	8 Textiles
9 Wearing apparel & fur	..	0.0	..	0.0	..	0.0	..	9 Articles d'habillement & fourrures
10 Leather products & footwear	..	0.4 [a,s]	..	0.2	..	0.2	..	10 Articles en cuir & chaussures
11 Wood, paper, printing, publishing	..	1.9 [a,s]	..	1.7	..	5.1	..	**11 Bois, papier, imprimerie, édition**
12 Wood & cork (not furniture)	..	0.8 [a,s]	..	0.1	..	4.9	..	12 Bois & liège (sauf meubles)
13 Pulp, paper & paper products	..	1.0 [a,s]	..	0.0	..	0.1	..	13 Carton, papier & articles
14 Publ., print. & repro. of rec. media	..	0.0 [a,s]	..	1.6	..	0.0	..	14 Edition, impress. & repro. enregis.
15 Coke, petroleum, nuclear fuel, chemicals, rubber plastics	..	2.5 [a,o,s]	..	5.3 [o]	..	3.8 [o]	..	**15 Coke, pétrole, combust. nucl., prod. chimi., caout. & mat. plast.**
16 Coke, ref. petrol. prod. & nuclear fuel	..	.. [n]	..	.. [n]	..	.. [n]	..	16 Coke, raffin. pétr. & combust. nucl.
17 Chemicals & chemical products	..	.. [n]	..	.. [n]	..	.. [n]	..	17 Produits chimiques
18 *Chemical prod.(less pharmaceuticals)*	..	.. [n]	..	.. [n]	..	.. [n]	..	18 *Prod. chimi. (moins prod. pharmaceutiques)*
19 *Pharmaceuticals*	..	.. [n]	..	.. [n]	..	.. [n]	..	19 *Produits pharmaceutiques*
20 Rubber & plastic products	..	.. [n]	..	.. [n]	..	.. [n]	..	20 Caoutchouc et matières plastiques
21 Non-metallic mineral products	..	0.0 [a,s]	..	0.2	..	0.2	..	**21 Produits minéraux non métalliques**
22 Basic metals	..	0.0 [a,s]	..	0.0 [o]	..	0.2 [o]	..	**22 Produits métallurgiques de base**
23 Basic metals, ferrous	..	..	..	.. [n]	..	.. [n]	..	23 Produits métallur. de base, ferreux
24 Basic metals, non-ferrous	..	..	..	.. [n]	..	.. [n]	..	24 Prod. métallur. de base, non ferreux
25 Fabricated metal products	..	1.5 [a,s]	..	0.1	..	2.1	..	**25 Ouvrages en métaux (sauf mach.& mat.)**
26 Machinery equipment, instruments and transport equipment	..	7.4 [a,s]	..	16.6	..	8.2	..	**26 Machines & matériel, instruments & mat. de transport**
27 Machinery n.e.c.	..	0.7 [a,s]	..	1.6	..	2.9	..	27 Machines, n.c.a.
28 Office, account. & computing machinery	..	0.0 [a,s]	..	0.1	..	1.2	..	28 Mach. de bur., compta. & trait. inform.
29 Electrical machinery	..	0.3 [a,s]	..	1.2	..	1.4	..	29 Machines électriques
30 Electro. equip. (radio, TV & commun.)	..	3.0 [a,o,s]	..	1.6 [o]	..	1.2 [o]	..	30 Equip. électro. (radio, tv & communic.)
31 *Electro. comp. (inc. semi-conductors)*	..	.. [n]	..	.. [n]	..	.. [n]	..	31 *Compos. électro. (semi-cond. incl.)*
32 *TV, radio & communications equip.*	..	.. [n]	..	.. [n]	..	.. [n]	..	32 *Appareils radio, tv et de communi.*
33 Instruments	..	0.2 [a,s]	..	0.2	..	0.9	..	33 Instruments
34 Motor vehicles	..	2.4 [a,s]	..	0.1	..	0.5	..	34 Véhicules automobiles
35 Other transport equipment	..	0.8 [a,s]	..	11.7 [o]	..	0.2 [o]	..	35 Autres matériels de transport
36 *Ships*	..	.. [n]	..	.. [n]	..	.. [n]	..	36 *Navires*
37 *Aerospace*	..	0.0	..	0.0	..	.. [n]	..	37 *Aérospatiale*
38 *Other transport equipment n.e.c.*	..	.. [n]	..	.. [n]	..	.. [n]	..	38 *Autres matériels de transport n.c.a.*
39 Furniture, other manufacturing n.e.c.	..	0.0 [a,o,s]	..	0.1 [o]	..	0.0 [o]	..	**39 Meubles, autres activ. de fabricat. n.c.a.**
40 Furniture	..	.. [n]	..	.. [n]	..	.. [n]	..	40 Meubles
41 Other manufacturing n.e.c.	..	.. [n]	..	.. [n]	..	.. [n]	..	41 Autres activités de fabrication n.c.a.
42 Recycling	..	0.0 [a,s]	..	0.0	..	0.0	..	**42 Récupération**
43 ELECTRICITY, GAS & WATER SUPPLY	..	0.5 [a,s]	..	2.8	..	0.1	..	**43 DISTRIBUTION D'ELECTRICITE, GAZ & EAU**
44 CONSTRUCTION	..	0.0 [a,s]	..	0.0	..	0.5	..	**44 CONSTRUCTION**
45 SERVICES SECTOR	..	3.8 [a,s]	..	14.4	..	12.7	..	**45 SECTEUR DES SERVICES**
46 Wholesale, retail trade & repair of motor vehicles, etc.	..	0.2 [a,s]	..	0.3	..	1.5	..	**46 Commerce (gros & détail), réparation de véhicules, etc.**
47 Hotels & restaurants	..	0.0	..	0.0	..	0.0	..	**47 Hôtels et restaurants**
48 Transport & storage	..	0.0 [a,s]	..	4.0	..	0.9	..	**48 Transports, entreposage**
49 Communications	..	2.3 [a,o,s]	..	2.9 [o]	..	0.0 [o]	..	**49 Communications**
50 Post	..	.. [n]	..	.. [n]	..	.. [n]	..	50 Postes
51 Telecommunications	..	.. [n]	..	.. [n]	..	.. [n]	..	51 Télécommunications
52 Financial intermediation (inc. insur.)	..	0.2 [a,s]	..	1.4	..	4.4	..	**52 Intermédiation financ. (assurances inc.)**
53 Real estate, renting & busin. activity	..	0.9 [a,s]	..	5.7	..	5.2	..	**53 Immobil., locat. & serv. aux entreprises**
54 Computer & related activities	..	0.2 [a,o,s]	..	0.6 [o]	..	1.0 [o]	..	54 Informatique et activités connexes
55 *Software consultancy*	..	.. [n]	..	.. [n]	..	.. [n]	..	55 *Production de logiciels*
56 *Other computer services n.e.c.*	..	.. [n]	..	.. [n]	..	.. [n]	..	56 *Autres activités informatiques*
57 Research & development	..	0.5 [a,s]	..	4.4	..	0.8	..	57 Recherche & développement expérimental
58 Other business activities n.e.c.	..	0.2 [a,s]	..	0.7	..	3.4	..	58 Autres activ. & serv. aux entreprises
59 Comm., soc. & pers. serv. activ., etc.	..	0.1 [a,s]	..	0.2	..	0.7	..	**59 Adm. pub., serv. soc. & collec., etc**
TOTAL BUSINESS ENTERPRISE SECTOR	**..**	**17.9** [s]	**..**	**42.8**	**..**	**34.3**	**..**	**TOTAL SECTEUR DES ENTREPRISES**

Source: OECD, R&D database (DSTI/EAS Division), January 2002.

Source : OCDE, base de données R-D (DSTI/Division AES), janvier 2002.

Total business enterprise R&D personnel (BEMP) - Personnel total de R-D du secteur des entreprises (PRDE)

by industry - par industrie

In full time equivalent / *En équivalence plein-temps*

	1994	1995	1996	1997	1998	1999	2000	
1 AGRICULTURE, HUNTING & FORESTRY	..	9 [a]	..	20	..	37	..	**1 AGRICULTURE, CHASSE & SYLVICULTURE**
2 MINING	..	9 [a]	..	2	..	8	..	**2 ACTIVITES EXTRACTIVES**
3 MANUFACTURING	..	1 356 [a]	..	1 340	..	1 920	..	**3 ACTIVITES DE FABRICATION**
4 Food, beverages & tobacco	..	71 [a]	..	76	..	71	..	**4 Produits alimentaires, boissons & tabac**
5 Food products & beverages	..	..	..	..	..	..	..	5 Produits alimentaires & boissons
6 Tobacco products	..	..	..	..	..	..	..	6 Produits à base de tabac
7 Textiles, wear app., fur & leather	..	103 [a]	..	71	..	129	..	**7 Textiles, habill., fourrures & cuirs**
8 Textiles	..	88 [a]	..	63	..	107	..	8 Textiles
9 Wearing apparel & fur	..	0	..	2	..	5	..	9 Articles d'habillement & fourrures
10 Leather products & footwear	..	15 [a]	..	6	..	17	..	10 Articles en cuir & chaussures
11 Wood, paper, printing, publishing	..	70 [a]	..	45	..	88	..	**11 Bois, papier, imprimerie, édition**
12 Wood & cork (not furniture)	..	17 [a]	..	14	..	56	..	12 Bois & liège (sauf meubles)
13 Pulp, paper & paper products	..	36 [a]	..	21	..	21	..	13 Carton, papier & articles
14 Publ., print. & repro. of rec. media	..	18 [a]	..	10	..	11	..	14 Edition, impress. & repro. enregis.
15 Coke, petroleum, nuclear fuel, chemicals, rubber plastics	..	276 [a]	..	321	..	339	..	**15 Coke, pétrole, combust. nucl., prod. chimi., caout. & mat. plast.**
16 Coke, ref. petrol. prod. & nuclear fuel	..	..	..	..	..	..	..	16 Coke, raffin. pétr. & combust. nucl.
17 Chemicals & chemical products	..	..	..	..	..	..	..	17 Produits chimiques
18 *Chemical prod.(less pharmaceuticals)*	..	..	..	..	..	..	..	18 *Prod. chimi. (moins prod. pharmaceutiques)*
19 *Pharmaceuticals*	..	..	..	..	..	..	..	19 *Produits pharmaceutiques*
20 Rubber & plastic products	..	..	..	..	..	..	..	20 Caoutchouc et matières plastiques
21 Non-metallic mineral products	..	22 [a]	..	19	..	37	..	**21 Produits minéraux non métalliques**
22 Basic metals	..	0 [a]	..	8	..	13	..	**22 Produits métallurgiques de base**
23 Basic metals, ferrous	..	..	..	..	..	..	..	23 Produits métallur. de base, ferreux
24 Basic metals, non-ferrous	..	..	..	..	..	..	..	24 Prod. métallur. de base, non ferreux
25 Fabricated metal products	..	19 [a]	..	39	..	46	..	**25 Ouvrages en métaux (sauf mach.& mat.)**
26 Machinery equipment, instruments and transport equipment	..	783 [a]	..	752	..	1 173	..	**26 Machines & matériel, instruments & mat. de transport**
27 Machinery n.e.c.	..	132 [a]	..	177	..	224	..	27 Machines, n.c.a.
28 Office, account. & computing machinery	..	3 [a]	..	34	..	58	..	28 Mach. de bur., compta. & trait. inform.
29 Electrical machinery	..	278 [a]	..	166	..	255	..	29 Machines électriques
30 Electro. equip. (radio, TV & commun.)	..	253 [a]	..	285	..	416	..	30 Equip. électro. (radio, tv & communic.)
31 *Electro. comp. (inc. semi-conductors)*	..	..	..	..	..	..	..	31 *Compos. électro. (semi-cond. incl.)*
32 *TV, radio & communications equip.*	..	..	..	..	..	..	..	32 *Appareils radio, tv et de communi.*
33 Instruments	..	44 [a]	..	42	..	112	..	33 Instruments
34 Motor vehicles	..	43 [a]	..	21	..	78	..	34 Véhicules automobiles
35 Other transport equipment	..	31 [a]	..	26	..	30	..	35 Autres matériels de transport
36 *Ships*	..	..	..	..	..	..	..	36 *Navires*
37 *Aerospace*	..	0	..	0	..	..	..	37 *Aérospatiale*
38 *Other transport equipment n.e.c.*	..	..	..	..	..	..	..	38 *Autres matériels de transport n.c.a.*
39 Furniture, other manufacturing n.e.c.	..	11 [a]	..	5	..	22	..	**39 Meubles, autres activ. de fabricat. n.c.a.**
40 Furniture	..	..	..	..	..	..	..	40 Meubles
41 Other manufacturing n.e.c.	..	..	..	..	..	..	..	41 Autres activités de fabrication n.c.a.
42 Recycling	..	2 [a]	..	5	..	2	..	**42 Récupération**
43 ELECTRICITY, GAS & WATER SUPPLY	..	71 [a]	..	40	..	30	..	**43 DISTRIBUTION D'ELECTRICITE, GAZ & EAU**
44 CONSTRUCTION	..	4 [a]	..	11	..	21	..	**44 CONSTRUCTION**
45 SERVICES SECTOR	..	469 [a]	..	568	..	1 244	..	**45 SECTEUR DES SERVICES**
46 Wholesale, retail trade & repair of motor vehicles, etc.	..	24 [a]	..	63	..	134	..	**46 Commerce (gros & détail), réparation de véhicules, etc.**
47 Hotels & restaurants	..	0	..	0	..	0	..	**47 Hôtels et restaurants**
48 Transport & storage	..	39 [a]	..	22	..	35	..	**48 Transports, entreposage**
49 Communications	..	133 [a]	..	97	..	2	..	**49 Communications**
50 Post	..	..	..	..	..	..	..	50 Postes
51 Telecommunications	..	..	..	..	..	..	..	51 Télécommunications
52 Financial intermediation (inc. insur.)	..	30 [a]	..	13	..	132	..	**52 Intermédiation financ. (assurances inc.)**
53 Real estate, renting & busin. activity	..	228 [a]	..	331	..	893	..	**53 Immobil., locat. & serv. aux entreprises**
54 Computer & related activities	..	42 [a]	..	89	..	480	..	54 Informatique et activités connexes
55 *Software consultancy*	..	..	..	..	..	..	..	55 *Production de logiciels*
56 *Other computer services n.e.c.*	..	..	..	..	..	..	..	56 *Autres activités informatiques*
57 Research & development	..	74 [a]	..	131	..	126	..	57 Recherche & développement expérimental
58 Other business activities n.e.c.	..	111 [a]	..	112	..	287	..	58 Autres activ. & serv. aux entreprises
59 Comm., soc. & pers. serv. activ., etc.	..	15 [a]	..	41	..	48	..	**59 Adm. pub., serv. soc. & collec., etc**
TOTAL BUSINESS ENTERPRISE SECTOR	**..**	**1 917 [a]**	**..**	**1 981**	**..**	**3 260**	**..**	**TOTAL SECTEUR DES ENTREPRISES**

Source: OECD, R&D database (DSTI/EAS Division), January 2002.

Source : OCDE, base de données R-D (DSTI/Division AES), janvier 2002.

PORTUGAL	Table 13 - Tableau 13	PORTUGAL

RSE in the business enterprise sector - Chercheurs dans le secteur des entreprises

by industry - par industrie

In full time equivalent — *En équivalence plein-temps*

	1994	1995	1996	1997	1998	1999	2000	
1 AGRICULTURE, HUNTING & FORESTRY	..	4 [a]	..	7	..	18	..	**1 AGRICULTURE, CHASSE & SYLVICULTURE**
2 MINING	..	3 [a]	..	2	..	5	..	**2 ACTIVITES EXTRACTIVES**
3 MANUFACTURING	..	650 [a]	..	754	..	1 117	..	**3 ACTIVITES DE FABRICATION**
4 Food, beverages & tobacco	..	38 [a]	..	51	..	41	..	**4 Produits alimentaires, boissons & tabac**
5 Food products & beverages	..	..	..	..	..	..	..	5 Produits alimentaires & boissons
6 Tobacco products	..	..	..	..	..	..	..	6 Produits à base de tabac
7 Textiles, wear app., fur & leather	..	21 [a]	..	29	..	33	..	**7 Textiles, habill., fourrures & cuirs**
8 Textiles	..	20 [a]	..	25	..	27	..	8 Textiles
9 Wearing apparel & fur	..	0	..	1	..	2	..	9 Articles d'habillement & fourrures
10 Leather products & footwear	..	1 [a]	..	3	..	4	..	10 Articles en cuir & chaussures
11 Wood, paper, printing, publishing	..	33 [a]	..	23	..	44	..	**11 Bois, papier, imprimerie, édition**
12 Wood & cork (not furniture)	..	3 [a]	..	6	..	25	..	12 Bois & liège (sauf meubles)
13 Pulp, paper & paper products	..	24 [a]	..	12	..	13	..	13 Carton, papier & articles
14 Publ., print. & repro. of rec. media	..	7 [a]	..	5	..	6	..	14 Edition, impress. & repro. enregis.
15 Coke, petroleum, nuclear fuel, chemicals, rubber plastics	..	113 [a]	..	159	..	186	..	**15 Coke, pétrole, combust. nucl., prod. chimi., caout. & mat. plast.**
16 Coke, ref. petrol. prod. & nuclear fuel	..	..	..	..	..	..	..	16 Coke, raffin. pétr. & combust. nucl.
17 Chemicals & chemical products	..	..	..	..	..	..	..	17 Produits chimiques
18 *Chemical prod.(less pharmaceuticals)*	..	..	..	..	..	..	..	18 *Prod. chimi. (moins prod. pharmaceutiques)*
19 *Pharmaceuticals*	..	..	..	..	..	..	..	19 *Produits pharmaceutiques*
20 Rubber & plastic products	..	..	..	..	..	..	..	20 Caoutchouc et matières plastiques
21 Non-metallic mineral products	..	8 [a]	..	9	..	21	..	**21 Produits minéraux non métalliques**
22 Basic metals	..	0 [a]	..	3	..	6	..	**22 Produits métallurgiques de base**
23 Basic metals, ferrous	..	..	..	..	..	..	..	23 Produits métallur. de base, ferreux
24 Basic metals, non-ferrous	..	..	..	..	..	..	..	24 Prod. métallur. de base, non ferreux
25 Fabricated metal products	..	7 [a]	..	10	..	18	..	**25 Ouvrages en métaux (sauf mach.& mat.)**
26 Machinery equipment, instruments and transport equipment	..	423 [a]	..	465	..	759	..	**26 Machines & matériel, instruments & mat. de transport**
27 Machinery n.e.c.	..	44 [a]	..	68	..	87	..	27 Machines, n.c.a.
28 Office, account. & computing machinery	..	3 [a]	..	22	..	34	..	28 Mach. de bur., compta. & trait. inform.
29 Electrical machinery	..	114 [a]	..	87	..	138	..	29 Machines électriques
30 Electro. equip. (radio, TV & commun.)	..	200 [a]	..	216	..	352	..	30 Equip. électro. (radio, tv & communic.)
31 *Electro. comp. (inc. semi-conductors)*	..	..	..	..	..	..	..	31 *Compos. électro. (semi-cond. incl.)*
32 *TV, radio & communications equip.*	..	..	..	..	..	..	..	32 *Appareils radio, tv et de communi.*
33 Instruments	..	31 [a]	..	38	..	90	..	33 Instruments
34 Motor vehicles	..	5 [a]	..	10	..	34	..	34 Véhicules automobiles
35 Other transport equipment	..	27 [a]	..	23	..	24	..	35 Autres matériels de transport
36 *Ships*	..	..	..	..	..	..	..	36 *Navires*
37 *Aerospace*	..	0	..	0	..	..	..	37 *Aérospatiale*
38 *Other transport equipment n.e.c.*	..	..	..	..	..	..	..	38 *Autres matériels de transport n.c.a.*
39 Furniture, other manufacturing n.e.c.	..	5 [a]	..	4	..	8	..	**39 Meubles, autres activ. de fabricat. n.c.a.**
40 Furniture	..	..	..	..	..	..	..	40 Meubles
41 Other manufacturing n.e.c.	..	3	..	..	..	..	..	41 Autres activités de fabrication n.c.a.
42 Recycling	..	2 [a]	..	1	..	1	..	**42 Récupération**
43 ELECTRICITY, GAS & WATER SUPPLY	..	52 [a]	..	36	..	28	..	**43 DISTRIBUTION D'ELECTRICITE, GAZ & EAU**
44 CONSTRUCTION	..	2 [a]	..	5	..	7	..	**44 CONSTRUCTION**
45 SERVICES SECTOR	..	364 [a]	..	389	..	819	..	**45 SECTEUR DES SERVICES**
46 Wholesale, retail trade & repair of motor vehicles, etc.	..	17 [a]	..	47	..	65	..	**46 Commerce (gros & détail), réparation de véhicules, etc.**
47 Hotels & restaurants	..	0	..	0	..	0	..	**47 Hôtels et restaurants**
48 Transport & storage	..	18 [a]	..	17	..	21	..	**48 Transports, entreposage**
49 Communications	..	120 [a]	..	90	..	2	..	**49 Communications**
50 Post	..	..	..	..	..	..	..	50 Postes
51 Telecommunications	..	..	..	..	..	..	..	51 Télécommunications
52 Financial intermediation (inc. insur.)	..	18 [a]	..	12	..	119	..	**52 Intermédiation financ. (assurances inc.)**
53 Real estate, renting & busin. activity	..	178 [a]	..	196	..	580	..	**53 Immobil., locat. & serv. aux entreprises**
54 Computer & related activities	..	31 [a]	..	67	..	294	..	54 Informatique et activités connexes
55 *Software consultancy*	..	..	..	..	..	..	..	55 *Production de logiciels*
56 *Other computer services n.e.c.*	..	..	..	..	..	..	..	56 *Autres activités informatiques*
57 Research & development	..	56 [a]	..	62	..	64	..	57 Recherche & développement expérimental
58 Other business activities n.e.c.	..	91 [a]	..	67	..	222	..	58 Autres activ. & serv. aux entreprises
59 Comm., soc. & pers. serv. activ., etc.	..	13 [a]	..	27	..	32	..	**59 Adm. pub., serv. soc. & collec., etc**
TOTAL BUSINESS ENTERPRISE SECTOR	**..**	**1 076 [a]**	**..**	**1 193**	**..**	**1 994**	**..**	**TOTAL SECTEUR DES ENTREPRISES**

Source: OECD, R&D database (DSTI/EAS Division), January 2002.

Source : OCDE, base de données R-D (DSTI/Division AES), janvier 2002.

Table 14 - Tableau 14

PORTUGAL

Higher education intramural expenditure on R&D - Dépenses intra-muros de R-D de l'enseignement supérieur
(HERD) by field of science - (DIRDES) par discipline scientifique

	1994	1995	1996	1997	1998	1999	2000	
	14.1 Million EUR (1999 PTE euro, EUR from 1999) - Millions EUR (1999 PTE euro, EUR à partir de 1999)							
Natural sciences	..	54.7 [s]	..	77.3 [a]	..	92.3	..	Sciences exactes et naturelles
Engineering	..	31.6 [s]	..	45.9 [a]	..	64.7	..	Sciences de l'ingénieur
Medical sciences	..	14.6 [s]	..	20.6 [a]	..	31.1	..	Sciences médicales
Agricultural sciences	..	14.2 [s]	..	18.7 [a]	..	25.9	..	Sciences agricoles
Sub-total NSE	..	115.0 [s]	..	162.5 [a]	..	214.1	..	Sous-total SENI
Social sciences	..	..	..	..	..	..	..	Sciences sociales
Humanities	..	..	..	..	..	..	..	Sciences humaines
Sub-total SSH	..	39.5 [s]	..	68.5 [a]	..	100.3	..	Sous-total SSH
Total NSE+SSH	..	170.4 [a]	..	231.0	..	314.4	..	*Total SENI+SSH*
	14.2 Million constant USD (1995 prices and PPPs) - Millions de USD constants (prix et PPA de 1995)							
Natural sciences	..	92.1 [s]	..	121.8 [a]	..	135.7	..	Sciences exactes et naturelles
Engineering	..	53.1 [s]	..	72.3 [a]	..	95.2	..	Sciences de l'ingénieur
Medical sciences	..	24.5 [s]	..	32.5 [a]	..	45.7	..	Sciences médicales
Agricultural sciences	..	23.9 [s]	..	29.5 [a]	..	38.0	..	Sciences agricoles
Sub-total NSE	..	193.5 [s]	..	256.2 [a]	..	314.7	..	Sous-total SENI
Social sciences	..	..	..	..	..	..	..	Sciences sociales
Humanities	..	..	..	..	..	..	..	Sciences humaines
Sub-total SSH	..	66.4 [s]	..	107.9 [a]	..	147.5	..	Sous-total SSH
Total NSE+SSH	..	286.9 [a]	..	364.0	..	462.1	..	*Total SENI+SSH*

Table 15 - Tableau 15

Higher education intramural expenditure on R&D - Dépenses intra-muros de R-D de l'enseignement supérieur
(HERD) by field of science and type of costs - (DIRDES) par discipline scientifique et par type de coûts

Million EUR (1999 PTE euro, EUR from 1999) — *Millions EUR (1999 PTE euro, EUR à partir de 1999)*

	1994	1995	1996	1997	1998	1999	2000	
Current expenditure								**Dépenses courantes**
Natural sciences	..	..	..	71.9	..	85.6	..	Sciences exactes et naturelles
Engineering	..	..	..	42.4	..	58.9	..	Sciences de l'ingénieur
Medical sciences	..	..	..	19.5	..	29.2	..	Sciences médicales
Agricultural sciences	..	..	..	17.2	..	22.9	..	Sciences agricoles
Sub-total NSE	..	..	..	150.9	..	196.6	..	Sous-total SENI
Social sciences	..	..	..	..	..	..	..	Sciences sociales
Humanities	..	..	..	..	..	..	..	Sciences humaines
Sub-total SSH	..	..	..	65.4	..	95.0	..	Sous-total SSH
Total NSE+SSH	..	144.8 [s]	..	216.3	..	291.5	..	*Total SENI+SSH*
Capital expenditure								**Dépenses en capital**
Natural sciences	..	..	..	5.4	..	6.8	..	Sciences exactes et naturelles
Engineering	..	..	..	3.5	..	5.9	..	Sciences de l'ingénieur
Medical sciences	..	..	..	1.1	..	1.9	..	Sciences médicales
Agricultural sciences	..	..	..	1.5	..	3.0	..	Sciences agricoles
Sub-total NSE	..	..	..	11.7	..	17.5	..	Sous-total SENI
Social sciences	..	..	..	..	..	..	..	Sciences sociales
Humanities	..	..	..	..	..	..	..	Sciences humaines
Sub-total SSH	..	..	..	3.0	..	5.4	..	Sous-total SSH
Total NSE+SSH	..	9.7 [s]	..	14.7	..	22.8	..	*Total SENI+SSH*

Source: OECD, R&D database (DSTI/EAS Division), January 2002.

Source : OCDE, base de données R-D (DSTI/Division AES), janvier 2002.

Higher education intramural expenditure on R&D - Dépenses intra-muros de R-D de l'enseignement supérieur

(HERD) by field of science and source of funds - (DIRDES) par discipline scientifique et par source de financement

Million EUR (1999 PTE euro, EUR from 1999) *Millions EUR (1999 PTE euro, EUR à partir de 1999)*

	1994	1995	1996	1997	1998	1999	2000	
Direct government								**État direct**
Natural sciences	..	..	..	..	..	..	..	Sciences exactes et naturelles
Engineering	..	..	..	..	..	..	..	Sciences de l'ingénieur
Medical sciences	..	..	..	..	..	..	..	Sciences médicales
Agricultural sciences	..	..	..	..	..	..	..	Sciences agricoles
Sub-total NSE	..	..	..	..	..	..	..	Sous-total SENI
Social sciences	..	..	..	..	..	..	..	Sciences sociales
Humanities	..	..	..	..	..	..	..	Sciences humaines
Sub-total SSH	..	..	..	..	..	..	..	Sous-total SSH
Total NSE+SSH	..	..	..	..	..	..	..	*Total SENI+SSH*
General university funds								**Fonds généraux des universités**
Natural sciences	..	..	..	..	..	..	..	Sciences exactes et naturelles
Engineering	..	..	..	..	..	..	..	Sciences de l'ingénieur
Medical sciences	..	..	..	..	..	..	..	Sciences médicales
Agricultural sciences	..	..	..	..	..	..	..	Sciences agricoles
Sub-total NSE	..	..	..	..	..	..	..	Sous-total SENI
Social sciences	..	..	..	..	..	..	..	Sciences sociales
Humanities	..	..	..	..	..	..	..	Sciences humaines
Sub-total SSH	..	..	..	..	..	..	..	Sous-total SSH
Total NSE+SSH	..	..	..	..	..	..	..	*Total SENI+SSH*
Sub-total government								**Sous-total État**
Natural sciences	..	..	..	72.0 [a]	..	85.5	..	Sciences exactes et naturelles
Engineering	..	..	..	40.0 [a]	..	57.0	..	Sciences de l'ingénieur
Medical sciences	..	..	..	18.5 [a]	..	28.5	..	Sciences médicales
Agricultural sciences	..	..	..	17.5 [a]	..	23.6	..	Sciences agricoles
Sub-total NSE	..	..	..	148.0 [a]	..	194.6	..	Sous-total SENI
Social sciences	..	..	..	..	..	..	..	Sciences sociales
Humanities	..	..	..	..	..	..	..	Sciences humaines
Sub-total SSH	..	..	..	59.4 [a]	..	84.4	..	Sous-total SSH
Total NSE+SSH	..	147.1 [a]	..	207.5 [a]	..	279.0	..	*Total SENI+SSH*
Higher education								**Enseignement supérieur**
Natural sciences	..	..	..	1.2 [a]	..	1.2	..	Sciences exactes et naturelles
Engineering	..	..	..	1.7 [a]	..	2.0	..	Sciences de l'ingénieur
Medical sciences	..	..	..	0.5 [a]	..	0.8	..	Sciences médicales
Agricultural sciences	..	..	..	0.4 [a]	..	0.6	..	Sciences agricoles
Sub-total NSE	..	..	..	3.9 [a]	..	4.5	..	Sous-total SENI
Social sciences	..	..	..	..	..	..	..	Sciences sociales
Humanities	..	..	..	..	..	..	..	Sciences humaines
Sub-total SSH	..	..	..	5.0 [a]	..	6.2	..	Sous-total SSH
Total NSE+SSH	..	5.6 [a]	..	8.9	..	10.7	..	*Total SENI+SSH*
Private non-profit								**Institutions sans but lucratif**
Natural sciences	..	..	..	0.8 [a]	..	0.8	..	Sciences exactes et naturelles
Engineering	..	..	..	0.6 [a]	..	0.6	..	Sciences de l'ingénieur
Medical sciences	..	..	..	0.6 [a]	..	0.7	..	Sciences médicales
Agricultural sciences	..	..	..	0.0 [a]	..	0.1	..	Sciences agricoles
Sub-total NSE	..	..	..	2.0 [a]	..	2.2	..	Sous-total SENI
Social sciences	..	..	..	..	..	..	..	Sciences sociales
Humanities	..	..	..	..	..	..	..	Sciences humaines
Sub-total SSH	..	..	..	1.1 [a]	..	8.1	..	Sous-total SSH
Total NSE+SSH	..	3.1 [a]	..	3.1	..	10.3	..	*Total SENI+SSH*
Business entreprise								**Entreprises**
Natural sciences	..	..	..	0.8 [a]	..	1.3	..	Sciences exactes et naturelles
Engineering	..	..	..	0.9 [a]	..	1.5	..	Sciences de l'ingénieur
Medical sciences	..	..	..	0.1 [a]	..	0.4	..	Sciences médicales
Agricultural sciences	..	..	..	0.2 [a]	..	0.1	..	Sciences agricoles
Sub-total NSE	..	..	..	2.0 [a]	..	3.4	..	Sous-total SENI
Social sciences	..	..	..	..	..	..	..	Sciences sociales
Humanities	..	..	..	..	..	..	..	Sciences humaines
Sub-total SSH	..	..	..	2.0 [a]	..	0.5	..	Sous-total SSH
Total NSE+SSH	..	1.6 [a]	..	4.0	..	3.9	..	*Total SENI+SSH*
Funds from abroad								**Fonds de l'étranger**
Natural sciences	..	..	..	2.5 [a]	..	3.6	..	Sciences exactes et naturelles
Engineering	..	..	..	2.7 [a]	..	3.6	..	Sciences de l'ingénieur
Medical sciences	..	..	..	0.9 [a]	..	0.8	..	Sciences médicales
Agricultural sciences	..	..	..	0.6 [a]	..	1.5	..	Sciences agricoles
Sub-total NSE	..	..	..	6.6 [a]	..	9.4	..	Sous-total SENI
Social sciences	..	..	..	..	..	..	..	Sciences sociales
Humanities	..	..	..	..	..	..	..	Sciences humaines
Sub-total SSH	..	..	..	0.9 [a]	..	1.1	..	Sous-total SSH
Total NSE+SSH	..	13.0 [a]	..	7.5 [a]	..	10.5	..	*Total SENI+SSH*

Source: OECD, R&D database (DSTI/EAS Division), January 2002.

Source : OCDE, base de données R-D (DSTI/Division AES), janvier 2002.

PORTUGAL Table 17 - Tableau 17 **PORTUGAL**

Higher education R&D personnel (HEMP) - Personnel de R-D du secteur de l'enseignement supérieur

by field of science and occupation - par discipline scientifique et par profession

In full-time equivalent — *En équivalence plein-temps*

	1994	1995	1996	1997	1998	1999	2000	
Total R&D personnel								**Personnel total de R-D**
Natural sciences	..	2 076 [s]	..	..	..	..	..	Sciences exactes et naturelles
Engineering	..	1 109 [s]	..	..	..	..	..	Sciences de l'ingénieur
Medical sciences	..	621 [s]	..	..	..	..	..	Sciences médicales
Agricultural sciences	..	595 [s]	..	..	..	..	..	Sciences agricoles
Sub-total NSE	..	4 401 [s]	..	..	..	..	..	Sous-total SENI
Social sciences	..	..	..	..	..	..	..	Sciences sociales
Humanities	..	..	..	..	..	..	..	Sciences humaines
Sub-total SSH	..	1 493 [s]	..	..	..	..	..	Sous-total SSH
Total NSE+SSH	..	6 484 [a]	..	8 442	..	9 187	..	*Total SENI+SSH*
Researchers								**Chercheurs**
Natural sciences	..	..	..	2 519	..	2 734	..	Sciences exactes et naturelles
Engineering	..	..	..	1 472	..	1 669	..	Sciences de l'ingénieur
Medical sciences	..	..	..	699	..	624	..	Sciences médicales
Agricultural sciences	..	..	..	506	..	571	..	Sciences agricoles
Sub-total NSE	..	..	..	5 196	..	5 598	..	Sous-total SENI
Social sciences	..	..	..	..	..	..	..	Sciences sociales
Humanities	..	..	..	..	..	..	..	Sciences humaines
Sub-total SSH	..	..	..	2 237	..	2 606	..	Sous-total SSH
Total NSE+SSH	..	5 850 [a]	..	7 475	..	8 243	..	*Total SENI+SSH*
Technicians								**Technicians**
Natural sciences	..	..	..	..	..	..	..	Sciences exactes et naturelles
Engineering	..	..	..	..	..	..	..	Sciences de l'ingénieur
Medical sciences	..	..	..	..	..	..	..	Sciences médicales
Agricultural sciences	..	..	..	..	..	..	..	Sciences agricoles
Sub-total NSE	..	..	..	..	..	..	..	Sous-total SENI
Social sciences	..	..	..	..	..	..	..	Sciences sociales
Humanities	..	..	..	..	..	..	..	Sciences humaines
Sub-total SSH	..	..	..	..	..	..	..	Sous-total SSH
Total NSE+SSH	..	297 [a]	..	405	..	467	..	*Total SENI+SSH*
Other								**Autres**
Natural sciences	..	..	..	..	..	..	..	Sciences exactes et naturelles
Engineering	..	..	..	..	..	..	..	Sciences de l'ingénieur
Medical sciences	..	..	..	..	..	..	..	Sciences médicales
Agricultural sciences	..	..	..	..	..	..	..	Sciences agricoles
Sub-total NSE	..	..	..	..	..	..	..	Sous-total SENI
Social sciences	..	..	..	..	..	..	..	Sciences sociales
Humanities	..	..	..	..	..	..	..	Sciences humaines
Sub-total SSH	..	..	..	..	..	..	..	Sous-total SSH
Total NSE+SSH	..	338 [a]	..	562	..	477	..	*Total SENI+SSH*

Table 18 - Tableau 18

Government budget appropriations or outlays for R&D (GBAORD) - Crédits budgétaires publics de R-D (CBPRD)

by socio-economic objective - par objectif socio-économique

Million EUR (1999 PTE euro, EUR from 1999) — *Millions EUR (1999 PTE euro, EUR à partir de 1999)*

	1995	1996	1997	1998	1999	2000	2001	
Agriculture, forestry and fishing	39.1	49.3	62.8	69.8	77.9	92.7	104.0 [p]	Agriculture, sylviculture et pêche
Industrial development	47.3	63.9	55.3	81.7	99.7	93.0	79.7 [p]	Développement industriel
Energy	1.9	2.1	6.6	7.4	5.5	6.2	7.3 [p]	Energie
Transport & telecommunications	..	..	..	..	..	..	..	*Transports & télécommunications*
Urban & rural planning	..	..	..	..	..	..	..	*Aménagement urbain et rural*
ST Infrastructure	9.5	10.9	13.4	14.9	47.5	53.9	63.4 [p]	ST Infrastructure
Prevention of pollution	..	..	..	..	..	..	..	*Prévention de la pollution*
Identificat. & treatment of pollution	..	..	..	..	..	..	..	*Détect. pollut. & mesures antipollut.*
ST Environment	16.1	18.4	20.6	23.0	27.7	31.4	36.9 [p]	ST Environnement
Health	18.1	20.7	25.2	28.0	41.8	47.3	55.7 [p]	Santé
Social development & services	16.2	18.5	20.4	22.8	21.6	24.5	28.8 [p]	Développement et services sociaux
Earth and atmosphere	14.4	16.5	12.9	14.3	10.2	11.5	13.5 [p]	Milieu terrestre et atmosphère
Advancement of research	38.0	43.4	35.6	39.6	51.5	58.4	68.7 [p]	*Promotion de la recherche*
General university funds	150.0	178.0	195.4	208.4	225.5	257.9	277.1 [p]	*Fonds généraux des universités*
ST Advancement of knowledge	188.0	221.5	230.9	248.0	276.9	316.3	345.7 [p]	ST promotion gén. des connaiss.
Civil space	0.0	0.0	1.5	1.7	3.4	3.9	4.6 [p]	Espace civil
Defence	9.5	9.4	6.6	7.2	10.3	8.4	9.8 [p]	Défense
Not elsewhere classified	4.1	4.7	20.3	22.7	21.4	24.2	28.5 [p]	Non classé
Total	**363.9**	**435.8**	**476.7**	**541.4**	**643.8**	**713.4**	**777.8 [p]**	**Total**

Source: OECD, R&D database (DSTI/EAS Division), January 2002.

Source : OCDE, base de données R-D (DSTI/Division AES), janvier 2002.

PORTUGAL Table 19 - Tableau 19 **PORTUGAL**

Technology balance of payments - Balance des paiements technologiques

by type of transfer - selon la nature des transferts

Million EUR (1999 PTE euro, EUR from 1999) *Millions EUR (1999 PTE euro, EUR à partir de 1999)*

	1983	1984	1985	1986	1987	1988	1989	1990	1991	
Receipts										**Recettes**
Related firms	..	..	..	..	..	..	..	..	..	Firmes affiliées
Unrelated firms	..	..	..	..	..	..	..	..	..	Firmes non affiliées
Other	..	..	..	..	..	..	..	..	..	Autre
Total	2	2	3	..	..	..	..	..	..	*Total*
Payments										**Paiements**
Related firms	..	..	..	..	..	..	..	..	..	Firmes affiliées
Unrelated firms	..	..	..	..	..	..	..	..	..	Firmes non affiliées
Other	..	..	..	..	..	..	..	..	..	Autre
Total	21	21	28	..	..	..	..	..	..	*Total*

	1992	1993	1994	1995	1996	1997	1998	1999	2000	
Receipts										**Recettes**
Related firms	..	..	..	..	..	..	..	..	..	Firmes affiliées
Unrelated firms	..	..	..	..	..	..	..	..	..	Firmes non affiliées
Other	..	..	..	..	..	..	..	..	..	Autre
Total	..	102	119	105	138	192	256	292 p	..	*Total*
Payments										**Paiements**
Related firms	..	..	..	..	..	..	..	..	..	Firmes affiliées
Unrelated firms	..	..	..	..	..	..	..	..	..	Firmes non affiliées
Other	..	..	..	..	..	..	..	..	..	Autre
Total	..	305	451	405	410	533	687	759 p	..	*Total*

Source: OECD, TBPdatabase (DSTI/EAS Division), Januay 2002.

Source : OCDE, base de données TBP (DSTI/Division AES), janvier 2002.

SLOVAK REPUBLIC — Table 1 - Tableau 1 — RÉPUBLIQUE SLOVAQUE

Gross domestic expenditure on R&D (GERD) - Dépenses intérieures brutes de R-D (DIRD)

by sector of performance and source of funds - par secteur d'exécution et par source de financement

	1994	1995	1996	1997	1998	1999	2000	
	1.1 Million SKK - Millions SKK							
Business enterprise								**Entreprises**
financed by								*financé par*
Business enterprise	1 985.0 [a]	2 540.0	2 768.0	4 787.0 [a]	3 019.0	2 564.0	3 118.0	Entreprises
Direct government	324.0 [a,d]	312.0 [d]	406.0 [d]	967.0 [a]	910.0	848.0	824.0	État direct
Higher education	0.0	0.0	0.0	0.0	0.0	0.0	0.0	Enseignement supérieur
Private non-profit	5.0	0.0	0.0	3.0	2.0	0.0	0.0	Institutions sans but lucratif
Funds from abroad	44.0	45.0	122.0	97.0	119.0	61.0	63.0	Fonds de l'étranger
Total	2 358.0 [a,d]	2 897.0 [d]	3 296.0 [d]	5 854.0 [a]	4 050.0	3 473.0	4 005.0	*Total*
Government								**État**
financed by								*financé par*
Business enterprise	693.0 [a]	704.0	618.0	132.0 [a]	166.0	199.0	192.0	Entreprises
Direct government	1 189.0 [a,d]	1 415.0 [d]	1 639.0 [d]	1 204.0 [a]	1 315.0	1 281.0	1 239.0	État direct
Higher education	0.0	0.0	0.0	0.0	0.0	0.0	13.0	Enseignement supérieur
Private non-profit	2.0	7.0	5.0	1.0	3.0	1.0	3.0	Institutions sans but lucratif
Funds from abroad	12.0	36.0	45.0	33.0	40.0	47.0	55.0	Fonds de l'étranger
Total	1 896.0 [a,d]	2 162.0 [d]	2 307.0 [d]	1 370.0 [a]	1 524.0	1 528.0	1 502.0	*Total*
Higher education								**Enseignement supérieur**
financed by								*financé par*
Business enterprise	2.0 [a,o]	3.0 [o]	2.0 [o]	2.0 [o]	3.0 [o]	5.0 [o]	2.0 [a]	Entreprises
Direct government	..	..	..	..	..	..	..	État direct
General university funds	..	..	..	..	..	..	..	Fonds généraux des universités
Sub-total government	214.0 [a]	306.0	289.0	504.0	565.0	529.0	530.0	Sous-Total État
Higher education	.. [n]	.. [n]	.. [n]	.. [n]	.. [n]	.. [n]	26.0	Enseignement supérieur
Private non-profit	0.0	0.0	0.0	0.0	0.0	0.0	0.0	Institutions sans but lucratif
Funds from abroad	3.0	6.0	11.0	14.0	12.0	17.0	21.0	Fonds de l'étranger
Total	219.0 [a]	315.0	302.0	520.0	580.0	551.0	579.0	*Total*
Private non-profit								**Institutions sans but lucratif**
financed by								*financé par*
Business enterprise	..	..	..	..	..	..	..	Entreprises
Direct government	..	..	..	..	..	..	..	État direct
Higher education	..	..	..	..	..	..	..	Enseignement supérieur
Private non-profit	..	..	..	..	..	..	..	Institutions sans but lucratif
Funds from abroad	..	..	..	..	..	..	..	Fonds de l'étranger
Total	..	..	..	0.0	0.0	0.0	0.0	*Total*
GERD								**DIRD**
financed by								*financé par*
Business enterprise	2 680.0 [a]	3 247.0	3 388.0	4 921.0	3 188.0	2 768.0	3 312.0	Entreprises
Direct government	..	..	..	..	..	..	..	État direct
General university funds	..	..	..	..	..	..	..	Fonds généraux des universités
Sub-total government	1 727.0 [a,d]	2 033.0 [d]	2 334.0 [d]	2 675.0 [a]	2 790.0	2 658.0	2 593.0	Sous-Total État
Higher education	0.0	0.0	0.0	0.0	0.0	0.0	39.0	Enseignement supérieur
Private non-profit	7.0	7.0	5.0	4.0	5.0	1.0	3.0	Institutions sans but lucratif
Funds from abroad	59.0	87.0	178.0	144.0	171.0	125.0	139.0	Fonds de l'étranger
Total	4 473.0 [a,d]	5 374.0 [d]	5 905.0 [d]	7 744.0 [a]	6 154.0	5 552.0	6 086.0	*Total*
	1.2 Million constant USD (1995 prices and PPPs) - Millions de USD constants (prix et PPA de 1995)							
Business enterprise								**Entreprises**
financed by								*financé par*
Business enterprise	183.1 [a,b]	213.5 [b]	222.6 [b]	361.2 [a,b]	216.7 [b]	172.7 [b]	197.3 [b]	Entreprises
Direct government	29.9 [a,b,d]	26.2 [b,d]	32.7 [b,d]	73.0 [a,b]	65.3 [b]	57.1 [b]	52.1 [b]	État direct
Higher education	0.0 [b]	0.0 [b]	0.0 [b]	0.0 [b]	0.0 [b]	0.0 [b]	0.0 [b]	Enseignement supérieur
Private non-profit	0.5 [b]	0.0 [b]	0.0 [b]	0.2 [b]	0.1 [b]	0.0 [b]	0.0 [b]	Institutions sans but lucratif
Funds from abroad	4.1 [b]	3.8 [b]	9.8 [b]	7.3 [b]	8.5 [b]	4.1 [b]	4.0 [b]	Fonds de l'étranger
Total	217.5 [a,b,d]	243.5 [b,d]	265.1 [b,d]	441.7 [a,b]	290.7 [b]	233.9 [b]	253.4 [b]	*Total*
Government								**État**
financed by								*financé par*
Business enterprise	63.9 [a,b]	59.2 [b]	49.7 [b]	10.0 [a,b]	11.9 [b]	13.4 [b]	12.1 [b]	Entreprises
Direct government	109.7 [a,b,d]	118.9 [b,d]	131.8 [b,d]	90.9 [a,b]	94.4 [b]	86.3 [b]	78.4 [b]	État direct
Higher education	0.0 [b]	0.0 [b]	0.0 [b]	0.0 [b]	0.0 [b]	0.0 [b]	0.8 [b]	Enseignement supérieur
Private non-profit	0.2 [b]	0.6 [b]	0.4 [b]	0.1 [b]	0.2 [b]	0.1 [b]	0.2 [b]	Institutions sans but lucratif
Funds from abroad	1.1 [b]	3.0 [b]	3.6 [b]	2.5 [b]	2.9 [b]	3.2 [b]	3.5 [b]	Fonds de l'étranger
Total	174.9 [a,b,d]	181.7 [b,d]	185.5 [b,d]	103.4 [a,b]	109.4 [b]	102.9 [b]	95.0 [b]	*Total*
Higher education								**Enseignement supérieur**
financed by								*financé par*
Business enterprise	0.2 [a,b,o]	0.3 [b,o]	0.2 [b,o]	0.2 [b,o]	0.2 [b,o]	0.3 [b,o]	0.1 [a,b]	Entreprises
Direct government	..	..	..	..	..	..	..	État direct
General university funds	..	..	..	..	..	..	..	Fonds généraux des universités
Sub-total government	19.7 [a,b]	25.7 [b]	23.2 [b]	38.0 [b]	40.6 [b]	35.6 [b]	33.5 [b]	Sous-Total État
Higher education	.. [b,n]	.. [b,n]	.. [b,n]	.. [b,n]	.. [b,n]	.. [b,n]	1.6 [b]	Enseignement supérieur
Private non-profit	0.0 [b]	0.0 [b]	0.0 [b]	0.0 [b]	0.0 [b]	0.0 [b]	0.0 [b]	Institutions sans but lucratif
Funds from abroad	0.3 [b]	0.5 [b]	0.9 [b]	1.1 [b]	0.9 [b]	1.1 [b]	1.3 [b]	Fonds de l'étranger
Total	20.2 [a,b]	26.5 [b]	24.3 [b]	39.2 [b]	41.6 [b]	37.1 [b]	36.6 [b]	*Total*
Private non-profit								**Institutions sans but lucratif**
financed by								*financé par*
Business enterprise	..	..	..	..	..	..	..	Entreprises
Direct government	..	..	..	..	..	..	..	État direct
Higher education	..	..	..	..	..	..	..	Enseignement supérieur
Private non-profit	..	..	..	..	..	..	..	Institutions sans but lucratif
Funds from abroad	..	..	..	..	..	..	..	Fonds de l'étranger
Total	..	..	..	0.0 [b]	0.0 [b]	0.0 [b]	0.0 [b]	*Total*
GERD								**DIRD**
financed by								*financé par*
Business enterprise	247.2 [a,b]	272.9 [b]	272.5 [b]	371.3 [b]	228.8 [b]	186.4 [b]	209.5 [b]	Entreprises
Direct government	..	..	..	..	..	..	..	État direct
General university funds	..	..	..	..	..	..	..	Fonds généraux des universités
Sub-total government	159.3 [a,b,d]	170.9 [b,d]	187.7 [b,d]	201.8 [a,b]	200.3 [b]	179.0 [b]	164.0 [b]	Sous-Total État
Higher education	0.0 [b]	0.0 [b]	0.0 [b]	0.0 [b]	0.0 [b]	0.0 [b]	2.5 [b]	Enseignement supérieur
Private non-profit	0.6 [b]	0.6 [b]	0.4 [b]	0.3 [b]	0.4 [b]	0.1 [b]	0.2 [b]	Institutions sans but lucratif
Funds from abroad	5.4 [b]	7.3 [b]	14.3 [b]	10.9 [b]	12.3 [b]	8.4 [b]	8.8 [b]	Fonds de l'étranger
Total	412.6 [a,b,d]	451.7 [b,d]	474.9 [b,d]	584.3 [a,b]	441.8 [b]	373.9 [b]	385.0 [b]	*Total*

Source: OECD, R&D database (DSTI/EAS Division), January 2002.

Source : OCDE, base de données R-D (DSTI/Division AES), janvier 2002.

SLOVAK REPUBLIC — Table 2 - Tableau 2 — RÉPUBLIQUE SLOVAQUE

Gross domestic expenditure on R&D (GERD) - Dépenses intérieures brutes de R-D (DIRD)

by sector of performance and type of costs - par secteur d'exécution et par type de coûts

Million SKK — *Millions SKK*

	1994	1995	1996	1997	1998	1999	2000	
Business enterprise								**Entreprises**
Labour costs	650.0 [a,d]	850.0 [d]	909.0 [d]	1 678.0 [a]	1 348.0	1 233.0	..	Coûts du personnel
Other current costs	1 569.0 [a,d]	1 651.0 [d]	1 924.0 [d]	3 620.0 [a]	2 343.0	1 973.0	..	Autres coûts courants
Total current expenditure	2 219.0 [a,d]	2 501.0 [d]	2 833.0 [d]	5 298.0 [a]	3 691.0	3 206.0	..	*Total dépenses courantes*
Land and buildings	..	..	102.0 [d]	241.0 [a]	182.0	137.0	..	Terrains et constructions
Instruments and equipment	..	..	361.0 [d]	315.0 [a]	177.0	130.0	..	Instruments et équipements
Total capital expenditure	139.0 [a,d]	396.0 [d]	463.0 [d]	556.0 [a]	359.0	267.0	..	*Total dépenses en capital*
Total	2 358.0 [a,d]	2 897.0 [d]	3 296.0 [d]	5 854.0 [a]	4 050.0	3 473.0	4 005.0	*Total*
Government								**État**
Labour costs	701.0 [a,d]	958.0 [d]	1 060.0 [d]	671.0 [a]	808.0	795.0	..	Coûts du personnel
Other current costs	973.0 [a,d]	948.0 [d]	918.0 [d]	395.0 [a]	441.0	431.0	..	Autres coûts courants
Total current expenditure	1 674.0 [a,d]	1 906.0 [d]	1 978.0 [d]	1 066.0 [a]	1 249.0	1 226.0	..	*Total dépenses courantes*
Land and buildings	..	..	151.0 [d]	147.0 [a]	187.0	156.0	..	Terrains et constructions
Instruments and equipment	..	..	178.0 [d]	157.0 [a]	88.0	146.0	..	Instruments et équipements
Total capital expenditure	222.0 [a,d]	256.0 [d]	329.0 [d]	304.0 [a]	275.0	302.0	..	*Total dépenses en capital*
Total	1 896.0 [a,d]	2 162.0 [d]	2 307.0 [d]	1 370.0 [a]	1 524.0	1 528.0	1 502.0	*Total*
Higher education								**Enseignement supérieur**
Labour costs	144.0 [a]	198.0	192.0	302.0	332.0	351.0	..	Coûts du personnel
Other current costs	51.0 [a]	88.0	76.0	160.0	182.0	149.0	..	Autres coûts courants
Total current expenditure	195.0 [a]	286.0	268.0	462.0	514.0	500.0	..	*Total dépenses courantes*
Land and buildings	..	..	0.0	0.0	0.0	0.0	..	Terrains et constructions
Instruments and equipment	..	..	34.0	58.0	66.0	51.0	..	Instruments et équipements
Total capital expenditure	24.0 [a]	29.0	34.0	58.0	66.0	51.0	..	*Total dépenses en capital*
Total	219.0 [a]	315.0	302.0	520.0	580.0	551.0	579.0	*Total*
Private non-profit								**Institutions sans but lucratif**
Labour costs	..	..	..	..	..	..	..	Coûts du personnel
Other current costs	..	..	..	..	..	..	..	Autres coûts courants
Total current expenditure	..	..	..	..	..	..	..	*Total dépenses courantes*
Land and buildings	..	..	..	..	..	..	..	Terrains et constructions
Instruments and equipment	..	..	..	..	..	..	..	Instruments et équipements
Total capital expenditure	..	..	..	..	..	..	..	*Total dépenses en capital*
Total	..	..	..	0.0	0.0	0.0	0.0	*Total*
GERD								**DIRD**
Labour costs	1 495.0 [a,d]	2 006.0 [d]	2 161.0 [d]	2 651.0 [a]	2 488.0	2 379.0	..	Coûts du personnel
Other current costs	2 593.0 [a,d]	2 687.0 [d]	2 918.0 [d]	4 175.0 [a]	2 966.0	2 553.0	..	Autres coûts courants
Total current expenditure	4 088.0 [a,d]	4 693.0 [d]	5 079.0 [d]	6 826.0 [a]	5 454.0	4 932.0	..	*Total dépenses courantes*
Land and buildings	..	..	253.0 [d]	388.0 [a]	369.0	293.0	..	Terrains et constructions
Instruments and equipment	..	..	573.0 [d]	530.0 [a]	331.0	327.0	..	Instruments et équipements
Total capital expenditure	385.0 [a,d]	681.0 [d]	826.0 [d]	918.0 [a]	700.0	620.0	..	*Total dépenses en capital*
Total	4 473.0 [a,d]	5 374.0 [d]	5 905.0 [d]	7 744.0 [a]	6 154.0	5 552.0	6 086.0	*Total*

Table 3 - Tableau 3

Current domestic expenditure on R&D - Dépenses intérieures courantes de R-D

by sector of performance and type of activity - par secteur d'exécution et par type d'activités

Million SKK — *Millions SKK*

	1994	1995	1996	1997	1998	1999	2000	
Business enterprise								**Entreprises**
Basic research	175.0 [d]	162.0 [d]	105.0 [d]	235.0 [a]	133.0	243.0	160.0	Recherche fondamentale
Applied research	1 307.0 [d]	1 593.0 [d]	1 489.0 [d]	3 651.0 [a]	2 290.0	1 843.0	2 343.0	Recherche appliquée
Experimental development	737.0 [d]	746.0 [d]	1 239.0 [d]	1 412.0 [a]	1 268.0	1 120.0	1 237.0	Développement expérimental
Not specified	0.0	0.0	0.0	0.0	0.0	0.0	0.0	Non spécifié
Total	2 219.0 [a,d]	2 501.0 [d]	2 833.0 [d]	5 298.0 [a]	3 691.0	3 206.0	3 740.0	*Total*
Government								**État**
Basic research	670.0 [a,d]	728.0 [d]	826.0 [d]	699.0 [a]	766.0	735.0	780.0	Recherche fondamentale
Applied research	880.0 [a,d]	990.0 [d]	1 050.0 [d]	352.0 [a]	459.0	477.0	510.0	Recherche appliquée
Experimental development	124.0 [a,d]	188.0 [d]	102.0 [d]	15.0 [a]	24.0	14.0	17.0	Développement expérimental
Not specified	0.0	0.0	0.0	0.0	0.0	0.0	0.0	Non spécifié
Total	1 674.0 [a,d]	1 906.0 [d]	1 978.0 [d]	1 066.0 [a]	1 249.0	1 226.0	1 307.0	*Total*
Higher education								**Enseignement supérieur**
Basic research	164.0	249.0	234.0	399.0	452.0	444.0	446.0	Recherche fondamentale
Applied research	22.0	30.0	29.0	54.0	50.0	42.0	60.0	Recherche appliquée
Experimental development	9.0	7.0	5.0	9.0	12.0	14.0	19.0	Développement expérimental
Not specified	0.0	0.0	0.0	0.0	0.0	0.0	0.0	Non spécifié
Total	195.0 [a]	286.0	268.0	462.0	514.0	500.0	525.0	*Total*
Private non-profit								**Institutions sans but lucratif**
Basic research	..	..	..	..	..	..	..	Recherche fondamentale
Applied research	..	..	..	..	..	..	..	Recherche appliquée
Experimental development	..	..	..	..	..	..	..	Développement expérimental
Not specified	..	..	..	..	..	..	..	Non spécifié
Total	..	..	..	0.0	0.0	0.0	0.0	*Total*
GERD								**DIRD**
Basic research	1 009.0 [d]	1 139.0 [d]	1 165.0 [d]	1 333.0 [a]	1 351.0	1 422.0	1 386.0	Recherche fondamentale
Applied research	2 209.0 [d]	2 613.0 [d]	2 568.0 [d]	4 057.0 [a]	2 799.0	2 362.0	2 913.0	Recherche appliquée
Experimental development	870.0 [d]	941.0 [d]	1 346.0 [d]	1 436.0 [a]	1 304.0	1 148.0	1 273.0	Développement expérimental
Not specified	0.0	0.0	0.0	0.0	0.0	0.0	0.0	Non spécifié
Total	4 088.0 [a,d]	4 693.0 [d]	5 079.0 [d]	6 826.0 [a]	5 454.0	4 932.0	5 572.0	*Total*

Source: OECD, R&D database (DSTI/EAS Division), January 2002.

Source : OCDE, base de données R-D (DSTI/Division AES), janvier 2002.

SLOVAK REPUBLI — Table 4 - Tableau 4 — RÉPUBLIQUE SLOVAQUE

Gross domestic expenditure on R&D (GERD) - Dépenses intérieures brutes de R-D (DIRD)

by sector of performance and socio-economic objective - par secteur d'exécution et par objectif socio-économique

Million SKK — *Millions SKK*

	1994	1995	1996	1997	1998	1999	2000	
Business enterprise								**Entreprises**
Agriculture, forestry and fishing	..	..	..	708.0	519.0	570.0	..	Agriculture, sylviculture et pêche
Industrial development	..	..	..	1 969.0	2 146.0	1 817.0	..	Développement industriel
Energy	..	..	..	1 730.0	181.0	142.0	..	Energie
Transport and telecommunications	..	..	..	248.0	264.0	321.0	..	Transports et télécommunications
Urban and rural planning	..	..	..	3.0	0.0	5.0	..	Aménagement urbain et rural
Prevention of pollution	..	..	..	96.0	51.0	21.0	..	Prévention de la pollution
Identification & treatment of pollution	..	..	..	0.0	0.0	0.0	..	Détection pollu. & mesures antipollu.
Health	..	..	..	153.0	163.0	207.0	..	Santé
Social development and services	..	..	..	6.0	2.0	5.0	..	Développement et services sociaux
Earth and atmosphere	..	..	..	160.0	114.0	118.0	..	Milieu terrestre et atmosphère
Advancement of knowledge	..	..	..	212.0	153.0	163.0	..	Promotion générale des connaissances
Civil space	..	..	..	0.0	0.0	0.0	..	Espace civil
Defence	..	..	..	66.0	61.0	104.0	..	Défense
Not elsewhere classified	2 358.0 [a,d]	2 897.0 [d]	3 296.0 [d]	503.0	396.0	0.0	4 005.0	Non classé
Total	2 358.0 [a,d]	2 897.0 [d]	3 296.0 [d]	5 854.0 [a]	4 050.0	3 473.0	4 005.0	*Total*
Government								**État**
Agriculture, forestry and fishing	..	..	..	135.0	243.0	254.0	..	Agriculture, sylviculture et pêche
Industrial development	..	..	..	167.0	125.0	27.0	..	Développement industriel
Energy	..	..	..	56.0	69.0	0.0	..	Energie
Transport and telecommunications	..	..	..	19.0	23.0	0.0	..	Transports et télécommunications
Urban and rural planning	..	..	..	0.0	0.0	0.0	..	Aménagement urbain et rural
Prevention of pollution	..	..	..	45.0	33.0	36.0	..	Prévention de la pollution
Identification & treatment of pollution	..	..	..	0.0	0.0	0.0	..	Détection pollu. & mesures antipollu.
Health	..	..	..	219.0	103.0	206.0	..	Santé
Social development and services	..	..	..	54.0	141.0	33.0	..	Développement et services sociaux
Earth and atmosphere	..	..	..	10.0	18.0	13.0	..	Milieu terrestre et atmosphère
Advancement of knowledge	..	..	..	455.0	609.0	952.0	..	Promotion générale des connaissances
Civil space	..	..	..	0.0	2.0	5.0	..	Espace civil
Defence	..	..	..	5.0	4.0	2.0	..	Défense
Not elsewhere classified	1 896.0 [a,d]	2 162.0 [d]	2 307.0 [d]	205.0 [a]	154.0	0.0	1 502.0	Non classé
Total	1 896.0 [a,d]	2 162.0 [d]	2 307.0 [d]	1 370.0 [a]	1 524.0	1 528.0	1 502.0	*Total*
Higher education								**Enseignement supérieur**
Agriculture, forestry and fishing	..	..	..	91.0	107.0	44.0	..	Agriculture, sylviculture et pêche
Industrial development	..	..	..	30.0	32.0	34.0	..	Développement industriel
Energy	..	..	..	6.0	6.0	7.0	..	Energie
Transport and telecommunications	..	..	..	9.0	8.0	9.0	..	Transports et télécommunications
Urban and rural planning	..	..	..	1.0	2.0	2.0	..	Aménagement urbain et rural
Prevention of pollution	..	..	..	7.0	8.0	9.0	..	Prévention de la pollution
Identification & treatment of pollution	..	..	..	0.0	0.0	0.0	..	Détection pollu. & mesures antipollu.
Health	..	..	..	10.0	10.0	41.0	..	Santé
Social development and services	..	..	..	8.0	2.0	8.0	..	Développement et services sociaux
Earth and atmosphere	..	..	..	0.0	5.0	1.0	..	Milieu terrestre et atmosphère
Advancement of knowledge	..	..	..	299.0	361.0	396.0	..	Promotion générale des connaissances
Civil space	..	..	..	0.0	0.0	0.0	..	Espace civil
Defence	..	..	..	0.0	0.0	0.0	..	Défense
Not elsewhere classified	219.0 [a]	315.0	302.0	59.0	39.0	0.0	579.0	Non classé
Total	219.0 [a]	315.0	302.0	520.0	580.0	551.0	579.0	*Total*
Private non-profit								**Institutions sans but lucratif**
Agriculture, forestry and fishing	..	..	..	..	..	..	..	Agriculture, sylviculture et pêche
Industrial development	..	..	..	..	..	..	..	Développement industriel
Energy	..	..	..	..	..	..	..	Energie
Transport and telecommunications	..	..	..	..	..	..	..	Transports et télécommunications
Urban and rural planning	..	..	..	..	..	..	..	Aménagement urbain et rural
Prevention of pollution	..	..	..	..	..	..	..	Prévention de la pollution
Identification & treatment of pollution	..	..	..	..	..	..	..	Détection pollu. & mesures antipollu.
Health	..	..	..	..	..	..	..	Santé
Social development and services	..	..	..	..	..	..	..	Développement et services sociaux
Earth and atmosphere	..	..	..	..	..	..	..	Milieu terrestre et atmosphère
Advancement of knowledge	..	..	..	..	..	..	..	Promotion générale des connaissances
Civil space	..	..	..	..	..	..	..	Espace civil
Defence	..	..	..	..	..	..	..	Défense
Not elsewhere classified	..	..	..	..	..	..	..	Non classé
Total	..	..	..	0.0	0.0	0.0	0.0	*Total*
GERD								**DIRD**
Agriculture, forestry and fishing	..	..	..	934.0	869.0	868.0	..	Agriculture, sylviculture et pêche
Industrial development	..	..	..	2 166.0	2 303.0	1 878.0	..	Développement industriel
Energy	..	..	..	1 792.0	256.0	149.0	..	Energie
Transport and telecommunications	..	..	..	276.0	295.0	330.0	..	Transports et télécommunications
Urban and rural planning	..	..	..	4.0	2.0	7.0	..	Aménagement urbain et rural
Prevention of pollution	..	..	..	148.0	92.0	66.0	..	Prévention de la pollution
Identification & treatment of pollution	..	..	..	0.0	0.0	0.0	..	Détection pollu. & mesures antipollu.
Health	..	..	..	382.0	276.0	454.0	..	Santé
Social development and services	..	..	..	68.0	145.0	46.0	..	Développement et services sociaux
Earth and atmosphere	..	..	..	170.0	137.0	132.0	..	Milieu terrestre et atmosphère
Advancement of knowledge	..	..	..	966.0	1 123.0	1 511.0	..	Promotion générale des connaissances
Civil space	..	..	..	0.0	2.0	5.0	..	Espace civil
Defence	..	..	..	71.0	65.0	106.0	..	Défense
Not elsewhere classified	4 473.0 [a,d]	5 374.0 [d]	5 905.0 [d]	767.0 [a]	589.0	0.0	6 086.0	Non classé
Total	4 473.0 [a,d]	5 374.0 [d]	5 905.0 [d]	7 744.0 [a]	6 154.0	5 552.0	6 086.0	*Total*

Source: OECD, R&D database (DSTI/EAS Division), January 2002.

Source : OCDE, base de données R-D (DSTI/Division AES), janvier 2002.

Table 5 - Tableau 5

Total R&D personnel - Personnel total de R-D

by sector of employment and occupation - par secteur d'emploi et par profession

In full-time equivalent / *En équivalence plein-temps*

	1994	1995	1996	1997	1998	1999	2000	
Business enterprise								**Entreprises**
RSE	2 648 [a]	2 103 [a]	2 259 [a]	3 387 [a]	2 903 [a]	2 522 [a]	2 420 [a]	Chercheurs
Technicians	1 721 [a]	1 611 [a]	1 812 [a]	2 546 [a]	2 477 [a]	2 094 [a]	1 817 [a]	Techniciens
Other occupations	1 326 [a]	1 145 [a]	1 224 [a]	1 475 [a]	1 328 [a]	1 076 [a]	935 [a]	Autres professions
Total	5 695 [a]	4 859 [a]	5 295 [a]	7 408 [a]	6 708 [a]	5 692 [a]	5 172 [a]	*Total*
Government								**État**
RSE	3 903 [a]	3 654 [a]	3 851 [a]	2 468 [a]	2 557 [a]	2 428 [a]	2 526 [a]	Chercheurs
Technicians	2 013 [a]	1 947 [a]	1 892 [a]	996 [a]	1 108 [a]	1 049 [a]	1 063 [a]	Techniciens
Other occupations	1 360 [a]	1 179 [a]	1 054 [a]	452 [a]	574 [a]	617 [a]	600 [a]	Autres professions
Total	7 276 [a]	6 780 [a]	6 797 [a]	3 916 [a]	4 239 [a]	4 094 [a]	4 189 [a]	*Total*
Higher education								**Enseignement supérieur**
RSE	3 698	3 954	3 900	4 138	4 685	4 254	5 009	Chercheurs
Technicians	510	523	540	776	746	715	717	Techniciens
Other occupations	77	66	81	127	83	94	134	Autres professions
Total	4 285	4 543	4 521	5 041	5 514	5 063	5 860	*Total*
Private non-profit								**Institutions sans but lucratif**
RSE	..	..	..	..	..	..	..	Chercheurs
Technicians	..	..	..	..	..	..	..	Techniciens
Other occupations	..	..	..	..	..	..	..	Autres professions
Total	..	..	..	0	0	0	0	*Total*
TOTAL								**TOTAL**
RSE	10 249 [a]	9 711 [a]	10 010 [a]	9 993 [a]	10 145 [a]	9 204 [a]	9 955 [a]	Chercheurs
Technicians	4 244 [a]	4 081 [a]	4 244 [a]	4 318 [a]	4 331 [a]	3 858 [a]	3 597 [a]	Techniciens
Other occupations	2 763 [a]	2 390 [a]	2 359 [a]	2 054 [a]	1 985 [a]	1 787 [a]	1 669 [a]	Autres professions
Total	17 256 [a]	16 182 [a]	16 613 [a]	16 365 [a]	16 461 [a]	14 849 [a]	15 221 [a]	*Total*

Table 6 - Tableau 6

Total R&D personnel - Personnel total de R-D

by sector of employment and formal qualification - par secteur d'emploi et par niveau de qualification

In full-time equivalent / *En équivalence plein-temps*

	1994	1995	1996	1997	1998	1999	2000	
Business enterprise								**Entreprises**
University PhD level degrees	2 681 [a]	2 278 [a]	2 566 [a]	3 619 [a]	3 288 [a]	2 909 [a]	2 772 [a]	Diplômes univ. de niv. doctorat
Other university degrees	3 [a]	1 [a]	1 [a]	10 [a]	9 [a]	8 [a]	13 [a]	Autres diplômes universitaires
Sub-total university degrees	2 684 [a]	2 279 [a]	2 567 [a]	3 629 [a]	3 297 [a]	2 917 [a]	2 785 [a]	Sous-total Universitaires
Other post-secondary diplomas	0	0	0	0	0	0	0	Autres diplômes post-secondaires
Secondary diplomas	2 649 [a]	2 327 [a]	2 472 [a]	3 407 [a]	3 117 [a]	2 577 [a]	2 141 [a]	Diplômes de fin d'études secondaires
Other qualifications	362 [a]	253 [a]	256 [a]	372 [a]	294 [a]	198 [a]	246 [a]	Autres qualifications
Total	5 695 [a]	4 859 [a]	5 295 [a]	7 408 [a]	6 708 [a]	5 692 [a]	5 172 [a]	*Total*
Government								**État**
University PhD level degrees	4 177 [a]	3 866 [a]	3 982 [a]	2 582 [a]	2 679 [a]	2 568 [a]	2 712 [a]	Diplômes univ. de niv. doctorat
Other university degrees	9 [a]	5 [a]	16 [a]	13 [a]	4 [a]	2 [a]	9 [a]	Autres diplômes universitaires
Sub-total university degrees	4 186 [a]	3 871 [a]	3 998 [a]	2 595 [a]	2 683 [a]	2 570 [a]	2 721 [a]	Sous-total Universitaires
Other post-secondary diplomas	0	0	0	0	0	0	0	Autres diplômes post-secondaires
Secondary diplomas	2 703 [a]	2 601 [a]	2 499 [a]	1 207 [a]	1 362 [a]	1 363 [a]	1 254 [a]	Diplômes de fin d'études secondaires
Other qualifications	387 [a]	308 [a]	300 [a]	114 [a]	194 [a]	161 [a]	214 [a]	Autres qualifications
Total	7 276 [a]	6 780 [a]	6 797 [a]	3 916 [a]	4 239 [a]	4 094 [a]	4 189 [a]	*Total*
Higher education								**Enseignement supérieur**
University PhD level degrees	3 750	4 091	4 044	4 323	4 818	4 397	5 106	Diplômes univ. de niv. doctorat
Other university degrees	23	6	4	2	3	1	3	Autres diplômes universitaires
Sub-total university degrees	3 773	4 097	4 048	4 325	4 821	4 398	5 109	Sous-total Universitaires
Other post-secondary diplomas	0	0	0	0	0	0	0	Autres diplômes post-secondaires
Secondary diplomas	510	442	472	701	662	651	731	Diplômes de fin d'études secondaires
Other qualifications	2	4	1	15	31	14	20	Autres qualifications
Total	4 285	4 543	4 521	5 041	5 514	5 063	5 860	*Total*
Private non-profit								**Institutions sans but lucratif**
University PhD level degrees	..	..	..	..	..	..	..	Diplômes univ. de niv. doctorat
Other university degrees	..	..	..	..	..	..	..	Autres diplômes universitaires
Sub-total university degrees	..	..	..	..	..	..	..	Sous-total Universitaires
Other post-secondary diplomas	..	..	..	..	..	..	..	Autres diplômes post-secondaires
Secondary diplomas	..	..	..	..	..	..	..	Diplômes de fin d'études secondaires
Other qualifications	..	..	..	..	..	..	..	Autres qualifications
Total	..	..	..	0	0	0	0	*Total*
TOTAL								**TOTAL**
University PhD level degrees	10 608 [a]	10 235 [a]	10 592 [a]	10 524 [a]	10 785 [a]	9 874 [a]	10 590 [a]	Diplômes univ. de niv. doctorat
Other university degrees	35 [a]	12 [a]	21 [a]	25 [a]	16 [a]	11 [a]	25 [a]	Autres diplômes universitaires
Sub-total university degrees	10 643 [a]	10 247 [a]	10 613 [a]	10 549 [a]	10 801 [a]	9 885 [a]	10 615 [a]	Sous-total Universitaires
Other post-secondary diplomas	0	0	0	0	0	0	0	Autres diplômes post-secondaires
Secondary diplomas	5 862 [a]	5 370 [a]	5 443 [a]	5 315 [a]	5 141 [a]	4 591 [a]	4 126 [a]	Diplômes de fin d'études secondaires
Other qualifications	751 [a]	565 [a]	557 [a]	501 [a]	519 [a]	373 [a]	480 [a]	Autres qualifications
Total	17 256 [a]	16 182 [a]	16 613 [a]	16 365 [a]	16 461 [a]	14 849 [a]	15 221 [a]	*Total*

Source: OECD, R&D database (DSTI/EAS Division), January 2002.

Source : OCDE, base de données R-D (DSTI/Division AES), janvier 2002.

Gross domestic expenditure on R&D (GERD) - Dépenses intérieures brutes de R-D (DIRD)

by sector of performance and main scientific field - par secteur d'exécution et par grand domaine scientifique

Million SKK *Millions SKK*

	1994	1995	1996	1997	1998	1999	2000	
Business enterprise								**Entreprises**
Natural sciences and engineering	..	..	3 286.0 [d]	5 838.0 [a]	4 039.0	3 417.0	..	Sciences exactes, nat. et de l'ingénieur
Social sciences and humanities	..	..	10.0 [d]	16.0 [a]	11.0	56.0	..	Sciences sociales et humaines
All fields of science	2 358.0 [a,d]	2 897.0 [d]	3 296.0 [d]	5 854.0 [a]	4 050.0	3 473.0	4 005.0	*Tous domaines scientifiques*
Government								**État**
Natural sciences and engineering	..	..	2 135.0 [d]	1 148.0 [a]	1 276.0	1 204.0	..	Sciences exactes, nat. et de l'ingénieur
Social sciences and humanities	..	..	172.0 [d]	222.0 [a]	248.0	324.0	..	Sciences sociales et humaines
All fields of science	1 896.0 [a,d]	2 162.0 [d]	2 307.0 [d]	1 370.0 [a]	1 524.0	1 528.0	1 502.0	*Tous domaines scientifiques*
Higher education								**Enseignement supérieur**
Natural sciences and engineering	..	..	275.0	436.0	499.0	468.0	..	Sciences exactes, nat. et de l'ingénieur
Social sciences and humanities	..	..	27.0	84.0	81.0	83.0	..	Sciences sociales et humaines
All fields of science	219.0 [a]	315.0	302.0	520.0	580.0	551.0	579.0	*Tous domaines scientifiques*
Private non-profit								**Institutions sans but lucratif**
Natural sciences and engineering	..	..	..	..	..	..	..	Sciences exactes, nat. et de l'ingénieur
Social sciences and humanities	..	..	..	..	..	..	..	Sciences sociales et humaines
All fields of science	..	..	..	0.0	0.0	0.0	0.0	*Tous domaines scientifiques*
TOTAL								**TOTAL**
Natural sciences and engineering	..	..	5 696.0 [d]	7 422.0 [a]	5 814.0	5 089.0	..	Sciences exactes, nat. et de l'ingénieur
Social sciences and humanities	..	..	209.0 [d]	322.0 [a]	340.0	463.0	..	Sciences sociales et humaines
All fields of science	4 473.0 [a,d]	5 374.0 [d]	5 905.0 [d]	7 744.0 [a]	6 154.0	5 552.0	6 086.0	*Tous domaines scientifiques*

Table 8 - Tableau 8

Total R&D personnel - Personnel total de R-D

by sector of employment and main scientific field - par secteur d'emploi et par grand domaine scientifique

In full-time equivalent *En équivalence plein-temps*

	1994	1995	1996	1997	1998	1999	2000	
Business enterprise								**Entreprises**
Natural sciences and engineering	..	..	5 262 [d]	7 360 [d]	6 674 [d]	5 574 [d]	..	Sciences exactes, nat. et de l'ingénieur
Social sciences and humanities	..	..	33 [d]	48 [d]	34 [d]	118 [d]	..	Sciences sociales et humaines
All fields of science	5 695 [d]	4 859 [d]	5 295 [d]	7 408 [d]	6 708 [d]	5 692 [d]	5 172 [d]	*Tous domaines scientifiques*
Government								**État**
Natural sciences and engineering	..	..	5 993 [d]	3 154 [d]	3 376 [d]	3 195 [d]	..	Sciences exactes, nat. et de l'ingénieur
Social sciences and humanities	..	..	804 [d]	762 [d]	863 [d]	899 [d]	..	Sciences sociales et humaines
All fields of science	7 276 [d]	6 780 [d]	6 797 [d]	3 916 [d]	4 239 [d]	4 094 [d]	4 189 [d]	*Tous domaines scientifiques*
Higher education								**Enseignement supérieur**
Natural sciences and engineering	..	..	3 864	4 048	4 460	4 087	..	Sciences exactes, nat. et de l'ingénieur
Social sciences and humanities	..	..	657	993	1 054	976	..	Sciences sociales et humaines
All fields of science	4 285	4 543	4 521	5 041	5 514	5 063	5 860	*Tous domaines scientifiques*
Private non-profit								**Institutions sans but lucratif**
Natural sciences and engineering	..	..	..	..	..	..	..	Sciences exactes, nat. et de l'ingénieur
Social sciences and humanities	..	..	..	..	..	..	..	Sciences sociales et humaines
All fields of science	..	..	..	0	0	0	0	*Tous domaines scientifiques*
TOTAL								**TOTAL**
Natural sciences and engineering	..	..	15 119 [d]	14 562 [d]	14 510 [d]	12 856 [d]	..	Sciences exactes, nat. et de l'ingénieur
Social sciences and humanities	..	..	1 494 [d]	1 803 [d]	1 951 [d]	1 993 [d]	..	Sciences sociales et humaines
All fields of science	17 256 [d]	16 182 [d]	16 613 [d]	16 365 [d]	16 461 [d]	14 849 [d]	15 221 [d]	*Tous domaines scientifiques*

Source: OECD, R&D database (DSTI/EAS Division), January 2002.

Source : OCDE, base de données R-D (DSTI/Division AES), janvier 2002.

Total business enterprise intramural expenditure on R&D - Dépenses intra-muros totales de R-D des entreprises

(BERD) by industry - (DIRDE) par industrie

	1994	1995	1996	1997	1998	1999	2000	
	9.1 Million SKK - Millions SKK							
1 AGRICULTURE, HUNTING & FORESTRY	..	..	..	24.0	392.0 [a]	465.0	442.0	**1 AGRICULTURE, CHASSE & SYLVICULTURE**
2 MINING	..	0.0	0.0	..	..	..	..	**2 ACTIVITES EXTRACTIVES**
3 MANUFACTURING	1 279.0 [a]	1 503.0	1 757.0	1 332.0	2 477.0 [a]	2 016.0	2 308.0	**3 ACTIVITES DE FABRICATION**
4 Food, beverages & tobacco	1.0 [a]	2.0	3.0	4.0	4.0 [a]	..	..	**4 Produits alimentaires, boissons & tabac**
5 Food products & beverages	1.0	2.0	3.0	4.0	4.0 [a]	..	..	5 Produits alimentaires & boissons
6 Tobacco products	0.0	0.0	0.0	0.0	0.0	0.0	0.0	6 Produits à base de tabac
7 Textiles, wear app., fur & leather	0.0	..	..	..	9.0 [a]	7.0	9.0	**7 Textiles, habill., fourrures & cuirs**
8 Textiles	..	..	..	..	9.0 [a]	7.0	9.0	8 Textiles
9 Wearing apparel & fur	0.0	0.0	0.0	0.0	0.0	0.0	0.0	9 Articles d'habillement & fourrures
10 Leather products & footwear	0.0	0.0	0.0	0.0	0.0	0.0	0.0	10 Articles en cuir & chaussures
11 Wood, paper, printing, publishing	5.0 [a]	..	..	..	..	..	..	**11 Bois, papier, imprimerie, édition**
12 Wood & cork (not furniture)	..	..	0.0	0.0	..	..	0.0	12 Bois & liège (sauf meubles)
13 Pulp, paper & paper products	..	..	..	..	..	..	..	13 Carton, papier & articles
14 Publ., print. & repro. of rec. media	0.0	0.0	0.0	0.0	0.0	0.0	0.0	14 Edition, impress. & repro. enregis.
15 Coke, petroleum, nuclear fuel, chemicals, rubber plastics	468.0 [a]	668.0	692.0	596.0	886.0 [a]	938.0	754.0	**15 Coke, pétrole, combust. nucl., prod. chimi., caout. & mat. plast.**
16 Coke, ref. petrol. prod. & nuclear fuel	..	..	..	..	..	..	0.0	16 Coke, raffin. pétr. & combust. nucl.
17 Chemicals & chemical products	..	..	..	..	..	..	446.0	17 Produits chimiques
18 *Chemical prod.(less pharmaceuticals)*	..	..	..	..	..	..	..	18 *Prod. chimi. (moins prod. pharmaceutiques)*
19 *Pharmaceuticals*	0.0	..	..	..	..	..	..	19 *Produits pharmaceutiques*
20 Rubber & plastic products	152.0 [a]	325.0	366.0	232.0	264.0 [a]	323.0	307.0	20 Caoutchouc et matières plastiques
21 Non-metallic mineral products	14.0 [a]	20.0	10.0	..	..	..	..	**21 Produits minéraux non métalliques**
22 Basic metals	124.0 [a]	135.0	140.0	134.0	131.0 [a]	115.0	..	**22 Produits métallurgiques de base**
23 Basic metals, ferrous	..	..	..	..	..	..	..	23 Produits métallur. de base, ferreux
24 Basic metals, non-ferrous	..	..	..	..	..	0.0	0.0	24 Prod. métallur. de base, non ferreux
25 Fabricated metal products	5.0 [a]	4.0	4.0	6.0	33.0 [a]	174.0	..	**25 Ouvrages en métaux (sauf mach.& mat.)**
26 Machinery equipment, instruments and transport equipment	658.0 [a]	660.0	898.0	569.0	1 187.0 [a]	494.0	1 185.0	**26 Machines & matériel, instruments & mat. de transport**
27 Machinery n.e.c.	253.0 [a]	242.0	322.0	132.0	906.0 [a]	191.0	381.0	27 Machines, n.c.a.
28 Office, account. & computing machinery	..	..	..	0.0	0.0	..	..	28 Mach. de bur., compta. & trait. inform.
29 Electrical machinery	34.0	27.0	50.0	64.0	111.0 [a]	53.0	157.0	29 Machines électriques
30 Electro. equip. (radio, TV & commun.)	92.0 [a]	98.0	109.0	81.0	..	70.0	513.0	30 Equip. électro. (radio, tv & communic.)
31 *Electro. comp. (inc. semi-conductors)*	0.0	..	0.0	0.0	0.0	0.0	0.0	31 *Compos. électro. (semi-cond. incl.)*
32 *TV, radio & communications equip.*	92.0	..	109.0	0.0	0.0	70.0	513.0	32 *Appareils radio, tv et de communi.*
33 Instruments	53.0 [a]	55.0	93.0	98.0	49.0 [a]	50.0	25.0	33 Instruments
34 Motor vehicles	102.0 [a]	105.0	94.0	81.0	17.0 [a]	..	..	34 Véhicules automobiles
35 Other transport equipment	..	..	..	113.0	..	..	..	35 Autres matériels de transport
36 *Ships*	0.0	0.0	0.0	0.0	0.0	0.0	0.0	36 *Navires*
37 *Aerospace*	0.0	..	..	..	0.0	0.0	0.0	37 *Aérospatiale*
38 *Other transport equipment n.e.c.*	0.0	0.0	0.0	..	..	..	0.0	38 *Autres matériels de transport n.c.a.*
39 Furniture, other manufacturing n.e.c.	..	..	..	..	..	..	71.0	**39 Meubles, autres activ. de fabricat. n.c.a.**
40 Furniture	0.0	..	..	0.0	0.0	0.0	0.0	40 Meubles
41 Other manufacturing n.e.c.	0.0	0.0	0.0	..	..	..	0.0	41 Autres activités de fabrication n.c.a.
42 Recycling	0.0	0.0	0.0	0.0	0.0	0.0	0.0	**42 Récupération**
43 ELECTRICITY, GAS & WATER SUPPLY	0.0	0.0	..	..	..	..	..	**43 DISTRIBUTION D'ELECTRICITE, GAZ & EAU**
44 CONSTRUCTION	11.0 [a]	..	19.0	13.0	34.0 [a]	29.0	15.0	**44 CONSTRUCTION**
45 SERVICES SECTOR	1 066.0 [a]	1 376.0	1 488.0	4 453.0 [a]	926.0 [a]	812.0	1 174.0	**45 SECTEUR DES SERVICES**
46 Wholesale, retail trade & repair of motor vehicles, etc.	0.0	0.0	0.0	0.0	0.0	0.0	..	**46 Commerce (gros & détail), réparation de véhicules, etc.**
47 Hotels & restaurants	0.0	0.0	0.0	0.0	0.0	0.0	0.0	**47 Hôtels et restaurants**
48 Transport & storage	0.0	0.0	..	..	..	..	..	**48 Transports, entreposage**
49 Communications	0.0	0.0	..	..	..	..	..	**49 Communications**
50 Post	0.0	0.0	0.0	0.0	0.0	0.0	0.0	50 Postes
51 Telecommunications	0.0	0.0	0.0	..	..	0.0	0.0	51 Télécommunications
52 Financial intermediation (inc. insur.)	0.0	0.0	0.0	0.0	0.0	0.0	0.0	**52 Intermédiation financ. (assurances inc.)**
53 Real estate, renting & busin. activity	1 066.0	1 376.0	1 457.0	4 421.0 [a]	795.0 [a]	703.0	1 153.0	**53 Immobil., locat. & serv. aux entreprises**
54 Computer & related activities	50.0	..	0.0	..	..	..	..	54 Informatique et activités connexes
55 *Software consultancy*	0.0	..	0.0	0.0	0.0	..	0.0	55 *Production de logiciels*
56 *Other computer services n.e.c.*	0.0	..	0.0	0.0	0.0	0.0	0.0	56 *Autres activités informatiques*
57 Research & development	1 016.0 [a,o]	1 372.0 [o]	1 457.0 [o]	4 412.0 [a,o]	719.0 [a]	679.0	1 139.0	57 Recherche & développement expérimental
58 Other business activities n.e.c.	0.0	0.0	0.0	..	..	..	..	58 Autres activ. & serv. aux entreprises
59 Comm., soc. & pers. serv. activ., etc.	0.0	..	0.0	0.0	0.0	0.0	..	**59 Adm. pub., serv. soc. & collec., etc**
TOTAL BUSINESS ENTERPRISE SECTOR	**2 358.0** [a,d]	**2 897.0** [d]	**3 296.0** [d]	**5 854.0** [a]	**4 050.0**	**3 473.0**	**4 005.0**	**TOTAL SECTEUR DES ENTREPRISES**

Source: OECD, R&D database (DSTI/EAS Division), January 2002.

Source : OCDE, base de données R-D (DSTI/Division AES), janvier 2002.

SLOVAK REPUBLIC — Table 9 - Tableau 9 — RÉPUBLIQUE SLOVAQUE

Total business enterprise intramural expenditure on R&D - Dépenses intra-muros totales de R-D des entreprises (BERD) by industry - (DIRDE) par industrie

	1994	1995	1996	1997	1998	1999	2000	
	9.2 Million constant USD(1995 prices and PPPs) - Millions de USD constants (prix et PPA de 1995)							
1 AGRICULTURE, HUNTING & FORESTRY	..	..	..	1.8	28.1 [a]	31.3	28.0	**1 AGRICULTURE, CHASSE & SYLVICULTURE**
2 MINING	..	0.0	0.0	..	..	..	..	**2 ACTIVITES EXTRACTIVES**
3 MANUFACTURING	118.0 [a,b]	126.3	141.3	100.5	177.8 [a]	135.8	146.0	**3 ACTIVITES DE FABRICATION**
4 Food, beverages & tobacco	0.1 [a,b]	0.2	0.2	0.3	0.3 [a]	..	..	**4 Produits alimentaires, boissons & tabac**
5 Food products & beverages	0.1 [b]	0.2	0.2	0.3	0.3 [a]	..	..	5 Produits alimentaires & boissons
6 Tobacco products	0.0 [b]	0.0	0.0	0.0	0.0	0.0	0.0	6 Produits à base de tabac
7 Textiles, wear app., fur & leather	0.0 [b]	..	..	..	0.6 [a]	0.5	0.6	**7 Textiles, habill., fourrures & cuirs**
8 Textiles	..	..	..	..	0.6 [a]	0.5	0.6	8 Textiles
9 Wearing apparel & fur	0.0 [b]	0.0	0.0	0.0	0.0	0.0	0.0	9 Articles d'habillement & fourrures
10 Leather products & footwear	0.0 [b]	0.0	0.0	0.0	0.0	0.0	0.0	10 Articles en cuir & chaussures
11 Wood, paper, printing, publishing	0.5 [a,b]	..	..	..	..	..	..	**11 Bois, papier, imprimerie, édition**
12 Wood & cork (not furniture)	..	..	0.0	0.0	..	..	0.0	12 Bois & liège (sauf meubles)
13 Pulp, paper & paper products	..	..	..	..	..	..	..	13 Carton, papier & articles
14 Publ., print. & repro. of rec. media	0.0 [b]	0.0	0.0	0.0	0.0	0.0	0.0	14 Edition, impress. & repro. enregis.
15 Coke, petroleum, nuclear fuel, chemicals, rubber plastics	43.2 [a,b]	56.2	55.7	45.0	63.6 [a]	63.2	47.7	**15 Coke, pétrole, combust. nucl., prod. chimi., caout. & mat. plast.**
16 Coke, ref. petrol. prod. & nuclear fuel	..	..	..	..	..	..	0.0	16 Coke, raffin. pétr. & combust. nucl.
17 Chemicals & chemical products	..	..	..	..	..	..	28.2	17 Produits chimiques
18 *Chemical prod.(less pharmaceuticals)*	..	..	..	..	..	..	..	18 *Prod. chimi. (moins prod. pharmaceutiques)*
19 *Pharmaceuticals*	0.0 [b]	..	..	..	..	..	..	19 *Produits pharmaceutiques*
20 Rubber & plastic products	14.0 [a,b]	27.3	29.4	17.5	19.0 [a]	21.8	19.4	20 Caoutchouc et matières plastiques
21 Non-metallic mineral products	1.3 [a,b]	1.7	0.8	..	..	..	..	**21 Produits minéraux non métalliques**
22 Basic metals	11.4 [a,b]	11.3	11.3	10.1	9.4 [a]	7.7	..	**22 Produits métallurgiques de base**
23 Basic metals, ferrous	..	..	..	..	..	..	..	23 Produits métallur. de base, ferreux
24 Basic metals, non-ferrous	..	..	..	..	..	0.0	0.0	24 Prod. métallur. de base, non ferreux
25 Fabricated metal products	0.5 [a,b]	0.3	0.3	0.5	2.4 [a]	11.7	..	**25 Ouvrages en métaux (sauf mach.& mat.)**
26 Machinery equipment, instruments and transport equipment	60.7 [a,b]	55.5	72.2	42.9	85.2 [a]	33.3	75.0	**26 Machines & matériel, instruments & mat. de transport**
27 Machinery n.e.c.	23.3 [a,b]	20.3	25.9	10.0	65.0 [a]	12.9	24.1	27 Machines, n.c.a.
28 Office, account. & computing machinery	..	..	..	0.0	0.0	..	..	28 Mach. de bur., compta. & trait. inform.
29 Electrical machinery	3.1 [b]	2.3	4.0	4.8	8.0 [a]	3.6	9.9	29 Machines électriques
30 Electro. equip. (radio, TV & commun.)	8.5 [a,b]	8.2	8.8	6.1	..	4.7	32.5	30 Equip. électro. (radio, tv & communic.)
31 *Electro. comp. (inc. semi-conductors)*	0.0 [b]	..	0.0	0.0	0.0	0.0	0.0	31 *Compos. électro. (semi-cond. incl.)*
32 *TV, radio & communications equip.*	8.5 [b]	..	8.8	0.0	0.0	4.7	32.5	32 *Appareils radio, tv et de communi.*
33 Instruments	4.9 [a,b]	4.6	7.5	7.4	3.5 [a]	3.4	1.6	33 Instruments
34 Motor vehicles	9.4 [a,b]	8.8	7.6	6.1	1.2 [a]	..	..	34 Véhicules automobiles
35 Other transport equipment	..	..	..	8.5	..	..	..	35 Autres matériels de transport
36 *Ships*	0.0 [b]	0.0	0.0	0.0	0.0	0.0	0.0	36 *Navires*
37 *Aerospace*	0.0 [b]	..	..	..	0.0	0.0	0.0	37 *Aérospatiale*
38 *Other transport equipment n.e.c.*	0.0 [b]	0.0	0.0	..	..	..	0.0	38 *Autres matériels de transport n.c.a.*
39 Furniture, other manufacturing n.e.c.	..	..	..	..	..	..	4.5	**39 Meubles, autres activ. de fabricat. n.c.a.**
40 Furniture	0.0 [b]	..	..	0.0	0.0	0.0	0.0	40 Meubles
41 Other manufacturing n.e.c.	0.0 [b]	0.0	0.0	..	..	..	0.0	41 Autres activités de fabrication n.c.a.
42 Recycling	0.0 [b]	0.0	0.0	0.0	0.0	0.0	0.0	**42 Récupération**
43 ELECTRICITY, GAS & WATER SUPPLY	0.0 [b]	0.0	..	..	..	..	..	**43 DISTRIBUTION D'ELECTRICITE, GAZ & EAU**
44 CONSTRUCTION	1.0 [a,b]	..	1.5	1.0	2.4 [a]	2.0	0.9	**44 CONSTRUCTION**
45 SERVICES SECTOR	98.3 [a,b]	115.7	119.7	336.0 [a]	66.5 [a]	54.7	74.3	**45 SECTEUR DES SERVICES**
46 Wholesale, retail trade & repair of motor vehicles, etc.	0.0 [b]	0.0	0.0	0.0	0.0	0.0	..	**46 Commerce (gros & détail), réparation de véhicules, etc.**
47 Hotels & restaurants	0.0 [b]	0.0	0.0	0.0	0.0	0.0	0.0	**47 Hôtels et restaurants**
48 Transport & storage	0.0 [b]	0.0	..	..	..	..	..	**48 Transports, entreposage**
49 Communications	0.0 [b]	0.0	..	..	..	..	..	**49 Communications**
50 Post	0.0 [b]	0.0	0.0	0.0	0.0	0.0	0.0	50 Postes
51 Telecommunications	0.0 [b]	0.0	0.0	..	..	0.0	0.0	51 Télécommunications
52 Financial intermediation (inc. insur.)	0.0 [b]	0.0	0.0	0.0	0.0	0.0	0.0	**52 Intermédiation financ. (assurances inc.)**
53 Real estate, renting & busin. activity	98.3 [b]	115.7	117.2	333.6 [a]	57.1 [a]	47.4	72.9	**53 Immobil., locat. & serv. aux entreprises**
54 Computer & related activities	4.6 [b]	..	0.0	..	..	..	..	54 Informatique et activités connexes
55 *Software consultancy*	0.0 [b]	..	0.0	0.0	0.0	..	0.0	55 *Production de logiciels*
56 *Other computer services n.e.c.*	0.0 [b]	..	0.0	0.0	0.0	0.0	0.0	56 *Autres activités informatiques*
57 Research & development	93.7 [a,b,]	115.3 [o]	117.2 [o]	332.9 [a,o]	51.6 [a]	45.7	72.1	57 Recherche & développement expérimental
58 Other business activities n.e.c.	0.0 [b]	0.0	0.0	..	..	..	..	58 Autres activ. & serv. aux entreprises
59 Comm., soc. & pers. serv. activ., etc.	0.0 [b]	..	0.0	0.0	0.0	0.0	..	**59 Adm. pub., serv. soc. & collec., etc**
TOTAL BUSINESS ENTERPRISE SECTOR	**217.5** [a,b,]	**243.5** [d]	**265.1** [d]	**441.7** [a]	**290.7**	**233.9**	**253.4**	**TOTAL SECTEUR DES ENTREPRISES**

Source: OECD, R&D database (DSTI/EAS Division), January 2002.

Source : OCDE, base de données R-D (DSTI/Division AES), janvier 2002.

Business enterprise intramural expenditure on R&D (BERD) - Dépenses intra-muros de R-D des entreprises (DIRDE)

by industry and source of funds - par industrie et par source de financement

Million SKK | *Millions SKK*

	1994	1995	1996	1997	1998	1999	2000	
	10.1 Business enterprise - Entreprises							
1 AGRICULTURE, HUNTING & FORESTRY	..	..	..	..	116.0 [a]	137.0	..	**1 AGRICULTURE, CHASSE & SYLVICULTURE**
2 MINING	..	0.0	0.0	..	..	..	..	**2 ACTIVITES EXTRACTIVES**
3 MANUFACTURING	..	1 367.0	1 615.0	1 252.0	2 211.0 [a]	1 797.0	..	**3 ACTIVITES DE FABRICATION**
4 Food, beverages & tobacco	..	2.0	3.0	4.0	4.0 [a]	..	..	**4 Produits alimentaires, boissons & tabac**
5 Food products & beverages	..	2.0	3.0	4.0	4.0 [a]	..	..	5 Produits alimentaires & boissons
6 Tobacco products	..	0.0	0.0	0.0	0.0	0.0	..	6 Produits à base de tabac
7 Textiles, wear app., fur & leather	..	..	..	..	7.0 [a]	6.0	..	**7 Textiles, habill., fourrures & cuirs**
8 Textiles	..	..	..	..	7.0 [a]	6.0	..	8 Textiles
9 Wearing apparel & fur	..	0.0	0.0	0.0	0.0	0.0	..	9 Articles d'habillement & fourrures
10 Leather products & footwear	..	0.0	0.0	0.0	0.0	0.0	..	10 Articles en cuir & chaussures
11 Wood, paper, printing, publishing	..	..	..	..	..	..	..	**11 Bois, papier, imprimerie, édition**
12 Wood & cork (not furniture)	..	..	0.0	0.0	..	..	..	12 Bois & liège (sauf meubles)
13 Pulp, paper & paper products	..	..	..	..	..	..	..	13 Carton, papier & articles
14 Publ., print. & repro. of rec. media	..	0.0	0.0	0.0	0.0	0.0	..	14 Edition, impress. & repro. enregis.
15 Coke, petroleum, nuclear fuel, chemicals, rubber plastics	..	658.0	647.0	583.0	809.0 [a]	831.0	..	**15 Coke, pétrole, combust. nucl., prod. chimi., caout. & mat. plast.**
16 Coke, ref. petrol. prod. & nuclear fuel	..	..	..	..	0.0	..	..	16 Coke, raffin. pétr. & combust. nucl.
17 Chemicals & chemical products	..	..	..	..	..	..	..	17 Produits chimiques
18 *Chemical prod.(less pharmaceuticals)*	..	..	..	..	..	..	..	18 *Prod. chimi. (moins prod. pharmaceutiques)*
19 *Pharmaceuticals*	..	..	0.0	..	..	..	..	19 *Produits pharmaceutiques*
20 Rubber & plastic products	..	325.0	323.0	224.0	258.0 [a]	293.0	..	20 Caoutchouc et matières plastiques
21 Non-metallic mineral products	..	20.0	10.0	..	0.0	..	..	**21 Produits minéraux non métalliques**
22 Basic metals	..	135.0	140.0	134.0	131.0 [a]	115.0	..	**22 Produits métallurgiques de base**
23 Basic metals, ferrous	..	..	..	..	..	..	..	23 Produits métallur. de base, ferreux
24 Basic metals, non-ferrous	..	..	0.0	..	..	0.0	..	24 Prod. métallur. de base, non ferreux
25 Fabricated metal products	..	4.0	2.0	6.0	33.0 [a]	154.0	..	**25 Ouvrages en métaux (sauf mach.& mat.)**
26 Machinery equipment, instruments and transport equipment	..	537.0	803.0	502.0	1 016.0 [a]	417.0	..	**26 Machines & matériel, instruments & mat. de transport**
27 Machinery n.e.c.	..	171.0	248.0	106.0	747.0 [a]	144.0	..	27 Machines, n.c.a.
28 Office, account. & computing machinery	..	..	..	0.0	0.0	..	..	28 Mach. de bur., compta. & trait. inform.
29 Electrical machinery	..	21.0	42.0	..	..	47.0	..	29 Machines électriques
30 Electro. equip. (radio, TV & commun.)	..	98.0	109.0	81.0	..	70.0	..	30 Equip. électro. (radio, tv & communic.)
31 *Electro. comp. (inc. semi-conductors)*	..	..	0.0	0.0	0.0	0.0	..	31 *Compos. électro. (semi-cond. incl.)*
32 *TV, radio & communications equip.*	..	..	109.0	0.0	0.0	70.0	..	32 *Appareils radio, tv et de communi.*
33 Instruments	..	37.0	89.0	..	49.0 [a]	40.0	..	33 Instruments
34 Motor vehicles	..	105.0	93.0	73.0	17.0 [a]	..	..	34 Véhicules automobiles
35 Other transport equipment	..	..	..	..	..	..	..	35 Autres matériels de transport
36 *Ships*	..	0.0	0.0	0.0	0.0	0.0	..	36 *Navires*
37 *Aerospace*	..	..	..	..	0.0	0.0	..	37 *Aérospatiale*
38 *Other transport equipment n.e.c.*	..	0.0	0.0	..	..	..	..	38 *Autres matériels de transport n.c.a.*
39 Furniture, other manufacturing n.e.c.	..	..	..	..	..	..	..	**39 Meubles, autres activ. de fabricat. n.c.a.**
40 Furniture	..	..	..	0.0	0.0	0.0	..	40 Meubles
41 Other manufacturing n.e.c.	..	0.0	0.0	..	..	..	..	41 Autres activités de fabrication n.c.a.
42 Recycling	..	0.0	0.0	0.0	0.0	0.0	..	**42 Récupération**
43 ELECTRICITY, GAS & WATER SUPPLY	..	0.0	..	..	..	..	..	**43 DISTRIBUTION D'ELECTRICITE, GAZ & EAU**
44 CONSTRUCTION	..	..	19.0	13.0	27.0 [a]	21.0	..	**44 CONSTRUCTION**
45 SERVICES SECTOR	..	1 159.0	1 118.0	3 491.0 [a]	531.0 [a]	513.0	..	**45 SECTEUR DES SERVICES**
46 Wholesale, retail trade & repair of motor vehicles, etc.	..	0.0	0.0	0.0	0.0	0.0	..	**46 Commerce (gros & détail), réparation de véhicules, etc.**
47 Hotels & restaurants	..	0.0	0.0	0.0	0.0	0.0	..	**47 Hôtels et restaurants**
48 Transport & storage	..	0.0	..	..	..	..	..	**48 Transports, entreposage**
49 Communications	..	0.0	..	..	..	..	..	**49 Communications**
50 Post	..	0.0	0.0	0.0	0.0	0.0	..	50 Postes
51 Telecommunications	..	0.0	0.0	..	..	0.0	..	51 Télécommunications
52 Financial intermediation (inc. insur.)	..	0.0	0.0	0.0	0.0	0.0	..	**52 Intermédiation financ. (assurances inc.)**
53 Real estate, renting & busin. activity	..	1 159.0	1 087.0	3 459.0 [a]	438.0 [a]	439.0	..	**53 Immobil., locat. & serv. aux entreprises**
54 Computer & related activities	..	..	0.0	..	..	..	..	54 Informatique et activités connexes
55 *Software consultancy*	..	..	0.0	0.0	0.0	..	..	55 *Production de logiciels*
56 *Other computer services n.e.c.*	..	0.0	0.0	0.0	0.0	0.0	..	56 *Autres activités informatiques*
57 Research & development	..	1 155.0 [o]	1 087.0 [o]	3 450.0 [a,o]	377.0 [a]	431.0	..	57 Recherche & développement expérimental
58 Other business activities n.e.c.	..	..	0.0	..	..	..	..	58 Autres activ. & serv. aux entreprises
59 Comm., soc. & pers. serv. activ., etc.	..	0.0	0.0	0.0	0.0	0.0	..	**59 Adm. pub., serv. soc. & collec., etc**
TOTAL BUSINESS ENTERPRISE SECTOR	**1 985.0** [a]	**2 540.0**	**2 768.0**	**4 787.0** [a]	**3 019.0**	**2 564.0**	**3 118.0**	**TOTAL SECTEUR DES ENTREPRISES**

Source: OECD, R&D database (DSTI/EAS Division), January 2002.

Source : OCDE, base de données R-D (DSTI/Division AES), janvier 2002.

SLOVAK REPUBLIC — Table 10 - Tableau 10 — RÉPUBLIQUE SLOVAQUE

Business enterprise intramural expenditure on R&D (BERD) - Dépenses intra-muros de R-D des entreprises (DIRDE)

by industry and source of funds - par industrie et par source de financement

Million SKK — *Millions SKK*

	1994	1995	1996	1997	1998	1999	2000	
				10.2 Government - État				
1 AGRICULTURE, HUNTING & FORESTRY	..	..	..	..	266.0 [a]	316.0	..	**1 AGRICULTURE, CHASSE & SYLVICULTURE**
2 MINING	..	0.0	0.0	0.0	0.0	0.0	..	**2 ACTIVITES EXTRACTIVES**
3 MANUFACTURING	..	136.0	104.0	80.0	194.0 [a]	198.0	..	**3 ACTIVITES DE FABRICATION**
4 Food, beverages & tobacco	..	0.0	0.0	0.0	0.0	0.0	..	**4 Produits alimentaires, boissons & tabac**
5 Food products & beverages	..	0.0	0.0	0.0	0.0	0.0	..	5 Produits alimentaires & boissons
6 Tobacco products	..	0.0	0.0	0.0	0.0	0.0	..	6 Produits à base de tabac
7 Textiles, wear app., fur & leather	..	0.0	0.0	0.0	2.0 [a]	0.0	..	**7 Textiles, habill., fourrures & cuirs**
8 Textiles	..	0.0	0.0	0.0	2.0 [a]	0.0	..	8 Textiles
9 Wearing apparel & fur	..	0.0	0.0	0.0	0.0	0.0	..	9 Articles d'habillement & fourrures
10 Leather products & footwear	..	0.0	0.0	0.0	0.0	0.0	..	10 Articles en cuir & chaussures
11 Wood, paper, printing, publishing	..	0.0	0.0	0.0	..	..	..	**11 Bois, papier, imprimerie, édition**
12 Wood & cork (not furniture)	..	0.0	0.0	0.0	..	..	..	12 Bois & liège (sauf meubles)
13 Pulp, paper & paper products	..	0.0	0.0	0.0	0.0	0.0	..	13 Carton, papier & articles
14 Publ., print. & repro. of rec. media	..	0.0	0.0	0.0	0.0	0.0	..	14 Edition, impress. & repro. enregis.
15 Coke, petroleum, nuclear fuel, chemicals, rubber plastics	..	10.0	..	13.0	71.0 [a]	91.0	..	**15 Coke, pétrole, combust. nucl., prod. chimi., caout. & mat. plast.**
16 Coke, ref. petrol. prod. & nuclear fuel	..	0.0	0.0	0.0	..	..	..	16 Coke, raffin. pétr. & combust. nucl.
17 Chemicals & chemical products	..	..	..	..	..	..	..	17 Produits chimiques
18 *Chemical prod.(less pharmaceuticals)*	..	0.0	0.0	..	..	..	..	18 *Prod. chimi. (moins prod. pharmaceutiques)*
19 *Pharmaceuticals*	..	..	0.0	0.0	..	..	..	19 *Produits pharmaceutiques*
20 Rubber & plastic products	..	..	..	8.0	5.0 [a]	30.0	..	20 Caoutchouc et matières plastiques
21 Non-metallic mineral products	..	0.0	0.0	0.0	0.0	0.0	..	**21 Produits minéraux non métalliques**
22 Basic metals	..	0.0	0.0	0.0	0.0	0.0	..	**22 Produits métallurgiques de base**
23 Basic metals, ferrous	..	0.0	0.0	0.0	0.0	0.0	..	23 Produits métallur. de base, ferreux
24 Basic metals, non-ferrous	..	0.0	0.0	0.0	0.0	0.0	..	24 Prod. métallur. de base, non ferreux
25 Fabricated metal products	..	0.0	..	0.0	0.0	..	..	**25 Ouvrages en métaux (sauf mach.& mat.)**
26 Machinery equipment, instruments and transport equipment	..	123.0	95.0	67.0	107.0 [a]	77.0	..	**26 Machines & matériel, instruments & mat. de transport**
27 Machinery n.e.c.	..	71.0	74.0	26.0	95.0 [a]	47.0	..	27 Machines, n.c.a.
28 Office, account. & computing machinery	..	0.0	0.0	0.0	0.0	0.0	..	28 Mach. de bur., compta. & trait. inform.
29 Electrical machinery	..	6.0	8.0	..	..	..	..	29 Machines électriques
30 Electro. equip. (radio, TV & commun.)	..	0.0	0.0	0.0	0.0	0.0	..	30 Equip. électro. (radio, tv & communic.)
31 *Electro. comp. (inc. semi-conductors)*	..	0.0	0.0	0.0	0.0	0.0	..	31 *Compos. électro. (semi-cond. incl.)*
32 *TV, radio & communications equip.*	..	0.0	0.0	0.0	0.0	0.0	..	32 *Appareils radio, tv et de communi.*
33 Instruments	..	18.0	..	..	0.0	..	..	33 Instruments
34 Motor vehicles	..	0.0	..	8.0	0.0	..	..	34 Véhicules automobiles
35 Other transport equipment	..	..	..	..	0.0	0.0	..	35 Autres matériels de transport
36 *Ships*	..	0.0	0.0	0.0	0.0	0.0	..	36 *Navires*
37 *Aerospace*	..	..	..	..	0.0	0.0	..	37 *Aérospatiale*
38 *Other transport equipment n.e.c.*	..	0.0	0.0	0.0	0.0	0.0	..	38 *Autres matériels de transport n.c.a.*
39 Furniture, other manufacturing n.e.c.	..	..	0.0	0.0	0.0	0.0	..	**39 Meubles, autres activ. de fabricat. n.c.a.**
40 Furniture	..	0.0	0.0	0.0	0.0	0.0	..	40 Meubles
41 Other manufacturing n.e.c.	..	0.0	0.0	0.0	0.0	0.0	..	41 Autres activités de fabrication n.c.a.
42 Recycling	..	0.0	0.0	0.0	0.0	0.0	..	**42 Récupération**
43 ELECTRICITY, GAS & WATER SUPPLY	..	0.0	..	..	..	..	..	**43 DISTRIBUTION D'ELECTRICITE, GAZ & EAU**
44 CONSTRUCTION	..	0.0	0.0	0.0	7.0 [a]	8.0	..	**44 CONSTRUCTION**
45 SERVICES SECTOR	..	172.0	286.0	862.0 [a]	355.0 [a]	273.0	..	**45 SECTEUR DES SERVICES**
46 Wholesale, retail trade & repair of motor vehicles, etc.	..	0.0	0.0	0.0	0.0	0.0	..	**46 Commerce (gros & détail), réparation de véhicules, etc.**
47 Hotels & restaurants	..	0.0	0.0	0.0	0.0	0.0	..	**47 Hôtels et restaurants**
48 Transport & storage	..	0.0	0.0	0.0	..	..	..	**48 Transports, entreposage**
49 Communications	..	0.0	0.0	0.0	..	..	..	**49 Communications**
50 Post	..	0.0	0.0	0.0	0.0	0.0	..	50 Postes
51 Telecommunications	..	0.0	0.0	0.0	0.0	0.0	..	51 Télécommunications
52 Financial intermediation (inc. insur.)	..	0.0	0.0	0.0	0.0	0.0	..	**52 Intermédiation financ. (assurances inc.)**
53 Real estate, renting & busin. activity	..	172.0	286.0	862.0 [a]	317.0 [a]	238.0	..	**53 Immobil., locat. & serv. aux entreprises**
54 Computer & related activities	..	0.0	0.0	0.0	..	0.0	..	54 Informatique et activités connexes
55 *Software consultancy*	..	0.0	0.0	0.0	0.0	0.0	..	55 *Production de logiciels*
56 *Other computer services n.e.c.*	..	0.0	0.0	0.0	0.0	0.0	..	56 *Autres activités informatiques*
57 Research & development	..	172.0 [o]	286.0 [o]	862.0 [a,o]	304.0 [a]	231.0	..	57 Recherche & développement expérimental
58 Other business activities n.e.c.	..	0.0	0.0	0.0	..	..	..	58 Autres activ. & serv. aux entreprises
59 Comm., soc. & pers. serv. activ., etc.	..	..	0.0	0.0	0.0	0.0	..	**59 Adm. pub., serv. soc. & collec., etc**
TOTAL BUSINESS ENTERPRISE SECTOR	**324.0 [a,d]**	**312.0 [d]**	**406.0 [d]**	**967.0 [a]**	**910.0**	**848.0**	**824.0**	**TOTAL SECTEUR DES ENTREPRISES**

Source: OECD, R&D database (DSTI/EAS Division), January 2002.

Source : OCDE, base de données R-D (DSTI/Division AES), janvier 2002.

Business enterprise intramural expenditure on R&D (BERD) - Dépenses intra-muros de R-D des entreprises (DIRDE)

by industry and source of funds - par industrie et par source de financement

Million SKK — *Millions SKK*

10.3 Other national funds - Autres sources nationales

	1994	1995	1996	1997	1998	1999	2000	
1 AGRICULTURE, HUNTING & FORESTRY	..	..	..	0.0	2.0 [a]	..	..	**1 AGRICULTURE, CHASSE & SYLVICULTURE**
2 MINING	..	..	..	0.0	0.0	..	..	**2 ACTIVITES EXTRACTIVES**
3 MANUFACTURING	..	..	..	0.0	0.0	..	..	**3 ACTIVITES DE FABRICATION**
4 Food, beverages & tobacco	..	..	..	0.0	0.0	..	..	**4 Produits alimentaires, boissons & tabac**
5 Food products & beverages	..	..	..	0.0	0.0	..	..	5 Produits alimentaires & boissons
6 Tobacco products	..	..	..	0.0	0.0	..	..	6 Produits à base de tabac
7 Textiles, wear app., fur & leather	..	..	..	0.0	0.0	..	..	**7 Textiles, habill., fourrures & cuirs**
8 Textiles	..	..	..	0.0	0.0	..	..	8 Textiles
9 Wearing apparel & fur	..	..	..	0.0	0.0	..	..	9 Articles d'habillement & fourrures
10 Leather products & footwear	..	..	..	0.0	0.0	..	..	10 Articles en cuir & chaussures
11 Wood, paper, printing, publishing	..	..	..	0.0	0.0	..	..	**11 Bois, papier, imprimerie, édition**
12 Wood & cork (not furniture)	..	..	..	0.0	0.0	..	..	12 Bois & liège (sauf meubles)
13 Pulp, paper & paper products	..	..	..	0.0	0.0	..	..	13 Carton, papier & articles
14 Publ., print. & repro. of rec. media	..	..	..	0.0	0.0	..	..	14 Edition, impress. & repro. enregis.
15 Coke, petroleum, nuclear fuel, chemicals, rubber plastics	..	..	..	0.0	0.0	..	..	**15 Coke, pétrole, combust. nucl., prod. chimi., caout. & mat. plast.**
16 Coke, ref. petrol. prod. & nuclear fuel	..	..	..	0.0	0.0	..	..	16 Coke, raffin. pétr. & combust. nucl.
17 Chemicals & chemical products	..	..	..	0.0	0.0	..	..	17 Produits chimiques
18 *Chemical prod.(less pharmaceuticals)*	..	..	..	0.0	0.0	..	..	18 *Prod. chimi. (moins prod. pharmaceutiques)*
19 *Pharmaceuticals*	..	..	..	0.0	0.0	..	..	19 *Produits pharmaceutiques*
20 Rubber & plastic products	..	..	..	0.0	0.0	..	..	20 Caoutchouc et matières plastiques
21 Non-metallic mineral products	..	..	..	0.0	0.0	..	..	**21 Produits minéraux non métalliques**
22 Basic metals	..	..	..	0.0	0.0	..	..	**22 Produits métallurgiques de base**
23 Basic metals, ferrous	..	..	..	0.0	0.0	..	..	23 Produits métallur. de base, ferreux
24 Basic metals, non-ferrous	..	..	..	0.0	0.0	..	..	24 Prod. métallur. de base, non ferreux
25 Fabricated metal products	..	..	..	0.0	0.0	..	..	**25 Ouvrages en métaux (sauf mach.& mat.)**
26 Machinery equipment, instruments and transport equipment	..	..	..	0.0	0.0	..	..	**26 Machines & matériel, instruments & mat. de transport**
27 Machinery n.e.c.	..	..	..	0.0	0.0	..	..	27 Machines, n.c.a.
28 Office, account. & computing machinery	..	..	..	0.0	0.0	..	..	28 Mach. de bur., compta. & trait. inform.
29 Electrical machinery	..	..	..	0.0	..	..	..	29 Machines électriques
30 Electro. equip. (radio, TV & commun.)	..	..	..	0.0	0.0	..	..	30 Equip. électro. (radio, tv & communic.)
31 *Electro. comp. (inc. semi-conductors)*	..	..	..	0.0	0.0	..	..	31 *Compos. électro. (semi-cond. incl.)*
32 *TV, radio & communications equip.*	..	..	..	0.0	0.0	..	..	32 *Appareils radio, tv et de communi.*
33 Instruments	..	..	..	0.0	0.0	..	..	33 Instruments
34 Motor vehicles	..	..	..	0.0	0.0	..	..	34 Véhicules automobiles
35 Other transport equipment	..	..	..	0.0	0.0	..	..	35 Autres matériels de transport
36 *Ships*	..	..	..	0.0	0.0	..	..	36 *Navires*
37 *Aerospace*	..	..	..	0.0	0.0	..	..	37 *Aérospatiale*
38 *Other transport equipment n.e.c.*	..	..	..	0.0	0.0	..	..	38 *Autres matériels de transport n.c.a.*
39 Furniture, other manufacturing n.e.c.	..	..	..	0.0	0.0	..	..	**39 Meubles, autres activ. de fabricat. n.c.a.**
40 Furniture	..	..	..	0.0	0.0	..	..	40 Meubles
41 Other manufacturing n.e.c.	..	..	..	0.0	0.0	..	..	41 Autres activités de fabrication n.c.a.
42 Recycling	..	..	..	0.0	0.0	..	..	**42 Récupération**
43 ELECTRICITY, GAS & WATER SUPPLY	..	..	..	0.0	0.0	..	..	**43 DISTRIBUTION D'ELECTRICITE, GAZ & EAU**
44 CONSTRUCTION	..	..	..	0.0	0.0	..	..	**44 CONSTRUCTION**
45 SERVICES SECTOR	..	..	..	3.0	0.0	..	..	**45 SECTEUR DES SERVICES**
46 Wholesale, retail trade & repair of motor vehicles, etc.	..	..	..	0.0	0.0	..	..	**46 Commerce (gros & détail), réparation de véhicules, etc.**
47 Hotels & restaurants	..	..	..	0.0	0.0	..	..	**47 Hôtels et restaurants**
48 Transport & storage	..	..	..	0.0	0.0	..	..	**48 Transports, entreposage**
49 Communications	..	..	..	0.0	0.0	..	..	**49 Communications**
50 Post	..	..	..	0.0	0.0	..	..	50 Postes
51 Telecommunications	..	..	..	0.0	0.0	..	..	51 Télécommunications
52 Financial intermediation (inc. insur.)	..	..	..	0.0	0.0	..	..	**52 Intermédiation financ. (assurances inc.)**
53 Real estate, renting & busin. activity	..	..	..	3.0	0.0 [a]	..	..	**53 Immobil., locat. & serv. aux entreprises**
54 Computer & related activities	..	..	..	0.0	0.0	..	..	54 Informatique et activités connexes
55 *Software consultancy*	..	..	..	0.0	0.0	..	..	55 *Production de logiciels*
56 *Other computer services n.e.c.*	..	..	..	0.0	0.0	..	..	56 *Autres activités informatiques*
57 Research & development	..	..	..	3.0	0.0 [a]	..	..	57 Recherche & développement expérimental
58 Other business activities n.e.c.	..	..	..	0.0	0.0	..	..	58 Autres activ. & serv. aux entreprises
59 Comm., soc. & pers. serv. activ., etc.	..	..	..	0.0	0.0	..	..	**59 Adm. pub., serv. soc. & collec., etc**
TOTAL BUSINESS ENTERPRISE SECTOR	**5.0**	**0.0**	**0.0**	**3.0**	**2.0**	**0.0**	**0.0**	**TOTAL SECTEUR DES ENTREPRISES**

Source: OECD, R&D database (DSTI/EAS Division), January 2002.

Source : OCDE, base de données R-D (DSTI/Division AES), janvier 2002.

Business enterprise intramural expenditure on R&D (BERD) - Dépenses intra-muros de R-D des entreprises (DIRDE)

by industry and source of funds - par industrie et par source de financement

Million SKK / *Millions SKK*

10.4 Funds from abroad - Fonds de l'étranger

	1994	1995	1996	1997	1998	1999	2000	
1 AGRICULTURE, HUNTING & FORESTRY	..	0.0	0.0	0.0	8.0 [a]	12.0	..	**1 AGRICULTURE, CHASSE & SYLVICULTURE**
2 MINING	..	0.0	0.0	0.0	0.0	0.0	..	**2 ACTIVITES EXTRACTIVES**
3 MANUFACTURING	..	0.0	38.0	0.0	72.0 [a]	21.0	..	**3 ACTIVITES DE FABRICATION**
4 Food, beverages & tobacco	..	0.0	0.0	0.0	0.0	0.0	..	**4 Produits alimentaires, boissons & tabac**
5 Food products & beverages	..	0.0	0.0	0.0	0.0	0.0	..	5 Produits alimentaires & boissons
6 Tobacco products	..	0.0	0.0	0.0	0.0	0.0	..	6 Produits à base de tabac
7 Textiles, wear app., fur & leather	..	0.0	0.0	0.0	0.0	1.0	..	**7 Textiles, habill., fourrures & cuirs**
8 Textiles	..	0.0	0.0	0.0	0.0	1.0	..	8 Textiles
9 Wearing apparel & fur	..	0.0	0.0	0.0	0.0	0.0	..	9 Articles d'habillement & fourrures
10 Leather products & footwear	..	0.0	0.0	0.0	0.0	0.0	..	10 Articles en cuir & chaussures
11 Wood, paper, printing, publishing	..	0.0	0.0	0.0	..	..	..	**11 Bois, papier, imprimerie, édition**
12 Wood & cork (not furniture)	..	0.0	0.0	0.0	..	..	..	12 Bois & liège (sauf meubles)
13 Pulp, paper & paper products	..	0.0	0.0	0.0	0.0	0.0	..	13 Carton, papier & articles
14 Publ., print. & repro. of rec. media	..	0.0	0.0	0.0	0.0	0.0	..	14 Edition, impress. & repro. enregis.
15 Coke, petroleum, nuclear fuel, chemicals, rubber plastics	..	0.0	..	0.0	6.0 [a]	16.0	..	**15 Coke, pétrole, combust. nucl., prod. chimi., caout. & mat. plast.**
16 Coke, ref. petrol. prod. & nuclear fuel	..	0.0	0.0	0.0	0.0	0.0	..	16 Coke, raffin. pétr. & combust. nucl.
17 Chemicals & chemical products	..	0.0	0.0	0.0	..	..	..	17 Produits chimiques
18 *Chemical prod.(less pharmaceuticals)*	..	0.0	0.0	0.0	..	..	..	18 *Prod. chimi. (moins prod. pharmaceutiques)*
19 *Pharmaceuticals*	..	0.0	0.0	0.0	1.0 [a]	0.0	..	19 *Produits pharmaceutiques*
20 Rubber & plastic products	..	0.0	..	0.0	0.0	0.0	..	20 Caoutchouc et matières plastiques
21 Non-metallic mineral products	..	0.0	0.0	0.0	0.0	0.0	..	**21 Produits minéraux non métalliques**
22 Basic metals	..	0.0	0.0	0.0	0.0	0.0	..	**22 Produits métallurgiques de base**
23 Basic metals, ferrous	..	0.0	0.0	0.0	0.0	0.0	..	23 Produits métallur. de base, ferreux
24 Basic metals, non-ferrous	..	0.0	0.0	0.0	0.0	0.0	..	24 Prod. métallur. de base, non ferreux
25 Fabricated metal products	..	0.0	0.0	0.0	0.0	..	..	**25 Ouvrages en métaux (sauf mach.& mat.)**
26 Machinery equipment, instruments and transport equipment	..	0.0	0.0	0.0	64.0 [a]	..	..	**26 Machines & matériel, instruments & mat. de transport**
27 Machinery n.e.c.	..	0.0	0.0	0.0	64.0 [a]	..	..	27 Machines, n.c.a.
28 Office, account. & computing machinery	..	0.0	0.0	0.0	0.0	0.0	..	28 Mach. de bur., compta. & trait. inform.
29 Electrical machinery	..	0.0	..	0.0	..	0.0	..	29 Machines électriques
30 Electro. equip. (radio, TV & commun.)	..	0.0	0.0	0.0	0.0	0.0	..	30 Equip. électro. (radio, tv & communic.)
31 *Electro. comp. (inc. semi-conductors)*	..	0.0	0.0	0.0	0.0	0.0	..	31 *Compos. électro. (semi-cond. incl.)*
32 *TV, radio & communications equip.*	..	0.0	0.0	0.0	0.0	0.0	..	32 *Appareils radio, tv et de communi.*
33 Instruments	..	0.0	0.0	0.0	0.0	0.0	..	33 Instruments
34 Motor vehicles	..	0.0	0.0	0.0	0.0	0.0	..	34 Véhicules automobiles
35 Other transport equipment	..	0.0	0.0	0.0	0.0	0.0	..	35 Autres matériels de transport
36 *Ships*	..	0.0	0.0	0.0	0.0	0.0	..	36 *Navires*
37 *Aerospace*	..	0.0	0.0	0.0	0.0	0.0	..	37 *Aérospatiale*
38 *Other transport equipment n.e.c.*	..	0.0	0.0	0.0	0.0	0.0	..	38 *Autres matériels de transport n.c.a.*
39 Furniture, other manufacturing n.e.c.	..	0.0	0.0	0.0	0.0	0.0	..	**39 Meubles, autres activ. de fabricat. n.c.a.**
40 Furniture	..	0.0	0.0	0.0	0.0	0.0	..	40 Meubles
41 Other manufacturing n.e.c.	..	0.0	0.0	0.0	0.0	0.0	..	41 Autres activités de fabrication n.c.a.
42 Recycling	..	0.0	0.0	0.0	0.0	0.0	..	**42 Récupération**
43 ELECTRICITY, GAS & WATER SUPPLY	..	0.0	0.0	0.0	..	..	..	**43 DISTRIBUTION D'ELECTRICITE, GAZ & EAU**
44 CONSTRUCTION	..	0.0	0.0	0.0	0.0	0.0	..	**44 CONSTRUCTION**
45 SERVICES SECTOR	..	45.0	84.0	97.0	40.0 [a]	26.0	..	**45 SECTEUR DES SERVICES**
46 Wholesale, retail trade & repair of motor vehicles, etc.	..	0.0	0.0	0.0	0.0	0.0	..	**46 Commerce (gros & détail), réparation de véhicules, etc.**
47 Hotels & restaurants	..	0.0	0.0	0.0	0.0	0.0	..	**47 Hôtels et restaurants**
48 Transport & storage	..	0.0	0.0	0.0	0.0	0.0	..	**48 Transports, entreposage**
49 Communications	..	0.0	0.0	0.0	0.0	0.0	..	**49 Communications**
50 Post	..	0.0	0.0	0.0	0.0	0.0	..	50 Postes
51 Telecommunications	..	0.0	0.0	0.0	0.0	0.0	..	51 Télécommunications
52 Financial intermediation (inc. insur.)	..	0.0	0.0	0.0	0.0	0.0	..	**52 Intermédiation financ. (assurances inc.)**
53 Real estate, renting & busin. activity	..	45.0	84.0	97.0	40.0 [a]	26.0	..	**53 Immobil., locat. & serv. aux entreprises**
54 Computer & related activities	..	0.0	0.0	0.0	..	..	..	54 Informatique et activités connexes
55 *Software consultancy*	..	0.0	0.0	0.0	0.0	..	..	55 *Production de logiciels*
56 *Other computer services n.e.c.*	..	0.0	0.0	0.0	0.0	0.0	..	56 *Autres activités informatiques*
57 Research & development	..	45.0 [o]	84.0 [o]	97.0 [o]	38.0 [a]	17.0	..	57 Recherche & développement expérimental
58 Other business activities n.e.c.	..	0.0	0.0	0.0	..	0.0	..	58 Autres activ. & serv. aux entreprises
59 Comm., soc. & pers. serv. activ., etc.	..	0.0	0.0	0.0	0.0	0.0	..	**59 Adm. pub., serv. soc. & collec., etc**
TOTAL BUSINESS ENTERPRISE SECTOR	**44.0**	**45.0**	**122.0**	**97.0**	**119.0**	**61.0**	**63.0**	**TOTAL SECTEUR DES ENTREPRISES**

Source: OECD, R&D database (DSTI/EAS Division), January 2002.

Source : OCDE, base de données R-D (DSTI/Division AES), janvier 2002.

SLOVAK REPUBLIC — Table 11 - Tableau 11 — RÉPUBLIQUE SLOVAQUE

Business enterprise intramural expenditure on R&D (BERD) - Dépenses intra-muros de R-D des entreprises (DIRDE)

by industry and type of costs - par industrie et par type de coûts

Million SKK — *Millions SKK*

	1994	1995	1996	1997	1998	1999	2000	
	11.1 Current expenditure - Dépenses courantes							
1 AGRICULTURE, HUNTING & FORESTRY	..	..	..	21.0	356.0 [a]	439.0	..	**1 AGRICULTURE, CHASSE & SYLVICULTURE**
2 MINING	..	0.0	0.0	..	..	..	..	**2 ACTIVITES EXTRACTIVES**
3 MANUFACTURING	..	1 273.0	1 491.0	1 245.0	2 244.0 [a]	1 821.0	..	**3 ACTIVITES DE FABRICATION**
4 Food, beverages & tobacco	..	2.0	3.0	4.0	4.0 [a]	..	..	**4 Produits alimentaires, boissons & tabac**
5 Food products & beverages	..	2.0	3.0	4.0	4.0 [a]	..	..	5 Produits alimentaires & boissons
6 Tobacco products	..	0.0	0.0	0.0	0.0	0.0	..	6 Produits à base de tabac
7 Textiles, wear app., fur & leather	..	..	..	..	..	..	..	**7 Textiles, habill., fourrures & cuirs**
8 Textiles	..	..	..	..	..	..	..	8 Textiles
9 Wearing apparel & fur	..	0.0	0.0	0.0	0.0	0.0	..	9 Articles d'habillement & fourrures
10 Leather products & footwear	..	0.0	0.0	0.0	0.0	0.0	..	10 Articles en cuir & chaussures
11 Wood, paper, printing, publishing	..	..	..	..	..	..	..	**11 Bois, papier, imprimerie, édition**
12 Wood & cork (not furniture)	..	..	0.0	0.0	..	..	..	12 Bois & liège (sauf meubles)
13 Pulp, paper & paper products	..	..	..	..	..	..	..	13 Carton, papier & articles
14 Publ., print. & repro. of rec. media	..	0.0	0.0	0.0	0.0	0.0	..	14 Edition, impress. & repro. enregis.
15 Coke, petroleum, nuclear fuel, chemicals, rubber plastics	..	535.0	549.0	555.0	839.0 [a]	869.0	..	**15 Coke, pétrole, combust. nucl., prod. chimi., caout. & mat. plast.**
16 Coke, ref. petrol. prod. & nuclear fuel	..	..	..	..	..	..	..	16 Coke, raffin. pétr. & combust. nucl.
17 Chemicals & chemical products	..	..	..	..	..	..	..	17 Produits chimiques
18 *Chemical prod.(less pharmaceuticals)*	..	..	..	..	..	..	..	18 *Prod. chimi. (moins prod. pharmaceutiques)*
19 *Pharmaceuticals*	..	..	..	..	..	..	..	19 *Produits pharmaceutiques*
20 Rubber & plastic products	..	211.0	240.0	225.0	254.0 [a]	311.0	..	20 Caoutchouc et matières plastiques
21 Non-metallic mineral products	..	20.0	10.0	..	..	..	..	**21 Produits minéraux non métalliques**
22 Basic metals	..	133.0	137.0	134.0	131.0 [a]	115.0	..	**22 Produits métallurgiques de base**
23 Basic metals, ferrous	..	..	..	..	..	..	..	23 Produits métallur. de base, ferreux
24 Basic metals, non-ferrous	..	..	..	..	..	..	..	24 Prod. métallur. de base, non ferreux
25 Fabricated metal products	..	4.0	..	6.0	33.0 [a]	167.0	..	**25 Ouvrages en métaux (sauf mach.& mat.)**
26 Machinery equipment, instruments and transport equipment	..	568.0	779.0	527.0	1 076.0 [a]	458.0	..	**26 Machines & matériel, instruments & mat. de transport**
27 Machinery n.e.c.	..	167.0	317.0	100.0	809.0 [a]	186.0	..	27 Machines, n.c.a.
28 Office, account. & computing machinery	..	..	..	0.0	0.0	..	..	28 Mach. de bur., compta. & trait. inform.
29 Electrical machinery	..	26.0	49.0	64.0	104.0 [a]	49.0	..	29 Machines électriques
30 Electro. equip. (radio, TV & commun.)	..	94.0	102.0	75.0	..	64.0	..	30 Equip. électro. (radio, tv & communic.)
31 *Electro. comp. (inc. semi-conductors)*	..	..	0.0	0.0	0.0	0.0	..	31 *Compos. électro. (semi-cond. incl.)*
32 *TV, radio & communications equip.*	..	..	102.0	0.0	0.0	64.0	..	32 *Appareils radio, tv et de communi.*
33 Instruments	..	54.0	91.0	96.0	..	50.0	..	33 Instruments
34 Motor vehicles	..	105.0	86.0	79.0	..	..	..	34 Véhicules automobiles
35 Other transport equipment	..	..	..	113.0	..	..	..	35 Autres matériels de transport
36 *Ships*	..	0.0	0.0	0.0	0.0	0.0	..	36 *Navires*
37 *Aerospace*	..	..	..	..	0.0	0.0	..	37 *Aérospatiale*
38 *Other transport equipment n.e.c.*	..	0.0	0.0	..	..	..	..	38 *Autres matériels de transport n.c.a.*
39 Furniture, other manufacturing n.e.c.	..	..	..	..	..	..	..	**39 Meubles, autres activ. de fabricat. n.c.a.**
40 Furniture	..	..	..	0.0	0.0	0.0	..	40 Meubles
41 Other manufacturing n.e.c.	..	0.0	0.0	..	..	..	..	41 Autres activités de fabrication n.c.a.
42 Recycling	..	0.0	0.0	0.0	0.0	0.0	..	**42 Récupération**
43 ELECTRICITY, GAS & WATER SUPPLY	..	0.0	..	..	..	..	..	**43 DISTRIBUTION D'ELECTRICITE, GAZ & EAU**
44 CONSTRUCTION	..	..	17.0	13.0	32.0 [a]	28.0	..	**44 CONSTRUCTION**
45 SERVICES SECTOR	..	1 209.0	1 297.0	3 995.0 [a]	864.0 [a]	772.0	..	**45 SECTEUR DES SERVICES**
46 Wholesale, retail trade & repair of motor vehicles, etc.	..	0.0	0.0	0.0	0.0	0.0	..	**46 Commerce (gros & détail), réparation de véhicules, etc.**
47 Hotels & restaurants	..	0.0	0.0	0.0	0.0	0.0	..	**47 Hôtels et restaurants**
48 Transport & storage	..	0.0	..	..	..	..	..	**48 Transports, entreposage**
49 Communications	..	0.0	..	..	..	..	..	**49 Communications**
50 Post	..	0.0	0.0	0.0	0.0	0.0	..	50 Postes
51 Telecommunications	..	0.0	0.0	..	..	0.0	..	51 Télécommunications
52 Financial intermediation (inc. insur.)	..	0.0	0.0	0.0	0.0	0.0	..	**52 Intermédiation financ. (assurances inc.)**
53 Real estate, renting & busin. activity	..	1 209.0	1 266.0	3 967.0 [a]	748.0 [a]	671.0	..	**53 Immobil., locat. & serv. aux entreprises**
54 Computer & related activities	..	..	0.0	..	..	..	..	54 Informatique et activités connexes
55 *Software consultancy*	..	..	0.0	0.0	0.0	..	..	55 *Production de logiciels*
56 *Other computer services n.e.c.*	..	..	0.0	0.0	0.0	0.0	..	56 *Autres activités informatiques*
57 Research & development	..	1 205.0 [o]	1 266.0 [o]	3 959.0 [a,o]	673.0 [a]	649.0	..	57 Recherche & développement expérimental
58 Other business activities n.e.c.	..	0.0	0.0	..	..	..	..	58 Autres activ. & serv. aux entreprises
59 Comm., soc. & pers. serv. activ., etc.	..	..	0.0	0.0	0.0	0.0	..	**59 Adm. pub., serv. soc. & collec., etc**
TOTAL BUSINESS ENTERPRISE SECTOR	**2 219.0** [a,d]	**2 501.0** [d]	**2 833.0** [d]	**5 298.0** [a]	**3 691.0**	**3 206.0**	..	**TOTAL SECTEUR DES ENTREPRISES**

Source: OECD, R&D database (DSTI/EAS Division), January 2002.

Source : OCDE, base de données R-D (DSTI/Division AES), janvier 2002.

Business enterprise intramural expenditure on R&D (BERD) - Dépenses intra-muros de R-D des entreprises (DIRDE)

by industry and type of costs - par industrie et par type de coûts

Million SKK — *Millions SKK*

11.2 Capital expenditure - Dépenses en capital

	1994	1995	1996	1997	1998	1999	2000	
1 AGRICULTURE, HUNTING & FORESTRY	..	0.0	..	3.0	36.0 [a]	26.0	..	**1 AGRICULTURE, CHASSE & SYLVICULTURE**
2 MINING	..	0.0	0.0	0.0	0.0	..	..	**2 ACTIVITES EXTRACTIVES**
3 MANUFACTURING	..	230.0	266.0	87.0	233.0 [a]	195.0	..	**3 ACTIVITES DE FABRICATION**
4 Food, beverages & tobacco	..	0.0	0.0	0.0	0.0	..	..	**4 Produits alimentaires, boissons & tabac**
5 Food products & beverages	..	0.0	0.0	0.0	0.0	..	..	5 Produits alimentaires & boissons
6 Tobacco products	..	0.0	0.0	0.0	0.0	0.0	..	6 Produits à base de tabac
7 Textiles, wear app., fur & leather	..	0.0	0.0	0.0	..	..	..	**7 Textiles, habill., fourrures & cuirs**
8 Textiles	..	0.0	0.0	0.0	..	..	..	8 Textiles
9 Wearing apparel & fur	..	0.0	0.0	0.0	0.0	0.0	..	9 Articles d'habillement & fourrures
10 Leather products & footwear	..	0.0	0.0	0.0	0.0	0.0	..	10 Articles en cuir & chaussures
11 Wood, paper, printing, publishing	..	..	..	..	..	..	..	**11 Bois, papier, imprimerie, édition**
12 Wood & cork (not furniture)	..	0.0	0.0	0.0	..	..	..	12 Bois & liège (sauf meubles)
13 Pulp, paper & paper products	..	..	..	..	..	..	..	13 Carton, papier & articles
14 Publ., print. & repro. of rec. media	..	0.0	0.0	0.0	0.0	0.0	..	14 Edition, impress. & repro. enregis.
15 Coke, petroleum, nuclear fuel, chemicals, rubber plastics	..	133.0	143.0	41.0	47.0 [a]	69.0	..	**15 Coke, pétrole, combust. nucl., prod. chimi., caout. & mat. plast.**
16 Coke, ref. petrol. prod. & nuclear fuel	..	0.0	0.0	0.0	0.0	0.0	..	16 Coke, raffin. pétr. & combust. nucl.
17 Chemicals & chemical products	..	..	..	..	..	..	..	17 Produits chimiques
18 *Chemical prod.(less pharmaceuticals)*	..	..	..	..	..	..	..	18 *Prod. chimi. (moins prod. pharmaceutiques)*
19 *Pharmaceuticals*	..	..	..	..	..	..	..	19 *Produits pharmaceutiques*
20 Rubber & plastic products	..	114.0	126.0	7.0	10.0 [a]	12.0	..	20 Caoutchouc et matières plastiques
21 Non-metallic mineral products	..	0.0	0.0	0.0	0.0	..	..	**21 Produits minéraux non métalliques**
22 Basic metals	..	2.0	3.0	0.0	0.0	0.0	..	**22 Produits métallurgiques de base**
23 Basic metals, ferrous	..	..	..	0.0	0.0	..	..	23 Produits métallur. de base, ferreux
24 Basic metals, non-ferrous	..	..	..	0.0	0.0	..	..	24 Prod. métallur. de base, non ferreux
25 Fabricated metal products	..	0.0	..	0.0	0.0	7.0	..	**25 Ouvrages en métaux (sauf mach.& mat.)**
26 Machinery equipment, instruments and transport equipment	..	92.0	119.0	42.0	111.0 [a]	36.0	..	**26 Machines & matériel, instruments & mat. de transport**
27 Machinery n.e.c.	..	75.0	5.0	32.0	97.0 [a]	5.0	..	27 Machines, n.c.a.
28 Office, account. & computing machinery	..	..	..	0.0	0.0	..	..	28 Mach. de bur., compta. & trait. inform.
29 Electrical machinery	..	1.0	1.0	0.0	7.0 [a]	4.0	..	29 Machines électriques
30 Electro. equip. (radio, TV & commun.)	..	4.0	7.0	6.0	..	6.0	..	30 Equip. électro. (radio, tv & communic.)
31 *Electro. comp. (inc. semi-conductors)*	..	..	0.0	0.0	0.0	0.0	..	31 *Compos. électro. (semi-cond. incl.)*
32 *TV, radio & communications equip.*	..	..	7.0	0.0	0.0	6.0	..	32 *Appareils radio, tv et de communi.*
33 Instruments	..	1.0	2.0	2.0	..	..	..	33 Instruments
34 Motor vehicles	..	0.0	8.0	2.0	..	..	..	34 Véhicules automobiles
35 Other transport equipment	..	..	..	0.0	..	..	..	35 Autres matériels de transport
36 *Ships*	..	0.0	0.0	0.0	0.0	0.0	..	36 *Navires*
37 *Aerospace*	..	..	..	..	0.0	0.0	..	37 *Aérospatiale*
38 *Other transport equipment n.e.c.*	..	0.0	0.0	..	..	..	..	38 *Autres matériels de transport n.c.a.*
39 Furniture, other manufacturing n.e.c.	..	0.0	0.0	..	..	..	..	**39 Meubles, autres activ. de fabricat. n.c.a.**
40 Furniture	..	0.0	0.0	0.0	0.0	0.0	..	40 Meubles
41 Other manufacturing n.e.c.	..	0.0	0.0	..	..	..	..	41 Autres activités de fabrication n.c.a.
42 Recycling	..	0.0	0.0	0.0	0.0	0.0	..	**42 Récupération**
43 ELECTRICITY, GAS & WATER SUPPLY	..	0.0	..	..	..	..	..	**43 DISTRIBUTION D'ELECTRICITE, GAZ & EAU**
44 CONSTRUCTION	..	0.0	2.0	0.0	2.0 [a]	1.0	..	**44 CONSTRUCTION**
45 SERVICES SECTOR	..	167.0	191.0	458.0 [a]	62.0 [a]	40.0	..	**45 SECTEUR DES SERVICES**
46 Wholesale, retail trade & repair of motor vehicles, etc.	..	0.0	0.0	0.0	0.0	0.0	..	**46 Commerce (gros & détail), réparation de véhicules, etc.**
47 Hotels & restaurants	..	0.0	0.0	0.0	0.0	0.0	..	**47 Hôtels et restaurants**
48 Transport & storage	..	0.0	0.0	..	..	..	..	**48 Transports, entreposage**
49 Communications	..	0.0	0.0	..	..	..	..	**49 Communications**
50 Post	..	0.0	0.0	0.0	0.0	0.0	..	50 Postes
51 Telecommunications	..	0.0	0.0	..	..	0.0	..	51 Télécommunications
52 Financial intermediation (inc. insur.)	..	0.0	0.0	0.0	0.0	0.0	..	**52 Intermédiation financ. (assurances inc.)**
53 Real estate, renting & busin. activity	..	167.0	191.0	454.0 [a]	47.0 [a]	32.0	..	**53 Immobil., locat. & serv. aux entreprises**
54 Computer & related activities	..	0.0	0.0	..	..	..	..	54 Informatique et activités connexes
55 *Software consultancy*	..	0.0	0.0	0.0	0.0	..	..	55 *Production de logiciels*
56 *Other computer services n.e.c.*	..	0.0	0.0	0.0	0.0	0.0	..	56 *Autres activités informatiques*
57 Research & development	..	167.0 [o]	191.0 [o]	453.0 [a,o]	46.0 [a]	30.0	..	57 Recherche & développement expérimental
58 Other business activities n.e.c.	..	0.0	0.0	..	..	..	..	58 Autres activ. & serv. aux entreprises
59 Comm., soc. & pers. serv. activ., etc.	..	0.0	0.0	0.0	0.0	0.0	..	**59 Adm. pub., serv. soc. & collec., etc**
TOTAL BUSINESS ENTERPRISE SECTOR	**139.0** [a,d]	**396.0** [d]	**463.0** [d]	**556.0** [a]	**359.0**	**267.0**	..	**TOTAL SECTEUR DES ENTREPRISES**

Source: OECD, R&D database (DSTI/EAS Division), January 2002.

Source : OCDE, base de données R-D (DSTI/Division AES), janvier 2002.

SLOVAK REPUBLIC — Table 12 - Tableau 12 — RÉPUBLIQUE SLOVAQUE

Total business enterprise R&D personnel (BEMP) - Personnel total de R-D du secteur des entreprises (PRDE)

by industry - par industrie

In full time equivalent — *En équivalence plein-temps*

	1994	1995	1996	1997	1998	1999	2000	
1 **AGRICULTURE, HUNTING & FORESTRY**	..	..	..	86	929 [a]	1 119	..	1 **AGRICULTURE, CHASSE & SYLVICULTURE**
2 **MINING**	..	0	0	..	..	..	..	2 **ACTIVITES EXTRACTIVES**
3 **MANUFACTURING**	2 688	2 637	2 591	2 161	3 264 [a]	2 478	..	3 **ACTIVITES DE FABRICATION**
4 **Food, beverages & tobacco**	6	6	10	10	8 [a]	..	..	4 **Produits alimentaires, boissons & tabac**
5 Food products & beverages	6	6	10	10	8 [a]	..	..	5 Produits alimentaires & boissons
6 Tobacco products	0	0	0	0	0	0	..	6 Produits à base de tabac
7 **Textiles, wear app., fur & leather**	..	..	..	..	91 [a]	81	..	7 **Textiles, habill., fourrures & cuirs**
8 Textiles	..	..	..	..	91 [a]	81	..	8 Textiles
9 Wearing apparel & fur	0	0	0	0	0	0	..	9 Articles d'habillement & fourrures
10 Leather products & footwear	0	0	0	0	0	0	..	10 Articles en cuir & chaussures
11 **Wood, paper, printing, publishing**	25	..	..	..	..	..	..	11 **Bois, papier, imprimerie, édition**
12 Wood & cork (not furniture)	..	..	0	0	..	..	..	12 Bois & liège (sauf meubles)
13 Pulp, paper & paper products	..	..	..	..	..	..	..	13 Carton, papier & articles
14 Publ., print. & repro. of rec. media	0	0	0	0	0	0	..	14 Edition, impress. & repro. enregis.
15 **Coke, petroleum, nuclear fuel, chemicals, rubber plastics**	847	791	706	658	1 143 [a]	1 099	..	15 **Coke, pétrole, combust. nucl., prod. chimi., caout. & mat. plast.**
16 Coke, ref. petrol. prod. & nuclear fuel	..	..	..	..	..	..	..	16 Coke, raffin. pétr. & combust. nucl.
17 Chemicals & chemical products	..	..	..	..	..	..	..	17 Produits chimiques
18 *Chemical prod.(less pharmaceuticals)*	..	..	..	..	..	..	..	18 *Prod. chimi. (moins prod. pharmaceutiques)*
19 *Pharmaceuticals*	0	..	..	..	..	..	..	19 *Produits pharmaceutiques*
20 Rubber & plastic products	308	215	244	218	257 [a]	300	..	20 Caoutchouc et matières plastiques
21 **Non-metallic mineral products**	66	49	25	..	..	..	..	21 **Produits minéraux non métalliques**
22 **Basic metals**	212	181	186	164	168 [a]	130	..	22 **Produits métallurgiques de base**
23 Basic metals, ferrous	..	..	..	..	..	..	..	23 Produits métallur. de base, ferreux
24 Basic metals, non-ferrous	..	..	..	..	..	0	..	24 Prod. métallur. de base, non ferreux
25 **Fabricated metal products**	18	13	27	7	55 [a]	188	..	25 **Ouvrages en métaux (sauf mach.& mat.)**
26 **Machinery equipment, instruments and transport equipment**	1 492	1 534	1 591	1 274	1 563 [a]	772	..	26 **Machines & matériel, instruments & mat. de transport**
27 Machinery n.e.c.	574	512	510	308	1 057 [a]	224	..	27 Machines, n.c.a.
28 Office, account. & computing machinery	..	..	..	0	0	..	..	28 Mach. de bur., compta. & trait. inform.
29 Electrical machinery	76	69	114	135	130 [a]	108	..	29 Machines électriques
30 Electro. equip. (radio, TV & commun.)	307	309	281	245	..	153	..	30 Equip. électro. (radio, tv & communic.)
31 *Electro. comp. (inc. semi-conductors)*	..	..	0	0	0	0	..	31 *Compos. électro. (semi-cond. incl.)*
32 *TV, radio & communications equip.*	..	..	281	0	0	153	..	32 *Appareils radio, tv et de communi.*
33 Instruments	176	184	264	270	158 [a]	130	..	33 Instruments
34 Motor vehicles	234	208	176	120	29 [a]	..	..	34 Véhicules automobiles
35 Other transport equipment	..	..	..	196	..	..	..	35 Autres matériels de transport
36 *Ships*	0	0	0	0	0	0	..	36 *Navires*
37 *Aerospace*	..	..	..	..	0	0	..	37 *Aérospatiale*
38 *Other transport equipment n.e.c.*	0	0	0	..	..	..	..	38 *Autres matériels de transport n.c.a.*
39 **Furniture, other manufacturing n.e.c.**	..	..	..	..	..	..	..	39 **Meubles, autres activ. de fabricat. n.c.a.**
40 Furniture	..	..	..	0	0	0	..	40 Meubles
41 Other manufacturing n.e.c.	0	0	0	..	..	..	..	41 Autres activités de fabrication n.c.a.
42 **Recycling**	0	0	0	0	0	0	..	42 **Récupération**
43 **ELECTRICITY, GAS & WATER SUPPLY**	0	0	..	..	..	..	..	43 **DISTRIBUTION D'ELECTRICITE, GAZ & EAU**
44 **CONSTRUCTION**	26	..	25	22	77 [a]	39	..	44 **CONSTRUCTION**
45 **SERVICES SECTOR**	2 971	2 173	2 581 [a]	5 086 [a]	1 784 [a]	1 422	..	45 **SECTEUR DES SERVICES**
46 **Wholesale, retail trade & repair of motor vehicles, etc.**	0	0	0	0	0	0	..	46 **Commerce (gros & détail), réparation de véhicules, etc.**
47 **Hotels & restaurants**	0	0	0	0	0	0	..	47 **Hôtels et restaurants**
48 **Transport & storage**	0	0	..	..	..	..	..	48 **Transports, entreposage**
49 **Communications**	0	0	..	..	..	..	..	49 **Communications**
50 Post	0	0	0	0	0	0	..	50 Postes
51 Telecommunications	0	0	0	..	..	0	..	51 Télécommunications
52 **Financial intermediation (inc. insur.)**	0	0	0	0	0	0	..	52 **Intermédiation financ. (assurances inc.)**
53 **Real estate, renting & busin. activity**	2 971	2 173	2 542 [a]	5 022 [a]	1 530 [a]	1 204	..	53 **Immobil., locat. & serv. aux entreprises**
54 Computer & related activities	235	..	0	..	..	..	..	54 Informatique et activités connexes
55 *Software consultancy*	..	..	0	0	0	..	..	55 *Production de logiciels*
56 *Other computer services n.e.c.*	..	..	0	0	0	0	..	56 *Autres activités informatiques*
57 Research & development	2 736 [o]	2 164 [o]	2 542 [a,o]	5 010 [a,o]	1 288 [a]	1 143	..	57 Recherche & développement expérimental
58 Other business activities n.e.c.	0	0	0	..	..	..	..	58 Autres activ. & serv. aux entreprises
59 **Comm., soc. & pers. serv. activ., etc.**	0	..	0	0	0	0	..	59 **Adm. pub., serv. soc. & collec., etc**
TOTAL BUSINESS ENTERPRISE SECTOR	**5 695** [d]	**4 859** [d]	**5 295** [d]	**7 408** [a,d]	**6 708** [d]	**5 692** [d]	**5 172** [d]	**TOTAL SECTEUR DES ENTREPRISES**

Source: OECD, R&D database (DSTI/EAS Division), January 2002.

Source : OCDE, base de données R-D (DSTI/Division AES), janvier 2002.

RSE in the business enterprise sector - Chercheurs dans le secteur des entreprises

by industry - par industrie

In full time equivalent / *En équivalence plein-temps*

	1994	1995	1996	1997	1998	1999	2000	
1 **AGRICULTURE, HUNTING & FORESTRY**	..	..	..	30	363 [a]	446	..	1 **AGRICULTURE, CHASSE & SYLVICULTURE**
2 **MINING**	..	0	0	..	..	..	..	2 **ACTIVITES EXTRACTIVES**
3 **MANUFACTURING**	1 079	1 078	1 108	899	1 383 [a]	1 054	..	3 **ACTIVITES DE FABRICATION**
4 **Food, beverages & tobacco**	3	3	5	8	6 [a]	..	..	4 **Produits alimentaires, boissons & tabac**
5 Food products & beverages	3	3	5	8	6 [a]	..	..	5 Produits alimentaires & boissons
6 Tobacco products	0	0	0	0	0	0	..	6 Produits à base de tabac
7 **Textiles, wear app., fur & leather**	..	..	0	..	35 [a]	33	..	7 **Textiles, habill., fourrures & cuirs**
8 Textiles	..	..	..	..	35 [a]	33	..	8 Textiles
9 Wearing apparel & fur	0	0	0	0	0	0	..	9 Articles d'habillement & fourrures
10 Leather products & footwear	0	0	0	0	0	0	..	10 Articles en cuir & chaussures
11 **Wood, paper, printing, publishing**	15	..	..	..	..	..	..	11 **Bois, papier, imprimerie, édition**
12 Wood & cork (not furniture)	..	..	0	0	..	..	..	12 Bois & liège (sauf meubles)
13 Pulp, paper & paper products	..	..	..	..	..	..	..	13 Carton, papier & articles
14 Publ., print. & repro. of rec. media	0	0	0	0	0	0	..	14 Edition, impress. & repro. enregis.
15 **Coke, petroleum, nuclear fuel, chemicals, rubber plastics**	402	392	333	346	526 [a]	509	..	15 **Coke, pétrole, combust. nucl., prod. chimi., caout. & mat. plast.**
16 Coke, ref. petrol. prod. & nuclear fuel	..	..	..	..	..	..	..	16 Coke, raffin. pétr. & combust. nucl.
17 Chemicals & chemical products	..	..	..	..	..	..	..	17 Produits chimiques
18 *Chemical prod.(less pharmaceuticals)*	..	..	..	..	..	..	..	18 *Prod. chimi. (moins prod. pharmaceutiques)*
19 *Pharmaceuticals*	0	..	..	..	..	..	..	19 *Produits pharmaceutiques*
20 Rubber & plastic products	100	114	132	117	125 [a]	110	..	20 Caoutchouc et matières plastiques
21 **Non-metallic mineral products**	22	14	..	..	..	..	..	21 **Produits minéraux non métalliques**
22 **Basic metals**	144	96	96	86	79 [a]	82	..	22 **Produits métallurgiques de base**
23 Basic metals, ferrous	..	..	..	..	..	..	..	23 Produits métallur. de base, ferreux
24 Basic metals, non-ferrous	..	..	0	..	..	0	..	24 Prod. métallur. de base, non ferreux
25 **Fabricated metal products**	3	4	..	..	14 [a]	59	..	25 **Ouvrages en métaux (sauf mach.& mat.)**
26 **Machinery equipment, instruments and transport equipment**	487	548	645	431	614 [a]	289	..	26 **Machines & matériel, instruments & mat. de transport**
27 Machinery n.e.c.	134	120	145	77	413 [a]	93	..	27 Machines, n.c.a.
28 Office, account. & computing machinery	..	..	..	0	0	..	..	28 Mach. de bur., compta. & trait. inform.
29 Electrical machinery	25	28	52	75	75 [a]	60	..	29 Machines électriques
30 Electro. equip. (radio, TV & commun.)	103	105	96	73	..	49	..	30 Equip. électro. (radio, tv & communic.)
31 *Electro. comp. (inc. semi-conductors)*	..	..	0	0	0	0	..	31 *Compos. électro. (semi-cond. incl.)*
32 *TV, radio & communications equip.*	..	..	96	0	0	49	..	32 *Appareils radio, tv et de communi.*
33 Instruments	85	88	151	155	55 [a]	53	..	33 Instruments
34 Motor vehicles	65	47	48	14	11 [a]	..	..	34 Véhicules automobiles
35 Other transport equipment	..	..	..	37	..	..	..	35 Autres matériels de transport
36 *Ships*	0	0	0	0	0	0	..	36 *Navires*
37 *Aerospace*	..	..	..	..	0	0	..	37 *Aérospatiale*
38 *Other transport equipment n.e.c.*	0	0	0	..	..	..	..	38 *Autres matériels de transport n.c.a.*
39 **Furniture, other manufacturing n.e.c.**	0	..	..	..	..	..	..	39 **Meubles, autres activ. de fabricat. n.c.a.**
40 Furniture	0	..	0	0	0	0	..	40 Meubles
41 Other manufacturing n.e.c.	0	0	0	..	..	..	..	41 Autres activités de fabrication n.c.a.
42 **Recycling**	0	0	0	0	0	0	..	42 **Récupération**
43 **ELECTRICITY, GAS & WATER SUPPLY**	0	0	..	..	..	..	..	43 **DISTRIBUTION D'ELECTRICITE, GAZ & EAU**
44 **CONSTRUCTION**	17	..	22	20	51 [a]	27	..	44 **CONSTRUCTION**
45 **SERVICES SECTOR**	1 548	996	1 078	2 400 [a]	837 [a]	743	..	45 **SECTEUR DES SERVICES**
46 **Wholesale, retail trade & repair of motor vehicles, etc.**	0	0	0	0	0	0	..	46 **Commerce (gros & détail), réparation de véhicules, etc.**
47 **Hotels & restaurants**	0	0	0	0	0	0	..	47 **Hôtels et restaurants**
48 **Transport & storage**	0	0	..	..	..	..	..	48 **Transports, entreposage**
49 **Communications**	0	0	..	..	..	..	..	49 **Communications**
50 Post	0	0	0	0	0	0	..	50 Postes
51 Telecommunications	0	0	0	..	..	0	..	51 Télécommunications
52 **Financial intermediation (inc. insur.)**	0	0	0	0	0	0	..	52 **Intermédiation financ. (assurances inc.)**
53 **Real estate, renting & busin. activity**	1 548	996	1 045	2 356 [a]	677 [a]	606	..	53 **Immobil., locat. & serv. aux entreprises**
54 Computer & related activities	233	..	0	..	..	..	..	54 Informatique et activités connexes
55 *Software consultancy*	..	..	0	0	0	..	..	55 *Production de logiciels*
56 *Other computer services n.e.c.*	..	..	0	0	0	0	..	56 *Autres activités informatiques*
57 Research & development	1 315 [o]	989 [o]	1 045 [o]	2 350 [a,o]	592 [a]	576	..	57 Recherche & développement expérimental
58 Other business activities n.e.c.	0	0	0	..	..	..	..	58 Autres activ. & serv. aux entreprises
59 **Comm., soc. & pers. serv. activ., etc.**	0	..	0	0	0	0	..	59 **Adm. pub., serv. soc. & collec., etc**
TOTAL BUSINESS ENTERPRISE SECTOR	**2 648** [d]	**2 103** [d]	**2 259** [d]	**3 387** [a,d]	**2 903** [d]	**2 522** [d]	**2 420** [d]	**TOTAL SECTEUR DES ENTREPRISES**

Source: OECD, R&D database (DSTI/EAS Division), January 2002.

Source : OCDE, base de données R-D (DSTI/Division AES), janvier 2002.

Higher education intramural expenditure on R&D - Dépenses intra-muros de R-D de l'enseignement supérieur

(HERD) by field of science - (DIRDES) par discipline scientifique

	1994	1995	1996	1997	1998	1999	2000	
	14.1 Million SKK - Millions SKK							
Natural sciences	..	..	101.1	101.7	111.9	120.1	..	Sciences exactes et naturelles
Engineering	..	..	122.5	203.2	228.9	234.8	..	Sciences de l'ingénieur
Medical sciences	..	..	..	64.8	67.8	50.6	..	Sciences médicales
Agricultural sciences	..	..	..	66.6	90.5	62.6	..	Sciences agricoles
Sub-total NSE	..	..	275.4	436.3	499.1	468.1	..	Sous-total SENI
Social sciences	..	..	..	..	..	57.0	..	Sciences sociales
Humanities	..	..	..	..	..	25.8	..	Sciences humaines
Sub-total SSH	..	..	26.9	83.6	81.1	82.8	..	Sous-total SSH
Total NSE+SSH	219.0 [a]	315.0	302.0	520.0	580.0	551.0	579.0	*Total SENI+SSH*
	14.2 Million constant USD (1995 prices and PPPs) - Millions de USD constants (prix et PPA de 1995)							
Natural sciences	..	..	8.1 [b]	7.7 [b]	8.0 [b]	8.1 [b]	..	Sciences exactes et naturelles
Engineering	..	..	9.9 [b]	15.3 [b]	16.4 [b]	15.8 [b]	..	Sciences de l'ingénieur
Medical sciences	..	..	..	4.9 [b]	4.9 [b]	3.4 [b]	..	Sciences médicales
Agricultural sciences	..	..	..	5.0 [b]	6.5 [b]	4.2 [b]	..	Sciences agricoles
Sub-total NSE	..	..	22.2 [b]	32.9 [b]	35.8 [b]	31.5 [b]	..	Sous-total SENI
Social sciences	..	..	..	..	..	3.8 [b]	..	Sciences sociales
Humanities	..	..	..	..	..	1.7 [b]	..	Sciences humaines
Sub-total SSH	..	..	2.2 [b]	6.3 [b]	5.8 [b]	5.6 [b]	..	Sous-total SSH
Total NSE+SSH	20.2 [a,b]	26.5 [b]	24.3 [b]	39.2 [b]	41.6 [b]	37.1 [b]	36.6 [b]	*Total SENI+SSH*

Table 15 - Tableau 15

Higher education intramural expenditure on R&D - Dépenses intra-muros de R-D de l'enseignement supérieur

(HERD) by field of science and type of costs - (DIRDES) par discipline scientifique et par type de coûts

Million SKK *Millions SKK*

	1994	1995	1996	1997	1998	1999	2000	
Current expenditure								**Dépenses courantes**
Natural sciences	..	..	90.7	91.4	103.4	111.3	..	Sciences exactes et naturelles
Engineering	..	..	107.6	178.3	198.5	208.6	..	Sciences de l'ingénieur
Medical sciences	..	..	..	56.5	59.3	45.4	..	Sciences médicales
Agricultural sciences	..	..	..	60.8	80.4	57.4	..	Sciences agricoles
Sub-total NSE	..	..	244.5	387.1	441.6	422.7	..	Sous-total SENI
Social sciences	..	..	..	..	..	52.6	..	Sciences sociales
Humanities	..	..	..	..	..	24.6	..	Sciences humaines
Sub-total SSH	..	..	23.5	74.4	72.2	77.2	..	Sous-total SSH
Total NSE+SSH	195.0 [a]	286.0	268.0	462.0	514.0	500.0	..	*Total SENI+SSH*
Capital expenditure								**Dépenses en capital**
Natural sciences	..	..	10.4	10.2	8.5	8.8	..	Sciences exactes et naturelles
Engineering	..	..	14.9	24.9	30.4	26.2	..	Sciences de l'ingénieur
Medical sciences	..	..	..	8.3	8.5	5.2	..	Sciences médicales
Agricultural sciences	..	..	..	5.8	10.1	5.2	..	Sciences agricoles
Sub-total NSE	..	..	30.9	49.2	57.5	45.4	..	Sous-total SENI
Social sciences	..	..	..	..	..	4.4	..	Sciences sociales
Humanities	..	..	..	..	..	1.2	..	Sciences humaines
Sub-total SSH	..	..	3.4	9.2	8.9	5.6	..	Sous-total SSH
Total NSE+SSH	24.0 [a]	29.0	34.0	58.0	66.0	51.0	..	*Total SENI+SSH*

Source: OECD, R&D database (DSTI/EAS Division), January 2002.

Source : OCDE, base de données R-D (DSTI/Division AES), janvier 2002.

SLOVAK REPUBLIC — Table 16 - Tableau 16 — RÉPUBLIQUE SLOVAQUE

Higher education intramural expenditure on R&D - Dépenses intra-muros de R-D de l'enseignement supérieur

(HERD) by field of science and source of funds - (DIRDES) par discipline scientifique et par source de financement

Million SKK — *Millions SKK*

	1994	1995	1996	1997	1998	1999	2000	
Direct government								**État direct**
Natural sciences	..	..	..	..	..	..	..	Sciences exactes et naturelles
Engineering	..	..	..	..	..	..	..	Sciences de l'ingénieur
Medical sciences	..	..	..	..	..	..	..	Sciences médicales
Agricultural sciences	..	..	..	..	..	..	..	Sciences agricoles
Sub-total NSE	..	..	..	..	..	..	..	Sous-total SENI
Social sciences	..	..	..	..	..	..	..	Sciences sociales
Humanities	..	..	..	..	..	..	..	Sciences humaines
Sub-total SSH	..	..	..	..	..	..	..	Sous-total SSH
Total NSE+SSH	..	..	..	..	..	..	..	*Total SENI+SSH*
General university funds								**Fonds généraux des universités**
Natural sciences	..	..	..	..	..	..	..	Sciences exactes et naturelles
Engineering	..	..	..	..	..	..	..	Sciences de l'ingénieur
Medical sciences	..	..	..	..	..	..	..	Sciences médicales
Agricultural sciences	..	..	..	..	..	..	..	Sciences agricoles
Sub-total NSE	..	..	..	..	..	..	..	Sous-total SENI
Social sciences	..	..	..	..	..	..	..	Sciences sociales
Humanities	..	..	..	..	..	..	..	Sciences humaines
Sub-total SSH	..	..	..	..	..	..	..	Sous-total SSH
Total NSE+SSH	..	..	..	..	..	..	..	*Total SENI+SSH*
Sub-total government								**Sous-total État**
Natural sciences	..	..	93.9	101.3	111.2	119.1	..	Sciences exactes et naturelles
Engineering	..	..	117.0	192.0	221.5	221.4	..	Sciences de l'ingénieur
Medical sciences	..	..	..	61.0	64.1	48.9	..	Sciences médicales
Agricultural sciences	..	..	..	66.6	90.5	61.5	..	Sciences agricoles
Sub-total NSE	..	..	262.4	420.8	487.3	450.9	..	Sous-total SENI
Social sciences	..	..	..	..	..	51.8	..	Sciences sociales
Humanities	..	..	..	..	..	25.8	..	Sciences humaines
Sub-total SSH	..	..	26.7	82.9	78.2	77.6	..	Sous-total SSH
Total NSE+SSH	214.0 [a]	306.0	289.0	504.0	565.0	529.0	530.0	*Total SENI+SSH*
Higher education								**Enseignement supérieur**
Natural sciences	..	..	.. [n]	..	..	.. [n]	..	Sciences exactes et naturelles
Engineering	..	..	..	..	..	.. [n]	..	Sciences de l'ingénieur
Medical sciences	..	..	..	.. [n]	.. [n]	.. [n]	..	Sciences médicales
Agricultural sciences	..	..	..	..	..	.. [n]	..	Sciences agricoles
Sub-total NSE	..	..	.. [n]	.. [n]	.. [n]	.. [n]	..	Sous-total SENI
Social sciences	..	..	..	..	..	.. [n]	..	Sciences sociales
Humanities	..	..	..	..	..	..	..	Sciences humaines
Sub-total SSH	..	..	.. [n]	.. [n]	.. [n]	.. [n]	..	Sous-total SSH
Total NSE+SSH	.. [n]	.. [n]	.. [n]	.. [n]	.. [n]	.. [n]	26.0	*Total SENI+SSH*
Private non-profit								**Institutions sans but lucratif**
Natural sciences	..	..	..	..	..	..	..	Sciences exactes et naturelles
Engineering	..	..	..	..	..	..	..	Sciences de l'ingénieur
Medical sciences	..	..	..	..	..	..	..	Sciences médicales
Agricultural sciences	..	..	..	..	..	..	..	Sciences agricoles
Sub-total NSE	..	..	..	..	..	..	..	Sous-total SENI
Social sciences	..	..	..	..	..	..	..	Sciences sociales
Humanities	..	..	..	..	..	..	..	Sciences humaines
Sub-total SSH	..	..	..	..	..	..	..	Sous-total SSH
Total NSE+SSH	..	..	..	0.0	0.0	0.0	0.0	*Total SENI+SSH*
Business entreprise								**Entreprises**
Natural sciences	..	..	1.6 [o]	0.0	0.0	0.7 [o]	..	Sciences exactes et naturelles
Engineering	..	..	..	0.0	0.0	1.6 [o]	..	Sciences de l'ingénieur
Medical sciences	..	..	..	2.0 [o]	2.0 [o]	1.5 [o]	..	Sciences médicales
Agricultural sciences	..	..	..	0.0	0.0	0.6 [o]	..	Sciences agricoles
Sub-total NSE	..	..	1.8 [o]	2.0 [o]	2.0 [o]	4.3 [o]	..	Sous-total SENI
Social sciences	..	..	..	..	..	0.7 [o]	..	Sciences sociales
Humanities	..	..	..	..	..	0.0	..	Sciences humaines
Sub-total SSH	..	..	0.2 [o]	0.1 [o]	0.7 [o]	0.7 [o]	..	Sous-total SSH
Total NSE+SSH	2.0 [a,o]	3.0 [o]	2.0 [o]	2.1 [o]	3.0 [o]	5.0 [o]	2.0	*Total SENI+SSH*
Funds from abroad								**Fonds de l'étranger**
Natural sciences	..	..	5.0	0.4	0.7	0.0	..	Sciences exactes et naturelles
Engineering	..	..	5.5	11.2	7.4	11.8	..	Sciences de l'ingénieur
Medical sciences	..	..	..	1.9	1.7	0.2	..	Sciences médicales
Agricultural sciences	..	..	..	0.0	0.0	0.5	..	Sciences agricoles
Sub-total NSE	..	..	10.6	13.4	9.8	12.5	..	Sous-total SENI
Social sciences	..	..	0.0	..	..	4.5	..	Sciences sociales
Humanities	..	..	0.0	..	..	0.0	..	Sciences humaines
Sub-total SSH	..	..	0.0	0.5	2.2	4.5	..	Sous-total SSH
Total NSE+SSH	3.0	6.0	11.0	14.0	12.0	17.0	21.0	*Total SENI+SSH*

Source: OECD, R&D database (DSTI/EAS Division), January 2002.

Source : OCDE, base de données R-D (DSTI/Division AES), janvier 2002.

SLOVAK REPUBLIC — Table 17 - Tableau 17 — RÉPUBLIQUE SLOVAQUE

Higher education R&D personnel (HEMP) - Personnel de R-D du secteur de l'enseignement supérieur

by field of science and occupation - par discipline scientifique et par profession

In full-time equivalent — *En équivalence plein-temps*

	1994	1995	1996	1997	1998	1999	2000	
Total R&D personnel								**Personnel total de R-D**
Natural sciences	..	..	1 108	1 155	949	891	..	Sciences exactes et naturelles
Engineering	..	..	2 055	1 680	1 877	1 952	..	Sciences de l'ingénieur
Medical sciences	..	..	..	645	908	758	..	Sciences médicales
Agricultural sciences	..	..	..	568	725	485	..	Sciences agricoles
Sub-total NSE	..	..	3 864	4 048	4 459	4 086	..	Sous-total SENI
Social sciences	..	..	..	..	..	755	..	Sciences sociales
Humanities	..	..	..	..	..	222	..	Sciences humaines
Sub-total SSH	..	..	657	993	1 055	977	..	Sous-total SSH
Total NSE+SSH	4 285	4 543	4 521	5 041	5 514	5 063	5 860	*Total SENI+SSH*
Researchers								**Chercheurs**
Natural sciences	..	..	919	920	814	712	..	Sciences exactes et naturelles
Engineering	..	..	1 772	1 443	1 644	1 688	..	Sciences de l'ingénieur
Medical sciences	..	..	..	551	738	594	..	Sciences médicales
Agricultural sciences	..	..	..	379	514	347	..	Sciences agricoles
Sub-total NSE	..	..	3 323	3 293	3 710	3 341	..	Sous-total SENI
Social sciences	..	..	..	..	..	713	..	Sciences sociales
Humanities	..	..	..	..	..	200	..	Sciences humaines
Sub-total SSH	..	..	577	845	975	913	..	Sous-total SSH
Total NSE+SSH	3 698	3 954	3 900	4 138	4 685	4 254	5 009	*Total SENI+SSH*
Technicians								**Technicians**
Natural sciences	..	..	184	183	127	170	..	Sciences exactes et naturelles
Engineering	..	..	230	192	190	222	..	Sciences de l'ingénieur
Medical sciences	..	..	..	88	163	139	..	Sciences médicales
Agricultural sciences	..	..	..	183	204	133	..	Sciences agricoles
Sub-total NSE	..	..	480	646	684	664	..	Sous-total SENI
Social sciences	..	..	..	..	..	36	..	Sciences sociales
Humanities	..	..	..	..	..	15	..	Sciences humaines
Sub-total SSH	..	..	60	131	62	51	..	Sous-total SSH
Total NSE+SSH	510	523	540	776	746	715	717	*Total SENI+SSH*
Other								**Autres**
Natural sciences	..	..	5	52	8	9	..	Sciences exactes et naturelles
Engineering	..	..	53	45	43	42	..	Sciences de l'ingénieur
Medical sciences	..	..	..	6	7	25	..	Sciences médicales
Agricultural sciences	..	..	..	6	7	5	..	Sciences agricoles
Sub-total NSE	..	..	61	109	65	81	..	Sous-total SENI
Social sciences	..	..	..	..	..	6	..	Sciences sociales
Humanities	..	..	..	..	..	7	..	Sciences humaines
Sub-total SSH	..	..	20	17	18	13	..	Sous-total SSH
Total NSE+SSH	77	66	81	127	83	94	134	*Total SENI+SSH*

Table 18 - Tableau 18

Government budget appropriations or outlays for R&D (GBAORD) - Crédits budgétaires publics de R-D (CBPRD)

by socio-economic objective - par objectif socio-économique

Million SKK — *Millions SKK*

	1995	1996	1997	1998	1999	2000	2001	
Agriculture, forestry and fishing	448.0	481.0	478.0	460.0	450.0	443.0	459.0	Agriculture, sylviculture et pêche
Industrial development	342.0	450.0	380.0	363.0	454.0	402.0	506.0	Développement industriel
Energy	16.0	86.0	92.0	122.0	62.0	60.0	43.0	Energie
Transport & telecommunications	55.0	64.0	44.0	39.0	35.0	37.0	34.0	*Transports & télécommunications*
Urban & rural planning	2.0	6.0	4.0	4.0	4.0	8.0	0.0	*Aménagement urbain et rural*
ST Infrastructure	57.0	70.0	48.0	43.0	39.0	45.0	34.0	ST Infrastructure
Prevention of pollution	44.0	65.0	58.0	51.0	49.0	44.0	45.0	*Prévention de la pollution*
Identificat. & treatment of pollution	.. n	.. n	.. n	.. n	.. n	.. n	.. n	*Détect. pollut. & mesures antipollut.*
ST Environment	44.0	65.0	58.0	51.0	49.0	44.0	45.0	ST Environnement
Health	161.0	192.0	194.0	185.0	183.0	177.0	190.0	Santé
Social development & services	125.0	147.0	197.0	172.0	252.0	346.0	372.0	Développement et services sociaux
Earth and atmosphere	0.0	0.0	0.0	0.0	0.0	0.0	0.0	Milieu terrestre et atmosphère
Advancement of research	746.0 o	839.0 o	927.0 o	1 052.0 o	1 028.0 o	970.0 o	841.0 o	*Promotion de la recherche*
General university funds	303.0	345.0	542.0	584.0	624.0	597.0	648.0	*Fonds généraux des universités*
ST Advancement of knowledge	1 049.0 o	1 184.0 o	1 469.0 o	1 636.0 o	1 652.0 o	1 567.0 o	1 489.0 o	ST promotion gén. des connaiss.
Civil space	.. n	.. n	.. n	10.0	.. n	.. n	.. n	Espace civil
Defence	..	..	..	..	..	..	..	Défense
Not elsewhere classified	0.0	0.0	0.0	217.0	243.0	262.0	351.0	Non classé
Total	**2 242.0**	**2 675.0**	**2 916.0**	**3 259.0**	**3 384.0**	**3 346.0**	**3 489.0**	**Total**

Source: OECD, R&D database (DSTI/EAS Division), January 2002.

Source : OCDE, base de données R-D (DSTI/Division AES), janvier 2002.

Table 19 - Tableau 19

Technology balance of payments - Balance des paiements technologiques

by type of transfer - selon la nature des transferts

Million SKK | *Millions SKK*

	1983	1984	1985	1986	1987	1988	1989	1990	1991	
Receipts										**Recettes**
Related firms	..	..	..	..	..	..	..	..	..	Firmes affiliées
Unrelated firms	..	..	..	..	..	..	..	..	..	Firmes non affiliées
Other	..	..	..	..	..	..	..	..	..	Autre
Total	..	..	..	..	..	..	..	..	..	*Total*
Payments										**Paiements**
Related firms	..	..	..	..	..	..	..	..	..	Firmes affiliées
Unrelated firms	..	..	..	..	..	..	..	..	..	Firmes non affiliées
Other	..	..	..	..	..	..	..	..	..	Autre
Total	..	..	..	..	..	..	..	..	..	*Total*

	1992	1993	1994	1995	1996	1997	1998	1999	2000	
Receipts										**Recettes**
Related firms	..	..	..	..	..	..	..	..	..	Firmes affiliées
Unrelated firms	..	..	..	..	..	..	..	..	..	Firmes non affiliées
Other	..	..	..	..	..	..	..	..	..	Autre
Total	..	..	..	280 *l*	417 *l*	483 *l*	724 *l*	645 *l*	1 080 *l*	*Total*
Payments										**Paiements**
Related firms	..	..	..	..	..	..	..	..	..	Firmes affiliées
Unrelated firms	..	..	..	..	..	..	..	..	..	Firmes non affiliées
Other	..	..	..	..	..	..	..	..	..	Autre
Total	..	..	..	793 *l*	1 364 *l*	2 011 *l*	2 119 *l*	2 573 *l*	2 970 *l*	*Total*

Source: OECD, TBPdatabase (DSTI/EAS Division), Januay 2002.

Source : OCDE, base de données TBP (DSTI/Division AES), janvier 2002.

SPAIN — Table 1 - Tableau 1 — ESPAGNE

Gross domestic expenditure on R&D (GERD) - Dépenses intérieures brutes de R-D (DIRD)

by sector of performance and source of funds - par secteur d'exécution et par source de financement

	1994	1995	1996	1997	1998	1999	2000	
	1.1 Million EUR (1999 ESP euro, EUR from 1999) - Millions EUR (1999 ESP euro, EUR à partir de 1999)							
Business enterprise								**Entreprises**
financed by								*financé par*
Business enterprise	1 234.1 [c]	1 444.6	1 614.6	1 672.3	2 188.9	2 251.8	..	Entreprises
Direct government	163.4 [c]	156.9	146.7	171.8	161.3	222.4	..	État direct
Higher education	..	0.1	0.4	0.3	1.5	3.8	..	Enseignement supérieur
Private non-profit	2.2 [c]	0.8	3.9	1.7	2.4	1.6	..	Institutions sans but lucratif
Funds from abroad	140.7 [c]	109.8	97.0	124.7	103.1	117.5	..	Fonds de l'étranger
Total	1 540.5	1 712.2	1 862.6	1 970.9	2 457.2	2 597.1	2 893.7 [c]	*Total*
Government								**État**
financed by								*financé par*
Business enterprise	28.7 [c]	35.2	37.7	36.4	45.6	61.2	..	Entreprises
Direct government	620.5 [c]	576.3	618.0	599.4	639.1	705.8	..	État direct
Higher education	..	1.4	1.8	1.9	2.2	2.5	..	Enseignement supérieur
Private non-profit	0.6 [c]	1.8	2.0	2.6	1.9	2.7	..	Institutions sans but lucratif
Funds from abroad	32.0 [c]	46.5	45.5	61.3	78.6	71.1	..	Fonds de l'étranger
Total	681.8	661.1	704.9	701.5	767.3	843.3	902.3 [c]	*Total*
Higher education								**Enseignement supérieur**
financed by								*financé par*
Business enterprise	61.6 [c]	94.7	92.9	85.3	100.3	116.1	..	Entreprises
Direct government	209.0 [c]	342.2	160.3	198.6	211.4	252.4	..	État direct
General university funds	726.8 [c,o]	458.6 [a]	753.9	777.0	795.4	844.0	..	Fonds généraux des universités
Sub-total government	935.8 [c,o]	800.8 [a]	914.2	975.6	1 006.8	1 096.4	..	Sous-Total État
Higher education	.. [n]	155.6	154.0	167.4	186.2	189.3	..	Enseignement supérieur
Private non-profit	6.5 [c]	5.7	11.0	10.1	14.5	15.0	..	Institutions sans but lucratif
Funds from abroad	36.3 [c]	80.1	70.7	83.6	130.8	87.8	..	Fonds de l'étranger
Total	1 040.3	1 136.9	1 242.7	1 321.9	1 438.7	1 504.6	1 609.9 [c]	*Total*
Private non-profit								**Institutions sans but lucratif**
financed by								*financé par*
Business enterprise	2.6 [c]	6.3	6.7 [c]	11.6	13.6 [c]	11.4	..	Entreprises
Direct government	5.9 [c]	12.5	13.3 [c]	15.4	17.9 [c]	14.3	..	État direct
Higher education	..	0.2	0.2 [c]	0.0	0.0 [c]	0.3	..	Enseignement supérieur
Private non-profit	23.0 [c]	19.8	21.0 [c]	14.8	17.2 [c]	20.2	..	Institutions sans but lucratif
Funds from abroad	0.3 [c]	1.1	1.1 [c]	2.7	3.1 [c]	4.3	..	Fonds de l'étranger
Total	31.9 [c]	39.8	42.4 [c]	44.6	51.9 [c]	50.4	53.9 [c]	*Total*
GERD								**DIRD**
financed by								*financé par*
Business enterprise	1 327.0 [c]	1 580.8	1 751.9 [c]	1 805.7	2 348.3 [c]	2 440.5	..	Entreprises
Direct government	998.8 [c]	1 087.9	938.3 [c]	985.2	1 029.7 [c]	1 194.8	..	État direct
General university funds	726.8 [c,o]	458.6 [a]	753.9	777.0	795.4	844.0	..	Fonds généraux des universités
Sub-total government	1 725.7 [c]	1 546.6 [a]	1 692.2 [c]	1 762.2	1 825.1 [c]	2 038.8	..	Sous-Total État
Higher education	.. [n]	157.2	156.3 [c]	169.6	190.0 [c]	195.9	..	Enseignement supérieur
Private non-profit	32.4 [c]	28.1	37.9 [c]	29.3	36.1 [c]	39.5	..	Institutions sans but lucratif
Funds from abroad	209.4 [c]	237.5	214.3 [c]	272.2	315.5 [c]	280.7	..	Fonds de l'étranger
Total	3 294.5 [c]	3 550.1	3 852.6 [c]	4 038.9	4 715.0 [c]	4 995.4	5 459.8 [c]	*Total*
	1.2 Million constant USD (1995 prices and PPPs) - Millions de USD constants (prix et PPA de 1995)							
Business enterprise								**Entreprises**
financed by								*financé par*
Business enterprise	1 763.1 [b]	1 968.9	2 125.8	2 154.3	2 757.1	2 756.8	..	Entreprises
Direct government	233.5 [b]	213.9	193.2	221.3	203.1	272.3	..	État direct
Higher education	..	0.1	0.5	0.4	1.9	4.7	..	Enseignement supérieur
Private non-profit	3.2 [b]	1.1	5.2	2.2	3.1	2.0	..	Institutions sans but lucratif
Funds from abroad	201.0 [b]	149.7	127.7	160.6	129.8	143.8	..	Fonds de l'étranger
Total	2 200.8 [b]	2 333.7	2 452.3	2 538.8	3 095.0	3 179.5	3 424.3 [c]	*Total*
Government								**État**
financed by								*financé par*
Business enterprise	41.0 [b]	48.0	49.7	46.8	57.4	74.9	..	Entreprises
Direct government	886.5 [b]	785.4	813.6	772.1	805.0	864.1	..	État direct
Higher education	..	1.9	2.3	2.4	2.7	3.0	..	Enseignement supérieur
Private non-profit	0.8 [b]	2.4	2.6	3.3	2.4	3.3	..	Institutions sans but lucratif
Funds from abroad	45.7 [b]	63.3	59.9	79.0	99.0	87.0	..	Fonds de l'étranger
Total	974.1 [b]	901.1	928.1	903.7	966.5	1 032.4	1 067.7 [c]	*Total*
Higher education								**Enseignement supérieur**
financed by								*financé par*
Business enterprise	88.1 [b]	129.1	122.3	109.9	126.3	142.2	..	Entreprises
Direct government	298.5 [b]	466.4	211.1	255.8	266.3	309.0	..	État direct
General university funds	1 038.4 [b,o]	625.1 [a]	992.5	1 000.9	1 001.9	1 033.3	..	Fonds généraux des universités
Sub-total government	1 336.9 [b,o]	1 091.5 [a]	1 203.6	1 256.7	1 268.2	1 342.2	..	Sous-Total État
Higher education	.. [b,n]	212.1	202.7	215.6	234.6	231.8	..	Enseignement supérieur
Private non-profit	9.3 [b]	7.8	14.5	13.1	18.3	18.3	..	Institutions sans but lucratif
Funds from abroad	51.9 [b]	109.1	93.1	107.7	164.7	107.5	..	Fonds de l'étranger
Total	1 486.2 [b]	1 549.6	1 636.1	1 702.9	1 812.1	1 842.0	1 905.1 [c]	*Total*
Private non-profit								**Institutions sans but lucratif**
financed by								*financé par*
Business enterprise	3.7 [b]	8.5	8.8 [c]	15.0	17.1 [c]	13.9	..	Entreprises
Direct government	8.4 [b]	17.1	17.5 [c]	19.9	22.6 [c]	17.5	..	État direct
Higher education	..	0.3	0.3 [c]	0.0	0.0 [c]	0.3	..	Enseignement supérieur
Private non-profit	32.9 [b]	26.9	27.7 [c]	19.1	21.7 [c]	24.7	..	Institutions sans but lucratif
Funds from abroad	0.4 [b]	1.5	1.5 [c]	3.4	3.9 [c]	5.3	..	Fonds de l'étranger
Total	45.5 [b]	54.3	55.8 [c]	57.4	65.3 [c]	61.7	63.8 [c]	*Total*
GERD								**DIRD**
financed by								*financé par*
Business enterprise	1 895.8 [b]	2 154.6	2 306.5 [c]	2 326.0	2 957.8 [c]	2 987.8	..	Entreprises
Direct government	1 427.0 [b]	1 482.8	1 235.4 [c]	1 269.0	1 297.0 [c]	1 462.8	..	État direct
General university funds	1 038.4 [b,o]	625.1 [a]	992.5	1 000.9	1 001.9	1 033.3	..	Fonds généraux des universités
Sub-total government	2 465.3 [b]	2 107.9 [a]	2 227.9 [c]	2 269.9	2 298.9 [c]	2 496.0	..	Sous-Total État
Higher education	.. [b,n]	214.3	205.8 [c]	218.4	239.3 [c]	239.8	..	Enseignement supérieur
Private non-profit	46.3 [b]	38.2	49.9 [c]	37.7	45.4 [c]	48.3	..	Institutions sans but lucratif
Funds from abroad	299.1 [b]	323.7	282.2 [c]	350.7	397.5 [c]	343.6	..	Fonds de l'étranger
Total	4 706.6 [b]	4 838.6	5 072.3 [c]	5 202.8	5 938.9 [c]	6 115.6	6 461.0 [c]	*Total*

Source: OECD, R&D database (DSTI/EAS Division), January 2002. *Source* : OCDE, base de données R-D (DSTI/Division AES), janvier 2002.

Table 2 - Tableau 2

Gross domestic expenditure on R&D (GERD) - Dépenses intérieures brutes de R-D (DIRD)

by sector of performance and type of costs - par secteur d'exécution et par type de coûts

Million EUR (1999 ESP euro, EUR from 1999) — *Millions EUR (1999 ESP euro, EUR à partir de 1999)*

	1994	1995	1996	1997	1998	1999	2000	
Business enterprise								**Entreprises**
Labour costs	869.6	930.8	1 050.3 c	1 049.0	1 198.7	1 355.5	..	Coûts du personnel
Other current costs	480.1	492.7	555.9 c	564.2	752.6	723.0	..	Autres coûts courants
Total current expenditure	1 349.7	1 423.5	1 606.3	1 613.1	1 951.2	2 078.5	..	*Total dépenses courantes*
Land and buildings	..	37.9	..	41.1	63.3	40.8	..	Terrains et constructions
Instruments and equipment	..	250.9	..	316.6	442.6	477.9	..	Instruments et équipements
Total capital expenditure	190.8	288.8	256.4	357.7	505.9	518.6	..	*Total dépenses en capital*
Total	1 540.5	1 712.2	1 862.6	1 970.9	2 457.2	2 597.1	2 893.7 c	*Total*
Government								**État**
Labour costs	397.5	359.2	374.5 c	394.3	427.5	461.1	..	Coûts du personnel
Other current costs	127.3	177.4	185.0 c	189.3	224.8	244.1	..	Autres coûts courants
Total current expenditure	524.8	536.6	559.5	583.6	652.3	705.2	..	*Total dépenses courantes*
Land and buildings	..	21.7	..	29.0	26.5	36.2	..	Terrains et constructions
Instruments and equipment	..	102.8	..	88.9	88.5	101.9	..	Instruments et équipements
Total capital expenditure	157.0	124.5	145.5	118.0	115.0	138.1	..	*Total dépenses en capital*
Total	681.8	661.1	704.9	701.5	767.3	843.3	902.3 c	*Total*
Higher education								**Enseignement supérieur**
Labour costs	683.7	726.7	796.4	829.1	908.8	940.1	..	Coûts du personnel
Other current costs	158.9	197.3	194.0	185.1	219.5	235.5	..	Autres coûts courants
Total current expenditure	842.6	923.9	990.5	1 014.3	1 128.3	1 175.5	..	*Total dépenses courantes*
Land and buildings	..	70.0	92.0	126.2	138.6	138.0	..	Terrains et constructions
Instruments and equipment	..	143.0	160.3	181.4	171.8	191.1	..	Instruments et équipements
Total capital expenditure	197.7	213.0	252.2	307.7	310.4	329.1	..	*Total dépenses en capital*
Total	1 040.3	1 136.9	1 242.7	1 321.9	1 438.7	1 504.6	1 609.9 c	*Total*
Private non-profit								**Institutions sans but lucratif**
Labour costs	23.6 c	22.7	24.2 c	23.8	27.7 c	27.2	..	Coûts du personnel
Other current costs	5.7 c	9.2	9.7 c	14.1	16.4 c	14.2	..	Autres coûts courants
Total current expenditure	29.4 c	31.9	33.9	37.9	44.1 c	41.4	..	*Total dépenses courantes*
Land and buildings	..	0.4	..	0.8	..	1.3	..	Terrains et constructions
Instruments and equipment	..	7.5	..	5.9	..	7.7	..	Instruments et équipements
Total capital expenditure	2.5 c	8.0	8.5 c	6.7	7.8 c	9.0	..	*Total dépenses en capital*
Total	31.9 c	39.8	42.4 c	44.6	51.9 c	50.4	53.9 c	*Total*
GERD								**DIRD**
Labour costs	1 974.4 c	2 039.4	2 245.5 c	2 296.1	2 562.7 c	2 783.9	..	Coûts du personnel
Other current costs	772.1 c	876.5	944.6 c	952.7	1 213.2 c	1 216.7	..	Autres coûts courants
Total current expenditure	2 746.5 c	2 915.9	3 190.1 c	3 248.8	3 776.0 c	4 000.6	..	*Total dépenses courantes*
Land and buildings	..	130.0	..	197.2	..	216.3	..	Terrains et constructions
Instruments and equipment	..	504.1	..	592.9	..	778.5	..	Instruments et équipements
Total capital expenditure	548.0 c	634.2	662.5 c	790.1	939.1 c	994.8	..	*Total dépenses en capital*
Total	3 294.5 c	3 550.1	3 852.6 c	4 038.9	4 715.0 c	4 995.4	5 459.8 c	*Total*

Table 3 - Tableau 3

Current domestic expenditure on R&D - Dépenses intérieures courantes de R-D

by sector of performance and type of activity - par secteur d'exécution et par type d'activités

Million EUR (1999 ESP euro, EUR from 1999) — *Millions EUR (1999 ESP euro, EUR à partir de 1999)*

	1994	1995	1996	1997	1998	1999	2000	
Business enterprise								**Entreprises**
Basic research	..	100.6	..	93.8	..	145.1	..	Recherche fondamentale
Applied research	..	461.8	..	550.2	..	651.8	..	Recherche appliquée
Experimental development	..	861.1	..	969.1	..	1 281.6	..	Développement expérimental
Not specified	1 349.7	0.0	1 606.3	0.0	1 951.2	0.0	..	Non spécifié
Total	1 349.7	1 423.5	1 606.3	1 613.1	1 951.2	2 078.5	..	*Total*
Government								**État**
Basic research	..	148.0	..	133.0	..	150.6	..	Recherche fondamentale
Applied research	..	306.4	..	353.6	..	404.7	..	Recherche appliquée
Experimental development	..	82.3	..	97.1	..	150.0	..	Développement expérimental
Not specified	524.8	0.0	559.5	0.0	652.3	0.0	..	Non spécifié
Total	524.8	536.6	559.5	583.6	652.3	705.2	..	*Total*
Higher education								**Enseignement supérieur**
Basic research	..	472.4	..	500.6	..	576.0	..	Recherche fondamentale
Applied research	..	301.0	..	339.7	..	395.2	..	Recherche appliquée
Experimental development	..	150.6	..	174.0	..	204.4	..	Développement expérimental
Not specified	842.6	0.0	990.5	0.0	1 128.3	0.0	..	Non spécifié
Total	842.6	923.9	990.5	1 014.3	1 128.3	1 175.5	..	*Total*
Private non-profit								**Institutions sans but lucratif**
Basic research	..	17.4	..	13.6	..	8.6	..	Recherche fondamentale
Applied research	..	9.7	..	17.4	..	24.1	..	Recherche appliquée
Experimental development	..	4.8	..	6.8	..	8.7	..	Développement expérimental
Not specified	29.4 c	0.0	33.9	0.0	44.1 c	0.0	..	Non spécifié
Total	29.4 c	31.9	33.9	37.9	44.1 c	41.4	..	*Total*
GERD								**DIRD**
Basic research	..	738.3	..	740.9	..	880.3	..	Recherche fondamentale
Applied research	..	1 078.9	..	1 260.9	..	1 475.7	..	Recherche appliquée
Experimental development	..	1 098.7	..	1 247.0	..	1 644.6	..	Développement expérimental
Not specified	2 746.5 c	0.0	3 190.1 c	0.0	3 776.0 c	0.0	..	Non spécifié
Total	2 746.5 c	2 915.9	3 190.1 c	3 248.8	3 776.0 c	4 000.6	..	*Total*

Source: OECD, R&D database (DSTI/EAS Division), January 2002.

Source : OCDE, base de données R-D (DSTI/Division AES), janvier 2002.

Gross domestic expenditure on R&D (GERD) - Dépenses intérieures brutes de R-D (DIRD)

by sector of performance and socio-economic objective - par secteur d'exécution et par objectif socio-économique

Million EUR (1999 ESP euro, EUR from 1999) — *Millions EUR (1999 ESP euro, EUR à partir de 1999)*

	1994	1995	1996	1997	1998	1999	2000	
Business enterprise								**Entreprises**
Agriculture, forestry and fishing	..	54.2	..	67.1	..	78.0	..	Agriculture, sylviculture et pêche
Industrial development	..	732.5	..	828.8	..	1 153.5	..	Développement industriel
Energy	..	132.3	..	123.1	..	120.6	..	Energie
Transport and telecommunications	..	402.2	..	494.5	..	607.8	..	Transports et télécommunications
Urban and rural planning	..	1.6	..	3.9	..	4.1	..	Aménagement urbain et rural
Prevention of pollution	..	43.4	..	34.4	..	51.9	..	Prévention de la pollution
Identification & treatment of pollution	..	23.7	..	19.2	..	37.0	..	Détection pollu. & mesures antipollu.
Health	..	189.5	..	238.9	..	279.1	..	Santé
Social development and services	..	5.7	..	8.7	..	15.5	..	Développement et services sociaux
Earth and atmosphere	..	7.8	..	6.8	..	7.9	..	Milieu terrestre et atmosphère
Advancement of knowledge	..	8.4	..	15.3	..	20.5	..	Promotion générale des connaissances
Civil space	..	33.4	..	37.5	..	67.6	..	Espace civil
Defence	..	77.7	..	92.6	..	153.7	..	Défense
Not elsewhere classified	1 540.5	0.0	1 862.6	0.0	2 457.2	0.0	2 893.7 [c]	Non classé
Total	1 540.5	1 712.2	1 862.6	1 970.9	2 457.2	2 597.1	2 893.7 [c]	*Total*
Government								**État**
Agriculture, forestry and fishing	..	144.4	..	162.9	..	189.5	..	Agriculture, sylviculture et pêche
Industrial development	..	51.8	..	51.2	..	73.4	..	Développement industriel
Energy	..	56.6	..	67.4	..	62.3	..	Energie
Transport and telecommunications	..	15.8	..	29.3	..	37.7	..	Transports et télécommunications
Urban and rural planning	..	2.3	..	2.0	..	3.8	..	Aménagement urbain et rural
Prevention of pollution	..	14.2	..	18.0	..	17.0	..	Prévention de la pollution
Identification & treatment of pollution	..	27.7	..	41.1	..	47.3	..	Détection pollu. & mesures antipollu.
Health	..	87.1	..	121.0	..	167.7	..	Santé
Social development and services	..	19.6	..	20.8	..	22.4	..	Développement et services sociaux
Earth and atmosphere	..	45.3	..	44.6	..	50.6	..	Milieu terrestre et atmosphère
Advancement of knowledge	..	114.7	..	92.3	..	107.0	..	Promotion générale des connaissances
Civil space	..	23.0	..	25.2	..	27.0	..	Espace civil
Defence	..	58.5	..	25.8	..	37.5	..	Défense
Not elsewhere classified	681.8	0.0	704.9	0.0	767.3	0.0	902.3 [c]	Non classé
Total	681.8	661.1	704.9	701.5	767.3	843.3	902.3 [c]	*Total*
Higher education								**Enseignement supérieur**
Agriculture, forestry and fishing	..	57.8	..	87.8	..	112.4	..	Agriculture, sylviculture et pêche
Industrial development	..	169.0	..	198.7	..	230.8	..	Développement industriel
Energy	..	42.9	..	47.8	..	36.3	..	Energie
Transport and telecommunications	..	59.0	..	76.8	..	94.8	..	Transports et télécommunications
Urban and rural planning	..	30.3	..	35.9	..	41.2	..	Aménagement urbain et rural
Prevention of pollution	..	30.4	..	38.0	..	39.5	..	Prévention de la pollution
Identification & treatment of pollution	..	45.9	..	57.1	..	66.3	..	Détection pollu. & mesures antipollu.
Health	..	126.1	..	143.3	..	184.2	..	Santé
Social development and services	..	103.1	..	99.4	..	111.1	..	Développement et services sociaux
Earth and atmosphere	..	32.7	..	46.0	..	57.8	..	Milieu terrestre et atmosphère
Advancement of knowledge	..	427.9	..	477.6	..	513.2	..	Promotion générale des connaissances
Civil space	..	9.6	..	8.8	..	12.6	..	Espace civil
Defence	..	2.2	..	4.5	..	4.5	..	Défense
Not elsewhere classified	1 040.3	0.0	1 242.7	0.0	1 438.7	0.0	1 609.9 [c]	Non classé
Total	1 040.3	1 136.9	1 242.7	1 321.9	1 438.7	1 504.6	1 609.9 [c]	*Total*
Private non-profit								**Institutions sans but lucratif**
Agriculture, forestry and fishing	..	0.4	..	0.8	..	1.5	..	Agriculture, sylviculture et pêche
Industrial development	..	18.4	..	7.6	..	15.6	..	Développement industriel
Energy	..	0.1	..	0.3	..	0.3	..	Energie
Transport and telecommunications	..	1.0	..	3.9	..	6.2	..	Transports et télécommunications
Urban and rural planning	..	0.1	..	0.4	..	0.3	..	Aménagement urbain et rural
Prevention of pollution	..	0.4	..	0.3	..	0.3	..	Prévention de la pollution
Identification & treatment of pollution	..	0.4	..	1.6	..	3.0	..	Détection pollu. & mesures antipollu.
Health	..	8.0	..	16.2	..	8.2	..	Santé
Social development and services	..	3.4	..	4.2	..	5.5	..	Développement et services sociaux
Earth and atmosphere	..	2.0	..	3.5	..	3.5	..	Milieu terrestre et atmosphère
Advancement of knowledge	..	5.6	..	5.9	..	5.9	..	Promotion générale des connaissances
Civil space	..	0.1	..	0.0	..	0.0	..	Espace civil
Defence	..	0.0	..	0.0	..	0.0	..	Défense
Not elsewhere classified	31.9 [c]	0.0	42.4 [c]	0.0	51.9 [c]	0.0	53.9 [c]	Non classé
Total	31.9 [c]	39.8	42.4 [c]	44.6	51.9 [c]	50.4	53.9 [c]	*Total*
GERD								**DIRD**
Agriculture, forestry and fishing	..	256.8	..	318.6	..	381.4	..	Agriculture, sylviculture et pêche
Industrial development	..	971.7	..	1 086.3	..	1 473.3	..	Développement industriel
Energy	..	231.9	..	238.7	..	219.6	..	Energie
Transport and telecommunications	..	478.0	..	604.5	..	746.5	..	Transports et télécommunications
Urban and rural planning	..	34.3	..	42.1	..	49.4	..	Aménagement urbain et rural
Prevention of pollution	..	88.3	..	90.7	..	108.6	..	Prévention de la pollution
Identification & treatment of pollution	..	97.7	..	119.0	..	153.6	..	Détection pollu. & mesures antipollu.
Health	..	410.6	..	519.4	..	639.2	..	Santé
Social development and services	..	131.9	..	133.1	..	154.6	..	Développement et services sociaux
Earth and atmosphere	..	87.8	..	100.9	..	119.7	..	Milieu terrestre et atmosphère
Advancement of knowledge	..	556.6	..	591.1	..	646.6	..	Promotion générale des connaissances
Civil space	..	66.1	..	71.5	..	107.1	..	Espace civil
Defence	..	138.5	..	123.0	..	195.7	..	Défense
Not elsewhere classified	3 294.5 [c]	0.0	3 852.6 [c]	0.0	4 715.0 [c]	0.0	5 459.8 [c]	Non classé
Total	3 294.5 [c]	3 550.1	3 852.6 [c]	4 038.9	4 715.0 [c]	4 995.4	5 459.8 [c]	*Total*

Source: OECD, R&D database (DSTI/EAS Division), January 2002.

Source : OCDE, base de données R-D (DSTI/Division AES), janvier 2002.

Table 5 - Tableau 5
Total R&D personnel - Personnel total de R-D
by sector of employment and occupation - par secteur d'emploi et par profession

In full-time equivalent *En équivalence plein-temps*

	1994	1995	1996	1997	1998	1999	2000	
Business enterprise								**Entreprises**
RSE	11 070 [c]	10 803	11 100	12 009	13 902	15 178	..	Chercheurs
Technicians	9 133 [c]	9 992	10 932 [c]	11 178	12 516	15 103	..	Techniciens
Other occupations	7 118 [c]	6 763	7 399 [c]	6 836	8 249	8 042	..	Autres professions
Total	27 321	27 558	29 431	30 023	34 667	38 323	38 706 [c]	*Total*
Government								**État**
RSE	7 820 [c]	8 359	9 126	10 490	11 021	11 934	..	Chercheurs
Technicians	4 005 [c]	3 364	3 343 [c]	3 826	4 087	4 879	..	Techniciens
Other occupations	5 721 [c]	5 430	5 397 [c]	4 873	5 062	5 470	..	Autres professions
Total	17 546	17 153	17 866	19 189	20 170	22 283	22 506 [c]	*Total*
Higher education								**Enseignement supérieur**
RSE	28 591	27 666	30 858	30 649	34 524	33 840	..	Chercheurs
Technicians	..	2 198	2 671 [c]	2 720	2 593	3 390	..	Techniciens
Other occupations	..	4 466	5 426 [c]	3 474	3 923	3 396	..	Autres professions
Total	34 642 [c]	34 330	38 955	36 843	41 040	40 626	41 032 [c]	*Total*
Private non-profit								**Institutions sans but lucratif**
RSE	386 [c]	514	549 [c]	735	822 [c]	616	..	Chercheurs
Technicians	394 [c]	314	335 [c]	201	218 [c]	221	..	Techniciens
Other occupations	110 [c]	119	127 [c]	159	181 [c]	168	..	Autres professions
Total	890 [c]	947	1 011 [c]	1 095	1 221 [c]	1 005	1 015 [c]	*Total*
TOTAL								**TOTAL**
RSE	47 867 [c]	47 342	51 633 [c]	53 883	60 269 [c]	61 568	..	Chercheurs
Technicians	13 532 [c,m]	15 868	17 281 [c]	17 925	19 414 [c]	23 593	..	Techniciens
Other occupations	12 949 [c,m]	16 778	18 349 [c]	15 342	17 415 [c]	17 076	..	Autres professions
Total	80 399 [c]	79 988	87 263 [c]	87 150	97 098 [c]	102 237	103 259 [c]	*Total*

Table 6 - Tableau 6
Total R&D personnel - Personnel total de R-D
by sector of employment and formal qualification - par secteur d'emploi et par niveau de qualification

In full-time equivalent *En équivalence plein-temps*

	1994	1995	1996	1997	1998	1999	2000	
Business enterprise								**Entreprises**
University PhD level degrees	..	1 601	..	1 576	..	1 825	..	Diplômes univ. de niv. doctorat
Other university degrees	..	15 539	..	17 641	..	23 689	..	Autres diplômes universitaires
Sub-total university degrees	..	17 140	..	19 217	..	25 514	..	Sous-total Universitaires
Other post-secondary diplomas	..	0	..	0	..	0	..	Autres diplômes post-secondaires
Secondary diplomas	..	8 117	..	8 665	..	10 310	..	Diplômes de fin d'études secondaires
Other qualifications	..	2 301	..	2 141	..	2 499	..	Autres qualifications
Total	27 321	27 558	29 431	30 023	34 667	38 323	38 706 [c]	*Total*
Government								**État**
University PhD level degrees	..	3 709	..	4 767	..	5 480	..	Diplômes univ. de niv. doctorat
Other university degrees	..	7 582	..	8 366	..	9 627	..	Autres diplômes universitaires
Sub-total university degrees	..	11 291	..	13 133	..	15 107	..	Sous-total Universitaires
Other post-secondary diplomas	..	0	..	0	..	0	..	Autres diplômes post-secondaires
Secondary diplomas	..	4 155	..	4 190	..	4 622	..	Diplômes de fin d'études secondaires
Other qualifications	..	1 707	..	1 866	..	2 554	..	Autres qualifications
Total	17 546	17 153	17 866	19 189	20 170	22 283	22 506 [c]	*Total*
Higher education								**Enseignement supérieur**
University PhD level degrees	..	13 603	..	16 331	..	18 722	..	Diplômes univ. de niv. doctorat
Other university degrees	..	16 178	..	16 916	..	17 856	..	Autres diplômes universitaires
Sub-total university degrees	28 591 [c]	29 781	..	33 247	..	36 578	..	Sous-total Universitaires
Other post-secondary diplomas	..	0	..	0	..	0	..	Autres diplômes post-secondaires
Secondary diplomas	..	2 991	..	2 412	..	2 768	..	Diplômes de fin d'études secondaires
Other qualifications	..	1 558	..	1 184	..	1 280	..	Autres qualifications
Total	34 642 [c]	34 330	38 955	36 843	41 040	40 626	41 032 [c]	*Total*
Private non-profit								**Institutions sans but lucratif**
University PhD level degrees	..	201	..	251	..	220	..	Diplômes univ. de niv. doctorat
Other university degrees	..	558	..	652	..	595	..	Autres diplômes universitaires
Sub-total university degrees	..	759	..	903	..	815	..	Sous-total Universitaires
Other post-secondary diplomas	..	0	..	0	..	0	..	Autres diplômes post-secondaires
Secondary diplomas	..	155	..	158	..	159	..	Diplômes de fin d'études secondaires
Other qualifications	..	33	..	34	..	31	..	Autres qualifications
Total	890 [c]	947	1 011 [c]	1 095	1 221 [c]	1 005	1 015 [c]	*Total*
TOTAL								**TOTAL**
University PhD level degrees	..	19 114	..	22 925	..	26 247	..	Diplômes univ. de niv. doctorat
Other university degrees	..	39 857	..	43 575	..	51 767	..	Autres diplômes universitaires
Sub-total university degrees	..	58 971	..	66 500	..	78 014	..	Sous-total Universitaires
Other post-secondary diplomas	..	0	..	0	..	0	..	Autres diplômes post-secondaires
Secondary diplomas	..	15 418	..	15 425	..	17 859	..	Diplômes de fin d'études secondaires
Other qualifications	..	5 599	..	5 225	..	6 364	..	Autres qualifications
Total	80 399 [c]	79 988	87 263 [c]	87 150	97 098 [c]	102 237	103 259 [c]	*Total*

Source: OECD, R&D database (DSTI/EAS Division), January 2002. *Source* : OCDE, base de données R-D (DSTI/Division AES), janvier 2002.

Table 7 - Tableau 7

Gross domestic expenditure on R&D (GERD) - Dépenses intérieures brutes de R-D (DIRD)

by sector of performance and main scientific field - par secteur d'exécution et par grand domaine scientifique

Million EUR (1999 ESP euro, EUR from 1999) — *Millions EUR (1999 ESP euro, EUR à partir de 1999)*

	1994	1995	1996	1997	1998	1999	2000	
Business enterprise								**Entreprises**
Natural sciences and engineering	1 540.5	1 710.8	..	1 965.0	..	2 590.5	..	Sciences exactes, nat. et de l'ingénieur
Social sciences and humanities	..	1.4	..	5.9	..	6.6	..	Sciences sociales et humaines
All fields of science	1 540.5	1 712.2	1 862.6	1 970.9	2 457.2	2 597.1	2 893.7 c	*Tous domaines scientifiques*
Government								**État**
Natural sciences and engineering	642.6	620.0	666.0	660.7	723.2	795.1	..	Sciences exactes, nat. et de l'ingénieur
Social sciences and humanities	39.2	41.1	38.9	40.8	44.1	48.2	..	Sciences sociales et humaines
All fields of science	681.8	661.1	704.9	701.5	767.3	843.3	902.3 c	*Tous domaines scientifiques*
Higher education								**Enseignement supérieur**
Natural sciences and engineering	781.2	884.0	967.1	1 032.6	1 121.2	1 172.6	..	Sciences exactes, nat. et de l'ingénieur
Social sciences and humanities	259.1	252.9	275.6	289.3	317.5	332.0	..	Sciences sociales et humaines
All fields of science	1 040.3	1 136.9	1 242.7	1 321.9	1 438.7	1 504.6	1 609.9 c	*Tous domaines scientifiques*
Private non-profit								**Institutions sans but lucratif**
Natural sciences and engineering	..	34.0	..	37.6	..	39.0	..	Sciences exactes, nat. et de l'ingénieur
Social sciences and humanities	..	5.9	..	7.0	..	11.4	..	Sciences sociales et humaines
All fields of science	31.9 c	39.8	42.4 c	44.6	51.9 c	50.4	53.9 c	*Tous domaines scientifiques*
TOTAL								**TOTAL**
Natural sciences and engineering	2 987.0	3 248.8	3 530.3	3 695.9	..	4 597.2	..	Sciences exactes, nat. et de l'ingénieur
Social sciences and humanities	307.4	301.3	322.4	343.0	..	398.2	..	Sciences sociales et humaines
All fields of science	3 294.5 c	3 550.1	3 852.6 c	4 038.9	4 715.0 c	4 995.4	5 459.8 c	*Tous domaines scientifiques*

Table 8 - Tableau 8

Total R&D personnel - Personnel total de R-D

by sector of employment and main scientific field - par secteur d'emploi et par grand domaine scientifique

In full-time equivalent — *En équivalence plein-temps*

	1994	1995	1996	1997	1998	1999	2000	
Business enterprise								**Entreprises**
Natural sciences and engineering	..	27 519	..	29 895	..	38 156	..	Sciences exactes, nat. et de l'ingénieur
Social sciences and humanities	..	39	..	128	..	167	..	Sciences sociales et humaines
All fields of science	27 321	27 558	29 431	30 023	34 667	38 323	38 706 c	*Tous domaines scientifiques*
Government								**État**
Natural sciences and engineering	16 565	15 978	16 784	17 947	18 897	20 895	..	Sciences exactes, nat. et de l'ingénieur
Social sciences and humanities	982	1 175	1 082	1 242	1 273	1 388	..	Sciences sociales et humaines
All fields of science	17 546	17 153	17 866	19 189	20 170	22 283	22 506 c	*Tous domaines scientifiques*
Higher education								**Enseignement supérieur**
Natural sciences and engineering	21 438 c	26 776	30 202	28 637	31 899	31 306	..	Sciences exactes, nat. et de l'ingénieur
Social sciences and humanities	13 204 c	7 554	8 753	8 206	9 142	9 320	..	Sciences sociales et humaines
All fields of science	34 642 c	34 330	38 955	36 843	41 040	40 626	41 032 c	*Tous domaines scientifiques*
Private non-profit								**Institutions sans but lucratif**
Natural sciences and engineering	..	777	..	908	..	760	..	Sciences exactes, nat. et de l'ingénieur
Social sciences and humanities	..	170	..	187	..	246	..	Sciences sociales et humaines
All fields of science	890 c	947	1 011 c	1 095	1 221 c	1 005	1 015 c	*Tous domaines scientifiques*
TOTAL								**TOTAL**
Natural sciences and engineering	..	71 050	..	77 387	..	91 117	..	Sciences exactes, nat. et de l'ingénieur
Social sciences and humanities	..	8 938	..	9 763	..	11 121	..	Sciences sociales et humaines
All fields of science	80 399 c	79 988	87 263 c	87 150	97 098 c	102 237	103 259 c	*Tous domaines scientifiques*

Source: OECD, R&D database (DSTI/EAS Division), January 2002. — *Source* : OCDE, base de données R-D (DSTI/Division AES), janvier 2002.

Total business enterprise intramural expenditure on R&D - Dépenses intra-muros totales de R-D des entreprises
(BERD) by industry - (DIRDE) par industrie

	1994	1995	1996	1997	1998	1999	2000	
	9.1 Million EUR (1999 ESP euro, EUR from 1999) - Millions EUR (1999 ESP euro, EUR à partir de 1999)							
1 AGRICULTURE, HUNTING & FORESTRY	22.5	30.8	30.5	37.6	66.8	43.5	..	**1 AGRICULTURE, CHASSE & SYLVICULTURE**
2 MINING	6.2	10.9	11.4	7.6	6.1	8.8	..	**2 ACTIVITES EXTRACTIVES**
3 MANUFACTURING	1 214.6	1 339.0	1 534.4	1 610.5	1 968.3	2 019.8	..	**3 ACTIVITES DE FABRICATION**
4 Food, beverages & tobacco	45.3	52.3	63.3	61.8	76.1	74.3	..	**4 Produits alimentaires, boissons & tabac**
5 Food products & beverages	38.2	48.3	58.8	57.0	71.2	70.0	..	5 Produits alimentaires & boissons
6 Tobacco products	7.0	4.0	4.5	4.8	4.9	4.4	..	6 Produits à base de tabac
7 Textiles, wear app., fur & leather	27.0	19.2	30.4	27.2	46.8	44.7	..	**7 Textiles, habill., fourrures & cuirs**
8 Textiles	12.4	9.9	18.6	16.4	25.9	31.3	..	8 Textiles
9 Wearing apparel & fur	9.4	5.1	5.3	6.2	10.4	5.6	..	9 Articles d'habillement & fourrures
10 Leather products & footwear	5.2	4.2	6.5	4.6	10.4	7.8	..	10 Articles en cuir & chaussures
11 Wood, paper, printing, publishing	10.8	14.0	21.4	20.5	66.6	32.5	..	**11 Bois, papier, imprimerie, édition**
12 Wood & cork (not furniture)	1.2	2.1	1.5	1.6	2.2	2.4	..	12 Bois & liège (sauf meubles)
13 Pulp, paper & paper products	6.9	7.6	11.3	13.9	56.3	24.0	..	13 Carton, papier & articles
14 Publ., print. & repro. of rec. media	2.7	4.3	8.6	5.0	8.2	6.0	..	14 Edition, impress. & repro. enregis.
15 Coke, petroleum, nuclear fuel, chemicals, rubber plastics	288.9	342.4	398.3	414.6	478.0	488.4	..	**15 Coke, pétrole, combust. nucl., prod. chimi., caout. & mat. plast.**
16 Coke, ref. petrol. prod. & nuclear fuel	22.7	27.5	26.2	32.2	29.7	30.7	..	16 Coke, raffin. pétr. & combust. nucl.
17 Chemicals & chemical products	226.0	272.3	320.6	319.1	380.4	389.7	..	17 Produits chimiques
18 *Chemical prod.(less pharmaceuticals)*	79.4	83.8	98.5	97.5	139.9	142.6	..	18 *Prod. chimi. (moins prod. pharmaceutiques)*
19 *Pharmaceuticals*	146.6	188.5	222.1	221.6	240.5	247.1	..	19 *Produits pharmaceutiques*
20 Rubber & plastic products	40.1	42.6	51.5	63.3	67.9	67.9	..	20 Caoutchouc et matières plastiques
21 Non-metallic mineral products	30.8	25.2	32.1	34.9	37.4	43.7	..	**21 Produits minéraux non métalliques**
22 Basic metals	29.2	28.6	39.6	35.9	40.2	41.2	..	**22 Produits métallurgiques de base**
23 Basic metals, ferrous	24.3	24.2	32.3	30.9	33.3	33.6	..	23 Produits métallur. de base, ferreux
24 Basic metals, non-ferrous	5.0	4.3	7.3	5.1	6.9	7.6	..	24 Prod. métallur. de base, non ferreux
25 Fabricated metal products	26.6	26.7	34.0	40.0	54.7	51.9	..	**25 Ouvrages en métaux (sauf mach.& mat.)**
26 Machinery equipment, instruments and transport equipment	738.6	812.6	895.2	949.5	1 125.9	1 199.4	..	**26 Machines & matériel, instruments & mat. de transport**
27 Machinery n.e.c.	98.2	116.9	138.0	139.8	178.7	184.1	..	27 Machines, n.c.a.
28 Office, account. & computing machinery	18.9	28.7	35.2	47.1	51.0	55.9	..	28 Mach. de bur., compta. & trait. inform.
29 Electrical machinery	47.4	56.7	69.5	92.4	113.9	117.2	..	29 Machines électriques
30 Electro. equip. (radio, TV & commun.)	207.4	203.9	217.1	222.4	248.9	240.8	..	30 Equip. électro. (radio, tv & communic.)
31 *Electro. comp. (inc. semi-conductors)*	13.7	18.6	29.1	29.5	29.4	36.1	..	31 *Compos. électro. (semi-cond. incl.)*
32 *TV, radio & communications equip.*	193.7	185.3	188.1	192.9	219.5	204.8	..	32 *Appareils radio, tv et de communi.*
33 Instruments	31.1	43.6	36.1	52.2	54.5	51.6	..	33 Instruments
34 Motor vehicles	166.0	169.9	183.0	172.6	187.8	243.9	..	34 Véhicules automobiles
35 Other transport equipment	169.4	192.9	216.3	222.8	291.2	305.8	..	35 Autres matériels de transport
36 *Ships*	21.0	25.7	32.6	32.1	64.5	66.0	..	36 *Navires*
37 *Aerospace*	131.8	149.6	157.5	167.0	197.4	199.3	..	37 *Aérospatiale*
38 *Other transport equipment n.e.c.*	16.5	17.7	26.1	23.7	29.3	40.6	..	38 *Autres matériels de transport n.c.a.*
39 Furniture, other manufacturing n.e.c.	17.3	18.0 [o]	20.2 [o]	26.1 [o]	41.0	40.3	..	**39 Meubles, autres activ. de fabricat. n.c.a.**
40 Furniture	7.7	7.6	7.5	11.4	13.1	16.9	..	40 Meubles
41 Other manufacturing n.e.c.	9.6	10.4 [o]	12.7 [o]	14.7 [o]	27.9	23.4	..	41 Autres activités de fabrication n.c.a.
42 Recycling	0.1	.. [n]	.. [n]	.. [n]	1.6	3.4	..	**42 Récupération**
43 ELECTRICITY, GAS & WATER SUPPLY	54.3	105.3	45.5	64.2	55.9	42.5	..	**43 DISTRIBUTION D'ELECTRICITE, GAZ & EAU**
44 CONSTRUCTION	4.6	5.9	6.0	6.3	4.1	6.5	..	**44 CONSTRUCTION**
45 SERVICES SECTOR	238.3	220.3	234.8	244.7	355.9	475.9	..	**45 SECTEUR DES SERVICES**
46 Wholesale, retail trade & repair of motor vehicles, etc.	0.8	2.2	0.9	2.6	3.5	18.3	..	**46 Commerce (gros & détail), réparation de véhicules, etc.**
47 Hotels & restaurants	0.0	0.0	0.0	0.0	0.0	0.0	..	**47 Hôtels et restaurants**
48 Transport & storage	6.7	.. [n]	.. [n]	1.5	4.7	1.7	..	**48 Transports, entreposage**
49 Communications	59.7	66.1 [o]	82.3 [o]	79.4	165.9	225.5	..	**49 Communications**
50 Post	0.0	0.0	0.0	0.0	0.0	0.0	..	50 Postes
51 Telecommunications	59.7	66.1 [o]	82.3 [o]	79.4	165.9	225.5	..	51 Télécommunications
52 Financial intermediation (inc. insur.)	0.0	.. [n]	.. [n]	0.0	.. [n]	.. [n]	..	**52 Intermédiation financ. (assurances inc.)**
53 Real estate, renting & busin. activity	165.1	144.2 [o]	143.9 [o]	146.5	170.7 [o]	215.1 [o]	..	**53 Immobil., locat. & serv. aux entreprises**
54 Computer & related activities	38.1	53.5	52.4	52.4	67.8	91.6	..	54 Informatique et activités connexes
55 *Software consultancy*	32.8	46.4	46.1	45.8	60.4	78.1	..	55 *Production de logiciels*
56 *Other computer services n.e.c.*	5.3	7.1	6.2	6.6	7.3	13.5	..	56 *Autres activités informatiques*
57 Research & development	40.9	3.1 [a,o]	3.7 [o]	4.3	15.1 [o]	20.1 [o]	..	57 Recherche & développement expérimental
58 Other business activities n.e.c.	86.1	87.6	87.9	89.7	87.9	103.5	..	58 Autres activ. & serv. aux entreprises
59 Comm., soc. & pers. serv. activ., etc.	5.9	7.8	7.8	14.7	11.1	15.3	..	**59 Adm. pub., serv. soc. & collec., etc**
TOTAL BUSINESS ENTERPRISE SECTOR	**1 540.5**	**1 712.2**	**1 862.6**	**1 970.9**	**2 457.2**	**2 597.1**	**2 893.7** [c]	**TOTAL SECTEUR DES ENTREPRISES**

Source: OECD, R&D database (DSTI/EAS Division), January 2002.

Source : OCDE, base de données R-D (DSTI/Division AES), janvier 2002.

Total business enterprise intramural expenditure on R&D - Dépenses intra-muros totales de R-D des entreprises
(BERD) by industry - (DIRDE) par industrie

	1994	1995	1996	1997	1998	1999	2000	
	9.2 Million constant USD(1995 prices and PPPs) - Millions de USD constants (prix et PPA de 1995)							
1 AGRICULTURE, HUNTING & FORESTRY	32.1 [b]	42.0	40.1	48.4	84.2	53.3	..	**1 AGRICULTURE, CHASSE & SYLVICULTURE**
2 MINING	8.9 [b]	14.8	15.0	9.8	7.7	10.8	..	**2 ACTIVITES EXTRACTIVES**
3 MANUFACTURING	1 735.2 [b]	1 825.0	2 020.2	2 074.6	2 479.3	2 472.8	..	**3 ACTIVITES DE FABRICATION**
4 Food, beverages & tobacco	64.7 [b]	71.3	83.3	79.6	95.8	91.0	..	**4 Produits alimentaires, boissons & tabac**
5 Food products & beverages	54.6 [b]	65.8	77.4	73.4	89.6	85.7	..	5 Produits alimentaires & boissons
6 Tobacco products	10.0 [b]	5.5	6.0	6.2	6.2	5.3	..	6 Produits à base de tabac
7 Textiles, wear app., fur & leather	38.6 [b]	26.1	40.0	35.1	58.9	54.7	..	**7 Textiles, habill., fourrures & cuirs**
8 Textiles	17.8 [b]	13.5	24.4	21.2	32.7	38.3	..	8 Textiles
9 Wearing apparel & fur	13.4 [b]	6.9	7.0	8.0	13.1	6.8	..	9 Articles d'habillement & fourrures
10 Leather products & footwear	7.4 [b]	5.7	8.6	5.9	13.1	9.6	..	10 Articles en cuir & chaussures
11 Wood, paper, printing, publishing	15.5 [b]	19.1	28.1	26.5	83.9	39.7	..	**11 Bois, papier, imprimerie, édition**
12 Wood & cork (not furniture)	1.7 [b]	2.9	2.0	2.1	2.7	3.0	..	12 Bois & liège (sauf meubles)
13 Pulp, paper & paper products	9.9 [b]	10.4	14.8	18.0	70.9	29.4	..	13 Carton, papier & articles
14 Publ., print. & repro. of rec. media	3.9 [b]	5.9	11.3	6.4	10.4	7.4	..	14 Edition, impress. & repro. enregis.
15 Coke, petroleum, nuclear fuel, chemicals, rubber plastics	412.7 [b]	466.7	524.3	534.1	602.1	597.9	..	**15 Coke, pétrole, combust. nucl., prod. chimi., caout. & mat. plast.**
16 Coke, ref. petrol. prod. & nuclear fuel	32.5 [b]	37.5	34.5	41.4	37.4	37.6	..	16 Coke, raffin. pétr. & combust. nucl.
17 Chemicals & chemical products	322.9 [b]	371.1	422.1	411.1	479.2	477.1	..	17 Produits chimiques
18 *Chemical prod.(less pharmaceuticals)*	113.4 [b]	114.2	129.7	125.6	176.3	174.6	..	18 *Prod. chimi. (moins prod. pharmaceutiques)*
19 *Pharmaceuticals*	209.5 [b]	256.9	292.4	285.5	302.9	302.5	..	19 *Produits pharmaceutiques*
20 Rubber & plastic products	57.3 [b]	58.1	67.7	81.6	85.5	83.1	..	20 Caoutchouc et matières plastiques
21 Non-metallic mineral products	44.0 [b]	34.4	42.3	44.9	47.1	53.5	..	**21 Produits minéraux non métalliques**
22 Basic metals	41.8 [b]	39.0	52.1	46.3	50.6	50.5	..	**22 Produits métallurgiques de base**
23 Basic metals, ferrous	34.7 [b]	33.0	42.5	39.8	41.9	41.2	..	23 Produits métallur. de base, ferreux
24 Basic metals, non-ferrous	7.1 [b]	5.9	9.6	6.5	8.7	9.3	..	24 Prod. métallur. de base, non ferreux
25 Fabricated metal products	38.0 [b]	36.4	44.8	51.5	68.9	63.5	..	**25 Ouvrages en métaux (sauf mach.& mat.)**
26 Machinery equipment, instruments and transport equipment	1 055.1 [b]	1 107.6	1 178.6	1 223.1	1 418.2	1 468.4	..	**26 Machines & matériel, instruments & mat. de transport**
27 Machinery n.e.c.	140.3 [b]	159.4	181.8	180.1	225.0	225.3	..	27 Machines, n.c.a.
28 Office, account. & computing machinery	27.1 [b]	39.1	46.4	60.7	64.3	68.5	..	28 Mach. de bur., compta. & trait. inform.
29 Electrical machinery	67.8 [b]	77.3	91.5	119.1	143.4	143.5	..	29 Machines électriques
30 Electro. equip. (radio, TV & commun.)	296.4 [b]	277.9	285.9	286.5	313.5	294.8	..	30 Equip. électro. (radio, tv & communic.)
31 *Electro. comp. (inc. semi-conductors)*	19.6 [b]	25.3	38.3	38.0	37.1	44.2	..	31 *Compos. électro. (semi-cond. incl.)*
32 *TV, radio & communications equip.*	276.8 [b]	252.6	247.6	248.5	276.4	250.7	..	32 *Appareils radio, tv et de communi.*
33 Instruments	44.5 [b]	59.4	47.5	67.3	68.7	63.2	..	33 Instruments
34 Motor vehicles	237.2 [b]	231.6	241.0	222.3	236.5	298.6	..	34 Véhicules automobiles
35 Other transport equipment	242.0 [b]	262.9	284.7	287.0	366.7	374.4	..	35 Autres matériels de transport
36 *Ships*	30.1 [b]	35.0	42.9	41.3	81.2	80.7	..	36 *Navires*
37 *Aerospace*	188.4 [b]	203.9	207.4	215.2	248.6	244.0	..	37 *Aérospatiale*
38 *Other transport equipment n.e.c.*	23.6 [b]	24.1	34.4	30.5	36.9	49.7	..	38 *Autres matériels de transport n.c.a.*
39 Furniture, other manufacturing n.e.c.	24.7 [b]	24.6 [o]	26.6 [o]	33.6 [o]	51.6	49.3	..	**39 Meubles, autres activ. de fabricat. n.c.a.**
40 Furniture	11.0 [b]	10.4	9.9	14.7	16.5	20.7	..	40 Meubles
41 Other manufacturing n.e.c.	13.7 [b]	14.2 [o]	16.7 [o]	18.9 [o]	35.2	28.6	..	41 Autres activités de fabrication n.c.a.
42 Recycling	0.2 [b]	.. [n]	.. [n]	.. [n]	2.0	4.2	..	**42 Récupération**
43 ELECTRICITY, GAS & WATER SUPPLY	77.6 [b]	143.5	60.0	82.6	70.4	52.1	..	**43 DISTRIBUTION D'ELECTRICITE, GAZ & EAU**
44 CONSTRUCTION	6.6 [b]	8.0	7.9	8.1	5.2	8.0	..	**44 CONSTRUCTION**
45 SERVICES SECTOR	340.4 [b]	300.3	309.2	315.2	448.2	582.6	..	**45 SECTEUR DES SERVICES**
46 Wholesale, retail trade & repair of motor vehicles, etc.	1.1 [b]	3.1	1.2	3.4	4.4	22.4	..	**46 Commerce (gros & détail), réparation de véhicules, etc.**
47 Hotels & restaurants	0.0 [b]	0.0	0.0	0.0	0.0	0.0	..	**47 Hôtels et restaurants**
48 Transport & storage	9.6 [b]	.. [n]	.. [n]	1.9	6.0	2.1	..	**48 Transports, entreposage**
49 Communications	85.3 [b]	90.1 [o]	108.4 [o]	102.3	208.9	276.0	..	**49 Communications**
50 Post	0.0 [b]	0.0	0.0	0.0	0.0	0.0	..	50 Postes
51 Telecommunications	85.3 [b]	90.1 [o]	108.4 [o]	102.3	208.9	276.0	..	51 Télécommunications
52 Financial intermediation (inc. insur.)	0.0 [b]	.. [n]	.. [n]	0.0	.. [n]	.. [n]	..	**52 Intermédiation financ. (assurances inc.)**
53 Real estate, renting & busin. activity	235.9 [b]	196.5 [o]	189.5 [o]	188.7	215.1 [o]	263.4 [o]	..	**53 Immobil., locat. & serv. aux entreprises**
54 Computer & related activities	54.5 [b]	73.0	68.9	67.5	85.4	112.2	..	54 Informatique et activités connexes
55 *Software consultancy*	46.9 [b]	63.3	60.8	59.0	76.1	95.6	..	55 *Production de logiciels*
56 *Other computer services n.e.c.*	7.5 [b]	9.7	8.2	8.5	9.2	16.6	..	56 *Autres activités informatiques*
57 Research & development	58.5 [b]	4.2 [a,o]	4.8 [o]	5.5	19.0 [o]	24.6 [o]	..	57 Recherche & développement expérimental
58 Other business activities n.e.c.	123.0 [b]	119.4	115.7	115.6	110.7	126.7	..	58 Autres activ. & serv. aux entreprises
59 Comm., soc. & pers. serv. activ., etc.	8.4 [b]	10.6	10.2	18.9	14.0	18.7	..	**59 Adm. pub., serv. soc. & collec., etc**
TOTAL BUSINESS ENTERPRISE SECTOR	**2 200.8** [b]	**2 333.7**	**2 452.3**	**2 538.8**	**3 095.0**	**3 179.5**	**3 424.3** [c]	**TOTAL SECTEUR DES ENTREPRISES**

Source: OECD, R&D database (DSTI/EAS Division), January 2002.

Source : OCDE, base de données R-D (DSTI/Division AES), janvier 2002.

Business enterprise intramural expenditure on R&D (BERD) - Dépenses intra-muros de R-D des entreprises (DIRDE)

by industry and source of funds - par industrie et par source de financement

Million EUR (1999 ESP euro, EUR from 1999) — *Millions EUR (1999 ESP euro, EUR à partir de 1999)*

10.1 Business enterprise - Entreprises

	1994	1995	1996	1997	1998	1999	2000	
1 AGRICULTURE, HUNTING & FORESTRY	..	15.8	17.2	20.8	50.7	20.4	..	**1 AGRICULTURE, CHASSE & SYLVICULTURE**
2 MINING	..	8.3	8.9	6.0	3.0	7.3	..	**2 ACTIVITES EXTRACTIVES**
3 MANUFACTURING	..	1 156.5	1 356.2	1 391.7	1 780.2	1 777.0	..	**3 ACTIVITES DE FABRICATION**
4 Food, beverages & tobacco	..	46.8	57.9	56.2	71.0	65.9	..	**4 Produits alimentaires, boissons & tabac**
5 Food products & beverages	..	42.8	53.4	51.5	66.3	61.7	..	5 Produits alimentaires & boissons
6 Tobacco products	..	4.0	4.5	4.7	4.8	4.2	..	6 Produits à base de tabac
7 Textiles, wear app., fur & leather	..	15.6	27.5	23.7	42.2	41.2	..	**7 Textiles, habill., fourrures & cuirs**
8 Textiles	..	8.8	17.6	14.9	24.0	30.2	..	8 Textiles
9 Wearing apparel & fur	..	5.1	5.3	6.1	10.3	5.6	..	9 Articles d'habillement & fourrures
10 Leather products & footwear	..	1.7	4.6	2.7	7.8	5.4	..	10 Articles en cuir & chaussures
11 Wood, paper, printing, publishing	..	11.9	19.7	16.0	62.6	29.2	..	**11 Bois, papier, imprimerie, édition**
12 Wood & cork (not furniture)	..	1.0	1.1	1.1	2.1	1.7	..	12 Bois & liège (sauf meubles)
13 Pulp, paper & paper products	..	7.2	10.8	11.0	54.5	23.0	..	13 Carton, papier & articles
14 Publ., print. & repro. of rec. media	..	3.7	7.8	4.0	6.1	4.4	..	14 Edition, impress. & repro. enregis.
15 Coke, petroleum, nuclear fuel, chemicals, rubber plastics	..	325.9	385.3	391.6	451.1	438.7	..	**15 Coke, pétrole, combust. nucl., prod. chimi., caout. & mat. plast.**
16 Coke, ref. petrol. prod. & nuclear fuel	..	25.5	24.5	30.2	26.5	29.7	..	16 Coke, raffin. pétr. & combust. nucl.
17 Chemicals & chemical products	..	259.3	310.1	300.5	360.8	346.3	..	17 Produits chimiques
18 *Chemical prod.(less pharmaceuticals)*	..	80.3	95.1	92.6	134.8	136.4	..	18 *Prod. chimi. (moins prod. pharmaceutiques)*
19 *Pharmaceuticals*	..	179.0	215.0	207.9	226.0	210.0	..	19 *Produits pharmaceutiques*
20 Rubber & plastic products	..	41.1	50.7	60.9	63.8	62.7	..	20 Caoutchouc et matières plastiques
21 Non-metallic mineral products	..	24.0	30.7	32.5	34.9	39.7	..	**21 Produits minéraux non métalliques**
22 Basic metals	..	19.8	30.2	25.8	29.0	35.2	..	**22 Produits métallurgiques de base**
23 Basic metals, ferrous	..	15.9	22.9	22.2	22.4	28.0	..	23 Produits métallur. de base, ferreux
24 Basic metals, non-ferrous	..	3.8	7.3	3.6	6.6	7.2	..	24 Prod. métallur. de base, non ferreux
25 Fabricated metal products	..	22.9	30.3	34.2	49.0	41.8	..	**25 Ouvrages en métaux (sauf mach.& mat.)**
26 Machinery equipment, instruments and transport equipment	..	673.4	757.3	789.8	1 001.4	1 045.9	..	**26 Machines & matériel, instruments & mat. de transport**
27 Machinery n.e.c.	..	99.9	122.1	119.0	160.2	157.5	..	27 Machines, n.c.a.
28 Office, account. & computing machinery	..	28.5	33.7	44.6	50.0	54.7	..	28 Mach. de bur., compta. & trait. inform.
29 Electrical machinery	..	53.8	66.5	83.6	102.5	102.8	..	29 Machines électriques
30 Electro. equip. (radio, TV & commun.)	..	174.9	184.3	185.8	235.7	224.7	..	30 Equip. électro. (radio, tv & communic.)
31 *Electro. comp. (inc. semi-conductors)*	..	16.0	25.3	23.6	23.1	29.7	..	31 *Compos. électro. (semi-cond. incl.)*
32 *TV, radio & communications equip.*	..	158.9	159.0	162.2	212.6	195.0	..	32 *Appareils radio, tv et de communi.*
33 Instruments	..	37.8	30.5	42.5	44.8	44.3	..	33 Instruments
34 Motor vehicles	..	167.3	177.3	168.8	182.5	229.8	..	34 Véhicules automobiles
35 Other transport equipment	..	111.3	142.9	145.6	225.7	232.1	..	35 Autres matériels de transport
36 *Ships*	..	8.4	14.5	18.5	46.6	52.6	..	36 *Navires*
37 *Aerospace*	..	85.6	104.3	107.5	151.0	141.4	..	37 *Aérospatiale*
38 *Other transport equipment n.e.c.*	..	17.3	24.1	19.5	28.1	38.1	..	38 *Autres matériels de transport n.c.a.*
39 Furniture, other manufacturing n.e.c.	..	16.1 [o]	17.2 [o]	21.8 [o]	37.5	36.4	..	**39 Meubles, autres activ. de fabricat. n.c.a.**
40 Furniture	..	7.0	5.7	8.0	10.9	14.5	..	40 Meubles
41 Other manufacturing n.e.c.	..	9.1 [o]	11.4 [o]	13.8 [o]	26.6	21.9	..	41 Autres activités de fabrication n.c.a.
42 Recycling	..	.. [n]	.. [n]	.. [n]	1.5	3.1	..	**42 Récupération**
43 ELECTRICITY, GAS & WATER SUPPLY	..	91.9	34.4	53.5	43.1	35.3	..	**43 DISTRIBUTION D'ELECTRICITE, GAZ & EAU**
44 CONSTRUCTION	..	5.4	5.8	5.7	3.2	5.7	..	**44 CONSTRUCTION**
45 SERVICES SECTOR	..	166.7	192.2	194.8	308.7	406.2	..	**45 SECTEUR DES SERVICES**
46 Wholesale, retail trade & repair of motor vehicles, etc.	..	2.2	0.4	2.3	3.0	18.0	..	**46 Commerce (gros & détail), réparation de véhicules, etc.**
47 Hotels & restaurants	..	0.0	0.0	0.0	0.0	0.0	..	**47 Hôtels et restaurants**
48 Transport & storage	..	.. [n]	.. [n]	0.6	4.2	1.7	..	**48 Transports, entreposage**
49 Communications	..	58.4 [o]	75.6 [o]	74.5	159.3	219.2	..	**49 Communications**
50 Post	..	0.0	0.0	0.0	0.0	0.0	..	50 Postes
51 Telecommunications	..	58.4 [o]	75.6 [o]	74.5	159.3	219.2	..	51 Télécommunications
52 Financial intermediation (inc. insur.)	..	.. [n]	.. [n]	0.0	.. [n]	.. [n]	..	**52 Intermédiation financ. (assurances inc.)**
53 Real estate, renting & busin. activity	..	100.7 [o]	110.8 [o]	106.1	135.2 [o]	155.9 [o]	..	**53 Immobil., locat. & serv. aux entreprises**
54 Computer & related activities	..	46.0	44.4	37.9	52.8	74.6	..	54 Informatique et activités connexes
55 *Software consultancy*	..	41.1	40.4	33.1	47.4	63.6	..	55 *Production de logiciels*
56 *Other computer services n.e.c.*	..	4.9	3.9	4.7	5.4	11.0	..	56 *Autres activités informatiques*
57 Research & development	..	2.5 [a,o]	2.6 [o]	3.2	10.8 [o]	13.6 [o]	..	57 Recherche & développement expérimental
58 Other business activities n.e.c.	..	52.1	63.8	65.1	71.6	67.7	..	58 Autres activ. & serv. aux entreprises
59 Comm., soc. & pers. serv. activ., etc.	..	5.5	5.4	11.3	7.0	11.4	..	**59 Adm. pub., serv. soc. & collec., etc**
TOTAL BUSINESS ENTERPRISE SECTOR	**1 234.1** [c]	**1 444.6**	**1 614.6**	**1 672.3**	**2 188.9**	**2 251.8**	**..**	**TOTAL SECTEUR DES ENTREPRISES**

Source: OECD, R&D database (DSTI/EAS Division), January 2002.

Source : OCDE, base de données R-D (DSTI/Division AES), janvier 2002.

Business enterprise intramural expenditure on R&D (BERD) - Dépenses intra-muros de R-D des entreprises (DIRDE)

by industry and source of funds - par industrie et par source de financement

Million EUR (1999 ESP euro, EUR from 1999) — *Millions EUR (1999 ESP euro, EUR à partir de 1999)*

10.2 Government - État

	1994	1995	1996	1997	1998	1999	2000	
1 AGRICULTURE, HUNTING & FORESTRY	..	13.1	12.3	13.8	14.1	17.9	..	**1 AGRICULTURE, CHASSE & SYLVICULTURE**
2 MINING	..	0.6	0.6	1.1	0.4	1.4	..	**2 ACTIVITES EXTRACTIVES**
3 MANUFACTURING	..	109.4	104.3	130.2	115.2	164.6	..	**3 ACTIVITES DE FABRICATION**
4 Food, beverages & tobacco	..	4.1	4.6	5.1	4.3	7.0	..	**4 Produits alimentaires, boissons & tabac**
5 Food products & beverages	..	4.1	4.6	5.1	4.2	7.0	..	5 Produits alimentaires & boissons
6 Tobacco products	..	0.0	0.0	0.0	0.0	0.0	..	6 Produits à base de tabac
7 Textiles, wear app., fur & leather	..	2.6	1.8	2.5	3.4	2.5	..	**7 Textiles, habill., fourrures & cuirs**
8 Textiles	..	1.0	0.5	1.2	1.3	0.6	..	8 Textiles
9 Wearing apparel & fur	..	0.0	0.0	0.1	0.1	0.0	..	9 Articles d'habillement & fourrures
10 Leather products & footwear	..	1.6	1.3	1.2	1.9	1.8	..	10 Articles en cuir & chaussures
11 Wood, paper, printing, publishing	..	1.8	1.4	4.2	3.5	3.0	..	**11 Bois, papier, imprimerie, édition**
12 Wood & cork (not furniture)	..	1.1	0.4	0.6	0.1	0.7	..	12 Bois & liège (sauf meubles)
13 Pulp, paper & paper products	..	0.2	0.2	2.9	1.6	0.8	..	13 Carton, papier & articles
14 Publ., print. & repro. of rec. media	..	0.5	0.8	0.8	1.8	1.5	..	14 Edition, impress. & repro. enregis.
15 Coke, petroleum, nuclear fuel, chemicals, rubber plastics	..	8.7	7.7	14.1	19.2	39.4	..	**15 Coke, pétrole, combust. nucl., prod. chimi., caout. & mat. plast.**
16 Coke, ref. petrol. prod. & nuclear fuel	..	0.2	0.6	0.7	0.6	0.5	..	16 Coke, raffin. pétr. & combust. nucl.
17 Chemicals & chemical products	..	7.3	6.4	11.4	15.1	34.1	..	17 Produits chimiques
18 *Chemical prod.(less pharmaceuticals)*	..	2.6	2.4	4.2	4.5	4.3	..	18 *Prod. chimi. (moins prod. pharmaceutiques)*
19 *Pharmaceuticals*	..	4.7	4.0	7.2	10.6	29.8	..	19 *Produits pharmaceutiques*
20 Rubber & plastic products	..	1.2	0.7	2.0	3.5	4.7	..	20 Caoutchouc et matières plastiques
21 Non-metallic mineral products	..	1.1	1.3	2.2	1.9	3.4	..	**21 Produits minéraux non métalliques**
22 Basic metals	..	6.6	5.5	8.5	8.0	3.6	..	**22 Produits métallurgiques de base**
23 Basic metals, ferrous	..	6.2	5.5	7.2	7.8	3.2	..	23 Produits métallur. de base, ferreux
24 Basic metals, non-ferrous	..	0.4	0.0	1.3	0.2	0.4	..	24 Prod. métallur. de base, non ferreux
25 Fabricated metal products	..	2.2	2.6	4.1	4.5	8.6	..	**25 Ouvrages en métaux (sauf mach.& mat.)**
26 Machinery equipment, instruments and transport equipment	..	80.4	77.2	87.2	68.0	94.0	..	**26 Machines & matériel, instruments & mat. de transport**
27 Machinery n.e.c.	..	14.1	12.9	15.6	12.2	21.8	..	27 Machines, n.c.a.
28 Office, account. & computing machinery	..	0.2	0.9	1.7	1.0	0.7	..	28 Mach. de bur., compta. & trait. inform.
29 Electrical machinery	..	2.2	2.3	7.0	7.5	10.0	..	29 Machines électriques
30 Electro. equip. (radio, TV & commun.)	..	5.9	7.8	10.1	10.9	10.7	..	30 Equip. électro. (radio, tv & communic.)
31 *Electro. comp. (inc. semi-conductors)*	..	2.3	3.8	5.2	5.1	4.6	..	31 *Compos. électro. (semi-cond. incl.)*
32 *TV, radio & communications equip.*	..	3.6	4.0	4.9	5.8	6.1	..	32 *Appareils radio, tv et de communi.*
33 Instruments	..	4.9	4.0	7.3	6.7	5.8	..	33 Instruments
34 Motor vehicles	..	2.5	5.1	3.6	4.4	7.9	..	34 Véhicules automobiles
35 Other transport equipment	..	50.6	44.2	41.9	25.4	37.2	..	35 Autres matériels de transport
36 *Ships*	..	13.3	13.9	12.2	9.5	6.2	..	36 *Navires*
37 *Aerospace*	..	36.9	28.4	27.1	15.0	28.9	..	37 *Aérospatiale*
38 *Other transport equipment n.e.c.*	..	0.4	1.9	2.6	0.9	2.2	..	38 *Autres matériels de transport n.c.a.*
39 Furniture, other manufacturing n.e.c.	..	1.8 [o]	2.3 [o]	2.2 [o]	2.4	2.9	..	**39 Meubles, autres activ. de fabricat. n.c.a.**
40 Furniture	..	0.6	1.2	1.5	1.2	1.7	..	40 Meubles
41 Other manufacturing n.e.c.	..	1.3 [o]	1.1 [o]	0.7 [o]	1.1	1.2	..	41 Autres activités de fabrication n.c.a.
42 Recycling	..	.. [n]	.. [n]	.. [n]	0.1	0.3	..	**42 Récupération**
43 ELECTRICITY, GAS & WATER SUPPLY	..	1.9	6.0	0.6	2.7	5.0	..	**43 DISTRIBUTION D'ELECTRICITE, GAZ & EAU**
44 CONSTRUCTION	..	0.4	0.1	0.4	0.5	0.6	..	**44 CONSTRUCTION**
45 SERVICES SECTOR	..	31.5	23.4	25.7	28.5	32.7	..	**45 SECTEUR DES SERVICES**
46 Wholesale, retail trade & repair of motor vehicles, etc.	..	0.1	0.3	0.3	0.4	0.3	..	**46 Commerce (gros & détail), réparation de véhicules, etc.**
47 Hotels & restaurants	..	0.0	0.0	0.0	0.0	0.0	..	**47 Hôtels et restaurants**
48 Transport & storage	..	.. [n]	.. [n]	0.8	0.3	0.0	..	**48 Transports, entreposage**
49 Communications	..	6.6 [o]	5.2 [o]	4.3	3.9	4.6	..	**49 Communications**
50 Post	..	0.0	0.0	0.0	0.0	0.0	..	50 Postes
51 Telecommunications	..	6.6 [o]	5.2 [o]	4.3	3.9	4.6	..	51 Télécommunications
52 Financial intermediation (inc. insur.)	..	.. [n]	.. [n]	0.0	.. [n]	.. [n]	..	**52 Intermédiation financ. (assurances inc.)**
53 Real estate, renting & busin. activity	..	22.9 [o]	15.8 [o]	17.5	21.4 [o]	25.3 [o]	..	**53 Immobil., locat. & serv. aux entreprises**
54 Computer & related activities	..	2.8	3.7	5.0	7.2	9.8	..	54 Informatique et activités connexes
55 *Software consultancy*	..	2.2	3.1	4.5	6.4	7.5	..	55 *Production de logiciels*
56 *Other computer services n.e.c.*	..	0.6	0.6	0.4	0.8	2.3	..	56 *Autres activités informatiques*
57 Research & development	..	0.5 [a,o]	0.4 [o]	1.1	3.1 [o]	5.3 [o]	..	57 Recherche & développement expérimental
58 Other business activities n.e.c.	..	19.6	11.7	11.4	11.1	10.2	..	58 Autres activ. & serv. aux entreprises
59 Comm., soc. & pers. serv. activ., etc.	..	1.9	2.1	2.8	2.4	2.5	..	**59 Adm. pub., serv. soc. & collec., etc**
TOTAL BUSINESS ENTERPRISE SECTOR	**163.4 [c]**	**156.9**	**146.7**	**171.8**	**161.3**	**222.4**	**..**	**TOTAL SECTEUR DES ENTREPRISES**

Source: OECD, R&D database (DSTI/EAS Division), January 2002.

Source : OCDE, base de données R-D (DSTI/Division AES), janvier 2002.

SPAIN Table 10 - Tableau 10 **ESPAGNE**

Business enterprise intramural expenditure on R&D (BERD) - Dépenses intra-muros de R-D des entreprises (DIRDE)

by industry and source of funds - par industrie et par source de financement

Million EUR (1999 ESP euro, EUR from 1999) *Millions EUR (1999 ESP euro, EUR à partir de 1999)*

10.3 Other national funds - Autres sources nationales

	1994	1995	1996	1997	1998	1999	2000	
1 AGRICULTURE, HUNTING & FORESTRY	..	0.1	0.1	0.1	0.1	0.9	..	**1 AGRICULTURE, CHASSE & SYLVICULTURE**
2 MINING	..	0.0	0.2	0.0	0.0	0.0	..	**2 ACTIVITES EXTRACTIVES**
3 MANUFACTURING	..	0.4	0.8	0.7	1.2	3.1	..	**3 ACTIVITES DE FABRICATION**
4 Food, beverages & tobacco	..	0.0	0.1	0.0	0.0	0.1	..	**4 Produits alimentaires, boissons & tabac**
5 Food products & beverages	..	0.0	0.1	0.0	0.0	0.1	..	5 Produits alimentaires & boissons
6 Tobacco products	..	0.0	0.0	0.0	0.0	0.0	..	6 Produits à base de tabac
7 Textiles, wear app., fur & leather	..	0.1	0.3	0.0	0.0	0.0	..	**7 Textiles, habill., fourrures & cuirs**
8 Textiles	..	0.0	0.2	0.0	0.0	0.0	..	8 Textiles
9 Wearing apparel & fur	..	0.0	0.0	0.0	0.0	0.0	..	9 Articles d'habillement & fourrures
10 Leather products & footwear	..	0.1	0.0	0.0	0.0	0.0	..	10 Articles en cuir & chaussures
11 Wood, paper, printing, publishing	..	0.0	0.0	0.0	0.0	0.0	..	**11 Bois, papier, imprimerie, édition**
12 Wood & cork (not furniture)	..	0.0	0.0	0.0	0.0	0.0	..	12 Bois & liège (sauf meubles)
13 Pulp, paper & paper products	..	0.0	0.0	0.0	0.0	0.0	..	13 Carton, papier & articles
14 Publ., print. & repro. of rec. media	..	0.0	0.0	0.0	0.0	0.0	..	14 Edition, impress. & repro. enregis.
15 Coke, petroleum, nuclear fuel, chemicals, rubber plastics	..	0.1	0.0	0.1	0.0	1.1	..	**15 Coke, pétrole, combust. nucl., prod. chimi., caout. & mat. plast.**
16 Coke, ref. petrol. prod. & nuclear fuel	..	0.1	0.0	0.0	0.0	0.0	..	16 Coke, raffin. pétr. & combust. nucl.
17 Chemicals & chemical products	..	0.0	0.0	0.1	0.0	1.1	..	17 Produits chimiques
18 *Chemical prod.(less pharmaceuticals)*	..	0.0	0.0	0.0	0.0	0.0	..	18 *Prod. chimi. (moins prod. pharmaceutiques)*
19 *Pharmaceuticals*	..	0.0	0.0	0.1	0.0	1.1	..	19 *Produits pharmaceutiques*
20 Rubber & plastic products	..	0.0	0.0	0.0	0.0	0.0	..	20 Caoutchouc et matières plastiques
21 Non-metallic mineral products	..	0.0	0.0	0.0	0.1	0.0	..	**21 Produits minéraux non métalliques**
22 Basic metals	..	0.1	0.2	0.2	0.1	0.0	..	**22 Produits métallurgiques de base**
23 Basic metals, ferrous	..	0.1	0.2	0.0	0.1	0.0	..	23 Produits métallur. de base, ferreux
24 Basic metals, non-ferrous	..	0.0	0.0	0.2	0.0	0.0	..	24 Prod. métallur. de base, non ferreux
25 Fabricated metal products	..	0.0	0.0	0.0	0.2	0.0	..	**25 Ouvrages en métaux (sauf mach.& mat.)**
26 Machinery equipment, instruments and transport equipment	..	0.2	0.3	0.3	0.8	1.9	..	**26 Machines & matériel, instruments & mat. de transport**
27 Machinery n.e.c.	..	0.2	0.1	0.3	0.4	0.3	..	27 Machines, n.c.a.
28 Office, account. & computing machinery	..	0.0	0.0	0.0	0.0	0.0	..	28 Mach. de bur., compta. & trait. inform.
29 Electrical machinery	..	0.0	0.1	0.0	0.4	0.1	..	29 Machines électriques
30 Electro. equip. (radio, TV & commun.)	..	0.0	0.0	0.0	0.0	1.3	..	30 Equip. électro. (radio, tv & communic.)
31 *Electro. comp. (inc. semi-conductors)*	..	0.0	0.0	0.0	0.0	1.3	..	31 *Compos. électro. (semi-cond. incl.)*
32 *TV, radio & communications equip.*	..	0.0	0.0	0.0	0.0	0.0	..	32 *Appareils radio, tv et de communi.*
33 Instruments	..	0.0	0.0	0.0	0.0	0.1	..	33 Instruments
34 Motor vehicles	..	0.0	0.0	0.0	0.0	0.0	..	34 Véhicules automobiles
35 Other transport equipment	..	0.0	0.0	0.0	0.0	0.2	..	35 Autres matériels de transport
36 *Ships*	..	0.0	0.0	0.0	0.0	0.0	..	36 *Navires*
37 *Aerospace*	..	0.0	0.0	0.0	0.0	0.2	..	37 *Aérospatiale*
38 *Other transport equipment n.e.c.*	..	0.0	0.0	0.0	0.0	0.0	..	38 *Autres matériels de transport n.c.a.*
39 Furniture, other manufacturing n.e.c.	..	0.0	0.1 [o]	0.0 [o]	0.0	0.0	..	**39 Meubles, autres activ. de fabricat. n.c.a.**
40 Furniture	..	0.0	0.0	0.0	0.0	0.0	..	40 Meubles
41 Other manufacturing n.e.c.	..	0.0	0.0 [o]	0.0 [o]	0.0	0.0	..	41 Autres activités de fabrication n.c.a.
42 Recycling	..	..	.. [n]	.. [n]	0.0	0.0	..	**42 Récupération**
43 ELECTRICITY, GAS & WATER SUPPLY	..	0.0	2.4	0.7	0.1	0.3	..	**43 DISTRIBUTION D'ELECTRICITE, GAZ & EAU**
44 CONSTRUCTION	..	0.0	0.0	0.0	0.0	0.0	..	**44 CONSTRUCTION**
45 SERVICES SECTOR	..	0.3	0.8	0.5	2.6	1.2	..	**45 SECTEUR DES SERVICES**
46 Wholesale, retail trade & repair of motor vehicles, etc.	..	0.0	0.0	0.0	0.0	0.0	..	**46 Commerce (gros & détail), réparation de véhicules, etc.**
47 Hotels & restaurants	..	0.0	0.0	0.0	0.0	0.0	..	**47 Hôtels et restaurants**
48 Transport & storage	..	..	.. [n]	0.0	0.0	0.0	..	**48 Transports, entreposage**
49 Communications	..	0.0	0.0 [o]	0.0	0.0	0.0	..	**49 Communications**
50 Post	..	0.0	0.0	0.0	0.0	0.0	..	50 Postes
51 Telecommunications	..	0.0	0.0 [o]	0.0	0.0	0.0	..	51 Télécommunications
52 Financial intermediation (inc. insur.)	..	.. [n]	.. [n]	0.0	.. [n]	.. [n]	..	**52 Intermédiation financ. (assurances inc.)**
53 Real estate, renting & busin. activity	..	0.1 [o]	0.5 [o]	0.0	1.2 [o]	0.2 [o]	..	**53 Immobil., locat. & serv. aux entreprises**
54 Computer & related activities	..	0.1	0.0	0.0	0.5	0.0	..	54 Informatique et activités connexes
55 *Software consultancy*	..	0.0	0.0	0.0	0.5	0.0	..	55 *Production de logiciels*
56 *Other computer services n.e.c.*	..	0.0	0.0	0.0	0.0	0.0	..	56 *Autres activités informatiques*
57 Research & development	..	0.0 [a]	0.5 [o]	0.0	0.2 [o]	0.1 [o]	..	57 Recherche & développement expérimental
58 Other business activities n.e.c.	..	0.0	0.0	0.0	0.4	0.1	..	58 Autres activ. & serv. aux entreprises
59 Comm., soc. & pers. serv. activ., etc.	..	0.2	0.3	0.5	1.5	0.9	..	**59 Adm. pub., serv. soc. & collec., etc**
TOTAL BUSINESS ENTERPRISE SECTOR	**2.2 [c]**	**0.9**	**4.3**	**2.0**	**4.0**	**5.4**	**..**	**TOTAL SECTEUR DES ENTREPRISES**

Source: OECD, R&D database (DSTI/EAS Division), January 2002.

Source : OCDE, base de données R-D (DSTI/Division AES), janvier 2002.

Business enterprise intramural expenditure on R&D (BERD) - Dépenses intra-muros de R-D des entreprises (DIRDE)

by industry and source of funds - par industrie et par source de financement

Million EUR (1999 ESP euro, EUR from 1999) — *Millions EUR (1999 ESP euro, EUR à partir de 1999)*

10.4 Funds from abroad - Fonds de l'étranger

	1994	1995	1996	1997	1998	1999	2000	
1 AGRICULTURE, HUNTING & FORESTRY	..	1.8	0.9	2.9	2.0	4.3	..	**1 AGRICULTURE, CHASSE & SYLVICULTURE**
2 MINING	..	1.9	1.7	0.5	2.8	0.1	..	**2 ACTIVITES EXTRACTIVES**
3 MANUFACTURING	..	72.7	73.1	87.9	71.9	75.1	..	**3 ACTIVITES DE FABRICATION**
4 Food, beverages & tobacco	..	1.3	0.7	0.5	0.8	1.4	..	**4 Produits alimentaires, boissons & tabac**
5 Food products & beverages	..	1.3	0.7	0.3	0.6	1.2	..	5 Produits alimentaires & boissons
6 Tobacco products	..	0.0	0.0	0.2	0.2	0.1	..	6 Produits à base de tabac
7 Textiles, wear app., fur & leather	..	0.9	0.8	1.0	1.2	1.0	..	**7 Textiles, habill., fourrures & cuirs**
8 Textiles	..	0.1	0.2	0.3	0.6	0.4	..	8 Textiles
9 Wearing apparel & fur	..	0.0	0.0	0.0	0.0	0.0	..	9 Articles d'habillement & fourrures
10 Leather products & footwear	..	0.8	0.6	0.7	0.6	0.6	..	10 Articles en cuir & chaussures
11 Wood, paper, printing, publishing	..	0.2	0.3	0.3	0.6	0.3	..	**11 Bois, papier, imprimerie, édition**
12 Wood & cork (not furniture)	..	0.0	0.0	0.0	0.0	0.0	..	12 Bois & liège (sauf meubles)
13 Pulp, paper & paper products	..	0.2	0.2	0.1	0.2	0.2	..	13 Carton, papier & articles
14 Publ., print. & repro. of rec. media	..	0.0	0.0	0.2	0.3	0.2	..	14 Edition, impress. & repro. enregis.
15 Coke, petroleum, nuclear fuel, chemicals, rubber plastics	..	7.7	5.2	8.7	7.7	9.2	..	**15 Coke, pétrole, combust. nucl., prod. chimi., caout. & mat. plast.**
16 Coke, ref. petrol. prod. & nuclear fuel	..	1.8	1.2	1.2	2.6	0.5	..	16 Coke, raffin. pétr. & combust. nucl.
17 Chemicals & chemical products	..	5.7	4.0	7.1	4.6	8.2	..	17 Produits chimiques
18 *Chemical prod.(less pharmaceuticals)*	..	0.9	1.0	0.7	0.7	1.9	..	18 *Prod. chimi. (moins prod. pharmaceutiques)*
19 *Pharmaceuticals*	..	4.7	3.0	6.5	3.9	6.3	..	19 *Produits pharmaceutiques*
20 Rubber & plastic products	..	0.3	0.0	0.4	0.6	0.5	..	20 Caoutchouc et matières plastiques
21 Non-metallic mineral products	..	0.2	0.1	0.2	0.5	0.7	..	**21 Produits minéraux non métalliques**
22 Basic metals	..	2.1	3.7	1.4	3.1	2.5	..	**22 Produits métallurgiques de base**
23 Basic metals, ferrous	..	2.0	3.7	1.4	3.0	2.5	..	23 Produits métallur. de base, ferreux
24 Basic metals, non-ferrous	..	0.1	0.0	0.0	0.1	0.0	..	24 Prod. métallur. de base, non ferreux
25 Fabricated metal products	..	1.5	1.0	1.6	1.0	1.4	..	**25 Ouvrages en métaux (sauf mach.& mat.)**
26 Machinery equipment, instruments and transport equipment	..	58.6	60.5	72.2	55.8	57.7	..	**26 Machines & matériel, instruments & mat. de transport**
27 Machinery n.e.c.	..	2.7	2.9	4.9	5.9	4.6	..	27 Machines, n.c.a.
28 Office, account. & computing machinery	..	0.0	0.6	0.8	0.0	0.6	..	28 Mach. de bur., compta. & trait. inform.
29 Electrical machinery	..	0.7	0.6	1.9	3.5	4.4	..	29 Machines électriques
30 Electro. equip. (radio, TV & commun.)	..	23.1	25.0	26.5	2.3	4.1	..	30 Equip. électro. (radio, tv & communic.)
31 *Electro. comp. (inc. semi-conductors)*	..	0.3	0.0	0.8	1.2	0.4	..	31 *Compos. électro. (semi-cond. incl.)*
32 *TV, radio & communications equip.*	..	22.8	25.0	25.8	1.1	3.6	..	32 *Appareils radio, tv et de communi.*
33 Instruments	..	0.9	1.6	2.5	3.1	1.4	..	33 Instruments
34 Motor vehicles	..	0.2	0.7	0.2	0.9	6.2	..	34 Véhicules automobiles
35 Other transport equipment	..	31.0	29.2	35.4	40.1	36.4	..	35 Autres matériels de transport
36 *Ships*	..	3.9	4.3	1.4	8.4	7.2	..	36 *Navires*
37 *Aerospace*	..	27.1	24.9	32.3	31.4	28.8	..	37 *Aérospatiale*
38 *Other transport equipment n.e.c.*	..	0.0	0.1	1.6	0.3	0.4	..	38 *Autres matériels de transport n.c.a.*
39 Furniture, other manufacturing n.e.c.	..	0.1 [o]	0.7 [o]	2.1 [o]	1.2	0.9	..	**39 Meubles, autres activ. de fabricat. n.c.a.**
40 Furniture	..	0.0	0.6	1.9	1.0	0.7	..	40 Meubles
41 Other manufacturing n.e.c.	..	0.0 [o]	0.1 [o]	0.2 [o]	0.2	0.3	..	41 Autres activités de fabrication n.c.a.
42 Recycling	..	.. [n]	.. [n]	.. [n]	0.0	0.0	..	**42 Récupération**
43 ELECTRICITY, GAS & WATER SUPPLY	..	11.5	2.7	9.4	10.0	1.9	..	**43 DISTRIBUTION D'ELECTRICITE, GAZ & EAU**
44 CONSTRUCTION	..	0.1	0.2	0.2	0.4	0.2	..	**44 CONSTRUCTION**
45 SERVICES SECTOR	..	21.9	18.4	23.7	16.1	35.8	..	**45 SECTEUR DES SERVICES**
46 Wholesale, retail trade & repair of motor vehicles, etc.	..	0.0	0.2	0.0	0.0	0.0	..	**46 Commerce (gros & détail), réparation de véhicules, etc.**
47 Hotels & restaurants	..	0.0	0.0	0.0	0.0	0.0	..	**47 Hôtels et restaurants**
48 Transport & storage	..	.. [n]	.. [n]	0.1	0.2	0.0	..	**48 Transports, entreposage**
49 Communications	..	1.2 [o]	1.5 [o]	0.6	2.7	1.6	..	**49 Communications**
50 Post	..	0.0	0.0	0.0	0.0	0.0	..	50 Postes
51 Telecommunications	..	1.2 [o]	1.5 [o]	0.6	2.7	1.6	..	51 Télécommunications
52 Financial intermediation (inc. insur.)	..	.. [n]	.. [n]	0.0	.. [n]	.. [n]	..	**52 Intermédiation financ. (assurances inc.)**
53 Real estate, renting & busin. activity	..	20.5 [o]	16.7 [o]	22.8	13.0 [o]	33.7 [o]	..	**53 Immobil., locat. & serv. aux entreprises**
54 Computer & related activities	..	4.7	4.3	9.6	7.2	7.2	..	54 Informatique et activités connexes
55 *Software consultancy*	..	3.1	2.6	8.2	6.1	6.9	..	55 *Production de logiciels*
56 *Other computer services n.e.c.*	..	1.5	1.7	1.4	1.1	0.3	..	56 *Autres activités informatiques*
57 Research & development	..	0.0 [a]	0.1 [o]	0.0	1.0 [o]	1.1 [o]	..	57 Recherche & développement expérimental
58 Other business activities n.e.c.	..	15.9	12.3	13.2	4.8	25.4	..	58 Autres activ. & serv. aux entreprises
59 Comm., soc. & pers. serv. activ., etc.	..	0.1	0.0	0.2	0.2	0.4	..	**59 Adm. pub., serv. soc. & collec., etc**
TOTAL BUSINESS ENTERPRISE SECTOR	**140.7 [c]**	**109.8**	**97.0**	**124.7**	**103.1**	**117.5**	..	**TOTAL SECTEUR DES ENTREPRISES**

Source: OECD, R&D database (DSTI/EAS Division), January 2002.

Source : OCDE, base de données R-D (DSTI/Division AES), janvier 2002.

Business enterprise intramural expenditure on R&D (BERD) - Dépenses intra-muros de R-D des entreprises (DIRDE)

by industry and type of costs - par industrie et par type de coûts

Million EUR (1999 ESP euro, EUR from 1999) — *Millions EUR (1999 ESP euro, EUR à partir de 1999)*

11.1 Current expenditure - Dépenses courantes

	1994	1995	1996	1997	1998	1999	2000	
1 AGRICULTURE, HUNTING & FORESTRY	..	23.5	25.8	30.4	54.9	37.5	..	**1 AGRICULTURE, CHASSE & SYLVICULTURE**
2 MINING	..	7.3	9.1	5.6	6.0	6.9	..	**2 ACTIVITES EXTRACTIVES**
3 MANUFACTURING	..	1 177.0	1 323.3	1 361.2	1 629.2	1 697.1	..	**3 ACTIVITES DE FABRICATION**
4 Food, beverages & tobacco	..	41.7	51.5	47.2	57.0	59.9	..	**4 Produits alimentaires, boissons & tabac**
5 Food products & beverages	..	37.8	47.0	42.8	52.3	55.7	..	5 Produits alimentaires & boissons
6 Tobacco products	..	4.0	4.5	4.3	4.7	4.2	..	6 Produits à base de tabac
7 Textiles, wear app., fur & leather	..	15.0	24.6	19.6	37.8	28.2	..	**7 Textiles, habill., fourrures & cuirs**
8 Textiles	..	7.4	14.4	11.4	20.9	17.9	..	8 Textiles
9 Wearing apparel & fur	..	4.2	4.9	4.4	8.7	5.0	..	9 Articles d'habillement & fourrures
10 Leather products & footwear	..	3.4	5.3	3.8	8.3	5.3	..	10 Articles en cuir & chaussures
11 Wood, paper, printing, publishing	..	12.2	19.3	15.0	21.4	22.0	..	**11 Bois, papier, imprimerie, édition**
12 Wood & cork (not furniture)	..	1.4	1.2	1.5	1.3	1.6	..	12 Bois & liège (sauf meubles)
13 Pulp, paper & paper products	..	7.1	9.9	9.4	14.3	15.3	..	13 Carton, papier & articles
14 Publ., print. & repro. of rec. media	..	3.7	8.2	4.1	5.9	5.2	..	14 Edition, impress. & repro. enregis.
15 Coke, petroleum, nuclear fuel, chemicals, rubber plastics	..	298.1	352.1	363.9	400.3	402.0	..	**15 Coke, pétrole, combust. nucl., prod. chimi., caout. & mat. plast.**
16 Coke, ref. petrol. prod. & nuclear fuel	..	24.3	23.7	27.9	24.6	26.8	..	16 Coke, raffin. pétr. & combust. nucl.
17 Chemicals & chemical products	..	239.8	290.7	288.0	319.3	320.0	..	17 Produits chimiques
18 *Chemical prod.(less pharmaceuticals)*	..	73.3	87.4	82.8	101.6	103.0	..	18 *Prod. chimi. (moins prod. pharmaceutiques)*
19 *Pharmaceuticals*	..	166.5	203.3	205.2	217.8	216.9	..	19 *Produits pharmaceutiques*
20 Rubber & plastic products	..	34.0	37.7	48.0	56.3	55.3	..	20 Caoutchouc et matières plastiques
21 Non-metallic mineral products	..	22.1	26.3	27.5	28.5	33.1	..	**21 Produits minéraux non métalliques**
22 Basic metals	..	19.9	30.4	23.5	30.5	33.6	..	**22 Produits métallurgiques de base**
23 Basic metals, ferrous	..	16.6	23.6	19.2	24.2	27.2	..	23 Produits métallur. de base, ferreux
24 Basic metals, non-ferrous	..	3.3	6.8	4.2	6.3	6.4	..	24 Prod. métallur. de base, non ferreux
25 Fabricated metal products	..	19.5	27.9	32.1	41.4	36.8	..	**25 Ouvrages en métaux (sauf mach.& mat.)**
26 Machinery equipment, instruments and transport equipment	..	732.7	774.2	810.6	984.0	1 051.6	..	**26 Machines & matériel, instruments & mat. de transport**
27 Machinery n.e.c.	..	99.8	111.7	121.5	151.9	159.6	..	27 Machines, n.c.a.
28 Office, account. & computing machinery	..	23.0	26.4	34.2	31.7	35.9	..	28 Mach. de bur., compta. & trait. inform.
29 Electrical machinery	..	49.5	61.2	79.2	97.5	94.2	..	29 Machines électriques
30 Electro. equip. (radio, TV & commun.)	..	189.9	194.4	197.8	226.8	219.5	..	30 Equip. électro. (radio, tv & communic.)
31 *Electro. comp. (inc. semi-conductors)*	..	13.8	16.7	18.0	25.4	30.3	..	31 *Compos. électro. (semi-cond. incl.)*
32 *TV, radio & communications equip.*	..	176.1	177.7	179.8	201.4	189.2	..	32 *Appareils radio, tv et de communi.*
33 Instruments	..	39.1	33.5	46.7	44.6	46.6	..	33 Instruments
34 Motor vehicles	..	148.8	161.4	148.0	155.0	206.7	..	34 Véhicules automobiles
35 Other transport equipment	..	182.7	185.5	183.2	276.5	289.1	..	35 Autres matériels de transport
36 *Ships*	..	24.1	28.0	25.1	61.9	62.5	..	36 *Navires*
37 *Aerospace*	..	142.3	134.7	136.7	188.3	189.0	..	37 *Aérospatiale*
38 *Other transport equipment n.e.c.*	..	16.3	22.8	21.4	26.2	37.6	..	38 *Autres matériels de transport n.c.a.*
39 Furniture, other manufacturing n.e.c.	..	15.8 [o]	17.2 [o]	21.9 [o]	27.1	27.6	..	**39 Meubles, autres activ. de fabricat. n.c.a.**
40 Furniture	..	6.7	6.2	9.5	11.2	14.2	..	40 Meubles
41 Other manufacturing n.e.c.	..	9.1 [o]	11.1 [o]	12.3 [o]	15.9	13.3	..	41 Autres activités de fabrication n.c.a.
42 Recycling	..	.. [n]	.. [n]	.. [n]	1.3	2.4	..	**42 Récupération**
43 ELECTRICITY, GAS & WATER SUPPLY	..	43.8	41.8	23.6	23.7	26.8	..	**43 DISTRIBUTION D'ELECTRICITE, GAZ & EAU**
44 CONSTRUCTION	..	5.4	3.9	5.7	3.7	6.2	..	**44 CONSTRUCTION**
45 SERVICES SECTOR	..	166.4	202.4	186.7	233.7	304.0	..	**45 SECTEUR DES SERVICES**
46 Wholesale, retail trade & repair of motor vehicles, etc.	..	1.3	0.5	2.4	2.3	16.7	..	**46 Commerce (gros & détail), réparation de véhicules, etc.**
47 Hotels & restaurants	..	0.0	0.0	0.0	0.0	0.0	..	**47 Hôtels et restaurants**
48 Transport & storage	..	.. [n]	.. [n]	0.7	2.9	1.0	..	**48 Transports, entreposage**
49 Communications	..	43.8 [o]	75.6 [o]	51.2	71.6	87.1	..	**49 Communications**
50 Post	..	0.0	0.0	0.0	0.0	0.0	..	50 Postes
51 Telecommunications	..	43.8 [o]	75.6 [o]	51.2	71.6	87.1	..	51 Télécommunications
52 Financial intermediation (inc. insur.)	..	.. [n]	0.0 [n]	0.0	.. [n]	.. [n]	..	**52 Intermédiation financ. (assurances inc.)**
53 Real estate, renting & busin. activity	..	115.0 [o]	119.3 [o]	120.4	147.1 [o]	186.6 [o]	..	**53 Immobil., locat. & serv. aux entreprises**
54 Computer & related activities	..	48.7	45.4	45.3	61.0	84.0	..	54 Informatique et activités connexes
55 *Software consultancy*	..	42.4	39.4	39.2	54.1	72.0	..	55 *Production de logiciels*
56 *Other computer services n.e.c.*	..	6.4	5.9	6.2	6.9	12.1	..	56 *Autres activités informatiques*
57 Research & development	..	3.0 [a,o]	3.6 [o]	3.9	12.8 [o]	16.6 [o]	..	57 Recherche & développement expérimental
58 Other business activities n.e.c.	..	63.3	70.3	71.2	73.3	86.0	..	58 Autres activ. & serv. aux entreprises
59 Comm., soc. & pers. serv. activ., etc.	..	6.3	7.0	11.9	9.8	12.7	..	**59 Adm. pub., serv. soc. & collec., etc**
TOTAL BUSINESS ENTERPRISE SECTOR	**1 349.7**	**1 423.5**	**1 606.3**	**1 613.1**	**1 951.2**	**2 078.5**	..	**TOTAL SECTEUR DES ENTREPRISES**

Source: OECD, R&D database (DSTI/EAS Division), January 2002.

Source : OCDE, base de données R-D (DSTI/Division AES), janvier 2002.

Business enterprise intramural expenditure on R&D (BERD) - Dépenses intra-muros de R-D des entreprises (DIRDE)

by industry and type of costs - par industrie et par type de coûts

Million EUR (1999 ESP euro, EUR from 1999) — Millions EUR (1999 ESP euro, EUR à partir de 1999)

11.2 Capital expenditure - Dépenses en capital

	1994	1995	1996	1997	1998	1999	2000	
1 AGRICULTURE, HUNTING & FORESTRY	..	7.3	4.7	7.2	12.0	6.0	..	1 AGRICULTURE, CHASSE & SYLVICULTURE
2 MINING	..	3.6	2.3	2.0	0.1	1.9	..	2 ACTIVITES EXTRACTIVES
3 MANUFACTURING	..	162.0	211.1	249.3	339.1	322.7	..	3 ACTIVITES DE FABRICATION
4 Food, beverages & tobacco	..	10.6	11.8	14.7	19.1	14.5	..	**4 Produits alimentaires, boissons & tabac**
5 Food products & beverages	..	10.5	11.7	14.1	18.9	14.3	..	5 Produits alimentaires & boissons
6 Tobacco products	..	0.1	0.1	0.5	0.2	0.1	..	6 Produits à base de tabac
7 Textiles, wear app., fur & leather	..	4.2	5.8	7.6	8.9	16.5	..	**7 Textiles, habill., fourrures & cuirs**
8 Textiles	..	2.5	4.2	5.0	5.1	13.3	..	8 Textiles
9 Wearing apparel & fur	..	0.9	0.4	1.8	1.7	0.6	..	9 Articles d'habillement & fourrures
10 Leather products & footwear	..	0.7	1.2	0.8	2.1	2.5	..	10 Articles en cuir & chaussures
11 Wood, paper, printing, publishing	..	1.8	2.1	5.5	45.3	10.5	..	**11 Bois, papier, imprimerie, édition**
12 Wood & cork (not furniture)	..	0.7	0.3	0.2	0.9	0.8	..	12 Bois & liège (sauf meubles)
13 Pulp, paper & paper products	..	0.5	1.3	4.5	42.0	8.7	..	13 Carton, papier & articles
14 Publ., print. & repro. of rec. media	..	0.6	0.4	0.8	2.4	0.9	..	14 Edition, impress. & repro. enregis.
15 Coke, petroleum, nuclear fuel, chemicals, rubber plastics	..	44.3	46.2	50.7	77.7	86.4	..	**15 Coke, pétrole, combust. nucl., prod. chimi., caout. & mat. plast.**
16 Coke, ref. petrol. prod. & nuclear fuel	..	3.2	2.5	4.3	5.1	4.0	..	16 Coke, raffin. pétr. & combust. nucl.
17 Chemicals & chemical products	..	32.4	29.9	31.1	61.1	69.7	..	17 Produits chimiques
18 *Chemical prod.(less pharmaceuticals)*	..	10.5	11.1	14.7	38.4	39.6	..	18 *Prod. chimi. (moins prod. pharmaceutiques)*
19 *Pharmaceuticals*	..	22.0	18.7	16.4	22.7	30.2	..	19 *Produits pharmaceutiques*
20 Rubber & plastic products	..	8.6	13.8	15.3	11.5	12.7	..	20 Caoutchouc et matières plastiques
21 Non-metallic mineral products	..	3.1	5.8	7.3	8.9	10.6	..	**21 Produits minéraux non métalliques**
22 Basic metals	..	8.7	9.2	12.5	9.7	7.6	..	**22 Produits métallurgiques de base**
23 Basic metals, ferrous	..	7.7	8.8	11.6	9.1	6.4	..	23 Produits métallur. de base, ferreux
24 Basic metals, non-ferrous	..	1.0	0.5	0.9	0.6	1.2	..	24 Prod. métallur. de base, non ferreux
25 Fabricated metal products	..	7.2	6.1	7.9	13.3	15.1	..	**25 Ouvrages en métaux (sauf mach.& mat.)**
26 Machinery equipment, instruments and transport equipment	..	80.0	121.1	138.9	141.9	147.8	..	**26 Machines & matériel, instruments & mat. de transport**
27 Machinery n.e.c.	..	17.2	26.3	18.4	26.8	24.5	..	27 Machines, n.c.a.
28 Office, account. & computing machinery	..	5.6	8.8	12.9	19.3	20.0	..	28 Mach. de bur., compta. & trait. inform.
29 Electrical machinery	..	7.2	8.3	13.2	16.3	23.0	..	29 Machines électriques
30 Electro. equip. (radio, TV & commun.)	..	14.0	22.8	24.7	22.1	21.3	..	30 Equip. électro. (radio, tv & communic.)
31 *Electro. comp. (inc. semi-conductors)*	..	4.8	12.4	11.5	4.0	5.8	..	31 *Compos. électro. (semi-cond. incl.)*
32 *TV, radio & communications equip.*	..	9.3	10.4	13.1	18.1	15.5	..	32 *Appareils radio, tv et de communi.*
33 Instruments	..	4.5	2.5	5.5	10.0	5.0	..	33 Instruments
34 Motor vehicles	..	21.2	21.6	24.6	32.8	37.2	..	34 Véhicules automobiles
35 Other transport equipment	..	10.3	30.8	39.6	14.7	16.8	..	35 Autres matériels de transport
36 *Ships*	..	1.6	4.6	7.0	2.6	3.5	..	36 *Navires*
37 *Aerospace*	..	7.3	22.8	30.3	9.1	10.3	..	37 *Aérospatiale*
38 *Other transport equipment n.e.c.*	..	1.4	3.3	2.3	3.0	3.0	..	38 *Autres matériels de transport n.c.a.*
39 Furniture, other manufacturing n.e.c.	..	2.2 [o]	3.0 [o]	4.2 [o]	13.9	12.7	..	**39 Meubles, autres activ. de fabricat. n.c.a.**
40 Furniture	..	0.9	1.4	1.8	1.9	2.7	..	40 Meubles
41 Other manufacturing n.e.c.	..	1.3 [o]	1.6 [o]	2.4 [o]	12.0	10.1	..	41 Autres activités de fabrication n.c.a.
42 Recycling	..	.. [n]	.. [n]	.. [n]	0.3	1.1	..	**42 Récupération**
43 ELECTRICITY, GAS & WATER SUPPLY	..	61.5	3.8	40.6	32.2	15.8	..	43 DISTRIBUTION D'ELECTRICITE, GAZ & EAU
44 CONSTRUCTION	..	0.4	2.1	0.6	0.4	0.3	..	44 CONSTRUCTION
45 SERVICES SECTOR	..	54.0	32.4	58.0	122.2	171.8	..	45 SECTEUR DES SERVICES
46 Wholesale, retail trade & repair of motor vehicles, etc.	..	1.0	0.4	0.3	1.2	1.6	..	**46 Commerce (gros & détail), réparation de véhicules, etc.**
47 Hotels & restaurants	..	0.0	0.0	0.0	0.0	0.0	..	**47 Hôtels et restaurants**
48 Transport & storage	..	.. [n]	.. [n]	0.8	1.8	0.7	..	**48 Transports, entreposage**
49 Communications	..	22.3 [o]	6.7 [o]	28.2	94.2	138.4	..	**49 Communications**
50 Post	..	0.0	0.0	0.0	0.0	0.0	..	50 Postes
51 Telecommunications	..	22.3 [o]	6.7 [o]	28.2	94.2	138.4	..	51 Télécommunications
52 Financial intermediation (inc. insur.)	..	.. [n]	.. [n]	0.0	.. [n]	.. [n]	..	**52 Intermédiation financ. (assurances inc.)**
53 Real estate, renting & busin. activity	..	29.2 [o]	24.6 [o]	26.0	23.7 [o]	28.5 [o]	..	**53 Immobil., locat. & serv. aux entreprises**
54 Computer & related activities	..	4.8	7.0	7.1	6.8	7.6	..	54 Informatique et activités connexes
55 *Software consultancy*	..	4.1	6.7	6.6	6.3	6.1	..	55 *Production de logiciels*
56 *Other computer services n.e.c.*	..	0.8	0.3	0.4	0.5	1.5	..	56 *Autres activités informatiques*
57 Research & development	..	0.1 [a,o]	0.1 [o]	0.4	2.3 [o]	3.5 [o]	..	57 Recherche & développement expérimental
58 Other business activities n.e.c.	..	24.3	17.6	18.5	14.6	17.5	..	58 Autres activ. & serv. aux entreprises
59 Comm., soc. & pers. serv. activ., etc.	..	1.5	0.8	2.7	1.3	2.7	..	**59 Adm. pub., serv. soc. & collec., etc**
TOTAL BUSINESS ENTERPRISE SECTOR	**190.8**	**288.8**	**256.4**	**357.7**	**505.9**	**518.6**	**..**	**TOTAL SECTEUR DES ENTREPRISES**

Source: OECD, R&D database (DSTI/EAS Division), January 2002.

Source : OCDE, base de données R-D (DSTI/Division AES), janvier 2002.

Total business enterprise R&D personnel (BEMP) - Personnel total de R-D du secteur des entreprises (PRDE)

by industry - par industrie

In full time equivalent — *En équivalence plein-temps*

	1994	1995	1996	1997	1998	1999	2000	
1 AGRICULTURE, HUNTING & FORESTRY	..	698	677	859	827	908	..	1 AGRICULTURE, CHASSE & SYLVICULTURE
2 MINING	..	189	165	115	67	101	..	2 ACTIVITES EXTRACTIVES
3 MANUFACTURING	..	22 558	24 708	25 236	28 753	30 642	..	3 ACTIVITES DE FABRICATION
4 Food, beverages & tobacco	..	914	1 041	1 017	1 214	1 330	..	**4 Produits alimentaires, boissons & tabac**
5 Food products & beverages	..	841	985	928	1 137	1 234	..	5 Produits alimentaires & boissons
6 Tobacco products	..	73	56	89	78	95	..	6 Produits à base de tabac
7 Textiles, wear app., fur & leather	..	376	578	493	1 025	838	..	**7 Textiles, habill., fourrures & cuirs**
8 Textiles	..	185	336	277	494	459	..	8 Textiles
9 Wearing apparel & fur	..	107	102	125	292	215	..	9 Articles d'habillement & fourrures
10 Leather products & footwear	..	85	140	91	239	163	..	10 Articles en cuir & chaussures
11 Wood, paper, printing, publishing	..	276	429	353	478	419	..	**11 Bois, papier, imprimerie, édition**
12 Wood & cork (not furniture)	..	39	38	35	38	50	..	12 Bois & liège (sauf meubles)
13 Pulp, paper & paper products	..	155	171	200	320	232	..	13 Carton, papier & articles
14 Publ., print. & repro. of rec. media	..	82	220	117	120	137	..	14 Edition, impress. & repro. enregis.
15 Coke, petroleum, nuclear fuel, chemicals, rubber plastics	..	5 742	6 157	6 375	6 894	7 288	..	**15 Coke, pétrole, combust. nucl., prod. chimi., caout. & mat. plast.**
16 Coke, ref. petrol. prod. & nuclear fuel	..	387	356	369	379	498	..	16 Coke, raffin. pétr. & combust. nucl.
17 Chemicals & chemical products	..	4 627	5 046	5 060	5 396	5 724	..	17 Produits chimiques
18 *Chemical prod.(less pharmaceuticals)*	..	1 804	2 038	2 025	2 300	2 393	..	18 *Prod. chimi. (moins prod. pharmaceutiques)*
19 *Pharmaceuticals*	..	2 822	3 008	3 035	3 096	3 331	..	19 *Produits pharmaceutiques*
20 Rubber & plastic products	..	728	756	946	1 119	1 066	..	20 Caoutchouc et matières plastiques
21 Non-metallic mineral products	..	519	601	591	631	711	..	**21 Produits minéraux non métalliques**
22 Basic metals	..	403	506	341	526	628	..	**22 Produits métallurgiques de base**
23 Basic metals, ferrous	..	335	427	272	401	519	..	23 Produits métallur. de base, ferreux
24 Basic metals, non-ferrous	..	68	79	69	125	108	..	24 Prod. métallur. de base, non ferreux
25 Fabricated metal products	..	443	668	691	913	882	..	**25 Ouvrages en métaux (sauf mach.& mat.)**
26 Machinery equipment, instruments and transport equipment	..	13 444	14 269	14 862	16 371	17 734	..	**26 Machines & matériel, instruments & mat. de transport**
27 Machinery n.e.c.	..	2 305	2 569	2 449	3 064	3 344	..	27 Machines, n.c.a.
28 Office, account. & computing machinery	..	281	394	475	391	362	..	28 Mach. de bur., compta. & trait. inform.
29 Electrical machinery	..	1 008	1 323	1 556	1 907	2 047	..	29 Machines électriques
30 Electro. equip. (radio, TV & commun.)	..	2 945	2 878	3 169	3 412	3 712	..	30 Equip. électro. (radio, tv & communic.)
31 *Electro. comp. (inc. semi-conductors)*	..	355	433	515	612	777	..	31 *Compos. électro. (semi-cond. incl.)*
32 *TV, radio & communications equip.*	..	2 589	2 445	2 654	2 800	2 935	..	32 *Appareils radio, tv et de communi.*
33 Instruments	..	918	695	922	984	1 041	..	33 Instruments
34 Motor vehicles	..	2 812	3 093	3 059	2 974	3 354	..	34 Véhicules automobiles
35 Other transport equipment	..	3 176	3 319	3 232	3 639	3 874	..	35 Autres matériels de transport
36 *Ships*	..	604	679	553	963	896	..	36 *Navires*
37 *Aerospace*	..	2 148	2 118	2 273	2 155	2 339	..	37 *Aérospatiale*
38 *Other transport equipment n.e.c.*	..	423	522	406	521	640	..	38 *Autres matériels de transport n.c.a.*
39 Furniture, other manufacturing n.e.c.	..	442 [o]	459 [o]	513 [o]	686	782	..	**39 Meubles, autres activ. de fabricat. n.c.a.**
40 Furniture	..	194	155	221	320	376	..	40 Meubles
41 Other manufacturing n.e.c.	..	248 [o]	304 [o]	292 [o]	366	406	..	41 Autres activités de fabrication n.c.a.
42 Recycling	..	.. [n]	.. [n]	.. [n]	15	30	..	**42 Récupération**
43 ELECTRICITY, GAS & WATER SUPPLY	..	662	540	338	256	350	..	43 DISTRIBUTION D'ELECTRICITE, GAZ & EAU
44 CONSTRUCTION	..	84	88	87	81	149	..	44 CONSTRUCTION
45 SERVICES SECTOR	..	3 367	3 252	3 388	4 682	6 174	..	45 SECTEUR DES SERVICES
46 Wholesale, retail trade & repair of motor vehicles, etc.	..	41	11	52	56	197	..	**46 Commerce (gros & détail), réparation de véhicules, etc.**
47 Hotels & restaurants	..	0	0	0	0	0	..	**47 Hôtels et restaurants**
48 Transport & storage	..	.. [n]	.. [n]	22	52	22	..	**48 Transports, entreposage**
49 Communications	..	782 [o]	856 [o]	877	1 267	1 318	..	**49 Communications**
50 Post	..	0	0	0	0	0	..	50 Postes
51 Telecommunications	..	782 [o]	856 [o]	877	1 267	1 318	..	51 Télécommunications
52 Financial intermediation (inc. insur.)	..	.. [n]	..	0	.. [n]	.. [n]	..	**52 Intermédiation financ. (assurances inc.)**
53 Real estate, renting & busin. activity	..	2 339 [o]	2 195	2 129	3 042 [o]	4 311 [o]	..	**53 Immobil., locat. & serv. aux entreprises**
54 Computer & related activities	..	1 041	860	902	1 458	2 214	..	54 Informatique et activités connexes
55 *Software consultancy*	..	892	723	751	1 293	1 900	..	55 *Production de logiciels*
56 *Other computer services n.e.c.*	..	148	137	151	165	314	..	56 *Autres activités informatiques*
57 Research & development	..	89 [o]	61 [o]	86	222 [o]	386 [o]	..	57 Recherche & développement expérimental
58 Other business activities n.e.c.	..	1 210	1 274	1 142	1 363	1 711	..	58 Autres activ. & serv. aux entreprises
59 Comm., soc. & pers. serv. activ., etc.	..	204	190	308	265	326	..	**59 Adm. pub., serv. soc. & collec., etc**
TOTAL BUSINESS ENTERPRISE SECTOR	**27 321**	**27 558**	**29 431**	**30 023**	**34 667**	**38 323**	**38 706** [c]	**TOTAL SECTEUR DES ENTREPRISES**

Source: OECD, R&D database (DSTI/EAS Division), January 2002.

Source : OCDE, base de données R-D (DSTI/Division AES), janvier 2002.

SPAIN | Table 13 - Tableau 13 | ESPAGNE

RSE in the business enterprise sector - Chercheurs dans le secteur des entreprises

by industry - par industrie

In full time equivalent | *En équivalence plein-temps*

	1994	1995	1996	1997	1998	1999	2000	
1 AGRICULTURE, HUNTING & FORESTRY	..	216	216	278	236	309	..	**1 AGRICULTURE, CHASSE & SYLVICULTURE**
2 MINING	..	70	51	38	19	49	..	**2 ACTIVITES EXTRACTIVES**
3 MANUFACTURING	..	8 351	8 900	9 665	10 960	11 771	..	**3 ACTIVITES DE FABRICATION**
4 Food, beverages & tobacco	..	353	389	393	483	456	..	**4 Produits alimentaires, boissons & tabac**
5 Food products & beverages	..	336	365	375	468	441	..	5 Produits alimentaires & boissons
6 Tobacco products	..	17	23	18	15	15	..	6 Produits à base de tabac
7 Textiles, wear app., fur & leather	..	132	189	159	450	248	..	**7 Textiles, habill., fourrures & cuirs**
8 Textiles	..	60	99	77	253	134	..	8 Textiles
9 Wearing apparel & fur	..	21	30	34	112	44	..	9 Articles d'habillement & fourrures
10 Leather products & footwear	..	52	60	49	85	70	..	10 Articles en cuir & chaussures
11 Wood, paper, printing, publishing	..	82	121	148	157	166	..	**11 Bois, papier, imprimerie, édition**
12 Wood & cork (not furniture)	..	16	14	17	17	27	..	12 Bois & liège (sauf meubles)
13 Pulp, paper & paper products	..	49	68	103	89	101	..	13 Carton, papier & articles
14 Publ., print. & repro. of rec. media	..	17	39	28	51	38	..	14 Edition, impress. & repro. enregis.
15 Coke, petroleum, nuclear fuel, chemicals, rubber plastics	..	2 194	2 383	2 532	2 763	2 797	..	**15 Coke, pétrole, combust. nucl., prod. chimi., caout. & mat. plast.**
16 Coke, ref. petrol. prod. & nuclear fuel	..	196	159	168	147	173	..	16 Coke, raffin. pétr. & combust. nucl.
17 Chemicals & chemical products	..	1 842	2 040	2 127	2 260	2 365	..	17 Produits chimiques
18 *Chemical prod.(less pharmaceuticals)*	..	687	770	779	995	961	..	18 *Prod. chimi. (moins prod. pharmaceutiques)*
19 *Pharmaceuticals*	..	1 155	1 270	1 348	1 264	1 404	..	19 *Produits pharmaceutiques*
20 Rubber & plastic products	..	156	185	238	357	259	..	20 Caoutchouc et matières plastiques
21 Non-metallic mineral products	..	131	169	168	171	189	..	**21 Produits minéraux non métalliques**
22 Basic metals	..	223	209	184	231	288	..	**22 Produits métallurgiques de base**
23 Basic metals, ferrous	..	192	187	149	201	249	..	23 Produits métallur. de base, ferreux
24 Basic metals, non-ferrous	..	30	23	35	30	40	..	24 Prod. métallur. de base, non ferreux
25 Fabricated metal products	..	150	265	249	398	333	..	**25 Ouvrages en métaux (sauf mach.& mat.)**
26 Machinery equipment, instruments and transport equipment	..	4 937	5 007	5 666	6 103	7 046	..	**26 Machines & matériel, instruments & mat. de transport**
27 Machinery n.e.c.	..	764	854	892	1 170	1 286	..	27 Machines, n.c.a.
28 Office, account. & computing machinery	..	207	166	244	240	254	..	28 Mach. de bur., compta. & trait. inform.
29 Electrical machinery	..	385	422	618	799	797	..	29 Machines électriques
30 Electro. equip. (radio, TV & commun.)	..	1 502	1 471	1 633	1 523	2 125	..	30 Equip. électro. (radio, tv & communic.)
31 *Electro. comp. (inc. semi-conductors)*	..	124	171	156	183	263	..	31 *Compos. électro. (semi-cond. incl.)*
32 *TV, radio & communications equip.*	..	1 377	1 300	1 477	1 340	1 862	..	32 *Appareils radio, tv et de communi.*
33 Instruments	..	425	272	462	514	551	..	33 Instruments
34 Motor vehicles	..	603	739	673	645	713	..	34 Véhicules automobiles
35 Other transport equipment	..	1 051	1 081	1 145	1 212	1 322	..	35 Autres matériels de transport
36 *Ships*	..	186	62	135	179	196	..	36 *Navires*
37 *Aerospace*	..	822	940	960	921	990	..	37 *Aérospatiale*
38 *Other transport equipment n.e.c.*	..	43	79	50	112	136	..	38 *Autres matériels de transport n.c.a.*
39 Furniture, other manufacturing n.e.c.	..	149 [o]	167 [o]	166 [o]	198	236	..	**39 Meubles, autres activ. de fabricat. n.c.a.**
40 Furniture	..	69	59	81	110	120	..	40 Meubles
41 Other manufacturing n.e.c.	..	80 [o]	109 [o]	85 [o]	87	116	..	41 Autres activités de fabrication n.c.a.
42 Recycling	..	.. [n]	.. [n]	.. [n]	7	13	..	**42 Récupération**
43 ELECTRICITY, GAS & WATER SUPPLY	..	323	262	142	98	113	..	**43 DISTRIBUTION D'ELECTRICITE, GAZ & EAU**
44 CONSTRUCTION	..	35	35	42	23	30	..	**44 CONSTRUCTION**
45 SERVICES SECTOR	..	1 808	1 636	1 845	2 566	2 907	..	**45 SECTEUR DES SERVICES**
46 Wholesale, retail trade & repair of motor vehicles, etc.	..	18	7	27	26	36	..	**46 Commerce (gros & détail), réparation de véhicules, etc.**
47 Hotels & restaurants	..	0	0	0	0	0	..	**47 Hôtels et restaurants**
48 Transport & storage	..	.. [n]	.. [n]	16	29	8	..	**48 Transports, entreposage**
49 Communications	..	588 [o]	514 [o]	691	976	1 030	..	**49 Communications**
50 Post	..	0	0	0	0	0	..	50 Postes
51 Telecommunications	..	588 [o]	514 [o]	691	976	1 030	..	51 Télécommunications
52 Financial intermediation (inc. insur.)	..	.. [n]	.. [n]	0	.. [n]	.. [n]	..	**52 Intermédiation financ. (assurances inc.)**
53 Real estate, renting & busin. activity	..	1 094 [o]	1 018 [o]	952	1 396 [o]	1 673 [o]	..	**53 Immobil., locat. & serv. aux entreprises**
54 Computer & related activities	..	457	372	327	498	670	..	54 Informatique et activités connexes
55 *Software consultancy*	..	398	290	274	434	544	..	55 *Production de logiciels*
56 *Other computer services n.e.c.*	..	59	82	52	64	127	..	56 *Autres activités informatiques*
57 Research & development	..	19 [o]	30 [o]	35	153 [o]	273 [o]	..	57 Recherche & développement expérimental
58 Other business activities n.e.c.	..	617	616	590	745	730	..	58 Autres activ. & serv. aux entreprises
59 Comm., soc. & pers. serv. activ., etc.	..	109	97	160	140	161	..	**59 Adm. pub., serv. soc. & collec., etc**
TOTAL BUSINESS ENTERPRISE SECTOR	**11 070 [c]**	**10 803**	**11 100**	**12 009**	**13 902**	**15 178**	..	**TOTAL SECTEUR DES ENTREPRISES**

Source: OECD, R&D database (DSTI/EAS Division), January 2002.

Source : OCDE, base de données R-D (DSTI/Division AES), janvier 2002.

Table 14 - Tableau 14

Higher education intramural expenditure on R&D - Dépenses intra-muros de R-D de l'enseignement supérieur
(HERD) by field of science - (DIRDES) par discipline scientifique

	1994	1995	1996	1997	1998	1999	2000	
	14.1 Million EUR (1999 ESP euro, EUR from 1999) - Millions EUR (1999 ESP euro, EUR à partir de 1999)							
Natural sciences	391.4	455.8	511.0	550.9	587.6	593.5	..	Sciences exactes et naturelles
Engineering	182.2	209.0	220.4	252.5	259.5	281.4	..	Sciences de l'ingénieur
Medical sciences	154.5	165.1	173.2	168.5	200.0	213.2	..	Sciences médicales
Agricultural sciences	53.1	54.1	62.5	60.7	74.2	84.5	..	Sciences agricoles
Sub-total NSE	781.2	884.0	967.1	1 032.6	1 121.2	1 172.6	..	Sous-total SENI
Social sciences	155.8	157.3	171.8	186.9	204.7	222.0	..	Sciences sociales
Humanities	103.3	95.6	103.8	102.4	112.8	110.0	..	Sciences humaines
Sub-total SSH	259.1	252.9	275.6	289.3	317.5	332.0	..	Sous-total SSH
Total NSE+SSH	1 040.3	1 136.9	1 242.7	1 321.9	1 438.7	1 504.6	1 609.9 c	*Total SENI+SSH*
	14.2 Million constant USD (1995 prices and PPPs) - Millions de USD constants (prix et PPA de 1995)							
Natural sciences	559.1 b	621.3	672.7	709.7	740.1	726.6	..	Sciences exactes et naturelles
Engineering	260.4 b	284.9	290.2	325.3	326.8	344.5	..	Sciences de l'ingénieur
Medical sciences	220.7 b	225.0	228.0	217.0	251.9	261.0	..	Sciences médicales
Agricultural sciences	75.9 b	73.7	82.3	78.3	93.4	103.4	..	Sciences agricoles
Sub-total NSE	1 116.0 b	1 204.8	1 273.2	1 330.2	1 412.2	1 435.6	..	Sous-total SENI
Social sciences	222.6 b	214.4	226.2	240.7	257.9	271.8	..	Sciences sociales
Humanities	147.6 b	130.3	136.7	132.0	142.0	134.6	..	Sciences humaines
Sub-total SSH	370.2 b	344.7	362.9	372.7	399.9	406.4	..	Sous-total SSH
Total NSE+SSH	1 486.2 b	1 549.6	1 636.1	1 702.9	1 812.1	1 842.0	1 905.1 c	*Total SENI+SSH*

Table 15 - Tableau 15

Higher education intramural expenditure on R&D - Dépenses intra-muros de R-D de l'enseignement supérieur
(HERD) by field of science and type of costs - (DIRDES) par discipline scientifique et par type de coûts

Million EUR (1999 ESP euro, EUR from 1999) *Millions EUR (1999 ESP euro, EUR à partir de 1999)*

	1994	1995	1996	1997	1998	1999	2000	
Current expenditure								**Dépenses courantes**
Natural sciences	..	369.4	..	423.6	463.7	469.2	..	Sciences exactes et naturelles
Engineering	..	170.6	..	192.1	198.8	212.4	..	Sciences de l'ingénieur
Medical sciences	..	132.5	..	132.6	163.4	170.8	..	Sciences médicales
Agricultural sciences	..	46.4	..	44.6	55.8	62.4	..	Sciences agricoles
Sub-total NSE	..	718.8	..	792.9	881.6	914.7	..	Sous-total SENI
Social sciences	..	127.4	..	141.9	157.6	173.7	..	Sciences sociales
Humanities	..	77.8	..	79.5	89.1	87.2	..	Sciences humaines
Sub-total SSH	..	205.1	..	221.4	246.7	260.8	..	Sous-total SSH
Total NSE+SSH	842.6	923.9	990.5	1 014.3	1 128.3	1 175.5	..	*Total SENI+SSH*
Capital expenditure								**Dépenses en capital**
Natural sciences	..	86.5	..	127.3	123.8	124.4	..	Sciences exactes et naturelles
Engineering	..	38.5	..	60.4	60.7	69.0	..	Sciences de l'ingénieur
Medical sciences	..	32.6	..	35.9	36.7	42.4	..	Sciences médicales
Agricultural sciences	..	7.7	..	16.1	18.4	22.1	..	Sciences agricoles
Sub-total NSE	..	165.2	..	239.8	239.6	257.9	..	Sous-total SENI
Social sciences	..	30.0	..	44.9	47.1	48.3	..	Sciences sociales
Humanities	..	17.9	..	23.0	23.7	22.8	..	Sciences humaines
Sub-total SSH	..	47.8	..	67.9	70.8	71.2	..	Sous-total SSH
Total NSE+SSH	197.7	213.0	252.2	307.7	310.4	329.1	..	*Total SENI+SSH*

Source: OECD, R&D database (DSTI/EAS Division), January 2002.

Source : OCDE, base de données R-D (DSTI/Division AES), janvier 2002.

Higher education intramural expenditure on R&D - Dépenses intra-muros de R-D de l'enseignement supérieur

(HERD) by field of science and source of funds - (DIRDES) par discipline scientifique et par source de financement

Million EUR (1999 ESP euro, EUR from 1999) — *Millions EUR (1999 ESP euro, EUR à partir de 1999)*

	1994	1995	1996	1997	1998	1999	2000	
Direct government								**État direct**
Natural sciences	..	143.7	..	84.7	87.3	103.6	..	Sciences exactes et naturelles
Engineering	..	44.0	..	34.3	41.4	50.4	..	Sciences de l'ingénieur
Medical sciences	..	53.3	..	26.2	26.7	33.6	..	Sciences médicales
Agricultural sciences	..	19.5	..	13.1	15.2	19.3	..	Sciences agricoles
Sub-total NSE	..	260.5	..	158.2	170.6	206.8	..	Sous-total SENI
Social sciences	..	51.0	..	25.1	24.6	29.0	..	Sciences sociales
Humanities	..	30.8	..	15.2	16.3	16.5	..	Sciences humaines
Sub-total SSH	..	81.8	..	40.3	40.9	45.5	..	Sous-total SSH
Total NSE+SSH	209.0 [c]	342.2	160.3	198.6	211.4	252.4	..	*Total SENI+SSH*
General university funds								**Fonds généraux des universités**
Natural sciences	..	189.5 [a]	..	335.4	340.2	345.2	..	Sciences exactes et naturelles
Engineering	..	96.3 [a]	..	136.3	128.6	142.6	..	Sciences de l'ingénieur
Medical sciences	..	61.7 [a]	..	104.5	116.0	124.6	..	Sciences médicales
Agricultural sciences	..	15.8 [a]	..	33.3	40.3	48.3	..	Sciences agricoles
Sub-total NSE	..	363.3 [a]	..	609.6	625.1	660.7	..	Sous-total SENI
Social sciences	..	59.2 [a]	..	108.0	109.5	120.0	..	Sciences sociales
Humanities	..	36.2 [a]	..	59.4	60.8	63.3	..	Sciences humaines
Sub-total SSH	..	95.3 [a]	..	167.4	170.3	183.3	..	Sous-total SSH
Total NSE+SSH	726.8 [c,o]	458.6 [a]	753.9	777.0	795.4	844.0	..	*Total SENI+SSH*
Sub-total government								**Sous-total État**
Natural sciences	..	333.2 [a]	..	420.1	427.5	448.8	..	Sciences exactes et naturelles
Engineering	..	140.2 [a]	..	170.6	170.0	193.0	..	Sciences de l'ingénieur
Medical sciences	..	115.0 [a]	..	130.7	142.7	158.2	..	Sciences médicales
Agricultural sciences	..	35.2 [a]	..	46.4	55.5	67.6	..	Sciences agricoles
Sub-total NSE	..	623.8 [a]	..	767.8	795.7	867.5	..	Sous-total SENI
Social sciences	..	110.2 [a]	..	133.1	134.1	149.1	..	Sciences sociales
Humanities	..	66.9 [a]	..	74.7	77.1	79.8	..	Sciences humaines
Sub-total SSH	..	177.1 [a]	..	207.8	211.2	228.8	..	Sous-total SSH
Total NSE+SSH	935.8 [c,o]	800.8 [a]	914.2	975.6	1 006.8	1 096.4	..	*Total SENI+SSH*
Higher education								**Enseignement supérieur**
Natural sciences	..	59.6	..	67.9	70.8	67.7	..	Sciences exactes et naturelles
Engineering	..	19.2	..	29.3	24.4	30.1	..	Sciences de l'ingénieur
Medical sciences	..	26.4	..	17.8	26.2	26.0	..	Sciences médicales
Agricultural sciences	..	11.2	..	6.3	7.6	6.8	..	Sciences agricoles
Sub-total NSE	..	116.4	..	121.3	129.1	130.6	..	Sous-total SENI
Social sciences	..	23.9	..	30.4	39.1	42.4	..	Sciences sociales
Humanities	..	15.3	..	15.6	18.0	16.3	..	Sciences humaines
Sub-total SSH	..	39.2	..	46.0	57.1	58.7	..	Sous-total SSH
Total NSE+SSH	.. [n]	155.6	154.0	167.4	186.2	189.3	..	*Total SENI+SSH*
Private non-profit								**Institutions sans but lucratif**
Natural sciences	..	1.2	..	2.4	3.1	4.1	..	Sciences exactes et naturelles
Engineering	..	2.0	..	2.6	4.0	3.5	..	Sciences de l'ingénieur
Medical sciences	..	1.2	..	1.6	3.9	2.7	..	Sciences médicales
Agricultural sciences	..	0.1	..	0.5	0.5	1.0	..	Sciences agricoles
Sub-total NSE	..	4.5	..	7.1	11.4	11.3	..	Sous-total SENI
Social sciences	..	0.8	..	1.6	2.0	2.5	..	Sciences sociales
Humanities	..	0.4	..	1.4	1.2	1.1	..	Sciences humaines
Sub-total SSH	..	1.2	..	3.0	3.1	3.6	..	Sous-total SSH
Total NSE+SSH	6.5 [c]	5.7	11.0	10.1	14.5	15.0	..	*Total SENI+SSH*
Business entreprise								**Entreprises**
Natural sciences	..	29.5	..	28.3	34.0	38.7	..	Sciences exactes et naturelles
Engineering	..	32.3	..	28.5	29.9	34.8	..	Sciences de l'ingénieur
Medical sciences	..	12.2	..	8.8	12.4	14.5	..	Sciences médicales
Agricultural sciences	..	4.0	..	3.8	4.2	5.3	..	Sciences agricoles
Sub-total NSE	..	78.0	..	69.4	80.6	93.3	..	Sous-total SENI
Social sciences	..	10.8	..	10.8	13.2	15.8	..	Sciences sociales
Humanities	..	5.9	..	5.2	6.6	7.0	..	Sciences humaines
Sub-total SSH	..	16.7	..	16.0	19.7	22.8	..	Sous-total SSH
Total NSE+SSH	61.6 [c]	94.7	92.9	85.3	100.3	116.1	..	*Total SENI+SSH*
Funds from abroad								**Fonds de l'étranger**
Natural sciences	..	32.3	..	32.2	52.1	34.2	..	Sciences exactes et naturelles
Engineering	..	15.3	..	21.6	31.1	20.0	..	Sciences de l'ingénieur
Medical sciences	..	10.3	..	9.5	14.8	11.9	..	Sciences médicales
Agricultural sciences	..	3.5	..	3.8	6.4	3.7	..	Sciences agricoles
Sub-total NSE	..	61.4	..	67.0	104.5	69.8	..	Sous-total SENI
Social sciences	..	11.6	..	10.9	16.4	12.2	..	Sciences sociales
Humanities	..	7.1	..	5.6	9.9	5.8	..	Sciences humaines
Sub-total SSH	..	18.7	..	16.5	26.3	18.0	..	Sous-total SSH
Total NSE+SSH	36.4 [c]	80.1	70.7	83.6	130.8	87.8	..	*Total SENI+SSH*

Source: OECD, R&D database (DSTI/EAS Division), January 2002.

Source : OCDE, base de données R-D (DSTI/Division AES), janvier 2002.

Table 17 - Tableau 17

Higher education R&D personnel (HEMP) - Personnel de R-D du secteur de l'enseignement supérieur

by field of science and occupation - par discipline scientifique et par profession

In full-time equivalent — *En équivalence plein-temps*

	1994	1995	1996	1997	1998	1999	2000	
Total R&D personnel								**Personnel total de R-D**
Natural sciences	..	13 714	..	15 202	16 365	16 153	..	Sciences exactes et naturelles
Engineering	..	6 487	..	6 370	7 051	6 789	..	Sciences de l'ingénieur
Medical sciences	..	4 859	..	5 165	6 282	6 020	..	Sciences médicales
Agricultural sciences	..	1 716	..	1 900	2 201	2 344	..	Sciences agricoles
Sub-total NSE	21 438 [c]	26 776	30 202	28 637	31 899	31 306	..	Sous-total SENI
Social sciences	..	4 719	..	5 157	5 706	6 106	..	Sciences sociales
Humanities	..	2 835	..	3 049	3 436	3 214	..	Sciences humaines
Sub-total SSH	13 204 [c]	7 554	8 753	8 206	9 142	9 320	..	Sous-total SSH
Total NSE+SSH	34 642 [c]	34 330	38 955	36 843	41 040	40 626	41 032 [c]	*Total SENI+SSH*
Researchers								**Chercheurs**
Natural sciences	..	11 088	..	12 606	13 606	13 339	..	Sciences exactes et naturelles
Engineering	..	5 128	..	5 240	6 119	5 668	..	Sciences de l'ingénieur
Medical sciences	..	3 934	..	4 333	5 326	5 069	..	Sciences médicales
Agricultural sciences	..	1 444	..	1 622	1 847	1 920	..	Sciences agricoles
Sub-total NSE	..	21 594	..	23 801	26 898	25 996	..	Sous-total SENI
Social sciences	..	3 778	..	4 266	4 714	5 084	..	Sciences sociales
Humanities	..	2 294	..	2 582	2 912	2 760	..	Sciences humaines
Sub-total SSH	..	6 072	..	6 848	7 626	7 844	..	Sous-total SSH
Total NSE+SSH	28 591	27 666	30 858	30 649	34 524	33 840	..	*Total SENI+SSH*
Technicians								**Technicians**
Natural sciences	..	840	..	1 108	1 097	1 433	..	Sciences exactes et naturelles
Engineering	..	485	..	566	369	627	..	Sciences de l'ingénieur
Medical sciences	..	260	..	340	372	418	..	Sciences médicales
Agricultural sciences	..	97	..	148	176	266	..	Sciences agricoles
Sub-total NSE	..	1 682	..	2 162	2 014	2 744	..	Sous-total SENI
Social sciences	..	337	..	365	367	435	..	Sciences sociales
Humanities	..	179	..	193	212	211	..	Sciences humaines
Sub-total SSH	..	516	..	558	579	646	..	Sous-total SSH
Total NSE+SSH	..	2 198	2 671 [c]	2 720	2 593	3 390	..	*Total SENI+SSH*
Other								**Autres**
Natural sciences	..	1 786	..	1 488	1 662	1 381	..	Sciences exactes et naturelles
Engineering	..	874	..	564	563	494	..	Sciences de l'ingénieur
Medical sciences	..	665	..	492	584	533	..	Sciences médicales
Agricultural sciences	..	175	..	130	178	158	..	Sciences agricoles
Sub-total NSE	..	3 500	..	2 674	2 987	2 566	..	Sous-total SENI
Social sciences	..	604	..	526	625	587	..	Sciences sociales
Humanities	..	362	..	274	312	243	..	Sciences humaines
Sub-total SSH	..	966	..	800	937	830	..	Sous-total SSH
Total NSE+SSH	..	4 466	5 426 [c]	3 474	3 923	3 396	..	*Total SENI+SSH*

Table 18 - Tableau 18

Government budget appropriations or outlays for R&D (GBAORD) - Crédits budgétaires publics de R-D (CBPRD)

by socio-economic objective - par objectif socio-économique

Million EUR (1999 ESP euro, EUR from 1999) — *Millions EUR (1999 ESP euro, EUR à partir de 1999)*

	1995	1996	1997	1998	1999	2000	2001	
Agriculture, forestry and fishing	91.3	89.3	98.0	119.4	112.9	..	..	Agriculture, sylviculture et pêche
Industrial development	429.7	451.3	423.0	447.2	593.8	..	..	Développement industriel
Energy	57.0	56.6	106.4 [a]	102.7	122.6	..	..	Energie
Transport & telecommunications	5.2	7.5	7.1	16.2	31.2	..	..	*Transports & télécommunications*
Urban & rural planning	8.0	6.6	2.7	2.2	3.0	..	..	*Aménagement urbain et rural*
ST Infrastructure	13.2	14.1	9.8	18.4	34.2	..	..	ST Infrastructure
Prevention of pollution	..	..	..	33.3	32.9	..	..	*Prévention de la pollution*
Identificat. & treatment of pollution	..	..	..	42.1	50.6	..	..	*Détect. pollut. & mesures antipollut.*
ST Environment	56.0	59.0	52.8	75.4	83.5	..	..	ST Environnement
Health	125.2	130.4	120.6	136.2	162.6	..	..	Santé
Social development & services	20.6	23.2	17.3	16.3	28.1	..	..	Développement et services sociaux
Earth and atmosphere	51.9	55.0	53.0	55.8	61.4	..	..	Milieu terrestre et atmosphère
Advancement of research	179.5	205.4	190.1	207.3	171.5	..	..	*Promotion de la recherche*
General university funds	687.7	687.7	701.4	726.0	832.7	..	..	*Fonds généraux des universités*
ST Advancement of knowledge	867.1	893.0	891.5	933.2	1 004.2	..	..	ST promotion gén. des connaiss.
Civil space	169.5	161.5	158.8	156.2	154.7	..	..	Espace civil
Defence	222.5	238.2	478.0	853.7	846.5	..	..	Défense
Not elsewhere classified	34.9	38.0	32.6	34.5	32.3	..	..	Non classé
Total	**2 139.0**	**2 209.4**	**2 441.6** [a]	**2 948.8**	**3 236.6**	..	..	**Total**

Source: OECD, R&D database (DSTI/EAS Division), January 2002.

Source : OCDE, base de données R-D (DSTI/Division AES), janvier 2002.

Table 19 - Tableau 19

Technology balance of payments - Balance des paiements technologiques

by type of transfer - selon la nature des transferts

Million EUR (1999 ESP euro, EUR from 1999) *Millions EUR (1999 ESP euro, EUR à partir de 1999)*

	1983	1984	1985	1986	1987	1988	1989	1990	1991	
Receipts										**Recettes**
Related firms	..	..	..	..	..	..	..	..	..	Firmes affiliées
Unrelated firms	..	..	..	..	..	..	..	..	..	Firmes non affiliées
Other	..	..	..	..	..	..	..	..	..	Autre
Total	110	125	141	150	123	128	206	245	400	*Total*
Payments										**Paiements**
Related firms	..	..	..	..	..	..	..	..	..	Firmes affiliées
Unrelated firms	..	..	..	..	..	..	..	..	..	Firmes non affiliées
Other	..	..	..	..	..	..	..	..	..	Autre
Total	531	509	564	624	680	966	1 145	1 333	1 422	*Total*

	1992	1993	1994	1995	1996	1997	1998	1999	2000	
Receipts										**Recettes**
Related firms	..	..	..	..	..	..	..	..	..	Firmes affiliées
Unrelated firms	..	..	..	..	..	..	..	..	..	Firmes non affiliées
Other	..	..	..	..	..	..	..	..	..	Autre
Total	487	..	75 [a]	60	67	142	171	..	..	*Total*
Payments										**Paiements**
Related firms	..	..	..	..	..	..	..	..	..	Firmes affiliées
Unrelated firms	..	..	..	..	..	..	..	..	..	Firmes non affiliées
Other	..	..	..	..	..	..	..	..	..	Autre
Total	1 951	..	770 [a]	832	805	945	921	..	..	*Total*

Source: OECD, TBPdatabase (DSTI/EAS Division), Januay 2002.

Source : OCDE, base de données TBP (DSTI/Division AES), janvier 2002.

SWEDEN

SUEDE

Table 1 - Tableau 1

Gross domestic expenditure on R&D (GERD) - Dépenses intérieures brutes de R-D (DIRD)

by sector of performance and source of funds - par secteur d'exécution et par source de financement

	1994	1995	1996	1997	1998	1999	2000	
				1.1 Million SEK - Millions SEK				
Business enterprise								**Entreprises**
financed by								*financé par*
Business enterprise	..	38 194.6 [a,m]	..	44 628.7 [m]	..	50 634.2 [m]	..	Entreprises
Direct government	..	4 166.8 [a,m]	..	3 826.5 [a,m]	..	4 439.8 [m]	..	État direct
Higher education	..	.. [n]	..	.. [n]	..	.. [n]	..	Enseignement supérieur
Private non-profit	..	31.9 [a,m,o]	..	33.7 [a,m,o]	..	36.5 [m,o]	..	Institutions sans but lucratif
Funds from abroad	..	1 635.1 [a]	..	1 662.2 [m]	..	1 843.0 [m]	..	Fonds de l'étranger
Total	..	44 029.0 [a,m]	..	50 151.0 [m]	..	56 953.5 [m]	..	*Total*
Government								**État**
financed by								*financé par*
Business enterprise	..	64.0 [h]	..	68.0 [h]	..	97.0 [h]	..	Entreprises
Direct government	..	2 041.0 [h]	..	2 234.3 [a,h]	..	2 318.0 [h]	..	État direct
Higher education	..	9.0 [h]	..	4.0 [h]	..	11.0 [h]	..	Enseignement supérieur
Private non-profit	..	2.0 [h]	..	9.0 [a,h]	..	25.0 [h]	..	Institutions sans but lucratif
Funds from abroad	..	56.0 [h]	..	57.0 [h]	..	96.0 [h]	..	Fonds de l'étranger
Total	..	2 171.0 [h]	..	2 372.3 [h]	..	2 547.0 [h]	..	*Total*
Higher education								**Enseignement supérieur**
financed by								*financé par*
Business enterprise	..	598.0 [a,j]	..	689.2 [a]	..	629.0	..	Entreprises
Direct government	..	3 485.0 [a,j]	..	3 796.9 [a]	..	4 089.0	..	État direct
General university funds	..	7 390.0 [a,j]	..	7 425.6 [a]	..	7 676.0	..	Fonds généraux des universités
Sub-total government	..	10 875.0 [a,j]	..	11 222.5 [a]	..	11 765.0	..	Sous-Total État
Higher education	..	394.0 [a,j]	..	225.9 [a]	..	224.0	..	Enseignement supérieur
Private non-profit	..	792.0 [a,j]	..	1 594.8 [a]	..	2 863.0	..	Institutions sans but lucratif
Funds from abroad	..	344.0 [a,j]	..	613.9 [a]	..	745.0	..	Fonds de l'étranger
Total	..	13 004.0 [a,j]	..	14 346.3 [a]	..	16 226.0	..	*Total*
Private non-profit								**Institutions sans but lucratif**
financed by								*financé par*
Business enterprise	..	..	..	17.3	..	13.0	..	Entreprises
Direct government	..	..	..	5.0 [a]	..	38.1	..	État direct
Higher education	..	..	..	0.5	..	3.5	..	Enseignement supérieur
Private non-profit	..	93.0 [a]	..	22.0 [a]	..	31.0	..	Institutions sans but lucratif
Funds from abroad	..	..	..	6.3	..	1.6	..	Fonds de l'étranger
Total	..	93.0 [a]	..	51.1	..	87.2	..	*Total*
GERD								**DIRD**
financed by								*financé par*
Business enterprise	..	38 857.3 [m]	..	45 403.2 [m]	..	51 373.2 [m]	..	Entreprises
Direct government	..	9 692.8 [a,m]	..	9 862.7 [a,m]	..	10 884.9 [m]	..	État direct
General university funds	..	7 390.0 [a,m]	..	7 425.6 [m]	..	7 676.0 [m]	..	Fonds généraux des universités
Sub-total government	..	17 082.8 [a,m]	..	17 288.3 [a,m]	..	18 560.9 [m]	..	Sous-Total État
Higher education	..	403.0 [m]	..	230.4 [m]	..	238.5 [m]	..	Enseignement supérieur
Private non-profit	..	918.9 [m]	..	1 659.5 [a,m]	..	2 955.5 [m]	..	Institutions sans but lucratif
Funds from abroad	..	2 035.1 [m]	..	2 339.4 [m]	..	2 685.6 [m]	..	Fonds de l'étranger
Total	..	59 297.0 [a,m]	..	66 920.7 [m]	..	75 813.5 [m]	..	*Total*
				1.2 Million constant USD (1995 prices and PPPs) - Millions de USD constants (prix et PPA de 1995)				
Business enterprise								**Entreprises**
financed by								*financé par*
Business enterprise	..	3 926.2 [a,m]	..	4 446.3 [m]	..	4 974.6 [m]	..	Entreprises
Direct government	..	428.3 [a,m]	..	381.2 [a,m]	..	436.2 [m]	..	État direct
Higher education	..	.. [n]	..	.. [n]	..	.. [n]	..	Enseignement supérieur
Private non-profit	..	3.3 [a,m,o]	..	3.4 [a,m,o]	..	3.6 [m,o]	..	Institutions sans but lucratif
Funds from abroad	..	168.1 [a]	..	165.6 [m]	..	181.1 [m]	..	Fonds de l'étranger
Total	..	4 526.0 [a,m]	..	4 996.5 [m]	..	5 595.5 [m]	..	*Total*
Government								**État**
financed by								*financé par*
Business enterprise	..	6.6 [h]	..	6.8 [h]	..	9.5 [h]	..	Entreprises
Direct government	..	209.8 [h]	..	222.6 [a,h]	..	227.7 [h]	..	État direct
Higher education	..	0.9 [h]	..	0.4 [h]	..	1.1 [h]	..	Enseignement supérieur
Private non-profit	..	0.2 [h]	..	0.9 [a,h]	..	2.5 [h]	..	Institutions sans but lucratif
Funds from abroad	..	5.8 [h]	..	5.7 [h]	..	9.4 [h]	..	Fonds de l'étranger
Total	..	223.2 [h]	..	236.4 [h]	..	250.2 [h]	..	*Total*
Higher education								**Enseignement supérieur**
financed by								*financé par*
Business enterprise	..	61.5 [a,j]	..	68.7 [a]	..	61.8	..	Entreprises
Direct government	..	358.2 [a,j]	..	378.3 [a]	..	401.7	..	État direct
General university funds	..	759.7 [a,j]	..	739.8 [a]	..	754.1	..	Fonds généraux des universités
Sub-total government	..	1 117.9 [a,j]	..	1 118.1 [a]	..	1 155.9	..	Sous-Total État
Higher education	..	40.5 [a,j]	..	22.5 [a]	..	22.0	..	Enseignement supérieur
Private non-profit	..	81.4 [a,j]	..	158.9 [a]	..	281.3	..	Institutions sans but lucratif
Funds from abroad	..	35.4 [a,j]	..	61.2 [a]	..	73.2	..	Fonds de l'étranger
Total	..	1 336.7 [a,j]	..	1 429.3 [a]	..	1 594.1	..	*Total*
Private non-profit								**Institutions sans but lucratif**
financed by								*financé par*
Business enterprise	..	..	..	1.7	..	1.3	..	Entreprises
Direct government	..	..	..	0.5 [a]	..	3.7	..	État direct
Higher education	..	..	..	0.1	..	0.3	..	Enseignement supérieur
Private non-profit	..	9.6 [a]	..	2.2 [a]	..	3.0	..	Institutions sans but lucratif
Funds from abroad	..	..	..	0.6	..	0.2	..	Fonds de l'étranger
Total	..	9.6 [a]	..	5.1	..	8.6	..	*Total*
GERD								**DIRD**
financed by								*financé par*
Business enterprise	..	3 994.3 [m]	..	4 523.5 [m]	..	5 047.2 [m]	..	Entreprises
Direct government	..	996.4 [a,m]	..	982.6 [a,m]	..	1 069.4 [m]	..	État direct
General university funds	..	759.7 [a,m]	..	739.8 [m]	..	754.1 [m]	..	Fonds généraux des universités
Sub-total government	..	1 756.0 [a,m]	..	1 722.4 [a,m]	..	1 823.5 [m]	..	Sous-Total État
Higher education	..	41.4 [m]	..	23.0 [m]	..	23.4 [m]	..	Enseignement supérieur
Private non-profit	..	94.5 [m]	..	165.3 [a,m]	..	290.4 [m]	..	Institutions sans but lucratif
Funds from abroad	..	209.2 [m]	..	233.1 [m]	..	263.9 [m]	..	Fonds de l'étranger
Total	..	6 095.4 [a,m]	..	6 667.3 [m]	..	7 448.4 [m]	..	*Total*

Source: OECD, R&D database (DSTI/EAS Division), January 2002.

Source : OCDE, base de données R-D (DSTI/Division AES), janvier 2002.

Table 2 - Tableau 2

Gross domestic expenditure on R&D (GERD) - Dépenses intérieures brutes de R-D (DIRD)

by sector of performance and type of costs - par secteur d'exécution et par type de coûts

Million SEK *Millions SEK*

	1994	1995	1996	1997	1998	1999	2000	
Business enterprise								**Entreprises**
Labour costs	..	18 251.7 [a]	..	20 988.7	..	23 439.3	..	Coûts du personnel
Other current costs	..	21 789.3 [a]	..	25 126.2	..	29 576.1	..	Autres coûts courants
Total current expenditure	..	40 041.1 [a]	..	46 114.9	..	53 015.4	..	*Total dépenses courantes*
Land and buildings	..	710.7 [a]	..	670.7	..	813.3	..	Terrains et constructions
Instruments and equipment	..	3 277.2 [a]	..	3 365.4	..	3 124.9	..	Instruments et équipements
Total capital expenditure	..	3 987.9 [a]	..	4 036.2	..	3 938.2	..	*Total dépenses en capital*
Total	..	44 029.0 [a,m]	..	50 151.0 [m]	..	56 953.5 [m]	..	*Total*
Government								**État**
Labour costs	..	1 397.0 [h]	..	1 440.4 [h]	..	1 450.0 [h]	..	Coûts du personnel
Other current costs	..	713.0 [h]	..	791.2 [h]	..	967.0 [h]	..	Autres coûts courants
Total current expenditure	..	2 111.0 [h]	..	2 231.6 [h]	..	2 417.0 [h]	..	*Total dépenses courantes*
Land and buildings	..	..	..	38.0 [h]	..	..	..	Terrains et constructions
Instruments and equipment	..	..	..	103.0 [h]	..	..	..	Instruments et équipements
Total capital expenditure	..	60.0 [h]	..	141.0 [h]	..	130.0 [h]	..	*Total dépenses en capital*
Total	..	2 171.0 [h]	..	2 372.3 [h]	..	2 547.0 [h]	..	*Total*
Higher education								**Enseignement supérieur**
Labour costs	..	..	..	..	..	..	..	Coûts du personnel
Other current costs	..	..	..	..	..	..	..	Autres coûts courants
Total current expenditure	..	13 004.0	..	13 411.8	..	15 158.0	..	*Total dépenses courantes*
Land and buildings	..	..	..	52.4	..	81.0	..	Terrains et constructions
Instruments and equipment	..	..	..	882.1	..	986.0	..	Instruments et équipements
Total capital expenditure	..	..	..	934.5	..	1 068.0	..	*Total dépenses en capital*
Total	..	13 004.0 [a,j]	..	14 346.3 [a]	..	16 226.0	..	*Total*
Private non-profit								**Institutions sans but lucratif**
Labour costs	..	..	..	37.2	..	60.0	..	Coûts du personnel
Other current costs	..	..	..	12.5	..	24.0	..	Autres coûts courants
Total current expenditure	..	93.0 [a]	..	49.7	..	84.0	..	*Total dépenses courantes*
Land and buildings	..	..	..	..	..	..	..	Terrains et constructions
Instruments and equipment	..	..	..	..	..	..	..	Instruments et équipements
Total capital expenditure	..	..	..	1.4	..	3.0	..	*Total dépenses en capital*
Total	..	93.0 [a]	..	51.1	..	87.2	..	*Total*
GERD								**DIRD**
Labour costs	..	..	..	..	..	..	..	Coûts du personnel
Other current costs	..	..	..	..	..	..	..	Autres coûts courants
Total current expenditure	..	55 249.1	..	61 808.0	..	70 674.4	..	*Total dépenses courantes*
Land and buildings	..	..	..	..	..	..	..	Terrains et constructions
Instruments and equipment	..	..	..	..	..	..	..	Instruments et équipements
Total capital expenditure	..	4 047.9	..	5 113.1	..	5 139.2	..	*Total dépenses en capital*
Total	..	59 297.0 [a,m]	..	66 920.7 [m]	..	75 813.5 [m]	..	*Total*

Table 3 - Tableau 3

Current domestic expenditure on R&D - Dépenses intérieures courantes de R-D

by sector of performance and type of activity - par secteur d'exécution et par type d'activités

Million SEK *Millions SEK*

	1994	1995	1996	1997	1998	1999	2000	
Business enterprise								**Entreprises**
Basic research	..	..	..	..	..	..	..	Recherche fondamentale
Applied research	..	..	..	..	..	..	..	Recherche appliquée
Experimental development	..	..	..	..	..	..	..	Développement expérimental
Not specified	..	40 041.1 [a]	..	46 114.9	..	53 015.4	..	Non spécifié
Total	..	40 041.1 [a]	..	46 114.9	..	53 015.4	..	*Total*
Government								**État**
Basic research	..	1 351.0 [h]	..	1 392.4 [h]	..	1 499.0	..	Recherche fondamentale
Applied research	..	..	..	..	..	..	..	Recherche appliquée
Experimental development	..	759.0 [h]	..	832.2 [h]	..	918.0	..	Développement expérimental
Not specified	..	..	..	7.0 [h]	..	..	..	Non spécifié
Total	..	2 111.0 [h]	..	2 231.6 [h]	..	2 417.0	..	*Total*
Higher education								**Enseignement supérieur**
Basic research	..	..	..	..	..	..	..	Recherche fondamentale
Applied research	..	..	..	..	..	..	..	Recherche appliquée
Experimental development	..	..	..	..	..	..	..	Développement expérimental
Not specified	..	13 004.0 [a,j]	..	13 411.8	..	15 158.0	..	Non spécifié
Total	..	13 004.0 [a,j]	..	13 411.8	..	15 158.0	..	*Total*
Private non-profit								**Institutions sans but lucratif**
Basic research	..	..	..	..	..	..	..	Recherche fondamentale
Applied research	..	..	..	..	..	..	..	Recherche appliquée
Experimental development	..	..	..	..	..	..	..	Développement expérimental
Not specified	..	93.0 [a]	..	49.7	..	84.0	..	Non spécifié
Total	..	93.0 [a]	..	49.7	..	84.0	..	*Total*
GERD								**DIRD**
Basic research	..	..	..	..	..	..	..	Recherche fondamentale
Applied research	..	..	..	..	..	..	..	Recherche appliquée
Experimental development	..	..	..	..	..	..	..	Développement expérimental
Not specified	..	55 249.1	..	61 808.0	..	70 674.4	..	Non spécifié
Total	..	55 249.1	..	61 808.0	..	70 674.4	..	*Total*

Source: OECD, R&D database (DSTI/EAS Division), January 2002.

Source : OCDE, base de données R-D (DSTI/Division AES), janvier 2002.

SWEDEN — Table 4 - Tableau 4 — SUEDE

Gross domestic expenditure on R&D (GERD) - Dépenses intérieures brutes de R-D (DIRD)

by sector of performance and socio-economic objective - par secteur d'exécution et par objectif socio-économique

Million SEK — *Millions SEK*

	1994	1995	1996	1997	1998	1999	2000	
Business enterprise								**Entreprises**
Agriculture, forestry and fishing	..	194.0 [a,j]	..	177.3 [j]	..	127.7 [j]	..	Agriculture, sylviculture et pêche
Industrial development	..	37 263.0 [a,j]	..	42 960.9 [j]	..	48 992.2 [j]	..	Développement industriel
Energy	..	..	..	145.3 [j]	..	99.8 [j]	..	Energie
Transport and telecommunications	..	..	..	101.6 [j]	..	375.1 [j]	..	Transports et télécommunications
Urban and rural planning	..	..	..	..	..	71.8 [j]	..	Aménagement urbain et rural
Prevention of pollution	..	..	..	101.2 [j]	..	..	..	Prévention de la pollution
Identification & treatment of pollution	..	..	..	..	..	35.7 [j]	..	Détection pollu. & mesures antipollu.
Health	..	..	..	..	..	104.9 [j]	..	Santé
Social development and services	..	..	..	..	..	..	..	Développement et services sociaux
Earth and atmosphere	..	..	..	..	..	..	..	Milieu terrestre et atmosphère
Advancement of knowledge	..	..	..	..	..	36.9 [j]	..	Promotion générale des connaissances
Civil space	..	..	..	..	..	..	..	Espace civil
Defence	..	2 121.0 [a,j]	..	2 437.3 [j]	..	3 101.7 [j]	..	Défense
Not elsewhere classified	..	463.0 [a,j]	..	191.2 [j]	..	69.6 [j]	..	Non classé
Total	..	44 029.0 [a,m]	..	50 151.0 [m]	..	56 953.5 [m]	..	*Total*
Government								**État**
Agriculture, forestry and fishing	..	96.0 [a,h]	..	126.2 [a,h,j]	..	45.0 [h,j]	..	Agriculture, sylviculture et pêche
Industrial development	..	36.0 [a,h]	..	41.1 [a,h,j]	..	59.0 [h,j]	..	Développement industriel
Energy	..	5.0 [a,h]	..	5.8 [a,h,j]	..	9.0 [h,j]	..	Energie
Transport and telecommunications	..	300.0 [a,h]	..	221.5 [a,h,j]	..	218.0 [h,j]	..	Transports et télécommunications
Urban and rural planning	..	47.0 [a,h]	..	50.1 [a,h,j]	..	61.0 [h,j]	..	Aménagement urbain et rural
Prevention of pollution	..	29.0 [a,h]	..	20.0 [a,h,j]	..	39.0 [h,j]	..	Prévention de la pollution
Identification & treatment of pollution	..	..	..	..	..	..	..	Détection pollu. & mesures antipollu.
Health	..	122.0 [a,h]	..	125.4 [a,h,j]	..	171.0 [h,j]	..	Santé
Social development and services	..	396.0 [a,h]	..	485.0 [a,h,j]	..	415.0 [h,j]	..	Développement et services sociaux
Earth and atmosphere	..	79.0 [a,h]	..	62.0 [a,h,j]	..	30.0 [h,j]	..	Milieu terrestre et atmosphère
Advancement of knowledge	..	72.0 [a,h]	..	122.0 [a,h,j]	..	79.0 [h,j]	..	Promotion générale des connaissances
Civil space	..	0.0 [a,h]	..	..	..	1.0 [h,j]	..	Espace civil
Defence	..	871.0 [a,h]	..	971.3 [a,h,j]	..	1 053.0 [h,j]	..	Défense
Not elsewhere classified	..	130.0 [a,h]	..	1.7 [a,h,j]	..	236.0 [h,j]	..	Non classé
Total	..	2 171.0 [h]	..	2 372.3 [h]	..	2 547.0 [h]	..	*Total*
Higher education								**Enseignement supérieur**
Agriculture, forestry and fishing	..	..	..	..	..	..	..	Agriculture, sylviculture et pêche
Industrial development	..	..	..	..	..	..	..	Développement industriel
Energy	..	..	..	..	..	..	..	Energie
Transport and telecommunications	..	..	..	..	..	..	..	Transports et télécommunications
Urban and rural planning	..	..	..	..	..	..	..	Aménagement urbain et rural
Prevention of pollution	..	..	..	..	..	..	..	Prévention de la pollution
Identification & treatment of pollution	..	..	..	..	..	..	..	Détection pollu. & mesures antipollu.
Health	..	..	..	..	..	..	..	Santé
Social development and services	..	..	..	..	..	..	..	Développement et services sociaux
Earth and atmosphere	..	..	..	..	..	..	..	Milieu terrestre et atmosphère
Advancement of knowledge	..	..	..	..	..	..	..	Promotion générale des connaissances
Civil space	..	..	..	..	..	..	..	Espace civil
Defence	..	..	..	..	..	..	..	Défense
Not elsewhere classified	..	13 004.0 [a,j]	..	14 346.3 [a]	..	16 226.0	..	Non classé
Total	..	13 004.0 [a,j]	..	14 346.3 [a]	..	16 226.0	..	*Total*
Private non-profit								**Institutions sans but lucratif**
Agriculture, forestry and fishing	..	..	..	..	..	..	..	Agriculture, sylviculture et pêche
Industrial development	..	..	..	11.0	..	10.0 [j]	..	Développement industriel
Energy	..	..	..	..	..	..	..	Energie
Transport and telecommunications	..	..	..	..	..	..	..	Transports et télécommunications
Urban and rural planning	..	..	..	..	..	12.0 [j]	..	Aménagement urbain et rural
Prevention of pollution	..	..	..	..	..	..	..	Prévention de la pollution
Identification & treatment of pollution	..	..	..	..	..	..	..	Détection pollu. & mesures antipollu.
Health	..	..	..	5.9 [j]	..	22.0 [j]	..	Santé
Social development and services	..	..	..	14.8 [j]	..	..	..	Développement et services sociaux
Earth and atmosphere	..	..	..	..	..	..	..	Milieu terrestre et atmosphère
Advancement of knowledge	..	..	..	17.9 [j]	..	15.0 [j]	..	Promotion générale des connaissances
Civil space	..	..	..	..	..	..	..	Espace civil
Defence	..	..	..	..	..	..	..	Défense
Not elsewhere classified	..	93.0 [a]	..	..	..	26.0 [j]	..	Non classé
Total	..	93.0 [a]	..	51.1	..	87.2	..	*Total*
GERD								**DIRD**
Agriculture, forestry and fishing	..	..	..	..	..	..	..	Agriculture, sylviculture et pêche
Industrial development	..	..	..	..	..	..	..	Développement industriel
Energy	..	..	..	..	..	..	..	Energie
Transport and telecommunications	..	..	..	..	..	..	..	Transports et télécommunications
Urban and rural planning	..	..	..	..	..	..	..	Aménagement urbain et rural
Prevention of pollution	..	..	..	..	..	..	..	Prévention de la pollution
Identification & treatment of pollution	..	..	..	..	..	..	..	Détection pollu. & mesures antipollu.
Health	..	..	..	..	..	..	..	Santé
Social development and services	..	..	..	..	..	..	..	Développement et services sociaux
Earth and atmosphere	..	..	..	..	..	..	..	Milieu terrestre et atmosphère
Advancement of knowledge	..	..	..	..	..	..	..	Promotion générale des connaissances
Civil space	..	..	..	..	..	..	..	Espace civil
Defence	..	..	..	..	..	..	..	Défense
Not elsewhere classified	..	59 297.0 [a,m]	..	66 920.7 [m]	..	75 813.5 [m]	..	Non classé
Total	..	59 297.0 [a,m]	..	66 920.7 [m]	..	75 813.5 [m]	..	*Total*

Source: OECD, R&D database (DSTI/EAS Division), January 2002.

Source : OCDE, base de données R-D (DSTI/Division AES), janvier 2002.

Table 5 - Tableau 5

Total R&D personnel - Personnel total de R-D

by sector of employment and occupation - par secteur d'emploi et par profession

In full-time equivalent | *En équivalence plein-temps*

	1994	1995	1996	1997	1998	1999	2000	
Business enterprise								**Entreprises**
RSE	..	19 054 [a]	..	20 924	..	22 822	..	Chercheurs
Technicians	..	..	..	..	..	..	..	Techniciens
Other occupations	..	..	..	..	..	..	..	Autres professions
Total	..	41 636 [a]	..	43 881	..	44 171	..	*Total*
Government								**État**
RSE	..	2 738	..	2 439	..	2 423	..	Chercheurs
Technicians	..	..	..	..	..	..	..	Techniciens
Other occupations	..	..	..	..	..	..	..	Autres professions
Total	..	3 518	..	3 334	..	3 195	..	*Total*
Higher education								**Enseignement supérieur**
RSE	..	11 873	..	13 515	..	14 623	..	Chercheurs
Technicians	..	3 974	..	3 576	..	3 393	..	Techniciens
Other occupations	..	1 454	..	1 107	..	1 159	..	Autres professions
Total	..	17 301	..	18 197	..	19 175	..	*Total*
Private non-profit								**Institutions sans but lucratif**
RSE	..	..	..	..	..	52	..	Chercheurs
Technicians	..	..	..	..	..	..	..	Techniciens
Other occupations	..	..	..	..	..	..	..	Autres professions
Total	..	180 [a]	..	83 [a]	..	133	..	*Total*
TOTAL								**TOTAL**
RSE	..	33 665	..	36 878	..	39 921	..	Chercheurs
Technicians	..	..	..	..	..	..	..	Techniciens
Other occupations	..	..	..	..	..	..	..	Autres professions
Total	..	62 635	..	65 495	..	66 674	..	*Total*

Table 6 - Tableau 6

Total R&D personnel - Personnel total de R-D

by sector of employment and formal qualification - par secteur d'emploi et par niveau de qualification

In full-time equivalent | *En équivalence plein-temps*

	1994	1995	1996	1997	1998	1999	2000	
Business enterprise								**Entreprises**
University PhD level degrees	..	2 592 [a]	..	3 242	..	3 598	..	Diplômes univ. de niv. doctorat
Other university degrees	..	16 462 [a]	..	17 682	..	19 225	..	Autres diplômes universitaires
Sub-total university degrees	..	19 054 [a]	..	20 924	..	22 822	..	Sous-total Universitaires
Other post-secondary diplomas	..	6 331 [a]	..	6 153	..	6 119	..	Autres diplômes post-secondaires
Secondary diplomas	..	12 105 [a]	..	12 013	..	11 188	..	Diplômes de fin d'études secondaires
Other qualifications	..	4 146 [a]	..	4 791	..	4 042	..	Autres qualifications
Total	..	41 636 [a]	..	43 881	..	44 171	..	*Total*
Government								**État**
University PhD level degrees	..	753	..	660	..	684	..	Diplômes univ. de niv. doctorat
Other university degrees	..	1 985	..	1 779	..	1 739	..	Autres diplômes universitaires
Sub-total university degrees	..	2 738	..	2 439	..	2 423	..	Sous-total Universitaires
Other post-secondary diplomas	..	190	..	274	..	197	..	Autres diplômes post-secondaires
Secondary diplomas	..	430	..	437	..	419	..	Diplômes de fin d'études secondaires
Other qualifications	..	159	..	184	..	155	..	Autres qualifications
Total	..	3 518	..	3 334	..	3 195	..	*Total*
Higher education								**Enseignement supérieur**
University PhD level degrees	..	5 313	..	5 450	..	6 070	..	Diplômes univ. de niv. doctorat
Other university degrees	..	5 573	..	5 700	..	6 435	..	Autres diplômes universitaires
Sub-total university degrees	..	10 886 [v]	..	11 150 [v]	..	12 505 [v]	..	Sous-total Universitaires
Other post-secondary diplomas	..	2 350 [v]	..	2 322 [v]	..	2 501 [v]	..	Autres diplômes post-secondaires
Secondary diplomas	..	1 828 [v]	..	1 504 [v]	..	1 337 [v]	..	Diplômes de fin d'études secondaires
Other qualifications	..	320 [v]	..	278 [v]	..	257 [v]	..	Autres qualifications
Total	..	17 301	..	18 197	..	19 175	..	*Total*
Private non-profit								**Institutions sans but lucratif**
University PhD level degrees	..	..	..	..	..	52	..	Diplômes univ. de niv. doctorat
Other university degrees	..	..	..	..	..	61	..	Autres diplômes universitaires
Sub-total university degrees	..	..	..	..	..	113	..	Sous-total Universitaires
Other post-secondary diplomas	..	..	..	..	..	9	..	Autres diplômes post-secondaires
Secondary diplomas	..	..	..	..	..	11	..	Diplômes de fin d'études secondaires
Other qualifications	..	..	..	..	..	..	..	Autres qualifications
Total	..	180 [a]	..	83 [a]	..	133	..	*Total*
TOTAL								**TOTAL**
University PhD level degrees	..	8 658	..	9 352	..	10 404	..	Diplômes univ. de niv. doctorat
Other university degrees	..	24 021	..	25 161	..	27 460	..	Autres diplômes universitaires
Sub-total university degrees	..	32 678 [v]	..	34 513 [v]	..	37 863 [v]	..	Sous-total Universitaires
Other post-secondary diplomas	..	8 871 [v]	..	8 749 [v]	..	8 825 [v]	..	Autres diplômes post-secondaires
Secondary diplomas	..	14 363 [v]	..	13 954 [v]	..	12 954 [v]	..	Diplômes de fin d'études secondaires
Other qualifications	..	4 624 [v]	..	5 253 [v]	..	4 454 [v]	..	Autres qualifications
Total	..	62 635	..	65 495	..	66 674	..	*Total*

Source: OECD, R&D database (DSTI/EAS Division), January 2002.

Source : OCDE, base de données R-D (DSTI/Division AES), janvier 2002.

Table 7 - Tableau 7

Gross domestic expenditure on R&D (GERD) - Dépenses intérieures brutes de R-D (DIRD)

by sector of performance and main scientific field - par secteur d'exécution et par grand domaine scientifique

Million SEK — *Millions SEK*

	1994	1995	1996	1997	1998	1999	2000	
Business enterprise								**Entreprises**
Natural sciences and engineering	..	44 029.0 [a,o]	..	50 151.0 [o]	..	56 953.5 [o]	..	Sciences exactes, nat. et de l'ingénieur
Social sciences and humanities	..	.. [n]	..	.. [n]	..	.. [n]	..	Sciences sociales et humaines
All fields of science	..	44 029.0 [a,m]	..	50 151.0 [m]	..	56 953.5 [m]	..	*Tous domaines scientifiques*
Government								**État**
Natural sciences and engineering	..	1 863.0 [h]	..	..	..	..	..	Sciences exactes, nat. et de l'ingénieur
Social sciences and humanities	..	308.0 [h]	..	..	..	..	..	Sciences sociales et humaines
All fields of science	..	2 171.0 [h]	..	2 372.3 [h]	..	2 547.0 [h,j]	..	*Tous domaines scientifiques*
Higher education								**Enseignement supérieur**
Natural sciences and engineering	..	10 086.0 [a,j,v]	..	11 299.0 [a]	..	12 381.0 [l]	..	Sciences exactes, nat. et de l'ingénieur
Social sciences and humanities	..	2 134.0 [a,j,v]	..	2 620.6 [a]	..	3 171.0 [l]	..	Sciences sociales et humaines
All fields of science	..	13 004.0 [a,j]	..	14 346.3 [a]	..	16 226.0	..	*Tous domaines scientifiques*
Private non-profit								**Institutions sans but lucratif**
Natural sciences and engineering	..	..	..	..	..	..	..	Sciences exactes, nat. et de l'ingénieur
Social sciences and humanities	..	..	..	..	..	..	..	Sciences sociales et humaines
All fields of science	..	93.0 [a]	..	51.1	..	87.2	..	*Tous domaines scientifiques*
TOTAL								**TOTAL**
Natural sciences and engineering	..	55 978.0 [v]	..	..	..	..	..	Sciences exactes, nat. et de l'ingénieur
Social sciences and humanities	..	2 442.0 [v]	..	..	..	..	..	Sciences sociales et humaines
All fields of science	..	59 297.0 [a,m]	..	66 920.7 [m]	..	75 813.5 [m]	..	*Tous domaines scientifiques*

Table 8 - Tableau 8

Total R&D personnel - Personnel total de R-D

by sector of employment and main scientific field - par secteur d'emploi et par grand domaine scientifique

In full-time equivalent — *En équivalence plein-temps*

	1994	1995	1996	1997	1998	1999	2000	
Business enterprise								**Entreprises**
Natural sciences and engineering	..	41 636 [a,o]	..	43 881 [o]	..	44 171 [o]	..	Sciences exactes, nat. et de l'ingénieur
Social sciences and humanities	..	.. [n]	..	.. [n]	..	.. [n]	..	Sciences sociales et humaines
All fields of science	..	41 636 [a]	..	43 881	..	44 171	..	*Tous domaines scientifiques*
Government								**État**
Natural sciences and engineering	..	3 018	..	..	..	..	..	Sciences exactes, nat. et de l'ingénieur
Social sciences and humanities	..	500	..	..	..	..	..	Sciences sociales et humaines
All fields of science	..	3 518	..	3 334	..	3 195	..	*Tous domaines scientifiques*
Higher education								**Enseignement supérieur**
Natural sciences and engineering	..	12 583 [v]	..	12 633 [v]	..	13 186 [v]	..	Sciences exactes, nat. et de l'ingénieur
Social sciences and humanities	..	4 224 [v]	..	4 783 [v]	..	4 657 [v]	..	Sciences sociales et humaines
All fields of science	..	17 301	..	18 197	..	19 175	..	*Tous domaines scientifiques*
Private non-profit								**Institutions sans but lucratif**
Natural sciences and engineering	..	..	..	..	..	..	..	Sciences exactes, nat. et de l'ingénieur
Social sciences and humanities	..	..	..	..	..	..	..	Sciences sociales et humaines
All fields of science	..	180 [a]	..	83 [a]	..	133	..	*Tous domaines scientifiques*
TOTAL								**TOTAL**
Natural sciences and engineering	..	57 237 [v]	..	..	..	..	..	Sciences exactes, nat. et de l'ingénieur
Social sciences and humanities	..	4 724 [v]	..	..	..	..	..	Sciences sociales et humaines
All fields of science	..	62 635	..	65 495	..	66 674	..	*Tous domaines scientifiques*

Source: OECD, R&D database (DSTI/EAS Division), January 2002.

Source : OCDE, base de données R-D (DSTI/Division AES), janvier 2002.

SWEDEN Table 9 - Tableau 9 **SUEDE**

Total business enterprise intramural expenditure on R&D - Dépenses intra-muros totales de R-D des entreprises
(BERD) by industry - (DIRDE) par industrie

9.1 Million SEK - Millions SEK

	1994	1995	1996	1997	1998	1999	2000	
1 AGRICULTURE, HUNTING & FORESTRY	..	126.3	..	143.4	..	224.9	..	**1 AGRICULTURE, CHASSE & SYLVICULTURE**
2 MINING	..	100.0	..	155.7	..	117.9	..	**2 ACTIVITES EXTRACTIVES**
3 MANUFACTURING	..	36 981.8	..	40 528.0	..	46 323.8	..	**3 ACTIVITES DE FABRICATION**
4 Food, beverages & tobacco	..	257.7	..	307.6	..	328.3	..	**4 Produits alimentaires, boissons & tabac**
5 Food products & beverages	..	..	..	..	..	.. [n]	..	5 Produits alimentaires & boissons
6 Tobacco products	..	..	..	..	..	.. [n]	..	6 Produits à base de tabac
7 Textiles, wear app., fur & leather	..	66.6	..	57.6	..	69.2	..	**7 Textiles, habill., fourrures & cuirs**
8 Textiles	..	..	..	..	..	.. [n]	..	8 Textiles
9 Wearing apparel & fur	..	..	..	..	..	.. [n]	..	9 Articles d'habillement & fourrures
10 Leather products & footwear	..	..	..	..	..	.. [n]	..	10 Articles en cuir & chaussures
11 Wood, paper, printing, publishing	..	1 072.9	..	1 425.7	..	781.4	..	**11 Bois, papier, imprimerie, édition**
12 Wood & cork (not furniture)	..	45.0	..	66.5 [o]	..	34.7 [o]	..	12 Bois & liège (sauf meubles)
13 Pulp, paper & paper products	..	1 010.6	..	1 359.2	..	723.4	..	13 Carton, papier & articles
14 Publ., print. & repro. of rec. media	..	17.3	..	.. [n]	..	23.3 [n]	..	14 Edition, impress. & repro. enregis.
15 Coke, petroleum, nuclear fuel, chemicals, rubber plastics	..	7 364.8	..	8 369.5	..	10 573.2	..	**15 Coke, pétrole, combust. nucl., prod. chimi., caout. & mat. plast.**
16 Coke, ref. petrol. prod. & nuclear fuel	..	.. [n]	..	.. [n]	..	.. [n]	..	16 Coke, raffin. pétr. & combust. nucl.
17 Chemicals & chemical products	..	6 930.6 [o]	..	8 195.6 [o]	..	10 341.1 [o]	..	17 Produits chimiques
18 *Chemical prod.(less pharmaceuticals)*	..	761.8 [o]	..	1 016.9 [o]	..	1 273.6 [o]	..	18 *Prod. chimi. (moins prod. pharmaceutiques)*
19 *Pharmaceuticals*	..	6 168.8	..	7 178.7	..	9 067.5	..	19 *Produits pharmaceutiques*
20 Rubber & plastic products	..	434.2	..	173.8	..	232.1	..	20 Caoutchouc et matières plastiques
21 Non-metallic mineral products	..	144.9	..	201.9	..	124.3	..	**21 Produits minéraux non métalliques**
22 Basic metals	..	348.7	..	416.3	..	1 000.0	..	**22 Produits métallurgiques de base**
23 Basic metals, ferrous	..	311.1	..	342.9	..	..	..	23 Produits métallur. de base, ferreux
24 Basic metals, non-ferrous	..	37.6	..	73.4	..	..	..	24 Prod. métallur. de base, non ferreux
25 Fabricated metal products	..	629.0	..	732.4	..	310.3	..	**25 Ouvrages en métaux (sauf mach.& mat.)**
26 Machinery equipment, instruments and transport equipment	..	27 004.0	..	28 930.6	..	33 052.8	..	**26 Machines & matériel, instruments & mat. de transport**
27 Machinery n.e.c.	..	5 129.7	..	5 080.0	..	4 873.3	..	27 Machines, n.c.a.
28 Office, account. & computing machinery	..	989.5	..	770.2	..	1 258.3	..	28 Mach. de bur., compta. & trait. inform.
29 Electrical machinery	..	793.7	..	1 819.1	..	..	..	29 Machines électriques
30 Electro. equip. (radio, TV & commun.)	..	7 373.9	..	8 214.6	..	11 265.9	..	30 Equip. électro. (radio, tv & communic.)
31 *Electro. comp. (inc. semi-conductors)*	..	..	..	..	..	..	..	31 *Compos. électro. (semi-cond. incl.)*
32 *TV, radio & communications equip.*	..	..	..	..	..	..	..	32 *Appareils radio, tv et de communi.*
33 Instruments	..	3 188.8	..	2 659.6	..	4 118.6	..	33 Instruments
34 Motor vehicles	..	9 528.4 [a,o]	..	10 387.1 [o]	..	11 536.7 [o]	..	34 Véhicules automobiles
35 Other transport equipment	..	.. [n]	..	.. [n]	..	.. [n]	..	35 Autres matériels de transport
36 *Ships*	..	.. [n]	..	.. [n]	..	.. [n]	..	36 *Navires*
37 *Aerospace*	..	.. [n]	..	.. [n]	..	.. [n]	..	37 *Aérospatiale*
38 *Other transport equipment n.e.c.*	..	.. [n]	..	.. [n]	..	.. [n]	..	38 *Autres matériels de transport n.c.a.*
39 Furniture, other manufacturing n.e.c.	..	93.1 [o]	..	86.5 [o]	..	84.2 [o]	..	**39 Meubles, autres activ. de fabricat. n.c.a.**
40 Furniture	..	60.2	..	..	..	.. [n]	..	40 Meubles
41 Other manufacturing n.e.c.	..	32.9 [o]	..	..	..	.. [n]	..	41 Autres activités de fabrication n.c.a.
42 Recycling	..	.. [n]	..	.. [n]	..	.. [n]	..	**42 Récupération**
43 ELECTRICITY, GAS & WATER SUPPLY	..	.. [n]	..	527.9 [o]	..	405.9 [o]	..	**43 DISTRIBUTION D'ELECTRICITE, GAZ & EAU**
44 CONSTRUCTION	..	.. [n]	..	.. [n]	..	.. [n]	..	**44 CONSTRUCTION**
45 SERVICES SECTOR	..	6 820.9 [a,o]	..	8 796.1 [a]	..	9 881.0	..	**45 SECTEUR DES SERVICES**
46 Wholesale, retail trade & repair of motor vehicles, etc.	..	495.5 [a,o]	..	594.3 [o]	..	542.8 [o]	..	**46 Commerce (gros & détail), réparation de véhicules, etc.**
47 Hotels & restaurants	..	.. [n]	..	.. [n]	..	.. [n]	..	**47 Hôtels et restaurants**
48 Transport & storage	..	94.7	..	79.2	..	32.2	..	**48 Transports, entreposage**
49 Communications	..	1 068.9 [o]	..	996.1 [a]	..	.. [n]	..	**49 Communications**
50 Post	..	..	..	..	..	..	..	50 Postes
51 Telecommunications	..	..	..	..	..	..	..	51 Télécommunications
52 Financial intermediation (inc. insur.)	..	..	..	..	..	..	..	**52 Intermédiation financ. (assurances inc.)**
53 Real estate, renting & busin. activity	..	5 085.5	..	7 068.7	..	9 237.0	..	**53 Immobil., locat. & serv. aux entreprises**
54 Computer & related activities	..	870.8	..	1 624.3	..	3 644.8 [o]	..	54 Informatique et activités connexes
55 *Software consultancy*	..	..	..	..	..	..	..	55 *Production de logiciels*
56 *Other computer services n.e.c.*	..	..	..	..	..	..	..	56 *Autres activités informatiques*
57 Research & development	..	3 314.3	..	4 870.0	..	4 861.3	..	57 Recherche & développement expérimental
58 Other business activities n.e.c.	..	900.4	..	574.4	..	730.9	..	58 Autres activ. & serv. aux entreprises
59 Comm., soc. & pers. serv. activ., etc.	..	76.2	..	57.8	..	69.0	..	**59 Adm. pub., serv. soc. & collec., etc**
TOTAL BUSINESS ENTERPRISE SECTOR	..	**44 029.0** [a]	..	**50 151.0**	..	**56 953.5**	..	**TOTAL SECTEUR DES ENTREPRISES**

Source: OECD, R&D database (DSTI/EAS Division), January 2002.

Source : OCDE, base de données R-D (DSTI/Division AES), janvier 2002.

SWEDEN Table 9 - Tableau 9 **SUEDE**

Total business enterprise intramural expenditure on R&D - Dépenses intra-muros totales de R-D des entreprises

(BERD) by industry - (DIRDE) par industrie

9.2 Million constant USD(1995 prices and PPPs) - Millions de USD constants (prix et PPA de 1995)

	1994	1995	1996	1997	1998	1999	2000	
1 AGRICULTURE, HUNTING & FORESTRY	..	13.0	..	14.3	..	22.1	..	1 AGRICULTURE, CHASSE & SYLVICULTURE
2 MINING	..	10.3	..	15.5	..	11.6	..	2 ACTIVITES EXTRACTIVES
3 MANUFACTURING	..	3 801.5	..	4 037.8	..	4 551.1	..	3 ACTIVITES DE FABRICATION
4 Food, beverages & tobacco	..	26.5	..	30.6	..	32.3	..	**4 Produits alimentaires, boissons & tabac**
5 Food products & beverages	..	..	..	..	..	.. [n]	..	5 Produits alimentaires & boissons
6 Tobacco products	..	..	..	..	..	.. [n]	..	6 Produits à base de tabac
7 Textiles, wear app., fur & leather	..	6.9	..	5.7	..	6.8	..	**7 Textiles, habill., fourrures & cuirs**
8 Textiles	..	..	..	..	..	.. [n]	..	8 Textiles
9 Wearing apparel & fur	..	..	..	..	..	.. [n]	..	9 Articles d'habillement & fourrures
10 Leather products & footwear	..	..	..	..	..	.. [n]	..	10 Articles en cuir & chaussures
11 Wood, paper, printing, publishing	..	110.3	..	142.0	..	76.8	..	**11 Bois, papier, imprimerie, édition**
12 Wood & cork (not furniture)	..	4.6	..	6.6 [o]	..	3.4 [o]	..	12 Bois & liège (sauf meubles)
13 Pulp, paper & paper products	..	103.9	..	135.4	..	71.1	..	13 Carton, papier & articles
14 Publ., print. & repro. of rec. media	..	1.8	..	.. [n]	..	2.3 [n]	..	14 Edition, impress. & repro. enregis.
15 Coke, petroleum, nuclear fuel, chemicals, rubber plastics	..	757.1	..	833.8	..	1 038.8	..	**15 Coke, pétrole, combust. nucl., prod. chimi., caout. & mat. plast.**
16 Coke, ref. petrol. prod. & nuclear fuel	..	.. [n]	..	.. [n]	..	.. [n]	..	16 Coke, raffin. pétr. & combust. nucl.
17 Chemicals & chemical products	..	712.4 [o]	..	816.5 [o]	..	1 016.0 [o]	..	17 Produits chimiques
18 *Chemical prod.(less pharmaceuticals)*	..	78.3 [o]	..	101.3 [o]	..	125.1 [o]	..	18 *Prod. chimi. (moins prod. pharmaceutiques)*
19 *Pharmaceuticals*	..	634.1	..	715.2	..	890.8	..	19 *Produits pharmaceutiques*
20 Rubber & plastic products	..	44.6	..	17.3	..	22.8	..	20 Caoutchouc et matières plastiques
21 Non-metallic mineral products	..	14.9	..	20.1	..	12.2	..	**21 Produits minéraux non métalliques**
22 Basic metals	..	35.8	..	41.5	..	98.3	..	**22 Produits métallurgiques de base**
23 Basic metals, ferrous	..	32.0	..	34.2	..	..	..	23 Produits métallur. de base, ferreux
24 Basic metals, non-ferrous	..	3.9	..	7.3	..	..	..	24 Prod. métallur. de base, non ferreux
25 Fabricated metal products	..	64.7	..	73.0	..	30.5	..	**25 Ouvrages en métaux (sauf mach.& mat.)**
26 Machinery equipment, instruments and transport equipment	..	2 775.9	..	2 882.3	..	3 247.3	..	**26 Machines & matériel, instruments & mat. de transport**
27 Machinery n.e.c.	..	527.3	..	506.1	..	478.8	..	27 Machines, n.c.a.
28 Office, account. & computing machinery	..	101.7	..	76.7	..	123.6	..	28 Mach. de bur., compta. & trait. inform.
29 Electrical machinery	..	81.6	..	181.2	..	..	..	29 Machines électriques
30 Electro. equip. (radio, TV & commun.)	..	758.0	..	818.4	..	1 106.8	..	30 Equip. électro. (radio, tv & communic.)
31 *Electro. comp. (inc. semi-conductors)*	..	..	..	..	..	..	..	31 *Compos. électro. (semi-cond. incl.)*
32 *TV, radio & communications equip.*	..	..	..	..	..	..	..	32 *Appareils radio, tv et de communi.*
33 Instruments	..	327.8	..	265.0	..	404.6	..	33 Instruments
34 Motor vehicles	..	979.5 [a,o]	..	1 034.9 [o]	..	1 133.4 [o]	..	34 Véhicules automobiles
35 Other transport equipment	..	.. [n]	..	.. [n]	..	.. [n]	..	35 Autres matériels de transport
36 *Ships*	..	.. [n]	..	.. [n]	..	.. [n]	..	36 *Navires*
37 *Aerospace*	..	.. [n]	..	.. [n]	..	.. [n]	..	37 *Aérospatiale*
38 *Other transport equipment n.e.c.*	..	.. [n]	..	.. [n]	..	.. [n]	..	38 *Autres matériels de transport n.c.a.*
39 Furniture, other manufacturing n.e.c.	..	9.6 [o]	..	8.6 [o]	..	8.3 [o]	..	**39 Meubles, autres activ. de fabricat. n.c.a.**
40 Furniture	..	6.2	..	..	..	.. [n]	..	40 Meubles
41 Other manufacturing n.e.c.	..	3.4 [o]	..	..	..	.. [n]	..	41 Autres activités de fabrication n.c.a.
42 Recycling	..	.. [n]	..	.. [n]	..	.. [n]	..	**42 Récupération**
43 ELECTRICITY, GAS & WATER SUPPLY	..	.. [n]	..	52.6 [o]	..	39.9 [o]	..	43 DISTRIBUTION D'ELECTRICITE, GAZ & EAU
44 CONSTRUCTION	..	.. [n]	..	.. [n]	..	.. [n]	..	44 CONSTRUCTION
45 SERVICES SECTOR	..	701.2 [a,o]	..	876.3 [a]	..	970.8	..	45 SECTEUR DES SERVICES
46 Wholesale, retail trade & repair of motor vehicles, etc.	..	50.9 [a,o]	..	59.2 [o]	..	53.3 [o]	..	**46 Commerce (gros & détail), réparation de véhicules, etc.**
47 Hotels & restaurants	..	.. [n]	..	.. [n]	..	.. [n]	..	**47 Hôtels et restaurants**
48 Transport & storage	..	9.7	..	7.9	..	3.2	..	**48 Transports, entreposage**
49 Communications	..	109.9 [o]	..	99.2 [a]	..	.. [n]	..	**49 Communications**
50 Post	..	..	..	..	..	..	..	50 Postes
51 Telecommunications	..	..	..	..	..	..	..	51 Télécommunications
52 Financial intermediation (inc. insur.)	..	..	..	..	..	..	..	**52 Intermédiation financ. (assurances inc.)**
53 Real estate, renting & busin. activity	..	522.8	..	704.2	..	907.5	..	**53 Immobil., locat. & serv. aux entreprises**
54 Computer & related activities	..	89.5	..	161.8	..	358.1 [o]	..	54 Informatique et activités connexes
55 *Software consultancy*	..	..	..	..	..	..	..	55 *Production de logiciels*
56 *Other computer services n.e.c.*	..	..	..	..	..	..	..	56 *Autres activités informatiques*
57 Research & development	..	340.7	..	485.2	..	477.6	..	57 Recherche & développement expérimental
58 Other business activities n.e.c.	..	92.6	..	57.2	..	71.8	..	58 Autres activ. & serv. aux entreprises
59 Comm., soc. & pers. serv. activ., etc.	..	7.8	..	5.8	..	6.8	..	**59 Adm. pub., serv. soc. & collec., etc**
TOTAL BUSINESS ENTERPRISE SECTOR	..	**4 526.0** [a]	..	**4 996.5**	..	**5 595.5**	..	**TOTAL SECTEUR DES ENTREPRISES**

Source: OECD, R&D database (DSTI/EAS Division), January 2002.

Source : OCDE, base de données R-D (DSTI/Division AES), janvier 2002.

Business enterprise intramural expenditure on R&D (BERD) - Dépenses intra-muros de R-D des entreprises (DIRDE)

by industry and source of funds - par industrie et par source de financement

Million SEK | Millions SEK

10.1 Business enterprise - Entreprises

	1994	1995	1996	1997	1998	1999	2000	
1 **AGRICULTURE, HUNTING & FORESTRY**	..	103.3	..	116.9	..	200.2	..	1 **AGRICULTURE, CHASSE & SYLVICULTURE**
2 **MINING**	..	99.1	..	153.2	..	113.6	..	2 **ACTIVITES EXTRACTIVES**
3 **MANUFACTURING**	..	32 148.0	..	36 669.3	..	41 901.6	..	3 **ACTIVITES DE FABRICATION**
4 **Food, beverages & tobacco**	..	252.2	..	302.3	..	322.2	..	4 **Produits alimentaires, boissons & tabac**
5 Food products & beverages	..	..	..	..	..	..[n]	..	5 Produits alimentaires & boissons
6 Tobacco products	..	..	..	..	..	..[n]	..	6 Produits à base de tabac
7 **Textiles, wear app., fur & leather**	..	66.6	..	57.6	..	68.7	..	7 **Textiles, habill., fourrures & cuirs**
8 Textiles	..	..	..	..	..	0.0[n]	..	8 Textiles
9 Wearing apparel & fur	..	..	..	..	..	0.0[n]	..	9 Articles d'habillement & fourrures
10 Leather products & footwear	..	..	..	..	..	0.0[n]	..	10 Articles en cuir & chaussures
11 **Wood, paper, printing, publishing**	..	976.8	..	1 413.4	..	760.8	..	11 **Bois, papier, imprimerie, édition**
12 Wood & cork (not furniture)	..	43.9	..	65.9[o]	..	33.3	..	12 Bois & liège (sauf meubles)
13 Pulp, paper & paper products	..	916.0	..	1 347.5	..	704.2	..	13 Carton, papier & articles
14 Publ., print. & repro. of rec. media	..	16.9	..	..[n]	..	23.3	..	14 Edition, impress. & repro. enregis.
15 **Coke, petroleum, nuclear fuel, chemicals, rubber plastics**	..	7 259.0	..	8 309.3	..	10 415.7	..	15 **Coke, pétrole, combust. nucl., prod. chimi., caout. & mat. plast.**
16 Coke, ref. petrol. prod. & nuclear fuel	..	..[n]	..	..[n]	..	0.0[n]	..	16 Coke, raffin. pétr. & combust. nucl.
17 Chemicals & chemical products	..	6 825.3[o]	..	8 135.8[o]	..	10 189.5[o]	..	17 Produits chimiques
18 *Chemical prod.(less pharmaceuticals)*	..	656.4[o]	..	957.1[o]	..	1 163.0[o]	..	18 *Prod. chimi. (moins prod. pharmaceutiques)*
19 *Pharmaceuticals*	..	6 168.8	..	7 178.7	..	9 026.5	..	19 *Produits pharmaceutiques*
20 Rubber & plastic products	..	433.7	..	173.5	..	226.3	..	20 Caoutchouc et matières plastiques
21 **Non-metallic mineral products**	..	142.9	..	198.7	..	123.3	..	21 **Produits minéraux non métalliques**
22 **Basic metals**	..	344.5	..	409.2	..	994.6	..	22 **Produits métallurgiques de base**
23 Basic metals, ferrous	..	308.7	..	335.8	..	..	..	23 Produits métallur. de base, ferreux
24 Basic metals, non-ferrous	..	35.8	..	73.4	..	..	..	24 Prod. métallur. de base, non ferreux
25 **Fabricated metal products**	..	619.7	..	727.5	..	253.9	..	25 **Ouvrages en métaux (sauf mach.& mat.)**
26 **Machinery equipment, instruments and transport equipment**	..	22 393.1	..	25 164.8	..	28 878.1	..	26 **Machines & matériel, instruments & mat. de transport**
27 Machinery n.e.c.	..	4 588.3	..	4 546.9	..	4 131.5	..	27 Machines, n.c.a.
28 Office, account. & computing machinery	..	289.7	..	240.6	..	885.8[o]	..	28 Mach. de bur., compta. & trait. inform.
29 Electrical machinery	..	758.9	..	1 655.0	..	0.0[n]	..	29 Machines électriques
30 Electro. equip. (radio, TV & commun.)	..	6 948.4	..	7 768.5	..	11 257.2	..	30 Equip. électro. (radio, tv & communic.)
31 *Electro. comp. (inc. semi-conductors)*	..	..	..	..	..	..	..	31 *Compos. électro. (semi-cond. incl.)*
32 *TV, radio & communications equip.*	..	..	..	..	..	..	..	32 *Appareils radio, tv et de communi.*
33 Instruments	..	1 748.8	..	1 980.8	..	3 025.0[o]	..	33 Instruments
34 Motor vehicles	..	8 059.1[a,o]	..	8 973.0[o]	..	9 578.5[n]	..	34 Véhicules automobiles
35 Other transport equipment	..	..[n]	..	..[n]	..	..[n]	..	35 Autres matériels de transport
36 *Ships*	..	..[n]	..	..[n]	..	..[n]	..	36 *Navires*
37 *Aerospace*	..	..[n]	..	..[n]	..	..[n]	..	37 *Aérospatiale*
38 *Other transport equipment n.e.c.*	..	..[n]	..	..[n]	..	..[n]	..	38 *Autres matériels de transport n.c.a.*
39 **Furniture, other manufacturing n.e.c.**	..	93.1[o]	..	86.5[o]	..	84.2[o]	..	39 **Meubles, autres activ. de fabricat. n.c.a.**
40 Furniture	..	60.2	..	..[n]	..	..[n]	..	40 Meubles
41 Other manufacturing n.e.c.	..	32.9[o]	..	..[n]	..	..[n]	..	41 Autres activités de fabrication n.c.a.
42 **Recycling**	..	..[n]	..	..[n]	..	..[n]	..	42 **Récupération**
43 **ELECTRICITY, GAS & WATER SUPPLY**	..	..[n]	..	480.9[o]	..	352.0[o]	..	43 **DISTRIBUTION D'ELECTRICITE, GAZ & EAU**
44 **CONSTRUCTION**	..	..[n]	..	..[n]	..	..[n]	..	44 **CONSTRUCTION**
45 **SERVICES SECTOR**	..	5 844.2[a,o]	..	7 208.4[a]	..	8 066.7	..	45 **SECTEUR DES SERVICES**
46 **Wholesale, retail trade & repair of motor vehicles, etc.**	..	474.9[a,o]	..	324.5[o]	..	247.5[o]	..	46 **Commerce (gros & détail), réparation de véhicules, etc.**
47 **Hotels & restaurants**	..	..[n]	..	..[n]	..	..[n]	..	47 **Hôtels et restaurants**
48 **Transport & storage**	..	94.7	..	67.1	..	25.8	..	48 **Transports, entreposage**
49 **Communications**	..	1 046.6[o]	..	992.7[a]	..	..[n]	..	49 **Communications**
50 Post	..	..	..	..	..	..	..	50 Postes
51 Telecommunications	..	..	..	..	..	..	..	51 Télécommunications
52 **Financial intermediation (inc. insur.)**	..	..	..	..	..	..	..	52 **Intermédiation financ. (assurances inc.)**
53 **Real estate, renting & busin. activity**	..	4 168.0	..	5 781.9	..	7 743.1	..	53 **Immobil., locat. & serv. aux entreprises**
54 Computer & related activities	..	778.1	..	1 602.8	..	3 418.0[o]	..	54 Informatique et activités connexes
55 *Software consultancy*	..	..	..	..	..	..	..	55 *Production de logiciels*
56 *Other computer services n.e.c.*	..	..	..	..	..	..	..	56 *Autres activités informatiques*
57 Research & development	..	2 606.7	..	3 789.8	..	3 834.7	..	57 Recherche & développement expérimental
58 Other business activities n.e.c.	..	783.3	..	389.3	..	490.4	..	58 Autres activ. & serv. aux entreprises
59 **Comm., soc. & pers. serv. activ., etc.**	..	60.0	..	42.2	..	50.3	..	59 **Adm. pub., serv. soc. & collec., etc**
TOTAL BUSINESS ENTERPRISE SECTOR	..	**38 194.5**[m]	..	**44 628.7**[m]	..	**50 634.2**[m]	..	**TOTAL SECTEUR DES ENTREPRISES**

Source: OECD, R&D database (DSTI/EAS Division), January 2002.

Source : OCDE, base de données R-D (DSTI/Division AES), janvier 2002.

Business enterprise intramural expenditure on R&D (BERD) - Dépenses intra-muros de R-D des entreprises (DIRDE)

by industry and source of funds - par industrie et par source de financement

Million SEK *Millions SEK*

10.2 Government - État

	1994	1995	1996	1997	1998	1999	2000	
1 AGRICULTURE, HUNTING & FORESTRY	..	23.0	..	23.7	..	17.8	..	**1 AGRICULTURE, CHASSE & SYLVICULTURE**
2 MINING	..	0.8	..	0.2	..	0.0	..	**2 ACTIVITES EXTRACTIVES**
3 MANUFACTURING	..	3 570.3	..	3 131.8	..	3 472.8	..	**3 ACTIVITES DE FABRICATION**
4 Food, beverages & tobacco	..	0.8	..	0.6	..	1.2	..	**4 Produits alimentaires, boissons & tabac**
5 Food products & beverages	..	..	..	..	..	.. [n]	..	5 Produits alimentaires & boissons
6 Tobacco products	..	..	..	..	..	.. [n]	..	6 Produits à base de tabac
7 Textiles, wear app., fur & leather	..	..	..	0.0	..	0.5	..	**7 Textiles, habill., fourrures & cuirs**
8 Textiles	..	..	..	..	..	.. [n]	..	8 Textiles
9 Wearing apparel & fur	..	..	..	..	..	.. [n]	..	9 Articles d'habillement & fourrures
10 Leather products & footwear	..	..	..	..	..	.. [n]	..	10 Articles en cuir & chaussures
11 Wood, paper, printing, publishing	..	3.4	..	2.5	..	2.1	..	**11 Bois, papier, imprimerie, édition**
12 Wood & cork (not furniture)	..	1.1	..	0.4 [o]	..	0.7	..	12 Bois & liège (sauf meubles)
13 Pulp, paper & paper products	..	2.0	..	2.1	..	1.4	..	13 Carton, papier & articles
14 Publ., print. & repro. of rec. media	..	0.4	..	.. [n]	..	0.0	..	14 Edition, impress. & repro. enregis.
15 Coke, petroleum, nuclear fuel, chemicals, rubber plastics	..	11.2	..	6.0	..	9.4	..	**15 Coke, pétrole, combust. nucl., prod. chimi., caout. & mat. plast.**
16 Coke, ref. petrol. prod. & nuclear fuel	..	.. [n]	..	.. [n]	..	.. [n]	..	16 Coke, raffin. pétr. & combust. nucl.
17 Chemicals & chemical products	..	11.1 [o]	..	6.0 [o]	..	4.0 [o]	..	17 Produits chimiques
18 *Chemical prod.(less pharmaceuticals)*	..	11.1 [o]	..	6.0 [o]	..	4.0 [o]	..	18 *Prod. chimi. (moins prod. pharmaceutiques)*
19 *Pharmaceuticals*	..	..	..	0.0	..	0.0	..	19 *Produits pharmaceutiques*
20 Rubber & plastic products	..	0.2	..	0.0	..	5.4	..	20 Caoutchouc et matières plastiques
21 Non-metallic mineral products	..	1.8	..	3.2	..	0.4	..	**21 Produits minéraux non métalliques**
22 Basic metals	..	2.7	..	2.5	..	0.2	..	**22 Produits métallurgiques de base**
23 Basic metals, ferrous	..	1.3	..	2.5	..	..	..	23 Produits métallur. de base, ferreux
24 Basic metals, non-ferrous	..	1.4	..	0.0	..	..	..	24 Prod. métallur. de base, non ferreux
25 Fabricated metal products	..	9.3	..	4.7	..	40.5	..	**25 Ouvrages en métaux (sauf mach.& mat.)**
26 Machinery equipment, instruments and transport equipment	..	3 541.1	..	3 112.3	..	3 418.5	..	**26 Machines & matériel, instruments & mat. de transport**
27 Machinery n.e.c.	..	473.4	..	479.8	..	522.8	..	27 Machines, n.c.a.
28 Office, account. & computing machinery	..	302.3	..	382.6	..	38.2 [o]	..	28 Mach. de bur., compta. & trait. inform.
29 Electrical machinery	..	..	..	0.0	..	.. [n]	..	29 Machines électriques
30 Electro. equip. (radio, TV & commun.)	..	2.2	..	363.7	..	1.1	..	30 Equip. électro. (radio, tv & communic.)
31 *Electro. comp. (inc. semi-conductors)*	..	..	..	..	..	..	..	31 *Compos. électro. (semi-cond. incl.)*
32 *TV, radio & communications equip.*	..	..	..	..	..	..	..	32 *Appareils radio, tv et de communi.*
33 Instruments	..	1 336.3	..	676.1	..	1 089.1 [o]	..	33 Instruments
34 Motor vehicles	..	1 426.9 [a,o]	..	1 210.1 [o]	..	1 767.3 [n]	..	34 Véhicules automobiles
35 Other transport equipment	..	.. [n]	..	.. [n]	..	.. [n]	..	35 Autres matériels de transport
36 *Ships*	..	.. [n]	..	.. [n]	..	.. [n]	..	36 *Navires*
37 *Aerospace*	..	.. [n]	..	.. [n]	..	.. [n]	..	37 *Aérospatiale*
38 *Other transport equipment n.e.c.*	..	.. [n]	..	.. [n]	..	.. [n]	..	38 *Autres matériels de transport n.c.a.*
39 Furniture, other manufacturing n.e.c.	..	..	..	0.0	..	0.0 [o]	..	**39 Meubles, autres activ. de fabricat. n.c.a.**
40 Furniture	..	..	..	..	..	.. [n]	..	40 Meubles
41 Other manufacturing n.e.c.	..	..	..	..	..	.. [n]	..	41 Autres activités de fabrication n.c.a.
42 Recycling	..	..	..	..	..	.. [n]	..	**42 Récupération**
43 ELECTRICITY, GAS & WATER SUPPLY	..	.. [n]	..	33.6 [o]	..	36.0 [o]	..	**43 DISTRIBUTION D'ELECTRICITE, GAZ & EAU**
44 CONSTRUCTION	..	.. [n]	..	.. [n]	..	.. [n]	..	**44 CONSTRUCTION**
45 SERVICES SECTOR	..	573.1 [a,o]	..	637.1 [a]	..	913.3	..	**45 SECTEUR DES SERVICES**
46 Wholesale, retail trade & repair of motor vehicles, etc.	..	1.8 [a,o]	..	0.0 [o]	..	5.5 [o]	..	**46 Commerce (gros & détail), réparation de véhicules, etc.**
47 Hotels & restaurants	..	.. [n]	..	.. [n]	..	.. [n]	..	**47 Hôtels et restaurants**
48 Transport & storage	..	..	..	1.5	..	0.0	..	**48 Transports, entreposage**
49 Communications	..	22.3 [o]	..	0.0 [a]	..	.. [n]	..	**49 Communications**
50 Post	..	..	..	..	..	..	..	50 Postes
51 Telecommunications	..	..	..	..	..	..	..	51 Télécommunications
52 Financial intermediation (inc. insur.)	..	..	..	..	..	..	..	**52 Intermédiation financ. (assurances inc.)**
53 Real estate, renting & busin. activity	..	536.8	..	624.3	..	893.6	..	**53 Immobil., locat. & serv. aux entreprises**
54 Computer & related activities	..	10.8	..	11.5	..	214.3 [o]	..	54 Informatique et activités connexes
55 *Software consultancy*	..	..	..	..	..	..	..	55 *Production de logiciels*
56 *Other computer services n.e.c.*	..	..	..	..	..	..	..	56 *Autres activités informatiques*
57 Research & development	..	443.1	..	519.9	..	553.0	..	57 Recherche & développement expérimental
58 Other business activities n.e.c.	..	82.9	..	92.9	..	126.2	..	58 Autres activ. & serv. aux entreprises
59 Comm., soc. & pers. serv. activ., etc.	..	12.2	..	11.2	..	14.2	..	**59 Adm. pub., serv. soc. & collec., etc**
TOTAL BUSINESS ENTERPRISE SECTOR	..	**4 166.8** [a,m]	..	**3 826.5** [a,m]	..	**4 439.8** [m]	..	**TOTAL SECTEUR DES ENTREPRISES**

Source: OECD, R&D database (DSTI/EAS Division), January 2002.

Source : OCDE, base de données R-D (DSTI/Division AES), janvier 2002.

Business enterprise intramural expenditure on R&D (BERD) - Dépenses intra-muros de R-D des entreprises (DIRDE)

by industry and source of funds - par industrie et par source de financement

Million SEK — *Millions SEK*

10.3 Other national funds - Autres sources nationales

	1994	1995	1996	1997	1998	1999	2000	
1 AGRICULTURE, HUNTING & FORESTRY	..	..	..	0.1	..	0.0	..	**1 AGRICULTURE, CHASSE & SYLVICULTURE**
2 MINING	..	..	..	0.0	..	0.0	..	**2 ACTIVITES EXTRACTIVES**
3 MANUFACTURING	..	4.6	..	5.1	..	4.9	..	**3 ACTIVITES DE FABRICATION**
4 Food, beverages & tobacco	..	4.6	..	4.4	..	4.9	..	**4 Produits alimentaires, boissons & tabac**
5 Food products & beverages	..	..	..	..	..	.. [n]	..	5 Produits alimentaires & boissons
6 Tobacco products	..	..	..	..	..	.. [n]	..	6 Produits à base de tabac
7 Textiles, wear app., fur & leather	..	..	..	0.0	..	0.0	..	**7 Textiles, habill., fourrures & cuirs**
8 Textiles	..	..	..	..	..	.. [n]	..	8 Textiles
9 Wearing apparel & fur	..	..	..	..	..	.. [n]	..	9 Articles d'habillement & fourrures
10 Leather products & footwear	..	..	..	..	..	.. [n]	..	10 Articles en cuir & chaussures
11 Wood, paper, printing, publishing	..	..	..	0.2	..	0.0	..	**11 Bois, papier, imprimerie, édition**
12 Wood & cork (not furniture)	..	..	..	0.0	..	0.0	..	12 Bois & liège (sauf meubles)
13 Pulp, paper & paper products	..	..	..	0.2	..	0.0	..	13 Carton, papier & articles
14 Publ., print. & repro. of rec. media	..	..	..	..	..	0.0	..	14 Edition, impress. & repro. enregis.
15 Coke, petroleum, nuclear fuel, chemicals, rubber plastics	..	..	..	0.0	..	0.0	..	**15 Coke, pétrole, combust. nucl., prod. chimi., caout. & mat. plast.**
16 Coke, ref. petrol. prod. & nuclear fuel	..	..	..	..	..	.. [n]	..	16 Coke, raffin. pétr. & combust. nucl.
17 Chemicals & chemical products	..	..	..	0.0	..	0.0 [o]	..	17 Produits chimiques
18 *Chemical prod.(less pharmaceuticals)*	..	..	..	0.0	..	0.0 [o]	..	18 *Prod. chimi. (moins prod. pharmaceutiques)*
19 *Pharmaceuticals*	..	..	..	0.0	..	0.0	..	19 *Produits pharmaceutiques*
20 Rubber & plastic products	..	..	..	0.0	..	0.0	..	20 Caoutchouc et matières plastiques
21 Non-metallic mineral products	..	..	..	0.0	..	0.0	..	**21 Produits minéraux non métalliques**
22 Basic metals	..	..	..	0.0	..	0.0	..	**22 Produits métallurgiques de base**
23 Basic metals, ferrous	..	..	..	0.0	..	..	..	23 Produits métallur. de base, ferreux
24 Basic metals, non-ferrous	..	..	..	0.0	..	..	..	24 Prod. métallur. de base, non ferreux
25 Fabricated metal products	..	..	..	0.0	..	0.0	..	**25 Ouvrages en métaux (sauf mach.& mat.)**
26 Machinery equipment, instruments and transport equipment	..	..	..	0.5	..	0.0	..	**26 Machines & matériel, instruments & mat. de transport**
27 Machinery n.e.c.	..	..	..	0.4	..	0.0	..	27 Machines, n.c.a.
28 Office, account. & computing machinery	..	..	..	0.0	..	0.0 [o]	..	28 Mach. de bur., compta. & trait. inform.
29 Electrical machinery	..	..	..	0.0	..	.. [n]	..	29 Machines électriques
30 Electro. equip. (radio, TV & commun.)	..	..	..	0.1	..	0.0	..	30 Equip. électro. (radio, tv & communic.)
31 *Electro. comp. (inc. semi-conductors)*	..	..	..	..	..	..	..	31 *Compos. électro. (semi-cond. incl.)*
32 *TV, radio & communications equip.*	..	..	..	..	..	..	..	32 *Appareils radio, tv et de communi.*
33 Instruments	..	..	..	0.0	..	0.0 [o]	..	33 Instruments
34 Motor vehicles	..	..	..	0.0	..	0.0 [n]	..	34 Véhicules automobiles
35 Other transport equipment	..	..	..	..	..	.. [n]	..	35 Autres matériels de transport
36 *Ships*	..	..	..	..	..	.. [n]	..	36 *Navires*
37 *Aerospace*	..	..	..	..	..	.. [n]	..	37 *Aérospatiale*
38 *Other transport equipment n.e.c.*	..	..	..	..	..	.. [n]	..	38 *Autres matériels de transport n.c.a.*
39 Furniture, other manufacturing n.e.c.	..	..	..	0.0	..	0.0 [o]	..	**39 Meubles, autres activ. de fabricat. n.c.a.**
40 Furniture	..	..	..	..	..	.. [n]	..	40 Meubles
41 Other manufacturing n.e.c.	..	..	..	..	..	.. [n]	..	41 Autres activités de fabrication n.c.a.
42 Recycling	..	..	..	..	..	.. [n]	..	**42 Récupération**
43 ELECTRICITY, GAS & WATER SUPPLY	..	..	..	0.0	..	0.0 [o]	..	**43 DISTRIBUTION D'ELECTRICITE, GAZ & EAU**
44 CONSTRUCTION	..	..	..	..	..	.. [n]	..	**44 CONSTRUCTION**
45 SERVICES SECTOR	..	27.2 [a]	..	28.5	..	31.6	..	**45 SECTEUR DES SERVICES**
46 Wholesale, retail trade & repair of motor vehicles, etc.	..	..	..	0.0 [o]	..	0.0 [o]	..	**46 Commerce (gros & détail), réparation de véhicules, etc.**
47 Hotels & restaurants	..	..	..	.. [n]	..	.. [n]	..	**47 Hôtels et restaurants**
48 Transport & storage	..	..	..	0.0	..	0.0	..	**48 Transports, entreposage**
49 Communications	..	..	..	0.0	..	.. [n]	..	**49 Communications**
50 Post	..	..	..	..	..	..	..	50 Postes
51 Telecommunications	..	..	..	..	..	..	..	51 Télécommunications
52 Financial intermediation (inc. insur.)	..	..	..	..	..	..	..	**52 Intermédiation financ. (assurances inc.)**
53 Real estate, renting & busin. activity	..	25.1	..	26.4	..	31.1	..	**53 Immobil., locat. & serv. aux entreprises**
54 Computer & related activities	..	..	..	0.0	..	0.0 [o]	..	54 Informatique et activités connexes
55 *Software consultancy*	..	..	..	..	..	..	..	55 *Production de logiciels*
56 *Other computer services n.e.c.*	..	..	..	..	..	..	..	56 *Autres activités informatiques*
57 Research & development	..	14.8	..	24.1	..	31.1	..	57 Recherche & développement expérimental
58 Other business activities n.e.c.	..	10.3	..	2.3	..	0.0	..	58 Autres activ. & serv. aux entreprises
59 Comm., soc. & pers. serv. activ., etc.	..	2.2	..	2.1	..	0.5	..	**59 Adm. pub., serv. soc. & collec., etc**
TOTAL BUSINESS ENTERPRISE SECTOR	..	**31.9** [a,m]	..	**33.7** [a,m]	..	**36.5** [m]	..	**TOTAL SECTEUR DES ENTREPRISES**

Source: OECD, R&D database (DSTI/EAS Division), January 2002.

Source : OCDE, base de données R-D (DSTI/Division AES), janvier 2002.

Business enterprise intramural expenditure on R&D (BERD) - Dépenses intra-muros de R-D des entreprises (DIRDE)

by industry and source of funds - par industrie et par source de financement

Million SEK — *Millions SEK*

10.4 Funds from abroad - Fonds de l'étranger

	1994	1995	1996	1997	1998	1999	2000	
1 AGRICULTURE, HUNTING & FORESTRY	..	..	..	2.7	..	7.0	..	**1 AGRICULTURE, CHASSE & SYLVICULTURE**
2 MINING	..	..	..	2.3	..	4.3	..	**2 ACTIVITES EXTRACTIVES**
3 MANUFACTURING	..	1 258.9	..	721.8	..	944.5	..	**3 ACTIVITES DE FABRICATION**
4 Food, beverages & tobacco	..	0.1	..	0.3	..	0.0	..	**4 Produits alimentaires, boissons & tabac**
5 Food products & beverages	..	..	..	..	..	.. [n]	..	5 Produits alimentaires & boissons
6 Tobacco products	..	..	..	..	..	.. [n]	..	6 Produits à base de tabac
7 Textiles, wear app., fur & leather	..	..	..	0.0	..	0.0	..	**7 Textiles, habill., fourrures & cuirs**
8 Textiles	..	..	..	..	..	.. [n]	..	8 Textiles
9 Wearing apparel & fur	..	..	..	..	..	.. [n]	..	9 Articles d'habillement & fourrures
10 Leather products & footwear	..	..	..	..	..	.. [n]	..	10 Articles en cuir & chaussures
11 Wood, paper, printing, publishing	..	92.6	..	9.5	..	18.5	..	**11 Bois, papier, imprimerie, édition**
12 Wood & cork (not furniture)	..	..	..	0.2 [o]	..	0.7	..	12 Bois & liège (sauf meubles)
13 Pulp, paper & paper products	..	92.6	..	9.3	..	17.8	..	13 Carton, papier & articles
14 Publ., print. & repro. of rec. media	..	..	..	.. [n]	..	0.0	..	14 Edition, impress. & repro. enregis.
15 Coke, petroleum, nuclear fuel, chemicals, rubber plastics	..	94.6	..	54.1	..	148.0	..	**15 Coke, pétrole, combust. nucl., prod. chimi., caout. & mat. plast.**
16 Coke, ref. petrol. prod. & nuclear fuel	..	.. [n]	..	.. [n]	..	.. [n]	..	16 Coke, raffin. pétr. & combust. nucl.
17 Chemicals & chemical products	..	94.3 [o]	..	53.8 [o]	..	147.6 [o]	..	17 Produits chimiques
18 *Chemical prod.(less pharmaceuticals)*	..	94.3 [o]	..	53.8 [o]	..	106.6 [o]	..	18 *Prod. chimi. (moins prod. pharmaceutiques)*
19 *Pharmaceuticals*	..	..	..	0.0	..	41.0	..	19 *Produits pharmaceutiques*
20 Rubber & plastic products	..	0.3	..	0.3	..	0.4	..	20 Caoutchouc et matières plastiques
21 Non-metallic mineral products	..	0.2	..	0.0	..	0.6	..	**21 Produits minéraux non métalliques**
22 Basic metals	..	1.5	..	4.6	..	5.2	..	**22 Produits métallurgiques de base**
23 Basic metals, ferrous	..	1.0	..	4.6	..	..	..	23 Produits métallur. de base, ferreux
24 Basic metals, non-ferrous	..	0.5	..	0.0	..	..	..	24 Prod. métallur. de base, non ferreux
25 Fabricated metal products	..	..	..	0.2	..	15.9	..	**25 Ouvrages en métaux (sauf mach.& mat.)**
26 Machinery equipment, instruments and transport equipment	..	1 069.8	..	653.0	..	756.2	..	**26 Machines & matériel, instruments & mat. de transport**
27 Machinery n.e.c.	..	68.0	..	52.8	..	219.0	..	27 Machines, n.c.a.
28 Office, account. & computing machinery	..	397.6	..	147.0	..	334.2 [o]	..	28 Mach. de bur., compta. & trait. inform.
29 Electrical machinery	..	34.7	..	164.2	..	.. [n]	..	29 Machines électriques
30 Electro. equip. (radio, TV & commun.)	..	423.3	..	82.3	..	7.7	..	30 Equip. électro. (radio, tv & communic.)
31 *Electro. comp. (inc. semi-conductors)*	..	..	..	..	..	..	..	31 *Compos. électro. (semi-cond. incl.)*
32 *TV, radio & communications equip.*	..	..	..	..	..	..	..	32 *Appareils radio, tv et de communi.*
33 Instruments	..	103.8	..	2.7	..	4.4 [o]	..	33 Instruments
34 Motor vehicles	..	42.4 [a,o]	..	204.0 [o]	..	190.9 [n]	..	34 Véhicules automobiles
35 Other transport equipment	..	.. [n]	..	.. [n]	..	.. [n]	..	35 Autres matériels de transport
36 *Ships*	..	.. [n]	..	.. [n]	..	.. [n]	..	36 *Navires*
37 *Aerospace*	..	.. [n]	..	.. [n]	..	.. [n]	..	37 *Aérospatiale*
38 *Other transport equipment n.e.c.*	..	.. [n]	..	.. [n]	..	.. [n]	..	38 *Autres matériels de transport n.c.a.*
39 Furniture, other manufacturing n.e.c.	..	..	..	0.0	..	0.0 [o]	..	**39 Meubles, autres activ. de fabricat. n.c.a.**
40 Furniture	..	..	..	..	..	.. [n]	..	40 Meubles
41 Other manufacturing n.e.c.	..	..	..	..	..	.. [n]	..	41 Autres activités de fabrication n.c.a.
42 Recycling	..	..	..	..	..	.. [n]	..	**42 Récupération**
43 ELECTRICITY, GAS & WATER SUPPLY	..	.. [n]	..	13.3 [o]	..	17.9 [n]	..	**43 DISTRIBUTION D'ELECTRICITE, GAZ & EAU**
44 CONSTRUCTION	..	.. [n]	..	.. [n]	..	..	..	**44 CONSTRUCTION**
45 SERVICES SECTOR	..	376.2 [a,o]	..	922.1 [a]	..	869.4	..	**45 SECTEUR DES SERVICES**
46 Wholesale, retail trade & repair of motor vehicles, etc.	..	18.8 [a,o]	..	269.8 [o]	..	289.9 [o]	..	**46 Commerce (gros & détail), réparation de véhicules, etc.**
47 Hotels & restaurants	..	.. [n]	..	.. [n]	..	.. [n]	..	**47 Hôtels et restaurants**
48 Transport & storage	..	..	..	10.6	..	6.3	..	**48 Transports, entreposage**
49 Communications	..	0.1 [o]	..	3.4 [a]	..	.. [n]	..	**49 Communications**
50 Post	..	..	..	..	..	..	..	50 Postes
51 Telecommunications	..	..	..	..	..	..	..	51 Télécommunications
52 Financial intermediation (inc. insur.)	..	..	..	..	..	..	..	**52 Intermédiation financ. (assurances inc.)**
53 Real estate, renting & busin. activity	..	355.5	..	636.1	..	569.2	..	**53 Immobil., locat. & serv. aux entreprises**
54 Computer & related activities	..	81.9	..	10.0	..	12.4 [o]	..	54 Informatique et activités connexes
55 *Software consultancy*	..	..	..	..	..	..	..	55 *Production de logiciels*
56 *Other computer services n.e.c.*	..	..	..	..	..	..	..	56 *Autres activités informatiques*
57 Research & development	..	249.7	..	536.2	..	442.5	..	57 Recherche & développement expérimental
58 Other business activities n.e.c.	..	23.9	..	89.8	..	114.3	..	58 Autres activ. & serv. aux entreprises
59 Comm., soc. & pers. serv. activ., etc.	..	1.9	..	2.3	..	4.0	..	**59 Adm. pub., serv. soc. & collec., etc**
TOTAL BUSINESS ENTERPRISE SECTOR	**..**	**1 635.1** [a,m]	**..**	**1 662.2** [m]	**..**	**1 843.0** [m]	**..**	**TOTAL SECTEUR DES ENTREPRISES**

Source: OECD, R&D database (DSTI/EAS Division), January 2002.

Source : OCDE, base de données R-D (DSTI/Division AES), janvier 2002.

Business enterprise intramural expenditure on R&D (BERD) - Dépenses intra-muros de R-D des entreprises (DIRDE)

by industry and type of costs - par industrie et par type de coûts

Million SEK *Millions SEK*

	1994	1995	1996	1997	1998	1999	2000	
				11.1 Current expenditure - Dépenses courantes				
1 AGRICULTURE, HUNTING & FORESTRY	..	119.9	..	140.4	..	206.3	..	**1 AGRICULTURE, CHASSE & SYLVICULTURE**
2 MINING	..	80.3	..	94.1	..	109.8	..	**2 ACTIVITES EXTRACTIVES**
3 MANUFACTURING	..	33 494.8	..	37 167.9	..	42 935.8	..	**3 ACTIVITES DE FABRICATION**
4 Food, beverages & tobacco	..	239.1	..	293.6	..	308.4	..	**4 Produits alimentaires, boissons & tabac**
5 Food products & beverages	..	..	..	..	..	.. [n]	..	5 Produits alimentaires & boissons
6 Tobacco products	..	..	..	..	..	.. [n]	..	6 Produits à base de tabac
7 Textiles, wear app., fur & leather	..	62.3	..	56.0	..	65.3	..	**7 Textiles, habill., fourrures & cuirs**
8 Textiles	..	..	..	..	..	.. [n]	..	8 Textiles
9 Wearing apparel & fur	..	..	..	..	..	.. [n]	..	9 Articles d'habillement & fourrures
10 Leather products & footwear	..	..	..	..	..	.. [n]	..	10 Articles en cuir & chaussures
11 Wood, paper, printing, publishing	..	965.0	..	1 191.5	..	713.2	..	**11 Bois, papier, imprimerie, édition**
12 Wood & cork (not furniture)	..	43.9	..	62.9 [o]	..	34.0	..	12 Bois & liège (sauf meubles)
13 Pulp, paper & paper products	..	904.0	..	1 128.6	..	657.6	..	13 Carton, papier & articles
14 Publ., print. & repro. of rec. media	..	17.1	..	.. [n]	..	21.6	..	14 Edition, impress. & repro. enregis.
15 Coke, petroleum, nuclear fuel, chemicals, rubber plastics	..	6 368.9	..	7 245.4	..	9 208.3	..	**15 Coke, pétrole, combust. nucl., prod. chimi., caout. & mat. plast.**
16 Coke, ref. petrol. prod. & nuclear fuel	..	.. [n]	..	.. [n]	..	.. [n]	..	16 Coke, raffin. pétr. & combust. nucl.
17 Chemicals & chemical products	..	5 999.9 [o]	..	7 085.9 [o]	..	9 001.3 [o]	..	17 Produits chimiques
18 *Chemical prod.(less pharmaceuticals)*	..	730.4 [o]	..	929.2 [o]	..	1 197.9 [o]	..	18 *Prod. chimi. (moins prod. pharmaceutiques)*
19 *Pharmaceuticals*	..	5 269.5	..	6 156.7	..	7 803.4	..	19 *Produits pharmaceutiques*
20 Rubber & plastic products	..	368.9	..	159.4	..	207.0	..	20 Caoutchouc et matières plastiques
21 Non-metallic mineral products	..	134.1	..	201.3	..	120.1	..	**21 Produits minéraux non métalliques**
22 Basic metals	..	308.9	..	392.1	..	948.9	..	**22 Produits métallurgiques de base**
23 Basic metals, ferrous	..	280.8	..	319.4	..	..	..	23 Produits métallur. de base, ferreux
24 Basic metals, non-ferrous	..	28.1	..	72.7	..	..	..	24 Prod. métallur. de base, non ferreux
25 Fabricated metal products	..	571.9	..	695.5	..	294.4	..	**25 Ouvrages en métaux (sauf mach.& mat.)**
26 Machinery equipment, instruments and transport equipment	..	24 758.5	..	27 007.8	..	31 196.6	..	**26 Machines & matériel, instruments & mat. de transport**
27 Machinery n.e.c.	..	4 713.9	..	4 747.0	..	4 604.8	..	27 Machines, n.c.a.
28 Office, account. & computing machinery	..	941.1	..	739.5	..	1 166.3 [o]	..	28 Mach. de bur., compta. & trait. inform.
29 Electrical machinery	..	696.4	..	1 609.3	..	.. [n]	..	29 Machines électriques
30 Electro. equip. (radio, TV & commun.)	..	6 311.6	..	7 691.2	..	10 564.6	..	30 Equip. électro. (radio, tv & communic.)
31 *Electro. comp. (inc. semi-conductors)*	..	..	..	..	..	..	..	31 *Compos. électro. (semi-cond. incl.)*
32 *TV, radio & communications equip.*	..	..	..	..	..	..	..	32 *Appareils radio, tv et de communi.*
33 Instruments	..	2 992.4	..	2 486.2	..	3 973.9 [o]	..	33 Instruments
34 Motor vehicles	..	9 103.0 [a,o]	..	9 734.7 [o]	..	10 887.1 [n]	..	34 Véhicules automobiles
35 Other transport equipment	..	.. [n]	..	.. [n]	..	.. [n]	..	35 Autres matériels de transport
36 *Ships*	..	.. [n]	..	.. [n]	..	.. [n]	..	36 *Navires*
37 *Aerospace*	..	.. [n]	..	.. [n]	..	.. [n]	..	37 *Aérospatiale*
38 *Other transport equipment n.e.c.*	..	.. [n]	..	.. [n]	..	.. [n]	..	38 *Autres matériels de transport n.c.a.*
39 Furniture, other manufacturing n.e.c.	..	86.1 [o]	..	84.8 [o]	..	80.5 [o]	..	**39 Meubles, autres activ. de fabricat. n.c.a.**
40 Furniture	..	55.6	..	..	..	.. [n]	..	40 Meubles
41 Other manufacturing n.e.c.	..	30.6 [o]	..	..	..	.. [n]	..	41 Autres activités de fabrication n.c.a.
42 Recycling	..	.. [n]	..	.. [n]	..	.. [n]	..	**42 Récupération**
43 ELECTRICITY, GAS & WATER SUPPLY	..	.. [n]	..	443.9 [o]	..	384.7 [o]	..	**43 DISTRIBUTION D'ELECTRICITE, GAZ & EAU**
44 CONSTRUCTION	..	.. [n]	..	.. [n]	..	.. [n]	..	**44 CONSTRUCTION**
45 SERVICES SECTOR	..	6 346.1 [a,o]	..	8 268.5 [a]	..	9 378.7	..	**45 SECTEUR DES SERVICES**
46 Wholesale, retail trade & repair of motor vehicles, etc.	..	479.5 [a,o]	..	573.4 [o]	..	515.9 [o]	..	**46 Commerce (gros & détail), réparation de véhicules, etc.**
47 Hotels & restaurants	..	.. [n]	..	.. [n]	..	.. [n]	..	**47 Hôtels et restaurants**
48 Transport & storage	..	90.8	..	75.6	..	32.2	..	**48 Transports, entreposage**
49 Communications	..	899.0 [o]	..	876.3 [a]	..	.. [n]	..	**49 Communications**
50 Post	..	..	..	..	..	..	..	50 Postes
51 Telecommunications	..	..	..	..	..	..	..	51 Télécommunications
52 Financial intermediation (inc. insur.)	..	..	..	..	..	..	..	**52 Intermédiation financ. (assurances inc.)**
53 Real estate, renting & busin. activity	..	4 803.1	..	6 687.8	..	8 764.2	..	**53 Immobil., locat. & serv. aux entreprises**
54 Computer & related activities	..	830.2	..	1 546.0	..	3 442.5 [o]	..	54 Informatique et activités connexes
55 *Software consultancy*	..	..	..	..	..	..	..	55 *Production de logiciels*
56 *Other computer services n.e.c.*	..	..	..	..	..	..	..	56 *Autres activités informatiques*
57 Research & development	..	3 114.1	..	4 595.2	..	4 624.3	..	57 Recherche & développement expérimental
58 Other business activities n.e.c.	..	858.8	..	546.6	..	697.4	..	58 Autres activ. & serv. aux entreprises
59 Comm., soc. & pers. serv. activ., etc.	..	73.8	..	55.3	..	66.4	..	**59 Adm. pub., serv. soc. & collec., etc**
TOTAL BUSINESS ENTERPRISE SECTOR	**..**	**40 041.1** [a]	**..**	**46 114.9**	**..**	**53 015.4**	**..**	**TOTAL SECTEUR DES ENTREPRISES**

Source: OECD, R&D database (DSTI/EAS Division), January 2002.

Source : OCDE, base de données R-D (DSTI/Division AES), janvier 2002.

Business enterprise intramural expenditure on R&D (BERD) - Dépenses intra-muros de R-D des entreprises (DIRDE)

by industry and type of costs - par industrie et par type de coûts

Million SEK — *Millions SEK*

11.2 Capital expenditure - Dépenses en capital

	1994	1995	1996	1997	1998	1999	2000	
1 **AGRICULTURE, HUNTING & FORESTRY**	..	6.4	..	3.0	..	18.6	..	1 **AGRICULTURE, CHASSE & SYLVICULTURE**
2 **MINING**	..	19.7	..	61.6	..	8.1	..	2 **ACTIVITES EXTRACTIVES**
3 **MANUFACTURING**	..	3 487.0	..	3 360.1	..	3 388.0	..	3 **ACTIVITES DE FABRICATION**
4 **Food, beverages & tobacco**	..	18.6	..	14.0	..	19.9	..	4 **Produits alimentaires, boissons & tabac**
5 Food products & beverages	..	..	..	..	..	.. [n]	..	5 Produits alimentaires & boissons
6 Tobacco products	..	..	..	..	..	.. [n]	..	6 Produits à base de tabac
7 **Textiles, wear app., fur & leather**	..	4.3	..	1.6	..	3.9	..	7 **Textiles, habill., fourrures & cuirs**
8 Textiles	..	..	..	..	..	.. [n]	..	8 Textiles
9 Wearing apparel & fur	..	..	..	..	..	.. [n]	..	9 Articles d'habillement & fourrures
10 Leather products & footwear	..	..	..	..	..	.. [n]	..	10 Articles en cuir & chaussures
11 **Wood, paper, printing, publishing**	..	108.0	..	234.2	..	68.1	..	11 **Bois, papier, imprimerie, édition**
12 Wood & cork (not furniture)	..	1.1	..	3.6 [o]	..	0.6	..	12 Bois & liège (sauf meubles)
13 Pulp, paper & paper products	..	106.7	..	230.6	..	65.8	..	13 Carton, papier & articles
14 Publ., print. & repro. of rec. media	..	0.2	..	.. [n]	..	1.7	..	14 Edition, impress. & repro. enregis.
15 **Coke, petroleum, nuclear fuel, chemicals, rubber plastics**	..	995.9	..	1 124.1	..	1 364.9	..	15 **Coke, pétrole, combust. nucl., prod. chimi., caout. & mat. plast.**
16 Coke, ref. petrol. prod. & nuclear fuel	..	.. [n]	..	.. [n]	..	.. [n]	..	16 Coke, raffin. pétr. & combust. nucl.
17 Chemicals & chemical products	..	930.7 [o]	..	1 109.7 [o]	..	1 339.8 [o]	..	17 Produits chimiques
18 *Chemical prod.(less pharmaceuticals)*	..	31.3 [o]	..	87.7 [o]	..	75.7 [o]	..	18 *Prod. chimi. (moins prod. pharmaceutiques)*
19 *Pharmaceuticals*	..	899.3	..	1 022.0	..	1 264.1	..	19 *Produits pharmaceutiques*
20 Rubber & plastic products	..	65.2	..	14.4	..	25.1	..	20 Caoutchouc et matières plastiques
21 **Non-metallic mineral products**	..	10.8	..	0.6	..	4.2	..	21 **Produits minéraux non métalliques**
22 **Basic metals**	..	39.8	..	24.2	..	51.1	..	22 **Produits métallurgiques de base**
23 Basic metals, ferrous	..	30.3	..	23.5	..	..	..	23 Produits métallur. de base, ferreux
24 Basic metals, non-ferrous	..	9.5	..	0.7	..	..	..	24 Prod. métallur. de base, non ferreux
25 **Fabricated metal products**	..	57.1	..	36.9	..	15.9	..	25 **Ouvrages en métaux (sauf mach.& mat.)**
26 **Machinery equipment, instruments and transport equipment**	..	2 245.5	..	1 922.8	..	1 856.2	..	26 **Machines & matériel, instruments & mat. de transport**
27 Machinery n.e.c.	..	415.8	..	333.0	..	268.5	..	27 Machines, n.c.a.
28 Office, account. & computing machinery	..	48.5	..	30.7	..	92.0 [o]	..	28 Mach. de bur., compta. & trait. inform.
29 Electrical machinery	..	97.3	..	209.8	..	.. [n]	..	29 Machines électriques
30 Electro. equip. (radio, TV & commun.)	..	1 062.3	..	523.4	..	701.3	..	30 Equip. électro. (radio, tv & communic.)
31 *Electro. comp. (inc. semi-conductors)*	..	..	..	..	..	..	..	31 *Compos. électro. (semi-cond. incl.)*
32 *TV, radio & communications equip.*	..	..	..	..	..	..	..	32 *Appareils radio, tv et de communi.*
33 Instruments	..	196.4	..	173.5	..	144.8 [o]	..	33 Instruments
34 Motor vehicles	..	425.3 [a,o]	..	652.4 [o]	..	649.6 [n]	..	34 Véhicules automobiles
35 Other transport equipment	..	.. [n]	..	.. [n]	..	.. [n]	..	35 Autres matériels de transport
36 *Ships*	..	.. [n]	..	.. [n]	..	.. [n]	..	36 *Navires*
37 *Aerospace*	..	.. [n]	..	.. [n]	..	.. [n]	..	37 *Aérospatiale*
38 *Other transport equipment n.e.c.*	..	.. [n]	..	.. [n]	..	.. [n]	..	38 *Autres matériels de transport n.c.a.*
39 **Furniture, other manufacturing n.e.c.**	..	6.9 [o]	..	1.8 [o]	..	3.7 [o]	..	39 **Meubles, autres activ. de fabricat. n.c.a.**
40 Furniture	..	4.6	..	..	..	.. [n]	..	40 Meubles
41 Other manufacturing n.e.c.	..	2.3 [o]	..	..	..	.. [n]	..	41 Autres activités de fabrication n.c.a.
42 **Recycling**	..	.. [n]	..	.. [n]	..	.. [n]	..	42 **Récupération**
43 **ELECTRICITY, GAS & WATER SUPPLY**	..	.. [n]	..	84.0 [o]	..	21.2 [o]	..	43 **DISTRIBUTION D'ELECTRICITE, GAZ & EAU**
44 **CONSTRUCTION**	..	.. [n]	..	.. [n]	..	.. [n]	..	44 **CONSTRUCTION**
45 **SERVICES SECTOR**	..	474.8 [a,o]	..	527.6 [a]	..	502.3	..	45 **SECTEUR DES SERVICES**
46 **Wholesale, retail trade & repair of motor vehicles, etc.**	..	16.1 [a,o]	..	20.9 [o]	..	26.9 [o]	..	46 **Commerce (gros & détail), réparation de véhicules, etc.**
47 **Hotels & restaurants**	..	.. [n]	..	.. [n]	..	.. [n]	..	47 **Hôtels et restaurants**
48 **Transport & storage**	..	3.9	..	3.6	..	0.0	..	48 **Transports, entreposage**
49 **Communications**	..	170.0 [o]	..	119.7 [a]	..	.. [n]	..	49 **Communications**
50 Post	..	..	..	..	..	..	..	50 Postes
51 Telecommunications	..	..	..	..	..	..	..	51 Télécommunications
52 **Financial intermediation (inc. insur.)**	..	..	..	..	..	..	..	52 **Intermédiation financ. (assurances inc.)**
53 **Real estate, renting & busin. activity**	..	282.5	..	380.9	..	472.8	..	53 **Immobil., locat. & serv. aux entreprises**
54 Computer & related activities	..	40.6	..	78.3	..	202.3 [o]	..	54 Informatique et activités connexes
55 *Software consultancy*	..	..	..	..	..	..	..	55 *Production de logiciels*
56 *Other computer services n.e.c.*	..	..	..	..	..	..	..	56 *Autres activités informatiques*
57 Research & development	..	200.2	..	274.8	..	237.0	..	57 Recherche & développement expérimental
58 Other business activities n.e.c.	..	41.7	..	27.8	..	33.4	..	58 Autres activ. & serv. aux entreprises
59 **Comm., soc. & pers. serv. activ., etc.**	..	2.4	..	2.5	..	2.6	..	59 **Adm. pub., serv. soc. & collec., etc**
TOTAL BUSINESS ENTERPRISE SECTOR	**..**	**3 987.9** [a]	**..**	**4 036.2**	**..**	**3 938.2**	**..**	**TOTAL SECTEUR DES ENTREPRISES**

Source: OECD, R&D database (DSTI/EAS Division), January 2002.

Source : OCDE, base de données R-D (DSTI/Division AES), janvier 2002.

SWEDEN Table 12 - Tableau 12 **SUEDE**

Total business enterprise R&D personnel (BEMP) - Personnel total de R-D du secteur des entreprises (PRDE)

by industry - par industrie

In full time equivalent | *En équivalence plein-temps*

	1994	1995	1996	1997	1998	1999	2000	
1 AGRICULTURE, HUNTING & FORESTRY	..	208	..	179	..	216	..	**1 AGRICULTURE, CHASSE & SYLVICULTURE**
2 MINING	..	108	..	105	..	129	..	**2 ACTIVITES EXTRACTIVES**
3 MANUFACTURING	..	34 180	..	35 229	..	35 252	..	**3 ACTIVITES DE FABRICATION**
4 Food, beverages & tobacco	..	360	..	419	..	402	..	**4 Produits alimentaires, boissons & tabac**
5 Food products & beverages	..	..	..	..	..	.. [n]	..	5 Produits alimentaires & boissons
6 Tobacco products	..	..	..	..	..	.. [n]	..	6 Produits à base de tabac
7 Textiles, wear app., fur & leather	..	108	..	71	..	74	..	**7 Textiles, habill., fourrures & cuirs**
8 Textiles	..	..	..	..	..	.. [n]	..	8 Textiles
9 Wearing apparel & fur	..	..	..	..	..	.. [n]	..	9 Articles d'habillement & fourrures
10 Leather products & footwear	..	..	..	..	..	.. [n]	..	10 Articles en cuir & chaussures
11 Wood, paper, printing, publishing	..	1 252	..	1 380	..	834	..	**11 Bois, papier, imprimerie, édition**
12 Wood & cork (not furniture)	..	71	..	104 [o]	..	48	..	12 Bois & liège (sauf meubles)
13 Pulp, paper & paper products	..	1 158	..	1 276	..	770	..	13 Carton, papier & articles
14 Publ., print. & repro. of rec. media	..	23	..	.. [n]	..	17	..	14 Edition, impress. & repro. enregis.
15 Coke, petroleum, nuclear fuel, chemicals, rubber plastics	..	5 435	..	6 492	..	6 422	..	**15 Coke, pétrole, combust. nucl., prod. chimi., caout. & mat. plast.**
16 Coke, ref. petrol. prod. & nuclear fuel	..	.. [n]	..	.. [n]	..	.. [n]	..	16 Coke, raffin. pétr. & combust. nucl.
17 Chemicals & chemical products	..	5 007 [o]	..	6 274 [o]	..	6 195 [o]	..	17 Produits chimiques
18 *Chemical prod.(less pharmaceuticals)*	..	981 [o]	..	1 162 [o]	..	1 206 [o]	..	18 *Prod. chimi. (moins prod. pharmaceutiques)*
19 *Pharmaceuticals*	..	4 026	..	5 112	..	4 989	..	19 *Produits pharmaceutiques*
20 Rubber & plastic products	..	428	..	217	..	226	..	20 Caoutchouc et matières plastiques
21 Non-metallic mineral products	..	166	..	193	..	125	..	**21 Produits minéraux non métalliques**
22 Basic metals	..	405	..	502	..	1 107	..	**22 Produits métallurgiques de base**
23 Basic metals, ferrous	..	373	..	433	..	..	..	23 Produits métallur. de base, ferreux
24 Basic metals, non-ferrous	..	32	..	69	..	..	..	24 Prod. métallur. de base, non ferreux
25 Fabricated metal products	..	869	..	874	..	296	..	**25 Ouvrages en métaux (sauf mach.& mat.)**
26 Machinery equipment, instruments and transport equipment	..	25 460	..	25 225	..	25 897	..	**26 Machines & matériel, instruments & mat. de transport**
27 Machinery n.e.c.	..	5 283	..	4 861	..	4 496	..	27 Machines, n.c.a.
28 Office, account. & computing machinery	..	821	..	721	..	1 176 [o]	..	28 Mach. de bur., compta. & trait. inform.
29 Electrical machinery	..	714	..	1 229	..	.. [n]	..	29 Machines électriques
30 Electro. equip. (radio, TV & commun.)	..	6 610	..	6 921	..	6 440	..	30 Equip. électro. (radio, tv & communic.)
31 *Electro. comp. (inc. semi-conductors)*	..	..	..	..	..	..	..	31 *Compos. électro. (semi-cond. incl.)*
32 *TV, radio & communications equip.*	..	..	..	..	..	..	..	32 *Appareils radio, tv et de communi.*
33 Instruments	..	3 318	..	2 860	..	3 681 [o]	..	33 Instruments
34 Motor vehicles	..	8 714 [a,o]	..	8 633 [o]	..	10 104 [n]	..	34 Véhicules automobiles
35 Other transport equipment	..	.. [n]	..	.. [n]	..	.. [n]	..	35 Autres matériels de transport
36 *Ships*	..	.. [n]	..	.. [n]	..	.. [n]	..	36 *Navires*
37 *Aerospace*	..	.. [n]	..	.. [n]	..	.. [n]	..	37 *Aérospatiale*
38 *Other transport equipment n.e.c.*	..	.. [n]	..	.. [n]	..	.. [n]	..	38 *Autres matériels de transport n.c.a.*
39 Furniture, other manufacturing n.e.c.	..	125 [o]	..	74 [o]	..	95 [o]	..	**39 Meubles, autres activ. de fabricat. n.c.a.**
40 Furniture	..	85	..	..	..	.. [n]	..	40 Meubles
41 Other manufacturing n.e.c.	..	39 [o]	..	..	..	.. [n]	..	41 Autres activités de fabrication n.c.a.
42 Recycling	..	.. [n]	..	.. [n]	..	.. [n]	..	**42 Récupération**
43 ELECTRICITY, GAS & WATER SUPPLY	..	.. [n]	..	390 [o]	..	.. [n]	..	**43 DISTRIBUTION D'ELECTRICITE, GAZ & EAU**
44 CONSTRUCTION	..	.. [n]	..	.. [n]	..	284 [o]	..	**44 CONSTRUCTION**
45 SERVICES SECTOR	..	7 140 [a,o]	..	7 978 [a]	..	8 291	..	**45 SECTEUR DES SERVICES**
46 Wholesale, retail trade & repair of motor vehicles, etc.	..	531 [a,o]	..	472 [o]	..	442 [o]	..	**46 Commerce (gros & détail), réparation de véhicules, etc.**
47 Hotels & restaurants	..	.. [n]	..	.. [n]	..	.. [n]	..	**47 Hôtels et restaurants**
48 Transport & storage	..	85	..	42	..	21	..	**48 Transports, entreposage**
49 Communications	..	1 120 [o]	..	890 [a]	..	..	..	**49 Communications**
50 Post	..	..	..	..	..	..	..	50 Postes
51 Telecommunications	..	..	..	..	..	.. [n]	..	51 Télécommunications
52 Financial intermediation (inc. insur.)	..	..	..	..	..	..	..	**52 Intermédiation financ. (assurances inc.)**
53 Real estate, renting & busin. activity	..	5 304	..	6 491	..	7 746	..	**53 Immobil., locat. & serv. aux entreprises**
54 Computer & related activities	..	884	..	1 602	..	2 921 [o]	..	54 Informatique et activités connexes
55 *Software consultancy*	..	..	..	..	..	..	..	55 *Production de logiciels*
56 *Other computer services n.e.c.*	..	..	..	..	..	..	..	56 *Autres activités informatiques*
57 Research & development	..	3 285	..	4 336	..	4 129	..	57 Recherche & développement expérimental
58 Other business activities n.e.c.	..	1 135	..	554	..	696	..	58 Autres activ. & serv. aux entreprises
59 Comm., soc. & pers. serv. activ., etc.	..	100	..	82	..	82	..	**59 Adm. pub., serv. soc. & collec., etc**
TOTAL BUSINESS ENTERPRISE SECTOR	**..**	**41 636 [a]**	**..**	**43 881**	**..**	**44 171**	**..**	**TOTAL SECTEUR DES ENTREPRISES**

Source: OECD, R&D database (DSTI/EAS Division), January 2002.

Source : OCDE, base de données R-D (DSTI/Division AES), janvier 2002.

Business enterprise university graduates - Diplômés universitaires des entreprises
by industry - par industrie

In full time equivalent — *En équivalence plein-temps*

	1994	1995	1996	1997	1998	1999	2000	
1 AGRICULTURE, HUNTING & FORESTRY	..	76	..	73	..	90	..	**1 AGRICULTURE, CHASSE & SYLVICULTURE**
2 MINING	..	39	..	33	..	62	..	**2 ACTIVITES EXTRACTIVES**
3 MANUFACTURING	..	14 690	..	15 558	..	16 811	..	**3 ACTIVITES DE FABRICATION**
4 Food, beverages & tobacco	..	172	..	183	..	203	..	**4 Produits alimentaires, boissons & tabac**
5 Food products & beverages	..	..	..	..	..	.. [n]	..	5 Produits alimentaires & boissons
6 Tobacco products	..	..	..	..	..	.. [n]	..	6 Produits à base de tabac
7 Textiles, wear app., fur & leather	..	29	..	21	..	30	..	**7 Textiles, habill., fourrures & cuirs**
8 Textiles	..	..	..	..	..	.. [n]	..	8 Textiles
9 Wearing apparel & fur	..	..	..	..	..	.. [n]	..	9 Articles d'habillement & fourrures
10 Leather products & footwear	..	..	..	..	..	.. [n]	..	10 Articles en cuir & chaussures
11 Wood, paper, printing, publishing	..	551	..	665	..	437	..	**11 Bois, papier, imprimerie, édition**
12 Wood & cork (not furniture)	..	20	..	31 [o]	..	21	..	12 Bois & liège (sauf meubles)
13 Pulp, paper & paper products	..	518	..	634	..	410	..	13 Carton, papier & articles
14 Publ., print. & repro. of rec. media	..	13	..	.. [n]	..	6	..	14 Edition, impress. & repro. enregis.
15 Coke, petroleum, nuclear fuel, chemicals, rubber plastics	..	2 913	..	3 572	..	3 681	..	**15 Coke, pétrole, combust. nucl., prod. chimi., caout. & mat. plast.**
16 Coke, ref. petrol. prod. & nuclear fuel	..	.. [n]	..	.. [n]	..	.. [n]	..	16 Coke, raffin. pétr. & combust. nucl.
17 Chemicals & chemical products	..	2 747 [o]	..	3 497 [o]	..	3 589 [o]	..	17 Produits chimiques
18 *Chemical prod.(less pharmaceuticals)*	..	484 [o]	..	525 [o]	..	647 [o]	..	18 *Prod. chimi. (moins prod. pharmaceutiques)*
19 *Pharmaceuticals*	..	2 263	..	2 972	..	2 942	..	19 *Produits pharmaceutiques*
20 Rubber & plastic products	..	165	..	75	..	92	..	20 Caoutchouc et matières plastiques
21 Non-metallic mineral products	..	61	..	110	..	50	..	**21 Produits minéraux non métalliques**
22 Basic metals	..	219	..	276	..	493	..	**22 Produits métallurgiques de base**
23 Basic metals, ferrous	..	204	..	229	..	..	..	23 Produits métallur. de base, ferreux
24 Basic metals, non-ferrous	..	15	..	47	..	..	..	24 Prod. métallur. de base, non ferreux
25 Fabricated metal products	..	333	..	325	..	107	..	**25 Ouvrages en métaux (sauf mach.& mat.)**
26 Machinery equipment, instruments and transport equipment	..	10 369	..	10 393	..	11 794	..	**26 Machines & matériel, instruments & mat. de transport**
27 Machinery n.e.c.	..	2 099	..	2 008	..	2 049	..	27 Machines, n.c.a.
28 Office, account. & computing machinery	..	564	..	494	..	807 [o]	..	28 Mach. de bur., compta. & trait. inform.
29 Electrical machinery	..	409	..	687	..	.. [n]	..	29 Machines électriques
30 Electro. equip. (radio, TV & commun.)	..	3 060	..	2 917	..	2 961	..	30 Equip. électro. (radio, tv & communic.)
31 *Electro. comp. (inc. semi-conductors)*	..	..	..	..	..	..	..	31 *Compos. électro. (semi-cond. incl.)*
32 *TV, radio & communications equip.*	..	..	..	..	..	..	..	32 *Appareils radio, tv et de communi.*
33 Instruments	..	1 756	..	1 764	..	2 374 [o]	..	33 Instruments
34 Motor vehicles	..	2 480 [a,o]	..	2 523 [o]	..	3 605 [n]	..	34 Véhicules automobiles
35 Other transport equipment	..	.. [n]	..	.. [n]	..	.. [n]	..	35 Autres matériels de transport
36 *Ships*	..	.. [n]	..	.. [n]	..	.. [n]	..	36 *Navires*
37 *Aerospace*	..	.. [n]	..	.. [n]	..	.. [n]	..	37 *Aérospatiale*
38 *Other transport equipment n.e.c.*	..	.. [n]	..	.. [n]	..	.. [n]	..	38 *Autres matériels de transport n.c.a.*
39 Furniture, other manufacturing n.e.c.	..	43 [o]	..	13 [o]	..	16 [o]	..	**39 Meubles, autres activ. de fabricat. n.c.a.**
40 Furniture	..	27	..	..	..	.. [n]	..	40 Meubles
41 Other manufacturing n.e.c.	..	16 [o]	..	..	..	.. [n]	..	41 Autres activités de fabrication n.c.a.
42 Recycling	..	.. [n]	..	.. [n]	..	.. [n]	..	**42 Récupération**
43 ELECTRICITY, GAS & WATER SUPPLY	..	.. [n]	..	224 [o]	..	.. [n]	..	**43 DISTRIBUTION D'ELECTRICITE, GAZ & EAU**
44 CONSTRUCTION	..	.. [n]	..	.. [n]	..	214 [o]	..	**44 CONSTRUCTION**
45 SERVICES SECTOR	..	4 250 [a,o]	..	5 036 [a]	..	5 646	..	**45 SECTEUR DES SERVICES**
46 Wholesale, retail trade & repair of motor vehicles, etc.	..	315 [a,o]	..	301 [o]	..	318 [o]	..	**46 Commerce (gros & détail), réparation de véhicules, etc.**
47 Hotels & restaurants	..	.. [n]	..	.. [n]	..	.. [n]	..	**47 Hôtels et restaurants**
48 Transport & storage	..	41	..	33	..	17	..	**48 Transports, entreposage**
49 Communications	..	680 [o]	..	704 [a]	..	..	..	**49 Communications**
50 Post	..	..	..	..	..	..	..	50 Postes
51 Telecommunications	..	..	..	..	..	.. [n]	..	51 Télécommunications
52 Financial intermediation (inc. insur.)	..	..	..	..	..	..	..	**52 Intermédiation financ. (assurances inc.)**
53 Real estate, renting & busin. activity	..	3 150	..	3 933	..	5 243	..	**53 Immobil., locat. & serv. aux entreprises**
54 Computer & related activities	..	567	..	1 122	..	2 261 [o]	..	54 Informatique et activités connexes
55 *Software consultancy*	..	..	..	..	..	..	..	55 *Production de logiciels*
56 *Other computer services n.e.c.*	..	..	..	..	..	..	..	56 *Autres activités informatiques*
57 Research & development	..	1 916	..	2 477	..	2 514	..	57 Recherche & développement expérimental
58 Other business activities n.e.c.	..	667	..	334	..	467	..	58 Autres activ. & serv. aux entreprises
59 Comm., soc. & pers. serv. activ., etc.	..	64	..	65	..	68	..	**59 Adm. pub., serv. soc. & collec., etc**
TOTAL BUSINESS ENTERPRISE SECTOR	**..**	**19 054 [a]**	**..**	**20 924**	**..**	**22 822**	**..**	**TOTAL SECTEUR DES ENTREPRISES**

Source: OECD, R&D database (DSTI/EAS Division), January 2002.

Source : OCDE, base de données R-D (DSTI/Division AES), janvier 2002.

Higher education intramural expenditure on R&D - Dépenses intra-muros de R-D de l'enseignement supérieur

(HERD) by field of science - (DIRDES) par discipline scientifique

	1994	1995	1996	1997	1998	1999	2000	
				14.1 Million SEK - Millions SEK				
Natural sciences	..	2 622.0 [a,j]	..	3 107.7 [a]	..	3 406.0	..	Sciences exactes et naturelles
Engineering	..	2 769.0 [a,j]	..	3 286.0 [a]	..	3 552.0	..	Sciences de l'ingénieur
Medical sciences	..	3 792.0 [a,j]	..	3 968.8 [a]	..	4 440.0	..	Sciences médicales
Agricultural sciences	..	903.0 [a,j]	..	936.4 [a]	..	982.0	..	Sciences agricoles
Sub-total NSE	..	10 086.0 [a,j,v]	..	11 299.0 [a]	..	12 381.0	..	Sous-total SENI
Social sciences	..	1 406.0 [a,j]	..	1 768.8 [a]	..	1 864.0	..	Sciences sociales
Humanities	..	729.0 [a,j]	..	851.8 [a]	..	992.0	..	Sciences humaines
Sub-total SSH	..	2 134.0 [a,j,v]	..	2 620.6 [a]	..	2 855.0	..	Sous-total SSH
Total NSE+SSH	..	13 004.0 [a,j]	..	14 346.3 [a]	..	16 226.0	..	*Total SENI+SSH*
				14.2 Million constant USD (1995 prices and PPPs) - Millions de USD constants (prix et PPA de 1995)				
Natural sciences	..	269.5 [a,j]	..	309.6 [a]	..	334.6	..	Sciences exactes et naturelles
Engineering	..	284.6 [a,j]	..	327.4 [a]	..	349.0	..	Sciences de l'ingénieur
Medical sciences	..	389.8 [a,j]	..	395.4 [a]	..	436.2	..	Sciences médicales
Agricultural sciences	..	92.8 [a,j]	..	93.3 [a]	..	96.5	..	Sciences agricoles
Sub-total NSE	..	1 036.8 [a,j,v]	..	1 125.7 [a]	..	1 216.4	..	Sous-total SENI
Social sciences	..	144.5 [a,j]	..	176.2 [a]	..	183.1	..	Sciences sociales
Humanities	..	74.9 [a,j]	..	84.9 [a]	..	97.5	..	Sciences humaines
Sub-total SSH	..	219.4 [a,j,v]	..	261.1 [a]	..	280.5	..	Sous-total SSH
Total NSE+SSH	..	1 336.8 [a,j]	..	1 429.3 [a]	..	1 594.1	..	*Total SENI+SSH*

Table 15 - Tableau 15

Higher education intramural expenditure on R&D - Dépenses intra-muros de R-D de l'enseignement supérieur

(HERD) by field of science and type of costs - (DIRDES) par discipline scientifique et par type de coûts

Million SEK *Millions SEK*

	1994	1995	1996	1997	1998	1999	2000	
Current expenditure								**Dépenses courantes**
Natural sciences	..	2 622.0	..	2 905.3	..	3 182.0	..	Sciences exactes et naturelles
Engineering	..	2 769.0	..	3 072.0	..	3 318.0	..	Sciences de l'ingénieur
Medical sciences	..	3 792.0	..	3 778.6	..	4 148.0	..	Sciences médicales
Agricultural sciences	..	903.0	..	875.5	..	918.0	..	Sciences agricoles
Sub-total NSE	..	10 086.0 [v]	..	10 631.4	..	11 566.0	..	Sous-total SENI
Social sciences	..	1 406.0	..	1 585.3	..	1 741.0	..	Sciences sociales
Humanities	..	729.0	..	796.3	..	926.0	..	Sciences humaines
Sub-total SSH	..	2 134.0 [v]	..	2 381.5	..	2 667.0	..	Sous-total SSH
Total NSE+SSH	..	13 004.0	..	13 411.8	..	15 158.0	..	*Total SENI+SSH*
Capital expenditure								**Dépenses en capital**
Natural sciences	..	..	..	202.4	..	224.0	..	Sciences exactes et naturelles
Engineering	..	..	..	214.1	..	234.0	..	Sciences de l'ingénieur
Medical sciences	..	..	..	258.5	..	292.0	..	Sciences médicales
Agricultural sciences	..	..	..	61.0	..	64.0	..	Sciences agricoles
Sub-total NSE	..	..	..	736.0	..	815.0	..	Sous-total SENI
Social sciences	..	..	..	115.2	..	123.0	..	Sciences sociales
Humanities	..	..	..	55.5	..	66.0	..	Sciences humaines
Sub-total SSH	..	..	..	170.7	..	188.0	..	Sous-total SSH
Total NSE+SSH	..	..	..	934.5	..	1 068.0	..	*Total SENI+SSH*

Source: OECD, R&D database (DSTI/EAS Division), January 2002.

Source : OCDE, base de données R-D (DSTI/Division AES), janvier 2002.

Higher education intramural expenditure on R&D - Dépenses intra-muros de R-D de l'enseignement supérieur

(HERD) by field of science and source of funds - (DIRDES) par discipline scientifique et par source de financement

Million SEK *Millions SEK*

	1994	1995	1996	1997	1998	1999	2000	
Direct government								**État direct**
Natural sciences	..	850.0 [a,j]	..	936.1 [a]	..	1 022.0	..	Sciences exactes et naturelles
Engineering	..	1 053.0 [a,j]	..	1 096.6 [a]	..	1 034.0	..	Sciences de l'ingénieur
Medical sciences	..	648.0 [a,j]	..	711.6 [a]	..	819.0	..	Sciences médicales
Agricultural sciences	..	191.0 [a,j]	..	250.6 [a]	..	226.0	..	Sciences agricoles
Sub-total NSE	..	2 742.0 [a,j]	..	2 994.9 [a]	..	3 102.0	..	Sous-total SENI
Social sciences	..	536.0 [a,j]	..	628.6 [a]	..	627.0	..	Sciences sociales
Humanities	..	135.0 [a,j]	..	148.6 [a]	..	171.0	..	Sciences humaines
Sub-total SSH	..	671.0 [a,j]	..	777.1 [a]	..	798.0	..	Sous-total SSH
Total NSE+SSH	..	3 485.0 [a,j]	..	3 796.9 [a]	..	4 089.0	..	*Total SENI+SSH*
General university funds								**Fonds généraux des universités**
Natural sciences	..	1 493.0 [a,j]	..	1 678.9 [a]	..	1 695.0	..	Sciences exactes et naturelles
Engineering	..	1 177.0 [a,j]	..	1 255.1 [a]	..	1 234.0	..	Sciences de l'ingénieur
Medical sciences	..	2 239.0 [a,j]	..	2 115.9 [a]	..	2 161.0	..	Sciences médicales
Agricultural sciences	..	570.0 [a,j]	..	553.2 [a]	..	476.0	..	Sciences agricoles
Sub-total NSE	..	5 479.0 [a,j]	..	5 603.1 [a]	..	5 566.0	..	Sous-total SENI
Social sciences	..	715.0 [a,j]	..	878.2 [a]	..	893.0	..	Sciences sociales
Humanities	..	539.0 [a,j]	..	595.5 [a]	..	665.0	..	Sciences humaines
Sub-total SSH	..	1 254.0 [a,j]	..	1 473.7 [a]	..	1 557.0	..	Sous-total SSH
Total NSE+SSH	..	7 390.0 [a,j]	..	7 425.6 [a]	..	7 676.0	..	*Total SENI+SSH*
Sub-total government								**Sous-total État**
Natural sciences	..	2 343.0 [a,j]	..	2 615.0 [a]	..	2 718.0	..	Sciences exactes et naturelles
Engineering	..	2 230.0 [a,j]	..	2 351.7 [a]	..	2 268.0	..	Sciences de l'ingénieur
Medical sciences	..	2 887.0 [a,j]	..	2 827.6 [a]	..	2 980.0	..	Sciences médicales
Agricultural sciences	..	761.0 [a,j]	..	803.7 [a]	..	703.0	..	Sciences agricoles
Sub-total NSE	..	8 221.0 [a,j]	..	8 598.0 [a]	..	8 668.0	..	Sous-total SENI
Social sciences	..	1 251.0 [a,j]	..	1 506.8 [a]	..	1 520.0	..	Sciences sociales
Humanities	..	674.0 [a,j]	..	744.1 [a]	..	836.0	..	Sciences humaines
Sub-total SSH	..	1 925.0 [a,j]	..	2 250.9 [a]	..	2 356.0	..	Sous-total SSH
Total NSE+SSH	..	10 875.0 [a,j]	..	11 222.5 [a]	..	11 765.0	..	*Total SENI+SSH*
Higher education								**Enseignement supérieur**
Natural sciences	..	23.0 [a,j]	..	28.1 [,a]	..	23.0	..	Sciences exactes et naturelles
Engineering	..	55.0 [a,j]	..	39.4 [,a]	..	15.0	..	Sciences de l'ingénieur
Medical sciences	..	257.0 [a,j]	..	105.9 [,a]	..	130.0	..	Sciences médicales
Agricultural sciences	..	3.0 [a,j]	..	- 3.0 [,a]	..	1.0	..	Sciences agricoles
Sub-total NSE	..	339.0 [a,j]	..	170.3 [,a]	..	169.0	..	Sous-total SENI
Social sciences	..	30.0 [a,j]	..	27.8 [,a]	..	34.0	..	Sciences sociales
Humanities	..	11.0 [a,j]	..	13.9 [,a]	..	15.0	..	Sciences humaines
Sub-total SSH	..	41.0 [a,j]	..	41.7 [,a]	..	49.0	..	Sous-total SSH
Total NSE+SSH	..	394.0 [a,j]	..	225.9 [,a]	..	224.0	..	*Total SENI+SSH*
Private non-profit								**Institutions sans but lucratif**
Natural sciences	..	137.0 [a,j]	..	277.5 [a]	..	422.0	..	Sciences exactes et naturelles
Engineering	..	134.0 [a,j]	..	378.3 [a]	..	819.0	..	Sciences de l'ingénieur
Medical sciences	..	309.0 [a,j]	..	618.3 [a]	..	848.0	..	Sciences médicales
Agricultural sciences	..	90.0 [a,j]	..	88.9 [a]	..	224.0	..	Sciences agricoles
Sub-total NSE	..	669.0 [a,j]	..	1 362.9 [a]	..	2 312.0	..	Sous-total SENI
Social sciences	..	65.0 [a,j]	..	126.9 [a]	..	216.0	..	Sciences sociales
Humanities	..	32.0 [a,j]	..	75.0 [a]	..	121.0	..	Sciences humaines
Sub-total SSH	..	97.0 [a,j]	..	201.8 [a]	..	337.0	..	Sous-total SSH
Total NSE+SSH	..	792.0 [a,j]	..	1 594.8 [a]	..	2 863.0	..	*Total SENI+SSH*
Business entreprise								**Entreprises**
Natural sciences	..	54.0 [a,j]	..	57.6 [a]	..	99.0	..	Sciences exactes et naturelles
Engineering	..	267.0 [a,j]	..	330.6 [a]	..	268.0	..	Sciences de l'ingénieur
Medical sciences	..	194.0 [a,j]	..	202.8 [a]	..	188.0	..	Sciences médicales
Agricultural sciences	..	28.0 [a,j]	..	22.1 [a]	..	22.0	..	Sciences agricoles
Sub-total NSE	..	543.0 [a,j]	..	613.2 [a]	..	576.0	..	Sous-total SENI
Social sciences	..	41.0 [a,j]	..	69.0 [a]	..	36.0	..	Sciences sociales
Humanities	..	3.0 [a,j]	..	4.1 [a]	..	1.0	..	Sciences humaines
Sub-total SSH	..	45.0 [a,j]	..	73.1 [a]	..	37.0	..	Sous-total SSH
Total NSE+SSH	..	598.0 [a,j]	..	689.2 [a]	..	629.0	..	*Total SENI+SSH*
Funds from abroad								**Fonds de l'étranger**
Natural sciences	..	66.0 [a,j]	..	129.6 [a]	..	145.0	..	Sciences exactes et naturelles
Engineering	..	83.0 [a,j]	..	186.0 [a]	..	183.0	..	Sciences de l'ingénieur
Medical sciences	..	145.0 [a,j]	..	214.3 [a]	..	294.0	..	Sciences médicales
Agricultural sciences	..	20.0 [a,j]	..	24.7 [a]	..	34.0	..	Sciences agricoles
Sub-total NSE	..	314.0 [a,j]	..	554.6 [a]	..	656.0	..	Sous-total SENI
Social sciences	..	18.0 [a,j]	..	38.4 [a]	..	58.0	..	Sciences sociales
Humanities	..	8.0 [a,j]	..	14.8 [a]	..	18.0	..	Sciences humaines
Sub-total SSH	..	25.0 [a,j]	..	53.1 [a]	..	76.0	..	Sous-total SSH
Total NSE+SSH	..	344.0 [a,j]	..	613.9 [a]	..	745.0	..	*Total SENI+SSH*

Source: OECD, R&D database (DSTI/EAS Division), January 2002.

Source : OCDE, base de données R-D (DSTI/Division AES), janvier 2002.

Higher education R&D personnel (HEMP) - Personnel de R-D du secteur de l'enseignement supérieur

by field of science and occupation - par discipline scientifique et par profession

In full-time equivalent — *En équivalence plein-temps*

	1994	1995	1996	1997	1998	1999	2000	
Total R&D personnel								**Personnel total de R-D**
Natural sciences	..	2 779	..	2 857	..	2 906	..	Sciences exactes et naturelles
Engineering	..	3 585	..	3 786	..	4 222	..	Sciences de l'ingénieur
Medical sciences	..	4 475	..	4 331	..	4 570	..	Sciences médicales
Agricultural sciences	..	1 745	..	1 659	..	1 487	..	Sciences agricoles
Sub-total NSE	..	12 583 v	..	12 633 v	..	13 186 v	..	Sous-total SENI
Social sciences	..	2 826	..	3 189	..	3 011	..	Sciences sociales
Humanities	..	1 398	..	1 594	..	1 646	..	Sciences humaines
Sub-total SSH	..	4 224 v	..	4 783 v	..	4 657 v	..	Sous-total SSH
Total NSE+SSH	..	17 301	..	18 197	..	19 175	..	*Total SENI+SSH*
Researchers								**Chercheurs**
Natural sciences	..	2 038	..	2 161	..	2 327	..	Sciences exactes et naturelles
Engineering	..	2 603	..	2 802	..	3 104	..	Sciences de l'ingénieur
Medical sciences	..	2 394	..	2 584	..	2 951	..	Sciences médicales
Agricultural sciences	..	920	..	1 082	..	1 043	..	Sciences agricoles
Sub-total NSE	..	7 955 v	..	8 629 v	..	9 425 v	..	Sous-total SENI
Social sciences	..	2 264	..	2 751	..	2 599	..	Sciences sociales
Humanities	..	1 230	..	1 448	..	1 473	..	Sciences humaines
Sub-total SSH	..	3 495 v	..	4 198 v	..	4 072 v	..	Sous-total SSH
Total NSE+SSH	..	11 873	..	13 515	..	14 623	..	*Total SENI+SSH*
Technicians								**Technicians**
Natural sciences	..	594	..	597	..	453	..	Sciences exactes et naturelles
Engineering	..	761	..	759	..	842	..	Sciences de l'ingénieur
Medical sciences	..	1 621	..	1 435	..	1 317	..	Sciences médicales
Agricultural sciences	..	679	..	448	..	379	..	Sciences agricoles
Sub-total NSE	..	3 656 v	..	3 238 v	..	2 990 v	..	Sous-total SENI
Social sciences	..	226	..	194	..	176	..	Sciences sociales
Humanities	..	43	..	64	..	78	..	Sciences humaines
Sub-total SSH	..	269 v	..	258 v	..	254 v	..	Sous-total SSH
Total NSE+SSH	..	3 974	..	3 576	..	3 393	..	*Total SENI+SSH*
Other								**Autres**
Natural sciences	..	146	..	99	..	126	..	Sciences exactes et naturelles
Engineering	..	221	..	226	..	276	..	Sciences de l'ingénieur
Medical sciences	..	460	..	313	..	303	..	Sciences médicales
Agricultural sciences	..	145	..	129	..	65	..	Sciences agricoles
Sub-total NSE	..	973 v	..	767 v	..	770 v	..	Sous-total SENI
Social sciences	..	336	..	244	..	236	..	Sciences sociales
Humanities	..	125	..	83	..	95	..	Sciences humaines
Sub-total SSH	..	460 v	..	327 v	..	331 v	..	Sous-total SSH
Total NSE+SSH	..	1 454	..	1 107	..	1 159	..	*Total SENI+SSH*

Table 18 - Tableau 18

Government budget appropriations or outlays for R&D (GBAORD) - Crédits budgétaires publics de R-D (CBPRD)

by socio-economic objective - par objectif socio-économique

Million SEK — *Millions SEK*

	1995	1996	1997	1998	1999	2000	2001	
Agriculture, forestry and fishing	337.0 c	337.0 c	..	275.0 a	291.0	292.0	471.0	Agriculture, sylviculture et pêche
Industrial development	1 165.0 c	1 165.0 c	..	769.0 a	612.0	861.0	520.0	Développement industriel
Energy	551.0 c	551.0 c	..	666.0 a	898.0	912.0	590.0	Energie
Transport & telecommunications	..	..	..	..	..	..	652.0	*Transports & télécommunications*
Urban & rural planning	..	..	..	..	..	..	95.0	*Aménagement urbain et rural*
ST Infrastructure	1 117.0 c	1 117.0 c	..	912.0 a	907.0	644.0	747.0	ST Infrastructure
Prevention of pollution	..	..	..	..	..	..	180.0	*Prévention de la pollution*
Identificat. & treatment of pollution	..	..	..	..	..	..	..	*Détect. pollut. & mesures antipollut.*
ST Environment	459.0 c	459.0 c	..	130.0 a	243.0	216.0	180.0	ST Environnement
Health	294.0 c	294.0 c	..	144.0 a	267.0	211.0	142.0	Santé
Social development & services	1 306.0 c	1 306.0 c	..	1 061.0 a	958.0	891.0	1 670.0	Développement et services sociaux
Earth and atmosphere	65.0 c	65.0 c	..	191.0 a	297.0	264.0	70.0	Milieu terrestre et atmosphère
Advancement of research	2 263.0 c	2 263.0 c	..	2 026.0 a	1 377.0	1 815.0	3 191.0	*Promotion de la recherche*
General university funds	7 651.0 c	7 651.0 c	..	7 689.3 a	7 724.0	8 050.0	8 249.0	*Fonds généraux des universités*
ST Advancement of knowledge	9 914.0 c	9 914.0 c	..	9 715.3 a	9 101.0	9 865.0	11 440.0	ST promotion gén. des connaiss.
Civil space	281.0 c	281.0 c	..	372.0 a	500.0	532.0	511.0	Espace civil
Defence	4 089.0 c	4 089.0 c	..	1 122.0 a	1 117.0	1 126.0	2 800.0	Défense
Not elsewhere classified	..	..	..	..	..	..	..	Non classé
Total	**19 578.0 c**	**19 578.0 c**	..	**15 357.0 a**	**15 191.0**	**15 813.0**	**19 141.0**	**Total**

Source: OECD, R&D database (DSTI/EAS Division), January 2002.

Source : OCDE, base de données R-D (DSTI/Division AES), janvier 2002.

Table 19 - Tableau 19

Technology balance of payments - Balance des paiements technologiques

by type of transfer - selon la nature des transferts

Million SEK *Millions SEK*

	1983	1984	1985	1986	1987	1988	1989	1990	1991	
Receipts										**Recettes**
Related firms	..	..	..	..	..	..	..	..	..	Firmes affiliées
Unrelated firms	..	..	..	..	..	..	..	..	..	Firmes non affiliées
Other	..	..	..	..	..	..	..	..	..	Autre
Total	632	..	752	..	861	..	1 282	..	1 315 [a]	*Total*
Payments										**Paiements**
Related firms	..	..	..	..	..	..	..	..	..	Firmes affiliées
Unrelated firms	..	..	..	..	..	..	..	..	..	Firmes non affiliées
Other	..	..	..	..	..	..	..	..	..	Autre
Total	289	..	424	..	259	..	224	..	701 [a]	*Total*

	1992	1993	1994	1995	1996	1997	1998	1999	2000	
Receipts										**Recettes**
Related firms	..	..	..	..	..	..	..	..	..	Firmes affiliées
Unrelated firms	..	..	..	..	..	..	..	..	..	Firmes non affiliées
Other	..	..	..	..	..	..	..	..	..	Autre
Total	..	3 092	..	..	..	..	..	..	..	*Total*
Payments										**Paiements**
Related firms	..	..	..	..	..	..	..	..	..	Firmes affiliées
Unrelated firms	..	..	..	..	..	..	..	..	..	Firmes non affiliées
Other	..	..	..	..	..	..	..	..	..	Autre
Total	..	348	..	..	..	..	..	..	..	*Total*

Source: OECD, TBPdatabase (DSTI/EAS Division), Januay 2002.

Source : OCDE, base de données TBP (DSTI/Division AES), janvier 2002.

Table 1 - Tableau 1

Gross domestic expenditure on R&D (GERD) - Dépenses intérieures brutes de R-D (DIRD)

by sector of performance and source of funds - par secteur d'exécution et par source de financement

	1994	1995	1996	1997	1998	1999	2000	
	1.1 Million CHF - Millions CHF							
Business enterprise								**Entreprises**
financed by								*financé par*
Business enterprise	..	..	6 530	..	..	..	..	Entreprises
Direct government	150 [h]	..	170 [h]	..	130 [h]	..	..	État direct
Higher education	..	..	0	..	..	..	..	Enseignement supérieur
Private non-profit	..	..	50	..	..	..	..	Institutions sans but lucratif
Funds from abroad	..	..	310	..	..	..	..	Fonds de l'étranger
Total	..	..	7 060	..	..	..	..	*Total*
Government								**État**
financed by								*financé par*
Business enterprise	..	..	..	..	..	..	..	Entreprises
Direct government	270 [h]	..	250 [h]	..	200 [h]	..	..	État direct
Higher education	..	..	..	..	..	..	..	Enseignement supérieur
Private non-profit	..	..	..	..	..	..	..	Institutions sans but lucratif
Funds from abroad	..	..	..	..	..	..	..	Fonds de l'étranger
Total	270 [h]	..	250 [h]	..	200 [h]	..	..	*Total*
Higher education								**Enseignement supérieur**
financed by								*financé par*
Business enterprise	40	..	150	..	170	..	..	Entreprises
Direct government	440 [h]	..	370 [h]	..	400 [h]	..	..	État direct
General university funds	1 700	..	1 780	..	1 520	..	..	Fonds généraux des universités
Sub-total government	2 140	..	2 150	..	1 920	..	..	Sous-Total État
Higher education	80	..	110	..	210	..	..	Enseignement supérieur
Private non-profit	70	..	20	..	90	..	..	Institutions sans but lucratif
Funds from abroad	..	..	..	..	..	..	..	Fonds de l'étranger
Total	2 330	..	2 430	..	2 390	..	..	*Total*
Private non-profit								**Institutions sans but lucratif**
financed by								*financé par*
Business enterprise	..	..	60	..	..	..	..	Entreprises
Direct government	..	..	120 [h]	..	110 [h]	..	..	État direct
Higher education	..	..	20	..	..	..	..	Enseignement supérieur
Private non-profit	..	..	50	..	..	..	..	Institutions sans but lucratif
Funds from abroad	..	..	..	..	..	..	..	Fonds de l'étranger
Total	..	..	250	..	..	..	..	*Total*
GERD								**DIRD**
financed by								*financé par*
Business enterprise	..	..	6 740	..	..	..	..	Entreprises
Direct government	..	..	910 [h]	..	..	..	..	État direct
General university funds	..	..	1 780	..	..	..	..	Fonds généraux des universités
Sub-total government	..	..	2 690	..	..	..	..	Sous-Total État
Higher education	..	..	130	..	..	..	..	Enseignement supérieur
Private non-profit	..	..	120	..	..	..	..	Institutions sans but lucratif
Funds from abroad	..	..	310	..	..	..	..	Fonds de l'étranger
Total	..	..	9 990	..	..	..	..	*Total*
	1.2 Million constant USD (1995 prices and PPPs) - Millions de USD constants (prix et PPA de 1995)							
Business enterprise								**Entreprises**
financed by								*financé par*
Business enterprise	..	..	3 235.4	..	..	..	..	Entreprises
Direct government	75.4 [h]	..	84.2 [h]	..	64.4 [h]	..	..	État direct
Higher education	..	..	0.0	..	..	..	..	Enseignement supérieur
Private non-profit	..	..	24.8	..	..	..	..	Institutions sans but lucratif
Funds from abroad	..	..	153.6	..	..	..	..	Fonds de l'étranger
Total	..	..	3 498.0	..	..	..	..	*Total*
Government								**État**
financed by								*financé par*
Business enterprise	..	..	..	..	..	..	..	Entreprises
Direct government	135.8 [h]	..	123.9 [h]	..	99.1 [h]	..	..	État direct
Higher education	..	..	..	..	..	..	..	Enseignement supérieur
Private non-profit	..	..	..	..	..	..	..	Institutions sans but lucratif
Funds from abroad	..	..	..	..	..	..	..	Fonds de l'étranger
Total	135.8 [h]	..	123.9 [h]	..	99.1 [h]	..	..	*Total*
Higher education								**Enseignement supérieur**
financed by								*financé par*
Business enterprise	20.1	..	74.3	..	84.2	..	..	Entreprises
Direct government	221.3 [h]	..	183.3 [h]	..	198.1 [h]	..	..	État direct
General university funds	855.0	..	881.9	..	752.9	..	..	Fonds généraux des universités
Sub-total government	1 076.2	..	1 065.3	..	951.0	..	..	Sous-Total État
Higher education	40.2	..	54.5	..	104.0	..	..	Enseignement supérieur
Private non-profit	35.2	..	9.9	..	44.6	..	..	Institutions sans but lucratif
Funds from abroad	..	..	..	..	..	..	..	Fonds de l'étranger
Total	1 171.8	..	1 204.0	..	1 183.8	..	..	*Total*
Private non-profit								**Institutions sans but lucratif**
financed by								*financé par*
Business enterprise	..	..	29.7	..	..	..	..	Entreprises
Direct government	..	..	59.5 [h]	..	54.5	..	..	État direct
Higher education	..	..	9.9	..	..	..	..	Enseignement supérieur
Private non-profit	..	..	24.8	..	..	..	..	Institutions sans but lucratif
Funds from abroad	..	..	..	..	..	..	..	Fonds de l'étranger
Total	..	..	123.9	..	..	..	..	*Total*
GERD								**DIRD**
financed by								*financé par*
Business enterprise	..	..	3 339.4	..	..	..	..	Entreprises
Direct government	..	..	450.9 [h]	..	..	..	..	État direct
General university funds	..	..	881.9	..	..	..	..	Fonds généraux des universités
Sub-total government	..	..	1 332.8	..	..	..	..	Sous-Total État
Higher education	..	..	64.4	..	..	..	..	Enseignement supérieur
Private non-profit	..	..	59.5	..	..	..	..	Institutions sans but lucratif
Funds from abroad	..	..	153.6	..	..	..	..	Fonds de l'étranger
Total	..	..	4 949.7	..	..	..	..	*Total*

Source: OECD, R&D database (DSTI/EAS Division), January 2002.

Source : OCDE, base de données R-D (DSTI/Division AES), janvier 2002.

Table 2 - Tableau 2

Gross domestic expenditure on R&D (GERD) - Dépenses intérieures brutes de R-D (DIRD)

by sector of performance and type of costs - par secteur d'exécution et par type de coûts

Million CHF | *Millions CHF*

	1994	1995	1996	1997	1998	1999	2000	
Business enterprise								**Entreprises**
Labour costs	..	..	3 980	..	..	..	..	Coûts du personnel
Other current costs	..	..	2 600	..	..	..	..	Autres coûts courants
Total current expenditure	..	..	6 580	..	..	..	..	*Total dépenses courantes*
Land and buildings	..	..	..	..	..	..	..	Terrains et constructions
Instruments and equipment	..	..	..	..	..	..	..	Instruments et équipements
Total capital expenditure	..	..	480	..	..	..	..	*Total dépenses en capital*
Total	..	..	7 060	..	..	..	..	*Total*
Government								**État**
Labour costs	..	..	170 [h]	..	150 [h]	..	..	Coûts du personnel
Other current costs	..	..	50 [h]	..	35 [h]	..	..	Autres coûts courants
Total current expenditure	..	..	220 [h]	..	185 [h]	..	..	*Total dépenses courantes*
Land and buildings	..	..	..	..	..	..	..	Terrains et constructions
Instruments and equipment	..	..	..	..	..	..	..	Instruments et équipements
Total capital expenditure	..	..	30 [h]	..	15 [h]	..	..	*Total dépenses en capital*
Total	270 [h]	..	250 [h]	..	200 [h]	..	..	*Total*
Higher education								**Enseignement supérieur**
Labour costs	..	..	1 380	..	1 490	..	..	Coûts du personnel
Other current costs	..	..	870	..	720	..	..	Autres coûts courants
Total current expenditure	..	..	2 250	..	2 210	..	..	*Total dépenses courantes*
Land and buildings	..	..	..	..	..	..	..	Terrains et constructions
Instruments and equipment	..	..	..	..	..	..	..	Instruments et équipements
Total capital expenditure	..	..	180	..	180	..	..	*Total dépenses en capital*
Total	2 330	..	2 430	..	2 390	..	..	*Total*
Private non-profit								**Institutions sans but lucratif**
Labour costs	..	..	140	..	..	..	..	Coûts du personnel
Other current costs	..	..	90	..	..	..	..	Autres coûts courants
Total current expenditure	..	..	230	..	..	..	..	*Total dépenses courantes*
Land and buildings	..	..	..	..	..	..	..	Terrains et constructions
Instruments and equipment	..	..	..	..	..	..	..	Instruments et équipements
Total capital expenditure	..	..	20	..	..	..	..	*Total dépenses en capital*
Total	..	..	250	..	..	..	..	*Total*
GERD								**DIRD**
Labour costs	..	..	5 670	..	..	..	..	Coûts du personnel
Other current costs	..	..	3 610	..	..	..	..	Autres coûts courants
Total current expenditure	..	..	9 280	..	..	..	..	*Total dépenses courantes*
Land and buildings	..	..	..	..	..	..	..	Terrains et constructions
Instruments and equipment	..	..	..	..	..	..	..	Instruments et équipements
Total capital expenditure	..	..	710	..	..	..	..	*Total dépenses en capital*
Total	..	..	9 990	..	..	..	..	*Total*

Table 3 - Tableau 3

Current domestic expenditure on R&D - Dépenses intérieures courantes de R-D

by sector of performance and type of activity - par secteur d'exécution et par type d'activités

Million CHF | *Millions CHF*

	1994	1995	1996	1997	1998	1999	2000	
Business enterprise								**Entreprises**
Basic research	..	..	705 [k]	..	..	..	..	Recherche fondamentale
Applied research	..	..	2 825 [k]	..	..	..	..	Recherche appliquée
Experimental development	..	..	3 530 [k]	..	..	..	..	Développement expérimental
Not specified	..	..	..	..	..	..	..	Non spécifié
Total	..	..	7 060 [k]	..	..	..	..	*Total*
Government								**État**
Basic research	..	..	17 [h,k]	..	10 [h]	..	..	Recherche fondamentale
Applied research	..	..	125 [h,k]	..	130 [h]	..	..	Recherche appliquée
Experimental development	..	..	108 [h,k]	..	60 [h]	..	..	Développement expérimental
Not specified	270 [h,k]	..	..	..	..	..	..	Non spécifié
Total	270 [h,k]	..	250 [h,k]	..	200 [h]	..	..	*Total*
Higher education								**Enseignement supérieur**
Basic research	..	..	2 070 [k]	..	2 010 [k]	..	..	Recherche fondamentale
Applied research	..	..	250 [k]	..	250 [k]	..	..	Recherche appliquée
Experimental development	..	..	110 [k]	..	130 [k]	..	..	Développement expérimental
Not specified	2 330 [k]	..	..	..	..	..	..	Non spécifié
Total	2 330 [k]	..	2 430 [k]	..	2 390 [k]	..	..	*Total*
Private non-profit								**Institutions sans but lucratif**
Basic research	..	..	..	..	..	..	..	Recherche fondamentale
Applied research	..	..	..	..	..	..	..	Recherche appliquée
Experimental development	..	..	..	..	..	..	..	Développement expérimental
Not specified	..	..	250 [k]	..	..	..	..	Non spécifié
Total	..	..	250 [k]	..	..	..	..	*Total*
GERD								**DIRD**
Basic research	..	..	2 792 [k]	..	..	..	..	Recherche fondamentale
Applied research	..	..	3 200 [k]	..	..	..	..	Recherche appliquée
Experimental development	..	..	3 748 [k]	..	..	..	..	Développement expérimental
Not specified	..	..	250 [k]	..	..	..	..	Non spécifié
Total	..	..	9 990 [k]	..	..	..	..	*Total*

Source: OECD, R&D database (DSTI/EAS Division), January 2002.

Source : OCDE, base de données R-D (DSTI/Division AES), janvier 2002.

Gross domestic expenditure on R&D (GERD) - Dépenses intérieures brutes de R-D (DIRD)

by sector of performance and socio-economic objective - par secteur d'exécution et par objectif socio-économique

Million CHF — *Millions CHF*

	1994	1995	1996	1997	1998	1999	2000	
Business enterprise								**Entreprises**
Agriculture, forestry and fishing	..	..	410	..	..	..	..	Agriculture, sylviculture et pêche
Industrial development	..	..	3 320	..	..	..	..	Développement industriel
Energy	..	..	540	..	..	..	..	Energie
Transport and telecommunications	..	..	.. [n]	..	..	..	..	Transports et télécommunications
Urban and rural planning	..	..	470 [o]	..	..	..	..	Aménagement urbain et rural
Prevention of pollution	..	..	140 [o]	..	..	..	..	Prévention de la pollution
Identification & treatment of pollution	..	..	140 [n]	..	..	..	..	Détection pollu. & mesures antipollu.
Health	..	..	1 550	..	..	..	..	Santé
Social development and services	..	..	50	..	..	..	..	Développement et services sociaux
Earth and atmosphere	..	..	..	..	..	..	..	Milieu terrestre et atmosphère
Advancement of knowledge	..	..	110	..	..	..	..	Promotion générale des connaissances
Civil space	..	..	..	..	..	..	..	Espace civil
Defence	..	..	70	..	..	..	..	Défense
Not elsewhere classified	..	..	400	..	..	..	..	Non classé
Total	..	..	7 060	..	..	..	..	*Total*
Government								**État**
Agriculture, forestry and fishing	..	..	40 [h]	..	70 [a,h]	..	..	Agriculture, sylviculture et pêche
Industrial development	..	..	15 [h]	..	15 [h]	..	..	Développement industriel
Energy	..	..	10 [h]	..	5 [h]	..	..	Energie
Transport and telecommunications	..	..	.. [n]	..	.. [n]	..	..	Transports et télécommunications
Urban and rural planning	..	..	20 [h,o]	..	5 [h,o]	..	..	Aménagement urbain et rural
Prevention of pollution	..	..	25 [h,o]	..	5 [h,o]	..	..	Prévention de la pollution
Identification & treatment of pollution	..	..	.. [n]	..	.. [n]	..	..	Détection pollu. & mesures antipollu.
Health	..	..	20 [h]	..	10 [a,h]	..	..	Santé
Social development and services	..	..	45 [h]	..	25 [h]	..	..	Développement et services sociaux
Earth and atmosphere	..	..	..	..	5 [h]	..	..	Milieu terrestre et atmosphère
Advancement of knowledge	..	..	0 [h]	..	..	..	..	Promotion générale des connaissances
Civil space	..	..	..	..	..	..	..	Espace civil
Defence	..	..	55 [h]	..	50 [h]	..	..	Défense
Not elsewhere classified	270 [h]	..	20 [h]	..	15 [h]	..	..	Non classé
Total	270 [h]	..	250 [h]	..	200 [h]	..	..	*Total*
Higher education								**Enseignement supérieur**
Agriculture, forestry and fishing	..	..	..	..	..	..	..	Agriculture, sylviculture et pêche
Industrial development	..	..	..	..	..	..	..	Développement industriel
Energy	..	..	..	..	..	..	..	Energie
Transport and telecommunications	..	..	..	..	..	..	..	Transports et télécommunications
Urban and rural planning	..	..	..	..	..	..	..	Aménagement urbain et rural
Prevention of pollution	..	..	..	..	..	..	..	Prévention de la pollution
Identification & treatment of pollution	..	..	..	..	..	..	..	Détection pollu. & mesures antipollu.
Health	..	..	..	..	..	..	..	Santé
Social development and services	..	..	..	..	..	..	..	Développement et services sociaux
Earth and atmosphere	..	..	..	..	..	..	..	Milieu terrestre et atmosphère
Advancement of knowledge	..	..	..	..	..	..	..	Promotion générale des connaissances
Civil space	..	..	..	..	..	..	..	Espace civil
Defence	..	..	..	..	..	..	..	Défense
Not elsewhere classified	2 330	..	2 430	..	2 390	..	..	Non classé
Total	2 330	..	2 430	..	2 390	..	..	*Total*
Private non-profit								**Institutions sans but lucratif**
Agriculture, forestry and fishing	..	..	..	..	..	..	..	Agriculture, sylviculture et pêche
Industrial development	..	..	..	..	..	..	..	Développement industriel
Energy	..	..	..	..	..	..	..	Energie
Transport and telecommunications	..	..	..	..	..	..	..	Transports et télécommunications
Urban and rural planning	..	..	..	..	..	..	..	Aménagement urbain et rural
Prevention of pollution	..	..	..	..	..	..	..	Prévention de la pollution
Identification & treatment of pollution	..	..	..	..	..	..	..	Détection pollu. & mesures antipollu.
Health	..	..	..	..	..	..	..	Santé
Social development and services	..	..	..	..	..	..	..	Développement et services sociaux
Earth and atmosphere	..	..	..	..	..	..	..	Milieu terrestre et atmosphère
Advancement of knowledge	..	..	..	..	..	..	..	Promotion générale des connaissances
Civil space	..	..	..	..	..	..	..	Espace civil
Defence	..	..	..	..	..	..	..	Défense
Not elsewhere classified	..	..	250	..	..	..	..	Non classé
Total	..	..	250	..	..	..	..	*Total*
GERD								**DIRD**
Agriculture, forestry and fishing	..	..	..	..	..	..	..	Agriculture, sylviculture et pêche
Industrial development	..	..	..	..	..	..	..	Développement industriel
Energy	..	..	..	..	..	..	..	Energie
Transport and telecommunications	..	..	..	..	..	..	..	Transports et télécommunications
Urban and rural planning	..	..	..	..	..	..	..	Aménagement urbain et rural
Prevention of pollution	..	..	..	..	..	..	..	Prévention de la pollution
Identification & treatment of pollution	..	..	..	..	..	..	..	Détection pollu. & mesures antipollu.
Health	..	..	..	..	..	..	..	Santé
Social development and services	..	..	..	..	..	..	..	Développement et services sociaux
Earth and atmosphere	..	..	..	..	..	..	..	Milieu terrestre et atmosphère
Advancement of knowledge	..	..	..	..	..	..	..	Promotion générale des connaissances
Civil space	..	..	..	..	..	..	..	Espace civil
Defence	..	..	..	..	..	..	..	Défense
Not elsewhere classified	..	..	9 990	..	..	..	..	Non classé
Total	..	..	9 990	..	..	..	..	*Total*

Source: OECD, R&D database (DSTI/EAS Division), January 2002.

Source : OCDE, base de données R-D (DSTI/Division AES), janvier 2002.

Table 5 - Tableau 5

Total R&D personnel - Personnel total de R-D

by sector of employment and occupation - par secteur d'emploi et par profession

In full-time equivalent — *En équivalence plein-temps*

	1994	1995	1996	1997	1998	1999	2000	
Business enterprise								**Entreprises**
RSE	..	..	12 500	..	..	..	..	Chercheurs
Technicians	..	..	17 950	..	..	..	..	Techniciens
Other occupations	..	..	4 000	..	..	..	..	Autres professions
Total	..	..	34 450	..	..	..	..	*Total*
Government								**État**
RSE	550 [h]	..	555 [h]	..	440 [h]	..	..	Chercheurs
Technicians	750 [h]	..	640 [h]	..	470 [h]	..	..	Techniciens
Other occupations	300 [h]	..	190 [h]	..	205 [h]	..	..	Autres professions
Total	1 600 [h]	..	1 385 [h]	..	1 115 [h]	..	..	*Total*
Higher education								**Enseignement supérieur**
RSE	7 680	..	8 580	..	8 970	..	..	Chercheurs
Technicians	4 170	..	4 960	..	4 770	..	..	Techniciens
Other occupations	690	..	890	..	810	..	..	Autres professions
Total	12 540	..	14 430	..	14 550	..	..	*Total*
Private non-profit								**Institutions sans but lucratif**
RSE	..	..	..	..	..	..	..	Chercheurs
Technicians	..	..	..	..	..	..	..	Techniciens
Other occupations	..	..	..	..	..	..	..	Autres professions
Total	..	..	..	..	..	..	..	*Total*
TOTAL								**TOTAL**
RSE	..	..	21 635	..	..	..	..	Chercheurs
Technicians	..	..	23 550	..	..	..	..	Techniciens
Other occupations	..	..	5 080	..	..	..	..	Autres professions
Total	..	..	50 265	..	..	..	..	*Total*

Table 6 - Tableau 6

Total R&D personnel - Personnel total de R-D

by sector of employment and formal qualification - par secteur d'emploi et par niveau de qualification

In full-time equivalent — *En équivalence plein-temps*

	1994	1995	1996	1997	1998	1999	2000	
Business enterprise								**Entreprises**
University PhD level degrees	..	..	..	..	..	..	..	Diplômes univ. de niv. doctorat
Other university degrees	..	..	..	..	..	..	..	Autres diplômes universitaires
Sub-total university degrees	..	..	10 240	..	..	..	..	Sous-total Universitaires
Other post-secondary diplomas	..	..	7 255	..	..	..	..	Autres diplômes post-secondaires
Secondary diplomas	..	..	..	..	..	..	..	Diplômes de fin d'études secondaires
Other qualifications	..	..	16 955	..	..	..	..	Autres qualifications
Total	..	..	34 450	..	..	..	..	*Total*
Government								**État**
University PhD level degrees	..	..	..	..	..	..	..	Diplômes univ. de niv. doctorat
Other university degrees	..	..	..	..	..	..	..	Autres diplômes universitaires
Sub-total university degrees	570 [h]	..	535 [h]	..	450 [h]	..	..	Sous-total Universitaires
Other post-secondary diplomas	320 [h]	..	275 [h]	..	215 [h]	..	..	Autres diplômes post-secondaires
Secondary diplomas	..	..	..	..	..	..	..	Diplômes de fin d'études secondaires
Other qualifications	710 [h]	..	575 [h]	..	450 [h]	..	..	Autres qualifications
Total	1 600 [h]	..	1 385 [h]	..	1 115 [h]	..	..	*Total*
Higher education								**Enseignement supérieur**
University PhD level degrees	..	..	..	..	..	..	..	Diplômes univ. de niv. doctorat
Other university degrees	..	..	..	..	..	..	..	Autres diplômes universitaires
Sub-total university degrees	7 680	..	8 580	..	8 760	..	..	Sous-total Universitaires
Other post-secondary diplomas	200	..	390	..	380	..	..	Autres diplômes post-secondaires
Secondary diplomas	..	..	..	..	..	..	..	Diplômes de fin d'études secondaires
Other qualifications	3 970	..	4 570	..	4 690	..	..	Autres qualifications
Total	12 540	..	14 430	..	14 550	..	..	*Total*
Private non-profit								**Institutions sans but lucratif**
University PhD level degrees	..	..	..	..	..	..	..	Diplômes univ. de niv. doctorat
Other university degrees	..	..	..	..	..	..	..	Autres diplômes universitaires
Sub-total university degrees	..	..	..	..	..	..	..	Sous-total Universitaires
Other post-secondary diplomas	..	..	..	..	..	..	..	Autres diplômes post-secondaires
Secondary diplomas	..	..	..	..	..	..	..	Diplômes de fin d'études secondaires
Other qualifications	..	..	..	..	..	..	..	Autres qualifications
Total	..	..	..	..	..	..	..	*Total*
TOTAL								**TOTAL**
University PhD level degrees	..	..	..	..	..	..	..	Diplômes univ. de niv. doctorat
Other university degrees	..	..	..	..	..	..	..	Autres diplômes universitaires
Sub-total university degrees	..	..	19 355	..	..	..	..	Sous-total Universitaires
Other post-secondary diplomas	..	..	7 920	..	..	..	..	Autres diplômes post-secondaires
Secondary diplomas	..	..	..	..	..	..	..	Diplômes de fin d'études secondaires
Other qualifications	..	..	22 100	..	..	..	..	Autres qualifications
Total	..	..	50 265	..	..	..	..	*Total*

Source: OECD, R&D database (DSTI/EAS Division), January 2002.

Source : OCDE, base de données R-D (DSTI/Division AES), janvier 2002.

Table 7 - Tableau 7

Gross domestic expenditure on R&D (GERD) - Dépenses intérieures brutes de R-D (DIRD)

by sector of performance and main scientific field - par secteur d'exécution et par grand domaine scientifique

Million CHF / *Millions CHF*

	1994	1995	1996	1997	1998	1999	2000	
Business enterprise								**Entreprises**
Natural sciences and engineering	..	..	..	..	..	..	..	Sciences exactes, nat. et de l'ingénieur
Social sciences and humanities	..	..	..	..	..	..	..	Sciences sociales et humaines
All fields of science	..	..	7 060	..	..	..	..	*Tous domaines scientifiques*
Government								**État**
Natural sciences and engineering	..	..	..	..	..	..	..	Sciences exactes, nat. et de l'ingénieur
Social sciences and humanities	..	..	..	..	..	..	..	Sciences sociales et humaines
All fields of science	270 [h]	..	250 [h]	..	200 [h]	..	..	*Tous domaines scientifiques*
Higher education								**Enseignement supérieur**
Natural sciences and engineering	..	..	2 020	..	1 590	..	..	Sciences exactes, nat. et de l'ingénieur
Social sciences and humanities	..	..	410	..	440	..	..	Sciences sociales et humaines
All fields of science	2 330	..	2 430	..	2 390	..	..	*Tous domaines scientifiques*
Private non-profit								**Institutions sans but lucratif**
Natural sciences and engineering	..	..	..	..	..	..	..	Sciences exactes, nat. et de l'ingénieur
Social sciences and humanities	..	..	..	..	..	..	..	Sciences sociales et humaines
All fields of science	..	..	250	..	..	..	..	*Tous domaines scientifiques*
TOTAL								**TOTAL**
Natural sciences and engineering	..	..	..	..	..	..	..	Sciences exactes, nat. et de l'ingénieur
Social sciences and humanities	..	..	..	..	..	..	..	Sciences sociales et humaines
All fields of science	..	..	9 990	..	..	..	..	*Tous domaines scientifiques*

Table 8 - Tableau 8

Total R&D personnel - Personnel total de R-D

by sector of employment and main scientific field - par secteur d'emploi et par grand domaine scientifique

In full-time equivalent / *En équivalence plein-temps*

	1994	1995	1996	1997	1998	1999	2000	
Business enterprise								**Entreprises**
Natural sciences and engineering	..	..	..	..	..	..	..	Sciences exactes, nat. et de l'ingénieur
Social sciences and humanities	..	..	..	..	..	..	..	Sciences sociales et humaines
All fields of science	..	..	34 450	..	..	..	..	*Tous domaines scientifiques*
Government								**État**
Natural sciences and engineering	..	..	..	..	..	..	..	Sciences exactes, nat. et de l'ingénieur
Social sciences and humanities	..	..	..	..	..	..	..	Sciences sociales et humaines
All fields of science	1 600 [h]	..	1 385 [h]	..	1 115 [h]	..	..	*Tous domaines scientifiques*
Higher education								**Enseignement supérieur**
Natural sciences and engineering	10 350	..	..	..	8 970	..	..	Sciences exactes, nat. et de l'ingénieur
Social sciences and humanities	1 800	..	..	..	2 140	..	..	Sciences sociales et humaines
All fields of science	12 540	..	14 430	..	14 550	..	..	*Tous domaines scientifiques*
Private non-profit								**Institutions sans but lucratif**
Natural sciences and engineering	..	..	..	..	..	..	..	Sciences exactes, nat. et de l'ingénieur
Social sciences and humanities	..	..	..	..	..	..	..	Sciences sociales et humaines
All fields of science	..	..	..	..	..	..	..	*Tous domaines scientifiques*
TOTAL								**TOTAL**
Natural sciences and engineering	..	..	..	..	..	..	..	Sciences exactes, nat. et de l'ingénieur
Social sciences and humanities	..	..	..	..	..	..	..	Sciences sociales et humaines
All fields of science	..	..	50 265	..	..	..	..	*Tous domaines scientifiques*

Source: OECD, R&D database (DSTI/EAS Division), January 2002.

Source : OCDE, base de données R-D (DSTI/Division AES), janvier 2002.

SWITZERLAND — Table 9 - Tableau 9 — SUISSE

Total business enterprise intramural expenditure on R&D - Dépenses intra-muros totales de R-D des entreprises
(BERD) by industry - (DIRDE) par industrie

	1994	1995	1996	1997	1998	1999	2000	
	9.1 Million CHF - Millions CHF							
1 AGRICULTURE, HUNTING & FORESTRY	..	..	..	..	..	..	..	**1 AGRICULTURE, CHASSE & SYLVICULTURE**
2 MINING	..	..	..	..	..	..	..	**2 ACTIVITES EXTRACTIVES**
3 MANUFACTURING	..	..	5 659	..	..	..	..	**3 ACTIVITES DE FABRICATION**
4 Food, beverages & tobacco	..	..	360	..	..	..	..	**4 Produits alimentaires, boissons & tabac**
5 Food products & beverages	..	..	..	..	..	..	..	5 Produits alimentaires & boissons
6 Tobacco products	..	..	..	..	..	..	..	6 Produits à base de tabac
7 Textiles, wear app., fur & leather	..	..	..	..	..	..	..	**7 Textiles, habill., fourrures & cuirs**
8 Textiles	..	..	..	..	..	..	..	8 Textiles
9 Wearing apparel & fur	..	..	..	..	..	..	..	9 Articles d'habillement & fourrures
10 Leather products & footwear	..	..	..	..	..	..	..	10 Articles en cuir & chaussures
11 Wood, paper, printing, publishing	..	..	..	..	..	..	..	**11 Bois, papier, imprimerie, édition**
12 Wood & cork (not furniture)	..	..	..	..	..	..	..	12 Bois & liège (sauf meubles)
13 Pulp, paper & paper products	..	..	..	..	..	..	..	13 Carton, papier & articles
14 Publ., print. & repro. of rec. media	..	..	..	..	..	..	..	14 Edition, impress. & repro. enregis.
15 Coke, petroleum, nuclear fuel, chemicals, rubber plastics	..	..	2 627	..	..	..	..	**15 Coke, pétrole, combust. nucl., prod. chimi., caout. & mat. plast.**
16 Coke, ref. petrol. prod. & nuclear fuel	..	..	..	..	..	..	..	16 Coke, raffin. pétr. & combust. nucl.
17 Chemicals & chemical products	..	..	..	..	..	..	..	17 Produits chimiques
18 *Chemical prod.(less pharmaceuticals)*	..	..	..	..	..	..	..	18 *Prod. chimi. (moins prod. pharmaceutiques)*
19 *Pharmaceuticals*	..	..	..	..	..	..	..	19 *Produits pharmaceutiques*
20 Rubber & plastic products	..	..	..	..	..	..	..	20 Caoutchouc et matières plastiques
21 Non-metallic mineral products	..	..	..	..	..	..	..	**21 Produits minéraux non métalliques**
22 Basic metals	..	..	..	..	..	..	..	**22 Produits métallurgiques de base**
23 Basic metals, ferrous	..	..	..	..	..	..	..	23 Produits métallur. de base, ferreux
24 Basic metals, non-ferrous	..	..	..	..	..	..	..	24 Prod. métallur. de base, non ferreux
25 Fabricated metal products	..	..	81	..	..	..	..	**25 Ouvrages en métaux (sauf mach.& mat.)**
26 Machinery equipment, instruments and transport equipment	..	..	2 293	..	..	..	..	**26 Machines & matériel, instruments & mat. de transport**
27 Machinery n.e.c.	..	..	..	..	..	..	..	27 Machines, n.c.a.
28 Office, account. & computing machinery	..	..	..	..	..	..	..	28 Mach. de bur., compta. & trait. inform.
29 Electrical machinery	..	..	..	..	..	..	..	29 Machines électriques
30 Electro. equip. (radio, TV & commun.)	..	..	..	..	..	..	..	30 Equip. électro. (radio, tv & communic.)
31 *Electro. comp. (inc. semi-conductors)*	..	..	..	..	..	..	..	31 *Compos. électro. (semi-cond. incl.)*
32 *TV, radio & communications equip.*	..	..	..	..	..	..	..	32 *Appareils radio, tv et de communi.*
33 Instruments	..	..	..	..	..	..	..	33 Instruments
34 Motor vehicles	..	..	..	..	..	..	..	34 Véhicules automobiles
35 Other transport equipment	..	..	..	..	..	..	..	35 Autres matériels de transport
36 *Ships*	..	..	..	..	..	..	..	36 *Navires*
37 *Aerospace*	..	..	..	..	..	..	..	37 *Aérospatiale*
38 *Other transport equipment n.e.c.*	..	..	..	..	..	..	..	38 *Autres matériels de transport n.c.a.*
39 Furniture, other manufacturing n.e.c.	..	..	..	..	..	..	..	**39 Meubles, autres activ. de fabricat. n.c.a.**
40 Furniture	..	..	..	..	..	..	..	40 Meubles
41 Other manufacturing n.e.c.	..	..	..	..	..	..	..	41 Autres activités de fabrication n.c.a.
42 Recycling	..	..	..	..	..	..	..	**42 Récupération**
43 ELECTRICITY, GAS & WATER SUPPLY	..	..	..	..	..	..	..	**43 DISTRIBUTION D'ELECTRICITE, GAZ & EAU**
44 CONSTRUCTION	..	..	..	..	..	..	..	**44 CONSTRUCTION**
45 SERVICES SECTOR	..	..	1 350	..	..	..	..	**45 SECTEUR DES SERVICES**
46 Wholesale, retail trade & repair of motor vehicles, etc.	..	..	..	..	..	..	..	**46 Commerce (gros & détail), réparation de véhicules, etc.**
47 Hotels & restaurants	..	..	..	..	..	..	..	**47 Hôtels et restaurants**
48 Transport & storage	..	..	..	..	..	..	..	**48 Transports, entreposage**
49 Communications	..	..	..	..	..	..	..	**49 Communications**
50 Post	..	..	..	..	..	..	..	50 Postes
51 Telecommunications	..	..	..	..	..	..	..	51 Télécommunications
52 Financial intermediation (inc. insur.)	..	..	290	..	..	..	..	**52 Intermédiation financ. (assurances inc.)**
53 Real estate, renting & busin. activity	..	..	..	..	..	..	..	**53 Immobil., locat. & serv. aux entreprises**
54 Computer & related activities	..	..	..	..	..	..	..	54 Informatique et activités connexes
55 *Software consultancy*	..	..	..	..	..	..	..	55 *Production de logiciels*
56 *Other computer services n.e.c.*	..	..	..	..	..	..	..	56 *Autres activités informatiques*
57 Research & development	..	..	395	..	..	..	..	57 Recherche & développement expérimental
58 Other business activities n.e.c.	..	..	595	..	..	..	..	58 Autres activ. & serv. aux entreprises
59 Comm., soc. & pers. serv. activ., etc.	..	..	..	..	..	..	..	**59 Adm. pub., serv. soc. & collec., etc**
TOTAL BUSINESS ENTERPRISE SECTOR	**..**	**..**	**7 060**	**..**	**..**	**..**	**..**	**TOTAL SECTEUR DES ENTREPRISES**

Source: OECD, R&D database (DSTI/EAS Division), January 2002.

Source : OCDE, base de données R-D (DSTI/Division AES), janvier 2002.

Total business enterprise intramural expenditure on R&D - Dépenses intra-muros totales de R-D des entreprises (BERD) by industry - (DIRDE) par industrie

9.2 Million constant USD(1995 prices and PPPs) - Millions de USD constants (prix et PPA de 1995)

	1994	1995	1996	1997	1998	1999	2000	
1 **AGRICULTURE, HUNTING & FORESTRY**	..	..	..	..	..	..	..	1 **AGRICULTURE, CHASSE & SYLVICULTURE**
2 **MINING**	..	..	..	..	..	..	..	2 **ACTIVITES EXTRACTIVES**
3 **MANUFACTURING**	..	..	2 803.8	..	..	..	..	3 **ACTIVITES DE FABRICATION**
4 **Food, beverages & tobacco**	..	..	178.4	..	..	..	..	4 **Produits alimentaires, boissons & tabac**
5 Food products & beverages	..	..	..	..	..	..	..	5 Produits alimentaires & boissons
6 Tobacco products	..	..	..	..	..	..	..	6 Produits à base de tabac
7 **Textiles, wear app., fur & leather**	..	..	..	..	..	..	..	7 **Textiles, habill., fourrures & cuirs**
8 Textiles	..	..	..	..	..	..	..	8 Textiles
9 Wearing apparel & fur	..	..	..	..	..	..	..	9 Articles d'habillement & fourrures
10 Leather products & footwear	..	..	..	..	..	..	..	10 Articles en cuir & chaussures
11 **Wood, paper, printing, publishing**	..	..	..	..	..	..	..	11 **Bois, papier, imprimerie, édition**
12 Wood & cork (not furniture)	..	..	..	..	..	..	..	12 Bois & liège (sauf meubles)
13 Pulp, paper & paper products	..	..	..	..	..	..	..	13 Carton, papier & articles
14 Publ., print. & repro. of rec. media	..	..	..	..	..	..	..	14 Edition, impress. & repro. enregis.
15 **Coke, petroleum, nuclear fuel, chemicals, rubber plastics**	..	..	1 301.6	..	..	..	..	15 **Coke, pétrole, combust. nucl., prod. chimi., caout. & mat. plast.**
16 Coke, ref. petrol. prod. & nuclear fuel	..	..	..	..	..	..	..	16 Coke, raffin. pétr. & combust. nucl.
17 Chemicals & chemical products	..	..	..	..	..	..	..	17 Produits chimiques
18 *Chemical prod.(less pharmaceuticals)*	..	..	..	..	..	..	..	18 *Prod. chimi. (moins prod. pharmaceutiques)*
19 *Pharmaceuticals*	..	..	..	..	..	..	..	19 *Produits pharmaceutiques*
20 Rubber & plastic products	..	..	..	..	..	..	..	20 Caoutchouc et matières plastiques
21 **Non-metallic mineral products**	..	..	..	..	..	..	..	21 **Produits minéraux non métalliques**
22 **Basic metals**	..	..	..	..	..	..	..	22 **Produits métallurgiques de base**
23 Basic metals, ferrous	..	..	..	..	..	..	..	23 Produits métallur. de base, ferreux
24 Basic metals, non-ferrous	..	..	..	..	..	..	..	24 Prod. métallur. de base, non ferreux
25 **Fabricated metal products**	..	..	40.1	..	..	..	..	25 **Ouvrages en métaux (sauf mach.& mat.)**
26 **Machinery equipment, instruments and transport equipment**	..	..	1 136.1	..	..	..	..	26 **Machines & matériel, instruments & mat. de transport**
27 Machinery n.e.c.	..	..	..	..	..	..	..	27 Machines, n.c.a.
28 Office, account. & computing machinery	..	..	..	..	..	..	..	28 Mach. de bur., compta. & trait. inform.
29 Electrical machinery	..	..	..	..	..	..	..	29 Machines électriques
30 Electro. equip. (radio, TV & commun.)	..	..	..	..	..	..	..	30 Equip. électro. (radio, tv & communic.)
31 *Electro. comp. (inc. semi-conductors)*	..	..	..	..	..	..	..	31 *Compos. électro. (semi-cond. incl.)*
32 *TV, radio & communications equip.*	..	..	..	..	..	..	..	32 *Appareils radio, tv et de communi.*
33 Instruments	..	..	..	..	..	..	..	33 Instruments
34 Motor vehicles	..	..	..	..	..	..	..	34 Véhicules automobiles
35 Other transport equipment	..	..	..	..	..	..	..	35 Autres matériels de transport
36 *Ships*	..	..	..	..	..	..	..	36 *Navires*
37 *Aerospace*	..	..	..	..	..	..	..	37 *Aérospatiale*
38 *Other transport equipment n.e.c.*	..	..	..	..	..	..	..	38 *Autres matériels de transport n.c.a.*
39 **Furniture, other manufacturing n.e.c.**	..	..	..	..	..	..	..	39 **Meubles, autres activ. de fabricat. n.c.a.**
40 Furniture	..	..	..	..	..	..	..	40 Meubles
41 Other manufacturing n.e.c.	..	..	..	..	..	..	..	41 Autres activités de fabrication n.c.a.
42 **Recycling**	..	..	..	..	..	..	..	42 **Récupération**
43 **ELECTRICITY, GAS & WATER SUPPLY**	..	..	..	..	..	..	..	43 **DISTRIBUTION D'ELECTRICITE, GAZ & EAU**
44 **CONSTRUCTION**	..	..	..	..	..	..	..	44 **CONSTRUCTION**
45 **SERVICES SECTOR**	..	..	668.9	..	..	..	..	45 **SECTEUR DES SERVICES**
46 **Wholesale, retail trade & repair of motor vehicles, etc.**	..	..	..	..	..	..	..	46 **Commerce (gros & détail), réparation de véhicules, etc.**
47 **Hotels & restaurants**	..	..	..	..	..	..	..	47 **Hôtels et restaurants**
48 **Transport & storage**	..	..	..	..	..	..	..	48 **Transports, entreposage**
49 **Communications**	..	..	..	..	..	..	..	49 **Communications**
50 Post	..	..	..	..	..	..	..	50 Postes
51 Telecommunications	..	..	..	..	..	..	..	51 Télécommunications
52 **Financial intermediation (inc. insur.)**	..	..	143.7	..	..	..	..	52 **Intermédiation financ. (assurances inc.)**
53 **Real estate, renting & busin. activity**	..	..	..	..	..	..	..	53 **Immobil., locat. & serv. aux entreprises**
54 Computer & related activities	..	..	..	..	..	..	..	54 Informatique et activités connexes
55 *Software consultancy*	..	..	..	..	..	..	..	55 *Production de logiciels*
56 *Other computer services n.e.c.*	..	..	..	..	..	..	..	56 *Autres activités informatiques*
57 Research & development	..	..	195.7	..	..	..	..	57 Recherche & développement expérimental
58 Other business activities n.e.c.	..	..	294.8	..	..	..	..	58 Autres activ. & serv. aux entreprises
59 **Comm., soc. & pers. serv. activ., etc.**	..	..	..	..	..	..	..	59 **Adm. pub., serv. soc. & collec., etc**
TOTAL BUSINESS ENTERPRISE SECTOR	**..**	**..**	**3 498.0**	**..**	**..**	**..**	**..**	**TOTAL SECTEUR DES ENTREPRISES**

Source: OECD, R&D database (DSTI/EAS Division), January 2002.

Source : OCDE, base de données R-D (DSTI/Division AES), janvier 2002.

Business enterprise intramural expenditure on R&D (BERD) - Dépenses intra-muros de R-D des entreprises (DIRDE)

by industry and source of funds - par industrie et par source de financement

Million CHF — *Millions CHF*

10.1 Business enterprise - Entreprises

	1994	1995	1996	1997	1998	1999	2000	
1 AGRICULTURE, HUNTING & FORESTRY	..	..	..	..	..	..	..	**1 AGRICULTURE, CHASSE & SYLVICULTURE**
2 MINING	..	..	..	..	..	..	..	**2 ACTIVITES EXTRACTIVES**
3 MANUFACTURING	..	..	5 413	..	..	..	..	**3 ACTIVITES DE FABRICATION**
4 Food, beverages & tobacco	..	..	355	..	..	..	..	**4 Produits alimentaires, boissons & tabac**
5 Food products & beverages	..	..	..	..	..	..	..	5 Produits alimentaires & boissons
6 Tobacco products	..	..	..	..	..	..	..	6 Produits à base de tabac
7 Textiles, wear app., fur & leather	..	..	..	..	..	..	..	**7 Textiles, habill., fourrures & cuirs**
8 Textiles	..	..	..	..	..	..	..	8 Textiles
9 Wearing apparel & fur	..	..	..	..	..	..	..	9 Articles d'habillement & fourrures
10 Leather products & footwear	..	..	..	..	..	..	..	10 Articles en cuir & chaussures
11 Wood, paper, printing, publishing	..	..	..	..	..	..	..	**11 Bois, papier, imprimerie, édition**
12 Wood & cork (not furniture)	..	..	..	..	..	..	..	12 Bois & liège (sauf meubles)
13 Pulp, paper & paper products	..	..	..	..	..	..	..	13 Carton, papier & articles
14 Publ., print. & repro. of rec. media	..	..	..	..	..	..	..	14 Edition, impress. & repro. enregis.
15 Coke, petroleum, nuclear fuel, chemicals, rubber plastics	..	..	2 613	..	..	..	..	**15 Coke, pétrole, combust. nucl., prod. chimi., caout. & mat. plast.**
16 Coke, ref. petrol. prod. & nuclear fuel	..	..	..	..	..	..	..	16 Coke, raffin. pétr. & combust. nucl.
17 Chemicals & chemical products	..	..	..	..	..	..	..	17 Produits chimiques
18 *Chemical prod.(less pharmaceuticals)*	..	..	..	..	..	..	..	18 *Prod. chimi. (moins prod. pharmaceutiques)*
19 *Pharmaceuticals*	..	..	..	..	..	..	..	19 *Produits pharmaceutiques*
20 Rubber & plastic products	..	..	..	..	..	..	..	20 Caoutchouc et matières plastiques
21 Non-metallic mineral products	..	..	..	..	..	..	..	**21 Produits minéraux non métalliques**
22 Basic metals	..	..	..	..	..	..	..	**22 Produits métallurgiques de base**
23 Basic metals, ferrous	..	..	..	..	..	..	..	23 Produits métallur. de base, ferreux
24 Basic metals, non-ferrous	..	..	..	..	..	..	..	24 Prod. métallur. de base, non ferreux
25 Fabricated metal products	..	..	78	..	..	..	..	**25 Ouvrages en métaux (sauf mach.& mat.)**
26 Machinery equipment, instruments and transport equipment	..	..	2 071	..	..	..	..	**26 Machines & matériel, instruments & mat. de transport**
27 Machinery n.e.c.	..	..	..	..	..	..	..	27 Machines, n.c.a.
28 Office, account. & computing machinery	..	..	..	..	..	..	..	28 Mach. de bur., compta. & trait. inform.
29 Electrical machinery	..	..	..	..	..	..	..	29 Machines électriques
30 Electro. equip. (radio, TV & commun.)	..	..	..	..	..	..	..	30 Equip. électro. (radio, tv & communic.)
31 *Electro. comp. (inc. semi-conductors)*	..	..	..	..	..	..	..	31 *Compos. électro. (semi-cond. incl.)*
32 *TV, radio & communications equip.*	..	..	..	..	..	..	..	32 *Appareils radio, tv et de communi.*
33 Instruments	..	..	..	..	..	..	..	33 Instruments
34 Motor vehicles	..	..	..	..	..	..	..	34 Véhicules automobiles
35 Other transport equipment	..	..	..	..	..	..	..	35 Autres matériels de transport
36 *Ships*	..	..	..	..	..	..	..	36 *Navires*
37 *Aerospace*	..	..	..	..	..	..	..	37 *Aérospatiale*
38 *Other transport equipment n.e.c.*	..	..	..	..	..	..	..	38 *Autres matériels de transport n.c.a.*
39 Furniture, other manufacturing n.e.c.	..	..	..	..	..	..	..	**39 Meubles, autres activ. de fabricat. n.c.a.**
40 Furniture	..	..	..	..	..	..	..	40 Meubles
41 Other manufacturing n.e.c.	..	..	..	..	..	..	..	41 Autres activités de fabrication n.c.a.
42 Recycling	..	..	..	..	..	..	..	**42 Récupération**
43 ELECTRICITY, GAS & WATER SUPPLY	..	..	..	..	..	..	..	**43 DISTRIBUTION D'ELECTRICITE, GAZ & EAU**
44 CONSTRUCTION	..	..	..	..	..	..	..	**44 CONSTRUCTION**
45 SERVICES SECTOR	..	..	1 086	..	..	..	..	**45 SECTEUR DES SERVICES**
46 Wholesale, retail trade & repair of motor vehicles, etc.	..	..	..	..	..	..	..	**46 Commerce (gros & détail), réparation de véhicules, etc.**
47 Hotels & restaurants	..	..	..	..	..	..	..	**47 Hôtels et restaurants**
48 Transport & storage	..	..	..	..	..	..	..	**48 Transports, entreposage**
49 Communications	..	..	..	..	..	..	..	**49 Communications**
50 Post	..	..	..	..	..	..	..	50 Postes
51 Telecommunications	..	..	..	..	..	..	..	51 Télécommunications
52 Financial intermediation (inc. insur.)	..	..	290	..	..	..	..	**52 Intermédiation financ. (assurances inc.)**
53 Real estate, renting & busin. activity	..	..	..	..	..	..	..	**53 Immobil., locat. & serv. aux entreprises**
54 Computer & related activities	..	..	..	..	..	..	..	54 Informatique et activités connexes
55 *Software consultancy*	..	..	..	..	..	..	..	55 *Production de logiciels*
56 *Other computer services n.e.c.*	..	..	..	..	..	..	..	56 *Autres activités informatiques*
57 Research & development	..	..	257	..	..	..	..	57 Recherche & développement expérimental
58 Other business activities n.e.c.	..	..	496	..	..	..	..	58 Autres activ. & serv. aux entreprises
59 Comm., soc. & pers. serv. activ., etc.	..	..	..	..	..	..	..	**59 Adm. pub., serv. soc. & collec., etc**
TOTAL BUSINESS ENTERPRISE SECTOR	**..**	**..**	**6 530**	**..**	**..**	**..**	**..**	**TOTAL SECTEUR DES ENTREPRISES**

Source: OECD, R&D database (DSTI/EAS Division), January 2002.

Source : OCDE, base de données R-D (DSTI/Division AES), janvier 2002.

SWITZERLAND — Table 10 - Tableau 10 — SUISSE

Business enterprise intramural expenditure on R&D (BERD) - Dépenses intra-muros de R-D des entreprises (DIRDE)

by industry and source of funds - par industrie et par source de financement

Million CHF — *Millions CHF*

10.2 Government - État

	1994	1995	1996	1997	1998	1999	2000	
1 **AGRICULTURE, HUNTING & FORESTRY**	..	..	..	..	..	..	..	1 **AGRICULTURE, CHASSE & SYLVICULTURE**
2 **MINING**	..	..	..	..	..	..	..	2 **ACTIVITES EXTRACTIVES**
3 **MANUFACTURING**	..	..	34	..	..	..	..	3 **ACTIVITES DE FABRICATION**
4 **Food, beverages & tobacco**	..	..	4	..	..	..	..	4 **Produits alimentaires, boissons & tabac**
5 Food products & beverages	..	..	..	..	..	..	..	5 Produits alimentaires & boissons
6 Tobacco products	..	..	..	..	..	..	..	6 Produits à base de tabac
7 **Textiles, wear app., fur & leather**	..	..	..	..	..	..	..	7 **Textiles, habill., fourrures & cuirs**
8 Textiles	..	..	..	..	..	..	..	8 Textiles
9 Wearing apparel & fur	..	..	..	..	..	..	..	9 Articles d'habillement & fourrures
10 Leather products & footwear	..	..	..	..	..	..	..	10 Articles en cuir & chaussures
11 **Wood, paper, printing, publishing**	..	..	..	..	..	..	..	11 **Bois, papier, imprimerie, édition**
12 Wood & cork (not furniture)	..	..	..	..	..	..	..	12 Bois & liège (sauf meubles)
13 Pulp, paper & paper products	..	..	..	..	..	..	..	13 Carton, papier & articles
14 Publ., print. & repro. of rec. media	..	..	..	..	..	..	..	14 Edition, impress. & repro. enregis.
15 **Coke, petroleum, nuclear fuel, chemicals, rubber plastics**	..	..	7	..	..	..	..	15 **Coke, pétrole, combust. nucl., prod. chimi., caout. & mat. plast.**
16 Coke, ref. petrol. prod. & nuclear fuel	..	..	..	..	..	..	..	16 Coke, raffin. pétr. & combust. nucl.
17 Chemicals & chemical products	..	..	..	..	..	..	..	17 Produits chimiques
18 *Chemical prod.(less pharmaceuticals)*	..	..	..	..	..	..	..	18 *Prod. chimi. (moins prod. pharmaceutiques)*
19 *Pharmaceuticals*	..	..	..	..	..	..	..	19 *Produits pharmaceutiques*
20 Rubber & plastic products	..	..	..	..	..	..	..	20 Caoutchouc et matières plastiques
21 **Non-metallic mineral products**	..	..	..	..	..	..	..	21 **Produits minéraux non métalliques**
22 **Basic metals**	..	..	..	..	..	..	..	22 **Produits métallurgiques de base**
23 Basic metals, ferrous	..	..	..	..	..	..	..	23 Produits métallur. de base, ferreux
24 Basic metals, non-ferrous	..	..	..	..	..	..	..	24 Prod. métallur. de base, non ferreux
25 **Fabricated metal products**	..	..	..	..	..	..	..	25 **Ouvrages en métaux (sauf mach.& mat.)**
26 **Machinery equipment, instruments and transport equipment**	..	..	22	..	..	..	..	26 **Machines & matériel, instruments & mat. de transport**
27 Machinery n.e.c.	..	..	..	..	..	..	..	27 Machines, n.c.a.
28 Office, account. & computing machinery	..	..	..	..	..	..	..	28 Mach. de bur., compta. & trait. inform.
29 Electrical machinery	..	..	..	..	..	..	..	29 Machines électriques
30 Electro. equip. (radio, TV & commun.)	..	..	..	..	..	..	..	30 Equip. électro. (radio, tv & communic.)
31 *Electro. comp. (inc. semi-conductors)*	..	..	..	..	..	..	..	31 *Compos. électro. (semi-cond. incl.)*
32 *TV, radio & communications equip.*	..	..	..	..	..	..	..	32 *Appareils radio, tv et de communi.*
33 Instruments	..	..	..	..	..	..	..	33 Instruments
34 Motor vehicles	..	..	..	..	..	..	..	34 Véhicules automobiles
35 Other transport equipment	..	..	..	..	..	..	..	35 Autres matériels de transport
36 *Ships*	..	..	..	..	..	..	..	36 *Navires*
37 *Aerospace*	..	..	..	..	..	..	..	37 *Aérospatiale*
38 *Other transport equipment n.e.c.*	..	..	..	..	..	..	..	38 *Autres matériels de transport n.c.a.*
39 **Furniture, other manufacturing n.e.c.**	..	..	..	..	..	..	..	39 **Meubles, autres activ. de fabricat. n.c.a.**
40 Furniture	..	..	..	..	..	..	..	40 Meubles
41 Other manufacturing n.e.c.	..	..	..	..	..	..	..	41 Autres activités de fabrication n.c.a.
42 **Recycling**	..	..	..	..	..	..	..	42 **Récupération**
43 **ELECTRICITY, GAS & WATER SUPPLY**	..	..	..	..	..	..	..	43 **DISTRIBUTION D'ELECTRICITE, GAZ & EAU**
44 **CONSTRUCTION**	..	..	..	..	..	..	..	44 **CONSTRUCTION**
45 **SERVICES SECTOR**	..	..	131	..	..	..	..	45 **SECTEUR DES SERVICES**
46 **Wholesale, retail trade & repair of motor vehicles, etc.**	..	..	..	..	..	..	..	46 **Commerce (gros & détail), réparation de véhicules, etc.**
47 **Hotels & restaurants**	..	..	..	..	..	..	..	47 **Hôtels et restaurants**
48 **Transport & storage**	..	..	..	..	..	..	..	48 **Transports, entreposage**
49 **Communications**	..	..	..	..	..	..	..	49 **Communications**
50 Post	..	..	..	..	..	..	..	50 Postes
51 Telecommunications	..	..	..	..	..	..	..	51 Télécommunications
52 **Financial intermediation (inc. insur.)**	..	..	..	..	..	..	..	52 **Intermédiation financ. (assurances inc.)**
53 **Real estate, renting & busin. activity**	..	..	..	..	..	..	..	53 **Immobil., locat. & serv. aux entreprises**
54 Computer & related activities	..	..	..	..	..	..	..	54 Informatique et activités connexes
55 *Software consultancy*	..	..	..	..	..	..	..	55 *Production de logiciels*
56 *Other computer services n.e.c.*	..	..	..	..	..	..	..	56 *Autres activités informatiques*
57 Research & development	..	..	24	..	..	..	..	57 Recherche & développement expérimental
58 Other business activities n.e.c.	..	..	66	..	..	..	..	58 Autres activ. & serv. aux entreprises
59 **Comm., soc. & pers. serv. activ., etc.**	..	..	..	..	..	..	..	59 **Adm. pub., serv. soc. & collec., etc**
TOTAL BUSINESS ENTERPRISE SECTOR	150 h	..	170 h	..	130 h	..	..	**TOTAL SECTEUR DES ENTREPRISES**

Source: OECD, R&D database (DSTI/EAS Division), January 2002.

Source : OCDE, base de données R-D (DSTI/Division AES), janvier 2002.

SWITZERLAND — Table 10 - Tableau 10 — SUISSE

Business enterprise intramural expenditure on R&D (BERD) - Dépenses intra-muros de R-D des entreprises (DIRDE)

by industry and source of funds - par industrie et par source de financement

Million CHF — *Millions CHF*

10.3 Other national funds - Autres sources nationales

	1994	1995	1996	1997	1998	1999	2000	
1 AGRICULTURE, HUNTING & FORESTRY	..	..	..	..	..	..	..	**1 AGRICULTURE, CHASSE & SYLVICULTURE**
2 MINING	..	..	..	..	..	..	..	**2 ACTIVITES EXTRACTIVES**
3 MANUFACTURING	..	..	2	..	..	..	..	**3 ACTIVITES DE FABRICATION**
4 Food, beverages & tobacco	..	..	..	..	..	..	..	**4 Produits alimentaires, boissons & tabac**
5 Food products & beverages	..	..	..	..	..	..	..	5 Produits alimentaires & boissons
6 Tobacco products	..	..	..	..	..	..	..	6 Produits à base de tabac
7 Textiles, wear app., fur & leather	..	..	..	..	..	..	..	**7 Textiles, habill., fourrures & cuirs**
8 Textiles	..	..	..	..	..	..	..	8 Textiles
9 Wearing apparel & fur	..	..	..	..	..	..	..	9 Articles d'habillement & fourrures
10 Leather products & footwear	..	..	..	..	..	..	..	10 Articles en cuir & chaussures
11 Wood, paper, printing, publishing	..	..	..	..	..	..	..	**11 Bois, papier, imprimerie, édition**
12 Wood & cork (not furniture)	..	..	..	..	..	..	..	12 Bois & liège (sauf meubles)
13 Pulp, paper & paper products	..	..	..	..	..	..	..	13 Carton, papier & articles
14 Publ., print. & repro. of rec. media	..	..	..	..	..	..	..	14 Edition, impress. & repro. enregis.
15 Coke, petroleum, nuclear fuel, chemicals, rubber plastics	..	..	..	..	..	..	..	**15 Coke, pétrole, combust. nucl., prod. chimi., caout. & mat. plast.**
16 Coke, ref. petrol. prod. & nuclear fuel	..	..	..	..	..	..	..	16 Coke, raffin. pétr. & combust. nucl.
17 Chemicals & chemical products	..	..	..	..	..	..	..	17 Produits chimiques
18 *Chemical prod.(less pharmaceuticals)*	..	..	..	..	..	..	..	18 *Prod. chimi. (moins prod. pharmaceutiques)*
19 *Pharmaceuticals*	..	..	..	..	..	..	..	19 *Produits pharmaceutiques*
20 Rubber & plastic products	..	..	..	..	..	..	..	20 Caoutchouc et matières plastiques
21 Non-metallic mineral products	..	..	..	..	..	..	..	**21 Produits minéraux non métalliques**
22 Basic metals	..	..	..	..	..	..	..	**22 Produits métallurgiques de base**
23 Basic metals, ferrous	..	..	..	..	..	..	..	23 Produits métallur. de base, ferreux
24 Basic metals, non-ferrous	..	..	..	..	..	..	..	24 Prod. métallur. de base, non ferreux
25 Fabricated metal products	..	..	..	..	..	..	..	**25 Ouvrages en métaux (sauf mach.& mat.)**
26 Machinery equipment, instruments and transport equipment	..	..	..	..	..	..	..	**26 Machines & matériel, instruments & mat. de transport**
27 Machinery n.e.c.	..	..	..	..	..	..	..	27 Machines, n.c.a.
28 Office, account. & computing machinery	..	..	..	..	..	..	..	28 Mach. de bur., compta. & trait. inform.
29 Electrical machinery	..	..	..	..	..	..	..	29 Machines électriques
30 Electro. equip. (radio, TV & commun.)	..	..	..	..	..	..	..	30 Equip. électro. (radio, tv & communic.)
31 *Electro. comp. (inc. semi-conductors)*	..	..	..	..	..	..	..	31 *Compos. électro. (semi-cond. incl.)*
32 *TV, radio & communications equip.*	..	..	..	..	..	..	..	32 *Appareils radio, tv et de communi.*
33 Instruments	..	..	..	..	..	..	..	33 Instruments
34 Motor vehicles	..	..	..	..	..	..	..	34 Véhicules automobiles
35 Other transport equipment	..	..	..	..	..	..	..	35 Autres matériels de transport
36 *Ships*	..	..	..	..	..	..	..	36 *Navires*
37 *Aerospace*	..	..	..	..	..	..	..	37 *Aérospatiale*
38 *Other transport equipment n.e.c.*	..	..	..	..	..	..	..	38 *Autres matériels de transport n.c.a.*
39 Furniture, other manufacturing n.e.c.	..	..	..	..	..	..	..	**39 Meubles, autres activ. de fabricat. n.c.a.**
40 Furniture	..	..	..	..	..	..	..	40 Meubles
41 Other manufacturing n.e.c.	..	..	..	..	..	..	..	41 Autres activités de fabrication n.c.a.
42 Recycling	..	..	..	..	..	..	..	**42 Récupération**
43 ELECTRICITY, GAS & WATER SUPPLY	..	..	..	..	..	..	..	**43 DISTRIBUTION D'ELECTRICITE, GAZ & EAU**
44 CONSTRUCTION	..	..	..	..	..	..	..	**44 CONSTRUCTION**
45 SERVICES SECTOR	..	..	48	..	..	..	..	**45 SECTEUR DES SERVICES**
46 Wholesale, retail trade & repair of motor vehicles, etc.	..	..	..	..	..	..	..	**46 Commerce (gros & détail), réparation de véhicules, etc.**
47 Hotels & restaurants	..	..	..	..	..	..	..	**47 Hôtels et restaurants**
48 Transport & storage	..	..	..	..	..	..	..	**48 Transports, entreposage**
49 Communications	..	..	..	..	..	..	..	**49 Communications**
50 Post	..	..	..	..	..	..	..	50 Postes
51 Telecommunications	..	..	..	..	..	..	..	51 Télécommunications
52 Financial intermediation (inc. insur.)	..	..	..	..	..	..	..	**52 Intermédiation financ. (assurances inc.)**
53 Real estate, renting & busin. activity	..	..	..	..	..	..	..	**53 Immobil., locat. & serv. aux entreprises**
54 Computer & related activities	..	..	..	..	..	..	..	54 Informatique et activités connexes
55 *Software consultancy*	..	..	..	..	..	..	..	55 *Production de logiciels*
56 *Other computer services n.e.c.*	..	..	..	..	..	..	..	56 *Autres activités informatiques*
57 Research & development	..	..	28	..	..	..	..	57 Recherche & développement expérimental
58 Other business activities n.e.c.	..	..	14	..	..	..	..	58 Autres activ. & serv. aux entreprises
59 Comm., soc. & pers. serv. activ., etc.	..	..	..	..	..	..	..	**59 Adm. pub., serv. soc. & collec., etc**
TOTAL BUSINESS ENTERPRISE SECTOR	**..**	**..**	**50**	**..**	**..**	**..**	**..**	**TOTAL SECTEUR DES ENTREPRISES**

Source: OECD, R&D database (DSTI/EAS Division), January 2002.

Source : OCDE, base de données R-D (DSTI/Division AES), janvier 2002.

Business enterprise intramural expenditure on R&D (BERD) - Dépenses intra-muros de R-D des entreprises (DIRDE)

by industry and source of funds - par industrie et par source de financement

Million CHF — *Millions CHF*

10.4 Funds from abroad - Fonds de l'étranger

	1994	1995	1996	1997	1998	1999	2000	
1 AGRICULTURE, HUNTING & FORESTRY	..	..	..	..	..	..	..	1 AGRICULTURE, CHASSE & SYLVICULTURE
2 MINING	..	..	..	..	..	..	..	2 ACTIVITES EXTRACTIVES
3 MANUFACTURING	..	..	209	..	..	..	..	3 ACTIVITES DE FABRICATION
4 Food, beverages & tobacco	..	..	..	..	..	..	..	**4 Produits alimentaires, boissons & tabac**
5 Food products & beverages	..	..	..	..	..	..	..	5 Produits alimentaires & boissons
6 Tobacco products	..	..	..	..	..	..	..	6 Produits à base de tabac
7 Textiles, wear app., fur & leather	..	..	..	..	..	..	..	**7 Textiles, habill., fourrures & cuirs**
8 Textiles	..	..	..	..	..	..	..	8 Textiles
9 Wearing apparel & fur	..	..	..	..	..	..	..	9 Articles d'habillement & fourrures
10 Leather products & footwear	..	..	..	..	..	..	..	10 Articles en cuir & chaussures
11 Wood, paper, printing, publishing	..	..	..	..	..	..	..	**11 Bois, papier, imprimerie, édition**
12 Wood & cork (not furniture)	..	..	..	..	..	..	..	12 Bois & liège (sauf meubles)
13 Pulp, paper & paper products	..	..	..	..	..	..	..	13 Carton, papier & articles
14 Publ., print. & repro. of rec. media	..	..	..	..	..	..	..	14 Edition, impress. & repro. enregis.
15 Coke, petroleum, nuclear fuel, chemicals, rubber plastics	..	..	6	..	..	..	..	**15 Coke, pétrole, combust. nucl., prod. chimi., caout. & mat. plast.**
16 Coke, ref. petrol. prod. & nuclear fuel	..	..	..	..	..	..	..	16 Coke, raffin. pétr. & combust. nucl.
17 Chemicals & chemical products	..	..	..	..	..	..	..	17 Produits chimiques
18 *Chemical prod.(less pharmaceuticals)*	..	..	..	..	..	..	..	18 *Prod. chimi. (moins prod. pharmaceutiques)*
19 *Pharmaceuticals*	..	..	..	..	..	..	..	19 *Produits pharmaceutiques*
20 Rubber & plastic products	..	..	..	..	..	..	..	20 Caoutchouc et matières plastiques
21 Non-metallic mineral products	..	..	..	..	..	..	..	**21 Produits minéraux non métalliques**
22 Basic metals	..	..	..	..	..	..	..	**22 Produits métallurgiques de base**
23 Basic metals, ferrous	..	..	..	..	..	..	..	23 Produits métallur. de base, ferreux
24 Basic metals, non-ferrous	..	..	..	..	..	..	..	24 Prod. métallur. de base, non ferreux
25 Fabricated metal products	..	..	..	..	..	..	..	**25 Ouvrages en métaux (sauf mach.& mat.)**
26 Machinery equipment, instruments and transport equipment	..	..	199	..	..	..	..	**26 Machines & matériel, instruments & mat. de transport**
27 Machinery n.e.c.	..	..	..	..	..	..	..	27 Machines, n.c.a.
28 Office, account. & computing machinery	..	..	..	..	..	..	..	28 Mach. de bur., compta. & trait. inform.
29 Electrical machinery	..	..	..	..	..	..	..	29 Machines électriques
30 Electro. equip. (radio, TV & commun.)	..	..	..	..	..	..	..	30 Equip. électro. (radio, tv & communic.)
31 *Electro. comp. (inc. semi-conductors)*	..	..	..	..	..	..	..	31 *Compos. électro. (semi-cond. incl.)*
32 *TV, radio & communications equip.*	..	..	..	..	..	..	..	32 *Appareils radio, tv et de communi.*
33 Instruments	..	..	..	..	..	..	..	33 Instruments
34 Motor vehicles	..	..	..	..	..	..	..	34 Véhicules automobiles
35 Other transport equipment	..	..	..	..	..	..	..	35 Autres matériels de transport
36 *Ships*	..	..	..	..	..	..	..	36 *Navires*
37 *Aerospace*	..	..	..	..	..	..	..	37 *Aérospatiale*
38 *Other transport equipment n.e.c.*	..	..	..	..	..	..	..	38 *Autres matériels de transport n.c.a.*
39 Furniture, other manufacturing n.e.c.	..	..	..	..	..	..	..	**39 Meubles, autres activ. de fabricat. n.c.a.**
40 Furniture	..	..	..	..	..	..	..	40 Meubles
41 Other manufacturing n.e.c.	..	..	..	..	..	..	..	41 Autres activités de fabrication n.c.a.
42 Recycling	..	..	..	..	..	..	..	**42 Récupération**
43 ELECTRICITY, GAS & WATER SUPPLY	..	..	..	..	..	..	..	43 DISTRIBUTION D'ELECTRICITE, GAZ & EAU
44 CONSTRUCTION	..	..	..	..	..	..	..	44 CONSTRUCTION
45 SERVICES SECTOR	..	..	106	..	..	..	..	45 SECTEUR DES SERVICES
46 Wholesale, retail trade & repair of motor vehicles, etc.	..	..	..	..	..	..	..	**46 Commerce (gros & détail), réparation de véhicules, etc.**
47 Hotels & restaurants	..	..	..	..	..	..	..	**47 Hôtels et restaurants**
48 Transport & storage	..	..	..	..	..	..	..	**48 Transports, entreposage**
49 Communications	..	..	..	..	..	..	..	**49 Communications**
50 Post	..	..	..	..	..	..	..	50 Postes
51 Telecommunications	..	..	..	..	..	..	..	51 Télécommunications
52 Financial intermediation (inc. insur.)	..	..	..	..	..	..	..	**52 Intermédiation financ. (assurances inc.)**
53 Real estate, renting & busin. activity	..	..	..	..	..	..	..	**53 Immobil., locat. & serv. aux entreprises**
54 Computer & related activities	..	..	..	..	..	..	..	54 Informatique et activités connexes
55 *Software consultancy*	..	..	..	..	..	..	..	55 *Production de logiciels*
56 *Other computer services n.e.c.*	..	..	..	..	..	..	..	56 *Autres activités informatiques*
57 Research & development	..	..	86	..	..	..	..	57 Recherche & développement expérimental
58 Other business activities n.e.c.	..	..	18	..	..	..	..	58 Autres activ. & serv. aux entreprises
59 Comm., soc. & pers. serv. activ., etc.	..	..	..	..	..	..	..	**59 Adm. pub., serv. soc. & collec., etc**
TOTAL BUSINESS ENTERPRISE SECTOR	**..**	**..**	**310**	**..**	**..**	**..**	**..**	**TOTAL SECTEUR DES ENTREPRISES**

Source: OECD, R&D database (DSTI/EAS Division), January 2002.

Source : OCDE, base de données R-D (DSTI/Division AES), janvier 2002.

SWITZERLAND Table 11 - Tableau 11 **SUISSE**

Business enterprise intramural expenditure on R&D (BERD) - Dépenses intra-muros de R-D des entreprises (DIRDE)

by industry and type of costs - par industrie et par type de coûts

Million CHF *Millions CHF*

11.1 Current expenditure - Dépenses courantes

	1994	1995	1996	1997	1998	1999	2000	
1 AGRICULTURE, HUNTING & FORESTRY	..	..	..	..	..	..	..	**1 AGRICULTURE, CHASSE & SYLVICULTURE**
2 MINING	..	..	..	..	..	..	..	**2 ACTIVITES EXTRACTIVES**
3 MANUFACTURING	..	..	5 273	..	..	..	..	**3 ACTIVITES DE FABRICATION**
4 Food, beverages & tobacco	..	..	309	..	..	..	..	**4 Produits alimentaires, boissons & tabac**
5 Food products & beverages	..	..	..	..	..	..	..	5 Produits alimentaires & boissons
6 Tobacco products	..	..	..	..	..	..	..	6 Produits à base de tabac
7 Textiles, wear app., fur & leather	..	..	..	..	..	..	..	**7 Textiles, habill., fourrures & cuirs**
8 Textiles	..	..	..	..	..	..	..	8 Textiles
9 Wearing apparel & fur	..	..	..	..	..	..	..	9 Articles d'habillement & fourrures
10 Leather products & footwear	..	..	..	..	..	..	..	10 Articles en cuir & chaussures
11 Wood, paper, printing, publishing	..	..	..	..	..	..	..	**11 Bois, papier, imprimerie, édition**
12 Wood & cork (not furniture)	..	..	..	..	..	..	..	12 Bois & liège (sauf meubles)
13 Pulp, paper & paper products	..	..	..	..	..	..	..	13 Carton, papier & articles
14 Publ., print. & repro. of rec. media	..	..	..	..	..	..	..	14 Edition, impress. & repro. enregis.
15 Coke, petroleum, nuclear fuel, chemicals, rubber plastics	..	..	2 456	..	..	..	..	**15 Coke, pétrole, combust. nucl., prod. chimi., caout. & mat. plast.**
16 Coke, ref. petrol. prod. & nuclear fuel	..	..	..	..	..	..	..	16 Coke, raffin. pétr. & combust. nucl.
17 Chemicals & chemical products	..	..	..	..	..	..	..	17 Produits chimiques
18 *Chemical prod.(less pharmaceuticals)*	..	..	..	..	..	..	..	18 *Prod. chimi. (moins prod. pharmaceutiques)*
19 *Pharmaceuticals*	..	..	..	..	..	..	..	19 *Produits pharmaceutiques*
20 Rubber & plastic products	..	..	..	..	..	..	..	20 Caoutchouc et matières plastiques
21 Non-metallic mineral products	..	..	..	..	..	..	..	**21 Produits minéraux non métalliques**
22 Basic metals	..	..	..	..	..	..	..	**22 Produits métallurgiques de base**
23 Basic metals, ferrous	..	..	..	..	..	..	..	23 Produits métallur. de base, ferreux
24 Basic metals, non-ferrous	..	..	..	..	..	..	..	24 Prod. métallur. de base, non ferreux
25 Fabricated metal products	..	..	76	..	..	..	..	**25 Ouvrages en métaux (sauf mach.& mat.)**
26 Machinery equipment, instruments and transport equipment	..	..	2 147	..	..	..	..	**26 Machines & matériel, instruments & mat. de transport**
27 Machinery n.e.c.	..	..	..	..	..	..	..	27 Machines, n.c.a.
28 Office, account. & computing machinery	..	..	..	..	..	..	..	28 Mach. de bur., compta. & trait. inform.
29 Electrical machinery	..	..	..	..	..	..	..	29 Machines électriques
30 Electro. equip. (radio, TV & commun.)	..	..	..	..	..	..	..	30 Equip. électro. (radio, tv & communic.)
31 *Electro. comp. (inc. semi-conductors)*	..	..	..	..	..	..	..	31 *Compos. électro. (semi-cond. incl.)*
32 *TV, radio & communications equip.*	..	..	..	..	..	..	..	32 *Appareils radio, tv et de communi.*
33 Instruments	..	..	..	..	..	..	..	33 Instruments
34 Motor vehicles	..	..	..	..	..	..	..	34 Véhicules automobiles
35 Other transport equipment	..	..	..	..	..	..	..	35 Autres matériels de transport
36 *Ships*	..	..	..	..	..	..	..	36 *Navires*
37 *Aerospace*	..	..	..	..	..	..	..	37 *Aérospatiale*
38 *Other transport equipment n.e.c.*	..	..	..	..	..	..	..	38 *Autres matériels de transport n.c.a.*
39 Furniture, other manufacturing n.e.c.	..	..	..	..	..	..	..	**39 Meubles, autres activ. de fabricat. n.c.a.**
40 Furniture	..	..	..	..	..	..	..	40 Meubles
41 Other manufacturing n.e.c.	..	..	..	..	..	..	..	41 Autres activités de fabrication n.c.a.
42 Recycling	..	..	..	..	..	..	..	**42 Récupération**
43 ELECTRICITY, GAS & WATER SUPPLY	..	..	..	..	..	..	..	**43 DISTRIBUTION D'ELECTRICITE, GAZ & EAU**
44 CONSTRUCTION	..	..	..	..	..	..	..	**44 CONSTRUCTION**
45 SERVICES SECTOR	..	..	1 276	..	..	..	..	**45 SECTEUR DES SERVICES**
46 Wholesale, retail trade & repair of motor vehicles, etc.	..	..	..	..	..	..	..	**46 Commerce (gros & détail), réparation de véhicules, etc.**
47 Hotels & restaurants	..	..	..	..	..	..	..	**47 Hôtels et restaurants**
48 Transport & storage	..	..	..	..	..	..	..	**48 Transports, entreposage**
49 Communications	..	..	..	..	..	..	..	**49 Communications**
50 Post	..	..	..	..	..	..	..	50 Postes
51 Telecommunications	..	..	..	..	..	..	..	51 Télécommunications
52 Financial intermediation (inc. insur.)	..	..	271	..	..	..	..	**52 Intermédiation financ. (assurances inc.)**
53 Real estate, renting & busin. activity	..	..	..	..	..	..	..	**53 Immobil., locat. & serv. aux entreprises**
54 Computer & related activities	..	..	..	..	..	..	..	54 Informatique et activités connexes
55 *Software consultancy*	..	..	..	..	..	..	..	55 *Production de logiciels*
56 *Other computer services n.e.c.*	..	..	..	..	..	..	..	56 *Autres activités informatiques*
57 Research & development	..	..	365	..	..	..	..	57 Recherche & développement expérimental
58 Other business activities n.e.c.	..	..	552	..	..	..	..	58 Autres activ. & serv. aux entreprises
59 Comm., soc. & pers. serv. activ., etc.	..	..	..	..	..	..	..	**59 Adm. pub., serv. soc. & collec., etc**
TOTAL BUSINESS ENTERPRISE SECTOR	**..**	**..**	**6 580**	**..**	**..**	**..**	**..**	**TOTAL SECTEUR DES ENTREPRISES**

Source: OECD, R&D database (DSTI/EAS Division), January 2002.

Source : OCDE, base de données R-D (DSTI/Division AES), janvier 2002.

Business enterprise intramural expenditure on R&D (BERD) - Dépenses intra-muros de R-D des entreprises (DIRDE)

by industry and type of costs - par industrie et par type de coûts

Million CHF — *Millions CHF*

11.2 Capital expenditure - Dépenses en capital

	1994	1995	1996	1997	1998	1999	2000	
1 AGRICULTURE, HUNTING & FORESTRY	..	..	..	..	..	..	..	**1 AGRICULTURE, CHASSE & SYLVICULTURE**
2 MINING	..	..	..	..	..	..	..	**2 ACTIVITES EXTRACTIVES**
3 MANUFACTURING	..	..	386	..	..	..	..	**3 ACTIVITES DE FABRICATION**
4 Food, beverages & tobacco	..	..	51	..	..	..	..	**4 Produits alimentaires, boissons & tabac**
5 Food products & beverages	..	..	..	..	..	..	..	5 Produits alimentaires & boissons
6 Tobacco products	..	..	..	..	..	..	..	6 Produits à base de tabac
7 Textiles, wear app., fur & leather	..	..	..	..	..	..	..	**7 Textiles, habill., fourrures & cuirs**
8 Textiles	..	..	..	..	..	..	..	8 Textiles
9 Wearing apparel & fur	..	..	..	..	..	..	..	9 Articles d'habillement & fourrures
10 Leather products & footwear	..	..	..	..	..	..	..	10 Articles en cuir & chaussures
11 Wood, paper, printing, publishing	..	..	..	..	..	..	..	**11 Bois, papier, imprimerie, édition**
12 Wood & cork (not furniture)	..	..	..	..	..	..	..	12 Bois & liège (sauf meubles)
13 Pulp, paper & paper products	..	..	..	..	..	..	..	13 Carton, papier & articles
14 Publ., print. & repro. of rec. media	..	..	..	..	..	..	..	14 Edition, impress. & repro. enregis.
15 Coke, petroleum, nuclear fuel, chemicals, rubber plastics	..	..	171	..	..	..	..	**15 Coke, pétrole, combust. nucl., prod. chimi., caout. & mat. plast.**
16 Coke, ref. petrol. prod. & nuclear fuel	..	..	..	..	..	..	..	16 Coke, raffin. pétr. & combust. nucl.
17 Chemicals & chemical products	..	..	..	..	..	..	..	17 Produits chimiques
18 *Chemical prod.(less pharmaceuticals)*	..	..	..	..	..	..	..	18 *Prod. chimi. (moins prod. pharmaceutiques)*
19 *Pharmaceuticals*	..	..	..	..	..	..	..	19 *Produits pharmaceutiques*
20 Rubber & plastic products	..	..	..	..	..	..	..	20 Caoutchouc et matières plastiques
21 Non-metallic mineral products	..	..	..	..	..	..	..	**21 Produits minéraux non métalliques**
22 Basic metals	..	..	..	..	..	..	..	**22 Produits métallurgiques de base**
23 Basic metals, ferrous	..	..	..	..	..	..	..	23 Produits métallur. de base, ferreux
24 Basic metals, non-ferrous	..	..	..	..	..	..	..	24 Prod. métallur. de base, non ferreux
25 Fabricated metal products	..	..	5	..	..	..	..	**25 Ouvrages en métaux (sauf mach.& mat.)**
26 Machinery equipment, instruments and transport equipment	..	..	146	..	..	..	..	**26 Machines & matériel, instruments & mat. de transport**
27 Machinery n.e.c.	..	..	..	..	..	..	..	27 Machines, n.c.a.
28 Office, account. & computing machinery	..	..	..	..	..	..	..	28 Mach. de bur., compta. & trait. inform.
29 Electrical machinery	..	..	..	..	..	..	..	29 Machines électriques
30 Electro. equip. (radio, TV & commun.)	..	..	..	..	..	..	..	30 Equip. électro. (radio, tv & communic.)
31 *Electro. comp. (inc. semi-conductors)*	..	..	..	..	..	..	..	31 *Compos. électro. (semi-cond. incl.)*
32 *TV, radio & communications equip.*	..	..	..	..	..	..	..	32 *Appareils radio, tv et de communi.*
33 Instruments	..	..	..	..	..	..	..	33 Instruments
34 Motor vehicles	..	..	..	..	..	..	..	34 Véhicules automobiles
35 Other transport equipment	..	..	..	..	..	..	..	35 Autres matériels de transport
36 *Ships*	..	..	..	..	..	..	..	36 *Navires*
37 *Aerospace*	..	..	..	..	..	..	..	37 *Aérospatiale*
38 *Other transport equipment n.e.c.*	..	..	..	..	..	..	..	38 *Autres matériels de transport n.c.a.*
39 Furniture, other manufacturing n.e.c.	..	..	..	..	..	..	..	**39 Meubles, autres activ. de fabricat. n.c.a.**
40 Furniture	..	..	..	..	..	..	..	40 Meubles
41 Other manufacturing n.e.c.	..	..	..	..	..	..	..	41 Autres activités de fabrication n.c.a.
42 Recycling	..	..	..	..	..	..	..	**42 Récupération**
43 ELECTRICITY, GAS & WATER SUPPLY	..	..	..	..	..	..	..	**43 DISTRIBUTION D'ELECTRICITE, GAZ & EAU**
44 CONSTRUCTION	..	..	..	..	..	..	..	**44 CONSTRUCTION**
45 SERVICES SECTOR	..	..	74	..	..	..	..	**45 SECTEUR DES SERVICES**
46 Wholesale, retail trade & repair of motor vehicles, etc.	..	..	..	..	..	..	..	**46 Commerce (gros & détail), réparation de véhicules, etc.**
47 Hotels & restaurants	..	..	..	..	..	..	..	**47 Hôtels et restaurants**
48 Transport & storage	..	..	..	..	..	..	..	**48 Transports, entreposage**
49 Communications	..	..	..	..	..	..	..	**49 Communications**
50 Post	..	..	..	..	..	..	..	50 Postes
51 Telecommunications	..	..	..	..	..	..	..	51 Télécommunications
52 Financial intermediation (inc. insur.)	..	..	19	..	..	..	..	**52 Intermédiation financ. (assurances inc.)**
53 Real estate, renting & busin. activity	..	..	..	..	..	..	..	**53 Immobil., locat. & serv. aux entreprises**
54 Computer & related activities	..	..	..	..	..	..	..	54 Informatique et activités connexes
55 *Software consultancy*	..	..	..	..	..	..	..	55 *Production de logiciels*
56 *Other computer services n.e.c.*	..	..	..	..	..	..	..	56 *Autres activités informatiques*
57 Research & development	..	..	30	..	..	..	..	57 Recherche & développement expérimental
58 Other business activities n.e.c.	..	..	43	..	..	..	..	58 Autres activ. & serv. aux entreprises
59 Comm., soc. & pers. serv. activ., etc.	..	..	..	..	..	..	..	**59 Adm. pub., serv. soc. & collec., etc**
TOTAL BUSINESS ENTERPRISE SECTOR	**..**	**..**	**480**	**..**	**..**	**..**	**..**	**TOTAL SECTEUR DES ENTREPRISES**

Source: OECD, R&D database (DSTI/EAS Division), January 2002.

Source : OCDE, base de données R-D (DSTI/Division AES), janvier 2002.

SWITZERLAND — Table 12 - Tableau 12 — SUISSE

Total business enterprise R&D personnel (BEMP) - Personnel total de R-D du secteur des entreprises (PRDE)

by industry - par industrie

In full time equivalent — *En équivalence plein-temps*

	1994	1995	1996	1997	1998	1999	2000	
1 AGRICULTURE, HUNTING & FORESTRY	..	..	..	..	..	..	..	**1 AGRICULTURE, CHASSE & SYLVICULTURE**
2 MINING	..	..	..	..	..	..	..	**2 ACTIVITES EXTRACTIVES**
3 MANUFACTURING	..	..	28 480	..	..	..	..	**3 ACTIVITES DE FABRICATION**
4 Food, beverages & tobacco	..	..	1 470	..	..	..	..	**4 Produits alimentaires, boissons & tabac**
5 Food products & beverages	..	..	..	..	..	..	..	5 Produits alimentaires & boissons
6 Tobacco products	..	..	..	..	..	..	..	6 Produits à base de tabac
7 Textiles, wear app., fur & leather	..	..	..	..	..	..	..	**7 Textiles, habill., fourrures & cuirs**
8 Textiles	..	..	..	..	..	..	..	8 Textiles
9 Wearing apparel & fur	..	..	..	..	..	..	..	9 Articles d'habillement & fourrures
10 Leather products & footwear	..	..	..	..	..	..	..	10 Articles en cuir & chaussures
11 Wood, paper, printing, publishing	..	..	..	..	..	..	..	**11 Bois, papier, imprimerie, édition**
12 Wood & cork (not furniture)	..	..	..	..	..	..	..	12 Bois & liège (sauf meubles)
13 Pulp, paper & paper products	..	..	..	..	..	..	..	13 Carton, papier & articles
14 Publ., print. & repro. of rec. media	..	..	..	..	..	..	..	14 Edition, impress. & repro. enregis.
15 Coke, petroleum, nuclear fuel, chemicals, rubber plastics	..	..	11 110	..	..	..	..	**15 Coke, pétrole, combust. nucl., prod. chimi., caout. & mat. plast.**
16 Coke, ref. petrol. prod. & nuclear fuel	..	..	..	..	..	..	..	16 Coke, raffin. pétr. & combust. nucl.
17 Chemicals & chemical products	..	..	..	..	..	..	..	17 Produits chimiques
18 *Chemical prod.(less pharmaceuticals)*	..	..	..	..	..	..	..	18 *Prod. chimi. (moins prod. pharmaceutiques)*
19 *Pharmaceuticals*	..	..	..	..	..	..	..	19 *Produits pharmaceutiques*
20 Rubber & plastic products	..	..	..	..	..	..	..	20 Caoutchouc et matières plastiques
21 Non-metallic mineral products	..	..	..	..	..	..	..	**21 Produits minéraux non métalliques**
22 Basic metals	..	..	..	..	..	..	..	**22 Produits métallurgiques de base**
23 Basic metals, ferrous	..	..	..	..	..	..	..	23 Produits métallur. de base, ferreux
24 Basic metals, non-ferrous	..	..	..	..	..	..	..	24 Prod. métallur. de base, non ferreux
25 Fabricated metal products	..	..	540	..	..	..	..	**25 Ouvrages en métaux (sauf mach.& mat.)**
26 Machinery equipment, instruments and transport equipment	..	..	13 560	..	..	..	..	**26 Machines & matériel, instruments & mat. de transport**
27 Machinery n.e.c.	..	..	..	..	..	..	..	27 Machines, n.c.a.
28 Office, account. & computing machinery	..	..	..	..	..	..	..	28 Mach. de bur., compta. & trait. inform.
29 Electrical machinery	..	..	..	..	..	..	..	29 Machines électriques
30 Electro. equip. (radio, TV & commun.)	..	..	..	..	..	..	..	30 Equip. électro. (radio, tv & communic.)
31 *Electro. comp. (inc. semi-conductors)*	..	..	..	..	..	..	..	31 *Compos. électro. (semi-cond. incl.)*
32 *TV, radio & communications equip.*	..	..	..	..	..	..	..	32 *Appareils radio, tv et de communi.*
33 Instruments	..	..	..	..	..	..	..	33 Instruments
34 Motor vehicles	..	..	..	..	..	..	..	34 Véhicules automobiles
35 Other transport equipment	..	..	..	..	..	..	..	35 Autres matériels de transport
36 *Ships*	..	..	..	..	..	..	..	36 *Navires*
37 *Aerospace*	..	..	..	..	..	..	..	37 *Aérospatiale*
38 *Other transport equipment n.e.c.*	..	..	..	..	..	..	..	38 *Autres matériels de transport n.c.a.*
39 Furniture, other manufacturing n.e.c.	..	..	..	..	..	..	..	**39 Meubles, autres activ. de fabricat. n.c.a.**
40 Furniture	..	..	..	..	..	..	..	40 Meubles
41 Other manufacturing n.e.c.	..	..	..	..	..	..	..	41 Autres activités de fabrication n.c.a.
42 Recycling	..	..	..	..	..	..	..	**42 Récupération**
43 ELECTRICITY, GAS & WATER SUPPLY	..	..	..	..	..	..	..	**43 DISTRIBUTION D'ELECTRICITE, GAZ & EAU**
44 CONSTRUCTION	..	..	..	..	..	..	..	**44 CONSTRUCTION**
45 SERVICES SECTOR	..	..	5 560	..	..	..	..	**45 SECTEUR DES SERVICES**
46 Wholesale, retail trade & repair of motor vehicles, etc.	..	..	..	..	..	..	..	**46 Commerce (gros & détail), réparation de véhicules, etc.**
47 Hotels & restaurants	..	..	..	..	..	..	..	**47 Hôtels et restaurants**
48 Transport & storage	..	..	..	..	..	..	..	**48 Transports, entreposage**
49 Communications	..	..	..	..	..	..	..	**49 Communications**
50 Post	..	..	..	..	..	..	..	50 Postes
51 Telecommunications	..	..	..	..	..	..	..	51 Télécommunications
52 Financial intermediation (inc. insur.)	..	..	..	..	..	..	..	**52 Intermédiation financ. (assurances inc.)**
53 Real estate, renting & busin. activity	..	..	..	..	..	..	..	**53 Immobil., locat. & serv. aux entreprises**
54 Computer & related activities	..	..	..	..	..	..	..	54 Informatique et activités connexes
55 *Software consultancy*	..	..	..	..	..	..	..	55 *Production de logiciels*
56 *Other computer services n.e.c.*	..	..	..	..	..	..	..	56 *Autres activités informatiques*
57 Research & development	..	..	2 140	..	..	..	..	57 Recherche & développement expérimental
58 Other business activities n.e.c.	..	..	2 530	..	..	..	..	58 Autres activ. & serv. aux entreprises
59 Comm., soc. & pers. serv. activ., etc.	..	..	..	..	..	..	..	**59 Adm. pub., serv. soc. & collec., etc**
TOTAL BUSINESS ENTERPRISE SECTOR	**..**	**..**	**34 450**	**..**	**..**	**..**	**..**	**TOTAL SECTEUR DES ENTREPRISES**

Source: OECD, R&D database (DSTI/EAS Division), January 2002.

Source : OCDE, base de données R-D (DSTI/Division AES), janvier 2002.

RSE in the business enterprise sector - Chercheurs dans le secteur des entreprises
by industry - par industrie

In full time equivalent | *En équivalence plein-temps*

	1994	1995	1996	1997	1998	1999	2000	
1 AGRICULTURE, HUNTING & FORESTRY	..	..	..	..	..	..	..	**1 AGRICULTURE, CHASSE & SYLVICULTURE**
2 MINING	..	..	..	..	..	..	..	**2 ACTIVITES EXTRACTIVES**
3 MANUFACTURING	..	..	9 880	..	..	..	..	**3 ACTIVITES DE FABRICATION**
4 Food, beverages & tobacco	..	..	870	..	..	..	..	**4 Produits alimentaires, boissons & tabac**
5 Food products & beverages	..	..	..	..	..	..	..	5 Produits alimentaires & boissons
6 Tobacco products	..	..	..	..	..	..	..	6 Produits à base de tabac
7 Textiles, wear app., fur & leather	..	..	..	..	..	..	..	**7 Textiles, habill., fourrures & cuirs**
8 Textiles	..	..	..	..	..	..	..	8 Textiles
9 Wearing apparel & fur	..	..	..	..	..	..	..	9 Articles d'habillement & fourrures
10 Leather products & footwear	..	..	..	..	..	..	..	10 Articles en cuir & chaussures
11 Wood, paper, printing, publishing	..	..	..	..	..	..	..	**11 Bois, papier, imprimerie, édition**
12 Wood & cork (not furniture)	..	..	..	..	..	..	..	12 Bois & liège (sauf meubles)
13 Pulp, paper & paper products	..	..	..	..	..	..	..	13 Carton, papier & articles
14 Publ., print. & repro. of rec. media	..	..	..	..	..	..	..	14 Edition, impress. & repro. enregis.
15 Coke, petroleum, nuclear fuel, chemicals, rubber plastics	..	..	4 270	..	..	..	..	**15 Coke, pétrole, combust. nucl., prod. chimi., caout. & mat. plast.**
16 Coke, ref. petrol. prod. & nuclear fuel	..	..	..	..	..	..	..	16 Coke, raffin. pétr. & combust. nucl.
17 Chemicals & chemical products	..	..	..	..	..	..	..	17 Produits chimiques
18 *Chemical prod.(less pharmaceuticals)*	..	..	..	..	..	..	..	18 *Prod. chimi. (moins prod. pharmaceutiques)*
19 *Pharmaceuticals*	..	..	..	..	..	..	..	19 *Produits pharmaceutiques*
20 Rubber & plastic products	..	..	..	..	..	..	..	20 Caoutchouc et matières plastiques
21 Non-metallic mineral products	..	..	..	..	..	..	..	**21 Produits minéraux non métalliques**
22 Basic metals	..	..	..	..	..	..	..	**22 Produits métallurgiques de base**
23 Basic metals, ferrous	..	..	..	..	..	..	..	23 Produits métallur. de base, ferreux
24 Basic metals, non-ferrous	..	..	..	..	..	..	..	24 Prod. métallur. de base, non ferreux
25 Fabricated metal products	..	..	80	..	..	..	..	**25 Ouvrages en métaux (sauf mach.& mat.)**
26 Machinery equipment, instruments and transport equipment	..	..	3 900	..	..	..	..	**26 Machines & matériel, instruments & mat. de transport**
27 Machinery n.e.c.	..	..	..	..	..	..	..	27 Machines, n.c.a.
28 Office, account. & computing machinery	..	..	..	..	..	..	..	28 Mach. de bur., compta. & trait. inform.
29 Electrical machinery	..	..	..	..	..	..	..	29 Machines électriques
30 Electro. equip. (radio, TV & commun.)	..	..	..	..	..	..	..	30 Equip. électro. (radio, tv & communic.)
31 *Electro. comp. (inc. semi-conductors)*	..	..	..	..	..	..	..	31 *Compos. électro. (semi-cond. incl.)*
32 *TV, radio & communications equip.*	..	..	..	..	..	..	..	32 *Appareils radio, tv et de communi.*
33 Instruments	..	..	..	..	..	..	..	33 Instruments
34 Motor vehicles	..	..	..	..	..	..	..	34 Véhicules automobiles
35 Other transport equipment	..	..	..	..	..	..	..	35 Autres matériels de transport
36 *Ships*	..	..	..	..	..	..	..	36 *Navires*
37 *Aerospace*	..	..	..	..	..	..	..	37 *Aérospatiale*
38 *Other transport equipment n.e.c.*	..	..	..	..	..	..	..	38 *Autres matériels de transport n.c.a.*
39 Furniture, other manufacturing n.e.c.	..	..	..	..	..	..	..	**39 Meubles, autres activ. de fabricat. n.c.a.**
40 Furniture	..	..	..	..	..	..	..	40 Meubles
41 Other manufacturing n.e.c.	..	..	..	..	..	..	..	41 Autres activités de fabrication n.c.a.
42 Recycling	..	..	..	..	..	..	..	**42 Récupération**
43 ELECTRICITY, GAS & WATER SUPPLY	..	..	..	..	..	..	..	**43 DISTRIBUTION D'ELECTRICITE, GAZ & EAU**
44 CONSTRUCTION	..	..	..	..	..	..	..	**44 CONSTRUCTION**
45 SERVICES SECTOR	..	..	2 580	..	..	..	..	**45 SECTEUR DES SERVICES**
46 Wholesale, retail trade & repair of motor vehicles, etc.	..	..	..	..	..	..	..	**46 Commerce (gros & détail), réparation de véhicules, etc.**
47 Hotels & restaurants	..	..	..	..	..	..	..	**47 Hôtels et restaurants**
48 Transport & storage	..	..	..	..	..	..	..	**48 Transports, entreposage**
49 Communications	..	..	..	..	..	..	..	**49 Communications**
50 Post	..	..	..	..	..	..	..	50 Postes
51 Telecommunications	..	..	..	..	..	..	..	51 Télécommunications
52 Financial intermediation (inc. insur.)	..	..	..	..	..	..	..	**52 Intermédiation financ. (assurances inc.)**
53 Real estate, renting & busin. activity	..	..	..	..	..	..	..	**53 Immobil., locat. & serv. aux entreprises**
54 Computer & related activities	..	..	..	..	..	..	..	54 Informatique et activités connexes
55 *Software consultancy*	..	..	..	..	..	..	..	55 *Production de logiciels*
56 *Other computer services n.e.c.*	..	..	..	..	..	..	..	56 *Autres activités informatiques*
57 Research & development	..	..	950	..	..	..	..	57 Recherche & développement expérimental
58 Other business activities n.e.c.	..	..	910	..	..	..	..	58 Autres activ. & serv. aux entreprises
59 Comm., soc. & pers. serv. activ., etc.	..	..	..	..	..	..	..	**59 Adm. pub., serv. soc. & collec., etc**
TOTAL BUSINESS ENTERPRISE SECTOR	..	..	**12 500**	..	..	..	..	**TOTAL SECTEUR DES ENTREPRISES**

Source: OECD, R&D database (DSTI/EAS Division), January 2002.

Source : OCDE, base de données R-D (DSTI/Division AES), janvier 2002.

Higher education R&D personnel (HEMP) - Personnel de R-D du secteur de l'enseignement supérieur

by field of science and occupation - par discipline scientifique et par profession

In full-time equivalent — *En équivalence plein-temps*

	1994	1995	1996	1997	1998	1999	2000	
Total R&D personnel								**Personnel total de R-D**
Natural sciences	..	..	..	..	..	..	..	Sciences exactes et naturelles
Engineering	..	..	..	..	..	..	..	Sciences de l'ingénieur
Medical sciences	..	..	..	..	..	..	..	Sciences médicales
Agricultural sciences	..	..	..	..	..	..	..	Sciences agricoles
Sub-total NSE	..	..	..	..	..	..	..	Sous-total SENI
Social sciences	..	..	..	..	..	..	..	Sciences sociales
Humanities	..	..	..	..	..	..	..	Sciences humaines
Sub-total SSH	..	..	..	..	..	..	..	Sous-total SSH
Total NSE+SSH	12 540	..	14 430	..	14 550	..	..	*Total SENI+SSH*
Researchers								**Chercheurs**
Natural sciences	..	..	..	..	..	..	..	Sciences exactes et naturelles
Engineering	..	..	..	..	..	..	..	Sciences de l'ingénieur
Medical sciences	..	..	..	..	..	..	..	Sciences médicales
Agricultural sciences	..	..	..	..	..	..	..	Sciences agricoles
Sub-total NSE	..	..	..	..	..	..	..	Sous-total SENI
Social sciences	..	..	..	..	..	..	..	Sciences sociales
Humanities	..	..	..	..	..	..	..	Sciences humaines
Sub-total SSH	..	..	..	..	..	..	..	Sous-total SSH
Total NSE+SSH	7 680	..	8 580	..	8 970	..	..	*Total SENI+SSH*
Technicians								**Technicians**
Natural sciences	..	..	..	..	..	..	..	Sciences exactes et naturelles
Engineering	..	..	..	..	..	..	..	Sciences de l'ingénieur
Medical sciences	..	..	..	..	..	..	..	Sciences médicales
Agricultural sciences	..	..	..	..	..	..	..	Sciences agricoles
Sub-total NSE	..	..	..	..	..	..	..	Sous-total SENI
Social sciences	..	..	..	..	..	..	..	Sciences sociales
Humanities	..	..	..	..	..	..	..	Sciences humaines
Sub-total SSH	..	..	..	..	..	..	..	Sous-total SSH
Total NSE+SSH	4 170	..	4 960	..	4 770	..	..	*Total SENI+SSH*
Other								**Autres**
Natural sciences	..	..	..	..	..	..	..	Sciences exactes et naturelles
Engineering	..	..	..	..	..	..	..	Sciences de l'ingénieur
Medical sciences	..	..	..	..	..	..	..	Sciences médicales
Agricultural sciences	..	..	..	..	..	..	..	Sciences agricoles
Sub-total NSE	..	..	..	..	..	..	..	Sous-total SENI
Social sciences	..	..	..	..	..	..	..	Sciences sociales
Humanities	..	..	..	..	..	..	..	Sciences humaines
Sub-total SSH	..	..	..	..	..	..	..	Sous-total SSH
Total NSE+SSH	690	..	890	..	810	..	..	*Total SENI+SSH*

Table 18 - Tableau 18

Government budget appropriations or outlays for R&D (GBAORD) - Crédits budgétaires publics de R-D (CBPRD)

by socio-economic objective - par objectif socio-économique

Million CHF — *Millions CHF*

	1995	1996	1997	1998	1999	2000	2001	
Agriculture, forestry and fishing	..	35.0 [m]	..	70.0 [a]	..	..	..	Agriculture, sylviculture et pêche
Industrial development	..	15.0 [m]	..	15.0 [m]	..	..	..	Développement industriel
Energy	..	10.0 [m]	..	5.0 [m]	..	..	..	Energie
Transport & telecommunications	..	15.0 [m]	..	..	..	..	..	*Transports & télécommunications*
Urban & rural planning	..	5.0 [m]	..	..	..	..	..	*Aménagement urbain et rural*
ST Infrastructure	..	20.0 [m]	..	5.0 [a,m]	..	..	..	ST Infrastructure
Prevention of pollution	..	..	..	..	..	..	..	*Prévention de la pollution*
Identificat. & treatment of pollution	..	..	..	..	..	..	..	*Détect. pollut. & mesures antipollut.*
ST Environment	..	25.0 [m]	..	5.0 [m]	..	..	..	ST Environnement
Health	..	20.0 [m]	..	10.0 [a,m]	..	..	..	Santé
Social development & services	..	45.0 [m]	..	25.0 [m]	..	..	..	Développement et services sociaux
Earth and atmosphere	..	5.0 [m]	..	0.0 [m]	..	..	..	Milieu terrestre et atmosphère
Advancement of research	..	..	..	..	..	..	..	*Promotion de la recherche*
General university funds	..	..	..	..	..	..	..	*Fonds généraux des universités*
ST Advancement of knowledge	..	1 780.0 [m]	..	1 520.0 [m]	..	..	..	ST promotion gén. des connaiss.
Civil space	..	..	..	..	..	..	..	Espace civil
Defence	..	55.0 [m]	..	50.0 [m]	..	..	..	Défense
Not elsewhere classified	..	920.0	..	915.0	..	..	..	Non classé
Total	**..**	**2 930.0**	**..**	**2 620.0**	**..**	**..**	**..**	**Total**

Source: OECD, R&D database (DSTI/EAS Division), January 2002.

Source : OCDE, base de données R-D (DSTI/Division AES), janvier 2002.

SWITZERLAND Table 19 - Tableau 19 **SUISSE**

Technology balance of payments - Balance des paiements technologiques

by type of transfer - selon la nature des transferts

Million CHF *Millions CHF*

	1983	1984	1985	1986	1987	1988	1989	1990	1991	
Receipts										**Recettes**
Related firms	..	..	..	..	..	..	..	..	..	Firmes affiliées
Unrelated firms	..	..	..	..	..	..	..	..	..	Firmes non affiliées
Other	..	..	..	..	..	..	..	..	..	Autre
Total	..	..	2 138	2 077	2 037	2 335	2 617	2 594	2 784	*Total*
Payments										**Paiements**
Related firms	..	..	..	..	..	..	..	..	..	Firmes affiliées
Unrelated firms	..	..	..	..	..	..	..	..	..	Firmes non affiliées
Other	..	..	..	..	..	..	..	..	..	Autre
Total	..	..	572	602	586	628	871	1 019	1 069	*Total*

	1992	1993	1994	1995	1996	1997	1998	1999	2000	
Receipts										**Recettes**
Related firms	..	..	..	..	..	..	..	..	..	Firmes affiliées
Unrelated firms	..	..	..	..	..	..	..	..	..	Firmes non affiliées
Other	..	..	..	..	..	..	..	..	..	Autre
Total	3 047	3 432	3 493	3 285	3 341	4 072	4 328	2 287	..	*Total*
Payments										**Paiements**
Related firms	..	..	..	..	..	..	..	..	..	Firmes affiliées
Unrelated firms	..	..	..	..	..	..	..	..	..	Firmes non affiliées
Other	..	..	..	..	..	..	..	..	..	Autre
Total	1 260	1 273	1 509	1 492	1 768	1 662	1 940	2 914	..	*Total*

Source: OECD, TBPdatabase (DSTI/EAS Division), Januay 2002.

Source : OCDE, base de données TBP (DSTI/Division AES), janvier 2002.

TURKEY — TURQUIE

Table 1 - Tableau 1

Gross domestic expenditure on R&D (GERD) - Dépenses intérieures brutes de R-D (DIRD)

by sector of performance and source of funds - par secteur d'exécution et par source de financement

	1994	1995	1996	1997	1998	1999	2000	
	1.1 Million TRL - Millions TRL							
Business enterprise								**Entreprises**
financed by								*financé par*
Business enterprise	3 221 563	6 360 104	16 897 729	43 612 968	79 890 934	160 082 422	..	Entreprises
Direct government	54 875	117 853	323 495	932 978	1 241 265	2 887 296	..	État direct
Higher education	.. n	.. n	.. n	.. n	.. n	.. n	..	Enseignement supérieur
Private non-profit	66 038 o	99 791 o	108 522 o	904 894 o	1 003 100 o	765 100 o	..	Institutions sans but lucratif
Funds from abroad	112 463	389 200	4 273	304 454	64 237	22 380 808	..	Fonds de l'étranger
Total	3 454 939	6 966 948	17 334 019	45 755 294	82 199 536	186 115 626	..	*Total*
Government								**État**
financed by								*financé par*
Business enterprise	25 908	64 261	223 978	291 324	710 226	1 706 000	..	Entreprises
Direct government	1 052 355	1 877 790	6 465 446	12 167 264	17 417 108	29 376 927	..	État direct
Higher education	.. n	.. n	.. n	.. n	.. n	.. n	..	Enseignement supérieur
Private non-profit	12 058 o	17 717 o	9 566 o	260 810 o	92 310 o	457 704 o	..	Institutions sans but lucratif
Funds from abroad	120 044	211 975	1 230 300	2 220 790	813 587	1 096 440	..	Fonds de l'étranger
Total	1 210 366	2 171 743	7 929 290	14 940 188	19 033 231	32 637 071	..	*Total*
Higher education								**Enseignement supérieur**
financed by								*financé par*
Business enterprise	1 368 085	3 282 788	7 443 413	15 368 296	28 167 399	50 106 235	..	Entreprises
Direct government	3 191 039	8 877 823	19 711 414	20 532 675	54 559 995	100 495 737	..	État direct
General university funds	4 508 191	7 526 265	11 259 403	42 551 894	65 715 989	100 627 170	..	Fonds généraux des universités
Sub-total government	7 699 230	16 404 088	30 970 817	63 084 569	120 275 984	201 122 907	..	Sous-Total État
Higher education	.. n	.. n	.. n	.. n	.. n	.. n	..	Enseignement supérieur
Private non-profit	258 650 o	683 828 o	3 050 444 o	2 586 580 o	10 713 604 o	19 181 043 o	..	Institutions sans but lucratif
Funds from abroad	..	..	..	46 738	32 383	0	..	Fonds de l'étranger
Total	9 325 965	20 370 704	41 464 674	81 086 183	159 189 370	270 410 185	..	*Total*
Private non-profit								**Institutions sans but lucratif**
financed by								*financé par*
Business enterprise	..	..	..	..	..	..	..	Entreprises
Direct government	..	..	..	..	..	..	..	État direct
Higher education	..	..	..	..	..	..	..	Enseignement supérieur
Private non-profit	..	..	..	..	..	..	..	Institutions sans but lucratif
Funds from abroad	..	..	..	..	..	..	..	Fonds de l'étranger
Total	..	..	..	..	..	..	..	*Total*
GERD								**DIRD**
financed by								*financé par*
Business enterprise	4 615 556	9 707 153	24 565 120	59 272 588	108 768 559	211 894 657	..	Entreprises
Direct government	4 298 269	10 873 466	26 500 355	33 632 917	73 218 368	132 759 960	..	État direct
General university funds	4 508 191	7 526 265	11 259 403	42 551 894	65 715 989	100 627 170	..	Fonds généraux des universités
Sub-total government	8 806 460	18 399 731	37 759 758	76 184 811	138 934 357	233 387 130	..	Sous-Total État
Higher education	.. n	.. n	.. n	.. n	.. n	.. n	..	Enseignement supérieur
Private non-profit	336 747 o	801 335 o	3 168 532 o	3 752 284 o	11 809 014 o	20 403 847 o	..	Institutions sans but lucratif
Funds from abroad	232 507	601 175	1 234 573	2 571 982	910 207	23 477 248	..	Fonds de l'étranger
Total	13 991 270	29 509 395	66 727 983	141 781 665	260 422 137	489 162 882	..	*Total*
	1.2 Million constant USD (1995 prices and PPPs) - Millions de USD constants (prix et PPA de 1995)							
Business enterprise								**Entreprises**
financed by								*financé par*
Business enterprise	270.0	284.8	425.4	604.8	630.7	812.4	..	Entreprises
Direct government	4.6	5.3	8.1	12.9	9.8	14.7	..	État direct
Higher education	.. n	.. n	.. n	.. n	.. n	.. n	..	Enseignement supérieur
Private non-profit	5.5 o	4.5 o	2.7 o	12.5 o	7.9 o	3.9 o	..	Institutions sans but lucratif
Funds from abroad	9.4	17.4	0.1	4.2	0.5	113.6	..	Fonds de l'étranger
Total	289.6	311.9	436.4	634.5	648.9	944.5	..	*Total*
Government								**État**
financed by								*financé par*
Business enterprise	2.2	2.9	5.6	4.0	5.6	8.7	..	Entreprises
Direct government	88.2	84.1	162.8	168.7	137.5	149.1	..	État direct
Higher education	.. n	.. n	.. n	.. n	.. n	.. n	..	Enseignement supérieur
Private non-profit	1.0 o	0.8 o	0.2 o	3.6 o	0.7 o	2.3 o	..	Institutions sans but lucratif
Funds from abroad	10.1	9.5	31.0	30.8	6.4	5.6	..	Fonds de l'étranger
Total	101.4	97.2	199.6	207.2	150.3	165.6	..	*Total*
Higher education								**Enseignement supérieur**
financed by								*financé par*
Business enterprise	114.7	147.0	187.4	213.1	222.4	254.3	..	Entreprises
Direct government	267.5	397.5	496.3	284.8	430.7	510.0	..	État direct
General university funds	377.9	337.0	283.5	590.1	518.8	510.6	..	Fonds généraux des universités
Sub-total government	645.3	734.5	779.7	874.9	949.5	1 020.6	..	Sous-Total État
Higher education	.. n	.. n	.. n	.. n	.. n	.. n	..	Enseignement supérieur
Private non-profit	21.7 o	30.6 o	76.8 o	35.9 o	84.6 o	97.3 o	..	Institutions sans but lucratif
Funds from abroad	..	..	..	0.6	0.3	0.0	..	Fonds de l'étranger
Total	781.7	912.1	1 043.9	1 124.5	1 256.7	1 372.2	..	*Total*
Private non-profit								**Institutions sans but lucratif**
financed by								*financé par*
Business enterprise	..	..	..	..	..	..	..	Entreprises
Direct government	..	..	..	..	..	..	..	État direct
Higher education	..	..	..	..	..	..	..	Enseignement supérieur
Private non-profit	..	..	..	..	..	..	..	Institutions sans but lucratif
Funds from abroad	..	..	..	..	..	..	..	Fonds de l'étranger
Total	..	..	..	..	..	..	..	*Total*
GERD								**DIRD**
financed by								*financé par*
Business enterprise	386.9	434.6	618.5	822.0	858.6	1 075.3	..	Entreprises
Direct government	360.3	486.9	667.2	466.4	578.0	673.7	..	État direct
General university funds	377.9	337.0	283.5	590.1	518.8	510.6	..	Fonds généraux des universités
Sub-total government	738.1	823.8	950.6	1 056.5	1 096.8	1 184.4	..	Sous-Total État
Higher education	.. n	.. n	.. n	.. n	.. n	.. n	..	Enseignement supérieur
Private non-profit	28.2 o	35.9 o	79.8 o	52.0 o	93.2 o	103.5 o	..	Institutions sans but lucratif
Funds from abroad	19.5	26.9	31.1	35.7	7.2	119.1	..	Fonds de l'étranger
Total	1 172.7	1 321.3	1 680.0	1 966.3	2 055.8	2 482.3	..	*Total*

Source: OECD, R&D database (DSTI/EAS Division), January 2002.

Source : OCDE, base de données R-D (DSTI/Division AES), janvier 2002.

Table 2 - Tableau 2

Gross domestic expenditure on R&D (GERD) - Dépenses intérieures brutes de R-D (DIRD)

by sector of performance and type of costs - par secteur d'exécution et par type de coûts

Million TRL | *Millions TRL*

	1994	1995	1996	1997	1998	1999	2000	
Business enterprise								**Entreprises**
Labour costs	1 115 469	2 382 926	5 302 902	..	..	..	..	Coûts du personnel
Other current costs	1 092 637	2 235 166	5 929 395	..	..	..	..	Autres coûts courants
Total current expenditure	2 208 106	4 618 092	11 232 297	..	..	..	..	*Total dépenses courantes*
Land and buildings	308 481	705 339	2 683 500	..	..	..	..	Terrains et constructions
Instruments and equipment	938 353	1 643 517	3 418 222	..	..	..	..	Instruments et équipements
Total capital expenditure	1 246 834	2 348 856	6 101 722	..	..	..	..	*Total dépenses en capital*
Total	3 454 939	6 966 948	17 334 019	45 755 294	82 199 536	186 115 626	..	*Total*
Government								**État**
Labour costs	667 942	1 277 840	3 131 414	..	..	..	..	Coûts du personnel
Other current costs	143 070	337 239	1 176 320	..	..	..	..	Autres coûts courants
Total current expenditure	811 012	1 615 079	4 307 734	..	..	..	..	*Total dépenses courantes*
Land and buildings	74 177	49 942	830 132	..	..	..	..	Terrains et constructions
Instruments and equipment	325 177	506 722	2 791 424	..	..	..	..	Instruments et équipements
Total capital expenditure	399 354	556 664	3 621 556	..	..	..	..	*Total dépenses en capital*
Total	1 210 366	2 171 743	7 929 290	14 940 188	19 033 231	32 637 071	..	*Total*
Higher education								**Enseignement supérieur**
Labour costs	3 601 788	9 358 434	20 333 128	..	..	..	..	Coûts du personnel
Other current costs	3 255 878	6 802 237	16 429 473	..	..	..	..	Autres coûts courants
Total current expenditure	6 857 666	16 160 671	36 762 601	..	..	..	..	*Total dépenses courantes*
Land and buildings	1 041 983	1 658 609	2 837 315	..	..	..	..	Terrains et constructions
Instruments and equipment	1 426 317	2 551 424	1 864 758	..	..	..	..	Instruments et équipements
Total capital expenditure	2 468 300	4 210 033	4 702 073	..	..	..	..	*Total dépenses en capital*
Total	9 325 965	20 370 704	41 464 674	81 086 183	159 189 370	270 410 185	..	*Total*
Private non-profit								**Institutions sans but lucratif**
Labour costs	..	..	..	..	..	..	..	Coûts du personnel
Other current costs	..	..	..	..	..	..	..	Autres coûts courants
Total current expenditure	..	..	..	..	..	..	..	*Total dépenses courantes*
Land and buildings	..	..	..	..	..	..	..	Terrains et constructions
Instruments and equipment	..	..	..	..	..	..	..	Instruments et équipements
Total capital expenditure	..	..	..	..	..	..	..	*Total dépenses en capital*
Total	..	..	..	..	..	..	..	*Total*
GERD								**DIRD**
Labour costs	5 385 198	13 019 200	28 767 444	..	..	..	..	Coûts du personnel
Other current costs	4 491 585	9 374 642	23 535 188	..	..	..	..	Autres coûts courants
Total current expenditure	9 876 783	22 393 842	52 302 632	..	..	..	..	*Total dépenses courantes*
Land and buildings	1 424 640	2 413 890	6 350 947	..	..	..	..	Terrains et constructions
Instruments and equipment	2 689 847	4 701 663	8 074 404	..	..	..	..	Instruments et équipements
Total capital expenditure	4 114 488	7 115 553	14 425 351	..	..	..	..	*Total dépenses en capital*
Total	13 991 270	29 509 395	66 727 983	141 781 665	260 422 137	489 162 882	..	*Total*

Table 3 - Tableau 3

Current domestic expenditure on R&D - Dépenses intérieures courantes de R-D

by sector of performance and type of activity - par secteur d'exécution et par type d'activités

Million TRL | *Millions TRL*

	1994	1995	1996	1997	1998	1999	2000	
Business enterprise								**Entreprises**
Basic research	231 221 [a,k]	494 600 [k]	1 713 187 [k]	..	4 321 963 [k]	6 083 193 [k]	..	Recherche fondamentale
Applied research	1 415 146 [a,k]	2 620 155 [k]	5 082 436 [k]	..	26 735 954 [k]	56 531 041 [k]	..	Recherche appliquée
Experimental development	1 808 572 [a,k]	3 852 192 [k]	10 538 395 [k]	..	51 141 619 [k]	123 501 392 [k]	..	Développement expérimental
Not specified	..	..	..	45 755 294 [k]	..	..	..	Non spécifié
Total	3 454 939 [k]	6 966 948 [k]	17 334 019 [k]	45 755 294 [k]	82 199 536 [k]	186 115 626 [k]	..	*Total*
Government								**État**
Basic research	191 277 [a,k]	333 713 [k]	1 789 273 [k]	..	3 734 805 [k]	5 641 315 [k]	..	Recherche fondamentale
Applied research	747 797 [a,k]	1 347 818 [k]	4 772 082 [k]	..	10 019 856 [k]	18 992 831 [k]	..	Recherche appliquée
Experimental development	271 292 [a,k]	490 213 [k]	1 367 936 [k]	..	5 278 570 [k]	8 002 924 [k]	..	Développement expérimental
Not specified	..	..	..	14 940 188 [k]	..	..	..	Non spécifié
Total	1 210 366 [k]	2 171 743 [k]	7 929 290 [k]	14 940 188 [k]	19 033 231 [k]	32 637 071 [k]	..	*Total*
Higher education								**Enseignement supérieur**
Basic research	..	..	..	..	..	..	..	Recherche fondamentale
Applied research	..	..	..	..	..	..	..	Recherche appliquée
Experimental development	..	..	..	..	..	..	..	Développement expérimental
Not specified	9 325 965 [k]	20 370 704 [k]	41 464 674 [k]	81 086 183 [k]	159 189 370 [k]	270 410 185 [k]	..	Non spécifié
Total	9 325 965 [k]	20 370 704 [k]	41 464 674 [k]	81 086 183 [k]	159 189 370 [k]	270 410 185 [k]	..	*Total*
Private non-profit								**Institutions sans but lucratif**
Basic research	..	..	..	..	..	..	..	Recherche fondamentale
Applied research	..	..	..	..	..	..	..	Recherche appliquée
Experimental development	..	..	..	..	..	..	..	Développement expérimental
Not specified	..	..	..	..	..	..	..	Non spécifié
Total	..	..	..	..	..	..	..	*Total*
GERD								**DIRD**
Basic research	..	..	..	..	..	..	..	Recherche fondamentale
Applied research	..	..	..	..	..	..	..	Recherche appliquée
Experimental development	..	..	..	..	..	..	..	Développement expérimental
Not specified	13 991 270 [k]	29 509 395 [k]	66 727 983 [k]	141 781 665 [k]	260 422 137 [k]	489 162 882 [k]	..	Non spécifié
Total	13 991 270 [k]	29 509 395 [k]	66 727 983 [k]	141 781 665 [k]	260 422 137 [k]	489 162 882 [k]	..	*Total*

Source: OECD, R&D database (DSTI/EAS Division), January 2002. | *Source* : OCDE, base de données R-D (DSTI/Division AES), janvier 2002.

Gross domestic expenditure on R&D (GERD) - Dépenses intérieures brutes de R-D (DIRD)

by sector of performance and socio-economic objective - par secteur d'exécution et par objectif socio-économique

Million TRL — *Millions TRL*

	1994	1995	1996	1997	1998	1999	2000	
Business enterprise								**Entreprises**
Agriculture, forestry and fishing	..	..	..	..	..	..	..	Agriculture, sylviculture et pêche
Industrial development	..	..	..	..	..	..	..	Développement industriel
Energy	..	..	..	..	..	..	..	Energie
Transport and telecommunications	..	..	..	..	..	..	..	Transports et télécommunications
Urban and rural planning	..	..	..	..	..	..	..	Aménagement urbain et rural
Prevention of pollution	..	..	..	..	..	..	..	Prévention de la pollution
Identification & treatment of pollution	..	..	..	..	..	..	..	Détection pollu. & mesures antipollu.
Health	..	..	..	..	..	..	..	Santé
Social development and services	..	..	..	..	..	..	..	Développement et services sociaux
Earth and atmosphere	..	..	..	..	..	..	..	Milieu terrestre et atmosphère
Advancement of knowledge	..	..	..	..	..	..	..	Promotion générale des connaissances
Civil space	..	..	..	..	..	..	..	Espace civil
Defence	..	..	..	..	..	..	..	Défense
Not elsewhere classified	3 454 939	6 966 948	17 334 019	45 755 294	82 199 536	186 115 626	..	Non classé
Total	3 454 939	6 966 948	17 334 019	45 755 294	82 199 536	186 115 626	..	*Total*
Government								**État**
Agriculture, forestry and fishing	463 384	869 906	2 050 114	..	..	..	..	Agriculture, sylviculture et pêche
Industrial development	250 537	372 818	1 045 884	..	..	..	..	Développement industriel
Energy	37 279	69 249	296 506	..	..	..	..	Energie
Transport and telecommunications	.. *n*	.. *n*	.. *n*	..	..	..	..	Transports et télécommunications
Urban and rural planning	31 823 *o*	39 964 *o*	227 234 *o*	..	..	..	..	Aménagement urbain et rural
Prevention of pollution	.. *n*	.. *n*	.. *n*	..	..	..	..	Prévention de la pollution
Identification & treatment of pollution	68 537 *o*	107 293 *o*	381 838 *o*	..	..	..	..	Détection pollu. & mesures antipollu.
Health	38 200	72 819	295 392	..	..	..	..	Santé
Social development and services	..	..	29 812	..	..	..	..	Développement et services sociaux
Earth and atmosphere	27 298	45 974	129 525	..	..	..	..	Milieu terrestre et atmosphère
Advancement of knowledge	146 925	295 883	2 458 640	..	..	..	..	Promotion générale des connaissances
Civil space	1 287	2 853	..	..	..	..	..	Espace civil
Defence	145 096	294 984	1 014 345	..	..	..	..	Défense
Not elsewhere classified	..	..	..	14 940 188	19 033 231	32 637 071	..	Non classé
Total	1 210 366	2 171 743	7 929 290	14 940 188	19 033 231	32 637 071	..	*Total*
Higher education								**Enseignement supérieur**
Agriculture, forestry and fishing	..	..	..	..	..	..	..	Agriculture, sylviculture et pêche
Industrial development	..	..	..	..	..	..	..	Développement industriel
Energy	..	..	..	..	..	..	..	Energie
Transport and telecommunications	..	..	..	..	..	..	..	Transports et télécommunications
Urban and rural planning	..	..	..	..	..	..	..	Aménagement urbain et rural
Prevention of pollution	..	..	..	..	..	..	..	Prévention de la pollution
Identification & treatment of pollution	..	..	..	..	..	..	..	Détection pollu. & mesures antipollu.
Health	..	..	..	..	..	..	..	Santé
Social development and services	..	..	..	..	..	..	..	Développement et services sociaux
Earth and atmosphere	..	..	..	..	..	..	..	Milieu terrestre et atmosphère
Advancement of knowledge	..	..	..	..	..	..	..	Promotion générale des connaissances
Civil space	..	..	..	..	..	..	..	Espace civil
Defence	..	..	..	..	..	..	..	Défense
Not elsewhere classified	9 325 965	20 370 704	41 464 674	81 086 183	159 189 370	270 410 185	..	Non classé
Total	9 325 965	20 370 704	41 464 674	81 086 183	159 189 370	270 410 185	..	*Total*
Private non-profit								**Institutions sans but lucratif**
Agriculture, forestry and fishing	..	..	..	..	..	..	..	Agriculture, sylviculture et pêche
Industrial development	..	..	..	..	..	..	..	Développement industriel
Energy	..	..	..	..	..	..	..	Energie
Transport and telecommunications	..	..	..	..	..	..	..	Transports et télécommunications
Urban and rural planning	..	..	..	..	..	..	..	Aménagement urbain et rural
Prevention of pollution	..	..	..	..	..	..	..	Prévention de la pollution
Identification & treatment of pollution	..	..	..	..	..	..	..	Détection pollu. & mesures antipollu.
Health	..	..	..	..	..	..	..	Santé
Social development and services	..	..	..	..	..	..	..	Développement et services sociaux
Earth and atmosphere	..	..	..	..	..	..	..	Milieu terrestre et atmosphère
Advancement of knowledge	..	..	..	..	..	..	..	Promotion générale des connaissances
Civil space	..	..	..	..	..	..	..	Espace civil
Defence	..	..	..	..	..	..	..	Défense
Not elsewhere classified	..	..	..	..	..	..	..	Non classé
Total	..	..	..	..	..	..	..	*Total*
GERD								**DIRD**
Agriculture, forestry and fishing	..	..	..	..	..	..	..	Agriculture, sylviculture et pêche
Industrial development	..	..	..	..	..	..	..	Développement industriel
Energy	..	..	..	..	..	..	..	Energie
Transport and telecommunications	..	..	..	..	..	..	..	Transports et télécommunications
Urban and rural planning	..	..	..	..	..	..	..	Aménagement urbain et rural
Prevention of pollution	..	..	..	..	..	..	..	Prévention de la pollution
Identification & treatment of pollution	..	..	..	..	..	..	..	Détection pollu. & mesures antipollu.
Health	..	..	..	..	..	..	..	Santé
Social development and services	..	..	..	..	..	..	..	Développement et services sociaux
Earth and atmosphere	..	..	..	..	..	..	..	Milieu terrestre et atmosphère
Advancement of knowledge	..	..	..	..	..	..	..	Promotion générale des connaissances
Civil space	..	..	..	..	..	..	..	Espace civil
Defence	..	..	..	..	..	..	..	Défense
Not elsewhere classified	13 991 270	29 509 395	66 727 983	141 781 665	260 422 137	489 162 882	..	Non classé
Total	13 991 270	29 509 395	66 727 983	141 781 665	260 422 137	489 162 882	..	*Total*

Source: OECD, R&D database (DSTI/EAS Division), January 2002.

Source : OCDE, base de données R-D (DSTI/Division AES), janvier 2002.

Table 5 - Tableau 5

Total R&D personnel - Personnel total de R-D

by sector of employment and occupation - par secteur d'emploi et par profession

In full-time equivalent — *En équivalence plein-temps*

	1994	1995	1996	1997	1998	1999	2000	
Business enterprise								**Entreprises**
RSE	1 992	2 211	2 456	3 232	3 070	3 247	..	Chercheurs
Technicians	882	1 023	1 283	1 658	1 441	1 633	..	Techniciens
Other occupations	358	400	554	742	712	734	..	Autres professions
Total	3 232	3 634	4 293	5 632	5 223	5 614	..	*Total*
Government								**État**
RSE	1 756	1 859	2 349	2 245	2 065	2 197	..	Chercheurs
Technicians	562	583	754	715	715	738	..	Techniciens
Other occupations	637	638	1 312	1 409	1 099	1 097	..	Autres professions
Total	2 955	3 080	4 415	4 369	3 879	4 032	..	*Total*
Higher education								**Enseignement supérieur**
RSE	10 712	11 784	13 287	13 431	13 790	14 621	..	Chercheurs
Technicians	..	..	..	..	..	..	..	Techniciens
Other occupations	..	..	..	..	..	..	..	Autres professions
Total	..	..	..	..	..	..	..	*Total*
Private non-profit								**Institutions sans but lucratif**
RSE	..	..	..	..	..	..	..	Chercheurs
Technicians	..	..	..	..	..	..	..	Techniciens
Other occupations	..	..	..	..	..	..	..	Autres professions
Total	..	..	..	..	..	..	..	*Total*
TOTAL								**TOTAL**
RSE	14 460	15 854	18 092	18 908	18 925	20 065	..	Chercheurs
Technicians	1 444 [m]	1 606 [m]	2 037 [m]	2 373 [m]	2 156 [m]	2 371 [m]	..	Techniciens
Other occupations	995 [m]	1 038 [m]	1 866 [m]	2 151 [m]	1 811 [m]	1 831 [m]	..	Autres professions
Total	16 899 [m]	18 498 [m]	21 995 [m]	23 432 [m]	22 892 [m]	24 267 [m]	..	*Total*

Table 6 - Tableau 6

Total R&D personnel - Personnel total de R-D

by sector of employment and formal qualification - par secteur d'emploi et par niveau de qualification

In full-time equivalent — *En équivalence plein-temps*

	1994	1995	1996	1997	1998	1999	2000	
Business enterprise								**Entreprises**
University PhD level degrees	117	130	153	165	199	215	..	Diplômes univ. de niv. doctorat
Other university degrees	2 063	2 323	2 581	3 006	3 018	3 246	..	Autres diplômes universitaires
Sub-total university degrees	2 180	2 453	2 734	3 171	3 217	3 461	..	Sous-total Universitaires
Other post-secondary diplomas	286	325	416	631	640	646	..	Autres diplômes post-secondaires
Secondary diplomas	581	661	830	1 128	881	1 038	..	Diplômes de fin d'études secondaires
Other qualifications	185	195	313	702	485	469	..	Autres qualifications
Total	3 232	3 634	4 293	5 632	5 223	5 614	..	*Total*
Government								**État**
University PhD level degrees	452	490	528	498	484	521	..	Diplômes univ. de niv. doctorat
Other university degrees	1 454	1 523	1 993	2 177	1 874	1 952	..	Autres diplômes universitaires
Sub-total university degrees	1 906	2 013	2 521	2 675	2 358	2 473	..	Sous-total Universitaires
Other post-secondary diplomas	75	91	190	187	228	250	..	Autres diplômes post-secondaires
Secondary diplomas	450	472	776	633	544	558	..	Diplômes de fin d'études secondaires
Other qualifications	524	504	928	874	749	751	..	Autres qualifications
Total	2 955	3 080	4 415	4 369	3 879	4 032	..	*Total*
Higher education								**Enseignement supérieur**
University PhD level degrees	5 019	5 417	5 826	6 161	6 591	6 994	..	Diplômes univ. de niv. doctorat
Other university degrees	5 693	6 367	7 461	7 270	7 199	7 627	..	Autres diplômes universitaires
Sub-total university degrees	10 712	11 784	13 287	13 431	13 790	14 621	..	Sous-total Universitaires
Other post-secondary diplomas	..	..	..	..	..	..	..	Autres diplômes post-secondaires
Secondary diplomas	..	..	..	..	..	..	..	Diplômes de fin d'études secondaires
Other qualifications	..	..	..	..	..	..	..	Autres qualifications
Total	..	..	..	..	..	..	..	*Total*
Private non-profit								**Institutions sans but lucratif**
University PhD level degrees	..	..	..	..	..	..	..	Diplômes univ. de niv. doctorat
Other university degrees	..	..	..	..	..	..	..	Autres diplômes universitaires
Sub-total university degrees	..	..	..	..	..	..	..	Sous-total Universitaires
Other post-secondary diplomas	..	..	..	..	..	..	..	Autres diplômes post-secondaires
Secondary diplomas	..	..	..	..	..	..	..	Diplômes de fin d'études secondaires
Other qualifications	..	..	..	..	..	..	..	Autres qualifications
Total	..	..	..	..	..	..	..	*Total*
TOTAL								**TOTAL**
University PhD level degrees	5 588	6 037	6 507	6 824	7 274	7 730	..	Diplômes univ. de niv. doctorat
Other university degrees	9 210	10 213	12 035	12 453	12 091	12 825	..	Autres diplômes universitaires
Sub-total university degrees	14 798	16 250	18 542	19 277	19 365	20 555	..	Sous-total Universitaires
Other post-secondary diplomas	361 [m]	416 [m]	606 [m]	818 [m]	868 [m]	896 [m]	..	Autres diplômes post-secondaires
Secondary diplomas	1 031 [m]	1 133 [m]	1 606 [m]	1 761 [m]	1 425 [m]	1 596 [m]	..	Diplômes de fin d'études secondaires
Other qualifications	709 [m]	699 [m]	1 241 [m]	1 576 [m]	1 234 [m]	1 220 [m]	..	Autres qualifications
Total	16 899 [m]	18 498 [m]	21 995 [m]	23 432 [m]	22 892 [m]	24 267 [m]	..	*Total*

Source: OECD, R&D database (DSTI/EAS Division), January 2002.

Source : OCDE, base de données R-D (DSTI/Division AES), janvier 2002.

TURKEY — TURQUIE

Table 7 - Tableau 7

Gross domestic expenditure on R&D (GERD) - Dépenses intérieures brutes de R-D (DIRD)

by sector of performance and main scientific field - par secteur d'exécution et par grand domaine scientifique

Million TRL — *Millions TRL*

	1994	1995	1996	1997	1998	1999	2000	
Business enterprise								**Entreprises**
Natural sciences and engineering	3 454 939	6 966 948	17 334 019	45 755 294	82 199 536	186 115 626	..	Sciences exactes, nat. et de l'ingénieur
Social sciences and humanities	..	..	..	..	..	..	..	Sciences sociales et humaines
All fields of science	3 454 939	6 966 948	17 334 019	45 755 294	82 199 536	186 115 626	..	*Tous domaines scientifiques*
Government								**État**
Natural sciences and engineering	..	..	..	..	..	..	..	Sciences exactes, nat. et de l'ingénieur
Social sciences and humanities	..	..	..	..	..	..	..	Sciences sociales et humaines
All fields of science	1 210 366	2 171 743	7 929 290	14 940 188	19 033 231	32 637 071	..	*Tous domaines scientifiques*
Higher education								**Enseignement supérieur**
Natural sciences and engineering	7 892 878	16 111 092	35 158 747	..	..	..	..	Sciences exactes, nat. et de l'ingénieur
Social sciences and humanities	1 433 087	4 259 612	6 305 927	..	..	..	..	Sciences sociales et humaines
All fields of science	9 325 965	20 370 704	41 464 674	81 086 183	159 189 370	270 410 185	..	*Tous domaines scientifiques*
Private non-profit								**Institutions sans but lucratif**
Natural sciences and engineering	..	..	..	..	..	..	..	Sciences exactes, nat. et de l'ingénieur
Social sciences and humanities	..	..	..	..	..	..	..	Sciences sociales et humaines
All fields of science	..	..	..	..	..	..	..	*Tous domaines scientifiques*
TOTAL								**TOTAL**
Natural sciences and engineering	..	..	..	..	..	..	..	Sciences exactes, nat. et de l'ingénieur
Social sciences and humanities	..	..	..	..	..	..	..	Sciences sociales et humaines
All fields of science	13 991 270	29 509 395	66 727 983	141 781 665	260 422 137	489 162 882	..	*Tous domaines scientifiques*

Table 8 - Tableau 8

Total R&D personnel - Personnel total de R-D

by sector of employment and main scientific field - par secteur d'emploi et par grand domaine scientifique

In full-time equivalent — *En équivalence plein-temps*

	1994	1995	1996	1997	1998	1999	2000	
Business enterprise								**Entreprises**
Natural sciences and engineering	3 232	3 634	4 293	5 632	5 223	5 614	..	Sciences exactes, nat. et de l'ingénieur
Social sciences and humanities	..	..	..	..	..	..	..	Sciences sociales et humaines
All fields of science	3 232	3 634	4 293	5 632	5 223	5 614	..	*Tous domaines scientifiques*
Government								**État**
Natural sciences and engineering	..	..	..	..	..	..	..	Sciences exactes, nat. et de l'ingénieur
Social sciences and humanities	..	..	..	..	..	..	..	Sciences sociales et humaines
All fields of science	2 955	3 080	4 415	4 369	3 879	4 032	..	*Tous domaines scientifiques*
Higher education								**Enseignement supérieur**
Natural sciences and engineering	..	..	..	..	..	..	..	Sciences exactes, nat. et de l'ingénieur
Social sciences and humanities	..	..	..	..	..	..	..	Sciences sociales et humaines
All fields of science	..	..	..	..	..	..	..	*Tous domaines scientifiques*
Private non-profit								**Institutions sans but lucratif**
Natural sciences and engineering	..	..	..	..	..	..	..	Sciences exactes, nat. et de l'ingénieur
Social sciences and humanities	..	..	..	..	..	..	..	Sciences sociales et humaines
All fields of science	..	..	..	..	..	..	..	*Tous domaines scientifiques*
TOTAL								**TOTAL**
Natural sciences and engineering	..	..	..	..	..	..	..	Sciences exactes, nat. et de l'ingénieur
Social sciences and humanities	..	..	..	..	..	..	..	Sciences sociales et humaines
All fields of science	16 899 m	18 498 m	21 995 m	23 432 m	22 892 m	24 267 m	..	*Tous domaines scientifiques*

Source: OECD, R&D database (DSTI/EAS Division), January 2002.

Source : OCDE, base de données R-D (DSTI/Division AES), janvier 2002.

Total business enterprise intramural expenditure on R&D - Dépenses intra-muros totales de R-D des entreprises
(BERD) by industry - (DIRDE) par industrie

	1994	1995	1996	1997	1998	1999	2000	
				9.1 Million TRL - Millions TRL				
1 AGRICULTURE, HUNTING & FORESTRY	24 895	118 233	35 512	5 144	28 987	27 153	..	**1 AGRICULTURE, CHASSE & SYLVICULTURE**
2 MINING	35 717	46 159	102 721	261 496	355 464	553 376	..	**2 ACTIVITES EXTRACTIVES**
3 MANUFACTURING	3 182 836	6 403 492	16 305 223	36 529 905	68 329 593	162 018 204	..	**3 ACTIVITES DE FABRICATION**
4 Food, beverages & tobacco	169 076	303 086	655 931	1 317 757	1 781 840	3 193 898	..	**4 Produits alimentaires, boissons & tabac**
5 Food products & beverages	161 327	285 156	597 300	1 257 425	1 659 367	2 901 697	..	5 Produits alimentaires & boissons
6 Tobacco products	7 749	17 930	58 631	60 332	122 473	292 201	..	6 Produits à base de tabac
7 Textiles, wear app., fur & leather	52 176	101 377	240 938	1 243 861	1 870 924	3 179 397	..	**7 Textiles, habill., fourrures & cuirs**
8 Textiles	52 176	101 377	232 888	1 208 091	1 870 924	3 179 397	..	8 Textiles
9 Wearing apparel & fur	0	0	3 717	0	..	..	..	9 Articles d'habillement & fourrures
10 Leather products & footwear	0	0	4 333	35 770	..	..	..	10 Articles en cuir & chaussures
11 Wood, paper, printing, publishing	7 331	13 872	130 714	548 581	385 378	576 967	..	**11 Bois, papier, imprimerie, édition**
12 Wood & cork (not furniture)	0	0	..	59 600	91 400	127 400	..	12 Bois & liège (sauf meubles)
13 Pulp, paper & paper products	7 331	13 872	121 120	488 981	260 378	346 367	..	13 Carton, papier & articles
14 Publ., print. & repro. of rec. media	..	..	9 594	0	33 600	103 200	..	14 Edition, impress. & repro. enregis.
15 Coke, petroleum, nuclear fuel, chemicals, rubber plastics	364 037	662 910	1 530 072	4 447 054	5 751 225	9 631 772	..	**15 Coke, pétrole, combust. nucl., prod. chimi., caout. & mat. plast.**
16 Coke, ref. petrol. prod. & nuclear fuel	22 395	29 736	12 763	0	..	..	..	16 Coke, raffin. pétr. & combust. nucl.
17 Chemicals & chemical products	263 814	478 011	1 066 688	3 307 918	4 476 426	8 052 196	..	17 Produits chimiques
18 *Chemical prod.(less pharmaceuticals)*	230 709	385 845	772 255	2 616 336	3 341 823	6 236 393	..	18 *Prod. chimi. (moins prod. pharmaceutiques)*
19 *Pharmaceuticals*	33 105	92 166	294 433	691 582	1 134 603	1 815 803	..	19 *Produits pharmaceutiques*
20 Rubber & plastic products	77 828	155 163	450 621	1 139 136	1 274 799	1 579 576	..	20 Caoutchouc et matières plastiques
21 Non-metallic mineral products	263 259	400 950	1 049 591	2 310 329	2 174 214	3 765 842	..	**21 Produits minéraux non métalliques**
22 Basic metals	189 816	401 939	2 364 398	1 964 014	3 626 622	4 076 686	..	**22 Produits métallurgiques de base**
23 Basic metals, ferrous	181 277	320 325	2 292 945	1 611 886	3 060 091	3 313 451	..	23 Produits métallur. de base, ferreux
24 Basic metals, non-ferrous	8 539	81 614	71 453	352 128	566 531	763 235	..	24 Prod. métallur. de base, non ferreux
25 Fabricated metal products	19 120	95 983	315 037	767 509	815 022	1 452 890	..	**25 Ouvrages en métaux (sauf mach.& mat.)**
26 Machinery equipment, instruments and transport equipment	2 118 021	4 423 375	9 845 834	23 824 438	51 858 678	136 047 744	..	**26 Machines & matériel, instruments & mat. de transport**
27 Machinery n.e.c.	658 478	1 323 905	2 196 771	5 640 455	8 044 209	13 080 654	..	27 Machines, n.c.a.
28 Office, account. & computing machinery	0	0	294 870	19 896	26 800	44 600	..	28 Mach. de bur., compta. & trait. inform.
29 Electrical machinery	121 530	330 055	943 531	2 514 071	2 583 374	2 251 527	..	29 Machines électriques
30 Electro. equip. (radio, TV & commun.)	1 064 446	2 136 984	4 078 453	6 085 968	9 804 766	17 512 778	..	30 Equip. électro. (radio, tv & communic.)
31 *Electro. comp. (inc. semi-conductors)*	1 615	2 898	7 921	102 758	146 828	228 742	..	31 *Compos. électro. (semi-cond. incl.)*
32 *TV, radio & communications equip.*	1 062 831	2 134 086	4 070 532	5 983 210	9 657 938	17 284 036	..	32 *Appareils radio, tv et de communi.*
33 Instruments	4 771	11 533	141 348	182 478	238 535	305 422	..	33 Instruments
34 Motor vehicles	230 952	522 345	1 930 897	8 623 714	29 035 511	99 790 240	..	34 Véhicules automobiles
35 Other transport equipment	37 844	98 553	259 964	757 856	2 125 483	3 062 523	..	35 Autres matériels de transport
36 *Ships*	0	0	..	0	..	..	..	36 *Navires*
37 *Aerospace*	20 015	50 923	176 763	647 507	1 918 067	2 780 183	..	37 *Aérospatiale*
38 *Other transport equipment n.e.c.*	17 829	47 630	83 201	110 349	207 416	282 340	..	38 *Autres matériels de transport n.c.a.*
39 Furniture, other manufacturing n.e.c.	0	0	172 708	106 362	65 690	93 008	..	**39 Meubles, autres activ. de fabricat. n.c.a.**
40 Furniture	0	0	..	106 362	65 690	93 008	..	40 Meubles
41 Other manufacturing n.e.c.	0	0	172 708	0	..	..	..	41 Autres activités de fabrication n.c.a.
42 Recycling	0	0	..	0	..	..	..	**42 Récupération**
43 ELECTRICITY, GAS & WATER SUPPLY	97 200	181 753	388 337	1 328 030	1 608 120	2 222 716	..	**43 DISTRIBUTION D'ELECTRICITE, GAZ & EAU**
44 CONSTRUCTION	0	0	..	3 848	17 010	41 929	..	**44 CONSTRUCTION**
45 SERVICES SECTOR	114 291	217 311	502 226	7 626 871	11 860 362	21 252 248	..	**45 SECTEUR DES SERVICES**
46 Wholesale, retail trade & repair of motor vehicles, etc.	0	0	..	82 240	..	..	..	**46 Commerce (gros & détail), réparation de véhicules, etc.**
47 Hotels & restaurants	0	0	..	0	..	..	..	**47 Hôtels et restaurants**
48 Transport & storage	2 034	3 295	8 635	40 476	56 940	93 276	..	**48 Transports, entreposage**
49 Communications	44 841	53 518	97 251	2 403 582	3 552 188	4 333 052	..	**49 Communications**
50 Post	0	0	..	0	..	..	..	50 Postes
51 Telecommunications	44 841	53 518	97 251	2 403 582	3 552 188	4 333 052	..	51 Télécommunications
52 Financial intermediation (inc. insur.)	..	..	..	3 669 709	6 258 814	13 505 039	..	**52 Intermédiation financ. (assurances inc.)**
53 Real estate, renting & busin. activity	67 416	160 498	396 340	1 430 864	1 888 449	3 153 182	..	**53 Immobil., locat. & serv. aux entreprises**
54 Computer & related activities	67 416	160 498	396 340	1 321 199	1 525 292	2 675 138	..	54 Informatique et activités connexes
55 *Software consultancy*	67 416	160 498	..	..	..	..	..	55 *Production de logiciels*
56 *Other computer services n.e.c.*	..	..	..	..	..	..	..	56 *Autres activités informatiques*
57 Research & development	..	..	..	..	..	..	..	57 Recherche & développement expérimental
58 Other business activities n.e.c.	0	0	..	109 665	363 157	478 044	..	58 Autres activ. & serv. aux entreprises
59 Comm., soc. & pers. serv. activ., etc.	0	0	..	0	103 971	167 699	..	**59 Adm. pub., serv. soc. & collec., etc**
TOTAL BUSINESS ENTERPRISE SECTOR	**3 454 939**	**6 966 948**	**17 334 019**	**45 755 294**	**82 199 536**	**186 115 626**	**..**	**TOTAL SECTEUR DES ENTREPRISES**

Source: OECD, R&D database (DSTI/EAS Division), January 2002.

Source : OCDE, base de données R-D (DSTI/Division AES), janvier 2002.

TURKEY Table 9 - Tableau 9 **TURQUIE**

Total business enterprise intramural expenditure on R&D - Dépenses intra-muros totales de R-D des entreprises
(BERD) by industry - (DIRDE) par industrie

9.2 Million constant USD(1995 prices and PPPs) - Millions de USD constants (prix et PPA de 1995)

	1994	1995	1996	1997	1998	1999	2000	
1 AGRICULTURE, HUNTING & FORESTRY	2.1	5.3	0.9	0.1	0.2	0.1	..	**1 AGRICULTURE, CHASSE & SYLVICULTURE**
2 MINING	3.0	2.1	2.6	3.6	2.8	2.8	..	**2 ACTIVITES EXTRACTIVES**
3 MANUFACTURING	266.8	286.7	410.5	506.6	539.4	822.2	..	**3 ACTIVITES DE FABRICATION**
4 Food, beverages & tobacco	14.2	13.6	16.5	18.3	14.1	16.2	..	**4 Produits alimentaires, boissons & tabac**
5 Food products & beverages	13.5	12.8	15.0	17.4	13.1	14.7	..	5 Produits alimentaires & boissons
6 Tobacco products	0.6	0.8	1.5	0.8	1.0	1.5	..	6 Produits à base de tabac
7 Textiles, wear app., fur & leather	4.4	4.5	6.1	17.3	14.8	16.1	..	**7 Textiles, habill., fourrures & cuirs**
8 Textiles	4.4	4.5	5.9	16.8	14.8	16.1	..	8 Textiles
9 Wearing apparel & fur	0.0	0.0	0.1	0.0	..	..	..	9 Articles d'habillement & fourrures
10 Leather products & footwear	0.0	0.0	0.1	0.5	..	..	..	10 Articles en cuir & chaussures
11 Wood, paper, printing, publishing	0.6	0.6	3.3	7.6	3.0	2.9	..	**11 Bois, papier, imprimerie, édition**
12 Wood & cork (not furniture)	0.0	0.0	..	0.8	0.7	0.6	..	12 Bois & liège (sauf meubles)
13 Pulp, paper & paper products	0.6	0.6	3.0	6.8	2.1	1.8	..	13 Carton, papier & articles
14 Publ., print. & repro. of rec. media	..	..	0.2	0.0	0.3	0.5	..	14 Edition, impress. & repro. enregis.
15 Coke, petroleum, nuclear fuel, chemicals, rubber plastics	30.5	29.7	38.5	61.7	45.4	48.9	..	**15 Coke, pétrole, combust. nucl., prod. chimi., caout. & mat. plast.**
16 Coke, ref. petrol. prod. & nuclear fuel	1.9	1.3	0.3	0.0	..	..	..	16 Coke, raffin. pétr. & combust. nucl.
17 Chemicals & chemical products	22.1	21.4	26.9	45.9	35.3	40.9	..	17 Produits chimiques
18 *Chemical prod.(less pharmaceuticals)*	19.3	17.3	19.4	36.3	26.4	31.6	..	18 *Prod. chimi. (moins prod. pharmaceutiques)*
19 *Pharmaceuticals*	2.8	4.1	7.4	9.6	9.0	9.2	..	19 *Produits pharmaceutiques*
20 Rubber & plastic products	6.5	6.9	11.3	15.8	10.1	8.0	..	20 Caoutchouc et matières plastiques
21 Non-metallic mineral products	22.1	18.0	26.4	32.0	17.2	19.1	..	**21 Produits minéraux non métalliques**
22 Basic metals	15.9	18.0	59.5	27.2	28.6	20.7	..	**22 Produits métallurgiques de base**
23 Basic metals, ferrous	15.2	14.3	57.7	22.4	24.2	16.8	..	23 Produits métallur. de base, ferreux
24 Basic metals, non-ferrous	0.7	3.7	1.8	4.9	4.5	3.9	..	24 Prod. métallur. de base, non ferreux
25 Fabricated metal products	1.6	4.3	7.9	10.6	6.4	7.4	..	**25 Ouvrages en métaux (sauf mach.& mat.)**
26 Machinery equipment, instruments and transport equipment	177.5	198.1	247.9	330.4	409.4	690.4	..	**26 Machines & matériel, instruments & mat. de transport**
27 Machinery n.e.c.	55.2	59.3	55.3	78.2	63.5	66.4	..	27 Machines, n.c.a.
28 Office, account. & computing machinery	0.0	0.0	7.4	0.3	0.2	0.2	..	28 Mach. de bur., compta. & trait. inform.
29 Electrical machinery	10.2	14.8	23.8	34.9	20.4	11.4	..	29 Machines électriques
30 Electro. equip. (radio, TV & commun.)	89.2	95.7	102.7	84.4	77.4	88.9	..	30 Equip. électro. (radio, tv & communic.)
31 *Electro. comp. (inc. semi-conductors)*	0.1	0.1	0.2	1.4	1.2	1.2	..	31 *Compos. électro. (semi-cond. incl.)*
32 *TV, radio & communications equip.*	89.1	95.6	102.5	83.0	76.2	87.7	..	32 *Appareils radio, tv et de communi.*
33 Instruments	0.4	0.5	3.6	2.5	1.9	1.6	..	33 Instruments
34 Motor vehicles	19.4	23.4	48.6	119.6	229.2	506.4	..	34 Véhicules automobiles
35 Other transport equipment	3.2	4.4	6.5	10.5	16.8	15.5	..	35 Autres matériels de transport
36 *Ships*	0.0	0.0	..	0.0	..	..	..	36 *Navires*
37 *Aerospace*	1.7	2.3	4.5	9.0	15.1	14.1	..	37 *Aérospatiale*
38 *Other transport equipment n.e.c.*	1.5	2.1	2.1	1.5	1.6	1.4	..	38 *Autres matériels de transport n.c.a.*
39 Furniture, other manufacturing n.e.c.	0.0	0.0	4.3	1.5	0.5	0.5	..	**39 Meubles, autres activ. de fabricat. n.c.a.**
40 Furniture	0.0	0.0	..	1.5	0.5	0.5	..	40 Meubles
41 Other manufacturing n.e.c.	0.0	0.0	4.3	0.0	..	..	..	41 Autres activités de fabrication n.c.a.
42 Recycling	0.0	0.0	..	0.0	..	..	..	**42 Récupération**
43 ELECTRICITY, GAS & WATER SUPPLY	8.1	8.1	9.8	18.4	12.7	11.3	..	**43 DISTRIBUTION D'ELECTRICITE, GAZ & EAU**
44 CONSTRUCTION	0.0	0.0	..	0.1	0.1	0.2	..	**44 CONSTRUCTION**
45 SERVICES SECTOR	9.6	9.7	12.6	105.8	93.6	107.8	..	**45 SECTEUR DES SERVICES**
46 Wholesale, retail trade & repair of motor vehicles, etc.	0.0	0.0	..	1.1	..	..	..	**46 Commerce (gros & détail), réparation de véhicules, etc.**
47 Hotels & restaurants	0.0	0.0	..	0.0	..	..	..	**47 Hôtels et restaurants**
48 Transport & storage	0.2	0.1	0.2	0.6	0.4	0.5	..	**48 Transports, entreposage**
49 Communications	3.8	2.4	2.4	33.3	28.0	22.0	..	**49 Communications**
50 Post	0.0	0.0	..	0.0	..	..	..	50 Postes
51 Telecommunications	3.8	2.4	2.4	33.3	28.0	22.0	..	51 Télécommunications
52 Financial intermediation (inc. insur.)	..	..	..	50.9	49.4	68.5	..	**52 Intermédiation financ. (assurances inc.)**
53 Real estate, renting & busin. activity	5.7	7.2	10.0	19.8	14.9	16.0	..	**53 Immobil., locat. & serv. aux entreprises**
54 Computer & related activities	5.7	7.2	10.0	18.3	12.0	13.6	..	54 Informatique et activités connexes
55 *Software consultancy*	5.7	7.2	..	..	..	..	..	55 *Production de logiciels*
56 *Other computer services n.e.c.*	..	..	..	..	..	..	..	56 *Autres activités informatiques*
57 Research & development	..	..	..	0.0	..	..	..	57 Recherche & développement expérimental
58 Other business activities n.e.c.	0.0	0.0	..	1.5	2.9	2.4	..	58 Autres activ. & serv. aux entreprises
59 Comm., soc. & pers. serv. activ., etc.	0.0	0.0	..	0.0	0.8	0.9	..	**59 Adm. pub., serv. soc. & collec., etc**
TOTAL BUSINESS ENTERPRISE SECTOR	**289.6**	**311.9**	**436.4**	**634.5**	**648.9**	**944.5**	**..**	**TOTAL SECTEUR DES ENTREPRISES**

Source: OECD, R&D database (DSTI/EAS Division), January 2002.

Source : OCDE, base de données R-D (DSTI/Division AES), janvier 2002.

Business enterprise intramural expenditure on R&D (BERD) - Dépenses intra-muros de R-D des entreprises (DIRDE)

by industry and source of funds - par industrie et par source de financement

Million TRL *Millions TRL*

	1994	1995	1996	1997	1998	1999	2000	
	10.1 Business enterprise - Entreprises							
1 AGRICULTURE, HUNTING & FORESTRY	11 155	56 403	34 860	..	..	..	..	**1 AGRICULTURE, CHASSE & SYLVICULTURE**
2 MINING	35 717	46 159	102 721	..	..	..	..	**2 ACTIVITES EXTRACTIVES**
3 MANUFACTURING	2 963 560	5 859 028	15 880 989	..	..	..	..	**3 ACTIVITES DE FABRICATION**
4 Food, beverages & tobacco	169 076	303 086	652 658	..	..	..	..	**4 Produits alimentaires, boissons & tabac**
5 Food products & beverages	161 327	285 156	594 027	..	..	..	..	5 Produits alimentaires & boissons
6 Tobacco products	7 749	17 930	58 631	..	..	..	..	6 Produits à base de tabac
7 Textiles, wear app., fur & leather	52 176	101 377	240 938	..	..	..	..	**7 Textiles, habill., fourrures & cuirs**
8 Textiles	52 176	101 377	232 888	..	..	..	..	8 Textiles
9 Wearing apparel & fur	..	..	3 717	..	..	..	..	9 Articles d'habillement & fourrures
10 Leather products & footwear	..	..	4 333	..	..	..	..	10 Articles en cuir & chaussures
11 Wood, paper, printing, publishing	7 331	13 872	128 714	..	..	..	..	**11 Bois, papier, imprimerie, édition**
12 Wood & cork (not furniture)	..	..	..	..	..	..	..	12 Bois & liège (sauf meubles)
13 Pulp, paper & paper products	7 331	13 872	121 120	..	..	..	..	13 Carton, papier & articles
14 Publ., print. & repro. of rec. media	..	..	7 594	..	..	..	..	14 Edition, impress. & repro. enregis.
15 Coke, petroleum, nuclear fuel, chemicals, rubber plastics	349 751	643 761	1 474 495	..	..	..	..	**15 Coke, pétrole, combust. nucl., prod. chimi., caout. & mat. plast.**
16 Coke, ref. petrol. prod. & nuclear fuel	22 395	29 736	12 763	..	..	..	..	16 Coke, raffin. pétr. & combust. nucl.
17 Chemicals & chemical products	249 528	458 862	1 012 611	..	..	..	..	17 Produits chimiques
18 *Chemical prod.(less pharmaceuticals)*	216 423	366 696	742 751	..	..	..	..	18 *Prod. chimi. (moins prod. pharmaceutiques)*
19 *Pharmaceuticals*	33 105	92 166	269 860	..	..	..	..	19 *Produits pharmaceutiques*
20 Rubber & plastic products	77 828	155 163	449 121	..	..	..	..	20 Caoutchouc et matières plastiques
21 Non-metallic mineral products	263 259	400 950	1 049 591	..	..	..	..	**21 Produits minéraux non métalliques**
22 Basic metals	189 816	383 070	2 282 899	..	..	..	..	**22 Produits métallurgiques de base**
23 Basic metals, ferrous	181 277	320 325	2 216 886	..	..	..	..	23 Produits métallur. de base, ferreux
24 Basic metals, non-ferrous	8 539	62 745	66 013	..	..	..	..	24 Prod. métallur. de base, non ferreux
25 Fabricated metal products	19 120	84 120	309 782	..	..	..	..	**25 Ouvrages en métaux (sauf mach.& mat.)**
26 Machinery equipment, instruments and transport equipment	1 913 031	3 928 792	9 569 204	..	..	..	..	**26 Machines & matériel, instruments & mat. de transport**
27 Machinery n.e.c.	560 215	1 161 461	2 161 595	..	..	..	..	27 Machines, n.c.a.
28 Office, account. & computing machinery	..	..	294 870	..	..	..	..	28 Mach. de bur., compta. & trait. inform.
29 Electrical machinery	105 256	305 103	880 168	..	..	..	..	29 Machines électriques
30 Electro. equip. (radio, TV & commun.)	979 269	1 841 754	4 035 704	..	..	..	..	30 Equip. électro. (radio, tv & communic.)
31 *Electro. comp. (inc. semi-conductors)*	1 615	2 898	7 921	..	..	..	..	31 *Compos. électro. (semi-cond. incl.)*
32 *TV, radio & communications equip.*	977 654	1 838 856	4 027 783	..	..	..	..	32 *Appareils radio, tv et de communi.*
33 Instruments	3 687	10 033	134 642	..	..	..	..	33 Instruments
34 Motor vehicles	226 760	518 888	1 930 897	..	..	..	..	34 Véhicules automobiles
35 Other transport equipment	37 844	91 553	131 328	..	..	..	..	35 Autres matériels de transport
36 *Ships*	..	..	..	..	..	..	..	36 *Navires*
37 *Aerospace*	20 015	43 923	48 127	..	..	..	..	37 *Aérospatiale*
38 *Other transport equipment n.e.c.*	17 829	47 630	83 201	..	..	..	..	38 *Autres matériels de transport n.c.a.*
39 Furniture, other manufacturing n.e.c.	..	..	172 708	..	..	..	..	**39 Meubles, autres activ. de fabricat. n.c.a.**
40 Furniture	..	..	..	..	..	..	..	40 Meubles
41 Other manufacturing n.e.c.	..	..	172 708	..	..	..	..	41 Autres activités de fabrication n.c.a.
42 Recycling	..	..	..	..	..	..	..	**42 Récupération**
43 ELECTRICITY, GAS & WATER SUPPLY	97 200	181 753	388 337	..	..	..	..	**43 DISTRIBUTION D'ELECTRICITE, GAZ & EAU**
44 CONSTRUCTION	..	..	..	..	..	..	..	**44 CONSTRUCTION**
45 SERVICES SECTOR	113 931	216 761	490 822	..	..	..	..	**45 SECTEUR DES SERVICES**
46 Wholesale, retail trade & repair of motor vehicles, etc.	..	..	..	..	..	..	..	**46 Commerce (gros & détail), réparation de véhicules, etc.**
47 Hotels & restaurants	..	..	..	..	..	..	..	**47 Hôtels et restaurants**
48 Transport & storage	2 034	3 295	8 635	..	..	..	..	**48 Transports, entreposage**
49 Communications	44 841	53 518	90 462	..	..	..	..	**49 Communications**
50 Post	..	..	..	..	..	..	..	50 Postes
51 Telecommunications	44 841	53 518	90 462	..	..	..	..	51 Télécommunications
52 Financial intermediation (inc. insur.)	..	..	..	..	..	..	..	**52 Intermédiation financ. (assurances inc.)**
53 Real estate, renting & busin. activity	67 056	159 948	391 725	..	..	..	..	**53 Immobil., locat. & serv. aux entreprises**
54 Computer & related activities	67 056	159 948	391 725	..	..	..	..	54 Informatique et activités connexes
55 *Software consultancy*	..	..	..	..	..	..	..	55 *Production de logiciels*
56 *Other computer services n.e.c.*	..	..	..	..	..	..	..	56 *Autres activités informatiques*
57 Research & development	..	..	..	..	..	..	..	57 Recherche & développement expérimental
58 Other business activities n.e.c.	..	..	..	..	..	..	..	58 Autres activ. & serv. aux entreprises
59 Comm., soc. & pers. serv. activ., etc.	..	..	..	..	..	..	..	**59 Adm. pub., serv. soc. & collec., etc**
TOTAL BUSINESS ENTERPRISE SECTOR	**3 221 563**	**6 360 104**	**16 897 729**	**43 612 968**	**79 890 934**	**160 082 422**	**..**	**TOTAL SECTEUR DES ENTREPRISES**

Source: OECD, R&D database (DSTI/EAS Division), January 2002.

Source : OCDE, base de données R-D (DSTI/Division AES), janvier 2002.

TURKEY — Table 10 - Tableau 10 — TURQUIE

Business enterprise intramural expenditure on R&D (BERD) - Dépenses intra-muros de R-D des entreprises (DIRDE)

by industry and source of funds - par industrie et par source de financement

Million TRL / *Millions TRL*

10.2 Government - État

	1994	1995	1996	1997	1998	1999	2000	
1 AGRICULTURE, HUNTING & FORESTRY	..	..	652	..	..	..	..	**1 AGRICULTURE, CHASSE & SYLVICULTURE**
2 MINING	..	..	..	..	..	..	..	**2 ACTIVITES EXTRACTIVES**
3 MANUFACTURING	54 515	117 303	314 254	..	..	..	..	**3 ACTIVITES DE FABRICATION**
4 Food, beverages & tobacco	..	..	400	..	..	..	..	**4 Produits alimentaires, boissons & tabac**
5 Food products & beverages	..	..	400	..	..	..	..	5 Produits alimentaires & boissons
6 Tobacco products	..	..	..	..	..	..	..	6 Produits à base de tabac
7 Textiles, wear app., fur & leather	..	..	..	..	..	..	..	**7 Textiles, habill., fourrures & cuirs**
8 Textiles	..	..	..	..	..	..	..	8 Textiles
9 Wearing apparel & fur	..	..	..	..	..	..	..	9 Articles d'habillement & fourrures
10 Leather products & footwear	..	..	..	..	..	..	..	10 Articles en cuir & chaussures
11 Wood, paper, printing, publishing	..	..	2 000	..	..	..	..	**11 Bois, papier, imprimerie, édition**
12 Wood & cork (not furniture)	..	..	..	..	..	..	..	12 Bois & liège (sauf meubles)
13 Pulp, paper & paper products	..	..	..	..	..	..	..	13 Carton, papier & articles
14 Publ., print. & repro. of rec. media	..	..	2 000	..	..	..	..	14 Edition, impress. & repro. enregis.
15 Coke, petroleum, nuclear fuel, chemicals, rubber plastics	14 286	19 149	38 044	..	..	..	..	**15 Coke, pétrole, combust. nucl., prod. chimi., caout. & mat. plast.**
16 Coke, ref. petrol. prod. & nuclear fuel	..	..	..	..	..	..	..	16 Coke, raffin. pétr. & combust. nucl.
17 Chemicals & chemical products	14 286	19 149	36 544	..	..	..	..	17 Produits chimiques
18 *Chemical prod.(less pharmaceuticals)*	14 286	19 149	11 971	..	..	..	..	18 *Prod. chimi. (moins prod. pharmaceutiques)*
19 *Pharmaceuticals*	..	..	24 573	..	..	..	..	19 *Produits pharmaceutiques*
20 Rubber & plastic products	..	..	1 500	..	..	..	..	20 Caoutchouc et matières plastiques
21 Non-metallic mineral products	..	..	..	..	..	..	..	**21 Produits minéraux non métalliques**
22 Basic metals	..	18 869	76 059	..	..	..	..	**22 Produits métallurgiques de base**
23 Basic metals, ferrous	..	..	76 059	..	..	..	..	23 Produits métallur. de base, ferreux
24 Basic metals, non-ferrous	..	18 869	..	..	..	..	..	24 Prod. métallur. de base, non ferreux
25 Fabricated metal products	..	200	5 255	..	..	..	..	**25 Ouvrages en métaux (sauf mach.& mat.)**
26 Machinery equipment, instruments and transport equipment	40 229	79 085	192 496	..	..	..	..	**26 Machines & matériel, instruments & mat. de transport**
27 Machinery n.e.c.	26 263	47 393	29 368	..	..	..	..	27 Machines, n.c.a.
28 Office, account. & computing machinery	..	..	..	..	..	..	..	28 Mach. de bur., compta. & trait. inform.
29 Electrical machinery	12 174	21 527	63 363	..	..	..	..	29 Machines électriques
30 Electro. equip. (radio, TV & commun.)	..	..	42 749	..	..	..	..	30 Equip. électro. (radio, tv & communic.)
31 *Electro. comp. (inc. semi-conductors)*	..	..	..	..	..	..	..	31 *Compos. électro. (semi-cond. incl.)*
32 *TV, radio & communications equip.*	..	..	42 749	..	..	..	..	32 *Appareils radio, tv et de communi.*
33 Instruments	63	127	255	..	..	..	..	33 Instruments
34 Motor vehicles	1 729	3 038	..	..	..	..	..	34 Véhicules automobiles
35 Other transport equipment	..	7 000	56 761	..	..	..	..	35 Autres matériels de transport
36 *Ships*	..	..	..	..	..	..	..	36 *Navires*
37 *Aerospace*	..	7 000	56 761	..	..	..	..	37 *Aérospatiale*
38 *Other transport equipment n.e.c.*	..	..	..	..	..	..	..	38 *Autres matériels de transport n.c.a.*
39 Furniture, other manufacturing n.e.c.	..	..	..	..	..	..	..	**39 Meubles, autres activ. de fabricat. n.c.a.**
40 Furniture	..	..	..	..	..	..	..	40 Meubles
41 Other manufacturing n.e.c.	..	..	..	..	..	..	..	41 Autres activités de fabrication n.c.a.
42 Recycling	..	..	..	..	..	..	..	**42 Récupération**
43 ELECTRICITY, GAS & WATER SUPPLY	..	..	..	..	..	..	..	**43 DISTRIBUTION D'ELECTRICITE, GAZ & EAU**
44 CONSTRUCTION	..	..	..	..	..	..	..	**44 CONSTRUCTION**
45 SERVICES SECTOR	360	550	8 589	..	..	..	..	**45 SECTEUR DES SERVICES**
46 Wholesale, retail trade & repair of motor vehicles, etc.	..	..	..	..	..	..	..	**46 Commerce (gros & détail), réparation de véhicules, etc.**
47 Hotels & restaurants	..	..	..	..	..	..	..	**47 Hôtels et restaurants**
48 Transport & storage	..	..	..	..	..	..	..	**48 Transports, entreposage**
49 Communications	..	..	6 789	..	..	..	..	**49 Communications**
50 Post	..	..	..	..	..	..	..	50 Postes
51 Telecommunications	..	..	6 789	..	..	..	..	51 Télécommunications
52 Financial intermediation (inc. insur.)	..	..	..	..	..	..	..	**52 Intermédiation financ. (assurances inc.)**
53 Real estate, renting & busin. activity	360	550	1 800	..	..	..	..	**53 Immobil., locat. & serv. aux entreprises**
54 Computer & related activities	360	550	1 800	..	..	..	..	54 Informatique et activités connexes
55 *Software consultancy*	..	..	..	..	..	..	..	55 *Production de logiciels*
56 *Other computer services n.e.c.*	..	..	..	..	..	..	..	56 *Autres activités informatiques*
57 Research & development	..	..	..	..	..	..	..	57 Recherche & développement expérimental
58 Other business activities n.e.c.	..	..	..	..	..	..	..	58 Autres activ. & serv. aux entreprises
59 Comm., soc. & pers. serv. activ., etc.	..	..	..	..	..	..	..	**59 Adm. pub., serv. soc. & collec., etc**
TOTAL BUSINESS ENTERPRISE SECTOR	**54 875**	**117 853**	**323 495**	**932 978**	**1 241 265**	**2 887 296**	..	**TOTAL SECTEUR DES ENTREPRISES**

Source: OECD, R&D database (DSTI/EAS Division), January 2002.

Source : OCDE, base de données R-D (DSTI/Division AES), janvier 2002.

Business enterprise intramural expenditure on R&D (BERD) - Dépenses intra-muros de R-D des entreprises (DIRDE)

by industry and source of funds - par industrie et par source de financement

Million TRL *Millions TRL*

	1994	1995	1996	1997	1998	1999	2000	
	10.3 Other national funds - Autres sources nationales							
1 AGRICULTURE, HUNTING & FORESTRY	13 740	61 830	..	..	..	..	..	**1 AGRICULTURE, CHASSE & SYLVICULTURE**
2 MINING	..	..	..	..	..	..	..	**2 ACTIVITES EXTRACTIVES**
3 MANUFACTURING	52 298	37 961	105 707	..	..	..	..	**3 ACTIVITES DE FABRICATION**
4 Food, beverages & tobacco	..	..	..	..	..	..	..	**4 Produits alimentaires, boissons & tabac**
5 Food products & beverages	..	..	..	..	..	..	..	5 Produits alimentaires & boissons
6 Tobacco products	..	..	..	..	..	..	..	6 Produits à base de tabac
7 Textiles, wear app., fur & leather	..	..	..	..	..	..	..	**7 Textiles, habill., fourrures & cuirs**
8 Textiles	..	..	..	..	..	..	..	8 Textiles
9 Wearing apparel & fur	..	..	..	..	..	..	..	9 Articles d'habillement & fourrures
10 Leather products & footwear	..	..	..	..	..	..	..	10 Articles en cuir & chaussures
11 Wood, paper, printing, publishing	..	..	..	..	..	..	..	**11 Bois, papier, imprimerie, édition**
12 Wood & cork (not furniture)	..	..	..	..	..	..	..	12 Bois & liège (sauf meubles)
13 Pulp, paper & paper products	..	..	..	..	..	..	..	13 Carton, papier & articles
14 Publ., print. & repro. of rec. media	..	..	..	..	..	..	..	14 Edition, impress. & repro. enregis.
15 Coke, petroleum, nuclear fuel, chemicals, rubber plastics	..	..	17 533	..	..	..	..	**15 Coke, pétrole, combust. nucl., prod. chimi., caout. & mat. plast.**
16 Coke, ref. petrol. prod. & nuclear fuel	..	..	..	..	..	..	..	16 Coke, raffin. pétr. & combust. nucl.
17 Chemicals & chemical products	..	..	17 533	..	..	..	..	17 Produits chimiques
18 *Chemical prod.(less pharmaceuticals)*	..	..	17 533	..	..	..	..	18 *Prod. chimi. (moins prod. pharmaceutiques)*
19 *Pharmaceuticals*	..	..	..	..	..	..	..	19 *Produits pharmaceutiques*
20 Rubber & plastic products	..	..	..	..	..	..	..	20 Caoutchouc et matières plastiques
21 Non-metallic mineral products	..	..	..	..	..	..	..	**21 Produits minéraux non métalliques**
22 Basic metals	..	..	5 440	..	..	..	..	**22 Produits métallurgiques de base**
23 Basic metals, ferrous	..	..	..	..	..	..	..	23 Produits métallur. de base, ferreux
24 Basic metals, non-ferrous	..	..	5 440	..	..	..	..	24 Prod. métallur. de base, non ferreux
25 Fabricated metal products	..	753	..	..	..	..	..	**25 Ouvrages en métaux (sauf mach.& mat.)**
26 Machinery equipment, instruments and transport equipment	52 298	37 208	82 734	..	..	..	..	**26 Machines & matériel, instruments & mat. de transport**
27 Machinery n.e.c.	45 000	30 180	5 808	..	..	..	..	27 Machines, n.c.a.
28 Office, account. & computing machinery	..	..	..	..	..	..	..	28 Mach. de bur., compta. & trait. inform.
29 Electrical machinery	4 100	3 425	..	..	..	..	..	29 Machines électriques
30 Electro. equip. (radio, TV & commun.)	2 177	2 230	..	..	..	..	..	30 Equip. électro. (radio, tv & communic.)
31 *Electro. comp. (inc. semi-conductors)*	..	..	..	..	..	..	..	31 *Compos. électro. (semi-cond. incl.)*
32 *TV, radio & communications equip.*	2 177	2 230	..	..	..	..	..	32 *Appareils radio, tv et de communi.*
33 Instruments	1 021	1 373	5 051	..	..	..	..	33 Instruments
34 Motor vehicles	..	..	..	..	..	..	..	34 Véhicules automobiles
35 Other transport equipment	..	..	71 875	..	..	..	..	35 Autres matériels de transport
36 *Ships*	..	..	..	..	..	..	..	36 *Navires*
37 *Aerospace*	..	..	71 875	..	..	..	..	37 *Aérospatiale*
38 *Other transport equipment n.e.c.*	..	..	..	..	..	..	..	38 *Autres matériels de transport n.c.a.*
39 Furniture, other manufacturing n.e.c.	..	..	..	..	..	..	..	**39 Meubles, autres activ. de fabricat. n.c.a.**
40 Furniture	..	..	..	..	..	..	..	40 Meubles
41 Other manufacturing n.e.c.	..	..	..	..	..	..	..	41 Autres activités de fabrication n.c.a.
42 Recycling	..	..	..	..	..	..	..	**42 Récupération**
43 ELECTRICITY, GAS & WATER SUPPLY	..	..	..	..	..	..	..	**43 DISTRIBUTION D'ELECTRICITE, GAZ & EAU**
44 CONSTRUCTION	..	..	..	..	..	..	..	**44 CONSTRUCTION**
45 SERVICES SECTOR	..	..	2 815	..	..	..	..	**45 SECTEUR DES SERVICES**
46 Wholesale, retail trade & repair of motor vehicles, etc.	..	..	..	..	..	..	..	**46 Commerce (gros & détail), réparation de véhicules, etc.**
47 Hotels & restaurants	..	..	..	..	..	..	..	**47 Hôtels et restaurants**
48 Transport & storage	..	..	..	..	..	..	..	**48 Transports, entreposage**
49 Communications	..	..	..	..	..	..	..	**49 Communications**
50 Post	..	..	..	..	..	..	..	50 Postes
51 Telecommunications	..	..	..	..	..	..	..	51 Télécommunications
52 Financial intermediation (inc. insur.)	..	..	..	..	..	..	..	**52 Intermédiation financ. (assurances inc.)**
53 Real estate, renting & busin. activity	..	..	2 815	..	..	..	..	**53 Immobil., locat. & serv. aux entreprises**
54 Computer & related activities	..	..	2 815	..	..	..	..	54 Informatique et activités connexes
55 *Software consultancy*	..	..	..	..	..	..	..	55 *Production de logiciels*
56 *Other computer services n.e.c.*	..	..	..	..	..	..	..	56 *Autres activités informatiques*
57 Research & development	..	..	..	..	..	..	..	57 Recherche & développement expérimental
58 Other business activities n.e.c.	..	..	..	..	..	..	..	58 Autres activ. & serv. aux entreprises
59 Comm., soc. & pers. serv. activ., etc.	..	..	..	..	..	..	..	**59 Adm. pub., serv. soc. & collec., etc**
TOTAL BUSINESS ENTERPRISE SECTOR	**66 038**	**99 791**	**108 522**	**904 894**	**1 003 100**	**765 100**	**..**	**TOTAL SECTEUR DES ENTREPRISES**

Source: OECD, R&D database (DSTI/EAS Division), January 2002.

Source : OCDE, base de données R-D (DSTI/Division AES), janvier 2002.

Business enterprise intramural expenditure on R&D (BERD) - Dépenses intra-muros de R-D des entreprises (DIRDE)

by industry and source of funds - par industrie et par source de financement

Million TRL — *Millions TRL*

	1994	1995	1996	1997	1998	1999	2000	
	10.4 Funds from abroad - Fonds de l'étranger							
1 AGRICULTURE, HUNTING & FORESTRY	..	..	..	..	..	..	..	**1 AGRICULTURE, CHASSE & SYLVICULTURE**
2 MINING	..	..	..	..	..	..	..	**2 ACTIVITES EXTRACTIVES**
3 MANUFACTURING	112 463	389 200	4 273	..	..	..	..	**3 ACTIVITES DE FABRICATION**
4 Food, beverages & tobacco	..	..	2 873	..	..	..	..	**4 Produits alimentaires, boissons & tabac**
5 Food products & beverages	..	..	2 873	..	..	..	..	5 Produits alimentaires & boissons
6 Tobacco products	..	..	..	..	..	..	..	6 Produits à base de tabac
7 Textiles, wear app., fur & leather	..	..	..	..	..	..	..	**7 Textiles, habill., fourrures & cuirs**
8 Textiles	..	..	..	..	..	..	..	8 Textiles
9 Wearing apparel & fur	..	..	..	..	..	..	..	9 Articles d'habillement & fourrures
10 Leather products & footwear	..	..	..	..	..	..	..	10 Articles en cuir & chaussures
11 Wood, paper, printing, publishing	..	..	..	..	..	..	..	**11 Bois, papier, imprimerie, édition**
12 Wood & cork (not furniture)	..	..	..	..	..	..	..	12 Bois & liège (sauf meubles)
13 Pulp, paper & paper products	..	..	..	..	..	..	..	13 Carton, papier & articles
14 Publ., print. & repro. of rec. media	..	..	..	..	..	..	..	14 Edition, impress. & repro. enregis.
15 Coke, petroleum, nuclear fuel, chemicals, rubber plastics	..	..	..	..	..	..	..	**15 Coke, pétrole, combust. nucl., prod. chimi., caout. & mat. plast.**
16 Coke, ref. petrol. prod. & nuclear fuel	..	..	..	..	..	..	..	16 Coke, raffin. pétr. & combust. nucl.
17 Chemicals & chemical products	..	..	..	..	..	..	..	17 Produits chimiques
18 *Chemical prod.(less pharmaceuticals)*	..	..	..	..	..	..	..	18 *Prod. chimi. (moins prod. pharmaceutiques)*
19 *Pharmaceuticals*	..	..	..	..	..	..	..	19 *Produits pharmaceutiques*
20 Rubber & plastic products	..	..	..	..	..	..	..	20 Caoutchouc et matières plastiques
21 Non-metallic mineral products	..	..	..	..	..	..	..	**21 Produits minéraux non métalliques**
22 Basic metals	..	..	..	..	..	..	..	**22 Produits métallurgiques de base**
23 Basic metals, ferrous	..	..	..	..	..	..	..	23 Produits métallur. de base, ferreux
24 Basic metals, non-ferrous	..	..	..	..	..	..	..	24 Prod. métallur. de base, non ferreux
25 Fabricated metal products	..	10 910	..	..	..	..	..	**25 Ouvrages en métaux (sauf mach.& mat.)**
26 Machinery equipment, instruments and transport equipment	112 463	378 290	1 400	..	..	..	..	**26 Machines & matériel, instruments & mat. de transport**
27 Machinery n.e.c.	27 000	84 871	..	..	..	..	..	27 Machines, n.c.a.
28 Office, account. & computing machinery	..	..	..	..	..	..	..	28 Mach. de bur., compta. & trait. inform.
29 Electrical machinery	..	..	..	..	..	..	..	29 Machines électriques
30 Electro. equip. (radio, TV & commun.)	83 000	293 000	..	..	..	..	..	30 Equip. électro. (radio, tv & communic.)
31 *Electro. comp. (inc. semi-conductors)*	..	..	..	..	..	..	..	31 *Compos. électro. (semi-cond. incl.)*
32 *TV, radio & communications equip.*	83 000	293 000	..	..	..	..	..	32 *Appareils radio, tv et de communi.*
33 Instruments	..	..	1 400	..	..	..	..	33 Instruments
34 Motor vehicles	2 463	419	..	..	..	..	..	34 Véhicules automobiles
35 Other transport equipment	..	..	..	..	..	..	..	35 Autres matériels de transport
36 *Ships*	..	..	..	..	..	..	..	36 *Navires*
37 *Aerospace*	..	..	..	..	..	..	..	37 *Aérospatiale*
38 *Other transport equipment n.e.c.*	..	..	..	..	..	..	..	38 *Autres matériels de transport n.c.a.*
39 Furniture, other manufacturing n.e.c.	..	..	..	..	..	..	..	**39 Meubles, autres activ. de fabricat. n.c.a.**
40 Furniture	..	..	..	..	..	..	..	40 Meubles
41 Other manufacturing n.e.c.	..	..	..	..	..	..	..	41 Autres activités de fabrication n.c.a.
42 Recycling	..	..	..	..	..	..	..	**42 Récupération**
43 ELECTRICITY, GAS & WATER SUPPLY	..	..	..	..	..	..	..	**43 DISTRIBUTION D'ELECTRICITE, GAZ & EAU**
44 CONSTRUCTION	..	..	..	..	..	..	..	**44 CONSTRUCTION**
45 SERVICES SECTOR	..	..	..	..	..	..	..	**45 SECTEUR DES SERVICES**
46 Wholesale, retail trade & repair of motor vehicles, etc.	..	..	..	..	..	..	..	**46 Commerce (gros & détail), réparation de véhicules, etc.**
47 Hotels & restaurants	..	..	..	..	..	..	..	**47 Hôtels et restaurants**
48 Transport & storage	..	..	..	..	..	..	..	**48 Transports, entreposage**
49 Communications	..	..	..	..	..	..	..	**49 Communications**
50 Post	..	..	..	..	..	..	..	50 Postes
51 Telecommunications	..	..	..	..	..	..	..	51 Télécommunications
52 Financial intermediation (inc. insur.)	..	..	..	..	..	..	..	**52 Intermédiation financ. (assurances inc.)**
53 Real estate, renting & busin. activity	..	..	..	..	..	..	..	**53 Immobil., locat. & serv. aux entreprises**
54 Computer & related activities	..	..	..	..	..	..	..	54 Informatique et activités connexes
55 *Software consultancy*	..	..	..	..	..	..	..	55 *Production de logiciels*
56 *Other computer services n.e.c.*	..	..	..	..	..	..	..	56 *Autres activités informatiques*
57 Research & development	..	..	..	..	..	..	..	57 Recherche & développement expérimental
58 Other business activities n.e.c.	..	..	..	..	..	..	..	58 Autres activ. & serv. aux entreprises
59 Comm., soc. & pers. serv. activ., etc.	..	..	..	..	..	..	..	**59 Adm. pub., serv. soc. & collec., etc**
TOTAL BUSINESS ENTERPRISE SECTOR	**112 463**	**389 200**	**4 273**	**304 454**	**64 237**	**22 380 808**	..	**TOTAL SECTEUR DES ENTREPRISES**

Source: OECD, R&D database (DSTI/EAS Division), January 2002.

Source : OCDE, base de données R-D (DSTI/Division AES), janvier 2002.

Business enterprise intramural expenditure on R&D (BERD) - Dépenses intra-muros de R-D des entreprises (DIRDE)

by industry and type of costs - par industrie et par type de coûts

Million TRL *Millions TRL*

	1994	1995	1996	1997	1998	1999	2000	
	11.1 Current expenditure - Dépenses courantes							
1 **AGRICULTURE, HUNTING & FORESTRY**	11 155	56 403	34 200	..	..	..	..	1 **AGRICULTURE, CHASSE & SYLVICULTURE**
2 **MINING**	32 874	40 715	94 300	..	..	..	..	2 **ACTIVITES EXTRACTIVES**
3 **MANUFACTURING**	2 015 737	4 223 846	10 499 568	..	..	..	..	3 **ACTIVITES DE FABRICATION**
4 **Food, beverages & tobacco**	159 108	278 894	569 303	..	..	..	..	4 **Produits alimentaires, boissons & tabac**
5 Food products & beverages	151 400	261 555	512 672	..	..	..	..	5 Produits alimentaires & boissons
6 Tobacco products	7 708	17 339	56 631	..	..	..	..	6 Produits à base de tabac
7 **Textiles, wear app., fur & leather**	51 558	95 751	178 460	..	..	..	..	7 **Textiles, habill., fourrures & cuirs**
8 Textiles	51 558	95 751	170 410	..	..	..	..	8 Textiles
9 Wearing apparel & fur	..	..	3 717	..	..	..	..	9 Articles d'habillement & fourrures
10 Leather products & footwear	..	..	4 333	..	..	..	..	10 Articles en cuir & chaussures
11 **Wood, paper, printing, publishing**	6 981	13 172	116 251	..	..	..	..	11 **Bois, papier, imprimerie, édition**
12 Wood & cork (not furniture)	..	..	..	..	..	..	..	12 Bois & liège (sauf meubles)
13 Pulp, paper & paper products	6 981	13 172	111 751	..	..	..	..	13 Carton, papier & articles
14 Publ., print. & repro. of rec. media	..	..	4 500	..	..	..	..	14 Edition, impress. & repro. enregis.
15 **Coke, petroleum, nuclear fuel, chemicals, rubber plastics**	226 648	493 252	1 183 087	..	..	..	..	15 **Coke, pétrole, combust. nucl., prod. chimi., caout. & mat. plast.**
16 Coke, ref. petrol. prod. & nuclear fuel	4 185	6 229	12 713	..	..	..	..	16 Coke, raffin. pétr. & combust. nucl.
17 Chemicals & chemical products	149 135	344 021	898 672	..	..	..	..	17 Produits chimiques
18 *Chemical prod.(less pharmaceuticals)*	119 900	255 914	626 889	..	..	..	..	18 *Prod. chimi. (moins prod. pharmaceutiques)*
19 *Pharmaceuticals*	29 235	88 107	271 783	..	..	..	..	19 *Produits pharmaceutiques*
20 Rubber & plastic products	73 328	143 002	271 702	..	..	..	..	20 Caoutchouc et matières plastiques
21 **Non-metallic mineral products**	64 504	150 126	726 330	..	..	..	..	21 **Produits minéraux non métalliques**
22 **Basic metals**	89 633	142 007	278 517	..	..	..	..	22 **Produits métallurgiques de base**
23 Basic metals, ferrous	81 222	120 073	226 861	..	..	..	..	23 Produits métallur. de base, ferreux
24 Basic metals, non-ferrous	8 411	21 934	51 656	..	..	..	..	24 Prod. métallur. de base, non ferreux
25 **Fabricated metal products**	15 210	38 532	169 226	..	..	..	..	25 **Ouvrages en métaux (sauf mach.& mat.)**
26 **Machinery equipment, instruments and transport equipment**	1 402 095	3 012 112	7 108 139	..	..	..	..	26 **Machines & matériel, instruments & mat. de transport**
27 Machinery n.e.c.	260 970	521 652	1 318 714	..	..	..	..	27 Machines, n.c.a.
28 Office, account. & computing machinery	..	..	114 870	..	..	..	..	28 Mach. de bur., compta. & trait. inform.
29 Electrical machinery	92 999	234 550	692 113	..	..	..	..	29 Machines électriques
30 Electro. equip. (radio, TV & commun.)	876 050	1 766 736	3 464 340	..	..	..	..	30 Equip. électro. (radio, tv & communic.)
31 *Electro. comp. (inc. semi-conductors)*	1 515	2 721	7 721	..	..	..	..	31 *Compos. électro. (semi-cond. incl.)*
32 *TV, radio & communications equip.*	874 535	1 764 015	3 456 619	..	..	..	..	32 *Appareils radio, tv et de communi.*
33 Instruments	2 871	9 233	59 807	..	..	..	..	33 Instruments
34 Motor vehicles	135 931	397 800	1 251 578	..	..	..	..	34 Véhicules automobiles
35 Other transport equipment	33 274	82 141	206 717	..	..	..	..	35 Autres matériels de transport
36 *Ships*	..	..	..	..	..	..	..	36 *Navires*
37 *Aerospace*	16 038	36 004	130 891	..	..	..	..	37 *Aérospatiale*
38 *Other transport equipment n.e.c.*	17 236	46 137	75 826	..	..	..	..	38 *Autres matériels de transport n.c.a.*
39 **Furniture, other manufacturing n.e.c.**	..	..	170 255	..	..	..	..	39 **Meubles, autres activ. de fabricat. n.c.a.**
40 Furniture	..	..	..	..	..	..	..	40 Meubles
41 Other manufacturing n.e.c.	..	..	170 255	..	..	..	..	41 Autres activités de fabrication n.c.a.
42 **Recycling**	..	..	..	..	..	..	..	42 **Récupération**
43 **ELECTRICITY, GAS & WATER SUPPLY**	85 064	142 801	248 163	..	..	..	..	43 **DISTRIBUTION D'ELECTRICITE, GAZ & EAU**
44 **CONSTRUCTION**	..	..	..	..	..	..	..	44 **CONSTRUCTION**
45 **SERVICES SECTOR**	63 276	154 327	356 066	..	..	..	..	45 **SECTEUR DES SERVICES**
46 **Wholesale, retail trade & repair of motor vehicles, etc.**	..	..	..	..	..	..	..	46 **Commerce (gros & détail), réparation de véhicules, etc.**
47 **Hotels & restaurants**	..	..	..	..	..	..	..	47 **Hôtels et restaurants**
48 **Transport & storage**	2 034	3 295	8 635	..	..	..	..	48 **Transports, entreposage**
49 **Communications**	19 998	31 298	82 631	..	..	..	..	49 **Communications**
50 Post	..	..	..	..	..	..	..	50 Postes
51 Telecommunications	19 998	31 298	82 631	..	..	..	..	51 Télécommunications
52 **Financial intermediation (inc. insur.)**	..	..	..	..	..	..	..	52 **Intermédiation financ. (assurances inc.)**
53 **Real estate, renting & busin. activity**	41 244	119 734	264 800	..	..	..	..	53 **Immobil., locat. & serv. aux entreprises**
54 Computer & related activities	41 244	119 734	264 800	..	..	..	..	54 Informatique et activités connexes
55 *Software consultancy*	..	..	..	..	..	..	..	55 *Production de logiciels*
56 *Other computer services n.e.c.*	..	..	..	..	..	..	..	56 *Autres activités informatiques*
57 Research & development	..	..	..	..	..	..	..	57 Recherche & développement expérimental
58 Other business activities n.e.c.	..	..	..	..	..	..	..	58 Autres activ. & serv. aux entreprises
59 **Comm., soc. & pers. serv. activ., etc.**	..	..	..	..	..	..	..	59 **Adm. pub., serv. soc. & collec., etc**
TOTAL BUSINESS ENTERPRISE SECTOR	**2 208 106**	**4 618 092**	**11 232 297**	**..**	**..**	**..**	**..**	**TOTAL SECTEUR DES ENTREPRISES**

Source: OECD, R&D database (DSTI/EAS Division), January 2002.

Source : OCDE, base de données R-D (DSTI/Division AES), janvier 2002.

TURKEY Table 11 - Tableau 11 **TURQUIE**

Business enterprise intramural expenditure on R&D (BERD) - Dépenses intra-muros de R-D des entreprises (DIRDE)

by industry and type of costs - par industrie et par type de coûts

Million TRL — *Millions TRL*

11.2 Capital expenditure - Dépenses en capital

	1994	1995	1996	1997	1998	1999	2000	
1 AGRICULTURE, HUNTING & FORESTRY	13 740	61 830	1 312	..	..	..	..	**1 AGRICULTURE, CHASSE & SYLVICULTURE**
2 MINING	2 843	5 444	8 421	..	..	..	..	**2 ACTIVITES EXTRACTIVES**
3 MANUFACTURING	1 167 099	2 179 645	5 805 655	..	..	..	..	**3 ACTIVITES DE FABRICATION**
4 Food, beverages & tobacco	9 968	24 192	86 628	..	..	..	..	**4 Produits alimentaires, boissons & tabac**
5 Food products & beverages	9 927	23 601	84 628	..	..	..	..	5 Produits alimentaires & boissons
6 Tobacco products	41	591	2 000	..	..	..	..	6 Produits à base de tabac
7 Textiles, wear app., fur & leather	618	5 626	62 478	..	..	..	..	**7 Textiles, habill., fourrures & cuirs**
8 Textiles	618	5 626	62 478	..	..	..	..	8 Textiles
9 Wearing apparel & fur	..	..	..	..	..	..	..	9 Articles d'habillement & fourrures
10 Leather products & footwear	..	..	..	..	..	..	..	10 Articles en cuir & chaussures
11 Wood, paper, printing, publishing	350	700	14 463	..	..	..	..	**11 Bois, papier, imprimerie, édition**
12 Wood & cork (not furniture)	..	..	..	..	..	..	..	12 Bois & liège (sauf meubles)
13 Pulp, paper & paper products	350	700	9 369	..	..	..	..	13 Carton, papier & articles
14 Publ., print. & repro. of rec. media	..	..	5 094	..	..	..	..	14 Edition, impress. & repro. enregis.
15 Coke, petroleum, nuclear fuel, chemicals, rubber plastics	137 389	169 657	346 985	..	..	..	..	**15 Coke, pétrole, combust. nucl., prod. chimi., caout. & mat. plast.**
16 Coke, ref. petrol. prod. & nuclear fuel	18 210	23 507	50	..	..	..	..	16 Coke, raffin. pétr. & combust. nucl.
17 Chemicals & chemical products	114 679	133 990	168 016	..	..	..	..	17 Produits chimiques
18 *Chemical prod.(less pharmaceuticals)*	110 809	129 930	145 366	..	..	..	..	18 *Prod. chimi. (moins prod. pharmaceutiques)*
19 *Pharmaceuticals*	3 870	4 059	22 650	..	..	..	..	19 *Produits pharmaceutiques*
20 Rubber & plastic products	4 500	12 161	178 919	..	..	..	..	20 Caoutchouc et matières plastiques
21 Non-metallic mineral products	198 755	250 824	323 261	..	..	..	..	**21 Produits minéraux non métalliques**
22 Basic metals	100 183	259 932	2 085 881	..	..	..	..	**22 Produits métallurgiques de base**
23 Basic metals, ferrous	100 055	200 252	2 066 084	..	..	..	..	23 Produits métallur. de base, ferreux
24 Basic metals, non-ferrous	128	59 680	19 797	..	..	..	..	24 Prod. métallur. de base, non ferreux
25 Fabricated metal products	3 910	57 451	145 811	..	..	..	..	**25 Ouvrages en métaux (sauf mach.& mat.)**
26 Machinery equipment, instruments and transport equipment	715 926	1 411 263	2 737 695	..	..	..	..	**26 Machines & matériel, instruments & mat. de transport**
27 Machinery n.e.c.	397 508	802 253	878 057	..	..	..	..	27 Machines, n.c.a.
28 Office, account. & computing machinery	..	..	180 000	..	..	..	..	28 Mach. de bur., compta. & trait. inform.
29 Electrical machinery	28 531	95 505	251 418	..	..	..	..	29 Machines électriques
30 Electro. equip. (radio, TV & commun.)	188 396	370 248	614 113	..	..	..	..	30 Equip. électro. (radio, tv & communic.)
31 *Electro. comp. (inc. semi-conductors)*	100	177	200	..	..	..	..	31 *Compos. électro. (semi-cond. incl.)*
32 *TV, radio & communications equip.*	188 296	370 071	613 913	..	..	..	..	32 *Appareils radio, tv et de communi.*
33 Instruments	1 900	2 300	81 541	..	..	..	..	33 Instruments
34 Motor vehicles	95 021	124 545	679 319	..	..	..	..	34 Véhicules automobiles
35 Other transport equipment	4 570	16 412	53 247	..	..	..	..	35 Autres matériels de transport
36 *Ships*	..	..	..	..	..	..	..	36 *Navires*
37 *Aerospace*	3 977	14 919	45 872	..	..	..	..	37 *Aérospatiale*
38 *Other transport equipment n.e.c.*	593	1 493	7 375	..	..	..	..	38 *Autres matériels de transport n.c.a.*
39 Furniture, other manufacturing n.e.c.	..	..	2 453	..	..	..	..	**39 Meubles, autres activ. de fabricat. n.c.a.**
40 Furniture	..	..	..	..	..	..	..	40 Meubles
41 Other manufacturing n.e.c.	..	..	2 453	..	..	..	..	41 Autres activités de fabrication n.c.a.
42 Recycling	..	..	..	..	..	..	..	**42 Récupération**
43 ELECTRICITY, GAS & WATER SUPPLY	12 136	38 952	140 174	..	..	..	..	**43 DISTRIBUTION D'ELECTRICITE, GAZ & EAU**
44 CONSTRUCTION	..	..	..	..	..	..	..	**44 CONSTRUCTION**
45 SERVICES SECTOR	51 015	62 984	146 160	..	..	..	..	**45 SECTEUR DES SERVICES**
46 Wholesale, retail trade & repair of motor vehicles, etc.	..	..	..	..	..	..	..	**46 Commerce (gros & détail), réparation de véhicules, etc.**
47 Hotels & restaurants	..	..	..	..	..	..	..	**47 Hôtels et restaurants**
48 Transport & storage	..	..	..	..	..	..	..	**48 Transports, entreposage**
49 Communications	24 843	22 220	14 620	..	..	..	..	**49 Communications**
50 Post	..	..	..	..	..	..	..	50 Postes
51 Telecommunications	24 843	22 220	14 620	..	..	..	..	51 Télécommunications
52 Financial intermediation (inc. insur.)	..	..	..	..	..	..	..	**52 Intermédiation financ. (assurances inc.)**
53 Real estate, renting & busin. activity	26 172	40 764	131 540	..	..	..	..	**53 Immobil., locat. & serv. aux entreprises**
54 Computer & related activities	26 172	40 764	131 540	..	..	..	..	54 Informatique et activités connexes
55 *Software consultancy*	..	..	..	..	..	..	..	55 *Production de logiciels*
56 *Other computer services n.e.c.*	..	..	..	..	..	..	..	56 *Autres activités informatiques*
57 Research & development	..	..	..	..	..	..	..	57 Recherche & développement expérimental
58 Other business activities n.e.c.	..	..	..	..	..	..	..	58 Autres activ. & serv. aux entreprises
59 Comm., soc. & pers. serv. activ., etc.	..	..	..	..	..	..	..	**59 Adm. pub., serv. soc. & collec., etc**
TOTAL BUSINESS ENTERPRISE SECTOR	**1 246 834**	**2 348 856**	**6 101 722**	**..**	**..**	**..**	**..**	**TOTAL SECTEUR DES ENTREPRISES**

Source: OECD, R&D database (DSTI/EAS Division), January 2002.

Source : OCDE, base de données R-D (DSTI/Division AES), janvier 2002.

Total business enterprise R&D personnel (BEMP) - Personnel total de R-D du secteur des entreprises (PRDE)

by industry - par industrie

In full time equivalent — *En équivalence plein-temps*

	1994	1995	1996	1997	1998	1999	2000	
1 AGRICULTURE, HUNTING & FORESTRY	28	28	39	..	..	..	..	**1 AGRICULTURE, CHASSE & SYLVICULTURE**
2 MINING	48	49	43	..	..	..	..	**2 ACTIVITES EXTRACTIVES**
3 MANUFACTURING	2 736	3 143	3 721	..	..	..	..	**3 ACTIVITES DE FABRICATION**
4 Food, beverages & tobacco	182	198	221	..	..	..	..	**4 Produits alimentaires, boissons & tabac**
5 Food products & beverages	165	180	209	..	..	..	..	5 Produits alimentaires & boissons
6 Tobacco products	17	18	12	..	..	..	..	6 Produits à base de tabac
7 Textiles, wear app., fur & leather	113	115	97	..	..	..	..	**7 Textiles, habill., fourrures & cuirs**
8 Textiles	113	115	86	..	..	..	..	8 Textiles
9 Wearing apparel & fur	..	..	5	..	..	..	..	9 Articles d'habillement & fourrures
10 Leather products & footwear	..	..	6	..	..	..	..	10 Articles en cuir & chaussures
11 Wood, paper, printing, publishing	30	28	45	..	..	..	..	**11 Bois, papier, imprimerie, édition**
12 Wood & cork (not furniture)	..	..	..	..	..	..	..	12 Bois & liège (sauf meubles)
13 Pulp, paper & paper products	30	28	37	..	..	..	..	13 Carton, papier & articles
14 Publ., print. & repro. of rec. media	..	..	8	..	..	..	..	14 Edition, impress. & repro. enregis.
15 Coke, petroleum, nuclear fuel, chemicals, rubber plastics	342	386	528	..	..	..	..	**15 Coke, pétrole, combust. nucl., prod. chimi., caout. & mat. plast.**
16 Coke, ref. petrol. prod. & nuclear fuel	6	6	3	..	..	..	..	16 Coke, raffin. pétr. & combust. nucl.
17 Chemicals & chemical products	258	293	409	..	..	..	..	17 Produits chimiques
18 *Chemical prod.(less pharmaceuticals)*	216	243	330	..	..	..	..	18 *Prod. chimi. (moins prod. pharmaceutiques)*
19 *Pharmaceuticals*	42	50	79	..	..	..	..	19 *Produits pharmaceutiques*
20 Rubber & plastic products	78	87	116	..	..	..	..	20 Caoutchouc et matières plastiques
21 Non-metallic mineral products	115	140	342	..	..	..	..	**21 Produits minéraux non métalliques**
22 Basic metals	176	212	170	..	..	..	..	**22 Produits métallurgiques de base**
23 Basic metals, ferrous	161	177	127	..	..	..	..	23 Produits métallur. de base, ferreux
24 Basic metals, non-ferrous	15	35	43	..	..	..	..	24 Prod. métallur. de base, non ferreux
25 Fabricated metal products	23	39	115	..	..	..	..	**25 Ouvrages en métaux (sauf mach.& mat.)**
26 Machinery equipment, instruments and transport equipment	1 755	2 025	2 200	..	..	..	..	**26 Machines & matériel, instruments & mat. de transport**
27 Machinery n.e.c.	346	444	487	..	..	..	..	27 Machines, n.c.a.
28 Office, account. & computing machinery	..	..	45	..	..	..	..	28 Mach. de bur., compta. & trait. inform.
29 Electrical machinery	107	137	246	..	..	..	..	29 Machines électriques
30 Electro. equip. (radio, TV & commun.)	1 012	1 071	976	..	..	..	..	30 Equip. électro. (radio, tv & communic.)
31 *Electro. comp. (inc. semi-conductors)*	6	6	6	..	..	..	..	31 *Compos. électro. (semi-cond. incl.)*
32 *TV, radio & communications equip.*	1 006	1 065	970	..	..	..	..	32 *Appareils radio, tv et de communi.*
33 Instruments	5	11	25	..	..	..	..	33 Instruments
34 Motor vehicles	228	270	315	..	..	..	..	34 Véhicules automobiles
35 Other transport equipment	57	92	106	..	..	..	..	35 Autres matériels de transport
36 *Ships*	..	..	..	..	..	..	..	36 *Navires*
37 *Aerospace*	16	20	35	..	..	..	..	37 *Aérospatiale*
38 *Other transport equipment n.e.c.*	41	72	71	..	..	..	..	38 *Autres matériels de transport n.c.a.*
39 Furniture, other manufacturing n.e.c.	..	..	3	..	..	..	..	**39 Meubles, autres activ. de fabricat. n.c.a.**
40 Furniture	..	..	..	..	..	..	..	40 Meubles
41 Other manufacturing n.e.c.	..	..	3	..	..	..	..	41 Autres activités de fabrication n.c.a.
42 Recycling	..	..	..	..	..	..	..	**42 Récupération**
43 ELECTRICITY, GAS & WATER SUPPLY	265	257	276	..	..	..	..	**43 DISTRIBUTION D'ELECTRICITE, GAZ & EAU**
44 CONSTRUCTION	..	..	..	..	..	..	..	**44 CONSTRUCTION**
45 SERVICES SECTOR	155	157	214	..	..	..	..	**45 SECTEUR DES SERVICES**
46 Wholesale, retail trade & repair of motor vehicles, etc.	..	..	..	..	..	..	..	**46 Commerce (gros & détail), réparation de véhicules, etc.**
47 Hotels & restaurants	..	..	..	..	..	..	..	**47 Hôtels et restaurants**
48 Transport & storage	6	6	10	..	..	..	..	**48 Transports, entreposage**
49 Communications	75	62	50	..	..	..	..	**49 Communications**
50 Post	..	..	..	..	..	..	..	50 Postes
51 Telecommunications	75	62	50	..	..	..	..	51 Télécommunications
52 Financial intermediation (inc. insur.)	..	..	..	..	..	..	..	**52 Intermédiation financ. (assurances inc.)**
53 Real estate, renting & busin. activity	74	89	154	..	..	..	..	**53 Immobil., locat. & serv. aux entreprises**
54 Computer & related activities	74	89	154	..	..	..	..	54 Informatique et activités connexes
55 *Software consultancy*	..	..	..	..	..	..	..	55 *Production de logiciels*
56 *Other computer services n.e.c.*	..	..	..	..	..	..	..	56 *Autres activités informatiques*
57 Research & development	..	..	..	..	..	..	..	57 Recherche & développement expérimental
58 Other business activities n.e.c.	..	..	..	..	..	..	..	58 Autres activ. & serv. aux entreprises
59 Comm., soc. & pers. serv. activ., etc.	..	..	..	..	..	..	..	**59 Adm. pub., serv. soc. & collec., etc**
TOTAL BUSINESS ENTERPRISE SECTOR	**3 232**	**3 634**	**4 293**	**5 632**	**5 223**	**5 614**	**..**	**TOTAL SECTEUR DES ENTREPRISES**

Source: OECD, R&D database (DSTI/EAS Division), January 2002.

Source : OCDE, base de données R-D (DSTI/Division AES), janvier 2002.

TURKEY | Table 13 - Tableau 13 | **TURQUIE**

RSE in the business enterprise sector - Chercheurs dans le secteur des entreprises

by industry - par industrie

In full time equivalent | *En équivalence plein-temps*

	1994	1995	1996	1997	1998	1999	2000	
1 AGRICULTURE, HUNTING & FORESTRY	22	22	22	..	..	..	..	**1 AGRICULTURE, CHASSE & SYLVICULTURE**
2 MINING	4	3	16	..	..	..	..	**2 ACTIVITES EXTRACTIVES**
3 MANUFACTURING	1 641	1 866	2 086	..	..	..	..	**3 ACTIVITES DE FABRICATION**
4 Food, beverages & tobacco	108	120	97	..	..	..	..	**4 Produits alimentaires, boissons & tabac**
5 Food products & beverages	96	107	90	..	..	..	..	5 Produits alimentaires & boissons
6 Tobacco products	12	13	7	..	..	..	..	6 Produits à base de tabac
7 Textiles, wear app., fur & leather	54	52	55	..	..	..	..	**7 Textiles, habill., fourrures & cuirs**
8 Textiles	54	52	47	..	..	..	..	8 Textiles
9 Wearing apparel & fur	..	..	2	..	..	..	..	9 Articles d'habillement & fourrures
10 Leather products & footwear	..	..	6	..	..	..	..	10 Articles en cuir & chaussures
11 Wood, paper, printing, publishing	19	15	30	..	..	..	..	**11 Bois, papier, imprimerie, édition**
12 Wood & cork (not furniture)	..	..	..	..	..	..	..	12 Bois & liège (sauf meubles)
13 Pulp, paper & paper products	19	15	22	..	..	..	..	13 Carton, papier & articles
14 Publ., print. & repro. of rec. media	..	..	8	..	..	..	..	14 Edition, impress. & repro. enregis.
15 Coke, petroleum, nuclear fuel, chemicals, rubber plastics	166	195	261	..	..	..	..	**15 Coke, pétrole, combust. nucl., prod. chimi., caout. & mat. plast.**
16 Coke, ref. petrol. prod. & nuclear fuel	1	1	2	..	..	..	..	16 Coke, raffin. pétr. & combust. nucl.
17 Chemicals & chemical products	130	154	209	..	..	..	..	17 Produits chimiques
18 *Chemical prod.(less pharmaceuticals)*	100	117	161	..	..	..	..	18 *Prod. chimi. (moins prod. pharmaceutiques)*
19 *Pharmaceuticals*	30	37	48	..	..	..	..	19 *Produits pharmaceutiques*
20 Rubber & plastic products	35	40	50	..	..	..	..	20 Caoutchouc et matières plastiques
21 Non-metallic mineral products	42	55	148	..	..	..	..	**21 Produits minéraux non métalliques**
22 Basic metals	107	129	87	..	..	..	..	**22 Produits métallurgiques de base**
23 Basic metals, ferrous	98	105	65	..	..	..	..	23 Produits métallur. de base, ferreux
24 Basic metals, non-ferrous	9	24	22	..	..	..	..	24 Prod. métallur. de base, non ferreux
25 Fabricated metal products	11	22	64	..	..	..	..	**25 Ouvrages en métaux (sauf mach.& mat.)**
26 Machinery equipment, instruments and transport equipment	1 134	1 278	1 344	..	..	..	..	**26 Machines & matériel, instruments & mat. de transport**
27 Machinery n.e.c.	216	266	302	..	..	..	..	27 Machines, n.c.a.
28 Office, account. & computing machinery	..	..	31	..	..	..	..	28 Mach. de bur., compta. & trait. inform.
29 Electrical machinery	41	63	152	..	..	..	..	29 Machines électriques
30 Electro. equip. (radio, TV & commun.)	731	764	653	..	..	..	..	30 Equip. électro. (radio, tv & communic.)
31 *Electro. comp. (inc. semi-conductors)*	2	2	2	..	..	..	..	31 *Compos. électro. (semi-cond. incl.)*
32 *TV, radio & communications equip.*	729	762	651	..	..	..	..	32 *Appareils radio, tv et de communi.*
33 Instruments	4	6	16	..	..	..	..	33 Instruments
34 Motor vehicles	122	145	148	..	..	..	..	34 Véhicules automobiles
35 Other transport equipment	20	34	42	..	..	..	..	35 Autres matériels de transport
36 *Ships*	..	..	..	..	..	..	..	36 *Navires*
37 *Aerospace*	15	18	30	..	..	..	..	37 *Aérospatiale*
38 *Other transport equipment n.e.c.*	5	16	12	..	..	..	..	38 *Autres matériels de transport n.c.a.*
39 Furniture, other manufacturing n.e.c.	..	..	..	..	..	..	..	**39 Meubles, autres activ. de fabricat. n.c.a.**
40 Furniture	..	..	..	..	..	..	..	40 Meubles
41 Other manufacturing n.e.c.	..	..	..	..	..	..	..	41 Autres activités de fabrication n.c.a.
42 Recycling	..	..	..	..	..	..	..	**42 Récupération**
43 ELECTRICITY, GAS & WATER SUPPLY	221	215	227	..	..	..	..	**43 DISTRIBUTION D'ELECTRICITE, GAZ & EAU**
44 CONSTRUCTION	..	..	..	..	..	..	..	**44 CONSTRUCTION**
45 SERVICES SECTOR	104	105	105	..	..	..	..	**45 SECTEUR DES SERVICES**
46 Wholesale, retail trade & repair of motor vehicles, etc.	..	..	..	..	..	..	..	**46 Commerce (gros & détail), réparation de véhicules, etc.**
47 Hotels & restaurants	..	..	..	..	..	..	..	**47 Hôtels et restaurants**
48 Transport & storage	6	6	8	..	..	..	..	**48 Transports, entreposage**
49 Communications	47	39	27	..	..	..	..	**49 Communications**
50 Post	..	..	..	..	..	..	..	50 Postes
51 Telecommunications	47	39	27	..	..	..	..	51 Télécommunications
52 Financial intermediation (inc. insur.)	..	..	..	..	..	..	..	**52 Intermédiation financ. (assurances inc.)**
53 Real estate, renting & busin. activity	51	60	70	..	..	..	..	**53 Immobil., locat. & serv. aux entreprises**
54 Computer & related activities	51	60	70	..	..	..	..	54 Informatique et activités connexes
55 *Software consultancy*	..	..	..	..	..	..	..	55 *Production de logiciels*
56 *Other computer services n.e.c.*	..	..	..	..	..	..	..	56 *Autres activités informatiques*
57 Research & development	..	..	..	..	..	..	..	57 Recherche & développement expérimental
58 Other business activities n.e.c.	..	..	..	..	..	..	..	58 Autres activ. & serv. aux entreprises
59 Comm., soc. & pers. serv. activ., etc.	..	..	..	..	..	..	..	**59 Adm. pub., serv. soc. & collec., etc**
TOTAL BUSINESS ENTERPRISE SECTOR	**1 992**	**2 211**	**2 456**	**3 232**	**3 070**	**3 247**	**..**	**TOTAL SECTEUR DES ENTREPRISES**

Source: OECD, R&D database (DSTI/EAS Division), January 2002.

Source : OCDE, base de données R-D (DSTI/Division AES), janvier 2002.

Table 14 - Tableau 14

Higher education intramural expenditure on R&D - Dépenses intra-muros de R-D de l'enseignement supérieur
(HERD) by field of science - (DIRDES) par discipline scientifique

	1994	1995	1996	1997	1998	1999	2000	
	14.1 Million TRL - Millions TRL							
Natural sciences	..	816 954	1 734 835	..	..	..	..	Sciences exactes et naturelles
Engineering	..	2 468 396	3 834 940	..	..	..	..	Sciences de l'ingénieur
Medical sciences	..	11 451 870	25 572 323	..	..	..	..	Sciences médicales
Agricultural sciences	..	1 373 872	4 016 649	..	..	..	..	Sciences agricoles
Sub-total NSE	7 892 878	16 111 092	35 158 747	..	..	..	..	Sous-total SENI
Social sciences	..	3 375 562	4 420 645	..	..	..	..	Sciences sociales
Humanities	..	884 050	1 885 282	..	..	..	..	Sciences humaines
Sub-total SSH	1 433 087	4 259 612	6 305 927	..	..	..	..	Sous-total SSH
Total NSE+SSH	9 325 965	20 370 704	41 464 674	81 086 183	159 189 370	270 410 185	..	*Total SENI+SSH*
	14.2 Million constant USD (1995 prices and PPPs) - Millions de USD constants (prix et PPA de 1995)							
Natural sciences	..	36.6	43.7	..	..	..	..	Sciences exactes et naturelles
Engineering	..	110.5	96.6	..	..	..	..	Sciences de l'ingénieur
Medical sciences	..	512.8	643.8	..	..	..	..	Sciences médicales
Agricultural sciences	..	61.5	101.1	..	..	..	..	Sciences agricoles
Sub-total NSE	661.6	721.4	885.2	..	..	..	..	Sous-total SENI
Social sciences	..	151.1	111.3	..	..	..	..	Sciences sociales
Humanities	..	39.6	47.5	..	..	..	..	Sciences humaines
Sub-total SSH	120.1	190.7	158.8	..	..	..	..	Sous-total SSH
Total NSE+SSH	781.7	912.1	1 043.9	1 124.5	1 256.7	1 372.2	..	*Total SENI+SSH*

Table 15 - Tableau 15

Higher education intramural expenditure on R&D - Dépenses intra-muros de R-D de l'enseignement supérieur
(HERD) by field of science and type of costs - (DIRDES) par discipline scientifique et par type de coûts

Million TRL *Millions TRL*

	1994	1995	1996	1997	1998	1999	2000	
Current expenditure								**Dépenses courantes**
Natural sciences	..	649 672	1 414 180	..	..	..	..	Sciences exactes et naturelles
Engineering	..	2 075 744	3 084 481	..	..	..	..	Sciences de l'ingénieur
Medical sciences	..	8 806 970	23 938 704	..	..	..	..	Sciences médicales
Agricultural sciences	..	1 046 487	3 784 978	..	..	..	..	Sciences agricoles
Sub-total NSE	..	12 578 873	32 222 343	..	..	..	..	Sous-total SENI
Social sciences	..	2 906 510	2 991 821	..	..	..	..	Sciences sociales
Humanities	..	675 288	1 548 437	..	..	..	..	Sciences humaines
Sub-total SSH	..	3 581 798	4 540 258	..	..	..	..	Sous-total SSH
Total NSE+SSH	6 857 666	16 160 671	36 762 601	..	..	..	..	*Total SENI+SSH*
Capital expenditure								**Dépenses en capital**
Natural sciences	..	167 282	320 655	..	..	..	..	Sciences exactes et naturelles
Engineering	..	392 652	750 459	..	..	..	..	Sciences de l'ingénieur
Medical sciences	..	2 644 900	1 633 619	..	..	..	..	Sciences médicales
Agricultural sciences	..	327 385	231 671	..	..	..	..	Sciences agricoles
Sub-total NSE	..	3 532 219	2 936 404	..	..	..	..	Sous-total SENI
Social sciences	..	469 052	1 428 824	..	..	..	..	Sciences sociales
Humanities	..	208 762	336 845	..	..	..	..	Sciences humaines
Sub-total SSH	..	677 814	1 765 669	..	..	..	..	Sous-total SSH
Total NSE+SSH	2 468 300	4 210 033	4 702 073	..	..	..	..	*Total SENI+SSH*

Source: OECD, R&D database (DSTI/EAS Division), January 2002.

Source : OCDE, base de données R-D (DSTI/Division AES), janvier 2002.

Higher education intramural expenditure on R&D - Dépenses intra-muros de R-D de l'enseignement supérieur

(HERD) by field of science and source of funds - (DIRDES) par discipline scientifique et par source de financement

Million TRL *Millions TRL*

	1994	1995	1996	1997	1998	1999	2000	
Direct government								**État direct**
Natural sciences	..	82 337	203 363	..	..	..	..	Sciences exactes et naturelles
Engineering	..	1 021 240	1 071 014	..	..	..	..	Sciences de l'ingénieur
Medical sciences	..	5 532 643	16 614 577	..	..	..	..	Sciences médicales
Agricultural sciences	..	373 860	652 197	..	..	..	..	Sciences agricoles
Sub-total NSE	..	7 010 080	18 541 151	..	..	..	..	Sous-total SENI
Social sciences	..	1 760 699	881 275	..	..	..	..	Sciences sociales
Humanities	..	107 044	288 988	..	..	..	..	Sciences humaines
Sub-total SSH	..	1 867 743	1 170 263	..	..	..	..	Sous-total SSH
Total NSE+SSH	3 191 039	8 877 823	19 711 414	20 532 675	54 559 995	100 495 737	..	*Total SENI+SSH*
General university funds								**Fonds généraux des universités**
Natural sciences	..	629 478	1 141 625	..	..	..	..	Sciences exactes et naturelles
Engineering	..	1 176 255	2 105 296	..	..	..	..	Sciences de l'ingénieur
Medical sciences	..	3 323 676	4 012 907	..	..	..	..	Sciences médicales
Agricultural sciences	..	492 233	784 111	..	..	..	..	Sciences agricoles
Sub-total NSE	..	5 621 642	8 043 939	..	..	..	..	Sous-total SENI
Social sciences	..	1 256 537	2 188 430	..	..	..	..	Sciences sociales
Humanities	..	648 086	1 027 034	..	..	..	..	Sciences humaines
Sub-total SSH	..	1 904 623	3 215 464	..	..	..	..	Sous-total SSH
Total NSE+SSH	4 508 191	7 526 265	11 259 403	42 551 894	65 715 989	100 627 170	..	*Total SENI+SSH*
Sub-total government								**Sous-total État**
Natural sciences	..	711 815	1 344 988	..	..	..	..	Sciences exactes et naturelles
Engineering	..	2 197 495	3 176 310	..	..	..	..	Sciences de l'ingénieur
Medical sciences	..	8 856 319	20 627 484	..	..	..	..	Sciences médicales
Agricultural sciences	..	866 093	1 436 308	..	..	..	..	Sciences agricoles
Sub-total NSE	..	12 631 722	26 585 090	..	..	..	..	Sous-total SENI
Social sciences	..	3 017 236	3 069 705	..	..	..	..	Sciences sociales
Humanities	..	755 130	1 316 022	..	..	..	..	Sciences humaines
Sub-total SSH	..	3 772 366	4 385 727	..	..	..	..	Sous-total SSH
Total NSE+SSH	7 699 230	16 404 088	30 970 817	63 084 569	120 275 984	201 122 907	..	*Total SENI+SSH*
Higher education								**Enseignement supérieur**
Natural sciences	..	.. [n]	.. [n]	..	..	..	..	Sciences exactes et naturelles
Engineering	..	.. [n]	.. [n]	..	..	..	..	Sciences de l'ingénieur
Medical sciences	..	.. [n]	.. [n]	..	..	..	..	Sciences médicales
Agricultural sciences	..	.. [n]	.. [n]	..	..	..	..	Sciences agricoles
Sub-total NSE	..	.. [n]	.. [n]	..	..	..	..	Sous-total SENI
Social sciences	..	.. [n]	.. [n]	..	..	..	..	Sciences sociales
Humanities	..	.. [n]	.. [n]	..	..	..	..	Sciences humaines
Sub-total SSH	..	.. [n]	.. [n]	..	..	..	..	Sous-total SSH
Total NSE+SSH	.. [n]	.. [n]	.. [n]	.. [n]	.. [n]	.. [n]	..	*Total SENI+SSH*
Private non-profit								**Institutions sans but lucratif**
Natural sciences	..	18 017 [o]	50 541 [o]	..	..	..	..	Sciences exactes et naturelles
Engineering	..	40 085 [o]	103 301 [o]	..	..	..	..	Sciences de l'ingénieur
Medical sciences	..	469 814 [o]	2 019 827 [o]	..	..	..	..	Sciences médicales
Agricultural sciences	..	57 816 [o]	245 139 [o]	..	..	..	..	Sciences agricoles
Sub-total NSE	..	585 732 [o]	2 418 808 [o]	..	..	..	..	Sous-total SENI
Social sciences	..	75 041 [o]	571 350 [o]	..	..	..	..	Sciences sociales
Humanities	..	23 055 [o]	60 286 [o]	..	..	..	..	Sciences humaines
Sub-total SSH	..	98 096 [o]	631 636 [o]	..	..	..	..	Sous-total SSH
Total NSE+SSH	258 650 [o]	683 828 [o]	3 050 444 [o]	2 586 580 [o]	10 713 604 [o]	19 181 043 [o]	..	*Total SENI+SSH*
Business entreprise								**Entreprises**
Natural sciences	..	87 122	339 306	..	..	..	..	Sciences exactes et naturelles
Engineering	..	230 816	555 329	..	..	..	..	Sciences de l'ingénieur
Medical sciences	..	2 125 737	2 925 012	..	..	..	..	Sciences médicales
Agricultural sciences	..	449 963	2 335 202	..	..	..	..	Sciences agricoles
Sub-total NSE	..	2 893 638	6 154 849	..	..	..	..	Sous-total SENI
Social sciences	..	283 285	779 590	..	..	..	..	Sciences sociales
Humanities	..	105 865	508 974	..	..	..	..	Sciences humaines
Sub-total SSH	..	389 150	1 288 564	..	..	..	..	Sous-total SSH
Total NSE+SSH	1 368 085	3 282 788	7 443 413	15 368 296	28 167 399	50 106 235	..	*Total SENI+SSH*
Funds from abroad								**Fonds de l'étranger**
Natural sciences	..	..	..	..	..	..	..	Sciences exactes et naturelles
Engineering	..	..	..	..	..	..	..	Sciences de l'ingénieur
Medical sciences	..	..	..	..	..	..	..	Sciences médicales
Agricultural sciences	..	..	..	..	..	..	..	Sciences agricoles
Sub-total NSE	..	..	..	..	..	..	..	Sous-total SENI
Social sciences	..	..	..	..	..	..	..	Sciences sociales
Humanities	..	..	..	..	..	..	..	Sciences humaines
Sub-total SSH	..	..	..	..	..	..	..	Sous-total SSH
Total NSE+SSH	..	..	..	46 738	32 383	..	..	*Total SENI+SSH*

Source: OECD, R&D database (DSTI/EAS Division), January 2002.

Source : OCDE, base de données R-D (DSTI/Division AES), janvier 2002.

Table 17 - Tableau 17

Higher education R&D personnel (HEMP) - Personnel de R-D du secteur de l'enseignement supérieur

by field of science and occupation - par discipline scientifique et par profession

In full-time equivalent — *En équivalence plein-temps*

	1994	1995	1996	1997	1998	1999	2000	
Total R&D personnel								**Personnel total de R-D**
Natural sciences	..	..	..	..	..	..	..	Sciences exactes et naturelles
Engineering	..	..	..	..	..	..	..	Sciences de l'ingénieur
Medical sciences	..	..	..	..	..	..	..	Sciences médicales
Agricultural sciences	..	..	..	..	..	..	..	Sciences agricoles
Sub-total NSE	..	..	..	..	..	..	..	Sous-total SENI
Social sciences	..	..	..	..	..	..	..	Sciences sociales
Humanities	..	..	..	..	..	..	..	Sciences humaines
Sub-total SSH	..	..	..	..	..	..	..	Sous-total SSH
Total NSE+SSH	..	..	..	..	..	..	..	*Total SENI+SSH*
Researchers								**Chercheurs**
Natural sciences	..	1 385	1 499	..	..	..	..	Sciences exactes et naturelles
Engineering	..	2 300	2 495	..	..	..	..	Sciences de l'ingénieur
Medical sciences	..	3 426	4 095	..	..	..	..	Sciences médicales
Agricultural sciences	..	884	973	..	..	..	..	Sciences agricoles
Sub-total NSE	7 262	7 995	9 062	..	..	..	..	Sous-total SENI
Social sciences	..	2 552	2 827	..	..	..	..	Sciences sociales
Humanities	..	1 238	1 398	..	..	..	..	Sciences humaines
Sub-total SSH	3 450	3 790	4 225	..	..	..	..	Sous-total SSH
Total NSE+SSH	10 712	11 784	13 287	13 431	13 790	14 621	..	*Total SENI+SSH*
Technicians								**Technicians**
Natural sciences	..	..	..	..	..	..	..	Sciences exactes et naturelles
Engineering	..	..	..	..	..	..	..	Sciences de l'ingénieur
Medical sciences	..	..	..	..	..	..	..	Sciences médicales
Agricultural sciences	..	..	..	..	..	..	..	Sciences agricoles
Sub-total NSE	..	..	..	..	..	..	..	Sous-total SENI
Social sciences	..	..	..	..	..	..	..	Sciences sociales
Humanities	..	..	..	..	..	..	..	Sciences humaines
Sub-total SSH	..	..	..	..	..	..	..	Sous-total SSH
Total NSE+SSH	..	..	..	..	..	..	..	*Total SENI+SSH*
Other								**Autres**
Natural sciences	..	..	..	..	..	..	..	Sciences exactes et naturelles
Engineering	..	..	..	..	..	..	..	Sciences de l'ingénieur
Medical sciences	..	..	..	..	..	..	..	Sciences médicales
Agricultural sciences	..	..	..	..	..	..	..	Sciences agricoles
Sub-total NSE	..	..	..	..	..	..	..	Sous-total SENI
Social sciences	..	..	..	..	..	..	..	Sciences sociales
Humanities	..	..	..	..	..	..	..	Sciences humaines
Sub-total SSH	..	..	..	..	..	..	..	Sous-total SSH
Total NSE+SSH	..	..	..	..	..	..	..	*Total SENI+SSH*

Table 18 - Tableau 18

Government budget appropriations or outlays for R&D (GBAORD) - Crédits budgétaires publics de R-D (CBPRD)

by socio-economic objective - par objectif socio-économique

Million TRL — *Millions TRL*

	1995	1996	1997	1998	1999	2000	2001	
Agriculture, forestry and fishing	..	..	..	..	..	..	..	Agriculture, sylviculture et pêche
Industrial development	..	..	..	..	..	..	..	Développement industriel
Energy	..	..	..	..	..	..	..	Energie
Transport & telecommunications	..	..	..	..	..	..	..	*Transports & télécommunications*
Urban & rural planning	..	..	..	..	..	..	..	*Aménagement urbain et rural*
ST Infrastructure	..	..	..	..	..	..	..	ST Infrastructure
Prevention of pollution	..	..	..	..	..	..	..	*Prévention de la pollution*
Identificat. & treatment of pollution	..	..	..	..	..	..	..	*Détect. pollut. & mesures antipollut.*
ST Environment	..	..	..	..	..	..	..	ST Environnement
Health	..	..	..	..	..	..	..	Santé
Social development & services	..	..	..	..	..	..	..	Développement et services sociaux
Earth and atmosphere	..	..	..	..	..	..	..	Milieu terrestre et atmosphère
Advancement of research	..	..	..	..	..	..	..	*Promotion de la recherche*
General university funds	..	..	..	..	..	..	..	*Fonds généraux des universités*
ST Advancement of knowledge	..	..	..	..	..	..	..	ST promotion gén. des connaiss.
Civil space	..	..	..	..	..	..	..	Espace civil
Defence	..	..	..	..	..	..	..	Défense
Not elsewhere classified	..	..	..	..	..	..	..	Non classé
Total	..	..	..	..	..	..	..	**Total**

Source: OECD, R&D database (DSTI/EAS Division), January 2002.

Source : OCDE, base de données R-D (DSTI/Division AES), janvier 2002.

UNITED KINGDOM — Table 1 - Tableau 1 — ROYAUME-UNI

Gross domestic expenditure on R&D (GERD) - Dépenses intérieures brutes de R-D (DIRD)

by sector of performance and source of funds - par secteur d'exécution et par source de financement

	1994	1995	1996	1997	1998	1999	2000	
	1.1 Million GBP - Millions GBP							
Business enterprise								**Entreprises**
financed by								*financé par*
Business enterprise	6 503.0 o	6 423.0 o	6 433.0 o	6 840.0 o	6 800.0 o	7 574.0 o	8 023.0 o	Entreprises
Direct government	910.0	953.0	842.0	915.0	1 094.0	1 157.0	1 013.0	État direct
Higher education	.. n	.. n	.. n	.. n	.. n	.. n	.. n	Enseignement supérieur
Private non-profit	19.0	3.0	5.0	0.0	0.0	1.0	3.0	Institutions sans but lucratif
Funds from abroad	1 410.0	1 738.0	2 018.0	1 800.0	2 238.0	2 570.0	2 470.0	Fonds de l'étranger
Total	8 842.0	9 116.0	9 297.0	9 556.0	10 133.0	11 302.0	11 510.0	*Total*
Government								**État**
financed by								*financé par*
Business enterprise	196.0	140.8	164.2	240.3	297.0	377.0	..	Entreprises
Direct government	1 742.1	1 799.0	1 796.3	1 670.5	1 652.7	1 283.0	..	État direct
Higher education	1.9	3.1	3.1	3.6	6.4	6.0	..	Enseignement supérieur
Private non-profit	47.1	33.1	35.5	34.0	45.5	47.0	..	Institutions sans but lucratif
Funds from abroad	64.0	67.9	70.5	68.8	77.0	75.0	..	Fonds de l'étranger
Total	2 051.0	2 043.8	2 069.7	2 017.1	2 078.6	1 788.0	..	*Total*
Higher education								**Enseignement supérieur**
financed by								*financé par*
Business enterprise	156.5	169.8	188.2	205.0	221.2	242.0	..	Entreprises
Direct government	783.6	812.1	828.2	852.2	874.0	1 016.0	..	État direct
General university funds	1 017.0	1 017.7	1 027.5	1 033.0	1 085.1	1 157.0	..	Fonds généraux des universités
Sub-total government	1 800.6	1 829.8	1 855.7	1 885.2	1 959.1	2 173.0	..	Sous-Total État
Higher education	112.4	114.2	115.8	117.0	122.3	136.0	..	Enseignement supérieur
Private non-profit	354.1	376.4	406.1	438.0	462.7	525.0	..	Institutions sans but lucratif
Funds from abroad	199.4	206.2	226.2	248.0	274.7	265.0	..	Fonds de l'étranger
Total	2 623.0	2 696.4	2 792.1	2 893.3	3 040.0	3 341.0	..	*Total*
Private non-profit								**Institutions sans but lucratif**
financed by								*financé par*
Business enterprise	30.6	32.3	32.2	35.0	37.9	42.0	..	Entreprises
Direct government	26.1	27.5	27.5	28.0	30.2	41.0	..	État direct
Higher education	1.3	1.4	1.4	1.0	1.5	2.0	..	Enseignement supérieur
Private non-profit	94.1	99.2	99.1	107.0	112.9	128.0	..	Institutions sans but lucratif
Funds from abroad	16.1	17.0	17.0	19.0	20.6	20.0	..	Fonds de l'étranger
Total	168.2	177.3	177.2	190.0	203.0	233.0	..	*Total*
GERD								**DIRD**
financed by								*financé par*
Business enterprise	6 886.1	6 765.9	6 817.6	7 320.3	7 356.1	8 235.0	..	Entreprises
Direct government	3 461.8	3 591.6	3 494.0	3 465.8	3 650.9	3 498.0	..	État direct
General university funds	1 017.0	1 017.7	1 027.5	1 033.0	1 085.1	1 157.0	..	Fonds généraux des universités
Sub-total government	4 478.8	4 609.3	4 521.5	4 498.8	4 736.0	4 655.0	..	Sous-Total État
Higher education	115.6	118.8	120.3	121.6	130.2	143.0	..	Enseignement supérieur
Private non-profit	514.3	511.7	545.7	579.0	621.0	701.0	..	Institutions sans but lucratif
Funds from abroad	1 689.5	2 029.1	2 331.7	2 135.8	2 610.0	2 930.0	..	Fonds de l'étranger
Total	13 684.2	14 033.5	14 336.0	14 656.4	15 454.6	16 664.0	..	*Total*
	1.2 Million constant USD (1995 prices and PPPs) - Millions de USD constants (prix et PPA de 1995)							
Business enterprise								**Entreprises**
financed by								*financé par*
Business enterprise	10 194.6 o	9 822.4 o	9 527.3 o	9 843.5 o	9 498.8 o	10 344.3 o	10 763.6 o	Entreprises
Direct government	1 426.6	1 457.4	1 247.0	1 316.8	1 528.2	1 580.2	1 359.0	État direct
Higher education	.. n	.. n	.. n	.. n	.. n	.. n	.. n	Enseignement supérieur
Private non-profit	29.8	4.6	7.4	0.0	0.0	1.4	4.0	Institutions sans but lucratif
Funds from abroad	2 210.4	2 657.9	2 988.7	2 590.4	3 126.2	3 510.0	3 313.7	Fonds de l'étranger
Total	13 861.4	13 940.7	13 768.9	13 752.1	14 154.7	15 435.9	15 441.7	*Total*
Government								**État**
financed by								*financé par*
Business enterprise	307.3	215.3	243.2	345.8	414.9	514.9	..	Entreprises
Direct government	2 731.0	2 751.1	2 660.3	2 404.0	2 308.6	1 752.3	..	État direct
Higher education	3.0	4.7	4.6	5.2	8.9	8.2	..	Enseignement supérieur
Private non-profit	73.8	50.6	52.6	48.9	63.6	64.2	..	Institutions sans but lucratif
Funds from abroad	100.3	103.8	104.4	99.0	107.6	102.4	..	Fonds de l'étranger
Total	3 215.3	3 125.5	3 065.2	2 902.8	2 903.6	2 442.0	..	*Total*
Higher education								**Enseignement supérieur**
financed by								*financé par*
Business enterprise	245.3	259.7	278.7	295.0	309.0	330.5	..	Entreprises
Direct government	1 228.4	1 241.9	1 226.6	1 226.5	1 220.9	1 387.6	..	État direct
General university funds	1 594.3	1 556.3	1 521.7	1 486.6	1 515.8	1 580.2	..	Fonds généraux des universités
Sub-total government	2 822.8	2 798.2	2 748.3	2 713.0	2 736.6	2 967.8	..	Sous-Total État
Higher education	176.2	174.6	171.5	168.4	170.8	185.7	..	Enseignement supérieur
Private non-profit	555.1	575.6	601.4	630.3	646.3	717.0	..	Institutions sans but lucratif
Funds from abroad	312.6	315.3	335.0	356.9	383.7	361.9	..	Fonds de l'étranger
Total	4 112.0	4 123.5	4 135.1	4 163.8	4 246.5	4 563.0	..	*Total*
Private non-profit								**Institutions sans but lucratif**
financed by								*financé par*
Business enterprise	48.0	49.4	47.7	50.4	52.9	57.4	..	Entreprises
Direct government	40.9	42.1	40.7	40.3	42.2	56.0	..	État direct
Higher education	2.0	2.1	2.1	1.4	2.1	2.7	..	Enseignement supérieur
Private non-profit	147.5	151.7	146.8	154.0	157.7	174.8	..	Institutions sans but lucratif
Funds from abroad	25.2	26.0	25.2	27.3	28.8	27.3	..	Fonds de l'étranger
Total	263.7	271.1	262.4	273.4	283.6	318.2	..	*Total*
GERD								**DIRD**
financed by								*financé par*
Business enterprise	10 795.2	10 346.8	10 096.9	10 534.6	10 275.7	11 247.1	..	Entreprises
Direct government	5 427.0	5 492.5	5 174.6	4 987.6	5 099.9	4 777.5	..	État direct
General university funds	1 594.3	1 556.3	1 521.7	1 486.6	1 515.8	1 580.2	..	Fonds généraux des universités
Sub-total government	7 021.3	7 048.8	6 696.4	6 474.2	6 615.7	6 357.7	..	Sous-Total État
Higher education	181.2	181.7	178.2	175.0	181.9	195.3	..	Enseignement supérieur
Private non-profit	806.3	782.5	808.2	833.2	867.5	957.4	..	Institutions sans but lucratif
Funds from abroad	2 648.6	3 103.0	3 453.3	3 073.6	3 645.9	4 001.7	..	Fonds de l'étranger
Total	21 452.4	21 460.8	21 231.7	21 092.1	21 588.4	22 759.2	..	*Total*

Source: OECD, R&D database (DSTI/EAS Division), January 2002.

Source : OCDE, base de données R-D (DSTI/Division AES), janvier 2002.

Table 2 - Tableau 2

Gross domestic expenditure on R&D (GERD) - Dépenses intérieures brutes de R-D (DIRD)

by sector of performance and type of costs - par secteur d'exécution et par type de coûts

Million GBP — *Millions GBP*

	1994	1995	1996	1997	1998	1999	2000	
Business enterprise								**Entreprises**
Labour costs	3 810.0	3 585.0	3 710.0	3 683.0	4 012.0	4 491.0	4 625.0	Coûts du personnel
Other current costs	4 012.0	4 520.0	4 627.0	4 769.0	5 082.0	5 586.0	5 706.0	Autres coûts courants
Total current expenditure	7 821.0	8 105.0	8 337.0	8 452.0	9 094.0	10 077.0	10 331.0	*Total dépenses courantes*
Land and buildings	381.0	243.0	255.0	338.0	332.0	374.0	398.0	Terrains et constructions
Instruments and equipment	640.0	768.0	705.0	766.0	707.0	851.0	781.0	Instruments et équipements
Total capital expenditure	1 021.0	1 011.0	960.0	1 104.0	1 039.0	1 225.0	1 179.0	*Total dépenses en capital*
Total	8 842.0	9 116.0	9 297.0	9 556.0	10 133.0	11 302.0	11 510.0	*Total*
Government								**État**
Labour costs	..	..	..	..	..	..	..	Coûts du personnel
Other current costs	..	..	..	..	..	..	..	Autres coûts courants
Total current expenditure	1 589.0	1 857.0	1 886.0	1 482.0	1 547.9	1 298.5	..	*Total dépenses courantes*
Land and buildings	..	..	..	..	..	..	..	Terrains et constructions
Instruments and equipment	..	..	..	..	..	..	..	Instruments et équipements
Total capital expenditure	462.0	186.4	183.7	535.1	530.0	489.5	..	*Total dépenses en capital*
Total	2 051.0	2 043.8	2 069.7	2 017.1	2 078.6	1 788.0	..	*Total*
Higher education								**Enseignement supérieur**
Labour costs	..	..	..	..	..	..	..	Coûts du personnel
Other current costs	..	..	..	..	..	..	..	Autres coûts courants
Total current expenditure	..	..	..	..	..	..	..	*Total dépenses courantes*
Land and buildings	..	..	..	..	..	..	..	Terrains et constructions
Instruments and equipment	..	..	..	..	..	..	..	Instruments et équipements
Total capital expenditure	..	..	..	..	..	..	..	*Total dépenses en capital*
Total	2 623.0	2 696.4	2 792.1	2 893.3	3 040.0	3 341.0	..	*Total*
Private non-profit								**Institutions sans but lucratif**
Labour costs	..	..	..	..	..	..	..	Coûts du personnel
Other current costs	..	..	..	..	..	..	..	Autres coûts courants
Total current expenditure	..	..	..	..	..	..	..	*Total dépenses courantes*
Land and buildings	..	..	..	..	..	..	..	Terrains et constructions
Instruments and equipment	..	..	..	..	..	..	..	Instruments et équipements
Total capital expenditure	..	..	..	..	..	..	..	*Total dépenses en capital*
Total	168.2	177.3	177.2	190.0	203.0	233.0	..	*Total*
GERD								**DIRD**
Labour costs	..	..	..	..	..	..	..	Coûts du personnel
Other current costs	..	..	..	..	..	..	..	Autres coûts courants
Total current expenditure	..	..	..	..	..	..	..	*Total dépenses courantes*
Land and buildings	..	..	..	..	..	..	..	Terrains et constructions
Instruments and equipment	..	..	..	..	..	..	..	Instruments et équipements
Total capital expenditure	..	..	..	..	..	..	..	*Total dépenses en capital*
Total	13 684.2	14 033.5	14 336.0	14 656.4	15 454.6	16 664.0	..	*Total*

Table 3 - Tableau 3

Current domestic expenditure on R&D - Dépenses intérieures courantes de R-D

by sector of performance and type of activity - par secteur d'exécution et par type d'activités

Million GBP — *Millions GBP*

	1994	1995	1996	1997	1998	1999	2000	
Business enterprise								**Entreprises**
Basic research	458.0 s	351.0 s	396.0 s	412.0 s	439.0 s	500.0	498.0	Recherche fondamentale
Applied research	2 531.0 s	2 721.0 s	2 951.0 s	3 069.0 s	3 564.0 s	3 931.0	3 542.0	Recherche appliquée
Experimental development	5 185.0 s	5 169.0 s	5 123.0 s	5 094.0 s	5 217.0 s	5 646.0	6 291.0	Développement expérimental
Not specified	..	..	..	..	..	..	..	Non spécifié
Total	7 821.0	8 105.0	8 337.0	8 452.0	9 094.0	10 077.0	10 331.0	*Total*
Government								**État**
Basic research	254.2	259.6	268.3	263.9	261.4	276.0	..	Recherche fondamentale
Applied research	1 000.7	1 295.7	1 290.2	885.7	908.2	914.9	..	Recherche appliquée
Experimental development	334.3	301.5	327.7	332.7	378.3	107.6	..	Développement expérimental
Not specified	..	..	..	..	..	..	..	Non spécifié
Total	1 589.1	1 857.0	1 886.0	1 482.0	1 547.9	1 298.5	..	*Total*
Higher education								**Enseignement supérieur**
Basic research	..	..	..	..	..	..	..	Recherche fondamentale
Applied research	..	..	..	..	..	..	..	Recherche appliquée
Experimental development	..	..	..	..	..	..	..	Développement expérimental
Not specified	..	..	..	..	..	..	..	Non spécifié
Total	..	..	..	..	..	..	..	*Total*
Private non-profit								**Institutions sans but lucratif**
Basic research	..	..	..	..	..	..	..	Recherche fondamentale
Applied research	..	..	..	..	..	..	..	Recherche appliquée
Experimental development	..	..	..	..	..	..	..	Développement expérimental
Not specified	..	..	..	..	..	..	..	Non spécifié
Total	..	..	..	..	..	..	..	*Total*
GERD								**DIRD**
Basic research	..	..	..	..	..	..	..	Recherche fondamentale
Applied research	..	..	..	..	..	..	..	Recherche appliquée
Experimental development	..	..	..	..	..	..	..	Développement expérimental
Not specified	..	..	..	..	..	..	..	Non spécifié
Total	..	..	..	..	..	..	..	*Total*

Source: OECD, R&D database (DSTI/EAS Division), January 2002.

Source : OCDE, base de données R-D (DSTI/Division AES), janvier 2002.

Gross domestic expenditure on R&D (GERD) - Dépenses intérieures brutes de R-D (DIRD)

by sector of performance and socio-economic objective - par secteur d'exécution et par objectif socio-économique

Million GBP / *Millions GBP*

	1994	1995	1996	1997	1998	1999	2000	
Business enterprise								**Entreprises**
Agriculture, forestry and fishing	..	..	..	..	..	..	..	Agriculture, sylviculture et pêche
Industrial development	..	..	..	..	..	..	..	Développement industriel
Energy	..	..	..	..	..	..	..	Energie
Transport and telecommunications	..	..	..	..	..	..	..	Transports et télécommunications
Urban and rural planning	..	..	..	..	..	..	..	Aménagement urbain et rural
Prevention of pollution	..	..	..	..	..	..	..	Prévention de la pollution
Identification & treatment of pollution	..	..	..	..	..	..	..	Détection pollu. & mesures antipollu.
Health	..	..	..	..	..	..	..	Santé
Social development and services	..	..	..	..	..	..	..	Développement et services sociaux
Earth and atmosphere	..	..	..	..	..	..	..	Milieu terrestre et atmosphère
Advancement of knowledge	..	..	..	..	..	..	..	Promotion générale des connaissances
Civil space	..	..	..	..	..	..	..	Espace civil
Defence	1 433.0	1 391.0	1 360.0	1 443.0	1 533.0	1 675.0	1 671.0	Défense
Not elsewhere classified	7 409.0	7 725.0	7 937.0	8 113.0	8 600.0	9 626.0	9 838.0	Non classé
Total	8 842.0	9 116.0	9 297.0	9 556.0	10 133.0	11 302.0	11 510.0	*Total*
Government								**État**
Agriculture, forestry and fishing	..	..	..	..	..	..	..	Agriculture, sylviculture et pêche
Industrial development	..	..	..	..	..	..	..	Développement industriel
Energy	..	..	..	..	..	..	..	Energie
Transport and telecommunications	..	..	..	..	..	..	..	Transports et télécommunications
Urban and rural planning	..	..	..	..	..	..	..	Aménagement urbain et rural
Prevention of pollution	..	..	..	..	..	..	..	Prévention de la pollution
Identification & treatment of pollution	..	..	..	..	..	..	..	Détection pollu. & mesures antipollu.
Health	..	..	..	..	..	..	..	Santé
Social development and services	..	..	..	..	..	..	..	Développement et services sociaux
Earth and atmosphere	..	..	..	..	..	..	..	Milieu terrestre et atmosphère
Advancement of knowledge	..	..	..	..	..	..	..	Promotion générale des connaissances
Civil space	..	..	..	..	..	..	..	Espace civil
Defence	647.7	680.0	732.0	713.0	753.0	479.0	..	Défense
Not elsewhere classified	1 403.3	1 363.8	1 337.7	1 304.1	1 325.0	1 309.0	..	Non classé
Total	2 051.0	2 043.8	2 069.7	2 017.1	2 078.6	1 788.0	..	*Total*
Higher education								**Enseignement supérieur**
Agriculture, forestry and fishing	..	..	..	..	..	..	..	Agriculture, sylviculture et pêche
Industrial development	..	..	..	..	..	..	..	Développement industriel
Energy	..	..	..	..	..	..	..	Energie
Transport and telecommunications	..	..	..	..	..	..	..	Transports et télécommunications
Urban and rural planning	..	..	..	..	..	..	..	Aménagement urbain et rural
Prevention of pollution	..	..	..	..	..	..	..	Prévention de la pollution
Identification & treatment of pollution	..	..	..	..	..	..	..	Détection pollu. & mesures antipollu.
Health	..	..	..	..	..	..	..	Santé
Social development and services	..	..	..	..	..	..	..	Développement et services sociaux
Earth and atmosphere	..	..	..	..	..	..	..	Milieu terrestre et atmosphère
Advancement of knowledge	..	..	..	..	..	..	..	Promotion générale des connaissances
Civil space	..	..	..	..	..	..	..	Espace civil
Defence	50.2	61.0	60.0	61.0	60.0	62.0	..	Défense
Not elsewhere classified	2 572.8	2 635.4	2 732.1	2 832.3	2 980.0	3 279.0	..	Non classé
Total	2 623.0	2 696.4	2 792.1	2 893.3	3 040.0	3 341.0	..	*Total*
Private non-profit								**Institutions sans but lucratif**
Agriculture, forestry and fishing	..	..	..	..	..	..	..	Agriculture, sylviculture et pêche
Industrial development	..	..	..	..	..	..	..	Développement industriel
Energy	..	..	..	..	..	..	..	Energie
Transport and telecommunications	..	..	..	..	..	..	..	Transports et télécommunications
Urban and rural planning	..	..	..	..	..	..	..	Aménagement urbain et rural
Prevention of pollution	..	..	..	..	..	..	..	Prévention de la pollution
Identification & treatment of pollution	..	..	..	..	..	..	..	Détection pollu. & mesures antipollu.
Health	..	..	..	..	..	..	..	Santé
Social development and services	..	..	..	..	..	..	..	Développement et services sociaux
Earth and atmosphere	..	..	..	..	..	..	..	Milieu terrestre et atmosphère
Advancement of knowledge	..	..	..	..	..	..	..	Promotion générale des connaissances
Civil space	..	..	..	..	..	..	..	Espace civil
Defence	9.1	0.0	0.0	0.0	0.0	0.0	..	Défense
Not elsewhere classified	159.1	177.3	177.2	190.0	203.0	233.0	..	Non classé
Total	168.2	177.3	177.2	190.0	203.0	233.0	..	*Total*
GERD								**DIRD**
Agriculture, forestry and fishing	..	..	..	..	..	..	..	Agriculture, sylviculture et pêche
Industrial development	..	..	..	..	..	..	..	Développement industriel
Energy	..	..	..	..	..	..	..	Energie
Transport and telecommunications	..	..	..	..	..	..	..	Transports et télécommunications
Urban and rural planning	..	..	..	..	..	..	..	Aménagement urbain et rural
Prevention of pollution	..	..	..	..	..	..	..	Prévention de la pollution
Identification & treatment of pollution	..	..	..	..	..	..	..	Détection pollu. & mesures antipollu.
Health	..	..	..	..	..	..	..	Santé
Social development and services	..	..	..	..	..	..	..	Développement et services sociaux
Earth and atmosphere	..	..	..	..	..	..	..	Milieu terrestre et atmosphère
Advancement of knowledge	..	..	..	..	..	..	..	Promotion générale des connaissances
Civil space	..	..	..	..	..	..	..	Espace civil
Defence	2 140.0	2 133.0	2 152.0	2 218.0	2 346.0	2 216.0	..	Défense
Not elsewhere classified	11 544.2	11 900.5	12 184.0	12 438.4	13 108.6	14 448.0	..	Non classé
Total	13 684.2	14 033.5	14 336.0	14 656.4	15 454.6	16 664.0	..	*Total*

Source: OECD, R&D database (DSTI/EAS Division), January 2002.

Source : OCDE, base de données R-D (DSTI/Division AES), janvier 2002.

Table 5 - Tableau 5

Total R&D personnel - Personnel total de R-D

by sector of employment and occupation - par secteur d'emploi et par profession

In full-time equivalent — *En équivalence plein-temps*

	1994	1995	1996	1997	1998	1999	2000	
Business enterprise								**Entreprises**
RSE	75 000	82 000	82 000	83 000	91 000	92 000	86 000	Chercheurs
Technicians	40 000	33 000	33 000	30 000	32 000	33 000	30 000	Techniciens
Other occupations	28 000	29 000	27 000	24 000	24 000	28 000	30 000	Autres professions
Total	143 000	154 000	142 000	137 000	148 000	153 000	145 000	*Total*
Government								**État**
RSE	14 000	13 673	13 021	12 496	14 368	14 958	..	Chercheurs
Technicians	6 000	5 978	6 032	5 507	6 676	6 726	..	Techniciens
Other occupations	12 000	9 268	8 433	7 895	8 153	7 988	..	Autres professions
Total	32 000	28 919	27 486	25 897	29 197	29 672	..	*Total*
Higher education								**Enseignement supérieur**
RSE	42 000 [a]	47 000	47 000	47 651	49 023	..	..	Chercheurs
Technicians	..	..	..	..	..	..	..	Techniciens
Other occupations	..	..	..	..	..	..	..	Autres professions
Total	..	..	..	..	..	..	..	*Total*
Private non-profit								**Institutions sans but lucratif**
RSE	3 000	3 000	2 595	2 799	3 000	3 450	..	Chercheurs
Technicians	1 000	1 000	794	857	918	1 056	..	Techniciens
Other occupations	2 000	2 000	1 906	2 041	2 205	2 536	..	Autres professions
Total	6 000	6 000	5 295	5 697	6 123	7 041	..	*Total*
TOTAL								**TOTAL**
RSE	134 000 [a]	145 673	144 616	145 946	157 391	..	..	Chercheurs
Technicians	..	..	..	..	..	..	..	Techniciens
Other occupations	..	..	..	..	..	..	..	Autres professions
Total	..	..	..	..	..	..	..	*Total*

Table 6 - Tableau 6

Total R&D personnel - Personnel total de R-D

by sector of employment and formal qualification - par secteur d'emploi et par niveau de qualification

In full-time equivalent — *En équivalence plein-temps*

	1994	1995	1996	1997	1998	1999	2000	
Business enterprise								**Entreprises**
University PhD level degrees	..	..	..	..	..	..	..	Diplômes univ. de niv. doctorat
Other university degrees	..	..	..	..	..	..	..	Autres diplômes universitaires
Sub-total university degrees	..	..	..	..	..	..	..	Sous-total Universitaires
Other post-secondary diplomas	..	..	..	..	..	..	..	Autres diplômes post-secondaires
Secondary diplomas	..	..	..	..	..	..	..	Diplômes de fin d'études secondaires
Other qualifications	..	..	..	..	..	..	..	Autres qualifications
Total	143 000	154 000	142 000	137 000	148 000	153 000	145 000	*Total*
Government								**État**
University PhD level degrees	..	..	..	..	..	..	..	Diplômes univ. de niv. doctorat
Other university degrees	..	..	..	..	..	..	..	Autres diplômes universitaires
Sub-total university degrees	..	..	..	..	..	..	..	Sous-total Universitaires
Other post-secondary diplomas	..	..	..	..	..	..	..	Autres diplômes post-secondaires
Secondary diplomas	..	..	..	..	..	..	..	Diplômes de fin d'études secondaires
Other qualifications	..	..	..	..	..	..	..	Autres qualifications
Total	32 000	28 919	27 486	25 897	29 197	29 672	..	*Total*
Higher education								**Enseignement supérieur**
University PhD level degrees	..	..	..	..	..	..	..	Diplômes univ. de niv. doctorat
Other university degrees	..	..	..	..	..	..	..	Autres diplômes universitaires
Sub-total university degrees	..	..	..	..	..	..	..	Sous-total Universitaires
Other post-secondary diplomas	..	..	..	..	..	..	..	Autres diplômes post-secondaires
Secondary diplomas	..	..	..	..	..	..	..	Diplômes de fin d'études secondaires
Other qualifications	..	..	..	..	..	..	..	Autres qualifications
Total	..	..	..	..	..	..	..	*Total*
Private non-profit								**Institutions sans but lucratif**
University PhD level degrees	..	..	..	..	..	..	..	Diplômes univ. de niv. doctorat
Other university degrees	..	..	..	..	..	..	..	Autres diplômes universitaires
Sub-total university degrees	..	..	..	..	..	..	..	Sous-total Universitaires
Other post-secondary diplomas	..	..	..	..	..	..	..	Autres diplômes post-secondaires
Secondary diplomas	..	..	..	..	..	..	..	Diplômes de fin d'études secondaires
Other qualifications	..	..	..	..	..	..	..	Autres qualifications
Total	6 000	6 000	5 295	5 697	6 123	7 041	..	*Total*
TOTAL								**TOTAL**
University PhD level degrees	..	..	..	..	..	..	..	Diplômes univ. de niv. doctorat
Other university degrees	..	..	..	..	..	..	..	Autres diplômes universitaires
Sub-total university degrees	..	..	..	..	..	..	..	Sous-total Universitaires
Other post-secondary diplomas	..	..	..	..	..	..	..	Autres diplômes post-secondaires
Secondary diplomas	..	..	..	..	..	..	..	Diplômes de fin d'études secondaires
Other qualifications	..	..	..	..	..	..	..	Autres qualifications
Total	..	..	..	..	..	..	..	*Total*

Source: OECD, R&D database (DSTI/EAS Division), January 2002.

Source : OCDE, base de données R-D (DSTI/Division AES), janvier 2002.

Table 7 - Tableau 7

Gross domestic expenditure on R&D (GERD) - Dépenses intérieures brutes de R-D (DIRD)

by sector of performance and main scientific field - par secteur d'exécution et par grand domaine scientifique

Million GBP — *Millions GBP*

	1994	1995	1996	1997	1998	1999	2000	
Business enterprise								**Entreprises**
Natural sciences and engineering	8 842.0	9 116.0	9 297.0	9 556.0	10 133.0	11 302.0	11 510.0	Sciences exactes, nat. et de l'ingénieur
Social sciences and humanities	..	..	..	..	..	..	..	Sciences sociales et humaines
All fields of science	8 842.0	9 116.0	9 297.0	9 556.0	10 133.0	11 302.0	11 510.0	*Tous domaines scientifiques*
Government								**État**
Natural sciences and engineering	..	..	..	..	..	..	..	Sciences exactes, nat. et de l'ingénieur
Social sciences and humanities	..	..	..	..	..	..	..	Sciences sociales et humaines
All fields of science	2 051.0	2 043.8	2 069.7	2 017.1	2 078.6	1 788.0	..	*Tous domaines scientifiques*
Higher education								**Enseignement supérieur**
Natural sciences and engineering	..	..	..	..	..	..	..	Sciences exactes, nat. et de l'ingénieur
Social sciences and humanities	..	..	..	..	..	..	..	Sciences sociales et humaines
All fields of science	2 623.0	2 696.4	2 792.1	2 893.3	3 040.0	3 341.0	..	*Tous domaines scientifiques*
Private non-profit								**Institutions sans but lucratif**
Natural sciences and engineering	..	..	..	..	..	..	..	Sciences exactes, nat. et de l'ingénieur
Social sciences and humanities	..	..	..	..	..	..	..	Sciences sociales et humaines
All fields of science	168.2	177.3	177.2	190.0	203.0	233.0	..	*Tous domaines scientifiques*
TOTAL								**TOTAL**
Natural sciences and engineering	..	..	..	..	..	..	..	Sciences exactes, nat. et de l'ingénieur
Social sciences and humanities	..	..	..	..	..	..	..	Sciences sociales et humaines
All fields of science	13 684.2	14 033.5	14 336.0	14 656.4	15 454.6	16 664.0	..	*Tous domaines scientifiques*

Table 8 - Tableau 8

Total R&D personnel - Personnel total de R-D

by sector of employment and main scientific field - par secteur d'emploi et par grand domaine scientifique

In full-time equivalent — *En équivalence plein-temps*

	1994	1995	1996	1997	1998	1999	2000	
Business enterprise								**Entreprises**
Natural sciences and engineering	143 000	154 000	142 000	137 000	148 000	153 000	145 000	Sciences exactes, nat. et de l'ingénieur
Social sciences and humanities	..	..	..	..	..	..	..	Sciences sociales et humaines
All fields of science	143 000	154 000	142 000	137 000	148 000	153 000	145 000	*Tous domaines scientifiques*
Government								**État**
Natural sciences and engineering	31 000	27 863	26 558	24 781	27 862 s	28 150	..	Sciences exactes, nat. et de l'ingénieur
Social sciences and humanities	1 000	1 057	928	1 117	1 341 s	1 521	..	Sciences sociales et humaines
All fields of science	32 000	28 919	27 486	25 897	29 197	29 672	..	*Tous domaines scientifiques*
Higher education								**Enseignement supérieur**
Natural sciences and engineering	..	..	..	..	..	..	..	Sciences exactes, nat. et de l'ingénieur
Social sciences and humanities	..	..	..	..	..	..	..	Sciences sociales et humaines
All fields of science	..	..	..	..	..	..	..	*Tous domaines scientifiques*
Private non-profit								**Institutions sans but lucratif**
Natural sciences and engineering	..	..	..	..	..	..	..	Sciences exactes, nat. et de l'ingénieur
Social sciences and humanities	..	..	..	..	..	..	..	Sciences sociales et humaines
All fields of science	6 000	6 000	5 295	5 697	6 123	7 041	..	*Tous domaines scientifiques*
TOTAL								**TOTAL**
Natural sciences and engineering	..	..	..	..	..	..	..	Sciences exactes, nat. et de l'ingénieur
Social sciences and humanities	..	..	..	..	..	..	..	Sciences sociales et humaines
All fields of science	..	..	..	..	..	..	..	*Tous domaines scientifiques*

Source: OECD, R&D database (DSTI/EAS Division), January 2002.

Source : OCDE, base de données R-D (DSTI/Division AES), janvier 2002.

Total business enterprise intramural expenditure on R&D - Dépenses intra-muros totales de R-D des entreprises
(BERD) by industry - (DIRDE) par industrie

9.1 Million GBP - Millions GBP

	1994	1995	1996	1997	1998	1999	2000	
1 AGRICULTURE, HUNTING & FORESTRY	80.0	..	76.0	84.0	102.0	115.0	135.0	**1 AGRICULTURE, CHASSE & SYLVICULTURE**
2 MINING	66.0	65.0	64.0	44.0	41.0	42.0	46.0	**2 ACTIVITES EXTRACTIVES**
3 MANUFACTURING	7 051.0	7 157.0	7 265.0	7 608.0	8 143.0	8 995.0	9 231.0	**3 ACTIVITES DE FABRICATION**
4 Food, beverages & tobacco	228.0	189.0	198.0	180.0	242.0	237.0	264.0	**4 Produits alimentaires, boissons & tabac**
5 Food products & beverages	..	..	..	..	..	..	..	5 Produits alimentaires & boissons
6 Tobacco products	..	..	..	..	..	..	..	6 Produits à base de tabac
7 Textiles, wear app., fur & leather	22.0	23.0	27.0	33.0	33.0	28.0	29.0	**7 Textiles, habill., fourrures & cuirs**
8 Textiles	..	..	..	..	..	..	..	8 Textiles
9 Wearing apparel & fur	..	..	..	..	..	..	..	9 Articles d'habillement & fourrures
10 Leather products & footwear	..	..	..	..	..	..	..	10 Articles en cuir & chaussures
11 Wood, paper, printing, publishing	44.0	39.0	57.0	44.0	49.0	45.0	38.0	**11 Bois, papier, imprimerie, édition**
12 Wood & cork (not furniture)	..	..	..	..	..	..	..	12 Bois & liège (sauf meubles)
13 Pulp, paper & paper products	..	..	..	..	..	..	..	13 Carton, papier & articles
14 Publ., print. & repro. of rec. media	..	..	..	..	..	..	..	14 Edition, impress. & repro. enregis.
15 Coke, petroleum, nuclear fuel, chemicals, rubber plastics	2 784.0	2 813.0	2 776.0	3 116.0	3 226.0	3 537.0	3 818.0	**15 Coke, pétrole, combust. nucl., prod. chimi., caout. & mat. plast.**
16 Coke, ref. petrol. prod. & nuclear fuel	203.0	239.0	230.0	225.0	234.0	212.0	182.0	16 Coke, raffin. pétr. & combust. nucl.
17 Chemicals & chemical products	2 509.0	2 514.0	2 479.0	2 831.0	2 926.0	3 253.0	3 528.0	17 Produits chimiques
18 *Chemical prod.(less pharmaceuticals)*	689.0	701.0	627.0	680.0	688.0	718.0	682.0	18 *Prod. chimi. (moins prod. pharmaceutiques)*
19 *Pharmaceuticals*	1 820.0	1 813.0	1 852.0	2 151.0	2 238.0	2 535.0	2 846.0	19 *Produits pharmaceutiques*
20 Rubber & plastic products	72.0	60.0	67.0	60.0	66.0	72.0	54.0	20 Caoutchouc et matières plastiques
21 Non-metallic mineral products	56.0	54.0	60.0	47.0	56.0	59.0	..	**21 Produits minéraux non métalliques**
22 Basic metals	66.0	66.0	54.0	54.0	67.0	63.0	..	**22 Produits métallurgiques de base**
23 Basic metals, ferrous	51.0	46.0	39.0	39.0	47.0	41.0	..	23 Produits métallur. de base, ferreux
24 Basic metals, non-ferrous	15.0	20.0	15.0	15.0	20.0	22.0	21.0	24 Prod. métallur. de base, non ferreux
25 Fabricated metal products	72.0	100.0	91.0	88.0	90.0	70.0	73.0	**25 Ouvrages en métaux (sauf mach.& mat.)**
26 Machinery equipment, instruments and transport equipment	3 750.0	3 851.0	3 985.0	4 021.0	4 360.0	4 922.0	4 928.0	**26 Machines & matériel, instruments & mat. de transport**
27 Machinery n.e.c.	689.0	583.0	577.0	622.0	640.0	642.0	703.0	27 Machines, n.c.a.
28 Office, account. & computing machinery	134.0	150.0	161.0	102.0	125.0	111.0	113.0	28 Mach. de bur., compta. & trait. inform.
29 Electrical machinery	567.0	494.0	490.0	424.0	423.0	357.0	422.0	29 Machines électriques
30 Electro. equip. (radio, TV & commun.)	517.0	602.0	662.0	655.0	772.0	867.0	1 024.0	30 Equip. électro. (radio, tv & communic.)
31 *Electro. comp. (inc. semi-conductors)*	..	..	..	..	..	..	..	31 *Compos. électro. (semi-cond. incl.)*
32 *TV, radio & communications equip.*	..	..	..	..	..	..	..	32 *Appareils radio, tv et de communi.*
33 Instruments	273.0	303.0	307.0	336.0	340.0	473.0	480.0	33 Instruments
34 Motor vehicles	669.0	795.0	926.0	924.0	913.0	1 060.0	864.0	34 Véhicules automobiles
35 Other transport equipment	901.0	924.0	862.0	958.0	1 147.0	1 412.0	1 321.0	35 Autres matériels de transport
36 *Ships*	17.0	20.0	20.0	15.0	36.0	76.0	72.0	36 *Navires*
37 *Aerospace*	860.0	886.0	812.0	893.0	1 039.0	1 237.0	1 091.0	37 *Aérospatiale*
38 *Other transport equipment n.e.c.*	24.0	18.0	30.0	50.0	72.0	99.0	158.0	38 *Autres matériels de transport n.c.a.*
39 Furniture, other manufacturing n.e.c.	28.0	21.0	16.0	25.0	20.0	33.0	27.0	**39 Meubles, autres activ. de fabricat. n.c.a.**
40 Furniture	..	..	..	..	..	..	..	40 Meubles
41 Other manufacturing n.e.c.	..	..	..	..	..	..	..	41 Autres activités de fabrication n.c.a.
42 Recycling	1.0	..	1.0	0.0	0.0	1.0	1.0	**42 Récupération**
43 ELECTRICITY, GAS & WATER SUPPLY	177.0	168.0	148.0	130.0	140.0	137.0	160.0	**43 DISTRIBUTION D'ELECTRICITE, GAZ & EAU**
44 CONSTRUCTION	11.0	8.0	8.0	38.0	39.0	41.0	34.0	**44 CONSTRUCTION**
45 SERVICES SECTOR	1 458.0	..	1 736.0	1 652.0	1 668.0	1 972.0	1 905.0	**45 SECTEUR DES SERVICES**
46 Wholesale, retail trade & repair of motor vehicles, etc.	..	..	..	..	..	..	..	**46 Commerce (gros & détail), réparation de véhicules, etc.**
47 Hotels & restaurants	..	..	..	..	..	..	..	**47 Hôtels et restaurants**
48 Transport & storage	..	..	..	..	..	..	..	**48 Transports, entreposage**
49 Communications	408.0	414.0	455.0	496.0	449.0	565.0	674.0	**49 Communications**
50 Post	..	..	..	..	..	..	..	50 Postes
51 Telecommunications	..	..	..	..	..	..	..	51 Télécommunications
52 Financial intermediation (inc. insur.)	104.0	..	141.0	142.0	157.0	196.0	131.0	**52 Intermédiation financ. (assurances inc.)**
53 Real estate, renting & busin. activity	1 055.0 [a]	..	1 118.0	993.0	1 034.0	1 161.0	1 039.0	**53 Immobil., locat. & serv. aux entreprises**
54 Computer & related activities	744.0	675.0	749.0	680.0	688.0	713.0	611.0	54 Informatique et activités connexes
55 *Software consultancy*	..	..	..	..	..	..	..	55 *Production de logiciels*
56 *Other computer services n.e.c.*	..	..	..	..	..	..	..	56 *Autres activités informatiques*
57 Research & development	178.0	247.0	369.0	313.0	346.0	448.0	428.0	57 Recherche & développement expérimental
58 Other business activities n.e.c.	..	..	..	..	..	..	..	58 Autres activ. & serv. aux entreprises
59 Comm., soc. & pers. serv. activ., etc.	10.0	14.0	10.0	6.0	8.0	11.0	12.0	**59 Adm. pub., serv. soc. & collec., etc**
TOTAL BUSINESS ENTERPRISE SECTOR	**8 842.0**	**9 116.0**	**9 297.0**	**9 556.0**	**10 133.0**	**11 302.0**	**11 510.0**	**TOTAL SECTEUR DES ENTREPRISES**

Source: OECD, R&D database (DSTI/EAS Division), January 2002.

Source : OCDE, base de données R-D (DSTI/Division AES), janvier 2002.

UNITED KINGDOM Table 9 - Tableau 9 **ROYAUME-UNI**

Total business enterprise intramural expenditure on R&D - Dépenses intra-muros totales de R-D des entreprises

(BERD) by industry - (DIRDE) par industrie

	1994	1995	1996	1997	1998	1999	2000	
	9.2 Million constant USD(1995 prices and PPPs) - Millions de USD constants (prix et PPA de 1995)							
1 AGRICULTURE, HUNTING & FORESTRY	125.4	..	112.6	120.9	142.5	157.1	181.1	**1 AGRICULTURE, CHASSE & SYLVICULTURE**
2 MINING	103.5	99.4	94.8	63.3	57.3	57.4	61.7	**2 ACTIVITES EXTRACTIVES**
3 MANUFACTURING	11 053.7	10 944.9	10 759.5	10 948.7	11 374.9	12 285.1	12 384.2	**3 ACTIVITES DE FABRICATION**
4 Food, beverages & tobacco	357.4	289.0	293.2	259.0	338.0	323.7	354.2	**4 Produits alimentaires, boissons & tabac**
5 Food products & beverages	..	..	..	..	..	..	..	5 Produits alimentaires & boissons
6 Tobacco products	..	..	..	..	..	..	..	6 Produits à base de tabac
7 Textiles, wear app., fur & leather	34.5	35.2	40.0	47.5	46.1	38.2	38.9	**7 Textiles, habill., fourrures & cuirs**
8 Textiles	..	..	..	..	..	..	..	8 Textiles
9 Wearing apparel & fur	..	..	..	..	..	..	..	9 Articles d'habillement & fourrures
10 Leather products & footwear	..	..	..	..	..	..	..	10 Articles en cuir & chaussures
11 Wood, paper, printing, publishing	69.0	59.6	84.4	63.3	68.4	61.5	51.0	**11 Bois, papier, imprimerie, édition**
12 Wood & cork (not furniture)	..	..	..	..	..	..	..	12 Bois & liège (sauf meubles)
13 Pulp, paper & paper products	..	..	..	..	..	..	..	13 Carton, papier & articles
14 Publ., print. & repro. of rec. media	..	..	..	..	..	..	..	14 Edition, impress. & repro. enregis.
15 Coke, petroleum, nuclear fuel, chemicals, rubber plastics	4 364.4	4 301.8	4 111.3	4 484.2	4 506.4	4 830.7	5 122.2	**15 Coke, pétrole, combust. nucl., prod. chimi., caout. & mat. plast.**
16 Coke, ref. petrol. prod. & nuclear fuel	318.2	365.5	340.6	323.8	326.9	289.5	244.2	16 Coke, raffin. pétr. & combust. nucl.
17 Chemicals & chemical products	3 933.3	3 844.6	3 671.4	4 074.1	4 087.3	4 442.8	4 733.1	17 Produits chimiques
18 *Chemical prod.(less pharmaceuticals)*	1 080.1	1 072.0	928.6	978.6	961.1	980.6	915.0	18 *Prod. chimi. (moins prod. pharmaceutiques)*
19 *Pharmaceuticals*	2 853.2	2 772.5	2 742.8	3 095.5	3 126.2	3 462.2	3 818.2	19 *Produits pharmaceutiques*
20 Rubber & plastic products	112.9	91.8	99.2	86.3	92.2	98.3	72.4	20 Caoutchouc et matières plastiques
21 Non-metallic mineral products	87.8	82.6	88.9	67.6	78.2	80.6	..	**21 Produits minéraux non métalliques**
22 Basic metals	103.5	100.9	80.0	77.7	93.6	86.0	..	**22 Produits métallurgiques de base**
23 Basic metals, ferrous	80.0	70.3	57.8	56.1	65.7	56.0	..	23 Produits métallur. de base, ferreux
24 Basic metals, non-ferrous	23.5	30.6	22.2	21.6	27.9	30.0	28.2	24 Prod. métallur. de base, non ferreux
25 Fabricated metal products	112.9	152.9	134.8	126.6	125.7	95.6	97.9	**25 Ouvrages en métaux (sauf mach.& mat.)**
26 Machinery equipment, instruments and transport equipment	5 878.8	5 889.2	5 901.8	5 786.6	6 090.4	6 722.3	6 611.4	**26 Machines & matériel, instruments & mat. de transport**
27 Machinery n.e.c.	1 080.1	891.6	854.5	895.1	894.0	876.8	943.1	27 Machines, n.c.a.
28 Office, account. & computing machinery	210.1	229.4	238.4	146.8	174.6	151.6	151.6	28 Mach. de bur., compta. & trait. inform.
29 Electrical machinery	888.9	755.5	725.7	610.2	590.9	487.6	566.2	29 Machines électriques
30 Electro. equip. (radio, TV & commun.)	810.5	920.6	980.4	942.6	1 078.4	1 184.1	1 373.8	30 Equip. électro. (radio, tv & communic.)
31 *Electro. comp. (inc. semi-conductors)*	..	..	..	..	..	..	..	31 *Compos. électro. (semi-cond. incl.)*
32 *TV, radio & communications equip.*	..	..	..	..	..	..	..	32 *Appareils radio, tv et de communi.*
33 Instruments	428.0	463.4	454.7	483.5	474.9	646.0	644.0	33 Instruments
34 Motor vehicles	1 048.8	1 215.8	1 371.4	1 329.7	1 275.4	1 447.7	1 159.1	34 Véhicules automobiles
35 Other transport equipment	1 412.5	1 413.0	1 276.6	1 378.7	1 602.2	1 928.5	1 772.2	35 Autres matériels de transport
36 *Ships*	26.7	30.6	29.6	21.6	50.3	103.8	96.6	36 *Navires*
37 *Aerospace*	1 348.2	1 354.9	1 202.6	1 285.1	1 451.4	1 689.5	1 463.7	37 *Aérospatiale*
38 *Other transport equipment n.e.c.*	37.6	27.5	44.4	72.0	100.6	135.2	212.0	38 *Autres matériels de transport n.c.a.*
39 Furniture, other manufacturing n.e.c.	43.9	32.1	23.7	36.0	27.9	45.1	36.2	**39 Meubles, autres activ. de fabricat. n.c.a.**
40 Furniture	..	..	..	..	..	..	..	40 Meubles
41 Other manufacturing n.e.c.	..	..	..	..	..	..	..	41 Autres activités de fabrication n.c.a.
42 Recycling	1.6	..	1.5	0.0	0.0	1.4	1.3	**42 Récupération**
43 ELECTRICITY, GAS & WATER SUPPLY	277.5	256.9	219.2	187.1	195.6	187.1	214.7	**43 DISTRIBUTION D'ELECTRICITE, GAZ & EAU**
44 CONSTRUCTION	17.2	12.2	11.8	54.7	54.5	56.0	45.6	**44 CONSTRUCTION**
45 SERVICES SECTOR	2 285.7	..	2 571.0	2 377.4	2 330.0	2 693.3	2 555.7	**45 SECTEUR DES SERVICES**
46 Wholesale, retail trade & repair of motor vehicles, etc.	..	..	..	..	..	..	..	**46 Commerce (gros & détail), réparation de véhicules, etc.**
47 Hotels & restaurants	..	..	..	..	..	..	..	**47 Hôtels et restaurants**
48 Transport & storage	..	..	..	..	..	..	..	**48 Transports, entreposage**
49 Communications	639.6	633.1	673.9	713.8	627.2	771.7	904.2	**49 Communications**
50 Post	..	..	..	..	..	..	..	50 Postes
51 Telecommunications	..	..	..	..	..	..	..	51 Télécommunications
52 Financial intermediation (inc. insur.)	163.0	..	208.8	204.4	219.3	267.7	175.7	**52 Intermédiation financ. (assurances inc.)**
53 Real estate, renting & busin. activity	1 653.9 [a]	..	1 655.8	1 429.0	1 444.4	1 585.7	1 393.9	**53 Immobil., locat. & serv. aux entreprises**
54 Computer & related activities	1 166.4	1 032.2	1 109.3	978.6	961.1	973.8	819.7	54 Informatique et activités connexes
55 *Software consultancy*	..	..	..	..	..	..	..	55 *Production de logiciels*
56 *Other computer services n.e.c.*	..	..	..	..	..	..	..	56 *Autres activités informatiques*
57 Research & development	279.0	377.7	546.5	450.4	483.3	611.9	574.2	57 Recherche & développement expérimental
58 Other business activities n.e.c.	..	..	..	..	..	..	..	58 Autres activ. & serv. aux entreprises
59 Comm., soc. & pers. serv. activ., etc.	15.7	21.4	14.8	8.6	11.2	15.0	16.1	**59 Adm. pub., serv. soc. & collec., etc**
TOTAL BUSINESS ENTERPRISE SECTOR	**13 861.4**	**13 940.7**	**13 768.9**	**13 752.1**	**14 154.7**	**15 435.9**	**15 441.7**	**TOTAL SECTEUR DES ENTREPRISES**

Source: OECD, R&D database (DSTI/EAS Division), January 2002.

Source : OCDE, base de données R-D (DSTI/Division AES), janvier 2002.

UNITED KINGDOM — Table 10 - Tableau 10 — ROYAUME-UNI

Business enterprise intramural expenditure on R&D (BERD) - Dépenses intra-muros de R-D des entreprises (DIRDE)

by industry and source of funds - par industrie et par source de financement

Million GBP — *Millions GBP*

	1994	1995	1996	1997	1998	1999	2000	
	10.1 Business enterprise - Entreprises							
1 **AGRICULTURE, HUNTING & FORESTRY**	74.0 [s]	..	..	75.0 [s]	81.0	96.0	..	1 **AGRICULTURE, CHASSE & SYLVICULTURE**
2 **MINING**	41.0 [s]	42.0	..	26.0	26.0	29.0	..	2 **ACTIVITES EXTRACTIVES**
3 **MANUFACTURING**	..	..	..	5 567.0 [s]	5 547.0 [s]	..	..	3 **ACTIVITES DE FABRICATION**
4 **Food, beverages & tobacco**	209.0 [s]	174.0	..	166.0 [s]	215.0	208.0	..	4 **Produits alimentaires, boissons & tabac**
5 Food products & beverages	..	..	..	..	..	..	..	5 Produits alimentaires & boissons
6 Tobacco products	..	..	..	..	..	..	..	6 Produits à base de tabac
7 **Textiles, wear app., fur & leather**	21.0 [s]	17.0	..	10.0 [s]	9.0	8.0	..	7 **Textiles, habill., fourrures & cuirs**
8 Textiles	..	..	..	..	..	..	..	8 Textiles
9 Wearing apparel & fur	..	..	..	..	..	..	..	9 Articles d'habillement & fourrures
10 Leather products & footwear	..	..	..	..	..	..	..	10 Articles en cuir & chaussures
11 **Wood, paper, printing, publishing**	42.0 [o,s]	35.0	..	39.0 [s]	42.0	38.0	..	11 **Bois, papier, imprimerie, édition**
12 Wood & cork (not furniture)	11.0 [o,s]	..	..	..	..	..	..	12 Bois & liège (sauf meubles)
13 Pulp, paper & paper products	31.0 [o,s]	..	..	..	..	..	..	13 Carton, papier & articles
14 Publ., print. & repro. of rec. media	.. [n]	..	..	..	..	..	..	14 Edition, impress. & repro. enregis.
15 **Coke, petroleum, nuclear fuel, chemicals, rubber plastics**	2 380.0 [s]	2 343.0 [s]	..	2 580.0 [s]	2 515.0 [s]	2 733.0	..	15 **Coke, pétrole, combust. nucl., prod. chimi., caout. & mat. plast.**
16 Coke, ref. petrol. prod. & nuclear fuel	13.0 [m,s]	209.0 [s]	..	180.0 [s]	182.0 [s]	143.0	..	16 Coke, raffin. pétr. & combust. nucl.
17 Chemicals & chemical products	2 308.0 [l,s]	2 099.0	..	2 358.0 [s]	2 282.0	2 528.0	..	17 Produits chimiques
18 *Chemical prod.(less pharmaceuticals)*	..	581.0	..	598.0 [s]	590.0	622.0	..	18 *Prod. chimi. (moins prod. pharmaceutiques)*
19 *Pharmaceuticals*	1 580.0 [s]	1 518.0	..	1 760.0 [s]	1 692.0	1 906.0	..	19 *Produits pharmaceutiques*
20 Rubber & plastic products	59.0 [s]	35.0	..	42.0 [s]	51.0	62.0	..	20 Caoutchouc et matières plastiques
21 **Non-metallic mineral products**	39.0 [s]	44.0	..	31.0 [s]	40.0	43.0	..	21 **Produits minéraux non métalliques**
22 **Basic metals**	..	47.0	..	37.0 [s]	45.0	39.0	..	22 **Produits métallurgiques de base**
23 Basic metals, ferrous	44.0 [s]	41.0	..	36.0 [s]	44.0	39.0	..	23 Produits métallur. de base, ferreux
24 Basic metals, non-ferrous	..	6.0	..	1.0 [s]	1.0	..	..	24 Prod. métallur. de base, non ferreux
25 **Fabricated metal products**	54.0 [s]	70.0	..	69.0 [s]	74.0	65.0	..	25 **Ouvrages en métaux (sauf mach.& mat.)**
26 **Machinery equipment, instruments and transport equipment**	..	2 318.0	..	2 610.0 [s]	2 587.0	..	..	26 **Machines & matériel, instruments & mat. de transport**
27 Machinery n.e.c.	386.0 [s]	333.0	..	383.0 [s]	396.0	397.0	..	27 Machines, n.c.a.
28 Office, account. & computing machinery	107.0 [s]	108.0	..	102.0 [s]	125.0	111.0	..	28 Mach. de bur., compta. & trait. inform.
29 Electrical machinery	309.0 [s]	242.0	..	217.0 [s]	214.0	185.0	..	29 Machines électriques
30 Electro. equip. (radio, TV & commun.)	366.0 [s]	382.0	..	397.0 [s]	415.0	523.0	..	30 Equip. électro. (radio, tv & communic.)
31 *Electro. comp. (inc. semi-conductors)*	..	..	..	..	..	..	..	31 *Compos. électro. (semi-cond. incl.)*
32 *TV, radio & communications equip.*	366.0 [s]	..	..	..	..	..	..	32 *Appareils radio, tv et de communi.*
33 Instruments	158.0 [s]	155.0	..	203.0 [s]	197.0	242.0	..	33 Instruments
34 Motor vehicles	573.0 [s]	620.0	..	748.0 [s]	712.0	826.0	..	34 Véhicules automobiles
35 Other transport equipment	..	478.0	..	560.0 [s]	528.0	515.0	..	35 Autres matériels de transport
36 *Ships*	..	12.0	..	3.0 [s]	6.0	5.0	..	36 *Navires*
37 *Aerospace*	422.0 [s]	452.0	..	509.0 [s]	458.0	510.0	..	37 *Aérospatiale*
38 *Other transport equipment n.e.c.*	21.0 [s]	14.0	..	48.0 [s]	64.0	..	..	38 *Autres matériels de transport n.c.a.*
39 **Furniture, other manufacturing n.e.c.**	18.0 [m,o,]	17.0	..	25.0 [s]	20.0	33.0	..	39 **Meubles, autres activ. de fabricat. n.c.a.**
40 Furniture	.. [n]	..	..	..	..	..	..	40 Meubles
41 Other manufacturing n.e.c.	18.0 [o,s]	..	..	..	..	..	..	41 Autres activités de fabrication n.c.a.
42 **Recycling**	.. [n]	..	..	..	..	1.0	..	42 **Récupération**
43 **ELECTRICITY, GAS & WATER SUPPLY**	168.0 [s]	..	..	124.0 [s]	134.0	132.0	..	43 **DISTRIBUTION D'ELECTRICITE, GAZ & EAU**
44 **CONSTRUCTION**	10.0 [s]	3.0	..	5.0	5.0	8.0	..	44 **CONSTRUCTION**
45 **SERVICES SECTOR**	..	..	..	1 066.0 [s]	1 035.0 [s]	1 249.0	..	45 **SECTEUR DES SERVICES**
46 **Wholesale, retail trade & repair of motor vehicles, etc.**	4.0 [s]	8.0 [o,s]	..	5.0 [o,s]	8.0 [o,s]	25.0	..	46 **Commerce (gros & détail), réparation de véhicules, etc.**
47 **Hotels & restaurants**	..	.. [n]	.. [n]	.. [n]	.. [n]	..	..	47 **Hôtels et restaurants**
48 **Transport & storage**	2.0 [s]	15.0 [s]	..	9.0 [s]	17.0	12.0	..	48 **Transports, entreposage**
49 **Communications**	391.0 [s]	379.0	..	457.0 [s]	423.0	528.0	..	49 **Communications**
50 Post	..	..	..	..	..	..	..	50 Postes
51 Telecommunications	..	..	..	..	..	..	..	51 Télécommunications
52 **Financial intermediation (inc. insur.)**	.. [n]	.. [n]	..	68.0	57.0 [o]	91.0	..	52 **Intermédiation financ. (assurances inc.)**
53 **Real estate, renting & busin. activity**	..	..	..	..	..	..	..	53 **Immobil., locat. & serv. aux entreprises**
54 Computer & related activities	532.0 [s]	386.0	..	428.0 [s]	419.0	446.0	..	54 Informatique et activités connexes
55 *Software consultancy*	..	..	..	..	..	..	..	55 *Production de logiciels*
56 *Other computer services n.e.c.*	..	..	..	..	..	..	..	56 *Autres activités informatiques*
57 Research & development	363.0 [s]	120.0	..	93.0 [s]	103.0	137.0	..	57 Recherche & développement expérimental
58 Other business activities n.e.c.	..	.. [n]	..	.. [n]	.. [n]	..	..	58 Autres activ. & serv. aux entreprises
59 **Comm., soc. & pers. serv. activ., etc.**	4.0 [s]	..	..	6.0 [s]	8.0	10.0	..	59 **Adm. pub., serv. soc. & collec., etc**
TOTAL BUSINESS ENTERPRISE SECTOR	**6 503.0** [o]	**6 423.0** [o]	**6 433.0** [o]	**6 840.0** [o]	**6 800.0** [o]	**7 574.0** [o]	**8 023.0** [o]	**TOTAL SECTEUR DES ENTREPRISES**

Source: OECD, R&D database (DSTI/EAS Division), January 2002.

Source : OCDE, base de données R-D (DSTI/Division AES), janvier 2002.

Business enterprise intramural expenditure on R&D (BERD) - Dépenses intra-muros de R-D des entreprises (DIRDE)

by industry and source of funds - par industrie et par source de financement

Million GBP — *Millions GBP*

10.2 Government - État

	1994	1995	1996	1997	1998	1999	2000	
1 AGRICULTURE, HUNTING & FORESTRY	..	..	..	..	14.0	..	..	1 AGRICULTURE, CHASSE & SYLVICULTURE
2 MINING	..	..	..	..	..	..	..	2 ACTIVITES EXTRACTIVES
3 MANUFACTURING	..	..	..	767.0 s	..	..	..	3 ACTIVITES DE FABRICATION
4 Food, beverages & tobacco	..	..	..	2.0	2.0	2.0	..	**4 Produits alimentaires, boissons & tabac**
5 Food products & beverages	..	..	..	..	..	..	..	5 Produits alimentaires & boissons
6 Tobacco products	..	..	..	..	..	..	..	6 Produits à base de tabac
7 Textiles, wear app., fur & leather	..	..	..	6.0	3.0	2.0	..	**7 Textiles, habill., fourrures & cuirs**
8 Textiles	..	..	..	..	..	..	..	8 Textiles
9 Wearing apparel & fur	..	..	..	..	..	..	..	9 Articles d'habillement & fourrures
10 Leather products & footwear	..	..	..	..	..	..	..	10 Articles en cuir & chaussures
11 Wood, paper, printing, publishing	..	..	..	..	..	..	..	**11 Bois, papier, imprimerie, édition**
12 Wood & cork (not furniture)	..	..	..	..	..	..	..	12 Bois & liège (sauf meubles)
13 Pulp, paper & paper products	..	..	..	..	..	..	..	13 Carton, papier & articles
14 Publ., print. & repro. of rec. media	..	..	..	..	..	..	..	14 Edition, impress. & repro. enregis.
15 Coke, petroleum, nuclear fuel, chemicals, rubber plastics	..	..	..	94.0 s	100.0 s	3.0	..	**15 Coke, pétrole, combust. nucl., prod. chimi., caout. & mat. plast.**
16 Coke, ref. petrol. prod. & nuclear fuel	..	..	..	91.0 s	96.0 s	..	..	16 Coke, raffin. pétr. & combust. nucl.
17 Chemicals & chemical products	..	..	..	3.0	4.0	3.0	..	17 Produits chimiques
18 *Chemical prod.(less pharmaceuticals)*	..	..	..	3.0	2.0	1.0	..	18 *Prod. chimi. (moins prod. pharmaceutiques)*
19 *Pharmaceuticals*	..	..	..	..	2.0	2.0	..	19 *Produits pharmaceutiques*
20 Rubber & plastic products	..	..	..	..	..	..	..	20 Caoutchouc et matières plastiques
21 Non-metallic mineral products	..	..	..	1.0	1.0	1.0	..	**21 Produits minéraux non métalliques**
22 Basic metals	..	..	..	..	..	..	..	**22 Produits métallurgiques de base**
23 Basic metals, ferrous	..	..	..	..	..	..	..	23 Produits métallur. de base, ferreux
24 Basic metals, non-ferrous	..	..	..	..	..	..	..	24 Prod. métallur. de base, non ferreux
25 Fabricated metal products	..	..	..	14.0	10.0	3.0	..	**25 Ouvrages en métaux (sauf mach.& mat.)**
26 Machinery equipment, instruments and transport equipment	..	..	..	650.0	..	..	..	**26 Machines & matériel, instruments & mat. de transport**
27 Machinery n.e.c.	..	..	..	194.0	187.0	208.0	..	27 Machines, n.c.a.
28 Office, account. & computing machinery	..	..	..	..	..	..	..	28 Mach. de bur., compta. & trait. inform.
29 Electrical machinery	..	..	..	129.0	149.0	127.0	..	29 Machines électriques
30 Electro. equip. (radio, TV & commun.)	..	..	..	65.0	85.0	55.0	..	30 Equip. électro. (radio, tv & communic.)
31 *Electro. comp. (inc. semi-conductors)*	..	..	..	..	..	..	..	31 *Compos. électro. (semi-cond. incl.)*
32 *TV, radio & communications equip.*	..	..	..	..	..	..	..	32 *Appareils radio, tv et de communi.*
33 Instruments	..	..	..	74.0	73.0	146.0	..	33 Instruments
34 Motor vehicles	..	..	..	1.0	..	..	..	34 Véhicules automobiles
35 Other transport equipment	..	..	..	187.0	..	..	..	35 Autres matériels de transport
36 *Ships*	..	..	..	7.0	24.0	70.0	..	36 *Navires*
37 *Aerospace*	..	..	..	180.0	265.0	279.0	..	37 *Aérospatiale*
38 *Other transport equipment n.e.c.*	..	..	..	..	..	..	..	38 *Autres matériels de transport n.c.a.*
39 Furniture, other manufacturing n.e.c.	..	..	..	..	..	..	..	**39 Meubles, autres activ. de fabricat. n.c.a.**
40 Furniture	..	..	..	..	..	..	..	40 Meubles
41 Other manufacturing n.e.c.	..	..	..	..	..	..	..	41 Autres activités de fabrication n.c.a.
42 Recycling	..	..	..	..	..	..	..	**42 Récupération**
43 ELECTRICITY, GAS & WATER SUPPLY	..	..	..	2.0	2.0	3.0	..	43 DISTRIBUTION D'ELECTRICITE, GAZ & EAU
44 CONSTRUCTION	..	..	..	30.0	30.0	30.0	..	44 CONSTRUCTION
45 SERVICES SECTOR	..	..	..	204.0 s	210.0 s	..	..	45 SECTEUR DES SERVICES
46 Wholesale, retail trade & repair of motor vehicles, etc.	..	..	..	..	..	..	..	**46 Commerce (gros & détail), réparation de véhicules, etc.**
47 Hotels & restaurants	..	..	..	..	..	..	..	**47 Hôtels et restaurants**
48 Transport & storage	..	..	..	..	..	..	..	**48 Transports, entreposage**
49 Communications	..	..	..	5.0	5.0	5.0	..	**49 Communications**
50 Post	..	..	..	..	..	..	..	50 Postes
51 Telecommunications	..	..	..	..	..	..	..	51 Télécommunications
52 Financial intermediation (inc. insur.)	..	..	..	40.0	40.0	11.0	..	**52 Intermédiation financ. (assurances inc.)**
53 Real estate, renting & busin. activity	..	..	..	..	..	..	..	**53 Immobil., locat. & serv. aux entreprises**
54 Computer & related activities	..	..	..	58.0	58.0	..	..	54 Informatique et activités connexes
55 *Software consultancy*	..	..	..	..	..	..	..	55 *Production de logiciels*
56 *Other computer services n.e.c.*	..	..	..	..	..	..	..	56 *Autres activités informatiques*
57 Research & development	..	..	..	101.0	107.0	134.0	..	57 Recherche & développement expérimental
58 Other business activities n.e.c.	..	..	..	..	..	..	..	58 Autres activ. & serv. aux entreprises
59 Comm., soc. & pers. serv. activ., etc.	..	..	..	..	..	..	..	**59 Adm. pub., serv. soc. & collec., etc**
TOTAL BUSINESS ENTERPRISE SECTOR	**910.0**	**953.0**	**842.0**	**915.0**	**1 094.0**	**1 157.0**	**1 013.0**	**TOTAL SECTEUR DES ENTREPRISES**

Source: OECD, R&D database (DSTI/EAS Division), January 2002.

Source : OCDE, base de données R-D (DSTI/Division AES), janvier 2002.

UNITED KINGDOM | Table 10 - Tableau 10 | ROYAUME-UNI

Business enterprise intramural expenditure on R&D (BERD) - Dépenses intra-muros de R-D des entreprises (DIRDE)

by industry and source of funds - par industrie et par source de financement

Million GBP | *Millions GBP*

	1994	1995	1996	1997	1998	1999	2000	
	10.4 Funds from abroad - Fonds de l'étranger							
1 AGRICULTURE, HUNTING & FORESTRY	..	..	..	7.0	7.0	..	..	**1 AGRICULTURE, CHASSE & SYLVICULTURE**
2 MINING	..	..	..	16.0	15.0	13.0	..	**2 ACTIVITES EXTRACTIVES**
3 MANUFACTURING	..	..	..	1 400.0 [s]	1 788.0 [s]	..	..	**3 ACTIVITES DE FABRICATION**
4 Food, beverages & tobacco	..	..	..	11.0	24.0	28.0	..	**4 Produits alimentaires, boissons & tabac**
5 Food products & beverages	..	..	..	..	..	..	..	5 Produits alimentaires & boissons
6 Tobacco products	..	..	..	..	..	..	..	6 Produits à base de tabac
7 Textiles, wear app., fur & leather	..	..	..	16.0	21.0	17.0	..	**7 Textiles, habill., fourrures & cuirs**
8 Textiles	..	..	..	..	..	..	..	8 Textiles
9 Wearing apparel & fur	..	..	..	..	..	..	..	9 Articles d'habillement & fourrures
10 Leather products & footwear	..	..	..	..	..	..	..	10 Articles en cuir & chaussures
11 Wood, paper, printing, publishing	..	..	..	5.0	6.0	7.0	..	**11 Bois, papier, imprimerie, édition**
12 Wood & cork (not furniture)	..	..	..	..	..	..	..	12 Bois & liège (sauf meubles)
13 Pulp, paper & paper products	..	..	..	..	..	..	..	13 Carton, papier & articles
14 Publ., print. & repro. of rec. media	..	..	..	..	..	..	..	14 Edition, impress. & repro. enregis.
15 Coke, petroleum, nuclear fuel, chemicals, rubber plastics	..	..	..	565.0 [s]	737.0 [s]	801.0	..	**15 Coke, pétrole, combust. nucl., prod. chimi., caout. & mat. plast.**
16 Coke, ref. petrol. prod. & nuclear fuel	..	..	..	78.0 [s]	83.0 [s]	70.0	..	16 Coke, raffin. pétr. & combust. nucl.
17 Chemicals & chemical products	..	..	..	469.0	639.0	721.0	..	17 Produits chimiques
18 *Chemical prod.(less pharmaceuticals)*	..	..	..	79.0	96.0	95.0	..	18 *Prod. chimi. (moins prod. pharmaceutiques)*
19 *Pharmaceuticals*	..	..	..	390.0	543.0	626.0	..	19 *Produits pharmaceutiques*
20 Rubber & plastic products	..	..	..	18.0	15.0	10.0	..	20 Caoutchouc et matières plastiques
21 Non-metallic mineral products	..	..	..	15.0	16.0	14.0	..	**21 Produits minéraux non métalliques**
22 Basic metals	..	..	..	18.0	23.0	..	..	**22 Produits métallurgiques de base**
23 Basic metals, ferrous	..	..	..	4.0	4.0	3.0	..	23 Produits métallur. de base, ferreux
24 Basic metals, non-ferrous	..	..	..	14.0	19.0	..	..	24 Prod. métallur. de base, non ferreux
25 Fabricated metal products	..	..	..	5.0	5.0	2.0	..	**25 Ouvrages en métaux (sauf mach.& mat.)**
26 Machinery equipment, instruments and transport equipment	..	..	..	765.0	956.0	..	..	**26 Machines & matériel, instruments & mat. de transport**
27 Machinery n.e.c.	..	..	..	45.0	58.0	37.0	..	27 Machines, n.c.a.
28 Office, account. & computing machinery	..	..	..	..	..	..	..	28 Mach. de bur., compta. & trait. inform.
29 Electrical machinery	..	..	..	78.0	59.0	45.0	..	29 Machines électriques
30 Electro. equip. (radio, TV & commun.)	..	..	..	193.0	272.0	288.0	..	30 Equip. électro. (radio, tv & communic.)
31 *Electro. comp. (inc. semi-conductors)*	..	..	..	..	..	..	..	31 *Compos. électro. (semi-cond. incl.)*
32 *TV, radio & communications equip.*	..	..	..	..	..	..	..	32 *Appareils radio, tv et de communi.*
33 Instruments	..	..	..	59.0	70.0	84.0	..	33 Instruments
34 Motor vehicles	..	..	..	176.0	173.0	..	..	34 Véhicules automobiles
35 Other transport equipment	..	..	..	214.0	324.0	..	..	35 Autres matériels de transport
36 *Ships*	..	..	..	6.0	6.0	1.0	..	36 *Navires*
37 *Aerospace*	..	..	..	205.0	316.0	448.0	..	37 *Aérospatiale*
38 *Other transport equipment n.e.c.*	..	..	..	3.0	2.0	..	..	38 *Autres matériels de transport n.c.a.*
39 Furniture, other manufacturing n.e.c.	..	..	..	..	..	..	..	**39 Meubles, autres activ. de fabricat. n.c.a.**
40 Furniture	..	..	..	..	..	..	..	40 Meubles
41 Other manufacturing n.e.c.	..	..	..	..	..	..	..	41 Autres activités de fabrication n.c.a.
42 Recycling	..	..	..	..	..	..	..	**42 Récupération**
43 ELECTRICITY, GAS & WATER SUPPLY	..	..	..	4.0	5.0	3.0	..	**43 DISTRIBUTION D'ELECTRICITE, GAZ & EAU**
44 CONSTRUCTION	..	..	..	3.0	3.0	3.0	..	**44 CONSTRUCTION**
45 SERVICES SECTOR	..	..	..	382.0 [s]	428.0 [s]	..	..	**45 SECTEUR DES SERVICES**
46 Wholesale, retail trade & repair of motor vehicles, etc.	..	..	..	..	..	..	..	**46 Commerce (gros & détail), réparation de véhicules, etc.**
47 Hotels & restaurants	..	..	..	..	..	..	..	**47 Hôtels et restaurants**
48 Transport & storage	..	..	..	2.0	3.0	1.0	..	**48 Transports, entreposage**
49 Communications	..	..	..	33.0	21.0	32.0	..	**49 Communications**
50 Post	..	..	..	..	..	..	..	50 Postes
51 Telecommunications	..	..	..	..	..	..	..	51 Télécommunications
52 Financial intermediation (inc. insur.)	..	..	..	34.0	59.0	95.0	..	**52 Intermédiation financ. (assurances inc.)**
53 Real estate, renting & busin. activity	..	..	..	..	..	..	..	**53 Immobil., locat. & serv. aux entreprises**
54 Computer & related activities	..	..	..	195.0	210.0	..	..	54 Informatique et activités connexes
55 *Software consultancy*	..	..	..	..	..	..	..	55 *Production de logiciels*
56 *Other computer services n.e.c.*	..	..	..	..	..	..	..	56 *Autres activités informatiques*
57 Research & development	..	..	..	118.0	135.0	177.0	..	57 Recherche & développement expérimental
58 Other business activities n.e.c.	..	..	..	..	..	..	..	58 Autres activ. & serv. aux entreprises
59 Comm., soc. & pers. serv. activ., etc.	..	..	..	..	..	..	..	**59 Adm. pub., serv. soc. & collec., etc**
TOTAL BUSINESS ENTERPRISE SECTOR	**1 410.0**	**1 738.0**	**2 018.0**	**1 800.0**	**2 238.0**	**2 570.0**	**2 470.0**	**TOTAL SECTEUR DES ENTREPRISES**

Source: OECD, R&D database (DSTI/EAS Division), January 2002.

Source : OCDE, base de données R-D (DSTI/Division AES), janvier 2002.

Table 17 - Tableau 17

Higher education R&D personnel (HEMP) - Personnel de R-D du secteur de l'enseignement supérieur

by field of science and occupation - par discipline scientifique et par profession

In full-time equivalent — *En équivalence plein-temps*

	1994	1995	1996	1997	1998	1999	2000	
Total R&D personnel								**Personnel total de R-D**
Natural sciences	..	..	..	..	..	..	..	Sciences exactes et naturelles
Engineering	..	..	..	..	..	..	..	Sciences de l'ingénieur
Medical sciences	..	..	..	..	..	..	..	Sciences médicales
Agricultural sciences	..	..	..	..	..	..	..	Sciences agricoles
Sub-total NSE	..	..	..	..	..	..	..	Sous-total SENI
Social sciences	..	..	..	..	..	..	..	Sciences sociales
Humanities	..	..	..	..	..	..	..	Sciences humaines
Sub-total SSH	..	..	..	..	..	..	..	Sous-total SSH
Total NSE+SSH	..	..	..	..	..	..	..	*Total SENI+SSH*
Researchers								**Chercheurs**
Natural sciences	..	..	..	..	..	..	..	Sciences exactes et naturelles
Engineering	..	..	..	..	..	..	..	Sciences de l'ingénieur
Medical sciences	..	..	..	..	..	..	..	Sciences médicales
Agricultural sciences	..	..	..	..	..	..	..	Sciences agricoles
Sub-total NSE	..	..	..	..	..	..	..	Sous-total SENI
Social sciences	..	..	..	..	..	..	..	Sciences sociales
Humanities	..	..	..	..	..	..	..	Sciences humaines
Sub-total SSH	..	..	..	..	..	..	..	Sous-total SSH
Total NSE+SSH	42 000 [a]	47 000	47 000	47 651	49 023	..	..	*Total SENI+SSH*
Technicians								**Techniciens**
Natural sciences	..	..	..	..	..	..	..	Sciences exactes et naturelles
Engineering	..	..	..	..	..	..	..	Sciences de l'ingénieur
Medical sciences	..	..	..	..	..	..	..	Sciences médicales
Agricultural sciences	..	..	..	..	..	..	..	Sciences agricoles
Sub-total NSE	..	..	..	..	..	..	..	Sous-total SENI
Social sciences	..	..	..	..	..	..	..	Sciences sociales
Humanities	..	..	..	..	..	..	..	Sciences humaines
Sub-total SSH	..	..	..	..	..	..	..	Sous-total SSH
Total NSE+SSH	..	..	..	..	..	..	..	*Total SENI+SSH*
Other								**Autres**
Natural sciences	..	..	..	..	..	..	..	Sciences exactes et naturelles
Engineering	..	..	..	..	..	..	..	Sciences de l'ingénieur
Medical sciences	..	..	..	..	..	..	..	Sciences médicales
Agricultural sciences	..	..	..	..	..	..	..	Sciences agricoles
Sub-total NSE	..	..	..	..	..	..	..	Sous-total SENI
Social sciences	..	..	..	..	..	..	..	Sciences sociales
Humanities	..	..	..	..	..	..	..	Sciences humaines
Sub-total SSH	..	..	..	..	..	..	..	Sous-total SSH
Total NSE+SSH	..	..	..	..	..	..	..	*Total SENI+SSH*

Table 18 - Tableau 18

Government budget appropriations or outlays for R&D (GBAORD) - Crédits budgétaires publics de R-D (CBPRD)

by socio-economic objective - par objectif socio-économique

Million GBP — *Millions GBP*

	1995	1996	1997	1998	1999	2000	2001	
Agriculture, forestry and fishing	281.9	257.0	268.9	255.5	259.8	..	..	Agriculture, sylviculture et pêche
Industrial development	165.8	144.6	102.8	61.6	37.4	..	..	Développement industriel
Energy	52.3	43.2	41.0	28.0	28.9	..	..	Energie
Transport & telecommunications	17.9	18.5	19.4	24.6	27.1	..	..	*Transports & télécommunications*
Urban & rural planning	76.3	80.3	79.5	78.9	76.9	..	..	*Aménagement urbain et rural*
ST Infrastructure	94.1	98.8	98.9	103.5	104.0	..	..	ST Infrastructure
Prevention of pollution	..	..	..	..	..	..	..	*Prévention de la pollution*
Identificat. & treatment of pollution	..	..	..	..	..	..	..	*Détect. pollut. & mesures antipollut.*
ST Environment	131.8	128.7	136.2	142.8	147.0	..	..	ST Environnement
Health	762.5 [a]	835.0	845.9	853.0	934.2	..	..	Santé
Social development & services	136.9	120.7	114.7	154.7	216.2	..	..	Développement et services sociaux
Earth and atmosphere	105.2	95.4	81.3	78.5	79.4	..	..	Milieu terrestre et atmosphère
Advancement of research	653.5	680.5	671.0	677.0	700.4	..	..	*Promotion de la recherche*
General university funds	1 018.6	1 027.5	1 033.3	1 085.1	1 157.1	..	..	*Fonds généraux des universités*
ST Advancement of knowledge	1 672.1	1 708.0	1 704.3	1 762.1	1 857.5	..	..	ST promotion gén. des connaiss.
Civil space	153.0	164.1	164.4	142.5	142.7	..	..	Espace civil
Defence	2 061.7	2 143.5	2 312.1	2 099.5	2 346.8	..	..	Défense
Not elsewhere classified	24.7	20.5	21.6	25.8	21.0	..	..	Non classé
Total	**5 642.0**	**5 759.3**	**5 892.0**	**5 707.4**	**6 174.9**	**..**	**..**	**Total**

Source: OECD, R&D database (DSTI/EAS Division), January 2002.

Source : OCDE, base de données R-D (DSTI/Division AES), janvier 2002.

Table 19 - Tableau 19

Technology balance of payments - Balance des paiements technologiques

by type of transfer - selon la nature des transferts

Million GBP *Millions GBP*

	1983	1984	1985	1986	1987	1988	1989	1990	1991	
Receipts										**Recettes**
Related firms	290	312	..	..	..	..	..	..	..	Firmes affiliées
Unrelated firms	325	334	..	..	..	..	..	..	..	Firmes non affiliées
Other	..	..	..	..	..	..	..	..	..	Autre
Total	615	766 [a]	809	719	861 [a]	966	1 152	1 162	1 323	*Total*
Payments										**Paiements**
Related firms	409	501	..	..	..	..	..	..	..	Firmes affiliées
Unrelated firms	73	91	..	..	..	..	..	..	..	Firmes non affiliées
Other	..	..	..	..	..	..	..	..	..	Autre
Total	482	731 [a]	719	653	951 [a]	1 054	1 264	1 536	1 305	*Total*

	1992	1993	1994	1995	1996	1997	1998	1999	2000	
Receipts										**Recettes**
Related firms	..	..	..	..	..	..	..	..	..	Firmes affiliées
Unrelated firms	..	..	..	..	..	..	..	..	..	Firmes non affiliées
Other	..	..	..	..	..	..	..	..	..	Autre
Total	1 799	1 972	2 437	2 673	2 407	3 559	3 673	4 020	..	*Total*
Payments										**Paiements**
Related firms	..	..	..	..	..	..	..	..	..	Firmes affiliées
Unrelated firms	..	..	..	..	..	..	..	..	..	Firmes non affiliées
Other	..	..	..	..	..	..	..	..	..	Autre
Total	1 663	1 767	2 075	2 237	2 108	2 099	1 916	2 027	..	*Total*

Source: OECD, TBPdatabase (DSTI/EAS Division), Januay 2002.

Source : OCDE, base de données TBP (DSTI/Division AES), janvier 2002.

UNITED STATES Table 1 - Tableau 1 **ÉTATS-UNIS**

Gross domestic expenditure on R&D (GERD) - Dépenses intérieures brutes de R-D (DIRD)

by sector of performance and source of funds - par secteur d'exécution et par source de financement

	1994	1995	1996	1997	1998	1999	2000	
	1.1 Million USD - Millions USD							
Business enterprise								**Entreprises**
financed by								*financé par*
Business enterprise	97 131 [l]	108 652 [l]	121 015 [l]	133 611 [l]	145 016 [l]	160 288 [l]	177 645 [l,p]	Entreprises
Direct government	22 463 [l]	23 451 [l]	23 653 [l]	23 928 [l]	24 164 [l]	22 535 [l]	22 210 [l,p]	État direct
Higher education	0	0	0	0	0	0	0	Enseignement supérieur
Private non-profit	0	0	0	0	0	0	0	Institutions sans but lucratif
Funds from abroad	0	0	0	0	0	0	0	Fonds de l'étranger
Total	119 594 [l]	132 103 [l]	144 668 [l]	157 539 [l]	169 180 [l]	182 823 [l]	199 855 [l,p]	*Total*
Government								**État**
financed by								*financé par*
Business enterprise	0	0	0	0	0	0	0	Entreprises
Direct government	16 769 [h]	17 365 [h]	17 041 [h]	17 387 [h]	17 819 [h]	18 891 [h]	19 844 [h,p]	État direct
Higher education	0	0	0	0	0	0	0	Enseignement supérieur
Private non-profit	0	0	0	0	0	0	0	Institutions sans but lucratif
Funds from abroad	0	0	0	0	0	0	0	Fonds de l'étranger
Total	16 769 [h]	17 365 [h]	17 041 [h]	17 387 [h]	17 819 [h]	18 891 [h]	19 844 [h,p]	*Total*
Higher education								**Enseignement supérieur**
financed by								*financé par*
Business enterprise	1 455 [l]	1 547 [l]	1 667 [l]	1 812 [l]	1 971 [l]	2 133 [l]	2 310 [l,p]	Entreprises
Direct government	19 923 [l]	20 702 [l]	21 335 [l]	22 128 [l]	23 165 [l]	24 299 [l]	25 473 [l,p]	État direct
General university funds	..	..	..	..	..	..	..	Fonds généraux des universités
Sub-total government	19 923 [l]	20 702 [l]	21 335 [l]	22 128 [l]	23 165 [l]	24 299 [l]	25 473 [l,p]	Sous-Total État
Higher education	3 936 [l]	4 108 [l]	4 430 [l]	4 846 [l]	5 183 [l]	5 562 [l]	5 969 [l,p]	Enseignement supérieur
Private non-profit	1 594 [l]	1 616 [l]	1 665 [l]	1 790 [l]	1 934 [l]	2 066 [l]	2 203 [l,p]	Institutions sans but lucratif
Funds from abroad	0	0	0	0	0	0	0	Fonds de l'étranger
Total	26 908 [l]	27 973 [l]	29 097 [l]	30 576 [l]	32 253 [l]	34 060 [l]	35 955 [l,p]	*Total*
Private non-profit								**Institutions sans but lucratif**
financed by								*financé par*
Business enterprise	617 [l]	671 [l]	730 [l]	809 [l]	880 [l]	976 [l]	1 085 [l,p]	Entreprises
Direct government	3 669 [l]	3 655 [l]	3 678 [l]	3 835 [l]	4 124 [l]	4 627 [l]	4 997 [l,p]	État direct
Higher education	0	0	0	0	0	0	0	Enseignement supérieur
Private non-profit	2 071 [l]	2 308 [l]	2 574 [l]	2 804 [l]	3 073 [l]	3 323 [l]	3 586 [l,p]	Institutions sans but lucratif
Funds from abroad	0	0	0	0	0	0	0	Fonds de l'étranger
Total	6 357 [l]	6 634 [l]	6 982 [l]	7 448 [l]	8 077 [l]	8 926 [l]	9 668 [l,p]	*Total*
GERD								**DIRD**
financed by								*financé par*
Business enterprise	99 203 [l]	110 870 [l]	123 412 [l]	136 232 [l]	147 867 [l]	163 397 [l]	181 040 [l,p]	Entreprises
Direct government	62 824 [l]	65 173 [l]	65 707 [l]	67 278 [l]	69 272 [l]	70 352 [l]	72 524 [l,p]	État direct
General university funds	..	..	..	..	..	..	..	Fonds généraux des universités
Sub-total government	62 824 [l]	65 173 [l]	65 707 [l]	67 278 [l]	69 272 [l]	70 352 [l]	72 524 [l,p]	Sous-Total État
Higher education	3 936 [l]	4 108 [l]	4 430 [l]	4 846 [l]	5 183 [l]	5 562 [l]	5 969 [l,p]	Enseignement supérieur
Private non-profit	3 665 [l]	3 924 [l]	4 239 [l]	4 594 [l]	5 007 [l]	5 389 [l]	5 789 [l,p]	Institutions sans but lucratif
Funds from abroad	0	0	0	0	0	0	0	Fonds de l'étranger
Total	169 628 [l]	184 075 [l]	197 788 [l]	212 950 [l]	227 329 [l]	244 700 [l]	265 322 [l,p]	*Total*
	1.2 Million constant USD (1995 prices and PPPs) - Millions de USD constants (prix et PPA de 1995)							
Business enterprise								**Entreprises**
financed by								*financé par*
Business enterprise	99 244.9 [l]	108 652.0 [l]	118 702.8 [l]	128 537.8 [l]	137 767.5 [l]	150 022.1 [l]	162 513.7 [b,l,p]	Entreprises
Direct government	22 951.9 [l]	23 451.0 [l]	23 201.1 [l]	23 019.5 [l]	22 956.2 [l]	21 091.7 [l]	20 318.2 [b,l,p]	État direct
Higher education	0.0	0.0	0.0	0.0	0.0	0.0	0.0 [b]	Enseignement supérieur
Private non-profit	0.0	0.0	0.0	0.0	0.0	0.0	0.0 [b]	Institutions sans but lucratif
Funds from abroad	0.0	0.0	0.0	0.0	0.0	0.0	0.0 [b]	Fonds de l'étranger
Total	122 196.8 [l]	132 103.0 [l]	141 903.8 [l]	151 557.2 [l]	160 723.7 [l]	171 113.8 [l]	182 831.9 [b,l,p]	*Total*
Government								**État**
financed by								*financé par*
Business enterprise	0.0	0.0	0.0	0.0	0.0	0.0	0.0 [b]	Entreprises
Direct government	17 134.3 [h]	17 365.1 [h]	16 715.3 [h]	16 727.1 [h]	16 928.4 [h]	17 681.5 [h]	18 154.1 [b,h,p]	État direct
Higher education	0.0	0.0	0.0	0.0	0.0	0.0	0.0 [b]	Enseignement supérieur
Private non-profit	0.0	0.0	0.0	0.0	0.0	0.0	0.0 [b]	Institutions sans but lucratif
Funds from abroad	0.0	0.0	0.0	0.0	0.0	0.0	0.0 [b]	Fonds de l'étranger
Total	17 134.3 [h]	17 365.1 [h]	16 715.3 [h]	16 727.1 [h]	16 928.4 [h]	17 681.5 [h]	18 154.1 [b,h,p]	*Total*
Higher education								**Enseignement supérieur**
financed by								*financé par*
Business enterprise	1 486.7 [l]	1 547.0 [l]	1 635.1 [l]	1 743.2 [l]	1 872.5 [l]	1 996.4 [l]	2 113.2 [b,l,p]	Entreprises
Direct government	20 356.6 [l]	20 702.0 [l]	20 927.4 [l]	21 287.8 [l]	22 007.1 [l]	22 742.7 [l]	23 303.3 [b,l,p]	État direct
General university funds	..	..	..	..	..	..	..	Fonds généraux des universités
Sub-total government	20 356.6 [l]	20 702.0 [l]	20 927.4 [l]	21 287.8 [l]	22 007.1 [l]	22 742.7 [l]	23 303.3 [b,l,p]	Sous-Total État
Higher education	4 021.7 [l]	4 108.0 [l]	4 345.4 [l]	4 662.0 [l]	4 923.9 [l]	5 205.8 [l]	5 460.6 [b,l,p]	Enseignement supérieur
Private non-profit	1 628.7 [l]	1 616.0 [l]	1 633.2 [l]	1 722.0 [l]	1 837.3 [l]	1 933.7 [l]	2 015.4 [b,l,p]	Institutions sans but lucratif
Funds from abroad	0.0	0.0	0.0	0.0	0.0	0.0	0.0 [b]	Fonds de l'étranger
Total	27 493.6 [l]	27 973.0 [l]	28 541.0 [l]	29 415.0 [l]	30 640.9 [l]	31 878.6 [l]	32 892.4 [b,l,p]	*Total*
Private non-profit								**Institutions sans but lucratif**
financed by								*financé par*
Business enterprise	630.4 [l]	671.0 [l]	716.1 [l]	778.3 [l]	836.0 [l]	913.5 [l]	992.6 [b,l,p]	Entreprises
Direct government	3 748.9 [l]	3 655.0 [l]	3 607.7 [l]	3 689.4 [l]	3 917.9 [l]	4 330.7 [l]	4 571.4 [b,l,p]	État direct
Higher education	0.0	0.0	0.0	0.0	0.0	0.0	0.0 [b]	Enseignement supérieur
Private non-profit	2 116.1 [l]	2 308.0 [l]	2 524.8 [l]	2 697.5 [l]	2 919.4 [l]	3 110.2 [l]	3 280.6 [b,l,p]	Institutions sans but lucratif
Funds from abroad	0.0	0.0	0.0	0.0	0.0	0.0	0.0 [b]	Fonds de l'étranger
Total	6 495.4 [l]	6 634.0 [l]	6 848.6 [l]	7 165.2 [l]	7 673.3 [l]	8 354.3 [l]	8 844.5 [b,l,p]	*Total*
GERD								**DIRD**
financed by								*financé par*
Business enterprise	101 362.0 [l]	110 870.0 [l]	121 054.0 [l]	131 059.3 [l]	140 476.0 [l]	152 932.0 [l]	165 619.5 [b,l,p]	Entreprises
Direct government	64 191.7 [l]	65 173.1 [l]	64 451.4 [l]	64 723.7 [l]	65 809.6 [l]	65 846.6 [l]	66 346.9 [b,l,p]	État direct
General university funds	..	..	..	..	..	..	..	Fonds généraux des universités
Sub-total government	64 191.7 [l]	65 173.1 [l]	64 451.4 [l]	64 723.7 [l]	65 809.6 [l]	65 846.6 [l]	66 346.9 [b,l,p]	Sous-Total État
Higher education	4 021.7 [l]	4 108.0 [l]	4 345.4 [l]	4 662.0 [l]	4 923.9 [l]	5 205.8 [l]	5 460.6 [b,l,p]	Enseignement supérieur
Private non-profit	3 744.8 [l]	3 924.0 [l]	4 158.0 [l]	4 419.6 [l]	4 756.7 [l]	5 043.9 [l]	5 295.9 [b,l,p]	Institutions sans but lucratif
Funds from abroad	0.0	0.0	0.0	0.0	0.0	0.0	0.0 [b]	Fonds de l'étranger
Total	173 320.1 [l]	184 075.1 [l]	194 008.8 [l]	204 864.6 [l]	215 966.3 [l]	229 028.1 [l]	242 722.9 [b,l,p]	*Total*

Source: OECD, R&D database (DSTI/EAS Division), January 2002.

Source : OCDE, base de données R-D (DSTI/Division AES), janvier 2002.

Table 2 - Tableau 2

Gross domestic expenditure on R&D (GERD) - Dépenses intérieures brutes de R-D (DIRD)

by sector of performance and type of costs - par secteur d'exécution et par type de coûts

Million USD — *Millions USD*

	1994	1995	1996	1997	1998	1999	2000	
Business enterprise								**Entreprises**
Labour costs	..	59 578	..	74 043	81 206	81 905	..	Coûts du personnel
Other current costs	..	72 525	..	83 496	87 974	100 918	..	Autres coûts courants
Total current expenditure	119 594	132 103	144 668	157 539	169 180	182 823	199 855	*Total dépenses courantes*
Land and buildings	..	..	..	..	..	..	..	Terrains et constructions
Instruments and equipment	..	..	..	..	..	..	..	Instruments et équipements
Total capital expenditure	..	..	..	..	..	..	..	*Total dépenses en capital*
Total	119 594 [l]	132 103 [l]	144 668 [l]	157 539 [l]	169 180 [l]	182 823 [l]	199 855 [l,p]	*Total*
Government								**État**
Labour costs	8 018 [h]	8 028 [h]	7 629 [h]	7 364 [h]	7 599 [h]	7 434 [h]	7 066 [h,p]	Coûts du personnel
Other current costs	8 337 [h]	8 875 [h]	8 956 [h]	9 456 [h]	9 764 [h]	10 898 [h]	12 077 [h,p]	Autres coûts courants
Total current expenditure	16 355 [h]	16 903 [h]	16 585 [h]	16 820 [h]	17 363 [h]	18 332 [h]	19 143 [h,p]	*Total dépenses courantes*
Land and buildings	..	..	..	..	..	..	..	Terrains et constructions
Instruments and equipment	..	..	..	..	..	..	..	Instruments et équipements
Total capital expenditure	414 [h]	462 [h]	456 [h]	567 [h]	456 [h]	559 [h]	701 [h]	*Total dépenses en capital*
Total	16 769 [h]	17 365 [h]	17 041 [h]	17 387 [h]	17 819 [h]	18 891 [h]	19 844 [h,p]	*Total*
Higher education								**Enseignement supérieur**
Labour costs	..	..	..	..	..	..	..	Coûts du personnel
Other current costs	..	..	..	..	..	..	..	Autres coûts courants
Total current expenditure	26 908	27 973	29 097	30 576	32 253	34 060	35 955 [p]	*Total dépenses courantes*
Land and buildings	..	..	..	..	..	..	..	Terrains et constructions
Instruments and equipment	..	..	..	..	..	..	..	Instruments et équipements
Total capital expenditure	..	..	..	..	..	..	..	*Total dépenses en capital*
Total	26 908 [l]	27 973 [l]	29 097 [l]	30 576 [l]	32 253 [l]	34 060 [l]	35 955 [l,p]	*Total*
Private non-profit								**Institutions sans but lucratif**
Labour costs	..	..	..	..	..	..	..	Coûts du personnel
Other current costs	..	..	..	..	..	..	..	Autres coûts courants
Total current expenditure	6 357	6 634	6 982	7 448	8 077	8 926	9 668	*Total dépenses courantes*
Land and buildings	..	..	..	..	..	..	..	Terrains et constructions
Instruments and equipment	..	..	..	..	..	..	..	Instruments et équipements
Total capital expenditure	..	..	..	..	..	..	..	*Total dépenses en capital*
Total	6 357 [l]	6 634 [l]	6 982 [l]	7 448 [l]	8 077 [l]	8 926 [l]	9 668 [l,p]	*Total*
GERD								**DIRD**
Labour costs	..	..	..	..	..	..	..	Coûts du personnel
Other current costs	..	..	..	..	..	..	..	Autres coûts courants
Total current expenditure	169 214	183 613	197 332	212 383	226 873	244 141	264 621 [p]	*Total dépenses courantes*
Land and buildings	..	..	..	..	..	..	..	Terrains et constructions
Instruments and equipment	..	..	..	..	..	..	..	Instruments et équipements
Total capital expenditure	..	..	..	..	..	..	..	*Total dépenses en capital*
Total	169 628 [l]	184 075 [l]	197 788 [l]	212 950 [l]	227 329 [l]	244 700 [l]	265 322 [l,p]	*Total*

Table 3 - Tableau 3

Current domestic expenditure on R&D - Dépenses intérieures courantes de R-D

by sector of performance and type of activity - par secteur d'exécution et par type d'activités

Million USD — *Millions USD*

	1994	1995	1996	1997	1998	1999	2000	
Business enterprise								**Entreprises**
Basic research	7 017	6 099	8 206	10 420	13 595	14 673	16 082 [p]	Recherche fondamentale
Applied research	23 491	27 454	29 241	32 643	30 571	35 641	37 933 [p]	Recherche appliquée
Experimental development	89 086	98 550	107 221	114 476	125 014	132 510	145 840 [p]	Développement expérimental
Not specified	..	..	..	..	..	..	..	Non spécifié
Total	119 594	132 103	144 668	157 539	169 180	182 823	199 855 [p]	*Total*
Government								**État**
Basic research	2 547 [h]	2 689 [h]	2 680 [h]	2 746 [h]	3 003 [h]	3 312 [h]	3 525 [h,p]	Recherche fondamentale
Applied research	4 985 [h]	4 952 [h]	4 872 [h]	4 997 [h]	5 146 [h]	5 503 [h]	5 826 [h,p]	Recherche appliquée
Experimental development	8 823 [h]	9 262 [h]	9 033 [h]	9 077 [h]	9 214 [h]	9 517 [h]	9 792 [h,p]	Développement expérimental
Not specified	..	..	..	..	..	..	..	Non spécifié
Total	16 355 [h]	16 903 [h]	16 585 [h]	16 820 [h]	17 363 [h]	18 332 [h]	19 143 [h,p]	*Total*
Higher education								**Enseignement supérieur**
Basic research	17 262	17 798	18 661	19 675	20 828	22 198	23 465 [p]	Recherche fondamentale
Applied research	6 366	6 772	7 153	7 516	7 801	8 090	8 661 [p]	Recherche appliquée
Experimental development	3 280	3 401	3 282	3 383	3 623	3 773	3 830 [p]	Développement expérimental
Not specified	..	..	..	..	..	..	..	Non spécifié
Total	26 908	27 973	29 097	30 576	32 253	34 060	35 955 [p]	*Total*
Private non-profit								**Institutions sans but lucratif**
Basic research	2 753	2 974	3 266	3 430	3 869	4 443	4 831 [p]	Recherche fondamentale
Applied research	1 770	1 821	1 903	2 054	2 185	2 398	2 621 [p]	Recherche appliquée
Experimental development	1 834	1 839	1 812	1 963	2 023	2 086	2 217 [p]	Développement expérimental
Not specified	..	..	..	..	..	..	..	Non spécifié
Total	6 357	6 634	6 982	7 448	8 077	8 926	9 668 [p]	*Total*
GERD								**DIRD**
Basic research	29 579	29 560	32 813	36 271	41 295	44 626	47 903 [p]	Recherche fondamentale
Applied research	36 612	40 999	43 169	47 210	45 703	51 632	55 041 [p]	Recherche appliquée
Experimental development	103 023	113 052	121 348	128 899	139 874	147 886	161 679 [p]	Développement expérimental
Not specified	..	..	..	..	..	..	..	Non spécifié
Total	169 214	183 611.0	197 330	212 380	226 872	244 144	264 623 [p]	*Total*

Source: OECD, R&D database (DSTI/EAS Division), January 2002.

Source : OCDE, base de données R-D (DSTI/Division AES), janvier 2002.

Table 5 - Tableau 5

Total R&D personnel - Personnel total de R-D

by sector of employment and occupation - par secteur d'emploi et par profession

In full-time equivalent *En équivalence plein-temps*

	1994	1995	1996	1997	1998	1999	2000	
Business enterprise								**Entreprises**
RSE	757 300	789 400	859 300	918 600	974 600 [p]	1 015 700 [p]	..	Chercheurs
Technicians	..	..	..	..	..	..	..	Techniciens
Other occupations	..	..	..	..	..	..	..	Autres professions
Total	..	..	..	..	..	..	..	*Total*
Government								**État**
RSE	..	53 900 [d,h,n]	52 100 [d,h,n]	49 800 [d,h,n]	..	..	..	Chercheurs
Technicians	..	..	..	..	..	..	..	Techniciens
Other occupations	..	..	..	..	..	..	..	Autres professions
Total	..	..	..	..	..	..	..	*Total*
Higher education								**Enseignement supérieur**
RSE	..	134 300	..	135 800	..	..	..	Chercheurs
Technicians	..	..	..	..	..	..	..	Techniciens
Other occupations	..	..	..	..	..	..	..	Autres professions
Total	..	..	..	..	..	..	..	*Total*
Private non-profit								**Institutions sans but lucratif**
RSE	..	10 000 [m]	..	10 000 [m]	..	..	..	Chercheurs
Technicians	..	..	..	..	..	..	..	Techniciens
Other occupations	..	..	..	..	..	..	..	Autres professions
Total	..	..	..	..	..	..	..	*Total*
TOTAL								**TOTAL**
RSE	..	987 700 [m]	..	1 114 100 [m]	..	..	..	Chercheurs
Technicians	..	..	..	..	..	..	..	Techniciens
Other occupations	..	..	..	..	..	..	..	Autres professions
Total	..	..	..	..	..	..	..	*Total*

Table 6 - Tableau 6

Total R&D personnel - Personnel total de R-D

by sector of employment and formal qualification - par secteur d'emploi et par niveau de qualification

In full-time equivalent *En équivalence plein-temps*

	1994	1995	1996	1997	1998	1999	2000	
Business enterprise								**Entreprises**
University PhD level degrees	..	..	..	..	..	..	..	Diplômes univ. de niv. doctorat
Other university degrees	..	..	..	..	..	..	..	Autres diplômes universitaires
Sub-total university degrees	..	..	..	..	..	..	..	Sous-total Universitaires
Other post-secondary diplomas	..	..	..	..	..	..	..	Autres diplômes post-secondaires
Secondary diplomas	..	..	..	..	..	..	..	Diplômes de fin d'études secondaires
Other qualifications	..	..	..	..	..	..	..	Autres qualifications
Total	..	..	..	..	..	..	..	*Total*
Government								**État**
University PhD level degrees	..	..	..	..	..	..	..	Diplômes univ. de niv. doctorat
Other university degrees	..	..	..	..	..	..	..	Autres diplômes universitaires
Sub-total university degrees	..	..	..	..	..	..	..	Sous-total Universitaires
Other post-secondary diplomas	..	..	..	..	..	..	..	Autres diplômes post-secondaires
Secondary diplomas	..	..	..	..	..	..	..	Diplômes de fin d'études secondaires
Other qualifications	..	..	..	..	..	..	..	Autres qualifications
Total	..	..	..	..	..	..	..	*Total*
Higher education								**Enseignement supérieur**
University PhD level degrees	..	89 300	..	91 800	..	..	..	Diplômes univ. de niv. doctorat
Other university degrees	..	45 000	..	44 000	..	..	..	Autres diplômes universitaires
Sub-total university degrees	..	134 300	..	135 800	..	..	..	Sous-total Universitaires
Other post-secondary diplomas	..	..	..	..	..	..	..	Autres diplômes post-secondaires
Secondary diplomas	..	..	..	..	..	..	..	Diplômes de fin d'études secondaires
Other qualifications	..	..	..	..	..	..	..	Autres qualifications
Total	..	..	..	..	..	..	..	*Total*
Private non-profit								**Institutions sans but lucratif**
University PhD level degrees	..	10 000	..	10 000	..	..	..	Diplômes univ. de niv. doctorat
Other university degrees	..	..	..	..	..	..	..	Autres diplômes universitaires
Sub-total university degrees	..	..	..	..	..	..	..	Sous-total Universitaires
Other post-secondary diplomas	..	..	..	..	..	..	..	Autres diplômes post-secondaires
Secondary diplomas	..	..	..	..	..	..	..	Diplômes de fin d'études secondaires
Other qualifications	..	..	..	..	..	..	..	Autres qualifications
Total	..	..	..	..	..	..	..	*Total*
TOTAL								**TOTAL**
University PhD level degrees	..	..	..	..	..	..	..	Diplômes univ. de niv. doctorat
Other university degrees	..	..	..	..	..	..	..	Autres diplômes universitaires
Sub-total university degrees	..	..	..	..	..	..	..	Sous-total Universitaires
Other post-secondary diplomas	..	..	..	..	..	..	..	Autres diplômes post-secondaires
Secondary diplomas	..	..	..	..	..	..	..	Diplômes de fin d'études secondaires
Other qualifications	..	..	..	..	..	..	..	Autres qualifications
Total	..	..	..	..	..	..	..	*Total*

Source: OECD, R&D database (DSTI/EAS Division), January 2002.

Source : OCDE, base de données R-D (DSTI/Division AES), janvier 2002.

UNITED STATES — Table 7 - Tableau 7 — ÉTATS-UNIS

Gross domestic expenditure on R&D (GERD) - Dépenses intérieures brutes de R-D (DIRD)

by sector of performance and main scientific field - par secteur d'exécution et par grand domaine scientifique

Million USD — *Millions USD*

	1994	1995	1996	1997	1998	1999	2000	
Business enterprise								**Entreprises**
Natural sciences and engineering	119 594 [j]	132 103 [j]	144 668 [j]	157 539 [j]	169 180 [j]	182 823 [j]	199 855 [j,p]	Sciences exactes, nat. et de l'ingénieur
Social sciences and humanities	..	..	..	..	..	..	..	Sciences sociales et humaines
All fields of science	119 594 [j]	132 103 [j]	144 668 [j]	157 539 [j]	169 180 [j]	182 823 [j]	199 855 [j,p]	*Tous domaines scientifiques*
Government								**État**
Natural sciences and engineering	..	..	..	..	..	..	..	Sciences exactes, nat. et de l'ingénieur
Social sciences and humanities	..	..	..	..	..	..	..	Sciences sociales et humaines
All fields of science	16 769 [h]	17 365 [h]	17 041 [h]	17 387 [h]	17 819 [h]	18 891 [h]	19 844 [h,p]	*Tous domaines scientifiques*
Higher education								**Enseignement supérieur**
Natural sciences and engineering	19 708 [j,v]	20 773 [j,v]	21 558 [j,v]	22 824 [j,v]	24 262 [j,v]	25 762 [j,v]	..	Sciences exactes, nat. et de l'ingénieur
Social sciences and humanities	..	..	..	..	..	..	..	Sciences sociales et humaines
All fields of science	26 908 [j]	27 973 [j]	29 097 [j]	30 576 [j]	32 253 [j]	34 060 [j]	35 955 [j,p]	*Tous domaines scientifiques*
Private non-profit								**Institutions sans but lucratif**
Natural sciences and engineering	..	..	..	..	..	..	..	Sciences exactes, nat. et de l'ingénieur
Social sciences and humanities	..	..	..	..	..	..	..	Sciences sociales et humaines
All fields of science	6 357 [j]	6 634 [j]	6 982 [j]	7 448 [j]	8 077 [j]	8 926 [j]	9 668 [j,p]	*Tous domaines scientifiques*
TOTAL								**TOTAL**
Natural sciences and engineering	..	..	..	..	..	..	..	Sciences exactes, nat. et de l'ingénieur
Social sciences and humanities	..	..	..	..	..	..	..	Sciences sociales et humaines
All fields of science	169 628 [j]	184 075 [j]	197 788 [j]	212 950 [j]	227 329 [j]	244 700 [j]	265 322 [j,p]	*Tous domaines scientifiques*

Table 8 - Tableau 8

Total R&D personnel - Personnel total de R-D

by sector of employment and main scientific field - par secteur d'emploi et par grand domaine scientifique

In full-time equivalent — *En équivalence plein-temps*

	1994	1995	1996	1997	1998	1999	2000	
Business enterprise								**Entreprises**
Natural sciences and engineering	..	..	..	..	..	..	..	Sciences exactes, nat. et de l'ingénieur
Social sciences and humanities	..	..	..	..	..	..	..	Sciences sociales et humaines
All fields of science	..	..	..	..	..	..	..	*Tous domaines scientifiques*
Government								**État**
Natural sciences and engineering	..	..	..	..	..	..	..	Sciences exactes, nat. et de l'ingénieur
Social sciences and humanities	..	..	..	..	..	..	..	Sciences sociales et humaines
All fields of science	..	..	..	..	..	..	..	*Tous domaines scientifiques*
Higher education								**Enseignement supérieur**
Natural sciences and engineering	..	..	..	..	..	..	..	Sciences exactes, nat. et de l'ingénieur
Social sciences and humanities	..	..	..	..	..	..	..	Sciences sociales et humaines
All fields of science	..	..	..	..	..	..	..	*Tous domaines scientifiques*
Private non-profit								**Institutions sans but lucratif**
Natural sciences and engineering	..	..	..	..	..	..	..	Sciences exactes, nat. et de l'ingénieur
Social sciences and humanities	..	..	..	..	..	..	..	Sciences sociales et humaines
All fields of science	..	..	..	..	..	..	..	*Tous domaines scientifiques*
TOTAL								**TOTAL**
Natural sciences and engineering	..	..	..	..	..	..	..	Sciences exactes, nat. et de l'ingénieur
Social sciences and humanities	..	..	..	..	..	..	..	Sciences sociales et humaines
All fields of science	..	..	..	..	..	..	..	*Tous domaines scientifiques*

Source: OECD, R&D database (DSTI/EAS Division), January 2002. — *Source* : OCDE, base de données R-D (DSTI/Division AES), janvier 2002.

UNITED STATES — Table 9 - Tableau 9 — ÉTATS-UNIS

Total business enterprise intramural expenditure on R&D - Dépenses intra-muros totales de R-D des entreprises
(BERD) by industry - (DIRDE) par industrie

9.1 Million USD - Millions USD

	1994	1995	1996	1997	1998	1999	2000	
1 AGRICULTURE, HUNTING & FORESTRY	.. [n]	.. [n]	44 [j]	7 [j]	.. [n]	..	..	**1 AGRICULTURE, CHASSE & SYLVICULTURE**
2 MINING	.. [n]	.. [n]	77 [j]	.. [n]	.. [n]	.. [n]	..	**2 ACTIVITES EXTRACTIVES**
3 MANUFACTURING	90 749 [j]	100 067 [j]	111 864 [j]	121 025 [j]	120 401 [j]	117 292 [a,j]	..	**3 ACTIVITES DE FABRICATION**
4 Food, beverages & tobacco	1 476 [j]	1 566 [j]	..	..	..	..	..	**4 Produits alimentaires, boissons & tabac**
5 Food products & beverages	..	..	..	1 808 [j]	1 728 [j]	..	..	5 Produits alimentaires & boissons
6 Tobacco products	..	..	..	..	..	..	..	6 Produits à base de tabac
7 Textiles, wear app., fur & leather	..	..	..	..	577 [j]	337 [a,j]	..	**7 Textiles, habill., fourrures & cuirs**
8 Textiles	..	..	..	..	257 [j]	..	..	8 Textiles
9 Wearing apparel & fur	..	..	..	..	294 [j]	..	..	9 Articles d'habillement & fourrures
10 Leather products & footwear	40 [j]	..	..	42 [j]	26 [j]	..	..	10 Articles en cuir & chaussures
11 Wood, paper, printing, publishing	..	..	..	..	..	..	..	**11 Bois, papier, imprimerie, édition**
12 Wood & cork (not furniture)	46 [j]	..	..	60 [j]	280 [j]	70 [a,j]	..	12 Bois & liège (sauf meubles)
13 Pulp, paper & paper products	..	..	..	..	..	..	..	13 Carton, papier & articles
14 Publ., print. & repro. of rec. media	318 [j]	..	..	367 [j]	332 [j]	..	..	14 Edition, impress. & repro. enregis.
15 Coke, petroleum, nuclear fuel, chemicals, rubber plastics	..	..	..	..	..	22 832 [a,j]	..	**15 Coke, pétrole, combust. nucl., prod. chimi., caout. & mat. plast.**
16 Coke, ref. petrol. prod. & nuclear fuel	1 950 [j]	1 760 [j]	1 662 [j]	..	1 808 [j]	615 [a,j]	..	16 Coke, raffin. pétr. & combust. nucl.
17 Chemicals & chemical products	..	17 547 [j]	..	..	21 930 [j]	20 372 [a,j]	..	17 Produits chimiques
18 *Chemical prod.(less pharmaceuticals)*	..	7 332 [j]	..	..	9 193 [j]	..	..	18 *Prod. chimi. (moins prod. pharmaceutiques)*
19 *Pharmaceuticals*	9 633 [j]	10 215 [j]	9 773 [j]	11 589 [j]	12 571 [j]	..	..	19 *Produits pharmaceutiques*
20 Rubber & plastic products	..	..	..	..	..	1 845 [a,j]	..	20 Caoutchouc et matières plastiques
21 Non-metallic mineral products	591 [j]	448 [j]	517 [j]	619 [j]	727 [j]	..	..	**21 Produits minéraux non métalliques**
22 Basic metals	690 [j]	593 [j]	..	988 [j]	..	470 [a,j]	..	**22 Produits métallurgiques de base**
23 Basic metals, ferrous	..	..	..	..	..	..	..	23 Produits métallur. de base, ferreux
24 Basic metals, non-ferrous	..	..	..	..	..	..	..	24 Prod. métallur. de base, non ferreux
25 Fabricated metal products	1 111 [j]	1 023 [j]	..	1 818 [j]	1 643 [j]	1 704 [a,j]	..	**25 Ouvrages en métaux (sauf mach.& mat.)**
26 Machinery equipment, instruments and transport equipment	62 976 [j]	..	82 412 [j]	90 054 [j]	..	..	..	**26 Machines & matériel, instruments & mat. de transport**
27 Machinery n.e.c.	4 004 [j]	..	..	5 659 [j]	6 014 [j]	6 327 [a,j]	..	27 Machines, n.c.a.
28 Office, account. & computing machinery	4 106 [j]	..	..	12 840 [j]	8 905 [j]	..	..	28 Mach. de bur., compta. & trait. inform.
29 Electrical machinery	..	..	..	4 909 [j]	4 811 [j]	..	..	29 Machines électriques
30 Electro. equip. (radio, TV & commun.)	..	..	..	..	..	..	..	30 Equip. électro. (radio, tv & communic.)
31 *Electro. comp. (inc. semi-conductors)*	6 032 [j]	..	..	..	..	10 827 [a,j]	..	31 *Compos. électro. (semi-cond. incl.)*
32 *TV, radio & communications equip.*	..	..	..	..	..	..	..	32 *Appareils radio, tv et de communi.*
33 Instruments	11 441 [j]	11 976 [j]	13 091 [j]	13 835 [j]	..	19 566 [a,j]	..	33 Instruments
34 Motor vehicles	..	..	..	..	..	18 274 [a,j]	..	34 Véhicules automobiles
35 Other transport equipment	..	..	..	..	..	15 785 [a,j]	..	35 Autres matériels de transport
36 *Ships*	..	..	..	..	..	..	..	36 *Navires*
37 *Aerospace*	14 260 [j]	16 951 [j]	16 224 [j]	16 296 [j]	14 449 [j]	14 425 [a,j]	..	37 *Aérospatiale*
38 *Other transport equipment n.e.c.*	..	..	..	..	..	1 359 [a,j]	..	38 *Autres matériels de transport n.c.a.*
39 Furniture, other manufacturing n.e.c.	..	..	..	..	..	..	..	**39 Meubles, autres activ. de fabricat. n.c.a.**
40 Furniture	..	..	..	314 [j]	..	251 [a,j]	..	40 Meubles
41 Other manufacturing n.e.c.	..	..	..	..	..	..	..	41 Autres activités de fabrication n.c.a.
42 Recycling	..	..	..	..	..	..	..	**42 Récupération**
43 ELECTRICITY, GAS & WATER SUPPLY	403 [j]	440 [j]	352 [j]	.. [n]	.. [n]	142 [a,j]	..	**43 DISTRIBUTION D'ELECTRICITE, GAZ & EAU**
44 CONSTRUCTION	.. [n]	.. [n]	283 [j]	261 [j]	.. [n]	699 [a,j]	..	**44 CONSTRUCTION**
45 SERVICES SECTOR	28 443 [j,o]	31 596 [j,o]	32 047 [a,j]	36 246 [a,j,o]	48 780 [a,j,o]	64 690 [a,j,o]	..	**45 SECTEUR DES SERVICES**
46 Wholesale, retail trade & repair of motor vehicles, etc.	..	..	6 364 [j]	..	..	19 960 [a,j]	..	**46 Commerce (gros & détail), réparation de véhicules, etc.**
47 Hotels & restaurants	..	..	272 [j]	155 [j]	..	..	..	**47 Hôtels et restaurants**
48 Transport & storage	..	..	223 [j]	..	..	466 [a,j,o]	..	**48 Transports, entreposage**
49 Communications	..	..	4 103 [j]	..	..	..	..	**49 Communications**
50 Post	..	..	..	..	..	.. [n]	..	50 Postes
51 Telecommunications	..	..	4 103 [j]	..	..	..	..	51 Télécommunications
52 Financial intermediation (inc. insur.)	..	..	1 297 [j]	..	..	..	..	**52 Intermédiation financ. (assurances inc.)**
53 Real estate, renting & busin. activity	..	..	..	..	..	..	..	**53 Immobil., locat. & serv. aux entreprises**
54 Computer & related activities	..	11 992 [j]	..	13 745 [j,m]	..	..	..	54 Informatique et activités connexes
55 *Software consultancy*	..	..	..	..	..	10 964 [a,j]	..	55 *Production de logiciels*
56 *Other computer services n.e.c.*	..	..	..	..	..	..	..	56 *Autres activités informatiques*
57 Research & development	3 765 [j]	..	5 484 [j,m]	..	9 169 [j,m]	11 264 [a,j,m]	..	57 Recherche & développement expérimental
58 Other business activities n.e.c.	..	..	920 [j]	..	2 299 [j]	..	..	58 Autres activ. & serv. aux entreprises
59 Comm., soc. & pers. serv. activ., etc.	..	..	..	..	..	..	..	**59 Adm. pub., serv. soc. & collec., etc**
TOTAL BUSINESS ENTERPRISE SECTOR	**119 595** [j]	**132 103** [j]	**144 668** [j]	**157 539** [j]	**169 180** [j]	**182 823** [j]	**199 855** [j,p]	**TOTAL SECTEUR DES ENTREPRISES**

Source: OECD, R&D database (DSTI/EAS Division), January 2002.

Source : OCDE, base de données R-D (DSTI/Division AES), janvier 2002.

Total business enterprise intramural expenditure on R&D - Dépenses intra-muros totales de R-D des entreprises
(BERD) by industry - (DIRDE) par industrie

	1994	1995	1996	1997	1998	1999	2000	
	9.2 Million constant USD(1995 prices and PPPs) - Millions de USD constants (prix et PPA de 1995)							
1 AGRICULTURE, HUNTING & FORESTRY	.. [n]	.. [n]	43.2 [j]	6.7 [j]	.. [n]	..	..	**1 AGRICULTURE, CHASSE & SYLVICULTURE**
2 MINING	.. [n]	.. [n]	75.5 [j]	.. [n]	.. [n]	.. [n]	..	**2 ACTIVITES EXTRACTIVES**
3 MANUFACTURING	92 724.0 [j]	100 067.0 [j]	109 726.6 [j]	116 429.7 [j]	114 382.9 [j]	109 779.8 [a,j]	..	**3 ACTIVITES DE FABRICATION**
4 Food, beverages & tobacco	1 508.1 [j]	1 566.0 [j]	..	..	..	..	..	**4 Produits alimentaires, boissons & tabac**
5 Food products & beverages	..	..	..	1 739.4 [j]	1 641.6 [j]	..	..	5 Produits alimentaires & boissons
6 Tobacco products	..	..	..	..	..	..	..	6 Produits à base de tabac
7 Textiles, wear app., fur & leather	..	..	..	..	548.2 [j]	315.4 [a,j]	..	**7 Textiles, habill., fourrures & cuirs**
8 Textiles	..	..	..	..	244.2 [j]	..	..	8 Textiles
9 Wearing apparel & fur	..	..	..	..	279.3 [j]	..	..	9 Articles d'habillement & fourrures
10 Leather products & footwear	40.9 [j]	..	..	40.4 [j]	24.7 [j]	..	..	10 Articles en cuir & chaussures
11 Wood, paper, printing, publishing	..	..	..	..	..	..	..	**11 Bois, papier, imprimerie, édition**
12 Wood & cork (not furniture)	47.0 [j]	..	..	57.7 [j]	266.0 [j]	65.5 [a,j]	..	12 Bois & liège (sauf meubles)
13 Pulp, paper & paper products	..	..	..	..	..	..	..	13 Carton, papier & articles
14 Publ., print. & repro. of rec. media	324.9 [j]	..	..	353.1 [j]	315.4 [j]	..	..	14 Edition, impress. & repro. enregis.
15 Coke, petroleum, nuclear fuel, chemicals, rubber plastics	..	..	..	..	..	21 369.7 [a,j]	..	**15 Coke, pétrole, combust. nucl., prod. chimi., caout. & mat. plast.**
16 Coke, ref. petrol. prod. & nuclear fuel	1 992.4 [j]	1 760.0 [j]	1 630.2 [j]	..	1 717.6 [j]	575.6 [a,j]	..	16 Coke, raffin. pétr. & combust. nucl.
17 Chemicals & chemical products	..	17 547.0 [j]	..	..	20 833.9 [j]	19 067.2 [a,j]	..	17 Produits chimiques
18 *Chemical prod.(less pharmaceuticals)*	..	7 332.0 [j]	..	..	8 733.5 [j,v]	..	..	18 *Prod. chimi. (moins prod. pharmaceutiques)*
19 *Pharmaceuticals*	9 842.7 [j]	10 215.0 [j]	9 586.3 [j]	11 149.0 [j]	11 942.7 [j,v]	..	..	19 *Produits pharmaceutiques*
20 Rubber & plastic products	..	..	..	..	..	1 726.8 [a,j]	..	20 Caoutchouc et matières plastiques
21 Non-metallic mineral products	603.9 [j]	448.0 [j]	507.1 [j]	595.5 [j]	690.7 [j]	..	..	**21 Produits minéraux non métalliques**
22 Basic metals	705.0 [j]	593.0 [j]	..	950.5 [j]	..	439.9 [a,j]	..	**22 Produits métallurgiques de base**
23 Basic metals, ferrous	..	..	..	..	..	..	..	23 Produits métallur. de base, ferreux
24 Basic metals, non-ferrous	..	..	..	..	..	..	..	24 Prod. métallur. de base, non ferreux
25 Fabricated metal products	1 135.2 [j]	1 023.0 [j]	..	1 749.0 [j]	1 560.9 [j]	1 594.9 [a,j]	..	**25 Ouvrages en métaux (sauf mach.& mat.)**
26 Machinery equipment, instruments and transport equipment	64 346.6 [j]	..	80 837.4 [j]	86 634.6 [j]	..	..	..	**26 Machines & matériel, instruments & mat. de transport**
27 Machinery n.e.c.	4 091.1 [j]	..	..	5 444.1 [j]	5 713.4 [j]	5 921.8 [a,j]	..	27 Machines, n.c.a.
28 Office, account. & computing machinery	4 195.4 [j]	..	..	12 352.5 [j]	8 459.9 [j]	..	..	28 Mach. de bur., compta. & trait. inform.
29 Electrical machinery	..	..	..	4 722.6 [j]	4 570.5 [j]	..	..	29 Machines électriques
30 Electro. equip. (radio, TV & commun.)	..	..	..	..	..	..	..	30 Equip. électro. (radio, tv & communic.)
31 *Electro. comp. (inc. semi-conductors)*	6 163.3 [j]	..	..	..	..	10 133.6 [a,j]	..	31 *Compos. électro. (semi-cond. incl.)*
32 *TV, radio & communications equip.*	..	..	..	..	..	..	..	32 *Appareils radio, tv et de communi.*
33 Instruments	11 690.0 [j]	11 976.0 [j]	12 840.9 [j]	13 309.7 [j]	..	18 312.9 [a,j]	..	33 Instruments
34 Motor vehicles	..	..	..	..	..	17 103.6 [a,j]	..	34 Véhicules automobiles
35 Other transport equipment	..	..	..	..	..	14 774.0 [a,j]	..	35 Autres matériels de transport
36 *Ships*	..	..	..	..	..	..	..	36 *Navires*
37 *Aerospace*	14 570.4 [j]	16 951.0 [j]	15 914.0 [j]	15 677.2 [j]	13 726.8 [j]	13 501.1 [a,j]	..	37 *Aérospatiale*
38 *Other transport equipment n.e.c.*	..	..	..	..	..	1 272.0 [a,j]	..	38 *Autres matériels de transport n.c.a.*
39 Furniture, other manufacturing n.e.c.	..	..	..	..	..	..	..	**39 Meubles, autres activ. de fabricat. n.c.a.**
40 Furniture	..	..	..	302.1 [j]	..	234.9 [a,j]	..	40 Meubles
41 Other manufacturing n.e.c.	..	..	..	..	..	..	..	41 Autres activités de fabrication n.c.a.
42 Recycling	..	..	..	..	..	..	..	**42 Récupération**
43 ELECTRICITY, GAS & WATER SUPPLY	411.8 [j]	440.0 [j]	345.3 [j]	.. [n]	.. [n]	132.9 [a,j]	..	**43 DISTRIBUTION D'ELECTRICITE, GAZ & EAU**
44 CONSTRUCTION	.. [n]	.. [n]	277.6 [j]	251.1 [j]	.. [n]	654.2 [a,j]	..	**44 CONSTRUCTION**
45 SERVICES SECTOR	29 062.0 [j,o]	31 596.0 [j,o]	31 434.7 [a,j]	34 869.7 [a,j,o]	46 341.8 [a,j,o]	60 546.8 [a,j,o]	..	**45 SECTEUR DES SERVICES**
46 Wholesale, retail trade & repair of motor vehicles, etc.	..	..	6 242.4 [j]	..	..	18 681.6 [a,j]	..	**46 Commerce (gros & détail), réparation de véhicules, etc.**
47 Hotels & restaurants	..	..	266.8 [j]	149.1 [j]	..	..	..	**47 Hôtels et restaurants**
48 Transport & storage	..	..	218.7 [j]	..	..	436.2 [a,j,o]	..	**48 Transports, entreposage**
49 Communications	..	..	4 024.6 [j]	..	..	..	..	**49 Communications**
50 Post	..	..	..	..	..	.. [n]	..	50 Postes
51 Telecommunications	..	..	4 024.6 [j]	..	..	..	..	51 Télécommunications
52 Financial intermediation (inc. insur.)	..	..	1 272.2 [j]	..	..	..	..	**52 Intermédiation financ. (assurances inc.)**
53 Real estate, renting & busin. activity	..	..	..	..	..	..	..	**53 Immobil., locat. & serv. aux entreprises**
54 Computer & related activities	..	11 992.0 [j]	..	13 223.1 [j,m]	..	..	..	54 Informatique et activités connexes
55 *Software consultancy*	..	..	..	..	..	10 261.8 [a,j]	..	55 *Production de logiciels*
56 *Other computer services n.e.c.*	..	..	..	..	..	..	..	56 *Autres activités informatiques*
57 Research & development	3 846.9 [j]	..	5 379.2 [j,m]	..	8 710.7 [j,m]	10 542.6 [a,j,m]	..	57 Recherche & développement expérimental
58 Other business activities n.e.c.	..	..	902.4 [j]	..	2 184.1 [j]	..	..	58 Autres activ. & serv. aux entreprises
59 Comm., soc. & pers. serv. activ., etc.	..	..	..	..	..	..	..	**59 Adm. pub., serv. soc. & collec., etc**
TOTAL BUSINESS ENTERPRISE SECTOR	**122 197.8** [j]	**132 103.0** [j]	**141 903.8** [j]	**151 557.2** [j]	**160 723.7** [j]	**171 113.8** [j]	**182 831.9** [b,j,p]	**TOTAL SECTEUR DES ENTREPRISES**

Source: OECD, R&D database (DSTI/EAS Division), January 2002.

Source : OCDE, base de données R-D (DSTI/Division AES), janvier 2002.

UNITED STATES — Table 10 - Tableau 10 — ÉTATS-UNIS

Business enterprise intramural expenditure on R&D (BERD) - Dépenses intra-muros de R-D des entreprises (DIRDE)

by industry and source of funds - par industrie et par source de financement

Million USD — *Millions USD*

	1994	1995	1996	1997	1998	1999	2000	
				10.1 Business enterprise - Entreprises				
1 AGRICULTURE, HUNTING & FORESTRY	.. [n]	.. [n]	44 [j]	7 [j]	.. [n]	..	..	**1 AGRICULTURE, CHASSE & SYLVICULTURE**
2 MINING	.. [n]	.. [n]	68 [j]	97 [j]	73 [j]	2 352 [a,j]	..	**2 ACTIVITES EXTRACTIVES**
3 MANUFACTURING	73 375 [j]	81 236 [j]	91 845 [j]	101 202 [j]	102 211 [j]	100 236 [a,j]	..	**3 ACTIVITES DE FABRICATION**
4 Food, beverages & tobacco	1 476 [j]	1 566 [j]	..	..	..	..	..	**4 Produits alimentaires, boissons & tabac**
5 Food products & beverages	..	..	..	1 808 [j]	1 728 [j]	..	..	5 Produits alimentaires & boissons
6 Tobacco products	..	..	..	..	..	..	..	6 Produits à base de tabac
7 Textiles, wear app., fur & leather	356 [j]	422 [j]	457 [j]	..	..	337 [a,j]	..	**7 Textiles, habill., fourrures & cuirs**
8 Textiles	242 [j]	276 [j]	303 [j]	265 [j]	257 [j]	..	..	8 Textiles
9 Wearing apparel & fur	74 [j]	106 [j]	113 [j]	242 [j]	..	..	..	9 Articles d'habillement & fourrures
10 Leather products & footwear	40 [j]	40 [j]	41 [j]	..	26 [j]	..	..	10 Articles en cuir & chaussures
11 Wood, paper, printing, publishing	1 629 [j]	1 794 [j]	1 938 [j]	1 918 [j]	2 188 [j]	2 937 [a,j]	..	**11 Bois, papier, imprimerie, édition**
12 Wood & cork (not furniture)	48 [j]	84 [j]	55 [j]	60 [j]	275 [j]	70 [a,j]	..	12 Bois & liège (sauf meubles)
13 Pulp, paper & paper products	1 263 [j]	1 404 [j]	1 558 [j]	1 491 [j]	1 583 [j]	..	..	13 Carton, papier & articles
14 Publ., print. & repro. of rec. media	318 [j]	306 [j]	325 [j]	367 [j]	330 [j]	..	..	14 Edition, impress. & repro. enregis.
15 Coke, petroleum, nuclear fuel, chemicals, rubber plastics	19 930 [j]	20 301 [j]	20 670 [j]	21 750 [j]	24 944 [j]	..	..	**15 Coke, pétrole, combust. nucl., prod. chimi., caout. & mat. plast.**
16 Coke, ref. petrol. prod. & nuclear fuel	1 939 [j]	1 754 [j]	1 638 [j]	1 651 [j]	1 802 [j]	..	..	16 Coke, raffin. pétr. & combust. nucl.
17 Chemicals & chemical products	16 559 [j]	17 337 [j]	17 698 [j]	18 693 [j]	21 448 [j]	20 178 [a,j]	..	17 Produits chimiques
18 *Chemical prod.(less pharmaceuticals)*	6 934 [j]	7 135 [j]	7 751 [j]	7 042 [j]	8 716 [j]	7 942 [a,j]	..	18 *Prod. chimi. (moins prod. pharmaceutiques)*
19 *Pharmaceuticals*	9 625 [j]	10 202 [j]	9 769 [j]	11 586 [j]	12 566 [j]	12 236 [a,j]	..	19 *Produits pharmaceutiques*
20 Rubber & plastic products	1 432 [j]	1 210 [j]	1 334 [j]	1 406 [j]	1 694 [j]	1 845 [a,j]	..	20 Caoutchouc et matières plastiques
21 Non-metallic mineral products	553 [j]	441 [j]	512 [j]	617 [j]	625 [j]	611 [a,j]	..	**21 Produits minéraux non métalliques**
22 Basic metals	672 [j]	580 [j]	637 [j]	767 [j]	722 [j]	457 [a,j]	..	**22 Produits métallurgiques de base**
23 Basic metals, ferrous	241 [j]	217 [j]	215 [j]	414 [j]	405 [j]	..	..	23 Produits métallur. de base, ferreux
24 Basic metals, non-ferrous	431 [j]	363 [j]	422 [j]	353 [j]	315 [j]	..	..	24 Prod. métallur. de base, non ferreux
25 Fabricated metal products	868 [j]	937 [j]	1 460 [j]	1 689 [j]	1 599 [j]	1 658 [a,j]	..	**25 Ouvrages en métaux (sauf mach.& mat.)**
26 Machinery equipment, instruments and transport equipment	47 301 [j]	54 563 [j]	63 523 [j]	71 215 [j]	68 607 [j]	69 209 [a,j]	..	**26 Machines & matériel, instruments & mat. de transport**
27 Machinery n.e.c.	3 933 [j]	4 977 [j]	5 206 [j]	5 606 [j]	5 956 [j]	5 916 [a,j]	..	27 Machines, n.c.a.
28 Office, account. & computing machinery	4 078 [j]	4 699 [j]	8 132 [j]	12 787 [j]	8 890 [j]	4 126 [a,j]	..	28 Mach. de bur., compta. & trait. inform.
29 Electrical machinery	2 664 [j]	3 473 [j]	3 360 [j]	4 432 [j]	4 273 [j]	3 967 [a,j]	..	29 Machines électriques
30 Electro. equip. (radio, TV & commun.)	10 873 [j]	13 587 [j]	16 996 [j]	18 315 [j]	20 104 [j]	17 385 [a,j]	..	30 Equip. électro. (radio, tv & communic.)
31 *Electro. comp. (inc. semi-conductors)*	5 870 [j]	9 628 [j]	12 497 [j]	10 786 [j]	9 776 [j]	10 750 [a,j]	..	31 *Compos. électro. (semi-cond. incl.)*
32 *TV, radio & communications equip.*	5 003 [j]	3 959 [j]	4 499 [j]	7 529 [j]	10 328 [j]	6 635 [a,j]	..	32 *Appareils radio, tv et de communi.*
33 Instruments	8 058 [j]	8 516 [j]	8 698 [j]	9 315 [j]	9 987 [j]	13 830 [a,j]	..	33 Instruments
34 Motor vehicles	11 950 [j]	13 590 [j]	14 528 [j]	13 758 [j]	13 502 [j]	18 033 [a,j]	..	34 Véhicules automobiles
35 Other transport equipment	5 745 [j]	5 721 [j]	6 007 [j]	5 984 [j]	5 446 [j]	5 952 [a,j]	..	35 Autres matériels de transport
36 *Ships*	..	..	..	..	..	..	..	36 *Navires*
37 *Aerospace*	5 466 [j]	5 489 [j]	5 710 [j]	5 677 [j]	5 108 [j]	5 309 [a,j]	..	37 *Aérospatiale*
38 *Other transport equipment n.e.c.*	..	..	..	..	..	643 [a,j]	..	38 *Autres matériels de transport n.c.a.*
39 Furniture, other manufacturing n.e.c.	593	..	..	..	..	..	..	**39 Meubles, autres activ. de fabricat. n.c.a.**
40 Furniture	155 [j]	145 [j]	598 [j]	314 [j]	289 [j]	251 [a,j]	..	40 Meubles
41 Other manufacturing n.e.c.	438	..	..	..	..	..	..	41 Autres activités de fabrication n.c.a.
42 Recycling	..	..	..	..	..	..	..	**42 Récupération**
43 ELECTRICITY, GAS & WATER SUPPLY	321 [j]	347 [j]	311 [j]	258 [j]	182 [j]	126 [a,j]	..	**43 DISTRIBUTION D'ELECTRICITE, GAZ & EAU**
44 CONSTRUCTION	.. [n]	.. [n]	283 [j]	260 [j]	443 [j]	697 [a,j]	..	**44 CONSTRUCTION**
45 SERVICES SECTOR	23 435 [j,o]	27 068 [j,o]	28 464 [j]	31 787 [j]	42 107 [j,o]	56 877 [a,j]	..	**45 SECTEUR DES SERVICES**
46 Wholesale, retail trade & repair of motor vehicles, etc.	..	..	6 328 [j]	..	..	19 864 [a,j]	..	**46 Commerce (gros & détail), réparation de véhicules, etc.**
47 Hotels & restaurants	..	..	271 [j]	155 [j]	113 [j]	..	..	**47 Hôtels et restaurants**
48 Transport & storage	..	80 [j]	211 [j,m]	670 [j,m]	269 [j,m]	466 [a,j]	..	**48 Transports, entreposage**
49 Communications	4 177 [j]	4 756 [j]	3 970 [j]	1 884 [j]	1 768 [j]	1 393 [a,j]	..	**49 Communications**
50 Post	..	..	..	..	..	..	..	50 Postes
51 Telecommunications	4 177 [j]	4 756 [j]	3 970 [j]	1 884 [j]	1 768 [j]	1 393 [a,j]	..	51 Télécommunications
52 Financial intermediation (inc. insur.)	..	..	1 294 [j]	..	..	..	..	**52 Intermédiation financ. (assurances inc.)**
53 Real estate, renting & busin. activity	..	..	..	..	..	..	..	**53 Immobil., locat. & serv. aux entreprises**
54 Computer & related activities	6 459	9 595 [j]	10 692 [j,m]	12 099 [j,m]	15 132 [j,m]	15 665 [a,j,m]	..	54 Informatique et activités connexes
55 *Software consultancy*	..	..	..	..	..	10 915 [a,j]	..	55 *Production de logiciels*
56 *Other computer services n.e.c.*	..	..	..	..	..	4 750 [a,j]	..	56 *Autres activités informatiques*
57 Research & development	1 794 [j]	2 829 [j]	3 776 [j,m]	4 782 [j,m]	6 780 [j,m]	8 022 [a,j,m]	..	57 Recherche & développement expérimental
58 Other business activities n.e.c.	..	132 [j]	129 [j]	346 [j]	567 [j]	..	..	58 Autres activ. & serv. aux entreprises
59 Comm., soc. & pers. serv. activ., etc.	..	..	..	..	..	..	..	**59 Adm. pub., serv. soc. & collec., etc**
TOTAL BUSINESS ENTERPRISE SECTOR	**97 131** [j]	**108 652** [j]	**121 015** [j]	**133 611** [j]	**145 016** [j]	**160 288** [j]	**177 645** [j,p]	**TOTAL SECTEUR DES ENTREPRISES**

Source: OECD, R&D database (DSTI/EAS Division), January 2002.

Source : OCDE, base de données R-D (DSTI/Division AES), janvier 2002.

UNITED STATES — Table 10 - Tableau 10 — ÉTATS-UNIS

Business enterprise intramural expenditure on R&D (BERD) - Dépenses intra-muros de R-D des entreprises (DIRDE)

by industry and source of funds - par industrie et par source de financement

Million USD — Millions USD

10.2 Government - État

	1994	1995	1996	1997	1998	1999	2000	
1 **AGRICULTURE, HUNTING & FORESTRY**	.. [n]	.. [n]	0 [j]	0 [j]	0 [j]	..	..	1 **AGRICULTURE, CHASSE & SYLVICULTURE**
2 **MINING**	.. [n]	.. [n]	8 [j]	.. [n]	.. [n]	..	..	2 **ACTIVITES EXTRACTIVES**
3 **MANUFACTURING**	17 373 [j]	18 831 [j]	20 020 [j]	19 823 [j]	18 189 [j]	17 055 [a,j]	..	3 **ACTIVITES DE FABRICATION**
4 **Food, beverages & tobacco**	0 [j]	0 [j]	1 [j]	0 [j]	0 [j]	0 [a,j]	..	4 **Produits alimentaires, boissons & tabac**
5 Food products & beverages	0 [j]	0 [j]	1 [j]	0 [j]	0 [j]	..	..	5 Produits alimentaires & boissons
6 Tobacco products	0 [j]	0 [j]	0 [j]	0 [j]	0 [j]	..	..	6 Produits à base de tabac
7 **Textiles, wear app., fur & leather**	..	..	..	..	0 [j]	0 [a,j]	..	7 **Textiles, habill., fourrures & cuirs**
8 Textiles	..	..	..	..	0 [j]	..	..	8 Textiles
9 Wearing apparel & fur	..	..	..	..	..	..	..	9 Articles d'habillement & fourrures
10 Leather products & footwear	0 [j]	..	..	..	0 [j]	..	..	10 Articles en cuir & chaussures
11 **Wood, paper, printing, publishing**	..	..	..	..	..	..	..	11 **Bois, papier, imprimerie, édition**
12 Wood & cork (not furniture)	0 [j]	..	..	0 [j]	5 [j]	0 [a,j]	..	12 Bois & liège (sauf meubles)
13 Pulp, paper & paper products	..	..	..	..	..	..	..	13 Carton, papier & articles
14 Publ., print. & repro. of rec. media	0 [j]	..	..	0 [j]	2 [j]	..	..	14 Edition, impress. & repro. enregis.
15 **Coke, petroleum, nuclear fuel, chemicals, rubber plastics**	..	..	..	..	..	..	..	15 **Coke, pétrole, combust. nucl., prod. chimi., caout. & mat. plast.**
16 Coke, ref. petrol. prod. & nuclear fuel	10 [j]	6 [j]	24 [j]	..	6 [j]	..	..	16 Coke, raffin. pétr. & combust. nucl.
17 Chemicals & chemical products	..	..	..	..	483	194 [a,j]	..	17 Produits chimiques
18 *Chemical prod.(less pharmaceuticals)*	..	196 [j]	..	..	478 [j]	..	..	18 *Prod. chimi. (moins prod. pharmaceutiques)*
19 *Pharmaceuticals*	8 [j]	14 [j]	3 [j]	..	..	..	..	19 *Produits pharmaceutiques*
20 Rubber & plastic products	..	..	..	..	..	0 [a,j]	..	20 Caoutchouc et matières plastiques
21 **Non-metallic mineral products**	38 [j]	6 [j]	5 [j]	2 [j]	101 [j]	..	..	21 **Produits minéraux non métalliques**
22 **Basic metals**	17 [j]	13 [j]	..	221 [j]	..	12 [a,j]	..	22 **Produits métallurgiques de base**
23 Basic metals, ferrous	..	..	..	..	..	..	..	23 Produits métallur. de base, ferreux
24 Basic metals, non-ferrous	..	..	..	..	..	..	..	24 Prod. métallur. de base, non ferreux
25 **Fabricated metal products**	243 [j]	86 [j]	..	129 [j]	43 [j]	46 [a,j]	..	25 **Ouvrages en métaux (sauf mach.& mat.)**
26 **Machinery equipment, instruments and transport equipment**	15 676 [l]	..	18 889 [l]	18 840 [l]	..	..	..	26 **Machines & matériel, instruments & mat. de transport**
27 Machinery n.e.c.	71 [j]	..	..	53 [j]	58 [j]	411 [a,j]	..	27 Machines, n.c.a.
28 Office, account. & computing machinery	28 [j]	..	..	53 [j]	15 [j]	..	..	28 Mach. de bur., compta. & trait. inform.
29 Electrical machinery	..	..	..	477 [j]	538 [j]	..	..	29 Machines électriques
30 Electro. equip. (radio, TV & commun.)	..	..	..	..	..	..	..	30 Equip. électro. (radio, tv & communic.)
31 *Electro. comp. (inc. semi-conductors)*	162 [j]	..	..	..	..	77 [a,j]	..	31 *Compos. électro. (semi-cond. incl.)*
32 *TV, radio & communications equip.*	..	..	..	..	..	..	..	32 *Appareils radio, tv et de communi.*
33 Instruments	3 384 [j]	3 460 [j]	4 393 [j]	4 520 [j]	..	5 736 [a,j]	..	33 Instruments
34 Motor vehicles	..	..	..	..	..	241 [a,j]	..	34 Véhicules automobiles
35 Other transport equipment	..	..	..	..	..	9 833 [a,j]	..	35 Autres matériels de transport
36 *Ships*	..	..	..	..	..	..	..	36 *Navires*
37 *Aerospace*	8 794 [j]	11 462 [j]	10 515 [j]	10 619 [j]	9 341 [j]	9 117 [a,j]	..	37 *Aérospatiale*
38 *Other transport equipment n.e.c.*	..	..	..	..	..	716 [a,j]	..	38 *Autres matériels de transport n.c.a.*
39 **Furniture, other manufacturing n.e.c.**	..	..	..	..	..	..	..	39 **Meubles, autres activ. de fabricat. n.c.a.**
40 Furniture	..	..	..	0 [j]	..	0 [a,j]	..	40 Meubles
41 Other manufacturing n.e.c.	..	..	..	..	..	..	..	41 Autres activités de fabrication n.c.a.
42 **Recycling**	..	..	..	..	..	..	..	42 **Récupération**
43 **ELECTRICITY, GAS & WATER SUPPLY**	82 [j]	93 [j]	42 [j]	.. [n]	.. [n]	17 [a,j]	..	43 **DISTRIBUTION D'ELECTRICITE, GAZ & EAU**
44 **CONSTRUCTION**	.. [n]	.. [n]	0 [j]	1 [j]	.. [n]	2 [a,j]	..	44 **CONSTRUCTION**
45 **SERVICES SECTOR**	5 008 [j,o]	4 527 [j,o]	3 583 [j,o]	4 104 [j,o]	5 975 [j,o]	5 460 [a,j]	..	45 **SECTEUR DES SERVICES**
46 **Wholesale, retail trade & repair of motor vehicles, etc.**	..	..	37 [l]	..	..	96 [a,j]	..	46 **Commerce (gros & détail), réparation de véhicules, etc.**
47 **Hotels & restaurants**	..	..	0 [j]	0 [j]	..	..	..	47 **Hôtels et restaurants**
48 **Transport & storage**	..	..	12 [j]	..	0 [j]	0 [a,j]	..	48 **Transports, entreposage**
49 **Communications**	..	..	133 [j]	..	..	..	..	49 **Communications**
50 Post	..	..	..	..	..	..	..	50 Postes
51 Telecommunications	..	..	133 [j]	..	..	..	..	51 Télécommunications
52 **Financial intermediation (inc. insur.)**	..	..	2 [j]	..	..	..	..	52 **Intermédiation financ. (assurances inc.)**
53 **Real estate, renting & busin. activity**	..	..	..	..	..	..	..	53 **Immobil., locat. & serv. aux entreprises**
54 Computer & related activities	..	2 397 [j]	..	1 647 [j,m]	..	..	..	54 Informatique et activités connexes
55 *Software consultancy*	..	..	..	..	..	49 [a,j]	..	55 *Production de logiciels*
56 *Other computer services n.e.c.*	..	..	..	..	..	..	..	56 *Autres activités informatiques*
57 Research & development	1 470 [j]	..	1 708 [j,m]	..	2 388 [j,m]	3 242 [a,j,n]	..	57 Recherche & développement expérimental
58 Other business activities n.e.c.	..	..	44 [j]	..	..	..	..	58 Autres activ. & serv. aux entreprises
59 **Comm., soc. & pers. serv. activ., etc.**	..	..	..	..	..	..	..	59 **Adm. pub., serv. soc. & collec., etc**
TOTAL BUSINESS ENTERPRISE SECTOR	**22 464** [j]	**23 451** [j]	**23 653** [j]	**23 928** [j]	**24 164** [j]	**22 535** [j]	**22 210** [j]	**TOTAL SECTEUR DES ENTREPRISES**

Source: OECD, R&D database (DSTI/EAS Division), January 2002.

Source : OCDE, base de données R-D (DSTI/Division AES), janvier 2002.

RSE in the business enterprise sector - Chercheurs dans le secteur des entreprises

by industry - par industrie

In full time equivalent — *En équivalence plein-temps*

	1994	1995	1996	1997	1998	1999	2000	
1 AGRICULTURE, HUNTING & FORESTRY	..	..	400	100	..	..	..	**1 AGRICULTURE, CHASSE & SYLVICULTURE**
2 MINING	..	..	600	700	600	5 600 [a]	..	**2 ACTIVITES EXTRACTIVES**
3 MANUFACTURING	540 700	606 200	639 900	682 100	659 000	599 900 [a]	..	**3 ACTIVITES DE FABRICATION**
4 Food, beverages & tobacco	9 000	..	..	..	..	10 000 [a]	..	**4 Produits alimentaires, boissons & tabac**
5 Food products & beverages	..	..	..	11 800	10 700	..	..	5 Produits alimentaires & boissons
6 Tobacco products	..	..	..	..	..	..	..	6 Produits à base de tabac
7 Textiles, wear app., fur & leather	3 400	4 100	4 100	4 300	3 200	11 100 [a]	..	**7 Textiles, habill., fourrures & cuirs**
8 Textiles	2 200	2 200	2 600	2 400	1 900	..	..	8 Textiles
9 Wearing apparel & fur	900	1 600	1 200	1 700	1 100	..	..	9 Articles d'habillement & fourrures
10 Leather products & footwear	300	300	300	200	200	..	..	10 Articles en cuir & chaussures
11 Wood, paper, printing, publishing	14 600	15 000	15 510	..	18 300	17 400 [a]	..	**11 Bois, papier, imprimerie, édition**
12 Wood & cork (not furniture)	500	700	610	..	2 800	700 [a]	..	12 Bois & liège (sauf meubles)
13 Pulp, paper & paper products	10 600	11 000	11 500	11 100	12 500	..	..	13 Carton, papier & articles
14 Publ., print. & repro. of rec. media	3 500	3 300	3 400	2 400	3 000	..	..	14 Edition, impress. & repro. enregis.
15 Coke, petroleum, nuclear fuel, chemicals, rubber plastics	109 300	112 900	112 400	111 200	115 500	101 900 [a]	..	**15 Coke, pétrole, combust. nucl., prod. chimi., caout. & mat. plast.**
16 Coke, ref. petrol. prod. & nuclear fuel	9 000	8 400	9 100	8 000	8 000	3 000 [a]	..	16 Coke, raffin. pétr. & combust. nucl.
17 Chemicals & chemical products	91 500	94 300	91 800	91 200	93 500	84 900 [a]	..	17 Produits chimiques
18 *Chemical prod.(less pharmaceuticals)*	49 300 [v]	45 300 [v]	45 900 [v]	44 200 [v]	40 700 [v]	43 600 [a]	..	18 *Prod. chimi. (moins prod. pharmaceutiques)*
19 *Pharmaceuticals*	50 300 [v]	49 000 [v]	43 100 [v]	45 300 [v]	50 000 [v]	41 300 [a]	..	19 *Produits pharmaceutiques*
20 Rubber & plastic products	8 800	10 200	11 500	12 000	14 000	14 000 [a]	..	20 Caoutchouc et matières plastiques
21 Non-metallic mineral products	4 300	3 900	4 430	4 000	4 700	3 800 [a]	..	**21 Produits minéraux non métalliques**
22 Basic metals	6 700	4 100	5 500	6 400	6 100	5 000 [a]	..	**22 Produits métallurgiques de base**
23 Basic metals, ferrous	1 900 [v]	1 600 [v]	1 600 [v]	3 900 [v]	3 900 [v]	..	..	23 Produits métallur. de base, ferreux
24 Basic metals, non-ferrous	4 600 [v]	2 500 [v]	3 900 [v]	2 500 [v]	2 200 [v]	..	..	24 Prod. métallur. de base, non ferreux
25 Fabricated metal products	9 500	9 100	11 000	13 200	13 200	10 500 [a]	..	**25 Ouvrages en métaux (sauf mach.& mat.)**
26 Machinery equipment, instruments and transport equipment	378 200	440 600	468 500	509 000	478 600	432 600 [a]	..	**26 Machines & matériel, instruments & mat. de transport**
27 Machinery n.e.c.	36 800 [v]	53 600 [v]	43 400 [v]	46 000 [v]	50 500 [v]	56 000 [a]	..	27 Machines, n.c.a.
28 Office, account. & computing machinery	31 800 [v]	33 700 [v]	45 300 [v]	66 100 [v]	45 600 [v]	21 300 [a]	..	28 Mach. de bur., compta. & trait. inform.
29 Electrical machinery	20 000 [v]	25 300 [v]	21 400 [v]	34 600 [v]	26 900 [v]	25 500 [a]	..	29 Machines électriques
30 Electro. equip. (radio, TV & commun.)	83 400 [v]	100 600 [v]	114 500 [v]	137 000 [v]	146 900 [v]	105 200 [a]	..	30 Equip. électro. (radio, tv & communic.)
31 *Electro. comp. (inc. semi-conductors)*	40 800 [v]	64 600 [v]	75 300 [v]	66 800 [v]	53 100 [v]	53 800 [a]	..	31 *Compos. électro. (semi-cond. incl.)*
32 *TV, radio & communications equip.*	42 600 [v]	36 000 [v]	39 200 [v]	70 200 [v]	93 800 [v]	51 400 [a]	..	32 *Appareils radio, tv et de communi.*
33 Instruments	90 600	70 900	72 900	71 600	64 500	85 200 [a]	..	33 Instruments
34 Motor vehicles	51 100 [v]	57 000 [v]	63 800 [v]	64 100 [v]	62 800 [v]	76 000 [a]	..	34 Véhicules automobiles
35 Other transport equipment	69 700 [v]	99 600 [v]	98 000 [v]	80 000 [v]	70 200 [v]	63 400 [a]	..	35 Autres matériels de transport
36 *Ships*	..	..	..	..	..	..	..	36 *Navires*
37 *Aerospace*	63 500 [v]	95 500 [v]	94 600 [v]	77 000 [v]	66 400 [v]	55 300 [a]	..	37 *Aérospatiale*
38 *Other transport equipment n.e.c.*	..	..	..	..	..	8 100 [a]	..	38 *Autres matériels de transport n.c.a.*
39 Furniture, other manufacturing n.e.c.	..	..	..	..	..	..	..	**39 Meubles, autres activ. de fabricat. n.c.a.**
40 Furniture	1 800	2 200	3 000	3 200	2 800	2 700 [a]	..	40 Meubles
41 Other manufacturing n.e.c.	..	..	..	..	..	..	..	41 Autres activités de fabrication n.c.a.
42 Recycling	..	..	..	..	..	..	..	**42 Récupération**
43 ELECTRICITY, GAS & WATER SUPPLY	1 900	3 200	4 000	1 100	1 000	700 [a]	..	**43 DISTRIBUTION D'ELECTRICITE, GAZ & EAU**
44 CONSTRUCTION	.. [n]	.. [n]	3 300	2 600	8 800	8 300 [a]	..	**44 CONSTRUCTION**
45 SERVICES SECTOR	206 600 [v]	223 400 [v]	237 500 [v]	265 000 [v]	328 300 [v]	419 300 [a]	..	**45 SECTEUR DES SERVICES**
46 Wholesale, retail trade & repair of motor vehicles, etc.	..	..	..	47 100	68 300	125 200 [a]	..	**46 Commerce (gros & détail), réparation de véhicules, etc.**
47 Hotels & restaurants	..	..	900	1 100	500	..	..	**47 Hôtels et restaurants**
48 Transport & storage	..	..	..	..	..	4 800 [a]	..	**48 Transports, entreposage**
49 Communications	30 100	25 400	24 700	8 400	14 100	15 700 [a]	..	**49 Communications**
50 Post	..	..	..	..	..	..	..	50 Postes
51 Telecommunications	30 100	25 400	24 700	8 400	14 100	15 700 [a]	..	51 Télécommunications
52 Financial intermediation (inc. insur.)	..	..	..	..	..	..	..	**52 Intermédiation financ. (assurances inc.)**
53 Real estate, renting & busin. activity	104 500	..	..	..	..	..	..	**53 Immobil., locat. & serv. aux entreprises**
54 Computer & related activities	76 500 [v]	100 100 [v]	67 100 [v]	81 700 [v]	106 300 [v]	122 600 [a]	..	54 Informatique et activités connexes
55 *Software consultancy*	..	..	..	..	..	76 500 [a]	..	55 *Production de logiciels*
56 *Other computer services n.e.c.*	..	..	..	..	..	46 100 [a]	..	56 *Autres activités informatiques*
57 Research & development	28 000 [v]	28 800 [v]	24 900 [v]	28 300 [v]	34 800 [v]	51 900 [a]	..	57 Recherche & développement expérimental
58 Other business activities n.e.c.	.. [v]	1 300 [v]	7 800 [v]	11 000 [v]	23 300 [v]	..	..	58 Autres activ. & serv. aux entreprises
59 Comm., soc. & pers. serv. activ., etc.	1 700	..	..	..	..	..	..	**59 Adm. pub., serv. soc. & collec., etc**
TOTAL BUSINESS ENTERPRISE SECTOR	**749 200**	**832 800**	**885 700**	**951 500**	**998 000**	**1 033 700** [a]	**..**	**TOTAL SECTEUR DES ENTREPRISES**

Source: OECD, R&D database (DSTI/EAS Division), January 2002.

Source : OCDE, base de données R-D (DSTI/Division AES), janvier 2002.

Table 14 - Tableau 14

Higher education intramural expenditure on R&D - Dépenses intra-muros de R-D de l'enseignement supérieur
(HERD) by field of science - (DIRDES) par discipline scientifique

	1994	1995	1996	1997	1998	1999	2000	
	14.1 Million USD - Millions USD							
Natural sciences	9 053 [i,v]	9 374 [i,v]	9 550 [i,v]	10 103 [i,v]	10 732 [i,v]	11 483 [i,v]	..	Sciences exactes et naturelles
Engineering	3 355 [i,v]	3 515 [i,v]	3 707 [i,v]	3 847 [i,v]	4 069 [i,v]	4 257 [i,v]	..	Sciences de l'ingénieur
Medical sciences	5 639 [i,v]	6 070 [i,v]	6 391 [i,v]	6 908 [i,v]	7 466 [i,v]	7 991 [i,v]	..	Sciences médicales
Agricultural sciences	1 661 [i,v]	1 814 [i,v]	1 910 [i,v]	1 966 [i,v]	1 995 [i,v]	2 031 [i,v]	..	Sciences agricoles
Sub-total NSE	19 708 [i,v]	20 773 [i,v]	21 558 [i,v]	22 824 [i,v]	24 262 [i,v]	25 762 [i,v]	..	Sous-total SENI
Social sciences	1 312 [i,v]	1 389 [i,v]	1 477 [i,v]	1 515 [i,v]	1 575 [i,v]	1 727 [i,v]	..	Sciences sociales
Humanities	..	..	..	..	..	..	..	Sciences humaines
Sub-total SSH	..	..	..	..	..	..	..	Sous-total SSH
Total NSE+SSH	26 908 [i]	27 973 [i]	29 097 [i]	30 576 [i]	32 253 [i]	34 060 [i]	35 955 [i,p]	*Total SENI+SSH*
	14.2 Million constant USD (1995 prices and PPPs) - Millions de USD constants (prix et PPA de 1995)							
Natural sciences	9 250.4 [i,v]	9 373.6 [i,v]	9 367.8 [i,v]	9 719.0 [i,v]	10 195.3 [i,v]	10 747.4 [i,v]	..	Sciences exactes et naturelles
Engineering	3 427.9 [i,v]	3 514.9 [i,v]	3 636.1 [i,v]	3 701.2 [i,v]	3 865.3 [i,v]	3 984.5 [i,v]	..	Sciences de l'ingénieur
Medical sciences	5 761.3 [i,v]	6 070.0 [i,v]	6 268.6 [i,v]	6 645.4 [i,v]	7 092.7 [i,v]	7 479.1 [i,v]	..	Sciences médicales
Agricultural sciences	1 697.5 [i,v]	1 814.3 [i,v]	1 873.6 [i,v]	1 891.4 [i,v]	1 895.7 [i,v]	1 901.3 [i,v]	..	Sciences agricoles
Sub-total NSE	20 137.0 [i,v]	20 772.7 [i,v]	21 146.1 [i,v]	21 957.0 [i,v]	23 049.0 [i,v]	24 112.2 [i,v]	..	Sous-total SENI
Social sciences	1 340.1 [i,v]	1 388.8 [i,v]	1 449.1 [i,v]	1 457.1 [i,v]	1 496.7 [i,v]	1 616.3 [i,v]	..	Sciences sociales
Humanities	..	..	..	..	..	..	..	Sciences humaines
Sub-total SSH	..	..	..	..	..	..	..	Sous-total SSH
Total NSE+SSH	27 493.6 [i]	27 973.0 [i]	28 541.0 [i]	29 415.0 [i]	30 640.9 [i]	31 878.6 [i]	32 892.5 [b,i,p]	*Total SENI+SSH*

Table 15 - Tableau 15

Higher education intramural expenditure on R&D - Dépenses intra-muros de R-D de l'enseignement supérieur
(HERD) by field of science and type of costs - (DIRDES) par discipline scientifique et par type de coûts

Million USD | *Millions USD*

	1994	1995	1996	1997	1998	1999	2000	
Current expenditure								**Dépenses courantes**
Natural sciences	..	..	..	..	..	..	..	Sciences exactes et naturelles
Engineering	..	..	..	..	..	..	..	Sciences de l'ingénieur
Medical sciences	..	..	..	..	..	..	..	Sciences médicales
Agricultural sciences	..	..	..	..	..	..	..	Sciences agricoles
Sub-total NSE	..	..	..	..	..	..	..	Sous-total SENI
Social sciences	..	..	..	..	..	..	..	Sciences sociales
Humanities	..	..	..	..	..	..	..	Sciences humaines
Sub-total SSH	..	..	..	..	..	..	..	Sous-total SSH
Total NSE+SSH	26 908.0	27 973.0	29 097.0	30 576.0	32 253.0	34 060.0	35 955.0 [p]	*Total SENI+SSH*
Capital expenditure								**Dépenses en capital**
Natural sciences	..	..	..	..	..	..	..	Sciences exactes et naturelles
Engineering	..	..	..	..	..	..	..	Sciences de l'ingénieur
Medical sciences	..	..	..	..	..	..	..	Sciences médicales
Agricultural sciences	..	..	..	..	..	..	..	Sciences agricoles
Sub-total NSE	..	..	..	..	..	..	..	Sous-total SENI
Social sciences	..	..	..	..	..	..	..	Sciences sociales
Humanities	..	..	..	..	..	..	..	Sciences humaines
Sub-total SSH	..	..	..	..	..	..	..	Sous-total SSH
Total NSE+SSH	..	..	..	..	..	..	..	*Total SENI+SSH*

Source: OECD, R&D database (DSTI/EAS Division), January 2002.

Source : OCDE, base de données R-D (DSTI/Division AES), janvier 2002.

Table 17 - Tableau 17

Higher education R&D personnel (HEMP) - Personnel de R-D du secteur de l'enseignement supérieur

by field of science and occupation - par discipline scientifique et par profession

In full-time equivalent / *En équivalence plein-temps*

	1994	1995	1996	1997	1998	1999	2000	
Total R&D personnel								**Personnel total de R-D**
Natural sciences	..	..	..	..	..	..	..	Sciences exactes et naturelles
Engineering	..	..	..	..	..	..	..	Sciences de l'ingénieur
Medical sciences	..	..	..	..	..	..	..	Sciences médicales
Agricultural sciences	..	..	..	..	..	..	..	Sciences agricoles
Sub-total NSE	..	..	..	..	..	..	..	Sous-total SENI
Social sciences	..	..	..	..	..	..	..	Sciences sociales
Humanities	..	..	..	..	..	..	..	Sciences humaines
Sub-total SSH	..	..	..	..	..	..	..	Sous-total SSH
Total NSE+SSH	..	..	..	..	..	..	..	*Total SENI+SSH*
Researchers								**Chercheurs**
Natural sciences	..	..	..	..	..	..	..	Sciences exactes et naturelles
Engineering	..	..	..	..	..	..	..	Sciences de l'ingénieur
Medical sciences	..	..	..	..	..	..	..	Sciences médicales
Agricultural sciences	..	..	..	..	..	..	..	Sciences agricoles
Sub-total NSE	..	..	..	..	..	..	..	Sous-total SENI
Social sciences	..	..	..	..	..	..	..	Sciences sociales
Humanities	..	..	..	..	..	..	..	Sciences humaines
Sub-total SSH	..	..	..	..	..	..	..	Sous-total SSH
Total NSE+SSH	..	134 300	..	135 800	..	..	..	*Total SENI+SSH*
Technicians								**Technicians**
Natural sciences	..	..	..	..	..	..	..	Sciences exactes et naturelles
Engineering	..	..	..	..	..	..	..	Sciences de l'ingénieur
Medical sciences	..	..	..	..	..	..	..	Sciences médicales
Agricultural sciences	..	..	..	..	..	..	..	Sciences agricoles
Sub-total NSE	..	..	..	..	..	..	..	Sous-total SENI
Social sciences	..	..	..	..	..	..	..	Sciences sociales
Humanities	..	..	..	..	..	..	..	Sciences humaines
Sub-total SSH	..	..	..	..	..	..	..	Sous-total SSH
Total NSE+SSH	..	..	..	..	..	..	..	*Total SENI+SSH*
Other								**Autres**
Natural sciences	..	..	..	..	..	..	..	Sciences exactes et naturelles
Engineering	..	..	..	..	..	..	..	Sciences de l'ingénieur
Medical sciences	..	..	..	..	..	..	..	Sciences médicales
Agricultural sciences	..	..	..	..	..	..	..	Sciences agricoles
Sub-total NSE	..	..	..	..	..	..	..	Sous-total SENI
Social sciences	..	..	..	..	..	..	..	Sciences sociales
Humanities	..	..	..	..	..	..	..	Sciences humaines
Sub-total SSH	..	..	..	..	..	..	..	Sous-total SSH
Total NSE+SSH	..	..	..	..	..	..	..	*Total SENI+SSH*

Table 18 - Tableau 18

Government budget appropriations or outlays for R&D (GBAORD) - Crédits budgétaires publics de R-D (CBPRD)

by socio-economic objective - par objectif socio-économique

Million USD / *Millions USD*

	1995	1996	1997	1998	1999	2000	2001	
Agriculture, forestry and fishing	1 731.0 [h,j]	1 681.0 [h,j]	1 706.2 [h,j]	1 546.5 [h,j]	1 648.3 [h,j]	1 744.9 [h,j,p]	1 829.8 [h,j,p]	Agriculture, sylviculture et pêche
Industrial development	525.0 [h,j]	432.0 [h,j]	408.5 [h,j]	397.9 [h,j]	432.0 [h,j]	373.1 [h,j,p]	516.4 [h,j,p]	Développement industriel
Energy	2 844.0 [h,j]	2 521.0 [h,j]	2 371.8 [h,j]	947.9 [a,h,j]	1 131.5 [h,j]	1 112.3 [h,j,p]	1 183.8 [h,j,p]	Energie
Transport & telecommunications	1 833.0 [h,j]	1 795.0 [h,j]	1 785.4 [h,j]	1 833.4 [h,j]	1 725.1 [h,j]	1 569.8 [h,j,p]	1 664.5 [h,j,p]	*Transports & télécommunications*
Urban & rural planning	70.0 [h,j]	50.0 [h,j]	48.0 [h,j]	42.0 [h,j]	59.0 [h,j]	46.0 [h,j,p]	63.0 [h,j,p]	*Aménagement urbain et rural*
ST Infrastructure	1 903.0 [h,j]	1 845.0 [h,j]	1 833.4 [h,j]	1 875.4 [h,j]	1 784.1 [h,j]	1 615.8 [h,j,p]	1 727.5 [h,j,p]	ST Infrastructure
Prevention of pollution	..	..	..	..	..	..	..	*Prévention de la pollution*
Identificat. & treatment of pollution	..	..	..	..	..	..	..	*Détect. pollut. & mesures antipollut.*
ST Environment	549.0 [h,j]	482.0 [h,j]	570.0 [h,j]	571.1 [h,j]	552.8 [h,j]	536.8 [h,j,p]	530.6 [h,j,p]	ST Environnement
Health	11 664.0 [h,j]	12 126.0 [h,j]	12 937.4 [h,j]	14 163.1 [h,j]	16 197.3 [h,j]	18 417.2 [h,j,p]	19 513.4 [h,j,p]	Santé
Social development & services	759.0 [h,j]	657.0 [h,j]	633.0 [h,j]	699.0 [h,j]	722.0 [h,j]	726.0 [h,j,p]	733.0 [h,j,p]	Développement et services sociaux
Earth and atmosphere	902.0 [h,j]	814.0 [h,j]	813.1 [h,j]	986.3 [h,j]	928.7 [h,j]	1 029.0 [h,j,p]	1 021.8 [h,j,p]	Milieu terrestre et atmosphère
Advancement of research	2 794.0 [h,j]	2 846.0 [h,j]	2 944.3 [h,j]	4 360.4 [a,h,j]	4 689.7 [h,j]	4 885.4 [h,j,p]	5 529.1 [h,j,p]	*Promotion de la recherche*
General university funds	..	..	..	..	..	..	..	*Fonds généraux des universités*
ST Advancement of knowledge	2 794.0 [h,i,j]	2 846.0 [h,i,j]	2 944.3 [h,i,j]	4 360.4 [a,h,i,j]	4 689.7 [h,j]	4 885.4 [h,j,p]	5 529.1 [h,j,p]	ST promotion gén. des connaiss.
Civil space	7 916.0 [h,j]	7 844.0 [h,j]	7 844.1 [h,j]	8 198.0 [h,j]	8 245.1 [h,j]	8 447.3 [h,j,p]	8 731.5 [h,j,p]	Espace civil
Defence	37 204.0 [h,j]	37 801.0 [h,j]	39 591.5 [h,j]	39 823.2 [h,j]	41 305.6 [h,j]	41 845.5 [h,j,p]	41 413.6 [h,j,p]	Défense
Not elsewhere classified	..	..	..	..	..	..	..	Non classé
Total	**68 791.0** [h,i,j]	**69 049.0** [h,i,j]	**71 653.3** [h,i,j]	**73 568.7** [h,i,j]	**77 637.1** [h,i,j]	**80 733.4** [h,i,j,p]	**82 730.5** [h,i,j,p]	**Total**

Source: OECD, R&D database (DSTI/EAS Division), January 2002.

Source : OCDE, base de données R-D (DSTI/Division AES), janvier 2002.

Table 19 - Tableau 19

Technology balance of payments - Balance des paiements technologiques

by type of transfer - selon la nature des transferts

Million USD — *Millions USD*

	1983	1984	1985	1986	1987	1988	1989	1990	1991	
Receipts										**Recettes**
Related firms	..	..	..	6 174	7 888	9 493	10 961	13 250	14 106	Firmes affiliées
Unrelated firms	..	..	..	1 939	2 286	2 646	2 857	3 384	3 712	Firmes non affiliées
Other	..	..	..	..	..	..	..	..	..	Autre
Total	5 778	6 177	6 678	8 113	10 174	12 139	13 818	16 634	17 819	*Total*
Payments										**Paiements**
Related firms	..	..	..	916	1 309	1 426	1 704	2 206	2 955	Firmes affiliées
Unrelated firms	..	..	..	484	547	1 175	824	929	1 080	Firmes non affiliées
Other	..	..	..	..	..	..	..	..	..	Autre
Total	943	1 168	1 170	1 401	1 857	2 601	2 528	3 135	4 035	*Total*

	1992	1993	1994	1995	1996	1997	1998	1999	2000	
Receipts										**Recettes**
Related firms	15 659	15 688	20 275	22 859	24 556	24 876 [s]	26 809 [s]	26 307	..	Firmes affiliées
Unrelated firms	5 183	6 007	6 437	7 430	7 914	8 763 [s]	9 388 [s]	10 160	..	Firmes non affiliées
Other	..	..	..	..	..	..	..	..	..	Autre
Total	20 841	21 695	26 712	30 289	32 470	33 639	36 197	36 467	..	*Total*
Payments										**Paiements**
Related firms	3 396	3 386	3 933	5 256	5 406	7 202 [s]	8 754 [s]	10 208	..	Firmes affiliées
Unrelated firms	1 766	1 646	1 919	1 663	2 431	2 412 [s]	2 959 [s]	3 067	..	Firmes non affiliées
Other	..	..	..	..	..	..	..	..	..	Autre
Total	5 161	5 032	5 852	6 919	7 837	9 614	11 713	13 275	..	*Total*

Source: OECD, TBPdatabase (DSTI/EAS Division), Januay 2002.

Source : OCDE, base de données TBP (DSTI/Division AES), janvier 2002.

GENERAL METHODOLOGY

1. R&D DATA

(Tables 1 to 18)

1.1. Definitions and coverage

OECD standards

The data in this publication have been collected and presented in line with the standard OECD methodology for R&D statistics entitled *The Measurement of Scientific and Technological Activities: Proposed Standard Practice for Surveys of Research and Experimental Development -- Frascati Manual 1993* (OECD) (and its 1989 supplement dealing with R&D statistics and output measurement in the Higher Education sector).

The two types of data

Most R&D data are derived from retrospective surveys of the units actually carrying out or "performing" R&D projects. Thus, the statistics in Tables 1 to 17 are based on the sum of performers' reports of their R&D expenditure and personnel on national territory (i.e. excluding payments to international organisations and other performers abroad). Personnel data are expressed in full-time equivalent (FTE) on R&D (i.e. a person working half-time on R&D counts as 0.5 person-years). Because of the complexity of the surveys, it is difficult to obtain very up-to-date series. In the present volume, 2000 data are still provisional and most data for later years are national estimates or projections (these data have been annotated "c" or "p").

More up-to-date information on government support for R&D can be derived from budgetary sources. The statistics in Table 18 are based on government budget appropriations or outlays for R&D as reported by the funding ministry or agency, and include payments to international organisations and other performers abroad.

The specification of data on the national effort financed by government in Table 1 differs significantly from the data on total government budget appropriations for R&D (Table 18); combining these two sets of data would give erroneous estimates for civil and defence GERD.

Fields of science

In general, the data cover R&D in all fields of science (NSE+SSH), i.e. natural sciences (including medical and agricultural sciences) and engineering (NSE) and social sciences and humanities (SSH). A large number of countries collect data on R&D activities in the Business Enterprise sector for NSE only.

Sectors of performance and sources of funds

Domestic R&D efforts (expenditures or personnel) are divided into *four sectors of performance* for statistical purposes, Business Enterprise (industry), Higher Education, Government and Private Non Profit institutions (PNP). This publication provides data on each of the four sector totals and detailed information for the first two sectors only.

R&D expenditure is subdivided into *five sources of funds*, from Business Enterprise, from Government, from Higher Education, from PNPs and from Abroad. Since the amounts financed by the Higher Education and PNP sectors are small, they have been combined as "other national sources" in Table 10.

R&D in the Business Enterprise sector (Tables 9 to 13)

These data are intended to provide internationally comparable indicators of resources devoted to R&D and trends at industry level. The Business Enterprise sector covers private and public enterprises and institutes serving such enterprises.

The classification is based on the International Standard Industrial Classification (ISIC, rev. 3) (see table on the following page).

The breakdown between industries is, in principle, made at the enterprise level, although some countries are able to break down R&D data for multi-product enterprises between their main lines of business. National statistical regulations prevent publication of results where there are very few firms in the given category, hence the many gaps in the tables. In principle R&D institutes serving enterprises are classified with the industry concerned. When this is not the case (e.g. Norway), the percentage of BERD performed by non-manufacturing industry is overestimated compared with other countries (see "National specifications" section for a summary of the main national characteristics affecting the comparability of data series over time or between countries).

R&D in the Higher Education sector (Tables 14 to 17)

These data are mainly intended to describe the characteristics and trends of R&D in the Higher Education sector by fields of science (see *Frascati Manual 1993*, section 3.7) because of the importance of the role played in R&D by universities and similar establishments. The Higher Education sector includes all universities, colleges of technology and other institutes of post-secondary education and also all research institutes, experimental stations and clinics operating under the direct control of, or administrated by, or associated with higher education establishments.

The basic classification comprises the six major fields of science and technology suggested in the UNESCO *Recommendation Concerning the International Standardisation of Statistics on Science and Technology.*

INDUSTRIAL CLASSIFICATION FOR RESOURCES DEVOTED TO R&D IN THE BUSINESS ENTERPRISE SECTOR IN THE *OECD R&D SURVEY* AND CORRESPONDENCE WITH ISIC Rev.3, NACE Rev.1 AND APPROXIMATE CORRESPONDENCE WITH ISIC Rev.2

	Title	ISIC Rev.3 Division/Group/Class	Corresponding NACE Rev. 1 Div./Group/Class	**Approximate** correspondence ISIC Rev. 2 Div./Group/Class
1.	**AGRICULTURE, HUNTING & FORESTRY**	**01+02+05**	**01+02+05**	**1**
2.	**MINING**	**10 thro' 14**	**10 thro' 14**	**2**
3.	**MANUFACTURING**	**15 thro' 37**	**15 thro' 37**	**3**
4.	**Food, beverages & tobacco**	**15+16**	**15+16**	**31**
5.	Food products & beverage	15	15	311 thro' 313
6.	Tobacco products	16	16	314
7.	**Textiles, wearing apparel, fur & leather**	**17 thro' 19**	**17 thro' 19**	**32**
8.	Textiles	17	17	321
9.	Wearing apparel & fur	18	18	(
10.	Leather products & footwear	19	19	(322 thro' 324
11.	**Wood, paper, printing, publishing**	**20 thro' 22**	**20 thro' 22**	**331+34+3832 (part)**
12.	Wood & cork (not furniture)	20	20	331
13.	Pulp, paper & paper products	21	21	341
14.	Publishing, printing & reproduction of recorded media	22	**22**	342+3832 (part)
15.	**Coke, petroleum, nuclear fuel, chemicals & products, rubber & plastics**	**23 thro' 25**	**23 thro'25**	**35**
16.	Coke, refined petroleum products & nuclear fuel	23	23	353+354
17.	Chemicals & chemical products	24	24	351+352
18.	*Chemical products less pharmaceuticals*	24 less 2423	24 less 24.4	351+352 less 3522
19.	*Pharmaceuticals*	2423	24.4	3522
20.	Rubber & plastic products	25	25	355+356
21.	**Non-metallic mineral products ("Stone, clay & glass")**	**26**	**26**	**36**
22.	**Basic metals**	**27**	**27**	**37**
23.	Basic metals, ferrous	271+2731	27.1 thro' 27.3 + 27.51/52	371
24.	Basic metals, non-ferrous	272+2732	27.4 + 27.53/54	372
25.	**Fabricated metal products (except Machinery & equipment)**	**28**	**28**	**381**
26.	**Machinery equipment, instruments & transport equipment**	**29 thro' 35**	**29 thro' 35**	**38 less 381 & 3832 (part)**
27.	Machinery, n.e.c.	29	29	382 less 3825+3829 (part)
28.	Office, accounting & computing machinery	30	30	3825
29.	Electrical machinery	31	31	383 less 3832
30.	Electronic equipment (radio, TV & communications)	32	32	3832 (part)
31.	*Electronic components (includes semiconductors)*	321	32.1	
32.	*Television, radio & communications equipment*	32 less 321	32 less 32.1	
33.	Medical, precision & optical instruments, watches clocks (instruments)	33	33	385
34.	Motor vehicles	34	34	3843
35.	Other transport equipment	35	35	384 (part)+3829(part)
36.	*Ships*	351	35.1	3841
37.	*Aerospace*	353	35.3	3845+3829(part)
38	*Other transport n.e.c.*	352+359	35.2+35.4+35.5	3842+3844+3849
39.	**Furniture, other manufacturing n.e.c.**	**36**	**36**	**332+39**
40.	Furniture	361	36.1	332
41.	Other manufacturing n.e.c.	369	36.2 thro' 36.6	39
42.	**Recycling**	**37**	**37**	**NA**
43.	**ELECTRICITY, GAS & WATER SUPPLY (UTILITIES)**	**40+41**	**40+41**	**4**
44.	**CONSTRUCTION**	**45**	**45**	**5**
45.	**SERVICES SECTOR**	**50 thro' 99**	**50 thro' 99**	**6 thro' 9**
46.	**Wholesale, retail trade & motor vehicle etc. repair**	**50 thro' 52**	**50 thro' 52**	**61+62+6(part)**
47.	**Hotels & restaurants**	**55**	**55**	**63**
48.	**Transport & storage**	**60 thro' 63**	**60 thro' 63**	**71**
49.	**Communications**	**64**	**64**	**72**
50.	Post	641	64.1	
51.	Telecommunications	642	64.2	
52.	**Financial intermediation (including insurance)**	**65 thro' 67**	**65 thro' 67**	**81+82**
53.	**Real estate, renting & business activities**	**70 thro' 74**	**70 thro' 74**	**83+932**
54.	Computer & related activities	72	72	8323
55.	*Software consultancy*	722	72.2	
56.	*Other computer services n.e.c.*	72 less 722	72 less 72.2	
57.	Research & development	73	73	932
58.	Other business activities n.e.c.	70+71+74	70+71+74	83(part)
59	**Community, social & personal service activ., etc.[a]**	**75 thro' 99**	**75 thro' 99**	**9 less 932**
60	**GRAND TOTAL**	**01 thro' 99**	**01 thro' 99**	**1 thro' 9**

a. Activities carried out in these industries by the business enterprise sector only. Figures are expected to be negligible: the heading is included as an aide-memoire.

Government budgetary appropriations or outlays for R&D (GBAORD) (Table 18)

These data are assembled by national authorities using statistics collected for budgets. This essentially involves identifying all the budget items involving R&D and measuring or estimating their R&D content. The series generally cover federal or central government only. These estimates, based on funders' reports, are less accurate than the "performer-reported" data in Tables 1 to 17, but as they are derived from the budget, they can be linked back to policy issues by means of a classification by "objectives" or "goals". Programmes are allocated between socio-economic objectives on the basis of *intentions* at the time the funds are committed and not the actual content of the projects concerned. These breakdowns reflect *policies* at a given moment in time.

The classification used is one specially developed at OECD for R&D analysis (see *Frascati Manual 1993*, sections 8.7.3 & 8.7.4). It is compatible with similar classifications used by the EC and by the Nordic countries.

Expenditures in national currency

National currency data is expressed in euro for the Euro-zone countries. Beginning with the year of entry into the Economic and Monetary Union (EMU), data are in EUR. For years prior to the year of entry into EMU, data have been converted from the former national currency using the appropriate irrevocable conversion rate. This presentation facilitates comparisons within a country over time and ensures that the historical evolution is preserved. Please note, however, that ***pre-EMU euro are a notional unit and should not be used to form area aggregates or to carry out cross-country comparisons***.

In the present publication, series stretch across the accession date of EMU member countries and thereby comprise both pre-accession data in the original national currency converted with the irrevocable conversion rate, and post-accession data in true EUR. Austrian national currency data, for example, is presented in 1999 ATS euro up through 1998, and beginning with the 1999 data, in EUR; similarly, Greek data is presented in 2001 GRD euro up through 2000, and beginning with the 2001 data, in EUR.

The Euro-zone countries, their irrevocable EUR/national currency exchange rate and year of EMU accession are the following:

Country	*1 EUR =*	*Year of accession to EMU*
Austria	13.7603 ATS	1999
Belgium	40.3399 BEF	1999
Finland	5.94573 FIM	1999
France	6.55957 FRF	1999
Germany	1.95583 DEM	1999
Greece	340.750 GRD	2001
Ireland	0.787564 IEP	1999
Italy	1936.27 ITL	1999
Luxembourg	40.3399 LUF	1999
Netherlands	2.20371 NLG	1999
Portugal	200.482 PTE	1999
Spain	166.386 ESP	1999

Expenditures in constant dollars

R&D expenditure series have been deflated using the implicit GDP deflator taken from the OECD National Accounts database updated for the most recent years by Secretariat forecasts of changes in the GDP deflator, as published twice a year in the *OECD Economic Outlook* (except in the case of Norway where a deflator excluding trends in petroleum prices has been used). These national currency data at 1995 prices have been converted to US$ using 1995 purchasing power parities (PPPs). The PPPs are those developed by the OECD National Accounts Division (for further details see sections VII and VIII of *National Accounts, Volume 1, 1988-1999*, OECD, Paris, 2001), updated for the most recent years by comparing the growth in prices (implicit

GDP deflator) in each country with that in the United States. These estimated parities are footnoted "*b*" in the series as are any data converted to current dollars using them.

International comparability

Though all OECD countries generally collect and report R&D in line with the *Frascati Manual*, some detailed national specifications may vary from OECD standards. The main exceptions are shown in the "National specifications" section below. They mainly deal with:

- breaks in series;
- whether or not the data cover all fields of science (NSE and SSH);
- difficulties encountered by national authorities in measuring the amount of resources devoted to R&D in the Higher Education sector because of the close relationship between teaching and research activities;
- differences in the coverage of the Higher Education and Government sectors due to institutions at the "borderline" of the two;
- whether or not R&D institutes serving enterprises are included and whether or not they are classified according to the industry concerned.

1.2. National specifications

A summary, by country, of the main national departures from OECD norms is given below for years 1981 to present. For more information concerning the differences in industrial classification (Tables 9 to 13), please consult the publication *Research and Development Expenditure in Industry 1974-1995*, OECD, Paris, 1997.

Sometimes retrospective revisions are received for the global data without the corresponding revision of the detailed breakdown (for example, a revision of the data covering expenditure by source of funds for total BERD and GERD in Table 1 without the corresponding revision of the BERD data by industry in Tables 9 and 10). Such data have been footnoted "s " in the tables concerned.

- **Australia**

  In the breakdown of R&D expenditure by type of cost in the business enterprise sector, the labour costs of maintenance staff and assistants not directly engaged in R&D are included in the cost of R&D personnel instead of being included in “other current costs” as the *Frascati Manual* recommends.

  From 1999, Australia has prepared its Federal Budget details according to the principles of accrual accounting, leading to a break in the series for GBAORD data. In 1999, the change from cash accounting to accrual accounting increased the estimate of GBAORD by AUD 232 million.

- **Austria**

  An estimate of the R&D expenditures of provincial hospitals was prepared on the basis of the annual financial reports of the Bundesländer (federal provinces) (AS 637.7 million for 1989) and is included in the expenditure tables. No corresponding estimate of personnel data is available.

In 1993:

- Because several institutions in the PNP sector were reclassified in the government sector, most of the 1993 data for these two sectors are not comparable with the data for earlier years.
- The total for gross domestic expenditure on R&D (GERD) includes government funding which was not identified by the R&D surveys. It has not been possible to break this funding down by type of cost or type of activity (Tables 2, 3, 11 and 15): it totalled AS 903.1 million in 1993: 750.5 million for current R&D expenditure and 152.6 million for capital expenditure. Similarly, in the business enterprise sector, it has not been broken down by industry and source of funds (Tables 9 and 10): AS 481 million in 1993.

- **Belgium**

As from 1993 (1992 for the business enterprise sector), the methodology for R&D surveys changed: data are based on full surveys and no longer on a combination of budget figures and survey findings. There is accordingly a significant break in the series, and data for these years cannot be compared with those for earlier years.

In 1989 the methods of evaluating the R&D effort changed (thereafter the surveys included the resources of communities and regions, which had become very significant following the reforms of Belgium's institutions of government in 1988), and from then on the data are not comparable with those for earlier years.

In 1987 and 1988, gross domestic expenditure on R&D (GERD) and GERD financed by government (Tables 1-4, 7) are underestimated as R&D financed by federative authorities (about 2-4 % of GERD and 7-15 % of government-financed GERD) is excluded. As a breakdown of this sum by sector of performance is not available, the impact on the other detailed R&D expenditure tables by sector of performance cannot be gauged precisely (business enterprise sector : Tables 9-11, and higher education: Tables 14-16) and the data there are underestimates.

As from 1983, data for GERD are not comparable with those for earlier years, because of changes in the methods of assessing R&D effort in Belgium.

As from 1981, resources devoted to R&D in the business enterprise sector are underestimated in all the detailed industry tables (Tables 9 to 13) since the resources of joint R&D centres of enterprises are not included (their breakdown by industry is no longer available). At the end of the 1970s the contribution of these centres represented approximately one-sixth of the R&D expenditure of non-manufacturing industries.

- **Canada**

The sources of funds for the higher education sector for General University Funds (GUF) have been adjusted at the national total level (Table 1) but not for the detailed data by scientific discipline (Table 16).

As from 1989, General University Funds from non-federal sources are included in GBAORD.

As from 1988, included in the R&D expenditure of the higher education sector are the estimated values for R&D in hospitals not covered by University reports and not previously included.

In 1987, modifications in the number of categories collected in the survey resulted in a break in series.

- **Czech Republic**

As from 1998:

- General University Funds (GUF) (not available separately) are included in the sub-total Government financing.

- reclassification of some funding sources (Ministry of Education and Grants agency, etc.) from higher education to Government.

As from 1995, the survey methodology was amended in order to collect data complying with OECD standards:

- Ancillary activities are excluded.
- All sources of funding are available, except General University Funds (GUF).
- Defence-related R&D is included.
- Personnel data by occupation are expressed in R&D FTE.

Until 1994 inclusive the data are not in line with the *Frascati Manual*:

- The data relate to all the activities of units making up the "R&D base", corresponding to scientific and technical activity rather than R&D proper. The national authorities consider that around 10% of all expenditure is devoted to activities other than R&D.
- The data on expenditure cover only expenditure financed by government.
- Defence-related R&D is excluded.
- The staff data cover all personnel in the units and are not expressed in R&D FTE. People working part-time are recorded in FTE for all their activities, whether R&D or not.

- **Denmark**

The HE-sector R&D expenditure is underestimated as all R&D carried out in hospital departments at the university-hospitals is included in the Government sector.

In 2001, 1993, 1988 and 1983 changes in the method of assessing government budget appropriations or outlays for R&D (GBAORD) by socio-economic objective (Table 18) led to breaks in the series.

In 2000, GBAORD coverage was extended to include provincial funding of R&D in hospitals as well as funds from The Danish National Research Foundation and the VaekstFonden.

In 1999, the breakdown of national total R&D expenditure by type of activity (Basic Research, Applied Research and Experimental Development) is based on the sum of current intramural R&D expenditure for the BE sector and total intramural R&D expenditure for the three other sectors.

- **Finland**

In 1998, due to a greater number of responses on the group level, the questionnaire category funds from other foreign enterprises of the group have been merged with business enterprise funds (own funds) thus reducing the share of funds coming from abroad.

As from 1997, the higher education sector covers central university hospitals, as do GBAORD data which cover appropriations for R&D for central university hospitals.

As from 1995, funds from external sources of the State research institutes are excluded from Government appropriations for R&D (GBAORD).

As from 1994, PNP institutions are included in the government sector in non-survey years, meaning a break in series for the data relating to the government sector for those years. In 1993 R&D by PNPs represented a little over 3 % of total government plus PNP R&D expenditure, around 4 % of R&D personnel and 5.5 % of researchers.

In 1991:

- The method of evaluating R&D expenditure by the government and higher education sectors changed; in particular, the coefficient for calculating pension costs was raised from 15 to 25 %

of salary. As a result, R&D expenditure by these sectors in 1991, and R&D financing by government and total national GERD, are not comparable with those for previous years.

- Similarly, data on GBAORD have been revised back to 1991 because of changes in R&D coefficients for certain research institutes and there was an upward adjustment in the total in 1991, due to the inclusion of pension costs.

In 1987:

- It would appear that the estimated expenditure on buildings and equipment, and hence capital expenditure, in the government and higher education sectors is not comparable with data for earlier years.
- The breakdown of R&D personnel by formal qualification was changed: other non-technical post-secondary qualifications, previously included under "other", are now contained in "other post-secondary".

In 1983, changes to the methodologies limit comparability for the detailed breakdowns for the higher education and government sectors.

In 1981 some data were retrospectively revised by the national authorities, and from that year the sectoral totals for government and higher education are no longer comparable with data for earlier years. The corresponding revisions in the more detailed breakdowns by category of cost or personnel have not been supplied.

- **France**

The National Centre for Scientific Research (CNRS) is included in the higher education sector whereas in other countries, such as Italy, this type of organisation is classified in the government sector. This obviously affects the breakdown of the R&D effort by sector of performance.

As from 1997:

- the method used to evaluate R&D personnel in the Government, higher education and PNP sectors has changed
- estimates for R&D expenditure and personnel are significantly revised due to:
  - changes in the methods used to evaluate domestic expenditure on defence for the 1998 survey;
  - Reevaluation of R&D activities of very large enterprises;
  - Reevaluation of clinical research and reevaluation of R&D effort of one civil organisation.

Consequently, 1997 R&D personnel data for the Government, higher education and PNP sectors are not comparable with those of previous years and 1997 data on intramural R&D of BE sector and funding of R&D by Government and BE sectors are not comparable with those of previous years. This break in series for 1997 also relates to GBAORD data.

As from 1992:

- Data for the enterprise and government sectors are not comparable to the corresponding 1991 data because France Télécom and GIAT Industries were transferred from government to business enterprise following a change in their legal status.
- Repayments of reimbursable assistance have been taken into account in the breakdown of R&D expenditure by source of funds.
- The survey method for the PNP sector changed.

As from 1981, following a re-evaluation of the proportion of time that teaching staff devote to R&D, the resources which the higher education sector devotes to R&D, and hence the national figures as well, are not comparable with those for earlier years.

- **Germany**

The total of the business enterprise gross domestic expenditure on R&D (BERD) includes government funding which was not identified by the enterprise survey but taken from government extramural expenditure for R&D performed by enterprises. It has not been possible to break this funding down by type of cost or type of activity (Tables 2 and 3) or by industry (Tables 9 and 10). In 1997 this amount was negligible; in 1995 it totalled DM 386 and DM 515 million in 1993.

Similarly, in the higher education sector it has not been possible to include some funding in the breakdown of R&D expenditure by type of cost or type of activity (Tables 2 and 3) or by main scientific field (Table 7), or in some detailed breakdowns by field of science (Tables 14 and 15).

Higher education's "own funds" are combined with government funding.

For 1997, the methodology of assessing GBAORD by socio-economic objective changed. The 1997 total budget figure of the Federal Ministry of Education, Science, Research and Technology was reduced, but the global reduction was not available by socio-economic objective. Therefore, total GBAORD reflects the adjusted budget figure, and the sum of the breakdown does not add to the total.

In 1995 the method of evaluating R&D resources in the higher education sector changed and was applied retroactively to 1981 to make the 1981 to 1994 data compatible with the more recent data.

As from 1993, R&D expenditure in the government sector, and consequently total GERD, includes R&D performed in German research institutions located abroad. This amounts to between 0.6 and 0.7 % of R&D expenditure in the government sector and less than 0.1 % of GERD.

In 1992:

- The restructuring that took place after unification created over 100 non-university research units in the government sector.
- The methodology of the survey on resources devoted to R&D in the government sector was altered.
- From that year on, resources devoted to R&D by the PNP sector have been included in the government sector.

As from 1991:

- R&D data are for unified Germany, until 1990 the data covered western Germany only.
- Graduate students conducting research and receiving grants for that purpose were included for the first time among higher education researchers (close to 10 000 FTE). The relevant grants paid to them in 1991 were included in higher education R&D expenditure (HERD), so total grants paid to higher education in 1991 were well above the earlier figures, because only grants paid directly by government had been included up till then; the outcome was an increase in total government funding for R&D performed by higher education in that year.

In 1987 a new survey for the PNP sector.

In 1985 several changes concerning the government sector.

In 1981 and 1983: downwards adjustment of R&D performed by the government sector and of government funding to higher education.

- **Greece**

In 1995 following the revision of research coefficients used to evaluate resources devoted to R&D by higher education, R&D expenditure and personnel for this sector and for national totals are not comparable to those for previous years.

As from 1993, the PNP sector is included in the survey.

In 1991 some activities included until then under "office machinery and computers" (and hence in the corresponding sub-total "machinery, equipment, instruments and transport equipment") in manufacturing were reclassified in services under "computer and related activities" (and hence in the corresponding sub-total "real estate, renting and business activities").

In 1989 the "research coefficients" used to evaluate resources devoted to R&D by higher education were revised.

In 1983:

- The methodology for evaluating resources devoted to R&D by higher education changed. The "research coefficients" used for the calculations were established on the basis of a survey of higher education institutions.
- The methodology for identifying R&D personnel was brought into line with the recommendations of the *Frascati Manual*, resulting in changes to personnel costs.

- **Hungary**

Only defence-related R&D expenditure performed in the civil sector is covered; R&D expenditure by the armed forces is excluded.

All breakdowns of R&D expenditure by sector of performance, by source of funds, by type of costs, by type of activity or by main scientific field (Tables 1 to 3, Table 7, and also Table 9 and Tables 14 to 16) are incomplete.

Since 1998 personnel dealing with activities related to safety and warehouse operations in R&D units are excluded, in accordance with the *FM*.

As from 1994:

- Payments for licences and know-how, previously included in business enterprise expenditure on R&D (BERD), were removed.
- The Central Technology Fund (KMÜFA), previously included in the business enterprise sector, was reclassified in the government sector.

- **Iceland**

In 1993, Iceland revised its methods for collection and processing budget data for GBAORD by socio-economic objective, resulting in a break in series.

- **Ireland**

Back to 1992 inclusive, the Government sector data was revised, as some expenditure is no longer classified as R&D.

Back to 1985 inclusive, HERD time series have been revised following results of the 1993 full survey of R&D in the higher education sector, showing that coefficient of research time derived from the 1984 survey had become seriously out-dated.

As from1983, non-government monies are no longer included in the GBAORD series.

- **Italy**

As from 1991, extramural expenditure is collected separately in Italian R&D surveys and can thus be excluded from most data.

Until 1990, the data presented in Table 1 – Gross domestic expenditure on R&D by sector of performance and source of funds – were based on all intramural and extramural R&D expenditure (including extramural expenditure abroad). In relation to the *Frascati Manual* recommendations, the inclusion of extramural R&D expenditure leads to gross domestic expenditure on R&D

(GERD) being overestimated. This relates most particularly to R&D expenditure by the business enterprise sector, but concerns the government sector as well.

Extramural expenditure represented:

- At national level, over 6 % of GERD in 1991 and some 15 % in 1993. In 1991, over 20 % of such extramural expenditure represented financing abroad; in 1993, over 26 %.
- For the business enterprise sector, 11 % in 1991 and around 15 % in 1993 of its R&D expenditure (BERD). Over 13 % in 1991 and around 18 % in 1993 of R&D financed by the business enterprise sector was extramural expenditure (a quarter of the sector's extramural expenditure represented payments abroad).
- For the government sector, only 1 % of its R&D expenditure (GOVERD).

As from 1987, the breakdown of R&D expenditure by type of activity is based on total expenditure and not current expenditure on R&D.

In 1984, part of the data for the ENEA (National Committee for R&D of Nuclear Energy and of Alternative Sources of Energy) were transferred from the objective "Environment" to "Health", reducing the former by half and increasing the latter accordingly (Table 18 - GBAORD).

As from 1981, data for defence R&D funding are not comparable with earlier years owing to an upward revision resulting following an improvement in the survey (Table 18 - GBAORD).

- **Japan**

Until and including 1995, data provided for R&D personnel and consequently labour cost data are overestimated by international standards. Data for researchers are expressed in number of persons regularly employed in R&D rather than in full-time equivalent. Studies by some Japanese authorities suggest that in order to reach FTE, the number of researchers might be reduced by perhaps 40 % in the higher education sector and by about 30 % for the national total. That would reduce HERD by about 25 % and GERD by about 15 %. OECD calculated, through 1995, the adjusted series for both expenditure and researchers for the higher education sector and the national total, and these data appear in the OECD publications Main Science and Technology Indicators, as well as various studies and analytical reports.

The breakdown of R&D expenditure in the higher education sector by source of funds has been adjusted by the Secretariat, at the national total level (Table 1), for funds from the government and higher education sectors, in particular for an estimate of GUF. The corresponding adjustment has not been made to the detailed data by scientific discipline (Table 16). For 1990, the breakdown of higher education source of funds was adjusted by the Japanese authorities using a different method, meaning a break in series in that year, mainly for direct financing by government. Reverting to adjustment by the Secretariat in 1991 resulted in a further break in series for direct financing by government.

The budget data on government outlays for R&D by socio-economic objective (GBAORD, Table 18) are underestimated because GUF excludes social sciences and humanities and the R&D portion of military procurement contracts is excluded.

- **Korea**

Social sciences and humanities are excluded from the R&D data, R&D is underestimated notably in the Government, Higher education and PNP sectors.

- **Mexico**

1993 is the first year for which a complete set of internationally comparable data on R&D in Mexico is available.

1993 and 1994: the data on resources devoted to R&D in the government sector and by PNPs come from the first national R&D surveys of those sectors based on the *Frascati Manual*; they are not comparable with the estimates available for previous years for the government sector.

1992 and 1993: the data on resources devoted to R&D in the business enterprise sector and in higher education come from the first national R&D surveys based on the *Frascati Manual*; they are not comparable with the estimates available for those sectors for earlier years.

- **Netherlands**

From 1997 onwards, a new definition of GUF is used for GBAORD series.

In 1996, firms with 10 to 49 employees in the Service sector were included in the scope of the business enterprise sector. Following this enlargement of the scope of the survey in the business enterprise sector, 1996 data on R&D expenditure and personnel in the Service sector, for the BE sector total, as well as the national total, are not directly comparable with those for earlier years.

In 1994:

- firms with 10 to 49 employees in the Manufacturing sector were included in the scope of the business enterprise sector. Following this enlargement of the scope of the survey in the business enterprise sector, 1994 data on R&D expenditure and personnel in the Manufacturing industries, the BE sector total, as well as the national total, are not directly comparable with those for earlier years.
- Following the reclassification of certain institutes from PNP to government, the data for those two sectors are not comparable with data for earlier years.

As from 1991 business enterprise SSH R&D, corresponding to R&D by institutes serving business enterprise only, was included in the PNP sector.

As from 1990 the methodology of the survey on HERD expenditure changed, and one consequence was an apparent increase of the government contribution to R&D performed by higher education.

As from 1985 the breakdown of R&D expenditure by type of activity is based on total expenditure rather than current intramural expenditure on R&D.

In 1982, as a result of a change in the method of estimating data on R&D by higher education, the data for 1982 (or 1983, depending on availability) for R&D in the higher education sector and the national total are not comparable with those for earlier years.

- **New Zealand**

Time comparisons are occasionally restricted because of changes to the method of collecting and estimating data in 1984 (higher education and national total), 1987 (BERD by industry) and 1992 (GUF for higher education and hence the national total).

- **Norway**

In the breakdown of gross domestic expenditure on R&D (GERD) by sector of performance and socio-economic objective (Table 4), capital expenditure has not been allocated by objective but shown under “not elsewhere classified”.

The method for compiling GBAORD data changed in 1996. The series have been revised back to 1983 to exclude contract research, state enterprises and payments to the European Commission.

In 1995 the survey of the business enterprise sector was considerably enlarged to include a sample of units with 10 or more employees (previously the lower limit was 50 employees) in branches traditionally covered by the survey; several services were also brought in. As a result, the 1995 data are not comparable with those for earlier years.

As from 1991 personnel in central administration units of higher education are not included in the data, but the cost of such personnel is included in other current R&D expenditure (in line with the 1993 *Frascati Manual*).

As from 1989 R&D performed by PNPs is included in the government sector.

As from 1987:

- The BE survey was expanded to include Bank and Insurance Services and more Engineering Services.
- The "own funds" of commercial State enterprises (included as executive units in the business enterprise sector), previously included in government funds, were reclassified as own funds of business enterprises.

In 1984 the coverage of the survey was expanded to include more private firms, notably within the branches of Transport, Commercial and Engineering Services and Other activities (ISIC 71, 8323/24).

As from 1981:

- at industrial branch level, all R&D performed by private and public institutes serving industry are classified under ISIC rev. 2 branches 8324 (Engineering, technical and architectural services n.e.c.) or 9320 (Research and scientific institutes n.e.c.). Accordingly, data by industry in the manufacturing sector include only R&D performed by firms (private and public), not institutes, and are not comparable with those for earlier years.
- The "own funds" of the mining sector (i.e. oil producers) include funds from their foreign parent companies during the eighties.

- **Poland**

Current R&D expenditure by type of activity (Table 3) is overestimated because it includes a portion of capital expenditure relating to special equipment for research.

Before 1995, defence R&D was mostly excluded.

- **Portugal**

In 1997, due to a new accounting method for structural funds from the European Commission, funds from Abroad and direct Government financing are not comparable with those of earlier years.

As from 1995:

- the classification by NACE Rev 1 of R&D activities in the BE sector was introduced and the data have been revised back to 1995.
- units of the PNP sector mainly controlled/financed by the BE and the HE sectors were reclassified under these sectors of performance (following the guidelines in Annex-6 of the 1993 *Frascati Manual*).

As from 1992 the data for RSE and technicians (Tables 5, 13 and 17) are not comparable with those for earlier years because the classification of R&D personnel by occupation was revised for the survey relating to 1995 and retrospectively back to 1992.

As from 1988 R&D personnel in the business enterprise sector is broken down by occupation and the data relating to RSE and technicians are not comparable with those for earlier years where breakdown by occupation was based on formal qualifications with data for researchers covering university graduates or holders of other post-secondary qualifications.

Until and including 1986, data for researchers in the business enterprise sector cover university graduates or holders of other post-secondary qualifications.

GBAORD data by socio-economic objective have been revised back to 1986 according to a new methodological exercise.

- **Slovak Republic**

As from 1997, some Defence R&D is separately available in the socio-economic breakdown of GERD. Nevertheless, it seems that Defence R&D is still largely underestimated in GERD data when comparing with information on Defence R&D from the Ministry of Defence. The former for years 1997 to 1999 represents respectively 28%, 31% and 44.5% of the later.

Before 1997 Defence R&D was excluded.

In 1994 *Frascati Manual* definitions have been adopted in the national R&D survey.

Before 1994 data referred to "Reseach and Development Base" (RDB) and cover the whole activity of institutions and not only R&D; personnel were not expressed in FTE and non R&D personnel were included in the "other" occupation class.

- **Spain**

In 1997:

- The defence objective in GBAORD almost doubled in magnitude due to an exceptional contribution by the Ministry for Industry and Energy.
- The incorporation of the Spanish contribution to CERN has involved substantial changes in the "Energy" category.

In 1995:

- Higher education's own funds were separated from GUF, in which they had earlier been included. As a result the 1995 data for GUF, the sub-total of government finance for R&D in the higher education sector and the national level are not comparable with those for previous years.
- The marked increase in funding from abroad in the higher education sector is chiefly due to funds from the European Commission, which have to some extent taken the place of government funding.

As from 1994, due to the availability of the breakdown of support personnel between Technician and Other in the higher education sector, the corresponding categories for total R&D personnel are no longer underestimated.

As from 1992:

- There was an upward re-estimation of GUF. This caused a break in series of HERD and GERD and of total government financing of HERD and GERD (Tables 1, 2, 3, 7, 14, 15 and 16). It should be noted that there was no equivalent reestimation for the data on budget appropriations for R&D (Table 18), where the source and method of estimation are both different for GUF.
- Labour costs and current intramural expenditure are no more underestimated as payroll costs for technicians and other support personnel in the higher education sector, previously not included, are incorporated in HERD.
- The method of estimating current and capital R&D expenditure by the higher education sector changed and the relevant data are now collected by a questionnaire sent to universities, which provide estimates.

Prior to 1989, the data for total R&D personnel, technicians and other supporting staff (Tables 5, 6 and 8) are underestimated because those for the higher education sector only cover RSE. As a result, total R&D personnel in Spain may be underestimated for these years by between 10 and

15 %. Totals for technicians and other supporting staff are still underestimated until 1994, as separately available data are missing for the higher education sector.

The series for R&D expenditure and personnel in the higher education sector were revised upwards, retrospectively, back to 1980 to allow for a change in the method of evaluating the 1988 data following the 1989-90 survey on the allocation of teachers' working time.

- **Sweden**

The data are systematically underestimated (by some 9 % for GERD in 1993) for two reasons:

- R&D performed in the government sector covers central government units only. Resources devoted to R&D by this sector could double if regional and local government units were included.
- Full coverage of SMEs could increase BERD by 7 %.

Furthermore, until 1991, the data for the business enterprise, government and PNP sectors, and consequently for national totals, (Tables 1 to 13) are further underestimated as the data for R&D performed by these sectors exclude the social sciences and humanities.

However, R&D expenditure by the higher education sector includes some non-R&D expenditure for education at ISCED level 7 (included in GUF).

As from 1998:

- GBAORD data are on a calendar year basis (January-December) instead of the period July-June which has been used until 1994. Budget appropriations for 1995 and 1996 are estimates based on the period July 1995-December 1996.
- GBAORD data exclude funding from *Public Research Foundations*.

In 1997:

- Due to a change in statute, the R&D funding from *Public Research Foundations* previously considered as coming from the PNP sector has been reclassified as coming from the government sector. Consequently, 1997 data on R&D funding from Government and PNP sectors are not comparable to those for previous years.
- Thc sharp decrease in R&D personnel in the PNP sector is partially due to modification of the national questionnaire.

In 1995:

- Capital expenditure for R&D in higher education is excluded for that year; consequently all 1995 data concerning HERD, total GERD and government-financed GERD are underestimated and not comparable to corresponding data for previous and following years.
- A number of institutions in the PNP sector have been reclassified mainly in the business enterprise sector.

As from 1993:

- SSH R&D was also included in the government, PNP and business enterprise sectors (in the latter case, this may affect service industries essentially); 1993 R&D data for Government and PNP sectors are not comparable with those for previous years.
- In the higher education sector, R&D personnel employed in hospitals administered by local government were not included (1 900 FTE in 1989 and 1991). But graduate students performing R&D funded from outside were included for the first time (1 322 FTE in 1993).
- The breakdown of R&D personnel by formal qualification in the higher education sector and consequently for national totals is incomplete.

- The breakdown of R&D expenditure and personnel by field of science in the higher education sector is incomplete.
- Public utilities of central government are excluded from the GBAORD data (Table 18).

In 1983, the changes in the industrial classification occasionally reduced the comparability of business enterprises sector data for 1983 and those for earlier years at individual branch level.

As from 1981:

- For the higher education sector, estimates of SSH R&D were made (SKr 580 million in 1981); the estimates of higher education R&D for 1979 related to NSE only. The SSH amount represented about 14 % of total higher education R&D expenditure in 1981, or 4 % of Sweden's GERD (which further covered R&D in NSE only for the other three sectors of performance in that year).
- Inclusion of total government-funded R&D expenditure in university hospitals (previously only R&D expenditure funded by medical faculties had been covered) brought an increase of around 18 % in 1981 in higher education R&D expenditure, or 5 % of GERD;
- The increased coverage of firms whose main activity is R&D (essentially ISIC rev.2, item 9320) brought an increase of some 4 to 5 % in the business enterprise sector total for 1981, compared with 1979.

- **Switzerland**

The business enterprise sector comprises private enterprises only. Public enterprises (Federal State Railways, Federal Alcohol Administration, Federal Post, Swiss National Bank, National Insurance Company) are included in the government sector.

As from 1998:

- The Federal Office of Agriculture (and its research institutes), no longer break down their R&D by socio-economic objective but group all under "Agriculture". This results, for the Government R&D expenditure by socio-economic objective (Table 4) and for the GBAORD series (Table 18), in a break in series both in the Agriculture and Health objectives, where half of the funds previously declared under Health are now declared under Agriculture.

- The field of telecommunications within the Federal Post has become the private enterprise Swisscom and is no longer considered in the GBAORD objective Sub-total Infrastructure (Table 18).

In 1992, R&D personnel data for the HE are not comparable with those for previous years following modifications in the evaluation method.

Comparisons over time are extremely limited as a result of various changes in the coverage of data and classifications in 1983, 1986, 1988 and 1989.

- **Turkey**

Only the number of researchers (in FTE for R&D) is available for higher education R&D personnel; as a result, total R&D personnel and the totals for technicians and other support staff (Tables 5, 6 and 8) for Turkey are underestimated.

As from 1994, R&D expenditure by type of activity (Basic research, Applied research, Experimental development) is based on total instead of current R&D expenditure, as was the case for previous years.

As from 1993, the survey was extended to include Financial Intermediation, Software consultancy and Other computer services. Consequently, 1993 data for Sub-Total Services are not comparable with those for previous years.

- **United Kingdom**

  R&D performed abroad as an integral part of United Kingdom R&D programmes is included in the national R&D effort.

  In the business enterprise sector, R&D funding by business also includes funds that may accrue from other national sources such as higher education or PNPs.

  All government R&D expenditure corresponds to expenditure by public services plus an estimate with three components: a small amount of R&D not identified by the government sector survey; R&D performed by local authorities; and as from 1991 an estimate of the portion of R&D performed by NHS hospitals (for the 1991 break in series, see below). This estimate supplementing government R&D expenditure is not broken down by type of cost or by type of activity (Tables 2 and 3).

  Following work to enhance the estimates of R&D by PNPs undertaken in 1996, estimates for R&D performed by the PNP sector were substantially revised downwards, in particular the funding from the business enterprise sector. The data for previous years were revised accordingly.

  As from 1995 for the GBAORD data (Table 18), the Health objective has been broadened to include the total net costs to *National Health Service* trusts of their involvement in R&D.

  In 1993:

  - The fall in government R&D personnel is chiefly due to the exclusion of staff in the *Atomic Weapons Establishment*, transferred from government to the enterprise sector.
  - A new method of estimating government-funded R&D in the higher education sector, introduced in 1994, was applied retrospectively back to 1993.

  In 1992 the increase in R&D personnel in the government sector is chiefly due to the inclusion of staff of the *Scottish Agricultural Research Institute*, the *Scottish Agricultural College* and *Scottish Natural Heritage*, transferred from the PNP sector to government.

  As from 1991:

  - Overall data for the government sector (Table 1) include an estimate of R&D support costs in NHS hospitals; those costs are excluded in the breakdown of R&D expenditure by type of cost or type of activity (Tables 2 and 3). R&D personnel in NHS hospitals are not included in the data for government R&D personnel as this information was not available.
  - Data concerning researchers and technicians in the government sector follow the *Frascati Manual* definitions; prior to 1991, RSE covered graduate scientists and engineers, and technicians covered non-graduate scientists and engineers.

  Between 1985 (85/86) and 1986 (86/87), the *United Kingdom Atomic Energy Authority* (UKAEA) was transferred from government to the enterprise sector.

  Between 1985 and 1986 R&D expenditure by a group of research institutes grant-aided by the *Agriculture and Food Research Council* in England and Wales seems to have been included with expenditure by the Research Council and transferred to the government sector; the institutes' staff appear to have been transferred over a longer period.

  Fresh estimates were made for R&D performed in the PNP sector, in particular regarding the proportion of expenditure funded by business. Downward revisions were made retrospectively back to 1985.

- **United States**

  Depreciation is reported in place of capital expenditure in the business enterprise sector.

The data for Tables 1 to 17 are probably somewhat underestimated:

- R&D performed in the government sector covers only federal government activities. State and local government establishments are excluded.
- In the higher education sector, R&D in the humanities is excluded as is capital expenditure.
- R&D expenditure in the PNP sector covers current expenditure only.

As far as R&D personnel data are concerned, only data covering research scientists and engineers (RSE) are collected, hence no data are available for total R&D personnel in the United States.

GERD data have been extensively revised so that they are on a calendar year basis and consistently tabulated. This affected mainly the Government and the higher education sectors.

As from 1999, Agriculture R&D is not surveyed in the BE sector.

As from 1998, as concerns the GBAORD data (Table 18), some components previously included under Energy have been transferred to Advancement of knowledge, resulting in a break in series for these two objectives.

For 1996, figures are separately available for R&D in Agriculture, Mining and Construction and these sectors are not included in the Services sub-total for that year.

Prior to 1996, the sub-total R&D expenditure for the Services sector in BE, which is calculated as a residual, included Agriculture, Mining and Construction.

As from 1985:

- The methodology for estimating the number of researchers was changed. In the government, higher education and PNP sectors the latest data refer to employed doctoral scientists and engineers who report their primary work activity as research, experimental development or the management of R&D, plus the number of full-time equivalent graduate students with research assistantships averaging an estimated 50 % of their time engaged in R&D activities.
- Researchers in the government sector exclude military personnel.

1.3. Sources

The data are derived from national R&D surveys and budgets and are mainly supplied to the Secretariat via the International Survey of Resources Devoted to R&D by OECD Member countries. This is a long "shuttle" questionnaire returned at least once every two years, with, at the beginning, the first section dealing with the "Main results" returned every six months for the OECD biannual *Main Science and Technology Indicators* publication. Some data on government budget appropriations or outlays on R&D have been provided by the Statistical Office of the European Communities (Eurostat).

Where countries do not regularly participate in these OECD surveys, the Secretariat may have made its own estimates based on data available from national publications. These estimated data have been footnoted "b".

2. TECHNOLOGY BALANCE OF PAYMENTS (TBP)
(Table 19)

2.1. Definitions and coverage

The TBP registers the commercial transactions related to international technology and know-how transfers. It consists of money paid or received for the acquisition or use of patents, licences, know-how, trademarks, patterns, designs, technical services (including technical assistance) and for industrial R&D carried out abroad, etc. The coverage may vary from country to country and the TBP data should be considered as only partial measures of international technology flows.

Data broken down by type of firm (affiliated or not affiliated companies) are shown whenever available.

2.2. National specifications

For certain countries, notably Canada, France and Japan, at least two national series of very different data are available. The series which appeared the most representative from the technological point of view was selected.

- **Austria**

  As from 1992, the data cover royalties, license fees and technology-related services.

  Until 1991 inclusive, data cover only royalties and license fees.

- **Belgium**

  In 1995, the balance of payments was prepared using a new method according to the OECD, IMF and Eurostat Manuals. 1995 data are not comparable with those for previous years. Data cover royalties, copyright and registration expenses of patents, manufacturing licences and trademarks, as well as technical assistance from BLEU.

- **France**

  Data cover commercial, financial, judicial and managerial assistance. R&D performed abroad is excluded.

- **Germany**

  The information given by type of firm is underestimated: it covers only patents and licences for firms of which foreign ownership is more than 20% (25% until 1989 inclusive).

As from 1986, the data cover transactions concerning patents, licences, trademarks, patterns and designs technical and also services and industrial R&D.

Until 1985 inclusive, the data cover only transactions concerning patents, licences, trademarks, patterns and designs.

- **Italy**

  Data relate to all foreign transactions over 20 million lira.

  As from 1992, data includes Research and Development, Technical studies, Staff training, Formation du personnel, Secondment of techniciens and experts and Other transactions in technolgy.

  Until 1991 inclusive, R&D performed abroad is excluded.

- **Japan**

  Transactions relating to trademarks, design and models are excluded. Also excluded is R&D performed abroad.

- **Mexico**

  Data are based on two surveys of firms benefiting from direct investment. Covered are international transactions on royalties and technical assistance and exclude R&D performed abroad. Receipts may be overestimated as they include copyrights and software.

- **Netherlands**

  As from 1982 technical services have been included in the data.

- **Norway**

  Before 1988, the data cover only patent licensing and are consequently underestimated.

- **Spain**

  In 1994, there is a break in series in the Technology Balance of Payments data due to the change in the data collection procedure. As from 1994 data are collected by survey and only refer to the exchange of non-incorporated technology in companies, excluding any non-industrial components.

  In 1993, the balance of payments was prepared using a new method according to the IMF Manual, 5th edition. 1993 data are not comparable with those for previous years. 1993 data include the following items:

  - Hardware services (which are part of "computer and Information services" in the balance of payments).
  - Royalties and licence fees (except copyrights), including the acquisition/disposal thereof.
  - Research and development.
  - Architectural, engineering and other technical services.
  - Agriculture, mining and on-site processing.

- **Switzerland**

  The TBP statistics are drawn from the Swiss balance of payments under current transactions balance, account: technological services. Not included are the financial transactions concerning purchase and sale of patents and licences but only those transactions relating to the taxes and rights resulting from the use of patents and licences. The trade of patents and licences is taken into account in the capital transactions balance.

- **United Kingdom**

  From 1996 onwards, the TBP survey was modified according to the recommendations of the IMF Manual, 5th edition. Excluded is R&D performed abroad. The data cover sale of patents and inventions, patent licensing, trademark patterns and designs, technology-related services and R&D.

  As from 1987, the coverage of the survey has been expanded.

  As from 1984, Oil company operations have been included in the data.

- **United States**

  The data cover only "royalties and licence fees" which are internationally relatively more comparable. Transactions concerning "other private services" have been excluded.

  As from 1992, payments and receipts relating to software are included.

2.3. Sources

The TBP data have been assembled from national sources (central banks, government institutions, etc.). The data have been updated from the first part (Main results) of the OECD International Survey of Resources devoted to R&D by Member countries, returned every six months and from the annual OECD survey on TBP.

MÉTHODOLOGIE GÉNÉRALE

1. RESSOURCES CONSACRÉES À LA R-D

(Tableaux 1 à 18)

1.1. Définitions et champ couvert

Normes OCDE

Les données de cette publication ont été recueillies et présentées selon les normes de la méthodologie de l'OCDE pour les statistiques de R-D publiées dans *La mesure des activités scientifiques et technologiques : méthode type proposée pour les enquêtes sur la recherche et le développement expérimental -- Manuel de Frascati 1993* (OCDE) (et son supplément paru en 1989 concernant les statistiques de R-D et la mesure des résultats dans l'enseignement supérieur).

Les deux principales catégories de données

La plupart des données sur la R-D sont tirées d'enquêtes rétrospectives effectuées auprès des unités qui réalisent ou "exécutent" des projets de R-D. Ainsi, les statistiques qui figurent dans les tableaux 1 à 17 sont fondées sur la somme des déclarations des exécutants en ce qui concerne leurs dépenses de R-D et leur personnel de R-D sur le territoire national (c'est-à-dire à l'exclusion des versements aux organisations internationales et autres exécutants à l'étranger). Les données sur le personnel de R-D sont exprimées en équivalence plein-temps (EPT) à des activités de R-D (c'est-à-dire qu'une personne qui travaille à mi-temps sur la R-D compte pour 0.5 année d'une personne). En raison de la complexité des enquêtes, il est difficile d'obtenir des séries très récentes. Dans ce volume, les données concernant 2000 sont encore provisoires et la plupart des données pour les années postérieures sont des estimations ou des projections nationales (ces données ont été annotées “c” ou “p”).

Des informations plus récentes concernant le soutien de l'État à la R-D peuvent être tirées de sources budgétaires. Les statistiques qui figurent dans le tableau 18 sont fondées sur les crédits budgétaires publics de R-D tels qu'ils ont été déclarés comme étant destinés à la R-D par les ministères ou organismes de l'État qui financent ; ces données comprennent les paiements aux organisations internationales et aux autres exécutants à l'étranger.

Les spécifications des données sur la R-D financée par l'État au tableau 1 diffèrent de façon significative de celles des données budgétaires utilisées pour les crédits budgétaires publics de R-D (tableau 18) ; la combinaison de ces deux séries fournirait une estimation erronée de la contribution civile et militaire aux efforts nationaux de R-D.

Disciplines scientifiques

Les données dans les différents tableaux couvrent généralement tous les disciplines scientifiques (SENI+SSH), c'est à dire l'ensemble des sciences exactes, naturelles (comprenant les sciences médicales et agricoles) et de l'ingénieur (SENI) et des sciences sociales et humaines (SSH). Beaucoup de pays recueillent des données détaillées sur les activités de R-D dans le secteur des entreprises pour les SENI seulement.

Secteurs d'exécution et sources de financement

Pour des besoins statistiques, l'effort national de R-D (dépenses et personnel) est divisé en *quatre secteurs d'exécution* : les entreprises (l'industrie), l'enseignement supérieur, l'État et les institutions sans but lucratif (ISBL). Cette publication fournit des renseignements sur le total de chacun de ces quatre secteurs et ne donne de renseignements détaillés que pour les deux premiers.

Les dépenses de R-D sont subdivisées en *cinq sources de financement* : les entreprises, l'État, l'enseignement supérieur, les institutions sans but lucratif et l'étranger. La somme des financements des secteurs de l'enseignement supérieur et des ISBL étant peu importante, elle a été présentée sous la rubrique "autres sources nationales" dans le tableau 10.

R-D dans le secteur des entreprises (tableaux 9 à 13)

Ces données sont destinées à fournir des mesures internationalement comparables des ressources consacrées à la R-D et leurs évolutions au niveau des industries. Le secteur des entreprises comprend les entreprises privées et publiques et les institutions à leur service.

La classification utilisée est fondée sur la Classification internationale type par industrie (CITI, 3ème révision) (voir tableau page suivante).

La distribution par industries est normalement établie au niveau de l'entreprise, mais certains pays répartissent les informations sur la R-D des entreprises multiproduits entre leurs principales lignes de produits. Dans certains pays, la réglementation statistique nationale interdit la publication de résultats lorsqu'il y a très peu dc compagnies dans une catégorie donnée, ce qui entraîne un manque d'informations dans certains tableaux. En principe, les institutions de R-D au service des entreprises sont classées avec l'industrie concernée. Lorsque ce n'est pas le cas (par exemple en Norvège), la part de la DIRDE exécutée par les entreprises non manufacturières est surestimée par rapport aux autres pays (on trouvera au point "Particularités nationales" ci-après, un résumé des principales caractéristiques nationales qui peuvent réduire la comparabilité dans le temps ou entre pays).

CLASSIFICATION PAR INDUSTRIE UTILISÉE POUR LES RESSOURCES CONSACRÉES À LA R-D DANS LE SECTEUR DES ENTREPRISES DANS L'*ENQUÊTE SUR LA R-D DE L'OCDE* ET CORRESPONDANCE AVEC LA CITI Rév.3, LA NACE Rév.1 ET CORRESPONDANCE APPROXIMATIVE AVEC LA CITI Rév.2

	CITI Rév.3 Division/ Groupe/Classe	Correspondance **NACE Rév.1** Division/ Groupe/Classe	Correspond. **approximative CITI Rév.2** Division/ Groupe/Classe
1.AGRICULTURE, CHASSE ET SYLVICULTURE	**01+02+05**	**01+02+05**	**1**
2.ACTIVITÉS EXTRACTIVES	**10 à 14**	**10 à 14**	**2**
3.ACTIVITÉS DE FABRICATION	**15 à 37**	**15 à 37**	**3**
4. Produits alimentaires, boissons et tabac	**15+16**	**15+16**	**31**
5. Produits alimentaires et boissons	15	15	311 à 313
6. Produits à base de tabac	16	16	314
7. Textiles, habillement, fourrures et cuirs	**17 à 19**	**17 à 19**	**32**
8. Textiles	17	17	321
9. Articles d'habillement et fourrures	18	18	}
10. Articles en cuir et chaussures	19	19	}322 à 324
11. Bois, papier, imprimerie, édition	**20 à 22**	**20 à 22**	**331+34+3832(part)**
12. Bois et liège (sauf meubles)	20	20	331
13. Carton, papier et articles	21	21	341
14. Édition, impression et reproduction d'enregistrements	22	22	342+3832(part)
15. Coke, pétrole, combustibles nucléaires, produits chimiques, caoutchouc et matières plastiques	**23 à 25**	**23 à 25**	**35**
16. Coke, raffinerie pétrole et combustibles nucléaires	23	23	353+354
17. Produits chimiques	24	24	351+352
18. *Produits chimiques moins pharmaceutiques*	24 moins 2423	24 moins 24.4	351+352 moins 3522
19. *Produits pharmaceutiques*	2423	24.4	3522
20. Caoutchouc et matières plastiques	25	25	355+356
21. Produits minéraux non métalliques	**26**	**26**	**36**
22. Produits métallurgiques de base	**27**	**27**	**37**
23 Produits métallurgiques de base ferreux	271+2731	27.1 à 27.3+27.51/52	371
24. Produits métallurgiques de base non ferreux	272+2732	27.4+27.53/54	372
25. Ouvrages en métaux (sauf machines et matériels)	**28**	**28**	**381**
26. Machines et matériels, instruments et matériels de transport	**29 à 35**	**29 à 35**	**38 moins 381 et 3832(part)**
27. Machines, n.c.a.	29	29	382 moins 3825+3829(part)
28. Machines de bureau, comptabilité et traitement informatique	30	30	3825
29. Machines électriques	31	31	383 moins 3832
30. Équipements électroniques (Radio, TV et communications)	32	32	3832(part)
31. *Composants électroniques (semi-conducteurs inclus)*	321	32.1	
32. *Appareils radio, TV et de communication*	32 moins 321	32 moins 32.1	
33. Instruments médicaux, de précision et d'optique et horlogerie	33	33	385
34. Véhicules automobiles	34	34	3843
35. Autres matériels de transport	35	35	384(part)+3829(part)
36. *Navires*	351	35.1	3841
37. *Aérospatiale*	353	35.3	3845+3829(part)
38. *Autres équipements de transport*	352+359	35.2+35.4+35.5	3842+3844+3849
39. Meubles, autres activités de fabrication n.c.a.	**36**	**36**	**332+39**
40. Meubles	361	36.1	332
41. Autres activités de fabrication n.c.a.	369	36.2 à 36.6	39
42. Récupération	**37**	**37**	**ND**
43.DISTRIBUTION D'ÉLECTRICITÉ, GAZ ET EAU	**40+41**	**40+41**	**4**
44.CONSTRUCTION	**45**	**45**	**5**
45.SECTEUR DES SERVICES	**50 à 99**	**50 à 99**	**6 à 9**
46. Commerce (gros et détail), réparation véhicules, etc.	**50 à 52**	**50 à 52**	**61+62+6(part)**
47. Hôtels et restaurants	**55**	**55**	**63**
48. Transports, entreposage	**60 à 63**	**60 à 63**	**71**
49. Communications	**64**	**64**	**72**
50. Postes	641	64.1	
51. Télécommunications	642	64.2	
52. Intermédiation financière (assurances incluses)	**65 à 67**	**65 à 67**	**81+82**
53. Immobilier, location, services aux entreprises	**70 à 74**	**70 à 74**	**83+932**
54. Informatique et activités connexes	72	72	8323
55 *Production de logiciels*	722	72.2	
56. *Autres activités informatiques*	72 moins 722	72 moins 72.2	
57. Recherche-développement expérimental	73	73	932
58. Autres activités et services aux entreprises	70+71+74	70+71+74	83(part)
59. Administration publique, services sociaux et collectifs, etc. [a]	**75 à 99**	**75 à 99**	**9 moins 932**
60.TOTAL GÉNÉRAL	**01 à 99**	**01 à 99**	**1 à 9**

a. Activités réalisées dans ces branches uniquement par le secteur des entreprises. Les chiffres sont probablement négligeables : la rubrique est reprise à titre d'aide-mémoire.

R-D dans le secteur de l'enseignement supérieur (tableaux 14 à 17)

Ces données sont principalement destinées à décrire les caractéristiques et l'évolution des travaux de R-D du secteur de l'enseignement supérieur selon les principales disciplines scientifiques (se reporter aussi au *"Manuel de Frascati 1993"*, section 3.7), en raison de l'importance du rôle joué par les universités et établissements analogues dans l'exécution de la R-D. Le secteur de l'enseignement supérieur comprend toutes les universités, grandes écoles, instituts de technologie et autres établissements post-secondaires, ainsi que les instituts de recherche, les stations d'essais et les cliniques travaillant sous le contrôle direct d'établissements d'enseignement supérieur ou administrés par, ou rattachés à ces derniers.

La classification de base est constituée des six disciplines scientifiques et technologiques principales suggérées par l'UNESCO dans la *Recommandation concernant la normalisation internationale des statistiques relatives à la science et à la technologie.*

Crédits budgétaires publics pour la R-D (CBPRD) (tableau 18)

Les autorités nationales se servent des statistiques budgétaires pour obtenir ces données. Celles-ci comprennent essentiellement l'identification de toutes les rubriques budgétaires concernant la R-D et le calcul ou l'estimation de leur contenu de R-D. Les séries ne concernent en général que le gouvernement fédéral ou central. Les estimations fondées sur les rapports des "financeurs" sont moins précises que les données fournies par ceux qui exécutent la R-D dans les tableaux 1 à 17. Néanmoins, ces données étant fondées sur le budget, elles peuvent être reliées aux politiques nationales au moyen d'une classification par "objectif" ou "but". Les programmes sont distribués entre les objectifs socio-économiques sur la base des *intentions* qui existent au moment de l'allocation des fonds et non sur le contenu véritable des projets R-D concernés. Cette distribution reflète les *politiques* nationales à un moment donné.

La classification utilisée a été spécialement développée à l'OCDE pour l'analyse de la R-D (se reporter au *"Manuel de Frascati 1993"*, section 8.7.3 & 8.7.4). Elle est compatible avec les classifications semblables utilisées par la CE et les pays nordiques.

Dépenses en monnaie nationale

Les données en monnaie nationale sont exprimées en euros pour les pays de la zone Euro. Les données sont en EUR à compter de la date d'entrée dans l'Union Économique et Monétaire (UEM). Les données avant l'entrée dans l'UEM ont été converties à partir de l'ancienne monnaie nationale et du taux de conversion irrévocable approprié. Cette présentation facilite les comparaisons entre périodes pour un pays donné et assure que l'évolution historique est respectée. Cependant, ***l'euro avant l'entrée dans l'UEM est une unité fictive qui ne se prête ni à des agrégats de pays ni à des comparaisons internationales***.

Dans cette publication, les séries temporelles s'étendent sur une période chevauchant la date d'accession des pays à l'UEM et comprennent donc des données antérieures à cette date exprimées dans l'ancienne monnaie nationale converties en utilisant le taux de conversion irrévocable et des données postérieures à la date d'accession exprimées en véritables EUR. Les données de l'Autriche en monnaie nationale, par exemple, sont présentées en euros ATS 1999 jusqu'à 1998 et en EUR à partir de 1999 ; de même, les données de la Grèce sont présentées en euros GRD 2001 jusqu'à 2000 et en EUR à partir de 2001.

Les pays de la zone Euro, leur taux de change irrévocable EUR/monnaie nationale et leur année d'accession à l'UEM sont présentés ci-dessous :

Pays	*1 EUR =*	*L'année d'accession à l'UEM*
Allemagne	1.95583 DEM	1999
Autriche	13.7603 ATS	1999
Belgique	40.3399 BEF	1999
Espagne	166.386 ESP	1999
Finlande	5.94573 FIM	1999
France	6.55957 FRF	1999
Grèce	340.750 GRD	2001
Irlande	0.787564 IEP	1999
Italie	1936.27 ITL	1999
Luxembourg	40.3399 LUF	1999
Pays-Bas	2.20371 NLG	1999
Portugal	200.482 PTE	1999

Dépenses en dollars constants

Certaines séries des dépenses de R-D ont été déflatées à l'aide des déflateurs implicites du PIB, tirés de la banque de données des Comptes nationaux de l'OCDE et mis à jour par le Secrétariat pour les années les plus récentes en utilisant les projections des taux d'augmentation de prix du PIB, tels qu'ils sont publiés deux fois par an dans les *Perspectives économiques de l'OCDE* (sauf dans le cas de la Norvège où c'est un déflateur ne tenant pas compte des fluctuations du prix du pétrole qui a été utilisé). Ces données en monnaie nationale aux prix de 1995, ont été converties en $US en utilisant les parités de pouvoir d'achat (PPA) de 1995. Les PPA ont été établies par la Division des Comptes nationaux de l'OCDE (pour plus de détails, on peut se reporter aux parties VII et VIII des *Comptes nationaux, vol. 1, 1988-1999*, OCDE, Paris, 2001), et mises à jour pour les années les plus récentes en comparant le taux de croissance des prix (déflateur implicite du PIB) dans chaque pays avec celui des États-Unis. Ces parités estimées et les données dérivées en dollars courants sont notées par la lettre "*b*" dans les séries.

Comparabilité internationale

Bien que tous les pays de l'OCDE recueillent et fournissent les données sur la R-D selon les recommandations du "*Manuel de Frascati*", certaines spécifications nationales peuvent différer des normes OCDE dans le détail. Les principales exceptions sont indiquées au point "Particularités nationales" ci-après. Elles se rapportent principalement :

- aux ruptures de séries ;
- à la couverture ou non de tous les domaines scientifiques (SENI et SSH) par les données ;
- aux difficultés que connaissent les pays lorsqu'ils s'efforcent de mesurer les montants des ressources consacrées à la R-D dans le secteur de l'enseignement supérieur, difficultés qui tiennent à la relation étroite qui existe entre les activités d'enseignement et les activités de recherche ;
- aux différences de champ couvert par les secteurs de l'enseignement supérieur et de l'État dues aux institutions "frontières" entre ces deux secteurs ;

– à l'inclusion ou non des instituts de R-D au service des entreprises, et à leur répartition ou non selon l'industrie qu'ils servent.

1.2. Particularités nationales

On trouvera dans ce qui suit, pour la période 1981 jusqu'à présent, un résumé par pays des principales divergences nationales par rapport aux normes OCDE. Pour les détails sur les différences de classification industrielle (tableaux 9 à 13), on se reportera à la publication *"Les dépenses en Recherche et développement dans l'industrie, 1974-95"*, OCDE, Paris 1997.

Il arrive que des révisions rétrospectives aient été communiquées pour des données globales sans les révisions pour les sous-ventilations correspondantes (par exemple une révision des sources de financement de la DIRDE totale et de la DIRD au tableau 1, mais pas les révisions correspondantes de la DIRDE par industrie aux tableaux 9 et 10). Ces dernières données non révisées ont été annotées par une note *"s"* dans les tableaux concernés.

- **Allemagne**

Dans le total des dépenses intra-muros de R-D du secteur des entreprises (DIRDE) sont inclus des fonds venant de l'État qui n'ont pas été saisis par l'enquête auprès des entreprises, mais sont tirés des dépenses extra-muros de R-D de l'État vers les entreprises. Ces fonds n'ont pas pu être ventilés par type de coûts, ou par type d'activités (tableaux 2 et 3), ou par industrie (tableaux 9 et 10). En 1997 le montant de ces fonds était négligeable ; en 1995 ces fonds s'élevaient à 386 millions de DM et `a 515 millions de DM en 1993).

Semblablement dans le secteur de l'enseignement supérieur, certains fonds n'ont pas pu être inclus dans la ventilation des dépenses de R-D par types de coûts ou par types d'activités (tableaux 2 et 3), mais aussi pour les grands domaines scientifiques (tableaux 7) ainsi que certaines ventilations détaillées par discipline scientifique (tableaux 14 et 15).

Les fonds propres de l'enseignement supérieur sont inclus avec les fonds venant de l'État.

En 1997, la méthode d'évaluation des CBPRD par objectif socio-économique a changée. Le montant total du budget du Ministère fédéral pour l'enseignement, la science, la recherche et la technologie pour 1997 a été réduit, mais la répartition de la réduction globale par objectif socio-économique n'était pas disponible. Il s'ensuit que le total des CBPRD reflète la donnée ajustée pour le budget, mais la somme des objectifs n'ajoute plus à ce total.

En 1995, la méthode d'évaluation des ressources consacrées à la R-D dans le secteur de l'enseignement supérieur a changé et ce système a été appliqué rétroactivement jusqu'en 1981 pour rendre les données de 1981 à 1994 compatibles aux donnés les plus récentes.

À partir de 1993, dans les dépenses de R-D du secteur de l'État, et en conséquence dans les dépenses totales de R-D (DIRD), sont comprises des dépenses pour des activités de R-D réalisées par des instituts de recherche allemands situés à l'étranger. Ces montants représentent environ 0.6% à 0.7% des dépenses de R-D du secteur de l'État et moins de 0.1% de la DIRD.

En 1992 :

– La restructuration faisant suite à la réunification a entraîné la création de plus de 100 instituts de recherche non-universitaires dans le secteur de l'État.

– La méthodologie de l'enquête sur les ressources consacrées à la R-D dans le secteur de l'État a été modifiée.

– Depuis cette date les ressources consacrées à la R-D par le secteur des ISBL sont incluses dans le secteur de l'État.

À partir de 1991:

- Les données se rapportent à l'Allemagne réunifiée, alors que jusqu'en 1990 elles ne concernent que l'Allemagne de l'Ouest.
- Les étudiants diplômés poursuivant des travaux de recherche et ayant reçu une subvention à cet effet ont été inclus pour la première fois dans les chercheurs de l'enseignement supérieur (près de 10 000 EPT). Les subventions correspondantes qui leurs ont été versées en 1991 ont été comprises dans la dépense de R-D de l'enseignement supérieur (HERD), ainsi le montant des subventions versées à l'enseignement supérieur en 1991 est bien plus élevé que précédemment car seules étaient incluses jusque là les subventions directement versées par l'État, ce qui a contribué à l'accroissement du financement total de l'État à la R-D de l'enseignement supérieur cette année-là.

1987 : une nouvelle enquête sur le secteur des ISBL.

1985 : plusieurs modifications concernant les secteurs de l'État et de l'enseignement supérieur.

1981 et 1983 ; ajustements à la baisse de la R-D exécutée par le secteur de l'État et des fonds de l'État vers l'enseignement supérieur.

- **Australie**

Dans la répartition de la dépense de R-D du secteur des entreprises par type de coûts, les charges salariales du personnel d'aide et d'entretien ne travaillant pas directement à la R-D sont comprises dans le coût du personnel de R-D au lieu d'être incluses dans les "autres coûts courants" comme cela est recommandé dans le *Manuel de Frascati.*

À partir de 1999, l'Australie prépare les détails du budget fédéral selon les principes de la comptabilité patrimoniale, menant à une rupture de séries pour des données des CBPRD. En 1999, ce changement de la comptabilité de caisse vers la comptabilité patrimoniale a augmenté l'évaluation des CBPRD de 232 millions de dollars australiens.

- **Autriche**

Une estimation des dépenses de R-D des hôpitaux provinciaux, fondée sur les rapports financiers annuels des provinces fédérales, a été incluse dans les tableaux sur les dépenses (pour 1989 cette estimation s'est élevée à 637.7 millions de schillings autrichiens). On ne dispose pas d'estimation correspondante pour le personnel de R-D.

En 1993 :

- Par suite du reclassement de plusieurs institutions du secteur des ISBL dans le secteur de l'État, la plupart des données concernant ces deux secteurs en 1993 ne sont pas comparables à celles des années antérieures.
- Dans le total des dépenses intra-muros de R-D (DIRD) sont inclus des fonds venant de l'État qui n'ont pas été saisis par les enquêtes sur la R-D. Ces fonds qui ont été inclus dans les totaux n'ont pas pu être ventilés par type de coûts ou type d'activités (tableaux 2, 3, 11 et 15) (ces fonds s'élevaient à 903.1 millions de SA en 1993, dont 750.5 millions pour des dépenses courantes de R-D et 152.6 millions pour des dépenses en capital). De même, ils n'ont pas pu être ventilés par industrie et source de financement (tableaux 9 et 10) (481 millions de SA en 1993).

- **Belgique**

À partir de 1993 (1992 pour le secteur des entreprises), la méthodologie des enquêtes sur la R-D a été modifiée et les données sont uniquement basées sur les résultats de l'enquête et ne sont plus des estimations fondées sur une combinaison de résultats d'enquêtes et de données budgétaires. Il y a donc une rupture de série, et les données pour ces années là ne sont pas comparables à celles des années précédentes.

En 1989, les méthodes d'évaluation de l'effort de R-D ont changées (enquêtes incluant désormais les ressources des communautés et régions devenues très importantes à la suite de la réforme de l'État belge de 1988) et les donnés ne sont plus comparables à celles des années antérieures à partir de cette date-là.

En 1987 et 1988 la dépense intérieure brute de R-D (DIRD) ainsi que la DIRD financée par l'État (pouvoirs publics) (tableaux 1 à 4 et 7) sont sous-estimées car la R-D financée par les Autorités fédérées est exclue (environ 2 à 4 % de la DIRD et 7 à 15 % de la DIRD financée par l'État). La distribution de ces fonds par secteur d'exécution n'étant pas disponible, il est impossible d'estimer avec précision l'impact sur les autres tableaux plus détaillés par secteurs d'exécution (entreprises : tableaux 9 à 11 ; et enseignement supérieur : tableaux 14 à 16) dont les données sont sous-estimées.

À partir de 1983, les données se rapportant à la DIRD ne sont plus comparables à celles des années antérieures par suite de changements dans les méthodes d'évaluation de l'effort de R-D en Belgique.

À partir de 1981, les ressources consacrées à la R-D dans le secteur des entreprises sont sous-estimées dans tous les tableaux détaillés par industrie (tableaux 9 à 13) parce que les ressources des centres communs de R-D des entreprises ne sont pas comprises (leur ventilation par industrie n'étant plus disponible). À la fin des années 70, la contribution de ces centres représentait environ un sixième des dépenses de R-D des industries non-manufacturières.

- **Canada**

Les sources de financement de l'enseignement supérieur ont été ajustées en ce qui concerne les FGU (fonds généraux des universités) au niveau des totaux nationaux (tableau 1), mais cet ajustement n'a pas pu être opéré au niveau des données détaillées par discipline scientifique dans le secteur de l'enseignement supérieur (tableau 16).

À partir de 1989, les fonds généraux des universités provenant de sources non fédérales sont inclus dans CBPRD.

À partir de 1988, on a inclus dans la dépense de R-D de l'enseignement supérieur les montants estimés au titre de la R-D par les hôpitaux universitaires, non couverts par les universités, non couverts par les données antérieures.

En 1987, une modification du nombre de catégories dans l'enquête a eu pour résultat une rupture de séries.

- **Corée**

Les sciences sociales et humaines sont exclues des données sur la R-D. Les ressources consacrées à la R-D sont donc sous-estimées notamment dans les secteurs de l'État, de l'enseignement supérieur et des ISBL.

- **Danemark**

La dépense de R-D du secteur de l'enseignement supérieur est sous-estimée puisque toute la R-D menée dans les départements hospitaliers des hôpitaux universitaires est comprise dans le secteur de l'État.

En 2001, 1993, 1988 et 1983, des changements de méthode d'évaluation du financement public destiné à la R-D (CBPRD) par objectif socio-économique (tableau 18) ont entraîné des ruptures de série.

En 2000, le champ couvert par les CBPRD a été élargi pour inclure le financement provincial de la R-D dans les hôpitaux ainsi que les fonds provenant de la Fondation nationale danoise de la research et de la VaekstFonden.

En 1999, la répartition de la dépense nationale de R-D par type d'activité (Recherche fondamentale, Recherche appliquée et Développement expérimental) est fondée sur la somme des dépenses courantes intra-muros de R&D de secteur des entreprises et des dépenses totales intra-muros de R-D des trois autres secteurs.

- **Espagne**

En 1997 :

- L'objectif défense dans les CBPRD a quasiment doublé à la suite d'une contribution exceptionnelle du Ministère de l'industrie et de l'énergie
- L'incorporation de la contribution de l'Espagne au CERN a entraîné un changement substantiel pour l'objectif énergie.

En 1995 :

- Les fonds propres de l'enseignement supérieur ont été séparés des fonds généraux des universités (FGU) dans lesquels ils étaient inclus auparavant. Il s'ensuit que les données de 1995 se rapportant aux FGU, au sous-total du financement de la R-D par l'État dans le secteur de l'enseignement supérieur ainsi qu'au niveau national ne sont pas comparables à celles des années antérieures.
- La forte croissance des fonds venant de l'étranger dans le secteur de l'enseignement supérieur est principalement due à des fonds provenant de la Commission européenne qui se sont en quelque sorte substitués aux fonds venant de l'État.

A partir de 1994, par suite de la disponibilité de la répartition du personnel de soutien de l'ES entre "techniciens" et "autres", les catégories correspondantes pour le personnel national total de R-D ne sont plus sous-évaluées.

À partir de 1992 :

- Il y a eu une réévaluation à la hausse des Fonds généraux des universités, ce qui a provoqué une rupture de série dans la dépense de R-D de l'enseignement supérieur et son financement par l'État, ainsi que pour la DIRD totale et son financement par l'État (tableaux 1, 2, 3, 7, 14, 15 et 16). Il faut noter qu'il n'y a pas eu de réévaluation équivalente pour les données sur les crédits budgétaires destinés à la R-D (tableau 18), qui ont une source et une méthode d'estimation différentes pour les FGU.
- Le coût du personnel et le sous-total des dépenses courantes ne sont plus sous-estimés par suite de l'inclusion des coûts relatifs aux techniciens et autres personnels de soutien du secteur de l'enseignement supérieur, qui n'étaient pas inclus jusque là, dans la dépense de R-D de l'enseignement supérieur.
- La méthode d'évaluation des dépenses courantes et des dépenses en capital pour la R-D de l'enseignement supérieur a changé et ces informations sont dorénavant recueillies au moyen d'un questionnaire auprès des universités qui en font les estimations.

Avant 1989, seul le nombre de chercheurs (en EPT à la R-D) est disponible pour le personnel de R-D de l'enseignement supérieur, et jusqu'en 1994 les données sur les techniciens et autre personnel de soutien ne sont pas non plus disponibles séparément pour ce secteur. Il s'ensuit qu'avant 1989 le personnel total de R-D, ainsi que les nombres totaux de techniciens et autres personnels de soutien jusqu'en 1994 (tableaux 5, 6 et 8), sont sous-estimés pour l'Espagne (de 10 à 15 % pour le personnel total).

Les séries des dépenses et du personnel de R-D du secteur de l'enseignement supérieur ont été révisées rétrospectivement à la hausse jusqu'en 1980 pour tenir compte d'une modification de la méthode d'évaluation des données de 1988 à la suite de l'enquête sur l'emploi du temps des professeurs menée en 1989-90.

- **États-Unis**

Les dépenses de R-D dans le secteur des entreprises comprennent les amortissements à la place des dépenses en capital.

Les données des tableaux 1 à 17 sont probablement légèrement sous-estimées :

- La R-D du secteur de l'État ne comprend que les activités du Gouvernement fédéral et pas les activités des établissements des États et des gouvernements locaux.
- Dans le secteur de l'enseignement supérieur, la R-D ne couvre pas les sciences humaines ni les dépenses en capital.
- Les dépenses de R-D du secteur des ISBL ne comprennent que les dépenses courantes.

En ce qui concerne le personnel de recherche, seules les données concernant les chercheurs sont recueillies, on ne dispose donc pas d'information sur le personnel total de R-D aux États-Unis.

Les données se rapportant à la DIRD par source de financement ont été largement révisées et correspondent à des années civiles. Ces révisions affectent principalement le secteur de l'État et celui de l'enseignement supérieur.

À partir de 1999, l'agriculture n'est plus incluse dans l'enquête sur la R-D du secteur des entreprises.

À partir de 1998, il y a eu quelques reclassifications dans les données se rapportant aux CBPRD (tableau 18), certains composants précédemment inclus sous Énergie ont été transférés sous Promotion de la science, ce qui a introduit une rupture de série pour ces deux objectifs.

En 1996, les données sont disponibles séparément pour l'agriculture, les industries minières et la construction et ne sont plus comprises dans le sous-total du secteur des services, qui pour 1996 n'est donc pas comparable à celui des années précédentes.

Avant 1996, le sous-total des dépenses de R-D dans les industries des services du secteur des entreprises était calculé comme un résiduel, comprenant l'agriculture, les industries minières et la construction.

À partir de 1985 :

- La méthode d'estimation du nombre de chercheurs a changé. Dans les secteurs de l'État, de l'enseignement supérieur et des institutions sans but lucratif les données se rapportent aux titulaires d'un doctorat dont l'activité principale est la recherche, le développement expérimental ou la gestion de projets de R-D, auxquels est ajouté, pour le secteur de l'enseignement supérieur, le nombre exprimé en équivalence plein temps des étudiants diplômés ayant une bourse de recherche, dont on a estimé qu'ils consacrent en moyenne 50 % de leur temps à des travaux de R-D.
- Les chercheurs de l'État ne comprennent plus les chercheurs des forces armées.

- **Finlande**

En 1998, en raison d'un plus grand taux de réponses au niveau des groupes, la catégorie "financement par d'autres entreprises étrangères du groupe" a été fusionnée avec celle du financement par les entreprises (fonds propres), réduisant par conséquent la part du financement de l'étranger.

À partir de 1997, les hôpitaux universitaires sont inclus dans le secteur de l'enseignement supérieur. De même, les CBPRD comprennent les crédits pour la R-D de ces centres hospitaliers universitaires.

À partir de 1995, les fonds provenant de sources extérieures aux instituts de recherche d'État sont exclus des crédits budgétaires publics pour la R-D (CBPRD).

À partir de 1994, toutes les institutions auparavant incluses dans le secteur des ISBL ont été reclassées dans le secteur de l'État pour les années où il n'y a pas d'enquête entraînant une rupture de série pour les données se rapportant au secteur de l'État ces années-là. En 1993, l'effort de R-D des ISBL représentait un peu plus de 3% de la dépense de R-D de l'ensemble État+ISBL, près de 4% du personnel de R-D et 5.5% des chercheurs.

En 1991 :

- La méthode d'évaluation des dépenses de R-D des secteurs de l'État et de l'enseignement supérieur a été modifiée, notamment le coefficient de calcul des coûts des pensions est passé de 15 à 25 % des salaires. Il s'ensuit que les dépenses de R-D de ces secteurs en 1991 ainsi que le financement de la R D par l'État et le total national DIRD ne sont pas comparables à celles des années précédentes.
- De même, les données se rapportant aux CBPRD ont été révisées jusqu'en 1991 par suite du changement des coefficients pour certains instituts de recherche et il y a eu un ajustement à la hausse en 1991 pour le total, du à l'inclusion des coûts de pensions.

En 1987 :

- Il semble que l'estimation des dépenses de constructions et équipements, et par conséquent le sous-total des dépenses en capital, dans les secteurs de l'État et de l'enseignement supérieur ne soient pas comparables aux données des années précédentes.
- La répartition du personnel de R-D par niveau de qualification a été modifiée : les autres qualifications post-secondaires non-techniques, qui étaient auparavant incluses dans la catégorie "autres diplômes" sont dorénavant comprises avec les "autres qualifications post-secondaires".

En 1983, les modifications apportées aux méthodologies utilisées pour établir les données réduisent la comparabilité pour les secteurs de l'enseignement Supérieur et de l'État en ce qui concerne les sous-ventilations.

En 1981 certaines données ont été révisés rétrospectivement par les autorités nationales, et les totaux par secteur pour l'État et l'enseignement supérieur pour 1981 et ne sont plus comparables aux données des années précédentes. Les révisions correspondantes dans les ventilations plus détaillées par catégories de dépenses ou de personnel n'ont pas été fournies.

- **France**

Le Centre national de la recherche scientifique (CNRS) est inclus dans le secteur de l'enseignement supérieur, alors que ce genre d'organisme est classé dans le secteur de l'État dans d'autres pays comme l'Italie; ceci a une influence sur la répartition de l'effort de R-D par secteur d'exécution.

À partir de 1997 :

- la méthode utilisée pour évaluer le personnel de R-D dans les secteurs de l'État, de l'enseignement supérieur et des ISBL a changé ;
- des révisions significatives ont été apportées aux données sur les dépenses et le personnel de R-D par suite des résultats obtenus dans les enquêtes de 1998 avec :
  - de nouvelles méthodes d'évaluation des dépenses intérieures pour la défense ;
  - la réévaluation des activités de R-D de très grandes entreprises;

- la réévaluation de la recherche clinique et la réévaluation de l'effort de R-D d'une organisation civile.

En conséquences, les données de 1997 se rapportant au personnel de R-D des secteurs de l'État, de l'enseignement supérieur et des ISBL ne sont pas comparables à celles des années précédentes de même que les données de 1997 se rapportant à la R-D intra-muros du secteur des entreprises ainsi que celles du financement de la R-D par les secteurs des entreprises et de l'État ne sont pas comparables à celles des années précédentes. La rupture de série de 1997 s'applique aussi aux données des CBPRD par objectif socio-économique.

À partir de 1992 :

– Les données du secteur des entreprises et celles du secteur de l'État ne sont pas comparables aux données correspondantes de 1991 par suite du transfert d'organismes ("France Télécom" et GIAT industries) du secteur de l'État dans le secteur des entreprises, conformément à leur changement de statut juridique.

– Il a été tenu compte du remboursement des aides remboursables dans la répartition de la dépense de R-D par source de financement.

– La méthode d'enquête sur le secteur des ISBL a été modifiée.

À partir 1981, à la suite de la réévaluation de la proportion du temps que les enseignants consacrent à la R-D, les ressources consacrées par le secteur de l'enseignement supérieur à la R-D et en conséquence au niveau national ne sont pas comparables à celles des années antérieures.

- **Grèce**

En 1995, à la suite de la révision des coefficients utilisés pour évaluer les ressources consacrées à la R-D par l'enseignement supérieur, les données se rapportant aux dépenses et au personnel de R-D du secteur de l'enseignement supérieur de même que pour les totaux nationaux ne sont pas comparables aux données des années précédentes.

À partir de 1993, le secteur des ISBL est inclus dans l'enquête.

En 1991, certaines activités comprises jusque là sous "machines de bureaux et ordinateurs" (et donc dans le sous-total correspondant "machines et matériels, instruments et transports) dans l'industrie manufacturière, ont été reclassées dans les services sous "informatique et activités connexes" (et donc dans le sous-total correspondant "immobilier, location, services aux entreprises").

En 1989 les "coefficients recherche" utilisés pour évaluer les ressources consacrées à la R-D par l'enseignement supérieur ont été révisés.

En 1983 :

– La méthodologie pour évaluer les ressources consacrées à la R-D par l'enseignement supérieur a été changée. Les "coefficients recherche" utilisés pour les calculs ont été établis au moyen d'une enquête dans les établissements de l'enseignement supérieur.

– La méthodologie pour recenser le personnel de R-D a été modifiée pour être conforme aux recommandations du *Manuel de Frascati* et ceci a entraîné une modification des coûts du personnel.

- **Hongrie**

Seules les dépenses de R-D liées à la défenses qui sont exécutées dans le secteur civil sont couvertes, celles des forces armées sont exclues.

Toutes les répartitions des dépenses de R-D par secteur d'exécution, par source de financement, par type de coût, par type d'activité ou par principal domaine scientifique (tableaux 1 à 3, tableau 7, et aussi tableau 9 et tableaux 14 à 16) sont incomplètes.

À partir de 1998, le personnel ayant des activités de sécurité ou d'entreposage dans les unités de R-D est exclu, en accord avec le *Manuel de Frascati*.

À partir de 1994 :

- Les achats de licences et savoir faire, auparavant inclus dans les dépenses de R-D du secteur des entreprises (DIRDE), en ont été exclus.
- La classification du Fond central de technologie (KMÜFA), auparavant compris dans le secteur des entreprises, a été inclus dans le secteur de l'État.

- **Irlande**

Jusqu'en 1992 compris, les données du secteur de l'État ont été révisées car certaines dépenses ne sont plus considérées comme des dépenses de R-D.

Rétrospectivement jusqu'à 1985 inclus, les séries sur la dépense de R-D de l'enseignement supérieur ont été révisées après les résultats obtenus dans l'enquête exhaustive de 1993 sur la R-D du secteur de l'enseignement supérieur, qui ont clairement indiqué que le coefficient du temps passé à la recherche avait sérieusement changé depuis l'enquête de 1984.

Depuis 1983, il n'y a plus de financement non-public dans les séries CBPRD.

- **Islande**

En 1993, les méthodes de collecte et de compilation des données sur les crédits budgétaires ont été révisées, entraînant à une rupture dans les séries des CBPRD.

- **Italie**

À partir de 1991, les dépenses extra-muros sont recueillies séparément dans les enquêtes italiennes sur la R-D et ont pu ainsi être exclues de la plupart des données.

Jusqu'en 1990, les données présentées au tableau 1 - Dépenses intérieure brute de R-D par secteur exécution et source de financement - sont fondées sur l'ensemble des dépenses intra plus extra-muros de R-D (y compris les dépenses extra-muros vers l'étranger). Par rapport aux normes du *Manuel de Frascati*, l'inclusion des dépenses extra-muros de R-D entraîne une surestimation de la dépense intérieure brute de R-D (DIRD). Ceci concerne plus particulièrement de la dépense de R-D du secteur des entreprises mais également celle du secteur de l'État.

Les dépenses extra-muros représentaient :

- Au niveau national, plus de 6 % de la DIRD en 1991 et près de 15 % en 1993. En 1991, plus de 20 % de ces dépenses extra-muros représentaient des financements vers l'étranger et plus de 26 % en 1993.
- Pour le secteur des entreprises, 11 % en 1991 et près de 15 % en 1993 de la dépense de R-D exécutée par ce secteur (DIRDE). Plus de 13 % en 1991 et près de 18 % en 1993 de la R-D financée par le secteur des entreprises étaient des dépenses extra-muros (le quart des dépenses extra-muros de ce secteur représente des fonds vers l'étranger).
- Pour le secteur de l'État, seulement 1 % de la dépense de R-D exécutée par ce secteur (DIRDET).

À partir de 1987, la répartition des dépenses de R-D par type d'activité est fondée sur les dépenses totales et non les dépenses courantes de R-D.

En 1984, une partie des ressources de l'ENEA (Comité National de R-D pour l'énergie nucléaire et l'énergie alternative) a été transféré de l'objectif « Environnement » à la « Santé », ce qui a réduit le premier de moitié et augmenté le second en conséquences (Tableau 18 - CBPRD).

À partir de 1981, les données sur le financement de la R-D pour la « Défense » ne sont plus comparables à celles des années précédentes à la suite d'une révision à la hausse due à une amélioration de l'enquête (Tableau 18 - CBPRD).

- **Japon**

Jusqu'en 1995, les données concernant le personnel de R-D sont surestimées, de même que les coûts du personnel selon les normes internationales. Les données sur les chercheurs sont exprimées en nombre de personnes physiques travaillant principalement à la R-D plutôt qu'en équivalence plein-temps (EPT). Selon des études effectuées par certaines autorités japonaises, le nombre de chercheurs en EPT à des activités de R-D serait inférieur d'environ 40 % dans le secteur de l'enseignement supérieur et de 30 % pour le total national. En conséquence, les dépenses de R-D devraient être réduites d'environ 25 % dans le secteur de l'enseignement supérieur et d'environ 15 % pour la DIRD totale. L'OCDE a calculé jusqu'en 1995, des séries ajustées aussi bien pour les dépenses que pour le personnel de R-D pour le secteur de l'enseignement supérieur ainsi qu'au niveau national ; ces données sont disponibles dans la publication de l'OCDE Principaux indicateurs de la science et de la technologie, ainsi que dans diverses études et rapports analytiques.

La répartition par source de financement de la dépense de R-D du secteur de l'enseignement supérieur, au niveau des données nationales globales (tableau 1), a été ajustée par le Secrétariat entre les fonds venant de l'État et ceux venant de l'enseignement supérieur, notamment pour une estimation des FGU. Cet ajustement n'a pas pu être opéré sur les données correspondantes détaillées par discipline scientifique du secteur de l'enseignement supérieur (tableau 16). Pour 1990, l'ajustement de la répartition des sources de financement du secteur de l'enseignement supérieur a été effectué par les autorités japonaises selon une méthode différente, ce qui a introduit une rupture de série cette année là principalement pour le financement direct par l'État. La reprise de l'ajustement par le Secrétariat en 1991 a entraîné une autre rupture de série pour le financement direct par l'État.

Les données budgétaires sur le financement public pour la R-D par objectif socio-économique (CPBRD, tableau 18) sont sous-estimées car les FGU ne comprennent pas la R-D en sciences sociales et humaines et la part R-D des commandes militaires est exclue.

- **Mexique**

1993 est la première année pour laquelle on dispose d'un ensemble complet de données sur la R-D au Mexique qui soit internationalement comparables.

1993 et 1994 : les données sur les ressources consacrées à la R-D dans le secteur de l'État ainsi que dans le secteur des ISBL proviennent des premières enquêtes nationales sur les ressources consacrées à la R-D fondées sur le *Manuel de Frascati* dans ces secteurs, elles ne sont pas comparables aux estimations disponibles pour les années précédentes pour le secteur de l'État.

1992 et 1993 : les données sur les ressources consacrées à la R-D dans le secteur des entreprises ainsi que dans le secteur de l'enseignement supérieur proviennent des premières enquêtes nationales sur les ressources consacrées à la R-D fondées sur le *Manuel de Frascati* dans ces secteurs, elles ne sont pas comparables aux estimations disponibles pour les années précédentes pour ces secteurs.

- **Norvège**

Dans la répartition de la dépense brute de R-D (DIRD) par secteur d'exécution et objectif socio-économique (tableau 4) les dépenses en capital n'ont pas été réparties par objectif mais regroupées sous "Non classé".

Une nouvelle méthode de collecte des données sur les CBPRD est appliquée depuis 1996 ; les séries ont été révisées rétrospectivement jusqu'en 1983 pour exclure la R-D en sous-traitance, les entreprises publiques et les paiements vers la Commission européenne.

En 1995, l'enquête sur le secteur des entreprises a été considérablement élargie pour inclure un échantillon d'unités ayant au minimum 10 salariés (auparavant la limite inférieure était de 50 salariés) dans des branches traditionnellement déjà couvertes par l'enquête ; plusieurs services ont été également inclus. Les données de 1995 ne sont en conséquence pas comparables à celles des années antérieures.

À partir de 1991, le personnel des unités d'administration centrales de l'enseignement supérieur n'est pas compris dans les données, mais le coût de ce personnel est inclus dans les autres dépenses courantes de R-D (en accord avec le *Manuel de Frascati* 1993).

À partir de 1989 la R-D exécutée par les institutions sans but lucratif a été incluse dans le secteur de l'État.

À partir de 1987 :

- le champs institutionnel couvert par le secteur des entreprises a été élargi pour inclure les banques et services d'assurances, ainsi que d'autres services d'ingénierie.

- les "fonds propres" des entreprises commerciales d'État (comprises en tant qu'exécutantes dans le secteur des entreprises), qui étaient auparavant inclus dans les fonds venant de l'État, ont été reclassés dans les fonds propres des entreprises.

En 1984 le champs institutionnel couvert par le secteur des entreprises a été considérablement élargi pour inclure d'autres firmes privés, notamment dans les branches des transports, services commerciaux et d 'ingénierie et autres activités (CITI rév. 2, 71 et 8323/24).

À partir de 1981 :

- Les instituts privés et publics au service de l'industrie ont été classés dans les branches 8324 de la CITI (Services techniques d'engineering, d'architecture n.c.a.) ou 9320 (Instituts scientifiques et de recherche n.c.a.). En conséquence, les données par industrie du secteur manufacturier ne comprennent que la R-D exécutée par les firmes (privées et publiques) et elles ne sont pas comparables à celles des années précédentes quand chaque institut de R-D était inclus dans l'industrie qu'il servait principalement.
- Les "fonds propres" des Industries extractives (c.à.d. les compagnies d'extraction pétrolière) comprennent les fonds en provenance de leur société-mère étrangère pendant les années 80.

- **Nouvelle-Zélande**

Les comparaisons dans le temps sont parfois limitées par suite de modifications dans la méthode de recueil et estimation des données intervenues en 1984 (enseignement supérieur et total national), 1987 (DIRDE par industrie) et 1992 (FGU pour l'enseignement supérieur et en conséquence total national).

- **Pays-Bas**

Depuis 1997, une nouvelle définition des fonds généraux des universités est utilisée pour les CBPRD.

En 1996, dans le secteur des services, les entreprises de 10 à 49 employés ont été inclus dans le champs de l'enquête du secteur des entreprises. Les données de 1996 sur les dépenses et le personnel de R-D dans le secteur des services, pour le total du secteur des entreprises ainsi que pour les totaux nationaux ne sont pas directement comparables à celles des années précédentes par suite de cet élargissement du champ de l'enquête du secteur des entreprises.

En 1994 :

- Le champ couvert par l'enquête sur le secteur des entreprises a été élargi pour inclure les entreprises de 10 à 49 employés dans le secteur manufacturier. Les données de 1994 relatives à la dépense et au personnel de R-D dans le secteur manufacturier, le total du secteur des entreprises ainsi que pour le total national ne sont pas comparables à celles des années antérieures.
- Par suite de la reclassification de certains instituts du secteur des ISBL dans le secteur de l'État, les données de ces deux secteurs ne sont pas comparables à celles des années antérieures.

À partir de 1991 la R-D en SSH des entreprises, qui correspond à la R-D des instituts au service des entreprises seulement, a été incluse dans le secteur des ISBL.

À partir de 1990 la méthodologie de l'enquête sur les dépenses de R-D du secteur de l'enseignement supérieur a été modifiée, une des conséquences en a été l'augmentation apparente de la contribution de l'État à la R-D exécutée par l'enseignement supérieur.

À partir de 1985 la ventilation des dépenses de R-D par type d'activités est fondée sur les dépenses totales au lieu des dépenses courantes intra-muros de R-D.

En 1982, par suite de la modification de la méthode d'estimation des données sur la R-D de l'enseignement supérieur, les données de 1982 (ou 1983 selon la disponibilité des données) sur la R-D dans l'enseignement supérieur et pour le total national ne sont pas comparables à celles des années précédentes.

- **Pologne**

Les dépenses courantes de R-D par type d'activités (tableau 3) sont surestimées car elles comprennent une partie des dépenses en capital se rapportant à des équipements spéciaux pour la recherche.

Avant 1995, la R-D pour la défense était principalement exclue.

- **Portugal**

En 1997, une nouvelle méthode a été utilisée pour comptabiliser les fonds structurels venant de la Commission européenne, et il s'ensuit que les données se rapportant au financement venant de l'étranger et celle du financement par l'État ne sont pas comparables aux données correspondantes des années précédentes.

À partir de 1995, à la suite de nouvelles procédures méthodologiques utilisées pour l'enquête de 1997 :

- La classification NACE rév. 1 a été utilisée pour répartir les activités de R-D du secteur des entreprises par industrie rétrospectivement jusqu'en 1995.
- Des unités du secteur des ISBL principalement contrôlées/financées par le secteur des entreprises ou le secteur de l'enseignement supérieur ont été reclassées dans ces secteurs d'exécution en accord avec les recommandations de l'Annexe 6 du *Manuel de Frascati 1993*.

À partir de 1992, les données se rapportant aux chercheurs et aux techniciens (tableaux 5, 13 et 17) ne sont pas comparables à celles des années précédentes par suite de la révision de la classification du personnel de R-D par profession effectuée pour l'enquête se rapportant à 1995 et rétrospectivement jusqu'en 1992.

À partir de 1988, la répartition du personnel de R-D du secteur des entreprises correspond à une ventilation par profession et les données se rapportant aux chercheurs et aux techniciens ne sont pas comparables à celles des années précédentes.

Jusqu'en 1986, les données se rapportant aux chercheurs dans le secteur des entreprises correspondent aux titulaires de diplôme universitaire ou d'un autre diplôme post-secondaire.

Les données sur les crédits budgétaires publics pour la R-D – CBPRD – par objectif socio-économique ont été révisées rétrospectivement jusqu'en 1986 suite à un nouvel exercice méthodologique.

- **République slovaque**

À partir de 1997, des données sur la R-D pour la « Défense » sont séparément disponibles dans la répartition de la DIRD par secteur d'exécution et objectif socio-économique. Néanmoins, il semble que la R-D de défense est encore largement sous-estimée dans les données de la DIRD si on les compare aux informations fournies par le Ministère de la Défense ; en effet celles de la DIRD pour les années 1997 à 1999 correspondent respectivement à 28%, 31% et 44.5% de celles du Ministère.

Avant 1997, la R-D pour la défense était exclue.

En 1994 les définitions du Manuel de Frascati ont été adoptées dans l'enquête nationale sur la R-D.

Avant 1994 les données se rapportaient à la « Base de Recherche et Développement » (RDB) et couvraient l'ensemble des activités des institutions et pas seulement la R-D, le personnel n'était pas exprimé en EPT et du personnel n'ayant pas d'activité de R-D était inclus dans la classe "Autres" professions.

- **République tchèque**

À partir de 1998 :

- Les fonds généraux des universités (qui ne sont pas disponible séparément) sont inclus dans le sous total du Financement par l'État.
- Des sources de financement (Ministère de l'enseignement et agence de subventions, etc.) ont été reclassées du secteur de l'enseignement supérieur à celui de l'État.

À partir de 1995, la méthodologie de l'enquête a changé afin de collecter des données compatibles avec les normes de l'OCDE:

- Les activités connexes ont été exclues.
- Toutes les sources de financement sont disponibles séparément sauf les fonds généraux des universités (FGU).
- Les données sur le personnel par profession sont exprimées en équivalence plein-temps à la R-D.
- La R-D en matière de défense est comprise.

Jusqu'en 1994 compris les données ne correspondent pas aux normes du Manuel de Frascati :

- Les données se rapportent à l'ensemble des activités d'unités constituant la "base R-D", ce qui correspond plutôt aux activités scientifiques et techniques qu'à la R-D proprement dite. Les autorités nationales considèrent qu'environ 10 % de l'ensemble des dépenses sont consacrés à d'autres activités que la R-D.

- Les données sur les dépenses ne comprennent que les dépenses financées par l'État.
- La R-D liée à la défense est exclue.
- Les données sur le personnel comprennent tout le personnel des unités et ne sont pas exprimées en équivalence plein temps à la R-D. Les personnes travaillant à temps partiel ont été comptabilisées en équivalence plein temps pour l'ensemble de leurs activités qu'il s'agisse de travaux de R-D ou non.

- **Royaume-Uni**

La R-D exécutée à l'étranger en tant que partie intégrante des programmes de R-D du Royaume-Uni est incluse dans l'effort national de R-D.

Dans le secteur des entreprises, le financement de la R-D par les entreprises comprend aussi des fonds pouvant provenir d'autres sources nationales telles que l'enseignement supérieur ou les ISBL.

L'ensemble de la dépense de R-D du secteur de l'État correspond aux dépenses des services publics auxquelles est ajoutée une estimation ayant trois composantes : un petit montant de R-D non disponible dans l'enquête sur le secteur de l'État ; la R-D exécutée par les autorités locales ; et à partir de 1991 une estimation de la part R-D réalisée par les hôpitaux des services de santé publique (*NHS hospitals*) (voir la rupture de série en 1991 ci-après). Cette estimation complétant le montant des dépenses de R-D de l'État n'a pas pu être ventilé par type de coûts ni par type d'activités (tableaux 2 et 3).

Suite au projet d'amélioration de l'estimation de l'effort de R-D du secteur des ISBL entrepris en 1996, il y a eu une révision substantielle à la baisse des estimations de la R-D exécutée par ce secteur, notamment concernant le financement venant du secteur des entreprises. Les données des années antérieures ont été révisées en conséquence.

A partir de 1995 pour les données des CBPRD (tableau 18), l'objectif socio-économique Santé a été élargi pour inclure le total des coûts nets de R-D des "National Health Service trusts".

En 1993 :

- La diminution du personnel de R-D du secteur de l'État est principalement due à l'exclusion du personnel de l'*Atomic Weapons Establishment* reclassé du secteur de l'État vers le secteur des entreprises.
- Une nouvelle méthode d'estimation de la R-D financée par l'État dans le secteur de l'enseignement supérieur introduite en 1994 a été appliquée rétrospectivement jusqu'en 1993.

En 1992 l'augmentation du personnel de R-D du secteur de l'État est principalement due à l'inclusion du personnel des Scottish Agricultural Research Institutes, Scottish Agricultural College et Scottish Natural Heritage reclassés du secteur des ISBL vers le secteur de l'État.

À partir de 1991 :

- Les données globales du secteur de l'État (tableau 1) comprennent une estimation des coûts de soutien à la R-D dans les hôpitaux des services de santé publics (*NHS hospitals*) ; ces coûts sont exclus dans la ventilation des dépenses de R-D par type de coûts ou type d'activités (tableaux 2 et 3). Le personnel de R-D de ces hôpitaux n'a pas pu être inclus dans les données du personnel de R-D de l'État car l'information n'est pas disponible.
- Les données concernant les chercheurs et les techniciens du secteur de l'État suivent les définitions du *Manuel de Frascati* ; avant 1991, les chercheurs se rapportaient aux scientifiques et ingénieurs diplômés, et les techniciens aux scientifiques et ingénieurs non-diplômés.

Entre 1985 (85/86) et 1986 (86/87), la United Kingdom Atomic Energy Authority (UKAEA) a été reclassé du secteur de l'État dans le secteur des entreprises.

Entre 1985 et 1986, il semble que les dépenses de R-D d'un groupe d'instituts de recherches subventionnées par le Conseil de recherches pour l'agriculture et l'alimentation en Angleterre et au Pays de Galles aient été intégrées avec celles de ce Conseil de recherches et transférées dans le secteur de l'État, le personnel de ces instituts semble avoir été transféré sur une période plus longue.

De nouvelles estimations ont été faites pour la dépense de R-D exécutée dans le secteur des ISBL, en particulier concernant la part financée par les entreprises. Ces révisions à la baisse ont été effectuées rétrospectivement jusqu'en 1985.

- **Suède**

Les données sont systématiquement sous-estimées (de près de 9 % pour la DIRD en 1993) pour deux raisons :

- La R-D exécutée dans le secteur de l'État couvre uniquement les unités administratives centrales. Les ressources consacrées à la R-D par ce secteur pourraient être multipliées par deux si l'on incluait les unités administratives départementales et locales.
- Une couverture complète des PME pourrait augmenter la DIRDE de 7 %.

De plus, jusqu'en 1991, les données concernant les secteurs des entreprises, de l'État et des ISBL et en conséquence les totaux nationaux (tableaux 1 à 13), ont été en plus sous-estimées du fait de l'exclusion des travaux de R-D en SSH dans les secteurs des entreprises, de l'État et des ISBL.

Cependant, dans les dépenses de R-D du secteur de l'enseignement supérieur sont incluses quelques dépenses non-R-D pour l'enseignement au niveau 7 de la CITE (comprises dans les FGU).

À partir de 1998 :

- les données des CBPRD sont basées sur l'année civile (janvier-décembre) (jusqu'en 1994 les données se référaient à la période juillet-juin) ; les CBPRD de 1995 et 1996 sont des estimations fondées sur le budget de juillet 1995 à décembre 1996.
- les données des CBPRD n'incluent pas le financement des Fondations Publiques de Recherche.

En 1997 :

- À la suite de changement de statut, le financement de la R-D par les Fondations Publiques de Recherche, précédemment considéré comme provenant du secteur des ISBL a été reclassé comme provenant du secteur de l'État. En conséquences, les données se rapportant au financement de la R-D par les secteurs de l'État ou des ISBL ne sont pas comparables à celles des années précédentes.
- La diminution prononcée du personnel de R-D du secteur des ISBL est partiellement due à une modification du questionnaire national.

En 1995 :

- Les dépenses en capital pour la R-D dans l'enseignement supérieur sont exclues cette année là ; en conséquences, les données se rapportant aux dépenses de R-D de l'ES, à la DIRD ainsi qu'aux sources de financement de la R-D sont sous-estimées et ne sont pas comparables à celles des années antérieures ou suivantes.
- Un certain nombre d'institutions du secteur des ISBL a été reclassé principalement dans le secteur des entreprises.

À partir de 1993 :

– La R-D en SSH a été elle aussi incluse dans le secteur de l'État, dans le secteur des ISBL et dans le secteur des entreprises (ce qui, pour ce dernier, peut affecter les industries du secteur des services essentiellement). Les données sur la R-D en 1993 dans les secteurs de l'État et des ISBL ne sont pas comparables à celles des années précédentes.

– Dans le secteur de l'enseignement supérieur, le personnel de R-D employé dans les hôpitaux administrés par le gouvernement local n'a pas été inclus (1 900 EPT en 1989 et en 1991). Par contre les étudiants diplômés poursuivant des travaux de R-D financés sur fonds extérieurs ont été inclus pour la première fois (1 322 EPT en 1993).

– La répartition du personnel de R-D par niveau de qualification dans le secteur de l'enseignement supérieur et en conséquence au niveau national est incomplète.

– La répartition des dépenses et du personnel de R-D par domaine scientifique dans le secteur de l'enseignement supérieur est incomplète.

– Les services publics du gouvernement central sont exclus des CBPRD (tableau18).

En 1983, les changements dans la classification industrielle ont eu pour conséquence de réduire parfois la comparabilité des données du secteur des entreprises de 1983 à celles des années précédentes au niveau des branches individuelles.

À partir de 1981 :

– Pour le secteur de l'enseignement supérieur, des estimations de la R-D en SSH ont été effectuées (580 millions de KrS en 1981), alors que les estimations de l'effort de R-D de l'enseignement supérieur pour 1979 ne se rapportaient qu'aux SENI seulement. Ce montant représente environ 14 % de la dépenses de R-D totale de l'enseignement supérieur en 1981, soit 4 % de la DIRD de la Suède (DIRD qui ne comprend que la R-D dans les SENI pour les trois autres secteurs d'exécution cette année là).

– L'inclusion des dépenses totales de R-D financées par l'État dans les hôpitaux universitaires (alors que précédemment seule les dépenses de R-D financées par les facultés de médecine étaient incluses) a causé en 1981 un accroissement d'environ 18 % des dépenses de R-D du secteur de l'enseignement supérieur, soit 5 % de la DIRD.

– L'augmentation de la couverture des firmes dont l'activité principale est la R-D (essentiellement CITI rév.2, 9320) a entraîné une augmentation de quelques 4 à 5 % au total pour le secteur en 1981 comparé à 1979.

- **Suisse**

  Le secteur des entreprises ne comprend que les entreprises privées. Les entreprises publiques (Chemins de fer fédéraux, Régie fédérale des alcools, Entreprises des PTT, Banque nationale suisse, Caisse nationale d'assurance) sont incluses dans le secteur de l'État.

  À partir de 1998 :

– L'Office fédéral de l'agriculture, ainsi que ses instituts de recherche, ne répartissent plus leurs activités de R-D par objectif socio-économique, mais ils les regroupent toutes sous « Agriculture ». Cela entraîne une rupture de série dans les dépenses de R-D par objectif socio-économique pour le secteur de l'État (tableau 4), ainsi que pour les CBPRD (tableau 18) pour les objectifs « Agriculture » et « Santé », où la moitié des fonds déclarés précédemment pour la « Santé » sont reportés sous « Agriculture ».

– Le domaine des télécommunications de la Poste fédérale est devenu l'entreprise privée Swisscom qui n'est désormais plus comprise dans l'objectif « Sous-total Infrastructure » des CBPRD (tableau 18).

En 1992, les données sur le personnel de R-D de l'enseignement supérieur ne sont pas comparables à celles des années antérieures par suite d'un changement dans la méthode dévaluation.

Les comparaisons dans le temps sont très limitées par suite de plusieurs modifications du champ couvert par les données et des classifications intervenues en 1983, 1986, 1988 et 1989.

- **Turquie**

  Seul le nombre de chercheurs (en EPT à la R-D) est disponible pour le personnel de R-D de l'enseignement supérieur ; il s'ensuit que le personnel total de R-D ainsi que les nombres totaux de techniciens et autres personnels de soutien (tableaux 5, 6 et 8) pour la Turquie sont sous-estimés.

  À partir de 1994, la ventilation des dépenses de R-D par type d'activité (recherche fondamentale, recherche appliquée, développement expérimental) est fondée sur les dépenses totales au lieu des dépenses courantes de R-D, comme c'était le cas pour les années précédentes.

  À partir de 1993, le champs de l'enquête a été élargi pour inclure l'Intermédiation financière, la Production de logiciels et les Autres activités informatiques. En conséquences, les données de 1993 pour l'ensemble des services ne sont pas comparables à celles des années précédentes.

1.3. Sources

Les données sont tirées des enquêtes nationales sur la R-D et des budgets nationaux. Elles ont été principalement fournies au Secrétariat via l'enquête internationale sur les ressources consacrées à la R-D par les pays Membres de l'OCDE. Celle-ci comprend le questionnaire navette complet, retourné au moins une fois tous les deux ans, avec, en première partie, une section concernant les "Principaux résultats" à retourner deux fois par an aux fins de la publication semestrielle *Principaux indicateurs de la science et de la technologie* de l'OCDE. Certaines données sur les crédits budgétaires publics de R-D ont été fournies par l'Office statistique des Communautés européennes (Eurostat).

Dans le cas de pays ne participant pas régulièrement aux enquêtes de l'OCDE, le Secrétariat a effectué des estimations fondées sur les données disponibles dans les publications nationales. Ces données sont annotées d'un "*b*" dans les tableaux.

2. BALANCE DES PAIEMENTS TECHNOLOGIQUES (BPT) (TABLEAU 19)

2.1. Définitions et champ couvert

La BPT mesure les opérations d'achat et de vente de savoir-faire et d'informations technologiques entre pays. Elle enregistre la contrepartie financière versée ou reçue pour l'acquisition ou l'utilisation des brevets, licences, savoir-faire, marques, modèles, dessins, services à contenu technique (y compris l'assistance technique), ainsi que la R-D industrielle réalisée à l'étranger, etc. L'étendue des opérations enregistrées peut varier d'un pays à l'autre et les données de la BPT devraient plutôt être considérées comme des mesures approximatives des flux internationaux de technologie.

Bien que rares, les données ventilées par type de firmes (affiliées et non affiliées) sont fournies chaque fois que l'information est disponible.

2.2. Particularités nationales

Pour certains pays, notamment le Canada, la France, le Japon, pour lesquels deux séries nationales au moins sont disponibles, mais très différentes entre elles, on a choisi la série qui a semblé la plus représentative du point de vue technologique.

- **Allemagne**

  Les informations par type de firme sont sous-estimées, elles se rapportent seulement aux recettes et dépenses au titre des brevets et licences pour les entreprises dans lesquelles la participation étrangère est de plus de 20 % (25 % jusqu'en 1989).

  À partir de 1986, les données couvrent les transactions se rapportant aux brevets, licences, marques, modèles et dessins ainsi que les services techniques et la R-D industrielle.

  Jusqu'à 1985 compris, les données ne couvraient que les transactions se rapportant aux brevets, licences, marques, modèles et dessins.

- **Autriche**

  Depuis 1992 les services à contenu technique sont inclus en plus des royalties et transactions relatives aux licences déjà incluses précédemment.

  Jusqu'en 1991, les données de la BPT couvraient uniquement les royalties et les transactions relatives aux licences.

- **Belgique**

  Les données de 1995 ne sont pas comparables à celles des années précédentes par suite de la nouvelle méthodologie utilisée pour la collecte des données sur la BPT conformément aux manuels de l'OCDE, du FMI et Eurostat. Les données couvrent les royalties, droits et frais d'enregistrement des brevets, de licences de fabrication et de marques, ainsi que les opérations d'assistance technique au sein de l'UEBL.

- **Espagne**

En 1994, suite au changement de procédure de collecte des données il y a une rupture de série dans les données de la balance des paiements technologiques. A partir de 1994, les données sont collectées au moyen d'une enquête et concernent uniquement les échanges de technologie non incorporée des firmes. Les composantes non industrielles sont donc exclues.

En 1993, la balance des paiements a été préparée en utilisant une nouvelle méthodologie conforme au 5ème Manuel du FMI. Les données de 1993 ne sont donc pas comparables à celles des années antérieures. Les données de 1993 comprennent les types d'opérations suivants :

- Services "hardware" (qui font partie des "services informatiques et d'information" dans la balance des paiements).
- Redevances et honoraires de licences (sauf droits d'auteur), y compris leur acquisition/vente.
- Recherche et développement.
- Architecture, engineering et autres services techniques.
- Agriculture, industries extractives et "*onsite processing*".

- **États-Unis**

Les données correspondent aux redevances et honoraires de licences ("royalties and licence fees") seulement, qui sont relativement plus comparables internationalement. Les transactions se rapportant aux « autres services privés ». ont été exclues.

À partir de 1992, les recettes et paiements relatifs aux logiciels sont inclus.

- **France**

Les données excluent la R-D exécutée à l'étranger et comprennent l'assistance commerciale, financière, juridique et de gestion.

- **Italie**

Les données concernent toutes les transactions avec l'étranger dont le montant dépasse 20 millions de lires.

À partir de 1992, les données incluent la R-D, les Études techniques et d'ingénierie, la Formation du personnel, le Déplacement de techniciens et d'experts et Autres transactions en technologie.

Jusqu'au 1991 inclus, la R-D exécutée à l'étranger est exclue.

- **Japon**

Les transactions relatives aux marques, dessins et modèles et à la R-D exécutée à l'étranger sont exclues.

- **Mexique**

Les données sont recueillies au moyen de deux enquêtes réalisées auprès des entreprises bénéficiaires d'investissement directs. Elles concernent les transactions internationales sur les royalties et l'assistance technique, et excluent la R-D exécutée à l'étranger. Les recettes peuvent être surestimés car elles incluent les droits d'auteur et les logiciels.

- **Norvège**

Avant 1988, les données de la BPT couvrent uniquement les concessions de licences et sont par conséquent sous-estimées.

- **Pays-Bas**

  À partir de 1982 les services à contenu technologique ont été inclus.

- **Royaume-Uni**

  En 1996, l'enquête sur la BPT a été modifiée pour prendre en compte les recommandations du 5ème manuel du FMI. La R-D exécutée à l'étranger n'est pas couverte. Sont couvertes les transactions relatives aux brevets et inventions, concessions de licences, marques, dessins, services à contenu technique et R-D.

  En 1987, la couverture de l'enquête sur la BPT a été étendue.

  À partir de 1984 les opérations des entreprises pétrolières sont incluses dans les données.

- **Suisse**

  Les statistiques de la BPT sont tirées de la balance suisse des paiements et se trouvent sous : balance des transactions courantes, compte : service à contenu technologique. Ces chiffres ne comprennent pas les transactions financières relevant de l'achat et de la vente de brevets et licences mais uniquement celles relatives aux taxes et droits découlant de l'utilisation de brevets et licences. Le commerce des brevets et licences est pris en compte dans la balance des capitaux.

2.3. Sources

Les données de la BPT ont été rassemblées à partir de sources nationales (banques centrales, organismes publics, etc.). Les données ont été mises à jour à partir des réponses semestrielles à la première section (Principaux résultats) de l'enquête internationale de l'OCDE sur les ressources consacrées à la R-D par les pays Membres ainsi qu'à partir de l'enquête annuelle de l'OCDE sur la BPT.

OECD PUBLICATIONS, 2, rue André-Pascal, 75775 PARIS CEDEX 16
PRINTED IN FRANCE
(92 2002 06 3 P) ISBN 92-64-09748-1 – No. 52446 2002

H. Fritsch · A. Lienemann · E. Brenner
B. Ludwikowski

Clinical Anatomy of the Pelvic Floor

With 42 Figures and 1 Table

Springer

Prof. Dr. Helga Fritsch
Institut für Anatomie,
Histologie und Embryologie
Leopold-Franzens-Universität
Müllerstr. 59
6010 Innsbruck
Austria

Dr. Erich Brenner
Institut für Anatomie
Histologie und Embryologie
Leopold-Franzens-Universität
Müllerstr. 59
6010 Innsbruck
Austria

Dr. Andreas Lienemann
Ludwig Maximilians Universität
Medizinische Klinik,
Radiologie
Ziemsenstrasse 1
80336 München
Germany

Dr. Barbara Ludwikowski
St Johanns Spital
Landesklinik für Kinderchirurgie
Müllner Hauptstrasse 48
5020 Salzburg
Austria

ISSN 0301-5556
ISBN 3-540-20525-X Springer-Verlag Berlin Heidelberg New York

Library of Congress Cataloging-in-Publication Data
Clinical anatomy of the pelvic floor / H. Fritsch ... [et al.].
p. ; cm. – (Advances in anatomy, embryology, and cell biology ; v. 175)
Includes bibliographical references and index.
ISBN 3-540-20525-X (alk. paper)
1. Pelvic floor–Anatomy. 2. Pelvic floor–Histology. 3. Pelvic floor–Magnetic resonance imaging. I. Fritsch, H. (Helga), 1957- II. Series.
[DNLM: 1. Pelvic Floor–anatomy & histology. 2. Magnetic Resonance Imaging. WP 155 C641 2004]

Springer-Verlag is a part of Springer Science+Business Media

springeronline.com

Printed in Germany

Typesetting: Stürtz, Würzburg

Printed on acid-free paper 27/3150/AG – 5 4 3 2 1 0

[redacted]
Embryology and Cell Biology

Vol. 175

Springer

Berlin
Heidelberg
New York
Hong Kong
London
Milan
Paris
Tokyo

Acknowledgements

This work has been initiated by Prof. Dr. Dr. h.c. K. Fleischhauer and Prof. Dr. Dr. h.c. F. Stelzner (Surgeon) at the Anatomical Institute of the University of Bonn; it has been continued at the Institute of Anatomy at the University of Lübeck with the support of Prof. Dr. Dr. h.c. mult. W. Kühnel. During the time in Lübeck, it has been supported by the DFG (Fr. 829/1–1, 829/1–2). It is still carried on with cooperations of urologists, surgeons and radiologists at the Department of Clinical and Functional Anatomy, Institute of Anatomy, Histology and Cell Biology, at the University of Innsbruck.

Our work is dedicated to our friends Andres and Harald who lived and worked with us and who died too early.

Contents

Abbreviations

ac	Anal canal
b	Bladder
bp	Bulbus penis
cu	Cervix uteri
dtm	Deep transverse perineal muscle
eas	External anal sphincter
if	Ischioanal fossa
ilc	Iliococcygeal muscle
isc	Ischiococcygeal muscle
is	Internal sphincter
lam	Levator ani muscle
lm	Longitudinal muscular layer
md	Müllerian duct
nvp	Nerve vessel plate
oi	Obturator internus muscle
p	Prostate
pbo	Pubic bone
pc	Pubococcygeal muscle
pr	Puborectalis muscle
r	Rectum
rup	Recto-uterine pouch
rvp	Recto-vesical pouch
u	Urethra
v	Vagina
vup	Vesico-uterine pouch

1 Introduction

The pelvic floor constitutes the caudal border of the human's visceral cavity. It is characterized by a complex morphology because different functional systems join here. Since a clear understanding of the pelvic floor region is crucial for both male and female pelvic surgery and for fundamental mechanisms of urogenital dysfunction and treatment, we have focused on a morphological concept of this region that is in accordance with actual clinical concepts.

The tissues comprising the pelvic floor region are striated muscles closely correlated to the smooth muscular layers of the pelvic organs as well as different kinds of connective tissue. The anatomy and physiology of all components of the pelvic floor are currently of clinical interest: Uterine or genital prolapse as well as sphincter defects, e.g. caused by damages during vaginal delivery, may be the reason for urinary or fecal incontinence in elderly female (Sultan et al. 1993). A keen knowledge of the morphological relationships of the prostate region is essential to prevent impotence and urinary incontinence, which frequently follow radical prostatectomy in the male (Herschhorn 1999). Due to the amount of important structures within the narrow cavity of the lesser pelvis, it is extremely difficult for all the pelvic surgeons to perform minimal invasive techniques. Nevertheless, in the near future these techniques have to be established quite similar to any other region.

Modern imaging techniques are used for the diagnosis of pelvic floor or sphincter disorders. Furthermore, they are employed to determine the extent of pelvic diseases and the staging of pelvic tumors. In order to be able to recognize the structures seen on ultrasound, CT, and MRI as well as on dynamic sonography and MRI, a detailed knowledge of the relationship of the anatomical entities within the pelvic floor region is required.

Most morphological descriptions of the pelvic floor region have been derived from dissectional or operative studies in the adult (DeLancey 1994; Gasparri and Brizzi 1961; Hafferl 1969; von Hayek 1969; Pernkopf 1941; Richter and Frick 1985). The results of these studies mainly emphasize the common idea of supportive connective tissue structures for each pelvic organ, of a common visceral or "endopelvic" fascia and of markable sexual differences relating the subdivision into connective tissue compartments (Fig. 1A, B). These studies do not provide any evidence for the developmental differentiation of the pelvic connective tissue nor of the pelvic floor muscles. As we suppose that the knowledge of the development of all these structures is necessary for the understanding of the subdivision of the adult pelvic floor region, we studied both human fetal and adult specimens. By means of this study we are able to test the following hypotheses: (1) In general the subdivision of male and female pelvic connective tissue is identical, (2) an endopelvic visceral fascia does not exist,

Fig. 1A, B Schematic drawings of the pelvic connective tissue, its subdivision, and compartments. **A** Ligaments and compartments in the female pelvis according to Tandler (1923). **B** Ligaments and compartments in the male pelvis according to Waldeyer (1899). (For definitions of abbreviations used in the legends, see list of abbreviations on page IX)

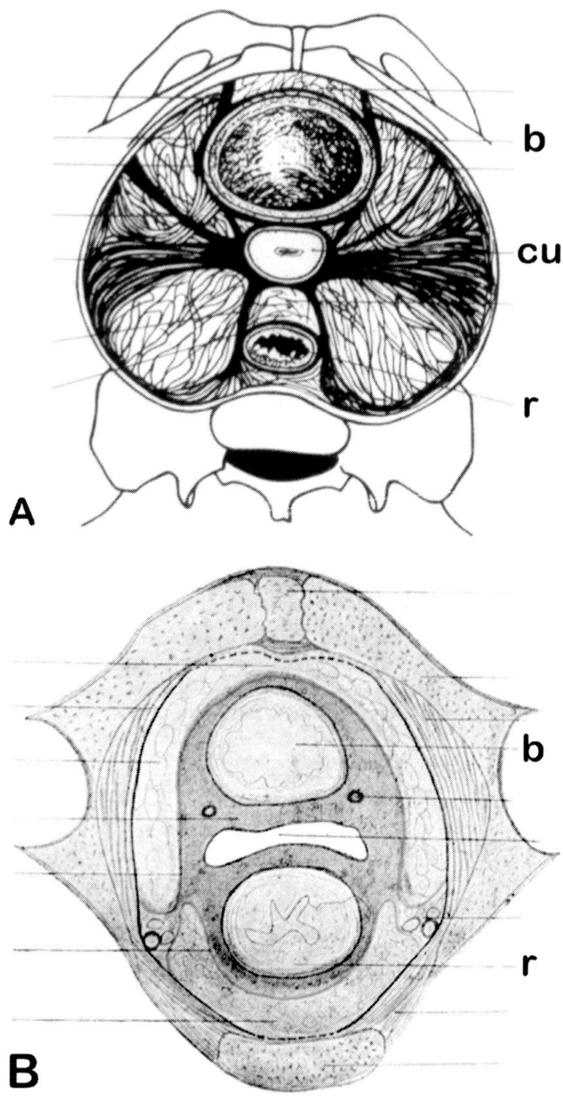

(3) striated muscles are supportive structures of the pelvic floor, and (4) sexual differences are found in the morphology of these muscles.

The *Terminologia Anatomica* (1998) contains a mixture of old and new terms describing the different structures of the pelvic floor. Throughout our review we will try to use actual anatomical terms as far as possible, we will compare them with clinical terms and, most importantly, we will define them exactly (see Table 1).

2 Morphological Approaches for a Spatial Insight into the Anatomy of the Pelvic Floor

In order to compare the results of our morphological studies with the planes of modern imaging techniques, both fetal and adult specimens were studied in undisturbed transparent sections of the entire pelvis. To gain the spatial insight into the pelvic floor anatomy that is most important for the operating surgeon, we produced computer-assisted three-dimensional reconstructions of our sections and thus accomplished dissectional impressions.

2.1 Macroscopic Dissections

Lateral dissection of the pelvis according to a protocol derived from Pernkopf (1941) was carried out in the pelves of two fetuses (23-week-old female and 25-week-old male) and two newborn children. The preparations were used to demonstrate the position of the inferior hypogastric plexus, the fat pad within the anterior compartment, as well as the so-called puboprostatic or pubovesical ligaments. The external anal sphincter and the levator ani muscle were carefully dissected in four embalmed specimens of adult pelves from outside.

2.2 Sections of Fetal Pelves

Pelvic floor anatomy was studied in seven newborn children and 79 fetuses, which were all obtained following legal abortion or miscarriage, and which showed no signs of maceration or macroscopic abnormality. The fetuses and newborn children had been fixed by immersion and stored in a 4% formaldehyde solution for at least 3 months. The crown-rump length (CRL) of the fetuses ranged between 34 and 351 mm, corresponding to a gestational age of 9–37 weeks after conception (Patten 1968).

The fetal pelves and the pelves of the newborn children were subjected to plastination histology (Fritsch 1988), and had therefore been impregnated with the epoxy resin BIODUR E 12 (Heidelberg, Germany). The embedded blocks were serially sectioned with a diamond-wire saw (Well, Mannheim, Germany) either in the transverse, coronal (= frontal) or sagittal plane. The thickness of these sections ranged from 300–700 μm.

After mounting and polishing, the sections were stained with azure II/methylene blue and counterstained with basic fuchsin (Fritsch 1989). The stained sections were examined and photographed with a macroscope (Wild/Leica, Stuttgart, Germany) at magnifications of ×4–80.

2.3 Sections of Adult Pelves

The sectional anatomy of the adult pelves was studied in 11 (6 female and 5 male) specimens from persons who were 58–81 years old. The pelves were frozen at −80°C and then serially cut into sections in the transverse, sagittal, or coronal plane with a bandsaw; the thickness of the sheets ranged between 3–5 mm. The sawdust was carefully removed from the sheets, which were then dehydrated with acetone at −25°C for 5 weeks, defatted in methylene chloride at room temperature for 2 weeks, and then impregnated with an epoxy resin mixture (von Hagens et al. 1987). Finally, the sheets were put into a chamber composed of glass plates, filled with epoxy resin, and cured at 50°C.

2.4 Three-Dimensional Reconstructions

Specimens subjected to plastination histology were scanned directly at a resolution of 1,200 dpi at 256 colors (HP ScanJet 5100 with transparency device; Hewlett Packard). The images were stored in an uncompressed TIFF-format. For further processing in Paint Shop Pro 5.03 (Jasc Software Inc.), the images were stacked and aligned manually using different layers in one file. After trimming of the whole layered image, each layer was saved to a separate file again. Each individual file was processed by isolating and coloring the structures, the spatial arrangement of which are to be shown. For color calibration, a common color table was used. These files were then imported to Spyglass Slicer 1.0 (Spyglass, USA). This program was used to create three-dimensional reconstruction, where the covering (hiding) structures were set nearly transparent (alpha 0.10), others semi-transparent (alpha 0.50) and decisive structures opaque.

2.5 Magnetic Resonance Imaging

In order to obtain morphological and functional images of the pelvic floor, 52 healthy volunteers (44 female, 8 male; mean age 26.9 years; range of age 20–51 years) were examined using Magnetic Resonance Imaging (MRI).

Functional cine MRI of the pelvic floor was performed on two 1.5-T superconductive magnet units (Magnetom Vision, Magnetom Sonata; Siemens Corp., Erlangen, Germany) according to a previously published protocol (Lienemann et al. 1997).

The examination was performed with the patient lying head first, supine, with slightly opened legs. We used a body-array-surface coil. Run-out was prevented with absorbent tissues. No premedication was utilized.

Pulse sequences for the acquisition of morphological images included T2-weighted turbo spin-echo sequences [TR 3500–3800 ms, TE 99 ms, matrix 308×512, two acquisitions, FOV (field of view) 370–250 mm] of the pelvis in axial, coronal, and sagittal orientation. With the exception of the sagittal orientation (3 mm), the slice thickness was 5 mm. In six individuals axial images of the anal sphincter complex were acquired with a slice thickness of 1 mm.

For the functional part, the bladder was emptied prior to the examination. Opacification of the vagina was achieved by using sonography gel. The rectum was filled with approximately 200 ml sonography gel until the patient expressed an urge to relieve the bowels.

For the functional examination, the urethral complex as seen on the axial image was used as a reference point for the sagittal single-slice True-FISP sequence (TR 5.8 ms, TE 2.5 ms, flip angle 70°, matrix 197×256, FOV field view 270 mm, a total of 30 measurements with one image every 1.2 s, in-plane resolution 1.02 mm). In addition, the protocol included an axial single slice True FISP-sequence at the level of the inferior rim with the pubic bone.

During the examination the volunteers were asked to relax the pelvic floor muscles, contract them slowly, and then relax again. Then the patient was asked to increase the intra-abdominal pressure by straining. Defecation was attempted in all cases. This cycle was repeated twice to a maximum of four times.

The functional imaging sequences were evaluated in a cine loop. The overall time of examination varied between 20 and 30 min.

3 Subdivision

The anatomy of both the male and female pelvis and perineum shows a lack of conceptual clarity. These regions are best understood when they are clearly described and subdivided according to functional and clinical requirements: The actual clinical subdivision discerns an *anterior*, a *middle*, and a *posterior* compartment. Whereas an anterior and posterior compartment may be found in the male as well as in the female, a middle compartment can only be found in the latter. The term "compartment" is routinely used by radiologists and all surgeons operating on the pelvic floor. This term is not identical with the term "space." According to former literature, a lot of spaces are supposed to be arranged in the region of the pelvic floor: retrorectal, pararectal, rectoprostatic, rectovaginal, retropubic, paravesical, etc. (Lierse 1984; Pernkopf 1941; Waldeyer 1899). From the point of view of the surgeon, "spaces" are empty (Richter and Frick 1985). They are only filled with loose connective tissue and neither contain large vessels nor nerves. Some years ago, we proposed to drop the term "space" and to speak of compartments instead, taking into account that a compartment may be filled by different tissue components (Fritsch 1994).

In this review, we first present the posterior compartment and then the anterior one. This is in accordance with the viewpoint of the radiologists and with the course of the vessels and nerves. An "extra" middle compartment that is characteristic for the female is presented in detail at the end of this review.

What is our common knowledge about the borders of the different pelvic compartments and what do we know about their content?

3.1 Posterior Compartment

The borders of the posterior compartment are the skeletal elements of the sacrum and the coccyx dorsally. They are completed by the anococcygeal body (see Table 1/1.) dorsocaudally and by the components of the levator ani muscle laterally and caudally. The rectoprostatic or rectovaginal fascia constitutes an incomplete border ventrocranially. The ventrocaudal border is composed by the perineal body (see Table 1/2.). The only organ of the posterior compartment is the anorectum (see Table 1/4.).

3.2 Anterior Compartment

The borders of the anterior compartment are the pubic symphysis ventrally, the components of the levator ani muscle laterally, and the perineal membrane (see Table 1/3.) caudally. In the female, a distinct border between the anterior and middle compartment cannot be found dorsally. In the male, however, the border between the anterior and posterior compartments is composed of the rectoprostatic fascia. The contents of the anterior compartment are different in the male and in the female. Whereas bladder, prostate with urethra, seminal glands, and the pelvic parts of the ductus deferens are situated in the male's anterior compartment, it only contains bladder and urethra in the female.

3.3 Middle Compartment

The borders are the components of the levator ani muscle laterally and the perineal body caudally. No distinct borders can be described ventrally, whereas the rectovaginal fascia constitutes the dorsal border. The middle compartment contains the female genital organs that are arranged in a more or less coronal plane. In more detail, the ovaries, the uterine tubes, the uterus, and the vagina are situated in this compartment.

3.4 Perineal Body

The perineal body is part of the perineum. It is situated between the genital organs and the anus and may be considered as a central or meeting point because a number of different structures join here.

4 Compartments

4.1 Posterior Compartment

4.1.1 Connective Tissue Structures

In macroscopic dissection of embalmed cadavers it is nearly impossible to distinguish subcompartments within the connective tissue of the posterior compartment. Our comparative study of adult and fetal pelves shows that two subcompartments can be distinguished within the posterior compartment:

A small presacral subcompartment (see Table 1/5.) is situated in front of the sacrum and coccyx. It is bordered by the caudal segments of the vertebral column dorsally and ventrolaterally; it is clearly demarcated by the pelvic parietal fascia (Fig. 2A–D), which is called presacral fascia (see Table 1/6.) at this position. In fetuses, the presacral compartment contains loose connective tissue that is replaced by adipose tissue in the adult. The subcompartment is predominated by large presacral veins.

The major part of the posterior pelvic compartment is filled by the anorectum and its accompanying tissues, constituting the perirectal "sub"compartment (see Table 1/7.). As we have shown (Fritsch 1990), this perirectal tissue is identical with the rectal adventitial tissue, which develops along the superior rectal vessels. In the adult, it mainly consists of adipose tissue subdivided by several connective tissue septa (Fig. 3A). Within this perirectal tissue the supplying structures of the rectum are enclosed: the superior rectal vessels, stems and branches, the branches of the variable medial rectal vessels, rectal nerves and rectal lymphatics, vessels, and nerves. The localization of these lymphatic nodes is strikingly different from that of the other lymph nodes of the posterior compartment that are situated laterally in the neighborhood of the iliac vessels (Nobis 1988; Stelzner 1998).

The rectal adventitia develops from a layer of condensed mesenchymal tissue, which—later on—forms a dense connective tissue in fetuses (Fig. 3B, C). In the newborn child it is remodeled by small fat lobules occurring between the connective tissue lamellae. The outer lamella covers the perirectal subcompartment and is called "rectal fascia" (Fritsch 1990) or "Grenzlamelle" (Stelzner 1989, 1998) (see Table 1/8.). It constitutes the morphological border of the perirectal subcompartment (Fig. 3D). The craniocaudal extent of the perirectal subcompartment depends on the branching pattern of the superior rectal vessels; thus, the perirectal compartment is broad laterally and dorsally and it is rather thin ventrally where it is only composed of some

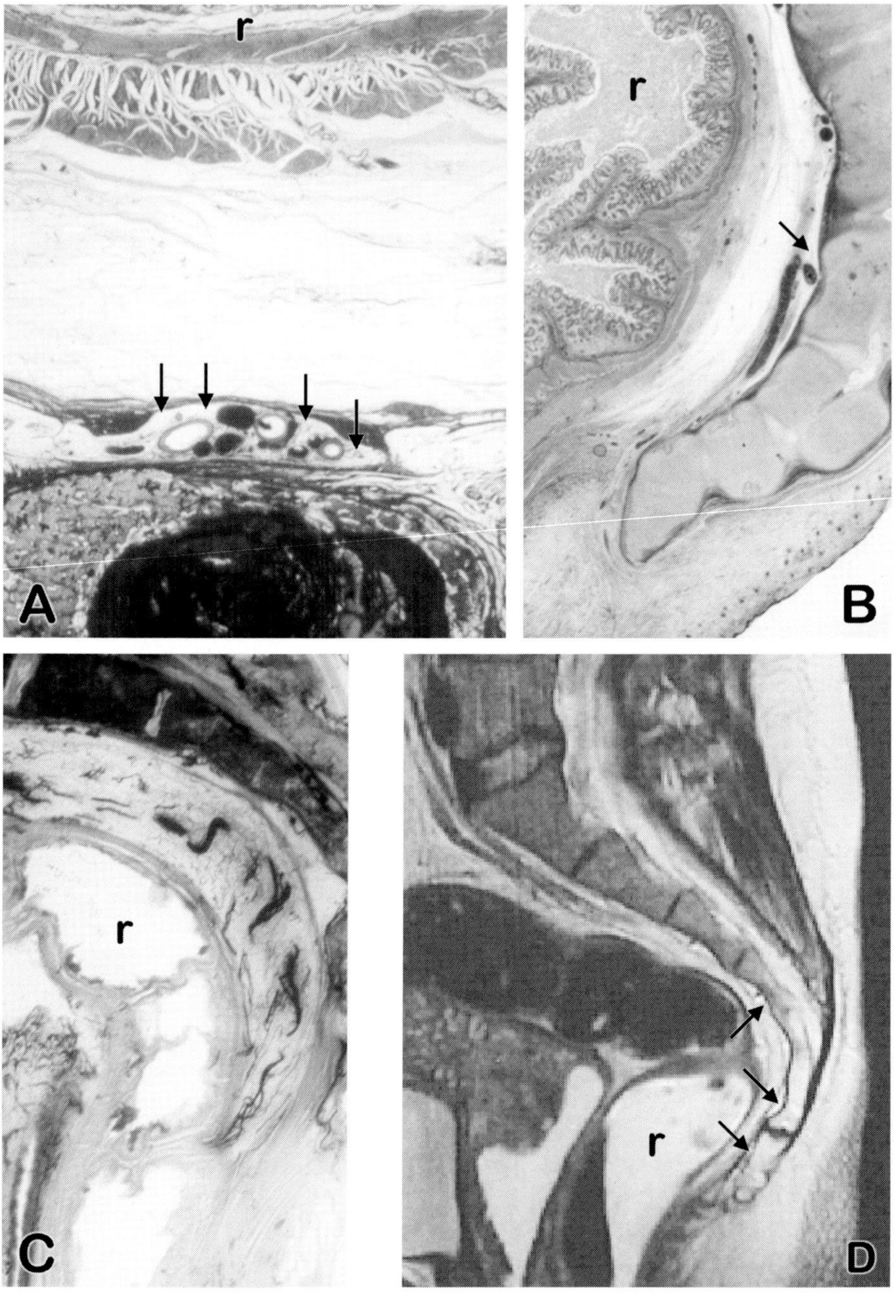

Fig. 2A–D Presacral space (*arrows*). **A** Axial section (500 μm) of an adult male. ×4. **B** Sagittal section (400 μm) of a 24-week-old female fetus. ×9. **C** Sagittal section (5 mm) of an adult female. ×0.45. **D** Mid-sagittal MR image of an adult female

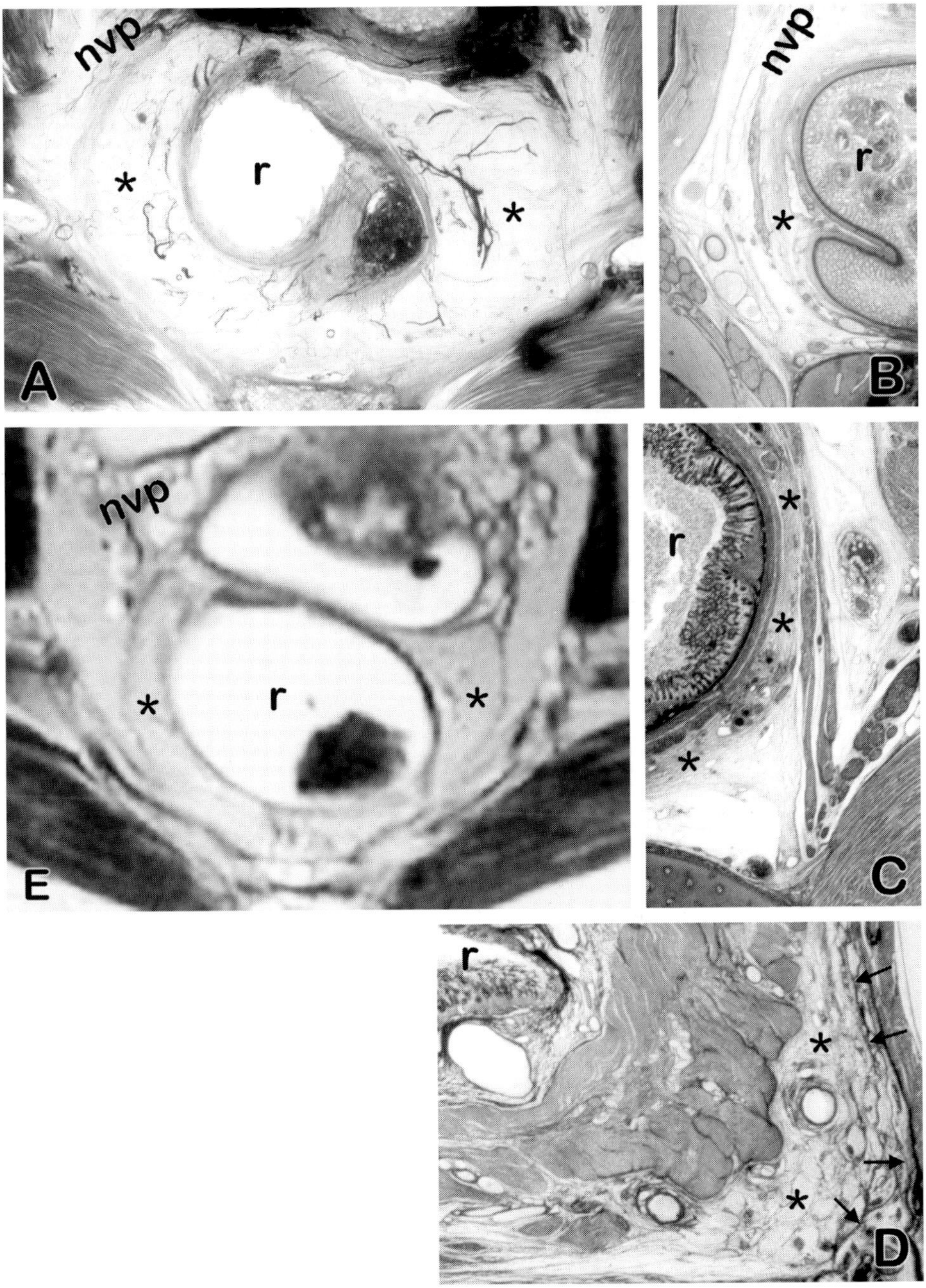

Fig. 3A–E Perirectal tissue (*asterisks*). **A** Axial section (5 mm) of an adult female. ×0.45. **B** Axial section (500 μm) of a 24-week-old male fetus. ×5. **C** Axial section (400 μm) of a 24-week old female fetus. ×5. **D** Axial section (400 μm) of a newborn male specimen. The *arrows* point to the rectal fascia. ×10. **E** Axial MR image of an adult female

connective tissue lamellae. As can be seen in sagittal sections, the extent of the perirectal subcompartment decreases in size in a craniocaudal direction (Fig. 2C).

What is situated outside the rectal fascia and therefore outside the perirectal subcompartment? Dorsally, the presacral subcompartment is loosely attached to the perirectal compartment (see Sect. 1). Laterally, the supplying structures (autonomic nerves and branches of the iliac vessels) of the urogenital organs constitute a nerve—vessel plate (Fig. 3B,C). The latter is accompanied by connective tissue and fills the remaining space between the perirectal compartment and the lateral pelvic wall. In the male, the nerves of the inferior hypogastric plexus (see Table 1/9.) are directly attached to the rectal fascia laterally (Fritsch 1989), whereas in the female the uterosacral ligament (see Table 1/10.) is situated between the rectal fascia and the inferior hypogastric plexus (Fig. 3A, E; Fritsch 1992).

Ventrally, the border of the perirectal compartment differs in a craniocaudal direction. In the male, the perirectal subcompartment borders on the anterior compartment, i.e., on the peritoneum of the rectovesical pouch and on the rectovesical fascia covering the seminal glands, and on the prostate. There is only a small distance between the ventral rectal wall and the seminal glands or the prostate. It is commonly accepted that a rectoprostatic fascia or fascia of Denonvilliers (see Table 1/11.) (Denonvilliers 1836, 1837 according to citation in Wesson 1922) is situated between the organs of the posterior compartment and of the anterior one (Fig. 4A). The fascia develops locally by differentiation of mesenchyme and is composed of collagenous fibers and smooth muscle bundles in the adult. Thus, the rectoprostatic fascia represents a kind of circumscribed septum between rectum and prostate that incompletely separates the two compartments (Fig. 4B, C). Laterally, the septum is neighbored to the inferior hypogastric plexus, but it does not have any connections to the lateral pelvic wall. A rectoprostatic fascia can already be found in middle-aged fetuses where several thin connective tissue lamellae can be discerned within the fascia: One part of them belongs to the prostate, the other to the rectal adventitia where smooth muscle bundles and nerves are intermingled. In adult specimens these lamellae seem to have fused and thus it is quite difficult for the surgeon to find a plane of cleavage (Stelzner 1998) between them.

In the female the perirectal subcompartment borders on the middle compartment, i.e., on the peritoneum of the recto-uterine pouch level with the cervix uteri and the fornix vaginae and on the posterior wall of the vagina more caudally. As we have recently shown (Ludwikowski et al. 2002) a two-layered rectovaginal fascia (see Table 1/12.) develops in the female and is identical to the male's rectoprostatic fascia (Tobin and Benjamin 1953). Level with the anorectal flexure, additional bundles of longitudinal smooth muscles are situated at the anterior rectal wall and directly borders on the rectovaginal fascia ventrally (Fig. 4D). The smooth muscle bundles are accompanied by nerves, some of them crossing the midline. They are found in the male as well as in the female (Fig. 4A, D), they are connected to the smooth muscle layer of the rectal wall. Caudally these additional smooth muscle bundles are attached to the connective tissue of the perineal body (Fig. 4E).

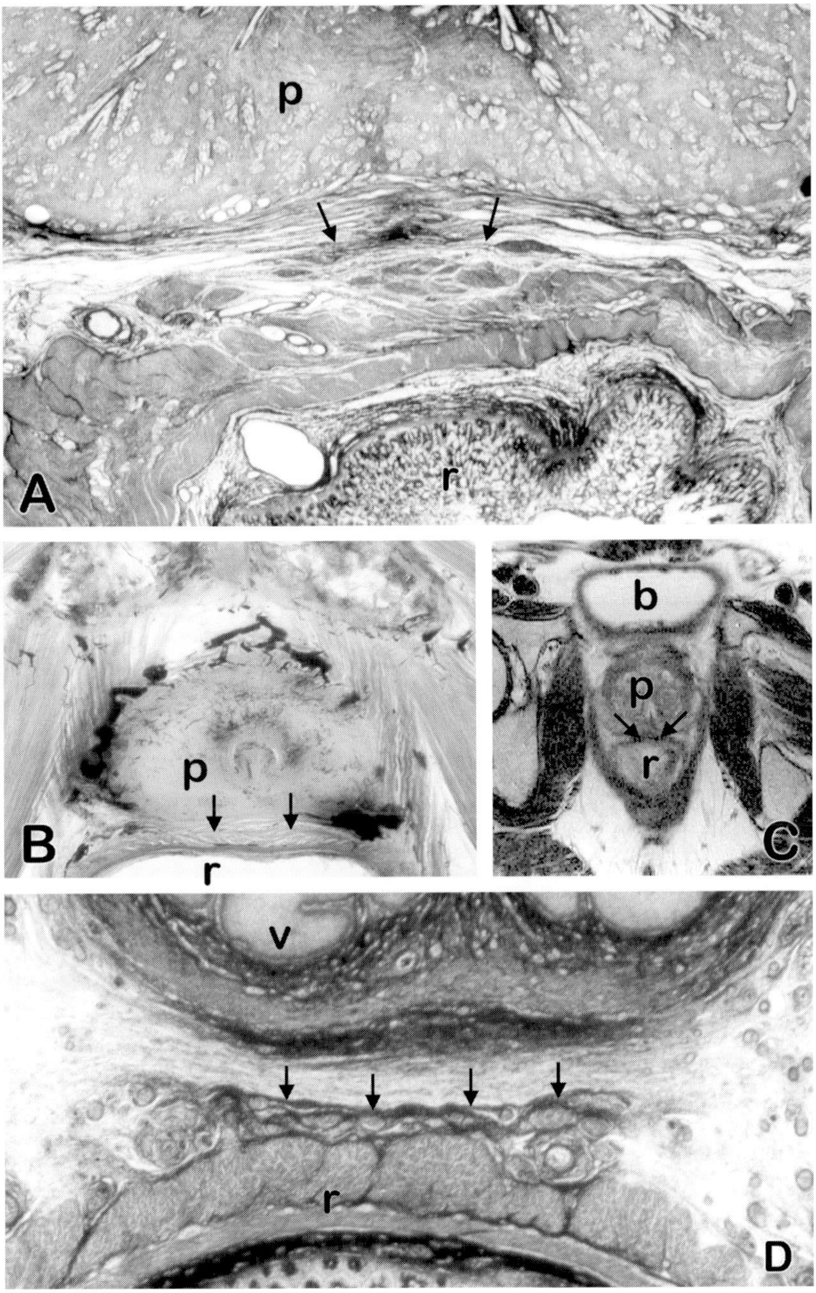

Fig. 4A–D Rectoprostatic (**A–C**) and rectovaginal (**D**) fascia (*arrows*). **A** Axial section (400 μm) of an adult male. ×10. **B** Axial section (3 mm) of an adult male. ×1.1. **C** Axial MR image of an adult male. **D** Axial section (400 μm) of a 24-week-old female fetus. ×28

4.1.2
Muscles

Within the posterior pelvic compartment all components of the levator ani muscle are to be found: The pubococcygeus muscles and the iliococcygeus muscles constitute an irregular plate and insert into the coccyx, where they overlap each other in a staggered arrangement (Fig. 5A–C). The inferior component, the puborectalis muscles, do not insert into any skeletal structure. Behind the rectal wall the fiber bundles of each puborectalis muscle criss-cross, thus constituting a muscular sling around the anorectal flexure (Fig. 6A, B). In the craniocaudal direction, the pubococcygeus muscle and the puborectalis muscle are more or less continuous. In sectional anatomy they can be differentiated by the different directions of their fiber bundles, those of the pubococcygeus taking a slightly descending course, those of the puborectalis exclusively situated in the horizontal plane. The different components of the levator ani muscle can already be distinguished in early fetal life (Fritsch and Fröhlich 1994). Sexual differences found in the levator ani muscle of the adult are already marked in late fetal life: The levator ani constitutes a thick and well-developed muscle in the male fetus whereas it is thinner and already intermingled with connective tissue in the female fetus (Fig. 6C, D). This is particularly true for its puborectalis portion.

The puborectalis muscle is continuous with the external anal sphincter caudally (Fig. 7A, B). The macroscopic distinction between both muscles is provided by the anococcygeal body. The puborectalis has no skeletal attachment, but the deep portion of the sphincter ani externus is indirectly fastened to the coccyx by the anococcygeal body.

The sphincter ani externus is the outer part of the anal sphincter complex (see Table 1/13.), the other components of which are the smooth internal sphincter and the longitudinal muscle layer of the anorectum; the latter is interposed between the sphincters. Whereas macroscopically the external anal sphincter presents itself as a continuous sheet covering the anal canal (Fig. 8A), it can be subdivided into a deep, anorectal portion and a superficial, subcutaneous portion in sectional anatomy (Fig. 8B). This deep portion is a clearly demarcated layer of circularly arranged striated muscle fibers; the superficial portion is characterized by an intermingling of the striated muscles fibers with the smooth longitudinal muscle. The form of the external anal sphincter can best be studied in three-dimensional reconstructions of histological or anatomical orthogonal sections (Fritsch et al. 2002): At a level above the perineum where the external anal sphincter is continuous with the puborectalis muscle dorsally (Fig. 8C), it is missing in the midline ventrally, but it is thickened ventrolaterally where it becomes part of the anterior compartment in the male and the middle compartment in the female. At the level of the perineum the external anal sphincter is complete ventrally (Fig. 8D), but it turns inwards and forms a muscular continuum with the smooth internal sphincter and the longitudinal muscle dorsally (Fig. 8E). As can be seen from the fetal sections, sexual differences in the anal sphincter complex are already present prenatally: The sphincter complex as a whole is thicker in the male than in the female, the anterior portion is thick in the female and thinner and more elongated in the male.

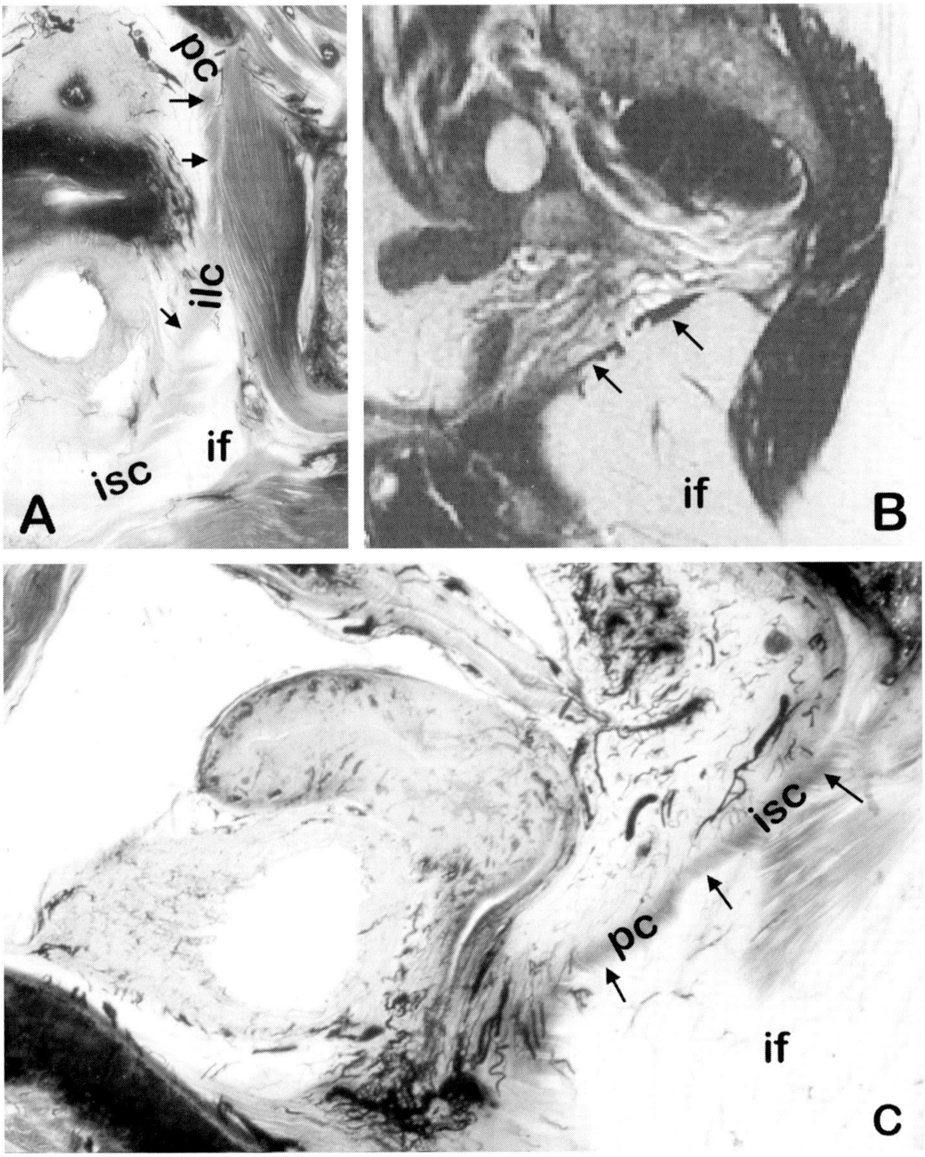

Fig. 5A–C Levator ani muscle (*arrows*). **A** Axial section (5 mm) of an adult female. ×0.6. **B** Parasagittal MR image of an adult female. **C** Sagittal section (5 mm) of an adult female. ×1.0. *ilc*, Iliococcygeal muscle; *isc*, ischiococcygeal muscle; *pc*, ubococcygeal muscle; *if*, ischioanal fossa

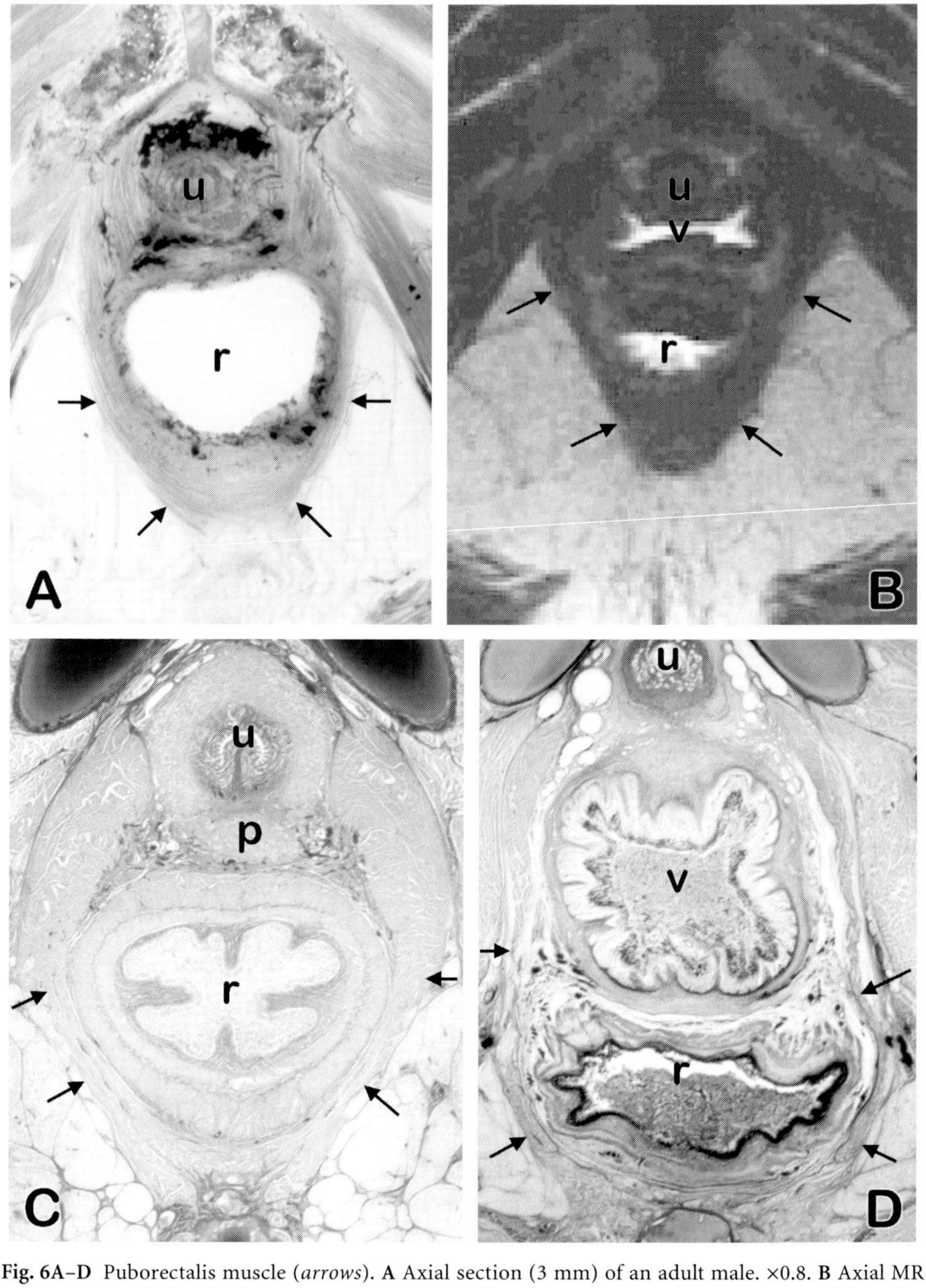

Fig. 6A–D Puborectalis muscle (*arrows*). **A** Axial section (3 mm) of an adult male. ×0.8. **B** Axial MR image of an adult female. **C** Axial section (400 μm) of a male newborn specimen. ×5. **D** Axial section (400 μm) of a female newborn specimen. ×4

Fig. 7A, B Computer-assisted reconstructions of a female fetus. A Oblique ventrolateral view. B Descending dorsoventral view

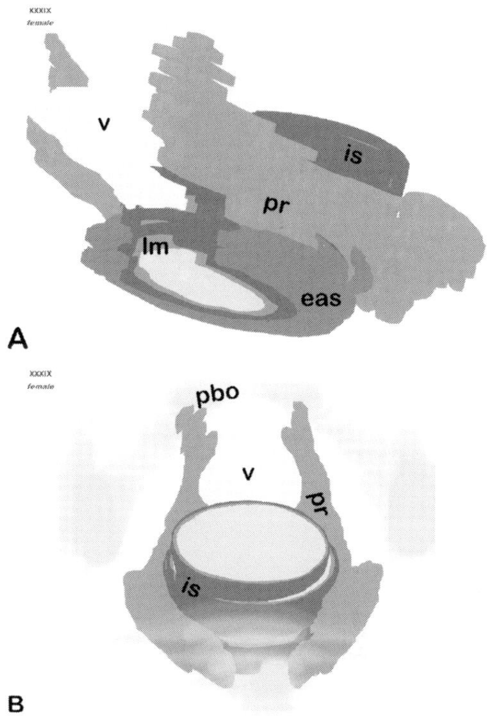

4.1.3
Reinterpreted Anatomy, Function and Fundamentals in Neuroanatomy

The posterior compartment is predominated by the rectum and its surrounding connective tissue. The morphological demarcation of this compartment is formed by the rectal fascia. In CT the rectal fascia may be discriminated as a slightly hyperdense sheath (Grabbe et al. 1982; Roediger and Tucker 1986) and in MRI it is visible as a thin, hypointense structure. According to our results the macroscopic borders of the perirectal compartment are clearly demarcated only in the adult female, where the sacrouterine ligaments constitute the lateral borders and where the posterior border is marked by the pelvic parietal fascia. The perirectal adipose tissue constitutes functional fat that adapts to the different filling volumes of the rectum and constitutes a gliding sheath for the movements of that organ. In contrast to prior literature (Richter 1998; Pernkopf 1941) we did not find any ligament or even ligamentous structures bounding the rectum to the lateral pelvic wall. Thus, there is neither a "rectal stalk" nor a dense "paraproctium."

From a morphological point of view, the surrounding tissue of the rectum has to be called rectal adventitia and not mesorectum (Heald 1995). Under the assumption that a "meso" is a duplication of the peritoneum, the term "mesorectum" is falsely used by the surgeons, because a "meso" cannot be found in the neighborhood of the mostly extraperitonally situated rectum in the adult.

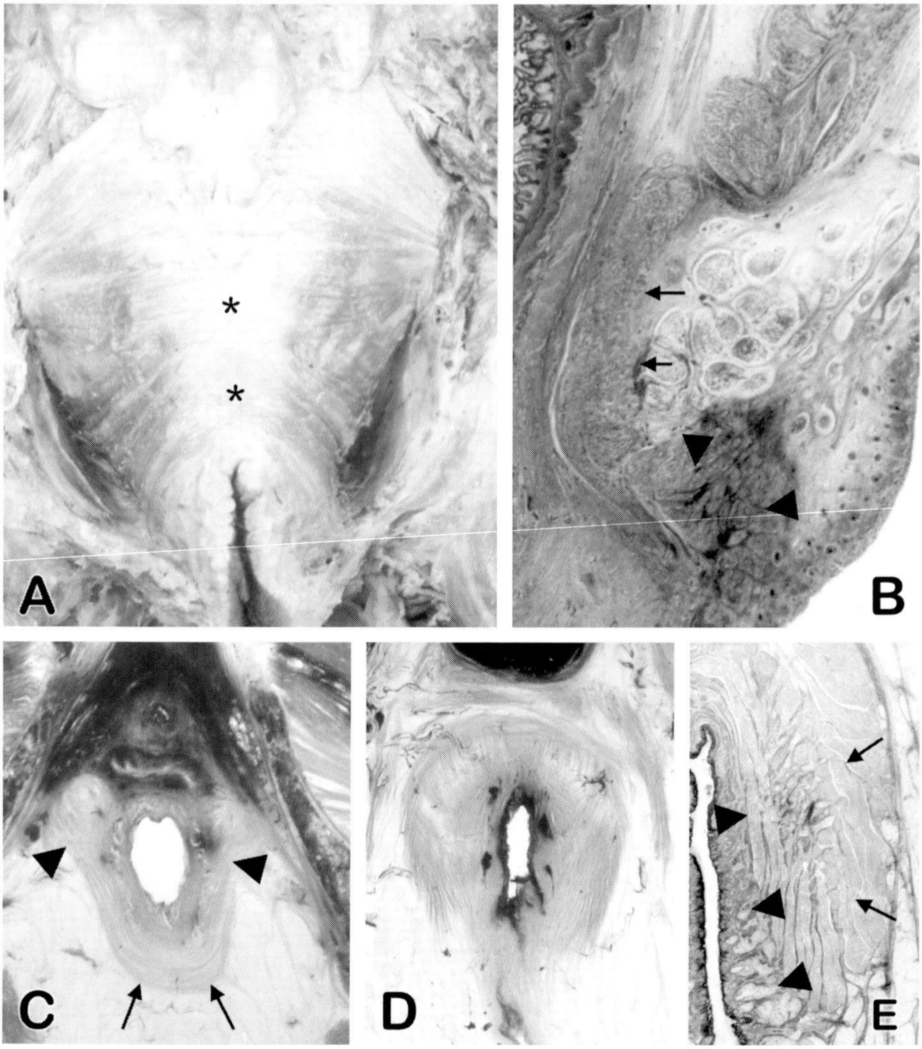

Fig. 8A–E Anal sphincter complex. **A** Macroscopic preparation of an adult female with anococcygeal body (*asterisks*). **B** Sagittal section (500 μm) of a 20-week-old female fetus with deep (*arrows*) and superficial (*arrowheads*) portion. ×10. **C** Axial section (5 mm) of an adult female, fusion of the external sphincter (*arrow*) and the puborectalis (*arrowhead*). ×0.6. **D** Axial section (3 mm) of an adult male, with the ventrally circular portion at the perineum. ×1.1. **E** Axial section (400 μm) of a male newborn specimen with the muscular continuum of external (*arrows*) and internal sphincter (*arrowheads*). ×130

It is obvious from former studies that there are similarities as to the source of innervation between fecal and urinary continence system. The external anal sphincter is innervated by the pudendal nerve and the puborectalis muscle is innervated by both the pudendal nerve and by branches of the sacral plexus (Roberts et al. 1988). According to Schroeder (1981), the nucleus, which is present in the ventral horn of the segments S2 -and S3 and called Onuf's nucleus (see Table 1/14.) (Onufrowicz 1901), is a good candidate as a source for the innervation of the striated anal and urethral sphincters in man. Both the internal anal sphincter and the longitudinal muscle layer are innervated by the autonomic nervous system. In contrast to the well-examined innervation of the smooth muscle system of the urinary tract, little is known about this part of the anal sphincter complex. Efferent fibers of the sympathetic division (Th6–L2) and efferent fibers of the parasympathetic division (S2–S4) are involved and approach the anorectum via inferior hypogastric plexus. Yamamoto et al. (1978) have found groups of neurons within Onuf's nucleus in animals that are responsible for the innervation of the smooth muscle sphincter. They have shown that it is difficult to distinguish those parts of Onuf's nucleus responsible for the smooth sphincter from the parts that are responsible for the striated sphincters. In contrast to the detailed knowledge of the extrinsic and intrinsic innervation of the colon (Wedel et al. 1999), little is known about the interaction of extrinsic and intrinsic innervation of the anorectum, especially on the initial process responsible for defecation and refilling. We suppose that the additional smooth muscle bundles and their nerves that are situated at the anterior rectal wall include receptors for the filling status of the rectum. It would make sense that such structures are situated at the anterior rectal wall because during filling the rectum can easily be widened in a lateral and dorsal direction but not in a ventral direction. As these bundles are connected to the longitudinal muscle layer of the rectal wall, they may be part of a dilatator similar to the puboprostatic or pubovesical ligaments in front of the urethra (see Sect. 4.2.3).

As to the anorectal continence, we have learned to distinguish between the maintenance of continence at rest where type I or slow fibers of the external sphincter are involved and the maintenance of continence during stress conditions where type II or fast fibers of the external anal sphincter are involved (Lierse et al. 1993). The external anal sphincter has capacity for both functions but the distribution differs during the developmental process. According to Beersiek et al. (1979), type I fibers predominate in the external anal sphincter and in the puborectalis of the adult. In newborn infants, type I fibers constitute only 40% of these muscles (unpublished data of J Steinfeld and W Lierse 1990; in: R Schäfer and P Enck 1997). According to our results, the external sphincter, the internal sphincter and the longitudinal muscle bundles form a morphologically intimately connected sphincter complex: Level with the perineum the external sphincter turns in to become continuous with the internal sphincter and both are enclosed by the longitudinal muscle bundles; furthermore, all parts of the sphincter complex are innervated by the pudendal nerve. Thus, our results support the physiological considerations of Schäfer and Enck (1997) who proposed that the type I fibers of the external sphincter intimately cooperate with the internal sphincter muscle.

4.1.4
Surgical and Obstetric Relevance

Anal sphincter-saving operations are obviously preferred to permanent stomas after resections for rectal cancer or mucosal ulcerative colitis. In this context, the clinical relevance of our findings concerning the surrounding tissue of the rectum is enormous (Stelzner 1989, 1998). As we have pointed out regarding the developmental origin of the rectal adventitia and fascia as well as the structures enclosed herein, we have found the explanation for the actual surgical procedure: The rectal fascia is the most important structure during operation of the rectum (Fritsch 1990, 1996; Heald 1995). It is possible for the surgeon to dissect the rectal fascia and thus to peel off the rectum and the surrounding adipose tissue with all lymph nodes and vessels (Stelzner 1998). There are three critical entities neighboring the rectal fascia: Dorsally, there is the presacral compartment with its large veins. A blood-sparing technique is necessary in this region to prevent copious bleeding; it would be better, if possible, not to touch this compartment. Ventrally, level with the rectovaginal or rectoprostatic fascia, it is important for the surgeon to separate the lamellae of this fascia and not to cut the smooth muscle bundles and the nerves of the inferior hypogastric plexus. Laterally, the nerves of the inferior hypogastric plexus and the iliac vessels are directly attached to the rectal fascia. If blood supply and/or innervation is disturbed, patients may have problems with rectal and urinary continence, constipation by stenosis, as well as erection in the male (Burnett and Wesselmann 1999; Heald 1988; Lepor et al. 1985; Long and Bernstein 1959; Maas et al. 1998; Stelzner 1977; Stelzner et al. 1989; Walsh and Schlegel 1988). This is the reason why—if possible—a small part of the ventral rectal wall should be left in cases of rectal excision (Stelzner et al. 1989).

In order to get the best functional results, it is very important for the surgeon to be well informed about the exact anatomy of the anal sphincter complex. Sphincter lesions are described after the use of staplers (Jehle et al. 1995). As this instrument is often used in anorectal operations, it should be pointed out that according to our results both the internal as well as the in-turned external sphincter can be damaged (Fritsch et al. 2002).

The most common reason for fecal incontinence in woman is childbirth with injury to the sphincter. External sphincter injuries occur in 6%–30% of woman (Sultan et al. 1993). It should be differentiated between complete or incomplete sphincter disruptions. Incomplete disruptions can be treated by biofeedback training. The results obtained by this treatment are sufficient. Complete sphincter disruptions are seldom apparent. Undoubtedly they need surgical intervention. Different techniques are used. Sphincteroplasty (suturing the muscular defect) is performed isolated or combined with surgery for urinary continence (Halverson et al. 2001). The same authors propagating this method showed one year later that years after sphincter repair (overlapping anal sphincter repair) more than half of the patients are incontinent to liquid or solid stool (Halverson and Hull 2002). In severe fecal incontinence a new prosthetic device, the Action artificial anal sphincter, was implanted. Continence was improved in 75%, but early infection, rectal erosion, and difficulties evacuating with developing obstructed defecation are major concerns in this operation (Altomare et al. 2001).

Gracilis plasties (a new circular muscle is placed around the anal canal) have been performed for 50 years. Though Pickrell's gracilis muscle transposition often results in building up a scar and never results in normal continence, acceptable continence can be achieved in selected patients (Han et al. 1995).

In order to improve the results, the gracilis plasty it is combined with a neuromuscular stimulator. Continence should be maintained by active muscle contraction and perfectly controllable sphincter function is described in a case report (Baeten et al. 1988). Nevertheless, the literature is unanimous. Cases with good results are always combined with intensive biofeedback. An explanation for this might be the redevelopment of the correct composition of type I and type II fibers and that the plasty is incomplete, because the smooth sphincter muscle is not included.

Despite different etiologies and surgical approaches, anterior levatorplasty and sphincteroplasty yielded similarly successful results in patients with fecal incontinence. Although a marked symptomatic improvement was seen in both groups, no associated physiological alterations could be detected. The reason for the improvement is thus unclear, but it may result from a stenosing effect in the anal canal.

Up to now, in children the artificial rectal sphincter and sphinctero-plasties have not been recommended. In children with anal atresia the so-called posterior anorectoplasty according to Pena (Pena 1997) is recommended. During this operation the distal part of the gut is torn through more or less incomplete sphincter anlagen. In doing so, the author recognized that the external sphincter is not a complete circular muscle. Thus he recommended a strictly midline preparation and an intraoperative electro-stimulation to identify the muscle fibers and to achieve the best functional outcome. Our morphological investigation (Fritsch et al. 2002) supports the fact that the external anal sphincter is not a totally circular muscle. We have thoroughly described the parts of the sphincter complex in order to help the surgeon to identify these structures and if possible, to reconstruct them in a meticulous way.

Until now, no plastic operation for rectal sphincter repair can really be recommended. An alternative method is daily enemas, which are performed by rectal enemas or using the Malone antegrade continence enema (Malone et al.1990; Griffiths and Malone 1995).

Rectoceles are hernial protrusions of the anterior rectal wall and the posterior vaginal wall into the vagina and/or throughout the vaginal introitus. It occurs with laxity of the connective tissue in advancing years, multiparity, poor bowel habits, perineal relaxation and increased intra-abdominal pressure in constipation (Khubchandani et al. 1983; Zbar et al. 2003). Rectoceles are usually associated with scarring and shortening of the anterior anal sphincter mechanism, and the perineal body is thinned. This may be the source of constipation, anal pain, and bleeding. The size of the rectocele does not correlate with symptoms and it is often diagnosed in a population without symptoms. Trauma or obstetrical injuries weaken the rectovaginal fascia. Pelvic denervation, secondary to a vaginal approach for hysterectomy, may contribute to formation of rectoceles and paradoxical sphincter contraction with elevated mean resting rectal pressure at rest. The etiology of rectoceles is different and should be clearly defined before treating. Pelvic floor outlet obstructions should be treated before a rectocele repair is done. Associated cystoceles or enteroceles, sphincter defects, should be corrected in the same operation. For this, different approaches are described: transvaginal, transrectal, transperineal, transabdominal, or combinations of these, but all of them result in a change of the rectovaginal fascia. In the successful repair of a rectocele, the rectovaginal fascia and its repair seem to be the key structure (Richardson 1993; Cundiff et al. 1998). The outcome of the operations depends on the surgeon's familiarity with the operation and associated defects and symptoms (Lawler and Fleshman 2002).

4.1.5
Important Vessels, Nerves and Lymphatics

1. Superior rectal artery
2. Rectal nerves
3. Rectal lymph nodes
4. Inferior hypogastric plexus
5. Superior hypogastric plexus
6. Common iliac artery
7. Internal iliac artery

(Veins have a corresponding course)

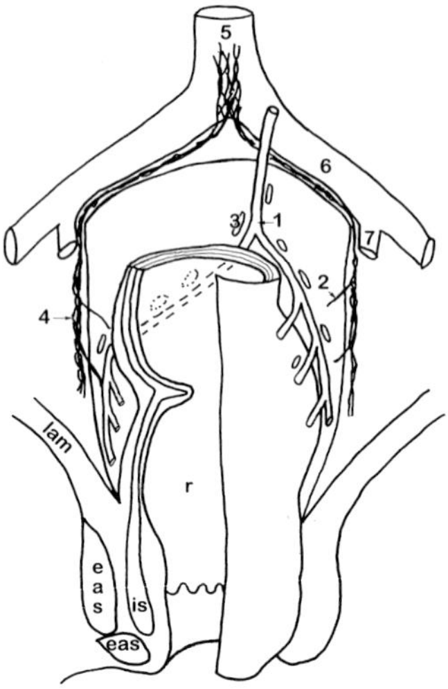

4.2
Anterior Compartment

4.2.1
Connective Tissue Structures

When dissecting along the lateral and ventral wall in embalmed cadavers, it is easy to isolate the bladder, including the embedding tissues and all the adjacent structures. During dissection, no lateral stalks are found that might be responsible for the fixation of the bladder or the urethra. Ventrally, a cord can be identified. It takes an ascending course from the pubic bone to the neck of the bladder and is thicker in the male, where it is usually called puboprostatic ligament (see Table 1/15.) and thinner in the female, where it is called pubovesical ligament (see Table 1/16.) (Fig. 9A, B). It is connected to the tendinous arch of the pelvic fascia (see Table 1/18.). Together, both structures incompletely subdivide the retropubic region into a prevesical subcompartment and a preurethral subcompartment. From the comparative sectional study of fetal and adult pelves we learned the detailed composition of the connective tissue structures within the anterior compartment:

With the exception of its neck and its posterior wall, the bladder is covered by adipose tissue (Fig. 9C). The latter constitutes a semicircular pad that fills the gap between the lateral pelvic wall and the ventral and lateral wall of the bladder. The fat pad is not subdivided by ligaments or any other dense connective tissue septa, but sometimes may be crossed by variable branches from the obturator vessels. It develops in situ (Fig. 9D) from a large paravisceral fat pad (see Table 1/19.) in human fetuses (Fritsch and Kühnel 1992) and neither contains large vessels, nerves, nor lym-

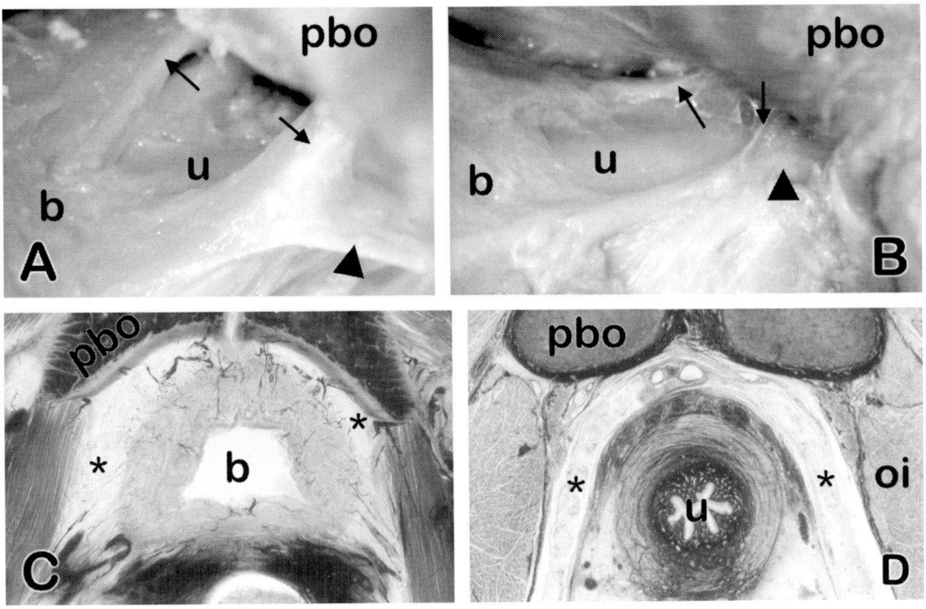

Fig. 9A–D Anterior compartment. **A** Macroscopic preparation of a male newborn specimen with the puboprostatic ligament (*arrow*) and the tendinous arch (*arrowhead*). ×6. **B** Macroscopic preparation of a 23-week-old female fetus with the pubovesical ligament (*arrow*) and the tendinous arch (*arrowhead*). ×9. **C** Axial section (5 mm) of an adult female with the paravisceral fat pad (*asterisks*). **D** Axial section (400 μm) of a 24-week-old female fetus with the developing paravisceral fat pad (*asterisks*). ×8

phatics. The latter derive from the internal iliac vessels and join the dorsolateral edge of the bladder, the seminal glands, and the prostate. Their branches, which are always accompanied by a sheath of dense connective tissue, embrace the bladder, urethra, and prostate. Thus nerves, vessels, and lymphatics are directly situated at the lateral and dorsal wall of the bladder and medially to the fat pad. Ventrocranially, both fat pads join in the midline. Their dorsal edge abuts at the perirectal compartment and their caudal border abuts the levator ani laterally and the pubovesical or puboprostatic ligament ventrally. Thus, they are not part of the preurethral subcompartment that is filled by connective tissue accompanying the deep dorsal vessels of the penis or the clitoris.

Within the anterior compartment two structures are found that are composed of dense connective tissue: the tendinous arch of the pelvic fascia that originates from the pubic bone and that is connected to the pelvic parietal fascia covering the levator ani muscle on its visceral side (superior fascia of the pelvic diaphragm); the semicircular fibrous sheath that covers the ventral and lateral wall of the bladder, the urethra, and the prostate. As the sheath is strong ventrally it can be considered as an incomplete ventral vesical, prostatic or urethral fascia. Whereas the ventral vesical fascia has absolutely no fixation to the lateral pelvic wall, at a level of the urogenital hiatus the ventral urethral fascia, but not the urethra (Ludwikowski et al. 2001), is attached to the fascia of the levator ani muscle laterally (Fig. 10A). Thus, within the

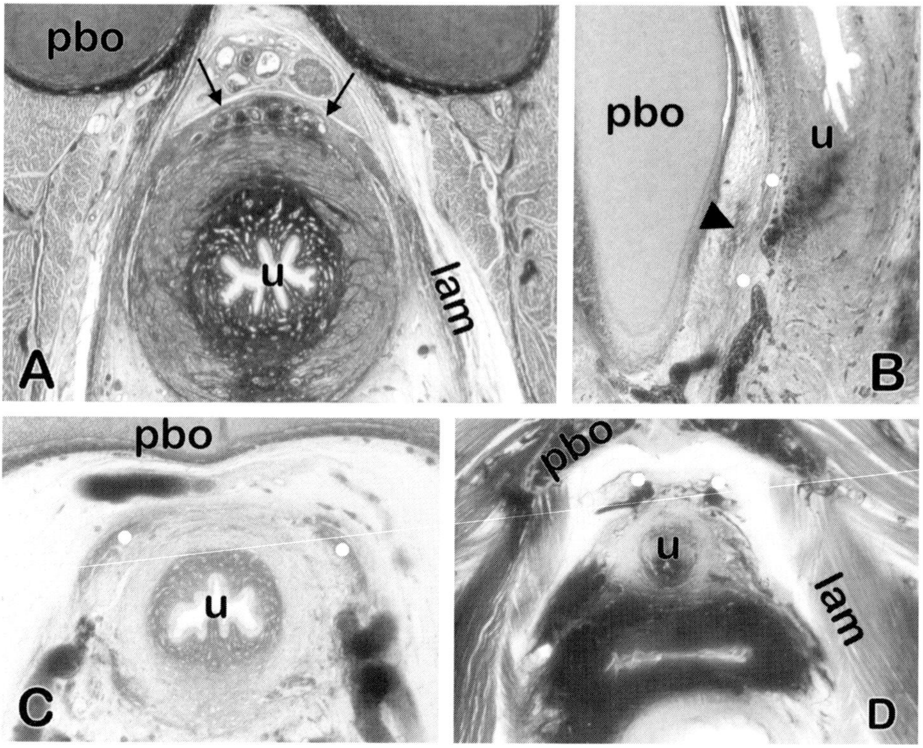

Fig. 10A–D Anterior compartment and the so-called ligaments of the urethra. **A** Axial section (400 μm) of a 24-week-old female fetus with the semicircular urethral sheath (*arrows*). ×12. **B** Sagittal section (500 μm) of a 13–14-week-old female fetus with the pubovesical ligament (*white spots*) and the origin of the tendinous arch (*arrowhead*). ×25. **C** Axial section (400 μm) of a 17-week-old female fetus with the pubovesical ligaments (*white dots*). ×12. **D** Axial section (5 mm) of an adult female with the pubovesical ligaments (*white dots*). ×7.5

hiatus a fibrous bridge connects the fasciae of the levator ani muscles of both sides. To summarize: the fibrous structures of the anterior compartment build up a hammock-like (DeLancey 1994) construction for bladder, prostate, and urethra. These findings can most clearly be shown in fetuses and are matching but not so evident in the adult. It is important to know that there is absolutely no kind of a lateral bony fixation for bladder, prostate, or urethra. In a dorsocranial direction, the ventral fascia of bladder, prostate, and urethra is continuous with the connective tissue sheath of the internal iliac vessels. Ventrally, the hammock-like construction is indirectly fixed to the pubic bone by means of the tendinous arch and by the so-called puboprostatic or pubovesical ligament (Fig. 10B–D). The latter is composed of cholinergic innervated smooth muscle cells (Wilson et al. 1983) and is connected to the vesical neck cranially (see Sect. 1).

An additional fibrous structure can be found to close the hiatus ventrally: A plate of dense connective tissue fills the space between pubic bone and urethral sphincter, thus constituting the perineal membrane (Fig. 11A).

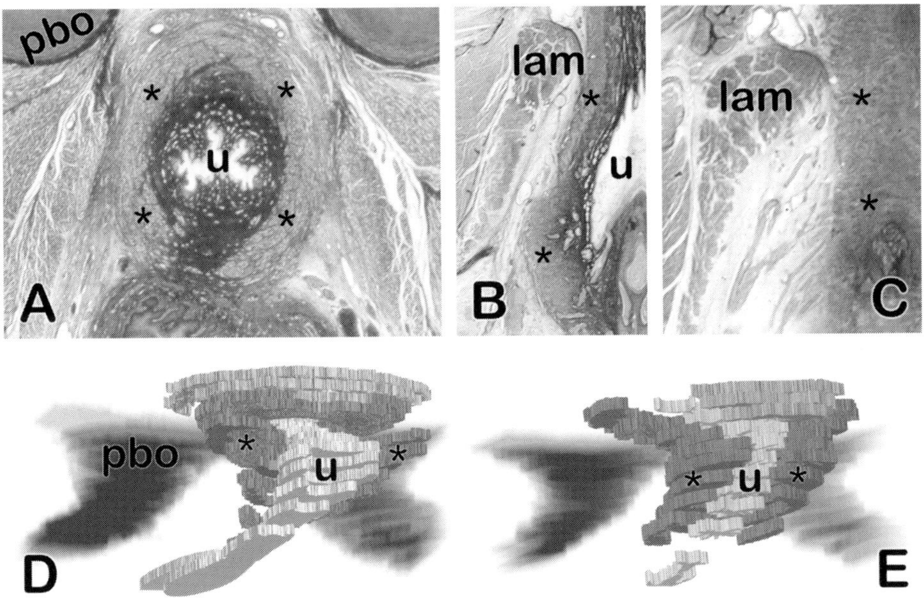

Fig. 11A–E External urethral sphincter (*asterisks*). **A** Axial section (400 μm) of a 24-week-old female fetus, embedded in the transverse perineal membrane. ×9. **B** Axial section (500 μm) of a 20-week-old male fetus just before leaving the urogenital hiatus. ×22.5. **C** Axial section of the same fetus (*b*) when leaving the urogenital hiatus. ×12. **D** Computer-assisted three-dimensional reconstruction of a male fetus. **E** Computer-assisted three-dimensional reconstruction of a female fetus

The male's cavernous nerves are clinically important structures that are in close contact to the connective tissue structures and the muscles of the anterior compartment. As has already been reported some years ago (Fritsch 1989) there are two pathways for these nerves to leave the pelvic cavity: Some of them run between the fascia of the levator ani muscle and the urethra, whereas others enter into the urethral external sphincter (Fig. 11B).

4.2.2
Muscles

The striated muscles of the anterior compartment are the ventral parts of the levator ani muscle (see Table 1/17.), i.e., the pubococcygeus and puborectalis muscle of each side. As they are covered by the superior fascia of the pelvic diaphragm, they are clearly separated by the adjacent organs (Figs. 10A, D, and 11A, B) and the external urethral sphincter. As has been reported previously (Ludwikowski et al. 2001), this muscle is horseshoe- or omega-shaped during fetal development and incompletely covers urethra or prostate (Fig. 11C, D). The dorsal ends of this muscle are connected by a plate of dense connective tissue that is broad in the male and small in the female, where it is firmly attached to the ventral wall of the vagina (Figs. 10D, 11A). Whereas most of the fibers of the external urethral sphincter run semicircular, the most caudal fibers nearly run in a transverse plane. This portion predominates in the male and therefore has been considered as the male's deep transverse perineal muscle (Oelrich 1980).

As has been described above, smooth muscles are found outside the walls of the urogenital organs constituting parts of the so-called puboprostatic or pubovesical ligament in front of the ventral wall of the urethra.

4.2.3 Reinterpreted Anatomy, Function and Fundamentals in Neuroanatomy

The extent of the fat pad described here is identical with the anterior portion of the paravisceral space as reported by Gasparri and Brizzi (1961). These authors described a large paravisceral compartment along the urogenital organs that is separated by the uterine vessels in the female. It is true that in the female the uterine vessels constitute the border between the anterior and posterior compartment, but nevertheless they separate the paravisceral fat pad and the perirectal subcompartment. The difference between our opinion and that of Gasparri and Brizzi (1961), though obviously slight, is of great clinical importance with respect to lymphatic and surgical pathways that are quite different within the two compartments. This issue demonstrates well that the interpretation regarding the subdivision of the pelvic connective tissue is incomplete without knowing its developmental and functional differentiation. It is obvious that the main function of the semicircular, paravisceral fat pad is to constitute a gliding pad for the bladder (Kux and Fritsch 2001). The fat pad accompanies the bladder whenever it is moving.

Function and innervation of the external urethral sphincter, the pubovesical and puboprostatic ligaments have been extensively described by Dorschner et al. (2001). According to their studies, the external urethral sphincter is composed of both a striated and a smooth muscular part and thus forms an urethral sphincter complex similar to what we have pointed out for the anal sphincter complex. There is no doubt about the function of the external urethral sphincter that is part of the urinary continence system. Furthermore, it is generally accepted that, similar to the anal sphincter complex, there are different fiber types within the striated part of the external urethral sphincter (Ho et al. 1997; Schroeder and Reske-Nielsen 1983), but as to the data about distribution of fast and low twitch fibers we are cautious when interpreting these data because they are taken from different specimens, from different places within the muscle itself, and from muscles of different age and of different sex.

Dorschner et al. (2001) pointed out the fact that the smooth muscle bundles of the puboprostatic or pubovesical ligaments are continuous with longitudinal muscle fibers of the neck of the bladder, which they call dilatator urethrae. Maybe again there is a similarity to the anorectum, where we also found smooth muscle bundles and autonomic nerves outside the ventral wall, which we think work in functional coactivity to the longitudinal internal bundles. Nevertheless, it seems certain that the function of the so-called puboprostatic or pubovesical ligaments which receive a presumptive cholinergic innervation (Wilson et al. 1983) is not fixing the urethra to the pubic bone but maintaining its position relative to the bone during micturition (Gosling et al. 1999). In contrast, the contraction of the levator ani muscle and the external urethral sphincter leads to a narrowing of the preurethral space and to an ascending movement of the urethra as can be seen in dynamic MRI (Fielding et al. 1998; Sprenger et al. 2000).

Whereas Oelrich (1980) pointed out the existence of a deep transverse perineal muscle in the male, Dorschner et al. (2001) completely deny the existence of a deep transverse perineal muscle and thus of the urogenital diaphragm as a similar portion does not exist in the female.

In order to clarify, we propose to compare the male and female situation in our preparations: The short female urethra is covered by the external urethral sphincter, which ends level with its orifice where it still consists of semicircular running bundles. The intrapelvic part of the male's urethra is also covered by the external urethral sphincter. Level with the perineum the bundles of this muscle take a more and more transverse course, running from left to right and thus constituting the so-called deep transverse perineal muscle. It seems to be easy to explain this sexual difference by the different function of the urethra at this level; whereas the female urethra has to be opened and closed at this level, the male urethra has to be secured to the bones additionally, as it changes its direction afterwards and proceeds into the penis.

4.2.4
Urological and Obstetric Relevance

As the supplying structures for the female's urological organs and the male's urogenital organs are situated at the medial border of the paravisceral fat pad, the latter constitutes an ideal surgical pathway. It must be taken into account, however, that whenever the fat pad is damaged, the prevesical space may undergo fibrous obliteration and thus distension and retraction of the urinary bladder may severely be impaired and ejaculation may become painful in the male (Kux and Fritsch 2001).

Due to our results that in principle support the hammock hypothesis of DeLancey (1994), an operative "refixation" of the urethra and the bladder neck should take into account an ascending dorsocranial traction (nerve guiding plate) as well as a descending ventrocaudal traction (tendinous arch of the pelvic fascia). Though there are innovative ideas as to the surgical reconstruction of the female urinary tract (Ulmsten 2001) they are not performed according to the morphological needs, because they mostly consider only one part of the so-called fixation system. Different sling techniques are performed by using different kinds of materials. In most techniques the urethra is torn in a dorsocranial direction. In the first years after operation patients are continent, but some years later the operation results in an urethral obstruction.

In cases of urogenital sinus malformation, a secure procedure can be performed using a perineal approach. The procedure is called total urogenital sinus mobilization (TUM), because the urethra is not fixed to the pubic bone. Level with the bladder neck, the pubovesical ligament should be protected to obtain urinary continence (Pena 1997; Ludwikowski et al. 1999).

To avoid postoperative impotence due to nerve damage of the cavernous nerves in the male, an effort is made to preserve these nerves during pelvic, urethral, or penile surgery. As their topography was not clear in the adult, it has extensively been studied in human fetuses (Walsh and Donker 1982; Fritsch 1989); the results of these studies have been compared with the relationships in the adult male (Lepor et al. 1985; Paick et al. 1993) and finally they have been used to improve operation techniques in male pelvis.

4.2.4.1
Important Vessels, Nerves and Lymphatics in the Male

1. Inferior vesical artery
2. Artery to the deferent duct
3. Superior vesical artery
4. Vesical lymph nodes
5. Internal iliac lymph nodes
6. Internal iliac artery
7. Inferior hypogastric plexus
8. Paravesical fat pad
9. Seminal gland

(Veins have a corresponding course)

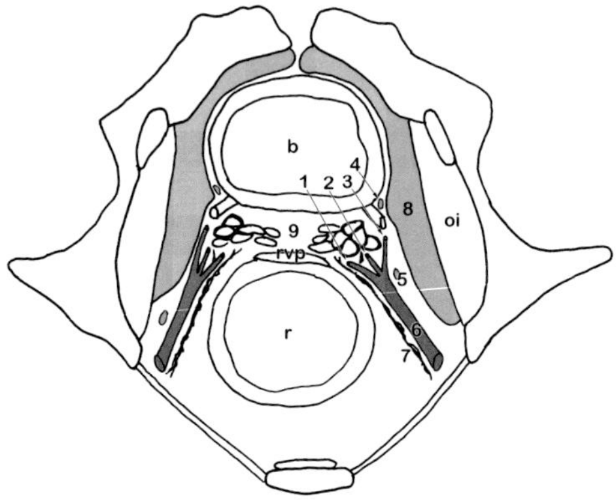

4.2.4.2
Important Vessels, Nerves and Lymphatics in the Female

1. Inferior vesical artery
2. Branches to the ureter
3. Superior vesical artery
4. Vesical lymph nodes
5. Internal iliac lymph nodes
6. Internal iliac artery
7. Inferior hypogastric plexus
8. Paravesical fat pad

(Veins have a corresponding course)

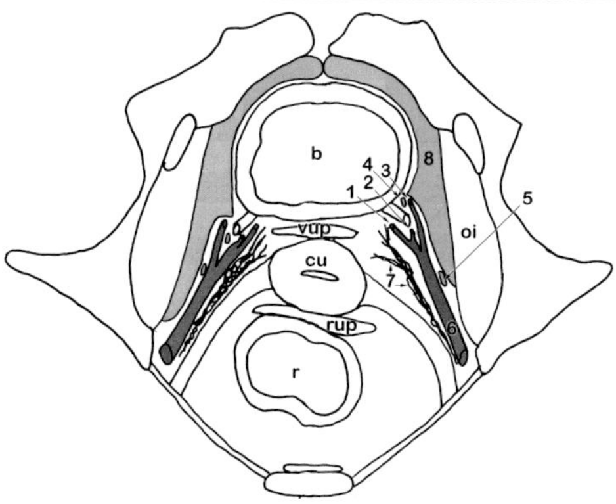

4.3
Middle Compartment

4.3.1
Connective Tissue Structures

In macroscopic dissections of the adult female pelvis it is impossible to isolate ligaments fastening the cervix uteri or the vagina to the lateral pelvic wall and thus separating the middle compartment from the anterior or the posterior one laterally. In a refined macroscopic dissection performed with a binocular dissecting microscope it is possible—as well as in any other part of the pelvic subperitoneal tissue—to isolate connective tissue septa within the adipose tissue surrounding uterus and vagina (DeBlok 1982). Our study of female fetal and adult pelvic sections reveals the true nature of the connective tissue structures surrounding uterus and vagina. The only connective tissue belonging to the middle compartment accompanies the vessels of uterus and vagina, thus running parallel to the lateral walls of these organs. In fetuses, the connective tissue is still loose and without a differentiated structure; in the adult it mainly consists of adipose tissue with regular connective tissue septa (Fig. 12A–D) and it is continuous with the broad ligaments (see Table 1/20.) laterally. As can clearly be seen in fetuses, the paracervical connective tissue abuts the paravesical adipose tissue laterally and the paravaginal connective tissue abuts the pelvic parietal fascia caudally (Fig. 12A, B). As shown in fetuses (Fritsch 1992), the broad ligaments themselves are part of the recto-uterine and the vesico-uterine fold (see Table 1/21., 1/23.) that tangentially cover the anterior and posterior uterine walls. Apart from dense subperitoneal connective tissue that covers the recto-uterine pouch (see Table 1/22., Fig. 12E) and mainly consists of collagenous fibers, no supportive ligaments are found for the female fetal uterus. In the adult, this condensation of subperitoneal connective tissue has developed to the uterosacral ligaments (see Table 1/10.). They are visible in the transparent sections as well as on MRI and form semicircular cords varying in thickness individually. They originate from the lateral margin of the cervix uteri and the vaginal vault and are directed dorsocranially, where they are connected to the pelvic parietal fascia covering the sacrospinous ligaments. As they are part of the recto-uterine ligaments, they cover the perirectal tissue laterally. Our study undoubtedly confirmed the existence of the round ligaments as well as their course and their components. However, ligamentous structures constituting cardinal or transverse ligaments (see Table 1/25.; Kocks 1880; Mackenrodt 1895) that are supposed to fasten the cervix uteri and the vaginal vault with the lateral pelvic wall can not be found in the adult pelvis. Our findings that have been taken from anatomic sections of elder specimens unrestrictedly correlate with the results of the MRI taken from young adult female pelves (Fig. 13A, B).

Subperitoneally, the middle compartment and its organs abut the anterior compartment ventrally. This area is predominated by the dense connective tissue bridge intimately connecting the ventral vaginal wall with the dorsal urethral wall (Fig. 12B; see also anterior compartment).

Dorsomedially, the middle compartment abuts the posterior compartment. The border between these compartments is demarcated by the rectovaginal fascia (see also posterior compartment), that is composed of dense connective tissue, elastic fi-

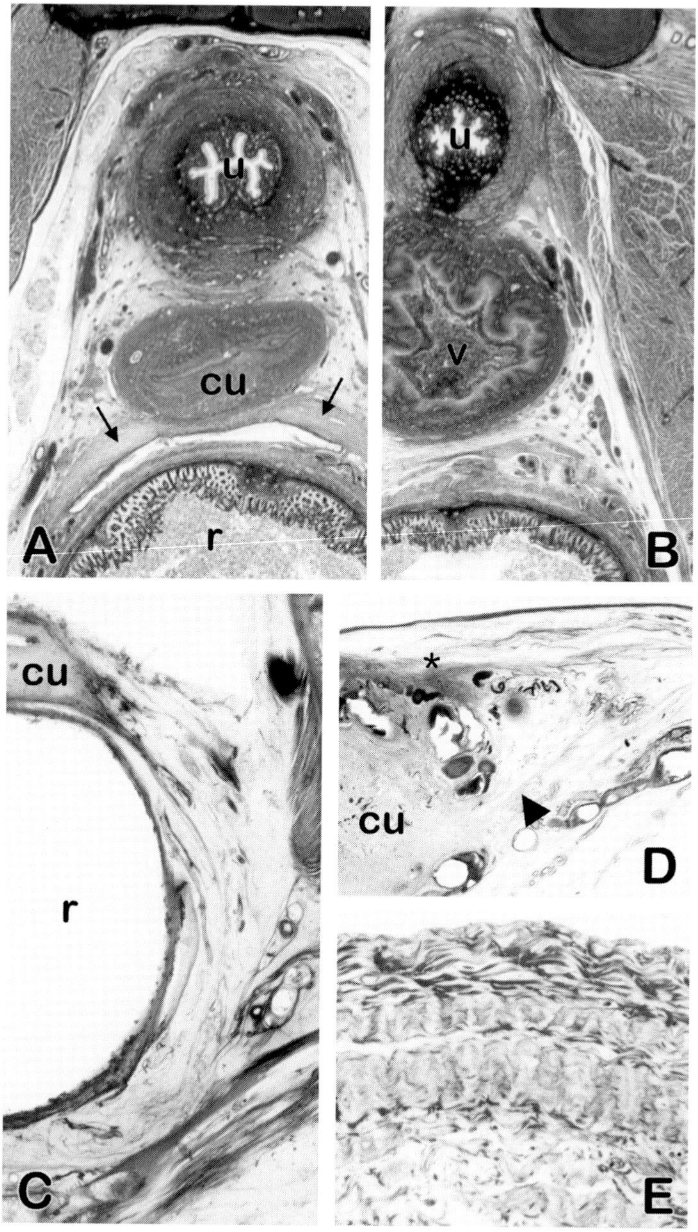

Fig. 12A–E Paracervical and paravaginal tissue. **A** Axial section (400 μm) of a 24-week-old-female fetus at a level with the recto-uterine pouch covered by dense connective tissue (*arrow*). ×8. **B** Axial section (400 μm) of the same fetus at a level with the vagina embedded in loose paravaginal tissue. Vagina and urethra are intimately connected. ×8. **C** Axial section (3 mm) of an adult female with the paracervical tissue. ×0.8. **D** Enlargement of an axial section (3 mm) of the same specimen with origin of the round ligament (*asterisk*) and the uterosacral ligament (*arrowhead*). ×3.5. **E** Enlargement of (*a*) with parallel oriented connective tissue fibers constituting the subperitoneal part of the utero-sacral ligament. ×40

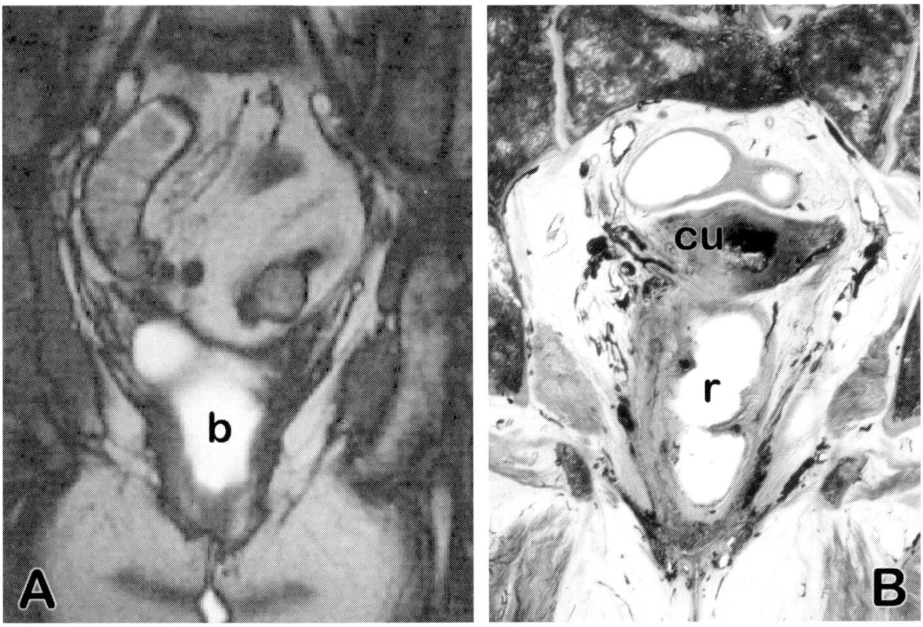

Fig. 13A, B Subperitoneal connective tissue and nerve vessel guiding plate. **A** Coronal section (3 mm) of an adult female with pararectal and paracervical tissue. ×0.4. **B** Coronal MR image of an adult female with paravesical and paracervical tissue

bers (Richardson 1993), and smooth muscle cells that belong to the longitudinal layer of the rectal wall.

4.3.2 Muscles

The middle compartment does not have any specific striated muscles. The lateral vaginal wall comes in close contact to the puborectalis portion of the levator ani muscle. Both structures are always separated by the superior fascia of this muscle (Fig. 6B).

4.3.3 Reinterpreted Anatomy, Function and Fundamentals of Neuroanatomy

Surgical techniques for the fixation of uterus and vagina are numerous. They all depend on the idea that there are sheath-like condensations within the pelvic cavity that are commonly called fascia. Moreover, these fasciae are thought to act as supportive structures for the uterus and vagina and thus they need to be reconstructed during operation. We think this point is one of the most critical to discuss within this review:

Our reinterpreted anatomy of the connective tissue surrounding uterus and vagina is:

- In accordance with some papers (Berglas and Rubin 1953; Koster 1933; Uhlenhut and Nolley 1957), we do *not* find any visceral fascia covering uterus and vagina. Both organs are accompanied by adventitial connective tissue. The rectovaginal fascia develops in situ (Ludwikowski et al. 2002) and is connected to the uterosacral ligaments, to the longitudinal muscular layer of the rectum and to the perineum (see posterior compartment and perineal body).

- As has clearly been summarized by Bastian and Lassau (1982) various ligaments are supposed to exist in the pelvis of the adult female. Our results show that—apart from the uterosacral and the round ligaments—no ligaments of the uterus can be found in conventional anatomical specimens, sections or by MRI. We showed, however, that the paracervical and paravaginal region contains adipose tissue, numerous vessels, nerves, and connective tissue septa. *All together these components may be confounded with a ligamentous structure* especially in the elder female. The connective tissue septa have carefully been described by new morphological approaches (De Blok 1982b; DeLancey 1996) but they have been overinterpreted as to their functional meaning. There is no doubt that some of these connective tissue septa are connected to the fascia of the levator ani muscle and the contraction of this muscle is directly transferred to the septa and thus to the vagina, too. However, due to their morphological characteristics they are not supposed to act as supportive structures.

Our results are still in disagreement with the classical descriptions in clinical and anatomical textbooks. We are aware of the fact that the variability of nomenclature is misleading, too. Nevertheless, it is morphologically true that if at all we speak of a fixation of the uterus it is only fulfilled by the uterosacral ligaments running in a dorsocranial direction. These ligaments are connected to the pelvic parietal fascia level with the sacrospinous ligaments, thus producing an upward traction for the whole uterovaginal complex.

4.3.4 Obstetric Relevance

There are various surgical procedures to reconstruct the so-called supportive ligaments in patients with genital prolapse. Due to our morphological data, in the case of prolapse it only makes sense to carry out a sacrospinous fixation of the uterovaginal complex (Niemen and Heinonen 2001; Thakar and Stanton 2002), taking into account that the pudendal vessels and the pudendal nerve are not injured during operation (Occelli et al. 2001). We are not able to answer the question whether such a fixation should be performed at the time of transvaginal hysterectomy (Cruikshank 1991), but we propose to not confine this technique to patients with an intact pelvic floor function. Based upon our morphological data, we would prefer a laparoscopic approach that bears a tremendous potential for the treatment of all aspects of urogenital organ and pelvic floor pathology by the same route without disturbing the pelvic floor itself (Wattiez et al. 2001).

In case of cervical carcinoma, a new technique has been developed, total mesometrial resection (TMMR), that allows nerve preservation and thus preservation of the adjacent organs and the pelvic floor (Höckel et al. in press).

4.3.5
Important Vessels, Nerves and Lymphatics

1 Uterine artery
2 Inferior hypogastric plexus

(Veins have a corresponding course)

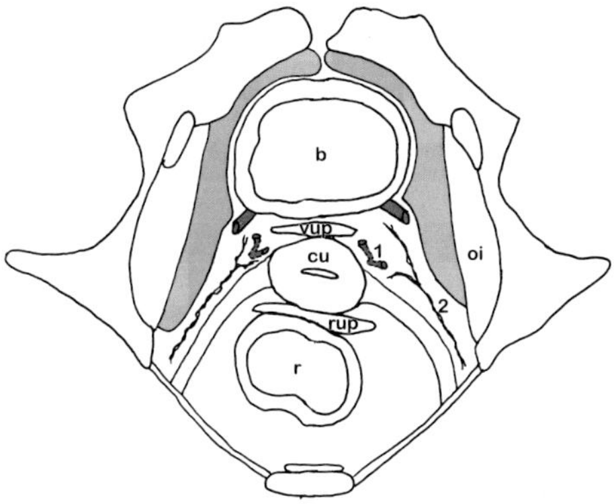

5 Perineal Body

5.1
Connective Tissue Structures and Muscles in the Male

The perineal body separates urogenital and anal hiatus. In the male it is situated between the rectum and urethra (prostate), i.e., between the posterior and anterior compartments. Within the region of the perineal body, the skin is firmly attached to the underlying connective tissue. This becomes obvious in macroscopic dissection as well as in histological sections. As can already be seen in early fetal life, the male's perineal body consists of dense connective tissue (Fig. 14A). It does not possess its own musculature, but numerous muscles originate or insert within the dense perineal body. The external anal sphincter is attached to it dorsally and the deep transverse perineal muscle, including the bulbourethral glands, abut the ventral portion of the perineal body (Fig. 14B, C). As has already been pointed out above (see posterior compartment), the additional smooth rectal muscle bundles that are situated in the rectoprostatic fascia are integrated and attached to the connective tissue of the perineal body (Fig. 14D). Apart from muscular structures, the cavernous nerves come in close contact to the perineal body. They pass just above its intrapelvic side (Fig. 14E) when leaving the pelvic cavity beside the membranous part of the urethra.

5.2
Connective Tissue Structures and Muscles in the Female

In the female, the perineal body is situated between the rectum and vagina, i.e., between the posterior and middle compartments. The female's perineal body is smaller than the male's, but it also consists of dense connective tissue. Apart from the deep transverse perineal muscle that does not exist in the female (Oelrich 1983), the muscular structures that are attached to the female's perineal body are the same as in the male. As the region of the female's perineal body is of high clinical interest with regard to birth damage and/or episiotomies (Woodman and Graney 2002), it will thoroughly be described according to the gynecologist's point of view, i.e., from outside (inferior) to the inside (superior): Level below the orifice of the vagina the external anal sphincter is attached to the perineal body (Fig. 15A), whereas level with the orifice of the vagina and above the internal sphincter abuts the perineal body and thus indirectly to the dorsal wall of the vagina (Fig. 15B). At these levels the external sphincter embraces the anal canal, the perineal body, and the dorsal wall of the vagina laterally. Cranially it is connected to the puborectalis muscle (Fig. 15C, D).

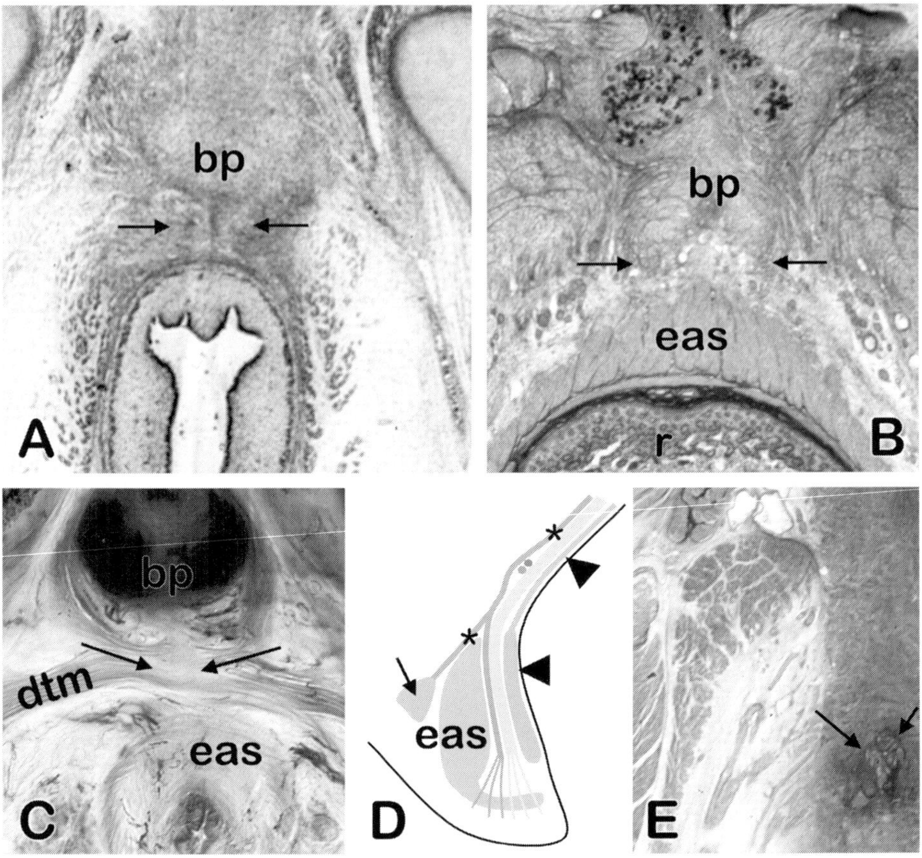

Fig. 14A–E Perineal body (*arrows*). **A** Axial section (300 μm) of a 9-week-old male fetus. ×36. **B** Axial section (400 μm) of a male newborn specimen. ×8. **C** Axial section (3 mm) of an adult male. ×2. **D** Schematic drawing of the sagittal plane pointing out the ventral anorectal wall (*arrowheads*) and the different muscle layers including the additional longitudinal muscle cells (*asterisks*). **E** Axial section (500 μm) of a 20-week-old male fetus just before leaving the urogenital hiatus. ×12

The infralevatoric side of the perineal body is connected with connective tissue septa of the ischioanal fossa (DeBlok and DeJong 1980) which are also connected to the inferior fascia of the levator ani muscle (Janssen et al. 2001)

5.3 Reinterpreted Anatomy, Function and Fundamentals of Neuroanatomy

A detailed knowledge concerning the anatomy of the perineal body has become of interest since transperineal or even dynamic transperineal ultrasound (Beer-Gabel et al. 2002) are carried out. With the help of this technique, the infralevatoric viscera, the soft tissues, and the puborectalis can be viewed and defined.

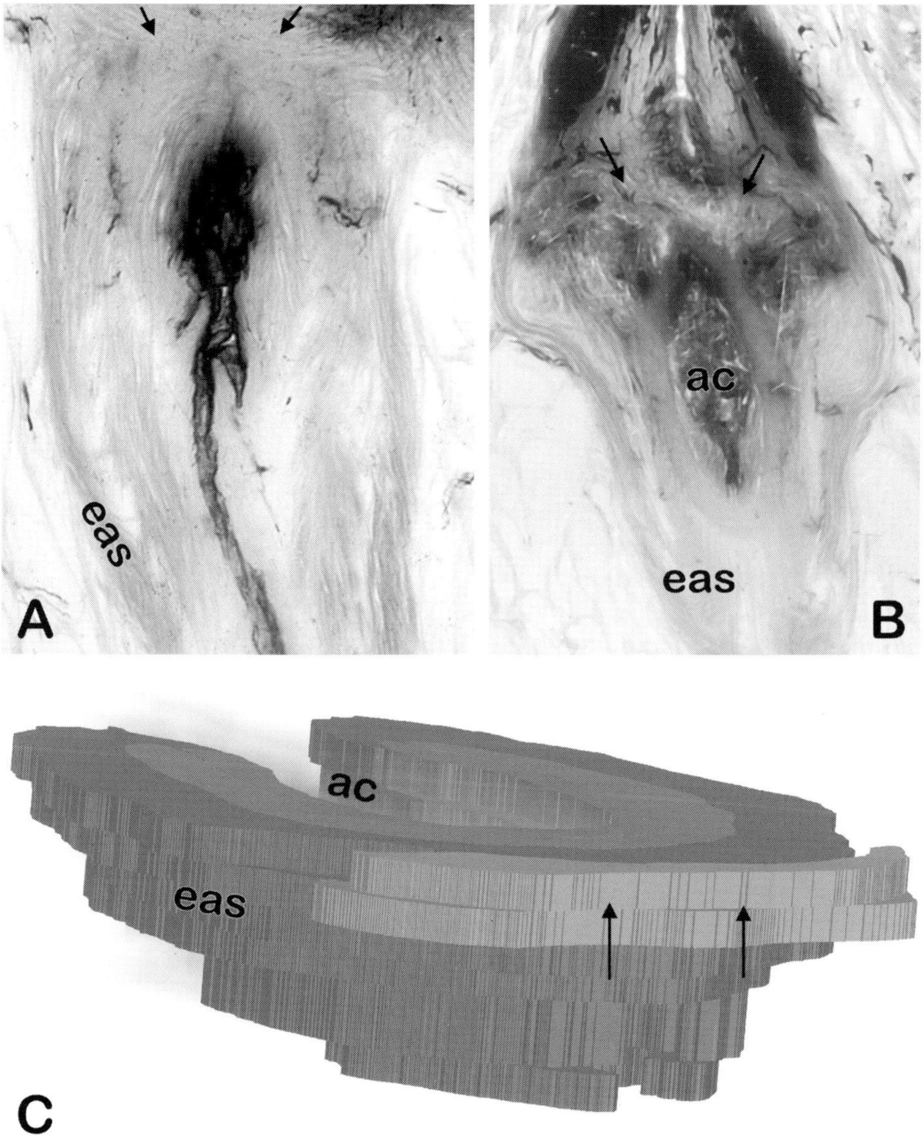

Fig. 15A–C Perineal body (*arrows*) and attached muscles. **A** Axial section (5 mm) of an adult female at a level with the anal cleft. ×2.2. **B** Axial section of the same specimen (**A**) at a level with the vaginal hiatus. ×1.2. **C** Computer-assisted reconstruction in an oblique view from the right side

For a long time there has been no doubt as to the fibrous components of this region. As, however, defined in the current *Terminologia Anatomica* (1998) the perineal body should be a fibromuscular rather than a tendinous structure. We do absolutely not agree with this opinion. The perineal body itself is a fibrous structure, but it is intermingled with all originating and inserting muscles. It has to be considered

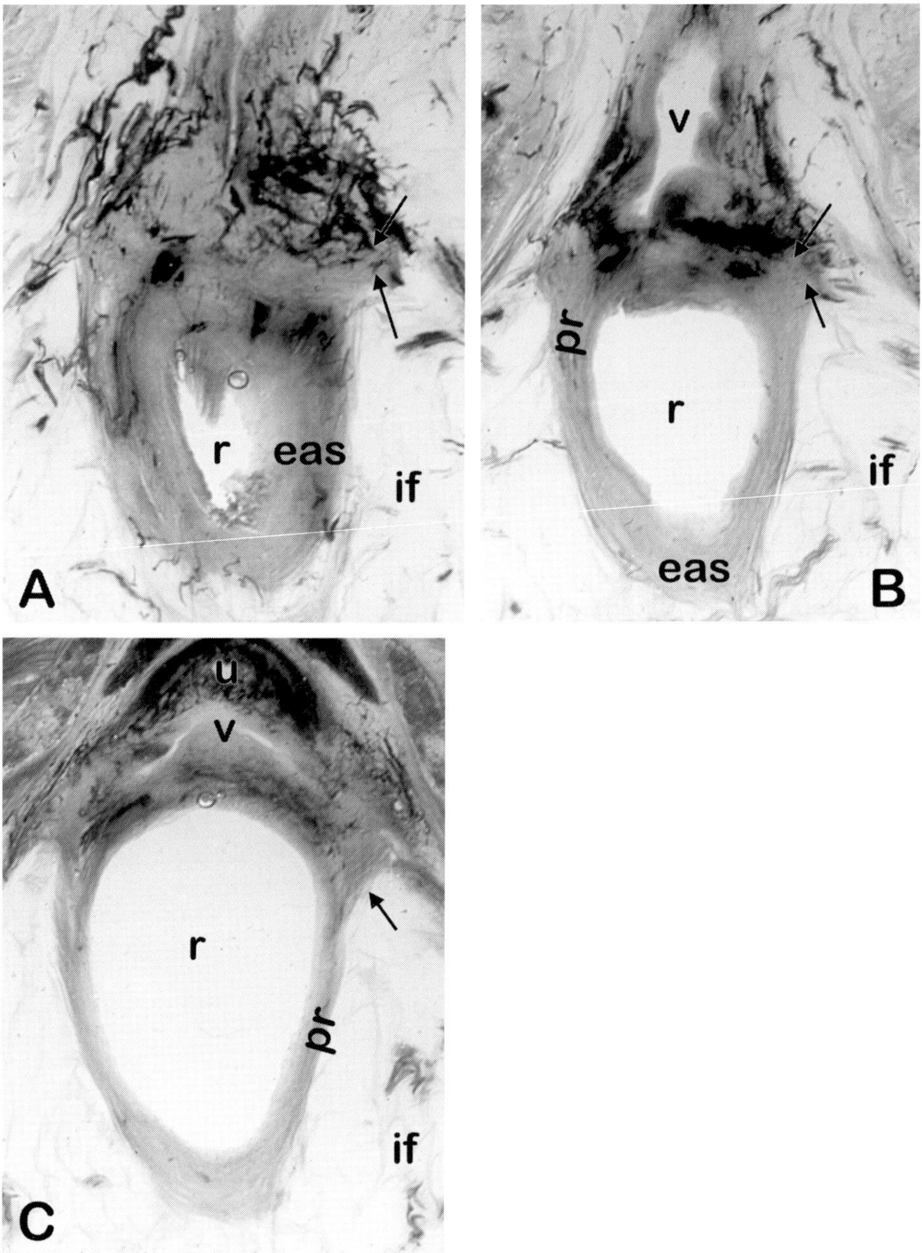

Fig. 16A–C Scar (*arrows*) of an old perineal rupture in axial sections (4 mm) of an adult female. **A** At a level with the perineum. ×0.8. **B** At a level with the fusion of external anal sphincter and puborectalis muscle. ×0. **C** At a level with the rectal ampulla. ×0.8

as a tendinous center for all the muscles that do not have a bony origin or attachment. There is no doubt that it is an important region for absorbing part of the intrapelvic (intra-abdominal) pressure. A stretched or even destroyed perineal body may be a cause for urogenital or rectal prolapse (Zbar et al. 2003).

5.4 Surgical, Urological, and Obstetric Relevance

From a morphological as well as from a functional point of view there is a need for discussion if and how a surgical approach through an intact perineal body should be performed.

Twenty years ago, an appeal was made by urologists to be careful when operating at the perineum, after they learned to avoid injury at the cavernous nerves passing in the neighborhood of the perineal body and the membranous part of the urethra (Lepor et al. 1985; Walsh et al. 1983; Fritsch 1989).

The discussion of pelvic floor damage during vaginal delivery and/or after episiotomies has been fueled by the remarkable statistics of Sultan et al. (1993), who showed that episiotomies do not prevent tearing. We think that the indication for episiotomies should be clearly defined by an international committee and it should be restricted to special cases, because perineal damage may occur not only spontaneously but also iatrogenically through the execution of an episiotomy. It is absolutely not old-fashioned to protect the perineum during vaginal delivery by hands-on methods.

We recommend not carrying out median and lateral episiotomies and being careful with the mediolateral ones: As can be seen from a pathological specimen in Fig. 16A–C, a perineal tear and/or a lateral episiotomy has led to a scar of the perineal body and the external anal sphincter. The connective tissue septa of the ischioanal fossa are irregular (Fig. 16A). At the border between the infralevatoric and levatoric levels, it becomes visible that the vaginal wall is slightly displaced, the puborectalis is rather thin, and the ischioanal fossa is not symmetric with the contralateral side (Fig. 16B), a diagnosis that still remains on supralevatoric levels (Fig. 16C). A refined and functional surgical treatment of perineal tears seems to be necessary to avoid such situations. As modern imaging techniques allow a fast and reliable control, it is the gynecologist's task to improve their surgical treatment.

6 Summary and Conclusions

The study presented here comparing cross-sectional anatomy of the fetal and the adult pelvic connective tissue with the results of modern imaging techniques and actual surgical techniques shows that the classical concepts concerning the subdivision of the pelvic connective tissue and muscles need to be revised.

According to clinical requirements, the subdivision of the pelvic cavity into anterior, posterior, and middle compartments is feasible. Predominating connecting tissue structures within the different compartments are:

Paravisceral fat pad within the anterior compartment (Fig. 17, I), *rectal adventitia or perirectal tissue* within the posterior compartment (Fig. 17, II), and *uterosacral ligaments* within the middle compartment. The *nerve-vessel guiding plate* can be found in all of these compartments; it starts within the posterior compartment and it ends within the anterior one. It constitutes the morphological border between the anterior and posterior compartments in the male. This border is supplied by the *uterosacral ligaments* in the female. Whereas in gross anatomy no further border is discernable between anterior and posterior or middle compartment, the *rectal fascia* (hardly visible in embalmed cadavers) demarcates the rectal adventitia and is one of the most important pelvic structures for the surgeon.

In principle, the outlined subdivision of the pelvic connective tissue is identical in the male and in the female; facts that become clear from early human life and that

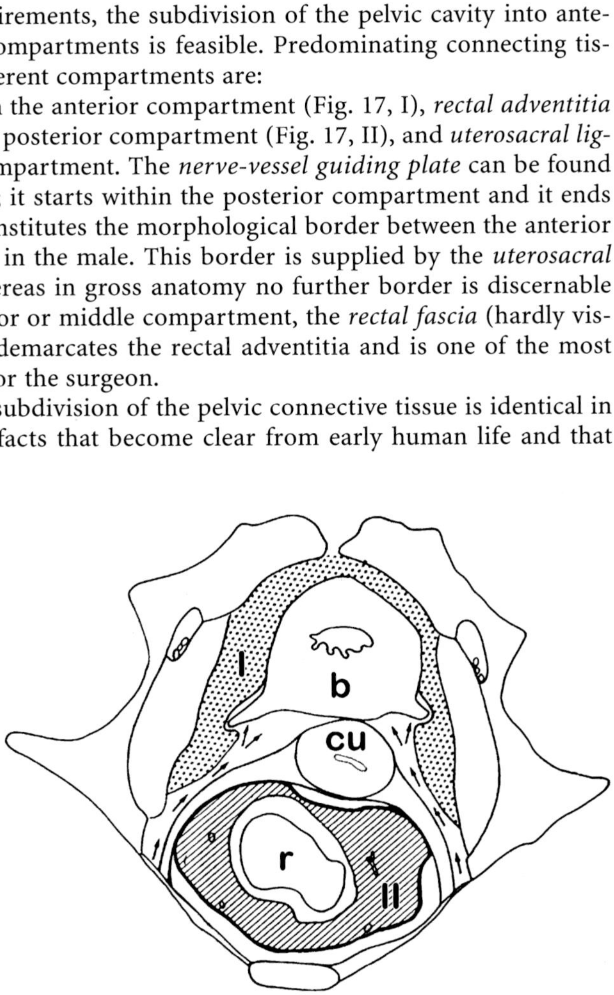

Fig. 17 Pelvic connective tissue compartments in a schematic drawing of the female pelvis: The large fat pad (*I*) fills the anterior compartment, the perirectal tissue (*II*) predominates the posterior one. The perirectal tissue has completely been removed in case of rectal cancer

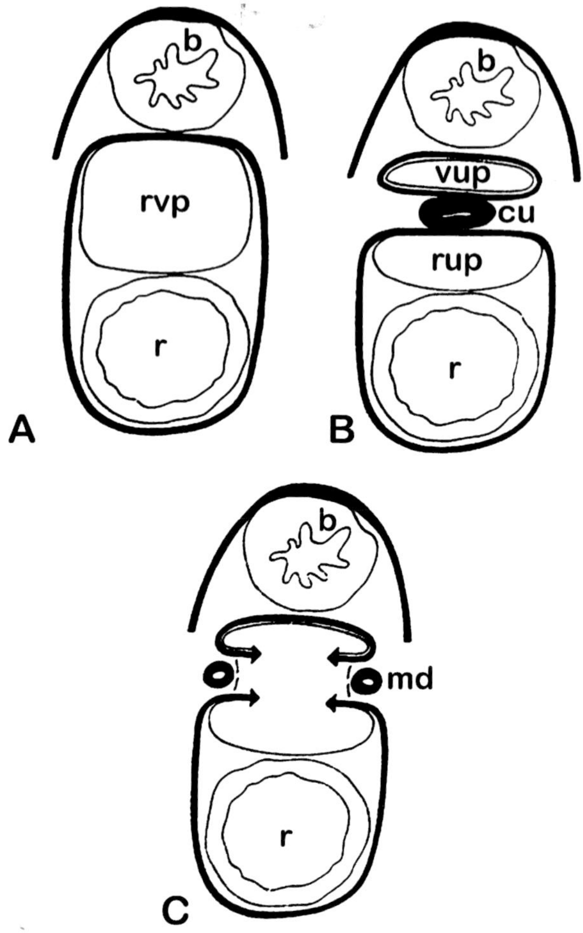

Fig. 18 Hypothesis on the development of the two peritoneal pouches in the female pelvis in correlation to the situation in the male

are already established during this period (Fig. 18). The uterus is interposed between the bladder and rectum and subdivides the pelvic peritoneum into two pouches thus establishing the only real difference between male and female pelvic cavity. The preferential direction of the pelvic connective tissue fibers is not changed by the interposition of the uterovaginal complex.

The pelvic floor muscles are composed of the portions of the *levator ani muscle, the muscles of the cavernous organs* and *the deep transverse perineal muscle in the male*. The latter does *not exist* in the female. We have clearly shown that the different muscles can already be found in early human life and that they are *never* intermingled with the muscular walls of the pelvic organs. The levator ani muscle of the female, however, is intermingled with connective tissue long before the female sexual hormones exert influence. We have also shown that the distinct sexual differences within the pelvic floor muscles as well as within the sphincter muscles can already be found in early human life. Both the external urethral and the external anal sphincter muscles *are not completely circular*. The external anal sphincter is intimately connected with the internal sphincter as well as with the longitudinal muscle. Whereas the innervation and function of the urethral sphincter muscles are mostly clear, cloacal development, innervation, and function of all parts of anal sphincter complex are not completely clarified.

As to the support of the pelvic viscera, we believe that intact pelvic floor muscles, an undisturbed topography of the pelvic organs, and an undisturbed perineum are of more importance than the so-called pelvic ligaments. Our hypothesis points to the fact that *the support of pelvic viscera is multistructural.* Thus in pelvic surgery, a lot of techniques have to be revised with the aim to preserve or to reconstruct all the structures mentioned. This is a multidisciplinary task that can only be solved by cooperation of morphologists, urologists, gynecologists, and coloproctologic surgeons or by creating a multidisciplinary pelvic floor specialist.

Table 1 Synonyms used for the sensory circumventricular organs

Term	Figure	*Terminologia Anatomica* (TA)		Clinical term	Definition	Renaming (according to our results)	Existence
		English	Latin				
1. Anococcygeal body	19	Anococcygeal body; anococcygeal ligament	Corpus anococcygeum; lig. anococcygeum	–	*TA*: The term *corpus*, rather than *ligamentum*, is used in TA because it is a stratified nonligamentous structure in which fleshy muscle attachments underlie a tendon	Not necessary	+
2. Perineal body	20	Perineal body	Corpus perineale; centrum perinei	–	*TA*: The perineal body is fibromuscular rather than tendinous and quite unlike the *centrum tendineum* of the diaphragm *Our opinion*: The perineal body itself is tendinous, nevertheless it can not be compared with the flat centrum tendineum of the diaphragm	Though tendinous, not necessary	+
3. Perineal membrane	21	Perineal membrane	Membrana perinei	–	Dense connective tissue between external urethral sphincter (and transverse perineal muscle in male) and pubic bone	Not necessary	+
4. Ano-rectum	22	Rectum and anal canal	Rectum et canalis analis	Ano-rectum	*Our opinion*: The clinical term includes both, the rectum and the anal canal not taking into account that they are of different origin	Necessary to pick up in TA	+
5. Presacral compartment	23	–	–	–	*Our opinion*: Small space between presacral fascia and sacral and coccygeal vertebrae containing vessels	Necessary to pick up in TA	+
6. Presacral fascia	23	Presacral fascia	Fascia presacralis	–	Caudal part of the parietal pelvic fascia		+
7. Perirectal compartment	24	–	–	Mesorectum	*Our opinion*: Compartment filled by the rectal adventitia including nerves, vessels, lymph nodes	Necessary to pick up in TA	+

Table 1 (continued)

Term	Figure	*Terminologia Anatomica* (TA) English	*Terminologia Anatomica* (TA) Latin	Clinical term	Definition	Renaming (according to our results)	Existence
8. Rectal fascia, or "Grenzlamelle"	25	–	–	–	*Our opinion*: Outer connective tissue lamella of the rectal adventitia, bordering the perirectal compartment	Necessary to pick up in TA	+
9. Inferior hypogastric plexus	26	Inferior hypogastric plexus; pelvic plexus	Plexus hypogastricus inferior; plexus pelvicus	Pelvic plexus	Autonomic nerve plexus within the recto-uterine or recto-vesical fold	Exclusively into the old and clinical term: *pelvic plexus*	+
10. Uterosacral ligament	27	Uterosacral ligament or recto-uterine ligament	Lig. rectouterinum	–	Dense connective tissue running from the edges of the cervix uteri to the region of the sacrospinous ligament	Exclusively into *uterosacral ligament*	+
11. Rectoprostatic fascia	28	Rectoprostatic fascia; rectovesical septum (male)	Fascia rectoprostatica; septum rectovesicale	–	*Our opinion*: Plate of dense connective tissue, smooth muscle cells and nerves, locally arranged between rectum and prostate	Exclusively into the term *rectoprostatic septum*	+
12. Rectovaginal fascia	28	Rectovaginal fascia; rectovaginal septum (female)	Fascia rectovaginalis; septum rectovaginale	–	*Our opinion*: Plate of dense connective tissue, smooth muscle cells and nerves, locally arranged between rectum and vagina	Exclusively into the term *rectovaginal septum*	+
13. Anal sphincter complex	29	–	–	–	Includes all muscle layers of the anal canal: internal (smooth) sphincter, longitudinal (smooth) muscle, external (striated) sphincter	Necessary to pick up in TA	+

Table 1 (continued)

Term	Fig-ure	*Terminologia Anatomica* (TA)		Clinical term	Definition	Renaming (according to our results)	Exis-tence
		English	Latin				
14. Onuf's nucleus	30	–	–	–	Onuf's nucleus (Onufrowicz, 1901): a – sexually dimorphic – longitudinally oriented group of small somatic motor neurons in the ventral horn of the sacral spinal cord (S2–S4), slightly medially to the lateral motor nucleus; axons from Onuf's nucleus are known to innervate the rhabdosphincter of the bladder, the bulbocavernosus muscle, and also the vagina in female. The neurotransmitters serotonin and norepinephrine are in high concentration in Onuf's nucleus		+
15. Puboprostatic ligament	31	Pubovesical ligament, medial puboprostatic ligament, pubovesicalis, puboprostatic ligament, lateral puboprostatic ligament; puboprostaticus	Lig. pubovesicale, lig. mediale puboprostaticum, m. pubovesicalis, lig. puboprostaticum, lig. laterale puboprostaticum; m. puboprostaticus	–	Most confusing structure! *Our opinion*: There is only one structure, running from the pubic bone to the vesical neck. It mainly consists of smooth muscle cells intermingled with strands of dense connective tissue	Exclusively into the term *pubovesical muscle*	+
16. Pubovesical ligament	31	Medial pubovesical ligament, pubovesicalis, lateral pubovesical ligament	Lig. mediale pubovesi-cale, m. pubovesicalis, lig. laterale pubovesicalis	–	Most confusing structure! *Our opinion*: there is only one structure, running from the pubic bone to the vesical neck. It mainly consists of smooth muscle cells intermingled with strands of dense connective tissue	Exclusively into the term *pubovesical muscle*	+

Table 1 (continued)

Term	Figure	*Terminologia Anatomica* (TA)		Clinical term	Definition	Renaming (according to our results)	Existence
		English	Latin				
17. Levator ani muscle	32	Levator ani	M. levator ani	–	Muscle that constitutes the main part of the pelvic diaphragm and is composed of the Mm. pubococcygei, iliococcygei, and puborectales of each side		+
18. Tendinous arch of the pelvic fascia	33	Tendinous arch of the pelvic fascia	Arcus tendineus fasciae pelvis	–	*Our opinion*: This structures originates from the pubic bone laterally, it is connected with the superior fascia of the pelvic diaphragm laterally and with the pubovesical ligament medially. It may falsely be called Lig. laterale puboprostaticum or Lig. laterale pubovesicale		+
19. Paravisceral fat pad	34	–	–	–	*Our opinion*: Fat pad at the lateral side of the bladder that develops in situ. Functionally necessary for the movements of bladder	Necessary to pick up in TA	+
20. Broad ligament	35	Broad ligament of the uterus	Lig. latum uteri	–	Peritoneal fold between the uterus and the lateral wall of the pelvis		+
21. Recto-uterine fold	36	Recto-uterine fold	Plica recto-uterina	–	Peritoneal fold passing from the cervix uteri on each side of the rectum to the posterior pelvic wall		+
22. Recto-uterine pouch	36	Recto-uterine pouch	Excavatio rectouterina	Space of Douglas	Deep peritoneal pouch situated between the recto-uterine folds of each side		+
23. Vesico-uterine fold	37	Vesico-uterine fold	Plica vesicouterina	–	Peritoneal fold between bladder and uterus on each side		+
24. Vesico-uterine pouch	37	Vesico-uterine pouch	Excavatio vesicouterina	–	Slight peritoneal pouch between the vesico-uterine folds of each side		+

Table 1 (continued)

Term	Figure	*Terminologia Anatomica* (TA)		Clinical term	Definition	Renaming (according to our results)	Existence
		English	Latin				
25. Transverse cervical ligament or cardinal ligament	38	Transverse cervical ligament, Cardinal ligament	Lig. transversum cervicis, lig. Cardinale	Cardinal ligament	Connective tissue structures that should extend from the side of the cervix to the lateral pelvic wall *Our opinion*: The cardinal ligament does not exist	Necessary to omit	∅

Fig. 19 See Table 1/1.

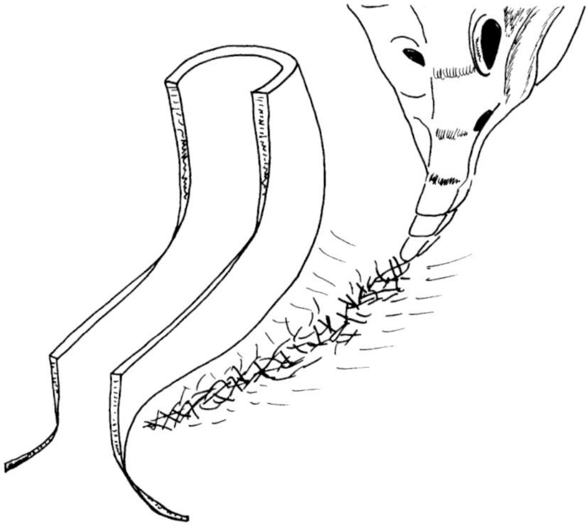

Fig. 20 See Table 1/2.

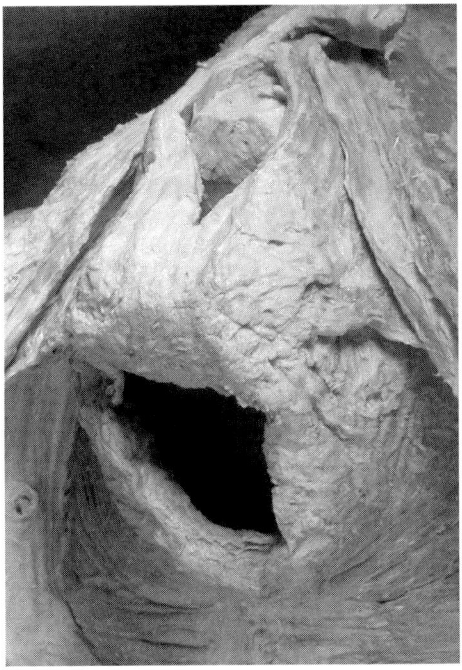

Fig. 21 See Table 1/3.

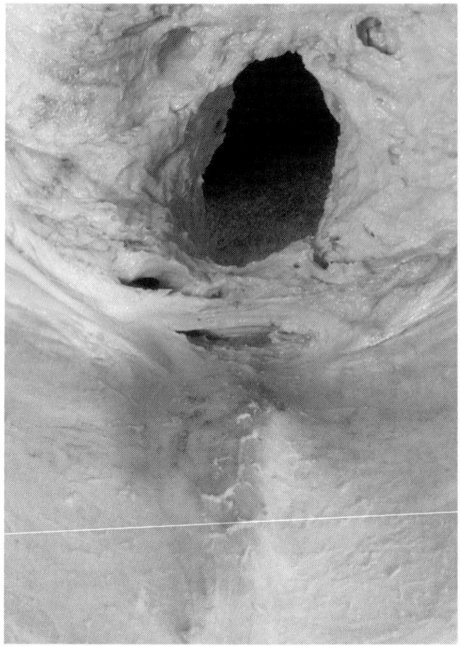

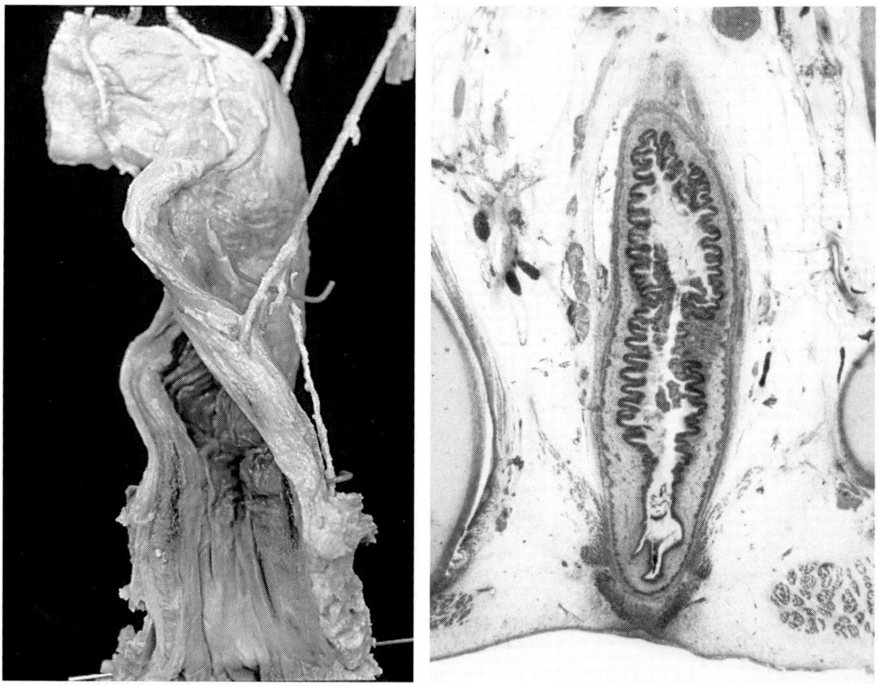

Fig. 22 See Table 1/4.

Fig. 23 See Table 1/5., 6.

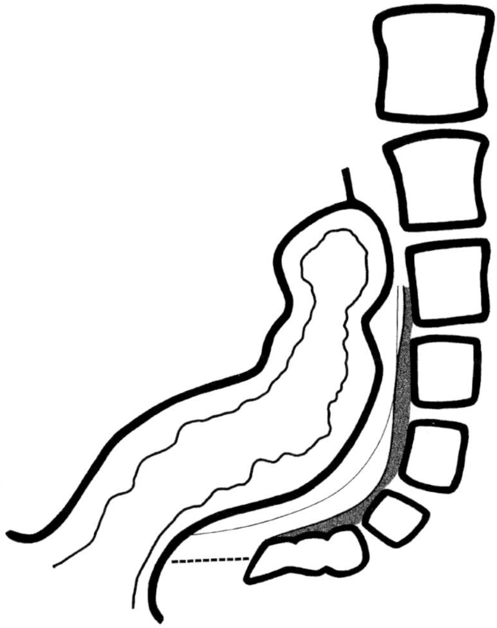

Fig. 24 See Table 1/7.

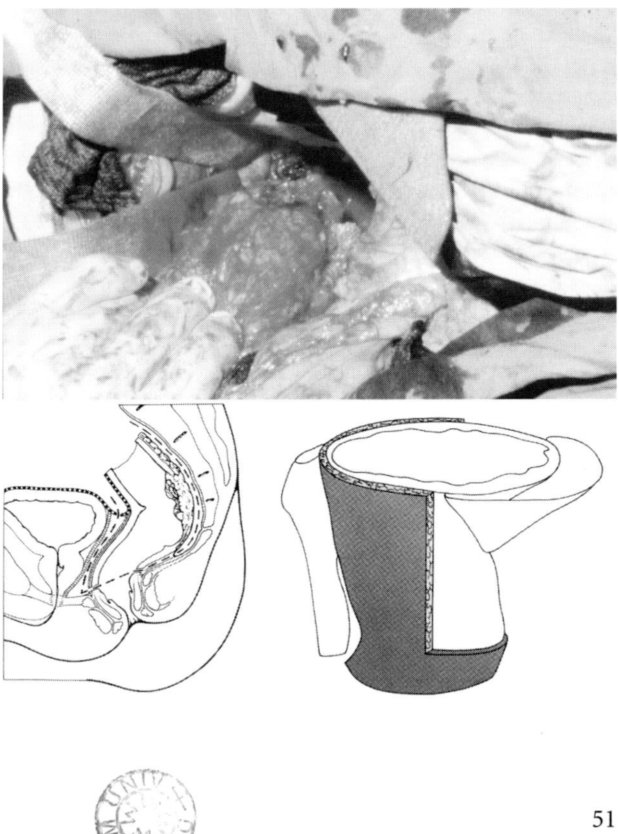

Fig. 25 See Table 1/8.

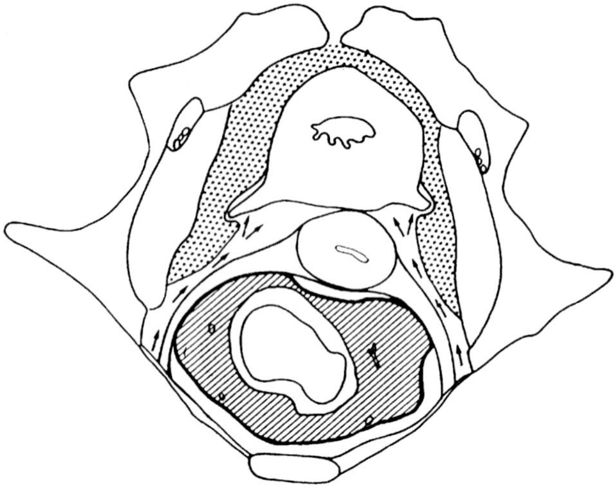

Fig. 26 See Table 1/9.

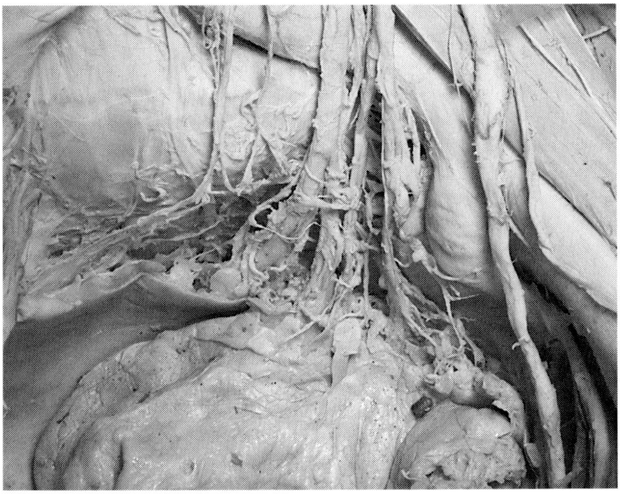

Fig. 27 See Table 1/10.

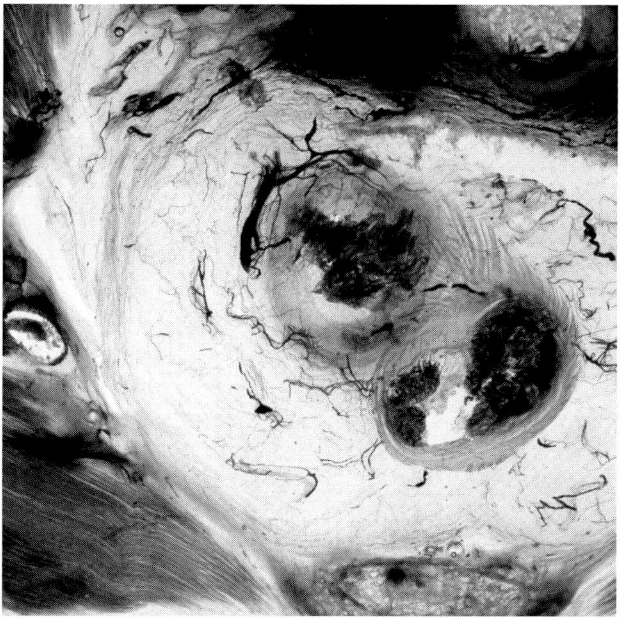

Fig. 28 See Table 1/11., 12.

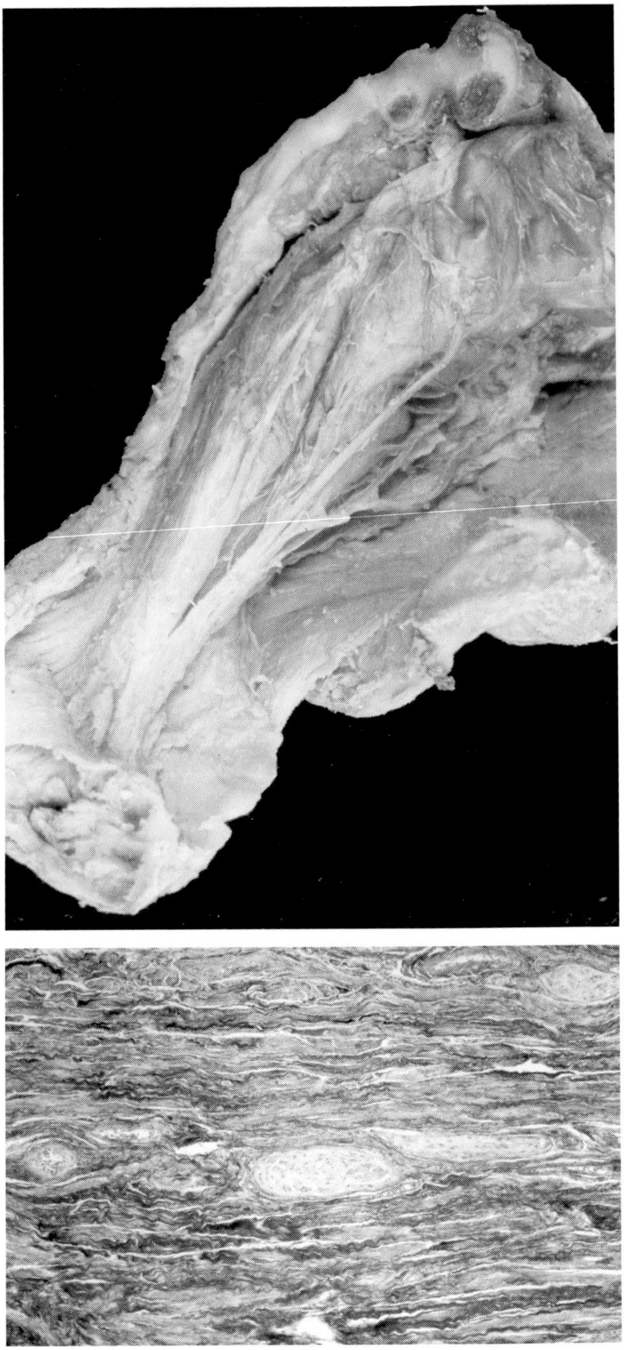

Fig. 29 See Table 1/13.

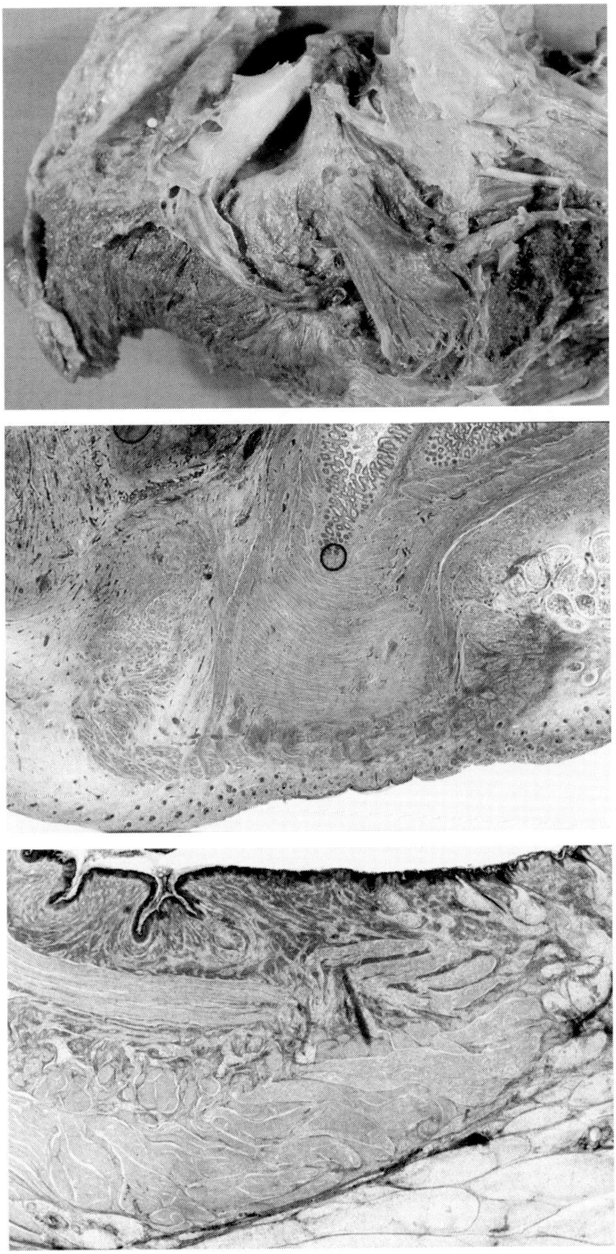

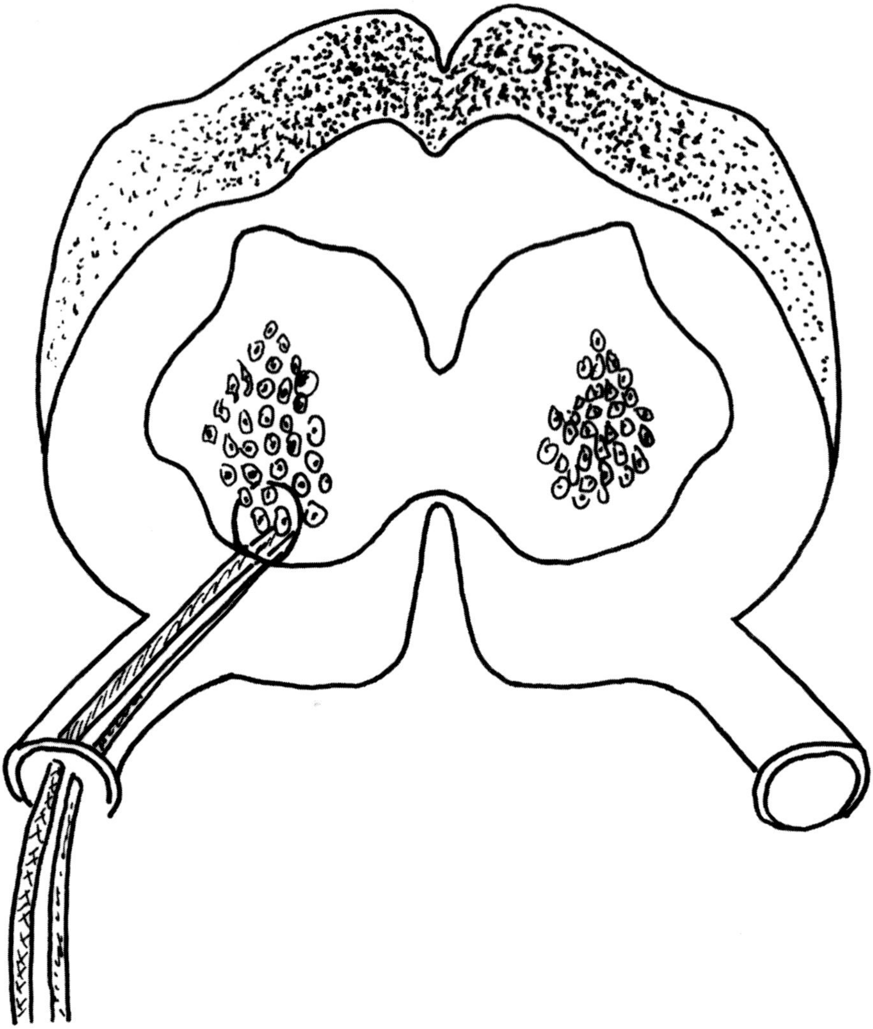

Fig. 30 See Table 1/14.

Fig. 31 See Table 1/15.

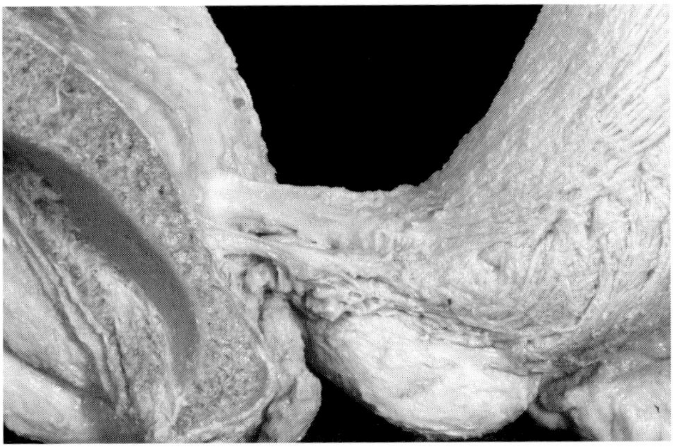

Fig. 32 See Table 1/17.

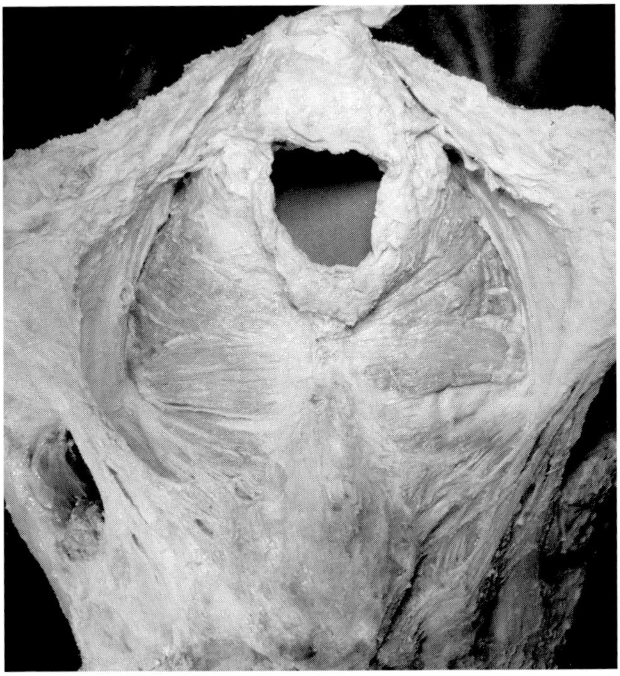

Fig. 33 See Table 1/18.

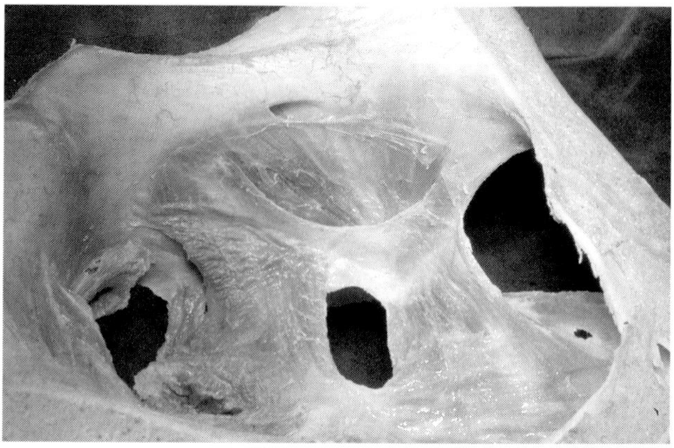

Fig. 34 See Table 1/19.

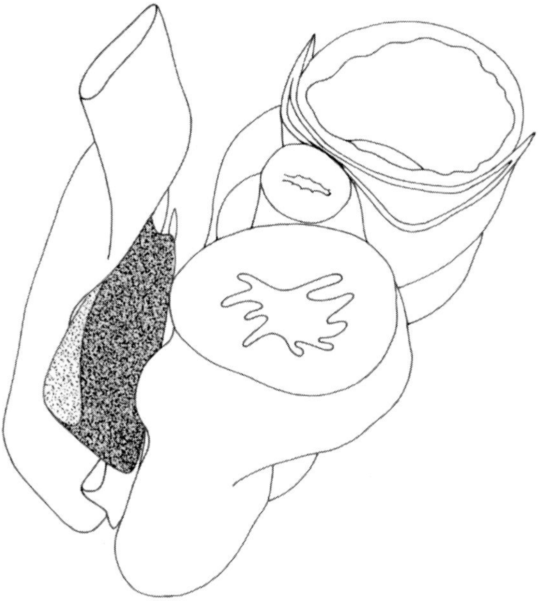

Fig. 35 See Table 1/20.

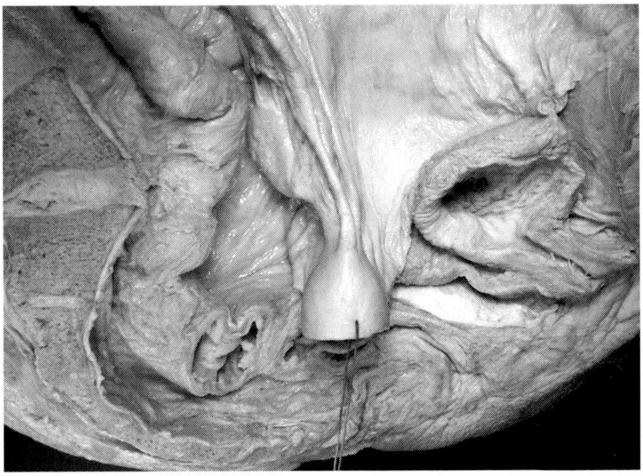

Fig. 36 See Table 1/21., 22.

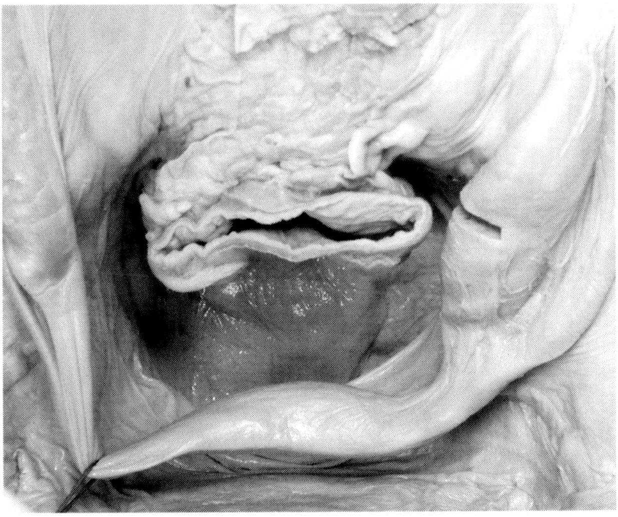

Fig. 37 See Table 1/23., 24.

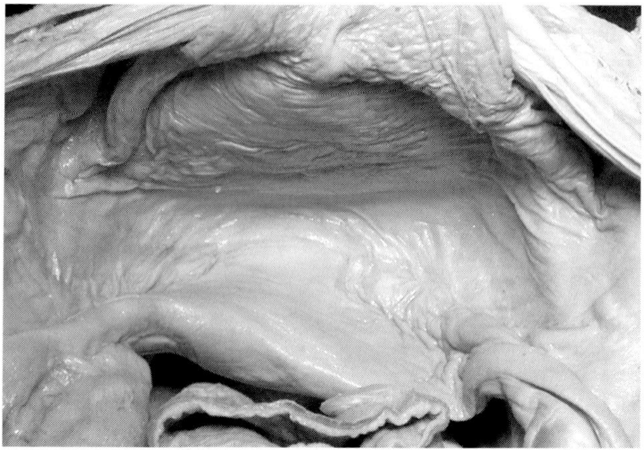

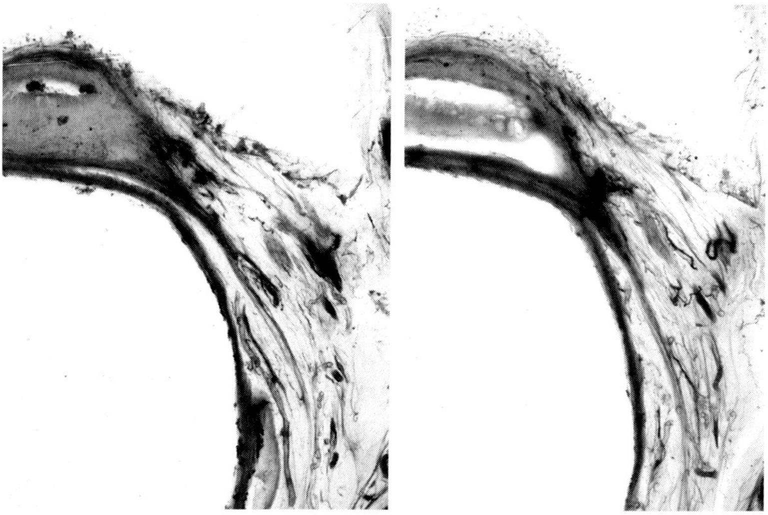

Fig. 38 See Table 1/25.

References

Altomare DF, Dodi G, LaTorre F, Romano G, Melega E, Rinaldi M (2001) Multicentre retrospective analysis of the outcome of artificial anal sphincter implantation for severe fecal incontinence. Br J Surg 88:1481–1486

Baeten C, Spaans F, Fluks A (1988) An implanted neuromuscular stimulator for fecal continence following previously implanted gracilis muscle. Report of a case. Dis Colon Rectum 31:134–7

Bastian D, Lassau JP (1982) The suspensory mechanism of the uterus. Surg Radiol Anat 4:147–160

Beer-Gabel M, Teshler M, Barzilai N, LurieY, Malnick S, Bass D, Zbar A (2002) Dynamic transperineal ultrasound in the diagnosis of pelvic floor disorders. Dis Colon Rectum 45:239–248

Beersiek F, Parks AG, Swash M (1979) Pathogenesis of ano-rectal incontinence. J Neurol Sci 42:111–127

Berglas B, Rubin IC (1953) Histologic study of the pelvic connective tissue. Surg Gynecol Obstet 97:277–289

Burnett AL, Wesselmann U (1999) History of the neurobiology of the pelvis. Urol 53:1082–1089

Cruikshank SH (1991) sacrospinous fixation – should this be performed at the time of vaginal hysterectomy? Am J Obstet Gynecol 1064:1072–1076

Cundiff GW, Weidner AC, Visco AG, Addison A, Bump RC (1998) An anatomic and functional assessment of the discrete defect rectocele repair. Am J Obstet Gynecol 179:1451–1457

DeBlok S (1982a) The connective tissue of the female fetal pelvic region. Acta Morphol Neerl Scand 20:65–92

DeBlok S (1982b) The connective tissue of the adult female pelvic region. Acta Morphol Neerl Scand 20:325–346

DeBlok S, DeJong E (1980) The fibrous tissue architecture of the female perineal region. Acta Morphol Neerl Scand 18:181–194

DeLancey JO (1994) Structural support of the urethra as it relates to stress urinary incontinence: the hammock hypothesis. Am J Obstet Gynecol 170:1713–1723

DeLancey JO (1996) Standing anatomy of the pelvic floor. J Pelvic Surg 2:260–263

Denonvilliers C (1836) Anatomie du perinee. Bull Soc Anat 2:105–105, (citation in Wesson MB 1922)

Denonvilliers C (1837) Propositions et observations d̀anatomie, de physiologie et pathologie. These x 285, Paris (citation in Wesson MB 1922)

Dorschner W, Stolzenburg JV, Neuhaus J (2001) Structure and function of the bladder neck. Adv Anat Embryol Cell Biol 159:1–109

Federative committee on anatomical terminology (1998) Terminologia Anatomica, International Anatomical Terminology. Georg Thieme Verlag, Stuttgart

Fielding JR, Griffiths DJ, Versi E, Mulkern RV, Lee ML, Jolesz FA (1998) MR imaging of pelvic floor continence mechanisms in the supine and sitting positions. Am J Roentgenol 171:1607–1610

Fritsch H (1988) Developmental changes in the retrorectal region of the human fetus. Anat Embryol 177:513–522

Fritsch H (1989) Staining of different tissues in thick epoxy resin impregnated sections of human fetuses. Stain Technol 64:75–79

Fritsch H (1989) Topography of the pelvic autonomic nerves in human fetuses between 21–29 weeks of gestation. Anat Embryol 180:57–64

Fritsch H (1990) Development of the rectal fascia. Anat Anz 170:273–280

Fritsch H (1992) The connective tissue sheath of uterus and vagina in the human female fetus. Ann Anat 174:261–266

Fritsch H (1994) Topography and subdivision of the pelvic connective tissue. Surg Radiol Anat 16:259–265

Fritsch H, Brenner E, Lienemann A, Ludwikowski B (2002) Anal sphincter complex. Dis Colon Rectum 45:188–194

Fritsch H, Fröhlich B (1994) Development of the levator ani muscle in human fetuses. Early Hum Dev 37:15–25

Fritsch H, Kühnel W (1992) Development and distribution of adipose tissue in the pelvis. Early Hum Dev 28:79–88
Fritsch H, Kühnel W, Stelzner F (1996) Entwicklung und klinische Anatomie der Adventitia recti. Langenbecks Arch Chir 381:237–243
Gasparri F, Brizzi E (1961) Significato anatomo-chirurgico delle formazioni connecttivali del piccolo bacino. Arch Ital Anat Embriol 66:151–169
Gosling J (1999) Gross anatomy of the lower urinary tract. In: Abrams P, Khoury S, Wein AJ (eds) Incontinence. Plymbridge Distributors Ltd, Plymouth, pp 21–56
Grabbe E, Lierse W, Winkler R (1982). Die Hüllfascien des Rektums. Fortsch Röntgenstr 136:653–659
Griffiths DM, Malone PS (1995) The Malone antegrade continence enema. J Pediatr Surg 30:68–71
Hafferl A (1969) Lehrbuch der topographischen Anatomie. Thiel W (ed) Springer Verlag, Berlin
Halverson AL, Hull TL, Flotura C (2001) Outcome of sphincteroplasty combined with surgery for urinary incontinence and pelvic organ prolapse. Dis Colon Rectum 44:1421–1426
Halverson AL, Hull TL (2002) Long-term outcome of overlapping anal sphincter repair. Dis Colon Rectum 45:345–348
Han SJ, Park HJ, Kim CB, Hwang EH (1995) Long-term follow-up of gracilis muscle transposition in children. Yonsei Med J 36:372–377
Heald RJ (1988) The holy plane of rectal surgery. J R Soc Med 81:503–508
Heald RJ (1995) Total mesorectal excision is optimal surgery for rectal cancer. Br J Surg 82:1297–1299
Herschhorn S (1999) Surgical treatment of urinary incontinence in men. In: Abrams P, Khoury S, Wein AJ (eds) Incontinence. Plymbridge Distributors Ltd, Plymouth, pp 691–729
Ho KMT, McMurray G, Mills IW Brading AF, Noble JG (1997) The human membranous striated sphincter: does it consist of a homogeneous population of fibres? Neurourol Urodynamics 16:453–454
Höckel M, Horn L-C, Hentschel B, Höckel S, Naumann G Total mesometrial resection (TMMR): high resolution extended hysterectomy with autonomic nerve preservation. Int J Gynecol Cancer (in press)
Janssen U, Lienemann A, Fritsch H (2001) Die Bedeutung des M. levator ani–Fossa ischioanalis-Glutaeus maximus (LFG)–Komplexes für den weiblichen Beckenboden. Ann Anat Suppl 183:11
Jehle EC, Haehnel T, Stralinger MJ, Becker HD (1995) Level of the anastomosis does not influence functional outcome after anterior rectal resection for rectal cancer. Am J Surg 169:147–152
Khubchandani IT, Sheets JA, Stasik JJ, Hakki AR (1983) Endorectal repair of rectocele. Dis Colon Rectum 26:792–796
Kocks J (1880) Normale und pathologische Lage und Gestalt des Uterus sowie deren Mechanik. Cohen, Bonn, pp 1–60
Koster H (1933) On the supports of the uterus. Am J Obstet Gynecol 25:67–74
Kux M, Fritsch H (2002) On the extraperitonal origin of hernia. Hernia 4:259–263
Lawler LP, Fleshman JW (2002) Solitary rectal ulcer, rectocele, hemorrhoids and pelvic pain. In: Pemberton JH, Swash M, Henry MM (eds) The pelvic floor. Its function and disorders. WB Saunders, London, pp 362–371
Lepor H, Gregerman M, Crosby R, Mostofi FK, Walsh PC (1985) Precise localization of the autonomic nerves from the pelvic plexus to the corpora cavernosa: A detailed anatomical study of the adult male pelvis. J Urol 133:207–212
Lienemann A, Anthuber C, Baron A, Kohz P, Reiser M (1997) Dynamic MR colocystorectography assessing pelvic floor descent. Eur Radiol 7:1309–1317
Lierse W (1984) Becken. In: von Lanz T, Wachsmuth W (eds) Praktische Anatomie, Bd 2, Teil 8A. Springer Verlag, Berlin Heidelberg New York Tokio
Lierse W, Holschneider AM, Steinfeld J (1993) The relative proportions of type I and type II fibers in the external sphincter ani muscle at different ages and stages of development – observations on the development of continence. Eur J Pediatr Surg 3:28–32
Long DM, Bernstein WC (1959) Sexual dysfunction as a complication of abdomino-perineal resection of the rectum in the male: an anatomic and physiologic study. Dis Colon Rectum 2:540–548

Ludwikowski B, Oesch-Hayward I, Fritsch H (2002) Rectovaginal fascia: An important structure in pelvic visceral surgery? About its development, structure, and function. J Pediatr Surg 37:634–638

Ludwikowski B, Oesch-Hayward I, Brenner E, Fritsch H (2001) The development of the external urethral sphincter in humans. BJU Int 87:565–568

Ludwikowski B, Oesch-Hayward I, Gonzalez R (1999) Total urogenital sinus mobilization: expanded applications. BJU Int 83:820–822

Maas CP, Moriya Y, Steup WH, Kiebert GM, Kranenbarg WM, van de Velde CJ (1998) Radical nerve preserving surgery for rectal cancer in the Netherlands: a prospective study on morbidity and functional outcome. Br J Surg 85:92–97

Mackenrodt A (1895) Ueber die Ursachen der normalen und pathologischen Lage des Uterus. Arch Gynaekol 48:393–421

Malone PS, Ransley PG, Kiely EM (1990) Preliminary report: The antegrade continence enema. Lancet 336:1217–1218

Niemen K, Heinonen PK (2001) Sacrospinous ligament fixation for massive genital prolapse in women aged over 80 years. BJOG 108:817–821

Nobis A (1988) Untersuchungen zur feineren Struktur des retrorektalen Raumes beim Menschen. Inaugural Dissertation, Bonn

Occelli B, Narducci F, Hautefeuille J, Francke JP, Querleu D, Crepin G, Cosson M (2001) Anatomic study of arcus tendineus fasciae pelvis. Eur J Obstet Gynecol Reprod Biol 97:213–219

Oelrich TM (1980) The urethral sphincter muscle in the male. Am J Anat 158:229–246

Oelrich TM (1983) The striated urogenital sphincter in the female. Anat Rec 205:223–232

Onufrowicz B (1901) On the arrangement and function of the cell groups of the sacral region of the spinal cord in man. Arch Neurol Psychopathol 3:387–412

Paick JS, Donatucci CF, Lue TF (1993) Anatomy of the cavernous nerves distal to the prostate: microdissection study in adult male cadavers. Urol 42:145–149

Patten BM (1968) Human embryology, 3rd edn. McGraw-Hill Book Company, New York Toronto Sydney London

Pena A (1997) Total urogenital mobilization – an easier way to repair cloacas. J Pediatr Surg 32:263–268

Pernkopf E (1941) Topographische Anatomie des Menschen. Urban & Schwarzenberg, Berlin, Bd 2, Teil 1: Bd 2, Teil 2

Richardson AC (1993) The rectovaginal septum revisited: its relationship to rectocele and its importance in rectocele repair. Clin Obstet Gynecol 36:976–983

Richter K, Frick H (1985) Die Anatomie der Fascia pelvic visceralis aus didaktischer Sicht. Geburtshilfe Frauenheilkd 45:282–287

Richter K (1998) Gynäkologische Chirurgie des Beckenbodens. Heinz F, Terruhn V (eds) Georg Thieme Verlag, Stuttgart New York

Roberts WH, Harrison CW, Mitchel DA, Fischer AF (1988) The levator ani muscle and the nerve supply of its puborectalis component. Clin Anat 1:267–283

Roediger WEW, Tucker WG (1986) Thickening of the pelvic fascia in carcinoma of the rectum. Dis Colon Rectum 29:117–119

Schäfer R, Enck P (1997) Physiologie und Pathophysiologie der Defäkation. In: Jost WH (ed) Neurologie des Beckenbodens. Chapman & Hall, Weinheim, pp 15–22

Schroeder HD (1981) Onuf's nucleus X, a morphological study of a human spinal nucleus. Anat Embryol 161:443–453

Schroeder HD, Reske-Nielsen E (1983) Fiber types in the striated urethral and canal sphincters. Acta Neuropathol 60:278–282

Sprenger D, Lienemann A, Anthuber C, Reiser M (2000) Funktionelle MRT des Beckenbodens: Normale anatomische und pathologische Befunde. Radiologe 40:458–464

Stelzner F (1977) Über Potenzstörungen nach Amputation und Kontinenzresektion des Rektums. Zentralbl Chir 102:212–219

Stelzner F (1989) Die Begründung, die Technik und die Ergebnisse der knappen transabdominalen Kontinenzresektion. Langenbecks Arch Chir 374:303–314

Stelzner F (1998) Chirurgie an vizeralen Abschlusssystemen. Georg Thieme, Stuttgart New York

Stelzner F, Fritsch H, Fleischhauer K (1989) Die chirurgische Anatomie der Genitalnerven des Mannes und ihre Schonung bei der Exzision des Rektums. Chirurg 60:228–234

Sultan AH, Kamm MA, Hudson CN, Thomas JM, Bartram CI (1993) Anal sphincter disruption during vaginal delivery. N Engl J Med 329:1905–1911

Tandler J (1923) Lehrbuch der systematischen Anatomie. Die Eingeweide. Bd. 2 Vogel, Leipzig

Thakar R, Stanton S (2002) Management of genital prolapse. BMJ 324:1258–1262

Tobin CE, Benjamin JA (1953) Anatomical and surgical study of Denonvilliers fascia. Surg Gynecol Obstet 80:373–388

Uhlenhuth E, Nolley GW (1957) Vaginal fascia, a myth? Obstet Gynecol 10:349–358

Ulmsten U (2001) The basic understanding and clinical results of tension-free vaginal tape for stress urinary incontinence. Urologe 40:269–273

von Hagens G, Tiedemann K, Kriz W (1987) The current potential of plastination. Anat Embryol 175:411–421

von Hayek H (1969) Das Bindegewebe und die glatte Muskulatur des Beckenbodens. In: Alken CE, Dix VW, Goodwin WE, Wildbolz EL (eds) Handbuch der Urologie. Springer Verlag, Berlin Heidelberg New York, pp 289–313

Waldeyer W (1899) Das Becken. Cohen, Bonn

Walsh PC, Donker PJ (1982) Impotence following radical prostatectomy: insight into etiology and prevention. J Urol 128:492–497

Walsh PC, Lepor U, Eggleston JCC (1983) Radical prostatectomy with preservation of sexual function: anatomical and pathological considerations. Prostate 4:473–485

Walsh PC, Schlegel PN (1988) Radical pelvic surgery with preservation of sexual function. Ann Surg 208:391–400

Wattiez A, Canis M, Mage G, Pouly JL, Bruhat MA (2001) Promontofixation for the treatment of prolapse. Urol Clin North Am 28:151–157

Wedel T, Roblick U, Gleiss J, Schiedeck T, Bruch HP, Kühnel W, Krammer HJ (1999) Organization of enteric nervous system in the human colon demonstrated by wholemount immunhistochemistry with special reference to the submucous plexus. Ann Anat181:327–337

Wesson MB (1922) The development and surgical importance of the rectourethralis muscle and Denonvilliers' fascia. J Urol 8:339–359

Wilson PD, Dixon JS, Brown ADG, Gosling JA (1983) Posterior pubo-urethral ligaments in normal and genuine stress incontinent women. J Urol 130:802–805

Woodmann PJ, Graney DO (2002) Anatomy and physiology of the female perineal body with relevance to obstetrical injury and repair. Clin Anat 15:321–334

Yamamoto T, Samoti H, Ise H (1978) Sacral spinal innervations of the rectal and vesical smooth muscle and sphincter striated muscles as demonstrated by the horseradish peroxidase method. Neurosci Lett 7: 41–47

Zbar AP, Lienemann A, Fritsch H, Beer-Gabel M, Pescatori M (2003) Rectocele: pathogenesis and surgical management. Int J Colorectal Dis 29:1–11

Subject Index